Textbook of
MEDICAL
PHYSIOLOGY

Eighth Edition

Arthur C. Guyton, M.D.

Professor
Department of Physiology and Biophysics
University of Mississippi School of Medicine

1991
W.B. SAUNDERS COMPANY
Harcourt Brace Jovanovich, Inc.
Philadelphia ■ London ■ Toronto ■ Montreal ■ Sydney ■ Tokyo

W. B. Saunders Company
Harcourt Brace Jovanovich, Inc.

The Curtis Center
Independence Square West
Philadelphia, PA 19106

Library of Congress Cataloging-in-Publication Data

Guyton, Arthur C.
    Textbook of medical physiology/Arthur C. Guyton. —8th ed.
        p.   cm.
    Includes bibliographical references.
    Includes index.
    ISBN 0-7216-3087-1
    1. Human physiology.   2. Physiology, Pathological.   I. Title.
    [DNLM: 1. Physiology.   QT 104 G992-]
QP34.5.G9   1991
612—dc20   90-8653
                                                               CIP

Listed here are the latest translated editions of this book together with the language of each translation and the publisher.

Japanese (*4th Edition*) — Hirokawa Publishing Compay, Tokyo, Japan
Serbo-Croatian (*4th Edition*) — Medicinska Knjiga, Belgrade, Yugoslavia
Italian (*4th Edition*) — Piccin Editore, Padova, Italy
Spanish (*7th Edition*) — Nueva Editorial Interamericana S.A. de C.V., Mexico City, Mexico
Persian (*7th Edition*) — Sherkat Sahami "TCHEHR," Tehran, Iran
Portuguese (*6th Edition*) — Editora Interamericana Ltda.,
                            Rio de Janeiro, Brazil

Editor: Martin J. Wonsiewicz
Developmental Editor: Rosanne Hallowell
Designer: Joan Wendt
Production Manager: Peter Faber
Manuscript Editor: Jessie Raymond
Mechanical Artist: Risa Clow
Illustration Coordinator: Brett MacNaughton
Indexer: Ann Kassar
Cover Designer: Michelle Maloney

TEXTBOOK OF MEDICAL PHYSIOLOGY, Eighth Edition   ISBN 0-7216-3087-1

Printed in the United States of America.

Last digit is the print number: 9   8   7   6   5   4   3   2

Dedicated to

MY FATHER
for the uncompromising principles that guided his life

MY MOTHER
for leading her children into intellectual pursuits

MY WIFE
for her magnificent devotion to her family

MY CHILDREN
for making everything worthwhile

Preface

Each time I revise this *Textbook of Medical Physiology,* I think that someday physiology will become a completely mature subject without change from year to year. However, this always proves to be far from the truth. Physiology is a vast discipline, and only now are we beginning to make inroads into many of its fundamental secrets. Most importantly, within the past few years many new techniques for learning cellular and molecular physiology have become available. Therefore, more and more can we present physiological principles in molecular and physical terms, rather than merely as a series of separate biological phenomena. This change we all welcome, but it is also a change that makes necessary the revision of virtually every chapter in this text with each new edition.

The beauty of physiology is that it attempts to integrate the individual functions of all the body's different cells and organs into a total functional whole, the human or animal body. Indeed, the life of the human being relies upon this total function, not on the functions of the single parts in isolation from the others.

This brings us to another entirely different subject: How are the separate organs and systems controlled in a way that no one system overfunctions while others fail to provide their share? Fortunately, the body is endowed with a vast network of feedback mechanisms that account for the necessary balances, without which we would not be able to live. Physiologists call this high level of internal bodily control *homeostasis.* In disease conditions, more often than not the functional balances become seriously disturbed — that is, homeostasis becomes very poor. When the disturbance is too great, the total body can no longer live. Therefore, one of the principal goals of any medical physiology text is to explain and emphasize the effectiveness and beauty of the body's homeostatic mechanisms, as well as to discuss their aberrant function in disease.

Another goal of this eighth edition has been to make the text as accurate as possible. To help attain this, suggestions and critiques from many physiologists, students, and clinicians throughout the United States and other parts of the world have been received and utilized in checking the factual accuracy of the text. Yet, even so, because of the likelihood of error in sorting through many thousands of bits of information, I wish to issue an invitation — in fact, much more than merely an invitation, actually a request — to all readers to send along notices of error or inaccuracy. Indeed, physiologists understand how important feedback is to proper function of the human body; so, too, is feedback equally important for progressive development of a textbook of medical physiology. I hope also that those many persons who have helped already will accept my sincerest appreciation for their efforts.

A word of explanation is needed about two features of the text — first, the references, and, second, the two print sizes. The references have been chosen primarily for their presentation of physiological principles and for the quality of their own bibliographies. Use of these references as well as of cross-references from them can give the student almost complete coverage of the entire field of physiology.

The print is set in two sizes. The material presented in small print is of several different kinds; first, anatomical, chemical, and other information that is needed for the immediate discussion but that most students will learn in more detail in other courses; second, information that is of special importance to certain clinical fields of medicine but that is not necessary to understanding the body's basic physiologic mechanisms; and, third, information that will be of value to those students who wish to pursue a subject more deeply than the average medical student. In contrast, the material in large print constitutes the fundamental physiological information that students will require in their medical studies and that they will not obtain in other courses. Those teachers who would like to present a limited course on physiology can direct students' study primarily to the large type.

Again, I wish to express my deepest appreciation to many others who have helped in the preparation of this book. I am particularly grateful to Ivadelle Osberg Heidke and Gwendolyn Robbins for their excellent secretarial services, to Tomiko Mita, Michael Schenk, Tina Burnham, and Diane Fleming for their superb work and helpfulness on the illustrations, and to Martin J. Wonsiewicz, Rosanne Hallowell, Jessie Raymond, Peter Faber, Brett MacNaughton, and the staff of the W. B. Saunders Company for continued editorial and production excellence.

ARTHUR C. GUYTON

Contents

CHAPTER 12   *Electrocardiographic Interpretation of Cardiac Muscle and
Coronary Muscle and Coronary Abnormalities: Vectorial
Analysis* 124

UNIT IX   *THE NERVOUS SYSTEM: A. GENERAL PRINCIPLES AND SENSORY PHYSIOLOGY* 477

UNIT I

INTRODUCTION TO PHYSIOLOGY: THE CELL AND GENERAL PHYSIOLOGY

1 ● Functional Organization of the Human Body and Control of the "Internal Environment"

2 ● The Cell and Its Function

3 ● Genetic Control of Protein Synthesis, Cell Function, and Cell Reproduction

CHAPTER 1

• Functional Organization of the Human Body and Control of the "Internal Environment" •

Physiology attempts to explain the physical and chemical factors that are responsible for the origin, development, and progression of life. Each type of life, from the very simple virus up to the largest tree or to the complicated human being, has its own functional characteristics. Therefore, the vast field of physiology can be divided into *viral physiology, bacterial physiology, cellular physiology, plant physiology, human physiology,* and many more subdivisions.

Human Physiology. In human physiology, we are concerned with the specific characteristics and mechanisms of the human body that make it a living being. The very fact that we remain alive is almost beyond our own control, for hunger makes us seek food and fear makes us seek refuge. Sensations of cold make us provide warmth, and other forces cause us to seek fellowship and to reproduce. Thus, the human being is actually an automaton, and the fact that we are sensing, feeling, and knowledgeable beings is part of this automatic sequence of life; these special attributes allow us to exist under widely varying conditions that otherwise would make life impossible.

● CELLS AS THE LIVING UNITS OF THE BODY

The basic living unit of the body is the cell, and each organ is an aggregate of many different cells held together by intercellular supporting structures. Each type of cell is specially adapted to perform one particular function. For instance, the red blood cells, 25 trillion in all, transport oxygen from the lungs to the tissues. Although this type of cell is perhaps the most abundant, there are perhaps another 75 trillion cells.

The entire body, then, contains about 100 trillion cells.

Although the many cells of the body often differ markedly from each other, all of them have certain basic characteristics that are alike. For instance, in all cells oxygen combines with carbohydrate, fat, or protein to release the energy required for cell function. Furthermore, the general mechanisms for changing nutrients into energy are basically the same in all cells, and all the cells also deliver the end-products of their chemical reactions into the surrounding fluids.

Almost all cells also have the ability to reproduce, and whenever cells of a particular type are destroyed from one cause or another, the remaining cells of this type often regenerate new cells until the appropriate number is replenished.

● THE EXTRACELLULAR FLUID—THE INTERNAL ENVIRONMENT

About 56 per cent of the adult human body is fluid. Although most of this fluid is inside the cells and is called *intracellular fluid,* about one third is in the spaces outside the cells and is called *extracellular fluid.* This extracellular fluid is in constant motion throughout the body. It is rapidly transported in the circulating blood and then mixed between the blood and the tissue fluids by diffusion through the capillary walls. In the extracellular fluid are the ions and nutrients needed by the cells for maintenance of cellular life. Therefore, all cells live in essentially the same environment, the extracellular fluid, for which reason the extracellular fluid is called the *internal environment* of the body, or the *milieu intérieur,* a term introduced a hundred years ago by the great 19th century French physiologist Claude Bernard.

Cells are capable of living, growing, and performing their special functions so long as the proper concentrations of oxygen, glucose, different ions, amino acids, fatty substances, and other constituents are available in this internal environment.

Differences Between Extracellular and Intracellular Fluids. The extracellular fluid contains large amounts of *sodium, chloride,* and *bicarbonate ions,* plus nutrients for the cells, such as *oxygen, glucose, fatty acids,* and *amino acids.* It also contains carbon dioxide that is being transported from the cells to the lungs to be excreted and other cellular products that are being transported to the kidneys for excretion.

The intracellular fluid differs significantly from the extracellular fluid; particularly, it contains large amounts of *potassium, magnesium,* and *phosphate ions* instead of the sodium and chloride ions found in the extracellular fluid. Special mechanisms for transporting ions through the cell membranes maintain these differences. These mechanisms are discussed in Chapter 4.

● "HOMEOSTATIC" MECHANISMS OF THE MAJOR FUNCTIONAL SYSTEMS

HOMEOSTASIS

The term *homeostasis* is used by physiologists to mean *maintenance of static, or constant, conditions in the internal environment.* Essentially all the organs and tissues of the body perform functions that help to maintain these constant conditions. For instance, the lungs provide oxygen to the extracellular fluid to replenish continually the oxygen that is being used by the cells, the kidneys maintain constant ion concentrations, and the gastrointestinal system provides nutrients. A large segment of this text is concerned with the manner in which each organ or tissue contributes to homeostasis. To begin this discussion, the different functional systems of the body and their homeostatic mechanisms are outlined briefly; then the basic theory of the control systems that cause the functional systems to operate in harmony with each other is discussed.

THE EXTRACELLULAR FLUID TRANSPORT SYSTEM—THE CIRCULATORY SYSTEM

Extracellular fluid is transported through all parts of the body in two different stages. The first stage entails movement of blood around and around the circulatory system, and the second, movement of fluid between the blood capillaries and the cells. Figure 1–1

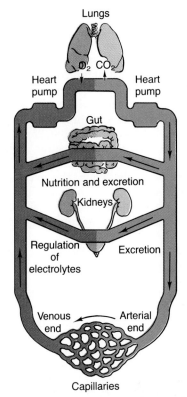

Figure 1–1. General organization of the circulatory system.

illustrates the overall circulation of blood. All the blood in the circulation traverses the entire circuit of the circulation an average of once each minute at rest and as many as six times each minute when a person becomes extremely active.

As blood passes through the capillaries, continual exchange of extracellular fluid occurs between the plasma portion of the blood and the interstitial fluid that fills the spaces between the cells, the *intercellular spaces.* This process is illustrated in Figure 1–2. Note that the capillaries are porous so that large amounts of

Figure 1–2. Diffusion of fluids through the capillary walls and through the interstitial spaces.

fluid and its dissolved constituents can *diffuse* back and forth between the blood and the tissue spaces, as illustrated by the arrows. This process of diffusion is caused by kinetic motion of the molecules in both the plasma and the interstitial fluid. That is, the fluid and dissolved molecules are continually moving and bouncing in all directions within the fluid itself and also through the pores and through the tissue spaces. Almost no cell is located more than 25 to 50 micrometers from a capillary, which insures diffusion of almost any substance from the capillary to the cell within a few seconds. Thus, the extracellular fluid everywhere in the body, both that of the plasma and that in interstitial spaces, is continually being mixed, thereby maintaining almost complete homogeneity.

ORIGIN OF NUTRIENTS IN THE EXTRACELLULAR FLUID

The Respiratory System. Figure 1–1 shows that each time the blood passes through the body it also flows through the lungs. The blood picks up oxygen in the alveoli, thus acquiring the *oxygen* needed by the cells. The membrane between the alveoli and the lumen of the pulmonary capillaries is only 0.4 to 2.0 micrometers in thickness, and oxygen diffuses through this membrane into the blood in exactly the same manner that water and ions diffuse through the tissue capillaries.

The Gastrointestinal Tract. A large portion of the blood pumped by the heart passes also through the walls of the gastrointestinal organs. Here different dissolved nutrients, including *carbohydrates, fatty acids, amino acids,* and others, are absorbed into the extracellular fluid.

The Liver and Other Organs That Perform Primarily Metabolic Functions. Not all substances absorbed from the gastrointestinal tract can be used in their absorbed form by the cells. The liver changes the chemical compositions of many of these to more usable forms, and other tissues of the body — the fat cells, the gastrointestinal mucosa, the kidneys, and the endocrine glands — help to modify the absorbed substances or store them until they are needed at a later time.

The Musculoskeletal System. Sometimes the question is asked: How does the musculoskeletal system fit into the homeostatic functions of the body? The answer to this is obvious and simple: Were it not for this system, the body could not move to the appropriate place at the appropriate time to obtain the foods required for nutrition. The musculoskeletal system also provides motility for protection against adverse surroundings, without which the entire body, and along with it all the homeostatic mechanisms, could be destroyed instantaneously.

REMOVAL OF METABOLIC END PRODUCTS

Removal of Carbon Dioxide by the Lungs. At the same time that blood picks up oxygen in the lungs, *carbon dioxide* is released from the blood into the alveoli, and the respiratory movement of air into and out of the alveoli carries the carbon dioxide to the atmosphere. Carbon dioxide is the most abundant of all the end products of metabolism.

The Kidneys. Passage of the blood through the kidneys removes most substances from the plasma that are not needed by the cells. These substances include especially different end products of cellular metabolism and excesses of ions and water that might have accumulated in the extracellular fluid. The kidneys perform their function by, first, filtering large quantities of plasma through the glomeruli into the tubules and then reabsorbing into the blood those substances needed by the body, such as glucose, amino acids, appropriate amounts of water, and many of the ions. However, most of the substances not needed by the body, especially the metabolic end products such as urea, are poorly reabsorbed and, instead, pass on through the renal tubules into the urine.

REGULATION OF BODY FUNCTIONS

The Nervous System. The nervous system is composed of three major parts: the *sensory portion,* the *central nervous system* (or *integrative portion*), and the *motor portion.* Sensory receptors detect the state of the body or the state of the surroundings. For instance, receptors present everywhere in the skin apprise one every time an object touches that person at any point. The eyes are sensory organs that give one a visual image of the surrounding area. The ears also are sensory organs. The central nervous system is composed of the brain and spinal cord. The brain can store information, generate thoughts, create ambition, and determine reactions the body performs in response to the sensations. Appropriate signals are then transmitted through the motor portion of the nervous system to carry out the person's desires.

A large segment of the nervous system is called the *autonomic system.* It operates at a subconscious level and controls many functions of the internal organs, including the action of the heart, the movements of the gastrointestinal tract, and the secretion by different glands.

The Hormonal System of Regulation. Located in the body are eight major endocrine glands that secrete chemical substances, the *hormones.* Hormones are transported in the extracellular fluid to all parts of the body to help regulate cellular function. For instance, thyroid hormone increases the rates of most chemical reactions in all cells. In this way thyroid hormone helps to set the tempo of bodily activity. Insulin controls glucose metabolism, adrenocortical hormones control ion and protein metabolism, and parathyroid hormone controls bone metabolism. Thus, the hormones are a system of regulation that complements the nervous system. The nervous system, in general, regulates mainly muscular and secretory activities of the body, whereas the hormonal system regulates mainly the metabolic functions.

REPRODUCTION

Reproduction sometimes is not considered to be a homeostatic function. However, reproduction does help to maintain static conditions by generating new beings to take the place of ones that are dying. This perhaps sounds like a permissive usage of the term homeostasis, but it does illustrate that, in the final analysis, essentially all structures of the body are so organized that they help maintain the automaticity and the continuity of life.

● THE CONTROL SYSTEMS OF THE BODY

The human body has literally thousands of control systems in it. The most intricate of all these are the genetic control systems that operate in all cells to control intracellular function and all extracellular functions as well. This subject is discussed in Chapter 3. Many other control systems operate within the organs to control functions of the individual parts of the organs; others operate throughout the entire body to control the interrelationships between the organs. For instance, the respiratory system, operating in association with the nervous system, regulates the concentration of carbon dioxide in the extracellular fluid. The liver and the pancreas regulate the concentration of glucose in the extracellular fluid. The kidneys regulate the concentrations of hydrogen, sodium, potassium, phosphate, and other ions in the extracellular fluid.

EXAMPLES OF CONTROL MECHANISMS

Regulation of Oxygen and Carbon Dioxide Concentrations in the Extracellular Fluid. Because oxygen is one of the major substances required for chemical reactions in the cells, it is fortunate that the body has a special control mechanism to maintain an almost exact and constant oxygen concentration in the extracellular fluid. This mechanism depends principally on the chemical characteristics of *hemoglobin*, which is present in all the red blood cells. Hemoglobin combines with oxygen as the blood passes through the lungs. Then, as the blood passes through the tissue capillaries, the hemoglobin will not release oxygen into the tissue fluid if too much oxygen is already there, but if the oxygen concentration is too little, sufficient oxygen will be released to re-establish an adequate tissue oxygen concentration. Thus, the regulation of oxygen concentration in the tissues is vested principally in the chemical characteristics of hemoglobin itself. This regulation is called the *oxygen-buffering function of hemoglobin*.

Carbon dioxide concentration in the extracellular fluid is regulated in quite a different way. Carbon dioxide is one of the major end-products of the oxidative reactions in cells. If all the carbon dioxide formed in the cells should continue to accumulate in the tissue fluids, the mass action of the carbon dioxide itself would soon halt all the energy-giving reactions of the cells. Fortunately, a nervous mechanism controls the expiration of carbon dioxide through the lungs and in this way maintains a constant and reasonable concentration of carbon dioxide in the extracellular fluid. That is, a high carbon dioxide concentration *excites the respiratory center,* causing the person to breathe rapidly and deeply. This increases the expiration of carbon dioxide and therefore increases its removal from the blood and the extracellular fluid, and the process continues until the concentration returns to normal.

Regulation of Arterial Pressure. Several different systems contribute to the regulation of arterial pressure. One of these, the *baroreceptor system,* is a very simple and excellent example of a control mechanism. In the walls of most of the great arteries of the upper body, especially in the bifurcation region of the carotids and the arch of the aorta, are many nerve receptors, called *baroreceptors,* which are stimulated by stretch of the arterial wall. When the arterial pressure becomes great, these baroreceptors are stimulated excessively, and impulses are transmitted to the medulla of the brain. Here the impulses inhibit the *vasomotor center,* which in turn decreases the number of impulses transmitted through the sympathetic nervous system to the heart and blood vessels. Lack of these impulses causes diminished pumping activity by the heart and increased ease of blood flow through the peripheral vessels, both of which lower the arterial pressure back toward normal. Conversely, a fall in arterial pressure relaxes the stretch receptors, allowing the vasomotor center to become more active than usual and thereby causing the arterial pressure to rise back toward normal.

Normal Ranges of Important Extracellular Fluid Constituents

Table 1–1 lists the more important constituents and physical characteristics of extracellular fluid along with their normal values, normal ranges, and maximum limits without causing death for short periods of time. Note especially the narrowness of the normal range for each one of these. Values outside these ranges are usually the cause of or the result of illness.

Even more important are the limits beyond which abnormalities can cause death. For instance, an increase in the body temperature of only 10° to 12°F (6° to 7°C) above normal can often lead to a vicious circle of increasing cellular metabolism that literally destroys the cells. Note also the very narrow range for the acid-base balance of the body, with a normal pH value of 7.4 and lethal values only about 0.5 on either side of the normal value. Another especially important factor is potassium ion, for whenever its concentration falls to less than one third normal, the person is likely to be paralyzed because of inability of the nerves to carry nerve signals, and if ever it rises to two

Table 1–1 SOME IMPORTANT CONSTITUENTS AND PHYSICAL CHARACTERISTICS OF THE EXTRACELLULAR FLUID, THE NORMAL RANGE OF CONTROL, AND THE APPROXIMATE NONLETHAL LIMITS

	Normal Value	Normal Range	Approximate Nonlethal Limits	Units
Oxygen	40	35–45	10–1000	mm Hg
Carbon dioxide	40	35–45	5–80	mm Hg
Sodium ion	142	138–146	115–175	mmol/L
Potassium ion	4.2	3.8–5.0	1.5–9.0	mmol/L
Calcium ion	1.2	1.0–1.4	0.5–2.0	mmol/L
Chloride ion	108	103–112	70–130	mmol/L
Bicarbonate ion	28	24–32	8–45	mmol/L
Glucose	85	75–95	20–1500	mg/dl
Body temperature	98.4 (37.0)	98–98.8 (37.0)	65–110 (18.3–43.3)	F° (C°)
Acid-base	7.4	7.3–7.5	6.9–8.0	pH

or more times normal, the heart muscle is likely to be depressed severely. Also, when the calcium ion concentration falls below about one half normal, the person is likely to experience tetanic contraction of muscles throughout the body because of spontaneous generation of nerve impulses in the peripheral nerves. When the glucose concentration falls below one half normal, the person frequently develops extreme mental irritability and sometimes even convulsions.

Thus, consideration of these examples should give one an extreme appreciation of the value and even necessity of the vast numbers of control systems that keep the body operating in health; in absence of any one of these controls, serious illness or death can result.

CHARACTERISTICS OF CONTROL SYSTEMS

The aforementioned examples of homeostatic control mechanisms are only a few of the many hundreds to thousands in the body, all of which have certain characteristics in common. These are explained in the following pages.

Negative Feedback Nature of Most Control Systems

Most control systems of the body act by a process of *negative feedback,* which can be explained best by reviewing some of the homeostatic control systems mentioned earlier. In the regulation of carbon dioxide concentration, a high concentration of carbon dioxide in the extracellular fluid causes increased pulmonary ventilation, and this in turn causes decreased carbon dioxide concentration because the lungs then excrete greater amounts of carbon dioxide out of the body. In other words, the high concentration causes a decreased concentration, which is *negative* to the initiating stimulus. Conversely, if the carbon dioxide concentration falls too low, this causes a feedback increase in the concentration. This response also is negative to the initiating stimulus.

In the arterial pressure–regulating mechanisms, a high pressure causes a series of reactions that promote a lowered pressure, or a low pressure causes a series of reactions that promote an elevated pressure. In both instances these effects are negative with respect to the initiating stimulus.

Therefore, in general, if some factor becomes excessive or too little, a control system initiates *negative feedback,* which consists of a series of changes that return the factor toward a certain mean value, thus maintaining homeostasis.

"Gain" of a Control System. The degree of effectiveness with which a control system maintains constant conditions is determined by the *gain* of the negative feedback. For instance, let us assume that a large volume of blood is transfused into a person whose baroreceptor pressure control system is not functioning, and the arterial pressure rises from the normal level of 100 mm Hg up to 175 mm Hg. Then, assume the same volume of blood is injected into the same person when the baroreceptor system is functioning, and this time the pressure rises only 25 mm Hg. Thus, the feedback control system has caused a "correction" of −50 mm Hg—that is, from 175 mm Hg to 125 mm Hg. Yet, there still remains an increase in pressure of +25 mm Hg, called the "error," which means that the control system is not 100 per cent effective in preventing change. The gain of the system is then calculated by the following formula:

$$\text{Gain} = \frac{\text{Correction}}{\text{Error}}$$

Thus, in the previous example, the correction is −50 mm Hg, and the error still persisting is +25 mm Hg. Therefore, the gain of the person's baroreceptor system for control of arterial pressure is −50 divided by +25, or −2. That is, an extraneous factor that tends to increase or decrease the arterial pressure does so only one third as much as would occur if this control system were not present.

The gains of some other physiological control systems are much greater than that of the baroreceptor system. For instance, the gain of the system controlling body temperature is approximately −33. Therefore, one can see that the temperature control system is much more effective than the baroreceptor pressure control system.

Positive Feedback — Vicious Circles and Death Caused by Positive Feedback

One might ask the question: Why do essentially all control systems of the body operate by negative feedback rather than positive feedback? However, if you will consider the nature of positive feedback, you will immediately see that positive feedback does not lead to stability but to instability and often to death.

Figure 1–3 illustrates an instance in which death can ensue from positive feedback. This figure depicts the pumping effectiveness of the heart, showing that the heart of the normal human being pumps about 5 liters of blood per minute. However, if the person is suddenly bled 2 liters, the amount of blood in the body is decreased to such a low level that not enough is available for the heart to pump effectively. As a result, the arterial pressure falls, and the flow of blood to the heart muscle through the coronary vessels also diminishes. This results in weakening of the heart, further diminished pumping, further decrease in coronary blood flow, and still more weakness of the heart; the cycle repeats itself again and again until death. Note that each cycle in the feedback results in further weakening of the heart. In other words, the initiating stimulus causes more of the same, which is *positive feedback*.

Positive feedback is better known as a "vicious circle," but actually a mild degree of positive feedback can be overcome by the negative feedback control mechanisms of the body, and a vicious circle will fail to develop. For instance, if the person in the aforementioned example were bled only 1 liter instead of 2 liters, the normal negative feedback mechanisms for controlling cardiac output and arterial pressure would overbalance the positive feedback, and the person would recover, as shown by the dashed curve of Figure 1–3.

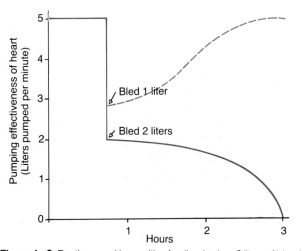

Figure 1–3. Death caused by positive feedback when 2 liters of blood are removed from the circulation.

Positive Feedback Can Sometimes Be Useful. In rare instances, the body has learned to use positive feedback to its advantage, as follows.

Blood clotting is an example of a very valuable use of positive feedback. When a blood vessel is ruptured and a clot begins to form, multiple enzymes called *clotting factors* are activated within the clot itself. Some of these enzymes act on other yet unactivated enzymes of the immediately adjacent blood, activating them and causing still more clot. This process continues until the hole in the vessel is plugged and bleeding no longer occurs. Unfortunately, on occasion, this mechanism can itself get out of hand and cause the formation of unwanted clots. In fact, this is what initiates most acute heart attacks, which are caused by a clot beginning on an atherosclerotic plaque in a coronary artery and then growing until the artery is completely blocked.

Childbirth is another instance in which positive feedback plays a role. When uterine contractions become strong enough for the baby's head to begin pushing through the cervix, stretch of the cervix sends signals through the uterine muscle back to the body of the uterus, causing even more powerful contractions. Thus, the uterine contractions stretch the cervix, and the cervical stretch causes more contractions. When this process becomes powerful enough, the baby is born. If it is not powerful enough, the contractions usually die out and wait until a few days later before beginning again.

Finally, another important use of positive feedback is for the generation of nerve signals. That is, when the membrane of a nerve fiber is stimulated, this causes slight leakage of sodium ions through sodium channels in the nerve membrane to the interior of the fiber. The sodium ions entering the fiber then change the membrane potential, which in turn causes more opening of channels, more change of potential, still more opening of channels, and so forth. Thus, from a slight beginning, there is an explosion of sodium leakage that creates the nerve action potential. This action potential in turn excites the nerve fiber still further along its length, with the process continuing until the nerve signal goes all the way to all ends of the nerve fiber.

Yet, we shall learn that in each one of these cases in which positive feedback is useful, the positive feedback itself is part of an overall negative feedback process. For instance, in the case of blood clotting, the positive feedback clotting process is a negative feedback process for maintenance of normal blood volume. And the positive feedback that causes nerve signals allows the nerves to participate in literally thousands of negative feedback nervous control systems.

Some More Complex Types of Control Systems — Adaptive Control System

Later in this text when we study the nervous system, we shall see that this system contains a morass of interconnected control mechanisms. Some of these are simple feedback systems similar to those that we have already discussed. However, many are not. For instance, some movements of the body occur so rapidly that there simply is not enough time for nerve signals to travel from the peripheral parts of the body all the way to the brain and then back to the periphery again in time to control the movements. Therefore, the brain uses a principle called *feed-forward control* to cause the required muscle contractions. Then, sensory nerve signals from the moving parts apprise the brain in retrospect whether or not the appropriate movement as envisaged by the brain actually has been performed correctly. If not, the brain corrects the feed-forward signals that it sends to the muscles the *next* time the movement is required. Then, once again, if still further correction needs to be made, this too will be done for subsequent movements. This is called *adaptive control.* Obviously, adaptive control, in a sense, is delayed negative feedback.

Thus, one can see how complex some of the feedback control systems of the body can be. The life of the person literally depends on all these. Therefore, a major share of this text is devoted to discussing these life-giving mechanisms.

● SUMMARY—AUTOMATICITY OF THE BODY

The purpose of this chapter has been to point out, first, the overall organization of the body and, second, the means by which the different parts of the body operate in harmony. To summarize, the body is actually a *social order of about 100 trillion cells* organized into different functional structures, some of which are called *organs.* Each functional structure provides its share in the maintenance of homeostatic conditions in the extracellular fluid, which is called the *internal environment.* As long as normal conditions are maintained in the internal environment, the cells of the body continue to live and function properly. Thus, each cell benefits from homeostasis, and in turn each cell contributes its share toward the maintenance of homeostasis. This reciprocal interplay provides continuous automaticity of the body until one or more functional systems lose their ability to contribute their share of function. When this happens, all the cells of the body suffer. Extreme dysfunction leads to death, whereas moderate dysfunction leads to sickness.

REFERENCES

Adolph, E. F.: Physiological integrations in action. The Physiologist, 25:(Suppl.) 1, 1982.
Adolph, E. F.: Physiological adaptations: Hypertrophies and superfunctions. Am. Sci., 60:608, 1972.
Bernard, C.: Lectures on the Phenomena of Life Common to Animals and Plants. Springfield, Ill., Charles C Thomas, 1974.
Brown, J. H. U. (ed.): Engineering Principles in Physiology. Vols. 1 and 2. New York, Academic Press, 1973.
Bruni, C., et al. (eds.): Systems Theory in Immunology. New York, Springer-Verlag, 1979.
Bryant, P. J., and Simpson, P.: Intrinsic and extrinsic control of growth in developing organs. Q. Rev. Biol., 59:387, 1984.
Burattini, R., and Borgdorff, P.: Closed-loop baroreflex control of total peripheral resistance in the cat: Identification of gains by aid of a model. Cardiovasc. Res., 18:715, 1984.
Cannon, W. B.: The Wisdom of the Body. New York, W. W. Norton & Co., 1932.
Frisancho, A. R.: Human Adaptation. St. Louis, C. V. Mosby Co., 1979.
Gann, D. S., et al.: Neural interaction in control of adrenocorticotropin. Fed. Proc., 44:161, 1985.
Guyton, A. C.: Arterial Pressure and Hypertension. Philadelphia, W. B. Saunders Co., 1980.
Guyton, A. C., and Coleman, T. G.: Quantitative analysis of the pathophysiology of hypertension. Circ. Res., 14:I-1, 1969.
Guyton, A. C., et al.: Dynamics and Control of the Body Fluids. Philadelphia, W. B. Saunders Co., 1975.
Huffaker, C. B. (ed.): Biological Control. New York, Plenum Press, 1974.
Iberall, A. S., and Guyton, A. C. (eds.): Proc. Int. Symp. on Dynamics and Controls in Physiological Systems. Regulation and Control in Physiological Systems. ISA, Pittsburgh, 1973.
Jones, R. W.: Principles of Biological Regulation: An Introduction to Feedback Systems. New York, Academic Press, 1973.
Klevecz, R. R., et al.: Cellular clocks and oscillators. Int. Rev. Cytol., 86: 97, 1984.
Krieger, D. T., and Aschoff, J.: Endocrine and other biological rhythms. In DeGroot, L. H., et al. (eds.): Endocrinology, Vol. 3. New York, Grune & Stratton, 1979, p. 2079.
McIntosh, J. E. A., and McIntosh, R. P.: Mathematical Modeling and Computers in Endocrinology. New York, Springer-Verlag, 1980.
Milhorn, H. T.: The Application of Control Theory to Physiological Systems. Philadelphia, W. B. Saunders Co., 1966.
Miller, S. L., and Orgel, L. E.: The Origins of Life on the Earth. Englewood Cliffs, N.J., Prentice-Hall, 1974.
Piva, F., et al.: Regulation of hypothalamic and pituitary function: Long, short, and ultrashort feedback loops. In DeGroot, L. J., et al. (eds.): Endocrinology. Vol. 1. New York, Grune & Stratton, 1979, p. 21.
Randall, J. E.: Microcomputers and Physiological Simulation. 2d Ed. New York, Raven Press, 1987.
Randall, J. E., Microcomputers and Physiological Simulation. Reading, Mass., Addison-Wesley Publishing Co., 1980.
Reeve, E. B., and Guyton, A. C.: Physical Bases of Circulatory Transport: Regulation and Exchange. Philadelphia, W. B. Saunders Co., 1967.
Rusak, B., and Zucker, I.: Neural regulation of circadian rhythms. Physiol. Rev., 59:449, 1979.
Stein, J. F.: Role of the cerebellum in the visual guidance of movement. Nature 323:217, 1986.
Sweetser, W.: Human Life (Aging and Old Age). New York, Arno Press, 1979.
Thompson, R. F.: The neurobiology of learning and memory. Science, 233:941, 1986.
Toates, F. M.: Control Theory in Biology and Experimental Psychology. London, Hutchinson Education Ltd., 1975.
Weston, L.: Body Rhythm: The Circadian Rhythms Within You. New York, Harcourt Brace Jovanovich, 1979.
Yates, F. E. (ed.): Self-Organizing Systems. New York, Plenum Publishing Corp., 1987.

CHAPTER 2

• *The Cell and Its Function* •

Each of the 75 to 100 trillion cells in the human being is a living structure that can survive indefinitely and, in most instances, can even reproduce itself, provided its surrounding fluids contain appropriate nutrients. To understand the function of organs and other structures of the body, it is essential that we first understand the basic organization of the cell and the functions of its component parts.

● ORGANIZATION OF THE CELL

A typical cell, as seen by the light microscope, is illustrated in Figure 2–1. Its two major parts are the *nucleus* and the *cytoplasm*. The nucleus is separated from the cytoplasm by a *nuclear membrane,* and the cytoplasm is separated from the surrounding fluids by a *cell membrane.*

The different substances that make up the cell are collectively called *protoplasm.* Protoplasm is composed mainly of five basic substances: water, electrolytes, proteins, lipids, and carbohydrates.

Water. The principal fluid medium of the cell is water, which is present in a concentration of between 70 and 85 per cent. Many cellular chemicals are dissolved in the water, whereas others are suspended in particulate or membranous form. Chemical reactions take place among the dissolved chemicals or at the surface boundaries between the suspended particles or membranes and the water.

Electrolytes. The most important electrolytes in the cell are *potassium, magnesium, phosphate, sulfate, bicarbonate,* and small quantities of *sodium, chloride,* and *calcium.* These are discussed in much greater detail in Chapter 4, which considers the interrelationships between the intracellular and extracellular fluids.

The electrolytes provide inorganic chemicals for cellular reactions. Also, they are necessary for operation of some of the cellular control mechanisms. For instance, electrolytes acting at the cell membrane allow transmission of electrochemical impulses in nerve and muscle fibers, and the intracellular electrolytes determine the activity of different enzymatically catalyzed reactions that are necessary for cellular metabolism.

Proteins. Next to water, the most abundant substance in most cells is proteins, which normally constitute 10 to 20 per cent of the cell mass. These can be divided into two different types, *structural proteins* and *globular proteins* that are mainly *enzymes.*

To get an idea of what is meant by *structural proteins,* one needs only to note that leather is composed principally of structural proteins and that hair is almost entirely a structural protein. Proteins of this type are present in the cell in the form of long thin filaments that themselves are polymers of many protein molecules. The most prominent use of such intracellular filaments is to provide the contractile mechanism of all muscles. However, filaments are also organized into microtubules that provide the "cytoskeletons" of such organelles as cilia and the mitotic spindles of mitosing cells. Extracellularly, fibrillar proteins are found especially in the collagen and elastin fibers of connective tissue, blood vessels, tendons, ligaments, and so forth.

The *globular proteins,* on the other hand, are an entirely different type of protein, composed usually of individual protein molecules or at most aggregates of a few molecules in a globular form rather than in a fibrillar form. These proteins are mainly the enzymes of the cell and, in contrast to the fibrillar proteins, are often soluble in the fluid of the cell or are integral parts of or adherent to membranous structures inside the cell. The enzymes come into direct contact with other substances inside the cell and catalyze chemical reactions. For instance, the chemical reactions that split glucose into its component parts and then combine these with oxygen to form carbon dioxide and water, while at the same time providing energy for cellular function, are catalyzed by a series of protein enzymes.

Lipids. Lipids are several different types of substances that are grouped together because of their common property of being soluble in fat solvents. The most important lipids in most cells are *phospholipids* and *cholesterol,* which constitute about 2 per cent of the total cell mass. The special importance of phospholipids and cholesterol is that they are mainly insoluble in water and therefore are used to form membranous barriers that separate the different intracellular compartments.

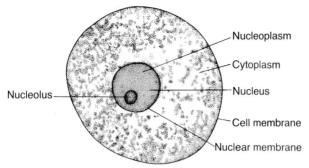

Figure 2–1. Structure of the cell as seen with the light microscope.

In addition to phospholipids and cholesterol, some cells contain large quantities of *triglycerides,* also called *neutral fat.* In the so-called fat cells, triglycerides often account for as much as 95 per cent of the cell mass. The fat stored in these cells represents the body's main storehouse of energy-giving nutrient that can later be dissoluted and used for energy wherever in the body it is needed.

Carbohydrates. In general, carbohydrates have very little structural function in the cell except as part of glycoprotein molecules, but they play a major role in nutrition of the cell. Most human cells do not maintain large stores of carbohydrates, usually averaging about 1 per cent of their total mass. However, carbohydrate, in the form of glucose, is always present in the surrounding extracellular fluid so that it is readily available to the cell. A small amount of carbohydrate is usually stored in the cells in the form of *glycogen,* which is an insoluble polymer of glucose and can be used rapidly to supply the cells' energy needs.

● PHYSICAL STRUCTURE OF THE CELL

The cell is not merely a bag of fluid, enzymes, and chemicals; it also contains highly organized physical structures, many of which are called *organelles,* and the physical nature of each of these is equally as important to the function of the cell as the cell's chemical constituents. For instance, without one of the organelles, the *mitochondria,* more than 95 per cent of the energy supply of the cell would cease immediately. Some principal organelles or structures of the cell are illustrated in Figure 2–2, including the *cell membrane, nuclear membrane, endoplasmic reticulum, Golgi apparatus, mitochondria, lysosomes,* and *centrioles.*

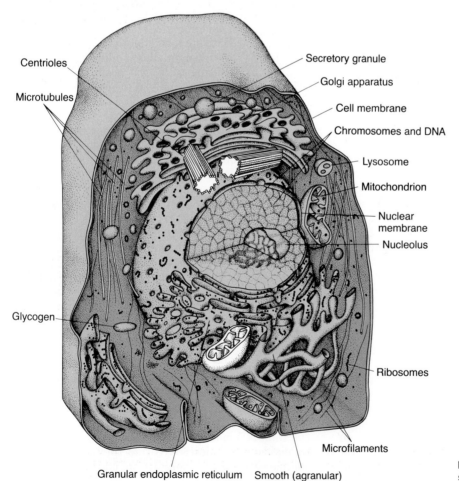

Figure 2–2. Reconstruction of a typical cell, showing the internal organelles in the cytoplasm and in the nucleus.

THE MEMBRANOUS STRUCTURES OF THE CELL

Essentially all organelles of the cell are lined by membranes composed primarily of lipids and proteins. These membranes include the *cell membrane,* the *nuclear membrane,* the *membrane of the endoplasmic reticulum,* and the *membranes of the mitochondria, lysosomes, Golgi apparatus* as well as still others. The lipids of the membranes provide a barrier that prevents free movement of water and water-soluble substances from one cell compartment to the other. The protein molecules in the membrane, on the other hand, often penetrate all the way through the membrane, thus interrupting the continuity of the lipid barrier, and therefore provide pathways for passage of specific substances through the membrane. Also, many of the membrane proteins are enzymes that catalyze a multitude of different chemical reactions, which are the subjects of further discussions in this and subsequent chapters.

The Cell Membrane

The cell membrane, which completely envelops the cell, is a very thin, elastic structure only 7.5 to 10 nanometers thick. It is composed almost entirely of proteins and lipids. The approximate composition is proteins, 55 per cent; phospholipids, 25 per cent; cholesterol, 13 per cent; other lipids, 4 per cent; and carbohydrates, 3 per cent.

The Lipid Barrier of the Cell Membrane. Figure 2–3 illustrates the cell membrane. Its basic structure is a *lipid bilayer,* which is a thin film of lipids only 2 molecules thick that is continuous over the entire cell surface. Interspersed in this lipid film are large globular protein molecules.

The lipid bilayer is composed almost entirely of phospholipids and cholesterol. One part of the phospholipid and the cholesterol molecules is soluble in water, that is, *hydrophilic,* whereas the other part is soluble only in fats, that is, *hydrophobic.* The phosphate radical of the phospholipid is hydrophilic, and the fatty acid radicals are hydrophobic. The cholesterol has a hydroxyl radical that is water soluble and a steroid nucleus that is fat soluble. Because the hydrophobic portions of both these molecules are repelled by water but are mutually attracted to each other, they have a natural tendency to line up, as illustrated in Figure 2–3, the fatty portions occupying the center of the membrane and the hydrophilic portions projecting to the two surfaces in contact with the surrounding water.

The membrane lipid bilayer is a major barrier impermeable to the usual water-soluble substances, such as ions, glucose, urea, and others. On the other hand, fat-soluble substances, such as oxygen, carbon dioxide, and alcohol, can penetrate this portion of the membrane with ease.

A special feature of the lipid bilayer is that it is a *fluid* and not a solid. Therefore, portions of the membrane can literally flow from one point to another along the surface of the membrane. Proteins or other substances dissolved in or floating in the lipid bilayer tend to diffuse to all areas of the cell membrane.

The Cell Membrane Proteins. Figure 2–3 illustrates globular masses floating in the lipid bilayer. These are membrane proteins, most of which are *glycoproteins.* Two types of proteins occur: the *integral proteins* that protrude all the way through the membrane and the *peripheral proteins* that are attached only to the surface of the membrane and do not penetrate.

Many of the integral proteins provide structural *channels* (or *pores*) through which water-soluble sub-

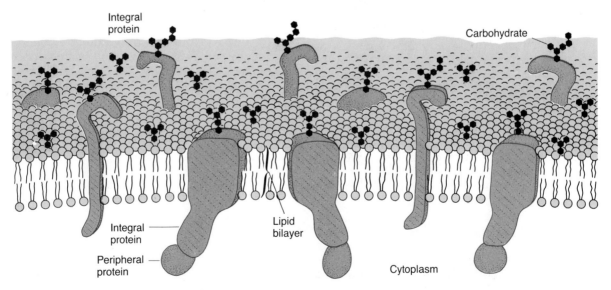

Figure 2–3. Structure of the cell membrane, showing that it is composed mainly from a lipid bilayer but with large numbers of protein molecules protruding through the layer. Also, carbohydrate moieties are attached to the protein molecules on the outside of the membrane and additional protein molecules on the inside. (From Lodish and Rothman: The assembly of cell membranes. *Sci. Am., 240:*48, 1979. ©1979 by Scientific American, Inc. All rights reserved.)

stances, especially the ions, can diffuse between the extracellular and intracellular fluid. However, these proteins have selective properties that cause preferential diffusion of some substances more than others. Others of the integral proteins act as *carrier proteins* for transporting substances in the direction opposite to their natural direction of diffusion, which is called "active transport." Still others act as *enzymes.*

The peripheral proteins occur either entirely or almost entirely on the inside of the membrane, and they are normally attached to one of the integral proteins. These peripheral proteins function almost entirely as enzymes.

The Membrane Carbohydrates—The Cell "Glycocalyx." The membrane carbohydrates occur almost invariably in combination with proteins and lipids in the form of *glycoproteins* and *glycolipids.* In fact, most of the integral proteins are glycoproteins, and about one tenth of the lipid molecules are glycolipids. The "glyco" portions of these molecules almost invariably protrude to the outside of the cell, dangling outward from the cell surface. Many other carbohydrate compounds, called *proteoglycans,* which are mainly carbohydrate substances bound together by small protein cores, are often loosely attached to the outer surface of the cell as well. Thus, the entire surface of the cell often has a loose carbohydrate coat called the *glycocalyx.*

The carbohydrate moieties attached to the outer surface of the cell have several important functions: (1) Many of them are negatively charged, which gives most cells an overall negative surface charge that repels other negative objects. (2) The glycocalyx of some cells attaches to the glycocalyx of other cells, thus attaching the cells to each other as well. (3) Many of the carbohydrates act as *receptor substances* for binding hormones like insulin and in doing so activate attached integral proteins that in turn also activate a cascade of intracellular enzymes. (4) Some enter into immune reactions, as we discuss in Chapter 34.

THE CYTOPLASM AND ITS ORGANELLES

The cytoplasm is filled with both minute and large dispersed particles and organelles, ranging in size from a few nanometers to many micrometers. The clear fluid portion of the cytoplasm in which the particles are dispersed is called *cytosol;* this contains mainly dissolved proteins, electrolytes, glucose, and minute quantities of lipid compounds.

The portion of the cytoplasm immediately beneath the cell membrane frequently contains an interwoven mat of microfilaments composed mainly of actin fibrillae. These provide a semisolid gel-like support for the cell membrane. This zone of the cytoplasm is called the *cortex,* or *ectoplasm.* The cytoplasm between the cortex and the nuclear membrane is liquefied and is called the *endoplasm.*

Dispersed in the cytoplasm are neutral fat globules, glycogen granules, ribosomes, secretory granules, and five especially important organelles: the *endoplasmic*

reticulum, the *Golgi apparatus,* the *mitochondria,* the *lysosomes,* and the *peroxisomes.*

The Endoplasmic Reticulum

Figure 2–2 illustrates in the cytoplasm a network of tubular and flat vesicular structures called the *endoplasmic reticulum.* The tubules and vesicles all interconnect with each other. Also, their walls are constructed of lipid bilayer membranes containing large amounts of proteins, similar to the cell membrane. The total surface area of this structure in some cells —the liver cells, for instance—can be as much as 30 to 40 times as great as the cell membrane area.

The detailed structure of a small portion of endoplasmic reticulum is illustrated in Figure 2–4. The space inside the tubules and vesicles is filled with *endoplasmic matrix,* a fluid medium that is different from the fluid outside the endoplasmic reticulum. Electron micrographs show that the space inside the endoplasmic reticulum is connected with the space between the two membranes of the double nuclear membrane.

Substances formed in some parts of the cell enter the space of the endoplasmic reticulum and are then conducted to other parts of the cell. Also, the vast surface area of the reticulum and multiple enzyme systems attached to its membranes provide the machinery for a major share of the metabolic functions of the cell.

Ribosomes and the Granular Endoplasmic Reticulum. Attached to the outer surfaces of many parts of the endoplasmic reticulum are large numbers of small granular particles called *ribosomes.* Where these are present, the reticulum is frequently called the *granular endoplasmic reticulum.* The ribosomes are composed of a mixture of ribonucleic acid (RNA) and proteins, and they function in the synthesis of protein in the cells, as discussed later in this and the following chapter.

The Agranular Endoplasmic Reticulum. Part of the endoplasmic reticulum has no attached ribosomes. This part is called the *agranular,* or *smooth, endoplasmic reticulum.* The agranular reticulum func-

Figure 2–4. Structure of the endoplasmic reticulum. (Modified from De Robertis, Saez, and De Robertis: Cell Biology. 6th ed. Philadelphia, W. B. Saunders Company, 1975.)

tions in the synthesis of lipid substances and in many other enzymatic processes of the cell.

Golgi Apparatus

The Golgi apparatus, illustrated in Figure 2–5, is closely related to the endoplasmic reticulum. It has membranes similar to those of the agranular endoplasmic reticulum. It is usually composed of four or more stacked layers of thin, flat enclosed vesicles lying near the nucleus. This apparatus is very prominent in secretory cells; in these it is located on the side of the cell from which the secretory substances are extruded.

The Golgi apparatus functions in association with the endoplasmic reticulum. As illustrated in Figure 2–5, small "transport vesicles," also called endoplasmic reticulum vesicles or simply _ER vesicles,_ continually pinch off from the endoplasmic reticulum and shortly thereafter fuse with the Golgi apparatus. In this way, substances are transported from the endoplasmic reticulum to the Golgi apparatus. The transported substances are then processed in the Golgi apparatus to form _lysosomes, secretory vesicles,_ or other cytoplasmic components to be discussed later in the chapter.

The Lysosomes

Lysosomes are vesicular organelles formed by the Golgi apparatus that then become dispersed throughout the cytoplasm. The lysosomes provide an intracellular digestive system that allows the cell to digest and thereby remove unwanted substances and structures, especially damaged or foreign structures, such as bacteria. The lysosome, illustrated in Figure 2–2, is quite different from one cell to another, but it is usually 250 to 750 nanometers in diameter. It is surrounded by a typical lipid bilayer membrane and is filled with large numbers of small granules 5 to 8 nanometers in diameter, which are protein aggregates of hydrolytic (digestive) enzymes. A hydrolytic enzyme is capable of splitting an organic compound into two or more parts by combining hydrogen from a water molecule with part of the compound and by combining the hydroxyl portion of the water molecule with the other part of the compound. For instance, protein is hydrolyzed to form amino acids, and glycogen is hydrolyzed to form glucose. More than 50 different _acid hydrolases_ have been found in lysosomes, and the principal substances that they digest are proteins, nucleic acids, mucopolysaccharides, lipids, and glycogen.

Ordinarily, the membrane surrounding the lysosome prevents the enclosed hydrolytic enzymes from coming in contact with other substances in the cell. However, many different conditions of the cell will break the membranes of some of the lysosomes, allowing release of the enzymes. These enzymes then split the organic substances with which they come in contact into small, highly diffusible substances, such as amino acids and glucose. Some of the more specific functions of lysosomes are discussed later in the chapter.

The Peroxisomes

Peroxisomes are similar physically to lysosomes, but they are different in two important ways: First, they are believed to be formed by budding off from the smooth endoplasmic reticulum rather than from the Golgi apparatus. Second, they contain oxidases rather than hydrolases. Several of the oxidases are capable of combining oxygen with hydrogen ions from different intracellular chemicals to form hydrogen peroxide (H_2O_2). The hydrogen peroxide in turn is itself a highly oxidizing substance, and this is used in association with _catalase,_ another oxidase enzyme present in large quantities in peroxisomes, to oxidize many substances that might otherwise be poisonous to the cell. For instance, most alcohol that a person drinks is detoxified by the peroxisomes of the liver cells in this manner. The hydrogen peroxide-catalase–oxidizing mechanism is also used for specific functional purposes in the cell, such as catalyzing the breakdown of fatty acids into acetyl-CoA that is then used for energy by the cell.

Secretory Vesicles

One of the important functions of many cells is secretion of special substances. Almost all such secretory substances are formed by the endoplasmic reticulum–Golgi apparatus system and are then released from the Golgi apparatus into the cytoplasm inside storage vesicles called _secretory vesicles_ or _secretory granules._ Figure 2–6 illustrates typical secretory vesicles inside pancreatic acinar cells storing protein proenzymes (enzymes that are not yet activated); the proenzymes are secreted later through the outer cell membrane into the pancreatic duct and thence into the duodenum, where they are activated and perform their digestive functions.

Golgi vesicles

Golgi apparatus

ER vesicles

Endoplasmic reticulum

Figure 2–5. A typical Golgi apparatus and its relationship to the endoplasmic reticulum and the nucleus.

Figure 2–6. Secretory granules in acinar cells of the pancreas.

The Mitochondria

The mitochondria are called the "powerhouses" of the cell. Without them, the cells would be unable to extract significant amounts of energy from the nutrients and oxygen, and as a consequence essentially all cellular functions would cease. As illustrated in the cell of Figure 2–2, these organelles are present in essentially all portions of the cytoplasm, but the total number per cell varies from less than a hundred up to several thousand, depending on the amount of energy required by each cell. Furthermore, the mitochondria are concentrated in those portions of the cell that are responsible for the major share of its energy metabolism. Mitochondria are also very variable in size and shape; some are only a few hundred nanometers in diameter and globular in shape, whereas others are as large as 1 micrometer in diameter, as long as 7 micrometers, or branching and filamentous in shape.

The basic structure of the mitochondrion is illustrated in Figure 2–7, which shows it to be composed mainly of two lipid bilayer–protein membranes: an *outer membrane* and an *inner membrane*. Many infoldings of the inner membrane form *shelves* onto which oxidative enzymes are attached. In addition, the inner cavity of the mitochondrion is filled with a *matrix* containing large quantities of dissolved en-

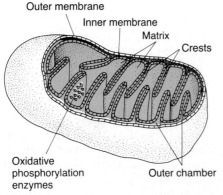

Figure 2–7. Structure of a mitochondrion. (Modified from De Robertis, Saez, and De Robertis: Cell Biology. 6th ed. Philadelphia, W. B. Saunders Company, 1975.)

zymes that are necessary for extracting energy from nutrients. These enzymes operate in association with the oxidative enzymes on the shelves to cause oxidation of the nutrients, thereby forming carbon dioxide and water. The liberated energy is used to synthesize a high-energy substance called *adenosine triphosphate (ATP)*. ATP is then transported out of the mitochondrion, and it diffuses throughout the cell to release its energy wherever it is needed for performing cellular functions. The details of ATP formation by the mitochondrion are given in Chapter 67, and some of the important functions of ATP in the cell are introduced later in this chapter.

Mitochondria are self-replicative, which means that one mitochondrion can form a second one, a third one, and so on, whenever there is need in the cell for increased amounts of ATP. Indeed, the mitochondria contain *deoxyribonucleic acid (DNA)* similar to that found in the nucleus. In the following chapter we see that DNA is the basic substance of the nucleus that controls replication of the cell. This substance plays a similar role in the mitochondrion but not entirely so because in the mitochondrion replication process, many proteins and lipids that have already been formed in the cytoplasm are incorporated into the mitochondria as they enlarge and bud off to form new mitochondria.

Filament and Tubular Structures of the Cell

The fibrillar proteins of the cell are usually organized into filaments or tubules. These originate as precursor protein molecules synthesized by ribosomes that at first are present in dissolved form in the cytoplasm. Then, they polymerize to form *filaments*. We have already pointed out that large numbers of actin filaments frequently occur in the outer zone of the cytoplasm, the zone called the *ectoplasm*, to form an elastic support for the cell membrane. Also, in muscle cells filaments are organized into a special contractile machine that is the basis of muscle contraction throughout the body, as is discussed in detail in Chapter 6.

A special type of filament composed of polymerized *tubulin* molecules is used in all cells to construct tubular structures, the *microtubules*. Almost invariably these contain 13 tubulin *protofilaments* lying parallel to each other in a circle to form a long hollow cylinder about 25 nanometers in diameter and 1 to many micrometers in length. These cylinders are often arranged in bundles that gives them, *en masse*, considerable structural strength. However, microtubules are stiff structures that break if bent too severely. Figure 2–8 illustrates typical microtubules that were teased from the flagellum of a sperm.

Another example of microtubules is the tubular mechanical structure of cilia that gives them structural strength, radiating upward from the cell cytoplasm to the tip of the cilium. Also, the *centrioles* and

Figure 2–8. Microtubules teased from the flagellum of a sperm. (From Porter: Ciba Foundation Symposium: Principles of Biomolecular Organization. Boston, Little, Brown & Company, 1966.)

the *mitotic spindle* of the mitosing cell are both composed of stiff microtubules.

Thus, a primary function of microtubules is to act as a *cytoskeleton,* providing rigid physical structures for certain parts of cells. Also, the cytoplasm often *streams* (flows) in the vicinity of microtubules, which might result from movement of arms that project outward from the microtubules.

THE NUCLEUS

The nucleus is the control center of the cell. Briefly, the nucleus contains large quantities of *DNA,* which we have called *genes* for many years. The genes determine the characteristics of the protein enzymes of the cytoplasm and in this way control cytoplasmic activities. They also control reproduction; the genes first reproduce themselves, and after this the cell splits by a special process called *mitosis* to form two daughter cells, each of which receives one of the two sets of genes. All these activities of the nucleus are considered in detail in the following chapter.

The appearance of the nucleus under the microscope does not give much of a clue to the mechanisms by which it performs its control activities. Figure 2–9 illustrates the light microscopic appearance of the interphase nucleus (period between mitoses), showing darkly staining *chromatin material* throughout the nucleoplasm. During mitosis, the chromatin material becomes readily identifiable as the highly structured *chromosomes,* which can be seen easily with the light microscope, as illustrated in the next chapter.

The Nuclear Envelope

The nuclear envelope is frequently called the *nuclear membrane.* However, it is actually two separate membranes, one inside the other. The outer membrane is continuous with the endoplasmic reticulum, and the space between the two nuclear membranes is also continuous with the compartment inside the endoplasmic reticulum.

The nuclear envelope is penetrated by several thousand *nuclear pores.* These are very large, almost 100 nanometers in diameter. However, large complexes of protein molecules are attached around the edges of the pores so that the central area of the pore is only about 9 nanometers in diameter. Even this size is large enough to allow some molecules up to 44,000 molecular weight to pass through with relative ease and molecules with molecular weight less than 15,000 to pass extremely rapidly.

Nucleoli

The nuclei of most cells contain one or more lightly staining structures called nucleoli. The nucleolus, unlike most of the organelles that we have discussed,

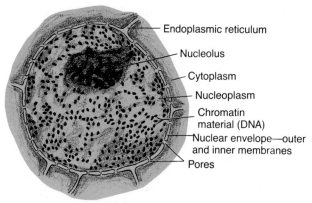

— Endoplasmic reticulum

— Nucleolus

— Cytoplasm

— Nucleoplasm

— Chromatin material (DNA)

—Nuclear envelope—outer and inner membranes

—Pores

Figure 2–9. Structure of the nucleus.

does not have a limiting membrane. Instead, it is simply a structure that contains a large amount of *RNA* and proteins of the types found in ribosomes. The nucleolus becomes considerably enlarged when a cell is actively synthesizing proteins. The genes of five separate chromosome pairs synthesize the RNA and then store it in the nucleolus, beginning with a loose fibrillar RNA that later condenses to form granular "subunits" of ribosomes. These in turn are transported through the nuclear membrane pores into the cytoplasm, where they assemble together to form the "mature" ribosomes that play an essential role for the formation of proteins, either in the cytosol or in association with the endoplasmic reticulum, as we discuss more fully in the following chapter.

● COMPARISON OF THE ANIMAL CELL WITH PRECELLULAR FORMS OF LIFE

Many of us think of the cell as the lowest level of life. However, the cell is a very complicated organism, which required many hundreds of million years to develop after the earliest form of life, an organism similar to the present-day *virus,* first appeared on earth. Figure 2–10 illustrates the relative sizes of the smallest known virus, a large virus, a *rickettsia,* a *bacterium,* and a nucleated cell, showing that the cell has a diameter about 1000 times that of the smallest virus and, therefore, a volume about 1 billion times that of the smallest virus. Correspondingly, the functions and anatomical organization of the cell are also far more complex than those of the virus.

The essential life-giving constituent of the very small virus is a *nucleic acid* embedded in a coat of protein. This nucleic acid is composed of the same basic constituents (DNA and RNA) as found in mammalian cells, and it is capable of reproducing itself if appropriate conditions are available. Thus, the virus is capable of propagating its lineage from generation to generation and, therefore, is a living structure in the same way that the cell and the human being are living structures.

As life evolved, other chemicals besides nucleic acid and simple proteins became integral parts of the organism, and specialized functions began to develop in different parts of the virus. A membrane formed around the virus, and inside the membrane a fluid matrix appeared. Specialized chemicals developed inside the matrix to perform special functions; many protein enzymes appeared, which were capable of catalyzing chemical reactions and, therefore, of determining the organism's activities.

In still later stages, particularly in the rickettsial and bacterial stages, *organelles* developed inside the organism, representing physical structures of chemical aggregates that perform functions in a more efficient manner than can be achieved by dispersed chemicals throughout the fluid matrix. Finally, in the nucleated cell, still more complex organelles developed, the most important of which is the *nucleus* itself. The nucleus distinguishes this type of cell from all lower forms of life; this structure provides a control center for all cellular activities, and it provides for very exact reproduction of new cells generation after generation, each new cell having essentially the same structure as its progenitor.

● FUNCTIONAL SYSTEMS OF THE CELL

In the remainder of this chapter, we discuss several representative functional systems of the cell that make it a living organism.

INGESTION BY THE CELL— ENDOCYTOSIS

If a cell is to live and grow, it must obtain nutrients and other substances from the surrounding fluids. Most substances pass through the cell membrane by *diffusion* and *active transport,* which are considered in detail in Chapter 4. However, large particles enter the cell by a specialized function of the cell membrane called *endocytosis.* The two principal forms of endocytosis are *pinocytosis* and *phagocytosis.* Pinocytosis means the ingestion of extremely small vesicles containing extracellular fluid. Phagocytosis means ingestion of large particles, such as bacteria, cells, or portions of degenerating tissue.

Pinocytosis. Pinocytosis occurs continually at the cell membranes of most cells but especially rapidly in some cells. For instance, it occurs so rapidly in macrophages that about 3 per cent of the total macrophage membrane is engulfed in the form of vesicles each minute. Even so, the pinocytic vesicles are so small, usually only 100 to 200 nanometers in diameter, that most of them can be seen only with the electron microscope.

Pinocytosis is the only means by which some very large macromolecules, such as most protein molecules, can enter cells. In fact, the rate at which pinocytic vesicles form usually is enhanced when such macromolecules attach to the cell membrane.

Figure 2–11 illustrates the successive steps of pinocytosis, showing three molecules of protein attaching to the membrane. These molecules usually attach to *receptors* on the surface of the membrane that are specific for the types of proteins that are to be absorbed. The receptors generally are concentrated in

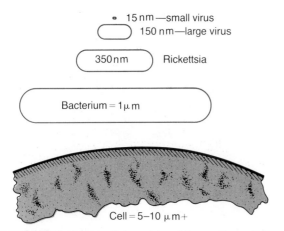

15 nm —small virus
150 nm —large virus
350 nm — Rickettsia
Bacterium = 1 μm
Cell = 5–10 μm +

Figure 2–10. Comparison of sizes of precellular organisms with that of the average cell in the human body.

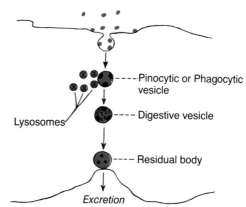

Figure 2–11. Mechanism of pinocytosis.

small pits in the cell membrane, called *coated pits*. On the inside of the cell membrane beneath these pits is a latticework of a fibrillar protein called *clathrin* as well as additional contractile filaments of *actin* and *myosin*. Once the protein molecules have bound with the receptors, the surface properties of the membrane change in such a way that the entire pit invaginates inward and the contractile proteins cause its borders to close over the attached proteins as well as over a small amount of extracellular fluid. Immediately thereafter, the invaginated portion of the membrane breaks away from the surface of the cell, forming a *pinocytic vesicle.*

What causes the cell membrane to go through the necessary contortions for forming pinocytic vesicles remains mainly a mystery. However, this process requires energy from within the cell; this is supplied by ATP, a high energy substance that is discussed later in the chapter. Also, it requires the presence of calcium ions in the extracellular fluid, which probably react with the contractile filaments beneath the coated pits to provide the force for pinching the vesicles away from the cell membrane.

Phagocytosis. Phagocytosis occurs in much the same way as pinocytosis except that it involves large particles rather than molecules. Only certain cells have the capability of phagocytosis, most notably the tissue macrophages and some of the white blood cells.

Phagocytosis is initiated when proteins or large polysaccharides on the surface of the particle that is to be phagocytized — such as a bacterium, a dead cell, or other tissue debris — bind with receptors on the surface of the phagocyte. In the case of bacteria, these usually are attached to specific antibodies, and the antibodies in turn attach to the phagocyte receptors. This intermediation of antibodies is called *opsonization,* which is discussed in Chapters 33 and 34.

Phagocytosis occurs in the following steps:

1. The cell membrane receptors attach to the surface ligands of the particle.

2. The edges of the membrane around the points of attachment evaginate outward within a fraction of a second to surround the particle; then, progressively more and more membrane receptors attach to the particle ligands, all this occurring suddenly in a zipperlike fashion.

3. Actin and other contractile fibrils in the cytoplasm surround the engulfed particle and contract around its outer edge, pushing the object further to the interior.

4. The contractile proteins then pinch the *phagocytic vesicle* off, leaving it in the cell interior in the same way that pinocytic vesicles are formed.

DIGESTION OF FOREIGN SUBSTANCES IN THE CELL—FUNCTION OF THE LYSOSOMES

Almost immediately after a pinocytic or phagocytic vesicle appears inside a cell, one or more lysosomes become attached to the vesicle and empty their acid hydrolases into the vesicle, as illustrated in Figure 2–12. Thus, a *digestive vesicle* is formed in which the hydrolases begin hydrolyzing the proteins, glycogen, nucleic acids, mucopolysaccharides, and other substances in the vesicle. The products of digestion are small molecules of amino acids, glucose, phosphates, and so forth that can then diffuse through the membrane of the vesicle into the cytoplasm. What is left of the digestive vesicle, called the *residual body,* represents the undigestible substances. In most instances this is finally excreted through the cell membrane by a process called *exocytosis,* which is essentially the opposite of endocytosis.

Figure 2–12. Digestion of substances in pinocytic vesicles by enzymes derived from lysosomes.

Thus, the lysosomes may be called the *digestive organs* of the cells.

Regression of Tissues and Autolysis of Cells. Often, tissues of the body regress to smaller size. For instance, this occurs in the uterus following pregnancy, in muscles during long periods of inactivity, and in mammary glands at the end of the period of lactation. Lysosomes are responsible for much of this regression. However, the mechanism by which lack of activity in a tissue causes the lysosomes to increase their activity is still unknown.

Another very special role of the lysosomes is the removal of damaged cells or damaged portions of cells from tissues — cells damaged by heat, cold, trauma, chemicals, or any other factor. Damage to the cell causes lysosomes to rupture, and the released hydrolases begin immediately to digest the surrounding organic substances. If the damage is slight, only a portion of the cell will be removed, followed by repair of the cell. However, if the damage is severe, the entire cell will be digested, a process called *autolysis.* In this way, the cell is completely removed, and a new cell of the same type ordinarily is formed by mitotic reproduction of an adjacent cell to take the place of the old one.

The lysosomes also contain bactericidal agents that can kill phagocytized bacteria before they can cause cellular damage. These agents include *lysozyme* that dissolves the bacterial cell membrane, *lysoferrin* that binds iron and other metals that are essential for bacterial growth, and acid at a pH of about 5.0 that activates the hydrolases and also inactivates some of the bacterial metabolic systems.

Also in lysosomes are stored enzymes that can begin the digestion of lipid aggregates and glycogen granules, making the lipid and glycogen available for use elsewhere in the cell or elsewhere in the body. In the absence of these enzymes, which results from occasional genetic disorders, extreme quantities of lipids or of glycogen often accumulate in the cells of many organs, especially the liver, and lead to early death.

SYNTHESIS AND FORMATION OF CELLULAR STRUCTURES BY THE ENDOPLASMIC RETICULUM AND THE GOLGI APPARATUS

The extensiveness of the endoplasmic reticulum and the Golgi apparatus, especially in secretory cells, has already been emphasized. These two structures are formed primarily of lipid bilayer membranes, and their walls are literally loaded with protein enzymes that catalyze the synthesis of many of the substances required by the cell.

In general, most of the synthesis begins in the endoplasmic reticulum, but most of the products that are formed are then passed on to the Golgi apparatus, where they are further processed prior to release into the cytoplasm. But, first, let us note the specific products that are synthesized in the specific portions of the endoplasmic reticulum and the Golgi apparatus.

Formation of Proteins by the Granular Endoplasmic Reticulum. The granular endoplasmic reticulum is characterized by the presence of large numbers of ribosomes attached to the outer surfaces of the reticulum membrane. As we discuss in the following chapter, protein molecules are synthesized within the structures of the ribosomes. Furthermore, the ribosomes extrude many of the synthesized protein molecules not into the cytosol but instead through the endoplasmic reticulum wall into the endoplasmic matrix.

Almost as rapidly as the protein molecules enter the endoplasmic matrix, enzymes in the endoplasmic reticulum wall cause rapid changes in these molecules. First, almost all of them are immediately glycosylated, that is, conjugated with carbohydrate moieties, to form *glycoproteins.* Therefore, essentially all the endoplasmic proteins are glycoproteins, in contrast to the proteins that are formed by the ribosomes in the cytosol, which are mainly free proteins. Second, the proteins are cross-linked and folded to form more compact molecules.

Synthesis of Lipids by the Endoplasmic Reticulum, Especially by the Smooth Endoplasmic Reticulum. The endoplasmic reticulum also synthesizes lipids, especially phospholipids and cholesterol. These are rapidly incorporated into the lipid bilayer of the endoplasmic reticulum itself, thus causing the endoplasmic reticulum to grow continually. This occurs mainly in the smooth portion of the endoplasmic reticulum.

To keep the endoplasmic reticulum from growing beyond the limits of the cell, however, small vesicles called *endoplasmic reticulum vesicles,* or *transport vesicles,* continually break away from the smooth reticulum; we see later that most of these vesicles migrate rapidly to the Golgi apparatus.

Other Functions of the Endoplasmic Reticulum. Other significant functions of the endoplasmic reticulum, again especially the smooth reticulum, are:

1. It contains the enzymes that control glycogen breakdown when glycogen is to be used for energy.

2. It contains a vast number of enzymes that are capable of detoxifying substances that are damaging to the cell, such as drugs. It achieves this by coagulation, oxidation, hydrolysis, conjugation with glycuronic acid, and in other ways.

Synthetic Functions of the Golgi Apparatus. Although the major function of the Golgi apparatus is to process substances already formed in the endoplasmic reticulum, it also has the capability of synthesizing certain carbohydrates that cannot be formed in the endoplasmic reticulum. This is especially true of sialic acid and galactose. In addition, the Golgi apparatus can cause the formation of very large saccharide polymers bound with only small amounts of protein; the most important of these are hyaluronic acid and chondroitin sulfate. A few of the many functions of hyaluronic acid and chondroitin sulfate in the body are as follows: (1) They are the major components of proteoglycans secreted in mucus and other glandular secretions. (2) They are the major components of the ground substance in the interstitial spaces, acting as a filler between collagen fibers and cells. (3) They are principal components of the organic matrix in both cartilage and bone.

Processing of Endoplasmic Secretions by the Golgi Apparatus — Formation of Vesicles. Figure 2–13 summarizes the major functions of the endoplasmic reticulum and Golgi apparatus. As substances are formed in the endoplasmic reticulum, especially the proteins, they are transported through the tubules toward the portions of the smooth endoplasmic reticulum that lie nearest the Golgi apparatus. At this point, small transport vesicles of smooth endoplasmic reticulum continually break away and diffuse to the *deeper layers* of the Golgi apparatus. Inside these vesicles are the synthesized proteins and other products. These vesicles instantly fuse with the Golgi apparatus and empty their contained substances into the vesicular spaces of the Golgi apparatus. Here, additional carbohydrate moieties are added to the secretions. Also, a most important function of the Golgi apparatus is to compact the endoplasmic reticular secretions into highly concentrated packets. As the secretions pass toward the outermost layers of the Golgi apparatus, the compaction and processing proceed; finally, both small and large vesicles continually break away from the Golgi apparatus, carrying with them the compacted secretory substances, and they then diffuse throughout the cell.

To give one an idea of the timing of these processes: When a glandular cell is bathed in radioactive amino acids, newly formed radioactive protein molecules can be detected in the granular endoplasmic reticulum within 3 to 5 minutes. Within 20 minutes, the newly formed proteins are present in the Golgi apparatus, and within 1 or 2 hours, radioactive proteins are secreted from the surface of the cell.

Types of Vesicles Formed by the Golgi Apparatus — Secretory Vesicles and Lysosomes. In a highly secretory cell, the vesicles that are formed by the Golgi apparatus are mainly *secretory vesicles,* containing especially the protein substances that are to be secreted through the surface of the cell.

These vesicles diffuse to the cell membrane then fuse with it and empty their substances to the exterior by a mechanism called *exocytosis,* which is essentially the opposite of endocytosis. Exocytosis, in most cases, is stimulated by entry of calcium ions into the cell; the calcium ions interact with the vesicular membrane, in some way not understood, to cause its fusion with the cell membrane.

On the other hand, some of the vesicles are destined for intracellular use. For instance, specialized portions of the Golgi apparatus form the *lysosomes* that have already been discussed. It is believed that the membranes of these specialized portions contain chemical receptors that cause acid hydrolases to attach. In this way these enzymes are concentrated and then released from the Golgi apparatus in the form of the lysosomal vesicles.

Another type of vesicle formed in the same way as the lysosomes are the *peroxisomes.* However, it is believed that these are formed by the smooth endoplasmic reticulum, along with formation of the transport vesicles, rather than by the Golgi apparatus. Here again, special receptors in the endoplasmic reticulum membrane probably attract and attach the oxidative enzymes that are to be released in concentrated form in the peroxisomes.

Use of Intracellular Vesicles to Replenish Cellular Membranes. Many of the vesicles finally fuse with the cell membrane or with the membranes of other intracellular structures, such as the mitochondria and even the endoplasmic reticulum itself. This obviously increases the expanse of these membranes and thereby replenishes these membranes as they themselves are destroyed. For instance, the cell membrane loses much of its substance every time it forms a phagocytic or pinocytic vesicle, and it is vesicles from the Golgi apparatus that continually replenish the cell membrane.

Thus, in summary, the membranous system of the endoplasmic reticulum and the Golgi apparatus represents a highly metabolic organ capable of forming both new cellular structures and secretory substances to be extruded from the cell.

Figure 2–13. Formation of proteins, lipids, and cellular vesicles by the endoplasmic reticulum and Golgi apparatus.

EXTRACTION OF ENERGY FROM NUTRIENTS — FUNCTION OF THE MITOCHONDRIA

The principal substances from which cells extract energy are oxygen and one or more of the foodstuffs — carbohydrates, fats, and proteins. In the human body, essentially all carbohydrates are converted into *glucose* before they reach the cell, the proteins are converted into *amino acids,* and the fats are converted into *fatty acids.* Figure 2–14 shows oxygen and the foodstuffs — glucose, fatty acids, and amino acids — all entering the cell. Inside the cell, the foodstuffs react chemically with the oxygen under the influence of various enzymes that control their rates of reactions and channel the energy that is released in the proper direction.

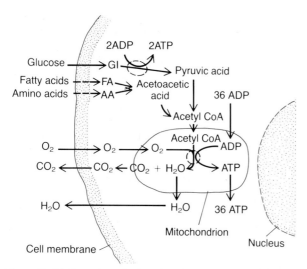

Figure 2–14. Formation of adenosine triphosphate (ATP) in the cell, showing that most of the ATP is formed in the mitochondria.

Almost all these oxidative reactions occur inside the mitochondria, and the energy that is released is used to form mainly *ATP*. Then, the ATP, not the original foodstuffs themselves, is used throughout the cell to energize almost all the intracellular metabolic reactions.

Functional Characteristics of ATP

The formula for ATP is:

Adenosine triphosphate

ATP is a nucleotide composed of the nitrogenous base *adenine,* the pentose sugar *ribose,* and three *phosphate radicals.* The last two phosphate radicals are connected with the remainder of the molecule by so-called *high-energy phosphate bonds,* which are represented by the symbol ~. Each of these bonds contains about 12,000 calories of energy per mole of ATP under the physical conditions of the body (7300 calories under standard conditions), which is much greater than the energy stored in the average chemical

bond of other organic compounds, thus giving rise to the term "high-energy" bond. Furthermore, the high-energy phosphate bond is very labile so that it can be split instantly on demand whenever energy is required to promote other cellular reactions.

When ATP releases its energy, a phosphoric acid radical is split away, and *adenosine diphosphate (ADP)* is formed. Then, energy derived from the cellular nutrients causes the ADP and phosphoric acid to recombine to form new ATP, the entire process continuing over and over again. For these reasons, ATP has been called the *energy currency* of the cell, for it can be spent and remade again and again, usually having a turnover time of only a few minutes at most.

Chemical Processes in the Formation of ATP —Role of the Mitochondria. On entry into the cells, glucose is subjected to enzymes in the cytoplasm that convert it into *pyruvic acid* (a process called *glycolysis*). A small amount of ADP is changed into ATP by energy released during this conversion, but this amount accounts for less than 5 per cent of the overall energy metabolism of the cell.

By far the major portion of the ATP formed in the cell is formed in the mitochondria. The pyruvic and fatty acids and most of the amino acids are all converted into the compound *acetyl-CoA* in the matrix of the mitochondrion. This substance, in turn, is acted upon by another series of enzymes in the mitochondrion matrix, undergoing dissolution in a sequence of chemical reactions called the *citric acid cycle,* or *Krebs cycle.* These chemical reactions are explained in detail in Chapter 67.

In the citric acid cycle, acetyl-CoA is split into its component parts, *hydrogen atoms* and *carbon dioxide.* The carbon dioxide, in turn, diffuses out of the mitochondria and eventually out of the cell.

The hydrogen atoms, on the other hand, are very highly reactive, and they eventually combine with oxygen that has diffused into the mitochondria. This releases a tremendous amount of energy, which is used by the mitochondria to convert large amounts of ADP to ATP. The processes of these reactions are very complex, requiring the participation of large numbers of protein enzymes that are integral parts of the mitochondrial *membranous shelves* that protrude into the mitochondrial matrix. The initial event is removal of an electron from the hydrogen atom, thus converting it to a hydrogen ion. The terminal event is movement of these ions through large globular proteins called *ATP synthetase* that protrude like knobs through the membranes of the mitochondrial shelves. ATP synthetase is an enzyme that uses the energy from the movement of the hydrogen ions to cause the conversion of ADP to ATP, while at the same time the hydrogen ions combine with oxygen to form water. Finally, the newly formed ATP is transported out of the mitochondria into all parts of the cytoplasm and nucleoplasm, where its energy is used to energize the functions of the cell.

This overall process for the formation of ATP is called the *chemiosmotic mechanism* of ATP formation. The chemical and physical details of this mechanism are presented in Chapter 67, and many of the metabolic functions of ATP in the body are presented in Chapters 67 through 71.

Uses of ATP for Cellular Function. ATP is used to promote three major categories of cellular functions: (1) *membrane transport,* (2) *synthesis of chemical compounds* throughout the cell, and (3) *mechanical work.* These three different uses of ATP are illustrated by examples in Figure 2–15: (1) to supply energy for the transport of sodium through the cell membrane, (2) to promote protein synthesis by the ribosomes, and (3) to supply the energy needed during muscle contraction.

In addition to membrane transport of sodium, energy from ATP is required either directly or indirectly for transport of potassium ions, calcium ions, magnesium ions, phosphate ions, chloride ions, urate ions, hydrogen ions, and still many other ions and various organic substances. Membrane transport is so important to cellular function that some cells, the renal tubular cells for instance, use as much as 80 per cent of the ATP formed in the cells for this purpose alone.

In addition to synthesizing proteins, cells also synthesize phospholipids, cholesterol, purines, pyrimidines, and a great host of other substances. Synthesis of almost any chemical compound requires energy. For instance, a single protein molecule might be composed of as many as several thousand amino acids attached to each other by peptide linkages; the formation of each of these linkages requires the breakdown of four high-energy bonds; thus, many thousand ATP molecules (or comparable guanosine triphosphate [GTP] molecules) must release their energy as each protein molecule is formed. Indeed, some cells use as much as 75 per cent of all the ATP formed in the cell simply to synthesize new chemical compounds; this is particularly true during the growth phase of cells.

The final major use of ATP is to supply energy for

special cells to perform mechanical work. We see in Chapter 6 that each contraction of a muscle fiber requires expenditure of tremendous quantities of ATP. Other cells perform mechanical work in additional ways, especially by *ciliary* and *ameboid motion,* which are described later in this chapter. The source of energy for all these types of mechanical work is ATP.

In summary, therefore, ATP is always available to release its energy rapidly and almost explosively wherever in the cell it is needed. To replace the ATP used by the cell, other much slower chemical reactions break down carbohydrates, fats, and proteins, and use the energy derived from these to form new ATP. Over 95 per cent of this ATP is formed in the mitochondria, which accounts for the name given to the mitochondria, the "powerhouses" of the cell.

AMEBOID LOCOMOTION OF CELLS

By far the most important type of cell movement that occurs in the body is that of the specialized muscle cells in skeletal, cardiac, and smooth muscle, which constitute almost 50 per cent of the entire body mass. The specialized functions of these cells are discussed in Chapters 6 through 9. However, two other types of movement occur in other cells, *ameboid locomotion* and *ciliary movement.*

Ameboid locomotion means movement of an entire cell in relation to its surroundings, such as the movement of white blood cells through tissues. However, it has received its name from the fact that amebae move in this manner and have provided an excellent tool for studying the phenomenon. Typically, ameboid locomotion begins with protrusion of a *pseudopodium* from one end of the cell. The pseudopodium projects far out away from the cell body then attaches to the new tissue area, and finally the remainder of the cell moves toward the pseudopodium. Figure 2–16 illustrates this process, showing an elongated cell, the right-hand end of which is a protruding pseudopodium. The membrane of this end of the cell is continually moving forward, and the membrane at the left-hand end of the cell is continually following along as the cell moves.

Mechanism of Ameboid Locomotion. Figure 2–16 illustrates the general principle of ameboid motion. Basically, it results from continual exocytosis that forms new cell membrane at the leading edge of the pseudopodium and continual endocytosis of the membrane in mid and rear portions of the cell. Also, one other effect is essential for forward movement of the cell. This effect is attachment of the pseudopodium to the surrounding tissues so that it becomes fixed in its leading position, while the remainder of

Figure 2–15. Use of adenosine triphosphate to provide energy for three major cellular functions: (1) membrane transport, (2) protein synthesis, and (3) muscle contraction.

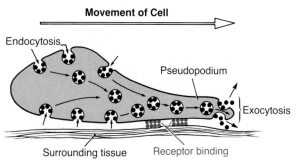

Figure 2–16. Ameboid motion by a cell.

the cell body is pulled forward toward the point of attachment. This attachment is effected by receptor proteins that line the insides of the exocytotic vesicles. When the vesicles become part of the pseudopodial membrane, they open so that their insides evert to the outside and the receptors now protrude to the outside and contact ligands in the surrounding tissues. An especially important ligand is a protein called *fibrinectin* bound to the collagen fibers of the tissue.

At the opposite end of the cell, the endocytotic activity pulls the receptors away from their ligands to form endocytotic vesicles. Then, inside the cell, these vesicles stream toward the pseudopodial end of the cell, where they are used to form still new membrane for the pseudopodium.

One of the unknowns in the process of ameboid motion is the source of the energy that causes the vesicular streaming from the endocytotic end of the cell toward the tip of the pseudopodium. Part of this might result from contraction of actin and myosin filaments in the ectoplasm of the cell, contracting the cell at its rear end and literally pushing the vesicles and cytoplasm toward the pseudopodial end.

Types of Cells That Exhibit Ameboid Locomotion. The most common cells exhibiting ameboid locomotion in the human body are the *white blood cells* moving out of the blood into the tissues in the form of *tissue macrophages* or *microphages.* However, many other types of cells can move by ameboid locomotion under certain circumstances. For instance, fibroblasts move into any damaged area to help repair the damage, and even some of the germinal cells of the skin, though ordinarily completely sessile cells, move toward a cut area to repair the rent. Finally, cell locomotion is especially important in the development of the fetus, for embryonic cells often migrate long distances from the primordial sites of origin to new areas during the development of special structures.

Control of Ameboid Locomotion—"Chemotaxis." The most important factor that usually initiates ameboid locomotion is the process called *chemotaxis.* This results from the appearance of certain chemical substances in the tissues. The chemical substance that causes chemotaxis to occur is called a *chemotactic substance.* Most cells that exhibit ameboid locomotion move toward the source of the chemotactic substance—that is, from an area of lower concentration toward an area of higher concentration—which is called *positive chemotaxis.* However, some cells move away from the source, which is called *negative chemotaxis.*

But how does chemotaxis control the direction of ameboid locomotion? Although the answer is not certain, it is known that the side of the cell most exposed to the chemotactic substance develops membrane changes that affect pseudopodial protrusion.

CILIA AND CILIARY MOVEMENTS

A second type of cellular motion, *ciliary movement,* is a whiplike movement of cilia on the surfaces of cells. This occurs in only two places in the human body: on the inside surfaces of the respiratory airways and on the inside surfaces of the uterine tubes (fallopian tubes) of the reproductive tract. In the nasal cavity and lower respiratory airways, the whiplike motion of the cilia causes a layer of mucus to move at a rate of about 1 cm/min toward the pharynx, in this way continually clearing these passageways of mucus and any particles that have become entrapped in the mucus. In the uterine tubes, the cilia cause slow movement of fluid from the ostium of the uterine tube toward the uterine cavity; it is this movement of fluid that transports the ovum from the ovary to the uterus.

As illustrated in Figure 2–17, a cilium has the appearance of a sharp-pointed curved hair that projects 2 to 4 micrometers from the surface of the cell. Many cilia project from each single cell—for instance, as many as 200 cilia on the surface of each epithelial cell in the respiratory tract. The cilium is covered by an outcropping of the cell membrane, and it is supported by 11 microtubules, 9 double tubules located around the periphery of the cilium, and 2 single tubules down the center, as shown in the cross-section illustrated in the figure. Each cilium is an outgrowth of a structure that lies immediately beneath the cell membrane, called the *basal body* of the cilium.

The *flagellum of a sperm* is also similar to a cilium; in fact, it has much the same type of structure and same type of contractile mechanism. However, the flagellum is much longer and moves in quasisinusoidal waves instead of with whiplike movements.

In the inset of Figure 2–17 movement of the cilium is illustrated. The cilium moves forward with a sudden rapid stroke, 10 to 20 times per second, bending sharply where it projects from the surface of the cell. Then it moves backward very slowly in a whiplike manner. The rapid forward movement pushes the fluid lying adjacent to the cell in the direction that the cilium moves, then the slow whiplike movement in the other direction has almost no effect on the fluid. As a result, fluid is continually propelled in the direc-

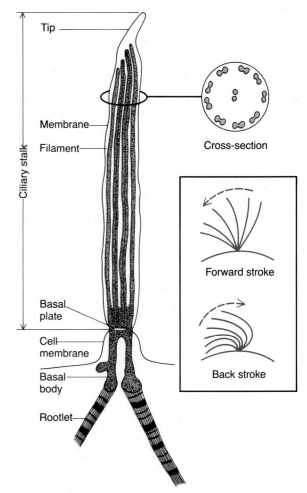

Figure 2–17. Structure and function of the cilium. (Modified from Satir: Cilia. *Sci. Am., 204:*108, 1961. ©1961 by Scientific American, Inc. All rights reserved.)

tion of the fast forward stroke. Since most ciliated cells have large numbers of cilia on their surfaces, and since all the cilia are oriented in the same direction, this is a very effective means for moving fluids from one part of the surface to another.

Mechanism of Ciliary Movement. Although all aspects of ciliary movement are not yet clear, we do know the following: First, the nine double tubules and the two single tubules are all linked to each other by a complex of protein cross-linkages; this total complex of tubules and cross-linkages is called the *axoneme*. Second, even after removal of the membrane and destruction of other elements of the cilium besides the axoneme, the cilium can still beat under appropriate conditions. Third, there are two necessary conditions for continued beating of the axoneme after removal of the other structures of the cilium: (1) the presence of ATP and (2) appropriate ionic conditions, including especially appropriate concentrations of magnesium and calcium. Fourth, during forward motion of the cilium, the tubules on the front edge of the cilium slide outward toward the tip of the cilium while the tubules on the back edge remain in place. Fifth, three protein arms, composed of the protein *dynein* that has ATPase activity, project from each set of peripheral tubules toward the next set.

Given this basic information, it has been postulated that the release of energy from ATP in contact with the ATPase dynein arms causes the arms to "crawl" along the surface of the adjacent pair of tubules. If the front tubules crawl outward while the back tubules remain stationary, this obviously will cause bending.

The way in which cilia contraction is controlled is not understood. However, the cilia of some genetically abnormal cells do not have the two central single tubules, and these cilia fail to beat at all. Therefore, it is presumed that some signal, perhaps an electrochemical signal, is transmitted along these two tubules to activate the dynein arms.

REFERENCES

Almers, W., and Stirling, C.: Distribution of transport proteins over animal cell membranes. J. Membr. Biol., 77:169, 1984.

Baker, P. F., and Knight, D. E.: Chemiosmotic hypotheses of exocytosis: A critique. Biosci. Rep., 4:285, 1984.

Balaban, R. S.: The application of nuclear magnetic resonance to the study of cellular physiology. Am. J. Physiol., 246:C10, 1984.

Bershadsky, A. D., and Vasiliev, J. M.: Cytoskeleton. New York, Plenum Publishing Corp., 1988.

Bettger, W. J., and McKeehan, W. L.: Mechanisms of cellular nutrition. Physiol. Rev., 66:1, 1986.

Blachley, J. E., et al.: The harmful effects of ethanol on ion transport and cellular respiration. Am. J. Med. Sci., 289:22, 1985.

Bohr, D. F.: Cell membrane in hypertension. News in Physiol. Sci., 4:85, 1989.

Boyd, A., and Simon, M.: Bacterial chemotaxis. Annu. Rev. Physiol., 44:501, 1982.

Brazy, P. C., and Mandel, L. J.: Does availability of inorganic phosphate regulate cellular oxidative metabolism? News Physiol. Sci., 1:100, 1986.

Case, R. M.: The role of Ca^{2+} stores in secretion. Cell-Calcium, 5:89, 1984.

Cereijido, M., et al.: Tight junction: Barrier between higher organisms and environment. New Physiol. Sci., 4:72, 1989.

Chien, S. (ed.): Molecular Biology in Physiology. New York, Raven Press, 1988.

Correia, J. J., and Williams, R. C., Jr.: Mechanisms of assembly and disassembly of microtubules. Annu. Rev. Biophys. Bioeng., 12:211, 1983.

Dean, R. T., et al.: Effects of exogenous amines on mammalian cells, with particular reference to membrane flow. Biochem. J., 217:27, 1984.

DeMello, W. C. (ed.): Cell-to-Cell Communication. New York, Plenum Publishing Corp., 1987.

DeRobertis, E. D. P., and DeRobertis, E. M. F., Jr.: Cell and Molecular Biology. 8th Ed. Philadelphia, Lea & Febiger, 1987.

Dowben, R. M. (ed.): Cell and Muscle Motility. New York, Plenum Publishing Corp. 1983.

Fawcett, D. W.: Bloom & Fawcett: A Textbook of Histology. 11th Ed. Philadelphia, W. B. Saunders Co., 1986.

Fawcett, D. W.: The Cell. 2nd Ed. Philadelphia, W. B. Saunders Co., 1981.

Fine, R. E., and Ockleford, C. D.: Supramolecular cytology of coated vesicles. Int. Rev. Cytol., 91:1, 1984.

Fitzgerald, P. G.: Gap junction heterogeneity in liver, heart, and lens. News in Physiol. Sci., 3:206, 1988.

Frankel, R. B.: Magnetic guidance of organisms. Annu. Rev. Biophys. Bioeng., 13:85, 1984.

Goldstein, D. B.: The effects of drugs on membrane fluidity. Annu. Rev. Pharmacol. Toxicol., 24:43, 1984.

Green, H., and Barrandon, Y.: Cultured epidermal cells and their use in the generation of epidermis. News in Physiol. Sci., 3:53, 1988.

Hoffman, E. K., and Simonsen, L. O.: Membrane mechanisms in volume and pH regulation in vertebrate cells. Physiol. Rev., 69:315, 1989.

Holtzman, E.: Lysosomes, New York, Plenum Publishing Corp., 1989.

Hubbard, A. L., et al.: Biogenesis of endogenous plasma membrane proteins in epithelian cells. Ann. Rev. Physiol., 51:755, 1989.

Kornfeld, S.: Trafficking of lysosomal enzymes. FASEB J., 1:462, 1987.

Kudlow, J. E., et al. (eds.): Biology of Growth Factors. New York, Plenum Publishing Corp., 1988.

Lane, M., et al.: The mitochondrion updated. Science, 234:526, 1986.

Lemasters, J. J., et al. (eds.): Integration of Mitochondrial Function. New York, Plenum Publishing Corp., 1988.

Low, M. G.: Glycosyl-phosphatidylinositol: A versatile anchor for cell surface proteins. FASEB J., 3:1600, 1989.

Machlin, L. J., and Bendich, A.: Free radical tissue damage: Protective role of antioxidant nutrients. FASEB J., 1:441, 1987.

Macnab, R. M., and Aizawa, S.-I.: Bacterial motility and the bacterial flagellar motor. Annu. Rev. Biophys. Bioeng., 13:51, 1984.

Malhotra, S. K.: The Plasma Membrane. New York, John Wiley & Sons, 1983.

McCloskey, M., and Poo, M. M.: Protein diffusion in cell membranes: Some biological implications. Int. Rev. Cytol., 87:19, 1984.

Moore, A. L., and Beechey, R. B. (eds.): Plant Mitochondria. New York, Plenum Publishing Corp., 1987.

Porter, K. R., and Bonneville, M. A.: Fine Structure of Cells and Tissues. 4th Ed. Philadelphia, Lea & Febiger, 1973.

Rodriguez-Boulan, E., and Salas, P. J. I.: External and internal signals for epithelial cell surface polarization. Ann. Rev. Physiol., 51:741, 1989.

Schachter, D.: Fluidity and function of hepatocyte plasma membranes. Hepatology, 4:140, 1984.

Sowers, A. E. (ed.): Cell Fusion. New York, Plenum Publishing Corp., 1987.

Thomas, K. A.: Fibroblast growth factors. FASEB J., 1:434, 1987.

van der Laarse, W. J., et al.: Energetics at the single cell level. News in Physiol. Sci., 4:91, 1989.

Wade, J. B.: Role of membrane fusion in hormonal regulation of epithelial transport. Ann. Rev. Physiol., 48:213, 1986.

Wall, D. A., and Maack, T.: Endocytic uptake, transport, and catabolism of proteins by epithelial cells. Am. J. Physiol., 248:C12, 1985.

Wheatley, D. N.: On the possible importance of an intracellular circulation. Life-Sci., 36:299, 1985.

Willingham, M. C., and Pastan, I.: Endocytosis and exocytosis: Current concepts of vesicle traffic in animal cells. Int. Rev. Cytol., 92:51, 1984.

CHAPTER 3

• *Genetic Control of Protein Synthesis, Cell Function, and Cell Reproduction* •

Virtually everyone knows that the genes control heredity from parents to children, but most persons do not realize that the same genes control the reproduction of and the day-by-day function of all cells. The genes control cell function by determining what substances will be synthesized within the cell—what structures, what enzymes, what chemicals.

Figure 3–1 illustrates the general schema of genetic control. Each gene, which is a nucleic acid called *deoxyribonucleic acid (DNA),* automatically controls the formation of another nucleic acid, *ribonucleic acid (RNA),* which spreads throughout the cell and controls the formation of a specific protein. Some proteins are *structural proteins,* which, in association with various lipids and carbohydrates, form the structures of the various organelles that are discussed in Chapter 2. By far the majority of the proteins are *enzymes* that catalyze the different chemical reactions in the cells. For instance, enzymes promote all the oxidative reactions that supply energy to the cell, and they promote the synthesis of various chemicals, such as lipids, glycogen, adenosine triphosphate (ATP), and so on.

For the formation of each cellular protein, there is usually only a single gene pair in each cell. It is estimated that cells of the human being have over 100,000 such gene pairs, which means that as many as 100,000 different proteins are formed in different cells, though not all of these in the same cell, for reasons that we discuss later in the chapter.

THE GENES

Large numbers of genes attached end on end are contained in extremely long double-stranded, helical molecules of *DNA* having molecular weights measured in the billions. A very short segment of such a molecule is illustrated in Figure 3–2. This molecule is composed of several simple chemical compounds arranged in a regular pattern explained in the following few paragraphs.

The Basic Building Blocks of DNA. Figure 3–3 illustrates the basic chemical compounds involved in the formation of DNA. These include (1) *phosphoric acid,* (2) a sugar called *deoxyribose,* and (3) four nitrogenous *bases* (two purines, *adenine* and *guanine,* and two pyrimidines, *thymine* and *cytosine*). The phosphoric acid and deoxyribose form

the two helical strands that are the backbone of the DNA molecule, and the bases lie between the two strands and connect them.

The Nucleotides. The first stage in the formation of DNA is the combination of one molecule of phosphoric acid, one molecule of deoxyribose, and one of the four bases to form a nucleotide. Four separate nucleotides are thus formed, one for each of the four bases: *deoxyadenylic, deoxythymidylic, deoxyguanylic,* and *deoxycytidylic acids.* Figure 3–4 illustrates the chemical structure of adenylic acid, and Figure 3–5 illustrates simple symbols for all the four basic nucleotides that form DNA.

Organization of the Nucleotides to Form DNA. Figure 3–6 illustrates the manner in which multiple numbers of nucleotides are bound together to form DNA. Note that these are combined in such a way that phosphoric acid and deoxyribose alternate with each other in the two separate strands, and these strands are held together by loose bonds between the purine and pyrimidine bases. But note carefully that

1. The purine base *adenine* always bonds with the pyrimidine base *thymine* and

2. The purine base *guanine* always bonds with the pyrimidine base *cytosine.*

Thus, in Figure 3–6 the sequence of complementary pairs of bases is CG, CG, GC, TA, CG, TA, GC, AT, and AT. However, the bases are bound together by very loose *hydrogen bonding,* represented in the figure by dashed lines. Because of the looseness of these bonds, the two strands can pull apart with ease, and they do so many times during the course of their function in the cell.

Now, to put the DNA of Figure 3–6 into its proper physical perspective, one needs merely to pick up the two ends and twist them into a helix. Ten pairs of nucleotides are present in each full turn of the helix in the DNA molecule, as is illustrated in Figure 3–2.

THE GENETIC CODE

The importance of DNA lies in its ability to control the formation of other substances in the cell. It does this by means of the so-called *genetic code.* When the two strands of a DNA molecule are split apart, this exposes the purine and

24

Gene (DNA)

↓

RNA formation

↓

Protein formation

↙   ↘

Cell structure   Cell enzymes

↘   ↙

Cell function

Figure 3 – 1. General schema by which the genes control cell function.

Phosphoric acid

Deoxyribose

Bases

Adenine   Thymine

Guanine   Cytosine

Purines | **Pyrimidines**

Figure 3 – 3. The basic building blocks of deoxyribonucleic acid (DNA).

pyrimidine bases projecting to the side of each strand. It is these projecting bases that form the code.

Research studies in the past few years have demonstrated that the genetic code consists of successive "triplets" of bases—that is, each three successive bases is a code word. The successive triplets eventually control the sequence of amino acids in a protein molecule during its synthesis in the cell. Note in Figure 3–6 that each of the two strands of the DNA molecule carries its own genetic code. For instance, the top strand, reading from left to right, has the genetic code GGC, AGA, CTT, the triplets being separated from each other by the arrows. As we follow this genetic code through Figures 3–7 and 3–8, we see that these three respective triplets are responsible for successive placement of the three amino acids, *proline, serine,* and *glutamic acid,* in a molecule of protein.

● **RNA—THE PROCESS OF TRANSCRIPTION**

Because almost all DNA is located in the nucleus of the cell and yet most of the functions of the cell are carried out in the cytoplasm, some means must be available for the genes of the nucleus to control the chemical reactions of the cytoplasm. This is achieved through the intermediary of another type of nucleic acid, *RNA,* the formation of which is controlled by the DNA of the nucleus. In this process the code is transferred to the RNA, which is called *transcription.* The RNA then diffuses from the nucleus through the

nuclear pores into the cytoplasmic compartment, where it controls protein synthesis.

Synthesis of RNA

During synthesis of RNA, the two strands of the DNA molecule separate temporarily; one of these strands is then used as a template for synthesis of the RNA molecules. The code triplets in the DNA cause the formation of *complementary* code triplets (called *codons*) in the RNA; these codons in turn control the sequence of amino acids in a protein to be synthesized later in the cytoplasm. When one strand of DNA is used in this manner to cause the formation of RNA, the opposite strand remains inactive. Each DNA strand in

Adenine

Phosphate   Deoxyribose

Figure 3 – 4. Deoxyadenylic acid, one of the nucleotides that make up DNA.

Figure 3 – 2. The helical, double-stranded structure of the gene. The outside strands are composed of phosphoric acid and the sugar deoxyribose. The internal molecules connecting the two strands of the helix are purine and pyrimidine bases; these determine the "code" of the gene.

Figure 3–5. Combinations of the basic building blocks of DNA to form nucleotides. (*P*, phosphoric acid; *D*, deoxyribose.) The four nucleotide bases are *A*, adenine; *T*, thymine; *G*, guanine; and *C*, cytosine. These four different types of nucleotides make up DNA.

each chromosome is such a large molecule that it carries the code for an average of about 4000 genes.

The Basic Building Blocks of RNA. The basic building blocks of RNA are almost the same as those of DNA except for two differences. First, the sugar deoxyribose is not used in the formation of RNA. In its place is another sugar of very slightly different composition, *ribose*. Second, thymine is replaced by another pyrimidine, *uracil*.

Formation of RNA Nucleotides. The basic building blocks of RNA first form nucleotides exactly as previously described for the synthesis of DNA. Here again, four separate nucleotides are used in the formation of RNA. These nucleotides contain the bases *adenine, guanine, cytosine, and uracil*. Note that these are the same as the bases in DNA except for one of them; the uracil replaces the thymine in DNA.

Activation of the Nucleotides. The next step in the synthesis of RNA is activation of the nucleotides. This occurs by addition to each nucleotide of two phosphate radicals to form triphosphates. These last two phosphates are combined with the nucleotide by *high-energy phosphate bonds* derived from the ATP of the cell.

The result of this activation process is that large quantities of energy are made available to each of the nucleotides, and this energy is used in promoting the subsequent chemical reactions that eventuate in the formation of the RNA chain.

Assembly of the RNA Molecule From Activated Nucleotides Using the DNA Strand as a Template — The Process of "Transcription"

Assembly of the RNA molecule is accomplished in the manner illustrated in Figure 3–7 under the influence of the enzyme *RNA polymerase*. This is a very large enzyme that has many functional properties necessary for the formation of the RNA molecule. These are as follows:

1. In the DNA strand immediately ahead of the initial gene is a sequence of nucleotides called the *promoter*. The RNA polymerase has an appropriate complementary structure that recognizes this promoter and becomes attached to it. This is the essential step in initiating the formation of the RNA molecule.

2. Once the RNA polymerase attaches to the promoter, it causes unwinding of about two turns of the DNA helix and separation of the unwound portions of the two strands.

3. Then the polymerase moves along the DNA strand, temporarily unwinding and separating the two DNA strands at each stage of its movement. As it moves along, it forms the RNA molecule by the following steps:

4. First, it causes hydrogen bonds between the successive bases of the DNA strand and the bases of complementary RNA nucleotides in the nucleoplasm.

5. Then, one at a time, the RNA polymerase breaks two of the three phosphate radicals away from each of these RNA nucleotides, liberating large amounts of energy from the broken high-energy phosphate bonds; this energy is used to cause covalent linkage of the remaining phosphate on the nucleotide with the ribose on the end of the growing RNA molecule.

6. When the RNA polymerase reaches the end of the gene or sequence of genes, it encounters a new sequence of DNA nucleotides called the *chain-terminating sequence*; this causes the polymerase to break away from the DNA strand. Then the polymerase can attach somewhere else on the same or another DNA strand to be used again and again for forming new RNA molecules.

7. As the new RNA strand is formed, its hydrogen bonds with the DNA template break away because the complementary strand of DNA has a higher energy of bonding that forces the new RNA strand loose and rebinds the two DNA strands. In this way, the RNA molecule is released into the nucleoplasm.

It should be remembered that there are four different types of DNA bases and also four different types of RNA nucleotide bases. Furthermore, these always combine with each other in specific combinations. Therefore, the code that is present in the DNA strand is transmitted in *complementary* form to the RNA molecule. The ribose nucleotide bases always combine with the deoxyribose bases in the following combinations:

DNA base	RNA base
guanine	cytosine
cytosine	guanine
adenine	uracil
thymine	adenine

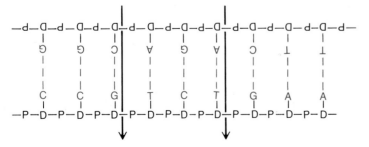

Figure 3–6. Arrangement of deoxyribose nucleotides in DNA.

Figure 3–7. Combination of ribose nucleotides with a strand of DNA to form a molecule of ribonucleic acid (RNA) that carries the DNA code from the gene to the cytoplasm. The RNA polymerase moves along the DNA strand and builds the RNA molecule.

After the RNA molecules have been released into the nucleoplasm, they still must be processed further before entering the cytoplasm. The reason for this is that the newly transcribed RNA contains many unwanted sequences of RNA nucleotides. Some of these occur at both ends of the RNA strand, and many others occur in the middle of the strand; this unwanted material probably constitutes more than 90 per cent of the total strand. Fortunately, a series of enzymes within the nucleoplasm has the capability of cutting out these unwanted sequences and then resplicing the RNA strand, a process called *RNA splicing*. The RNA is now ready to be used for the formation of proteins.

There are three separate types of RNA, each of which plays an independent and entirely different role in protein formation. These are

1. *Messenger RNA,* which carries the genetic code to the cytoplasm for controlling the formation of the proteins;

2. *Transfer RNA,* which transports activated amino acids to the ribosomes to be used in assembling the protein molecules; and

3. *Ribosomal RNA,* which, along with about 75 different proteins, forms the *ribosomes,* the physical and chemical structures on which protein molecules are actually assembled.

MESSENGER RNA—THE "CODONS"

Messenger RNA molecules are long single RNA strands that are suspended in the cytoplasm. These molecules are composed of several hundred to several thousand nucleotides in unpaired strands, and they contain *codons* that are exactly complementary to the code triplets of the genes. Figure 3–8 illustrates a small segment of a molecule of messenger RNA. Its codons are CCG, UCU, and GAA. These are the codons for the amino acids proline, serine, and glutamic acid. The transcription of these codons from the DNA molecule is demonstrated in Figure 3–7.

RNA Codons for the Different Amino Acids. Table 3–1 gives the RNA codons for the 20 common amino acids found in protein molecules. Note that most of the amino

acids are represented by more than one codon; also, one codon represents the signal "start manufacturing a protein molecule," and three codons represent "stop manufacturing a protein molecule." In Table 3–1, these two types of codons are designated CI for "chain-initiating" and CT for "chain-terminating."

TRANSFER RNA—THE "ANTICODONS"

Another type of RNA that plays an essential role in protein synthesis is called *transfer RNA* because it transfers amino acid molecules to protein molecules as the protein is synthesized. Each type of transfer RNA combines specifically with one of the 20 amino acids that are to be incorporated into proteins. The transfer RNA then acts as a *carrier* to transport its specific type of amino acid to the ribosomes, where protein molecules are forming. In the ribosomes, each specific type of transfer RNA recognizes a particular codon on the messenger RNA, as is described subsequently, and thereby delivers the appropriate amino acid to the appropriate place in the chain of the newly forming protein molecule.

Transfer RNA, containing only about 80 nucleotides, is a

Figure 3–8. Portion of a ribonucleic acid molecule, showing three "code" words, CCG, UCU, and GAA, which represent the three amino acids *proline, serine,* and *glutamic acid.*

Table 3–1 RNA CODONS FOR THE DIFFERENT AMINO ACIDS AND FOR START AND STOP

Amino Acid	RNA Codons					
Alanine	GCU	GCC	GCA	GCG		
Arginine	CGU	CGC	CGA	CGG	AGA	AGG
Asparagine	AAU	AAC				
Aspartic acid	GAU	GAC				
Cysteine	UGU	UGC				
Glutamic acid	GAA	GAG				
Glutamine	CAA	CAG				
Glycine	GGU	GGC	GGA	GGG		
Histidine	CAU	CAC				
Isoleucine	AUU	AUC	AUA			
Leucine	CUU	CUC	CUA	CUG	UUA	UUG
Lysine	AAA	AAG				
Methionine	AUG					
Phenylalanine	UUU	UUC				
Proline	CCU	CCC	CCA	CCG		
Serine	UCU	UCC	UCA	UCG	AGC	AGU
Threonine	ACU	ACC	ACA	ACG		
Tryptophan	UGG					
Tyrosine	UAU	UAC				
Valine	GUU	GUC	GUA	GUG		
Start (CI)	AUG					
Stop (CT)	UAA	UAG	UGA			

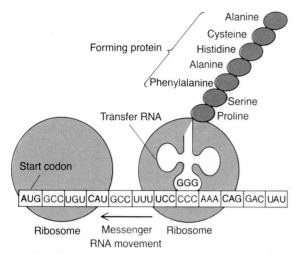

Figure 3 – 9. Mechanism by which a protein molecule is formed in ribosomes in association with messenger RNA and transfer RNA.

relatively small molecule in comparison with messenger RNA. It is a folded chain of nucleotides with a cloverleaf appearance similar to that illustrated in Figure 3–9. At one end of the molecule is always an adenylic acid; it is to this that the transported amino acid attaches to a hydroxyl group of the ribose in the adenylic acid. A specific enzyme causes this attachment for each specific type of transfer RNA; this enzyme also determines the type of amino acid that will attach to the respective type of transfer RNA.

Since the function of transfer RNA is to cause attachment of a specific amino acid to a forming protein chain, it is essential that each type of transfer RNA also have specificity for a particular codon in the messenger RNA. The specific code in the transfer RNA that allows it to recognize a specific codon is again a triplet of nucleotide bases and is called an *anticodon.* This is located approximately in the middle of the transfer RNA molecule (at the bottom of the cloverleaf configuration illustrated in Figure 3–9). During formation of a protein molecule, the anticodon bases combine loosely by hydrogen bonding with the codon bases of the messenger RNA. In this way the respective amino acids are lined up one after another along the messenger RNA chain, thus establishing the appropriate sequence of amino acids in the protein molecule.

RIBOSOMAL RNA

The third type of RNA in the cell is ribosomal RNA; it constitutes about 60 per cent of the *ribosome.* The remainder of the ribosome is protein, containing about 75 different types of proteins that are both structural proteins and enzymes needed in the manufacture of protein molecules.

The ribosome is the physical structure in the cytoplasm on which protein molecules are actually synthesized. However, it always functions in association with both the other types of RNA as well: transfer RNA transports amino acids to the ribosome for incorporation into the developing protein molecule, whereas messenger RNA provides the information necessary for sequencing the amino acids in proper order for each specific type of protein to be manufactured.

The ribosomes of nucleated cells are composed of two physical subunits, called the *small subunit,* containing 1 RNA molecule and 33 proteins, and the *large subunit,* con-

taining 3 RNAs and more than 40 proteins. Although we have only partial knowledge of the mechanism of protein manufacture in the ribosome, it is known that messenger RNA and transfer RNA first complex with the small subunit. Then the large subunit provides most of the enzymes that promote peptide linkages between the successive amino acids. Thus, the ribosome acts as a manufacturing plant in which the protein molecules are formed.

Formation of Ribosomes in the Nucleolus. The DNA genes for formation of ribosomal RNA are located in five different chromosomal pairs of the nucleus, and each chromosome contains many duplicates of these genes because of the large amount of ribosomal RNA required for cellular function.

As the ribosomal RNA forms, it collects in the *nucleolus,* a specialized structure lying adjacent to the chromosomes. When large amounts of ribosomal RNA are being synthesized, as occurs in cells that manufacture large amounts of protein, the nucleolus is a very large structure, whereas in cells that synthesize very little protein, the nucleolus may not be seen at all. The ribosomal RNA is specially processed in the nucleolus and combined with "ribosomal proteins" to form granular condensation products that are primordial subunits of the ribosomes. These subunits are then released from the nucleolus and transported through the large pores of the nuclear envelope to almost all parts of the cytoplasm. Only after the subunits enter the cytoplasm are they assembled to form mature, functional ribosomes. Therefore, proteins are not formed in the nucleus because the nucleus does not contain mature ribosomes.

FORMATION OF PROTEINS ON THE RIBOSOMES — THE PROCESS OF "TRANSLATION"

When a molecule of messenger RNA comes in contact with a ribosome, it travels along the ribosome, beginning at a predetermined end of the RNA molecule specified by an appropriate sequence of RNA bases. Then, as illustrated in Figure 3–9, while the messenger RNA travels along the ribosome, a protein molecule is formed — a process called *translation.* Thus, the ribosome reads the code of the messenger RNA in much the same way that a tape is "read" as it passes through the playback head of a tape recorder. Then, when a "stop" (or "chain-terminating") codon slips past the ribosome, the end of a protein molecule is signaled, and the protein molecule is freed into the cytoplasm.

Polyribosomes. A single messenger RNA molecule can form protein molecules in several different ribosomes at the same time, the RNA strand passing to a successive ribosome as it leaves the first, as shown in Figure 3–9. The protein molecules obviously are in different stages of development in each ribosome. As a result, clusters of ribosomes frequently occur, 3 to 10 ribosomes being attached by a single messenger RNA at the same time. These clusters are called *polyribosomes.*

It is especially important to note that a messenger RNA can cause the formation of a protein molecule in any ribosome, that there is no specificity of ribosomes for given types of protein. The ribosome is simply the structure in which the chemical reactions take place.

Attachment of Ribosomes to the Endoplasmic Reticulum. In the previous chapter it was noted that many ribosomes become attached to the endoplasmic reticulum. This does not occur until the ribosomes begin to form the protein molecules. It occurs because the initial ends of some

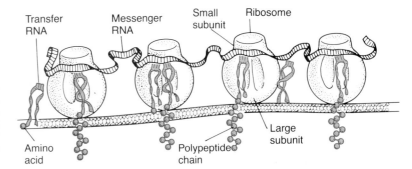

Figure 3–10. An artist's concept of the physical structure of the ribosomes as well as their functional relationship to messenger RNA, transfer RNA, and the endoplasmic reticulum during the formation of protein molecules. (From Bloom and Fawcett: A Textbook of Histology. 10th ed. Philadelphia, W. B. Saunders Company, 1975.)

protein molecules have amino acid sequences that immediately attach to specific receptor sites on the endoplasmic reticulum; this causes these molecules to penetrate the reticulum wall and enter the endoplasmic reticulum matrix. This occurs while the protein molecule is still being formed by the ribosome, which pulls the ribosome against the endoplasmic reticulum, thereby giving the "granular" appearance to the reticulum.

Figure 3–10 shows the functional relationship of messenger RNA to the ribosomes and also the manner in which the ribosomes attach to the membrane of the endoplasmic reticulum. Note the process of translation occurring in several ribosomes at the same time in response to the same strand of messenger RNA. And note also the newly forming polypeptide chains passing through the endoplasmic reticulum membrane into the endoplasmic matrix.

Yet, it should be noted that except in glandular cells that form large amounts of protein-containing secretory vesicles, most proteins formed by the ribosomes are released directly into the cytosol. These are the enzymes and internal structural proteins of the cell.

Chemical Steps in Protein Synthesis. Some of the chemical events that occur in synthesis of a protein molecule are illustrated in Figure 3–11. This figure shows representative reactions for three separate amino acids, AA_1, AA_2, and AA_{20}. The stages of the reactions are the following: (1) Each amino acid is *activated* by a chemical process in which ATP combines with the amino acid to form an

adenosine monophosphate complex with the amino acid, giving up two high-energy phosphate bonds in the process. (2) The activated amino acid, having an excess of energy, then *combines with its specific transfer RNA to form an amino acid-tRNA complex* and, at the same time, releases the adenosine monophosphate. (3) The transfer RNA carrying the amino acid complex then comes in contact with the messenger RNA molecule in the ribosome, where the anticodon of the transfer RNA attaches temporarily to its specific codon of the messenger RNA, thus lining up the amino acids in appropriate sequence to form a protein molecule. Then, under the influence of the enzyme *peptidyl transferase,* one of the proteins in the ribosome, *peptide bonds* are formed between the successive amino acids, thus adding progressively to the protein chain. These chemical events require the energy from two additional high-energy phosphate bonds, making a total of four high-energy bonds used for each amino acid added to the protein chain. Thus, the synthesis of proteins is one of the most energy-consuming processes of the cell.

Peptide Linkage. The successive amino acids in the protein chain combine with each other according to the following typical reaction:

$$\underset{R-\overset{\overset{NH_2}{|}}{C}-\overset{\overset{O}{\|}}{C}-OH}{} + \underset{H-\overset{\overset{H}{|}}{N}-\overset{\overset{R}{|}}{C}-COOH}{} \longrightarrow$$

$$R-\overset{\overset{NH_2}{|}}{C}-\overset{\overset{O}{\|}}{C}-\overset{\overset{H}{|}}{N}-\overset{\overset{R}{|}}{C}-COOH + H_2O$$

In this chemical reaction, a hydroxyl radical is removed from the COOH portion of one amino acid, while a hydrogen of the NH_2 portion of the other amino acid is removed. These combine to form water, and the two reactive sites left on the two successive amino acids bond with each other, resulting in a single molecule. This process is called *peptide linkage.*

Amino acid	AA_1	AA_2	AA_{20}
	+	+	+
	ATP	ATP	ATP
	↓	↓	↓
Activated amino acid	AMP – AA_1	AMP – AA_2	AMP – AA_{20}
	+	+	+
	$_tRNA_1$	$_tRNA_2$	$_tRNA_{20}$
	↓	↓	↓
RNA-amino acyl complex	$_tRNA_1 - AA_1$	$_tRNA_2 - AA_2$	$_tRNA_{20} - AA_{20}$
	+	+	+

Messenger RNA	GCC	UGU	AAU	CAU	CGU	AUG	GUU
	↓	↓	↓	↓	↓	↓	↓
	GCC	UGU	AAU	CAU	CGU	AUG	GUU

Complex between $_tRNA$, messenger RNA, and amino acid:

$_tRNA_1 - AA_1$ $_tRNA_5 - AA_5$ $_tRNA_3 - AA_3$ $_tRNA_9 - AA_9$ $_tRNA_2 - AA_2$ $_tRNA_{13} - AA_{13}$ $_tRNA_{20} - AA_{20}$

↓GTP ↓ GTP↓ GTP ↓ GTP↓GTP↓ GTP ↓ GTP

AA_1 AA_5 AA_3 AA_9 AA_2 AA_{13} AA_{20}

Figure 3–11. Chemical events in the formation of a protein molecule.

● **SYNTHESIS OF OTHER SUBSTANCES IN THE CELL**

Many thousand protein enzymes formed in the manner just described control essentially all the other chemical reactions that take place in cells. These enzymes promote synthesis of lipids, glycogen, purines, pyrimidines, and hundreds of other substances. We discuss many of these synthetic processes in relation to carbohydrate, lipid, and protein metabolism in Chapters 67 through 69. It is by means of all these different substances that the many functions of the cells are performed.

● **CONTROL OF GENETIC FUNCTION AND BIOCHEMICAL ACTIVITY IN CELLS**

From our discussion thus far, it is clear that the genes control both the physical and chemical functions of the cells. However, the activation of the genes themselves must be controlled as well; otherwise, some parts of the cell might overgrow or some chemical reactions might overact until they should kill the cell. Fortunately, each cell has powerful internal feedback control mechanisms that keep the various functional operations of the cell in step with each other. For each gene or for each small group of genes (100,000 in all), there is at least one such feedback mechanism.

The smallest viruses have contributed immensely to our understanding of cellular function because they are simple enough for biologists to work out almost the exact details of function, molecule by molecule. At a slightly higher level, the bacteria have also been very valuable, especially the bacterium *Escherichia coli,* which is immensely abundant in feces. Most of what we discuss in the next few pages has been learned from studies in these lower forms of life. Unfortunately, though, the nucleated cell is so complex that we are only now barely beginning to understand the special control mechanisms invented by this higher form of life. To illustrate the difference in complexity of the nucleated cell (called a *eukaryote*) and the non-nucleated cell (called a *prokaryote*), the eukaryote of the human being has almost 1000 times as much DNA as each *E. coli* bacterium.

There are basically two different methods by which the biochemical activities in the cell are controlled. One of these is called *genetic regulation,* in which the activities of the genes themselves are controlled, and the other *enzyme regulation,* in which the levels of activity of the enzymes within the cell are controlled.

GENETIC REGULATION

The Operon of the Prokaryote and Its Control of Biochemical Synthesis—Function of the "Promoter." The synthesis of a cellular biochemical product usually requires a series of reactions, and each of these reactions is catalyzed by a special protein enzyme. Formation of all the enzymes needed for the synthetic process is often controlled by a sequence of genes located in series one after the other on the same chromosomal DNA strand. This area of the DNA strand is called an *operon,* and the genes responsible for forming the respective enzymes are called the *structural genes*. In Figure 3–12, three respective structural genes are illustrated in an operon, and it is shown that they control the formation of three respective enzymes used in a particular biochemical synthetic process.

Now note in the figure the segment on the DNA strand called the *promoter.* This is a series of nucleotides that has specific affinity for RNA polymerase, as already discussed. The polymerase must bind with this promoter before it can begin traveling along the DNA strand to synthesize RNA. Therefore, the promoter is the essential element in activating the operon.

Control of the Operon by a "Repressor Protein"— The "Repressor Operator." Also note in Figure 3–12 an additional band of nucleotides lying in the middle of the promoter. This area is called a *repressor operator* because a "regulatory" protein can bind here and prevent attachment of RNA polymerase to the promoter, thereby blocking transcription of the genes. Such a regulatory protein is called a *repressor protein*. However, each regulatory repressor pro-

Figure 3–12. Function of the operon to control biosynthesis. Note that the synthesized product exerts a negative feedback to inhibit function of the operon, in this way automatically controlling the concentration of the product itself.

tein generally exists in two different allosteric forms, one that can bind with the operator and repress transcription and the other that does not bind. That is, one of its forms might be a straight protein molecule and the other the same molecule bent at an angle in the middle. Only one of these forms can repress the operator. In turn, various nonprotein substances in the cell, such as some of the cell metabolites, can bind with the repressor protein to change its state. A substance that causes it to change so that it will bind with the operator and stop transcription is called a *repressor substance* or *inhibitor substance*. On the other hand, a substance that changes the repressor protein so that it will break its bond with the operator is called an *activator substance* or *inducer substance* because it activates, or induces, the transcription process by removing the repressor protein.

To illustrate the control of gene transcription by a repressor protein, let us give an example. The saccharide lactose usually is not available to *E. coli* cells as a food substrate. Therefore, the bacterium normally will not synthesize the required enzymes for metabolic breakdown of lactose. However, when lactose does become available, it induces an allosteric conformational change in a repressor protein, causing it to break away from a repressor operator on the operon that transcribes for the necessary metabolic enzymes. Therefore, the operon becomes derepressed, and within a few minutes the appropriate enzymes are present in the bacterium to cause breakdown of the lactose. Then, as the lactose begins to disappear from within the cell, the rate of synthesis of the enzymes decreases back to that level only required for the amount of lactose that is available. Thus, one can see the logic of having such regulatory systems in the cell.

Control of the Operon by an "Activator Protein" —The "Activator Operator." Now note in Figure 3–12 another operator, called the *activator operator,* that lies adjacent to but ahead of the promoter. When a regulatory protein binds to this operator, it helps to attract the RNA polymerase to the promoter, in this way activating the operon. Therefore, a regulatory protein of this type is called an *activator protein*. The operon can be activated or inhibited through the activator operator in ways exactly opposite to the control of the repressor operator.

Negative Feedback Control of the Operon. Finally, note in Figure 3–12 that the presence of a critical amount of a synthesized product in the cell can cause negative feedback inhibition of the operon that is responsible for its synthesis. It can do this either by causing a regulatory repressor protein to bind at the repressor operator or by causing a regulatory activator protein to break its bond with the activator operator. In either case, the operon becomes inhibited. Therefore, once the required synthesized product has become abundant enough, the operon becomes dormant. On the other hand, when the synthesized product becomes degraded in the cell and its concentration falls, the operon once again becomes active. In this way, the concentration of the product is automatically controlled.

Other Mechanisms for Control of Transcription by the Operon. Variations in the basic mechanism for control of the operon have been discovered with rapidity in the last several years. Without giving details, let us list some of these:

1. An operon is frequently controlled by a *regulatory gene* located elsewhere in the genetic complex of the nucleus. That is, the regulatory gene causes the formation of a regulatory protein that in turn acts either as an activator or as a repressor substance to control the operon.

2. Occasionally, many different operons are controlled at the same time by the same regulatory protein. In some instances, the same regulatory protein functions as an activator for one operon and as a repressor for another operon. When multiple operons are controlled simultaneously in this manner, all the operons that function together are called a *regulon.*

3. Some operons are controlled not at the starting point of transcription on the DNA strand but instead farther along the strand. Sometimes, the control is not even at the DNA strand itself but is instead during the processing of the RNA molecules in the nucleus before they are released into the cytoplasm; or, rarely, control might occur at the level of RNA translation by the ribosomes.

4. In eukaryotes, the nuclear DNA is packaged in specific structural units, the *chromosomes.* And within each chromosome, the DNA is wound around small proteins called *histones,* which in turn are compacted tightly together by still other proteins. As long as the DNA is in this compacted state, it cannot function to form RNA. However, multiple control mechanisms are beginning to be discovered that can cause selected areas of the chromosomes to become decompacted one part at a time so that RNA transcription can occur. Thus, in the eukaryote, still higher orders of control are employed for establishing proper cell function. In addition, signals from outside the cell, such as some of the hormones, can activate specific chromosomal areas, thus providing the chemical machinery for specific functions.

Because there are as many as 100,000 different genes in each human cell, the large number of different ways in which genetic activity can be controlled is not surprising. The genetic control systems are especially important for controlling the intracellular concentrations of amino acids, amino acid derivatives, and the intermediate substrates of carbohydrate, lipid, and protein metabolism.

CONTROL OF ENZYME ACTIVITY

In addition to control of the genetic regulatory system, some of the intracellular enzymes themselves can be directly controlled by intracellular inhibitors or activators. This, therefore, represents a second category of mechanisms by which cellular biochemical functions can be controlled.

Enzyme Inhibition. Some of the chemical substances formed in the cell have a direct feedback effect in inhibiting the respective enzyme systems that synthesize them. Al-

most always the synthesized product acts on the first enzyme in a sequence, rather than on the subsequent enzymes, usually binding directly with the enzyme and causing an allosteric conformational change that inactivates it. One can readily recognize the importance of inactivating this first enzyme: It prevents buildup of intermediary products that will not be used.

This process of enzyme inhibition is another example of negative feedback control; it is responsible for controlling the intracellular concentrations of some of the amino acids, purines, pyrimidines, vitamins, and other substances.

Enzyme Activation. Enzymes either that are normally inactive or that have been inactivated by some inhibitor substance often can be activated. An example of this is the action of cyclic adenosine monophosphate (cAMP) in causing glycogen to split and release glucose molecules to form high-energy ATP, as discussed in the previous chapter. When most of the ATP has been depleted in a cell, a considerable amount of cAMP begins to be formed as a breakdown product of the ATP; the presence of this cAMP indicates that the cellular reserves of ATP have approached a low ebb. However, the cAMP immediately activates the glycogen-splitting enzyme phosphorylase, liberating glucose molecules that are rapidly metabolized and their energy used for replenishment of the ATP stores. Thus, in this case the cAMP acts as an enzyme activator and thereby helps control intracellular ATP concentration.

Another interesting instance of both enzyme inhibition and enzyme activation occurs in the formation of the purines and pyrimidines. These substances are needed by the cell in approximately equal quantities for formation of DNA and RNA. When purines are formed, they *inhibit* the enzymes that are required for formation of additional purines. However, they *activate* the enzymes for formation of the pyrimidines. Conversely, the pyrimidines inhibit their own enzymes but activate the purine enzymes. In this way there is continual cross-feed between the synthesizing systems for these two substances, resulting in almost exactly equal amounts of the two substances in the cells at all times.

In summary, there are two principal methods by which the cells control proper proportions and proper quantities of different cellular constituents: (1) the mechanism of genetic regulation and (2) the mechanism of enzyme regulation. The genes can be either activated or inhibited, and, likewise, the enzyme systems can be either activated or inhibited. Most often, these regulatory mechanisms function as feedback control systems that continually monitor the cell's biochemical composition and make corrections as needed. But, on occasion, substances from without the cell (especially some of the hormones that will be discussed later in this text) also control the intracellular biochemical reactions by activating or inhibiting one or more of the intracellular control systems.

● CELL REPRODUCTION

Cell reproduction is another example of the pervading, ubiquitous role that the DNA-genetic system plays in all life processes. The genes and their regulatory mechanisms determine the growth characteristics of the cells and also when or whether these cells divide to form new cells. In this way, the all-important genetic system controls each stage in the development of the human being from the single-cell fertilized ovum to the whole functioning body. Thus, if there is any central theme to life, it is the DNA-genetic system.

Figure 3–13. Stages in the reproduction of the cell. *A, B,* and *C,* prophase; *D,* prometaphase; *E,* metaphase; *F,* anaphase; *G* and *H,* telophase. (Redrawn from Mazia: How cells divide. *Sci. Am., 205*:102, 1961. © by Scientific American, Inc. All rights reserved.)

The Life Cycle of the Cell. The life cycle of a cell is the period of time from cell reproduction to reproduction. When mammalian cells are not inhibited and are reproducing as rapidly as they can, this life cycle lasts for 10 to 30 hours. It is terminated by a series of distinct physical events called *mitosis* that cause division of the cell into two new daughter cells. The events of mitosis are illustrated in Figure 3–13 and are described later. The actual stage of mitosis, however, lasts for only about 30 minutes, so that more than 95 per cent of the life cycle even of rapidly reproducing cells is represented by the interval between mitosis, called *interphase*. In fact, except in special conditions of rapid cellular reproduction, inhibitory factors almost always slow or stop the uninhibited life cycle of the cell. Therefore, different cells of the body have life cycle periods that vary from as little as 10 hours for stimulated bone marrow cells to an entire lifetime of the human body for nerve and striated muscle cells.

REPLICATION OF THE DNA, THE INITIAL STEP IN CELL REPRODUCTION

As is true of almost all other important events in the cell, reproduction begins in the nucleus itself. The first step is *replication (duplication) of all DNA in the chromosomes.* Only after this has occurred can mitosis take place.

The DNA begins to be duplicated some 5 to 10 hours before mitosis, and this is completed in 4 to 8 hours. The DNA is duplicated only once, so that the net result is two exact *replicates* of all DNA. These replicates, in turn, be-

come the DNA in the two new daughter cells that will be formed at mitosis. Following replication of the DNA, there is another period of 1 to 2 hours before mitosis begins abruptly. However, even during this period, preliminary changes are already beginning to take place that will lead to the mitotic process.

Chemical and Physical Events of DNA Replication. DNA is replicated in much the same way that RNA is transcribed by DNA except for a few important differences:

1. Both strands of the DNA in each chromosome are replicated, not simply one of them.

2. Both entire strands of the DNA helix are replicated from end to end rather than small portions of them, as occurs in the transcription of RNA by the genes.

3. The principal enzymes for replicating DNA are a complex of several enzymes called *DNA polymerase,* which is comparable to RNA polymerase. It attaches to and moves along the DNA template strand, while another enzyme, *DNA ligase,* causes bonding of the successive nucleotides to each other, using high-energy phosphate bonds to energize these attachments.

4. Formation of each new DNA strand occurs simultaneously in hundreds of segments along each of the two strands of the helix until the entire strand is replicated. Then the ends of the subunits are joined together by the DNA ligase enzyme.

5. Each newly formed strand of DNA remains attached by loose hydrogen bonding to the original DNA strand that is used as its template. Therefore, two new DNA helixes are formed that are exact duplicates of each other and that are still coiled together.

6. Because the DNA helixes in each chromosome are about 6 centimeters in length and have millions of turns in each helix, it would be impossible for the two newly formed DNA helixes to uncoil from each other were it not for some special mechanism. This is achieved by enzymes that periodically cut each helix along its entire length, rotate each segment enough to cause separation, and then resplice the helix. Thus, the two new helixes become uncoiled.

DNA Repair and DNA "Proofreading." During the hour or so between DNA replication and the beginning of mitosis, there is a period of very active repair and "proofreading" of the DNA strands. That is, wherever inappropriate DNA nucleotides have been matched up with the nucleotides of the original template strand, special enzymes cut out the defective areas and replace these with the appropriate complementary nucleotides. This is achieved by the same DNA polymerases and DNA ligase that are used in the process of replication. This repair process is referred to as *DNA proofreading.*

Because of repair and proofreading, the transcription process almost never makes a mistake. But when a mistake is made, this is called a *mutation;* it in turn will cause the formation of some abnormal protein in the cell, often leading to abnormal cellular function and sometimes even to death. Yet, because of the precision of the transcription process, it has been calculated that each human gene mutates only about once in every 200,000 years of human life. Nevertheless, when one realizes that there are 100,000 or more genes in the human genome and that the period from one generation to another is about 30 years, one would still expect as many as 10 or more mutations in the passage of the genome from parent to child. Fortunately, though, each human genome is represented by two separate sets of chromosomes with almost identical genes, so that one functional gene of each pair almost always is still available to the child despite mutations.

THE CHROMOSOMES AND THEIR REPLICATION

The DNA helixes of the nucleus are packaged in chromosomes. The human cell contains 46 chromosomes arranged in 23 pairs. Most of the genes in the two chromosomes of each pair are identical or almost identical with each other, so that it is usually stated that the different genes also exist in pairs, although occasionally this is not the case.

Aside from the DNA in the chromosome, there is also a large amount of protein, composed mainly of many small molecules of electropositively charged *histones*. The histones are organized into vast numbers of small, bobbinlike cores. Successive segments of each DNA helix are coiled sequentially around one core after another. Then, during mitosis the successive cores are packed one against the other, thus allowing the tremendously long DNA molecule, having a linear length of 6 centimeters and a molecular weight of about 60 billion, to be packaged in a coiled and folded arrangement of the *mitotic chromosome* that is only a few micrometers in length, 1/10,000 the stretched-out length of the DNA itself.

The histone cores probably play an important role in the regulation of DNA activity because as long as the DNA is packaged tightly, it cannot function as a template either for the formation of RNA or for the replication of new DNA. Furthermore, some of the regulatory proteins have been shown to *decondense* the histone packaging of the DNA and to allow small segments at a time to form RNA. Thus, this is still a higher order of regulation than some of the types of regulation discussed earlier.

Several nonhistone proteins are also major components of chromosomes, functioning both as chromosomal structural proteins and, in connection with the genetic regulatory machinery, as activators, inhibitors, and enzymes.

Replication of the chromosomes in their entirety follows during the next few minutes after replication of the DNA helixes; the new DNA helixes collect new protein molecules as needed. At this stage, the two newly formed chromosomes are called *chromatids*. They remain temporarily attached to each other (until the time for mitosis) at a point called the *centromere* located near the center of each of the chromatids.

MITOSIS

The actual process by which the cell splits into two new cells is called *mitosis*. Once each chromosome has been replicated to form the two chromatids, mitosis follows automatically within an hour or two.

The Mitotic Apparatus. One of the first events of mitosis takes place in the cytoplasm, occurring during the latter part of interphase and early part of prophase, in or around the small structures called *centrioles*. As illustrated in Figure 3–13, two pairs of centrioles lie close to each other near one pole of the nucleus. (These centrioles, like the DNA and chromosomes, had also been replicated during interphase, usually shortly before the replication of the DNA.) Each centriole is a small cylindrical body about 0.4 micrometer long and about 0.15 micrometer in diameter, consisting mainly of nine parallel tubular structures arranged in the form of a cylinder. The two centrioles of each pair lie at right angles to each other.

Shortly before mitosis is to take place, the two pairs of centrioles begin to move apart from each other. This is caused by successive polymerization of protein microtubules growing between the respective centriole pairs and actually pushing them apart. At the same time, other microtubules grow radially away from each of the centriole pairs, forming a spiny star, called the *aster*, in each end of the cell. Some of the spines penetrate the nucleus and play a role in separating the two sets of chromatids during mitosis. The complex of microtubules connecting the two centriole pairs is called the *spindle*, and the entire set of microtubules plus the two pairs of centrioles is called the *mitotic apparatus*.

Prophase. The first stage of mitosis, called *prophase*, is shown in Figure 3–13A, B, and C. While the spindle is forming, the chromosomes of the nucleus, which in interphase consist of loosely coiled strands, become condensed into well-defined chromosomes.

Prometaphase. During this stage (Fig. 3–13D), the nuclear envelope fragments. At the same time, a new set of microtubules begins to extend outward from a small condensed portion of each chromatid called the *kinetochore*, located on the outside of the centromere, where the two chromatids are attached to each other. These new microtubules, in turn, either become attached to or interact with the microtubules from the two asters of the mitotic apparatus, with one chromatid connecting to the aster at one pole of the cell and the other chromatid connecting to the opposite aster.

Metaphase. During metaphase (Fig. 3–13E), the two asters of the mitotic apparatus are pushed farther apart by additional growth of the mitotic spindle. Simultaneously the chromatids are pulled tightly by the attached microtubules to the very center of the cell, lining up to form the *equatorial plate* of the mitotic spindle.

Anaphase. During this phase (Fig. 3–13F), the two chromatids of each chromosome are pulled apart at the centromere. Exactly how the microtubular system causes this is not known, except that it is known that the microtubules contain *actin* in addition to tubulin; actin is one of the contractile proteins of muscle. Therefore, it has been presumed that the microtubules might contract or that the chromosomal tubules might interact in a sliding way with the microtubules of the aster to cause the pulling force. Regardless of the mechanism, all 46 pairs of chromatids are separated, forming two separate sets of 46 *daughter chromosomes*. One of these sets is pulled toward one mitotic aster and the other toward the other aster at the two respective poles of the dividing cell.

Telophase. In telophase (Fig. 3–13G and H), the two sets of daughter chromosomes are now pulled completely apart. Then the mitotic apparatus dissolutes, and a new nuclear membrane develops around each set of chromosomes, this membrane being formed from portions of the endoplasmic reticulum that are already present in the cytoplasm. Shortly thereafter, the cell pinches in two midway between the two nuclei. This is caused by a contractile ring of *microfilaments* composed of *actin* and probably *myosin*, the two contractile proteins of muscle, forming at the juncture of the newly developing cells and pinching them off from each other.

CONTROL OF CELL GROWTH AND REPRODUCTION

We all know that certain cells grow and reproduce all the time, such as the blood-forming cells of the bone marrow, the germinal layers of the skin, and the epithelium of the gut. However, many other cells, such as smooth muscle cells, may not reproduce for many years. A few cells, such as the neurons and most striated muscle cells, do not reproduce during the entire life of the person.

In certain tissues, an insufficiency of some types of cells

causes these to grow and reproduce very rapidly until appropriate numbers of them are again available. For instance, seven eighths of the liver can be removed surgically, and the cells of the remaining one eighth will grow and divide until the liver mass returns almost to normal. The same occurs for almost all glandular cells, cells of the bone marrow, subcutaneous tissue, intestinal epithelium, and almost any other tissue except highly differentiated cells, such as nerve and muscle cells.

We know very little about the mechanisms that maintain proper numbers of the different types of cells in the body. However, experiments have shown at least three ways in which growth can be controlled. First, growth is often controlled by *growth factors* that come from other parts of the body. Some of these circulate in the blood, but others originate in adjacent tissues. For instance, the epithelial cells of some glands, such as the pancreas, will fail to grow without a growth factor from the sublying connective tissue of the gland. Second, most normal cells will stop growing when they have run out of space for growth. This occurs when cells are grown in tissue culture; the cells grow until they contact a solid object and then growth stops. Third, cells grown in tissue culture often stop growing when minute amounts of their own secretions are allowed to collect in the culture medium. This, too, could provide a means for negative feedback control of growth.

Regulation of Cell Size. Cell size is determined almost entirely by the amount of functioning DNA in the nucleus. If replication of the DNA does not occur, the cell grows to a certain size and thereafter remains at that size. On the other hand, it is possible, by use of the chemical *colchicine,* to prevent formation of the mitotic spindle and therefore to prevent mitosis, even though replication of the DNA continues. In this event, the nucleus then contains far greater quantities of DNA than normally, and the cell grows proportionately larger. It is assumed that this results simply from increased production of RNA and cell proteins, which in turn cause the cell to grow larger.

● CELL DIFFERENTIATION

A special characteristic of cell growth and cell division is *cell differentiation,* which means changes in physical and functional properties of cells as they proliferate in the embryo to form the different bodily structures.

The earliest and simplest theory for explaining differentiation was that the genetic composition of the nucleus undergoes changes during successive generations of cells in such a way that one daughter cell inherits a different set of genes from that of the other daughter cell.

However, this theory has now been completely disproved in many ways but especially well illustrated by the following simple experiment. The nucleus from an intestinal mucosal cell of a frog, when surgically implanted into a frog ovum from which the original nucleus has been removed, will often cause the formation of a completely normal frog. This demonstrates that even the intestinal mucosal cell, which is a reasonably well-differentiated cell, still carries all the necessary genetic information for development of all structures required in the frog's body.

Therefore, it has become clear that differentiation results not from loss of genes but instead from selective repression of different genetic operons. In fact, electron micrographs suggest that some segments of DNA helixes wound around histone cores become so condensed that they will no longer uncoil to form RNA molecules. One suggestion for the cause of this effect is the following: It has been supposed that a regulatory gene in the genome begins at a certain stage of cell differentiation to produce a regulatory protein that causes positive feedback activation of this same regulatory gene. This positive feedback causes continued production of the regulatory protein forever thereafter, but this same regulatory protein represses another select group of genes. Therefore, the repressed genes will never function again. Regardless of the mechanism, most mature cells of the human being produce about 8000 to 10,000 proteins rather than the potential 100,000 or more if all genes were active.

Embryological experiments show also that certain cells in an embryo control the differentiation of adjacent cells. For instance, the *primordial chordamesoderm* is called the *primary organizer* of the the embryo because it forms a focus around which the rest of the embryo develops. It differentiates into a *mesodermal axis* containing segmentally arranged *somites* and, as a result of *inductions* in the surrounding tissues, causes formation of essentially all the organs of the body.

Another instance of induction occurs when the developing eye vesicles come in contact with the ectoderm of the head and cause it to thicken into a lens plate that folds inward to form the lens of the eye. Therefore, a large share of the embryo develops as a result of such inductions, one part of the body affecting another part, and this part affecting still other parts.

Thus, although our understanding of cell differentiation is still hazy, we know many different control mechanisms by which differentiation *could* occur.

● CANCER

Cancer is caused in all or almost all instances by *mutation* or *abnormal activation* of cellular genes that control cell growth and cell mitosis. The abnormal genes are called *oncogenes.*

Only a minute fraction of the cells that mutate in the body ever lead to cancer. There are several reasons for this. First, most mutated cells have less survival capability than normal cells and therefore simply die. Second, only a few of the mutated cells that do survive lose the normal feedback controls that prevent excessive growth. Third, those cells that are potentially cancerous are often, if not usually, destroyed by the body's immune system before they grow into a cancer. This occurs in the following way: Most mutated cells form abnormal proteins within their cell bodies because of their altered genes, and these proteins then stimulate the body's immune system, causing it to form antibodies or sensitized lymphocytes against the cancerous cells, in this way destroying them. In support of this is the fact that in persons whose immune systems have been suppressed, such as those who are taking immunosuppressant drugs following transplantation of a kidney or a heart, the probability of developing a cancer is multiplied severalfold.

But what is it that causes the altered genes? When one realizes that many trillions of new cells are formed each year in the human being, this question should probably better be asked in the following form: Why is it that we do not develop literally millions or billions of mutant cancerous cells? The answer is the incredible precision with which DNA chromosomal strands are replicated in each cell before mitosis takes place and also because the "proofreading" process cuts and repairs any abnormal DNA strand before the mitotic process is allowed to proceed. Yet, despite all these precautions, probably one newly formed cell in every few million still has significant mutant characteristics.

Thus, chance alone is all that is required for mutations to

take place, so we may suppose that a very large number of cancers are merely the result of an unlucky occurrence.

Yet, the probability of mutations can be increased many-fold when a person is exposed to certain chemical, physical, or biological factors. Some of these are the following:

1. It is well known that *ionizing radiation,* such as x-rays, gamma rays, and particle radiations from radioactive substances, and even ultraviolet light can predispose to cancer. Ions formed in tissue cells under the influence of such radiation are highly reactive and can rupture DNA strands, thus causing many mutations.

2. *Chemical substances* of certain types also have a high propensity for causing mutations. Historically, it was long ago discovered that various aniline dye derivatives are very likely to cause cancer, so that workers in chemical plants producing such substances, if unprotected, have a special predisposition to cancer. Chemical substances that can cause mutation are called *carcinogens.* The carcinogens that cause by far the greatest number of deaths in our present-day society are those in cigarette smoke. These cause about one quarter of all cancer deaths.

3. *Physical irritants* can also lead to cancer, such as continued abrasion of the linings of the intestinal tract by some types of food. The damage to the tissues leads to rapid mitotic replacement of the cells. The more rapid the mitosis, the greater the chance for mutation.

4. In many families there is a strong *hereditary tendency to cancer.* This probably results from the fact that most cancers require not one mutation but two or more mutations before cancer occurs. In those families that are particularly predisposed to cancer, it is presumed that one or more of the genes are already mutated in the inherited genome. Therefore, far fewer additional mutations must take place in such a person before a cancer begins to grow.

5. In experimental animals certain types of viruses can cause some kinds of cancer, including leukemia. This occasionally results by either of two ways. First, in the case of DNA viruses, the DNA strand of the virus can insert itself directly into one of the chromosomes and thereby cause the mutation that leads to cancer. In the case of RNA viruses, some of these carry with them an enzyme called *reverse transcriptase* that causes DNA to be transcribed from the RNA. Then the transcribed DNA inserts itself into the animal cell genome, thus leading to cancer. Yet, despite the demonstration that viral cancer sometimes occurs in animals, it has not yet been proved that cancer spreads this way in human beings nor that cancer is contagious from one person to another.

Invasive Characteristic of the Cancer Cell. The three major differences between the cancer cell and the normal cell are: (1) The cancer cell does not respect usual cellular growth limits; the reason for this is that the cells presumably do not require the growth factors that are necessary to cause growth of normal cells. (2) Cancer cells are far less adhesive to each other than are normal cells. Therefore, they have a tendency to wander through the tissues, to enter the blood stream, and to be transported all through the body, where they form nidi for numerous new cancerous growths. (3) Some cancers also produce *angiogenic factors* that cause many new blood vessels to grow into the cancer, thus supplying the nutrients required for cancer growth.

Why Do Cancer Cells Kill? The answer to this question is usually very simple. Cancer tissue competes with normal tissues for nutrients. Because cancer cells continue to proliferate indefinitely, their number multiplying day by day, one can readily understand that the cancer cells will soon demand essentially all the nutrition available to the body. As a result, the normal tissues gradually suffer nutritive death.

REFERENCES

Bajaj, M., and Blundell, T.: Evolution and the tertiary structure of proteins. Annu. Rev. Biophys. Bioeng., 13:453, 1984.
Beers, R. F. (ed): Cell Fusion. Gene Transfer and Transformation. New York, Raven Press, 1984.
Bownes, M.: Differentiation of Cells. New York, Methuen, Inc., 1985.
Bradshaw, R. A., and Prentis, S.: Oncogenes and Growth Factors. New York, Elsevier Science Publishing Co., 1987.
Bryant, P. J., and Simpson, P.: Intrinsic and extrinsic control of growth in developing organs. Q. Rev. Biol., 59:387, 1984.
Busch, H. (ed.): The Cell Nucleus, Nuclear Particles. New York, Academic Press, 1981.
Cohn, W. E. (ed.): Progress in Nucleic Acid Research and Molecular Biology. DNA: Multiprotein Interactions. New York, Academic Press, 1981.
Davidson, R.: Somatic Cell Genetics. New York, Van Nostrand Reinhold, 1984.
DeRecondo, A. M. (ed.): New Approaches in Eukaryotic DNA Replication. New York, Plenum Publishing Corp., 1983.
Echols, H.: Multiple DNA-protein interactions governing high-precision DNA transactions. Science, 233:1050, 1986.
Fox, J. L.: Contemplating the human genome. BioScience, 37:457, 1987.
Garvey, E. P., and Santi, D. V.: Stable amplified DNA in drug-resistant *Leishmania* exists as extrachromosomal circles. Science, 233:535, 1986.
Go, N.: Theoretical studies of protein folding. Annu. Rev. Biophys. Bioeng., 12:183, 1983.
Hall, A.: Oncogenes—implications for human cancer: A review. J. R. Soc. Med., 77:410, 1984.
Horton, J. D. (ed.): Development and Differentiation of Vertebrates. New York, Elsevier/North-Holland, 1980.
Hunt, T. (ed.): DNA Makes RNA Makes Protein. New York, Elsevier Science Publishing Co., 1983.
Klevecz, R. R., et al.: Cellular clocks and oscillators. Int. Rev. Cytol., 86:97, 1984.
Kornberg, A.: DNA Replication. San Francisco, W. H. Freeman, 1980.
Kucherlapati, R., and Skoultchi, A. I.: Introduction of purified genes into mammalian cells. CRC Crit. Rev. Biochem., 16:349, 1984.
Kulaev, J. S. (ed.): Environmental Regulation of Microbial Metabolism. New York, Academic Press, 1985.
Macara, I. G.: Oncogenes and cellular signal transduction. Physiol. Rev., 69:797, 1989.
Margalit, H., et al.: Initiation of DNA replication in bacteria: Analysis of an autorepressor control model. J. Theor. Biol., 111:183, 1984.
Marx, J. L.: Zipping up DNA binding proteins. Science, 240:1732, 1988.
Nomura, M., et al.: Regulation of the synthesis of ribosomes and ribosomal components. Annu. Rev. Biochem., 53:75, 1984.
Nora, J. J., and Fraser, F. C.: Medical Genetics: Principles and Practice. 3rd Ed. Philadelphia, Lea & Febiger, 1989.
Ohlendorf, D. H., and Matthews, B. W.: Structural studies of protein-nucleic acid interactions. Annu. Rev. Biophys. Bioeng., 12:259, 1983.
Orcutt, B. D., et al.: Protein and nucleic acid sequence database systems. Annu. Rev. Biophys. Bioeng., 12:419, 1983.
Pabo, C. O., and Sauer, R. T.: Protein-DNA recognition. Annu. Rev. Biochem., 53:293, 1984.
Raff, E. C.: Genetics of microtubule systems. J. Cell. Biol., 99:1, 1984.
Rigler, R., and Wintermeyer, W.: Dynamics of tRNA. Annu. Rev. Biophys. Bioeng., 12:475, 1983.
Schimke, R. T.: Gene amplification in cultured animal cells. Cell, 37:705, 1984.
Schleif, R.: DNA looping. Science, 240:127, 1988.
Schlesinger, D. H. (ed.): Macromolecular Sequencing and Synthesis: Selected Methods and Applications. New York, Alan R. Liss, Inc., 1988.
Schopf, J. W.: The evolution of the earliest cells. Sci. Am., 239:110, 1978.
Scriver, C. R., et al. (eds.): The Metabolic Basis of Inherited Disease. 6th Ed. New York, McGraw-Hill Book Co., 1989.
Smith, H. S., et al.: The biology of breast cancer at the cellular level. Biochem. Biophys. Acta, 738:103, 1984.
Stoffler, G., and Stoffler-Meilicke, M.: Immunoelectron microscopy of ribosomes. Annu. Rev. Biophys. Bioeng., 13:303, 1984.
Thomas, K. A.: Fibroblast growth factors. FASEB J. 1:434, 1987.
Thompson, M. W., et al.: Genetics in Medicine. 5th Ed. Philadelphia, W. B. Saunders Co., 1989.
Varmus, H. E.: Oncogenes and transcriptional control. Science, 238:1337, 1987.
Weiss, L., et al.: Interactions of cancer cells with the microvasculature during metastasis. FASEB J., 2:12, 1988.

UNIT II

MEMBRANE PHYSIOLOGY, NERVE, AND MUSCLE

CHAPTER 4

• *Transport of Ions and Molecules Through the Cell Membrane* •

Figure 4–1 gives the approximate compositions of the *extracellular fluid,* which lies outside the cell membranes, and the *intracellular fluid,* inside the cells. Note that the extracellular fluid contains large quantities of *sodium* but only small quantities of *potassium.* Exactly the opposite is true of the intracellular fluid. Also, the extracellular fluid contains large quantities of chloride, whereas the intracellular fluid contains very little. But the concentrations of phosphates, essentially all of which are organic metabolic intermediates, and proteins in the intracellular fluid are considerably greater than in the extracellular fluid. These many differences are extremely important to the life of the cell. It is the purpose of this chapter to explain how these differences are brought about by the transport mechanisms of the cell membranes.

The Lipid Barrier and the Transport Proteins of the Cell Membrane

The structure of the cell membrane is discussed in Chapter 2 and shown in Figure 2–3. It consists almost entirely of a *lipid bilayer* with large numbers of protein molecules floating in the lipid, many penetrating all the way through, as illustrated in Figure 4–2.

The lipid bilayer is not miscible with either the extracellular fluid or the intracellular fluid. Therefore, it constitutes a barrier for the movement of most water molecules and water-soluble substances between the extracellular and intracellular fluid compartments. However, as illustrated by the left-hand arrow of Figure 4–2, a few substances can penetrate this bilayer and can either enter the cell or leave it, passing directly through the lipid substance itself.

The protein molecules, on the other hand, have entirely different transport properties. Their molecular structures interrupt the continuity of the lipid bilayer and therefore constitute an alternate pathway through the cell membrane. Most of these penetrating proteins, therefore, are *transport proteins.* Different proteins function differently. Some have watery spaces all the way through the molecule and allow free movement of certain ions or molecules; these are called *channel proteins.* Others, called *carrier proteins,* bind with substances that are to be transported, and conformational changes in the protein molecules then move the substances through the interstices of the molecules to the other side of the membrane. Both the channel proteins and the carrier proteins are highly selective in the type or types of molecules or ions that are allowed to cross the membrane.

Diffusion Versus Active Transport. Transport through the cell membrane, either directly through the lipid bilayer or through the proteins, occurs by one of two basic processes, *diffusion* (which is also called "passive transport") or *active transport.* Although there are many different variations of these two basic mechanisms, as we see later in this chapter, diffusion means random molecular movement of substances molecule by molecule either through intermolecular spaces in the membrane or in combination with a carrier protein. The energy that causes diffusion is the energy of the normal kinetic motion of matter. By contrast, active transport means movement of ions or other substances across the membrane in combination with a carrier protein but additionally *against an energy gradient,* such as from a low concentration state to a high concentration state, a process that requires an additional source of energy besides kinetic energy to cause the movement. Let us explain in more detail the basic physics and physical chemistry of these two separate processes.

● DIFFUSION

All molecules and ions in the body fluids, including both water molecules and dissolved substances, are in constant motion, each particle moving its own separate way. Motion of these particles is what physicists call heat—the greater the motion, the higher the

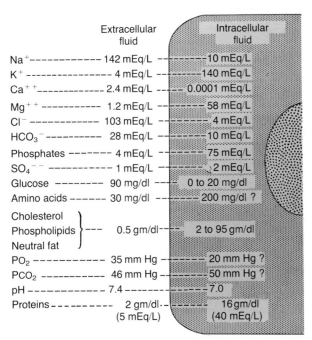

	Extracellular fluid	Intracellular fluid
Na$^+$	142 mEq/L	10 mEq/L
K$^+$	4 mEq/L	140 mEq/L
Ca^{++}	2.4 mEq/L	0.0001 mEq/L
Mg^{++}	1.2 mEq/L	58 mEq/L
Cl$^-$	103 mEq/L	4 mEq/L
HCO$_3^-$	28 mEq/L	10 mEq/L
Phosphates	4 mEq/L	75 mEq/L
SO$_4^{--}$	1 mEq/L	2 mEq/L
Glucose	90 mg/dl	0 to 20 mg/dl
Amino acids	30 mg/dl	200 mg/dl ?
Cholesterol Phospholipids Neutral fat	0.5 gm/dl	2 to 95 gm/dl
PO$_2$	35 mm Hg	20 mm Hg ?
PCO$_2$	46 mm Hg	50 mm Hg ?
pH	7.4	7.0
Proteins	2 gm/dl (5 mEq/L)	16 gm/dl (40 mEq/L)

Figure 4–1. Chemical compositions of extracellular and intracellular fluids.

temperature—and motion never ceases under any conditions except at absolute zero temperature. When a moving molecule, A, approaches a stationary molecule, B, the electrostatic and internuclear forces of molecule A repel molecule B, transferring some of the energy of motion to molecule B. Consequently, molecule B gains kinetic energy of motion, while molecule A slows down, losing some of its kinetic energy. Thus, as shown in Figure 4–3, a single molecule in solution bounces among the other molecules first in one direction, then another, then another, and so forth, bouncing randomly billions of times each second.

This continual movement of molecules among each other in liquids, or in gases, is called *diffusion*. Ions diffuse in exactly the same manner as whole molecules, and even suspended colloid particles diffuse in a similar manner except that they diffuse far less rapidly than molecular substances because of their very large sizes.

Figure 4–3. Diffusion of a fluid molecule during a billionth of a second.

DIFFUSION THROUGH THE CELL MEMBRANE

Diffusion through the cell membrane is divided into two separate subtypes called *simple diffusion* and *facilitated diffusion*. Simple diffusion means the molecular kinetic movement of molecules or ions through a membrane opening or intermolecular spaces without the necessity of binding with carrier proteins in the membrane. The rate of diffusion is determined by the amount of substance available, by the velocity of kinetic motion, and by the number of openings in the cell membrane through which the molecules or ions can move. On the other hand, facilitated diffusion requires the interaction of the molecules or ions with a carrier protein that aids its passage through the membrane, probably by binding chemically with it and shuttling it through the membrane in this form.

Simple diffusion can occur through the cell membrane by two pathways: through the interstices of the lipid bilayer and through watery channels in some of the transport proteins, as illustrated to the left in Figure 4–2.

Simple Diffusion Through the Lipid Bilayer

Diffusion of Lipid-Soluble Substances. In experimental studies, the lipids of cells have been separated from the proteins and then reconstituted as artificial membranes consisting of a lipid bilayer but

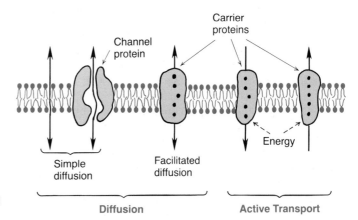

Figure 4–2. Transport pathways through the cell membrane, and the basic mechanisms of transport.

without any transport proteins. Using such an artificial membrane, the transport properties of the lipid bilayer by itself have been determined.

One of the most important factors that determines how rapidly a substance will move through the lipid bilayer is the *lipid solubility* of the substance. For instance, the lipid solubilities of oxygen, nitrogen, carbon dioxide, and alcohols are very high, so that all these can dissolve directly in the lipid bilayer and diffuse through the cell membrane in exactly the same manner that diffusion occurs in a watery solution. For obvious reasons, the rate of diffusion of these substances through the membrane is directly proportional to their lipid solubility. Especially large quantities of oxygen can be transported in this way; therefore, oxygen is delivered to the interior of the cell almost as though the cell membrane did not exist.

Transport of Water and Other Lipid-Insoluble Molecules. Even though water is highly insoluble in the membrane lipids, nevertheless it penetrates the cell membrane very readily, much of it passing directly through the lipid bilayer and still more passing through protein channels. The rapidity with which water molecules can penetrate the cell membrane is astounding. As an example, the total amount of water that diffuses in each direction through the red cell membrane during each second is approximately 100 times as great as the volume of the red cell itself.

The reason for the large amount of diffusion of water through the lipid bilayer is still not certain, but it is believed that water molecules are small enough and their kinetic energy great enough that they can simply penetrate like bullets through the lipid portion of the membrane before the "hydrophobic" character of the lipids can stop them.

Other lipid-insoluble molecules also can pass through the lipid bilayer in the same way as water molecules if they are small enough. However, as they become larger, their penetration falls off extremely rapidly. For instance, the diameter of the urea molecule is only 20 per cent greater than that of water. Yet its penetration through the cell membrane is about a thousand times less than that of water. Even so, remembering the astonishing rate of water penetration, this amount of penetration still allows rapid transport of urea through the cell membrane. The glucose molecule, which has a diameter only three times that of the water molecule, penetrates the lipid bilayer 100,000 times less rapidly than water, thus illustrating that the only lipid-insoluble molecules that can penetrate the lipid bilayer are the very small ones.

Failure of Ions to Diffuse Through the Lipid Bilayer. Even though water and other very small uncharged molecules diffuse easily through the lipid bilayer, ions — even small ones, such as hydrogen ions, sodium ions, potassium ions, and so forth — penetrate the lipid bilayer about one million times less rapidly than does water. Therefore, any significant transport of these through the cell membrane must occur through channels in the proteins, as we discuss shortly.

The reason for the impenetrability of the lipid bi-

Table 4–1 RELATIONSHIP OF EFFECTIVE DIAMETERS OF DIFFERENT SUBSTANCES TO THEIR LIPID BILAYER PERMEABILITIES

Substance	Diameter (nm)	Relative Permeability
Water molecule	0.3	1.0
Urea molecule	0.36	0.0006
Hydrated chloride ion	0.386	0.00000001
Hydrated potassium ion	0.396	0.0000000006
Hydrated sodium ion	0.512	0.0000000002
Glycerol	0.62	0.0006
Glucose	0.86	0.000009

layer to ions is the electrical charge of the ions; this impedes ionic movement in two separate ways: (1) The electrical charge of these ions causes multiple molecules of water to become bonded to the ions, forming so-called *hydrated ions.* This greatly increases the sizes of ions, which alone impedes penetration of the lipid bilayer. (2) Even more important, the electrical charge of the ion also interacts with the charges of the lipid bilayer in the following way. It will be recalled that each half of the bilayer is composed of "polar" lipids that have an excess of negative charge facing toward the surfaces of the membrane. Therefore, when a charged ion tries to penetrate either the negative or the positive electrical barrier, it is instantaneously repulsed.

To summarize, Table 4–1 gives the relative permeabilities of the lipid bilayer to a number of different molecules or ions of different diameters. Note especially the *extremely poor permeance of the ions* because of their electrical charges and the *poor permeance of glucose* because of its molecular diameter. Note also that glycerol penetrates the membrane almost as easily as urea even though its diameter is almost twice as great. The reason for this is a slight degree of lipid solubility.

Simple Diffusion Through Protein Channels and "Gating" of These Channels

The protein channels are believed to be watery pathways through the interstices of the protein molecules. In fact, computerized three-dimensional reconstructions of some of these proteins have demonstrated tube-shaped channels from the extracellular to the intracellular ends. Therefore, substances can diffuse directly through these channels from one side of the membrane to the other. However, the protein channels are distinguished by two important characteristics: (1) They are often selectively permeable to certain substances. (2) Many of the channels can be opened or closed by *gates.*

Selective Permeability of Different Protein Channels. Most, but not all, protein channels are highly selective for the transport of one or more specific ions or molecules. This results from the characteristics of the channel itself, such as its diameter, its shape, and the nature of the electrical charges along its inside surfaces. To give an example, one of the most important of the protein channels, the so-called *sodium channels,* calculate to be only 0.3 by 0.5 nanometer in size, but more importantly the inner surfaces of these channels are *strongly negatively charged,* as illustrated by the negative signs inside the channel proteins in the top panel of Figure 4–4. These strong negative charges are postulated to pull sodium ions more than they pull other physiologically important ions into the channels because of the smaller ionic diameter of the dehydrated sodium ions than for the others. Once in the channel, the sodium ions then diffuse in either direction according to the usual laws of diffusion. Thus, the sodium channel is specifically selective for the passage of sodium ions.

On the other hand, another set of protein channels is selective for potassium transport, illustrated in the lower panel of Figure 4–4. These channels calculate to be slightly smaller than the sodium channels, only 0.3 by 0.3 nanometer, but *they are not negatively charged.* Therefore, no strong attractive force is pulling ions into the channels, and the ions are not pulled away from the water molecules that hydrate them. The hydrated form of the potassium ion is considerably smaller than the hydrated form of sodium because the sodium ion has one whole orbital set of electrons less than the potassium ion, which allows the sodium nucleus to attract far more water molecules than can the potassium. Therefore, the smaller hydrated potassium ions can pass easily through this smaller channel, whereas sodium ions are mainly rejected, thus once again providing selective permeability for a specific ion.

Gating of Protein Channels. Gating of protein channels provides a means for controlling the perme-

ability of the channels. This is illustrated in both the upper and the lower panels of Figure 4–4 for the sodium and the potassium ion. It is believed that the gates are actual gatelike extensions of the transport protein molecule, which can close over the opening of the channel or can be lifted away from the opening by a conformational change in the shape of the protein molecule itself. In the case of the sodium channels, this gate opens and closes on the outer surface of the cell membrane, whereas for the potassium channels it opens and closes on the inner surface.

The opening and closing of gates are controlled in two principal ways:

1. *Voltage gating.* In this instance, the molecular conformation of the gate responds to the electrical potential across the cell membrane. For instance, when there is a strong negative charge on the inside of the cell membrane, the sodium gates remain tightly closed; on the other hand, when the inside of the membrane loses its negative charge, these gates open suddenly and allow tremendous quantities of sodium to pass inward through the sodium pores (until still another set of gates at the cytoplasmic ends of the channels close, as explained in Chapter 5). This is the basic cause of action potentials in nerves that are responsible for nerve signals. The potassium gates also open when the inside of the cell membrane becomes positively charged, but this response is much slower than that for the sodium gates. These events are discussed in the following chapter.

2. *Ligand gating.* Some protein channel gates are opened by the binding of another molecule with the protein; this causes a conformational change in the protein molecule that opens or closes the gate. This is called *ligand gating,* and the substance that binds is the *ligand.* One of the most important instances of ligand gating is the effect of acetylcholine on the so-called *acetylcholine channel.* This opens the gate of this channel, providing a pore about 0.65 nanometer in diameter that allows all molecules and positive ions smaller than this diameter to pass through. This gate is exceedingly important in the transmission of signals from one nerve cell to another (Chapter 45) and from nerve cells to muscle cells (Chapter 7).

The Open-State, Closed-State of Gated Channels. Figure 4–5 illustrates an especially interesting characteristic of voltage-gated channels. This figure shows two recordings of electrical current flowing through a single sodium channel when there was an approximate 25 millivolt potential gradient across the membrane. Note that the channel conducts current either all or none. That is, the gate of the channel snaps open and then snaps closed, each snapping event occurring within a few millionths of a second. This illustrates the rapidity with which conformational changes can occur in the shape of the protein molecular gates. At one voltage potential the channel may remain closed all the time or almost all the time, whereas at another voltage level it may remain open either all or most of the time. However, at in-between voltages, the gates tend to snap open and closed intermittently, as illustrated in the upper recording, giving

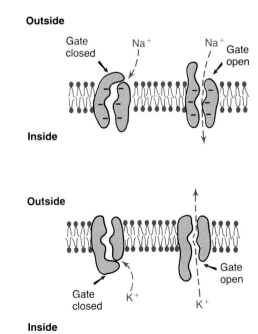

Figure 4–4. Transport of sodium and potassium ions through protein channels. Also shown are conformational changes of the channel protein molecules that open or close the "gates" guarding the channels.

an average current flow somewhere between the minimum and the maximum.

The Patch-Clamp Method for Recording Ion Current Flow Through Single Channels. One might wonder how it is technically possible to record ion current flow through single channels as shown in Figure 4–5A. This has been achieved by using the "patch-clamp" method illustrated in Figure 4–5B. Very simply, a micropipette, having a tip diameter of only 1 or 2 micrometers, is abutted against the outside of a cell membrane. Then suction is applied inside the pipette to pull the membrane slightly into the tip of the pipette. This creates a seal where the edges of the pipette touch the cell membrane. The result is a minute "patch" at the tip of the pipette through which current flow can be recorded.

Alternatively, as shown to the right in Figure 4–5B, the small cell membrane patch at the end of the pipette can be torn away from the cell. The pipette with its sealed patch is then inserted into a free solution. This allows the concentration of the ions both inside the micropipette and in the outside solution to be altered as desired. Also, the voltage between the two sides of the membrane can be set at will — that is, "clamped" to a given voltage.

Fortunately, it has been possible to make such patches small enough that one often finds only a single channel protein in the membrane patch that is being studied. By varying the concentrations of different ions and the voltage across the membrane, one can determine the transport characteristics of the channel as well as its gating properties.

Facilitated Diffusion

Facilitated diffusion is also called *carrier-mediated diffusion* because a substance transported in this manner usually cannot pass through the membrane without a specific carrier protein helping it. That is, the carrier *facilitates* the diffusion of the substance to the other side.

Facilitated diffusion differs from simple diffusion through an open channel in the following very important way: Although the rate of diffusion through an open channel increases proportionately with the concentration of the diffusing substance, in facilitated diffusion the rate of diffusion approaches a maximum, called V_{max}, as the concentration of the substance increases. This difference between simple diffusion and facilitated diffusion is illustrated in Figure 4–6, showing that as the concentration of the substance

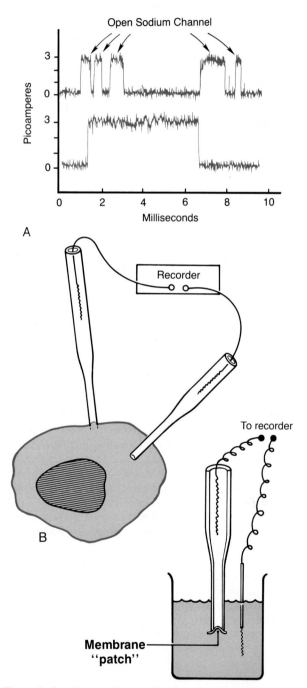

Figure 4–5. *A,* Record of current flow through a single voltage-gated sodium channel, demonstrating the all-or-none principle for opening of the channel. *B,* The "patch-clamp" method for recording current flow through a single protein channel. To the left, recording is performed from a "patch" of a living cell membrane. To the right, recording is from a membrane patch that has been torn away from the cell.

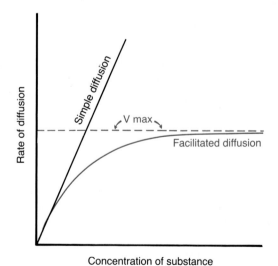

Figure 4–6. Effect of concentration of a substance on rate of diffusion through a membrane in which there is simple diffusion and a membrane in which there is facilitated diffusion. This illustrates that facilitated diffusion approaches a maximum rate called the V_{max}.

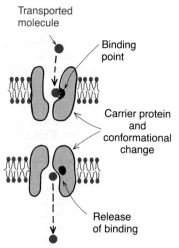

Figure 4-7. A postulated mechanism for facilitated diffusion.

increases, the rate of simple diffusion continues to increase proportionately but showing also the limitation of facilitated diffusion to the V_{max} level.

What is it that limits the rate of facilitated diffusion? A probable answer is the mechanism illustrated in Figure 4-7. This shows a carrier protein with a channel large enough to transport a specific molecule part way but not all the way through the membrane. It also shows a binding "receptor" on the protein carrier. The molecule to be transported enters the channel and becomes bound. Then in a fraction of a second a conformational change occurs in the carrier protein, so that the channel now opens to the opposite side of the membrane. Because the binding force of the receptor is weak, the thermal motion of the attached molecule causes it to break away and to be released on the opposite side. Obviously, the rate at which molecules can be transported by this mechanism can never be greater than the rate at which the carrier protein molecule can undergo conformational change back and forth between its two states. Note specifically that this mechanism allows the transported molecule to "diffuse" in either direction through the membrane.

Among the most important substances that cross cell membranes by facilitated diffusion are *glucose* and most of the *amino acids.* In the case of glucose, the carrier molecule is known to have a molecular weight of about 45,000; it can also transport several other monosaccharides that have structures similar to that of glucose, including mannose, galactose, xylose, and arabinose. Also, insulin can increase the rate of facilitated diffusion of glucose as much as 10-fold to 20-fold. This is the principal mechanism by which insulin controls glucose use in the body, as we discuss in Chapter 78.

FACTORS THAT AFFECT NET RATE OF DIFFUSION

By now it is evident that many different substances can diffuse either through the lipid bilayer of the cell membrane or through protein channels. However, please understand clearly that substances that diffuse in one direction can also diffuse in the opposite direction. Usually, what is important to the cell is not the total substance diffusing in both directions but the difference between these two, which is the *net rate of diffusion* in one direction. The factors that affect this are (1) the permeability of the membrane, (2) the difference in concentration of the diffusing substance between the two sides of the membrane, (3) the pressure difference across the membrane, and (4) in the case of ions, the electrical potential difference between the two sides of the membrane.

Permeability of the Membrane. The permeability of a membrane for a given substance is expressed as the *net* rate of diffusion of the substance through each unit area of the membrane for a unit concentration difference (when there are no electrical or pressure differences). The different factors that affect cell membrane permeability are

1. Thickness of the membrane — the greater the thickness, the less the rate of diffusion.

2. Lipid solubility — the greater the solubility of the substance in the cell membrane lipids, the greater the quantity of substances that dissolves in the membrane and therefore that can pass through.

3. Number of protein channels through which the substance can pass — the rate of diffusion is directly related to the number of channels per unit area.

4. The temperature — the greater the temperature, the greater is the thermal motion of molecules and ions in a solution, so that diffusion increases directly in proportion to temperature.

5. The molecular weight of the diffusing substance — this has a complex effect; the velocity of thermal motion of a dissolved substance is proportional to the square root of its molecular weight. On the other hand, as the molecular diameter approaches the diameter of a channel, the resistance increases tremendously, so that a membrane is frequently hundreds to millions of times as permeable to very small molecules as to very large ones, as is evident from the relative values given in Table 4-1.

The Diffusion Coefficient. Another factor that affects the total rate of diffusion is the area of the membrane. Therefore, to determine the total permeability of a cell membrane, of a capillary, or of any other complete membrane, one must multiply its permeability (which measures the movement of substances through a *unit* membrane area) by the *total* area of the membrane. This total permeability is expressed as the *diffusion coefficient;* its relationship to permeability is:

$$D = P \times A$$

where D is the diffusion coefficient, P is the permeability, and A is the total area.

Effect of a Concentration Difference. Figure 4-8A illustrates a cell membrane with a substance in high concentration on the outside and low concentration on the inside. The rate at which the substance diffuses *inward* is proportional to the concentration of molecules on the *outside,* for this concentration determines how many molecules strike the outside of the channels each second. On the other hand, the rate at which molecules diffuse *outward* is proportional to their concentration *inside* the membrane. Obviously, therefore, the rate of net diffusion into the cell is proportional to the concentration on the outside *minus* the concentration on the inside or

$$\text{Net diffusion} \propto D\ (C_o - C_i)$$

in which C_o is the concentration on the outside, C_i is the concentration on the inside, and D is the diffusion coefficient of the membrane for the substance.

Effect of an Electrical Potential on the Diffusion of Ions. If an electrical potential is applied across the membrane as shown in Figure 4–8B, because of their electrical charges, ions will move through the membrane even though no concentration difference exists to cause their movement. Thus, in the left panel of the figure, the concentrations of negative ions are exactly the same on both sides of the membrane, but a positive charge has been applied to the right side of the membrane and a negative charge to the left, creating an electrical gradient across the membrane. The positive charge attracts the negative ions, while the negative charge repels them. There-

fore, net diffusion occurs from left to right. After much time, large quantities of negative ions will have moved to the right (if we neglect, for the time being, the disturbing effects of the positive ions of the solution), creating the condition illustrated in the right panel of Figure 4–8B, in which a concentration difference of the same ions has developed in the direction opposite to the electrical potential difference. Obviously, the concentration difference is now tending to move the ions to the left, while the electrical difference is tending to move them to the right. When the concentration difference rises high enough, the two effects exactly balance each other. At normal body temperature (37°C), the electrical difference that will exactly balance a given concentration difference of *univalent* ions, such as sodium (Na^+), potassium (K^+), or chloride (Cl^-), can be determined from the following formula called the *Nernst equation:*

$$\text{EMF (in millivolts)} = \pm 61 \log \frac{C_1}{C_2}$$

in which EMF is the electromotive force (voltage) between side 1 and side 2 of the membrane, C_1 is the concentration on side 1, and C_2 is the concentration on side 2. The polarity of the voltage on side 1 in the above equation is $+$ for negative ions and $-$ for positive ions. This relationship is extremely important in understanding the transmission of nerve impulses, for which reason it is discussed in even greater detail in Chapter 5.

Effect of a Pressure Difference. At times considerable pressure difference develops between the two sides of a membrane. This occurs, for instance, at the capillary membrane, which has a pressure approximately 20 mm Hg greater inside the capillary than outside. Pressure actually means the sum of all the forces of the different molecules striking a unit surface area at a given instant. Therefore, when the pressure is higher on one side of a membrane than the other, this means that the sum of all the forces of the molecules striking the channels on that side of the membrane is greater than on the other side. This can result either from greater numbers of molecules striking the membrane per second or from greater kinetic energy of the average molecule striking the membrane. In either event, increased amounts of energy are available to cause net movement of molecules from the high pressure side toward the low pressure side. This effect is illustrated in Figure 4–8C, which shows a piston developing high pressure on one side of a cell membrane, thereby causing net diffusion through the membrane to the other side.

OSMOSIS ACROSS SELECTIVELY PERMEABLE MEMBRANES—NET DIFFUSION OF WATER

By far the most abundant substance to diffuse through the cell membrane is water. It should be recalled that enough water ordinarily diffuses in each

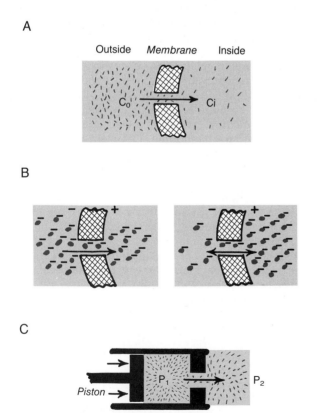

Figure 4–8. Effect of (A) concentration difference, (B) electrical difference, and (C) pressure difference on net diffusion of molecules and ions through a cell membrane.

direction through the red cell membrane per second to equal about *100 times the volume of the cell itself.* Yet, *normally,* the amount that diffuses in the two directions is so precisely balanced that not even the slightest *net* movement of water occurs. Therefore, the volume of the cell remains constant. However, under certain conditions, a *concentration difference for water* can develop across a membrane, just as concentration differences for other substances can also occur. When this happens, net movement of water does occur across the cell membrane, causing the cell either to swell or to shrink, depending on the direction of the net movement. This process of net movement of water caused by a concentration difference of water is called *osmosis.*

To give an example of osmosis, let us assume the conditions shown in Figure 4–9, with pure water on one side of the cell membrane and a solution of sodium chloride on the other side. Referring back to Table 4–1, we see that water molecules pass through the cell membrane with ease, whereas sodium and chloride ions pass through only with extreme difficulty. Therefore, sodium chloride solution is actually a mixture of permeant water molecules and nonpermeant sodium and chloride ions, and the membrane is said to be *selectively permeable* (or "semipermeable") to water but not to sodium and chloride ions. Yet the presence of the sodium and chloride has displaced some of the water molecules and therefore has reduced the concentration of water molecules to less than that of pure water. As a result, in the example of Figure 4–9, more water molecules strike the channels on the left side, where there is pure water, than on the right side, where the water concentration has been reduced. Thus, net movement of water occurs from left to right —that is, osmosis occurs from the pure water into the sodium chloride solution.

Osmotic Pressure

If in Figure 4–9 pressure were applied to the sodium chloride solution, osmosis of water into this solution would be slowed, stopped, or even reversed. The amount of pressure required exactly to stop osmosis is called the *osmotic pressure* of the sodium chloride solution.

The principle of a pressure difference opposing osmosis is illustrated in Figure 4–10, which shows a selectively permeable membrane separating two columns of fluid, one containing water and the other containing a solution of water and any solute that will not penetrate the membrane. Osmosis of water from chamber B into chamber A causes the levels of the fluid columns to become farther and farther apart, until eventually a pressure difference is developed that is great enough to oppose the osmotic effect. The pressure difference across the membrane at this point is the osmotic pressure of the solution containing the nondiffusible solute.

Importance of Numbers of Osmotic Particles (or of Molar Concentration) in Determining Osmotic Pressure. The osmotic pressure exerted by particles in a solution, whether they be molecules or ions, is determined by the *numbers* of particles per unit volume of fluid and not the mass of the particles. The reason for this is that each particle in a solution, regardless of its mass, exerts, on the average, the same amount of pressure against the membrane. That is, all particles are bouncing among each other with, on the average, equal energy. If some particles have greater kinetic energy of movement than others, their impact with the low-energy particles will impart part of their energy to these, thus decreasing the energy level of the high-energy particles, while increasing the energy level of the others until the energy level of all of them, averaged over a period of time, is the same. Therefore, the large particles, which have greater mass (m) than the small particles, move at lower velocities (v), while the small particles move at higher velocities in such a

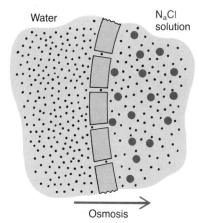

Figure 4–9. Osmosis at a cell membrane when a sodium chloride solution is placed on one side of the membrane and water on the other side.

Water   N$_a$Cl solution

Osmosis

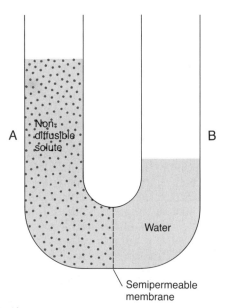

Figure 4–10. Demonstration of osmotic pressure on the two sides of a semipermeable membrane.

A   Non-diffusible solute   B

Water

Semipermeable membrane

way that their average kinetic energies (k), determined by the equation

$$k = \frac{mv^2}{2},$$

become equal to each other. Therefore, on the average, the kinetic energy of each molecule or ion that strikes a membrane is approximately the same regardless of its molecular size. Consequently, the factor that determines the osmotic pressure of a solution is the concentration of the solution in terms of numbers of particles (which is the same as its *molar concentration* if it is a nondissociated molecule) and not in terms of mass of the solute.

"Osmolality"—The Osmole. Because the amount of osmotic pressure exerted by a solute is proportional to the concentration of the solute in numbers of molecules or ions, expressing the solute concentration in terms of mass is of no value in determining osmotic pressure. To express the concentration in terms of numbers of particles, the unit called the *osmole* is used in place of grams.

One osmole is the number of molecules in 1 gram molecular weight of undissociated solute. Thus, 180 grams of glucose, which is 1 gram molecular weight of glucose, is equal to 1 osmole of glucose because glucose does not dissociate. On the other hand, if the solute dissociates into two ions, 1 gram molecular weight of the solute equals 2 osmoles because the number of osmotically active particles is now twice as great as is the case in the undissociated solute. Therefore, 1 gram molecular weight of sodium chloride, 58.5 grams, is equal to 2 osmoles.

A solution that has *1 osmole of solute dissolved in each kilogram of water* is said to have an *osmolality of 1 osmole per kilogram,* and a solution that has ½₀₀₀ osmole dissolved per kilogram has an osmolality of 1 milliosmole per kilogram. The normal osmolality of the extracellular and intracellular fluids is about 300 milliosmoles per kilogram.

Relationship of Osmolality to Osmotic Pressure. At normal body temperature, 37°C, a concentration of 1 osmole per liter will cause *19,300 mm Hg* osmotic pressure in the solution. Likewise, 1 milliosmole per liter concentration is equivalent to *19.3 mm Hg* osmotic pressure. Multiplying this value times the 300 milliosmolar concentration of the body fluids gives a total calculated osmotic pressure of these fluids of 5790 mm Hg. The measured value for this, however, averages only about 5500 mm Hg. The reason for this difference is that many of the ions in the body fluids, such as the sodium and the chloride ions, are highly attracted to each other; consequently, they cannot move totally unrestrained in the fluids and create their full osmotic potential. Therefore, on the average, the actual osmotic pressure of the body fluids is about 0.93 times the calculated value.

The Term "Osmolarity." Because of the difficulty of measuring kilograms of water in a solution, which is required to determine "osmolality," one usually uses another term, "osmolarity," which is the osmolar concentration expressed as *osmoles per liter of solution* rather than osmoles per kilogram of water. Although, strictly speaking, it is osmoles per kilogram of water (osmolality) that determines the osmotic pressure, nevertheless, for dilute solutions, such as those in the body, the quantitative differences between osmolarity and osmolality are less than 1 per cent. Since it is far more practical to measure osmolarity than osmolality, this is the usual practice in almost all physiological studies.

● ACTIVE TRANSPORT

From the discussion thus far, it is evident that *no substances can diffuse against an "electrochemical gradient,"* which is the sum of all the diffusion forces acting at the membrane—the forces caused by concentration difference, electrical difference, and pressure difference. That is, it is often said that substances cannot diffuse "uphill."

Yet, at times a large concentration of a substance is required in the intracellular fluid even though the extracellular fluid contains only a minute concentration. This is true, for instance, for potassium ions. Conversely, it is important to keep the concentrations of other ions very low inside the cell even though their concentrations in the extracellular fluid are very great. This is especially true for sodium ions. Obviously, neither of these two effects could occur by the process of simple diffusion, for simple diffusion tends always to equilibrate the concentrations on the two sides of the membrane. Instead, some energy source must cause movement of potassium ions "uphill" to the inside of cells and cause movement of sodium ions also "uphill" but in this instance to the outside of the cell. When a cell membrane moves molecules or ions uphill against a concentration gradient (or uphill against an electrical or pressure gradient), the process is called *active transport.*

Among the different substances that are actively transported through cell membranes are sodium ions, potassium ions, calcium ions, iron ions, hydrogen ions, chloride ions, iodide ions, urate ions, several different sugars, and most of the amino acids.

Primary Active Transport and Secondary Active Transport. Active transport is divided into two types according to the source of the energy used to cause the transport. These are called *primary active transport* and *secondary active transport.* In primary active transport, the energy is derived directly from the breakdown of adenosine triphosphate (ATP) or some other high-energy phosphate compound. In secondary active transport, the energy is derived secondarily from ionic concentration gradients that have been created in the first place by primary active transport. In both instances, transport depends on *carrier proteins* that penetrate through the membrane, the same as is true for facilitated diffusion. However, in active transport, the carrier protein functions differently from the carrier in facilitated diffusion, for it is capable of imparting energy to the transported substance to move it against an electrochemical gradient. Let us give some examples of primary active transport and secondary active transport and explain their principles of function more fully.

Primary Active Transport — The Sodium-Potassium "Pump"

Among the substances that are transported by primary active transport are sodium, potassium, calcium, hydrogen, chloride, and a few other ions. However, not all of these substances are transported by the membranes of all cells. Furthermore, some of the pumps function at intracellular membranes rather than (or in addition to) the surface membrane of the cell, such as at the membrane of the muscle sarcoplasmic reticulum or at one of the two membranes of the mitochondria. Nevertheless, they all operate by essentially the same basic mechanism.

The active transport mechanism that has been studied in greatest detail is the *sodium-potassium pump,* a transport process that pumps sodium ions outward through the cell membrane and at the same time pumps potassium ions from the outside to the inside. This pump is present in all cells of the body, and it is responsible for maintaining the sodium and potassium concentration differences across the cell membrane as well as for establishing a negative electrical potential inside the cells. Indeed, we see in the next chapter that this pump is the basis of nerve function to transmit nerve signals throughout the nervous system.

Figure 4–11 illustrates the basic components of the Na^+-K^+ pump. The *carrier protein* is a complex of two separate globular proteins, a larger one with a molecular weight of about 100,000 and a smaller one with a molecular weight of 55,000. Although the function of the smaller protein is not known, the larger protein has three specific features that are important for function of the pump:

1. It has three *receptor sites for binding sodium ions* on the portion of the protein that protrudes to the interior of the cell.

2. It has two *receptor sites for potassium ions* on the outside.

3. The inside portion of this protein adjacent to or near to the sodium binding sites has ATPase activity.

Now to put the pump into perspective: When three sodium ions bind on the inside of the carrier protein and two potassium ions on the outside, the ATPase function of the protein becomes activated. This then cleaves one molecule of ATP, splitting it to adenosine diphosphate (ADP) and liberating a high-energy phosphate bond of energy. This energy is then believed to cause a conformational change in the protein carrier molecule, extruding the sodium ions to the outside and the potassium ions to the inside. Unfortunately, the precise mechanism of the conformational change of the carrier is still to be discovered.

Importance of the Na^+-K^+ Pump in Controlling Cell Volume. One of the most important functions of the Na^+-K^+ pump is to control the volume of the cells. Without function of this pump, most cells of the body would swell until they burst. The mechanism for controlling the volume is the following: Inside the cell are large numbers of proteins and other organic compounds that cannot escape from the cell. Most of these are negatively charged and therefore collect around them large numbers of positive ions as well. All these substances then tend to cause osmosis of water to the interior of the cell; unless this is checked, the cell will swell indefinitely until it bursts. However, the normal mechanism for preventing this is the Na^+-K^+ pump. Note again that this device pumps three Na^+ ions to the outside of the cell for every two K^+ ions pumped to the interior. Also, the membrane is far less permeable to sodium ions than to potassium ions, so that once the sodium ions are on the outside they have a strong tendency to stay there. Thus, this represents a continual net loss of ionic substances out of the cell, which initiates an opposite osmotic tendency to move water out of the cell. Furthermore, when the cell begins to swell, this automatically activates the Na^+-K^+ pump, moving still more ions to the exterior and carrying water with them. Therefore, the Na^+-K^+ pump performs a continual surveillance role in maintaining normal cell volume.

The Electrogenic Nature of the Na^+-K^+ Pump. The fact that the Na^+-K^+ pump moves three sodium ions to the exterior for every two potassium ions to the interior means that a net of one positive charge is moved from the interior of the cell to the exterior for each revolution of the pump. This obviously creates positivity outside the cell but leaves a deficit of positive ions inside the cell; that is, it causes negativity on the inside. Therefore, the Na^+-K^+ pump is said to be *electrogenic* because it creates an electrical potential across the cell membrane as it pumps.

The Calcium Pump

Another very important primary active transport mechanism is the calcium pump. Calcium ions are normally maintained at extremely low concentration in the intracellular cytosol, at a concentration about 10,000 times less than that in the extracellular fluid.

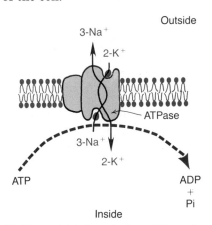

Figure 4–11. The postulated mechanism of the sodium-potassium pump.

This is achieved by two calcium pumps. One is in the cell membrane and pumps calcium to the outside of the cell. The other pumps calcium ions into one or more of the internal vesicular organelles of the cell, such as into the sarcoplasmic reticulum of muscle cells and into the mitochondria in all cells. In both instances, the carrier protein penetrates the membrane from side to side and also serves as an ATPase having the same capability to cleave ATP as the ATPase sodium carrier protein. The difference is that this protein has a binding site for calcium instead of sodium.

Saturation of Active Transport

Active transport saturates in exactly the same way that facilitated diffusion saturates, as illustrated in Figure 4–6. When the concentration of the substance to be transported is small, the rate of transport increases approximately in proportion to the increase in concentration. However, at large concentrations, transport approaches a maximum, which is called the V_{max}, as is also true for facilitated diffusion. The saturation is caused by limitation of the rates at which the chemical reactions of binding, release, and carrier conformational changes can occur.

Energetics of Active Transport

The amount of energy required to transport a substance actively through a membrane (aside from energy lost as heat in the chemical reactions) is determined by the degree that the substance is concentrated during transport. Compared with the energy required to concentrate a substance tenfold, to concentrate it hundredfold requires twice as much energy and to concentrate it thousandfold requires three times as much. In other words, the energy required is proportional to the *logarithm* of the degree that the substance is concentrated, as expressed by the following formula:

$$\text{Energy (in calories per osmole)} = 1400 \log \frac{C_1}{C_2}$$

That is, in terms of calories, the amount of energy required to concentrate 1 osmole of substance tenfold is about 1400 calories, or hundredfold, 2800 calories. One can see that the energy expenditure for concentrating substances in cells or for removing substances from cells against a concentration gradient can be tremendous. Some cells, such as those lining the renal tubules as well as many glandular cells, expend as much as 90 per cent of their energy for this purpose alone.

Secondary Active Transport — Co-Transport and Counter-Transport

When sodium ions are transported out of cells by primary active transport, a very large concentration gradient of sodium usually develops — very high con-

centration outside the cell and very low concentration inside. This gradient represents a storehouse of energy because the excess sodium outside the cell membrane is always attempting to diffuse to the interior. Under the appropriate conditions, this diffusion energy of sodium can literally pull other substances along with the sodium through the cell membrane. This phenomenon is called *co-transport;* it is one form of secondary active transport.

For sodium to pull another substance along with it, a coupling mechanism is required. This is achieved by means of still another carrier protein in the cell membrane. The carrier in this instance serves as an attachment point for both the sodium ion and the substance to be co-transported. Once they both are attached, a conformational change occurs in the carrier protein, and the energy gradient of the sodium ion causes both the sodium ion and the other substance to be transported together to the interior of the cell.

In *counter-transport,* sodium ions again attempt to diffuse to the interior of the cell because of their large concentration gradient. However, this time, the substance to be transported is on the inside of the cell and must be transported to the outside. Therefore, the sodium ion binds to the carrier protein where it projects through the exterior surface of the membrane, while the substance to be counter-transported binds to the interior projection of the carrier protein. Once both have bound, a conformational change occurs again, with the energy of the sodium ion moving to the interior causing the other substance to move to the exterior.

Sodium Co-Transport of Glucose and Amino Acids. Glucose and many amino acids are transported into most cells against very large concentration gradients; the mechanism of this is entirely by the co-transport mechanism illustrated in Figure 4–12. Note that the transport carrier protein has two binding sites on its exterior side, one for sodium and one for glucose. Also, the concentration of sodium ions is very high on the outside and very low inside, which provides the energy for the transport. A special property of the transport protein is that the conformational change to allow sodium movement to the interior will not occur until a glucose molecule also attaches. But when they are both attached, the conformational change takes place automatically, and both the sodium and the glucose are transported to

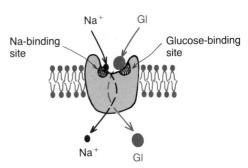

Figure 4–12. A postulated mechanism for sodium co-transport of glucose.

the inside of the cell at the same time. Hence, this is a *sodium-glucose co-transport* mechanism.

Sodium co-transport of the amino acids occurs in the same manner as for glucose except that it uses a different set of transport proteins. Five separate *amino acid transport proteins* have been identified, each of which is responsible for transporting one subset of amino acids with specific molecular characteristics.

Sodium co-transport of glucose and amino acids occurs especially in the epithelial cells of the intestinal tract and renal tubules to aid in the absorption of these substances into the blood, as we discuss in later chapters.

Two other important co-transport mechanisms are (1) a *sodium-potassium-two chloride* co-transporter that allows two chloride ions to be carried into cells along with one sodium and one potassium ion, all moving in the same direction, and (2) a *potassium-chloride co-transporter* that allows potassium and chloride ions to be transported from inside cells to the exterior together. Still other co-transport mechanisms into at least some cells include co-transport of iodine ions, iron ions, and urate ions.

Sodium Counter-Transport of Calcium and Hydrogen Ions. Two especially important counter-transport mechanisms are *sodium-calcium counter-transport* and *sodium-hydrogen counter-transport*. Calcium counter-transport occurs in all or almost all cell membranes, with sodium ions moving to the interior and calcium ions to the exterior, both bound to the same transport protein in a counter-transport mode. This is in addition to primary active transport of calcium that occurs in some cells.

Sodium-hydrogen counter-transport occurs in several tissues. An especially important example is in the proximal tubules of the kidneys, where sodium ions move from the lumen of the tubule to the interior of the tubular cells, while hydrogen ions are counter-transported into the lumen. This mechanism is a key to hydrogen ion control in the body fluids, as is discussed in detail in Chapter 30.

Other counter-transport mechanisms include cation exchanges of calcium or sodium ions on one side of the membrane for magnesium or potassium ions on the other side and anion exchanges of chloride ions moving in one direction for bicarbonate ions or sulfate ions moving in the other direction.

ACTIVE TRANSPORT THROUGH CELLULAR SHEETS

At many places in the body, substances must be transported all the way through a cellular sheet instead of simply through the cell membrane. Transport of this type occurs through the intestinal epithelium, the epithelium of the renal tubules, the epithelium of all exocrine glands, the epithelium of the gallbladder, the membrane of the choroid plexus of the brain, and many other membranes.

The basic mechanism for transport of a substance through a cellular sheet is (1) to provide *active transport* through the cell membrane *on one side* of the cell

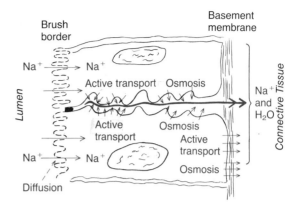

Figure 4–13. Basic mechanism of active transport all the way through a layer of cells.

and then (2) to provide for either *simple diffusion* or *facilitated diffusion* through the membrane *on the opposite side* of the cell.

Figure 4–13 illustrates the mechanism for transport of sodium ions through the epithelium of the intestines, the gallbladder, and the renal tubules. This figure shows that the epithelial cells are connected together at the luminal pole by means of tight junctions that mainly block sodium ion diffusion through the spaces between the cells. However, the luminal surfaces of the cells are very permeable to both sodium ions and to water. Therefore, sodium and water diffuse readily to the interior of the cell. Then, at the basal and lateral membranes of the cells, the sodium ions are actively transported into the extracellular fluid. This creates a high sodium ion concentration gradient across these membranes, which in turn causes osmosis of water as well. Thus, the active transport of sodium ions at the basolateral sides of the epithelial cells results in the transport not only of sodium ion but of water also.

Furthermore, any other substances that might be linked by co-transport with the sodium can be transported in this same way. For instance, the positive charges of the sodium ions usually pull negatively charged chloride ions along with the sodium. Likewise, when glucose (or amino acids) is co-transported along with sodium through the luminal surface of the cell, the concentration of the glucose increases inside the cell. Then, this glucose is transported by facilitated diffusion through the basolateral borders of the cell, finally entering the extracellular fluid along with the sodium ions, the chloride ions, and the water.

These are the mechanisms by which almost all the nutrients, the ions, and other substances are absorbed into the blood from the intestine; they are also the way in which these same substances are reabsorbed from the glomerular filtrate by the renal tubules.

REFERENCES

Agnew, W. S.: Voltage-regulated sodium channel molecules. Annu. Rev. Physiol., 46:517, 1984.
Almers, W., and Stirling, C.: Distribution of transport proteins over animal membranes. J. Membr. Biol., 77:169, 1984.
Andreoli, T. E., et al. (eds.): Physiology of Membrane Disorders. 2nd Ed. New York, Plenum Publishing Corp., 1986.

Auerbach, A., and Sachs, F.: Patch clamp studies of single ionic channels. Annu. Rev. Biophys. Bioeng., 13:269, 1984.

Barry, P. H., and Diamond, J. M.: Effects of unstirred layers on membrane phenomena. Physiol. Rev., 64:763, 1984.

Biggio, G., and Costa, E. (eds.): Chloride Channels and Their Modulation by Neurotransmitters and Drugs. New York, Raven Press, 1988.

Blachley, J. E., et al.: The harmful effects of ethanol on ion transport and cellular respiration. Am. J. Med. Sci., 289:22, 1985.

Bohr, D. F.: Cell membrane in hypertension. News Physiol. Sci., 4:85, 1989.

Bretag, A. H.: Muscle chloride channels. Physiol. Rev., 67:618, 1987.

Cereijido, M., et al.: Tight junction: Barrier between higher organisms and environment. News Physiol. Sci., 4:72, 1989.

Chasis, J. A., and Shohet, S. B.: Red cell biochemical anatomy and membrane properties. Annu. Rev. Physiol., 49:237, 1987.

Deuticke, B., and Haest, C. W. M.: Lipid modulation of transport proteins in vertebrate cell membranes. Annu. Rev. Physiol., 49:221, 1987.

Dinno, M. A., and Armstrong, W. M. (eds.): Membrane Biophysics III: Biological Transport. New York, Alan R. Liss, Inc., 1988.

DiPolo, R., and Beauge, L.: The calcium pump and sodium-calcium exchange in squid axons. Annu. Rev. Physiol., 45:313, 1983.

Donowitz, M., and Welsh, M. J.: Ca^{2+} and cyclic AMP in regulation of intestinal Na, K, and Cl transport. Annu. Rev. Physiol., 48:135, 1986.

Ellis, D.: Na-Ca exchange in cardiac tissues. Adv. Myocardiol., 5:295, 1985.

Finkelstein, A., et al.: Osmotic swelling of vesicles. Annu. Rev. Physiol., 48:135, 1986.

Flatman, P. W.: Magnesium transport across cell membranes. J. Membr. Biol., 80:1, 1984.

Forgac, M.: Structure and function of vacuolar class of ATP-driven proton pumps. Physiol. Rev., 69:765, 1989.

Gadsby, D. C.: The Na/K pump of cardiac cells. Annu. Rev. Biophys. Bioeng., 13:373, 1984.

Haas, M.: Properties and diversity of Na-K-Cl cotransporters. Annu. Rev. Physiol., 51:443, 1989.

Haynes, D. H., and Mandveno, A.: Computer modeling of Ca^{2+} pump function of Ca^{2+}-Mg^{2+}-ATPase of sarcoplasmic reticulum. Physiol. Rev., 67:244, 1987.

Hidalgo, C. (ed.): Physical Properties of Biological Membranes and Their Functional Implications. New York, Plenum Publishing Corp., 1988.

Hoffmann, E. K., and Simonsen, L. O.: Membrane mechanisms in volume and pH regulation in vertebrate cells. Physiol. Rev., 69:315, 1989.

Ives, H. E., and Rector, F. C., Jr.: Proton transport and cell function. J. Clin. Invest., 73:285, 1984.

Jacobson, K., et al.: Lateral diffusion of proteins in membranes. Annu. Rev. Physiol., 49:163, 1987.

Jennings, M. L.: Kinetics and mechanism of anion transport in red blood cells. Annu. Rev. Physiol., 47:519, 1985.

Kaplan, J. H.: Ion movements through the sodium pump. Annu. Rev. Physiol., 47:535, 1985.

Kim, C. H., and Tedeschi, H. (eds.): Advances in Membrane Biochemistry and Bioenergetics. New York, Plenum Publishing Corp., 1987.

Latorre, R., et al.: Varieties of calcium-activated potassium channels. Annu. Rev. Physiol., 51:385, 1989.

Lauger, P.: Dynamics of ion transport systems in membranes. Physiol. Rev., 67:1296, 1987.

Lear, J. D., et al.: Synthetic amphiphilic peptide models for protein ion channels. Science, 240:1177, 1988.

Lerner, J.: Cell membrane amino acid transport processes in the domestic fowl *(Gallus domesticus)*. Comp. Biochem. Physiol., 78:205, 1984.

Lindermann, B.: Fluctuation analysis of sodium channels in epthelia. Annu. Rev. Physiol., 46:497, 1984.

Macey, R. I.: Transport of water and urea in red blood cells. Am. J. Physiol., 246:C195, 1984.

Malhotra, S. K.: The Plasma Membrane. New York, John Wiley & Sons, 1983.

Miller, C.: Ion channels in liposomes. Annu. Rev. Physiol., 46:549, 1984.

Miller, C.: Integral membrane channels: Studies in model membranes. Physiol. Rev., 63:1209, 1983.

Narahashi, T.: Ion Channels. New York, Plenum Publishing Corp., 1988.

Parker, J. D., and Berkowitz, L. R.: Physiologically instructive genetic variants involving the human red cell membrane. Physiol. Rev., 63:261, 1983.

Petersen, O. H.: Potassium channels and fluid secretion. News Physiol. Sci., 1:92, 1986.

Petersen, O. H., and Petersen, C. C. H.: The patch-clamp technique: Recording ionic currents through single pores in the cell membrane. News Physiol. Sci., 1:5, 1986.

Reuter, H.: Modulation of ion channels by phosphorylation and second messengers. News Physiol. Sci., 2:168, 1987.

Reuter, H.: Ion channels in cardiac cell membranes. Annu. Rev. Physiol., 46:473, 1984.

Rudel, R., and Lehmann-Horn, F.: Membrane changes in cells from myotonia patients. Physiol. Rev., 65:310, 1985.

Sakmann, B., and Neher, E.: Patch clamp techniques for studying ionic channels in excitable membranes. Annu. Rev. Physiol., 46:455, 1984.

Salkoff, L. B., and Tonouye, M. A.: Genetics of ion channels. Physiol. Rev., 66:301, 1986.

Schatzmann, H. J.: The calcium pump of the surface membrane and of the sarcoplasmic reticulum. Annu. Rev. Physiol., 51:473, 1989.

Schatzmann, H. J.: The red cell calcium pump. Annu. Rev. Physiol., 445:303, 1983.

Schultz, S. G.: A cellular model for active sodium absorption by mammalian colon. Annu. Rev. Physiol., 46:435, 1984.

Schwartz, G. J., and Al-Awqati, Q.: Regulation of transepithelia H^+ transport by exocytosis and endocytosis. Annu. Rev. Physiol., 48:153, 1986.

Smith, P. L., and McCabe, R. D.: Mechanism and regulation of transcellular potassium transport by the colon. Am. J. Physiol., 247:G445, 1984.

Stein, W. D. (ed.): The Ion Pumps: Structure, Function, and Regulation. New York, Alan R. Liss, Inc., 1988.

Trimmer, J. S., and Agnew, W. S.: Molecular diversity of voltage-sensitive Na channels. Annu. Rev. Physiol., 51:401, 1989.

Verner, K., and Schatz, G.: Protein translocution across membranes. Science, 241:1307, 1988.

Wallace, B. A., and Ravikumar, K.: The gramicidin pore: Crystal structure of a cesium complex. Science, 241:182, 1988.

Welsh, M. J.: Electrolyte transport by airway epithelia. Physiol. Rev., 67:1143, 1987.

Wright, E. M.: Electrophysiology of plasma membrane vesicles. Am. J. Physiol., 246:F363, 1984.

CHAPTER 5

• *Membrane Potentials and Action Potentials* •

Electrical potentials exist across the membranes of essentially all cells of the body, and some cells, such as nerve and muscle cells, are "excitable"—that is, capable of self-generation of electrochemical impulses at their membranes and, in most instances, employment of these impulses to transmit signals along the membranes. In still other types of cells, such as glandular cells, macrophages, and ciliated cells, other types of changes in membrane potentials probably play significant roles in controlling many of the cell's functions. However, the present discussion is concerned with membrane potentials generated both at rest and during action by nerve and muscle cells.

● BASIC PHYSICS OF MEMBRANE POTENTIALS

MEMBRANE POTENTIALS CAUSED BY DIFFUSION

Figure 5–1A and B illustrates a nerve fiber when there is no active transport of either sodium or potassium ions. In Figure 5–1A, the potassium concentration is very great inside the membrane, whereas that outside is very low. Let us also assume that the membrane in this instance is very permeable to the potassium ions but not to any other ions. Because of the large potassium concentration gradient from the inside toward the outside, there is a strong tendency for potassium ions to diffuse outward. As they do so, they carry positive charges to the outside, thus creating a state of electropositivity outside the membrane and electronegativity on the inside because of the negative anions that remain behind, which do not diffuse outward along with the potassium. This new potential difference repels the positively charged potassium ions in a backward direction from the outside to the inside. Within a millisecond or so, the potential change becomes great enough to block further net diffusion of potassium ions to the exterior despite the high potassium ion concentration gradient. In the normal large mammalian nerve fiber, the potential

difference required is about 94 millivolts, with negativity inside the fiber membrane.

Figure 5–1B illustrates the same phenomenon as that in Figure 5–1A but this time with a high concentration of sodium ions outside the membrane and a low sodium concentration inside. These ions are also positively charged, and this time the membrane is highly permeable to the sodium ions but impermeable to all other ions. Diffusion of the sodium ions to the inside creates a membrane potential now of opposite polarity, with negativity outside and positivity inside. Again, the membrane potential rises high enough within milliseconds to block further net diffusion of the sodium ions to the inside; however, this time, for the large mammalian nerve fiber, the potential is about 61 millivolts and with positivity inside the fiber.

Thus, in both parts of Figure 5–1 we see that a concentration difference of ions across a selectively permeable membrane can, under appropriate conditions, cause the creation of a membrane potential. In later sections of this chapter, we see that many of the rapid changes in membrane potentials observed during the course of nerve and muscle impulse transmission result from the occurrence of rapidly changing diffusion potentials of this nature.

Relationship of the Diffusion Potential to the Concentration Difference—The Nernst Equation. The potential level across the membrane that will exactly prevent net diffusion of an ion in either direction through the membrane is called the *Nernst potential* for that ion. The magnitude of this potential is determined by the *ratio* of the ion concentrations on the two sides of the membrane—the greater this ratio, the greater the tendency for the ions to diffuse in one direction, and therefore the greater is the Nernst potential. The following equation, called the *Nernst equation,* can be used to calculate the Nernst potential for any univalent ion at normal body temperature of 37°C:

$$\text{EMF (millivolts)} = \pm 61 \log \frac{\text{Conc. inside}}{\text{Conc. outside}}$$

Diffusion Potentials

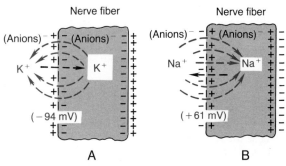

Nerve fiber   Nerve fiber

A   B

Figure 5–1. *A,* Establishment of a diffusion potential across a cell membrane, caused by potassium ions diffusing from inside the cell to the outside through a membrane that is selectively permeable only to potassium. *B,* Establishment of a diffusion potential when the membrane is permeable only to sodium ions. Note that the internal membrane potential is negative when potassium ions diffuse and positive when sodium ions diffuse because of opposite concentration gradients of these two ions.

When using this formula, it is assumed that the potential outside the membrane always remains at exactly zero potential, and the Nernst potential that is calculated is the potential inside the membrane. Also, the sign of the potential is positive (+) when the ion under consideration is a negative ion and negative (−) when it is a positive ion.

Thus, when the concentration of a positive ion (potassium ions, for instance) on the inside is 10 times that on the outside, the log of 10 is 1, so that the Nernst potential calculates to be −61 millivolts inside the membrane.

Calculation of the Diffusion Potential When the Membrane Is Permeable to Several Different Ions

When a membrane is permeable to several different ions, the diffusion potential that develops depends on three factors: (1) the polarity of the electrical charge of each ion, (2) the permeability of the membrane *(P)* to each ion, and (3) the concentrations *(C)* of the respective ions on the inside *(i)* and outside *(o)* of the membrane. Thus, the following formula, called the *Goldman equation,* or the *Goldman-Hodgkin-Katz equation,* gives the calculated membrane potential on the *inside* of the membrane when two univalent positive ions, sodium (Na⁺) and potassium (K⁺), and one univalent negative ion, chloride (Cl⁻), are involved.

$$\text{EMF (millivolts)} = -61 \cdot \log \frac{C_{Na^+_i}P_{Na^+} + C_{K^+_i}P_{K^+} + C_{Cl^-_o}P_{Cl^-}}{C_{Na^+_o}P_{Na^+} + C_{K^+_o}P_{K^+} + C_{Cl^-_i}P_{Cl^-}}$$

Now, let us study the importance and the meaning of this equation. First, sodium, potassium, and chloride ions are the ions most importantly involved in the development of membrane potentials in nerve and muscle fibers as well as in the neuronal cells in the central nervous system. The concentration gradient of each of these ions across the membrane helps determine the voltage of the membrane potential.

Second, the degree of importance of each of the ions in determining the voltage is proportional to the membrane permeability for that particular ion. Thus, if the membrane is impermeable to both potassium and chloride ions, the membrane potential becomes entirely dominated by the concentration gradient of sodium ions alone, and the resulting potential will be exactly equal to the Nernst potential for sodium. The same principle holds for each of the other two ions if the membrane should become selectively permeable for either one of them alone.

Third, a positive ion concentration gradient from *inside* the membrane *to the outside* causes electronegativity inside the membrane. The reason for this is that positive ions diffuse to the outside when their concentration is higher inside than outside. This carries positive charges to the outside but leaves the nondiffusible negative anions on the inside. Exactly the opposite effect occurs when there is a negative ion gradient. That is, a chloride ion gradient from the *outside to the inside* causes negativity inside the cell because negatively charged chloride ions then diffuse to the inside, while leaving the positive ions on the outside.

Fourth, we see later that the permeabilities of the sodium and potassium channels undergo very rapid changes during conduction of the nerve impulse, whereas the permeability of the chloride channels does not change greatly during this process. Therefore, the changes in the sodium and potassium permeabilities are primarily responsible for signal transmission in the nerves, which is the subject of most of the remainder of this chapter.

MEASURING THE MEMBRANE POTENTIAL

The method for measuring the membrane potential is simple in theory but often very difficult in practice because of the small sizes of many of the fibers. Figure 5–2 illustrates a small pipette, filled with a very strong electrolyte solution (KCl), that is impaled through the cell membrane to the interior of the fiber. Then another electrode, called the "indifferent electrode," is placed in the interstitial fluids, and the potential difference between the inside and

Silver-silver chloride electrode

Figure 5–2. Measurement of the membrane potential of the nerve fiber using a microelectrode.

outside of the fiber is measured using an appropriate volt-meter. This is a highly sophisticated electronic apparatus that is capable of measuring very small voltages despite extremely high resistance to electrical flow through the tip of the micropipette, which has a diameter usually less than 1 micrometer and a resistance often as great as a billion ohms. For recording rapid *changes* in the membrane potential during the transmission of nerve impulses, the microelectrode is connected to an oscilloscope, as explained later in the chapter.

THE CELL MEMBRANE AS AN ELECTRICAL CAPACITOR

In each of the figures shown thus far, the negative and positive ionic charges that cause the membrane potential have been shown to be lined up against the membrane, and we have not spoken of the arrangement of the charges elsewhere in the fluids, either inside the nerve fiber or on the outside in the interstitial fluid. However, Figure 5–3 illustrates this, showing that everywhere except adjacent to the surfaces of the cell membrane itself, the negative and positive charges are exactly equal. This is called the principle of *electrical neutrality;* that is, for every positive ion there is a negative ion nearby to neutralize it, or otherwise electrical potentials of billions of volts would appear within the fluids.

When positive charges are pumped to the outside of the membrane, these positive charges line up along the outside of the membrane, and on the inside the anions line up that have been left behind. This creates a *dipole layer* of positive and negative charges between the outside and inside of the membrane, but it still leaves equal numbers of negative and positive charges everywhere else within the fluids. This is the same effect that occurs when the plates of an electrical capacitor become electrically charged—that is, lining up of negative and positive charges on the opposite sides of the dielectric membrane of the capacitor. Therefore, the lipid bilayer of the cell membrane actually functions as a *dielectric* of a cell membrane capacitor, much as mica, paper, and Mylar function as dielectrics in electrical capacitors.

Because of the extreme thinness of the cell membrane (only 7 to 10 nanometers), its capacitance is tremendous for its area—about *1 microfarad per square centimeter.*

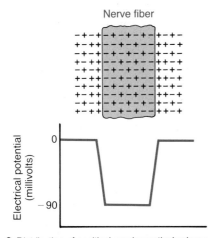

Figure 5–3. Distribution of positively and negatively charged ions in the interstitial fluid surrounding a nerve fiber and in the fluid inside the fiber; note the dipolar alignment of negative charges along the inside surface of the membrane and positive charges along the outside surface. In the lower panel are illustrated the abrupt changes in membrane potential that occur at the membranes on the two sides of the fiber.

The lower part of Figure 5–3 illustrates the electrical potential that will be recorded at each point in or near the nerve fiber membrane, beginning at the left side of the figure and passing to the right. As long as the electrode is outside the nerve membrane, the potential that is recorded is zero, which is the potential of the extracellular fluid. Then, as the recording electrode passes through the electrical dipole layer at the cell membrane, the potential decreases immediately to −90 millivolts. Again, the electrical potential remains at a steady level as the electrode passes across the interior of the fiber but reverses back to zero the instant it passes through the opposite side of the membrane.

The fact that the nerve membrane functions as a capacitor has one especially important point of significance: To create a negative potential inside the membrane, only enough positive ions must be transported outward to develop the electrical dipole layer at the membrane itself. All the remaining ions inside the nerve fiber can still be both positive and negative ions. Therefore, an incredibly small number of ions needs to be transferred through the membrane to establish the normal potential of −90 millivolts inside the nerve fiber — only about $\frac{1}{3,000,000}$ to $\frac{1}{100,000,000}$ of the total positive charges inside the fiber need be so transferred. Also, an equally small number of positive ions moving from outside to the inside of the fiber can reverse the potential from −90 millivolts to as much as +35 millivolts within as little as $\frac{1}{10,000}$ of a second. This rapid shifting of ions in this manner causes the nerve signals that we discuss in the subsequent sections of this chapter.

● THE RESTING MEMBRANE POTENTIAL OF NERVES

The membrane potential of large nerve fibers when they are not transmitting nerve signals is about −90 millivolts (mV). That is, the potential *inside the fiber* is 90 mV more negative than the potential in the interstitial fluid on the outside of the fiber. In the next few paragraphs, we explain all the factors that determine the level of this potential, but before doing so we must describe the transport properties of the resting nerve membrane for sodium and potassium.

Active Transport of Sodium and Potassium Ions Through the Membrane—The Sodium-Potassium Pump. First, let us recall from the discussions of the previous chapter that all cell membranes of the body have a powerful sodium-potassium pump and that this continually pumps sodium to the outside of the fiber and potassium to the inside. Further, let us remember that this is an *electrogenic pump* because more positive charges are pumped to the outside than to the inside (three Na^+ ions to the outside for each two K^+ ions to the inside), leaving a net deficit of positive ions on the inside; this is the same as causing a negative charge inside the cell membrane.

This sodium-potassium pump also causes the tremendous concentration gradients for sodium and potassium across the resting nerve membrane. These gradients are the following:

Na^+ (outside):	142 mEq/L
Na^+ (inside):	14 mEq/L

Outside

Figure 5–4. The functional characteristics of the Na⁺-K⁺ pump and also of the potassium-sodium "leak" channels.

| K⁺ (outside): | 4 mEq/L |
| K⁺ (inside): | 140 mEq/L |

The ratios of these two respective ions from the inside to the outside are

$$Na^+_{inside}/Na^+_{outside} = 0.1$$
$$K^+_{inside}/K^+_{outside} = 35.0$$

Leakage of Potassium and Sodium Through the Nerve Membrane. To the right in Figure 5–4 is illustrated a channel protein in the cell membrane, through which potassium and sodium ions can leak, called a *potassium-sodium "leak" channel.* There are actually multiple different proteins of this type with different leak characteristics. However, the emphasis is on potassium leakage because on the average the channels are far more permeable to potassium than to sodium, normally about 100 times as permeable. We see later that this differential in permeability is exceedingly important in determining the level of the normal resting membrane potential.

ORIGIN OF THE NORMAL RESTING MEMBRANE POTENTIAL

Figure 5–5 illustrates the important factors in the establishment of the normal resting membrane potential of −90 mV. These are as follows.

Contribution of the Potassium Diffusion Potential. In Figure 5–5A, we make the assumption that the only movement of ions through the membrane is the diffusion of potassium ions, as illustrated by the open channels between the potassium inside the membrane and the outside. Because of the high ratio of potassium ions inside to outside, 35 to 1, the Nernst potential corresponding to this ratio is −94 mV, for the logarithm of 35 is 1.54, and this times −61 mV is −94 mV. Therefore, if potassium ions were the only factor causing the resting potential, this resting potential would also be equal to −94 mV, as illustrated in the figure.

Contribution of Sodium Diffusion Through the Nerve Membrane. Figure 5–5B illustrates the addition of very slight permeability of the nerve mem-

Figure 5–5. Establishment of resting membrane potentials in nerve fibers under three separate conditions: *A*, when the membrane potential is caused entirely by potassium diffusion alone; *B*, when the membrane potential is caused by diffusion of both sodium and potassium ions; and *C*, when the membrane potential is caused by diffusion of both sodium and potassium ions plus pumping of both these ions by the Na⁺–K⁺ pump.

brane to sodium ions, caused by the minute diffusion of the sodium ions through the K⁺-Na⁺ leak channels. The ratio of sodium ions from inside to outside the membrane is 0.1, and this gives a calculated Nernst potential for the inside of the membrane of +61 mV. But also shown in Figure 5–5B is the Nernst potential for potassium diffusion of −94 mV. How do these interact with each other, and what will be the summated potential? This can be answered by using the Goldman equation described earlier. However, intuitively, one can see that if the membrane is highly permeable to potassium but only very slightly permeable to sodium, it is logical that the diffusion of potassium will contribute far more to the membrane potential than will the diffusion of sodium. In the normal nerve fiber, the permeability of the membrane to potassium is about 100 times as great as to sodium. Using this value in the Goldman equation gives an internal membrane potential of −86 mV, as shown to the right in the figure.

Contribution of the Na⁺-K⁺ Pump. Finally, in Figure 5–5C an additional contribution of the Na⁺-K⁺ pump is illustrated. In this figure, there is continuous pumping of three sodium ions to the outside for each two potassium ions pumped to the inside of the membrane. The fact that more sodium ions are being pumped to the outside than potassium to the inside causes a continual loss of positive charges from inside the membrane; this creates an additional degree of negativity (about −4 mV additional) on the inside beyond that which can be accounted for by diffusion alone. Therefore, as illustrated in Figure 5–5C, the net membrane potential with all these factors operative at the same time is −90 mV.

In summary, the diffusion potentials alone caused by potassium and sodium diffusion would give a membrane potential of approximately −86 mV, almost all of this being determined by potassium diffusion. Then, an additional −4 mV is contributed to the membrane potential by the electrogenic Na⁺-K⁺ pump, giving a net resting membrane potential of −90 mV.

The resting membrane potential in large skeletal muscle fibers is approximately the same as that in large nerve fibers, also −90 mV. However, in both small nerve fibers and small muscle fibers—smooth muscle, for instance—as well as in many of the neurons of the central nervous system, the membrane potential is often as little as −40 to −60 mV instead of −90 mV.

● THE NERVE ACTION POTENTIAL

Nerve signals are transmitted by *action potentials,* which are rapid changes in the membrane potential. Each action potential begins with a sudden change from the normal resting negative potential to a positive membrane potential and then ends with an almost equally rapid change back again to the negative potential. To conduct a nerve signal, the action potential moves along the nerve fiber until it comes to the fiber's end. The upper panel of Figure 5–6 shows the disturbances that occur at the membrane during the action potential, with transfer of positive charges to the interior of the fiber at its onset and return of positive charges to the exterior at its end. The lower panel illustrates graphically the successive changes in the membrane potential over a few 10,000ths of a second, illustrating the explosive onset of the action potential and the almost equally as rapid recovery.

The successive stages of the action potential are as follows:

Resting Stage. This is the resting membrane potential before the action potential occurs. The membrane is said to be "polarized" during this stage because of the very large negative membrane potential that is present.

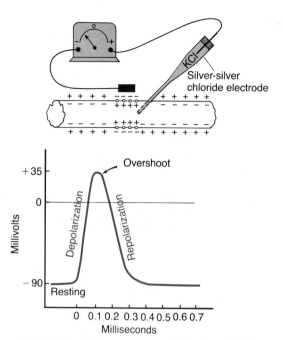

Figure 5–6. A typical action potential recorded by the method illustrated in the upper panel of the figure.

Depolarization Stage. At this time, the membrane suddenly becomes very permeable to sodium ions, allowing tremendous numbers of sodium ions to flow to the interior of the axon. The normal "polarized" state of −90 mV is lost, with the potential rising rapidly in the positive direction. This is called *depolarization.* In large nerve fibers, the membrane potential actually "overshoots" beyond the zero level and becomes somewhat positive, but in some smaller fibers as well as many central nervous system neurons, the potential merely approaches the zero level and does not overshoot to the positive state.

Repolarization Stage. Within a few 10,000ths of a second after the membrane becomes highly permeable to sodium ions, the sodium channels begin to close, and the potassium channels open more than normally. Then, rapid diffusion of potassium ions to the exterior re-establishes the normal negative resting membrane potential. This is called *repolarization* of the membrane.

To explain more fully the factors that cause both the depolarization and the repolarization processes, we need now to describe the special characteristics of yet two other types of transport channels through the nerve membrane: the voltage-gated sodium and potassium channels.

THE VOLTAGE-GATED SODIUM AND POTASSIUM CHANNELS

The necessary actor in causing both depolarization and repolarization of the nerve membrane during the action potential is the *voltage-gated sodium channel.* However, the *voltage-gated potassium channel* also plays an important role in increasing the rapidity of repolarization of the membrane. *These two voltage-gated channels are in addition to the Na⁺-K⁺ pump and also in addition to the Na⁺-K⁺ leak channels.*

The Voltage-Gated Sodium Channel — "Activation" and "Inactivation" of the Channel

The upper panel of Figure 5–7 illustrates the voltage-gated sodium channel in three separate states. This channel has two *gates,* one near the outside of the channel called the *activation gate* and another near the inside called the *inactivation gate.* To the left is shown the state of these two gates in the normal resting membrane when the membrane potential is −90 mV. In this state, the activation gate is closed, which prevents any entry of sodium ions to the interior of the fiber through these sodium channels. On the other hand, the inactivation gate is open and does not at this time constitute any barrier to the movement of the sodium ions.

Activation of the Sodium Channel. When the membrane potential becomes less negative than during the resting state, rising from −90 mV toward zero, it finally reaches a voltage, usually somewhere between −70 and −50 mV, that causes a sudden conformational change in the activation gate, flipping it to the open position. This is called the *activated state;* during this state, sodium ions can literally pour inward through the channel, increasing the sodium permeability of the membrane as much as 500-fold to 5000-fold.

Figure 5–7. Characteristics of the voltage-gated sodium and potassium channels, showing both activation and inactivation of the sodium channels but activation of the potassium channels only when the membrane potential is changed from the normal resting negative value to a positive value.

Inactivation of the Sodium Channel. To the far right in the upper panel of Figure 5–7 is illustrated a third state of the sodium channel. The same increase in voltage that opens the activation gate also closes the inactivation gate. However, closure of the inactivation gate occurs a few 10,000ths of a second after the activation gate opens. That is, the conformational change that flips the inactivation gate to the closed state is a slower process, whereas the conformational change that opens the activation gate is a very rapid process. Therefore, after the sodium channel has remained open for a few 10,000ths of a second, it closes, and sodium ions can no longer pour to the inside of the membrane. At this point the membrane potential begins to recover back toward the resting membrane state, which is the repolarization process.

A very important characteristic of the sodium channel inactivation process is that *the inactivation gate will not reopen again until the membrane potential returns either to or nearly to the original resting membrane potential level.* Therefore, it is not possible for the sodium channels to open again without the nerve fiber first repolarizing.

The Voltage-Gated Potassium Channels and Their Activation

The lower panel of Figure 5–7 illustrates the voltage-gated potassium channel in two separate states: during the resting state and toward the end of the action potential. During the resting state, the gate of the potassium channel is closed, as illustrated to the left in the figure, and potassium ions are prevented from passing through this channel to the exterior. When the membrane potential rises from −90 mV toward zero, this voltage change causes a slow conformational opening of the gate and allows increased potassium diffusion outward through the channel. However, because of the slowness of opening of these potassium channels, they mainly open just at the same time that the sodium channels are beginning to become inactivated and therefore are closing. Thus, the decrease in sodium entry to the cell and simultaneous increase in potassium exit from the cell greatly speeds the repolarization process, leading within a few 10,000ths of a second to full recovery of the resting membrane potential.

The Research Method for Measuring the Effect of Voltage on Opening and Closing of the Voltage-Gated Channels—The "Voltage Clamp." The original research that led to our quantitative understanding of the sodium and potassium channels was so ingenious that it led to Nobel prizes for the scientists responsible, Hodgkin and Huxley. The essence of these studies is illustrated in Figures 5–8 and 5–9.

Figure 5–8 illustrates the experimental apparatus called the *voltage clamp,* which is used to measure the flow of ions through the different channels. In using this apparatus, two electrodes are inserted into the nerve fiber. One of these is for the purpose of measuring the voltage of the membrane potential. The other is to conduct electrical current either into or out of the nerve fiber. This apparatus is used in the

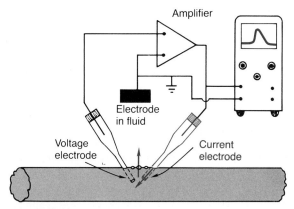

Figure 5–8. The "voltage clamp" method for studying flow of ions through specific channels.

following way: The investigator decides what voltage he or she wishes to establish inside the nerve fiber. He or she then adjusts the electronic portion of the apparatus to the desired voltage, and this automatically injects either positive or negative electricity through the current electrode at whatever rate is required to hold the voltage, as measured by the voltage electrode, at the level set by the operator. For instance, when the membrane potential is suddenly increased by this voltage clamp from −90 mV to zero, the voltage-gated sodium and potassium channels open, and sodium and potassium ions begin to pour through the channels. To counterbalance the effect of these ion movements on the set potential, electrical current is injected automatically through the current electrode of the voltage clamp to maintain the intracellular voltage at the zero level. To achieve this, the current injected must be exactly equal to but of opposite polarity to the net current flow through the membrane channels. To measure how much current flow is occurring at each instant, the current electrode is connected to an oscilloscope that records the current flow, as illustrated on the screen of the oscilloscope in the figure. Finally, the investigator adjusts the concentrations of the ions to desired levels both inside and outside the nerve fiber and repeats the study. This can be done very easily when using very large nerve fibers removed from some crustaceans, especially the giant squid axon that is sometimes as large as 1 millimeter in diameter. When sodium is the only permeant ion in the solutions inside and outside the squid axon, the

Figure 5–9. Typical changes in conductance of sodium and potassium ion channels when the membrane potential is suddenly increased from the normal resting value of −90 mV to a positive value of +10 mV for 2 milliseconds. This figure illustrates that the sodium channels open (activate) and then close (inactivate) before the end of the 2 milliseconds, whereas the potassium channels only open (activate).

voltage clamp measures current flow only through the sodium channels. When potassium is the only permeant ion, current flow only through the potassium channels is measured.

Another means for studying the flow of ions through individual channels is to block one type of channel at a time. For instance, the sodium channels can be blocked by a toxin called *tetrodotoxin* by applying this toxin to the outside of the cell membrane where the sodium activation gates are located. Conversely, *tetraethylammonium ion* blocks the potassium pores when it is applied to the interior of the nerve fiber.

Figure 5–9 illustrates the typical changes in conductance of the voltage-gated sodium and potassium channels when the membrane potential is suddenly changed by use of the voltage clamp from −90 mV to +10 mV and 2 milliseconds later back again to −90 mV. Note the sudden opening of the sodium channels (the activation stage) within a very small fraction of a millisecond after the membrane potential is increased to the positive value. However, during approximately the next millisecond or so, the sodium channels automatically close (the inactivation stage).

Now, note the opening (activation) of the potassium channels. These open slowly and reach the full open state only after the sodium channels have already become almost completely closed. Furthermore, once the potassium channels open, they remain open for the entire duration of the positive membrane potential and do not close again until after the membrane potential is decreased back to a very negative value.

Finally, let us recall that the voltage-gated channels normally flip to the open state or the closed state very suddenly, which was illustrated in Figure 4–5 in the previous chapter. Therefore, how is it that the curves in Figure 5–9 are so smooth? The answer is that these curves represent the flow of sodium and potassium ions through literally thousands of channels at the same time. Some open at one voltage level, others at another voltage, and so forth. Likewise, some become inactivated at different points in the cycle from the others. Thus, the illustrated curves represent summated current flows through the many channels.

SUMMARY OF THE EVENTS THAT CAUSE THE ACTION POTENTIAL

Figure 5–10 illustrates in summary form the sequential events that occur during and shortly after the action potential. These are as follows:

At the bottom of the figure are shown the changes in membrane conductances for sodium and potassium ions. During the resting state, before the action potential begins, the conductance for potassium ions is shown to be 50 to 100 times as great as the conductance for sodium ions. This is caused by much greater leakage of potassium ions than sodium ions through the leak channels. However, at the onset of the action potential, the sodium channels instantaneously become activated and allow an up to 5000-fold increase in sodium conductance. Then the inactivation process closes the sodium channels within another few fractions of a millisecond. The onset of the action potential also causes voltage gating of the potassium channels, causing them to begin opening a fraction of a millisecond after the sodium channels open. And at

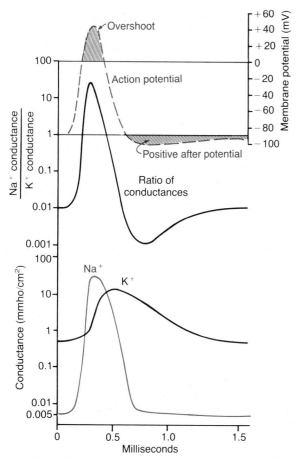

Figure 5–10. Changes in sodium and potassium conductances during the course of the action potential. Note that sodium conductance increases several thousandfold during the early stages of the action potential, whereas potassium conductance increases only about 30-fold during the latter stages of the action potential and for a short period thereafter. (Curves constructed from data in Hodgkin and Huxley papers but transposed from squid axon to apply to the membrane potentials of large mammalian nerve fibers.)

the end of the action potential, the return of the membrane potential to the negative state causes the potassium channels to close back to their original status, but again only after a short delay.

In the middle portion of Figure 5–10 is shown the ratio of sodium conductance to potassium conductance at each instant during the action potential, and above this is shown the action potential itself. During the early portion of the action potential, this ratio increases more than a thousandfold. Therefore, far more sodium ions now flow to the interior of the fiber than do potassium ions to the exterior. This is what causes the membrane potential to become positive. Then the sodium channels begin to become inactivated, and at the same time the potassium channels open, so that the ratio of conductance now shifts far in favor of high potassium conductance but low sodium conductance. This allows extremely rapid loss of potassium ions to the exterior, while essentially no sodium ions flow to the interior. Consequently, the action potential quickly returns to its baseline level.

The "Positive" Afterpotential

Note also in Figure 5–10 that the membrane potential becomes even more negative than the original resting membrane potential for a few milliseconds after the action potential is over. Strangely enough, this is called the *"positive" afterpotential,* which is a misnomer because the positive afterpotential is actually even more negative than the resting potential. The reason for calling it "positive" is that, historically, the first potential measurements were made on the outside of the nerve fiber membrane rather than inside, and when measured on the outside this potential causes a positive record on the recording meter rather than a negative one.

The cause of the positive afterpotential is mainly that many potassium channels remain open for several milliseconds after the repolarization process of the membrane is complete. This allows excess potassium ions to diffuse out of the nerve fiber, leaving an extra deficit of positive ions on the inside, which means more negativity.

ROLES OF OTHER IONS DURING THE ACTION POTENTIAL

Thus far, we have considered only the roles of sodium and potassium ions in the generation of the action potential. However, at least three other types of ions must be considered. These are as follows:

The Impermeant Negatively Charged Ions (Anions) Inside the Axon. Inside the axon are many negatively charged ions that cannot go through the membrane channels. These include protein molecules, many organic phosphate compounds, sulfate compounds, and so forth. Because these cannot leave the interior of the axon, any deficit of positive ions inside the membrane leaves an excess of the impermeant negative ions. Therefore, these impermeant negative ions are responsible for the negative charge inside the fiber when there is a deficit of the positively charged potassium ions and other positive ions.

Calcium Ions. The cell membranes of almost all, if not all, cells of the body have a calcium pump similar to the sodium pump. Like the sodium pump, this device pumps calcium ions from the interior to the exterior of the cell membrane (or into the endoplasmic reticulum), creating a calcium ion gradient of about 10,000-fold, leaving an internal concentration of calcium ions of about 10^{-7} molar in contrast to an external concentration of about 10^{-3} molar.

In addition, there are also voltage-gated calcium channels. These channels are slightly permeable to sodium ions as well as to calcium ions; when they open, both calcium and sodium ions flow to the interior of the fiber. Therefore, these channels are sometimes also called Ca^{++}-Na^{+} *channels.* The calcium channels are very slow to become activated, requiring 10 to 20 times as long for activation as the sodium channels. Therefore, they are also frequently called *slow channels,* in contrast to the sodium channels that are called *fast channels.*

Calcium channels are very numerous in both cardiac muscle and smooth muscle. In fact, in some types of smooth muscle, the fast sodium channels are hardly present at all, so the action potentials then are caused almost entirely by activation of the slow calcium channels.

Increased Permeability of the Sodium Channels When There Is a Deficit of Calcium Ions. The concentration of calcium ions in the interstitial fluid also has a profound effect on the voltage level at which the sodium channels become activated. When there is a deficit of calcium ions, the sodium channels are activated (opened) by very little increase of the membrane potential above the normal resting level. Therefore, the nerve fiber becomes highly excitable, sometimes discharging repetitively without any provocation rather than remaining in the resting state. In fact, the calcium ion concentration needs to fall only 30 to 50 per cent below normal before spontaneous discharge occurs in many peripheral nerves, often causing muscle "tetany" that can actually be lethal because of tetanic contraction of the respiratory muscles.

The probable way in which calcium ions affect the sodium channels is the following: These ions appear to bind to the exterior surfaces of the sodium channel protein molecule. The positive charges of these calcium ions, in turn, alter the electrical state of the channel protein itself, in this way increasing the voltage level required to open the gate.

Chloride Ions. Chloride ions leak through the resting membrane in the same way that small quantities of potassium and sodium ions leak through. In the usual nerve fiber, the rate of chloride diffusion through the membrane is about one half as great as the diffusion of potassium ions. Therefore, the question must be asked: Why have we not considered the chloride ions in our explanation of the action potential? The answer is that the chloride ions function passively in this process. Also, the permeability of the chloride leak channels does not change significantly during the action potential.

In the normal resting state of the nerve fiber, the −90 mV inside the fiber repels most of the chloride ions from the fiber. Therefore, the concentration of chloride ions inside the fiber is only 3 to 4 mEq/liter, whereas the concentration outside the fiber is about 103 mEq/liter. The Nernst potential for this ratio of chloride ions is exactly equal to the −90 mV membrane potential, which is what one would expect for an ion that is not actively pumped.

During the action potential, small quantities of chloride ions do diffuse into the nerve fiber because of the temporary loss of the internal negativity. This movement of chloride ions serves to alter slightly the timing of the successive voltage changes during the action potential, but it does not alter the fundamental process.

INITIATION OF THE ACTION POTENTIAL

Up to this point, we have explained the changing sodium and potassium permeabilities of the membrane as well as the development of the action potential itself, but we have not explained what initiates the action potential. The answer to this, as follows, is really quite simple:

A Positive Feedback Opens the Sodium Channels. First, as long as the membrane of the nerve fiber remains totally undisturbed, no action potential occurs in the normal nerve. However, if any event at all causes enough initial rise in the membrane potential from −90 mV up toward the zero level, the rising voltage itself will cause many voltage-gated sodium channels to begin opening. This allows rapid inflow of sodium ions, which causes still further rise of the membrane potential, thus opening still more voltage-gated sodium channels and more streaming of sodium ions to the interior of the fiber. Obviously, this process is a positive-feedback vicious circle that, once the feedback is strong enough, will continue until all the voltage-gated sodium channels have become totally

activated (opened). Then, within another fraction of a millisecond, the rising membrane potential causes beginning inactivation of the sodium channels as well as opening of potassium channels, and the action potential soon terminates.

Threshold for Initiation of the Action Potential. An action potential will not occur until the initial rise in membrane potential is great enough to create the vicious circle described in the last paragraph. Usually, a sudden rise in membrane potential of 15 to 30 mV is required. Therefore, a sudden increase in the membrane potential in a large nerve fiber of from −90 mV up to about −65 mV will usually cause the explosive development of the action potential. This level of −65 mV, therefore, is said to be the *threshold* for stimulation.

Accommodation of the Membrane—Failure to Fire Despite Rising Voltage. If the membrane potential rises very slowly—over many milliseconds instead of a fraction of a millisecond—the slow, inactivating gates of the sodium channels will have time to close at the same time that the activating gates are opening. Consequently, the opening of the activating gates will not be as effective in increasing the flow of sodium ions as normally. Therefore, a slow increase in the internal potential of a nerve fiber either requires a higher threshold voltage than normal to cause firing or prevents firing entirely at times, even with a voltage rise all the way to zero or even to positive voltage. This phenomenon is called *accommodation* of the membrane to the stimulus.

● PROPAGATION OF THE ACTION POTENTIAL

In the preceding paragraphs we have discussed the action potential as it occurs at one spot on the membrane. However, an action potential elicited at any one point on an excitable membrane usually excites adjacent portions of the membrane, resulting in propagation of the action potential. The mechanism of this is illustrated in Figure 5–11. Figure 5–11A shows a normal resting nerve fiber, and Figure 5–11B shows a nerve fiber that has been excited in its midportion—that is, the midportion has suddenly developed increased permeability to sodium. The arrows illustrate a "local circuit" of current flow between the depolarized areas of the membrane and the adjacent resting membrane areas; positive electrical charges carried by the inward diffusing sodium ions flow inward through the depolarized membrane and then for several millimeters along the core of the axon. These positive charges increase the voltage for a distance of 1 to 3 millimeters inside large fibers to above the threshold voltage value for initiating an action potential. Therefore, the sodium channels in these new areas immediately activate, and, as illustrated in Figure 5–11C and D, the explosive action potential spreads. Then these newly depolarized areas cause local circuits of current flow still further along the membrane, causing progressively more and more depolarization. Thus, the depolarization process travels along the entire extent of the fiber. The transmission of the depolarization

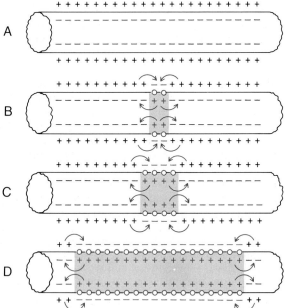

Figure 5–11. Propagation of action potentials in both directions along a conductive fiber.

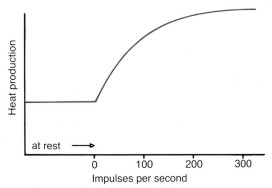

Figure 5–12. Heat production in a nerve fiber at rest and at progressively increasing rates of stimulation.

process along a nerve or muscle fiber is called a *nerve* or *muscle impulse.*

Direction of Propagation. It is obvious, as illustrated in Figure 5–11, that an excitable membrane has no single direction of propagation, but that the action potential can travel in both directions away from the stimulus—and even along all branches of a nerve fiber—until the entire membrane has become depolarized.

The All-or-Nothing Principle. It is equally obvious that, once an action potential has been elicited at any point on the membrane of a normal fiber, the depolarization process will travel over the entire membrane if conditions are right, or it might not travel at all if conditions are not right. This is called the all-or-nothing principle, and it applies to all normal excitable tissues. Occasionally, though, the action potential will reach a point on the membrane at which it does not generate sufficient voltage to stimulate the next area of the membrane. When this occurs, the spread of depolarization stops. Therefore, for continued propagation of an impulse to occur, the ratio of action potential to threshold for excitation must at all times be greater than 1. This is called the *safety factor* for propagation.

● RE-ESTABLISHING SODIUM AND POTASSIUM IONIC GRADIENTS AFTER ACTION POTENTIALS— IMPORTANCE OF ENERGY METABOLISM

Transmission of each impulse along the nerve fiber reduces infinitesimally the concentration differences of sodium and potassium between the inside and out-

side of the membrane because of diffusion of sodium ions to the inside during depolarization and diffusion of potassium ions to the outside during repolarization. For a single action potential, this effect is so minute that it cannot be measured. Indeed, 100,000 to 50,000,000 impulses can be transmitted by nerve fibers, the number depending on the size of the fiber and several other factors, before the concentration differences have run down to the point that action potential conduction ceases. Yet, even so, with time it becomes necessary to re-establish the sodium and potassium membrane concentration differences. This is achieved by the action of the Na^+-K^+ pump in exactly the same way as that described earlier in the chapter for original establishment of the resting potential. That is, the sodium ions that have diffused to the interior of the cell during the action potentials and the potassium ions that have diffused to the exterior are returned to their original state by the Na^+-K^+ pump. Since this pump requires energy for operation, this process of "recharging" the nerve fiber is an active metabolic one, using energy derived from the adenosine triphosphate energy system of the cell. Figure 5–12 shows that the nerve fiber produces excess heat, which is a measure of its energy expenditure, when the impulse frequency increases.

A special feature of the sodium-potassium ATPase pump is that its degree of activity is strongly stimulated when excess sodium ions accumulate inside the cell membrane. In fact, the pumping activity increases approximately in proportion to the third power of the sodium concentration. That is, as the internal sodium concentration rises from 10 to 20 mEq/liter, the activity of the pump does not merely double but instead increases approximately eightfold. Therefore, it can easily be understood how the "recharging" process of the nerve fiber can rapidly be set into motion whenever the concentration differences of sodium and potassium ions across the membrane begin to "run down."

● PLATEAU IN SOME ACTION POTENTIALS

In some instances the excitable membrane does not repolarize immediately after depolarization, but, in-

Figure 5-13. An action potential from a Purkinje fiber of the heart, showing a "plateau."

stead, the potential remains on a plateau near the peak of the spike for many milliseconds before repolarization begins. Such a plateau is illustrated in Figure 5-13; one can readily see that the plateau greatly prolongs the period of depolarization. This type of action potential occurs in heart muscle fibers, where the plateau lasts for as long as two to three tenths second and causes contraction of the heart muscle also to last for this same long period of time.

The cause of the plateau is a combination of several different factors. First, in heart muscle, two separate types of channels enter into the depolarization process: (1) the usual voltage-activated sodium channels, called *fast channels,* and (2) voltage-activated calcium channels, which are slow to be activated and therefore are called *slow channels* — these channels allow diffusion mainly of calcium ions but of some sodium ions as well. Activation of the fast channels causes the spike portion of the action potential, whereas the slow but prolonged activation of the slow channels is mainly responsible for the plateau portion of the action potential.

A second factor partly responsible for the plateau is that the voltage-gated potassium channels are slow to be activated in some instances, often not opening until the very end of the plateau. This delays the return of the membrane potential toward the resting value. But, then, this opening of the potassium channels at the same time that the slow channels begin to close causes rapid return of the action potential from its plateau level back to the negative resting level, accounting for the rapid downslope at the end of the action potential.

● RHYTHMICITY OF CERTAIN EXCITABLE TISSUES — REPETITIVE DISCHARGE

Repetitive self-induced discharges, or *rhythmicity,* occur normally in the heart, in most smooth muscle,

and in many of the neurons of the central nervous system. It is these rhythmical discharges that cause the heart rhythm, that cause peristalsis, and that cause such neuronal events as the rhythmical control of breathing.

Also, all other excitable tissues can discharge repetitively if the threshold for stimulation is reduced low enough. For instance, even nerve fibers and skeletal muscle fibers, which normally are highly stable, discharge repetitively when they are placed in a solution containing the drug veratrine or when the calcium ion concentration falls below a critical value.

The Re-Excitation Process Necessary for Rhythmicity. For rhythmicity to occur, the membrane, even in its natural state, must already be permeable enough to sodium ions (or to calcium and sodium ions through the slow calcium channels) to allow automatic membrane depolarization. Thus, Figure 5-14 shows that the "resting" membrane potential is only −60 to −70 mV. This is not enough negative voltage to keep the sodium and calcium channels closed. That is, (1) sodium and calcium ions flow inward, (2) this further increases the membrane permeability, (3) still more ions flow inward, (4) the permeability increases more, and so forth, thus eliciting the regenerative process of sodium and calcium channel openings until an action potential is generated. Then, at the end of the action potential, the membrane repolarizes. But shortly thereafter, the depolarization process begins again, and a new action potential occurs spontaneously. This cycle continues again and again and causes self-induced rhythmical excitation of the excitable tissue.

Yet, why does the membrane not depolarize immediately after it has become repolarized rather than delaying for nearly a second before the onset of the next action potential? The answer to this can be found by referring back to Figure 5-10, which shows that toward the end of all action potentials, and continuing for a short period thereafter, the membrane becomes excessively permeable to potassium. The excessive outflow of potassium ions carries tremendous numbers of positive charges to the outside of the membrane, creating inside the fiber considerably more negativity than would otherwise occur for a short period after the preceding action potential is

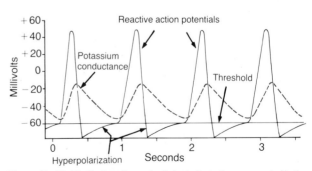

Figure 5-14. Rhythmic action potentials similar to those recorded in the rhythmical control center of the heart. Note their relationship to potassium conductance and to the state of hyperpolarization.

over, thus drawing the membrane potential nearer to the potassium Nernst potential. This is a state called *hyperpolarization,* which is illustrated in Figure 5–14. As long as this state exists, re-excitation will not occur; but gradually the excess potassium conductance (and the state of hyperpolarization) disappears, as shown in the figure, thereby allowing the membrane potential to increase until it reaches the *threshold* for excitation; then suddenly a new action potential results, the process occurring again and again.

SPECIAL ASPECTS OF SIGNAL TRANSMISSION IN NERVE TRUNKS

Myelinated and Unmyelinated Nerve Fibers. Figure 5–15 illustrates a cross-section of a typical small nerve trunk, showing many large nerve fibers that compose most of the cross-sectional area. However, look very carefully and you will see many more small fibers lying between the large ones. The large fibers are *myelinated,* and the small ones are *unmyelinated.* The average nerve trunk contains about twice as many unmyelinated fibers as myelinated fibers.

Figure 5–16 illustrates a typical myelinated fiber. The central core of the fiber is the *axon,* and the membrane of the axon is the actual *conductive membrane.* The axon is filled in its center with *axoplasm,* which is a viscid intracellular fluid. Surrounding the axon is a *myelin sheath* that is often thicker than the axon itself, and about once every 1 to 3 millimeters along the length of the axon the myelin sheath is interrupted by a *node of Ranvier.*

The myelin sheath is deposited around the axon by Schwann cells in the following manner: The membrane of a Schwann cell first envelops the axon. Then the cell rotates around the axon many times, laying down multiple layers of

cellular membrane containing the lipid substance *sphingomyelin.* This substance is an excellent insulator that decreases the ion flow through the membrane approximately 5000-fold and also decreases the membrane capacitance as much as 50-fold. However, at the juncture between each two successive Schwann cells along the axon, a small, uninsulated area remains, only 2 to 3 *micro*meters in length, where ions can still flow with ease between the extracellular fluid and the axon. This area is the node of Ranvier.

"Saltatory" Conduction in Myelinated Fibers from Node to Node. Even though ions cannot flow significantly through the thick myelin sheaths of myelinated nerves, they can flow with considerable ease through the nodes of Ranvier. Therefore, action potentials can occur *only at the nodes.* Yet, the action potentials are conducted from node to node, as illustrated in Figure 5–17; this is called *saltatory conduction.* That is, electrical current flows through the surrounding extracellular fluids and also through the axoplasm from node to node, exciting successive nodes one after another. Thus, the nerve impulse jumps down the fiber, which is the origin of the term "saltatory."

Saltatory conduction is of value for two reasons. First, by causing the depolarization process to jump long intervals along the axis of the nerve fiber, this mechanism increases the velocity of nerve transmission in myelinated fibers as much as 5-fold to 50-fold. Second, saltatory conduction conserves energy for the axon, for only the nodes depolarize, allowing perhaps a hundred times smaller loss of ions than would otherwise be necessary and therefore requiring little extra metabolism for re-establishing the sodium and potassium concentration differences across the membrane after a series of nerve impulses.

Still another feature of saltatory conduction in large myelinated fibers is the following: The excellent insulation afforded by the myelin membrane and the 50-fold decrease in membrane capacitance allows the repolarization process to occur with very little transfer of ions. Therefore, at the end of the action potential when the sodium channels begin to

Figure 5–15. Cross-section of a small nerve trunk containing myelinated and unmyelinated fibers.

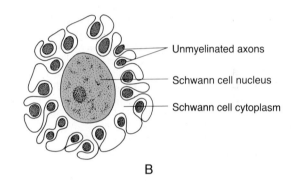

Figure 5–16. Function of the Schwann cell to insulate nerve fibers. *A,* The wrapping of a Schwann cell membrane around a large axon to form the myelin sheath of the myelinated nerve fiber. (Modified from Leeson and Leeson: Histology. Philadelphia, W. B. Saunders Company, 1979.) *B,* Evagination of the membrane and cytoplasm of a Schwann cell around multiple unmyelinated nerve fibers.

close, repolarization occurs so rapidly that the potassium channels usually have not yet begun to open significantly. Therefore, conduction of the nerve impulse in the myelinated nerve fiber is accomplished almost entirely by the sequential changes in the voltage-gated sodium channels, with very little contribution by the potassium channels.

VELOCITY OF CONDUCTION IN NERVE FIBERS

The velocity of conduction in nerve fibers varies from as little as 0.5 m/sec in very small unmyelinated fibers to as high as 100 m/sec (the length of a football field in 1 second) in very large myelinated fibers. The velocity increases ap-

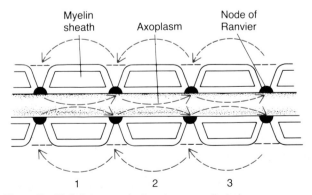

Figure 5–17. Saltatory conduction along a myelinated axon.

proximately with the fiber diameter in myelinated nerve fibers and approximately with the square root of fiber diameter in unmyelinated fibers.

● EXCITATION—THE PROCESS OF ELICITING THE ACTION POTENTIAL

Basically, any factor that causes sodium ions to begin to diffuse inward through the membrane in sufficient numbers will set off the automatic, regenerative opening of the sodium channels. This can result from simple *mechanical* disturbance of the membrane, *chemical* effects on the membrane, or passage of *electricity* through the membrane. All these are used at different points in the body to elicit nerve or muscle action potentials: mechanical pressure to excite sensory nerve endings in the skin, chemical neurotransmitters to transmit signals from one neuron to the next in the brain, and electrical current to transmit signals between muscle cells in the heart and intestine. For the purpose of understanding the excitation process, let us begin by discussing the principles of electrical stimulation.

Excitation of a Nerve Fiber by a Negatively Charged Metal Electrode. The usual means for exciting a nerve or muscle in the experimental laboratory is to apply electricity at the nerve or muscle surface through two small electrodes, one of which is negatively charged and the other positively charged. When this is done, one finds that the excitable membrane becomes stimulated at the negative electrode.

The cause of these effects is the following: Remember that the action potential is initiated by the opening of voltage-gated sodium channels. Furthermore, these channels are opened by a decrease in the electrical voltage across the membrane. The negative current from the negative electrode reduces the voltage immediately outside the membrane, drawing this voltage nearer to the voltage of the negative membrane potential inside the fiber. This decreases the electrical voltage across the membrane and allows activation of the sodium channels, thus resulting in an action potential. Conversely, at the anode, the injection of positive charges on the outside of the nerve membrane heightens the voltage difference across the membrane rather than lessening it. And this causes a state of "hyperpolarization," which decreases the excitability of the fiber.

Threshold for Excitation and "Acute Local Potentials." A weak electrical stimulus may not be able to excite a fiber. However, when the stimulus is progressively increased, there comes a point at which excitation does take place. Figure 5–18 illustrates the effects of successively applied stimuli of progressing strength. A very weak stimulus at point A causes the membrane potential to change from −90 to −85 mV, but this is not sufficient change for the automatic regenerative processes of the action potential to develop. At point B, the stimulus is greater, but, here again, the intensity still is not enough. Nevertheless, the stimulus does disturb the membrane potential locally for as long as a millisecond or more after both of these weak stimuli. These local potential changes are called *acute local potentials,* and when they fail to elicit an action potential, they are called *acute subthreshold potentials.*

At point C in Figure 5–18, the stimulus is even stronger. Now the local potential has reached barely the level to elicit an action potential, called the threshold level, but this occurs only after a short "latent period." At point D, the stimulus is still stronger, the acute local potential is also stronger, and the action potential occurs after less of a latent period.

Thus, this figure shows that even a very weak stimulus always causes a local potential change at the membrane, but the intensity of the local potential must rise to a *threshold level* before the action potential will be set off.

The "Refractory Period" During Which New Stimuli Cannot Be Elicited

A new action potential cannot occur in an excitable fiber as long as the membrane is still depolarized from the preceding action potential. The reason for this is that shortly after the action potential is initiated the sodium channels (or calcium channels, or both) become inactivated, and any amount of excitatory signal applied to these channels at this point will not open the inactivation gates. The only condition that will reopen them is for the membrane potential to return either to or almost to the original resting membrane potential level. Then, within another small fraction of a second, the inactivation gates of the channels open, and a new action potential can then be initiated.

The period of time during which a second action potential cannot be elicited, even with a very strong stimulus, is called the *absolute refractory period.* This period for large myelinated nerve fibers is about ½500 second. Therefore, one can readily calculate that such a fiber can carry a maximum of about 2500 impulses per second.

Following the absolute refractory period is a *relative refractory period,* lasting about one quarter to one half as long as the absolute period. During this time, stronger than normal stimuli can excite the fiber. The cause of this relative refractoriness is twofold: (1) During this time some of the sodium channels still have not been reversed from their inactivation state, and (2) the potassium channels are usually wide open at this time, causing a state of hyperpolarization that makes it more difficult to stimulate the fiber.

INHIBITION OF EXCITABILITY— "STABILIZERS" AND LOCAL ANESTHETICS

In contrast to the factors that increase nerve excitability, still others, called *membrane-stabilizing factors, can decrease excitability. For instance, a high extracellular fluid calcium ion concentration* decreases the membrane permeability and simultaneously reduces its excitability. Therefore, calcium ions are said to be a "stabilizer." Also, *low potassium ion concentration* in the extracellular fluids, because it has a direct effect of decreasing the permeability of the potassium channels, likewise acts as a stabilizer and reduces membrane excitability. Indeed, in a hereditary disease known as *familial periodic paralysis,* the extracellular potassium ion concentration is often so much reduced that the person actually becomes paralyzed but reverts to normal instantly after intravenous administration of potassium.

Local Anesthetics. Among the most important stabilizers are the many substances used clinically as local anesthetics, including *procaine, tetracaine,* and many other drugs. Most of these act directly on the activation gates of the sodium channels, making it much more difficult for these gates to open and thereby reducing the membrane excitability. When the excitability has been reduced so low that the ratio of *action potential strength to excitability threshold* (called the "safety factor") is reduced below 1.0, a nerve impulse fails to pass through the anesthetized area.

● RECORDING MEMBRANE POTENTIALS AND ACTION POTENTIALS

The Cathode Ray Oscilloscope. Earlier in this chapter we noted that the membrane potential changes occur very rapidly throughout the course of an action potential. Indeed, most of the action potential complex of large nerve fibers takes place in less than 1/1000 second. In some figures of this chapter an electrical meter has been shown recording these potential changes. However, it must be understood

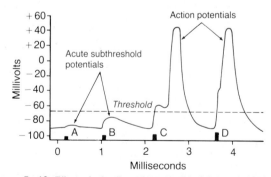

Figure 5–18. Effect of stimuli on the potential of the excitable membrane, showing the development of "acute subthreshold potentials" when the stimuli are below the threshold value required for eliciting an action potential.

that any meter capable of recording them must be capable of responding extremely rapidly. For practical purposes, the only common type of meter that is capable of responding accurately to the very rapid membrane potential changes is the cathode ray oscilloscope.

Figure 5–19 illustrates the basic components of a cathode ray oscilloscope. The cathode ray tube itself is composed basically of an *electron gun* and a *fluorescent surface* against which electrons are fired. Where the electrons hit the surface, the fluorescent material glows. If the electron beam is moved across the surface, the spot of glowing light also moves and draws a fluorescent line on the screen.

In addition to the electron gun and fluorescent surface, the cathode ray tube is provided with two sets of electrically charged plates, one set positioned on either side of the electron beam and the other set positioned above and below. Appropriate electronic control circuits change the voltage on these plates so that the electron beam can be bent up or down in response to electrical signals coming from recording electrodes on nerves. The beam of electrons is also swept horizontally across the screen at a constant rate. This gives the record illustrated on the face of the cathode ray tube, giving a time base horizontally and the voltage changes at the nerve electrodes vertically. Note at the left end of the record a small *stimulus artifact* caused by the electrical stimulus used to elicit the action potential. Then further to the right is the recorded action potential itself.

Recording the Monophasic Action Potential. Throughout this chapter "monophasic" action potentials have been shown in the different diagrams. To record these, a micropipette electrode, illustrated earlier in the chapter in Figure 5–2, is inserted into the interior of the fiber. Then, as the action potential spreads down the fiber, the changes in the potential inside the fiber are recorded, as illustrated earlier in the chapter in Figures 5–6, 5–10, and 5–13.

Recording a Biphasic Action Potential. When one wishes to record impulses from a whole nerve trunk, it is not feasible to place electrodes inside the nerve fibers. Therefore, the usual method of recording is to place two electrodes on the outside of fibers. However, the record that is obtained is then biphasic for the following reasons: When an action potential moving down the nerve fiber reaches the first electrode, it becomes charged negatively, while the sec-

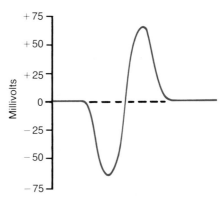

Figure 5–20. Recording of a biphasic action potential.

ond electrode is still unaffected. This causes the oscilloscope to record in the negative direction. Then as the action potential proceeds still farther down the nerve, there comes a point when the membrane beneath the first electrode becomes repolarized, while the second electrode is negative, and the oscilloscope records in the opposite direction. Thus, a graphic record such as that illustrated in Figure 5–20 is recorded by the oscilloscope, showing a potential change first in one direction and then in the opposite direction.

REFERENCES

Agnew, W. S.: Voltage-regulated sodium channel molecules. Annu. Rev. Physiol., 45:517, 1984.

Armstrong, C. M.: Sodium channels and gating currents. Physiol. Rev., 61:644, 1981.

Auerbach, A., and Sachs, F.: Patch clamp studies of single ionic channels. Annu. Rev. Biophys. Bioeng., 13:269, 1984.

Biggio, G., and Costa, E.: Chloride Channels and Their Modulation by Neurotransmitters and Drugs. New York, Raven Press, 1988.

Bretag, A. H.: Muscle chloride channels. Physiol. Rev., 67:1987.

Byrne, J. H., and Schultz, S. G.: An Introduction to Membrane Transport and Bioelectricity. New York, Raven Press, 1988.

Clausen, T.: Regulation of active Na$^+$-K$^+$ transport in skeletal muscle. Physiol. Rev., 66:542, 1986.

Cole, K. S.: Electrodiffusion models for the membrane of squid giant axon. Physiol. Rev., 45:340, 1965.

Cooper, S. A.: New peripherally-acting oral analgesic agents. Annu. Rev. Pharmacol. Toxicol., 23:617, 1983.

DeWeer, P., et al.: Voltage dependence of the Na-K pump. Annu. Rev. Physiol., 50:225, 1988.

DiFrancesco, D., and Noble, D.: A model of cardiac electrical activity incorporating ionic pumps and concentration changes. Phil. Trans. R. Soc. Lond. (Biol.), 307:353, 1985.

DiPolo, R., and Beauge, L.: The calcium pump and sodium-calcium exchange in squid axons. Annu. Rev. Physiol., 45:313, 1983.

French, R. J., and Horn, R.: Sodium channel gating: Models, mimics, and modifiers. Annu. Rev. Biophys. Bioeng., 12:319, 1983.

Garty, H., and Benos, D. J.: Characteristics and regulatory mechanisms of the amiloride-blockable Na$^+$ channel. Physiol. Rev., 68:309, 1988.

Grinnell, A. D., et al. (eds.): Calcium and Ion Channel Modulation. New York, Plenum Publishing Corp., 1988.

Hille, B.: Gating in sodium channels of nerve. Annu. Rev. Physiol., 38:139, 1976.

Hodgkin, A. L.: The Conduction of the Nervous Impulse. Springfield, Ill., Charles C Thomas, 1963.

Hodgkin, A. L., and Horowicz, P.: The effect of sudden changes in ionic concentrations on the membrane potential of single muscle fibers. J. Physiol. (Lond.), 153:370, 1960.

Hodgkin, A. L., and Huxley, A. F.: Movement of sodium and potassium ions during nervous activity. Cold Spr. Harb. Symp. Quant. Biol., 17:43, 1952.

Hodgkin, A. L., and Huxley, A. F.: Quantitative description of membrane current and its application to conduction and excitation in nerve. J. Physiol. (Lond.), 117:500, 1952.

Kaplan, J. H.: Ion movements through the sodium pump. Annu. Rev. Physiol., 47:535, 1985.

Katz, B.: Nerve, Muscle, and Synapse. New York, McGraw-Hill, 1968.

Keynes, R. D.: Ion channels in the nerve-cell membrane. Sci. Am., 240:126, 1979.

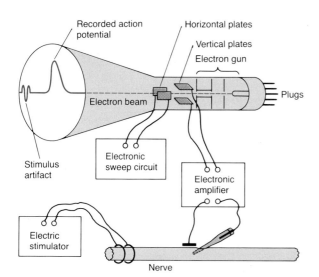

Figure 5–19. The cathode ray oscilloscope for recording transient action potentials.

Kostyuk, P. G.: Intracellular perfusion of nerve cells and its effects on membrane currents. Physiol. Rev., 64:435, 1984.

Krueger, B. K.: Toward an understanding of structure and function of ion channels. FASEB J., 3:1906, 1989.

Latorre, R., and Alvarez, O.: Voltage-dependent channels in lipid bilayer membranes. Physiol. Rev., 61:77, 1981.

Latorre, R., et al.: K^+ channels gated by voltage and ions. Annu. Rev. Physiol., 46:485, 1984.

Levitan, I. B.: Modulation of ion channels in neutrons and other cells. Annu. Rev. Neurosci., 11:119, 1988.

Malhotra, S. K.: The Plasma Membrane. New York, John Wiley & Sons, 1983.

Miller, R. J.: Multiple calcium channels and neuronal function. Science, 235:46, 1987.

Moody, W., Jr.: Effects of intracellular H^+ on the electrical properties of excitable cells. Annu. Rev. Neurosci., 7:257, 1984.

Naftalin, R. J.: The thermostatics and thermodynamics of cotransport. Biochem. Biophys. Acta., 778:155, 1984.

Narahashi, T.: Ion Channels. New York, Plenum Publishing Corp., 1988.

Requena, J.: Calcium transport and regulation in nerve fibers. Annu. Rev. Biophys. Bioeng., 12:237, 1983.

Reuter, H.: Modulation of ion channels by phosphorylation and second messengers. News Physiol. Sci., 2:168, 1987.

Rogart, R.: Sodium channels in nerve and muscle membrane. Annu. Rev. Physiol., 43:711, 1981.

Ross, W. N.: Changes in intracellular calcium during neuron activity. Annu. Rev. Physiol., 51:491, 1989.

Sakmann, B., and Neher, E.: Patch clamp techniques for studying ionic channels in excitable membranes. Annu. Rev. Physiol., 46:455, 1984.

Schubert, D.: Developmental Biology of Cultured Nerve, Muscle and Glia. New York, John Wiley & Sons, 1984.

Schwartz, W., and Passow, H.: Ca^{2+}-activated K^+ channels in erythrocytes and excitable cells. Annu. Rev. Physiol., 45:359, 1983.

Shepherd, G. M.: Neurobiology. New York, Oxford University Press, 1987.

Sjodi, R. A.: Ion Transport in Skeletal Muscle. New York, John Wiley & Sons, 1982.

Skene, J. H. P.: Axonal growth-associated proteins. Ann. Rev. Neurosci., 12:127, 1989.

Snell, R. M. (ed.): Transcellular Membrane Potentials and Ionic Fluxes. New York, Gordon Press Pubs., 1984.

Sperelakis, N.: Hormonal and neurotransmitter regulation of Ca^{++} influx through voltage-dependent slow channels in cardiac muscle membrane. Membr. Biochem., 5:131, 1984.

Stefani, E., and Chiarandini, D. J.: Ionic channels in skeletal muscle. Annu. Rev. Physiol., 44:357, 1982.

Swadlow, H. A., et al.: Modulation of impulse conduction along the axonal tree. Annu. Rev. Biophys. Bioeng., 9:143, 1980.

Trimmer, J. A., and Agnew, W. S.: Molecular diversity of voltage-sensitive Na channels. Annu. Rev. Physiol., 51:401, 1989.

Tsien, R. W.: Calcium channels in excitable cell membranes. Annu. Rev. Physiol., 45:341, 1983.

Ulbricht, W.: Kinetics of drug action and equilibrium results at the node of Ranvier. Physiol. Rev., 61:785, 1981.

Vinores, S., and Guroff, G.: Nerve growth factor: Mechanicsms of action. Annu. Rev. Biophys. Bioeng., 9:223, 1980.

Weiss, D. C. (ed.): Axioplasmic Transport in Physiology and Pathology. New York, Springer-Verlag, 1982.

Windhager, E. E., and Taylor, A.: Regulatory role of intracellular calcium ions in epithelial Na transport. Annu. Rev. Physiol., 45:519, 1983.

Wright, E. M.: Electrophysiology of plasma membrane vesicles. Am. J. Physiol., 246:F363, 1984.

Zigmond, R. E., and Bowers, C. W.: Influence of nerve activity on the macromolecular content of neurons and their effector organs. Annu. Rev. Physiol., 43:673, 1981.

CHAPTER 6

• *Contraction of Skeletal Muscle* •

Approximately 40 per cent of the body is skeletal muscle, and almost another 10 per cent is smooth and cardiac muscle. Many of the same principles of contraction apply to all these different types of muscle, but in the present chapter the function of skeletal muscle is considered mainly; the specialized functions of smooth muscle are discussed in Chapter 8 and cardiac muscle in Chapter 9.

● PHYSIOLOGIC ANATOMY OF SKELETAL MUSCLE

THE SKELETAL MUSCLE FIBER

Figure 6–1 illustrates the organization of skeletal muscle, showing that all skeletal muscles are made of numerous fibers ranging between 10 and 80 micrometers in diameter. Each of these fibers in turn is made up of successively smaller subunits, also illustrated in Figure 6–1, that are described in subsequent paragraphs.

In most muscles the fibers extend the entire length of the muscle; except for about 2 per cent of the fibers, each is innervated by only one nerve ending, located near the middle of the fiber.

The Sarcolemma. The sarcolemma is the cell membrane of the muscle fiber. However, the sarcolemma consists of a true cell membrane, called the *plasma membrane,* and an outer coat made up of a thin layer of polysaccharide material containing numerous thin collagen fibrils. At the end of the muscle fiber, this surface layer of the sarcolemma fuses with a tendon fiber, and the tendon fibers in turn collect into bundles to form the muscle tendons and thence insert into the bones.

Myofibrils; Actin and Myosin Filaments. Each muscle fiber contains several hundred to several thousand *myofibrils,* which are illustrated by the many small open dots in the cross-sectional view of Figure 6–1C. Each myofibril (Figure 6–1D and E) in turn has, lying side-by-side, about 1500 *myosin filaments* and 3000 *actin filaments,* which are large polymerized protein molecules that are responsible for muscle contraction. These can be seen in longitudinal view in the electron micrograph of Figure 6–2 and are represented diagrammatically in Figure 6–1, parts E through L. The thick filaments in the diagrams are *myosin,* and the thin filaments are *actin.*

Note that the myosin and actin filaments partially interdigitate and thus cause the myofibrils to have alternate light and dark bands. The light bands contain only actin filaments and are called *I bands* because they are *isotropic* to polarized light. The dark bands contain the myosin filaments as well as the ends of the actin filaments where they overlap the myosin and are called *A bands* because they are *anisotropic* to polarized light. Note also the small projections from the sides of the myosin filaments. These are called *cross-bridges.* They protrude from the surfaces of the myosin filaments along the entire extent of the filament except in the very center. It is interaction between these cross-bridges and the actin filaments that causes contraction.

Figure 6–1E also shows that the ends of the actin filaments are attached to a so-called *Z disc.* From this disc, these filaments extend in both directions to interdigitate with the myosin filaments. The Z disc, which itself is composed of filamentous proteins different from the actin and myosin filaments, passes from myofibril to myofibril, attaching the myofibrils to each other all the way across the muscle fiber. Therefore, the entire muscle fiber has light and dark bands, as do the individual myofibrils. These bands give skeletal and cardiac muscle their "striated" appearance.

The portion of a myofibril (or of the whole muscle fiber) that lies between two successive Z discs is called a *sarcomere.* When the muscle fiber is at its normal, fully stretched resting length, the length of the sarcomere is about 2 micrometers. At this length, the actin filaments completely overlap the myosin filaments and are just beginning to overlap each other. We see later that at this length the sarcomere also is capable of generating its greatest force of contraction

The Sarcoplasm. The mofibrils are suspended inside the muscle fiber in a matrix called *sarcoplasm,* which is composed of usual intracellular constituents. The fluid of the sarcoplasm contains large quantities of potassium, magnesium, phosphate, and protein enzymes. Also present are tremendous numbers of *mitochondria* that lie between and parallel to the myofi-

Skeletal Muscle

Figure 6–1. Organization of skeletal muscle, from the gross to the molecular level. *F, G, H.* and *I* are cross-sections at the levels indicated. (Drawing by Sylvia Colard Keene. Modified from Fawcett: Bloom and Fawcett: A Textbook of Histology. Philadelphia, W. B. Saunders Company, 1986.)

brils, a condition that is indicative of the great need of the contracting myofibrils for large amounts of adenosine triphosphate (ATP) formed by the mitochondria.

The Sarcoplasmic Reticulum. Also in the sarcoplasm is an extensive endoplasmic reticulum, which in the muscle fiber is called the *sarcoplasmic reticulum.* This reticulum has a special organization that is extremely important in the control of muscle contraction, which is discussed in the following chapter. The electron micrograph of Figure 6–3 illustrates the arrangement of this sarcoplasmic reticulum and shows how extensive it can be. The more rapidly contracting types of muscle have especially extensive sarcoplasmic reticula, indicating that this structure is important in causing rapid muscle contraction, as is also discussed later.

● THE GENERAL MECHANISM OF MUSCLE CONTRACTION

The initiation and execution of muscle contraction occurs in the following sequential steps:

1. An action potential travels along a motor nerve to its endings on muscle fibers.

2. At each ending, the nerve secretes a small amount of the neurotransmitter substance called *acetylcholine.*

3. The acetylcholine acts on a local area of the muscle fiber membrane to open multiple acetylcholine-gated protein channels in the muscle fiber membrane.

4. Opening of the acetylcholine channels allows large quantities of sodium ions to flow to the interior of the muscle fiber membrane at the point of the nerve terminal. This initiates an action potential in the muscle fiber.

5. The action potential travels along the muscle fiber membrane in the same way that action potentials travel along nerve membranes.

6. The action potential depolarizes the muscle fiber membrane and also travels deeply within the muscle fiber. Here it causes the sarcoplasmic reticulum to release into the myofibrils large quantities of calcium ions that have been stored within the reticulum.

7. The calcium ions initiate attractive forces be-

Figure 6-2. Electron micrograph of muscle myofibrils, showing the detailed organization of actin and myosin filaments. Note the mitochondria lying between the myofibrils. (From Fawcett: The Cell. Philadelphia, W. B. Saunders Company, 1981.)

tween the actin and myosin filaments, causing them to slide together, which is the contractile process.

8. After a fraction of a second, the calcium ions are pumped back into the sarcoplasmic reticulum, where they remain stored until a new muscle action potential comes along; muscle contraction ceases.

We now describe the machinery of the contractile process but return to the details of muscle excitation in the following chapter.

● MOLECULAR MECHANISM OF MUSCLE CONTRACTION

Sliding Mechanism of Contraction. Figure 6-4 illustrates the basic mechanism of muscle contraction. It shows the relaxed state of a sarcomere (above) and the contracted state (below). In the relaxed state, the ends of the actin filaments derived from two successive Z discs barely begin to overlap each other,

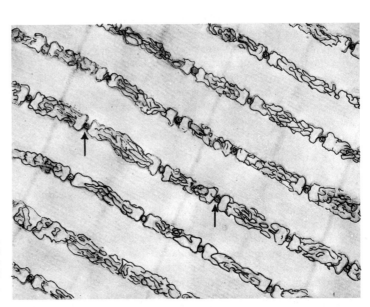

Figure 6-3. Sarcoplasmic reticulum surrounding the myofibrils, showing the longitudinal system paralleling the myofibrils. Also shown in cross-section are the T tubules (arrows) that lead to the exterior of the fiber membrane and that contain extracellular fluid (From Fawcett: The Cell. Philadelphia, W. B. Saunders Company, 1981.)

Figure 6–4. The relaxed and contracted states of a myofibril, showing sliding of the actin filaments (black) into the spaces between the myosin filaments (red).

while at the same time completely overlapping the myosin filaments. On the other hand, in the contracted state these actin filaments have been pulled inward among the myosin filaments, so that they now overlap each other to a major extent. Also, the Z discs have been pulled by the actin filaments up to the ends of the myosin filaments. Indeed, the actin filaments can be pulled together so tightly that the ends of the myosin filaments actually buckle during very intense contraction. Thus, muscle contraction occurs by a *sliding filament mechanism.*

But what causes the actin filaments to slide inward among the myosin filaments? This is caused by mechanical forces generated by the interaction of the cross-bridges of the myosin filaments with the actin filaments, as we discuss in the following sections. Under resting conditions, these forces are inhibited, but when an action potential travels over the muscle fiber membrane, this causes the release of large quantities of calcium ions into the sarcoplasm surrounding the myofibrils. These calcium ions in turn activate the forces between the filaments, and contraction begins. but energy is also needed for the contractile process to proceed. This energy is derived from the high-energy bonds of ATP, which is degraded to adenosine diphosphate (ADP) to liberate the energy required.

In the next few sections, we describe what is known about the details of the molecular processes of contraction. To begin this discussion, however, we must first characterize in detail the myosin and actin filaments.

MOLECULAR CHARACTERISTICS OF THE CONTRACTILE FILAMENTS

The Myosin Filament. The myosin filament is composed of multiple myosin molecules, each having a molecular weight of about 480,000. Figure 6–5A illustrates an individual molecule; section B illustrates the organization of the molecules to form a myosin filament as well as its interaction on one side with the ends of two actin filaments.

The *myosin molecule* is composed of six polypeptide

chains, two *heavy chains* each with a molecular weight of about 200,000 and four *light chains* with molecular weights of about 20,000 each. The two heavy chains wrap spirally around each other to form a double helix. However, one end of each of these chains is folded into a globular protein mass called the myosin *head.* Thus, there are two free heads lying side by side at one end of the double helix myosin molecule; the elongated portion of the coiled helix is called the *tail.* The four light chains are also parts of the myosin heads, two to each head. These light chains help control the function of the head during the process of muscle contraction.

The *myosin filament* is made up of 200 or more individual myosin molecules. The central portion of one of these filaments is illustrated in Figure 6–5B, showing the tails of the myosin molecules bundled together to form the *body* of the filament, while many heads of the molecules hang outward to the sides of the body. Also, part of the helix portion of each myosin molecule extends to the side along with the head, thus providing an *arm* that extends the head outward from the body, as shown in the figure. The protruding arms and heads together are called *cross-bridges.* Each cross-bridge is believed to be flexible at two points called *hinges,* one where the arm leaves the body of the myosin filament and the other where the two heads attach to the arm. The hinged arms allow the heads either to be extended far outward from the body of the myosin filament or to be brought close to the body. The hinged heads are believed to participate in the actual contraction process, as we discuss in the following sections.

The total length of each myosin filament is very uniform, almost exactly 1.6 micrometers. However, note that there are no cross-bridge heads in the very

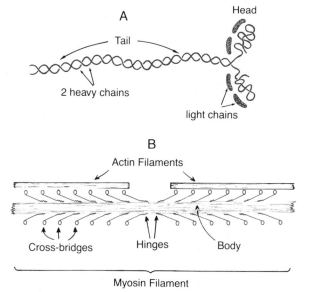

Figure 6–5. *A,* The myosin molecule. *B,* Combination of many myosin molecules to form a myosin filament. Also shown are the cross-bridges and the interaction between the heads of the cross-bridges and adjacent actin filaments.

center of the myosin filament for a distance of about 0.2 micrometer because the hinged arms extend toward both ends of the myosin filament away from the center; therefore, in the center there are only tails of the myosin molecules and no heads.

Now, to complete the picture, the myosin filament itself is twisted so that each successive set of cross-bridges is axially displaced from the previous set by 120 degrees. This insures that the cross-bridges extend in all directions around the filament.

ATPase Activity of the Myosin Head. Another feature of the myosin head that is essential for muscle contraction is that it functions as an ATPase enzyme. As we see later, this property allows the head to cleave ATP and to use the energy derived from the ATP's high-energy phosphate bond to energize the contraction process.

The Actin Filament. The actin filament is also complex. It is composed of three different protein components: *actin, tropomyosin,* and *troponin.*

The backbone of the actin filament is a double-stranded F-actin protein molecule, illustrated by the two lighter-colored strands in Figure 6–6. The two strands are wound in a helix in the same manner as the myosin molecule but with a complete revolution every 70 nanometers.

Each strand of the double F-actin helix is composed of polymerized G-actin molecules, each having a molecular weight of about 42,000. There are approximately 13 of these molecules in each revolution of each strand of helix. Attached to each one of the G-actin molecules is one molecule of ADP. It is believed that these ADP molecules are the active sites on the actin filaments with which the cross-bridges of the myosin filaments interact to cause muscle contraction. The active sites on the two F-actin strands of the double helix are staggered, giving one active site on the overall actin filament approximately every 2.7 nanometers.

Each actin filament is approximately 1 micrometer long. The bases of the actin filaments are inserted strongly into the Z discs, while the other ends protrude in both directions into the adjacent sarcomeres to lie in the spaces between the myosin molecules, as illustrated in Figure 6–4.

Tropomyosin Molecules. The actin filament also contains another protein, *tropomyosin.* Each molecule of tropomyosin has a molecular weight of 70,000 and a length of 40 nanometers. These mole-cules are connected loosely with the F-actin strands, wrapped spirally around the sides of the F-actin helix. In the resting state, the tropomyosin molecules are believed to lie on top of the active sites of the actin strands, so that attraction cannot occur between the actin and myosin filaments to cause contraction. Each tropomyosin molecule covers about seven of these active sites.

Troponin and Its Role in Muscle Contraction. Attached near one end of each tropomyosin molecule is still another protein molecule called *troponin.* This is actually a complex of three loosely bound protein subunits, each of which plays a specific role in the control of muscular contraction. One of the subunits (troponin I) has a strong affinity for actin, another (troponin T) for tropomyosin, and a third (troponin C) for calcium ions. This complex is believed to attach the tropomyosin to the actin. The strong affinity of the troponin for calcium ions is believed to initiate the contraction process, as is explained in the following section.

Interaction of Myosin, Actin Filaments, and Calcium Ions to Cause Contraction

Inhibition of the Actin Filament by the Troponin-Tropomyosin Complex; Activation by Calcium Ions. A pure actin filament without the presence of the troponin-tropomyosin complex binds strongly with myosin molecules in the presence of magnesium ions and ATP, both of which are normally abundant in the myofibril. However, if the troponin-tropomyosin complex is added to the actin filament, this binding does not take place. Therefore, it is believed that the active sites on the normal actin filament of the relaxed muscle are inhibited or actually physically covered by the troponin-tropomyosin complex. Consequently, the sites cannot attach to the myosin filaments to cause contraction. Before contraction can take place, the inhibitory effect of the troponin-tropomyosin complex must itself be inhibited.

Now, let us discuss the role of the calcium ions. In the presence of large amounts of calcium ions, the inhibitory effect of the troponin-tropomyosin on the actin filaments is itself inhibited. The mechanism of this is not known, but one suggestion is the following: When calcium ions combine with troponin C, each molecule of which can bind strongly with up to four calcium ions even when these are present in minute quantities, the troponin complex supposedly undergoes a conformational change that in some way tugs on the tropomyosin molecule and supposedly moves it deeper into the groove between the two actin strands. This "uncovers" the active sites of the actin, thus allowing contraction to proceed. Although this is a hypothetical mechanism, nevertheless it does emphasize that the normal relationship between the tropomyosin-troponin complex and actin is altered by calcium ions, producing a new condition that leads to contraction.

Figure 6–6. The actin filament, composed of two helical strands of F-actin and tropomyosin molecules that fit loosely in the grooves between the actin strands. Attached to one end of each tropomyosin molecule is a troponin complex that initiates contraction.

Interaction Between the "Activated" Actin Filament and the Myosin Cross-Bridges — The "Walk-Along" Theory of Contraction. As soon as the actin filament becomes activated by the calcium ions, the heads of the cross-bridges from the myosin filaments immediately become attracted to the active sites of the actin filament, and this in some way causes contraction to occur. Although the precise manner by which this interaction between the cross-bridges and the actin causes contraction is still unknown, a suggested hypothesis for which considerable evidence exists is the "walk-along" theory (or *"ratchet" theory) of contraction.*

Figure 6–7 illustrates the postulated walk-along mechanism for contraction. This figure shows the heads of two cross-bridges attaching to and disengaging from the active sites of an actin filament. It is postulated that when the head attaches to an active site, this attachment simultaneously causes profound changes in the intramolecular forces between the head and arm of the cross-bridge. The new alignment of forces causes the head to tilt toward the arm and to drag the actin filament along with it. This tilt of the head is called the *power stroke.* Then, immediately after tilting, the head automatically breaks away from the active site. Next, the head returns to its normal perpendicular direction. In this position it combines with a new active site farther down along the actin filament; then, the head tilts again to cause a new power stroke, and the actin filament moves another step. Thus, the heads of the cross-bridges bend back and forth and step by step walk along the actin filament, pulling the ends of the actin filaments toward the center of the myosin filament.

Each one of the cross-bridges is believed to operate independently of all others, each attaching and pulling in a continuous but random cycle. Therefore, the greater the number of cross-bridges in contact with the actin filament at any given time, the greater, theoretically, is the force of contraction.

ATP as the Source of Energy for Contraction — Chemical Events in the Motion of the Myosin Heads. When a muscle contracts against a load, work is performed and energy is required. Large amounts of ATP are cleaved to form ADP during the contraction process. Furthermore, the greater the amount of work performed by the muscle, the greater the amount of ATP that is cleaved, which is called the *Fenn effect.*

Although it is still not known exactly how ATP is used to provide the energy for contraction, the following is a sequence of events that has been suggested as the means by which this occurs:

1. Before contraction begins, the heads of the cross-bridges bind with ATP. The ATPase activity of the myosin head immediately cleaves the ATP but leaves the cleavage products, ADP plus Pi, bound to the head. In this state, the conformation of the head is such that it extends perpendicularly toward the actin filament but is not yet attached to the actin.

2. Next, when the inhibitory effect of the troponin-tropomyosin complex is itself inhibited by calcium ions, active sites on the actin filament are uncovered, and the myosin heads do then bind with these, as illustrated in Figure 6–7.

3. The bond between the head of the cross-bridge and the active site of the actin filament causes a conformational change in the head, prompting the head to tilt toward the arm of the cross-bridge. This provides the *power stroke* for pulling the actin filament. The energy that activates the power stroke is the energy already stored, like a "cocked" spring, by the conformational change in the head when the ATP molecule had been cleaved.

4. Once the head of the cross-bridge is tilted, this allows release of the ADP and Pi that were previously attached to the head; at the site of release of the ADP, a new molecule of ATP binds. This binding in turn causes detachment of the head from the actin.

5. After the head has detached from the actin, the new molecule of ATP is also cleaved, and the energy again "cocks" the head back to its perpendicular condition ready to begin a new power stroke cycle.

6. Then, when the cocked head, with its stored energy derived from the cleaved ATP, binds with a new active site on the actin filament, it becomes uncocked and once again provides the power stroke.

7. Thus, the process proceeds again and again until the actin filament pulls the Z membrane up against the ends of the myosin filaments or until the load on the muscle becomes too great for further pulling to occur.

DEGREE OF ACTIN AND MYOSIN FILAMENT OVERLAP — EFFECT ON TENSION DEVELOPED BY THE CONTRACTING MUSCLE

Figure 6–8 illustrates the effect of sarcomere length and of myosin-actin filament overlap on the active tension developed by a contracting muscle fiber. To the right are illustrated different degrees of overlap of the myosin and actin filaments at different sarcomere lengths. At point D on the diagram, the actin filament has pulled all the way out to the end of the myosin filament with no overlap at all. At this point, the tension developed by the activated muscle is zero. Then, as the sarcomere shortens and the actin filament begins to overlap the myosin filament, the tension increases progressively until the sarcomere

Figure 6–7. The "walk-along" mechanism for contraction of the muscle.

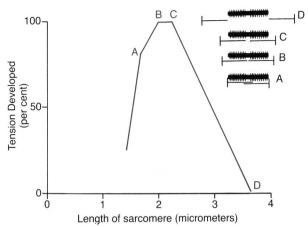

Figure 6–8. Length-tension diagram for a single sarcomere, illustrating maximum strength of contraction when the sarcomere is 2.0 to 2.2 micrometers in length. At the upper right are shown the relative positions of the actin and myosin filaments at different sarcomere lengths from point A to point D. (Modified from Gordon, Huxley, and Julian: The length-tension diagram of single vertebrate striated muscle fibers. *J. Physiol.*, 171:28P, 1964.)

length decreases to about 2.2 micrometers. At this point, the actin filament has already overlapped all the cross-bridges of the myosin filament but has not yet reached the center of the myosin filament. Upon further shortening, the sarcomere maintains full tension until point B at a sarcomere length of approximately 2.0 micrometers. At this point, the ends of the two actin filaments begin to overlap each other, in addition to overlapping the myosin filaments. As the sarcomere length falls from 2 micrometers down to about 1.65 micrometers, at point A, the strength of contraction decreases. It is at this point that the two Z discs of the sarcomere abut the ends of the myosin filaments. Then, as contraction proceeds to still shorter sarcomere lengths, the ends of the myosin filaments are actually crumpled, and as illustrated in the figure, the strength of contraction also decreases precipitously.

This diagram illustrates that maximum contraction occurs when there is maximum overlap between the actin filaments and the cross-bridges of the myosin filaments, and it supports the idea that the greater the number of cross-bridges pulling the actin filaments, the greater the strength of contraction.

Effect of Muscle Length on Force of Contraction in the Intact Muscle. The upper curve of Figure 6–9 is similar to that in Figure 6–8, but this illustrates the intact, whole muscle rather than a single muscle fiber. The whole muscle has a large amount of connective tissue in it; also, the sarcomeres in different parts of the muscle do not necessarily contract exactly in unison. Therefore, the curve has somewhat different dimensions from those illustrated for the individual muscle fiber, but it nevertheless exhibits the same form.

Note in Figure 6–9 that when the muscle is at its normal resting length, which is at a sarcomere length of about 2 micrometers, it contracts with maximum force of contraction. If the muscle is stretched to

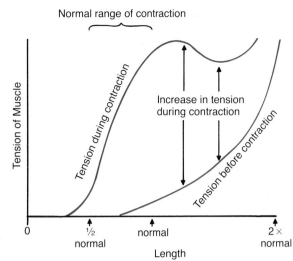

Figure 6–9. Relation of muscle length to force of contraction.

much greater than normal length prior to contraction, a large amount of *resting tension* develops in the muscle even before contraction takes place; this tension results from the elastic forces of the connective tissue, the sarcolemma, the blood vessels, the nerves, and so forth. However, the *increase* in tension during contraction, called *active tension*, decreases as the muscle is stretched much beyond its normal length—that is, to a sarcomere length greater than about 2.2 micrometers. This is demonstrated by the decrease in the arrow length in the figure.

RELATION OF VELOCITY OF CONTRACTION TO LOAD

A muscle contracts extremely rapidly when it contracts against no load—to a state of full contraction in approximately 0.1 second for the average muscle. However, when loads are applied, the velocity of contraction becomes progressively less as the load increases, as illustrated in Figure 6–10. When the load increases to equal the maximum force that the muscle can exert, the velocity of contraction becomes zero and no contraction at all results, despite activation of the muscle fiber.

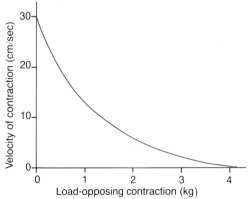

Figure 6–10. Relation of load to velocity of contraction in a skeletal muscle 8 centimeters long.

This decreasing velocity with load is caused by the fact that a load on a contracting muscle is a reverse force that opposes the contractile force caused by muscle contraction. Therefore, the net force that is available to cause velocity of shortening is correspondingly reduced.

● ENERGETICS OF MUSCLE CONTRACTION

Work Output During Muscle Contraction

When a muscle contracts against a load, it performs *work*. This means that *energy* is transferred from the muscle to the external load, for example, to lift an object to a greater height or to overcome resistance to movement.

In mathematical terms, work is defined by the following equation:

$$W = L \times D$$

in which W is the work output, L is the load, and D is the distance of movement against the load. The energy required to perform the work is derived from the chemical reactions in the muscle cells during contraction, as we describe in the following sections.

Sources of Energy for Muscle Contraction

We have already seen that muscle contraction depends on energy supplied by ATP. Most of this energy is required to actuate the walk-along mechanism by which the cross-bridges pull the actin filaments, but small amounts are required for (1) pumping calcium from the sarcoplasm into the sarcoplasmic reticulum after the contraction is over and (2) pumping sodium and potassium ions through the muscle fiber membrane to maintain an appropriate ionic environment for the propagation of action potentials.

However, the concentration of ATP present in the muscle fiber, about 4 millimolar, is sufficient to maintain full contraction for only 1 to 2 seconds at most. Fortunately, after the ATP is split into ADP, as described in Chapter 2, the ADP is rephosphorylated to form new ATP within a fraction of a second. There are several sources of the energy for this rephosphorylation.

The first source of energy that is used to reconstitute the ATP is the substance *phosphocreatine,* which carries a high-energy phosphate bond similar to those of ATP. The high-energy phosphate bond of the phosphocreatine has a slightly higher amount of free energy than that of the ATP bond, as we discuss more fully in Chapters 67 and 72. Therefore, phosphocreatine is instantly cleaved, and the released energy causes bonding of a new phosphate ion to ADP to reconstitute the ATP. However, the total amount of phosphocreatine is also very little—only about five times as great as the ATP. Therefore, the combined energy of both the stored ATP and the phosphocrea-

tine in the muscle is still capable of causing maximal muscle contraction for no longer than 7 to 8 seconds.

The next important source of energy, which is used to reconstitute both ATP and phosphocreatine, is *glycogen* previously stored in the muscle cells. Rapid enzymatic breakdown of the glycogen to pyruvic acid and lactic acid liberates energy that is used to convert ADP to ATP, and the ATP can then be used directly to energize muscular contraction or to re-form the stores of phosphocreatine. The importance of this "glycolysis" mechanism is twofold. First, the glycolytic reactions occur even in the absence of oxygen, so that muscle contraction can be sustained for a short time when oxygen is not available. Second, the rate of formation of ATP by the glycolytic process is about two and one half times as rapid as ATP formation when the cellular foodstuffs react with oxygen. Unfortunately, though, so many end-products of glycolysis accumulate in the muscle cells that glycolysis alone can sustain maximum muscle contraction for only about 1 minute.

The final source of energy is the process of *oxidative metabolism.* This means the combining of oxygen with the various cellular foodstuffs to liberate ATP. Over 95 per cent of all energy used by the muscles for sustained, long-term contraction is derived from this source. The foodstuffs that are consumed are carbohydrates, fats, and protein. For extremely long-term muscle activity—over a period of hours — by far the greatest proportion of energy comes from fats.

The detailed mechanisms of these energetic processes are discussed in Chapters 67 through 72. In addition, the importance of the different mechanisms of energy release in different sports is discussed in Chapter 84 on sports physiology.

Efficiency of Muscle Contraction. The "efficiency" of an engine or a motor is calculated as the percentage of energy input that is converted into work instead of heat. The percentage of the input energy to the muscle (the chemical energy in the nutrients) that can be converted into work is less than 20 to 25 per cent, the remainder becoming heat. The reason for this low efficiency is that about half of the energy in the foodstuffs is lost during the formation of ATP, and even then only 40 to 45 per cent of the energy in the ATP itself can later be converted into work.

Maximum efficiency can be realized only when the muscle contracts at a moderate velocity. If the muscle contracts very slowly or without any movement at all, large amounts of *maintenance heat* are released during the process of contraction even though little or no work is being performed, thereby decreasing the efficiency. On the other hand, if contraction is too rapid, large proportions of the energy are used to overcome the viscous friction within the muscle itself, and this, too, reduces the efficiency of contraction. Ordinarily, maximum efficiency is developed when the velocity of contraction is about 30 per cent of maximum.

● CHARACTERISTICS OF WHOLE MUSCLE CONTRACTION

Many features of muscle contraction can be especially well demonstrated by eliciting single *muscle twitches.* This can be accomplished by instantaneously exciting the nerve to a muscle or by passing a short electrical stimulus through the muscle itself, giving rise to a single, sudden contraction lasting for a fraction of a second.

Isometric Versus Isotonic Contraction. Muscle contraction is said to be *isometric* when the muscle does not shorten during contraction and *isotonic* when it shortens with the tension on the muscle remaining constant. Systems for recording the two types of muscle contraction are illustrated in Figure 6–11.

In the isometric system, the muscle contracts against a force transducer without decreasing the muscle length, as illustrated to the right in Figure 6–11. In the isotonic system, the muscle shortens against a fixed load; this is illustrated to the left in the figure, showing a muscle lifting a pan of weights. Obviously, the characteristics of isotonic contraction depend on the load against which the muscle contracts as well as on the inertia of the load. On the other hand, the isometric system records strictly changes in force of muscle contraction itself. Therefore, the isometric system is most often used when comparing the functional characteristics of different muscle types.

The Series Elastic Component of Muscle Contraction. When muscle fibers contract against a load, those portions of the muscle that do not contract—the tendons, the sarcolemmal ends of the muscle fibers where they attach to the tendons, and perhaps even the hinged arms of the cross-bridges—stretch slightly as the tension increases. Consequently, the muscle must shorten an extra 3 to 5 per cent to make up for the stretch of these elements. The elements of the muscle that stretch during contraction are called the *series elastic component* of the muscle.

Characteristics of Isometric Twitches Recorded from Different Muscles

The body has many different sizes of skeletal muscles—from the very small stapedius muscle in the middle ear of only a few millimeters length and a millimeter or so in diameter up to the very large quadriceps muscle, a half million times as large as the stapedius. Furthermore, the fibers may be as small as 10 micrometers in diameter or as large as 80 micrometers. Finally, the energetics of muscle contraction vary considerably from one muscle to another. Therefore, it is no wonder that the characteristics of muscle contraction differ among muscles.

Figure 6–12 illustrates isometric contractions of three different types of skeletal muscles: an ocular muscle, which has a duration of contraction of less than $\frac{1}{40}$ second; the gastrocnemius muscle, which has a duration of contraction of about $\frac{1}{15}$ second; and the soleus muscle, which has a

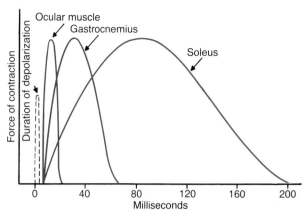

Figure 6–12. Duration of isometric contractions of different types of mammalian muscles, showing also a latent period between the action potential and muscle contraction.

duration of contraction of about $\frac{1}{5}$ second. It is interesting that these durations of contractions are adapted to the function of each of the respective muscles. Ocular movements must be extremely rapid to maintain fixation of the eyes on specific objects, and the gastrocnemius muscle must contract moderately rapidly to provide sufficient velocity of limb movement for running and jumping, whereas the soleus muscle is concerned principally with slow contraction for continual support of the body against gravity.

Fast Versus Slow Muscle Fibers. As we discuss more fully in Chapter 84 on sports physiology, every muscle of the body is composed of a mixture of so-called *fast* and *slow* muscle fibers, with still other fibers gradated between these two extremes. The muscles that react very rapidly are composed mainly of the fast fibers with only small numbers of the slow variety. Conversely, the muscles that respond slowly but with prolonged contraction are composed mainly of slow fibers. The differences between these two types of fibers are the following:

Fast fibers: (1) Much larger fibers for great strength of contraction. (2) Extensive sarcoplasmic reticulum for rapid release of calcium ions to initiate contraction. (3) Large amounts of glycolytic enzymes for rapid release of energy by the glycolytic process. (4) Less extensive blood supply because oxidative metabolism is of secondary importance. (5) Fewer mitochondria, also because oxidative metabolism is secondary.

Slow fibers: (1) Smaller fibers. (2) Also innervated by smaller nerve fibers. (3) More extensive blood vessel system and capillaries to supply extra amounts of oxygen. (4) Greatly increased numbers of mitochondria, also to support high levels of oxidative metabolism. (5) Fibers contain large amounts of myoglobin, an iron-containing protein similar to hemoglobin in red blood cells. Myoglobin combines with oxygen and stores it until needed, and it greatly speeds oxygen transport to the mitochondria. The myoglobin gives the slow muscle a reddish appearance and the name *red muscle*, whereas the lack of red in fast muscle gives the name *white muscle*.

From these descriptions, one can see that the fast fibers are adapted for very rapid and very powerful muscle contractions, such as for jumping or for short-distance powerful running. The slow fibers are adapted for prolonged, continued muscle activity, such as support of the body against gravity, and long-continuing athletic events, such as marathon races.

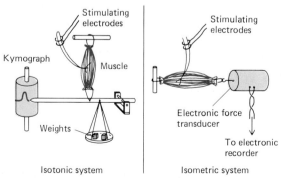

Figure 6–11. Isotonic and isometric recording systems.

MECHANICS OF SKELETAL MUSCLE CONTRACTION

The Motor Unit

Each motoneuron that leaves the spinal cord innervates many different muscle fibers, the number depending on the type of muscle. All the muscle fibers innervated by a single motor nerve fiber are called a *motor unit.* In general, small muscles that react rapidly and whose control must be exact have few muscle fibers (as few as two to three in some of the laryngeal muscles) in each motor unit. On the other hand, the large muscles that do not require very fine control, such as the gastrocnemius muscle, may have several hundred muscle fibers in a motor unit. An average figure for all the muscles of the body can be considered to be about 100 muscle fibers to the motor unit.

The muscle fibers in each motor unit are not all bunched together in a muscle but instead are spread out in the muscle in microbundles of 3 to 15 fibers. Therefore, these lie among similar microbundles of other motor units. This interdigitation allows the separate motor units to contract in support of each other rather than entirely as individual segments.

Muscle Contractions of Different Force — Force Summation

Summation means the adding together of individual twitch contractions to increase the intensity of overall muscle contraction. Summation occurs in two different ways: (1) by increasing the number of motor units contracting simultaneously, which is called *multiple fiber summation,* and (2) by increasing the frequency of contraction, which is called *frequency summation* or *tetanization.*

Multiple Fiber Summation. When the central nervous system sends a weak signal to contract a muscle, the motor units in the muscle that contain the smallest and fewest muscle fibers are stimulated in preference to the larger motor units. Then as the strength of the signal increases, larger and larger motor units begin to be excited as well, with the largest motor units often having as much as 50 times as much contractile force as the smallest units. This is called the *size principle.* It is important because it allows the gradations of muscle force during weak contraction to occur in very small steps, while the steps become progressively greater when large amounts of force are required. The cause of this size principle is that the smaller motor units are driven by small motor nerve fibers, and the small motoneurons in the spinal cord are far more excitable than the larger ones, so that they naturally are excited first.

Another important feature of multiple fiber summation is that the different motor units are driven asynchronously by the spinal cord, so that contraction alternates among motor units one after the other, thus providing smooth contraction even at low frequencies of nerve signals.

Frequency Summation and Tetanization. Figure 6–13 illustrates the principles of frequency summation and tetanization. To the left are illustrated individual twitch contractions occurring one after another at low frequency of stimulation. Then as the frequency increases, there comes a point when each new contraction occurs before the preceding one is over. As a result, the second contraction is added partially to the first, so that the total strength of contraction rises progressively with increasing frequency. When the frequency reaches a critical level, the successive contractions are so rapid that they literally fuse together, and the contraction appears to be completely smooth and continuous, as illustrated in the figure. This is called *tetaniza-*

Figure 6–13. Frequency summation and tetanization.

tion. At a still higher frequency, the strength of contraction reaches its maximum, so that additional increase in frequency beyond that point will have no further effect in increasing contractile force. This occurs because enough calcium ions are then maintained in the muscle sarcoplasm even between action potentials, so that the full contractile state is sustained without allowing any relaxation between the action potentials.

Maximum Strength of Contraction. The maximum strength of tetanic contraction of a muscle operating at a normal muscle length averages between 3 and 4 kilograms per square centimeter of muscle, or 50 pounds per square inch. Because a quadriceps muscle can at times have as much as 16 square inches of muscle belly, as much as 800 pounds of tension may at times be applied to the patellar tendon. One can readily understand, therefore, how it is possible for muscles sometimes to pull their tendons out of the insertions in bones.

Changes in Muscle Strength at the Onset of Contraction — The Staircase Effect (Treppe). When a muscle begins to contract after a long period of rest, its initial strength of contraction may be as little as one half its strength 10 to 50 muscle twitches later. That is, the strength of contraction increases to a plateau, a phenomenon called the *staircase effect* or *treppe.*

Although all the possible causes of the staircase effect are not yet known, it is believed primarily to be caused by increase of calcium ions in the cytosol because of release of these ions from the sarcoplasmic reticulum with each muscle action potential and failure to recapture the ions immediately.

Skeletal Muscle Tone

Even when muscles are at rest, a certain amount of tautness usually remains. This is called *muscle tone.* Since skeletal muscle fibers do not contract without an actual action potential to stimulate the fibers except in certain pathological conditions, skeletal muscle tone results entirely from nerve impulses coming from the spinal cord. These in turn are controlled partly by impulses transmitted from the brain to the appropriate anterior motoneurons and partly by impulses that originate in *muscle spindles* located in the muscle itself. Both of these are discussed in relation to muscle spindle and spinal cord function in Chapter 54.

Muscle Fatigue

Prolonged and strong contraction of a muscle leads to the well-known state of muscle fatigue. Studies in athletes have shown that muscle fatigue increases in almost direct proportion to the rate of depletion of muscle glycogen. Therefore, most fatigue probably results simply from inability of the contractile and metabolic processes of the muscle fibers to continue supplying the same work output. However, experiments have also shown that transmission of the nerve signal through the neuromuscular junction, which is discussed in the following chapter, can occasionally diminish following prolonged muscle activity, thus further diminishing muscle contraction.

Interruption of blood flow through a contracting muscle leads to almost complete muscle fatigue in a minute or more because of the obvious loss of nutrient supply — especially loss of oxygen.

The Lever Systems of the Body

Muscles obviously operate by applying tension to their points of insertion into bones, and the bones in turn form various types of lever systems. Figure 6–14 illustrates the lever system activated by the biceps muscle to lift the forearm. If we assume that a large biceps muscle has a cross-sectional area of 6 square inches, the maximum force of contraction would be about 300 pounds. When the forearm is exactly at right angles with the upper arm, the tendon attachment of the biceps is about 2 inches anterior to the fulcrum at the elbow, and the total length of the forearm lever is about 14 inches. Therefore, the amount of lifting power that the biceps would have at the hand would be only one seventh of the 300 pounds force, or about 43 pounds. When the arm is in the fully extended position, the attachment of the biceps is much less than 2 inches anterior to the fulcrum, and the force with which the forearm can be brought forward is much less than 43 pounds.

In short, an analysis of the lever systems of the body depends on (1) a discrete knowledge of the point of muscle insertion and (2) its distance from the fulcrum of the lever as well as (3) the length of the lever arm and (4) the position of the lever. Obviously, many different types of movement are required in the body, some of which need great strength and others large distances of movement. For this reason there are all varieties of muscles; some are long and contract a long distance, and some are short but have large cross-sectional areas and therefore can provide extreme strengths of contraction over short distances. The study of different types of muscles, lever systems, and their movements is called *kinesiology* and is a very important phase of human physioanatomy.

Figure 6–14. The lever system activated by the biceps muscle.

"Positioning" of a Body Part by Contraction of Antagonist Muscles on Opposite Sides of a Joint — "Coactivation" of Antagonist Muscles

Virtually all movements of the body are caused by simultaneous contraction of antagonist muscles on opposite sides of joints. This is called *coactivation* of the antagonist muscles, and it is controlled by the motor mechanisms of the brain and spinal cord.

The position of each separate part of the body, such as a limb, is determined by the relative degrees of contraction of the antagonist sets of muscles. For instance, let us assume that a limb is to be placed in a mid-range of motion. To achieve this, antagonist muscles are excited approximately equally. Remember that an elongated muscle contracts with more force than a shortened muscle, which was illustrated in Figure 6–9, showing maximum strength of contraction at full muscle length and almost no strength of contraction at half normal length. Therefore, the longer antagonist muscle contracts with far greater force than the shorter muscle. As the limb moves toward its mid position, the strength of the longer muscle decreases, while the strength of the shorter muscle increases until the two strengths exactly equal each other. It is at this point that movement of the limb stops. Thus, by varying the ratios of the degree of activation of the antagonist muscles, the nervous system directs the positioning of the limb.

We learn in Chapter 54 that the motor nervous system also has additional very important mechanisms to compensate for different muscle loads when directing this positioning process.

REMODELING OF MUSCLE TO MATCH FUNCTION

All the muscles of the body are continually being remodeled to match the functions that are required of them. Their diameters are altered, their lengths are altered, their strengths are altered, their vascular supplies are altered, and even the types of muscle fibers are altered at least to a slight extent. This remodeling process is often quite rapid, within a few weeks. Indeed, experiments have shown that even normally the muscle contractile proteins can be totally replaced once every 2 weeks.

Muscle Hypertrophy and Muscle Atrophy

When the total mass of a muscle enlarges, this is called *muscle hypertrophy.* When it decreases, the process is called *muscle atrophy.*

Virtually all muscle hypertrophy results from hypertrophy of the individual muscle fibers, which is called simply *fiber hypertrophy.* This usually occurs in response to contraction of a muscle at maximal or almost maximal force. Hypertrophy occurs to a much greater extent when the muscle is simultaneously stretched during the contractile process. Only a few such strong contractions each day are required to cause almost maximum hypertrophy within 6 to 10 weeks.

Unfortunately, the manner in which forceful contraction leads to hypertrophy is not known. Yet it is known that the rate of synthesis of muscle contractile proteins is far greater during developing hypertrophy than their rate of decay, leading to progressively greater numbers of both actin and myosin filaments in the myofibrils. In turn, the myofibrils themselves split within each muscle fiber to form new myofibrils. Thus, it is mainly this great increase in numbers of additional myofibrils that causes muscle fibers to hypertrophy.

Along with the increasing numbers of myofibrils, the enzyme systems that provide energy also increase. This is especially true of the enzymes for glycolysis, allowing a rapid supply of energy during short-term forceful muscle contraction.

When a muscle remains unused for long periods of time, the rate of decay of the contractile proteins as well as the numbers of myofibrils occurs more rapidly than the rate of replacement. Therefore, muscle atrophy occurs.

Adjustment of Muscle Length. Another type of hypertrophy occurs when muscles are stretched to a greater-than-normal length. This causes new sarcomeres to be added at the ends of the muscle fibers where they attach to the tendons. In fact, new sarcomeres can be added as rapidly as several per minute, illustrating the rapidity of this type of hypertrophy.

Conversely, when a muscle remains shortened continually to less than its normal length, sarcomeres at the ends of the muscle fibers disappear approximately equally as rapidly. It is by these processes that muscles are continually remolded to have the appropriate length for proper muscle contraction.

Hyperplasia of Muscle Fibers. Under very rare conditions of extreme muscle force generation, the actual numbers of muscle fibers have been observed to increase, but by only a few percent, in addition to the fiber hypertrophy process. This increase in fiber numbers is called *fiber hyperplasia.* When it does occur, the mechanism is linear splitting of previously enlarged fibers.

Remodeling of "Slow" Muscles in Marathon Runners

The very fast, spring-type muscles, such as the gastrocnemius, can maintain a high level of contractile strength for only short periods of continuous contraction. Therefore, the so-called slow muscles, such as the soleus, are used for prolonged activity, as in marathon running. These muscles do not hypertrophy nearly so much as do the fast muscles. Instead, they become remodeled in another way. Prolonged activity over a period of many hours each day causes, in addition to mild to moderate hypertrophy of the fibers, the following changes that increase the ability of the fibers to use nutrients:

1. Increased myoglobin in each fiber to transport oxygen to the mitochondria.

2. Greatly increased numbers of mitochondria to form increased quantities of ATP.

3. Increased amounts of oxidative enzymes in the mitochondria to cause increased rate of oxidative metabolism, thus further increasing the production of ATP.

4. Increased growth of capillaries in the muscle itself, causing closer spacing of the capillaries among the muscle fibers, so that oxygen and other nutrients can be delivered rapidly and easily during prolonged periods of activity.

Effects of Muscle Denervation

When a muscle loses its nerve supply, it no longer receives the contractile signals that are required to maintain normal muscle size. Therefore, atrophy begins almost immediately. After about 2 months, degenerative changes also begin to appear in the muscle fibers themselves. If the nerve supply grows back to the muscle, full return of function will usually occur up to about 3 months, but from that time onward the capability of functional return becomes less and less, with no return of function after 1 to 2 years.

In the final stage of denervation atrophy, most of the muscle fibers are completely destroyed and replaced by fibrous and fatty tissue. Those fibers that do remain are composed of a long cell membrane with a line-up of muscle cell nuclei but with no contractile properties and with no capability of regenerating myofibrils if a nerve regrows.

Unfortunately, the fibrous tissue that replaces the muscle fibers during denervation atrophy has a tendency to continue shortening for many months, which is called *contracture.* Therefore, one of the most important problems in the practice of physical therapy is to keep atrophying muscles from developing debilitating and disfiguring contracture. This is achieved by daily stretching of the muscles or use of appliances that keep the muscles stretched during the atrophying process.

Recovery of Muscle Contraction in Poliomyelitis: Development of Macromotor Units. When some nerve fibers to a muscle are destroyed but not all of them, as often occurs in poliomyelitis, the remaining nerve fibers sprout forth new axons to form many new branches that then innervate many of the paralyzed muscle fibers. This causes very large motor units called *macromotor units,* containing as many as five times the normal number of muscle fibers for each motoneuron in the spinal cord. This obviously decreases the fineness of control that one has over the muscles but, nevertheless, allows the muscles to regain strength.

RIGOR MORTIS

Several hours after death all the muscles of the body go into a state of *contracture* called rigor mortis; that is, the muscle contracts and becomes rigid even without action potentials. This rigidity is caused by loss of all the ATP, which is required to cause separation of the cross-bridges from the actin filaments during the relaxation process. The muscles remain in rigor until the muscle proteins are destroyed, which usually results from autolysis caused by enzymes released from the lysosomes some 15 to 25 hours later, the process occurring more rapidly at higher temperature.

REFERENCES

Clausen, T.: Regulation of active Na^+-K^+ transport in skeletal muscle. Physiol. Rev., 66:542, 1986.
DiDonato, S., et al. (eds.): Molecular Genetics of Neurological and Neuromuscular Disease. New York, Raven Press, 1988.
Engel, A. G., and Banker, B. Q. (eds.): Myology. New York, McGraw-Hill Book Co., 1986.
Freund, H.-J.: Motor unit and muscle activity in voluntary motor control. Physiol. Rev., 63:387, 1983.
Goldman, Y. E.: Kinetics of the actomyosin ATPase in muscle fibers. Annu. Rev. Physiol., 49:637, 1987.
Goldman, Y. E., and Brenner, B. (eds.): General introduction. Annu. Rev. Physiol., 49:629, 1987.
Gowitzke, B. A., et al.: Scientific Bases of Human Movement. Baltimore, Williams & Wilkins, 1988.
Hasselbach, W., and Oetliker, H.: Energetics and electrogenicity of the sarcoplasmic reticulum calcium pump. Annu. Rev. Physiol., 45:325, 1983.
Haynes, D. H., and Mandveno, A.: Computer modeling of Ca^{2+}-Mg^{2+}-ATPase of sarcoplasmic reticulum. Physiol. Rev., 67:244, 1987.
Homsher, E.: Muscle enthalpy production and its relationship to actomyosin ATPase. Annu. Rev. Physiol., 49:673, 1987.
Huang, C. L. H.: Intramembrane charge movements in skeletal muscle. Physiol. Rev., 68:1197, 1988.

Huxley, A. F.: Muscular Contraction. Annu. Rev. Physiol., 50:1, 1988.

Huxley, A. F., and Gordon, A. M.: Striation patterns in active and passive shortening of muscle. Nature (Lond.), 193:280, 1962.

Huxley, H. E., and Faruqi, A. R.: Time-resolved x-ray diffraction studies on vertebrate striated muscle. Annu. Rev. Biophys. Bioeng., 12:381, 1983.

Ikemoto, N.: Structure and function of the calcium pump protein of sarcoplasmic reticulum. Annu. Rev. Physiol., 44:297, 1982.

Johnson, E. W. (ed.): Practical Electromyography. Baltimore, Williams & Wilkins, 1988.

Kodama, T.: Thermodynamic analysis of muscle ATPase mechanisms. Physiol. Rev., 65:476, 1985.

Kolata, G.: Metabolic catch-22 of exercise regimens. Science, 236:146, 1987.

Korn, E. D.: Actin polymerization and its regulation by proteins from non-muscle cells. Physiol. Rev., 62:672, 1982.

Korn, E. D., et al.: Actin polymerization and ATP hydrolysis. Science, 238:638, 1987.

Laufer, R., et al.: Regulation of acetylcholine receptor biosynthesis during motor endplate morphogenesis. News Physiol. Sci., 4:5, 1989.

Martonosi, A. N.: Mechanisms of Ca^{2+} release from sarcoplasmic reticulum of skeletal muscle. Physiol. Rev., 64:1240, 1984.

Morgan, D. L., and Proiske, U.: Vertebrate slow muscle: Its structure, pattern of innervation, and mechanical properties. Physiol. Rev., 64:103, 1984.

Morkin, E.: Chronic adaptations in contractile proteins: Genetic regulation. Annu. Rev. Physiol., 49:545, 1987.

Oho, S. J.: Electromyography: Neuromuscular Transmission Studies. Baltimore, Williams & Wilkins, 1988.

Partridge, L. D., and Benton, L. A.: Muscle, the motor. In Brooks, V. B. (ed.): Handbook of Physiology. Sec. 1, Vol. II. Bethesda, American Physiological Society, 1981, p. 43.

Pollack, G. H.: The cross-bridge theory. Physiol. Rev., 63:1049, 1983.

Rash, J.: Neuromuscular Atlas. New York, Prager Publishers, 1984.

Ringel, S. P.: Neuromuscular Disorders: A Guide for Patient and Family. New York, Raven Press, 1987.

Rios, E., and Pizarro, G.: Voltage sensors and calcium channels of excitation-contraction coupling. News Physiol. Sci., 3:223, 1988.

Rogart, R.: Sodium channels in nerve and muscle membrane. Annu. Rev. Physiol., 43:711, 1981.

Rowland, L. P., et al. (eds.): Molecular Genetics in Diseases of Brain, Nerve, and Muscle. New York, Oxford University Press, 1989.

Rudel, R., and Lehmann-Horn, F.: Membrane changes in cells from myotonia patients. Physiol. Rev., 65:310, 1985.

Schafer, R. C.: Clinical Biomechanics: Musculoskeletal Actions and Reactions. Baltimore, Williams & Wilkins, 1986.

Schubert, D.: Developmental Biology of Cultured Nerve, Muscle and Glia. New York, John Wiley & Sons, 1984.

Sjodin, R. A.: Ion Transport in Skeletal Muscle. New York, John Wiley & Sons, 1982.

Soderberg, G. L.: Kinesiology. Baltimore, Williams & Wilkins, 1986.

Stefani, E., and Chiarandini, D. J.: Ionic channels in skeletal muscle. Annu. Rev. Physiol., 44:357, 1982.

Steinbach, J. H.: Structural and functional diversity in vertebrate skeletal muscle nicotinic acetylcholine receptors. Annu. Rev. Physiol., 51:353, 1989.

Sugi, H., and Pollack, G. H. (eds.): Molecular Mechanism of Muscle Contraction. New York, Plenum Publishing Corp., 1988.

Swynghedauw, B.: Developmental and functional adaptation of contractile proteins in cardiac and skeletal muscles. Physiol. Rev., 66:710, 1986.

Thomas, D. D.: Spectroscopic probes of muscle cross-bridge rotation. Annu. Rev. Physiol., 49:691, 1987.

Vergara, J., and Asotra, K.: The chemical transmission mechanism of excitation-contraction coupling in skeletal muscle. News Physiol. Sci., 2:182, 1987.

Wingrad, S.: Regulation of cardiac contractile proteins. Correlations between physiology and biochemistry. Circ. Res., 55:565, 1984.

CHAPTER 7

• *Excitation of Skeletal Muscle Contraction: Neuromuscular Transmission and Excitation-Contraction Coupling* •

● TRANSMISSION OF IMPULSES FROM NERVES TO SKELETAL MUSCLE FIBERS: THE NEUROMUSCULAR JUNCTION

The skeletal muscle fibers are innervated by large, myelinated nerve fibers that originate in the large motoneurons of the anterior horns of the spinal cord. As pointed out in the previous chapter, each nerve fiber normally branches many times and stimulates from three to several hundred skeletal muscle fibers. The nerve ending makes a junction, called the *neuromuscular junction,* with the muscle fiber near the fiber's midpoint, and the action potential in the fiber travels in both directions toward the muscle fiber ends. With the exception of about 2 per cent of the muscle fibers, there is only one such junction per muscle fiber.

Physiologic Anatomy of the Neuromuscular Junction — The "Motor End-Plate." Figure 7–1, parts A and B, illustrates the neuromuscular junction between a large, myelinated nerve fiber and a skeletal muscle fiber. The nerve fiber branches at its end to form a complex of branching nerve *terminals,* which invaginate into the muscle fiber but lie entirely outside the muscle fiber plasma membrane. The entire structure is called the *motor end-plate.* It is covered by one or more Schwann cells that insulate it from the surrounding fluids.

Figure 7–1C shows an electron micrographic sketch of the junction between a single-branch axon terminal and the muscle fiber membrane. The invagination of the membrane is called the *synaptic gutter* or *synaptic trough,* and the space between the terminal and the fiber membrane is called the *synaptic cleft.* The synaptic cleft is 20 to 30 nanometers wide and is occupied by a basal lamina, which is a thin layer of spongy reticular fibers through which diffuses extracellular fluid. At the bottom of the gutter are numerous smaller *folds* of the muscle membrane called *subneural clefts,* which greatly increase the surface area at which the synaptic transmitter can act.

In the axon terminal are many mitochondria that supply energy mainly for synthesis of the excitatory transmitter *acetylcholine* that, in turn, excites the muscle fiber. The acetylcholine is synthesized in the cytoplasm of the terminal but is rapidly absorbed into many small synaptic vesicles, approximately 300,000 of which are normally in the terminals of a single end-plate. Attached to the matrix of the basal lamina are large quantities of the enzyme *acetylcholinesterase,* which is capable of destroying acetylcholine, to be explained in further detail.

Secretion of Acetylcholine by the Nerve Terminals

When a nerve impulse reaches the neuromuscular junction, about 300 vesicles of acetylcholine are released from the terminals into the synaptic trough. Figure 7–2 illustrates some of the details of this

A

Myofibrils

Teloglial cell

Terminal nerve branches

Myelin sheath

Axon

Muscle nuclei

B

C

Synaptic vesicles

Axon terminal in synaptic trough

Subneural clefts

Figure 7 – 1. Different views of the motor end-plate. *A,* Longitudinal section through the end-plate. *B,* Surface view of the end-plate. *C,* Electron micrographic appearance of the contact point between one of the axon terminals and the muscle fiber membrane, representing the rectangular area shown in *A.* (From Fawcett, as modified from R. Couteaux: Bloom and Fawcett: A Textbook of Histology. Philadelphia, W. B. Saunders Company, 1986.)

mechanism, showing an expanded view of a synaptic trough with the neural membrane above and the muscle membrane and its subneural clefts below.

On the inside surface of the neural membrane are linear *dense bars,* shown in cross-section in Figure 7 – 2. To each side of each dense bar are protein particles that penetrate the membrane, believed to be voltage-gated calcium channels. When the action potential spreads over the terminal, these channels open and allow large quantities of calcium to diffuse to the interior of the terminal. The calcium ions in turn exert an attractive influence on the acetylcholine vesicles, drawing them to the neural membrane adjacent to the dense bars. Some of the vesicles fuse with the neural membrane and empty their acetylcholine into the synaptic trough by the process of exocytosis.

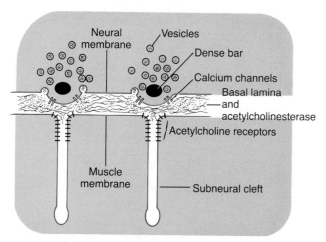

Neural membrane

Vesicles

Dense bar

Calcium channels

Basal lamina and acetylcholinesterase

Acetylcholine receptors

Muscle membrane

Subneural cleft

Figure 7 – 2. Release of acetylcholine from synaptic vesicles at the neural membrane of the neuromuscular junction. Note the close proximity of the release sites to the acetylcholine receptors at the mouths of the subneural clefts.

Although some of the aforementioned details are still speculative, it is known that the effective stimulus for causing acetylcholine release from the vesicles is entry of calcium ions. Furthermore, the vesicles are emptied through the membrane adjacent to the dense bars.

Effect of Acetylcholine to Open Acetylcholine-Gated Ion Channels. Figure 7 – 2 shows many acetylcholine receptors in the muscle membrane; these are actually acetylcholine-gated ion channels, located almost entirely near the mouths of the subneural clefts lying immediately below the dense bar areas, where the acetylcholine vesicles empty into the synaptic trough.

Each receptor is a large protein complex having a total molecular weight of 275,000. The complex is composed of five subunit proteins, which penetrate all the way through the membrane lying side by side in a circle to form a tubular channel. The channel remains constricted until acetylcholine attaches to one of the subunits. This causes a conformational change that opens the channel, as illustrated in Figure 7 – 3; the upper channel is closed, while the bottom one has been opened by attachment of an acetylcholine molecule.

The acetylcholine channel has a diameter when open of about 0.65 nanometer, which is large enough to allow all the important positive ions — sodium (Na^+), potassium (K^+), and calcium (Ca^{++}) — to move easily through the opening. On the other hand, negative ions, such as chloride ions, do not pass through because of strong negative charges in the mouth of the channel.

However, in practice, far more sodium ions flow through the acetylcholine channels than any other ions for two reasons. First, there are only two positive ions in great enough concentration to matter greatly,

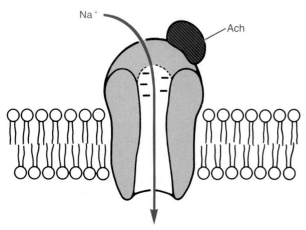

Figure 7–3. The *acetylcholine channel:* Above, while in the closed state. Below, after acetylcholine has become attached and a conformational change has opened the channel, allowing excess sodium to enter the muscle fiber and excite contraction. Note the negative charges at the channel mouth that prevent passage of negative ions.

sodium ions in the extracellular fluid and potassium ions in the intracellular fluid. Second, the very negative potential on the inside of the muscle membrane, −80 to −90 millivolts, pulls the positively charged sodium ions to the inside of the fiber, while at the same time preventing the efflux of the potassium ions when they attempt to pass outward.

Therefore, as illustrated in the lower panel of Figure 7–3, the net effect of opening the acetylcholine-gated channels is to allow large numbers of sodium ions to pour to the inside of the fiber, carrying with them large numbers of positive charges. This creates a local potential inside the fiber called the *end-plate potential* that initiates an action potential at the muscle membrane and thus causes muscle contraction.

Destruction of the Released Acetylcholine by Acetylcholinesterase. The acetylcholine, once released into the synaptic trough, continues to activate the acetylcholine receptors as long as it persists in the trough. However, it is rapidly removed by two means: (1) Most of the acetylcholine is destroyed by the en-

zyme *acetylcholinesterase* that is attached mainly to the *basal lamina,* a spongy layer of fine connective tissue that fills the synaptic trough between the presynaptic terminal and the postsynaptic muscle membrane. (2) A small amount diffuses out of the synaptic trough and is then no longer available to act on the muscle fiber membrane.

Yet, the very short period of time that the acetylcholine remains in the synaptic trough — a few milliseconds at most — is almost always sufficient to excite the muscle fiber. Then the rapid removal of the acetylcholine prevents muscle re-excitation after the fiber has recovered from the first action potential.

The "End-Plate Potential" and Excitation of the Skeletal Muscle Fiber. The sudden insurgence of sodium ions into the muscle fiber when the acetylcholine channels open causes the membrane potential in the *local area of the end-plate* to increase in the positive direction as much as 50 to 75 millivolts, creating a *local potential* called the *end-plate potential.* If one recalls from Chapter 5 that a sudden increase in membrane potential of more than 15 to 30 millivolts is sufficient to initiate the positive feedback effect of sodium channel activation, one can understand that the end-plate potential created by the acetylcholine stimulation is normally far greater than enough to initiate an action potential in the muscle fiber.

Figure 7–4 illustrates the principle of an end-plate potential initiating the action potential. In this figure are shown three separate end-plate potentials. End-plate potentials A and C are too weak to elicit an action potential, but they do nevertheless give the weak local potentials recorded in the figure. In contrast, end-plate potential B is much stronger and causes sodium channels to activate so that the self-regenerative effect of more and more sodium ions flowing to the interior of the fiber initiated an action potential. The weak end-plate potential at point A was caused by poisoning the muscle fiber with *curare,* a drug that blocks the gating action of acetylcholine on the acetylcholine channels by competing with the acetylcholine for the acetylcholine receptor site. The weak end-plate potential at point C resulted from the effect of botulinum toxin, a bacterial toxin that decreases the release of acetylcholine by the nerve terminals.

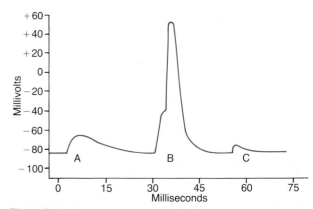

Figure 7–4. End-plate potentials. *A,* A weakened end-plate potential recorded in a curarized muscle, too weak to elicit an action potential; *B,* normal end-plate potential eliciting a muscle action potential; and *C,* weakened end-plate potential caused by botulinum toxin that decreases end-plate release of acetylcholine, again too weak to elicit a muscle action potential.

"Safety Factor" for Transmission at the Neuromuscular Junction; Fatigue of the Junction. Ordinarily, each impulse that arrives at the neuromuscular junction causes about three to four times as much end-plate potential as that required to stimulate the muscle fiber. Therefore, the normal neuromuscular junction is said to have a very high *safety factor.* However, artificial stimulation of the nerve fiber at rates greater than 100 times per second for several minutes often diminishes the number of vesicles of acetylcholine released with each impulse so much that impulses then fail to pass into the muscle fiber. This is called *fatigue* of the neuromuscular junction, and it is analogous to fatigue of the synapse in the central nervous system. Under normal functioning conditions, fatigue of the neuromuscular junction probably occurs very rarely and even then only at the most exhausting levels of muscular activity.

Molecular Biology of Acetylcholine Formation and Release

Because the neuromuscular junction is large enough to be easily studied, it is one of the few synapses of the nervous system at which most of the details of chemical transmission have been worked out. The formation and release of acetylcholine at this junction occurs in the following stages:

1. Very small vesicles, about 40 nanometers in size, are formed by the Golgi apparatus in the cell body of the motoneuron in the spinal cord. These vesicles are then transported by "streaming" of the axoplasm through the core of the axon from the central cell body to the neuromuscular junction at the tips of the nerve fibers. About 300,000 of these small vesicles collect in the nerve terminals of a single end-plate.

2. Acetylcholine is synthesized in the cytosol of the terminal nerve fibers but is then transported through the membranes of the vesicles to their interior, where it is stored in highly concentrated form, with about 10,000 molecules of acetylcholine in each vesicle.

3. Under resting conditions, an occasional vesicle fuses with the surface membrane of the nerve terminal and releases its acetylcholine into the synaptic gutter. When this occurs, a so-called *miniature end-plate* potential, about 1 millivolt in intensity and lasting for a few milliseconds, occurs in the local area of the muscle fiber because of the action of this "packet" of acetylcholine.

4. When an action potential arrives at the nerve terminal, this opens many calcium channels in the membrane of the terminal because this terminal has an abundance of voltage-gated calcium channels. As a result, the calcium ion concentration in the terminal increases about 100-fold, which in turn increases the rate of fusion of the acetylcholine vesicles with the terminal membrane about 10,000-fold. As each vesicle fuses, its outer surface ruptures through the cell membrane, thus causing *exocytosis* of acetylcholine into the synaptic cleft. Usually about 200 to 300 vesicles rupture with each action potential. The acetylcholine is then split by acetylcholinesterase into acetate ion and choline, and the choline is actively reabsorbed back into the neural terminal to be reused in forming new acetylcholine. This entire sequence of events occurs in 5 to 10 milliseconds.

5. After each vesicle has released its acetylcholine, the membrane of the vesicle becomes part of the cell membrane. However, the number of vesicles available in the nerve ending is sufficient to allow transmission of only a few thousand nerve impulses. Therefore, for continued function of the neuromuscular junction, the vesicles need to be retrieved from the nerve membrane. Retrieval is achieved by the process of *endocytosis* that was explained in Chapter 2. Within a few seconds after the action potential is over, "coated pits" appear on the surface of the terminal nerve membrane, caused by contractile proteins of the cytosol, especially the protein *cathrin,* attaching underneath the membrane in the areas of the original vesicles. Within about 20 seconds, the proteins contract and cause the pits to break away to the interior of the membrane, thus forming new vesicles. Within another few seconds, acetylcholine is transported to the interior of these vesicles, and they are then ready for a new cycle of acetylcholine release.

Drugs That Affect Transmission at the Neuromuscular Junction

Drugs That Stimulate the Muscle Fiber by Acetylcholine-like Action. Many different compounds, including *methacholine, carbachol,* and *nicotine,* have the same effect on the muscle fiber as does acetylcholine. The difference between these drugs and acetylcholine is that they are not destroyed by cholinesterase or are destroyed very slowly, so that when once applied to the muscle fiber the action persists for many minutes to several hours. These drugs work by causing localized areas of depolarization at the motor end-plate, where the acetylcholine receptors are located. Then, every time the muscle fiber becomes repolarized elsewhere, these depolarized areas, by virtue of their leaking ions, cause new action potentials, thereby causing a state of spasm.

Drugs That Block Transmission at the Neuromuscular Junction. A group of drugs, known as the *curariform drugs,* can prevent passage of impulses from the end-plate into the muscle. Thus, D-tubocurarine affects the membrane by competing with acetylcholine for the receptor sites of the membrane, so that the acetylcholine cannot increase the permeability of the acetylcholine channels sufficiently to initiate a depolarization wave.

Drugs That Stimulate the Neuromuscular Junction by Inactivating Acetylcholinesterase. Three particularly well-known drugs, *neostigmine, physostigmine,* and *diisopropyl fluorophosphate,* inactivate acetylcholinesterase so that the cholinesterase normally in the synapses will not hydrolyze the acetylcholine released at the end-plate. As a result, acetylcholine increases in quantity with successive nerve impulses so that extreme amounts of acetylcholine can accumulate and then repetitively stimulate the muscle fiber. This causes *muscular spasm* when even a few nerve impulses reach the muscle; this can cause death due to laryngeal spasm, which smothers the person.

Neostigmine and physostigmine combine with acetylcholinesterase to inactivate it for up to several hours, after which they are displaced from the acetylcholinesterase so that it once again becomes active. On the other hand, diisopropyl fluorophosphate, which has military potential as a very powerful "nerve" gas, actually inactivates acetylcholinesterase for weeks, which makes this a particularly lethal drug.

Myasthenia Gravis

The disease *myasthenia gravis,* which occurs in about one of every 20,000 persons, causes the person to become paralyzed because of inability of the neuromuscular junctions to transmit signals from the nerve fibers to the muscle fibers. Pathologically, antibodies that attack the acetylcholine-gated transport proteins have been demonstrated in the blood of most of these patients. Therefore, it is believed that myasthenia gravis in most instances is an autoimmune disease in which patients have developed antibodies against their own acetylcholine-activated ion channels.

Regardless of the cause, the end-plate potentials developed in the muscle fibers are too weak to stimulate the muscle fibers adequately. If the disease is intense enough, the patient dies of paralysis — in particular, of paralysis of the respiratory muscles. However, the disease can usually be ameliorated by administering *neostigmine* or some other anticholinesterase drug. This allows for more acetylcholine to accumulate in the synaptic cleft. Within minutes, some of these paralyzed persons can begin to function almost normally.

● THE MUSCLE ACTION POTENTIAL

Almost everything discussed in Chapter 5 regarding initiation and conduction of action potentials in nerve fibers applies equally well to skeletal muscle fibers except for quantitative differences. Some of the quantitative aspects of muscle potentials are the following:

1. Resting membrane potential: approximately −80 to −90 millivolts in skeletal fibers — the same as in large myelinated nerve fibers.

2. Duration of action potential: 1 to 5 milliseconds in skeletal muscle — about five times as long as large myelinated nerves.

3. Velocity of conduction: 3 to 5 m/sec — about ⅟₁₈ the velocity of conduction in the large myelinated nerve fibers that excite skeletal muscle.

Spread of the Action Potential to the Interior of the Muscle Fiber by Way of the Transverse Tubule System

The skeletal muscle fiber is so large that action potentials spreading along its surface membrane cause almost no current flow deep within the fiber. Yet, to cause contraction, these electrical currents must penetrate to the vicinity of all the separate myofibrils. This is achieved by transmission of the action potentials along *transverse tubules* (T tubules) that penetrate all the way through the muscle fiber from one side to the other. The T tubule action potentials in turn cause the sarcoplasmic reticulum to release calcium ions in the immediate vicinity of all the myofibrils, and these calcium ions then cause contraction. This overall process is called *excitation-contraction* coupling. Now, let us describe this in much greater detail.

● EXCITATION-CONTRACTION COUPLING

The Transverse Tubule – Sarcoplasmic Reticulum System

Figure 7–5 illustrates several myofibrils surrounded by the transverse tubule – sarcoplasmic reticulum system. The transverse tubules are very small and run transverse to the myofibrils. They begin at the cell membrane and penetrate all the way from one side of the muscle fiber to the opposite side. Not shown in the figure is the fact that these tubules branch among themselves so that they form entire *planes* of T tubules interlacing among all the separate myofibrils. Also, it should be noted that *where the T tubules originate from the cell membrane they are open to the exterior.* Therefore, they communicate with the fluid surrounding the muscle fiber and contain extracellular fluid in their lumens. In other words, the T tubules are internal extensions of the cell membrane. Therefore, when an action potential spreads over a muscle fiber membrane, it spreads along the T tubules to the deep interior of the muscle fiber as well. The action potential currents surrounding these transverse tubules then elicit the muscle contraction.

Figure 7–5 shows the *sarcoplasmic reticulum* as well, shown in red. This is composed of two major parts: (1) long *longitudinal tubules* that run parallel to the myofibrils and terminate in (2) large chambers called *terminal cisternae* that abut the transverse tubules. When the muscle fiber is sectioned longitudinally and electron micrographs are made, one sees this abutting of the cisternae against the transverse tubule, which gives the appearance of a *triad* with a small central tubule and a large cisterna on either side. This is illustrated in Figure 7–5 and is also seen in the electron micrograph of Figure 6–3 in the previous chapter.

In the muscle of lower animals, such as the frog, there is a single T tubule network for each sarcomere, located at the level of the Z disc, as illustrated in Figure 7–5. Cardiac muscle also has this type of T tubule system. However, in mammalian skeletal muscle there are two T tubule networks for each sarcomere located near the two ends of the myosin filaments, which are the points where the actual mechanical forces of muscle contraction are created. Thus, mammalian skeletal muscle is optimally organized for rapid excitation of muscle contraction.

RELEASE OF CALCIUM IONS BY THE SARCOPLASMIC RETICULUM

One of the special features of the sarcoplasmic reticulum is that it contains calcium ions in very high concentration, and many of these ions are released when the adjacent T tubule is excited.

Figure 7–6 shows that the action potential of the T tubule causes current flow through the tips of the cisternae that abut the T tubule. At these points, each cisterna projects *junctional feet* that attach to the membrane of the T tubule, presumably facilitating passage of some signal from the T tubule to the cisterna. Possibly this signal is electrical current of the action potential itself. However, there are also reasons to believe that it could be some chemical or mechanical signal. Whatever the signal, it causes rapid opening of large numbers of calcium channels through the membranes of the cisternae and their

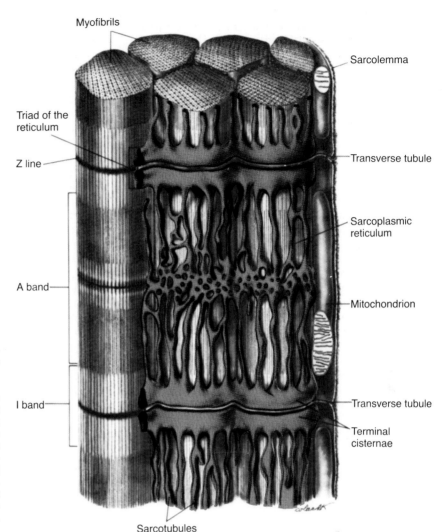

Myofibrils

Sarcolemma

Triad of the reticulum

Z line

Transverse tubule

A band

Sarcoplasmic reticulum

I band

Mitochondrion

Transverse tubule

Terminal cisternae

Sarcotubules

Figure 7-5. The transverse tubule–sarcoplasmic reticulum system. Note the *longitudinal tubules* that terminate in large *cisternae*. The cisternae in turn abut the transverse tubules. Note also that the transverse tubules communicate with the outside of the cell membrane. This illustration was drawn from frog muscle, which has one transverse tubule per sarcomere, located at the Z line. A similar arrangement is found in mammalian heart muscle, but mammalian skeletal muscle has two transverse tubules per sarcomere, located at the A-I junctions. (From Fawcett: Bloom and Fawcett: A Textbook of Histology. Philadelphia, W. B. Saunders Company, 1986. Modified after Peachey: *J. Cell Biol.* 25:209, 1965. Drawn by Sylvia Colard Keene.)

attached longitudinal tubules of the sarcoplasmic reticulum. These channels remain open for a few milliseconds; during this time the calcium ions responsible for muscle contraction are released into the sarcoplasm surrounding the myofibrils.

The calcium ions that are thus released from the sarcoplasmic reticulum diffuse to the adjacent myofibrils, where they bind strongly with troponin C, as discussed in the previous chapter, and this in turn elicits the muscle contraction.

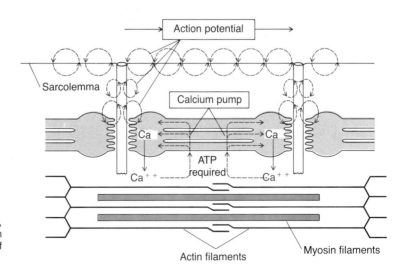

Action potential

Sarcolemma

Calcium pump

Ca

Ca

ATP required

Ca⁺⁺

Ca⁺⁺

Actin filaments

Myosin filaments

Figure 7-6. Excitation-contraction coupling in the muscle, showing an action potential that causes release of calcium ions from the sarcoplasmic reticulum and then reuptake of the calcium ions by a calcium pump.

The Calcium Pump for Removing Calcium Ions from the Sarcoplasmic Fluid. Once the calcium ions have been released from the sarcoplasmic tubules and have diffused to the myofibrils, muscle contraction will continue as long as the calcium ions remain in high concentration in the sarcoplasmic fluid. However, a continually active calcium pump located in the walls of the sarcoplasmic reticulum pumps calcium ions out of the sarcoplasmic fluid back into the sarcoplasmic tubules. This pump can concentrate the calcium ions about 10,000-fold inside the sarcoplasmic reticulum. In addition, inside the reticulum is a protein called *calsequestrin* that can bind over 40 times as much calcium as that in the ionic state, thus providing another 40-fold increase in the storage of calcium. Thus, this massive transfer of calcium into the sarcoplasmic reticulum causes virtual total depletion of calcium ions in the fluid of the myofibrils. Therefore, except immediately after an action potential, the calcium ion concentration in the myofibrils is kept at an extremely low level.

The Excitatory "Pulse" of Calcium Ions. The normal concentration (less than 10^{-7} molar) of calcium ions in the cytosol that bathes the myofibrils is too little to elicit contraction. Therefore, in the resting state, the troponin-tropomyosin complex keeps the actin filaments inhibited and maintains a relaxed state of the muscle.

On the other hand, full excitation of the T tubule–sarcoplasmic reticulum system causes enough release of calcium ions to increase the concentration in the myofibrillar fluid to as high as 2×10^{-4} molar concentration, which is about 10 times the level required to cause maximum muscle contraction (about 2×10^{-5} molar). Immediately thereafter, the calcium pump depletes the calcium ions again. The total duration of this calcium "pulse" in the usual skeletal muscle fiber lasts about 1/20 of a second, though it may last several times as long as this in some skeletal muscle fibers and be several times shorter in others (in heart muscle, the calcium pulse lasts for as long as 1/3 second because of the long duration of the cardiac action potential). It is during this calcium pulse that muscle contraction occurs. If the contraction is to continue without interruption for longer intervals, a series of such pulses must be initiated by a continuous series of repetitive action potentials, as discussed in the previous chapter.

REFERENCES

See references for Chapters 5 and 6.

CHAPTER 8

• Contraction and Excitation of Smooth Muscle •

● CONTRACTION OF SMOOTH MUSCLE

In the previous two chapters, the discussion has been concerned with skeletal muscle. We now turn to smooth muscle, which is composed of far smaller fibers — usually 2 to 5 micrometers in diameter and only 20 to 500 micrometers in length — in contrast to the skeletal muscle fibers that are as much as 20 times as large (in diameter) and thousands of times as long. Nevertheless, many of the principles of contraction apply to smooth muscle the same as to skeletal muscle. Most important, essentially the same attractive forces between myosin and actin filaments cause contraction in smooth muscle as in skeletal muscle, but the internal physical arrangement of smooth muscle fibers is entirely different, as we see subsequently.

TYPES OF SMOOTH MUSCLE

The smooth muscle of each organ is distinctive from that of most other organs in several different ways: physical dimensions, organization into bundles or sheets, response to different types of stimuli, characteristics of innervation, and function. Yet, for the sake of simplicity, smooth muscle can generally be divided into two major types, which are illustrated in Figure 8–1: *multiunit smooth muscle* and *single-unit smooth muscle.*

Multiunit Smooth Muscle. This type of smooth muscle is composed of discrete smooth muscle fibers. Each fiber operates entirely independently of the others and is often innervated by a single nerve ending, as occurs for skeletal muscle fibers. Furthermore, the outer surfaces of these fibers, like those of skeletal muscle fibers, are covered by a thin layer of "basement membrane–like" substance, a mixture of fine collagen and glycoprotein fibrillae that helps insulate the separate fibers from each other.

The most important characteristic of multiunit smooth muscle fibers is that each fiber can contract independently of the others, and their control is exerted mainly by nerve signals. This is in contrast to a major share of the control of visceral smooth muscle by non-nervous stimuli. An additional characteristic is that they rarely exhibit spontaneous contractions.

Some examples of multiunit smooth muscle found in the body are the smooth muscle fibers of the ciliary muscle of the eye, the iris of the eye, the nictitating membrane that covers the eyes in some lower animals, and the piloerector muscles that cause erection of the hairs when stimulated by the sympathetic nervous system.

Single-Unit Smooth Muscle. The term "single-unit" is confusing because it does not mean single muscle fibers. Instead, it means a whole mass of hundreds to millions of muscle fibers that contract together as a single unit. The fibers are usually aggregated into sheets or bundles, and their cell membranes are adherent to each other at multiple points so that force generated in one muscle fiber can be transmitted to the next. In addition, the cell membranes are joined by many *gap junctions* through which ions can flow freely from one cell to the next so that action potentials travel from one fiber to the next and cause the muscle fibers all to contract together. This type of smooth muscle is also known as *syncytial smooth muscle* because of its interconnections among fibers. Because such muscle is found in the walls of most viscera of the body — including the gut, the bile ducts, the ureters, the uterus, and many blood vessels — it is also often called *visceral smooth muscle.*

THE CONTRACTILE PROCESS IN SMOOTH MUSCLE

The Chemical Basis for Smooth Muscle Contraction

Smooth muscle contains both *actin* and *myosin filaments,* having chemical characteristics similar to but not exactly the same as those of the actin and myosin filaments in skeletal muscle. However, it does not contain troponin, so that the mechanism for control of contraction is entirely different. This is discussed in detail in a subsequent section of this chapter.

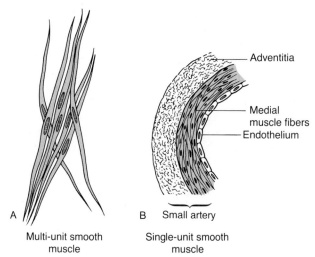

Figure 8–1. Multi-unit and single-unit smooth muscle.

Chemical studies have shown that actin and myosin derived from smooth muscle interact with each other in much the same way that this occurs for actin and myosin derived from skeletal muscle. Furthermore, the contractile process is activated by calcium ions, and adenosine triphosphate (ATP) is degraded to adenosine diphosphate (ADP) to provide the energy for contraction.

On the other hand, there are major differences between the physical organization of smooth muscle and that of skeletal muscle as well as differences in excitation-contraction coupling, control of the contractile process by calcium ions, duration of contraction, and amount of energy required for the contractile process.

The Physical Basis for Smooth Muscle Contraction

Smooth muscle does not have the same striated arrangement of the actin and myosin filaments as that found in skeletal muscle. For a long time, it was impossible to discern even in electron micrographs any specific organization in the smooth muscle cell that could account for contraction. However, recent special electron micrographic techniques suggest the physical organization illustrated in Figure 8–2. This shows large numbers of actin filaments attached to so-called *dense bodies*. Some of these bodies are attached to the cell membrane. Others are dispersed inside the cell and are held in place by a scaffold of structural proteins linking one dense body to another. Note in Figure 8–2 that some of the membrane dense bodies of adjacent cells are also bonded together by intracellular protein bridges. It is mainly through these bonds that the force of contraction is transmitted from one cell to the next.

Interspersed among the many actin filaments in the muscle fiber are a few myosin filaments. These have a diameter over two times as great as that of the actin filaments. When seen in electron micrographic cross-section, one usually finds about 15 times as many

actin filaments as myosin filaments. Part of this difference is caused by the fact that the ratio of actin filament length to myosin filament length in smooth muscle is much greater than in skeletal muscle. Therefore, the probability of seeing excess actin filaments is increased. Nevertheless, one is impressed by the relative sparsity of myosin filaments with respect to actin filaments.

To the right in Figure 8–2 is the postulated structure of individual contractile units within smooth muscle cells, showing large numbers of actin filaments radiating from two dense bodies; these filaments overlap a single myosin filament located midway between the dense bodies. Obviously, this contractile unit is similar to the contractile unit of skeletal muscle but without the regularity of the skeletal muscle structure; in fact, the dense bodies of

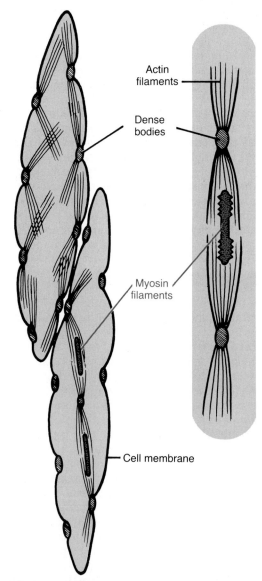

Figure 8–2. Physical structure of smooth muscle. The upper left-hand fiber shows actin filaments radiating from "dense bodies." The lower fiber in the right-hand insert demonstrates the relationship of myosin filaments to the actin filaments.

smooth muscle serve the same role as the Z discs in skeletal muscle.

Comparison of Smooth Muscle Contraction with Skeletal Muscle Contraction

Although most skeletal muscle contracts rapidly, most smooth muscle contraction provides prolonged tonic contraction, often lasting hours or even days. Therefore, it is to be expected that both the physical and the chemical characteristics of smooth muscle versus skeletal muscle contraction would differ. The following are some of the differences:

Slow Cycling of the Cross-Bridges. The rapidity of cycling of the cross-bridges in smooth muscle — that is, their attachment to actin, then release from the actin, and attachment again for the next cycle — is much, much slower in smooth muscle than in skeletal muscle, in fact as little as $1/10$ to $1/300$ the frequency in skeletal muscle. Yet, the *fraction of time* that the cross-bridges remain attached to the actin filaments, which is the major factor that determines the force of contraction, is believed to be very greatly increased in smooth muscle. A possible reason for the slow cycling is that the cross-bridge heads have far less ATPase activity than that in skeletal muscle, so that degradation of the ATP that energizes the movements of the heads is greatly reduced, with corresponding slowing of the rate of cycling.

Energy Required to Sustain Smooth Muscle Contraction. Only $1/10$ to $1/300$ as much energy is required to sustain the same tension of contraction in smooth muscle as in skeletal muscle. This, too, is believed to result because of the very slow attachment cycling of the cross-bridges and because only one molecule of ATP is required for each cycle regardless of its duration.

This economy of energy utilization by smooth muscle is exceedingly important to the overall energy economy of the body because organs, such as the intestines, the urinary bladder, the gallbladder, and other viscera must maintain tonic muscle contraction on a daily basis.

Slowness of Onset of Contraction and Relaxation of Smooth Muscle. A typical smooth muscle tissue begins to contract 50 to 100 milliseconds after it is excited, reaches full contraction about ½ second later, and then declines in contractile force in another 1 to 2 seconds, giving a total contraction time of 1 to 3 seconds. This is about 30 times as long as a single contraction of an average skeletal muscle. However, because of the many different types of smooth muscle, contraction of some types can be as short as 0.2 second or as long as 30 seconds.

The slow onset of contraction in smooth muscle as well as the prolonged contraction is probably caused by the slowness of attachment and detachment of the cross-bridges. In addition, the initiation of contraction in response to calcium ions, called the excitation-contraction coupling mechanism, is much slower than in skeletal muscle, as we discuss later.

Force of Muscle Contraction. Despite the relatively few myosin filaments in smooth muscle and despite the slow cycling time of the cross-bridges, the maximum force of contraction of smooth muscle is often even greater than that of skeletal muscle — as great as 4 to 6 kg/cm² cross-sectional area for smooth muscle in comparison with 3 to 4 kilograms for skeletal muscle. This great force of attraction is postulated to result from the prolonged period of attachment of the myosin cross-bridges to the actin filaments.

Percentage Shortening of Smooth Muscle During Contraction. A characteristic of smooth muscle that is different from skeletal muscle is its ability to shorten a far greater percentage of its length than can skeletal muscle while still maintaining almost full force of contraction. Skeletal muscle has a useful distance of contraction of only about one third its stretched length, whereas smooth muscle can often contract quite effectively more than two thirds its stretched length. This allows smooth muscle to perform especially important functions in the hollow viscera, allowing the gut, the bladder, the blood vessels, and other internal bodily structures to change their lumen diameters from very large down to almost zero.

Why this difference between smooth muscle and skeletal muscle? The answer to this is not completely known, but there appear to be two probable reasons. First, it is likely that some contractile units of smooth muscle have optimal overlapping of their actin and myosin filaments at one length of the muscle and others at other lengths, rather than all of these being synchronized together, as usually occurs in skeletal muscle. Therefore, a greater distance of contraction can be achieved. Second, the actin filaments in smooth muscle are much longer than those in skeletal muscle. Therefore, these filaments can be pulled along the myosin filaments for a much greater distance during smooth muscle contraction than occurs in skeletal muscle contraction.

The "Latch" Mechanism for Prolonged Holding Contractions of Smooth Muscle. Once smooth muscle has developed full contraction, the degree of activation of the muscle can usually be reduced to far less than the initial level and yet the muscle will still maintain its full strength of contraction. Furthermore, the energy consumed to maintain contraction is often miniscule, sometimes as little as $1/300$ the energy required for comparable skeletal muscle continuous contraction. This is called the "latch" mechanism. This same effect occurs to a slight extent in skeletal muscle but to an extent many times less than that in smooth muscle.

The importance of the latch mechanism is that it can maintain prolonged tonic contraction in smooth muscle for hours and hours with very little use of energy. Also, very little excitatory signal is required from nerve fibers or hormonal sources.

The cause of the latch phenomenon is undoubtedly related to the prolonged attachment of the myosin cross-bridges to the actin filaments. However, why this mechanism is more evident in some smooth muscle than in others is not understood, and even then the degree of the effect can change at different times.

Stress-Relaxation of Smooth Muscle. Another important characteristic of smooth muscle, especially of the visceral type of smooth muscle in many hollow organs, is its ability to return nearly to its original *force* of contraction seconds or minutes after it has been elongated or shortened. For example, a sudden increase in volume of fluid in the urinary bladder causes an immediate large increase in pressure in the bladder. However, during the next 15 seconds to a minute or so, despite continued stretch of the bladder wall, the pressure returns almost exactly back to the original level. Then, when the volume is increased by another step, the same effect occurs again. When the volume is suddenly decreased, the pressure falls very low at first but then returns in another few seconds or minutes back to the original level. This phenomenon is called *stress-relaxation*. Its obvious importance is that it allows a hollow organ to maintain approximately the same amount of pressure inside its lumen regardless of length of the muscle fibers.

The stress-relaxation phenomenon is probably closely related to the latch phenomenon. When the muscle is first stretched, the latch phenomenon resists change in length. Yet, with successive recycling of the myosin heads during the succeeding seconds to minutes, the heads release and attach further along the actin filaments. Therefore, eventually, the length of the muscle changes, while the tension in the muscle returns very near to its original value because the number of myosin cross-bridges causing contractile force is still very nearly the same as had been true previously.

REGULATION OF CONTRACTION BY CALCIUM IONS

As is true for skeletal muscle, the initiating event in most smooth muscle contraction is an increase in intracellular calcium ions. This increase can be caused by nerve stimulation of the smooth muscle fiber, hormonal stimulation, stretch of the fiber, or even changes in the chemical environment of the fiber.

Yet, smooth muscle does not contain troponin, the regulatory protein that is activated by calcium ions to cause skeletal muscle contraction. Instead, smooth muscle contraction is activated by an entirely different mechanism, as follows:

Combination of Calcium Ions with "Calmodulin"—Activation of Myosin Kinase and Phosphorylation of the Myosin Head. In place of troponin, smooth muscle cells contain large quantities of another regulatory protein called *calmodulin*. Although this protein is similar to troponin in that it reacts with four calcium ions, it is different in the manner in which it initiates the contraction. Calmodulin does this by activating the myosin cross-bridges.

This activation and subsequent contraction occurs in the following sequence:

1. The calcium ions bind with calmodulin.

2. The calmodulin-calcium combination then joins with and activates myosin kinase, a phosphorylating enzyme.

3. One of the light chains of each myosin head, called the *regulatory chain,* becomes phosphorylated in response to the myosin kinase. When this chain is not phosphorylated, the attachment-detachment cycling of the head will not occur. But, when the regulatory chain is phosphorylated, the head has the capability of binding with the actin filament and proceeding through the entire cycling process, thus causing muscle contraction.

Cessation of Contraction — Role of "Myosin Phosphatase." When the calcium ion concentration falls below a critical level, the aforementioned processes all automatically reverse except for the phosphorylation of the myosin head. Reversal of this requires another enzyme, *myosin phosphatase,* which splits the phosphate from the regulatory light chain. Then, the cycling stops and the contraction ceases. The time required for relaxation of muscle contraction, therefore, is determined to a great extent by the amount of active myosin phosphatase in the cell.

A Possible Mechanism for Regulation of the Latch Phenomenon

Because of the importance of the latch phenomenon in smooth muscle and because it is this phenomenon that allows long-term maintenance of tone in many smooth muscle organs, many attempts have been made to explain the latch phenomenon. Among many of the mechanisms that have been postulated, one of the simplest is the following:

When the myosin kinase and myosin phosphatase enzymes are both strongly activated, the cycling frequency of the myosin heads is very great, and the velocity of contraction is also great. Then, as the activation of the enzymes decreases, the cycling frequency decreases, but at the same time the lower activation also causes the myosin heads to remain attached to the actin filament for a longer and longer proportion of the cycling period. Therefore, the number of heads attached to the actin filament at any given time still remains very large. Since it is the number of heads attached to the actin that determines the force of contraction, tension is still maintained, or "latched"; yet, very little energy is used by the muscle because ATP is not degraded to ADP except on the rare occasions when a head detaches.

● NEURAL AND HORMONAL CONTROL OF SMOOTH MUSCLE CONTRACTION

Although skeletal muscle is activated exclusively by the nervous system, smooth muscle can be stimulated to contract by multiple types of signals: by nervous signals, by hormonal stimulation, and in several other ways. The principal reason for the difference is that the smooth muscle membrane contains many different types of receptor proteins that can initiate the contractile process. Still other receptor proteins inhibit smooth muscle contraction, which is another difference from skeletal muscle. Therefore, in this section, we discuss, first, neural control of smooth muscle contraction, followed by hormonal control and other means of control.

NEUROMUSCULAR JUNCTIONS OF SMOOTH MUSCLE

Physiologic Anatomy of Smooth Muscle Neuromuscular Junctions. Neuromuscular junctions of the type found on skeletal muscle fibers do not occur in smooth muscle. Instead, the *autonomic nerve fibers* that innervate smooth muscle generally branch diffusely on top of a sheet of muscle fibers, as illustrated in Figure 8–3. In most instances, these fibers do not make direct contact with the smooth muscle fibers at all but instead form so-called *diffuse junctions* that secrete their transmitter substance into the interstitial fluid from a few nanometers to a few microns away from the muscle cells; the transmitter substance then diffuses to the cells. Furthermore, where there are many layers of muscle cells, the nerve fibers often innervate only the outer layer, and the muscle excitation then travels from this outer layer to the inner layers by action potential conduction in the muscle mass or by subsequent diffusion of the transmitter substance.

The axons innervating smooth muscle fibers also do not have typical branching end-feet of the type in the motor end-plate on skeletal muscle fibers. Instead, most of the fine terminal axons have multiple *varicosities* distributed along their axes. At these points the Schwann cells are interrupted so that transmitter substance can be secreted through the walls of the varicosities. In the varicosities are vesicles similar to those in the skeletal muscle end-plate containing transmitter substance. However, in contrast to the vesicles of skeletal muscle junctions that contain only acetylcholine, the vesicles of the autonomic nerve fiber endings contain *acetylcholine* in some fibers and *norepinephrine* in others.

In a few instances, particularly in the multiunit type of smooth muscle, the varicosities lie directly on the muscle fiber membrane with a separation from this membrane of as little as 20 to 30 nanometers — the same width as the synaptic cleft that occurs in the skeletal muscle junction. These *contact junctions* function in much the same way as the skeletal muscle neuromuscular junction, and the latent period of contraction of these smooth muscle fibers is considerably shorter than of fibers stimulated by the diffuse junctions.

Excitatory and Inhibitory Transmitter Substances at the Smooth Muscle Neuromuscular Junction. Two different transmitter substances known to be secreted by the autonomic nerves innervating smooth muscle are *acetylcholine* and *norepinephrine*. Acetylcholine is an excitatory transmitter substance for smooth muscle fibers in some organs but an inhibitory substance for smooth muscle in other organs. When acetylcholine excites a muscle fiber, norepinephrine ordinarily inhibits it. Conversely, when acetylcholine inhibits a fiber, norepinephrine usually excites it.

But why these different responses? The answer is that both acetylcholine and norepinephrine excite or inhibit smooth muscle by first binding with a *receptor protein* on the surface of the muscle cell membrane. This receptor in turn controls the opening or closing of ion channels or controls some other means for activating or inhibiting the smooth muscle fiber. Furthermore, some of the receptor proteins are *excitatory receptors*, whereas others are *inhibitory receptors*. Thus, it is the type of receptor that determines whether the smooth muscle will be inhibited or excited and also determines which of the two transmitters, acetylcholine or norepinephrine, will be effective in causing the excitation or inhibition. These receptors are discussed in more detail in Chapter 60 in relation to the function of the autonomic nervous system.

MEMBRANE POTENTIALS AND ACTION POTENTIALS IN SMOOTH MUSCLE

Membrane Potentials in Smooth Muscle. The quantitative value of the membrane potential of smooth muscle is variable from one type of smooth muscle to another, and it also depends on the momentary condition of the muscle. However, in the normal resting state, the membrane potential is usually about -50 to -60 millivolts, or about 30 millivolts less negative than in skeletal muscle.

Action Potentials in Single-Unit Smooth Muscle. Action potentials occur in single-unit smooth muscle in the same way that they occur in skeletal muscle. However, action potentials do not normally occur in many if not most multiunit types of smooth muscle, as is discussed in a subsequent section.

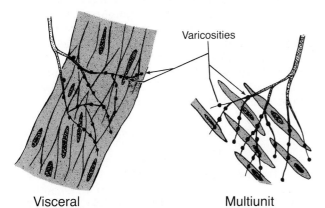

Varicosities

Visceral   Multiunit

Figure 8–3. Innervation of smooth muscle.

The action potentials of visceral smooth muscle occur in two different forms: (1) spike potentials and (2) action potentials with plateaus.

Spike Potentials. Typical spike action potentials, such as those seen in skeletal muscle, occur in most types of single-unit smooth muscle. The duration of this type of action potential is 10 to 50 milliseconds, as illustrated in Figure 8–4A and B. Such action potentials can be elicited in many ways, such as by electrical stimulation, by the action of hormones on the smooth muscle, by the action of transmitter substances from nerve fibers, or as a result of spontaneous generation in the muscle fiber itself, as discussed subsequently.

Action Potentials With Plateaus. Figure 8–4C illustrates an action potential with a plateau. The onset of this action potential is similar to that of the typical spike potential. However, instead of rapid repolarization of the muscle fiber membrane, the repolarization is delayed for several hundred to several thousand milliseconds. The importance of the plateau is that it can account for the prolonged periods of contraction that occur in some types of smooth muscle, such as the ureter, the uterus under some conditions, and some types of vascular smooth muscle. (Also, this is the type of action potential seen in cardiac muscle fibers that have a prolonged period of contraction, as we discuss in the next two chapters.)

Importance of Calcium Channels in Generating the Smooth Muscle Action Potential. The smooth muscle cell membrane has far more voltage-gated calcium channels than does skeletal muscle but very, very few voltage-gated sodium channels. Therefore, sodium participates very little if any in the generation of the action potential in most smooth muscle. Instead, the flow of calcium ions to the interior of the fiber is mainly responsible for the action potential. This occurs in the same self-regenerative way as occurs for the sodium channels in nerve fibers and in skeletal muscle fibers. However, calcium channels open many times more slowly than do sodium channels. This accounts in large measure for the slow action potentials of smooth muscle fibers.

Another important feature of calcium entry into the cells during the action potential is that this same calcium acts directly on the smooth muscle contractile mechanism to cause contraction, as discussed earlier. Thus, the calcium performs two tasks at once.

Slow Wave Potentials in Single-Unit Smooth Muscle and Spontaneous Generation of Action Potentials. Some smooth muscle is self-excitatory. That is, action potentials arise within the smooth muscle itself without an extrinsic stimulus. This is usually associated with a basic *slow wave rhythm* of the membrane potential. A typical slow wave of this type in the visceral smooth muscle of the gut is illustrated in Figure 8–4B. The slow wave itself is not an action potential. It is not a self-regenerative process that spreads progressively over the membranes of the muscle fibers. Instead, it is a local property of the smooth muscle fibers that make up the muscle mass.

The cause of the slow wave rhythm is unknown; one suggestion is that the slow waves are caused by waxing and waning of the pumping of sodium ions outward through the muscle fiber membrane; the membrane potential becomes more negative when sodium is pumped rapidly and less negative when the sodium pump becomes less active. Another suggestion is that the conductances of the ion channels increase and decrease rhythmically.

The importance of the slow waves lies in the fact that they can initiate action potentials. The slow waves themselves cannot cause muscle contraction, but when the potential of the slow wave rises above the level of approximately −35 millivolts (the approximate threshold for eliciting action potentials in most visceral smooth muscle), an action potential develops and spreads over the muscle mass, and then contraction does occur. Figure 8–4B illustrates this effect, showing that at each peak of the slow wave, one or more action potentials occur. This effect can obviously promote a series of rhythmical contractions of the smooth muscle mass. Therefore, the slow waves are frequently called *pacemaker waves*. In Chapter 62 we see that this type of activity controls the rhythmical contractions of the gut.

Figure 8–4. *A,* Typical smooth muscle action potential (spike potential) elicited by an external stimulus. *B,* Repetitive spike potentials elicited by slow rhythmical electrical waves that occur spontaneously in the smooth muscle of the intestinal wall. *C,* An action potential with a plateau recorded from a smooth muscle fiber of the uterus.

Excitation of Visceral Smooth Muscle by Stretch. When visceral (single-unit) smooth muscle is stretched sufficiently, spontaneous action potentials are usually generated. These result from a combination of the normal slow wave potentials plus a decrease in the negativity of the membrane potential caused by the stretch itself. This response to stretch allows a hollow organ that is excessively stretched to contract automatically and therefore to resist the stretch. For instance, when the gut is overstretched by intestinal contents, a local automatic contraction often sets up a peristaltic wave that moves the contents away from the excessively stretched intestine.

Depolarization of Multiunit Smooth Muscle Without Action Potentials. The smooth muscle fibers of multiunit smooth muscle normally contract mainly in response to nerve stimuli. The nerve endings secrete acetylcholine in the case of some multiunit smooth muscles and norepinephrine in the case of others. In both instances, these transmitter substances cause depolarization of the smooth muscle membrane, and this response in turn elicits the contraction. However, action potentials most often do not develop. The reason for this is that the fibers are too small to generate an action potential. (When action potentials are elicited in visceral single-unit smooth muscle, as many as 30 to 40 smooth muscle fibers must depolarize simultaneously before a self-propagating action potential ensues.) Yet, even without an action potential in the multiunit smooth muscle fibers, the local depolarization, called the "junctional potential," caused by the nerve transmitter substance itself spreads "electrotonically" over the entire fiber and is all that is needed to cause the muscle contraction.

SMOOTH MUSCLE CONTRACTION WITHOUT ACTION POTENTIALS — EFFECT OF LOCAL TISSUE FACTORS AND HORMONES

Probably half or more of all smooth muscle contraction is initiated not by action potentials but by stimulatory factors acting directly on the smooth muscle contractile machinery. The two types of nonnervous and nonaction potential stimulating factors most often involved are (1) local tissue factors and (2) various hormones.

Smooth Muscle Contraction in Response to Local Tissue Factors. In Chapter 17 we discuss the control of contraction of the arterioles, meta-arterioles, and precapillary sphincters. The smaller of these vessels have little or no nervous supply. Yet, the smooth muscle is highly contractile, responding rapidly to changes in local conditions in the surrounding interstitial fluid. In this way, a powerful local feedback control system controls the blood flow to the local tissue area. Some of the specific control factors are as follows:

1. Lack of oxygen in the local tissues causes smooth muscle relaxation and therefore vasodilatation.

2. Excess carbon dioxide causes vasodilatation.

3. Increased hydrogen ion concentration also causes increased vasodilatation.

Such factors as adenosine, lactic acid, increased potassium ions, diminished calcium ion concentration, and decreased body temperature also cause local vasodilatation.

Effects of Hormones on Smooth Muscle Contraction. Most of the circulating hormones in the body affect smooth muscle contraction at least to some degree, and some have very profound effects. Some of the more important blood-borne hormones that affect contraction are *norepinephrine, epinephrine, acetylcholine, angiotensin, vasopressin, oxytocin, serotonin,* and *histamine.*

A hormone causes contraction of smooth muscle when the muscle cell membrane contains *hormone-gated excitatory receptors* for the respective hormone. However, the hormone causes inhibition instead of contraction if the membrane contains *inhibitory receptors* rather than excitatory receptors.

Smooth Muscle Excitation or Inhibition Caused by Hormones or Local Tissue Factors. Some hormone receptors in the smooth muscle membrane open sodium or calcium ion channels and depolarize the membrane the same as following nerve stimulation. Occasionally, but not always, action potentials result, or rhythmical action potentials that are already occurring may be enhanced. However, in many instances depolarization occurs without action potentials; nevertheless, even this depolarization is associated with calcium ion entry into the cell that promotes contraction.

Activation of other membrane receptors inhibits contraction. This is achieved by closing sodium and calcium channels to prevent the entry of these positive ions or by opening potassium channels to allow positive potassium ions to flow to the exterior, in both instances increasing the degree of negativity inside the muscle cell, a state called *hyperpolarization.*

Sometimes contraction or inhibition is initiated by hormones without causing any change at all in the membrane potential. In these instances, the hormone usually activates a membrane receptor that does not open any ion channels but instead causes an internal change in the muscle fiber, such as release of calcium ions from the sarcoplasmic reticulum; the calcium then induces contraction. Or, to inhibit contraction, other receptor mechanisms are known to activate the enzyme *adenylate cyclase* or *guanylate cyclase* in the cell membrane; a portion of the enzyme protrudes to the interior of the cell and causes the formation of *cyclic AMP* or *cyclic GMP,* so-called *second messengers.* The cyclic AMP or GMP in turn has many effects, one of which is to change the degree of phosphorylation of several enzymes that indirectly inhibit contraction. Especially, the pump that pumps calcium ions from the sarcoplasm into the sarcoplasmic reticulum is activated as well as the cell membrane pump that pumps calcium ions out of the cell itself;

these effects reduce the intracellular calcium ion concentration, thereby inhibiting contraction.

Unfortunately, it is not known how most other local tissue factors besides the hormones, such as oxygen lack, excess carbon dioxide, or changes in hydrogen ion concentration, either excite or inhibit smooth muscle contraction. However, possible mechanisms include changes in the cell membrane potential, changes in the permeability of the membrane to calcium ions, changes in the intracellular contractile machinery, or some combination.

SOURCE OF CALCIUM IONS THAT CAUSE CONTRACTION BOTH THROUGH THE CELL MEMBRANE AND FROM THE SARCOPLASMIC RETICULUM

Although the contractile process in smooth muscle, as in skeletal muscle, is activated by calcium ions, the source of the calcium ions differs at least partly in smooth muscle; the difference is that the sarcoplasmic reticulum, from which virtually all the calcium ions are derived in skeletal muscle contraction, is often only rudimentary in most smooth muscle. Instead, in most types of smooth muscle, almost all the calcium ions that cause contraction enter the muscle cell from the extracellular fluid at the time of the action potential. There is a reasonably high concentration of calcium ions in the extracellular fluid, greater than 10^{-3} molar in comparison with less than 10^{-7} molar in the cell sarcoplasm, and as was pointed out earlier, the smooth muscle action potential is caused mainly by influx of calcium ions into the muscle fiber. Because the smooth muscle fibers are extremely small (in contrast to the sizes of the skeletal muscle fibers), these calcium ions can diffuse to all parts of the smooth muscle and elicit the contractile process. The time required for this diffusion to occur is usually 200 to 300 milliseconds and is called the *latent period* before the contraction begins; this latent period is some 50 times as great as that for skeletal muscle contraction.

Still additional calcium can enter the smooth muscle fiber via *hormone-activated calcium channels*; these, too, cause contraction. Usually, the opening of these channels does not cause an action potential, and sometimes not much change in the resting membrane potential either, because the sodium pump in the cell membrane pumps enough sodium ions to the exterior to maintain an almost normal membrane potential. Even so, contraction continues as long as these calcium channels remain open because it is calcium ions, not a change in membrane potential, that cause contraction. This is one means by which contraction is achieved in smooth muscle without significant change in cell membrane potential.

Role of the Sarcoplasmic Reticulum. Some smooth muscle contains a moderately developed sarcoplasmic reticulum. Figure 8–5 shows an example, illustrating separate sarcoplasmic tubules that lie near the cell membrane. Small invaginations of the

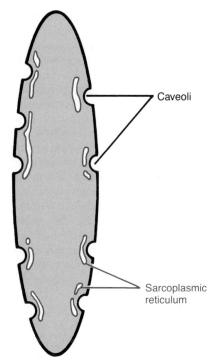

Figure 8–5. Sarcoplasmic tubules in a smooth muscle fiber, showing their relationship to invaginations in the cell membrane called *caveoli*.

membrane, called *caveoli*, abut against the surfaces of these tubules. The caveoli are believed to represent a rudimentary analog of the T tubule system of skeletal muscle. When an action potential is transmitted into the caveoli invaginations, this seems to excite calcium ion release from the sarcoplasmic tubules in the same way that action potentials in skeletal muscle T tubules also cause release of calcium ions.

In general, the more extensive the sarcoplasmic reticulum in the smooth muscle fiber, the more rapidly it contracts, presumably because calcium entry through the cell membrane is much slower than internal release of calcium ions from the sarcoplasmic reticulum.

Effect of Extracellular Calcium Ion Concentration on Smooth Muscle Contraction. Although the extracellular fluid calcium ion concentration has almost no effect on the force of contraction of skeletal muscle, this is not true for most smooth muscle. When the extracellular fluid calcium ion concentration falls to a low level, smooth muscle contraction usually almost ceases. In fact, after several minutes of being immersed in a low calcium medium, even the sarcoplasmic reticulum of smooth muscle fibers loses its calcium supply. Therefore, the force of contraction of smooth muscle is highly dependent on the extracellular fluid calcium ion concentration. We see in the following chapter that this is also true for cardiac muscle.

The Calcium Pump. To cause relaxation of the smooth muscle contractile elements, it is necessary to remove the calcium ions. This removal is achieved by calcium pumps that pump the calcium ions out of the smooth muscle fiber back into the extracellular fluid or pump the calcium ions into the sarcoplasmic reticulum. However, these pumps are very slow-acting in comparison with the fast-acting sarcoplasmic reticulum pump in skeletal muscle. Therefore, the duration of smooth muscle contraction is often in the order of seconds rather than hundredths to tenths of a second, as occurs for skeletal muscle.

REFERENCES

Bennett, M. R.: Development of neuromuscular synapses. Physiol. Rev., 63:915, 1983.

Blaustein, M. P., and Hamlyn, J. M.: Sodium transport inhibition, cell calcium, and hypertension. The natriuretic hormone/Na^+-Ca^{2+} exchange/hypertension hypothesis. Am. J. Med., 77:45, 1984.

Bohr, D. F., and Webb, R. C.: Vascular smooth muscle function and its changes in hypertension. Am. J. Med., 77:3, 1984.

Bolton, T. B.: Mechanisms of action of transmitters and other substances on smooth muscle. Physiol. Rev., 59:606, 1979.

Borgstrom, P., et al.: An evaluation of the metabolic interaction with myogenic vascular reactivity during blood flow autoregulation. Acta Physiol. Scand., 122:275, 1984.

Bulbring, E. (ed.): Smooth Muscle. An Assessment of Current Knowledge. Austin, University of Texas Press, 1981.

Butler, T. M., and Siegman, M. J.: High-energy phosphate metabolism in vascular smooth muscle. Annu. Rev. Physiol., 47:629, 1985.

Campbell, J. H., and Campbell, G. R.: Endothelial cell influences on vascular smooth muscle phenotype. Annu. Rev. Physiol., 48:295, 1986.

Dowben, R. M. (ed.): Cell and Muscle Motility. New York, Plenum Publishing Corp., 1983.

Eisenberg, E., and Greene, L. E.: The relation of muscle biochemistry to muscle physiology. Annu. Rev. Physiol., 42:293, 1980.

Fleming, W. W.: The electromagnetic Na^+, K^+ -pump in smooth muscle: Physiologic and pharmacologic significance. Annu. Rev. Pharmacol. Toxicol., 20:129, 1980.

Furchgott, R. F.: The role of endothelium in the responses of vascular smooth muscle to drugs. Annu. Rev. Pharmacol. Toxicol., 24:175, 1984.

Gabella, G.: Structural apparatus for force transmission in smooth muscle. Physiol. Rev., 64:455, 1984.

Hai, C.-M., and Murphy, R. A.: Ca^{2+}, crossbridge phosphorylation, and contraction. Annu. Rev. Physiol., 51:285, 1989.

Hartshorne, D. J., and Gorecka, A.: Biochemistry of the contractile proteins of smooth muscle. In Bohr, D. F., et al. (eds.): Handbook of Physiology. Sec. 2, Vol. II. Baltimore, Williams & Wilkins, 1980, p. 83.

Hartshorne, D. J., and Siemankowski, R. R.: Regulation of smooth muscle actomyosin. Annu. Rev. Physiol., 43:519, 1981.

Hertzberg, E. L., et al.: Gap junctional communication. Annu. Rev. Physiol., 43:479, 1981.

Hess, G. P., et al.: Acetylcholine receptor–controlled ion translocation: Chemical kinetic investigations of the mechanism. Annu. Rev. Biophys. Bioeng., 12:443, 1983.

Hirst, G. D. S., and Edwards, F. R.: Sympathetic neuroeffector transmission in arteries and arterioles. Physiol. Rev., 69:546, 1989.

Homsher, E.: Muscle enthalpy production and its relationship to actomyosin ATPase. Annu. Rev. Physiol., 49:673, 1987.

Johansson, B., and Somlyo, A. P.: Electrophysiology and excitation-contraction coupling. In Bohr, D. F., et al. (eds.): Handbook of Physiology. Sec. 2, Vol. II. Baltimore, Williams & Wilkins, 1980, p. 301.

Kamm, K. E., and Stull, J. T.: Regulation of smooth muscle contractile elements by second messengers. Annu. Rev. Physiol., 51:299, 1989.

Kito, S., et al. (eds.): Neuroreceptors and Signal Transduction. New York, Plenum Publishing Corp., 1988.

Lambert, J. J., et al.: Drug-induced modification of ionic conductance at the neuromuscular junction. Annu. Rev. Pharmacol. Toxicol., 23:505, 1983.

Lowenstein, W. R.: Junctional intercellular communication: The cell-to-cell membrane channel. Physiol. Rev., 61:829, 1981.

McKinney, M., and Richelson, E.: The coupling of neuronal muscarinic receptor to responses. Annu. Rev. Pharmacol. Toxicol., 24:121, 1984.

Morgan, D. L., and Proske, U.: Vertebrate slow muscle: Its structure, pattern of innervation, and mechanical properties. Physiol. Rev., 64:103, 1984.

Murphy, R. A.: Muscle cells of hollow organs. News Physiol. Sci., 3:124, 1988.

Paul, R. J.: Smooth muscle energetics. Annu. Rev. Physiol., 51:331, 1989.

Paul, R. J.: Chemical energetics of vascular smooth muscle. *In* Bohr, D. F., et al. (eds.): Handbook of Physiology. Sec. 2, Vol. II. Baltimore, Williams & Wilkins, 1980, p. 201.

Prosser, C. L.: Evolution and diversity of nonstriated muscles. In Bohr, D. F., et al. (eds.): Handbook of Physiology. Sec. 2, Vol. II. Baltimore, Williams & Wilkins, 1980, p. 635.

Purves, D., and Lichtman, J. W.: Specific connections between nerve cells. Annu. Rev. Physiol., 45:553, 1983.

Putney, J. W., Jr., et al.: How do inositol phosphates regulate calcium signaling? FASEB J., 3:1899, 1989.

Rasmussen, H., et al.: Protein kinase C in the regulation of smooth muscle contraction. FASEB J., 1:177, 1987.

Rosenthal, W., et al.: Control of voltage-dependent Ca^{2+} channels by G protein-coupled receptors. FASEB J., 2:2784, 1988.

Rowland, L. P., et al. (eds.): Molecular Genetics in Diseases of Brain, Nerve, and Muscle. New York, Oxford University Press, 1989.

Schneider, M. F.: Membrane charge movement and depolarization-contraction coupling. Annu. Rev. Physiol., 43:507, 1981.

Seidel, C. L., and Schildmeyer, L. A.: Vascular smooth muscle adaptation to increased load. Annu. Rev. Physiol., 49:489, 1987.

Somlyo, A. P.: Ultrastructure of vascular smooth muscle. In Bohr, D. F., et al. (eds.): Handbook of Physiology. Sec. 2, Vol. II. Baltimore, Williams & Wilkins, 1980, p. 33.

Spray, D. C., and Bennett, M. V. L.: Physiology and pharmacology of gap junctions. Annu. Rev. Physiol., 47:281, 1985.

van Breemen, C., and Saida, K.: Cellular mechanisms regulating $[Ca^{2+}]_i$ smooth muscle. Annu. Rev. Physiol., 51:315, 1989.

Vanhoutte, P. M.: Calcium-entry blockers, vascular smooth muscle and systemic hypertension. Am. J. Cardiol., 55:17B, 1985.

(See also Chapters 5 and 6.)

UNIT III

THE HEART

CHAPTER 9

• *Heart Muscle; The Heart as a Pump* •

With this chapter we begin discussion of the heart and circulatory system. The heart, illustrated in Figure 9–1, is actually two separate pumps: a *right heart* that pumps the blood through the lungs and a *left heart* that pumps the blood through the peripheral organs. In turn, each of these two separate hearts is a pulsatile two-chamber pump composed of an *atrium* and a *ventricle*. The atrium functions principally as a blood reservoir and as an entryway to the ventricle, but it also pumps weakly to help move the blood into the ventricle. The ventricle in turn supplies the main force that propels the blood through either the pulmonary or the peripheral circulation.

Special mechanisms in the heart maintain cardiac rhythmicity and transmit action potentials throughout the heart muscle to cause the heart's rhythmical beat. This rhythmical control system is explained in the following chapter. In the present chapter, we explain how the heart operates as a pump, beginning with the special features of heart muscle itself.

● PHYSIOLOGY OF CARDIAC MUSCLE

The heart is composed of three major types of cardiac muscle: atrial muscle, ventricular muscle, and specialized excitatory and conductive muscle fibers. The atrial and ventricular types of muscle contract in much the same way as skeletal muscle except that the duration of contraction is much longer. On the other hand, the specialized excitatory and conductive fibers contract only feebly because they contain few contractile fibrils; instead, they exhibit rhythmicity and varying rates of conduction, providing an excitatory system for the heart and a transmission system for controlled conduction of the cardiac excitatory signal throughout the heart.

PHYSIOLOGIC ANATOMY OF CARDIAC MUSCLE

Figure 9–2 illustrates a typical histological picture of cardiac muscle, showing the cardiac muscle fibers arranged in a latticework, the fibers dividing, then recombining, and then

spreading again. One notes immediately from this figure that cardiac muscle is *striated* in the same manner as typical skeletal muscle. Furthermore, cardiac muscle has typical myofibrils that contain *actin* and *myosin filaments* almost identical to those found in skeletal muscle, and these filaments interdigitate and slide along each other during the process of contraction in the same manner as occurs in skeletal muscle. (See Chapter 6.)

Cardiac Muscle as a Syncytium. The angulated dark areas crossing the cardiac muscle fibers in Figure 9–2 are called *intercalated discs;* they are actually cell membranes that separate individual cardiac muscle cells from each other. That is, cardiac muscle fibers are made up of many individual cells connected in series with each other. Yet electrical resistance through the intercalated disc is only ¼₀₀ the resistance through the outside membrane of the cardiac muscle fiber because the cell membranes fuse with each other and form very permeable "communicating" junctions (gap junctions) that allow relatively free diffusion of ions. Therefore, from a functional point of view, ions move with ease along the axes of the cardiac muscle fibers, so that action potentials travel from one cardiac muscle cell to another, past the intercalated discs, with only slight hindrance. Therefore, cardiac muscle is a *syncytium* of many heart muscle cells, in which the cardiac cells are so interconnected that when one of these cells becomes excited, the action potential spreads to all of them, spreading from cell to cell and spreading throughout the latticework interconnections.

The heart is composed of two separate syncytiums: the *atrial syncytium* that constitutes the walls of the two atria and the *ventricular syncytium* that constitutes the walls of the two ventricles. The atria are separated from the ventricles by fibrous tissue that surrounds the valvular openings between the atria and ventricles. Normally, action potentials can be conducted from the atrial syncytium into the ventricular syncytium only by way of a specialized conductive system, the *A-V bundle,* which is discussed in detail in the following chapter. This division of the muscle mass of the heart into two separate functional syncytiums allows the atria to contract a short time ahead of ventricular contraction, which is important for the effectiveness of heart pumping.

ACTION POTENTIALS IN CARDIAC MUSCLE

The *resting membrane potential* of normal cardiac muscle is approximately −85 to −95 millivolts (mV)

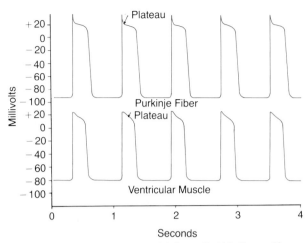

Figure 9-3. Rhythmic action potentials from a Purkinje fiber and from a ventricular muscle fiber, recorded by means of microelectrodes.

15 times as long in cardiac muscle as in skeletal muscle.

At this point, we must ask the question: Why is the action potential of cardiac muscle so long, and why does it have a plateau, whereas that of skeletal muscle does not? The basic biophysical answers to these questions are presented in Chapter 5, but they merit summarizing again.

At least two major differences between the membrane properties of cardiac and skeletal muscle account for the prolonged action potential and the plateau in cardiac muscle.

First, the action potential of skeletal muscle is caused almost entirely by sudden opening of large numbers of so-called *fast sodium channels* that allow tremendous numbers of sodium ions to enter the skeletal muscle fiber. These channels are called "fast" channels because they remain open for only a few 10,000ths of a second and then abruptly close. At the end of this closure, the process of repolarization occurs, and the action potential is over within another 10,000th of a second or so. In cardiac muscle, on the other hand, the action potential is caused by the opening of two types of channels: (1) the same *fast sodium channels* as those in skeletal muscle and (2) another entire population of so-called *slow calcium channels*, also called *calcium-sodium channels*. This second population of channels differs from the fast sodium channels in being slower to open, but, more importantly, they remain open for several tenths of a second. During this time, large amounts of both calcium and sodium ions flow through these channels to the interior of the cardiac muscle fiber, and this maintains a prolonged period of depolarization, causing the plateau in the action potential. Furthermore, the calcium ions that enter the muscle during this action potential play an important role in helping excite the muscle contractile process, which is another difference between cardiac muscle and skeletal muscle, as we discuss later in this chapter.

The second major functional difference between

and approximately −90 to −100 mV in the specialized conductive fibers, the Purkinje fibers, which are discussed in the following chapter.

The *action potential* recorded in ventricular muscle, shown by the bottom record of Figure 9-3, is 105 mV, which means that the membrane potential rises from its normally very negative value to a slightly positive value of about +20 mV. The positive portion is called the *overshoot potential*. Then, after the initial *spike,* the membrane remains depolarized for about 0.2 second in atrial muscle and about 0.3 second in ventricular muscle, exhibiting a *plateau* as illustrated in Figure 9-3, followed at the end of the plateau by abrupt repolarization. The presence of this plateau in the action potential causes muscle contraction to last 3 to

Figure 9-1. Structure of the heart and course of blood flow through the heart chambers.

Figure 9-2. The "syncytial," interconnecting nature of cardiac muscle.

cardiac muscle and skeletal muscle that helps account for both the prolonged action potential and its plateau is this: Immediately after the onset of the action potential, the permeability of the cardiac muscle membrane for potassium *decreases* about fivefold, an effect that does not occur in skeletal muscle. It is possible that this decreased potassium permeability is caused in some way by the excess calcium influx through the calcium channels just noted. However, regardless of the cause, the decreased potassium permeability greatly decreases the outflux of potassium ions during the action potential plateau and thereby prevents early recovery. When the slow calcium-sodium channels close at the end of 0.2 to 0.3 second and the influx of calcium and sodium ions ceases, the membrane permeability for potassium increases very rapidly, and the rapid loss of potassium from the fiber returns the membrane potential to its resting level, thus ending the action potential.

Velocity of Conduction in Cardiac Muscle. The velocity of conduction of the action potential in both atrial and ventricular muscle fibers is about 0.3 to 0.5 meter per second, or about $\frac{1}{250}$ the velocity in very large nerve fibers and about $\frac{1}{10}$ the velocity in skeletal muscle fibers. The velocity of conduction in the specialized conductive system varies from 0.02 to 4 meters per second in different parts of the system, as is explained in the following chapter.

Refractory Period of Cardiac Muscle. Cardiac muscle, like all excitable tissue, is refractory to restimulation during the action potential. Therefore, the refractory period of the heart is the interval of time, as shown to the left in Figure 9-4, during which a normal cardiac impulse cannot re-excite an already excited area of cardiac muscle. The normal refractory period of the ventricle is 0.25 to 0.3 second, which is approximately the duration of the action potential. There is an additional *relative refractory period* of about 0.05 second during which the muscle is more difficult than normal to excite but nevertheless can be excited, as illustrated by the early premature contraction in the second example of Figure 9-4.

The refractory period of atrial muscle is much shorter than that for the ventricles (about 0.15 second), and the relative refractory period is another 0.03 second. Therefore, the rhythmical rate of contraction of the atria can be much faster than that of the ventricles.

CONTRACTION OF CARDIAC MUSCLE

Excitation-Contraction Coupling—Function of Calcium Ions and of the T Tubules. The term "excitation-contraction coupling" means the mechanism by which the action potential causes the myofibrils of muscle to contract. This is discussed for skeletal muscle in Chapter 7. However, once again there are differences in this mechanism in cardiac muscle that have important effects on the characteristics of cardiac muscle contraction.

As is true for skeletal muscle, when an action potential passes over the cardiac muscle membrane, the action potential also spreads to the interior of the cardiac muscle fiber along the membranes of the T tubules. The T tubule action potentials in turn act on the membranes of the longitudinal sarcoplasmic tubules to cause instantaneous release of very large quantities of calcium ions into the muscle sarcoplasm from the sarcoplasmic reticulum. In another few thousandths of a second, these calcium ions diffuse into the myofibrils and catalyze the chemical reactions that promote sliding of the actin and myosin filaments along each other; this in turn produces the muscle contraction.

Thus far, this mechanism of excitation-contraction coupling is the same as that for skeletal muscle, but there is a second effect that is quite different. In addition to the calcium ions released into the sarcoplasm from the cisternae of the sarcoplasmic reticulum, large quantities of extra calcium ions also diffuse into the sarcoplasm from the T tubules at the time of the action potential. Indeed, without this extra calcium from the T tubules, the strength of cardiac muscle contraction would be considerably reduced because the sarcoplasmic reticulum of cardiac muscle is less well developed than that of skeletal muscle and does not store enough calcium to provide full contraction. On the other hand, the T tubules of cardiac muscle have a diameter 5 times as great as that of the skeletal muscle tubules and a volume 25 times as great; also, inside the T tubules is a large quantity of mucopolysaccharides that are electronegatively charged and bind an abundant store of calcium ions, keeping this always available for diffusion to the interior of the cardiac muscle fiber when the T tubule action potential occurs.

The strength of contraction of cardiac muscle depends to a great extent on the concentration of calcium ions in the extracellular fluids. The reason for this is that the ends of the T tubules open directly to the outside of the cardiac muscle fibers, allowing the

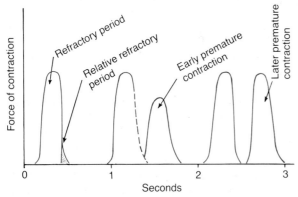

Figure 9-4. Contraction of the heart, showing the durations of the refractory period and the relative refractory period, the effect of an early premature contraction, and the effect of a later premature contraction. Note that the premature contractions do not cause wave summation, as occurs in skeletal muscle.

same extracellular fluid that is in the cardiac muscle interstitium to percolate through the T tubules as well. Consequently, the quantity of calcium ions in the T tubule system as well as the availability of calcium ions to cause cardiac muscle contraction depends directly on the extracellular fluid calcium ion concentration.

By way of contrast, the strength of skeletal muscle contraction is hardly affected by the extracellular fluid calcium concentration because its contraction is caused almost entirely by calcium ions released from the sarcoplasmic reticulum inside the skeletal muscle fiber itself.

At the end of the plateau of the action potential, the influx of calcium ions to the interior of the muscle fiber is suddenly cut off, and the calcium ions in the sarcoplasm are rapidly pumped back into both the sarcoplasmic reticulum and the T tubules. As a result, the contraction ceases until a new action potential occurs.

Duration of Contraction. Cardiac muscle begins to contract a few milliseconds after the action potential begins and continues to contract for a few milliseconds after the action potential ends. Therefore, the duration of contraction of cardiac muscle is mainly a function of the duration of the action potential — about 0.2 second in atrial muscle and 0.3 second in ventricular muscle.

Effect of Heart Rate on Duration of Contraction. When the heart rate increases, the duration of each total cycle of the heart, including both the contraction phase and the relaxation phase, obviously decreases. The duration of the action potential and the period of contraction (systole) also decrease but not by as great a percentage as does the relaxation phase (diastole). At a normal heart rate of 72 beats per minute, the period of contraction is about 0.40 of the entire cycle. At three times normal heart rate, this period is about 0.65 of the entire cycle, which means that the heart beating at a very fast rate does not remain relaxed long enough to allow complete filling of the cardiac chambers prior to the next contraction.

● THE CARDIAC CYCLE

The period from the beginning of one heartbeat to the beginning of the next is called the *cardiac cycle.* Each cycle is initiated by spontaneous generation of an action potential in the sinus node, as is explained in the following chapter. This node is located in the superior lateral wall of the right atrium near the opening of the superior vena cava, and the action potential travels rapidly through both atria and thence through the A-V bundle into the ventricles. However, because of a special arrangement of the conducting system from the atria into the ventricles, there is a delay of more than 1/10 second between passage of the cardiac impulse from the atria into the ventricles. This allows the atria to contract ahead of the ventricles, thereby pumping blood into the ventricles prior to the very strong ventricular contraction. Thus, the atria act as *primer pumps* for the ventricles, and the

ventricles then provide the major source of power for moving blood through the vascular system.

SYSTOLE AND DIASTOLE

The cardiac cycle consists of a period of relaxation called *diastole,* during which the heart fills with blood, followed by a period of contraction called *systole.*

Figure 9–5 illustrates the different events during the cardiac cycle. The top three curves show the pressure changes in the aorta, the left ventricle, and the left atrium, respectively. The fourth curve depicts the changes in ventricular volume, the fifth the electrocardiogram, and the sixth a phonocardiogram, which is a recording of the sounds produced by the heart — mainly by the heart valves — as it pumps. It is especially important that the reader study in detail the diagram of this figure and understand the causes of all the events illustrated. These are explained as follows:

RELATIONSHIP OF THE ELECTROCARDIOGRAM TO THE CARDIAC CYCLE

The electrocardiogram in Figure 9–5 shows the *P, Q, R, S,* and *T waves,* which are discussed in Chapters 11, 12, and 13. These are electrical voltages generated by the heart and recorded by the electrocardiograph from the surface of the body. The *P wave* is caused by the *spread of depolarization* through the atria, and this is followed by atrial contraction, which causes a slight rise in the atrial pressure curve immediately after the P wave. Approximately 0.16 second after the onset of the P wave, the *QRS waves* appear as a result of depolarization of the ventricles, which initiates contraction of the ventricles and causes the ventricular pressure to begin rising, as also illustrated in the figure. Therefore, the QRS complex begins slightly before the onset of ventricular systole.

Finally, one observes the *ventricular T wave* in the electrocardiogram. This represents the stage of repolarization of the ventricles at which time the ventricular muscle fibers begin to relax. Therefore, the T wave occurs slightly prior to the end of ventricular contraction.

Function of the Atria as Pumps. Blood normally flows continually from the great veins into the atria; approximately 75 per cent of blood flows directly through the atria into the ventricles even before the atria contract. Then, atrial contraction usually causes an additional 25 per cent filling of the ventricles. Therefore, the atria simply function as primer pumps that increase the ventricular pumping effectiveness as much as 25 per cent. Yet, the heart can continue to operate quite satisfactorily under normal resting conditions even without this extra 25 per cent effectiveness because it normally has the capability of pumping 300 to 400 per cent more blood than is required by the body anyway. Therefore, when the atria fail to function, the difference is unlikely to be noticed unless a person exercises; then acute signs of heart fail-

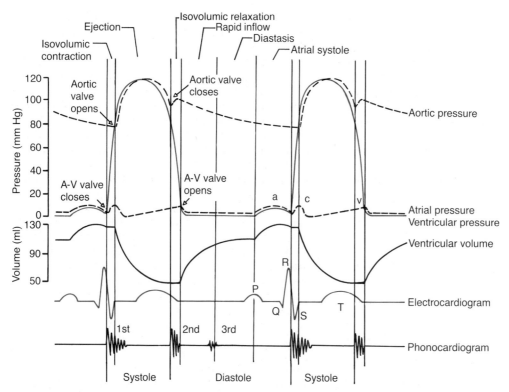

Figure 9–5. The events of the cardiac cycle, showing changes in left atrial pressure, left ventricular pressure, aortic pressure, ventricular volume, the electrocardiogram, and the phonocardiogram.

ure occasionally develop, especially shortness of breath.

Pressure Changes in the Atria—The a, c, and v Waves. In the atrial pressure curve of Figure 9–5, three major pressure elevations, called the *a, c,* and *v atrial pressure waves,* can be noted.

The *a wave* is caused by atrial contraction. Ordinarily, the *right* atrial pressure rises 4 to 6 mm Hg during atrial contraction, whereas the *left* atrial pressure rises about 7 to 8 mm Hg.

The *c wave* occurs when the ventricles begin to contract; it is caused partly by slight backflow of blood into atria at the onset of ventricular contraction but probably mainly by bulging of the A-V valves backward toward the atria because of increasing pressure in the ventricles.

The *v wave* occurs toward the end of ventricular contraction; it results from slow buildup of blood in the atria while the A-V valves are closed during ventricular contraction. Then, when ventricular contraction is over, the A-V valves open, allowing blood to flow rapidly into the ventricles and causing the v wave to disappear.

FUNCTION OF THE VENTRICLES AS PUMPS

Filling of the Ventricles. During ventricular systole, large amounts of blood accumulate in the atria because of the closed A-V valves. Therefore, just as soon as systole is over and the ventricular pressures fall again to their low diastolic values, the high pressures in the atria immediately push the A-V valves open and allow blood to flow rapidly into the ventricles, as shown by the rise of the *ventricular volume curve* in Figure 9–5. This is called the *period of rapid*

filling of the ventricles. The atrial pressures fall to within a fraction of a millimeter of the ventricular pressures because the normal A-V valve openings are so large that they offer almost no resistance to blood flow.

The period of rapid filling lasts approximately the first third of diastole. During the middle third of diastole only a small amount of blood normally flows into the ventricles; this is blood that continues to empty into the atria from the veins and passes on through the atria directly into the ventricles.

During the latter third of diastole, the atria contract and give an additional thrust to the inflow of blood into the ventricles; this accounts for approximately 25 per cent of the filling of the ventricles during each heart cycle.

Emptying of the Ventricles During Systole. *Period of Isovolumic (Isometric) Contraction.* Immediately after ventricular contraction begins, the ventricular pressure abruptly rises, as shown in Figure 9–5, causing the A-V valves to close. Then an additional 0.02 to 0.03 second is required for the ventricle to build up sufficient pressure to push the semilunar (aortic and pulmonary) valves open against the pressures in the aorta and pulmonary artery. Therefore, during this period of time, contraction is occurring in the ventricles, but there is no emptying. This period is called the period of *isovolumic* or *isometric contraction,* meaning by these terms that tension is increasing in the muscle, but no shortening of the muscle fibers is occurring. (This is not strictly true because there is apex-to-base shortening and circumferential elongation.)

Period of Ejection. When the left ventricular pressure rises slightly above 80 mm Hg (and the right ventricular pressure slightly above 8 mm Hg), the ventricular pressures now push the semilunar valves open. Immediately, blood begins to pour out of the ventricles, with about 70 per cent of the emptying occurring during the first third of the period of ejection and the remaining 30 per cent during the next two thirds. Therefore, the first third is called the *period of rapid ejection* and the last two thirds the *period of slow ejection.*

For a very peculiar reason, the ventricular pressure falls to a value slightly *below* that in the aorta during the period of slow ejection, despite the fact that some blood is still leaving the left ventricle. The reason is that the blood flowing out of the ventricle has built up momentum. As this momentum decreases during the latter part of systole, the kinetic energy of the momentum is converted into pressure in the aorta, which makes the arterial pressure slightly greater than the pressure inside the ventricle.

Period of Isovolumic (Isometric) Relaxation. At the end of systole, ventricular relaxation begins suddenly, allowing the intraventricular pressures to fall rapidly. The elevated pressures in the distended large arteries immediately push blood back toward the ventricles, which snaps the aortic and pulmonary valves closed. For another 0.03 to 0.06 second, the ventricular muscle continues to relax, even though the ventricular volume does not change, giving rise to the period of *isovolumic* or *isometric relaxation.* During this period, the intraventricular pressures fall rapidly back to their very low diastolic levels. Then the A-V valves open to begin a new cycle of ventricular pumping.

End-Diastolic Volume, End-Systolic Volume, and Stroke Volume Output. During diastole, filling of the ventricles normally increases the volume of each ventricle to about 110 to 120 milliliters (ml). This volume is known as the *end-diastolic volume.* Then, as the ventricles empty during systole, the volume decreases about 70 ml, which is called the *stroke volume output.* The remaining volume in each ventricle, about 40 to 50 ml, is called the *end-systolic volume.* The fraction of the end-diastolic volume that is ejected is called the *ejection fraction* — usually equal to about 60 per cent.

When the heart contracts strongly, the end-systolic volume can fall to as little as 10 to 20 ml. On the other hand, when large amounts of blood flow into the ventricles during diastole, their end-diastolic volumes can become as great as 150 to 180 ml in the normal heart. And by both increasing the end-diastolic volume and decreasing the end-systolic volume, the stroke volume output can at times be increased to about double normal.

FUNCTION OF THE VALVES

The Atrioventricular Valves. The *A-V valves* (the *tricuspid* and the *mitral* valves) prevent backflow

of blood from the ventricles to the atria during systole, and the *semilunar valves* (the *aortic* and *pulmonary* valves) prevent backflow from the aorta and pulmonary arteries into the ventricles during diastole. All these valves, which are illustrated in Figure 9-6, close and open *passively.* That is, they close when a backward pressure gradient pushes blood backward, and they open when a forward pressure gradient forces blood in the forward direction. For obvious anatomical reasons, the thin, filmy A-V valves require almost no backflow to cause closure, whereas the much heavier semilunar valves require rather strong backflow for a few milliseconds.

Function of the Papillary Muscles. Figure 9-6 also illustrates the papillary muscles that attach to the vanes of the A-V valves by the *chordae tendineae.* The papillary muscles contract when the ventricular walls contract, but, contrary to what might be expected, they *do not* help the valves to close. Instead, they pull the vanes of the valves inward toward the ventricles to prevent their bulging too far backward toward the atria during ventricular contraction. If a chorda tendinea becomes ruptured or if one of the papillary muscles becomes paralyzed, the valve bulges far backward, sometimes so far that it leaks severely and results in severe or even lethal cardiac incapacity.

The Aortic and Pulmonary Valves. There are differences between the operation of the aortic and pulmonary valves and that of the A-V valves. First, the high pressures in the arteries at the end of systole cause the semilunar valves to snap to the closed position in comparison with a much softer closure of the A-V valves. Second, because of smaller openings, the velocity of blood ejection through the aortic and pulmonary valves is far greater than that through the much larger A-V valves. Also, because of the rapid closure and rapid ejection, the edges of the semilunar valves are subjected to much greater mechanical abrasion than are the A-V valves, which also are supported by the chordae tendineae. It is obvious from

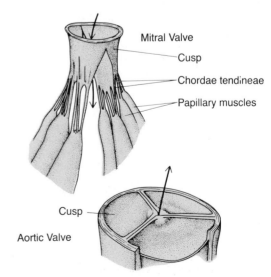

Figure 9-6. Mitral and aortic valves.

the anatomy of the aortic and pulmonary valves, as illustrated in Figure 9–6, that they are well adapted to withstand this extra physical trauma.

THE AORTIC PRESSURE CURVE

When the left ventricle contracts, the ventricular pressure rises rapidly until the aortic valve opens. Then the pressure in the ventricle rises much less thereafter, as illustrated in Figure 9–5, because blood immediately flows out of the ventricle into the aorta.

The entry of blood into the arteries causes the walls of these arteries to stretch and the pressure to rise. Then, at the end of systole, after the left ventricle stops ejecting blood and the aortic valve closes, the elastic recoil of the arteries maintains a high pressure in the arteries even during diastole.

A so-called *incisura* occurs in the aortic pressure curve when the aortic valve closes. This is caused by a short period of backward flow of blood immediately prior to closure of the valve, followed then by sudden cessation of the backflow.

After the aortic valve has closed, pressure in the aorta falls slowly throughout diastole because blood stored in the distended elastic arteries flows continually through the peripheral vessels back to the veins. Before the ventricle contracts again, the aortic pressure usually falls to approximately 80 mm Hg (diastolic pressure), which is two thirds the maximal pressure of 120 mm Hg (systolic pressure) occurring in the aorta during ventricular contraction.

The pressure curve in the pulmonary artery is similar to that in the aorta except that the pressures are only about one sixth as great, as is discussed in Chapter 14.

RELATIONSHIP OF THE HEART SOUNDS TO HEART PUMPING

When listening to the heart with a stethoscope, one does not hear the opening of the valves, for this is a relatively slowly developing process that makes no noise. However, when the valves close, the vanes of the valves and the surrounding fluids vibrate under the influence of the sudden pressure differentials that develop, giving off sound that travels in all directions through the chest.

When the ventricles first contract, one hears a sound that is caused by closure of the A-V valves. The vibration is low in pitch and relatively long continued and is known as the *first heart sound*. When the aortic and pulmonary valves close, one hears a relatively rapid snap, for these valves close extremely rapidly, and the surroundings vibrate for only a short period of time. This sound is known as the *second heart sound*.

Occasionally, one can hear an *atrial sound* when the atria beat because of vibrations associated with the flow of blood into the ventricles. Also, a *third heart sound* sometimes occurs at about the end of the first third of diastole, believed to be caused by blood flowing with a rumbling motion into the almost-filled ventricles. The precise causes of the heart sounds are discussed more fully in Chapter 23, in relation to auscultation.

WORK OUTPUT OF THE HEART

Stroke Work Output and Minute Work Output. The *stroke work output* of the heart is the amount of energy that the heart converts to work during each heartbeat while pumping blood into the arteries. *Minute work output* is the total amount of energy converted in a period of 1 minute; obviously, this is equal to the stroke work output times the heart rate.

Work output of the heart is in two forms. First, by far the major proportion is used to move the blood from the low pressure veins to the high pressure arteries. This is called *volume-pressure work* or *external work*. Second, a minor proportion of the energy is used to accelerate the blood to its velocity of ejection through the aortic and pulmonary valves. This is the *kinetic energy of blood flow* component of the work output.

External Work (Volume-Pressure Work). The work performed by the left ventricle to raise the pressure of the blood during each heartbeat (the left ventricular stroke external work output) is equal to the *stroke volume* times *(left ventricular mean ejection pressure* minus *mean left ventricular input pressure during diastolic filling)*. When pressure is expressed in dynes per square centimeter and stroke volume in milliliters, the external work output is in ergs.

Right ventricular external work output is normally about one sixth the work output of the left ventricle because of the difference in systolic pressure against which the two ventricles must pump.

The Kinetic Energy of Blood Flow. The additional work output of each ventricle required to create kinetic energy of blood flow is proportional to the mass of blood ejected times the square of velocity of ejection. That is,

$$\text{Kinetic energy} = \frac{mv^2}{2}$$

When the mass is expressed in *grams* of blood ejected and the velocity in *centimeters per second*, the work output is in *ergs.*

Ordinarily, the work output of the left ventricle required to create kinetic energy of blood flow is only about 1 per cent of the total work output of the ventricle and therefore is ignored in the calculation of the total stroke work output. In certain abnormal conditions, such as aortic stenosis, in which the blood flows with great velocity through the stenosed valve, more than 50 per cent of the total work output may be required to create kinetic energy of blood flow.

Graphical Analysis of Ventricular Pumping

Figure 9–7 illustrates a graphical diagram that is especially useful in explaining the pumping mechanics of the left ventricle. The two most important components of the diagram are the two heavy black curves labeled "diastolic pressure" and "systolic pressure." These two curves are volume-pressure curves.

The diastolic pressure curve is determined by filling the heart with progressively greater quantities of blood and then measuring the diastolic pressure immediately before ventricular contraction occurs, which is the *end-diastolic pressure* of the ventricle.

The systolic pressure curve is determined by preventing any outflow of blood from the heart and measuring the maximum systolic pressure that is achieved during ventricular contraction at each volume of filling.

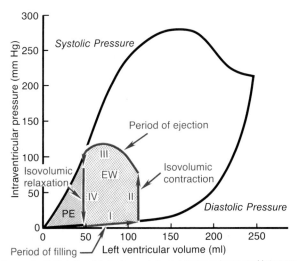

Figure 9-7. Relationship between left ventricular volume and intraventricular pressure during diastole and during systole. Also shown by the heavy red lines is the "volume-pressure diagram" that illustrates the changes in intraventricular volume and pressure during the cardiac cycle.

It is very clear that until the volume of the ventricle rises above about 150 ml, the diastolic pressure does not increase greatly. Therefore, up to this volume, blood can flow easily into the ventricle from the atrium. Above 150 ml, the diastolic pressure does increase rapidly, partly because of fibrous tissue in the heart that will stretch no more and partly because the pericardium that surrounds the heart becomes stretched nearly to its limit. During ventricular contraction, the systolic pressure increases rapidly at progressively greater ventricular volumes but reaches a maximum at a ventricular volume of 150 to 170 ml. Then, as the volume increases still further, the systolic pressure actually decreases under some conditions, as illustrated by the falling systolic pressure curve, because at these very great volumes the actin and myosin filaments of the cardiac muscle fibers are actually pulled apart enough for the strength of cardiac fiber contraction to become less than optimal.

Note especially in the figure that the maximum systolic pressure for the normal unstimulated left ventricle is between 250 and 300 mm Hg, but this varies widely with strength of the heart. For the normal right ventricle, it is between 60 and 80 mm Hg.

The "Volume-Pressure Diagram" During the Cardiac Cycle; Cardiac Work Output. The red curves in Figure 9-7 form a loop called the *volume-pressure diagram* of the cardiac cycle for the left ventricle. It is divided into four separate phases:

Phase I: *Period of filling.* This phase in the volume-pressure diagram begins at a ventricular volume of about 45 ml and a diastolic pressure very near 0 mm Hg. Forty-five milliliters is the amount of blood that remains in the ventricle after the previous heartbeat and is called the *end-systolic volume.* As pulmonary venous blood flows into the ventricle from the atrium, the volume normally increases to about 115 ml, called the *end-diastolic volume,* an increase of 70 ml. Therefore, the volume-pressure diagram during phase I extends along the line labeled "I," with the volume increasing to 115 ml and the diastolic pressure rising to about 5 mm Hg.

Phase II: *Period of isovolumic contraction.* During isovolumic contraction, the volume of the ventricle does not change. However, the pressure inside the ventricle rises to

equal the pressure in the aorta, a pressure value of about 80 mm Hg, as depicted by the line labeled "II."

Phase III: *Period of ejection.* During ejection, the systolic pressure rises even higher because of still more contraction of the heart. At the same time, the volume of the ventricle decreases because blood now flows out of the ventricle into the aorta. Therefore, the curve labeled "III" traces the changes in volume and systolic pressure during this period of ejection.

Phase IV: *Period of isovolumic relaxation.* At the end of the period of ejection, the semilunar valves of the ventricles close, and the ventricular pressure falls back to the diastolic pressure level. The line labeled "IV" traces this decrease in intraventricular pressure without any change in volume. Thus, the ventricle returns to its starting point, with about 45 ml of blood left in the ventricle and at an atrial pressure usually very close to 0 mm Hg.

Work Output Calculated From the Volume-Pressure Diagram. Readers well trained in the basic principles of physics should recognize that the area subtended by this volume-pressure diagram, the right-hand portion of the shaded area, labeled EW, is equal to the *net external work output* of the ventricle during its contraction cycle. In experimental studies of cardiac contraction, this diagram is therefore used for calculating cardiac work output.

When the heart pumps large quantities of blood, the work diagram becomes much larger in area. That is, it extends far to the right because the ventricle now fills with more blood during diastole, it rises much higher because the ventricle contracts with greater pressure, and it usually extends farther to the left because the ventricle contracts to a smaller volume—especially if the ventricle is stimulated to increased activity by the sympathetic nervous system.

The Concepts of "Preload" and "Afterload." In assessing the contractile properties of muscle, it is important to specify the degree of tension on the muscle when it begins to contract, which is called the *preload,* and also to specify the load against which the muscle exerts its contractile force, which is called the *afterload.*

For cardiac contraction, the preload is usually considered to be the volume of blood in the ventricle at the end of diastole, that is, the end-diastolic volume. However, sometimes this preload is expressed as the end-diastolic pressure when the ventricle has become filled.

The afterload of the ventricle is the pressure in the artery leading from the ventricle. In Figure 9-7, this corresponds to the systolic pressure described by the Phase III curve of the volume-pressure diagram. (Sometimes, the afterload is very loosely considered to be the resistance in the circulation rather than the pressure.)

The importance of the concepts of preload and afterload is that in many abnormal functional states of the heart or circulation, the degree of filling of the ventricle (the preload), the arterial pressure against which the ventricle must contract (the afterload), or both are severely altered from the normal.

THE CHEMICAL ENERGY FOR CARDIAC CONTRACTION: OXYGEN UTILIZATION BY THE HEART

Heart muscle, like skeletal muscle, uses chemical energy to provide the work of contraction. This energy is derived mainly from oxidative metabolism of fatty acids and to a lesser extent of other nutrients, especially lactate and glucose. Therefore, the rate of oxygen consumption by the heart is an excellent measure of the chemical energy liberated while the heart performs its work. The different reactions that liberate this energy are discussed in Chapters 67 and 68.

Experimental studies on isolated hearts have shown that the oxygen consumption of the heart, and therefore the chemical energy expended during contraction, is directly related to the total shaded area of Figure 9-7. This shaded portion consists of the *external work*, EW, as explained earlier, and an additional portion called the *potential energy*, labeled PE. The potential energy represents additional work output that could be accomplished by contraction of the ventricle if the ventricle should empty completely all the blood in its chamber with each contraction.

Unfortunately, it is impossible to measure the total shaded area of Figure 9-7 in the living animal or human. Instead, it has also been found experimentally that oxygen consumption is nearly proportional to the *tension* that occurs in the heart muscle during contraction *times* the *duration of time* that the contraction persists, called the *tension-time index*. Because tension is very high when the systolic pressure is high, correspondingly more oxygen is used. Also, much more chemical energy is expended even at normal systolic pressures when the ventricle is abnormally dilated because the heart muscle tension during contraction is proportional to pressure *times* the diameter of the ventricle. This is especially important in heart failure, for the ventricle is then dilated, and paradoxically the amount of chemical energy required for a given amount of work output must be greater than ever even though the heart is already failing.

Efficiency of Cardiac Contraction. During muscular contraction, most of the chemical energy is converted into heat and a much smaller portion into work output. The ratio of work output to chemical energy expenditure is called the efficiency of cardiac contraction, or simply *efficiency of the heart*. The maximum efficiency of the normal heart is between 20 and 25 per cent. In heart failure, this may fall as low as 5 to 10 per cent.

● REGULATION OF HEART PUMPING

When a person is at rest, the heart pumps only 4 to 6 liters of blood each minute. However, during severe exercise, the heart may be required to pump as much as four to seven times this amount. The present section discusses the means by which the heart can adapt to such extreme increases in cardiac output.

The two basic means by which the volume pumped by the heart is regulated are (1) intrinsic cardiac regulation of pumping in response to changes in volume of blood flowing into the heart and (2) control of the heart by the autonomic nervous system.

INTRINSIC REGULATION OF HEART PUMPING—THE FRANK-STARLING MECHANISM

In Chapter 20 we see that the amount of blood pumped by the heart each minute is determined by the rate of blood flow into the heart from the veins, which is called *venous return*. That is, each peripheral tissue

of the body controls its own blood flow, and the total of all the local blood flows through all the peripheral tissues returns by way of the veins to the right atrium. The heart in turn automatically pumps this incoming blood on into the systemic arteries, so that it can flow around the circuit again.

This intrinsic ability of the heart to adapt to changing volumes of inflowing blood is called the *Frank-Starling mechanism of the heart*, in honor of Frank and Starling, two great physiologists of nearly a century ago. Basically, the Frank-Starling mechanism means that the greater the heart is filled during diastole, the greater will be the quantity of blood pumped into the aorta. Or another way to express this is: *Within physiological limits, the heart pumps all the blood that comes to it without allowing excessive damming of blood in the veins.*

What Is the Explanation of the Frank-Starling Mechanism? When an extra amount of blood flows into the ventricles, the cardiac muscle itself is stretched to a greater length. This in turn causes the muscle to contract with increased force because the actin and myosin filaments are then brought to a more nearly optimal degree of interdigitation for force generation. Therefore, the ventricle, because of its increased pumping, automatically pumps the extra blood into the arteries. This ability of stretched muscle, up to an optimal length, to contract with increased force is characteristic of all striated muscle, as explained in Chapter 6, not simply of cardiac muscle.

In addition to the important effect of stretching the heart muscle, still another factor increases heart pumping when its volume is increased. Stretch of the right atrial wall directly increases the heart rate by as much as 10 to 20 per cent; this too helps increase the amount of blood pumped each minute, though its contribution is much less than that of the Frank-Starling mechanism.

Lack of Effect on Cardiac Output of Changes in Arterial Pressure Load. One of the most important consequences of the Frank-Starling mechanism of the heart is that, within reasonable limits, changes in the arterial pressure against which the heart pumps have almost no effect on the rate at which blood is

Figure 9-8. Constancy of cardiac output even in the face of wide changes in arterial pressure. Only when the arterial pressure rises above the normal operating pressure range does the pressure load cause the heart to begin to fail.

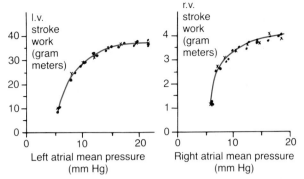

Figure 9-9. Left and right ventricular function curves in a dog, depicting ventricular stroke work output as a function of left and right mean atrial pressures. (Curves reconstructed from data in Sarnoff: *Physiol. Rev.* 35:107, 1955.)

pumped by the heart each minute (the cardiac output). This effect is illustrated in Figure 9-8, which is a curve extrapolated to the human from data in dogs in which the arterial pressure was progressively changed by constricting the aorta, while the cardiac output was measured simultaneously. The significance of this effect is the following: Regardless of the arterial pressure load up to a reasonable limit, the important factor determining the amount of blood pumped by the heart is still the rate of entry of blood into the heart.

Ventricular Function Curves

One of the best ways to express the functional ability of the ventricles to pump blood is by ventricular function curves, as shown in Figures 9-9 and 9-10. Figure 9-9 illustrates a type of ventricular function curve called the *stroke work output curve*. Note that as the atrial pressures increase, the stroke work output also increases until it reaches the limit of the heart's ability.

Figure 9-10 illustrates another type of ventricular function curve called the *minute ventricular output curve*. These two curves represent function of the two ventricles of the human heart based on data extrapolated from lower animals. As each atrial pressure rises, the respective ventricular volume output per minute also increases.

Thus, ventricular function curves are another way of expressing the Frank-Starling mechanism of the heart. That is, as the ventricles fill to higher atrial pressures, the ventricular volume and strength of cardiac contraction increase, causing the heart to pump increased quantities of blood into the arteries.

CONTROL OF THE HEART BY THE SYMPATHETIC AND PARASYMPATHETIC NERVES

The pumping effectiveness of the heart is highly controlled by the *sympathetic* and *parasympathetic* (vagus) nerves, which abundantly supply the heart, as illustrated in Figure 9-11. The amount of blood pumped by the heart each minute, the *cardiac output,* can often be increased more than 100 per cent by sympathetic stimulation. By contrast, it can be decreased to as low as zero or almost zero by vagal (parasympathetic) stimulation.

Excitation of the Heart by the Sympathetic Nerves. Strong sympathetic stimulation can increase the heart rate in the human to as high as 200 and rarely even 250 beats per minute in young people. Also, sympathetic stimulation increases the force with which the heart muscle contracts, therefore also increasing the volume of blood pumped as well as increasing the ejection pressure. Thus, sympathetic stimulation can often increase the cardiac output as much as twofold to threefold.

Inhibition of the sympathetic nervous system can be used to decrease cardiac pumping to a moderate extent in the following way: Under normal conditions, the sympathetic nerve fibers to the heart discharge continuously at a slow rate that maintains pumping at about 30 per cent above that with no sympathetic stimulation at all. Therefore, when the activity of the sympathetic nervous system is depressed below normal, this decreases both the heart rate and the strength of ventricular contraction, thereby decreasing the level of cardiac pumping to as much as 30 per cent below normal.

Figure 9-10. Approximate normal right and left ventricular output curves for the human heart as extrapolated from data obtained in dogs.

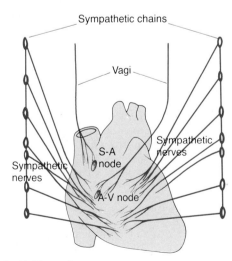

Figure 9-11. The cardiac nerves.

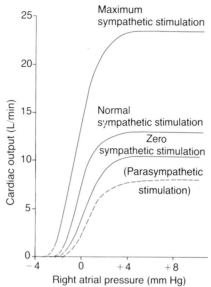

Figure 9–12. Effect on the cardiac output curve of different degrees of sympathetic and parasympathetic stimulation.

Parasympathetic (Vagal) Stimulation of the Heart. Strong vagal stimulation of the heart can actually stop the heartbeat for a few seconds, but then the heart usually "escapes" and beats at a rate of 20 to 30 beats per minute thereafter. In addition, strong parasympathetic stimulation decreases the strength of heart contraction by as much as 20 to 30 per cent. This is not a great decrease because the vagal fibers are distributed mainly to the atria and not much to the ventricles, where the power contraction of the heart occurs. Nevertheless, the great decrease in heart rate combined with a slight decrease in heart contraction can decrease ventricular pumping as much as 50 or more per cent — especially so when the heart is working under great workload.

Effect of Sympathetic or Parasympathetic Stimulation on the Cardiac Function Curve. Figure 9–12 illustrates four separate cardiac function curves. These are much the same as the ventricular function curves of Figures 9–9 and 9–10 except that they represent function of the entire heart rather than of a single ventricle; they show the relationship between the right atrial pressure at the input of the heart and cardiac output into the aorta.

The curves of Figure 9–12 demonstrate that at any given right atrial pressure, the cardiac output increases with increasing sympathetic stimulation and decreases with increasing parasympathetic stimulation. It should be remembered that the changes in output caused by nerve stimulation are brought about by both *changes in heart rate* and *changes in contractile strength of the heart,* for both of these affect cardiac output.

EFFECT OF HEART RATE ON FUNCTION OF THE HEART AS A PUMP

In general, the more times the heart beats per minute, the more blood it can pump, but there are impor-

tant limitations to this effect. For instance, once the heart rate rises above a critical level, the heart strength itself decreases, presumably because of overuse of metabolic substrates in the cardiac muscle. In addition, the period of diastole between the contractions becomes so reduced that blood does not have time to flow adequately from the atria into the ventricles. For these reasons, when the heart rate is increased artificially by *electrical stimulation,* the heart has its peak ability to pump large quantities of blood at a heart rate between 100 and 150 beats per minute. On the other hand, when its rate is increased by *sympathetic stimulation,* it reaches its peak ability to pump blood at a heart rate between 170 and 220 beats per minute. The reason for this difference is that sympathetic stimulation not only increases the heart rate but also increases heart strength. At the same time it also decreases the duration of systolic contraction and allows more time for filling during diastole.

ASSESSMENT OF CARDIAC CONTRACTILITY

Although it is very easy to determine the heart rate by simply timing the pulse, it has always been difficult to determine the strength of contraction of the heart, commonly called *cardiac contractility.* Very often the change in contractility is exactly opposite to the change in heart rate. Indeed, this effect occurs almost invariably in heart-debilitating diseases.

One of the ways in which cardiac contractility can be determined with great precision is to record one or more of the cardiac function curves. However, this can be done easily only in experimental animals. Therefore, many physiologists and clinicians have searched for methods to assess the cardiac contractility in a simple way. One of these methods is to determine the so-called dP/dt.

dP/dt as a Measure of Cardiac Contractility. dP/dt means the *rate of change of the ventricular pressure* with respect to time. The dP/dt record is generated by an electronic computer that *differentiates* the ventricular pressure wave, thus giving a record of the rate of change of the ventricular pressure. Figure 9–13 illustrates two separate recordings of the ventricular pressure wave as well as simultaneous recordings (in color) of the dP/dt. In the upper part of the figure, the heart was beating normally, and in the lower part, the heart had been stimulated by isoproterenol, a drug that has essentially the same effect on the heart as sympathetic stimulation.

Note in the upper record that at the same time that the ventricular pressure is increasing at its most rapid rate, the recording of the dP/dt record also reaches its greatest height. On the other hand, at the time that the ventricular pressure is falling most rapidly, the dP/dt record reaches its lowest level. When the ventricular pressure is neither rising nor falling, the dP/dt record is at zero value.

Experimental studies have shown that the rate of rise of ventricular pressure, the dP/dt, in general correlates very well with the strength of contraction of the ventricle. This effect is illustrated by a comparison of the dP/dt record in the upper part of Figure 9–13 with that in the lower part. Thus, the *peak* dP/dt is often used as a means for comparing the contractilities of hearts in different functional states.

Unfortunately, the quantitative value for peak dP/dt is also affected by other factors that are not related to cardiac contractility. For instance the value is increased by both

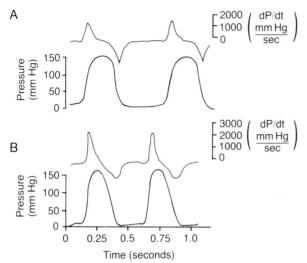

Figure 9-13. Simultaneous recordings of ventricular pressure and dP/dt. *A* shows results from a normal heart, and *B* shows results from a heart stimulated by isoproterenol. (Modified from Mason et al., in Sodeman and Sodeman (eds.): Pathologic Physiology. 6th ed. Philadelphia, W. B. Saunders Company, 1979.)

increased input pressure to the left ventricle (the end-diastolic ventricular pressure), which is the *preload* of the ventricle, and the pressure in the aorta against which the heart is pumping the blood, called the *afterload*. Therefore, it is often difficult to use dP/dt as a measure of contractility in comparing hearts from one person to another because one of these factors may differ. For this reason other quantitative measures have also been used in attempts to assess cardiac contractility. One of these has been to use dP/dt divided by the instantaneous pressure in the ventricle, or (dP/dt)/P.

EFFECT OF POTASSIUM AND CALCIUM IONS ON HEART FUNCTION

In the discussion of membrane potentials in Chapter 5, it was pointed out that potassium ions have a marked effect on membrane potentials and action potentials, and in Chapter 6 it was noted that calcium ions play an especially important role in initiating the muscle contractile process. Therefore, it is to be expected that the concentrations of these two ions in the extracellular fluids will also have important effects on cardiac pumping.

Effect of Potassium Ions. Excess potassium in the extracellular fluids causes the heart to become extremely dilated and flaccid and slows the heart rate. Very large quantities can also block conduction of the cardiac impulse from the atria to the ventricles through the A-V bundle. Elevation of potassium concentration to only 9 to 12 mEq/liter—two to three times the normal value—can cause such weakness of the heart and abnormal rhythm that this can cause death.

These effects are caused partially by the fact that a high potassium concentration in the extracellular fluids causes a decreased resting membrane potential in the cardiac muscle fibers, as explained in Chapter 5. As the membrane potential decreases, the intensity of the action potential also decreases, which makes the contraction of the heart progressively weaker.

Effect of Calcium Ions. An excess of calcium ions causes effects almost exactly opposite of those of potassium ions, causing the heart to go into spastic contraction. This is caused by the direct effect of calcium ions in exciting the cardiac contractile process, as explained earlier in the chapter. Conversely, a deficiency of calcium ions causes cardiac flaccidity, similar to the effect of potassium. However, because the calcium ion levels in the blood are normally regulated within narrow ranges, it is rare that these cardiac effects of abnormal calcium concentrations are of clinical concern.

EFFECT OF TEMPERATURE ON THE HEART

Increased temperature, as occurs when one has fever, causes greatly increased heart rate, sometimes to as great as double normal. Decreased temperature causes greatly decreased rate, falling to as low as a few beats per minute when a person is near death from hypothermia in the range of 60° to 70°F (15.5° to 21.1°C). These effects presumably result from the heat causing increased permeability of the muscle membrane to the ions, resulting in acceleration of the self-excitation process.

Contractile strength of the heart is often enhanced temporarily by a moderate increase in temperature, but prolonged elevation of the temperature exhausts the metabolic systems of the heart and causes weakness.

REFERENCES

Brunwald, E., and Ross, J., Jr.: Control of cardiac performance. In Berne, R. M., et al. (eds.): Handbook of Physiology. Sec. 2, Vol. I. Baltimore, Williams & Wilkins, 1979, p. 533.

Cooper, G., IV: Cardiocyte adaptation to chronically altered load. Annu. Rev. Physiol., 49:501, 1987.

Cowley, A. W., Jr., and Guyton, A. C.: Heart rate as a determinant of cardiac output in dogs with arteriovenous fistula. Am. J. Cardiol., 28:321, 1971.

Ellis, D.: Na-Ca exchange in cardiac tissue. Adv. Myocardiol., 5:295, 1985.

Fabiato, A., and Fabiato, F.: Calcium and cardiac excitation-contraction coupling. Annu. Rev. Physiol., 41:473, 1979.

FitzGerald, P. G.: Gap junction heterogeneity in liver, heart, and lens. News Physiol. Sci., 3:206, 1988.

Fozzard, H. A., et al.: The Heart and Cardiovascular System: Scientific Foundations. New York, Raven Press, 1986.

Gadsby, D. C.: The Na/K pump of cardiac cells. Annu. Rev. Biophys. Bioeng., 13:373, 1984.

Gevers, W.: Protein metabolism of the heart. J. Mol. Cell. Cardiol., 16:3, 1984.

Gilmour, R. F., Jr., and Zipes, D. P.: Slow inward current and cardiac arrhythmias. Am. J. Cardiol., 55:89B, 1985.

Guyton, A. C.: Determination of cardiac output by equating venous return curves with cardiac response curves. Physiol. Rev., 35:123, 1955.

Guyton, A. C., et al.: Circulatory Physiology: Cardiac Output and Its Regulation. 2nd Ed. Philadelphia, W. B. Saunders Co., 1973.

Hoffman, B. B., and Lefkowitz, R. J.: Adrenergic receptors in the heart. Annu. Rev. Physiol., 44:475, 1982.

Hurst, J. W., et al.: The Heart. New York, McGraw-Hill Book Co., 1990.

Jacobus, W. E.: Respiratory control and the integration of heart high-energy phosphate metabolism by mitochondrial creatine kinase. Annu. Rev. Physiol., 47:707, 1985.

Johnson, E. A.: Force-interval relationship of cardiac muscle. In Berne, R. M., et al. (eds.): Handbook of Physiology. Sec. 2, Vol. I. Baltimore, Williams & Wilkins, 1979, p. 475.

Manfredi, J. P., and Holmes, E. W.: Purine salvage pathways in myocardium. Annu. Rev. Physiol., 47:691, 1985.

Matsubara, I.: X-ray diffraction studies of the heart. Annu. Rev. Biophys. Bioeng., 9:81, 1980.

Mela-Riker, L. M., and Bukoski, R. D.: Regulation of mitochondrial activity in cardiac cells. Annu. Rev. Physiol., 47:645, 1985.

Morgan, H. E., et al.: Biochemical mechanisms of cardiac hypertrophy. Annu. Rev. Physiol., 49:533, 1987.

Nozawa, T., et al.: Relation between oxygen consumption and pressure-volume area of in situ dog heart. Am. J. Physiol., 253:H31, 1987.

Page, E., and Shibata, Y.: Permeable junctions between cardiac cells. Annu. Rev. Physiol., 43:431, 1981.

Parmley, W. W., and Talbot, W.: Heart as a pump. In Berne, R. M., et al. (eds.): Handbook of Physiology. Sec. 2, Vol. I. Baltimore, Williams & Wilkins, 1979, p. 429.

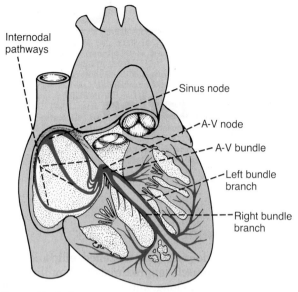

Figure 10–1. The sinus node and the Purkinje system of the heart, showing also the A-V node, the atrial internodal pathways, and the ventricular bundle branches.

slow calcium-sodium channels, and (3) the *potassium channels.* The opening of the fast sodium channels for a few 10,000ths of a second is responsible for the very rapid spikelike onset of the action potential observed in ventricular muscle because of rapid influx of positive sodium ions to the interior of the fiber. Then the plateau of the ventricular action potential is caused primarily by slower opening of the slow calcium-sodium channels, which lasts for a few tenths of a second. Finally, increased opening of the potassium channels and diffusion of large amounts of positive potassium ions out of the fiber return the membrane potential to its resting level.

But there is a difference in the function of these channels in the sinus nodal fiber because of the much lesser negativity of the "resting" potential—only −55 millivolts. At this level of negativity, the fast sodium channels have mainly become "inactivated," which means that they have become blocked. The cause of this is that any time the membrane potential

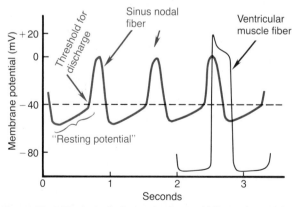

Figure 10–2. Rhythmic discharge of a sinus nodal fiber and comparison of the sinus nodal action potential with that of a ventricular muscle fiber.

remains less negative than about −60 millivolts for more than a few milliseconds, the inactivation gates on the inside of the cell membrane that close these channels become closed and remain so. Therefore, only the slow calcium-sodium channels can open (that is, can become "activated") and thereby cause the action potential. Therefore, the action potential is slower to develop than that of the ventricular muscle and also recovers with a slow decrement of the potential rather than the abrupt recovery that occurs for the ventricular fiber.

Self-Excitation of Sinus Nodal Fibers. Sodium ions naturally tend to leak to the interior of the sinus nodal fibers through multiple membrane channels, and this influx of positive charges also causes a rising membrane potential. Thus, as illustrated in Figure 10–2, the "resting" potential gradually rises between each two heartbeats. When it reaches a *threshold voltage* of about −40 millivolts, the calcium-sodium channels become activated, leading to very rapid entry of both calcium and sodium ions, thus causing the action potential. Therefore, basically, the inherent leakiness of the sinus nodal fibers to sodium ions causes their self-excitation.

Why does this leakiness to sodium ions not cause the sinus nodal fibers to remain depolarized all the time? The answer is that two events occur during the course of the action potential. First, the calcium-sodium channels become inactivated (that is, they close) within about 100 to 150 milliseconds after opening, and second, at about the same time greatly increased numbers of potassium channels open. Therefore, now the influx of calcium and sodium ions through the calcium-sodium channels ceases simultaneously, while large quantities of positive potassium ions diffuse *out* of the fiber, thus terminating the action potential. Furthermore, the potassium channels remain open for another few tenths of a second, carrying a great excess of positive potassium charges out of the cell, which temporarily causes considerable excess negativity inside the fiber; this is called *hyperpolarization.* This hyperpolarization initially carries the "resting" membrane potential down to about −55 to −60 millivolts at the termination of the action potential.

Last, we must explain why the state of hyperpolarization also is not maintained forever. The reason is that during the next few tenths of a second after the action potential is over, progressively more and more of the potassium channels begin to close. Now the inward-leaking sodium ions once again overbalance the outward flux of potassium ions, which causes the "resting" potential to drift upward, finally reaching the threshold level for discharge at a potential of about −40 millivolts. Then the entire process begins again: self-excitation, recovery from the action potential, hyperpolarization after the action potential is over, upward drift of the "resting" potential, then re-excitation still again to elicit another cycle. This process continues indefinitely throughout the life of the person.

INTERNODAL PATHWAYS AND TRANSMISSION OF THE CARDIAC IMPULSE THROUGH THE ATRIA

The ends of the sinus nodal fibers fuse with the surrounding atrial muscle fibers, and action potentials originating in the sinus node travel outward into these fibers. In this way, the action potential spreads through the entire atrial muscle mass and eventually also to the A-V node. The velocity of conduction in the atrial muscle is approximately 0.3 m/sec. However, conduction is somewhat more rapid in several small bundles of atrial muscle fibers. One of these, called the *anterior interatrial band,* passes through the anterior walls of the atria to the left atrium and conducts the cardiac impulse at a velocity of about 1 m/sec. In addition, three other small bundles curve through the atrial walls and terminate in the A-V node, also conducting the cardiac impulse at this rapid velocity. These three small bundles, illustrated in Figure 10–1, are called, respectively, the *anterior, middle,* and *posterior internodal pathways.* The cause of the more rapid velocity of conduction in these bundles is the presence of a number of specialized conduction fibers mixed with the atrial muscle. These fibers are similar to the very rapidly conducting Purkinje fibers of the ventricles, which are discussed subsequently.

THE A-V NODE AND DELAY IN IMPULSE CONDUCTION

Fortunately, the conductive system is organized so that the cardiac impulse will not travel from the atria into the ventricles too rapidly; this allows time for the atria to empty their contents into the ventricles before ventricular contraction begins. It is primarily the A-V node and its associated conductive fibers that delay this transmission of the cardiac impulse from the atria into the ventricles.

The A-V node is located in the posterior septal wall of the right atrium immediately behind the tricuspid valve and adjacent to the opening of the coronary sinus, as illustrated in Figure 10–1. Figure 10–3 shows diagrammatically the different parts of this node and its connections with the atrial internodal pathway fibers and the A-V bundle. The figure also shows the approximate intervals of time in fractions of a second between the genesis of the cardiac impulse in the sinus node and its appearance at different points in the A-V nodal system. Note that the impulse, after traveling through the internodal pathway, reaches the A-V node approximately 0.03 second after its origin in the sinus node. Then there is a further delay of 0.09 second in the A-V node itself before the impulse enters the *penetrating portion of the A-V bundle.* A final delay of another 0.04 second occurs mainly in this penetrating A-V bundle, which is composed of multiple small fascicles passing through the fibrous tissue separating the atria from the ventricles.

Thus, the total delay in the A-V nodal and A-V bundle system is approximately 0.13 second. About a quarter of this time lapse occurs in the *transitional*

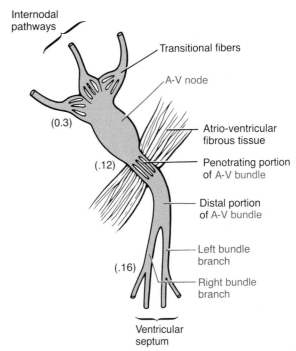

Figure 10–3. Organization of the A-V node. The numbers represent the interval of time from the origin of the impulse in the sinus node. The values have been extrapolated to the human being.

fibers, which are very small fibers that connect the fibers of the atrial internodal pathways with the A-V node (see Fig. 10–3). The velocity of conduction in these fibers is as little as 0.02 to 0.05 m/sec (about 1/12 that in normal cardiac muscle), which greatly delays entrance of the impulse into the A-V node. After entering the node proper, the velocity of conduction in the nodal fibers is still quite low, only 0.05 m/sec, about one-eighth the conduction velocity in normal cardiac muscle. This low velocity of conduction is also approximately true for the penetrating portion of the A-V bundle.

Cause of the Slow Conduction. The cause of the extremely slow conduction in the transitional as well as the nodal and penetrating A-V bundle fibers is partly that their sizes are considerably smaller than the sizes of the normal atrial muscle fibers. However, most of the slow conduction is probably caused by two entirely different factors. First, all these fibers have resting membrane potentials that are much less negative than the normal resting potential of other cardiac muscle. Second, very few gap junctions connect the successive fibers in the pathway, so that there is great resistance to the conduction of excitatory ions from one fiber to the next. Thus, with both low voltage to drive the ions and great resistance to the movement of the ions, it is easy to see why each succeeding fiber is slow to be excited.

TRANSMISSION IN THE PURKINJE SYSTEM

The *Purkinje fibers* lead from the A-V node through the A-V bundle into the ventricles. Except for the initial portion of these fibers where they penetrate the atrioventricular fibrous barrier, they have functional characteristics quite the opposite of those of the A-V nodal fibers; they are very large fibers, even larger than the normal ventricular muscle fibers, and they transmit action potentials at a velocity of 1.5 to 4.0 m/sec, a velocity about 6 times that in the usual cardiac muscle and 150 times that in some transitional fibers. This allows almost immediate transmission of the cardiac impulse throughout the entire ventricular system.

The very rapid transmission of action potentials by Purkinje fibers is believed to be caused by increased permeability of the gap junctions at the intercalated discs between the successive cardiac cells that make up the Purkinje fibers. At these discs, ions are transmitted easily from one cell to the next, thus enhancing the velocity of transmission. The Purkinje fibers also have very few myofibrils, which means that they barely contract during the course of impulse transmission.

One-Way Conduction Through the A-V Bundle. A special characteristic of the A-V bundle is the inability except in abnormal states of action potentials to travel backward in the bundle from the ventricles to the atria. This prevents re-entry of cardiac impulses by this route from the ventricles into the atria, allowing only forward conduction from the atria to the ventricles.

Furthermore, it should be recalled that the atrial muscle is separated from the ventricular muscle by a continuous fibrous barrier, a portion of which is illustrated in Figure 10–3. This barrier normally acts as an insulator to prevent passage of the cardiac impulse between the atria and the ventricles through any other route besides forward conduction through the A-V bundle itself. (However, in rare instances an abnormal muscle bridge does penetrate through the fibrous barrier elsewhere besides at the A-V bundle. Under such conditions, the cardiac impulse can then re-enter the atria from the ventricles and cause serious cardiac arrhythmia.)

Distribution of the Purkinje Fibers in the Ventricles. After penetrating through the atrioventricular fibrous tissue, the distal portion of the A-V bundle passes downward in the ventricular septum for 5 to 15 millimeters toward the apex of the heart, as shown in Figure 10–1 and 10–3. Then the bundle divides into the *left* and *right bundle branches* that lie beneath the endocardium of the two respective sides of the septum. Each branch spreads downward to the apex of the ventricle, dividing into smaller branches that course around each ventricular chamber and back toward the base of the heart. The terminal Purkinje fibers penetrate about one third of the way into the muscle mass and then become continuous with the cardiac muscle fibers.

From the time that the cardiac impulse first enters the bundle branches until it reaches the terminations of the Purkinje fibers, the total time that elapses averages only 0.03 second; therefore, once the cardiac impulse enters the Purkinje system, it spreads almost immediately to the entire endocardial surface of the ventricular muscle.

TRANSMISSION OF THE CARDIAC IMPULSE IN THE VENTRICULAR MUSCLE

Once the impulse has reached the ends of the Purkinje fibers, it is then transmitted through the ventricular muscle mass by the ventricular muscle fibers themselves. The velocity of transmission is now only 0.3 to 0.5 m/sec, one sixth that in the Purkinje fibers.

The cardiac muscle wraps around the heart in a double spiral with fibrous septa between the spiraling layers; therefore, the cardiac impulse does not necessarily travel directly outward toward the surface of the heart but instead angulates toward the surface along the directions of the spirals. Because of this, transmission from the endocardial surface to the epicardial surface of the ventricle requires as much as another 0.03 second, approximately equal to the time required for transmission through the entire Purkinje system. Thus, the total time for transmission of the cardiac impulse from the initial bundle branches to the last of the ventricular muscle fibers in the normal heart is about 0.06 second.

SUMMARY OF THE SPREAD OF THE CARDIAC IMPULSE THROUGH THE HEART

Figure 10–4 illustrates in summary form the transmission of the cardiac impulse through the human heart. The numbers on the figure represent the intervals of time in hundredths of a second that lapse be-

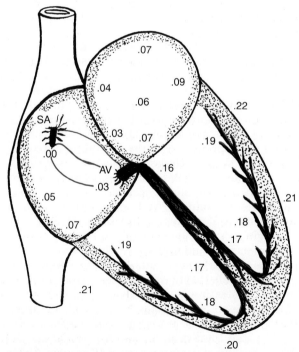

Figure 10–4. Transmission of the cardiac impulse through the heart, showing the time of appearance (in fractions of a second) of the impulse in different parts of the heart.

tween the origin of the cardiac impulse in the sinus node and its appearance at each respective point in the heart. Note that the impulse spreads at moderate velocity through the atria but is delayed more than 0.1 second in the A-V nodal region before appearing in the ventricular septal A-V bundle. Once it has entered this bundle, it spreads rapidly through the Purkinje fibers to the entire endocardial surfaces of the ventricles. Then the impulse once again spreads slowly through the ventricular muscle to the epicardial surfaces.

It is extremely important that the reader learn in detail the course of the cardiac impulse through the heart and the times of its appearance in each separate part of the heart, for a quantitative knowledge of this process is essential to the understanding of electrocardiography, which is discussed in the following three chapters.

● CONTROL OF EXCITATION AND CONDUCTION IN THE HEART

THE SINUS NODE AS THE PACEMAKER OF THE HEART

In the previous discussion of the genesis and transmission of the cardiac impulse through the heart, we have noted that the impulse normally arises in the sinus node. However, this often is not the case under abnormal conditions, for other parts of the heart can exhibit rhythmic contraction in the same way that the sinus nodal fibers can; this is particularly true of the A-V nodal and Purkinje fibers.

The A-V nodal fibers, when not stimulated from some outside source, discharge at an intrinsic rhythmic rate of 40 to 60 times per minute, and the Purkinje fibers discharge at a rate of somewhere between 15 and 40 times per minute. These rates are in contrast to the normal rate of the sinus node of 70 to 80 times per minute.

Therefore, the question that we must ask is: Why does the sinus node control the heart's rhythmicity rather than the A-V node or the Purkinje fibers? The answer to this derives from the fact that the discharge rate of the sinus node is considerably greater than that of either the A-V node or the Purkinje fibers. Each time the sinus node discharges, its impulse is conducted into both the A-V node and the Purkinje fibers, discharging their excitable membranes. Then these tissues as well as the sinus node recover from the action potential and become hyperpolarized. But the sinus node loses its hyperpolarization much more rapidly than does either of the other two and emits a new impulse before either one of them can reach its own threshold for self-excitation. The new impulse again discharges both the A-V node and Purkinje fibers. This process continues on and on, the sinus node always exciting these other potentially self-excitatory tissues before self-excitation can actually occur.

Thus, the sinus node controls the beat of the heart because its rate of rhythmic discharge is greater than that of any other part of the heart. Therefore, the sinus node is the normal *pacemaker* of the heart.

Abnormal Pacemakers — The Ectopic Pacemaker. Occasionally some other part of the heart develops a rhythmic discharge rate that is more rapid than that of the sinus node. For instance, this often occurs in the A-V node or in the Purkinje fibers. In either of these cases, the pacemaker of the heart shifts from the sinus node to the A-V node or to the excitable Purkinje fibers. Under more rare conditions, a point in the atrial or ventricular muscle develops excessive excitability and becomes the pacemaker.

A pacemaker elsewhere than the sinus node is called an *ectopic pacemaker*. Obviously, an ectopic pacemaker causes an abnormal sequence of contraction of the different parts of the heart.

Another cause of shift of the pacemaker is blockage of transmission of the impulses from the sinus node to the other parts of the heart, this occurring most frequently at the A-V node or in the penetrating portion of the A-V bundle on the way to the ventricles. When A-V block occurs, the atria continue to beat at the normal rate of rhythm of the sinus node, while a new pacemaker develops in the Purkinje system of the ventricles and drives the ventricular muscle at a new rate somewhere between 15 and 40 beats per minute. However, after a sudden block, the Purkinje system does not begin to emit its rhythmical impulses until 5 to 30 seconds later because up to that point it has been "overdriven" by the rapid sinus impulses and consequently is in a suppressed state. During this 5 to 30 seconds, the ventricles fail to pump any blood, and the person faints after the first 4 to 5 seconds because of lack of blood flow to the brain. This delayed pickup of the heartbeat is called *Stokes-Adams syndrome*. If the period is too long, it can lead to death.

ROLE OF THE PURKINJE SYSTEM IN CAUSING SYNCHRONOUS CONTRACTION OF THE VENTRICULAR MUSCLE

It is clear from the previous description of the Purkinje system that the cardiac impulse arrives at almost all portions of the ventricles within a very narrow span of time, exciting the first ventricular muscle fiber only 0.06 second ahead of excitation of the last ventricular muscle fiber. This causes all portions of the ventricular muscle in both ventricles to begin contracting at almost exactly the same time. Effective pumping by the two ventricular chambers requires this synchronous type of contraction. If the cardiac impulse traveled through the ventricular muscle very slowly, much of the ventricular mass would contract prior to contraction of the remainder, in which case the overall pumping effect would be greatly depressed. Indeed, in some types of cardiac debilities, some of which are discussed in Chapters 12 and 13, slow transmission does occur, and the pumping effectiveness of the ventricles is decreased perhaps as much as 20 to 30 per cent.

CONTROL OF HEART RHYTHMICITY AND CONDUCTION BY THE SYMPATHETIC AND PARASYMPATHETIC NERVES

The heart is supplied with both sympathetic and parasympathetic nerves, as illustrated in Figure 9–11 of the previous chapter. The parasympathetic nerves (the vagi) are distributed mainly to the sinus and A-V nodes, to a lesser extent to the muscle of the two atria, and even less to the ventricular muscle. The sympathetic nerves, on the other hand, are distributed to all parts of the heart, with a strong representation to the ventricular muscle as well as to all the other areas.

Effect of Parasympathetic (Vagal) Stimulation on Cardiac Rhythm and Conduction — Ventricular Escape. Stimulation of the parasympathetic nerves to the heart (the vagi) causes the hormone *acetylcholine* to be released at the vagal endings. This hormone has two major effects on the heart. First, it decreases the rate of rhythm of the sinus node, and, second, it decreases the excitability of the A-V junctional fibers between the atrial musculature and the A-V node, thereby slowing transmission of the cardiac impulse into the ventricles. Very strong stimulation of the vagi can completely stop the rhythmic contraction of the sinus node or completely block transmission of the cardiac impulse through the A-V junction. In either case, rhythmic impulses are no longer transmitted into the ventricles. The ventricles stop beating for 4 to 10 seconds, but then some point in the Purkinje fibers, usually in the ventricular septal portion of the A-V bundle, develops a rhythm of its own and causes ventricular contraction at a rate of 15 to 40 beats per minute. This phenomenon is called *ventricular escape.*

Mechanism of the Vagal Effects. The acetylcholine released at the vagal nerve endings greatly increases the permeability of the fiber membranes to potassium, which allows rapid leakage of potassium to the exterior. This causes increased negativity inside the fibers, an effect called *hyperpolarization,* which makes excitable tissue much less excitable, as was explained in Chapter 5.

In the sinus node, the state of hyperpolarization decreases the "resting" membrane potential of the sinus nodal fibers to a level considerably more negative than the normal value, to a level as low as − 65 to − 75 millivolts rather than the normal level of − 55 to − 60 millivolts. Therefore, the upward drift of the resting membrane potential caused by sodium leakage requires much longer to reach the threshold potential for excitation. Obviously, this greatly slows the rate of rhythmicity of these nodal fibers. And, if the vagal stimulation is strong enough, it is possible to stop completely the rhythmical self-excitation of this node.

In the A-V node, the state of hyperpolarization makes it difficult for the minute junctional fibers, which can generate only small quantities of current during the action potential, to excite the nodal fibers. Therefore, the *safety factor* for transmission of the cardiac impulse through the junctional fibers and into the nodal fibers decreases. A moderate decrease in this simply delays conduction of the impulse, but a decrease in safety factor below unity (which means so low that the action potential of a fiber cannot cause an action potential in the successive portion of the fiber) completely blocks conduction.

Effect of Sympathetic Stimulation on Cardiac Rhythm and Conduction. Sympathetic stimulation causes essentially the opposite effects on the heart to those caused by vagal stimulation as follows: First, it increases the rate of sinus nodal discharge. Second, it increases the rate of conduction as well as the level of excitability in all portions of the heart. Third, it increases greatly the force of contraction of all the cardiac musculature, both atrial and ventricular, as discussed in the previous chapter.

In short, sympathetic stimulation increases the overall activity of the heart. Maximal stimulation can almost triple the rate of heartbeat and can increase the strength of heart contraction as much as twofold.

Mechanism of the Sympathetic Effect. Stimulation of the sympathetic nerves releases the hormone *norepinephrine* at the sympathetic nerve endings. The precise mechanism by which this hormone acts on cardiac muscle fibers is still somewhat doubtful, but the present belief is that it increases the permeability of the fiber membrane to sodium and calcium. In the sinus node, an increase of sodium permeability causes a more positive resting potential and an increased rate of upward drift of the membrane potential to the threshold level for self-excitation, both of which obviously accelerate the onset of self-excitation and therefore increase the heart rate.

In the A-V node, increased sodium permeability makes it easier for the action potential to excite the succeeding portion of the conducting fiber, thereby decreasing the conduction time from the atria to the ventricles.

The increase in permeability to calcium ions is at least partially responsible for the increase in contractile strength of the cardiac muscle under the influence of sympathetic stimulation because calcium ions play a powerful role in exciting the contractile process of the myofibrils.

REFERENCES

Akera, T., and Brody, T. M.: Myocardial membranes: Regulation and function of the sodium pump. Annu. Rev. Physiol., 44:375, 1982.

Brown, H. F.: Electrophysiology of the sinoatrial node. Physiol. Rev., 62:505, 1982.

Brutsaert, D. L., and Paulus, W. J.: Contraction and relaxation of the heart as muscle and pump. In Guyton, A. C., and Young, D. B. (eds.): International Review of Physiology: Cardiovascular Physiology III. Vol. 18. Baltimore, University Park Press, 1979, p. 1.

DiFrancesco, D., and Noble, D.: A model of cardiac electrical activity incorporating ionic pumps and concentration changes. Phil. Trans. R. Soc. Lond. (Biol.), 307:353, 1985.

Ellis, D.: Na-Ca exchange in cardiac tissues. Adv. Myocardiol., 5:295, 1985.

Farah, A. E., et al.: Positive inotropic agents. Annu. Rev. Pharmacol. Toxicol., 24:275, 1985.

Fozzard, H. A.: Heart: Excitation-contraction coupling. Annu. Rev. Physiol., 39:201, 1977.

Fozzard, H. A., et al.: The Heart and Cardiovascular System: Scientific Foundations. New York, Raven Press, 1986.

Geddes, L. A.: Cardiovascular Medical Devices. New York, John Wiley & Sons, 1984.

Gilmour, R. F., Jr., and Zipes, D. P.: Slow inward current and cardiac arrhythmias. Am. J. Cardiol., 55:89B, 1985.

Glitsch, H. G.: Electrogenic Na pumping in the heart. Annu. Rev. Physiol., 44:389, 1982.

Gravanis, M. B. (ed.): Cardiovascular Pathophysiology. New York, McGraw-Hill Book Co., 1987.

Guyton, A. C., and Satterfield, J.: Factors concerned in electrical defibrillation of the heart, particularly through the unopened chest. Am. J. Physiol., 167:81, 1951.

Herbette, L., et al.: The interaction of drugs with the sarcoplasmic reticulum. Annu. Rev. Pharmacol. Toxicol., 22:413, 1982.

Hondeghem, L. M., and Katzung, B. G.: Antiarrhythmic agents: The modulated receptor mechanism of action of sodium and calcium channel-blocking drugs. Annu. Rev. Pharmacol. Toxicol., 24:387, 1984.

Irisawa, H.: Comparative physiology of the cardiac pacemaker mechanism. Physiol. Rev., 58:461, 1984.

Josephson, M. E., and Singh, B. N.: Use of calcium antagonists in ventricular dysfunction. Am. J. Cardiol., 55:81B, 1985.

Langer, G. A.: Sodium-calcium exchange in the heart. Annu. Rev. Physiol., 44:435, 1982.

Latorre, R., et al.: K$^+$ channels gated by voltage and ions. Annu. Rev. Physiol., 46:485, 1984.

Lazdunski, M., and Renaud, J. F.: The action of cardiotoxins on cardiac plasma membranes. Annu. Rev. Physiol., 44:463, 1982.

Levy, M. N., and Martin, P. J.: Neural control of the heart. In Berne, R. M., et al. (eds.): Handbook of Physiology. Sec. 2, Vol. I. Baltimore, Williams & Wilkins, 1979, p. 581.

Levy, M. N., et al.: Neural regulation of the heart beat. Annu. Rev. Physiol., 43:443, 1981.

Loewenstein, W. R.: Junctional intercellular communication: the cell-to-cell membrane channel. Physiol. Rev., 61:829, 1981.

Mazzanti, M., and DeFelice, L. J.: K channel kinetics during the spontaneous heart beat in embryonic chick ventricle cells. Biophys. J., 54:1139, 1988.

Mazzanti, M., and DeFelice, L. J.: Na channel kinetics during the spontaneous heart beat in embryonic chick ventricle cells. Biophys. J., 52:95, 1987.

Mazzanti, M., and DeFelice, L. J.: Regulation of the Na-conducting Ca channel during the cardiac action potential. Biophys. J., 51:115, 1987.

McAnulty, J., and Rahimtoola, S.: Prognosis in bundle branch block. Annu. Rev. Med., 32:499, 1981.

McDonald, T. F.: The slow inward calcium current in the heart. Annu. Rev. Physiol., 44:425, 1982.

Meijler, F. L.: Atrioventricular conduction versus heart size from mouse to whale. J. Am. Coll. Cardiol., 5:363, 1985.

Meijler, F. L., and Janse, M. J.: Morphology and electrophysiology of the mammalian atrioventricular node. Physiol. Rev., 68:608, 1988.

Orrego F.: Calcium and the mechanism of action of digitalis. Gen. Pharmacol., 15:273, 1984.

Reuter, H.: Ion channels in cardiac cell membranes. Annu. Rev. Physiol., 44:473, 1984.

Sheridan, J. D., and Atkinson, M. M.: Physiological roles of permeable junctions: Some possibilities. Annu. Rev. Physiol., 47:337, 1985.

Spear, J. F., and Moore, E. N.: Mechanisms of cardiac arrhythmias. Annu. Rev. Physiol., 44:485, 1982.

Sperelakis, N.: Hormonal and neurotransmitter regulation of Ca^{++} influx through voltage-dependent slow channels in cardiac muscle membrane. Membr. Biochem., 5:131, 1984.

Vasselle, M.: Electrogenesis of the plateau and pacemaker potential. Annu. Rev. Physiol., 41:425, 1979.

Verrier, R. L., and Lown, B.: Behavioral stress and cardiac arrhythmias. Annu. Rev. Physiol., 46:155, 1984.

(See also Chapter 9.)

CHAPTER 11

• *The Normal Electrocardiogram* •

As the cardiac impulse passes through the heart, electrical currents spread into the tissues surrounding the heart, and a small proportion of these spreads all the way to the surface of the body. If electrodes are placed on the skin on opposite sides of the heart, electrical potentials generated by these currents can be recorded; the recording is known as an *electrocardiogram.* A normal electrocardiogram for two beats of the heart is illustrated in Figure 11–1.

● CHARACTERISTICS OF THE NORMAL ELECTROCARDIOGRAM

The normal electrocardiogram (Fig. 11–1) is composed of a P wave, a "QRS complex," and a T wave. The QRS complex is often three separate waves, the Q wave, the R wave, and the S wave.

The P wave is caused by electrical potentials generated as the atria depolarize prior to contraction. The QRS complex is caused by potentials generated when the ventricles depolarize prior to contraction, that is, as the depolarization wave spreads through the ventricles. Therefore, both the P wave and the components of the QRS complex are *depolarization waves.*

The T wave is caused by potentials generated as the ventricles recover from the state of depolarization. This process occurs in ventricular muscle 0.25 to 0.35 second after depolarization, and this wave is known as a *repolarization wave.*

Thus, the electrocardiogram is composed of both depolarization and repolarization waves. The principles of depolarization and repolarization are discussed in Chapter 5. However, the distinction between depolarization waves and repolarization waves is so important in electrocardiography that further clarification is needed, as follows:

DEPOLARIZATION WAVES VERSUS REPOLARIZATION WAVES

Figure 11–2 illustrates a muscle fiber in four different stages of depolarization and repolarization. During the process of depolarization the normal negative potential inside the fiber is lost and the membrane potential actually reverses; that is, it becomes slightly positive inside and negative outside.

In Figure 11–2A, the process of depolarization, illustrated by the red positive charges inside and negative charges outside, is traveling from left to right, and the first half of the fiber has already depolarized, while the remaining half is still polarized. Therefore, the left electrode on the fiber is in an area of negativity where it touches the outside of the fiber, while the right electrode is in an area of positivity; this causes the meter to record positively. To the right of the muscle fiber is illustrated a record of the potential between the electrodes as recorded by a high-speed recording meter at this particular stage of depolarization. Note that when depolarization has reached this halfway mark, the record has risen to a maximum positive value.

In Figure 11–2B, depolarization has extended over the entire muscle fiber, and the recording to the right has returned to the zero baseline because both electrodes are now in areas of equal negativity. The completed wave is a *depolarization wave* because it results from spread of depolarization along the entire extent of the muscle fiber.

Figure 11–2C illustrates the repolarization process in the muscle fiber, which has proceeded halfway along the fiber from left to right. At this point, the left electrode is in an area of positivity, while the right electrode is in an area of negativity. This is opposite to the polarity in Figure 11–2A. Consequently, the recording, as illustrated to the right, becomes negative.

In Figure 11–2D, the muscle fiber has completely repolarized, and both electrodes are now in areas of positivity, so that no potential is recorded between them. Thus, in the recording to the right, the potential returns once more to the zero level. This completed negative wave is a *repolarization wave* because it results from spread of the repolarization process over the muscle fiber.

Relationship of the Monophasic Action Potential of Cardiac Muscle to the QRS and T Waves. The monophasic action potential of ventricular muscle, which is discussed in the preceding chapter, normally lasts between 0.25 and 0.35 second. The top part of Figure 11–3 illustrates a monophasic action potential recorded from a microelectrode inserted to the inside of a single ventricular muscle fiber. The upsweep of this action potential is caused by *depolarization,* and the return of the potential to the baseline is caused by *repolarization.*

Note below in the figure the simultaneous recording of the electrocardiogram from this same ventricle, which shows the QRS wave appearing at the beginning of the monophasic action potential and the T wave appearing at the end. Note especially that *no potential at all is recorded in the electrocardiogram when the ventricular muscle is either completely polarized or completely depolarized.* It is only when the muscle is partly polarized and partly depolarized that current flows from one part of the ventricles to another part and therefore also flows to the surface of the body to cause the electrocardiogram.

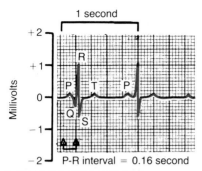

Figure 11-1. The normal electrocardiogram.

RELATIONSHIP OF ATRIAL AND VENTRICULAR CONTRACTION TO THE WAVES OF THE ELECTROCARDIOGRAM

Before contraction of muscle can occur, depolarization must spread through the muscle to initiate the chemical processes of contraction. Therefore, the P wave occurs immediately before the *beginning of contraction of the atria,* and the QRS wave occurs immediately before the *beginning of contraction of the ventricles.* The ventricles remain contracted until a few milliseconds after repolarization has occurred, that is, until after the end of the T wave.

The ventricular repolarization wave is the T wave of the normal electrocardiogram. Ordinarily, ventricular muscle begins to repolarize in some fibers approximately 0.20 second after the beginning of the depolarization wave but in many others not until as long as 0.35 second. Thus, the process of repolarization extends over a long period of time, about 0.15 second. For this reason, the T wave in the normal electrocardiogram is often a prolonged wave, but the voltage of the T wave is considerably less than the voltage of the QRS complex, partly because of its prolonged length.

The atria repolarize approximately 0.15 to 0.20 second after the P wave. However, this occurs just at the moment

Figure 11-2. Recording the *depolarization wave* and the *repolarization wave* from a cardiac muscle fiber.

Figure 11-3. *Above:* Monophasic action potential from a ventricular muscle fiber during normal cardiac function, showing rapid depolarization and then repolarization occurring slowly during the plateau stage but very rapidly toward the end. *Below:* Electrocardiogram recorded simultaneously.

that the QRS wave is being recorded in the electrocardiogram. Therefore, the atrial repolarization wave, known as the *atrial T wave,* is usually totally obscured by the much larger QRS wave. For this reason, an atrial T wave is rarely observed in the electrocardiogram.

VOLTAGE AND TIME CALIBRATION OF THE ELECTROCARDIOGRAM

All recordings of electrocardiograms are made with appropriate calibration lines on the recording paper. Either these calibration lines are already ruled on the paper, as is the case when a pen recorder is used, or they are recorded on the paper at the same time that the electrocardiogram is recorded, which is the case with the photographic types of electrocardiographs.

As illustrated in Figure 11-1, the horizontal calibration lines are arranged so that 10 small divisions upward or downward in the standard electrocardiogram represent 1 millivolt (mV) with positivity in the upward direction and negativity in the downward direction.

The vertical lines on the electrocardiogram are time calibration lines. Each inch in the horizontal direction is 1 second, and each inch in turn is usually broken into five segments by dark vertical lines; the intervals between these lines represent 0.20 second. These intervals are then broken into five smaller intervals by thin lines, and each of these represents 0.04 second.

Normal Voltages in the Electrocardiogram. The voltages of the waves in the normal electrocardiogram depend on the manner in which the electrodes are applied to the surface of the body. When one electrode is placed directly over the heart and the second electrode is placed elsewhere on the body, the voltage of the QRS complex may be as great as 3 to 4 mV. Even this voltage is very small in comparison with the monophasic action potential of 110 mV recorded directly at the heart muscle membrane. When electrocardiograms are recorded from electrodes on the two arms or on one arm and one leg, the voltage of the QRS complex usually is approximately 1 mV from the top of the R wave to the bottom of the S wave; the voltage of the P wave, between 0.1 and 0.3 mV; and that of the T wave, between 0.2 and 0.3 mV.

The P-Q or P-R Interval. The duration of time between the beginning of the P wave and the beginning of the

QRS wave is the interval between the beginning of contraction of the atrium and the beginning of contraction of the ventricle. This period of time is called the P-Q interval. The normal P-Q interval is approximately 0.16 second. This interval is sometimes also called the P-R interval because the Q wave is frequently absent.

The Q-T Interval. Contraction of the ventricle lasts almost from the beginning of the Q wave to the end of the T wave. This interval of time is called the Q-T interval and ordinarily is approximately 0.35 second.

The Rate of the Heart as Determined From Electrocardiograms. The rate of heartbeat can be determined easily from electrocardiograms because the time interval between two successive beats is the reciprocal of the heart rate. If the interval between two beats as determined from the time calibration lines is 1 second, the heart rate is 60 beats per minute. The normal interval between two successive QRS complexes is approximately 0.83 second. This is a heart rate of 60/0.83 times per minute, or 72 beats per minute.

● **METHODS FOR RECORDING ELECTROCARDIOGRAMS**

The electrical currents generated by the cardiac muscle during each beat of the heart sometimes change potentials and polarity in less than 0.01 second. Therefore, it is essential that any apparatus for recording electrocardiograms be capable of responding rapidly to these changes in electrical potentials. In general, two different types of recording apparatuses are used for this purpose, as follows:

THE PEN RECORDER

Most modern clinical electrocardiographic apparatus employ a direct pen writing recorder that writes the electrocardiogram with a pen directly on a moving sheet of paper. The pen is often a thin tube connected at one end to an inkwell, and its recording end is connected to a powerful electromagnet system that is capable of moving the pen back and forth at high speed. As the paper moves forward, the pen records the electrocardiogram. The movement of the pen in turn is controlled by means of appropriate electronic amplifiers connected to electrocardiographic electrodes on the patient.

Other pen recording systems use special paper that does not require ink in the recording stylus. One such paper turns black when it is exposed to heat; the stylus itself is made very hot by electrical current flowing through its tip. Another type turns black when electrical current flows from the tip of the stylus through the paper to an electrode at its back. This leaves a black line everywhere on the paper that the stylus touches.

● **FLOW OF CURRENT AROUND THE HEART DURING THE CARDIAC CYCLE**

RECORDING ELECTRICAL POTENTIALS FROM A PARTIALLY DEPOLARIZED MASS OF SYNCYTIAL CARDIAC MUSCLE

Figure 11–4 illustrates a syncytial mass of cardiac muscle that has been stimulated at its centralmost point. Prior to stimulation, all the exteriors of the muscle cells had been

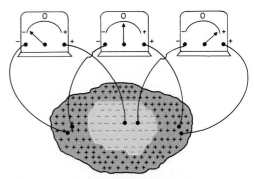

Figure 11–4. Instantaneous potentials developed on the surface of a cardiac muscle mass that has been depolarized in its center.

positive and the interiors negative. However, for reasons presented in Chapter 5 in the discussion of membrane potentials, as soon as an area of the cardiac syncytium becomes depolarized, negative charges leak to the outsides of the depolarized muscle fibers, making this surface area electronegative, as represented by the negative signs in the figure, with respect to the remaining surface of the heart, which is still polarized in the normal manner, as represented by the positive signs. Therefore, a meter connected with its negative terminal on the area of depolarization and its positive terminal on one of the still-polarized areas, as illustrated to the right in the figure, records positively.

Two other possible electrode placements and meter readings are also illustrated in Figure 11–4. *These should be studied carefully, and the reader should be able to explain the causes of the respective meter readings.* Obviously, since the process of depolarization spreads in all directions through the heart, the potential differences shown in the figure last for only a few milliseconds, and the actual voltage measurements can be accomplished only with a high-speed recording apparatus.

FLOW OF ELECTRICAL CURRENTS AROUND THE HEART IN THE CHEST

Figure 11–5 illustrates the ventricular muscle lying within the chest. Even the lungs, though mostly filled with

Figure 11–5. Flow of current in the chest around a partially depolarized heart.

air, conduct electricity to a surprising extent, and fluids of the other tissues surrounding the heart conduct electricity even more easily. Therefore, the heart is actually suspended in a conductive medium. When one portion of the ventricles becomes electronegative with respect to the remainder, electrical current flows from the depolarized area to the polarized area in large circuitous routes, as noted in the figure.

It should be recalled from the discussion of the Purkinje system in Chapter 10 that the cardiac impulse first arrives in the ventricles in the septum and shortly thereafter reaches the endocardial surfaces of the remainder of the ventricles, as shown by the colored areas and the negative signs in Figure 11–5. This provides electronegativity on the insides of the ventricles and electropositivity on the outer walls of the ventricles, and current flows through the fluids surrounding the ventricles along elliptical paths, as illustrated in the figure. If one algebraically averages all the lines of current flow (the elliptical lines), one finds that the average current flow, *in the negative to positive direction,* is from the base of the heart toward the apex. During most of the remainder of the depolarization process, the current continues to flow in this direction as the depolarization spreads from the endocardial surface outward through the ventricular muscle. However, immediately before the depolarization has completed its course through the ventricles, the direction of current flow reverses for about ¹⁄₁₀₀ second, flowing then from the apex toward the base because the very last part of the heart to become depolarized is the outer walls of the ventricles near the base of the heart.

Thus, in the normal heart current flows primarily in the direction from the base toward the apex during almost the entire cycle of depolarization except at the very end. Therefore, if a meter is connected to the surface of the body, as shown in Figure 11–5, the electrode nearer the base will be negative, whereas the electrode nearer the apex will be positive, and the recording meter will show a positive recording in the electrocardiogram.

● ELECTROCARDIOGRAPHIC LEADS

THE THREE BIPOLAR LIMB LEADS

Figure 11–6 illustrates electrical connections between the limbs and the electrocardiograph for recording electrocardiograms from the so-called standard bipolar limb leads. The term "bipolar" means that the electrocardiogram is recorded from two specific electrodes on the body, in this case on the limbs. Thus, a "lead" is not a single wire connecting from the body but a combination of two wires and their electrodes to make a complete circuit with the electrocardiograph. The electrocardiograph in each instance is illustrated by mechanical meters in the diagram, though the actual electrocardiograph is a high-speed recording meter with a moving paper.

Lead I. In recording limb lead I, the *negative terminal of the electrocardiograph is connected to the right arm* and the *positive terminal to the left arm.* Therefore, when the point on the chest where the right arm connects to the chest is electronegative with respect to the point where the left arm connects, the electrocardiograph records positively—that is, above the zero voltage line in the electrocardiogram. When the opposite is true, the electrocardiograph records below the line.

Lead II. In recording limb lead II, the *negative terminal of the electrocardiograph is connected to the right arm* and

Figure 11–6. Conventional arrangement of electrodes for recording the standard electrocardiographic leads. Einthoven's triangle is superimposed on the chest.

the *positive terminal to the left leg.* Therefore, when the right arm is negative with respect to the left leg, the electrocardiograph records positively.

Lead III. In recording limb lead III, the *negative terminal of the electrocardiograph is connected to the left arm* and the *positive terminal to the left leg.* This means that the electrocardiograph records positively when the left arm is negative with respect to the left leg.

Einthoven's Triangle. In Figure 11–6, a triangle, called *Einthoven's triangle,* is drawn around the area of the heart. This is a diagrammatic means of illustrating that the two arms and the left leg form apices of a triangle surrounding the heart. The two apices at the upper part of the triangle represent the points at which the two arms connect electrically with the fluids around the heart, and the lower apex is the point at which the left leg connects with the fluids.

Einthoven's Law. Einthoven's law states simply that if the electrical potentials of any two of the three bipolar limb electrocardiographic leads are known at any given instant, the third one can be determined mathematically from the first two by simply summing the first two (but note that the positive and negative signs of the different leads must be observed when making this summation).

For instance, let us assume that momentarily, as noted in Figure 11–6, the right arm is 0.2 mV negative with respect to the average potential in the body, the left arm is 0.3 mV positive, and the left leg is 1.0 mV positive. Observing the meters in the figure, it can be seen that lead I records a positive potential of 0.5 mV because this is the difference between the −0.2 mV on the right arm and the +0.3 mV on the left arm. Similarly, lead III records a positive potential of 0.7 mV, and lead II records a positive potential of 1.2 mV

Figure 11–7. Normal electrocardiograms recorded from the three standard electrocardiographic leads.

because these are the instantaneous potential differences between the respective pairs of limbs.

Now, note that *the sum of the voltages in leads I and III equals the voltage in lead II.* That is, 0.5 plus 0.7 equals 1.2. Mathematically, this principle, called Einthoven's law, holds true at any given instant while the electrocardiogram is being recorded.

Normal Electrocardiograms Recorded From the Three Bipolar Limb Leads. Figure 11–7 illustrates simultaneous recordings of the electrocardiogram in leads I, II, and III. It is obvious from this figure that the electrocardiograms in these three leads are very similar to each other, for they all record positive *P* waves and positive *T* waves, and the major portion of the QRS complex is also positive in each electrocardiogram.

On analysis of the three electrocardiograms, it can be shown with careful measurements that at any given instant the sum of the potentials in leads I and III equals the potential in lead II, thus illustrating the validity of Einthoven's law.

Because the recordings from all the bipolar limb leads are

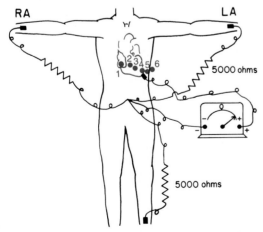

Figure 11–8. Connections of the body with the electrocardiograph for recording chest leads.

Figure 11–9. Normal electrocardiograms recorded from the six standard chest leads.

similar to each other, it does not matter greatly which lead is recorded when one wishes to diagnose the different cardiac arrhythmias, for diagnosis of arrhythmias depends mainly on the time relationships between the different waves of the cardiac cycle. On the other hand, when one wishes to diagnose damage in the ventricular or atrial muscle or in the conducting system, it does matter greatly which leads are recorded, for abnormalities of the cardiac muscle change the patterns of the electrocardiograms markedly in some leads and yet may not affect other leads.

Electrocardiographic interpretation of these two types of conditions—cardiac myopathies and cardiac arrhythmias—is discussed separately in the following two chapters.

CHEST LEADS (PRECORDIAL LEADS)

Often electrocardiograms are recorded with an electrode placed on the anterior surface of the chest over the heart at one of the six separate red points in Figure 11–8. This electrode is connected to the positive terminal of the electrocardiograph, while the negative electrode, called the *indifferent electrode,* is normally connected through electrical resistances to the right arm, left arm, and left leg all at the same time, as also shown in the figure. Usually six different standard chest leads are recorded from the anterior chest wall, the chest electrode being placed respectively at the six points illustrated in the diagram. The different recordings recorded by the method illustrated in Figure 11–8 are known as leads V_1, V_2, V_3, V_4, V_5, and V_6.

Figure 11–9 illustrates the electrocardiograms of the normal heart as recorded from these six standard chest leads. Because the heart surfaces are close to the chest wall, each chest lead records mainly the electrical potential of the cardiac musculature immediately beneath the electrode. Therefore, relatively minute abnormalities in the ventricles, particularly in the anterior ventricular wall, frequently cause marked changes in the electrocardiograms recorded from chest leads.

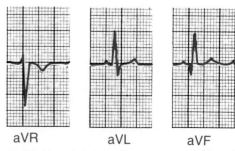

Figure 11–10. Normal electrocardiograms recorded from the three augmented unipolar limb leads.

In leads V_1 and V_2, the QRS recordings of the normal heart are mainly negative because, as illustrated in Figure 11–8, the chest electrode in these leads is nearer the base of the heart than the apex, which is the direction of electronegativity during most of the ventricular depolarization process. On the other hand, the QRS complexes in leads V_4, V_5, and V_6 are mainly positive because the chest electrode in these leads is nearer the apex, which is the direction of electropositivity during most of depolarization.

AUGMENTED UNIPOLAR LIMB LEADS

Another system of leads in wide use is the "augmented unipolar limb lead." In this type of recording, two of the limbs are connected through electrical resistances to the negative terminal of the electrocardiograph, while the third limb is connected to the positive terminal. When the positive terminal is on the right arm, the lead is known as the aV_R lead; when on the left arm, the aV_L lead; and when on the left leg, the aV_F lead.

Normal recordings of the augmented unipolar limb leads are shown in Figure 11–10. These are all similar to the standard limb lead recordings except that the recording from the aV_R lead is inverted.

REFERENCES

See references for Chapter 13.

CHAPTER 12

• Electrocardiographic Interpretation of Cardiac Muscle and Coronary Abnormalities: Vectorial Analysis •

From the discussion in Chapter 10 of impulse transmission through the heart, it is obvious that any change in the pattern of this transmission can cause abnormal electrical potentials around the heart and, consequently, can alter the shapes of the waves in the electrocardiogram. For this reason, almost all serious abnormalities of the heart muscle can be detected by analyzing the contours of the different waves in the different electrocardiographic leads. This is the subject of the present chapter.

● PRINCIPLES OF VECTORIAL ANALYSIS OF ELECTROCARDIOGRAMS

USE OF VECTORS TO REPRESENT ELECTRICAL POTENTIALS

Before it is possible to understand how cardiac abnormalities affect the contours of waves in the electrocardiogram, one must first become familiar with the concept of vectors and vectorial analysis as applied to electrical potentials in and around the heart.

Several times in the preceding chapter it is pointed out that heart current flows in a particular direction at a given instant in the cardiac cycle. A vector is an arrow that points in the direction of the electrical potential generated by the current flow *with the arrowhead in the positive direction.* Also, by convention, the length of the arrow is drawn *proportional to the voltage of the potential.*

The "Resultant" Vector in the Heart at Any Given Instant. Figure 12–1 shows, via the shaded area and the negative signs, depolarization of the ventricular septum and parts of the lateral endocardial walls of the two ventricles. Electrical current flows between these depolarized areas inside the heart and the nondepolarized areas on the outside of the heart as indicated by the elliptical arrows. Current also flows inside the heart chambers directly from the depolarized areas toward the polarized areas. Even though a small amount of current flows upward inside the heart, a considerably greater quantity flows downward on the out-

side of the ventricles toward the apex. Therefore, the summated vector of the generated potential at this particular instant, called the *instantaneous mean vector,* is drawn through the center of the ventricles in a direction from the base of the heart toward the apex. Furthermore, because these currents are considerable in quantity, the potential is large so that the vector is relatively long.

DENOTING THE DIRECTION OF A VECTOR IN TERMS OF DEGREES

When a vector is horizontal and directed toward the subject's left side, the vector is said to extend in the direction of 0 degrees, as is illustrated in Figure 12–2. From this zero reference point, the scale of vectors rotates clockwise; when the vector extends from above downward, it has a direction of 90 degrees; when it extends from the subject's left to the right, it has a direction of 180 degrees; and, when it extends upward, it has a direction of −90 or +270 degrees.

In a normal heart, the average direction of the vector of the heart during spread of the depolarization wave through the ventricles, called the *mean QRS vector,* is approximately 59 degrees, which is illustrated by vector A drawn through the center of Figure 12–2 in the 59 degree direction. This means that during most of the depolarization wave, the apex of the heart remains positive with respect to the base of the heart, as is discussed later in the chapter.

"AXIS" OF EACH OF THE BIPOLAR AND UNIPOLAR LIMB LEADS

In the preceding chapter the three bipolar and three unipolar limb leads are described. Each lead is actually a pair of electrodes connected to the body on opposite sides of the heart, and the direction from the negative to the positive electrode is called the *axis of the lead.* Lead I is recorded from two electrodes placed respectively on the two arms. Since the electrodes lie in the horizontal direction with the positive electrode to the left, the axis of lead I is 0 degrees.

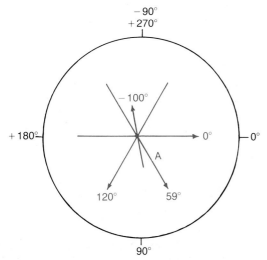

Figure 12-1. A mean vector through the partially depolarized heart.

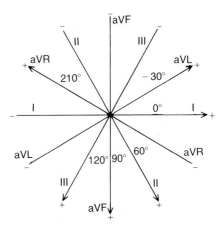

Figure 12-3. Axes of the three bipolar and three unipolar leads.

In recording lead II, electrodes are placed on the right arm and left leg. The right arm connects to the torso in the upper right-hand corner and the left leg to the lower left-hand corner. Therefore, the direction of this lead is approximately 60 degrees.

By a similar analysis it can be seen that lead III has an axis of approximately 120 degrees; lead aV_R, 210 degrees; aV_F, 90 degrees; and aV_L, −30 degrees. The directions of the axes of all these different leads are shown in the diagram of Figure 12-3, which is known as the *hexagonal reference system*. The polarities of the electrodes are illustrated by the plus and minus signs. *The reader must learn these axes and their polarities, particularly for the bipolar limb leads I, II, and III, in order to understand the remainder of this chapter.*

VECTORIAL ANALYSIS OF POTENTIALS RECORDED IN DIFFERENT LEADS

Now that we have discussed, first, the conventions for representing potentials across the heart by means of vectors and, second, the axes of the leads, it becomes possible to put these two together to determine the potential that will be recorded in each lead for a given vector in the heart.

Figure 12-4 illustrates a partially depolarized heart; vector *A* represents the instantaneous mean direction of current flow in the heart and its potential. In this instance the direction of the potential is 55 degrees, and the voltage of the potential will be assumed to be 2 millivolts (mV). Through the base of vector *A* is drawn the axis of lead I in the 0 degree direction. To determine how much of the voltage in vector *A* will be recorded in lead I, a perpendicular is dropped from the tip of vector *A* to the lead I axis, and a so-called *projected vector (B)* is drawn along the axis. The head of this vector points toward the positive end of the lead I axis, which means that the record momentarily being recorded in the electrocardiogram of lead I will be positive. The voltage recorded will be equal to the length of *B* divided by the length of *A* times 2 mV, or approximately 1 mV.

Figure 12-5 illustrates another example of vectorial analysis. In this example, vector *A* represents the electrical potential at a given instant during ventricular depolarization in another heart, in which the left side of the heart becomes depolarized somewhat more rapidly than the right. In this instance the vector has a direction of 100 degrees, and the voltage is again 2 mV. To determine the potential actually recorded in lead I, we drop a perpendicular to the lead I axis and find the projected vector *B*. Vector *B* is very short and this time in the negative direction, indicating that at this particular instant the recording in lead I will be negative (below the zero line), and the voltage recorded will be slight. This figure illustrates that *when the vector in the heart is in a direction almost perpendicular to the axis of the lead, the voltage recorded in the electrocardiogram of this lead is very low.* On the other hand, *when the heart vector has*

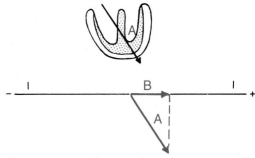

Figure 12-4. Determination of a projected vector B along the axis of lead I when vector A represents the instantaneous potential in the ventricles.

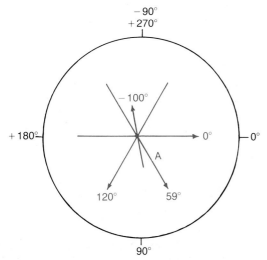

Figure 12-2. Vectors drawn to represent directions of potentials for several different hearts.

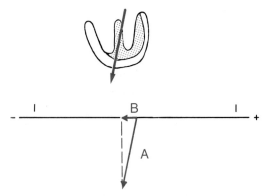

Figure 12-5. Determination of the projected vector B along the axis of lead I when vector A represents the instantaneous potential in the ventricles.

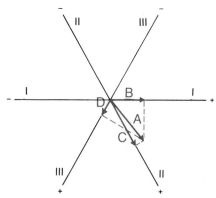

Figure 12-6. Determination of projected vectors in leads I, II, and III when vector A represents the instantaneous potential in the ventricles.

almost exactly the same axis as the lead, essentially the entire voltage of the vector will be recorded.

Vectorial Analysis of Potentials in the Three Standard Bipolar Limb Leads. In Figure 12-6, vector *A* depicts the instantaneous electrical potential of a partially depolarized ventricle. To determine the potential recorded at this instant in the electrocardiogram of each one of the three standard bipolar limb leads, perpendiculars are dropped to all the lines representing the different leads, as illustrated in the figure. The projected vector *B* depicts the potential recorded at that instant in lead I, projected vector *C* depicts the potential in lead II, and projected vector *D* depicts the potential in lead III. In each of these the record in the electrocardiogram is positive — that is, above the zero line — because the projected vectors point in the positive directions along the axes of all the leads. The potential in lead I is approximately one half that of the actual potential in the heart as depicted by vector *A;* in lead II, it is almost exactly equal to that in the heart, and in lead III, it is about one third that in the heart.

An identical analysis can be used to determine potentials recorded in augmented limb leads except that the respective axes of these leads (see Fig. 12-3) are used in place of the standard bipolar limb lead axes used in Figure 12-6.

● VECTORIAL ANALYSIS OF THE NORMAL ELECTROCARDIOGRAM

VECTORS OCCURRING DURING DEPOLARIZATION OF THE VENTRICLES — THE QRS COMPLEX

When the cardiac impulse enters the ventricles through the A-V bundle, the first part of the ventricles to become depolarized is the left endocardial surface of the septum. This depolarization spreads rapidly to involve both endocardial surfaces of the septum, as illustrated by the shaded portion of the ventricle in Figure 12-7A. Then the depolarization spreads along the endocardial surfaces of the two ventricles, as shown in Figure 12-7B and C. Finally, it spreads through the ventricular muscle to the outside of the heart, as shown progressively in Figure 12-7C, D, and E.

At each stage of depolarization of the ventricles in Figure 12-7, parts A to E, the instantaneous electrical potential is represented by a vector superimposed on the ventricle in each figure. Each of these vectors is analyzed by the method described in the preceding section to determine the voltages

that will be recorded at each instant in each of the three standard electrocardiographic leads. To the right in each stage of the figure is shown the progressive development of the QRS complex. *Keep in mind that a positive vector in a lead will cause the recording in the electrocardiogram to be above the zero line, whereas a negative vector will cause the recording to be below the line.*

Before proceeding with any further consideration of vectorial analysis, it is essential that this analysis of the successive normal vectors presented in Figure 12-7 be understood. Each of these analyses should be studied in detail by the procedure given previously. A short summary of this sequence is the following:

In Figure 12-7A, the ventricular muscle has just begun to be depolarized, representing an instant about 0.01 second after the onset of depolarization. At this time, the vector is short because only a small portion of the ventricles — the septum — is depolarized. Therefore, all electrocardiographic voltages are low, as recorded to the right of the ventricular muscle for each of the leads. The voltage in lead II is greater than the voltages in leads I and III because the heart vector extends mainly in the same direction as the axis of lead II.

In Figure 12-7B, which represents approximately 0.02 second after onset of depolarization, the heart vector is long because much of the ventricles has now become depolarized. Therefore, the voltages in all electrocardiographic leads have increased.

In Figure 12-7C, about 0.035 second after onset of depolarization, the heart vector is becoming shorter and the recorded electrocardiographic voltages are less because the outside of the apex of the heart is now electronegative, neutralizing much of the negativity on the endocardial surfaces of the heart. Also, the axis of the vector is shifting toward the left side of the chest because the left ventricle is slightly slower to depolarize than the right. The ratio of the voltage in lead I to that in lead III is increasing.

In Figure 12-7D, about 0.05 second after onset of depolarization, the heart vector points toward the base of the left ventricle, and it is short because only a minute portion of the ventricular muscle is still polarized. Because of the direction of the vector at this time, the voltages recorded in leads II and III are both negative — that is, below the line.

In Figure 12-7E, about 0.06 second after onset of depolarization, the entire ventricular muscle mass is depolarized, so that no current flows around the heart at all and no electrical potential is generated. The vector becomes zero, and the voltages in all leads become zero.

Thus, the QRS complexes are completed in the three

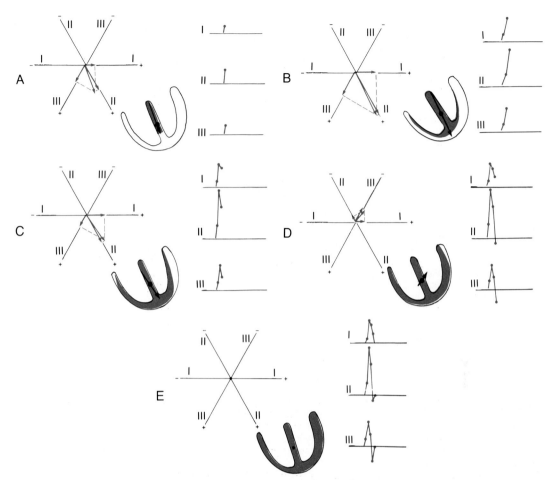

Figure 12–7. *A,* The ventricular vectors and the QRS complexes 0.01 second after onset of ventricular depolarization. *B,* 0.02 second after onset of depolarization. *C,* 0.035 second after onset of depolarization. *D,* 0.05 second after onset of depolarization. *E,* After depolarization of the ventricles is complete, 0.06 second after onset.

standard bipolar limb leads. The QRS complex sometimes has a slight negative depression at its beginning in one or more of the leads, which is not shown in Figure 12–7; this is the Q wave. When it occurs, it is caused by initial depolarization of the left side of the septum before the right side, which creates a weak vector from left to right for a fraction of a second before the usual base-to-apex vector occurs. The major positive deflection shown in Figure 12–7 is the R wave, and the final negative deflection is the S wave.

THE ELECTROCARDIOGRAM DURING REPOLARIZATION—THE T WAVE

Once the ventricular muscle has become depolarized, approximately 0.15 second elapses before sufficient repolarization begins for it to be observed in the electrocardiogram; then repolarization proceeds throughout the ventricular muscle until it is complete at about 0.35 second after the onset of the QRS complex. It is this process of repolarization that causes the T wave in the electrocardiogram.

Because the septum and other endocardial areas of the ventricular muscle depolarize first, it seems logical that these areas should repolarize first as well, but this is not the usual case because the septum and the other endocardial areas have a longer period of contraction and therefore are slower to repolarize than the external surfaces of the heart. Therefore, *the greatest portion of ventricular muscle to repo-*

larize first is that located over the entire outer surface of the ventricles and especially near the apex of the heart. The endocardial areas, on the other hand, normally repolarize last. The reason for this abnormal sequence of repolarization is believed to be that high pressure in the ventricles during contraction greatly reduces coronary blood flow to the endocardium, thereby slowing the repolarization process in the endocardial areas.

Because the outer and apical surfaces of the ventricles repolarize before the inner and basal surfaces, the positive end of the heart vector during repolarization is toward the apex of the heart. Thus, *the predominant direction of the vector through the heart during* repolarization *of the ventricles is from base to apex, which is also the predominant direction of the vector during* depolarization. *As a result, the T wave in the normal bipolar limb leads is positive, which is also the polarity of most of the normal QRS complex.*

In Figure 12–8, five stages of repolarization of the ventricles are denoted by the progressive increase of the white areas—the repolarized areas. At each stage, the vector extends from the base toward the apex until it disappears in the last stage. At first the vector is relatively small because the area of repolarization is small. Later the vector becomes stronger and stronger because of greater degrees of repolarization. Finally, the vector becomes weaker again because the areas of depolarization still persisting become so slight that the total quantity of current flow begins to decrease.

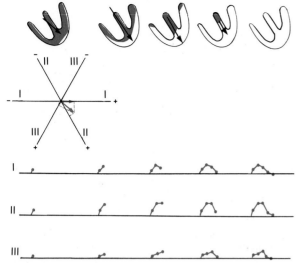

Figure 12-8. Generation of the T wave during repolarization of the ventricles, showing a vectorial analysis of the first stage of repolarization. The total time from the beginning of the T wave to the end is about 0.15 second.

These changes illustrate that the vector is greatest when approximately half the heart is in the polarized state and approximately half is depolarized.

The changes in the electrocardiograms of the three standard limb leads during the process of repolarization are noted under each of the ventricles, depicting the progressive stages of repolarization. Over approximately 0.15 second, the period required for the whole process to take place, the T wave of the electrocardiogram is generated.

DEPOLARIZATION OF THE ATRIA— THE P WAVE

Depolarization of the atria begins in the sinus node and spreads in all directions over the atria. Therefore, the point of original electronegativity in the atria is approximately at the point of entry of the superior vena cava where the sinus node lies, and the direction of the electrical potential in the atrium at the beginning of depolarization is in the direction noted in Figure 12-9. Furthermore, the vector remains gen-

erally in this direction throughout the process of atrial depolarization.

Thus, the vector of current flow during depolarization in the atria points in almost the same direction as that in the ventricles. And, because this direction is in the directions of the axes of the standard bipolar limb leads I, II, and III, the electrocardiograms recorded from the atria during the process of depolarization usually are positive in all three of these leads, as illustrated in Figure 12-9. The record of atrial depolarization is known as the P wave.

Repolarization of the Atria—The Atrial T Wave. Spread of the depolarization wave through the atrial muscle is *much slower than in the ventricles.* Therefore, the musculature around the sinus node becomes depolarized a long time before the musculature in distal parts of the atria. Because of this, *the area in the atria that becomes repolarized first is the sinus nodal region, the area that had originally become depolarized first,* which is an entirely different situation from that in the ventricles. Thus, when repolarization begins, the region around the sinus node becomes positive with respect to the rest of the atria. Therefore, the atrial repolarization vector is *backward to the vector of depolarization.* (Note again that this is opposite to the effect that occurs in the ventricles.) Therefore, as noted to the right in Figure 12-9, the so-called atrial T wave follows about 0.15 second after the atrial P wave, but this T wave is on the opposite side of the zero reference line from the P wave— that is, it is normally negative rather than positive in the three standard bipolar limb leads. In the normal electrocardiogram this T wave appears at approximately the same time that the QRS complex of the ventricle appears. Therefore, it is almost always totally obscured by the larger QRS complex, though in some abnormal states it does play a role in the recorded electrocardiogram.

THE VECTORCARDIOGRAM

It has been noted in the preceding discussions that the vector of current flow through the heart changes rapidly as the impulse spreads through the myocardium. It changes in two aspects: First, the vector increases and decreases in length because of the increasing and decreasing voltage of the vector. Second, the vector changes its direction because of changes in the average direction of the electrical potential of the heart. The so-called *vectorcardiogram* depicts these changes in the vectors at the different times during the cardiac cycle, as illustrated in Figure 12-10.

In the vectorcardiogram of Figure 12-10, point 5 is the

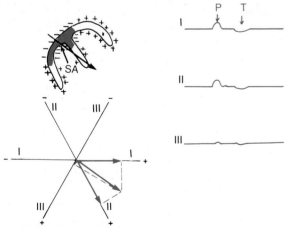

Figure 12-9. Depolarization of the atria and generation of the P wave, showing the vector through the atria and the resultant vectors in the three standard leads. At right are the atrial P and T waves.

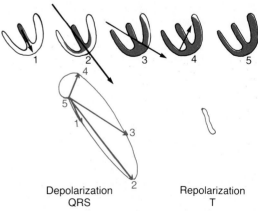

Depolarization   Repolarization
QRS   T

Figure 12-10. The QRS and T vectorcardiograms.

zero reference point, and this point is the negative end of all the vectors. While the heart is quiescent, the positive end of the vector also remains at the zero point because there is no electrical potential. However, just as soon as current begins to flow through the heart, the positive end of the vector leaves the zero reference point.

When the septum first becomes depolarized, the vector extends downward toward the apex of the heart, but it is relatively weak, thus generating the first portion of the vectorcardiogram, as illustrated by the positive end of vector 1. As more of the heart becomes depolarized, the vector becomes stronger and stronger, usually swinging slightly to one side. Thus, vector 2 of Figure 12–10 represents the state of depolarization of the heart about 0.02 second after vector 1. After another 0.02 second, vector 3 represents the potential of the heart, and vector 4 occurs in still another 0.01 second. Finally, the heart becomes totally depolarized, and the vector becomes zero once again, as shown at point 5.

The elliptical figure generated by the positive ends of the vectors is called the *QRS vectorcardiogram.*

Vectorcardiograms can be recorded instantaneously on an oscilloscope by connecting electrodes from above and below the heart to the vertical plates of the oscilloscope and connecting electrodes from each side of the heart to the horizontal plates. When the vector changes, the spot of light on the oscilloscope follows the course of the positive end of the changing vector, thus inscribing the vectorcardiogram on the oscilloscopic screen.

The "T" Vectorcardiogram. Changing vectors in the heart occur not only during the depolarization process, for vectors depicting current flow around the ventricles also reappear during repolarization. Therefore, a second, smaller vectorcardiogram—the *T vectorcardiogram*—is inscribed during repolarization of the muscular mass; this is depicted to the right in Figure 12–10. Also, a still smaller *P vectorcardiogram* is inscribed during atrial depolarization.

● THE MEAN ELECTRICAL AXIS OF THE VENTRICULAR QRS

The vectorcardiogram of the ventricular depolarization wave (the QRS vectorcardiogram) shown in Figure 12–10 is that of a normal heart. Note from this vectorcardiogram that the preponderant direction of the vectors of the ventricles is normally toward the apex of the heart—that is, during most of the cycle of ventricular depolarization, the direction of the electrical potential is from the base of the ventricles toward the apex. This preponderant direction of the potential during depolarization is called the *mean electrical axis of the ventricles* or the *mean QRS vector.* The mean electrical axis of the normal ventricles is 59 degrees. However, in certain pathological conditions of the heart, this direction is changed markedly—sometimes even to opposite poles of the heart.

DETERMINING THE ELECTRICAL AXIS FROM STANDARD LEAD ELECTROCARDIOGRAMS

Clinically, the electrical axis of the heart is usually determined from the standard bipolar limb lead electrocardiograms rather than from the vectorcardiogram. Figure 12–11 illustrates a method for doing this. After recording the standard leads, one determines the maximum potential and polarity of the recording in two of the leads. In lead I of the

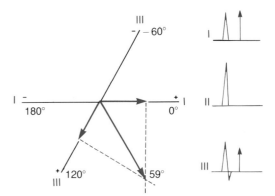

Figure 12–11. Plotting the mean electrical axis of the heart from two electrocardiographic leads.

figure, the recording is positive, and in lead III, the recording is mainly positive but negative during part of the cycle. If any part of the recording is negative, *this negative potential is subtracted from the positive potential* to determine the net potential for that lead, as illustrated by the arrows to the right of the QRS complexes of leads I and III. (To be even more accurate, some cardiologists subtract the *area* of the negative wave from the *area* of the positive wave.) After subtracting the negative portion of the QRS wave in lead III from the positive portion, each net potential is plotted on the axes of the respective leads, with the base of the potential at the point of intersection of the axes, as illustrated in Figure 12–11.

If the net potential of lead I is positive, it is plotted in a positive direction along the line depicting lead I. On the other hand, if this potential is negative, it is plotted in a negative direction. Also, for lead III, the net potential is placed with its base at the point of intersection, and, if positive, it is plotted in the positive direction along the line depicting lead III. If it is negative, it is plotted in the negative direction.

To determine the actual vector of the ventricular mean electrical potential, one draws perpendicular lines from the apices of the two net potentials of leads I and III, respectively. The point of intersection of these two perpendicular lines represents, by vectorial analysis, the apex of the actual mean QRS vector in the ventricles, and the point of intersection of the two lead axes represents the negative end of the actual vector. Therefore, the *mean QRS vector* is drawn between these two points. The approximate average potential generated by the ventricles during depolarization is represented by the length of the vector, and the mean electrical axis is represented by the direction of the vector. Thus, the orientation of the mean electrical axis of the normal ventricles, as determined in Figure 12–11, is 59 degrees.

ABNORMAL VENTRICULAR CONDITIONS THAT CAUSE AXIS DEVIATION

Although the mean electrical axis of the ventricles averages approximately 59 degrees, this axis can swing to the left even in the normal heart to approximately 20 degrees or to the right to approximately 100 degrees. The causes of the normal variations are mainly anatomical differences in the Purkinje system distribution or in the musculature itself of different hearts. Yet, a number of conditions can cause axis deviation even beyond these normal limits, as follows:

Change in the Position of the Heart. Obviously, if the heart itself is angulated to the left, the mean electrical axis of the heart will also shift to the left. Such shift occurs (1) during expiration, (2) when a person lies down because the abdominal contents press upward against the diaphragm, and (3) quite frequently in stocky, fat persons whose diaphragms normally press upward against the heart all the time.

Likewise, angulation of the heart to the right causes the mean electrical axis of the ventricles to shift to the right. This condition occurs (1) during inspiration, (2) when a person stands up, and (3) normally in tall, lanky persons whose hearts hang downward.

Hypertrophy of One Ventricle. When one ventricle greatly hypertrophies, the axis of the heart shifts toward the hypertrophied ventricle for two reasons. First, far greater quantity of muscle exists on the hypertrophied side of the heart than on the other side, and this allows excess generation of electrical potential on that side. Second, more time is required for the depolarization wave to travel through the hypertrophied ventricle than through the normal ventricle. Consequently, the normal ventricle becomes depolarized— that is, becomes negative—considerably in advance of the hypertrophied ventricle, and this causes a strong vector from the normal side of the heart toward the hypertrophied side, which still remains positively charged. Thus, the axis deviates toward the hypertrophied ventricle.

Vectorial Analysis of Left Axis Deviation Resulting from Hypertrophy of the Left Ventricle. Figure 12–12 illustrates the three standard bipolar limb leads of an electrocardiogram in which an analysis of the axis direction shows left axis deviation with the mean electrical axis pointing in the −15 degree direction. This is a typical electrocardiogram resulting from increased muscular mass of the left ventricle. In this instance, the axis deviation was caused by *hypertension* (high blood pressure), which caused the left ventricle to hypertrophy in order to pump blood against the elevated systemic arterial pressure. However, a similar picture of left axis deviation occurs when the left ventricle hypertrophies as a result of *aortic valvular stenosis, aortic valvular regurgitation,* or any of a number of con-

genital heart conditions in which the left ventricle enlarges while the right side of the heart remains relatively normal in size.

Vectorial Analysis of Right Axis Deviation Resulting from Hypertrophy of the Right Ventricle. The electrocardiogram of Figure 12–13 illustrates intense right axis deviation with an electrical axis of approximately 170 degrees, which is 111 degrees to the right of the normal 59 degree mean electrical axis of the ventricles. The right axis deviation illustrated in this figure was caused by hypertrophy of the right ventricle as a result of *pulmonary stenosis.* However, right axis deviation may also occur in other congenital heart conditions that cause hypertrophy of the right ventricle, such as *tetralogy of Fallot* or *interventricular septal defect.* Also, hypertrophy of the right ventricle as a result of *increased pulmonary vascular resistance* can cause right axis deviation.

Bundle Branch Block. Ordinarily, the two lateral walls of the ventricles depolarize at almost the same time because both the left and right bundle branches of the Purkinje system transmit the cardiac impulse to the endocardial surfaces of the two ventricular walls at almost the same instant. As a result, the potentials generated by the two ventricles almost neutralize each other. However, if one of the major bundle branches is blocked, the cardiac impulse spreads through the normal ventricle long before it spreads through the other. Therefore, depolarization of the two ventricles does not occur even nearly simultaneously, and the depolarization potentials do not neutralize each other. As a result, axis deviation occurs as follows:

Vectorial Analysis of Left Axis Deviation in Left Bundle Branch Block. When the left bundle branch is blocked, cardiac depolarization spreads through the right

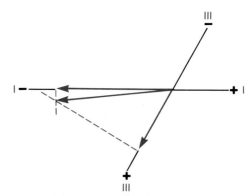

Figure 12–13. High-voltage electrocardiogram in pulmonary stenosis with right ventricular hypertrophy. Intense right axis deviation is also seen and a slightly prolonged QRS complex as well.

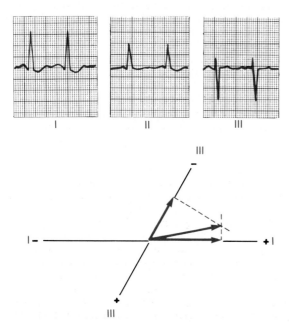

Figure 12–12. Left axis deviation in hypertensive heart disease. Note the slightly prolonged QRS complex as well.

Figure 12-14. Left axis deviation due to left bundle branch block. Note the greatly prolonged QRS complex.

ventricle two to three times as rapidly as through the left ventricle. Consequently, much of the left ventricle remains polarized for a long time after the right ventricle has become totally depolarized. Thus, the right ventricle becomes electronegative while the left ventricle remains electropositive during most of the depolarization process, and a very strong vector projects from the right ventricle toward the left ventricle. In other words, there is intense left axis deviation because the positive end of the vector points toward the left ventricle. This is illustrated in Figure 12-14, which shows typical left axis deviation resulting from left bundle branch block. Note that the axis is approximately −50 degrees.

Because of slowness of impulse conduction when the Purkinje system is blocked, axis deviation resulting from bundle branch block also greatly increases the duration of the QRS complex, which one can see by observing the excessive widths of the QRS waves in Figure 12-14. This is discussed in greater detail later in the chapter. This prolonged QRS complex differentiates this condition from axis deviation caused by hypertrophy.

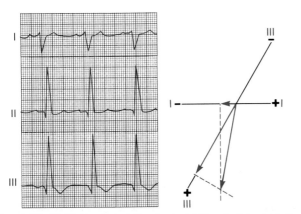

Figure 12-15. Right axis deviation due to right bundle branch block. Note the greatly prolonged QRS complex.

Vectorial Analysis of Right Axis Deviation in Right Bundle Branch Block. When the right bundle branch is blocked, the left ventricle depolarizes far more rapidly than the right ventricle, so that the left becomes electronegative while the right remains electropositive. Therefore, a very strong vector develops with its negative end toward the left ventricle and its positive end toward the right ventricle. In other words, intense right axis deviation occurs.

Right axis deviation caused by right bundle branch block is illustrated and its vector analyzed in Figure 12-15, which shows an axis of approximately 105 degrees and a prolonged QRS complex because of blocked conduction. Note also the great prolongation of the QRS complex.

● CONDITIONS THAT CAUSE ABNORMAL VOLTAGES OF THE QRS COMPLEX

INCREASED VOLTAGE IN THE STANDARD BIPOLAR LIMB LEADS

Normally, the voltages in the three standard bipolar limb leads, as measured from the peak of the R wave to the bottom of the S wave, vary between 0.5 and 2.0 mV, with lead III usually recording the lowest voltage and lead II the highest. However, these relationships are not invariably true even in the normal heart. In general, when the sum of the voltages of all the QRS complexes of the three standard leads is greater than 4 mV, one considers that the patient has a high-voltage electrocardiogram.

The cause of high-voltage QRS complexes is most often increased muscular mass of the heart, which ordinarily results from *hypertrophy of the muscle* in response to excessive load on one part of the heart or the other. For instance, the right ventricle hypertrophies when it must pump blood through a stenotic pulmonary valve, and the left ventricle hypertrophies when a person has high blood pressure. The increased quantity of muscle allows generation of increased quantities of electricity around the heart. As a result, the electrical potentials recorded in the electrocardiographic leads are considerably greater than normal, as shown in Figures 12-12 and 12-13.

DECREASED VOLTAGE OF THE ELECTROCARDIOGRAM

There are three major causes of decreased voltage in the electrocardiogram. These are, first, abnormalities of the cardiac muscle itself that prevent generation of large quantities of current; second, abnormal conditions around the heart, so that current cannot be conducted from the heart to the surface of the body with ease; and, third, rotation of the apex of the heart to point toward the anterior chest wall, so that the electrical current of the heart flows mainly anteroposteriorly in the chest rather than in the frontal plane of the body, which decreases the voltages in the limb leads.

Decreased Voltage Caused by Cardiac Myopathies. One of the most usual causes of decreased voltage of the QRS complex is a series of *old myocardial infarctions* with resultant *diminished muscle mass*. This also causes the depolarization wave to move through the ventricles slowly and prevents major portions of the heart from becoming massively depolarized all at once. Consequently, this condition causes moderate prolongation of the QRS complex along

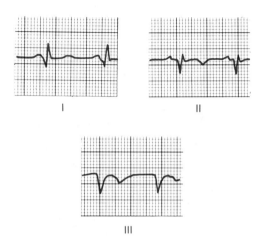

Figure 12–16. Low-voltage electrocardiogram, with evidence of local damage throughout the ventricles caused by old myocardial infarction.

with the decreased voltage. Figure 12–16 illustrates the typical low-voltage electrocardiogram with prolongation of the QRS complex that one often finds after multiple small infarctions of the heart that have resulted in local blocks and loss of muscle mass throughout the ventricles.

Decreased Voltage Caused by Conditions Surrounding the Heart. One of the most important causes of decreased voltage in the electrocardiographic leads is *fluid in the pericardium*. Because extracellular fluid conducts electrical currents with great ease, a large portion of electricity flowing out of the heart is conducted from one part of the heart to another through the pericardial effusion. Thus, this effusion effectively "short-circuits" the electrical potentials generated by the heart. *Pleural effusion* to a lesser extent can also "short" the electricity around the heart so that the voltages at the surface of the body and in the electrocardiograms are decreased.

Pulmonary emphysema can decrease the electrocardiographic potentials but by a different method from that of pericardial effusion. In pulmonary emphysema, conduction of electrical current through the lungs is considerably depressed because of the excessive quantity of air in the lungs. Also, the chest cavity enlarges, and the lungs tend to envelop the heart to a greater extent than normally. Therefore, the lungs act as an insulator to prevent spread of electrical voltage from the heart to the surface of the body, and this results in decreased electrocardiographic potentials in the various leads.

● PROLONGED AND BIZARRE PATTERNS OF THE QRS COMPLEX

PROLONGED QRS COMPLEX AS A RESULT OF CARDIAC HYPERTROPHY OR DILATATION

The QRS complex lasts as long as the process of depolarization continues to spread through the ventricles—that is, as long as part of the ventricles is depolarized and part is still polarized. Therefore, the cause of a prolonged QRS complex is always *prolonged conduction* of the impulse through the ventricles. Such prolongation often occurs when one or both ventricles are hypertrophied or dilated, owing to the longer pathway that the impulse must then travel. The normal QRS complex lasts 0.06 to 0.09 second, whereas in hypertrophy or dilatation of the left or right ventricle, the QRS complex may be prolonged to 0.10 to 0.12 second.

PROLONGED QRS COMPLEX RESULTING FROM PURKINJE SYSTEM BLOCKS

When the Purkinje fibers are blocked, the cardiac impulse must be conducted by the ventricular muscle instead, thereby decreasing the velocity of impulse conduction to approximately one-third to one-fourth normal. Therefore, if complete block of one of the bundle branches occurs, the duration of the QRS complex is usually increased to 0.14 second or greater.

In general, a QRS complex is considered to be abnormally long when it lasts more than 0.09 second, and when it lasts more than 0.12 second, the prolongation is almost certain to be caused by pathological block of the conduction system somewhere in the ventricles, as illustrated by the electrocardiograms for bundle branch block in Figures 12–14 and 12–15.

CONDITIONS CAUSING BIZARRE QRS COMPLEXES

Bizarre patterns of the QRS complex are most frequently caused by two conditions: first, destruction of cardiac muscle in various areas throughout the ventricular system with replacement of this muscle by scar tissue and, second, local blocks in the conduction of impulses by the Purkinje system.

Sometimes local blocks occur at multiple points in the ventricles. As a result, cardiac impulse conduction becomes very irregular, causing rapid shifts between left and right axis deviations. This often causes double or even triple peaks in some of the electrocardiographic leads, such as those illustrated in Figure 12–14.

● CURRENT OF INJURY

Many different cardiac abnormalities, especially those that damage the heart muscle itself, often cause part of the heart to remain partially or totally *depolarized all the time*. When this occurs, current flows between the pathologically depolarized and the normally polarized areas. This is called a *current of injury*. Note especially that *the injured part of the heart is negative because it is this part that is depolarized and emits negative charges into the surrounding fluids while the remainder of the heart is positive.*

Some of the abnormalities that can cause current of injury are: (1) *mechanical trauma*, which makes the membranes remain so permeable that full repolarization cannot take place; (2) *infectious processes* that damage the muscle membranes; and (3) *ischemia of local areas of muscle caused by coronary occlusion*, which is by far the most common cause of current of injury in the heart. During ischemia enough nutrients from the coronary blood supply simply are not available to the heart muscle to maintain normal function.

EFFECT OF CURRENT OF INJURY ON THE QRS COMPLEX

In Figure 12–17, a shaded area in the base of the left ventricle is newly infarcted. Therefore, during the T-P interval—that is, when the normal ventricular muscle is polarized—*negative* current flows from the base of the left ventricle toward the rest of the ventricles. The potential vector of this "current of injury," as illustrated in the first and the last hearts in the figure, is in a direction of approxi-

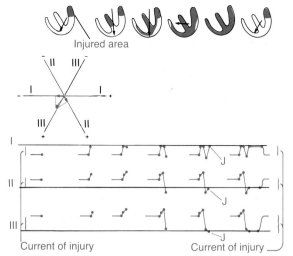

Figure 12–17. Effect of a current of injury on the electrocardiogram.

mately 125 degrees, with the base of the vector, the *negative end,* toward the injured muscle. As illustrated in the lower portions of the figure, even before the QRS complex begins *this vector causes an initial record in lead I below the zero potential line* because the projected vector of the current of injury in lead I points toward the negative end of the lead I axis. In lead II, the record is above the line because the projected vector points toward the positive terminal of lead II. In lead III, the projected vector of the current flow is also in the same direction as the polarity of lead III, so that the record is positive. Furthermore, because the vector of the current of injury lies almost exactly along the axis of lead III, the potential of the current of injury in lead III is much greater than in either of the other two records.

As the heart then proceeds through its normal process of depolarization, the septum first becomes depolarized, and the depolarization spreads down to the apex and back toward the bases of the ventricles. The last portion of the ventricles to become totally depolarized is the base of the right ventricle, because the base of the left ventricle is already totally and permanently depolarized. By vectorial analysis, as illustrated in the figure, the electrocardiogram generated by the depolarization wave traveling through the ventricles may be constructed graphically, as shown in Figure 12–17.

When the heart becomes totally depolarized at the end of the depolarization process, as noted by the next to last stage in Figure 12–17, all of the ventricular muscle is in a negative state. Therefore, at this instant in the electrocardiogram, absolutely no current flows around the musculature of the ventricles because now both the injured heart muscle and the contracting muscle are completely depolarized.

As repolarization then takes place, all of the heart finally repolarizes except the area of permanent depolarization in the injured base of the left ventricle. Thus, the repolarization causes a return of the current of injury in each lead, as noted at the far right in Figure 12–17.

THE J POINT—THE ZERO REFERENCE POTENTIAL FOR ANALYZING CURRENT OF INJURY

One would think that the electrocardiograph machines for recording electrocardiograms could determine when no current is flowing around the heart. However, many stray currents exist in the body, such as currents resulting from

"skin potentials" and from differences in ionic concentrations in different parts of the body. Therefore, when two electrodes are connected between the arms or between an arm and a leg, these stray currents make it impossible for one to predetermine the exact zero reference level in the electrocardiogram. For these reasons, the following procedure must be used to determine the zero potential level: First, one notes *the exact point at which the wave of depolarization just completes its passage through the heart,* which occurs at the very end of the QRS complex. At exactly this point, all parts of the ventricles are depolarized, so that no current is flowing around the heart. Even the current of injury disappears at this point. Therefore, the potential of the electrocardiogram at this instant is at exactly zero voltage. This point is known as the "J point" in the electrocardiogram, as illustrated in Figures 12–17 and 12–18.

For analysis of the electrical axis of the injury potential caused by a current of injury, a horizontal line is drawn through the electrocardiogram at the level of the J point, and this horizontal line is the zero potential line in the electrocardiogram from which all potentials caused by currents of injury must be measured.

Use of the J Point in Plotting the Axis of an Injury Potential. Figure 12–18 illustrates electrocardiograms recorded from leads I and III, both of which show currents of injury. In other words, the J point of each of these two electrocardiograms is not on the same line as the T-P segment. A horizontal line has been drawn through the J point to represent the zero potential level in each of the two recordings. The potential of the current of injury in each lead is the difference between the level of the T-P segment of the electrocardiogram (which is recorded between heartbeats when there is a current of injury) and the zero potential line, as illustrated by the arrows. In lead I, the recorded potential caused by the current of injury is above the zero potential line and is, therefore, positive. On the other hand, in lead III

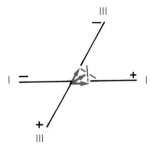

Figure 12–18. The "J" point as the zero reference voltage of the electrocardiogram. Also, method for plotting the axis of a current of injury is shown (*lower*).

the T-P segment is below the zero potential line; therefore, the potential of the current of injury in lead III is negative.

At the bottom in Figure 12–18, the potentials of the current of injury in leads I and III are plotted on the coordinates of these leads, and the resultant vector of the injury potential for the whole ventricular mass is determined by the method already described. In this instance, the vector of the current of injury extends from the right side of the ventricles toward the left and slightly upward, with an axis of approximately −30 degrees.

If one places the vector of the current of injury directly over the ventricles, *the negative end of the vector points toward the permanently depolarized, "injured" area of the ventricles.* In the instance illustrated in Figure 12–18, the injured area would be in the lateral wall of the right ventricle.

The Phenomenon of S-T Segment Shift. The portion of the electrocardiogram that occurs between the end of the QRS complex and the beginning of the T wave is called the *S-T segment.* The J point lies at the very beginning of this segment. Therefore, every time a current of injury occurs in one of the electrocardiographic leads, one also finds that the S-T segment and the T-P segments of the electrocardiogram are not at the same potential levels in the record. Actually, it is the T-P segment and not the S-T segment that is shifted away from the zero axis. However, most persons are conditioned to consider the T-P segment of the electrocardiogram as the reference potential level rather than the J point. Therefore, when a current of injury is evident in an electrocardiogram, it appears that the S-T segment is shifted from its normal level in the electrocardiogram, and this is called an *S-T segment shift.* Obviously, when one sees an S-T segment shift in an electrocardiogram, he or she knows immediately that the electrocardiogram shows the characteristics of a current of injury. In fact, most electrocardiographers do not speak of current of injury at all but simply speak of S-T segment shift, which means the same thing.

CORONARY ISCHEMIA AS A CAUSE OF CURRENT OF INJURY

Insufficient blood flow to the cardiac muscle depresses metabolism of the muscle for three different reasons: oxygen lack, excess accumulation of carbon dioxide, and lack of sufficient nutrients. Consequently, repolarization of the membranes cannot occur in areas of severe myocardial ischemia. Often, the heart muscle does not die because the blood flow is sufficient to maintain life of the muscle even though it is not sufficient to cause repolarization of the membranes. As long as this state exists, a current of injury continues to flow during diastole.

Extreme ischemia of the cardiac muscle occurs following coronary occlusion, and a strong current of injury flows from the infarcted area of the ventricles during the T-P interval between heartbeats, as is illustrated in Figures 12–19 and 12–20. Therefore, one of the most important diagnostic features of electrocardiograms recorded following acute coronary thrombosis is the current of injury.

Acute Anterior Wall Infarction. Figure 12–19 illustrates the electrocardiogram in the three standard bipolar limb leads and in one chest lead recorded from a patient with acute anterior wall cardiac infarction. The most important diagnostic feature of this electrocardiogram is the intense current of injury in the chest lead. If one draws a zero potential line through the J point of this electrocardio-

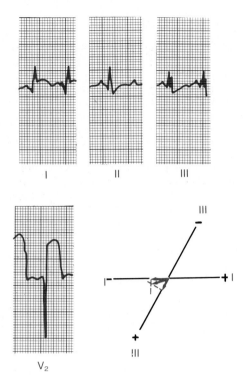

Figure 12–19. Current of injury in acute anterior wall infarction. Note the intense current of injury in lead V₂.

gram, a strong *negative* injury potential during the T-P interval is found, which means that the chest electrode over the front of the heart is in an area of strongly negative potential. In other words, the negative end of the injury potential vector is against the chest wall. This means that the current of injury is emanating from the anterior wall of the ventricles, which diagnoses this condition as anterior wall infarction.

Analyzing the currents of injury in leads I and III, one finds a negative potential caused by the current of injury in

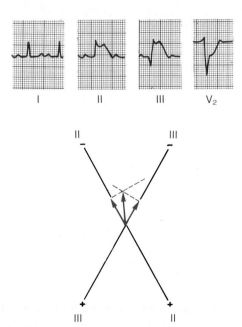

Figure 12–20. Current of injury in acute posterior wall, apical infarction.

lead I and a positive potential for the current of injury in lead III. This means that the resultant vector of the current of injury in the heart is approximately +150 degrees, with the negative end of the vector pointing toward the left ventricle and the positive end pointing toward the right ventricle. Thus, in this particular electrocardiogram the current of injury appears to be coming mainly from the left ventricle as well as from the anterior wall of the heart. Therefore, one would suspect that this anterior wall infarction is probably caused by thrombosis of the anterior descending limb of the left coronary artery.

Posterior Wall Infarction. Figure 12–20 illustrates the three standard bipolar limb leads and one chest lead from a patient with posterior wall infarction. The major diagnostic feature of this electrocardiogram is also in the chest lead. If a zero potential reference line is drawn through the J point of this lead, it is readily apparent that during the T-P interval the potential of the current of injury is positive. This means that the positive end of the vector is at the chest wall and the negative end (injured end) is away from the chest wall. In other words, the current of injury is coming from the side of the heart opposite that portion adjacent to the chest wall, which is the reason why this type of electrocardiogram is the basis for diagnosing posterior wall infarction.

If one analyzes the currents of injury in leads II and III of Figure 12–20, it is readily apparent that the injury potential is negative in both leads. By vectorial analysis as shown in the figure, one finds that the vector of the injury potential is approximately −95 degrees, with the negative end of the vector pointing downward and the positive end upward. Thus, because the infarct, as indicated by the chest lead, is on the posterior wall of the heart and, as indicated by the currents of injury in leads II and III, is in the apical portion of the heart, one would suspect that this infarct is close to the apex on the posterior wall of the left ventricle.

Infarction in Other Parts of the Heart. By the same procedures as those illustrated in the preceding two discussions of anterior and posterior wall infarctions, it is possible to determine the locus of any infarcted area emitting a current of injury regardless of which part of the heart is involved. In making such vectorial analyses, it must always be remembered that *the positive end of the injury potential vector points toward the normal cardiac muscle, and the negative end points toward the abnormal portion of the heart that is emitting the current of injury.*

Recovery from Acute Coronary Thrombosis. Figure 12–21 illustrates a V$_3$ chest lead from a patient with acute posterior infarction, showing the change in the electrocardiogram of this lead from the day of the attack to 1 week

later, then 3 weeks later, and finally 1 year later. From this electrocardiogram it can be seen that the current of injury is strong immediately after the acute attack (T-P segment displaced positively from the J point and S-T segment), but after approximately 1 week the current of injury has diminished considerably, and after 3 weeks it is completely gone. After that, the electrocardiogram does not change greatly during the following year. This is the usual recovery pattern following acute cardiac infarction of moderate degree when the collateral coronary blood flow is sufficient to re-establish appropriate nutrition to most of the infarcted area.

On the other hand, when all the coronary vessels throughout the heart are fairly well sclerosed, it may not be possible for the adjacent coronary vessels to supply sufficient blood to the infarcted area for recovery. Therefore, in some patients with coronary infarction, the infarcted area never redevelops an adequate coronary blood supply, some of the heart muscle dies, and relative coronary insufficiency persists in this area of the heart indefinitely. If the muscle does not die and become replaced by scar tissue, it continually emits a current of injury as long as the relative ischemia exists, particularly during bouts of exercise when the heart is overloaded.

Old Recovered Myocardial Infarction. Figure 12–22 illustrates leads I and III following anterior infarction and posterior infarction as these leads appear approximately a year after the acute episode. These are what might be called the "ideal" configurations of the QRS complex in these types of recovered myocardial infarction. Usually a Q wave develops at the beginning of the QRS complex in lead I in anterior infarction because of loss of muscle mass in the anterior wall of the left ventricle, whereas in posterior infarction a Q wave develops at the beginning of the QRS complex in lead III because of loss of muscle in the posterior apical part of the ventricle.

These configurations are certainly not those found in all cases of old anterior and posterior cardiac infarction. Local loss of muscle and local areas of conduction block can cause the following abnormalities of the QRS complex: bizarre patterns (the prominent Q waves, for instance), decreased voltage, and prolongation.

Current of Injury in Angina Pectoris. "Angina pectoris" means simply pain in the pectoral regions of the upper chest, this pain usually radiating into the neck and down the left arm. The pain is caused by *relative* ischemia of the heart. No pain is usually felt as long as the person is perfectly quiet, but, just as soon as the patient overworks the heart, the pain appears.

A current of injury often occurs during an attack of severe angina pectoris, for the relative coronary insufficiency then occasionally becomes great enough to prevent adequate repolarization of the membranes in some areas of the heart during diastole.

Same day   1 week   3 weeks   1 year

Figure 12–21. Recovery of the myocardium following moderate posterior wall infarction, illustrating disappearance of the current of injury (lead V$_3$).

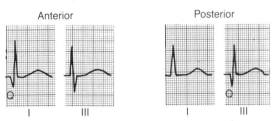

Anterior   Posterior

I   III   I   III

Figure 12–22. Electrocardiograms of old anterior and posterior wall infarctions, illustrating the Q wave in lead I in old anterior wall infarction and the Q wave in lead III in old posterior wall infarction.

● ABNORMALITIES IN THE T WAVE

Earlier in the chapter it was pointed out that the T wave is normally positive in all the standard bipolar limb leads and that this is caused by repolarization of the apex and outer surfaces of the ventricles ahead of the endocardial surfaces. This direction in which repolarization spreads over the heart is backward to the direction in which depolarization takes place. (If the basic principles of the upright T wave in the standard leads have not been understood by now, the reader should become familiar with the earlier, more detailed discussion of this before proceeding to the following few sections.)

The T wave becomes abnormal when the normal sequence of repolarization does not occur. Several factors can change this sequence of repolarization, as follows:

EFFECT OF SLOW CONDUCTION OF THE DEPOLARIZATION WAVE ON THE T WAVE

Referring back to Figure 12–14, note that the QRS complex is considerably prolonged. The reason for this prolongation is delayed conduction in the left ventricle as a result of left bundle branch block. The left ventricle becomes depolarized approximately 0.08 second after depolarization of the right ventricle, which gives a strong mean QRS vector *to the left.* The refractory periods of the right and left ventricular muscle masses are not greatly different from each other. Therefore, the right ventricle begins to repolarize long before the left ventricle; this causes positivity in the right ventricle and negativity in the left ventricle. In other words, the mean axis of the T wave is deviated *to the right,* which is opposite the mean electrical axis of the QRS complex in this same electrocardiogram. Thus, when conduction of the impulse through the ventricles is greatly delayed, the T wave is almost always of opposite polarity to that of the QRS complex.

In Figure 12–15 and in several figures in the following chapter, conduction also does not occur through the Purkinje system. As a result, the rate of conduction is greatly slowed, and in each instance the T wave is of opposite polarity to that of the QRS complex, whether the condition causing this delayed conduction happens to be left bundle branch block, right bundle branch block, premature ventricular contraction, or otherwise.

PROLONGED DEPOLARIZATION IN PORTIONS OF THE VENTRICULAR MUSCLE AS A CAUSE OF ABNORMALITIES IN THE T WAVE

If the apex of the ventricles should have an abnormally long period of depolarization, that is, a prolonged action potential, repolarization of the ventricles would not begin at the apex as it normally does. Instead, the base of the ventricles would repolarize ahead of the apex, and the vector of repolarization would point from the apex toward the base of the heart, opposite to the usual vector of repolarization. Consequently, the T wave in all three standard leads would be negative rather than the usual positive. Thus, the simple fact that the apical muscle of the heart had a prolonged period of depolarization would be sufficient to cause marked changes in the T wave, even to the extent of changing the entire polarity, as is illustrated in Figure 12–23.

Mild ischemia is by far the most common cause of increased duration of depolarization of cardiac muscle, and,

Figure 12–23. Inverted T wave resulting from mild ischemia of the apex of the ventricles.

when the ischemia occurs in only one area of the heart, the depolarization period of this area increases out of proportion to that in other portions. As a result, definite changes in the T wave may take place. The ischemia may result from chronic, progressive coronary occlusion, acute coronary occlusion, or relative coronary insufficiency occurring during exercise.

One means for detecting mild coronary insufficiency is to have the patient exercise and then to record the electrocardiogram immediately thereafter, noting whether or not changes occur in the T waves. The changes in the T waves need not be specific, for any change in the T wave in any lead—inversion, for instance, or a biphasic wave—is often evidence enough that some portion of the ventricular muscle has increased its period of depolarization out of proportion to the rest of the heart, and this is probably caused by relative coronary insufficiency.

All the other conditions that can cause currents of injury, including pericarditis, myocarditis, and mechanical trauma of the heart, can also cause changes in the T wave. A current of injury occurs when the period of depolarization of some muscle is so long that the muscle fails to repolarize completely before the next cardiac cycle begins. Therefore, a current of injury is actually an exacerbated form of abnormal T wave, for both of these result from an increased depolarization period of one or more portions of cardiac muscle, and the difference is only a matter of degree.

Effect of Digitalis on the T Wave. As discussed in Chapter 22, digitalis is a drug that can be used during relative coronary insufficiency to increase the strength of cardiac muscular contraction. However, digitalis also increases the period of depolarization of cardiac muscle. It usually increases this period by approximately the same proportion in all or most of the ventricular muscle, but when overdosages of digitalis are given, the depolarization period of one part of the heart may be increased out of porportion to that of other parts. As a result, nonspecific changes, such as T wave inversion or biphasic T waves, may occur in one or more of the electrocardiographic leads. A biphasic T wave caused by excessive administration of digitalis is illustrated in Figure 12–24. There is a slight amount of current of injury, too. This probably results from continuous depolarization of part of the ventricular muscle.

Changes in the T wave during digitalis administration are

Figure 12–24. Biphasic T wave caused by digitalis toxicity.

the earliest signs of digitalis toxicity. If still more digitalis is given to the patient, strong currents of injury may develop. Also, digitalis can block conduction of the cardiac impulse to various portions of the heart, so that various arrhythmias can result. It is desirable clinically to keep the effects of digitalis from going beyond the stage of mild T wave abnor-malities. Therefore, the electrocardiograph is used routinely in following digitalized patients.

REFERENCES

See references for Chapter 13.

CHAPTER 13

• Cardiac Arrhythmias and Their Electrocardiographic Interpretation •

Some of the most distressing types of heart malfunction occur not as a result of abnormal heart muscle but instead abnormal rhythm of the heart. Sometimes the heart rate is too fast or too slow to pump proper amounts of blood; sometimes the interval between heartbeats is too short for the ventricles to fill; and sometimes the beat of the atria is totally uncoordinated with the beat of the ventricles, so that the atria no longer function as primers for the ventricles.

The purpose of the present chapter is to discuss the common cardiac arrhythmias and their effects on heart pumping as well as their diagnosis by electrocardiography. The causes of the cardiac arrhythmias are usually one or a combination of the following abnormalities in the rhythmicity-conduction system of the heart:

1. Abnormal rhythmicity of the pacemaker.
2. Shift of the pacemaker from the sinus node to other parts of the heart.
3. Blocks at different points in the transmission of the impulse through the heart.
4. Abnormal pathways of impulse transmission through the heart.
5. Spontaneous generation of abnormal impulses in almost any part of the heart.

● ABNORMAL SINUS RHYTHMS

TACHYCARDIA

The term "tachycardia" means fast heart rate, usually defined as faster than 100 beats per minute. An electrocardiogram recorded from a patient with tachycardia is illustrated in Figure 13-1. This electrocardiogram is normal except that the rate of heartbeat, as determined from the time intervals between QRS complexes, is approximately 150 per minute instead of the normal 72 per minute.

The three general causes of tachycardia are *increased body temperature, stimulation of the heart by the sympathetic nerves,* and *toxic conditions of the heart.*

The rate of the heart increases approximately 10 beats per minute for each degree Fahrenheit (18 beats per degree

Celsius) increase in body temperature up to a body temperature of about 105°; beyond this the heart rate may actually decrease, owing to progressive weakening of the heart muscle as a result of the fever. Fever causes tachycardia because increased temperature increases the rate of metabolism of the sinus node, which in turn directly increases its excitability and rate of rhythm.

Many factors can cause the sympathetic nervous system to excite the heart, as we discuss at multiple points in this text. For instance, when a patient loses blood and passes into a state of shock or semishock, reflex stimulation of the heart increases its rate to as high as 150 to 180 beats per minute. Also, simple weakening of the myocardium usually increases the heart rate because the weakened heart does not pump blood into the arterial tree to a normal extent, and this elicits reflexes to increase the rate of the heart.

BRADYCARDIA

The term "bradycardia" means a slow heart rate, usually defined as less than 60 beats per minute. Bradycardia is illustrated by the electrocardiogram in Figure 13-2.

Bradycardia in Athletes. The athlete's heart is considerably stronger than that of a normal person, a fact that allows the athlete's heart to pump a greater stroke volume output per beat. The excessive quantities of blood pumped into the arterial tree with each beat initiate circulatory reflexes or other effects to cause the bradycardia.

Vagal Stimulation as a Cause of Bradycardia. Any circulatory reflex that stimulates the vagus nerve can cause the heart rate to decrease considerably because of the inhibitory effect that parasympathetic nervous signals have on heart function. Perhaps the most striking example of this occurs in patients with the *carotid sinus syndrome*. In these patients, an arteriosclerotic process in the carotid sinus region of the carotid artery causes excessive sensitivity of the pressure receptors (baroreceptors) located in the arterial wall; as a result, mild pressure on the neck elicits a strong baroreceptor reflex, causing intense vagal stimulation of the heart and extreme bradycardia. Indeed, sometimes this reflex is so powerful that it actually stops the heart.

Figure 13–1. Sinus tachycardia (lead I).

Figure 13–3. Sinus arrhythmia as detected by a cardiotachometer. To the left is the recording taken when the subject was breathing normally; to the right, when breathing deeply.

SINUS ARRHYTHMIA

Figure 13–3 illustrates a *cardiotachometer* recording of the heart rate during normal and deep respiration. A cardio-tachometer is an instrument that records by the height of successive spikes the duration of the interval between each two QRS complexes in the electrocardiogram. Note from this record that the heart rate increases and decreases approximately 5 per cent during the various phases of the quiet respiratory cycle. However, during deep respiration, as shown to the right in Figure 13–3, the heart rate even normally increases and decreases with each respiratory cycle by as much as 30 per cent.

Sinus arrhythmia can result from any one of many circulatory reflexes or other nervous effects that alter the strength of the sympathetic and parasympathetic nerve signals to the sinus node. In the respiratory type of sinus arrhythmia illustrated in Figure 13–3, this results mainly from "spillover" of signals from the medullary respiratory center into the vasomotor center during the inspiratory and expiratory cycles of respiration. The spillover signals cause alternate increase and decrease in the number of impulses transmitted to the heart through the sympathetic and vagus nerves.

● ABNORMAL RHYTHMS RESULTING FROM IMPULSE CONDUCTION BLOCK

SINOATRIAL BLOCK

In rare instances the impulse from the sinus node is blocked before it enters the atrial muscle. This phenomenon is illustrated in Figure 13–4, which shows the sudden cessation of P waves with resultant standstill of the atrium. However, the ventricle picks up a new rhythm, the impulse usually originating in the A-V node, so that the ventricular QRS-T complex is not altered.

ATRIOVENTRICULAR BLOCK

The only means by which impulses can ordinarily pass from the atria into the ventricles is through the *A-V bundle,* which is also known as the *bundle of His.* The different conditions that can either decrease the rate of conduction of the impulse through this bundle or totally block the impulse are:

1. *Ischemia of the A-V nodal or A-V bundle fibers* often delays or blocks conduction from the atria to the ventricles. Coronary insufficiency can cause ischemia of the A-V node

and bundle in the same manner that it can cause ischemia of the myocardium.

2. *Compression of the A-V bundle* by scar tissue or by calcified portions of the heart can depress or block conduction from the atria to the ventricles.

3. *Inflammation of the A-V node or A-V bundle* can depress conductivity between the atria and the ventricles. Inflammation results frequently from different types of myocarditis, such as occur in diphtheria and rheumatic fever.

4. *Extreme stimulation of the heart by the vagus nerves* in rare instances blocks impulse conduction through the A-V node. Such vagal excitation occasionally results from strong stimulation of the baroreceptors in persons with the *carotid sinus syndrome,* which was discussed earlier in relation to bradycardia.

Incomplete Heart Block. ***Prolonged P-R (or P-Q) Interval — "First Degree Block."*** The normal lapse of time between the *beginning* of the P wave and the *beginning* of the QRS complex is approximately 0.16 second when the heart is beating at a normal rate. This P-R interval usually decreases in length with faster heartbeat and increases with slower heartbeat. In general, when the P-R interval increases above a value of approximately 0.20 second in a heart beating at normal rate, the P-R interval is said to be prolonged, and the patient is said to have *first degree incomplete heart block.* Figure 13–5 illustrates an electrocardiogram with a prolonged P-R interval, the interval in this instance being approximately 0.30 second. Thus, first degree block is defined as a delay of conduction from the atria to the ventricles but not actual blockage of conduction.

The P-R interval rarely increases above 0.35 to 0.45 second, for by the time the rate of conduction through the A-V node and bundle is depressed to such an extent, conduction stops entirely. Thus, when a patient's P-R interval is approaching these limits, additional slight increase in the severity of the condition will block impulse conduction rather than simply delay conduction further.

One of the means for determining the severity of some heart diseases — acute rheumatic fever, for instance — is to measure the P-R interval.

Second Degree Block. When conduction through the A-V junction is slowed until the P-R interval is 0.25 to 0.45 second, the action potentials traveling through the A-V node sometimes are strong enough to pass on through the A-V node and at other times are not strong enough. Often the impulse passes into the ventricles following one atrial contraction and fails to pass following the next one or two atrial contractions, thus alternating between conduction

Figure 13–2. Sinus bradycardia (lead III).

Figure 13–4. S-A nodal block with A-V nodal rhythm (lead III).

Figure 13–5. Prolonged P-R interval (lead II).

Figure 13–7. Complete atrioventricular block (lead II).

and nonconduction. In this instance, the atria beat at a considerably faster rate than the ventricles, and it is said that there are "dropped beats" of the ventricles. This condition is called *second degree incomplete heart block.*

Figure 13–6 illustrates P-R intervals of 0.30 second, and it also illustrates one dropped beat as a result of failure of conduction from the atria to the ventricles.

At times every other beat of the ventricles is dropped, so that a "2:1 rhythm" develops in the heart, with the atria beating twice for every single beat of the ventricles. Sometimes other rhythms, such as 3:2 or 3:1, also develop.

Complete A-V Block (Third Degree Block). When the condition causing poor conduction in the A-V node or bundle becomes extremely severe, complete block of the impulse from the atria into the ventricles occurs. In this instance the P waves become completely dissociated from the QRS-T complexes, as illustrated in Figure 13–7. Note that the rate of rhythm of the atria in this electrocardiogram is approximately 100 beats per minute, whereas the rate of ventricular beat is less than 40 per minute. Furthermore, there is no relationship whatsoever between the rhythm of the atria and that of the ventricles, for the ventricles have "escaped" from control by the atria, and they are beating at their own natural rate.

Stokes-Adams Syndrome — Ventricular Escape. In some patients with A-V block, the total block comes and goes — that is, impulses are conducted from the atria into the ventricles for a period of time, and then suddenly no impulses at all are transmitted. The duration of total block may be a few seconds, a few minutes, or a few hours, or it may be weeks or even longer before conduction returns. In particular, this condition occurs in hearts with borderline ischemia.

Immediately after A-V conduction is first blocked, the ventricles stop contracting entirely for 5 to 30 seconds because of the phenomenon called *overdrive suppression,* which means that ventricular excitability has been suppressed because the ventricles have been driven by the atria at a rate greater than their natural rate of rhythm. Finally, some part of the Purkinje system beyond the block, usually in the distal part of the A-V node beyond the blocked point in the node or in the A-V bundle, begins discharging rhythmically at a rate of 15 to 40 times per minute and acting as the pacemaker of the ventricles. This is called *ventricular escape.*

Because the brain cannot remain active for more than 3 to 5 seconds without blood supply, patients usually faint a few seconds after complete block occurs because the heart then does not pump any blood for 5 to 30 seconds until the ventricles "escape." After escape, even the slowly beating

ventricles usually pump enough blood to allow rapid recovery from the faint and then to sustain the person. These periodic fainting spells are known as the Stokes-Adams syndrome.

Occasionally the interval of ventricular standstill at the onset of complete block is so long that it either becomes detrimental to the health of the patient or causes death. Consequently, most of these patients are now provided with an *artificial pacemaker,* which is a small battery-operated electrical stimulator planted beneath the skin, the electrodes from which are usually connected to the right ventricle. This pacemaker provides continued rhythmic impulses that take over control of the ventricles. The batteries are replaced about once every 5 years.

INCOMPLETE INTRAVENTRICULAR BLOCK —ELECTRICAL ALTERNANS

Most of the same factors that can cause A-V block can also block impulse conduction in peripheral portions of the ventricular Purkinje system. At times, *incomplete* block occurs so that the impulse is sometimes transmitted and sometimes not transmitted, causing block of the impulse during some heart cycles and not during others. The QRS complex may be considerably abnormal during those cycles in which the impulse is blocked. Figure 13–8 illustrates the condition known as *electrical alternans,* which results from partial intraventricular block every other heartbeat. This electrocardiogram also illustrates tachycardia, which is probably the reason the block has occurred, for when the rate of the heart is very rapid it may be impossible for portions of the Purkinje system to recover from the refractory period quickly enough to respond during each succeeding heartbeat. Also, many conditions that depress the heart, such as ischemia, myocarditis, and digitalis toxicity, can cause incomplete intraventricular block with resultant electrical alternans.

● PREMATURE CONTRACTIONS

A premature contraction is a contraction of the heart prior to the time that normal contraction would have been expected. This condition is also frequently called *extrasystole, premature beat,* or *ectopic beat.*

Causes of Premature Contractions. Most premature contractions result from *ectopic foci* in the heart, which emit abnormal impulses at odd times during the cardiac rhythm.

Dropped beat

Figure 13–6. Second degree incomplete atrioventricular block (lead V₃).

Figure 13–8. Partial intraventricular block — electrical alternans (lead III).

Among the possible causes of ectopic foci are (1) local areas of ischemia; (2) small calcified plaques at different points in the heart, which press against the adjacent cardiac muscle so that some of the fibers are irritated; and (3) toxic irritation of the A-V node, Purkinje system, or myocardium caused by drugs, nicotine, or caffeine. Mechanical initiation of premature contractions is also frequent during cardiac catheterization, large numbers of premature contractions often occurring when the catheter enters the right ventricle and presses against the endocardium.

In persons with ischemic heart disease, the ectopic rhythm is, on occasion, believed to be caused by a re-entrant signal as follows: The normal heartbeat excites an area of ischemic tissue in which the cardiac impulse travels extremely slowly. Then, after the contraction of the normal heart muscle is over, the slowly traveling signal in the ischemic tissue escapes back into the normal cardiac muscle, in this way causing a second contraction at a later time in the heartbeat.

PREMATURE ATRIAL CONTRACTIONS

Figure 13–9 is an electrocardiogram showing a single premature atrial contraction. The P wave of this beat occurs too soon in the heart cycle, and the P-R interval is shortened, indicating that the ectopic origin of the beat is near the A-V node. Also, the interval between the premature contraction and the next succeeding contraction is slightly prolonged, which is called a *compensatory pause*. The reason for this is that the premature contraction originated in the atrium some distance from the sinus node, and the impulse had to travel through a considerable amount of atrial muscle before it discharged the sinus node. Consequently, the sinus node discharged very late in the premature cycle, and this made the succeeding heartbeat also late in appearing.

Premature atrial contractions occur frequently in healthy persons and, indeed, are often found in athletes or others whose hearts are certain to be in healthy condition. Yet mild toxic conditions resulting from such factors as excess smoking, lack of sleep, ingestion of too much coffee, alcoholism, and use of various drugs can also initiate such contractions.

Pulse Deficit. When the heart contracts ahead of schedule, the ventricles will not have filled with blood normally, and the stroke volume output during that contraction is depressed or sometimes almost absent. Therefore, the pulse wave passing to the periphery following a premature contraction may be so weak that the pulse cannot be felt at all in the radial artery. Thus, a deficit in the number of pulses felt in the radial pulse occurs in relation to the number of contractions of the heart.

Bigeminal Pulse. At times every other beat of the heart may be a premature contraction. This causes the patient to have a bigeminal pulse—that is, two pulses close together, then a longer diastolic interval, then two again, and so on.

Figure 13–10. A-V nodal premature contraction (lead III).

A-V NODAL OR A-V BUNDLE PREMATURE CONTRACTIONS

Figure 13–10 illustrates a premature contraction originating either in the A-V node or in the A-V bundle. The P wave is missing from the record of the premature contraction. Instead, the P wave is superimposed onto the QRS-T complex of the premature contraction because the cardiac impulse travels backward into the atria at the same time that it travels forward into the ventricles; this P wave distorts the complex, but the P wave itself cannot be discerned as such.

In general, A-V nodal premature contractions have the same significance and causes as atrial premature contractions.

PREMATURE VENTRICULAR CONTRACTIONS

The electrocardiogram of Figure 13–11 illustrates a series of premature ventricular contractions (PVCs) alternating with normal contractions. Most PVCs of this type probably result from a re-entrant signal from an ischemic area of ventricular muscle, as described earlier. This causes several effects in the electrocardiogram, as follows:

1. The QRS complex is usually considerably prolonged. The reason is that the impulse is conducted mainly through the slowly conducting muscle of the ventricle rather than through the Purkinje system.

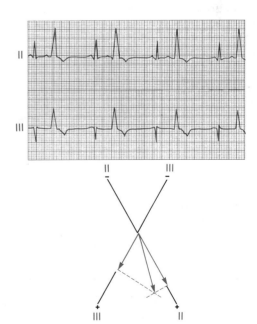

Figure 13–11. Premature ventricular contractions (PVCs) illustrated by the large abnormal QRS-T complexes (leads II and III). Axis of the premature contractions is plotted in accord with the principles of vectorial analysis explained in Chapter 12.

Figure 13–9. Atrial premature contraction (lead I).

2. The QRS complex has a very high voltage for the following reason: When the normal impulse passes through the heart, it passes through both ventricles approximately simultaneously; consequently, the depolarization waves of the two sides of the heart partially neutralize each other. However, when a PVC occurs, the impulse travels in only one direction, so that there is no such neutralization effect, and one entire side of the heart is depolarized, while the other entire side is still polarized; this causes intense electrical potentials.

3. Following almost all PVCs the T wave has a potential opposite to that of the QRS complex because the *slow conduction of the impulse* through the cardiac muscle causes the area first depolarized also to repolarize first. As a result, the direction of current flow in the heart during repolarization is opposite to that during depolarization, and the potential of the T wave is reversed to that of the QRS complex. This is not true of the normal T wave, as explained in Chapter 11.

Some PVCs are relatively benign in their origin and result from simple factors, such as cigarettes, coffee, lack of sleep, various mild toxic states, and even emotional irritability. On the other hand, a large share of PVCs results from stray impulses or re-entrant signals originating around the borders of infarcted or ischemic areas of the heart. Therefore, the presence of such PVCs is not to be taken lightly. Statistics show that persons with significant numbers of PVCs have a much higher than normal chance of developing spontaneous lethal ventricular fibrillation, presumably initiated by one of the PVCs itself. This is especially true when the PVCs occur during the vulnerable period for causing fibrillation, just at the end of the T wave when the ventricles are coming out of refractoriness, as is explained later in the chapter.

Vector Analysis of the Origin of an Ectopic Premature Ventricular Contraction. In Chapter 12 the principles of vectorial analysis were explained. Applying these principles, one can determine from the electrocardiogram in Figure 13–11 the point of origin of the PVC as follows: Note that the potentials of the premature contractions in leads II and III are both strongly positive. Plotting these potentials on the axes of leads II and III and solving by vectorial analysis for the mean QRS vector in the heart, one finds that this vector of the premature contraction has its negative end (origin) at the base of the heart and its positive end toward the apex. Thus, the first portion of the heart to become depolarized during the premature contractions lies near the base of the heart, which therefore is the locus of the ectopic focus.

● PAROXYSMAL TACHYCARDIA

Abnormalities in any portion of the heart, including the atria, the Purkinje system, and the ventricles, can sometimes cause rapid rhythmic discharge of impulses that spread in all directions throughout the heart. This is believed to be caused most frequently by re-entrant pathways that set up local repeated self–re-excitation. Because of the rapid rhythm in the irritable focus, this focus becomes the pacemaker of the heart.

The term "paroxysmal" means that the heart rate usually becomes very rapid in paroxysms, with the paroxysms beginning suddenly and lasting for a few seconds, a few minutes, a few hours, or sometimes much longer. Then the paroxysms usually end as suddenly as they had begun, the pacemaker of the heart then shifting back to the sinus node.

Paroxysmal tachycardia can often be stopped by eliciting a vagal reflex. A strange type of vagal reflex sometimes elicited for this purpose is one that occurs when painful pressure is applied to the eyes. Also, pressure on the carotid sinuses can sometimes elicit enough of a vagal reflex to stop the tachycardia. Various drugs may also be used. Two drugs frequently used are quinidine and lidocaine, both of which depress the normal increase in sodium permeability of the cardiac muscle membrane during the generation of the action potential, thereby often blocking the rhythmic discharge of the focal region that is causing the paroxysmal attack.

ATRIAL PAROXYSMAL TACHYCARDIA

Figure 13–12 illustrates in the middle of the record a sudden increase in rate of heartbeat from approximately 95 to approximately 150 beats per minute. On close observation of the electrocardiogram it can be seen that an inverted P wave occurs before each of the QRS-T complexes during the paroxysm of rapid heartbeat, and this P wave is partially superimposed on the normal T wave of the preceding beat. This indicates that the origin of this paroxysmal tachycardia is in the atrium, but, because the P wave is abnormal, the origin is not near the sinus node.

A-V Nodal Paroxysmal Tachycardia. Paroxysmal tachycardia very often results from an aberrant rhythm that involves the A-V node. One of the postulated causes of this type of tachycardia is a local re-entrant signal entering the A-V node in the following manner: It is believed that some of the A-V nodal fibers are much more conductive than others. Therefore, a signal is conducted to the ventricles by way of some of the fibers but without exciting all of them. Then, after the cardiac impulse has entered the ventricles, a re-entrant signal travels backward through the previously unexcited fibers, thus initiating still a new impulse from the A-V node. Thus, a new ventricular contraction occurs, followed by re-entry once again through the rapidly recovering A-V nodal fibers and continuation of the repetitive cycle.

Regardless of the cause of the A-V nodal tachycardia, it usually causes almost normal QRS-T complexes but missing or obscured P waves.

Atrial or A-V nodal paroxysmal tachycardia, both of which are called *supraventricular tachycardias,* usually occurs in young, otherwise healthy persons, and such persons generally grow out of the predisposition to the tachycardias after adolescence. In general, supraventricular tachycardia frightens the individual tremendously and may cause weakness during the paroxysms but only rarely does permanent harm come from the attacks.

VENTRICULAR PAROXYSMAL TACHYCARDIA

Figure 13–13 illustrates a typical short paroxysm of ventricular tachycardia. The electrocardiogram of ventricular paroxysmal tachycardia has the appearance of a series of

Figure 13–12. Atrial paroxysmal tachycardia—onset in middle of record (lead I).

Figure 13-13. Ventricular paroxysmal tachycardia (lead III).

ventricular premature beats occurring one after another without any normal beats interspersed.

Ventricular paroxysmal tachycardia is usually a serious condition for two reasons. First, this type of tachycardia usually does not occur unless considerable ischemic damage is present in the ventricles. Second, ventricular tachycardia very frequently initiates ventricular fibrillation because of rapid repeated stimulation of the ventricular muscle, as we discuss in the following section.

Digitalis intoxication sometimes causes irritable foci that lead to ventricular tachycardia. On the other hand, quinidine, which increases the refractory period of cardiac muscle and also increases its threshold for excitation, may be used to block irritable foci causing ventricular tachycardia.

● VENTRICULAR FIBRILLATION

The most serious of all cardiac arrhythmias is *ventricular fibrillation,* which, if not treated instantly, is almost invariably fatal.

Ventricular fibrillation results from cardiac impulses that have gone berserk within the ventricular muscle mass, stimulating first one portion of the ventricular muscle, then another portion, then another, and eventually feeding back onto itself to re-excite the same ventricular muscle over and over again — never stopping. When this happens, many small portions of the ventricular muscle will be contracting at the same time, while equally as many other portions will be relaxing. Thus, there is never a coordinate contraction of all of the heart muscle at once, which is required for a pumping cycle of the heart. Therefore, despite massive flow of stimulatory signals throughout the ventricles, the ventricular chamber neither enlarges nor contracts but remains in an indeterminate stage of partial contraction, pumping either no blood at all or negligible amounts. Therefore, after fibrillation begins, unconsciousness occurs within 4 to 5 seconds for lack of blood flow to the brain, and irretrievable death of the tissues occurs throughout the body within a few minutes.

Multiple factors can spark the beginning of ventricular fibrillation — with a normal heartbeat one second and a second later the ventricles in fibrillation. Especially likely to initiate fibrillation are (1) sudden electrical shock of the heart or (2) ischemia of the heart muscle, its specialized conducting system, or both. In either instance, an instantaneous pattern of re-entry signals can be established, so that the contractile impulses travel around and around through the heart muscle. This phenomenon is also frequently called a *circus movement.*

THE PHENOMENON OF "RE-ENTRY" — CIRCUS MOVEMENTS AS THE BASIS FOR VENTRICULAR FIBRILLATION

When the *normal* cardiac impulse has traveled through the entire extent of the ventricles, it then has no place else to go because all of the ventricular muscle is at that time refractory and cannot conduct the impulse further. Therefore, that impulse dies, and the heart awaits a new signal to begin in the sinus node.

However, under some circumstances this normal sequence of events does not occur. Therefore, let us explain more fully the background conditions that can initiate re-entry and lead to the circus movements of ventricular fibrillation.

Figure 13-14 illustrates several small cardiac muscle strips cut in the form of circles. If such a strip is stimulated at the 12 o'clock position *so that the impulse travels in only one direction,* the impulse spreads progressively around the circle until it returns to the 12 o'clock position. If the originally stimulated muscle fibers are still in a refractory state, the impulse then dies out, for refractory muscle cannot transmit a second impulse. However, there are three different conditions that can cause this impulse to continue to travel around the circle, that is, to cause "re-entry" of the impulse into muscle that has already been excited.

First, if the *length of the pathway around the circle is long,* by the time the impulse returns to the 12 o'clock position the originally stimulated muscle will no longer be refractory, and the impulse will continue around the circle again and again.

Second, if the length of the pathway remains constant but the *velocity of conduction becomes decreased* enough, an increased interval of time will elapse before the impulse returns to the 12 o'clock position. By this time the originally stimulated muscle might be out of the refractory state, and the impulse can continue around the circle again and again.

Third, *the refractory period of the muscle might become greatly shortened.* In this case, the impulse could also continue around and around the circle.

All three of these conditions occur in different pathological states of the human heart as follows: (1) A long pathway frequently occurs in dilated hearts. (2) Decreased rate of conduction frequently results from blockage of the Purkinje system, ischemia of the muscle, high blood potassium, and many other factors. (3) A shortened refractory period frequently occurs in response to various drugs, such as epinephrine, or following repetitive electrical stimulation. Thus, in many different cardiac disturbances re-entry can cause abnormal patterns of cardiac contraction or abnormal cardiac rhythms that completely ignore the pace-setting effects of the sinus node.

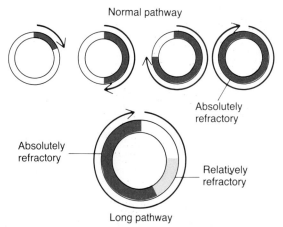

Figure 13-14. The circus movement, showing annihilation of the impulse in the short pathway and continued propagation of the impulse in the long pathway.

The "Chain Reaction" Mechanism of Fibrillation

In ventricular fibrillation one sees many separate and small contractile waves spreading at the same time in different directions over the cardiac muscle. Obviously, then, the re-entrant impulses in fibrillation are not simply a single impulse moving in a circle, as illustrated in Figure 13–14. Instead, they have degenerated into a series of multiple wave fronts that have the appearance of a "chain reaction." One of the best ways to explain this process in fibrillation is to describe the initiation of fibrillation by electric shock caused by 60 cycle alternating electrical current.

Fibrillation Caused by 60 Cycle Alternating Current. At a central point in the ventricles of heart A in Figure 13–15, a 60 cycle electrical stimulus is applied through a stimulating electrode. The first cycle of the electrical stimulus causes a depolarization wave to spread in all directions, leaving all the muscle beneath the electrode in a refractory state. After about 0.25 second, this muscle begins to come out of the refractory state. However, some portions of the muscle come out of refractoriness prior to other portions. This state of events is depicted in heart A by many light patches, which represent excitable cardiac muscle, and dark patches, which represent still refractory muscle. New stimuli from the electrode can now cause impulses to travel in certain directions through the heart but not in all directions. Thus, in heart A certain impulses travel for short distances until they reach refractory areas of the heart and then are blocked. Other impulses, however, pass between the refractory areas and continue to travel in the excitable patches of muscle. Now, several events transpire in rapid succession, all occurring simultaneously and eventuating in a state of fibrillation. These are as follows:

First, block of the impulses in some directions but successful transmission in other directions creates one of the necessary conditions for a re-entrant signal to develop — that is, *transmission of some of the depolarization waves around the heart in only one direction.* As a result, these waves do not run into waves traveling in the opposite direction and therefore do not annihilate themselves on the opposite side of the heart but can continue around and around the ventricles.

Second, the rapid stimulation of the heart causes two changes in the cardiac muscle itself, both of which predispose to circus movement: (1) the *velocity of conduction through the heart becomes decreased,* which allows a longer time interval for the impulses to travel around the heart. (2) The *refractory period of the muscle becomes shortened,* allowing re-entry of the impulse into previously excited heart muscle within a much shorter period of time than normally.

Third, one of the most important features of fibrillation is the *division of impulses,* as illustrated in heart A. When a depolarization wave reaches a refractory area in the heart, it travels to both sides around the area. Thus, a single impulse becomes two impulses. Then when each of these reaches another refractory area it, too, divides to form still two more impulses. In this way many different new wave fronts are continually being formed in the heart by a progressive *chain reaction* until, finally, there are many small depolarization waves traveling in many different directions at the same time. Furthermore, this irregular pattern of impulse travel causes a *circuitous route for the impulses to travel, greatly lengthening the conductive pathway, which is one of the conditions that sustains the fibrillation.* It also results in a continual irregular pattern of patchy refractory areas in the heart. One can readily see that a vicious cycle has been initiated: More and more impulses are formed, these cause more and more patches of refractory muscle, and the refractory patches cause more and more division of the impulses. Therefore, any time a single area of cardiac muscle comes out of refractoriness, an impulse is always close at hand to re-enter the area.

Heart B in Figure 13–15 illustrates the final state that develops in fibrillation. Here one can see many impulses traveling in all directions, some dividing and increasing the number of impulses, while others are blocked entirely by refractory areas.

Vulnerable Period for Causing Ventriculation Fibrillation. The time period during the heart cycle when there are likely to be simultaneous areas of refractoriness and nonrefractoriness in the heart muscle is at the very moment that the heart is recovering from the previous cardiac cycle — that is, just at the end of the cardiac contraction. Therefore, this instant in the cycle is said to be the *vulnerable period* of the ventricles for fibrillation. In fact, a single electric shock during this vulnerable period frequently can lead to the odd pattern of impulses spreading unidirectionally around refractory areas of muscle, thus initiating fibrillation.

The Electrocardiogram in Ventricular Fibrillation

In ventricular fibrillation the electrocardiogram is extremely bizarre, as seen in Figure 13–16, and ordinarily shows no tendency toward a regular rhythm of any type. In the early phases of ventricular fibrillation, relatively large masses of muscle contract simultaneously, and this causes strong though irregular waves in the electrocardiogram. However, after only a few seconds the coarse contractions of the ventricles disappear, and the electrocardiogram changes into a new pattern of low-voltage, very irregular

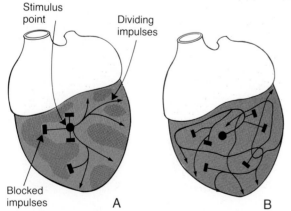

Stimulus point

Dividing impulses

Blocked impulses

A   B

Figure 13–15. *A,* Initiation of fibrillation in a heart when patches of refractory musculature are present. *B,* Continued propagation of *fibrillatory impulses* in the fibrillating ventricle.

Figure 13–16. Ventricular fibrillation (lead II).

waves. Thus, no repetitive electrocardiographic pattern can be ascribed to ventricular fibrillation except that the electrical potentials constantly and spasmodically change because the currents in the heart flow first in one direction, then in another, and rarely repeat any specific cycle.

The voltages of the waves in the electrocardiogram in ventricular fibrillation are usually about 0.5 mV when ventricular fibrillation first begins, but these decay rapidly so that after 20 to 30 seconds they are usually only 0.2 to 0.3 mV. Minute voltages of 0.1 mV or less may be recorded for 10 minutes or longer after ventricular fibrillation begins. As already pointed out, ventricular fibrillation is lethal unless it is stopped by some heroic therapy, such as immediate electroshock through the heart, as explained in the following section.

Electroshock Defibrillation of the Ventricles

Although a weak alternating current almost invariably throws the ventricles into fibrillation, a very strong electrical current passed through the ventricles for a short interval of time can stop fibrillation by throwing all the ventricular muscle into refractoriness simultaneously. This is accomplished by passing intense current through electrodes placed on two sides of the heart. The current penetrates most of the fibers of the ventricles, thus stimulating essentially all parts of the ventricles simultaneously and causing them to become refractory. All impulses stop, and the heart then remains quiescent for 3 to 5 seconds, after which it begins to beat again, usually with the sinus node or some other part of the heart becoming the pacemaker. Occasionally, however, the same re-entrant focus that had originally thrown the ventricles into fibrillation is still present, and fibrillation begins again immediately.

When electrodes are applied directly to the two sides of the heart, fibrillation can usually be stopped with 110 volts of 60 cycle alternating current applied for 0.1 second, or 1000 volts direct current applied for a few thousandths of a second. When applied through the chest wall, as illustrated in Figure 13–17, the usual procedure is to charge a large electrical capacitor up to several thousand volts and then to cause the capacitor to discharge in a few thousandths of a second through the electrodes and the heart. In our laboratory, the heart of a single anesthetized dog was defibrillated 130 times through the chest wall, and the animal remained in perfectly normal condition.

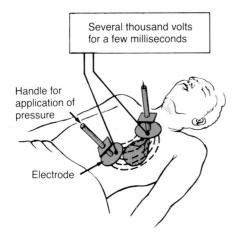

Figure 13–17. Application of electrical current to the chest to stop ventricular fibrillation.

Hand Pumping of the Heart (Cardiopulmonary Resuscitation) As an Aid to Defibrillation

Unless defibrillated within 1 minute after fibrillation begins, the heart is usually too weak to be revived by defibrillation alone because of lack of nutrition from coronary blood flow. However, it is still possible to revive the heart by preliminarily pumping it by hand and then defibrillating it later. In this way small quantities of blood are delivered into the aorta, and a renewed coronary blood supply develops. Then, after a few minutes, electrical defibrillation often becomes possible. Indeed, fibrillating hearts have been pumped by hand as long as 90 minutes before defibrillation.

A technique of pumping the heart without opening the chest consists of intermittent, very powerful thrusts of pressure on the chest wall, along with artificial respiration as well. This is called *cardiopulmonary resuscitation,* or simply CPR.

Lack of blood flow to the brain for more than 5 to 10 minutes usually results in permanent mental impairment or even total destruction of the brain. Even though the heart should be revived, the person might die from the effects of brain damage or live with permanent mental impairment.

● ATRIAL FIBRILLATION

Remember that, except for the connection through the A-V bundle, the atrial muscle mass is entirely separated from the ventricular muscle mass, insulated from each other by fibrous tissue. Therefore, ventricular fibrillation often occurs entirely independently of atrial fibrillation. Likewise, fibrillation often occurs in the atria independently of ventricular fibrillation and is called *atrial fibrillation;* this is illustrated to the right in Figure 13–18.

The mechanism of atrial fibrillation is identical with that of ventricular fibrillation except that the process occurs in the atrial muscle mass instead of the ventricular mass. A very frequent cause of atrial fibrillation is atrial enlargement resulting from heart valve lesions that prevent the atria from emptying adequately into the ventricles or resulting from ventricular failure with excess damming of blood in the atria. The dilated atrial walls provide the ideal conditions of a long conductive pathway as well as slow conduction, both of which predispose to atrial fibrillation.

Pumping Characteristics of the Atria During Atrial Fibrillation. For the same reasons that the ventricles will not pump blood during ventricular fibrillation, neither do the atria pump blood in atrial fibrillation. Therefore, the atria become useless as primer pumps for the

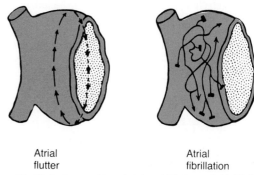

Atrial
flutter

Atrial
fibrillation

Figure 13–18. Pathways of impulses in atrial flutter and atrial fibrillation.

ventricles. Even so, blood flows passively through the atria into the ventricles, and the efficiency of ventricular pumping is decreased only 20 to 30 per cent. Therefore, in contrast to the lethality of ventricular fibrillation, a person can live for months or even years with atrial fibrillation, though at a reduced efficiency of overall heart pumping.

The Electrocardiogram in Atrial Fibrillation. Figure 13–19 illustrates the electrocardiogram during atrial fibrillation. Numerous small depolarization waves spread in all directions through the atria during atrial fibrillation. Because the waves are weak and because many of them are of opposite polarity at any given time, they usually almost completely neutralize each other. Therefore, in the electrocardiogram one can see either no P waves from the atria or a fine, high frequency, very low-voltage wavy record. On the other hand, the QRS-T complexes are completely normal unless there is some pathology of the ventricles, but their timing is very irregular for the following reasons:

Irregularity of the Ventricular Rhythm During Atrial Fibrillation. When the atria are fibrillating, impulses arrive at the A-V node rapidly but also irregularly. Because the A-V node will not pass a second impulse for approximately 0.35 second after a previous one, at least 0.35 second must elapse between one ventricular contraction and the next, and an additional but variable interval of 0 to 0.6 second occurs before one of the irregular fibrillatory impulses happens to arrive at the A-V node. Thus, the interval between successive ventricular contractions varies from a minimum of about 0.35 second to a maximum of about 0.95 second, causing a very irregular heartbeat. In fact, this irregularity, illustrated by the very variable spacing of the heartbeats in the electrocardiogram of Figure 13–18, is one of the clinical findings used to diagnose the condition. Also, because of the rapid rate of the fibrillatory impulses in the atria, the ventricle is usually driven at a fast heart rate, usually between 125 and 150 beats per minute.

Electroshock Treatment of Atrial Fibrillation. In the same manner that ventricular fibrillation can be converted back to a normal rhythm by electroshock, so also can atrial fibrillation be converted by electrical shock. The procedure is essentially the same as for ventricular conversion, namely passage of a single strong electric shock through the atria, which throws the entire heart into refractoriness for a few seconds; usually a normal rhythm will follow if the heart is capable of this.

ATRIAL FLUTTER

Atrial flutter is another condition caused by a circus movement in the atria. However, it is different from atrial fibrillation in that the electrical signal travels as a single large wave front always in one direction around and around the atrial muscle mass. As illustrated to the left in Figure 13–18, this wave front usually travels from top to bottom around the openings of the superior and inferior vena cava.

Atrial flutter causes a very rapid rate of contraction of the atria, usually between 200 and 350 beats per minute. How-

Figure 13–19. Atrial fibrillation (lead I).

Figure 13–20. Atrial flutter—2:1 and 3:1 rhythm (lead I).

ever, because one side of the atria is contracting while the other side is relaxing, the amount of blood pumped by the atria is very slight. Furthermore, the signals reach the A-V node too rapidly for all of these to be passed into the ventricles because the refractory period of the A-V node and A-V bundle is too long to pass more than a fraction of the atrial signals. Therefore, there are usually two to three beats of the atria for every single beat of the ventricles.

Figure 13–20 illustrates the typical electrocardiogram in atrial flutter. The P waves are strong because of contraction of semicoordinate masses of muscle. However, note in the record that a QRS-T complex follows an atrial P wave only once for every two to three beats of the atria, giving a 2:1 and a 3:1 rhythm.

● CARDIAC ARREST

A final serious abnormality of the cardiac rhythmicity-conduction system is cardiac arrest. This results from cessation of all rhythmic impulses of the heart. That is, no spontaneous rhythm at all remains.

Cardiac arrest is especially likely to occur during deep anesthesia when patients often develop severe hypoxia because of inadequate respiration. The hypoxia prevents the muscle fibers and conductive fibers from maintaining normal electrolyte concentration differentials across their membranes, and their excitability may be so affected that the automatic rhythmicity disappears.

Following temporary cardiac arrest, cardiopulmonary resuscitation is usually quite successful in re-establishing a normal heart rhythm. However, in some patients, severe myocardial disease leads to permanent or semipermanent cardiac arrest, which obviously can cause immediate death. In many cases, however, rhythmic electrical impulses from an implanted electronic cardiac pacemaker have been used successfully to keep patients alive for many years.

REFERENCES

Brown, H. F.: Electrophysiology of the sinoatrial node. Physiol. Rev., 62:505, 1982.

Chung, E. K.: Electrocardiography: Self-Assessment. East Norwalk, Conn., Appleton & Lange, 1988.

Chung, E. K.: Principles of Cardiac Arrhythmias. 4th Ed. Baltimore, Williams & Wilkins, 1988.

Chung, E. K.: Electrocardiography: Practical Applications With Vectorial Principles. 2nd Ed. Hagerstown, Md., Harper & Row, 1980.

Gilmour, R. F., Jr., and Zipes, D. P.: Slow inward current and cardiac arrhythmias. Am. J. Cardiol., 55:89B, 1985.

Goldman, M. J.: Principles of Clinical Electrocardiography. 12th Ed. East Norwalk, Conn., Appleton & Lange, 1986.

Goldschlager, N., and Goldman, M. J.: Principles of Clinical Electrocardiography. 13th Ed. East Norwalk, Conn., Appleton & Lange, 1986.

Guyton, A. C., and Crowell, J. W.: A stereovectorcardiograph. J. Lab. Clin. Med., 40:726, 1952.

Hurst, J. W., et al. (eds.): The Heart. 7th Ed. New York, McGraw-Hill Book Co., 1990.

Johnson, R., and Swartz, M. H.: A Simplified Approach to Electrocardiography. Philadelphia, W. B. Saunders Co., 1986.

Josephson, M. E.: Clinical Cardiac Electrophysiology: Techniques and Interpretations. 2nd Ed. Philadelphia, Lea & Febiger, 1989.

Kennedy, H. L.: Ambulatory Electrocardiography and Its Technology. 2nd Ed. Philadelphia, Lea & Febiger, 1989.

Laiken, N., et al.: Interpretation of Electrocardiograms: A Self-Instructional Approach. 2nd Ed. New York, Raven Press, 1988.

Lipman, B.: Clinical Electrocardiography. Chicago, Year Book Medical Publishers, 1984.

Marriott, H. J. L., and Conover, M. B.: Advanced Concepts in Arrhythmias. 2nd Ed. St. Louis, C. V. Mosby Co., 1989.

Marriott, H. J. L., and Wagner, G.: Practical Electrocardiography. 8th Ed. Baltimore, Williams & Wilkins, 1988.

McAnulty, J., and Rahimtoola, S.: Prognosis in bundle branch block. Annu. Rev. Med., 32:499, 1981.

Meijler, F. L.: Atrioventricular conduction versus heart size from mouse to whale. J. Am. Coll. Cardiol., 5:363, 1985.

Morganroth, J.: Ambulatory Holter electrocardiography: Choice of technologies and clinical uses. Ann. Intern. Med., 102:73, 1985.

Pick, A., and Langendorf, R.: Interpretation of Complex Arrhythmias. Philadelphia, Lea & Febiger, 1980.

Saksena, S., and Goldschlager, N. F. (ed.): Electrical Therapy for Cardiac Arrhythmias. Philadelphia, W. B. Saunders Co., 1989.

Scher, A. M., and Spach, M. S.: Cardiac depolarization and repolarization and the electrocardiogram. In Berne, R. M., et al. (eds.): Handbook of Physiology. Sec. 2, Vol. I. Baltimore, Williams & Wilkins, 1979, p. 357.

Stein, E.: Interpretation of Arrhythmias: A Self-Study Program. Philadelphia, Lea & Febiger, 1988.

Summerall, C. P.: Monitoring Heart Rhythm. New York, John Wiley & Sons, 1982.

Tamargo, J., et al.: Modulated receptor hypothesis: Selectivity and interactions of antiarrhythmic drugs. News Physiol. Sci., 4:88, 1989.

UNIT IV

THE CIRCULATION

CHAPTER 14

• *Overview of the Circulation, and Medical Physics of Pressure, Flow, and Resistance* •

The function of the circulation is to service the needs of the tissues — to transport nutrients to the tissues, to transport waste products away, to conduct hormones from one part of the body to another, and in general to maintain an appropriate environment in all the tissue fluids for optimal survival and function of the cells.

Yet, it is sometimes difficult to understand how blood flow is controlled in relation to the tissue's needs and to understand how the heart and circulation are controlled to provide the necessary cardiac output and arterial pressure to propel the flowing blood. Also, what are the mechanisms for controlling blood volume, and how does this relate to all the other functions of the circulation? These are some of the questions that we propose to answer in the following chapters on the circulation.

● PHYSICAL CHARACTERISTICS OF THE CIRCULATION

The circulation, illustrated in Figure 14–1, is divided into the *systemic circulation* and the *pulmonary circulation.* Because the systemic circulation supplies all the tissues of the body except the lungs with blood flow, it is also frequently called the *greater circulation* or *peripheral circulation.*

Although the vascular system in each separate tissue of the body has its own special characteristics, some general principles of vascular function nevertheless apply in all parts of the system. It is the purpose of the present chapter to discuss these general principles.

The Functional Parts of the Circulation. Before attempting to discuss the details of function in the circulation, it is important to understand the overall role of each of its parts, as follows:

The function of the *arteries* is to transport blood *under high pressure* to the tissues. For this reason, the arteries have strong vascular walls, and blood flows rapidly in the arteries.

The *arterioles* are the last small branches of the arterial system, and they act as *control valves* through which blood is released into the capillaries. The arteriole has a strong muscular wall that is capable of closing the arteriole completely or of allowing it to be dilated severalfold, thus having the capability of vastly altering blood flow to the capillaries in response to the needs of the tissues.

The function of the *capillaries* is to exchange fluid, nutrients, electrolytes, hormones, and other substances between the blood and the interstitial fluid. For this role, the capillary walls are very thin and are permeable to small molecular substances.

The *venules* collect blood from the capillaries; they gradually coalesce into progressively larger veins.

The *veins* function as conduits for transport of blood from the tissues back to the heart, but, equally important, they serve as a major reservoir of blood. Because the pressure in the venous system is very low, the venous walls are thin. Even so, they are muscular, and this allows them to contract or expand and thereby act as a reservoir for extra blood, either a small or a large amount, depending on the needs of the body.

Volumes of Blood in the Different Parts of the Circulation. By far the greater proportion of the blood in the circulation is contained in the systemic veins. Figure 14–1 shows this, illustrating that approximately 84 per cent of the entire blood volume of the body is in the systemic circulation, with 64 per cent in the veins, 13 per cent in the arteries, and 7 per cent in the systemic arterioles and capillaries. The heart contains 7 per cent of the blood and the pulmonary vessels 9 per cent. Most surprising is the very low blood volume in the capillaries of the systemic circu-

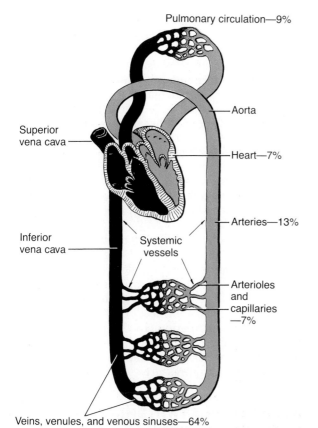

Figure **14 – 1.** Distribution of blood volume in the different portions of the circulatory system.

	cm^2
Aorta	2.5
Small arteries	20
Arterioles	40
Capillaries	2500
Venules	250
Small veins	80
Venae cavae	8

Note particularly the much larger cross-sectional areas of the veins than of the arteries, averaging about four times those of the corresponding arteries. This explains the very large storage of blood in the venous system in comparison with that in the arterial system.

Because the same volume of blood flows through each segment of the circulation each minute, the velocity of blood flow is inversely proportional to its cross-sectional area. Thus, under resting conditions, the velocity averages 33 cm/sec in the aorta but 1/1000 of this in the capillaries, or about 0.3 mm/sec. However, since the capillaries have a typical length of only 0.3 to 1 mm, the blood remains in the capillaries for only 1 to 3 seconds — a very surprising fact, since all diffusion that takes place through the capillary walls must occur in this exceedingly short time.

Pressures in the Various Portions of the Circulation. Because the heart pumps blood continually into the aorta, the pressure in the aorta is obviously high, averaging approximately 100 mm Hg. Also, because the pumping by the heart is pulsatile, the arterial pressure fluctuates between a *systolic level* of 120 mm Hg and a *diastolic level* of 80 mm Hg, as illustrated in Figure 14 – 2. As the blood flows through the systemic circulation, its pressure falls progressively to approximately 0 mm Hg by the time it reaches the termination of the venae cavae in the right atrium.

The pressure in the systemic capillaries varies from as high as 35 mm Hg near the arteriolar ends to as low as 10 mm Hg near their venous ends, but their average "functional" pressure in most vascular beds is about 17 mm Hg, a pressure low enough that very little of

lation. Yet, it is here that the most important function of the systemic circulation occurs, diffusion of substances back and forth between the blood and the tissues. This function is so important that it is discussed in detail in Chapter 16.

Cross-Sectional Areas and Velocities of Blood Flow. If all the systemic vessels of each type were put side by side, their total cross-sectional areas would be:

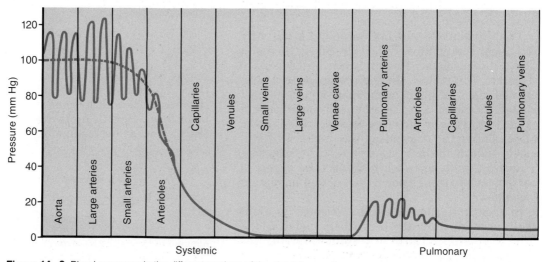

Figure **14 – 2.** Blood pressures in the different portions of the circulatory system.

the plasma leaks out of the porous capillaries even though nutrients can diffuse easily to the tissue cells.

Note to the far right in Figure 14–2 the respective pressures in the different parts of the pulmonary circulation. In the pulmonary arteries, the pressure is pulsatile, just as in the aorta, but the pressure level is far less, at a systolic pressure of about 25 mm Hg and a diastolic pressure of 8 mm Hg, with a mean pulmonary arterial pressure of only 16 mm Hg. The pulmonary capillary pressure averages only 7 mm Hg. Yet, the total blood flow through the lungs each minute is the same as that through the systemic circulation. The low pressures of the pulmonary system are in accord with the needs of the lungs, for all that is required is to expose the blood in the pulmonary capillaries to the oxygen and other gases in the pulmonary alveoli, and the distances that the blood must flow before returning to the heart are short.

● THE BASIC THEORY OF CIRCULATORY FUNCTION

Although the details of circulatory function are often complex, there are three basic principles that underlie all functions of the system. These are as follows:

The Blood Flow to Each Tissue of the Body Is Almost Always Precisely Controlled in Relation to the Tissue Needs. When tissues are active, they need much more blood flow than when at rest, occasionally as much as 20 to 30 times the resting level. Yet, the heart normally cannot increase its cardiac output more than four to seven times. Therefore, it is not possible simply to increase the blood flow everywhere in the body when a particular tissue demands increased flow. Instead, the microvessels of each tissue continuously monitor the tissue needs, such as the availability of nutrition and the accumulation of tissue waste products, and these in turn control the local blood flow very precisely to the level required for the tissue activity. Also, nervous control of the circulation provides additional specific attributes to tissue blood flow control.

The Cardiac Output Is Controlled Mainly by the Local Tissue Flow. When blood flows through a tissue, it immediately returns by way of the veins to the heart. Fortunately, the heart responds to this increased inflow of blood by pumping almost all of it immediately back into the arteries from whence it came. In this sense, the heart acts as an automaton, responding to the demands of the tissues. Unfortunately, though, the heart is not perfect in its response. Therefore, it often needs help in the form of special nerve signals to make it pump the required amounts of blood flow.

In General, the Arterial Pressure Is Controlled Independently of Either Local Blood Flow Control or Cardiac Output Control. The circulatory system is provided with a very extensive system for controlling the arterial pressure. For instance, if at any time the pressure falls significantly below its normal mean level of about 100 mm Hg, a barrage of nervous reflexes within seconds elicits a series of circulatory changes to raise the pressure back to normal, including increased force of heart pumping, contraction of the large venous reservoirs to provide more blood for the heart, and generalized constriction of most of the arterioles throughout the body, so that more blood will accumulate in the arterial tree. Then, over more prolonged periods of time, hours and days, the kidneys play an additional major role in pressure control both by secreting pressure-controlling hormones and by regulating the blood volume.

The importance of pressure control is that it prevents changes in blood flow in one area of the body from significantly affecting flow elsewhere in the body because the common pressure head for both areas is not allowed to change greatly.

Thus, in summary the needs of the local tissues are served by the circulation. In the remainder of this chapter, we begin to discuss the basic details of the management of blood flow and the control of cardiac output and arterial pressure.

● INTERRELATIONSHIPS AMONG PRESSURE, FLOW, AND RESISTANCE

Flow through a blood vessel is determined entirely by two factors: (1) the *pressure difference* between the two ends of the vessel, which is the force that pushes the blood through the vessel, and (2) the impediment to blood flow through the vessel, which is called vascular *resistance*. Figure 14–3 illustrates these relationships, showing a blood vessel segment located anywhere in the circulatory system.

P_1 represents the pressure at the origin of the vessel; at the other end the pressure is P_2. The flow through the vessel can be calculated by the following formula, which is called *Ohm's law:*

$$Q = \frac{\Delta P}{R} \qquad (1)$$

in which Q is blood flow, ΔP is the pressure difference $(P_1 - P_2)$ between the two ends of the vessel, and R is the resistance. This formula states, in effect, that the blood flow is directly proportional to the pressure difference but inversely proportional to the resistance.

Figure 14–3. Relationships among pressure, resistance, and blood flow.

It should be noted especially that it is the *difference* in pressure between the two ends of the vessel, not the absolute pressure in the vessel, that determines the rate of flow. For instance, if the pressure at both ends of the segment were 100 mm Hg and yet no difference existed between the two ends, there would be no flow despite the presence of 100 mm Hg pressure.

Ohm's law expresses the most important of all the relationships that the reader needs to understand to comprehend the hemodynamics of the circulation. Because of the extreme importance of this formula, the reader should also become familiar with its other two algebraic forms:

$$\Delta P = Q \times R \qquad (2)$$

$$R = \frac{\Delta P}{Q} \qquad (3)$$

BLOOD FLOW

Blood flow means simply the quantity of blood that passes a given point in the circulation in a given period of time. Ordinarily, blood flow is expressed in *milliliters* or *liters per minute,* but it can be expressed in milliliters per second or in any other unit of flow.

The overall blood flow in the circulation of an adult person at rest is about 5000 ml per minute. This is called the *cardiac output* because it is the amount of blood pumped by the heart in a unit period of time.

Methods for Measuring Blood Flow. Many different mechanical or mechanoelectrical devices can be inserted in series with a blood vessel or in some instances applied to the outside of the vessel to measure flow. These are called simply *flowmeters.*

The Electromagnetic Flowmeter. One of the most important devices for measuring blood flow *without opening the vessel* is the electromagnetic flowmeter, the principles of which are illustrated in Figure 14–4. Figure 14–4A shows generation of electromotive force in a wire that is moved rapidly through a magnetic field. This is the well-known principle for production of electricity by the electric generator. Figure 14–4B shows that exactly the same principle applies for generation of electromotive force in blood when it moves through a magnetic field. In this case, a blood vessel is placed between the poles of a strong magnet, and electrodes are placed on the two sides of the vessel perpendicular to the magnetic lines of force. When blood flows through the vessel, electrical voltage proportional to the rate of flow is generated between the two electrodes, and this is recorded using an appropriate meter or electronic apparatus. Figure 14–4C illustrates an actual "probe" that is placed on a large blood vessel to record its blood flow. This probe contains both the strong magnet and the electrodes.

A special advantage of the electromagnetic flowmeter is that it can record changes in flow that occur in less than 0.01 second, allowing accurate recording of pulsatile changes in flow as well as steady flow.

The Ultrasonic Doppler Flowmeter. Another type of flowmeter that can be applied to the outside of the vessel and that has many of the same advantages as the electromagnetic flowmeter is the ultrasonic Doppler flowmeter, illustrated in Figure 14–5. A minute piezoelectric crystal is mounted in the wall of the device. This crystal, when energized with an appropriate electronic apparatus, transmits sound at a frequency of several million cycles per second downstream along the flowing blood. A portion of the sound is reflected by the flowing red blood cells, so that reflected sound waves travel backward from the blood toward the crystal. However, these reflected waves have a lower frequency than the transmitted wave because the red cells are moving away from the transmitter crystal. This is called the Doppler effect. (It is the same effect that one experiences when a train approaches and passes by while blowing the whistle. Once the whistle has passed by the person, the pitch of the sound from the whistle suddenly becomes much lower than when the train is approaching.) The transmitted wave

A   B

C

Figure 14–4. A flowmeter of the electromagnetic type, showing *A,* generation of an electromotive force in a wire as it passes through an electromagnetic field; *B,* generation of an electromotive force in electrodes on a blood vessel when the vessel is placed in a strong magnetic field and blood flows through the vessel; and *C,* a modern electromagnetic flowmeter "probe" for chronic implantation around blood vessels.

Figure 14–5. An ultrasonic Doppler flowmeter.

is intermittently cut off, and the reflected wave is received back onto the crystal then amplified greatly by the electronic apparatus. Another portion of the apparatus determines the frequency difference between the transmitted wave and the reflected wave, thus also determining the velocity of blood flow.

Like the electromagnetic flowmeter, the ultrasonic Doppler flowmeter is capable of recording very rapid, pulsatile changes in flow as well as steady flow.

Laminar Flow of Blood in Vessels

When blood flows at a steady rate through a long, smooth vessel, it flows in *streamlines,* with each layer of blood remaining the same distance from the wall. Also, the central portion of the blood stays in the center of the vessel. This type of flow is called *laminar flow* or *streamline flow,* and it is opposite to *turbulent flow,* which is blood flowing in all directions in the vessel and continually mixing within the vessel, as is discussed subsequently.

Parabolic Velocity Profile During Laminar Flow. When laminar flow occurs, the velocity of flow in the center of the vessel is far greater than that toward the outer edges. This is illustrated by the experiment shown in Figure 14–6. In vessel A are two different fluids, the one to the left colored by a dye and the one to the right a clear fluid, but there is no flow in the vessel. Then the fluids are made to flow; a parabolic interface develops between the two fluids, as shown 1 second later in vessel B, illustrating that the portion of fluid adjacent to the vessel wall has hardly moved at all, the portion slightly away from the wall has moved a small distance, and the portion in the center of the vessel has moved a long distance. This effect is called the parabolic profile for the velocity of blood flow.

The cause of the parabolic profile is the following: The fluid molecules touching the wall hardly move because of adherence to the vessel wall. The next layer of molecules slips over these, the third layer over the second, the fourth layer over the third, and so forth. Therefore, the fluid in the middle of the vessel can move rapidly because many layers of slipping molecules exist between the middle of the vessel and the vessel wall, all of these capable of slipping over each other, whereas those portions of fluid near the wall do not have this advantage.

Turbulent Flow of Blood Under Some Conditions. When the rate of blood flow becomes too great, when it

Figure 14–6. An experiment illustrating laminar blood flow. *A,* Two separate fluids before flow begins; *B,* the same fluids 1 second after flow begins.

passes an obstruction in a vessel, when it makes a sharp turn, or when it passes over a rough surface, the flow may then become *turbulent* rather than streamline. Turbulent flow means that the blood flows crosswise in the vessel as well as along the vessel, usually forming whorls in the blood called *eddy currents.* These are similar to the whirlpools that one frequently sees in a rapidly flowing river at a point of obstruction.

When eddy currents are present, blood flows with much greater resistance than when the flow is streamline because the eddies add tremendously to the overall friction of flow in the vessel.

The tendency for turbulent flow increases in direct proportion to the velocity of blood flow, in direct proportion to the diameter of the blood vessel, and inversely proportional to the viscosity of the blood divided by its density in accordance with the following equation:

$$\text{Re} = \frac{v \cdot d}{\dfrac{\eta}{\rho}} \qquad (4)$$

in which Re is *Reynolds' number* and is the measure of the tendency for turbulence to occur, v is the velocity of blood flow (in centimeters/second), η is the viscosity (in poises), and ρ is the density. When Reynolds' number rises above 200 to 400, turbulent flow will occur at some branches of vessels but will die out along the smooth portions of the vessels. However, when Reynolds' number rises above approximately 2000, turbulence will usually occur even in a straight, smooth vessel. Reynolds' number even normally rises to 200 to 2000 in large arteries; as a result there is almost always some turbulence of flow in the root of the aorta and at the major arterial branches.

BLOOD PRESSURE

The Standard Units of Pressure. Blood pressure is almost always measured in *millimeters of mer-*

Figure 14–7. Recording arterial pressure with a mercury manometer, a method that has been used in the manner shown above for recording pressure throughout the history of physiology.

cury *(mm Hg)* because the mercury manometer (shown in Figure 14–7) has been used since antiquity as the standard reference for measuring blood pressure. Actually, blood pressure means the *force exerted by the blood against any unit area of the vessel wall.* When one says that the pressure in a vessel is 50 mm Hg, this means that the force exerted is sufficient to push a column of mercury up to a level 50 mm high. If the pressure is 100 mm Hg, it will push the column of mercury up to 100 millimeters.

Occasionally, pressure is measured in *centimeters of water.* A pressure of 10 centimeters of water means a pressure sufficient to raise a column of water to a height of 10 centimeters. *One millimeter of mercury equals 1.36 centimeter of water* because the specific gravity of mercury is 13.6 times that of water, and 1 centimeter is 10 times as great as 1 millimeter. Dividing 13.6 by 10, we derive the factor 1.36.

High-Fidelity Methods for Measuring Blood Pressure. Unfortunately, the mercury in the mercury manometer has so much *inertia* that it cannot rise and fall rapidly. For this reason the mercury manometer, though excellent for recording steady pressures, cannot respond to pressure changes that occur more rapidly than approximately one cycle every 2 to 3 seconds. Whenever it is desired to record rapidly changing pressures, some other type of pressure recorder is needed. Figure 14–8 demonstrates the basic principles of three electronic pressure *transducers* commonly used for converting pressure into electrical signals and then recording the pressure on a high-speed electrical recorder. Each of these transducers employs a very thin and highly stretched metal membrane that forms one wall of the fluid chamber. The fluid chamber in turn is connected through a

needle or catheter with the vessel in which the pressure is to be measured. Pressure variations in the vessel cause changes of pressure in the chamber beneath the membrane. When the pressure is high, the membrane bulges outward slightly, and when low, it returns toward its resting position.

In Figure 14–8A, a simple metal plate is placed a few thousandths of an inch above the membrane. When the membrane bulges outward, the *capacitance* between the plate and membrane increases, and this change in capacitance can be recorded by an appropriate electronic system.

In Figure 14–8B, a small iron slug rests on the membrane, and this can be displaced upward into a coil. Movement of the iron changes the *inductance* of the coil, and this, too, can be recorded electronically.

Finally, in Figure 14–8C, a very thin, stretched resistance wire is connected to the membrane. When this wire is greatly stretched, its resistance increases, and when less stretched, the resistance decreases. These changes also can be recorded by means of an electronic system.

With some of these high-fidelity types of recording systems, pressure cycles up to 500 cycles per second have been recorded accurately. In common use are recorders capable of registering pressure changes occurring as rapidly as 20 to 100 cycles per second.

RESISTANCE TO BLOOD FLOW

Units of Resistance. Resistance is the impediment to blood flow in a vessel, but it cannot be measured by any direct means. Instead, resistance must be calculated from measurements of blood flow and pressure difference in the vessel. If the pressure difference between two points in a vessel is 1 mm Hg and the flow is 1 ml/sec, the resistance is said to be 1 *peripheral resistance unit,* usually abbreviated *PRU.*

Expression of Resistance in CGS Units. Occasionally, a basic physical unit called the CGS (centimeters, grams, seconds) unit is used to express resistance. This unit is *dyne seconds/centimeters⁵*. Resistance in these units can be calculated by the following formula:

$$R\left(in\ \frac{dyne\ sec}{cm^5}\right) = \frac{1333 \times mm\ Hg}{ml/sec} \qquad (5)$$

Total Peripheral Resistance and Total Pulmonary Resistance. The rate of blood flow through the circulatory system when a person is at rest is close to 100 ml/sec, and the pressure difference from the systemic arteries to the systemic veins is about 100 mm Hg. Therefore, in round figures the resistance of the entire systemic circulation, called the *total peripheral resistance,* is approximately 100/100 or 1 PRU. In some conditions in which all the blood vessels throughout the body become strongly constricted, the total peripheral resistance rises to as high as 4 PRU, and when the vessels become greatly dilated, it can fall to as little as 0.2 PRU.

In the pulmonary system the mean arterial pressure averages 16 mm Hg, and the mean left atrial pressure averages 2 mm Hg, giving a net pressure difference of 14 mm. Therefore, in round figures the *total pulmonary resistance* at rest calculates to be about 0.14 PRU.

Figure 14–8. Principles of three different types of electronic transducers for recording rapidly changing blood pressures.

"Conductance" of Blood in a Vessel and Its Relationship to Resistance. Conductance is a measure of the blood flow through a vessel for a given pressure difference. This is generally expressed in terms of ml/sec/mm Hg pressure, but it can also be expressed in terms of liters/sec/mm Hg or in any other units of blood flow and pressure.

It is immediately evident that conductance is the reciprocal of resistance in accord with the following equation:

$$\text{Conductance} = \frac{1}{\text{Resistance}} \qquad (6)$$

Effect of Vascular Diameter on Conductance. Slight changes in the diameter of a vessel cause tremendous changes in its ability to conduct blood when the blood flow is streamline. This is illustrated forcefully by the experiment in Figure 14–9A, which shows three separate vessels with relative diameters of 1, 2, and 4 but with the same pressure difference of 100 mm Hg between the two ends of the vessels. Although the diameters of these vessels increase only 4-fold, the respective flows are 1, 16, and 256 ml/mm, which is a 256-fold increase in flow. Thus, the conductance of the vessel increases in proportion to the *fourth power of the diameter,* in accord with the following formula:

$$\text{Conductance} \propto \text{Diameter}^4 \qquad (7)$$

Poiseuille's Law. The cause of this great increase in conductance with an increase in diameter can be explained by referring to Figure 14–9B. This illustrates cross-sections of a large and a small vessel. The concentric rings inside these vessels indicate that the velocity of flow in each ring is different from that in the other rings because of *laminar* flow, which is discussed earlier in the chapter. That is, the blood in the ring touching the wall of the vessel is flowing hardly at all because of its adherence to the vascular endothelium. The next ring of blood slips past the first ring and, therefore, flows at a more rapid velocity. The third, fourth, fifth, and sixth rings likewise flow at progressively increasing velocities. Thus, the blood that is very near the wall of the vessel flows extremely slowly, whereas that in the middle of the vessel flows extremely rapidly.

In the small vessel, essentially all the blood is very near the wall, so that the extremely rapidly flowing central stream of blood simply does not exist.

By integrating the velocities of all the concentric rings of flowing blood and multiplying these by the areas of the rings, one can derive the following formula known as Poiseuille's law:

$$Q = \frac{\pi \Delta P r^4}{8 \eta l} \qquad (8)$$

in which Q is the rate of blood flow, ΔP is the pressure difference between the ends of the vessel, r is the radius of the vessel, l is the length of the vessel, and η is the viscosity of the blood.

Note particularly in this equation that the rate of blood flow is directly proportional to the *fourth power of the radius* of the vessel, which illustrates once again that the diameter of a blood vessel plays by far the greatest role of all factors in determining the rate of blood flow through the vessel.

Importance of the Vessel Diameter "Fourth Power Law" in Determining Arteriolar Resistance. In the systemic circulation, about two thirds of the resistance is in the small arterioles. The internal diameters of these range from as little as 8 micrometers up to as great as 30 micrometers. However, their strong vascular walls allow the internal diameters to change tremendously, often as much as fourfold. From the fourth power law discussed earlier that relates blood flow to diameter of the vessel, one can see that a 4-fold increase in vessel diameter theoretically could increase the flow as much as 256-fold. Thus, this fourth power law makes it possible for the arterioles, responding with only small changes in diameter to nervous signals or local tissue signals, either to turn off almost completely the blood flow to the tissue or at other times to cause a vast increase in flow. Indeed, ranges of blood flow of more than a hundredfold have actually been recorded between the limits of maximum arteriolar constriction and maximum arteriolar dilatation.

Effect of Blood Hematocrit and Viscosity on Vascular Resistance and Blood Flow

Note especially that one of the important factors in Poiseuille's law is the viscosity of the blood. The greater the viscosity, the less the flow in a vessel if all other factors are constant. Furthermore, the viscosity of normal blood is about three times as great as the viscosity of water.

But what is it that makes the blood so viscous? It is mainly the large numbers of suspended red cells in the blood, each of which exerts frictional drag against adjacent cells and against the wall of the blood vessel. The per cent of cells in the blood is called the *hematocrit.*

The Hematocrit. The per cent of the blood that is cells is called the hematocrit. Thus, if a person has a

Figure 14–9. *A,* Demonstration of the effect of vessel diameter on blood flow. *B,* Concentric rings of blood flowing at different velocities; the farther away from the vessel wall, the faster the flow.

hematocrit of 40, 40 per cent of the blood volume is cells, and the remainder is plasma. The hematocrit of normal men averages about 42, whereas that of normal women averages about 38. These values vary tremendously, depending on whether or not the person has anemia, the degree of bodily activity, and the altitude at which the person resides. These effects are discussed in relation to the red blood cells and their function in Chapter 32.

Blood hematocrit is determined by centrifuging blood in a calibrated tube, such as that shown in Figure 14–10. The calibration allows direct reading of the per cent of cells.

Effect of Hematocrit on Blood Viscosity. The greater the percentage of cells in the blood—that is, the greater the hematocrit—the more friction there is between successive layers of blood, and this friction determines viscosity. Therefore, the viscosity of blood increases drastically as the hematocrit increases, as illustrated in Figure 14–11. If we consider the viscosity of whole blood at normal hematocrit to be about 3, this means that three times as much pressure is required to force whole blood as to force water through the same tube. Note that when the hematocrit rises to 60 or 70, which it often does in polycythemia, the blood viscosity can become as great as 10 times that of water, and its flow through blood vessels is greatly retarded.

Another factor that affects blood viscosity is the concentration and types of proteins in the plasma, but these effects are so much less important than the effect of hematocrit that they are not significant considerations in most hemodynamic studies. The viscosity of blood plasma is about 1.5 times that of water.

Blood Viscosity in the Microcirculation. Because most resistance in the circulatory system occurs in the very small blood vessels, it is especially important to know how

Figure 14–11. Effect of hematocrit on viscosity.

blood viscosity affects blood flow in these minute vessels. At least three additional factors besides hematocrit and plasma proteins affect blood viscosity in these vessels:

1. Blood flow in very minute tubes exhibits far less viscous effect than it does in large vessels. This is called the *Fahraeus-Lindqvist effect.* It begins to appear when the vessel diameter falls below approximately 1.5 mm, and in tubes as small as capillaries the viscosity of whole blood is as little as one half that in large vessels. The Fahraeus-Lindqvist effect is caused by alignment of the red cells as they pass through the vessels. That is, the red cells, instead of moving randomly, line up and move through the vessels as a single plug, thus eliminating the viscous resistance that occurs internally in the blood itself. The Fahraeus-Lindqvist effect, however, is probably more than offset by the following two effects under most conditions.

2. The viscosity of blood increases tremendously as its velocity of flow decreases. Since the velocity of blood flow in the small vessels is extremely minute, often less than 1 mm per second, blood viscosity can increase as much as tenfold from this factor alone. This effect is partly caused by adherence of the red cells to each other (formation of rouleaux and larger aggregates) and to the vessel walls.

3. Cells also often become stuck at constrictions in small blood vessels; this happens especially in capillaries where the nuclei of endothelial cells protrude into the capillary lumen. When this occurs, blood flow can become totally blocked for a fraction of a second, for several seconds, or for much longer periods of time, thus giving an apparent effect of greatly increased viscosity.

Because of these special effects that occur in the minute vessels of the circulatory system, it has been impossible to derive an exact mathematical relationship that describes how the hematocrit affects viscosity in the minute vessels, which is the place in the circulatory system where viscosity plays its most important role. Nevertheless, because some of these effects tend to decrease viscosity and others tend to increase viscosity, it is reasonable to assume that the overall viscous effects in the small vessels are approximately equivalent to those that occur in the larger vessels.

EFFECTS OF PRESSURE ON VASCULAR RESISTANCE AND TISSUE BLOOD FLOW

From the discussions thus far, one would expect that an increase in arterial pressure would cause a

Figure 14–10. Hematocrits in the normal person and in patients with anemia and polycythemia.

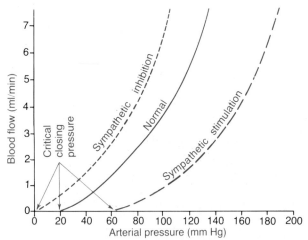

Figure 14–12. Effect of arterial pressure on blood flow through a blood vessel at different degrees of vascular tone caused by increased or decreased sympathetic stimulation of the vessels.

proportionate increase in blood flow through the various tissues of the body. However, the effect of pressure on blood flow is far greater than one would expect, as illustrated in Figure 14–12. The reason for this is that an increase in arterial pressure not only increases the force that tends to push blood through the vessels but also distends the vessels at the same time, which decreases their resistance. Thus, increased pressure increases the flow in two different ways, and for most tissues, blood flow at 100 mm Hg arterial pressure is usually about four times as great as the blood flow at 50 mm Hg.

Note also in Figure 14–12 the large changes in blood flow that can be caused by either increased or decreased sympathetic stimulation of the peripheral blood vessels. Thus, inhibition of sympathetic stimulation greatly dilates the vessels and therefore increases the blood flow, sometimes as much as twofold or more. Conversely, very strong sympathetic stimulation can constrict the vessels so much that blood flow can be decreased sometimes to as low as zero for short periods of time despite high arterial pressure.

REFERENCES

See references for Chapter 15.

• Vascular Distensibility and Functions of the Arterial and Venous Systems •

• VASCULAR DISTENSIBILITY

A valuable characteristic of the vascular system is that all blood vessels are distensible. We have seen one example of this in the previous chapter, when the pressure in the arterioles is increased; this dilates the arterioles and therefore decreases their resistance. The result is increased blood flow not only because of increased pressure, but also because of decreased resistance, usually giving at least twice as much flow increase as one might expect.

Vascular distensibility also plays other important roles in circulatory function. For example, the distensible nature of the arteries allows these to accommodate the pulsatile output of the heart and to average out the pressure pulsations. This provides an almost completely smooth, continuous flow of blood through the tissues.

The most distensible by far of all the vessels are the veins. Even slight increases in pressure cause the veins to store as much as a half a liter to a liter of extra blood. Therefore, the veins provide a reservoir function for storing large quantities of blood that can be called into use whenever required elsewhere in the circulation.

Units of Vascular Distensibility. Vascular distensibility is normally expressed as the fractional increase in volume for each millimeter mercury rise in pressure in accordance with the following formula:

$$\text{Vascular distensibility} = \frac{\text{Increase in volume}}{\text{Increase in pressure} \times \text{Original volume}} \quad (1)$$

That is, if 1 mm Hg causes a vessel originally containing 10 ml of blood to increase its volume by 1 ml, the distensibility would be 0.1 per mm Hg or 10 per cent per mm Hg.

Difference in Distensibility of the Arteries and the Veins. Anatomically, the walls of arteries are far stronger than those of veins. Consequently, the veins, on the average, are about 6 to 10 times as distensible as the arteries. That is, a given rise in pressure causes about 6 to 10 times as much extra blood to fill a vein as to fill an artery of comparable size.

In the pulmonary circulation, the veins are very similar to those of the systemic circulation. However, the pulmonary arteries, which normally operate under pressures about one sixth those in the systemic arterial system, have distensibilities about one half those of veins, rather than one eighth as is true of the systemic arteries.

VASCULAR COMPLIANCE (OR CAPACITANCE)

Usually in hemodynamic studies it is much more important to know the *total quantity of blood* that can be stored in a given portion of the circulation for each mm Hg pressure rise than to know the distensibility of the individual vessels. This value is called the *compliance* or *capacitance* of the respective vascular bed. That is,

$$\text{Vascular compliance} = \frac{\text{Increase in volume}}{\text{Increase in pressure}} \quad (2)$$

Compliance and distensibility are quite different. A highly distensible vessel that has a very slight volume may have far less compliance than a much less distensible vessel that has a very large volume, for *compliance is equal to distensibility times volume.*

The compliance of a vein is about 24 times that of its corresponding artery because it is about 8 times as distensible, and it has a volume about 3 times as great $(8 \times 3 = 24)$.

VOLUME-PRESSURE CURVES OF THE ARTERIAL AND VENOUS CIRCULATIONS

A convenient method for expressing the relationship of pressure to volume in a vessel or in a large portion of the circulation is the so-called *volume-pressure curve* (also frequently called the *pressure-volume curve*). The two solid curves of Figure 15–1 represent respectively the volume-pressure curves of the normal arterial and venous systems, showing that when the arterial system, including the larger arteries, small arteries, and arterioles, is filled with approximately 750 ml of blood, the mean arterial pressure is 100 mm Hg, but when filled with only 500 milliliters, the pressure falls to zero.

In the entire venous system, on the other hand, the volume of blood normally is about 2500 milliliters, and tremendous changes in this volume are required to change the venous pressure only a few millimeters of mercury.

Effect of Sympathetic Stimulation or Sympathetic Inhibition on the Volume-Pressure Relationships of the Arterial and Venous Systems. Also shown in Figure 15–1 are the effects of sympathetic stimulation and sympathetic inhibition on the volume-pressure curves. It is evident that the increase in vascular smooth muscle tone caused by sympathetic stimulation increases the pressure at each volume of the arteries or veins, whereas sympathetic inhibition decreases the pressure at each volume. Obviously, control of the vessels in this manner by the sympathetics is a valuable means for diminishing the dimensions of one segment of the circulation, thus transferring blood to other segments. For instance, an increase in vascular tone throughout the systemic circulation often causes large volumes of blood to shift into the heart, which is a major way in which pumping by the heart is increased.

Sympathetic control of vascular capacity is also especially important during hemorrhage. Enhancement of the sympathetic tone of the vessels, especially of the veins, reduces the dimensions of the circulatory system, and the circulation continues to operate almost normally even when as much as 25 per cent of the total blood volume has been lost.

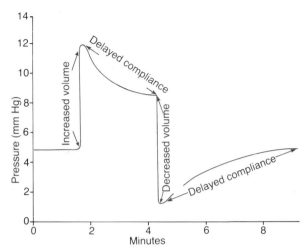

Figure 15–2. Effect on the intravascular pressure of injecting a small volume of blood into a venous segment, illustrating the principles of delayed compliance.

DELAYED COMPLIANCE (STRESS-RELAXATION) OF VESSELS

The term "delayed compliance" means that a vessel exposed to increased volume will at first exhibit a large increase in pressure, but delayed stretch of the vessel wall will allow the pressure to return back toward normal. This effect is illustrated in Figure 15–2.

In this figure, the pressure is being recorded in a small segment of a vein that is occluded at both ends. Then, an extra volume of blood is suddenly injected until the pressure rises from 5 to 12 mm Hg. Even though none of the blood is removed after it is injected, the pressure nevertheless begins to fall immediately and approaches approximately 9 mm Hg after several minutes. In other words, the volume of blood injected caused immediate *elastic* distension of the vein, but then the smooth muscle fibers of the vein began to "creep" to longer lengths, and their tensions correspondingly decreased. This effect is a characteristic of all smooth muscle tissue and is called *stress-relaxation,* which is explained in Chapter 8.

After the delayed increase in compliance had taken place in the experiment illustrated in Figure 15–2, the extra blood volume was suddenly removed, and the pressure immediately fell to a very low value. Subsequently, the smooth muscle fibers began to readjust their tensions back to their initial values, and after a number of minutes the normal vascular pressure of 5 mm Hg returned.

Delayed compliance is a valuable mechanism by which the circulation can accommodate much extra blood when necessary, such as following too large a transfusion. Also, delayed compliance in the reverse direction is one of the ways in which the circulation automatically adjusts itself over a period of minutes or hours to diminished blood volume after serious hemorrhage.

● THE ARTERIAL PRESSURE PULSATIONS

With each beat of the heart a new surge of blood fills the arteries. Were it not for the distensibility of the

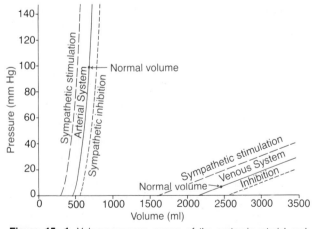

Figure 15–1. Volume-pressure curves of the systemic arterial and venous systems, showing also the effects of sympathetic stimulation and sympathetic inhibition.

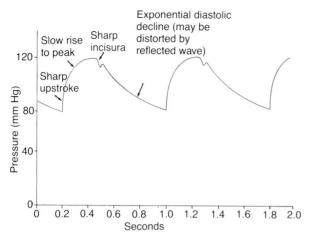

Figure 15-3. A normal pressure pulse contour recorded from the ascending *aorta*. (From Opdyke: *Fed. Proc., 11*:734, 1952.)

arterial system, blood flow through the tissues would occur only during cardiac systole, and no blood flow at all would occur during diastole. Fortunately, the combination of distensibility of the arteries and their resistance reduces the pressure pulsations almost to zero by the time the blood reaches the capillaries; therefore, tissue blood flow is hardly affected by the pulsatile nature of heart pumping.

A typical record of the *pressure pulsations* at the root of the aorta is illustrated in Figure 15-3. In the normal young adult, the pressure at the height of each pulse, the *systolic pressure,* is about 120 mm Hg and at its lowest point, the *diastolic pressure,* is about 80 mm Hg. The difference between these two pressures, about 40 mm Hg, is called the *pulse pressure.*

Two major factors affect the pulse pressure: (1) the *stroke volume output* of the heart and (2) the *compliance (total distensibility)* of the arterial tree. A third, less important factor is the character of ejection from the heart during systole.

In general, the greater the stroke volume output, the greater is the amount of blood that must be ac-

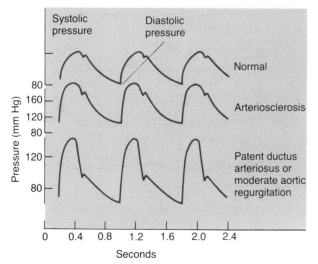

Figure 15-4. Pressure pulse contours in arteriosclerosis, patent ductus arteriosus, and moderate aortic regurgitation.

commodated in the arterial tree with each heartbeat, and, therefore, the greater is the pressure rise and fall during systole and diastole, thus causing a greater pulse pressure.

On the other hand, the less the compliance of the arterial system, the greater will be the rise in pressure for a given stroke volume of blood pumped into the arteries. For instance, as illustrated by the middle curve in Figure 15-4, the pulse pressure sometimes rises to as much as two times normal in old age because the arteries become hardened with *arteriosclerosis* and therefore noncompliant.

In effect, then, the pulse pressure is determined approximately by the *ratio of stroke volume output to compliance of the arterial tree.* Therefore, any condition of the circulation that affects either of these two factors will also affect the pulse pressure.

Abnormal Pressure Pulse Contours

Some conditions of the circulation also cause abnormal contours to the pressure pulse wave in addition to altering the pulse pressure. Especially distinctive among these are patent ductus arteriosus and aortic regurgitation.

Patent Ductus Arteriosus. The ductus arteriosus is a special artery that carries blood from the pulmonary artery to the aorta during fetal life, so that the blood flow can bypass the inactive fetal lungs. Normally, it closes within hours after birth, but in a few persons the ductus arteriosus remains patent (open) indefinitely. Therefore, after birth the blood flows backward from the aorta through the open ductus into the pulmonary artery, allowing very rapid runoff of blood from the arterial tree after each heartbeat and a greatly decreased diastolic pressure. However, this is compensated for by a much greater than normal stroke volume output because the blood flowing through the ductus passes rapidly through the lungs a second time and re-enters the left heart as extra inflowing blood. Therefore, the left heart beats with extra force and volume, and the systolic pressure rises much higher than normal. These effects give the pressure pulse contour shown by the lowest curve in Figure 15-4. Here one finds an elevated systolic pressure, a greatly depressed diastolic pressure, and a greatly increased pulse pressure.

Aortic Regurgitation. Aortic regurgitation causes an abnormal pressure pulse contour similar to that of patent ductus arteriosis but for a different reason. Regurgitation means backward flow of blood through the aortic valve because of failure of the valve to close. This usually results from heart diseases that destroy the valve, especially rheumatic fever. In aortic regurgitation, much of the blood that is pumped into the aorta during systole flows back into the left ventricle during diastole, thus giving a low diastolic pressure. However, this backflow overfills the ventricle before its next beat; therefore, the ventricle pumps a much greater than normal stroke volume output during systole, which causes a high systolic pressure. Thus, the pressure pulse contour is very similar to that of patent ductus arteriosus (the lower curve of Figure 15-4) but is not identical, for in aortic regurgitation, the valve sometimes fails to close at all. When this is true, the incisura (the "notch" in the curve that occurs when the valve closes) is entirely absent.

THE RADIAL PULSE

Clinically, it has been the habit for many years for a physician to feel the radial pulse of each patient. This is done to determine the rate of the heartbeat or, frequently, because of the psychic contact that it gives the physician with the patient. Under certain circumstances, however, the character of the pulse can also be of value in the diagnosis of circulatory diseases.

Weak Pulse. A weak pulse at the radial artery usually indicates either (1) greatly decreased central pulse pressure, such as occurs when the stroke volume output is low, or (2) increased "damping" of the pulse wave caused by vascular spasm; the latter occurs when the sympathetic nervous system becomes overly active following blood loss or when a person is having a chill.

Pulsus Paradoxus. Occasionally, the strength of the pulse becomes strong, then weak, then strong, occurring in synchrony with the phases of respiration. This is caused by alternate increase and decrease in cardiac output with each respiration. During inspiration, all the blood vessels of the lungs increase in size because of increased negative pressure in the thorax. Therefore, blood collects in the lungs, and the stroke volume output and pulse strength decrease. During expiration, opposite effects occur. This is a normal phenomenon in all persons, but it becomes extremely distinct in some conditions, such as in very deep breathing or in cardiac tamponade (compression of the heart from the outside by fluid in the pericardial sac or by a constricted pericardium).

Pulse Deficit. The rhythm of the heart is very irregular in atrial fibrillation or in the case of premature heartbeats. In these arrhythmias, which are discussed in Chapter 13, two beats of the heart often come so close together that the second beat pumps no blood or very little blood because the left ventricle has too little time to fill between the beats. In this circumstance, one can hear the second beat of the heart with a stethoscope applied directly over the heart but cannot feel a pulse in the radial artery, an effect called a *pulse deficit*. The greater the pulse deficit each minute, the more serious, ordinarily, is the arrhythmia.

TRANSMISSION OF PRESSURE PULSES TO THE PERIPHERAL ARTERIES

When the heart ejects blood into the aorta during systole, at first only the proximal portion of the aorta becomes distended because the inertia of the blood prevents sudden movement all the way to the periphery. However, the rising pressure in the central aorta rapidly overcomes this inertia and the wave front of distension spreads further and further along the aorta, as illustrated in Figure 15–5. This is called *transmission of the pressure pulse* in the arteries.

The velocity of pressure pulse transmission in the normal aorta is 3 to 5 meters per second, in the large arterial branches 7 to 10 meters per second, and in the small arteries 15 to 35 meters per second. In general, the greater the compliance of each vascular segment, the slower is the velocity, which explains the slow transmission in the aorta and the much faster transmission in the much less compliant small distal arteries.

It should also be recognized that the velocity of transmission of the pressure pulse is 15 or more times the velocity of blood flow in the aorta, for the pressure

Figure 15–5. Progressive stages in the transmission of the pressure pulse along the aorta.

pulse is simply a moving wave of *pressure* that involves very little forward movement of blood volume.

Damping of the Pressure Pulses in the Smaller Arteries, Arterioles, and Capillaries. Figure 15–6 shows typical changes in the contours of the pressure pulse as the pulse travels into the peripheral vessels. Note especially in the three lower curves that the intensity of pulsation becomes progressively

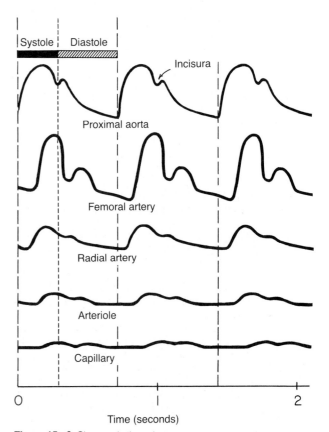

Figure 15–6. Changes in the pulse pressure contour as the pulse wave travels toward the smaller vessels.

less in the smaller arteries, the arterioles, and especially the capillaries. In fact, only when the aortic pulsations are extremely large or when the arterioles are greatly dilated can pulsations be observed at all in the capillaries.

This progressive diminishment of the pulsations in the periphery is called *damping* of the pressure pulses. The cause of this is twofold: (1) the resistance to blood movement in the vessels and (2) the compliance of the vessels. The resistance damps the pulsations because a small amount of blood must flow forward at the pulse wave front to distend the next segment of the vessel; the greater the resistance, the more difficult is it for this to occur. The compliance damps the pulsations because the more compliant a vessel, the more the quantity of blood flow at the pulse wave front must be to cause the rise in pressure. Therefore, in effect, *the degree of damping is almost directly proportional to the product of resistance times compliance.*

Increase in Intensity of Some Peripheral Pulses Caused by Reflected Pulse Waves. Observe carefully the second curve in Figure 15–6, which shows the pulse pressure contour in the femoral artery. Note that the systolic pressure is slightly greater in this peripheral artery than at the origin of the aorta, and the diastolic pressure is distinctly lower than the diastolic pressure at the origin of the aorta. In addition, the contour is considerably altered, showing greatly enhanced secondary waves after the initial peak. All these effects are caused by reflection of the pulse wave from distal points in the arterial tree, such as from the legs where the arteries become narrowed and from intersections where side arteries leave the central aortic tree. These reflections are analogous to the reflections of ocean waves from a steep shoreline. After a wave washes onto the shore, it then reverses and rolls back (is reflected) toward the sea; there it meets still new oncoming waves, and the impact of the opposing smaller waves causes great surging waves with bizarre characteristics — much the same as occurs in the femoral artery.

CLINICAL METHODS FOR MEASURING SYSTOLIC AND DIASTOLIC PRESSURES

Obviously, it is impossible to use the various pressure recorders that require needle insertion into an artery, as described earlier in the chapter, for making routine pressure measurements in human patients, although they are used on occasion when special studies are necessary. Instead, the clinician determines systolic and diastolic pressures by indirect means, most usually by the auscultatory method.

The Auscultatory Method. Figure 15–7 illustrates the auscultatory method for determining systolic and diastolic arterial pressures. A stethoscope is placed over the antecubital artery, while a blood pressure cuff is inflated around the upper arm. As long as the cuff presses against the arm with so little pressure that the artery remains distended with blood, no sounds whatsoever are heard by the stethoscope, despite the fact that the blood within the artery is pulsating. When the cuff pressure is great enough to close the artery during part of the arterial pressure cycle, a sound is then heard with each pulsation. These sounds are called *Korotkoff sounds.*

The exact cause of Korotkoff sounds is still debated, but they are believed to be caused by blood jetting through the

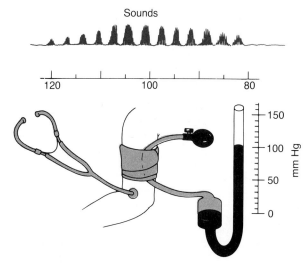

Figure 15–7. The auscultatory method of measuring systolic and diastolic pressures.

partly occluded vessel. The jet causes turbulence in the open vessel beyond the cuff, and this sets up the vibrations heard through the stethoscope.

In determining blood pressure by the auscultatory method, the pressure in the cuff is first elevated well above arterial systolic pressure. As long as this pressure is higher than systolic pressure, the brachial artery remains collapsed and no blood whatsoever flows into the lower artery during any part of the pressure cycle. Therefore, no Korotkoff sounds are heard in the lower artery. But then the cuff pressure is gradually reduced. Just as soon as the pressure in the cuff falls below systolic pressure, blood slips through the artery beneath the cuff during the peak of systolic pressure, and one begins to hear *tapping* sounds in the antecubital artery in synchrony with the heartbeat. As soon as these sounds are heard, the pressure level indicated by the manometer connected to the cuff is approximately equal to the systolic pressure.

As the pressure in the cuff is lowered still more, the Korotkoff sounds change in quality, having less of the tapping quality but more of a rhythmic, harsher quality. Then, finally, when the pressure in the cuff falls to equal diastolic pressure, the artery no longer closes during diastole, which means that the basic factor causing the sounds (the jetting of blood through a squeezed artery) is no longer present. Therefore, the sounds suddenly change to a muffled quality and then usually disappear entirely after another 5 to 10 mm drop in cuff pressure. One notes the manometer pressure when the Korotkoff sounds change to the muffled quality, and this pressure is approximately equal to the diastolic pressure.

The auscultatory method for determining systolic and diastolic pressures is not entirely accurate, but it usually gives values within 10 per cent of those determined by direct measurement from the arteries.

Oscillometric Method for Estimating Arterial Pressure. Arterial blood pressure can also be estimated by recording the pulsation in the lower arm with an *oscillometer,* while a cuff is inflated over the upper arm. An oscillometer consists of a recording apparatus that can register pulsations in a lightly inflated pressure cuff around the forearm. The importance of this method is that many of the automatic blood pressure recording devices employ this principle.

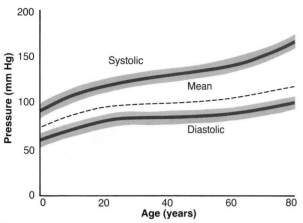

Figure 15–8. Changes in systolic, diastolic, and mean arterial pressures with age. The shaded areas show the normal range.

Normal Arterial Pressures as Measured by the Auscultatory Method. Figure 15–8 illustrates the normal ranges of systolic and diastolic pressures at different ages. The progressive increase in pressure with age results from the effects of aging on the long-term blood pressure control mechanisms. We see in Chapter 19 that the kidneys primarily are responsible for this long-term regulation of arterial pressure; it is well known that the kidneys do indeed exhibit definitive changes with age, especially after the age of 50.

The exceptional rise in *systolic* pressure beyond the age of 60 results from hardening of the arteries, which itself is an end-stage result of *atherosclerosis*. This causes a bounding systolic pressure and also considerable increase in the pulse pressure.

The Mean Arterial Pressure. The mean arterial pressure is the average of all the pressures measured millisecond by millisecond over a period of time. It is not equal to the average of systolic and diastolic pressure because the pressure remains nearer to diastolic pressure than to systolic pressure during the greater part of the cardiac cycle. Therefore, the mean arterial pressure is determined about 60 per cent by the diastolic pressure and 40 per cent by the systolic pressure. Note in Figure 15–8 that the mean pressure at all ages is nearer to the diastolic pressure than to the systolic pressure, especially so at older ages.

● THE VEINS AND THEIR FUNCTIONS

For years the veins have been considered to be nothing more than passageways for flow of blood into the heart, but it is rapidly becoming apparent that they perform many functions that are necessary to the operation of the circulation. Especially important, they are capable of constricting and enlarging, of storing large quantities of blood and making this blood available when it is required by the remainder of the circulation, of actually propelling blood forward by means of a so-called venous pump, and even of helping regulate cardiac output, an exceedingly important function that is described in Chapter 20.

VENOUS PRESSURES — RIGHT ATRIAL PRESSURE (CENTRAL VENOUS PRESSURE) AND PERIPHERAL PRESSURES

To understand the various functions of the veins, it is first necessary to know something about the pressures in the veins and how they are regulated. Blood from all the systemic veins flows into the right atrium; therefore, the pressure in the right atrium is frequently called the *central venous pressure.* Obviously, anything that affects right atrial pressure usually affects venous pressure everywhere in the body.

Right atrial pressure is regulated by a balance between the ability of the heart to pump blood out of the right atrium and, second, *the tendency for blood to flow from the peripheral vessels back into the right atrium.*

If the heart is pumping strongly, the right atrial pressure tends to decrease. On the other hand, weakness of the heart tends to elevate the right atrial pressure. Likewise, any effect that causes rapid inflow of blood into the right atrium from the veins tends to elevate the right atrial pressure. Some of the factors that increase this tendency for venous return (and also tend to increase the right atrial pressure) are (1) increased blood volume, (2) increased large vessel tone throughout the body with resultant increased peripheral venous pressures, and (3) dilatation of the arterioles, which decreases the peripheral resistance and allows rapid flow of blood from the arteries to the veins.

The same factors that regulate right atrial pressure also enter into the regulation of cardiac output, for the amount of blood pumped by the heart depends both on the ability of the heart to pump and the tendency for blood to flow into the heart from the peripheral vessels. Therefore, we discuss the regulation of right atrial pressure in more depth in Chapter 20 in connection with the regulation of cardiac output.

The *normal right atrial pressure* is approximately 0 mm Hg, which is about equal to the atmospheric pressure around the body. However, it can rise to as high as 20 to 30 mm Hg under very abnormal conditions, such as (1) serious heart failure or (2) following massive transfusion of blood, which will cause excessive quantities of blood to attempt to flow into the heart from the peripheral vessels.

The lower limit to the right atrial pressure is usually about −3 to −5 mm Hg, which is the pressure in the chest cavity that surrounds the heart. The right atrial pressure approaches these very low values when the heart pumps with exceptional vigor or when the flow of blood into the heart from the peripheral vessels is greatly depressed, such as following severe hemorrhage.

Venous Resistance and Peripheral Venous Pressure

Large veins have almost no resistance *when they are distended.* However, as illustrated in Figure 15–9,

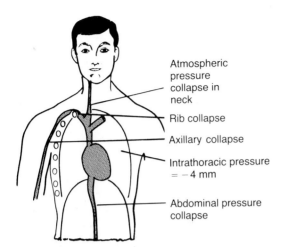

Figure 15–9. Factors tending to collapse the veins entering the thorax.

most of the large veins entering the thorax are compressed at many points by the surrounding tissues, so that blood flow is impeded. For instance, the veins from the arms are compressed by their sharp angulation over the first rib. Second, the pressure in the neck veins often falls so low that the atmospheric pressure on the outside of the neck causes them to collapse. Finally, veins coursing through the abdomen are often compressed by different organs and by the intra-abdominal pressure, so that usually they are at least partially collapsed to an ovoid or slitlike state. For these reasons, the *large veins do usually offer considerable resistance to blood flow,* and because of this the pressure in the peripheral veins is usually 4 to 7 mm Hg greater than the right atrial pressure.

Effect of High Right Atrial Pressure on Peripheral Venous Pressure. When the right atrial pressure rises above its normal value of 0 mm Hg, blood begins to back up in the large veins and to open them up. The pressures in the peripheral veins do not rise until all the collapsed points between the peripheral veins and the large veins have opened up. This usually occurs when the right atrial pressure rises to about +4 to 6 mm Hg. Then, as the right atrial pressure rises still further, the additional increase in pressure is reflected by a corresponding rise in peripheral venous pressure. Because the heart must be greatly weakened to cause a rise in right atrial pressure to as high as 4 to 6 mm Hg, one often finds that the peripheral venous pressure is not elevated in the early stages of heart failure.

Effect of Abdominal Pressure on Venous Pressures of the Leg. The normal pressure in the peritoneal cavity averages about 2 mm Hg, but at times it can rise to as high as 15 to 20 mm Hg as a result of pregnancy, large tumors, or excessive fluid (called "ascites") in the peritoneal cavity. When this happens, the pressure in the veins of the legs must rise *above* the abdominal pressure before the abdominal veins will open and allow the blood to flow from the legs to the heart. Thus, if the intra-abdominal pres-

sure is 20 mm Hg, the lowest possible pressure in the femoral veins is 20 mm Hg.

Effect of "Hydrostatic" Pressure on Venous Pressure

In any body of water, the pressure at the surface of the water is equal to atmospheric pressure, but the pressure rises 1 mm Hg for each 13.6 mm distance below the surface. This pressure results from the weight of the water and therefore is called *hydrostatic pressure.*

Hydrostatic pressure also occurs in the vascular system of the human being because of the weight of the blood in the vessels, as illustrated in Figure 15–10. When a person is standing, the pressure in the right atrium remains approximately 0 mm Hg because the heart pumps into the arteries any excess blood that attempts to accumulate at this point. However, in an adult *who is standing absolutely still* the pressure in the veins of the feet is approximately +90 mm Hg simply because of the weight of the blood in the veins between the heart and the feet. The venous pressures at other levels of the body lie proportionately between 0 and 90 mm Hg.

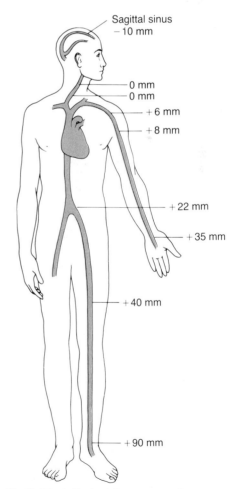

Figure 15–10. Effect of hydrostatic pressure on the venous pressures throughout the body.

In the arm veins, the pressure at the level of the top rib is usually about +6 mm Hg because of compression of the subclavian vein as it passes over this rib. The hydrostatic pressure down the length of the arm is then determined by the distance below the level of this rib. Thus, if the hydrostatic difference between the level of the rib and the hand is 29 mm Hg, this hydrostatic pressure is added to the 6 mm Hg pressure caused by compression of the vein as it crosses the rib, making a total of 35 mm Hg pressure in the veins of the hand.

The neck veins collapse almost completely all the way to the skull owing to atmospheric pressure on the outside of the neck. This collapse causes the pressure in these veins to remain zero along their entire extent. The reason for this is that any tendency for the pressure to rise above this level opens the veins and allows the pressure to fall back to zero, and any tendency for the pressure to fall below this level collapses the veins still more, which increases their resistance and again returns the pressure back to zero.

The veins inside the skull, however, are in a noncollapsible chamber, and they will not collapse. Consequently, *negative pressure can exist in the dural sinuses of the head;* in the standing position, the venous pressure in the sagittal sinus is approximately −10 mm Hg because of the hydrostatic "suction" between the top of the skull and the base of the skull. Therefore, if the sagittal sinus is opened during surgery, air can be sucked immediately into this vein; the air may even pass downward to cause air embolism in the heart, so that the heart valves will not function satisfactorily, and death can ensue.

Effect of the Hydrostatic Factor on Arterial and Other Pressures. The hydrostatic factor also affects the peripheral pressures in the arteries and capillaries as well as in the veins. For instance, a standing person who has an arterial pressure of 100 mm Hg at the level of the heart has an arterial pressure in the feet of about 190 mm Hg. Therefore, any time one states that the arterial pressure is 100 mm Hg, it generally means that this is the pressure at the hydrostatic level of the heart.

Venous Valves, the "Venous Pump," and the Venous Pressure

Were it not for valves in the veins, the hydrostatic pressure effect would cause the venous pressure in the feet always to be about +90 mm Hg in a standing adult. However, every time one moves the legs one tightens the muscles and compresses the veins either in the muscles or adjacent to them, and this squeezes the blood out of the veins. The valves in the veins, illustrated in Figure 15–11, are arranged so that the direction of blood flow can be only toward the heart. Consequently, every time a person moves the legs or even tenses the muscles, a certain amount of blood is propelled toward the heart, and the pressure in the veins is lowered. This pumping system is known as the "venous pump" or "muscle pump," and it is efficient

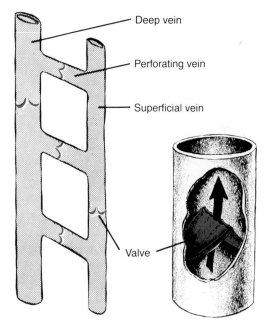

Figure 15–11. The venous valves of the leg.

enough that under ordinary circumstances the venous pressure in the feet of a walking adult remains less than 25 mm Hg.

If the human being stands perfectly still, the venous pump does not work, and the venous pressures in the lower part of the leg will rise to the full hydrostatic value of 90 mm Hg in about 30 seconds. The pressures in the capillaries also increase greatly, causing fluid to leak from the circulatory system into the tissue spaces. As a result, the legs swell, and the blood volume diminishes. Indeed, as much as 15 to 20 per cent of the blood volume is frequently lost from the circulatory system within the first 15 minutes of standing absolutely still, as often occurs when a soldier is made to stand at absolute attention.

Venous Valve Incompetence and Varicose Veins. The valves of the venous system frequently become "incompetent" or sometimes are even destroyed. This is especially true when the veins have been overstretched by excess venous pressure lasting weeks or months, as occurs in pregnancy or when one stands most of the time. Stretching the veins increases their cross-sectional areas, but the valves do not increase in size. Therefore, the veins of the valves no longer close completely. When this develops, the pressure in the veins of the legs increases still more owing to failure of the venous pump; this further increases the size of the veins and finally destroys the function of the valves entirely. Thus, the person develops "varicose veins," which are characterized by large, bulbous protrusions of the veins beneath the skin of the entire leg and particularly of the lower leg. The venous and capillary pressures become very high, and leakage of fluid from the capillaries causes constant edema in the legs whenever these persons stand for more than a few minutes. The edema in turn prevents adequate diffusion of nutritional materials from the capillaries to the muscle and skin cells, so that the muscles become painful and weak, and the skin frequently becomes gangrenous and ulcerates. Obviously, the best treatment for such a condition is continual elevation of the legs to a level at least as high as the heart, but tight binders on the legs are also of considerable aid in preventing the edema and its sequelae.

Clinical Estimation of Venous Pressure. The venous pressure can often be estimated by simply observing the degree of distension of the peripheral veins — especially the neck veins. For instance, in the sitting position, the neck veins are never distended in the normal person. However, when the right atrial pressure becomes increased to as much as 10 mm Hg, the lower veins of the neck begin to protrude, and at 15 mm Hg essentially all the veins in the neck become distended.

Direct Measurement of Venous Pressure and Right Atrial Pressure. Venous pressure can be measured with ease by inserting a syringe needle directly into a vein and connecting it to a pressure recorder.

The only means by which *right atrial pressure* can be measured accurately is by inserting a catheter through the veins into the right atrium. Pressures measured through such "central venous catheters" are used almost routinely in hospitalized cardiac patients to provide constant assessment of heart pumping ability.

Pressure Reference Level for Measuring Venous and Other Circulatory Pressures

In discussions up to this point we have often spoken of right atrial pressure as being 0 mm Hg and arterial pressure as being 100 mm Hg, but we have not stated the hydrostatic level in the circulatory system to which this pressure is referred. There is one point in the circulatory system at which hydrostatic pressure factors caused by changes in body position usually do not affect the pressure measurement by more than 1 mm Hg. This is at the level of the tricuspid valve, as shown by the crossed axes in Figure 15–12. Therefore, all pressure measurements discussed in this text are referred to this level, which is called the *reference level for pressure measurement.*

The reason for lack of hydrostatic effects at the tricuspid valve is that the heart automatically prevents significant hydrostatic changes in pressure at this point in the following way:

If the pressure at the tricuspid valve rises slightly above normal, the right ventricle fills to a greater extent than usual, causing the heart to pump blood more rapidly and therefore to decrease the pressure at the tricuspid valve back toward the normal mean value. On the other hand, if

the pressure falls, the right ventricle fails to fill adequately, its pumping decreases, and blood dams up in the venous system until the tricuspid pressure again rises to a normal value. In other words, *the heart acts as a feedback regulator of pressure* at the tricuspid valve.

When a human is lying on his or her back, the tricuspid valve is located almost exactly 60 per cent of the chest thickness in front of the back. Therefore, this is the *zero pressure reference level.*

BLOOD RESERVOIR FUNCTION OF THE VEINS

It was pointed out in the previous chapter that over 60 per cent of all the blood in the circulatory system is in the veins. For this reason, and also because the veins are so compliant, it is frequently said that the venous system serves as a *blood reservoir* for the circulation.

When blood is lost from the body and the arterial pressure begins to fall, pressure reflexes are elicited from the carotid sinuses and other pressure-sensitive areas of the circulation, as is discussed in Chapter 18; these in turn send sympathetic nerve signals to the veins, causing them to constrict, and this takes up much of the slack in the circulatory system caused by the lost blood. Indeed, even after as much as 20 per cent of the total blood volume has been lost, the circulatory system often functions almost normally because of this variable reservoir system of the veins.

Specific Blood Reservoirs. Certain portions of the circulatory system are so extensive and so compliant that they are called specific "blood reservoirs." These include (1) the *spleen,* which can sometimes decrease in size sufficiently to release as much as 100 milliliters of blood into other areas of the circulation; (2) the *liver,* the sinuses of which can release several hundred milliliters of blood into the remainder of the circulation; (3) the *large abdominal veins,* which can contribute as much as 300 milliliters; and (4) the *venous plexus beneath the skin,* which can also contribute several hundred milliliters. The *heart* and the *lungs,* though not parts of the systemic venous reservoir system, must also be considered to be blood reservoirs. The heart, for instance, becomes reduced in size during sympathetic stimulation

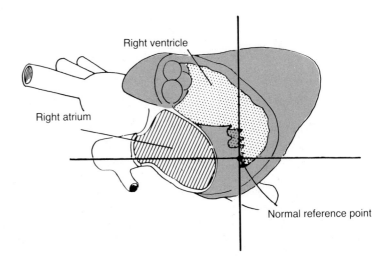

Figure 15–12. Location of the reference point for pressure measurement at the tricuspid valve.

and in this way can contribute some 50 to 100 milliliters of blood, and the lungs can contribute another 100 to 200 milliliters when the pulmonary pressures fall to low values.

The Spleen as a Reservoir for Storing Red Blood Cells. Figure 15–13 illustrates that the spleen has two separate areas for storing blood: the *venous sinuses* and the *pulp.* Small vessels flow directly into the venous sinuses, and these can swell the same as any other part of the venous system and store whole blood.

In the splenic pulp, the capillaries are so permeable that whole blood oozes through the capillary walls into a trabecular mesh forming the *red pulp.* The red cells are trapped by the trabeculae, while the plasma returns into the venous sinuses and then into the general circulation. As a consequence, the red pulp of the splenic pulp is a *special reservoir containing large quantities of extra red blood cells* that are expelled into the general circulation when the sympathetic nervous system is excited and contracts the spleen or its vessels. In lower animals, this extra storage of red blood cells is much greater than in humans, but even in the human possibly as much as 50 milliliters of concentrated red blood cells can be released into the circulation, raising the hematocrit as much as 1 to 2 per cent.

In other areas of the splenic pulp are islands of white blood cells, which collectively are called the *white pulp.* Here lymphoid cells are formed similar to those formed in the lymph nodes. These are part of the body's immune system, which is described in Chapter 34.

The Blood-Cleansing Function of the Spleen—Removal of Old Cells. Blood passing through the splenic pulp before it enters the sinuses undergoes thorough squeezing. Therefore, it is to be expected that fragile red blood cells would not withstand the trauma. For this reason, many of the red blood cells destroyed in the body have their final demise in the spleen. After the cells rupture, the released hemoglobin and the cell stroma are ingested by the reticuloendothelial cells of the spleen.

Reticuloendothelial Cells of the Spleen. The pulp of the spleen contains many large phagocytic reticuloendothelial cells, and the venous sinuses are lined with similar cells. These cells act as a cleansing system for the blood, acting in concert with a similar system in the venous sinuses of the liver. When the blood is invaded by infectious agents, the reticuloendothelial cells of the spleen rapidly remove debris, bacteria, parasites, and so forth. Also, in many infectious processes the spleen enlarges in the same manner that

lymph glands enlarge and then performs its cleansing function even more adequately.

Much of the spleen is also filled with *white pulp,* which contains large aggregates of lymphocytes and plasma cells. These function in exactly the same way in the spleen as in the lymph glands to cause either humoral or lymphocytic immunity against toxins, bacteria, and so forth, as described in Chapter 34.

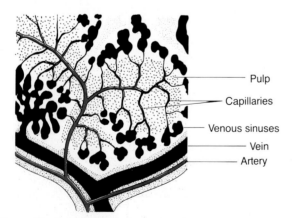

Figure 15–13. The functional structures of the spleen. (Modified from Bloom and Fawcett: A Textbook of Histology. 10th ed. Philadelphia, W. B. Saunders Company, 1975).

REFERENCES

American Physiological Society: Peripheral Circulation and Organ Blood Flow. Washington, D.C., American Physiological Society, 1983.

Caro, C. G., and Lever, M. J.: Factors influencing arterial wall mass transport. Biorheology, 21:197, 1984.

Chien, S.: Red cell deformability and its relevance to blood flow. Annu. Rev. Physiol., 49:177, 1987.

Chien, S., et al.: Blood flow in small tubes. In Renkin, E. M., and Michel, C. C. (eds.): Handbook of Physiology. Sec. 2, Vol. IV. Bethesda, Md., American Physiological Society, 1984, p. 217.

Dobrin, P. B.: Vascular mechanics. In Shepard, J. T., and Abboud, F. M. (eds.): Handbook of Physiology. Sec. 2, Vol. III. Bethesda, Md., American Physiological Society, 1983, p. 65.

Donald, D. E.: Splanchnic circulation. In Shepherd, J. T., and Abboud, F. M. (eds.): Handbook of Physiology. Sec. 2, Vol. III. Bethesda, Md., American Physiological Society, 1983, p. 219.

Fozzard, H. A., et al.: The Heart and Cardiovascular System: Scientific Foundations. New York, Raven Press, 1986.

Fronek, A. (ed.): Noninvasive Diagnostics in Vascular Disease. New York, McGraw-Hill Book Co., 1989.

Fukada, E., et al.: Methods to study rheological properties of blood during clotting. Biorheology, 1:(Suppl.)9, 1984.

Goerke, J., and Mines, A. H.: Cardiovascular Physiology. New York, Raven Press, 1988.

Goodman, A. H., et al.: A television method for measuring capillary red cell velocities. J. Appl. Physiol., 37:126, 1974.

Gow, B. S.: Circulation correlates: Vascular impedance, resistance, and capacity. In Bohr, D. F., et al. (eds.): Handbook of Physiology. Sec. 2, Vol. II. Baltimore, Williams & Wilkins, 1980, p. 353.

Green, H. D.: Circulation: Physical principles. In Glasser, O. (ed.): Medical Physics. Chicago, Year Book Medical Publishers, 1944.

Gross, J. F., and Popel, A. (eds): Mathematics of Microcirculation Phenomena. New York, Raven Press, 1980.

Guyton, A. C.: Arterial Pressure and Hypertension. Philadelphia, W. B. Saunders Co., 1980.

Guyton, A. C.: Peripheral circulation. Annu. Rev. Physiol., 21:239, 1959.

Guyton, A. C.: The venous system and its role in the circulation. Mod. Conc. Cardiov. Dis., 17:483, 1958.

Guyton, A. C., and Greganti, F. P.: A physiologic reference point for measuring circulation pressures in the dog—particularly venous pressure. Am. J. Physiol., 185:137, 1956.

Guyton, A. C., and Jones, C. E.: Central venous pressure: Physiological significance and clinical implications. Am. Heart J., 86:431, 1973.

Guyton, A. C., et al.: Cardiac Output and Its Regulation. Philadelphia, W. B. Saunders Co., 1973.

Guyton, A. C., et al.: Evidence for tissue oxygen demand as the major factor causing autoregulation. Circ. Res., 14:60, 1964.

Guyton, A. C., et al.: Pressure-volume curves of the entire arterial and venous systems in the living animal. Am. J. Physiol., 184:253, 1956.

Guyton, J. R.: Mechanical control of smooth muscle growth. In Seidel, C. L., and Weisbrodt, N. W.: Hypertrophic Response in Smooth Muscle. Boca Raton, Fla., CRC Press, 1987, p. 121.

Hirakawa, S., et al.: Volume-pressure relations of the human pulmonary "venous" system studied by radionuclide angiocardiography and passive leg elevation, with special reference to the effect of nitroglycerin. Jpn. Circ. J., 50:303, 1986.

Hirakawa, S., et al.: The role of alpha and beta adrenergic receptors in constriction and dilation of the systemic capacitance vessels: A study with measurements of the mean circulatory pressure in dogs. Jpn. Circ. J., 48:620, 1984.

Hochmuth, R. M., and Waugh, R. E.: Erythrocyte membrane elasticity and viscosity. Annu. Rev. Physiol., 49:209, 1987.

Kenner, T. (ed.): Cardiovascular System Dynamics. Models and Measurements. New York, Plenum Publishing Corp., 1982.

Lassen, N. A., et al.: Indicator methods for measurement of organ and tissue blood flow. In Shepherd, J. T., and Abboud, F. M. (eds.): Handbook of Physiology. Sec. 2, Vol. III. Bethesda, Md., American Physiological Society, 1983, p. 21.

Mellander, S.: Systemic circulation: Local control. Annu. Rev. Physiol., 32:313, 1970.

Milnor, W. R.: Hemodynamics. 2nd Ed. Baltimore, Williams & Wilkins, 1989.

Neil, E.: Peripheral circulation: Historical aspects. In Shepherd, J. T., and Abboud, F. M. (eds.): Handbook of Physiology. Sec. 2, Vol. III. Bethesda, Md., American Physiological Society, 1983, p. 1.

O'Rourke, M. F.: Vascular impedance in studies of arterial and cardiac function. Physiol. Rev., 62:570, 1982.

Peterson, N. S., and Campbell, K. B.: Teaching cardiovascular integrations with computer laboratories. The Physiologist, 28:159, 1985.

Piene, H.: Pulmonary arterial impedance and right ventricular function. Physiol. Rev., 66:606, 1986.

Rodkiewicz, C. M.: Arteries and Arterial Blood Flow. New York, Springer-Verlag, 1983.

Rothe, C. F.: Reflex control of veins and vascular capacitance. Physiol. Rev., 63:1281, 1983.

Rothe, C. F.: Venous system: Physiology of the capacitance vessels. In Shepherd, J. T., and Abboud, F. M. (eds.): Handbook of Physiology. Sec. 2, Vol. III. Bethesda, Md., American Physiological Society, 1983, p. 397.

Rowan: Physics and the Circulation. Philadelphia, Heyden & Sons, Inc., 1981.

Schmid-Schonbein, G. W.: Granulocyte: Friend and foe. News Physiol. Sci., 3:144, 1988.

Schmid-Schonbein, H. J.: Microrheology of erythrocytes, blood viscosity, and the distribution of blood flow in the microcirculation. Int. Rev. Physiol., 9:1, 1976.

Schmidt-Nielsen, K., and Pennycuik, P.: Capillary density in mammals in relation to body size and oxygen consumption. Am. J. Physiol., 200:746, 1961.

Schneck, D. J., and Vawter, D. L. (eds.): Biofluid Mechanics. New York, Plenum Press, 1980.

Sigel, B., et al: Ultrasonic detection of red cell aggregation immediately preceding blood clotting. Invest. Radiol., 19:458, 1984.

Vanhoutte, P. M.: Vasodilation: Vascular Smooth Muscle, Peptides, Autonomic Nerves, and Endothelium. New York, Raven Press, 1988.

Vanhoutte, P. M., and Leusen, I. (eds.): Mechanics of Vasodilatation. New York, S. Karger, 1978.

Vanhoutte, P. M., et al.: Local modulation of the adrenergic neuroeffector interaction in the blood vessel wall. Physiol. Rev., 61:151, 1981.

Wiedeman, M. P.: Dimensions of blood vessels from distributing artery to collecting vein. Circ. Res., 12:375, 1963.

Ziegler, M. G.: Postural hypotension. Annu. Rev. Med., 31:239, 1980.

Zweifach, B. W., and Lipowsky, H. H.: Pressure-flow relations in blood and lymph microcirculation. In Renkin, E. M., and Michel, C. C. (eds.): Handbook of Physiology. Sec. 2, Vol. IV. Bethesda, Md., American Physiological Society, 1984, p. 251.

CHAPTER 16

• The Microcirculation and the Lymphatic System: Capillary Fluid Exchange, Interstitial Fluid, and Lymph Flow •

In the microcirculation the most purposeful function of the circulation occurs, transport of nutrients to the tissues and removal of cellular excreta. The small arterioles control the blood flow to each tissue area, and local conditions in the tissues themselves control the diameters of the arterioles in turn. Thus, each tissue in most instances controls its own blood flow in relation to its needs, a subject that is discussed in detail in the following chapter.

The capillaries are extremely thin structures with walls of a single layer of highly permeable endothelial cells. Here interchange of nutrients and cellular excreta occurs between the tissues and the circulating blood. About 10 billion capillaries with a total surface area estimated to be 500 to 700 square meters (approximately the surface area of a football field) provide this function for the whole body. Indeed, it is rare that any single functional cell of the body is more than 20 to 30 micrometers away from a capillary.

The purpose of this chapter is to discuss the transfer of substances between the blood and interstitial fluid and especially to discuss the factors that affect the transfer of fluid volume itself between the circulating blood and the interstitial fluid.

STRUCTURE OF THE MICROCIRCULATION AND CAPILLARY SYSTEM

The microcirculation of each organ is specifically organized to serve that organ's own special needs. In general, each nutrient artery entering an organ branches six to eight times before the arteries become small enough to be called "arterioles," which have internal diameters generally less than 20 micrometers. Then the arterioles themselves branch two to five times, reaching diameters of 5 to 9 micrometers at their ends where they supply blood to the capillaries.

Figure 16–1 illustrates the structure of a representative capillary bed as seen in the mesentery, showing that blood enters the capillaries through an *arteriole* and leaves by way of a *venule*. Blood from the arteriole passes into a series of *metarterioles,* which are called by some physiologists *terminal arterioles* and which have a structure midway between that of arterioles and capillaries. After leaving the metarteriole, the blood enters the *capillaries,* some of which are large and are called *preferential channels* and others of which are small and are *true capillaries.* After passing through the capillaries, the blood enters the venule and returns to the general circulation.

The arterioles are highly muscular, and their diameters can change manyfold. The metarterioles (the terminal arterioles) do not have a continuous muscular coat, but smooth muscle fibers encircle the vessel at intermediate points, as illustrated in Figure 16–1 by the large black dots to the sides of the metarteriole.

At the point where the true capillaries originate from the metarterioles a smooth muscle fiber usually encircles the capillary. This is called the *precapillary sphincter.* This sphincter can open and close the entrance to the capillary.

The venules are considerably larger than the arterioles and have a much weaker muscular coat. Yet, it must be remembered that the pressure in the venules is much less than that in the arterioles, so that the venules can still contract considerably.

This typical arrangement of the capillary bed is not found in all parts of the body; however, some similar arrangement serves the same purposes. Most important of all, the metarterioles (and the precapillary sphincters when they also exist) are in extremely close contact with the tissues that they serve. Therefore, the local conditions of the tissues — the concentrations of nutrients, end products of metabolism, hydrogen ions, and so forth — can cause direct effects on these in controlling the local blood flow in each minute tissue area.

Structure of the Capillary Wall. Figure 16–2 illustrates the ultramicroscopic structure of a typical capillary wall as found in most organs of the body,

Figure 16 – 1. Structure of the mesenteric capillary bed. (From Zweifach: Factors Regulating Blood Pressure. New York, Josiah Macy, Jr., Foundation, 1950.)

especially in the muscles and connective tissue. Note that the wall is composed of a unicellular layer of endothelial cells and is surrounded by a basement membrane on the outside. The total thickness of the wall is about 0.5 micrometer.

The diameter of the capillary is 4 to 9 micrometers, barely large enough for red blood cells and other blood cells to squeeze through.

"Pores" in the Capillary Membrane. Studying Figure 16–2, one sees two minute passageways connecting the interior of the capillary with the exterior. One of these is the *intercellular cleft,* which is the thin slit that lies between adjacent endothelial cells. Each of these clefts is interrupted periodically by short ridges of protein attachments that hold the endothelial cells together, but each ridge in turn is broken after a short distance, so that in between them fluid can percolate freely through the cleft. The cleft normally has a very uniform spacing with a width of approximately 6 to 7 nanometers (60 to 70 angstroms), slightly smaller than the diameter of an albumin protein molecule.

Because the intercellular clefts are located only at the edges of the endothelial cells, they usually represent no more than 1/1000 of the total surface area of the capillary. Nevertheless, the rate of thermal motion of water molecules as well as most other water-soluble ions and small solutes is so rapid that all of these

diffuse with ease between the interior and exterior of the capillaries through these "slit-pores," the intercellular clefts.

Also present in the endothelial cells are many minute *plasmalemmal vesicles.* These form at one surface of the cell by imbibing small packets of plasma or extracellular fluid. They can then move slowly through the endothelial cell, and it has been postulated that they transport significant amounts of substances through the capillary wall. It has also been postulated that some of these vesicles coalesce to form vesicular channels all the way through the membrane, thus allowing another pathway for fluid and large molecule movement. Yet, careful experiments in which physical and chemical means have been used to prevent the formation of or movement of plasmalemmal vesicles have shown that normal movement through the cell membrane of even very large macromolecules still occurs, indicating that these vesicles are probably of only minor importance for most capillary wall transport.

The "pores" in the capillaries of some organs have special characteristics to meet the peculiar needs of the organs. Some of these are:

1. In the *brain,* the junctions between the capillary endothelial cells are mainly "tight" junctions that will allow only very small molecules to pass into the brain tissues. This is called the *blood-brain barrier* and is discussed further in Chapter 61.

2. In the *liver,* exactly the opposite is true. The clefts between the capillary endothelial cells are very wide open, so that almost all dissolved substances of the plasma, including even the plasma proteins, can pass from the blood into the liver tissues. The pores of the intestinal membranes are midway between those of the muscles and the liver.

3. In the *glomerular tufts of the kidney,* numerous small oval windows called *fenestrae* penetrate directly through the very middle of the endothelial cells, so that tremendous amounts of substances can be filtered through the glomeruli without having to pass through the clefts between the endothelial cells.

● FLOW OF BLOOD IN THE CAPILLARIES— VASOMOTION

Blood usually does not flow continuously through the capillaries. Instead, it flows intermittently, turning on and off every few seconds or minutes. The cause of this intermittency is the phenomenon called *vasomotion,* which means intermittent contraction of the metarterioles and precapillary sphincters.

Regulation of Vasomotion. The most important factor found thus far to affect the degree of opening and closing of the metarterioles and precapillary sphincters is the concentration of *oxygen* in the tissues. When the rate of oxygen usage is great, the intermittent periods of blood flow occur more often, and the duration of each period of flow lasts for a longer time, thereby allowing the blood to carry increased quantities of oxygen (as well as other nutrients) to the tissues. This effect as well as multiple other factors that con-

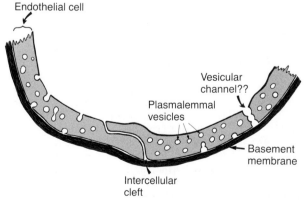

Figure 16 – 2. Structure of the capillary wall. Note especially the *intercellular cleft* at the junction between adjacent endothelial cells; it is believed that most water-soluble substances diffuse through the capillary membrane along this cleft.

trol local tissue blood flow is discussed in the following chapter.

AVERAGE FUNCTION OF THE CAPILLARY SYSTEM

Despite the fact that blood flow through each capillary is intermittent, so many capillaries are present in the tissues that their overall function becomes averaged. That is, there is an *average rate of blood flow* through each tissue capillary bed, an *average capillary pressure* within the capillaries, and an *average rate of transfer of substances* between the blood of the capillaries and the surrounding interstitial fluid. In the remainder of this chapter, we are concerned with these averages, though one must remember that the average functions are in reality the functions of literally billions of individual capillaries, each operating intermittently in response to the local conditions in the tissues.

● EXCHANGE OF NUTRIENTS AND OTHER SUBSTANCES BETWEEN THE BLOOD AND INTERSTITIAL FLUID

DIFFUSION THROUGH THE CAPILLARY MEMBRANE

By far the most important means by which substances are transferred between the plasma and interstitial fluid is by *diffusion*. Figure 16–3 illustrates this process, showing that as the blood traverses the capillary, tremendous numbers of water molecules and dissolved particles diffuse back and forth through the capillary wall, providing continual mixing between the interstitial fluid and the plasma. Diffusion results from thermal motion of the water molecules and the dissolved substances in the fluid, the different particles moving first in one direction, then another, moving randomly in every direction.

Diffusion of Lipid-Soluble Substances Through the Capillary Membrane. If a substance is lipid-soluble, it can diffuse directly through the cell membranes of the capillary without having to go through the pores. Such substances include especially oxygen and carbon dioxide. Because these can permeate all areas of the capillary membrane, their rates of transport through the capillary membrane are many times the rates for most lipid-insoluble substances, such as sodium ions, glucose, and so forth.

Diffusion of Water-Soluble Substances Through the Capillary Membrane. Many substances needed by the tissues are soluble in water but cannot pass through the lipid membranes of the endothelial cells; such substances include water molecules themselves, sodium ions, chloride ions, and glucose. Despite the fact that not over 1/1000 of the surface area of the capillaries is represented by the intercellular junctions, the velocity of thermal molecular motion is so great that even this small area is sufficient to allow tremendous diffusion of water and water-soluble substances through these pores. To give one an idea of the extreme rapidity with which these substances diffuse, *the rate at which water molecules diffuse through the capillary membrane is approximately 80 times as great as the rate at which plasma itself flows linearly along the capillary.* That is, the water of the plasma is exchanged with the water of the interstitial fluid 80 times before the plasma can go the entire distance through the capillary.

Effect of Molecular Size on Passage Through the Pores. The width of the capillary intercellular slit-pores, 6 to 7 nanometers, is about 20 times the diameter of the water molecule, which is the smallest molecule that normally passes through the capillary pores. On the other hand, the diameters of plasma protein molecules are slightly greater than the width of the pores. Other substances, such as sodium ions, chloride ions, glucose, and urea, have intermediate diameters. Therefore, it is obvious that the permeability of the capillary pores for different substances varies according to their molecular diameters.

Table 16–1 gives the relative permeabilities of the capillary pores in muscle for substances commonly

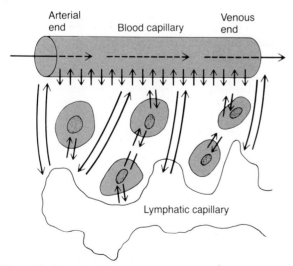

Figure 16–3. Diffusion of fluid and dissolved substances between the capillary and interstitial fluid spaces.

Table 16–1 RELATIVE PERMEABILITY OF MUSCLE CAPILLARY PORES TO DIFFERENT-SIZED MOLECULES

Substance	Molecular Weight	Permeability
Water	18	1.00
NaCl	58.5	0.96
Urea	60	0.8
Glucose	180	0.6
Sucrose	342	0.4
Inulin	5000	0.2
Myoglobin	17,600	0.03
Hemoglobin	68,000	0.01
Albumin	69,000	.0001

Modified from Pappenheimer.

encountered by the capillary membrane, illustrating, for instance, that the permeability for glucose molecules is 0.6 times that for water molecules, whereas the permeability for albumin molecules is very, very slight.

However, a word of caution must be issued at this point. The capillaries in different tissues have extreme differences in their permeabilities. For instance, the membrane of the liver capillary sinusoids is so permeable that even plasma proteins pass freely through these walls almost as easily as water and other substances. Also, the permeability of the renal glomerular membrane for water and electrolytes is about 500 times the permeabilty of the muscle capillaries, but the glomerular and muscle permeabilities for protein are about the same. When we study these different organs later in this text, it should become clear why some tissues require greater degrees of capillary permeability than others—the liver, for instance, to transfer tremendous amounts of nutrients between the blood and the liver parenchymal cells and the kidneys to allow filtration of large quantities of fluid for the formation of urine.

Effect of Concentration Difference on Net Rate of Diffusion Through the Capillary Membrane. The "net" rate of diffusion of a substance through any membrane is proportional to the *concentration difference* between the two sides of the membrane. That is, the greater the difference between the concentrations of any given substance on the two sides of the capillary membrane, the greater will be the net movement of the substance through the membrane. Thus, the concentration of oxygen in the blood is normally greater than that in the interstitial fluid. Therefore, large quantities of oxygen normally move from the blood toward the tissues. Conversely, the concentration of carbon dioxide is greater in the tissues than in the blood, which obviously causes carbon dioxide to move into the blood and to be carried away from the tissues.

Fortunately, the rates of diffusion through the capillary membranes of most nutritionally important substances are so great that only slight concentration differences suffice to cause more than adequate transport between the plasma and interstitial fluid. For instance, the concentration of oxygen in the interstitial fluid immediately outside the capillary is probably no more than 1 per cent less than the concentration in the blood, and yet this 1 per cent difference causes enough oxygen to move from the blood into the interstitial spaces to provide all the oxygen required for tissue metabolism.

● THE INTERSTITIUM AND THE INTERSTITIAL FLUID

Approximately one sixth of the body consists of spaces between cells, which collectively are called the *interstitium*. The fluid in these spaces is the *interstitial fluid*.

The structure of the interstitium is illustrated in Figure 16–4. It has two major types of solid structures: (1) collagen fiber bundles and (2) proteoglycan filaments. The collagen fiber bundles extend long distances in the interstitium. They are extremely strong and therefore provide most of the tensional strength of the tissues. The proteoglycan filaments, on the other hand, are extremely thin, coiled molecules composed of about 98 per cent hyaluronic acid and 2 per cent protein. These molecules are so thin that they can never be seen with a light microscope and are very difficult to demonstrate even with the electron microscope. Nevertheless, they form a mat of very fine reticular filaments aptly described as a "brush pile."

"Gel" in the Interstitium. The fluid in the interstitium is derived by filtration from the capillaries. It contains almost the same constituents as plasma except for much lower concentrations of proteins because these do not filter out of the capillaries with ease. This fluid is mainly entrapped in the minute spaces among the proteoglycan filaments. This combination of the proteoglycan filaments and the fluid entrapped within them has the characteristics of a *gel* and therefore is often called the *tissue gel.*

Because of the large number of proteoglycan filaments, fluid *flows* through the tissue gel only very poorly. Instead, it mainly *diffuses* through the gel; that is, it moves molecule by molecule from one place to another by the process of kinetic motion rather than by large numbers of molecules moving together.

Fortunately, diffusion through the gel occurs about 95 to 99 per cent as rapidly as it does through free fluid. For the short distances between the capillaries and the tissue cells, this diffusion allows rapid transport through the interstitium not only of the water molecules, but also electrolytes, nutrients, cellular excreta, oxygen, carbon dioxide, and so forth.

"Free" Fluid in the Interstitium. Although almost all the fluid in the interstitium is normally entrapped within the tissue gel, occasionally small *rivu-*

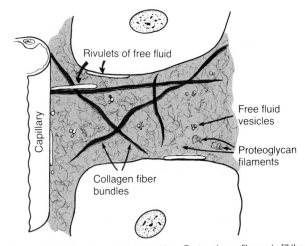

Figure 16–4. Structure of the interstitium. Proteoglycan filaments fill the spaces between the collagen fiber bundles. Free fluid vesicles are seen and small amounts of free fluid in the form of rivulets.

lets of "free" fluid and small *free fluid vesicles* are also present, which means fluid that is free of the proteoglycan molecules and therefore can flow freely. When a dye is injected into the circulating blood, it can often be seen to flow through the interstitium in the small rivulets, usually coursing along the surfaces of collagen fibers or surfaces of cells. However, the amount of "free" fluid present in *normal* tissues is very slight, usually much less than 1 per cent. On the other hand, when the tissues develop *edema, these small pockets and rivulets of free fluid expand tremendously.*

Relationship of the Free Fluid to the Gel Fluid. The free fluid and the gel fluid are continually interchanging with each other by diffusion. At the surface between the free fluid and the gel, there is a slight but uncertain osmotic pressure difference that probably makes the pressure of the fluid in the gel about 2 mm Hg more positive than that in the free fluid.

Because of the uncertainty of the quantitative value of the pressures in the gel, in most of the discussions of this chapter we speak mainly of the free fluid pressure rather than the intragel fluid pressure.

● DISTRIBUTION OF FLUID VOLUME BETWEEN THE PLASMA AND THE INTERSTITIAL FLUID

The pressure in the capillaries tends to force fluid and its dissolved substances through the capillary pores into the interstitial spaces. In contrast, osmotic pressure caused by the plasma proteins (called *colloid* osmotic pressure) tends to cause fluid movement by osmosis from the interstitial spaces into the blood; it is this osmotic pressure that prevents significant loss of fluid volume from the blood into the interstitial spaces. Yet, also important is the lymphatic system to return back to the circulation the small amounts of protein and fluid that do leak into the interstitial spaces. In the following few paragraphs we discuss the factors that determine the movement, or lack of movement, of fluid volume through the capillary membrane, and later in the chapter, we discuss the role of the lymphatic system in this overall mechanism.

The Four Primary Forces That Determine Fluid Movement Through the Capillary Membrane. Figure 16–5 illustrates the four primary forces that determine whether fluid will move out of the blood into the interstitial fluid or in the opposite direction; these, called "Starling forces" in honor of the physiologist who first demonstrated their importance, are:

1. The *capillary pressure* (Pc), which tends to force fluid outward through the capillary membrane.

2. The *interstitial fluid pressure* (Pif), which tends to force fluid inward through the capillary membrane when Pif is positive but outward when Pif is negative.

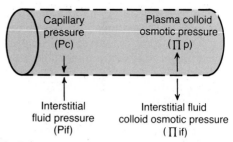

Figure 16–5. Forces operative at the capillary membrane tending to move fluid either outward or inward through the membrane.

3. The *plasma colloid osmotic pressure* (Πp), which tends to cause osmosis of fluid inward through the membrane.

4. The *interstitial fluid colloid osmotic pressure* (Πif), which tends to cause osmosis of fluid outward through the membrane.

CAPILLARY PRESSURE

Two different methods have been used to estimate the capillary pressure: (1) *direct cannulation of the capillaries,* which has given an average mean capillary pressure of about 25 mm Hg, and (2) *indirect functional measurement of the capillary pressure,* which has given a capillary pressure averaging about 17 mm Hg. These methods are the following:

Micropipet Method for Measuring Capillary Pressure. To measure pressure in a capillary by cannulation, a microscopic glass pipet is thrust directly into the capillary, and the pressure is measured by an appropriate micromanometer system. Using this method, capillary pressures have been measured in capillaries of exposed tissues of lower animals and in large capillary loops of the eponychium at the base of the fingernail in humans. These measurements have given pressures of 30 to 40 mm Hg in the arterial ends of the capillaries, 10 to 15 mm Hg in the venous ends, and about 25 mm Hg in the middle.

Isogravimetric Method for Indirectly Measuring "Functional" Capillary Pressure. Figure 16–6 illustrates an *isogravimetric* method for estimating capillary pressure. This figure shows a section of gut held by one arm of a gravimetric balance. Blood is perfused through the gut. When the arterial pressure is decreased, the resulting decrease in capillary pressure allows the osmotic pressure of the plasma proteins to cause absorption of fluid out of the gut wall and makes the weight of the gut decrease. This immediately causes displacement of the balance arm. However, to prevent this weight decrease, the venous pressure is raised an amount sufficient to overcome the effect of decreasing the arterial pressure. In other words, the capillary pressure is kept constant while decreasing the arterial pressure but raising the venous pressure.

In the lower part of the figure, the changes in arterial and venous pressures that exactly nullify all weight changes are illustrated. The arterial and

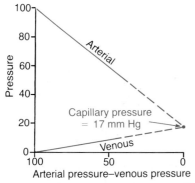

Figure 16–6. Isogravimetric method for measuring capillary pressure (explained in the text).

venous lines meet each other at a value of 17 mm Hg. Therefore, the capillary pressure must have remained at this same level of 17 mm Hg throughout these maneuvers, or otherwise filtration or absorption of fluid through the capillary walls would have occurred. Thus, in a roundabout way, the "functional" capillary pressure is measured to be about 17 mm Hg.

The Functional Capillary Pressure. It is clear that the aforementioned two methods do not give the same capillary pressure. However, the isogravimetric method determines the capillary pressure that exactly balances all the other forces tending to move fluid into or out of the capillaries. Since such a balance of forces is the normal state, the average functional capillary pressure must be very close to the pressure measured by the isogravimetric method. Therefore, one is justified in believing that the true functional capillary pressure averages about 17 mm Hg.

It is very easy to explain why the cannulation methods give higher pressure values. The most important reason is that these measurements are usually made in capillaries whose arterial ends are open and blood is actively flowing into the capillary. However, it should be recalled from the earlier discussion of capillary vasomotion that the metarterioles and precapillary sphincters are normally closed during a very large part of the vasomotion cycle. When closed, the pressure in the capillaries beyond the closures should be almost equal to the pressure at the venous ends of the capillaries, about 10 mm Hg. Therefore, when averaged over a period of time, one would expect the *functional* mean capillary pressure to be much nearer to the pressure in the venous ends of the capillaries than to the pressure in the arterial ends.

Also, there are two other reasons why the functional capillary pressure is less than the values measured by cannulation. One of these is that there are far more venous capillaries than arterial capillaries. Second, the venous capillaries are several times as permeable as the arterial capillaries. Both of these effects further weight the functional capillary pressure to a lower value.

INTERSTITIAL FLUID PRESSURE

As is true for the measurement of capillary pressure, there have also been several different methods for measuring interstitial fluid pressure, and each of these gives slightly different values but usually slightly *negative*, less than atmospheric pressure. The three methods most widely used have been (1) direct cannulation of the tissues with a micropipet, (2) measurement of the pressure from implanted perforated capsules, and (3) measurement of the pressure from a cotton wick inserted into the tissue.

Measurement of Interstitial Fluid Pressure Using the Micropipet. The same type of micropipet used for measuring capillary pressure can also be used in some tissues for measuring interstitial fluid pressure. The tip of the micropipet is about 1 micron in diameter, but even this is 20 or more times larger than the sizes of the spaces between the proteoglycan filaments of the interstitium. Therefore, the pressure that is measured is probably the pressure in a free fluid pocket.

The first pressures measured using the micropipet method ranged from −1 to +2 mm Hg but usually slightly positive. However, with experience in making such measurements in additional types of tissues, the more recent pressures have averaged about −1 mm Hg and sometimes as low as −3 mm Hg, giving average pressure values in *loose* tissues that are slightly less than atmospheric pressure.

Measurement of Interstitial Free Fluid Pressure in Implanted Perforated Hollow Capsules. Figure 16–7 illustrates an indirect method for measuring interstitial fluid pressure that may be explained as follows: A small hollow plastic capsule perforated by several hundred small holes is implanted in the tissues, and the surgical wound is allowed to heal for approximately 1 month. At the end of that time, tissue will have grown inward through the holes to line the inner surface of the sphere. Furthermore, the cavity is filled with fluid that flows freely through the perforations back and forth between the fluid in the interstitial spaces and the fluid in the cavity. Therefore, the pressure in the cavity should equal the free fluid pressure in the interstitial fluid spaces. A needle is inserted through the skin and through one of the perforations to the interior of the cavity, and the pressure is measured by use of an appropriate manometer.

Interstitial free fluid pressure measured by this method in normal *loose* subcutaneous tissue averages about −6 mm Hg. That is, the pressure is *less than atmospheric pressure*—in other words, is a semivacuum or a suction.

Instead of capsules with large perforations, capsules of porous plastic material with pore sizes too small for tissues to grow into the cavity have also been used. The pressures that have been measured have averaged −3 to −5 mm Hg.

Measurement of Interstitial Free Fluid Pressure by Means of a Cotton Wick. Still another method is to

Skin   Implanted capsule   Blood vessels

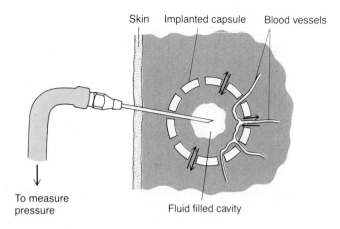

To measure pressure

Fluid filled cavity

Figure 16–7. The perforated capsule method for measuring interstitial fluid pressure.

insert into a tissue a small Teflon tube with about eight cotton fibers protruding from its end. The cotton fibers form a "wick" that makes excellent contact with the tissue fluids and transmits interstitial fluid pressure into the Teflon tube; the pressure can then be measured from the tube by usual manometric means. Pressures measured by this technique in *loose* subcutaneous tissue have also been negative, usually measuring −1 to −3 mm Hg.

Interstitial Fluid Pressures in Tightly Encased Tissues

Some tissues of the body are surrounded by tight encasements, such as the cranial vault around the brain, the strong fibrous capsule around the kidney, the fibrous sheaths around the muscles, and the sclera around the eye. In most of these, regardless of the method used for measurement, the interstitial fluid pressures are usually positive. However, even so, these interstitial fluid pressures almost invariably are still less than the pressures exerted on the outsides of the tissues by their encasements. For instance, the cerebrospinal fluid pressure surrounding the brain averages about +10 mm Hg, whereas the brain interstitial fluid pressure averages about +4 to +6 mm Hg. In the kidneys, the capsular pressure surrounding the kidney averages about +13 mm Hg, whereas the reported interstitial fluid pressures have averaged about +6 mm Hg.

Thus, if one remembers that the pressure exerted on the skin is atmospheric pressure, which is considered to be zero pressure, one might formulate a general rule that the normal interstitial fluid pressure is usually several millimeters of mercury negative with respect to the pressure that surrounds each tissue.

Is the True Interstitial Fluid Pressure in Subcutaneous Tissue Subatmospheric?

The concept that the interstitial fluid pressure is subatmospheric in many tissues of the body began with clinical observations that could not be explained by the previously held concept that interstitial fluid pressure was always positive. Some of the pertinent observations are the following:

1. When a skin graft is placed on a concave surface of the body, such as in an eye socket after removal of the eye, before the skin becomes attached to the sublying socket, fluid tends to collect underneath the graft. Also, the skin attempts to shorten, which tries

to pull it away from the concavity. Nevertheless, some negative force underneath the skin causes absorption of the fluid and usually will literally pull the skin back into the concavity.

2. Less than 1 mm Hg of positive pressure is required to inject tremendous volumes of fluid into very loose subcutaneous tissues, such as beneath the lower eyelid, in the axillary space, and in the scrotum. Amounts of fluid calculated to be more than 100 times the amount of fluid normally in the interstitial space, when injected into these areas, will cause no more than about 2 mm Hg of positive pressure. The importance of these observations is that they show that such tissues do not have strong fibers that can prevent the accumulation of fluid. Therefore, some other mechanism must be available to prevent such fluid accumulation.

3. In most natural cavities of the body where there is free fluid in dynamic equilibrium with the surrounding interstitial fluids, the pressures that have been measured have been negative. Some of these are the following:

> Intrapleural space: −8 mm Hg.
> Joint synovial spaces: −4 to −6 mm Hg.
> Epidural space: −4 to −6 mm Hg.

4. The implanted capsule method for measuring the interstitial fluid pressure can be used to record dynamic changes in this pressure. The changes are approximately those that one would calculate to occur when (1) the arterial pressure is increased or decreased, (2) fluid is injected into the surrounding tissue spaces, or (3) a highly concentrated colloid osmotic agent is injected into the blood to absorb fluid from the tissue spaces. It is not likely that these dynamic changes could be measured unless the capsule pressure closely approximated the true interstitial pressure.

Summary — An Average Value for Interstitial Fluid Pressure in Loose Subcutaneous Tissue. Although the aforementioned different methods give slightly different values for interstitial fluid pressure, there is now a general belief among most physiologists that the true interstitial fluid pressure in *loose* subcutaneous tissue is slightly less than atmospheric pressure. A pressure value that many are beginning to accept is an average value of about −3 mm Hg. However, in all tissues with tight fibrous or fascial coverings that hold the tissues together, the pressures underneath these constrictive coverings are usually more positive.

Pumping by the Lymphatic System Is the Basic Cause of the Negative Pressure

The lymphatic system is discussed later in the chapter, but we need to understand here the basic role that this system plays in determining interstitial fluid pressure. The lymphatic system is a "scavenger" system that removes excess fluid, debris, and other matter from the tissue spaces. When fluid enters the terminal lymphatic capillaries, they contract and propel the lymph forward through the lymphatic system, eventually emptying back into the circulation. In this way, any time any free fluid accumulates in the tissues, it is simply pumped away. When the amount of fluid leaking from the blood capillaries is very slight, as is true for most tissues, research evidence suggests that the lymphatic capillaries can actually pump a slight intermittent negative pressure that gives the average negativity observed in the loose tissues. The details of this lymphatic pumping system are discussed later in the chapter.

PLASMA COLLOID OSMOTIC PRESSURE

Colloid Osmotic Pressure Caused by Proteins. The proteins are the only dissolved substances in the plasma and interstitial fluid that do not diffuse readily through the capillary membrane. Furthermore, when small quantities of protein do diffuse into the interstitial fluid, these are soon removed from the interstitial spaces by way of the lymph vessels. Therefore, the concentration of protein in the plasma averages about three times as much as that in the interstitial fluid, 7.3 gm/dl in the plasma versus 2 to 3 gm/dl in the interstitial fluid.

In the basic discussion of osmotic pressure in Chapter 4, it was pointed out that only those substances that fail to pass through the pores of a semipermeable membrane exert osmotic pressure. Since the proteins are the only dissolved constituents that do not readily penetrate the pores of the capillary membrane, it is the dissolved proteins of the plasma and interstitial fluids that are responsible for the osmotic pressure at the capillary membrane. To distinguish this osmotic pressure from that which occurs at the cell membrane, it is called either *colloid osmotic pressure* or *oncotic pressure*. The term "colloid" osmotic pressure is derived from the fact that a protein solution resembles a colloidal solution despite the fact that it is actually a true solution. (The osmotic pressure that results at the cell membrane is often called *total osmotic pressure* to distinguish it from the colloid osmotic pressure because essentially all dissolved substances of the

body fluids exert osmotic pressure at the cell membrane.)

The Donnan Effect on the Colloid Osmotic Pressure. The so-called *Donnan effect* causes the colloid osmotic pressure of the plasma to be about 50 per cent greater than that caused by the proteins alone. This results from the fact that the proteins are negative ions, and to balance these negative ions a large number of positively charged ions (cations), mainly sodium ions, are held near to, but usually not bound to, the electronegative charges of the proteins. These extra cations, therefore, increase the number of osmotically active substances wherever the proteins occur, and they increase the osmotic pressure. Even more important, however (for mathematical reasons that cannot be explained here), the Donnan equilibrium effect becomes progressively more significant the higher the concentration of proteins. This means, as illustrated in Figure 16–8, that the initial few grams of protein in each 100 millimeters of plasma or interstitial fluid have much less colloid osmotic effect than do the next few grams.

Normal Values for Plasma Colloid Osmotic Pressure. The colloid osmotic pressure of normal human plasma averages approximately 28 mm Hg; 19 mm of this is caused by the dissolved protein and 9 mm by the cations held in the plasma by the Donnan effect of the proteins, as just discussed.

Note particularly that the 28 mm Hg colloid osmotic pressure that can develop at the capillary mem-

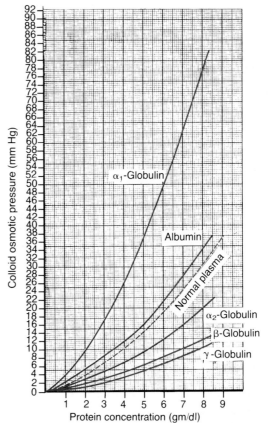

Figure 16–8. The osmotic pressure of five fractions of the plasma proteins at different concentrations. Also, the dashed line shows the osmotic pressure of normal plasma proteins, which are a mixture of the others. (Modified from Ott: *Klin. Wschr.,* 34 : 1079, 1956.)

brane is only ½₀₀ the total osmotic pressure that would develop at a *cell* membrane if normal interstitial fluid were on one side of the cell membrane and pure water on the other side. Thus, the colloid osmotic pressure of the plasma is actually a weak osmotic force; but, even so, it plays an exceedingly important role in the maintenance of normal blood and interstitial fluid volumes.

Diminishment of the Colloid Osmotic Pressure Caused by Protein Molecules Leaking Through the Capillary Pores—The "Reflection Coefficient." Only when the protein molecules are unable to pass through the capillary pores will they cause osmotic pressure. In other words, those molecules that do not pass through when they come to a pore are said to be "reflected" from the pore rather than filtered through it; osmotic pressure is created as a product of this reflection process. When all of the molecules are reflected, the pores are said to have a "reflection coefficient" for the protein molecules of 1.0. When none are reflected but instead pass on through the pores, the reflection coefficient is said to be 0.0.

One can easily see that when all of the protein molecules are reflected, they all exert full osmotic pressure, but when only one half are reflected—a "reflection coefficient" of 0.5—only half of the protein molecules then exert osmotic pressure. Thus, for the brain capillaries the reflection coefficient is almost exactly 1, and for muscle capillaries it approaches 1; therefore, either all or almost all of the plasma colloid osmotic force is exerted. On the other hand, in the liver sinusoids, the reflection coefficient approaches 0, and the plasma proteins there exert almost no osmotic pressure.

Effect of the Different Plasma Proteins on Colloid Osmotic Pressure. The plasma proteins are a mixture of proteins that contains albumin, with an average molecular weight of 69,000; globulins, 140,000; and fibrinogen, 400,000. Thus, 1 gram of globulin contains only half as many molecules as 1 gram of albumin, and 1 gram of fibrinogen contains only one sixth as many molecules as 1 gram of albumin. (It should be recalled from the discussion of osmotic pressure in Chapter 4 that the osmotic pressure is determined by the *number of molecules* dissolved in a fluid rather than by the weight of these molecules.) The average relative concentrations of the different types of proteins in the plasma and their respective colloid osmotic pressures are the following:

	gm/dl	*IIp (mm Hg)*
Albumin	4.5	21.8
Globulins	2.5	6.0
Fibrinogen	0.3	0.2
Total	7.3	28.0

Thus, about 75 per cent of the total colloid osmotic pressure of the plasma results from the albumin fraction, 25 per cent from the globulins, and almost none from the fibrinogen. Therefore, from the point of view of capillary dynamics, it is mainly albumin that is important.

Figure 16–8 illustrates graphically the colloid osmotic pressures exerted by different concentrations of albumin and four different fractions of globulins.

INTERSTITIAL FLUID COLLOID OSMOTIC PRESSURE

Although the size of the usual capillary pore is smaller than the molecular sizes of the plasma pro-

teins, this is not true of all the pores. Therefore, small amounts of plasma proteins do leak through the pores into the interstitial spaces.

The total quantity of protein in the entire 12 liters of interstitial fluid of the body is actually greater than the total quantity of protein in the plasma itself, but since this volume is four times the volume of plasma the average protein *concentration* of the interstitial fluid is usually about 40 per cent of that in plasma, or approximately 3 gm/dl. Referring to the diagram in Figure 16–8, one finds that the average colloid osmotic pressure for this concentration of proteins in the interstitial fluids is approximately 8 mm Hg when the reflection coefficient of the capillary membranes is 1.0.

EXCHANGE OF FLUID VOLUME THROUGH THE CAPILLARY MEMBRANE

Now that the different factors affecting capillary membrane dynamics have been discussed, it is possible to put all these together to see how normal capillaries function.

The average capillary pressure at the arterial ends of the capillaries is 15 to 25 mm Hg greater than at the venous ends. Because of this difference, fluid "filters" out of the capillaries at their arterial ends and then is reabsorbed into the capillaries at their venous ends. Thus, a small amount of fluid actually "flows" through the tissues from the arterial ends of the capillaries to the venous ends. The dynamics of this flow are the following:

Analysis of the Forces Causing Filtration at the Arterial End of the Capillary. The approximate average forces operative at the arterial end of the capillary that cause movement through the capillary membrane are:

	mm Hg
Forces tending to move fluid outward:	
Capillary pressure	30
Negative interstitial free fluid pressure	3
Interstitial fluid colloid osmotic pressure	8
TOTAL OUTWARD FORCE	41
Forces tending to move fluid inward:	
Plasma colloid osmotic pressure	28
TOTAL INWARD FORCE	28
Summation of forces:	
Outward	41
Inward	28
NET OUTWARD FORCE	13

Thus, the summation of forces at the arterial end of the capillary shows a net *filtration pressure* of 13 mm Hg, tending to move fluid in the outward direction.

This 13 mm Hg filtration pressure causes, on the average, about 0.5 per cent of the plasma to filter out of the arterial end of the capillaries into the interstitial spaces.

Analysis of Reabsorption at the Venous End of the Capillary. The low pressure at the venous end of the capillary changes the balance of forces in favor of absorption as follows:

	mm Hg
Forces tending to move fluid inward:	
Plasma colloid osmotic pressure	28
TOTAL INWARD FORCE	28
Forces tending to move fluid outward:	
Capillary pressure	10
Negative interstitial free fluid pressure	3
Insterstitial fluid colloid osmotic pressure	8
TOTAL OUTWARD FORCE	21
Summation of forces:	
Inward	28
Outward	21
NET INWARD FORCE	7

Thus, the force that causes fluid to move into the capillary, 28 mm Hg, is greater than that opposing reabsorption, 21 mm Hg. The difference, 7 mm Hg, is the *reabsorption pressure*. This reabsorption pressure is considerably less than the filtration pressure, but remember that the venous capillaries are more numerous and more permeable than the arterial capillaries, so that less pressure is required to cause the inward movement of fluid.

The reabsorption pressure causes about nine tenths of the fluid that has filtered out of the arterial ends of the capillaries to be reabsorbed at the venous ends. The other one tenth flows into the lymph vessels.

Flow of Fluid Through the Interstitial Spaces

The 0.5 per cent of the plasma fluid that filters out of the arterial ends of the capillaries *flows* through the tissue spaces to the venous ends of the capillaries, where all but about one tenth of it is reabsorbed, and the remainder *flows* into the lymphatic capillaries.

Distinction Between Filtration and Diffusion. It is especially important to distinguish between *filtration* and *diffusion* through the capillary membrane. Diffusion occurs in both directions, whereas filtration is the *net* movement of fluid out of the capillaries at the arterial ends. The rate of diffusion of water through all the capillary membranes of the entire body is about 240,000 ml/min, whereas the normal rate of filtration at the arterial ends of all the capillaries is only 16 ml/min, a difference of about 15,000-fold. Thus, the quantitative rate of diffusion of water and nutrients back and forth between the capillaries and the interstitial spaces is tremendous in comparison with the minute rate of "flow" of fluid through the tissues.

THE STARLING EQUILIBRIUM FOR CAPILLARY EXCHANGE

E. H. Starling pointed out almost a century ago that under normal conditions a state of near equilibrium exists at the capillary membrane, whereby the amount of fluid filtering outward from some capillaries equals almost exactly (but not quite) that quantity of fluid that is returned to the circulation by absorption through other capillaries. The very slight disequilibrium that does occur accounts for the small amount of fluid that is eventually returned by way of the lymphatics. Thus, the following chart shows the normal mean dynamics of the capillary system; however, note this time that we assume the mean *functional* capillary pressure to be 17.3 mm Hg, a slightly higher value than assumed in the earlier analyses, this time to emphasize the very slight disequilibrium:

	mm Hg
Mean forces tending to move fluid outward:	
Mean capillary pressure	17.3
Negative interstitial free fluid pressure	3.0
Insterstitial fluid colloid osmotic pressure	8.0
TOTAL OUTWARD FORCE	28.3
Mean force tending to move fluid inward:	
Plasma colloid osmotic pressure	28.0
TOTAL INWARD FORCE	28.0
Summation of mean forces:	
Outward	28.3
Inward	28.0
NET OUTWARD FORCE	0.3

Thus, we find a near-equilibrium but nevertheless a slight imbalance of forces, 0.3 mm Hg, that causes slightly more filtration of fluid into the interstitial spaces than reabsorption. This slight excess of filtration is called the *net filtration,* and it is balanced by fluid return to the circulation through the lymphatics. The normal rate of net filtration in the entire body is only about 2 ml/min.

The Filtration Coefficient. In the previous example an average net imbalance of forces at the capillary membranes of 0.3 mm Hg causes a net rate of fluid filtration in the entire body of 2 ml/min. Expressing this per millimeter of mercury, one finds a net filtration rate of 6.67 ml fluid per minute per millimeter of mercury for the entire body. This expression is the *filtration coefficient.*

The filtration coefficient can also be expressed for different areas of the body in terms of the rate of filtration per minute per millimeter of mercury per 100 grams of tissue. On this basis the filtration coefficient of the average tissue is about 0.01 ml/min/mm Hg/100 gm of tissue. However, because of extreme differences in permeabilities of the capillary systems in different tissues, this coefficient varies more than a hundredfold among the different tissues. It is very small in both brain and muscle, moderately great in subcutaneous tissue, large in the intestine, and extreme in the liver and the glomerulus of the kidney, where the pores are either numerous or wide open. By the same token, the permeation of proteins through the capillary membranes varies greatly as well. The concentration of protein in the interstitial fluid of muscles is about 1.5 gm/dl, in subcutaneous tissue 2 gm/dl, in intestine 4 gm/dl, and in liver 6 gm/dl.

Effect of Abnormal Imbalance of Forces at the Capillary Membrane

If the mean capillary pressure rises above 17 mm Hg, the net force tending to cause filtration of fluid into the tissue spaces obviously rises. Thus, a 20 mm Hg rise in mean capillary pressure causes an increase in the net filtration pressure from 0.3 mm Hg to 20.3 mm Hg, which results in 68 times as much net filtration of fluid into the interstitial spaces as normally occurs, and this would require also 68 times the normal flow of fluid into the lymphatic system, an amount that is usually too much for the lymphatics to carry away. As a result, fluid begins to accumulate in the interstitial spaces, and edema results.

Conversely, if the capillary pressure falls very low, net reabsorption of fluid into the capillaries occurs instead of net filtration, and the blood volume increases at the expense of the interstitial fluid volume. The effects of these imbalances at the capillary membrane are discussed in Chapter 25 in relation to the formation of edema.

● THE LYMPHATIC SYSTEM

The lymphatic system represents an accessory route by which fluids can flow from the interstitial spaces into the blood. And, most important of all, the lymphatics can carry proteins and large particulate matter away from the tissue spaces, neither of which can be removed by absorption directly into the blood capillary. This removal of proteins from the interstitial spaces is an absolutely essential function, without which we would die within about 24 hours.

THE LYMPH CHANNELS OF THE BODY

With the exception of a very few, almost all tissues of the body have lymphatic channels that drain excess fluid directly from the interstitial spaces. The exceptions include the superficial portions of the skin, the central nervous system, deeper portions of peripheral nerves, the endomysium of muscles, and the bones. However, even these tissues have minute interstitial channels called *prelymphatics* through which interstitial fluid can flow; eventually this fluid flows into lymphatic vessels or, in the case of the brain, flows into the cerebrospinal fluid and thence directly back into the blood.

Essentially all the lymph from the lower part of the body

—even most of that from the legs—flows up the *thoracic duct* and empties into the venous system at the juncture of the *left* internal jugular vein and subclavian vein, as illustrated in Figure 16–9.

Lymph from the left side of the head, the left arm, and parts of the chest region also enters the thoracic duct before it empties into the veins. Lymph from the right side of the neck and head, from the right arm, and from parts of the thorax enters the *right lymph duct*, which then empties into the venous system at the juncture of the *right* subclavian vein and internal jugular vein.

The Lymphatic Capillaries and Their Permeability. Most of the fluid filtering from the arterial capillaries flows among the cells and is finally reabsorbed back into the *venous ends* of the *blood capillaries;* but about *one tenth* of the fluid enters the *lymphatic capillaries* instead and returns to the blood through the lymphatic system rather than through the venous capillaries.

The minute quantity of fluid that returns to the circulation by way of the lymphatics is extremely important because substances of high molecular weight, such as proteins, cannot be reabsorbed into the venous capillaries, but they can enter the lymphatic capillaries almost completely unimpeded. The reason for this is a special structure of the lymphatic capillaries, illustrated in Figure 16–10. This figure shows the endothelial cells of the capillary attached by *anchoring filaments* to the surrounding connective tissue. At the junctions of adjacent endothelial cells, the edge of one endothelial cell usually overlaps the edge of the adjacent one in such a way that the overlapping edge is free to flap inward, thus forming a minute valve that opens to the interior of the capillary. Interstitial fluid, along with its suspended particles, can push the valve open and flow directly into the lymphatic capillary. But this fluid has difficulty leaving the capillary once it has entered because any backflow will close the flap valve. Thus, the lymphatics have valves at the very tips of the terminal lymphatic capillaries as well as valves along their larger vessels up to the point where they empty into the blood circulation.

FORMATION OF LYMPH

Lymph is derived from interstitial fluid that flows into the lymphatics. Therefore, lymph as it first flows from each tissue has almost the same composition as the interstitial fluid.

The protein concentration in the interstitial fluid of most tissues averages about 2 gm/dl, and the protein concentration of lymph flowing from these tissues is near this value. On the other hand, lymph formed in the liver has a protein concentration as high as 6 gm/dl, and lymph formed in the intestines has a protein concentration as high as 3 to 4 gm/dl. Because about two thirds of all lymph is derived from the liver and intestines, the thoracic lymph, which is a mixture of lymph from all areas of the body, usually has a protein concentration of 3 to 5 gm/dl.

The lymphatic system is also one of the major routes for absorption of nutrients from the gastroin-

Figure 16–9. The lymphatic system.

testinal tract, being responsible principally for the absorption of fats, as is discussed in Chapter 65. Indeed, after a fatty meal, thoracic duct lymph sometimes contains as much as 1 to 2 per cent fat.

Finally, even large particles, such as bacteria, can push their way between the endothelial cells of the lymphatic capillaries and in this way enter the lymph. As the lymph passes through the lymph nodes, these particles are removed and destroyed, as discussed in Chapter 33.

RATE OF LYMPH FLOW

Approximately 100 milliliters of lymph flows through the *thoracic duct* of a resting human per hour, and perhaps another 20 milliliters flows into the circulation each hour through other channels, making a total estimated lymph flow of perhaps 120 ml/hr. This is about $\frac{1}{120,000}$ of the calculated rate of fluid *diffusion* back and forth through the capillary membranes, and it is only one tenth the rate of fluid *filtration* from the arterial ends of the capillaries into the tissue spaces in the entire body. These facts illustrate that the flow of lymph is relatively small in comparison with the total exchange of fluid between the plasma and the interstitial fluid.

The rate of lymph flow is determined mainly by two factors: (1) the interstitial fluid pressure and (2) the degree of activity of the lymphatic pump.

Effect of Interstitial Fluid Pressure on Lymph Flow. Figure 16–11 illustrates the effect of

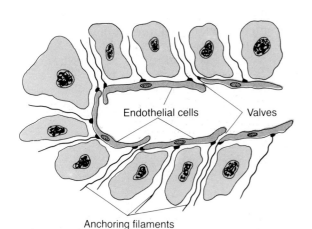

Figure 16–10. Special structure of the lymphatic capillaries that permits passage of substances of high molecular weight back into the circulation.

Figure 16–11. Relationship between interstitial fluid pressure and lymph flow. Note that lymph flow reaches a maximum as the interstitial pressure rises slightly above atmospheric pressure (0 mm Hg). (Courtesy of Drs. Harry Gibson and Aubrey Taylor.)

different levels of interstitial fluid pressure on lymph flow as measured in dog legs. Note that the lymph flow is very slight at interstitial fluid pressures more negative than −6 mm Hg. Then, as the pressure rises up to values slightly greater than 0 mm Hg (atmospheric pressure), the flow increases more than 20-fold. Therefore, any factor that increases interstitial fluid pressure will normally also increase the lymph flow. Such factors include

Elevated capillary pressure.
Decreased plasma colloid osmotic pressure.
Increased interstitial fluid protein.
Increased permeability of the capillaries.

All of these cause the balance of fluid exchange at the blood capillary membrane to favor fluid movement into the interstitium, thus increasing interstitial fluid volume, interstitial fluid pressure, and lymph flow all at the same time.

However, note that when the interstitial fluid pressure reaches a millimeter or two greater than atmospheric pressure (0 mm Hg), lymph flow fails to rise further at still higher pressures. This probably results from the fact that the increasing tissue pressure not only increases entry of fluid into the lymphatic capillaries, but also compresses the outside surfaces of the larger lymphatics, thus impeding lymph flow. These two factors appear to balance each other almost exactly. A similar phenomenon is observed in the flow of air from the alveoli of the lungs during expiration; it is called the "maximum expiratory rate," the basic principles of which are explained in detail in Chapter 37.

Effect of the Lymphatic Pump. Valves exist in all lymph channels; typical valves are illustrated in Figure 16–12 in *collecting lymphatics* into which the lymphatic capillaries empty. In the large lymphatics, valves exist every few millimeters and in the smaller lymphatics much closer than this.

Intrinsic Pumping by the Lymph Vessel. Motion pictures of exposed lymph vessels, both in animals and humans, show that when a lymph vessel becomes stretched with fluid, the smooth muscle in the wall of the vessel automatically contracts. Furthermore, each segment of the lymph vessel between successive valves functions as a separate automatic pump. That is, filling of a segment causes it to contract, and the fluid is pumped through the valve into the following lymphatic segment. This fills the subsequent segment, and a few seconds later it too contracts, the process continuing all along the lymph vessel until the fluid is finally emptied. In a large lymph vessel, this lymphatic pump can generate pressures as high as 25 to 50 mm Hg if the outflow from the vessel becomes blocked.

Pumping Caused by External Compression of the Lymphatics. In addition to the pumping caused by intrinsic contraction of the lymph vessel walls, any external factor that compresses the lymph vessel can also cause pumping. In order of their importance, such factors are

Contraction of muscles.
Movement of the parts of the body.
Arterial pulsations.
Compression of the tissues by objects outside the body.

Figure 16–12. Structure of lymphatic capillaries and a collecting lymphatic, showing also the lymphatic valves.

Obviously, the lymphatic pump becomes very active during exercise, often increasing lymph flow as much as 10-fold to 30-fold. On the other hand, during periods of rest, lymph flow is very sluggish.

The Lymphatic Capillary Pump. Many physiologists believe that the lymphatic capillary is also capable of pumping lymph, in addition to the lymphatic pump of the larger lymph vessels. As explained earlier in the chapter, the walls of the lymphatic capillaries are tightly adherent to the surrounding tissue cells by means of their anchoring filaments. Therefore, each time excess fluid enters the tissue and causes the tissue to swell, the anchoring filaments pull the lymphatic capillary open, and fluid flows into the capillary through the junctions between the endothelial cells. Then, when the tissue is compressed or when the lymphatic capillary contracts, the pressure inside the capillary increases and causes the overlapping edges of the endothelial cells to close like valves. Therefore, the pressure pushes the lymph forward into the collecting lymphatic instead of backwards through the cell junctions.

The lymphatic capillary endothelial cells contain contractile actomyosin filaments. In some animals these have been observed to cause rhythmical contraction of the lymphatic capillaries in the same way that many of the small blood vessels contract rhythmically. Therefore, it is possible that at least part of the lymph pumping results from lymph capillary contraction as well as contraction of the larger lymphatics.

Summary of Factors That Determine Lymph Flow. From the previous discussion one can see that the two primary factors that determine lymph flow are the interstitial fluid pressure and the activity of the lymphatic pump. Therefore, one can state that, roughly, *the rate of lymph flow is determined by the product of interstitial fluid pressure and the activity of the lymphatic pump.*

ROLE OF THE LYMPHATIC SYSTEM IN CONTROLLING INTERSTITIAL FLUID PROTEIN CONCENTRATION, INTERSTITIAL FLUID VOLUME, AND INTERSTITIAL FLUID PRESSURE

It is already clear that the lymphatic system functions as an "overflow mechanism" to return to the circulation excess proteins and excess fluid volume from the tissue spaces. Therefore, the lymphatic system also plays a central role in controlling (1) the concentration of proteins in the interstitial fluids, (2) the volume of interstitial fluid, and (3) the interstitial fluid pressure. Let us explain how these different factors interact.

First, remember that small amounts of proteins leak continuously out of the blood capillaries into the interstitium. Only very minute amounts of the leaked proteins return to the circulation by way of the venous ends of the blood capillaries. Therefore, these proteins tend to accumulate in the interstitial fluid, and

this in turn increases the colloid osmotic pressure of the interstitial fluids.

Second, the increasing colloid osmotic pressure in the interstitial fluid shifts the balance of forces at the blood capillary membranes in favor of fluid filtration into the interstitium. Therefore, fluid is pulled osmotically by these proteins into the interstitium, thus increasing both the interstitial fluid volume and the interstitial fluid pressure.

Third, the increasing interstitial fluid pressure greatly increases the rate of lymph flow, as explained earlier. This in turn carries away the excess volume and excess protein that has accumulated in the spaces.

Thus, once the interstitial fluid protein concentration reaches a certain level and causes a comparable increase in interstitial fluid volume and interstitial fluid pressure, the return of protein and fluid by way of the lymphatic system becomes great enough to balance exactly the rate of leakage of these from the blood capillaries. Therefore, the values of all these factors reach a steady state; they will remain exactly balanced at these levels until something changes the rate of leakage of proteins and fluid from the blood capillaries.

Significance of Negative Interstitial Fluid Pressure as a Means for Holding the Body Tissues Together

In the past it has been assumed that the different tissues of the body are held together entirely by connective tissue fibers. However, at many places in the body, connective tissue fibers are absent. This occurs particularly at points were tissues slide over each other. Yet, even at these places, the tissues are held together by the negative interstitial fluid pressure, which is actually a partial vacuum. When the tissues lose their negative pressure, fluid accumulates in the spaces, and the condition known as *edema* occurs, which is discussed in Chapter 25.

REFERENCES

Arendshorst, W. J, and Gottschalk, C. W.: Glomerular ultrafiltration dynamics: Historical perspective. Am. J. Physiol., 248:F163, 1985.

Barry, P. H, and Diamond, J. M.: Effects of unstirred layers on membrane phenomena. Physiol. Rev., 64:763, 1984.

Bassingthwaighte, J. B., and Sparks, H. V., Jr.: Indicator dilution estimation of capillary endothelial transport. Annu. Rev. Physiol., 48:321, 1986.

Brace, R. A., and Guyton, A. C.: Interaction of transcapillary Starling forces in the isolated dog forelimb. Am. J. Physiol., 233:H136, 1977.

Brace, R. A., and Guyton, A. C.: Effect of hindlimb isolation procedure on isogravimetric capillary pressure and transcapillary fluid dynamics in dogs. Circ. Res., 38:192, 1976.

Bundgaard, M.: Transport pathways in capillaries—in search of pores. Annu. Rev. Physiol., 42:325, 1980.

Chien, S. (ed.): Vascular Endothelium in Health and Disease. New York, Plenum Publishing Corp., 1988.

Comper, W. D., and Laurent, T. C.: Physiological function of connective tissue polysaccharides. Physiol. Rev., 58:255, 1978.

Crone, C.: The Malpighi lecture. From "Porositates carnis" to cellular microcirculation. Int. J. Mirocir. Clin. Exp., 6:101, 1987.

Crone, C.: When capillary permeability increases. News Physiol. Sci., 2:16, 1987.

Crone, C.: Modulation of solute permeability in microvascular endothelium. Fed. Proc., 45:77, 1986.

Crone, C., and Levitt, D. G.: Capillary permeability to small solutes. In Renkin, E. M., and Michel, C. C. (eds.): Handbook of Physiology. Sec. 2, Vol. IV. Bethesda, Md., American Physiological Society, 1984, p. 411.

Duling, B. R., and Desjardins, C.: Capillary hematocrit—What does it mean? News Physiol. Sci., 2:66, 1987.

Folkman, J., and Klagsbrun, M.: Angiogenic factors. Science, 235:442, 1987.

Gore, R. W., and McDonagh, P. E.: Fluid exchange across single capillaries. Annu. Rev. Physiol., 42:337, 1980.

Gross, J. F., and Popel, A. (eds.): Mathematics of Microcirculation Phenomena. New York, Raven Press, 1980.

Guyton, A. C.: Interstitial fluid pressure: II. Pressure-volume curves of interstitial space. Circ. Res., 16:452, 1965.

Guyton, A. C.: Concept of negative interstitial pressure based on pressures in implanted perforated capsules. Circ. Res., 12:399, 1963.

Guyton, A. C., et al.: Circlatory Physiology II. Dynamics and Control of the Body Fluids. Philadelphia, W. B. Saunders Co., 1975.

Guyton, A. C., et al.: Interstitial fluid pressure. Physiol. Rev., 51:527, 1971.

Guyton, A. C., et al.: Interstitial fluid pressure: III. Its effect on resistance to tissue fluid mobility. Circ. Res., 19:412, 1966.

Guyton, A. C., et al.: Interstitial fluid pressure: IV. Its effect on fluid movement through the capillary wall. Circ. Res., 19:1022, 1966.

Handley, D. A., and Crone, C.: Capillary permeability. News Physiol. Sci., 2:151, 1987.

Haraldsson, B.: Physiological studies of macromolecular transport across capillary walls. Acta Physiol. Scand., 128:1, 1986.

Haraldsson, B., and Rippe, B.: Serum factors other than albumin are needed for the maintenance of normal capillary permselectivity in rat hindlimb muscle. Acta Physiol. Scand., 123:427, 1985.

Hoppeler, H., and Kayar, S. R.: Capillarity and oxidative capacity of muscles. News Physiol. Sci., 3:113, 1988.

Intaglietta, M.: Transcapillary exchange of fluid in single microvessels. In Kaley, G., and Altura, B. M. (eds.): Microcirculation. Vol. 1. Baltimore, University Park Press, 1977, p. 197.

Intaglietta, M.: The measurement of pressure and flow in the microcirculation. Microvasc. Res., 5:357, 1973.

Landis, E. M.: Capillary pressure and capillary permeability. Physiol. Rev., 14:404, 1934.

Landis, E. M., and Pappenheimer, J. R.: Exchange of substances through the capillary walls. In Hamilton, W. F. (ed.): Handbook of Physiology. Sec. 2, Vol. 2. Baltimore, Williams & Wilkins, 1963, p. 961.

Marx, J. L.: Angiogenesis research comes of age. Res. News, 237:23, 1987.

Michel, C. C.: Fluid movements through capillary walls. In Renkin, E. M., and Michel, C. C. (eds.): Handbook of Physiology. Sec. 2, Vol. IV. Bethesda, Md., American Physiological Society, 1984, p. 375.

Nicoll, P. A., and Taylor, A. E.: Lymph formation and flow. Annu. Rev. Physiol., 39:73, 1977.

Olesen, S. -P., and Crone, C.: Substances that rapidly augment ionic conductance of endothelium in cerebral venules. Acta Physiol. Scand., 127: 233, 1986.

Pappenheimer, J. R.: Contributions to microvascular research of Jean Leonard Marie Poiseuille. In Renkin, E. M., and Michel, C. C. (eds.): Handbook of Physiology. Sec. 2, Vol. IV. Bethesda, Md., American Physiological Society, 1984, p. 1.

Pappenheimer, J. R.: Passage of molecules through capillary walls. Physiol. Rev., 33:387, 1953.

Rhodin, J. A. G.: Architecture of the vessel wall. In Bohr, D. F., et al. (eds.): Handbook of Physiology. Sec. 2, Vol. II. Baltimore, Williams & Wilkins, 1980, p. 1.

Rippe, B., and Haraldsson, B.: How are macromolecules transported across the capillary wall? New Physiol. Sci., 2:135, 1987.

Rothschild, M. A., et al.: Albumin synthesis. In Javitt, N. B. (ed.): International Review of Physiology: Liver and Biliary Tract Physiology I. Vol. 21. Baltimore, University Park Press, 1980, p. 249.

Shepro, D., and D'Amore, P. A.: Physiology and biochemistry of the vascular wall endothelium. In Renkin, E. M., and Michel, C. C. (eds.): Handbook of Physiology. Sec. 2, Vol. IV. Bethesda, Md., American Physiological Society, 1984, p. 103.

Simionescu, M., and Simionescu, N.: Ultrastructure of the microvascular wall: Functional correlations. In Renkin, E. M., and Michel, C. C. (eds.): Handbook of Physiology. Sec. 2, Vol. IV. Bethesda, Md., American Physiological Society, 1984, p. 41.

Simionescu, N.: Cellular aspects of transcapillary exchange. Physiol. Rev., 63:1536, 1983.

Stromme, S. B., et al.: Interstitial fluid pressure in terrestrial and semiterrestrial animals. J. Appl. Physiol., 27:123,1969.

Taylor, A. E., and Granger, D. N.: Exchange of macromolecules across the microcirculation. In Renkin, E. M., and Michel, C. C. (eds.): Handbook of Physiology. Sec. 2, Vol. IV. Bethesda, Md., American Physiological Society, 1984, p. 467.

Taylor, A. E., and Townsley, M. I.: Evaluation of the Starling fluid flux equation. News Physiol. Sci., 2:48, 1987.

Vanhoutte, P. M.: Vasodilation: Vascular Smooth Muscle, Peptides, Autonomic Nerves, and Endothelium. New York, Raven Press, 1988.

Wiedeman, M. P.: Architecture. In Renkin, E. M., and Michel, C. C. (eds.): Handbook of Physiology. Sec. 2, Vol. IV. Bethesda, Md., American Physiological Society, 1984, p. 11.

CHAPTER 17

• *Local Control of Blood Flow by the Tissues, and Humoral Regulation* •

● LOCAL CONTROL OF BLOOD FLOW IN RESPONSE TO TISSUE NEED

One of the most fundamental and important characteristics of the circulation is the ability of each tissue to control its own local blood flow in proportion to its need. Furthermore, as the need for blood flow changes, the flow follows the change.

What are some of the specific needs of the tissues for blood flow? The answer to this is manyfold, including the following:

1. Delivery of oxygen to the tissues.
2. Delivery of other nutrients, such as glucose, amino acids, fatty acids, and so forth.
3. Removal of carbon dioxide from the tissues.
4. Removal of hydrogen ions from the tissues.
5. Maintenance of proper concentrations of other ions in the tissues.
6. Transport of various hormones and other specific substances to the different tissues.

Also, certain organs have special requirements. For instance, blood flow to the skin determines heat loss from the body and in this way helps control the body temperature. Delivery of adequate quantities of blood plasma to the kidneys allows the kidneys to excrete the waste products of the body.

We see later that most of these factors exert extreme degrees of control over local blood flow.

Variations in Blood Flow in Different Tissues and Organs. In general, the greater the degree of metabolism in an organ, the greater its blood flow. Note, for instance, in Table 17–1 the very large blood flows in the various glandular organs—for example, several hundred ml/min/100 gm of thyroid or adrenal gland tissue and a blood flow of 95 ml/min/100 gm of liver.

Also note the extremely large blood flow through the kidneys, 360 ml/min/100 gm. This extreme amount of flow is required for the kidneys to perform their function of cleansing the blood of waste products.

On the other hand, most surprising is the low blood flow to the resting muscles of the body, even though they constitute between 30 and 40 per cent of the total body mass. However, in the resting state the metabolic activity of the muscles is very low, and so also is the blood flow, only 4 ml/min/100 gm. Yet, during very heavy exercise, the metabolic activity can increase as much as 50-fold and the blood flow as much as 20-fold, rising to as high as 80 ml/min/100 gm.

The Importance of Blood Flow Control by the Local Tissues. One might ask the simple question: Why not simply allow a very large blood flow through every tissue of the body, always enough to supply the tissue's needs whether the activity of the tissue is very little or very great? The answer to this is equally simple: To do this would require many times the amount of blood flow that the heart can pump.

Experiments have shown that the blood flow to each tissue is usually regulated at the minimal level that will supply its requirements, no more, no less. For instance, in tissues for which the most important requirement is delivery of oxygen, the blood flow is only slightly more than the amount required to maintain the tissues at full oxygenation but no more than this. By controlling the local blood flow in such an exact way, the tissues never suffer from nutritional deficiency, and yet the workload on the heart is kept at a minimum.

MECHANISMS OF BLOOD FLOW CONTROL

Local blood flow control can be divided into two different phases: (1) acute control and (2) long-term control. Acute control is achieved by rapid changes in local constriction of the arterioles, metarterioles, and precapillary sphincters, occurring within seconds to minutes to provide a rapid means for maintaining appropriate local tissue blood flow. Long-term con-

Table 17–1 BLOOD FLOW TO DIFFERENT ORGANS AND TISSUES UNDER BASAL CONDITIONS

	Per cent	Ml/min	Ml/min/ 100 gm
Brain	14	700	50
Heart	4	200	70
Bronchi	2	100	25
Kidneys	22	1100	360
Liver	27	1350	95
Portal	(21)	(1050)	
Arterial	(6)	(300)	
Muscle (inactive state)	15	750	4
Bone	5	250	3
Skin (cool weather)	6	300	3
Thyroid gland	1	50	160
Adrenal glands	0.5	25	300
Other tissues	3.5	175	1.3
Total	100.0	5000	—

Based mainly on data compiled by Dr. L. A. Sapirstein.

trol, on the other hand, means slow changes in flow over a period of days, weeks, or even months. In general, the long-term changes come about as a result of an increase or decrease in the physical sizes and numbers of actual blood vessels supplying the tissues.

Acute Control of Local Blood Flow

Effect of Tissue Metabolism on Local Blood Flow. Figure 17–1 illustrates the approximate quantitative effect on blood flow of increasing the rate of metabolism in a local tissue, such as muscle. Note that an increase in metabolism up to eight times normal increases the blood flow about fourfold. The increase in flow at first is less than the increase in metabolism. However, once the metabolism rises high enough to remove most of the nutrients from the blood, further increase in metabolism can occur only with a concomitant increase in blood flow to supply the required nutrients.

Local Blood Flow Regulation When Oxygen Availability Changes. One of the most necessary of the nutrients is oxygen. Whenever the availability of oxygen to the tissues decreases, such as at high altitude, in pneumonia, in carbon monoxide poisoning (which poisons the ability of hemoglobin to transport oxygen), or in cyanide poisoning (which poisons the ability of the tissues to use oxygen), the blood flow through the tissues increases markedly. Figure 17–2 shows that as the arterial oxygen saturation falls to about 25 per cent of normal, the blood flow through an isolated leg increases about threefold; that is, the blood flow increases almost enough, but not quite, to make up for the decreased amount of oxygen in the blood, thus automatically maintaining an almost constant supply of oxygen to the tissues. Cyanide poisoning of local tissue areas can cause a local blood flow increase as much as sevenfold, thus illustrating the extreme effect of oxygen deficiency in increasing blood flow.

There are two basic theories for the regulation of local blood flow when either the rate of tissue metabolism changes or the availability of oxygen changes. These are (1) the *vasodilator theory* and (2) the *oxygen demand theory*.

The Vasodilator Theory for Local Blood Flow Regulation. According to this theory, the greater the rate of metabolism, the less the blood flow, or the less the availability of oxygen and other nutrients to a tissue, the greater becomes the rate of formation of a *vasodilator substance*. The vasodilator substance then is believed to diffuse back to the precapillary sphincters, metarterioles, and arterioles to cause dilatation. Some of the different vasodilator substances that have been suggested are *adenosine, carbon dioxide, lactic acid, adenosine phosphate compounds, histamine, potassium ions,* and *hydrogen ions*.

Most of the vasodilator theories assume that the vasodilator substance is released from the tissue mainly in response to oxygen deficiency. For instance, experiments have shown that decreased availability of oxygen can cause both adenosine and lactic acid to be released from the tissues; these are substances that can cause intense vasodilation and therefore could be responsible, or partially responsible, for the local blood flow regulation.

Many physiologists have suggested that the sub-

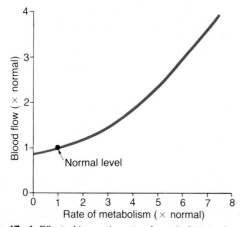

Figure 17–1. Effect of increasing rate of metabolism on tissue blood flow.

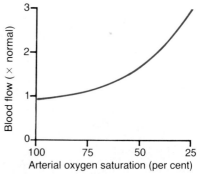

Figure 17–2. Effect of arterial oxygen saturation on blood flow through an isolated dog leg.

stance *adenosine* is perhaps the most important of the local vasodilators for controlling local blood flow. For instance, minute quantities of adenosine are released from heart muscle cells whenever coronary blood flow becomes too little, and it is believed that this causes local vasodilation in the heart and thereby returns the blood flow back toward normal. Also, whenever the heart becomes overly active and the heart's metabolism increases, this too causes excessive utilization of oxygen, followed by decreased oxygen concentration in the heart muscle with consequent degradation of adenosine triphosphate (ATP) and increased formation of adenosine. Here again it is believed that some of this adenosine leaks out of the cells to cause coronary vasodilation, with increased coronary blood flow to supply the nutrient demands of the active heart.

Although the research evidence is less clear, some physiologists have also suggested that the same adenosine mechanism might control blood flow in skeletal muscle and other tissues of the body as well as in the heart.

The problem with the different vasodilator theories of local blood flow regulation has been the following: It has been difficult to prove that sufficient quantities of any single vasodilator substance are indeed formed in the tissues to cause all of the measured increase in blood flow either in states of increased tissue metabolic demand or in decreased tissue oxygen. On the other hand, perhaps a combination of all the different vasodilators could increase the blood flow sufficiently.

The Oxygen Demand Theory for Local Blood Flow Control. Although the vasodilator theory is accepted by most physiologists, several critical facts have made a few physiologists favor still another theory, which can be called either the oxygen demand theory or, more accurately, the *nutrient demand theory* (because probably other nutrients besides oxygen are involved). Oxygen (and other nutrients as well) is required to maintain vascular muscle contraction. Therefore, in the absence of an adequate supply of oxygen and other nutrients, it is reasonable to believe that the blood vessels would naturally dilate. Also, increased utilization of oxygen in the tissues as a result of increased metabolism would theoretically decrease the local tissue oxygen availability, and this too would cause local vasodilation.

A mechanism by which the oxygen demand theory could operate is illustrated in Figure 17–3. This figure shows what might be called a "tissue unit," consisting of a metarteriole, a single capillary, and its surrounding tissue. At the origin of the capillary is a *precapillary sphincter*, and around the metarteriole are several smooth muscle fibers. Observing under a microscope a thin tissue, such as a bat's wing, one sees that the precapillary sphincters are normally either completely open or completely closed, and the degree of constriction of the metarteriole also varies with time. The number of precapillary sphincters that are open at any given time is approximately proportional to the requirements of the tissue for nutrition. In addition,

Precapillary sphincter

Figure 17–3. Diagram of a tissue unit area for explanation of local feedback control of blood flow.

the precapillary sphincters and metarterioles often open and close cyclically several times per minute, with the duration of the open phases approximately proportional to the metabolic needs of the tissues. The cyclic opening and closing is called *vasomotion.*

Now, let us explain how oxygen concentration in the local tissue could regulate blood flow through the area. Since smooth muscle requires oxygen (or other nutrients or both) to remain contracted, one might assume that the strength of contraction of the sphincters would increase with an increase in oxygen concentration. Consequently, when the oxygen concentration in the tissue rises above a certain level, the precapillary and metarteriole sphincters presumably close and remain closed until the tissue cells consume the excess oxygen. When the oxygen concentration falls low enough, the sphincters open once more to begin the cycle again.

The evidence against the oxygen demand theory is that under some conditions vascular smooth muscle can remain contracted for long periods of time in the presence of extremely minute concentrations of oxygen—concentrations below those normally found in the tissues. A possible answer to this is that the smooth muscle in the microvessels might be genetically more sensitive to oxygen lack than are the types of smooth muscle that have thus far been studied. Indeed, in very small, isolated, perfused arteries (with internal diameters of approximately 0.5 millimeters), *marked vasodilation does occur at the low oxygen concentrations found even normally in the tissues.*

Thus, on the basis of presently available data, either a vasodilator theory or an oxygen demand theory could explain local blood flow regulation in response to the metabolic needs of the tissues. Perhaps the truth lies in a combination of the two mechanisms.

Possible Role of Other Nutrients Besides Oxygen in the Control of Local Blood Flow. Under special conditions, it has been shown that lack of glucose in the perfusing blood for longer than a few minutes can cause local tissue vasodilation. It is possible that this same effect occurs when other nutrients, such as amino acids or fatty acids, are also deficient, though this has not been studied adequately. Also, vasodilation occurs in the vitamin deficiency disease *beriberi,* in which the patient usually has deficiencies of thiamine, niacin, and riboflavin. In this disease, the peripheral vascular blood flow increases as much as twofold to threefold. Since these vitamins are all concerned with the oxidative phosphorylation mechanism for generating ATP in the local tissues, one could suspect that deficiency of these vitamins leads to diminished smooth muscle contractile ability and therefore to the local vasodilation.

Special Examples of "Metabolic" Control of Local Blood Flow

The mechanisms that we have described thus far for local blood flow control are frequently called "metabolic mechanisms" because all of them function in response to the metabolic needs of the tissues. Two additional special instances of metabolic control of local blood flow are *reactive hyperemia and active hyperemia.*

Reactive Hyperemia. When the blood supply to a tissue is blocked for a few seconds to several hours and then is unblocked, the flow through the tissue usually increases to about five times normal; the increased flow will continue for a few seconds if the block has lasted only a few seconds but sometimes for as long as many hours if the blood flow has been stopped for an hour or more. This phenomenon is called *reactive hyperemia.* Reactive hyperemia is almost certainly another manifestation of the local "metabolic" blood flow regulation mechanism; that is, lack of flow sets into motion all of those factors that cause vasodilation. Following short periods of vascular occlusion, the extra blood flow during the reactive hyperemia phase lasts long enough to repay almost exactly the tissue oxygen deficit that has accrued during the period of occlusion. This mechanism emphasizes the close connection between local blood flow regulation and delivery of nutrients to the tissues.

Active Hyperemia. When any tissue becomes highly active, such as a muscle during exercise, a gastrointestinal gland during a hypersecretory period, or even the brain during rapid mental activity, the rate of blood flow through the tissue increases. Here again, by simply applying the basic principles of local blood flow control, one can easily understand this so-called *active hyperemia.* The increase in local metabolism causes the cells to devour the tissue fluid nutrients extremely rapidly and also to release large quantities of vasodilator substances. The result obviously would be to dilate the local blood vessels and, therefore, to increase local blood flow. In this way, the active tissue will receive the additional nutrients required to sustain its new level of function.

As pointed out earlier, active hyperemia in skeletal muscle can increase the local blood flow as much as 20-fold during intense exercise.

"Autoregulation" of Blood Flow When the Arterial Pressure Changes From Normal —"Metabolic" Versus "Myogenic" Mechanisms

In most tissues of the body, an acute increase or a decrease in arterial pressure will cause an immediate increase or decrease in blood flow. However, within less than a minute, the blood flow usually returns most of the way back toward the normal level. Therefore, if a few minutes are allowed for the local blood flow control mechanisms to function properly, the blood flow will be related to arterial pressure approximately in accord with the solid curve labeled "acute" in Figure 17–4. Note that between an arterial pressure of approximately 70 mm Hg and 175 mm Hg the blood flow increases only to 30 per cent. This maintenance of the flow at relatively normal levels despite marked changes in arterial pressure is called *autoregulation of blood flow.* For almost a century, two different views have been proposed to explain the acute autoregulation mechanism. These have been called (1) the metabolic theory and (2) the myogenic theory.

The *metabolic theory* can be understood very easily by simply applying the basic principles of local blood flow regulation already discussed in earlier sections. Thus, when the arterial pressure becomes too great, the excess flow provides too many nutrients to the tissues and also flushes vasodilator substances out of the tissues; both these effects will then cause the blood vessels to constrict and the flow to return nearly to normal despite the increased pressure.

The *myogenic theory,* on the other hand, suggests that still another mechanism not at all related to tissue metabolism explains the phenomenon of autoregulation. This theory is based on the observation that sudden stretch of small blood vessels will cause the smooth muscle of the vessel wall to contract. Therefore, it is believed that when high arterial pressure stretches the vessel, this in turn causes vascular con-

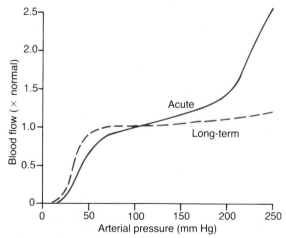

Figure 17–4. Effect of increasing arterial pressure on blood flow through a muscle. The solid curve shows the effect if the arterial pressure is raised over a period of a few minutes. The dashed curve shows the effect if the arterial pressure is raised extremely slowly over a period of many weeks.

striction and reduces the blood flow nearly back to normal. Conversely, at low pressures, the degree of stretch of the vessel is less, so that the smooth muscle relaxes and allows increased flow.

Certainly, some experiments are difficult to explain without invoking the myogenic theory of autoregulation. Yet it is doubtful that myogenic autoregulation is a very powerful functional mechanism in most of the body for a very clear reason. A strong myogenic mechanism everywhere in the body could rapidly lead to death in the following way: An increase in pressure would stretch the blood vessels, which would then cause vasoconstriction. The increased constriction would increase the peripheral resistance and raise the blood pressure still more. This secondary increase in pressure would cause another cycle of stretch followed by still more vasoconstriction and then still more increase in pressure. Thus, a vicious circle would ensue, and if this should occur strongly in all the body at once, the arterial pressure would rise suddenly to such a high level that the heart would fail. Even so, in certain isolated tissues as well as in many intermediate-size arteries, it is difficult to explain some of the phenomena observed without invoking the myogenic theory.

It has been suggested especially that the myogenic mechanism protects the capillaries from excessively high blood pressures. That is, if the pressure in the small arteries and arterioles rises too high, these vessels would simply constrict and prevent this high pressure from being transmitted into the capillaries, which are so weak that excessive pressure could rupture them.

Special Mechanisms for Controlling Local Blood Flow in Specific Tissues

Although the general mechanisms for local blood flow control discussed thus far are present in almost all tissues of the body, distinctly different mechanisms operate in a few special areas. These are all discussed in relation to specific organs, but two notable ones are as follows:

1. In the *kidneys,* blood flow control is vested mainly in a mechanism called *tubuloglomerular feedback,* in which the composition of the fluid in the early distal tubule is detected where this tubule abuts against the afferent arteriole at the *juxtaglomerular apparatus.* When too much fluid filters from the blood through the glomerulus into the tubular system, appropriate feedback signals cause changes in constriction of both the afferent and efferent arterioles, in this way reducing both renal blood flow and glomerular filtration rate. The details of this mechanism are discussed in Chapter 26.

2. In the *brain,* in addition to control of blood flow by tissue oxygen concentration, the concentrations of carbon dioxide and hydrogen ions also play very prominent roles. An increase of either of these dilates the cerebral vessels and allows rapid washout of the excess carbon dioxide or hydrogen ions. This is important because the excitatory function of the brain itself is highly dependent on very exact control of both the carbon dioxide concentration and the hydrogen ion concentration. This special mechanism for cerebral blood flow control is presented in Chapter 61.

A Mechanism for Secondary Dilatation of the Larger Arteries When Microvascular Blood Flow Increases — The Endothelial-Derived Relaxing Factor

The local mechanisms for controlling tissue blood flow can dilate only the very small microvessels located in the immediate tissue itself because local feedback caused by vasodilator substances or oxygen deficiency can affect only these vessels, not the larger arteries back upstream. Yet, when blood flow through the microvascular portion of the circulation increases, this is believed to entrain secondarily another mechanism that does dilate the larger vessels as well. This mechanism is the following:

The endothelial cells lining the arterioles and small arteries synthesize several different substances that, when released, can affect the degree of contraction of the arterial wall. The most important of these is a vasodilator substance called *endothelial-derived relaxing factor (EDRF).* Rapid flow of blood through the arteries causes "sheer-stress" on the endothelial cells because of viscous drag of the blood against the vascular walls. This stress contorts the endothelial cells in the direction of flow and causes greatly increased release of EDRF. The EDRF then relaxes the arterial wall, causing it to dilate.

This is a fortunate mechanism because it causes a secondary increase in the dimensions of the larger blood vessels whenever the microvascular blood flow increases. Without such a response, the effectiveness of local blood flow control would be greatly compromised because much of the resistance to blood flow is in the upstream arterioles and small arteries.

Long-Term Vascular Dilatation of Upstream Vessels. When the blood flow through an arteriole, small artery, or even larger artery continues at excessive velocity for days, weeks, or months, the structural dimensions of the vessel enlarge — not merely relaxation of the smooth muscle in the vessel wall. This is caused either by some effect of the EDRF mechanism or perhaps by the prolonged physical overdilation itself of the vessel wall. The sizes of all the different arteries throughout the body appear to be continually adjusted and readjusted throughout the lifetime of the person, so that the velocity of blood flow is never great enough to cause an inordinate amount of blood flow resistance.

LONG-TERM BLOOD FLOW REGULATION

Thus far, most of the mechanisms for local blood flow regulation that we have discussed act within a few seconds to a few minutes after the local tissue conditions have changed. Yet, even after full function of these acute mechanisms, the blood flow usually is adjusted only about three quarters of the way to the exact requirement of the tissues. For instance, when the arterial pressure is suddenly increased from 100 mm Hg to 150 mm Hg, the blood flow increases almost instantaneously about 100 per cent. Then within 30 seconds to 2 minutes the flow decreases back to about 15 per cent above the original control value. This illustrates the rapidity of the acute type of local regulation, but at the same time it also demonstrates that the regulation is still very incomplete because there remains an excess 15 per cent increase in blood flow.

However, over a period of hours, days, and weeks a long-term type of local blood flow regulation develops in addition to the acute regulation. This long-term regulation gives far more complete regulation than the acute mechanism. For instance, in the above example, if the arterial pressure remains at 150 mm Hg indefinitely, within a few weeks the blood flow through the tissues will gradually reapproach almost exactly the normal value. Figure 17–4 illustrates by the dashed curve the extreme effectiveness of this long-term local blood flow regulation. Note that once the long-term regulation has had time to occur, changes in arterial pressure between 50 and 250 mm Hg have very little effect on the rate of local blood flow.

Long-term regulation also occurs when the metabolic demands of a tissue change. Thus, if a tissue becomes chronically overactive and therefore requires chronically increased quantities of nutrients, the blood supply gradually increases to match the needs of the tissue.

Mechanism of Long-Term Regulation — Change in Tissue Vascularity

The mechanism of long-term local blood flow regulation is a change in the degree of vascularity of the tissues. That is, if the arterial pressure falls to 60 mm Hg and remains at this level for many weeks, the physical structural sizes of the vessels in the tissue increase, and under some conditions even the numbers of vessels increase as well; if the pressure then rises to a very high level, the number and sizes of vessels decrease. Likewise, if the metabolism in a given tissue is increased for a prolonged period of time, vascularity increases; if the metabolism is decreased, vascularity decreases.

Thus, there is reconstruction of the tissue vasculature to meet the needs of the tissues. This reconstruction occurs very rapidly (within days) in extremely young animals. It also occurs rapidly in new growth of tissue, such as in scar tissue and in cancerous tissue; on the other hand, it occurs very slowly in old, well-established tissues. Therefore, the time required for long-term regulation to take place may be only a few days in the neonate or as long as months or even years in the elderly person. Furthermore, the final degree of response is much greater in younger tissues than in older, so that in the neonate the vascularity will adjust to match almost exactly the needs of the tissue for blood flow, whereas in older tissues, vascularity frequently lags far behind the needs of the tissues.

Role of Oxygen in Long-Term Regulation. A probable stimulus for increased or decreased vascularity in many instances is need of the tissue for oxygen. One reason for believing this is that increased vascularity occurs in the tissues of many animals that live at high altitudes, where the atmospheric oxygen is low. A second reason is that fetal chicks hatched in low oxygen have up to two times as much vascular

conductivity as is normally true. This effect has also been demonstrated dramatically in newborn premature humans who are put into an oxygen tent for therapeutic purposes. The excess oxygen causes almost immediate cessation of new vascular growth in the retina of the eye and even causes degeneration of some of the capillaries that have already formed. Then when the baby is taken out of the oxygen, there is explosive overgrowth of new vessels to make up for the sudden decrease in available oxygen; indeed, there is often so much overgrowth that the vessels grow into the eye's vitreous humor and eventually cause blindness. (This condition is called *retrolental fibroplasia.*)

Growth of New Vessels — Angiogenesis and Angiogenic Factors

The term "angiogenesis" means growth of new blood vessels. Angiogenesis occurs mainly in response to the presence of angiogenic factors released from (1) ischemic tissues, (2) tissues that are growing rapidly, or (3) tissues that have excessively high metabolic rates.

A dozen or more such angiogenic factors have been found, almost all of which are small peptides. Three of those that have been best characterized are *endothelial cell growth factor (ECGF), fibroblast growth factor (FGF),* and *angiogenin,* each of which has been isolated either from tumors or from other tissues that generally have inadequate blood supply. Presumably it is the deficiency of tissue oxygen, other nutrients, or both that leads to the formation of the angiogenic factors.

Essentially all the angiogenic factors promote new vessel growth in the same manner. They cause new vessels to sprout from either small venules or occasionally capillaries. The first step is dissolution of the basement membrane of the endothelial cells. This is followed by rapid reproduction of new endothelial cells that then stream out of the vessel wall in extended cords directed toward the source of the angiogenic factor. The cells in each cord continue to divide and eventually fold over into a tube. Next, the tube connects up with another tube budding from another donor vessel and forms a capillary loop through which blood begins to flow. If the flow is great enough, smooth muscle cells eventually invade the wall, so that some of the new vessels eventually grow to be small arterioles or perhaps even larger arteries. Thus, this process of angiogenesis explains the manner in which metabolic factors in local tissues can cause the growth of new vessels.

However, it should also be noted that certain other substances, such as some steroid hormones, have exactly the opposite effect on blood vessels, occasionally even causing dissolution of vascular cells and disappearance of vessels. Therefore, blood vessels can increase when needed or at other times disappear as well.

DEVELOPMENT OF COLLATERAL CIRCULATION—A PHENOMENON OF LONG-TERM LOCAL BLOOD FLOW REGULATION

When either an artery or a vein is blocked, a new vascular channel usually develops around the blockage and allows at least partial resupply of blood to the affected tissue. The first stage in this process is dilatation of many minute vascular loops that are already present in the tissues and connect to the vessel above and below the blockage. This dilatation occurs within the first minute or two, indicating that it is simply a metabolic relaxation of the muscular fibers of the small vessels involved. After this initial opening of these small collateral vessels, the blood flow usually will still be less than one quarter that needed to supply the tissue needs. However, further opening occurs within the ensuing hours, so that within one day as much as half of the tissue needs may be met and within a few days often all of the tissue needs. The collateral vessels continue to grow for many months thereafter, almost always forming multiple small collateral channels rather than one single large vessel. Under resting conditions, the blood flow usually returns to normal, but the new channels rarely ever become large enough to supply maximum blood flow during strenuous tissue activity.

Thus, the development of collateral vessels follows the usual principles of both acute and long-term local blood flow control, the acute control being rapid metabolic dilatation, followed by growth and enlargement of the vessels manyfold over a period of weeks and months.

The most important example of the development of collateral blood vessels occurs following thrombosis of one of the coronary arteries. Almost all persons by the age of 60 will have had at least one coronary vessel to close. Yet most persons will not know that this has happened because collaterals will have developed rapidly enough to prevent any myocardial damage. It is in those instances in which thrombosis occurs too rapidly for simultaneous development of collaterals that serious heart attacks occur.

● HUMORAL REGULATION OF THE CIRCULATION

Humoral regulation of the circulation means regulation by substances secreted into or absorbed into the body fluids, such as by hormones, ions, or so forth. Some of these substances are formed by special glands and then transported in the blood throughout the entire body. Others are formed in local tissue areas and cause only local circulatory effects. Among the most important of the humoral factors that affect circulatory function are the following:

Vasoconstrictor Agents

Norepinephrine and Epinephrine. Norepinephrine is an especially powerful vasoconstrictor hormone; epinephrine is less so and in some instances even causes mild vasodilation, which occasionally occurs in the heart to dilate the coronary arteries during increased heart activity. When the sympathetic nervous system is generally stimulated during stress or exercise, the sympathetic nerves directly release norepinephrine that excites the heart, the veins, and the arterioles. They also cause the adrenal medullae to secrete both norepinephrine and epinephrine. These then circulate in the blood and cause almost the same excitatory effects on the circulation as the direct stimulation, thus providing a dual system of control.

Angiotensin. Angiotensin is one of the most powerful vasoconstrictor substances known. As little as *one millionth* of a gram can increase the arterial pressure of a human as much as 50 or more mm Hg.

The effect of angiotensin is to constrict very powerfully the small arterioles. If this occurs in an isolated tissue area, the blood flow to the area can be severely depressed. However, the real importance of angiotensin is that it normally acts simultaneously on all the arterioles of the body to increase the *total* peripheral resistance, thereby increasing the arterial pressure. Because of this plus several renal and adrenocortical stimulatory effects of angiotensin, this hormone plays an integral role in the regulation of the arterial pressure, as is discussed in detail in Chapter 19.

Vasopressin. Vasopressin, also called *antidiuretic hormone,* is formed in the hypothalamus (see Chapter 75) but is transported down the center of nerve axons to the posterior pituitary gland, where it is eventually secreted into the blood. Vasopressin is even more powerful than angiotensin as a vasoconstrictor, thus making it perhaps the body's most potent constrictor substance. Therefore, it is clear that vasopressin could have very intense effects on circulatory function. Yet, normally, only very minute amounts of vasopressin are secreted, so that most physiologists have thought that vasopressin plays little role in vascular control. On the other hand, experiments have shown that the concentration of circulating vasopressin during severe hemorrhage can rise enough to increase the arterial pressure as much as 60 mm Hg, and in many instances this can by itself bring the arterial pressure almost back up to normal.

Also, vasopressin has an *all-important* function in controlling water reabsorption in the renal tubules, which is discussed in Chapter 28, and therefore to help control body fluid volume. That is why this hormone is also called antidiuretic hormone.

Vasodilator Agents

Bradykinin. Several substances called *kinins* that can cause powerful vasodilation are frequently formed in the blood and tissue fluids. One of these substances is *bradykinin.*

The kinins are small polypeptides that are split away by proteolytic enzymes from alpha$_2$-globulins in the plasma or tissue fluids. An enzyme of particular importance is *kallikrein,* which is present in the blood and tissue fluids in an inactive form. Kallikrein is activated by maceration of the blood, tissue inflammation, and other similar chemical and physical effects on the blood. As kallikrein becomes activated, it acts immediately on the alpha$_2$-globulin to release a kinin called *kallidin* that is then converted by tissue enzymes into *bradykinin*. Once formed, the bradykinin persists for only a few minutes because it is digested by the enzyme *carboxypeptidase* or by *converting enzyme*, an enzyme that also plays an essential role in activating angiotensin, as discussed in Chapter 19. The activated kallikrein enzyme is destroyed by a kallikrein inhibitor also present in the body fluids.

Bradykinin causes very powerful *arteriolar dilatation* and also *increased capillary permeability*. For instance, injection of 1 *microgram* of bradykinin into the brachial artery of a person increases the blood flow through the arm as much as sixfold, and even smaller amounts injected locally into tissues can cause marked edema because of the increase in capillary pore size.

There is reason to believe that kinins play special roles in regulating blood flow and capillary leakage of fluids in inflamed tissues. It is also believed that bradykinin plays a role in regulating blood flow in the skin and also in the salivary and gastrointestinal glands.

Serotonin. Serotonin (5-hydroxytryptamine) is present in large concentrations in the chromaffin tissue of the intestine and other abdominal structures. Also, it is present in high concentration in the platelets. Serotonin can have either a vasodilator or a vasoconstrictor effect, depending on the condition or the area of the circulation. Even though these effects can sometimes be powerful, the functions of serotonin in regulation of the circulation are almost entirely unknown. Occasionally, tumors composed of chromaffin tissue develop, called *carcinoid tumors*. These secrete tremendous quantities of serotonin and cause mottled areas of vasodilation in the skin; but the very fact that these tremendous quantities of serotonin do not drastically disturb the circulation makes it doubtful that serotonin plays a widespread general role in regulation of circulatory function.

Histamine. Histamine is released in essentially every tissue of the body when it becomes damaged, inflamed, or is the subject of an allergic reaction. Most of the histamine is derived from mast cells in the damaged tissues and from basophils in the blood.

Histamine has a powerful vasodilator effect on the arterioles and, like bradykinin, also has the ability to greatly increase capillary porosity, allowing leakage of both fluid and plasma protein into the tissues. In many pathological conditions, the intense arteriolar dilatation and increased capillary porosity caused by histamine cause tremendous quantities of fluid to leak out of the circulation into the tissues, inducing edema. The local vasodilatory and edema-producing effects of histamine are especially prominent in allergic reactions and are discussed in Chapter 34.

Prostaglandins. Almost every tissue of the body contains small to moderate amounts of several chemically related substances called prostaglandins. These substances probably have important intracellular effects, but some of them are also released into the local tissue fluids and into the circulating blood under both physiological and pathological conditions. Although some of the prostaglandins cause vasoconstriction, most of the more important ones seem to be mainly vasodilator agents. Thus far, no specific pattern of function of the prostaglandins in circulatory control has been found. However, their widespread prevalence in the tissues and their myriad effects on the circulation make them ideal candidates for special roles in circulatory control, especially for control in local vascular areas. For this reason, these substances are presently under intensive research investigation, though unequivocal conclusions about their roles in circulatory regulation are still evasive.

Effects of Different Ions and Other Chemical Factors on Vascular Control

Many different ions and other chemical factors can either dilate or constrict local blood vessels. Most of them have little function in the overall *regulation* of the circulation, but their specific effects can be listed as follows:

An increase in *calcium ion* concentration causes vasoconstriction. This results from the general effect of calcium to stimulate smooth muscle contraction, as discussed in Chapter 8.

An increase in *potassium ion* concentration causes vasodilation. This results from the ability of potassium ions to inhibit smooth muscle contraction.

An increase in *magnesium ion* concentration causes powerful vasodilation, for magnesium ions inhibit smooth muscle generally.

Increased *sodium ion* concentration causes mild arteriolar dilatation. This results mainly from an increase in osmolality of the fluids rather than from a specific effect of sodium ion itself. *Increased osmolality* of the blood caused by increased quantities of *glucose* or other nonvasoactive substances also causes arteriolar dilatation. Decreased osmolality causes arteriolar constriction.

The only anions to have significant effects on blood vessels are *acetate* and *citrate,* both of which cause mild degrees of vasodilation.

An *increase in hydrogen ion* concentration (decrease in pH) causes dilatation of the arterioles. A slight *decrease in hydrogen ion* concentration causes arteriolar constriction, but an intense decrease causes dilatation, which is the same effect as that which occurs with increased hydrogen ion concentration.

An increase in carbon dioxide concentration causes moderate vasodilation in most tissues and marked vasodilation in the brain. However, carbon dioxide, acting on the vasomotor center, has an extremely powerful indirect vasoconstrictor effect that is transmitted through the sympathetic vasoconstrictor system.

REFERENCES

Antonaccio, M. J. (ed.): Cardiovascular Pharmacology. New York, Raven Press, 1984.

Banchero, N.: Cardiovascular responses to chronic hypoxia. Annu. Rev. Physiol., 49:465, 1987.

Borgstrom, P., et al.: An evaluation of the metabolic interaction with myogenic vascular reactivity during blood flow autoregulation. Acta Physiol. Scand., 122:275, 1984.

Brenner, B. M., and Laragh, J. H.: Advances in Atrial Peptide Research (American Society of Hypertension Symposium Series, Vol. 2). New York, Raven Press, 1988.

Buckley, J. P., and Ferrario, C. M.: Brain Peptides and Catecholamines in Cardiovascular Regulation. New York, Raven Press, 1987.

Burton, H. W., and Faulkner, J. A.: Angiogenesis in skeletal muscle. News Physiol. Sci., 1:160, 1986.

Chien, S. (ed.): Vascular Endothelium in Health and Disease. New York, Plenum Publishing Corp., 1988.

Cowley, A. W., and Guyton, A. C.: Quantification of intermediate steps in the renin-angiotensin-vasoconstrictor feedback loop in the dog. Circ. Res., 30:557, 1972.

Cowley, A. W., Jr., et al: Vasopressin: Cellular and Integrative Functions. New York, Raven Press, 1988.

Davies, P. F.: How do vascular endothelial cells respond to flow? News Physiol. Sci., 4:22, 1989.

Duling, B.: Oxygen, metabolism, and microcirculatory regulation. In Kaley, G., and Altura, B. M. (eds.): Microcirculation. Vol. II. Baltimore, University Park Press, 1977.

Duling, B. R., and Klitzman, B.: Local control of microvascular function: Role in tissue oxygen supply. Annu. Rev. Physiol., 42:373, 1980.

Folkman, J.: Angiogenesis: What makes blood vessels grow? News Physiol. Sci., 1:199, 1986.

Folkman, J.: The vascularization of tumors. Sci. Am., 234(5):58, 1978.

Fridovich, I., et al.: Endothelium-derived relaxing factor: In search of the endogenous nitroglycerin. News Physiol. Sci., 2:61, 1987.

Furchgott, R. F., and Vanhoutte, P. M.: Endothelium-derived relaxing and contracting factors. FASEB J., 3:1007, 1989.

Granger, H. J., and Guyton, A. C.: Autoregulation of the total systemic circulation following destruction of the central nervous system in the dog. Circ. Res., 25:379, 1969.

Guyton, A. C.: Integrative hemodynamics. In Sodeman, W. A., Jr., and Sodeman, T. M. (eds.): Pathologic Physiology: Mechanisms of Disease. 6th Ed. Philadelphia, W. B. Saunders Co., 1979, p. 169.

Guyton, A. C., et al.: Cardiac Output and Its Regulation. Philadelphia, W. B. Saunders Co., 1973.

Guyton, A. C., et al.: Circulation: Overall regulation. Annu. Rev. Physiol., 34:13, 1972.

Haddy, F. J.: The role of a humoral Na+,K+-ATPase inhibitor in regulating precapillary vessel tone. J. Cardiovasc. Pharmacol., 6:(Suppl. 2)S439, 1984.

Harris, P. D.: Movement of oxygen in skeletal muscle. News Physiol. Sci., 1:147, 1986.

Hegyeli, R. J. (ed.): Prostaglandins and Cardiovascular Disease. New York, Raven Press, 1981.

Hudlicka, O.: Development of microcirculation: Capillary growth and adaptation. In Renkin, E. M., and Michel, C. C. (eds.): Handbook of Physiology. Sec. 2, Vol. IV. Bethesda, Md., American Physiological Society, 1984, p. 165.

Johansson, B.: Vascular smooth muscle reactivity. Annu. Rev. Physiol., 43:359, 1981.

Johnson, P. C.: The myogenic response. In Bohr, D. F., et al. (eds.): Handbook of Physiology. Sec. 2. Vol. II. Baltimore, Williams & Wilkins, 1980, p. 409.

Kreye, V. A.: Direct vasodilators with unknown modes of action: The nitrocompounds and hydralazine. J. Cardiovasc. Pharmacol., 6:(Suppl. 4)S646, 1984.

Manfredi, J. P., and Holmes, E. W.: Purine salvage pathways in myocardium. Annu. Rev. Physiol., 47:691, 1985.

Olsson, R. A.: Local factors regulating cardiac and skeletal muscle blood flow. Annu. Rev. Physiol., 43:385, 1981.

Pearson, J. D., and Gordon, J. L.: Nucleotide metabolism by endothelium. Annu. Rev. Physiol., 47:617, 1985.

Peroutka, S. J.: 5-Hydroxytryptamine receptor subtypes. Annu. Rev. Neurosci., 11:45, 1988.

Regoli, D.: Neurohumoral regulation of precapillary vessels: The kallikrein-kinin system. J. Cardiovasc. Pharmacol., 6:(Suppl. 2)S401, 1984.

Renkin, E. M.: Control of microcirculation and blood-tissue exchange. In Renkin, E. M., and Michel, C. C. (eds.): Handbook of Physiology. Sec. 2, Vol. IV. Bethesda, Md., American Physiological Society, 1984, p. 627.

Rosenthal, D. L., and Guyton, A. C.: Hemodynamics of collateral vasodilatation following femoral artery occlusion in anesthetized dogs. Circ. Res., 23:239, 1968.

Samuelsson, B., et al. (eds.): Prostaglandins and Related Compounds: Sixth International Conference, Florence, Italy (Advances in Prostaglandin, Thromboxane, and Leukotriene Research, Vols. 17A and 17B). New York, Raven Press, 1987.

Schmid-Schonbein G. W.: Granulocyte: Friend and foe. News Physiol. Sci., 3:144, 1988.

Seidel, C. L., and Schildmeyer, L. A.: Vascular smooth muscle adaptation to increased load. Annu. Rev. Physiol., 49:489, 1987.

Silverman, W. A.: The lesson of retrolental fibroplasia. Sci. Am., 236(6):100, 1977.

Smith, W. L.: Prostaglandin biosynthesis and its compartmentation in vascular smooth muscle and endothelial cells. Annu. Rev. Physiol., 48:251, 1986.

Sparks, H. V., Jr.: Effect of local metabolic factors on vascular smooth muscle. In Bohr, D. F., et al. (eds.): Handbook of Physiology. Sec. 2, Vol. II. Baltimore, Williams & Wilkins, 1980, p. 475.

Steranka, L. R., et al.: Antagonists of B₂ bradykinin receptors. FASEB J., 3:2019, 1989.

Vallee, B. L., et al.: Tumor derived angiogenesis factors from rat Walker 256 carcinoma: An experimental investigation and review. Experientia, 41:1, 1985.

Vanhoutte, P. M.: Vasodilation: Vascular Smooth Muscle, Peptides, Autonomic Nerves, and Endothelium. New York, Raven Press, 1988.

Vanhoutte, P. M.: Endothelium and the control of vascular tissue. News Physiol. Sci., 2:18, 1987.

Vanhoutte, P. M., et al.: Local modulation of the adrenergic neuroeffector interaction in the blood vessel wall. Physiol. Rev., 61:151, 1981.

Wyse, G. D., et al.: Responses of human peripheral arteries to agonists: Similarities and differences. Fed. Proc., 44:331, 1985.

CHAPTER 18

•Nervous Regulation of the Circulation, and Rapid Control of Arterial Pressure •

● NERVOUS REGULATION OF THE CIRCULATION

As well as the all-important regulation of blood flow by each local tissue as discussed in the previous chapter, the nervous system provides additional powerful control of the circulation but usually quite a different kind of control. Nervous control normally has little to do with adjustment of blood flow tissue by tissue — this is the function of the local tissue blood flow control. Instead, nervous control mainly affects more global functions, such as redistributing the blood flow to different areas of the body, increasing the pumping activity of the heart, and especially providing very rapid control of arterial pressure.

The means by which the nervous system controls the circulation is almost entirely through the autonomic nervous system. The total function of this system is presented in Chapter 60. Yet, its specific anatomical and functional characteristics relating to circulatory control require special attention here as well.

THE AUTONOMIC NERVOUS SYSTEM

By far the most important part of the autonomic nervous system for regulation of the circulation is the *sympathetic nervous system.* However, the *parasympathetic nervous system* is also important for its regulation of heart function, as we see later in the chapter.

The Sympathetic Nervous System. Figure 18–1 illustrates the anatomy of sympathetic nervous control of the circulation. Sympathetic vasomotor nerve fibers leave the spinal cord through all the thoracic and the first one to two lumbar spinal nerves. These pass into the sympathetic chain and thence by two routes to the circulation: (1) through specific *sympathetic nerves* that innervate mainly the vasculature of the internal viscera and the heart and (2) through the *spinal nerves* that innervate mainly the vasculature of the peripheral areas. The precise pathways of these fibers in the spinal cord and in the sympathetic chains are discussed in Chapter 60.

Sympathetic Innervation of the Blood Vessels. Figure 18–2 illustrates the distribution of sympathetic nerve fibers to the blood vessels, showing that all the vessels except the capillaries, precapillary sphincters, and most of the metarterioles are innervated.

The innervation of the *small arteries* and *arterioles* allows sympathetic stimulation to increase the *resistance* and thereby to change the rate of blood flow through the tissues.

The innervation of large vessels, particularly of the *veins,* makes it possible for sympathetic stimulation to change the volume of these vessels and thereby to alter the volume of the peripheral circulatory system. This can translocate blood into the heart and thereby play a major role in the regulation of cardiovascular function, as we see later in this and subsequent chapters.

Sympathetic Nerve Fibers to the Heart. In addition to sympathetic nerve fibers supplying the blood vessels, other sympathetic fibers go to the heart, as was discussed in Chapter 9. It should be recalled that sympathetic stimulation markedly increases the activity of the heart, increasing the heart rate and enhancing its strength of pumping.

Parasympathetic Control of Heart Function, Especially Heart Rate. Although the parasympathetic nervous system is exceedingly important for many other autonomic functions of the body, it plays only a minor role in regulation of the circulation. Its only really important circulatory effect is its control of heart rate by way of parasympathetic fibers carried to the heart in the *vagus nerves,* shown in Figure 18–1 by the dashed nerve from the medulla directly to the heart.

The effects of parasympathetic stimulation on heart function are discussed in detail in Chapter 9. Principally, parasympathetic stimulation causes a marked *decrease* in heart rate and a slight decrease in contractility.

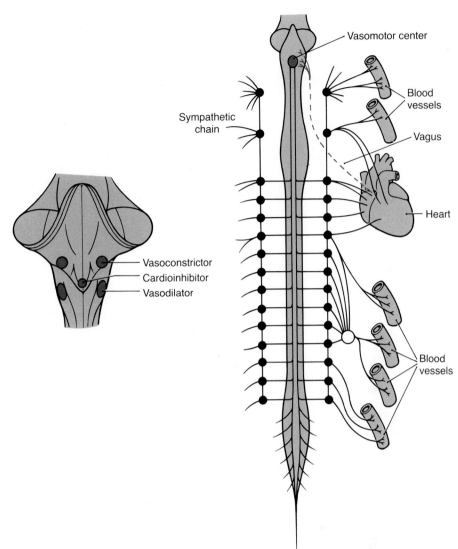

Figure 18–1. Anatomy of sympathetic nervous control of the circulation.

The Sympathetic Vasoconstrictor System and Its Control by the Central Nervous System

The sympathetic nerves carry tremendous numbers of *vasoconstrictor fibers* and only a very few vasodilator fibers. The vasoconstrictor fibers are distrib-

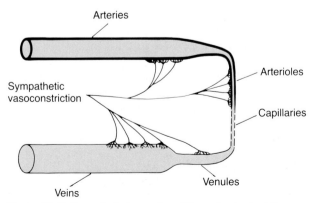

Figure 18–2. Sympathetic innervation of the systemic circulation.

uted to essentially all segments of the circulation. However, this distribution is greater in some tissues than in others. It is especially powerful in the kidneys, the gut, the spleen, and the skin but less potent in skeletal muscle and in the brain.

The Vasomotor Center and Its Control of the Vasoconstrictor System. Located bilaterally in the reticular substance of the medulla and lower third of the pons, illustrated in Figure 18–3, is an area called the *vasomotor center.* The center transmits impulses downward through the cord and thence through the sympathetic vasoconstrictor fibers to all or almost all the blood vessels of the body.

Although the total organization of the vasomotor center is still unclear, experiments have made it possible to identify certain important areas in the center, as follows:

1. A *vasoconstrictor area,* called area "C-1," located bilaterally in the anterolateral portions of the upper medulla. The neurons in this area secrete *norepinephrine;* their fibers are distributed throughout the cord, where they excite the vasoconstrictor neurons of the sympathetic nervous system.

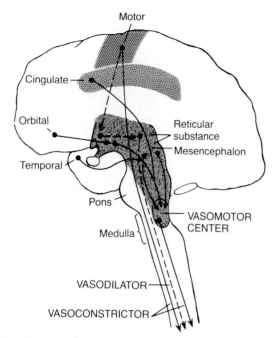

Figure 18–3. Areas of the brain that play important roles in the nervous regulation of the circulation. The dashed lines represent inhibitory pathways.

2. A *vasodilator area,* called area "A-1," located bilaterally in the anterolateral portions of the lower half of the medulla. The fibers from these neurons project upward to the vasoconstrictor area (C-1) and inhibit the vasoconstrictor activity of that area, thus causing vasodilation.

3. A *sensory area,* area "A-2," located bilaterally in the *tractus solitarius* in the posterolateral portions of the medulla and lower pons. The neurons of this area receive sensory nerve signals mainly from the vagus and glossopharyngeal nerves, and the output signals from this sensory area then help to control the activi-

ties of both the vasoconstrictor and vasodilator areas, thus providing "reflex" control of many circulatory functions. An example is the baroreceptor reflex for controlling arterial pressure, which we describe later in the chapter.

Continuous Partial Constriction of the Blood Vessels Caused by Sympathetic Vasoconstrictor Tone. Under normal conditions, the vasoconstrictor area of the vasomotor center transmits signals continuously to the sympathetic vasoconstrictor nerve fibers, causing continuous slow firing of these fibers at a rate of about one half to two impulses per second. This continual firing is called *sympathetic vasoconstrictor tone.* These impulses maintain a partial state of contraction in the blood vessels, a state called *vasomotor tone.*

Figure 18–4 demonstrates the significance of vasoconstrictor tone. In the experiment of this figure, total spinal anesthesia was administered to an animal, which completely blocked all transmission of nerve impulses from the central nervous system to the periphery. As a result, the arterial pressure fell from 100 to 50 mm Hg, illustrating the effect of loss of vasoconstrictor tone throughout the body. A few minutes later a small amount of the hormone norepinephrine was injected intravenously—norepinephrine is the substance secreted at the endings of sympathetic nerve fibers throughout the body. As this hormone was transported in the blood to all the blood vessels, the vessels once again became constricted, and the arterial pressure rose to a level even greater than normal for a minute or two until the norepinephrine was destroyed.

Control of Heart Activity by the Vasomotor Center. At the same time that the vasomotor center is controlling the degree of vascular constriction, it also controls heart activity. The lateral portions of the vasomotor center transmit excitatory impulses through the sympathetic nerve fibers to the heart to

Figure 18–4. Effect of total spinal anesthesia on the arterial pressure, showing a marked fall in pressure resulting from loss of vasomotor tone.

increase heart rate and contractility, whereas the medial portion of the vasomotor center, which lies in immediate apposition to the *dorsal motor nucleus of the vagus nerve*, transmits impulses through the vagus nerve to the heart to decrease heart rate. Therefore, the vasomotor center can either increase or decrease heart activity, this ordinarily increasing at the same time that vasoconstriction occurs throughout the body and ordinarily decreasing at the same time that vasoconstriction is inhibited.

Control of the Vasomotor Center by Higher Nervous Centers. Large numbers of areas throughout the *reticular substance* of the *pons, mesencephalon*, and *diencephalon* can either excite or inhibit the vasomotor center. This reticular substance is illustrated in Figure 18–3 by the diffuse shaded area. In general, the more lateral and superior portions of the reticular substance cause excitation, whereas the more medial and inferior portions cause inhibition.

The *hypothalamus* plays a special role in the control of the vasoconstrictor system, for it can exert powerful excitatory or inhibitory effects on the vasomotor center. The *posterolateral portions* of the hypothalamus cause mainly excitation, whereas the *anterior part* can cause mild excitation or inhibition, depending on the precise part of the anterior hypothalamus stimulated.

Many different parts of the *cerebral cortex* can also excite or inhibit the vasomotor center. Stimulation of the *motor cortex*, for instance, excites the vasomotor center because of impulses transmitted downward into the hypothalamus and thence to the vasomotor center. Also, stimulation of the *anterior temporal lobe*, the *orbital areas of the frontal cortex*, the *anterior part of the cingulate gyrus*, the *amygdala*, the *septum*, and the *hippocampus* can all either excite or inhibit the vasomotor center, depending on the precise portion of these areas that is stimulated and on the intensity of the stimulus.

Thus, widespread areas of the brain can have profound effects on the cardiovascular function.

Norepinephrine—The Sympathetic Vasoconstrictor Transmitter Substance. The substance secreted at the endings of the vasoconstrictor nerves is norepinephrine. Norepinephrine acts directly on the so-called alpha receptors of the vascular smooth muscle to cause vasoconstriction, as is discussed in Chapter 60.

The Adrenal Medullae and Their Relationship to the Sympathetic Vasoconstrictor System. Sympathetic impulses are transmitted to the adrenal medullae at the same time that they are transmitted to all the blood vessels. These cause the medullae to secrete both epinephrine and norepinephrine into the circulating blood. These two hormones are carried in the bloodstream to all parts of the body, where they act directly on the blood vessels usually to cause vasoconstriction, but sometimes the epinephrine causes vasodilation because it has a potent "beta" receptor stimulatory effect, which often dilates vessels, as is discussed in Chapter 60.

The Sympathetic Vasodilator System and Its Control by the Central Nervous System

The sympathetic nerves to skeletal muscles carry sympathetic vasodilator fibers as well as constrictor fibers. In lower animals, such as the cat, these fibers release *acetylcholine*, not norepinephrine, at their endings, though in primates the vasodilator effect might be caused by epinephrine exciting the beta receptors in the muscle vasculature.

The pathways for central nervous system control of the vasodilator system are illustrated by the dashed lines in Figure 18–3. The principal area of the brain controlling this system is the *anterior hypothalamus*.

Importance of the Sympathetic Vasodilator System. It is doubtful that the sympathetic vasodilator system plays an important role in the control of the circulation in the human because complete block of the sympathetic nerves to the muscles hardly affects the ability of the muscles to control their own blood flow in response to their needs. Yet, it is possible that at the onset of exercise the sympathetic vasodilator system causes initial vasodilation in the skeletal muscles to allow an *anticipatory increase in blood flow* even before the muscles require increased nutrients.

Emotional Fainting—Vasovagal Syncopy. A particularly interesting vasodilatory reaction occurs in persons who experience intense emotional disturbances that cause fainting. The muscle vasodilator system becomes powerfully activated, and at the same time the vagal cardioinhibitory center transmits strong signals to the heart to slow markedly the heart rate. The arterial pressure falls instantly, which also reduces blood flow to the brain and causes the person to lose consciousness. This overall effect is called *vasovagal syncopy*. Emotional fainting obviously begins with disturbing thoughts in the cerebral cortex. The pathway probably then goes to the vasodilatory center of the anterior hypothalamus, thence to the vagal centers of the medulla, and finally through the cord to the vasodilator nerves of the muscles.

● ROLE OF THE NERVOUS SYSTEM FOR RAPID CONTROL OF ARTERIAL PRESSURE

One of the most important functions of nervous control of the circulation is its capability to cause very rapid increases in arterial pressure. For this purpose, the entire vasoconstrictor and cardioaccelerator functions of the sympathetic nervous system are stimulated as a unit. At the same time there is reciprocal inhibition of the normal parasympathetic vagal inhibitory signals to the heart. In consequence, three major changes occur simultaneously, each of which helps to increase the arterial pressure. These are as follows:

1. *Almost all arterioles of the body are constricted.* This greatly increases the total peripheral resistance, impeding the run-off of blood from the arteries and thereby increasing the arterial pressure.

2. *The veins especially but the other large vessels of the circulation as well are strongly constricted.* This displaces blood out of the circulation toward the heart, thus increasing the volume of blood in the heart

chambers. This then causes the heart to beat with far greater force and therefore to pump increased quantities of blood. This, too, increases the arterial pressure.

3. Finally, *the heart itself is directly stimulated by the autonomic nervous system, further enhancing cardiac pumping.* Much of this is caused by an increase in the heart rate sometimes to as great as three times normal. In addition, sympathetic nervous signals have a direct effect to increase the contractile force of the heart muscle, this too increasing the capability of the heart to pump larger volumes of blood. Therefore, under strong sympathetic stimulation, the heart can pump for several minutes at least two times as much blood as under normal conditions. Thus, this too contributes still more to the rise in arterial pressure.

Rapidity of Nervous Control of Arterial Pressure. An especially important characteristic of nervous control of arterial pressure is its rapidity of response, beginning within seconds and often increasing the pressure to two times normal within 5 to 15 seconds. Conversely, sudden inhibition of nervous stimulation can decrease the arterial pressure to as little as one half normal within 10 to 40 seconds. Therefore, nervous control of arterial pressure is by far the most rapid of all our mechanisms for pressure control.

INCREASE IN ARTERIAL PRESSURE DURING MUSCLE EXERCISE AND OTHER TYPES OF STRESS

An important example of the ability of the nervous system to increase the arterial pressure is the increased pressure during muscle exercise. During heavy exercise, the muscles require greatly increased blood flow. Part of this increase results from local vasodilation of the muscle vasculature caused by increased metabolism of the muscle cells, as explained in the previous chapter. However, still additional increase results from simultaneous elevation of arterial pressure during the exercise. In most heavy exercise, the arterial pressure rises about 30 to 40 per cent, which will increase blood flow by approximately an additional twofold.

The increase in arterial pressure during exercise is believed to result mainly from the following effect: At the same time that the motor areas of the nervous system become activated to cause exercise, most of the reticular activating system of the brain stem is also activated, which includes greatly increased stimulation of the vasoconstrictor and cardioacceleratory areas of the vasomotor center. These raise the arterial pressure instantaneously to keep pace with the increase in muscle activity.

In many other types of stress besides muscle exercise, a similar rise in pressure can also take place. For instance, during extreme fright, the arterial pressure often rises to as high as double normal within a few seconds. This is called the *alarm reaction,* and it obviously provides a head of pressure that can immediately supply blood to any or all muscles of the body that might wish to respond instantly to cause flight from danger.

THE REFLEX MECHANISMS FOR MAINTAINING NORMAL ARTERIAL PRESSURE

Aside from the exercise and stress functions of the autonomic nervous system to raise the arterial pressure, there are also multiple subconscious nervous mechanisms for maintaining the arterial pressure at or near its normal operating level. Almost all of these are *negative feedback reflex mechanisms,* which we explain in the following sections.

The Arterial Baroreceptor Control System — Baroreceptor Reflexes

By far the best known of the mechanisms for arterial pressure control is the *baroreceptor reflex.* Basically, this reflex is initiated by stretch receptors, called either *baroreceptors* or *pressoreceptors,* which are located in the walls of the large systemic arteries. A rise in pressure stretches the baroreceptors and causes them to transmit signals into the central nervous system, and "feedback" signals are then sent back through the autonomic nervous system to the circulation to reduce arterial pressure downward toward the normal level.

Physiologic Anatomy of the Baroreceptors, and Their Innervation. Baroreceptors are spray-type nerve endings lying in the walls of the arteries; they are stimulated when stretched. A few baroreceptors are located in the wall of almost every large artery of the thoracic and neck regions; but, as illustrated in Figure 18–5, baroreceptors are extremely abundant, in (1) the wall of each internal carotid artery slightly above the carotid bifurcation, an area known as the *carotid sinus,* and (2) the wall of the aortic arch.

Figure 18–5 also shows that signals are transmitted from each carotid sinus through the very small *Hering's nerve* to the glossopharyngeal nerve and thence to the *tractus solitarius* in the medullary area of the brain stem. Signals from the arch of the aorta are transmitted through the vagus nerves also into this area of the medulla.

Response of the Baroreceptors to Pressure. Figure 18–6 illustrates the effect of different arterial pressures on the rate of impulse transmission in a Hering's nerve. Note that the carotid sinus baroreceptors are not stimulated at all by pressures between 0 and 60 mm Hg, but above 60 mm Hg they respond progressively more and more rapidly and reach a maximum at about 180 mm Hg. The responses of the aortic baroreceptors are similar to those of the carotid receptors except that they operate, in general, at pressure levels about 30 mm Hg higher.

Note especially that in the normal operating range of arterial pressure, around 100 mm Hg, even a slight change in pressure causes strong autonomic reflexes to readjust the arterial pressure back toward normal.

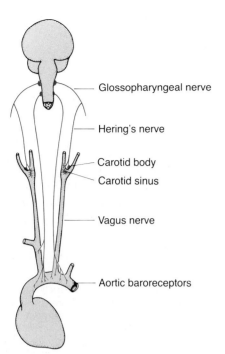

Figure 18–5. The baroreceptor system.

- Glossopharyngeal nerve
- Hering's nerve
- Carotid body
- Carotid sinus
- Vagus nerve
- Aortic baroreceptors

is falling, the rate might be as little as one quarter that for the stationary pressure.

The Reflex Initiated by the Baroreceptors. After the baroreceptor signals have entered the tractus solitarius of the medulla, secondary signals *inhibit the vasoconstrictor center* of the medulla and *excite the vagal center*. The net effects are (1) *vasodilation* of the veins and the arterioles throughout the peripheral circulatory system and (2) *decreased heart rate* and *strength of heart contraction*. Therefore, excitation of the baroreceptors by pressure in the arteries reflexly *causes the arterial pressure to decrease* because of both a decrease in peripheral resistance and a decrease in cardiac output. Conversely, low pressure has opposite effects, reflexly causing the pressure to rise back toward normal.

Figure 18–7 illustrates a typical reflex change in arterial pressure caused by occluding the common carotid arteries. This reduces the carotid sinus pressure; as a result, the baroreceptors become inactive and lose their inhibitory effect on the vasomotor center. The vasomotor center then becomes much more active than usual, causing the arterial pressure to rise and to remain elevated during the 10 minutes that the carotids are occluded. Removal of the occlusion allows the pressure to fall immediately to slightly below normal as a momentary overcompensation and then to return to normal in another minute or so.

Function of the Baroreceptors During Changes in Body Posture. The ability of the baroreceptors to maintain relatively constant arterial pressure is extremely important when a person sits or stands after having been lying down. Immediately upon standing, the arterial pressure in the head and upper part of the body obviously tends to fall, and marked reduction of this pressure can cause loss of consciousness. Fortunately, however, the falling pressure at the baroreceptors elicits an immediate reflex, resulting in strong sympathetic discharge throughout the body, and this minimizes the decrease in pressure in the head and upper body.

Thus, the baroreceptor feedback mechanism functions most effectively in the very pressure range where it is most needed.

The baroreceptors respond extremely rapidly to changes in arterial pressure; in fact, the rate of impulse firing increases during systole and decreases again during diastole. Furthermore, the baroreceptors *respond much more to a rapidly changing pressure* than to a stationary pressure. That is, if the mean arterial pressure is 150 mm Hg but at that moment is rising rapidly, the rate of impulse transmission may be as much as twice that when the pressure is stationary at 150 mm Hg. On the other hand, if the pressure

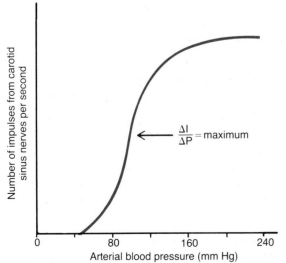

Figure 18–6. Response of the baroreceptors at different levels of arterial pressure.

$\frac{\Delta I}{\Delta P}$ = maximum

Figure 18–7. Typical carotid sinus reflex effect on arterial pressure caused by clamping both common carotids (after the two vagus nerves have been cut).

The "Buffer" Function of the Baroreceptor Control System. Because the baroreceptor system opposes either increases or decreases in arterial pressure, it is often called a *pressure buffer system,* and the nerves from the baroreceptors are called *buffer nerves.*

Figure 18–8 illustrates the importance of this buffer function of the baroreceptors. The upper record in this figure shows an arterial pressure recording for 2 hours from a normal dog and the lower record from a dog whose baroreceptor nerves from both the carotid sinuses and the aorta had previously been removed. Note the extreme variability of pressure in the denervated dog caused by simple events of the day, such as lying down, standing, excitement, eating, defecation, noises, and so forth.

Figure 18–9 illustrates the frequency distributions of the arterial pressures recorded for a full 24-hour day in both the normal dog and the denervated dog. Note that when the baroreceptors were normal, the arterial pressure remained throughout the day within a narrow range between 85 and 115 mm Hg — indeed, during most of the day at almost exactly 100 mm Hg. On the other hand, after denervation of the baroreceptors, the frequency distribution curve became the broad, low curve of the figure, showing that the pressure range increased 2.5-fold, frequently falling to as low as 50 mm Hg or rising to over 160 mm Hg. Thus, one can see the extreme variability of pressure in the absence of the arterial baroreceptor system.

Figure 18–9. Frequency distribution curves of the arterial pressure for a 24-hour period in a normal dog and in the same dog several weeks after the baroreceptors had been denervated. (From Cowley, Liard, and Guyton: *Circ. Res., 32:*564, 1973. By permission of the American Heart Association, Inc.)

In summary, a primary purpose of the arterial baroreceptor system is to reduce the daily variation in arterial pressure to about one half to one third that which would occur were the baroreceptor system not present.

Unimportance of the Baroreceptor System for Long-Term Regulation of Arterial Pressure — "Resetting" of the Baroreceptors. The baroreceptor control system is probably of little or no importance in long-term regulation of arterial pressure for a very simple reason: The baroreceptors themselves reset in 1 to 2 days to whatever pressure level they are exposed. That is, if the pressure rises from the normal value of 100 mm Hg to 200 mm Hg, extreme numbers of baroreceptor impulses are at first transmitted. During the next few seconds, the rate of firing diminishes considerably; then it diminishes much more slowly during the next 1 to 2 days, at the end of which time the rate will have returned essentially to the normal level despite the fact that the arterial pressure now remains 200 mm Hg. Conversely, when the arterial pressure falls to a very low level, the baroreceptors at first transmit no impulses at all, but gradually over a day or two the rate of baroreceptor firing returns again to the original control level.

This "resetting" of the baroreceptors obviously prevents the baroreceptor reflex from functioning as a control system for arterial pressure changes that last longer than a few days at a time. In fact, referring again to Figure 18–8 and 18–9, one can see that the average arterial pressure over any prolonged period of time is almost exactly the same whether the baroreceptors are present or not. This illustrates the *unimportance of the baroreceptor system for long-term regulation of the arterial pressure* even though it is a potent

Figure 18–8. Two-hour records of arterial pressure in a normal dog (above) and in the same dog (below) several weeks after the baroreceptors had been denervated. (From Cowley, Liard, and Guyton: *Circ. Res., 32:*564, 1973. By permission of the American Heart Association, Inc.)

mechanism for preventing the rapid changes of arterial pressure that occur moment by moment or hour by hour. Prolonged regulation of arterial pressure requires other control systems, principally the renal-body fluid-pressure control system (along with its associated hormonal mechanisms), discussed in the following chapter.

Control of Arterial Pressure by the Carotid and Aortic Chemoreceptors — Effect of Oxygen Lack on Arterial Pressure

Closely associated with the baroreceptor pressure control system is a chemoreceptor reflex that operates in much the same way as the baroreceptor reflex except that instead of stretch receptors initiating the response, chemoreceptors do this.

The chemoreceptors are chemosensitive cells sensitive to oxygen lack, carbon dioxide excess, or hydrogen ion excess. They are located in several small organs 1 to 2 millimeters in size: two *carotid bodies,* one of which lies in the bifurcation of each common carotid artery, and several *aortic bodies* adjacent to the aorta. The chemoreceptors excite nerve fibers that pass along with the baroreceptor fibers through Hering's nerves and the vagus nerves into the vasomotor center.

Each carotid or aortic body is supplied with an abundant blood flow through a small nutrient artery, so that the chemoreceptors are always in close contact with the arterial blood. Whenever the arterial pressure falls below a critical level, the chemoreceptors become stimulated because of diminished blood flow to the bodies and therefore diminished availability of oxygen and excess buildup of carbon dioxide and hydrogen ions that are not removed by the slow flow of blood.

The signals transmitted from the chemoreceptors into the vasomotor center *excite* the vasomotor center, and this elevates the arterial pressure. Obviously, this reflex helps to return the arterial pressure back toward the normal level whenever it falls too low.

However, the chemoreceptor reflex is not a powerful arterial pressure controller in the normal arterial pressure range because the chemoreceptors themselves are not stimulated strongly until the arterial pressure falls below 80 mm Hg. Therefore, it is at the lower pressures that this reflex becomes important to help prevent still further fall in pressure.

The chemoreceptors are discussed in much more detail in Chapter 41 in relation to respiratory control, in which they play an even more important role than in pressure control.

Atrial and Pulmonary Artery Reflexes that Help Regulate Arterial Pressure and Other Circulatory Factors

Both the atria and the pulmonary arteries have stretch receptors, called *low pressure receptors,* in their walls similar to the baroreceptor stretch receptors of the large systemic arteries. These low pressure receptors play an important role to minimize arterial pressure changes in response to changes in blood volume. To give an example, if 300 milliliters of blood are infused into a dog with all the receptors intact, the arterial pressure will rise only 15 mm Hg. With the arterial baroreceptors denervated, the pressure will rise 50 mm Hg. If the low pressure receptors are also denervated, the pressure will rise 120 mm Hg.

Thus, one can see that even though the low pressure receptors in the pulmonary artery and in the atria cannot detect the systemic arterial pressure, these receptors nevertheless do detect simultaneous increases in pressure in the low pressure areas of the circulation caused by an increase in volume, and they elicit reflexes parallel to the baroreceptor reflexes to make the total reflex system much more potent for control of arterial pressure.

Atrial Reflexes to the Kidneys — The Volume Reflex. Stretch of the atria also causes reflex dilatation of the afferent arterioles in the kidneys, which is the same reflex effect that occurs in other peripheral arterioles but is unusually potent in the kidneys. Also, signals are transmitted simultaneously to the hypothalamus to decrease the secretion of antidiuretic hormone, thereby indirectly affecting kidney function. The decreased afferent arteriolar resistance causes the glomerular capillary pressure to rise, with resultant increase in filtration of fluid into the kidney tubules. The diminution of antidiuretic hormone diminishes the reabsorption of water from the tubules. The combination of these two effects therefore causes rapid loss of fluid into the urine, which serves as a powerful means to return the blood volume back toward normal.

Obviously, all these mechanisms that tend to return the blood volume back toward normal following a volume overload act indirectly as pressure controllers as well as volume controllers because excess volume drives the heart to greater cardiac output and leads therefore to greater arterial pressure. This volume reflex mechanism is discussed again in Chapter 29, along with discussion of mechanisms of blood volume control.

Atrial Reflex Control of Heart Rate (the Bainbridge Reflex). An increase in atrial pressure also causes an increase in heart rate, sometimes increasing the heart rate as much as 75 per cent. A small part of this increase is caused by a direct effect of the increased atrial volume to stretch the sinus node; it was pointed out in Chapter 10 that such direct stretch can increase the heart rate as much as 15 per cent. An additional 40 to 60 per cent increase in rate is caused by a reflex called the *Bainbridge reflex.* The stretch receptors of the atria that elicit the Bainbridge reflex transmit their afferent signals through the vagus nerves to the medulla of the brain. Then, efferent signals are transmitted back through both the vagal and sympathetic nerves to increase the heart's rate and presumably also strength of contraction. Thus, this reflex helps prevent damming of blood in the veins, the atria, and the pulmonary circulation. This reflex obviously has a different purpose from that of controlling arterial pressure.

THE CENTRAL NERVOUS SYSTEM ISCHEMIC RESPONSE — CONTROL OF ARTERIAL PRESSURE BY THE VASOMOTOR CENTER IN RESPONSE TO DIMINISHED BRAIN BLOOD FLOW

Normally, most nervous control of blood pressure is achieved by reflexes originating in the baroreceptors, the chemoreceptors, and the low pressure receptors, all of which are located in the peripheral circulation outside the brain. However, when blood flow to the vasomotor center in the lower brain stem becomes decreased enough to cause nutritional deficiency, that is, to cause *cerebral ischemia,* the neurons in the vasomotor center itself respond directly to the ischemia and become strongly excited. When this occurs, the systemic arterial pressure often rises to a level as high as the heart can possibly pump. This effect is believed to be caused by failure of the slowly flowing blood to carry carbon dioxide away from the vasomotor center; the local concentration of carbon dioxide then increases greatly and has an extremely potent effect in stimulating the sympathetic nervous system. It is possible that other factors, such as the buildup of lactic acid and other acidic substances, also contribute to the marked stimulation of the vasomotor center and to the elevation in pressure. This arterial pressure elevation in response to cerebral ischemia is known as the *central nervous system ischemic response* or simply *CNS ischemic response.*

The magnitude of the ischemic effect on vasomotor activity is tremendous; it can elevate the mean arterial pressure for as long as 10 minutes sometimes to as high as 250 mm Hg. *The degree of sympathetic vasoconstriction caused by intense cerebral ischemia is often so great that some of the peripheral vessels become totally or almost totally occluded.* The kidneys, for instance, often entirely cease their production of urine because of arteriolar constriction in response to the sympathetic discharge. Therefore, *the CNS ischemic response is one of the most powerful of all the activators of the sympathetic vasoconstrictor system.*

Importance of the CNS Ischemic Response as a Regulator of Arterial Pressure. Despite the extremely powerful nature of the CNS ischemic response, it does not become very active until the arterial pressure falls far below normal, down to 60 mm Hg and below, reaching its greatest degree of stimulation at a pressure of 15 to 20 mm Hg. Therefore, it is not one of the usual mechanisms for regulating normal arterial pressure. Instead, it operates principally as an *emergency arterial pressure control system that acts rapidly and extremely powerfully to prevent further decrease in arterial pressure whenever blood flow to the brain decreases dangerously close to the lethal level.* It is sometimes called the "last ditch stand" pressure control mechanism.

The Cushing Reaction. The so-called Cushing reaction is a special type of CNS ischemic response that results from increased pressure in the cranial vault. For instance, when the cerebrospinal fluid pressure rises to equal the arterial pressure, it compresses the arteries in the brain and cuts off the blood supply to the brain. Obviously, this initiates a CNS ischemic response, which causes the arterial pressure to rise. When the arterial pressure has risen to a level higher than the cerebrospinal fluid pressure, blood flows once again into the vessels of the brain to relieve the ischemia. Ordinarily, the blood pressure comes to a new equilibrium level slightly higher than the cerebrospinal fluid pressure,

thus allowing blood to continue flowing to the brain. A typical Cushing reaction is illustrated in Figure 18–10, caused in this instance by pumping fluid under pressure into the cranial vault around the brain.

The Cushing reaction helps protect the vital centers of the brain from loss of nutrition if ever the cerebrospinal fluid pressure rises high enough to compress the cerebral arteries.

Depressant Effect of Extreme, Prolonged Ischemia on the Vasomotor Center. If cerebral ischemia becomes so severe that maximum rise in mean arterial pressure still cannot relieve the ischemia, the neuronal cells begin to suffer metabolically, and within 3 to 10 minutes they become totally inactive. The arterial pressure then falls to about 40 to 50 mm Hg, which is the level to which the pressure falls when the vasomotor center loses all its control of the circulation, so that all tonic vasoconstrictor activity is lost. Therefore, it is fortunate that the ischemic response is extremely powerful, so that arterial pressure can usually rise high enough to correct brain ischemia before it causes nutritional depression and death of the neuronal cells.

● SPECIAL FEATURES OF NERVOUS CONTROL OF ARTERIAL PRESSURE

ROLE OF THE SKELETAL NERVES AND SKELETAL MUSCLES IN INCREASING CARDIAC OUTPUT AND ARTERIAL PRESSURE

Although most nervous control of the circulation is effected through the autonomic nervous system, at least two conditions in which the skeletal nerves and muscles also play major roles in circulatory responses are the following:

The Abdominal Compression Reflex. When a baroreceptor or chemoreceptor reflex is elicited, or whenever almost any other factor stimulates the sympathetic vasoconstrictor system, the vasomotor center and other areas of the reticular substance of the brain stem transmit impulses simultaneously through skeletal nerves to skeletal muscles of the body, particularly to the abdominal muscles. This increases the basal tone of these muscles, which compresses all the venous reservoirs of the abdomen, helping translocate blood out of the abdominal vascular reservoirs toward the heart. As a result, increased quantities of blood are made available to the heart to be pumped. This overall response is called the *abdominal compression reflex.* The resulting effect

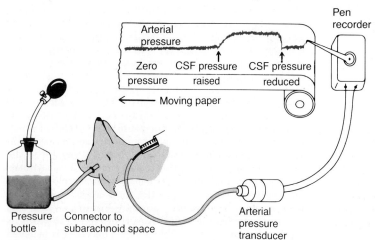

Figure 18–10. The Cushing reaction, showing a rise in arterial pressure resulting from increased cerebrospinal fluid (CSF) pressure.

on the circulation is the same as that caused by sympathetic vasoconstrictor impulses when they constrict the veins, namely, an increase in both cardiac output and arterial pressure.

The abdominal compression reflex is probably much more important than has been realized in the past, for it is well known that persons whose skeletal muscles have been paralyzed are considerably more prone to hypotensive episodes than are persons with normal skeletal muscles.

Increased Cardiac Output and Arterial Pressure Caused by Skeletal Muscle Contraction During Exercise. When the skeletal muscles contract during exercise, they compress blood vessels throughout the body. Even anticipation of exercise tightens the muscles, thereby compressing the vessels. The resulting effect is to translocate large quantities of blood from the peripheral vessels into the heart and lungs and therefore to increase the cardiac output. This is an essential effect in helping to cause the fivefold to sixfold increase in cardiac output that sometimes occurs in severe exercise. The increase in cardiac output in turn is an essential ingredient in increasing the arterial pressure during exercise, an increase usually in the range of 20 to 60 per cent.

RESPIRATORY WAVES IN THE ARTERIAL PRESSURE

With each cycle of respiration, the arterial pressure rises and falls usually 4 to 6 mm Hg in a wavelike fashion, giving rise to so-called *respiratory waves* in the arterial pressure. These result from several different effects, some of which are reflex in nature, as follows:

1. Many impulses arising in the respiratory center of the medulla "spill over" into the vasomotor center with each respiratory cycle.

2. Every time a person inspires, the pressure in the thoracic cavity becomes more negative than usual, causing the blood vessels in the chest to expand. This reduces the quantity of blood returning to the left side of the heart and thereby momentarily decreases the cardiac output and arterial pressure.

3. The pressure changes caused in the thoracic vessels by respiration can excite vascular and atrial stretch receptors.

Although it is difficult to analyze the exact relationship of all these factors in causing the respiratory pressure waves, the net result during normal respiration is usually an increase in arterial pressure during the early part of expiration and a fall in pressure during the remainder of the respiratory cycle. During deep respiration the blood pressure can rise and fall as much as 20 mm Hg with each respiratory cycle.

ARTERIAL PRESSURE "VASOMOTOR" WAVES — OSCILLATION OF THE PRESSURE REFLEX CONTROL SYSTEMS

Often in recording arterial pressure in an animal, in addition to the small pressure waves caused by the respiration some much larger waves are noted — as great as 20 to 40 mm Hg at times — that rise and fall more slowly than the respiratory waves. The duration of each cycle varies from 26 seconds in the anesthetized dog to as short as 7 to 10 seconds in the human. These waves are called *vasomotor waves* or sometimes "Mayer waves." Such records are illustrated in Figure 18–11, showing the cyclical rise and fall in arterial pressure.

Figure 18–11. *A,* Vasomotor waves caused by oscillation of the CNS ischemic response. *B,* Vasomotor waves caused by baroreceptor reflex oscillation.

The cause of vasomotor waves is usually oscillation of one or more nervous pressure control mechanisms, some of which are the following:

Oscillation of the Baroreceptor and Chemoreceptor Reflexes. The vasomotor waves of Figure 18–11B are the common type of vasomotor waves that are seen almost daily in experimental pressure recordings, though usually much less intense than shown in the figure. They are caused mainly by oscillation of the *baroreceptor reflex.* That is, a high pressure excites the baroreceptors; this then inhibits the sympathetic nervous system and lowers the pressure a few seconds later. The decreased pressure reduces the baroreceptor stimulation and allows the vasomotor center to become active once again, elevating the pressure to a high value, but again the response is not instantaneous; it is delayed until a few seconds later. This high pressure then initiates another cycle, and the oscillation continues on and on.

The *chemoreceptor reflex* can also oscillate to give the same type of waves. Usually, this reflex oscillates simultaneously with the baroreceptor reflex. In fact, it probably plays the major role in causing vasomotor waves when the arterial pressure is in the range of 40 to 80 mm Hg because in this range chemoreceptor control of the circulation becomes powerful, while baroreceptor control becomes weaker.

Oscillation of the CNS Ischemic Response. The record in Figure 18–11A resulted from oscillation of the CNS ischemic pressure control mechanism. In this experiment, the cerebrospinal fluid pressure was raised to 160 mm Hg, which compressed the cerebral vessels and initiated a CNS ischemic response. When the arterial pressure rose above 160 mm Hg, the ischemia was relieved, and the sympathetic nervous system became inactive. As a result, the arterial pressure fell rapidly back to a lower value, causing medullary ischemia once again. The ischemia then initiated another rise in pressure. Again the ischemia was relieved, and the pressure fell. This repeated itself cyclically as long as the cerebrospinal fluid pressure remained elevated.

Thus, any reflex pressure control mechanism can oscillate if the intensity of "feedback" is strong enough and there is also a delay between excitation of the pressure receptor and the subsequent pressure response. The vasomotor waves are of considerable theoretical importance because they show that the nervous reflexes that control arterial pressure obey identically the same principles as those applicable to mechanical and electrical control systems. For instance, if the feedback "gain" is too great in the guiding mechanism of an automatic pilot for an airplane and there is also a delay in response by the guiding mechanism, the plane will oscillate from side to side instead of following a straight course.

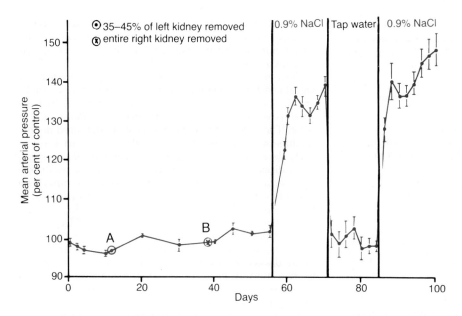

Figure 19–7. Effect on arterial pressure of drinking 0.9 per cent saline solution instead of water in four dogs with 70 per cent of their renal tissue removed. (From Langston, Guyton, Douglas, and Dorsett: *Circ. Res., 12*:508, 1963. By permission of The American Heart Association, Inc.)

The Sequential Changes in Circulatory Function During the Development of Volume-Loading Hypertension. It is especially instructive to study the sequential changes in circulatory function during the progressive development of volume-loading hypertension. Figure 19–8 illustrates these sequential changes. A week or so prior to the point labeled "0" days, the kidney mass had already been decreased to only 30 per cent of normal. Then at this point, the intake of salt and water was increased to about six times normal. The acute effect was increases in extracellular fluid volume, blood volume, and cardiac output to 20 to 40 per cent above normal. Simultaneously, the arterial pressure began to rise as well but not nearly so much at first as did the fluid volumes and cardiac output. The reason for this can be discerned by studying the total peripheral resistance curve, which shows an initial *decrease* in total peripheral resist-

ance. This decrease was caused by the baroreceptor mechanism, discussed in the previous chapter, which tried to prevent the rise in pressure. However, after a few days, the baroreceptors adapted (reset) and no longer opposed the rise in pressure. By this time, the arterial pressure had risen almost to its full height because of the increase in cardiac output, even though the total peripheral resistance was still almost at the normal level.

After these early acute changes in the circulatory variables had occurred, more prolonged secondary changes occurred during the next few days and weeks. Especially important was a *progressive increase in total peripheral resistance,* while at the same time *the cardiac output decreased almost all the way back to normal.* Multiple experiments have now shown that these changes were almost certainly caused mainly by *the long-term autoregulation*

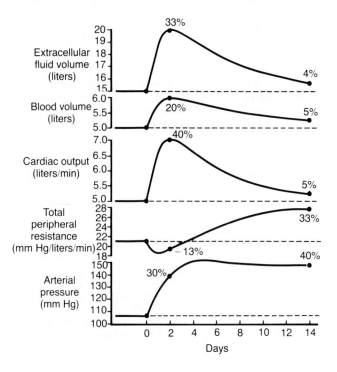

Figure 19–8. Progressive changes in important circulatory system variables during the first few weeks of *volume-loading hypertension.* Note especially the initial rise in cardiac output as the basic cause of the hypertension. Subsequently, the autoregulation mechanism returns the cardiac output almost to normal while at the same time causing a *secondary increase in total peripheral resistance.* (Modified from Guyton: Arterial Pressure and Hypertension. Philadelphia, W. B. Saunders Company, 1980.)

mechanism that is discussed in Chapter 17 as well as earlier in this chapter. That is, after the cardiac output had risen to a high level and had initiated the hypertension, the excess blood flow through the tissues then caused progressive constriction of the local arterioles, thus returning the cardiac output almost all the way back to normal but simultaneously causing a *secondary increase in total peripheral resistance.*

Note also that the extracellular fluid volume and blood volume returned almost all the way back to normal along with the decrease in cardiac output. This resulted from two factors: First, the increase in arteriolar resistance decreased the capillary pressure, which allowed the fluid in the tissue spaces to be absorbed back into the blood. Second, the elevated arterial pressure now caused the kidneys to excrete the excess volume of fluid that had initially accumulated in the body.

Finally, let us take stock of the final state of the circulation several weeks after the initial onset of volume loading. We find the following effects:

1. *Hypertension.*
2. *Marked increase in total peripheral resistance.*
3. *Almost complete return of the extracellular fluid volume, blood volume, and cardiac output back to normal.*

Therefore, we can divide volume-loading hypertension into two separate sequential stages:

The first stage results from increased fluid volumes and increased cardiac output. It is this increase in cardiac output that causes the hypertension.

The second stage in volume-loading hypertension is characterized by high blood pressure and high total peripheral resistance but return of the cardiac output so near to normal that the usual measuring techniques most frequently cannot detect an abnormally elevated output.

It should be especially noted that *the increased total peripheral resistance in volume-loading hypertension occurs* after *the hypertension has developed and therefore is* secondary *to the hypertension rather than being the cause of the hypertension.*

Volume-Loading Hypertension in Patients Who Have No Kidneys but Are Being Maintained on an Artificial Kidney

When a patient is maintained on an artificial kidney, it is especially important to keep the body fluid volume at a normal level — that is, to remove the appropriate amount of water and salt each time the patient is dialyzed. If this is not done and the extracellular fluid volume is allowed to increase, hypertension will develop in almost exactly the same way as that illustrated in Figure 19–8. That is, the cardiac output increases at first and causes hypertension. Then the autoregulation mechanism returns the cardiac output back toward normal while causing a secondary increase in total peripheral resistance. Therefore, in the end, the hypertension is a high peripheral resistance type of hypertension.

Hypertension Caused by Primary Aldosteronism

Another type of volume-loading hypertension is caused by excess aldosterone in the body or occasionally by excesses of other types of steroids. A small tumor in one of the adrenal glands occasionally secretes very large quantities of aldosterone, which is the condition called "primary aldosteronism." As is discussed in Chapter 29, aldosterone increases the rate of reabsorption of salt and water by the tubules of the kidneys, thereby reducing the loss of these in the urine while causing an increase in the extracellular fluid volume. Consequently, hypertension occurs; when salt intake is increased at the same time, the hypertension becomes greater. Furthermore, if the condition persists for months or years, the excess aldosterone seems to cause pathological changes in the kidneys that make the kidneys then retain still more salt and water in addition to that caused directly by the aldosterone. Therefore, the hypertension often finally becomes very severe.

Here again, in the early stages of this type of hypertension, the cardiac output is increased, but in the later stages the cardiac output generally returns almost entirely to normal, while the total peripheral resistance becomes secondarily elevated.

● THE RENIN-ANGIOTENSIN SYSTEM: ITS ROLE IN PRESSURE CONTROL AND IN HYPERTENSION

Aside from the capability to control arterial pressure through changes in extracellular fluid volume, the kidneys also have another powerful mechanism for controlling pressure. This is the renin-angiotensin system.

Renin is a small protein enzyme released by the kidneys when the arterial pressure falls too low. In turn, it raises the arterial pressure in several different ways, thus helping to correct the initial fall in pressure.

The Components of the Renin-Angiotensin System

Figure 19–9 illustrates the functional steps by which the renin-angiotensin system helps to regulate arterial pressure.

Renin is synthesized and stored in an inactive form called *prorenin* in the *juxtaglomerular cells* of the kidneys, which are modified smooth muscle cells located in the walls of the afferent arterioles immediately proximal to the glomeruli. When the arterial pressure falls, intrinsic reactions in the kidneys themselves cause many of these prorenin molecules to split and release *renin.* Most of the renin enters the blood and leaves the kidneys to circulate throughout the entire blood stream, although a small amount remains in the local fluids of the kidney and initiates several intrarenal functions.

Renin is an enzyme, not a vasoactive substance itself. Instead, as illustrated in the schema of Figure 19–9, it acts enzymatically on another plasma protein, a globulin called *renin substrate* (or *angiotensinogen*), to release a 10–amino acid peptide, *angiotensin I.* Angiotensin I has mild vasoconstrictor properties but not enough to cause significant functional changes in circulatory function. The renin persists in the blood for 30 minutes to an hour and continues to cause formation of angiotensin I during this entire time.

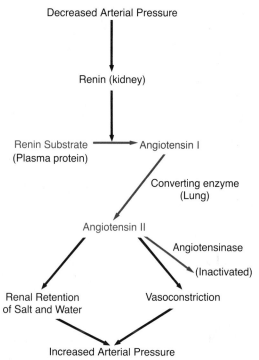

Figure 19-9. The renin-angiotensin-vasoconstrictor mechanism for arterial pressure control.

Within a few seconds after formation of the angiotensin I, two additional amino acids are split from it to form the 8–amino acid peptide *angiotensin II*. This conversion occurs almost entirely in the small vessels of the lungs, catalyzed by the enzyme *converting enzyme* that is present in the endothelium of the lung vessels. Angiotensin II is an extremely powerful vasoconstrictor and has other effects as well that affect the circulation. However, it persists in the blood only for a minute or two because it is rapidly inactivated by multiple blood and tissue enzymes collectively called *angiotensinase.*

During its persistence in the blood, angiotensin II has two principal effects that can elevate arterial pressure. The first of these, *vasoconstriction,* occurs very rapidly. Vasoconstriction occurs very intensely in the arterioles and to considerably less extent in the veins. Constriction of the arterioles increases the peripheral resistance, thereby raising the arterial pressure back toward normal, as illustrated at the bottom of the schema in Figure 19–9. Also, the mild constriction of the veins promotes increased venous return of blood to the heart, thereby helping the heart pump against the increasing pressure.

The second principal means by which angiotensin increases the arterial pressure is to act on the kidneys to *decrease the excretion of both salt and water.* This increases the extracellular fluid volume, which then increases the arterial pressure slowly over a period of hours and days. This long-term effect, acting through the extracellular fluid volume mechanism, is even more powerful than the acute vasoconstrictor mechanism in eventually returning the arterial pressure back to normal.

Rapidity and Intensity of the Vasoconstrictor Pressure Response to the Renin-Angiotensin System

Figure 19–10 illustrates a typical experiment showing the effect of hemorrhage on the arterial pressure under two separate conditions: (1) with the renin-angiotensin system functioning and (2) without the system functioning (the system was interrupted by a renin-blocking antibody). Note that following the hemorrhage, which caused an acute fall of the arterial pressure to 50 mm Hg, the arterial pressure rose back to 83 mm Hg when the renin-angiotensin system was functional. On the other hand, it rose to only 60 mm Hg when the renin-angiotensin system was blocked. This illustrates that the renin-angiotensin system is powerful enough to return the arterial pressure at least halfway back to normal following severe hemorrhage. Therefore, it can sometimes be of lifesaving service to the body, especially in circulatory shock.

Note that the renin-angiotensin vasoconstrictor system requires approximately 20 minutes to become fully active. Therefore, it is far slower to act than are the nervous reflexes and the sympathetic norepinephrine-epinephrine system.

Effect of Angiotensin to Cause Renal Retention of Salt and Water—An Especially Important Means for Long-Term Control of Arterial Pressure

Angiotensin causes the kidneys to retain both salt and water in two different ways:

1. Angiotensin *acts directly on the kidneys to cause salt and water retention.*

2. Angiotensin *causes the adrenal glands to secrete aldosterone, and the aldosterone in turn increases salt and water reabsorption by the kidney tubules.*

Obviously, to achieve balance between fluid intake and output, the arterial pressure must rise to a considerably increased level to overcome these two fluid-retaining effects of angiotensin. For this reason, whenever excess amounts of angiotensin circulate in the

Figure 19-10. Pressure-compensating effect of the renin-angiotensin-vasoconstrictor system following severe hemorrhage. (Drawn from experiments by Dr. Royce Brough.)

blood, the entire long-term renal–body fluid mechanism for arterial pressure control automatically becomes set to a higher than normal arterial pressure level.

Mechanisms of the Direct Renal Effects of Angiotensin to Cause Renal Retention of Salt and Water. Angiotensin has several intrarenal effects that make the kidneys retain salt and water. Probably the most important is to constrict the renal blood vessels, thereby diminishing blood flow through the kidneys. As a result, less fluid filters through the glomeruli into the tubules. Also, the slow flow of blood in the peritubular capillaries reduces their pressure, which allows rapid osmotic reabsorption of fluid from the tubules. Thus, for both of these reasons, less urine is excreted. In addition, angiotensin has a weak effect on the tubular cells themselves to increase tubular reabsorption of sodium and water. The total result of all these effects is very significant, sometimes decreasing urinary output as much as fourfold to sixfold.

Stimulation of Aldosterone Secretion by Angiotensin, and the Effect of Aldosterone in Increasing Salt and Water Retention by the Kidneys. Angiotensin is also one of the most powerful controllers of aldosterone secretion, as we discuss in relation to body fluid regulation in Chapter 29 and in relation to adrenal gland function in Chapter 77. Therefore, when the renin-angiotensin system becomes activated, the rate of aldosterone secretion usually increases at the same time. One of the most important functions of aldosterone is to cause marked increase in sodium reabsorption by the kidney tubules, thus increasing the extracellular fluid sodium. This then causes water retention, as already explained, thus increasing the extracellular fluid volume and leading secondarily to still more long-term elevation of the arterial pressure.

Both the direct effect of angiotensin on the kidney and its effect acting through aldosterone are important in long-term arterial pressure control. However, research has suggested that the direct effect on the kidneys is perhaps three or more times as potent as the indirect effect acting through aldosterone — even though the indirect effect is the one most widely known.

Quantitative Analysis of Arterial Pressure Changes Caused by Angiotensin. Figure 19–11 illustrates a quantitative analysis of the effect of angiotensin arterial pressure control. This figure shows two separate renal output curves and also illustrates the normal level of sodium intake. The left-hand renal output curve is approximately that measured in dogs whose renin-angiotensin system has been blocked by the drug captopril (which blocks the conversion of angiotensin I to angiotensin II). The right-hand curve was measured in dogs infused continuously with angiotensin II at a level about 2.5 times the normal rate of angiotensin formation in the blood. Note the obvious shift of the renal output curve toward higher pressure levels under the influence of angiotensin. This shift is caused by both the direct effect of angiotensin on the kidney and the indirect effect acting through aldosterone secretion, as explained previously.

Finally, note the two separate equilibrium points, one for

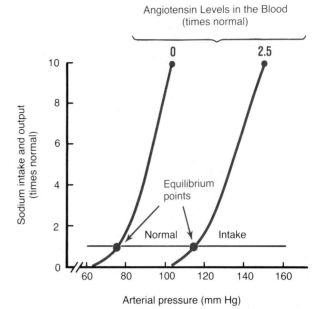

Figure 19–11. Effect of two different angiotensin levels on the renal output curve, showing regulation of the arterial pressure at an "equilibrium point" of 75 mm Hg when the angiotensin level is low and at 115 mm Hg when the angiotensin level is high.

zero angiotensin at an arterial pressure level of 75 mm Hg and one for elevated angiotensin at a pressure level of 115 mm Hg.

Therefore, even though the vasoconstrictor effect of angiotensin on all the *nonrenal* blood vessels of the body cannot cause a long-term increase in arterial pressure, the effects of angiotensin on renal retention of water and salt can have a very powerful effect in causing chronic elevation of the arterial pressure.

Role of the Renin-Angiotensin System in Maintaining a Normal Arterial Pressure Despite Wide Variations in Salt Intake

Probably the most important function of the renin-angiotensin system is to allow a person to eat either a very small or a very large amount of salt without causing great changes in either extracellular fluid volume or arterial pressure. This function is explained by the schema in Figure 19–12, which shows that the direct effect of increased salt intake is to elevate the extracellular fluid volume and this, in turn, to elevate the arterial pressure. Then, the increased arterial pressure causes increased blood flow through the kidneys, which reduces the rate of secretion of renin to a much lower level; this leads sequentially to decreased renal retention of salt and water, return of the extracellular fluid volume almost to normal, and finally return of the arterial pressure also almost to normal. Thus, the renin-angiotensin system is an automatic feedback mechanism that helps maintain the arterial pressure at or near the normal level even when the salt intake is increased. When the salt intake is decreased below normal, exactly the opposite effects take place.

To emphasize the importance of this effect of the renin-angiotensin system, the arterial pressure rises no more than 4 to 6 mm Hg in response to as much as a 50-fold increase in salt intake when the system functions normally. On the other hand, when the renin-angiotensin system is blocked,

Increased salt intake

↓

Increased extracellular volume

↓

Increased arterial pressure

↓

Decreased renin and angiotensin

↓

Decreased renal retention of salt and water

↓

Return of extracellular volume almost to normal

↓

Return of arterial pressure almost to normal

Figure 19–12. The sequential events by which increased salt intake tends to increase the arterial pressure. However, the schema also shows a feedback decrease in activity of the renin-angiotensin system that returns the arterial pressure back toward the normal level.

the same increase in salt intake will then cause the pressure to rise about 10 times as much, sometimes rising 50 to 60 mm Hg.

Graphical Analysis of the Combined Role of Sodium Intake and of the Renin-Angiotensin System in Arterial Pressure Control—The "Sodium-Loading Renal Function Curve." Figure 19–13 presents a graphical analysis of the automatic renin-angiotensin feedback mechanism for preventing excessive rise in arterial pressure when sodium intake increases. It shows four separate levels of salt intake, as illustrated by the four separate horizontal

Figure 19–13. The effect of different levels of salt intake on the arterial pressure. The lighter curves are renal output curves. The very dark red curve is the "sodium-loading renal function curve" for a person with normal kidneys. See text for explanation.

lines. To the right of each line is shown the approximate level of circulating angiotensin that one finds in the blood at each of these salt intake levels. Note that at very low salt intake, the concentration of angiotensin in the blood increases to about 2.5 times normal. On the other hand, when the sodium intake is 50 times greater, the level of circulating angiotensin becomes essentially zero.

Now, observe the four separate renal output curves (the black curves). These are the renal output curves caused by the four different levels of angiotensin that occur respectively at the different levels of salt intake. Thus, when the level of angiotensin in the blood is high, it is the renal output curve to the far right that depicts the relationship between arterial pressure and sodium output. Conversely, when the circulating level of angiotensin is zero, the far left renal output curve depicts this relationship.

Now, finally, match the appropriate levels of sodium intake with the appropriate renal output curves at the different levels of angiotensin and you will be able to determine the equilibrium point for these four separate pairs of intake and output curves. These equilibrium points are illustrated by the very heavy red points on the graph.

As a final maneuver, draw a heavy line connecting the four separate equilibrium points, as is illustrated in red in the figure. This line represents the relationship between long-term sodium intake and long-term level of arterial pressure, called the *salt-loading renal function curve.* In other words, this salt-loading renal function curve shows the relationship between arterial pressures and the renal sodium outputs that are recorded when the initial factor that is changed is the salt intake. Note that despite an approximate 50-fold change in sodium intake, as depicted in the figure, the arterial pressure change is only 4 mm Hg. These are the results that have been observed in animals when the sodium intake has been changed as much as 50-fold. Furthermore, the same experiment performed in humans has shown that changing the sodium intake from $\frac{1}{15}$ normal to 10 times normal—a total range of 150-fold—changes the arterial pressure only about 17 mm Hg.

TYPES OF HYPERTENSION IN WHICH ANGIOTENSIN IS INVOLVED: HYPERTENSION CAUSED BY SECRETION OF RENIN BY A RENIN-SECRETING TUMOR OR BY INFUSION OF ANGIOTENSIN II

Occasionally a tumor of the juxtaglomerular cells occurs and secretes tremendous quantities of renin; in turn, equally large quantities of angiotensin II are formed. In all patients in whom this has occurred, very severe hypertension has developed. Also, when large amounts of angiotensin are infused continuously for days or weeks on end into animals, similar severe hypertension develops.

We have already noted that angiotensin can increase the arterial pressure in two ways:

1. By constricting the arterioles throughout the entire body, thereby increasing the total peripheral resistance and arterial pressure; this effect occurs within seconds after one begins to infuse angiotensin.

2. By causing the kidneys to retain water and salt; over a period of days this, too, will cause hypertension.

Which of these causes the chronic hypertension? To answer this, let us suppose that angiotensin were infused into a person at a rate high enough to increase greatly the total peripheral resistance and therefore also cause an acute rise in arterial pressure. Let us further suppose that the angiotensin had no effect at all on the renal output curve—that is, the angiotensin did not in any way affect the normal capability of the kidneys to excrete salt and water. In such a theoretical instance, the marked rise in arterial pressure would cause very intense pressure diuresis and pressure natriuresis because the kidneys would still be normally responsive to the rising arterial pressure. Consequently, the blood volume would decrease rapidly, and the arterial pressure would eventually return *all the way* to normal. The reason it would return all the way to normal is that it is only at the normal arterial pressure that urinary output would decrease enough to come into balance with the salt and water intake. Therefore, the vasoconstriction in the nonrenal areas of the body, without affecting the kidneys, could not sustain a long-term increase in arterial pressure. Instead, to sustain the increased pressure, it is essential that the renal function curve be shifted to a high pressure level. Thus, once again, we see that the long-term level of arterial pressure can be changed to a new level only by changing one of the two determinants of arterial pressure—in this case shift of the renal function curve to a high pressure level.

To review, the two determinants of arterial pressure regulation are (1) the level of intake of salt and water and (2) the degree of shift of the renal function curve along the pressure axis. In angiotensin-induced hypertension, the angiotensin causes the renal function curve to shift toward higher pressure levels, thus also shifting the equilibrium point for long-term arterial pressure control to a higher level as well. This is the factor that is crucial in maintaining the pressure at a high level even though the initial rise in pressure is often caused by the acute effect of angiotensin to increase the total peripheral resistance.

Goldblatt Hypertension

"One-Kidney" Goldblatt Hypertension. When one kidney is removed and a constrictor is placed on the renal artery of the remaining kidney, as illustrated in Figure 19–14, the immediate effect is greatly reduced pressure in the renal artery beyond the constrictor, as shown by the dashed curve in the figure. Then, within a few minutes the systemic arterial pressure begins to rise and continues to rise for several days. The pressure usually rises rapidly for the first hour or so, and this is followed by a slower rise over a period of several days to a much higher pressure level. When the systemic arterial pressure reaches its new stable pressure level, the renal arterial pressure will have returned either all the way or almost all the way back to normal. The hypertension produced in this way is called *"one-kidney" Goldblatt hypertension* in honor of Goldblatt, who first studied the important quantitative features of hypertension caused by renal artery constriction.

The early rise in arterial pressure in Goldblatt hypertension is caused by the renin-angiotensin vasoconstrictor mechanism. Because of the poor blood flow through the kidney after acute reduction of renal arterial pressure, large quantities of renin are secreted by the kidneys, as illustrated by the lowermost curve in Figure 19–14, and this causes angiotensin to be formed in the blood, as described earlier in the chapter; the angiotensin in turn raises the arterial pressure acutely. The secretion of renin rises to a peak in a few hours but returns all the way back to normal within 5 to 7 days because the renal arterial pressure by that time has also risen back to normal, so that the kidney is no longer ischemic.

The second rise in arterial pressure is caused by fluid

Figure 19-14. Effect of placing a constricting clamp on the renal artery of one kidney after the other kidney has been removed. Note the changes in systemic arterial pressure, renal artery pressure distal to the clamp, and rate of renin secretion. The resulting hypertension is called "one-kidney Goldblatt hypertension."

retention; within 5 to 7 days the fluid volume has increased enough to raise the arterial pressure to its new sustained level. The quantitative value of this sustained pressure level is determined by the degree of constriction of the renal artery. That is, the aortic pressure must now rise to a much higher than normal level in order to make the renal arterial pressure distal to the constrictor rise high enough to cause normal urinary output.

Note especially that one-kidney Goldblatt hypertension has two phases. The first phase is a vasoconstrictor type of hypertension caused by angiotensin, but this is only transient. The second stage is a volume-loading type of hypertension. However, it is often very difficult to tell that this second stage is true volume-loading hypertension because neither the blood volume nor the cardiac output is significantly elevated. Instead, the total peripheral resistance is increased. To understand this, recall that in pure volume-loading hypertension blood volume and cardiac output are elevated only the first few days during the onset; after the first few days, volume-loading hypertension is a high resistance hypertension exactly as seen in the second stage of one-kidney Goldblatt hypertension. The transient increases in blood volume and cardiac output that are normally seen in pure volume-loading hypertension are obscured by the angiotensin vasoconstriction that occurs in Goldblatt hypertension during these early days.

"Two-Kidney" Goldblatt Hypertension. Hypertension also often results when the artery to one kidney is constricted, while the artery to the other kidney is still normal. This hypertension results from the following mechanism: The constricted kidney retains salt and water because of decreased renal arterial pressure in this kidney. Also, the "normal" kidney retains salt and water because of renin produced by the ischemic kidney. This renin causes the formation of angiotensin, which circulates to the opposite kidney to cause it also to retain salt and water. Thus, both kidneys, but for different reasons, become salt and water retainers. Consequently, hypertension develops.

Hypertension Caused by Diseased Kidneys That Secrete Renin Chronically. Often patchy areas of one or both kidneys are diseased and ischemic, while other areas of the kidneys are normal. When this occurs, the almost identical effects occur as in the two-kidney type of Goldblatt hypertension. That is, the patchy ischemic kidney tissue secretes renin, and this in turn, acting through the formation of angiotensin II, causes the remaining kidney mass also to retain salt and water. Indeed, one of the most common causes of renal hypertension is such patchy ischemic kidney disease.

OTHER TYPES OF HYPERTENSION CAUSED BY COMBINATIONS OF VOLUME-LOADING AND VASOCONSTRICTION

Hypertension in the Upper Part of the Body Caused by Coarctation of the Aorta

Out of every few thousand babies, one is born with a constricted aorta at a point beyond the arterial branches to the head and arms but proximal to the kidneys, a condition called coarctation of the aorta. When this occurs, much of the blood flow to the lower body is carried by collateral blood vessels in the body wall, with much vascular resistance between the upper aorta and the lower aorta. As a consequence, the arterial pressure in the upper part of the body averages 55 per cent higher than in the lower body.

The mechanism of this upper body hypertension is almost identical to that of one-kidney Goldblatt hypertension. That is, when a constrictor is placed on the aorta above the renal arteries, the blood pressure in both kidneys at first falls, renin is secreted, angiotensin is formed, and acute hypertension occurs in the upper body because of the vasoconstrictor effects of the angiotensin. Within a few days, salt and water retention occurs so that the arterial pressure in the lower body at the level of the kidneys rises approximately to normal, while very high pressure persists in the upper body. Now, the kidneys are no longer ischemic, so that the secretion of renin and formation of angiotensin return to normal. Likewise, in coarctation of the aorta, the arterial pressure in the lower body is usually almost normal, while the pressure in the upper body is far higher than normal.

Role of Autoregulation in the Hypertension of Coarctation. A very significant feature of hypertension caused by aortic coarctation is that the blood flow in the arms, where the pressure averages 55 per cent above normal, is almost exactly normal. Also, the blood flow in the legs, where the pressure is not elevated, is almost exactly normal. How could this be, with the pressure in the upper body 55 per cent greater than in the lower body? The answer is not that there are differences in vasoconstrictor substances in the bloods of the upper and lower body because

the same blood flows in both areas. Likewise, the nervous system innervates both areas of the circulation similarly, so that there is no reason to believe that there is a difference in nervous control of the blood vessels. Therefore, the only reasonable answer is that the process of long-term autoregulation develops so completely in aortic coarctation that the local blood flow mechanisms have compensated almost 100 per cent for the differences in pressure. The result is that both in the high pressure area and in the low pressure area the local blood flow is controlled exactly in accord with the needs of the tissues and not in accord with the level of the pressure.

One of the reasons why these observations are so important is that they illustrate how complete the long-term autoregulation process can be. These observations also illustrate how autoregulation can convert an initial high cardiac output hypertension into a high total peripheral resistance hypertension with normal cardiac output.

Hypertension in Toxemia of Pregnancy

During pregnancy many patients develop hypertension, which is one of the manifestations of the syndrome called *toxemia of pregnancy*. The principal pathological abnormality that causes this hypertension is believed to be thickening of the glomerular membranes (perhaps caused by an autoimmune process), which reduces the rate of fluid filtration from the glomeruli into the renal tubules. For obvious reasons, the pressure level of the renal output curve is elevated, and the long-term level of arterial pressure becomes correspondingly elevated. These patients are especially prone to hypertension when they eat large quantities of salt.

Neurogenic Hypertension

Acute hypertension can be caused by strong *stimulation of the sympathetic nervous system*. For instance, when a person becomes excited for any reason or at times during states of anxiety, the sympathetic system becomes excessively stimulated, peripheral vasoconstriction occurs everywhere in the body, and acute hypertension ensues.

Acute Neurogenic Hypertension Caused by Sectioning the Baroreceptor Nerves. Another type of acute neurogenic hypertension occurs when the nerves leading from the baroreceptors are cut or when the tractus solitarius is destroyed in each side of the medulla oblongata (these are the areas where the nerves from the baroreceptors connect in the brain stem). The sudden cessation of normal nerve signals from the baroreceptors has the same effect on the nervous pressure control mechanisms as a sudden reduction of the arterial pressure in the aorta and in the carotid arteries. That is, the vasomotor center suddenly becomes extremely active and the mean arterial pressure rises from 100 mm Hg to as high as 160 mm Hg. However, the pressure returns to normal within about 2 days because the response of the vasomotor center to the absent baroreceptor signal fades away, which is called central "resetting" of the baroreceptor pressure control mechanism. Therefore, the neurogenic hypertension caused by sectioning the baroreceptor nerves is only an acute type of hypertension, not a chronic type.

Possible Neurogenic Mechanisms in Chronic Hypertension. Despite the well-documented specific instances of *acute* hypertension caused by neurogenic mechanisms, it has been difficult to prove that abnormalities of the nervous system can cause chronic hypertension. Nevertheless, most clinicians have the "clinical impression" that prolonged nervous tension can maintain prolonged increase in sympathetic stimulation of the blood vessels and kidneys and therefore can lead to chronic elevation of the arterial pressure. Unfortunately, research studies thus far in hypertensive patients have failed to give positive proof of continuous excess sympathetic activity.

Probably more likely as a neurogenic mechanism of chronic hypertension is the following: Frequently, the sympathetic nervous system becomes extremely activated for short periods of time. During such periods, renal blood flow is often decreased to extremely low levels because of severe renal vascular vasoconstriction. Experiments have demonstrated that infusing norepinephrine, the sympathetic transmitter substance, directly into the renal arteries can cause permanent damage to the kidneys that can possibly lead to a prolonged renal type of hypertension. Thus, it is possible that episodes of extreme sympathetic activity, which are likely to occur in anyone's life, could under appropriate conditions lead to renal organic changes that could be the basis for permanent hypertension.

Spontaneous Hypertension In Lower Animals

Spontaneous, hereditary hypertension has been observed in a number of different strains of lower animals, including four different strains of rats, at least one strain of rabbits, and at least one strain of dogs. In the strain of rats that has been studied to the greatest extent, the Okamoto strain, there is evidence that in the early development of hypertension the sympathetic nervous system is considerably more active than in normal rats. However, in the late stages of this type of hypertension, two different structural changes have been observed in the nephrons of the kidneys: (1) increased preglomerular renal arterial resistance and (2) decreased permeability of the glomerular membranes. These structural changes could easily be the basis for the long-term continuance of the hypertension. In two other strains of hypertensive rats, changes in renal function also have been observed.

ESSENTIAL HYPERTENSION

Approximately 90 per cent of all persons who have hypertension are said to have "essential hypertension." This term means simply that *the hypertension is of unknown origin.* However, in most patients with essential hypertension, there is a very strong hereditary tendency, the same as occurs in strains of hypertensive lower animals.

Some of the characteristics of severe essential hypertension are the following:

1. The mean arterial pressure is increased 40 to 60 per cent.

2. The renal blood flow in the later stages is decreased to about one half normal.

3. The resistance to blood flow through the kidneys is increased twofold to fourfold.

4. Despite the great decrease in renal blood flow, the glomerular filtration rate is often very near normal. The reason for this is that the high arterial pressure still causes adequate filtration of fluid through the glomeruli into the renal tubules.

5. The cardiac output is approximately normal.

6. The total peripheral resistance is increased about 40 to 60 per cent, about the same amount that the arterial pressure is increased.

Finally, the most important finding of all in persons with essential hypertension is the following:

7. The kidneys will not excrete adequate amounts of salt and water unless the arterial pressure is high. In other words, if the mean arterial pressure in the essential hypertensive person is 150 mm Hg, reduction of the arterial pressure artificially to the normal value of 100 mm Hg (but without otherwise altering renal function except for the decreased pressure) will cause almost total anuria, and the person will retain salt and water until the pressure rises back to the elevated value of 150 mm Hg.

The exact reason for this failure of the kidneys of essential hypertensive persons to excrete salt and water at normal pressure levels is unknown. However, the very significant vascular changes in the kidneys listed earlier in this section suggest that decreased renal blood flow is the cause of this.

Graphical Analysis of Arterial Pressure Control in Essential Hypertension. Figure 19–15 is a graphical analysis of essential hypertension. The curves of this figure are not the same renal output curves that have been used in the early figures of this chapter. Instead, these curves are sodium-loading renal function curves of the type shown for the normal kidney by the heavy red curve in Figure 19–13. The reason for using this curve rather than the renal output curve is the following: It is very easy to determine the sodium-loading type of curve by simply increasing the level of sodium intake to a new level every few days, then waiting for the renal output of sodium to come into balance with the intake, and finally recording the changes in arterial pressure. When this procedure is employed in essential hypertensive patients, the two curves to the right in Figure 19–15 are recorded in two different groups of essential hypertensive patients, one called (1) *nonsalt-sensitive* patients, and the other (2) *salt-sensitive.* Note in both instances that the curves are shifted to a much higher pressure level than for normal persons. Now, let us plot on this same graph (1) a normal level of salt intake, and (2) a high level of salt intake representing 3.5 times the normal intake. In the case of the person with nonsalt-sensitive essential hypertension, the arterial pressure does not increase significantly when changing from normal salt intake to high salt intake. On the other hand, in those patients who have salt-sensitive essen-

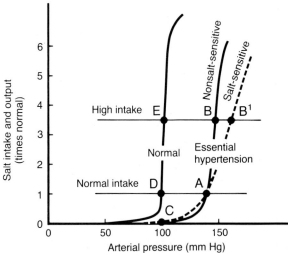

Figure 19–15. Analysis of arterial pressure regulation in "nonsalt-sensitive" essential hypertension and "salt-sensitive" essential hypertension. Explained in the text. (From Guyton et al.: *Annu. Rev. Med. 31:*15, 1980.)

tial hypertension, the high salt intake further exacerbates the hypertension.

The reason for the difference between nonsalt-sensitive essential hypertension and salt-sensitive hypertension is presumably structural or functional differences in the kidneys of these two types of hypertensive patients.

Treatment of Essential Hypertension. Essential hypertension generally can be treated by two different types of drugs: (1) drugs that increase renal blood flow or (2) drugs that decrease tubular reabsorption of salt and water.

Those drugs that increase renal blood flow are the various *vasodilator drugs*. Different ones act in one of the following three ways: (1) by inhibiting sympathetic nervous signals to the kidneys or by blocking the action of the sympathetic transmitter substance on the renal vasculature, (2) by directly paralyzing the smooth muscle of the renal vasculature, or (3) by blocking the action of the renin-angiotensin system on the renal vasculature.

Those drugs that reduce reabsorption of salt and water by the tubules include especially drugs that block active transport of sodium through the tubular wall; this blockage in turn also prevents the reabsorption of water as well, as explained earlier in the chapter. Such drugs are called *natriuretics* or *diuretics;* they are discussed in greater detail in Chapter 31.

SUMMARY OF THE INTEGRATED, MULTIFACETED SYSTEM FOR ARTERIAL PRESSURE REGULATION

By now, it is clear that arterial pressure is not regulated by a single pressure controlling system but instead by several interrelated systems each of which performs a specific function. For instance, when a person bleeds severely so that the pressure falls suddenly, two problems confront the pressure control system. The first is survival, to return the arterial pressure immediately to a high enough level that the person can live through the acute episode. The second is to return the blood volume eventually to its normal level, so that the circulatory system can re-establish full normality, including return of the arterial pressure all the way back to its normal value, not merely back to a pressure level required for survival.

In the previous chapter we saw that the first line of defense against acute changes in arterial pressure is the nervous control system. In the present chapter we have emphasized the role of the kidneys in long-term control of arterial pressure. However, there are a few other pieces to the puzzle as well. Figure 19–16 helps to put these together.

Figure 19–16 illustrates the approximate control responses, expressed as feedback gain, of eight different arterial pressure control mechanisms. These mechanisms can be divided into three separate groups: (1) those that react very rapidly, within seconds or minutes; (2) those that respond over an intermediate time period, minutes or hours; and (3) those that provide long-term arterial pressure regulation,

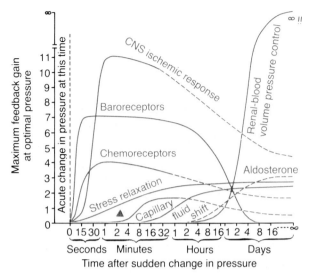

▲ Renin-angiotensin-vasoconstriction

Figure 19–16. Approximate potency of various arterial pressure control mechanisms at different time intervals after the onset of a disturbance to the arterial pressure. Note especially the infinite gain of the renal–body fluid pressure control mechanism that occurs after a few days' time. (Reprinted from Guyton: Arterial Pressure and Hypertension. Philadelphia, W. B. Saunders Company, 1980.)

days, months, and years. Let us see how these all fit together as a total, concerted system for pressure control.

The Rapidly Acting Pressure Control Mechanisms, Acting Within Seconds or Minutes. The rapidly acting pressure control mechanisms are almost entirely acute nervous reflexes or other nervous responses. Note in Figure 19–16 the three mechanisms that show responses within seconds. These are (1) the baroreceptor feedback mechanism, (2) the central nervous system ischemic mechanism, and (3) the chemoreceptor mechanism. These not only begin to react within seconds, but they are also very powerful. After any acute fall in pressure, as might be caused by severe hemorrhage, the nervous mechanisms combine to cause constriction of the veins to provide increased transfer of blood into the heart, increased heart rate and contractility of the heart to provide greater pumping capacity by the heart, and constriction of the arterioles to impede the flow of the blood out of the arteries; all these effects occur almost instantly to raise the arterial pressure back into a survival range.

When the pressure suddenly rises too high, as might occur in response to a drug or over-administration of a blood transfusion, these same mechanisms operate in the reverse direction, again returning the pressure back toward the normal range.

The Intermediate Time-Period Pressure Control Mechanisms. Several pressure control mechanisms exhibit significant responses only after a few minutes following an acute arterial pressure change. Three of these, illustrated in Figure 19–16, are (1) the renin-angiotensin vasoconstrictor mechanism, (2) stress-relaxation of the vasculature, and (3) shift of fluid through the capillary walls in and out of the circulation to readjust the blood volume as needed.

We have already described at length the role of the renin-angiotensin vasoconstrictor system to provide a semiacute means for increasing the arterial pressure when this is needed. The *stress-relaxation mechanism* is illustrated by the following example: When the pressure in the blood vessels becomes too high, they become stretched, and they keep on stretching more and more for minutes or hours; as a result, the pressure in the vessels falls toward normal. This continuing stretch of the vessels, called stress-relaxation, therefore, can serve as an intermediate-term pressure "buffer."

The *capillary fluid shift mechanism* means simply that any time the capillary pressure falls too low, fluid is absorbed by osmosis from the tissue, into the circulation, thus building up the blood volume and increasing the pressure in the circulation. Conversely, when the capillary pressure rises too high, fluid is lost out of the circulation into the tissues, thus reducing the blood volume and also all the pressures throughout the circulation.

These three intermediate mechanisms become mostly activated within 30 minutes to several hours. Their effect can last for long periods of time, days if necessary. During this time, the nervous mechanisms usually fatigue and become less and less effective, which explains the importance of these intermediate pressure control measures.

The Long-Term Mechanisms for Arterial Pressure Regulation. The goal of the present chapter has been to explain the role of the kidneys in long-term control of arterial pressure. To the far right in Figure 19–16 is shown the renal–blood volume pressure control mechanism (which is the same as the renal–body fluid pressure control mechanism), illustrating that it takes a few hours to show any significant response. Yet, it eventually develops a feedback gain for control of arterial pressure equal to infinity. This means that this mechanism can eventually return the arterial pressure *all the way* back to that pressure level that will provide normal output of salt and water by the kidneys. By now, the reader should be fully familiar with this concept, which has been the major point of the present chapter.

However, it must also be remembered that many different factors can affect the pressure regulating level of the renal–body fluid mechanism. One of these, illustrated in Figure 19–16, is aldosterone. Within a few hours after a sudden change in the arterial pressure, the rate of secretion of aldosterone also changes, and this plays an important role in modifying the pressure control characteristics of the renal–body fluid mechanism. Another factor especially important in the daily control of the arterial pressure is the interaction of the renin-angiotensin system with the aldosterone and renal fluid mechanism. For instance, a person's salt intake varies tremendously

from one day to another. We have seen in the present chapter that the salt intake can decrease to as little as 1/10 normal or can rise to as high as 10 to 15 times normal, and yet the regulated level of the mean arterial pressure will change only a few millimeters of mercury if the renin-angiotensin system is fully operative, as explained earlier in this chapter. *Without the renin-angiotensin-aldosterone system, the effect on arterial pressure of excessive salt intake is 10 times as great.*

Thus, arterial pressure control begins with the lifesaving measures of the nervous pressure controls, then continues with the sustaining characteristics of the intermediate pressure controls, and, finally is stabilized at the long-term pressure level by the renal–body fluid mechanism. This long-term mechanism, in turn, has multiple interactions with the renin-angiotensin-aldosterone system, the nervous system, and several other factors that provide special control capabilities for special purposes.

REFERENCES

Anderson, D. E.: Interactions of stress, salt, and blood pressure. Annu. Rev. Physiol., 46:143, 1984.

Beauchamp, G. K.: The human preference for execess salt. Am. Scientist, 75:27, 1987.

Bernstein, K. N., and O'Connor, D. T.: Antiadrenergic antihypertensive drugs: Their effect on renal function. Annu. Rev. Pharmacol. Toxicol., 24:105, 1984.

Bianchi, G., et al.: Sodium balance and peripheral resistance in arterial hypertension. J. Cardiovasc. Pharmacol., 6:(Suppl. 2)S457, 1984.

Blaustein, M. P.: Sodium chloride, extracellular fluid volume. Hypertension, 7:834, 1985.

Blaustein, M. P., and Hamlyn, J. M.: Sodium transport inhibition, cell calcium, and hypertension. The natriuretic hormone/Na^+-Ca^+ exchange/hypertension hypothesis. Am. J. Med., 77:45, 1984.

Blessing, W. W.: Central neurotransmitter pathways for baroreceptor-initiated secretion of vasopressin. News Physiol. Sci., 1:90, 1986.

Bohr, D. F.: Cell membrane in hypertension. News Physiol. Sci., 4:85, 1989.

Borgstrom, P., et al.: An evaluation of the metabolic interaction with myogenic vascular reactivity during blood flow autoregulation. Acta Physiol. Scand., 122:275, 1984.

Cauvin, C., et al.: Mechanisms of calcium antagonist-induced vasodilation. Annu. Rev. Pharmacol. Toxicol., 23:373, 1983.

Coleman, T. G., and Guyton, A. C.: The pressor role of angiotensin in salt deprivation and renal hypertension. Clin. Sci., 48:458, 1975.

Coleman, T. G., and Guyton, A. C.: Hypertension caused by salt loading in the dog. III. Onset transients of cardiac output and other circulatory variables. Circ. Res., 25:153, 1969.

Coleman, T. G., et al.: Experimental hypertension and the long-term control of arterial pressure. In MTP International Review of Science: Physiology. Baltimore, University Park Press, 1974, Vol. 1, p. 259.

Coleman, T. G., et al.: Regulation of arterial pressure in the anephric state. Circulation, 42:509, 1970.

Conway, J.: Hemodynamic aspects of essential hypertension in humans. Physiol. Rev., 64:617, 1984.

Cowley, A. W., Jr., et al.: Vasopressin: Cellular and Integrative Functions. New York, Raven Press, 1988.

Cowley, A. W., Jr., et al.: Open-loop analysis of the renin-angiotensin system in the dog. Circ. Res., 28:568, 1971.

DeClue, J. W., et al.: Subpressor angiotensin infusion, renal sodium handling, and salt-induced hypertension in the dog. Circ. Res., 43(4):503, 1978.

Estafanous, F. G., and Tarazi, R. C.: Systemic arterial hypertension associated with cardiac surgery. Am. J. Cardiol., 46:685, 1980.

Folkman, J.: Angiogenesis: What makes blood vessels grow? News Physiol. Sci., 1:199, 1986.

Folkow, B.: Physiological aspects of primary hypertension. Physiol. Rev., 62:347, 1982.

Freis, E. D.: Hemodynamics of hypertension. Physiol. Rev., 40:27, 1960.

Gibbons, G. H., et al.: Interaction of signals influencing renin release. Annu. Rev. Physiol., 46:291, 1984.

Glorioso, N., et al.: Renovascular Hypertension. New York, Raven Press, 1987.

WHAT IS THE ROLE OF THE NERVOUS SYSTEM IN CONTROLLING CARDIAC OUTPUT?

Importance of the Nervous System to Maintain the Arterial Pressure When the Cardiac Output Increases

Figure 20–5 illustrates an important difference in cardiac output control with and without a functioning autonomic nervous system. The solid curves show the effect in the normal animal of intense dilatation of the peripheral blood vessels caused by administering the drug dinitrophenol, which increased the metabolism of virtually all tissues of the body about fourfold. Note that with normal venous control, dilating all of the peripheral blood vessels caused almost no change in arterial pressure but increased the cardiac output almost fourfold. However, after autonomic control of the nervous system had been totally blocked, none of the normal circulatory reflexes for maintaining the arterial pressure could function, and vasodilation of the vessels with dinitrophenol (dashed curves) then caused a profound fall in arterial pressure to about one half normal, while the cardiac output rose only 1.6-fold instead of 4-fold.

Therefore, maintenance of a normal arterial pressure by the nervous reflexes, by mechanisms explained in Chapter 18, is essential to achieve high cardiac outputs when the peripheral tissues dilate their vessels to increase their local blood flow.

Effect of the Nervous System to *Increase* the Arterial Pressure During Exercise. During exercise, the nervous system not only prevents the arterial pressure from falling, but it also actually *increases* the pressure. The cause of this is mainly the following: The same brain activity that sends signals to the pe-ripheral muscles to cause exercise also sends simultaneous signals into the autonomic centers to excite circulatory activity, causing venous constriction, increased heart rate, and increased contractility of the heart; all these changes acting together increase the arterial pressure even above normal. This increased pressure often allows the cardiac output to increase during exercise at least an extra 50 per cent or more.

In summary, when the local tissue vessels dilate and attempt to increase the cardiac output above normal, the nervous system plays an exceedingly important role in preventing the arterial pressure from falling to disastrously low levels. In fact, in exercise, the nervous system goes even further, providing additional signals to raise the arterial pressure above normal, which serves to increase the cardiac output an extra 50 per cent or more.

● PATHOLOGICALLY HIGH AND PATHOLOGICALLY LOW CARDIAC OUTPUTS

In normal human beings, the cardiac outputs are surprisingly constant from one person to another. However, multiple clinical abnormalities can cause either high or low cardiac outputs. Some of the more important of these are illustrated in Figure 20–6.

High Cardiac Output — Virtually Always Caused by Reduced Total Peripheral Resistance

The left side of Figure 20–6 identifies conditions that commonly cause cardiac outputs higher than normal. One of the distinguishing features of these is that *they all result from chronically reduced total peripheral resistance.* None of them result from excessive excitation of the heart itself, which we explain subsequently. For the present, let us look at some of the peripheral factors that can increase the cardiac output to above normal:

Figure 20–5. An experiment in a dog to demonstrate the importance of nervous control of arterial pressure as a prerequisite for cardiac output control. Note that with pressure control, the metabolic stimulant dinitrophenol increases cardiac output; without pressure control, the arterial pressure falls and the cardiac output rises very little. (Drawn from experiments by Dr. M. Banet.)

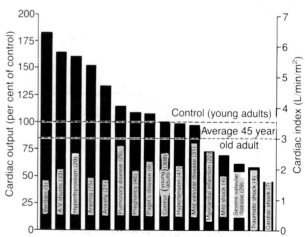

Figure 20–6. Cardiac output in different pathological conditions. The numbers in parentheses indicate number of patients studied in each condition. (From Guyton, Jones, and Coleman: Circulatory Physiology: Cardiac Output and Its Regulation. Philadelphia, W. B. Saunders Company, 1973.)

1. *Beriberi.* This is a disease caused by insufficient quantity of the vitamin *thiamine* in the diet. Lack of this vitamin causes diminished ability of the tissues to use cellular nutrients, and the local tissue blood flow mechanisms in turn cause marked peripheral vasodilation. The total peripheral resistance decreases sometimes to as little as one half normal. Consequently, the long-term level of cardiac output also often increases to as much as two times normal.

2. *Arteriovenous fistula (shunt).* Earlier in the chapter we pointed out that whenever a fistula (also called a shunt) occurs between a major artery and a major vein, tremendous amounts of blood will flow directly from the artery into the vein. This, too, greatly decreases the total peripheral resistance and likewise increases the venous return and cardiac output.

3. *Hyperthyroidism.* In hyperthyroidism, the metabolism of all the tissues of the body becomes greatly increased. Oxygen usage increases, and vasodilator products are released from the tissues. Therefore, the total peripheral resistance decreases markedly, again because of the local tissue blood flow control reactions throughout the body, and the cardiac output often increases to as much as 40 to 80 per cent above normal.

4. *Anemia.* In anemia, two peripheral effects greatly decrease the total peripheral resistance. One of these is reduced viscosity of the blood, resulting from the decreased concentration of red blood cells. The other is diminished delivery of oxygen to the tissues because of the decreased hemoglobin, which causes local vasodilation. As a consequence, the cardiac output increases greatly.

Any other factor that decreases the total peripheral resistance chronically will also increase the cardiac output.

Failure of Increased Cardiac Pumping to Cause Prolonged Increase of the Cardiac Output. If the heart is suddenly stimulated excessively, the cardiac output often increases as much as 20 to 30 per cent. However, even this small increase is maintained no longer than a few minutes, even though the heart continues to be strongly stimulated. There are two reasons for this: (1) Excess blood flow through the tissues causes automatic vasoconstriction of the blood vessels because of the autoregulation mechanism discussed in previous chapters, and this reduces the venous return and cardiac output back toward normal. (2) The slightly increased arterial pressure that results following acute cardiac stimulation raises the capillary pressure, and fluid filters out of the capillaries into the tissues, thereby decreasing the blood volume and also decreasing the venous return back toward normal. Over a period of hours and days, the increased pressure causes the kidneys to lose fluid volume as well until the arterial pressure and cardiac output return to normal.

Thus, all the known conditions that cause *chronic* elevation of the cardiac output result from decreased total peripheral resistance and not increased cardiac activity.

Low Output

Figure 20–6 illustrates to the far right the conditions that cause abnormally low cardiac output. These fall into two categories: (1) those abnormalities that cause the pumping effectiveness of the heart to fall too low and (2) those that cause venous return to fall too low.

Decreased Cardiac Output Caused by Cardiac Factors. Whenever the heart becomes severely damaged, from whatever cause, its limiting level of pumping may fall below that needed for adequate blood flow to the tissues. Some examples of this include *severe myocardial infarction, severe valvular heart disease, myocarditis, cardiac tamponade,* and certain *cardiac metabolic derangements.* The effect of several of these is illustrated on the right in Figure 20–6, showing the low cardiac outputs that result.

When the cardiac output falls so low that the tissues throughout the body begin to suffer nutritional deficiency, the condition is called *cardiac shock.* This is discussed fully in Chapter 22 in relation to cardiac failure.

Decrease in Cardiac Output Caused by Peripheral Factors — Decreased Venous Return. Any factor that interferes with venous return also can lead to decreased cardiac output. Some of these are

1. *Decreased blood volume.* By far the most common peripheral factor that leads to decreased cardiac output is decreased blood volume, resulting most often from hemorrhage. It is very clear why this decreases the cardiac output: Loss of blood decreases the filling of the vascular system to such a low level that there simply is not enough blood in the peripheral vessels to create the pressure required to push blood back to the heart.

2. *Acute venous dilatation.* On some occasions the peripheral veins become acutely vasodilated. This results most often when the sympathetic nervous system suddenly becomes inactive. For instance, fainting often results from sudden loss of sympathetic nervous system activity, which causes the peripheral capacitative vessels, especially the veins, to dilate markedly. This decreases the filling pressure of the vascular system because the blood volume can no longer create adequate pressure in the now flaccid peripheral blood vessels. As a result, the blood "pools" in the vessels and will not return to the heart.

3. *Obstruction of the large veins.* On rare occasions, the large veins leading into the heart become obstructed, so that the blood in the peripheral vessels cannot flow back into the heart. Consequently, the cardiac output falls markedly.

Regardless of the cause of low cardiac output, whether it be a peripheral factor or a cardiac factor, if ever the cardiac output falls below that level required for adequate nutrition of the tissues, the person is said to suffer *circulatory shock.* This can be lethal within a few minutes to a few hours. Circulatory shock is such an important clinical problem that it is discussed in detail in Chapter 24.

● A MORE QUANTITATIVE ANALYSIS OF CARDIAC OUTPUT REGULATION

Our discussion of cardiac output regulation thus far is adequate for understanding the factors that control cardiac output in most simple conditions. However, in especially stressful situations, such as the extremes of exercise, cardiac failure, and circulatory shock, a more complex quantitative analysis as presented in the following sections is required to give much deeper insight.

To perform the more quantitative analysis, it is necessary to distinguish separately the two primary factors concerned with cardiac output regulation: (1) the pumping ability of the heart, as represented by *cardiac output curves,* and (2) the peripheral factors that affect flow of blood from the veins into the heart, as represented by *venous return curves.* Then one can put these curves together in a quantitative way to show how they interact with each other to determine at the same time cardiac output, venous return, and right atrial pressure.

CARDIAC OUTPUT CURVES USED IN THE QUANTITATIVE ANALYSIS

Cardiac output curves used to depict the quantitative effects of different levels of heart effectiveness have already been illustrated in Figure 20–4. Still an additional set of curves is required to show the effect on cardiac output of changing the external pressure on the outside of the heart, as explained in the following section.

Effect of Extracardiac Pressure on Cardiac Output Curves. Figure 20–7 illustrates the effect of changes in extracardiac pressure on the cardiac output curve. The normal extracardiac pressure, which is also equal to the normal intrapleural pressure (the pressure in the chest cavity), is −4 mm Hg. Note in the figure that a rise in intrapleural pressure to −2 mm Hg shifts the entire curve to the right by this same amount. This shift occurs because an extra 2 mm Hg right atrial pressure is now required to overcome the increased pressure on the outside of the heart. Likewise, an increase in intrapleural pressure to +2 mm Hg requires a 6 mm Hg increase in right atrial pressure, which obviously shifts the entire curve 6 mm Hg to the right. In the same manner, a more negative intrapleural pressure shifts the curve to the left.

Some of the factors that can alter the intrapleural pressure and thereby shift the cardiac output curve include

1. *Cyclic changes during normal respiration,* which can be as much as ±3 mm Hg in normal expiration versus inspiration and as much as ±50 mm Hg in strenuous breathing.

2. *Breathing against a negative pressure,* which shifts the curve to a more negative atrial pressure (to the left).

3. *Positive pressure breathing,* which shifts the curve to the right.

4. *Opening the thoracic cage,* which increases the intrapleural pressure to 0 mm Hg and shifts the curve to the right 4 mm Hg.

5. *Cardiac tamponade,* which means an accumulation of large quantities of fluid in the pericardial cavity around the heart with resultant increase in extracardiac pressure and shifting of the curve to the right. Note in Figure 20–7 that cardiac tamponade shifts the upper parts of the curves farther to the right than the lower parts because the extracardiac pressure rises to higher values as the chambers of the heart fill to increased volumes during high cardiac output.

Combinations of Different Patterns of Cardiac Output Curves. Figure 20–8 illustrates that the cardiac

Figure 20–8. Combinations of two major patterns of cardiac output curves, showing the effect of alterations in both extracardiac pressure and effectiveness of the heart as a pump. (From Guyton, Jones, and Coleman: Circulatory Physiology: Cardiac Output and Its Regulation. Philadelphia, W. B. Saunders Company, 1973.)

output curve can change as a result of simultaneous changes in both extracardiac pressure and effectiveness of the heart as a pump. Thus, knowing what is happening to the extracardiac pressure and also the approximate capability of the heart as a pump, one can express the momentary ability of the heart to pump blood by a single cardiac output curve.

VENOUS RETURN CURVES

There still remains the entire systemic circulation that must be considered before a total analysis of cardiac regulation can be made. To analyze the function of the systemic circulation, we remove the heart and lungs from the circulation of an animal and replace these with a pump and artificial oxygenator system. Then, different factors, such as changes in blood volume, changes in vascular resistances, and changes in right atrial pressure, are altered to determine how the systemic circulation operates in different circulatory states. In these studies, one finds three principal factors that affect venous return to the heart from the systemic circulation. These are

1. The *right atrial pressure,* which exerts a backward force on the veins to impede flow of blood into the right atrium.

2. The degree of filling of the systemic circulation, measured by the *mean systemic filling pressure,* which forces the systemic blood toward the heart (this is the pressure measured everywhere in the systemic circulation when all flow of blood is stopped—we discuss this in detail later).

3. The *resistance to blood flow* between the peripheral vessels and the right atrium.

These three factors can all be expressed quantitatively by the *venous return curve,* as we explain in the following sections.

The Normal Venous Return Curve

In the same way that the cardiac output curve relates cardiac output to right atrial pressure, the *venous return curve relates venous return also to right atrial pressure.*

The curve in Figure 20–9 is the normal venous return

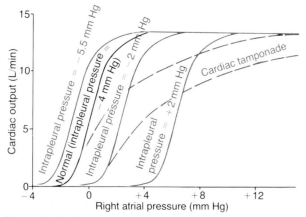

Figure 20–7. Cardiac output curves at different levels of intrapleural pressure and at different degrees of cardiac tamponade. (From Guyton, Jones, and Coleman: Circulatory Physiology: Cardiac Output and Its Regulation. Philadelphia, W. B. Saunders Company, 1973.)

Figure 20–9. The normal venous return curve. The *plateau* is caused by collapse of the large veins entering the chest when the right atrial pressure falls below atmospheric pressure. Note also that venous return becomes zero when the right atrial pressure rises to equal the mean systemic filling pressure.

curve. This curve shows that as the right atrial pressure increases, it causes back pressure on the systemic circulation and thereby decreases venous return of blood to the heart. *If all circulatory reflexes are prevented from acting,* venous return decreases to zero when the right atrial pressure rises to about 7 mm Hg. Such a slight rise in right atrial pressure causes a drastic decrease in venous return because the systemic circulation is a very distensible bag, so that any increase in back pressure causes blood to dam up in this bag instead of returning to the heart. Lack of venous flow into the heart decreases the cardiac output, thereby decreasing the arterial pressure. Thus, at the same time that the right atrial pressure is rising, the arterial pressure is falling, and these two pressures come to equilibrium when all flow in the systemic circulation ceases at a pressure of 7 mm Hg, which, by definition, is the *mean systemic filling pressure.*

Plateau in the Venous Return Curve at Negative Atrial Pressures Caused by Collapse of the Veins. When the right atrial pressure falls *below* zero—that is, below atmospheric pressure—venous return does not increase significantly. By the time the right atrial pressure has fallen to about −2 mm Hg, the venous return will have reached a plateau; then it remains at this plateau level even though the right atrial pressure falls as low as −20 to −50 mm Hg. This plateau is caused by collapse of the veins entering the chest. Low right atrial pressure sucks the walls of the veins together where they enter the chest, which prevents the negative pressure from sucking blood through the veins. Instead, the pressure in the veins immediately outside the chest remains almost exactly equal to atmospheric pressure (zero pressure) because that is the pressure on the outsides of the very flaccid veins causing them to collapse. Therefore, for all practical purposes, the venous pressure where the large veins empty into the chest never falls below a value of 0 mm Hg despite the fact that the right atrial pressure may fall to very negative values. Consequently, even very negative pressures in the right atrium cannot increase venous return significantly above that which exists at a normal atrial pressure of 0 mm Hg.

The Mean Circulatory Filling Pressure, the Mean Systemic Filling Pressure, and Their Effect on Venous Return

When the heart pumping is stopped by causing ventricular fibrillation or in any other way, the flow of blood everywhere in the circulation ceases a few seconds later. Without blood flow, the pressures everywhere in the circulation become exactly equal after about a minute. This equilibrated pressure level is called the *mean circulatory filling pressure.*

The Effect of Blood Volume on the Mean Circulatory Filling Pressure. Obviously, the greater the volume of blood in the circulation, the greater will be the mean circulatory filling pressure because extra blood volume stresses the walls of the vasculature. The *solid curve* in Figure 20–10 illustrates the approximate normal effect of different levels of blood volume on the mean circulatory filling pressure. Note that at a blood volume of about 4000 milliliters, the mean circulatory filling pressure is close to zero because this is the "unstressed volume" of the circulation, but at a volume of 5000 milliliters, the filling pressure is the normal value of 7 mm Hg. Similarly, at still higher volumes, the mean circulatory filling pressure increases almost linearly.

Effect of Sympathetic Stimulation of the Circulation on the Mean Circulatory Filling Pressure. The two *dashed curves* in Figure 20–10 illustrate the effects respectively of high and low levels of sympathetic stimulation on the mean circulatory filling pressure. Strong sympathetic stimulation constricts all of the systemic blood vessels as well as the larger pulmonary vessels and even the chambers of the heart. Therefore, the capacity of the system decreases, so that at each level of blood volume the mean

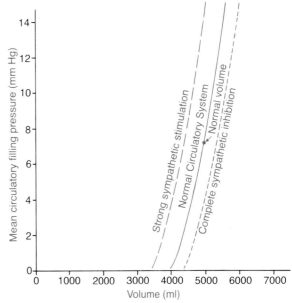

Figure 20–10. Volume-pressure curves of the entire circulation, illustrating the effect of strong sympathetic stimulation and complete sympathetic inhibition.

circulatory filling pressure is increased. At normal blood volume, maximal sympathetic stimulation will increase the mean circulatory filling pressure from 7 mm Hg to two and a half times that value, or about 17 mm Hg. Conversely, complete inhibition of the sympathetic nervous system relaxes the blood vessels as well as the heart, decreasing the mean circulatory filling pressure from the normal value of 7 mm Hg down to about 4 mm Hg.

Before leaving Figure 20–10, note specifically how steep the curves are. This means that even slight changes in blood volume or slight changes in the capacity of the system caused by various levels of sympathetic activity can have very large effects on the mean circulatory filling pressure.

The Mean "Systemic" Filling Pressure and Its Relationship to the Mean "Circulatory" Filling Pressure. The *mean systemic filling pressure* is slightly different from the mean circulatory filling pressure. It is the pressure measured everywhere in the systemic circulation after blood flow has been stopped by clamping the large blood vessels at the heart, so that the pressures in the systemic circulation can be measured independently from those in the pulmonary circulation. The mean systemic pressure, though almost impossible to measure in the living animal, is the important pressure for determining venous return. Fortunately, the mean systemic filling pressure is almost always very nearly equal to the mean circulatory filling pressure because the pulmonary circulation has less than one-eighth as much capacitance as the systemic circulation and only about one-tenth as much blood volume. Therefore, the pulmonary circulation contributes very little in determining the mean circulatory filling pressure, whereas the systemic system contributes almost all.

Effect on the Venous Return Curve of Changes in Mean Systemic Filling Pressure. Figure 20–11 illustrates the effects on the venous return curve caused by increasing or decreasing the mean systemic filling pressure *(Psf)*. Note in Figure 20–11 that the normal mean systemic filling pressure is 7 mm Hg. For the uppermost curve in the figure, the mean systemic filling pressure is 14 mm Hg, and for the lowermost curve it is 3.5 mm Hg. These curves demonstrate that the greater the mean systemic filling pressure, which also means the greater the "tightness" with which the circulatory system is filled with blood, the more the venous return curve shifts *upward* and *to the right.* Conversely, the lower the mean systemic filling pressure, the more the curve shifts *downward* and *to the left.* To express this another way, the greater the system is filled, the easier it is for blood to

flow into the heart. And the less the filling, the more difficult it is for blood to flow into the heart.

Pressure Gradient for Venous Return. When the right atrial pressure rises to equal the mean systemic filling pressure, all other pressures in the systemic circulation also come to this same pressure. Therefore, there is no longer any pressure difference between the peripheral vessels and the right atrium. Consequently, there also can no longer be any flow from any peripheral vessel back to the right atrium. However, when the right atrial pressure falls progressively lower than the mean systemic filling pressure, the flow to the heart increases proportionately, as one can see by studying any of the venous return curves in Figure 20–11. That is, *the greater the difference between the mean systemic filling pressure and the right atrial pressure, the greater becomes the venous return.* Therefore, the difference between these two pressures is called the *pressure gradient for venous return.*

The Resistance to Venous Return

In the same way that the mean systemic filling pressure represents a pressure pushing the blood in the veins from the periphery toward the heart, there is also resistance to this venous flow of blood. This resistance is called the *resistance to venous return.*

Most of the resistance to venous return occurs in the veins, though some occurs in the arterioles and small arteries as well. Why is venous resistance so important in determining the resistance to venous return? The answer is that when the resistance in the veins increases, blood begins to be dammed up in all parts of the circulation, including the veins, but the venous pressure rises very little because the veins are highly distensible. Therefore, the rise in pressure upstream from the resistance (that is, in the upstream veins themselves) is not very effective in overcoming the resistance, and the venous return decreases drastically. On the other hand, when the arteriolar and small artery resistances increase, blood accumulates in the arteries, which have a capacitance only 1/30 as great as that of the veins. Therefore, even slight accumulation of blood in the arteries raises the pressure very greatly—30 times as much as in the veins, and this high pressure overcomes much of the increased resistance so that the venous return decreases very little. Therefore, mathematically, it turns out that about two thirds of the so-called resistance to venous return is determined by venous resistance and about one third by the arteriolar and small artery resistance.

Venous return can be calculated by the following formula:

$$VR = \frac{MSFP - RAP}{RVR}$$

in which VR is venous return, MSFP is mean systemic filling pressure, RAP is right atrial pressure, and RVR is resistance to venous return. In the normal human, the values for these are venous return equals 5 liters/min, mean systemic filling pressure 7 mm Hg, right atrial pressure 0 mm Hg, and resistance to venous return 1.4 mm Hg per liter of blood flow.

Effect of Resistance to Venous Return on the Venous Return Curve. Figure 20–12 illustrates the effect of different resistances to venous return on the venous return curve, showing that a *decrease* in this resistance to one half normal allows two times as much flow of blood and therefore *rotates the curve upward* to two times as great a

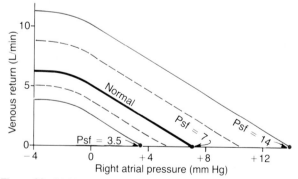

Figure 20–11. Venous return curves, showing the normal curve when the mean systemic filling pressure *(Psf)* is 7 mm Hg and showing the effect of altering the mean systemic filling pressure. (From Guyton, Jones, and Coleman: Circulatory Physiology: Cardiac Output and Its Regulation. Philadelphia, W. B. Saunders Company, 1973.)

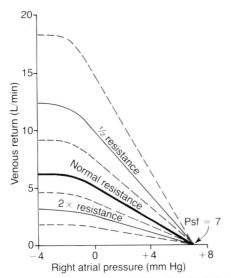

Figure 20–12. Venous return curves, depicting the effect of altering the "resistance to venous return." (From Guyton, Jones, and Coleman: Circulatory Physiology: Cardiac Output and Its Regulation. Phialdelphia, W. B. Saunders Company, 1973.)

ANALYSIS OF CARDIAC OUTPUT AND RIGHT ATRIAL PRESSURE, USING CARDIAC OUTPUT AND VENOUS RETURN CURVES

In the complete circulation, the heart and the systemic circulation must operate together. This means: (1) The venous return from the systemic circulation must equal the cardiac output from the heart. (2) The right atrial pressure is the same for both the heart and the systemic circulation.

Therefore, one can predict the cardiac output and right atrial pressure in the following way: (1) Determine the momentary pumping ability of the heart, and depict this in the form of a cardiac output curve; (2) determine the momentary state of the systemic circulation, and depict this in the form of a venous return curve; and (3) "equate" these two curves against each other, as shown in Figure 20–14.

The two black curves in the figure depict the *normal cardiac output curve* and the *normal venous return curve*. Obviously, there is only one point on the graph, point A, at which the venous return equals the cardiac output and at which the right atrial pressure is the same for both the heart and the systemic circulation. Therefore, in the normal circulation the right atrial pressure, cardiac output, and venous return are all depicted by point A, called the *equilibrium point.*

slope. Conversely, an *increase* in resistance to two times normal *rotates the curve downward* to one half as great a slope. However, note that when the right atrial pressure rises to equal the mean systemic filling pressure, venous return becomes zero at all levels of resistance to venous return because when there is no pressure gradient to cause flow of blood, it makes no difference what the resistance is in the circulation; the flow is still zero. Therefore, *the highest level to which the right atrial pressure can rise* regardless of how much the heart might fail is equal to the mean systemic filling pressure.

Combinations of Venous Return Curve Patterns. Figure 20–13 illustrates the effects on the venous return curve caused by simultaneous changes in mean systemic pressure and resistance to venous return, illustrating that both these factors can operate simultaneously.

Effect of Increased Blood Volume on Cardiac Output

A sudden increase in blood volume of about 20 per cent increases the cardiac output to about 2.5 to 3 times normal. An analysis of this effect is illustrated by the dashed curve of Figure 20–14. Immediately upon infusing the large quantity of extra blood, the increased filling of the system causes the mean systemic filling pressure to rise to 16 mm Hg, which shifts the venous return curve to the right. At the same time, the increased blood volume distends the blood vessels, thus reducing their resistance and thereby also reducing the resistance to venous return, which rotates the curve upward. As a result of these two effects, the venous return curve of Figure 20–14 changes from the solid curve

Figure 20–13. Combinations of the major patterns of venous return curves, illustrating the effects of simultaneous changes in mean systemic filling pressure *(Psf)* and in "resistance to venous return." (From Guyton, Jones, and Coleman: Circulatory Physiology: Cardiac Output and Its Regulation. Philadelphia, W. B. Saunders Company, 1973.)

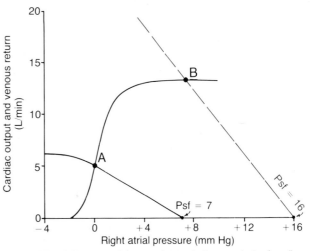

Figure 20–14. The two solid curves demonstrate an analysis of cardiac output and right atrial pressure when the cardiac output and venous return curves are normal. Transfusion of blood equal to 20 per cent of the blood volume causes the venous return curve to become the dashed curve; as a result, the cardiac output and right atrial pressure shift from point A to point B.

to the dashed curve. This new curve equates with the cardiac output curve at point B, showing that the cardiac output increases 2.5 to 3 times and that the right atrial pressure rises to about +8 mm Hg.

Compensatory Effects Initiated in Response to the Increased Blood Volume. The increased cardiac output caused by the increased blood volume lasts only a few minutes because several different compensatory effects immediately begin to occur: (1) The increased cardiac output increases the capillary pressure, so that fluid begins to transude out of the capillaries into the tissues, thereby returning the blood volume toward normal. (2) The increased pressure in the veins causes the veins to distend gradually by the mechanism called *stress-relaxation,* especially causing the venous blood reservoirs, such as the liver and spleen, to distend. (3) The excess blood flow through the peripheral tissues causes autoregulatory increase in the peripheral resistance. These factors cause the mean systemic filling pressure to return back toward normal and also cause the resistance vessels of the systemic circulation to constrict. Therefore, gradually over a period of 20 to 40 minutes, the cardiac output returns almost to normal.

Effect of Sympathetic Stimulation on Cardiac Output

Sympathetic stimulation affects both the heart and the systemic circulation: (1) It makes the heart a stronger pump. (2) In the systemic circulation, it increases the mean systemic filling pressure because of contraction of the peripheral vessels—especially the veins—and it also increases the resistance to venous return. In Figure 20–15, the normal cardiac output and venous return curves are depicted by the very dark lines; these equate with each other at point A, which represents a normal venous return and cardiac output of 5 liters/min and a right atrial pressure of 0 mm Hg. Moderate sympathetic stimulation is depicted by

the long-dashed curves and maximal sympathetic stimulation by the dot-dash red curves.

Note that maximal sympathetic stimulation (the red curves) increases the mean systemic filling pressure (depicted by the point at which the venous return curve reaches the zero venous return level) to 17 mm Hg, and it also increases the pumping effectiveness of the heart by nearly 100 per cent. As a result, the cardiac output rises from the normal value at equilibrium point A to approximately double normal at equilibrium point D—and yet *the right atrial pressure hardly changes.* Thus, different degrees of sympathetic stimulation can increase the cardiac output progressively to about two times normal—at least for short periods of time until compensatory effects occur.

Effect of Sympathetic Inhibition

The sympathetic nervous system can be blocked completely by inducing *total spinal anesthesia* or by using some drug, such as *hexamethonium,* that blocks transmission of nerve impulses through the autonomic ganglia. The two lowermost curves in Figure 20–15, the short-dashed curves, show the effect of total sympathetic inhibition caused by total spinal anesthesia, illustrating that (1) the mean systemic filling pressure falls to about 4 mm Hg, and (2) the effectiveness of the heart as a pump decreases to about 80 per cent of normal. The cardiac output falls from point A to point B, which is a decrease to about 60 per cent of normal.

Effect of Opening a Large Arteriovenous Fistula (A Direct Open Hole Between a Large Artery and Vein)

Figure 20–16 illustrates various stages of circulatory changes that occur after opening a large arteriovenous fistula, as follows:

1. The two curves crossing at point A illustrate the normal condition.

2. The curves crossing at point B illustrate the circulatory condition immediately after opening the large fistula. The principal effects are (1) a sudden and precipitous rotation of the venous return curve in the upward direction, caused by the very large decrease in resistance to venous return when blood is allowed to flow with almost no impediment directly from the large artery into the venous system, by-passing most of the resistance elements of the peripheral circulation, and (2) a slight increase in the level of the cardiac output curve because opening the fistula decreases the peripheral resistance and allows an acute fall in arterial pressure against which the heart must pump. The net result, depicted by point B, is an increase in cardiac output from 5 liters/min to 13 liters/min and an increase in right atrial pressure to about +3 mm Hg.

3. Point C represents the effects approximately 1 minute later after the sympathetic nerve reflexes have restored the arterial pressure almost to normal and have caused two other effects: (1) an increase in the mean systemic filling pressure because of constriction of all vessels, from 7 mm Hg to 9 mm Hg, thus shifting the venous return curve 2 mm Hg to the right, and (2) further elevation of the cardiac output curve because of sympathetic nervous excitation of the heart. Therefore, the cardiac output now rises to almost 16 liters/min and the right atrial pressure to about 4 mm Hg.

4. Point D illustrates the effect after several more weeks. By this time, the blood volume has increased because the

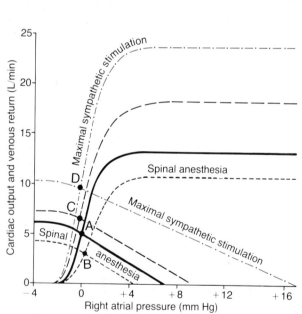

Figure 20–15. Analysis of the effect on cardiac output of (1) moderate sympathetic stimulation, (2) maximal sympathetic stimulation (colored curves), and (3) sympathetic inhibition caused by total spinal anesthesia. (From Guyton, Jones, and Coleman: Circulatory Physiology: Cardiac Output and Its Regulation. Philadelphia, W. B. Saunders Company, 1973.)

Figure 20–16. Analysis of successive changes in cardiac output and right atrial pressure after a large arteriovenous fistula is suddenly opened. The stages of the analysis, as illustrated by the equilibrium points, are: (A) normal conditions, (B) immediately after opening the A-V fistula, (C) a minute or so later after the sympathetic reflexes have become active, and (D) several weeks later after the blood volume has increased and the heart has begun to hypertrophy. (From Guyton, Jones, and Coleman: Circulatory Physiology: Cardiac Output and Its Regulation. Philadelphia, W. B. Saunders Company, 1973.)

slight reduction in arterial pressure and the sympathetic stimulation have both reduced kidney output of urine. Therefore, the mean systemic filling pressure has now risen to +12 mm Hg, shifting the venous return curve another 3 mm Hg to the right. Also, the prolonged increased workload on the heart has caused the heart muscle to hypertrophy, thus raising the level of the cardiac output curve still further. Therefore, point D illustrates a cardiac output now of almost 20 liters/min and a right atrial pressure of about 6 mm Hg.

Other Graphical Analyses

In the following chapter, a graphical analysis of cardiac output regulation during exercise is presented, and in Chapter 22 analyses of cardiac output regulation at various stages of congestive heart failure are shown.

● METHODS FOR MEASURING CARDIAC OUTPUT

In animal experiments, one can cannulate the aorta, the pulmonary artery, or the great veins entering the heart and measure the cardiac output using any type of flowmeter. For instance, an electromagnetic or ultrasonic flowmeter can be placed on the aorta or pulmonary artery to measure cardiac

output. However, except in rare instances, in the human cardiac output is measured by indirect methods that do not require surgery. Two of the methods commonly used are the *oxygen Fick method* and the *indicator dilution method.*

Pulsatile Output of the Heart as Measured by an Electromagnetic or Ultrasonic Flowmeter

Figure 20–17 illustrates a recording in a dog of blood flow in the root of the aorta made using an electromagnetic flowmeter. It demonstrates that the blood flow rises rapidly to a peak during systole and, then, at the end of systole, actually reverses for a small fraction of a second. It is this reverse flow that causes the aortic valve to close. A minute amount of reverse flow continues throughout diastole to supply blood to the coronary vessels.

Measurement of Cardiac Output by the Oxygen Fick Method

The Fick procedure is explained by Figure 20–18. This figure shows that 200 milliliters of oxygen are

Figure 20–17. Pulsatile blood flow in the root of the aorta recorded by an electromagnetic flowmeter.

Figure 20–18. The Fick principle for determining cardiac output.

being absorbed from the lungs into the pulmonary blood each minute. It also illustrates that the blood entering the right side of the heart has an oxygen concentration of 160 milliliters per liter of blood, whereas that leaving the left side has an oxygen concentration of 200 milliliters per liter of blood. From these data one can calculate that each liter of blood passing through the lungs picks up 40 milliliters of oxygen. Since the total quantity of oxygen absorbed into the blood from the lungs each minute is 200 milliliters, a total of five 1-liter portions of blood must pass through the pulmonary circulation each minute to absorb this amount of oxygen. Therefore, the quantity of blood flowing through the lungs each minute is 5 liters, which is also a measure of the cardiac output. Thus, the cardiac output can be calculated by the following formula:

Cardiac output (liters/min) =

$$\frac{O_2 \text{ absorbed per minute by the lungs (ml/min)}}{\text{Arteriovenous } O_2 \text{ difference (ml/liter of blood)}}$$

In applying the Fick procedure, mixed venous blood is usually obtained through a catheter inserted up the brachial vein of the forearm, through the subclavian vein, down to the right atrium, and finally into the right ventricle or pulmonary artery. Arterial blood can be obtained from any artery in the body, and the rate of oxygen absorption by the lungs is measured by the disappearance of oxygen from the respired air, using any of many types of oxygen meters.

THE INDICATOR DILUTION METHOD

In measuring the cardiac output by the indicator dilution method, a small amount of *indicator,* such as a dye, is injected into a large vein or preferably into the right side of the heart itself. This passes rapidly through the right heart, the lungs, the left heart, and finally into the arterial system. Then one records the concentration of the dye as it passes through one of the peripheral arteries, giving a curve such as one of the two red curves illustrated in Figure 20–19. In

each of these instances, 5 milligrams of Cardio-Green dye was injected at zero time. In the top recording, none of the dye passed into the arterial tree until approximately 3 seconds after the injection, but then the arterial concentration of the dye rose rapidly to a maximum in approximately 6 to 7 seconds. After that, the concentration fell rapidly. However, before the concentration reached the zero point, some of the dye had already circulated all the way through some of the peripheral vessels and returned through the heart for a second time. Consequently, the dye concentration in the artery began to rise again. For the purpose of calculation, however, it is necessary to extrapolate the early downslope of the curve to the zero point, as shown by the dashed portion of the curve. In this way, the *time-concentration curve* of the dye in an artery can be measured in its first portion and estimated reasonably accurately in its latter portion.

Once the time-concentration curve has been determined, one can then calculate the mean concentration of dye in the arterial blood for the duration of the curve. In Figure 20–19, this was done by measuring the area under the entire curve and then averaging the concentration of dye for the duration of the curve; one can see from the shaded rectangle straddling the upper curve of the figure that the average concentration of dye was approximately 0.25 mg/dl of blood and that the duration of this average value was 12 seconds. A total of 5 mg of dye was injected at the beginning of the experiment. In order for blood carrying only 0.25 mg of dye in each deciliter to carry the entire 5 mg of dye through the heart and lungs in 12 seconds, it would be necessary for a total of 20 1-deciliter portions of blood to pass through the heart during this time, which would be the same as a cardiac output of 2 liters per 12 seconds, or 10 liters/min.

We leave it to the reader to calculate the cardiac output from the bottom curve of Figure 20–19.

To summarize, the cardiac output can be determined using the following formula:

Cardiac output (ml/min) =

$$\frac{\text{Milligrams of dye injected} \times 60}{\left(\begin{array}{c}\text{Average concentration of dye}\\\text{in each milliliter of blood}\\\text{for the duration of the curve}\end{array}\right) \times \left(\begin{array}{c}\text{Duration of}\\\text{the curve}\\\text{in seconds}\end{array}\right)}$$

REFERENCES

Banet, M., and Guyton, A. C.: Effect of body metabolism on cardiac output: Role of the central nervous system. Am. J. Physiol., 220:662, 1971.

Bishop, V. S., and Stone, H. L.: Quantitative description of ventricular output curves in conscious dogs. Circ. Res., 20:581, 1967.

Bishop, V. S., et al.: Cardiac function curves in conscious dogs. Am. J. Physiol., 207:677, 1964.

Coleman, T. G., et al.: Control of cardiac output of regional blood flow distribution. Ann. Biomed. Eng., 2:149, 1974.

Dobbs, W. A., Jr., et al.: Relative importance of nervous control of cardiac output and arterial pressure. Am. J. Cardiol., 27:507, 1971.

Dodge, H. T., and Kennedy, J. W.: Cardiac output, cardiac performance, hypertrophy, dilatation, valvular disease, ischemic heart disease, and pericardial disease. In Sodeman, W. A., Jr., and Sodeman, T. M. (eds.): Pathologic Physiology: Mechanisms of Disease. 6th Ed. Philadelphia, W. B. Saunders Co., 1979, p. 271.

Donald, D. E., and Shepherd, J. T.: Response to exercise in dogs with cardiac denervation. Am. J. Physiol., 205:393, 1963.

Dow, P.: Estimations of cardiac output and central blood volume by dye dilution. Physiol. Rev., 36:77, 1956.

Fermoso, J. D., et al.: Mechanism of decrease in cardiac output caused by opening the chest. Am. J. Physiol., 207:1112, 1964.

Green, J. F.: Determinants of systemic blood flow. In Guyton, A. C., and Young, D. B. (eds.): International Review of Physiology: Cardiovascular Physiology, III. Vol. 18. Baltimore, University Park Press, 1979, p. 33.

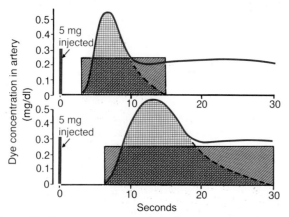

Figure 20–19. Dye concentration curves used to calculate the cardiac output by the dilution method. (The rectangular areas are the calculated average concentrations of dye in the arterial blood for the durations of the respective curves.)

Grodins, F. S.: Integrative cardiovascular physiology: A mathematical synthesis of cardiac and blood vessel hemodynamics. Q. Rev. Biol., 34:93, 1959.

Guyton, A. C.: Essential cardiovascular regulation — the control linkages between bodily needs and circulatory function. In Dickinson, C. J., and Marks, J. (eds.): Developments in Cardiovascular Medicine. Lancaster, England, MTP Press, 1978, p. 265.

Guyton, A. C.: An overall analysis of a cardiovascular regulation. Anesth. Analg., 56:761, 1977.

Guyton, A. C.: Regulation of cardiac output. N. Engl. J. Med., 277:805, 1967.

Guyton, A. C.: Venous return. In Hamilton, W. F. (ed.): Handbook of Physiology. Sec. 2, Vol. 2. Baltimore, Williams & Wilkins, 1963, p. 1099.

Guyton, A. C.: Determination of cardiac output by equating venous return curves with cardiac response curves. Physiol. Rev., 35:123, 1955.

Guyton, A. C., et al: Cardiac Output and Its Regulation. Philadelphia, W. B. Saunders Co., 1973.

Guyton, A. C., et al.: Circulation: Overall regulation. Annu. Rev. Physiol., 34:13, 1972.

Guyton, A. C., et al: Systems analysis of arterial pressure regulation and hypertension. Ann. Biomed. Eng., 1:254, 1972.

Guyton, A. C., et al.: Autoregulation of the total systemic circulation and its relation to control of cardiac output and arterial pressure. Circ. Res., 28:(Suppl. 1)93, 1971.

Guyton, A. C., et al.: Instantaneous increase in mean circulatory pressure and cardiac output at the onset of muscular activity. Circ. Res., 11:431, 1962.

Guyton, A. C., et al.: Relative importance of venous and arterial resistance in controlling venous return and cardiac output. Am. J. Physiol., 196:1008, 1959.

Guyton, A. C., et al.: Venous return at various right atrial pressures and the normal venous return curve. Am. J. Physiol., 189:609, 1957.

Jones, C. E., et al.: Cardiac output and physiological mechanisms in circulatory shock. In MTP International Review of Science: Physiology. Vol. 1. Baltimore, University Park Press, 1974, p. 233.

Kenner, T. (ed.): Cardiovascular System Dynamics. Models and Measurements. New York, Plenum Publishing Corp., 1982.

Levy, M. N., and Zieske, H.: A closed circulatory system model. Physiologist, 10:419, 1967.

Longhurst, J. C., and Mitchell, J. H.: Reflex control of the circulation by afferents from skeletal muscle. In Guyton, A. C., and Young, D. B. (eds.): International Review of Physiology: Cardiovascular Physiology III. Vol. 18. Baltimore, University Park Press, 1979, p. 125.

Lyons, K. P.: Cardiovascular Nuclear Medicine. East Norwalk, Conn., Appleton & Lange, 1988.

Mark, A. L., and Mancia, G.: Cardiopulmonary baroreflexes in humans. In Shepherd, J. T., and Abboud, F. M. (eds.): Handbook of Physiology. Sec. 2, Vol. III. Bethesda, Md., American Physiological Society, 1983, p. 795.

Mitchell, J. H., and Wildenthal, K.: Static (isometric) exercise and the heart: Physiological and clinical considerations. Annu. Rev. Med., 25:369, 1974.

Prather, J. W., et al.: Effect of blood volume, mean circulatory pressure and stress relaxation on cardiac output. Am. J. Physiol., 216:467, 1969.

Rothe, C. F.: Reflex control of veins and vascular capacitance. Physiol. Rev., 63:1281, 1983.

Rothe, C. F.: Venous system: Physiology of the capacitance vessels. In Shepherd, J. T., and Abboud, F. M. (eds.): Handbook of Physiology. Sec. 2, Vol. III. Bethesda, Md., American Physiological Society, 1983, p. 397.

Sarnoff, S., and Mitchell, J. H.: The regulation of the performance of the heart. Am. J. Med., 30:747, 1961.

Sugimoto, T., et al.: Effect of tachycardia on cardiac output during normal and increased venous return. Am. J. Physiol., 211:288, 1966.

Varat, M. A., et al.: Cardiovascular effects of anemia. Am. Heart J., 83:415, 1972.

Weisel, R. D., et al.: Current concepts measurement of cardiac output by thermodilution. N. Engl. J. Med., 292:682, 1975.

CHAPTER 21

• *Muscle Blood Flow and Cardiac Output During Exercise; the Coronary Circulation; and Ischemic Heart Disease* •

In the present chapter we consider blood flow to the skeletal muscles and coronary blood flow to the heart. Regulation of both of these is achieved mainly by local control of vascular resistance by tissue metabolic needs. In addition, many related subjects, such as cardiac output control during exercise, the characteristics of heart attacks, and the pain of angina pectoris, are discussed.

● BLOOD FLOW IN SKELETAL MUSCLES AND ITS REGULATION IN EXERCISE

Very strenuous exercise is the most stressful condition that the normal circulatory system faces. This is true because the blood flow in muscles can increase more than 20-fold (a greater increase than in any other tissue of the body) and also because there is such a very large mass of skeletal muscle in the body. The product of these two factors is so great that the total muscle blood flow in the normal young adult can increase during heavy exercise from the normal level of less than 1 liter/min to as great as 20 liters/min, high enough to increase the cardiac output to as much as five times normal and in the well-trained athlete to as much as six to seven times normal.

RATE OF BLOOD FLOW THROUGH THE MUSCLES

During rest, blood flow through skeletal muscle averages 3 to 4 ml/min per 100 grams of muscle. However, during extreme exercise, this rate can increase as much as 15-fold to 25-fold, rising to 50 to 80 milliliters per 100 grams of muscle.

Intermittent Flow During Muscle Contraction. Figure 21–1 illustrates a study of blood flow changes in the calf muscles of the human leg during strong rhythmic muscular exercise. Note that the flow increases and decreases with each muscle contraction, decreasing during the contraction phase and increasing between contractions. At the end of the rhythmic contractions, the blood flow remains very high for a few seconds longer but then gradually fades toward normal during the next few minutes.

The cause of the decreased flow during muscle contraction is compression of the blood vessels by the contracted muscle. During strong *tetanic* contraction, which causes sustained compression of the blood vessels, the blood flow can be almost totally stopped.

Opening of Muscle Capillaries During Exercise. During rest, only 20 to 25 per cent of the muscle capillaries have flowing blood. But during strenuous exercise, all the capillaries open up. This opening up of dormant capillaries also diminishes the distance that oxygen and other nutrients must diffuse from the capillaries to the muscle fibers and contributes a much increased surface area through which nutrients can diffuse from the blood.

CONTROL OF BLOOD FLOW THROUGH THE SKELETAL MUSCLES

Local Regulation. The tremendous increase in muscle blood flow that occurs during skeletal muscle activity is caused primarily by local effects in the muscles acting directly on the arterioles to cause vasodilation.

Figure 21–1. Effects of muscle exercise on blood flow in the calf of a leg during strong rhythmic contraction. The blood flow was much less during contraction than between contractions. (From Barcroft and Dornhorst: *J. Physiol., 109*:402, 1949.)

This local increase in blood flow during muscle contraction is probably caused by several different factors all operating at the same time. One of the most important of these is reduction of oxygen in the muscle tissues. That is, during muscle activity the muscle uses oxygen very rapidly, thereby decreasing the oxygen concentration in the tissue fluids. This in turn causes vasodilation either because the vessel walls cannot maintain contraction in the absence of oxygen or because oxygen deficiency causes release of vasodilator substances. The vasodilator substance that has been suggested most widely in recent years has been adenosine, but experiments have shown that even large amounts of adenosine infused directly into a muscle artery cannot cause long-sustained vasodilation in skeletal muscle. Furthermore, even after the muscle blood vessels have become insensitive to the vasodilator effects of adenosine, these same vessels will dilate fully in response to muscle activity.

Other vasodilator substances released during muscle contraction include potassium ions, acetylcholine, adenosine triphosphate, lactic acid, and carbon dioxide. Unfortunately, we still do not know quantitatively how great a role each of these plays in increasing muscle blood flow during muscle activity; this subject is discussed in more detail in Chapter 17.

Nervous Control of Muscle Blood Flow. In addition to the local tissue regulatory mechanism, the skeletal muscles are also provided with sympathetic vasoconstrictor nerves and, in some species of animals, sympathetic vasodilator nerves as well.

Sympathetic Vasoconstrictor Nerves. The sympathetic vasoconstrictor nerve fibers secrete norepinephrine and when maximally stimulated can decrease blood flow through the muscles to perhaps one half to one fourth normal. This represents rather poor vasoconstriction in comparison with that caused by sympathetic nerves in some other areas of the body in which blood flow can be almost completely blocked. Yet even this degree of vasoconstriction is of physiological importance in circulatory shock and during other periods of stress when it is desirable to reduce blood flow through the many muscles of the body.

In addition to the norepinephrine secreted at the sympathetic vasoconstrictor nerve endings, the adrenal medullae secrete large amounts of additional norepinephrine plus epinephrine into the circulating blood during strenuous exercise. The circulating norepinephrine acts on the muscle vessels to cause a vasoconstrictor effect similar to that caused by direct sympathetic nerve stimulation. The epinephrine, on the other hand, often has a slight vasodilator effect because epinephrine excites the *beta* receptors of the vessels, which are vasodilator receptors, in contrast to the *alpha* vasoconstrictor receptors excited by the norepinephrine. These receptors are discussed in Chapter 60.

Sympathetic Vasodilator Fibers. In the cat and some other lower animals, there are also sympathetic *vasodilator* fibers that secrete acetylcholine—this hormone, in turn, causing the vasodilation. However, such fibers have not yet been proved to occur in the human. Instead, as noted previously, circulating epinephrine from the adrenal medullae acting at beta receptors in the muscle arterioles sometimes seems to cause mild vasodilation, and this may subserve the same function as the vasodilator system of the lower animals.

CIRCULATORY READJUSTMENTS DURING EXERCISE

Three major effects occur during exercise that are essential for the circulatory system to supply the tremendous blood flow required by the muscles. These effects are (1) mass discharge of the sympathetic nervous system throughout the body with consequent stimulatory effects on the circulation, (2) increase in arterial pressure, and (3) increase in cardiac output.

Mass Sympathetic Discharge

At the onset of exercise, signals are transmitted not only from the brain to the muscle to cause muscle contraction, but also from the higher levels of the brain into the vasomotor center to initiate mass sympathetic discharge. Simultaneously, the parasympathetic signals to the heart are greatly attenuated. Therefore, three major circulatory effects result.

First, the *heart is stimulated* to greatly increased heart rate and pumping strength as a result of the sympathetic drive to the heart as well as release of the heart from the normal parasympathetic inhibition.

Second, most of the arterioles of the peripheral circulation are strongly contracted except the arterioles in the active muscles, which are strongly vasodilated by the local vasodilator effects in the muscles themselves. Thus, the heart is stimulated to supply the increased blood flow required by the muscles, and blood flow through most nonmuscular areas of the body is temporarily reduced, thereby temporarily "lending" their blood supply to the muscles. This effect accounts for as much as 2 liters of extra blood flow to the muscles, which is exceedingly important when one thinks of the wild animal running for its life because even a fractional increase in running speed may make the difference between life and death. However, two of the organ circulatory systems, the *coronary* and *cerebral systems, are spared this vasoconstrictor effect* because both these circulatory areas have very poor vasoconstrictor innervation — fortunately so because both the heart and the brain are as essential to exercise as are the skeletal muscles themselves.

Third, the muscular walls of the *veins and other capacitative areas of the circulation are contracted powerfully,* which greatly *increases the mean systemic filling pressure.* As we learned in the previous chapter, this is one of the most important factors in promoting venous return of blood to the heart and therefore in increasing the cardiac output as well.

A Muscle Reflex That May Further Stimulate the Sympathetic Nervous System. Aside from the sympathetic stimulation caused by direct signals from the brain, reflex signals from the contracting muscles are also believed to pass up the spinal cord to the vasomotor center and to excite the sympathetic nerves. These signals have been postulated to be initiated by metabolic end products acting on small sensory nerve endings in the muscle tissue.

Increase in Arterial Pressure — An Important Result of Increased Sympathetic Activity

One of the most important effects of increased sympathetic activity in exercise is to increase the arterial pressure. This results from multiple stimulatory effects, including (1) vasoconstriction of the arterioles and small arteries in most of the tissues of the body besides the active muscles, (2) increased pumping activity by the heart, and (3) a great increase in mean circulatory filling pressure caused mainly by venous contraction. These effects working together virtually always increase the arterial pressure during exercise. This increase can be as little as 20 or as great as 80 mm Hg, depending on the conditions under which the exercise is performed. When a person performs exercise under very tense conditions but uses only a few muscles, the sympathetic response still occurs everywhere in the body, but vasodilation occurs in only a few muscles. Therefore, the net effect is mainly one of vasoconstriction, often increasing the mean arterial pressure to as high as 170 mm Hg. Such a condition occurs in a person standing on a ladder and nailing with a hammer on the ceiling above. The tenseness of the situation is obvious, and yet the amount of muscle vasodilation is relatively slight.

On the other hand, when a person performs whole-body exercise, such as running or swimming, the increase in arterial pressure is often only 20 to 40 mm Hg. The lack of a tremendous rise in pressure results from the extreme vasodilation occurring in large masses of muscle.

In rare instances, persons are found in whom the sympathetic nervous system is absent, either congenitally or because of surgical removal. When such a person exercises, instead of the arterial pressure rising, the pressure actually falls — sometimes to as low as one-half normal, and the cardiac output rises only about one third as much as it does normally. Therefore, one can readily understand the major importance of increased sympathetic activity during exercise.

Why Is the Arterial Pressure Rise Important During Exercise? When muscles are stimulated maximally in a laboratory experiment, but without allowing the arterial pressure to rise, the blood flow will rarely rise more than about eightfold. Yet, we know from studies of marathon runners that the blood flow through the muscles can increase from as little as 1 liter/min for the whole body during rest to at least 20 liters/min during maximal activity. Therefore, it is clear that the muscle blood flow can increase much more than occurs in simple laboratory experiments. What is the difference? The difference is mainly that in physiologic exercise the arterial pressure rises at the same time. Let us assume, for instance, that the arterial pressure rises 30 per cent, a common rise during exercise. This 30 per cent increase causes 30 per cent more force to push the blood through the tissue vessels. However, this is not the only effect of the increasing pressure, for the pressure also dilates the blood vessels and allows about another 100 per cent increase in flow. Thus, a 30 per cent increase over the original 8-fold increase in blood flow would increase the flow 10.4-fold, and doubling this again would increase the flow more than 20-fold.

The Increase in Cardiac Output in Exercise

Many different physiological effects occurring together during exercise cause the cardiac output to increase almost in proportion to the degree of exercise. This increase in output, in turn, is essential to supply the large amounts of oxygen and other nutrients needed by the working muscles. In fact, the ability of the circulatory system to provide increased cardiac output in heavy exercise is equally as important as the strength of the muscles themselves in setting the limit for the performance of muscle work. For instance, those marathon runners who can increase their cardiac outputs the most are generally the same ones who have the record-breaking times.

Graphical Analysis of the Changes in Cardiac Output in Heavy Exercise. Figure 21–2 illustrates a graphical analysis of the very large increase in cardiac output that occurs in heavy exercise. The black curves, crossing at point A, give the analysis for the normal circulation, whereas the red curves and point B analyze heavy exercise. Note that the great increase in cardiac output requires very significant changes in both the cardiac output curve and the venous return curve, as follows:

The increased level of the cardiac output curve is very easy to understand. It results almost entirely from sympathetic stimulation of the heart that causes both increased heart rate, often up to rates as great as 170 to 190 beats per minute, and increased strength of contraction of the heart, often to as much as two times normal. Obviously, without the increased level of the output curve, the increase in cardiac output would be limited to the plateau level of the normal heart, which would mean a maximum increase of

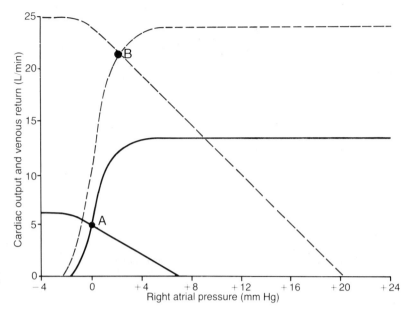

Figure 21-2. Graphical analysis of the changes in cardiac output and right atrial pressure with the onset of strenuous exercise.

cardiac output of only about 2.5-fold rather than the 4-fold that can commonly be achieved and the 7-fold that can be achieved in some marathon runners.

Now study the venous return curves. If no change occurred from the normal venous return curve, the cardiac output could hardly rise at all in exercise because the upper plateau level of the normal curve is only 6 liters/min. Yet, two very important changes do occur: (1) The mean systemic filling pressure rises tremendously at the very outset of heavy exercise. This results partly from sympathetic stimulation of the veins and other capacitative parts of the circulation. In addition, tensing of the abdominal and other muscles of the body compresses many of the internal vessels, thus providing still much more compression of the entire capacitative vascular system with still greater increase in the mean systemic filling pressure. In maximal exercise these two effects together can increase the mean systemic filling pressure from a normal level of 7 mm Hg to as high as 30 mm Hg. (2) The slope of the venous return curve rotates in the upward direction. This is caused by decreased resistance in virtually all the blood vessels in the active muscle tissue, which causes the resistance to venous return to decrease, thus increasing the slope of the venous return curve. The combination of increased mean systemic filling pressure and decreased resistance to venous return raises the entire level of the venous return curve.

Thus, in response to the changes in both the venous return curve and the cardiac output curve, the new equilibrium point in Figure 21-2 for arterial pressure and right atrial pressure is now point B, in contrast to the normal level at point A. Note especially that the right atrial pressure has hardly changed, having risen only 1½ mm Hg. In fact, in a person with a strong heart, the right atrial pressure often actually falls below normal because of the greatly increased sympathetic stimulation of the heart during exercise.

● THE CORONARY CIRCULATION

Approximately one third of all deaths in the affluent society of the Western world result from coronary artery disease, and almost all elderly persons have at least some impairment of the coronary artery circulation. For this reason, the normal and pathological physiology of the coronary circulation is one of the most important subjects in the entire field of medicine.

PHYSIOLOGIC ANATOMY OF THE CORONARY BLOOD SUPPLY

Figure 21-3 illustrates the heart with its coronary blood supply. Note that the main coronary arteries lie on the surface of the heart, and small arteries penetrate into the cardiac muscle mass. It is almost entirely through these arteries that the heart receives its nutritive blood supply. Only the inner 75 to 100 micrometers of the endocardial surface can obtain significant amounts of nutrition directly from the blood in the cardiac chambers.

The *left coronary artery* supplies mainly the anterior and lateral portions of the left ventricle, whereas the *right coronary artery* supplies most of the right ventricle as well as the posterior part of the left ventricle in 80 to 90 per cent of all persons. In about half of all human beings, more blood flows through the right coronary artery than through the left; in about 30 per cent, the arteries are about equal; and in 20 per cent, the left artery predominates.

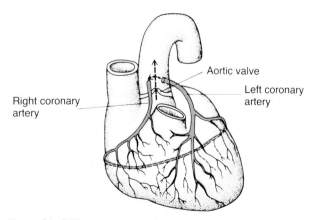

Figure 21-3. The coronary vessels.

Most of the venous blood flow from the left ventricle leaves by way of the *coronary sinus*—which is about 75 per cent of the total coronary blood flow—and most of the venous blood from the right ventricle flows through the small *anterior cardiac veins* directly into the right atrium and are not connected with the coronary sinus. A small amount of coronary blood flows back into the heart through very minute *thebesian veins,* which empty directly into all chambers of the heart.

NORMAL CORONARY BLOOD FLOW

The resting coronary blood flow in the human averages approximately 225 ml/min, which is about 0.7 to 0.8 milliliter per gram of heart muscle, or 4 to 5 per cent of the total cardiac output.

In strenuous exercise, the heart increases its cardiac output as much as fourfold to sevenfold, and it pumps this blood against a higher than normal arterial pressure. Consequently, the work output of the heart under severe conditions may increase as much as sixfold to eightfold. The coronary blood flow increases threefold to fourfold to supply the extra nutrients needed by the heart. Obviously, this increase is not as much as the increase in work load, which means that the ratio of coronary blood flow to energy expenditure by the heart decreases. However, the "efficiency" of cardiac utilization of energy increases to make up for this relative deficiency of blood supply.

Phasic Changes in Coronary Blood Flow— Effect of Cardiac Muscle Compression. Figure 21–4 illustrates the average blood flow *through the nutrient capillaries* of the left ventricular coronary system in milliliters per minute in the human heart during systole and diastole, as *calculated* from experiments in lower animals. Note from this diagram that the blood flow in the left ventricle falls to a low value during *systole,* which is opposite to the flow in other vascular beds of the body. The reason for this is the strong compression of the left ventricular muscle around the intramuscular vessels during systole.

During *diastole,* the cardiac muscle relaxes completely and no longer obstructs the blood flow through the left ventricular capillaries, so that blood now flows rapidly during all of diastole.

Blood flow through the coronary capillaries of the right ventricle also undergoes phasic changes during the cardiac cycle, but, because the force of contraction of the right ventricle is far less than that of the left ventricle, the inverse phasic changes are only partial in contrast to those in the left ventricle.

Epicardial Versus Subendocardial Blood Flow—Effect of Intramyocardial Pressure. During cardiac contraction, all the cardiac muscle squeezes toward the centers of the ventricles. That is, the ventricular muscle adjacent to the ventricular chambers (the subendocardial muscle) squeezes the blood in the ventricle, the muscle in the middle layer of the ventricle squeezes the blood in the ventricle *and also the subendocardial muscle,* and the outermost muscle squeezes both the middle and the subendocardial muscle as well as the blood in the ventricle. Therefore, during systole, a tissue pressure gradient develops within the heart muscle itself, with the pressure in the subendocardial muscle almost as great as the pressure inside the ventricle, while the pressure in the outer layer of the heart muscle is only slightly greater than atmospheric pressure. The importance of this pressure gradient is that *the intramyocardial pressure in the inner layers of the heart muscle is so much greater than in the outer layers that it compresses the subendocardial blood vessels far more than the outer vessels.*

Figure 21–5 illustrates the special arrangement of the coronary vessels at different depths in the heart, showing on the surface of the cardiac muscle the large epicardial coronary arteries that supply the heart. Smaller intramuscular arteries penetrate the muscle, supplying the needed nutrients en route to the endocardium. Then lying immediately beneath the endocardium is a plexus of subendocardial arteries. During systole, blood flow through the subendocardial plexus of the left ventricle, where the contractile force of the muscle is very great, falls almost to zero. To compensate for this almost total lack of flow during systole, the subendocardial arterial plexus is more extensive than the nutrient arteries in the middle and outer layers of the heart. Therefore, during diastole, blood flow in the subendocardial arteries is considerably greater than is blood flow in the outermost arteries.

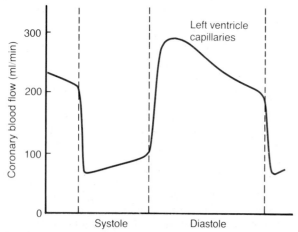

Figure 21–4. Phasic flow of blood through the coronary capillaries of the human left ventricle (as extrapolated from studies in dogs).

Figure 21–5. Diagram of the epicardial, intramuscular, and subendocardial coronary vasculature.

Later in the chapter, we see that this peculiar difference between blood flow in the epicardial and subendocardial arteries plays an important role in certain types of coronary ischemia.

CONTROL OF CORONARY BLOOD FLOW

Local Metabolism as the Primary Controller of Coronary Flow

Blood flow through the coronary system is regulated almost entirely by vascular response to the local needs of the cardiac musculature for nutrition. This mechanism works equally well when the nerves to the heart are intact or are removed. That is, whenever the vigor of contraction is increased, regardless of cause, the rate of coronary blood flow simultaneously increases, and, conversely, decreased activity is accompanied by decreased coronary flow. It is immediately obvious that this local regulation of blood flow is almost identical with that occurring in many other tissues, especially in the skeletal muscles of all the body.

Oxygen Demand as a Major Factor in Local Blood Flow Regulation. Blood flow in the coronaries is regulated almost exactly in proportion to the need of the cardiac musculature for oxygen. Even in the normal resting state, about 70 per cent of the oxygen in the arterial blood is removed as the blood passes through the heart; and, because not much oxygen is left, not much additional oxygen can be supplied to the heart musculature unless the blood flow increases. Fortunately, the blood flow does increase almost directly in proportion to the metabolic consumption of oxygen by the heart.

Yet, the exact means by which increased oxygen consumption causes coronary dilatation has not been determined. It is speculated that a decrease in the oxygen concentration in the heart causes vasodilator substances to be released from the muscle cells, and that these dilate the arterioles. The substance with the greatest vasodilator propensity is *adenosine*. In the presence of very low concentrations of oxygen in the muscle cells, a large proportion of the cell's adenosine triphosphate (ATP) degrades to adenosine monophosphate; then small portions of this are further degraded to release adenosine into the tissue fluids of the heart muscle. After the adenosine causes vasodilation, much of it is reabsorbed back into the cardiac cells to be reused.

Adenosine is not the only vasodilator product that has been identified. Others include adenosine phosphate compounds, potassium ions, hydrogen ions, carbon dioxide, bradykinin, and possibly prostaglandins.

Yet, difficulties still exist with the vasodilator hypothesis. For one, agents that block or partially block the vasodilator effect of adenosine do not prevent the coronary vasodilation caused by increased muscle activity. Second, studies in skeletal muscle have shown that adenosine will maintain vascular dilatation for only 1 to 3 hours, and yet muscle activity will still dilate the local blood vessels even when the adenosine can no longer dilate them.

Therefore, yet another theory to explain the coronary artery dilatation should be remembered until proved false: In the absence of adequate amounts of oxygen in the cardiac muscle, not only does the muscle itself suffer oxygen deficiency, but so do the arteriolar muscle walls as well. This could easily cause local vasodilation because of lack of the required energy to keep the coronary vessels contracted against the high arterial pressure. But this, too, has its problems because the coronary arteries require only minute amounts of oxygen to maintain full contraction.

Nervous Control of Coronary Blood Flow

Stimulation of the autonomic nerves to the heart can affect coronary blood flow in two ways — directly and indirectly. The direct effects result from direct action of the nervous transmitter substances, acetylcholine from the vagus nerves and norepinephrine from the sympathetic nerves, on the coronary vessels themselves. The indirect effects result from secondary changes in coronary blood flow caused by increased or decreased activity of the heart.

The indirect effects, which are mostly opposite to the direct effects, play by far the more important role in normal control of coronary blood flow. Thus, sympathetic stimulation, which releases norepinephrine, increases both heart rate and heart contractility as well as its rate of metabolism. In turn, the increased activity of the heart sets off local blood flow regulatory mechanisms for dilating the coronary vessels, and the blood flow increases approximately in proportion to the metabolic needs of the heart muscle. In contrast, vagal stimulation, with its release of acetylcholine, slows the heart and also has a slight depressive effect on cardiac contractility. Both these effects decrease cardiac oxygen consumption and therefore indirectly constrict the coronaries.

Direct Effects of Nervous Stimuli on the Coronary Vasculature. The distribution of parasympathetic (vagal) nerve fibers to the ventricular coronary system is so slight that parasympathetic stimulation has only a very slight direct effect to *dilate* the coronaries.

On the other hand, there is much more extensive sympathetic innervation of the coronary vessels. In Chapter 60 we see that the sympathetic transmitter substances norepinephrine and epinephrine can have either constrictor or dilator effects, depending on the presence or absence of specific receptors in the blood vessel walls. The constrictor receptors are called *alpha receptors,* and the dilator receptors are called *beta receptors.* Both alpha and beta receptors are known to exist in the coronary vessels. In general, the epicardial coronary vessels have a preponderance of alpha receptors, whereas the intramuscular arteries may have a preponderance of beta receptors. Therefore, sympathetic stimulation can at least theoretically cause either slight overall coronary constriction or dilatation but usually more constriction. Yet, in some persons, the alpha vasoconstrictor effects seem to be disproportionately severe, and these persons can have vasospastic myocardial ischemia during periods of excess sympathetic drive, often with resultant anginal pain.

It must be pointed out again, however, that metabolic factors — especially myocardial oxygen consumption — are the major controllers of myocardial blood flow. Therefore, whenever the direct effects of nervous stimulation alter the coronary blood flow, the metabolic factors usually override these direct coronary nervous effects and return the flow most of the way back toward normal within seconds or minutes.

SPECIAL FEATURES OF CARDIAC MUSCLE METABOLISM

The principles of cellular metabolism, which are discussed in Chapters 67 through 72, apply to cardiac muscle the same as for other tissues, but there are quantitative differences. Most importantly, under resting conditions cardiac muscle normally uses mainly fatty acids for its energy instead of carbohydrates, with approximately 70 per cent of the normal metabolism being derived from fatty acids. However, as is true of other tissues, under anaerobic or ischemic conditions cardiac metabolism must call on the anaerobic glycolysis mechanism for energy. This, unfortunately, can supply very little extra energy in relation to the large energy requirements of the heart. Also, glycolysis uses tremendous quantities of the blood glucose and at the same time forms large amounts of lactic acid in the cardiac tissue, which is probably one of the causes of cardiac pain in cardiac ischemic conditions, as is discussed later in the chapter.

As is true in other tissues, more than 95 per cent of the metabolic energy liberated from the foods is used to form ATP in the mitochondria. This ATP in turn supplies the energy for cellular function. In very severe coronary ischemia, the ATP degrades to adenosine diphosphate, adenosine monophosphate, and adenosine. Because the cell membrane is permeable to adenosine, much of this can be lost from the muscle cells into the circulating blood. This released adenosine is believed to be one of the substances that causes dilatation of the coronary arterioles during coronary hypoxia, as discussed earlier. However, the loss of adenosine also has a very serious cellular consequence. Within as little as half an hour of severe coronary ischemia, as occurs after a myocardial infarct or during cardiac arrest, about half of the adenine base can be lost from the cardiac muscle cells. Furthermore, this can be replaced by new synthesis of adenine at a rate of only 2 per cent per hour. Therefore, once a serious bout of ischemia has persisted for half an hour or so, relief of the coronary ischemia may be too late to save the lives of the cardiac cells. This almost certainly is one of the major causes of cardiac cellular death following myocardial ischemia and also one of the most important causes of cardiac debility in the late stages of circulatory shock, as is discussed in Chapter 24.

ISCHEMIC HEART DISEASE

The single most common cause of death in Western culture is ischemic heart disease, which results from insufficient coronary blood flow. Approximately 35 per cent of all humans in the United States die of this cause. Some deaths occur suddenly as a result of an acute coronary occlusion or of fibrillation of the heart, whereas others occur slowly over a period of weeks to years as a result of progressive weakening of the heart pumping process. In the present chapter, we

discuss the coronary ischemia problem itself as well as acute coronary occlusion and myocardial infarction. In the following chapter, we discuss congestive heart failure, the most frequent cause of which is progressive coronary ischemia.

Atherosclerosis as the Cause of Ischemic Heart Disease. The most frequent cause of diminished coronary blood flow is atherosclerosis. The atherosclerotic process is discussed in connection with lipid metabolism in Chapter 68; briefly, this process is the following: In certain persons who have a genetic predisposition to atherosclerosis or in persons who eat excessive quantities of cholesterol and other fats, large quantities of cholesterol gradually become deposited beneath the intima at many points in the arteries. Later, these areas of deposit become invaded by fibrous tissue, and they also frequently become calcified. The net result is the development of atherosclerotic plaques that protrude into the vessel lumens and either block or partially block blood flow.

A very common site for development of atherosclerotic plaques is the first few centimeters of the coronary arteries.

Acute Coronary Occlusion

Acute occlusion of a coronary artery frequently occurs in a person who already has serious underlying atherosclerotic coronary heart disease but almost never in a person with a normal coronary circulation. This condition can result from any one of several different effects, two of which are the following:

1. The atherosclerotic plaque can cause a local blood clot called a *thrombus,* which in turn occludes the artery. The thrombus usually begins where the plaque has grown so much that it has broken through the intima, thus coming in contact with the flowing blood. Because the plaque presents an unsmooth surface to the blood, platelets begin to adhere to it, fibrin begins to be deposited, and blood cells become entrapped and form a clot that grows until it occludes the vessel. Or, occasionally the clot breaks away from its attachment on the atherosclerotic plaque and flows to a more peripheral branch of the coronary arterial tree, where it blocks the artery at that point. A thrombus that flows along the artery in this way and occludes the vessel more distally is called an *embolus.*

2. It is believed by many clinicians that local spasm of a coronary artery can also cause sudden occlusion. The spasm might result from irritation of the smooth muscle of the arterial wall by the edges of an arteriosclerotic plaque, or it might result from nervous reflexes that cause the coronary contraction. The spasm may then lead to *secondary* thrombosis of the vessel.

Collateral Circulation in the Heart. The degree of damage to the heart caused either by slowly developing atherosclerotic constriction of the coronary arteries or by sudden occlusion is determined to a great extent by the degree of collateral circulation that is already developed or that can develop within a short period of time after the occlusion.

Unfortunately, in a normal heart, almost no communications exist among the larger coronary arteries. But many anastomoses do exist among the smaller arteries sized 20 to 250 micrometers in diameter, as shown in Figure 21–6.

When a sudden occlusion occurs in one of the larger coronary arteries, these small anastomoses dilate within a few seconds. The blood flow through these minute collaterals is usually less than one half that needed to keep alive the

Figure 21–6. Minute anastomoses of the coronary arterial system.

cardiac muscle that they supply; unfortunately, the diameters of the collateral vessels do not enlarge further for the next 8 to 24 hours. But then collateral flow does begin to increase, doubling by the second or third day and often reaching normal or almost normal coronary flow in the previously ischemic muscle within about 1 month; it is capable of increasing even further with increased metabolic loads. It is because of these developing collateral channels that a patient often recovers from various types of coronary occlusion when the area of muscle involved is not too great.

When atherosclerosis constricts the coronary arteries slowly over a period of many years rather than suddenly, collateral vessels can develop at the same time that the atherosclerosis does. Therefore, the person may never experience an acute episode of cardiac dysfunction. Eventually, however, the sclerotic process develops beyond the limits of even the collateral blood supply to provide the needed blood flow, and even the collaterals sometimes develop atherosclerosis. When this occurs, the heart muscle becomes severely limited in its work output, often so much so that the heart cannot pump even the normally required amounts of blood flow. This is the most common cause of cardiac failure and occurs in vast numbers of older persons.

Myocardial Infarction

Immediately after an acute coronary occlusion, blood flow ceases in the coronary vessels beyond the occlusion except for small amounts of collateral flow from surrounding vessels. The area of muscle that has either zero flow or so little flow that it cannot sustain cardiac muscle function is said to be *infarcted*. The overall process is called a *myocardial infarction*.

Soon after the onset of the infarction small amounts of collateral blood seep into the infarcted area, and this, combined with progressive dilatation of the local blood vessels, causes the area to become overfilled with stagnant blood. Simultaneously, the muscle fibers use the last vestiges of the oxygen in the blood, causing the hemoglobin to become totally reduced and very dark blue in color. Therefore, the infarcted area takes on a dark bluish-brown hue, and the

blood vessels of the area appear to be engorged despite the lack of blood flow. In later stages, the vessels become highly permeable and leak fluid, the tissue becomes edematous, and the cardiac muscle cells begin to swell because of diminished cellular metabolism. Within a few hours of almost no blood supply, the cells die.

Cardiac muscle requires approximately 1.3 milliliters of oxygen per 100 grams of muscle tissue per minute simply to remain alive. This is in comparison with approximately 8 milliliters of oxygen per 100 grams delivered to the normal resting left ventricle each minute. Therefore, if there is even as much as 15 to 30 per cent of normal resting coronary blood flow, the muscle will not die. In the central portion of a large infarct, however, the blood flow is usually less than this, so that this muscle does die.

Subendocardial Infarction. The subendocardial muscle very frequently becomes infarcted even when there is no evidence of infarction in the outer surface portions of the muscle. The reason for this is that the subendocardial muscle even normally has difficulty obtaining adequate blood flow because the blood vessels in the subendocardium are so intensely compressed by systolic contraction of the heart, as was explained earlier. Therefore, any condition that compromises blood flow to any area of the heart usually causes damage first in the subendocardial regions, and the damage then spreads outward toward the epicardium.

Furthermore, two specific conditions can make the problem of subendocardial infarction even much worse. First, any heart condition that causes a *low diastolic blood pressure,* such as patent ductus arteriosus or aortic regurgitation, will make the subendocardial blood flow especially poor because virtually all blood flow to the subendocardium occurs during diastole, when the heart muscle is not compressing the vessels. Second, conditions that cause *high diastolic pressures in the ventricles* will also compromise subendocardial blood flow because of direct compression of the subendocardial vessels by the intraventricular pressure during diastole. This occurs especially in severe heart failure, in which many persons develop very severe subendocardial muscular damage throughout the left ventricle as the heart becomes weaker and the diastolic pressure in the ventricle rises.

CAUSES OF DEATH FOLLOWING ACUTE CORONARY OCCLUSION

The four major causes of death following acute myocardial infarction are decreased cardiac output; damming of blood in the pulmonary or systemic veins with death resulting from edema, especially pulmonary edema; fibrillation of the heart; and, occasionally, rupture of the heart.

Decreased Cardiac Output — Cardiac Shock. When some of the cardiac muscle fibers are not functioning at all and others are too weak to contract with great force, the overall pumping ability of the affected ventricle is proportionately depressed. Indeed, the overall pumping strength of the heart is often decreased more than one might expect because of the phenomenon called *systolic stretch,* which is illustrated in Figure 21–7. When the normal portions of the ventricular muscle contract, the ischemic muscle, whether this be dead or simply nonfunctional, instead of contracting is actually forced outward by the pressure that develops inside the ventricle. Therefore, much of the pumping force of the ventricle is dissipated by bulging of the area of nonfunctional cardiac muscle.

When the heart becomes incapable of contracting with

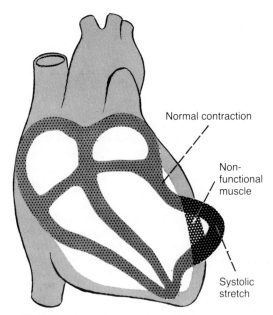

Normal contraction

Non-functional muscle

Systolic stretch

Figure 21-7. Systolic stretch in an area of ischemic cardiac muscle.

sufficient force to pump enough blood into the arterial tree, cardiac failure and death of the peripheral tissues ensue as a result of peripheral ischemia. This condition is called *coronary shock, cardiogenic shock, cardiac shock,* or *low cardiac output failure.* It is discussed in the following chapter. Cardiac shock almost always occurs when more than 40 per cent of the left ventricle is infarcted. Death occurs in about 85 per cent of those patients who develop cardiac shock.

Damming of Blood in the Venous System. When the heart is not pumping blood forward, it must be damming blood in the blood vessels of the lungs or in the systemic circulation. Obviously, this increases both the left and right atrial pressures and also leads to increased capillary pressures, particularly in the lungs. These effects often cause little difficulty during the first few hours after the myocardial infarction. Instead, the symptoms develop a few days later for the following reason: The diminished cardiac output leads to diminished blood flow to the kidneys, and, for reasons that are discussed in the following chapter, the kidneys retain large quantities of fluid. This adds progressively to the blood volume and therefore to the congestive symptoms. Consequently, many patients who seemingly are getting along well develop acute pulmonary edema several days after a myocardial infarction and often die within a few hours after appearance of the initial edema symptoms.

Rupture of the Infarcted Area. During the first day of an acute infarct there is little danger of rupture of the ischemic portion of the heart, but a few days after a large infarct occurs, the dead muscle fibers begin to degenerate, and the dead tissue is likely to become very thin. If this happens, the dead muscle bulges outward severely with each heart contraction, and the systolic stretch becomes greater and greater until finally the heart ruptures. In fact, one of the means used in assessing the progress of a severe myocardial infarction is to record by radiograph whether the degree of systolic stretch is getting worse.

When a ventricle does rupture, the loss of blood into the pericardial space causes rapid development of *cardiac tamponade*—that is, compression of the heart from the outside by blood collecting in the pericardial cavity. Because the heart is compressed, blood cannot flow into the right atrium, and the patient dies of suddenly decreased cardiac output.

Fibrillation of the Ventricles Following Myocardial Infarction. Many persons who die of coronary occlusion die because of ventricular fibrillation. The tendency to develop fibrillation is especially great following a large infarction, but fibrillation can occur following small occlusions as well. Indeed, some patients with coronary insufficiency die acutely from fibrillation without any infarction at all.

There are two especially dangerous periods during which fibrillation is most likely to occur. The first is during the first 10 minutes after the infarction occurs. Then there is a short period of relative safety, followed by a secondary period of cardiac irritability beginning an hour or so later and lasting for another few hours. However, fibrillation can occur for many days after the infarct.

At least four different factors enter into the tendency for the heart to fibrillate:

1. Acute loss of blood supply to the cardiac muscle causes rapid depletion of potassium from the ischemic musculature. This increases the potassium concentration in the extracellular fluids surrounding the cardiac muscle fibers. Experiments in which potassium has been injected into the coronary system have demonstrated that an elevated extracellular potassium concentration increases the irritability of the cardiac musculature.

2. Ischemia of the muscle causes an "injury current," which was described in Chapter 12 in relation to electrocardiograms in patients with acute myocardial infarction. The ischemic musculature cannot repolarize its membranes, so that the external surface of this muscle remains negative with respect to the normal cardiac muscle membrane elsewhere in the heart. Therefore, electrical current flows from this ischemic area of the heart to the normal area and can elicit abnormal impulses that can cause fibrillation.

3. Powerful sympathetic reflexes develop following massive infarction, principally because the heart does not pump an adequate volume of blood into the arterial tree. The sympathetic stimulation also increases the irritability of the cardiac muscle and thereby predisposes to fibrillation.

4. The myocardial infarction itself often causes the ventricle to dilate excessively. This increases the pathway length for impulse conduction in the heart and also frequently causes abnormal conduction pathways around the infarcted area of the cardiac muscle. Both of these effects predispose to development of circus movements. As discussed in Chapter 13, excess prolongation of the conduction pathway in the ventricles will allow an impulse to re-enter muscle that is already recovering from refractoriness, thereby initiating a subsequent cycle of excitation and causing the process to continue on and on.

THE STAGES OF RECOVERY FROM ACUTE MYOCARDIAL INFARCTION

The upper part of Figure 21-8 illustrates the effects of acute coronary occlusion, on the left, in a patient with a small area of muscle ischemia and, on the right, in a patient with a large area of ischemia. When the area of ischemia is small, little or no death of the muscle cells may occur, but part of the muscle often does become temporarily nonfunctional because of inadequate nutrition to support muscle contraction.

When the area of ischemia is large, some of the muscle fibers in the very center of the area die rapidly, within 1 to 3 hours in an area of total cessation of coronary blood supply. Immediately around the dead area is a nonfunctional area because of failure of contraction and usually also failure of impulse conduction. Then, extending circumferentially

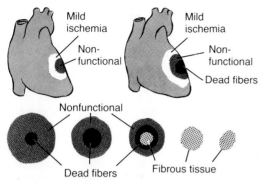

Figure 21–8. *Top,* Small and large areas of coronary ischemia. *Bottom,* Stages of recovery from myocardial infarction.

around the nonfunctional area is an area that is still contracting but weakly so because of mild ischemia.

Replacement of Dead Muscle by Scar Tissue. In the lower part of Figure 21–8, the various stages of recovery following a myocardial infarction are illustrated. Shortly after the occlusion, the muscle fibers die in the very center of the ischemic area. Then during the ensuing days, this area of dead fibers grows because many of the marginal fibers finally succumb to the prolonged ischemia. At the same time, owing to the enlargement of the collateral arterial channels growing into the outer rim of the infarcted area, the nonfunctional area of muscle becomes smaller and smaller. After a few days to 3 weeks, most of the nonfunctional area of muscle becomes functional again or dead — one or the other. In the meantime, fibrous tissue begins developing among the dead fibers, for ischemia stimulates growth of fibroblasts and promotes development of greater than normal quantities of fibrous tissue. Therefore, the dead muscular tissue is gradually replaced by fibrous tissue. Then, because it is a general property of fibrous tissue to undergo progressive elastomeric contraction and dissolution, the size of the fibrous scar may become smaller over a period of several months to a year.

Finally, the normal areas of the heart gradually hypertrophy to compensate at least partially for the lost cardiac musculature. By these means the heart recovers.

Value of Rest in Treating Myocardial Infarction. The degree of cellular death is determined by the *degree of ischemia × the degree of metabolism* of the heart muscle. When the metabolism of the heart muscle is greatly increased, such as during exercise, in severe emotional strain, or as a result of fatigue, the heart needs increased oxygen and other nutrients for sustaining its life. Furthermore, anastomotic blood vessels that supply blood to ischemic areas of the heart must also still supply the areas of the heart that they normally supply. When the heart becomes excessively active, the vessels of the normal musculature become greatly dilated, and this allows most of the blood flowing into the coronary vessels to flow through the normal muscle tissue, thus leaving little blood to flow through the small anastomotic channels into the ischemic area, so that the ischemic condition is worsened. This condition is called the "coronary steal" syndrome. Consequently, one of the most important factors in the treatment of a patient with myocardial infarction is observance of absolute rest.

FUNCTION OF THE HEART FOLLOWING RECOVERY FROM MYOCARDIAL INFARCTION

Occasionally, a heart that has recovered from a large myocardial infarction returns to full functional capability, but more frequently its pumping capability is permanently decreased below that of a normal heart. This does not mean that the person is necessarily a cardiac invalid or that the resting cardiac output is depressed below normal because the normal person's heart is capable of pumping about 300 per cent more blood per minute than the body requires — that is, a person has a "cardiac reserve" of 300 per cent. Even when the cardiac reserve is reduced to as little as 100 per cent, the person can still perform normal activity of a quiet, restful type but not strenuous exercise that would overload the heart.

PAIN IN CORONARY DISEASE

Normally, a person cannot "feel" his or her heart, but ischemic cardiac muscle does exhibit pain sensation. Exactly what causes this pain is not known, but it is believed that ischemia causes the muscle to release acidic substances, such as lactic acid, or other pain-promoting products, such as histamine, kinins, or cellular proteolytic enzymes, that are not removed rapidly enough by the slowly moving blood. The high concentrations of these abnormal products then stimulate the pain endings in the cardiac muscle, and pain impulses are conducted through the sympathetic afferent nerve fibers into the central nervous system.

Angina Pectoris

In most persons who develop progressive constriction of their coronary arteries, cardiac pain, called *angina pectoris,* begins to appear whenever the load on the heart becomes too great in relation to the coronary blood flow. This pain is usually felt beneath the upper sternum and is often also referred to surface areas of the body, most often to the left arm and left shoulder but also frequently to the neck and even to the side of the face or to the opposite arm and shoulder. The reason for this distribution of pain is that the heart originates during embryonic life in the neck, as do the arms. Therefore, both of these structures receive pain nerve fibers from the same spinal cord segments.

In general, most persons who have chronic angina pectoris feel the pain when they exercise, also when they experience emotions that increase metabolism of the heart or temporarily constrict the coronary vessels because of sympathetic vasoconstrictor nerve signals. Usually the pain lasts for only a few minutes. However, some patients have such severe and lasting ischemia that the pain is present all the time. The pain is frequently described as hot, pressing, and constricting; it is of such quality that it usually makes the patient stop all activity and come to a complete state of rest.

When a person has frequent attacks of angina, the likelihood of developing an acute coronary occlusion is generally very great.

Treatment With Drugs. Several vasodilator drugs, when administered during an acute anginal attack, will usually give immediate relief from the pain. Commonly used vasodilators are *nitroglycerin* and *other nitrate drugs.*

A second class of drugs that are used for prolonged treatment of angina pectoris is the *beta blockers,* such as propranolol. These block the sympathetic beta receptors, which prevents sympathetic stimulation of heart rate and cardiac metabolism during exercise or emotional episodes. Therefore, therapy with a beta blocker decreases the need of the heart for metabolic oxygen during stressful conditions.

For obvious reasons, this can greatly reduce the number of anginal attacks as well as their severity.

SURGICAL TREATMENT OF CORONARY DISEASE

Aortic-Coronary Bypass Surgery. In many patients with coronary ischemia, the constricted areas of the coronary vessels are located at only a few discrete points, and the coronary vessels beyond these points are normal or almost normal. A surgical procedure has been developed in the past 25 years, called *aortic-coronary bypass,* for anastomosing small vein grafts to the aorta and to the sides of the more peripheral coronary vessels. Usually, one to five such grafts are performed during the operation, each of which supplies a peripheral coronary artery beyond a block. The vein that is used for the graft is usually the long superficial saphenous vein removed from the leg of the patient.

The acute results from this type of surgery have been especially good, causing this to be the most common cardiac operation performed. Anginal pain is relieved in most patients. Also, in patients whose hearts have not become too severely damaged prior to the operation, the coronary bypass procedure often can provide the patient with a normal survival expectation. On the other hand, if the heart has already been severely damaged, the bypass procedure is likely to be of little value.

Coronary Angioplasty. During the past 10 years, a procedure has been employed to open partially blocked coronary vessels before they become totally occluded. This procedure, called *coronary artery angioplasty,* is the following: A very small balloon-tipped catheter, about 1 millimeter in diameter, is passed under radiographic guidance into the coronary system and pushed through the partially occluded artery until the balloon portion of the catheter straddles the partially occluded point. Then the balloon is inflated with several atmospheres of pressure, which stretches the diseased artery almost to the point of bursting. After this procedure is performed, the blood flow through the vessel often increases as much as threefold to fourfold, and more than three quarters of the patients who undergo the procedure are relieved of the coronary ischemic symptoms for at least several years, though many of the patients will still require coronary bypass surgery eventually.

REFERENCES

American Physiological Society: Peripheral Circulation and Organ Blood Flow. Washington, D.C., American Physiological Society, 1983.

Berne, R. M., and Rubio, R.: Coronary circulation. In Berne, R. M., et al. (eds.): Handbook of Physiology. Sec. 2. Vol. I. Baltimore, Williams & Wilkins, 1979, p. 873.

Braunwald, E.: Heart Disease. Philadelphia, W. B. Saunders Co., 1988.

Chapman, C. B.: The physiology of exercise. Sci. Am., 212:88, 1965.

Chilian, W. M., and Marcus, M. L.: Coronary vascular adaptations to myocardial hypertrophy. Annu. Rev. Physiol., 49:477, 1987.

Cohn, P. F.: Silent myocardial ischemia: Present status. Mod. Concepts Cardiovas. Dis., 56:1, 1987.

Conti, C. R.: Large vessel coronary vasospasm: Diagnosis, natural history and treatment. Am. J. Cardiol., 55:41B, 1985.

Epstein, S. E., et al.: Dynamic coronary obstruction as a cause of angina pectoris: Implications regarding therapy. Am. J. Cardiol., 55:61B, 1985.

Feigl, E. O.: Coronary physiology. Physiol. Rev., 63:1, 1983.

Fletcher, G. F., and Cantwell, J. D.: Exercise and Coronary Heart Disease: Role in Prevention, Diagnosis, Treatment. Springfield, Ill., Charles C Thomas, 1978.

Fox, E. L.: Sports Physiology. Philadelphia, W. B. Saunders Co., 1979.

Fuster, V., and Chesebro, J. A.: Symposium on thrombosis and antithrombotic therapy. Am. J. Cardiol., 8:1B, 1986.

Geddes, L. A.: Cardiovascular Medical Devices. New York, John Wiley & Sons, 1984.

Gensini, G. G. (ed.): Coronary Arteriography. Mt. Kisco, N.Y., Futura Publishing Co., 1985.

Gerard, R. D., and Meidell, R. S.: Regulation of tissue plasminogen activator expression. Annu. Rev. Physiol., 51:245, 1989.

Gibbs, C. L., and Chapman, J. B.: Cardiac energetics. In Berne, R. M., et al. (eds.): Handbook of Physiology. Sec. 2, Vol. I. Baltimore, Williams & Wilkins, 1979, p. 775.

Goldberger, E. (ed.): Treatment of Cardiac Emergencies. St. Louis, C. V. Mosby Co., 1989.

Gregg, D. E.: The natural history of coronary collateral development. Circ. Res., 35:335, 1974.

Gregg, D. E.: Coronary Circulation in Health and Disease. Philadelphia, Lea & Febiger, 1950.

Guyton, A. C., et al.: Cardiac Output and Its Regulation. Philadelphia, W. B. Saunders Co., 1973.

Guyton, R. A., and Daggett, W. M.: The evolution of myocardial infarction: Physiological basis for clinical intervention. Int. Rev. Physiol., 9:305, 1976.

Hurst, J. W. et al. (eds.): Atlas of the Heart. New York, McGraw-Hill Book Co., 1988.

Hutchinson: Prevention of Coronary Heart Disease. Chicago, Year Book Medical Publishers, 1985.

Jang, G. D. (ed.): Angioplasty. New York, McGraw-Hill Book Co., 1986.

Jennings, R. B., and Steenbergen, C., Jr.: Nucleotide metabolism and cellular damage in myocardial ischemia. Annu. Rev. Physiol., 47:727, 1985.

Juergens, J. L., et al. (eds.): Allen, Barker, Hines Peripheral Vascular Disease. 5th Ed. Philadelphia, W.B. Saunders Co., 1980.

Katz, A. M.: Basic cellular mechanisms of action of the calcium-channel blockers. Am. J. Cardiol., 55:2B, 1985.

Klocke, F. J., and Ellish, A. K.: Control of coronary blood flow. Annu. Rev. Med., 31:489, 1980.

Lawrie: Coronary Artery Bypass Surgery. Chicago, Year Book Medical Publishers, 1985.

Leon, M. B., et al.: Combination therapy with calcium-channel blockers and beta blockers for chronic stable angina pectoris. Am. J. Cardiol., 55:69B, 1985.

McAllister, R. G., Jr., et al.: Pharmacokinetics of calcium-entry blockers. Am. J. Cardiol., 55:30B, 1985.

Michell, J. H., et al.: The exercise pressor reflex: Its cardiovascular effects, afferent mechanisms, and control pathways. Annu. Rev. Physiol., 45:229, 1983.

Miller, A. J.: Diagnosis of Chest Pain. New York, Raven Press, 1988.

Najarian, J. S.: Advances in Vascular Surgery. Chicago, Year Book Medical Publishers, 1983.

Reeder, G. S., and Vliestra, R. E.: Coronary angioplasty: 1986. Mod. Concepts Cardiovas. Dis., 55:49, 1986.

Reimer, K. A., and Jennings, R. B.: Effects of calcium-channel blockers on myocardial preservation during experimental acute myocardial infarction. Am. J. Cardiol., 55:107B, 1985.

Reitz, B. A.: Myocardial protection during cardiac surgery. Annu. Rev. Med., 33:151, 1982.

Rigotti, N. A., et al.: Exercise and coronary heart disease. Annu. Rev. Med., 34:391, 1983.

Roberts, L.: Study bolsters case against cholesterol. Science, 237:28, 1987.

Rose, C. P., and Goresky, C. A.: Interactions between capillary exchange, cellular entry, and metabolic sequestration processes in the heart. In Renkin, E. M., and Michel, C. C. (eds.): Handbook of Physiology. Sec. 2, Vol. IV. Bethesda, Md., American Physiological Society, 1984, p. 781.

Sharma, B., et al.: Intracoronary prostaglandin E_1 plus streptokinase in acute myocardial infarction. Am. J. Cardiol., 58:1161, 1986.

Shepherd, J. T.: Circulation to skeletal muscle. In Shepherd, J. T., and Abboud, F. M. (eds.): Handbook of Physiology. Sec. 2, Vol. III. Bethesda, Md., American Physiological Society, 1983, p. 319.

Shepherd, J. T., and Vanhoutte, P. M.: Spasm of the coronary arteries: Causes and consequences (the scientist's viewpoint). Mayo Clin. Proc., 60:33, 1985.

Singh, B. N., et al.: Second-generation calcium antagonists: Search for greater selectivity and versatility. Am. J. Cardiol., 55:214B, 1985.

Stone, H. L.: Control of the coronary circulation during exercise. Annu. Rev. Physiol., 45:213, 1983.

Strauss, R. H. (ed.): Sports Medicine and Physiology. Philadelphia, W. B. Saunders Co., 1979.

Su, C.: Extracellular functions of nucleotides in heart and blood vessels. Annu. Rev. Physiol., 47:665, 1985.

Sugimoto, T., et al.: Quantitative effect of low coronary pressure on left ventricular performance. Jap. Heart J., 9:46, 1968.

Vedin, J. A., and Wilhelmsson, C. E.: Beta receptor blocking agents in the secondary prevention of coronary heart disease. Annu. Rev. Pharmacol. Toxicol., 23:29, 1983.

Virmani, R., and Forman, M. B.: Nonatherosclerotic Ischemic Heart Disease. New York, Raven Press, 1989.

Vogel, J. H. K., and King, S. B., III: Interventional Cardiology: Future Directions. St. Louis, C. V. Mosby Co., 1989.

Winniford, M. D., et al.: Calcium antagonists for acute ischemic heart disease. Am. J. Cardiol., 55:116B, 1985.

Zelis, R., and Flaim, S. F.: Calcium blocking drugs for angina pectoris. Annu. Rev. Med., 33:465, 1982.

CHAPTER 22

• *Cardiac Failure* •

Perhaps the most important ailment that must be treated by the physician is cardiac failure, which can result from any heart condition that reduces the ability of the heart to pump blood. Usually the cause is decreased contractility of the myocardium, resulting from diminished coronary blood flow, but failure to pump can also be caused by damage to the heart valves, external pressure around the heart, vitamin deficiency, primary cardiac muscle disease, or any other abnormality that makes the heart a hypoeffective pump. In the present chapter, we discuss primarily cardiac failure caused by ischemic heart disease. In the following chapter, we discuss valvular and congenital heart disease.

Definition of Cardiac Failure. The term "cardiac failure" means simply *failure of the heart, usually because of some malfunction of the heart itself, to pump enough blood to satisfy the needs of the body.* Cardiac failure may be manifest in either of two ways: (1) by a decrease in cardiac output or (2) by a damming of the blood in the veins behind the left or right heart — even though the cardiac output may be normal or at times even above normal, as we discuss later.

● DYNAMICS OF THE CIRCULATION IN CARDIAC FAILURE

ACUTE EFFECTS OF MODERATE CARDIAC FAILURE

If a heart suddenly becomes severely damaged, such as by myocardial infarction, the pumping ability of the heart is immediately depressed. As a result, two essential effects occur: (1) reduced cardiac output and (2) damming of blood in the veins, resulting in increased systemic venous pressure. These two effects are shown graphically in Figure 22–1. This figure illustrates, first, a normal cardiac output curve, and point A on this curve is the normal operating point, with a normal cardiac output under resting conditions of 5 liters/min and a right atrial pressure of 0 mm Hg.

Immediately after the heart becomes damaged, the cardiac output curve becomes greatly reduced, falling to the lower, short-dashed curve at the very bottom of the graph. Within a few seconds, a new circulatory state is established at point B rather than point A, showing that the cardiac output has fallen to 2 liters/min, about two-fifths normal, while the right atrial pressure has risen to 4 mm Hg because blood returning to the heart is dammed up in the right atrium. This low cardiac output is still sufficient to sustain life, but it is likely to be associated with fainting. Fortunately, this acute stage lasts for only a few seconds because sympathetic reflexes occur immediately that can compensate to a great extent for the damaged heart as follows:

Compensation for Acute Cardiac Failure by Sympathetic Reflexes. When the cardiac output falls precariously low, many of the different circulatory reflexes discussed in Chapter 18 are immediately activated. The best known of these is the baroreceptor reflex, which is activated by diminished arterial pressure. It is probable that the chemoreceptor reflex, the central nervous system ischemic response, and even reflexes originating in the damaged heart itself also contribute to the nervous response. But whatever all the reflexes might be, the sympathetics become strongly stimulated within a few seconds, and the parasympathetics become reciprocally inhibited at the same time.

Strong sympathetic stimulation has two major effects on the circulation: first, on the heart itself and, second, on the peripheral vasculature. If all the musculature is diffusely damaged but is still functional, sympathetic stimulation strengthens this damaged musculature. If part of the muscle is totally nonfunctional, while part of it is still normal, the normal muscle is strongly stimulated by sympathetic stimulation, in this way compensating for the nonfunctional muscle. Thus, *the heart one way or another becomes a stronger pump, often as much as 100 per cent stronger, under the influence of the sympathetic impulses.* This effect is also illustrated in Figure 22–1, which shows elevation of the cardiac output curve after sympathetic compensation (the dotted curve).

Sympathetic stimulation also increases the tendency for venous return, for it increases the tone of most of the blood vessels of the circulation, especially of the veins, *raising the mean systemic filling pressure* to 12 to 14 mm Hg, almost 100 per cent above normal. As should be recalled from the discussion in Chapter 20, this greatly increases the tendency for blood to flow back to the heart. Therefore, the damaged heart becomes primed with more inflowing blood than usual, and the right atrial pressure rises still further, which helps the heart to pump larger quantities of blood. Thus, in Figure 22–1 the new circulatory state is depicted by point C, showing a cardiac output of 4.2 liters/min and a right atrial pressure of 5 mm Hg.

The sympathetic reflexes become maximally developed in about 30 seconds. Therefore, a person who has a sudden moderate heart attack might experience nothing more than cardiac pain and a few seconds of fainting. Shortly thereafter, with the aid of the sympathetic reflex compensations as described previously, the cardiac output may return to a level entirely adequate to sustain the person who remains quiet, though the pain might persist.

THE CHRONIC STAGE OF FAILURE

After the first few minutes of an acute heart attack, a prolonged secondary state begins. This is characterized

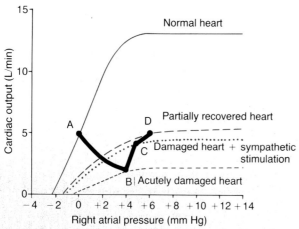

Figure 22 – 1. Progressive changes in the cardiac output curve following acute myocardial infarction. The cardiac output and right atrial pressure change progressively from point A to point D, as explained in the text.

mainly by two events: (1) retention of fluid by the kidneys and (2) usually progressive recovery of the heart itself over a period of several weeks to months, as discussed in the previous chapter.

Renal Retention of Fluid and Increase in Blood Volume

A low cardiac output has a profound effect on renal function, sometimes causing anuria when the cardiac output falls to as low as one-half to two-thirds normal. In general, the urinary output remains reduced as long as the cardiac output is significantly less than normal, and it usually does not return to normal after an acute heart attack until the cardiac output rises either all the way back to normal or almost to normal.

The Beneficial Effects of Moderate Fluid Retention in Cardiac Failure. Although many cardiologists formerly considered fluid retention always to have a detrimental effect in cardiac failure, it is now known that a moderate increase in body fluid and blood volume is actually a very important factor helping to compensate for the diminished pumping ability of the heart. It does this by increasing the tendency for venous return. That is, the increased blood volume increases the venous return in two ways: First, it increases the mean systemic filling pressure, which *increases the pressure gradient for causing flow of blood toward the heart.* Second, it distends the veins, which *reduces the venous resistance* and thereby allows increased ease of flow of blood to the heart.

If the heart is not too greatly damaged, this increased tendency for venous return can often fully compensate for the heart's diminished pumping ability—so much so, in fact, that if the heart's pumping ability is reduced to as little as 40 to 50 per cent of normal, still the increased venous return can often cause an entirely normal cardiac output.

On the other hand, when the heart's maximum pumping ability is reduced to less than 25 to 45 per cent of normal, the blood flow to the kidneys becomes too low for the urinary output of salt and water to equal the intake. Therefore, fluid retention begins and will continue indefinitely unless major therapeutic procedures are employed to prevent this. Furthermore, because the heart is already pumping at its maximum pumping capacity, this excess fluid no longer has a beneficial effect on the circulation. Instead, very severe

edema develops throughout the body, which can be very detrimental in itself and can lead to death. This problem of edema is discussed in more detail later in the chapter.

Detrimental Effects of Excess Fluid Retention in the Severe Stages of Cardiac Failure. In contrast to the beneficial effects of moderate fluid retention in cardiac failure, in severe failure with extreme excesses of fluid retention, the fluid then begins to have very serious physiological consequences, including overstretching of the heart, thus weakening the heart still more; filtration of fluid into the lungs to cause pulmonary edema and consequent deoxygenation of the blood; and, often, development of extensive edema in all of the peripheral tissues of the body as well. These detrimental effects of excessive fluid are discussed in subsequent sections of the chapter.

Recovery of the Myocardium Following Myocardial Infarction

After a heart becomes suddenly damaged as a result of myocardial infarction, the natural reparative processes of the body begin immediately to help restore normal cardiac function. Thus, a new collateral blood supply begins to penetrate the peripheral portions of the infarcted area, often causing much of the muscle in the fringe areas to become functional again. Also, the undamaged musculature hypertrophies, in this way offsetting much of the cardiac damage.

Obviously, the degree of recovery depends on the type of cardiac damage, and it varies from no recovery at all to almost complete recovery. Ordinarily, after myocardial infarction the heart recovers rapidly during the first few days and weeks and will have achieved most of its final state of recovery within 5 to 7 weeks, though some recovery continues for months.

Cardiac Output Curve After Partial Recovery. The long-dashed curve of Figure 22 – 1 illustrates function of the partially recovered heart a week or so after the acute myocardial infarction. By this time, considerable fluid also has been retained in the body, and the tendency for venous return has increased markedly; therefore, the right atrial pressure has risen even more. As a result, the state of the circulation is now changed from point C to point D, which represents a *normal* cardiac output of 5 liters per minute but a right atrial pressure elevated to 6 mm Hg.

Because the cardiac output has returned to normal, renal output also will have returned to normal, and no further fluid retention will occur. Therefore, except for the high right atrial pressure represented by point D in this figure, the person now has essentially normal cardiovascular dynamics *as long as he remains at rest.*

If the heart itself recovers to a significant extent and if adequate fluid retention occurs, the sympathetic stimulation gradually abates toward normal for the following reasons: The partial recovery of the heart can do the same thing for the cardiac output curve as sympathetic stimulation, and the increased blood volume can do the same thing for venous return as sympathetic contraction of the veins. Thus, as these two factors develop, the fast pulse rate, cold skin, and pallor resulting from sympathetic stimulation in the acute stage of cardiac failure gradually disappear.

SUMMARY OF THE CHANGES THAT OCCUR FOLLOWING ACUTE CARDIAC FAILURE— "COMPENSATED HEART FAILURE"

To summarize the events discussed in the past few sections describing the dynamics of circulatory changes following an acute, moderate heart attack, we may divide the stages into (1) the instantaneous effect of the cardiac damage; (2) compensation by the sympathetic nervous system, which occurs mainly within the first 30 seconds to 1 minute; and (3) chronic compensations resulting from partial cardiac recovery and renal retention of fluid. All these changes are shown graphically by the very heavy line in Figure 22–1. The progression of this line shows the normal state of the circulation (point A), the state a few seconds after the heart attack but before sympathetic reflexes have occurred (point B), the rise in cardiac output toward normal caused by sympathetic stimulation (point C), and final return of the cardiac output to normal following several days to several weeks of cardiac recovery and fluid retention (point D). This final state is called *compensated heart failure.*

Compensated Heart Failure. Note especially in Figure 22–1 that the pumping ability of the heart, as depicted by the cardiac output curve, is still depressed to less than one-half normal. This illustrates that factors that increase the right atrial pressure (principally the increased blood volume caused by retention of fluid) can maintain the cardiac output at a normal level despite continued weakness of the heart itself. However, one of the results of chronic cardiac weakness is this chronic increase in right atrial pressure itself; in Figure 22–1 it is shown to be 6 mm Hg. Many persons, especially in old age, have completely normal resting cardiac outputs but mildly to moderately elevated right atrial pressures because of compensated heart failure. These persons may not know that they have cardiac damage because the damage more often than not has occurred a little at a time, and the compensation has occurred concurrently with the progressive stages of damage.

When a person is in compensated heart failure, any attempt to perform heavy exercise will usually cause immediate return of the symptoms of acute failure because the heart simply is not able to increase its pumping capacity to the levels required to sustain the exercise. Therefore, it is said that the *cardiac reserve* is reduced in compensated heart failure. This concept of cardiac reserve is discussed more fully later in the chapter.

DYNAMICS OF SEVERE CARDIAC FAILURE —DECOMPENSATED HEART FAILURE

If the heart becomes severely damaged, no amount of compensation, either by sympathetic nervous reflexes or by fluid retention, can make this weakened heart pump a normal cardiac output. As a consequence, the cardiac output cannot rise to a high enough value to bring about return of normal renal function. Fluid continues to be retained, the person develops progressively more and more edema, and this state of events eventually leads to death. This is called *decompensated heart failure.*

Thus, the main basis of decompensated heart failure is *failure of the heart to pump sufficient blood to make the kidneys function adequately.*

Analysis of Decompensated Heart Failure. Figure 22–2 illustrates a greatly depressed cardiac output curve, depicting the function of a heart that has become extremely weakened and cannot be strengthened. Point A on this curve represents the approximate state of the circulation before any compensation has occurred, and point B the state after the first few minutes of acute compensation by sympathetic stimulation, as described previously. At point B, the cardiac output has risen to 4 liters/min and the right atrial pressure to 5 mm Hg. The person appears to be in reasonably good condition, but this state will not remain static for the following reason: The cardiac output has not risen quite high enough to cause adequate kidney excretion of fluid; therefore, fluid retention continues unabated and can eventually be the cause of death. These events can be explained quantitatively in the following way:

Note the dashed line at a cardiac output level of 5 liters in Figure 22–2. This is the critical cardiac output level that is required to allow the kidneys to re-establish normal fluid balance—that is, for the output of salt and water to be as great as the intake of these. At any cardiac output below this level, all the fluid-retaining mechanisms discussed in the earlier section remain in play, and the body fluid volumes increase progressively. Because of this progressive increase in fluid volume, the mean systemic filling pressure continues to rise; this forces progressively increasing quantities of blood into the right atrium, thus increasing the right atrial pressure. After a day or so, the state of the circulation changes in Figure 22–2 from point B to point C—the right atrial pressure rising to 7 mm Hg and the cardiac output to 4.2 liters/min. However, note again that the cardiac output is still not high enough to cause normal renal output of fluid; therefore, fluid continues to be retained, and after another day or so the right atrial pressure rises to 9 mm Hg, and the circulatory state becomes that depicted by point D. Still, the cardiac output is not enough to establish normal fluid balance.

After another few days of fluid retention, the right atrial pressure has risen still further, but by now the cardiac function curve is beginning to decline toward a lower level. This decline is caused by overstretch of the heart, edema of the heart muscle, and other factors that diminish the pumping performance of the heart. It is now clear that further retention of fluid will be more detrimental than beneficial to the circulation. Yet, the cardiac output still is not high enough to bring about normal renal function, and fluid retention not only continues but actually accelerates because of the falling cardiac output. Consequently, within a few days the state of the circulation has reached point F on the curve, with the cardiac output less than 2.5 liters/min and the

Figure 22–2. Greatly depressed cardiac output curve that indicates decompensated heart disease. Progressive fluid retention raises the right atrial pressure, and the cardiac output progresses from point A to point F, as explained in the text.

right atrial pressure 16 mm Hg. This is a state that now has approached or reached incompatibility with life, and the patient dies. This state of heart failure in which the failure continues to worsen is called *decompensated heart failure.*

Thus, one can see from this analysis that failure of the cardiac output ever to rise to the critical level required for normal renal function results in (1) progressive retention of fluid, which causes (2) progressive elevation of the mean systemic filling pressure, and (3) progressive elevation of the right atrial pressure until finally the heart is so overstretched or so edematous that it becomes unable to pump even moderate quantities of blood and, therefore, fails completely. Clinically, one detects this serious condition of decompensation principally by the progressive edema, especially edema of the lungs, which leads to bubbling rales in the lungs and dyspnea (air hunger). All clinicians know that lack of appropriate therapy when this state of events occurs leads to rapid death.

Treatment of Decompensation. The two ways in which the decompensation process can often be stopped are (1) by strengthening the heart in any one of several ways, especially by administration of a cardiotonic drug, such as digitalis, so that it can pump adequate quantities of blood to make the kidneys function normally again, or (2) by administering diuretics and reducing water and salt intake, which brings about a balance between fluid intake and output despite the low cardiac output.

Both methods stop the decompensation process by reestablishing normal fluid balance, so that at least as much fluid leaves the body as enters it.

Mechanism of Action of the Cardiotonic Drugs. Cardiotonic drugs, such as digitalis, when administered to a person with a normal heart, have little effect on increasing the contractile strength of the cardiac muscle. On the other hand, when administered to a person with a chronically failing heart, these same drugs can sometimes increase the strength of the failing myocardium as much as 50 to 100 per cent. Therefore, they are the mainstay of therapy in chronic heart failure.

The way in which digitalis and the other cardiotonic glycosides strengthen heart contraction is believed to be by increasing the quantity of calcium ions in the muscle fibers. In the failing heart muscle, the sarcoplasmic reticulum fails to accumulate normal quantities of calcium and therefore cannot release enough calcium to cause full contraction of the muscle. The effect of digitalis is to depress the calcium pump of the cardiac muscle cell membrane that normally pumps calcium ions out of the muscle cells. This mechanism is required to keep the concentration of calcium from rising too high in these cells because calcium ions have a natural tendency to leak inward through the membrane. However, in the case of the failing heart, a slight extra amount of calcium is desirable to increase the muscle contractile force. Therefore, it is usually beneficial to depress the calcium pumping mechanism a moderate amount using digitalis, thus allowing the intracellular calcium level to rise.

● UNILATERAL LEFT HEART FAILURE

In the discussions thus far in this chapter, we have considered failure of the heart as a whole. Yet in a large number of patients, especially those with early acute failure, left-sided failure predominates over right-sided failure, and only in rare instances will the right side fail without significant failure of the left side. Therefore, we need especially to discuss the special features of unilateral left heart failure.

When the left side of the heart fails without concomitant failure of the right side, blood continues to be pumped into the lungs with usual right heart vigor, whereas it is not pumped adequately out of the lungs into the systemic circulation. As a result, the *mean pulmonary filling pressure* rises because of shift of large volumes of blood from the systemic circulation into the pulmonary circulation.

As the volume of blood in the lungs increases, the pulmonary capillary pressure also increases, and if this rises above a value approximately equal to the colloid osmotic pressure of the plasma, about 28 mm Hg, fluid begins to filter out of the capillaries into the interstitial spaces and alveoli, resulting in pulmonary edema.

Thus, among the most important problems of left heart failure are *pulmonary vascular congestion* and *pulmonary edema.* In severe acute left heart failure, pulmonary edema occasionally occurs so rapidly that it causes death by suffocation in as little as 20 to 30 minutes, which we discuss more fully later in the chapter.

● "HIGH OUTPUT CARDIAC FAILURE" — A CONDITION OF OVERLOADING THE HEART

A condition called "high output cardiac failure" frequently occurs in persons who have cardiac outputs much higher than normal but who also have signs of cardiac failure because the left or right or both atrial pressures are much too high, and there usually also is accumulation of edema. The true problem is often not failure of the pumping ability of the heart but instead *overloading of the heart with too much venous return.* Most often this is caused by some circulatory abnormality that drastically decreases the total peripheral resistance, such as

1. *Arteriovenous fistula,* in which blood is shunted directly from the arteries to the veins. This greatly increases the venous return and frequently causes congestive symptoms similar to those of typical heart failure, even though the heart is normal and is actually pumping a tremendously increased cardiac output.

2. *Beriberi heart disease,* which is caused by lack of the B vitamins, especially of thiamine. This causes poor metabolic utilization of nutrients throughout the body. In turn, the tissue metabolic mechanisms controlling local blood flow dilate all the local blood vessels, profoundly reducing the total peripheral resistance and often doubling the venous return. The heart is usually weakened as well; yet, even with the weakness of the heart, the cardiac output is still as much as two times normal. Therefore, severe congestive symptoms occur both because of the excess venous return and because of the weakened heart, so that there is *both excess loading of the heart and true heart failure.*

3. *Thyrotoxicosis,* which, because of its stimulatory effect on body metabolism, causes generalized systemic vasodilation. Here again, the thyrotoxicosis frequently weakens the heart as well, so that there is often both overloading of the heart and true heart failure, occurring simultaneously.

● LOW OUTPUT CARDIAC FAILURE — "CARDIOGENIC SHOCK"

In many instances after acute heart attacks, and even sometimes after prolonged periods of slow progressive cardiac deterioration, the heart becomes incapable of pumping even the minimal amount of blood flow required to keep the body alive. Consequently, all of the body tissues begin to suffer and even to deteriorate, often leading to death within a few hours to a few days. The picture then is one of circulatory shock, as explained in Chapter 24. Even the cardiovascular system itself suffers from lack of nutrition, and it, too, along with the remainder of the body deteriorates, thus hastening the death. This circulatory shock syndrome caused by inadequate cardiac pumping is called *cardiogenic shock* or simply *cardiac shock*. Sometimes it is also called the *power failure syndrome*. Once a person develops cardiogenic shock, the survival rate even with the best of therapy is usually less than 15 per cent.

Vicious Circle of Cardiac Deterioration in Cardiogenic Shock. The discussion of circulatory shock in Chapter 24 emphasizes the tendency for the heart itself to become progressively more damaged when its coronary blood supply is reduced during the course of shock. That is, the low arterial pressure that occurs during shock reduces the coronary supply, which makes the heart still weaker, which makes the shock still worse, the process eventually becoming a vicious circle of cardiac deterioration. In cardiogenic shock caused by myocardial infarction, this problem is greatly compounded by the already existing coronary thrombosis. For instance, in a normal heart, the arterial pressure usually must be reduced below about 45 mm Hg before cardiac deterioration sets in. However, in a heart that already has a blocked major coronary vessel, deterioration will set in when the arterial pressure falls as low as 80 to 90 mm Hg. In other words, even the minutest amount of fall in arterial pressure can set off a vicious circle of cardiac deterioration following myocardial infarction. For this reason, in treating myocardial infarction it is extremely important to prevent even short periods of hypotension.

Physiology of Treatment. Often a patient dies of cardiogenic shock before the various compensatory processes can return the cardiac output to a life-sustaining level. Therefore, treatment of this condition is one of the most important problems in the management of acute heart attacks. Immediate digitalization of the heart is often employed for strengthening the heart if the remaining functioning ventricular muscle shows signs of deterioration. Also, infusion of whole blood, plasma, or a blood pressure–raising drug is used to sustain the arterial pressure. If the arterial pressure can be raised high enough, the coronary blood flow may be elevated to a high enough value to prevent the vicious circle of deterioration until appropriate compensatory mechanisms in the body can correct the shock. Even with the best therapy, however, once the shock syndrome has begun, with the arterial pressure remaining as much as 20 mm Hg below normal for as long as an hour, 85 per cent of the patients die.

Some success has been achieved in saving the lives of patients with cardiogenic shock using one of the following two procedures: (1) surgical removal of the clot in the coronary artery, often combined with coronary bypass graft, or (2) catheterizing the blocked coronary artery and infusing either *streptokinase* or *tissue type plasminogen activator (t-PA)* enzymes that cause dissolution of the clot. The results are occasionally astounding when the procedure is instituted within the first hour but of little if any benefit after 3 hours.

● EDEMA IN PATIENTS WITH CARDIAC FAILURE

Inability of *Acute* Cardiac Failure to Cause Peripheral Edema. Although acute left heart failure can cause terrific congestion of the lungs, with rapid development of pulmonary edema and even death within minutes to hours, *acute* heart failure of any type never causes immediate development of peripheral edema. This can be explained best by referring to Figure 22–3. When a previously normal heart acutely fails as a pump, the arterial pressure falls and the right atrial pressure rises. As the cardiac output approaches zero, these two pressures approach each other at an equilibrium value of about 13 mm Hg. It is obvious that capillary pressure must also fall from its normal value of 17 mm Hg to the equilibrium pressure of 13 mm Hg. Thus, *severe acute cardiac failure actually causes a fall in peripheral capillary pressure rather than a rise.* Animal experiments as well as experience in humans show that acute cardiac failure does not cause immediate development of peripheral edema.

Role of the Kidneys in Cardiac Edema

After the first day or so, peripheral edema does begin to occur, but principally *because of fluid retention by the kidneys.* The retention of fluid increases the mean systemic filling pressure, resulting in increased tendency for blood to return to the heart. This now elevates the right atrial pressure to a still higher value, and it also returns the arterial pressure to normal. Therefore, the rising right atrial pressure *now causes the capillary pressure to rise markedly,* thus causing loss of fluid into the tissues and development of severe edema.

There are three known causes of the reduced renal output of urine during cardiac failure, all of which are perhaps equally important but in different ways.

1. Decreased Glomerular Filtration. A decrease in cardiac output has a tendency to reduce the glomerular pressure in the kidneys because of (1) *reduced arterial pressure* and (2) *intense sympathetic constriction of the afferent arterioles of the kidney.* As a consequence, except in the mildest degrees of heart failure, the glomerular filtration rate becomes less than normal. It should become evident from the discussion of kidney function in Chapters 26 to 29 that *even a very slight decrease in glomerular filtration often decreases urine output markedly.* When the cardiac output falls to about one-half normal, this factor alone can result in almost complete anuria.

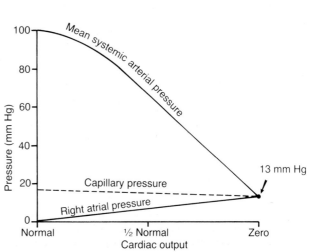

Figure 22–3. Progressive changes in mean systemic arterial pressure and right atrial pressure as the cardiac output falls from normal to zero. This figure also shows the effect of decreased cardiac output on capillary pressure.

2. Activation of the Renin-Angiotensin System and Increased Reabsorption of Water and Salt by the Renal Tubules. The reduced blood flow to the kidneys causes marked increase in renin output, and this in turn causes the formation of angiotensin by the mechanism described in Chapter 19. The angiotensin has a direct effect on the arterioles of the kidneys to decrease further the blood flow through the kidneys. This especially reduces the peritubular capillary pressure, thus promoting greatly increased reabsorption of both water and salt from the renal tubules. Therefore, the net loss of water and salt into the urine is greatly decreased, and the quantities of salt and water in the body fluids increase.

3. Increased Aldosterone Secretion. In the chronic stage of heart failure, large quantities of aldosterone are secreted by the adrenal cortex. This results mainly from the effect of angiotensin in stimulating aldosterone secretion. Also, some of the increase in aldosterone secretion might result from increased plasma potassium because excess potassium is the most powerful stimulus known for aldosterone secretion, and the potassium concentration rises in response to reduced renal function in cardiac failure. The elevated aldosterone further increases the reabsorption of sodium from the renal tubules, and this in turn leads to a secondary increase in water reabsorption for two reasons. First, as the sodium is reabsorbed, it reduces the osmotic pressure in the tubules while increasing the osmotic pressure in the renal interstitial fluids; these changes promote osmosis of water into the blood. Second, the absorbed sodium increases the osmotic concentration of the extracellular fluid and elicits *antidiuretic hormone* secretion by the hypothalamic–posterior pituitary system, which is discussed in Chapter 29. The antidiuretic hormone then promotes increased tubular reabsorption of water.

Possible Role of Atrial Natriuretic Factor to Delay the Onset of Cardiac Decompensation. *Atrial natriuretic factor (ANF)* is a hormone released by the atrial walls when they become stretched. Because heart failure almost always causes excessive increase in atrial pressures that stretch the atrial walls, the circulating levels of ANF in the blood increase as much as fivefold to tenfold in severe heart failure. The atrial natriuretic factor in turn has a direct effect on the kidneys to increase greatly their excretion of salt and water. Therefore, it has been suggested, but not proved, that ANF plays a natural role to help prevent the extreme congestive symptoms of cardiac failure. The renal effects of ANF are discussed in Chapter 29.

Acute Pulmonary Edema in Chronic Heart Failure — A Lethal Vicious Circle

A frequent cause of death is acute pulmonary edema occurring in patients who have had chronic heart failure for a long time. When this occurs in a person without new cardiac damage, it usually is set off by some temporary overload of the heart, such as might result from a bout of heavy exercise, some emotional experience, or even a severe cold. This acute pulmonary edema is believed to result from the following vicious circle:

1. A temporarily increased load on the already weak left ventricle results from increased venous return from the peripheral circulation. Because of the limited pumping capacity of the left heart, blood begins to dam up in the lungs.

2. The increased blood in the lungs elevates the pulmonary capillary pressure, and a small amount of fluid begins to transude into the lung tissues and alveoli.

3. The increased fluid in the lungs diminishes the degree of oxygenation of the blood.

4. The decreased oxygen in the blood further weakens the heart and also causes peripheral vasodilatation.

5. The peripheral vasodilatation increases venous return from the peripheral circulation still more.

6. The increased venous return further increases the damming of the blood in the lungs, leading to still more transudation of fluid, more arterial oxygen desaturation, more venous return, and so forth. Thus, a vicious circle has been established.

Once this vicious circle has proceeded beyond a certain critical point, it will continue until death of the patient unless heroic therapeutic measures are employed. The types of heroic therapeutic measures that can reverse the process and save the life of the patient include

1. Putting tourniquets on all four limbs to sequester much of the blood in the veins and therefore to decrease the workload on the left heart.

2. Bleeding the patient.

3. Giving a rapidly acting diuretic, such as furosemide, to cause rapid loss of fluid from the body.

4. Giving the patient pure oxygen to breathe to reverse the blood desaturation, the heart deterioration, and the peripheral vasodilatation.

5. Giving the patient a rapidly acting cardiotonic drug, such as ouabain, to strengthen the heart.

Unfortunately, this vicious circle of acute pulmonary edema can proceed so rapidly that death can occur within 20 minutes to an hour. Therefore, any procedure that is to be successful must be instituted immediately.

● HEART FAILURE IN SEVERE HYPERTENSION AND IN VALVULAR AND CONGENITAL HEART DISEASE

Thus far we have discussed heart failure mainly as it occurs in patients with acute or chronically progressive coronary artery disease. However, heart failure can result from other conditions that also reduce the heart's pumping effectiveness. For instance, malfunction of the heart valves or congenital anomalies of the heart, which are discussed in the following chapter, can both reduce heart pumping and lead to all the typical sequelae of heart failure.

In addition, serious hypertension is another common cause of chronic heart failure, particularly in older people whose heart muscle is already weakened by age as well as by various degrees of progressive coronary vascular sclerosis. Especially patients who develop the extreme state of hypertension called malignant hypertension almost always eventually develop both heart failure and kidney failure, and death follows shortly afterward because of the inability of the heart to cope with both the extra fluid and the extra workload caused by the high pressure.

Hypertensive heart failure can often be alleviated by simply treating the hypertension itself. This is achieved mainly by giving a diuretic to cause the kidneys to eliminate large amounts of fluid; by giving a converting enzyme inhibitor to block the renin-angiotensin system; by giving a vasodilator that dilates all the peripheral blood vessels, including dilating the renal arteries, thus aiding the kidneys as well; or by giving some combination of therapy.

● DETERIORATION OF THE HEART IN CHRONIC FAILURE — A VICIOUS CIRCLE

Regardless of the cause of chronic heart failure, the heart often goes into a final stage of rapid deterioration because the heart failure itself causes still more damage to the myocardium. This generally results from the following vicious circle: In chronic heart failure, not only is the blood flow to the peripheral tissues of the body compromised, but the blood flow in the coronary system is frequently poor as well. This is especially true of the blood flow to the subendocardial portions of the heart muscle. Gradually at first, and more rapidly later, many of the subendocardial heart muscle cells begin to die and are replaced by fibrous tissue. Thus, the heart becomes weaker and the failure becomes worse, instituting still another cycle of diminished coronary blood flow, additional death of the myocardium, and further degrees of failure. Thus, this progressive damage to the myocardium begins in the subendocardial muscle and spreads farther and farther in the outward direction through the heart wall. Because of this vicious circle of deterioration, cardiac failure beyond a critical point becomes almost impossible to treat.

● CARDIAC RESERVE

The maximum percentage that the cardiac output can increase above normal is called the *cardiac reserve.* Thus, in the normal young adult the cardiac reserve is 300 to 400 per cent. In the athletically trained person, it is occasionally as high as 500 to 600 per cent, whereas in the asthenic or elderly person, it may be as low as 200 per cent. In heart failure, there is no reserve at all. As an example, during severe exercise the cardiac output of the normal healthy young adult can rise to about five times normal; this is an increase above normal of 400 per cent—that is, a cardiac reserve of 400 per cent.

Any factor that prevents the heart from pumping blood satisfactorily decreases the cardiac reserve. This can result from ischemic heart disease, primary myocardial disease, vitamin deficiency, damage to the myocardium, valvular heart disease, and many other factors, some of which are illustrated in Figure 22–4.

Diagnosis of Low Cardiac Reserve — The Exercise Test. So long as people with low cardiac reserve remain in a state of rest, they usually will not know that they have heart disease. However, a diagnosis of low cardiac reserve can usually be made easily by requiring a person to exercise either on a treadmill or by walking up and down steps. The increased load on the heart rapidly uses up the small amount of reserve that is available, and the cardiac output fails to rise high enough to sustain the body's new level of activity. The acute effects are

1. Immediate and sometimes extreme shortness of breath (dyspnea) resulting from the heart's not pumping

sufficient blood to the tissues, thereby causing tissue ischemia and creating a sensation of air hunger.

2. Extreme muscle fatigue resulting from muscle ischemia, thus limiting the person's ability to continue with the exercise.

3. Excessive increase in heart rate because the nervous reflexes overreact in an attempt to overcome the inadequate cardiac output.

Therefore, exercise tests are part of the armamentarium of the cardiologist. These tests take the place of cardiac output measurements that unfortunately cannot be made with ease in most clinical settings.

● APPENDIX

A QUANTITATIVE GRAPHICAL METHOD FOR ANALYSIS OF CARDIAC FAILURE

Although it is possible to understand most of the general principles of cardiac failure using mainly qualitative logic, as we have done thus far in this chapter, one can grasp the importance of the different factors in cardiac failure with far greater depth by using quantitative approaches. One such approach is the graphical method for analysis of cardiac output regulation that was introduced in Chapter 20. In the following few paragraphs, we analyze several aspects of cardiac failure using this graphical technique.

Graphical Analysis of Acute Heart Failure and of Chronic Compensation

Figure 22–5 illustrates cardiac output and venous return curves for different states of the heart and of the peripheral circulation. The two *solid curves* are (1) the normal cardiac output curve and (2) the normal venous return curve. As pointed out in Chapter 20, there is only one point on each of these two curves at which the circulatory system can operate so long as these two curves remain normal. This point is where the two curves cross at point A. Therefore, the normal state of the circulation is a cardiac output and venous return of 5 liters/min and a right atrial pressure of 0 mm Hg.

Effect of Acute Heart Attack. During the first few seconds after a moderately severe heart attack, the cardiac output curve falls to the lowermost *long-dashed curve.* During this few seconds, the venous return curve still has not changed because the peripheral circulatory system is still operating normally. Therefore, the new state of the circulation is depicted by *point B*, where the new cardiac output

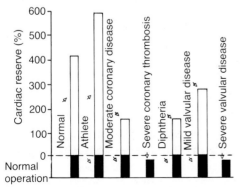

Figure 22–4. Cardiac reserve in different conditions.

Figure 22–5. The progressive changes in cardiac output and right atrial pressure during different stages of cardiac failure.

curve crosses the normal venous return curve. Thus, the right atrial pressure rises to 4 mm Hg, whereas the cardiac output falls to 2 liters/min.

Effect of Sympathetic Reflexes. Within the next 30 seconds, the sympathetic reflexes become very active. These affect both curves, raising them to the *short-dashed curves.* Sympathetic stimulation can increase the plateau level of the cardiac output curve as much as 100 per cent. It can also increase the mean systemic filling pressure (depicted by the point where the venous return curve crosses the zero venous return axis) by several millimeters of mercury — in this figure from a normal value of 7 mm Hg up to 10 mm Hg. This increase in mean systemic filling pressure shifts the entire venous return curve to the right and upward. The new cardiac output and venous return curves now equilibrate at point C, that is, at a right atrial pressure of 5 mm and at a cardiac output of 4 liters/min.

Compensation During the Next Few Days. During the ensuing week, the cardiac output and venous return curves rise to the *dot-dashed curves* because of (1) some recovery of the heart and (2) renal retention of salt and water, which raises the mean systemic filling pressure still further — this time up to 12 mm Hg. The two new curves, the dot-dashed curves, now equilibrate at *point D.* Thus, the cardiac output has now returned entirely to normal. The right atrial pressure, however, has risen still further to 6 mm Hg. Because the cardiac output is now normal, renal output is also normal, so that a new state of equilibrated fluid balance has been achieved. Therefore, the circulatory system will continue to function at point D and will remain stable, with a normal cardiac output and an elevated right atrial pressure, until some additional extrinsic factor changes either the cardiac output curve or the venous return curve.

Using this technique for analysis, one can see especially the importance of moderate fluid retention and how it eventually leads to a new stable state of the circulation; one can also see the interrelationship between the mean systemic filling pressure and cardiac pumping at various degrees of cardiac failure.

Note that the events described in Figure 22–5 are the same as those presented in Figure 22–1 but presented in a more quantitative manner.

Graphical Analysis of "Decompensated" Cardiac Failure

The cardiac output curve in Figure 22–6 is the same as that in Figure 22–2, a very low curve that in this case has already reached a degree of recovery as great as this heart can achieve. In this figure we have added venous return

curves that occur during successive days following the acute fall of the cardiac output curve to this very low level. At *point A* the curve equates with the normal venous return curve to give a cardiac output of approximately 3 liters/min. However, stimulation of the sympathetic nervous system, caused by this low cardiac output, increases the mean systemic filling pressure within 30 seconds from 7 to 10.5 mm Hg and shifts the venous return curve upward and to the right to give the curve labeled "autonomic compensation." Thus, the new venous return curve equates with the cardiac output curve at *point B.* The cardiac output has been improved up to a level of 4 liters/min but at the expense of an additional rise in right atrial pressure to 5 mm Hg.

The cardiac output of 4 liters per minute is still too low to cause the kidneys to function normally. Therefore, fluid is retained, the mean systemic filling pressure rises from 10.5 almost to 13 mm Hg, and the venous return curve becomes that labeled "second day." This equilibrates with the cardiac output curve at *point C,* and the cardiac output rises to 4.2 liters/min, while the right atrial pressure rises to 7 mm Hg.

During the succeeding days the cardiac output never rises high enough to re-establish normal renal function. Therefore, fluid continues to be retained, the mean systemic filling pressure continues to rise, the venous return curve continues to shift to the right, and the equilibrium point between the venous return curve and the cardiac output curve shifts progressively to *point D,* to *point E,* and finally to *point F.* The equilibration process is now on the downslope of the cardiac output curve, so that further retention of fluid causes only a severe cardiac edema and a detrimental effect on cardiac output. Therefore, the condition accelerates in a downhill direction until death occurs.

Thus, the process of "decompensation" results from the fact that the cardiac output curve never rises to the critical level of 5 liters/min required to re-establish normal balance between fluid input and output.

Treatment of Decompensated Heart Disease With Digitalis. Let us assume that the stage of decompensation has already reached point E in Figure 22–6, and let us proceed to the same point E in Figure 22–7. At this point, digitalis is given to strengthen the heart and to raise the cardiac output curve to the level of the heavy curve in Figure 22–7. This does not immediately change the venous return curve. Therefore, the new cardiac output curve equates with the venous return curve at *point G.* The cardiac output is now 5.7 liters/min, a value greater than the critical level of 5 liters required for normal fluid balance. Therefore, the kidneys eliminate far more fluid than normally, causing *diuresis,* a well-known therapeutic effect of digitalis.

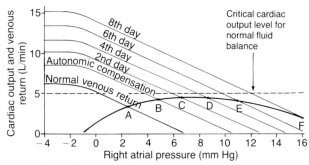

Figure 22–6. Graphical analysis of decompensated heart disease, showing progressive shift of the venous return curve to the right as a result of fluid retention.

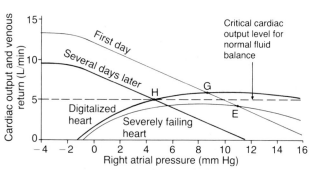

Figure 22–7. Treatment of decompensated heart disease, showing the effect of digitalis in elevating the cardiac output curve, this in turn causing progressive shift of the venous return curve to the left.

The progressive loss of fluid over a period of several days reduces the mean systemic filling pressure back down to 11.5 mm Hg, and the new venous return curve becomes the heavy curve labeled "several days later." This curve equates with the cardiac output curve of the digitalized heart at *point H,* at an output of 5 liters/min and a right atrial pressure of 4.6 mm Hg. This cardiac output is precisely that required for normal fluid balance. Therefore, no additional fluid will be lost, and none will be gained. Consequently, the circulatory system has now stabilized, or, in other words, the decompensation of the heart failure has been "compensated." To state this another way, the final steady-state condition of the circulation is defined by the crossing point of three different curves: the cardiac output curve, the venous return curve, and the critical level for normal fluid balance. The compensatory mechanisms automatically stabilize the circulation when all three curves cross at the same point.

Graphical Analysis of High Output Cardiac Failure

Figure 22–8 gives an analysis of two types of high output cardiac failure. One of these is caused by an arteriovenous fistula that overloads the heart because of excessive venous return, even though the pumping capability of the heart itself is not depressed. The other is caused by the disease beriberi, in which the venous return is greatly increased because of diminished systemic vascular resistance, but at the same time the pumping capability of the heart is also depressed.

Arteriovenous Fistula. The heavy black curves of the figure depict the normal cardiac output and normal venous return curve. These equate with each other at *point A,* which depicts a normal cardiac output of 5 liters/min and a normal right atrial pressure of 0 mm Hg.

Now let us assume that systemic resistance (the total peripheral resistance) becomes greatly decreased because of opening a very large arteriovenous fistula. The venous return curve rotates upward to give the curve labeled "A-V fistula." This venous return curve equates with the cardiac output curve at *point B,* with a cardiac output of 12.5 liters/min and a right atrial pressure of 3 mm Hg. Thus, the cardiac output has become greatly elevated, the right atrial pressure is slightly elevated, and there are mild signs of

peripheral congestion. If the person attempts to exercise, he or she will have very little cardiac reserve because the heart is already being used almost to maximum capacity simply to pump the extra blood through the arteriovenous fistula. Therefore, this condition resembles a failure condition and is called "high output failure," but in reality the heart is simply overloaded by excess venous return.

Beriberi. The dashed lines in Figure 22–8 illustrate the approximate changes in the cardiac output and venous return curves caused by beriberi. The decreased level of the cardiac output curve is caused by weakening of the heart because of the avitaminosis (mainly lack of thiamine) that causes the beriberi syndrome. The weakening of the heart has decreased the blood flow to the kidneys. Therefore, the kidneys have retained a large amount of fluid, which in turn has increased the mean systemic filling pressure (represented by the point where the venous return curve intersects with the zero cardiac output level) from the normal value of 7 mm Hg up to 11 mm Hg. This has shifted the venous return curve to the right. Finally, the venous return curve has rotated upward from the normal because the avitaminosis has dilated the peripheral blood vessels, as explained in Chapter 17.

Therefore, the two dashed curves intersect with each other at *point C,* which describes the circulatory condition in beriberi, with a right atrial pressure in this instance of 9 mm Hg and a cardiac output about 65 per cent above normal—this very high cardiac output occurs despite the weak heart, as illustrated by the depressed plateau level of the cardiac output curve.

REFERENCES

Barger, A. C.: The kidney in congestive heart failure. Circulation, 21:124, 1960.
Braunwald, E.: Heart Disease. Philadelphia, W. B. Saunders Co., 1988.
Braunwald, E.: Heart failure—pathophysiological considerations. In Dickinson, C. J., and Marks, J. (eds.): Developments in Cardiovascular Medicine. Lancaster, England, MTP Press, 1978, p. 213.
Braunwald, E., et al.: Mechanism of Contractility of the Normal and Failing Heart. Boston, Little, Brown, 1968.
Chatterjee, K., and Parmley, W. W.: Vasodilator therapy for chronic heart failure. Annu. Rev. Pharmacol. Toxicol., 20:475, 1980.
Fein, F. S., and Sonnenblick, E. H.: Diabetic cardiomyopathy. Prog. Cardiovasc. Dis., 27:255, 1985.
Francis, G. S.: Neurohumoral mechanisms involved in congestive heart failure. Am. J. Cardiol., 55:15A, 1985.
Gilbert, I., and Henning, R. J.: Adenocarcinoma of the lung presenting with pericardial tamponade: Report of a case and review of the literature. Heart-Lung, 14:83, 1985.
Goldberger, E.: Treatment of Cardiac Emergencies. St. Louis, C. V. Mosby Co., 1989.
Guyton, A. C.: The systemic venous system in cardiac failure. J. Chronic Dis., 9:465, 1959.
Guyton, A. C., et al.: Cardiac Output and Its Regulation. Philadelphia, W. B. Saunders Co., 1973.
Kones, R. J.: Cardiogenic Shock: Mechanisms and Management. New York, Futura Publishing Co., 1975.
Lange, L. G., and Sobel, B. E.: Pharmacological salvage of myocardium. Annu. Rev. Pharmacol. Toxicol., 22:115, 1982.
McManus: Cardiovascular Complications of Exercise. Chicago, Year Book Medical Publishers, 1985.
Miller, A. J.: Diagnosis of Chest Pain. New York, Raven Press, 1988.
Norris, R. M.: Myocardial Infarction. Its Presentation, Pathogenesis and Treatment. New York, Churchill Livingstone, Inc., 1982.
Orrego, F.: Calcium and the mechanism of action of digitalis. Gen. Pharmacol., 15:273, 1984.
Rapaport, E. (ed.): Current Controversies in Cardiovascular Disease. Philadelphia, W. B. Saunders Co., 1980.
Regan, T. J.: Congestive heart failure in the diabetic. Annu. Rev. Med., 34:161, 1983.
Samuel, P., et al.: The role of diet in the etiology and treatment of atherosclerosis. Annu. Rev. Med., 34:179, 1983.
Shoemaker, W. C., et al.: Society of Critical Care Medicine, The Textbook of Critical Care. 2nd Ed. Philadelphia, W. B. Saunders Co., 1988.

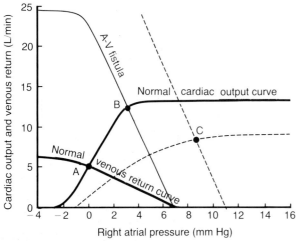

Figure 22–8. Graphical analysis of the two types of factors that can cause high output cardiac failure.

Silver, M. D.: Cardiovascular Pathology. New York, Churchill Livingstone, Inc., 1983.

Stallones, R. A.: Ischemic heart disease and lipids in blood and diet. Annu. Rev. Nutr., 3:155, 1983.

Sullivan, J. M.: Drug suppression of the angiotensin system in congestive heart failure. Annu. Rev. Med., 34:169, 1983.

Tintinalli, J. E., et al.: Emergency Medicine: A Comprehensive Study Guide. 2nd Ed. New York, McGraw-Hill Book Co., 1988.

Wasserman, K., et al.: Principles of Exercise Testing and Interpretation. Philadelphia, Lea & Febiger, 1987.

Weisfeldt, M. L.: The aging heart. Hosp. Pract., 20:115, 1985.

Winkelman, J. L.: Essentials of Basic Life Support. Minneapolis, Burgess, 1981.

Winwood, R. S.: Essentials of Clinical Diagnosis in Cardiology. Baltimore, E. Arnold, 1981.

Zelis, R., et al: Cardiocirculatory dynamics in the normal and failing heart. Annu. Rev. Physiol., 43:455, 1981.

(See also Chapter 21.)

CHAPTER 23

• *Heart Sounds; Dynamics of Valvular and Congenital Heart Defects* •

● THE HEART SOUNDS

The function of the heart valves was discussed in Chapter 9, and it was pointed out that closure of the valves is associated with audible sounds. On the other hand, no sounds occur when the valves open. The purpose of the first section of this chapter is to discuss the factors that cause the sounds in the heart, under both normal and abnormal conditions.

NORMAL HEART SOUNDS

Listening with a stethoscope to a normal heart, one hears a sound usually described as "lub, dub, lub, dub . . ." The "lub" is associated with closure of the A-V valves at the beginning of systole and the "dub" with closure of the semilunar valves at the end of systole. The "lub" sound is called the *first heart sound* and the "dub" the *second heart sound* because the normal cycle of the heart is considered to start with the beginning of systole.

Causes of the First and Second Heart Sounds. The earliest suggestion for the cause of the heart sounds was that the slapping together of the valve leaflets sets up vibrations, but this has been shown to cause little if any of the sound because of the cushioning effect of the blood. Instead, the cause is *vibration of the taut valves immediately after closure* along with *vibration of the adjacent blood, walls of the heart, and major vessels around the heart.* That is, in generating the first heart sound, contraction of the ventricles causes sudden backflow of blood against the A-V valves (the tricuspid and mitral valves), causing them to bulge toward the atria until the chordae tendineae abruptly stop the backbulging. The elastic tautness of the valves then causes the backsurging blood to bounce forward again into each respective ventricle. This sets the blood and the ventricular walls as well as

the valves into vibration and also causes vibrating turbulence in the blood. The vibrations then travel through the adjacent tissues to the chest wall, where they can be heard as sound by the stethoscope.

The second heart sound results from sudden closure of the semilunar valves (the aortic and pulmonary valves). When the semilunar valves close, they bulge backward toward the ventricles, and their elastic stretch recoils the blood back into the arteries, which causes a short period of reverberation of blood back and forth between the walls of the arteries and the valves and also between the valves and the ventricular walls. The vibrations set up in the arterial walls are then transmitted along the arteries. When the vibrations of the vessels or ventricles come into contact with a "sounding board," such as the chest wall, they create sound that can be heard.

Duration and Pitch of the First and Second Heart Sounds. The duration of each of the heart sounds is slightly more than 0.10 second—the first sound about 0.14 second and the second about 0.11 second. The reason for the shorter second sound is the extra tautness of the semilunar valves, so that they vibrate for a shorter period of time than do the A-V valves.

The audible range of frequencies (pitch) in the first and second heart sounds, as illustrated in Figure 23–1, begins at the very lowest frequency that the ear can detect and goes up to about 500 cycles per second. However, when special electronic apparatus is used to record these sounds, by far the largest proportion of the recorded sound is at frequencies below the audible range going down to as low as 3 or 4 cycles per second and peaking at about 20 cycles per second. For this reason, some heart sounds can be recorded electronically in phonocardiograms when they cannot be heard with a stethoscope.

The second heart sound normally has a higher fre-

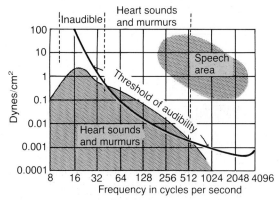

Figure 23-1. Amplitude of different frequency vibrations in the heart sounds and heart murmurs in relation to the threshold of audibility, showing that the range of sounds that can be heard is between about 40 and 500 cycles per second. (Modified from Butterworth, Chassin, and McGrath: Cardiac Auscultation. New York, Grune and Stratton.)

quency than the first heart sound for two reasons: (1) because of the tautness of the semilunar valves in comparison with much less taut A-V valves, and (2) because of greater elasticity coefficient of the arteries that provide the principal vibrating chambers for the second sound in comparison with the much looser ventricular chambers that provide the vibrating system of the first heart sound. The clinician uses these sound differences to distinguish special characteristics of the two respective sounds.

The Third Heart Sound. Occasionally a very weak rumbling third heart sound is heard at the beginning of the *middle third of diastole.* A logical, but yet unproved, explanation of this sound is oscillation of blood back and forth between the walls of the ventricles initiated by inrushing blood from the atria. This is analogous to running water from a faucet into a sack, the inrushing water reverberating back and forth between the walls of the sack to cause vibrations in the walls. The reason the third heart sound does not occur until the middle third of diastole is presumably that in the early part of diastole the heart is not filled sufficiently to create even the small amount of elastic tension in the ventricles necessary for reverberation. The frequency of this sound is usually so low that the ear cannot hear the sound; yet it can often be recorded in the phonocardiogram.

The Atrial Heart Sound (Fourth Heart Sound). An atrial heart sound can be recorded in many persons in the phonocardiogram, but it can almost never be heard with a stethoscope because of its low frequency—usually 20 cycles per second or less. This sound occurs when the atria contract, and presumably it is caused by inrush of blood into the ventricles, which initiates vibrations similar to those of the third heart sound.

AREAS FOR AUSCULTATION OF NORMAL HEART SOUNDS

Listening to the sounds of the body, usually with the aid of a stethoscope, is called *auscultation.* Figure 23-2 illus-

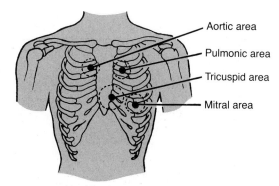

Figure 23-2. Chest areas from which each valve sound is best heard.

trates the areas of the chest wall from which the different valvular sounds can best be distinguished. Unfortunately, the sounds from all the valves can be heard from each of these areas, though the sound from the designated valve is as loud, *relative to the other sounds,* as it ever will be. The cardiologist distinguishes the sounds from the different valves by a process of elimination; that is, he or she moves the stethoscope from one area to another, noting the loudness of the sounds in different areas and gradually picking out the sound components from each valve.

The areas for listening to the different heart sounds are not directly over the valves themselves. The aortic area is upward along the aorta because of sound transmission up the aorta, the pulmonic area is upward along the pulmonary artery, the tricuspid area is over the right ventricle, and the mitral area is over the apex of the heart, which is the portion of the left ventricle nearest the surface of the chest because the heart is rotated so that most of the left ventricle lies hidden behind the right ventricle. In other words, the sounds caused by the A-V valves are transmitted to the chest wall through each respective ventricle, and the sounds from the semilunar valves are transmitted especially along the great vessels leading from the heart.

The Phonocardiogram

If a microphone specially designed to detect low frequency sound is placed on the chest, the heart sounds can be amplified and recorded by a high-speed recording appa-

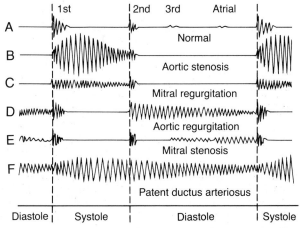

Figure 23-3. Phonocardiograms from normal and abnormal hearts.

ratus. The recording is called a *phonocardiogram,* and the heart sounds appear as waves, as illustrated schematically in Figure 23–3. Record A is a recording of normal heart sounds, showing the vibrations of the first, second, and third heart sounds and even the atrial sound. Note specifically that the third and atrial heart sounds are each a very low rumble. The third heart sound can be recorded in only one third to one half of all persons, and the atrial heart sound can be recorded in perhaps one fourth of all persons.

VALVULAR LESIONS

Rheumatic Valvular Lesions

By far the greatest number of valvular lesions results from rheumatic fever. Rheumatic fever is an autoimmune disease in which the heart valves are likely to be damaged or destroyed. It is initiated by streptococcal toxin in the following manner:

The sequence of events almost always begins with a preliminary streptococcal infection caused specifically by group A hemolytic streptococci, such as a sore throat, scarlet fever, or middle ear infection. The streptococci release several different proteins against which antibodies are formed, the most important of which seems to be a protein called the "M" antigen. The antibodies then react not only with this M antigen, but unfortunately also with many different tissues of the body, often causing severe immunological damage. These reactions continue to take place as long as the antibodies persist in the blood—6 months or more.

Rheumatic fever causes damage in many parts of the body but especially in certain very susceptible areas, such as the heart valves. The degree of heart valve damage is directly correlated with the titer and the persistence of these antibodies. Principles of immunity relating to this type of reaction are discussed in Chapter 34, and it is also noted in Chapter 31 that acute glomerular nephritis has a similar basis.

In rheumatic fever, large hemorrhagic, fibrinous, bulbous lesions grow along the inflamed edges of the heart valves. Because the mitral valve receives more trauma during valvular action than any of the other valves, this valve is the one most often seriously damaged, and the aortic valve is second most frequently damaged. The tricuspid and pulmonary valves are usually affected much less severely, probably because the stresses acting on these valves are slight compared with those on the left ventricular valves.

Scarring of the Valves. The lesions of acute rheumatic fever frequently occur on adjacent valve leaflets simultaneously, so that the edges of the leaflets become stuck together. Then, weeks, months, or years later, the lesions become scar tissue, permanently fusing portions of the leaflets. Also, the free edges of the leaflets, which are normally filmy and free-flapping, become solid, scarred masses.

A valve in which the leaflets adhere to each other so extensively that blood cannot flow through satisfactorily is said to be *stenosed.* On the other hand, when the valve edges are so destroyed by scar tissue that they cannot close when the ventricles contract, *regurgitation,* or backflow, of blood occurs when the valve should be closed. Stenosis usually does not occur without the coexistence of at least some degree of regurgitation, and vice versa.

Other Causes of Valvular Lesions. Stenosis or lack of one or more leaflets of a valve frequently occurs as a congenital defect. Complete lack of leaflets is rare, though stenosis is common, as is discussed later in this chapter.

● HEART MURMURS CAUSED BY VALVULAR LESIONS

As illustrated by the phonocardiograms in Figure 23–3, many abnormal heart sounds, known as "heart murmurs," occur when there are abnormalities of the valves, as follows:

The Murmur of Aortic Stenosis. In aortic stenosis, blood is ejected from the left ventricle through only a small opening of the aortic valve. Because of the resistance to ejection, the pressure in the left ventricle rises sometimes as high as 300 mm Hg, while the pressure in the aorta is still normal. Thus, a nozzle effect is created *during systole,* with blood jetting at tremendous velocity through the small opening of the valve. This causes *severe turbulence* of the blood in the root of the aorta. The turbulent blood impinging against the aortic walls causes intense vibration, and a loud murmur is transmitted throughout the upper aorta and even into the larger arteries of the neck. This sound is harsh and in severe stenosis occasionally is so loud that it can be heard several feet away from the patient. Also, the sound vibrations can often be felt with the hand on the upper chest and lower neck, a phenomenon known as a "thrill."

The Murmur of Aortic Regurgitation. In aortic regurgitation, no sound is heard during systole, but *during diastole* blood flows backward from the aorta into the left ventricle, causing a "blowing" murmur of relatively high pitch and with a swishing quality heard maximally over the left ventricle. This murmur results from *turbulence* of blood jetting backward into the blood already in the left ventricle.

The Murmur of Mitral Regurgitation. In mitral regurgitation, blood flows backward through the mitral valve *during systole.* This also causes a high frequency "blowing," swishing sound similar to that of aortic regurgitation, transmitted most strongly into the left atrium. However, the left atrium is so deep within the chest that it is difficult to hear this sound directly over the atrium. As a result, the sound of mitral regurgitation is transmitted to the chest wall mainly through the left ventricle instead, and it is usually heard best at the apex of the heart.

The Murmur of Mitral Stenosis. In mitral stenosis, blood passes with difficulty from the left atrium into the left ventricle, and, because the pressure in the left atrium rarely rises above 35 mm Hg except for short periods of time, a great pressure differential forcing blood from the left atrium into the left ventricle never develops. Consequently, the abnormal sounds heard in mitral stenosis are usually weak and of very low frequency, so that most of the sound spectrum is below the low frequency end of human hearing.

During the early part of diastole, the ventricle has so little blood in it and its walls are so flabby that blood does not reverberate back and forth between the walls of the ventricle. For this reason, even in severe mitral stenosis, no murmur at all might be heard during the first third of diastole. Then, after the first third the ventricle is stretched enough for blood to reverberate, and a low rumbling murmur often begins. In mild stenosis, the murmur lasts only during the first half of the middle third of diastole, but in severe stenosis it can begin earlier and persist for the whole remainder of diastole. Because of the extremely low frequency of the sound vibrations, one can often feel these, called a "thrill," over the apex of the heart even when the sound cannot be heard through the stethoscope.

Phonocardiograms of Valvular Murmurs. Phonocardiograms B, C, D, and E of Figure 23–3 illustrate, respectively, idealized records obtained from patients with aortic stenosis, mitral regurgitation, aortic regurgitation, and mitral stenosis. It is obvious from these phonocardio-

2. A *progressive stage,* in which the shock becomes steadily worse until death.

3. An *irreversible stage,* in which the shock has progressed to such an extent that all forms of known therapy will be inadequate to save the life of the person even though for the moment the person is still alive.

Now, let us discuss the different stages of circulatory shock caused by decreased blood volume, which will illustrate the basic principles. Then we can consider the special characteristics of other initiating causes of shock.

● SHOCK CAUSED BY HYPOVOLEMIA — HEMORRHAGIC SHOCK

Hypovolemia means diminished blood volume, and hemorrhage is the most common cause of hypovolemic shock.

Hemorrhage *decreases the mean systemic filling pressure* and as a consequence decreases venous return. As a result, the cardiac output falls below normal, and shock ensues. Obviously, all degrees of shock can result from hemorrhage, from the mildest diminishment of cardiac output to almost complete cessation of output.

RELATIONSHIP OF BLEEDING VOLUME TO CARDIAC OUTPUT AND ARTERIAL PRESSURE

Figure 24–1 illustrates the approximate effect on both cardiac output and arterial pressure of removing blood from the circulatory system over a period of about half an hour. Approximately 10 per cent of the total blood volume can be removed with no significant effect on either arterial pressure or cardiac output, but greater blood loss usually diminishes the cardiac output first and later the pressure, both of these falling to zero when about 35 to 45 per cent of the total blood volume has been removed.

Sympathetic Reflex Compensation in Shock. Fortunately, the decrease in arterial pressure as well as in the pressures in the low pressure areas of the thorax following hemorrhage initiates powerful sympathetic reflexes (initiated mainly by the baroreceptors and the low pressure stretch receptors). These reflexes stimulate the sympathetic vasoconstrictor system throughout the body, resulting in three important effects: (1) The arterioles constrict in most parts of the body, thereby greatly increasing the total peripheral resistance. (2) The veins and venous reservoirs constrict, thereby helping to maintain adequate venous re-

turn despite diminished blood volume. (3) Heart activity increases markedly, sometimes increasing the heart rate from the normal value of 72 beats per minute to as much as 170 to 200 beats per minute.

Value of the Reflexes. In the absence of the sympathetic reflexes, only 15 to 20 per cent of the blood volume can be removed over a period of half an hour before a person will die; this is in contrast to a 30 to 40 per cent loss of blood volume that a person can sustain when the reflexes are intact. Therefore, the reflexes extend the amount of blood loss that can occur without causing death to about two times that which is possible in their absence.

Greater Effect of the Reflexes in Maintaining Arterial Pressure Than in Maintaining Cardiac Output. Referring again to Figure 24–1, one sees that the arterial pressure is maintained at or near normal levels in the hemorrhaging person longer than is the cardiac output. The reason for this is that the sympathetic reflexes are geared more for maintaining arterial pressure than for maintaining output. They increase the arterial pressure mainly by increasing the total peripheral resistance, which has no beneficial effect on cardiac output; however, the sympathetic constriction of the veins is important to keep venous return and cardiac output from falling too much.

Especially interesting is the second plateau in the arterial pressure curve of Figure 24–1. This results from activation of the central nervous system (CNS) ischemic response, which causes extreme stimulation of the sympathetic nervous system, as discussed in Chapter 18, when the arterial pressure falls below 50 mm Hg. This effect of the CNS ischemic response can be called the "last ditch stand" of the sympathetic reflexes in their attempt to keep the arterial pressure from falling too low.

Protection of Coronary and Cerebral Blood Flow by the Reflexes. A special value of the maintenance of normal arterial pressure even in the face of decreasing cardiac output is protection of blood flow through the coronary and cerebral circulatory systems. Sympathetic stimulation does not cause significant constriction of either the cerebral or the cardiac vessels. In addition, in both these vascular beds local autoregulation is excellent, which prevents moderate decreases in arterial pressure from significantly affecting their blood flows. Therefore, blood flow through the heart and brain is maintained essentially at normal levels as long as the arterial pressure does not fall below about 70 mm Hg, despite the fact that blood flow in many other areas of the body might be decreased to as little as one-quarter normal by this time because of vasospasm.

NONPROGRESSIVE AND PROGRESSIVE HEMORRHAGIC SHOCK

Figure 24–2 illustrates an experiment that we performed in dogs to demonstrate the effects of different degrees of hemorrhage on the subsequent course of arterial pressure. The dogs were bled rapidly until their arterial pressures fell to different levels. Those dogs whose pressures fell immediately to no lower than 45 mm Hg (groups I, II, and III) all eventually recovered; the recovery occurred rapidly if the pressure fell only slightly (group I) but occurred slowly if it fell almost to the 45 mm Hg level (group III). On the other hand, when the arterial pressure fell below 45 mm Hg (groups IV, V, and VI), all the dogs died, though many of them hovered between life and death for many hours before the circulatory system began to deteriorate.

This experiment demonstrates that the circulatory sys-

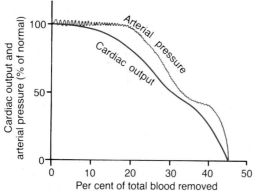

Figure 24–1. Effect of hemorrhage on cardiac output and arterial pressure.

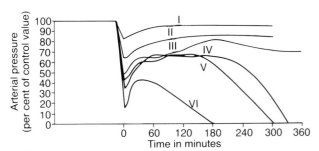

Figure 24–2. Course of arterial pressure in dogs after different degrees of acute hemorrhage. Each curve represents the average results from six dogs.

tem can recover as long as the degree of hemorrhage is no greater than a certain critical amount. However, crossing this critical amount by even a few milliliters of blood loss makes the eventual difference between life and death. Thus, hemorrhage beyond a certain critical level causes shock to become *progressive.* That is, *the shock itself causes still more shock,* the condition becoming a vicious circle that leads eventually to complete deterioration of the circulation and to death.

Nonprogressive Shock — Compensated Shock

If shock is not severe enough to cause its own progression, the person eventually recovers. Therefore, shock of this lesser degree can be called *nonprogressive shock.* It is also frequently called *compensated shock,* meaning that the sympathetic reflexes and other factors have compensated enough to prevent further deterioration of the circulation.

The factors that cause a person to recover from moderate degrees of shock are all the negative feedback control mechanisms of the circulation that attempt to return cardiac output and arterial pressure to normal levels. These include

1. The *baroreceptor reflexes,* which elicit powerful sympathetic stimulation of the circulation.

2. The *central nervous system ischemic response,* which elicits even more powerful sympathetic stimulation throughout the body but is not activated significantly until the arterial pressure falls below 50 mm Hg.

3. *Reverse stress-relaxation of the circulatory system,* which causes the blood vessels to contract down around the diminished blood volume, so that the blood volume that is available will more adequately fill the circulation.

4. *Formation of angiotensin,* which constricts the peripheral arteries and causes increased conservation of water and salt by the kidneys, both of which help prevent progression of the shock.

5. *Formation of vasopressin (antidiuretic hormone),* which constricts the peripheral arteries and veins and also greatly increases water retention by the kidneys.

6. *Compensation mechanisms that return the blood volume back toward normal,* including absorption of large quantities of fluid from the intestinal tract, absorption of fluid from the interstitial spaces of the body, conservation of water and salt by the kidneys, and increased thirst and increased appetite for salt, which make the person drink water and eat salty foods if able.

The sympathetic reflexes provide immediate help toward bringing about recovery, for they become maximally activated within 30 seconds after hemorrhage. The angiotensin and vasopressin mechanisms as well as the reverse stress-

relaxation that causes contraction of the blood vessels and venous reservoirs all require 10 minutes to an hour to occur completely, but, nevertheless, these aid greatly in increasing the arterial pressure or increasing the mean systemic filling pressure and thereby increasing the return of blood to the heart. Finally, the readjustment of blood volume by absorption of fluid from the interstitial spaces and from the intestinal tract as well as the ingestion and absorption of additional quantities of fluid and salt may require from 1 to 48 hours, but eventually recovery takes place, provided the shock does not become severe enough to enter the progressive stage.

Progressive Shock — The Vicious Circle of Cardiovascular Deterioration

Once shock has become severe enough, the structures of the circulatory system themselves begin to deteriorate, and various types of positive feedback develop that can cause a vicious circle of progressively decreasing cardiac output. Figure 24–3 illustrates some of these positive feedbacks that further depress the cardiac output in shock. These are the following:

Cardiac Depression. When the arterial pressure falls low enough, *coronary blood flow decreases below that required for adequate nutrition of the myocardium* itself. This obviously weakens the heart and thereby decreases the cardiac output still more. Thus, a positive feedback cycle has developed whereby the shock becomes more and more severe.

Figure 24–4 illustrates cardiac output curves extrapolated to the human heart from experiments in dogs, showing progressive deterioration of the heart at different times following the onset of shock. A dog was bled until the arterial pressure fell to 30 mm Hg, and the pressure was held at this level by further bleeding or retransfusion of blood as required. Note that there was little deterioration of the heart during the first two hours, but by four hours the heart had deteriorated about 40 per cent; then, rapidly, during the last hour of the experiment the heart deteriorated almost completely.

Thus, one of the important features of progressive shock, whether it be hemorrhagic in origin or caused in any other way, is eventual progressive deterioration of the heart. In the early stages of shock, this plays very little role in the condition of the person, partly because deterioration of the heart itself is not very severe during the first hour or so of shock but mainly because the heart has tremendous reserve that makes it normally capable of pumping 300 to 400 per cent more blood than is required by the body for adequate nutrition. However, in the very latest stages of shock, deterioration of the heart is probably the most important factor in the further progression of the shock.

Vasomotor Failure. In the early stages of shock, various circulatory reflexes cause intense activity of the sympathetic nervous system. This, as discussed previously, helps delay depression of the cardiac output and especially helps prevent decreased arterial pressure. However, there comes a point at which diminished blood flow to the vasomotor center itself so depresses the center that it becomes progressively less active and finally totally inactive. For instance, complete circulatory arrest to the brain causes, during the first 4 to 8 minutes, the most intense of all sympathetic discharges, resulting in a typical CNS ischemic response, but by the end of 10 to 15 minutes, the vasomotor center becomes so depressed that no evidence of sympa-

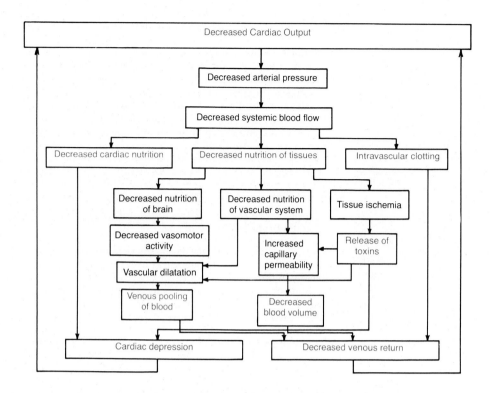

Figure 24-3. Different types of feedback that can lead to progression of shock.

thetic discharge at all can be demonstrated. Fortunately, though, the vasomotor center does not usually fail in the early stages of shock—only in the late stages.

Thrombosis of the Minute Vessels—Sludged Blood. Thrombosis occurring in many of the minute vessels in the circulatory system also often causes the shock to progress. The initiating cause of this is sluggish blood flow in the microvessels. Because tissue metabolism continues despite the low flow, large amounts of acid, either carbonic acid or lactic acid, continue to empty into the local blood vessels and to increase greatly the acidity of the blood. This acid, plus other deterioration products from the ischemic tissues, causes blood agglutination or actual blood clots, thus leading to minute plugs in the small vessels. Even if the vessels do not become plugged, the tendency for the cells to

stick to each other makes it more difficult for blood to flow through the microvasculature, giving rise to the term *sludged blood.*

Increased Capillary Permeability. After many hours of capillary hypoxia and lack of other nutrients as well, the permeability of the capillaries gradually increases, and large quantities of fluid begin to transude into the tissues. This further decreases the blood volume, with resultant further decrease in cardiac output, thus making the shock still more severe.

Fortunately, capillary hypoxia does not cause increased capillary permeability until the very late stages of prolonged shock.

Release of Toxins by Ischemic Tissues. Throughout the history of research in the field of shock, it has been suggested time and again that shock causes tissues to release toxic substances, such as histamine, serotonin, tissue enzymes, and so forth, that then cause further deterioration of the circulatory system. Quantitative studies have especially proved the significance of at least one toxin, *endotoxin,* in many types of shock.

Endotoxin. Endotoxin is a toxin released from the bodies of dead gram-negative bacteria in the intestines. Diminished blood flow to the intestines causes enhanced formation and absorption of this toxic substance, and it then causes extensive vascular dilatation, greatly increased cellular metabolism despite the inadequate nutrition of the cells, and cardiac depression. Although this toxin can play a major role in some types of shock, especially septic shock, discussed later in the chapter, it still is not clear how much endotoxin is released during hemorrhagic shock and whether or not it is an especially important factor in the progression of this type of shock.

Generalized Cellular Deterioration. As shock becomes very severe, many signs of generalized cellular deterioration occur throughout the body. One organ especially affected is the *liver,* primarily because of the lack of enough nutrients to support the normally high rate of metabolism

Figure 24-4. Cardiac output curves of the heart at different times after hemorrhagic shock begins. (These curves are extrapolated to the human heart from data obtained in dog experiments by Dr. J. W. Crowell.)

in liver cells, but also partly because of the extreme vascular exposure of the liver cells to any toxic or other abnormal metabolic factor in shock.

Among the different damaging cellular effects that are known to occur are the following:

1. Active transport of sodium and potassium through the cell membrane is greatly diminished. As a result, sodium and chloride accumulate in the cells and potassium is lost from the cells. In addition, the cells begin to swell.

2. Mitochondrial activity in the liver cells as well as in many other tissues of the body becomes severely depressed.

3. Lysosomes begin to split in widespread tissue areas, with intracellular release of hydrolases that cause further intracellular deterioration.

4. Cellular metabolism of nutrients, such as glucose, eventually becomes greatly depressed in the last stages of shock. The activities of some hormones are depressed as well, including as much as 200-fold depression in the action of insulin.

Obviously, all these effects contribute to further deterioration of many different organs of the body, including especially (1) the *liver,* with depression of its many metabolic and detoxification functions; (2) the lungs, with eventual development of pulmonary edema and poor ability to oxygenate the blood; and (3) the heart, thereby further depressing the contractility of the heart.

Tissue Necrosis in Severe Shock — Effect of Pattern of Vascular Blood Flow in Different Organs. Not all cells of the body are equally damaged by shock because some tissues have better blood supplies than others. For instance, the cells adjacent to the arterial ends of capillaries receive better nutrition than the cells adjacent to the venous ends of the same capillaries. Therefore, one would expect more nutritive deficiency around the venous ends of capillaries than elsewhere. This is precisely the effect that Crowell found in studying tissue areas in many parts of the body. For instance, Figure 24–5 illustrates necrosis in the center of a liver lobule, the portion of the lobule that is last to be bathed by the blood as it passes through the liver sinusoids.

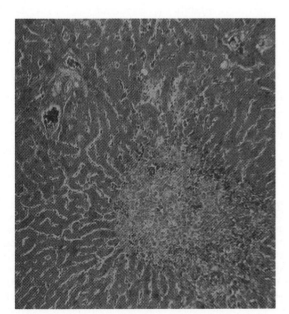

Figure 24–5. Necrosis of the central portion of a liver lobule in severe shock. (Courtesy of Dr. J. W. Crowell.)

Similar punctate lesions occur in heart muscle, though here a definite repetitive pattern, such as occurs in the liver, cannot be demonstrated. Nevertheless, the cardiac lesions probably play an important role in leading to the final irreversible stage of shock. Deteriorative lesions also occur in the kidneys, especially in the kidney tubules, leading to kidney failure and subsequent uremic death several days later. Deterioration of the lungs often leads to respiratory distress and death several days later — called the *shock lung syndrome.*

Acidosis in Shock. Most of the metabolic derangements that occur in shocked tissue can lead to acidosis. Especially important is the poor delivery of oxygen to the tissues, which greatly diminishes oxidative metabolism of the foodstuffs. When this occurs, the cells obtain their energy by the anaerobic process of glycolysis, which leads to tremendous quantities of *excess lactic acid* in the blood. In addition, the poor blood flow through the tissues prevents normal removal of carbon dioxide; the carbon dioxide reacts locally in the cells with water to form very high concentrations of intracellular carbonic acid; this in turn reacts with the various tissue buffers to form still other intracellular acidic substances.

Thus, another deteriorative effect of shock is both generalized and local tissue acidosis, leading to still further progression of the shock itself.

Positive Feedback and the Vicious Circle of Progressive Shock

All the factors just discussed that can lead to further progression of shock are types of *positive feedback.* That is, each increase in the degree of shock causes a further increase in the shock.

However, positive feedback does not always necessarily lead to a vicious circle. Whether or not a vicious circle develops depends on the intensity of the positive feedback. In mild degrees of shock, the negative feedback mechanisms — the sympathetic reflexes, the stress-relaxation mechanism of the blood reservoirs, the absorption of fluid into the blood from the interstitial spaces, and others — can easily overcome the positive feedback influences and can therefore cause recovery. In severe degrees of shock, however, the positive feedback mechanisms become more and more powerful, thus leading to such rapid deterioration of the circulation that the negative feedback systems cannot return the cardiac output to normal.

Thus, considering once again the principles of positive feedback and vicious circles as discussed in Chapter 1, one can readily understand why there is a critical cardiac output level above which a person in shock recovers and below which the person enters a vicious circle of circulatory deterioration.

IRREVERSIBLE SHOCK

After shock has progressed to a certain stage, transfusion or any other type of therapy becomes incapable of saving the life of the person. Therefore, the person is then said to be in the *irreversible stage of shock.* Ironically, even in this irreversible stage, therapy can on occasion still return the arterial pressure and even the cardiac output to normal for short periods of time, but the circulatory system nevertheless continues to deteriorate, and death ensues in another few minutes to few hours.

Figure 24–6 illustrates this effect, showing that transfu-

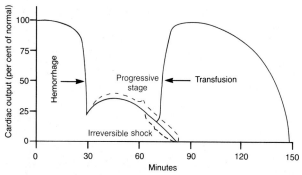

Figure 24-6. Failure of transfusion to prevent death in irreversible shock.

sion during the irreversible stage can sometimes cause the cardiac output (as well as the arterial pressure) to return to normal. However, the cardiac output soon begins to fall again, and subsequent transfusions have less and less effect. Thus, something has changed in the cells of the heart, other tissues, or both that may not necessarily affect the *immediate* ability of the heart to pump blood but over a long period of time does depress this ability and results in death. Now the question remains: What factor or factors lead to the eventual total deterioration of circulatory function?

The answer to this question seems to be, simply, that beyond a certain point so much tissue damage has occurred, so many destructive enzymes have been released into the body fluids, so much acidosis has developed, and so many other destructive factors are now in progress that even a normal cardiac output cannot reverse the continuing deterioration. Therefore, in severe shock a stage is eventually reached beyond which the person is destined to die even though vigorous therapy can still return the cardiac output to normal for short periods of time.

Depletion of Cellular High-Energy Phosphate Reserves in Irreversible Shock. The high-energy phosphate reserves in the tissues of the body, especially in the liver and in the heart, are greatly diminished in severe degrees of shock. Essentially all the creatine phosphate is degraded, and almost all of the adenosine triphosphate (ATP) has also been degraded to adenosine diphosphate (ADP), adenosine monophosphate (AMP), and eventually to adenosine. Then much of this adenosine diffuses out of the cells into the circulating blood and is converted into uric acid, a substance that cannot re-enter the cells to reconstitute the adenosine phosphate system. Unfortunately, new adenosine can be synthesized at a rate of only about 2 per cent an hour, meaning that, once depleted, the high-energy phosphate stores of the cells are difficult to replenish. Therefore, one of the most devastating end-results of deterioration in shock, and one that is perhaps the most significant of all in the development of the final state of irreversibility, is this cellular depletion of the high-energy compounds.

HYPOVOLEMIC SHOCK CAUSED BY PLASMA LOSS

Loss of plasma from the circulatory system, even without the loss of whole blood, can sometimes be severe enough to reduce the total blood volume markedly, in this way causing typical hypovolemic shock similar in almost all details to that caused by hemorrhage. Severe plasma loss occurs in the following conditions:

1. *Intestinal obstruction* is often a cause of reduced plasma volume. The resulting distention of the intestine causes fluid to leak from the intestinal capillaries into the intestinal walls and intestinal lumen. This loss of fluid might result from elevated capillary pressure caused by increased resistance in the stretched veins over the surface of the intestine, or it might be caused by direct damage to the capillaries themselves. Regardless of the cause, the lost fluid has a very high content of protein, thereby reducing the total plasma protein as well as the plasma volume.

2. In patients who have *severe burns* or other denuding conditions of the skin, so much plasma almost always is lost through the exposed areas that the plasma volume becomes markedly reduced.

The hypovolemic shock that results from plasma loss has almost the same characteristics as the shock caused by hemorrhage, except for one additional complicating factor — the blood viscosity increases greatly as a result of plasma loss, and this further exacerbates the sluggishness of blood flow.

Loss of fluid from all fluid compartments of the body is called *dehydration;* this, too, can reduce the blood volume and cause hypovolemic shock very similar to that resulting from hemorrhage. Some of the causes of this type of shock are: (1) excessive sweating; (2) fluid loss in severe diarrhea or vomiting; (3) excess loss of fluid by nephrotic kidneys; (4) inadequate intake of fluid and electrolytes; and (5) destruction of the adrenal cortices, with consequent failure of the kidneys to reabsorb sodium, chloride, and water because of the absence of the hormone aldosterone.

HYPOVOLEMIC SHOCK CAUSED BY TRAUMA

One of the most common causes of circulatory shock is trauma to the body. Often the shock results simply from hemorrhage caused by the trauma, but it can also occur even without hemorrhage, for contusion of the body can often damage the capillaries sufficiently to allow excessive loss of plasma into the tissues. This results in greatly reduced plasma volume with resultant hypovolemic shock.

The *pain* associated with serious trauma can be an additional aggravating factor in traumatic shock, for pain sometimes strongly inhibits the vasomotor center, thereby increasing the vascular capacitance and reducing the venous return. Various attempts have also been made to implicate toxic factors released by the traumatized tissues as one of the causes of shock following trauma. However, cross-transfusion experiments have failed to show significant toxic elements.

In summary, traumatic shock seems to result mainly from hypovolemia, although there may also be a moderate degree of concomitant neurogenic shock caused by the pain.

● NEUROGENIC SHOCK—INCREASED VASCULAR CAPACITY

Occasionally shock results without any loss of blood volume whatsoever. Instead, the *vascular capacity* increases so much that even the normal amount of blood becomes incapable of adequately filling the circulatory system. One of the major causes of this is *sudden loss of vasomotor tone* throughout the body causing especially massive dilatation of the veins. The resulting condition is known as *neurogenic shock.*

The relationship of vascular capacity to blood volume is discussed in Chapter 15, where it is pointed out that either an increase in vascular capacity or a decrease in blood volume *reduces the mean systemic filling pressure,* which in turn reduces the venous return to the heart. Diminished venous return caused by vascular dilatation is often called "venous pooling" of blood.

Causes of Neurogenic Shock. Some of the different factors that can cause loss of vasomotor tone include the following:

1. *Deep general anesthesia* often depresses the vasomotor center enough to cause vasomotor collapse, with resulting neurogenic shock.

2. *Spinal anesthesia,* especially when this extends all the way up the spinal cord, blocks the sympathetic outflow from the nervous system and is a common cause of neurogenic shock.

3. *Brain damage* is often a cause of vasomotor collapse. Many patients who have had brain concussion or contusion of the basal regions of the brain develop profound neurogenic shock. Also, even though short periods of medullary ischemia cause extreme vasomotor activity, prolonged ischemia can cause inactivation of the vasomotor neurons and can cause development of severe neurogenic shock.

Vasovagal Syncope — Emotional Fainting. The circulatory collapse that results from "emotional" fainting usually is not caused by vasomotor failure but instead by strong emotional activation of parasympathetic signals to slow the heart and sympathetic signals to dilate the peripheral vasculature, thereby decreasing cardiac output and arterial pressure. Therefore, the fainting that results from an emotional disturbance is called *vasovagal syncope* to differentiate it from other types of fainting that result from other causes of reduced cardiac output.

● ANAPHYLACTIC SHOCK

"Anaphylaxis" is an allergic condition in which the cardiac output and arterial pressure often fall drastically. This is discussed in Chapter 34. It results primarily from an antigen-antibody reaction that takes place immediately after an antigen to which the person is sensitive has entered the circulation. One of the principal effects is to cause the *basophils* in the blood and the *mast* cells in the pericapillary tissues to release *histamine* or a *histamine-like substance.* The histamine in turn causes (1) an increase in vascular capacity because of venous dilatation, (2) dilatation of the arterioles with resultant greatly reduced arterial pressure, and (3) greatly increased capillary permeability with rapid loss of fluid and protein into the tissue spaces. The net effect is a great reduction in venous return and often such serious shock that the person dies within minutes.

Intravenous injection of large amounts of histamine causes "histamine shock," which has characteristics almost identical with those of anaphylactic shock, though usually less severe.

● SEPTIC SHOCK

The condition that was formerly known by the popular name of "blood poisoning" is now called *septic shock* by most clinicians. This simply means widely disseminated infection to many areas of the body, with the infection being borne through the blood from one tissue to another and causing extensive damage. Actually, there are many different varieties of septic shock because of the many different types of bacterial infection that can cause it and also because infection in one part of the body will produce different effects from those caused by infection elsewhere in the body.

Septic shock is extremely important to the clinician because this type of shock, more frequently than any other kind of shock besides cardiogenic shock, causes patient death in the modern hospital. Some of the typical causes of septic shock include

1. Peritonitis caused by spread of infection from the uterus and fallopian tubes, frequently resulting from instrumental abortion.

2. Peritonitis resulting from rupture of the gut, sometimes caused by intestinal disease and sometimes by wounds.

3. Generalized infection resulting from spread of a simple skin infection, such as streptococcal or staphylococcal infection.

4. Generalized gangrenous infection resulting specifically from gas gangrene bacilli, spreading first through the tissues themselves and finally by way of the blood to the internal organs, especially to the liver.

5. Infection spreading into the blood from the kidney or urinary tract, often caused by colon bacilli.

Special Features of Septic Shock. Because of the multiple types of septic shock, it is difficult to categorize this condition. However, some features are

1. High fever.

2. Marked vasodilation throughout the body, especially in the infected tissues.

3. High cardiac output in perhaps half of the patients, caused by vasodilation in the infected tissues and also by high metabolic rate and vasodilation elsewhere in the body resulting from bacterial toxin stimulation of cellular metabolism and from the high body temperature.

4. Sludging of the blood, presumably caused by red cell agglutination in response to degenerating tissues.

5. Development of microclots in widespread areas of the body, a condition called *disseminated intravascular coagulation.* Also, this causes the clotting factors to be used up so that hemorrhages occur into many tissues, especially into the gut wall and into the intestinal tract.

In the early stages of septic shock, the patient usually does not have signs of circulatory collapse but, instead, only signs of the bacterial infection itself. However, as the infection becomes more severe, the circulatory system usually becomes involved either directly or as a secondary result of toxins from the bacteria, and *there finally comes a point at which deterioration of the circulation becomes progressive in the same way that progression occurs in all other types of shock. Therefore, the end-stages of septic shock are not greatly different from the end-stages of hemorrhagic shock,* even though the initiating factors are markedly different in the two conditions.

Endotoxin Shock. A special type of septic shock is known as endotoxin shock. It frequently occurs when a large segment of the gut becomes strangulated and loses most of its blood supply. The gut rapidly becomes gangrenous, and the bacteria in the gut multiply rapidly. Most of these bacteria are so-called "gram-negative" bacteria, mainly colon bacilli, that contain a toxin called *endotoxin.* Another condition that also frequently causes colon bacilli septicemia is extension of urinary tract infections into the blood.

On entering the circulation, endotoxin causes an effect very similar to that of anaphylaxis, often resulting in severe shock. Also, further compounding the circulatory depression is a direct effect of endotoxin on the heart to decrease myocardial contractility.

● EFFECTS OF SHOCK ON THE BODY

Decreased Metabolism and Cellular Deterioration in Hypovolemic Shock. In hypovolemic shock, the decreased cardiac output reduces the delivery of both oxygen and other nutrients to the tissues. This in turn reduces the metabolism of virtually all cells of the body, leading to multiple types of cellular damage, including (1) decreased ability of the mitochondria to synthesize ATP, (2) decreased ability of the cell's membrane pump to keep the sodium concentration low and the potassium concentration high inside the cell, (3) depressed processing of nutrients by the cell's metabolic machinery, and (4) eventual rupture of many lysosomes, releasing digestive enzymes within the cells that cause intracellular destruction and even death of cells. Obviously, all these effects lead to still greater cellular deterioration and further decrease in total bodily metabolism. Usually a person can continue to live for only a few hours if the cardiac output falls to as low as 40 per cent of normal.

Muscular Weakness. One of the earliest symptoms of shock is severe muscular weakness that is also associated with profound and rapid fatigue whenever patients attempt to use their muscles. This obviously results from the diminished supply of nutrients — especially oxygen — to the muscles.

Body Temperature. Because of the depressed metabolism in shock, the amount of heat liberated in the body is reduced (except in septic shock in which the infection may cause an opposite effect). As a result, the body temperature tends to decrease if the body is exposed to even the slightest cold.

Mental Function. In the early stages of shock the person is usually conscious, though signs of mental haziness may be noted. As the shock progresses, the person falls into a state of stupor, and in the last stages of shock even the subconscious mental functions, including vasomotor control and respiration, fail.

A person who recovers from shock usually exhibits no permanent impairment of mental functions.

Reduced Renal Function and Renal Deterioration. Even the slight decreases in cardiac output and arterial pressure that occur in the early stages of shock greatly diminish or even abolish urine output, because the glomerular pressure falls below the critical level required for filtration. Other renal effects are discussed in Chapters 26 through 29. This retention of fluid by the kidneys is usually helpful in shock, for it aids in preventing further diminishment of blood volume.

In the late stages of shock, however, the renal tubular epithelial cells deteriorate extremely rapidly because normally they have a very high metabolism and require large amounts of nutrients. The result is severe *tubular necrosis* with tubular cell death and sloughing and blockage of the tubules. Therefore, even though a person might survive the shock itself, the damage often subsequently causes renal shutdown, with death occurring a week or so later because of uremia.

● PHYSIOLOGY OF TREATMENT IN SHOCK

REPLACEMENT THERAPY

Blood and Plasma Transfusion. If a person is in shock caused by hemorrhage, the best possible therapy is usually transfusion of whole blood. If the shock is caused by plasma loss, the best therapy is administration of plasma; when dehydration is the cause, administration of the appropriate electrolytic solution can correct the shock.

Unfortunately, whole blood is not always available, such as under battlefield conditions. However, plasma can usually substitute adequately for whole blood because it increases the blood volume and restores normal hemodynamics. Plasma cannot restore a normal hematocrit, but the human can usually stand a decrease in hematocrit to about one-third normal before serious consequences result if the cardiac output is adequate. Therefore, in acute conditions it is reasonable to use plasma in place of whole blood for treatment of hemorrhagic and most other types of hypovolemic shock.

Sometimes plasma also is unavailable. In these instances, various *plasma substitutes* have been developed that perform almost exactly the same hemodynamic functions as plasma. One of these is the following:

Dextran Solution as a Plasma Substitute. The principal requirement of a truly effective plasma substitute is that it remain in the circulatory system — that is, not filter through the capillary pores into the tissue spaces. In addition, the solution must be nontoxic and must contain appropriate electrolytes to prevent derangement of the extracellular fluid electrolytes on administration. To remain in the circulation the plasma substitute must contain some substance that has a large enough molecular size to exert colloid osmotic pressure.

One of the most satisfactory substances that has been developed thus far for this purpose is dextran, a large polysaccharide polymer of glucose. Certain bacteria secrete dextran as a by-product of their growth, and commercial dextran is manufactured by a bacterial culture procedure. By varying the growth conditions of the bacteria, the molecular weight of the dextran can be controlled to the desired value. Dextrans of appropriate molecular size do not pass through the capillary pores and, therefore, can replace plasma proteins as colloid osmotic agents.

Fortunately, few toxic reactions have been observed when using dextran to provide colloid osmotic pressure; therefore, solutions of this substance have proved to be a satisfactory substitute for plasma in much fluid replacement therapy.

TREATMENT OF SHOCK WITH SYMPATHOMIMETIC DRUGS

A *sympathomimetic drug* is a drug that mimics sympathetic stimulation. These drugs include norepinephrine, epinephrine, and a large number of long-acting drugs that have the same effects as epinephrine and norepinephrine.

In two types of shock, sympathomimetic drugs have proved to be especially beneficial. The first of these is *neurogenic shock,* in which the sympathetic nervous system is severely depressed. Administering a sympathomimetic drug takes the place of the diminished sympathetic activity and can often restore full circulatory function.

The second type of shock in which sympathomimetic drugs are very valuable is *anaphylactic shock,* in which histamine plays a prominent role. The sympathomimetic drugs have a vasoconstrictor effect that opposes the vasodilating effect of histamine. Therefore, either norepinephrine or another sympathomimetic drug is often lifesaving.

Unfortunately, sympathomimetic drugs have not proved to be very valuable in hemorrhagic shock. The reason is that in this type of shock the sympathetic nervous system has almost always already become maximally activated by the circulatory reflexes, and so much norepinephrine and epinephrine are circulating in the blood that the sympathomimetic drugs have essentially no additional beneficial effect.

OTHER THERAPY

Treatment by the Head-Down Position. When the pressure falls too low in most types of shock, especially hemorrhagic and neurogenic shock, placing the patient with the head as much as a foot lower than the feet will help tremendously in promoting venous return and thereby increasing cardiac output. Therefore, this is the first essential in the treatment of many types of shock.

Oxygen Therapy. Since the major deleterious effect of most types of shock is too little delivery of oxygen to the tissues, giving the patient oxygen to breathe can be of benefit in some instances. However, this frequently is of far less value than one might expect because the problem usually is not inadequate oxygenation of the blood in the lungs, but inadequate transport of the blood after it is oxygenated.

Treatment With Glucocorticoids. Glucocorticoids are frequently given to patients in severe shock for several reasons: (1) Experiments have shown empirically that glucocorticoids frequently increase the strength of the heart in the late stages of shock; (2) glucocorticoids stabilize the lysosomal membranes and prevent release of lysosomal enzymes into the cytoplasm of the cells, thus preventing deterioration from this source; (3) glucocorticoids might also aid in the metabolism of glucose by the severely damaged cells.

● CIRCULATORY ARREST

A condition closely allied to circulatory shock is circulatory arrest, in which all blood flow completely stops. This occurs frequently on the surgical operating table as a result of *cardiac arrest* or of *ventricular fibrillation*.

Ventricular fibrillation can usually be stopped by strong electroshock of the heart, the basic principles of which were described in Chapter 13.

Cardiac arrest usually results from too little oxygen in the anesthetic gaseous mixture or from a depressant effect of the anesthesia itself. A normal cardiac rhythm can usually be restored by removing the anesthetic and then applying cardiopulmonary resuscitation procedures for a few minutes while supplying the patient's lungs with adequate quantities of ventilatory oxygen.

Effect of Circulatory Arrest on the Brain

A special problem in circulatory arrest is to prevent detrimental effects in the brain as a result of the arrest. In general, 4 to 5 minutes of circulatory arrest causes permanent brain damage in over half the patients. Circulatory arrest for as long as 10 minutes almost universally destroys most, if not all, of the mental powers.

For many years it was taught that this detrimental effect on the brain was caused by the cerebral hypoxia that occurs during circulatory arrest. However, experiments have shown that, if blood clots are prevented from occurring in the blood vessels of the brain, this will also prevent the very rapid deterioration of the brain during circulatory arrest. For instance, in one animal experiment, all of the animal's

blood was removed from the blood vessels at the beginning of circulatory arrest and then replaced at the end of the circulatory arrest, so that no intravascular blood clotting could occur. In these, the brain was able to withstand up to 30 minutes of circulatory arrest without any permanent brain damage. Also, administration of heparin or streptokinase prior to cardiac arrest was shown to increase the survivability of the brain two to four times as long as is usually the case. Therefore, it is likely that the severe brain damage that occurs following circulatory arrest results mainly from permanent blockage of many small or even large blood vessels by blood clots, thus causing prolonged ischemia and eventual death of the neurons.

REFERENCES

Achauer, B. M.: Management of the Burned Patient. East Norwalk, Conn., Appleton & Lange, 1987.
Bernton, E. W., et al: Opioids and neuropeptides: Mechanisms in circulatory shock. Fed. Proc., 44:190, 1985.
Bond, R. F., and Johnson, G., III: Vascular adrenergic interactions during hemorrhagic shock. Fed. Proc., 44:281, 1985.
Braunwald, E. (ed.): Heart Disease. A Textbook of Cardiovascular Medicine. Philadelphia, W. B. Saunders Co., 1988.
Crowell, J. W., and Guyton, A. C.: Evidence favoring a cardiac mechanism in irreversible hemorrhagic shock. Am. J. Physiol., 201:893, 1961.
Crowell, J. W., and Smith, E. E.: Oxygen deficit and irreversible hemorrhagic shock. Am. J. Physiol., 206:313, 1964.
Filkins, J. P.: Monokines and the metabolic pathophysiology of septic shock. Fed. Proc., 44:300, 1985.
Franklin, J., and Doelp, A.: Shocktrauma. New York, St. Martin's Press, 1980.
Guyton, A. C., and Crowell, J. W.: Dynamics of the heart in shock. Fed. Proc., 20:51, 1961.
Guyton, A. C., et al.: Cardiac Output and Its Regulation. Philadelphia, W. B. Saunders Co., 1973.
Heffernan, J. J., et al.: Clinical Problems in Acute Care Medicine. Philadelphia, W. B. Saunders Co., 1989.
Intensive Care of the Surgical Cardiopulmonary Patient. Chicago, Year Book Medical Publishers, 1983.
Jamieson, G. A., and Greenwalt, T. J. (eds.): Blood Substitutes and Plasma Expanders. New York, A. R. Liss, 1978.
Jones, C. E., et al.: Cardiac output and physiological mechanisms in circulatory shock. In MTP International Review of Science: Physiology, Vol. 1. Baltimore. University Park Press, 1974, p. 233.
Jones, C. E., et al.: A cause-effect relationship between oxygen deficit and irreversible hemorrhagic shock. Surgery, 127:93, 1968.
Jones, C. E., et al.: Significance of increased blood uric acid following extensive hemorrhage. Am. J. Physiol., 214:1374, 1968.
Kovach, A. G. B., and Sandor, P.: Cerebral blood flow and brain function during hypotension. Annu. Rev. Physiol., 38:571, 1976.
Lefer, A. M.: Eicosanoids as mediators of ischemia and shock. Fed. Proc., 44:275, 1985.
Lefer, A. M., and Martin, J.: Relationship of plasma peptides to the myocardial depressant factor in hemorrhagic shock in cats. Circ. Res., 26:59, 1970.
Luce, J. M., and Pierson, D. J.: Critical Care Medicine. Philadelphia, W. B. Saunders Co., 1988.
McCord, J. M.: Oxygen-deprived free radicals in postischemic tissue injury. N. Engl. J. Med., 312:159, 1985.
Movat, H. Z.: Microcirculation in disseminated intravascular coagulation induced by endotoxins. In Renkin, E. M., and Michel, C. C. (eds.): Handbook of Physiology. Sec. 2 Vol. IV. Bethesda, American Physiological Society, 1984, p. 1047.
Rothe, C. F., et al.: Control of total vascular resistance in hemorrhagic shock in the dog. Circ. Res., 12:667, 1963.
Schmid-Schönbein, H., and Teitel, P. (eds.): Basic Aspects of Blood Trauma. Hingham, Mass., Kluwer Boston, 1979.
Shoemaker, W. C., et al. (eds.): Society of Critical Care Medicine, The Textbook of Critical Care. Philadelphia, W. B. Saunders Co., 1988.
Sibbald, W. J. (ed.): Synopsis of Critical Care. Baltimore, Williams & Wilkins, 1988.
Smith, E. E., and Crowell, J. W.: Effect of hemorrhagic hypotension on oxygen consumption of dogs. Am. J. Physiol., 207:647, 1964.
Tintinalli, J. E.: Emergency Medicine: A Comprehensive Study Guide. 2nd Ed. New York, McGraw-Hill Book Co., 1988.
Winkelman, J. L.: Essentials of Basic Life Support. Minneapolis, Burgess, 1981.

UNIT V

THE KIDNEYS AND BODY FLUIDS

• *The Body Fluid Compartments: Extracellular and Intracellular Fluids; Interstitial Fluid and Edema* •

The body fluids are so important to the basic physiology of bodily function that we discuss them at several points in this text but each time in different contexts. Most importantly, it is pointed out in Chapter 4 that the body fluids can be divided into *extracellular* and *intracellular fluids;* the basic differences between these are discussed as well as the manner in which these differences come about as the result of cell membrane transport.

In the present chapter and in the following chapters on the kidneys, we are concerned with the body fluids from the total point of view, including regulation of body fluid volume, regulation of the constituents in the extracellular fluid, regulation of acid-base balance, and factors that govern gross interchange of fluid between extracellular and the intracellular compartments — especially the osmotic relationships between these compartments.

TOTAL BODY WATER

The total amount of water in a man of average weight (70 kilograms) is approximately 40 liters (Fig. 25–1), averaging 57 per cent of his total body weight. In a newborn infant, this may be as high as 75 per cent of the body weight, but it progressively decreases from birth to old age, most of the decrease occurring in the first 10 years of life. Also, obesity decreases the percentage of water in the body, sometimes to as low as 45 per cent.

INTAKE VERSUS OUTPUT OF WATER

Daily Intake of Water. Most of our daily intake of water enters by the oral route. Approximately two thirds is in the form of pure water or some other beverage, and the remainder is in the food that is eaten. A small amount is also synthesized in the body as the result of oxidation of hydrogen in the food; this quantity ranges between 150 and 250 ml/day, depending on the rate of metabolism. The normal intake of fluid, including that synthesized, averages about 2300 ml/day.

Daily Loss of Body Water. Table 25–1 shows the routes by which water is lost from the body under different conditions. Normally, at an atmospheric temperature of about 68°F (20°C), approximately 1400 of the 2300 milliliters of water intake is lost in the *urine,* 100 milliliters is lost in the *sweat,* and 100 milliliters in the *feces.* The remaining 700 milliliters is lost by *evaporation from the respiratory tract* or by *diffusion through the skin.*

Insensible Water Loss. Loss of water by diffusion through the skin and by evaporation from the respiratory tract is known as *insensible water loss* because we do not know that we are actually losing water at the time that it is leaving the body.

The average loss of water by diffusion through the skin is approximately 300 to 400 ml/day; this amount is lost even in a person who is born without sweat glands. In other words, the water molecules themselves actually diffuse through the cells of the skin. Fortunately, the cholesterol-filled, cornified layer of the skin acts as a protector against still much greater loss of water by diffusion. When the cornified layer becomes denuded, as after extensive burns, the rate of evaporation can increase to as much as 3 to 5 liters each day.

All air that enters the respiratory tract becomes totally saturated with moisture, to a vapor pressure of approximately 47 mm Hg, before it is expelled. Since the vapor pressure of the inspired atmospheric air is usually far less than 47 mm Hg, the average water loss through the lungs is about 300 to 400 ml/day. Because the atmospheric vapor pressure normally decreases with decreasing temperature, the loss is greatest in very cold weather and least in very warm weather. This explains the dry feeling in the respiratory passages in cold weather.

Loss of Water in Hot Weather and During Exercise. In very hot weather, water loss in sweat is occasionally increased to as much as 1.5 to 2.0 liters an hour, which

Total Body Fluids
(40 liters)

Figure 25–1. The body fluids diagrammed, showing the extracellular fluid volume, intracellular fluid volume, blood volume, and total body fluids.

Table 25–1 DAILY LOSS OF WATER (IN MILLILITERS)

	Normal Temperature	Hot Weather	Prolonged Heavy Exercise
Insensible loss:			
Skin	350	350	350
Respiratory tract	350	250	650
Urine	1400	1200	500
Sweat	100	1400	5000
Feces	100	100	100
TOTAL	2300	3300	6600

obviously can rapidly deplete the body fluids. Sweating is discussed in Chapter 73.

Exercise increases the loss of water in two ways. First, exercise increases the rate of respiration, which promotes increased water loss through the respiratory tract in proportion to the increased ventilatory rate. Second, and much more important, exercise increases the body heat and consequently is likely to result in excessive sweating.

● BODY FLUID COMPARTMENTS

THE INTRACELLULAR FLUID COMPARTMENT

About 25 of the 40 liters of fluid in the body are inside the approximately 75 trillion cells of the body and are collectively called the *intracellular fluid.* The fluid of each cell contains its own individual mixture of different constituents, but the concentrations of these constituents are reasonably similar from one cell to another. For this reason the intracellular fluid of all the different cells together is considered to be one large fluid compartment.

THE EXTRACELLULAR FLUID COMPARTMENT

All the fluids outside the cells are called *extracellular fluid,* and these fluids are constantly mixing, as explained in Chapter 1. The total amount of fluid in the extracellular compartment averages about 15 liters in a 70 kilogram adult.

The extracellular fluid can be divided into *interstitial fluid, plasma, cerebrospinal fluid, intraocular fluid, fluids of the gastrointestinal tract,* and *fluids of the potential spaces.*

Plasma. The plasma is the noncellular portion of blood. It is part of the extracellular fluid and communicates continually with the interstitial fluid through pores in the capillaries. The plasma volume averages 3 liters in the normal adult.

BLOOD VOLUME

Blood contains both extracellular fluid (the fluid of the plasma) and intracellular fluid (the fluid in the red blood cells). However, since blood is contained in a closed chamber all its own — the circulatory system — its volume and its special dynamics are exceedingly important.

The average blood volume of normal adult is almost exactly 5000 milliliters. On the average, approximately *3000 milliliters of this is plasma,* and the remainder, *2000 milliliters, is red blood cells.* However, these values vary greatly in different individuals; also, sex, weight, and many other factors affect the blood volume.

The Hematocrit. The hematocrit is the percentage of red blood cells in the blood as determined by centrifuging blood in a "hematocrit tube" until the cells become packed tightly in the bottom of the tube. This was explained in Chapter 14. Unfortunately, it is impossible for the red blood cells to be packed completely together; about 3 to 8 per cent of plasma remains entrapped among the cells. Therefore, the true hematocrit (H) averages about 96 per cent of the measured hematocrit (Hct); that is:

$$H = 0.96 \text{ Hct}$$

The true hematocrit (H) is approximately 40 for a normal man and 36 for a normal woman.

In severe anemia, the hematocrit may fall to as low as 10, but this small quantity of red blood cells is barely sufficient to sustain life. On the other hand, a few conditions cause excessive production of red blood cells, resulting in polycythemia. In these instances, the hematocrit often rises to 65. Obviously, there is an upper limit to the level of the hematocrit in polycythemic blood because excessive hematocrit causes the blood to become so viscous that death ensues as a result of multiple plugging of the peripheral vascular tree.

The Body Hematocrit. The hematocrit of the blood is *considerably less in the capillaries, arterioles, and other very small vessels* than in the large veins and arteries. Part of the cause of this is *axial streaming* of blood cells in blood vessels, which was explained in Chapter 14. In the very small vessels, red cells cannot flow near the walls of the vessels nearly so easily as can plasma. Therefore, the ratio of plasma to cells is far greater in the small vessels than in the large vessels.

If one averages the hematocrit in both the large and small vessels, a value called *body hematocrit* is determined. For normal man the body hematocrit (H_0) averages 91 per cent of the large vessel hematocrit; that is:

$$H_0 = 0.91 \, H = 0.87 \, Hct$$

MEASUREMENT OF BODY FLUID VOLUMES

The Dilution Principle for Measuring Fluid Volumes

The volume of any fluid compartment of the body can be measured by placing a substance in the compartment, allowing it to disperse evenly throughout the fluid, and then measuring the extent to which the substance has become diluted. Figure 25–2 illustrates this "dilution" principle for measuring the volume of any fluid compartment of the body. In this example, a small quantity of dye or other foreign substance is placed in fluid chamber A, and the substance is allowed to disperse throughout the chamber until it becomes mixed in equal concentrations in all areas, as shown in chamber B. Then a sample of the dispersed fluid is removed and the concentration of the substance is analyzed chemically, photoelectrically, or by any other means. The volume of the chamber can then be determined from the following formula:

Volume in milliliters

$$= \frac{\text{Quantity of test substance instilled}}{\text{Concentration per ml of dispersed fluid}}$$

Note that all one needs to know is (1) the *total quantity of the test substance* put into the chamber and (2) the *concentration in the fluid after dispersement*. As an example, if 10 milligrams of dye is dispersed in a chamber and the final concentration is 0.01 milligram for each milliliter of fluid, then the volume calculates to be 10 divided by 0.01, or 1000 milliliters.

Figure 25–2. Principles of the dilution method for measuring fluid volumes (explained in the text).

Determination of Volumes of Different Body Fluid Compartments

Fortunately, certain substances remain in specific fluid compartments without diffusing into the other compartments. Therefore, by choosing the appropriate "marker substance," the fluid volume of almost any compartment of the body can be measured using the dilution principle. Some of the substances used for measuring fluid volumes are as follows:

Total Body Water. Radioactive water or heavy water contains respectively radioactive *tritium* (H^3) or *deuterium* (H^2). These mix with the total body water within a few hours after being injected into the blood, and the dilution principle can be used to calculate the total body water.

Extracellular Fluid Volume. Multiple substances diffuse into most of the extracellular fluids, with only small amounts diffusing into the cells; these include *radioactive sodium, radioactive chloride, radioactive bromide, radioiothalamate, thiosulfate ion, thiocyanite ion, inulin,* and *sucrose.* When any one of these is injected into the blood, it will usually mix almost completely with the extracellular fluids within 30 minutes to an hour, and the dilution principle can then be used to calculate the extracellular fluid volume. However, some of these substances, such as radioactive sodium, diffuse more widely throughout the extracellular space than others and even slightly into cells. Therefore, one frequently speaks of the *sodium space,* the *inulin space,* or the *chloride space* instead of calling the measured volume the true extracellular fluid volume.

Calculation of Intracellular Volume. Obviously, by subtracting the extracellular fluid volume from the total body water, one can calculate the intracellular fluid volume.

Measurement of Plasma Volume. Any substance that attaches strongly to the plasma proteins will diffuse throughout the plasma with only small amounts leaking into the interstitium. Therefore, radioactive plasma protein, such as ^{131}I-protein, or plasma protein to which a so-called vital dye has been attached can be used to measure plasma volume. A vital dye often used is *T-1824,* also called *Evan's blue.*

Calculation of Interstitial Fluid Volume. To calculate interstitial fluid volume, the plasma volume is subtracted from the extracellular fluid volume.

Measurement of Blood Volume. Blood volume is usually measured by injecting into the circulation *radioactively labeled red cells.* After these mix in the circulation, the radioactivity of a sample of blood can be measured and the total blood volume calculated using the dilution principle. The substance most frequently used to label the red blood cells is *radioactive chromium* (^{51}Cr), which binds tightly with the red blood cells.

The total blood volume can also be calculated from the plasma volume by multiplying plasma volume × $100 \div (1 - \text{hematocrit})$.

Also, the red blood cell volume can be calculated by subtracting the plasma volume from the total blood volume.

● CONSTITUENTS OF EXTRACELLULAR AND INTRACELLULAR FLUIDS

The constituents of extracellular and intracellular fluids are illustrated diagrammatically and quantitatively in Figures 25–3 and 25–4 and in Table 25–2.

Figure 25–3. The compositions of plasma, interstitial fluid, and intracellular fluid. (Modified from Gamble: Chemical Anatomy, Physiology, and Pathology of Extracellular Fluid: A Lecture Syllabus. Cambridge, Mass.: Harvard University Press, 1954. Reprinted with permission.)

Figure 25–4. The nonelectrolytes of the plasma.

Figure 25–3 gives the compositions of plasma, interstitial fluid, and intracellular fluid in milliequivalents per liter. Figure 25–4 shows the concentrations of the nonelectrolytes in plasma, expressed as milligrams per deciliter of plasma. Finally, Table 25–2 gives the constituents of plasma, interstitial fluid, and intra-

cellular fluid in terms of their milliosmolar concentrations.

Important Constituents of the Extracellular Fluid. Referring again to Figure 25–3, one sees that extracellular fluid, both that of the blood plasma and of the interstitial fluid, contains large quantities of

Table 25–2 OSMOLAR SUBSTANCES IN EXTRACELLULAR AND INTRACELLULAR FLUIDS

	Plasma (mOsm/liter of H₂O)	Interstitial (mOsm/liter of H₂O)	Intracellular (mOsm/liter of H₂O)
Na^+	143	140	14
K^+	4.2	4.0	140
Ca^{++}	1.3	1.2	0
Mg^{++}	0.8	0.7	20
Cl^-	108	108	4
HCO_3^-	24	28.3	10
HPO_4^{--}, $H_2PO_4^-$	2	2	11
SO_4	0.5	0.5	1
Phosphocreatine			45
Carnosine			14
Amino acids	2	2	8
Creatine	0.2	0.2	9
Lactate	1.2	1.2	1.5
Adenosine triphosphate			5
Hexose monophosphate			3.7
Glucose	5.6	5.6	
Protein	1.2	0.2	4
Urea	4	4	4
Others	4.8	3.9	11
TOTAL mOsm/liter	302.8	301.8	302.2
Corrected osmolar activity (mOsm/liter)	282.5	281.3	281.3
Total osmotic pressure at 37°C (mm Hg)	5450	5430	5430

sodium and *chloride ions,* reasonably large quantities of *bicarbonate ion,* but only small quantities of potassium, calcium, magnesium, phosphate, sulfate, and organic acid ions. In addition, plasma contains a large amount of protein, whereas interstitial fluid contains much less. (The proteins in these fluids and their significance were discussed in Chapter 16.)

In Chapter 1, it is pointed out that the extracellular fluid is called the *internal environment* of the body and that its constituents are accurately regulated, mainly by the kidneys, so that the cells remain bathed continually in a fluid containing the proper electrolytes and nutrients for continued cellular function. The regulation of most of these constituents is presented in Chapter 29.

Important Constituents of the Intracellular Fluid. From Figure 25–3 it is also clear that the intracellular fluid contains only small quantities of sodium and chloride ions and almost no calcium ions; but it does contain large quantities of *potassium* and *phosphate* and moderate quantities of *magnesium* and *sulfate ions,* all of which are present in only small concentrations in the extracellular fluid. In addition, cells contain very large amounts of protein, approximately four times as much as in plasma.

● OSMOTIC EQUILIBRIA AND FLUID SHIFTS BETWEEN THE EXTRACELLULAR AND INTRACELLULAR FLUIDS

One of the most troublesome of all problems in clinical medicine is maintenance of adequate body fluids and proper balance between the extracellular and intracellular fluid volumes in seriously ill patients. The purpose of the following discussion, therefore, is to explain the interrelationships between extracellular and intracellular fluid volumes and the osmotic factors that cause shifts of fluid between the extracellular and intracellular compartments

BASIC PRINCIPLES OF OSMOSIS AND OSMOTIC PRESSURE

The basic principles of osmosis and osmotic pressure are presented in Chapter 4. The essence of these is reviewed here briefly.

Whenever a membrane between two fluid compartments is permeable to water but not to some of the dissolved solutes — that is, the membrane is selectively *permeable* — and the concentration of the solutes is greater on one side of the membrane than on the other, water passes through the membrane toward the side with the greater concentration of solutes. This phenomenon is called *osmosis.*

Osmosis results from the kinetic motion of the individual particles — both the molecules and the ions — in the solutions on the two sides of the membrane. This can be explained in the following way: When the temperature is the same in both solutions, the particles on both sides of the membrane, on the average, all have the same amount of kinetic energy of motion. However, the nonpermeant particles in the two solutions displace some of the water molecules. As a result, the so-called *chemical potential* of the water is reduced on each side of the membrane in direct proportion to the concentration of nonpermeant particles. Therefore, on the side of the membrane where there is the lower concentration of nonpermeant particles, the concentration of water molecules is the greater; that is, the chemical potential of the water is greater. This means that more water molecules will strike each pore of the membrane each second on the side of the membrane that has the lower concentration of nonpermeant solutes. As a result, net diffusion of water molecules will occur from this side toward the solution on the opposite side of the membrane, where the concentration of nonpermeant particles is greater. This net rate of diffusion is called the *rate of osmosis.*

Osmotic Pressure. Osmosis of water molecules can be opposed by applying a pressure across the selectively permeable membrane in the direction opposite that of the osmosis. The amount of pressure required to oppose the osmosis exactly is called the *osmotic pressure.*

Relationship of the Molecular Concentration of a Solution to Its Osmotic Pressure. Each nonpermeant molecule dissolved in water changes the chemical potential of the water by a given amount. Consequently, the tendency for this to cause osmosis of water is directly proportional to the concentration of the nonpermeant molecules. As a corollary, the osmotic pressure of the solution is also proportional to the concentration of the nonpermeant molecules in the solution. This relationship holds true for all nonpermeant molecules almost regardless of their molecular weight. For instance, one molecule of albumin with a molecular weight of 70,000 has the same osmotic effect as a molecule of glucose with a molecular weight of 180.

Osmotic Effect of Ions. Nonpermeant ions cause osmosis and osmotic pressure in exactly the same manner as do nonpermeant molecules. Furthermore, when a molecule dissociates into two or more ions, each of the ions then exerts osmotic pressure individually. Therefore, to determine the osmotic effect, all the nonpermeant ions must be added to all the nonpermeant molecules; but note that a bivalent ion, such as calcium, exerts no more osmotic pressure than does a univalent ion, such as sodium.

Osmoles. The ability of solutes to cause osmosis and osmotic pressure is measured in terms of "osmoles"; the osmole is a measure of the total number of particles. *One gram mole of nonpermeant and non-ionizable substance is equal to 1 osmole.* On the other hand, if a substance ionizes into two ions (sodium chloride into sodium and chloride ions, for instance), 0.5 gram mole of the substance equals 1 osmole. The obvious reason for using the osmole is that osmotic pressure is determined by the number of particles instead of the mass of the solute.

In general, the osmole is too large a unit for satisfactory use in expressing osmotic activity of solutes in the body. Therefore, the term *milliosmole*, which equals 1/1000 osmole, is commonly used.

Osmolality and Osmolarity. The osmolal concentration of a solution is called its *osmolality* when the concentration is expressed in osmoles per kilogram of water; it is called *osmolarity* when it is expressed as osmoles per liter of solution. The term "osmolality" is generally preferred because the osmotic pressure of a solution is considerably more closely related to its osmolality than to its osmolarity in very concentrated solutions. However, in the very dilute solutions of the normal human body, the differences are so slight that the terms are frequently used interchangeably. Furthermore, it is so much easier to express body fluid quantities in liters than in kilograms of water that almost all calculations are based on osmolarities rather than on osmolalities. It is common practice for many physiologists to speak in terms of osmolality even though they make their calculations in terms of osmolarities.

Relationship of Osmotic Pressure to Osmolality. The osmotic pressure of a solution *at body temperature of 98.6° F (37° C)* can be determined approximately from the following formula:

Osmotic pressure (mm Hg)
= 19.3 × Osmolality (milliosmole/kg water)

OSMOLALITY OF THE BODY FLUIDS

Table 25–2 lists the osmotically active substances in plasma, interstitial fluid, and intracellular fluid. The approximate milliosmoles of each of these per liter of water is given. Note especially that approximately *four fifths* of the total osmolality of the interstitial fluid and plasma is caused by sodium and chloride ions, whereas almost *half* of the intracellular osmolality is caused by potassium ions; the remainder is divided among the many other intracellular substances.

As noted at the bottom of Table 25–2, the total osmolality of each of the three compartments is approximately 300 mOsm/liter, with that of the plasma 1.3 milliosmoles greater than that of the interstitial and intracellular fluids. This slight difference between plasma and interstitial fluid is caused by the osmotic effect of the plasma proteins, which maintains about 20 mm Hg greater pressure in the capillaries than in the surrounding interstitial fluid spaces, as explained in Chapter 16.

Corrected Osmolar Activity of the Body Fluids. At the bottom of Table 25–2 are shown *corrected osmolar activities* of plasma, interstitial fluid, and intracellular fluid. The reason for these corrections is that molecules and ions in solution exert either *intermolecular attraction* from one solute molecule to the next or *intermolecular repulsion*, and these

two effects can cause, respectively, a decrease or an increase in the osmotic "activity" of the dissolved substance.

Total Osmotic Pressure Exerted by the Body Fluids. At the bottom of Table 25–2 is also shown the total *osmotic pressure* in millimeters of mercury that would be exerted by each of the different fluids if it were placed on one side of a cell membrane with pure water on the other side. Note that this total pressure averages about 5450 mm Hg for plasma, which is 19.3 times the corrected osmolality of 282.5 mOsm/kg for plasma.

MAINTENANCE OF OSMOTIC EQUILIBRIUM BETWEEN EXTRACELLULAR AND INTRACELLULAR FLUIDS

The tremendous osmotic pressure that can develop across the cell membrane when one side of the membrane is exposed to pure water—more than 5400 mm Hg—illustrates how much force can become available to move water molecules through the membrane when the solutions on the two sides of the membrane are not in osmotic equilibrium. For instance, in Figure 25–5A, a cell is placed in a solution that has an osmolality far less than that of the intracellular fluid. As a result, osmosis of water into the cell begins immediately, causing the cell to swell and diluting the intracellular fluid while concentrating the extracellular fluid. When the fluid inside the cell becomes diluted sufficiently to equal the osmolal concentration of the fluid on the outside, as shown in Figure 25–5B, further osmosis then ceases.

In Figure 25–5C, a cell is placed in a solution having a much higher concentration outside the cell than inside. This time, water passes by osmosis to the exte-

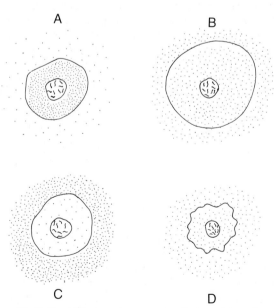

Figure 25–5. Establishment of osmotic equilibrium when cells are placed in a hypotonic or hypertonic solution.

rior, concentrating the intracellular fluid and diluting the extracellular fluid. This time the cell shrinks until the two concentrations become equal, as shown in Figure 25–5D.

Rapidity of Attaining Extracellular and Intracellular Osmotic Equilibrium. The transfer of water through the cell membrane by osmosis occurs so rapidly that any lack of osmotic equilibrium between the two fluid compartments in any given tissue is usually corrected within a few seconds and at most within a minute or so. However, this rapid transfer of water does not mean that complete equilibration occurs between the extracellular and intracellular compartments throughout the whole body within this same short period of time. The reason for this is that fluid usually enters the body through the gut and must then be transported by the blood to all tissues before complete equilibration can occur. In the normal person, it may take as long as 30 minutes to achieve reasonably good equilibration everywhere in the body after drinking water.

Isotonicity, Hypotonicity, and Hypertonicity. A fluid into which normal body cells can be placed without causing either swelling or shrinkage of the cells is said to be *isotonic* with the cells. A 0.9 per cent solution of sodium chloride or a 5 per cent glucose solution is approximately isotonic. Isotonic solutions are especially important in clinical medicine because they can be infused into the blood without danger of upsetting the osmotic equilibria between the fluids outside and inside the cells.

A solution that will cause the cells to swell is said to be *hypotonic;* any solution of sodium chloride with less than 0.9 per cent concentration is hypotonic.

A solution that will cause the cells to shrink is said to be *hypertonic;* sodium chloride solutions of greater than 0.9 per cent concentration are hypertonic.

● CHANGES IN THE VOLUMES AND OSMOLALITIES OF THE EXTRACELLULAR AND INTRACELLULAR FLUID COMPARTMENTS IN ABNORMAL STATES

CALCULATION OF FLUID SHIFTS BETWEEN THE EXTRACELLULAR AND INTRACELLULAR FLUID COMPARTMENTS

Among the different factors that can cause extracellular or intracellular volumes to change markedly are ingestion of water, dehydration, intravenous infusion of different types of solutions, loss of large quantities of fluids from the gastrointestinal tract, and loss of abnormal quantities of fluid by sweating or through the kidneys.

The changes in both extracellular and intracellular fluid volumes can be calculated easily if one keeps the following two basic principles in mind:

1. The osmolalities of the extracellular and intracellular fluids remain exactly equal to each other except for a few minutes after a change in one of the fluids occurs.

2. The number of osmoles of osmotically active substance in each compartment, in the extracellular fluid or in the intracellular fluid, remains constant unless one of the osmotically active substances moves through the cell membranes to the other compartment.

Using these two basic principles, we can now show by example how one can analyze the effects of different abnormal fluid conditions on extracellular and intracellular fluid volumes and osmolalities.

Effect of Adding Saline Solution to the Extracellular Fluid

If an *isotonic* saline solution is added to the extracellular fluid compartment, the osmolality of the extracellular fluid does not change, and no osmosis results. The only effect is an increase in extracellular fluid volume.

However, if a *hypertonic* solution is added to the extracellular fluid, the osmolality increases and causes osmosis of water out of the cells into the extracellular compartment.

Finally, if a *hypotonic* solution is added, the osmolality of the extracellular fluid decreases, and some of the extracellular fluid passes into the cells.

Table 25–3 calculates the sequential effects that would occur if 2 liters of 4.5 per cent (five times isotonic concentration) sodium chloride solution were added to the extracellular fluid compartment. The added solution contains a large total number of milliosmoles, 3000, which are added to the extracellular fluid compartment. The total of milliosmoles in the body becomes 15,000, and the volume of the total body water rises from 40 to 42 liters. Osmotic equilibration occurs within a few minutes. Dividing 42 into the new total milliosmoles, 15,000, one finds that the new milliosmolar concentration is 357. Now, dividing 357 into the total milliosmoles in the extracellular fluid, 7500, one finds that the new extracellular fluid volume is 21 liters. Dividing 357 into the total intracellular milliosmoles, 7500, the intracellular fluid is found also to be 21 liters.

The reader should become completely familiar with this method of calculation, for an understanding of the mathematical aspects of osmotic equilibria between the two compartments is essential to understanding almost all fluid problems of the body.

Table 25–3 EFFECT OF ADDING 2 LITERS OF 4.5 PER CENT SODIUM CHLORIDE SOLUTION

	Extracellular			Intracellular			Total Body Fluid		
	Volume (Liters)	Concentration (mOsm/liter)	Total mOsm	Volume (Liters)	Concentration (mOsm/liter)	Total mOsmole	Volume (Liters)	Concentration (mOsm/liter)	Total mOsm
Initial	15	300	4500	25	300	7500	40	300	12000
Solution added	2	1500	3000	0	0	0	2	1500	3000
Instantaneous effect	17	441	7500	25	300	7500	42	No equilibrium	15000
After osmotic equilibrium	21	357	7500	21	357	7500	42	357	15000

GLUCOSE AND OTHER SOLUTIONS ADMINISTERED FOR NUTRITIVE PURPOSES

Many different types of solutions are often administered intravenously to provide nutrition to patients who cannot otherwise take adequate amounts of food. Especially used are *glucose solutions* and to a lesser extent *amino acid* solutions. Rarely used are *homogenized fat solutions*. When these are administered, their concentrations of osmotically active substances are adjusted nearly to isotonicity, or they are given slowly enough that they do not upset the osmotic equilibria of the body fluids. However, after the glucose or other nutrient is metabolized, an excess of water often remains. Ordinarily, the kidneys excrete this in the form of very dilute urine. Thus, the net result is only addition of the nutrient to the body.

● EDEMA

Edema means the presence of excess fluid in the tissues of the body. In many instances edema occurs mainly in the extracellular fluid compartment, but it can involve the intracellular fluids as well. Two conditions are prone to cause intracellular swelling.

First, *depression of the metabolic systems of the tissues or lack of adequate nutrition to the cells* can cause serious intracellular edema. This especially occurs in any area of the body where the local blood flow is depressed and the delivery of oxygen and other nutrients is too low to maintain normal tissue metabolism, which depresses the cell membrane ionic pumps, especially the pump that removes sodium from the cells. Then, when sodium ions leak into the interior of the cells, the pump fails to remove these; instead, they cause osmosis of water into the cells as well. This can sometimes increase the intracellular volume of a tissue area, even an entire ischemic leg, for instance, to as much as two or more times normal. It is usually a prelude to death of the tissues.

Second, *intracellular edema also occurs in inflamed tissue areas.* The inflammation usually has a direct effect on the cell membranes to increase their permeability, allowing sodium and other ions to diffuse to the interior, with subsequent osmosis of water also into the cells.

EXTRACELLULAR FLUID EDEMA

Aside from the aforementioned instances that cause intracellular edema, a great host of conditions can cause extracellular edema. These can usually be classified into two different types: (1) edema caused by abnormal leakage of fluid from the blood capillaries or failure of the lymphatic system to return fluid from the interstitium and (2) edema caused by renal retention of salt and water.

Edema Caused by Abnormal Leakage of Fluid From the Capillaries or by Lymphatic Obstruction

Obviously, excess interstitial fluid will accumulate and cause extracellular edema if either too much fluid filters from the capillaries into the interstitium or lymphatic blockage prevents the return of interstitial fluid and protein to the circulation. The following is a list of the different conditions that can cause extracellular edema by one of these two methods.

I. Increased Capillary Pressure
  A. Excessive kidney retention of salt and water
  B. High venous pressure
    1. Heart failure
    2. Local venous block
    3. Failure of venous pumps
      (a) Paralysis of muscles
      (b) Immobilized parts of body
      (c) Failure of venous valves
  C. Decreased arteriolar resistance
    1. Excessive body heat
    2. Paralysis of sympathetic nervous system
    3. Effects of vasodilator drugs
II. Decreased Plasma Proteins
  A. Loss of proteins in urine (nephrosis)
  B. Loss of protein from denuded skin areas
    1. Burns
    2. Wounds
  C. Failure to produce proteins
    1. Liver disease
    2. Serious protein or caloric malnutrition
III. Increased Capillary Permeability
  A. Immune reactions that cause release of histamine and other immune products
  B. Toxins
  C. Bacterial infections
  D. Vitamin deficiency — especially vitamin C
  E. Prolonged ischemia
  F. Burns
IV. Blockage of Lymph Return
  A. Blockage of lymph nodes by cancer
  B. Blockage of lymph nodes by infection — especially with filaria nematodes
  C. Congenital absence of or abnormality of lymphatic vessels

From this outline it is clear than any one of three factors can increase the leakage of fluid and protein into the interstitium: (1) *increased capillary pressure;* (2) *decreased plasma proteins*, which causes diminished plasma colloid osmotic pressure; and (3) *increased capillary permeability*, which allows excessive leakage of both proteins and fluid through the pores.

A fourth factor leading to interstitial edema is *lymphatic blockage*. Figure 25–6 A to D illustrates the capillary, tissue, and lymphatic dynamics in edema caused by each of these factors. Note especially the very high interstitial colloid osmotic pressure shown in Figure 25–6C that occurs in lymphatic blockage edema. Also, note that the interstitial fluid pressure is positive, not the normal negative value, in all of these types of edema.

Two of the most prevalent causes of edema in the whole body are heart failure and the kidney disease nephrosis. In *heart failure*, edema occurs for several reasons. First, the heart fails to pump blood normally from the veins into the arteries; this causes the venous pressure and capillary pressure both to increase. Second, the arterial pressure tends to fall; this decreases the urinary output of water and salt by the kidneys. Third, diminished blood flow to the kidneys causes them to secrete renin; the renin causes the formation of angiotensin II in the blood, and this causes the adrenal glands to secrete aldosterone. Finally, both the angiotensin and the aldosterone have direct effects on the kidneys to cause additional retention of water and salt (see Chapter 19). Putting all these different factors together, serious generalized edema is very common in untreated heart failure.

In those heart failure patients in whom the left heart fails without significant failure of the right heart, blood is pumped into the lungs normally by the right heart but can-

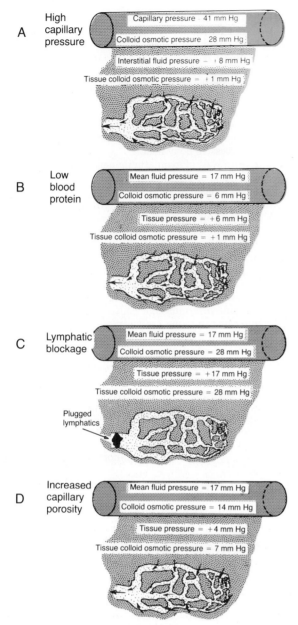

Figure 25-6. Various causes of edema.

not escape easily from the pulmonary veins through the left heart. Consequently, all the pulmonary pressures, including the pulmonary capillary pressure, rise far above normal and cause serious pulmonary edema. This is discussed in Chapter 38.

In *nephrosis*, the membranes of the renal glomeruli are damaged and become very leaky to plasma proteins, often allowing tremendous quantities of these proteins to pass into the urine. The loss of these causes the plasma colloid osmotic pressure to fall to a very low value, so low that it no longer prevents excessive filtration of fluid from the tissue capillaries into the interstitium. Very serious generalized edema usually occurs when the plasma protein concentration falls below 2.5 to 3 gm/dl.

Edema Caused by Renal Retention of Salt and Water

In the discussions earlier in this chapter, it has already become clear that sodium chloride normally remains almost

entirely in the extracellular fluid compartment, with only small amounts entering the cells. Also, we learn in the following chapters that most diseases that compromise renal function have an extreme tendency to reduce the excretion of salt and water into the urine. Therefore, the quantity of sodium chloride and water in the extracellular fluid often increases drastically. The effects of this are twofold: (1) widespread extracellular edema and (2) hypertension because of the increased blood volume (if all other aspects of the circulation are functioning normally), as explained in Chapter 19. To give an example, children who develop acute glomerulonephritis, in which the renal glomeruli are blocked by inflammation and therefore fail to filter adequate quantities of fluid, develop serious extracellular fluid edema in the entire body, and along with this they also develop hypertension, as we discuss in Chapter 31.

Relationship Between Interstitial Fluid Pressure and Interstitial Fluid Volume

In Chapter 16, we noted that the interstitial fluid pressure in most loose subcutaneous tissues of the body is slightly less than atmospheric pressure — that is, a slight *suction* in the tissues averaging about −3 mm Hg that helps to hold the tissues together. Figure 25-7 illustrates the approximate effects of different levels of interstitial fluid pressure on the interstitial fluid volume, as extrapolated to the human from animal studies. Note that at −3 mm Hg the interstitial fluid volume is approximately 12 liters. Furthermore, virtually all of this fluid is in a *gel* form. That is, the

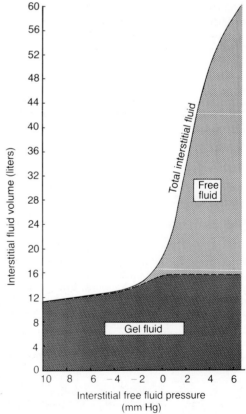

Figure 25-7. Effect of increasing interstitial fluid pressure on the volumes of total interstitial fluid, gel fluid, and free fluid. Note that significant amounts of free fluid occur only when the interstitial free fluid pressure becomes positive. (Modified from Guyton, Granger, and Taylor: *Physiol. Rev., 51*: 527, 1971.)

fluid is *bound in a proteoglycan meshwork,* so that there are virtually no "free" fluid spaces larger than a few hundredths of a micron diameter. The importance of the gel is that it prevents the fluid from flowing easily through the tissues because of the impediment of the "brushpile" of trillions of proteoglycan filaments. Also, when the interstitial fluid pressure falls to very negative values, the gel fluid does not contract greatly because the meshwork of proteoglycan filaments offers an elastic resistance to compression. Therefore, in the negative interstitial fluid pressure range, the interstitial fluid volume does not change greatly regardless of whether the degree of suction is only a millimeter or so of negative pressure or as much as 10 to 20 mm Hg negative pressure.

Effect of Positive Pressure on Interstitial Volume —"Free Fluid," and Pitting Edema. Note to the right in Figure 25–7 the sudden increase in total interstitial fluid volume when the interstitial fluid pressure rises above atmospheric pressure (above zero pressure level). Because the loose subcutaneous tissues of the body are highly distensible, even 2, 3, or 4 mm Hg of positive pressure can increase the interstitial fluid volume from its normal level of 12 liters up to two to three times normal. Furthermore, most of the extra fluid is "free fluid" that forms large free fluid spaces in the tissues that are no longer enmeshed tightly within a meshwork of the tissue gel. This fluid can now flow freely through the tissue spaces. When this occurs, the edema is said to be "pitting" edema because one can press the thumb against a tissue area and push the fluid out of that area into other areas. Then, when the thumb is removed, a pit is left in the skin for a few seconds, or sometimes as long as a minute, until the pit fills again with fluid that flows back from the surrounding tissues. This type of edema is distinguished from "nonpitting" edema, which occurs when the cells swell instead of the interstitium swelling or when the fluid in the interstitium is clotted with fibrinogen, so that it cannot move freely within the tissue spaces.

Figure 25–8 illustrates a physical model of the tissues constructed of a large rubber bag to represent the skin, filled with small balloons to represent the tissue cells, and with large amounts of cotton fibers between the balloons to represent collagen and proteoglycan fibers and filaments of the interstitium. When fluid is sucked out of the bag, as shown to the right in the figure, the cells crunch together but are

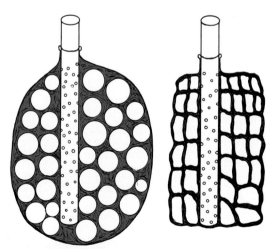

Figure 25–8. Physical model of the tissues constructed of a rubber bag, a perforated tube to simulate a capillary, balloons filled with water to simulate cells, and cotton between the balloons to simulate intercellular elements. *Left,* Edematous state. *Right,* Nonedematous state.

still held apart by the interstitial fibrillar network, still containing moderate amounts of interstitial fluid filling the interstices of the fibrils. To the left, on the other hand, is a typical edematous state, with large amounts of excess "free" fluid between the cells and free movement of this excess fluid within the spaces as well as rounding up of the cells, a typical condition that occurs in edema.

Importance of the Proteoglycan Filaments in the Interstitium

First, the proteoglycan filaments, along with collagen fibrils also in the interstitial spaces, act as a "spacer" in the spaces between the cells. Many nutrients and ions do not diffuse readily through cell membranes. Therefore, without adequate spacing between the cells, these nutrients, electrolytes, and cellular waste products could not be exchanged between blood capillaries and cells located at a distance from each other.

Second, without proteoglycan filaments, fluid flows extremely easily through the tissue spaces. For instance, when too much fluid occurs in the interstitium, as occurs in edema, so that there are not enough filaments to match the increased fluid, fluid can then create large channels and flow as much as 250,000 times as easily as when filaments are compacted in the interstitium. For instance, a woman lying on her side will often have edema in the lowermost breast, while the upper breast is nonedematous. Then, when she rolls to the opposite side, within minutes the breast that is now lowermost becomes edematous, while the first breast loses its fluid because of mobility of free fluid through the tissues from one breast to the other. In fact, even a standing person would almost certainly lose most of his or her interstitial fluid from the upper body to the lower body if it were not for the proteoglycan filaments.

Yet, even though fluid does not *flow* through the tissues easily in the presence of the compacted proteoglycan filaments, *the substances within the fluid do diffuse through the tissues at least 95 per cent as easily as normally.* Therefore, the usual diffusion of nutrients to the cells and excreta away from the cells is not compromised by the filamentous mesh.

The Concept of an "Edema Safety Factor"

Even though any one of many abnormalities can cause edema, usually the abnormality must become quite severe before edema actually occurs. The reason for this is that there is a so-called edema safety factor. This safety factor is made up of three different components:

1. Figure 25–7 illustrates that edema does not begin to appear until the interstitial fluid pressure rises above the atmospheric pressure level. Therefore, *all of the negative interstitial fluid pressure must be dissipated,* and the pressure must rise into the positive range before edema will occur. Obviously, this allows at least some degree of capillary filtration abnormality or even some degree of lymphatic obstruction before edema would occur.

2. *Lymph flow can increase as much as 10-fold to 50-fold.* This constitutes another safety factor because the lymph vessels can carry away large amounts of fluid as they are formed and help to prevent the interstitial fluid pressure from rising into the positive pressure range.

3. As the lymph flow increases, it carries protein away from the interstitium, thus reducing the interstitial fluid colloid osmotic pressure. This is called *"washdown" of the interstitial fluid protein.* As washdown occurs, the differential between plasma colloid osmotic pressure and the colloid

osmotic pressure in the interstitium increases, thus increasing the osmotic absorption of fluid from the interstitium into the capillaries. This, too, increases the safety factor against edema.

Putting all these factors together we find the following: The safety factor caused by the normal negative interstitial fluid pressure averages about 3 mm Hg, that caused by increased lymph flow averages about 7 mm Hg, and that caused by washdown of protein from the interstitial spaces averages as much as another 7 mm Hg. Thus, the total safety factor against edema is about 17 mm Hg. This means that the capillary pressure in a peripheral tissue could theoretically rise to as high as 34 mm Hg, to approximately double its normal value of 17 mm Hg, before edema would occur.

Thus, the concept of a safety factor against edema explains why we are not continuously edematous.

● FLUIDS IN THE POTENTIAL SPACES OF THE BODY

A "potential space" can best be described by listing several of the important potential spaces: the pleural cavity, the pericardial cavity, the peritoneal cavity, and the synovial cavities, including both the joint cavities and the bursae. Virtually all of these potential spaces have surfaces that almost touch each other with only a thin layer of fluid in between, and the surfaces slide over each other. To facilitate this sliding, a viscid fluid lubricates the surfaces.

Fluid Exchange Between the Capillaries and the Potential Spaces

The surface membrane of a potential space usually does not offer significant resistance to the passage of fluid, electrolytes, or even proteins, all moving back and forth between the space and the interstitial fluid in the surrounding tissue with relative ease. Therefore, each potential space is, in reality, a large tissue space. Consequently, fluid in the capillaries adjacent to a potential space membrane not only diffuses into the interstitial fluid but also into the potential space.

Function of the Lymphatic Vessels Draining the Potential Spaces. In the same manner that protein collects in the interstitium because of leakage out of the capillaries, so does it also collect in the potential spaces, and it must be removed through lymphatics or some other channel. Fortunately, each potential space is either directly or indirectly connected with lymph vessels. In fact, in some cases, such as for the pleural cavity and peritoneal cavity, large lymphatic vessels arise directly from the cavity itself.

Edema Fluid in the Potential Spaces. When edema occurs in the subcutaneous tissues adjacent to a potential space, edema fluid usually collects in the potential space as well and is called an *effusion*. Thus, lymphatic blockage or any one of the multiple abnormalities that causes excessive capillary filtration can cause effusion in the same way that interstitial edema is caused. The abdominal cavity is especially prone to collect effusion fluid, and in this instance it is called *ascites*. In very serious cases, the amount of ascites can become as great as 20 or more liters.

In addition, the other potential spaces, such as the pleural cavity, the pericardial cavity, the joint spaces, the bursae, and so forth, can all also become seriously swollen in generalized edema. Also, local infection in any one of the cavities often blocks the lymphatic drainage, and this too can lead to isolated swelling of any one of the cavities.

The dynamics of fluid exchange in the pleural cavity are discussed in detail in Chapter 38. These dynamics are mainly representative of all the other potential spaces as well. It is especially interesting that the pressure in most or all of the potential spaces in the nonedematous state is *negative* in the same way that the pressure is negative (subatmospheric) in the loose subcutaneous tissue. For instance, it measures about -7 to -8 mm Hg in the pleural cavity, -3 to -5 mm Hg in the joint spaces, and -5 to -6 mm Hg in the pericardial cavity.

REFERENCES

Adoph, E. F.: Physiology of Man in the Desert. New York, Interscience Publishers, 1947.

Anderson, D. H.: Compartmental Modeling and Tracer Kinetics. New York, Springer-Verlag, 1983.

Andreoli, T. E., et al. (eds.): Membrane Physiology. New York, Plenum Press, 1980.

Barry, P. H., and Diamond, J. M.: Effects of unstirred layers on membrane phenomena. Physiol. Rev., 64:763, 1984.

Borow, M.: Fundamentals of Homeostasis: A Clinical Approach to Fluid Electrolyte Acid-Base Energy Metabolism in Health and Disease. Flushing, N.Y., Medical Examination Publishing Co., 1977.

Borut, A., and Shkolnik, A.: Physiological adaptations to the desert environment. Int. Rev. Physiol., 7:185, 1974.

Dawson, K. G.: Endocrine physiology of electrolyte metabolism. Drugs, 28(Suppl.1):98, 1984.

Gilles, R. (ed): Mechanisms of Osmoregulation: Maintenance of Cell Volume. New York, John Wiley & Sons, 1979.

Gilmore, J. P.: Neural control of extracellular volume in the human and nonhuman primate. In Shepherd, J. T., and Abboud, F. M. (eds.): Handbook of Physiology. Sec. 2, Vol. III. Bethesda, Md., American Physiological Society, 1983, p. 885.

Goldberger, E.: A Primer of Water, Electrolyte, and Acid-Base Syndromes. Philadelphia, Lea & Febiger, 1980.

Grinstein, S., et al.: Responses of lymphocytes to anisotonic media: Volume-regulating behavior. Am. J. Physiol., 246:C204, 1984.

Guyton, A. C., et al.: Circulatory Physiology II: Dynamics and Control of Body Fluids. Philadelphia, W. B. Saunders Co., 1975.

Guyton, A. C., et al.: Interstitial fluid pressure. Physiol. Rev., 51:527, 1971.

Haas, M.: Properties and diversity of (Na-K-Cl) Cotransporters. Annu. Rev. Physiol., 51:443, 1989.

Jamieson, G. A., and Greenwalt, T. J. (eds.): Blood Substitutes and Plasma Expanders. New York, A. R. Liss, 1978.

Jennings, M. L.: Kinetics and mechanism of anion transport in red blood cells. Annu. Rev. Physiol., 47:519, 1985.

Kinmonth, J. G.: The Lymphatics. Surgery, Lymphography and Diseases of the Chyle and Lymph Systems. Baltimore, E. Arnold, 1982.

Klahr, S. (ed.): Cancer of the Prostate and Kidney. New York, Plenum Publishing Corp., 1983.

Kokko, J. P., and Tannen, R. L.: Fluids and Electrolytes. Philadelphia, W. B. Saunders Co., 1986.

Leak, L. V., and Burke, J. F.: Ultrastructural studies on the lymphatic anchoring of filaments. J. Cell Biol., 36:129, 1968.

Lebenthal, E.: Total Parenteral Nutrition: Indications, Utilization, Complications, and Pathophysiological Considerations. New York, Raven Press, 1986.

Levick, J. R.: Blood flow and mass transport in synovial joints. In Renkin, E. M., and Michel, C. C. (eds.): Handbook of Physiology. Sec. 2, Vol. IV. Bethesda, Md., American Physiological Society, 1984, p. 917.

Macey, R. I.: Transport of water and urea in red blood cells. Am. J. Physiol., 246:C195, 1984.

Maxwell, M. H., et al. (eds.): Clinical Disorders of Fluid and Electrolyte Metabolism, 4th Ed. New York, McGraw-Hill Book Co., 1987.

Nicoll, P. A., and Taylor, A. E.: Lymph formation and flow. Annu. Rev. Physiol., 39:73, 1977.

Nieuwenhuis, P. (ed.): In Vivo Immunology, Histophysiology of the Lymphoid System. New York, Plenum Publishing Corp., 1982.

Peters, T., and Sjoholm, I. (eds.): Albumin: Structure, Biosynthesis, Function. New York, Pergamon Press, 1978.

Post, R. L.: Seeds of sodium, potassium ATPase. Annu. Rev. Physiol., 51:1, 1989.

Renkin, E. M.: Lymph as a measure of the composition of interstitial fluid. In Fishman, A. P., and Renkin, E. M. (eds.): Pulmonary Edema. Baltimore, Waverly Press, 1979, p. 145.

Rose, B. D. (ed.): Clinical Physiology of Acid-Base and Electrolyte Disorders, 3rd Ed. New York, McGraw-Hill Book Co., 1989.

Schreiber, G., and Urban, J.: The synthesis and secretion of albumin. In Adrian, R. H., et al. (eds.): Reviews of Physiology, Biochemistry, and Pharmacology. New York, Springer-Verlag, 1978, p. 27.

Shoemaker, V. H., and Nagy, K. A.: Osmoregulation in amphibians and reptiles. Annu. Rev. Physiol., 39:449, 1977.

Smith, K.: Fluids and Electrolytes: A Conceptual Approach. New York, Churchill Livingstone, 1980.

Tinker, D., and Rucker, R. B.: Role of selected nutrients in synthesis, accu-

mulation, and chemical modification of connective tissue proteins. Physiol. Rev., 65:607, 1985.

Wolf, A. V., and Crowder, N. A.: Introduction to Body Fluid Metabolism. Baltimore, Williams & Wilkins, 1964.

Yet, M. G., et al.: Effects of the protein matrix on glycan processing in glycoproteins. FASEB J., 2:22, 1988.

Yoffey, J. M., and Courtice, F. C. (eds.): Lymphatics, Lymph and Lympho-myeloid Complex. New York, Academic Press, 1970.

CHAPTER 26

• *Formation of Urine by the Kidney: I. Renal Blood Flow, Glomerular Filtration, and Their Control* •

The kidneys perform two major functions. First, they excrete most of the end products of bodily metabolism, and second, they control the concentrations of most of the constituents of the body fluids. The purpose of this chapter is to discuss the basic principles of urine formation. Then, in the following chapters, we present detailed mechanisms for processing and control of excretion of the individual constituents in the urine.

PHYSIOLOGIC ANATOMY OF THE KIDNEY

The two kidneys together contain about 2,000,000 nephrons, and each nephron is capable of forming urine by itself. Therefore, in most instances, it is not necessary to discuss the entire kidney but merely the function of a single nephron to explain the function of the kidney. The nephron is composed basically of (1) a *glomerulus* through which fluid is filtered from the blood, and (2) a long *tubule* in which the filtered fluid is converted into urine on its way to the *pelvis* of the kidney.

Figure 26–1 shows the general organizational plan of the kidney, illustrating especially the distinction between the *cortex* of the kidney and the *medulla*. Figure 26–2A illustrates the basic anatomy of the nephron, which may be described as follows: Blood enters the glomerulus through the *afferent arteriole* and then leaves through the *efferent arteriole*. The glomerulus is a network of up to 50 parallel branching and anastomosing capillaries covered by epithelial cells and encased in *Bowman's capsule*. Pressure of the blood in the glomerulus causes fluid to filter into Bowman's capsule, and from here the fluid flows into the *proximal tubule* that lies in the *cortex* of the kidney along with the glomerulus.

From the proximal tubule the fluid passes into the *loop of Henle* that dips deeply into the kidney mass, some of the loops passing all the way to the bottom of the renal medulla. Each loop is divided into the *descending limb* and the *ascending limb*. The wall of the descending limb and the lower end of the ascending limb is very thin and therefore is called the *thin segment* of the loop of Henle. However, after the ascending limb of the loop has returned part way back in the cortical direction, its wall once again becomes thick like that of the other portions of the tubular system; this portion of the loop of Henle is called the *thick segment of the ascending limb*.

After passing through the loop of Henle, the fluid then enters the *distal tubule,* which, like the proximal tubule, lies in the renal cortex. Then, still in the cortex, as many as eight of the distal tubules coalesce to form the *cortical collecting duct* (also called *collecting tubule*), the end of which turns once again away from the cortex and passes downward into the medulla, where it becomes the *medullary collecting duct* but usually called simply the *collecting duct.* Successive generations of collecting ducts coalesce to form progressively larger collecting ducts that penetrate all the way through the medulla, parallel to the loops of Henle. The largest collecting ducts empty into the *renal pelvis* through the tips of the *renal papillae;* these are conical projections of the medulla that protrude into the *renal calyces,* which are themselves recesses of the *renal pelvis.* In each kidney there are about 250 of these very large collecting ducts, each of which transmits the urine from about 4000 nephrons.

As the glomerular filtrate flows through the tubules, over 99 per cent of its water and varying amounts of its solutes are normally reabsorbed into the vascular system, and small amounts of some substances are also *secreted* into the tubules. The re-

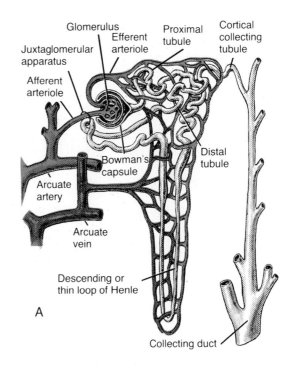

Figure 26-1. The general organizational plan of the urinary system.

maining tubular water and dissolved substances become the urine.

"Cortical" and "Juxtamedullary" Nephrons.
The characteristics of the nephrons are somewhat different, depending on how deep the nephrons lie within the kidney mass. Those nephrons of which the glomeruli lie close to the surface of the kidney are called *cortical nephrons,* as illustrated in Figure 26–2B. These have very short thin segments in their loops of Henle, and the loops penetrate only a very short distance into the outer portion of the medulla. Between one fifth and one third of the nephrons have glomeruli that lie deep in the renal cortex near the medulla; these are called *juxtamedullary nephrons.* These nephrons have very long loops of Henle and especially long thin segments of the loops that penetrate deeply into the inner zone of the medulla, many of them all the way to the tips of the renal papillae.

The Peritubular Capillary Network and the Vasa Recta. Surrounding the entire tubular system of the kidney is an extensive network of capillaries called the *peritubular capillary network.* This network is supplied with blood from the *efferent arterioles,* blood that has already passed through the glomerulus. Most of the peritubular capillary network lies in the renal cortex alongside the proximal tubules, distal tubules, and cortical collecting ducts. However, from the deeper portions of this peritubular network, long branching capillary loops called *vasa recta* extend downward into the medulla to lie side by side with the juxtaglomerular loops of Henle all the way to the renal papillae. Then, like the loops of Henle, they also loop back toward the cortex and empty into the cortical veins.

Functional Diagram of the Nephron. Figure 26–3 illustrates a simplified diagram of the "physiologic nephron." This diagram contains most of the

Figure 26-2. *A,* The nephron. (From Smith: The Kidney: Structure and Functions in Health and Disease. New York, Oxford University Press, 1951). *B,* Differences between a cortical and a juxtamedullary nephron. (From Pitts: Physiology of the Kidney and Body Fluids. Chicago, Year Book Medical Publishers, 1974.)

nephron's functional structures, and it is used in the present discussion to explain many aspects of renal function.

BASIC THEORY OF NEPHRON FUNCTION

The basic function of the nephron is to clean, or "clear," the blood plasma of unwanted substances as

Figure 26-3. The functional nephron.

it passes through the kidney. The substances that must be cleared include particularly the end products of metabolism, such as urea, creatinine, uric acid, and urates. In addition, many other substances, such as sodium ions, potassium ions, chloride ions, and hydrogen ions, tend to accumulate in the body in excess quantities; it is the function of the nephron also to clear the plasma of these excesses.

The principal mechanism by which the nephron clears the plasma of unwanted substances is as follows: (1) It filters a large proportion of the plasma in the flowing glomerular blood, usually about one fifth of it, through the glomerular membrane into the tubular system of the nephron. (2) Then, as this filtered fluid flows through the tubules, the *unwanted substances fail to be reabsorbed* while the *wanted substances, especially almost all of the water and many of the electrolytes, are reabsorbed* back into the plasma of the peritubular capillaries. In other words, the wanted portions of the tubular fluid are returned to the blood and the unwanted portions pass into the urine.

A second mechanism by which the nephron clears the plasma of other unwanted substances is by *secretion.* That is, substances are secreted from the plasma directly through the epithelial cells lining the tubules into the tubular fluid. Thus, the urine that is eventually formed is composed mainly of *filtered* substances but also of small amounts of *secreted* substances.

● RENAL BLOOD FLOW AND PRESSURES

BLOOD FLOW THROUGH THE KIDNEYS

The rate of blood flow through both kidneys of a 70 kilogram man is about 1200 ml/min.

The portion of the total cardiac output that passes through the kidneys is called the *renal fraction.* Since the normal cardiac output of a 70 kilogram adult man is about 5600 ml/min and the blood flow through both kidneys is about 1200 ml/min, one can calculate that the normal renal fraction is about 21 per cent. This can vary from as little as 12 per cent to as high as 30 per cent in the normal resting person.

Special Aspects of Blood Flow Through the Renal Vasculature. Note in Figure 26-3 that there are two capillary beds associated with the nephron: (1) the *glomerulus* and (2) the *peritubular capillaries.* The glomerular capillary bed receives its blood from the *afferent arteriole,* and from this bed the blood flows into the peritubular capillary bed through the *efferent arteriole,* which offers considerable resistance to blood flow. As a result, the glomerular capillary bed is a *high pressure bed,* whereas the peritubular capillary bed is a *low pressure bed.* Because of the high pressure in the glomerulus, it functions in much the same way as the usual arterial ends of the tissue capillaries, with fluid filtering continually out of the glomerulus into Bowman's capsule. On the other hand, the low pressure in the peritubular capillary system causes it to function in much the same way as the venous ends of the tissue capillaries, with fluid being absorbed continually into the capillaries.

Blood Flow in the Vasa Recta. A special portion of the peritubular capillary system is the vasa recta, which are a network of capillaries that descend into the medulla around the loops of Henle. These capillaries form loops in the medulla of the kidney and then return to the cortex before emptying into the veins. The vasa recta play a special role in the formation of concentrated urine, which is discussed in Chapter 28.

Only a small proportion of the total renal blood flow, about 1 to 2 per cent, flows through the vasa recta. In other words, blood flow through the medulla of the kidney is sluggish in contrast to the rapid blood flow in the cortex.

PRESSURES IN THE RENAL CIRCULATION

Figure 26-4 gives the approximate pressures in different parts of the renal circulation and tubules, showing an initial pressure of approximately 100 mm Hg in the large arcuate arteries and about 8 mm Hg in the veins into which the blood finally drains. The two major areas of resistance to blood flow through the nephron are (1) the *small renal arteries and afferent arteriole* and (2) the *efferent arteriole.* In the small arteries and afferent arteriole, the pressure falls from 100 mm Hg at its arterial end to an estimated mean pressure of about 60 mm Hg in the glomerulus. (This pressure is still in serious doubt, having been calculated to be as high as 70 mm Hg in the dog and measured to be as low as 45 mm Hg in the rat. Therefore, 60 mm Hg is only an average estimate.) As the blood flows through the efferent arterioles from

100 mm Hg

60 mm Hg

18 mm Hg

18 mm Hg

13 mm Hg

10 mm Hg

10 mm Hg

8 mm Hg

Interstitial fluid
pressure 6 mm Hg

0 mm Hg

Figure 26–4. Approximate pressures at different points in the vessels and tubules of the functional nephron and in the interstitial fluid.

the glomerulus to the peritubular capillary system, the pressure falls another 47 mm Hg to a mean peritubular capillary pressure of 13 mm Hg. Thus, the high pressure capillary bed in the glomerulus operates at a mean pressure of about 60 mm Hg and therefore causes rapid filtration of fluid, whereas the low pressure capillary bed in the peritubular capillary system operates at a mean capillary pressure of about 13 mm Hg, therefore allowing rapid absorption of fluid because of the high osmotic pressure of the plasma.

"INTRARENAL PRESSURE" AND RENAL INTERSTITIAL FLUID PRESSURE

The kidney is encased in a tight *fibrous capsule.* When a needle is inserted into the kidney and the pressure in the needle is gradually raised until fluid flows into the kidney tissue, the pressure at which flow begins is between 10 and 18 mm Hg, averaging perhaps 13 mm Hg. This "needle" pressure is called the *intrarenal pressure.* It is pointed out in Chapter 16 that such needle pressures do not measure the interstitial fluid pressure but instead measure the *total tissue pressure* (the pressure of fluid plus all solid tissue elements that tends to collapse blood vessels and tubules).

Attempts to measure the *interstitial fluid pressure* of the kidney using implanted perforated capsules as described in Chapter 16 have given an average mean value of +6 mm Hg, which at present is probably the best estimate of interstitial fluid pressure of the kidney.

FUNCTION OF THE PERITUBULAR CAPILLARIES

Tremendous quantities of fluid, about 180 liters each day, are filtered through all the glomeruli; all but the 1 to 1.5 liters of this that becomes urine is reabsorbed from the tubules into the renal interstitial spaces and thence into the peritubular capillaries. This represents about four times as much fluid as that reabsorbed at the venous ends of all the other capillaries of the entire body. Therefore, one can readily see that reabsorption of fluid into the peritubular capillaries presents a special problem. However, the peritubular capillaries are extremely porous in comparison with those in other body tissues, so that extremely rapid osmosis of fluid resulting from the colloid osmotic pressure of the plasma proteins can account for the rapid absorption that is required.

● GLOMERULAR FILTRATION AND THE GLOMERULAR FILTRATE

The Glomerular Membrane and Glomerular Permeability. The fluid that filters through the glomerulus into Bowman's capsule is called *glomerular filtrate,* and the membrane of the glomerular capillaries is called the *glomerular membrane.* Although, in general, this membrane is similar to that of other capillaries throughout the body, it has several differences, as illustrated in Figure 26–5. First, it has three major layers: (1) the endothelial layer of the capillary itself, (2) a basement membrane, and (3) a layer of epithelial cells that is illustrated on the outer surfaces of the glomerular capillaries in the figure. Yet, despite the number of layers, the permeability of the glomerular membrane is from 100 to 500 times as great as that of the usual capillary.

The tremendous permeability of the glomerular membrane is caused by its special structure. The capillary *endothelial cells* lining the glomerulus are perforated by literally thousands of small holes called *fenestrae.* Then, outside the endothelial cells is a *basement membrane,* composed mainly of a meshwork of collagen and proteoglycan fibrillae that also has large spaces through which fluid can filter. A final layer of the glomerular membrane is a layer of *epithelial cells* that line the outer surfaces of the glomerulus. However, these cells are not continuous but instead consist mainly of fingerlike projections that cover the basement membrane. Figure 26–6 is a scanning electron micrograph of the outer surfaces of the glomerular capillaries, illustrating thousands to millions of minute parallel "fingers" of the epithelial cells. These fingers form slits called *slit-pores* through which the glomerular filtrate filters.

Thus, the glomerular filtrate passes through three different layers before entering Bowman's capsule, but each of these layers is several hundred times as

porous as the usual capillary membrane, which accounts for the tremendous volume of glomerular filtrate that can be formed each minute. Yet, despite the tremendous permeability of the glomerular membrane, it has an extremely high degree of selectivity for the sizes of molecules that it allows to pass.

Overall, the permeability of the glomerular membrane to substances of different molecular weights (expressed as the ratio of concentration of the dissolved substance on the filtrate side of the membrane to its concentration on the plasma side) is approximately as follows:

Molecular Weight	Permeability	Example Substance
5,200	1.00	Inulin
30,000	0.5	Very small protein
69,000	0.005	Albumin

This means that at a molecular weight of 5,200, the dissolved substance filters 100 per cent as easily as water, but at the protein molecular weight of 69,000 only 0.5 per cent of the protein molecules filter. Note that the molecular weight of the smallest plasma protein, albumin, is 69,000. Therefore, for practical purposes the glomerular membrane is almost completely impermeable to all plasma proteins but is highly permeable to essentially all other dissolved substances in normal plasma.

There are two basic reasons for the high degree of molecular selectivity by the glomerular membrane. First is the sizes of the pores in the membrane itself. That is, the pores of the membrane are large enough to allow molecules with diameters up to about 8 nanometers (80 angstroms) to pass through. Yet, the molecular diameter of the plasma protein albumin molecule is only about 6 nanometers, which is somewhat smaller than the sizes of these large pores. Therefore, why do the protein molecules not pass through in great quantity? The answer to this is the second factor that determines the permeability of the membrane: The *basement membrane portions of the glomerular pores are lined with a complex of proteoglycans that have very strong negative electrical charges.* The plasma proteins also have strong negative electrical charges. Therefore, electrostatic repulsion of the molecules by the pore walls keeps virtually all protein molecules larger than 69,000 molecular weight from passing through.

Composition of the Glomerular Filtrate. The glomerular filtrate has almost exactly the same composition as the fluid that filters from the arterial ends of the capillaries into the interstitial fluids. It contains no red blood cells and about 0.03 per cent protein, or about 1/240 the protein in the plasma.

The electrolyte and other solute composition of glomerular filtrate is also similar to that of the interstitial fluid. However, because of the paucity of negatively charged protein ions in the filtrate, a Donnan equilibrium effect occurs that causes the concentra-

Capillary endothelial cell
Basement membrane
Epithelial cell

Plasma

Filtrate

Fenestra

Slit-pore

Figure 26–5. Functional structure of the glomerular membrane.

Figure 26-6. Scanning electron micrograph of a glomerulus from a normal rat kidney. The visceral epithelial cells, or podocytes (P), extend multiple processes outward from the main cell body, which wrap around the individual capillary loops. Note that immediately adjacent pedicels, or foot processes, arise from different podocytes. (Magnification, × 3300.) (From Brenner and Rector: The Kidney. Philadelphia, W. B. Saunders Company, 1981.)

tion of the nonprotein negative ions, including chloride and bicarbonate ions, to be about 5 per cent higher in the glomerular filtrate than in plasma; the concentration of positive ions is about 5 per cent lower.

To summarize: For all practical purposes, *glomerular filtrate is the same as plasma except that it has no significant amount of proteins.*

THE GLOMERULAR FILTRATION RATE

The quantity of glomerular filtrate formed each minute in all nephrons of both kidneys is called the *glomerular filtration rate.* In the normal person, this averages approximately 125 ml/min. To express this differently, the total quantity of glomerular filtrate formed each day averages about 180 liters, or more than two times the total weight of the body. Over 99 per cent of the filtrate is normally reabsorbed in the tubules, with the remaining small portion passing into the urine.

The Filtration Fraction. The filtration fraction is the fraction of the renal plasma flow that becomes glomerular filtrate. Since the normal plasma flow through both kidneys is 650 ml/min and the normal glomerular filtration rate in both kidneys is 125 milliliters, *the average filtration fraction is approximately ¹/₅ or 19 per cent.*

Dynamics of Filtration Through the Glomerular Membrane

The same forces that cause fluid to filter from any high pressure capillary also apply to filtration from the glomerulus into Bowman's capsule. Thus, the forces are as follows:

1. The *pressure inside the glomerular capillaries* promotes filtration through the glomerular membrane.

2. The *pressure in Bowman's capsule* outside the capillaries opposes filtration.

3. The *colloid osmotic pressure of the plasma proteins* also opposes filtration.

4. The *colloid osmotic pressure of the proteins in Bowman's capsule* promotes filtration; however, so little protein normally filters into the glomerular filtrate that this factor usually has no significant effect and is actually considered to be zero.

Glomerular Pressure. The glomerular pressure is the average pressure in the glomerular capillaries. This unfortunately has been measured directly only in one mammal, the rat, in which the average value is about 45 mm Hg. However, from various indirect measurements it has been calculated to be from 55 to 70 mm Hg in the dog. Because humans are large mammals, *a reasonable average value can be considered to be 60 mm Hg,* though, as noted later, this can increase or decrease considerably under varying conditions.

Pressure in Bowman's Capsule. In lower animals, pressure measurements actually have been made in Bowman's capsule and at different points along the renal tubules by inserting micropipets into the lumen. On the basis of these studies, *capsular pressure in the human being is estimated to be about 18 mm Hg.*

Colloid Osmotic Pressure in the Glomerular Capillaries. Because approximately one fifth of the plasma in the capillaries filters into the capsule, the protein concentration increases about 20 per cent as the blood passes from the arterial to the venous ends of the glomerular capillaries. If the normal colloid osmotic pressure of blood entering the capillaries is 28 mm Hg, it rises to approximately 36 mm Hg by the time the blood reaches the venous ends of the capillaries, and the average colloid osmotic pressure is about 32 mm Hg.

Filtration Pressure, Filtration Coefficient, and Glomerular Filtration Rate. The *filtration pressure* is the net pressure forcing fluid through the glomerular membrane, and this is equal to the *glomerular pressure* minus *the sum of glomerular colloid osmotic pressure and capsular pressure* (that is, FP = GP − GCP − CP). In Figure 26–7A, *the normal filtration pressure is shown to be about 10 mm Hg.*

The *filtration coefficient*, called K_f, is defined as the glomerular filtration rate in both kidneys per millimeter of mercury of filtration pressure. That is, the glomerular filtration rate is equal to the filtration pressure times the filtration coefficient, or

$$GFR = \text{Filtration pressure} \cdot K_f$$

The normal filtration coefficient is calculated to be about 12.5 ml/min/mm Hg of filtration pressure. Thus, at a normal mean filtration pressure of 10 mm Hg, the total filtration rate of both kidneys is 125 ml/min.

Factors That Affect the Glomerular Filtration Rate

It is clear from the previous discussion that, aside from the filtration coefficient, the other factors that determine the glomerular filtration rate are (1) glomerular pressure, (2) plasma colloid osmotic pressure, and (3) Bowman's capsule pressure. Some of the important conditions that affect these, and therefore also affect glomerular filtration rate, are the following:

Effect of Renal Blood Flow on Glomerular Filtration Rate. An increase in the rate of blood flow through the nephrons greatly elevates the glomerular filtration rate. One of the reasons for this is that the increasing flow elevates the glomerular pres-

sure, which obviously enhances filtration. However, a second reason that is not quite so easily understood is the following: At normal renal blood flow, about 20 per cent of the plasma is filtered through the glomerular membrane; therefore, the concentration of the plasma proteins becomes considerably increased before the blood leaves the glomerulus and, consequently, the colloid osmotic pressure increases. This rising colloid osmotic pressure exerts a strong influence on reducing further filtration. Now let us see what effect increased blood flow will have on this. As the flow increases, greater quantities of plasma enter the glomeruli, so that filtration of fluid from the plasma causes a smaller percentage increase in protein concentration and colloid osmotic pressure. Therefore, the colloid osmotic pressure rises much less and now exerts far less inhibitory influence on glomerular filtration. Consequently, even when the glomerular pressure remains constant, the greater the rate of blood flow into the glomerulus, the greater the glomerular filtration rate.

Effect of Afferent Arteriolar Constriction on Glomerular Filtration Rate. Afferent arteriolar constriction decreases the rate of blood flow into the glomerulus and also decreases the glomerular pressure; both of these effects decrease the filtration rate. This effect is illustrated in Figure 26–7B. Conversely, dilatation of the afferent arteriole increases the glomerular filtration rate.

Effect of Efferent Arteriolar Constriction on Glomerular Filtration Rate. Constriction of the efferent arteriole increases the resistance to outflow from the glomeruli. This obviously increases the glomerular pressure and at small increases in efferent resistance causes a slight increase in glomerular filtration rate, as illustrated in Figure 26–7C. *However,* the blood flow decreases at the same time, and if the increase in efferent arteriolar constriction is moderate or severe, the plasma will remain for a longer period of time in the glomerulus, and extra large portions of plasma will filter out. This will increase the plasma colloid osmotic pressure to excessive levels, which will cause a paradoxical decrease in the glomerular filtration rate despite the elevated glomerular pressure.

Figure 26–7. *A,* Normal pressures at different points in the nephron and the normal filtration pressure. *B,* Effect of afferent arteriolar constriction on pressures in the nephron and on filtration pressure. *C,* Effect of efferent arteriolar constriction on pressures in the nephron and on filtration pressure.

● CONTROL OF THE GLOMERULAR FILTRATION RATE AND RENAL BLOOD FLOW

For the most part, renal blood flow and glomerular filtration rate are controlled together by local feedback control mechanisms within the kidney that keep both flow and filtration rate at very constant levels. Therefore, it is said that glomerular filtration and blood flow are both "autoregulated." The mechanisms of these autoregulation processes are the following:

AUTOREGULATION OF THE GLOMERULAR FILTRATION RATE

The glomerular filtration rate normally remains very nearly constant hour after hour, usually varying very little either above or below the normal value of about 125 ml/min for the two kidneys. Even a change in arterial pressure from as little as 75 mm Hg to as high as 160 mm Hg hardly changes the glomerular filtration rate. This effect is illustrated in Figure 26–8, and it is called *autoregulation of glomerular filtration rate.*

Why Is It Important for the Glomerular Filtration Rate to Be Autoregulated? To understand the importance of maintaining a constant glomerular filtration rate, let us consider what would happen if the glomerular filtration rate should become, first, very slight or, second, very great.

At a very slight glomerular filtration rate, the tubular fluid would pass through the tubules so slowly that essentially all of it would be reabsorbed. Therefore, the kidney would fail to eliminate the necessary waste products.

At the other extreme, with a much too high glomerular filtration rate, the fluid would pass so rapidly through the tubules that they would be unable to reabsorb those substances that need to be conserved in the body.

Thus, one can readily see that the glomerular fil-

trate must flow into the tubular system at an appropriate rate (1) to allow the unwanted substances to pass on into the urine while (2) reabsorbing the wanted substances. However, it is often not appreciated how narrow the range of glomerular filtration rate must be if optimal function of the tubular system is to occur. To emphasize this narrow range, analyses of tubular function have shown that even a 5 per cent too great or too little rate of glomerular filtration can have considerable effects in causing either excess loss of solutes and water into the urine or, at the other extreme, too little of the necessary excretion of waste products.

Mechanism of Autoregulation of the Glomerular Filtration Rate— Tubuloglomerular Feedback

The precision with which glomerular filtration rate must be autoregulated demands that there be a highly efficient control system for controlling this filtration rate. Fortunately, each nephron is provided with not one but *two* special feedback mechanisms from the distal tubule to the periglomerular arterioles; these add together to provide the degree of glomerular filtration autoregulation that is required. These two mechanisms are (1) an *afferent arteriolar vasodilator feedback mechanism* and (2) an *efferent arteriolar vasoconstrictor feedback mechanism.* The combination of these is called *tubuloglomerular feedback.* The feedback process occurs either entirely or almost entirely at the *juxtaglomerular complex,* which has the following characteristics:

The Juxtaglomerular Complex. Figure 26–9 illustrates the juxtaglomerular complex, showing that the initial portion of the distal tubule, immediately after the upper end of the thick segment of the ascending limb of the loop of Henle, passes in the angle between the afferent and efferent arterioles, actually abutting each of these two arterioles. Furthermore, those epithelial cells of the distal tubule that come in contact with the arterioles are more dense than the other tubular cells and are collectively called the *macula densa.* The macula densa cells appear to secrete some substance toward the arterioles because the Golgi apparatus, an intracellular secretory organelle, is directed toward the arterioles and not toward the lumen of the tubule, in contrast to all the other tubular epithelial cells. Note also in Figure 26–9 that the smooth muscle cells of both the afferent and efferent arterioles are swollen and contain dark granules where they come in contact with the macula densa. These cells are called *juxtaglomerular cells,* and the granules are composed mainly of inactive *renin.* The whole complex of macula densa and juxtaglomerular cells is called the *juxtaglomerular complex.*

Thus, the anatomical structure of the juxtaglomerular apparatus suggests strongly that the fluid in the distal tubule in some way plays an important role in helping to control nephron function by providing feedback signals to both the afferent and the efferent arterioles.

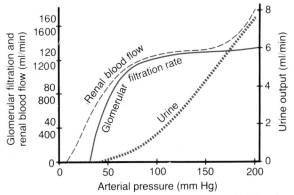

Figure 26–8. Autoregulation of glomerular filtration rate (GFR) and renal blood flow (RBF) when the arterial pressure is increased, but there is lack of autoregulation of urine flow.

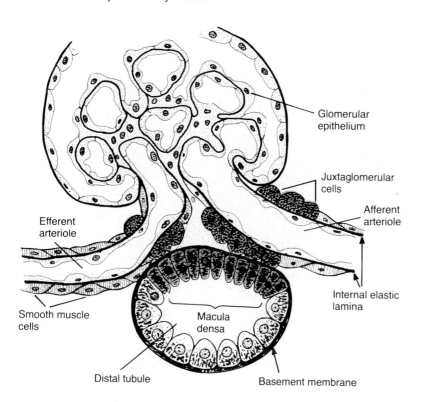

Glomerular
epithelium

Juxtaglomerular
cells

Afferent
arteriole

Efferent
arteriole

Internal elastic
lamina

Smooth muscle
cells

Macula
densa

Distal tubule

Basement membrane

Figure 26–9. Structure of the juxtaglomerular apparatus, illustrating its possible feedback role in the control of nephron function. (Modified from Ham: Histology. Philadelphia, J. B. Lippincott Company, 1971.)

The Afferent Arteriolar Vasodilator Feedback Mechanism

A low rate of glomerular filtration causes over-reabsorption of sodium and chloride ions in the ascending limb of the loop of Henle and therefore decreases the ionic concentration at the macula densa. This decrease in ions in turn initiates a signal from the macula densa to dilate the afferent arteriole. Putting these two facts together, the following is the postulated mechanism by which the afferent arteriolar vasodilator feedback mechanism controls glomerular filtration rate:

1. Too little flow of glomerular filtrate into the tubules causes decreased sodium and chloride ion concentration at the macula densa.

2. The decreased ionic concentration causes afferent arteriolar dilatation.

3. This in turn allows increased blood flow into the glomerulus, which increases the glomerular pressure.

4. The increased glomerular pressure as well as the increased glomerular blood flow increases the glomerular filtration rate back toward the required level.

Thus, this is a typical negative feedback mechanism for controlling the glomerular filtration rate at a steady rate. This mechanism also helps autoregulate the renal blood flow at the same time, as we discuss subsequently.

The Efferent Arteriolar Vasoconstrictor Feedback Mechanism

Too few sodium and chloride ions at the macula densa are believed also to cause the juxtaglomerular cells to release active renin, and this in turn causes

formation of angiotensin. The angiotensin then constricts mainly the efferent arteriole because it is highly sensitive to angiotensin II, much more so than is the afferent arteriole.

With these facts in mind, we can now describe the efferent arteriolar vasoconstrictor mechanism that helps maintain a constant glomerular filtration rate:

1. A too low glomerular filtration rate causes excess reabsorption of sodium and chloride ions in the ascending limb of the loop of Henle, reducing the ionic concentration at the macula densa.

2. The low concentration of ions then causes the juxtaglomerular cells to release renin from their granules.

3. The renin causes formation of angiotensin II.

4. The angiotensin II constricts the efferent arterioles, which causes the pressure in the glomerulus to rise.

5. The increased pressure then causes the glomerular filtration rate to return back toward normal.

Thus, this is still another negative feedback mechanism that helps to maintain a very constant glomerular filtration rate; it does so by constricting the efferent arterioles at the same time that the afferent vasodilator mechanism described previously dilates the afferent arterioles. When both these mechanisms function together, the glomerular filtration rate increases only a few per cent even though the arterial pressure changes between the limits of 75 and 160 mm Hg.

AUTOREGULATION OF RENAL BLOOD FLOW

Not only is the glomerular filtration rate autoregulated at a very constant level, but the renal blood flow also remains near the normal level despite considerable change in the arterial pressure; this is called *autoregulation of renal blood flow.* For instance, the renal blood flow of both kidneys remains very near the normal level of 1200 ml/min even when the arterial pressure falls as low as 75 mm Hg or rises to as high as 160 mm Hg. This, too, is illustrated in Figure 26–8.

The most important mechanism for autoregulating the renal blood flow is probably the first of the two mechanisms that autoregulates glomerular filtration rate, the *afferent arteriolar vasodilator mechanism.* That is, when the renal blood flow falls too low, this at first decreases the glomerular filtration rate. The falling glomerular filtration in turn causes a feedback effect at the juxtaglomerular complex to dilate the afferent arteriole, as explained earlier. This then allows increased blood flow into the glomerulus, thus returning both the renal blood flow and the glomerular filtration rate back toward normal. It should be recognized, however, that this mechanism is primarily concerned with autoregulating glomerular filtration rate, which is a necessary condition for normal function of the nephron. The autoregulation of renal blood flow is merely incidental to the autoregulation of the filtration.

Some physiologists believe that still another mechanism is important in autoregulating the renal blood flow. This is the so-called *myogenic mechanism,* which has been postulated as follows: When the arterial pressure rises, it stretches the wall of the arteriole, which in turn causes a secondary contraction of the arteriole. This decreases the renal blood flow back toward normal, thus opposing the effect of the rising arterial pressure to increase the flow. Unfortunately, this mechanism has not yet been proved, and there are many reasons for believing that it would be inherently unstable, so that some physiologists do not believe that such a mechanism truly does help in autoregulating the renal blood flow.

EFFECT OF ARTERIAL PRESSURE ON URINARY OUTPUT—THE PHENOMENON OF "PRESSURE DIURESIS"

Because of the autoregulation phenomenon, increasing the arterial pressure between the limits of 75 mm Hg and 160 mm Hg usually has only a slight effect on renal blood flow and glomerular filtration rate. However, the very slight increase in glomerular filtration rate that does occur is nevertheless enough to cause marked increase in urinary output, an effect that is illustrated by the dotted curve in Figure 26–8. This curve shows that a decrease in the mean arterial pressure from its normal value of about 100 mm Hg down to about 50 mm Hg causes complete cessation of urine output, whereas an increase in arterial pressure to double normal (to 200 mm Hg) increases the urine output as much as sevenfold to eightfold. As we study tubular reabsorptive mechanisms in the following chapters, it becomes clear that tubular reabsorption does not necessarily increase when the arterial pressure rises. Therefore, all or most of the increase in glomerular filtration becomes also an increase in urinary output. This large effect that arterial pressure has on urinary output is called *pressure diuresis.*

EFFECT OF SYMPATHETIC STIMULATION ON RENAL BLOOD FLOW AND THE GLOMERULAR FILTRATION RATE

The sympathetic nerves innervate both the afferent arterioles and the efferent arterioles and to a lesser extent some of the renal tubules as well. Even so, slight to moderate degrees of sympathetic stimulation usually have only mild effects on either renal blood flow or glomerular filtration. The reason for this is probably that the renal autoregulation mechanisms override the nervous stimuli.

On the other hand, very strong *acute* sympathetic stimulation can constrict the renal arterioles so greatly that renal blood flow decreases temporarily to as little as 10 to 30 per cent of normal, and urine output may decrease all the way to zero. Yet, if this strong sympathetic stimulation is continued, renal blood flow, glomerular filtration, and urine output all recover most of the way to normal within 20 to 30 minutes despite continued stimulation. Part of this recovery results from diminishing release of sympathetic neurotransmitter at the sympathetic nerve endings. However, internal adjustments of ions and other solutes in the kidneys, as well as hormonal changes discussed in subsequent chapters, probably also play significant roles in this.

● REABSORPTION OF FLUID BY THE PERITUBULAR CAPILLARIES

As the glomerular filtrate passes through the renal tubular system, the tubular epithelium reabsorbs more than 99 per cent of the water in the filtrate as well as large quantities of electrolytes and other substances. This absorbed fluid passes first into the interstitium and is absorbed from there into the peritubular capillaries, thus returning to the blood, as illustrated in Figure 26–10.

Note in the figure that the normal peritubular capillary pressure averages about 13 mm Hg, whereas the renal interstitial fluid pressure averages about 6 mm Hg. Thus, there is a positive pressure gradient from the peritubular capillary to the interstitial fluid of 7 mm Hg that opposes fluid reabsorption. However, this is more than overbalanced by the colloid osmotic pressures, with −32 mm Hg in the plasma and −15 mm Hg in the interstitium, giving a net −17 mm Hg osmotic force to reabsorb the interstitial fluid into the capillary blood. Combining the 7 mm Hg pressure gradient and the −17 mm Hg colloid osmotic pressure gradient, the result is −10 mm Hg net absorption pressure. It is this large absorption pressure that causes continual reabsorption by the peritubular capillaries of the large amounts of fluid entering the interstitium from the renal tubules.

Effect of Peritubular Absorption on Reabsorption Through the Renal Tubular Epithelium. Figure 26–10 illustrates *active* absorption of sodium plus *osmotic* absorption of water into the intercellular channels between the tubular epithelial cells. Once this fluid enters the intercellular channels, it can then move in either of two directions. First, it can continue toward the bases of the epithelial cells, then through the basement membrane into the interstitium and finally into the peritubular capillary. Second, because

Figure 26–10. Absorption of fluid at the peritubular capillary membrane and the effect of this on the absorption of water and sodium through the tubular epithelium.

the so-called tight junctions (zona occludens) between the epithelial cells are quite leaky, the fluid in the intercellular channels can diffuse or flow through these junctions backward into the tubular lumen. Which one of these directions the fluid will actually flow depends to a great extent on the absorptive forces of the peritubular capillary. This is especially true for the proximal tubules, where the so-called tight junctions are very, very leaky. When the capillary forces cause absorption of interstitial fluid into the peritubular capillaries, the interstitial pressure decreases, which promotes very rapid flow of fluid from the intercellular channels in the direction of the blood. Conversely, when the capillary forces do not favor absorption of fluid, the interstitial fluid pressure rises, and much of the fluid then returns in the backward direction through the "tight" junctions back into the tubular lumen from which it had just been absorbed. Therefore, *net* tubular reabsorption is greatly reduced anytime the fluid in the interstitium fails to be absorbed adequately into the peritubular capillaries.

Therefore, one of the important factors controlling fluid absorption from the renal tubules, especially from the proximal tubules, is the net absorption pressure at the peritubular capillary membrane. For instance, when the peritubular plasma protein concentration is high, increased quantities of fluid will be reabsorbed from the proximal tubules. But, conversely, when the peritubular capillary blood pressure is high, the forces shift in the opposite direction, and decreased reabsorption occurs.

REFERENCES

Andersson, B.: Regulation of body fluids. Annu. Rev. Physiol., 39:185, 1977.
Andreoli, T. E., and Schaefer, J. A.: Mass transport across cell membranes: The effects of antidiuretic hormone on water and solute flows in epithelia. Annu. Rev. Physiol., 38:451, 1976.
Andreoli, T. E., et al. (eds.): Clinical Disorders of Membrane Transport Processes. New York, Plenum Publishing Corp., 1986.
Andreoli, T. E., et al. (eds.): Membrane Transport Processes in Organized Systems. New York, Plenum Publishing Corp., 1986.
Andreoli, T. E., et al. (eds.): Physiology of Membrane Disorders, 2nd Ed. New York, Plenum Publishing Corp., 1986.
Aronson, P. S.: The renal proximal tubule: A model for diversity of anion exchangers and stilbene-sensitive anion transporters. Annu. Rev. Physiol., 51:419, 1989.
Barger, A. C., and Herd, J. A.: Renal vascular anatomy and distribution of blood flow. In Orloff, F., and Berliner, R. W. (eds.): Handbook of Physiology. Sec. 8. Baltimore, Williams & Wilkins, 1973, p. 249.
Baylis, C., and Blantz, R. C.: Glomerular hemodynamics. News in Physiol. Sci., 1:86, 1986.
Beeuwkes R., III: The vascular organization of the kidney. Annu. Rev. Physiol, 42:531 1980.
Berry, C. A.: Heterogeneity of tubular transport processes in the nephron. Annu. Rev. Physiol., 44:181, 1982.
Bertani, T., et al.: Drugs and Kidney. Vol. 33. New York, Raven Press, 1986.
Brenner, B. M., and Berliner, R. W.: Transport of potassium. In Orloff, F., and Berliner, R. W. (eds.): Handbook of Physiology. Sec. 8. Baltimore, Williams & Wilkins, 1973, p. 497.
Brenner, B. M., and Rector, F. C., Jr.: The Kidney, 3rd Ed. Philadelphia, W. B. Saunders Co., 1986.
Brown, D.: Vesicle recycling and cell-specific function in kidney epithelial cells. Annu. Rev. Physiol., 51:771, 1989.
Bulger, R. E., and Dobyan, D. C.: Recent advances in renal morphology. Annu. Rev. Physiol., 44:147, 1982.
Burg, M. B.: Origins of isolated tubule microperfusion methodology. News Physiol. Sci., 3:176, 1988.
Burg, M., and Good, D.: Sodium chloride coupled transport in mammalian nephrons. Annu. Rev. Physiol., 45:533, 1983.
Charlton, C. A.: The Urological System. New York, Churchill Livingstone, 1983.
Edwards, R. M.: Direct assessment of glomerular arteriole reactivity. News Physiol. Sci., 3:216, 1988.
Giebisch, G., and Windhager, E.: Electrolyte transport across renal tubular membranes. In Orloff, F., and Berliner, R. W. (eds.): Handbook of Physiology. Sec. 8. Baltimore, Williams & Wilkins, 1973, p. 315.
Glynn, I. M., and Krlish, S. J. D.: The sodium pump. Annu. Rev. Physiol., 37:13, 1975.
Gottschalk, C. W., and Lassiter, W. E.: Micropuncture methodology. In Orloff, F., and Berliner, R. W. (eds.): Handbook of Physiology. Sec. 8, Baltimore, Williams & Wilkins, 1973, p. 129.
Greger, R.: Chloride transport in thick ascending limb, distal convolution, and collecting duct. Annu. Rev. Physiol., 50:111, 1988.
Grimellec, C. L., et al.: Lipid asymmetry and transport function in renal epithelial cells. News Physiol. Sci., 3:227, 1988.
Guyton, A. C., et al.: Dynamics and Control of the Body Fluids. Philadelphia, W. B. Saunders Co., 1975.
Karniski, L. P., and Aronson, P. S.: Formate: A critical intermediate for chloride transport in the proximal tubule. News Physiol. Sci., 2:160, 1987.
Knox, F. G., and Granger, J. P.: Control of sodium excretion: The kidney produces under pressure. News Physiol. Sci., 2:26, 1987.
Knox, F. G., and Spieman, W. S.: Renal circulation. In Shepherd, J. T., and Abboud, F. M. (eds.): Handbook of Physiology. Sec. 2, Vol. III. Bethesda, Md., American Physiological Society, 1983, p. 183.
Koeppen, B. M., et al.: Electrophysiology of mammalian renal tubules: Inferences from intracellular microelectrode studies. Annu. Rev. Physiol., 45:497, 1983.
Kumar, R., et al.: Ca²⁺-Mg²⁺-ATPase calcium pumps in the kidney. News in Physiol. Sci., 3:219, 1988.
Latta, H.: Ultrastructure of the glomerulus and juxtaglomerular apparatus. In Orloff, F., and Berliner, R. W. (eds.): Handbook of Physiology. Sec. 8. Baltimore, Williams & Wilkins, 1973, p. 1.
Lauger, P.: Dynamics of ion transport systems in membranes. Physiol. Rev., 67:1296, 1987.
Mandel, L. J.: Metabolic substates, cellular energy production, and the regulation of proximal tubular transport. Annu. Rev. Physiol., 47:85, 1985.
Massry, S. G., and Gassock, R. J. (eds.): Textbook of Nephrology. Baltimore, Williams & Wilkins, 1988.
Morel, F., and Doucet A.: Hormonal control of kidney functions at the cell level. Physiol. Rev., 66:377, 1986.
Morel, R., et al.: Distribution of hormone-dependent adenylate cyclase in the nephron and its physiological significance. Annu. Rev. Physiol., 43:569, 1981.
Philipson, K. D.: Sodium-calcium exchange in plasma membrane vesicles. Annu. Rev. Physiol., 47:561, 1985.
Pinter, G. G.: Renal lymph: Vital for the kidney and valuable for the physiologist. News Physiol. Sci., 3:189, 1988.
Post, R. L.: Seeds of sodium, potassium ATPase. Annu. Rev. Physiol., 51:1, 1989.
Ross, C. R., and Holohan, P. D.: Transport of organic anions and cations in isolated renal plasma membranes. Annu. Rev. Pharmacol. Toxicol., 23:65, 1983.

Schafer, J. A., and Williams, J. C., Jr.: Transport of metabolic substrates by the proximal nephron. Annu. Rev. Physiol., 47:103, 1985.

Schafer, J. A.: Fluid absorption in the kidney proximal tubule. News in Physiol. Sci., 2:22, 1987.

Schild, L., et al.: Chloride transport in the proximal renal tubule. Annu. Rev. Physiol., 50:97, 1988.

Schoolwerth, A. C., and LaNoue, K. F.: Transport of metabolic substrates in renal mitochondria. Annu. Rev. Physiol., 47:143, 1985.

Silbernagl, S.: Tubular reabsorption of amino acids in the kidney. News Physiol. Sci., 1:167, 1986.

Silbernagl, S.: The renal handling of amino acids and oligopeptides. Physiol. Rev., 68:911, 1988.

Smith H. W.: The Kidney: Structure and Function in Health and Disease. New York, Oxford University Press, 1951.

Soltoff, S. P.: ATP and the regulation of renal cell function. Annu. Rev. Physiol., 48:9, 1986.

Ulrich, K. L.: Sugar, amino acids, and Na^+ cotransport in the proximal tubule. Annu. Rev. Physiol., 41:181, 1979.

Vander, A. J.: Renal Physiology. New York, McGraw-Hill, 1980.

Walker, L. A., and Valtin, H.: Biological importance of nephron heterogeneity. Annu. Rev. Physiol., 44:203, 1982.

Wirthensohn, G., and Guder, W. G.: Renal substrate metabolism. Physiol. Rev., 66:469, 1986.

CHAPTER 27

• *Formation of Urine by the Kidney: II. Processing of the Filtrate in the Tubules* •

REABSORPTION AND SECRETION IN THE TUBULES

The glomerular filtrate entering the tubules of the nephron flows (1) through the *proximal tubule,* (2) through the *loop of Henle,* (3) through the *distal tubule,* (4) through the *cortical collecting duct,* and (5) through the *collecting duct* into the pelvis of the kidney. Along this course, substances are selectively reabsorbed or secreted by the tubular epithelium, and the resultant fluid after this processing enters the renal pelvis as *urine.* Reabsorption plays a much greater role than does secretion in this formation of urine, but secretion is especially important in determining the amounts of potassium ions, hydrogen ions, and a few other substances in the urine, as discussed later.

Ordinarily, more than 99 per cent of the water in the glomerular filtrate is reabsorbed as it is processed in the tubules. Therefore, if some dissolved constituent of the glomerular filtrate is not reabsorbed at all along the entire course of the tubules, this reabsorption of water obviously concentrates the substance more than 99-fold. On the other hand, some constituents, such as glucose and amino acids, are reabsorbed almost entirely, so that their concentrations decrease almost to zero before the fluid becomes urine. In this way the tubules separate substances that are to be conserved in the body from those that are to be eliminated in the urine, and they do this without losing much water in the urine.

The basic mechanisms for transport through the tubular membrane are essentially the same as those discussed in Chapter 4 for transport through other membranes of the body. These can be divided into *active transport* and *passive transport.*

ACTIVE TRANSPORT THROUGH THE TUBULAR MEMBRANE

As explained in Chapter 4, there are two basic mechanisms of active transport, *primary active transport* and *secondary active transport.* These two types can best be explained by considering actual examples.

Primary Active Transport of Sodium Ions Through the Tubular Membrane — Function of Na+,K+-ATPase

The basic mechanism for transporting sodium ions through the tubular membrane, which always occurs in the direction from the tubular lumen to the interstitium, is illustrated in Figure 27–1A. On the basal and lateral surfaces of the tubular epithelial cell, the cell membrane contains an extensive Na+,K+-ATPase system that is capable of cleaving adenosine trisphosphate (ATP) and using the released energy to transport sodium ions out of the cell into the interstitium, while at the same time transporting potassium ions from the interstitium to the interior of the cell. From Chapter 4 it should also be recalled that this ATPase system pumps three sodium ions for every two potassium ions pumped. However, the basolateral sides of the tubular epithelial cell are so permeable to potassium that virtually all of the potassium immediately diffuses back out of the cell into the interstitium. Therefore, the net effect, illustrated in Figure 27–1A, is to pump so much sodium out that this reduces the sodium inside the cell down to a very low concentration. Also, because three positive electrical charges are pumped out of the cell along with the sodium ions, the interior of the cell is left with a very negative

Figure 27–1. *A,* Basic mechanism for active transport of sodium through the tubular epithelial cell. This figure shows active transport by the sodium-potassium pump, which pumps sodium out of the basolateral membrane of the cell and simultaneously creates a very low intracellular sodium concentration as well as a negative intracellular potential. The low intracellular sodium concentration and the negative potential then cause sodium ions to diffuse from the tubular lumen into the cell through the brush border. *B,* The net mechanism for active transport of sodium from the tubular lumen all the way into the peritubular capillary.

potential of about -70 millivolts. Therefore, two factors now cause sodium ions to *diffuse* through the cell's *luminal* membrane *from the tubular lumen to the interior* of the cell: (1) the very large sodium concentration gradient across the membrane, with high sodium concentration in the tubular lumen and low sodium concentration inside the cell, and (2) the attraction of the positive sodium ions in the tubular lumen to the interior of the cell by the -70 millivolts intracellular potential.

On the tubular border of the epithelial cell is an extensive brush border that multiplies the surface area of luminal exposure about 20-fold. In the membrane of this brush border are *sodium carrier proteins* that bind with the sodium ions on the luminal surface of the membrane and release them inside the cell, thus providing *facilitated diffusion* of the sodium to the interior of the cell. This ensures rapid diffusion of sodium through the luminal border of the epithelial cell at the same time that the sodium ion is being actively transported out of the cell at the basolateral borders. Sodium carrier proteins in the brush border also play a very important role in secondary active transport, as explained subsequently.

Thus, the net mechanism for transport of sodium through the tubular epithelial wall is illustrated in Figure 27–1B, which shows that the sodium pumped from the tubule is eventually absorbed into the peritubular capillary and carried away by the blood.

Although by far most of the primary active transport in the entire tubular system is to transport sodium, primary active transport also occurs for a few other substances as well. For instance, active *secretory* transport of hydrogen ions occurs in some of the distal portions of the tubular system, as we discuss later, and

active *absorptive* transport of calcium occurs in other portions of the tubular system.

Secondary Active Absorption From the Tubular Lumen

In secondary active transport, no energy is used directly from ATP or any other high-energy phosphate source. Instead, the movement of sodium ions from the tubular lumen to the interior of the cell energizes most of the secondary transport of other substances. This is achieved by multiple different types of sodium carrier proteins in the epithelial cell brush border. For instance, in Figure 27–2, the uppermost cell illustrates secondary active transport of glucose, and the second cell illustrates secondary active transport of amino acid ions. In each case, the carrier protein in the brush border membrane combines with both the substance to be transported and a sodium ion at the same time. As the sodium moves down its electrochemical gradient to the interior of the cell, it pulls glucose or the amino acid ion along with it. Usually, each type of carrier protein is specific for transport of each different substance or for a class of substances. This type of secondary active transport, in which the sodium ion pulls a substance along with it through the membrane, is called *co-transport.*

Glucose, amino acids, and several other organic compounds are especially strongly co-transported in the proximal tubules. Chloride ions are co-transported mainly in the thick segment of the ascending limb of the loop of Henle. Other substances also co-transported at some point in the tubular system include phosphate, calcium, magnesium, and hydrogen ions. We speak more about these later.

Co-transport

Counter Transport

Figure 27–2. Mechanisms of secondary active transport. The upper two cells show *co-transport* of glucose and amino acids through the brush border of the epithelial cells along with sodium ions, followed by facilitated diffusion through the basolateral membranes. The third cell shows *counter-transport* of hydrogen ions through the brush border into the tubular lumen; inward movement of sodium ions provides the energy for outward movement of the hydrogen ions.

Once the glucose, amino acid, or other substance has been co-transported from the tubular lumen into the epithelial cell, it then usually moves through the basolateral border of the cell by facilitated diffusion in conjunction with another carrier protein.

Secondary Active Secretion Into the Tubules

A few substances are secondarily actively *secreted* into the tubules. In general, the process is the reverse of that discussed in the previous paragraphs for secondary absorption. Some of the important substances that are secreted this way in specific portions of the tubules are hydrogen ions, potassium ions, and urate ions. As an example, the lowermost cell in Figure 27–2 illustrates secondary active secretion of hydrogen

ions into the proximal tubules. To achieve this, a hydrogen ion inside the epithelial cell binds with the carrier protein in the brush border, while a sodium ion in the tubular lumen binds with the other end of the same carrier protein. Then as the sodium ion moves inward, the hydrogen ion is forced outward in the opposite direction. For obvious reasons, this is called *counter-transport.*

To understand more fully the principles of primary and secondary active transport, the reader is also referred to more detailed discussions in Chapter 4.

PASSIVE ABSORPTION OF WATER: OSMOSIS THROUGH THE TUBULAR EPITHELIUM

When the different solutes are transported out of the tubule by either primary or secondary active transport, their total concentration decreases inside the tubular lumen but increases in the interstitium. This obviously creates a concentration difference that causes osmosis of water in the same direction that the solutes have been transported.

A large share of this osmosis occurs through the so-called tight junctions between the epithelial cells rather than through the cells themselves. The reason for this is that these junctions are not nearly so tight as their name would imply but instead allow rapid diffusion of water and many other small ions. This is especially true of the proximal tubules, where the "tight" junctions are very, very loose. As solutes are absorbed through the proximal tubular epithelial cells, this increases the osmolality in the interstitium and causes almost instantaneous osmosis of a commensurate volume of water to go along with the solutes. Therefore, in effect, the fluid absorbed from the proximal tubules is almost exactly isosmotic.

In the more distal parts of the tubular system, beginning in the loop of Henle and extending through the remaining tubules, the "tight" junctions are much tighter, and the epithelial cells also have much less extensive membrane surfaces. Therefore, in general, the latter portions of the tubular system are far less permeable than are the proximal tubules.

PASSIVE ABSORPTION OF CHLORIDE IONS, UREA, AND OTHER SOLUTES BY THE PROCESS OF DIFFUSION

When sodium ions are transported through the tubular epithelial cell, a negative ion, such as the *chloride ion,* is usually transported along with each sodium ion to maintain electrical neutrality. Earlier it was pointed out that in some segments of the tubules, chloride ions can be transported by secondary active transport. However, in most tubular segments, chloride ions are transported mainly by passive diffusion. This occurs especially through the "tight" junctions of the proximal tubules but also to a less extent through the "tight" junctions of the later portions of the tubular system.

Urea is another substance that is passively reabsorbed but to much less degree than chloride ions. Actually, one of the principal functional purposes of the kidneys is *not* to reabsorb urea but instead to allow as much as possible of this waste product of metabolism to pass on into the urine. Unfortunately, though, the urea molecule is very small, and the tubules are partially permeable to it. Therefore, usually, as water is reabsorbed from the tubules, about half of the urea in the glomerular filtrate is passively reabsorbed by diffusing along with the water, while the other half passes on into the urine.

Another waste product of metabolism is *creatinine*. However, its molecule is somewhat larger than that of urea, so that virtually none of it is reabsorbed; instead, virtually all creatinine that is filtered in the glomerular filtrate passes on through the tubular system and is excreted in the urine.

ABSORPTIVE CAPABILITIES OF DIFFERENT TUBULE SEGMENTS

In subsequent chapters the absorption and secretion of specific substances in different segments of the tubular system are discussed. However, it is important first to point out basic differences between the absorptive and secretory capabilities of the different tubular segments.

Proximal Tubular Epithelium. Figure 27–3 illustrates the cellular characteristics of the tubular membrane in (1) the proximal tubule, (2) the thin segment of the loop of Henle, (3) the distal tubule, and (4) the collecting duct. The proximal tubular cells have the appearance of being highly metabolic cells, having large numbers of mitochondria to support extremely rapid active transport processes; true enough, one finds that *about 65 per cent of the glomerular filtrate normally is reabsorbed before reaching the loops of Henle.* As already described in relation to sodium transport, the proximal tubular epithelial cells have an extensive brush border. They also have a far-ranging labyrinth of intercellular and basal channels that provide an extensive membrane area on the interstitial fluid side of the epithelium, the side where abundant active transport of sodium ions occurs.

The extensive membrane surface of the epithelial cell brush border, on the other hand, is literally loaded with protein carrier molecules that promote either *co-transport* for absorbing certain substances from the tubular lumen into the interstitium or *counter-transport* for secreting other substances into the tubules. The most important substances that are specifically absorbed by secondary active transport in the proximal tubules are *glucose* and the *amino acids*. The most important substance that is secreted by secondary active transport is *hydrogen ions.*

Thin Segment of the Loop of Henle. The epithelium of the thin segment of the loop of Henle, as the name implies, is very thin. The cells have no brush border and very few mitochondria, indicating a minimal level of metabolic activity. The *descending portion* of this thin segment is highly permeable to water and moderately permeable to urea, sodium, and most other ions. Therefore, it appears to be adapted primarily for simple diffusion of substances through its walls.

The ascending portion of the thin segment, on the other hand, is believed to be different in one very important characteristic; it is believed to be far less permeable to water than is the descending portion. This difference is important for explaining the mechanism for concentrating urine, as we discuss later.

Thick Segment of the Loop of Henle. The thick segment of the loop of Henle begins about halfway up the ascending limb of the loop, where the epithelial cells become grossly thickened, as illustrated in Figure 26–2. This segment then ascends all the way back to the same glomerulus from which the tubule originated and passes snugly through the angle between the afferent and efferent arterioles, forming a complex with these arterioles called the *juxtaglomerular complex,* discussed in the previous chapter. Beyond this point, the tubule becomes the distal tubule.

The epithelial cells of the thick segment of the loop of Henle are similar to those of the proximal tubules except that they have a rudimentary brush border, fewer basal channels, and much tighter "tight" junctions where the cells attach to each other. The cells are especially adapted for strong active transport of sodium and chloride ions, transporting these from the tubular lumen into the interstitial fluid.

On the other hand, this thick segment is almost entirely impermeable to both water and urea. Therefore, even though over three quarters of all the ions in the tubular fluid are transported out of the thick segment into the interstitium, almost all the water and urea remain in the tubule. Thus, the ascending limb tubular fluid becomes very dilute except that its con-

Proximal tubule   Loop of Henle thin segment   Distal tubule   Collecting duct

Basement Membrane

Figure 27–3. Characteristics of the epithelial cells in different tubular segments.

centration of urea is high. This thick segment plays an exceedingly important role under different conditions in the renal mechanisms for diluting or concentrating the urine that is eventually formed by the kidney.

The Distal Tubule. The distal tubule begins at the juxtaglomerular complex, continuing from the thick segment of the ascending limb of the loop of Henle. The early part of the distal tubule is highly convoluted and eventually coalesces with several other distal tubules to form the cortical collecting duct that was described earlier.

The distal tubule is divided into two important functional segments, the *diluting segment* and the *late distal tubule.*

The Diluting Segment. The first half of the distal tubule has almost the same characteristics as the thick segment of the ascending limb of the loop of Henle. It absorbs most of the ions avidly but is almost entirely impermeable to both water and urea. Therefore, this diluting segment, too, contributes to the dilution of the tubular fluid in the same way as the thick segment of the ascending limb of the loop of Henle.

The Late Distal Tubule and Cortical Collecting Duct. The functional characteristics of the late distal tubule and the cortical collecting duct are similar to each other; even their lining epithelial cells are similar. Several important characteristics of these tubular segments are the following:

1. The epithelium of both is almost entirely impermeable to urea, the same as in the diluting segment of the distal tubule, so that essentially all the urea passes on into the collecting duct eventually to be excreted in the urine.

2. These two segments reabsorb *sodium ions* avidly, but the rate of this reabsorption is controlled to a very great extent by aldosterone, as we discuss in more detail in the following chapter. Simultaneously with the pumping of sodium out of the tubular lumen into the peritubular interstitium, *potassium ions* are transported in the opposite direction into the tubular lumen; this too is controlled by aldosterone and by several other factors, including the concentration of potassium ions in the body fluids. Thus, potassium ions are actively secreted into these tubular segments, and *it is mainly by this means that the potassium ion concentration is controlled in the extracellular fluids of the body.*

3. The late distal tubule and the cortical collecting duct also contain a special type of epithelial cell called *intercalated cells* or "brown cells" that secretes hydrogen ions by primary active secretion. This is quite different from the secondary active secretion of hydrogen ions that occurs in the proximal tubules and in some other parts of the tubular system. The intercalated cells can secrete hydrogen ions against a concentration gradient of as much as a thousand to one, in contrast to only a fewfold gradient for secondary active hydrogen ion secretion. Therefore, these intercalated cells play an absolutely essential role in the final high degrees of acidification that can occur in the urine.

4. The late distal tubule and the cortical collecting duct differ from the diluting segment in still another very important aspect: They are *permeable to water in the presence of antidiuretic hormone* but *impermeable when this hormone is absent,* providing a method for controlling the degree of dilution of the urine, a subject that we discuss in detail later. The collecting duct also shares this responsiveness to antidiuretic hormone.

The Collecting Duct. The epithelial cells of the collecting duct are nearly cuboidal in shape, with smooth surfaces, and contain relatively few mitochondria. This epithelium has two especially important characteristics for renal function:

1. The permeability of the collecting duct to water is controlled mainly by the level of antidiuretic hormone in the circulating blood, as just mentioned. When excessive amounts of antidiuretic hormone are available, water is reabsorbed into the medullary interstitium with great avidity, thus reducing the volume of urine and concentrating most of the dissolved substances in the urine. The collecting duct epithelium is slightly permeable to urea as well. Therefore, some urea is reabsorbed into the medullary interstitium. Most of this then diffuses back into the loop of Henle and is returned through the distal tubule once again to the collecting duct and is finally excreted.

2. The second important characteristic of the collecting duct epithelium is that it too is capable of secreting hydrogen ions against a very high hydrogen ion gradient. Therefore, as explained in Chapter 30, the late distal tubule and the collecting duct system play an exceedingly important role in controlling the acid-base balance of the body fluids.

REABSORPTION OF WATER IN DIFFERENT SEGMENTS OF THE TUBULES

Water transport occurs entirely by osmotic diffusion. That is, whenever some solute in the glomerular filtrate is absorbed either by active reabsorption or by diffusion caused by an electrochemical gradient, the resulting decreased concentration of solute in the tubular fluid and increased concentration in the interstitial fluid causes osmosis of water out of the tubules. Therefore, the volume of tubular fluid progressively diminishes along the tubular system.

Figure 27–4 shows the volumes of fluid that flow per minute at different points in the tubular system. In the two kidneys of the human being, the total fluid volumes flowing into each segment each minute (under normal resting conditions) are the following:

	ml/min
Glomerular filtrate	125
Flowing into the loops of Henle	45
Flowing into the distal tubules	25
Flowing into the collecting tubules	12
Flowing into the urine	1

Figure 27–4. Volume flow of fluid in each segment of the tubular system per minute. Note that the flow is plotted on a semilogarithmic scale, illustrating the tremendous difference in flow between the earlier and later segments of the tubules.

Figure 27–5. Composite figure showing average concentrations of different substances at different points in the tubular system.

From this chart one can also deduce the approximate per cent of the glomerular filtrate water that is reabsorbed in each segment of the tubules, as follows:

	Per cent
Proximal tubules	65
Loops of Henle	15
Distal tubules	10
Collecting ducts	9.3
Passing into the urine	0.7

We see later in this and the following chapters that some of these values vary greatly under different operational conditions of the kidney, particularly when the kidney is forming very dilute or very concentrated urine.

Concentrations of Different Substances at Different Points in the Tubules

Whether or not a substance will become concentrated in the tubular fluid is determined by the *relative degree of reabsorption of that substance* versus *the reabsorption of water*. If a greater percentage of water is reabsorbed, the substance becomes more concentrated. If a greater percentage of the substance is reabsorbed, it becomes more diluted.

Figure 27–5 illustrates the degree of concentration of most of the important substances in the different tubular segments. All the values in this figure are relative concentrations, with each substance considered to have a normal concentration of 1 in the glomerular filtrate, as shown at the left-hand side of the figure. As the filtrate moves along the tubular system, the concentration rises to progressively greater than 1 if more water is reabsorbed than solute, or the concentration becomes progressively less than 1 if more solute is reabsorbed than water. Also, if a substance is secreted

by the tubular epithelium into the tubule, this too will increase its concentration.

Those substances represented at the top of Figure 27–5 are obviously the ones that become highly concentrated in the urine. In general, these substances are not needed by the body, and the kidneys have become adapted not to reabsorb them either at all or only very slightly or even to secrete them into the tubules, thereby excreting especially great quantities into the urine.

On the other hand, those substances represented toward the bottom of the figure are all strongly reabsorbed; these are all substances that the body needs to conserve, and almost none of them are lost in the urine.

Table 27–1 summarizes the concentrating ability of the tubular system for the many different substances excreted in the urine. It also gives the normal quantities of the substances entering the proximal tubules each minute in the glomerular filtrate. The quantity of each of the substances entering each minute is called the *tubular load* of that substance per minute.

Reabsorption of Specific Substances at Different Points Along the Tubular System

Reabsorption of Substances of Nutritional Value to the Body — Glucose, Proteins, Amino Acids, Acetoacetate Ions, and Vitamins. Five different substances in the glomerular filtrate of particular importance to bodily nutrition are *glucose, proteins, amino acids, acetoacetate ions,* and *the vitamins.*

Table 27-1 RELATIVE CONCENTRATIONS OF SUBSTANCES IN THE GLOMERULAR FILTRATE AND IN THE URINE

	Glomerular Filtrate (125 ml/min)		Urine (1 ml/min)		Concentration in Urine/ Concentration in Plasma (Plasma clearance per min)
	Quantity/min	*Concentration*	*Quantity/min*	*Concentration*	
Na^+	17.7 mEq	142 mEq/liter	0.128 mEq	128 mEq/liter	0.9
K^+	0.63	5	0.06	60	12
Ca^{++}	0.5	4	0.0048	4.8	1.2
Mg^{++}	0.38	3	0.015	15	5.0
Cl^-	12.9	103	0.134	134	1.3
HCO_3^-	3.5	28	0.014	14	0.5
$H_2PO_4^-$ HPO_4^{--}	0.25	2	0.05	50	25
SO_4^{--}	0.09	0.7	0.033	33	47
Glucose	125 mg	100 mg/dl	0 mg	0 mg/dl	0
Urea	33	26	18.2	1820	70
Uric acid	3.8	3	0.42	42	14
Creatinine	1.4	1.1	1.96	196	140
Inulin	—	—	—	—	125
PAH	—	—	—	—	585

Normally, all of these are completely or almost completely *reabsorbed by active processes in the proximal tubules* of the kidney. Thus, Figure 27-5 shows that the concentrations of glucose, protein, and amino acids all decrease to the vanishing point before the tubular fluid has passed through the proximal tubules. Therefore, almost none of these substances remain in the fluid entering the loop of Henle.

Special Mechanism for Absorption of Protein. As much as 30 grams of plasma protein filter into the glomerular filtrate each day. This would be a great metabolic drain on the body if the protein were not returned to the body fluids. Because the protein molecule is much too large to be transported by the usual transport processes, *protein is absorbed through the brush border of the proximal tubular epithelium by pinocytosis,* which means simply that the protein attaches itself to the membrane, and this portion of the membrane then invaginates to the interior of the cell. Once inside the cell the protein is digested into its constituent amino acids, which are then absorbed by facilitated diffusion through the base and sides of the cell into the interstitial fluid. Details of the pinocytosis mechanism are discussed in Chapter 4.

Poor Reabsorption of the Metabolic End Products: Urea, Creatinine, and Others. Figure 27-5 also illustrates the concentrations of two major metabolic end products in the different segments of the tubular system — urea and creatinine. Only small quantities of *urea* are reabsorbed during the entire course through the tubular system. Yet, about 99.3 per cent of the water is reabsorbed. The removal of all this water therefore concentrates the urea about 65-fold.

Creatinine is not reabsorbed in the tubules at all; indeed, small quantities of creatinine are actually secreted into the tubules by the proximal tubules, so that the concentration of creatinine increases about 140-fold.

The *urate ion* is another end product of metabolism. It is reabsorbed more than urea, but even so, large quantities of urate still remain in the fluid that finally issues into the urine. Several other end products, such as *sulfates, phosphates,* and *nitrates,* are transported in much the same way as urate ions. These, too, are normally reabsorbed to a far less extent than is water, so that their concentrations become greatly increased as they flow along the tubules. Yet *each is actively reabsorbed to some extent,* which keeps their concentrations in the extracellular fluid from ever falling too low.

Reabsorption of Inulin and Para-Aminohippuric Acid by the Tubules. Note also in Figure 27-5 that when the substance *inulin,* a large polysaccharide, is infused into the blood and is then filtered in the glomerular filtrate, its concentration increases 125-fold by the time it reaches the urine. The cause of this is simply that inulin is neither reabsorbed nor secreted in any segment of the tubules, whereas all but 1 milliliter of the 125 milliliters of water in the glomerular filtrate is reabsorbed.

Also, Figure 27-5 shows that when *para-aminohippuric acid* (PAH) is infused into the blood and then excreted by the kidneys, its concentration increases 585-fold as the tubular fluid passes through the tubular system. This is because large quantities of PAH are *secreted* into the tubular fluid by the proximal tubular epithelial cells, and it is not reabsorbed in any segment of the tubular system.

These two substances play an important role in experimental studies of tubular function, as is discussed later in the chapter.

Reabsorption of Different Ions by the Tubules — Sodium, Potassium, Chloride, Bicarbonate, Potassium, and Others. Finally, note in Figure 27-5 the concentration changes of several important ions — sodium, potassium, chloride, and bicarbonate. Some of the concentrations rise above 1.0, which means greater concentration than in the glomerular filtrate, whereas others fall below 1.0. The degree of concentration or dilution depends on various mechanisms that increase or decrease reabsorption of the different ions for the purpose of controlling their extracellular fluid concentrations. In the following chapters we are particularly concerned with several of these control systems.

To excrete enough potassium and hydrogen ions, it is necessary for both of these to be actively secreted into the tubular system; the amounts of secretion are precisely determined by the potassium and hydrogen ion concentrations in the extracellular fluids.

Bicarbonate ion is transported in a very peculiar way, by converting the bicarbonate ion into carbon dioxide and then allowing the carbon dioxide simply to diffuse through the tubular wall into the interstitial fluid. The method for converting the bicarbonate ion into carbon dioxide is first to secrete a hydrogen ion into the tubule; this then binds with the bicarbonate ion to form H_2CO_3. In turn, the H_2CO_3 dissociates into water and carbon dioxide. After the carbon dioxide diffuses through the tubular membrane, on the other side it recombines with water to form new bicarbonate ion.

Finally, both calcium and magnesium ions are actively reabsorbed in some of the tubules, and many negative ions, especially chloride ions, are reabsorbed mainly by passive diffusion as a result of the electrical gradient that develops across the tubular wall when positive ions are reabsorbed. In addition, some negative ions—urate, phosphates, sulfate, and nitrate—can be reabsorbed by active transport, occurring to the greatest extent in the proximal tubules.

● EFFECT OF "TUBULAR LOAD" AND "TUBULAR TRANSPORT MAXIMUM" ON URINE CONSTITUENTS

Tubular Load. The *tubular load* of a substance is the total amount of the substance that filters through the glomerular membrane into the tubules each minute. For instance, if 125 milliliters of glomerular filtrate is formed each minute with a glucose concentration of 100 mg/dl, the tubular load of glucose is 100 mg × 1.25, or 125 milligrams of glucose per minute. Similarly, the load of sodium that enters the tubules each minute is approximately 18 mEq/min, the load of chloride ion is about 13 mEq/min, the load of urea is approximately 33 mg/min, and so forth.

Maximum Rate of Transport of Actively Reabsorbed or Secreted Substances — The "Tubular Transport Maximum" (Tm)

Because each substance that is actively reabsorbed (or secreted) requires a specific transport system in the tubular epithelial cells, the maximum amount that can be reabsorbed often depends on the maximum rate at which the transport system itself can operate, and this in turn depends on the total amounts of carrier and specific enzymes available, as explained in Chapter 4. Consequently, for most actively reabsorbed substances there is a maximum rate at which each can be reabsorbed; this is called the *tubular transport maximum* for the substance, and the abbreviation is *Tm*. For instance, the Tm for glucose averages 320 mg/min for the adult human, and if the tubular load of glucose

becomes greater than 320 mg/min, the excess above this amount is not reabsorbed but instead passes on into the urine. Ordinarily, though, the tubular load of glucose is only 125 mg/min, so that for practical purposes all of it is reabsorbed.

Figure 27–6 demonstrates the relationship between tubular load of glucose, tubular transport maximum for glucose, and rate of glucose loss into the urine. Note that when the tubular load is at its normal level of 125 mg/min, there is no detectable loss of glucose into the urine. However, when the tubular load rises above about 220 mg/min, significant quantities of glucose begin to appear in the urine. Once the load has risen above approximately 400 mg/min, the loss into the urine thereafter is equal to the *tubular load minus 320 mg/min*. Thus, at 400 mg/min tubular load, the loss is 80 mg/min, and at 800 mg/min tubular load, it is 480 mg/min. In other words, 320 mg/min of the tubular load, which represents the tubular transport maximum for glucose, is reabsorbed, and all the remainder is lost into the urine.

Threshold for Substances That Have a Tubular Transport Maximum. Every substance that has a reabsorptive transport maximum also has a "threshold" concentration in the plasma below which none of it appears in the urine and above which progressively larger quantities appear.

Thus, Figure 27–6 shows that glucose begins to spill into the urine when its tubular load exceeds 220 mg/min. The threshold concentration of glucose in plasma that gives this tubular load is 180 m/dl when the kidneys are operating at their normal glomerular filtration rate of 125 ml/min.

Tubular Transport Maximums of Important Substances Absorbed From the Tubules. Some of the important tubular transport maximums for substances absorbed from the tubules are the following:

Glucose	320	mg/min
Phosphate	0.1	mM/min
Sulfate	0.06	mM/min
Amino acids	1.5	mM/min
Urate	15	mg/min
Plasma protein	30	mg/min
Hemoglobin	1	mg/min
Lactate	75	mg/min
Acetoacetate	variable	(about 30 mg/min)

Figure 27–6. Relationship of tubular load of glucose to loss of glucose into the urine.

Tubular Transport Maximums for Secretion. Substances that are actively secreted by the tubules also exhibit tubular transport maximums as follows:

	mg/min
Creatinine	16
PAH	80

Substances That Do Not Exhibit a Transport Maximum — Especially Sodium Ions in the Proximal Tubules

In the previous list of substances that exhibit transport maximums, it is notable that all of the substances that are reabsorbed by diffusion are missing. Instead, their rates of transport are determined by two factors: (1) the concentration gradient of the substance across the membrane, without any maximum, and (2) the time that the fluid containing the substance remains within the tubule. Therefore, transport of this type is called *gradient-time transport.*

It is notable that many of the rapidly absorbed actively transported substances, such as sodium ions, are likewise missing from the list of substances that exhibit transport maximums. The reason for this is that other factors besides the maximum rate of active transport are the limiting factors that determine the transport rate. For instance, in the proximal tubules the rate of sodium active transport by the basolateral membrane of the tubular epithelial cell is far greater than the rate at which sodium ions can diffuse from the tubular lumen into the cell through the brush border. Therefore, the maximum rate that the active transport mechanism of the basolateral membrane can transport sodium is virtually never the limiting factor in determining how rapidly sodium can be absorbed. And still another factor enters into the limitation of sodium reabsorption. This is the back-leakage of much of the transported sodium from the interstitium into the tubular lumen through the epithelial junctions.

Therefore, sodium transport in the proximal tubules also obeys mainly gradient-time transport principles rather than tubular maximum transport principles. This means simply that the greater the concentration of sodium in the proximal tubules, the greater is its reabsorption; also the greater the time that tubular fluid remains in the proximal tubules, the greater is sodium reabsorption.

In the more distal parts of the tubular system, in the *distal tubules* and beyond, the epithelial cells are quite different from those in the proximal tubules, having much tighter epithelial junctions and also transporting far less quantities of sodium. Because of these differences, sodium transport in these distal segments exhibits a transport maximum the same as most other actively transported substances. Furthermore, this transport maximum changes in response to the hormones aldosterone and angiotensin, thus providing a means for controlling the rate of sodium excretion into the urine.

Glomerulotubular Balance in the Proximal Tubules. An especially important discovery concerning the absorption of sodium and fluid in the proximal tubules is the effect called *glomerulotubular balance.* This means that under normal conditions, almost a constant percentage, about 65 per cent, of both of these is reabsorbed during passage through the proximal tubules regardless of the rate at which glomerular filtrate enters the tubular system. The reason for this is yet unclear. It may result from greater stretch of the tubules or some other physical effect at the

higher flow rates. But, regardless of the cause, when the glomerular filtration rate is 100 ml/min, the proximal tubular reabsorption is about 65 ml/min; an elevation in glomerular filtrate to 200 ml/min increases proximal reabsorption to about 130 ml/min, maintaining the proportional "balance" very near to 65 per cent.

The importance of glomerulotubular balance is that it helps to prevent overloading of the more distal segments of the tubular system when the glomerular filtration rate increases. In addition, the tubuloglomerular feedback mechanisms that control the glomerular filtration rate itself, as discussed in the previous chapter, also play an important role in preventing the distal tubular segments from becoming overloaded. It is essential that they not become overloaded because in these final segments of the tubular system, the amount of each substance excreted in the urine must be controlled. If the tubular load entering the distal tubules varies significantly up or down, the tubular control systems will not function properly. For example, aldosterone has only a limited range of control of sodium reabsorption; therefore, for it to be effective in controlling sodium excretion in the urine, the sodium load entering the distal tubules each minute must not vary more than a few per cent.

● THE CONCEPT OF "PLASMA CLEARANCE" — ITS USE IN ASSESSING RENAL FUNCTION

The term "plasma clearance" is used to express the ability of the kidneys to clean, or "clear," the plasma of various substances. Thus, if the plasma passing through the kidneys contains 0.1 gram of a substance in each deciliter, and 0.1 gram of this substance also passes into the urine each minute, 1 deciliter of the plasma is cleaned or "cleared" of the substance per minute.

Referring again to Table 27–1, note that the normal concentration of urea in each milliliter of plasma and glomerular filtrate is 0.26 milligram, and the quantity of urea that passes into the urine each minute is approximately 18.2 milligrams. Therefore, the equivalent quantity of plasma that completely loses its entire content of urea each minute can be calculated by dividing the quantity of urea entering the urine each minute by the quantity of urea in each milliliter of plasma. Thus $18.2 \div 0.26 = 70$; that is, 70 milliliters of plasma is *cleaned* or *cleared* of urea each minute. The amount that is cleared each minute is known as the *plasma clearance of urea.* Obviously, therefore, the plasma clearance of each substance is a measure of the effectiveness of the kidney to remove that substance from the extracellular fluid.

Plasma clearance for any substance can be calculated by the following formula:

$$\text{Plasma clearance (ml/min)} = \frac{\text{Urine flow (ml/min)} \times \text{Concentration in urine}}{\text{Concentration in plasma}}$$

The plasma clearances of the usual constituents of urine are shown in the last column of Table 27–1.

INULIN CLEARANCE AS A MEASURE OF GLOMERULAR FILTRATION RATE

Inulin is a polysaccharide that has the specific attributes of not being reabsorbed to a significant extent by the tubules of the nephron and yet being of small enough molecular weight (about 5200) that it passes through the glomerular membrane as freely as the crystalloids and water of the plasma. Also, inulin is not actively secreted even in the minutest amount by the tubules. Consequently, glomerular filtrate contains virtually the same concentration of inulin as does plasma, and as the filtrate flows down the tubules, all the filtered inulin continues on into the urine. Thus, *all the glomerular filtrate formed is cleared of inulin.* Therefore, the *plasma clearance per minute of inulin is equal to the glomerular filtration rate.*

As an example, let us assume that chemical analysis shows the inulin concentration in the plasma to be 0.001 gram in each milliliter and that 0.125 gram of inulin passes into the urine per minute. By dividing 0.001 into 0.125, one finds that 125 ml of glomerular filtrate must be formed each minute in order to deliver to the urine the analyzed quantity of inulin. In other words, by measuring the plasma clearance of inulin, we have determined that the glomerular filtration rate is 125 ml/min.

Inulin is not the only substance that can be used for determining the quantity of glomerular filtrate formed each minute, for the plasma clearance of any other substance that is totally diffusible through the glomerular membrane but is neither absorbed nor secreted by the tubular walls is equal to the glomerular filtration rate. *Mannitol* is a monosaccharide that is frequently used instead of inulin for such measurements, and *radioactive iothalamate* is still another substance frequently used because its radioactivity allows easy quantitative analysis.

PARA-AMINOHIPPURIC ACID CLEARANCE AS A MEASURE OF PLASMA FLOW THROUGH THE KIDNEYS

PAH, like inulin, passes through the glomerular membrane with perfect ease. However, it is different from inulin in that most of the PAH remaining in the plasma after the glomerular filtrate is formed is secreted from the peritubular capillaries into the tubules by the proximal tubular epithelium (if the plasma concentration of PAH is very low). Indeed, only about one-tenth of the original PAH remains in the plasma by the time the blood leaves the kidneys.

One can use the clearance of PAH for estimating the *flow of plasma* through the kidneys. As an example, let us assume that 1 milligram of PAH is present in each deciliter of plasma, and that 5.85 milligram of PAH passes into the urine per minute. Consequently, 585 ml of plasma is cleared of PAH each minute. Obviously, if this much plasma is cleared of PAH, *at least this much plasma must have passed through the kidneys* in this same period of time. And, since it is known that almost all the PAH is cleared from the blood as it passes through the kidneys, 585 milliliters would be a reasonable first approximation of the actual plasma flow per minute.

Yet, to be still more accurate, one can correct for the average amount of PAH that is still in the blood when it leaves the kidney. In different experiments the PAH clearance has been found to be about 91 per cent of the plasma load of PAH entering the kidneys; this percentage is known as the *extraction ratio* of PAH. Therefore, the 585 milliliters of plasma just calculated would be only 91 per cent of the total amount of plasma flowing through the kidneys. Dividing 585 by 0.91 gives a total plasma flow per minute of approximately 650 milliliters.

One can calculate the *total blood flow* through the kidneys each minute from the plasma flow and the hematocrit (the percentage of red blood cells in the blood). If the hematocrit is 45 per cent and the plasma flow 650 ml/min, the total blood flow through both kidneys is $650 \times 100 \div 55$, or 1182 ml/min.

CALCULATING THE FILTRATION FRACTION FROM PLASMA CLEARANCES

To calculate the filtration fraction — that is, the fraction of the plasma that filters through the glomerular membrane — one must determine (1) the plasma flow through the two kidneys (PAH clearance) and (2) the glomerular filtration rate per minute (inulin clearance). Using 650 milliliters plasma flow and 125 milliliters glomerular filtration rate as normal values, we find that the calculated filtration fraction is 125/650, or, to express this as a percentage, 19 per cent.

REFERENCES

See references for Chapter 26.

CHAPTER 28

• *Renal and Associated Mechanisms for Controlling Extracellular Fluid Osmolality and Sodium Concentration* •

In the previous two chapters, we have discussed the mechanisms by which glomerular filtrate is formed and the way in which it is then processed in the tubules to become urine. In the following few chapters, we present the mechanisms by which the kidneys employ negative feedback systems to control the composition of the extracellular fluid, including the volume of the fluid, its osmolality, and the concentrations of the various dissolved substances. These feedback controls frequently involve the nervous system, hormonal mechanisms, and even physical factors, such as the effects that changing levels of arterial pressure have on urinary output.

The present chapter discusses specifically (1) the mechanisms by which the kidneys can eliminate excess water by excreting a dilute urine or can conserve water by excreting a concentrated urine; (2) the nervous and hormonal mechanisms for controlling body fluid osmolality, by turning on or off the dilution or concentration mechanisms of the kidney; and (3) the thirst and salt appetite mechanisms for determining the intakes of water and salt, thus also helping to control extracellular fluid osmolality and sodium concentration.

● THE MECHANISM FOR EXCRETING EXCESS WATER: EXCRETION OF A DILUTE URINE

One of the most important functions of the kidney is to control the osmolality of the body fluids. When the osmolality falls too low — that is, when the fluids become too dilute — nervous and hormonal feedback mechanisms cause the kidneys to excrete a great excess of water in the urine. This obviously causes a dilute urine, but it also removes water from the body, in this way increasing the body fluid osmolality back toward normal. Conversely, when the osmolality of the body fluids is too great, the kidneys excrete an excess of solutes, thereby reducing the body fluid osmolality again back toward normal but at the same time excreting a concentrated urine.

Role of Antidiuretic Hormone in Controlling Urine Concentration. The signal that tells the kidney whether to excrete a dilute or a concentrated urine is the hormone called *antidiuretic hormone* (ADH) (also called "vasopressin") that is secreted by the posterior pituitary gland. When the body fluids are too concentrated, the posterior pituitary gland secretes large amounts of ADH, which allows the kidneys to continue excreting large amounts of solutes but to conserve water. Conversely, in the absence of ADH, the kidneys excrete a dilute urine, thus losing excess water from the body. The feedback mechanisms for controlling this system are described later in the chapter; for now, let us discuss the renal mechanisms for excreting a dilute or a concentrated urine.

The Renal Mechanism for Excreting a Dilute Urine. Figure 28–1 illustrates the mechanism for excreting a dilute urine. When the glomerular filtrate is initially formed by the glomerulus, its osmolality is almost exactly the same as that of the plasma, approximately 300 mOsm/liter. To excrete excess water, it is necessary to dilute the filtrate as it passes through the

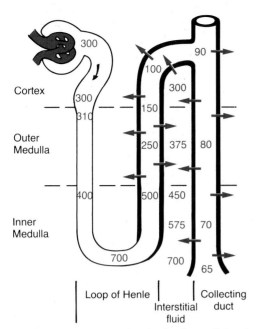

Figure 28–1. The renal mechanism for forming a dilute urine. The darkened walls of the distal portions of the tubular system indicate that these portions of the tubules are relatively impermeable to the reabsorption of water in the absence of antidiuretic hormone. The solid arrows indicate active processes for absorption of most of the solutes besides the urinary waste products. (Numerical values are in milliosmoles per liter.)

tubules. This is achieved by reabsorbing a higher proportion of solutes than water.

The colored arrows in Figure 28–1 represent reabsorption of most of the tubular solutes (besides the waste products) in all the distal segments of the tubular system. These same tubular segments are depicted in the figure with thickened, darkened walls to indicate that their epithelia are almost completely impermeable to water when the kidneys are excreting a dilute urine. The ascending limb of the loop of Henle and the diluting segment of the distal tubule even normally are very impermeable to water, whereas the late distal tubule, the cortical collecting duct, and the collecting duct also become almost completely impermeable to water *when ADH is not present* in the circulating body fluids.

The reabsorption of solutes in these distal segments of the tubular system is strong and active. In the thick segment of the ascending limb of the loop of Henle, active reabsorption of sodium, potassium, and chloride ions is especially powerful, and one can see from the numerical values in the figure that the osmolality of the fluid in the ascending limb of the loop of Henle decreases progressively to about 100 mOsm/liter by the time the fluid leaves this tubular segment. That is, most of the solutes besides the waste products are reabsorbed, but the water remains behind. Then, as this remaining dilute fluid passes through the distal tubule, cortical collecting duct, and collecting duct, some additional reabsorption of solutes, especially of sodium ions, causes still more dilution of the tubular fluid, often decreasing its osmolality to as little as 65 mOsm/liter and rarely to levels below 50 mOsm/liter

by the time it leaves the collecting duct to enter the urine.

To summarize, the process for excreting a dilute urine is very simple; it is one of absorbing solutes from the distal segments of the tubular system while leaving the water in the tubules. However, this failure of water to be reabsorbed occurs only in the absence of ADH.

● THE MECHANISM FOR EXCRETING EXCESS SOLUTES: THE COUNTERCURRENT MECHANISM FOR EXCRETING A CONCENTRATED URINE

The process for concentrating urine is not nearly so simple as for diluting it. Yet, at times it is exceedingly important to concentrate the urine as much as possible, so that excess solutes can be eliminated with as little loss of water from the body as possible — for instance, when one is exposed to desert conditions with an inadequate supply of water. Fortunately, the kidneys have developed a special though very complex mechanism for concentrating the urine called the *countercurrent mechanism.*

The countercurrent mechanism depends on a peculiar anatomical arrangement of the loops of Henle and the vasa recta in the renal medulla. In the human, the loops of Henle of one third to one fifth of the nephrons dip deep into the medulla and then return to the cortex; some dip all the way to the tips of the papillae that project into the renal pelvis. This group of nephrons with the long loops of Henle is called the *juxtamedullary nephrons.* Paralleling the long loops of Henle are loops of peritubular capillaries called the *vasa recta;* these also loop down into the medulla from the cortex and then back out to the cortex. These arrangements of the different parts of the juxtamedullary nephron and the vasa recta are diagrammed in Figure 28–2.

Hyperosmolality of the Medullary Interstitial Fluid, and Mechanisms for Achieving It

The first step in the excretion of excess solutes in the urine — that is, for excreting a concentrated urine — is to create a very high osmotic pressure (hyperosmolality) of the medullary interstitial fluid. As we see later, this in turn is necessary for concentrating the urine. But, first, let us explain the mechanism for creating this hyperosmolality.

The normal osmolality of the fluids in almost all parts of the body is about 300 mOsm/liter. However, as shown by the numbers in Figure 28–2, the osmolality of the interstitial fluid in the medulla of the kidney is much higher than this, and it becomes progressively greater the deeper one goes into the medulla, increasing from 300 mOsm/liter in the cortex to 1200 mOsm/

Figure 28–2. The countercurrent mechanism for concentrating the urine. (Numerical values are in milliosmoles per liter.)

liter (occasionally as high as 1400 mOsm/liter) in the pelvic tip of the medulla. Three different solute-concentrating mechanisms are responsible for this hyperosmolality; these are as follows:

First, the principal cause of the greatly increased medullary osmolality is active transport of sodium ions (plus co-transport of potassium, chloride, and other ions) out of the thick portion of the ascending limb of the loop of Henle into the interstitium. The large colored arrows shown in this tubular segment in Figure 28–2 illustrate this transport into the outer medullary interstitial fluid. The sodium and its associated ions become concentrated in this fluid. Also, they are carried downward into the inner medulla by the downward flowing blood in the descending limbs of the vasa recta and also by diffusing into the descending thin limb of the loop of Henle, as we see shortly.

Second, smaller quantities of ions are also transported into the medullary interstitial fluid from the collecting duct, mainly resulting from active transport of sodium ions and electrogenic passive absorption of chloride ions along with the sodium ions.

Third, when the concentration of antidiuretic hormone is high in the blood, large amounts of urea are also absorbed into the fluid of the inner medulla from the collecting duct. The reason is: The inner medullary portion of the collecting duct is moderately permeable to urea, and ADH increases this permeability still slightly more. Even more important, ADH causes the inner medullary collecting duct to become highly permeable to water. The fact that it becomes highly permeable to water allows the high osmotic pressure of the sodium in the interstitium to cause osmotic reabsorption of water out of the collecting duct; this greatly increases the concentration of urea in the duct. Now, because of this high urea concentration, the urea too diffuses through the collecting duct wall into the medullary interstitium. Consequently, the urea concentration in the medullary interstitial fluid rises almost to equal the concentration in the collecting duct. In the human, during maximal antidiuretic hormone stimulation this may be to as high as 400 to 500 mOsm/liter, which obviously adds greatly to the osmolality of the inner medullary interstitial fluid.

In summary, at least three different factors contribute to the marked increase in osmolality in the medullary interstitial fluid. These are (1) active transport of the ions into the interstitium by the thick portion of the ascending limb of the loop of Henle, (2) active transport of ions from the collecting duct into the interstitium, and (3) passive diffusion of large amounts of urea from the collecting duct into the interstitium. The net result is an increase in the osmolality of the medullary interstitial fluid, when adequate amounts of ADH are present, to as high as 1200 to 1400 mOsm/liter near the tips of the papillae.

Mechanism by Which Antidiuretic Hormone Increases Water Reabsorption. The final details of the mechanism by which ADH increases water reabsorption by the late distal tubules, cortical collecting ducts, and collecting ducts are still to be established. However, several facts about the mechanism are the following: ADH itself acts not on the luminal membrane of the tubular epithelial cells but instead on the basolateral membrane of these cells. It activates the enzyme *adenyl cyclase* in this membrane, which then causes formation of *cyclic adenosine monophosphate* (cyclic AMP) in the cell cytoplasm. This then diffuses to the luminal side of the cell and within a few minutes causes elongated vesicular structures in the cytoplasm to develop and fuse with the cell's luminal membrane. In this way the membranes of these vesicles become part of the luminal cell membrane, providing patches of membrane containing aggregates of protein particles that have very large water conducting channels. Therefore, the luminal membrane becomes highly water permeable in contrast to its normal state of almost total impermeability. As water diffuses to the inside of the epithelial cell, it continues to diffuse in the normal manner through the cell to the basolateral membrane and then into the interstitial fluid.

When ADH is no longer present, the vesicular structures detach from the luminal membrane within 10 to 15 minutes and return to their internal position in the cytoplasm; the tubules again become water impermeable.

A Mechanism for Enhancing Further Medullary Hyperosmolality — The "Countercurrent Multiplier"

In the previous discussion, we pointed out that when sodium and its associated chloride ions are transported from the lumen of the thick ascending limb of the loop of Henle into the medullary interstitium, much of it immediately diffuses into both the descending limbs of the vasa recta and the descending thin limbs of the loops of Henle. This sodium chloride is then carried by fluid flow down to the tip of the papilla. Once there, much of the sodium chloride diffuses into the papillary interstitium, further enhancing its osmolality. But the remaining sodium chloride is carried again back up the ascending limb of the loop of Henle, where the thick ascending segment retransports this sodium chloride once again into the papillary interstitium. This repetitive reabsorption of sodium chloride by the thick ascending segment of the loop of Henle, along with the continual inflow of new sodium chloride from the proximal tubule into the loop of Henle, is called the *countercurrent multiplier.* Obviously, the reabsorbed sodium chloride keeps adding to the newly arriving sodium chloride, thus "multiplying" its concentration in the medullary interstitium.

Countercurrent Exchange Mechanism in the Vasa Recta — A Mechanism for Holding Solutes in the Medulla

We have now discussed the mechanisms by which high concentrations of solutes are achieved in the medullary interstitium. However, without a special medullary vascular system as well, the flow of blood through the interstitium would rapidly remove the excess solutes and keep the concentration from rising very high. Fortunately, the medullary blood flow has two characteristics, both exceedingly important, for maintaining the high solute concentration in the medullary interstitial fluids. These are as follows:

First, the inner medullary blood flow is very slight in quantity, amounting to only 1 to 2 per cent of the total blood flow of the kidney. Because of this very sluggish blood flow, removal of solutes is minimized.

Second, the vasa recta function as a *countercurrent exchanger* that also minimizes the washout of solutes from the medulla. This can be explained as follows: A countercurrent fluid exchange mechanism is one in which fluid flows through a long U-tube, with the two arms of the U lying in close proximity to each other, so that fluid and solutes can exchange readily between the two arms. This obviously also requires that each of the two arms of the U be highly permeable, which is true for the vasa recta. When the fluids and solutes in the two parallel streams of flow can exchange rapidly, tremendous concentrations of solute can be maintained at the tip of the loop with negligible washout of solute.

Thus, in Figure 28-2, as blood flows down the descending limbs of the vasa recta, sodium chloride and urea diffuse into the blood from the interstitial fluid, while water diffuses outward into the interstitium. These two effects cause the osmolal concentration in the capillary blood to rise progressively higher, to a maximum concentration of 1200 mOsm/liter at the tips of the vasa recta. Then, as the blood flows back up the ascending limbs, the extreme diffusibility of all molecules through the capillary membrane allows almost all the extra sodium chloride and urea to diffuse back out of the blood into the interstitial fluid, while water diffuses back into the blood. Therefore, by the time the blood finally leaves the medulla, its osmolal concentration is only slightly greater than that of the blood that had initially entered the vasa recta. As a result, the blood flowing through the vasa recta carries only a minute amount of the medullary interstitial solutes away from the medulla.

Mechanism for Excreting a Concentrated Urine — Role of Antidiuretic Hormone

Now that we have explained how the kidney creates hyperosmolality in the medullary interstitium, it becomes a simple matter to explain the mechanism for excreting a concentrated urine.

When the concentration of ADH in the blood is high, the epithelium of the cortical collecting duct, the collecting duct, and in some species of animals the late distal tubule as well becomes highly permeable to water. This is illustrated in Figure 28-2 by the thin walls of these segments of the tubular system. Most importantly, as the tubular fluid flows through the collecting duct, water is pulled by osmosis into the highly concentrated fluid of the medullary interstitium. Thus, the collecting duct fluid also becomes highly concentrated, and it issues from the papilla into the pelvis of the kidney at a concentration of about 1200 mOsm/liter, almost exactly equal to the osmolal concentration of the solutes in the medullary interstitium near the papilla.

Summary of the Osmolal Concentration Changes in the Different Segments of the Tubules

Figure 28–3 illustrates the changes in osmolality of tubular fluid as it passes through the different segments of the tubules. In the proximal tubules, the tubular membranes are so highly permeable to water that whenever a solute is transported across the membrane an almost exact proportionate amount of water crosses the membrane by osmosis at the same time; therefore, the osmolality of the fluid remains almost exactly equal to that of the glomerular filtrate, 300 mOsm/liter, throughout the entire extent of the proximal tubule.

Next, in the loop of Henle, the osmolality rises rapidly because of the countercurrent mechanism explained in the previous few paragraphs. When a concentrated urine is being formed—that is, when the blood concentration of ADH is high—the loop of Henle osmolality rises much higher than when a dilute urine is being formed because of the large quantity of urea that is passively reabsorbed into the medullary interstitium from the collecting ducts.

Next, beginning in the thick portion of the ascending limb of the loop of Henle, the osmolality falls once again to a very low level, usually to about 100 mOsm/liter.

Finally, in the late distal tubule, cortical collecting duct, and collecting duct, the osmolality depends entirely on the presence or absence of ADH. In the absence of ADH, very little water is reabsorbed from these segments. Therefore, the osmolality remains less than 100 mOsm/liter and even falls still additional milliosmoles because of active transport of ions through the epithelium of these tubules. Therefore, a very dilute urine is formed.

But, in the presence of excess ADH, the cortical collecting duct, the collecting duct, and in some animal species the late distal tubule as well become highly permeable to water, so that most of the water is reabsorbed, thus producing a very concentrated urine.

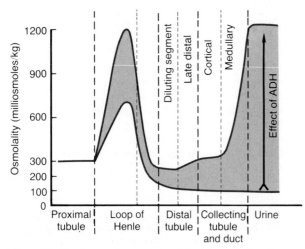

Figure 28–3. Changes in osmolality of the tubular fluid as it passes through the tubular system.

Note especially the shaded areas in Figure 28–3. These denote the ranges of concentration of the tubular fluid and the usual range of urine concentration between 65 and 1200 mOsm/liter in humans, depending on the blood concentration of ADH at any given time.

OSMOLAR CLEARANCE; FREE WATER CLEARANCE

One can calculate the clearance of osmolar substances (C_{osm}) in terms of the volume of plasma cleared per minute in the same way that the clearance of a single substance is calculated, using the following formula:

$$C_{osm} = \frac{\text{Osmoles entering urine per minute}}{\text{Plasma osmolar concentration}}$$

For instance, if the plasma osmolality is 300 mOsm/liter and the milliosmoles entering the urine per minute is 1.5, the osmolar clearance will be 1.5/300 liters/min, or 5 ml/min.

Free Water Clearance. When the kidney forms urine that is osmotically more dilute than plasma, it is obvious that a higher proportion of the water in the glomerular filtrate is excreted than of the osmolar substances. The excess water that is excreted is called *free water,* and the total plasma volume that is cleared of this excess water each minute is called the *free water clearance.* The free water clearance can be calculated by first determining the osmolar clearance and then subtracting this rate from the rate of urine flow per minute. Thus, the formula for free water clearance (C_{H_2O}) is:

$$C_{H_2O} = \text{Urine volume per minute} - C_{osm}$$

The free water clearance is important because it tells one how rapidly the kidneys are changing the body fluid osmolality. The free water clearance can be *positive,* in which case excess water is being removed, or it can be *negative,* in which case excess solutes are being removed.

● CONTROL OF EXTRACELLULAR FLUID OSMOLALITY AND SODIUM CONCENTRATION

Now that the mechanisms by which the kidney can excrete either a dilute or a concentrated urine have been discussed, in the next few pages we explain how these mechanisms are manipulated to control extracellular fluid osmolality and sodium concentration. Osmolality and sodium concentration are inextricably related to each other because the sodium ions in the extracellular fluids play the dominant role in determining the extracellular fluid osmolality, as we explain in the following section.

The osmolality of the extracellular fluid averages almost exactly 300 mOsm/liter, and the sodium ion concentration 142 mEq/liter. These rarely change more than ± 3 per cent day in and day out, which gives one an idea how tightly both of these are controlled.

Relationship Between Extracellular Fluid Osmolality and Sodium Concentration. The osmolality of the extracellular fluids is determined almost entirely by the extracellular fluid sodium concentration. The reason for this is that sodium is by far the most abundant positive ion of the extracellular fluid, representing over 90 per cent of these ions. Furthermore, every time a positive ion is reabsorbed by the renal tubules a negative ion also is reabsorbed. Therefore, in effect, the control of the positive ions controls the total ion concentration. Also, the glucose and urea, which are the most abundant of the nonionic osmolar solutes in the extracellular fluids, normally represent only 3 per cent of the total osmolality, and even then the urea exerts very little *effective* osmotic pressure because it penetrates cells too easily to cause significant osmotic results. Therefore, *the sodium ions of the extracellular fluid determine either directly or indirectly over 90 per cent of the osmotic pressure of the extracellular* fluid. Consequently, we can generally talk in terms of control of osmolality and control of sodium ion concentration at the same time.

Three separate control systems operate in close association to regulate extracellular osmolality and sodium concentration. These are (1) the osmoreceptor–antidiuretic hormone system, (2) the thirst mechanism, and (3) the salt appetite mechanism.

THE OSMORECEPTOR–ANTIDIURETIC HORMONE FEEDBACK CONTROL SYSTEM

Figure 28–4 illustrates the osmoreceptor–antidiuretic hormone system for control of extracellular fluid sodium concentration and osmolality. It is a typical feedback control system that operates by the following steps:

1. An increase in osmolality (mainly excess sodium and the negative ions that go with it) excites *osmoreceptors* located in the anterior hypothalamus near to the supraoptic nuclei.

2. Excitation of the osmoreceptors in turn stimu-

lates the supraoptic nuclei, which then cause the posterior pituitary gland to release ADH.

3. The ADH increases the permeability of the late distal tubules, the cortical collecting ducts, and the collecting ducts to water, as explained earlier, and therefore *causes increased conservation of water by the kidneys.*

4. The conservation of water but *loss of sodium and other osmolar substances in the urine* causes dilution of the sodium and other substances in the extracellular fluid, thus correcting the initial, excessively concentrated extracellular fluid.

Conversely, when the extracellular fluid becomes too dilute (hypo-osmotic), less ADH is formed, and excess water is lost in comparison with the extracellular fluid solutes, thus concentrating the body fluids back toward normal.

The Osmoreceptors (or Osmosodium Receptors)—The "AV3V Region" of the Brain. Figure 28–5 illustrates the hypothalamus and the pituitary gland. The hypothalamus contains two areas that are important in controlling ADH secretion and also in controlling thirst. One of these is the *supraoptic nuclei.* Here about five sixths of the ADH is formed in the cell bodies of large neuronal cells; the remaining one sixth is formed in the nearby *paraventricular nuclei.* This hormone is transported down the axons of the neurons to their tips, terminating in the posterior pituitary gland. When the supraoptic and paraventricular nuclei are stimulated, nerve impulses pass to these nerve endings and cause ADH to be released into the capillary blood of the posterior pituitary gland.

A second neuronal area important in controlling osmolality is a broad area located along the anteroventral border of the third ventricle, called the *AV3V region,* also illustrated in Figure 28–5. At the superior part of this area is a special structure called the *subfornical organ* and at the inferior part another called the *organum vasculosum of the lamina terminalis.* In between these two "organs" is the *median preoptic nucleus,* which has multiple nervous connections with the two organs and also with both the supraoptic nu-

Figure 28–4. Control of extracellular fluid osmolality and sodium ion concentration by the osmosodium receptor–antidiuretic hormone feedback control system.

Hypothalamus

Supraoptic nuclei

Excited by concentrated extracellular fluid (excess Na$^+$)

Antidiuretic hormone released by posterior pituitary gland

Excess reabsorption of water corrects excess concentration of extracellular fluid and excess Na$^+$

Urine decreased but concentrated

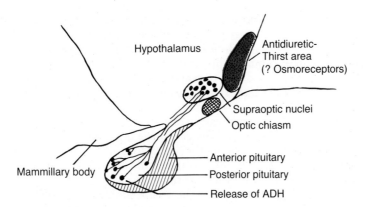

Figure 28–5. The supraopticopituitary antidiuretic system and its relationship to the thirst center in the hypothalamus.

clei and the blood pressure control centers in the medulla of the brain. Lesions in this AV3V region cause multiple deficits in the control of ADH secretion, thirst, sodium appetite, and blood pressure. Also, electrical stimulation as well as stimulation by the hormone angiotensin II can alter ADH secretion, thirst, and sodium appetite.

In the vicinity of the AV3V region and the supraoptic nuclei are other neuronal cells that are excited by very minute increases in extracellular fluid osmolality and, conversely, inhibited by decreases in osmolality. These neurons are called *osmoreceptors.* They in turn send nerve signals to the supraoptic nuclei to control the secretion of ADH. It is likely that they also induce thirst as well.

Both the subfornical organ and the organum vasculosum of the lamina terminalis have vascular supplies that lack the typical blood-brain barrier present elsewhere in the brain that impedes diffusion of most ions from the blood into the brain tissue. Therefore, this makes it possible for ions and other solutes to cross with ease between the blood and the local interstitial fluid. In this way the osmoreceptors respond rapidly to changes in the osmolality of the fluid in the blood, exerting powerful control over the secretion of ADH and probably over thirst as well.

Summary of the Antidiuretic Hormone Mechanism for Controlling Extracellular Fluid Osmolality and Extracellular Fluid Sodium Concentration. From these discussions, we can reiterate once again the importance of the ADH mechanism for controlling at the same time both extracellular fluid osmolality and extracellular fluid sodium concentration. That is, an increase in sodium concentration causes almost an exactly parallel increase in osmolality, which in turn excites the osmoreceptors of the hypothalamus. These receptors then cause the secretion of ADH, which markedly increases the reabsorption of water in the renal tubules. Consequently, very little water is lost into the urine, but the urinary solutes continue to be lost. Therefore, the relative proportion of water in the extracellular fluid increases, whereas the proportion of solutes decreases. In this way, the sodium ion concentration of the extracellular fluid, and the osmolality as well, decrease toward the normal level. This is a very powerful

mechanism for controlling both the extracellular fluid osmolality and the extracellular fluid sodium concentration.

Function of the Antidiuretic Hormone Mechanism in Special Conditions

Water Diuresis. When a person drinks a large amount of water, a phenomenon called *water diuresis* occurs; a typical record of this is shown in Figure 28–6. In this example, a man drank 1 liter of water, and approximately 45 minutes later his urine output had increased to eight times normal. It remained at this level for 2 hours—that is, until the osmolality of the extracellular fluid returned essentially to normal. The delay in onset of water diuresis is caused partly by delay in absorption of the water from the gastrointestinal tract but mainly by the time required for destruction of the ADH that had already been released by the pituitary gland prior to drinking the water.

Syndrome of Inappropriate Antidiuretic Hormone Secretion. Certain types of tumors, especially bronchogenic tumors of the lungs or tumors of the basal regions of the brain, occasionally secrete ADH or a similar hormone. This condition is called the *syndrome of inappropriate ADH secretion.* This excess ADH causes only a slight increase in extracellular fluid volume. Instead, its principal effect is *to decrease greatly the sodium concentration (and the osmolality) of the extracellular fluid.* The explanation of this effect is

Figure 28–6. Water diuresis in a human being following ingestion of 1000 ml of water. (Redrawn from Smith: The Kidney: Structure and Functions in Health and Disease. New York, Oxford University Press, 1951.)

the following: The ADH at first causes a decrease in urine output and a simultaneous slight increase in blood volume. This in turn causes a slight rise in arterial pressure, which then causes a secondary *increase* in urinary output. Yet, the urine that is excreted is still tremendously concentrated because the ADH is still causing excessive reabsorption of water by the kidney tubules. Consequently, the kidneys excrete extreme amounts of sodium and other ions into the urine but keep the water in the extracellular fluid. Therefore, the sodium concentration becomes seriously reduced, sometimes falling from a normal value of 142 mEq/liter to as low as 110 to 120 mEq/liter. At values this low, patients frequently die sudden deaths because of coma and convulsions.

This disease is especially instructive because it illustrates the extreme importance of the ADH mechanism for control of sodium concentration and extracellular osmolality and yet its relatively mild effect on control of body fluid volume.

THIRST AND ITS ROLE IN CONTROLLING EXTRACELLULAR FLUID OSMOLALITY AND SODIUM CONCENTRATION

The phenomenon of thirst is equally as important for regulating body water, osmolality, and sodium concentration as is the osmoreceptor-renal mechanism previously discussed because the amount of water in the body at any one time is determined by the balance between both *intake* and *output* of water. Thirst, the primary regulator of the intake of water, is defined as the *conscious desire for water.*

Neural Integration of Thirst — The "Thirst" Center

Referring again to Figure 28-5, note that the same area along the anteroventral wall of the third ventricle that promotes antidiuresis can also cause thirst. Also located anterolaterally in the preoptic hypothalamus are other small areas that, when stimulated electrically, will cause immediate onset of drinking that continues as long as the stimulation lasts. All these areas together are called the *thirst center.*

Injection of hypertonic salt solution into portions of the center causes osmosis of water out of the neuronal cells, and this causes drinking. Therefore, these cells function as *osmoreceptors* to activate the thirst mechanism. Probably these are the same osmoreceptors that activate the antidiuretic system as well.

In addition, an increase in osmotic pressure of the cerebrospinal fluid in the third ventricle has essentially the same effect to promote drinking. Some experiments suggest that the site of this effect is the *organum vasculosum of the lamina terminalis,* which lies immediately beneath the ventricular surface at the inferiormost end of the AV3V region. Neuronal cells found here have been shown to be excited by increased osmolality.

Basic Stimulus for Exciting Thirst — Intracellular Dehydration. Any factor that will cause *intracellular dehydration* will in general cause the sensation of thirst. The most common cause of this is increased osmolar concentration of the extracellular fluid, especially increased sodium concentration, which causes osmosis of fluid from the neuronal cells of the thirst center. However, another important cause is excessive loss of potassium from the body, which reduces the intracellular potassium of the thirst cells and therefore decreases their volume.

Temporary Relief of Thirst Caused by the Act of Drinking

A thirsty person receives relief from thirst immediately after drinking water, even before the water has been absorbed from the gastrointestinal tract. In fact, in persons who have an esophageal opening to the exterior so that the water is lost to the outside and never goes into the gastrointestinal tract, partial relief of thirst still occurs, but this relief is only temporary, and the thirst returns after 15 or more minutes. If the water does enter the stomach, distension of the stomach and other portions of the upper gastrointestinal tract provides still further temporary relief from thirst. Indeed, even simple inflation of a balloon in the stomach can relieve thirst for 5 to 30 minutes.

One might wonder what the value of this temporary relief from thirst could be, but there is good reason for its occurrence. After a person has drunk water, as long as ½ to 1 hour may be required for all the water to be absorbed and distributed throughout the body. Were the thirst sensation not temporarily relieved after drinking water, the person would continue to drink more and more. When all this water should finally become absorbed, the body fluids would be far more diluted than normal, and an abnormal condition opposite to that which the person was attempting to correct would have been created. It is well known that a thirsty animal almost never drinks more than the amount of water needed to relieve its state of dehydration. Indeed, it is uncanny that the animal usually drinks almost exactly the right amount.

Role of Thirst in Controlling Extracellular Fluid Osmolality and Sodium Concentration

Threshold for Drinking—The Tripping Mechanism. The kidneys are continually excreting fluid; also water is lost by evaporation from the skin and lungs. Therefore, a person is continually being dehydrated, causing the volume of extracellular fluid to decrease and its concentration of sodium and other osmolar elements to rise. When the sodium concentration rises approximately 2 mEq/liter above normal (or the osmolality rises approximately 4 mOsm/liter above normal), the drinking mechanism becomes "tripped"; that is, the person then reaches a level of thirst that is strong enough to activate the necessary motor effort to cause drinking. This is called the *threshold for drinking.* The person ordinarily drinks precisely the required amount of fluid to bring the extracellular fluids back to normal—that is, to a state of *satiety*. Then the process of dehydration and sodium concentration begins again, and after a period of time the drinking act is tripped again, the process continuing on and on indefinitely.

In this way, both the osmolality and the sodium concentration of the extracellular fluid are very precisely controlled.

COMBINED ROLES OF THE ANTIDIURETIC AND THIRST MECHANISMS FOR CONTROL OF EXTRACELLULAR FLUID OSMOLALITY AND SODIUM CONCENTRATION

When either the ADH mechanism or the thirst mechanism fails, the other ordinarily can still control both extracellular fluid osmolality and sodium concentration with reasonable effectiveness. On the other hand, if both of them fail simultaneously, neither sodium nor osmolality is then adequately controlled.

Figure 28–7 illustrates dramatically the overall capability of the ADH-thirst system to control extracellular fluid sodium concentration (and therefore osmolality as well). This figure demonstrates the ability of the same animals to control their extracellular fluid sodium concentrations in two different conditions: (1) in the normal state and (2) after both the ADH and thirst mechanisms had been blocked. Note that in the normal state (the solid curve), a sixfold increase in sodium intake caused the sodium concentration to change only two thirds of 1 per cent (from 142 mEq/liter to 143 mEq/liter)—an excellent degree of sodium concentration control. Now note the dashed curve of the figure, which shows the change in sodium concentration when the ADH-thirst system was blocked. In this case, the sodium concentration increased 10 per cent with only a fivefold increase in sodium intake (a change in sodium concentration from 137 mEq/liter to 151 mEq/liter), which is an extreme change in sodium concentration when one realizes that the normal sodium concentration rarely rises or falls more than 1 per cent from day to day.

Therefore, the major feedback mechanism for control of sodium concentration (and also for extracellular osmolality) is the ADH-thirst mechanism. In the absence of this dual mechanism there is no feedback mechanism that will cause the body to increase water ingestion or water conservation by the kidneys when excess sodium enters the body. Therefore, the sodium concentration simply increases.

Effect of Cardiovascular Reflexes on the ADH-Thirst Control System

Two cardiovascular reflexes also have powerful effects on the ADH-thirst mechanism: (1) the *arterial baroreceptor reflex* and (2) the *volume receptor reflex*, both of which are discussed in Chapter 18. When the blood volume falls, both of these cause increased secretion of ADH and increased thirst. That is, a decrease in blood volume causes the arterial pressure to fall and activates the arterial baroreceptor reflex. And the volume receptor reflex is activated when the pressures in the two atria, in the pulmonary artery, and in other low pressure areas of the lesser circulation fall below normal, all a common result of too little volume in the circulation. The net result is to activate the ADH-thirst system and thereby to increase the body fluid volume.

To compare the osmolality effects in activating the ADH system with the circulatory reflex effects, Fig-

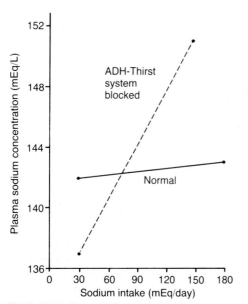

Figure 28–7. Effect on the extracellular fluid sodium concentration in dogs caused by tremendous changes in sodium intake (1) under normal conditions, and (2) after the antidiuretic hormone and thirst feedback systems had been blocked. This figure shows lack of sodium ion control in the absence of these systems. (Courtesy of Dr. David B. Young.)

Figure 28–8. Effect of changes in plasma osmolality or blood volume on the level of plasma antidiuretic hormone (ADH) (arginine vasopressin [AVP]). (From Dunn et al.: *J. Clin. Invest., 52*:3212, 1973. By copyright permission of the American Society for Clinical Investigation.)

ure 28–8 illustrates by the open circles the effect of increases in body fluid osmolality to cause ADH secretion and by the solid circles the effect of decreased blood volume.

● SODIUM EXCRETION AND ITS CONTROL BY ALDOSTERONE

The glomerular filtrate normally contains about 26,000 mEq of sodium each day, and yet the average intake of sodium per day is only 150 mEq. Therefore, the kidneys can be allowed to excrete only about 150 of the total 26,000 mEq; otherwise the body would become depleted of sodium. Consequently, the principal role of the tubular system in sodium excretion is to reabsorb sodium, not to excrete it.

Reabsorption of Most of the Tubular Sodium in the Proximal Tubules and Loops of Henle. By the time the tubular fluid reaches the distal tubules, all but about 8 per cent of the sodium has already been reabsorbed. About 65 per cent of this is reabsorbed in the proximal tubules because of the active transport of sodium by the proximal tubular epithelial cells. Also, as explained in the previous chapter, when the sodium is reabsorbed, the positive charge of the sodium ions causes passive diffusion or co-transport of negative ions, especially chloride ions, through the epithelium as well. Then the cumulative reabsorption of ions creates an osmotic pressure difference that moves water through the membrane, too. In fact, the epithelium is so permeable to water that almost the same identical proportion of water and sodium ions is reabsorbed.

In the thick portion of the ascending limb of the loop of Henle approximately another 27 per cent of the sodium is reabsorbed, leaving only 8 per cent to enter the distal tubules. However, reabsorption of sodium here is quite different from that in the proximal tubules because this thick segment of the loop is almost impermeable to water, as noted earlier in this chapter. Therefore, the concentration of sodium ions falls quite low in the ascending limb tubular fluid before it enters the distal tubules, down to as little as one quarter of its concentration in the plasma — to a concentration of 30 to 40 mEq/liter in contrast to the plasma level of 140 mEq/liter.

Thus, the proximal tubules and the loops of Henle are responsible for returning most of the sodium that enters the tubular system in the glomerular filtrate back to the plasma, thereby conserving the sodium for the body.

Variable Reabsorption of Sodium in the Late Distal Tubules and Cortical Collecting Ducts — Role of Aldosterone

Sodium reabsorption in the late distal tubules, and the cortical collecting ducts is highly variable. Its rate of reabsorption is controlled mainly by the concentration in the blood of *aldosterone,* a hormone secreted by the adrenal cortex. In the presence of large amounts of aldosterone, almost the last vestiges of tubular sodium are reabsorbed from these portions of the tubular system so that essentially none of the sodium enters the urine. On the other hand, in the absence of aldosterone, almost all of the sodium that enters the late distal tubules, about 800 mEq/day, is not reabsorbed and does then pass into the urine. Thus, the sodium excretion may be as little as 0.1 gram per day or as great as 20 grams per day, depending on the amount of aldosterone that is secreted.

Mechanism by Which Aldosterone Enhances Sodium Transport. Upon entering a tubular epithelial cell, aldosterone combines with a *receptor protein;* this combination diffuses within minutes into the nucleus, where it activates the DNA molecules to form one or more types of messenger RNA. The RNA is then believed to cause formation of carrier proteins or protein enzymes that are necessary for the sodium transport process. Various theories have suggested (1) that a specific protein increases the permeability of the luminal border of the cell to sodium, (2) that an increased quantity of Na^+,K^+-ATPase (which is the sodium pumping protein) develops in the basolateral membrane of the epithelial cell, or (3) that a protein enzyme or enzymes increase the availability of ATP to the ATPase so that it can function more actively. Unfortunately, the precise mechanism is yet unknown.

Ordinarily, aldosterone has no effect on sodium transport for the first 45 minutes after it is administered; after this time the specific proteins important for transport begin to appear in the epithelial cells, followed by progressive increase in transport during the ensuing few hours.

Control of Aldosterone Secretion. Aldosterone is secreted by the zona glomerulosa cells in the outer cortex of the adrenal glands, as is discussed in much greater detail in Chapter 77. Three different factors are known to stimulate this secretion of aldosterone. These are (1) increased angiotensin II in the blood, (2) increased extracellular fluid potassium ion concentration, and (3) decreased extracellular fluid sodium ion concentration. The second of these is important for the control of potassium ion concentration, as is discussed in the following chapter. The first and the third are especially important for the control of sodium excretion by the kidneys as well as extracellular volume.

When the extracellular fluid volume falls too low, the resulting effects on the circulatory system are to decrease the arterial pressure and to increase the reflex stimulation of the sympathetic nervous system. Both of these reduce blood flow through the kidneys and excite the secretion of renin, as is explained in Chapter 19. The renin in turn causes the formation of angiotensin I that is later converted to angiotensin II. Finally, the angiotensin II has a direct effect on the zona glomerulosa cells to increase the secretion of aldosterone.

In addition, a decrease in extracellular fluid sodium

ion concentration also seems to have a direct weak stimulatory effect on aldosterone secretion, though this effect is much, much less potent than the effect of angiotensin II.

Thus, whenever either the extracellular fluid volume or the extracellular fluid sodium ion concentration falls below normal, aldosterone is secreted and the renal tubules reabsorb extra quantities of sodium and water, leading to return of both extracellular fluid sodium and extracellular fluid volume back toward normal.

Relative Unimportance of the Aldosterone Feedback Mechanism for Determining Sodium Ion *Concentration* **Under Normal Conditions.** Although aldosterone increases the *quantity* of sodium in the extracellular fluid, the increased reabsorption of water along with the sodium usually prevents a rise in sodium *concentration* but instead mainly increases the total quantity of extracellular fluid. This is illustrated by the experiment of Figure 28-9. This figure shows the effect on sodium concentration of more than a sixfold increase in sodium intake in the same dog (1) under normal conditions and (2) after the aldosterone feedback system had been blocked—that is, the adrenal glands had been removed, and the animals were infused at a constant rate of aldosterone that could neither change upward nor downward. Note that the sodium concentration changed no more than 1 to 2 per cent in either case. In other words, even without a functional aldosterone feedback system (because the aldosterone could neither change upward nor downward) sodium concentration was still very well regulated.

Because of the great effect that aldosterone has on tubular sodium reabsorption, this lack of importance of aldosterone feedback control of sodium concentration seems to be a paradox, but it results from the following simple effect: When the aldosterone causes increased sodium reabsorption from the tubules, as discussed earlier, this causes a simultaneous reabsorption of water and an increase in extracellular fluid volume. An increase of only a few per cent in the extracellular fluid volume eventually leads to an increase in arterial pressure, and the increase in arterial pres-

sure leads to increased glomerular filtration rate, a well-known effect in the presence of excess aldosterone. The rapid flow of filtrate down the tubular system then overrides the excessive reabsorptive effect of the aldosterone and thereby almost completely nullifies the effect of aldosterone on extracellular fluid sodium concentration.

Furthermore, as was explained earlier, the ADH-thirst system is an extremely powerful controller of sodium concentration—much more powerful than the aldosterone feedback system—so that the ADH-thirst system greatly overshadows the aldosterone system for sodium concentration control under normal conditions. Indeed, even in patients who have *primary aldosteronism* (these patients secrete tremendous quantities of aldosterone), the sodium concentration still rises only 2 to 3 mEq/liter above normal.

CONTROL OF SODIUM INTAKE BY THE SALT APPETITE MECHANISM

Maintenance of normal extracellular sodium requires not only control of sodium excretion but also control of sodium intake. Fortunately, the body uses the salt appetite mechanism to control sodium intake, which is analogous to the thirst mechanism for control of water intake.

In the same way that two major stimuli elicit thirst, two major stimuli also excite salt appetite: (1) decreased sodium concentration in the extracellular fluid and (2) circulatory insufficiency, often caused by decreased blood volume. However, a major difference between thirst and salt appetite is that thirst is elicited almost immediately, whereas the craving for salt usually begins only after several hours then builds progressively more and more.

The neuronal mechanism for salt appetite also is analogous to that of the thirst mechanism. Some of the same neuronal centers in the AV3V region of the brain seem to be involved, for lesions in this region frequently affect both thirst and salt appetite together. Also, circulatory reflexes elicited by low blood pressure or decreased blood volume feed through the medulla of the brain to the AV3V region to affect thirst and salt appetite together.

The importance of salt craving is especially illustrated in persons who have Addison's disease. In persons with this disease virtually no aldosterone is secreted, which causes excessive loss of salt in the urine and leads to both low sodium ion concentration in the extracellular fluid and decreased blood volume; both of these strongly excite the desire for salt. Likewise, it is well known that animals living in the wild where little salt often is in the food will actively search out mineral deposits called "salt licks" that contain concentrated salt.

REFERENCES

Andersson, B., and Rundgren, M.: Thirst and its disorders. Annu. Rev. Med., 33:231, 1982.

Arendshorst, W. J.: Altered reactivity of tubuloglomerular feedback. Annu. Rev. Physiol., 49:295, 1987.

Baer, P. G., and McGiff, J. C.: Hormonal systems and renal hemodynamics. Annu. Rev. Physiol., 42:589, 1980.

Bell, P. D., et al.: Calcium as a mediator of tubuloglomerular feedback. Annu. Rev. Physiol., 49:275, 1987.

Blaustein, M. P., and Hamlyn, J. M.: Sodium transport inhibition, cell calcium, and hypertension. The natriuretic hormone/Na+−Ca2+ exchange/hypertension hypothesis. Am. J. Med., 77:45, 1984.

Brenner, B. M., and Rector F. C.: The Kidney, 3rd Ed. Philadelphia, W. B. Saunders Co., 1986.

Briggs, J. P., and Schnermann, J.: The tubuloglomerular feedback mecha-

Figure 28-9. Effect on extracellular fluid sodium concentration in dogs caused by tremendous changes in sodium intake (1) under normal conditions and (2) after the aldosterone feedback system had been blocked. Note that sodium is exceedingly well controlled with or without aldosterone feedback control. (Courtesy of Dr. David B. Young.)

nism: Functional and biochemical aspects. Annu. Rev. Physiol., 49:251, 1987.

Brownstein, M. J.: Biosynthesis of vasopressin and oxytocin. Annu. Rev. Physiol., 45:129, 1983.

Buckalew, B. M., Jr., and Gruber, K. A.: Natriuretic hormone. Annu. Rev. Physioll., 46:343, 1984.

Bulger, R. E.: Composition of renal medullary tissue. Kid. Int., 31:556, 1987.

Cowley, A. W. Jr., et al.: Vasopressin: Cellular and Integrative Functions. New York, Raven Press, 1988.

Dawson, K. G.: Endocrine physiology of electrolyte metabolism. Drugs, 28(Suppl. 1):98, 1984.

deRouffignac, C., et al.: Physiological control of the urinary concentrating mechanism by peptide hormones. Kid. Int., 31:611, 1987.

Doris, P. A.: Vasopressin and central integrative processes. Neuroendocrinology, 38:75, 1984.

Doucet, A.: Multiple hormonal control of kidney tubular functions. News in Physiol. Sci., 2:141, 1987.

Earley, L, E., and Schrier, R. W.: Intrarenal control of sodium excretion by hemodynamic and physical factors. In Orloff, F., and Berliner, R. W. (eds.): Handbook of Physiology. Sec. 8. Baltimore, Williams & Wilkins, 1973, p. 721.

Ehrlich, E. N.: Adrenocortical regulation of salt and water metabolism: Physiology, pathophysiology, and clinical syndromes. In DeGroot, I. J., et al. (eds.): Endocrinology. Vol. 3. New York, Grune & Stratton, 1979, p. 1883.

Epstein, M., et al.: β-Blockers and the kidney: Implications for renal function and renin release. Physiologist, 28:53, 1985.

Fanestil, D. D., and Park, C. S.: Steroid hormones and the kidney. Annu. Rev. Physiol., 43:637, 1981.

Fitzsimons, J. T.: Thirst. Physiol. Rev., 52:468, 1972.

Flatman, P. W.: Magnesium transport across cell membranes. J. Membr. Biol., 80:1, 1984.

Fregley, M. J., and Rowland, N. E.: Role of renin-angiotensin-aldosterone system in NaCl appetite of rats. Am. J. Physiol., 248:R1, 1985.

Gash, D. M., and Boer, G. J. (eds.): Vasopressin. New York, Plenum Publishing Corp., 1987.

Gibbons, G. H., et al.: Interaction of signals renin release. Annu. Rev. Physiol., 46:291, 1984.

Goetz, K. L., et al.: Atrial receptors and renal function. Physiol. Rev., 55:157, 1975.

Gottschalk, C. W.: History of the urinary concentrating mechanism. Kid. Int., 31:507, 1987.

Greenwald L., and Stetson, D.: Urine concentration and the length of the renal papilla. News Physiol. Sci., 3:46, 1988.

Greger, R., and Valazquez, H.: The cortical thick ascending limb and early distal convoluted tubule in the urinary concentration mechanism. Kid. Int., 31:590, 1987.

Gregor, R.: Ion transport mechanisms in thick ascending limb of Henle's loop of mammalian nephron. Physiol. Rev., 65:760, 1985.

Guyton, A. C., et al.: A system analysis of volume regulation. Alfred Benzon Symposium XI. Munksgaard, 1978.

Guyton, A. C., et al.: Dynamics and Control of the Body Fluids. Philadelphia, W. B. Saunders Co., 1975.

Guyton, A. C., et al.: Theory for renal autoregulation by feedback at the juxtaglomerular apparatus. Circ. Res., 14:187, 1964.

Hall, J. E., et al.: Control of glomerular filtration rate by the renin-angiotensin system. Am. J. Physiol., 233:F366. 1977.

Hall, J. E., et al.: Dissociation of renal blood flow and filtration rate autoregulation by renin depletion. Am. J. Physiol., 232:F215, 1977.

Hall, J. E., et al.: Intrarenal control of electrolyte excretion by angiotensin II. Am. J. Physiol., 232:F538, 1977.

Hall, J. E., et al.: Intrarenal role of angiotensin II and (des-Asp) angiotensin II. Am J. Physiol., 236:F252, 1979.

Handler, J. S., and Orloff, J.: Antidiuretic hormone. Annu. Rev. Physiol., 43:611, 1981.

Harrison, M. H.: Effects of thermal stress and exercise on blood volume in humans. Physiol. Rev., 65:149, 1985.

Hays, R. M., et al.: Effects of antidiuretic hormone on the collecting duct. Kid. Int., 31:530, 1987.

Hayward, J. N.: Neural control of the posterior pituitary. Annu. Rev. Physiol., 37:191, 1975.

Hebert, S. C., et al.: The medullary thick limb: Function and modulation of the single-effect multiplier. Kid. Int., 31:580, 1987.

Imai, M., et al.: Function of thin loops of Henle. Kid. Int., 31:565, 1987.

Jackson, T. E., et al.: Transient response of glomerular filtration rate and renal blood flow to step changes in arterial pressure. Am. J. Physiol., 233:F396, 1977.

Jamison, R. L.: Short and long loop nephrons. Kid. Int., 31:597, 1987.

Koepke, J. P., and DiBona, G. F.: Functions of the renal nerves. Physiologist, 28:47, 1985.

Kokko, J. P.: The role of the collecting duct in urinary concentration. Kid. Int., 31:606, 1987.

Marsh, D. J.: Functional transitions in the renal countercurrent system. Kid. Int., 31:668, 1987.

McKinley M. J.: An important region for osmoregulation: The anterior wall of the third ventricle. News Physiol. Sci., 2:13, 1987.

Menninger, R. P.: Current concepts of volume receptor regulation of vasopressin release. Fed. Proc., 44:55, 1985.

Miller, M.: Assessment of hormonal disorders of water metabolism. Clin. Lab. Med., 4:729, 1984.

Naftalin, R. J.: The thermostatics and thermodynamics of cotransport. Biochem. Biophys. Acta, 778:155, 1984.

Navar, L. G., et al.: Distal tubular feedback control of renal hemodynamics and autoregulation. Annu. Rev. Physiol., 42:557. 1980.

O'Connor, W. J.: Normal Renal Function. The Excretion of Water and Electrolytes Derived from Food and Drink. New York, Oxford University Press. 1982.

Rasch, R. W.: A microprocessor analog simulation of the operation of countercurrent multiplier systems. Physiologist, 27:58, 1984.

Sawyer, W. H., and Manning, M.: The use of antagonists of vasopressin in studies of its physiological functions. Fed. Proc., 44:78, 1985.

Seldin, D. W., and Giebisch, G.: The Regulation of Acid-Base Balance. New York, Raven Press, 1989.

Sklar, A. H., and Schrier, R. W.: Central nervous system mediators of vasopressin release. Physiol. Rev., 63:1243, 1983.

Stein, J. H., and Reineck, H. H.: Effect of alterations in extracellular fluid volume on segmental sodium transport. Physiol. Rev., 55:127, 1975.

Stephenson, J. L.: Models of the urinary concentrating mechanism. Kid. Int., 31:648, 1987.

Stricker, E. M., and Verbalis, J. G.: Hormones and behavior: The biology of thirst and sodium appetite. Am. Scientist, 76:261, 1988.

Sutton, R. A.: Diuretics and calcium metabolism. Am. J. Kidney Dis., 5:4. 1985.

Swanson. L. W., and Sawchenko. P. E.: Hypothalamic integration: Organization of the paraventricular and supraoptic nuclei. Annu. Rev. Neurosci., 6:269, 1983.

Thomsen, K., and Schou, M.: Lithium clearance: A new research area. News in Physiol. Sci., 1:126, 1986.

Valtin, H.: Renal Dysfunction: Mechanisms Involved in Fluid and Solute Imbalance. Boston, Little, Brown, 1979.

Verney, E. B.: Absorption and excretion of water: Antidiuretic hormone. Lancet, 2:739, 1946.

Young, D. B., et al.: Control of extracellular sodium concentration by antidiuretic hormone-thirst feedback mechanism. Am. J. Physiol., 232:R145, 1977.

Young, D. B., et al.: Effectiveness of the aldosterone-sodium and -potassium feedback control system. Am. J. Physiol., 231:945, 1976.

Zerbe, R. et al.: Vasopressin function in the syndrome of inappropriate antidiuresis. Annu. Rev. Med., 31:315, 2980.

CHAPTER 29

• *Renal Regulation of Blood Volume and Extracellular Fluid Volume; Excretion and Regulation of Urea, Potassium, and Other Substances* •

One mechanism above all others dominates the control of blood volume and extracellular fluid volume. This is the effect of blood volume on arterial pressure, on the one hand, and the effect, on the other hand, of arterial pressure on urinary excretion of sodium and water. This interplay between the kidneys and the circulatory system is so powerful in controlling blood volume—and secondarily extracellular fluid volume as well—that it dominates over the other mechanisms that help to control salt and water intake or salt and water excretion.

● CONTROL OF BLOOD VOLUME

Constancy of the Blood Volume. The extreme degree of precision with which the blood volume is controlled is illustrated in Figure 29–1. This shows that almost no change in blood volume occurs despite tremendous changes in daily intake of water and electrolytes, except when the intake becomes so low that it is not sufficient to make up for the fluid losses caused by evaporation or other inescapable losses.

Pressure Natriuresis and Pressure Diuresis as the Major Basis for Blood Volume Control

Although the hormonal factors discussed in the last two chapters—antidiuretic hormone (ADH), aldosterone, angiotensin, and atrial natriuretic factor (ANF)—all play roles in the control of blood volume,

320

by far the most important basis for blood volume control is a purely mechanical mechanism: the mechanical effect of increased arterial pressure to cause greatly increased fluid volume output by the kidneys. This is called *pressure diuresis*. Increased salt output also occurs along with the diuresis; this is called *pressure natriuresis*.

Figure 29–2 illustrates this effect of arterial pressure on urinary sodium or volume output. Note that an increase in arterial pressure up to double normal increases urinary output about eightfold. Fluid is then lost rapidly from the body until the blood volume becomes small enough to return the arterial pressure to its normal value. Conversely, when the arterial pressure falls from normal to about 50 mm Hg, urinary output falls essentially to zero. Then fluid builds up in the body, because of continued fluid intake, until the blood volume rises high enough to return the arterial pressure to its normal value.

Thus, this simple mechanical effect of pressure on urinary output provides a negative feedback mechanism for control of blood volume. Following are the specific steps in this feedback control system.

The Overall Mechanism for Blood Volume Control. The overall mechanism for blood volume control is illustrated in Figure 29–3. This is essentially the same as the basic mechanism for arterial pressure control that was presented in Chapter 19. It was pointed out in that chapter that extracellular fluid volume, blood volume, cardiac output, arterial pressure, and urine output are all controlled at the same time as separate parts of a common basic feed-

Figure 29-1. Approximate effect on blood volume caused by marked changes in daily fluid intake. Note the precision of blood volume control in the normal range.

back mechanism. There are six major steps in this mechanism as they relate to blood volume control:

1. An increase in *blood volume increases cardiac output.*

2. An increase in cardiac output increases *arterial pressure.*

3. An increase in *arterial pressure increases urinary output;* the solid, steep curve in this block indicates that the effect of pressure on the output is especially potent when the pressure remains elevated for days or weeks at a time.

4. This block shows *summation of the fluid intake minus the fluid losses* from the body, the losses including both urinary output and other fluid loss. The output of this block is the *rate of change of extracellular fluid volume.* If the intake is greater than the output, the rate of change will be positive; if the output is greater than the intake, the rate of change will be negative.

5. This block *integrates the rate of change of extracellular fluid volume*—that is, it shows the accumulation of greater or less fluid volume with time, depending on whether the rate of change is positive or negative. The output of block 5 is the actual extracellular fluid volume.

6. This block gives the *relationship between extra-cellular fluid volume and blood volume,* showing that, in general, *as the extracellular fluid volume increases, the blood volume also increases.*

Summary of the Basic Blood Volume Control Mechanisms. To summarize the principles illustrated in Figure 29-3, we can trace what happens when the blood volume becomes abnormal. When the blood volume becomes too great, the cardiac output also becomes too great and therefore increases the arterial pressure. This, in turn, has a profound effect on the kidneys, causing loss of fluid from the body and returning the blood volume to normal. Conversely, if the blood volume falls below normal, the cardiac output and arterial pressure decrease, the kidneys retain fluid, and progressive accumulation of the ingested fluid builds the blood volume eventually back to normal.

Usually, parallel processes will also occur to reconstitute red cell mass, plasma proteins, and so forth if these have become abnormal at the same time. However, if the red cell volume remains abnormal, the plasma volume will simply make up the difference, the volume becoming essentially normal despite the low red cell mass.

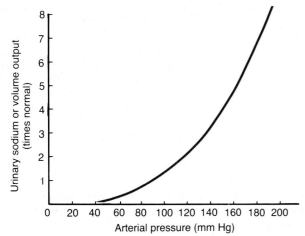

Figure 29-2. Effect of arterial pressure change on urinary output.

Figure 29-3. Basic feedback mechanism for control of blood volume and extracellular fluid volume (the points on each of the curves represent normal values).

Reason for the Precision of the Blood Volume —Regulating Mechanism. Studying the diagram of Figure 29–3 very carefully, one can see why the blood volume remains almost exactly constant despite extreme changes in daily fluid intake. The reason for this is the slopes of the curves in blocks 1, 2, and 3 are all very steep, meaning that (a) a slight change in blood volume causes a marked change in cardiac output, (b) a slight change in cardiac output causes a marked change in arterial pressure, and (c) a slight change in arterial pressure causes a marked change in urinary output. These factors all multiply together to give an extremely high gain for the feedback control of blood volume.

EFFECT OF NERVOUS AND HORMONAL FACTORS ON THE CONTROL OF BLOOD VOLUME

The discussions of the previous chapter made it clear that both nervous and hormonal factors can cause acute changes in salt and water excretion by the kidneys. These acute effects are often dramatic in short-term experiments; therefore, most physiology texts emphasize greatly these mechanisms of blood volume control. However, after full adjustment of the basic renal mechanism for blood volume control discussed above, the chronic changes in blood volume caused by any one of the nervous or hormonal factors are usually no more than 5 to 10 percent. Let us speak of each of these separately.

Arterial Baroreceptor and Low-Pressure Stretch Receptor Reflexes — The "Volume Reflex." When the blood volume becomes too great, the arterial pressure rises; also, the pressures in the pulmonary artery and other low-pressure regions of the thorax increase. The stretch of the arterial baroreceptors and of the low-pressure stretch receptors in turn cause reflex inhibition of the sympathetic nervous system, which acutely dilates the renal arterioles and often allows an immediate severalfold increase in urinary output. For this reason, this set of reflexes is sometimes called the *volume reflex.*

However, under chronic conditions, total sympathetic denervation is not associated with a significant change in blood volume. Also, in patients who have norepinephrine- and epinephrine-secreting tumors, with tremendous output of these sympathetic hormones, the blood volume has been reported to be either normal or decreased only 5 to 10 per cent. These hormones have the effect on the kidneys of causing retention of fluid, but they have the simultaneous vasoconstrictor effect on the blood vessels of decreasing the capacity of the circulatory system. It is the second of these that seems to prevail, so that the blood volume might decrease a few percentage points.

Effect of Atrial Natriuretic Factor. Overstretching the two atria as a result of excess blood volume causes the release of ANF into the blood. This acts on the kidneys acutely to increase sodium and volume excretion as much as 3- to 10-fold. However,

this extreme effect on urine output does not last, so that in the chronic state, there is only a suggestion of slightly reduced blood volume associated with a slight reduction in arterial pressure as well.

Effect of Aldosterone. Aldosterone causes very strong reabsorption of sodium in the distal segments of the renal tubular system. At first, almost all sodium entering the glomerular filtrate may be reabsorbed by the kidneys and retained in the blood, and this causes a secondary reabsorption of water as well. As a result, the blood volume increases during the first day or so as much as 10 to 20 per cent. Then, as the arterial pressure rises in response to the increasing blood volume, the phenomenon of "aldosterone escape" occurs, which means that the kidneys begin to excrete thereafter an amount of sodium equal to the daily intake despite the continued presence of aldosterone. The reason for the escape is mainly the pressure natriuresis and pressure diuresis that occurs when the arterial pressure rises. In experiments in which the renal arterial pressure has been prevented from rising along with the rise in arterial pressure, pressure natriuresis is prevented and escape also does not occur.

After aldosterone escape, the blood volume returns within a few days to only 5 to 10 per cent above normal, which is associated with continued moderate elevation of the arterial pressure.

Only in patients who secrete no aldosterone at all (patients with Addison's disease, who often excrete as much as 20 grams of sodium per day) is there a significant change in volume, because the sodium depletion in turn causes depletion of water as well. However, this goes far beyond the normal pattern of blood volume control. And even then, if the person is allowed to eat all the salt desired, there may be no more than a 5 per cent decrease in blood volume.

Role of Angiotensin. Neither a decrease in circulating angiotensin II to zero nor a great increase in angiotensin has more than a 5 to 10 per cent chronic effect on blood volume, despite the fact that angiotensin has both the direct renal effect of causing renal retention of salt and water and an indirect effect as well, because angiotensin stimulates aldosterone secretion.

Effect of Antidiuretic Hormone. Infusion of large amounts of ADH (also called *vasopressin*) into animals can cause so much renal retention of water that the blood volume increases 15 to 25 per cent during the first few days. However, the arterial pressure also rises and then causes excretion of the excess volume. After several weeks of continued infusion, the blood volume measures no more than 5 to 10 per cent above normal, and the arterial pressure also returns to within 10 mm Hg of normal. Thus, there is not a serious volume change. However, the animals do usually develop severe reduction in extracellular sodium ion concentration. The reason for this is that the water retained by the kidneys dilutes the extracellular sodium, and at the same time the slight pressure rise causes excess sodium loss in the urine.

In patients who have lost their ability to secrete

ADH because of destruction of the supraoptic nuclei, the urine output may be as great as 5 to 10 times normal day in and day out, but this is virtually always compensated for by ingestion of enough water to make up the difference. Therefore, the blood volume is rarely decreased more than 5 per cent as long as sufficient water is available.

Summary

In summary, within the normal ranges of operation of the nervous reflexes and the various hormonal factors that affect urinary sodium and fluid volume output, these rarely affect the blood volume more than plus or minus 5 to 10 per cent. However, even these slight changes in volume can sometimes have considerable effect on the long-term arterial pressure level.

The most important fundamental controller of blood volume, therefore, is the combined natriuresis and pressure diuresis mechanism, whereby changes in arterial pressure cause marked changes in rate of urinary output. This mechanism controls at the same time the arterial pressure, blood volume, extracellular fluid volume, and renal excretion of salt and water, all together in a common feedback loop. The importance of this mechanism for controlling of *arterial pressure* was also presented in Chapter 19.

Conditions That Cause Large Increases in Blood Volume

There are three conditions, none of them related to nervous or hormonal mechanisms, that often cause large increases in blood volume. All of these result from circulatory abnormalities as follows.

Increased Blood Volume Caused by Heart Disease. The blood volume is often increased as much as 15 to 20 per cent as a result of heart disease and occasionally as much as 30 to 40 per cent. The reason is that a weak heart often cannot attain a high enough arterial pressure to cause necessary urinary output of fluid. Therefore, the kidneys retain volume until either the person dies or the input pressure to the heart becomes great enough to overprime the heart, thus finally achieving the necessary arterial pressure to provide normal urine output. Therefore, in myocardial failure, in heart valvular disease, and in congenital abnormalities of the heart — in all of these — one of the most important circulatory compensations is increased blood volume. This allows even the weakened heart to pump a life-sustaining level of cardiac output.

Increased Blood Volume in Polycythemia. In persons who have a great excess of red blood cells (polycythemia), blood viscosity increases. This in turn greatly increases the peripheral resistance and even resistance to blood flow in the venules and small veins, which tends to reduce venous return to the heart. The compensation for this is that the kidneys retain fluid until the blood volume becomes great enough to provide normal venous return. When this is achieved, the arterial pressure is high enough to provide the necessary urinary output and to prevent further fluid retention. In *polycythemia vera*, a condition in which the bone marrow produces high numbers of red blood cells for an unknown reason (presumably a tumorous condition of the red cell production system), blood volume frequently increases to greater than 50 per cent above normal.

Increased Blood Volume Caused by Increased Capacity of the Circulation. Any condition that increases the total capacity of the circulatory system will also cause the blood volume to increase to fill up this extra capacity. For instance, in a pregnant mother, the extra vascular capacity of the uterus, the placenta, and other enlarged organs of the mother's body regularly increases the blood volume 15 to 25 per cent. In a similar manner, patients with very large varicose veins in the legs, which in rare instances may hold as much as a liter of extra blood, the blood volume simply increases to fill up the extra capacity. That is, salt and water are retained by the kidneys until the total vascular capacity is filled enough to raise the arterial pressure to the level required to balance renal output of fluid with daily intake of fluid.

● CONTROL OF EXTRACELLULAR FLUID VOLUME

It is clear from the foregoing discussion of the basic mechanisms for blood volume control that extracellular fluid volume is controlled at the same time. That is, fluid first goes into the blood, but it rapidly becomes distributed between the interstitial spaces and the plasma. Therefore, it is impossible to control blood volume to any given level without controlling the extracellular fluid volume at the same time. Yet the relative volumes of distribution between the interstitial spaces and the blood can vary greatly, depending on the physical characteristics of the circulatory system and of the interstitial spaces. Under normal conditions, the fluid in the interstitial spaces is bound in a gel-like matrix of proteoglycan molecules, and there is essentially no free fluid. At other times, however, abnormal conditions can cause edema to occur. These abnormal conditions were discussed in detail in Chapter 25. The principal factors that can cause edema are (1) increased capillary pressure, (2) decreased plasma colloid osmotic pressure, (3) increased tissue colloid osmotic pressure, and (4) increased permeability of the capillaries. Whenever any one of these conditions occurs, a very high proportion of the extracellular fluid becomes distributed to the interstitial spaces.

Normal Distribution of Fluid Volume Between the Interstitial Spaces and Vascular System. Figure 29–4 illustrates the approximate normal relationship between extracellular fluid volume and blood volume. When an extra amount of fluid accumulates in the extracellular fluid space, either as a

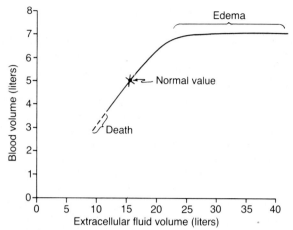

Figure 29-4. Approximate relationship between extracellular fluid volume and blood volume, showing a nearly linear relationship in the normal range but indicating failure of the blood volume to continue rising when the extracellular fluid volume becomes excessive.

result of intake of too much fluid or because of decreased kidney output of fluid, about one sixth to one third of the extra fluid normally stays in the blood and increases the blood volume. The remainder of the fluid is distributed to the interstitial spaces. However, when the extracellular fluid volume rises more than 30 to 50 per cent above normal, as shown in the figure, very little of the additional fluid then remains in the blood—almost all of it instead goes into the interstitial spaces. This occurs because the interstitial fluid pressure rises from its normal negative (subatmospheric) value to a positive value, which causes the loose tissues to swell rapidly and allows fluid literally to pour out of the capillaries. That is, the interstitial spaces literally become an "overflow" reservoir for excess fluid, sometimes increasing in volume as much as 10 to 30 liters. This obviously causes edema, as was explained in Chapter 25, but it also acts as an important overflow release valve for the circulatory system, a well-known phenomenon that is utilized daily by the clinician to allow administration of almost unlimited quantities of intravenous fluid and yet not force the heart into cardiac failure.

To summarize, extracellular fluid volume is controlled simultaneously with the control of blood volume, but the relative ratio of the extracellular fluid volume to blood volume depends on the physical properties of the circulation and of the interstitial spaces and also on the dynamics of fluid exchange through the capillary membranes.

● UREA EXCRETION

The body forms an average of 25 to 30 grams of urea each day—more than this in persons who eat a very high protein diet and less in persons who are on a low protein diet. All this urea must be excreted in the urine; otherwise, it will accumulate in the body fluids. Its normal concentration in plasma is approximately

26 mg/dl, but patients with renal insufficiency frequently have levels as high as 200 mg/dl, and it has been recorded as high as 800 mg/dl.

The two major factors that determine the rate of urea excretion are (1) the concentration of urea in the plasma, and (2) the glomerular filtration rate. These factors increase urea excretion mainly because the "load" of urea entering the proximal tubules is equal to the *product of the plasma urea concentration and the glomerular filtration rate.* In general, the quantity of urea that passes on through the tubules into the urine averages between 40 and 60 per cent of the urea load that enters the proximal tubules. Only when the function of kidneys is abnormal does this fraction of tubular load change much from this normal range.

Effect of Decreased Glomerular Filtration Rate on Urea Excretion and on Plasma Urea Concentration. In many renal diseases, the glomerular filtration rate of the two kidneys falls considerably below normal. Because urea excretion is directly related to the glomerular filtration of urea, there is decreased excretion of urea when the glomerular filtration decreases. However, the body continues to form large quantities of urea, which means that urea will progressively collect in the body fluids until its plasma concentration rises very high. Then the urea filtered in the glomerular filtrate, which is equal to the plasma urea concentration times the glomerular filtration rate, eventually will become great enough to allow excretion of the urea as rapidly as it is formed. Yet one will recognize that this occurs only because the plasma concentration of urea has risen very high, and this in itself is a very abnormal state that can be severely damaging to the body.

Therefore, probably the most important reason for forming large amounts of glomerular filtrate each day by the two kidneys is to excrete the necessary amounts of urea.

Many of the other waste products that must be excreted by the kidneys obey the same principles of excretion as urea, for their rates of excretion also depend highly upon the amount of glomerular filtrate formed each day. Such substances include creatinine, uric acid, and various other toxic waste products, which will be discussed more fully in Chapter 31.

The Mechanism for Excreting Large Quantities of Urea but Minimal Quantities of Water. Were it not for the ability of the tubular system to concentrate urea, the kidneys could excrete the necessary urea only by passing large amounts of water into the urine at the same time. Instead, urea is normally concentrated at least 50-fold and when water is being conserved by the body and a concentrated urine is excreted, sometimes as much as several hundred-fold. Figure 29-5 illustrates this progressive increase in urea concentration in the tubular fluid as it passes through the tubular system. The numbers given in the diagram are those recorded during moderate antidiuresis, which means formation of small amounts of concentrated urine. Note that the concentration of urea in the proximal tubule fluid is only 4.5 mOsm/

Figure 29-5. Recirculation of the urea absorbed from the collecting duct, this passing into the loop of Henle, through the distal tubule, and finally back to the collecting duct. (Numerical values are milliosmolalities of urea during antidiuresis caused by the presence of large amounts of antidiuretic hormone.)

liter, but this increases about 100 times (to 450 mOsm/liter) by the time the urea is excreted in the urine.

Recirculation of Urea From the Collecting Duct to the Loop of Henle as a Special Means for Concentrating the Urea. Note in Figure 29-5 the high concentration of urea in the lower collecting duct and also the reabsorption of urea from the medullary portion of this duct into the medullary interstitium, which increases the urea concentration there to high levels, to as high as 400 to 500 mOsm/liter. A moderate share of this urea is then reabsorbed into the thin limb of the loop of Henle so that it passes upward through the distal tubule, through the cortical collecting duct, and back down through the collecting duct again. In this way, urea can recirculate through these terminal portions of the tubular system several times before it is excreted. Each time around this circuit contributes to the high concentrations shown in the figure, so that very little water is excreted along with the urea.

Thus, this urea recirculation mechanism through the loop of Henle, the distal tubule, and the collecting tubule and duct is a way to concentrate urea in the medullary interstitium and in the urine at the same time. Since urea is the most abundant of the waste products that must be excreted by the kidneys, this mechanism for concentrating urea before it is excreted is essential to the economy of the body fluid when a person must live on short rations of water.

● POTASSIUM EXCRETION

The amount of potassium entering the glomerular filtrate each day is about 800 mEq, while the daily intake of potassium is only about 100 mEq. Therefore,

to maintain normal body potassium balance, one eighth of the total daily tubular load of potassium must be excreted. Furthermore, as is true for sodium excretion, the rate of potassium excretion must be carefully controlled so that it will exactly match the daily potassium intake.

Reabsorption of Large Amounts of Potassium in the Proximal Tubules and Loops of Henle. Large amounts of potassium are reabsorbed as a result of potassium co-transport with sodium in both the proximal tubules and the thick ascending limbs of the loops of Henle. Active transport by the proximal tubular epithelial cells reabsorbs about 65 per cent of all the filtered potassium. Then, approximately another 27 per cent is reabsorbed in the thick portion of the ascending limb of the loop of Henle, leaving about 8 per cent of the original filtered potassium to enter the distal tubules. This represents only 65 mEq of potassium per day, which is actually less than the average daily intake of 100 mEq of potassium for most people.

Additional Reabsorption of Potassium by the Distal Tubules and Cortical Collecting Ducts. In addition to the absorption of potassium by the proximal tubules and loops of Henle, a small amount of reabsorption of potassium occurs in the distal tubules and cortical collecting ducts. Although this reabsorption is slight, it is nevertheless still sufficient to remove almost all of the remaining tubular potassium. This is exactly what does happen when the extracellular fluid potassium concentration is very low, thus conserving potassium tenaciously in the body.

On the other hand, when the extracellular fluid potassium concentration is normal or high, large amounts of potassium need to be excreted into the urine simply to rid the body of the normal amounts of potassium ingested each day. For this purpose, the distal tubular segments have evolved a specific mechanism for secreting potassium, as follows.

Secretion of Potassium in the Late Distal Tubules and Cortical Collecting Ducts

In the *late distal tubules,* and even more so in the *cortical collecting ducts,* so-called *principal cells,* which make up about 90 per cent of the epithelial cells in these regions, have the special capability of secreting large quantities of potassium into the tubular lumen. However, this secretion occurs only when the extracellular fluid potassium concentration is higher than a critical level. The mechanism of this secretion is illustrated in Figure 29-6. It begins with the usual Na^+-K^+-adenosine triphosphatase (ATPase) pump in the basolateral membrane of the cell, which pumps sodium out of the cell into the interstitium and at the same time pumps potassium to the interior of the cell. In contrast to the epithelial cells elsewhere in the renal tubules, the luminal border of the principal cells is very permeable to potassium. Therefore, the great increase in potassium inside the cell promotes rapid diffusion of potassium also into the tubular lumen.

Figure 29–6. Mechanisms of Na$^+$ and K$^+$ transport through the distal tubular epithelium.

Thus, the basic driving force for the potassium secretory mechanism is the Na$^+$-K$^+$ pump in the basolateral membrane. For this pump to work, sodium ions must continually diffuse from the tubular lumen into the cell and then be exchanged at the basolateral membrane for potassium ion. Therefore, the greater the quantity of sodium that is available in the tubular lumen to diffuse into the cell, the greater also is the rate of potassium secretion. This is sometimes important clinically because persons on a low sodium diet often fail to excrete adequate amounts of potassium and therefore develop hyperkalemia.

Except when a person is on an extremely low intake of potassium, tubular secretion of potassium is usually necessary to prevent death. Cardiac arrhythmias often appear when the plasma potassium concentration rises from the normal value of 4.5 mEq/liter to a level of about 8 mEq/liter. Still higher potassium concentrations can end in cardiac arrest or fibrillation.

CONTROL OF POTASSIUM EXCRETION AND POTASSIUM CONCENTRATION IN THE EXTRACELLULAR FLUID

It is especially important to control the extracellular fluid potassium ion concentration because very slight changes in concentration can alter nervous and cardiac functions seriously, as discussed in Chapter 5 for peripheral nerves, in Chapter 45 for the central nervous system, and Chapter 10 for the heart. The normal concentration is about 4.5 mEq/liter, and this rarely rises or falls more than ±0.3 mEq.

Two major factors play important roles in controlling the potassium ion concentration: (1) the direct effect of increased extracellular fluid potassium concentration in causing increased secretion of potassium into the tubules, and (2) the effect of aldosterone in increasing potassium secretion.

Direct Effect of Extracellular Potassium Concentration on Potassium Secretion

The rate of secretion of potassium ions into the late distal tubules and cortical collecting ducts is directly stimulated by increased extracellular fluid potassium ion concentration, as illustrated by the *black curve* in Figure 29–7 — in fact, drastically so after the extracellular potassium concentration rises above about 4.1 mEq/liter. Therefore, this effect serves as one of the two very important feedback mechanisms for controlling the extracellular fluid potassium ion concentration.

A postulated mechanism to explain the increased potassium secretion when extracellular potassium increases is the following: The increased extracellular potassium increases the potassium gradient from the renal interstitial fluid to the interior of the epithelial cell. This decreases back-leakage of potassium ions from inside the cell through the basolateral membrane and also allows the very negative electrical potential inside the cell to attract more potassium ions inward through the basolateral membrane. These two effects together therefore increase the concentration of potassium ions inside the cell, which in turn increases the exchange of potassium for sodium across the luminal membrane. Regardless of the mechanism, it should be remembered that even a slight increase in the extracellular potassium concentration will cause a significant increase in potassium excretion into the urine.

Effect of Aldosterone in Controlling Potassium Ion Secretion

In the previous chapter we learned that active reabsorption of sodium ions by the principal cells in the late distal tubules and cortical collecting ducts is strongly controlled by the hormone aldosterone. Aldosterone also has the same powerful effect of con-

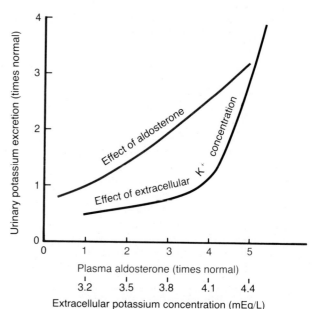

Figure 29–7. Effect of (1) plasma aldosterone concentration and (2) extracellular potassium ion concentration on the rate of urinary potassium excretion. These are the two most important factors regulating the rate of potassium excretion in the urine. (Drawn from data in Young: *Am. J. Physiol.* 244:F28, 1983.)

trolling *potassium ion secretion* by these same cells, for it activates the Na^+,K^+-ATPase pump that pumps sodium from the tubular fluid into the interstitial fluid and potassium in the opposite direction both at the same time. The *red curve* in Figure 29-7 illustrates this, showing a threefold increase in urinary excretion of potassium when aldosterone is increased.

Therefore, aldosterone is the second important factor that controls potassium excretion; therefore, it also helps greatly in controlling extracellular fluid potassium ion concentration, as we discuss more fully in the following sections.

Effect of Potassium Ion Concentration on the Rate of Aldosterone Secretion. In any properly functioning feedback control system, the factor that is controlled almost invariably has a feedback effect for control of the controller. In the case of the aldosterone-potassium control system, the rate of aldosterone secretion by the zona glomerulosa cells of the adrenal cortex is controlled very strongly by the extracellular fluid potassium concentration. Figure 29-8 illustrates this, showing that an increase in the potassium concentration of about 3 mEq/liter can increase the plasma aldosterone concentration from almost 0 to as high as 60 ng/dl of plasma, a concentration almost ten times normal.

Putting the effects illustrated in Figure 29-8 together with the fact that aldosterone greatly increases renal excretion of potassium (as shown in Figure 29-7), one can construct a very simple system for negative feedback control of potassium concentration, illustrated in Figure 29-9. That is, an increase in potassium concentration causes an increase in aldosterone concentration in the circulating blood (block 1). The increase in aldosterone then causes a marked increase in potassium excretion by the kidneys (block 2). The increased potassium excretion then decreases the extracellular fluid potassium concentration back toward normal (blocks 3 and 4).

Figure 29-9. Simplified schema of the aldosterone system for control of extracellular fluid potassium concentration.

Importance of the Aldosterone Feedback System for Control of Potassium Concentration. Without a functioning aldosterone feedback system, an animal can easily die of hyperpotassemia.

Figure 29-10 illustrates the potent effect of the aldosterone feedback system in controlling potassium concentration. In the experiment of this figure, dogs were subjected to an almost sevenfold increase in potassium intake in two different states: (1) the normal state and (2) after the aldosterone feedback system had been blocked by removing the adrenal glands and the animals given a fixed rate of aldosterone infusion.

Note that in the normal animal the sevenfold increase in potassium intake caused an increase in plasma potassium concentration of only 2.4 per cent—from a concentration of 4.2 mEq/liter to 4.3 mEq/liter. Thus, when the aldosterone feedback system

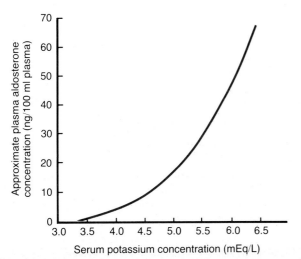

Figure 29-8. Effect on extracellular fluid aldosterone concentration of potassium ion concentration changes. Note the extreme change in aldosterone concentration for very minute changes in potassium concentration.

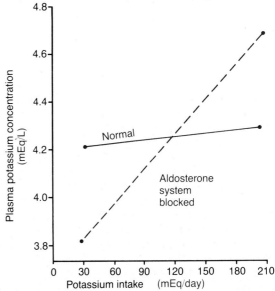

Figure 29-10. Effect on extracellular fluid potassium concentration of tremendous changes in potassium intake (1) under normal conditions and (2) after the aldosterone feedback system had been blocked. This figure demonstrates that potassium concentration is very poorly controlled after block of the aldosterone system. (Courtesy of Dr. David B. Young.)

was functioning normally, the potassium concentration remained very precisely controlled despite the tremendous change in potassium intake.

On the other hand, the *dashed curve* in the figure shows the effect after the aldosterone system had been blocked. Note that the same increase in potassium intake now caused a 26 per cent increase in potassium concentration! Thus, the control of potassium concentration in the normal animals was many times as effective as in the animals without an aldosterone feedback mechanism.

Effect of Primary Aldosteronism and Addison's Disease on Extracellular Fluid Potassium Concentration. *Primary aldosteronism* is caused by a tumor of the zona glomerulosa of one of the adrenal glands, the tumor secreting tremendous quantities of aldosterone. One of the most important effects of this disease is a marked increase in aldosterone-stimulated potassium excretion; this causes a severe decrease in extracellular fluid potassium concentration, so that many of these patients experience paralysis caused by failure of nerve transmission, as was explained in Chapter 5.

Conversely, in *Addison's disease* the adrenal glands have been destroyed, aldosterone secretion approaches zero, renal potassium secretion becomes very poor, and the extracellular fluid potassium concentration frequently rises to as high as double normal. This is often the cause of death in these patients, resulting in cardiac arrest.

● CONTROL OF THE EXTRACELLULAR CONCENTRATIONS OF OTHER IONS

Regulation of Calcium Ion Concentration

The role of calcium in the body and control of its concentration in the extracellular fluid are discussed in detail in Chapter 79 in relation to the endocrinology of parathyroid hormone, calcitonin, and bone. It is discussed here in brief.

The *day-to-day* calcium ion concentration remains within a few per cent of 2.4 mEq/liter and is *controlled principally by the effect of parathyroid hormone on bone reabsorption.* When the extracellular fluid concentration of calcium falls too low, the parathyroid glands are directly stimulated to promote increased secretion of parathyroid hormone. This hormone in turn acts directly on the bones to increase the reabsorption of bone salts, thus releasing large amounts of calcium into the extracellular fluid and elevating the calcium level back to normal. On the other hand, when the calcium concentration becomes too great, parathyroid hormone secretion becomes depressed so that almost no bone reabsorption then occurs. Yet the osteoblastic system for forming new bone does continue to deposit calcium, thus removing calcium from the extracellular fluid, in this way reducing the calcium ion concentration to its normal level.

However, the bones do not have an inexhaustible supply of calcium, and eventually they will run out of calcium. Therefore, *long-term control* of calcium ion concentration results from the effect of parathyroid hormone on reabsorption of calcium from the kidney tubules and absorption of calcium from the gut through the gastrointestinal mucosa, both of which effects are markedly increased by parathyroid hormone.

The effect of parathyroid hormone on calcium handling by the kidneys closely parallels the effect of aldosterone on sodium handling. That is, even in the absence of parathyroid hormone, much of the calcium is reabsorbed from the tubular fluid in the proximal tubules, the loop of Henle, and the diluting segment of the distal tubules, but about 10 per cent of the filtered load of calcium still remains to enter the late distal tubules. Then, if large amounts of parathyroid hormone are present in the body fluids, essentially all of the remaining calcium will be reabsorbed from the late distal tubules and cortical collecting ducts, thus conserving the calcium in the body. So the principal mechanism for *long-term control of calcium ion concentration* is as follows: low calcium in the extracellular fluid leads to parathyroid hormone secretion, and the parathyroid hormone then promotes intense *conservation of calcium by the kidneys and also greatly increased calcium absorption from the gastrointestinal tract.* These effects will be discussed in more detail in Chapter 79.

Active calcium reabsorption in the late distal tubules and cortical collecting ducts does not seem to occur secondary to sodium transport. Instead, it is believed that a separate primary active transport mechanism for calcium functions in the basolateral membranes of the epithelial cells in these tubular segments. On the other hand, in the earlier renal tubular segments, calcium transport, like that of so many other ions, probably occurs secondary to sodium transport, either as secondary active transport or as passive reabsorption of calcium down an electrochemical gradient.

Regulation of Magnesium Ion Concentration

Much less is known about the regulation of magnesium ion concentration than of calcium ion concentration. Magnesium ions are reabsorbed by all portions of the renal tubules. However, the magnesium ion itself directly affects the tubular epithelial cells by decreasing this reabsorption. Therefore, when the extracellular fluid magnesium concentration is high, excessive amounts of magnesium are excreted; conversely, when the magnesium concentration is low, magnesium is conserved.

Regulation of Phosphate Concentration

Phosphate concentration is regulated primarily by an *overflow* mechanism, which can be explained as follows: The renal tubules have a normal transport maximum for absorbing phosphate at about 0.1 mmol/min. When less than this "load" of phosphate is present in the glomerular filtrate, all of it is reabsorbed. When more than this amount is present, the excess is excreted. Therefore, normally phosphate ion spills into the urine when its concentration in the extracellular fluid is above a threshold value of approximately 0.8 mmol/liter, which gives a tubular load of phosphate of 0.1 mmol/min. Since most people ingest large quantities of phosphate day in and day out, either in milk or in meat, the concentration of phosphate is usually maintained at a level of about 1.0 mmol/liter, a level at which there is continual overflow of excess phosphate into the urine.

Role of Parathyroid Hormone in Phosphate Ion Regulation. Parathyroid hormone, which plays a major role in regulation of calcium ion concentration, as explained above, also affects phosphate ion concentration in two different ways. First, parathyroid hormone promotes bone reabsorption, thereby dumping large quantities of phosphate ions into the extracellular fluid from the bone salts. Second, this hormone decreases the transport maximum for phosphate by the renal tubules so that a greater proportion of the tubular phosphate is lost in the urine. The combination of these factors causes marked loss of phosphates in the urine. The interrelationships between phosphate, calcium, and parathyroid hormone control will be discussed in Chapter 79.

REFERENCES

Altura, B. M., and Altura, B. T.: Magnesium, electrolyte transport, and coronary vascular tone. Drugs, 28(Suppl. 1):120, 1984.

Anagnostopoulos, T., et al.: Transport of chlorine in the proximal tubule. Its effects on water-electrolyte absorption. J. Physiol. (Paris), 79:132, 1984.

Andersson, D., and Rundgren, M.: Thirst and it disorders. Annu. Rev. Med., 33:231, 1982.

Angus, Z. S., et al.: PTH, calcitonin, cyclic nucleotides, and the kidney. Annu. Rev. Physiol,. 43:583, 1981.

Arendshorst, W. J.: Altered reactivity of tubuloglomuerular feedback. Annu. Rev. Physiol., 49:295, 1987.

Baer, P. G., and McGiff, J. C.: Hormonal systems and renal hemodynamics. Annu. Rev. Physiol., 42:589, 1980.

Bell, P. D., et al.: Calcium as a mediator of tubuloglomerular feedback. Annu. Rev. Physiol., 49:275, 1987.

Boron, W. F., and Sackin, H.: Measurement of intracellular ionic composition and activities in renal tubules. Annu. Rev. Physiol., 45:483, 1983.

Brenner, B. M., and Rector, F. C., Jr.: The Kidney, 3rd Ed. Philadelphia, W. B. Saunders Co., 1986.

Bricker, N. S.: The Kidney: Diagnosis and Management. New York, John Wiley and Sons, 1984.

Briggs, J. P., and Schnermann, J.: The tubuloglomerular feedback mechanism: Functional and biochemical aspects. Annu. Rev. Physiol., 49:251, 1987.

Buckalew, V. M., Jr., and Gruber, K. A.: Natriuretic hormone. Annu. Rev. Physiol., 46:343, 1984.

Currie, M. G., and Needleman, P.: Renal arachidonic acid metabolism. Annu. Rev. Physiol., 46:327, 1984.

Dawson, K. G.: Endocrine physiology of electrolyte metabolism. Drugs, 28 (Suppl. 1):98, 1984.

Dennis, V. W., et al.: Renal handling of phosphate and calcium. Annu. Rev. Physiol., 41:257, 1979.

deWardener, H. E., and Clarkson, E. M.: Concept of natriuretic hormone. Physiol. Rev., 65:658, 1985.

Dunn, M. J. (ed.): Prostaglandins and the Kidney. Biochemistry, Physiology, Pharmacology, and Clinical Applications. New York, Plenum Publishing Corp., 1983.

Epstein, M., et al.: β-Blockers and the kidney: Implications for renal function and renin release. Physiologist, 28:53, 1985.

Flatman, P. W.: Magnesium transport across cell membranes. J. Membr. Biol., 80:1, 1984.

Friedman, P. A.: Renal calcium transport: Sites and insights. News Physiol. Sci., 3:17, 1988.

Gauer, O. H., and Henry, J. P.: Neurohormonal control of plasma volume. Int. Rev. Physiol., 9:145, 1976.

Gibbons, G. H., et al.: Interaction of signal influencing renin release. Annu. Rev. Physiol., 46:291, 1984.

Giebisch, G., and Stanton, B.: Potassium transport in the nephron. Annu. Rev. Physiol., 41:241, 1979.

Gmaj, P., and Murer, H.: Cellular mechanisms of inorganic phosphate transport in kidney. Physiol. Rev., 66:36, 1986.

Goetz, K. L., et al.: Atrial receptors and renal function. Physiol. Rev., 55:157, 1975.

Goldberg, M., et al.: Renal handling of calcium and phosphate. Int. Rev. Physiol., 11:211, 1976.

Guyton, A. C., et al.: A system analysis of volume regulation. Alfred Benxon Symposium XI. Munksgaard, 1978.

Guyton, A. C., et al.: Dynamics and Control of the Body Fluids. Philadelphia, W. B. Saunders Co., 1975.

Hall, J. E., et al.: Control of glomerular filtration rate by renin-angiotensin system. Am. J. Physiol., 233:F366, 1977.

Hall, J. E., et al.: Dissociation of renal blood flow and filtration rate autoregulation by renin depletion. Am. J. Physiol., 232:F215, 1977.

Hall, J. E., et al.: Intrarenal control of electrolyte excretion by angiotensin II. Am. J. Physiol., 232:F538, 1977.

Handler, J. S., and Orloff, J.: Antidiuretic hormone. Annu. Rev. Physiol., 43:611, 1981.

Harrison, M H.: Effects of thermal stress and exercise on blood volume in humans. Physiol. Rev., 65:149, 1985.

Hayward, J. N.: Neural control of the posterior pituitary. Annu. Rev. Physiol., 37:191, 1975.

Honig, A.: Salt and water metabolism in acute high-altitude hypoxia: Role of peripheral arterial chemoreceptors. News Physiol. Sci., 4:109, 1989.

Kanwar, Y. S.: Biophysiology of glomerular filtration and proteinuria. Lab. Invest., 51:7, 1984.

Koepke, J. P., and DiBona, G. F.: Functions of renal nerves. Physiologist, 28:47, 1985.

Kramer, H. J., and Kruck, F. (eds.): Natriuretic Hormone. New York, Springer-Verlag, 1978.

McKinley, M. J.: An important region for osmoregulation: The anterior wall of the third ventricle. News Physiol. Sci., 2:13, 1987.

Mizgala, C. L., and Quamme, G. A.: Renal handling of phosphate. Physiol. Rev., 65;431, 1985.

Murer, H., and Malmstrom K.: How renal phosphate transport is regulated. News Physiol. Sci., 2:45, 1987.

O'Connor, W. J.: Normal Renal Function. The Excretion of Water and Electrolytes Derived From Food and Drink. New York, Oxford University Press, 1982.

Salazar, F. J., et al.: Atrial natriuretic peptide levels during acute and chronic saline loading in conscious dogs. Am. J. Physiol., 251:R499, 1986.

Seldin, D. W., and Giebisch, G.: The regulation of potassium balance. New York, Raven Press, 1989.

Smith, W. L., and Garcia Perez, A.: Immunodissection: Use of monoclonal antibodies to isolate specific types of renal cells. Am. J. Physiol., 248:F1, 1985.

CHAPTER 30

•Regulation of Acid-Base Balance •

When one speaks of the regulation of acid-base balance, regulation of hydrogen ion concentration in the body fluids is actually meant. The hydrogen ion concentration in different solutions can vary from less than 10^{-14} Eq/liter to higher than 10^0, which means a total variation of more than a quadrillionfold. On a logarithmic basis, the hydrogen ion concentration in the human body is approximately midway between these two extremes.

Only slight changes in hydrogen ion concentration from the normal value can cause marked alterations in the rates of chemical reactions in the cells, some being depressed and others accelerated. For this reason the regulation of hydrogen ion concentration is one of the most important aspects of homeostasis. Later in the chapter the effects on the body of high hydrogen ion concentration (acidosis) and low hydrogen ion concentration (alkalosis) are discussed. In general, when people become acidotic, they are likely to die in a coma; when they become alkalotic, they may die of tetany or convulsions.

Acids and Bases — Their Definitions and Meanings

An acid is defined as a molecule or an ion that can function as a *proton donor*. A base is defined as a molecule or an ion that can function as a *proton acceptor*. If one will remember that a proton is actually a hydrogen ion, one can restate these definitions in the following ways:

An *acid* is a molecule or an ion that can contribute a hydrogen ion to a solution. Thus, HCl ionizes in water to form hydrogen ions and chloride ions and therefore is the acid called hydrochloric acid. Likewise, H_2CO_3 ionizes in water to form H^+ and HCO_3^-, thus also contributing hydrogen ions to the solution, and is the acid known as carbonic acid. Other important acids in the body are acetic acid, sodium acid phosphate, uric acid, acetoacetic acid, and so forth.

A *base* is a molecule or an ion that will combine with hydrogen ions to remove them from a solution. For instance, the bicarbonate ion, HCO_3^-, is a base because it can combine with hydrogen ions to form H_2CO_3. Likewise HPO_4^{--} is a base because it can combine with hydrogen ions to form $H_2PO_4^-$. The proteins of the body also function as bases because certain of the amino acids in the protein molecules function as negative ions that bind readily with excess hydrogen ions. In fact, the hemoglobin in the red blood cells and the other proteins in the other cells of the body are among the most important of the body's bases. Commonly, molecules such as sodium bicarbonate and sodium phosphate are called bases as well. However, the negative ions of these molecules are actually the real bases, according to the most-used definition of the word "base."

Finally, one needs to understand the relationship between a "base" and an "alkali." An alkali is the combination of one of the alkaline metals — sodium, potassium, or so forth — with a highly basic ion such as the hydroxyl ion (OH^-). The basic portions of these molecules react vigorously with hydrogen ions to remove them from the solutions and therefore are typical bases. Since alkalis are well known, the term alkali is frequently used synonymously with the term base. And, for similar reasons, the term "alkalosis" is used to mean the *opposite* of "acidosis," that is, excess removal of hydrogen ions from the solution, in contrast to excess addition of hydrogen ions (acidosis).

Strong and Weak Acids and Bases. A strong acid is one that has a very strong tendency to dissociate into ions and therefore to discharge its hydrogen ion into the solution. A typical example is hydrochloric acid. On the other hand, acids that release hydrogen ions with far less vigor are called weak acids. Examples are carbonic acid and sodium acid phosphate. A strong base is one that reacts very powerfully with hydrogen ions and therefore removes these with extreme avidity from the solution. A typical example is the hydroxyl ion (OH^-). A typical weak base is the bicarbonate ion (HCO_3^-), for it binds much more weakly with hydrogen ions.

Most of the acids and bases that are concerned with normal regulation of acid-base balance in the body are weak acids and bases, the most important of which are carbonic acid and bicarbonate base.

Hydrogen Ion Concentration and pH of Normal Body Fluids and in Acidosis and Alkalosis. The hydrogen ion concentration in the extracellular fluid is normally regulated at a constant value of approximately 4×10^{-8} Eq/liter; this value can vary from as low as 1×10^{-8} to as high as 1.6×10^{-7} without causing death.

From these values, it is already apparent that expressing hydrogen ion concentration in terms of actual concentrations is a cumbersome procedure. Therefore, the symbol *pH* has come into usage for expressing the concentration; pH is related to actual hydrogen ion concentration by the following formula (when H^+ concentration is expressed in equivalents per liter):

$$pH = \log \frac{1}{H^+ \text{ conc.}} = -\log H^+ \text{ conc.} \quad (1)$$

Note from this formula that a low pH corresponds to a high hydrogen ion concentration, which is called *acidosis;* and, conversely, a high pH corresponds to a low hydrogen ion concentration, which is called *alkalosis.*

The normal pH of arterial blood is 7.4, while the pH of venous blood and of interstitial fluids is about 7.35 because of extra quantities of carbon dioxide that form carbonic acid in these fluids.

Since the normal pH of the arterial blood is 7.4, a person is considered to have acidosis whenever the pH is below this value and to have alkalosis when it rises above 7.4. The lower limit at which a person can live more than a few hours is about 6.8, and the upper limit approximately 8.0.

Intracellular pH. On the basis of indirect measurements, it has been found that the intracellular pH usually ranges between 6.0 and 7.4 in different cells, perhaps averaging about 7.0. A *rapid rate of metabolism* in cells increases the rate of acid formation, especially carbonic acid (H_2CO_3), and consequently decreases pH. Also, *poor blood flow* to any tissue causes acid accumulation and a decrease in pH.

DEFENSE AGAINST CHANGES IN HYDROGEN ION CONCENTRATION

To prevent acidosis or alkalosis, several special control systems are available: (1) All the body fluids are supplied with acid-base *buffer systems* that immediately combine with any acid or base and thereby prevent excessive changes in hydrogen ion concentration. (2) If the hydrogen ion concentration does change measurably, the *respiratory center is immediately stimulated* to alter the rate of breathing. As a result, the rate of carbon dioxide removal from the body fluids is automatically changed; and, for reasons that will be presented later, this causes the hydrogen ion concentration to return toward normal. (3) When the hydrogen ion concentration changes from normal, *the kidneys excrete either an acid or alkaline urine,* thereby also helping to readjust the hydrogen ion concentration of the body fluids back to normal.

The buffer systems can act within a fraction of a second to prevent excessive changes in hydrogen ion concentration. On the other hand, it takes 1 to 12 minutes for the respiratory system to make acute adjustments and another day or so to make still additional chronic adjustments. Finally, the kidneys, although providing the most powerful of all the acid-base regulatory systems, require many hours to several days to readjust the hydrogen ion concentration.

● FUNCTION OF ACID-BASE BUFFERS

An acid-base buffer is a solution containing two or more chemical compounds that prevents marked changes in hydrogen ion concentration when either an acid or a base is added to the solution. As an example, if only a few drops of concentrated hydrochloric acid are added to a beaker of pure water, the pH of the water might immediately fall from a neutral 7.0 to as low as 1.0. However, if a satisfactory buffer system is present, the hydrochloric acid combines instantaneously with the buffer, and the pH falls only slightly. Perhaps the best way to explain the action of an acid-base buffer is to consider an actual simple buffer system, such as the bicarbonate buffer, which is extremely important in regulation of acid-base balance in the body.

THE BICARBONATE BUFFER SYSTEM

A typical bicarbonate buffer system consists of a *mixture* of carbonic acid (H_2CO_3) and sodium bicarbonate ($NaHCO_3$) in the same solution. It must first be noted that carbonic acid is a very weak acid for its degree of dissociation into hydrogen ions and bicarbonate ions is poor in comparison with that of many other acids.

When a strong acid, such as hydrochloric acid, is added to the bicarbonate buffer solution, the following reaction takes place:

$$HCl + NaHCO_3 \rightarrow H_2CO_3 + NaCl \quad (2)$$

From this equation it can be seen that the strong hydrochloric acid is converted into the very weak carbonic acid. Therefore, addition of the HCl lowers the pH of the solution only slightly.

Now let us see what happens when a strong base, such as sodium hydroxide, is added to a buffer solution containing carbonic acid; the following reaction takes place:

$$NaOH + H_2CO_3 \rightarrow NaHCO_3 + H_2O \quad (3)$$

This equation shows that the hydroxyl ion of the sodium hydroxide combines with a hydrogen ion from the carbonic acid to form water and that the other product formed is sodium bicarbonate. The net result is exchange of the strong base NaOH for the weak base $NaHCO_3$.

Quantitative Dynamics of Buffer Systems

Dissociation of Carbonic Acid. All acids are ionized to a certain extent, and the percentage of ionization is called the *degree of dissociation.* Equation 4 illustrates the reversible relationship between undissociated carbonic acid and the two ions that it forms, H^+ and HCO_3^-.

$$H_2CO_3 \rightleftarrows H^+ + HCO_3^- \quad (4)$$

A physicochemical law has been found to apply to the dissociation of all molecules; as applied in this case specifically to carbonic acid, it is expressed by the following formula:

$$\frac{H^+ \times HCO_3^-}{H_2CO_3} = K' \tag{5}$$

This formula states that in any given carbonic acid solution the concentration of hydrogen ions times the concentration of bicarbonate ions divided by the concentration of undissociated carbonic acid molecules is equal to a constant, K'.

However, it is almost impossible to measure the concentration of undissociated carbonic acid in a solution because it rapidly dissociates also into dissolved CO_2 and H_2O as well as into H^+ and HCO_3^-. On the other hand, the concentration of dissolved carbon dioxide is easily measured; and, since the amount of undissociated carbonic acid is proportional to the amount of dissolved carbon dioxide, equation 5 above can also be expressed as follows:

$$\frac{H^+ \times HCO_3^-}{CO_2} = K \tag{6}$$

The only real difference between the preceding two formulas is that the constant K is approximately ¹⁄₄₀₀ times the constant K' because the proportionality ratio between carbonic acid and carbon dioxide is 1 to 400.

Formula 6 can be changed into the following form:

$$H^+ = K \cdot \frac{CO_2}{HCO_3^-} \tag{7}$$

If we take the logarithm of each of the two sides of Equation 7, it becomes the following:

$$\log H^+ = \log K + \log \frac{CO_2}{HCO_3^-} \tag{8}$$

The signs of the log H^+ and of the log K are changed from positive to negative and the carbon dioxide and bicarbonate are inverted in the last term, which is the same as changing its sign also, giving the following formula:

$$-\log H^+ = -\log K + \log \frac{HCO_3^-}{CO_2} \tag{9}$$

It will be recalled from earlier in the chapter that $-\log H^+$ is equal to the pH of the solution. Likewise, $-\log K$ *is called the pK of a buffer*. Therefore, this formula can be changed still further to the following:

$$pH = pK + \log \frac{HCO_3^-}{CO_2} \tag{10}$$

The Henderson-Hasselbalch Equation. For the bicarbonate buffer system the pK is 6.1, and Equation 10 may be expressed as follows:

$$pH = 6.1 + \log \frac{HCO_3^-}{CO_2} \tag{11}$$

This is called the Henderson-Hasselbalch equation, and by using it one can calculate the pH of a solution with reasonable accuracy if the molar concentrations of bicarbonate ion and dissolved carbon dioxide are known. If the bicarbonate concentration is equal to the dissolved carbon dioxide concentration, the second member of the right-hand portion of the equation becomes log of 1, which is equal to zero. Therefore, *for equal concentrations the pH of the solution is equal to the pK.*

From the Henderson-Hasselbalch equation one can readily see that *an increase in bicarbonate ion concentration causes the pH to rise,* or, in other words, shifts the acid-base balance toward the alkaline side. On the other hand, *an increase in the concentration of dissolved carbon dioxide decreases the pH,* or shifts the acid-base balance toward the acid side.

It will be apparent later in this chapter that one can change the concentration of dissolved carbon dioxide in the body fluids by increasing or decreasing the rate of respiration. In this way the respiratory system can change the pH of the body fluids. On the other hand, the kidneys can increase or decrease the concentration of bicarbonate ion in the body fluids, in this way increasing or decreasing the pH. Thus, these two major mechanisms for regulating hydrogen ion concentration operate principally by altering one or the other of the two elements of the bicarbonate buffer system.

The "Titration Curve" of the Bicarbonate Buffer System. Figure 30-1 shows the changes in pH of the body fluids when the ratio of bicarbonate ion to carbon dioxide changes. Note that when the concentrations of the two elements of the buffer are equal, the pH of the solution is 6.1, which is equal to the pK of the bicarbonate buffer system. When base is added to the buffer a large proportion of the dissolved carbon dioxide is converted into bicarbonate ions, and the ratio is altered. As a result, the pH rises as indicated by the forward slope of the curve. On the other hand, when acid is added, a large proportion of the bicarbonate ion is converted into dissolved carbon dioxide so that the pH falls as illustrated by the downslope of the curve.

Buffering Power of the Bicarbonate Buffer System. Referring once again to Figure 30-1, note that at the central point of the curve addition of a slight amount of acid or base causes minimal change in pH. However, toward each end of the curve addition of a slight amount of acid or base causes the pH to change greatly. Thus, the so-called *buffer-*

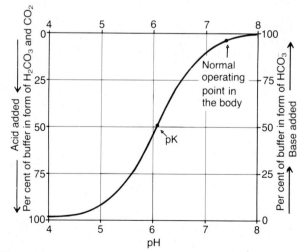

Figure 30-1. "Titration curve" for the bicarbonate buffer system.

ing power of the buffer system *is greatest when the pH is equal to the pK,* which is in the exact center of the curve. The buffering power is still reasonably effective until the ratio of one element of the buffer system to the other reaches as much as 8:1 or 1:8, but beyond these limits the buffering power diminishes rapidly. When all the carbon dioxide has been converted into bicarbonate ion or when all the bicarbonate ion has been converted into carbon dioxide, the system has no more buffering power at all.

A second factor that determines the buffering power is the concentration of the two elements of the buffer solution, the CO_2 and HCO_3^-. Obviously, if the concentrations are slight, only a small amount of acid or base added to the solution will change the pH considerably. Thus, *the buffering power of a buffer is also directly proportional to the concentrations of the buffer substances.*

THE BUFFER SYSTEMS OF THE BODY FLUIDS

The three major buffer systems of the body fluids are the *bicarbonate buffer,* which was described above, the *phosphate buffer,* and the *protein buffer.* Each of these performs major buffering functions under different conditions.

The Bicarbonate Buffer System. The bicarbonate system is not an exceedingly powerful buffer, for two reasons. First, the pH in the extracellular fluids is about 7.4, whereas the pK of the bicarbonate buffer system is 6.1. This means that approximately 20 times as much of the bicarbonate buffer is in the form of bicarbonate ion as in the form of dissolved carbon dioxide. For this reason, the system operates on a portion of its buffering curve where the buffering power is poor. Second, the concentrations of the two elements of the bicarbonate system, CO_2 and HCO_3^-, are not great.

Yet despite the fact that the bicarbonate buffer system is not especially powerful, it is actually more important than all the others in the body, because the concentration of each of the two elements of the bicarbonate system can be regulated, carbon dioxide by the respiratory system and the bicarbonate ion by the kidneys. As a result, the pH of the blood can be shifted up or down by the respiratory and renal regulatory systems.

The Phosphate Buffer System. The phosphate buffer system acts in an almost identical manner, but it is composed of the following two elements: $H_2PO_4^-$ and HPO_4^{--}. When a strong acid, such as hydrochloric acid, is added to a mixture of these two substances, the following reaction occurs:

$$HCl + Na_2HPO_4 \rightarrow NaH_2PO_4 + NaCl \quad (12)$$

The net result of this reaction is that the hydrochloric acid is removed, and in its place an additional quantity of NaH_2PO_4 is formed. NaH_2PO_4 is only weakly acidic, so that the added strong acid is immediately traded for a very weak acid, and the pH changes relatively slightly.

Conversely, if a strong base, such as sodium hydroxide, is added to the buffer system, the following reaction takes place:

$$NaOH + NaH_2PO_4 \rightarrow Na_2HPO_4 + H_2O \quad (13)$$

Here sodium hydroxide is decomposed to form water and Na_2HPO_4. That is, a strong base is traded for the very weak base Na_2HPO_4, allowing only a slight shift in pH toward the alkaline side.

The phosphate buffer system has a pK of 6.8, which is not far from the normal pH of 7.4 in the body fluids; this allows the phosphate system to operate near its maximum buffering power. However, despite the fact that this buffer system operates in a reasonably good portion of the buffer curve, its concentration in the extracellular fluid is only one-twelfth that of the bicarbonate buffer. Therefore, its total buffering power *in the extracellular fluid* is far less than that of the bicarbonate system.

On the other hand, the phosphate buffer is especially important in the tubular fluids of the kidneys, for two reasons: First, phosphate usually becomes greatly concentrated in the tubules, thereby also greatly increasing the buffering power of the phosphate system. Second, the tubular fluid usually becomes more acidic than the extracellular fluid, bringing the operating range of the buffer closer to the pK of the system.

The phosphate buffer is also very important in the intracellular fluids, because the concentration of phosphate in these fluids is many times that in the extracellular fluids and also because the pH of the intracellular fluids is usually closer to the pK of the phosphate buffer system than is the pH of the extracellular fluid.

The Protein Buffer System. The most plentiful buffer of the body is the proteins of the cells and plasma, mainly because of their very high concentrations. There is a slight amount of diffusion of hydrogen ions through the cell membrane; even more important, carbon dioxide can diffuse within seconds through cell membranes, and even bicarbonate ions can diffuse to some extent (the hydrogen and bicarbonate ions require several hours to come to equilibrium in most cells other than the red blood cells). This diffusion of hydrogen ions and of the two elements of the bicarbonate buffer system causes the pH in the intracellular fluids to change approximately in proportion to the changes in pH in the extracellular fluids. Thus, all the buffer systems inside the cells help buffer the extracellular fluids as well but may take several hours to do so. These include the extremely large amounts of proteins inside the cells. Indeed, experimental studies have shown that about three quarters of all the *chemical* buffering power of the body fluids is inside the cells, and most of this results from the intracellular proteins. However, except for the red blood cells, the slowness of movement of hydrogen and bicarbonate ions through the cell membranes often delays the ability of the intracellular buffers to buffer extracellular acid-base abnormalities for several hours.

The method by which the protein buffer system operates is precisely the same as that of the bicarbonate buffer system. It will be recalled that a protein is composed of amino acids bound together by peptide linkages, but some of the different amino acids, especially histidine, have free acidic radicals that can dissociate to form base plus H^+. Furthermore, the pKs of a few of these amino acid buffering systems are not far from 7.4. This, too, helps make the protein buffering systems the most powerful of the body.

THE ISOHYDRIC PRINCIPLE

Each of the above buffer systems has been discussed as if it could operate individually in the body fluids. However, they all actually work together, for the hydrogen ion is common to the chemical reactions of all the systems. Therefore, whenever any condition causes the hydrogen ion concentration to change, it causes the balance of all the buffer systems to change at the same time. This phenomenon, called the *isohydric principle,* is presented by the following formula:

$$H^+ = \frac{K_1 \times HA_1}{A_1^-} = \frac{K_2 \times HA_2}{A_2^-} = \frac{K_3 \times HA_3}{A_3^-} \quad (14)$$

K_1, K_2, and K_3 are the dissociation constants of three respective acids, HA_1, HA_2, and HA_3, and A_1^-, A_2^-, and A_3^- are the concentrations of the free negative ions that constitute the respective bases of the three buffer systems.

The important feature of this principle is that any condition that changes the balance of any one of the buffer systems also changes the balance of all the others, for *the buffer systems actually buffer each other by shifting hydrogen ions from one to the other.*

● RESPIRATORY REGULATION OF ACID-BASE BALANCE

In the discussion of the Henderson-Hasselbalch equation, it was noted that an *increase in carbon dioxide concentration in the body fluids decreases the pH toward the acidic side,* whereas a *decrease in carbon dioxide raises the pH toward the alkaline side.* It is on the basis of this effect that the respiratory system is capable of altering the pH either up or down.

Balance Between Metabolic Formation of Carbon Dioxide and Pulmonary Expiration of Carbon Dioxide. Carbon dioxide is continually being formed in the body by the different intracellular metabolic processes, the carbon in the foods being oxidized by oxygen to form carbon dioxide. This in turn diffuses into the interstitial fluids and blood and is transported to the lungs, where it diffuses into the alveoli and then is transferred to the atmosphere by pulmonary ventilation. However, several minutes are required for this passage of carbon dioxide from the cells to the atmosphere, so that an average of 1.2 mmol/liter of dissolved carbon dioxide is normally in the extracellular fluids at all times.

If the rate of metabolic formation of carbon dioxide becomes increased, its concentration in the extracellular fluids is likewise increased. Conversely, decreased metabolism decreases the carbon dioxide concentration.

On the other hand, if the rate of pulmonary ventilation is increased, carbon dioxide is blown off from the lungs and the amount of carbon dioxide in the extracellular fluids decreases.

Effect of Increasing or Decreasing Alveolar Ventilation on the pH of Extracellular Fluids

If we assume that the rate of metabolic formation of carbon dioxide remains constant, then the only factor that affects the carbon dioxide concentration in the body fluids is the rate of the alveolar ventilation, as expressed by the following formula:

$$CO_2 \propto \frac{1}{\text{Alveolar ventilation}} \quad (15)$$

Because an increase in carbon dioxide decreases the pH, changes in alveolar ventilation cause reciprocal changes in the hydrogen ion concentration.

Figure 30–2 illustrates the approximate change in pH in the blood that can be caused by increasing or decreasing the rate of alveolar ventilation. Note that an increase in alveolar ventilation to two times normal raises the pH of the extracellular fluids by about 0.23. This means that if the pH of the body fluids had been 7.4 with normal alveolar ventilation, doubling the ventilation raises the pH to 7.63. Conversely, a decrease in alveolar ventilation to one-quarter normal reduces the pH by 0.45. That is, if at normal alveolar ventilation the pH had been 7.4, reducing the ventilation to one quarter reduces the pH to 6.95. Because alveolar ventilation can be reduced to zero ventilation or increased to about 15 times normal, one can readily

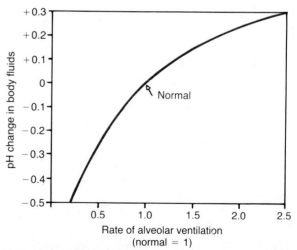

Figure 30–2. Approximate change in extracellular fluid pH caused by increased or decreased rate of alveolar ventilation.

understand how much the pH of the body fluids can be changed by alterations in the activity of the respiratory system.

Effect of Hydrogen Ion Concentration on Alveolar Ventilation

Not only does the rate of alveolar ventilation affect the hydrogen ion concentration of the body fluids, but, in turn, the hydrogen ion concentration affects the rate of alveolar ventilation. This results from a *direct action of hydrogen ions on the respiratory center in the medulla oblongata* that controls breathing, which will be discussed in detail in Chapter 41.

Figure 30–3 illustrates the changes in alveolar ventilation caused by changing the pH of arterial blood from 7.0 to 7.6. From this graph it is evident that a decrease in pH from the normal value of 7.4 to the strongly acidic level of 7.0 can increase the rate of alveolar ventilation to as much as four to five times normal, whereas an increase in pH into the alkaline range can decrease the rate of alveolar ventilation to only a fraction of the normal level.

Feedback Control of Hydrogen Ion Concentration by the Respiratory System. Because of the ability of the respiratory center to respond to hydrogen ion concentration, and because changes in alveolar ventilation in turn alter the hydrogen ion concentration in the body fluids, the respiratory system acts as a typical feedback controller of hydrogen ion concentration. That is, any time the hydrogen ion concentration becomes high, the respiratory system becomes more active, and alveolar ventilation increases. As a result, the carbon dioxide concentration in the extracellular fluids decreases, thus reducing the hydrogen concentration back toward normal. Conversely, if the hydrogen ion concentration falls too low, the respiratory center becomes depressed, alveolar ventilation also decreases, and the hydrogen ion concentration rises back toward normal.

Efficiency of Respiratory Control of Hydrogen Ion Concentration. Unfortunately, respiratory control cannot return the hydrogen ion concentration all the way to the normal value of 7.4 when some abnormality outside the respiratory system has altered the pH from normal. The reason for this is that as the pH returns toward normal, the stimulus that has been causing either increased or decreased respiration will itself begin to be lost. Ordinarily, the respiratory mechanism for controlling hydrogen ion concentration has a control effectiveness of between 50 and 75 per cent (a feedback gain of 1 to 3). That is, if the hydrogen ion concentration should suddenly be decreased from 7.4 to 7.0 by some extraneous factor, the respiratory system, in 3 to 12 minutes, returns the pH to a value of about 7.2 to 7.3.

Buffering Power of the Respiratory System. In effect, respiratory regulation of acid-base balance is a *physiological type of buffer system* having almost the same importance as the chemical buffering systems of the body discussed earlier in the chapter. The overall "buffering power" of the respiratory system is one to two times as great as that of all the chemical buffers combined. That is, one to two times as much acid or base normally can be buffered by this mechanism as by the chemical buffers.

● RENAL CONTROL OF HYDROGEN ION CONCENTRATION

The kidneys control extracellular fluid hydrogen ion concentration by excreting either an acidic urine or a basic urine. Excreting an acidic urine reduces the amount of acid in the extracellular fluids, whereas excreting a basic urine removes base from the extracellular fluids.

The body's means for determining whether the urine will be acidic or basic is as follows: Large numbers of bicarbonate ions are filtered continually into the glomerular filtrate, which removes base from the blood. On the other hand, large numbers of hydrogen

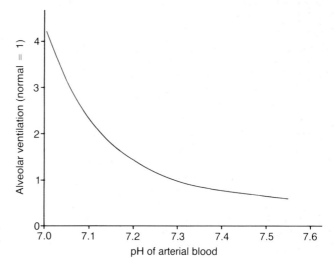

Figure 30–3. Effect of blood pH on the rate of alveolar ventilation. (Constructed from data obtained by Gray: Pulmonary Ventilation and Its Regulation. Springfield, Ill., Charles C Thomas.)

ions are secreted at the same time into the tubular lumens by the tubular epithelium, thus removing acid. If more hydrogen ions are secreted than bicarbonate ions are filtered, there will be net loss of acid from the extracellular fluids. Conversely, if more bicarbonate is filtered than hydrogen is secreted, there will be net loss of base. The following sections describe the different renal mechanisms that accomplish these events.

TUBULAR SECRETION OF HYDROGEN IONS

The epithelial cells everywhere in the tubular system, except in the thin limb of the loop of Henle, secrete hydrogen ions into the tubular fluid. However, in different tubular segments there are two quite different methods for doing this, each of which has separate characteristics and even separate purposes.

Secondary Active Transport of Hydrogen Ions in the Early Tubular Segments. The epithelial cells of the proximal tubule, the thick segment of the ascending limb of the loop of Henle, and the distal tubule all secrete hydrogen ions into the tubular fluid by secondary active transport; the mechanism of this is illustrated in Figure 30–4. Tremendous numbers of hydrogen ions are secreted in this manner, several thousand milliequivalents per day, but never against a very high hydrogen ion gradient, because the tubular fluid becomes very acidic only in the latter segments of the tubular system.

Figure 30–4 shows that the secretory process *begins with carbon dioxide* that either diffuses into or is formed by metabolism in the tubular epithelial cells. The carbon dioxide, under the influence of the enzyme *carbonic anhydrase* combines with water to form *carbonic acid*. This then dissociates into *bicarbonate ion* and *hydrogen ion*. Finally, the hydrogen ions are secreted into the tubule by a mechanism of Na^+-H^+ *countertransport*. That is, when sodium moves from the lumen of the tubule to the interior of the cell, it

first combines with a *carrier protein* in the luminal border of the cell membrane, and at the same time a hydrogen ion inside the cell combines with the opposite side of the same carrier protein. Then, because the concentration of sodium is much lower inside the cell than in the tubular lumen, this causes movement of the sodium down the concentration gradient to the interior, providing at the same time the energy for moving the hydrogen ion in the opposite direction (the "counter" direction) into the tubular lumen.

Primary Active Transport of Hydrogen Ions in the Late Tubular Segments. Beginning in the late distal tubules and then continuing all the way through the remainder of the tubular system to the renal pelvis, the tubules secrete hydrogen ions by *primary active transport*. The characteristics of this transport are quite different from the secondary active transport system in the earlier tubular segments. First, it normally accounts for less than 5 per cent of the total hydrogen ions excreted. On the other hand, it can concentrate hydrogen ions as much as 900-fold, in contrast to only 3- to 4-fold concentration that can be achieved in the proximal tubules and 10- to 15-fold concentration in the early distal tubules by the secondary transport mechanism. The 900-fold concentration of the hydrogen ions can decrease the pH of the tubular fluid to about 4.5, which therefore is the lower limit of pH that can be achieved in the excreted urine.

The mechanism of primary active hydrogen ion transport is illustrated in Figure 30–5. It occurs at the luminal membrane of the tubular cell, where hydrogen ions are transported directly by a specific transport protein, *a hydrogen-transporting adenosine triphosphatase (ATPase)*. The energy required for pumping the hydrogen ions against the 900-fold concentration gradient is derived from the breakdown of ATP to adenosine diphosphate (ADP).

Figure 30–4. Chemical reactions for (1) secondary active secretion of hydrogen ions into the tubule, (2) sodium ion reabsorption in exchange for the hydrogen ions secreted, and (3) combination of hydrogen ions with bicarbonate ions in the tubules to form carbon dioxide and water.

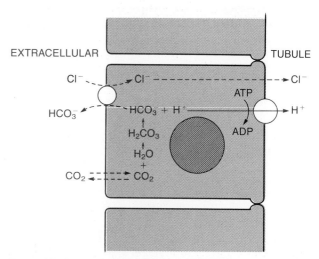

Figure 30–5. Primary active transport of hydrogen ions through the luminal membrane of the tubular epithelial cell. Note that one bicarbonate ion is absorbed for each hydrogen ion secreted, and a chloride ion is secreted passively along with the hydrogen ion.

The hydrogen ions pumped by this process are generated inside the cell in the following two steps: (1) Dissolved carbon dioxide combines with water inside the cell to form carbonic acid (H_2CO_3). (2) The carbonic acid then dissociates into bicarbonate ions (HCO_3^-) that are absorbed into the blood, plus hydrogen ions (H^+) that are secreted into the urine.

This primary active transport of hydrogen ions is believed to be a function of the special type of cell called the *intercalated cells.* These make their first appearance in the late distal tubules and then extend all the way to the end of the collecting duct system, reaching maximum numbers — about 10 per cent of the total epithelial cells — in the outer medullary collecting ducts. These cells are dark in appearance and therefore are frequently also called *dark cells.*

Regulation of Hydrogen Ion Secretion by the Hydrogen Ion Concentration in the Extracellular Fluids

The rate of hydrogen ion secretion into the tubules changes markedly in response to only slight changes in extracellular fluid hydrogen ion concentration. Thus, when the concentration of hydrogen ions is great (pH less than 7.4), hydrogen ion secretion can increase severalfold. Conversely, at an extracellular pH greater than 7.4, the hydrogen ion secretion diminishes equally as much. The control of these changes is achieved in the following way:

In acidosis, the *ratio* of carbon dioxide to bicarbonate ions in the extracellular fluid is above normal, as one can determine by referring again to the Henderson-Hasselbalch equation. Also, a similar ratio occurs inside the epithelial secretory cell, giving a high level of hydrogen ions and a correspondingly high rate of hydrogen ion secretion into the tubular lumen. In alkalosis, exactly the opposite occurs, with a resulting decrease in hydrogen ion secretion.

At normal extracellular fluid pH, the rate of hydrogen ion secretion is about 3.5 mM/min, but this rises or falls almost directly in proportion to the change in extracellular hydrogen ion concentration.

Interaction of Bicarbonate Ions with Hydrogen Ions in the Tubules — "Reabsorption" of Bicarbonate Ions

The renal tubules are not very permeable to the bicarbonate ion because it is a large ion and also is electrically charged. However, the bicarbonate ion can, in effect, be "reabsorbed" by a special process illustrated in Figure 30-4.

The reabsorption of bicarbonate ions is initiated by a reaction in the tubules between the bicarbonate ions filtered in the glomerular filtrate and the hydrogen ions secreted by the tubular cells, as illustrated in the figure. The carbonic acid then dissociates into carbon dioxide and water. The carbon dioxide, having the capability of diffusing extremely readily through all

cellular membranes, instantaneously diffuses into the tubular cell; and the water remains in the tubule.

If we now note in Figure 30-4 the chemical reactions that are responsible for formation of the hydrogen ions in the epithelial cells, we see that each time a hydrogen ion is formed, a bicarbonate ion also is formed inside these cells by the dissociation of H_2CO_3 into H^+ and HCO_3^-. This bicarbonate ion then diffuses into the extracellular fluid through the basolateral membrane.

Thus, the net effect of all these reactions is a mechanism for "reabsorption" of bicarbonate ions from the tubules, though the bicarbonate ions that enter the extracellular fluid are not the same bicarbonate ions that are removed from the tubular fluid.

Titration of Bicarbonate Ions Against Hydrogen Ions in the Tubules. Under normal conditions the *rate of hydrogen ion secretion is about 3.5 mmol/min* and the *rate of filtration of bicarbonate ions in the glomerular filtrate is about 3.46 mmol/min.* Thus, the quantities of the two ions entering the tubules are almost equal, and they combine with each other and actually annihilate each other, the end products being carbon dioxide and water. Therefore, it is said that the bicarbonate ions and hydrogen ions normally "titrate" each other in the tubules.

However, note also that this titration process is not quite exact, for usually a slight excess of hydrogen ions (the acidic component) remains in the tubules to be excreted in the urine. The reason for this is that under normal conditions a person's metabolic processes continually form a small amount of excess acid (about 60 mEq/day) that gives rise to the slight excess of hydrogen ions over bicarbonate ions in the tubules.

On rare occasions the bicarbonate ions are in excess, as we shall see in subsequent discussions. When this occurs, the titration process again is not quite complete; this time, excess bicarbonate ions (the basic component) are left in the tubules to pass into the urine.

Thus, the basic mechanism by which the kidney corrects either acidosis or alkalosis is by incomplete titration of hydrogen ions against bicarbonate ions, leaving one or the other of these to pass into the urine and therefore to be removed from the extracellular fluid.

RENAL CORRECTION OF ACIDOSIS — INCREASE OF BICARBONATE IONS IN THE EXTRACELLULAR FLUID

Now that we have described the mechanisms by which the renal tubules secrete hydrogen ions and reabsorb bicarbonate ions, we can explain the manner in which the kidneys readjust the pH of the extracellular fluids when it becomes abnormal.

First, let us consider *acidosis.* Referring again to Equation 11, the Henderson-Hasselbalch equation, we see that in *acidosis* the ratio of *carbon dioxide to bicarbonate ions* in the extracellular fluid increases. Therefore, the *rate of hydrogen ion secretion* rises to a level greater than the *rate of bicarbonate ion filtration* into the tubules. As a result, an excess of hydrogen ions is secreted into the tubules, while diminished quantities of bicarbonate ions enter the glomerular filtrate, so that now there are far too few bicarbonate ions to react with the hydrogen ions. The excess hydrogen ions combine with the buffers in the tubular fluid, as explained in later paragraphs, and are excreted into the urine.

Figure 30–4 shows that each time a hydrogen ion is secreted into the tubules two other effects occur simultaneously: first, a bicarbonate ion is formed in the tubular epithelial cell; and, second, a sodium ion is absorbed from the tubule into the epithelial cell. The sodium ion and bicarbonate ion then are both transported from the epithelial cell into the extracellular fluid.

Thus, *the net effect of secreting excess hydrogen ions into the tubules is to increase the quantity of bicarbonate ions in the extracellular fluid.* This increases the bicarbonate portion of the bicarbonate buffer system, which, in accordance with the Henderson-Hasselbalch equation and the isohydric principle, shifts all the buffers in the alkaline direction, increasing the pH in the process, and thereby correcting the acidosis.

RENAL CORRECTION OF ALKALOSIS— DECREASE OF BICARBONATE IONS IN THE EXTRACELLULAR FLUID

In *alkalosis,* the *ratio* of bicarbonate ions to dissolved carbon dioxide molecules increases. The effect of this on the titration process in the tubules is to increase the *ratio* of bicarbonate ions filtered into the tubules to hydrogen ions secreted. This increase occurs because the high extracellular bicarbonate ion concentration increases the bicarbonate ions filtered in the glomerular filtrate while at the same time the low carbon dioxide concentration in the extracellular fluid decreases the secretion of hydrogen ions. Therefore, the fine balance that normally exists in the tubules between the hydrogen and bicarbonate ions no longer occurs. Instead, far greater quantities of bicarbonate ions than hydrogen ions now enter the tubules.

Because almost no bicarbonate ions can be reabsorbed without first reacting with hydrogen ions, all the excess bicarbonate ions pass into the urine and carry with them sodium ions or other positive ions. Thus, in effect, sodium bicarbonate is removed from the extracellular fluid.

Loss of sodium bicarbonate from the extracellular fluid decreases the bicarbonate ion portion of the bicarbonate buffer system, and, in accordance with the Henderson-Hasselbalch equation, this shifts the pH of the body fluids back in the acid direction. Furthermore, because of the isohydric principle, all the other body buffers shift back in the acid direction too. Thus, the alkalosis is corrected.

COMBINATION OF EXCESS HYDROGEN IONS WITH TUBULAR BUFFERS AND TRANSPORT INTO THE URINE

When excess hydrogen ions are secreted into the tubules, only a small portion of these can be carried in the free form by the tubular fluid into the urine. The reason for this is that the maximum hydrogen ion concentration that the tubular system can achieve is $10^{-4.5}$ molar, which corresponds to a pH of 4.5. At normal daily urine flows, only 1 per cent of the daily excretion of excess hydrogen ions can be transported into the urine at this concentration.

Therefore, to carry the excess hydrogen ions into the urine the hydrogen ions must be carried in some form other than free hydrogen ions. This is achieved by the hydrogen ions' first combining with intratubular buffers and then being transported in this form.

The tubular fluids have two very important buffer systems that transport the excess hydrogen ions into the urine: (1) the phosphate buffer and (2) the ammonia buffer. In addition, there are a number of weak buffer systems such as urate and citrate that are of much less importance.

Transport of Excess Hydrogen Ions Into the Urine by the Phosphate Buffer. The phosphate buffer is composed of a mixture of HPO_4^{--} and $H_2PO_4^-$. Both become considerably concentrated in the tubular fluid because of their relatively poor reab-

Figure 30–6. Chemical reactions in the tubules involving hydrogen ions, sodium ions, and the phosphate buffer system.

sorption and because of removal of water from the tubular fluid. Therefore, even though the phosphate buffer is very weak in the blood, it is a much more powerful buffer in the tubular fluid.

Another factor that increases the importance of the phosphate buffer in the tubular fluids during acidosis is that the pK of this buffer is 6.8. When excess hydrogen ions are secreted, the tubular fluid normally begins at a pH of nearly 7.4 in the early proximal tubules and falls to an average of about 6.0 in the distal tubules and collecting ducts. Therefore, in these tubules the phosphate buffer functions in its most effective range, very near its pK value, as explained earlier in the chapter.

Figure 30–6 illustrates the manner in which hydrogen ions are removed from the tubular fluid by the phosphate buffer system and also how this functions in the total picture of renal acid-base control. Note that for each hydrogen ion bound by the phosphate buffer, a new bicarbonate ion is formed by the epithelial cell and transported into the blood. This contributes further to correction of acidosis when excess hydrogen ions are secreted.

Transport of Excess Hydrogen Ions Into the Urine by the Ammonia Buffer System. Another even more important, more complex, hydrogen ion buffer system of the tubular fluid is composed of ammonia (NH_3) and the ammonium ion (NH_4^+). The epithelial cells of all the tubules besides those of the thin segment of the loop of Henle continually synthesize ammonia, and this diffuses into the tubules. The ammonia then reacts with hydrogen ions, as illustrated in Figure 30–7, to form ammonium ions. These are then excreted into the urine in combination with chloride ions and other tubular anions. Note in the figure that the net effect of these reactions is, again, *to increase the bicarbonate concentration* in the extracellular fluid.

This ammonium ion mechanism for transport of excess hydrogen ions in the tubules is especially important, for two reasons: (1) Each time an ammonia molecule (NH_3) combines with a hydrogen ion to form an ammonium ion (NH_4^+), the concentration of ammonia in the tubular fluid becomes decreased, which causes still more ammonia to diffuse from the epithelial cells into the tubular fluid. Thus, the rate of ammonia secretion into the tubular fluid is actually controlled by the amount of excess hydrogen ions to be transported. (2) Most of the negative ions of the tubular fluid are chloride ions. Only a few hydrogen ions could be transported into the urine in direct combination with chloride, because hydrochloric acid is a very strong acid and the tubular pH would fall rapidly below the *critical value of 4.5, below which further hydrogen ion secretion ceases.* However, when hydrogen ions combine with ammonia and the resulting ammonium ions then combine with chloride, the pH does not fall significantly, because ammonium chloride is only very weakly acidic.

Sixty per cent of the ammonia secreted by the tubular epithelium is derived from *glutamine,* and the remaining 40 per cent from other amino acids or amines.

Enhancement of the Ammonia Buffer System in Chronic Acidosis. If the tubular fluids remain highly acidic for long periods of time, the formation of ammonia steadily increases during the first 2 to 3 days, rising as much as tenfold. For instance, immediately after acidosis begins, as little as 30 millimoles of ammonia might be secreted each day, but after several days as much as 300 to 450 millimoles can be secreted, illustrating that the ammonia-secreting mechanism can adapt readily to handle greatly increased loads of acid elimination. The principal cause of the increasing ammonia formation is that local acidosis of the tubular cells induces the production of large amounts of the enzyme *glutaminase* that is responsible for releasing ammonia from glutamine.

RAPIDITY OF ACID-BASE REGULATION BY THE KIDNEYS

The kidney mechanism for regulating acid-base balance cannot readjust the pH within seconds, as can the extracellular fluid buffer systems, nor within minutes, as can the respiratory compensatory mechanism; but it is different from these other two in that it can keep on functioning for hours or days until it brings the pH almost exactly back to normal. In other words, its ultimate capability to regulate the body fluid pH, though it is slow to act, is infinitely more powerful than that of the other two regulatory mechanisms. The following few paragraphs explain the quantitative importance of the kidneys for regulating the hydrogen ion concentration.

Figure 30–8 illustrates the effect of extracellular fluid pH on the rate at which bicarbonate ions are lost from or gained by the extracellular fluids each minute. For instance, at a pH of 7.0, approximately 2.3 millimoles of bicarbonate ions are gained each minute, but as the pH returns toward the normal value of 7.4, the rate of gain falls to zero. Then, when the pH becomes significantly greater than 7.4, bicarbonate ions are lost by the extracellular fluids. For instance, at a pH of

Figure 30–7. Secretion of ammonia by the tubular epithelial cells and reaction of the ammonia with hydrogen ions in the tubules.

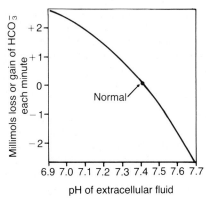

Figure 30 – 8. Effect of extracellular fluid pH on the rate of loss or gain of bicarbonate ions in the body fluids each minute.

7.6 about 1.5 millimoles of bicarbonate ions are lost each minute.

The total amount of buffers in the entire body (within the range of pH 7.0 to 7.8) is approximately 1000 millimoles. If all these should be suddenly shifted to the basic or acidic side by injecting a base or an acid, the kidneys would be able to return the pH of the body fluids back almost to normal in 1 to 3 days. But most important of all, this mechanism continues acting until the pH returns almost exactly to normal, rather than a certain percentage of the way. Therefore, the real value of the renal mechanism for regulating hydrogen ion concentration is not rapidity of action but instead its ability in the end to neutralize completely any excess acid or alkali that enters the body fluids, *unless the excess continues to enter.*

Ordinarily, the kidneys can remove up to about 500 millimoles of either acid or base each day. If greater quantities than this enter the body fluids, the kidneys are unable to cope with the extra load, and severe acidosis or alkalosis ensues.

Range of Urinary pH. In the process of adjusting the hydrogen ion concentration of the extracellular fluid, the kidneys often excrete urine at pHs as low as 4.5 or as high as 8.0. When acid is being excreted, the urine pH falls; and when alkali is being excreted, the pH rises. Even when the pH of the extracellular fluids is at the normal value of 7.4, a fraction of a millimole of acid is still lost each minute. The reason for this is that about 50 to 80 millimoles more acid than alkali are formed in the body each day, and this acid must be removed continually. Because of the presence of this excess acid in the urine, the normal urine pH averages about 6.0 instead of 7.4, the pH of the blood.

RENAL REGULATION OF PLASMA CHLORIDE CONCENTRATION — THE CHLORIDE-TO-BICARBONATE RATIO

In the preceding discussions we have emphasized the ability of the kidneys to conserve bicarbonate ion in the extracellular fluids whenever a state of acidosis develops, or to remove bicarbonate ions in a state of alkalosis. Thus, the bicarbonate ion is shuttled back and forth between high and low values as one of the principal means of adjusting the acid-base balance of the extracellular buffer systems and therefore also for adjusting the extracellular fluid pH.

However, in the process of juggling the extracellular fluid concentration of bicarbonate ion, it is essential to remove some other anion from the extracellular fluids each time the bicarbonate is increased, or to increase some other anion when the bicarbonate concentration is decreased. In general, the anion that is reciprocally juggled up or down with the bicarbonate ion is the chloride ion because this is the anion in greatest concentration in the extracellular fluid.

● CLINICAL ABNORMALITIES OF ACID-BASE BALANCE

RESPIRATORY ACIDOSIS AND ALKALOSIS

From the discussions earlier in the chapter it is obvious that any factor that decreases the rate of pulmonary ventilation increases the concentration of dissolved carbon dioxide in the extracellular fluid. And this in turn leads to increased carbonic acid and hydrogen ions, thus resulting in acidosis. Because this type of acidosis is caused by an abnormality of respiration, it is called *respiratory acidosis.*

On the other hand, excessive pulmonary ventilation reverses the process and decreases the hydrogen ion concentration, thus resulting in alkalosis; this condition is called *respiratory alkalosis.*

A person can cause respiratory acidosis in himself by simply holding his breath, which he can do until the pH of the body fluids falls to as low as perhaps 7.0. On the other hand, he can voluntarily overbreathe and cause alkalosis to a pH of about 7.9.

Respiratory acidosis frequently results from pathological conditions. For instance, damage to the respiratory center in the medulla oblongata that causes reduced breathing, obstruction of the passageways in the respiratory tract, pneumonia, decreased pulmonary membrane surface area, or any other factor that interferes with the exchange of gases between the blood and alveolar air can result in respiratory acidosis.

On the other hand, only rarely do pathological conditions cause *respiratory alkalosis.* However, occasionally a psychoneurosis causes overbreathing to the extent that a person becomes alkalotic. Also, a physiologic type of respiratory alkalosis occurs when a person ascends to a *high altitude.* The low oxygen content of the air stimulates respiration, which causes excess loss of carbon dioxide and development of mild respiratory alkalosis.

METABOLIC ACIDOSIS AND ALKALOSIS

The terms *metabolic acidosis* and *metabolic alkalosis* refer to all other abnormalities of acid-base balance besides those caused by excess or insufficient carbon dioxide in the body fluids. Use of the word "metabolic" in this instance is unfortunate, because carbon dioxide is also a metabolic product. Yet, by convention, carbonic acid resulting from dissolved carbon dioxide is called a *respiratory acid* while any other acid in the body, whether it be formed by metabolism or simply ingested by the person, is called a *metabolic acid,* or a *fixed acid.*

Causes of Metabolic Acidosis

Metabolic acidosis can result from (1) failure of the kidneys to excrete the metabolic acids normally formed in the body, (2) formation of excessive quantities of metabolic acids in the body, (3) intravenous administration of metabolic acids, or (4) addition of metabolic acids by absorption from the gastrointestinal tract. Metabolic acidosis can result also from (5) *loss* of base from the body fluids. Some of the specific conditions that cause metabolic acidosis are the following:

Diarrhea. Severe diarrhea is one of the most frequent causes of metabolic acidosis for the following reasons: The gastrointestinal secretions normally contain large amounts of sodium bicarbonate. Therefore, excessive loss of these secretions during a bout of diarrhea is exactly the same as excretion of large amounts of sodium bicarbonate into the urine. In accordance with the Henderson-Hasselbalch equation, this results in a shift of the bicarbonate buffer system toward the acid side and results in metabolic acidosis. In fact, acidosis resulting from severe diarrhea can be so serious that it is one of the most common causes of death in young children.

Vomiting. A second cause of metabolic acidosis is vomiting. Vomiting of gastric contents alone, which sometimes occurs, obviously would cause loss of acid because the stomach secretions are highly acidic and would lead to alkalosis. But vomiting of contents from deeper in the gastrointestinal tract, which often occurs in much more quantity than loss of the stomach contents, will cause loss of alkali and result in metabolic acidosis.

Uremia. A third common type of acidosis is uremic acidosis, which occurs in severe renal disease. The cause of this is failure of the kidneys to rid the body of even the normal amounts of acids formed each day by the metabolic processes of the body.

Diabetes Mellitus. A fourth and extremely important cause of metabolic acidosis is diabetes mellitus. In this condition, lack of insulin secretion by the pancreas prevents normal use of glucose for metabolism. Instead, some of the fats are split into acetoacetic acid, and this in turn is metabolized by the tissues for energy in place of glucose. Simultaneously, the concentration of acetoacetic acid in the extracellular fluids often rises very high and causes very severe acidosis. Also, large quantities of the acetoacetic acid are excreted in the urine, sometimes as much as 500 to 1000 mmol/day.

Causes of Metabolic Alkalosis

Metabolic alkalosis does not occur nearly as often as metabolic acidosis. However, there are several common causes of metabolic alkalosis.

Alkalosis Caused by Administering Diuretics (Except the Carbonic Anhydrase Inhibitors). All diuretics cause increased flow of fluid along the tubules, and this increase usually causes a great excess of sodium ions to flow into the distal and collecting tubules, leading also to rapid reabsorption of sodium ions from these tubules. This rapid reabsorption is coupled with enhanced hydrogen ion secretion because of Na^+-H^+ exchange mechanisms in the luminal membranes of the tubular cells that link hydrogen secretion to sodium absorption; hence excessive loss of hydrogen ions from the body and resultant extracellular fluid alkalosis.

Excessive Ingestion of Alkaline Drugs. Perhaps the second most common cause of alkalosis is excessive ingestion of alkaline drugs, such as sodium bicarbonate, for the treatment of gastritis or peptic ulcer.

Alkalosis Caused by Loss of Chloride Ions. Excessive vomiting of gastric contents *without* vomiting of lower gastrointestinal contents causes excessive loss of hydrochloric acid secreted by the stomach mucosa. The net result is loss of acid from the extracellular fluid and development of metabolic alkalosis. This type of alkalosis occurs in newborn children who have pyloric obstruction caused by a greatly hypertrophied pyloric sphincter muscle.

Alkalosis Caused by Excess Aldosterone. When excess quantities of aldosterone are secreted by the adrenal glands, the extracellular fluid becomes slightly alkalotic. This is caused in the following way: The aldosterone promotes extensive reabsorption of sodium ions from the distal segments of the tubular system, but coupled with this is increased secretion of hydrogen ions as well, which promotes alkalosis.

EFFECTS OF ACIDOSIS AND ALKALOSIS ON THE BODY

Acidosis. The major clinical effect of acidosis is depression of the *central nervous system.* When the pH of the blood falls below 7.0, the nervous system becomes so depressed that the person first becomes disoriented and later comatose. Therefore, patients dying of diabetic acidosis, uremic acidosis, and other types of acidosis usually die in a state of coma.

In metabolic acidosis the high hydrogen ion concentration causes increased rate and depth of respiration. Therefore, one of the diagnostic signs of *metabolic* acidosis is increased pulmonary ventilation. On the other hand, *in respiratory acidosis,* the cause of the acidosis is *depressed respiration,* which is opposite to the effect in metabolic acidosis.

Alkalosis. The major clinical effect of alkalosis is *overexcitability of the nervous system.* This occurs both in the central nervous system and in the peripheral nerves, but usually the peripheral nerves are affected before the central nervous system. The nerves sometimes become so excitable that they automatically and repetitively fire even when they are not stimulated by normal stimuli. As a result, the muscles go into a state of *tetany,* which means a state of tonic spasm. This tetany usually appears first in the muscles of the forearm, then spreads to the muscles of the face and finally all over the body. Extremely alkalotic patients may die of tetany of the respiratory muscles.

Occasionally an alkalotic person develops severe symptoms of central nervous system overexcitability. The symptoms may manifest themselves as extreme nervousness or, in susceptible persons, as convulsions. For instance, in persons who are predisposed to epileptic seizures, simply overbreathing often results in an attack. Indeed, this is one of the clinical methods for assessing one's degree of epileptic predisposition.

RESPIRATORY COMPENSATION OF METABOLIC ACIDOSIS OR ALKALOSIS

It was pointed out earlier that the high hydrogen ion concentration of metabolic acidosis causes increased pulmonary ventilation, which in turn results in rapid removal of carbon dioxide from the body fluids and reduces the hydrogen ion concentration back toward normal. Thus, this respiratory effect helps to compensate for the metabolic acidosis. However, this compensation occurs only partially. Ordinarily, the respiratory system compensates between 50 and 75 per cent. That is, if the metabolic factor makes the pH of the blood fall to 7.0 at normal pulmonary ventilation, the rate of pulmonary ventilation normally increases sufficiently to return the pH of the blood to 7.2 to 7.3, as was pointed out earlier in the chapter.

The opposite effect occurs in metabolic acidosis. That is, alkalosis diminishes the pulmonary ventilation, which in turn increases the hydrogen ion concentration back toward normal. Here again the compensation can take place to about 50 to 75 per cent.

RENAL COMPENSATION OF RESPIRATORY ACIDOSIS OR ALKALOSIS

If a person develops respiratory acidosis that continues for a long period of time, the kidneys will secrete an excess of hydrogen ions, resulting in an increase in sodium bicarbonate in the extracellular fluids. After 1 to 6 days, the pH of the body fluids will have returned about 65 to 75 per cent of the way to normal even though the person continues to breathe poorly.

Exactly the converse effect occurs in respiratory alkalosis. Large amounts of sodium bicarbonate are lost into the urine, decreasing the extracellular bicarbonate ion, thereby decreasing the pH also almost back toward normal.

PHYSIOLOGY OF TREATMENT IN ACIDOSIS OR ALKALOSIS

Obviously, the best treatment for acidosis or alkalosis is to remove the condition causing the abnormality, but, if this cannot be effected, different drugs can be used to neutralize the excess acid or alkali.

To neutralize excess acid, large amounts of *sodium bicarbonate* can be ingested by mouth. This is absorbed into the blood stream and increases the bicarbonate ion portion of the bicarbonate buffer, thereby shifting the pH toward the alkaline side. Sodium bicarbonate is occasionally used also for intravenous therapy, but this has such strong and often dangerous physiological effects that other substances are often used instead, such as *sodium lactate* or *sodium gluconate.* The lactate and gluconate portions of the molecules are metabolized in the body, leaving the sodium in the extracellular fluids in the form of sodium bicarbonate, and thereby shifting the pH of the fluids in the alkaline direction.

For treatment of alkalosis, ammonium chloride is often administered by mouth. When this is absorbed into the blood, the ammonia portion of the ammonium chloride is converted by the liver into urea; this reaction liberates hydrochloric acid, which immediately reacts with the buffers of the body fluids to shift the hydrogen ion concentration in the acid direction. Occasionally, ammonium chloride is infused intravenously, but the ammonium ion is highly toxic, and this procedure can be dangerous. Another substance occasionally used is *lysine monohydrochloride.*

CLINICAL MEASUREMENTS AND ANALYSIS OF ACID-BASE ABNORMALITIES

pH Measurements. In studying a patient with acidosis or alkalosis it is desirable to know the pH of the body fluids. This can be determined easily by measuring the pH of the plasma with a glass electrode pH meter. However, extreme care must be exercised in removing the plasma and in making the measurement, because even the slightest diffusion of carbon dioxide out of the plasma into the air shifts the bicarbonate buffer system in the alkaline direction, resulting in a much elevated pH measurement.

The pH-Bicarbonate Diagram. The so-called pH-bicarbonate diagram, which is illustrated in Figure 30–9, can be used to determine the *type* of acidosis or alkalosis a person has and its *severity.* Its use may be explained as follows:

The more vertical curves of the diagram depict different carbon dioxide concentrations. The normal carbon dioxide concentration of 1.2 mmol/liter is denoted by the colored line (this is equivalent to a P_{CO_2} of 40 mm Hg). The points along this line represent the possible combinations of bicarbonate concentration and pH that can exist in the body fluids when the carbon dioxide concentration is normal.

The more horizontal lines represent the concentrations of excess metabolic acids or bases in the body fluids. The colored line, denoted by the numeral zero, indicates a balance between both of these. That is, the points along this line represent the possible combinations of bicarbonate concentration and pH that can occur as long as the metabolic acids and bases in the body fluids are normal. The upper two horizontal lines indicate, respectively, additions of 5 and 10 mmol/liter of extra metabolic base to the body fluids, and the two lower horizontal lines indicate additions of an extra 5 and 10 mmol/liter of metabolic acid.

To use this diagram we simply find the pH of the blood and the bicarbonate concentration, then plot the appropri-

Figure 30–9. The pH-bicarbonate diagram for determining relative degrees of metabolic and respiratory acidosis or alkalosis in a patient. (Modified from Davenport: The ABC of Acid-Base Chemistry. Chicago, The University of Chicago Press. © 1947, 1949, 1950, 1958, 1969, 1974 by The University of Chicago. All rights reserved.)

ate point on the diagram. For instance, if the pH is found to be the normal value of 7.4 and bicarbonate concentration the normal value of 25 mmol/liter, we plot point A, which represents the normal condition.

Now using data obtained from another patient, we plot a new point with a pH of 7.63 and a bicarbonate concentration of 28 mmol/liter. This is point B on the diagram, which represents a carbon dioxide concentration of 0.8 mmol/liter and 7 mM/liter extra metabolic base. Thus, this person has *metabolic alkalosis* because of the considerable extra metabolic base in the body fluids, and the person also has *respiratory alkalosis* because of overventilating to an extent that the carbon dioxide concentration is considerably less than normal.

Similarly, from other patients, we plot points C, D, and E. Point C represents 6 mmol/liter of *metabolic acidosis* and sufficient *respiratory alkalosis* to reduce the carbon dioxide concentration to 0.7 mmol/liter. A person with such a finding as this could have respiratory alkalosis that has been partially compensated by metabolic acidosis produced by the kidneys.

Point D represents *mild metabolic acidosis,* 2 mmol/liter, combined with *severe respiratory acidosis.* A person could get into this state with severe primary respiratory acidosis and mild metabolic acidosis resulting from some other cause.

Point E represents *mild respiratory acidosis and severe metabolic alkalosis.* Presumably, in this case the metabolic alkalosis was primary, and respiratory compensation caused mild respiratory acidosis in an attempt to compensate for the metabolic alkalosis.

In summary, by using the pH-bicarbonate diagram, one can determine both the degree of metabolic acidosis or alkalosis and the degree of respiratory acidosis or alkalosis in a patient at the same time.

REFERENCES

Aickin, C. C.: Intracellular pH regulation by vertebrate muscle. Annu. Rev. Physiol., 48:349, 1986.

Aronson, P. S.: Kinetic properties of the plasma membrane Na^+-H^+ exchanger. Annu. Rev. Physiol., 47:545, 1985.

Brenner, B. M., and Rector, F. C., Jr.: The Kidney, 3rd Ed. Philadelphia, W. B. Saunders Co., 1986.

Catto, G. R.: Clinical Aspects of Renal Physiology. Philadelphia, W. B. Saunders Company, 1981.

de Roufignac, C., and Ealouf, J-M.: Hormonal regulation of chloride transport in the proximal and distal nephron. Annu. Rev. Physiol., 50:123, 1988.

DuBose, T. D., Jr., and Bidani, A.: Kinetics of CO_2 exchange in the kidney. Annu. Rev. Physiol., 50:653, 1988.

Galla, J. H., and Luke, R. G.: Chloride transport and disorders of acid-base balance. Annu. Rev. Physiol., 50:141. 1988.

Gamble, J. L., Jr.: Evolution of acid-base concept (1917–1984). Physiologist, 27, 375, 1984.

Gennari, F. J., et al.: High P_{CO_2} in rat kidney. News Physiol. Sci., 1:137, 1986.

Gottschalk, C. W., et al.: Localization of urine acidification in the mammalian kidney. Am. J. Physiol., 198:581, 1960.

Halperin, M. L., and Goldstein, M. B.: Fluid, Electrolyte, and Acid-Base Emergencies. Philadelphia, W. B. Saunders Co., 1988.

Halperin, M. L., et al.: Distal renal tubular acidosis syndromes: A pathophysiological approach. Am. J. Nephrol., 5:1, 1985.

Haussinger, D. (ed.): pH Homeostasis. San Diego, Calif., Academic Press, 1988.

Holliday, M. A., et al.: Pediatric Nephrology, 2nd Ed. Baltimore, Williams & Wilkins, 1987.

Ives, H. E., and Rector, F. C., Jr.: Proton transport and cell function. J. Clin. Invest., 73:285, 1984.

Jones, N. L.: Blood Bases and Acid-Base Physiology. New York, B. C. Decker, 1980.

Kazemi, H., and Johnson, D. C.: Regulation of cerebrospinal fluid acid-base balance. Physiol. Rev., 48:953, 1986.

Knepper, M. A., et al.: Ammonium transport in the kidney. Physiol. Rev., 69:179, 1989.

Kovacevic, A., and McGivan, J. D.: Mitochondrial metabolism of glutamine and glutamate and its physiological significance. Physiol. Rev., 63:547, 1983.

Maren, T. H.: The kinetics of HCO_3^- synthesis related to fluid secretion, pH control, and CO_2 elimination. Annu. Rev. Physiol., 50:695, 1988.

Massry, S. C., and Glassock, R. J.: Textbook of Nephrology, 2nd Ed. Baltimore, Williams & Wilkins, 1988.

Maude, D. L.: Acid-base disorders—A computer simulation. Physiologist, 28:118, 1985.

Murer, H., and Malmstrom, K.: How renal phosphate transport is regulated. News Physiol. Sci., 2:45, 1987.

Pitts, R. F.: Production and excretion of ammonia in relation to acid-base regulation. In Orloff, F., and Berliner, R. W. (eds.): Handbook of Physiology. Sec. 8. Baltimore, Williams & Wilkins, 1973, p. 455.

Quintero, J. A.: Acid-Base Balance: A Manual for Clinicians. St. Louis. W. H. Green, 1979.

Rahn, H.: Body temperature and acid-base regulation. Pneumonologie, 151:87, 1974.

Rector, F. C., Jr.: Acidification of urine. In Orloff, F., and Berliner, R. W. (eds.): Handbook of Physiology. Sec. 8. Baltimore, Williams & Wilkins, 1973, p. 431.

Rose, B. D. (ed.): Clinical Physiology of Acid-Base and Electrolyte Disorders, 3rd Ed. New York, McGraw-Hill Book Co., 1989.

Rosival, V.: The influence of blood hydrogen ion concentration on the level of consciousness in diabetic ketoacidosis. Ann. Clin. Res., 19:23, 1987.

Rudnick, G.: ATP-driven H^+ pumping into intracellular organelles. Annu. Rev. Physiol., 48:403, 1986.

Sasaki, S., and Marumo, F.: Mechanisms of transcellular Cl transport in mammalian renal proximal tubules. News Physiol. Sci., 4:18, 1989.

Schwartz, G. J., and Al-Awqati, Q.: Regulation of transepithelial H^+ transport by exocytosis and endocytosis. Annu. Rev. Physiol., 48:153, 1986.

Seldin, D. W., and Giebisch, G.: The Regulation of Acid-Base Balance. New York, Raven Press, 1989.

Steinmetz, P. R.: Cellular mechanisms of urinary acidification. Physiol. Rev., 54:890, 1974.

Tannen, R. L.: Control of acid secretion by the kidney. Annu. Rev. Med., 31:35, 1980.

Tanner, G. A.: Renal regulation of acid-base balance: Ammonia excretion. Physiologist, 27:95, 1984.

White, F. N., and Somero, G.: Acid-base regulation and phospholipid adaptations to temperature: Time courses and physiological significance of modifying the milieu for protein function. Physiol. Rev., 62:40, 1982.

Wright, E. M.: Transport of carboxylic acids by renal membrane vesicles. Annu. Rev. Physiol., 47:127, 1985.

CHAPTER 31

• Renal Disease, Diuresis, and Micturition •

● RENAL DISEASE

Renal disease can be classified into five different physiological categories: (1) *acute renal failure,* in which the kidneys stop working entirely or almost entirely, (2) *chronic renal failure,* in which progressively more nephrons are destroyed until the kidneys simply cannot perform all the necessary functions, (3) *hypertensive kidney disease,* in which vascular or glomerular lesions cause hypertension but not renal failure, (4) *nephrotic syndrome,* in which the glomeruli have become far more permeable than normal so that large amounts of protein are lost into the urine, and (5) *specific tubular abnormalities* that cause abnormal reabsorption or lack of reabsorption of certain substances by the tubules.

ACUTE RENAL FAILURE

Almost any condition that seriously interferes with kidney function can cause acute renal failure. Two of the most common causes are (1) acute glomerulonephritis and (2) acute damage and obstruction of the tubules.

Renal Failure Caused by Acute Glomerulonephritis. Acute glomerulonephritis is a disease caused by an abnormal immune reaction. In about 95 per cent of the patients, this occurs 1 to 3 weeks following an infection elsewhere in the body caused by certain types of group A beta streptococci. The infection may have been a streptococcal sore throat, streptococcal tonsillitis, or even streptococcal infection of the skin. It is not the infection itself that causes damage to the kidneys. Instead, as antibodies develop during the succeeding few weeks against the streptococcal antigen, it is believed that the antibodies and antigen react with each other to form an insoluble immune complex that becomes entrapped in the glomerulus, especially in the basement membrane portion of the glomerulus.

Once the immune complex has deposited in the glomeruli, all the cells of the glomerulus begin to proliferate, but mainly the epithelial cells and the mesangial cells that lie between the endothelium and epithelium. In addition, large numbers of white blood cells become entrapped in the glomeruli. Many of the glomeruli become blocked entirely by this inflammatory reaction, and those that are not blocked usually become excessively permeable, allowing both protein and red blood cells to leak into the glomerular filtrate. In a few of the severest cases, either total or almost total renal shutdown occurs.

The acute inflammation of the glomeruli usually subsides in 10 days to 2 weeks, and in most patients the kidneys return to normal function within the next few weeks to few months. Sometimes, however, many of the glomeruli are destroyed beyond repair, and in a small percentage of patients progressive renal deterioration continues indefinitely, similar to that described for chronic glomerulonephritis in a subsequent section.

Tubular Necrosis as a Cause of Acute Renal Failure. Another common cause of acute renal shutdown is *tubular necrosis,* which means destruction of epithelial cells in the tubules, as illustrated in Figure 31–1. Common causes of tubular necrosis are (1) *various poisons* that destroy the tubular epithelial cells and (2) *severe acute ischemia* of the kidneys.

Renal Poisons. Among the different renal poisons are *carbon tetrachloride* and the heavy metals such as the *mercuric ion.* These substances have specific nephrotoxic action on the tubular epithelial cells, causing death of many of them. As a result, the epithelial cells slough away from the basement membrane and plug the tubules. In some instances the basement membrane also is destroyed; but if not, new tubular epithelial cells can usually grow along the surface of the membrane, so that the tubule becomes repaired within 10 to 20 days.

Severe Acute Renal Ischemia. Severe ischemia of the kidney is likely to result from *severe circulatory shock.* In shock, the heart simply fails to pump sufficient amounts of blood to supply adequate nutrition to the different parts of the body, and renal blood flow is particularly likely to suffer because of sympathetic constriction of the renal vessels or because of vasoconstrictive substances in the blood of shock patients (Chapter 24). Therefore, lack of adequate nutrition often destroys many tubular epithelial cells and thereby plugs many of the nephrons.

Transfusion Reaction as a Cause of Acute Renal Failure. A severe transfusion reaction normally results in hemolysis of large amounts of red blood cells with release of hemoglobin into the plasma. The size of the hemoglobin molecule is slightly less than that of the pores in the glomerular membrane, so that much of the hemoglobin passes through the membrane into the glomerular filtrate. Therefore, after a transfusion reaction the tubular load of hemoglobin often becomes far greater than can be reabsorbed by the proximal tubules. The excess hemoglobin becomes so concentrated that it is likely to precipitate in the nephron and in this way cause blockage. In addition, hemolysis of red

344

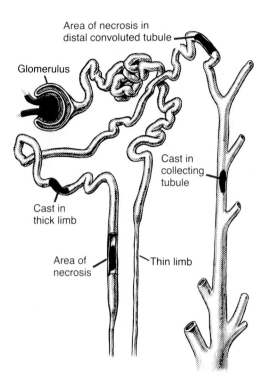

Figure 31–1. Damage to the distal tubules as a result of shock. (Modified from MacLean: Acute Renal Failure. Springfield, Ill., Charles C Thomas.)

blood cells probably also releases vasoconstrictor agents into the blood stream, and it is believed that the vasoconstriction can promote poor blood supply to the tubules, thus acting as a further cause of tubular damage.

Physiological Effects of Acute Renal Failure

When the degree of acute renal failure is moderate, the major physiological effect is retention of salt and water. At first the tissues become edematous, but the person has few other symptoms. However, within a few days the patient also develops hypertension, usually with a blood pressure increase of 30 to 40 mm Hg; this usually continues until the acute renal failure subsides.

In the severest cases, uremic retention of waste products ensues and acidosis soon develops. In complete renal shutdown without treatment, the patient dies within 8 to 14 days. Other effects of renal retention will be discussed in the following section in relation to chronic renal failure.

CHRONIC RENAL FAILURE—DECREASE IN NUMBER OF FUNCTIONAL NEPHRONS

In many types of kidney disease large numbers of nephrons are destroyed or damaged so severely that the remaining nephrons simply cannot perform the normal functions of the kidney. Some of the different causes of this are *chronic glomerulonephritis, traumatic loss of kidney tissue, congenital absence of kidney tissue, congenital polycystic disease* (in which large cysts develop in the kidneys and destroy surrounding nephrons by compression), *urinary tract obstruction* resulting from renal stones, *pyelonephritis,* and disease of the renal vasculature.

Chronic Glomerulonephritis. Chronic glomerulonephritis is caused by any one of many diseases that damage principally the glomeruli but often the tubules as well. The basic glomerular lesion is usually very similar to that which occurs in acute glomerulonephritis. It seems to begin with accumulation of precipitated antigen-antibody complex in the glomerular membrane, though in only a few instances is this caused by streptococcal infection. The result is inflammation of the glomeruli. The glomerular membrane becomes progressively thickened and is eventually invaded by fibrous tissue. In the later stages of the disease, the glomerular filtration coefficient becomes greatly reduced because of decreased numbers of filtering capillaries in the glomerular tufts and because of thickened glomerular membranes. In the final stages of the disease many of the glomeruli are completely replaced by fibrous tissue, and the function of these nephrons is thereafter lost forever.

Pyelonephritis. Pyelonephritis is an infectious and inflammatory process that usually begins in the renal pelvis and then extends progressively into the renal parenchyma. The infection can result from many different types of bacteria, but especially from the colon bacillus that originates from fecal contamination of the urinary tract. Invasion of the kidneys by these bacteria results in progressive destruction of renal tubules, glomeruli, and any other structures in the path of the invading organisms. Consequently, large portions of the functional renal tissue are lost.

A particularly interesting feature of pyelonephritis is that the invading infection usually affects the medulla of the kidney more than it affects the cortex. Because one of the primary functions of the medulla is to provide the countercurrent mechanism for concentrating the urine, patients with pyelonephritis frequently have reasonably normal renal function except for reduced ability to concentrate their urine.

Destruction of Nephrons by Renal Vascular Disease—Benign Nephrosclerosis. Many different types of vascular lesions can lead to renal ischemia and death of the renal tissue; the most common of these are (1) *atherosclerosis* of the larger renal arteries with progressive sclerotic constriction of the vessels; (2) *fibromuscular hyperplasia* of one or more of the large arteries, which also causes occlusion of these large vessels; and (3) *benign nephrosclerosis,* a very common condition caused by sclerotic lesions of the smaller arteries and arterioles.

Arteriosclerotic or hyperplastic lesions of the larger arteries frequently affect one kidney more than the other and therefore cause unilaterally diminished renal function.

Benign nephrosclerosis is believed to begin with leakage of plasma through the intimal membrane of the small arteries and arterioles. This causes a fibrinoid deposit to develop in the medial layers of these vessels, followed by progressive fibrous tissue invasion that eventually constricts the vessel —in many instances totally occluding it. Because there is essentially no collateral circulation among the smaller renal arteries, destruction of one of these also causes destruction of a comparable number of nephrons. Therefore, much of the kidney tissue becomes replaced by minute areas of fibrous tissue; the kidneys become greatly reduced in size and progressively develop a nodular surface. This process occurs at least to some extent in most persons in old age and causes progressive decrease in both renal blood flow and renal plasma clearance. Figure 31–2 illustrates the Diodrast clearance of the kidneys (which is a measure of renal plasma flow, like PAH clearance) of otherwise normal persons at different ages. Note that even in the "normal" person, kidney plasma flow decreases on the average to about 45 per

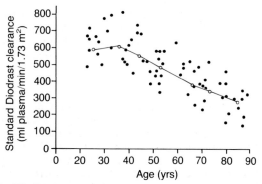

Figure 31–2. Effect of aging on the Diodrast clearance of the kidneys. (Modified from Wolstenholme et al.: Ciba Foundation Colloquia on Ageing. Boston, Little, Brown and Co.)

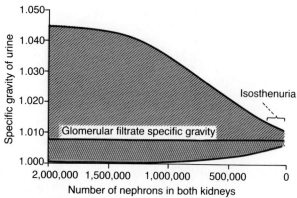

Figure 31–3. Development of *isosthenuria* in patients with decreased number of active nephrons.

cent of normal by the age of 80 years, and there is accompanying reduction of excretory function.

Abnormal Nephron Function in Chronic Renal Failure

Inability of Failing Kidneys to Maintain Normal Blood Constituents. Normally, only one third of the normal number of nephrons can eliminate essentially all the normal "load" of waste products from the body without serious accumulation of any of these in the body fluids. However, further reduction of numbers of nephrons leads to retention, especially of those waste products that depend on a high glomerular filtration rate for excretion. The most prominent of these are *urea* and *creatinine*. Death usually ensues when the number of nephrons falls below 5 to 20 per cent of normal.

Function of the Remaining Nephrons in Renal Failure—Paradoxical Increase in Urine Volume Output. In renal failure, the still-functioning nephrons usually become greatly overloaded in several different ways. First, for reasons not understood, the blood flow through the glomerulus and the amount of glomerular filtrate formed by each nephron often increase 50 to 100 per cent. Second, extra large quantities of excretory substances, such as urea, phosphates, sulfates, uric acid, and creatinine, accumulate in the extracellular fluid. These constitute markedly increased tubular loads of substances that are poorly reabsorbed, sometimes increasing as much as 1000 per cent per nephron. Therefore, only a small fraction of the tubular solutes is reabsorbed, and the remaining solutes act like an osmotic diuretic, resulting in rapid flushing of tubular fluid through the tubules. Consequently, the volume of urine formed by each nephron can rise to as much as 20 times normal, and a person will occasionally have *as much as two to three times normal total urine output* despite significant renal insufficiency. This paradoxical situation is caused by a greater increase in urine volume output per nephron than reduction in number of nephrons.

Isosthenuria. Another effect of rapid flushing of fluid through the tubules is that the normal concentrating and diluting mechanisms of the kidney do not function properly. This occurs mainly because of too rapid flow of tubular fluid through the collecting ducts for adequate water absorption to occur. Therefore, as progressively more nephrons are destroyed, the specific gravity of the urine approaches that of the glomerular filtrate, approximately 1.008. These effects are illustrated in Figure 31–3, which gives the approximate upper and lower limits of urine specific gravity as the number of nephrons decreases. Because the concentrating mechanism is impaired more than the diluting mechanism, an important renal function test is to determine how well the kidneys can concentrate urine when the person is dehydrated for 12 or more hours.

Effects of Renal Failure on the Body Fluids —Uremia

The effect of renal failure on the body fluids depends to a great extent on water and food intake. Assuming that the person continues to ingest moderate amounts of water and food after complete renal shutdown, the concentration changes of different substances in the extracellular fluid are approximately those shown in Figure 31–4. The most important effects are (1) *generalized edema* resulting from water and salt retention; (2) *acidosis* resulting from failure of the kidneys to rid the body of normal acidic products; (3) *high concentrations of the nonprotein nitrogens*, especially *urea, creatinine*, and *uric acid*, resulting from failure of the body to excrete the metabolic end products of proteins; and (4) *high concentration of other urinary retention products* including *phenols, guanidine bases, sulfates, phosphates*, and *potassium*. This condition is called *uremia* because of the high concentration of urea in the body fluids.

Water Retention and Edema. If treatment is begun to restrict water intake immediately after acute renal failure,

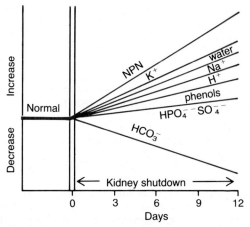

Figure 31–4. Effect of kidney shutdown on extracellular fluid constituents.

the total body fluid content may not change at all. However, when the patient drinks water in response to normal desire, the body fluids begin increasing immediately and rapidly. If the patient takes in no electrolytes at the same time, as much as one half of the water may enter the cells instead of the extracellular fluid.

Acidosis in Renal Failure. Each day the metabolic processes of the body normally produce 50 to 80 millimoles more metabolic acid than metabolic alkali. Therefore, any time the kidneys fail to function, acid begins to accumulate in the body fluids. Normally, the buffers of the fluids can buffer up to a total of 500 to 1000 millimoles of acid without lethal depression of the extracellular fluid pH, and the phosphate compounds in the bones can buffer an additional few thousand millimoles, but gradually this buffering power is used up so that the pH falls drastically. The patient becomes *comatose* at about this same time, caused mainly by the acidosis as is discussed subsequently.

Increase in Urea and Other Nonprotein Nitrogens (Azotemia) in Uremia. The nonprotein nitrogens include urea, uric acid, creatinine, and a few less important compounds. These, in general, are the end products of protein metabolism and must be removed from the body continually to ensure continued protein metabolism in the cells. The concentrations of these, particularly of urea, can rise to as high as ten times normal during 1 to 2 weeks of renal failure. However, even these high levels do not seem to affect physiological function nearly so much as do the high concentrations of hydrogen ions and some of the other less obvious substances such as very toxic guanidine bases, ammonium ions, and others. Yet one of the most important means of assessing the degree of renal failure is to measure the concentrations of urea and creatinine.

Uremic Coma. After a week or more of renal failure the sensorium becomes clouded, and the patient soon progresses into a state of coma. The acidosis is believed to be the principal factor responsible for the coma because acidosis caused by other conditions, such as severe diabetes mellitus, also causes coma. However, many other abnormalities could also be contributory—the generalized edema, the high potassium concentration, and possibly even the high nonprotein nitrogen concentration.

The respiration usually is deep and rapid in coma, which is a respiratory attempt to compensate for the metabolic acidosis. In addition to this, during the last day or so before death, the arterial pressure falls progressively, then rapidly in the last few hours. Death occurs usually when the pH of the blood falls to about 6.8.

Anemia in Chronic Renal Failure. A patient with chronic renal failure of severe degree almost always also develops severe *anemia*. The probable cause of this is the following: The kidneys normally secrete the substance *erythropoietin*, which in turn stimulates the bone marrow to produce red blood cells. Obviously, if the kidneys are seriously damaged, they are unable to form adequate quantities of erythropoietin, which leads to diminished red blood cell production and consequent anemia. However, other factors, such as the high plasma concentrations of urea, hydrogen ions, and other waste products, might also play important roles in causing the anemia.

Osteomalacia in Renal Failure. Prolonged renal failure also causes *osteomalacia*, a condition in which the bones are partially absorbed and therefore greatly weakened, as will be explained in relation to the physiology of bone in Chapter 79. The most important cause of this condition is the following: Vitamin D must be converted by a two-stage process, first in the liver and then in the kidney, into 1,25-dihydroxycholecalciferol before it is able to promote calcium absorption from the intestine. Therefore, serious damage to the kidneys greatly reduces the availability of calcium to the bones.

Dialysis of Patients with the Artificial Kidney

Artificial kidneys have now been used for almost 40 years to treat patients with severe renal failure. In certain types of acute renal failure, such as that following mercury poisoning or following circulatory shock, the artificial kidney is used simply to tide the patient over for a few weeks until the renal damage heals so that the kidneys can resume function. However, the artificial kidney has now been developed to the point that thousands of persons with permanent renal failure or even total kidney removal are being maintained in health for years at a time, their lives depending entirely on the artificial kidney.

The basic principle of the artificial kidney is to pass blood through very minute blood channels bounded by a thin membrane. On the other side of the membrane is a *dialyzing fluid* into which unwanted substances in the blood pass by diffusion.

Figure 31–5 illustrates the components of one type of artificial kidney in which blood flows continually between two thin membranes of cellophane; on the outside of the membranes is the dialyzing fluid. The cellophane is porous enough to allow all constituents of the plasma except the plasma proteins to diffuse in both directions—from plasma into the dialyzing fluid and from the dialyzing fluid back into the plasma. If the concentration of a substance is greater in the plasma than in the dialyzing fluid, there will be net transfer of the substance from the plasma into the dialyzing fluid. The amount of the substance that is transferred depends on (1) the permeability characteristics of the membrane as well as its surface area; (2) the difference between the concentrations on the two sides of the membrane; (3) molecular size, the smaller molecules diffusing more rapidly than larger ones; and (4) the length of time that the blood and the fluid remain in contact with the membrane.

In the normal operation of the artificial kidney, blood flows continually or intermittently back into a vein. The total amount of blood in the artificial kidney at any one time is usually less than 500 milliliters, the rate of flow may be several hundred milliliters per minute, and the total diffusing surface is usually between 0.6 and 2.5 square meters. To prevent coagulation of blood in the artificial kidney, a small amount of heparin is infused into the blood as it enters the "kidney."

The Dialyzing Fluid. Table 31–1 compares the constituents in a typical dialyzing fluid with those in normal plasma and uremic plasma. Note that the concentrations of the ions and other substances in the dialyzing fluid are not the same as the concentrations in normal plasma or in uremic plasma. Instead, they are adjusted to levels that are needed to cause appropriate movement of water and each solute through the membrane during the dialysis period.

Note also that there is no phosphate, urea, urate, sulfate, or creatinine in the dialyzing fluid; but they are present in high concentrations in the uremic blood. Therefore, when the uremic patient is dialyzed, these substances are lost in large quantities into the dialyzing fluid; thereby major proportions of them are removed from the plasma.

Effectiveness of the Artificial Kidney. The effectiveness of an artificial kidney is expressed in terms of the amount of plasma that can be cleared of different sub-

Figure 31–5. Principles of the artificial kidney.

stances each minute, which, as will be recalled from Chapter 27, is also the principal means for expressing the functional effectiveness of the kidneys themselves. Most artificial kidneys can clear 100 to 225 milliliters of plasma per minute of urea, which shows that, at least in the excretion of this substance, the artificial kidney can function about twice as rapidly as the two normal kidneys together, whose urea clearance is only 70 ml/min. Yet the artificial kidney is normally used for only 4 to 6 hours three times a week.

Table 31–1 COMPARISON OF DIALYZING FLUID WITH NORMAL AND UREMIC PLASMA

Constituent	Normal Plasma	Dialyzing Fluid	Uremic Plasma
Electrolytes (mEq/Liter)			
Na^+	142	133	142
K^+	5	1.0	7
Ca^{++}	3	3.0	2
Mg^{++}	1.5	1.5	1.5
Cl^-	107	105	107
HCO_3^-	27	35.7	14
$Lactate^-$	1.2	1.2	1.2
HPO_4^{--}	3	0	9
$Urate^-$	0.3	0	2
$Sulfate^{--}$	0.5	0	3
Nonelectrolytes (mg/dl)			
Glucose	100	125	100
Urea	26	0	200
Creatinine	1	0	6

Therefore, the overall plasma clearance is still considerably limited when the artificial kidney replaces the normal kidneys.

HYPERTENSIVE KIDNEY DISEASE

Many of the same types of kidney disease that lead to chronic renal failure can also cause hypertension. However, this is not invariably true, because damage to certain portions of the kidney is very prone to cause hypertension, whereas damage to other portions will cause uremia without hypertension. A classification of renal disease relative to its hypertensive or nonhypertensive effects is the following.

Hypertensive Renal Lesions. Essentially all renal lesions that cause *diminished blood flow* or *diminished glomerular filtration per nephron* will cause hypertension. Both these conditions tend to cause retention of salt and water, which eventually leads to hypertension (see Chapter 19). Once the hypertension has developed, the glomerular filtration rate may return entirely to normal. If all the tubules are normal, the filtrate is then processed normally in these so that urinary excretion may be completely normal, and there may be absolutely no sign of renal failure.

Renal Diseases That Lead to Renal Failure but Might Not Lead to Hypertension. Loss of large numbers of whole nephrons, such as occurs from loss of one kidney and part of another kidney, will always lead to renal failure if the amount of kidney tissue lost is great enough. However, if the remaining nephrons are completely normal, this condition frequently will not cause hypertension because even a slight rise in arterial pressure will increase the glomerular filtration rate sufficiently to promote rapid water and salt loss in the urine—even with few nephrons. On the other hand, a patient with this type of kidney abnormality who eats large quantities of salt will then develop very severe hypertension because the kidneys simply cannot clear adequate quantities of salt under these conditions.

Hypertension Caused by Renal Secretion of Renin. When one part of the kidney mass is ischemic and the remainder is nonischemic (such as occurs when one renal artery is severely constricted), the ischemic renal tissue secretes large quantities of renin. This secretion leads to the formation of angiotensin II, which in turn leads to hypertension. The most probable cause of the chronic hypertension, as was discussed in Chapter 19, is that (1) the ischemic kidney tissue itself excretes less than normal amounts of water and salt, and (2) the angiotensin affects the nonischemic kidney tissue, causing it also to retain water and salt. (Many physiologists have believed in the past that the hypertension resulted from peripheral vasoconstriction induced by the angiotensin. However, recent experiments have demonstrated almost conclusively that this is not the case, as was discussed in Chapter 19. In fact, the blood concentration of angiotensin is usually much too low to cause much peripheral vasoconstriction, but it can strongly affect renal excretion of salt.)

THE NEPHROTIC SYNDROME—INCREASED GLOMERULAR PERMEABILITY

Large numbers of patients with renal disease develop a so-called *nephrotic syndrome*, which is characterized especially by *loss of large quantities of plasma proteins into the urine*. In some instances this occurs without evidence of any other abnormality of renal function, but more often it is associated with some degree of renal failure.

The cause of the protein loss in the urine is increased permeability of the glomerular membrane. Therefore, any disease condition that can increase the permeability of this membrane can cause the nephrotic syndrome. Such diseases include *chronic glomerulonephritis* (in the previous discussion, it was noted that this disease primarily affects the glomeruli and often causes a greatly increased permeability of the glomerular membrane); *amyloidosis*, which results from deposition of an abnormal proteinoid substance in the walls of the blood vessels and seriously damages the basement membrane of the glomerulus; and *minimal change nephrotic syndrome*, a disease found mainly in young children.

"Minimal Change" Nephrotic Syndrome. In so-called minimal change nephrotic syndrome, one can rarely discern with the light microscope any abnormality of the glomerular membrane. However, with special study techniques, it has been found that the negative electrical charge normally exhibited by the glomerular membrane is reduced or lacking. Also, immunologic studies show abnormal immune reactions in some cases, suggesting that the loss of the negative charge might have resulted from antibody attack on the membrane.

Loss of the negative charge allows proteins, especially albumin, to pass through the glomerular membrane with ease, for it will be recalled that this negative charge normally repulses the negatively charged plasma protein molecules, which is a principal means of preventing leakage of proteins into the urine.

Minimal change nephrotic syndrome occurs mainly in children between the ages of 2 and 6 years, but it occurs occasionally in adults as well.

The greatly increased permeability of the glomerular membrane occasionally allows as much as 40 grams of plasma protein loss into the urine each day, which is an extreme amount for a young child. Therefore, the plasma proteins often fall below 2 gm/dl, and the colloid osmotic pressure falls from a normal of 28 to as low as 6 to 8 mm Hg. A consequence of this is that the blood capillaries all over the body leak tremendous quantities of fluid into all tissues, causing very severe *low protein edema*. Even in the very young child, as much as 10 liters of extra tissue fluid sometimes occurs, as well as an additional 10 liters of *ascites* in the abdomen. Also, the joints swell, and the pleural cavity and the pericardium can become partially filled with fluid.

Approximately 90 per cent of these children respond exceedingly well to the administration of glucocorticoid steroids, which are known to alter certain types of immunologic abnormalities. However, the cellular mechanism of the glucocorticoid effect is not understood.

In all types of nephrosis in which plasma albumin concentration falls very low, large quantities of lipids appear in the blood plasma, with especially greatly increased blood cholesterol. This is believed to be caused by a direct effect of low plasma albumin on the liver to increase the production of the plasma lipoproteins.

SPECIFIC TUBULAR DISORDERS

In the discussion of active reabsorption and secretion by the tubules in Chapter 27, it was pointed out that different transport mechanisms are responsible for transporting different individual substances. In Chapter 3 it was also pointed out that each cellular enzyme and each carrier protein is formed in response to a respective gene in the nucleus. If any required gene happens to be absent or abnormal, the tubules might be deficient in one of the appropriate enzymes or carriers. For this reason many different specific tubular disorders are known to occur for the transport of individual or special groups of substances through the tubular membrane. Essentially all of these are hereditary disorders. Some of the more important ones are discussed here.

Renal Glycosuria. In this condition the blood glucose concentration may be completely normal, but the transport mechanism for reabsorption of glucose is greatly limited or absent. Consequently, despite the normal blood glucose level, large quantities of glucose pass into the urine each day. Because one of the tests for diabetes mellitus (which results from lack of insulin secretion by the pancreas) is the presence of glucose in the urine, renal glycosuria (a benign condition which causes essentially no dysfunction of the body) must always be ruled out before making a diagnosis of diabetes mellitus.

Excess Proximal Tubular Reabsorption of Urate Ion—The Cause of Gout. Urate ions are one of the important waste products of cellular metabolism. Therefore, it is important to rid the body fluid of any excesses of this ion. However, for reasons that are not known, the renal tubules have the capability of both actively secreting urate ion and actively reabsorbing it. The active reabsorption process occurs in the proximal tubules. In a small subset of persons this reabsorption is overly active and causes persistently elevated urate ion in the extracellular fluid. Above a critical concentration, the urate precipitates in many tissues of the body in the form of uric acid crystals. Precipitation occurs especially in the joints, leading to the clinical syndrome known as *gout*, a type of inflammatory arthritis that can affect multiple joints at a time.

Nephrogenic Diabetes Insipidus. Occasionally the renal tubules do not respond to antidiuretic hormone secreted by the supraopticohypophysial system; as a consequence, large quantities of dilute urine are continually excreted. As long as the person is supplied with plenty of water, this condition rarely causes any severe difficulty. However, when adequate quantities of water are not available, the person rapidly becomes dehydrated.

Metabolic Acidosis Caused by Failure of the Tubules to Secrete Hydrogen Ions. In this condition the person is unable to secrete adequate quantities of hydrogen ions; as a result, large amounts of sodium bicarbonate are continually lost into the urine for reasons discussed in the preceding chapter. This causes a continual state of *metabolic acidosis.* However, adequate replacement therapy for continual administration of alkali can maintain normal bodily function.

Renal Hypophosphatemia. In renal hypophosphatemia the renal tubules fail to reabsorb adequate quantities of phosphate ions even when the phospate concentration of the body fluids falls very low. This condition does not cause any serious immediate abnormalities, because the phosphate level of the extracellular fluids can vary widely without significant cellular dysfunction. However, over a long period of time the low phosphate level results in diminished calcification of the bones and causes the person to develop rickets. Furthermore, this type of rickets is refractory to vitamin D therapy in contrast to the rapid response of the usual type of rickets, which is discussed in Chapter 79.

Aminoaciduria. Some amino acids share mutual transport systems for reabsorption, while other amino acids have distinct transport systems of their own. Rarely, a condition called *generalized aminoaciduria* results from deficient reabsorption of all amino acids; but more frequently, deficiencies of specific carrier systems may result in (1) *essential cystinuria,* in which large amounts of cystine fail to be reabsorbed and often crystallize in the urine to form renal stones; (2) simple *glycinuria,* in which glycine fails to be reabsorbed; or (3) *beta-amino-isobutyric aciduria,* which occurs in about 5 per cent of all people but apparently has little or no clinical significance.

● **RENAL FUNCTION TESTS**

The renal function tests can be divided into three categories: (1) determination of renal clearances, (2) measurement of substances in the blood that are normally excreted by the kidneys, and (3) chemical and physical analysis of the urine.

Renal Clearance Tests. Any of the renal clearance tests, including clearance of para-aminohippuric acid, inulin, mannitol, or other substances, as described in Chapter 27, can be used as renal function tests. Indeed, if all these are performed one can determine the glomerular filtration rate, the effective blood flow through the kidney per minute, the filtration fraction, and many other characteristics of renal function. However, many of these clearance tests are difficult to perform. Yet several special clearance tests in which radiopaque or radioactive substances are excreted from the blood into the renal pelvis are easily employed. Two of these are intravenous pyelography and radioactive clearance studies.

Intravenous Pyelography. Several substances containing large quantities of iodine in their molecules — Diodrast, Hippuran, and Iopax — are excreted into the urine both by glomerular filtration and by active tubular secretion. Consequently, their concentration in the urine becomes very high within a few minutes after intravenous injection. Also, the iodine in the compounds makes them relatively opaque to x-rays. Therefore, x-ray pictures can be made showing shadows of the renal pelves, of the ureters, and even of the urinary bladder. Ordinarily, a sufficient quantity is excreted — that is, is "cleared" — within 5 minutes after injection to give good shadows of the kidney pelves. Failure to show a distinct shadow within this time indicates depressed renal clearance.

Radioactive Clearance Studies. If any of the above substances (or many others) is prepared with *radioactive* iodine or some other radioactive nuclide, one can measure the radioactivity from both renal pelves by placing appropriate radioactivity counters over the kidneys. Only a trace amount of the substance need be injected intravenously and the degree of radioactivity recorded during the following few minutes to determine roughly the renal clearances.

A special value of both x-ray and radioactive pyelography is that they measure the functioning of each kidney independently of the other, rather than the total functioning of both kidneys together, as is measured by the other renal function tests.

Blood Analyses as Tests of Renal Function. One can also estimate how well the kidneys are performing their functions by measuring the concentrations of various substances in the blood. For instance, the normal concentration of *urea* in the blood is 26 mg/dl, but in severe cases of renal insufficiency this can rise to as high as 300 mg/dl. The normal concentration of *creatinine* in the blood is 1.1 mg/dl; but this, too, can increase as much as tenfold. To determine the degree of metabolic acidosis resulting from renal dysfunction, one can use the *pH-bicarbonate diagram* or some other similar procedure, as discussed and illustrated in the preceding chapter. Though these different tests are not as satisfactory as clearance tests for determining the functional capabilities of the kidneys, they are easy to perform, and they do tell the physician how seriously the internal environment has been disturbed.

Physical Measurements of the Urine as Renal Function Tests. Obviously, one of the most important urinary measurements is the *volume of urine* formed each day. In acute renal failure this can fall to zero, and in chronic renal failure it often is diminished. On the other hand, moderate renal failure may actually increase the urinary output, as was described earlier, because of vast over-diuresis of the remaining nephrons when the majority have been destroyed.

A second factor frequently measured is the *specific gravity* of the urine. Depending on the types of substances being cleared, this can vary tremendously: its upper extreme can go as high as 1.045, and it can fall to as little as 1.002. To test the ability of the kidneys to dilute the urine, the patient drinks large quantities of water, and measurements are made of the minimal specific gravity that can be attained. Then, at another time, the patient goes without water for 12 or more hours, and the maximum concentration of the urine is measured. Referring back to Figure 31–3, note that the concentrating ability of the kidneys is especially compromised as the number of nephrons decreases.

● **DIURETICS AND MECHANISMS OF THEIR ACTIONS**

A diuretic is a substance that increases the rate of urine output. Most diuretics act by decreasing the rate of reabsorption of fluid from the tubules.

The principal use of diuretics is to reduce the total amount of fluid in the body. They are especially important in treating edema and hypertension.

When one is using a diuretic it is usually important that the rate of sodium loss in the urine also be increased as well as the rate of water loss. The reason for this is the following: If water alone were removed from the body fluids, the fluids would become hypertonic and elicit an osmoreceptor response, followed by marked secretion of antidiuretic hormone. This hormone would then cause large amounts of water to be reabsorbed from the tubules, which would nullify the effect of the diuretic. However, if sodium is lost along with the water, this nullification will not result. Therefore, all valuable diuretics cause marked *natriuresis* (sodium loss) as well as diuresis.

Several significant types of diuretics are the following:

The Osmotic Diuretics. Injection into the blood stream of *urea, sucrose, mannitol,* or any other substance not easily reabsorbed by the tubules causes a great increase in the osmotically active substances in the tubules. The osmotic pressure of these then greatly decreases water reabsorption, so that large amounts of tubular fluid flush on into the urine.

The same effect occurs when the glucose concentration of the blood rises to very high levels in diabetes mellitus. Above a glucose concentration of about 250 mg/dl very little of the extra glucose is reabsorbed by the tubules; instead it acts as an osmotic diuretic and causes rapid loss of fluid into the urine. The name "diabetes" refers to the prolific urine flow.

Diuretics That Diminish Active Reabsorption. Any substance that inhibits carrier systems in the tubular epithelial cells and thereby diminishes active reabsorption of tubular solutes increases the tubular osmotic pressure and causes osmotic diuresis. Some of the most commonly used drugs of this type are the following:

The "Loop" Diuretics — Furosemide and Ethacrynic Acid. Furosemide and ethacrynic acid are the most powerful of the clinically used diuretics. They are called *loop diuretics* because they function mainly to decrease active reabsorption in the ascending limb of the loop of Henle, though also in the early portions of the distal tubule as well. Their mechanism of action is to block active sodium-chloride co-transport in the luminal membrane of the epithelial cells, which is the principal mechanism for reabsorbing chloride and sodium ions. This causes diuresis for two reasons: (1) It allows greatly increased quantities of solutes to be delivered into the distal portions of the nephrons, and these then act as osmotic agents to prevent water reabsorption as well. (2) Failure to absorb sodium and chloride ions from the loop of Henle into the medullary interstitium decreases the concentration of the medullary interstitial fluid. Consequently, the concentrating ability of the kidney is greatly decreased, so that reabsorption of fluid in the collecting ducts is further decreased. Because of these two effects, as much as 20 to 30 per cent of the glomerular filtrate may be delivered into the urine, causing under acute conditions urine outputs as great as 25 times normal for a period of a few minutes.

Chlorothiazide. Chlorothiazide and other thiazide derivatives act primarily on the distal tubules to prevent active sodium reabsorption; under favorable conditions they cause as much as 8 per cent of the glomerular filtrate to pass into the urine.

Carbonic Anhydrase Inhibitors — Acetazolamide. Acetazolamide (Diamox) and other carbonic anhydrase inhibitors primarily block reabsorption of bicarbonate ions from the proximal tubules. They do this by inhibiting the carbonic anhydrase that is attached to the luminal brush border of the tubular epithelial cells and normally catalyzes the dissociation of carbonic acid into water and carbon dioxide. Blockage of this enzyme-catalyzed reaction prevents removal of bicarbonate ions from the tubular fluid, and these remain in the tubules to act as an osmotic diuretic. However, use of this drug also causes some degree of acidosis because of excessive loss of bicarbonate ions in the urine.

Competitive Aldosterone Inhibitors — Spironolactone. Spironolactone and several other similar substances compete with aldosterone for receptor sites in the distal nephron epithelial cells and thereby block the sodium reabsorption–promoting effect of aldosterone. As a consequence, the sodium remains in the tubules and acts as an osmotic diuretic. These drugs also block the effect of aldosterone to promote potassium secretion into the tubules. Therefore, in some instances, the extracellular fluid potassium concentration becomes dangerously elevated.

● MICTURITION

Micturition is the process by which the urinary bladder empties when it becomes filled. Basically the bladder (1) progressively fills until the tension in its walls rises above a threshold value, at which time (2) a nervous reflex called the "micturition reflex" occurs that either causes micturition or, if it fails in this, at least causes a conscious desire to urinate.

PHYSIOLOGIC ANATOMY AND NERVOUS CONNECTIONS OF THE BLADDER

The urinary bladder, illustrated in Figure 31–6, is a smooth muscle chamber composed of two principal parts: (1) the *body,* which is the major part of the bladder, in which the urine collects, and (2) the *neck,* which is a funnel-shaped extension of the body, passing inferiorly and anteriorly into the urogenital triangle and connecting with the urethra. The lower part of the bladder neck is also called the *posterior urethra* because of its relationship to the urethra.

The smooth muscle of the bladder is known as the *detrusor muscle.* Its muscle fibers extend in all directions and, when contracted, can increase the pressure in the bladder sometimes to as high as 40 to 60 mm Hg. Thus, it is the detrusor muscle that empties the bladder. The smooth muscle cells of the detrusor muscle fuse with each other so that low resistance electrical pathways exist from one to the other. Therefore, an action potential can spread throughout the detrusor muscle to cause contraction of the entire bladder at once.

On the posterior wall of the bladder, lying immediately above the bladder neck, is a small triangular area called the *trigone.* At the lowermost apex of the trigone is the opening of the bladder through the *bladder neck* into the *posterior urethra,* and the two ureters enter the bladder at the uppermost angles of the trigone. The trigone can be identified by the fact that its mucosa is very smooth, in contrast to the remainder of the bladder mucosa, which is folded to form *rugae.* Where each ureter enters the bladder, it courses obliquely through the detrusor muscle and then passes still another 1 to 2 centimeters underneath the bladder mucosa before emptying into the bladder.

The Automatic Bladder. If the spinal cord is damaged above the sacral region but the sacral segments are still intact, typical micturition reflexes still occur. However, they are no longer controllable by the brain. During the first few days to several weeks after the damage to the cord has occurred, the micturition reflexes are completely suppressed because of the state of "spinal shock" caused by the sudden loss of facilitatory impulses from the brain stem and cerebrum. However, if the bladder is emptied periodically by catheterization to prevent physical bladder injury, the excitability of the micturition reflex gradually increases until typical micturition reflexes return.

It is especially interesting that stimulating the skin in the genital region can sometimes elicit a micturition reflex in this condition, thus providing a means by which some patients can still control urination.

The Uninhibited Neurogenic Bladder. Another common abnormality of micturition is the so-called uninhibited neurogenic bladder, which results in frequent and relatively uncontrollable micturition. This condition derives from partial damage in the spinal cord or brain stem that interrupts most of the inhibitory signals. Therefore, facilitatory impulses passing continually down the cord keep the sacral centers so excitable that even a very small quantity of urine will elicit an uncontrollable micturition reflex and thereby promote urination.

REFERENCES

Adlere, S. G., et al.: Hypersensitivity phenomena and the kidney: Role of drugs and environmental agents. Am. J. Kidney Dis., 5:75, 1985.

Avram, M. M. (ed.): Prevention of Kidney Disease and Long-Term Survival. New York, Plenum Publishing Corp., 1982.

Berstein, K. N., and O'Connor, D. T.: Antiadrenergic antihypertensive drugs: Their effect on renal function. Annu. Rev. Pharmacol. Toxicol., 24:105, 1984.

Brater, D. C.: Pharmacodynamic considerations in the use of diuretics. Annu. Rev. Pharmacol. Toxicol., 23:45, 1983.

Brenner, B. M., and Rector, F. C., Jr.: The Kidney, 3rd Ed. Philadelphia, W. B. Saunders Co., 1986.

Bricker, N. S.: The Kidney. Diagnosis and Management. New York, John Wiley & Sons, 1984.

Brun, C., and Olsen S.: Atlas of Renal Biopsy. Philadelphia, W. B. Saunders Company, 1981.

Bullock, M. L., et al.: The assessment of risk factors in 462 patients with acute renal failure. Am. J. Kidney Dis., 5:97, 1985.

Catto, B. E.: Clinical Aspects of Renal Physiology. Philadelphia, W. B. Saunders Co., 1981.

Charlton, C. A.: The Urological System. New York, Churchill Livingstone, 1983.

Churg, J: Renal Disease-Classification and Atlas of Glomerular Diseases. New York, Igaku-Shoin Medical Publishers, 1982.

Conger, J. D., and Schrier, R. W.: Renal hemodynamics in acute renal failure. Annu. Rev. Physiol., 42:603, 1980.

Cummings, N. B. (ed.): Immune Mechanisms in Renal Disease. New York, Plenum Publishing Corp., 1983.

DeFronzo, R. A., et al.: Clinical disorders of hyperkalemia. Annu. Rev. Med., 33:521, 1982.

Dirks, J. H., and Sutton, R. A.: Diuretics: Physiology, Pharmacology and Clinical Use. Philadelphia, W. B. Saunders Co., 1986.

Dirks, J. H., and Wong, N. L. M.: Acute functional adaptation to nephron loss: Micropuncture studies. Yale J. Biol. Med., 51(3):255, 1978.

Earley, L. E., and Gottschalk, C. W. (eds.): Strauss and Welt's Diseases of the Kidney, 3rd Ed. Boston, Little, Brown, 1979.

Epstein, M.: The Kidney in Liver Disease, 3rd Ed. Baltimore, Williams & Wilkins, 1988.

Fine, L. G., and Norman, J.: Cellular events in renal hypertrophy. Annu. Rev. Physiol., 51:19, 1989.

Ginzler, E. M., et al.: The natural history and responses to therapy of lupus nephritis. Annu. Rev. Med., 31:463, 1980.

Gorevic, P. D., and Franklin, E. C.: Amyloidosis. Annu. Rev. Med., 32:261, 1981.

Grantham, J. J.: Fluid secretion in the nephron: Relation to renal failure. Physiol. Rev., 56:248, 1976.

Halperin, M. L., and Goldstein, M. B.: Fluid, Electrolyte, and Acid-Base Emergencies. Philadelphia, W. B. Saunders Co., 1988.

Harrington, J. T., and Kassirer, J. P.: Renal vein thrombosis. Annu. Rev. Med., 33:255, 1982.

Herrin, J. T.: Management of children with progressive renal failure. Annu. Rev. Med., 34:21, 1983.

Holliday, M. A., et al.: Pediatric Nephrology, 2nd Ed. Baltimore, Williams & Wilkins, 1987.

Kahan, B. D., et al.: Clinical and experimental studies with cyclosporine in renal transplantation. Surgery, 97:125, 1985.

Kanwar, Y. S.: Biophysiology of glomerular filtration and proteinuria. Lab. Invest., 51:7, 1984.

Kincaid-Smith, P., et al. (eds.): Progress in Glomerulonephritis. New York, John Wiley & Sons, 1979.

Knox, F. G. (ed.): Textbook of Renal Pathophysiology. Hagerstown, Md., Harper & Row, 1978.

Leaf, A., and Cotran, R. S.: Renal Pathophysiology, 2nd Ed. New York, Oxford University Press, 1980.

Martinez, M. M.: Handbook of Renal Therapeutics. New York, Plenum Publishing Corp., 1983.

Massry, S. G., and Glassock, R. J.: Textbook of Nephrology, 2nd Ed. Baltimore, Williams & Wilkins, 1988.

Oplez, G., and Lenhard, V.: Immunological factors influencing renal graft survival. Annu. Rev. Med., 34:133, 1983.

Pavon, M. M. (ed.): Cancer of the Prostate and Kidney. New York, Plenum Publishing Corp., 1983.

Porter, B. (ed.): Nephrotoxic Mechanisms of Drugs and Environmental Toxins. New York, Plenum Publishing Corp., 1982.

Risdon, R. A.: Atlas of Renal Pathology. Philadelphia, J. B. Lippincott Co., 1981.

Rose, B. D. (ed.): Pathophysiology of Renal Disease, 2nd Ed. New York, McGraw-Hill Book Co., 1987.

Shah, S. V.: Oxidant mechanisms in glomerular injury. News Physiol. Sci., 3:254, 1988.

Sinniah, R., and Cohen, A. H.: Glomerular capillary aneurysms in light-chain nephropathy. An ultrastructural proposal of pathogenesis. Am. J. Pathol., 118:298, 1985.

Strauss, J. (ed.): Nephrotic Syndrome. New York, Garland Press, 1979.

Sutton, R. A.: Diuretics and calcium metabolism. Am. J. Kidney Dis., 5:4, 1985.

Tanagho, E. A., and McAninch, J. W.: Smith's General Urology. East Norwalk, Conn., Appleton & Lange, 1988.

Troyer, D. A., et al.: Renal metabolism during normoxia, hypoxia, and ischemic injury. Annu. Rev. Physiol., 48:33, 1986.

Wagner, R.: Nephrotic syndrome. Contemporary Patient Management. New Hyde Park, N. Y., Medical Examination Publishing Co., 1981.

Walser, M.: Nutrition in renal failure. Annu. Rev. Nutr., 3:125, 1983.

Warren, S., Jr. (ed.): Self-Assessment of Current Knowledge in Nephrology. New Hyde Park, N.Y., Medical Examination Publishing Co., 1981.

Williams, R. C., Jr.: Immune complexes in human diseases. Annu. Rev. Med., 32:13, 1981.

Wills, M. R.: Uremic toxins, and their effects on intermediary metabolism. Clin. Chem., 31:5, 1985.

Wood, R. F.: Renal Transplantation: A Clinical Handbook. Philadelphia, W. B. Saunders Co., 1984.

Zmijewski, C. M.: Human leukocyte antigen matching in renal transplantation: Review and current status. J. Surg. Res., 38:66, 1985.

UNIT VI

BLOOD CELLS, IMMUNITY, AND BLOOD CLOTTING

CHAPTER 32

• Red Blood Cells, Anemia, and Polycythemia •

With this chapter we begin a discussion of the blood cells and of other cells closely related to those of the blood: the cells of the macrophage system and of the lymphatic system. We first present the functions of red blood cells, which are the most abundant of all the cells of the body and are necessary for delivery of oxygen to the tissues.

● THE RED BLOOD CELLS (ERYTHROCYTES)

The major function of red blood cells, also known as erythrocytes, is to transport *hemoglobin,* which in turn carries oxygen from the lungs to the tissues. In some lower animals hemoglobin circulates as free protein in the plasma, not enclosed in red blood cells. However, when it is free in the plasma of the human being, approximately 3 per cent of it leaks through the capillary membrane into the tissue spaces or through the glomerular membrane of the kidney into the glomerular filtrate each time the blood passes through the capillaries. Therefore, for hemoglobin to remain in the blood stream, it must exist inside red blood cells.

The red blood cells have other functions besides simply transport of hemoglobin. For instance, they contain a large quantity of *carbonic anhydrase,* which catalyzes the reaction between carbon dioxide and water, increasing the rate of this reaction many thousandfold. The rapidity of this reaction makes it possible for the water in blood to react with large quantities of carbon dioxide and thereby transport it from the tissues to the lungs in the form of the bicarbonate ion (HCO_3^-). Also, the hemoglobin in the cells is an excellent *acid-base buffer* (as is true of most proteins), so that the red blood cells are responsible for most of the buffering power of whole blood.

The Shape and Size of Red Blood Cells. Normal red blood cells, illustrated in Figure 32–3, are biconcave discs having a mean diameter of approximately 7.5 micrometers and a thickness at the thickest point of 1.9 micrometer and in the center of 1 micrometer or less. The average volume of the red blood cell is 83 cubic micrometers.

The shapes of red blood cells can change remarkably as the cells pass through capillaries. Actually, the red blood cell is a "bag" that can be deformed into almost any shape. Furthermore, because the normal cell has a great excess of cell membrane for the quantity of material inside, deformation does not stretch the membrane greatly, and consequently does not rupture the cell, as would be the case with many other cells.

Concentration of Red Blood Cells in the Blood. In normal men the average number of red blood cells per cubic millimeter is 5,200,000 (±300,000) and in normal women 4,700,000 (±300,000). Also, the altitude at which the person lives affects the number of red blood cells; this is discussed later.

Quantity of Hemoglobin in the Cells. Red blood cells have the ability to concentrate hemoglobin in the cell fluid up to approximately 34 gm/dl of cells. The concentration never rises above this value, for this is a metabolic limit of the cell's hemoglobin-forming mechanism. Furthermore, in normal persons the percentage of hemoglobin is almost always near the maximum in each cell. However, when hemoglobin formation is deficient in the bone marrow, the percentage of hemoglobin in the cells may fall considerably below this value, and the volume of the red cell may decrease as well because of diminished hemoglobin to fill the cell.

When the hematocrit (the percentage of the blood that is cells—normally 40 to 45 per cent) and the quantity of hemoglobin in each respective cell are normal, the whole blood of men contains an average of 16 grams of hemoglobin per deciliter and of women an average of 14 gm/dl. As will be discussed in connection with the transport of oxygen in Chapter 40, each gram of pure hemoglobin is capable of combining with approximately 1.39 milliliters of oxygen. Therefore, in normal man, over 21 milliliters of oxygen can be carried in combination with hemoglobin in each deciliter of blood, and in normal woman 19 milliliters of oxygen can be carried.

PRODUCTION OF RED BLOOD CELLS

Areas of the Body That Produce Red Blood Cells. In the early few weeks of embryonic life, primitive, nucleated red blood cells are produced in the *yolk sac.* During the middle trimester of gestation the *liver* is the main organ for production of red blood cells, though reasonable numbers of red blood cells are also produced by the *spleen* and *lymph nodes.* Then, during the latter part of gestation and after birth, red blood cells are produced exclusively by the *bone marrow.*

As illustrated by Figure 32–1, the bone marrow of essentially all bones produces red blood cells until a person is 5 years old; but the marrow of the long bones, except for the proximal portions of the humeri and tibiae, becomes quite fatty and produces no more red blood cells after the age of about 20 years. Beyond this age most red cells are produced in the marrow of the membranous bones, such as the verte-

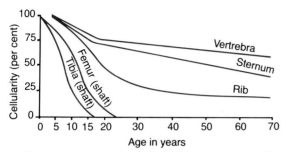

Figure 32–1. Relative rates of red blood cell production in the different bones at different ages.

brae, the sternum, the ribs, and the ilia. Even in these bones the marrow becomes less productive as age increases.

Genesis of Blood Cells

Pluripotential Hemopoietic Stem Cells, Growth Inducers, and Differentiation Inducers. In the bone marrow are cells called *pluripotential hemopoietic stem cells* (PHSCs), from which all the cells in the circulating blood are derived. Figure 32–2 illustrates the successive divisions of the pluripotential cells to form the different peripheral blood cells. As these cells reproduce, continuing throughout the life of the person, a portion of them are exactly like the original pluripotential cells and are retained in the bone marrow to maintain a supply of these, though their numbers do diminish with age. The larger portion of the reproduced stem cells, however, differentiate to form the other cells illustrated to the right in Figure 32–2. The early offspring still cannot be recognized as the different types of blood cells, even though they have already become committed to a particular line of cells and are called *committed stem cells.*

The different committed stem cells, when grown in culture, will produce colonies of specific types of blood cells. Therefore, a committed stem cell that produces erythrocytes is called a *colony-forming unit–erythrocyte,* and the abbreviation CFU-E is used to designate this type of stem cell. Likewise, colony-forming units that form granulocytes and monocytes have the designation CFU-GM, and so forth.

Growth and reproduction of the different stem cells are controlled by multiple proteins called *growth inducers.* Four major growth inducers have been described, each having different characteristics. One of these, *interleukin-3,* promotes growth and reproduction of virtually all the different types of stem cells, whereas the others induce growth of only specific types of committed stem cells.

The growth inducers promote growth but not differentiation of the cells. Instead, this is the function of still another set of proteins, called *differentiation inducers.* Each of these causes one type of stem cell to differentiate one or more steps toward a final type of adult blood cell.

Formation of the growth inducers and differentiation inducers is itself controlled by factors outside the bone marrow. For instance, in the case of red blood cells, exposure to low oxygen for a long period of time results in growth induction, differentiation, and production of greatly increased numbers of erythrocytes, as we shall discuss later in the chapter. In the case of some of the white blood cells, infectious diseases cause growth, differentiation, and eventual formation of the specific types of white blood cells that are needed to combat the infection.

Stages of Differentiation of Red Blood Cells

The first cell that can be identified as belonging to the red blood cell series is the *proerythroblast,* illustrated in Figure

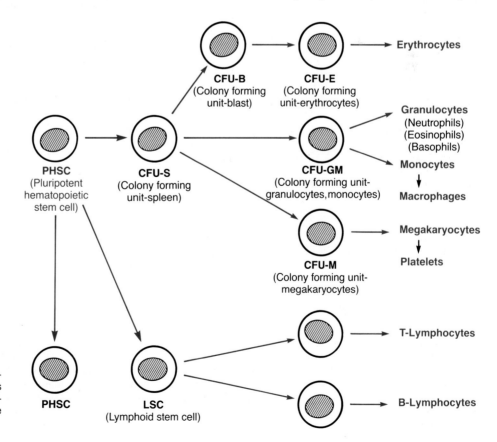

Figure 32–2. Formation of the multiple different peripheral blood cells from the original pluripotential hematopoietic stem cell (PHSC) in the bone marrow.

GENESIS OF RBC

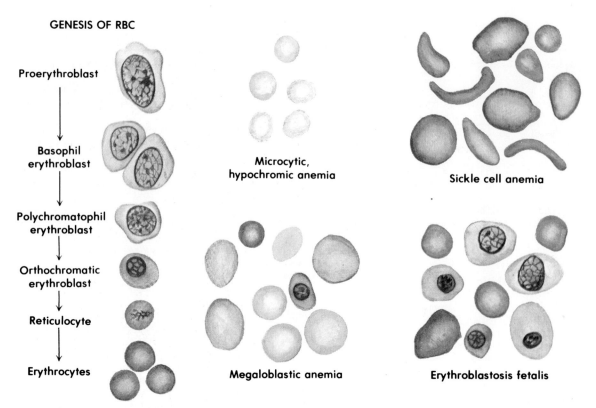

Figure 32–3. Genesis of red blood cells, and red blood cells in different types of anemias.

32–3. Under appropriate stimulation, large numbers of these cells are formed from the CFU-E stem cells.

Once the proerythroblast has been formed, it divides several more times, eventually forming many mature red blood cells. The first-generation cells are called *basophil erythroblasts* because they stain with basic dyes; the cell at this time has accumulated very little hemoglobin. However, in the succeeding generations, as illustrated in Figure 32–3, the cells become filled with hemoglobin to a concentration of approximately 34 per cent, the nucleus condenses to a small size, and its final remnant is extruded from the cell. At the same time the endoplasmic reticulum is reabsorbed. The cell at this stage is called a *reticulocyte* because it still contains a small amount of basophilic material, consisting of remnants of the Golgi apparatus, mitochondria, and a few other cytoplasmic organelles. It is during this reticulocyte stage that the cells pass into the blood capillaries by diapedesis (squeezing through the pores of the membrane).

The remaining basophilic material in the reticulocyte normally disappears within 1 to 2 days, and the cell is then the *mature erythrocyte*. Because of the short life of the reticulocytes, their concentration among all the red cells of the blood is normally slightly less than 1 per cent.

Regulation of Red Blood Cell Production— The Role of Erythropoietin

The total mass of red blood cells in the circulatory system is regulated within very narrow limits, so that an adequate number of red cells is always available to provide sufficient tissue oxygenation and yet so that the cells do not become so concentrated that they impede blood flow. What we know about this control mechanism is diagrammed in Figure 32–4 and is as follows.

Tissue Oxygenation as the Basic Regulator of Red Blood Cell Production. Any condition that causes the quantity of oxygen transported to the tissues to decrease ordinarily increases the rate of red blood cell production. Thus, when a person becomes extremely *anemic* as a result of hemorrhage or any other condition, the bone marrow immediately begins to produce large quantities of red blood cells. Also, destruction of major portions of the bone marrow by any means, especially x-ray therapy, causes hyperplasia of the remaining bone marrow, thereby attempting to supply the demand for red blood cells in the body.

Figure 32–4. Function of the erythropoietin mechanism to increase the production of red blood cells when various factors decrease tissue oxygenation.

At very *high altitudes,* where the quantity of oxygen in the air is greatly decreased, insufficient oxygen is transported to the tissues, and red cells are produced so rapidly that their number in the blood is considerably increased.

Therefore, it is obvious that it is not the concentration of red blood cells in the blood that controls the rate of red cell production, but instead it is the functional ability of the cells to transport oxygen to the tissues in relation to the tissue demand for oxygen.

Various diseases of the circulation that cause decreased blood flow through the peripheral vessels, and particularly those that cause failure of oxygen absorption by the blood as it passes through the lungs, can also increase the rate of red cell production. This is especially apparent in prolonged *cardiac failure* and in many *lung diseases,* for the tissue hypoxia resulting from these conditions increases the rate of red cell production, with resultant increase in the hematocrit and usually increase in the total blood volume as well.

Erythropoietin, Its Formation in Response to Hypoxia, and Its Function in Regulating Red Blood Cell Production. The principal factor that stimulates red blood cell production is a circulating hormone called *erythropoietin,* a glycoprotein with a molecular weight of about 34,000. In the absence of erythropoietin, hypoxia has either no effect or very little effect in stimulating red blood cell production. On the other hand, when the erythropoietin system is functional, hypoxia causes marked increase in erythropoietin production, and the erythropoietin in turn enhances red blood cell production until the hypoxia is relieved.

Role of the Kidneys in the Formation of Erythropoietin. In the normal person, 80 to 90 per cent of all erythropoietin is formed in the kidneys; the remainder, mainly in the liver. However, it is not known exactly where in the kidney the erythropoietin is formed. One possibility is the mesangial cells located at the pole of the glomerulus and extending also into the tuft of glomerular capillaries. These cells, when grown in tissue culture and subjected simultaneously to low oxygen concentrations, secrete large quantities of erythropoietin. Furthermore, their location adjacent to the glomerular capillaries ensures that they are bathed in arterial blood, which is important because it allows them to detect the *arterial* oxygen saturation, not venous saturation. This would explain why exposure of a human being to low oxygen at high altitude will lead to greatly increased production of red blood cells.

However, large amounts of erythropoietin and large numbers of red blood cells are also formed in anemic persons even when the arterial oxygen Po$_2$ is normal. Therefore, there must be some other mechanism for stimulating erythropoietin secretion besides simply response of the mesangial cells to low oxygen. One suggestion, based on other research, is that the renal tubular epithelial cells also secrete erythropoietin, in which case anemic blood would be unable to deliver enough oxygen from the peritubular capillaries to the tubular cells and therefore could stimulate red cell production.

Finally, at times hypoxia in other parts of the body but not in the kidneys will also stimulate erythropoietin secretion, which suggests that there might be some nonrenal sensor that sends an additional signal to the kidneys to produce this hormone. In particular, both norepinephrine and epinephrine and several of the prostaglandins stimulate erythropoietin production.

When both kidneys are removed from a person or when the kidneys are destroyed by renal disease, the person invariably becomes very anemic because the 10 to 20 per cent of the normal erythropoietin formed in other tissues (mainly in the liver) is sufficient to cause only one third to one half as much red blood cell formation as needed by the body.

Effect of Erythropoietin on Erythrogenesis. Upon placing an animal or person in an atmosphere of low oxygen, erythropoietin begins to be formed within minutes or hours, and it reaches maximum production within 24 hours. Yet almost no new red blood cells appear in the circulating blood until about 5 days later. From this fact, as well as still other studies, it has been determined that the important effect of erythropoietin is to stimulate the production of proerythroblasts from hemopoietic stem cells in the bone marrow. In addition, once the proerythroblasts are formed, the erythropoietin causes these cells also to pass more rapidly through the different erythroblastic stages than normally, also speeding up the production of new cells. The rapid production of cells continues as long as the person remains in the low oxygen state or until enough red blood cells are produced to carry adequate amounts of oxygen to the tissues despite the low oxygen; at this time the rate of erythropoietin production decreases to a level that will maintain the required number of red cells, but not an excess.

In the complete absence of erythropoietin, few red blood cells are formed by the bone marrow. At the other extreme, when very large quantities of erythropoietin are formed, the rate of red blood cell production can rise to as high as ten or more times normal. Therefore, the erythropoietin control mechanism for red blood cell production is a powerful one.

Maturation of Red Blood Cells — Requirement for Vitamin B$_{12}$ (Cyanocobalamin) and Folic Acid

Because of the continuing need to replenish red blood cells, the cells of the bone marrow are among the most rapidly growing and reproducing cells of the entire body. Therefore, as would be expected, their maturation and rate of production is affected greatly by a person's nutritional status.

Especially important for final maturation of the red blood cells are the two vitamins *vitamin B$_{12}$* and *folic acid.* Both of these are essential for the synthesis of DNA, for each in a different way is required for the formation of thymidine triphosphate, one of the essential building blocks of DNA. Therefore, lack of either vitamin B$_{12}$ or folic acid causes diminished DNA and consequently failure of nuclear maturation and division. Furthermore, the erythroblastic cells of the bone marrow, in addition to failing to proliferate rapidly, become larger than normal, developing into so-called *megaloblasts,* and the adult erythrocyte has a flimsy membrane and is often irregular, large, and oval instead of the usual biconcave disc. These poorly formed cells, after entering the circulating blood, are capable of carrying oxygen normally but their fragility causes them to have a short life, one half to one third of normal. Therefore, it is said that vitamin B$_{12}$ or folic acid deficiency causes *maturation failure* in the process of erythropoiesis.

The cause of the abnormal cells seems to be as follows: The inability of the cells to synthesize adequate quantities of DNA leads to slow reproduction of the cells but does not prevent formation of RNA by the DNA that is already available in each presently existing cell. Therefore, the quantity of RNA in each cell becomes much greater than normal, leading to excess production of cytoplasmic hemoglobin and other constituents, which causes the cells to enlarge. Yet because of abnormalities of some of the genes (the DNA), the structural components of the cell membrane and cytoskeleton are malformed, which leads to abnormal cell shapes and especially greatly increased cell membrane fragility.

Maturation Failure Caused by Poor Absorption of Vitamin B₁₂ — Pernicious Anemia. A common cause of maturation failure is failure to absorb vitamin B_{12} from the gastrointestinal tract. This often occurs in the disease called *pernicious anemia,* in which the basic abnormality is an *atrophic gastric mucosa* that fails to secrete normal gastric secretions. The parietal cells of the gastric glands secrete a glycoprotein called *intrinsic factor,* which combines with vitamin B_{12} of the food and makes the B_{12} available for absorption by the gut. It does this in the following way: (1) The intrinsic factor binds tightly with the vitamin B_{12}. In this bound state the B_{12} is protected from digestion by the gastrointestinal enzymes. (2) Still in the bound state, the intrinsic factor binds to specific receptor sites on the brush border membranes of the mucosal cells in the ileum. (3) Vitamin B_{12} is transported into the blood during the next few hours by the process of pinocytosis, carrying the intrinsic factor and the vitamin together through the membrane.

Lack of intrinsic factor, therefore, causes loss of much of the vitamin because of both enzyme action in the gut and failure of its absorption.

Once vitamin B_{12} has been absorbed from the gastrointestinal tract, it is stored in large quantities in the liver and then released slowly as needed to the bone marrow and other tissues of the body. The minimum amount of vitamin B_{12} required each day to maintain normal red cell maturation is only 1 to 3 micrograms, and the normal store in the liver and other body tissues is about 1000 times this amount. Therefore, as many as 4 to 5 years of defective B_{12} absorption are required to cause maturation failure anemia.

Failure of Maturation Caused by Deficiency of Absorption of Folic Acid (Pteroylglutamic Acid). Folic acid is a normal constituent of green vegetables, some fruits, liver, and other meats. However, it is easily destroyed during cooking. Also, persons with gastrointestinal absorption abnormalities, such as the common small intestinal disease called *sprue,* often have serious difficulty in absorbing both folic acid and vitamin B_{12}. Therefore, in many instances of maturation failure, the cause is deficiency of absorption of both folic acid and vitamin B_{12}.

FORMATION OF HEMOGLOBIN

Synthesis of hemoglobin begins in the proerythroblasts and continues slightly even into the reticulocyte stage, for when the reticulocytes leave the bone marrow and pass into the blood stream they continue to form minute quantities of hemoglobin for another day or so.

Figure 32-5 shows the basic chemical steps in the formation of hemoglobin. From tracer studies with isotopes it is known that the heme portion of hemoglobin is synthesized mainly from acetic acid and glycine and that most of this synthesis occurs in the mitochondria. The acetic acid is changed in the Krebs cycle, which is explained in Chapter 67, into succinyl-CoA, and then two molecules of this combine with two molecules of glycine to form a pyrrole compound. In turn, four pyrrole compounds combine to form a protoporphyrin compound. One of the protoporphyrins, known as protoporphyrin IX, then combines with iron to form the *heme* molecule. Finally, each heme molecule combines with a very long polypeptide chain, called a *globin,* synthesized by the ribosomes, forming a subunit of hemo-

I.   2 succinyl-CoA + 2 glycine ⟶ (pyrrole)
II. 4 pyrrole ⟶ protoporphyrin IX
III. protoporphyrin IX + Fe^{++} ⟶ heme
IV. heme + polypeptide ⟶ hemoglobin chain (α or β)
V.   2 α chains + 2 β chains ⟶ hemoglobin A

Figure 32-5. Formation of hemoglobin.

globin called a *hemoglobin chain* (Fig. 32-6). Each of these chains has a molecular weight of about 16,000; four of them in turn bind together loosely to form the whole hemoglobin molecule.

There are slight variations in the subunit hemoglobin chains, depending on the amino acid composition of the polypeptide portion. The different types of chains are designated *alpha chains, beta chains, gamma chains,* and *delta chains.* The most common form of hemoglobin in the adult human being, *hemoglobin A,* is a combination of *two alpha chains* and *two beta chains.*

Because each chain has a heme prosthetic group, there are four separate iron atoms in each hemoglobin molecule; each of these can bind with 1 molecule of oxygen, making a total of 4 molecules of oxygen (or 8 atoms) that can be transported by each hemoglobin molecule. Hemoglobin has a molecular weight of 64,458.

The nature of the hemoglobin chains determines the binding affinity of the hemoglobin for oxygen. Abnormalities of the chains can alter the physical characteristics of the hemoglobin molecule as well. For instance, in sickle cell anemia the amino acid valine is substituted for glutamic acid at one point in each of the two beta chains. When this type of hemoglobin is exposed to low oxygen, it forms elongated crystals inside the red blood cells that are sometimes 15 micrometers in length. These make it almost impossible

Figure 32-6. Basic structure of the hemoglobin molecule, showing one of the four heme complexes bound together to form the hemoglobin molecule.

for the cells to pass through the small capillaries, and the spiked ends of the crystals are very likely to rupture the cell membranes, thus leading to *sickle cell anemia.*

Combination of Hemoglobin with Oxygen. The most important feature of the hemoglobin molecule is its ability to combine loosely and reversibly with oxygen. This ability is discussed in detail in Chapter 40 in relation to respiration, for the primary function of hemoglobin in the body depends upon its ability to combine with oxygen in the lungs and then to release this oxygen readily in the tissue capillaries where the gaseous tension of oxygen is much lower than in the lungs.

Oxygen *does not* combine with the two positive bonds of the iron in the hemoglobin molecule. Instead, it binds loosely with one of the six "coordination" bonds of the iron atom. This is an extremely loose bond so that the combination is easily reversible. Furthermore, the oxygen does not become ionic oxygen but is carried as molecular oxygen to the tissues, where, because of the loose, readily reversible combination, it is released into the tissue fluids in the form of dissolved molecular oxygen, rather than ionic oxygen.

IRON METABOLISM

Because iron is important for formation of hemoglobin, myoglobin, and other substances such as the cytochromes, cytochrome oxidase, peroxidase, and catalase, it is essential to understand the means by which iron is utilized in the body.

The total quantity of iron in the body averages about 4 grams, approximately 65 per cent of which is present in the form of hemoglobin. About 4 per cent is present in the form of myoglobin, 1 per cent in the form of the various heme compounds that promote intracellular oxidation, 0.1 per cent combined with the protein transferrin in the blood plasma, and 15 to 30 per cent stored mainly in the reticuloendothelial system and liver parenchymal cells in the form of ferritin.

Transport and Storage of Iron. Transport, storage, and metabolism of iron in the body are illustrated in Figure 32–7, and may be explained as follows: When iron is absorbed from the small intestine, it immediately combines in the blood plasma with a beta globulin, *apotransferrin,* to form *transferrin,* which is then transported in the plasma. The iron is loosely combined with the globulin molecule and, consequently, can be released to any of the tissue cells

at any point in the body. Excess iron in the blood is deposited in all cells of the body *but especially* in the reticuloendothelial cells and liver hepatocytes. In the cell cytoplasm, it combines mainly with a protein, *apoferritin,* to form *ferritin.* Apoferritin has a molecular weight of approximately 460,000, and varying quantities of iron can combine in clusters of iron radicals with this large molecule; therefore, ferritin may contain only a small amount of iron or a large amount. This iron stored in ferritin is called *storage iron.*

Smaller quantities of the iron in the storage pool are stored in an extremely insoluble form called *hemosiderin.* This is especially true when the total quantity of iron in the body is more than the apoferritin storage pool can accommodate. Hemosiderin forms especially large clusters in the cells and consequently can be stained and observed microscopically as large particles in tissue slices by histologic techniques. Ferritin can also be stained, but the ferritin particles are so small and dispersed that they usually can be seen only with the electron microscope.

When the quantity of iron in the plasma falls very low, iron is removed from ferritin quite easily, but less easily from hemosiderin. The iron is then transported by the transferrin in the plasma to the portions of the body where it is needed.

A unique characteristic of the transferrin molecule is that it binds especially strongly with receptors in the cell membranes of erythroblasts in the bone marrow. Then, along with its bound iron, it is ingested into the erythroblasts by endocytosis. There the transferrin delivers the iron directly to the mitochondria, where heme is synthesized. In persons who do not have adequate quantities of transferrin in their blood, failure to transport iron to the erythroblasts in this manner can cause severe hypochromic anemia—that is, decreased numbers of red cells containing very little hemoglobin.

When red blood cells have lived their life span and are destroyed, the hemoglobin released from the cells is ingested by the cells of the monocyte-macrophage system. There free iron is liberated, and it can then be either stored in the ferritin pool or reused for formation of hemoglobin.

Daily Loss of Iron. About 1 milligram of iron is excreted each day by men, mainly into the feces. Additional quantities of iron are lost whenever bleeding occurs. In women, the menstrual loss of blood brings the average iron loss to a value of approximately 2 mg/day.

Obviously, the average quantity of iron derived from the diet each day must at least equal that lost from the body.

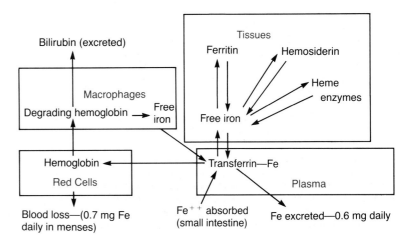

Figure 32–7. Iron transport and metabolism.

Absorption of Iron From the Gastrointestinal Tract

Iron is absorbed from all parts of the small intestine, mostly by the following mechanism. The liver secretes moderate amounts of *apotransferrin* into the bile that flows through the bile duct into the duodenum. In the small intestine, the apotransferrin binds with free iron and also with some iron compounds such as hemoglobin and myoglobin from meat, two of the most important sources of iron in the diet. This combination is called *transferrin*. It in turn is attracted to and binds with receptors in the membranes of the intestinal epithelial cells. Then, by the process of pinocytosis, the transferrin molecule, carrying with it its iron store, is absorbed into the epithelial cells and later is released on the blood side of these cells in the form of *plasma transferrin*.

The rate of iron absorption is extremely slow, with a maximum rate of only a few milligrams per day. This means that when tremendous quantities of iron are present in the food, only small proportions of this can be absorbed.

Regulation of Total Body Iron by Alteration of Rate of Absorption. When the body has become saturated with iron so that essentially all of the apoferritin in the iron storage areas is already combined with iron, the rate of absorption of iron from the intestinal tract becomes greatly decreased. On the other hand, when the iron stores have been depleted of iron, the rate of absorption becomes greatly accelerated, to as much as five or more times as great as when the iron stores are saturated. Thus, the total body iron is regulated to a great extent by altering the rate of absorption.

Feedback Mechanisms for Regulating Iron Absorption. Two mechanisms that play at least some role in regulating iron absorption are the following: (1) When essentially all the apoferritin in the body has become saturated with iron, it becomes difficult for transferrin to release iron to the tissues. As a consequence, the transferrin, which is normally only one-third saturated with iron, now becomes almost fully bound with iron, so that the transferrin accepts almost no new iron from the mucosal cells. Then, as a final stage of this process, the buildup of excess iron in the mucosal cells themselves depresses active absorption of iron from the intestinal lumen. (2) When the body already has excess stores of iron, the liver decreases its rate of formation of apotransferrin, thus reducing the concentration of this iron-transporting molecule in the plasma and also in the bile. Therefore, less iron is then absorbed by the intestinal apotransferrin mechanism, and less iron can also be transported away from the intestinal epithelial cells in the plasma by plasma transferrin.

Yet, despite these feedback control mechanisms for regulating iron absorption, when a person eats extremely large amounts of iron compounds, excess iron does enter the blood and can lead to massive deposition of hemosiderin in the reticuloendothelial cells throughout the body, which at times can be very damaging.

● DESTRUCTION OF RED BLOOD CELLS

When red blood cells are delivered from the bone marrow into the circulatory system, they normally circulate an average of 120 days before being destroyed. Even though mature red cells do not have a nucleus, mitochondria, or endoplasmic reticulum, they nevertheless have cytoplasmic enzymes that are capable of metabolizing glucose and form-

ing small amounts of adenosine triphosphate (ATP). The ATP in turn serves the red cell in several important ways: (1) maintaining the pliability of the cell membrane, (2) maintaining membrane transport of ions, (3) keeping the iron of the cell's hemoglobin in the ferrous form, rather than the ferric form (which causes the formation of methemoglobin that will not carry oxygen), and (4) preventing oxidation of the proteins in the red cell. However, these metabolic systems of the red cell become progressively less active with time, and the cells become more and more fragile, presumably because their life processes simply wear out.

Once the red cell membrane becomes very fragile, the cell ruptures during passage through some tight spot of the circulation. Many of the red cells fragment in the spleen, where they squeeze through the red pulp of the spleen. Here the spaces between the structural trabeculae of the pulp are only 3 micrometers wide, in comparison with the 8 micrometer diameter of the red cell. When the spleen is removed, the number of abnormal cells and old cells circulating in the blood increases considerably.

Destruction of Hemoglobin. The hemoglobin released from the cells when they burst is phagocytized almost immediately by macrophages in many parts of the body, but especially in the liver (the Kupffer cells), spleen, and bone marrow. During the next few hours to days, the macrophages release the iron from the hemoglobin back into the blood to be carried by transferrin either to the bone marrow for production of new red blood cells or to the liver and other tissues for storage in the form of ferritin. The porphyrin portion of the hemoglobin molecule is converted by the macrophages, through a series of stages, into the bile pigment *bilirubin*, which is released into the blood and later secreted by the liver into the bile; this will be discussed in relation to liver function in Chapter 70.

● THE ANEMIAS

Anemia means a deficiency of red blood cells, which can be caused either by too rapid loss or too slow production of red blood cells. Some types of anemia and their physiological causes are the following:

Blood Loss Anemia. After rapid hemorrhage the body replaces the plasma within 1 to 3 days, but this leaves a low concentration of red blood cells. If a second hemorrhage does not occur, the red blood cell concentration returns to normal within 3 to 4 weeks.

In chronic blood loss, the person frequently cannot absorb enough iron from the intestines to form hemoglobin as rapidly as it is lost. Therefore, red cells are then produced with too little hemoglobin inside them, giving rise to *microcytic hypochromic anemia*, which is illustrated in Figure 32–3.

Aplastic Anemia. *Bone marrow aplasia* means lack of a functioning bone marrow. For instance, the person exposed to gamma ray radiation from a nuclear bomb blast is likely to sustain complete destruction of bone marrow, followed in a few weeks by lethal anemia. Likewise, excessive x-ray treatment, certain industrial chemicals, and even drugs to which the person might be sensitive can cause the same effect.

Megaloblastic Anemia. From the earlier discussion in this chapter of vitamin B_{12}, folic acid, and intrinsic factor from the stomach mucosa, one can readily understand that loss of any one of these factors can lead to very slow reproduction of the erythroblasts in the bone marrow. As a result, these grow too large, with odd shapes, and are called *mega-*

loblasts. Thus, atrophy of the stomach mucosa, as occurs in *pernicious anemia,* or loss of the entire stomach as the result of total gastrectomy can lead to megaloblastic anemia. Also, patients who have intestinal sprue, in which folic acid, B$_{12}$, and other vitamin B compounds are poorly absorbed, often develop megaloblastic anemia. Because the erythroblasts cannot proliferate rapidly enough to form normal numbers of red blood cells, the cells that are formed are mostly oversized, of bizarre shapes, and have fragile membranes. Therefore, these cells rupture easily, leaving the person in dire need of an adequate number of red cells.

Hemolytic Anemia. Many different abnormalities of the red blood cells, many of which are hereditarily acquired, make the cells very fragile, so that they rupture easily as they go through the capillaries, especially through the spleen. Therefore, even though the number of red blood cells formed is completely normal, or even vastly excessive in some hemolytic diseases, the red cell life span is so short that serious anemia results. Some of these types of anemia are the following:

In *hereditary spherocytosis* the red cells are very small in size, and they are *spherical* in shape, rather than being biconcave discs. These cells cannot be compressed, because they do not have the normal loose, baglike cell membrane structure of the biconcave discs. Therefore, on passing through the splenic pulp they are easily ruptured by even slight compression.

In *sickle cell anemia,* which is present in 0.3 to 1.0 per cent of West African and American blacks, the cells contain an abnormal type of hemoglobin called *hemoglobin S,* caused by abnormal composition of the beta chains of the hemoglobin, as explained earlier in the chapter. When this hemoglobin is exposed to low concentrations of oxygen, it precipitates into long crystals inside the red blood cell. These crystals elongate the cell and give it the appearance of being a sickle, rather than a biconcave disc. The precipitated hemoglobin also damages the cell membrane so that the cells become highly fragile, leading to serious anemia. Such patients frequently go into a vicious circle called a sickle cell disease "crisis," in which low oxygen tension in the tissues causes sickling, which causes impediment of blood flow though the tissues, in turn causing still further decrease in oxygen tension. Thus, once the process starts, it progresses rapidly, leading to a serious decrease in red blood cell mass within a few hours and, often, to death.

In *erythroblastosis fetalis,* Rh positive red blood cells in the fetus are attacked by antibodies from an Rh negative mother. These antibodies make the cells fragile and cause the child to be born with serious anemia. This is discussed in Chapter 35 in relation to the Rh factor of blood. The extremely rapid formation of new red cells that occurs in erythroblastosis fetalis causes a large number of early *blast* forms of red cells to be released into the blood.

Hemolysis also occasionally results from transfusion reactions, from malaria, from reactions to certain drugs, and as an autoimmune process.

● EFFECTS OF ANEMIA ON THE CIRCULATORY SYSTEM

The viscosity of the blood, which was discussed in Chapter 14, depends almost entirely on the concentration of red blood cells. In severe anemia the blood viscosity may fall to as low as 1.5 times that of water rather than the normal value of approximately 3. This decreases the resistance to blood flow in the peripheral vessels so that far greater than

normal quantities of blood return to the heart. Moreover, hypoxia due to diminished transport of oxygen by the blood causes the tissue vessels to dilate, allowing still further increase in return of blood to the heart, increasing the cardiac output to a still higher level. Thus, one of the major effects of anemia is greatly *increased work load on the heart.*

The increased cardiac output in anemia partially offsets many of the effects of anemia; for even though each unit quantity of blood carries only small quantities of oxygen, the rate of blood flow may be increased to such an extent that almost normal quantities of oxygen are delivered to the tissues. However, when the anemic person begins to exercise, the heart is not capable of pumping much greater quantities of blood than it is already pumping. Consequently, during exercise, which greatly increases the tissue demand for oxygen, extreme tissue hypoxia results, and acute cardiac failure often ensues.

● POLYCYTHEMIA

Secondary Polycythemia. Whenever the tissues become hypoxic because of too little oxygen in the atmosphere such as at high altitudes, or because of failure of delivery of oxygen to the tissues as occurs in cardiac failure, the blood-forming organs automatically produce large quantities of red blood cells. This condition is called *secondary polycythemia,* and the red cell count commonly arises to as high as 6 to 8 million/mm^3.

A very common type of secondary polycythemia, called *physiologic polycythemia,* occurs in natives who live at altitudes of 14,000 to 17,000 feet. The blood count is generally 6 to 8 million/mm^3; this is associated with the ability of these persons to perform high levels of continuous work even in a rarefied atmosphere.

Polycythemia Vera (Erythremia). In addition to those persons who have physiologic polycythemia, others have a condition known as *polycythemia vera,* in which the red blood cell count may be as high as 7 to 8 million and the hematocrit as high as 60 to 70 per cent. Polycythemia vera is a tumorous condition of the organs that produce blood cells. It causes excess production of red blood cells in the same manner that a tumor of a breast causes excess production of a specific type of breast cell. It usually also causes excess production of white blood cells and platelets.

In polycythemia vera not only does the hematocrit increase, but the total blood volume also increases, rarely to almost twice normal. As a result, the entire vascular system becomes intensely engorged. In addition, many of the capillaries become plugged by the viscous blood, for the viscosity of the blood in polycythemia vera sometimes increases from the normal of three times the viscosity of water to ten times that of water.

EFFECT OF POLYCYTHEMIA ON THE CIRCULATORY SYSTEM

Because of the greatly increased viscosity of the blood in polycythemia, the flow of blood through the vessels is often very sluggish. In accordance with the factors that regulate the return of blood to the heart, as discussed in Chapter 20, it is obvious that increasing the viscosity tends to decrease the rate of venous return to the heart. On the other hand, the blood volume is greatly increased in polycythemia, which tends to increase the venous return. Actually, the cardiac output in polycythemia is not far from normal because these two factors more or less neutralize each other.

The arterial pressure is normal in most persons with polycythemia, though in approximately one third of them the pressure is elevated. This means that the blood pressure-regulating mechanisms can usually offset the tendency for increased blood viscosity to increase peripheral resistance and thereby to increase arterial pressure. Yet beyond certain limits, these regulations fail.

The color of the skin depends to a great extent on the quantity of blood in the subpapillary venous plexus. In polycythemia vera the quantity of blood in this plexus is greatly increased. Furthermore, because the blood passes sluggishly through the skin capillaries before entering the venous plexus, a larger than normal quantity of hemoglobin is deoxygenated before the blood enters the plexus. The blue color of this deoxygenated hemoglobin masks the red color of the oxygenated hemoglobin. Therefore, a person with polycythemia vera ordinarily has a ruddy complexion but often with a bluish (cyanotic) tint to the skin. (In secondary polycythemia cyanosis is also almost always evident because hypoxia is the usual cause of this type of polycythemia.)

REFERENCES

Barnes, D. M.: Blood-forming stem cells purified. Science, 241:24, 1988.

Bauer, C., and Kurtz, A.: Erythropoietin production in the kidney. News Physiol. Sci., 2:69, 1987.

Bauer, C., and Kurtz, A.: Oxygen sensing in the kidney and its relation to erythropoietin production. Annu. Rev. Physiol., 51:845, 1989.

Benz, E. H., Jr., and Forge, B. G.: The thalassemia syndromes: Models for the molecular analysis of human disease. Annu. Rev. Med., 33:363, 1982

Brain, M. C.: Current Therapy in Hematology-Oncology 1983–1984. St. Louis, C. V. Mosby Co., 1983.

Charache, S.: Treatment of sickle cell anemia. Annu. Rev. Med., 32:195, 1981.

Charlton, R. W., and Bothwell, T. H.: Iron absorption. Annu. Rev. Med., 34:55, 1983.

Chasis, J. A., and Shohet, S. B.: Red cell biochemical anatomy and membrane properties. Annu. Rev. Physiol., 49:237, 1987.

Clark, M. R.: Senescence of red blood cells: Progress and problems. Physiol. Rev., 68:503, 1988.

Eaves, A. C., and Eaves, C. J.: Erythropoiesis in culture. Clin. Haematol., 13:371, 1984.

Eaves, C. J., and Eaves, A. C.: Factors and hemopoiesis: Pandora's box revisited. Prog. Clin. Biol. Res., 148:83, 1984.

Ersley, A. H., and Gabuzda, T. G.: Basic Pathophysiology of Blood, 3rd Ed. Philadelphia, W. B. Saunders Co., 1985.

Fevery, J., and Heirwegh, K. P. M.: Bilirubin metabolism. In Javitt, N. B. (ed.): International Review of Physiology: Liver and Biliary Tract Physiology I. Vol. 21. Baltimore, University Park Press, 1980, p. 171.

Freedman, M. H.: Congential failure of hematopoiesis in the newborn infant. Clin. Perinatol., 11:417, 1984.

Golde, D. W., and Gasson, J. C.: Hormones that stimulate the growth of blood cells. Sci. Am., July 1988, p. 62.

Grignani, F., et al.: Genotypic, Phenotypic, and Functional Aspects of Haematopoiesis. New York, Raven Press, 1987.

Hansen, R. M., et al.: Failure to suspect and diagnose thalassemic syndromes. Interpretation of RBC indices by the nonhematologist. Arch. Intern. Med., 145:93, 1985.

Harrison, P. R.: Molecular analysis of erythropoiesis. A current appraisal. Exp. Cell. Res., 155:321, 1984.

Huebers, H. A., and Finch, C. A.: The physiology of transferrin and transferrin receptors. Physiol. Rev., 67:520, 1987.

Jacquez, J. A.: Red blood cell as glucose carrier: Significance for placental and cerebral glucose transfer. Am. J. Physiol., 246:R289, 1984.

Johannessen, J. V.: Electron microscopy in human medicine. In Cardiovascular System. Lymphoreticular and Hematopoietic System. Vol. 5. New York, McGraw-Hill Book Co., 1982.

Johnson, G. R.: Haemopoietic multipotential stem cells in culture. Clin. Haematol., 13:309, 1984.

Johnstone R. M., and Teng, K.: Membrane remodelling during reticulocyte maturation. News Physiol. Sci., 4:37, 1989.

Konigsberg, W.: Protein structure and molecular dysfunction: Hemoglobin. In Bondy, P. K., and Roseberg, L. E. (eds.): Metabolic Control and Disease, 8th Ed. Philadelphia, W. B. Saunders Co., 1980, p. 27.

Lee, G. R., et al.: Wintrobe's Clinical Hematology, 9th Ed. Philadelphia, Lea & Febiger, 1989.

Macey, R. I.: Transport of water and urea in red blood cells. Am. J. Physiol., 246:C195, 1984.

Mackinney, A. A. Jr.: The Pathophysiology of Blood. New York, John Wiley & Sons, 1984.

Metcalf, D., and Nicola, N. A.: The regulatory factors controlling murine erythropoiesis in vitro. Prog. Clin. Biol. Res., 148:93, 1984.

Miller, D. R., et al. (eds.): Blood Diseases of Infancy and Childhood. St. Louis, C. V. Mosby Co., 1989.

Ogawa, M., et al.: Differentiation and proliferation kinetics of hemopoietic stem cells in culture. Prog. Clin. Biol. Res., 148:35, 1984.

Peschle, C.: Erythropoiesis. Annu. Rev. Med., 31:303, 1980.

Petz, L. D.: Red cell transfusion problems in immunohematologic disease. Annu. Rev. Med., 33:355, 1982.

Plowman, P. N.: Hematology and Immunology. New York, Elsevier Science Publishing, 1987.

Propper, R., and Nathan, D.: Clinical removal of iron. Annu. Rev. Med., 33:509, 1982.

Schmid-Schonbein, H., et al.: Biology of red cells: Non-nucleated erythrocytes as fluid drop-like cell fragments. Int. J. Microcirc. Clin. Exp., 3:161, 1984.

Seiverd, C. E.: Hematology for Medical Technologists. Philadelphia, Lea & Febiger, 1983.

Spivak, J. L. (ed.): Fundamentals of Clinical Hematology. Hagerstown, Md., Harper & Row, 1980.

Stanley, E. T., and Jubinsky, P. T.: Factors affecting the growth and differentiation of haemopoietic cells in culture. Clin. Haematol., 13:329, 1984.

Thompson, R. B.: A Short Textbook of Haematology. Baltimore, Urban & Schwarzenberg, 1984.

Torok-Storb, B.: T-cell effects on in vitro erythropoiesis: Immune regulation and immune reactivity. Prog. Clin. Biol. Res., 148:163, 1984.

Tschudy, D. P., and Lamon, J. M.: Porphyrin metabolism and the porphyrias. In Bondy, P. K., and Rosenberg, L. E. (eds.): Metabolic Control and Disease, 8th Ed. Philadelphia, W. B. Saunders Co., 1980, p. 939.

Weatherall, D. J. (ed.): Advances in Red Blood Cell Biology. New York, Raven Press, 1982.

Williams, W. J., et al. (eds.): Hematology, 4th Ed. New York, McGraw-Hill Book Co., 1989.

Wood, W. G.: Haemoglobin switching. News Physiol. Sci., 3:33, 1988.

CHAPTER 33

• *Resistance of the Body to Infection: I. Leukocytes, Granulocytes, the Monocyte-Macrophage System, and Inflammation* •

Our bodies normally are exposed to bacteria, viruses, fungi, and parasites, which occur especially in the skin, the mouth, the respiratory passageways, the intestinal tract, the lining membranes of the eyes, and even the urinary tract. Many of these agents are capable of causing serious disease if they invade the deeper tissues. In addition, we are exposed intermittently to other highly infectious bacteria and viruses besides those that are normally present in our bodies, and these cause lethal diseases such as pneumonia, streptococcal infection, and typhoid fever.

Fortunately, our bodies have a special system for combating the different infectious and toxic agents. This is composed of the blood leukocytes (the white blood cells) and tissue cells originally derived from the leukocytes. These cells all work together in two different ways to prevent disease: (1) by actually destroying invading agents by the process of phagocytosis and (2) by forming antibodies and sensitized lymphocytes, one or both of which may destroy the invader. The present chapter is concerned with the first of these methods, and the following chapter is concerned with the second.

● THE LEUKOCYTES (WHITE BLOOD CELLS)

The leukocytes are the *mobile units* of the body's protective system. They are formed partially in the bone marrow (the *granulocytes* and *monocytes,* and a few *lymphocytes*) and partially in the lymph tissue (*lymphocytes* and *plasma cells*), but after formation they are transported in the blood to the different parts of the body where they are to be used. The real value of the white blood cells is that most of them are specifically transported to areas of serious inflamma-

tion, thereby providing a rapid and potent defense against any infectious agent that might be present. As we see later, the granulocytes and monocytes have a special capability to "seek out and destroy" any foreign invader.

GENERAL CHARACTERISTICS OF LEUKOCYTES

The Types of White Blood Cells. Six different types of white blood cells are normally found in the blood. These are *polymorphonuclear neutrophils, polymorphonuclear eosinophils, polymorphonuclear basophils, monocytes, lymphocytes,* and occasional *plasma cells.* In addition, there are large numbers of *platelets,* which are fragments of a seventh type of white cell found in the bone marrow, the *megakaryocyte.* The three types of polymorphonuclear cells have a granular appearance, as illustrated in Figure 33–1, for which reason they are called *granulocytes,* or, in clinical terminology, simply "polys."

The granulocytes and the monocytes protect the body against invading organisms by ingesting them — that is, by the process of *phagocytosis.* The lymphocytes and plasma cells function mainly in connection with the immune system; this is discussed in the following chapter. However, a function of certain lymphocytes is to attach themselves to specific invading organisms and destroy them, an action similar to those of the granulocytes and monocytes. Finally, the function of platelets is mainly to activate the blood-clotting mechanism, which is discussed in Chapter 36.

Concentrations of the Different White Blood Cells in the Blood. The adult human being has approximately 7000 white blood cells per microliter of blood. The normal percentages of the different types of white blood cells are approximately the following:

Genesis of Myelocytes

Genesis of Lymphocytes

Figure 33–1. Genesis of the white blood cells. The different cells of the myelogenous series are 1, myeloblast; 2, promyelocyte; 3, megakaryocyte; 4, neutrophil myelocyte; 5, young neutrophil metamyelocyte; 6, "band" neutrophil metamyelocyte; 7, polymorphonuclear neutrophil; 8, eosinophil myelocyte; 9, eosinophil metamyelocyte; 10, polymorphonuclear eosinophil; 11, basophil myelocyte; 12, polymorphonuclear basophil; 13–16, stages of monocyte formation.

Polymorphonuclear neutrophils	62.0%
Polymorphonuclear eosinophils	2.3%
Polymorphonuclear basophils	0.4%
Monocytes	5.3%
Lymphocytes	30.0%

The number of platelets, which are only cell fragments, in each microliter of blood is normally about 300,000.

GENESIS OF THE LEUKOCYTES

The early differentiation of the pluripotential hemopoietic stem cell into the different types of committed stem cells was illustrated in Figure 32–2 in the previous chapter. Aside from those cells committed to the formation of red blood cells, two major lineages of *white blood cells* are also formed, the myelocytic and the lymphocytic lineages. To the left in Figure 33–1 is illustrated the *myelocytic lineage* beginning with the *myeloblast;* and to the right is illustrated the *lymphocytic lineage* beginning with the *lymphoblast.*

The granulocytes and monocytes are formed only in the bone marrow. Lymphocytes and plasma cells are produced mainly in the various lymphogenous organs, including the lymph glands, the spleen, the thymus, the tonsils, and various lymphoid rests in the bone marrow, gut, and elsewhere.

The white blood cells formed in the bone marrow, especially the granulocytes, are stored within the marrow until they are needed in the circulatory system. Then when the need arises, various factors that are discussed later cause them to be released. Normally, about three times as many granulocytes are stored in the marrow as circulate in the entire blood. This represents about a 6 day supply of granulocytes.

As illustrated in Figure 33–1, megakaryocytes are also formed in the bone marrow and are part of the myelogenous group of bone marrow cells. These megakaryocytes fragment in the bone marrow, the small fragments known as *platelets* or *thrombocytes* passing then into the blood.

LIFE SPAN OF THE WHITE BLOOD CELLS

The main reason white blood cells are present in the blood is simply to be transported from the bone marrow or lymphoid tissue to the areas of the body where they are needed. The life of the granulocytes once released from the bone marrow is normally 4 to 8 hours circulating in the blood and another 4 to 5 days in the tissues. In times of serious tissue infection, this total life span is often shortened to only a few hours because the granulocytes then proceed rapidly to the infected area, perform their functions, and in the process are themselves destroyed.

The monocytes also have a short transit time, 10 to 20 hours, in the blood before wandering through the capillary membranes into the tissues. However, once in the tissues they swell to much larger sizes to become *tissue macrophages* and in this form can live for months or even years unless destroyed by performing phagocytic function. These tissue macrophages form the basis of the tissue macrophage system that provides a continuing defense in the tissues against infection, as we shall discuss later in the chapter.

Lymphocytes enter the circulatory system continually along with the drainage of lymph from the lymph nodes. Then, after a few hours, they pass back into the tissues by diapedesis, then re-enter the lymph and return to the blood again and again; thus, there is continual circulation of the lymphocytes through the tissues. The lymphocytes have life spans of months or even years, but this depends on the body's need for these cells.

The platelets in the blood are totally replaced approximately once every 10 days; in other words, about 30,000 platelets are formed each day for each microliter of blood.

● DEFENSIVE PROPERTIES OF NEUTROPHILS AND MONOCYTE-MACROPHAGES

It is mainly the neutrophils and the monocytes that attack and destroy invading bacteria, viruses, and other injurious agents. The neutrophils are mature cells that can attack and destroy bacteria and viruses even in the circulating blood. On the other hand, the blood monocytes are immature cells that have very little ability to fight infectious agents. However, once they enter the tissues, they begin to swell, sometimes increasing their diameters as much as fivefold, to as great as 80 micrometers, a size that can be seen with the naked eye. Also, extremely large numbers of lysosomes develop in the cytoplasm, giving the cytoplasm the appearance of a bag filled with granules. These cells are now called *macrophages,* and they are extremely capable of combating disease agents.

Diapedesis. Neutrophils and monocytes can squeeze through the pores of the blood vessels by the process of diapedesis. That is, even though a pore is much smaller than the size of the cell, a small portion of the cell slides through the pore at a time, the portion sliding through being momentarily constricted to the size of the pore, as illustrated in Figure 33–2.

Ameboid Motion. Both neutrophils and macrophages move through the tissues by ameboid motion, which was described in Chapter 2. Some of the cells can move at velocities as great as 40 μm/min, several times their own length each minute.

Chemotaxis. Many different chemical substances in the tissues cause both neutrophils and macrophages to move toward the source of the chemical. This phenomenon, illustrated in Figure 33–2, is known as *chemotaxis.* When a tissue becomes inflamed, at least a dozen different products are formed that can cause chemotaxis toward the inflamed area. These include (1) some of the bacterial toxins, (2) degenerative products of the inflamed tissues themselves, (3) several reaction products of the "complement complex" (discussed in the following chapter), and (4) several reaction products caused by plasma clotting in the inflamed area, as well as still other substances.

As illustrated in Figure 33–2, chemotaxis depends on a concentration gradient of the chemotactic substance. The concentration is greatest near the source, which causes the directional movement of the leukocytes. Chemotaxis is very effective up to 100 micrometers away from an inflamed tissue; since almost no tissue area is more than 30 to 50 micrometers away from a capillary, the chemotactic signal can easily move vast hordes of leukocytes from the capillaries into the inflamed area.

PHAGOCYTOSIS

The most important function of the neutrophils and macrophages is phagocytosis.

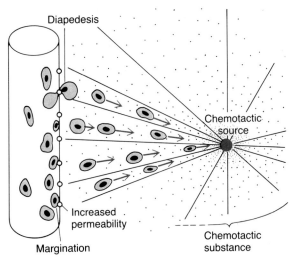

Figure 33–2. Movement of neutrophils by the process of *chemotaxis* toward an area of tissue damage.

Obviously, phagocytes must be selective of the material that is phagocytized, or otherwise some of the normal cells and structures of the body would be ingested. Whether or not phagocytosis will occur depends especially upon three selective procedures. First, if the surface of a particle is rough, the likelihood of phagocytosis is increased. Second, most natural substances of the body have protective protein coats that repel the phagocytes. On the other hand, dead tissues and foreign particles frequently have no protective coats, and many are also strongly electrically charged, which also makes them subject to phagocytosis. Third, the body has a specific means of recognizing certain foreign materials. This is a function of the immune system that is described in the following chapter. The immune system develops antibodies against infectious agents like bacteria. The variable portions of the antibodies then adhere to the bacterial membranes and thereby make the bacteria especially susceptible to phagocytosis. To do this, the constant portion of the antibody molecule combines with the C3 product of the *complement* cascade, which is an additional part of the immune system that is discussed in the following chapter. The C3 molecules then attach themselves to receptors on the phagocyte membrane, thus initiating phagocytosis. The entire process is called *opsonization.*

Phagocytosis by Neutrophils. The neutrophils entering the tissues are already mature cells that can immediately begin phagocytosis. On approaching a particle to be phagocytized, the neutrophil first attaches itself to the particle, then projects pseudopodia in all directions around the particle, and the pseudopodia meet each other on the opposite side and fuse. This creates an enclosed chamber containing the phagocytized particle. Then the chamber invaginates to the inside of the cytoplasmic cavity and breaks away from the outer cell membrane to form a free-floating *phagocytic vesicle* (also called a *phagosome*) inside the cytoplasm.

A neutrophil can usually phagocytize 5 to 20 bacteria before the neutrophil itself becomes inactivated and dies.

Phagocytosis by Macrophages. Macrophages, when activated by the immune system as described in the following chapter, are much more powerful phagocytes than the neutrophils, often capable of phagocytizing as many as 100 bacteria. They also have the ability to engulf much larger particles, even whole red blood cells, or malarial parasites; whereas neutrophils are not capable of phagocytizing particles much larger than bacteria. Also, macrophages, after digesting particles, can extrude the residual products and often survive many more months.

Enzymatic Digestion of the Phagocytized Particles. Once a foreign particle has been phagocytized, lysosomes and other cytoplasmic granules immediately come in contact with the phagocytic vesicle, and their membranes fuse with those of the vesicle, thereby dumping many digestive enzymes and bactericidal agents into the vesicle. Thus, the phagocytic vesicle now becomes a *digestive vesicle,* and digestion of the phagocytized particle begins immediately.

Neutrophils and macrophages both have an abundance of lysosomes filled with *proteolytic enzymes* especially geared for digesting bacteria and other foreign protein matter. The lysosomes of macrophages also contain large amounts of *lipases,* which digest the thick lipid membranes possessed by some bacteria.

The Ability of Neutrophils and Macrophages to Kill Bacteria. In addition to digestion of ingested bacteria in the phagosomes, neutrophils and macrophages also contain bactericidal agents that kill most bacteria even when the lysosomal enzymes fail to digest them. This is especially important because some bacteria have protective coats or other factors that prevent their destruction by the digestive enzymes. Much of the killing effect results from several powerful *oxidizing agents* formed by enzymes in the membrane of the phagosome or by the special organelle called the *peroxisome.* These oxidizing agents include large quantities of *superoxide* (O_2^-), *hydrogen peroxide* (H_2O_2), and *hydroxyl ions* ($—OH^-$), all of which are lethal to most bacteria even in very small quantities. Also, one of the lysosomal enzymes, myeloperoxidase, catalyzes the reaction between H_2O_2 and chloride ions to form hypochlorite, which is exceedingly bactericidal.

Another lysosomal substance that is lethal to many bacteria in the phagosome is *lysozyme,* a chemical that can cause dissolution of the lipid membranes of the bacteria.

Unfortunately, though, some bacteria, notably the tubercle bacillus, have coats that are resistant to lysosomal digestion and at the same time secrete substances that resist even the killing effects of the neutrophils and macrophages. Often these bacteria are responsible for many of the chronic diseases.

● THE MONOCYTE-MACROPHAGE SYSTEM AND THE RETICULOENDOTHELIAL SYSTEM

In the above paragraphs we have described the macrophages mainly as mobile cells that are capable of wandering through the tissues. However, a large portion of the monocytes, on entering the tissues and after becoming macrophages, become attached to the tissues and remain attached for months or perhaps even years unless they are called upon to perform specific protective functions. They have the same capabilities as the mobile macrophages to phagocytize large quantities of bacteria, viruses, necrotic tissue, or other foreign particles in the tissue. When appropriately stimulated, they can break away from their attachments and become mobile macrophages that respond to chemotaxis and all the other stimuli related to the inflammatory process.

The combination of monocytes, mobile macrophages, fixed tissue macrophages, and a few specialized endothelial cells in the bone marrow, spleen, and lymph nodes is called the *reticuloendothelial system.* The reason for categorizing all of these cells together as a single system is that they all exhibit similar phagocytic properties. Also, either all or almost all of these cells originate from monocytic stem cells; therefore, the reticuloendothelial system is almost synonymous with the monocyte-macrophage system. Yet because the term "reticuloendothelial system" is much better known in medical literature than the term "monocyte-macrophage system," it should be remembered as a generalized phagocytic system located in all tissues but especially so in those tissue areas where large quantities of particles, toxins, and other unwanted substances must be destroyed.

Tissue Macrophages in the Skin and Subcutaneous Tissues (Histiocytes). Though the skin is normally impregnable to infectious agents, this no longer holds true when the skin is broken. When infection does begin in the subcutaneous tissues and local inflammation ensues, the tissue macrophages can divide *in situ* and form more macrophages. Then they perform the usual functions of attacking and destroying the infectious agents, as described earlier.

Macrophages of the Lymph Nodes. Essentially no particulate matter that enters the tissues can be absorbed directly through the capillary membranes into the blood. Instead, if the particles are not destroyed locally in the tis-

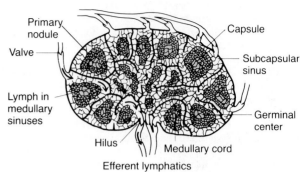

Figure 33–3. Functional diagram of a lymph node. (Redrawn from Ham: Histology. Philadelphia, J. B. Lippincott Company, 1971.)

sues, they enter the lymph and flow through the lymphatic vessels to the lymph nodes located intermittently along the course of the lymphatic. The foreign particles are trapped there in a meshwork of sinuses lined by *tissue macrophages.*

Figure 33–3 illustrates the general organization of the lymph node, showing lymph entering by way of the *afferent lymphatics,* flowing through the *medullary sinuses,* and finally passing out of the *hilus* into the *efferent lymphatics.* Large numbers of macrophages line the sinuses, and if any particles enter the sinuses the macrophages phagocytize them and prevent general dissemination throughout the body.

Alveolar Macrophages. Another route by which invading organisms frequently enter the body is through the respiratory system. Fortunately, large numbers of tissue macrophages are present as integral components of the alveolar walls. These can phagocytize particles that become entrapped in the alveoli. If the particles are digestible, the macrophages can also digest them and release the digestive products into the lymph. If the particle is not digestible, the macrophages often form a "giant cell" capsule around the particle until such time, if ever, that it can be slowly dissolved. Such capsules are frequently formed around tubercle bacilli, silica dust particles, and even carbon particles.

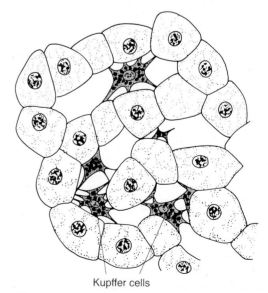

Figure 33–4. Kupffer cells lining the liver sinusoids, showing phagocytosis of India ink particles. (Redrawn from Copenhaver *et al.:* Bailey's Textbook of Histology. Baltimore, Williams & Wilkins, 1969.)

Figure 33 – 5. Functional structures of the spleen. (Modified from Bloom and Fawcett: Textbook of Histology. Philadelphia, W. B. Saunders Company, 1975.)

Tissue Macrophages (Kupffer Cells) in the Liver Sinuses. Still another favorite route by which bacteria invade the body is through the gastrointestinal tract. Large numbers of bacteria constantly pass through the gastrointestinal mucosa into the portal blood. However, before this blood enters the general circulation, it must pass through the sinuses of the liver; these sinuses are lined with tissue macrophages called *Kupffer cells,* illustrated in Figure 33 – 4. These cells form such an effective particulate filtration system that almost none of the bacteria from the gastrointestinal tract succeeds in passing from the portal blood into the general systemic circulation. Indeed, motion pictures of phagocytosis by Kupffer cells have demonstrated phagocytosis of a single bacterium in less than 0.01 second.

Macrophages of the Spleen and Bone Marrow. If an invading organism does succeed in entering the general circulation, there still remain other lines of defense by the tissue macrophage system, especially by macrophages of the spleen and bone marrow. In both these tissues, macrophages have become entrapped by the reticular meshwork of the two organs; and when foreign particles come in contact with these, they are phagocytized.

The spleen is similar to the lymph nodes, except that blood, instead of lymph, flows through the substance of the spleen. Figure 33 – 5 illustrates the spleen's general structure, showing a small peripheral segment. Note that a small artery penetrates from the splenic capsule into the *splenic pulp* and terminates in small capillaries. The capillaries are highly porous, allowing large numbers of whole blood cells to pass out of the capillaries into the *cords of the red pulp.* These cells then gradually *squeeze* through the trabecular meshwork of the cords and eventually return to the circulation through the endothelial walls of the *venous sinuses.* The trabeculae of the red pulp are lined with vast numbers of macrophages, and the venous sinuses are also lined with macrophages. This peculiar passage of blood through the cords of the red pulp provides an exceptional means of phagocytosis of unwanted debris in the blood, especially old and abnormal red blood cells.

● INFLAMMATION AND FUNCTION OF NEUTROPHILS AND MACROPHAGES

THE PROCESS OF INFLAMMATION

When tissue injury occurs, whether caused by bacteria, trauma, chemicals, heat, or any other phenomenon, multi-

ple substances that cause dramatic secondary changes in the tissues are released by the injured tissues. The entire complex of tissue changes is called *inflammation.* Inflammation is characterized by (1) vasodilatation of the local blood vessels with consequent excess local blood flow, (2) increased permeability of the capillaries with leakage of large quantities of fluid into the interstitial spaces, (3) often clotting of the fluid in the interstitial spaces because of excessive amounts of fibrinogen and other proteins leaking from the capillaries, (4) migration of large numbers of granulocytes and monocytes into the tissue, and (5) swelling of the tissue cells. Some of the many tissue products that cause these reactions are *histamine, bradykinin, serotonin, prostaglandins,* several different *reaction products of the complement system* (which are described in the following chapter), *reaction products of the blood-clotting system,* and multiple *hormonal substances called lymphokines* that are released by sensitized T cells (part of the immune system, as also discussed in the following chapter). Several of these substances strongly activate the macrophage system, and within a few hours the macrophages begin to devour the destroyed tissues — but at times the macrophages also further injure the still-living tissue cells.

The "Walling Off" Effect of Inflammation. One of the first results of inflammation is to "wall off" the area of injury from the remaining tissues. The tissue spaces and the lymphatics in the inflamed area are blocked by fibrinogen clots so that fluid barely flows through the spaces. Therefore, this walling-off process delays the spread of bacteria or toxic products.

The intensity of the inflammatory process is usually proportional to the degree of tissue injury. For instance, staphylococci invading the tissues liberate extremely lethal cellular toxins. As a result, the process of inflammation develops rapidly — indeed, much more rapidly than the staphylococci themselves can multiply and spread. Therefore, staphylococcal infection is characteristically walled off rapidly. On the other hand, streptococci do not cause such intense local tissue destruction. Therefore, the walling off process develops slowly while the streptococci reproduce and migrate. As a result, streptococci have a far greater tendency to spread through the body and cause death than do staphylococci, even though staphylococci are far more destructive to the tissues.

THE MACROPHAGE AND NEUTROPHIL RESPONSE TO INFLAMMATION

The Tissue Macrophage as the First Line of Defense. Within minutes after inflammation begins, those macrophages already present in the tissues, whether histiocytes in the subcutaneous tissues, alveolar macrophages in the lungs, microglia in the brain, or others, immediately begin their phagocytic actions. When activated by the products of inflammation, the first effect is rapid enlargement of each of these cells. Next, many of the previously sessile macrophages break loose from their attachments and become mobile, forming the first line of defense against infection during the first hour or so. Unfortunately, though, their numbers often are not very great.

Neutrophil Invasion of the Inflamed Area — The Second Line of Defense. Also within the first hour or so after inflammation begins, large numbers of neutrophils begin to invade the inflamed area from the blood. This is caused by products from the inflamed tissues that initiate the following reactions: (1) They alter the inside surface of the capillary endothelium, causing neutrophils to stick to the capillary walls in the inflamed area. This effect is called *margination* and is illustrated in Figure 32–2. (2) They cause the endothelial cells of the capillaries and small venules to separate, allowing openings large enough for neutrophils to pass by *diapedesis* into the tissue spaces. (3) Other products of the inflammation cause *chemotaxis* of the neutrophils toward the injured tissues, as explained in an earlier section.

Thus, within several hours after tissue damage begins, the area becomes well supplied with neutrophils. Because the blood neutrophils are already mature cells, they are ready to begin immediately their scavenger functions for removal of foreign matter.

Acute Increase in Neutrophils in the Blood—"Neutrophilia." Also within a few hours after the onset of acute, severe inflammation, the number of neutrophils in the blood sometimes increases as much as four- to fivefold —to as high as 15,000 to 25,000 per microliter. This is called *neutrophilia,* which means an increase in the number of neutrophils in the blood. Neutrophilia is caused by products of the inflammation that enter the blood stream, then are transported to the bone marrow, and there act on the marrow capillaries and on stored neutrophils to mobilize these immediately into the circulating blood. This obviously makes more neutrophils available to the inflamed tissue area.

Monocyte-Macrophage Invasion of the Inflamed Tissue—The Third Line of Defense. Along with the invasion of neutrophils, monocytes from the blood also enter the inflamed tissue. However, the number of monocytes in the circulating blood is low; also, the storage pool of monocytes in the bone marrow is much less than that of neutrophils. Therefore, the buildup of monocytes in the inflamed tissue area is much slower than that of neutrophils, requiring several days to become effective. Furthermore, even after invading the inflamed tissue, monocytes are still immature cells, requiring 8 hours or more to swell to much larger sizes and develop tremendous quantities of lysosomes, only then acquiring the full capacity for phagocytosis. Yet after several days to several weeks, the macrophages finally come to dominate the phagocytic cells of the inflamed area because of greatly increased bone marrow production of monocytes, as explained below.

It has already been pointed out that macrophages can phagocytize far more bacteria and far larger particles, including even neutrophils themselves and large quantities of necrotic tissue, than can neutrophils. Also, the macrophages play an important role in initiating the development of antibodies, as we discuss in the following chapter.

Increased Production of Granulocytes and Monocytes by the Bone Marrow—The Fourth Line of Defense. The fourth line of defense is greatly increased production of granulocytes and monocytes by the bone marrow. This results from stimulation of the committed granulocytic and monocytic stem cells. However, it takes 3 to 4 days before newly formed granulocytes and monocytes reach the stage of leaving the bone marrow. If the stimulus from the inflamed tissue continues, the bone marrow can continue to produce these cells in tremendous quantities for months and even for years, sometimes at rates of production as great as 50 times normal.

Feedback Control of the Macrophage and Neutrophil Responses

Though more than two dozen different factors have been implicated in the control of the macrophage-neutrophil re-

sponse to inflammation, five of these are believed to play dominant roles. These are illustrated in Figure 33–6. (1) *Tumor necrosis factor* (TNF), (2) *interleukin-1* (IL-1), (3) *granulocyte-monocyte colony stimulating factor* (GM-CSF), (4) *granulocyte colony stimulating factor* (G-CSF), and (5) *monocyte colony stimulating factor* (M-CSF).

The first two of these, TNF and IL-1, appear to initiate most of the other events. All five of the factors are probably formed mainly by activated macrophages in the inflamed tissues and in smaller quantities by other inflamed tissue cells as well.

The cause of the increased production of granulocytes and monocytes by the bone marrow is mainly the three colony stimulating factors, one of which, GM-CSF, stimulates both granulocyte and monocyte production, and the

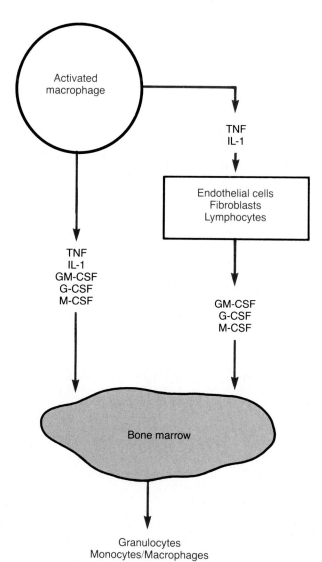

Figure 33–6. Control of bone marrow production of granulocytes and monocyte-macrophages in response to multiple growth factors released from activated macrophages in an inflamed tissue. (TNF, tissue necrosis factor; IL-1, interleukin-1; GM-CSF, granulocyte-monocyte colony stimulating factor; G-CSF, granulocyte colony stimulating factor; M-CSF, monocyte-macrophage colony stimulating factor.)

other two, G-CSF and M-CSF, granulocyte and monocyte production.

Thus, this combination of TNF, IL-1, and colony stimulating factors, along with still other less important factors, provides a powerful feedback mechanism that begins with tissue inflammation, then formation of defensive white blood cells, and finally removal of the cause of the inflammation.

Formation of Pus

When the neutrophils and macrophages engulf large numbers of bacteria and necrotic tissue, essentially all the neutrophils and many, if not most, of the macrophages themselves eventually die. After several days, a cavity is often excavated in the inflamed tissues containing varying portions of necrotic tissue, dead neutrophils, and dead macrophages. Such a mixture is commonly known as *pus*. After the infection has been suppressed, the dead cells and necrotic tissue in the pus gradually autolyze over a period of days, and the end products are usually absorbed into the surrounding tissues until most of the evidence of tissue damage is gone.

● THE EOSINOPHILS

The eosinophils normally comprise 2 to 3 per cent of all the blood leukocytes. Eosinophils are weak phagocytes, and they exhibit chemotaxis, but in comparison with the neutrophils it is doubtful that the eosinophils are of significant importance in protection against the usual types of infection.

On the other hand, eosinophils are often produced in very large numbers in persons with parasitic infections, and they migrate into tissues diseased by parasites. Though most of the parasites are too large to be phagocytized by the eosinophils or by any other phagocytic cells, nevertheless the eosinophils attach themselves to the parasites and release substances that kill many of them. For instance, one of the most widespread infections of the world is *schistosomiasis*, a parasitic infection found in as many as one third of the population of some tropical countries; the parasite invades literally any part of the body. Eosinophils attach themselves to the juvenile forms of the parasite and kill many of them. They do so in several ways: (1) by releasing hydrolytic enzymes from their granules, which are modified lysosomes, (2) probably by also releasing highly reactive forms of oxygen that are especially lethal, and (3) by releasing from the granules a highly larvacidal polypeptide called *major basic protein* (MBP). In our own country, another parasitic disease that causes eosinophilia is *trichinosis,* which results from invasion of the muscles by the *Trichinella* parasite ("pork worm") after a person eats uncooked pork.

Eosinophils also have a special propensity to collect in tissues in which allergic reactions have occurred, such as in the peribronchial tissues of the lungs in asthmatic persons, in the skin after allergic skin reactions, and so forth. This is caused at least partly by the fact that many mast cells and basophils participate in allergic reactions, as we discuss in the following section; and these release an *eosinophil chemotactic factor* that causes eosinophils to migrate toward the inflamed allergic tissue. The eosinophils are believed to detoxify some of the inflammation-inducing substances released by the mast cells and basophils, and probably also to phagocytize and destroy allergen-antibody complexes, thus preventing spread of the local inflammatory process.

● THE BASOPHILS

The basophils in the circulating blood are very similar to, though not identical with, the large *mast* cells located immediately outside many of the capillaries in the body. Both of these cells liberate *heparin* into the blood, a substance that can prevent blood coagulation and that can also speed the removal of fat particles from the blood after a fatty meal.

The mast cells and basophils also release histamine, as well as smaller quantities of bradykinin and serotonin. Indeed, it is mainly the mast cells in inflamed tissues that release these substances during inflammation.

The mast cells and basophils play an exceedingly important role in some types of allergic reactions because the type of antibody that causes allergic reactions, the IgE type (see Chapter 34), has a special propensity to become attached to mast cells and basophils. Then, when the specific antigen subsequently reacts with the antibody, the resulting attachment of the antigen to the antibody causes the mast cell or basophil to rupture and release exceedingly large quantities of histamine, bradykinin, serotonin, heparin, slow reacting substance of anaphylaxis, and a number of lysosomal enzymes. These in turn cause local vascular and tissue reactions that cause the allergic manifestations. These are discussed in greater detail in the following chapter.

● LEUKOPENIA

A clinical condition known as leukopenia or agranulocytosis occasionally occurs, in which the bone marrow stops producing white blood cells, leaving the body unprotected against bacteria and other agents that might invade the tissues.

Normally, the human body lives in symbiosis with many bacteria, for all the mucous membranes of the body are constantly exposed to large numbers of bacteria. The mouth almost always contains various spirochetal, pneumococcal, and streptococcal bacteria, and these same bacteria are present to a lesser extent in the entire respiratory tract. The gastrointestinal tract is especially loaded with colon bacilli. Furthermore, one can always find bacteria in the eyes, the urethra, and the vagina. Therefore, any decrease in the number of white blood cells immediately allows invasion of the tissues by the bacteria that are already present in the body. Within 2 days after the bone marrow stops producing white blood cells, ulcers may appear in the mouth and colon, or the person develops some form of severe respiratory infection. Bacteria from the ulcers then rapidly invade the surrounding tissues and the blood. Without treatment, death often ensues 3 to 6 days after acute total leukopenia begins.

Irradiation of the body by gamma rays caused by a nuclear explosion, or exposure to drugs and chemicals containing benzene or anthracene nuclei is quite likely to cause aplasia of the bone marrow. Indeed, some of the common drugs, such as chloramphenicol (an antibiotic), thiouracil (used to treat thyrotoxicosis), and even the various barbiturate hypnotics, on occasion cause agranulocytosis (or *bone marrow aplasia* in which no cells of any type—red cells included—are produced in the bone marrow), thus setting off the entire infective sequence of this malady.

After irradiation injury to the bone marrow, some stem cells, myeloblasts, and hemocytoblasts usually remain undestroyed and are capable of regenerating the bone marrow, provided sufficient time is available. Therefore, the patient properly treated with antibiotics and other drugs to ward off infection will usually develop enough new bone marrow within several weeks to several months for blood cell concentration to return to normal.

● THE LEUKEMIAS

Uncontrolled production of white blood cells is caused by cancerous mutation of a myelogenous or lymphogenous cell. This causes leukemia, which is usually characterized by greatly increased numbers of abnormal white blood cells in the circulating blood.

Types of Leukemia. Leukemias are divided into two general types: the *lymphogenous leukemias* and the *myelogenous leukemias.* The lymphogenous leukemias are caused by cancerous production of lymphoid cells, usually beginning in a lymph node or other lymphogenous tissue and then spreading to other areas of the body. The second type of leukemia, myelogenous leukemia, begins by cancerous production of young myelogenous cells in the bone marrow and then spreads throughout the body so that white blood cells are produced in many extramedullary organs.

In myelogenous leukemia, the cancerous process occasionally produces partially differentiated cells, resulting in what might be called *neutrophilic leukemia, eosinophilic leukemia, basophilic leukemia,* or *monocytic leukemia.* More frequently, however, the leukemia cells are bizarre and undifferentiated and not identical to any of the normal white blood cells. Usually the more undifferentiated the cells, the more *acute* is the leukemia, often leading to death within a few months if untreated. But with some of the more differentiated cells, the process can be quite *chronic,* sometimes developing slowly over a period of 10 to 20 years.

Leukemic cells, especially the very undifferentiated cells, are usually nonfunctional, so that they cannot provide the usual protection against infection associated with white blood cells.

EFFECTS OF LEUKEMIA ON THE BODY

The first effect of leukemia is metastatic growth of leukemic cells in abnormal areas of the body. The leukemic cells of the bone marrow may reproduce so greatly that they invade the surrounding bone, causing pain and eventually a tendency to easy fracture. Almost all leukemias spread to the spleen, the lymph nodes, the liver, and other especially vascular regions, regardless of whether the origin of the leukemia is in the bone marrow or in the lymph nodes. In each of these areas the rapidly growing cells invade the surrounding tissues, utilizing the metabolic elements of these tissues and consequently causing tissue destruction.

Very common effects in leukemia are the development of infections, severe anemia, and bleeding tendency caused by thrombocytopenia (lack of platelets). These effects result mainly from displacement of the normal bone marrow by the nonfunctional leukemic cells.

Finally, perhaps the most important effect of leukemia on the body is the excessive use of metabolic substrates by the growing cancerous cells. The leukemic tissues reproduce new cells so rapidly that tremendous demands are made on the body fluids for foodstuffs, especially the amino acids and vitamins. Consequently, the energy of the patient is greatly depleted, and the excessive utilization of amino acids causes rapid deterioration of the normal protein tissues of the body. Thus, while the leukemic tissues grow, the other tissues are debilitated. Obviously, after metabolic starvation has continued long enough, it alone is sufficient to cause death.

REFERENCES

Askonas, B. A., and Bancroft, B. J.: Interaction of African trypanosomes with the immune system. Phil. Trans. R. Soc. Lond. (Biol.), 307:41, 1984.
Barrett, K. E., and Metcalfe, D. D.: Mast cell heterogeneity: Evidence and implications. J. Clin. Immunol., 4:253, 1984.
Besemer, J., et al.: Colony stimulating factors. Triangle, 25:177, 1986.
Bocci, V.: What are the roles of interferons in physiological conditions? News Physiol. Sci., 3:201, 1988.
Brain, M. C.: Current Therapy in Hematology-Oncology 1983–1984. St. Louis, C. V. Mosby Co., 1983.
Dinarello, C. A.: Biology of interleukin 1. FASEB J., 2:108, 1988.
Feuerstein, G., and Hallenbeck, J. M.: Leukotrienes in health and disease. FASEB J., 1:186, 1987.
Ford-Hutchinson, A. W.: Leukotrienes: Their formation and role as inflammatory mediators. Fed. Proc., 44:25, 1985.
Forman, H. J., and Thomas, M. J.: Oxidant production and bactericidal activity in phagocytes. Annu. Rev. Physiol., 48:669, 1986.
Foucar, K., et al.: Nonendemic adult T-cell leukemia/lymphoma in the United States: Report of two cases and review of the literature. Am. J. Clin. Pathol., 83:18, 1985.
Freedman, M. H.: Congenital failure of hematopoiesis in the newborn infant. Clin. Perinatol., 11:417. 1984.
Froese, A.: Receptors for IgE on mast cells and basophils. Prog. Allergy, 34:142, 1984.
Gabig, T. G., and Babior, B. M.: The killing of pathogens by phagocytes. Annu. Rev. Med., 32:313, 1981.
Galli, S. J., et al.: Basophils and mast cells: Morphologic insights into their biology, secretory patterns, and function. Prog. Allergy, 34:1 1984.
Gallin, J. I., et al.: Inflammation: Basic Principles and Clinical Correlates. New York, Raven Press, 1988.
Goldstein, I. R. M., and Marder, S. R.: Infections and hypocomplementemia. Annu. Rev. Med., 34:47, 1983.
Graziano, F. M., and Lemanske, R. F., Jr. (eds.): Clinical Immunology. Baltimore, Williams & Wilkins, 1988.
Grinstein, S., and Dixon, S. J.: Ion transport, membrane potential, and cytoplasmic pH in lymphocytes: Changes during activation. Physiol. Rev., 69:417, 1989.
Henderson, E. S., and Lister, T. A.: Leukemia, 5th Ed. Philadelphia, W. B. Saunders Co., 1989.
Hogg, J. C.: Neutrophil kinetics and lung injury. Physiol. Rev., 67:1249, 1987.
Holleb, A. I. (ed.): Interferon: A Current Perspective. In Ca-A Cancer Journal for Clinicians, Vol. 38: No. 5. New York, American Cancer Society by H & W Publishing, 1988.
Howard, J. G., and Liew, F. Y.: Mechanisms of acquired immunity in leishmaniasis. Phil. Trans. R. Soc. Lond. (Biol.), 307:87, 1984.
Johnson, G. R.: Haemopoietic multipotential stem cells in culture. Clin. Haematol., 13:309, 1984.
Kawasaki, E. S., et al.: Molecular cloning of a complementary DNA encoding human macrophage-specific colony-stimulating factor (CSF-1). Science, 230:291, 1985.
Kolata, G.: Clinical promise with new hormones. Science, 236:517, 1987.
Lagunoff, D., et al.: Agents that release histamine from mast cells. Annu. Rev. Pharmacol. Toxicol., 23:331, 1983.
Mackinney, A. A., Jr.: The Pahtophysiology of Blood. New York, John Wiley & Sons, 1984.
Mahapatro, D., and Mahapatro, R. C.: Cutaneous basophil hypersensitivity. Am. J. Dermatopathol., 6:483, 1984.
May, M. E., and Waddell, C. C.: Basophils in peripheral blood and bone marrow: A retrospective review. Am. J. Med., 76:509, 1984.
Metcalf, J. A., et al.: Laboratory Manual of Neutrophil Function. New York, Raven Press, 1986.
Miller, D. R., et al. (eds.): Blood Diseases of Infancy and Childhood. St. Louis, C. V. Mosby Co., 1989.
Ogawa, M., et al.: Differentiation and proliferative kinetics of hemopoietic stem cells in culture. Prog Clin. Biol. Res., 148:35, 1984.
Olsen, E. G., and Spry, C. J.: Relation between eosinophilia and endomyocardial disease. Prog. Cardiovasc. Dis., 21:241, 1985.
Omann, G. M., et al.: Signal transduction and cytoskeletal activation in the neutrophil. Physiol. Rev., 67:285, 1987.
Pardoll, D. M., et al.: The unfolding story of T cell receptor c. FASEB J., 1:103, 1987.
Peterson, P. L., and Quie, P. G.: Bacterial surface components and the pathogenesis of infectious diseases. Annu. Rev. Med., 32:39, 1981.
Rowley, J. D.: Biological implications of consistent chromosome rearrangements in leukemia and lymphoma. Cancer Res., 44:3159, 1984.

Sachs, L.: The molecular control of blood cell development. Science, 238:1374, 1987.

Samuelsson, B., et al.: Leukotrienes and lipoxins: Structures, biosynthesis, and biological effects. Science, 237:1171, 1987.

Schmid-Schonbein. G. W.: Granulocyte: Friend and foe. News Physiol. Sci., 3:144, 1988.

Seiverd, C. E.: Hematology for Medical Technologist. Philadelphia, Lea & Febiger, 1983.

Shiffmann, E.: Leukocyte chemotaxis. Annu. Rev. Physiol., 44:553. 1982.

Stanley, E. R., and Jubinsky, P. T.: Factors affecting the growth and differentiation of haemopoietic cells in culture. Clin. Haematol., 13:329, 1984.

Thompson, R. B.: A Short Textbook of Haematology. Baltimore, Urban & Schwarzenberg, 1984.

Vane, J., and Botting, R.: Inflammation and the mechanism of action of anti-inflammatory drugs. FASEB J., 1:89, 1987.

Warnke, R. A., and Link, M. P.: Identification and significance of cell markers in leukemia and lymphoma. Annu. Rev. Med., 34:117, 1983.

Williams, W. J., et al. (eds.): Hematology, 4th Ed. New York, McGraw-Hill Book Co., 1990.

Young, J. D.: Killing of target cells by lymphocytes: A mechanistic view. Physiol. Rev., 69:250, 1989.

CHAPTER 34

•*Resistance of the Body to Infection: II. Immunity and Allergy* •

● INNATE IMMUNITY

The human body has the ability to resist almost all types of organisms or toxins that tend to damage the tissues and organs. This capacity is called *immunity*. Much of immunity is caused by a special immune system that forms antibodies and activated lymphocytes that attack and destroy the specific organisms or toxins. This type of immunity is *acquired immunity*. However, an additional portion of immunity results from general processes, rather than from processes directed at specific disease organisms. This is called *innate immunity*. It includes the following:

1. Phagocytosis of bacteria and other invaders by white blood cells and cells of the tissue macrophage system, as described in the previous chapter.

2. Destruction by the acid secretions of the stomach and by the digestive enzymes of organisms swallowed into the stomach.

3. Resistance of the skin to invasion by organisms.

4. Presence in the blood of certain chemical compounds that attach to foreign organisms or toxins and destroy them. Some of these are (1) *lysozyme,* a mucolytic polysaccharide that attacks bacteria and causes them to dissolute; (2) *basic polypeptides,* which react with and inactivate certain types of gram-positive bacteria; (3) the *complement complex* that will be described later, a system of about 20 proteins that can be activated in various ways to destroy bacteria; and (4) *natural killer lymphocytes* that can recognize and destroy foreign cells, tumor cells, and even some infected cells.

This innate immunity makes the human body resistant to such diseases as some paralytic virus infections of animals, hog cholera, cattle plague, and distemper—a viral disease that kills a large percentage of dogs that become afflicted with it. On the other hand, lower animals are resistant or even completely immune to many human diseases, such as poliomyelitis, mumps, human cholera, measles, and syphilis, which are very destructive or even lethal to the human being.

● ACQUIRED IMMUNITY

In addition to its innate immunity, the human body also has the ability to develop extremely powerful specific im-

munity against individual invading agents such as lethal bacteria, viruses, toxins, and even foreign tissues from other animals. This is called *acquired immunity*. It is with this immune mechanism and some of its associated reactions—especially the allergies—that the remainder of this chapter will be concerned.

Acquired immunity can often bestow extreme protection. For instance, certain toxins such as the paralytic toxin of botulinum or the tetanizing toxin of tetanus can be protected against in doses as high as 100,000 times the amount that would be lethal without immunity. This is the reason the process known as "vaccination" is so extremely important in protecting human beings against disease and against toxins, as is explained in the course of this chapter.

TWO BASIC TYPES OF ACQUIRED IMMUNITY

Two basic, but closely allied, types of acquired immunity occur in the body. In one of these the body develops circulating *antibodies,* which are globulin molecules that are capable of attacking the invading agent. This type of immunity is called *humoral immunity* or *B cell immunity*. The second type of acquired immunity is achieved through the formation of large numbers of *activated lymphocytes* that are specifically designed to destroy the foreign agent. This type of immunity is called *cell-mediated immunity* or *T cell immunity*.

We shall see shortly that both the antibodies and the activated lymphocytes are formed in the lymphoid tissue of the body. First, let us discuss the initiation of the immune process by *antigens*.

ANTIGENS

Because acquired immunity does not occur until after first invasion by a foreign organism or toxin, it is clear that the body must have some mechanism for recognizing the initial invasion. Each toxin or each type of organism almost always contains one or more specific chemical compounds in its makeup that are different from all other compounds. In general, these are proteins or large polysaccharides, and it is they that initiate the acquired immunity. These substances are called *antigens*.

For a substance to be antigenic it usually must have a high

374

molecular weight, 8000 or greater. Furthermore, the process of antigenicity usually depends upon regularly recurring molecular groups, called *epitopes,* on the surface of the large molecule, which explains why proteins and large polysaccharides are almost always antigenic, for they both have this type of stereochemical characteristic.

Haptens. Though substances with molecular weights of less than 8000 only rarely act as antigens, immunity can nevertheless be developed against substances of low molecular weight in a very special way, as follows: If the low molecular weight compound, which is called a *hapten,* first combines with a substance that is antigenic, such as a protein, then the combination will elicit an immune response. The antibodies or activated lymphocytes that develop against the combination can then react either against the protein or against the hapten. Therefore, on second exposure to the hapten, some of the antibodies or lymphocytes react with it before it can spread through the body and cause damage.

The haptens that elicit immune responses of this type are usually drugs, chemical constituents in dust, breakdown products of dandruff from animals, degenerative products of scaling skin, various industrial chemicals, the toxin of poison ivy, and so forth.

ROLE OF LYMPHOCYTES IN ACQUIRED IMMUNITY

Acquired immunity is the product of the body's lymphocyte system. In persons who have a genetic lack of lymphocytes or whose lymphocytes have been destroyed by radiation or by chemicals, no acquired immunity whatsoever can develop. And almost immediately after birth such a person dies of fulminating infection unless treated by heroic measures. Therefore, it is clear that the lymphocytes are essential to survival of the human being.

The lymphocytes are located most extensively in the *lymph nodes,* but they are also found in special lymphoid tissues such as the *spleen, submucosal areas of the gastrointestinal tract,* and the *bone marrow.* The lymphoid tissue is distributed very advantageously in the body to intercept the invading organisms or toxins before they can spread too

widely. For instance, the lymphoid tissue of the gastrointestinal tract is exposed immediately to antigens invading through the gut. The lymphoid tissue of the throat and pharynx (the tonsils and adenoids) is extremely well located to intercept antigens that enter by way of the upper respiratory tract. The lymphoid tissue in the lymph nodes is exposed to antigens that invade the peripheral tissues of the body. And, finally, the lymphoid tissue of the spleen and bone marrow plays the specific role of intercepting antigenic agents that have succeeded in reaching the circulating blood.

Two Types of Lymphocytes That Promote, Respectively, Cell-Mediated Immunity and Humoral Immunity — The T and the B Lymphocytes. Though most of the lymphocytes in normal lymphoid tissue look alike when studied under the microscope, these cells are distinctly divided into two major populations. One of the populations is responsible for forming the activated lymphocytes that provide cell-mediated immunity and the other for forming the antibodies that provide humoral immunity.

Both of these types of lymphocytes are derived originally in the embryo from *pluripotent hemopoietic stem cells* that differentiate and become committed to form lymphocytes. The lymphocytes that are formed eventually end up in the lymphoid tissue, but before doing so they are further differentiated or "preprocessed" in the following ways:

Those lymphocytes that are eventually destined to form activated lymphocytes first migrate to and are preprocessed in the *thymus* gland, for which reason they are called *T lymphocytes.* These are responsible for cell-mediated immunity.

The other population of lymphocytes — those that are destined to form antibodies — are preprocessed in the liver during midfetal life and in the bone marrow in late fetal life and after birth. However, this population of cells was first discovered in birds in which the preprocessing occurs in the *bursa of Fabricius,* a structure not found in mammals. For this reason these lymphocytes are called B lymphocytes, and they are responsible for humoral immunity.

Figure 34–1 illustrates the two separate lymphocyte systems for the formation, respectively, of the activated T lymphocytes and the antibodies.

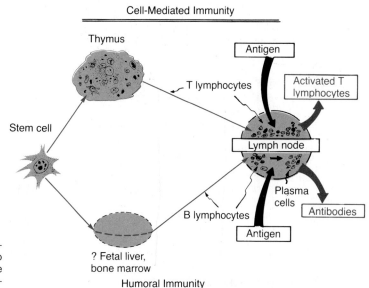

Figure 34–1. Formation of antibodies and sensitized lymphocytes by a lymph node in response to antigens. This figure also shows the origin of *thymic* (T) and *bursal* (B) lymphocytes that are responsible for the cell-mediated and humoral immune processes of the lymph nodes.

PREPROCESSING OF THE T AND B LYMPHOCYTES

Though all the lymphocytes of the body originate from the lymphocytic committed stem cells of the embryo, these stem cells are themselves incapable of forming either activated lymphocytes or antibodies. Before they can do so, they must be further differentiated in appropriate processing areas in the thymus or in the B cell processing area.

Role of the Thymus Gland in Preprocessing the T Lymphocytes. Most of the preprocessing of the T lymphocytes of the thymus gland occurs shortly before birth of the baby and for a few months after birth. Therefore, beyond this period of time, removal of the thymus gland usually will not seriously impair the T lymphocytic immune system, the system necessary for cell-mediated immunity. However, removal of the thymus several months before birth can completely prevent development of all cell-mediated immunity. Because it is this cellular type of immunity that is mainly responsible for rejection of transplanted organs such as hearts and kidneys, one can transplant organs with little likelihood of rejection if the thymus is removed from an animal a reasonable period of time before birth.

Thymic Hormone. In addition to preprocessing the T lymphocytes, the thymus is believed by some research workers to secrete one or more stimulatory factors, collectively called *thymic hormone*. This hormone supposedly spreads through the body fluids and increases the activity of the T lymphocytes that have already left the thymus gland and have migrated to the lymphoid tissue. This hormone is believed to cause further proliferation and increased activity of these lymphocytes. Otherwise, little is known about either the nature or the function of this hormone.

Role of the Bursa of Fabricius for Preprocessing B Lymphocytes in Birds. During the latter part of fetal life the bursa of Fabricius preprocesses the B lymphocytes and prepares them to manufacture antibodies. Here again, this preprocessing continues for a while after birth. In mammals, it is believed that the B cells are preprocessed during midfetal life in the liver but thereafter in the bone marrow.

Spread of Processed Lymphocytes to the Lymphoid Tissue. After formation of processed lymphocytes in both the thymus and the bursa, these first circulate freely in the blood for a few hours but then become entrapped in the lymphoid tissue. Thus, the lymphocytes do not originate primordially in the lymphoid tissue but, instead, are transported to this tissue by way of the preprocessing areas of the thymus and probably fetal liver and bone marrow.

SPECIFICITY OF T LYMPHOCYTES AND ANTIBODIES — ROLE OF LYMPHOCYTE CLONES

When a specific antigen comes in contact with the T and B lymphocytes in the lymphoid tissue, certain of the T lymphocytes become activated to form "T cells," and certain of the B lymphocytes form antibodies. The T cells and antibodies in turn then react highly specifically against the particular type of antigen that initiated their development. The mechanism of this specificity is the following:

Millions of Preformed Specific Lymphocytes in the Lymphoid Tissue. There are at least a million different types of preformed B lymphocytes and equally as many preformed T lymphocytes that are capable of forming the highly specific antibodies or T cells when stimulated by the appropriate antigens. Each one of these preformed lymphocytes is capable of forming only one type of antibody or one type of T cell with a single type of specificity. And only the specific type of antigen with which it can react can activate it. However, once the specific lymphocyte is activated by its antigen, it reproduces wildly, forming tremendous numbers of duplicate lymphocytes. If it is a B lymphocyte, its progeny will eventually secrete antibodies that then circulate throughout the body. If it is a T lymphocyte, its progeny are sensitized T cells that are released into the lymph and carried to the blood, then circulated through all the tissue fluids and back into the lymph, circulating around and around in this circuit sometimes for months or years.

All the different lymphocytes that are capable of forming one specificity of antibody or T cell are called a *clone of lymphocytes.* That is, the lymphocytes in each clone are alike and are derived originally from one or a few early lymphocytes of its specific type.

ORIGIN OF THE MANY CLONES OF LYMPHOCYTES

Only about a thousand genes code for the different types of antibodies and T lymphocytes. At first, it was a mystery how it was possible for so few genes to code for the million or more different specificities of antibody molecules or T cells that can be produced by the lymphoid tissue, especially when one considers that a single gene is usually necessary for the formation of each different type of protein. However, the answer to this has now been discovered. During the preprocessing stage, the genes cause the formulation of large RNA molecules containing many small functional segments, each of which codes for an individual portion of an antibody or of a T cell "marker" (the protein on the T cell surface that gives it specificity), but not for the entire antibody or T cell marker. Then, while still in the nucleus, the RNA molecules are cut into segments and respliced to form the final messenger RNA molecules, employing the cutting and splicing mechanism that was described in Chapter 3. However, during this cutting and splicing the individual segments of the original RNA molecules are spliced in different combinations in different ones of the processed lymphocytes. And when one realizes that there are a thousand or more different genes, each coding for multiple segments of antibodies or cell markers, one can understand what a vast number of different combinations can be formed, giving rise to literally millions of possible antibody and T cell specificities.

Mechanism for Activating a Clone of Lymphocytes

Each clone of lymphocytes is responsive to only a single type of antigen (or to similar antigens that have almost exactly the same stereochemical characteristics). The reason for this is the following: In the case of the B lymphocytes, each one has on the surface of its cell membrane about 100,000 antibody molecules that will react highly specifically with only the one specific type of antigen. Therefore, when the appropriate antigen comes along, it immediately attaches to the cell membrane; this leads to the activation process, which we will describe in more detail subsequently. In the case of the T lymphocytes, molecules very similar to antibodies, called *surface receptor proteins* (or T cell markers), are on the surface of the cell membrane, and these too are highly specific for the one specified activating antigen.

Role of Macrophages in the Activation Process.
Aside from the lymphocytes in lymphoid tissue, literally millions of macrophages are also present in the same tissue. These line the sinusoids of the lymph nodes, spleen, and other lymphoid tissue, and they lie in apposition to many of the lymph node lymphocytes. Most invading organisms are first phagocytized and partially digested by the macrophages, and the antigenic products are liberated into the macrophage cytosol. The macrophages then pass these antigens directly to the lymphocytes, thus leading to activation of the specified clones. The macrophages also secrete an activating substance that promotes the growth and reproduction of the specific lymphocytes. The substance is called *interleukin-1.*

Role of the T Cells in the Activation of the B Lymphocytes.
Most antigens activate both T lymphocytes and B lymphocytes at the same time. Some of the T cells that are formed, called *helper cells,* in turn secrete specific substances (collectively called lymphokines) that further activate the B lymphocytes. Indeed, without the help of these T cells, the quantity of antibodies formed by the B lymphocytes is usually very slight. We discuss this cooperative relationship between helper cells and B cells again after we have an opportunity to describe the mechanisms of the T cell system of immunity.

SPECIFIC ATTRIBUTES OF THE B LYMPHOCYTE SYSTEM—HUMORAL IMMUNITY AND THE ANTIBODIES

Formation of Antibodies by Plasma Cells.
Prior to exposure to a specific antigen, the clones of B lymphocytes remain dormant in the lymphoid tissue. However, upon entry of a foreign antigen, the lymphoid tissue macrophages phagocytize the antigen and then present it to the adjacent B lymphocytes. In addition, the antigen is also presented to T cells at the same time, and activated "helper" T cells then also contribute to the activation of the B lymphocytes, as we discuss more fully later. Those B lymphocytes specific for the antigen immediately enlarge and take on the appearance of *lymphoblasts.* Some of these then further differentiate to form *plasmablasts,* which are the precursors of *plasma cells.* In these cells the cytoplasm expands and the rough endoplasmic reticulum vastly proliferates. They then begin to divide at a rate of approximately once every 10 hours for about nine divisions, giving in 4 days a total population of about 500 cells for each original plasmablast. The mature plasma cell then produces gamma globulin antibodies at an extremely rapid rate—about 2000 molecules per second for each plasma cell. The antibodies are secreted into the lymph and are carried to the circulating blood. This process continues for several days or weeks until death of the plasma cells.

Formation of "Memory" Cells—Difference Between the Primary Response and the Secondary Response.
Some of the lymphoblasts formed by activation of a clone of B lymphocytes do not go on to form plasma cells, but, instead, form moderate numbers of new B lymphocytes similar to those of the original clone. In other words, the population of the specifically activated clone becomes greatly enhanced. And the new B lymphocytes are added to the original lymphocytes of the clone. They also circulate throughout the body to inhabit all the lymphoid tissue, but immunologically they remain dormant until activated once again by a new quantity of the same antigen. These are called memory cells. Obviously, subsequent exposure to the same antigen will then cause a much more rapid and much more potent antibody response, because there are many more memory cells than there were original lymphocytes of the specific clone.

Figure 34–2 illustrates the differences between the *primary response* that occurs on first exposure to a specific antigen and the *secondary response* that occurs after a second exposure to the same antigen. Note the delay in the appearance of the primary response, its weak potency, and its short life. The secondary response, by contrast, begins rapidly after exposure to the antigen, is far more potent, and forms antibodies for many months, rather than for only a few weeks.

The increased potency and duration of the secondary response explains why *vaccination* is usually accomplished by injecting antigen in multiple doses with periods of several weeks or several months between injections.

The Nature of the Antibodies

The antibodies are gamma globulins called *immunoglobulins,* and they have molecular weights between approximately 150,000 and 900,000. Usually they constitute about 20 per cent of all the plasma proteins.

All the immunoglobulins are composed of combinations of *light* and *heavy polypeptide chains;* most are a combination of two light and two heavy chains, as illustrated in Figure 34–3. Some of the immunoglobulins, though, have combinations of as many as ten heavy and ten light chains, which gives rise to the much larger molecular weight immunoglobulins. Yet in all immunoglobulins, each heavy chain is paralleled by a light chain at one of its ends, thus forming a heavy-light pair, and there are always at least two such pairs in each immunoglobulin molecule.

Figure 34–3 shows a designated end of each of the light and each of the heavy chains, called the *variable portion;* the remainder of each chain is called the *constant portion.* The variable portion is different for each specificity of antibody, and it is this portion that attaches specifically to a particular type of antigen. The constant portion of the antibody determines other properties of the antibody, establishing such factors as diffusivity of the antibody in the tissues, adherence of the antibody to specific structures within the tissues, attachment to the *complement complex,* the ease with which the antibodies pass through membranes, and other biological properties of the antibody.

Specificity of Antibodies.
Each antibody is specific for a particular antigen; this is caused by its unique structural

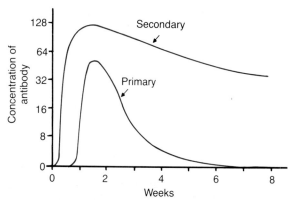

Figure 34–2. The time course of the antibody response to a *primary* injection of antigen and to a *secondary* injection several weeks later.

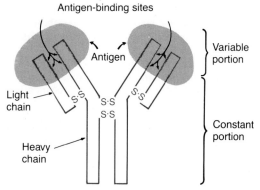

Figure 34–3. Structure of the typical IgG antibody, showing it to be composed of two heavy polypeptide chains and two light polypeptide chains. The antigen binds at two different sites on the variable portions of the chains.

organization of amino acids in the variable portions of both the light and heavy chains. These have a different steric shape for each antigen specificity, so that when an antigen comes in contact with it the prosthetic groups of the antigen fit as a mirror image with those of the antibody, thus allowing rapid bonding between the antibody and the antigen. The bonding is noncovalent, but when the antibody is highly specific there are so many bonding sites that the antibody-antigen coupling is nevertheless exceedingly strong, held together by (1) hydrophobic bonding, (2) hydrogen bonding, (3) ionic attractions, and (4) van der Waals forces. It also obeys the thermodynamic mass action law:

$$K_a = \frac{\text{concentration of bound antibody-antigen}}{\text{concentration of antibody} \times \text{concentration of antigen}}$$

K_a is called the *affinity constant* and is a measure of how tightly the antibody binds with the antigen.

Note, especially, in Figure 34–3 that there are two variable sites on the illustrated antibody for attachment of antigens, making this type of antibody *bivalent*. However, a small proportion of the antibodies, which consist of combinations of up to 10 light and 10 heavy chains, have as many as ten *binding* sites.

Classes of Antibodies. There are five general classes of antibodies, respectively named *IgM, IgG, IgA, IgD,* and *IgE.* Ig stands for immunoglobulin, and the other five respective letters simply designate the respective classes.

For the purpose of our present limited discussion, two of these classes of antibodies are of particular importance: IgG, which is a bivalent antibody and comprises about 75 per cent of the antibodies of the normal person; and IgE, which has 10 binding sites but constitutes only a small percentage of the antibodies and is especially involved in allergy. However, the IgM class is also interesting because a large share of the antibodies formed during the primary response are of this type. Their ten binding sites make them exceedingly effective in protecting the body against invaders, even though there are not many IgM antibodies.

Mechanisms of Action of Antibodies

Antibodies act mainly in two different ways to protect the body against invading agents: (1) by direct attack on the invader and (2) by activation of the complement system that then destroys the invader.

Direct Action of Antibodies on Invading Agents. Figure 34–4 illustrates antibodies (designated by the Y-shaped bars) reacting with antigens (designated by the darkened dumbbells). Because of the bivalent nature of the antibodies and the multiple antigen sites on most invading agents, the antibodies can inactivate the invading agent in one of several ways, as follows:

1. *Agglutination,* in which multiple large particles with antigens on their surfaces, such as bacteria or red cells, are bound together into a clump.

2. *Precipitation,* in which the molecular complex of soluble antigen (such as tetanus toxin) and antibody becomes so large that it is rendered insoluble and precipitates.

3. *Neutralization,* in which the antibodies cover the toxic sites of the antigenic agent.

4. *Lysis,* in which some very potent antibodies are occasionally capable of directly attacking membranes of cellular agents and thereby causing rupture of the cell.

However, the direct actions of antibodies attacking the antigenic invaders probably, under normal conditions, are not strong enough to play a major role in protecting the body against the invader. Most of the protection comes through the *amplifying* effects of the complement system described below.

The Complement System for Antibody Action

"Complement" is a collective term to describe a system of about 20 different proteins, many of which are enzyme precursors. The principal actors in this system are 11 proteins designated C1 through C9, B, and D, illustrated in Figure 34–5. All these are present normally among the plasma proteins and also among the plasma proteins that leak out of the capillaries into the tissue spaces. The enzyme precursors are normally inactive, but they can be activated in two separate ways called (1) the *classical pathway* and (2) the *alternate pathway.*

The Classical Pathway. The classical pathway is activated by an antigen-antibody reaction. That is, when an antibody binds with an antigen, a specific reactive site on the "constant" portion of the antibody becomes uncovered, or activated, and this in turn binds directly with the C1 molecule of the complement system, setting into motion a "cascade" of sequential reactions, illustrated in Figure 34–5, beginning with activation of the proenzyme C1 itself. Only a few antigen-antibody combinations are required to activate many molecules in the first stage of the complement system. The C1 enzymes thus formed *then activate*

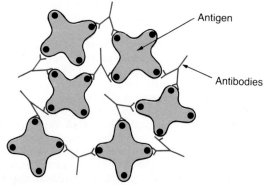

Figure 34–4. Binding of antigen molecules to each other by bivalent antibodies.

Figure 34–5. Cascade of reactions during activation of the classical pathway of complement. (Modified from Alexander and Good: Fundamentals of Clinical Immunology. Philadelphia, W. B. Saunders Company, 1977.)

successively increasing quantities of enzymes in the later stages of the system, so that from a very small beginning an extremely large "amplified" reaction occurs. Multiple end products are formed, as illustrated in the figure, and several of these cause important effects that help to prevent damage by the invading organism or toxin. Among the more important effects are the following:

1. *Opsonization* and *phagocytosis.* One of the products of the complement cascade, C3b, strongly activates phagocytosis by both neutrophils and macrophages, causing them to engulf the bacteria to which the antigen-antibody complexes are attached. This process is called *opsonization.* It often enhances the number of bacteria that can be destroyed by many hundredfold.

2. *Lysis.* One of the most important of all the products of the complement cascade is the *lytic complex,* which is a combination of multiple complement factors and is designated C5b6789. This has a direct effect of rupturing the cell membranes of bacteria or other invading organisms.

3. *Agglutination.* The complement products also change the surfaces of the invading organisms, causing them to adhere to each other, thus promoting agglutination.

4. *Neutralization of viruses.* The complement enzymes and other complement products can attack the structures of some viruses and thereby render them nonvirulent.

5. *Chemotaxis.* Fragment C5a causes chemotaxis of neutrophils and macrophages, thus causing large numbers of these phagocytes to migrate into the local region of the antigenic agent.

6. *Activation of mast cells and basophils.* Fragments C3a, C4a, and C5a all activate mast cells and basophils, causing them to release histamine and several other substances into the local fluids. These substances in turn cause increased local blood flow, increased leakage of fluid and plasma protein into the tissue, and other local tissue reactions that help inactivate or immobilize the antigenic agent. These same factors play a major role in inflammation, which was discussed in the preceding chapter, and also in allergy, as we discuss later.

7. *Inflammatory effects.* In addition to inflammatory effects caused by activation of the mast cells and basophils, several other complement products contribute further to local inflammation. These products cause the hyperemia to be increased, the capillary leakage of proteins to be in-

creased, and the proteins to coagulate in the tissue spaces, thus preventing movement of the invading organism through the tissues.

The Alternate Pathway. The complement system sometimes is activated without the intermediation of an antigen-antibody reaction. This occurs especially in response to large polysaccharide molecules in the cell membranes of some invading microorganisms. These substances react with complement factors B and D, forming an activation product that activates factor C3, setting off the remainder of the complement cascade beyond the C3 level. Thus, essentially all the same final products of the system are formed as in the classical pathway, and these cause the same effects as those just listed in protecting the body against the invader.

Because the alternate pathway does not involve an antigen-antibody reaction, it is one of the first lines of defense against invading microorganisms, capable of functioning even before a person becomes immunized against the organism.

SPECIAL ATTRIBUTES OF THE T LYMPHOCYTE SYSTEM—ACTIVATED T CELLS AND "CELL-MEDIATED IMMUNITY"

Release of Activated T Cells from Lymphoid Tissue, and Formation of Memory Cells. Upon exposure to the proper antigens, as presented by adjacent macrophages, the T lymphocytes of the lymphoid tissue proliferate and release large numbers of activated T cells in ways that parallel antibody release. The principal difference is that instead of releasing antibodies, whole activated T cells are formed and released into the lymph. These then pass into the circulation and are distributed throughout the body, passing through the capillary walls into the tissue spaces, back into the lymph and blood once again, and circulating again and again throughout the body, sometimes lasting for months or even years.

Also, *T lymphocyte memory cells* are formed in the same way that memory cells are formed in the humoral antibody system. That is, when a clone of T lymphocytes is activated by an antigen, many of the newly formed lymphocytes are preserved in the lymphoid tissue to become additional T lymphocytes of that specific clone; in fact, these memory cells even spread throughout the lymphoid tissue of the entire body. Therefore, on subsequent exposure to the same antigen, the release of activated T cells occurs far more rapidly and much more powerfully than in the first response.

Antigen Receptors on the T Lymphocytes. Antigens bind with *receptor molecules* on the surfaces of T cells in the same way that they bind with antibodies. These receptor molecules are composed of a variable unit similar to the variable portion of the humoral antibody, but its stem section is firmly bound to the cell membrane. Therefore, it is never secreted from the cell into the body fluids. There are as many as 100,000 receptor sites on a single T cell.

THE MANY TYPES OF T CELLS AND THEIR FUNCTIONS

In the last few years it has become clear that there are many different types of T cells. These are classified into three major groups: *(1) helper T cells, (2) cytotoxic T cells,* and *(3) suppressor T cells.* The functions of each of these are quite distinct.

The Helper T Cells — Their Role in Overall Regulation of Immunity

The helper T cells are by far the most numerous of the T cells, usually constituting more than three quarters of all of them. As their name implies, they help in the functions of the immune system, and in many ways. In fact, they serve as the major regulator of virtually all immune functions, as illustrated in Figure 34–6. They do this by forming a series of protein mediators, called *lymphokines,* that act on other cells of the immune system as well as on bone marrow cells. Among the important lymphokines secreted by the helper T cells are the following:

> Interleukin-2
> Interleukin-3
> Interleukin-4
> Interleukin-5
> Interleukin-6
> Granulocyte-monocyte colony stimulating factor
> Interferon-γ

Specific Regulatory Functions of the Lymphokines. In the absence of the lymphokines from the helper T cells, the remainder of the immune system is almost paralyzed. In fact, it is the helper T cells that are inactivated or destroyed by the acquired immunodeficiency syndrome (AIDS) virus, which leaves the body almost totally unprotected against infectious disease, therefore leading to the now well-known rapid lethal effects of AIDS. Some of the specific regulatory functions are the following:

Stimulation of Growth and Proliferation of Cytotoxic T Cells and Suppressor T Cells. In the absence of helper T cells, the clones for producing cytotoxic T cells and suppressor T cells are activated very little by most antigens. The lymphokine interleukin-2 has an especially strong stimulatory effect in causing growth and proliferation of both cytotoxic and suppressor T cells. In addition, several of the other lymphokines have less potent effects, especially interleukin-4 and interleukin-5.

Stimulation of B Cell Growth and Differentiation to Form Plasma Cells and Antibodies. The direct ac-

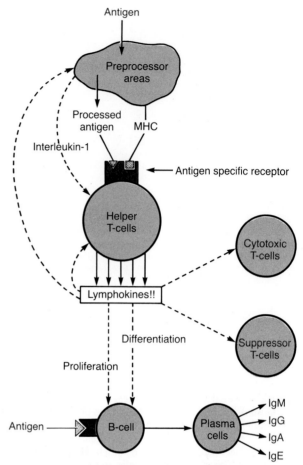

Figure 34 – 6. Regulation of the immune system, emphasizing the pivotal role of the helper T cells.

tions of antigen to cause B cell growth, proliferation, formation of plasma cells, and secretion of antibodies are also very slight without the "help" of the helper T cells. Almost all of the interleukins participate in the B cell response, but especially interleukins 4, 5, and 6. In fact, these three interleukins have such potent effects on the B cells that they have been called *B cell stimulating factors* or *B cell growth factors.*

Activation of the Macrophage System. The lymphokines also affect the macrophages. First, they slow or stop the migration of the macrophages after they have been chemotactically attracted into the inflamed tissue area, thus causing great accumulation of macrophages. Second, they activate the macrophages to cause far more efficient phagocytosis, allowing them to attack and destroy greatly increasing numbers of invading organisms.

Feedback Stimulatory Effect on the Helper Cells Themselves. Some of the lymphokines, especially interleukin-2, have a direct positive feedback effect in stimulating activation of the helper T cells themselves. This acts as an amplifier in further enhancing the helper cell response as well as the entire immune response to an invading antigen.

Cytotoxic T Cells

The cytotoxic T cell is a direct attack cell, capable of killing microorganisms and at times even some of the body's own cells. For this reason, these cells are frequently called *killer cells.* The receptor proteins on the surfaces of the cytotoxic cells cause them to bind tightly to those organisms or cells that contain their binding-specific antigen. Then, they kill the attacked cell in the manner illustrated in Figure 34–7. After binding, the cytotoxic T cell secretes *hole-forming proteins,* called perforins, that literally punch large round holes in the membrane of the attacked cell. Then the cytotoxic cell releases cytotoxic substances directly into the attacked cell. Almost immediately, the attacked cell becomes greatly swollen and usually dissolves entirely shortly thereafter.

Especially important is that these cytotoxic "killer" cells can pull away from the victim cells after they have punched holes and delivered cytotoxic substances, then move on to kill multiple more cells as well. Indeed, even after destruction of all the invaders, many of these cells still persist for months thereafter in the tissues.

The cytotoxic T cells are especially lethal to tissue cells that have been invaded by viruses because many virus particles become entrapped in the membranes of these cells and attract the T cells because of the viral antigenicity. The cytotoxic cells also play an important role in destroying cancer cells, heart transplant cells, or other types of cells that are "foreign" to the person's own body.

Suppressor T Cells

Much less is known about the suppressor T cells than about the others, but they are capable of suppressing the functions of both cytotoxic and helper T cells. It is believed that these suppressor functions serve the purpose of regulating the activities of the other cells, keeping them from causing excessive immune reactions that might be severely damaging to the body. For this reason, the suppressor cells are, along with the helper T cells, classified as *regulatory T cells.* One scenario for the function of the regulatory suppressor T cells is the following: the helper T cells activate the suppressor T cells; these cells in turn act as a negative feedback controller of the helper T cells; and this automatically sets the level of activity of the helper T cell system. It is also probable that the suppressor T cells play an important role in limiting the ability of the immune system to attack a person's own body tissues, called *immune tolerance,* as we discuss in the following section.

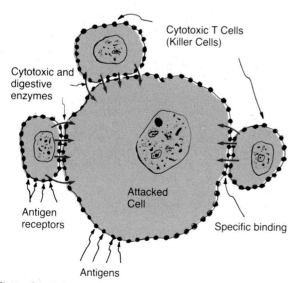

Figure 34–7. Direct destruction of an invading cell by sensitized lymphocytes.

Cytotoxic T Cells (Killer Cells)

Cytotoxic and digestive enzymes

Antigen receptors

Attacked Cell

Antigens

Specific binding

TOLERANCE OF THE ACQUIRED IMMUNITY SYSTEM TO ONE'S OWN TISSUES—ROLE OF PREPROCESSING IN THE THYMUS AND BONE MARROW

Obviously, if a person should become immune to his or her own tissues, the process of acquired immunity would destroy the individual's own body. Fortunately, the immune mechanism normally "recognizes" a person's own tissues as being completely distinctive from those of invaders, and his immunity system forms very few antibodies or activated T cells against his own antigens. This phenomenon is known as *self-tolerance* to the body's own tissues.

The Clone Selection Mechanism of Self-Tolerance. It is believed that most of the phenomenon of tolerance develops during the processing of the T lymphocytes in the thymus and the B lymphocytes in the B lymphocyte processing area—in the human being in the bone marrow. The reason for this belief is that injecting a strong antigen into a fetus at the time that the lymphocytes are being processed in these two areas will prevent the development of clones of lymphocytes in the lymphoid tissue that are specific for the injected antigen. Also, experiments have shown that specific immature lymphocytes in the thymus, when exposed to a strong antigen, become lymphoblastic, proliferate considerably, and then combine with the stimulating antigen—an effect that is believed to cause the cells themselves to be destroyed by the thymic epithelial cells before they can migrate to and colonize the lymphoid tissue.

Therefore, it is believed that during the processing of lymphocytes in the thymus and in the B lymphocyte processing area, all or most of those clones of lymphocytes that are specific for the body's own tissues are self-destroyed because of their continual exposure to the body's antigens.

Role of the Suppressor T Cells in the Development of Tolerance. The suppressor T cells are probably responsible for still another type of "self-tolerance." For instance, an autoimmune reaction sometimes occurs acutely against one of the body's tissues, but after a few days or a few weeks disappears even though autoimmune antibodies still persist in the circulating plasma. What has happened is that the number of suppressor T cells specifically sensitized to the offending self-antigen has greatly increased. It is believed that these suppressor T cells function to counteract the effects of the autoimmune antibodies as well as the sensitized helper cells and sensitized cytotoxic T cells, thus blocking the immune attack on the tissue. Unfortunately, though, this is not entirely clear.

Failure of the Tolerance Mechanism—Autoimmune Diseases. Unfortunately, people frequently lose some of their immune tolerance to their own tissues. This occurs to a greater extent the older a person becomes. It usually occurs after destruction of some of the body's tissues, which releases considerable quantities of antigens that circulate in the body and presumably cause acquired immunity in the form of either activated T cells or antibodies. Some of these antigens seem to combine with other proteins, such as proteins from bacteria or viruses, to form a new type of antigen that can then cause immunity. Then the resulting activated T cells and antibodies attack the body's own tissues. Also, it is believed that some of the proteins of the body are normally sequestered from the immune system during embryonic development of tolerance so that tolerance to these proteins never develops in the first place. For instance, the proteins of the cornea do not seem to circulate in the fluids of the fetus; this is also true of the thyroglobulin molecule of the thyroid; therefore, tolerance to these never develops. When damage occurs to either of these two tissues, these protein molecules can then elicit immunity, and the immunity in turn can attack the cornea in the first instance or the thyroid gland in the second instance to cause corneal opacity or thyroiditis.

Other diseases that result from autoimmunity include *rheumatic fever,* in which the bone becomes immunized against tissues in the joints and heart, especially the heart valves, following exposure to a specific type of streptococcal toxin that has an epitope in its molecular structure similar to the structure of some of the body's own self-antigens; one type of *glomerulonephritis,* in which the person becomes immunized against the basement membranes of glomeruli; *myasthenia gravis,* in which immunity develops against the acetylcholine receptor proteins of the neuromuscular junction, causing paralysis; and *lupus erythematosus,* in which the person becomes immunized against many different body tissues at the same time, a disease that causes extensive damage and often rapid death.

VACCINATION

The process of vaccination has been used for many years to cause acquired immunity against specific diseases. A person can be vaccinated by injecting dead organisms that are no longer capable of causing disease but which still have their chemical antigens. This type of vaccination is used to protect against typhoid fever, whooping cough, diphtheria, and many other types of bacterial diseases. Also, immunity can be achieved against toxins that have been treated with chemicals so that their toxic nature has been destroyed even though their antigens for causing immunity are still intact. This procedure is used in vaccinating against tetanus, botulism, and other similar toxic diseases. And, finally, a person can be vaccinated by infection with live organisms that have been "attenuated." That is, these organisms either have been grown in special culture media or have been passed through a series of animals until they have mutated enough that they will not cause disease but do still carry the specific antigens. This procedure is used to protect against poliomyelitis, yellow fever, measles, smallpox, and many other viral diseases.

PASSIVE IMMUNITY

Thus far, all the acquired immunity that we have discussed has been *active immunity.* That is, the person's body develops either antibodies or activated lymphocytes in response to invasion of the body by a foreign antigen. However, temporary immunity can be achieved in a person without injecting any antigen whatsoever. This is done by infusing antibodies, activated T cells, or both from someone else or from some other animal that has been actively immunized against the antigen. The antibodies will last for 2 to 3 weeks, and during that time the person is protected against the invading disease. Activated T cells will last for a few weeks if transfused from another person, and for a few hours to a few days if transfused from an animal. The transfusion of antibodies or lymphocytes to confer immunity is called *passive immunity.*

● ALLERGY

One of the important side effects of immunity is the development, under some conditions, of allergy. There are several different types of allergy, some of which can occur in any person, and others that occur only in persons who have a specific allergic tendency.

AN ALLERGY THAT OCCURS IN NORMAL PEOPLE: DELAYED-REACTION ALLERGY

This type of allergy frequently causes skin eruptions in response to certain drugs or chemicals, particularly some cosmetics and household chemicals, to which one's skin is often exposed. Another example of such an allergy is the skin eruption caused by exposure to poison ivy.

Delayed-reaction allergy is caused by activated T cells and not by antibodies. In the case of poison ivy, the toxin of poison ivy in itself does not cause much harm to the tissues. However, upon repeated exposure it does cause the formation of activated helper and cytotoxic T cells. Then, following subsequent exposure to the poison ivy toxin, within a day or so the activated T cells diffuse in sufficient numbers into the skin to respond to the poison ivy toxin and elicit a cell-mediated type of immune reaction. Remembering that this type of immunity can cause release of many toxic substances from the activated T cells as well as extensive invasion of the tissues by macrophages and their subsequent effects, one can well understand that the eventual result of some delayed-reaction allergies can be serious tissue damage.

ALLERGIES IN THE "ALLERGIC" PERSON

Some persons have an "allergic" tendency. Their allergies are called *atopic allergies* because they are caused by a nonordinary response of the immune system. The allergic tendency is genetically passed on from parent to child, and it is characterized by the presence of large quantities of IgE *antibodies.* These antibodies are called *reagins* or *sensitizing antibodies* to distinguish them from the more common IgG antibodies. When an *allergen* (defined as an antigen that reacts specifically with a specific type of IgE reagin antibody) enters the body, an allergen-reagin reaction takes place, and a subsequent allergic reaction takes place.

A special characteristic of the IgE antibodies (the reagins) is a strong propensity to attach to mast cells and basophils. Indeed, a single mast cell or basophil can bind as many as half a million molecules of IgE antibodies. Then, when an antigen (an allergen) that has multiple binding sites binds with several of the IgE antibodies attached to a mast cell or basophil, this causes an immediate change in the membrane of the cell, perhaps resulting from a simple physical effect of the antibody molecules being pulled together by the antigen. At any rate, many of the mast cells and basophils rupture; others release their granules without rupturing and also secrete additional substances not already preformed in the granules. Some of the many different substances that are either released immediately or secreted shortly thereafter include *histamine, slow-reacting substance of anaphylaxis* (which is a mixture of toxic substances called leukotrienes), *eosinophil chemotactic substance,* a *protease,* a *neutrophil chemotactic substance, heparin,* and *platelet activating factors.* These substances cause such phenomena as dilatation of the local blood vessels, attraction of eosinophils and neutrophils to the reactive site, damage to the local tissues by the protease, increased permeability of the capillaries and loss of fluid into the tissues, and contraction of local smooth muscle cells. Therefore, any number of different types of abnormal tissue responses can occur, depending upon the type of tissue in which the allergen-reagin reaction occurs. Among the different types of allergic reactions caused in this manner are the following:

Anaphylaxis. When a specific allergen is injected directly into the circulation, it can react in widespread areas of the body with the basophils of the blood and the mast cells located immediately outside the small blood vessels. Therefore, a widespread allergic reaction occurs throughout the vascular system and in closely associated tissues. This is called *anaphylaxis.* The *histamine* released into the circulation causes widespread peripheral vasodilation as well as increased permeability of the capillaries and marked loss of plasma from the circulation. Often, persons experiencing this reaction die of circulatory shock within a few minutes unless treated with norepinephrine to oppose the effects of the histamine. But also released from the cells is the mixture of leukotrienes called *slow-reacting substance of anaphylaxis,* which causes spasm of the smooth muscle of the bronchioles, eliciting an asthma-like attack and sometimes causing death by suffocation.

Urticaria. Urticaria results from antigen entering specific skin areas and causing localized anaphylactoid reactions. *Histamine* released locally causes (1) vasodilatation that induces an immediate *red flare* and (2) increased local permeability of the capillaries that leads to local circumscribed areas of swelling of the skin in another few minutes. The swellings are commonly called hives. Administration of antihistamine drugs to a person prior to exposure will prevent the hives.

Hay Fever. In hay fever, the allergen-reagin reaction occurs in the nose. *Histamine* released in response to this causes local vascular dilatation with resultant increased capillary pressure, and it also causes increased capillary permeability. Both of these effects cause rapid fluid leakage into the tissues of the nose, and the nasal linings become swollen and secretory. Here again, use of antihistamine drugs can prevent this swelling reaction. However, other products of the allergen-reagin reaction still cause irritation of the nose, still eliciting the typical sneezing syndrome despite drug therapy.

Asthma. In asthma, the allergen-reagin reaction occurs in the bronchioles of the lungs. Here, the most important product released from the mast cells seems to be the *slow-reacting substance of anaphylaxis,* which causes spasm of the bronchiolar smooth muscle. Consequently, the person has difficulty breathing until the reactive products of the allergic reaction have been removed. Unfortunately, administration of antihistaminics has little effect on the course of asthma, because histamine does not appear to be the major factor eliciting the asthmatic reaction.

REFERENCES

Atassi, M. S.: Antigenic structures of proteins. Eur. J. Biochem., 145:1, 1984.

Beer, D. J., et al.: The influence of histamine on immune and inflammatory responses. Adv. Immunol., 35:209, 1984.

Bierman, C. W., and Pearlman, D. S.: Allergic Diseases from Infancy to Adulthood, 2nd Ed. Philadelphia, W. B. Saunders Co., 1988.

Blalock, J. E.: A molecular basis for bidirectional communication between the immune and neuroendocrine systems. Physiol. Rev., 69:1, 1989.

Bloch, E. F., and Malveaux, F. J.: The significance of immunoglobulin E in resistance to parasitic infections. Annu. Allergy, 54:83, 1985.

Capron, A., et al.: Immunity to schistosomes: Progress toward vaccine. Science, 238:1065, 1987.

Castro, G. A.: Gut immunophysiology: Regulatory pathways within a common mucosal immune system. News Physiol. Sci., 4:59, 1989.

Caterson, B., et al.: Production and characterization of monoclonal antibod-ies directed against connective tissue proteoglycans. Fed. Proc., 44:386, 1985.

Cooper, J., et al.: Network regulation among T cells: Qualitative and quantitative studies on suppression in the non-immune state. Immunol. Rev., 79:63, 1984.

Cotman, C. W., et al. (eds.): The Neuro-Immune-Endocrine Connection. New York, Raven Press, 1987.

Coutinho, A., et al.: T-cell-dependent B cell activation. Immunol. Rev., 78:211, 1984.

Denning, P. J.: Computer models of AIDS epidemiology. Am. Sci., 75:347, 1987.

Doherty, P. C., et al.: Immunological surveillance of tumors in the context of major histocompatibility complex restriction of T cell formation. Adv. Cancer Res., 42:1, 1984.

Evans, A. S. (ed.): Viral Infections of Humans, 3rd Ed. New York, Plenum Publishing Corp., 1989.

Ford-Hutchinson, A. W.: Leukotrienes: Their formation and role as inflammatory mediators. Fed. Proc., 44:25, 1985.

Forman, J.: T cells, the MHC, and function. Immunol. Rev., 81:203, 1984.

Froese, A.: Receptors for IgE on mast cells and basophils. Prog. Allergy, 34:142, 1984.

Fudenberg, H. H., and Whitten, H. D.: Immunostimulation: Synthetic and biological modulators of immunity. Annu. Rev. Pharmacol. Toxicol., 24:147, 1984.

Gatenby, P. A., et al.: T cells, T cell subsets and immunoregulation. Aust. N. Z. J. Med., 14:89, 1984.

Gerard, J. W.: Food Allergy. Springfield, Ill., Charles C Thomas, 1980.

Glick, B.: Interrelation of the avian immune and neuroendocrine systems. J. Exp. Zool., 232:671, 1984.

Gmaj, P., and Murer, H.: Cellular mechanisms of inorganic phosphate transport in kidney. Physiol. Rev., 66:36, 1986.

Graziano, F. M., and Lemanske, R. F. Jr., (eds.): Clinical Immunology. Baltimore, Williams & Wilkins, 1988.

Heberman, R. B.: Adoptive therapy for cancer with interleukin-2-activated killer cells. Cancer Bull., 39:6, 1987.

Hobbs, J. R. (ed.): Thymic Factor Therapy. New York, Raven Press, 1985.

Horisberger, M. A., and Hochkeppel, H. K.: An interferon-induced mouse protein involved in the mechanism of resistance to influenza viruses. Its purification to homogeneity and characterization by polyclonal antibodies. J. Biol. Chem., 260:1730, 1985.

Kay, M. M.: Immunological aspects of aging: Early changes in thymic activity. Mech. Ageing Dev., 20:193, 1984.

Kelley, W. N., et al.: Textbook of Rheumatology, 3rd Ed. Philadelphia, W. B. Saunders Co., 1989.

Kendall, M. D.: Have we underestimated the importance of the thymus in man? Experientia, 40:1181, 1984.

Kolata G.: Immune boosters. Discover, September 1987, p. 68.

Kolitx, J. E., et al.: Interleukin-2: Basic and clinical studies. Cancer Bull. 39:13, 1987.

Koshland, D. E., Jr.: Frontiers in immunology. Science, 238:1023, 1987.

Kouttab, N. M., and Maizel, A. L.: Interleukin-2 — Its role in the regulation of T-cell proliferation. Cancer Bull., 39:51, 1987.

Kouttab, N. M., et al.: Lymphokines and monokines as regulators of human lymphoproliferation. Clin. Chem., 30:1539, 1984.

Lancki, D. W., et al.: Cell surface structures involved in T cell activation. Immunol. Rev., 81:65, 1984.

Langer, J. A., and Pestka, S.: Purification, bacterial expression, and biological activities of the human interferons. J. Invest. Dermatol., 83:(Suppl. 1):128s. 1984.

Lessof, M. H., et al.: Allergy: Immunological and Clinical Aspects, 2nd Ed. Baltimore, Williams & Wilkins, 1984.

Lockey, R. F., and Bukantz, S. C.: Principles of Immunology and Allergy. Philadelphia, W. B. Saunders Co., 1987.

Marrack, P., and Kappler, J.: The T cell receptor. Science, 238:1073, 1987.

Marx J. L.: What T cells see and how they see it. Science, 242:863, 1988.

Marx, J. L.: Antibody research garners Nobel Prize. Science, 238:484, 1987.

Marx, J. L.: How killer cells kill their targets. Science, 231:1367, 1986.

McNamara, M., et al.: T-cell helper circuits. Immunol. Rev., 79:87, 1984.

Metzger, H., and Kinet, J.-P.: How antibodies work: focus on Fc receptors. FASEB J., 2:3, 1988.

Mier, J. W.: Therapeutic uses of recombinant interleukin-2 in patients with cancer. Cancer Bull., 39:19, 1987.

Milstein, C.: From antibody structure to immunological diversification of immune response. Science, 231, 1986.

Miyajima, A., et al.: Coordinate regulation of immune and inflammatory responses by T cell-derived lymphokines. FASEB J., 2:2462, 1988.

Muraguchi, A., et al.: Regulation of human B-cell activation, proliferation, and differentiation by soluble factors. J. Clin. Immunol., 4:337, 1984.

Nossal, G. J. V.: Immunologic tolerance: Collaboration between antigen and lymphokines. Science, 245:147, 1989.

Paige, C. J., and Wu, G. E.: The B cell repertoire. FASEB J., 3:1818, 1989.

Parakevas, F.: Pathways of T cell suppression. CRC Crit. Rev. Immunol., 5:95, 1984.

Paul, W. E.: Interleukin 4/B cell stimulatory factor 1: One lymphokine, many functions. FASEB J., 1:456, 1987.

Piper, P. J.: Formation and actions of leukotrienes. Physiol. Rev., 64:744, 1984.

Rajewsky, K., et al.: Evolutionary and somatic selection of the antibody repertoire in the mouse. Science, 238:1088, 1987.

Schwartz, L. M. (ed.): Compendium of Immunology. New York, Van Nostrand Reinhold, 1983.

Sell, S.: Immunology, Immunopathology, and Immunity, 4th Ed. New York, Elsevier Science Publishing Co., 1987.

Shannon, K. M., and Ammann, A. J.: Acquired immune deficiency syndrome in childhood. J. Pediatr., 106:332, 1985.

Simonsen, M., et al.: Compound receptors in the cell membrane: Ruminations from the borderland of immunology and physiology. Prog. Allergy, 36:151, 1985.

Smith, K. A.: Interleukin-2: inception, impact, and implications. Science, 240:1169, 1988.

Smith, W. L., and Garcia-Perez, A.: Immunodissection: Use of monoclonal antibodies to isolate specific types of renal cells. Am. J. Physiol., 248:F1, 1985.

Stites, D. P., et al.: Basic and Clinical Immunology, 6th Ed. East Norwalk, Conn., Appleton & Lange, 1987.

Streilein, J. W.: Immune regulation and the eye: A dangerous compromise. FASEB J., 1:199, 1987.

Takemore, T., and Rajewsky, K.: Mechanism of neonatally induced idiotype suppression and its relevance for the acquisition of self-tolerance. Immunol. Rev., 70:103, 1984.

Taylor, D. W., and Siddiqui, W. A.: Recent advances in malarial immunity. Annu. Rev. Med., 33:69, 1982.

Unanue, E. R., and Allen, P. M.: The basis for the immunoregulatory role of macrophages and other accessory cells. Science, 236:551, 1987.

Vitetta, E. S., et al.: T-cell-derived lymphokines that induce IgM and IgG secretion in activated murine B cells. Immunol. Rev., 78:137, 1984.

Wybran, J.: Enkephalins and endorphins as modifiers of the immune system: Present and future. Fed. Proc., 44:92, 1985.

CHAPTER 35

• *Blood Groups; Transfusion; Tissue and Organ Transplantation* •

● ANTIGENICITY AND IMMUNE REACTIONS OF BLOOD

When blood transfusions from one person to another were first attempted, the transfusions were successful in some instances; but in many more, immediate or delayed agglutination and hemolysis of the red blood cells occurred, resulting in typical transfusion reactions that occasionally led to death. Soon it was discovered that the bloods of different persons usually have different antigenic and immune properties, so that antibodies in the plasma of one blood react with antigens on the surfaces of the red cells of another. For this reason, it is easy for blood from a donor to be mismatched with that of a recipient. Fortunately, if proper precautions are taken, one can determine ahead of time whether or not appropriate antibodies and antigens are present in the donor and recipient bloods to cause a reaction.

Multiplicity of Antigens in the Blood Cells. At least 30 commonly occurring antigens and hundreds of other rare antigens, each of which can at times cause antigen-antibody reactions, have been found in human blood cells, especially on the surfaces of the cell membranes. Most of these are weak and therefore are of importance principally for studying the inheritance of genes to establish parentage and so forth. However, two particular groups of antigens are more likely than the others to cause blood transfusion reactions. These are the so-called *O-A-B* system of antigens and the *Rh* system.

● O-A-B BLOOD GROUPS

THE A AND B ANTIGENS — AGGLUTINOGENS

Two related antigens — type A and type B — occur on the surfaces of the red blood cells in a large proportion of the population. Because of the way these antigens are inherited, people may have neither of them on their cells, they may have one, or they may have both simultaneously.

As will be discussed subsequently, strong antibodies that react specifically with either the type A or type B antigen almost always occur *in the plasmas of persons who do not have the antigens on their red cells.* These antibodies bind with the red cell antigens to cause agglutination of the red cells. Therefore, the type A and type B antigens are called *agglutinogens,* and the plasma antibodies that cause the agglutination are called *agglutinins.* It is on the basis of the presence or absence of the A and B agglutinogens in the red blood cells that blood is grouped (or "typed") for the purpose of transfusion.

The Four Major O-A-B Blood Types. In transfusing blood from one person to another, the bloods of donors and recipients are normally classified into four major O-A-B types, as illustrated in Table 35–1, depending on the presence or absence of the two agglutinogens. When neither A nor B agglutinogen is present, the blood group is *type O.* When only type A agglutinogen is present, the blood is *type A.* When only type B agglutinogen is present, the blood is *type B.* And when both A and B agglutinogens are present, the blood is *type AB.*

Relative Frequencies of the Different Blood Types. The prevalence of the different blood types among caucasoids is approximately as follows:

O	47%
A	41%
B	9%
AB	3%

It is obvious from these percentages that the O and A genes occur frequently, whereas the B gene is infrequent.

Genetic Determination of the Agglutinogens. Two genes, one on each of two paired chromosomes, determine the O-A-B blood groups. These two genes are allelomorphic genes that can be any one of three different types, but only one type on each chromosome: type O, type A, or type B. However, the type O gene is either functionless or almost functionless, so that it causes either no type O agglutinogen on the cells or such a weak agglutinogen that it is normally insignificant. On the other hand, the type A and type B genes do cause strong agglutinogens on the cells.

The six possible combinations of genes, as shown in Table 35–1, are OO, OA, OB, AA, BB, and AB. These different combinations of genes are known as the *genotypes,* and each person is one of the six different genotypes.

Table 35-1 THE BLOOD GROUPS WITH THEIR GENOTYPES AND THEIR CONSTITUENT AGGLUTINOGENS AND AGGLUTININS

Genotypes	Blood Types	Agglutinogens	Agglutinins
OO	O	—	Anti-A and Anti-B
OA or AA	A	A	Anti-B
OB or BB	B	B	Anti-A
AB	AB	A and B	—

One can observe from the table that a person with genotype OO produces no agglutinogens at all, and, therefore, the blood type is O. A person with either genotype OA or AA produces type A agglutinogens and, therefore, has blood type A. Genotypes OB and BB give type B blood, and genotype AB gives type AB blood.

THE AGGLUTININS

When type A agglutinogen *is not present* in a person's red blood cells, antibodies known as anti-A agglutinins develop in his plasma. Also, when type B agglutinogen *is not present* in the red blood cells, antibodies known as anti-B agglutinins develop in the plasma.

Thus, referring once again to Table 35-1, it will be observed that group O blood, though containing no agglutinogens, does contain both *anti-A* and *anti-B agglutinins;* group A blood contains type A agglutinogens and *anti-B agglutinins;* and group B blood contains type B agglutinogens and *anti-A agglutinins.* Finally, group AB blood contains both A and B agglutinogens but no agglutinins at all.

Titer of the Agglutinins at Different Ages. Immediately after birth the quantity of agglutinins in the plasma is almost zero. Two to 8 months after birth, the infant begins to produce agglutinins—anti-A agglutinins when type A agglutinogens are not present in the cells and anti-B agglutinins when type B agglutinogens are not in the cells. Figure 35-1 illustrates the changing titer of the anti-A and anti-B agglutinins at different ages. A maximum titer is usually reached at 8 to 10 years of age, and this gradually declines throughout the remaining years of life.

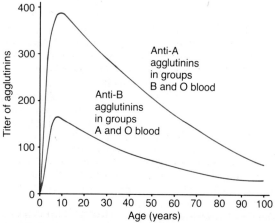

Figure 35-1. Average titers of anti-A and anti-B agglutinins in the blood of persons in group B and group A at different ages.

Origin of Agglutinins in the Plasma. The agglutinins are gamma globulins, as are other antibodies, and are produced by the same cells that produce antibodies to any other antigens. Most of them are IgM and IgG immunoglobulin molecules.

But why are these agglutinins produced in individuals who do not have the antigenic substances in their red blood cells? The answer to this seems to be that small amounts of group A and B antigens enter the body in the food, in bacteria, and in other ways, and these substances initiate the development of the anti-A or anti-B agglutinins. One of the reasons for believing this is that injection of group A or group B antigen into a recipient having another blood type causes a typical immune response with formation of greater quantities of agglutinins than ever. Also, the newborn baby has few if any agglutinins, showing that agglutinin formation occurs almost entirely after birth.

THE AGGLUTINATION PROCESS IN TRANSFUSION REACTIONS

When bloods are mismatched so that anti-A or anti-B plasma agglutinins are mixed with red blood cells containing A or B agglutinogens, respectively, the red cells agglutinate by the following process: The agglutinins attach themselves to the red blood cells. Because the agglutinins have two binding sites (IgG type) or ten sites (IgM type), a single agglutinin can attach to two or more different red blood cells at the same time, thereby causing the cells to adhere to each other. This causes the cells to clump. Then these clumps plug small blood vessels throughout the circulatory system. During the ensuing few hours to a few days, the phagocytic white blood cells and the reticuloendothelial system destroy the agglutinated cells, releasing hemoglobin into the plasma.

Hemolysis in Transfusion Reactions. Sometimes, when recipient and donor bloods are mismatched, immediate hemolysis of red cells occurs in the circulating blood. In this case the antibodies cause lysis of the red blood cells by activating the complement system. This in turn releases proteolytic enzymes (the *lytic complex)* that rupture the cell membranes, as was described in Chapter 34.

However, *immediate* intravascular hemolysis is far less common than agglutination followed by *delayed* hemolysis, because not only does there have to be a very high titer of antibodies for this to occur but also a different type of antibody seems to be required, mainly the IgM antibodies; these antibodies are called *hemolysins.*

BLOOD TYPING

Prior to giving a transfusion, it is necessary to determine the blood type of the recipient and the type of the donor blood so that the bloods will be appropriately matched. This is called *blood typing,* and it is performed in the following way: The red blood cells are first diluted with saline. One portion is then mixed with anti-A agglutinin, and another portion is mixed with anti-B agglutinin. After several minutes, the mixture is observed under a microscope. If the red blood cells have become clumped—that is, "agglutinated" —one knows that an antibody-antigen reaction has resulted.

Table 35-2 illustrates the presence (+) or absence (−) of agglutination with each of the four different types of blood. Type O red blood cells have no agglutinogens and therefore do not react with either the anti-A or the anti-B serum.

Table 35-2 BLOOD TYPING—SHOWING AGGLUTINATION OF CELLS OF THE DIFFERENT BLOOD TYPES WITH ANTI-A AND ANTI-B AGGLUTININS

Red Blood Cell Types	Sera	
	Anti-A	Anti-B
O	−	−
A	+	−
B	−	+
AB	+	+

Type A blood has A agglutinogens and therefore agglutinates with anti-A agglutinins. Type B blood has B agglutinogens and agglutinates with anti-B serum. Type AB blood has both A and B agglutinogens and agglutinates with both types of serum.

● THE Rh BLOOD TYPES

Along with the O-A-B blood group system, the most important other system in the transfusion of blood is the Rh system. The one major difference between the O-A-B system and the Rh system is the following: In the O-A-B system, the agglutinins responsible for causing transfusion reactions develop spontaneously, whereas in the Rh system spontaneous agglutinins almost never occur. Instead, the person must first be massively exposed to an Rh antigen, usually by transfusion of blood or by having a baby with the antigen, before enough agglutinins to cause a significant transfusion reaction will develop.

The Rh Antigens—"Rh Positive" and "Rh Negative" Persons. There are six common types of Rh antigens, each of which is called an Rh factor. These types are designated C, D, E, c, d, and e. A person who has a C antigen will not have the c antigen, but the person missing the C antigen will always have the c antigen. The same is true for the D-d and E-e antigens. Also, because of the manner of inheritance of these factors, each person will have one of each of the three pairs of antigens.

The type D antigen is widely prevalent in the population, and it is also considerably more antigenic than the other Rh antigens. Therefore, anyone who has this type of antigen is said to be *Rh positive*, whereas those persons who do not have type D antigen are said to be *Rh negative*. However, it must be noted that even in Rh negative persons some of the other Rh antigens can still cause transfusion reactions, though usually much milder.

Approximately 85 per cent of all whites are Rh positive and 15 per cent, Rh negative. In American blacks, the percentage of Rh positives is about 95, whereas in African blacks it is virtually 100 per cent.

THE Rh IMMUNE RESPONSE

Formation of Anti-Rh Agglutinins. When red blood cells containing Rh factor, or even protein breakdown products of such cells, are injected into a person without the factor—that is, into the Rh negative person—anti-Rh agglutinins develop very slowly, the maximum concentration of agglutinins occurring approximately 2 to 4 months later. This immune response occurs to a much greater extent in some people than in others. On multiple exposure to the Rh factor, the Rh negative person eventually becomes strongly "sensitized" to Rh factor.

Characteristics of Rh Transfusion Reactions. If an Rh negative person has never before been exposed to Rh positive blood, transfusion of Rh positive blood into him causes no immediate reaction at all. However, in some of these persons anti-Rh antibodies develop in sufficient quantities during the next 2 to 4 weeks to cause agglutination of the transfused cells that are still in the blood. These cells are then hemolyzed by the tissue macrophage system. Thus, a delayed transfusion reaction occurs, though it is usually mild. Yet on subsequent transfusion of Rh positive blood into the same person, who is now immunized against the Rh factor, the transfusion reaction is greatly enhanced and can be as severe as the reactions that occur with blood of types A and B.

Erythroblastosis Fetalis

Erythroblastosis fetalis is a disease of the fetus and newborn infant characterized by progressive agglutination and subsequent phagocytosis of the red blood cells. In most instances of erythroblastosis fetalis the mother is Rh negative and the father is Rh positive. The baby has inherited the Rh positive antigen from the father, and the mother has developed anti-Rh agglutinins that diffuse through the placenta into the fetus to cause red blood cell agglutination.

Prevalence of the Disease. An Rh negative mother having her first Rh positive child usually does not develop sufficient anti-Rh agglutinins to cause any harm. However, approximately 3 per cent of second Rh positive babies exhibit some signs of erythroblastosis fetalis; approximately 10 per cent of the third babies exhibit the disease; and the incidence rises progressively with subsequent pregnancies.

The Rh negative mother develops anti-Rh agglutinins only when the fetus is Rh positive. Many of the Rh positive fathers are heterozygous (about 55 per cent), causing about one fourth of the offspring to be Rh negative. Therefore, after an erythroblastotic child has been born, it is not certain that future children will also be erythroblastotic.

Effect of the Mother's Antibodies on the Fetus. After anti-Rh antibodies have formed in the mother, they diffuse slowly through the placental membrane into the fetus' blood. There they cause agglutination of the fetus' blood. The agglutinated red blood cells subsequently hemolyze, releasing hemoglobin into the blood. The macrophages then convert the hemoglobin into bilirubin, which causes yellowness (jaundice) of the skin. The antibodies probably also attack and damage some other cells of the body.

Clinical Picture of Erythroblastosis. The newborn, jaundiced, erythroblastotic baby is usually anemic at birth, and the anti-Rh agglutinins from the mother usually circulate in the baby's blood for 1 to 2 months after birth, destroying more and more red blood cells.

The hemopoietic tissues of the baby attempt to replace the hemolyzing red blood cells. The liver and the spleen become greatly enlarged and produce red blood cells in the same manner that they normally do during the middle of gestation. Because of the very rapid production of cells, many early forms, including many nucleated blastic forms, are emptied into the circulatory system, and it is because of the presence of these that the disease is called erythroblastosis fetalis.

Although the severe anemia of erythroblastosis fetalis is usually the cause of death, many children who barely survive the anemia exhibit permanent mental impairment or damage to motor areas of the brain because of precipitation of bilirubin in the neuronal cells, causing their destruction, a condition called *kernicterus*.

Treatment of the Erythroblastotic Baby. The usual treatment for erythroblastosis fetalis is to replace the newborn infant's blood with Rh negative blood. Approximately 400 ml of Rh negative blood is infused over a period of 1.5 or more hours while the baby's own Rh positive blood is being removed. This procedure may be repeated several times during the first few weeks of life, mainly to keep the bilirubin level low and thereby to prevent kernicterus. By the time these transfused Rh negative cells are replaced with the baby's own Rh positive cells, a process that requires 6 or more weeks, the anti-Rh agglutinins that had come from the mother will have been destroyed.

TRANSFUSION REACTIONS RESULTING FROM MISMATCHED BLOOD TYPES

If blood of one blood type is transfused to a recipient of another blood type, a transfusion reaction is likely to occur in which the red blood cells *of the donor blood* are agglutinated. It is very rare that the transfused blood causes agglutination *of the recipient's cells,* for the following reason: The plasma portion of the donor blood immediately becomes diluted by all the plasma of the recipient, thereby decreasing the titer of the infused agglutinins to a level too low to cause agglutination. On the other hand, the infused blood does not dilute the agglutinins in the recipient's plasma to a major extent. Therefore, the recipient's agglutinins can still agglutinate the donor cells.

As explained earlier, all transfusion reactions eventually cause hemolysis, either immediate hemolysis caused by hemolysins or phagocytosis of agglutinated cells. The hemoglobin released from the red cells is then converted by the phagocytes into bilirubin and later excreted into the bile by the liver, as discussed in Chapter 70. The concentration of bilirubin in the body fluids often rises high enough to cause *jaundice*—that is, the person's tissues become tinged with yellow bile pigment. But if liver function is normal, jaundice usually does not appear unless more than 300 to 500 milliliters of blood is hemolyzed in less than a day.

Acute Kidney Shutdown Following Transfusion Reactions. One of the most lethal effects of transfusion reactions is kidney shutdown, which can begin within a few minutes to a few hours and continue until the person dies of renal failure.

The kidney shutdown seems to result from three different causes: First, the antigen-antibody reaction of the transfusion reaction releases toxic substances from the hemolyzing blood that cause powerful renal vasoconstriction. Second, the loss of circulating red cells along with production of toxic substances from the hemolyzed cells and from the immune reaction often causes circulatory shock. The arterial blood pressure falls very low, and the renal blood flow and urinary output decrease. Third, if the total amount of free hemoglobin in the circulating blood is greater than that quantity which can bind with haptoglobin (a plasma protein that can bind small amounts of hemoglobin), much of the excess leaks through the glomerular membranes into the kidney tubules. If this amount is still slight, it can be reabsorbed through the tubular epithelium into the blood and will cause no harm; but if it is great, then only a small percentage is reabsorbed. Yet water continues to be reabsorbed, causing the tubular hemoglobin concentration to rise so high that it precipitates and blocks many of the

tubules; this is especially true if the urine is acidic. Thus, renal vasoconstriction, circulatory shock, and tubular blockage all add together to cause acute renal shutdown. If the shutdown is complete and fails to open up, the patient dies within a week to 12 days, as explained in Chapter 31, unless treated with the artificial kidney.

● TRANSPLANTATION OF TISSUES AND ORGANS

Many of the different antigens of red blood cells that cause transfusion reactions plus still many more are present in other cells of the body as well. Consequently, any foreign cells transplanted into a recipient can cause immune responses and immune reactions. In other words, most recipients are just as able to resist invasion by foreign tissue cells as to resist invasion by foreign bacteria or red cells.

Isografts, Allografts, and Xenografts. A transplant of a tissue or whole organ from one identical twin to another is called an *isograft.* A transplant from one human being to another or from any animal to another animal of the same species is called an *allograft.* Finally, a transplant from a lower animal to a human being or from an animal of one species to one of another species is called a *xenograft.*

Transplantation of Cellular Tissues. In the case of isografts, cells in the transplant will almost always live indefinitely if an adequate blood supply is provided; in the case of allografts and xenografts, immune reactions almost always occur, causing death of all the cells in the graft within 1 to 5 weeks after transplantation unless some specific therapy is used to prevent the immune reaction. The cells of allografts usually persist longer than those of xenografts because the antigenic makeup of the allograft is more nearly the same as that of the recipient's tissues than is true for the xenograft. When the tissues are properly "typed" and are very similar prior to transplant, successful allografts, even without drug therapy, occasionally result.

Some of the different cellular tissues and organs that have been transplanted either experimentally or for therapeutic benefit from one person to another are skin, kidney, heart, liver, glandular tissue, bone marrow, and lung. Many kidney allografts have been successful for as long as 5 to 15 years, and liver and heart transplants for 1 to 15 years.

ATTEMPTS TO OVERCOME THE IMMUNE REACTION IN TRANSPLANTED TISSUE

Because of the extreme potential importance of transplanting certain tissues and organs, serious attempts have been made to prevent the antigen-antibody reactions associated with transplants. The following specific procedures have met with certain degrees of clinical or experimental success.

Tissue Typing—The HLA Complex of Antigens

The most important antigens for causing graft rejection are a complex called the HLA antigens. These are subdivided into three different classes: class 1, which are present on the surfaces of all cells of the body; class 2, which are present specifically on the surfaces of lymphocytic cells associated with immunity; and class 3, which have to do with production of complement and other substances related to immunity.

Because the class 1 HLA antigens are present on almost all cells in the body, these are the ones that are mainly responsible for graft rejection. This is a group of about 150 different antigens located on the cell membrane surfaces of all cells in the body. Despite this large number of different antigens, they are coded for by only six separate genes, three located on each chromosome of a chromosomal pair. The genes are allelomorphic and therefore can code for six different HLA antigens in any one person. However, even this represents more than a trillion possible combinations. Consequently, it is virtually impossible for two different persons, except in the case of identical twins, to have the same six HLA antigens. Development of immunity against any one of these antigens can cause graft rejection.

The HLA antigens occur in the white blood cells as well as in the tissue cells. Therefore, tissue typing for these antigens is done in the membranes of lymphocytes that have been separated from the person's blood. The lymphocytes are mixed with appropriate antisera and complement; and after appropriate incubation, the cells are tested for membrane damage, usually by testing the rate of uptake by the lymphocytic cells of a supravital dye.

Fortunately, some of the HLA antigens are not severely antigenic, for which reason a precise match of some of the antigens between donor and recipient is not absolutely essential to allow allograft acceptance. Therefore, by obtaining the best possible match between donor and recipient, the grafting procedure has become far less hazardous. The best success has been with tissue-type matches between members of the same family. Of course, the match in identical twins is exact, so that transplants between identical twins are almost never rejected because of immune reactions.

Prevention of Graft Rejection by Suppressing the Immune System

Obviously, if the immune system were completely suppressed, graft rejection would not occur. In fact, in an occasional person who has serious depression of the immune system, grafts can be successful without the use of any therapy at all to prevent rejection. However, in the normal person, even with the best possible tissue typing, allografts will rarely resist rejection for more than a few weeks without the employment of some therapeutic procedure to suppress the immune system. Furthermore, because the T cells are mainly important for killing the cells of the grafts, their suppression is much more important than suppression of plasma antibodies. Some of the therapeutic agents that have been used for this purpose include the following:

1. *Glucocorticoid hormones (or drugs with glucocorticoid activity),* which suppress the growth of all lymphoid tissue and therefore decrease the formation of both antibodies and T cells.

2. *Various drugs that have a toxic effect on the lymphoid system* and therefore block the formation of antibodies and T cells, especially the drug azathioprine (Imuran).

3. *Cyclosporine,* which has a specific inhibitory effect on the formation of helper T cells and therefore is especially efficacious in blocking the T cell rejection reaction. This has proved to be the most valuable of all the drugs because it does not depress some other portions of the immune system.

Unfortunately, use of these agents leaves the person unprotected from disease; therefore, bacterial and viral infections sometimes become rampant. In addition, the incidence of cancer is several times as great in an immunosuppressed person, presumably because the immune system is important in destroying many early cancer cells before they can really begin to proliferate.

To summarize, transplantation of living tissues in human beings until the present has had limited but important success. On the other hand, when someone succeeds in blocking the immune response of the recipient to a donor organ without at the same time destroying the recipient's specific immunity for disease, the story will change overnight.

REFERENCES

Ascher, N. L., et al.: Cellular basis of allograft rejection. Immunol. Rev., 77:217, 1984

Baumgartner, W. A., and Reitz, B. A.: Heart and Heart-Lung Transplantation. Philadelphia, W. B. Saunders Co., 1989.

Chatterjee, S. N. (ed.): Renal Transplantation: A Multidisciplinary Approach. New York, Raven Press, 1980.

Claas, F. H., and van-Rood, J. J.: The interaction of drugs and endogenous substances with HLA class-I antigens. Prog. Allergy, 36:135, 1985.

Colvin, R. B.: Flow cytometric analysis of T cells: Diagnostic applications in transplantation. Ann. N.Y. Acad. Sci., 428:5, 1984.

Demetris, A. J., et al.: Pathology of hepatic transplantation: A review of 62 adult allograft recipients immunosuppressed with a cyclosporin/steroid regimen. Am. J. Pathol., 118:151, 1985.

Doherty, P. C., et al.: Immunological surveillance of tumors in the context of major histocompatibility complex restriction of T cell formation. Adv. Cancer Res., 42:1. 1984.

Fitch, F. W., et al.: Contributions of T cell clones to an understanding of the T cell receptor for antigen. Transplantation, 38:210, 1984.

Fowler, M. B., and Schroeder, J. S.: Current status of cardiac transplantation. Mod. Conc. Cardio. Dis., 55:37, 1986.

Goulmy, E.: Class-I-restricted human cytotoxic T lymphocytes directed against minor transplantation antigens and their possible role in organ transplantation. Prog. Allergy, 36:44, 1985.

Hancock, W. W.: Analysis of intragraft effector mechanisms associated with human renal allograft rejection: Immunohistological studies with monoclonal antibodies. Immunol. Rev., 77:61, 1984.

Hobbs, J. R. (ed.): Thymic Factor Therapy. New York, Raven Press, 1985.

Hunt, S. A., and Stinson, E. B.: Cardiac transplantation. Annu. Rev. Med., 32:213, 1982.

Kahan, B. D., et al.: Clinical and experimental studies with cyclosporine in renal transplantation. Surgery, 97:125, 1985.

Kaufman, J. F., et al: The class II molecules of the human and murine major histocompatibility complex. Cell, 36:1, 1984.

Mason, D. W., et al.: Mechanisms of allograft rejection: The roles of cytotoxic T-cells and delayed-typed hypersensitivity. Immunol. Rev., 77:167, 1984.

Miller, D. R., et al. (eds.): Blood Diseases of Infancy and Childhood. St. Louis, C. V. Mosby Co., 1989.

Morris, P. J.: Kidney Transplantation, 3rd Ed. Philadelphia, W. B. Saunders Co., 1988.

Rodger, J. C., and Drake B. L.: The enigma of the fetal graft. Am. Sci., 75:51, 1987.

Strom, T. B.: Toward more selective therapies to block graft rejection. AFK Nephrol. Letter, 4:13, 1987.

Williams, W. J., et al. (eds.): Hematology, 4th Ed. New York, McGraw-Hill Book Co., 1990.

Wood, R. F.: Renal Transplantation. A Clinical Handbook. Philadelphia, W. B. Saunders Co., 1984.

Zmijewski, C. M.: Human leukocyte antigen matching in renal transplantation: Review and current status. J. Surg. Res., 38:66, 1985.

CHAPTER 36

• *HEMOSTASIS AND BLOOD COAGULATION* •

● EVENTS IN HEMOSTASIS

The term *hemostasis* means prevention of blood loss. Whenever a vessel is severed or ruptured, hemostasis is achieved by several different mechanisms, including (1) vascular spasm, (2) formation of a platelet plug, (3) blood coagulation, and (4) eventual growth of fibrous tissue into the blood clot to close the hole in the vessel permanently.

VASCULAR CONSTRICTION

Immediately after a blood vessel is cut or ruptured, the stimulus of the traumatized vessel causes the wall of the vessel to contract; this instantaneously reduces the flow of blood from the vessel rupture. The contraction results from nervous reflexes, local myogenic spasm, and local humoral factors from the traumatized tissues and blood platelets. The nervous reflexes presumably are initiated by pain or other impulses originating from the traumatized vessel or from nearby tissues. However, much of the vasoconstriction probably results from local myogenic contraction of the blood vessels initiated by direct damage to the vascular wall. For the smaller vessels, the platelets are responsible for much of the vasoconstriction by releasing the vasoconstrictor substance *thromboxane A_2.* The more vessel traumatized, the greater the degree of spasm; this means that a sharply cut blood vessel usually bleeds much more than does a vessel ruptured by crushing. This local vascular spasm can last for many minutes or even hours, during which time the ensuing processes of platelet plugging and blood coagulation can take place.

The value of vascular spasm as a mean of hemostasis is illustrated by the fact that persons whose legs have been severed by crushing types of trauma sometimes have such intense spasm in vessels as large as the anterior tibial artery that there is no lethal loss of blood.

FORMATION OF THE PLATELET PLUG

If the rent in the blood vessel is very small — and many very small vascular holes do develop each day — it is often sealed by a *platelet plug,* rather than by a blood clot. To understand this it is important that we first discuss the nature of platelets themselves.

Physical and Chemical Characteristics of Platelets

Platelets are minute round or oval discs 2 to 4 micrometers in diameter. They are formed in the bone marrow from *megakaryocytes,* which are extremely large cells of the hemopoietic series in the bone marrow that fragment into platelets either in the bone marrow or soon after entering the blood, especially as they try to squeeze through the pulmonary capillaries. The normal concentration of platelets in the blood is between 150,000 and 300,000 per microliter.

Platelets have many functional characteristics of whole cells, even though they do not have nuclei and cannot reproduce. In their cytoplasm are such active factors as (1) *actin* and *myosin molecules,* similar to those found in muscle cells, as well as still another contractile protein, *thrombosthenin,* that can cause the platelets to contract; (2) residuals of both the *endoplasmic reticulum* and the *Golgi apparatus* that synthesize various enzymes and store large quantities of calcium ions; (3) mitochondria and enzyme systems that are capable of forming *adenosine triphosphate (ATP)* and *adenosine diphosphate (ADP);* (4) enzyme systems that synthesize *prostaglandins,* which are local hormones that cause many different types of vascular and other local tissue reactions; (5) an important protein called *fibrin-stabilizing factor,* which we will discuss later in relation to blood coagulation; and (6) a *growth factor* that can cause vascular endothelial cells, vascular smooth muscle cells, and fibroblasts to multiply and grow, thus causing cellular growth that helps repair damaged vascular walls.

The cell membrane of the platelets is also important. On its surface is a coat of *glycoproteins* that causes it to avoid adherence to normal endothelium and yet to adhere to *injured* areas of the vessel wall, especially to injured endothelial cells and even more so to any exposed collagen from deeper in the vessel wall. In addition, the membrane contains large amounts of *phospholipids* containing *platelet factor 3,* which plays several activating roles at multiple points in the blood clotting process, as we discuss later.

Thus, the platelet is a very active structure. It has a half-life in the blood of 8 to 12 days, at the end of which time its life processes run out. Then it is eliminated from the circulation mainly by the tissue macrophage system; more than half the platelets are removed by the macrophages in the spleen, where the blood passes through a latticework of tight trabeculae.

Mechanism of the Platelet Plug

Platelet repair of vascular openings is based on several important functions of the platelet itself: When platelets come in contact with a damaged vascular surface, such as the collagen fibers in the vascular wall or even damaged endothelial cells, they immediately change their characteristics drastically. They begin to swell; they assume irregular forms with numerous irradiating processes protruding from their surfaces; their contractile proteins contract forcefully and cause the release of granules containing multiple active factors; they become sticky so that they stick to the collagen fibers; they secrete large quantities of *ADP;* and their enzymes form *thromboxane A₂,* which is also secreted into the blood. The ADP and thromboxane, in turn, act on nearby platelets to activate them as well, and the stickiness of these additional platelets causes them to adhere to the originally activated platelets. Therefore, at the site of any rent in a vessel, the damaged vascular wall or extravascular tissues elicit a vicious cycle of activation of successively increasing numbers of platelets that themselves attract more and more additional platelets, thus forming a *platelet plug.* This is at first a fairly loose plug, but it is usually successful in blocking the blood loss if the vascular opening is small. Then, during the subsequent process of blood coagulation, to be described in subsequent paragraphs, *fibrin threads* form that attach to the platelets, thus forming a tight and unyielding plug.

Importance of the Platelet Method for Closing Vascular Holes. If the rent in a vessel is small, the platelet plug by itself can stop blood loss completely, but if there is a large hole, a blood clot in addition to the platelet plug is required to stop the bleeding.

The platelet-plugging mechanism is extremely important for closing the minute ruptures in very small blood vessels that occur hundreds of times daily. Indeed, multiple small holes through the endothelial cells themselves are often closed by platelets actually fusing with the endothelial cells to form additional endothelial cell membrane. A person who has very few platelets develops literally hundreds of small hemorrhagic areas under the skin and throughout the internal tissues, but this does not occur in the normal person.

BLOOD COAGULATION IN THE RUPTURED VESSEL

The third mechanism for hemostasis is formation of the blood clot. The clot begins to develop in 15 to 20 seconds if the trauma of the vascular wall has been severe and in 1 to 2 minutes if the trauma has been minor. Activator substances both from the traumatized vascular wall and from platelets and blood proteins adhering to the traumatized vascular wall initiate the clotting process. The physical events of this process are illustrated in Figure 36–1, and the chemical events will be discussed in detail later in the chapter.

Within 3 to 6 minutes after rupture of a vessel, if the vessel opening is not too large, the entire opening or broken end of the vessel is filled with clot. After 20 minutes to an hour, the clot retracts; this closes the vessel still further. Platelets play an important role in this clot retraction, as is also discussed later.

FIBROUS ORGANIZATION OR DISSOLUTION OF THE BLOOD CLOT

Once a blood clot has formed, it can follow two separate courses: it can become invaded by fibroblasts, which subse-

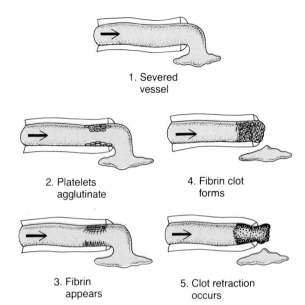

Figure 36–1. The clotting process in the traumatized blood vessel. (Modified from Seegers: Hemostatic Agents. Springfield, Ill., Charles C Thomas.)

quently form connective tissue all through the clot; or it can dissolve. The usual course for a clot that forms in a small hole of a vessel wall is invasion by fibroblasts, beginning within a few hours after the clot is formed (which is promoted at least partially by the *growth factor* secreted by platelets) and continuing to complete organization of the clot into fibrous tissue within approximately 1 to 2 weeks. On the other hand, when blood coagulates to form a larger clot, such as blood that has leaked into tissues, special substances within the clot itself become activated, and these then function as enzymes to dissolve the clot itself, as will be discussed later in the chapter.

● MECHANISM OF BLOOD COAGULATION

Basic Theory. Over 50 important substances that affect blood coagulation have been found in the blood and tissues, some promoting coagulation, called *procoagulants,* and others inhibiting coagulation, called *anticoagulants.* Whether or not the blood will coagulate depends on the degree of balance between these two groups of substances. Normally the anticoagulants predominate, and the blood does not coagulate; but when a vessel is ruptured, procoagulants in the area of damage become "activated" and override the anticoagulants, and then a clot does develop.

General Mechanism. All research workers in the field of blood coagulation agree that clotting takes place in three essential steps: (1) A complex of substances called *prothrombin activator* is formed in response to rupture of the vessel or damage to the blood itself. (2) The prothrombin activator catalyzes the conversion of *prothrombin* into *thrombin.* (3) The thrombin acts as an enzyme to convert *fibrinogen* into *fibrin threads* that enmesh platelets, blood cells, and plasma to form the clot itself.

Let us discuss first the mechanism by which the blood clot itself is formed, beginning with the conversion of prothrombin to thrombin; then we will come back to the initiating stages in the clotting process by which prothrombin activator is formed.

CONVERSION OF PROTHROMBIN TO THROMBIN

After prothrombin activator has been formed as a result of rupture of the blood vessel or as a result of damage to special activator substances in the blood, the prothrombin activator then causes conversion of prothrombin to thrombin (Fig. 36-2), which in turn causes polymerization of fibrinogen molecules into fibrin threads within another 10 to 15 seconds. Thus, the rate-limiting factor in causing blood coagulation is usually the formation of prothrombin activator and not the subsequent reactions beyond that point, for these normally occur very rapidly to form the clot itself.

Prothrombin and Thrombin. Prothrombin is a plasma protein, an alpha$_2$-globulin, having a molecular weight of 68,700. It is present in normal plasma in a concentration of about 15 mg/dl. It is an unstable protein that can split easily into smaller compounds, one of which is *thrombin,* which has a molecular weight of 33,700, almost exactly half that of prothrombin.

Prothrombin is formed continually by the liver, and it is continually being used throughout the body for blood clotting. If the liver fails to produce prothrombin, its concentration in the plasma falls too low to provide normal blood coagulation within 24 hours. Vitamin K is required by the liver for normal formation of prothrombin as well as three other clotting factors that we discuss later; therefore, either lack of vitamin K or the presence of liver disease that prevents normal prothrombin formation can often decrease the prothrombin level so low that a bleeding tendency results.

CONVERSION OF FIBRINOGEN TO FIBRIN— FORMATION OF THE CLOT

Fibrinogen. Fibrinogen is a high molecular weight protein (340,000) occurring in the plasma in quantities of 100 to 700 mg/dl. Fibrinogen is formed in the liver, and liver disease occasionally decreases the concentration of circulating fibrinogen, as it does the concentration of prothrombin, as was pointed out previously.

Because of its large molecular size, very little fibrinogen normally leaks into the interstitial fluids; and since it is one of the essential factors in the coagulation process, interstitial fluids ordinarily coagulate poorly if at all. Yet when the permeability of the capillaries becomes pathologically increased, fibrinogen does then leak into the tissue fluids in sufficient quantities to allow clotting in much the same way that plasma and whole blood clot.

Action of Thrombin on Fibrinogen to Form Fibrin. Thrombin is a protein *enzyme* with proteolytic capabilities. It acts on fibrinogen to remove four low molecular weight peptides from each molecule of fibrinogen, forming a molecule of *fibrin monomer* that has the automatic capability of polymerizing with other fibrin monomer molecules. Therefore, many fibrin monomer molecules polymerize within seconds into *long fibrin threads* that form the *reticulum* of the clot.

In the early stages of this polymerization, the fibrin monomer molecules are held together by weak noncovalent hydrogen bonding, and the threads also are not cross-linked with each other; therefore, the resultant clot is weak and can be broken apart with ease. However, still another process occurs during the following few minutes that greatly strengthens the fibrin reticulum. This involves a substance called *fibrin-stabilizing factor* that is normally present in small amounts in the plasma globulins but that is also released from platelets entrapped in the clot. Before fibrin-stabilizing factor can have an effect on the fibrin threads it must itself be activated. Fortunately, the same thrombin that causes fibrin formation also activates the fibrin-stabilizing factor. Then this activated substance operates as an enzyme to cause *covalent bonds* between the fibrin monomer molecules as well as multiple cross-linkages between the adjacent fibrin threads, thus adding tremendously to the three-dimensional strength of the fibrin meshwork.

The Blood Clot. The clot is composed of a meshwork of fibrin threads running in all directions and entrapping blood cells, platelets, and plasma. The fibrin threads adhere to damaged surfaces of blood vessels; therefore, the blood clot becomes adherent to any vascular opening and thereby prevents blood loss.

Clot Retraction — Serum. Within a few minutes after a clot is formed, it begins to contract and usually expresses most of the fluid from the clot within 20 to 60 minutes. The fluid expressed is called *serum,* because all its fibrinogen and most of the other clotting factors have been removed; in this way, serum differs from plasma. Serum obviously cannot clot because of lack of these factors.

Platelets are necessary for clot retraction to occur. Therefore, failure of clot retraction is an indication that the number of platelets in the circulating blood is low. Electron micrographs of platelets in blood clots show that they become attached to the fibrin threads in such a way that they actually bond different threads together. Furthermore, platelets entrapped in the clot continue to release procoagulant substances, one of which is fibrin-stabilizing factor, which causes more and more cross-linking bonds between the adjacent fibrin threads. In addition, the platelets themselves contribute directly to clot contraction by activating the platelet thrombosthenin, actin, and myosin molecules, which are contractile proteins and cause strong contraction of the platelet spicules attached to the fibrin. Obviously, this also helps compress the fibrin meshwork into a smaller mass. The contraction is probably activated by thrombin as well as by calcium ions released from the calcium stores in the mitochondria, endoplasmic reticulum, and Golgi apparatus of the platelets.

As the clot retracts, the edges of the broken blood vessel are pulled together, thus possibly or probably contributing to the ultimate state of hemostasis.

THE VICIOUS CIRCLE OF CLOT FORMATION

Once a blood clot has started to develop, it normally extends within minutes into the surrounding blood. That is, the clot itself initiates a vicious circle to promote more clotting. One of the most important causes of this is the fact that the proteolytic action of thrombin allows it to act on many of the other blood-clotting factors in addition to fibrinogen. For instance, thrombin has a direct proteolytic effect on prothrombin itself, tending to split this into still more thrombin, and it acts on some of the blood-clotting

Figure 36-2. Schema for conversion of prothrombin to thrombin and polymerization of fibrinogen to form fibrin threads.

factors responsible for the formation of prothrombin activator. (These effects, to be discussed in subsequent paragraphs, include (1) acceleration of the actions of Factors VIII, IX, X, XI, and XII; and (2) aggregation of platelets.) Once a critical amount of thrombin is formed, a vicious circle develops that causes still more blood clotting and more thrombin to be formed; thus, the blood clot continues to grow until something stops its growth.

INITIATION OF COAGULATION: FORMATION OF PROTHROMBIN ACTIVATOR COMPLEX

Now that we have discussed the clotting process initiated by the formation of thrombin from prothrombin, we must turn to the more complex mechanisms that activate the prothrombin and cause it to form thrombin. These mechanisms can be set into play by trauma to the vascular wall and adjacent tissues, trauma to the blood, or contact of the blood with damaged endothelial cells or with collagen and other tissue elements outside the blood vessel endothelium. In each instance, they lead to the formation of *prothrombin activator complex,* which then causes prothrombin conversion to thrombin.

Prothrombin activator is generally considered to be formed in two basic ways, although in reality these interact constantly with each other: (1) by the *extrinsic pathway* that begins with trauma to the vascular wall and surrounding tissues, and (2) by the *intrinsic pathway* that begins in the blood itself.

In both the extrinsic and intrinsic pathways a series of different plasma proteins, especially beta-globulins, play major roles. These, along with the other factors, already discussed that enter into the clotting process, are called *blood-clotting factors* and for the most part they are *inactive* forms of proteolytic enzymes. When converted to the active forms, their enzymatic actions cause the successive, cascading reactions of the clotting process.

Most of the clotting factors, which are listed in Table 36–1, are designated by Roman numerals.

The Extrinsic Mechanism for Initiating Clotting

The extrinsic mechanism for initiating the formation of prothrombin activator begins with a traumatized vascular wall or extravascular tissues and occurs according to the following three basic steps, as illustrated in Figure 36–3.

1. *Release of tissue thromboplastin.* Traumatized tissue releases a complex of several factors called *tissue thromboplastin.* This includes especially *phospholipids* from the membranes of the tissues and a *lipoprotein complex* containing an important glycoprotein that functions as a *proteolytic enzyme.*

2. *Activation of Factor X to form activated Factor X — role of Factor VII and tissue thromboplastin.* The lipoprotein complex of tissue thromboplastin further complexes with blood coagulation Factor VII, and, in the presence of tissue phospholipids and calcium ions acts enzymatically on factor X to form *activated Factor X.*

3. *Effect of activated Factor X to form prothrombin activator — role of Factor V.* The activated Factor X complexes immediately with the tissue phospholipids released as part of the tissue thromboplastin or released from platelets and also with Factor V to form the complex called *prothrombin activator.* Within a few seconds this splits prothrombin to form thrombin, and the clotting process

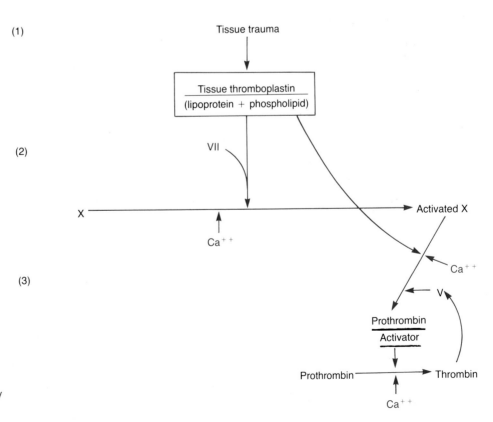

Extrinsic Pathway

(1)

(2)

(3)

Figure 36–3. The extrinsic pathway for initiating blood clotting.

Table 36–1 CLOTTING FACTORS IN THE BLOOD AND THEIR SYNONYMS

Clotting Factor	Synonyms
Fibrinogen	Factor I
Prothrombin	Factor II
Tissue thromboplastin	Factor III; tissue factor
Calcium	Factor IV
Factor V	Proaccelerin; labile factor; Ac-globulin (Ac-G)
Factor VII	Serum prothrombin conversion accelerator (SPCA); proconvertin; stable factor
Factor VIII	Antihemophilic factor (AHF); antihemophilic globulin (AHG); antihemophilic factor A
Factor IX	Plasma thromboplastin component (PTC); Christmas factor; antihemophilic factor B
Factor X	Stuart factor; Stuart-Prower factor
Factor XI	Plasma thromboplastin antecedent (PTA); antihemophilic factor C
Factor XII	Hageman factor
Factor XIII	Fibrin-stabilizing factor
Prekallikrein	Fletcher factor
High molecular weight kininogen	Fitzgerald factor; HMWK
Platelets	

proceeds as has already been explained. At first, the Factor V in the prothrombin activator complex is inactive; but once clotting begins and thrombin begins to be formed, the proteolytic action of thrombin activates Factor V; this then becomes an additional strong accelerator of prothrombin activation. Thus, in the final prothrombin activator complex, activated Factor X is the actual protease that causes splitting of prothrombin to thrombin, activated Factor V greatly accelerates this protease activity, and phospholipids act as a vehicle that further accelerates the process. Note especially the *positive feedback* effect of thrombin, acting through Factor V, in accelerating the entire process once it begins.

The Intrinsic Mechanism for Initiating Clotting

The second mechanism for initiating the formation of prothrombin activator, and therefore for initiating clotting, begins with trauma to the blood itself or exposure of the blood to collagen in a traumatized vascular wall, and then continues through the following series of cascading reactions illustrated in Figure 36–4:

1. *Activation of Factor XII and release of platelet phospholipids.* Trauma to the blood or exposure of the blood to vascular wall collagen alters two important clotting factors in the blood—Factor XII and the platelets. When Factor XII is disturbed, such as by coming into contact with collagen or with a wettable surface such as glass, it takes on a new configuration that converts it into a proteolytic enzyme called "activated Factor XII." Simultaneously, the blood trauma also damages the platelets, either because of adherence to collagen or to a wettable surface (or by damage in other ways), and this releases platelet phospholipid containing the lipoprotein called platelet factor 3, which also plays a role in subsequent clotting reactions.

2. *Activation of Factor XI.* The activated Factor XII acts enzymatically on Factor XI to activate this as well, which is the second step in the intrinsic pathway. This reaction also requires *HMW kininogen,* and it is accelerated by prekallikrein.

3. *Activation of Factor IX by activated Factor XI.* The activated factor XI then acts enzymatically on Factor IX to activate this factor, also.

Intrinsic Pathway

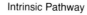

Figure 36–4. The intrinsic pathway for initiating blood clotting.

4. *Activation of Factor X—role of Factor VIII.* The activated factor IX, acting in concert with factor VIII and with the platelet phospholipids and factor 3 from the traumatized platelets, activates Factor X. It is clear that when either Factor VIII or platelets are in short supply, this step is deficient. Factor VIII is the factor that is missing in the person who has classic *hemophilia,* for which reason it is called *antihemophilic factor.* Platelets are the clotting factor that is lacking in the bleeding disease called *thrombocytopenia.*

5. *Action of activated Factor X to form prothrombin activator—role of Factor V.* This step in the intrinsic pathway is the same as the last step in the extrinsic pathway. That is, activated Factor X combines with Factor V and platelet or tissue phospholipids to form the complex called *prothrombin activator.* The prothrombin activator in turn initiates within seconds the cleavage of prothrombin to form thrombin, thereby setting into motion the final clotting process, as described earlier.

Role of Calcium Ions in the Intrinsic and Extrinsic Pathways

Except for the first two steps in the intrinsic pathway, calcium ions are required for promotion of all the reactions. Therefore, in the absence of calcium ions, blood clotting will not occur.

Fortunately, in the living body the calcium ion concentration rarely falls low enough to affect significantly the kinetics of blood clotting. On the other hand, when blood is removed from a person, it can be prevented from clotting by reducing the calcium ion concentration below the threshold level for clotting, either by deionizing the calcium by reacting it with substances such as *citrate ion* or by precipitating the calcium with substances such as *oxalate ion.*

Interaction Between the Extrinsic and Intrinsic Pathways—Summary of Blood Clotting Initiation

It is clear from the above schemas of the intrinsic and extrinsic systems that after blood vessels rupture clotting is initiated by both pathways simultaneously. The tissue thromboplastin initiates the extrinsic pathway, whereas contact of Factor XII and the platelets with collagen in the vascular wall initiates the intrinsic pathway.

Another important interaction also occurs between the two systems, caused by the proteolytic effects of thrombin to activate enzymes in the intrinsic pathway. For instance, note in Figure 36–4 that thrombin activates Factor VIII, and it also acts directly on platelets to cause them to aggregate and release their granules containing multiple other procoagulant factors. Therefore, in effect, initiation of the extrinsic pathway for clotting secondarily activates the intrinsic pathway as well.

When blood is removed from the body and held in a test tube, the intrinsic pathway alone must elicit the clotting. This usually results from contact of Factor XII and platelets with the wall of the vessel, which activates both of these and initiates the intrinsic mechanism. If the surface of the container is very "nonwettable," such as a siliconized surface, blood clotting is delayed for up to an hour or more.

Intravascular clotting sometimes results also from other factors that activate the intrinsic pathway. For instance, antigen-antibody reactions sometimes initiate the clotting process, and the same is true of some drugs or debris that might happen to enter the circulation.

An especially important difference between the extrinsic and intrinsic pathways is that *the extrinsic pathway* can be explosive in nature; once initiated, its speed of occurrence is limited only by the amount of tissue thromboplastin released from the traumatized tissues and by the quantities of Factors X, VII, and V in the blood. With severe tissue trauma, clotting can occur in as little as 15 seconds. On the other hand, the intrinsic pathway is much slower to proceed, usually requiring 1 to 6 minutes to cause clotting.

PREVENTION OF BLOOD CLOTTING IN THE NORMAL VASCULAR SYSTEM—THE INTRAVASCULAR ANTICOAGULANTS

Endothelial Surface Factors. Probably the most important factors for preventing clotting in the normal vascular system are (1) the *smoothness* of the endothelium, which prevents contact activation of the intrinsic clotting system; (2) a layer of *glycocalyx,* a mucopolysaccharide adsorbed to the inner surface of the endothelium, which repels the clotting factors and platelets, thereby preventing activation of clotting; and (3) a protein bound with the endothelial membrane, *thrombomodulin,* which binds thrombin. The binding of thrombomodulin with thrombin not only slows the clotting process by removing thrombin, but the combined thrombomodulin-thrombin complex also activates a plasma protein, *protein C,* that acts as an anticoagulant by inactivating activated Factors V and VIII.

When the endothelial wall is damaged, its smoothness and glycocalyx-thrombomodulin layer are both lost, which activates both Factor XII and the platelets, thus setting off the intrinsic pathway of clotting. If Factor XII and platelets come in contact with the subendothelial collagen, the activation is even more powerful.

The Antithrombin Action of Fibrin and of Antithrombin III. Among the most important anticoagulants in the blood itself are those that remove thrombin from the blood. The two most powerful of these are the *fibrin threads* formed during the process of clotting and an alpha-globulin called *antithrombin III* or *antithrombin-heparin cofactor.*

While a clot is forming, approximately 85 to 90 per cent of the thrombin formed from the prothrombin becomes adsorbed to the fibrin threads as they develop. This obviously helps prevent the spread of thrombin into the remaining blood and therefore prevents excessive spread of the clot.

The thrombin that does not adsorb to the fibrin threads soon combines with antithrombin III, which blocks the effect of the thrombin on the fibrinogen and then inactivates the bound thrombin during the next 12 to 20 minutes.

Heparin. Heparin is another powerful anticoagulant. Yet its concentration in the blood is normally very slight, so that only under some physiological conditions does it have significant anticoagulant effects. On the other hand, it is used very widely in medical practice to prevent intravascular clotting.

The heparin molecule is a highly negatively charged conjugated polysaccharide. By itself it has little or no anticoagulant property, but it combines with antithrombin III and increases as much as a hundred- to a thousandfold the effectiveness of antithrombin III in removing thrombin and thus acting as an anticoagulant. Therefore, in the presence of excess heparin the removal of thrombin from the circulating blood is almost instantaneous.

The complex of heparin and antithrombin III also removes several other activated coagulation factors in addition to thrombin, thus further enhancing the effectiveness of anticoagulation. These others include Factors XII, XI, IX, and X.

Heparin is produced by many different cells of the human body, but especially large quantities are formed by the basophilic *mast cells* located in the pericapillary connective tissue throughout the body. These cells continually secrete small quantities of heparin that diffuse into the circulatory system. The *basophil cells* of the blood, which are functionally almost identical with the mast cells, also release small quantities of heparin into the plasma.

Mast cells are extremely abundant in the tissue surrounding the capillaries of the lungs and to a lesser extent the capillaries of the liver. It is easy to understand why large quantities of heparin might be needed in these areas, for the capillaries of the lungs and liver receive many embolic clots formed in the slowly flowing venous blood; sufficient formation of heparin might prevent further growth of the clots.

Alpha₂-Macroglobulin. *Alpha₂-macroglobulin* is a very large globulin molecule having a molecular weight of 360,000. It is similar to antithrombin-heparin complex in that it combines with the proteolytic coagulation factors. However, its activity is not accelerated by heparin. Its function is mainly to act as a binding agent for several of the coagulation factors and prevent their proteolytic actions until they can be destroyed in various ways, but not by the alpha₂-macroglobulin itself. In this way it probably plays an important role even normally in preventing blood clotting.

LYSIS OF BLOOD CLOTS—PLASMIN

The plasma proteins contain a euglobulin called *plasminogen* or *profibrinolysin,* which, when activated, becomes a substance called *plasmin* or *fibrinolysin.* Plasmin is a proteolytic enzyme that resembles trypsin, the most important digestive enzyme of pancreatic secretion. It digests the fibrin threads and also digests other substances in the surrounding blood, such as fibrinogen, Factor V, Factor VIII, prothrombin, and Factor XII. Therefore, whenever plasmin is formed in a blood clot, it can cause lysis of the clot and also destruction of many of the clotting factors, thereby causing hypocoagulability of the blood.

Formation of Plasmin and Lysis of Clots. When a clot is formed, a large amount of plasminogen is trapped in the clot along with other plasma proteins. However, this will not become plasmin and will not cause lysis of the clot until it is activated. Fortunately, the injured tissues and vascular endothelium release a powerful activator called *tissue plasminogen activator* that a day or so later, after the clot has stopped the bleeding, eventually converts plasminogen to plasmin and removes the clot. In fact, many small blood vessels in which the blood flow has been stopped by clots are reopened by this mechanism.

A Plasmin Inhibitor, Alpha₂-Antiplasmin. Plasmin not only destroys fibrin threads but also functions as a proteolytic enzyme to digest fibrinogen and a number of other clotting factors as well. Small amounts of plasmin are formed in the blood all the time, which could seriously impede the activation of the clotting system were it not for the fact that the blood also contains another factor, *alpha₂-antiplasmin,* that binds with plasmin and inhibits it. Therefore, the rate of plasmin formation must rise above a certain critical level before it becomes effective.

Significance of the Plasmin System. The lysis of blood clots allows slow clearing (over a period of several days) of extraneous clotted blood in the tissues and sometimes allows reopening of clotted vessels. An especially important function of the plasmin system is to remove very minute clots from the millions of tiny peripheral vessels that eventually would all become occluded were there no way to cleanse them.

● CONDITIONS THAT CAUSE EXCESSIVE BLEEDING IN HUMAN BEINGS

Excessive bleeding can result from deficiency of any one of the many different blood clotting factors. Three particular types of bleeding tendencies that have been studied to the greatest extent will be discussed: (1) bleeding caused by vitamin K deficiency, (2) hemophilia, and (3) thrombocytopenia (platelet deficiency).

DECREASED PROTHROMBIN, FACTOR VII, FACTOR IX, AND FACTOR X CAUSED BY VITAMIN K DEFICIENCY

With few exceptions, almost all the blood clotting factors are formed by the liver. Therefore, diseases of the liver such as *hepatitis, cirrhosis,* and *acute yellow atrophy* all can depress the clotting system so greatly that the patient develops a severe tendency to bleed.

Another cause of depressed formation of clotting factors by the liver is vitamin K deficiency. Vitamin K is necessary for promotion of formation of four of the most important clotting factors, *prothrombin, Factor VII, Factor IX,* and *Factor X.* In the absence of vitamin K, insufficiency of these coagulation factors can also lead to a serious bleeding tendency.

Fortunately, vitamin K is continually synthesized in the intestinal tract by bacteria so that vitamin K deficiency rarely if ever occurs simply because of its absence from the diet, except in newborn children before they establish their intestinal bacterial flora. However, vitamin K deficiency does often occur as a result of poor absorption of fats from the gastrointestinal tract, because vitamin K is fat-soluble and ordinarily is absorbed into the blood along with the fats.

One of the most prevalent causes of vitamin K deficiency is failure of the liver to secrete bile into the gastrointestinal tract (which occurs either as a result of obstruction of the bile ducts or as a result of liver disease), for lack of bile prevents adequate fat digestion and absorption, and therefore depresses vitamin K absorption as well. Thus, liver disease often causes decreased production of prothrombin and the other factors both because of poor vitamin K absorption and because of the diseased liver cells as well. Because of this, vitamin K is injected into all patients with liver disease or obstructed bile ducts prior to performing any surgical procedure. Ordinarily, if vitamin K is given to a deficient patient 4 to 8 hours prior to operation, and the liver parenchymal cells are at least one-half normal in function, sufficient clotting factors will be produced to prevent excessive bleeding during the operation.

HEMOPHILIA

Hemophilia is a bleeding tendency that occurs almost exclusively in males. In 85 per cent of cases it is caused by *deficiency of Factor VIII;* this type of hemophilia is called *hemophilia A* or *classic hemophilia.* About one of every 10,000 males in the United States has classic hemophilia. In the other 15 per cent of the patients, the bleeding tendency is caused by deficiency of Factor IX. Both of these factors are transmitted genetically by way of the female chromosome as a recessive trait. Therefore, almost never will a woman have hemophilia because at least one of her two X chromosomes will have the appropriate genes. However, if one of her X chromosomes is deficient, she will be a *hemophilia carrier,* transmitting the disease to half her male offspring and transmitting the carrier state to half her female offspring.

The bleeding trait in hemophilia can have various degrees of severity, depending upon the severity of the genetic deficiency. Usually, bleeding will not occur except after trauma, but the degree of trauma required to cause severe and prolonged bleeding may be so mild that it is hardly noticeable. And bleeding can often last for literally weeks after extraction of a tooth.

Factor VIII is composed of two separate components, a very large component with a molecular weight in the millions and another smaller component with a molecular weight of about 230,000. This smaller component is most important in the intrinsic pathway for clotting, and it is deficiency of this part of Factor VIII that causes classic hemophilia. Another bleeding disease with somewhat different characteristics, called von Willebrand's disease, results from loss of the large component.

When a person with classic hemophilia develops severe, prolonged bleeding, almost the only therapy that is truly effective is injection of purified Factor VIII. Unfortunately, the cost of Factor VIII is very high; and its availability is limited because it can be gathered only from human blood and only in extremely small quantities. Fortunately, a genetically engineered Factor VIII should be available for human use soon.

THROMBOCYTOPENIA

Thrombocytopenia means the presence of a very low quantity of platelets in the circulatory system. Persons with thrombocytopenia have a tendency to bleed as do hemophiliacs, except that the bleeding is usually from many small venules or capillaries, rather than from larger vessels, as in hemophilia. As a result, small punctate hemorrhages occur throughout all the body tissues. The skin of such a person displays many small, purplish blotches, giving the disease the name *thrombocytopenic purpura.* It will be remembered that platelets are especially important for repair of minute breaks in capillaries and other small vessels.

Ordinarily, bleeding does not occur until the number of platelets in the blood falls below a value of approximately 50,000 per microliter, rather than the normal 150,000 to 300,000. Levels as low as 10,000 per microliter are frequently lethal.

Even without making specific platelet counts on the blood, one can sometimes suggest the existence of thrombocytopenia by simply noting whether or not a clot of the person's blood retracts, for, as pointed out earlier, clot retraction is normally dependent upon the presence of large numbers of platelets entrapped in the fibrin mesh of the clot.

Most persons with thrombocytopenia have the disease known as *idiopathic thrombocytopenia,* which means simply "thrombocytopenia of unknown cause." However, in the past few years it has been discovered that in most of these persons specific antibodies are destroying the platelets. Occasionally these have developed because of transfusions from other persons, but usually they result from development of autoimmunity to the person's own platelets, the cause of which, however, is not known.

Relief from bleeding for 1 to 4 days can often be effected in the thrombocytopenic patient by giving *fresh whole blood transfusions.* Also, *splenectomy* is usually very helpful, sometimes effecting almost a complete cure, because the spleen removes large numbers of platelets, particularly damaged ones, from the blood.

● THROMBOEMBOLIC CONDITIONS IN THE HUMAN BEING

Thrombi and Emboli. An abnormal clot that develops in a blood vessel is called a *thrombus.* Once a clot has developed, continued flow of blood past the clot is likely to break it away from its attachment, and such freely flowing clots are known as *emboli.* Emboli generally do not stop flowing until they come to a narrow point in the circulatory system. Thus, emboli originating in large arteries or in the left side of the heart eventually plug either smaller systemic arteries or arterioles—in the brain, the kidneys, or elsewhere. On the other hand, emboli originating in the venous system and in the right side of the heart flow into the vessels of the lung to cause pulmonary arterial embolism.

Causes of Thromboembolic Conditions. The causes of thromboembolic conditions in the human being are usually twofold: First, any *roughened endothelial surface of a vessel*—as may be caused by arteriosclerosis, infection, or trauma—is likely to initiate the clotting process. Second, blood often clots *when it flows very slowly* through blood vessels, for small quantities of thrombin and other procoagulants are always being formed. These are generally removed from the blood by the macrophage system, mainly the Kupffer cells of the liver. If the blood is flowing too slowly, the concentrations of the procoagulants in local areas often rise high enough to initiate clotting, but when the blood flows rapidly these are rapidly mixed with large quantities of blood and are removed during passage through the liver.

Use of Tissue Plasminogen Activator in Treating Intravascular Clots. Genetically engineered tissue plasminogen activator (TPA) is now available. It is very effective in dissolving intravascular clots when delivered directly to a thrombosed area through a catheter. For instance, if used within the first hour or so after thrombotic occlusion of a coronary artery, the heart is often spared serious damage.

FEMORAL THROMBOSIS AND MASSIVE PULMONARY EMBOLISM

Because clotting almost always occurs when blood flow is blocked for many hours in any vessel of the body, the immobility of bed patients plus the practice of propping the knees with underlying pillows often causes intravascular clotting because of blood stasis in one or more of the leg veins for hours at a time. Then the clot grows, mainly in the direction of the slowly moving blood, sometimes growing the entire length of the leg veins and occasionally even up into the common iliac vein and inferior vena cava. Then, about one time out of every ten, a large part of the clot disengages from its attachments to the vessel wall and flows freely with the venous blood into the right side of the heart and thence into the pulmonary arteries to cause *massive pulmonary embolism*. If the clot is large enough to occlude both the pulmonary arteries, immediate death ensues. If only one pulmonary artery or a smaller branch is blocked, death may not occur, or the embolism may lead to death a few hours to several days later because of further growth of the clot within the pulmonary vessels. But, again, TPA therapy can be a lifesaver.

DISSEMINATED INTRAVASCULAR COAGULATION

Occasionally, the clotting mechanism becomes activated in widespread areas of the circulation, giving rise to the condition called *disseminated intravascular coagulation*. This often results from the presence of large amounts of traumatized or dying tissue in the body that releases tissue thromboplastin into the blood. Frequently, the clots are small but numerous, and they plug a large share of the small peripheral blood vessels. This occurs especially in septicemic shock, in which either circulating bacteria or bacterial toxins—especially *endotoxins*—activate the clotting mechanisms. The plugging of the small peripheral vessels greatly diminishes the delivery of oxygen and other nutrients to the tissues—a situation that exacerbates the shock picture. It is partly for this reason that septicemic shock is lethal in 85 per cent or more of patients.

A peculiar effect of disseminated intravascular coagulation is that the patient frequently begins to bleed. The reason for this is that so many of the clotting factors are removed by the widespread clotting that too few procoagulants remain to allow normal hemostasis of the remaining blood.

● ANTICOAGULANTS FOR CLINICAL USE

In some thromboembolic conditions it is desirable to delay the coagulation process. Therefore, various anticoagulants have been developed for treatment of these conditions. The ones most useful clinically are heparin and the coumarins.

HEPARIN AS AN INTRAVENOUS ANTICOAGULANT

Commercial heparin is extracted from animal tissues from several organs and is prepared in almost pure form. Injection of relatively small quantities, approximately 0.5 to 1 mg per kilogram of body weight, causes the blood clotting time to increase from a normal of approximately 6 minutes to 30 or more minutes. Furthermore, this change in clotting time occurs instantaneously, thereby immediately preventing further development of the thromboembolic condition.

The action of heparin lasts approximately three to four hours. The injected heparin is destroyed by an enzyme in the blood known as *heparinase*.

In the treatment of a patient with heparin, too much heparin is sometimes given, and *serious* bleeding crises occur. In these instances, *protamine* acts specifically as an antiheparin, and the clotting mechanism can be reverted to normal by administering this substance. Protamine combines with heparin and inactivates it because it carries strong positive electrical charges, whereas heparin carries strong negative charges.

COUMARINS AS ANTICOAGULANTS

When a coumarin, such as *warfarin,* is given to a patient, the plasma levels of prothrombin and Factors VII, IX, and X, all formed by the liver, begin to fall, indicating that warfarin has a potent depressant effect on liver formation of all these compounds. Warfarin causes this effect by competing with vitamin K for reactive sites in the enzymatic processes for formation of prothrombin and the other three clotting factors, thereby blocking the action of vitamin K.

After administration of an effective dose of warfarin, the coagulant activity of the blood decreases to approximately 50 per cent of normal by the end of 12 hours and to approximately 20 per cent of normal by the end of 24 hours. In other words, the coagulation process is not blocked immediately, but must await the consumption of the prothrombin and other factors already present in the plasma. Normal coagulation returns one to three days after discontinuing therapy.

PREVENTION OF BLOOD COAGULATION OUTSIDE THE BODY

Although blood removed from the body and held in a glass test tube normally clots in about six minutes, blood collected in *siliconized containers* often does not clot for as long as an hour or more. The reason for this delay is that preparing the surfaces of the containers with silicone prevents contact activation of platelets as well as Factor XII, which initiate the intrinsic clotting mechanism. On the other hand, untreated glass containers allow contact activation and rapid development of clots.

Heparin can be used for preventing coagulation of blood outside the body as well as in the body. Heparin is especially used in all surgical procedures in which the blood is passed through a heart-lung machine or artificial kidney and then back into the person.

Various substances that *decrease the concentration of calcium ions* in the blood can be used for preventing blood coagulation outside the body. For instance, soluble *oxalate* compounds mixed in very small quantity with a sample of blood cause precipitation of calcium oxalate from the plasma and thereby decrease the ionic calcium levels so much that blood coagulation is blocked.

A second calcium deionizing agent used for preventing coagulation is *sodium, ammonium,* or *potassium citrate.* The citrate ion combines with calcium in the blood to cause an un-ionized calcium compound, and the lack of ionic calcium prevents coagulation. Citrate anticoagulants have a very important advantage over the oxalate anticoagulants, for oxalate is toxic to the body, whereas moderate quantities of citrate can be injected intravenously. After injection, the citrate ion is removed from the blood within a few minutes by the liver and is polymerized into glucose or metabolized directly for energy. Consequently, 500 ml of blood that has been rendered incoagulable by citrate can ordinarily be injected into a recipient within a few minutes without any dire consequences. If the liver is damaged or if large quantities of citrated blood or plasma are given too rapidly, the citrate ion may not be removed quickly enough, and the citrate can then greatly depress the level of calcium ion in the blood, which can result in tetany and convulsive death.

● BLOOD COAGULATION TESTS

BLEEDING TIME

When a sharp knife is used to pierce the tip of the finger or lobe of the ear, bleeding ordinarily lasts 1 to 6 minutes. However, the time depends largely on the depth of the wound and on the degree of hyperemia in the finger at the time of the test. Lack of several of the clotting factors can prolong the bleeding time, but this is especially prolonged by lack of platelets.

CLOTTING TIME

Many methods have been devised for determining clotting times. The one most widely used is to collect blood in a chemically clean glass test tube and then to tip the tube back and forth approximately every 30 seconds until the blood has clotted. By this method, the normal clotting time is about 6 to 10 minutes.

Procedures using multiple test tubes have been devised for determining clotting time more accurately. However, clotting times are also very dependent on the condition of the glass itself and even on the size of the tube, which makes a high degree of standardization necessary to obtain accurate results. A typical condition that causes a prolonged clotting time is hemophilia, but a deficiency of any of the factors in the intrinsic pathway for clotting can be the culprit.

PROTHROMBIN TIME

The prothrombin time gives an indication of the total quantity of prothrombin in the blood. Figure 36-5 shows the relationship of prothrombin concentration to prothrombin time. The means for determining prothrombin time is the following:

Blood removed from the patient is immediately oxalated so that none of the prothrombin can change into thrombin. Later, a large excess of calcium ion and tissue thromboplastin is suddenly mixed with the oxalated blood. The calcium nullifies the effect of the oxalate, and the tissue thromboplastin activates the prothrombin-to-thrombin reaction by means of the extrinsic clotting pathway. The time required for coagulation to take place is known as the *prothrombin time*. The normal prothrombin time is approximately 12 seconds, but this depends to a certain extent on the exact procedure employed. In each laboratory a curve relating prothrombin concentration to prothrombin time, such as that shown in Figure 36-5, is ordinarily drawn for the method used so that the prothrombin in the blood can be quantitated.

Tests similar to that for prothrombin time have been devised to determine the quantities of other clotting factors in the body. In each of these tests, excesses of calcium ions and all the other factors besides the one being tested are added to oxalated blood all at once, and then the time of coagulation is determined in the same manner as the usual prothrombin time. If the factor is deficient, the time will be considerably prolonged.

REFERENCES

Carpentier, A.: Low molecular weight (LMW) heparin derivatives in experimental extra-corporeal circulation (ECC). Haemostasis, 14:325, 1984.
Chavin, S. I.: Factor VIII: Structure and function in blood clotting. Am. J. Hematol., 16:297, 1984.
Coller, B. S. (ed.): Progress in Hemostasis and Thrombosis, Vol. 9. Philadelphia, W. B. Saunders Co., 1989.
Esmon, C. T.: The regulation of natural anticoagulant pathways. Science, 235:1348, 1987.
Frojmovic, M. M., and Milton, J. G.: Human platelet size, shape, and related functions in health and disease. Physiol. Rev., 62:185, 1982.
Fukada, E., et al.: Methods to study rheological properties of blood during clotting. Biorheology, Suppl., 1:9, 1984.
Gowans, J. L. (in honour of): Blood Cells and Vessel Walls: Functional Interactions. Princeton, N. J., Excerpta Medica, 1980.
Haber, E., et al.: Innovative approaches to plasminogen activator therapy. Science, 243:51, 1989.
Hassan, H. J., et al.: Intragenic Factor IX restriction site polymorphism in hemophilia B variants. Blood, 65:441. 1985.
Hilgartner, M. W., and Pochedly, C.: Hemophilia in the Child and Adult, 3rd Ed. New York, Raven Press, 1989.
Holmsen, H.: Nucleotide metabolism of platelets. Annu. Rev. Physiol., 47:677, 1985.
Kane, K. K.: Fibrinolysis: A review. Ann. Clin. Lab. Sci., 14:443, 1984.
Kline, D. L., and Reddy, K. N. N. (eds.): Fibrinolysis. Boca Raton, Fla., CRC Press, 1980.
Kumada, T., and Abika, Y.: Purification and some properties of rat alpha 2-plasmin inhibitor. Thromb. Res., 36:143, 1984.
Mann, K. G.: The biochemistry of coagulation. Clin. Lab. Med., 4:207, 1984.
Miller, D. R., et al. (eds.): Blood Diseases of Infancy and Childhood. St. Louis, C. V. Mosby Co., 1989.
Murano, G., and Bick, R. L. (eds): Basic Concepts of Hemostasis and Thrombosis: Clinical Laboratory Evaluation of Thrombohemorrhagic Phenomena. Boca Raton, Fla., CRC Press, 1980.
Mustard, J. F., et al.: Prostaglandins and platelets. Annu. Rev. Med., 31:89, 1980.
Nemerson, Y., and Nossel, H. L.: The biology of thrombosis. Annu. Rev. Med., 33:479, 1982.
Olson, R. E.: The function and metabolism of vitamin K. Annu. Rev. Nutr., 4:281, 1984.
Renck, H.: Bleeding and Thrombotic Disorders in Surgical Patients. East Norwalk, Conn., Appleton & Lange, 1988.
Rodgers, G. M.: Hemostatic properties of normal and perturbed vascular cells. FASEB J., 2:116, 1988.
Rosenberg, R. D.: Role of heparin and heparinlike molecules in thrombosis and athersclerosis. Fed. Proc., 44:404, 1985.
Siess, W.: Molecular mechanisms of platelet activation. Physiol. Rev., 69:58, 1989.
Sigel, B., et al.: Ultrasonic detection of red cell aggregation immediately preceding blood clotting. Invest. Radiol., 19:458, 1984.
Wall, R. T., and Harker, L. A.: The endothelium and thrombosis. Annu. Rev. Med., 31:361, 1980.
Williams, W. J., et al.: Hematology, 4th Ed. New York, McGraw-Hill Book Co., 1990.
Zwaal, R. F. A.: Scrambling membrane phospholipids and local control of blood clotting. News Physiol. Sci., 3:57, 1988.

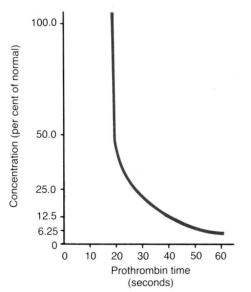

Figure 36-5. Relationship of prothrombin concentration in the blood to the prothrombin time.

UNIT VII

RESPIRATION

CHAPTER 37

• *Pulmonary Ventilation* •

The obvious goal of respiration is to provide oxygen to the tissues and to remove carbon dioxide. To achieve this, respiration can be divided into four major functional events: (1) pulmonary ventilation, which means the inflow and outflow of air between the atmosphere and the lung alveoli, (2) diffusion of oxygen and carbon dioxide between the alveoli and the blood, (3) transport of oxygen and carbon dioxide in the blood and body fluids to and from the cells, and (4) regulation of ventilation and other facets of respiration. The present chapter is a discussion of pulmonary ventilation, and the subsequent five chapters cover the other respiratory functions as well as the physiology of special respiratory problems.

● MECHANICS OF PULMONARY VENTILATION

THE MUSCLES THAT CAUSE LUNG EXPANSION AND CONTRACTION

The lungs can be expanded and contracted in two ways: (1) by downward and upward movement of the diaphragm to lengthen or shorten the chest cavity, and (2) by elevation and depression of the ribs to increase and decrease the anteroposterior diameter of the chest cavity. Figure 37–1 illustrates these two methods.

Normal quiet breathing is accomplished almost entirely by the first of the above two methods, that is, by movement of the diaphragm. During inspiration, contraction of the diaphragm pulls the lower surfaces of the lungs downward. Then, during expiration, the diaphragm simply relaxes; and the *elastic recoil* of the lungs, chest wall, and abdominal structures compresses the lungs. During heavy breathing, however, the elastic forces are not powerful enough to cause the necessary rapid expiration, so that the extra required force is achieved mainly by contraction of the *abdominal muscles,* which pushes the abdominal contents upward against the bottom of the diaphragm.

The second method for expanding the lungs is to raise the rib cage. This expands the lungs because, in the natural resting position, the ribs slant downward,

thus allowing the sternum to fall backward toward the spinal column. But when the rib cage is elevated, the ribs then project almost directly forward so that the sternum now also moves forward, away from the spine, making the anteroposterior thickness of the chest about 20 per cent greater during maximum inspiration than during expiration. Therefore, those muscles that elevate the chest cage can be classified as muscles of inspiration; and those muscles that depress the chest cage, as muscles of expiration. The most important muscles that raise the rib cage are the *external intercostals,* but others that help slightly are (1) the *sternocleidomastoid* muscles, which lift upward on the sternum; (2) the *anterior serrati,* which lift many of the ribs; and (3) the *scaleni,* which lift the first two ribs.

The muscles that pull the rib cage downward during expiration are (1) the *abdominal recti,* which have the powerful effect of pulling downward on the lower ribs at the same time that they and the other abdominal muscles also compress the abdominal contents upward toward the diaphragm, and (2) the *internal intercostals.*

Figure 37–1 also illustrates the mechanism by which the external and internal intercostals act to cause inspiration and expiration. To the left, the ribs during expiration are angled downward and the external intercostals are elongated in a forward and downward direction. As they contract, they pull the upper ribs forward in relation to the lower ribs, and this causes leverage on the ribs to raise them upward, thereby causing inspiration. The internal intercostals function exactly oppositely, functioning as expiratory muscles, for they angle between the ribs in the opposite direction and cause opposite leverage.

MOVEMENT OF AIR IN AND OUT OF THE LUNGS — AND THE PRESSURES THAT CAUSE IT

The lung is an elastic structure that will collapse like a balloon and expel all of its air through the trachea whenever there is no force to keep it inflated. Also, there are no attachments between the lung and the walls of the chest cage except where it is sus-

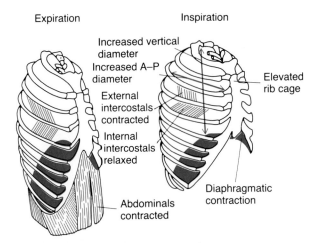

Expiration   Inspiration

Increased vertical diameter

Increased A–P diameter

Elevated rib cage

External intercostals contracted

Internal intercostals relaxed

Diaphragmatic contraction

Abdominals contracted

Figure 37–1. Expansion and contraction of the thoracic cage during expiration and inspiration, illustrating especially diaphragmatic contraction, elevation of the rib cage, and function of the intercostals.

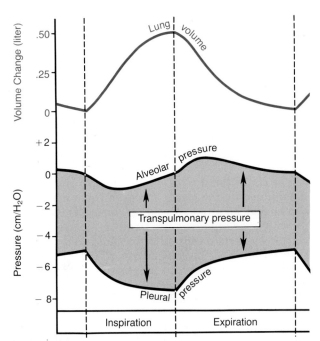

Figure 37–2. Changes in lung volume, alveolar pressure, pleural pressure, and transpulmonary pressure during normal breathing.

pended at its hilum from the mediastinum. Instead, the lung literally floats in the thoracic cavity, surrounded by a very thin layer of *pleural fluid* that lubricates the movements of the lungs within the cavity. Furthermore, continual pumping of this fluid into lymphatic channels maintains a slight suction between the visceral surface of the lung pleura and the parietal pleural surface of the thoracic cavity. Therefore, the two lungs are held to the thoracic wall as if glued there, except that they can slide freely, well lubricated, as the chest expands and contracts.

The Pleural Pressure and Its Changes During Respiration

Pleural pressure is the pressure in the narrow space between the lung pleura and the chest wall pleura. As noted above, this is normally a slight suction, which means a slightly negative pressure. The normal pleural pressure at the beginning of inspiration is approximately −5 centimeters of water, which is the amount of suction that is required to hold the lungs open to their resting level. Then, during normal inspiration, the expansion of the chest cage pulls the surface of the lungs with still greater force and creates a still more negative pressure down to an average of about −7.5 centimeters of water.

These relationships between pleural pressure and changing lung volume are illustrated in Figure 37–2, showing in the bottom panel the increasing negativity of the pleural pressure from −5 to −7.5 during inspiration and in the top panel an increase in lung volume of 0.5 liter. Then, during expiration, the events are essentially reversed.

Alveolar Pressure

Alveolar pressure is the pressure inside the lung alveoli. When the glottis is open and no air is flowing into or out of the lungs, the pressures in all parts of the

respiratory tree, all the way to the alveoli, are all exactly equal to atmospheric pressure, which is considered to be 0 centimeter water pressure. To cause inward flow of air during inspiration, the pressure in the alveoli must fall to a value slightly below atmospheric pressure. The second panel of Figure 37–2 illustrates in normal inspiration a decrease in alveolar pressure to about −1 centimeter of water. This very slight negative pressure is enough to move about 0.5 liter of air into the lungs in the 2 seconds required for inspiration.

During expiration, opposite changes occur: the alveolar pressure rises to about +1 centimeter of water, and this forces the 0.5 liter of inspired air out of the lungs during the 2 to 3 seconds of expiration.

The Transpulmonary Pressure. Finally, note in the third panel of Figure 37–2 the pressure difference between the alveolar pressure and the pleural pressure. This is called the *transpulmonary pressure.* It is the pressure difference between the alveoli and the outer surfaces of the lungs, and it is actually a measure of the elastic forces in the lungs that tend to collapse the lungs at each point of expansion, called the *recoil pressure.*

Compliance of the Lungs

The extent to which the lungs expand for each unit increase in transpulmonary pressure is called their *compliance.* The normal total compliance of both lungs together in the average adult human being is approximately 200 ml/cm of water pressure. That is, every time the transpulmonary pressure increases by 1 centimeter of water, the lungs expand 200 milliliters.

The Compliance Diagram of the Lungs. Figure 37–3 is a diagram relating lung volume changes to changes in transpulmonary pressure. Note that the relationship is different for inspiration and expiration. Each curve is recorded by changing the transpulmonary pressure in small steps and allowing the lung volume to come to a steady level between successive steps. The two curves are called the *inspiratory compliance curve* and the *expiratory compliance curve,* and the entire diagram is called the *compliance diagram of the lungs.*

The characteristics of the compliance diagram are determined by the elastic forces of the lungs. These can be divided into two separate parts: (1) the *elastic forces of the lung tissue* itself and (2) the *elastic force caused by surface tension of the fluid that lines the inside walls of the alveoli* and other lung air spaces.

The elastic forces of the lung tissue are determined mainly by the elastin and collagen fibers interwoven among the lung parenchyma. In the deflated lungs, these fibers are in a partially contracted and kinked state, and then when the lungs are expanded, the fibers are partly stretched and partly unkinked, thereby elongating.

The elastic forces caused by surface tension are much more complex. However, it accounts for about two thirds of the total elastic forces in the normal lungs. The significance of surface tension is illustrated in Figure 37–4, in which are compared the compliance of the lungs when filled with air and that when filled with saline. When filled with air, there is an interface between the alveolar lining fluid and the air in the alveoli. In the case of the saline-filled lung, there is no air-fluid interface, and therefore the surface tension effect is not present — only tissue elastic forces are operative in the saline-filled lung.

Note that the transpulmonary pressures required to expand the air-filled lungs are about three times as

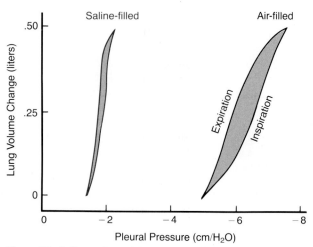

Figure 37–4. Comparison of the compliance diagrams of air-filled and saline-filled lungs.

great as the pressures required to expand the saline-filled lungs. Thus, one can conclude that the tissue elastic forces tending to cause collapse of the air-filled lung represent only about one third of the total lung elasticity, whereas the surface tension forces represent about two thirds.

The surface tension elastic forces of the lungs also change tremendously when the substance called "surfactant" is not present in the alveolar fluid. Therefore, let us discuss surfactant and its relationship to the surface tension forces.

"Surfactant," Surface Tension, and Collapse of the Lungs

The Principle of Surface Tension. When water forms a surface with air, the water molecules on the surface of the water have an extra strong attraction for each other. As a result, the water surface is always attempting to contract. This is what holds raindrops together; that is, there is a tight contractile membrane of water molecules around the entire surface of the raindrop. Now let us reverse these principles and see what happens on the inner surfaces of the alveoli and other air spaces. Here, the water surface is also attempting to contract, but this time the surface of the water lining the alveoli surrounds the alveolar air, always attempting to contract like a balloon. Obviously, this attempts to force the air out of the alveoli through the bronchi, and in doing so it causes the alveoli (and other air spaces in the lungs) to attempt to collapse. Since this occurs in all of the air spaces of the lungs, the net effect is to cause an elastic contractile force of the entire lungs, which is called the *surface tension elastic force.*

"Surfactant" and Its Effect on Surface Tension. Surfactant is a *surface active agent,* which means that when it spreads over the surface of a fluid it greatly reduces the surface tension. It is secreted by special surfactant-secreting epithelial cells that comprise about 10 per cent of the surface area of the al-

Figure 37–3. Compliance diagram in a normal person. This diagram shows the compliance of the lungs alone.

veoli. These cells are granular in nature, containing lipid inclusions; they are called *type II alveolar epithelial cells.*

Surfactant is a complex mixture of several phospholipids, proteins, and ions. The three most important components are the phospholipid *dipalmitoyl lecithin, surfactant apoproteins,* and *calcium ions.* The dipalmitoyl lecithin, along with several less important phospholipids, is responsible for reducing the surface tension. These will not dissolve in the fluid; instead, they spread over its surface, because one portion of each phospholipid molecule is hydrophilic and dissolves in the water lining the alveoli, whereas the lipid portion of the molecule is hydrophobic and is oriented toward the air, forming a lipid hydrophobic surface exposed to the air. This surface has from one-twelfth to one-half the surface tension of a pure water surface. The importance of the surfactant apoproteins and the calcium ions is that in their absence the dipalmitoyl lecithin spreads so slowly over the fluid surface that it cannot function effectively.

In quantitative terms, the surface tension of different watery fluids is approximately the following: pure water, 70 dynes/cm; normal fluids lining the alveoli but without surfactant, 50 dynes/cm; fluids lining the alveoli with surfactant included, between 5 and 30 dynes/cm.

The Collapse Pressure of Occluded Alveoli Caused by Surface Tension. If the air passages leading from the air spaces of the lungs are blocked, the surface tension tending to cause collapse of the spaces will create a positive pressure in the alveoli, attempting to push the air out. The amount of pressure generated in this way in a spherical air space can be calculated from the following formula:

$$\text{Pressure} = \frac{2 \times \text{Surface Tension}}{\text{Radius}}$$

For the average-sized alveolus with a radius of about 100 micrometers and lined with normal surfactant, this calculates to be about 4 centimeters of water pressure (3 mm Hg). However, if the alveoli were lined with pure water, it would calculate to be about 18 centimeters of water pressure. Thus, one sees how important surfactant is in reducing the amount of transpulmonary pressure required to keep the lungs expanded.

Effect of Size of the Alveoli on the Collapse Pressure Caused by Surface Tension. Note from the above formula that the collapse pressure generated in the alveoli is *inversely* affected by the radius of the alveolus, which means that the smaller the alveolus the greater becomes the collapse pressure. Thus, when the alveoli have half the normal radius, only 50 instead of 100 micrometers, the collapse pressures noted above are exactly doubled. This is specifically significant in very small premature babies who often have alveoli with radii less than one-quarter normal. Furthermore, surfactant does not normally begin to be secreted into the alveoli until between the sixth and seventh month of gestation and in some babies even later than that, so that many premature babies have little or no surfactant in the alveoli. Therefore, the lungs of these babies have extreme collapse tendencies, sometimes as much as 30 mm Hg or more, and this causes the condition called *respiratory distress syndrome of the newborn.* It is fatal if not treated with strong measures.

Role of Surfactant, "Interdependence," and Lung Fibrous Tissue in "Stabilizing" the Sizes of the Alveoli. Now, let us see what would happen if many of the alveoli in the lung were very small and others very large. The collapse tendency of the smaller alveoli would be far greater than that of the larger ones. Therefore, the smaller alveoli theoretically would tend to collapse, decreasing their volume in the lungs; and this loss of volume in a portion of the lungs would cause expansion of the larger alveoli. Then, as the smaller alveoli became still smaller, their collapse tendency would become even greater, whereas the collapse tendency of the expanding larger alveoli would become less. Therefore, theoretically, all of the smaller alveoli would collapse, and this would pull the larger alveoli to still larger sizes. This phenomenon is called *instability of the alveoli.*

However, in practice, this phenomenon of instability of the alveoli does not occur in the normal lung, though it does occur under special conditions such as when there is very little surfactant in the alveolar fluid and the lung volume is decreased at the same time. There are several reasons why instability does not occur in the normal lung. One of these is the phenomenon called *interdependence* between the adjacent alveoli, alveolar ducts, and other air spaces. That is, each of these spaces splints the other in such a way that a large alveolus usually cannot exist adjacent to a small alveolus because they share common septal walls. This is the interdependence phenomenon.

A second reason why instability does not occur is that the lung is constructed of about 50,000 functional units, each containing one or a few alveolar ducts and their associated alveoli. All of these are surrounded by fibrous septa that penetrate from the lung surface into the lung parenchyma. This fibrous tissue acts as an additional split.

Finally, one must not forget the role of surfactant in opposing instability. It does this in two separate ways: First, it reduces the total amount of surface tension, and this allows the interdependence phenomenon and the fibrous tissue to overcome the surface tension effects. Second, as an alveolus becomes smaller, the surfactant molecules on the alveolar surface become squeezed together, increasing their concentration, and this reduces the surface tension still further. Thus, the smaller the alveolus becomes, the less becomes the surface tension, which opposes the extra tendency of smaller alveoli to collapse. Conversely, as alveoli become larger, the surface concentration of surfactant decreases, and the surface tension becomes greater; and this opposes further enlargement of these already large alveoli.

EFFECT OF THE THORACIC CAGE ON LUNG EXPANSIBILITY

Thus far we have discussed the expansibility of the lungs alone without considering the thoracic cage. However, the thoracic cage also has its own elastic and viscous characteristics, similar to those of the lungs; and even if the lungs were not present in the thorax, considerable muscular effort would still be required to expand the thoracic cage.

Compliance of the Thorax and the Lungs Together

The compliance of the entire pulmonary system (the lungs and the thoracic cage together) is measured while expanding the lungs of a totally relaxed or paralyzed person. To do this, air is forced into the lungs a little at a time while recording the lung pressures and volumes. It is found that to respire this total pulmonary system almost twice as much pressure is needed as when respiring the same lungs after removal from the chest cage. Therefore, the compliance of the combined lung-thorax system is only slightly greater than one-half that of the lungs alone — 110 milliliters of volume per centimeter of water for the combined system, compared with 200 ml/cm for the lungs alone. Furthermore, when the lungs are expanded to very high volumes or compressed to very low volumes, the limitations of the chest become extreme; when near these limits, the compliance of the combined lung-thorax system can be as little as one-fifth that of the lungs alone.

THE "WORK" OF BREATHING

We have already pointed out that during normal quiet respiration respiratory muscle contraction occurs only during inspiration, whereas expiration is entirely a passive process caused by elastic recoil of the lung and chest cage structures. Thus, the respiratory muscles normally perform "work" only to cause inspiration and not at all to cause expiration.

The work of inspiration can be divided into three different fractions: (1) that required to expand the lungs against its elastic forces, called *compliance work* or *elastic work;* (2) that required to overcome the viscosity of the lung and chest wall structures, called *tissue resistance work;* and (3) that required to overcome airway resistance during the movement of air into the lungs, called *airway resistance work.* These three different types of work are illustrated graphically in Figure 37–5. In this diagram the curve labeled "inspiration" illustrates the progressive *change in pleural pressure* and *lung volume* during inspiration, and the total shaded area of the figure represents the total work per-

formed on the lungs by the inspiratory muscles during the act of inspiration. The shaded area, in turn, is divided into three different segments representing the three different types of work performed during inspiration. These can be explained as follows:

Compliance Work. The dotted area represents the compliance work that is required to expand the lungs against the elastic forces. This can be calculated by multiplying the volume of expansion times the average pressure required to cause the expansion, and this is equal to the area represented by the dots. That is,

$$\text{Compliance work} = \frac{\Delta V \cdot \Delta P}{2}$$

where ΔV is the increase in volume, and ΔP is the increase in intrapleural pressure.

Tissue Resistance Work. The area represented by the vertical bars is proportional to the work expenditure required to overcome the viscosity of the lungs and chest cage.

Airway Resistance Work. Finally, the area in Figure 37–5 illustrated by the horizontal bars represents the work required to overcome the resistance to airflow through the respiratory passageways.

Additional Work Required to Expand and Contract the Thoracic Cage. The work of breathing calculated in Figure 37–5 applies only to the lungs, not the thoracic cage. However, we have seen that the compliance of the total lung-thorax system is only slightly over half that of the lungs alone. Therefore, almost two times as much energy is required for normal expansion and contraction of the total lung-thorax system as for expansion of the lungs alone.

Comparison of the Different Types of Work. It is clear from Figure 37–5 that during normal quiet breathing most of the work performed by the respiratory muscles is used simply to expand the lungs. Normally only a small percentage of the total work is used to overcome tissue resistance (tissue viscosity), and somewhat more is used to overcome airway resistance. On the other hand, during very heavy breathing, when air must flow through the respiratory passageways at very high velocity, the greater proportion of the work is then used to overcome airway resistance.

In pulmonary disease, all three of the different types of work are frequently vastly increased. Compliance work and tissue resistance work are especially increased by diseases that cause fibrosis of the lungs, and airway resistance work is especially increased by diseases that obstruct the airways.

During normal quiet respiration, no muscle "work" is performed during expiration because expiration results from elastic recoil of the lungs and chest. But in heavy breathing or when airway resistance and tissue resistance are great, then expiratory work does of course occur and sometimes becomes even greater than inspiratory work. This is especially true in asthma, which often increases airway resistance manyfold during expiration but much less so during inspiration for reasons that will be explained later.

Energy Required for Respiration. During normal quiet respiration, only 3 to 5 per cent of the total energy expended by the body is required to energize the pulmonary ventilatory process. But during very heavy exercise, the amount of energy required can increase as much as 50-fold, especially if the person has any degree of increased airway resistance or decreased pulmonary compliance. Therefore, one of the major limitations in the intensity of exercise that a person can perform is his ability to provide muscle energy for the respiratory process alone.

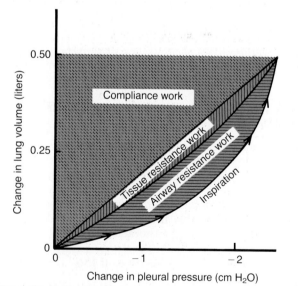

Figure 37–5. Graphical representation of the three different types of work accomplished during inspiration: (1) compliance work, (2) tissue resistance work, and (3) airway resistance work.

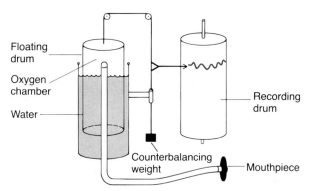

Figure 37–6. A spirometer.

● THE PULMONARY VOLUMES AND CAPACITIES

RECORDING CHANGES IN PULMONARY VOLUME — SPIROMETRY

A simple method for studying pulmonary ventilation is to record the volume movement of air into and out of the lungs, a process called *spirometry*. A typical spirometer is illustrated in Figure 37–6. This consists of a drum inverted over a chamber of water, with the drum counterbalanced by a weight. In the drum is a breathing mixture of gases, usually air or oxygen; a tube connects the mouth with the gas chamber. When one breathes in and out of the chamber the drum rises and falls, and an appropriate recording is made on a moving sheet of paper.

Figure 37–7 illustrates a spirogram showing changes in lung volume under different conditions of breathing. For ease in describing the events of pulmonary ventilation, the air in the lungs has been subdivided at different points on this diagram into four different *volumes* and four different *capacities*, which are the following:

THE PULMONARY "VOLUMES"

To the left in Figure 37–7 are listed four different pulmonary lung "volumes," which, when added together, equal the maximum volume to which the lungs can be expanded. The significance of each of these volumes is the following:

1. The *tidal volume* is the volume of air inspired or expired with each normal breath; it amounts to about 500 milliliters in the average young adult man.

2. The *inspiratory reserve volume* is the extra volume of air that can be inspired over and beyond the normal tidal volume; it is usually equal to approximately 3000 milliliters.

3. The *expiratory reserve volume* is the extra amount of air that can be expired by forceful expiration after the end of a normal tidal expiration; this normally amounts to about 1100 milliliters.

4. The *residual volume* is the volume of air still remaining in the lungs after the most forceful expiration. This volume averages about 1200 milliliters.

THE PULMONARY "CAPACITIES"

In describing events in the pulmonary cycle, it is sometimes desirable to consider two or more of the above volumes together. Such combinations are called *pulmonary capacities*. To the right in Figure 37–7 are listed the different pulmonary capacities, which can be described as follows:

1. The *inspiratory capacity* equals the *tidal volume* plus the *inspiratory reserve volume*. This is the amount of air (about 3500 milliliters) that a person can breathe beginning at the normal expiratory level and distending the lungs to the maximum amount.

2. The *functional residual capacity* equals the *expiratory reserve volume* plus the *residual volume*. This is the amount of air remaining in the lungs at the end of normal expiration (about 2300 milliliters).

3. The *vital capacity* equals the *inspiratory reserve*

Figure 37–7. Diagram showing respiratory excursions during normal breathing and during maximal inspiration and maximal expiration.

volume plus the *tidal volume* plus the *expiratory reserve volume.* This is the maximum amount of air that a person can expel from the lungs after first filling the lungs to their maximum extent and then expiring to the maximum extent (about 4600 milliliters).

4. The *total lung capacity* is the maximum volume to which the lungs can be expanded with the greatest possible inspiratory effort (about 5800 milliliters); it is equal to the vital capacity plus the residual volume.

All pulmonary volumes and capacities are about 20 to 25 per cent less in women than in men, and they obviously are greater in large and athletic persons than in small and asthenic persons.

ABBREVIATIONS AND SYMBOLS USED IN PULMONARY FUNCTION STUDIES

Spirometry is only one of many measurement procedures that the pulmonary physician uses daily. Furthermore, we shall see in subsequent discussions that many of the measurement procedures depend heavily on mathematical computations. To simplify these calculations as well as the presentation of pulmonary function data, a number of abbreviations and symbols have become standardized. The more important of these are given in Table 37–1. Using these symbols, we present here a few simple algebraic exercises showing some of the interrelationships among the pulmonary volumes and capacities; the student should think through and verify these interrelationships:

$$VC = IRV + V_T + ERV$$
$$VC = IC + ERV$$
$$TLC = VC + RV$$
$$TLC = IC + FRC$$
$$FRC = ERV + RV$$

DETERMINATION OF FUNCTIONAL RESIDUAL CAPACITY—THE HELIUM DILUTION METHOD

The functional residual capacity, which is the volume of air that normally remains in the lungs between breaths, is very important to lung function. Its value changes markedly in some types of pulmonary disease, for which reason it is often desirable to measure this capacity. Unfortunately, the spirometer cannot be used in a direct way to measure the functional residual capacity, because the air in the residual volume of the lungs cannot be expired into the spirometer, and this volume comprises about half of the functional residual capacity. Therefore, to measure functional residual capacity, the spirometer must be used in an indirect manner, usually by means of a helium dilution method, as follows:

A spirometer of known volume is filled with air mixed with helium at a known concentration. Before breathing from the spirometer, the person expires normally. At the end of this expiration the remaining volume in the lungs is exactly equal to the functional residual capacity. At this point the subject immediately begins to breathe from the spirometer, and the gases of the spirometer begin to mix with the gases of the lungs. As a result, the helium becomes diluted by the functional residual capacity gases, and the volume of the functional residual capacity can then be calculated from the degree of dilution of the helium with the use of the following formula:

$$FRC = \left(\frac{Ci_{He}}{Cf_{He}} - 1\right) Vi_{Spir}$$

in which

FRC is *functional residual capacity*
Ci_{He} is *initial concentration of helium in the spirometer*
Cf_{He} is *final concentration of helium in the spirometer*
Vi_{Spir} is *initial volume of the spirometer*

Table 37–1 LIST OF ABBREVIATIONS AND SYMBOLS FOR PULMONARY FUNCTION

V_T	tidal volume	P_B	atmospheric pressure
FRC	functional residual capacity	Palv	alveolar pressure
ERV	expiratory reserve volume	Ppl	pleural pressure
RV	residual volume	PO_2	partial pressure of oxygen
IC	inspiratory capacity	PCO_2	partial pressure of carbon dioxide
IRV	inspiratory reserve volume	PN_2	partial pressure of nitrogen
TLC	total lung capacity	Pa_{O_2}	partial pressure of oxygen in arterial blood
VC	vital capacity		
Raw	resistance of tracheobronchial tree to flow of air into the lung	Pa_{CO_2}	partial pressure of carbon dioxide in arterial blood
C	compliance	PA_{O_2}	partial pressure of oxygen in alveolar gas
V_D	volume of dead space gas		
V_A	volume of alveolar gas	PA_{CO_2}	partial pressure of carbon dioxide in alveolar gas
$\dot{V}_I$	inspired volume of ventilation per minute	PA_{H_2O}	partial pressure of water in alveolar gas
$\dot{V}_E$	expired volume of ventilation per minute	R	respiratory exchange ratio
$\dot{V}_A$	alveolar ventilation per minute	$\dot{Q}$	cardiac output
$\dot{V}_{O_2}$	rate of oxygen uptake per minute	$\dot{Q}s$	shunt flow
$\dot{V}_{CO_2}$	amount of carbon dioxide eliminated per minute	Ca_{O_2}	concentration of oxygen in arterial blood
$\dot{V}_{CO}$	rate of carbon monoxide uptake per minute	$C\bar{v}_{O_2}$	concentration of oxygen in mixed venous blood
DL_{O_2}	diffusing capacity of the lung for oxygen	SO_2	percentage saturation of hemoglobin with oxygen
DL_{CO}	diffusing capacity of the lung for carbon monoxide	Sa_{O_2}	percentage saturation of hemoglobin with oxygen in arterial blood

Once the functional residual capacity has been determined, the residual volume can then be determined by subtracting the expiratory reserve volume from the functional residual capacity. Also, the total lung capacity can be determined by adding the inspiratory capacity to the functional residual capacity. That is,

$$RV = FRC - ERV$$

and

$$TLC = FRC + IC$$

● THE MINUTE RESPIRATORY VOLUME — RESPIRATORY RATE TIMES TIDAL VOLUME

The *minute respiratory volume* is the total amount of new air moved into the respiratory passages each minute; this is equal to the *tidal volume* times the *respiratory rate*. The normal tidal volume is about 500 milliliters, and the normal respiratory rate is approximately 12 breaths per minute. Therefore, the *minute respiratory volume averages about 6 liters per minute.* A person can occasionally live for short periods of time with a minute respiratory volume as low as 1.5 liters per minute and with a respiratory rate as low as two to four breaths per minute.

The respiratory rate occasionally rises to as high as 40 to 50 per minute, and the tidal volume can become as great as the vital capacity, about 4600 milliliters in the young adult man. However, at rapid breathing rates a person usually cannot sustain a tidal volume greater than about one-half the vital capacity.

● ALVEOLAR VENTILATION

The ultimate importance of the pulmonary ventilatory system is to renew continually the air in the gas exchange areas of the lungs where the air is in close proximity to the pulmonary blood. These areas include the alveoli, the alveolar sacs, the alveolar ducts, and the respiratory bronchioles. The rate at which new air does reach these areas is called *alveolar ventilation.* Strangely, though, during normal quiet respiration the volume of air in the tidal air is only enough to fill the respiratory passageways down as far as the terminal bronchioles, with only a very small portion of the inspired air actually flowing all the way into the alveoli. Therefore, how does new air move this last short distance from the terminal bronchioles into the alveoli? The answer: by *diffusion.* Diffusion is caused by the kinetic motion of molecules, each gas molecule moving at high velocity among the other molecules. Fortunately, the velocity of movement of the molecules in the respiratory air is so great and the distances so short from the terminal bronchioles to the alveoli that the gases move this remaining distance in only a fraction of a second.

The Dead Space and Its Effect on Alveolar Ventilation

Unfortunately, some of the air that a person breathes never reaches the gas exchange areas but instead goes to fill respiratory passages where gas exchange does not occur. This air is called *dead space air* because it is not useful for the gas exchange process; the respiratory passages where no gas exchange takes place is called the *dead space.*

On expiration, the air in the dead space is expired first before any of the air from the alveoli reaches the atmosphere. Therefore, the dead space is equally disadvantageous for removal of the expiratory gases from the lungs.

Measurement of the Dead Space Volume. A simple method for measuring dead space volume is illustrated by Figure 37-8. In making this measurement the subject suddenly takes a deep breath of oxygen. This, obviously, fills the entire dead space with pure oxygen, and some oxygen also mixes with the alveolar air but does not completely replace this air. Then the person expires through a rapidly recording nitrogen meter, which makes the record shown in the figure. The first portion of the expired air comes from the dead space regions of the respiratory passageways where the air has been completely replaced by oxygen. Therefore, in the early part of the record only oxygen appears, and the nitrogen concentration is exactly zero. Then when alveolar air begins to reach the nitrogen meter, the nitrogen concentration rises rapidly because alveolar air containing large amounts of nitrogen begins to mix with the dead space air. After still more air has been expired, all of the dead space air has been washed from the passages, and then only alveolar air remains. Therefore, the recorded nitrogen concentration reaches a plateau level equal to its concentration in the alveoli as illustrated to the right in the figure. With a little thought, the student can see that the area covered by the dots represents the air that has no nitrogen in it; therefore, this area is a measure of the volume of dead space air. For exact quantification, the following equation is used:

$$V_D = \frac{\text{Area of dots} \times V_E}{\text{Area of hatching} + \text{Area of dots}}$$

where V_D is *dead space air,* and V_E is the *total volume of expired air.*

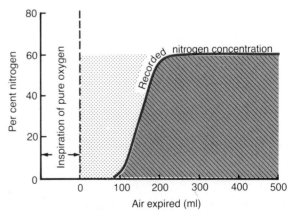

Figure 37-8. Continuous record of the changes in nitrogen concentration in the expired air following a previous inspiration of pure oxygen. This record can be used to calculate dead space, as discussed in the text.

Let us assume, for instance, that the area of the dots on the graph is 30 square centimeters, the area of the hatching is 70 square centimeters, and the total volume expired is 500 milliliters. The dead space then would be

$$\frac{30}{30 + 70} \times 500, \text{ or } 150 \text{ ml}$$

Normal Dead Space Volume. The normal dead space air in the young adult is about 150 milliliters. This increases slightly with age.

Anatomic Versus Physiologic Dead Space. The method just described for measuring the dead space measures the volume of all the space of the respiratory system besides the gas exchange areas, and this is called the *anatomic dead space.* On occasion, however, some of the alveoli themselves are nonfunctional or are only partially functional because of absent or poor blood flow through adjacent pulmonary capillaries. Therefore, from a functional point of view, these alveoli must also be considered to be dead space. When the alveolar dead space is included in the total measurement of dead space, this is then called *physiologic dead space,* in contradistinction to the anatomic dead space. In the normal person, the anatomic and the physiologic dead spaces are nearly equal because all alveoli are functional in the normal lung, but in persons with partially functional or nonfunctional alveoli in some parts of the lungs, the physiologic dead space is sometimes as much as ten times the anatomic dead space, or as much as 1 to 2 liters. These problems will be discussed further in Chapter 39 in relation to pulmonary gaseous exchange and in Chapter 42 in relation to certain pulmonary diseases.

RATE OF ALVEOLAR VENTILATION

Alveolar ventilation per minute is the total volume of new air entering the alveoli (and other adjacent gas exchange areas) each minute. It is equal to the respiratory rate times the amount of new air that enters the alveoli with each breath:

$$\dot{V}_A = \text{Freq} \cdot (V_T - V_D)$$

where $\dot{V}_A$ is the *volume of alveolar ventilation per minute,* Freq is the *frequency of respiration per minute,* V_T is the *tidal volume,* and V_D is the *dead space volume.*

Thus, with a normal tidal volume of 500 milliliters, a normal dead space of 150 milliliters, and a respiratory rate of 12 times per minute, alveolar ventilation equals $12 \times (500 - 150)$, or 4200 ml/min.

Alveolar ventilation is one of the major factors determining the concentrations of oxygen and carbon dioxide in the alveoli. Therefore, almost all discussions of gaseous exchange in the following chapters emphasize alveolar ventilation.

● FUNCTIONS OF THE RESPIRATORY PASSAGEWAYS

THE TRACHEA, BRONCHI, AND BRONCHIOLES

Figure 37–9 illustrates the respiratory system, showing especially the respiratory passageways. The air is distributed to the lungs by way of the trachea, bronchi, and bronchioles. The *trachea* is called the *first generation respiratory passageway,* and the two *main right* and *left bronchi* are the *second generation;* each division thereafter is an additional

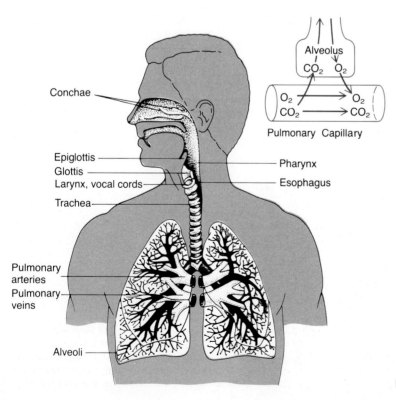

Figure 37–9. The respiratory passages.

generation. There are between 20 and 25 generations before the air finally reaches the alveoli.

One of the most important problems in all the respiratory passageways is to keep them open to allow easy passage of the air to and from the alveoli. To keep the trachea from collapsing, multiple cartilage rings extend about five sixths of the way around the trachea. In the walls of the bronchi are less extensive cartilage plates that also maintain a reasonable amount of rigidity, yet allow sufficient motion for the lungs to expand and contract. These plates become progressively less extensive in the later generations of bronchi and are completely gone in the bronchioles, which have diameters usually less than 1 to 1.5 millimeters. The bronchioles are not prevented from collapsing by any rigidity of their walls. Instead, they are expanded by the same transpulmonary pressures that expand the alveoli. That is, as the alveoli enlarge, so also do the bronchioles enlarge.

The Muscular Wall of the Bronchi and Bronchioles and Its Control. In all areas of the trachea and bronchi not occupied by cartilage plates, the walls are composed mainly of smooth muscle. Also, the walls of the bronchioles are almost entirely smooth muscle, with the exception of the most terminal bronchiole, called the *respiratory bronchiole,* that has only a few smooth muscle fibers. Many obstructive diseases of the lung cause narrowing of the smaller bronchi and the bronchioles, often because of excessive contraction of the smooth muscle itself.

Resistance to Air Flow in the Bronchial Tree. Under *normal respiratory conditions,* air flows through the respiratory passageways so easily that less than 1 centimeter of water pressure gradient from the alveoli to the atmosphere is sufficient to cause enough air flow for quiet breathing. The greatest amount of resistance to air flow occurs not in the very minute air passages of the bronchioles, but instead in some of the larger bronchi near to the trachea. The reason for this high resistance is that there are relatively few of these larger bronchi in comparison with about 65,000 parallel terminal bronchioles, through each of which only a very minute amount of air must pass.

Yet in disease conditions the smaller bronchioles very often do play a far greater role in determining air flow resistance, for two reasons: (1) because of their small size, they are easily occluded; (2) because they have a greater percentage of smooth muscle in their walls, they constrict very easily.

Nervous and Local Control of the Bronchiolar Musculature — Sympathetic Control. Direct control of the bronchioles by sympathetic nerve fibers is relatively weak because few of these fibers penetrate to the central portions of the lung. However, the bronchial tree is very much exposed to circulating norepinephrine and epinephrine released into the blood by sympathetic stimulation of the adrenal medullae. Both of these hormones, especially epinephrine, because of its greater stimulation of *beta receptors,* cause dilatation of the bronchial tree.

Parasympathetic Stimulation. A few parasympathetic nerve fibers derived from the vagus nerves also penetrate the lung parenchyma. These nerves secrete acetylcholine and, when activated, cause mild to moderate constriction of the bronchioles. When a disease process such as asthma has already caused some constriction, parasympathetic superimposed nervous stimulation often worsens the condition. When this occurs, administration of drugs that block the effects of acetylcholine, such as *atropine,* can sometimes relax the respiratory passages enough to relieve the obstruction.

The parasympathetic nerves are sometimes activated by reflexes that originate in the lungs. Most of these begin with irritation of the epithelial membrane of the respiratory passageways themselves, initiated by noxious gases, dust, cigarette smoke, or bronchial infection. Also, a bronchiolar constrictor reflex often occurs when microemboli occlude small pulmonary arteries.

Local Factors Affecting Bronchial Contraction. Several different substances formed in the lungs themselves are often quite active in causing bronchiolar constriction. Two of the most important of these are *histamine* and the substance called *slow reactive substance of anaphylaxis.* Both of these are released in the lung tissues by mast cells during allergic reactions, especially allergic reactions caused by pollen in the air. Therefore, they play key roles in causing the airway obstruction that occurs in allergic asthma; this is especially true of the slow reactive substance of anaphylaxis.

Also, the same irritants that cause parasympathetic vasoconstrictor reflexes of the airways — smoke, dust, sulfur dioxide, and some of the acidic elements in smog — can initiate local, non-nervous reactions that cause obstructive constriction of the airways.

The Mucous Coat of the Respiratory Passageways, and Action of Cilia to Clear the Passageways

All the respiratory passages, from the nose to the terminal bronchioles, are kept moist by a layer of mucus that coats the entire surface. This mucus is secreted partly by individual goblet cells in the epithelial lining of the passages and partly by small submucosal glands. In addition to keeping the surfaces moist, the mucus also traps small particles out of the inspired air and keeps most of these from ever reaching the alveoli. The mucus itself is removed from the passages in the following manner:

The entire surface of the respiratory passages, both in the nose and in the lower passages down as far as the terminal bronchioles, is lined with ciliated epithelium, with about 200 cilia on each epithelial cell. These beat continually at a rate of 10 to 20 times per second by the mechanism explained in Chapter 2, and the direction of their "power stroke" is always toward the pharynx. That is, the cilia in the lower respiratory passages beat upward while those in the nose beat downward. This continual beating causes the coat of mucus to flow slowly, at a velocity of about 1 cm/min, toward the pharynx. Then the mucus and its entrapped particles are either swallowed or coughed to the exterior.

The Cough Reflex

The bronchi and the trachea are so sensitive to light touch that excessive amounts of any foreign matter or any other cause of irritation initiates the cough reflex. The larynx and carina (the point where the trachea divides into the bronchi) are especially sensitive, and the terminal bronchioles and even the alveoli are very sensitive to corrosive chemical stimuli such as sulfur dioxide gas and chlorine. Afferent impulses pass from the respiratory passages mainly through the vagus nerves to the medulla. There, an automatic sequence of events is triggered by the neuronal circuits of the medulla, causing the following effects:

First, about 2.5 liters of air is inspired. Second, the epiglottis closes, and the vocal cords shut tightly to entrap the air within the lungs. Third, the abdominal muscles contract forcefully, pushing against the diaphragm while other expiratory muscles, such as the internal intercostals, also contract forcefully. Consequently, the pressure in the lungs rises to as high as 100 mm Hg or more. Fourth, the vocal cords and the epiglottis suddenly open widely so that air under pressure in the lungs *explodes* outward. Indeed, this air is sometimes expelled at velocities as high as 75 to 100 miles an hour. Furthermore, and very important, the strong compression of the lungs also collapses the bronchi and trachea by causing the noncartilaginous parts of these to invaginate inward so that the exploding air actually passes through *bronchial* and *tracheal slits*. The rapidly moving air usually carries with it any foreign matter that is present in the bronchi or trachea.

The Sneeze Reflex

The sneeze reflex is very much like the cough reflex except that it applies to the nasal passageways instead of to the lower respiratory passages. The initiating stimulus of the sneeze reflex is irritation in the nasal passageways, the afferent impulses passing in the fifth nerve to the medulla, where the reflex is triggered. A series of reactions similar to those for the cough reflex takes place; however, the uvula is depressed so that large amounts of air pass rapidly through the nose, thus helping clear the nasal passages of foreign matter.

RESPIRATORY FUNCTIONS OF THE NOSE

As air passes through the nose, three distinct functions are performed by the nasal cavities: (1) The *air is warmed* by the extensive surfaces of the conchae and septum, a total area of about 160 cm², illustrated in Figure 37–9. (2) The *air is almost completely humidified* even before it passes beyond the nose. (3) The *air is filtered*. These functions together are called the *air conditioning function* of the upper respiratory passageways. Ordinarily, the temperature of the inspired air rises to within 1°F of body temperature and within 2 to 3 per cent of full saturation with water vapor before it reaches the trachea. When a person breathes air through a tube directly into the trachea (as through a tracheostomy), the cooling and especially the drying effect in the lower lung can lead to serious lung crusting and infection.

Filtration Function of the Nose. The hairs at the entrance to the nostrils are important for removing large particles. Much more important, though, is the removal of particles by *turbulent precipitation*. That is, the air passing through the nasal passageways hits many obstructing vanes: the conchae (also called "turbinates" because they cause turbulence of the air), the septum, and the pharyngeal wall. Each time air hits one of these obstructions it must change its direction of movement; the particles suspended in the air, having far more mass and momentum than air, cannot change their direction of travel as rapidly as can the air. Therefore, they continue forward, striking the surfaces of the obstructions, and are entrapped in the mucous coating and transported by the cilia to the pharynx to be swallowed.

Size of Particles Entrapped in the Respiratory Passages. The nasal turbulence mechanism for removing particles from air is so effective that almost no particles larger than 4 to 6 micrometers in diameter enter the lungs through the nose. This size is smaller than the size of red blood cells.

Of the remaining particles, many of sizes between 1 and 5 micrometers *settle* out in the smaller bronchioles as a result of *gravitational precipitation*. For instance, terminal bronchiolar disease is very common in coal miners because of settled dust particles. Some of the still smaller particles (smaller than 1 micrometer in diameter) *diffuse* against the walls of the alveoli and adhere to the alveolar fluid. But many particles smaller than 0.5 micrometer in diameter remain suspended in the alveolar air and are later expelled by expiration. For instance, the particles of cigarette smoke have a particle size of approximately 0.3 micrometer. Almost none of these are precipitated in the respiratory passageways before they reach the alveoli. However, up to one third of them do precipitate in the alveoli by the diffusion process, with the balance remaining suspended and expelled in the expired air.

Many of the particles that become entrapped in the alveoli are removed by *alveolar macrophages,* as explained in Chapter 33, and others are carried away by the lung lymphatics. An excess of particles causes growth of fibrous tissue in the alveolar septa, leading to permanent debility.

VOCALIZATION

Speech involves not only the respiratory system but also (1) specific speech nervous control centers in the cerebral cortex, which will be discussed in Chapter 57, (2) respiratory control centers of the brain, and (3) the articulation and resonance structures of the mouth and nasal cavities. Basically, speech is composed of two separate mechanical functions: (1) *phonation,* which is achieved by the larynx, and (2) *articulation,* which is achieved by the structures of the mouth.

Phonation. The larynx is especially adapted to act as a vibrator. The vibrating element is the *vocal folds,* commonly called the *vocal cords.* The vocal folds protrude from the lateral walls of the larynx toward the center of the glottis; they are stretched and positioned by several specific muscles of the larynx itself.

Figure 37–10B illustrates the vocal folds as they are seen when looking into the glottis with a laryngoscope. During normal breathing, the folds are wide open to allow easy passage of air. During phonation, the folds close together so that passage of air between them will cause vibration. The pitch of the vibration is determined mainly by the degree of stretch of the folds but also by how tightly the folds are approximated to each other and by the mass of their edges.

Figure 37–10A shows a dissected view of the vocal folds after removing the mucous lining. Immediately inside each fold is a strong elastic ligament called the *vocal ligament.* This is attached anteriorly to the large *thyroid cartilage,* which is the cartilage that protrudes from the anterior surface of the neck and is called the "Adam's apple." Posteriorly, the vocal ligament is attached to the *vocal processes* of two *arytenoid cartilages.* Both the thyroid cartilage and the arytenoid cartilages in turn articulate from below with another cartilage not shown in Figure 37–10, the *cricoid cartilage.*

The vocal folds can be stretched by either forward rotation of the thyroid cartilage or posterior rotation of the arytenoid cartilages, activated by muscles from the thyroid cartilage and arytenoid cartilages to the cricoid cartilage. Muscles located within the vocal folds lateral to the vocal ligaments, the thyroarytenoid muscles, can pull the arytenoid cartilages toward the thyroid cartilage and therefore loosen the vocal folds. Also, slips of these muscles can

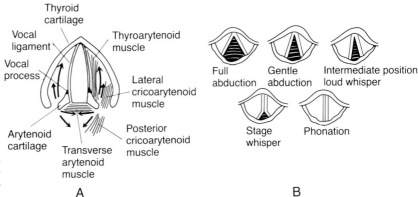

Figure 37–10. Laryngeal function in phonation. (Modified from Greene: The Voice and Its Disorders. 4th ed. Philadelphia, J. B. Lippincott Company, 1980.)

change the *shapes and masses of the vocal fold edges,* sharpening them to emit high-pitch sounds and blunting them for the more bass sounds.

Finally, several other sets of very small laryngeal muscles lie between the arytenoid cartilages and the cricoid cartilage and can rotate these cartilages inward or outward or pull their bases together or apart to give the various configurations of the vocal folds shown in Figure 37–10B.

Articulation and Resonance. The three major organs of articulation are the *lips,* the *tongue,* and the *soft palate.* These need not be discussed in detail because all of us are familiar with their movements during speech and other vocalizations.

The resonators include the *mouth,* the *nose and associated nasal sinuses,* the *pharynx,* and even the *chest cavity* itself. Here again we are all familiar with the resonating qualities of these different structures. For instance, the function of the nasal resonators is illustrated by the change in the quality of the voice when a person has a severe cold that blocks the air passages to these resonators.

REFERENCES

Agostoni, E.: Thickness and pressure of the pleural liquid. In Fishman, A. P., and Hecht, H. H. (eds.): The Pulmonary Circulation and Interstitial Space. Chicago, University of Chicago Press, 1969, p. 65.

Bailey, B. J., and Biller, H. F., Surgery of the Larynx. Philadelphia, W. B. Saunders Co., 1985.

Barnes, P. J.: Airway neuropeptides: Roles in fine tuning and in disease? News Physiol. Sci., 4:116, 1989.

Bartlett, D., Jr.: Respiratory functions of the larynx. Physiol. Rev., 69:33, 1989.

Bless, D. M. (ed.): Vocal Fold Physiology. Contemporary Research and Clinical Issues. San Diego, College-Hill Press, 1983.

Bourbon, J. R., and Rieutort, M.: Pulmonary surfactant: biochemistry, physiology, and pathology. News Physiol. Sci., 2:129, 1987.

Burri, P. H.: Fetal and postnatal development of the lung. Annu. Rev. Physiol., 46:617, 1984.

Coburn, R. F. (ed.): Airway Smooth Muscle in Health and Disease. New York, Plenum Publishing Corp., 1989.

Deslauriers, J., and Lacquet, L. K.: The Pleural Space, Vol. 6. In Delarue, N. C., and Eschapasse, H. (eds.), International Trends in General Thoracic Surgery Series. St. Louis, C. V. Mosby Co., 1989.

Drazen, J. M., et al.: High-frequency ventilation. Physiol. Rev., 64:505, 1984.

Fisher, A. B., and Chander, A.: Intracellular processing of surfactant lipids in the lung. Annu. Rev. Physiol., 47:789, 1985.

Fishman, A. P.: Assessment of Pulmonary Function. New York, McGraw-Hill, 1980.

Forster, R. E., II: Introduction to respiratory physiology. Annu. Rev. Physiol., 49:555, 1987.

Fujimura, O.: Vocal Physiology: Voice Production, Mechanisms and Function (Vocal Fold Physiology, Vol. 2). New York, Raven Press, 1988.

Gabella, G.: Innervation of airway smooth muscle: Fine structure. Annu. Rev. Physiol., 49:583, 1987.

Gil, J.: Organization of microcirculation in the lung. Annu. Rev. Physiol., 42:177, 1980.

Gil, J.: Histological preservation and ultrastructure of alveolar surfactant. Annu. Rev. Physiol., 47:753, 1985.

Gonzalez, N. C., and Fedde, M. R. (eds.): Oxygen Transfer from Atmosphere to Tissues. New York, Plenum Publishing Corp., 1988.

Green, M., and Moxham, J.: The respiratory muscles. Clin. Sci., 68:1, 1985.

Harada, R. N., and Repine, J. E.: Pulmonary host defense mechanisms. Chest, 87:247, 1985.

Harding, R.: Function of the larynx in the fetus and newborn. Annu. Rev. Physiol., 46:645, 1984.

Hodgkin, J. E., and Petty, T. L.: Chronic Obstructive Pulmonary Disease: Current Concepts. Philadelphia, W. B. Saunders Co., 1987.

King, R. J.: Composition and metabolism of the apolipoproteins of pulmonary surfactant. Annu. Rev. Physiol., 47:775, 1985.

Levitzky, M. G.: Pulmonary Physiology. New York, McGraw-Hill Book Co., 1982.

Lundberg, J. M., and Coburn, R. F.: Polypeptide-containing neurons in airway smooth muscle. Annu. Rev. Physiol., 49:557, 1987.

Morrow, P. E.: Factors determining hygroscopic aerosol deposition in airways. Physiol. Rev., 66:330, 1986.

Mortola, J. P.: Dynamics of breathing in newborn mammals. Physiol. Rev., 67:187, 1987.

Murray, J. F.: The Normal Lung, 2nd Ed. Philadelphia, W. B. Saunders Co., 1986.

Newman, J. D. (ed.): The Physiological Control of Mammalian Vocalization. New York, Plenum Publishing Corp., 1988.

Nicolosi, L., et al. Terminology of Communication Disorders: Speech-Language-Hearing, 3rd Ed. Baltimore, Williams & Wilkins, 1988.

Notter, R. H., and Finkelstein, J. N.: Pulmonary surfactant: An interdisciplinary approach. J. Appl. Physiol., 57:1613, 1984.

Perelman, R. H., et al.: Developmental aspects of lung lipids. Annu. Rev. Physiol., 47:803, 1985.

Rahn, H., et al.: The pressure-volume diagram of the thorax and lung. Am. J. Physiol., 146:161, 1946.

Ryan, J. W.: Processing of endogenous polypeptides by the lungs. Annu. Rev. Physiol., 44:241, 1982.

Ryan, U. S.: Metabolic activity of pulmonary endothelium. Annu. Rev. Physiol., 48:263, 1986.

Sant'Ambrogio, G., and Mathew, O. P.: Control of upper airway muscles. News Physiol. Sci., 3:167, 1988.

Schneiderman, C. R.: Basic Anatomy and Physiology in Speech and Hearing. San Diego, College-Hill Press, 1984.

Taylor, A. E., et al.: Clinical Respiratory Physiology. Philadelphia, W. B. Saunders Co., 1989.

van Golde, L. M. G., et al.: The pulmonary surfactant system: Biochemical aspects and functional significance. Physiol. Rev., 68:374, 1988.

van Golde, L. M. G.: Synthesis of surfactant lipids in the adult lung. Annu. Rev. Physiol., 47:765, 1985.

Wagner, P. D.: Ventilation-perfusion relationships. Annu. Rev. Physiol., 42:235, 1980.

Wagner, P. D.: The lungs during exercise. News Physiol. Sci., 2:6, 1987.

Weibel, E. R., and Bachofen, H.: How to stabilize the pulmonary alveoli: Surfactant or fibers? News Physiol. Sci., 2:72, 1987.

Weibel, E. R., and Bachofen, H.: Structural design of the alveolar septum and fluid exchange. In Fishman, A. P., and Renkin, E. M. (eds.): Pulmonary Edema. Baltimore, Waverly Press, 1979, p. 1.

Weibel, E. R.: Scaling of structural and functional variables in the respiratory system. Annu. Rev. Physiol., 49:147, 1987.

Welsh, M. J.: Electrolyte transport by airway epithelia. Physiol. Rev., 67:1143, 1987.

West, J. B.: Pulmonary Pathophysiology—The Essentials, 3rd Ed. Baltimore, Williams & Wilkins, 1987.

CHAPTER 38

• *Pulmonary Circulation; Pulmonary Edema; Pleural Fluid* •

The quantity of blood flowing through the lungs is essentially equal to that flowing through the systemic circulation. However, certain problems related to distribution of blood flow and other hemodynamics are special to the pulmonary circulation and are especially important to the gas exchange function of the lungs. Therefore, the present discussion is concerned specifically with these special features of the pulmonary circulation.

● PHYSIOLOGIC ANATOMY OF THE PULMONARY CIRCULATORY SYSTEM

The Pulmonary Vessels. The pulmonary artery extends only 5 centimeters beyond the apex of the right ventricle and then divides into the right and left main branches, which supply blood to the two respective lungs. The pulmonary artery is also thin, with a wall thickness approximately twice that of the venae cavae and one-third that of the aorta. The pulmonary arterial branches are all very short. However, all the pulmonary arteries, even the smaller arteries and arterioles, have much larger diameters than their counterpart systemic arteries. This, combined with the fact that the vessels are very thin and distensible, gives the pulmonary arterial tree a very large compliance, averaging almost 7 ml/mm Hg, which is similar to that of the systemic arterial tree. This large compliance allows the pulmonary arteries to accommodate about two thirds of the stroke volume output of the right ventricle.

The pulmonary veins, like the pulmonary arteries, are also short, but their distensibility characteristics are similar to those of the veins in the systemic circulation.

The Bronchial Vessels. Blood also flows to the lungs through several bronchial arteries, amounting to about 1 to 2 per cent of the total cardiac output. This bronchial arterial blood is *oxygenated* blood, in contrast to the partially deoxygenated blood in the pulmonary arteries. It supplies the supporting tissues of the lungs, including the connective tissue, the septa, and the large and small bronchi. After this bronchial arterial blood has passed through the supporting tissues, it empties into the pulmonary veins and *enters the left atrium,* rather than passing back to the right atrium. Therefore, the left ventricular output is slightly greater than the right ventricular output.

The Lymphatics. Lymphatics extend from all the supportive tissues of the lung, beginning in the connective tissue spaces surrounding the terminal bronchioles and coursing to the hilum of the lung and thence mainly into the right lymphatic duct. Particulate matter entering the alveoli is partly removed via these channels, and protein is also removed from the lung tissues, thereby helping to prevent edema.

● PRESSURES IN THE PULMONARY SYSTEM

The Pressure Pulse Curve in the Right Ventricle. The pressure pulse curves of the right ventricle and pulmonary artery are illustrated in the lower portion of Figure 38–1. These are contrasted with the much higher aortic pressure curve shown above. The systolic pressure in the right ventricle of the normal human being averages approximately 25 mm Hg, and the diastolic pressure averages about 0 to 1 mm Hg, values that are only one-fifth those for the left ventricle.

Pressures in the Pulmonary Artery. During *systole,* the pressure in the pulmonary artery is essentially equal to the pressure in the right ventricle, as also shown in Figure 38–1. However, after the pulmonary valve closes at the end of systole, the ventricular pressure falls precipitously, whereas the pulmonary arterial pressure falls slowly as blood flows through the capillaries of the lungs.

As shown in Figure 38–2, the *systolic pulmonary arterial pressure* averages approximately 25 mm Hg in the normal human being; the *diastolic pulmonary arterial pressure,* approximately 8 mm Hg; and the *mean pulmonary arterial pressure,* 15 mm Hg.

Pulmonary Arterial Pulse Pressure. The pulse pressure in the pulmonary arteries averages 17 mm Hg, which is two thirds as much as the systolic pressure. In previous discussions of pulse pressure in the systemic circulation, it has been pointed out that the less the compliance of an elastic reservoir that

Figure 38–1. Pressure pulse contours in the right ventricle, pulmonary artery, and aorta.

receives pulsatile injections of blood, the greater will be the pulse pressure. If it were not for the large compliance of the thin pulmonary arteries, the pulmonary arterial pulse pressure would be even greater than it is. However, another factor that keeps the pulse pressure from being higher is that about one-third the blood ejected by the right ventricle runs off from the pulmonary arteries through the capillaries into the pulmonary veins and left atrium at the same time that it is being ejected during systole.

Pulmonary Capillary Pressure. The mean pulmonary capillary pressure, as diagrammed in Figure 38–2, has been estimated by indirect means to be approximately 7 mm Hg. This will be discussed in detail later in the chapter in relation to fluid exchange functions of the capillary.

Left Atrial and Pulmonary Venous Pressure. The mean pressure in the left atrium and in the major pulmonary veins averages approximately 2 mm Hg in the recumbent human being, varying from as low as 1 mm Hg to as high as 5 mm Hg.

It usually is not reasonable to measure the left atrial pressure directly in the normal human being because it is difficult to pass a catheter through the heart chambers into the left atrium. However, the left atrial pressure often can be estimated closely by measuring the so-called *pulmonary wedge pressure*. This is achieved by inserting a catheter through the right side of the heart and the pulmonary artery into one of the small branches of the pulmonary arteries and then pushing the catheter until it wedges tightly in the artery. The pressure then measured through the catheter, called the "wedge pressure," is about 5 mm Hg. Since all blood flow has been stopped in the small artery, and the blood vessels extending from the artery make almost direct connection through the pulmonary capillaries with the blood in the pulmonary veins, this wedge pressure is usually only 2 to 3 mm Hg greater than the left atrial pressure. Also, when the left atrial pressure rises to high values, so also rises the pulmonary wedge pressure. Therefore, wedge pressure measurements are used frequently for studying changes in left atrial pressure in congestive heart failure.

● THE BLOOD VOLUME OF THE LUNGS

The blood volume of the lungs is approximately 450 milliliters, about 9 per cent of the total blood volume of the circulatory system. About 70 milliliters of this is in the capillaries, and the remainder is divided about equally between the arteries and veins.

The Lungs as a Blood Reservoir. Under different physiological and pathological conditions, the quantity of blood in the lungs can vary from as little as one-half up to two times normal. For instance, when a person blows air out so hard that high air pressure is built up in the lungs—such as when blowing a trumpet—as much as 250 ml of blood can be expelled from the pulmonary circulatory system into the systemic circulation. Also, loss of blood from the systemic circulation by hemorrhage can be partly compensated for by automatic shift of blood from the lungs into the systemic vessels.

Shift of Blood Between the Pulmonary and Systemic Circulatory Systems as a Result of Cardiac Pathology. Failure of the left side of the heart or increased resistance to blood flow through the mitral valve as a result of mitral stenosis or mitral regurgitation causes blood to dam up in the pulmonary circulation, sometimes increasing the pulmonary blood volume as much as 100 per cent, and also causing corresponding increases in the pulmonary vascular pressures.

Because the volume of the systemic circulation is about nine times that of the pulmonary system, a shift of blood from one system to the other affects the pulmonary system greatly but usually has only mild systemic effects.

● BLOOD FLOW THROUGH THE LUNGS AND ITS DISTRIBUTION

The blood flow through the lungs is essentially equal to the cardiac output. Therefore, the factors that control cardiac output—mainly peripheral fac-

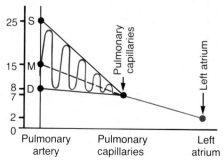

Figure 38–2. Pressures in the different vessels of the lungs.

tors, as discussed in Chapter 20 — also control pulmonary blood flow. Under most conditions, the pulmonary vessels act as passive, distensible tubes that enlarge with increasing pressure and narrow with decreasing pressure. However, for adequate aeration of the blood, it is important for the blood to be distributed to those segments of the lungs where the alveoli are best oxygenated. This is achieved by the following mechanism.

Effect of Diminished Alveolar Oxygen on Local Alveolar Blood Flow — Automatic Control of Pulmonary Blood Flow Distribution. When the concentration of oxygen in the alveoli decreases below normal, the adjacent blood vessels slowly constrict during the ensuing 3 to 10 minutes, the vascular resistance increasing as much as fivefold at extremely low oxygen levels. It should be noted specifically that this is *opposite the effect* normally observed in systemic vessels, which dilate, rather than constrict, in response to low oxygen. It is believed that the low oxygen concentration causes some yet undiscovered vasoconstrictor substance to be released from the lung tissue, this substance in turn promoting constriction of the small arteries. It has been suggested that this vasoconstrictor might be secreted by the alveolar epithelial cells when they become hypoxic.

This effect of a low oxygen level on pulmonary vascular resistance has an important function: to distribute blood flow where it is most effective. That is, when some of the alveoli are poorly ventilated so that the oxygen concentration in them becomes low, the local vessels constrict. This in turn causes most of the blood to flow through other areas of the lungs that are better aerated, thus providing an automatic control system for distributing blood flow to the different pulmonary areas in proportion to their degree of ventilation.

Paucity of Autonomic Nervous Control of Blood Flow in the Lungs. Although nerves innervate the lung tissues, it is doubtful that these have a major function in normal control of pulmonary blood flow. Normally, stimulation of the vagal fibers to the lungs causes a slight decrease in pulmonary vascular resistance, and stimulation of the sympathetics causes a slight increase in resistance, both effects perhaps too little to be more than marginally important.

However, research workers have described reflexes in the pulmonary vascular system that might be of clinical importance under certain conditions. For instance, it has been claimed that small emboli occluding the small pulmonary arteries cause a reflex that elicits sympathetic vasoconstriction throughout the lungs, leading to an increase in pulmonary arterial pressure. However, the significance of this reflex is still not certain.

In contrast to the very slight effect (almost none) that sympathetic stimulation has on the small resistance vessels of the lung, it does have a very great effect in constricting the large pulmonary capacitative vessels, especially the veins. This large vessel constriction provides a means by which sympathetic stimulation can displace much of the extra blood in the lungs into other segments of the circulation when needed elsewhere.

EFFECT OF HYDROSTATIC PRESSURE GRADIENTS IN THE LUNGS ON REGIONAL PULMONARY BLOOD FLOW

In Chapter 15 it was pointed out that the pressure in the foot of a standing person can be as much as 90 mm Hg greater than the pressure at the level of the heart. This is caused by *hydrostatic pressure* — that is, by the weight of the blood itself. The same effect, but to a lesser degree, occurs in the lungs. In the normal, upright adult person, the lowest point in the lungs is about 30 cm below the highest point. This represents a 23 mm Hg pressure difference, about 15 mm Hg of which are above the heart and 8 below. That is, the pulmonary arterial pressures in the uppermost portion of the lung of a standing person are about 15 mm Hg less than the pulmonary arterial pressure at the level of the heart, and the pressure in the lowest portion of the lungs is about 8 mm Hg greater. Such pressure differences have profound effects on blood flow through the different areas of the lungs. This is illustrated by the lower curve in Figure 38–3, which plots blood flow per unit of lung tissue against hydrostatic level in the lung. Note that in the standing position, at rest, there is little flow in the top of the lung but about five times this much flow in the lower lung. To help explain these differences, one often describes the lung as divided into three different zones, as illustrated in Figure 38–4, in which the patterns of blood flow are quite different. Let us explain these differences.

Zones 1, 2, and 3 of Pulmonary Blood Flow

The capillaries in the alveolar walls are distended by the blood pressure inside them, but simultaneously they are compressed by the alveolar pressure on their outsides. Therefore, any time the alveolar air pressure becomes greater than the capillary blood pressure, the capillaries close and there is no blood flow. Under different normal and pathological lung conditions one may find any one of three possible zones of pulmonary blood flow as follows:

Zone 1: No blood flow at all during any part of the cardiac cycle because the local capillary pressure in that area of the

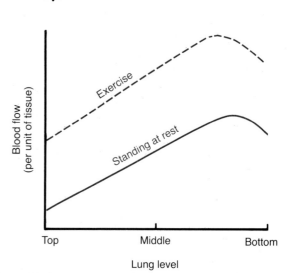

Figure 38–3. Blood flow at different levels in the lung of an upright person, both at rest and during exercise. Note that when the person is at rest, the blood flow is very low at the top of the lungs and most of the flow is through the lower lung.

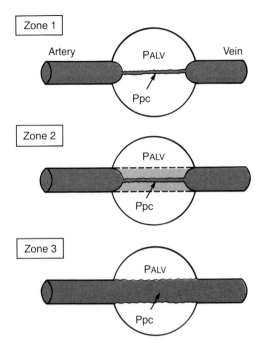

Figure 38–4. Mechanics of blood flow in the three different blood flow zones of the lung: *zone 1, no flow,* because alveolar pressure is greater than arterial pressure; *zone 2, intermittent flow,* because systolic arterial pressure rises higher than alveolar pressure but diastolic pressure falls below alveolar pressure; *zone 3, continuous flow,* because arterial pressure remains greater than alveolar pressure at all times.

lung never rises higher than the alveolar pressure during any part of the cardiac cycle.

Zone 2: Intermittent blood flow during the pulmonary arterial pressure peaks because the systolic pressure is greater than the alveolar pressure but the diastolic pressure is less than alveolar pressure.

Zone 3: Continuous blood flow, because the alveolar capillary pressure remains greater than alveolar pressure during the entire cardiac cycle.

Normally, the lungs have only zones 2 and 3 blood flow, zone 2 (intermittent flow) in the apices, and zone 3 (continuous flow) in all the lower areas. Let us explain. When a person is in the upright position, the pulmonary arterial pressure at the lung apex is about 15 mm Hg less than the pressure at the level of the heart. Therefore, the apical systolic pressure is only 10 mm Hg (25 mm Hg at heart level minus 15 mm Hg hydrostatic pressure difference). Obviously, this is greater than the zero alveolar pressure, so that blood flows through the pulmonary apical blood vessels during systole. On the other hand, during diastole, the 8 mm Hg diastolic pressure at the level of the heart is not sufficient to push the blood up the 15 mm Hg hydrostatic pressure gradient required to cause diastolic flow. Therefore, blood flow through the apical part of the lung is intermittent, with flow during systole but cessation of flow during diastole; this is called zone 2 blood flow. Zone 2 blood flow begins in the normal lungs approximately 10 centimeters above the level of the heart and extends from there to the top of the lungs.

In the lower regions of the lungs, from about 10 centimeters above the level of the heart all the way to the bottom, the pulmonary arterial pressure remains during both systole and diastole greater than the zero alveolar pressure. Therefore, there is continuous flow, which means zone 3 blood flow. Also, when a person is in a lying position, no part of the lung is more than a few centimeters above the level of the heart. Therefore, blood flow then, in a normal person, is always entirely zone 3 blood flow, including the apices.

Zone 1 Blood Flow Occurs Only Under Abnormal Conditions. Zone 1 blood flow, which is blood flow at no time during the cardiac cycle, occurs when either the pulmonary systolic arterial pressure is too low or the alveolar pressure is too high to allow flow. For instance, if an upright person is breathing against a positive air pressure so that the intra-alveolar pressure is 10 mm Hg or more greater than normal, but the pulmonary systolic pressure is normal, then one would expect zone 1 blood flow—no blood flow at all—at least in the lung apices. Another instance in which zone 1 blood flow occurs is in an upright person whose pulmonary systolic pressure is exceedingly low, as might occur in hypovolemic states.

Effect of Exercise on Blood Flow Through the Different Parts of the Lungs. Referring again to Figure 38–3, one sees that the blood flow in all parts of the lung increases during exercise. However, the increase in flow in the top of the lung may be as much as 700 to 800 per cent, while the increase in the lower part of the lung may be no more than 200 to 300 per cent. The reason for these effects is the considerably higher pulmonary vascular pressures that occur during exercise, which effectively convert the entire lung into a zone 3 pattern of flow.

EFFECT OF INCREASED CARDIAC OUTPUT ON THE PULMONARY CIRCULATION DURING HEAVY EXERCISE

During heavy exercise the blood flow through the lungs increases as much as four- to sevenfold. This extra flow is achieved in two ways: (1) by increasing the number of open capillaries, sometimes as much as threefold, and (2) by increasing the rate of flow through each capillary, often as much as twofold. Fortunately, in the normal person, these two factors together decrease the pulmonary vascular resistance so much that the pulmonary arterial pressure rises very little even during maximum exercise; this effect is illustrated in Figure 38–5.

This ability of the lungs to accommodate greatly increased blood flow during exercise obviously con-

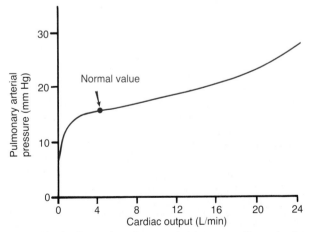

Figure 38–5. Effect on the pulmonary arterial pressure of increasing the cardiac output.

serves the energy of the right side of the heart, and it also prevents a significant rise in pulmonary capillary pressure and therefore prevents development of pulmonary edema during the increased cardiac output.

FUNCTION OF THE PULMONARY CIRCULATION WHEN THE LEFT ATRIAL PRESSURE RISES AS A RESULT OF LEFT HEART FAILURE

When the left side of the heart fails, blood begins to dam up in the left atrium. As a result, the left atrial pressure can rise on occasion from its normal value of 1 to 5 mm Hg to as high as 40 to 50 mm Hg. The initial pressure rise, up to about 7 mm Hg, has almost no effect on pulmonary circulatory function because this initial rise merely expands the venules and opens up more capillaries so that blood continues to flow with almost equal ease from the pulmonary arteries. Figure 38–6 illustrates this effect, showing almost no change in pulmonary arterial pressure at the lower left atrial pressures. Because the left atrial pressure in a normal person almost never rises above +6 mm Hg even during the most strenuous exercise, the changes in left atrial pressure have virtually no effect on pulmonary circulatory function except when the left side of the heart fails.

However, when the left atrial pressure rises to greater than 7 or 8 mm Hg, then further increases in left atrial pressure cause almost equally great increases in pulmonary arterial pressure, as also shown in Figure 38–6, with concomitant increased load on the right side of the heart as well.

It is also true that the initial rise in left atrial pressure up to about 7 or 8 mm Hg has virtually no effect on pulmonary capillary pressure either. But any increase in left atrial pressure beyond this increases the capillary pressure almost equally as much. When the left atrial pressure has risen above 25 to 30 mm Hg, causing similar increases in capillary pressure, pulmonary edema is then very likely to develop.

● PULMONARY CAPILLARY DYNAMICS

Exchange of gases between the alveolar air and the pulmonary capillary blood will be discussed in the next chapter. However, it is important for us to note here that the alveolar walls are lined with so many capillaries that in most places they almost touch each other. Therefore, it has often been said that the capillary blood flows in the alveolar walls as a "sheet," rather than in individual vessels.

Pulmonary Capillary Pressure. Unfortunately, no direct measurements of pulmonary capillary pressure have been made. However, "isogravimetric" measurement of pulmonary capillary pressure, using a technique described in Chapter 16, has given a value of 7 mm Hg. This is probably very nearly correct, because the mean left atrial pressure is about 2 mm

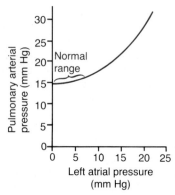

Figure 38–6. Effect of left atrial pressure on pulmonary arterial pressure.

Hg and the mean pulmonary arterial pressure is only 15 mm Hg, so that the mean pulmonary capillary pressure must lie somewhere between these two values.

Length of Time Blood Stays in the Capillaries. From histologic study of the total cross-sectional area of all the pulmonary capillaries, it can be calculated that when the cardiac output is normal, blood passes through the pulmonary capillaries in about 0.8 second. Increasing the cardiac output shortens this time sometimes to as little as 0.3 second, but the shortening would be much greater were it not for the fact that additional capillaries, which normally are collapsed, open up to accommodate the increased blood flow. Thus, in less than 1 second, blood passing through the capillaries becomes oxygenated and loses its excess carbon dioxide.

CAPILLARY EXCHANGE OF FLUID IN THE LUNGS, AND PULMONARY INTERSTITIAL FLUID DYNAMICS

The dynamics of fluid exchange through the lung capillaries are *qualitatively* the same as for peripheral tissues. However, *quantitatively*, there are important differences:

1. The pulmonary capillary pressure is very low, about 7 mm Hg, in comparison with a considerably higher functional capillary pressure in the peripheral tissues, about 17 mm Hg.

2. The interstitial fluid pressure in the lung is slightly more negative than in the peripheral subcutaneous tissue. (This has been measured in two ways: by a pipette inserted into the pulmonary interstitium giving a value of about −5 mm Hg, and by measuring the absorption pressure of fluid from the alveoli giving a value of about −8 mm Hg.)

3. The pulmonary capillaries are relatively leaky to protein molecules, so that the colloid osmotic pressure of the pulmonary interstitial fluids is probably about 14 mm Hg in comparison with less than half this in the peripheral tissues.

4. The alveolar walls are extremely thin, and the alveolar epithelium covering the alveolar surfaces is

so weak that it is ruptured by any positive pressure in the interstitial spaces greater than atmospheric pressure (0 mm Hg), which allows dumping of fluid from the interstitial spaces into the alveoli.

Now let us see how these quantitative differences affect pulmonary fluid dynamics.

Interrelationship Between Interstitial Fluid Pressure and Other Pressures in the Lung. Figure 38–7 illustrates a pulmonary capillary, a pulmonary alveolus, and a lymphatic capillary draining the interstitial space between the capillary and the alveolus. Note the balance of forces at the capillary membrane as follows:

	mm Hg
Forces tending to cause movement of fluid outward from the capillaries and into the pulmonary interstitium:	
Capillary pressure	7
Interstitial fluid colloid osmotic pressure	14
Negative interstitial fluid pressure	8
TOTAL OUTWARD FORCE	29
Forces tending to cause absorption of fluid into the capillaries:	
Plasma colloid osmotic pressure	28
TOTAL INWARD FORCE	28

Thus the normal outward forces are slightly greater than the inward forces. The *net mean filtration pressure* at the pulmonary capillary membrane can be calculated as follows:

	mm Hg
Total outward force	+29
Total inward force	−28
NET MEAN FILTRATION PRESSURE	+1

This net filtration pressure causes a slight continual flow of fluid from the pulmonary capillaries into the interstitial spaces; and, except for a small amount that evaporates in the alveoli, this fluid is pumped back to the circulation through the pulmonary lymphatic system.

Negative Interstitial Pressure and the Mechanism for Keeping the Alveoli "Dry." One of the most important problems in lung function is to understand why the alveoli do not fill up with fluid. One's first impulse is to say that the alveolar epithelium keeps fluid from leaking out of the interstitial spaces into the alveoli. However, this is not true, because there are always a small number of openings between the alveolar epithelial cells through which even large protein molecules as well as large quantities of water and electrolytes can pass.

However, if one remembers that the pulmonary capillaries and the pulmonary lymphatic system normally maintain a *negative pressure* in the interstitial

Pressures Causing Fluid Movement

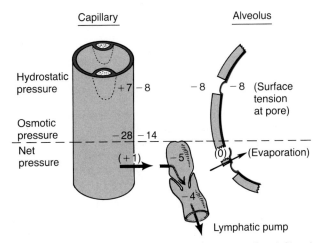

Figure 38–7. Hydrostatic and osmotic forces at the capillary (left) and alveolar membrane (right) of the lungs. Also shown is a lymphatic (center) that pumps fluid from the pulmonary interstitial spaces. (Modified from Guyton, Taylor, and Granger: Dynamics and Control of the Body Fluids. Philadelphia, W. B. Saunders Company, 1975.)

spaces, then it is clear that whenever extra fluid appears in the alveoli, it will simply be sucked mechanically into the lung interstitium through the small openings between the alveolar epithelial cells. Then the excess fluid is either carried away in the pulmonary lymphatics or is absorbed into the pulmonary capillaries. Thus, under normal conditions the alveoli are kept in a "dry" state except for a small amount of fluid that seeps from the epithelium onto the lining surfaces of the alveoli to keep them moist.

PULMONARY EDEMA

Pulmonary edema occurs in the same way that it occurs elsewhere in the body. Any factor that causes the pulmonary interstitial fluid pressure to rise from the negative range into the positive range will cause sudden filling of the pulmonary interstitial spaces and, in more severe cases, even the alveoli with large amounts of free fluid.

The usual causes of pulmonary edema are as follows:

1. Left heart failure or mitral valvular disease with consequent great increase in pulmonary capillary pressure and flooding of the interstitial spaces.

2. Damage to the pulmonary capillary membrane caused by infections such as pneumonia or by breathing noxious substances such as chlorine gas or sulfur dioxide gas, with resultant rapid leakage of both plasma proteins and fluid out of the capillaries.

Pulmonary "Interstitial Fluid" Edema Versus Pulmonary "Alveolar" Edema. The interstitial fluid volume of the lungs usually cannot increase more than about 50 per cent (representing less than 100 milliliters of fluid) before the alveolar epithelial membranes rupture and fluid begins to pour from the interstitial spaces into the alveoli. The cause of this is simply the very slight tensional strength of the pulmonary alveolar epithelium; that is, any positive pressure in the interstitial fluid spaces seems to cause immediate rupture of this epithelium.

Therefore, except in the mildest cases of pulmonary

edema, edema fluid always enters the alveoli; if this edema becomes severe enough, it can cause death by suffocation.

The "Pulmonary Edema Safety Factor." All the factors that tend to prevent edema in the peripheral tissues also tend to prevent edema in the lungs. That is, before positive interstitial fluid pressure can occur and cause edema, all the following factors must be overcome: (1) the normal negativity of the interstitial fluid pressure of the lungs, (2) the lymphatic pumping of fluid out of the interstitial spaces, and (3) the increased osmosis of fluid into the pulmonary capillaries caused by decreased protein in the interstitial fluid when the lymph flow increases.

In experiments in animals it has been found that the pulmonary capillary pressure normally must rise to a value at least equal to the plasma colloid osmotic pressure before significant pulmonary edema will occur. To give an example, Figure 38–8 illustrates the effect of different levels of left atrial pressure to increase the rate of edema formation in dogs. Remember that every time the left atrial pressure rises to high values, the pulmonary capillary pressure rises to a level 1 to 2 mm Hg greater than the left atrial pressure. In this figure, as soon as the left atrial pressure rose above 23 mm Hg (with the pulmonary capillary pressure above about 25 mm Hg), fluid began to accumulate in the lungs and increased extremely rapidly, with further increases in pressure. Yet below 25 mm Hg pulmonary capillary pressure, there was no significant increase in pulmonary fluid. The plasma colloid osmotic pressure in dogs is almost exactly equal to this 25 mm Hg critical pressure level. Therefore, in the human being, who normally has a plasma colloid osmotic pressure of 28 mm Hg, one can predict that the pulmonary capillary pressure must rise from the normal level of 7 mm Hg to over 28 mm Hg to cause pulmonary edema, giving a safety factor against pulmonary edema of about 21 mm Hg.

Safety Factor in Chronic Conditions. When the pulmonary capillary pressure remains elevated chronically (for at least 2 weeks), the lungs become even more resistant to pulmonary edema because the lymph vessels expand greatly, increasing their capability for carrying fluid away from the interstitial spaces as much as tenfold. Therefore, in a patient with chronic mitral stenosis a pulmonary capillary pressure as great as 40 to 45 mm Hg occasionally develops without significant pulmonary edema.

Thus, in chronic pulmonary edema, the safety factor against pulmonary edema can rise to as high as 35 to 45 mm Hg, in comparison with a normal value of 21 mm Hg under acute conditions.

Rapidity of Death in Acute Pulmonary Edema. When the pulmonary capillary pressure does rise even slightly above the safety factor level, lethal pulmonary edema can occur within hours, and even within 20 to 30 minutes if the capillary pressure rises as much as 25 to 30 mm Hg above the safety factor level. Thus, in acute left heart failure, in which the pulmonary capillary pressure occasionally rises to as high as 50 mm Hg, death frequently ensues in less than one-half hour from acute pulmonary edema.

● THE FLUIDS IN THE PLEURAL CAVITY

When the lungs expand and contract during normal breathing, they slide back and forth within the pleural cavity. To facilitate this, a very thin layer of fluid lies between the parietal and visceral pleurae.

Figure 38–9 illustrates the dynamics of fluid exchange in the pleural space. Each of the two pleurae is a very porous mesenchymal serous membrane through which small amounts of interstitial fluid transude continually into the pleural space. These fluids carry with them tissue proteins, giving the pleural fluid a mucoid characteristic, which is what allows extremely easy slippage of the moving lungs.

The total amount of fluid in each pleural cavity is very slight, only a few milliliters. Yet whenever the quantity becomes more than barely enough to separate the two plurae, the excess is pumped away by lymphatic vessels opening directly from the pleural cavity into (1) the mediastinum, (2) the superior surface of the diaphragm, and (3) the lateral surfaces of the parietal pleura. Therefore, the *pleural space* — the space between the visceral and parietal plurae — is called a *potential space* because normally it is so narrow that it is not obviously a physical space at all.

Negative Pressure in the Pleural Fluid. Because the recoil tendency of the lungs causes them to try to collapse, a negative force is always required on the outside of the lungs to keep them expanded. This is provided by a negative pressure in the normal pleural space. The basic cause of this

Figure 38–8. Rate of fluid loss into the lung tissues when the left atrial pressure (and also pulmonary capillary pressure) is increased. (From Guyton and Lindsey: *Circ. Res.*, 7:649, 1959. By permission of The American Heart Association, Inc.)

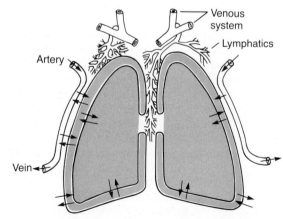

Figure 38–9. Dynamics of fluid exchange in the intrapleural spaces.

negative pressure is the pumping of fluid from the space by the lymphatics (which is also the basis of the negative pressure found in most tissue spaces of the body). Because the normal collapse tendency of the lungs is about −4 mm Hg (−5 or −6 centimeters of water) the pleural fluid pressure must always be at least as negative as −4 mm Hg simply to keep the lungs expanded. Actual measurements have shown it usually to be about −7 mm Hg, which is a few millimeters of mercury more negative than the collapse pressure of the lungs. Thus, the negativity of the pleural fluid keeps the lungs pulled tightly against the parietal pleura of the chest cavity except for the extremely thin layer of mucoid fluid that acts as a lubricant.

Pleural Effusion. Pleural effusion means the collection of large amounts of free fluid in the pleural space. It is analogous to edema fluid in the tissues and can actually be called edema of the pleural cavity. The possible causes of effusion are the same as the causes of edema in other tissues, including (1) blockage of lymphatic drainage from the pleural cavity; (2) cardiac failure, which causes excessively high peripheral and pulmonary capillary pressures, leading to excessive transudation of fluid into the pleural cavity; (3) greatly reduced plasma colloid osmotic pressure, thus also allowing excessive transudation of fluid; and (4) infection or any other cause of inflammation of the pleural surfaces of the pleural cavity, which breaks down the capillary membranes and allows rapid dumping of both plasma proteins and fluid into the cavity.

REFERENCES

Agostoni, E.: Mechanics of the pleural space. Physiol. Rev., 52:57, 1972.

Arakawa, M., et al.: Pulmonary blood volume and pulmonary extravascular water volume in men. Jpn. Circ. J., 49:475, 1985.

Bergofsky, E. H.: Humoral control of the pulmonary circulation. Annu. Rev. Physiol., 42:221, 1980.

Cournand, A.: Some aspects of the pulmonary circulation in normal man and in chronic cardiopulmonary diseases. Circulation, 2:641, 1950.

Culver, B. H., and Butler, J.: Mechanical influences on the pulmonary microcirculation. Annu. Rev. Physiol., 42:187, 1980.

Dawson, C. A.: Role of pulmonary vasomotion in physiology of the lung. Physiol. Rev., 64:544, 1984.

Deslauriers, J., and Lacquet, L. K.: The Pleural Space, Vol. 6. In Delarue, N. C., and Eschapasse, H. (eds.): International Trends in General Thoracic Surgery Series. St. Louis, C. V. Mosby Co., 1989.

Downing, S. E., and Lee, J. C.: Nervous control of the pulmonary circulation. Annu. Rev. Physiol., 42:199, 1980.

Effros, R. M.: Pulmonary microcirculation and exchange. In Renkin, E. M., and Michel, C. C. (eds.): Handbook of Physiology. Sec. 2, Vol. IV. Bethesda, Md., American Physiological Society, 1984, p. 865.

Fishman, A. P., and Pietra, G. G.: Hemodynamic pulmonary edema. In Fishman, A. P. and Renkin, E. M. (eds.): Pulmonary Edema. Baltimore, Waverly Press, 1979, p. 79.

Fishman, A. P., and Renkin, E. M. (eds.): Pulmonary Edema. Baltimore, American Physiological Society, 1979.

Fishman, A. P.: Vasomotor regulation of the pulmonary circulation. Annu. Rev. Physiol., 42:211, 1980.

Gaar, K. A., Jr., et al.: Effect of lung edema on pulmonary capillary pressure. Am. J. Physiol., 216:1370, 1969.

Gaar, K. A., Jr., et al.: Effects of capillary pressure and plasma protein on development of pulmonary edema. Am. J. Physiol., 213:79, 1967.

Gil, J.: Organization of microcirculation in the lung. Annu. Rev. Physiol., 42:177, 1980.

Grover, R. F., et al.: Pulmonary Circulation. In Shepherd, J. T., and Abboud, F. M. (eds.): Handbook of Physiology, Sec. 2, Vol. III. Bethesda, American Physiological Society, 1983, p. 103.

Guyton, A. C., and Lindsey, A. W.: Effect of elevated left atrial pressure and decreased plasma protein concentration on the development of pulmonary edema. Circ. Res., 7:649, 1959.

Guyton, A. C., et al.: Circulatory Physiology. II. Dynamics and Control of the Body Fluids. Philadelphia, W. B. Saunders Co., 1975.

Guyton, A. C., et al.: Forces governing water movement in the lung. In Fishman, A. P., and Renkin, E. M. (eds.): Pulmonary Edema. Baltimore, Waverly Press, 1979, p. 65.

Hills, B. A.: The pleural interface. Thorax, 40:1, 1985.

Iinuma, J., et al.: Fluid volume balance between pulmonary intravascular space and extravascular space in dogs. Jpn Circ. J., 50:818, 1986.

Laine, G. A., et al.: A new look at pulmonary edema. News Physiol. Sci., 1:150, 1986.

Malik, A. B.: Pulmonary microembolism. Physiol. Rev., 63:1114, 1983.

Mark, A. L., and Mancia, G.: Cardiopulmonary baroreflexes in humans. In Shepherd, J. T., and Abboud, F. M. (eds.): Handbook of Physiology. Sec. 2, Vol. III. Bethesda, American Physiological Society, 1983, p. 795.

Meyer, B. J., et al.: Interstitial fluid pressure V. Negative pressure in the lungs. Circ. Res., 22:263, 1968.

Notter, R. H., and Finkelstein, J. N.: Pulmonary surfactant: An interdisciplinary approach. J. Appl. Physiol., 57:1613, 1984.

Parker, J. C., et al.: Pulmonary interstitial and capillary pressures estimated from intra-alveolar fluid pressures. J. Appl. Physiol., 44(2):267, 1978.

Parker, J. C., et al.: Pulmonary transcapillary exchange and pulmonary edema. In Guyton, A. C., and Young, D. B. (eds.): International Review of Physiology: Cardiovascular Physiology III. Vol. 18. Baltimore, University Park Press, 1979, p. 261.

Piene, H.: Pulmonary arterial impedance and right ventricular function. Physiol. Rev., 66:606, 1986.

Ryan, U. S.: Metabolic activity of pulmonary endothelium. Annu. Rev. Physiol., 48:263, 1986.

Staub, N. C.: Pulmonary edema due to increased microvascular permeability. Annu. Rev. Med., 32:291, 1981.

Staub, N. C.: Pulmonary edema. Physiol. Rev., 54:678, 1974.

Takaya, T., et al.: Pulmonary vein blood flow velocity waveform. Jpn. Circ. J., 50:405, 1986.

Taylor, A. E., et al.: Na^{24} space, D$_2$O space, and blood volume in isolated dog lung. Am. J. Physiol., 211:66, 196.

Taylor, A. E., et al.: Permeability of the alveolar membrane to solutes. Circ. Res., 16:353, 1965.

Visscher, M. B., et al.: The physiology and pharmacology of lung edema. Pharm. Rev., 8:389, 1956.

Weibel, E. R., and Bachofen, H.: Structural design of the alveolar septum and fluid exchange. In Fishman, A. P., and Renkin, E. M. (eds.): Pulmonary Edema. Baltimore, Waverly Press, 1979, p. 1.

West, J. B.: Pulmonary Pathophysiology — The Essentials, 3rd Ed. Baltimore, Williams & Wilkins, 1987.

CHAPTER 39

•*Physical Principles of Gaseous Exchange; Diffusion of Oxygen and Carbon Dioxide Through the Respiratory Membrane* •

After the alveoli are ventilated with fresh air, the next step in the respiratory process is *diffusion* of oxygen from the alveoli into the pulmonary blood and diffusion of carbon dioxide in the opposite direction. The process of diffusion is simply random molecular motion of molecules intertwining their ways in both directions through the respiratory membrane. However, in respiratory physiology one is concerned not only with the basic mechanism by which diffusion occurs but also with the *rate* at which it occurs; and this is a much more complicated problem, requiring a deeper understanding of the physics of diffusion and gaseous exchange.

● PHYSICS OF DIFFUSION AND GAS PRESSURES

THE MOLECULAR BASIS OF GASEOUS DIFFUSION

All the gases that are of concern in respiratory physiology are simple molecules that are free to move among each other, which is the process called "diffusion." This is also true of the gases dissolved in the fluids and tissues of the body.

However, for diffusion to occur, there must be a source of energy. This is provided by the kinetic motion of the molecules themselves. That is, except at absolute zero temperature, all molecules of all matter are continually undergoing motion. For free molecules that are not physically attached to others, this means linear movement at high velocity until they strike other molecules. Then they bounce away in new directions and continue again until striking still other mole-

cules. In this way the molecules move rapidly among each other.

Net Diffusion of a Gas in One Direction — Effect of a Concentration Gradient. If a gas chamber or a solution has a high concentration of a gas at one end of the chamber and a low concentration at the other end, as illustrated in Figure 39–1, net diffusion of the gas will occur from the high concentration area toward the low area. The reason for this is obvious: There simply are far more molecules at end A of the chamber to diffuse toward end B than there are molecules to diffuse in the opposite direction. Therefore, the rates of diffusion in each of the two directions are proportionately different, as illustrated by the lengths of the two arrows.

GAS PRESSURES IN A MIXTURE OF GASES—PARTIAL PRESSURES OF INDIVIDUAL GASES

Pressure is caused by the constant impact of kinetically moving molecules against a surface. Therefore, the pressure of a gas acting on the surfaces of the respiratory passages and alveoli is proportional to the summated force of impact of all the molecules striking the surface at any given instant. This means that *the total pressure is directly proportional to the concentration of the gas molecules.*

However, in respiratory physiology, one deals with mixtures of gases, mainly of oxygen, nitrogen, and carbon dioxide. Furthermore, the rate of diffusion of each of these gases is directly proportional to the pressure caused by this gas alone, which is called the *partial pressure* of the gas. Therefore, let us explain the concept of partial pressure.

Consider air, which has an approximate composition of 79 per cent nitrogen and 21 per cent oxygen. The total pressure of this mixture is 760 mm Hg, and it is clear from the

Figure 39–1. Net diffusion of oxygen from one end of a chamber to the other.

above description of the molecular basis of pressure that each gas contributes to the total pressure in direct proportion to its concentration. Therefore, 79 per cent of the 760 mm Hg is caused by nitrogen (about 600 mm Hg) and 21 per cent by oxygen (about 160 mm Hg). Thus, the "partial pressure" of nitrogen in the mixture is 600 mm Hg, and the "partial pressure" of oxygen is 160 mm Hg; the total pressure is 760 mm Hg, the sum of the individual partial pressures.

The partial pressures of the individual gases in a mixture are designated by the symbols P_{O_2}, P_{CO_2}, P_{N_2}, P_{H_2O}, P_{He}, and so forth.

PRESSURES OF GASES IN WATER AND TISSUES

Gases dissolved in water or in the body tissues also exert pressure, because the dissolved molecules are moving randomly and have kinetic energy, as do the molecules in the gaseous phase. Furthermore, when the molecules of a gas dissolved in fluid encounter a surface like the membrane of a cell, they exert their own pressure in the same way that a gas in the gaseous phase exerts its own individual partial pressure. The pressures of the separate dissolved gases are designated similarly as for the partial pressures of the gases in the gaseous state, i.e., P_{O_2}, P_{CO_2}, P_{N_2}, P_{He}.

Factors That Determine the Pressure of a Gas Dissolved in a Fluid. The pressure of a gas in a solution is determined not only by its concentration but also by the *solubility coefficient* of the gas. That is, some types of molecules, especially carbon dioxide, are physically or chemically attracted to water molecules, whereas others are repelled. Obviously, when molecules are attracted, far more of them can then become dissolved without building up excess pressure within the solution. On the other hand, in the case of those that are repelled, excessive pressures will develop for many fewer dissolved molecules. These relationships can be expressed by the following formula, which is *Henry's law:*

$$\text{Pressure} = \frac{\text{Concentration of dissolved gas}}{\text{Solubility coefficient}}$$

When pressure is expressed in atmospheres and concentration is expressed in volume of gas dissolved in each volume of water, the solubility coefficients for important respiratory gases at body temperature are the following:

Oxygen	0.024
Carbon dioxide	0.57
Carbon monoxide	0.018
Nitrogen	0.012
Helium	0.008

From this table, one can see that carbon dioxide is more than 20 times as soluble as oxygen, and oxygen is moderately more soluble than any of the other three gases.

Diffusion of Gases Between the Gaseous Phase in the Alveoli and the Dissolved Phase in the Pulmonary Blood. The partial pressure of each gas in the respiratory gas mixture tends to force molecules of that gas into solution first in the alveolar membrane and then in the blood of the alveolar capillaries. Conversely, the molecules of the same gas that are already dissolved in the blood are bouncing randomly in the fluid of the blood, and some escape back into the alveoli. The rate at which they escape is directly proportional to their partial pressure in the blood. But in which direction will *net diffusion* of the gas occur? The answer to this is that net diffusion is determined by the difference between the two partial pressures. If the partial pressure is greater in the gaseous phase, as is normally true for oxygen, then more molecules will go into the blood than in the other direction. Alternately, if the pressure of the gas is greater in the blood, which is normally true for carbon dioxide, then net diffusion will occur toward the gaseous phase in the alveoli.

THE VAPOR PRESSURE OF WATER

When air enters the respiratory passageways, water immediately evaporates from the surfaces of these passages and humidifies the air. This results from the fact that water molecules, like the different dissolved gas molecules, are continually escaping from the water surface into the gas phase. The pressure that the water molecules exert to escape through the surface is called the *vapor pressure* of the water. At normal body temperature, 37°C, this vapor pressure is 47 mm Hg. Therefore, once the gas mixture has become fully humidified—that is, once it is in "equilibrium" with the surrounding water—the partial pressure of the water vapor in the gas mixture is also 47 mm Hg. This partial pressure, like the other partial pressures, is designated P_{H_2O}.

The vapor pressure of water depends entirely on the temperature of the water. The greater the temperature, the greater the kinetic activity of the molecules, and therefore the greater the likelihood that the water molecules will escape from the surface of the water into the gas phase. For instance, the water vapor pressure at 0°C is 5 mm Hg and at 100°C is 760 mm Hg. But the most important value to remember is the vapor pressure at body temperature, 47 mm Hg; this value will appear in many of our subsequent discussions.

DIFFUSION OF GASES THROUGH FLUIDS— THE PRESSURE DIFFERENCE FOR DIFFUSION

Now, let us return to the problem of diffusion. From the above discussion it is already clear that when the concentration, or pressure, of a gas is greater in one area than in another area, there will be net diffusion from the high pressure area toward the low pressure area. For instance, returning to Figure 39–1, one can readily see that the molecules in the area of high pressure, because of their greater number, have more statistical chance of moving randomly into the area of low pressure than do molecules attempting to go in the other direction. However, some molecules do bounce randomly from the area of low pressure toward the area of high pressure. Therefore, the *net diffusion* of gas from the area of high pressure to the area of low pressure is equal to the number of molecules bouncing in this direction *minus* the number bouncing in the opposite direction, and this in turn is proportional to the gas pressure difference between the two areas, called simply the *pressure difference for diffusion.*

Quantifying the Net Rate of Diffusion in Fluids. In addition to the pressure difference, several other factors affect the rate of gas diffusion in a fluid. These are (1) the solubility of the gas in the fluid, (2) the cross-sectional area of the fluid, (3) the distance through which the gas must diffuse, (4) the molecular weight of the gas, and (5) the temperature of the fluid. In the body, the last of these factors, the temperature, remains reasonably constant and usually need not be considered.

Obviously, the greater the solubility of the gas, the greater will be the number of molecules available to diffuse for any given pressure difference. Also, the greater the cross-sectional area of the chamber, the greater will be the total number of molecules to diffuse. On the other hand, the greater the distance that the molecules must diffuse, the longer will it take the molecules to diffuse the entire distance. Finally, the greater the velocity of kinetic movement of the molecules, which is inversely proportional to the square root of the molecular weight, the greater is the rate of diffusion of the gas. All of these factors can be expressed in a single formula, as follows:

$$D \propto \frac{\Delta P \times A \times S}{d \times \sqrt{MW}}$$

in which D is the diffusion rate, ΔP is the pressure difference between the two ends of the diffusion pathway, A is the cross-sectional area of the pathway, S is the solubility of the gas, d is the distance of diffusion, and MW is the molecular weight of the gas.

It is obvious from this formula that the characteristics of the gas itself determine two factors of the formula: solubility and molecular weight, and these together are called the *diffusion coefficient* of the gas. That is, the diffusion coefficient equals $S/\sqrt{MW}$; the relative rates at which different gases at the same pressure levels will diffuse are proportional to their diffusion coefficients. Considering the diffusion coefficient for oxygen to be 1, the *relative* diffusion coefficients for different gases of respiratory importance in the body fluids are as follows:

Oxygen	1.0
Carbon dioxide	20.3
Carbon monoxide	0.81
Nitrogen	0.53
Helium	0.95

DIFFUSION OF GASES THROUGH TISSUES

The gases that are of respiratory importance are highly soluble in lipids and, consequently, are also highly soluble in cell membranes. Because of this, these gases diffuse through the cell membranes with very little impediment. Instead, the major limitation to the movement of gases in tissues is the rate at which the gases can diffuse through the tissue water instead of through the cell membranes. Therefore, diffusion of gases through the tissues, including through the respiratory membrane, is almost equal to the diffusion of gases through water, as given in the above list.

● COMPOSITION OF ALVEOLAR AIR—ITS RELATION TO ATMOSPHERIC AIR

Alveolar air does not have the same concentrations of gases as atmospheric air by any means, which can readily be seen by comparing the alveolar air composition in column 5 of Table 39–1 with the composition of atmospheric air in column 1. There are several reasons for the differences. First, the alveolar air is only partially replaced by atmospheric air with each breath. Second, oxygen is constantly being absorbed from the alveolar air. Third, carbon dioxide is constantly diffusing from the pulmonary blood into the alveoli. And, fourth, dry atmospheric air that enters the respiratory passages is humidified even before it reaches the alveoli.

Humidification of the Air As It Enters the Respiratory Passages. Column 1 of Table 39–1 shows that atmospheric air is composed almost entirely of nitrogen and oxygen; it normally contains almost no carbon dioxide and little water vapor. However, as soon as the atmospheric air enters the respiratory passages, it is exposed to the fluids covering the respiratory surfaces. Even before the air enters the alveoli, it becomes totally humidified.

The partial pressure of water vapor at normal body temperature of 37°C is 47 mm Hg, which, therefore, is the partial pressure of water in the alveolar air. Because the total pressure in the alveoli cannot rise to more than the atmospheric pressure, this water vapor

Table 39–1 PARTIAL PRESSURES OF RESPIRATORY GASES AS THEY ENTER AND LEAVE THE LUNGS (AT SEA LEVEL)

	Atmospheric Air* (mm Hg)		Humidified Air (mm Hg)		Alveolar Air (mm Hg)		Expired Air (mm Hg)	
N_2	597.0	(78.62%)	563.4	(74.09%)	569.0	(74.9%)	566.0	(74.5%)
O_2	159.0	(20.84%)	149.3	(19.67%)	104.0	(13.6%)	120.0	(15.7%)
CO_2	0.3	(0.04%)	0.3	(0.04%)	40.0	(5.3%)	27.0	(3.6%)
H_2O	3.7	(0.50%)	47.0	(6.20%)	47.0	(6.2%)	47.0	(6.2%)
TOTAL	760.0	(100.00%)	760.0	(100.00%)	760.0	(100.0%)	760.0	(100.0%)

*On an average cool, clear day.

simply *dilutes* all the other gases in the inspired air. In column 3 of Table 39–1 it can be seen that humidification of the air dilutes the oxygen partial pressure at sea level from an average of 159 mm Hg in atmospheric air to 149 mm Hg in the humidified air, and it dilutes the nitrogen partial pressure from 597 to 563 mm Hg.

RATE AT WHICH ALVEOLAR AIR IS RENEWED BY ATMOSPHERIC AIR

In Chapter 37, it was pointed out that the *functional residual capacity* of the lungs, the amount of air remaining in the lungs at the end of normal expiration, measures approximately 2300 milliliters. Yet only 350 milliliters of new air is brought into the alveoli with each normal respiration, and this same amount of old alveolar air is expired. Therefore, the amount of alveolar air replaced by new atmospheric air with each breath is only one seventh of the total, so that many breaths are required to exchange most of the alveolar air. Figure 39–2 illustrates this slow rate of renewal of the alveolar air. In the first alveolus of the figure an excess amount of a gas is in all the alveoli, but note that even at the end of 16 breaths the excess gas still has not been completely removed from the alveoli.

Figure 39–3 illustrates graphically the rate at which an excess of a gas in the alveoli is normally removed, showing that with normal alveolar ventilation approximately half the gas is removed in 17 seconds. When a person's rate of alveolar ventilation is only half normal, half the gas is removed in 34 seconds, and, when the rate of ventilation is two times normal, half is removed in about 8 seconds.

Importance of the Slow Replacement of Alveolar Air. This slow replacement of alveolar air is of particular importance in preventing sudden changes in gaseous concentrations in the blood.

This makes the respiratory control mechanism much more stable than it would otherwise be and helps to prevent excessive increases and decreases in tissue oxygenation, tissue carbon dioxide concentration, and tissue pH when respiration is temporarily interrupted.

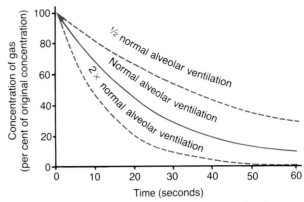

Figure 39–3. Rate of removal of excess gas from the alveoli.

OXYGEN CONCENTRATION AND PARTIAL PRESSURE IN THE ALVEOLI

Oxygen is continually being absorbed into the blood of the lungs, and new oxygen is continually being breathed into the alveoli from the atmosphere. The more rapidly oxygen is absorbed, the lower becomes its concentration in the alveoli; on the other hand, the more rapidly new oxygen is breathed into the alveoli from the atmosphere, the higher becomes its concentration. Therefore, oxygen concentration in the alveoli, and therefore its partial pressure as well, is controlled by, first, the rate of absorption of oxygen into the blood and, second, the rate of entry of new oxygen into the lungs by the ventilatory process.

Figure 39–4 illustrates the effect both of alveolar ventilation and of rate of oxygen absorption into the blood on the alveolar partial pressure of oxygen (PA_{O_2}). The solid curve represents oxygen absorption at a rate of 250 ml/min, and the dotted curve at 1000 ml/min. At a normal ventilatory rate of 4.2 liters/min and an oxygen consumption of 250 ml/min, the normal operating point in Figure 39–4 is point A. The figure also shows that when 1000 milliliters of oxygen is being absorbed each minute, as occurs during moderate exercise, the rate of alveolar ventilation must increase fourfold to maintain the alveolar PO_2 at the normal value of 104 mm Hg.

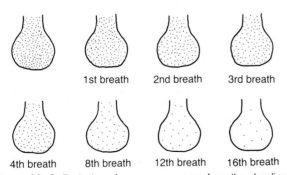

Figure 39–2. Expiration of a gaseous excess from the alveoli with successive breaths.

Figure 39–4. Effect of alveolar ventilation and of two rates of oxygen absorption, 250 ml/min and 1000 ml/min, from the alveoli on the alveolar PO_2.

Another effect illustrated in Figure 39–4 is that an extremely marked increase in alveolar ventilation can never increase the alveolar Po_2 above 149 mm Hg as long as the person is breathing normal atmospheric air, for this is the maximum Po_2 of oxygen in humidified air. However, if the person breathes gases containing partial pressures of oxygen higher than 149 mm Hg, the alveolar Po_2 can approach these higher pressures at very high rates of ventilation.

CO₂ CONCENTRATION AND PARTIAL PRESSURE IN THE ALVEOLI

Carbon dioxide is continually being formed in the body, then discharged into the alveoli; and it is continually being removed from the alveoli by ventilation. Figure 39–5 illustrates the effects on the alveolar Pco_2 of both alveolar ventilation and two rates of carbon dioxide excretion. The solid curve represents a normal rate of carbon dioxide excretion of 200 ml/min. At the normal rate of alveolar ventilation of 4.2 liters/min, the operating point for alveolar Pco_2 is at point A in Figure 39–5—that is, 40 mm Hg.

Two other facts are also evident from Figure 39–5: First, *the alveolar Pco_2 increases directly in proportion to the rate of carbon dioxide excretion*, as represented by the elevation of the dotted curve for 800 milliliters CO_2 excretion per minute. Second, *the alveolar Pco_2 decreases in inverse proportion to alveolar ventilation*. Therefore, the concentrations and partial pressures of both oxygen and carbon dioxide in the alveoli are determined by the rates of absorption or excretion of the two gases, and also by the alveolar ventilation.

EXPIRED AIR

Expired air is a combination of dead space air and alveolar air, and its overall composition is, therefore, determined by, first, the proportion of the expired air that is dead space air and the proportion that is alveolar air. Figure 39–6 shows the progressive changes in oxygen and carbon dioxide partial pressures in the expired air during the course of expiration. The very first portion of this air, the dead space

Figure 39–5. Effect on alveolar Pco_2 of alveolar ventilation and rate of carbon dioxide excretion from the blood.

Figure 39–6. Oxygen and carbon dioxide partial pressures in the various portions of normal expired air.

air, is typical humidified air as shown in column 3 of Table 39–1. Then, progressively more and more alveolar air becomes mixed with the dead space air until all the dead space air has finally been washed out and nothing but alveolar air is expired at the end of expiration. Therefore, the means for collecting alveolar air for study is simply to collect a sample of the last portion of expired air.

Normal expired air, containing both dead space air and alveolar air, has gaseous concentrations and partial pressures approximately as shown in column 7 of Table 39–1— that is, concentrations between those of alveolar air and humidified atmospheric air.

● DIFFUSION OF GASES THROUGH THE RESPIRATORY MEMBRANE

The Respiratory Unit. Figure 39–7 illustrates the respiratory unit, which is composed of a *respiratory bronchiole, alveolar ducts, atria,* and *alveoli* (of which there are about 300 million in the two lungs, each alveolus having an average diameter of about 0.2 millimeter). The alveolar walls are extremely thin, and within them is an almost solid network of interconnecting capillaries, illustrated in Figure 39–8. Indeed, because of the extensiveness of the capillary plexus, the flow of blood in the alveolar wall has been described as a "sheet" of flowing blood. Thus, it is obvious that the alveolar gases are in close proximity to the blood of the capillaries. Consequently, gaseous exchange between the alveolar air and the pulmonary blood occurs through the membranes of all the terminal portions of the lungs. These membranes are collectively known as the *respiratory membrane,* also called the *pulmonary membrane.*

The Respiratory Membrane. Figure 39–9 illustrates to the left the ultrastructure of the respiratory membrane shown in cross-section and to the right a red blood cell. It also shows the diffusion of oxygen from the alveolus into the red blood cell and diffusion

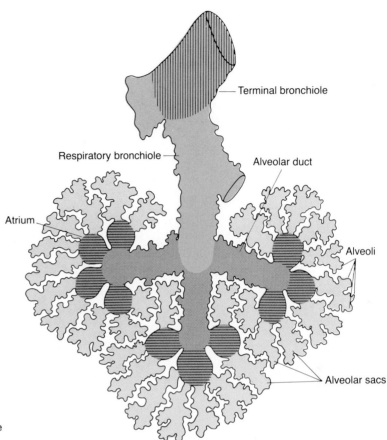

Figure 39-7. The respiratory lobule. (From Miller: The Lung. Springfield, Ill., Charles C Thomas.)

of carbon dioxide in the opposite direction. Note the following different layers of the respiratory membrane:

1. A layer of fluid lining the alveolus and containing surfactant that reduces the surface tension of the alveolar fluid.

2. The alveolar epithelium comprised of very thin epithelial cells.

3. An epithelial basement membrane.

4. A very thin interstitial space between the alveolar epithelium and the capillary membrane.

5. A capillary basement membrane that in many places fuses with the epithelial basement membrane.

6. The capillary endothelial membrane.

Despite the large number of layers, the overall thickness of the respiratory membrane in some areas is as little as 0.2 micrometer, and it averages about 0.6 micrometer.

From histologic studies it has been estimated that the total surface area of the respiratory membrane is approximately 50 to 100 square meters in the normal adult. This is equivalent to the floor area of a 25 by 30 foot room. The total quantity of blood in the capillaries of the lung at any given instant is 60 to 140 milliliters. Now imagine this small amount of blood spread over the entire surface of a 25 by 30 foot floor, and it is very easy to understand the rapidity of respiratory exchange of gases.

The average diameter of the pulmonary capillaries

is only about 5 micrometers, which means that red blood cells must actually squeeze through them. Therefore, the red blood cell membrane usually touches the capillary wall so that oxygen and carbon dioxide need not pass through significant amounts of plasma as they diffuse between the alveolus and the red cell. Obviously, this, too, increases the rapidity of diffusion.

FACTORS THAT AFFECT THE RATE OF GAS DIFFUSION THROUGH THE RESPIRATORY MEMBRANE

Referring to the earlier discussion of diffusion through water, one can apply the same principles and same formula to diffusion of gases through the respiratory membrane. Thus, the factors that determine how rapidly a gas will pass through the membrane are (1) the *thickness of the membrane;* (2) the *surface area of the membrane;* (3) the *diffusion coefficient* of the gas in the substance of the membrane — that is, in the water of the membrane; and (4) the *pressure difference* between the two sides of the membrane.

The *thickness of the respiratory membrane* occasionally increases, often as a result of edema fluid in the interstitial space of the membrane and in the alveoli, so that the respiratory gases must then diffuse not only through the membrane but also through this fluid. Also, some pulmonary diseases cause fibrosis of

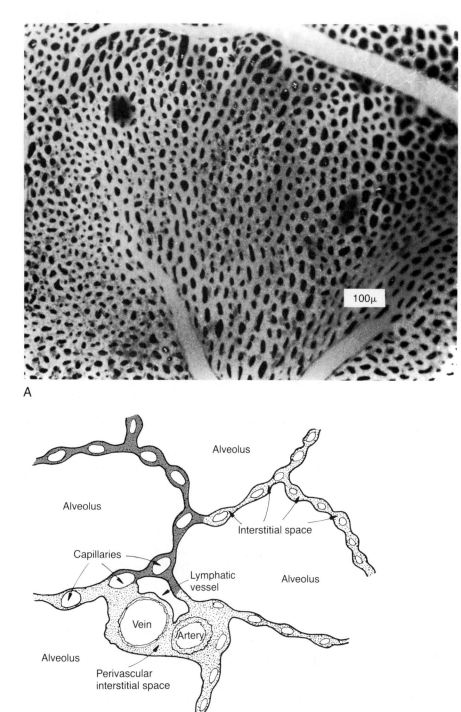

A

B

Figure 39–8. *A,* Surface view of capillaries in the alveolar wall. (From Maloney and Castle: *Resp. Physiol.,* 7:150, 1969. Reproduced by permission of ASP Biological and Medical Press. North-Holland Division.) *B,* Cross-sectional view of alveolar walls and their vascular supply.

the lungs, which can increase the thickness of some portions of the respiratory membrane. Because the rate of diffusion through the membrane is inversely proportional to the thickness of the membrane, any factor that increases the thickness to more than two to three times normal can interfere significantly with normal respiratory exchange of gases.

The *surface area of the respiratory membrane* can be greatly decreased by many different conditions. For instance, removal of an entire lung decreases the sur-

face area to half normal. Also, in *emphysema* many of the alveoli coalesce, with dissolution of many alveolar walls. Therefore, the new chambers are much larger than the original alveoli, but the total surface area of the respiratory membrane is considerably decreased because of loss of the alveolar walls. When the total surface area is decreased to approximately one-third to one-fourth normal, exchange of gases through the membrane is impeded to a significant degree *even under resting conditions.* During competitive sports

Figure 39–9. Ultrastructure of the respiratory membrane as shown in cross-section.

and other strenuous exercise, even the slightest decrease in surface area of the lungs can be a serious detriment to respiratory exchange of gases.

The *diffusion coefficient* for the transfer of each gas through the respiratory membrane depends on its *solubility* in the membrane and inversely on the *square root* of its *molecular weight*. The rate of diffusion in the respiratory membrane is almost exactly the same as that in water, for reasons explained earlier. Therefore, for a given pressure difference, carbon dioxide diffuses through the membrane about 20 times as rapidly as oxygen. Oxygen in turn diffuses about two times as rapidly as nitrogen.

The *pressure difference* across the respiratory membrane is the difference between the partial pressure of the gas in the alveoli and the pressure of the gas in the blood. The partial pressure represents a measure of the total number of molecules of a particular gas striking a unit area of the alveolar surface of the membrane in unit time, and the pressure of the gas in the blood represents the number of molecules attempting to escape from the blood in the opposite direction. Therefore, the difference between these two pressures is a measure of the *net tendency* for the gas to move through the membrane. Obviously, when the partial pressure of a gas in the alveoli is greater than the pressure of the gas in the blood, as is true for oxygen, net diffusion from the alveoli into the blood occurs; but when the pressure of the gas in the blood is greater than the partial pressure in the alveoli, as is true for carbon dioxide, net diffusion from the blood into the alveoli occurs.

DIFFUSING CAPACITY OF THE RESPIRATORY MEMBRANE

The ability of the respiratory membrane to exchange a gas between the alveoli and the pulmonary blood can be expressed in quantitative terms by its *diffusing capacity,* which is defined as the *volume of a gas that diffuses through the membrane each minute for a pressure difference of 1 mm Hg.*

Obviously, all the factors discussed above that affect diffusion through the respiratory membrane can affect the diffusing capacity.

The Diffusing Capacity for Oxygen. In the average young male adult the *diffusing capacity for oxygen* under resting conditions averages *21 ml/min/mm Hg.* In functional terms, what does this mean? The mean oxygen pressure difference across the respiratory membrane during normal, quiet breathing is approximately 11 mm Hg. Multiplication of this pressure by the diffusing capacity (11×21) gives a total of about 230 milliliters of oxygen diffusing through the respiratory membrane each minute; this is equal to the rate at which the body uses oxygen.

Change in Oxygen-Diffusing Capacity During Exercise. During strenuous exercise, or during other conditions that greatly increase pulmonary blood flow and alveolar ventilation, the diffusing capacity for oxygen increases in young male adults to a maximum of about 65 ml/min/mm Hg, which is three times the diffusing capacity under resting conditions. This increase is caused by several different factors, among which are (1) opening up of a number of previously dormant pulmonary capillaries, thereby increasing the surface area of the blood into which the oxygen can diffuse; and (2) a better match between the ventilation of the alveoli and the perfusion of the alveolar capillaries with blood, called the "ventilation-perfusion ratio," which will be explained in detail later in the chapter. Therefore, during exercise, the oxygenation of the blood is increased not only by increased alveolar ventilation but also by a greater capacity of the respiratory membrane for transmitting oxygen into the blood.

Diffusing Capacity for Carbon Dioxide. The diffusing capacity for carbon dioxide has never been measured because of the following technical difficulty: Carbon dioxide diffuses through the respiratory membrane so rapidly that the average P_{CO_2} in the pulmonary blood is not far different from the P_{CO_2} in the alveoli—the average difference is less than 1 mm Hg—and with the available techniques, this difference is too small to be measured.

Nevertheless, measurements of diffusion of other gases have shown that the diffusing capacity varies directly with the diffusion coefficient of the particular gas. Since the diffusion coefficient of carbon dioxide is

20 times that of oxygen, one would expect a diffusing capacity for carbon dioxide under resting conditions of about 400 to 450 ml/min/mm Hg and during exercise of about 1200 to 1300 ml/min/mm Hg.

Figure 39–10 compares the measured or calculated diffusing capacities of oxygen, carbon dioxide, and carbon monoxide at rest and during exercise, showing the extreme diffusing capacity of carbon dioxide and also the effect of exercise on the diffusing capacities of all these gases.

Measurement of Diffusing Capacity. The oxygen-diffusing capacity can be calculated from measurements of (1) alveolar P_{O_2}, (2) P_{O_2} in the pulmonary capillary blood, and (3) the rate of oxygen uptake by the blood. However, to measure the P_{O_2} in the pulmonary capillary blood is so difficult and so imprecise that it is not practical to measure oxygen-diffusing capacity by such a direct procedure except on an experimental basis.

To obviate the difficulties encountered in measuring oxygen-diffusing capacity directly, physiologists usually measure carbon monoxide–diffusing capacity instead and then calculate the oxygen-diffusing capacity from this. The principle of the carbon monoxide method is the following: A small amount of carbon monoxide is breathed into the alveoli, and the partial pressure of the carbon monoxide in the alveoli is measured from appropriate alveolar air samples. The carbon monoxide pressure in the blood is essentially zero because hemoglobin combines with this gas so rapidly that its pressure never has time to build up. Therefore, the pressure difference of carbon monoxide across the respiratory membrane is equal to its partial pressure in the alveoli. Then, by measuring the volume of carbon monoxide absorbed in a period of time and dividing this by the alveolar carbon monoxide partial pressure, one can determine accurately the carbon monoxide–diffusing capacity.

To convert carbon monoxide–diffusing capacity to oxygen-diffusing capacity, the value is multiplied by a factor of 1.23 because the diffusion coefficient for oxygen is 1.23 times that for carbon monoxide. Thus, the average diffusing capacity for carbon monoxide in young male adults at rest is 17 ml/min/mm Hg, and the diffusing capacity for oxygen is 1.23 times this, or 21 ml/min/mm Hg.

● EFFECT OF THE VENTILATION-PERFUSION RATIO ON ALVEOLAR GAS CONCENTRATION

In the early part of this chapter we learned that two factors determine the P_{O_2} and the P_{CO_2} in the aveoli: (1) the rate of alveolar ventilation and (2) the rate of transfer of oxygen and carbon dioxide through the respiratory membrane. However, these earlier discussions made the assumption that all the alveoli are ventilated equally and also that the blood flow through the alveolar capillaries is the same for each alveolus. However, even normally to some extent, and especially in many lung diseases, some areas of the lungs are well ventilated but have almost no blood flow, whereas other areas may have excellent blood flow but little or no ventilation. Obviously, in either of these conditions, gas exchange through the respiratory membrane is seriously impaired; and the person may suffer severe respiratory distress despite both normal *total* ventilation and also normal *total* pulmonary blood flow, but with the ventilation and blood flow going to different parts of the lungs. Therefore, a concept has been developed to help us understand respiratory exchange when there is imbalance between alveolar ventilation and alveolar blood flow. This concept is called the *ventilation-perfusion ratio.*

In quantitative terms, the ventilation-perfusion ratio is expressed as $\dot{V}_A/\dot{Q}$. When $\dot{V}_A$ (alveolar ventilation) is normal for a given alveolus and $\dot{Q}$ (blood flow) is also normal for the same alveolus, then the ventilation-perfusion ratio ($\dot{V}_A/\dot{Q}$) is also said to be normal. However, when the ventilation ($\dot{V}_A$) is zero and yet there is still perfusion ($\dot{Q}$) of the alveolus, the ventilation-perfusion ratio is zero. Or, at the other extreme, when there is adequate ventilation ($\dot{V}_A$) but zero perfusion ($\dot{Q}$), then the ratio is infinity. At a ratio of either zero or infinity, there will be no exchange of gases through the respiratory membrane of the affected alveoli, which explains the importance of this concept. Therefore, let us explain the respiratory consequences of these two extremes.

Alveolar Oxygen and Carbon Dioxide Partial Pressures When $\dot{V}_A/\dot{Q}$ Equals Zero. When $\dot{V}_A/\dot{Q}$ is equal to zero—that is, without any alveolar ventilation—the air in the alveolus comes to equilibrium with the blood oxygen and carbon dioxide, because these gases diffuse between the blood and the alveolar air. Since the blood that perfuses the capillaries is venous blood returning to the lungs from the systemic circulation, it is the gases in this blood that come to equilibrium with the alveolar gases. In the next chapter we shall see that the normal venous blood ($\bar{v}$) has a P_{O_2} of 40 mm Hg and a P_{CO_2} of 45 mm Hg. Therefore, these are also the normal partial pressures of these two gases in alveoli that have blood flow but no ventilation.

Alveolar Oxygen and Carbon Dioxide Partial Pressures When $\dot{V}_A/\dot{Q}$ Equals Infinity. The effect on the alveolar gas partial pressures is entirely different when $\dot{V}_A/\dot{Q}$ equals infinity from the effect when $\dot{V}_A/\dot{Q}$ equals zero, because now there is no capillary blood flow to carry oxygen away or to bring carbon dioxide to the alveoli. Therefore, instead of the alveolar gases coming to equilibrium with the venous blood, the alveolar air now becomes equal to the

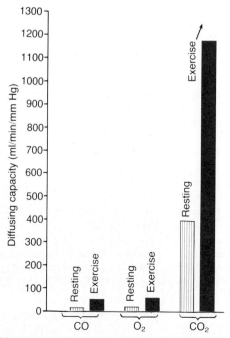

Figure 39–10. Lung diffusing capacities for carbon monoxide, oxygen, and carbon dioxide in the normal lungs.

humidified inspired air. That is, the air that is inspired loses no oxygen to the blood and it gains no CO_2. And, since normal inspired and humidified air has a Po_2 of 149 mm Hg and a Pco_2 of 0 mm Hg, these are the partial pressures of these two gases in the alveoli.

Gas Exchange, and Alveolar Partial Pressures When $\dot{V}A/\dot{Q}$ Is Normal. When there is both normal alveolar ventilation and also normal alveolar capillary blood flow (which means normal alveolar perfusion), exchange of oxygen and carbon dioxide through the respiratory membrane is nearly optimal, and alveolar Po_2 is normally at a level of 104 mm Hg, which lies between that of the inspired air (149 mm Hg) and that of venous blood (40 mm Hg). Likewise, alveolar Pco_2 lies between the two extremes; it is normally 40 mm Hg, in contrast to 45 mm Hg in venous blood and 0 mm Hg in the inspired air. Thus, under normal conditions, the alveolar air Po_2 averages 104 mm Hg and the Pco_2, 40 mm Hg.

The Po_2-Pco_2, $\dot{V}A/\dot{Q}$ Diagram

The concepts presented in the above sections can be shown very usefully in graphical form, as illustrated in Figure 39-11, in a diagram called the Po_2-Pco_2, $\dot{V}A/\dot{Q}$ diagram. The curve in the diagram represents all possible Po_2 and Pco_2 combinations between the limits of $\dot{V}A/\dot{Q}$ equals zero and $\dot{V}A/\dot{Q}$ equals infinity when the gaseous pressures in the venous blood are normal and the person is breathing air at sea level pressure. Thus, point $\bar{v}$ is the plot of Po_2 and Pco_2 when $\dot{V}A/\dot{Q}$ equals zero. At this point, the Po_2 is 40 mm Hg and the Pco_2 is 45 mm Hg, which are the values in venous blood.

At the other end of the curve, when $\dot{V}A/\dot{Q}$ equals infinity, point I represents inspired air, showing Po_2 to be 149 mm Hg while Pco_2 is zero.

Also plotted on the curve is the point representing normal alveolar air when $\dot{V}A/\dot{Q}$ is normal. At this point, Po_2 is 104 mm Hg and Pco_2 is 40 mm Hg.

The Concept of "Physiologic Shunt" (When $\dot{V}A/\dot{Q}$ Is Below Normal)

Whenever $\dot{V}A/\dot{Q}$ is below normal, there obviously is not ventilation enough to provide the oxygen needed to oxygenate the blood flowing through the alveolar capillaries. Therefore, a certain fraction of the venous blood passing through the pulmonary capillaries does not become oxygenated. This fraction is called *shunted blood*. Also, some additional blood flows through the bronchial vessels, rather than through the alveolar capillaries, normally about 2 per

cent of the cardiac output; this, too, is unoxygenated, shunted blood.

The total quantitative amount of shunted blood per minute is called the *physiologic shunt*. This physiologic shunt is measured in the clinical pulmonary function laboratory by analyzing the concentration of oxygen in both mixed venous blood and arterial blood along with simultaneous measurement of the cardiac output. From these values the physiologic shunt can then be calculated by means of the following equation:

$$\frac{\dot{Q}PS}{\dot{Q}T} = \frac{Ci_{O_2} - Ca_{O_2}}{Ci_{O_2} - C\bar{v}_{O_2}}$$

in which

$\dot{Q}PS$ is the physiologic shunt blood flow per minute
$\dot{Q}T$ is cardiac output per minute
Ci_{O_2} is the concentration of oxygen in the arterial blood if there is "ideal" ventilation-perfusion ratio
Ca_{O_2} is the measured concentration of oxygen in the arterial blood
$C\bar{v}_{O_2}$ is the measured concentration of oxygen in the mixed venous blood

Obviously, the greater the physiologic shunt, the greater the *amount of blood that fails to be oxygenated* as it passes through the lungs.

The Concept of "Physiologic Dead Space" (When $\dot{V}A/\dot{Q}$ Is Greater Than Normal)

When ventilation of some of the alveoli is great but the alveolar blood flow is low, there is then far more available oxygen in the alveoli than can be transported away from the alveoli by the flowing blood. Thus, the ventilation of these alveoli is said to be *wasted*. In addition, the ventilation of the anatomical dead space areas of the respiratory passageways is also wasted. The sum of these two types of wasted ventilation is called the *physiologic dead space*. This is measured in the clinical pulmonary function laboratory by making appropriate blood and expiratory gas measurements and using the following equation, called the Bohr equation:

$$\frac{VD_{phys}}{VT} = \frac{Pa_{CO_2} - P\bar{E}_{CO_2}}{Pa_{CO_2}}$$

in which

VD_{phys} is the physiologic dead space
VT is the tidal volume
Pa_{CO_2} is the partial pressure of carbon dioxide in the arterial blood
$P\bar{E}_{CO_2}$ is the average partial pressure of carbon dioxide in the entire expired air

Obviously, when the physiologic dead space is very great, much of the *work of ventilation* is wasted effort because so much of the ventilated air never reaches the blood.

Abnormalities of Ventilation-Perfusion Ratio

Abnormal $\dot{V}A/\dot{Q}$s in the Upper and Lower Normal Lung. In a normal person in the upright position, both

Figure 39-11. The normal Po_2-Pco_2, $\dot{V}A/\dot{Q}$ diagram.

blood flow and alveolar ventilation are considerably less in the upper part of the lung than in the lower part; however, blood flow is decreased considerably more than ventilation. Therefore, at the top of the lung, $\dot{V}_A/\dot{Q}$ is as much as 3.0 times as great as the ideal value, which causes a moderate degree of *physiologic dead space* in this area of the lung.

At the other extreme, in the bottom of the lung there is slightly too little ventilation in relation to blood flow, with $\dot{V}_A/\dot{Q}$ as low as 0.6 times the ideal value. Therefore, in this area a small fraction of the blood fails to become normally oxygenated, and this represents a *physiologic shunt.*

Therefore, in both extremes of the lung, inequalities of ventilation and perfusion decrease the effectiveness of the lung slightly for exchange of oxygen and carbon dioxide. However, during exercise the blood flow to the upper part of the lung increases markedly, so that far less physiologic dead space occurs and the effectiveness of gas exchange approaches optimum.

Abnormal $\dot{V}_A/\dot{Q}$ in Chronic Obstructive Lung Disease. Most persons who smoke for prolonged periods of time develop various degrees of bronchial obstruction; in a large share of them this eventually becomes so severe that they develop serious alveolar air trapping and resultant emphysema. The emphysema in turn causes many of the alveolar walls to be destroyed. Thus, two different abnormalities cause abnormal $\dot{V}_A/\dot{Q}$. First, because many of the small bronchioles are obstructed, the alveoli beyond the obstructions are unventilated, causing a $\dot{V}_A/\dot{Q}$ that approaches zero. Second, in those areas of the lung where the alveolar walls have been mainly destroyed but there is still alveolar ventilation, most of the ventilation is wasted because of inadequate blood flow to transport the blood gases.

Therefore, in chronic obstructive lung disease, some areas of the lung exhibit very *serious physiologic shunt* and other areas very *serious physiologic dead space.* Both of these tremendously decrease the effectiveness of the lungs as gas exchange organs, sometimes reducing the effectiveness to as little as one-tenth normal. In fact, this is the most prevalent cause of pulmonary disability today.

REFERENCES

Bauer, C., et al. (eds.): Biophysics and Physiology of Carbon Dioxide. New York, Springer-Verlag, 1980.

Bidani, A., and Crandall, E. D.: Velocity of CO_2 exchanges in the lungs. Annu. Rev. Physiol., 50:639, 1988.

Drazen, J. M., et al.: High-frequency ventilation. Physiol. Rev., 64:505, 1984.

Forster, R. E., and Crandall, E. D.: Pulmonary gas exchange. Annu. Rev. Physiol., 38:69, 1976.

Forster, R. E.: Diffusion of gases. In Fenn, W. O., and Rahn, H. (eds.): Handbook of Physiology. Sec. 3, Vol. 1. Baltimore, Williams & Wilkins, 1964, p. 839.

Forster, R. E.: Interpretation of measurements of pulmonary diffusing capacity. In Fenn, W. O., and Rahn, H. (eds.): Handbook of Physiology. Sec. 3, Vol. 2. Baltimore, Williams & Wilkins, 1965, p. 1435.

Glass, M. L., and Wood, S. C.: Gas exchange and control of breathing in reptiles. Physiol. Rev., 63:232, 1983.

Gonzales, N. C., and Fedde, M. R. (eds.): Oxygen Transfer from Atmosphere to Tissues. New York, Plenum Publishing Corp., 1988.

Guyton, A. C., et al.: An arteriovenous oxygen difference recorder. J. Appl. Physiol., 10:158, 1957.

Huch, A., et al. (eds.): Continuous Transcutaneous Monitoring. New York, Plenum Publishing Corp., 1987.

Hughes, J. M.: Proceedings: Regional differences in gas exchange. Proc. R. Soc. Med., 6:974, 1973.

Johansen, K.: Comparative physiology: Gas exchange and circulation in fishes. Annu. Rev. Physiol., 33:569, 1971.

Jones, N. L.: Blood Gases and Acid-Base Physiology. New York, B. C. Decker, 1980.

Klocke R. A.: Velocity of CO_2 exchange in blood. Annu. Rev. Physiol., 50:625, 1988.

Lane, E. E., and Walker, J. F.: Clinical Arterial Blood Gas Analysis. St. Louis, C. V. Mosby Co., 1987.

Levitsky, M. G.: Pulmonary Physiology. New York, McGraw-Hill Book Co., 1982.

Milhorn, H. T., Jr., and Pulley, P. E., Jr.: A theoretical study of pulmonary capillary gas exchange and venous admixture. Biophys. J., 8:337, 1968.

Mochizuki, M., et al. (eds.): Oxygen Transport to Tissue X. New York, Plenum Publishing Corp., 1988.

Olson, L. E.: Physiology laboratories quantifying gas exchange in health and disease. The Physiologist, 28:170, 1985.

Otis, A. B.: Quantitative relationships in steady-state gas exchange. In Fenn, W. O., and Rahn, H. (eds): Handbook of Physiology. Sec. 3, Vol. 1. Baltimore, Williams & Wilkins, 1964, p. 681.

Paiva, M., and Engel, L. A.: Theoretical studies of gas mixing and ventilation distribution in the lung. Physiol. Rev., 67:750, 1987.

Piper, J., and Scheid, P.: Comparative physiology of respiration: Functional analysis of gas exchange organs in vertebrates. Int. Rev. Physiol., 14:219, 1977.

Piper, J., and Scheid, P.: Respiration: Alveolar gas exchange. Annu. Rev. Physiol., 33:131, 1971.

Quintanilha, A., (ed.): Oxygen Radicals in Biological Systems. New York, Plenum Publishing Corp., 1988.

Rahn, H., and Farhi, E. E.: Ventilation, perfusion, and gas exchange—the Va/Q concept. In Fenn, W. O., and Rahn, H. (eds.): Handbook of Physiology. Sec. 3, Vol. 1. Baltimore, Williams & Wilkins, 1964, p. 125.

Rahn, H., and Fenn, W. O.: A Graphical Analysis of Respiratory Gas Exchange. Washington, D.C., American Physiological Society, 1955.

Rahn, H., et al.: How bird eggs breathe. Sci. Am., 240(2):46, 1979.

Riley, R. L., and Permutt, S.: The four-quadrant diagram for analyzing the distribution of gas and blood in the lung. In Fenn, W. O., and Rahn, H. (eds.): Handbook of Physiology. Sec. 3, Vol. 2. Baltimore, Williams & Wilkins, 1965, p. 1413.

Wagner, P. D.: Diffusion and chemical reaction in pulmonary gas exchange. Physiol. Rev., 57:257, 1977.

Wagner, P. D.: Ventilation-perfusion relationships. Annu. Rev. Physiol., 42:235, 1980.

Weibel, E. R.: Morphological basis of alveolar capillary gas exchange. Physiol. Rev., 53:419, 1973.

West, J. B.: Pulmonary gas exchange. Int. Rev. Physiol., 14:83, 1977.

CHAPTER 40

• *Transport of Oxygen and Carbon Dioxide in the Blood and Body Fluids* •

Once oxygen has diffused from the alveoli into the pulmonary blood, it is transported principally in combination with hemoglobin to the tissue capillaries, where it is released for use by the cells. The presence of hemoglobin in the red cells of the blood allows the blood to transport 30 to 100 times as much oxygen as could be transported simply in the form of dissolved oxygen in the water of the blood.

In the tissue cells oxygen reacts with various foodstuffs to form large quantities of carbon dioxide. This in turn enters the tissue capillaries and is transported back to the lungs. Carbon dioxide, like oxygen, also combines with chemical substances in the blood that increase carbon dioxide transport 15- to 20-fold.

The purpose of the present chapter, therefore, is to present both qualitatively and quantitatively the physical and chemical principles of oxygen and carbon dioxide transport in the blood and body fluids.

• PRESSURES OF OXYGEN AND CARBON DIOXIDE IN THE LUNGS, BLOOD, AND TISSUES

In the discussions of the preceding chapter it was pointed out that gases can move from one point to another by diffusion and that the cause of this movement is always a pressure difference from the first point to the other. Thus, oxygen diffuses from the alveoli into the pulmonary capillary blood because the oxygen pressure (Po_2) in the alveoli is greater than the Po_2 in the pulmonary blood. Then in the tissues a higher Po_2 in the capillary blood causes oxygen to diffuse to the cells.

Conversely, when oxygen is metabolized in the cells, the carbon dioxide pressure (Pco_2) rises to a high value, which causes carbon dioxide to diffuse into the tissue capillaries. Similarly, it diffuses out of the blood into the alveoli because the Pco_2 in the alveoli is lower than that in the pulmonary capillary blood.

Basically, then, the transport of oxygen and carbon dioxide by the blood depends on both diffusion and the movement of blood. We now need to consider quantitatively the factors responsible for these effects.

UPTAKE OF OXYGEN BY THE PULMONARY BLOOD

The top part of Figure 40-1 illustrates a pulmonary alveolus adjacent to a pulmonary capillary, showing diffusion of oxygen molecules between the alveolar air and the pulmonary blood. The Po_2 of the gaseous oxygen in the alveolus averages 104 mm Hg, whereas the Po_2 of the venous blood entering the capillary averages only 40 mm Hg because a large amount of oxygen has been removed from this blood as it has passed through the peripheral tissues. Therefore, the *initial* pressure difference that causes oxygen to diffuse into the pulmonary capillary is $104 - 40$, or 64 mm Hg. The curve below the capillary shows the progressive rise in blood Po_2 as the blood passes through the capillary, showing that the Po_2 rises essentially to equal that of the alveolar air by the time the blood has moved a third of the distance through the capillary, becoming almost 104 mm Hg.

Uptake of Oxygen by the Pulmonary Blood During Exercise. During strenuous exercise, a person's body may require as much as 20 times the normal amount of oxygen. Also, because of the increased cardiac output, the time that the blood remains in the capillary may be reduced to less than one-half normal despite the fact that additional capillaries open up. Therefore, oxygenation of the blood could suffer for two reasons. Yet because of the great *safety factor* for diffusion of oxygen through the pulmonary membrane, the blood is still *almost completely saturated* with oxygen when it leaves the pulmonary capillaries. The reasons for this are as follows:

First, it was pointed out in the previous chapter that the diffusing capacity for oxygen increases almost threefold during exercise; this results mainly from in-

Figure 40–1. Uptake of oxygen by the pulmonary capillary blood. (The curve in this figure was constructed from data in Milhorn and Pulley: *Biophys. J., 8*:337, 1968.)

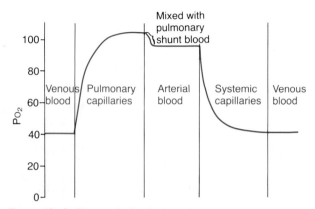

Figure 40–2. Changes in P_{O_2} in the pulmonary capillary blood, the arterial blood, and the systemic capillary blood, illustrating the effect of "venous admixture."

creased numbers of capillaries participating in the diffusion but also from more nearly ideal ventilation-perfusion ratio in the upper part of the lungs.

Second, note in Figure 40–1 that during normal pulmonary blood flow the blood becomes almost saturated with oxygen by the time it has passed through one third of the pulmonary capillary, and little additional oxygen enters the blood during the latter two thirds of its transit. That is, the blood normally stays in the lung capillaries about three times as long as necessary to cause full oxygenation. Therefore, even with the shortened time of exposure in exercise, the blood still can become either fully oxygenated or nearly so.

TRANSPORT OF OXYGEN IN THE ARTERIAL BLOOD

About 98 per cent of the blood that enters the left atrium from the lung has passed through the alveolar capillaries and has become fully oxygenated, that is, up to a P_{O_2} of approximately 104 mm Hg, as just explained. However, another 2 per cent of the blood has passed through the bronchial circulation, which supplies mainly the deep tissues of the lungs and is not exposed to the pulmonary air. Therefore, this blood flow represents "shunt" flow, meaning blood that is shunted past the gas exchange areas. On leaving the lungs, its P_{O_2} is approximately that of normal venous blood, about 40 mm Hg. This blood combines in the pulmonary veins with the oxygenated blood from the alveolar capillaries; this mixing of the bloods is called *venous admixture of blood,* and it causes the P_{O_2} of the blood pumped by the left heart into the aorta to fall to approximately 95 mm Hg. These changes in blood P_{O_2} at different points in the circulatory system are illustrated in Figure 40–2.

DIFFUSION OF OXYGEN FROM THE CAPILLARIES TO THE INTERSTITIAL FLUID

When the arterial blood reaches the peripheral tissues, its P_{O_2} is still 95 mm Hg. On the other hand, as shown in Figure 40–3, the P_{O_2} in the interstitial fluid averages only 40 mm Hg. Thus, there is a tremendous initial pressure difference that causes oxygen to diffuse very rapidly from the blood into the tissues, so rapidly that the capillary P_{O_2} falls almost to equal the 40 mm Hg pressure in the interstitium. Therefore, the P_{O_2} of the blood entering the veins from the tissue capillaries is also about 40 mm Hg.

Effect of Rate of Blood Flow on Interstitial Fluid P_{O_2}. If the blood flow through a particular tissue becomes increased, greater quantities of oxygen are transported into the tissue in a given period of time, and the tissue P_{O_2} becomes correspondingly increased. This effect is illustrated in Figure 40–4. Note that an increase in flow to 400 per cent of normal increases the P_{O_2} from 40 mm Hg at point A in the figure to 66 mm Hg at point B. However, the upper limit to which the P_{O_2} can rise, even with maximal blood flow, is 95 mm Hg, because this is the oxygen pressure in the arterial blood.

Effect of Rate of Tissue Metabolism on Interstitial Fluid P_{O_2}. If the cells utilize more oxygen for metabolism than normally, this tends to reduce the interstitial fluid P_{O_2}. Figure 40–4 also illustrates this effect, showing reduced interstitial fluid P_{O_2} when the oxygen consumption is increased and increased P_{O_2} when the consumption is decreased.

In summary, tissue P_{O_2} is determined by a balance between (1) the rate of oxygen transport to the tissues in the blood and (2) the rate at which the oxygen is utilized by the tissues.

Figure 40–3. Diffusion of oxygen from a tissue capillary to the cells.

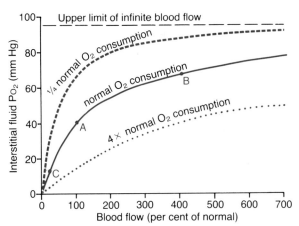

Figure 40–4. Effect of blood flow and rate of oxygen consumption on tissue PO$_2$.

Figure 40–5. Uptake of carbon dioxide by the blood in the capillaries.

DIFFUSION OF OXYGEN FROM THE CAPILLARIES TO THE CELLS

Oxygen is always being used by the cells. Therefore, the intracellular Po$_2$ remains lower than the Po$_2$ in the capillaries. Also, in many instances, there is considerable distance between the capillaries and the cells. Therefore, the normal intracellular Po$_2$ ranges from as low as 5 mm Hg to as high as 40 mm Hg, averaging (by direct measurement in lower animals) 23 mm Hg. Because only 1 to 3 mm Hg of oxygen pressure is normally required for full support of the metabolic processes of the cell, one can see that even this low cellular Po$_2$ of 23 mm Hg is more than adequate and actually provides a considerable safety factor.

DIFFUSION OF CARBON DIOXIDE FROM THE CELLS TO THE TISSUE CAPILLARIES, AND FROM THE PULMONARY CAPILLARIES TO THE ALVEOLI

When oxygen is used by the cells, most of it becomes carbon dioxide, and this increases the intracellular Pco$_2$. Therefore, carbon dioxide diffuses from the cells into the tissue capillaries and is then carried by the blood to the lungs, where it diffuses from the pulmonary capillaries into the alveoli. Thus, at each point in the gas transport chain, carbon dioxide diffuses in a direction exactly opposite that of the diffusion of oxygen. Yet there is one major difference between the diffusion of carbon dioxide and that of oxygen: *carbon dioxide diffuses about 20 times as rapidly as oxygen.* Therefore, the pressure differences that cause carbon dioxide diffusion are, in each instance, far less than the pressure differences required to cause oxygen diffusion. These pressures are the following:

1. Intracellular Pco$_2$, about 46 mm Hg; interstitial Pco$_2$, about 45 mm Hg; thus, there is only a 1 mm Hg pressure differential, as illustrated in Figure 40–5.

2. Pco$_2$ of the arterial blood entering the tissues, 40 mm Hg; Pco$_2$ of the venous blood leaving the tissues, about 45 mm Hg; thus, as also illustrated in Figure 40–5, the tissue capillary blood comes almost exactly to equilibrium with the interstitial Pco$_2$, also 45 mm Hg.

3. Pco$_2$ of the venous blood entering the pulmonary capillaries, 45 mm Hg; Pco$_2$ of the alveolar air, 40 mm Hg; thus, only a 5 mm Hg pressure difference causes all the required carbon dioxide diffusion out of the pulmonary capillaries into the alveoli. Furthermore, as illustrated in Figure 40–6, the Pco$_2$ of the pulmonary capillary blood falls almost exactly to equal the alveolar Pco$_2$ of 40 mm Hg before it has passed more than about one-third the distance through the capillaries. This is the same effect that was observed earlier for oxygen diffusion.

Effect of Tissue Metabolism and Blood Flow on Interstitial Pco$_2$. Tissue capillary blood flow and tissue metabolism affect the Pco$_2$ in a way exactly opposite the way in which they affect tissue Po$_2$. Figure 40–7 shows these effects, as follows:

1. A decrease in blood flow from normal, point A, to one-quarter normal, point B, increases the tissue Pco$_2$ from the normal value of 45 mm Hg up to the elevated level of 60 mm Hg. Conversely, increasing the blood flow to six times normal, point C, decreases the Pco$_2$ from the normal value of 45 mm Hg to 41 mm Hg, almost down to a level equal to the Pco$_2$ in the arterial blood, 40 mm Hg, entering the tissue capillaries.

2. Note also that a tenfold increase in metabolic rate greatly elevates the Pco$_2$ at all levels of blood

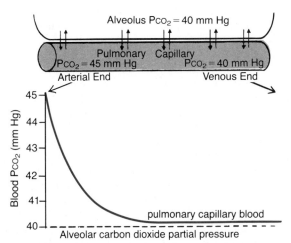

Figure 40–6. Diffusion of carbon dioxide from the pulmonary blood into the alveolus. (This curve was constructed from data in Milhorn and Pulley: *Biophys. J.,* 8:337, 1968.)

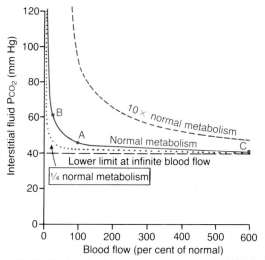

Figure 40-7. Effect of blood flow and metabolic rate on tissue P_{CO_2}.

flow, whereas decreasing the metabolism to one-quarter normal causes the interstitial fluid P_{CO_2} to fall to about 41 mm Hg, closely approaching that of the arterial blood, 40 mm Hg.

● TRANSPORT OF OXYGEN IN THE BLOOD

Normally, about 97 per cent of the oxygen transported from the lungs to the tissues is carried in chemical combination with hemoglobin in the red blood cells, and the remaining 3 per cent is carried in the dissolved state in the water of the plasma and cells. Thus, *under normal conditions* oxygen is carried to the tissues almost entirely by hemoglobin.

THE REVERSIBLE COMBINATION OF OXYGEN WITH HEMOGLOBIN

The chemistry of hemoglobin was presented in Chapter 32, where it was pointed out that the oxygen molecule combines loosely and reversibly with the heme portion of the hemoglobin. When the P_{O_2} is high, as in the pulmonary capillaries, oxygen binds with the hemoglobin, but when the P_{O_2} is low, as in the tissue capillaries, oxygen is released from the hemoglobin. This is the basis for almost all oxygen transport from the lungs to the tissues.

The Oxygen-Hemoglobin Dissociation Curve. Figure 40-8 illustrates the oxygen-hemoglobin dissociation curve, which shows a progressive increase in the percentage of the hemoglobin that is bound with oxygen as the P_{O_2} increases, which is called the *per cent saturation of the hemoglobin*. Because the blood in the arteries usually has a P_{O_2} of about 95 mm Hg, one can see from the dissociation curve that the *usual oxygen saturation of arterial blood is about 97 per cent.* On the other hand, in normal venous blood returning from the tissues, the P_{O_2} is about 40 mm Hg and *the saturation of the hemoglobin is about 75 per cent.*

Maximum Amount of Oxygen That Can Combine With the Hemoglobin of the Blood. The blood of a normal person contains approximately 15 grams of hemoglobin in each 100 milliliters of blood, and each gram of hemoglobin can bind with a maximum of about 1.34 milliliters of oxygen (1.39 milliliters when the hemoglobin is chemically pure, but this is reduced by impurities such as methemoglobin). Therefore, on the average, the hemoglobin in 100 milliliters of blood can combine with a total of almost exactly 20 milliliters of oxygen when the hemoglobin is 100 per cent saturated. This is usually expressed as 20 *volumes per cent.* The oxygen-hemoglobin dissociation curve for the normal person, therefore, can also be expressed in terms of volume per cent of oxygen, as shown by the right-most scale in Figure 40-8, instead of per cent saturation of hemoglobin.

Amount of Oxygen Released From the Hemoglobin in the Tissues. The total quantity of oxygen *bound with hemoglobin* in normal arterial blood, which is 97 per cent saturated, is approximately 19.4 milliliters per deciliter of blood. This is illustrated in Figure 40-9. On passing through the tissue capil-

Figure 40-8. The oxygen-hemoglobin dissociation curve.

Figure 40–9. Effect of blood P_{O_2} on the quantity of oxygen bound with hemoglobin in each 100 ml of blood.

laries, this amount is reduced, on the average, to 14.4 milliliters (P_{O_2} of 40 mm Hg, 75 per cent saturated hemoglobin). Thus, *under normal conditions about 5 milliliters of oxygen is transported to the tissues by each deciliter of blood.*

Transport of Oxygen During Strenuous Exercise. In heavy exercise the muscle cells utilize oxygen at a rapid rate, which causes the interstitial fluid P_{O_2} to fall to as low as 15 mm Hg. At this pressure only 4.4 milliliters of oxygen remains bound with the hemoglobin in each deciliter of blood, as also shown in Figure 40–9. Thus, 19.4 − 4.4, or 15 milliliters, is the quantity of oxygen then transported by each 100 milliliters of blood. Thus, three times as much oxygen is transported in each volume of blood that passes through the tissues as normally. And when one remembers that the cardiac output can increase to six to seven times normal in well-trained marathon runners, multiplying these two gives a 20-fold increase in oxygen transport to the tissues: this is about the limit that can be achieved.

The Utilization Coefficient. The percentage of the blood that gives up its oxygen as it passes through the tissue capillaries is called the *utilization coefficient.* The normal value for this is approximately 25 per cent, evident from the preceding discussion. However, during strenuous exercise the utilization coefficient in the entire body can increase to as much as 75 to 85 per cent. And, in local tissue areas where the blood flow is extremely slow or the metabolic rate very high, utilization coefficients approaching 100 per cent have been recorded—that is, essentially all the oxygen is removed.

THE EFFECT OF HEMOGLOBIN TO "BUFFER" THE TISSUE OXYGEN P_{O_2}

Although hemoglobin is necessary for transport of oxygen to the tissues, it performs still another major function essential to life. This is its function as a "tissue oxygen buffer" system. That is, the hemoglobin in the blood is mainly responsible for stabilizing the oxygen pressure in the tissues. This can be explained as follows:

Role of Hemoglobin in Maintaining Constant P_{O_2} in the Tissues. Under basal conditions the tissues require about 5 milliliters of oxygen from each deciliter of blood passing through the tissue capillaries. Referring again to the oxygen-hemoglobin dissociation curve in Figure 40–9, one can see that for the 5 milliliters of oxygen to be released, the P_{O_2} must fall to about 40 mm Hg. Therefore, the tissue P_{O_2} normally cannot rise above this 40 mm Hg level, for if such should occur, the oxygen needed by the tissues could not be released from the hemoglobin. In this way, the hemoglobin normally sets an upper limit on the gaseous pressure in the tissues at approximately 40 mm Hg.

On the other hand, in heavy exercise extra large amounts of oxygen must be delivered from the hemoglobin to the tissues. But this can be achieved with very little further decrease in tissue P_{O_2}—down to a level of 15 to 25 mm Hg—because of the steep slope of the dissociation curve; that is, a small fall in P_{O_2} causes large amounts of oxygen to be released.

It can be seen, then, that the hemoglobin in the blood automatically delivers oxygen to the tissues at a pressure that is held rather tightly between approximately 15 and 40 mm Hg.

Value of Hemoglobin for Maintaining Constant Tissue P_{O_2} When Atmospheric Oxygen Concentration Changes Markedly. The normal P_{O_2} in the alveoli is approximately 104 mm Hg; but as one ascends a mountain or ascends in an airplane, the P_{O_2} can easily fall to less than half this. Or when one enters areas of compressed air, such as deep in the sea or in pressurized chambers, the P_{O_2} may rise to ten times this level. Even so, the tissue P_{O_2} changes very little. Let us explain.

It will be seen from the oxygen-hemoglobin dissociation curve in Figure 40–8 that when the alveolar P_{O_2} is decreased to as low as 60 mm Hg, the arterial hemoglobin is still 89 per cent saturated, only 8 per cent below the normal saturation of 97 per cent. Furthermore, the tissues still remove approximately 5 milliliters of oxygen from each deciliter of blood passing through the tissues; to remove this oxygen, the P_{O_2} of the venous blood falls to 35 mm Hg, only 5 mm below the normal value. Thus, the tissue P_{O_2} hardly changes, despite the marked fall in alveolar P_{O_2} from 104 to 60 mm Hg.

On the other hand, when the alveolar P_{O_2} rises as high as 500 mm Hg, the maximum oxygen saturation of hemoglobin can never rise above 100 per cent, which is only 3 per cent above the normal level of 97 per cent. Only a small amount of additional oxygen dissolves in the fluid of the blood, as will be discussed subsequently. Then, when the blood passes through the tissue capillaries, it still loses several milliliters of oxygen to the tissues, which automatically reduces the P_{O_2} of the capillary blood to a value only a few millimeters greater than the normal 40 mm Hg.

Consequently, the level of alveolar oxygen may vary greatly—from 60 to more than 500 mm Hg P_{O_2}—and still the P_{O_2} in the tissue does not vary more than a few millimeters from normal, illustrating beautifully the tissue oxygen buffer function of the blood hemoglobin.

SHIFT OF THE OXYGEN-HEMOGLOBIN DISSOCIATION CURVE AND ITS SIGNIFICANCE

The oxygen-hemoglobin dissociation curves of Figures 40–8 and 40–9 are for normal, average blood. However, a number of different factors can displace the dissociation curve in one direction or the other in the manner illustrated in Figure 40–10. This figure shows that when the blood becomes slightly acidic, with the pH decreasing from the normal value of 7.4 to 7.2, the oxygen-hemoglobin dissociation curve shifts on the average about 15 per cent to the right. On the other hand, an increase in the pH to 7.6 shifts the curve a similar amount to the left.

In addition to pH changes, several other factors are also known to shift the curve. Three of these, all of which shift the curve to the right, are (1) increased carbon dioxide concentration, (2) increased blood temperature, and (3) increased 2,3-diphosphoglycerate, a phosphate compound normally present in the blood but in different concentrations under different conditions.

A condition that shifts the dissociation curve to the left is the presence in the blood of large quantities of *fetal hemoglobin,* a type of hemoglobin normally present in the fetus before birth that is different from normal hemoglobin called *adult hemoglobin.* Shift of the curve to the left when fetal hemoglobin is present is important for oxygen delivery to the fetal tissues under the hypoxic conditions in which the fetus exists. This is discussed further in Chapter 82.

Importance of the Effect of Carbon Dioxide and Hydrogen Ions in Shifting the Oxygen-Hemoglobin Dissociation Curve — The Bohr Effect. Shift of the oxygen-hemoglobin dissociation curve by changes in the blood CO_2 and hydrogen ions has a very significant effect in enhancing oxygenation of the blood in the lungs and then again in enhancing release of oxygen from the blood in the tissues. This is called the *Bohr effect,* and it can be explained as follows: As the blood passes through the lungs, carbon dioxide diffuses from the blood into the alveoli. This reduces the blood Pco_2 and also decreases the hydrogen ion concentration because of the resulting decrease in blood carbonic acid. Both these effects shift the oxygen-hemoglobin dissociation curve to the left and upward, as shown in Figure 40–10. Therefore, the quantity of oxygen that binds with the hemoglobin at any given alveolar Po_2 becomes considerably increased, thus allowing greater oxygen transport to the tissues. Then when the blood reaches the tissue capillaries, exactly the opposite effects occur. Carbon dioxide entering the blood from the tissues shifts the curve rightward, which displaces oxygen from the hemoglobin and therefore delivers oxygen to the tissues at a higher Po_2 than would otherwise occur.

Effect of 2,3-Diphosphoglycerate (DPG). The normal DPG in the blood keeps the oxygen-hemoglobin dissociation curve shifted slightly to the right all the time. In addition, in hypoxic conditions that last longer than a few hours, the quantity of DPG in the blood increases considerably, thus shifting the oxygen-hemoglobin dissociation curve even farther to the right. This causes oxygen to be released to the tissues at as much as 10 mm Hg higher oxygen pressure than would be the case without this increased DPG. Therefore, it was taught in the past that this might be an important mechanism for adaptation to hypoxia. However, the presence of the excess DPG also makes it difficult for the hemoglobin to combine with oxygen in the lungs when the alveolar Po_2 is reduced, thereby often causing as much harm as good. Therefore, it is questionable whether the DPG shift of the dissociation curve is as beneficial in hypoxia as had been believed.

Shift of the Dissociation Curve During Exercise. In exercise, several factors shift the dissociation curve considerably to the right. The exercising muscles release large quantities of carbon dioxide; and this, plus several acids released by the exercising muscle, increases the hydrogen ion concentration in the muscle capillary blood. In addition, the temperature of the muscle often rises as much as 2° to 3° C; and, finally, phosphate compounds are also released. All of these factors act together to shift the oxygen-hemoglobin dissociation curve *of the muscle capillary blood* considerably to the right. This right-hand shift of the hemoglobin curves sometimes allows oxygen to be released to the muscle at Po_2s as great as 40 mm Hg (the normal resting value) even when as much as 75 to 85 per cent of the oxygen is being removed from the hemoglobin. Then, in the lungs, the shift occurs in the opposite direction, thus allowing pickup of extra amounts of oxygen from the alveoli.

METABOLIC USE OF OXYGEN BY THE CELLS

Relationship Between Intracellular Po_2 and Rate of Oxygen Usage. Only a minute level of oxygen pressure is required in the cells for normal intracellular chemical reactions to take place. The reason for this is that the respiratory enzyme systems of the cell, which are discussed in Chapter 67, are geared so that when the cellular Po_2 is more than 1 to 3 mm Hg, oxygen availability is no longer a limiting factor in the rates of the chemical reactions. Instead, the main limiting factor then is the *concentration of adenosine diphosphate* (ADP) in the cells, as was explained in Chapter 3. This effect is illustrated in Figure 40–11, which shows the relationship between intracellular Po_2 and rate of oxygen usage. Note that whenever the intracellular Po_2 is above 1 to 3 mm Hg, the rate of

Figure 40–10. Shift of the oxygen-hemoglobin dissociation curve to the right by increases in (1) hydrogen ions, (2) CO_2, (3) temperature, or (4) 2,3-diphosphoglycerate (DPG).

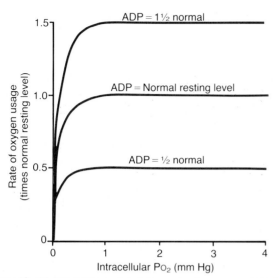

Figure 40 – 11. Effect of intracellular P_{O_2} on rate of oxygen usage by the cells. Note that increasing the intracellular concentration of adenosine diphosphate (ADP) increases the rate of oxygen usage.

oxygen usage becomes constant for any given concentration of ADP in the cell. On the other hand, when the ADP concentration is altered, the rate of oxygen usage changes in proportion to the change in ADP concentration.

It will be recalled from the discussion in Chapter 3 that when adenosine triphosphate (ATP) is utilized in the cells to provide energy, it is converted into ADP. The increasing concentration of ADP, in turn, increases the metabolic usage of both oxygen and the various nutrients that combine with the oxygen to release energy. This energy is used to reform the ATP. Therefore, *under normal operating conditions the rate of oxygen utilization by the cells is controlled ultimately by the rate of energy expenditure within the cells — that is, by the rate at which ADP is formed from ATP.* Only in very hypoxic states does the availability of oxygen become a limiting condition.

Effect of Diffusion Distance From the Capillary to the Cell on Oxygen Utilization. Cells are rarely more than 50 micrometers away from a capillary, and oxygen normally can diffuse readily enough from the capillary to the cell to supply all the required amounts of oxygen for metabolism. However, occasionally cells are located farther than this from the capillaries, and the rate of oxygen diffusion to these cells is so low that the intracellular P_{O_2} falls then below the critical level of 1 to 3 mm Hg required to maintain maximum intracellular metabolism. Thus, under these conditions, oxygen utilization by the cells is said to be *diffusion limited* and is no longer determined by the amount of ADP formed in the cells. Fortunately, this almost never occurs except in pathological states.

Effect of Blood Flow on Metabolic Use of Oxygen. The total amount of oxygen available each minute for use in any given tissue is determined by (1) the quantity of oxygen transported in each deciliter of

blood and (2) the rate of blood flow. If the rate of blood flow falls to zero, the amount of available oxygen obviously also falls to zero. Thus, there are times when the rate of blood flow through a tissue can be so low that the tissue P_{O_2} falls below the critical 1 to 3 mm Hg required for maximal intracellular metabolism. Under these conditions, the rate of tissue utilization of oxygen is *blood flow limited.* However, neither diffusion-limited nor blood flow–limited oxygen can continue for long, because the cells then often receive less oxygen than is required for continuation of the life of the cells themselves.

TRANSPORT OF OXYGEN IN THE DISSOLVED STATE

At the normal arterial P_{O_2} of 95 mm Hg, approximately 0.29 milliliter of oxygen is dissolved in every deciliter of water in the blood. Then when the P_{O_2} of the blood falls to 40 mm Hg in the tissue capillaries, only 0.12 milliliter of oxygen remains dissolved. In other words, 0.17 milliliter of oxygen is normally transported in the dissolved state to the tissues by each deciliter of blood water. This compares with almost 5.0 milliliters transported by the hemoglobin. Therefore, the amount of oxygen transported to the tissues in the dissolved state is normally slight, only about 3 per cent of the total, as compared with 97 per cent transported by the hemoglobin. During strenuous exercise, when hemoglobin release of oxygen to the tissues increases another threefold, the relative quantity then transported in the dissolved state falls to as little as 1.5 per cent. Yet if a person breathes oxygen at very high alveolar P_{O_2}s, the amount then transported in the dissolved state can become much greater, sometimes so much so that serious excesses of oxygen occur in the tissues and "oxygen poisoning" ensues. This often leads to convulsions and even death, as will be discussed in detail in Chapter 44 in relation to high pressure breathing.

COMBINATION OF HEMOGLOBIN WITH CARBON MONOXIDE — DISPLACEMENT OF OXYGEN

Carbon monoxide combines with hemoglobin at the same point on the hemoglobin molecule as does oxygen and therefore can displace oxygen from the hemoglobin. Furthermore, it binds with about 250 times as much tenacity as oxygen, which is illustrated by the carbon monoxide–hemoglobin dissociation curve in Figure 40–12. This curve is almost identical with the oxygen-hemoglobin dissociation curve, except that the pressures of the carbon monoxide, shown on the abscissa, are at a level 1/250 of those in the oxygen-hemoglobin dissociation curve of Figure 40–8. Therefore, a carbon monoxide pressure of only 0.4 mm Hg in the alveoli, 1/250 that of the alveolar oxygen, allows the carbon monoxide to compete equally with the oxygen for combination with the hemoglobin and causes half the hemoglobin in the blood to become bound with carbon monoxide instead of with oxygen. Therefore, a carbon monoxide pressure only a little greater than 0.4 mm Hg (about 0.7 mm Hg, or a concentration of about 0.1 per cent in the air) can be lethal.

A patient severely poisoned with carbon monoxide can be advantageously treated by administering pure oxygen, for oxygen at high alveolar pressures displaces carbon monoxide from its combination with hemoglobin far more rapidly

Figure 40–12. The carbon monoxide–hemoglobin dissociation curve. Note the extremely low pressures at which carbon monoxide combines with hemoglobin.

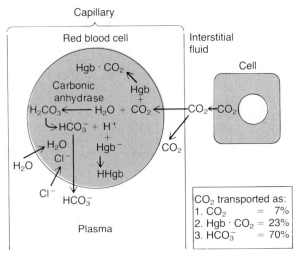

Figure 40–13. Transport of carbon dioxide in the blood.

than can oxygen at the low pressure of atmospheric oxygen.

The patient can also be benefited by simultaneous administration of a few per cent carbon dioxide because this strongly stimulates the respiratory center, which is discussed in the following chapter. This increases alveolar ventilation and reduces the alveolar carbon monoxide concentration, which allows increased carbon monoxide release from the blood.

With intensive oxygen and carbon dioxide therapy, carbon monoxide can be removed from the blood 10 to 20 times as rapidly as without therapy.

● TRANSPORT OF CARBON DIOXIDE IN THE BLOOD

Transport of carbon dioxide by the blood is not nearly so great a problem as transport of oxygen, because even in the most abnormal conditions carbon dioxide can usually be transported in far greater quantities than can oxygen. However, the amount of carbon dioxide in the blood does have much to do with acid-base balance of the body fluids, which was discussed in Chapter 30.

Under normal resting conditions *an average of 4 milliliters of carbon dioxide is transported from the tissues to the lungs in each deciliter of blood.*

CHEMICAL FORMS IN WHICH CARBON DIOXIDE IS TRANSPORTED

To begin the process of carbon dioxide transport, carbon dioxide diffuses out of the tissue cells in the dissolved molecular CO_2 form (but not to any significant extent in the bicarbonate form because the cell membrane is almost impermeable to bicarbonate ions). On entering the capillary, the carbon dioxide initiates a host of almost instantaneous physical and chemical reactions, illustrated in Figure 40–13, that are essential for CO_2 transport.

Transport of Carbon Dioxide in the Dissolved State. A small portion of the carbon dioxide is trans-

ported in the dissolved state to the lungs. It will be recalled that the Pco_2 of venous blood is 45 mm Hg and that of arterial blood is 40 mm Hg. The amount of carbon dioxide dissolved in the fluid of the blood at 45 mm Hg is about 2.7 ml/dl (2.7 volumes per cent). The amount dissolved at 40 mm Hg is about 2.4 milliliters, or a difference of 0.3 milliliter. Therefore, only about 0.3 milliliter of carbon dioxide is transported in the form of dissolved carbon dioxide by each deciliter of blood. This is about 7 per cent of all the carbon dioxide transported.

Transport of Carbon Dioxide in the Form of Bicarbonate Ion. *Reaction of Carbon Dioxide With Water in the Red Blood Cells — Effect of Carbonic Anhydrase.* The dissolved carbon dioxide in the blood reacts with water to form carbonic acid. However, this reaction would occur much too slowly to be of importance were it not for the fact that inside the red blood cells is an enzyme called *carbonic anhydrase,* which catalyzes the reaction between carbon dioxide and water, accelerating its rate about 5000-fold. Therefore, instead of requiring many seconds or minutes to occur, as is true in the plasma, the reaction occurs so rapidly in the red blood cells that it reaches almost complete equilibrium within a small fraction of a second. This allows tremendous amounts of carbon dioxide to react with the red cell water even before the blood leaves the tissue capillaries.

Dissociation of the Carbonic Acid Into Bicarbonate and Hydrogen Ions. In another small fraction of a second the carbonic acid formed in the red cells dissociates into *hydrogen* and *bicarbonate ions.* Most of the hydrogen ions then combine with the hemoglobin in the red blood cells because hemoglobin is a powerful acid-base buffer. In turn, many of the bicarbonate ions diffuse into the plasma while chloride ions diffuse into the red cells to take their place. This is made possible by the presence of a special *bicarbonate-chloride carrier protein* in the red cell membrane that shuttles these two ions in opposite directions at rapid velocities. Thus, the chloride content of venous red blood cells is greater than that of arterial cells, a phenomenon called the *chloride shift.*

The reversible combination of carbon dioxide with water in the red blood cells under the influence of carbonic anhydrase probably accounts for about 70 per cent of the carbon dioxide transported from the tissues to the lungs. Thus, this means of transporting carbon dioxide is by far the most important of all the methods for transport. Indeed, when a carbonic anhydrase inhibitor (acetazolamide) is administered to an animal to block the action of carbonic anhydrase in the red blood cells, carbon dioxide transport from the tissues becomes very poor — so poor that the tissue Pco_2 can be made to rise to as high as 80 mm Hg instead of the normal 45 mm Hg.

Transport of Carbon Dioxide in Combination with Hemoglobin and Plasma Proteins — Carbaminohemoglobin. In addition to reacting with water, carbon dioxide also reacts directly with hemoglobin to form the compound *carbaminohemoglobin* (CO_2HHb). This combination of carbon dioxide with the hemoglobin is a reversible reaction that occurs with a very loose bond, so that the carbon dioxide is easily released into the alveoli where the Pco_2 is lower than in the tissue capillaries. A small amount of carbon dioxide also reacts in this same way with the plasma proteins, but this is much less significant because the quantity of these proteins is only one fourth as great as the quantity of hemoglobin.

The *theoretical* quantity of carbon dioxide that can be carried from the tissues to the lungs in combination with hemoglobin and plasma proteins is approximately 30 per cent of the total quantity transported — that is, about 1.5 milliliters of carbon dioxide in each deciliter of blood. However, this reaction is much slower than the reaction of carbon dioxide with water inside the red blood cells. Therefore, it is doubtful that this mechanism provides transport of more than 15 to 25 per cent of the total quantity of carbon dioxide.

THE CARBON DIOXIDE DISSOCIATION CURVE

It is apparent that carbon dioxide can exist in the blood in many different forms: (1) as free carbon dioxide and (2) in chemical combinations with water, hemoglobin, and plasma protein. The total quantity of carbon dioxide combined with the blood in all these forms depends on the Pco_2. The curve shown in Figure 40–14 depicts this dependence of total blood CO_2 in all its forms on Pco_2; this curve is called the *carbon dioxide dissociation curve.*

Note that the normal blood Pco_2 ranges between the limits of 40 mm Hg in arterial blood and 45 mm Hg in venous blood, which is a very narrow range. Note also that the normal concentration of carbon dioxide in the blood is about 50 volumes per cent but that only 4 volumes per cent of this is actually exchanged during normal transport of carbon dioxide from the tissues to

Figure 40–14. The carbon dioxide dissociation curve.

the lungs. That is, the concentration rises to about 52 volumes per cent as the blood passes through the tissues and falls to about 48 volumes per cent as it passes through the lungs.

EFFECT OF THE OXYGEN-HEMOGLOBIN REACTION ON CARBON DIOXIDE TRANSPORT — THE HALDANE EFFECT

Earlier in the chapter it was pointed out that an increase in carbon dioxide in the blood will cause oxygen to be displaced from the hemoglobin and that this is an important factor in promoting oxygen transport. The reverse is also true: binding of oxygen with hemoglobin tends to displace carbon dioxide from the blood. Indeed, this effect, called the *Haldane effect,* is quantitatively far more important in promoting carbon dioxide transport than is the Bohr effect in promoting oxygen transport.

The Haldane effect results from the simple fact that combination of oxygen with hemoglobin causes the hemoglobin to become a stronger acid. This in turn displaces carbon dioxide from the blood in two ways: (1) The more highly acidic hemoglobin has less tendency to combine with carbon dioxide to form carbaminohemoglobin, thus displacing much of the carbon dioxide that is present in the carbamino form from the blood. (2) The increased acidity of the hemoglobin causes it to release an excess of hydrogen ions, and these in turn bind with bicarbonate ions to form carbonic acid; then this dissociates into water and carbon dioxide, and the carbon dioxide is released from the blood into the alveoli.

Therefore, in the tissue capillaries the Haldane effect causes increased pickup of carbon dioxide because of oxygen removal from the hemoglobin, and in the lungs it causes increased release of carbon dioxide because of oxygen pickup by the hemoglobin.

Figure 40–15 illustrates quantitatively the significance of the Haldane effect on the transport of carbon dioxide from the tissues to the lungs. This figure shows small portions of two separate carbon dioxide dissociation curves, the solid curve when the Po_2 is 100 mm Hg, which is the case in the lungs, and the dashed curve when the Po_2 is 40 mm Hg, which is the case in the tissue capillaries. Point A on the dashed curve shows that the normal Pco_2 of 45 mm Hg in the tissues causes 52 volumes per cent of carbon dioxide to combine with the blood. On entering the lungs the Pco_2 falls to 40 mm Hg while the Po_2 rises to 100 mm Hg. If the carbon dioxide dissociation curve did not shift because of the Haldane effect, the carbon dioxide content of the blood would

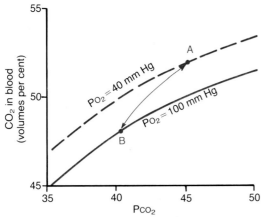

Figure 40–15. Portions of the carbon dioxide dissociation curves when the P_{O_2} is 100 mm Hg and 40 mm Hg, respectively. The arrow represents the Haldane effect on the transport of carbon dioxide, as explained in the text.

fall only to 50 volumes per cent, which would be a loss of only 2 volumes per cent of carbon dioxide. However, the increase in P_{O_2} in the lungs decreases the carbon dioxide dissociation curve from the dashed curve to the solid curve of the figure, so that the carbon dioxide content falls to 48 volumes per cent (point B). This represents an additional 2 volume per cent loss of carbon dioxide. Thus, the Haldane effect approximately doubles the amount of carbon dioxide released from the blood in the lungs and approximately doubles the pickup of CO_2 in the tissues.

CHANGE IN BLOOD ACIDITY DURING CARBON DIOXIDE TRANSPORT

The carbonic acid formed when carbon dioxide enters the blood in the tissues decreases the blood pH. Fortunately, though, the reaction of this acid with the buffers of the blood prevents the hydrogen ion concentration from rising greatly. Ordinarily, arterial blood has a pH of approximately 7.41, and, as the blood acquires carbon dioxide in the tissue capillaries, the pH falls to a venous value of approximately 7.37. In other words, a pH change of 0.04 unit takes place. The reverse occurs when carbon dioxide is released from the blood in the lungs, the pH rising to the arterial value once again. In exercise, or in other conditions of high metabolic activity, or when the blood flow through the tissues is sluggish, the decrease in pH in the tissue blood (and in the tissues themselves) can be as much as 0.50, or on occasion even more than this, thus causing severe tissue acidosis.

● THE RESPIRATORY EXCHANGE RATIO

The discerning student will have noted that normal transport of oxygen from the lungs to the tissues by each deciliter of blood is about 5 milliliters, whereas normal transport of carbon dioxide from the tissues to the lungs is about 4 milliliters. Thus, under normal resting conditions only about 80 per cent as much carbon dioxide is expired from the lungs as there is oxygen uptake by the lungs. The ratio of carbon dioxide output to oxygen uptake is called the *respiratory exchange ratio* (R). That is,

$$R = \frac{\text{Rate of carbon dioxide output}}{\text{Rate of oxygen uptake}}$$

The value for R changes under different metabolic conditions. When a person is utilizing entirely carbohydrates for body metabolism, R rises to 1.00. On the other hand, when the person is utilizing fats almost entirely for metabolic energy, the level falls to as low as 0.7. The reason for this difference is that when oxygen is metabolized with carbohydrates, one molecule of carbon dioxide is formed for each molecule of oxygen consumed; whereas when oxygen reacts with fats, a large share of the oxygen combines with hydrogen atoms from the fats to form water instead of carbon dioxide. In other words, the *respiratory quotient of the chemical reactions* in the tissues when fats are metabolized is about 0.70 instead of 1.00, which is the case when carbohydrates are being utilized. The tissue respiratory quotient will be discussed in Chapter 71.

For a person on a normal diet consuming average amounts of carbohydrates, fats, and proteins, the average value for R is considered to be 0.825.

REFERENCES

Adamson, J. W., and Finch, C. A.: Hemoglobin function, oxygen affinity, and erythropoietin. Annu. Rev. Physiol., 37:351, 1975.

Bartels, H., and Baumann, R.: Respiratory function of hemoglobin. Int. Rev. Physiol., 14:107, 1977.

Bauer, C., et al. (eds.): Biophysics and Physiology of Carbon Dioxide. New York, Springer-Verlag, 1980.

Caughey, W. S. (ed.): Oxygen: Biochemical and Clinical Aspects. New York, Academic Press, 1979.

Cherniack, N. S., and Logobardo, G. S.: Oxygen and carbon dioxide gas stores of the body. Physiol. Rev., 50:196, 1970.

Crowell, J. W., and Smith, E. E.: Determinants of the optimal hematocrit. J. Appl. Physiol., 22:501, 1967.

Forster, R. E.: CO₂: chemical, biochemical, and physiological aspects. Physiologist, 13:398, 1970.

Forster, R. E.: Pulmonary ventilation and blood gas exchange. In Sodeman, W. A., Jr., and Sodeman, W. A. (eds.): Pathologic Physiology: Mechanisms of Disease, 5th Ed. Philadelphia, W. B. Saunders Co., 1974, p. 371.

Gaar, K. A., Jr.: Oxygen transport: A simple model for study and examination. Physiologist, 28:412, 1985.

Gonzalez, N. C., and Fedde, M. R.: Oxygen Transfer from Atmosphere to Tissues. New York, Plenum Publishing Corp., 1988.

Grodins, F. S., and Yamashiro, S. M.: Optimization of mammalian respiratory gas transport system. Annu. Rev. Biophys. Bioeng., 4:115, 1973.

Haldane, J. S., and Priestley, J. G.: Respiration. New Haven, Yale University Press, 1935.

Hayaishi, O.: Molecular Oxygen in Biology. New York, American Elsevier Publishing Co., 1974.

Jobsis, F. F.: Intracellular metabolism of oxygen. Am. Rev. Resp. Dis., 110:58, 1974.

Jones. C. E., et al.: Determination of mean tissue oxygen tensions by implanted perforated capsules. J. Appl. Physiol., 26:630, 1969.

Jones, N. L.: Blood Gases and Acid-Base Physiology. New York, B. C. Decker, 1980.

Kessler, M., et al.: (eds.): Oxygen Supply. Baltimore, University Park Press, 1973.

Kilmartin, J. V., and Rossi-Bernardi, L.: Interaction of hemoglobin with hydrogen ions, carbon dioxide, and organic phosphates. Physiol. Rev., 53:836, 1973.

Konigsberg, W.: Protein structure and molecular dysfunction: Hemoglobin. In Bondy, P. K., and Rosenberg, L. E. (eds.): Metabolic Control and Disease, 8th Ed. Philadelphia, W. B. Saunders Co., 1980, p. 27.

Lane, E. E., and Walker, J. F.: Clinical Arterial Blood Gas Analysis. St. Louis, C. V. Mosby Co., 1987.

Lehman, H., and Huntsman, R. G.: Man's Hemoglobins, 2nd Ed. Philadelphia, J. B. Lippincott Co., 1974.

Michel, C. C.: The transport of oxygen and carbon dioxide by the blood. In MTP International Review of Science: Physiology. Vol. 2. Baltimore, University Park Press, 1974, p. 67.

Mochizuki, M., et al. (eds.): Oxygen Transport to Tissue X. New York, Plenum Publishing Corp., 1988.

Morson, B. C. (ed.): Hypoxia and Ischaemia. London, Journal of Clinical Pathology for the College, 1977.

Oxygen Transport in Birds at Altitude. Real and simulated microgravity alters cellular processes: Emergence of a new environmental frontier. News Physiol. Sci., 3:36, 1988.

Perutz, M. F.: Hemoglobin structure and respiratory transport. Sci. Am., 239(6):92, 1987.

Randall, D. J.: The Evolution of Air Breathing in Vertebrates. New York, Cambridge University Press, 1980.

Riggs, A. F.: The Bohr effect. Annu. Rev. Physiol., 50:181, 1988.

Stainsby, W. N., and Barclay, J. K.: Oxyen uptake by striated muscle. Muscle Biol., 1:273, 1972.

Tamoru, M., et al.: In vivo study of tissue oxygen metabolism using optical and nuclear magnetic resonance. Annu. Rev. Physiol., 51:813, 1989.

Wagner, P. D.: Diffusion and chemical reaction in pulmonary gas exchange. Physiol. Rev., 57:257, 1977.

Wagner, P. D.: The oxyhemoglobin dissociation curve and pulmonary gas exchange. Semin. Hematol., 11:405, 1974.

Whalen, W. J.: Intracellular PO_2 in heart and skeletal muscle. Physiologist, 14:69, 1971.

CHAPTER 41

•*Regulation of Respiration*•

The nervous system adjusts the rate of alveolar ventilation almost exactly to the demands of the body so that the arterial blood oxygen pressure (Po_2) and carbon dioxide pressure (Pco_2) are hardly altered even during strenuous exercise and most other types of respiratory stress.

The present chapter describes the operation of this neurogenic system for regulation of respiration.

● THE RESPIRATORY CENTER

The "respiratory center" is composed of several widely dispersed groups of neurons located *bilaterally* in the medulla oblongata and pons, as illustrated for one side in Figure 41–1. It is divided into three major collections of neurons: (1) a *dorsal respiratory group,* located in the dorsal portion of the medulla, which mainly causes inspiration, (2) a *ventral respiratory group,* located in the ventrolateral part of the medulla, which can cause either expiration or inspiration, depending upon which neurons in the group are stimulated, and (3) the *pneumotaxic center,* located dorsally in the superior portion of the pons, which helps control both the rate and pattern of breathing. The dorsal respiratory group of neurons plays the fundamental role in the control of respiration. Therefore, let us discuss its function first.

The Dorsal Respiratory Group of Neurons — Its Inspiratory and Rhythmical Functions

The dorsal respiratory group of neurons extends most of the length of the medulla. Either all or most of its neurons are located within the *nucleus of the tractus solitarius,* though additional neurons in the adjacent reticular substance of the medulla probably also play important roles in respiratory control. The nucleus of the tractus solitarius is also the sensory termination of both the vagal and glossopharyngeal nerves, which transmit sensory signals into the respiratory center from the peripheral chemoreceptors, the baroreceptors, and several different types of receptors in the lung. All the signals from these peripheral areas

help in the control of respiration, as we discuss in subsequent sections of this chapter.

Rhythmical Inspiratory Discharges from the Dorsal Respiratory Group. The basic rhythm of respiration is generated mainly in the dorsal respiratory group of neurons. Even when all the peripheral nerves entering the medulla are sectioned and the brain stem is transected both above and below the medulla, this group of neurons still emits repetitive bursts of *inspiratory* action potentials. Unfortunately, though, the basic cause of these repetitive discharges is still unknown. In primitive animals, neural networks have been found in which activity of one set of neurons excites a second set, which in turn inhibits the first. Then after a period of time the mechanism repeats itself, continuing throughout the life of the animal. Therefore, most respiratory physiologists believe that some similar network of neurons located entirely within the medulla, probably involving not only the dorsal respiratory group but adjacent areas of the medulla as well, is responsible for the basic rhythm of respiration.

The Inspiratory "Ramp" Signal. The nervous signal that is transmitted to the inspiratory muscles is not an instantaneous burst of action potentials. Instead, in normal respiration, it begins very weakly at first and increases steadily in a ramp fashion for about 2 seconds. It abruptly ceases for approximately the next 3 seconds, then begins again for still another cycle, and again and again. Thus, the inspiratory signal is said to be a *ramp signal.* The obvious advantage of this is that it causes a steady increase in the volume of the lungs during inspiration, rather than inspiratory gasps.

There are two ways in which the inspiratory ramp is controlled:

1. Control of the rate of increase of the ramp signal, so that during very active respiration the ramp increases rapidly and therefore fills the lungs rapidly as well.

2. Control of the limiting point at which the ramp suddenly ceases. This is the usual method for controlling the rate of respiration; that is, the earlier the ramp ceases, the shorter the duration of inspiration. For reasons not presently understood, this also

444

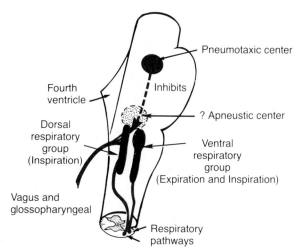

Figure 41–1. Organization of the respiratory center.

shortens the duration of expiration. Thus, the rate of respiration is increased.

The Pneumotaxic Center—Its Function in Limiting the Duration of Inspiration and Increasing Respiratory Rate

The pneumotaxic center, located dorsally in the *nucleus parabrachialis* of the upper pons, transmits impulses continuously to the inspiratory area. The primary effect of these is to control the "switch-off" point of the inspiratory ramp, thus controlling the duration of the filling phase of the lung cycle. When the pneumotaxic signals are strong, inspiration might last for as little as 0.5 second; but when weak, for as long as 5 or more seconds, thus filling the lungs with a great excess of air.

Therefore, the function of the pneumotaxic center is primarily to limit inspiration. However, this has a secondary effect of increasing the rate of breathing because limitation of inspiration also shortens expiration and the entire period of respiration. Thus, a strong pneumotaxic signal can increase the rate of breathing up to 30 to 40 breaths per minute, whereas a weak pneumotaxic signal may reduce the rate to only a few breaths per minute.

The Ventral Respiratory Group of Neurons—Its Function in Both Inspiration and Expiration

Located about 5 millimeters anterior and lateral to the dorsal respiratory group of neurons is the ventral respiratory group of neurons, found in the *nucleus ambiguus* rostrally and the *nucleus retroambiguus* caudally. The function of this neuronal group differs from that of the dorsal respiratory group in several important ways:

1. The neurons of the ventral respiratory group remain almost totally *inactive* during normal quiet respiration. Therefore, normal quiet breathing is caused only by repetitive inspiratory signals from the dorsal respiratory group transmitted mainly to the diaphragm, and expiration results from elastic recoil of the lungs and thoracic cage.

2. There is no evidence that the ventral respiratory neurons participate in the basic rhythmic oscillation that controls respiration.

3. When the respiratory drive for increased pulmonary ventilation becomes greater than normal, respiratory signals then spill over into the ventral respiratory neurons from the basic oscillating mechanism of the dorsal respiratory area. As a consequence, the ventral respiratory area then does contribute its share to the respiratory drive as well.

4. Electrical stimulation of some of the neurons in the ventral group causes inspiration, whereas stimulation of others causes expiration. Therefore, these neurons contribute to both inspiration and expiration. However, they are especially important in providing the powerful expiratory signals to the abdominal muscles during expiration. Thus, this area operates more or less as an overdrive mechanism when high levels of pulmonary ventilation are required.

Possibility of an "Apneustic Center" in the Lower Pons

To add confusion to our knowledge about respiratory center function, there is another strange center in the lower part of the pons, called the *apneustic center*. However, its function can be demonstrated only when the vagus nerves to the medulla have been sectioned and when the connections from the pneumotaxic center have also been blocked by transecting the pons in its midregion. Then the apneustic center of the lower pons sends signals to the dorsal respiratory group of neurons that prevent the "switch-off" of the inspiratory ramp signal. Therefore, the lungs become almost completely filled with air, and only occasional short expiratory gasps occur.

The function of the apneustic center is not understood, but it presumably operates in association with the pneumotaxic center to control the depth of inspiration.

Reflex Limitation of Inspiration by Lung Inflation Signals—The Hering-Breuer Inflation Reflex

In addition to the neural mechanisms operating entirely within the brain stem, reflex signals from the periphery also help to control respiration. Most important, located in the walls of the bronchi and bronchioles throughout the lungs are *stretch receptors* that transmit signals through the *vagi* into the dorsal respiratory group of neurons when the lungs become overstretched. These signals affect inspiration in much the same way as signals from the pneumotaxic center; that is, when the lungs become overly inflated, the stretch reflectors activate an appropriate feedback response that "switches off" the inspiratory ramp and thus stops further inspiration. This is called the *Hering-Breuer inflation reflex*. This reflex also increases the rate of respiration, the same as is true for signals from the pneumotoxic center.

However, in human beings, the Hering-Breuer reflex probably is not activated until the tidal volume increases to greater than approximately 1.5 liters. Therefore, this reflex appears to be mainly a protective mechanism for preventing excess lung inflation rather than an important ingredient in the normal control of ventilation.

CONTROL OF OVERALL RESPIRATORY CENTER ACTIVITY

Up to this point we have discussed the basic mechanisms for causing inspiration and expiration, but it is also important to know how the intensity of the respiratory control signals is increased or decreased to match the ventilatory needs of the body. For example, during very heavy exercise, the rates of oxygen utilization and carbon dioxide formation are often increased to as much as 20 times normal, requiring commensurate increases in pulmonary ventilation.

The major purpose of the remainder of this chapter is to discuss this control of ventilation in response to the needs of the body.

● CHEMICAL CONTROL OF RESPIRATION

The ultimate goal of respiration is to maintain proper concentrations of oxygen, carbon dioxide, and hydrogen ions in the tissues. It is fortunate, therefore, that respiratory activity is highly responsive to changes in each of these.

Excess carbon dioxide or hydrogen ions mainly stimulate the respiratory center itself, causing greatly increased strength of both the inspiratory and expiratory signals to the respiratory muscles.

Oxygen, on the other hand, does not have a significant *direct* effect on the respiratory center of the brain in controlling respiration. Instead, it acts almost entirely on peripheral chemoreceptors located in the carotid and aortic bodies, and these in turn transmit appropriate nervous signals to the respiratory center for control of respiration.

Let us discuss first the stimulation of the respiratory center itself by carbon dioxide and hydrogen ions.

DIRECT CHEMICAL CONTROL OF RESPIRATORY CENTER ACTIVITY BY CARBON DIOXIDE AND HYDROGEN IONS

The Chemosensitive Area of the Respiratory Center. We have discussed mainly three different areas of the respiratory center: the dorsal respiratory group of neurons, the ventral respiratory group, and the pneumotoxic center. However, it is believed that none of these are affected directly by changes in blood carbon dioxide concentration or hydrogen ion concentration. Instead, an additional neuronal area, a very sensitive *chemosensitive area*, illustrated in Figure 41–2, is located bilaterally lying less than 1 millimeter beneath the ventral surface of the medulla. This area is highly sensitive to changes in either blood P_{CO_2} or hydrogen ion concentration, and it in turn excites the other portions of the respiratory center.

Response of the Chemosensitive Neurons to Hydrogen Ions — The Primary Stimulus

The sensor neurons in the chemosensitive area are especially excited by hydrogen ions; in fact, it is believed that hydrogen ions are perhaps the only important direct stimulus for these neurons. Unfortunately, though, hydrogen ions do not easily cross either the blood-brain barrier or the blood–cerebrospinal fluid barrier. For this reason, changes in hydrogen ion concentration in the blood actually have considerably less effect in stimulating the chemosensitive neurons than do changes in carbon dioxide, even though carbon dioxide stimulates these neurons indirectly, as is explained below.

Effect of Blood Carbon Dioxide on Stimulating the Chemosensitive Area

Though carbon dioxide has very little direct effect to stimulate the neurons in the chemosensitive area, it does have a very potent indirect effect. It does this by reacting with the water of the tissues to form carbonic acid. This in turn dissociates into hydrogen and bicarbonate ions; the hydrogen ions then have a potent direct stimulatory effect. These reactions are illustrated in Figure 41–2.

Why is it that blood carbon dioxide has a more potent effect in stimulating the chemosensitive neurons than do blood hydrogen ions? The answer is that

Figure 41–2. Stimulation of the inspiratory area by the *chemosensitive area* located bilaterally in the medulla, lying only a few microns beneath the ventral medullary surface. Note also that hydrogen ions stimulate the chemosensitive area, whereas carbon dioxide in the fluid gives rise to most of the hydrogen ions.

the blood-brain barrier and the blood–cerebrospinal fluid barrier are both almost completely impermeable to hydrogen ions, whereas carbon dioxide passes through both these barriers almost as if they did not exist. Consequently, whenever the blood P_{CO_2} increases, so also does the P_{CO_2} of both the interstitial fluid of the medulla and of the cerebrospinal fluid. In both of these fluids the carbon dioxide immediately reacts with the water to form hydrogen ions. Thus, paradoxically, more hydrogen ions are released into the respiratory chemosensitive sensory area when the blood carbon dioxide concentration increases than when the blood hydrogen ion concentration increases. For this reason, respiratory center activity is affected considerably more by changes in blood carbon dioxide than by changes in blood hydrogen ions, a fact that we subsequently discuss quantitatively.

Importance of Cerebrospinal Fluid P_{CO_2} in Stimulating the Chemoreceptive Area. Changing the P_{CO_2} in the cerebrospinal fluid that bathes the surface of the brainstem chemoreceptive area will excite respiration in the same way that increased P_{CO_2} in the medullary interstitial fluids also excites respiration. However, the excitation occurs far more rapidly. The reason for this is believed to be that the cerebrospinal fluid has very little protein acid-base buffers. Therefore, the hydrogen ion concentration increases almost instantly when carbon dioxide enters the cerebrospinal fluid from the extensive arachnoid blood vessels. By contrast, the brain tissues have great quantities of protein buffers, so that the change there in hydrogen ion concentration in response to carbon dioxide is greatly delayed. Consequently, the initial rapid excitation of the respiratory system by carbon dioxide entering the cerebrospinal fluid occurs within seconds, in comparison with as long as a minute or more for the stimulation through the brain interstitial fluid.

Decreased Stimulatory Effect of Carbon Dioxide After the First 1 to 2 Days. The excitation of the respiratory center by carbon dioxide is very great the first few hours but then gradually declines over the next 1 to 2 days, decreasing to about one-fifth the initial effect. Part of this decline results from renal readjustment of the hydrogen ion concentration back toward normal after the carbon dioxide first increases the hydrogen concentration. The kidneys achieve this by increasing the blood bicarbonate. This binds with the hydrogen ions in the cerebrospinal fluid to reduce their concentration. Furthermore, over a period of hours, the bicarbonate ions also slowly diffuse through the blood-brain and blood–cerebrospinal fluid barriers and reduce the hydrogen ions around the respiratory neurons as well.

Therefore, a change in blood carbon dioxide concentration has a very potent *acute* effect on controlling respiratory but only a weak *chronic* effect after a few days' adaptation.

Quantitative Effects of Blood P_{CO_2} and Hydrogen Ion Concentration on Alveolar Ventilation

Figure 41–3 illustrates quantitatively the approximate effects of blood P_{CO_2} and blood pH (which is an inverse logarithmic measure of hydrogen ion concen-

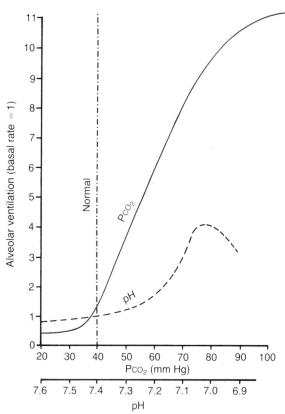

Figure 41–3. Effects of increased arterial P_{CO_2} and decreased arterial pH on the rate of alveolar ventilation.

tration) on alveolar ventilation. Note the marked increase in ventilation caused by the increase in P_{CO_2}. But note also the much smaller effect of increased hydrogen ion concentration (that is, decreased pH).

Finally, note the very great change in alveolar ventilation in the normal blood P_{CO_2} range between 35 and 60 mm Hg. This illustrates the tremendous effect that carbon dioxide changes have in controlling respiration. By contrast, the change in respiration in the normal pH range between 7.3 and 7.5 is more than ten times less. The likely reason for this very great difference is the slight permeability of the blood-brain barrier to hydrogen ions in comparison with its extreme permeability to carbon dioxide. After the carbon dioxide crosses the barrier, however, it reacts with water to form large numbers of hydrogen ions that then do stimulate respiration strongly. Yet the hydrogen ions already formed before crossing the barrier cannot cross in large enough numbers to be very effective.

Unimportance of Oxygen for Direct Control of the Respiratory Center

Changes in oxygen concentration have virtually no *direct* effect on the respiratory center itself to alter respiratory drive (although it does have an indirect effect, as explained in the next section). Nevertheless, the respiratory control system is quite poor in controlling the P_{O_2} in the arterial blood going from the lungs to the peripheral tissues. Yet we learned in Chapter 40 that the hemoglobin oxygen buffer system will deliver almost exactly normal amounts of oxygen to the tissues even when the pulmonary P_{O_2} changes from a value as low as 60 mm Hg up to as high as 1000 mm Hg. Therefore, except under special conditions, proper delivery of oxygen can occur despite changes in lung ventilation ranging from slightly below one-half normal to as high as 20 or more times normal. On the other hand, this is not true for carbon dioxide, for both the blood and tissue P_{CO_2} change almost exactly inversely with the rate of pulmonary ventilation; thus, evolution has made carbon dioxide the major controller of respiration, not oxygen.

Yet for those special conditions where the tissues do get into trouble for lack of oxygen, the body has a special mechanism for respiratory control located outside the brain respiratory center; this responds when the blood oxygen falls too low, as is explained in the following section.

● THE PERIPHERAL CHEMORECEPTOR SYSTEM FOR CONTROL OF RESPIRATORY ACTIVITY— ROLE OF OXYGEN IN RESPIRATORY CONTROL

Aside from the direct control of respiratory activity by the respiratory center itself, still another accessory mechanism is also available for controlling respiration. This is the *peripheral chemoreceptor system,* illustrated in Figure 41–4. Special nervous chemical receptors, called *chemoreceptors,* are located in several areas outside the brain and are especially important for detecting changes in oxygen in the blood, although they also respond to changes in carbon dioxide and hydrogen ion concentrations, too. The chemoreceptors in turn transmit nervous signals to the respiratory center to help regulate respiratory activity.

By far the largest number of chemoreceptors are located in the *carotid bodies.* However, a sizable number are in the *aortic bodies,* also illustrated in Figure 41–4; and a few are located elsewhere in association with other arteries of the thoracic and abdominal regions of the body. The *carotid bodies* are located bilaterally in the bifurcations of the common carotid arteries, and their afferent nerve fibers pass through Hering's nerves to the *glosso-pharyngeal nerves* and thence to the dorsal respiratory area of the medulla. The *aortic bodies* are located along the arch of the aorta; their afferent nerve fibers pass through the *vagi* also to the dorsal respiratory area. Each of these chemoreceptor bodies receives a special blood supply through a minute artery directly from the adjacent arterial trunk. Furthermore, the blood flow through these bodies is extreme, twenty times the weight of the bodies themselves each minute. Therefore, the percentage removal of oxygen is virtually zero. This means that *the chemoreceptors are exposed at all times to arterial blood,* not venous blood, and their P_{O_2}s are arterial P_{O_2}s.

Stimulation of the Chemoreceptors by Decreased Arterial Oxygen. Changes in arterial oxygen concentration have *no* direct stimulatory effect on the respiratory center itself, but when the oxygen concentration in the arterial blood falls below normal, the chemoreceptors become strongly stimulated. This effect is illustrated in Figure 41–5, which shows the effect of different levels of *arterial* P_{O_2} rate of nerve impulse transmission from a carotid body. Note that the impulse rate is particularly sensitive to changes in arterial P_{O_2} in the range between 60 and 30 mm Hg, the range in which the arterial hemoglobin saturation with oxygen decreases rapidly.

Effect of Carbon Dioxide and Hydrogen Ion Concentration on Chemoreceptor Activity. An increase in either carbon dioxide concentration or hydrogen ion concentration also excites the chemoreceptors and in this way indirectly increases respiratory activity. However, the direct effects of both these factors in the respiratory center itself are so much more powerful than their effects mediated through the chemoreceptors (about seven times as power-

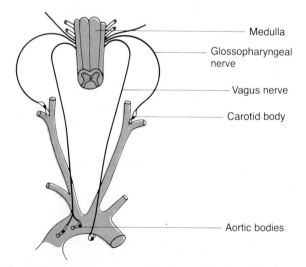

Figure 41–4. Respiratory control by the carotid and aortic bodies.

Figure 41–5. Effect of arterial P_{O_2} on impulse rate from the carotid body of a cat. (Curve drawn from data from several sources, but primarily from Von Euler.)

ful) that for most practical purposes the indirect effects through the chemoreceptors do not need to be considered. Yet there is one difference between the peripheral and central effects of carbon dioxide: the peripheral stimulation of the chemoreceptors occurs as much as five times as rapidly as central stimulation, so that the peripheral chemoreceptors might increase the rapidity of response to carbon dioxide at the onset of exercise.

Basic Mechanism of Stimulation of the Chemoreceptors by Oxygen Deficiency. The exact means by which low P_{O_2} excites the nerve endings in the carotid and aortic bodies is still unknown. However, these bodies have two different, highly characteristic glandular-like cells in them. For this reason, some investigators have suggested that these cells might function as the chemoreceptors and then in turn stimulate the nerve endings. However, other studies suggest that the nerve endings themselves are directly sensitive to the low P_{O_2}.

QUANTITATIVE EFFECT OF LOW ARTERIAL P_{O_2} ON ALVEOLAR VENTILATION

When a person breathes air that has too little oxygen, this obviously will decrease the blood P_{O_2} and excite the carotid and aortic chemoreceptors, thereby increasing respiration. However, the effect usually is much less than one would expect, because the increased respiration will remove carbon dioxide from the lungs and decrease both blood P_{CO_2} and hydrogen ion concentration. Both of these changes then severely depress the respiratory center, as discussed earlier, so that the final effect of the chemoreceptors in increasing respiration in response to low P_{O_2} is mostly suppressed; this is clearly evident in the experimental results shown in Figure 41–6.

Yet the effect of low arterial P_{O_2} on alveolar ventilation is far greater under some other conditions, two of which are (1) low arterial P_{O_2} when arterial carbon dioxide and hydrogen ion concentrations remain normal despite increased respiration and (2) breathing of oxygen at low concentrations for many days.

Effect of Low Arterial P_{O_2} When Arterial Carbon Dioxide and Hydrogen Ion Concentrations Remain Normal

Figure 41–7 illustrates the effect of low arterial P_{O_2} on alveolar ventilation when the P_{CO_2} and the hydrogen ion concentration remain constant at their normal levels. In other words, the ventilatory drive due to both the carbon dioxide and the hydrogen ions does not change; only the ventilatory drive due to the effect of low oxygen on the chemoreceptors is present. The figure shows almost no effect on ventilation as long as the arterial P_{O_2} remains greater than 100 mm Hg. However, at pressures lower than 100 mm Hg, ventilation approximately doubles when the arterial P_{O_2} falls to 60 mm Hg and rises almost sixfold when it falls to 20 mm Hg. Thus, under these conditions, low arterial P_{O_2} *can* drive the ventilatory process quite strongly.

Conditions in Which Ventilatory Drive by Low Arterial P_{O_2} Is Not Blocked by Decreases in P_{CO_2} and Hydrogen Ions. In pneumonia, emphysema, or almost any other pulmonary condition that prevents adequate exchange of gases through the pulmonary membrane, too little oxygen will be absorbed into the arterial blood, and at the same time the arterial P_{CO_2} and hydrogen ion concentration will usually remain near normal or at times even be increased because of poor transport of carbon dioxide through the membrane. In these instances, the ventilatory drive of the low P_{O_2} is not blocked by changes in blood P_{CO_2} and hydrogen ions. Therefore, the low oxygen is then very important in helping to increase respiration. In fact, if the person is given a high concentration of oxygen to breathe, he loses the stimulatory drive from the low arterial P_{O_2} and pulmonary ventilation often decreases enough to cause death because of excessive buildup of lethal P_{CO_2} and hydrogen ion concentration.

Also, in some instances of extremely heavy exercise, especially in persons with slightly compromised pulmonary systems, the arterial P_{CO_2} and hydrogen ion concentration will rise at the same time that the P_{O_2} falls. In such conditions the decreasing arterial P_{O_2} combines with the increasing carbon dioxide and increasing hydrogen ions to give extremely strong ventilatory drive.

Figure 41–6. The lower curve shows the effect of arterial P_{O_2} on alveolar ventilation when the ventilation was allowed to decrease the arterial P_{CO_2}. Note the far less stimulation of ventilation by low P_{O_2} than in Figure 41–7, because the fall in P_{CO_2} had a depressant effect on ventilation. (Drawn from data in Gray: Pulmonary Ventilation and Its Physiological Regulation. Springfield, Ill., Charles C Thomas.)

Figure 41–7. The lower curve illustrates the effect of different levels of arterial P_{O_2} on alveolar ventilation, showing a six-fold increase in ventilation as the P_{O_2} falls from the normal level of 100 mm Hg down to 20 mm Hg. The upper curve shows that the arterial P_{CO_2} was kept at a constant level during the measurements of this study; pH was also kept constant.

Effect of Chronic Breathing of Low Oxygen

Mountain climbers have found that when they ascend a mountain slowly over a period of days, rather than a period of hours, they can withstand far lower atmospheric oxygen concentrations than they can when they ascend rapidly. The reason for this is that the respiratory center in the brain stem loses within 2 to 3 days about four fifths of its sensitivity to changes in arterial Pco_2 and hydrogen ions. Therefore, the blow-off of carbon dioxide that normally would inhibit respiration fails to do so, and the low oxygen now can drive the respiratory system to a much higher level of alveolar ventilation than under acute low oxygen conditions. Instead of a 70 per cent increase in ventilation that might occur on acute exposure to low oxygen, the alveolar ventilation often increases as much as 400 to 500 per cent after 2 to 3 days of low oxygen; and this helps immensely in supplying additional oxygen to the climber. To give a practical example, even experienced mountain climbers can get into difficulty for lack of oxygen if they ascend within 1 day to an altitude of only 18,000 to 20,000 feet. Yet Mount Everest, at an altitude of over 29,000 feet, has been climbed to its peak without any supplemental oxygen; but the ascent was made in very slow stages, so that complete acclimatization of the respiratory drive to low Po_2 could be achieved.

COMPOSITE EFFECTS OF Pco_2, PH, AND Po_2 ON ALVEOLAR VENTILATION

Figure 41–8 gives a quick overview of the manner in which the three chemical factors Po_2, Pco_2, and pH — all together — affect alveolar ventilation. To understand this diagram, first observe the four solid curves. Each of these curves represents the effect of different levels of alveolar Pco_2 on alveolar ventilation, but each curve itself was measured at a different Po_2, beginning at a low Po_2 of 40 mm Hg and going to a high level of 100 mm Hg. Thus, this "family" of solid curves represents the combined effects of alveolar Pco_2 and Po_2 on ventilation.

Now observe the dashed curves. The solid curves were all measured at a blood pH of 7.4; the dashed curves, at a pH of

7.3. Therefore, we now have two families of curves representing the combined effects of Pco_2 and Po_2 on ventilation at two different pH values. Obviously, still different families of curves would be displaced to the right at higher pHs and displaced to the left at lower pHs.

Thus, using this diagram, one can predict the level of alveolar ventilation for most combinations of alveolar Pco_2, alveolar Po_2, and arterial pH.

● REGULATION OF RESPIRATION DURING EXERCISE

In strenuous exercise, oxygen consumption and carbon dioxide formation can increase as much as 20-fold. Yet alveolar ventilation ordinarily increases almost exactly in step with the increased level of metabolism, as illustrated by the relationship between oxygen consumption and ventilation in Figure 41–9. Therefore, the arterial Po_2, Pco_2, and pH all remain *almost exactly normal.*

In trying to analyze the factors that cause increased ventilation during exercise, one is tempted immediately to ascribe this to the chemical alterations in the body fluids during exercise, including increase of carbon dioxide, increase of hydrogen ions, and decrease of oxygen. However, this is not valid, for measurements of arterial Pco_2, pH, and Po_2 show that none of these usually changes significantly.

Therefore, the question must be asked: What is it during exercise that causes the intense ventilation? This question has not been answered, but at least two different effects seem to be predominantly concerned:

1. The brain, on transmitting impulses to the contracting muscles, is believed to transmit collateral impulses into the brain stem to excite the respiratory center. This is analogous to the stimulatory effect of the higher centers of the brain on the vasomotor center of the brain stem during exercise, causing a rise in arterial pressure as well as an increase in ventilation.

Figure 41–8. A composite diagram showing the interrelated effects of Pco_2, Po_2, and pH on alveolar ventilation. (Drawn from data presented in Cunningham and Lloyd: The Regulation of Human Respiration. Philadelphia, F. A. Davis Company.)

Figure 41–9. Effect of exercise on oxygen consumption and ventilatory rate. (From Gray: Pulmonary Ventilation and Its Physiological Regulation. Springfield, Ill., Charles C Thomas.)

2. During exercise, the body movements, especially of the limbs, are believed to increase pulmonary ventilation by exciting joint proprioceptors that then transmit excitatory impulses to the respiratory center. The reason for believing this is that even passive movements of the limbs often increase pulmonary ventilation severalfold.

It is possible that still other factors are also important in increasing pulmonary ventilation during exercise. For instance, some experiments even suggest that hypoxia developing in the muscles during exercise elicits afferent nerve signals to the respiratory center to excite respiration. However, because a large share of the total increase in ventilation begins immediately upon the initiation of exercise, most of the increase in respiration probably results from the two neurogenic factors noted above, namely, *stimulatory impulses from the higher centers of the brain and proprioceptive stimulatory reflexes.*

Interrelationship Between Chemical Factors and Nervous Factors in the Control of Respiration During Exercise. When a person exercises, usually the nervous factors stimulate the respiratory center almost exactly the proper amount to supply the extra oxygen requirements for the exercise and to blow off the extra carbon dioxide. Occasionally, however, the nervous signals are either too strong or too weak in their stimulation of the respiratory center. Then the chemical factors play a very significant role in bringing about the final adjustment in respiration required to keep the carbon dioxide and hydrogen ion concentrations of the body fluids as nearly normal as possible. This effect is illustrated in Figure 41–10,

Figure 41–10. Changes in alveolar ventilation and arterial PCO_2 during a 1-minute period of exercise and also following termination of the exercise. (Extrapolated to the human being from data in dogs; from Bainton: *J. Appl. Physiol., 33:778, 1972.*)

which shows by the lower curve the changes in the alveolar ventilation during a 1 minute period of exercise and by the upper curve the changes in PCO_2. Note that at the onset of the exercise, the alveolar ventilation increases instantaneously without an initial increase in arterial PCO_2. In fact, this onset increase in ventilation is usually great enough that it actually *decreases* arterial PCO_2 below normal, as shown in the figure. The reason the ventilation forges ahead of the buildup of blood carbon dioxide is that the brain provides an "anticipatory" stimulation of respiration at

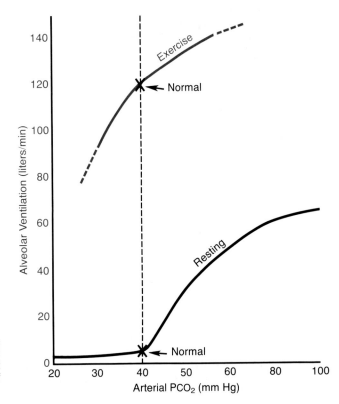

Figure 41–11. Approximate effect of maximum exercise to shift the alveolar PCO_2-ventilation response curve to a much higher than normal level. The shift, believed to be caused by neurogenic factors, is almost exactly the right amount to maintain the arterial PCO_2 at the normal level of 40 mm Hg both in the resting state and during very heavy exercise.

the onset of exercise, causing extra alveolar ventilation even before this is needed. However, after about 30 to 40 seconds, the amount of carbon dioxide released into the blood from the active muscles then approximately matches the increased rate of ventilation, and the arterial P_{CO_2} returns essentially to normal as shown toward the end of the 1 minute period of exercise in the figure.

Figure 41–11 summarizes the control of respiration in still another way, this time more quantitatively. The lower curve of this figure shows the effect of different levels of arterial P_{CO_2} on alveolar ventilation when the body is at rest — that is, not exercising. The upper curve shows the approximate shift of this ventilatory curve caused by the neurogenic drive to the respiratory center that occurs during very heavy exercise. The crosses on the two curves show the arterial P_{CO_2}s first in the resting state and then in the exercising state. Note in both instances that the P_{CO_2} is exactly at the normal level of 40 mm Hg. In other words, the neurogenic factor shifts the curve more than 20-fold in the upward direction so that the ventilation almost exactly matches the rate of oxygen consumption and rate of carbon dioxide release, thus keeping the arterial P_{O_2} and P_{CO_2} very near to their normal values.

Yet the upper curve of Figure 41–11 also illustrates that if the arterial P_{CO_2} does change from the normal value of 40 mm Hg, it has its usual stimulatory effect on ventilation at P_{CO_2}s greater than 40, and its usual depressant effect at P_{CO_2}s less than 40 mm Hg.

Possibility That the Neurogenic Factor for Control of Ventilation During Exercise Is a Learned Response. Many experiments suggest that the brain's ability to shift the ventilatory response curve during exercise, as illustrated in Figure 41–11, is mainly a *learned* response. That is, with repeated exercise, the brain becomes progressively more able to provide the proper amount of brain signal required to keep the blood chemical factors at their normal levels. Also, there is much reason to believe that some of the higher learning centers of the brain are important in this neurogenic respiratory control factor — probably even the cerebral cortex. One important reason for believing this is that when the cerebral cortex is anesthetized, the respiratory control system loses its special ability to maintain the arterial blood gases near normal during exercise.

● OTHER FACTORS THAT AFFECT RESPIRATION

Voluntary Control of Respiration. Thus far, we have discussed the involuntary system for control of respiration. However, we all know that respiration can be controlled voluntarily and that one can hyperventilate or hypoventilate to such an extent that serious derangements in P_{CO_2}, pH, and P_{O_2} can occur in the blood.

Voluntary control of respiration seems not to be mediated through the respiratory center of the medulla. Instead, the nervous pathway for voluntary control passes directly from the cortex and other higher centers downward through the cortico-spinal tract to the spinal neurons that drive the respiratory muscles.

Effect of Irritant Receptors in the Airways. The epithelium of the trachea, bronchi, and bronchioles is supplied with sensory nerve endings that are stimulated by irritants that enter the respiratory airways. These cause coughing and sneezing, as was discussed in Chapter 39. They possibly also cause bronchial constriction in such diseases as asthma and emphysema.

Function of Lung "J" Receptors. A few sensory nerve endings occur in the alveolar walls in *juxta*position to the pulmonary capillaries, from whence comes the name "J receptors." They are stimulated when irritant chemicals are injected into the pulmonary blood, and they are also excited when the pulmonary capillaries become engorged with blood or when pulmonary edema occurs in such conditions as congestive heart failure. Though the functional role of the J receptors is not known, their excitation perhaps does give the person a feeling of dyspnea.

Effect of Brain Edema. The activity of the respiratory center may be depressed or even totally inactivated by acute brain edema resulting from brain concussion. For instance, the head might be struck against some solid object, following which the damaged brain tissues swell, compressing the cerebral arteries against the cranial vault and thus totally or partially blocking the cerebral blood supply.

Occasionally, respiratory depression resulting from brain edema can be relieved temporarily by intravenous injection of hypertonic solutions such as highly concentrated mannitol solution. These solutions osmotically remove some of the fluids of the brain, thus relieving intracranial pressure and sometimes re-establishing respiration within a few minutes.

Anesthesia. Perhaps the most prevalent cause of respiratory depression and respiratory arrest is overdosage of anesthetics or narcotics. For instance, sodium pentobarbital is a poor anesthetic because it depresses the respiratory center considerably more than many other anesthetics such as halothane. At one time, morphine was used as an anesthetic, but this drug is now used only as an adjunct to anesthetics because it greatly depresses the respiratory center while having much less ability to anesthetize the cerebral cortex.

PERIODIC BREATHING

An abnormality of respiration called *periodic breathing* occurs in a number of different disease conditions. The person breathes deeply for a short interval of time and then breathes slightly or not at all for an additional interval, the cycle repeating itself over and over again.

The most common type of periodic breathing, *Cheyne-Stokes breathing,* is characterized by slowly waxing and waning respiration, occurring over and over again approximately every 40 to 60 seconds.

Basic Mechanism of Cheyne-Stokes Breathing. The basic cause of Cheyne-Stokes breathing is the following: When a person overbreathes, thus blowing off too much carbon dioxide from the pulmonary blood and also increasing the blood oxygen, it still takes several seconds before the changed pulmonary blood can be transported to the brain and inhibit ventilation. By this time, the person has already overventilated for an extra few seconds. Therefore, when the respiratory center does eventually respond, it becomes depressed, and now the opposite cycle begins. That is, carbon dioxide builds up and oxygen decreases in the pulmo-

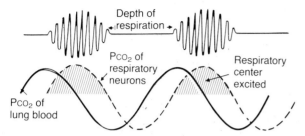

Figure 41–12. Cheyne-Stokes breathing, showing the changing P_{CO_2} in the pulmonary blood (solid line) and the delayed changes in P_{CO_2} of the fluids of the respiratory center (dashed line).

nary blood. Then again it is a few seconds before the brain can respond to the new changes. However, when the brain does respond, the person breathes hard once again. The cycle repeats itself again and again.

Thus, the basic cause of Cheyne-Stokes breathing is present in everyone. However, this mechanism is highly "damped." That is, the fluids of the blood and of the respiratory center have large amounts of stored and chemically bound carbon dioxide and oxygen. Therefore, normally the lungs cannot build up enough extra carbon dioxide or depress the oxygen sufficiently in a few seconds to cause the next cycle of the periodic breathing. Yet under two separate conditions the damping factors are overriden, and Cheyne-Stokes breathing does occur:

1. When there is a long delay in the transport of the blood from the lungs to the brain, the gas changes in the blood will continue for many more seconds than usual. Under these conditions, the storage capacities of the blood and tissues for the gases is exceeded; then the periodic respiratory drive becomes extreme, and Cheyne-Stokes breathing begins. This type of Cheyne-Stokes breathing occurs often in patients with severe cardiac failure, because the left heart is very large and also the blood flow is slow, thus delaying the transport of blood gases from the lungs to the brain. In fact, in patients with chronic heart failure, Cheyne-Stokes breathing is likely to occur on and off for months.

2. A second cause of Cheyne-Stokes breathing is increased negative feedback gain in the respiratory center. This means that a change in blood carbon dioxide or oxygen now causes far greater change in ventilation than normally. For instance, instead of the normal 2- to 3-fold increase in ventilation when the P_{CO_2} rises 3 mm Hg, this same 3 mm Hg might increase the ventilation as much as 10- to 20-fold. Therefore, the brain feedback tendency for periodic breathing is now strong enough to cause Cheyne-Stokes breathing without any blood flow delay between the lungs and brain. This type of Cheyne-Stokes breathing occurs mainly in patients with brain damage. The brain damage often turns off the respiratory drive entirely for a few seconds; then a very slight increase in blood carbon dioxide turns it back on with great force. Cheyne-Stokes breathing of this type is frequently a prelude to death.

Typical records of changes in pulmonary and respiratory center P_{CO_2}s during Cheyne-Stokes breathing are illustrated in Figure 41–12. Note that the P_{CO_2} of the pulmonary blood changes in advance of the P_{CO_2} of the respiratory

neurons. And the depth of respiration corresponds with the P_{CO_2} in the brain, not with the P_{CO_2} in the pulmonary blood where the ventilation is occurring.

REFERENCES

Acker, H.: P_{O_2} chemoreception in arterial chemoreceptors. Annu. Rev. Physiol., 51:835, 1989.

Bainton, C. R.: Effect of speed versus grade and shivering on ventilation in dogs during active exercise. J. Appl. Physiol., 33:778, 1972.

Cohen, M. I.: Central determinants of respiratory rhythm. Annu. Rev. Physiol., 43:91, 1981.

Cohen, M. I.: Neurogenesis of respiratory rhythm in the mammal. Physiol. Rev., 59:1105, 1979.

Cunningham, D. J. C.: Integrative aspects of the regulation of breathing: A personal view. In MTP International Review of Science: Physiology. Vol. 2. Baltimore, University Park Press, 1974, p. 303.

Eldridge, F. L., and Millhorn, D. E.: Central regulation of respiration by endogenous neurotransmitters and neuromodulators. Annu. Rev. Physiol., 43:121, 1981.

Eyzaguirre, C., et al.: Arterial chemoreceptors. In Shepherd, J. T., and Abboud, F. M. (eds.): Handbook of Physiology. Sec. 2, Vol. III. Bethesda, Md., American Physiological Society, 1983, p. 557.

Feldman, J. L., and Ellenberger, H. H.: Central coordination of respiratory and cardiovascular control in mammals. Annu. Rev. Physiol., 50:593, 1988.

Flenley, D. C., and Warren, P. M.: Ventilatory responses to O_2 and CO_2 during exercise. Annu. Rev. Physiol., 45:415, 1983.

Guyton, A. C., et al.: Basic oscillating mechanism of Cheyne-Stokes breathing. Am. J. Physiol., 187:395, 1956.

Haddad, G. G., and Mellins, R. B.: Hypoxia and respiratory control in early life. Annu. Rev. Physiol., 46:629, 1984.

Honig, A.: Salt and water metabolism in acute high-altitude hypoxia: Role of peripheral arterial chemoreceptors. News Physiol. Sci., 4:109, 1989.

Jansen, A. H., and Cherniack, V.: Development of respiratory control. Physiol. Rev., 63:437, 1983.

Kalia, M. P.: Anatomical organization of central respiratory neurons. Annu. Rev. Physiol., 43:105, 1981.

Karczewski, W. A., et al.: Control of Breathing During Sleep and Anesthesia. New York, Plenum Publishing Corp., 1988.

Masuyama, H., and Honda, Y.: Differences in overall "gain" of CO_2-feedback system between dead space and CO_2 ventilations in man. Bull. Eur. Physiopathol. Respir., 20:501, 1984.

Milhorn, H. T., Jr., and Guyton, A. C.: An analog computer analysis of Cheyne-Stokes breathing. J. Appl. Physiol., 20:328, 1965.

Milhorn, H. T., Jr., et al.: A mathematical model of the human respiratory control system. Biophys. J., 5:27, 1965.

Mitchell, G. S., et al.: Changes in the V_I-V_{CO_2} relationship during exercise in goats: Role of carotid bodies. J. Appl. Physiol., 57:1894, 1984.

Read, D. J., and Henderson-Smart, D. J.: Regulation of breathing in the newborn during different behavioral states. Annu. Rev. Physiol., 46:675, 1984.

Richter, D. W., et al.: How is the respiratory rhythm generated? A model. News Physiol. Sci., 1:109, 1986.

Rigatto, H.: Control of ventilation in the newborn. Annu. Rev. Physiol., 46:661, 1984.

Rowell, L. B., and Sheriff, D. D.: Are muscle "chemoreflexes" functionally important? News Physiol. Sci., 3:250, 1988.

Schlaefke, M. E. (ed.): Central Neurone Environment and the Control Systems of Breathing and Circulation. New York, Springer-Verlag, 1983.

Sinclair, J. D.: Respiratory drive in hypoxia: Carotid body and other mechanisms compared. News Physiol. Sci., 2:57, 1987.

Von Euler, C., and Lagercrantz, H.: Neurobiology of the Control of Breathing. New York, Raven Press, 1987.

Von Euler, C., and Langercrantz, H. (eds.): Central Nervous Control Mechanisms in Breathing. New York, Pergamon Press, 1980.

Walker, D. W.: Peripheral and central chemoreceptors in the fetus and newborn. Annu. Rev. Physiol., 46:687, 1984.

West, J. B.: Pulmonary Pathophysiology—The Essentials, 3rd Ed. Baltimore, Williams & Wilkins, 1987.

Whipp, B. J.: Ventilatory control during exercise in humans. Annu. Rev. Physiol., 45:393, 1983.

Williams, M. H. (ed.): Symposium on Disturbances of Respiratory Control. Philadelphia, W. B. Saunders Co., 1980.

CHAPTER 42

• Respiratory Insufficiency — Pathophysiology, Diagnosis, Oxygen Therapy •

The diagnosis and treatment of most respiratory disorders have come to depend highly on an understanding of the basic physiological principles of respiration and gas exchange. Some diseases of respiration result from inadequate ventilation, while others result from abnormalities of diffusion through the pulmonary membrane or of transport from the lungs to the tissues. In each of these instances, the therapy is often entirely different, so that it is no longer satisfactory simply to make a diagnosis of "respiratory insufficiency."

SOME USEFUL METHODS FOR STUDYING RESPIRATORY ABNORMALITIES

In the previous few chapters we have discussed a number of different methods for studying respiratory abnormalities, including measuring vital capacity, tidal air, functional residual capacity, dead space, physiologic shunt, and physiologic dead space. This array of measurements is only part of the armamentarium of the clinical pulmonary physiologist. Some other interesting tools available are described.

STUDY OF BLOOD GASES AND pH

Among the most fundamental of all tests of pulmonary performance are determinations of the blood P_{O_2}, P_{CO_2}, and pH. Indeed, it is often important to make these measurements extremely rapidly as an aid in determining appropriate therapy for acute respiratory distress or acute abnormalities of acid-base balance. Therefore, several simple and very rapid methods have been developed to make these measurements in only a few minutes, using no more than a few drops of blood. These are the following:

Determination of Blood pH. Blood pH is measured using a glass pH electrode of the type used in all chemical laboratories. However, the electrodes used for this purpose are miniaturized, so that no more than a drop or so of blood need be used. The voltage generated by the glass electrode is a direct measure of the pH, and this is generally read directly from a voltmeter scale, or it is recorded on a chart.

Determination of Blood P_{CO_2}. A glass electrode pH meter can be used to determine blood P_{CO_2} in the following way: When a weak solution of sodium bicarbonate is exposed to carbon dioxide, the carbon dioxide dissolves in the solution until an equilibrium state is established. In this equilibrium state, the pH of the solution is a function of the carbon dioxide and bicarbonate ion concentrations in accordance with the Henderson-Hasselbalch equation that was discussed in Chapter 30; that is,

$$pH = \log \frac{HCO_3^-}{CO_2} + 6.1$$

Therefore, with appropriate calibration of the pH meter, the carbon dioxide concentration (or P_{CO_2}) can be read directly from the meter.

When this apparatus is used to measure P_{CO_2} in blood, a miniature glass electrode is coated with a thin solution of sodium bicarbonate, and this is separated from the blood by a very thin plastic membrane that allows carbon dioxide to diffuse from the blood into the solution. Here again, only a drop or so of blood is required.

Determination of Blood P_{O_2}. The concentration of oxygen in a fluid can be measured by a technique called *polarography*. Electrical current is made to flow between a small negative electrode and the solution. If the voltage of the electrode is more than -0.6 volt different from the voltage of the solution, oxygen will deposit on the electrode. Furthermore, the rate of current flow through the electrode will be directly proportional to the concentration of oxygen (and therefore P_{O_2} as well). In practice, a negative platinum electrode with a surface area of about 1 square millimeter is used, and this is separated from the blood by a thin plastic membrane that allows diffusion of oxygen but not diffusion of proteins or other substances that will "poison" the electrode.

Often, all three of the measuring devices for pH, P_{CO_2}, and P_{O_2} are built into the same apparatus, and all these measurements can be made within a minute or so using a single, very small sample of blood. Thus, changes in the blood gases and pH can be followed almost moment by moment.

MEASUREMENT OF MAXIMUM EXPIRATORY FLOW

In many respiratory diseases, especially in asthma, the resistance to air flow becomes especially great during expiration, sometimes causing tremendous difficulty in breathing. This has led to the concept called the *maximum expiratory flow,* which can be defined as follows: When a person expires with great force, the expiratory air flow reaches a maximum flow beyond which the flow cannot be increased even with greatly increased additional force. This is the maximum expiratory flow. The maximum expiratory flow is much greater when the lungs are filled with a large volume of air than when they are almost empty. These principles can be understood by referring to Figure 42–1.

Figure 42–1A illustrates the effect of pressure applied to the outsides of the alveoli and the passageways caused by compressing the chest cage. The arrows indicate that the same amount of pressure is applied to the outsides of both the alveoli and the bronchioles. Therefore, not only does this pressure force air from the alveoli into the bronchioles, but it tends also to collapse the bronchioles at the same time which will oppose the movement of the air to the exterior. Once the bronchioles have become almost completely collapsed, further expiratory force can still greatly increase the alveolar pressure, but it also increases the airway resistance by an equal amount, thus preventing any further increase in flow. Therefore, beyond a critical degree of expiratory force, the maximum expiratory flow has been reached.

Figure 42–1B illustrates the effect of bronchiolar collapse on the maximum expiratory flow. The curve recorded in this section shows the maximum expiratory flow at all levels of lung volume when a normal person first inhales as much air as possible and then expires with maximum expiratory effort until he can expire no more. Note that he quickly reaches a *maximum expiratory air flow* of over 400 liters/min. But it does not matter how much additional expiratory effort he exerts; this is still the maximum flow that he can achieve.

Note also that as the lung volume becomes smaller this maximum expiratory flow also becomes less. The main reason for this is that in the enlarged lung the bronchi and bronchioles are held open partially via elastic pull on their outsides by lung structural elements; however, as the lung becomes smaller, these structures are relaxed so that the bronchi and bronchioles are collapsed more easily by external pressure.

Abnormalities of the Maximum Expiratory Flow-Volume Curve. Figure 42–2 illustrates once more the normal maximum expiratory flow-volume curve, along with two additional curves recorded in two types of lung diseases: (1) constricted lungs and (2) partial airway obstruction.

Note that the *constricted lungs* have both reduced total lung capacity (TLC) and reduced residual volume (RV). Furthermore, because the lung cannot expand to its normal volume, even with the greatest possible expiratory effort the maximal expiratory flow cannot rise to equal that of the normal curve. Constricted lung diseases include fibrotic diseases of the lung itself, such as *tuberculosis, silicosis,* and others, and also diseases that constrict the chest cage such as *kyphosis, scoliosis,* and *fibrotic pleurisy.*

In diseases with *airway obstruction,* it is usually much more difficult to expire than to inspire because the closing tendency of the airways is greatly increased by the positive pressure in the chest during expiration. By contrast, the negative pleural pressure of inspiration actually "pulls" the airways open at the same time that it expands the alveoli. Therefore, air tends to enter the lung easily but then become trapped in the lungs. Over a period of months or years, this effect increases both the total lung capacity and the residual volume, as illustrated in Figure 42–2. Also, because of the obstruction of the airways and because these collapse more easily than normal airways, the maximum expiratory flow is greatly reduced. The classic disease that causes severe airway obstruction is *asthma.* However, serious airway obstruction also occurs in some stages of *emphysema.*

FORCED EXPIRATORY VITAL CAPACITY AND FORCED EXPIRATORY VOLUME

Another exceedingly useful clinical pulmonary test, and one that is also very simple, is to make a record on a spirometer of the *forced expiratory vital capacity* (FVC). Such a record is illustrated in Figure 42–3A for a person with normal lungs and in Figure 42–3B for a person with airway

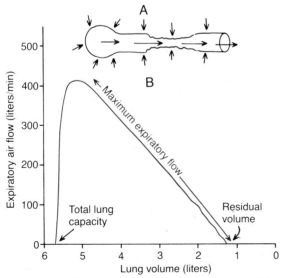

Figure 42–1. *A,* Collapse of the respiratory passageway during maximum expiratory effort, an effect that limits the expiratory flow rate. *B,* Effect of lung volume on the maximum expiratory air flow, showing decreasing maximum expiratory air flow as the lung volume becomes smaller.

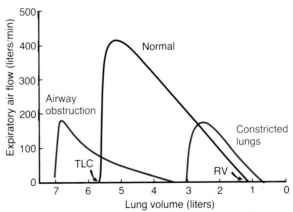

Figure 42–2. Effect of two different respiratory abnormalities — constricted lungs and airway obstruction — on the maximum expiratory flow-volume curve.

Figure 42–3. Recordings during a forced vital capacity maneuver: *A*, in the normal person, and *B*, in the person with airway obstruction.

obstruction. In performing the forced expiratory vital capacity maneuver the person first inspires maximally to the total lung capacity, then exhales into the spirometer with maximum expiratory effort as rapidly and as completely as possible. The downslope of the record as recorded against time represents the *forced vital capacity,* as illustrated in the figure.

Now, study the difference between the two records for normal lungs and airway obstruction. The total volume changes of the forced vital capacities are nearly equal, indicating only a moderate difference in basic lung volumes in the two persons. On the other hand, there is a major difference in the amounts of air that these persons can expire each second, especially during the first second. Therefore, it is customary to record the forced expiratory volume during the first second (FEV_1) and to use this comparison between the normal and the abnormal. In the normal person the percentage of the forced vital capacity that is expired in the first second (FEV_1/FVC%) is about 80 per cent. However, note in Figure 42–3B that with airway obstruction this value was decreased to only 47 per cent. In serious airway obstruction, as often occurs in acute asthma, this can decrease to less than 20 per cent.

● PHYSIOLOGIC PECULIARITIES OF SPECIFIC PULMONARY ABNORMALITIES

CHRONIC PULMONARY EMPHYSEMA

The term *pulmonary emphysema* literally means excess air in the lungs. However, when one speaks of chronic pulmonary emphysema, a complex obstructive and destructive process of the lungs generally is meant, and in most instances it is a consequence of long-term smoking. It results from three major pathophysiological events in the lungs:

1. *Chronic infection* caused by inhaling smoke or other substances that irritate the bronchi and bronchioles. The principal reason for the chronic infection is that the irritant seriously deranges the normal protective mechanisms of the

airways, including partial paralysis of the cilia of the respiratory epithelium by the effects of nicotine so that mucus cannot be moved easily out of the passageways; stimulation of excess mucus secretion, which further exacerbates the condition; and inhibition of the alveolar macrophages, so that they become less effective in combating infection.

2. The infection, the excess mucus, and inflammatory edema of the bronchiolar epithelium together cause *chronic obstruction* of many of the smaller airways.

3. The obstruction of the airways makes it especially difficult to expire, thus causing *entrapment of air in the alveoli* and overstretching them. This, combined with the lung infection, causes marked destruction of many of the alveolar walls. Therefore, the final picture of the emphysematous lung is that illustrated in Figures 42–4 and 42–5.

The physiological effects of chronic emphysema are extremely varied, depending on the severity of the disease and on the relative degree of bronchiolar obstruction versus lung parenchymal destruction. However, among the different abnormalities are the following:

1. The bronchiolar obstruction greatly *increases airway resistance* and results in greatly increased work of breathing. It is especially difficult for the person to move air through the bronchioles during expiration because the compressive force on the outside of the lung not only compresses the alveoli but also compresses the bronchioles, which further increases their resistance during expiration.

2. The marked loss of lung parenchyma greatly *decreases the diffusing capacity* of the lung, which reduces the ability of the lungs to oxygenate the blood and to remove carbon dioxide.

3. The obstructive process is frequently much worse in some parts of the lungs than in other parts so that some portions of the lungs are well ventilated while other portions are poorly ventilated. This often causes *extremely abnormal ventilation-perfusion ratios,* with a very low $\dot{V}a/\dot{Q}$ in some parts *(physiologic shunt)* resulting in poor aeration of the blood, and a very high $\dot{V}a/\dot{Q}$ in other parts *(physiologic dead space)* resulting in wasted ventilation, both effects occurring in the same lungs.

4. Loss of large portions of the lung parenchyma also decreases the number of pulmonary capillaries through which blood can pass. As a result, the pulmonary vascular resistance increases markedly, causing pulmonary hypertension. This in turn overloads the right side of the heart and frequently causes right-heart failure.

Chronic emphysema usually progresses slowly over many years. The person develops hypoxia and hypercapnia because of hypoventilation of many alveoli and because of loss of lung parenchyma. The net result of all of these effects is severe, prolonged, devastating air hunger that can last for years until the hypoxia and hypercapnia cause death—a very high penalty to pay for smoking.

PNEUMONIA

The term *pneumonia* describes any inflammatory condition of the lung in which some or all of the alveoli are filled with fluid and blood cells, as shown in Figure 42–5. A common type of pneumonia is *bacterial pneumonia,* caused most frequently by pneumococci. This disease begins with infection in the alveoli; the pulmonary membrane becomes inflamed and highly porous so that fluid and even red and white blood cells pass out of the blood into the alveoli. Thus, the infected alveoli become progressively filled with fluid and cells, and the infection spreads by extension of bacteria

Figure 42–4. Contrast of the emphysematous lung (above) with the normal lung (below), showing extensive alveolar destruction. (Reproduced with permission of Patricia Delaney and the Department of Anatomy, The Medical College of Wisconsin.)

from alveolus to alveolus. Eventually, large areas of the lungs, sometimes whole lobes or even a whole lung, become "consolidated," which means that they are filled with fluid and cellular debris.

The pulmonary function of the lungs during pneumonia changes in different stages of the disease. In the early stages, the pneumonia process might well be localized to only one lung, and alveolar ventilation may be reduced even

though blood flow through the lung continues normally. This results in two major pulmonary abnormalities: (1) reduction in the total available surface area of the respiratory membrane and (2) decreased ventilation-perfusion ratio. Both these effects cause reduced diffusing capacity, which results in *hypoxemia* (low blood oxygen) and *hypercapnia* (high blood carbon dioxide).

Figure 42–6 illustrates the effect of the decreased venti-

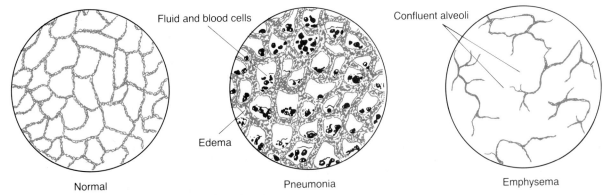

Figure 42–5. Pulmonary changes in pneumonia and emphysema.

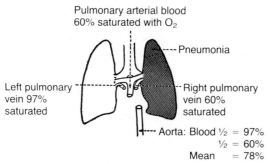

Figure 42–6. Effect of pneumonia on arterial blood oxygen saturation.

lation-perfusion ratio in pneumonia, showing that the blood passing through the aerated lung becomes 97 per cent saturated, whereas that passing through the unaerated lung remains only 60 per cent saturated, causing the mean saturation of the aortic blood to be about 78 per cent, which is far below normal.

ATELECTASIS

Atelectasis means collapse of the alveoli. It can occur in a localized area of a lung, in an entire lobe, or in an entire lung. Its most common causes are: (1) obstruction of the airway and (2) lack of surfactant in the fluids lining the alveoli.

Airway Obstruction. The airway obstruction type of atelectasis usually results from (1) blockage of many small bronchi with mucus or (2) obstruction of a major bronchus by either a large mucous plug or some solid object such as cancer. The air entrapped beyond the block is absorbed within minutes to hours by the blood flowing in the pulmonary capillaries. If the lung tissue is pliable enough, this will lead simply to collapse of the alveoli. However, if the lung tissue cannot collapse, absorption of air from the alveoli creates tremendously negative pressures within the alveoli and pulls fluid out of the pulmonary interstitium into the alveoli, thus causing the alveoli to fill completely with edema fluid. This almost always is the effect that occurs when an entire lung becomes atelectatic, a condition called *massive collapse* of the lung, because the solidity of the chest wall and of the mediastinum allows the lung to decrease only to about one-half normal size, rather than collapsing completely. The remainder of the space in the alveoli must then become filled with fluid.

The effects on overall pulmonary function caused by *massive collapse* (atelectasis) of an entire lung are illustrated in Figure 42–7. Collapse of the lung tissue not only occludes the alveoli but almost always also increases the

resistance to blood flow through the pulmonary vessels. This resistance increase occurs partially because of the collapse itself, which compresses and folds the vessels as the volume of the lung decreases. In addition, hypoxia in the collapsed alveoli causes additional vasoconstriction, as was explained in Chapter 38.

Because of the vascular constriction, blood flow through the atelectatic lung becomes slight. Most of the blood is routed through the ventilated lung and therefore becomes well aerated. In the situation shown in Figure 42–7, five sixths of the blood passes through the aerated lung and only one sixth through the unaerated lung. As a result, the overall ventilation-perfusion ratio is only moderately compromised, so that the aortic blood has only mild oxygen desaturation despite total loss of ventilation by an entire lung.

Lack of Surfactant. The secretion and function of surfactant in the alveoli was discussed in Chapter 37. It was pointed out that the substance surfactant is secreted by the alveolar epithelium into the fluids that line the alveoli. This substance decreases the surface tension in the alveoli two- to tenfold and plays a major role in preventing alveolar collapse. However, in a number of different conditions, such as in *hyaline membrane disease* (or *respiratory distress syndrome*), which often occurs in newborn premature babies, the quantity of surfactant secreted by the alveoli is greatly depressed. As a result, the surface tension of the alveolar fluid increases so much that it causes a serious tendency for the lungs of these babies to collapse or to become filled with fluid, as explained earlier; many of these infants die of suffocation as increasing portions of the lungs become atelectatic.

ASTHMA

Asthma is characterized by spastic contraction of the bronchiolar smooth muscle bronchioles, which causes extremely difficult breathing. It occurs in 3 to 5 per cent of all persons at some time in life. The usual cause is hypersensitivity of the bronchioles to foreign substances in the air. In younger patients, under the age of 30 years, the asthma in about 70 per cent is caused by allergic hypersensitivity, especially sensitivity to plant pollens. In older persons, the cause is almost always hypersensitivity to nonallergic types of irritants in the air, such as irritants in smog.

The allergic reaction that occurs in the allergic type of asthma is believed to occur in the following way: The typically allergic person has a tendency to form abnormally large amounts of IgE antibodies, and these antibodies cause allergic reactions when they react with their complementary antigens, as was explained in Chapter 34. In asthma, these antibodies mainly attach to mast cells that lie in the lung interstitium in close association with the bronchioles and small bronchi. When the person breathes in pollen to which he or she is sensitive (that is, to which the person has developed IgE antibodies), the pollen reacts with the mast cell–attached antibodies and causes these cells to release several different substances. Among them are *histamine, slow-reacting substance of anaphylaxis* (which is a mixture of leukotrienes), *eosinophilic chemotactic factor,* and *bradykinin.* The combined effects of all these factors, especially the slow-reacting substance of anaphylaxis (SRS-A), are to produce (1) localized edema in the walls of the small bronchioles as well as secretion of thick mucus into the bronchiolar lumens and (2) spasm of the bronchiolar smooth muscle. Obviously, therefore, the airway resistance increases greatly.

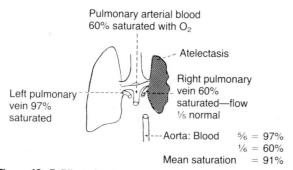

Figure 42–7. Effect of atelectasis on arterial blood oxygen saturation.

As discussed earlier in this chapter, the bronchiolar diameter becomes more reduced during expiration than during inspiration in asthma because the increased intrapulmonary pressure during expiratory effort compresses the outsides of the bronchioles. Because the bronchioles are already partially occluded, further occlusion resulting from the external pressure creates especially severe obstruction during expiration. Therefore, the asthmatic person usually can inspire quite adequately but has great difficulty expiring. Therefore, clinical measurements show greatly reduced maximum expiratory rate and timed expiratory volume. Also, this results in dyspnea, or "air hunger," which is discussed later in the chapter.

The functional residual capacity and the residual volume of the lung become greatly increased during the asthmatic attack because of the difficulty in expiring air from the lungs. Also, over a long period of time the chest cage becomes permanently enlarged, causing a "barrel chest," and the functional residual capacity and residual volume become permanently increased.

TUBERCULOSIS

In tuberculosis the tubercle bacilli cause a peculiar tissue reaction in the lungs, including (1) invasion of the infected region by macrophages and (2) walling off of the lesion by fibrous tissue to form the so-called tubercle. This walling-off process helps to limit further transmission of the tubercle bacilli in the lungs and therefore is part of the protective process against the infection. However, in approximately 3 per cent of all persons who contract tuberculosis, if untreated, the walling-off process fails, and tubercle bacilli spread throughout the lungs, often causing extreme destruction of lung tissue with formation of large abscess cavities. Thus, tuberculosis in its late stages causes many areas of fibrosis throughout the lungs and reduces the total amount of functional lung tissue. These effects cause (1) increased "work" on the part of the respiratory muscles to cause pulmonary ventilation and *reduced vital capacity and breathing capacity;* (2) *reduced total respiratory membrane surface area* and *increased thickness of the respiratory membrane,* these causing progressively diminished pulmonary diffusing capacity; and (3) *abnormal ventilation-perfusion ratio* in the lungs, further reducing the pulmonary diffusing capacity.

● HYPOXIA

Obviously, almost any of the conditions discussed in the past few sections of this chapter can cause serious degrees of cellular hypoxia. In some of these, oxygen therapy is of great value; in others it is of moderate value; in still others it is of almost no value. Therefore, it is important to classify the different types of hypoxia; then we can readily discuss the physiological principles of therapy. The following is a descriptive classification of the different causes of hypoxia:

1. Inadequate oxygenation of the lungs because of extrinsic reasons
   a. Deficiency of oxygen in atmosphere
   b. Hypoventilation (neuromuscular disorders)
2. Pulmonary disease
   a. Hypoventilation due to increased airway resistance or decreased pulmonary compliance
   b. Uneven alveolar ventilation-perfusion ratio (including increased physiologic dead space and physiologic shunt)
   c. Diminished respiratory membrane diffusion
3. Venous-to-arterial shunts ("right-to-left" cardiac shunts)
4. Inadequate transport and delivery of oxygen
   a. Anemia, abnormal hemoglobin
   b. General circulatory deficiency
   c. Localized circulatory deficiency (peripheral, cerebral, coronary vessels)
   d. Tissue edema
5. Inadequate tissue capability of using oxygen
   a. Poisoning of cellular enzymes
   b. Diminished cellular metabolic capacity because of toxicity, vitamin deficiency, or other factors.

This classification of the different types of hypoxia is mainly self-evident from the discussions earlier in the chapter. Only one of the types of hypoxia in the above classification needs further elaboration; this is the hypoxia caused by inadequate capability of the cells to use oxygen.

Inadequate Tissue Capability of Using Oxygen. The classic cause of inability of the tissues to use oxygen is *cyanide poisoning,* in which the action of cytochrome oxidase is completely blocked—to such an extent that the tissues simply cannot utilize the oxygen even though plenty is available. Also, deficiencies of oxidative enzymes or other elements in the tissue oxidative system can lead to this type of hypoxia. A special example of this occurs in the disease beriberi, in which several important steps in the tissue utilization of oxygen and the formation of carbon dioxide are compromised because of vitamin B deficiency.

Effects of Hypoxia on the Body. Hypoxia, if severe enough, can actually cause death of the cells, but in less severe degrees it results principally in (1) depressed mental activity, sometimes culminating in coma, and (2) reduced work capacity of the muscles. These effects are discussed in the following chapter in relation to high altitude physiology.

OXYGEN THERAPY IN THE DIFFERENT TYPES OF HYPOXIA

Oxygen can be administered by (1) placing the patient's head in a "tent" that contains air fortified with oxygen, (2) allowing the patient to breathe either pure oxygen or high concentrations of oxygen from a mask, or (3) administering oxygen through an intranasal tube.

Oxygen therapy is of great value in certain types of hypoxia but of almost no value at all in other types. However, recalling the basic physiological principles of the different types of hypoxia, one can readily decide when oxygen therapy is of value and, if so, how valuable. For instance:

In *atmospheric hypoxia,* oxygen therapy can obviously completely correct the depressed oxygen level in the inspired gases and, therefore, provide 100 per cent effective therapy.

In *hypoventilation hypoxia,* a person breathing 100 per cent oxygen can move five times as much oxygen into the alveoli with each breath as when breathing normal air. Therefore, here again oxygen therapy can be extremely beneficial. (However, this provides no benefit for the hypercapnia also caused by the hypoventilation.)

In *hypoxia caused by impaired diffusion,* essentially the same result occurs as in hypoventilation hypoxia, for oxygen therapy can increase the Po_2 in the lungs from a normal value of about 100 mm Hg to as high as 600 mm Hg. This raises the oxygen diffusion gradient between the alveoli and the blood from a normal value of 60 mm Hg to as high as 560 mm Hg, or an increase of more than 800 per cent. This highly beneficial effect of oxygen therapy in diffusion hypoxia is illustrated in Figure 42–8, which shows that the pulmonary blood in this patient with pulmonary edema picks up oxygen four times as rapidly as it would with no therapy.

In *hypoxia caused by anemia, abnormal hemoglobin transport, circulatory deficiency, or physiological shunt,* oxygen therapy is of much less value because plenty of oxygen is already available in the alveoli. The problem, instead, is that appropriate mechanisms for transporting the oxygen to the tissues are deficient. Even so, a small amount of extra oxygen, between 7 and 30 per cent, can be transported in the dissolved state in the blood even though the amount transported by the hemoglobin is hardly altered. This small amount of extra oxygen may be the difference between life and death.

In the different types of *hypoxia caused by inadequate tissue use of oxygen,* there is abnormality neither of oxygen pickup by the lungs nor of transport to the tissues. Instead, the tissue metabolic system is simply incapable of utilizing the oxygen that is delivered. Therefore, it is doubtful that oxygen therapy is of any benefit at all.

● HYPERCAPNIA

Hypercapnia means excess carbon dioxide in the body fluids.

One might suspect on first thought that any respiratory condition that causes hypoxia would also cause hypercapnia. However, hypercapnia usually occurs in association with hypoxia only when the hypoxia is caused by *hypoventilation* or by *circulatory deficiency.* The reasons for this are the following:

Obviously, hypoxia caused by *too little oxygen in the air,* by *too little hemoglobin,* or by *poisoning of the oxidative enzymes* has to do only with the availability of oxygen or use of oxygen by the tissues. Therefore, it is readily understandable that hypercapnia is *not* a concomitant of these types of hypoxia.

Also, in hypoxia resulting from poor diffusion through the pulmonary membrane or through the tissues, serious hypercapnia usually does not occur because carbon dioxide diffuses 20 times as rapidly as oxygen. Also, if hypercapnia does begin to occur this immediately stimulates pulmonary ventilation, which corrects the hypercapnia but not necessarily the hypoxia.

However, in hypoxia caused by hypoventilation, carbon dioxide transfer between the alveoli and the atmosphere is affected as much as is oxygen transfer. Therefore, hypercapnia always results along with the hypoxia. And in circulatory deficiency, diminished flow of blood decreases the removal of carbon dioxide from the tissues, resulting in tissue hypercapnia. However, the transport capacity of the blood for carbon dioxide is about three times that for oxygen, so that even here the tissue hypercapnia is much less than the tissue hypoxia.

When the alveolar P_{CO_2} rises above approximately 60 to 75 mm Hg, the person by then is breathing about as rapidly and deeply as he can, and "air hunger," also called *dyspnea,* becomes very severe. As the P_{CO_2} rises to 80 to 100 mm Hg, the person becomes lethargic and sometimes even semicomatose. Anesthesia and death can result when the P_{CO_2} rises to 100 to 150 mm Hg.

CYANOSIS

The term "cyanosis" means blueness of the skin, and its cause is excessive amounts of deoxygenated hemoglobin in the skin blood vessels, especially in the capillaries. This deoxygenated hemoglobin has an intense dark blue-purple color that is transmitted through the skin.

It is *not* the percentage deoxygenation of the hemoglobin that causes the bluish hue of the skin, but principally the *concentration of deoxygenated hemoglobin without regard to the amount of oxygenated hemoglobin.* The reason for this is that the red color of oxygenated blood is weak in comparison with the dark blue color of deoxygenated blood. Therefore, when the two are mixed together, the oxygenated blood has relatively little coloring effect in comparison with that of the deoxygenated blood.

In general, definite cyanosis appears whenever the *arterial blood* contains more than 5 grams deoxygenated hemoglobin in each deciliter of blood.

DYSPNEA

Dyspnea means a mental anguish associated with inability to ventilate enough to satisfy the demand for air. A common synonym is "air hunger."

At least three different factors often enter into the development of the sensation of dyspnea. These are (1) abnormality of the respiratory gases in the body fluids, especially hypercapnia (which will be discussed below) and to much less an extent hypoxia, (2) the amount of work that must be performed by the respiratory muscles to provide adequate ventilation, and (3) state of mind.

A person becomes very dyspneic especially from excess buildup of carbon dioxide in body fluids. At times, however, the levels of both carbon dioxide and oxygen in the body fluids are completely normal, but to attain this normality of the respiratory gases, the person has to breathe forcefully. In these instances the forceful activity of the respiratory muscles frequently gives the person a sensation of very severe dyspnea.

Finally, the person's respiratory functions may be completely normal, and still dyspnea may be experienced because of an abnormal state of mind. This is called *neurogenic dyspnea* or, sometimes, *emotional dyspnea.* For

Figure 42–8. Absorption of oxygen into the pulmonary capillary blood in pulmonary edema with and without oxygen therapy.

instance, almost anyone momentarily thinking about the act of breathing may suddenly start taking breaths a little more deeply than ordinarily because of a feeling of mild dyspnea. This feeling is greatly enhanced in persons who have a psychological fear of not being able to receive a sufficient quantity of air, such as on entering small or crowded rooms.

● ARTIFICIAL RESPIRATION

The Resuscitator. Many types of resuscitators are available, and each has its own characteristic principles of operation. Basically, the resuscitator illustrated in Figure 42–9A consists of a supply of oxygen or air; a mechanism for applying intermittent positive pressure and, with some machines, negative pressure as well; and a mask that fits over the face of the patient or a connector for connecting the equipment to an endotracheal tube. This apparatus forces air through the mask into the lungs of the patient during the positive pressure cycle and then usually allows the air to flow passively out of the lungs during the remainder of the cycle.

Earlier resuscitators often caused such severe damage to the lungs because of excessive positive pressure that their usage was at one time greatly decried. However, most resuscitators now have adjustable positive pressure limits that are commonly set at 12 to 15 cm water pressure for normal lungs but sometimes much higher for very noncompliant lungs.

The Tank Respirator. Figure 42–9B illustrates the tank respirator with a patient's body inside the tank and his head protruding through a flexible but airtight collar. At the end of the tank opposite the patient's head is a motor-driven leather diaphragm that moves back and forth with sufficient excursion to raise and lower the pressure inside the tank. As the leather diaphragm moves inward, positive pressure develops around the body and causes expiration; as the diaphragm moves outward, negative pressure causes inspiration. Check valves on the respirator control the positive and negative pressure. Ordinarily these pressures are adjusted so that the negative pressure that causes inspiration falls to −10 to −20 cm water and the positive pressure rises to 0 to +5 cm water.

Effect of the Resuscitator and the Tank Respirator on Venous Return. When air is forced into the lungs under positive pressure, or when the pressure around the patient's body is greatly reduced but with the trachea exposed to atmospheric pressure through the nose, as in the case of the tank respirator, the pressure inside the chest cavity is greater than the pressure everywhere else in the body. Therefore, the flow of blood into the chest from the peripheral veins becomes impeded. As a result, use of excessive pressures with either the resuscitator or the tank respirator can reduce the cardiac output—sometimes to lethal levels. A normal person can usually survive as much as 20 mm Hg continuous positive pressure in the lungs, but exposure for more than a few minutes to greater than 30 mm Hg may cause death.

REFERENCES

Bates, D. V.: Respiratory Function in Disease, 3rd Ed. Philadelphia, W. B. Saunders Co., 1989.
Berte, J. B.: Critical Care: The Lung, 2nd Ed. East Norwalk, Conn., Appleton & Lange, 1986.
Busse, W. W.: The precipitation of asthma by upper respiratory infections. Chest, 87:44S, 1985.
Crapo, J. D.: Morphologic changes in pulmonary oxygen toxicity. Annu. Rev. Physiol., 49:721, 1987.
Dempsey, J. A., and Forster, H. V.: Mediation of ventilatory adaptations. Physiol. Rev., 62:262, 1982.
Fishman, A. P. (ed.): Pulmonary Diseases and Disorders, 2nd Ed. New York, McGraw-Hill Book Co., 1988.
Fridovich, I., and Freeman, B.: Antioxidant defenses in the lung. Annu. Rev. Physiol., 49:693, 1987.

Mechanism for applying positive and negative pressures

A

Positive pressure valve

Negative pressure valve

Leather diaphragm

B

Figure 42–9. *A,* The resuscitator. *B,* Tank respirator.

Green, J. F.: Fundamental Cardiovascular and Pulmonary Physiology, 2nd Ed. Philadelphia, Lea & Febiger, 1987.

Hall, W. H., and Douglas, R. G., Jr.: Pulmonary function during and after common respiratory infections. Annu. Rev. Med., 31:233, 1980.

Hodgkin, J. E., and Petty, T. L.: Chronic Obstructive Pulmonary Disease: Current Concepts. Philadelphia, W. B. Saunders Co., 1987.

Hogg, J. C.: Neutrophil kinetics and lung injury. Physiol. Rev., 67:1249, 1987.

Huch, A., et al. (eds.): Continuous Transcutaneous Monitoring. New York, Plenum Publishing Corp., 1987.

Imlay, J. A., and Linn, S.: DNA damage and oxygen radical toxicity. Science, 240:1302, 1988.

Kelsey, N.: Respiratory Therapy Review: A Workbook and Study Guide. St. Louis, C. V. Mosby Co., 1982.

Killian, K. J., and Campbell, E. J. M.: Dyspnea and exercise. Annu. Rev. Physiol., 45:465, 1983.

Malik, A. B.: Pulmonary microembolism. Physiol. Rev., 63:1114, 1983.

Martin, L.: Pulmonary Physiology in Clinical Practice: The Essentials for Patient Care and Evaluation. St. Louis, C. V. Mosby Co., 1987.

Masuyama, H., and Honda, Y.: Differences in overall 'gain' of CO_2-feedback system between dead space and CO_2 ventilations in man. Bull. Eur. Physiopathol. Respir., 20:501, 1984.

McFadden, E. R., Jr., and Ingram, R. H., Jr.: Exercise-induced airway obstruction. Annu. Rev. Physiol., 45:453, 1983.

Miller, A. A. (ed.): Pulmonary Function Tests: A Guide for the Student and House Officer. Philadelphia, W. B. Saunders Co., 1987.

Morris, H. G.: Mechanisms of action and therapeutic role of corticosteroids in asthma. J. Allergy Clin. Immunol., 75:1, 1985.

Murray, J. F., and Nadel, J. A.: Textbook of Respiratory Medicine. Philadelphia, W. B. Saunders Co., 1988.

Notter, R. H., and Finkelstein, J. N.: Pulmonary surfactant: An interdisciplinary approach. J. Appl. Physiol., 57:1613, 1984.

Packer, M.: Therapeutic application of calcium-channel antagonists for pulmonary hypertension. Am. J. Cardiol., 55:196B, 1985.

Pryor, W. A.: Oxy-radicals and related species. Annu. Rev. Physiol., 49:657, 1987.

Quintanilha, A. (ed.): Oxygen Radicals in Biological Systems. New York, Plenum Publishing Corp., 1988.

Quintanilha, A. (ed.): Reactive Oxygen Species in Chemistry, Biology, and Medicine. New York, Plenum Publishing Corp., 1988.

Sant'Ambrogio, G.: Information arising from the tracheobronchial tree of mammals. Physiol. Rev., 62:531, 1982.

Sant'Ambrogio, G.: Nervous receptors of the tracheobronchial tree. Annu. Rev. Physiol., 49:611, 1987.

Smith, L. L.: The response of the lung to foreign compounds that produce free radicals. Annu. Rev. Physiol., 49:681, 1987.

Spector, S. L.: The use of corticosteroids in the treatment of asthma. Chest, 87:73S, 1985.

Staub, N. C.: Pulmonary edema. Physiol. Rev., 54:678, 1974.

Stewart, R. I.: Exercise in patients with chronic obstructive pulmonary disease. S. Afr. Med. J., 67:87, 1985.

Van Scott, M. R., and Boucher, R. C.: Current perspectives of Clara cell function. News Physiol. Sci., 3:13, 1988.

Way, J. L.: Cyanide intoxication and its mechanism of antagonism. Annu. Rev. Pharmacol. Toxicol., 24: 451, 1984.

West, J. B.: Pulmonary Pathophysiology — the Essentials, 3rd Ed. Baltimore, Williams & Wilkins, 1987.

Wolfe, W. G., and Sabiston, D. C.: Pulmonary Embolism. Philadelphia, W. B. Saunders Co., 1980.

AVIATION, SPACE, AND DEEP SEA DIVING PHYSIOLOGY

CHAPTER 43

• Aviation, High Altitude, and Space Physiology •

As man has ascended to higher and higher altitudes in aviation, in mountain climbing, and in space vehicles, it has become progressively more important to understand the effects of altitude and low gas pressures, as well as several other factors — acceleratory forces, weightlessness, and so forth — on the human body. The present chapter deals with all these problems.

● EFFECTS OF LOW OXYGEN PRESSURE ON THE BODY

Barometric Pressures at Different Altitudes. Table 43–1 gives the approximate *barometric* and *oxygen pressures* at different altitudes, showing that at sea level the barometric pressure is 760 mm Hg; at 10,000 feet, only 523 mm Hg; and at 50,000 feet, 87 mm Hg. This decrease in barometric pressure is the basic cause of all the hypoxia problems in high altitude physiology, because as the barometric pressure decreases, the oxygen partial pressure decreases proportionately, remaining at all times slightly less than 21 per cent of the total barometric pressure — at sea level approximately 159 mm Hg but at 50,000 feet only 18 mm Hg.

ALVEOLAR PO₂ AT DIFFERENT ELEVATIONS

Effect of Carbon Dioxide and Water Vapor on Alveolar Oxygen. Even at high altitudes carbon dioxide is continually excreted from the pulmonary blood into the alveoli. Also, water vaporizes into the inspired air from the respiratory surfaces. Therefore, these two gases dilute the oxygen in the alveoli, thus reducing the oxygen concentration.

Water vapor pressure remains 47 mm Hg as long as the body temperature is normal, regardless of altitude; and during exposure to very high altitudes the pressure of carbon dioxide falls in the acclimatized person, who increases his alveolar ventilation about fivefold, from the normal sea level value of 40 mm Hg to about 7 mm Hg, because of increased respiration.

Now let us see how the pressures of these two gases affect the alveolar oxygen. For instance, assume that the barometric pressure falls to 253 mm Hg, which is the measured value at the top of 29,028 foot Mount Everest. Forty-seven millimeters of mercury of this must be water vapor, leaving only

206 mm Hg for all the other gases. In the acclimatized person, 7 mm of the 206 mm Hg must be carbon dioxide, leaving only 199 mm Hg. If there were no use of oxygen by the body, one fifth of this 199 mm Hg would be oxygen and four fifths would be nitrogen; or the PO₂ in the alveoli would be 40 mm Hg. However, some of this remaining alveolar oxygen would be absorbed into the blood, leaving about 35 mm Hg oxygen pressure in the alveoli. Therefore, at the summit of Mount Everest, only the best of acclimatized persons can barely survive when breathing air. But the effect is very different when the person is breathing pure oxygen, as we see in the following discussions.

Alveolar PO₂ at Different Altitudes. The fifth column of Table 43–1 shows the PO₂s in the alveoli at different altitudes when one is breathing air, for both the unacclimatized and the acclimatized person. The alveolar PO₂ is 104 mm Hg at sea level; at 20,000 feet altitude it falls to approximately 40 mm Hg in the unacclimatized person but 53 mm Hg in the acclimatized. The difference between these two is that the alveolar ventilation increases only slightly in the unacclimatized but about six times as much in the acclimatized person, as we discuss later.

Saturation of Hemoglobin with Oxygen at Different Altitudes. Figure 43–1 illustrates arterial oxygen saturation at different altitudes in breathing air and in breathing oxygen. Up to an altitude of approximately 10,000 feet, even when air is breathed, the arterial oxygen saturation remains at least as high as 90 per cent. However, above 10,000 feet the arterial oxygen saturation falls progressively, as illustrated by the left-hand curve of the figure, until it is only 70 per cent at 20,000 feet and still very much less at higher altitudes.

EFFECT OF BREATHING PURE OXYGEN ON THE ALVEOLAR PO₂ AT DIFFERENT ALTITUDES

When a person breathes pure oxygen instead of air, most of the space in the alveoli formerly occupied by nitrogen now becomes occupied by oxygen instead. Therefore, at 30,000 feet the aviator could have an alveolar PO₂ of 139 mm Hg instead of the 18 mm Hg when breathing air.

The second curve of Figure 43–1 illustrates the arterial oxygen saturation at different altitudes when one is breathing pure oxygen. Note that the saturation remains above 90 per cent until the aviator ascends to approximately 39,000

464

Table 43–1 EFFECTS OF ACUTE EXPOSURE TO LOW ATMOSPHERIC PRESSURES ON ALVEOLAR GAS CONCENTRATIONS AND ON ARTERIAL OXYGEN SATURATION*

Altitude (ft)	Barometric Pressure (mm Hg)	PO_2 in Air (mm Hg)	Breathing Air			Breathing Pure Oxygen		
			PCO_2 in Alveoli (mm Hg)	PO_2 in Alveoli (mm Hg)	Arterial Oxygen Saturation (%)	PCO_2 in Alveoli (mm Hg)	PO_2 in Alveoli (mm Hg)	Arterial Oxygen Saturation (%)
0	760	159	40 (40)	104 (104)	97 (97)	40	673	100
10,000	523	110	36 (23)	67 (77)	90 (92)	40	436	100
20,000	349	73	24 (10)	40 (53)	73 (85)	40	262	100
30,000	226	47	24 (7)	18 (30)	24 (38)	40	139	99
40,000	141	29				36	58	84
50,000	87	18				24	16	15

* Numbers in parentheses are acclimatized values.

feet; then it falls rapidly to approximately 50 per cent at about 47,000 feet.

The "Ceiling" When Breathing Air and When Breathing Oxygen in an Unpressurized Airplane

Comparing the two arterial oxygen saturation curves in Figure 43–1, one notes that an aviator breathing oxygen can ascend to far higher altitudes than one not breathing oxygen. For instance, the arterial saturation at 47,000 feet when one is breathing oxygen is about 50 per cent and is equivalent to the arterial oxygen saturation at 23,000 feet when one is breathing air. In addition, because an unacclimatized person ordinarily can remain conscious until the arterial oxygen saturation falls to 40 to 50 per cent, the ceiling for an aviator in an unpressurized airplane when breathing air is approximately 23,000 feet and when breathing pure oxygen about 47,000 feet, provided the oxygen-supplying equipment operates perfectly.

ACUTE EFFECTS OF HYPOXIA

Some of the important acute effects of hypoxia, beginning at an altitude of approximately 12,000 feet, are drowsiness, lassitude, mental and muscle fatigue, sometimes headache,

Figure 43–1. Effect of low atmospheric pressure on arterial oxygen saturation when one is breathing air and when one is breathing pure oxygen.

occasionally nausea, and sometimes euphoria. All of these progress to a stage of twitchings or convulsions above 18,000 feet and end, above 23,000 feet in the unacclimatized person, in coma.

One of the most important effects of hypoxia is decreased mental proficiency, which decreases judgment, memory, and the performance of discrete motor movements. For instance, if an aviator stays at 15,000 feet for 1 hour, mental proficiency ordinarily will have fallen to approximately 50 per cent of normal and after 18 hours at this level to approximately 20 per cent of normal.

ACCLIMATIZATION TO LOW PO_2

A person remaining at high altitudes for days, weeks, or years becomes more and more acclimatized to the low PO_2, so that it causes fewer deleterious effects on the body and also so that it becomes possible for the person to work harder without hypoxic effects or to ascend to still higher altitudes. The five principal means by which acclimatization comes about are (1) a great increase in pulmonary ventilation, (2) increased red blood cells, (3) increased diffusing capacity of the lungs, (4) increased vascularity of the tissues, and (5) increased ability of the cells to utilize oxygen despite the low PO_2.

Increased Pulmonary Ventilation. On immediate exposure to very low PO_2, the hypoxic stimulation of the chemoreceptors increases alveolar ventilation to a maximum of about 65 per cent. This is an immediate compensation for the high altitude, and it alone allows the person to rise several thousand feet higher than would be possible without the increased ventilation. Then, if the person remains at a very high altitude for several days, the ventilation gradually increases to an average of about five times normal. The basic cause of this gradual increase is the following:

The immediate 65 per cent increase in pulmonary ventilation on rising to a high altitude blows off large quantities of carbon dioxide, reducing the PCO_2 and increasing the pH of the body fluids. Both these changes *inhibit* the respiratory center and thereby *oppose the chemoreceptor stimulation of the center.* However, during the ensuing 2 to 5 days, this inhibition fades away, allowing the respiratory center now to respond with full force to the chemoreceptor stimuli resulting from hypoxia, and the ventilation increases to about five times normal. The cause of this fading inhibition is believed to be mainly a reduction of bicarbonate ion concentration in the cerebrospinal fluid and probably also in the brain tissues. This, in turn, decreases the pH in the fluids surrounding the chemosensitive neurons of the respiratory center, thus increasing the activity of the center. (See Chapter 41 for a more detailed discussion of this effect.)

Increase in Red Blood Cells and Hemoglobin During Acclimatization. It will be recalled from Chapter 32 that hypoxia is the principal stimulus for causing an increase in red blood cell production. Ordinarily, in full acclimatization to low oxygen, the hematocrit rises from a normal value of 40 to 45 to an average of 60 to 65, with an average increase in hemoglobin concentration from a normal of 15 gm/dl to about 22 gm/dl.

In addition, the blood volume also increases, often by as much as 20 to 30 per cent, resulting in a total increase in circulating hemoglobin of as much as 50 to 90 per cent.

Unfortunately, this increase in hemoglobin and blood volume is a slow one, having almost no effect until after 2 to 3 weeks, reaching half development in a month or so, and becoming fully developed only after many months.

Increased Diffusing Capacity During Acclimatization. It will be recalled that the normal diffusing capacity for oxygen through the pulmonary membrane is approximately 21 ml/mm Hg/min, and this diffusing capacity can increase as much as threefold during exercise. A similar increase in diffusing capacity occurs at high altitude. Part of the increase probably results from greatly increased pulmonary capillary blood volume, which expands the capillaries and increases the surface through which oxygen can diffuse into the blood. Another part results from an increase in lung volume, which expands the surface area of the alveolar membrane. A final part results from an increase in pulmonary arterial pressure; this forces blood into greater numbers of alveolar capillaries than normally—especially in the upper parts of the lungs, which are poorly perfused under usual conditions.

The Circulatory System in Acclimatization— Increased Capillarity. The cardiac output often increases as much as 30 per cent immediately after a person ascends to high altitude but then decreases back toward normal with the *increase in blood hematocrit* so that the amount of oxygen transported to the tissues remains approximately normal—that is, unless the altitude becomes so high that severe hypoxia results.

Another circulatory adaptation is an increase in numbers of capillaries in the tissues, which is called *increased capillarity*. This occurs especially in animals born and bred at high altitudes, but less so for animals that later in life have become exposed to the altitude. In very active tissues exposed to chronic hypoxia, the increase in capillarity is especially marked. For instance, the capillary density in the right ventricular muscle increases many percentage points because of the combined effects of hypoxia and excess work load on the right ventricle caused by pulmonary hypertension at high altitude (the hypertension results from pulmonary vasoconstriction caused by low alveolar oxygen concentration).

Cellular Acclimatization. In animals native to altitudes of 13,000 to 17,000 feet, mitochondria and some cellular oxidative enzyme systems are slightly more plentiful than in sea level inhabitants. Therefore, it is presumed that acclimatized human beings as well as these animals can utilize oxygen more effectively than can their sea level counterparts, but this is still uncertain.

NATURAL ACCLIMATIZATION OF NATIVES LIVING AT HIGH ALTITUDES

Many natives in the Andes and in the Himalayas live at altitudes above 13,000 feet—one group in the Peruvian Andes actually lives at an altitude of 17,500 feet and works a mine at an altitude of 19,000 feet. Many of these natives are born at these altitudes and live there all their lives. In all the aspects of acclimatization the natives are superior to even the best-acclimatized lowlanders, even though the lowlanders might also have lived at high altitudes for 10 or more years. This process of acclimatization of the natives begins in infancy. The chest size, especially, is greatly increased, whereas the body size is somewhat decreased, giving a high ratio of ventilatory capacity to body mass. In addition, their hearts, particularly the right side of the heart, which provides a high pulmonary arterial pressure to pump blood through a greatly expanded pulmonary capillary system, are considerably larger than the hearts of lowlanders.

The delivery of oxygen by the blood to the tissues is also highly facilitated in these natives. For instance, Figure 43-2 shows the hemoglobin-oxygen dissociation curves for natives who live at sea level and for their counterparts who live at 15,000 feet. Note that the arterial oxygen P_{O_2} in the natives at high altitude is only 40 mm Hg, but because of the greater quantity of hemoglobin the quantity of oxygen in the arterial blood is actually greater than in the blood of the natives at the lower altitude. Note also that the venous P_{O_2} in the high altitude natives is only 15 mm Hg less than the venous P_{O_2} for the lowlanders, despite the very low arterial P_{O_2}, indicating that oxygen transport to the tissues is exceedingly effective in the naturally acclimatized high-altitude natives.

WORK CAPACITY AT HIGH ALTITUDES: THE EFFECT OF ACCLIMATIZATION

Aside from the mental depression caused by hypoxia, as discussed earlier, the work capacity of all the muscles is also greatly decreased in hypoxia. This includes not only the skeletal muscles but also the cardiac muscle, so that even the maximum level of cardiac output is reduced. In general, the work capacity is reduced in direct proportion to the decrease in maximum rate of oxygen uptake that the body can achieve.

To give an idea of the importance of acclimatization for work capacity, consider this: The work capacities in per cent of normal for unacclimatized and acclimatized persons

Figure 43-2. Oxygen-dissociation curves for blood of high-altitude and sea-level residents, showing the respective arterial and venous P_{O_2}s and oxygen contents as recorded in their native surroundings. (From Oxygen-dissociation curves for bloods of high-altitude and sea-level residents. PAHO Scientific Publication No. 140, Life at High Altitudes, 1966.)

at an altitude of 17,000 feet have been found to be the following:

	Per cent
Unacclimatized	50
Acclimatized for two months	68
Native living at 13,200 feet but working at 17,000 feet	87

Thus, naturally acclimatized natives can achieve a daily work output even at this high altitude almost equal to that of a normal person at sea level, but even well-acclimatized lowlanders can almost never achieve this result.

CHRONIC MOUNTAIN SICKNESS

Occasionally, a person who remains at high altitude too long develops chronic mountain sickness, in which the following effects occur: (1) the red cell mass and hematocrit become exceptionally high, (2) the pulmonary arterial pressure becomes elevated even more than the normal elevation that occurs during acclimatization, (3) the right side of the heart becomes greatly enlarged, (4) the peripheral arterial pressure begins to fall, (5) congestive heart failure ensues, and (6) death often follows unless the person is removed to a lower altitude. The cause of this sequence of events is probably threefold: First, the red cell mass becomes so great that the blood viscosity increases severalfold; this now actually *decreases* tissue blood flow so that oxygen delivery begins to decrease as well. Second, the pulmonary arterioles become vasospastic because of the lung hypoxia. This results from the hypoxic vascular constrictor effect that normally operates to divert blood flow from unaerated to aerated alveoli, as explained in Chapter 38. But in this instance all the arterioles become constricted, the pulmonary arterial pressure rises excessively, and the right side of the heart fails. Third, the pulmonary arteriolar spasm diverts much of the blood flow through nonalveolar pulmonary vessels, thus causing an excess of pulmonary shunt blood flow where the blood is not oxygenated; and this further compounds the problem. Fortunately, most of these persons recover within days or weeks when they are moved to a lower altitude.

ACUTE MOUNTAIN SICKNESS AND HIGH-ALTITUDE PULMONARY EDEMA

A small percentage of persons who ascend rapidly to high altitudes become acutely sick and can die if not given oxygen or removed to a low altitude. The sickness begins from a few hours up to about 2 days after the ascent. Two different events frequently occur:

1. *Acute cerebral edema.* This is believed to result from local vasodilatation of the cerebral blood vessels, caused by the hypoxia. Dilatation of the arterioles increases the capillary pressure, which in turn causes fluid to leak into the cerebral tissues. The cerebral edema can then lead to severe disorientation and other effects related to cerebral dysfunction.

2. *Acute pulmonary edema.* The cause of this is still unknown, but a suggested answer is the following: The severe hypoxia causes the pulmonary arterioles to constrict potently, but the constriction is much greater in some parts of the lungs than in other parts, so that more and more of the pulmonary blood flow is forced through fewer and fewer still unconstricted pulmonary vessels. The postulated result is that the capillary pressure in these areas of the lungs be-

comes especially high, and local edema occurs. Then extension of the process to progressively more and more areas of the lungs leads to severe pulmonary dysfunction that can be lethal. However, allowing the person to breathe oxygen will usually reverse the process within hours.

Some persons have extreme pulmonary vascular reactivity to hypoxia, many times the reactivity of the normal person; these people are especially susceptible to acute high-altitude pulmonary edema.

● EFFECTS OF ACCELERATORY FORCES ON THE BODY IN AVIATION AND SPACE PHYSIOLOGY

Because of rapid changes in velocity and direction of motion in airplanes and spacecraft, several types of acceleratory forces often affect the body during flight. At the beginning of flight, simple linear acceleration occurs; at the end of flight, deceleration; and every time the vehicle turns, angular and centrifugal acceleration.

CENTRIFUGAL ACCELERATORY FORCES

When an airplane makes a turn, the force of centrifugal acceleration is determined by the following relationship:

$$f = \frac{mv^2}{r}$$

in which f is the centrifugal acceleratory force, m is the mass of the object, v is the velocity of travel, and r is the radius of curvature of the turn. From this formula it is obvious that as the velocity increases, the force of centrifugal acceleration increases in proportion to the square of the velocity. It is also obvious that the force of acceleration is directly proportional to the sharpness of the turn (the less the radius).

Measurement of Acceleratory Force—"G." When a man is simply sitting in his seat, the force with which he is pressing against the seat results from the pull of gravity, and it is equal to his weight. The intensity of this force is said to be +1 G because it is equal to the pull of gravity. If the force with which he presses against the seat becomes five times his normal weight during pull-out from a dive, the force acting upon the seat is +5 G.

If the airplane goes through an outside loop so that the man is held down by his seat belt, *negative G* is applied to his body; and if the force with which he is thrown against his belt is equal to the weight of his body, the negative force is −1 G.

Effects of Centrifugal Acceleratory Force (Positive G) on the Body. *Effects on the Circulatory System.* The most important effect of centrifugal acceleration is on the circulatory system because blood is mobile and can be translocated by centrifugal forces.

When the aviator is subjected to *positive G,* the blood is centrifuged toward the lower part of the body. Thus, if the centrifugal acceleratory force is +5 G and the person is in an immobilized standing position, the hydrostatic pressure in the veins of the feet is five times normal, or approximately 450 mm Hg; even in the sitting position this pressure is nearly 300 mm Hg. As the pressure in the vessels of the lower part of the body increases, the vessels passively dilate, and a major proportion of the blood from the upper part of the body is translocated into these lower vessels. Because the heart cannot pump unless blood returns to it, the greater

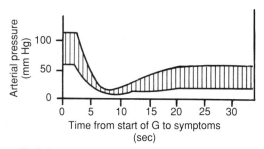

Figure 43–3. Changes in systolic and diastolic arterial pressures following abrupt and continuing exposure to an acceleratory force of +3.3 G. (Modified from Martin and Henry: *J. Aviation Med.,* 22:382, 1951.)

the quantity of blood "pooled" in the lower body the less becomes the cardiac output.

Figure 43–3 illustrates the changes in systolic and diastolic arterial pressures in the upper body when an acceleratory force of +3.3 G is suddenly applied to a sitting person. Note that both these pressures fall below 22 mm Hg for the first few seconds after the acceleration begins but then return to a systolic pressure of about 55 mm Hg and a diastolic pressure of 20 mm Hg within another 10 to 15 seconds. This secondary recovery is caused mainly by activation of the baroreceptor reflexes.

Acceleration greater than 4 to 6 G causes "blackout" of vision within a few seconds and unconsciousness shortly thereafter. If this degree of acceleration is continued, the person will die.

Effects on the Vertebrae. Extremely high acceleratory forces for even a fraction of a second can fracture the vertebrae. The degree of positive acceleration that the average person can withstand in the sitting position before vertebral fracture occurs is approximately 20 G.

Negative G. The effects of negative G on the body are less dramatic acutely but possibly more damaging permanently than the effects of positive G. An aviator can usually go through outside loops up to negative acceleratory forces of −4 to 5 G without harm except for intense momentary hyperemia of the head, though occasionally psychotic disturbances lasting for 15 to 20 minutes occur thereafter as a result of brain edema.

Occasionally negative G forces can be so great and centrifugation of the blood into the head so great that the cerebral blood pressure reaches 300 to 400 Hg, sometimes causing small vessels on the surface of the head and in the brain to rupture. However, the vessels inside the cranium show less tendency for rupture than would be expected, for the cerebrospinal fluid is centrifuged toward the head at the same time that blood is centrifuged toward the cranial vessels, and the greatly increased pressure of this fluid acts as a cushioning buffer on the outside of the brain to prevent vascular rupture.

Because the eyes are not protected by the cranium, intense hyperemia occurs in them during negative G. As a result, the eyes often become temporarily blinded with "redout."

Protection of the Body Against Centrifugal Acceleratory Forces. Specific procedures and apparatus have been developed to protect aviators against the circulatory collapse that occurs during positive G. First, if the aviator tightens the abdominal muscles to an extreme degree and leans forward to compress the abdomen, some of the pooling of blood in the large vessels of the abdomen can be prevented, thereby delaying the onset of blackout. Also, special "anti-G" suits have been devised to prevent pooling of blood

in the lower abdomen and legs. The simplest of these applies positive pressure to the legs and abdomen by inflating compression bags as the G increases. Theoretically, a pilot submerged in a tank or suit of water could withstand very high degrees of acceleration, both positive and negative, for the pressures developed in the water and pressing on the outside of the body during centrifugal acceleration almost exactly balance the forces acting on the body. Unfortunately, however, the presence of air in the lungs still allows displacement of the heart, the lung tissues, and the diaphragm into seriously abnormal positions despite submersion in water. Therefore, even if this procedure were used, the limits of safety would be about 15 to 20 G.

EFFECTS OF LINEAR ACCELERATORY FORCES ON THE BODY

Acceleratory Forces in Space Travel. Unlike an airplane, a spacecraft cannot make rapid turns; therefore, centrifugal acceleration is of little importance except when the spacecraft goes into abnormal gyrations. On the other hand, blast-off acceleration and landing deceleration might be tremendous; both of these are types of linear acceleration.

Figure 43–4 illustrates a typical profile of the acceleration during blast-off in a three-stage spacecraft, showing that the first-stage booster causes acceleration as high as 9 G and the second-stage booster, as high as 8 G. In the standing position the human body could not withstand this much acceleration, but in a semi-reclining position *transverse to the axis of acceleration,* this amount of acceleration can be withstood with ease despite the fact that the acceleratory forces continue for as long as 5 minutes at a time. Therefore, we see the reason for the reclining seats used by the astronauts.

Problems also occur during deceleration when the spacecraft re-enters the atmosphere. A person traveling at Mach 1 (the speed of sound and of fast airplanes) can be safely decelerated in a distance of approximately 0.12 mile, whereas a person traveling at a speed of Mach 100 (a speed possible in interplanetary space travel) requires a distance of about 10,000 miles for safe deceleration. The principal reason for this difference is that the total amount of energy that must be dispelled during deceleration is proportional to the *square* of the velocity, which alone increases the dis-

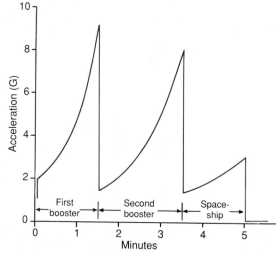

Figure 43–4. Acceleratory forces during the takeoff of a spacecraft.

tance 10,000-fold. But, in addition to this, a human being can withstand far less deceleration if the period of deceleration lasts for a long time than for a short time. Therefore, deceleration must be accomplished much more slowly from the very high velocities than is necessary at lower velocities.

Deceleratory Forces Associated With Parachute Jumps. When the parachuting aviator leaves the airplane, the velocity of fall is exactly 0 feet per second at first. However, because of the acceleratory force of gravity, within 1 second the velocity of fall is 32 feet per second (if there is no air resistance), in 2 seconds it is 64 feet per second, and so on. However, as the rate of fall increases, the air resistance tending to slow the fall also increases. Finally, the deceleratory force of the air resistance exactly balances the acceleratory force of gravity, so that after falling for approximately 12 seconds and a distance of 1400 feet, the person will be falling at a "terminal velocity" of 109 to 119 miles per hour (175 feet per second).

If the parachutist has already reached the terminal velocity of fall before opening the parachute, an "opening shock load" of approximately 1200 pounds occurs on the parachute shrouds.

The usual-sized parachute slows the fall of the parachutist to approximately one-ninth the terminal velocity. In other words, the speed of landing is approximately 20 feet per second, and the force of impact against the earth is approximately 1/81 the impact force without a parachute. Even so, the force of impact is still great enough to cause considerable damage to the body unless the parachutist is properly trained in landing. Actually, the force of impact with the earth is approximately the same as that which would be experienced by jumping from a height of about 6 feet. Unless forewarned, the parachutist will be tricked by the senses into striking the earth with extended legs, and this will result in tremendous deceleratory forces along the skeletal axis of the body, resulting in fracture of the pelvis, of vertebrae, or of a leg. Consequently, the trained parachutist strikes the earth with knees bent, but muscles taut, to cushion the shock of landing.

● "ARTIFICIAL CLIMATE" IN THE SEALED SPACECRAFT

Since there is no atmosphere in outer space, an artificial atmosphere and climate must be provided. Most important of all, the oxygen concentration must remain high enough and the carbon dioxide concentration low enough to prevent suffocation. In some of the earlier space missions a capsule atmosphere containing pure oxygen at about 260 mm Hg pressure has been used. In others, a gas mixture of 50 per cent oxygen and 50 per cent nitrogen at a total pressure of about 380 mm Hg has been used. The presence of nitrogen in the mixture greatly diminishes the likelihood of fire and explosion. It also protects against the development of local patches of atelectasis that often occur in breathing pure oxygen, because oxygen is absorbed very rapidly when small bronchi are temporarily blocked by mucous plugs.

For space travel lasting more than several months, it will be impractical to carry along an adequate oxygen supply and enough carbon dioxide absorbent. For this reason, "recycling techniques" have been proposed for use of the same oxygen over and over again. Some recycling processes depend on purely physical procedures, such as distillation, electrolysis of water to release oxygen, and so forth. Others depend on biological methods, such as use of algae with their large store of chlorophyll to generate foodstuffs and release oxygen from carbon dioxide at the same time by the process of photosynthesis. Unfortunately, a completely practical system for recycling is yet to be achieved.

● WEIGHTLESSNESS IN SPACE

A person in an orbiting satellite or in any nonpropelled spacecraft experiences *weightlessness*. That is, the person is not drawn toward the bottom, sides, or top of the spacecraft but simply floats inside its chambers. The cause of this is not failure of gravity to pull on the body, because gravity from any nearby heavenly body is still active. However, the gravity acts on both the spacecraft and the person at the same time, and since there is no resistance to movement in space, both are pulled with exactly the same acceleratory forces and in the same direction. For this reason, the person simply is not attracted toward any wall of the spacecraft.

Physiological Problems of Weightlessness. Fortunately, the physiological problems of weightlessness have not proved to be severe. Most of the problems that do occur appear to be related to three effects of the weightlessness: (1) motion sickness during the first few days of travel, (2) translocation of fluids within the body because of failure of gravity to cause hydrostatic pressures, and (3) diminishment of physical activity because no strength of muscle contraction is required to oppose the force of gravity.

Almost 50 per cent of astronauts experience motion sickness, with nausea and vomiting, during the first 2 to 5 days of space travel. This probably results from an unfamiliar pattern of motion signals arriving in the equilibrium centers of the brain but a total lack of gravitational signals.

The observed effects of prolonged stay in space are the following: (1) decrease in blood volume, (2) decrease in red cell mass, (3) decreased muscle strength and work capacity, (4) decrease in maximum cardiac output, and (5) loss of calcium and phosphate from the bones as well as loss of bone mass. Most of these same effects also occur in persons lying in bed for an extended period of time. For this reason extensive exercise programs are carried out during prolonged Space Laboratory missions, and most of the above effects are greatly reduced, except for in the case of bone loss. In previous Space Laboratory expeditions, in which the exercise program had been less vigorous, the astronauts had considerably lower work capacities for the first few days after returning to earth. They also had a tendency to faint (and still do to some extent) when they stood up during the first day or so after return to gravity because of translocation of their diminished blood volume toward their abdomens and legs.

The effects of weightlessness that do occur usually reach their maximum within the first few weeks after the astronaut enters the space environment and fortunately are not progressive thereafter, with one exception: the bone loss continues for many months, presumably because the stimulation of bone deposition requires the extra stresses of gravity. Nevertheless, it appears that with an appropriate exercise program the physiologic effects of weightlessness will not be a serious problem even during prolonged space voyages.

REFERENCES

American Physiological Society: High Altitude and Man. Washington, D.C., American Physiological Society, 1984.

Blomqvist, C. G., and Stone, H. L.: Cardiovascular adjustments to gravitational stress. In Shepard, J. T., and Abboud, F. M. (eds.): Handbook of Physiology. Sec. 2, Vol. III, Bethesda, Md., American Physiological Society, 1983, p. 1025.

Brendel, W. (ed.): High Altitude Physiology and Medicine. I. Physiology of Adaptation. New York, Springer-Verlag, 1982.

Burton, R. R. et al.: Man at high sustained +G acceleration: A review. Aerospace Med., 45:1115, 1974.

DeHart, R. L. (ed.): Fundamentals of Aerospace Medicine. Philadelphia, Lea & Febiger, 1985.

Grover, R. F., et al.: High-altitude pulmonary edema. In Fishman, A. P., and Renkin, E. M. (eds.): Pulmonary Edema. Baltimore, Waverly Press, 1979, p. 229.

Hochachka, P. W.: Living Without Oxygen: Closed and Open Systems of Hypoxia Tolerance. Cambridge, Mass., Harvard University Press, 1980.

Hock, R. J.: The physiology of high altitude. Sci. Am., 222:52, 1979.

Honig, A.: Salt and water metabolism in acute high-altitude hypoxia: Role of peripheral arterial chemoreceptors. News Physiol. Sci., 4:109, 1989.

Life Sciences Report, National Aeronautics and Space Administration, Washington, D.C., December 1987.

McCormack, P. D., et al. (eds.): Terrestrial Space Radiation and Its Biological Effects. New York, Plenum Publishing Corp., 1988.

Monge, C., et al.: Laying eggs at high altitude. News Physiol. Sci., 3:69, 1988.

Nicogossian, A. E., et al. (eds.): Space Physiology and Medicine, 2nd Ed. Philadelphia, Lea & Febiger, 1989.

Private Pilot's Medical Handbook. Blue Ridge Summit, Pa., TAB Books, 1982.

Proceedings of the Ninth Annual Meeting of the IUPS Commission on Gravitational Physiology. Physiologist (Suppl.) 31(1), 1988.

Quintanilha, A. (ed.): Reactive Oxygen Species in Chemistry, Biology, and Medicine. New York, Plenum Publishing Corp., 1988.

Read, K.: Aeromedicine for Aviators. New York, State Mutual Book and Periodical Services, 1982.

Sloan, A. W.: Man in Extreme Environments. Springfield, Ill., Charles C Thomas, 1979.

Smith, E. E., and Crowell, J. W.: Role of the hematocrit in altitude acclimatization. Aerospace Med., 38:39, 1966.

Smith, E. E., and Crowell, J. W.: Influence of hematocrit ratio on survival of unacclimatized dogs at simulated high altitude. Am. J. Physiol., 205:1172, 1963.

Sutton, J. R., et al.: Exercise at altitude. Annu. Rev. Physiol., 45:427, 1983.

Talbot, J. M., and Fisher, K. D.: Space sickness. Physiologist, 27:423, 1984.

West, J. B.: Climbing Mount Everest without oxygen. News Physiol. Sci., 1:25, 1986.

West, J. B.: Human physiology at extreme altitude on Mount Everest. Science, 223:784, 1984.

West, J. B.: Man in space. News Physiol. Sci., 1:198, 1986.

CHAPTER 44

• *Physiology of Deep Sea Diving and Other Hyperbaric Conditions* •

When human beings descend beneath the sea, the pressure around them increases tremendously. To keep the lungs from collapsing, air must be supplied also under high pressure, which exposes the blood in the lungs to extremely high alveolar gas pressure, a condition called *hyperbarism.* Beyond certain limits these high pressures can cause tremendous alterations in the body physiology.

Relationship of Sea Depth to Pressure. A column of sea water 33 feet deep exerts the same pressure at its bottom as all the atmosphere above the earth. Therefore, a person 33 feet beneath the ocean surface is exposed to a pressure of 2 atmospheres, 1 atmosphere of pressure caused by the air above the water and the second atmosphere by the weight of the water itself. At 66 feet the pressure is 3 atmospheres, and so forth, in accord with the table in Figure 44–1.

Effect of Depth on the Volume of Gases — Boyle's Law. Another important effect of depth is the compression of gases to smaller and smaller volumes. Figure 44–1 also illustrates a bell jar at sea level containing 1 liter of air. At 33 feet beneath the sea where the pressure is 2 atmospheres, the volume has been compressed to only one-half liter, and at 8 atmospheres (233 feet) to one-eighth liter. Thus, the volume to which a given quantity of gas is compressed is inversely proportional to the pressure. This is the physical principle called *Boyle's Law,* which is extremely important in diving because increased pressures can collapse air chambers of the diver's body, including the lungs, and often cause serious damage.

Many times in this chapter it is necessary to refer to *actual volume* versus *sea level volume.* For instance, we might speak of an actual volume of 1 liter at a depth of 300 feet; this is the same *quantity* of air as a sea level volume of 10 liters.

● EFFECT OF HIGH PARTIAL PRESSURES OF GASES ON THE BODY

The three gases to which a diver breathing air is normally exposed are nitrogen, oxygen, and carbon dioxide, and each of these at times can cause serious physiological effects at high pressures.

Nitrogen Narcosis at High Nitrogen Pressures. Approximately four fifths of the air is nitrogen. At sea level pressure the nitrogen has no known effect on bodily func-

tion, but at high pressures it can cause varying degrees of narcosis. When the diver remains beneath the sea for an hour or more and is breathing compressed air, the depth at which the first symptoms of mild narcosis appear is approximately 120 feet, at which level he begins to exhibit joviality and to lose many of his cares. At 150 to 200 feet, he becomes drowsy. At 200 to 250 feet, his strength wanes considerably, and he often becomes too clumsy to perform the work required of him. Beyond 250 feet (8.5 atmospheres pressure), the diver usually becomes almost useless as a result of nitrogen narcosis if he remains at these depths too long.

Nitrogen narcosis has characteristics very similar to those of alcohol intoxication, and for this reason it has frequently been called "raptures of the depths."

The mechanism of the narcotic effect is believed to be the same as that of essentially all the gas anesthetics. That is, nitrogen dissolves freely in the fats of the body, and it is presumed that it, like most other anesthetic gases, dissolves in the membranes of the neurons and, because of its *physical* effect on altering electrical conductance of the membranes, reduces their excitability.

Oxygen Toxicity at High Pressures. *Effect of Extremely High Po_2 on Blood Oxygen Transport.* When the Po_2 in the blood rises far above 100 mm Hg, the amount of oxygen dissolved in the water of the blood increases markedly. This is illustrated in Figure 44–2, which depicts the same oxygen-hemoglobin dissociation curve as that shown in Chapter 40 except that the alveolar Po_2 is now extended to over 3000 mm Hg. Also depicted is the volume of oxygen dissolved in the fluid of the blood at each Po_2 level. Note that in the normal range of alveolar Po_2, almost none of the total oxygen in the blood is accounted for by dissolved oxygen, but as the pressure rises progressively into the thousands of millimeters of mercury, a large portion of the total oxygen is then dissolved, rather than bound with hemoglobin.

Effect of High Alveolar Po_2 on Tissue Po_2. Let us assume that the Po_2 in the lungs is about 3000 mm Hg (4 atmospheres pressure). Referring to Figure 44–2, one finds that this represents a total oxygen content in each 100 milliliters of blood of about 29 volumes per cent, as illustrated by point A in the figure. As this blood passes through the tissue capillaries and the tissues utilize their normal amount of oxygen, about 5 milliliters from each 100 milliliters of blood, the oxygen content on leaving the tissue

471

Depth (feet)	Atmosphere (s)
Sea level	1
33	2
66	3
100	4
133	5
166	6
200	7
300	10
400	13
500	16

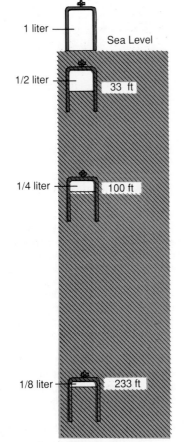

Figure 44–1. Effect of depth on gas volumes.

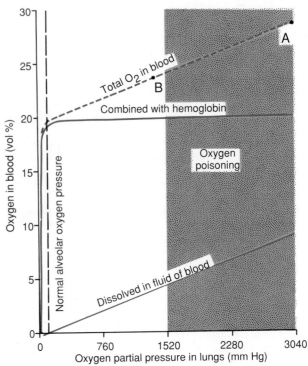

Figure 44–2. Quantity of oxygen dissolved in the fluid of the blood and in combination with hemoglobin at very high PO_2s.

capillaries is still 24 volumes per cent (point B in the figure). At this point the PO_2 is still about 1200 mm Hg, which means that oxygen is delivered to the tissues at this extremely high pressure instead of at the normal value of 40 mm Hg. Thus, once the alveolar PO_2 rises above a critical level, the hemoglobin-oxygen buffer mechanism (discussed in Chapter 40) is no longer capable of keeping the tissue PO_2 in the normal safe range between 20 and 60 mm Hg.

Acute Oxygen Poisoning. Because of the extremely high tissue PO_2 that occurs when oxygen is breathed at a very high alveolar oxygen pressure, one can readily understand that this can be very detrimental to many of the body's tissues. This is especially true of the brain. In fact, exposure to 3 atmospheres pressure of oxygen ($PO_2 = 2280$ mm Hg) will cause convulsions followed by coma in most persons after one to several hours. The convulsions often occur without any warning and, for obvious reasons, are likely to be lethal to divers submerged beneath the sea.

Other symptoms encountered in acute oxygen poisoning

include *nausea, muscle twitchings, dizziness, disturbances of vision, irritability,* and *disorientation.*

Exercise greatly increases a diver's susceptibility to oxygen toxicity, causing symptoms to appear much earlier and with far greater severity than in the resting person.

Excessive Intracellular Oxidation as the Cause of Nervous System Oxygen Toxicity — "Oxidizing Free Radicals." Molecular oxygen (O_2) has very little capability of oxidizing other chemical compounds. Instead, it must first be converted into an "active" form of oxygen. There are several different forms of active oxygen; these are called *oxygen free radicals.* One of the most important of these is the superoxide free radical O_2^-, and another is the peroxide radical in the form of hydrogen peroxide. Even when the tissue PO_2 is normal at the level of 40 mm Hg, small amounts of free radicals are continually being formed from the dissolved molecular oxygen. However, the tissues also contain multiple enzymes that rapidly remove these free radicals, including especially peroxidases, catalases, and superoxide dismutases. Therefore, so long as the hemoglobin-oxygen buffering mechanism functions properly and maintains a normal tissue PO_2, the oxidizing free radicals are removed so rapidly that they have little or no effect in the tissues.

On the other hand, above a critical alveolar PO_2 (above about 2 atmospheres PO_2), the hemoglobin-oxygen buffering mechanism fails, and the tissue PO_2 can then rise to hundreds or thousands of millimeters of mercury. Then, the amounts of oxidizing free radicals literally swamp the enzyme systems for removing them, and now they do have serious destructive and even lethal effects on the cells. One of the principal effects is to oxidize the polyunsaturated fatty acids that are essential components of many of the membranous structures of the cells, and another effect is to oxidize some of the cellular enzymes, thus damaging severely the cellular metabolic systems. The nervous tissues are especially susceptible because of their high lipid content. Therefore, most of the acute lethal effects of acute oxygen toxicity are related to brain dysfunction.

Chronic Oxygen Poisoning as a Cause of Pulmonary Disability. A person can be exposed to 100 per cent oxygen at normal atmospheric pressure almost indefinitely without developing the *acute* oxygen toxicity of the nervous system just described. However, after only 12 or so hours of 100 per cent oxygen exposure, *lung passageway congestion, pulmonary edema,* and *atelectasis* caused by damage to the linings of the bronchi and alveoli begin to develop. The reason for this effect in the lungs and not in the other tissues is that the air spaces of the lungs are directly exposed to the high oxygen pressure, whereas oxygen is delivered to the other tissues at almost normal Po_2 because of the hemoglobin-oxygen buffer system so long as the air Po_2 remains less than about 2 atmospheres.

Carbon Dioxide Toxicity at Great Depths. If the diving gear is properly designed and also functions properly, the diver has no problem due to carbon dioxide toxicity, for depth alone does not increase the carbon dioxide partial pressure in the alveoli. This is true because depth does not increase the rate of carbon dioxide production in the body; and as long as the diver continues to breathe a normal tidal volume, he continues to expire the carbon dioxide as it is formed, maintaining his alveolar carbon dioxide partial pressure at a normal value.

Unfortunately, though, in certain types of diving gear, such as the diving helmet and the different types of rebreathing apparatuses, carbon dioxide can frequently build up in the dead space air of the apparatus and be rebreathed by the diver. Up to an alveolar carbon dioxide pressure (Pco_2) of about 80 mm Hg, two times that of normal alveoli, the diver tolerates this buildup, his minute respiratory volume increasing up to a maximum of eight- to elevenfold to compensate for the increased carbon dioxide. However, beyond the 80 mm Hg level the situation becomes intolerable, and eventually the respiratory center begins to be depressed, rather than excited; the diver's respiration then begins to fail, rather than to compensate. In addition, the diver develops severe respiratory acidosis, and varying degrees of lethargy, narcosis, and finally anesthesia ensue, as discussed in Chapter 42.

DECOMPRESSION OF THE DIVER AFTER EXPOSURE TO HIGH PRESSURES

When a person breathes air under high pressure for a long time, the amount of nitrogen dissolved in the body fluids becomes great. The reason for this is the following: The blood flowing through the pulmonary capillaries becomes saturated with nitrogen to the same pressure as that in the breathing mixture. Over several hours, enough nitrogen is carried to all the tissues of the body to saturate the tissues also with dissolved nitrogen. Because nitrogen is not metabolized by the body, it remains dissolved until the nitrogen pressure in the lungs decreases, at which time the nitrogen is then removed by the reverse respiratory process.

Volume of Nitrogen Dissolved in the Body Fluids at Different Depths. At sea level almost 1 liter of nitrogen is dissolved in the entire body. A little less than half of this is dissolved in the water of the body and a little more than half in the fat of the body. This is true despite the fact that fat constitutes only 15 per cent of the normal body, because nitrogen is five times as soluble in fat as in water.

After the diver has become totally saturated with nitrogen, the *sea level volume of nitrogen* dissolved in the body fluids at the different depths is as follows:

Feet	Liters
33	2
100	4
200	7
300	10

However, several hours are required for the gas pressures of nitrogen in all the body tissues to come to equilibrium with the gas pressure of nitrogen in the alveoli, simply because the blood does not flow rapidly enough and the nitrogen does not diffuse rapidly enough to cause instantaneous equilibrium. The nitrogen dissolved in the water of the body comes to almost complete saturation in less than 1 hour, but the fat, requiring much more nitrogen for saturation and also having a relatively poor blood supply, reaches saturation only after several hours. For this reason, if a person remains at deep levels for only a few minutes, not much nitrogen dissolves in fluids and tissues; whereas if the person remains at a deep level for several hours, both the fluids and tissues become almost completely saturated with nitrogen.

Decompression Sickness (Synonyms: Bends, Compressed Air Sickness, Caisson Disease, Diver's Paralysis, Dysbarism). If a diver has been beneath the sea long enough that large amounts of nitrogen have dissolved in his body and then he suddenly comes back to the surface of the sea, significant quantities of nitrogen bubbles can develop in his body fluids either intracellularly or extracellularly, and these can cause minor or serious damage in almost any area of the body, depending on the number of bubbles formed; this is "decompression sickness."

The principles underlying bubble formation are shown in Figure 44–3. To the left, the diver's tissues have become equilibrated to a very high nitrogen pressure. However, as long as the diver remains deep beneath the sea, the pressure against the outside of his body (5000 mm Hg) compresses all the body tissues sufficiently to keep the dissolved gases in solution. But when the diver suddenly rises to sea level, the pressure on the outside of his body becomes only 1 atmosphere (760 mm Hg), whereas the pressure inside the body fluids is the sum of the pressures of water vapor, carbon

Figure 44–3. Gaseous pressure both inside and outside the body, showing at right the great excess of intrabody pressure that is responsible for bubble formation in the body tissues.

dioxide, oxygen, and nitrogen, or a total of 4065 mm Hg, which is far greater than the pressure on the outside of the body. Therefore, the gases can escape from the dissolved state and form actual bubbles both in the tissues and especially in the blood, where they plug the small blood vessels. However, the bubbles may not appear for many minutes to hours because the gases can remain dissolved in the "supersaturated" state sometimes for hours before bubbling.

Exercise hastens the formation of bubbles during decompression because of increased agitation of the tissues and fluids. This is an effect analogous to that of shaking an opened bottle of soda pop to release the bubbles.

Symptoms of Decompression Sickness. Most of the symptoms of decompression sickness are caused by gas bubbles blocking blood vessels in the different tissues. At first, only the smallest vessels are blocked by very minute bubbles, but as the bubbles coalesce, progressively larger vessels are affected. Obviously, tissue ischemia and sometimes tissue death are the result.

In most persons with decompression sickness, the symptoms are pain in the joints and muscles of the legs or arms, affecting about 89 per cent of those who develop decompression sickness. The joint pain accounts for the term "bends" that is often applied to this condition.

In 5 to 10 per cent of persons with decompression sickness, nervous system symptoms occur, ranging from dizziness in about 5 per cent to paralysis or collapse and unconsciousness in as many as 3 per cent. The paralysis may be temporary, but in some instances the damage is permanent.

Finally, about 2 per cent of persons with decompression sickness develop "the chokes," caused by massive numbers of microbubbles plugging the capillaries of the lungs; this is characterized by serious shortness of breath, often followed by severe pulmonary edema and occasionally death.

Rate of Nitrogen Elimination From the Body; Decompression Tables. Fortunately, if a diver is brought to the surface slowly, the dissolved nitrogen is eliminated through the lungs rapidly enough to prevent decompression sickness. Approximately two thirds of the total nitrogen is liberated in 1 hour and about 90 per cent in 6 hours.

Special decompression tables have been prepared by the U.S. Navy that detail procedures for safe decompression. To give the student an idea of the decompression process, a diver who has been breathing air and has been on the bottom for 60 minutes at a depth of 190 feet is decompressed according to the following schedule:

10 minutes at 50 feet depth
17 minutes at 40 feet depth
19 minutes at 30 feet depth
50 minutes at 20 feet depth
84 minutes at 10 feet depth

Thus, for a work period on the bottom of only 1 hour, the total time for decompression is about 3 hours.

Tank Decompression, and Treatment of Decompression Sickness. Another procedure widely used for decompression of professional divers is to put the diver into a pressurized tank and then to lower the pressure gradually back to normal atmospheric pressure, using essentially the same time schedule as noted above.

Tank decompression is even more important for treating persons in whom symptoms of decompression sickness develop minutes or even hours after they have returned to the surface. In this case, the diver is recompressed immediately to a deep level. Then decompression is carried out over a time period several times as long as the usual decompression period. For best results, the oxygen concentration in the gas mixture of the decompression tank is increased, sometimes even to pure oxygen. This reduces the nitrogen partial pressure in the alveoli to a very low value, which causes several times as rapid removal of the nitrogen.

"Saturation Diving" and Use of Helium-Oxygen Mixtures in Deep Dives. When divers must work at very deep levels—between 250 feet and nearly 1000 feet—they frequently live in a large compression tank for weeks at a time, remaining compressed at a pressure level near that at which they will be working. This keeps the tissues and fluids of the body saturated with the gases to which they will be exposed while diving. Then when they work and later return to the same tank after working, there are not significant changes in pressure, and so decompression bubbles do not occur.

In very deep dives, especially during saturation diving, helium is usually used in the gas mixture instead of nitrogen for three different reasons: (1) it has only about one-fifth the narcotic effect of nitrogen, (2) only about half as much volume of helium dissolves in the body tissues as nitrogen, and (3) the low density of helium (one-seventh the density of nitrogen) keeps the airway resistance for breathing at a minimum, which is extremely important, because highly compressed nitrogen is so dense that airway resistance can become extreme, sometimes making the work of breathing beyond endurance.

Finally, in very deep dives it is important to reduce the oxygen concentration in the gaseous mixture, for otherwise oxygen toxicity would result. For instance, at a depth of 700 feet (22 atmospheres of pressure) a 1 per cent oxygen mixture will provide all the oxygen required by the diver, while a 21 per cent mixture of oxygen (the percentage in air) delivers a Po_2 to the lungs of over 4 atmospheres, a level likely to cause convulsions in as little as 30 minutes.

● **SCUBA DIVING (SELF-CONTAINED UNDERWATER BREATHING APPARATUS)**

Prior to the 1940s, almost all diving was done using a diving helmet connected to a hose through which air was pumped to the diver from the surface. Then, in 1943, Jacques Cousteau developed and popularized the *self-contained underwater breathing apparatus,* popularly known simply as the SCUBA apparatus. The type of SCUBA apparatus used in over 99 per cent of all sports and commercial diving is the *open circuit demand system* illustrated in Figure 44–4. This system consists of the following components: (1) one or more tanks of compressed air or of some other breathing mixture, (2) a first-stage "reducing" valve for reducing the pressure from the tanks to a constant low pressure level, (3) a combination inhalation "demand" valve and exhalation valve that allows air to be pulled into the lungs with very slight negative pressure of breathing and then to be exhausted into the sea at very slight positive pressure, and (4) a mask and tube system with small "dead space."

Basically, the demand system operates as follows: The first-stage reducing valve reduces the pressure from the tanks usually to a pressure of about 140 PSI. However, the breathing mixture does not flow continually into the mask. Instead, with each inspiration, slight negative pressure in the mask pulls the diaphragm of the demand valve inward, and this automatically releases air from the hose into the mask and lungs. In this way only the amount of air needed

Figure 44–4. The open circuit demand type of SCUBA apparatus.

for inhalation enters the system. Then, on expiration, the air cannot go back into the tank, but instead is expired through the expiration valve.

The most important problem in use of the self-contained underwater breathing apparatus is the time limit that one can remain beneath the surface; for instance, only a few minutes are possible at a 200 foot depth. The reason for this is that tremendous airflow from the tanks is required to wash carbon dioxide out of the lungs—the greater the depth, the greater the airflow in terms of *quantity* of air that is required, because the *volumes* have been compressed to small sizes.

● SPECIAL PHYSIOLOGICAL PROBLEMS IN SUBMARINES

Escape From Submarines. Essentially the same problems as those of deep sea diving are often met in relation to submarines, especially when it is necessary to escape from a submerged submarine. Escape is possible from as deep as 300 feet even without using any special type of apparatus. Proper use of rebreathing devices, especially when using helium, theoretically can allow escape from as deep as 600 feet or perhaps more.

One of the major problems of escape is prevention of air embolism. As the person ascends, the gases in his lungs expand and sometimes rupture a pulmonary vessel, allowing the gases to enter into the pulmonary vascular system to cause embolism of the circulation. Therefore, as the person ascends, he must consciously exhale continually.

Exhalation of the expanding gases from the lungs during ascent, even without breathing, is often rapid enough to blow off the accumulating carbon dioxide in the lungs. This keeps the concentration of carbon dioxide from building up in the blood and keeps the person from having the desire to breathe. Therefore, he can hold his breath for an extra long time during ascent.

Health Problems in the Submarine Internal Environment. Except for escape, submarine medicine generally centers around several engineering problems to keep hazards out of the internal environment. In atomic submarines there exists the problem of radiation hazards; but with appropriate shielding, the amount of radiation received by the crew submerged beneath the sea has actually been less than the normal radiation received above the surface of the sea from cosmic rays.

Second, poisonous gases on occasion escape into the atmosphere of the submarine and must be controlled early. For instance, during several weeks' submergence, cigarette smoking by the crew can liberate sufficient amounts of carbon monoxide, if it is not removed from the air, to cause carbon monoxide poisoning, and on occasion even Freon gas has been found to diffuse through the tubes in refrigeration systems in sufficient quantity to cause toxicity.

● HYPERBARIC OXYGEN THERAPY

In recent years it has been learned that the intense oxidizing properties of high pressure oxygen *(hyperbaric oxygen)* can have very valuable therapeutic effects in several important clinical conditions. Therefore, large pressure tanks are now available in many medical centers into which patients can be placed and treated with hyperbaric oxygen. The oxygen is usually administered at Po_2s of 2 to 3 atmospheres of pressure through a mask or intratracheal tube, while the gas around the body is normal air compressed to the same high pressure level. It is believed that the same oxidizing free radicals responsible for oxygen toxicity are also responsible for the therapeutic benefits. Some of the conditions in which hyperbaric oxygen therapy has been especially beneficial are the following:

Probably the most successful use of hyperbaric oxygen has been in the treatment of *gas gangrene.* The bacteria that cause this condition, the *clostridial organisms,* grow best under anaerobic conditions and actually stop growing at oxygen pressures greater than about 70 mm Hg. Therefore, hyperbaric oxygenation of the tissues can frequently stop the infectious process entirely and thus convert a condition that formerly was almost 100 per cent fatal into one that is cured in most instances by hyperbaric therapy.

Recent results also suggest that hyperbaric oxygenation might have almost as dramatic an effect in curing *leprosy* as in curing gas gangrene—also because of the susceptibility of the leprosy bacillus to destruction by high oxygen pressures.

Other conditions in which hyperbaric oxygen therapy has been either valuable or possibly valuable include decompression sickness, arterial gas embolism, carbon monoxide poisoning, osteomyelitis, and myocardial infarction.

REFERENCES

Bennett, P. B., and Elliott, D. H.: The Physiology and Medicine of Diving and Compressed Air Work, 2nd Ed. Baltimore, Williams & Wilkins, 1975.
Brauer, R. W.: Problems of exposure to high pressures. News Physiol. Sci., 1:192, 1986.
Burakovsky, V. I.: Hyperbaric Oxygenation and Its Value in Cardiovascular Surgery. Chicago, Imported Publications, 1981.
Cole, T.: Deep duty. Popular Mechanics, November, 1986, p. 71.
Crapo, J. D.: Morphologic changes in pulmonary oxygen toxicity. Annu. Rev. Physiol., 48:721, 1986.
Elsner, R., and de Burgh Daly, M.: Coping with asphyxia: Lessons from seals. News Physiol. Sci., 3:65, 1988.
Fisher, A. B., et al.: Oxygen toxicity of the lung: Biochemical aspects. In Fishman, A. P., and Renkin, E. M. (eds.): Pulmonary Edema. Baltimore, Waverly Press, 1979, p. 207.
Fridovich, I., and Freeman, B.: Antioxidant defenses in the lung. Annu. Rev. Physiol., 48:693, 1986.

Fridovich, I.: Superoxide radical: An endogenous toxicant. Annu. Rev. Pharmacol. Toxicol., 23:239, 1983.

Gamarra, J. A.: Decompression Sickness. Hagerstown, Md., Harper & Row, 1974.

Gooden, B. A.: Drowning and the diving reflex in man. Med. J. Aust., 2:583, 1972.

Halsey, M. J.: Effects of high pressure on the central nervous system. Physiol. Rev., 62:1341, 1982.

Hochachka, B. W., and Murphy, B.: Metabolic status during diving and recovery in marine animals. In Robertshaw, D. (ed.): International Review of Physiology: Environmental Physiology III. Vol. 20, Baltimore, University Park Press, 1979, p. 253.

Hochachka, B. W., and Storey, K. B.: Metabolic consequences of diving in animals and man. Science, 187:613, 1975.

Jamieson, D., et al.: The relation of free radical production to hyperoxia. Annu. Rev. Physiol., 48:703, 1986.

Miller, J. W.: Vertical Excursions Breathing Air from Nitrogen-Oxygen or Air Saturation Exposures. Washington, D.C., U.S. Government Printing Office, 1976.

Oxygen Free Radicals and Tissue Damage. Ciba Foundation Symposium. New York, Excerpta Medica, 1979.

Pryor, W. A.: Oxy-radicals and related species. Annu. Rev. Physiol., 48:657, 1986.

Shilling, C. W., and Beckett, M. W. (eds.): Underwater Physiology IV. Bethesda, Md., Federation of American Societies for Experimental Biology, 1978.

Sloan, A. W.: Man in Extreme Environments. Springfield, Ill., Charles C Thomas, 1979.

Smith, L. L.: The response of the lung to foreign compounds that produce free radicals. Annu. Rev. Physiol., 48:681, 1986.

Undersea Medical Society: Glossary of Diving and Hyperbaric Terms. Bethesda, Md., Undersea Medical Society, 1978.

Zimmerman, A. M. (ed.): High Pressure Effects on Cellular Processes. New York, Academic Press, 1970.

UNIT IX

THE NERVOUS SYSTEM: A. GENERAL PRINCIPLES AND SENSORY PHYSIOLOGY

• Organization of the Nervous System; Basic Functions of Synapses and Transmitter Substances •

The nervous system, along with the endocrine system, provides most of the control functions for the body. In general, the nervous system controls the rapid activities of the body, such as muscular contractions, rapidly changing visceral events, and even the rates of secretion of some endocrine glands. The endocrine system, by contrast, regulates principally the metabolic functions of the body.

The nervous system is unique in the vast complexity of the control actions that it can perform. It receives literally millions of bits of information from the different sensory organs and then integrates all these to determine the response to be made by the body. The purpose of this chapter is to present, first, a general outline of the overall mechanisms by which the nervous system performs such functions. Then we will discuss the function of central nervous system synapses, the basic structures that control the passage of signals into, through, and out of the nervous system. In succeeding chapters we will analyze in detail the functions of the individual parts of the nervous system. Before beginning this discussion, however, the reader should refer to Chapters 5 and 7, which present, respectively, the principles of membrane potentials and transmission of signals in nerves and through neuromuscular junctions.

• GENERAL DESIGN OF THE NERVOUS SYSTEM

THE CENTRAL NERVOUS SYSTEM NEURON — THE BASIC FUNCTIONAL UNIT

The central nervous system is composed of more than 100 billion neurons. Figure 45–1 illustrates a typical neuron of the type found in the cerebral motor cortex. The incoming information enters the cell almost entirely through synapses on the neuronal dendrites or cell body; there may be only a few hundred or as many as 200,000 such synaptic connections from the input fibers. On the other hand, the output signal travels by way of a single axon, but this axon gives off many separate branches to other parts of the brain, the spinal cord, or the peripheral body. These terminals then provide synapses with the next order of neurons or with muscle cells or secretory cells.

A special feature of most synapses is that the signal normally passes only in the forward direction except under rare conditions. This allows the signals to be conducted in the required directions for performing necessary nervous functions. We shall also see that the neurons are organized into a great multitude of neural networks that determine the functions of the nervous system.

THE SENSORY DIVISION OF THE NERVOUS SYSTEM — SENSORY RECEPTORS

Most activities of the nervous system are initiated by sensory experience emanating from *sensory receptors,* whether visual receptors, auditory receptors, tactile receptors on the surface of the body, or other kinds of receptors. This sensory experience can cause an immediate reaction, or its memory can be stored in the brain for minutes, weeks, or years and then can help determine the bodily reactions at some future date.

Figure 45–2 illustrates a portion of the sensory system, the *somatic* portion, which transmits sensory information from the receptors of the entire surface of

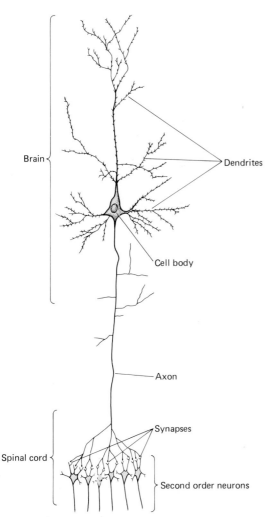

Figure 45–1. Structure of a large neuron of the brain, showing its important functional parts. (From Guyton: Basic Neuroscience: Anatomy and Physiology. Philadelphia, W. B. Saunders Company, 1987.)

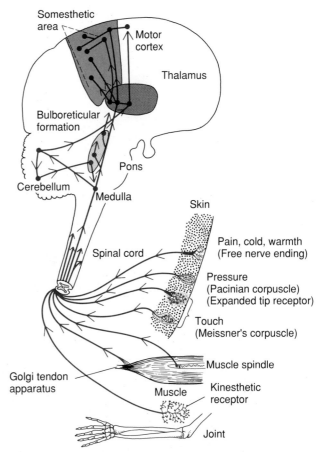

Figure 45–2. The somatic sensory axis of the nervous system.

the body and some deep structures. This information enters the central nervous system through the spinal nerves and is conducted to multiple "primary" sensory areas in (1) the spinal cord at all levels, (2) the reticular substance of the medulla, pons, and mesencephalon, (3) the cerebellum, (4) the thalamus, and (5) the somesthetic areas of the cerebral cortex. But in addition to these primary sensory areas, signals are then relayed to essentially all other parts of the nervous system as well.

THE MOTOR DIVISION— THE EFFECTORS

The most important ultimate role of the nervous system is to control the various bodily activities. This is achieved by controlling (1) contraction of skeletal muscles throughout the body, (2) contraction of smooth muscle in the internal organs, and (3) secretion by both exocrine and endocrine glands in many parts of the body. These activities are collectively called *motor functions* of the nervous system, and the

muscles and glands are called *effectors,* because they perform the functions dictated by the nerve signals.

Figure 45–3 illustrates the *motor axis* of the nervous system for controlling skeletal muscle contraction. Operating parallel to this axis is another similar system for control of the smooth muscles and glands called the *autonomic nervous system,* which is presented in Chapter 60. Note in Figure 45–3 that the skeletal muscles can be controlled from many different levels of the central nervous system, including (1) the spinal cord, (b) the reticular substance of the medulla, pons, and mesencephalon, (3) the basal ganglia, (4) the cerebellum, and (5) the motor cortex. Each of these different areas plays its own specific role in the control of body movements, the lower regions being concerned primarily with automatic, instantaneous responses of the body to sensory stimuli; and the higher regions, with deliberate movements controlled by the thought process of the cerebrum.

PROCESSING OF INFORMATION— "INTEGRATIVE" FUNCTION OF THE NERVOUS SYSTEM

The major function of the nervous system is to process incoming information in such a way that *appropriate* motor responses occur. More than 99 per cent of all sensory information is discarded by the brain as

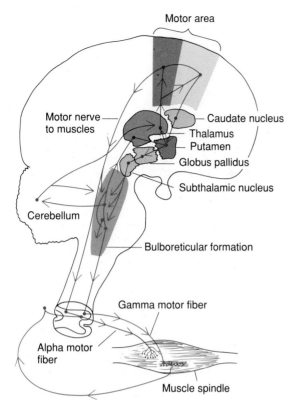

Figure 45–3. The motor axis of the nervous system.

irrelevant and unimportant. For instance, one is ordinarily totally unaware of the parts of the body that are in contact with clothing and is also unaware of the seat pressure when sitting. Likewise, attention is drawn only to an occasional object in one's field of vision, and even the perpetual noise of our surroundings is usually relegated to the background.

After the important sensory information has been selected, it is then channeled into proper motor regions of the brain to cause the desired responses. This channeling of information is called the *integrative function* of the nervous system. Thus, if a person places a hand on a hot stove, the desired response is to lift the hand. There are other associated responses, too, such as moving the entire body away from the stove and perhaps even shouting with pain. Yet even these responses represent activity by only a small fraction of the total motor system of the body.

Role of Synapses in Processing Information. The synapse is the junction point from one neuron to the next and, therefore, is an advantageous site for control of signal transmission. Later in this chapter we discuss the details of synaptic function. However, it is important to point out here that the synapses determine the directions that the nervous signals spread in the nervous system. Some synapses transmit signals from one neuron to the next with ease, and others transmit signals only with difficulty. Also, facilitatory and inhibitory signals from other areas in the nervous system can control synaptic activity, sometimes opening the synapses for transmission and at other times closing them. In addition, some post-

synaptic neurons respond with large numbers of impulses, and others respond with only a few. Thus, the synapses perform a selective action, often blocking the weak signals while allowing the strong signals to pass, often selecting and amplifying certain weak signals, and often channeling the signals in many different directions, rather than simply in one direction.

STORAGE OF INFORMATION—MEMORY

Only a small fraction of the important sensory information causes an immediate motor response. Much of the remainder is stored for future control of motor activities and for use in the thinking processes. Most of this storage occurs in the *cerebral cortex,* but not all, for even the basal regions of the brain and perhaps even the spinal cord can store small amounts of information.

The storage of information is the process we call *memory,* and this, too, is a function of the synapses. That is, each time certain types of sensory signals pass through sequences of synapses, these synapses become more capable of transmitting the same signals the next time, which process is called *facilitation.* After the sensory signals have passed through the synapses a large number of times, the synapses become so facilitated that signals generated within the brain itself can also cause transmission of impulses through the same sequences of synapses even though the sensory input has not been excited. This gives the person a perception of experiencing the original sensations, although in effect they are only memories of the sensations.

Unfortunately, we do not know the precise mechanism by which facilitation of synapses occurs in the memory process, but what is known about this and other details of the memory process will be discussed in Chapter 57.

Once memories have been stored in the nervous system, they become part of the processing mechanism. The thought processes of the brain compare new sensory experiences with the stored memories; the memories help to select the important new sensory information and to channel this into appropriate storage areas for future use or into motor areas to cause bodily responses.

● THE THREE MAJOR LEVELS OF CENTRAL NERVOUS SYSTEM FUNCTION

The human nervous system has inherited specific characteristics from each stage of evolutionary development. From this heritage, three major levels of the central nervous system have specific functional attributes: (1) the *spinal cord level,* (2) the *lower brain level,* and (3) the *higher brain* or *cortical level.*

The Spinal Cord Level

We often think of the spinal cord as being only a conduit for signals from the periphery of the body to the brain or in the opposite direction from the brain back to the body. However, this is far from the truth. Even after the spinal cord has been cut in the high neck region, many spinal cord functions still occur. For instance, neuronal circuits in the cord can cause (1) walking movements, (2) reflexes that withdraw portions of the body from objects, (3) reflexes that stiffen the legs to support the body against gravity, and (4) reflexes that control local blood vessels, gastrointestinal movements, and so forth, in addition to many other functions.

In fact, the upper levels of the nervous system often operate not by sending signals directly to the periphery of the body but instead by sending signals to the control centers of the cord, simply "commanding" the cord centers to perform their functions.

The Lower Brain Level

Many if not most of what we call subconscious activities of the body are controlled in the lower areas of the brain — in the medulla, pons, mesencephalon, hypothalamus, thalamus, cerebellum, and basal ganglia. Subconscious control of arterial pressure and respiration is achieved mainly in the medulla and pons. Control of equilibrium is a combined function of the older portions of the cerebellum and the reticular substance of the medulla, pons, and mesencephalon. Feeding reflexes, such as salivation in response to the taste of food and the licking of the lips, are controlled by areas in the medulla, pons, mesencephalon, amygdala, and hypothalamus; and many emotional patterns, such as anger, excitement, sexual activities, reaction to pain, or reaction of pleasure, can occur in animals without a cerebral cortex.

The Higher Brain or Cortical Level

After recounting all the nervous system functions that can occur at the cord and lower brain levels, what is left for the cerebral cortex to do? The answer to this is a complex one, but it begins with the fact that the cerebral cortex is an extremely large memory storehouse. The cortex never functions alone, but always in association with the lower centers of the nervous system.

Without the cerebral cortex, the functions of the lower brain centers are often very imprecise. The vast storehouse of cortical information usually converts these functions to very determinative and precision operations.

Finally, the cerebral cortex is essential for most of our thought processes even though it also cannot function alone in this. In fact, it is the lower centers that cause *wakefulness* in the cerebral cortex, thus opening its bank of memories to the thinking machinery of the brain.

Thus, each portion of the nervous system performs specific functions. Many integrative functions are well developed in the spinal cord, and many of the subconscious functions originate and are executed entirely in the lower regions of the brain. But it is the cortex that opens the world up for one's mind.

● COMPARISON OF THE NERVOUS SYSTEM WITH AN ELECTRONIC COMPUTER

When electronic computers were first developed in many different laboratories of the world by as many different scientists, it soon became apparent that all these machines have many features in common with the nervous system. First, they all have input circuits that are comparable to the sensory portion to the nervous system and output circuits that are comparable to the motor portion of the nervous system. In the conducting pathway between the inputs and the outputs are the mechanisms for performing the different types of computations.

In simple computers, the output signals are controlled directly by the input signals, operating in a manner similar to that of the simple reflexes of the spinal cord. However, in the more complex computers, the output is determined both by input signals and by information that has already been stored in memory in the computer, which is analogous to the more complex reflex and processing mechanisms of our higher nervous system. Furthermore, as the computers become even more complex, it is necessary to add still another unit, called the *central programming unit*, which determines the sequence of all operations. This unit is analogous to the mechanism in our brain that allows us to direct our attention first to one thought or sensation or motor activity, then to another, and so forth, until complex sequences of thought or action take place.

Figure 45–4 is a simple block diagram of a modern computer. Even a rapid study of this diagram will demonstrate its similarity to the nervous system. The fact that the basic components of the general purpose computer are analogous to those of the human nervous system demonstrates that the brain is basically a computer that continuously collects sensory information and uses this along with stored information to compute the daily course of bodily activity.

● THE CENTRAL NERVOUS SYSTEM SYNAPSES

Every medical student is aware that information is transmitted in the central nervous system mainly in the form of nerve impulses through a succession of neurons, one after another. However, it is not immediately apparent that each impulse (1) may be blocked in its transmission from one neuron to the next, (2) may be changed from a single impulse into repetitive impulses, or (3) may be integrated with impulses from

Figure 45–4. Block diagram of a general purpose electronic computer, showing the basic components and their interrelationships.

other neurons to cause highly intricate patterns of impulses in successive neurons. All these functions can be classified as *synaptic functions of neurons.*

Types of Synapses—Chemical and Electrical

Nerve signals are transmitted from one neuron to the next through interneuronal junctions called *synapses.* In the animal world there are basically two different types of synapses: (1) the *chemical synapse* and (2) the *electrical synapse.*

Almost all the synapses utilized for signal transmission in the central nervous system of the human being are *chemical synapses.* In these, the first neuron secretes a chemical substance called a *neurotransmitter* at the synapse, and this transmitter in turn acts on receptor proteins in the membrane of the next neuron to excite the neuron, to inhibit it, or to modify its sensitivity in some other way. Over 40 different transmitter substances have been discovered thus far. Some of the best known are acetylcholine, norepinephrine, histamine, gamma-aminobutyric acid (GABA), and glutamate.

Electrical synapses, on the other hand, are characterized by direct channels that conduct electricity from one cell to the next. Most of these consist of small protein tubular structures called *gap junctions* that allow free movements of ions from the interior of one cell to the next. These were discussed in Chapter 4. Only a few gap junctions have been found in the central nervous system, and their significance in general is not known. On the other hand, it is by way of gap junctions and other similar junctions that action potentials are transmitted from one smooth muscle fiber to the next in visceral smooth muscle (Chapter 8) and also from one cardiac muscle cell to the next in cardiac muscle (Chapter 10).

One-Way Conduction Through Chemical Synapses. Chemical synapses have one exceedingly important characteristic that makes them highly desirable as the form of transmission of nervous system signals: they always transmit the signals in one direction—that is, from the neuron that secretes the transmitter, called the *presynaptic neuron,* to the neuron on which the transmitter acts, called the *postsynaptic neuron.* This is the principle of *one-way conduction* through chemical synapses, and it is quite different from conduction through electrical synapses that can transmit signals in either direction.

Think for a moment about the extreme importance of the one-way conduction mechanism. It allows signals to be directed toward specific goals. Indeed, it is this specific transmission of signals to discrete and highly focused areas in the nervous system that allows the nervous system to perform its myriad functions of sensation, motor control, memory, and many others.

PHYSIOLOGIC ANATOMY OF THE SYNAPSE

Figure 45–5 illustrates a typical *motor neuron* in the anterior horn of the spinal cord. It is composed of

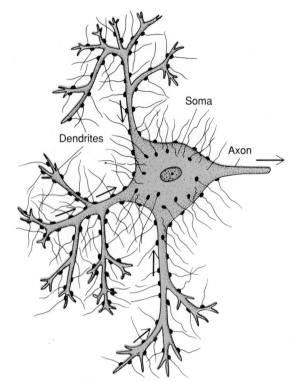

Figure 45–5. A typical motor neuron, showing presynaptic terminals on the neuronal soma and dendrites. Note also the single axon.

three major parts: the *soma,* which is the main body of the neuron; a single *axon,* which extends from the soma into the peripheral nerve; and the *dendrites,* which are thin projections of the soma that extend up to 1 mm into the surrounding areas of the cord.

Up to as many as 100,000 small knobs called *presynaptic terminals* lie on the surfaces of the dendrites and soma of the motor neuron, approximately 80 to 95 per cent of them on the dendrites and only 5 to 20 per cent on the soma. These terminals are the ends of nerve fibrils that originate in many other neurons; usually not more than a few of the terminals are derived from any single previous neuron. Later it will become evident that many of these presynaptic terminals are *excitatory* and secrete a substance that excites the postsynaptic neuron, but many others are *inhibitory* and secrete a substance that inhibits the neuron.

Neurons in other parts of the cord and brain differ markedly from the motor neuron in (1) the size of the cell body; (2) the length, size, and number of dendrites; ranging in length from almost none at all up to as long as many centimeters; (3) the length and size of the axon; and (4) the number of presynaptic terminals, which may range from only a few to several hundred thousand. These differences make neurons in different parts of the nervous system react differently to incoming signals and therefore perform different functions.

The Presynaptic Terminals. Electron microscopic studies of the presynaptic terminals show that these have varied anatomical forms, but most resemble small round or oval knobs and therefore are frequently called *terminal knobs, boutons, end-feet,* or *synaptic knobs.*

Figure 45–6 illustrates the basic structure of the presynaptic terminal. It is separated from the neuronal soma by a *synaptic cleft* having a width usually of 200 to 300 angstroms. The terminal has two internal structures important to the excitatory or inhibitory functions of the synapse: the *synaptic vesicles* and the *mitochondria.* The synaptic vesicles contain *transmitter substances* that, when released into the synaptic cleft, either *excite* or *inhibit* the postsynaptic neuron—excite if the neuronal membrane contains *excitatory receptors,* inhibit if it contains *inhibitory receptors.* The mitochondria provide adenosine triphosphate (ATP), which is required to synthesize new transmitter substance.

When an action potential spreads over a presynaptic terminal, the membrane depolarization causes emptying of a small number of vesicles into the cleft; and the released transmitter in turn causes an immediate change in the permeability characteristics of the postsynaptic neuronal membrane, which leads to excitation or inhibition of the neuron, depending on its receptor characteristics.

Mechanism by Which Action Potentials Cause Transmitter Release at the Presynaptic Terminals — Role of Calcium Ions

The synaptic membrane of the presynaptic terminals contains large numbers of *voltage-gated calcium channels.* This is quite different from the other areas of the nerve fiber, which contain very few of these channels. When the action potential depolarizes the terminal, large numbers of calcium ions, along with the sodium ions that cause most of the action potential, flow into the terminal. The quantity of transmitter substance that is released into the synaptic cleft is directly related to the number of calcium ions that enter the terminal. The precise mechanism by which

the calcium ions cause this release is not known but is believed to be the following:

When the calcium ions enter the synaptic terminal, it is believed that they bind with protein molecules on the inner surfaces of the synaptic membrane, called *release sites.* This in turn causes the transmitter vesicles in the local vicinity to bind also with the membrane and actually to fuse with it, and finally to open to the exterior by the process called *exocytosis,* which is described in Chapter 2. Usually, a few vesicles release their transmitter into the cleft following each single action potential. For the vesicles that store the neurotransmitter acetylcholine, between 2000 and 10,000 molecules of acetylcholine are present in each vesicle, and there are enough vesicles in the presynaptic terminal to transmit from a few hundred to more than 10,000 action potentials.

Action of the Transmitter Substance on the Postsynaptic Neuron — The Function of Receptors

At the synapse, the membrane of the postsynaptic neuron contains large numbers of *receptor proteins,* also illustrated in Figure 45–6. These receptors have two important components: (1) a *binding component* that protrudes outward from the membrane into the synaptic cleft—it binds with the neurotransmitter from the presynaptic terminal—and (2) an *ionophore component* that passes all the way through the membrane to the interior of the postsynaptic neuron. The ionophore in turn is one of two types: (1) a *chemically activated ion channel* or (2) an *enzyme that activates an internal metabolic change inside the cell.*

The Ion Channels. The chemically activated ion channels (also called ligand-activated channels) are usually of three types: (1) *sodium channels* that allow mainly sodium ions (but some potassium ions as well) to pass through, (2) *potassium channels* that allow mainly potassium ions to pass, and (3) *chloride channels* that allow chloride and a few other anions to pass. We learn later that opening the sodium channels excites the postsynaptic neuron. Therefore, a transmitter substance that opens the sodium channels is called an *excitatory transmitter.* On the other hand, opening of potassium and chloride channels inhibits the neuron, and transmitters that open either or both of these are called *inhibitory transmitters.*

The Enzyme Receptors. Activation of an enzymatic type of receptor causes other effects on the postsynaptic neuron. One effect is to *activate the metabolic machinery of the cell,* such as causing formation of cyclic adenosine monophosphate (AMP), which in turn excites many other intracellular activities. Another is to *activate cellular genes,* which in turn manufacture additional receptors for the postsynaptic membrane. Still a third effect is to *activate protein kinases,* which decrease the numbers of receptors. Changes such as these can alter the reactivity of the synapse for minutes, days, months, or even years. Therefore, transmitter substances that cause such ef-

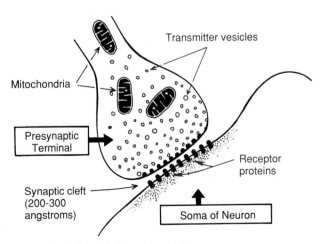

Figure 45–6. Physiologic anatomy of the synapse.

fects are sometimes called synaptic *modulators*. Recent experiments have demonstrated that such modulators are important in at least some of the memory processes, which we discuss in Chapter 57.

Excitatory and Inhibitory Receptors

Some postsynaptic receptors, when activated, cause excitation of the postsynaptic neuron and others cause inhibition. The importance of having inhibitory as well as excitatory types of receptors is that this gives an additional dimension to nervous function, allowing restraint of nervous action as well as excitation.

The different molecular and membrane mechanisms employed by the different receptors to cause excitation or inhibition include the following:

Excitation.

1. Opening of sodium channels to allow large numbers of positive electrical charges to flow to the interior of the postsynaptic cell. This raises the membrane potential in the positive direction up toward the threshold level for excitation. It is by far the most widely used means of causing excitation.

2. Depressed conduction through potassium or chloride channels, or both. This decreases the diffusion of positively charged potassium ions out of the postsynaptic neuron or decreases the diffusion of negatively charged chloride ions to the inside. In either instance, the effect is to make the internal membrane potential more positive than normally, which is excitatory.

3. Various changes in the internal metabolism of the cell to excite cell activity, or in some instances increase in the number of excitatory membrane receptors or decrease in the number of inhibitory membrane receptors.

Inhibition.

1. Opening of potassium ion channels through the receptor molecule. This allows rapid diffusion of positively charged potassium ions from inside the postsynaptic neuron to the outside, thereby carrying positive charges outward and increasing the negativity inside, which is inhibitory.

2. Increase in the conductance of chloride ions through the receptor. This allows negative chloride ions to diffuse to the interior, which is also inhibitory.

3. Activation of receptor enzymes that inhibit cellular metabolic functions or that increase the number of inhibitory synaptic receptors or decrease the number of excitatory receptors.

CHEMICAL SUBSTANCES THAT FUNCTION AS SYNAPTIC TRANSMITTERS

More than 40 different chemical substances have been proved or postulated to function as synaptic transmitters. Most of these are listed in Tables 45–1 and 45–2, which give two different groups of synaptic

Table 45–1
SMALL-MOLECULE, RAPIDLY ACTING TRANSMITTERS

Class I
  Acetylcholine
Class II: *The Amines*
  Norepinephrine
  Epinephrine
  Dopamine
  Serotonin
  Histamine
Class III: *Amino Acids*
  γ-Aminobutyric acid (GABA)
  Glycine
  Glutamate
  Aspartate

transmitters. One is comprised of small-molecule, rapidly acting transmitters. The other is a large number of neuropeptides of much larger molecular size and much more slowly acting.

The small-molecule, rapidly acting transmitters are the ones that cause most of the acute responses of the nervous system, such as transmission of sensory signals to and inside the brain and motor signals back to the muscles. The neuropeptides, on the other hand, usually cause more prolonged actions, such as long-term changes in numbers of receptors, long-term closure of certain ion channels, and possibly even long-term changes in numbers of synapses.

Table 45–2 NEUROPEPTIDE, SLOWLY ACTING TRANSMITTERS

A. *Hypothalamic-releasing hormones*
  Thyrotropin-releasing hormone
  Luteinizing hormone-releasing hormone
  Somatostatin (growth hormone-inhibitory factor)
B. *Pituitary peptides*
  ACTH
  β-Endorphin
  α-Melanocyte-stimulating hormone
  Prolactin
  Luteinizing hormone
  Thyrotropin
  Growth hormone
  Vasopressin
  Oxytocin
C. *Peptides that act on gut and brain*
  Leucine enkephalin
  Methionine enkephalin
  Substance P
  Gastrin
  Cholecystokinin
  Vasoactive intestinal polypeptide (VIP)
  Neurotensin
  Insulin
  Glucagon
D. *From other tissues*
  Angiotensin II
  Bradykinin
  Carnosine
  Sleep peptides
  Calcitonin

The Small-Molecule, Rapidly Acting Transmitters

Almost without exception, the small-molecule types of transmitters are synthesized in the cytosol of the presynaptic terminal and then are absorbed by active transport into the many transmitter vesicles in the terminal. Then, each time an action potential reaches the presynaptic terminal, a few vesicles at a time release their transmitter into the synaptic cleft, usually within a millisecond or less, by the mechanism described earlier. The subsequent action of the small-molecule type of transmitter on the postsynaptic membrane receptors usually also occurs within another millisecond or less. Most often the effect is to increase or decrease conductance through ion channels; an example is to increase sodium conductance, which causes excitation, or to increase potassium or chloride conductance, which causes inhibition. However, occasionally the small-molecule types of transmitters can stimulate receptor-activated enzymes instead of opening ion channels, thus changing the internal metabolic machinery of the cell.

Recycling of the Small-Molecule Types of Vesicles. The vesicles that store and release small-molecule transmitters are continually recycled, that is, used over and over again. After they fuse with the synaptic membrane and open to release their transmitters, the vesicle membrane at first simply becomes part of the synaptic membrane. However, within seconds to minutes, the vesicle portion of the membrane invaginates back to the inside of the presynaptic terminal and pinches off to form a new vesicle. It still contains the appropriate transport proteins required for concentrating new transmitter substance inside the vesicle.

Acetylcholine is a typical small-molecule transmitter that obeys the above principles of synthesis and release. It is synthesized in the presynaptic terminal from acetyl coenzyme A (acetyl-CoA) and choline in the presence of the enzyme *choline acetyltransferase.* Then it is transported into its specific vesicles. When the vesicles later release the acetylcholine into the synaptic cleft, the acetylcholine is rapidly split again to acetate and choline by the enzyme *cholinesterase,* which is bound to the proteoglycan reticulum that fills the space of the synaptic cleft. Then the vesicles are recycled, and choline also is actively transported back into the terminal to be used again for synthesis of new acetylcholine.

Characteristics of Some of the More Important Small-Molecule Transmitters. The most important of the small-molecule transmitters are the following:

Acetylcholine is secreted by neurons in many areas of the brain, but specifically by the large pyramidal cells of the motor cortex, by many different neurons in the basal ganglia, by the motor neurons that innervate the skeletal muscles, by the preganglionic neurons of the autonomic nervous system, by the postganglionic neurons of the parasympathetic nervous system, and by some of the postganglionic neurons of the sympathetic nervous system. In most instances acetylcholine has an excitatory effect; however, it is known to have inhibitory effects at some of the peripheral parasympathetic nerve endings, such as inhibition of the heart by the vagus nerves.

Norepinephrine is secreted by many neurons whose cell bodies are located in the brain stem and hypothalamus. Specifically, norepinephrine-secreting neurons located in the *locus ceruleus* in the pons send nerve fibers to widespread areas of the brain and help control the overall activity and mood of the mind. In most of these areas it probably activates excitatory receptors; but in a few areas, inhibitory receptors instead. Norepinephrine is also secreted by most of the postganglionic neurons of the sympathetic nervous system, where it excites some organs but inhibits others.

Dopamine is secreted by neurons that originate in the substantia nigra. The termination of these neurons are mainly in the striatal region of the basal ganglia. The effect of dopamine is usually inhibition.

Glycine is secreted mainly at synapses in the spinal cord. It probably always acts as an inhibitory transmitter.

Gamma-aminobutyric acid (GABA) is secreted by nerve terminals in the spinal cord, the cerebellum, the basal ganglia, and many areas of the cortex. It is believed always to cause inhibition.

Glutamate is probably secreted by the presynaptic terminals in many of the sensory pathways as well as in many areas of the cortex. It probably always causes excitation.

Serotonin is secreted by nuclei that originate in the median raphe of the brain stem and project to many brain areas, especially to the dorsal horns of the spinal cord and to the hypothalamus. Serotonin acts as an inhibitor of pain pathways in the cord, and it is also believed to help control the mood of the person, perhaps even to cause sleep.

The Neuropeptides

The neuropeptides are an entirely different group of transmitters that are synthesized differently and whose actions are usually slow and in other ways quite different from those of the small-molecule transmitters.

The neuropeptides are not synthesized in the cytosol of the presynaptic terminals. Instead, they are synthesized as integral parts of large protein molecules by the ribosomes in the neuronal cell body. The protein molecules are transported immediately into the endoplasmic reticulum of the cell body, and the endoplasmic reticulum and subsequently the Golgi apparatus function together to do two things: First, they enzymatically split the original protein into smaller fragments and thereby release either the neuropeptide itself or a precursor of it. Second, the Golgi apparatus packages the neuropeptide into minute transmitter vesicles that are released into the cytoplasm. Then the transmitter vesicles are transported all the way to the tips of the nerve fibers by *axonal streaming* of the axon cytoplasm, traveling at the slow rate of only a few centimeters per day. Finally, these vesicles release their transmitter in response to action potentials in the same manner as for small-molecule transmitters. However, the vesicle is autolyzed and is not reused.

Because of this laborious method of forming the neuropeptides, much smaller quantities of these are usually released than for the small-molecule transmitters. However, this is partly compensated for by the fact that the neuropeptides are generally a thousand or more times as potent as the small-molecule transmitters. Another important characteristic of the neuropeptides is that they usually cause much more prolonged actions. Some of these actions include prolonged closure of calcium pores, prolonged changes in the metabolic machinery of cells, prolonged changes in activation or deactivation of specific genes in the cell nucleus, and prolonged alterations in numbers of excitatory or inhibitory receptors. Some of these effects can last for days or perhaps even months or years. Unfortunately, our knowledge of the functions of the neuropeptides is only at an early beginning.

Release of Only a Single Small-Molecule Transmitter by Each Type of Neuron

Almost invariably, only a single small-molecule type of transmitter is released by each type of neuron. However, the terminals of the same neuron may also release one or more neuropeptides at the same time. Yet whatever small-molecule transmitter and neuropeptides are released at one terminal of the neuron, these same transmitters will be released at all other terminals of the same neuron, whether these are few in number or many thousand and also wherever these terminate within the nervous system or in peripheral organs.

Removal of the Transmitter Substance From the Synapse

After a transmitter is released at a nerve ending, it is either destroyed or removed in some other way to prevent continued action forever thereafter. In the case of the neuropeptides, these are removed mainly by diffusion into the surrounding tissues, followed by destruction within a few minutes to several hours by specific or nonspecific enzymes. For the small-molecule, rapidly acting transmitters, removal usually occurs within a few milliseconds. This is achieved in three different ways:

1. By *diffusion* of the transmitter out of the cleft into the surrounding fluids.

2. By *enzymatic destruction* within the cleft itself. For instance, in the case of acetylcholine, the enzyme *cholinesterase* is present in the cleft, bound in the proteoglycan matrix that fills the space. Each molecule of this enzyme can split as many as 10 molecules of acetylcholine each millisecond, thus inactivating this transmitter substance. Similar effects occur for other transmitters.

3. By *active transport back into the presynaptic terminal itself* and reuse. This is called *transmitter re-uptake*. It occurs especially prominently at the presynaptic terminals of the sympathetic nervous system for the re-uptake of norepinephrine, as we shall discuss in Chapter 60.

The degree to which each of these methods of removal is utilized is different for each type of transmitter.

ELECTRICAL EVENTS DURING NEURONAL EXCITATION

The electrical events in neuronal excitation have been studied especially in the large motor neurons of the anterior horns of the spinal cord. Therefore, the events to be described in the following few sections pertain essentially to these neurons. However, except for some quantitative differences, they apply to most other neurons of the nervous system as well.

The Resting Membrane Potential of the Neuronal Soma. Figure 45–7 illustrates the soma of a motor neuron showing the resting membrane potential to be about −65 millivolts. This is somewhat less than the −90 millivolts found in large peripheral nerve fibers and in skeletal muscle fibers; the lower voltage is important, however, because it allows both positive and negative control of the degree of excitability of the neuron. That is, decreasing the voltage to a less negative value makes the membrane of the neuron more excitable, whereas increasing this voltage to a more negative value makes the neuron less excitable. This is the basis of the two modes of function of the neuron—either excitation or inhibition—as is explained in detail in the following sections.

Concentration Differences of Ions Across the Neuronal Somal Membrane. Figure 45–7 also illustrates the concentration differences across the

Figure 45–7. Distribution of sodium, potassium, and chloride ions across the neuronal somal membrane; origin of the intrasomal membrane potential.

neuronal somal membrane of the three ions that are most important for neuronal function: sodium ions, potassium ions, and chloride ions.

At the top, the sodium ion concentration is shown to be very great in the extracellular fluid but low inside the neuron. This sodium concentration gradient is caused by a strong sodium pump that continually pumps sodium out of the neuron.

The figure also shows that the potassium ion concentration is large inside the neuronal soma but very low in the extracellular fluid. It illustrates that there is also a potassium pump (the other half of the Na^+-K^+ pump, as described in Chapter 4) that pumps potassium to the interior. However, potassium ions leak through the membrane ion channels at a rate sufficient to nullify much of the effectiveness of the potassium pump.

Figure 45–7 shows the chloride ion to be of high concentration in the extracellular fluid but low concentration inside the neuron. It also shows that the membrane is quite permeable to chloride ions and that there may be a weak chloride pump. Yet most of the reason for the low concentration of chloride ions inside the neuron is the −65 millivolts in the neuron. That is, this negative voltage repels the negatively charged chloride ions, forcing them outward through the pores until the concentration difference is much greater outside the membrane than inside.

Let us recall at this point what we learned in Chapters 4 and 5 about the relationship of ionic concentration differences to membrane potentials. It will be recalled that an electrical potential across the membrane can exactly oppose the movement of ions through a membrane despite concentration differences between the outside and inside of the membrane if the potential is of the proper polarity and magnitude. Such a potential that exactly opposes movement of each type of ion is called the Nernst potential for that ion; the equation for this is the following:

$$EMF(mV) = \pm 61 \times \log\left(\frac{\text{Concentration outside}}{\text{Concentration inside}}\right)$$

where EMF is the Nernst potential in millivolts on the *inside of the membrane*. The potential will be positive (+) for a positive ion and negative (−) for a negative ion.

Now, let us calculate the Nernst potential that will exactly oppose the movement of each of the three separate ions: sodium, potassium, and chloride.

For the sodium concentration difference shown in Figure 45–7, 142 mEq/liter on the exterior and 14 mEq/liter on the interior, the membrane potential that would exactly oppose sodium ion movement through the sodium channels would be +61 millivolts. However, the actual membrane potential is −65 millivolts, not +61 millivolts. Therefore, sodium ions normally diffuse inward through the sodium channels; however, not many sodium ions will diffuse because most of the sodium channels are normally

closed. Furthermore, those sodium ions that do diffuse to the interior are normally pumped immediately back to the exterior by the sodium pump.

For potassium ions, the concentration gradient is 120 mEq/liter inside the neuron and 4.5 mEq/liter outside. This gives a Nernst potential of −86 millivolts inside the neuron, which is more negative than the −65 that actually exists. Therefore, there is a tendency for potassium ions to diffuse to the outside of the neuron, but this is opposed by the continual pumping of these potassium ions back to the interior.

Finally, the chloride ion gradient, 107 mEq/liter outside and 8 mEq/liter inside, yields a Nernst potential of −70 millivolts inside the neuron, which is slightly more negative than the actual value measured. Therefore, chloride ions tend normally to leak to the interior of the neuron, whereas those that do diffuse are moved back to the exterior, perhaps by an active chloride pump.

Keep these three Nernst potentials in mind and also remember the direction in which the different ions tend to diffuse, for this information will be important in understanding both excitation and inhibition of the neuron by synaptic activation of receptor ion channels.

Origin of the Resting Membrane Potential of the Neuronal Soma. The basic cause of the −65 millivolt resting membrane potential of the neuronal soma is the sodium-potassium pump. This pump causes the extrusion of more positively charged sodium ions to the exterior than potassium to the interior—three sodium ions outward for each two potassium ions inward. Because there are large numbers of negatively charged ions inside the soma that cannot diffuse through the membrane—protein ions, phosphate ions, and many others—extrusion of the excess positive ions to the exterior leaves some of these nondiffusible negative ions inside the cell unbalanced by positive ions. Therefore, the interior of the neuron becomes negatively charged as the result of the sodium-potassium pump. This principle was discussed in more detail in Chapter 5 in relation to the resting membrane potential of nerve fibers. In addition, as also explained in Chapter 5, diffusion of potassium ions outward through the membrane is another cause of intracellular negativity.

Uniform Distribution of the Potential Inside the Soma. The interior of the neuronal soma contains a very highly conductive electrolytic solution, the intracellular fluid of the neuron. Furthermore, the diameter of the neuronal soma is very large (from 10 to 80 micrometers in diameter), causing there to be almost no resistance to conduction of electrical current from one part of the somal interior to another part. Therefore, any change in potential in any part of the intrasomal fluid causes an almost exactly equal change in potential at all other points inside the soma. This is an important principle because it plays a major role in the summation of signals entering the neuron from multiple sources, as we shall see in subsequent sections of this chapter.

Effect of Synaptic Excitation on the Postsynaptic Membrane — The Excitatory Postsynaptic Potential. Figure 45–8A illustrates the resting neuron with an unexcited presynaptic terminal resting upon its surface. The resting membrane potential everywhere in the soma is −65 millivolts.

Figure 45–8B illustrates a presynaptic terminal that has secreted a transmitter into the cleft between the terminal and the neuronal somal membrane. This transmitter acts on a membrane excitatory receptor *to increase the membrane's permeability to Na+.* Because of the large electrochemical gradient that tends to move sodium inward, this large increase in membrane conductance for sodium ions allows these ions to rush to the inside of the membrane.

The rapid influx of the positively charged sodium ions to the interior neutralizes part of the negativity of the resting membrane potential. Thus, in Figure 45–8B the resting membrane potential has increased from −65 millivolts to −45 millivolts. This increase in voltage above the normal resting neuronal potential — that is, to a less negative value — is called the *excitatory postsynaptic potential* (or EPSP) because if this potential rises high enough it will elicit an action potential in the neuron, thus exciting it. In this case the EPSP is +20 millivolts.

However, we must issue a word of warning at this point. Discharge of a single presynaptic terminal can never increase the neuronal potential from −65 millivolts up to −45 millivolts. Instead, an increase of this magnitude requires the simultaneous discharge of many terminals — about 40 to 80 for the usual anterior motor neuron — at the same time or in rapid suc-

cession. This occurs by a process called *summation,* which will be discussed in detail in the following sections.

Generation of Action Potentials at the Initial Segment of the Axon Leaving the Neuron — Threshold for Excitation. When the excitatory postsynaptic potential rises high enough, there comes a point at which this initiates an action potential in the neuron. However, the action potential does not begin on the somal membrane adjacent to the excitatory synapses. Instead, it begins in the initial segment of the axon leaving the neuronal soma. The main reason for this point of origin of the action potential is that the soma has relatively few voltage-gated sodium channels in its membrane, which makes it difficult to open the required number of channels to elicit an action potential. On the other hand, the membrane of the initial segment has seven times as great a concentration of voltage-gated sodium channels and therefore can generate an action potential with much greater ease than can the soma. The excitatory postsynaptic potential that will elicit an action potential at the initial segment is between +15 and +20 millivolts. This is in contrast to the +30 millivolts or more required on the soma.

Once the action potential begins, it travels both peripherally along the axon and often also backward over the soma. In some instances, it travels backward into the dendrites, too, but not into all of them, because they, like the neuronal soma, also have very few voltage-gated sodium channels and therefore frequently cannot generate action potentials at all.

Thus, in Figure 45–8B, it is shown that under normal conditions the *threshold* for excitation of the neuron is about −45 millivolts, which represents an excitatory postsynaptic potential of +20 millivolts — that is, 20 millivolts more positive than the normal resting neuronal potential of −65 millivolts.

ELECTRICAL EVENTS IN NEURONAL INHIBITION

Effect of Inhibitory Synapses on the Postsynaptic Membrane — The Inhibitory Postsynaptic Potential. The inhibitory synapses open the potassium or the chloride channels, or both, instead of sodium channels, allowing easy passage of one or both of these ions. Now, to understand how the inhibitory synapses inhibit the postsynaptic neuron, we must recall what we learned about the Nernst potentials for both the potassium ions and the chloride ions. We calculated this potential for potassium ions to be about −86 millivolts and for chloride ions about −70 millivolts. Both of these potentials are more negative than the −65 millivolts normally present inside the resting neuronal membrane. Therefore, opening the potassium channels will allow positively charged potassium ions to move to the exterior, which will make the membrane potential more negative than normal; and opening the chloride channels will allow negatively charged chloride ions to move to the interior,

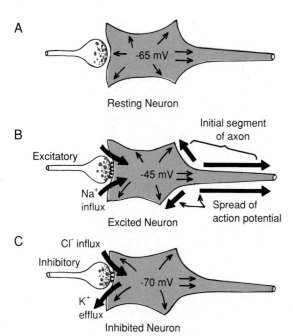

Figure 45–8. Three states of a neuron. *A,* A resting neuron. *B,* A neuron in an excited state, with more positive intraneuronal potential caused by sodium influx. *C,* A neuron in an inhibited state, with more negative intraneuronal membrane potential caused by potassium ion efflux, chloride ion influx, or both.

which also will make the membrane potential more negative than usual. This increases the degree of intracellular negativity, which is called *hyperpolarizaton.* It obviously inhibits the neuron because the membrane potential is now farther away than ever from the threshold for excitation. Therefore, an increase in negativity beyond the normal resting membrane potential level is called the *inhibitory postsynaptic potential* (IPSP).

Thus, Figure 45–8C illustrates the effect on the membrane potential caused by activation of inhibitory synapses, allowing chloride influx into the cell or potassium efflux from the cell, with the membrane potential decreasing from its normal value of −65 millivolts to the more negative value of −70 millivolts. This membrane potential that is 5 millivolts more negative is the inhibitory postsynaptic potential. Thus the IPSP in this instance is −5 millivolts.

Inhibition of Neurons Without Causing an Inhibitory Postsynaptic Potential — "Short Circuiting" of the Membrane. Sometimes activation of the inhibitory synapses causes little or no inhibitory postsynaptic potential but nevertheless still inhibits the neuron.

The reason that the potential often does not change is that in some neurons the concentration differences across the membrane for the potassium and chloride ions are only able to cause a Nernst equilibrium potential equal to the normal resting potential. Therefore, when the inhibitory channels open, there is no net flow of ions to cause an inhibitory postsynaptic potential. Yet the potassium or chloride ions, or both, do diffuse bidirectionally through the wide-open channels many times as rapidly as normally, and this high flux inhibits the neuron in the following way: When excitatory synapses cause sodium ions to flow into the neuron, the wide-open potassium or chloride channels cause far less excitatory postsynaptic potential than usual because any tendency for the membrane potential to change away from the resting potential is immediately opposed by rapid flux of potassium or chloride ions through the inhibitory channels to bring the potential back toward the negative Nernst equilibrium potential for these two ions. Therefore, the influx of sodium ions required to overcome the potassium or chloride flux and therefore cause excitation may be as much as 5 to 20 times normal.

This tendency for the potassium and chloride ions to maintain the membrane potential near the resting value when the inhibitory channels are wide open is called "short circuiting" of the membrane, thus making the sodium current flow caused by excitatory synapses ineffective in exciting the cell.

To express the phenomenon of short circuiting more mathematically, one needs to recall the Goldman equation from Chapter 5. This equation shows that the membrane potential is determined by summation of the tendencies for the different ions to carry electrical charges through the membrane in the two directions. The membrane potential will approach the Nernst equilibrium potentials for those ions that permeate the membrane to the greatest extent.

When the inhibitory channels are wide open, the chloride and potassium ions permeate the membrane greatly. Therefore, when the excitatory channels open, the summated effect of the inhibitory channels makes it difficult to raise the neuronal potential up to the threshold value for excitation.

Presynaptic Inhibition

In addition to the inhibition caused by inhibitory synapses operating at the neuronal membrane, which is called *postsynaptic inhibition,* another type of inhibition often occurs in the presynaptic terminals before the signal ever reaches the synapse. This type of inhibition, called *presynaptic inhibition,* is believed to occur in the following way:

In presynaptic inhibition, the inhibition is caused by "presynaptic" synapses that lie on the terminal nerve fibrils before they themselves terminate on the following neuron. It is believed that activation of these synapses on the presynaptic terminals decreases the ability of the calcium channels in the terminals to open. Because calcium ions must enter the presynaptic terminals before the vesicles can release transmitter at the neuronal synapse, the obvious result is to reduce neuronal excitation.

The cause of the reduced calcium entry into the presynaptic terminals is still unknown. One theory suggests that the presynaptic synapses release a transmitter that directly blocks calcium channels. Another theory proposes that the transmitter inhibits the opening of sodium channels, thus reducing the action potential in the terminal; because the voltage-activated calcium channels are very highly voltage-sensitive, any decrease in action potential greatly reduces calcium entry.

Presynaptic inhibition occurs in many of the sensory pathways in the nervous system. That is, the adjacent nerve fibers inhibit each other, which minimizes the sideways spread of signals from one fiber to the next. We discuss this phenomenon more fully in subsequent chapters.

Presynaptic inhibition is different from postsynaptic inhibition in its time sequence. It requires many milliseconds to develop; but once it does occur, it can last for as long as minutes or even hours. *Postsynaptic* inhibition, on the other hand, normally lasts for only a few milliseconds.

Summation of Postsynaptic Potentials

Time Course of Postsynaptic Potentials. When a synapse excites the anterior motor neuron the neuronal membrane remains highly permeable for only 1 to 2 milliseconds. During this time sodium ions diffuse rapidly to the interior of the cell to increase the intraneuronal potential, thus creating the *excitatory postsynaptic potential,* illustrated by the two lower curves of Figure 45–9. This potential then slowly dissipates over the next 15 milliseconds, because this is the time required for the positive charges to flow away from the excited synapses along the lengths of the dendrites and axon and for potassium ions to leak out

Figure 45–9. Excitatory postsynaptic potentials, showing that simultaneous firing of only a few synapses will not cause sufficient summated potential to elicit an action potential, but the simultaneous firing of many synapses will raise the summated potential to the threshold for excitation and cause a superimposed action potential.

or chloride ions to leak in to re-establish the normal resting membrane potential.

Precisely the opposite effect occurs for the inhibitory postsynaptic potential. That is, the inhibitory synapse increases the permeability of the membrane to potassium or chloride ions, or both, for 1 to 2 milliseconds, and this decreases the intraneuronal potential to a more negative value than normal, thereby creating the *inhibitory postsynaptic potential.* This potential also persists for about 15 milliseconds.

However, other types of transmitter substances acting on other neurons can excite or inhibit for hundreds of milliseconds or even for seconds, minutes, or hours.

Spatial Summation of the Postsynaptic Potentials — The Threshold for Firing

It has already been pointed out that excitation of a single presynaptic terminal on the surface of a neuron will almost never excite the neuron. The reason for this is that sufficient transmitter substance is released by a single terminal to cause an excitatory postsynaptic potential usually no more than 0.5 to 1 millivolt at most, instead of the required 10 to 20 millivolts to reach the usual threshold for excitation. However, during excitation in a neuronal pool of the nervous system, many presynaptic terminals are usually stimulated at the same time; and even though these terminals are spread over wide areas of the neuron, their effects can still summate. The reason for this is the following: It has already been pointed out that a change in the potential at any single point within the soma will cause the potential to change everywhere in the soma almost exactly equally. Therefore, for each excitatory synapse that discharges simultaneously, the intrasomal potential becomes more positive by as much as a fraction of a millivolt up to about 1 millivolt. When the excitatory postsynaptic potential becomes great enough, the *threshold for firing* will be reached, and an action potential will generate in the initial segment of the axon. This effect is illustrated in Figure 45–9, which shows several excitatory postsynaptic potentials. The bottom postsynaptic potential in the figure was caused by simultaneous stimulation of four synapses; the next higher potential was caused by stimulation of two

times as many synapses; finally, a still higher excitatory postsynaptic potential was caused by stimulation of four times as many synapses. This time an action potential was generated in the initial axon segment.

This effect of summing simultaneous postsynaptic potentials by excitation of multiple terminals on widely spaced areas of the membrane is called *spatial summation.*

Temporal Summation

Each time a terminal fires, the released transmitter substance opens the membrane channels for a millisecond or so. Since the postsynaptic potential lasts up to 15 milliseconds, a second opening of the same channels can increase the postsynaptic potential to a still greater level; therefore, the more rapid the rate of terminal stimulation, the greater the effective postsynaptic potential. Thus, successive postsynaptic potentials in a presynaptic terminal, if they occur rapidly enough, can summate in the same way that postsynaptic potentials can summate from widely distributed terminals over the surface of the neuron. This summation is called *temporal summation.*

Simultaneous Summation of Inhibitory and Excitatory Postsynaptic Potentials. Obviously, if an inhibitory postsynaptic potential is tending to decrease the membrane potential to a more negative value while an excitatory postsynaptic potential is tending to increase the potential at the same time, these two effects can either completely nullify each other or partially nullify each other. Also, inhibitory "short circuiting" of the membrane potential can nullify much of an excitatory potential. Thus, if a neuron is currently being excited by an excitatory postsynaptic potential, then an inhibitory signal from another source can easily reduce the postsynaptic potential to less than the threshold value for excitation, thus turning off the activity of the neuron.

Facilitation of Neurons. Often the summated postsynaptic potential is excitatory in nature but has not risen high enough to reach the threshold for excitation. When this happens the neuron is said to be *facilitated.* That is, its membrane potential is nearer the threshold for firing than normally but not yet to the firing level. Nevertheless, another signal entering the neuron from some other source can then excite the neuron very easily. Diffuse signals in the nervous system often facilitate large groups of neurons so that they can respond quickly and easily to signals arriving from second sources.

SPECIAL FUNCTIONS OF DENDRITES IN EXCITING NEURONS

The Large Spatial Field of Excitation of the Dendrites. The dendrites of the anterior motor neurons extend for 500 to 1000 micrometers in all directions from the neuronal soma. Therefore, these dendrites can receive signals from a large spatial area around the motor neuron. This provides vast opportunity for summation of signals from many separate presynaptic neurons.

It is also important that between 80 and 90 per cent of all the presynaptic terminals terminate on the dendrites of the anterior motor neuron, in contrast to only 10 to 20 per cent terminating on the neuronal soma. Therefore, the preponderant share of the excitation is provided by signals transmitted over the dendrites.

Many Dendrites Cannot Transmit Action Potentials — But They Can Transmit Signals by Electronic Conduction. Many dendrites fail to transmit action potentials because their membranes have relatively few voltage-gated sodium channels, so that their thresholds for excitation are too high for action potentials ever to occur. Yet they do transmit *electrotonic current* down the dendrites to the soma. Transmission of electrotonic current means the direct spread of current by electrical conduction in the fluids of the dendrites with no generation of action potentials. Stimulation of the neuron by this current has special characteristics, as follows:

Decrement of Electrotonic Conduction in the Dendrites — Greater Excitation by Synapses Near the Soma. In Figure 45–10 a number of excitatory and inhibitory synapses are shown stimulating the dendrites of a neuron. On the two dendrites to the left in the figure are shown excitatory effects near the ends of the dendrites; note the high levels of the excitatory postsynaptic potentials at these ends — that is, the less negative membrane potentials at these points. However, a large share of the excitatory postsynaptic potential is lost before it reaches the soma. The reason for this is that the dendrites are long and thin, and their membranes are also thin and excessively permeable to potassium and chloride ions, making them "leaky" to electrical current. Therefore, before the excitatory potentials can reach the soma, a large share of the potential is lost by leakage through the membrane. This decrease in membrane potential as it spreads electrotonically along dendrites toward the soma is called *decremental conduction.*

It is also obvious that the nearer the excitatory synapse is to the soma of the neuron, the less will be the decrement of conduction. Therefore, those synapses that lie near the soma have far more excitatory effect than those that lie far away from the soma.

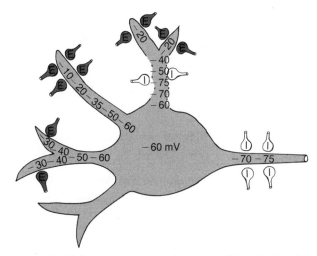

Figure 45–10. Stimulation of a neuron by presynaptic terminals located on dendrites, showing, especially, decremental conduction of excitatory electrotonic potentials in the two dendrites to the left and inhibition of dendritic excitation in the dendrite that is uppermost. A powerful effect of inhibitory synapses at the initial segment is also shown.

Rapid Re-excitation of the Neuron by the Dendrites After the Neuron Fires. When an action potential is generated in a neuron, this action potential usually spreads back over the soma but not always over the dendrites. Therefore, the excitatory postsynaptic potentials in the dendrites often are only partially disturbed by the action potential, so that just as soon as the action potential is over, the potentials still existing in the dendrites are ready and waiting to excite the neuron again. Thus, the dendrites have a "holding capacity" for the excitatory signal from presynaptic sources.

Summation of Excitation and Inhibition in Dendrites. The uppermost dendrite of Figure 45–10 is shown to be stimulated by both excitatory and inhibitory synapses. At the tip of the dendrite is a strong excitatory postsynaptic potential, but nearer to the soma are two inhibitory synapses acting on the same dendrite. These inhibitory synapses provide a hyperpolarizing voltage that completely nullifies the excitatory effect and indeed transmits a small amount of inhibition by electrotonic conduction toward the soma. Thus, dendrites can summate excitatory and inhibitory postsynaptic potentials in the same way that the soma can.

Also shown in the figure are several inhibitory synapses located directly on the axon hillock and initial axon segment. This location provides especially powerful inhibition because it has the direct effect of increasing the threshold for excitation at the very point where the action potential is normally generated.

RELATION OF STATE OF EXCITATION OF THE NEURON TO THE RATE OF FIRING

The "Excitatory State." The "excitatory state" of a neuron is defined as the degree of excitatory drive to the neuron. If there is a higher degree of excitation than inhibition of the neuron at any given instant, then it is said that there is an *excitatory state*. On the other hand, if there is more inhibition than excitation, then it is said that there is an *inhibitory state*.

When the excitatory state of a neuron rises above the threshold for excitation, then the neuron will fire repetitively as long as the excitatory state remains at this level. However, *the rate at which it will fire is determined by how much* the excitatory state is above threshold. To explain this, we must first consider what happens to the neuronal somal potential during and following the action potential.

Changes in Neuronal Somal Potential During and Following the Action Potential. The lower curve of Figure 45–11 illustrates an action potential spreading backward over the neuronal soma after being initiated in the initial axon segment by an excitatory postsynaptic potential. After the spike portion of the action potential, there is a very long state of "hyperpolarization" lasting for many milliseconds. During this interval the somal membrane potential falls below the normal resting membrane potential. This is caused at least partly by a high degree of permeability of the neuronal membrane to potassium ions that persists for many milliseconds after the action potential is over. The high membrane conductivity for potassium ions also "short circuits" excitatory potentials, as has already been explained.

The importance of this hyperpolarization as well as the short circuiting that occurs after the spike potential is over is that the neuron remains in an *inhibited state* during this period of time. Therefore, a far greater excitatory state is required during this time than normally to cause re-excitation of the neuron.

Relationship of Excitatory State to Frequency of Firing. The curve shown at the top of Figure 45–11, labeled "Excitatory state required for re-excitation," depicts the relative level of the excitatory state required at each instance after an action potential is over to re-excite the neuron. Note that very soon after an action potential is over, a very high excitatory state is required. That is, a very large num-

ber of excitatory synapses must be firing simultaneously. Then, after many milliseconds have passed and the state of hyperpolarization and the short circuiting of the neuron have begun to disappear, the excitatory state required becomes greatly reduced.

Therefore, when the excitatory state is high, a second action potential will appear very soon after the previous one. Then still a third action potential will appear soon after the second, and this process will continue indefinitely. Thus, at a very high excitatory state the frequency of firing of the neuron is great.

On the other hand, when the excitatory state is only barely above threshold, the neuron must recover almost completely from the hyperpolarization and short circuiting, which requires many milliseconds, before it will fire again. Therefore, the frequency of neuronal firing is low.

Response Characteristics of Different Neurons to Increasing Levels of Excitatory State. Histological study of the nervous system convinces one of the widely varying types of neurons in different parts of the nervous system. And, physiologically, the different types of neurons perform different functions. Therefore, as would be expected, the ability to respond to stimulation by the synapses varies from one type of neuron to another.

Figure 45–12 illustrates theoretical responses of three different types of neurons to varying levels of the excitatory state. Note that neuron 1 has a low threshold for excitation, whereas neuron 3 has a high threshold. But note also that neuron 2 has the lowest maximum frequency of discharge, whereas neuron 3 has the highest maximum frequency.

Some neurons in the central nervous system fire continuously because even the normal excitatory state is above the threshold level. Their frequency of firing can usually be increased still more by further increasing their excitatory state. Or the frequency may be decreased, or firing even be stopped, by superimposing an inhibitory state on the neuron.

Thus, different neurons respond differently, have different thresholds for excitation, and have widely

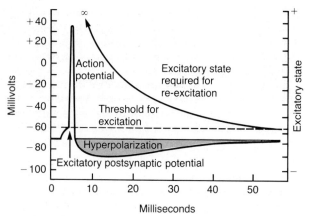

Figure 45–11. A neuronal action potential followed by a prolonged period of neuronal hyperpolarization. Also shown is the "excitatory state" required for re-excitation of the neuron at given intervals after the action potential is over.

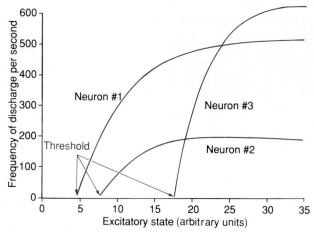

Figure 45–12. Response characteristics of different types of neurons to progressively increasing levels of excitatory state.

differing maximal frequencies of discharge. With a little imagination one can readily understand the importance of having neurons with many different types of response characteristics to perform the widely varying functions of the nervous system.

● SOME SPECIAL CHARACTERISTICS OF SYNAPTIC TRANSMISSION

Fatigue of Synaptic Transmission. When excitatory synapses are repetitively stimulated at a rapid rate, the number of discharges by the postsynaptic neuron is at first very great, but it becomes progressively less in succeeding milliseconds or seconds. This is called *fatigue* of synaptic transmission.

Fatigue is an exceedingly important characteristic of synaptic function, for when areas of the nervous system become overexcited, fatigue causes them to lose this excess excitability after a while. For example, fatigue is probably the most important means by which the excess excitability of the brain during an epileptic convulsion is finally subdued so that the convulsion ceases. Thus, the development of fatigue is a protective mechanism against excess neuronal activity. This is discussed further in the description of reverberating neuronal circuits in the following chapter.

The mechanism of fatigue is mainly exhaustion of the stores of transmitter substance in the synaptic terminals, particularly because it has been calculated that the excitatory terminals on most neurons can store enough excitatory transmitter for only 10,000 normal synaptic transmissions, so that the transmitter can be exhausted in only a few seconds to a few minutes of rapid stimulation. However, part of the fatigue process probably also results from two other factors as well: (1) progressive inactivation of many of the postsynaptic membrane receptors and (2) slow buildup of calcium ions inside the *postsynaptic* neuronal cell caused by the successive action potentials — these calcium ions in turn open calcium-activated potassium channels, which cause an inhibitory effect on the postsynaptic neuron.

Post-tetanic Facilitation. When a rapidly repetitive series of impulses stimulates an excitatory synapse for a period of time and then a rest period is allowed, the synapse will often become for a period of seconds or minutes even more responsive to subsequent stimulation than normally. This is called *post-tetanic facilitation.*

Experiments have shown that post-tetanic facilitation is caused mainly by the buildup of excess calcium ions in the *presynaptic* terminals because the calcium pump pumps too slowly to remove all of these immediately after each action potential. These accumulated calcium ions cause more and more vesicular release of transmitter substance, occasionally increasing the release of transmitter to a rate two times normal.

The physiological significance of post-tetanic facilitation is still very doubtful, and it may have no real significance at all. However, neurons could possibly store information by this mechanism. Therefore post-tetanic facilitation might well be a mechanism of "short-term" memory in the central nervous system.

Effect of Acidosis and Alkalosis on Synaptic Transmission. The neurons are highly responsive to changes in pH of the surrounding interstitial fluids. *Alkalosis greatly increases neuronal excitability.* For instance, a rise in arterial pH from the 7.4 norm to 7.8 to 8.0 often causes cerebral convulsions because of increased excitability of the neurons. This can be demonstrated especially well by having a person who is predisposed to epileptic convulsions overbreathe. The overbreathing elevates the pH of the blood only momentarily, but even this short interval can often precipitate an epileptic attack.

On the other hand, *acidosis greatly depresses neuronal activity;* a fall in pH from 7.4 below 7.0 usually causes a comatose state. For instance, in very severe diabetic or uremic acidosis, coma always develops.

Effect of Hypoxia on Synaptic Transmission. Neuronal excitability is also highly dependent on an adequate supply of oxygen. Cessation of oxygen for only a few seconds can cause complete inexcitability of the neurons. This is often seen when the cerebral circulation is temporarily interrupted, for within 3 to 5 seconds the person becomes unconscious.

Effect of Drugs on Synaptic Transmission. Many different drugs are known to increase the excitability of neurons, and others are known to decrease the excitability. For instance, caffeine, theophylline, and theobromine, which are found in coffee, tea, and cocoa, respectively, all increase neuronal excitability, presumably by reducing the threshold for excitation of the neurons. Also, strychnine is one of the best known of all the agents that increase the excitability of neurons. However, it does not reduce the threshold for excitation of the neurons at all; instead, it *inhibits the action of some of the inhibitory transmitters* on the neurons, especially the inhibitory effect of glycine in the spinal cord. In consequence, the effects of the excitatory transmitters become overwhelming, and the neurons become so excited that they go into rapidly repetitive discharge, resulting in severe tonic muscle spasms.

Most anesthetics increase the membrane threshold for excitation and thereby decrease synaptic transmission at many points in the nervous system. Because most of the anesthetics are lipid-soluble, it has been reasoned that they might change the physical characteristics of the neuronal membranes, making them less responsive to excitatory agents.

Synaptic Delay. In transmission of an action potential from a presynaptic neuron to a postsynaptic neuron, a certain amount of time is consumed in the process of (1) discharge of the transmitter substance by the presynaptic terminal, (2) diffusion of the transmitter to the postsynaptic neuronal membrane, (3) action of the transmitter on the membrane receptor, (4) action of the receptor to increase the membrane permeability, and (5) inward diffusion of sodium to raise the excitatory postsynaptic potential to a high enough value to elicit an action potential. The *minimal* period of time required for all these events to take place, even when large numbers of excitatory synapses are stimulated simultaneously, is about 0.5 millisecond. This is called the *synaptic delay.* It is important for the following reason: Neurophysiologists can measure the *minimal* delay time between an input volley of impulses and an output volley and from this can estimate the number of series neurons in the circuit.

REFERENCES

Bahill, A. T., and Hamm, T. M.: Using open-loop experiments to study physiological systems, with examples from the human eye-movement systems. News Physiol. Sci., 4:104, 1989.

Barchi, R. L.: Probing the molecular structure of the voltage-dependent sodium channel. Annu. Rev. Neurosci., 11:455, 1988.

Berg, D. K.: New neuronal growth factors. Annu. Rev. Neurosci., 7:149, 1984.

Bloom, F. E.: Neurotransmitters: past, present, and future directions. FASEB J., 2:32, 1988.

Bousfield, D. (ed.): Neurotransmitters in Action. New York, Elsevier Science Publishing Co., 1985.

Byrn, J. H., and Schultz, S. G.: An Introduction to Membrane Transport and Bioelectricity. New York, Raven Press, 1988.

Changeux, J.-P., et al.: Acetylcholine receptor: An allosteric protein. Science, 225:1335, 1984.

Cotman, C. W., et al.: Excitatory animo acid neurotransmission: NMDA receptors and Hebb-type synaptic plasticity. Annu. Rev. Neurosci., 11:61, 1988.

Eldefrawi, A. T., and Eldefrawi, M. E.: Receptors for γ-aminobutyric acid and voltage-dependent chloride channels as targets for drugs and toxicants. FASEB J., 1:262, 987.

Eyzaguirre, C.: Physiology of the Nervous System. Chicago, Year Book Medical Publishers, 1985.

Grinnell, A. D., et al. (eds.): Calcium and Ion Channel Modulation. New York, Plenum Publishing Corp., 1988.

Grinvald, A., et al.: Optical imaging of neuronal activity. Physiol. Rev., 68:1285, 1988.

Hanin, I. (ed.): Dynamics of Neurotransmitter Function. New York, Raven Press, 1984.

Hansen, A. J.: Disturbed ion gradients in brain anoxia. News Physiol. Sci., 2:54, 1987.

Heinemann, S., and Patrick, J. (eds.): Molecular Neurobiology. New York, Plenum Publishing Corp., 1987.

Ito, M.: Where are neurophysiologists going? News Physiol. Sci., 1:30, 1986.

Iversen, L. L., and Goodman, E. C. (eds.): Fast and Slow Chemical Signalling in the Nervous System. New York, Oxford University Press, 1986.

Iversen, L. L.: Nonopioid neuropeptides in mammalian CNS. Annu. Rev. Pharmacol. Toxicol., 23:1, 1983.

Johnson, R. G., Jr.: Accumulation of biological amines into chromaffin granules: A model for hormone and neurotransmitter transport. Physiol. Rev., 68:232, 1988.

Kito, S., et al. (eds.): Neuroreceptors and Signal Transduction. New York, Plenum Publishing Corp., 1988.

Kostyuk, P. G.: Intracellular perfusion of nerve cells and its effect on membrane currents. Physiol. Rev., 64:435, 1984.

Krnjevic, K.: Ephaptic interactions: A significant mode of communications in the brain. News Physiol. Sci., 1:28, 1986.

Laduron, P. M.: Presynaptic heteroreceptors in regulation of neuronal transmission. Biochem. Pharmacol., 34:467, 1985.

Marx, J. L.: NMDA receptors trigger excitement. Science, 239:254, 1988.

McGeer, P. L., et al.: Molecular Neurobiology of the Mammalian Brain, 2nd Ed. New York, Plenum Publishing Corp., 1987.

McKay, R. D. G.: Molecular approach to the nervous system. Annu. Rev. Neurosci., 6:527, 1983.

Millhorn, D. E., and Hokfelt, T.: Chemical messengers and their coexistence in individual neurons. News Physiol. Sci., 3:1, 1988.

Nakanishi S.: Substance P precursor and kininogen: Their structures, gene organizations, and regulation. Physiol. Rev., 67:1117, 1987.

Narahashi, T. (ed.): Ion Channels. New York, Plenum Publishing Corp., 1988.

Nicholl, R. A.: The coupling of neurotransmitter receptors to ion channels in the brain. Science, 241:545, 1988.

Phillips, M. I.: Functions of angiotensin in the central nervous system. Annu. Rev. Physiol., 49:413, 1987.

Pickering, P. T., et al. (eds.): Neurosecretion. New York, Plenum Publishing Corp., 1988.

Popot, J.-L., and Changeux, J.-P.: Nicotinic receptor of acetylcholine: Structure of an oligomeric integral membrane protein. Physiol. Rev., 64:1162, 1984.

Purves, D., and Lichtman, J. W.: Specific connections between nerve cells. Annu. Rev. Physiol., 45:553, 1983.

Purves, D., et al.: Nerve terminal remodeling visualized in living mice by repeated examination of the same neuron. Science, 238:1122, 1987.

Reichardt, L. F.: Immunological approaches to the nervous system. Science, 225:1294, 1984.

Robinson, M. B., and Coyle, J. T.: Glutamate and related acidic excitatory neurotransmitters: From basic science to clinical application. FASEB J., 1:446, 1987.

Schubert, D.: Developmental Biology of Cultured Nerve, Muscle and Glia. New York, John Wiley & Sons, 1984.

Skok, V. I., et al. (eds.): Neuronal Acetylcholine Receptors. New York, Plenum Publishing Corp., 1989.

Snyder, S. H.: Neuronal receptors. Annu. Rev. Physiol., 48:461, 1986.

Starke, K.: Presynaptic autoregulation: Does it play a role? News Physiol. Sci., 4:1, 1989.

Stein, J. F.: Introduction to Neurophysiology. St. Louis, C. V. Mosby, Co. 1982.

Su, C.: Purinergic neurotransmission and neuromodulation. Annu. Rev. Pharmacol. Toxicol., 23:397, 1983.

Thompson, R. F.: The neurobiology of learning and memory. Science, 233:941, 1986.

Tucek, S.: Regulation of acetylcholine synthesis in the brain. J. Neurochem., 44:11, 1985.

White, J. D., et al.: Biochemistry of peptide-secreting neurons. Physiol. Rev., 65:553, 1985.

Williams, R. W., and Herrup, K.: The control of neuron number. Annu. Rev. Neurosci., 11:423, 1988.

Wurtman, R. J.: Presynaptic control of release of amine neurotransmitters by precursor levels. News Physiol. Sci., 3:158, 1988.

Zucker, R. S.: Short-term synaptic plasticity. Annu. Rev. Physiol., 12:13, 1989.

CHAPTER 46

• Sensory Receptors; Neuronal Circuits for Processing Information •

Input to the nervous system is provided by the sensory receptors that detect such sensory stimuli as touch, sound, light, pain, cold, warmth, and so forth. The purpose of this chapter is to discuss the basic mechanisms by which these receptors change sensory stimuli into nerve signals and, also, how the information conveyed in the signals is processed in the nervous system.

● TYPES OF SENSORY RECEPTORS AND THE SENSORY STIMULI THEY DETECT

Table 46–1 gives a list and classification of most of the body's sensory receptors. This table shows that there are basically five different types of sensory receptors: (1) *mechanoreceptors,* which detect mechanical deformation of the receptor or of cells adjacent to the receptor; (2) *thermoreceptors,* which detect changes in temperature, some receptors detecting cold and others warmth; (3) *nociceptors* (pain receptors), which detect damage in the tissues, whether physical damage or chemical damage; (4) *electromagnetic receptors,* which detect light on the retina of the eye; and (5) *chemoreceptors,* which detect taste in the mouth, smell in the nose, oxygen level in the arterial blood, osmolality of the body fluids, carbon dioxide concentration, and perhaps other factors that make up the chemistry of the body.

In this chapter we discuss the function of a few specific types of receptors, primarily peripheral mechanoreceptors, to illustrate some of the basic principles by which receptors in general operate. Other receptors will be discussed in relation to the sensory systems that they subserve.

Figure 46–1 illustrates some of the different types of mechanoreceptors found in the skin or in the deep structures of the body, and Table 46–1 gives their respective sensory functions. All these receptors are discussed in the following chapters in relation to the respective sensory systems.

DIFFERENTIAL SENSITIVITY OF RECEPTORS

The first question that must be answered is, how do two types of sensory receptors detect different types of sensory stimuli? The answer is, by virtue of differential sensitivities. That is, each type of receptor is very highly sensitive to one type of stimulus for which it is designed, and yet is almost nonresponsive to normal intensities of the other types of sensory stimuli. Thus, the rods and cones are highly responsive to light but are almost completely nonresponsive to heat, cold, pressure on the eyeballs, or chemical changes in the blood. The osmoreceptors of the supraoptic nuclei in the hypothalamus detect minute changes in the osmolality of the body fluids but have never been known to respond to sound. Finally, pain receptors in the skin are almost never stimulated by usual touch or pressure stimuli but do become highly active the moment tactile stimuli become severe enough to damage the tissues.

Modality of Sensation — The "Labeled Line" Principle

Each of the principal types of sensation that we can experience — pain, touch, sight, sound, and so forth — is called a *modality* of sensation. Yet despite the fact that we experience these different modalities of sensation, nerve fibers transmit only impulses. Therefore, how is it that different nerve fibers transmit different modalities of sensation?

The answer to this is that each nerve tract terminates at a specific point in the central nervous system, and the type of sensation felt when a nerve fiber is stimulated is determined by the point in the nervous

Table 46–1 CLASSIFICATION OF SENSORY RECEPTORS

Mechanoreceptors
    Skin tactile sensibilities
        (epidermis and dermis)
        Free nerve endings
        Expanded tip endings
            Merkel's discs
            Plus several other variants
        Spray endings
            Ruffini's endings
        Encapsulated endings
            Meissner's corpuscles
            Krause's corpuscles
        Hair end-organs
    Deep tissue sensibilities
        Free nerve endings
        Expanded tip endings
        Spray endings
            Ruffini's endings
        Encapsulated endings
            Pacinian corpuscles
            Plus a few other variants
        Muscle endings
            Muscle spindles
            Golgi tendon receptors
    Hearing
        Sound receptors of cochlea
    Equilibrium
        Vestibular receptors
    Arterial pressure
        Baroreceptors of carotid
            sinuses and aorta
Thermoreceptors
    Cold
        Cold receptors
    Warmth
        Warm receptors
Nociceptors
    Pain
        Free nerve endings
Electromagnetic Receptors
    Vision
        Rods
        Cones
Chemoreceptors
    Taste
        Receptors of taste buds
    Smell
        Receptors of olfactory
            epithelium
    Arterial oxygen
        Receptors of aortic and
            carotid bodies
    Osmolality
        Probably neurons in or near
            supraoptic nuclei
    Blood CO_2
        Receptors in or on surface
            of medulla and in aortic and
            carotid bodies
    Blood glucose, amino acids,
        fatty acids
        Receptors in hypothalamus

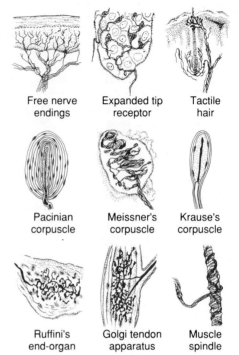

Figure 46–1. Several types of somatic sensory nerve endings.

(Labels in figure:)
Free nerve endings — Expanded tip receptor — Tactile hair
Pacinian corpuscle — Meissner's corpuscle — Krause's corpuscle
Ruffini's end-organ — Golgi tendon apparatus — Muscle spindle

cells. Yet the person still perceives pain. Likewise, if a touch fiber is stimulated by exciting a touch receptor electrically or in any other way, the person perceives touch because touch fibers lead to specific touch areas in the brain. Similarly, fibers from the retina of the eye terminate in the vision areas of the brain, fibers from the ear terminate in the auditory areas of the brain, and temperature fibers terminate in the temperature areas.

This specificity of nerve fibers for transmitting only one modality of sensation is called the *"labeled line"* principle.

● TRANSDUCTION OF SENSORY STIMULI INTO NERVE IMPULSES

LOCAL CURRENTS AT NERVE ENDINGS — RECEPTOR POTENTIALS

All sensory receptors have one feature in common. Whatever the type of stimulus that excites the receptor, its immediate effect is to change the membrane potential of the receptor. This change in potential is called a *receptor potential*.

Mechanisms of Receptor Potentials. Different receptors can be excited in several different ways to cause receptor potentials: (1) by mechanical deformation of the receptor, which stretches the membrane and opens ion channels; (2) by application of a chemical to the membrane, which also opens ion channels; (3) by change of the temperature of the membrane, which alters the permeability of the membrane; and

system to which the fiber leads. For instance, if a pain fiber is stimulated, the person perceives pain regardless of what type of stimulus excites the fiber. The stimulus can be electricity, heat, crushing, or stimulation of the pain nerve ending by damage to the tissue

(4) by the effects of electromagnetic radiation such as light on the receptor, which either directly or indirectly changes the membrane characteristics and allows ions to flow through membrane channels. It will be recognized that these four different means of exciting receptors correspond in general with the different types of known sensory receptors. In all instances, the basic cause of the change in membrane potential is a change in receptor membrane permeability, which allows ions to diffuse more or less readily through the membrane and thereby change the transmembrane potential.

The Receptor Potential Amplitude. The maximum amplitude of most sensory receptor potentials is around 100 millivolts. This is approximately the same maximum voltage recorded in action potentials and is also approximately the voltage when the membrane becomes maximally permeable to sodium ions.

Relationship of the Receptor Potential to Action Potentials. When the receptor potential rises above the *threshold* for eliciting action potentials in the nerve fiber attached to the receptor, then action potentials begin to appear. This is illustrated in Figure 46–2. Note also that the more the receptor potential rises above the threshold level, the greater becomes the action potential frequency. Thus, the receptor potential stimulates the sensory nerve fiber in the same way that the excitatory postsynaptic potential in the central nervous system neuron stimulates the neuron's axon.

The Receptor Potential of the Pacinian Corpuscle — An Illustrative Example of Receptor Function

The student should at this point restudy the anatomical structure of the pacinian corpuscle illustrated in Figure 46–1. Note that the corpuscle has a central nerve fiber extending through its core. Surrounding this are multiple concentric capsule layers, so that compression anywhere on the outside of the corpuscle will elongate, indent, or otherwise deform the central fiber.

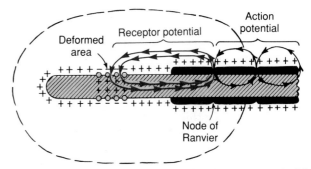

Figure 46–3. Excitation of a sensory nerve fiber by a receptor potential produced in a pacinian corpuscle. (Modified from Loëwenstein: *Ann. N.Y. Acad. Sci., 94*:510, 1961.)

Now study Figure 46–3, which illustrates only the central fiber of the pacinian corpuscle after all capsule layers have been removed by microdissection. The very tip of the central fiber is unmyelinated, but it becomes myelinated shortly before leaving the corpuscle to enter the peripheral sensory nerve.

The figure also illustrates the mechanism by which the receptor potential is produced in the pacinian corpuscle. Observe the small area of the terminal fiber that has been deformed by compression of the corpuscle, and note that ion channels have opened in the membrane, allowing positively charged sodium ions to diffuse to the interior of the fiber. This in turn creates increased positivity inside the fiber, which is the receptor potential. The receptor potential in turn induces a *local circuit* of current flow, illustrated by the red arrows, that spreads along the nerve fiber. At the first node of Ranvier, which itself lies inside the capsule of the pacinian corpuscle, the local current flow depolarizes the node, and this then sets off typical action potentials that are transmitted along the nerve fiber toward the central nervous system.

Relationship Between Stimulus Intensity and the Receptor Potential. Figure 46–4 illustrates the changing amplitude of the receptor potential caused by progressively stronger mechanical compression

Figure 46–2. Typical relationship between receptor potential and action potentials when the receptor potential rises above the threshold level.

Figure 46–4. Relationship of amplitude of receptor potential to strength of a stimulus applied to a pacinian corpuscle. (From Loëwenstein: *Ann. N.Y. Acad. Sci., 94*:510, 1961.)

applied experimentally to the central core of a pacinian corpuscle. Note that the amplitude increases rapidly at first but then progressively less rapidly at high stimulus strength.

In general, the frequency of repetitive action potentials transmitted from sensory receptors increases approximately in proportion to the increase in receptor potential. Putting this information together with the data in Figure 46–4, one can see that even though a very weak sensory stimulus can usually elicit at least some sensory signal, very intense stimulation of the receptor causes progressively less and less further increase in numbers of action potentials. This is an exceedingly important principle, employed by almost all sensory receptors. It allows the receptor to be very sensitive to weak sensory experience and yet not reach a maximum firing rate until the sensory experience is extreme. Obviously, this allows the receptor to have an extreme range of response, from very weak to very intense.

ADAPTATION OF RECEPTORS

A special characteristic of all sensory receptors is that they *adapt* either partially or completely to their stimuli after a period of time. That is, when a continuous sensory stimulus is applied, the receptors respond at a very high impulse rate at first, then at a progressively lower rate until finally many of them no longer respond at all.

Figure 46–5 illustrates typical adaptation of certain types of receptors. Note that the pacinian corpuscle adapts extremely rapidly and hair receptors adapt within a second or so, whereas joint capsule and muscle spindle receptors adapt very slowly.

Furthermore, some sensory receptors adapt to a far greater extent than others. For example, the pacinian corpuscles adapt to "extinction" within a few hundredths of a second, and the hair base receptors adapt to extinction within a second or more. It is probable that all other *mechanoreceptors* also adapt completely eventually, but some require hours or days to do so, for which reason they are frequently called "nonadapt-

ing" receptors. The longest measured time for complete adaptation of a mechanoreceptor is about two days for the carotid and aortic baroreceptors.

Some of the nonmechanoreceptors, the chemoreceptors and pain receptors for instance, probably never adapt completely.

Mechanisms by Which Receptors Adapt. Adaptation of receptors is an individual property of each type of receptor in much the same way that development of a receptor potential is an individual property. For instance, in the eye, the rods and cones adapt by changing the concentrations of their light-sensitive chemicals (which is discussed in Chapter 50).

In the case of the mechanoreceptors, the receptor that has been studied for adaptation in greatest detail is again the pacinian corpuscle. Adaptation occurs in this receptor in two ways. First, the pacinian corpuscle is a viscoelastic structure so that when a distorting force is suddenly applied to one side of the corpuscle, this force is instantly transmitted by the viscous component of the corpuscle directly to the same side of the central core, thus eliciting the receptor potential. However, within a few hundredths of a second the fluid within the corpuscle redistributes, so that the pressure becomes essentially equal all through the corpuscle; this now applies an even pressure on all sides of the central core fiber, so that the receptor potential is no longer elicited. Thus, the receptor potential appears at the onset of compression but then disappears within a small fraction of a second even though the compression continues.

Then, when the distorting force is removed from the corpuscle, essentially the reverse events occur. The sudden removal of the distortion from one side of the corpuscle allows rapid expansion on that side, and a corresponding distortion of the central core occurs once more. Again, within hundredths of a second, the pressure becomes equalized all through the corpuscle, and the stimulus is lost. Nevertheless, this disturbance of the central core fiber signals the offset of compression as well as signaling the onset of compression.

The second mechanism of adaptation of the pacinian corpuscle, but a much slower one, results from a process called *accommodation* that occurs in the nerve fiber itself. That is, even if by chance the central core fiber should continue to be distorted, as can be achieved after the capsule has been removed and the core compressed with a stylus, the tip of the nerve fiber itself gradually becomes "accommodated" to the stimulus. This probably results from "inactivation" of the sodium channels in the nerve fiber membrane, which means that the current flow itself through the channels in some way causes them gradually to close, as was explained in Chapter 5.

Presumably, these same two general mechanisms of adaptation apply also to the other types of mechanoreceptors. That is, part of the adaptation results from readjustments in the structure of the receptor itself, and part results from accommodation in the terminal nerve fibril.

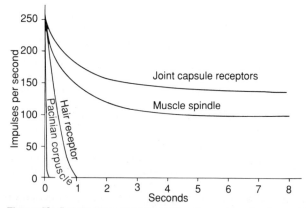

Figure 46–5. Adaptation of different types of receptors, showing rapid adaptation of some receptors and slow adaptation of others.

Function of the Slowly Adapting Receptors to Detect Continuous Stimulus Strength — The "Tonic" Receptors. The slowly adapting receptors continue to transmit impulses to the brain as long as the stimulus is present (or at least for many minutes or hours). Therefore, they keep the brain constantly apprised of the status of the body and its relation to its surroundings. For instance, impulses from the muscle spindles and Golgi tendon apparatus allow the central nervous system to know the status of muscle contraction and the load on the muscle tendon at each instant.

Other types of slowly adapting receptors include the receptors of the macula in the vestibular apparatus, the pain receptors, the baroreceptors of the arterial tree, the chemoreceptors of the carotid and aortic bodies, and some of the tactile receptors, such as Ruffini's endings and Merkel's discs.

Because the slowly adapting receptors can continue to transmit information for many hours, they are also called *tonic* receptors. Many of these slowly adapting receptors will adapt to extinction if the intensity of the stimulus remains absolutely constant for several hours or days. Fortunately, because of our continually changing bodily state, these receptors almost never reach a state of complete adaptation.

Function of the Rapidly Adapting Receptors to Detect Change in Stimulus Strength — The "Rate Receptors" or "Movement Receptors" or "Phasic Receptors." Obviously, receptors that adapt rapidly cannot be used to transmit a continuous signal because these receptors are stimulated only when the stimulus strength changes. Yet they react strongly *while a change is actually taking place.* Furthermore, the number of impulses transmitted is directly related to the *rate at which the change takes place.* Therefore, these receptors are called *rate* receptors, *movement* receptors, or *phasic* receptors. Thus, in the case of the pacinian corpuscle, sudden pressure applied to the skin excites this receptor for a few milliseconds, and then its excitation is over even though the pressure continues. But later it transmits a signal again when the pressure is released. In other words, the pacinian corpuscle is exceedingly important for transmitting information about rapid changes in pressure against the body, but it is useless for transmitting information about constant pressure applied to the body.

Importance of the Rate Receptors — Their Predictive Function. If one knows the rate at which some change in bodily status is taking place, one can predict the state of the body a few seconds or even a few minutes later. For instance, the receptors of the semicircular canals in the vestibular apparatus of the ear detect the rate at which the head begins to turn when one runs around a curve. Using this information, a person can predict how much he will turn within the next 2 seconds and can adjust the motion of the limbs *ahead of time* to keep from losing balance. Likewise, receptors located in or near the joints help detect the rates of movement of the different parts of the body. Therefore, when one is running, information from these receptors allows the nervous system to predict where the feet will be during any precise fraction of a second, and appropriate motor signals can be transmitted to the muscles of the legs to make any necessary anticipatory corrections in limb position so that the person will not fall. Loss of this predictive function makes it impossible for the person to run.

● THE NERVE FIBERS THAT TRANSMIT SIGNALS AND THEIR PHYSIOLOGICAL CLASSIFICATION

Some signals need to be transmitted to the central nervous system extremely rapidly; otherwise the information would be useless. An example of this is the sensory signals that apprise the brain of the momentary positions of the limbs at each fraction of a second during running. Another example is the motor signals sent back to the muscles from the brain. At the other extreme, some types of sensory information, such as that depicting prolonged, aching pain, do not need to be transmitted rapidly at all, so that very slowly conducting fibers will suffice. Fortunately, nerve fibers come in all sizes between 0.2 and 20 micrometers in diameter — the larger the diameter, the greater the conducting velocity. The range of conducting velocities is between 0.5 and 120 m/sec.

Figure 46-6 gives two different classifications of nerve fibers that are in general use. One of these is a general classification that includes both sensory and motor fibers, including the autonomic nerve fibers as well. The other is a classification of sensory nerve fibers that is used primarily by sensory neurophysiologists.

In the general classification, the fibers are divided into types A and C, and the type A fibers are further subdivided into α, β, γ, and δ fibers.

Type A fibers are the typical myelinated fibers of spinal nerves. Type C fibers are the very small, unmyelinated nerve fibers that conduct impulses at low velocities. These constitute more than half the sensory fibers in most peripheral nerves and also all of the postganglionic autonomic fibers.

The sizes, velocities of conduction, and functions of the different nerve fiber types are given in the figure. Note that a few very large fibers can transmit impulses at velocities as great as 120 m/sec, a distance in 1 second that is longer than a football field. On the other hand, the smallest fibers transmit impulses as slowly as 0.5 m/sec, requiring about 2 seconds to go from the big toe to the spinal cord.

Alternate Classification Used by Sensory Physiologists. Certain recording techniques have made it possible to separate the type Aα fibers into two subgroups; and yet these same recording techniques cannot distinguish easily between Aβ and Aγ fibers. Therefore, the following classification is frequently used by sensory physiologists:

Group Ia. Fibers from the annulospiral endings of muscle spindles (average about 17 microns in diameter; these are alpha type A fibers in the general classification).

Group Ib. Fibers from the Golgi tendon organs (average about 16 micrometers in diameter; these also are alpha type A fibers).

Group II. Fibers from the discrete cutaneous tactile receptors and also from the flower-spray endings of the muscle spindles (average about 8 micrometers in diameter;

Figure 46–6. Physiological classifications and functions of nerve fibers.

these are beta and gamma type A fibers in the other classification).

Group III. Fibers carrying temperature, crude touch, and pricking pain sensations (average about 3 micrometers in diameter; these are delta type A fibers in the other classification).

Group IV. Unmyelinated fibers carrying pain, itch, temperature, and crude touch sensations (0.5 to 2 microns in diameter; these are called type C fibers in the other classification).

● TRANSMISSION OF SIGNALS OF DIFFERENT INTENSITY IN NERVE TRACTS— SPATIAL AND TEMPORAL SUMMATION

One of the characteristics of each signal that always must be conveyed is its intensity, for instance, the intensity of pain. The different gradations of intensity can be transmitted either by utilizing increasing numbers of parallel fibers or by sending more impulses along a single fiber. These two mechanisms are called, respectively, spatial summation and temporal summation.

Figure 46–7 illustrates the phenomenon of *spatial summation,* whereby increasing signal strength is transmitted by using progressively greater numbers of fibers. This figure shows a section of skin innervated by a large number of parallel pain nerve fibers. Each of these arborizes into hundreds of minute *free nerve endings* that serve as pain receptors. The entire cluster of fibers from one pain fiber frequently covers an area of skin as large as 5 centimeters in diameter, and this area is called the *receptor field* of that fiber. The number of endings is large in the center of the field but diminishes toward the periphery. One can also see from the figure that the arborizing nerve fibrils overlap those from other pain fibers. Therefore, a pinprick of the skin usually stimulates endings from many different pain fibers simultaneously. When the pinprick is in the center of the receptive field of a particular pain fiber, however, the degree of stimulation of that fiber is far greater than when it is in the periphery of the field.

Thus, in the lower part of Figure 46–7 are shown three separate views of the cross section of the nerve bundle leading from the skin area. To the left is shown the effect of a weak stimulus, with only a single nerve fiber in the middle of the bundle stimulated very strongly (represented by the solid fiber), whereas several adjacent fibers are stimulated weakly (half-solid fibers). The other two views of the nerve cross section show the effect respectively of a moderate stimulus and a strong stimulus, with progressively more fibers being stimulated. Thus, the stronger signals spread to more and more fibers. This is the phenomenon of spatial summation.

A second means for transmitting signals of increasing strength is by increasing the *frequency* of nerve impulses in each fiber, which is called *temporal summation.* Figure 46–8 illustrates this, showing in the upper part a changing strength of signal and in the lower part the actual impulses transmitted by the nerve fiber.

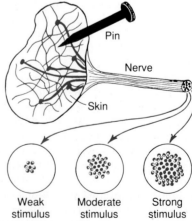

Figure 46–7. Pattern of stimulation of pain fibers in a nerve trunk leading from an area of skin pricked by a pin. This is an example of *spatial summation.*

Figure 46–8. Translation of signal strength into a frequency-modulated series of nerve impulses, showing *above* the strength of signal and *below* the separate nerve impulses. This is an example of *temporal summation.*

● TRANSMISSION AND PROCESSING OF SIGNALS IN NEURONAL POOLS

The central nervous system is made up of literally hundreds or even thousands of separate neuronal pools, some of which contain very few neurons while others hold vast numbers. For instance, the entire cerebral cortex could be considered to be a single large neuronal pool, or it could be considered to be a collection of smaller pools each observing separate functions. Other neuronal pools include the different basal ganglia, the specific nuclei in the thalamus, in the cerebellum, the mesencephalon, pons, and medulla. Also, the entire dorsal gray matter of the spinal cord could be considered to be one long pool of neurons, and the entire anterior gray matter another long neuronal pool.

Each pool has its own special characteristics of organization that cause it to process signals in its own special way, thus allowing these special characteristics to achieve the multitude of functions of the nervous system. Yet despite their differences in function, the pools also have many similar principles of function, described in the following pages.

RELAYING OF SIGNALS THROUGH NEURONAL POOLS

Organization of Neurons for Relaying Signals. Figure 46–9 is a schematic diagram of several neurons in a neuronal pool, showing "input" fibers to the left and "output" fibers to the right. Each input fiber divides hundreds to thousands of times, providing an average of a thousand or more terminal fibrils that spread over a large area in the pool to synapse with the dendrites or cell bodies of the neurons in the pool. The dendrites usually also arborize and spread for hundreds to thousands of micrometers in the pool. The neuronal area stimulated by each incoming nerve fiber is called its *stimulatory field.* Note that large numbers of the terminals from each input fiber lie on the centermost neuron in its "field," but progressively

fewer terminals lie on the neurons farther from the center of the field.

Threshold and Subthreshold Stimuli—Facilitation. From the discussion of synaptic function in the previous chapter, it will be recalled that discharge of a single excitatory presynaptic terminal almost never stimulates the postsynaptic neuron. Instead, large numbers of terminals must discharge on the same neuron either simultaneously or in rapid succession to cause excitation. For instance, in Figure 46–9, let us assume that six separate terminals must discharge simultaneously to excite any one of the neurons. If the student will count the number of terminals on each one of the neurons from each input fiber, he or she will see that input *fiber 1* has more than enough terminals to cause *neuron a* to discharge. Therefore, the stimulus from input fiber 1 to this neuron is said to be an *excitatory stimulus;* it is also called a *suprathreshold stimulus* because it is above the threshold required for excitation.

Input fiber 1 also contributes terminals to neurons b and c, but not enough to cause excitation. Nevertheless, discharge of these terminals makes both these neurons more excitable to signals arriving through other incoming nerve fibers. Therefore, the stimulus to these neurons is said to be *subthreshold,* and the neurons are said to be *facilitated.*

Similarly, for input *fiber 2,* the stimulus to *neuron d* is a suprathreshold stimulus; and the stimulus to *neurons b* and *c* is a subthreshold, but facilitating, stimulus.

It must be recognized that Figure 46–9 represents a highly condensed version of a neuronal pool, for each input nerve fiber usually provides terminals to hundreds or thousands of separate neurons in its distribution "field," as illustrated in Figure 46–10. In the

Figure 46–9. Basic organization of a neuronal pool.

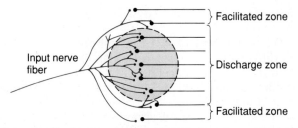

Figure 46-10. "Discharge" and "facilitated" zones of a neuronal pool.

central portion of the field, almost all the neurons are stimulated by the incoming fiber, designated in Figure 46–10 by the darkened circle area. Therefore, this is said to be the *discharge zone* of the incoming fiber, also called *excited zone* or *liminal zone*. To either side, the neurons are facilitated but not excited, and these areas are called the *facilitated zone*, also called *subthreshold zone* or *subliminal zone*.

Inhibition of a Neuronal Pool. We must also remember that some incoming fibers inhibit neurons, rather than exciting them. This is exactly the opposite of facilitation, and the entire field of the inhibitory branches is called the *inhibitory zone*. The degree of inhibition in the center of this zone obviously is very great because of large numbers of endings in the center; it becomes progressively less toward its edges.

Divergence of Signals Passing Through Neuronal Pools

Often it is important for signals entering a neuronal pool to excite far greater numbers of nerve fibers leaving the pool. This phenomenon is called *divergence*. Two major types of divergence occur and have entirely different purposes:

An *amplifying* type of divergence is illustrated in Figure 46–11A. This means simply that an input signal spreads to an increasing number of neurons as it passes through successive orders of neurons in its path. This type of divergence is characteristic of the corticospinal pathway in its control of skeletal muscles, with a single large pyramidal cell in the motor

cortex capable, under appropriate conditions, of exciting as many as 10,000 muscle fibers.

The second type of divergence, illustrated in Figure 46–11B, is *divergence into multiple tracts*. In this case, the signal is transmitted in two separate directions from the pool. For instance, information transmitted in the dorsal columns of the spinal cord takes two courses in the lower part of the brain: (1) into the cerebellum and (2) on through the lower regions of the brain to the thalamus and cerebral cortex. Likewise, in the thalamus almost all sensory information is relayed both into deep structures of the thalamus and to discrete regions of the cerebral cortex.

Convergence of Signals

"Convergence" means signals from multiple inputs converging to excite a single neuron. Figure 46–12A shows *convergence from a single source*. That is, multiple terminals from an incoming fiber tract terminate on the same neuron. The importance of this is that neurons are almost never excited by an action potential from a single input terminal. But action potentials from multiple input terminals will provide enough "spatial summation" to bring the neuron to the threshold required for discharge.

However, *convergence can also result from input signals* (excitatory or inhibitory) *from multiple sources*, as illustrated in Figure 46–12B. For instance, the interneurons of the spinal cord receive converging signals from (1) peripheral nerve fibers entering the cord, (2) propriospinal fibers passing from one segment of the cord to another, (3) corticospinal fibers from the cerebral cortex, and (4) several other long pathways descending from the brain into the spinal cord. Then the signals from the interneurons converge on the anterior motor neurons to control muscle function.

Such convergence allows summation of information from different sources, and the resulting response is a summated effect of all the different types of information. Obviously, therefore, convergence is

Figure 46-11. "Divergence" in neuronal pathways. *A,* Divergence within a pathway to cause "amplification" of the signal. *B,* Divergence into multiple tracts to transmit the signal to separate areas.

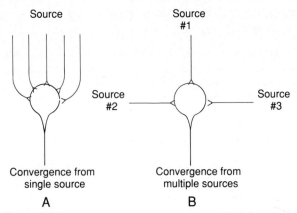

Figure 46-12. "Convergence" of multiple input fibers on a single neuron. *A,* Input fibers from a single source. *B,* Input fibers from multiple sources.

one of the important means by which the central nervous system correlates, summates, and sorts different types of information.

Neuronal Circuit Causing Both Excitatory and Inhibitory Output Signals

Sometimes an incoming signal to a neuronal pool causes an output excitatory signal going in one direction and at the same time an inhibitory signal going elsewhere. For instance, at the same time that an excitatory signal is transmitted by one set of neurons in the spinal cord to cause forward movement of a leg, an inhibitory signal is transmitted simultaneously through a separate set of neurons to inhibit the muscles on the back of the leg so that they will not oppose the forward movement. This type of circuit is characteristic of control of all antagonistic pairs of muscles, and it is called the *reciprocal inhibition circuit.*

Figure 46–13 illustrates the means by which the inhibition is achieved. The input fiber directly excites the excitatory output pathway, but it stimulates an intermediate *inhibitory neuron* (neuron 2) which then inhibits the second output pathway from the pool. This type of circuit is also important in preventing overactivity in many parts of the brain.

PROLONGATION OF A SIGNAL BY A NEURONAL POOL— "AFTERDISCHARGE"

Thus far, we have considered signals that are merely relayed through neuronal pools. However, in many instances, a signal entering a pool causes a prolonged output discharge, called *afterdischarge,* even after the incoming signal is over, and lasting from a few milliseconds to as long as many minutes. The two most important mechanisms by which afterdischarge occurs are the following:

Synaptic Afterdischarge. When excitatory synapses discharge on the surfaces of dendrites or the soma of a neuron, a postsynaptic potential develops in the neuron that lasts for many milliseconds, especially so when some of the long-acting synaptic transmitter substances are involved. As long as this potential lasts, it can continue to excite the neuron, causing it to transmit a continuous train of output impulses, as was explained in the previous chapter. Thus, as a result of this synaptic "afterdischarge" mechanism alone, it is possible for a single instantaneous input to cause a sustained signal output (a series of repetitive discharges) lasting for many milliseconds.

The Reverberatory (Oscillatory) Circuit as a Cause of Signal Prolongation. One of the most important of all circuits in the entire nervous system is the *reverberatory,* or *oscillatory, circuit.* Such circuits are caused by positive feedback within the neuronal network. That is, the output of a neuronal circuit feeds back to re-excite the input of the same circuit. Consequently, once stimulated, the circuit discharges repetitively for a long time.

Several different possible varieties of reverberatory circuits are illustrated in Figure 46–14, the simplest — in Figure 46–14A — involving only a single neuron. In this case, the output neuron simply sends a collateral nerve fiber back to its own dendrites or soma to restimulate itself; therefore, once the neuron discharged, the feedback stimuli could help keep the neuron discharging for a long time thereafter.

Figure 46–14B illustrates a few additional neurons in the feedback circuit, which would give a longer period of time between the initial discharge and the feedback signal. Figure 46–14C illustrates a still more complex system in which both facilitatory and inhibitory fibers impinge on the reverberating circuit. A facilitatory signal enhances the intensity and frequency of reverberation, whereas an inhibitory signal depresses or stops the reverberation.

Figure 46–14D illustrates that most reverberating pathways are constituted of many parallel fibers, and at each cell station the terminal fibrils diffuse widely.

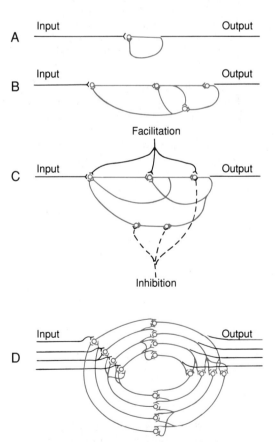

Figure 46–14. Reverberatory circuits of increasing complexity.

Figure 46–13. Inhibitory circuit. Neuron 2 is an inhibitory neuron.

In such a system the total reverberating signal can be either weak or strong, depending on how many parallel nerve fibers are momentarily involved in the reverberation.

Characteristics of Signal Prolongation From a Reverberatory Circuit. Figure 46–15 illustrates output signals from a typical reverberatory circuit. The input stimulus need last only 1 millisecond or so, and yet the output can last for many milliseconds or even minutes. The figure demonstrates that the intensity of the output signal usually increases to a high value early in the reverberation, then decreases to a critical point, at which it suddenly ceases entirely. The cause of this sudden cessation of reverberation is fatigue of one or more of the synaptic junctions in the circuit, for fatigue beyond a certain critical level lowers the stimulation of the next neuron in the circuit below threshold level so that the circuit is suddenly broken. Obviously, the duration of the signal before cessation can also be controlled by signals from other parts of the brain that inhibit or facilitate the circuit.

Almost these exact patterns of output signals are recorded from the motor nerves exciting a muscle involved in the flexor reflex following pain stimulation of the foot (as illustrated later in the chapter in Figure 46–18).

Continuous Signal Output From Neuronal Circuits

Some neuronal circuits emit output signals continuously even without excitatory input signals. At least two different mechanisms can cause this effect: (1) intrinsic neuronal discharge and (2) continuous reverberatory signals.

Continuous Discharge Caused by Intrinsic Neuronal Excitability. Neurons, like other excitable tissues, discharge repetitively if their membrane potentials rise above certain threshold levels. The membrane potentials of many neurons even normally are high enough to cause them to emit impulses continually. This occurs especially in large numbers of the neurons of the cerebellum as well as in most of the interneurons of the spinal cord. The rates at which these cells emit impulses can be increased by facilita-

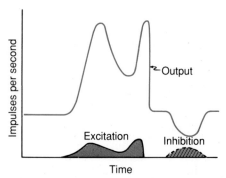

Figure 46–16. Continuous output from either a reverberating circuit or a pool of intrinsically discharging neurons. This figure also shows the effect of excitatory or inhibitory input signals.

tory signals or decreased by inhibitory signals; the latter can sometimes decrease the rate to extinction.

Continuous Signals Emitted From Reverberating Circuits as a Means for Transmitting Information. Obviously, a reverberating circuit that never fatigues to extinction can also be a source of continual impulses. And facilitatory impulses entering the reverberating pool can increase the output signal, while inhibition can decrease or even extinguish the signal.

Figure 46–16 illustrates a continual output signal from a pool of neurons, whether a pool emitting impulses because of intrinsic neuronal excitability or as a result of reverberation. Note that an excitatory (or facilitatory) input signal greatly increases the output signal, whereas an inhibitory input signal greatly decreases the output. Those students who are familiar with radio transmitters will recognize this to be a *carrier wave* type of information transmission. That is, the excitatory and inhibitory control signals are not the *cause* of the output signal, but they do *control* it. Note that this carrier wave system allows decrease in signal intensity as well as increase, whereas, up to this point, the types of information transmission that we have discussed have been only positive information, rather than negative information. This type of information transmission is used by the autonomic nervous system to control such functions as vascular tone, gut tone, degree of constriction of the iris, heart rate, and others.

RHYTHMIC SIGNAL OUTPUT

Many neuronal circuits emit rhythmic output signals—for instance, the rhythmic respiratory signal originating in the reticular substance of the medulla and pons. This repetitive rhythmic signal continues throughout life, while other rhythmic signals, such as those that cause scratching movements by the hind leg of a dog or the walking movements in an animal, require input stimuli into the respective circuits to initiate the signals.

Either all or almost all rhythmic signals that have been studied experimentally have been found to result from reverberating circuits or successive reverberat-

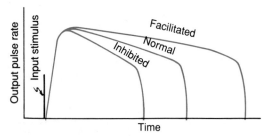

Figure 46–15. Typical pattern of the output signal from a reverberatory circuit following a single input stimulus, showing the effects of facilitation and inhibition.

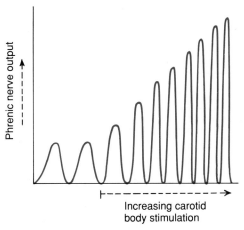

Figure 46-17. The rhythmical output from the respiratory center, showing that progressively increasing stimulation of the carotid body increases both the intensity and frequency of oscillation.

ing circuits that feed excitatory or inhibitory signals from one to the next.

Obviously, facilitatory or inhibitory signals can affect rhythmic signal output in the same way that they can affect continuous signal outputs. Figure 46-17, for instance, illustrates the rhythmic respiratory signal in the phrenic nerve. However, when the carotid body is stimulated by arterial oxygen deficiency, the frequency and amplitude of the rhythmic signal pattern increase progressively.

● INSTABILITY AND STABILITY OF NEURONAL CIRCUITS

Almost every part of the brain connects either directly or indirectly with every other part, and this creates a serious problem. If the first part excites the second, the second the third, the third the fourth, and so on until finally the signal re-excites the first part, it is clear that an excitatory signal entering any part of the brain would set off a continuous cycle of re-excitation of all parts. If this should occur, the brain would be inundated by a mass of uncontrolled reverberating signals — signals that would be transmitting no information but, nevertheless, would be consuming the circuits of the brain so that none of the informational signals could be transmitted. Such an effect actually occurs in widespread areas of the brain during *epileptic convulsions*.

How does the central nervous system prevent this from happening all the time? The answer seems to lie in two basic mechanisms that function throughout the central nervous system: (1) inhibitory circuits, and (2) fatigue of synapses.

Inhibitory Circuits as a Mechanism for Stabilizing Nervous System Function

Two types of inhibitory circuits in widespread areas of the brain help prevent excessive spread of signals:

(1) inhibitory feedback circuits that return from the termini of pathways back to the initial excitatory neurons of the same pathways — these are believed to occur in all the sensory nervous pathways, and they inhibit the input neurons when the termini become overly excited; and (2) some neuronal pools that exert gross inhibitory control over widespread areas of the brain — for instance, many of the basal ganglia exert inhibitory influences throughout the motor control system.

Synaptic Fatigue as a Means of Stabilizing the Nervous System

Synaptic fatigue means simply that synaptic transmission becomes progressively weaker the more prolonged the period of excitation. Figure 46-18 illustrates three successive records of a flexor reflex elicited in an animal caused by inflicting pain in the footpad of the paw. Note in each record that the strength of contraction progressively "decrements" — that is, its strength diminishes; this is believed to be caused by *fatigue* of the synapses in the flexor reflex circuit. Furthermore, the shorter the interval between the successive flexor reflexes, the less is the intensity of the subsequent reflex response. Thus, in most neuronal circuits that are overused, the sensitivities of the circuits become depressed.

Automatic Short-term Adjustment of Pathway Sensitivity by the Fatigue Mechanism. Now let us apply this phenomenon of fatigue to multiple pathways in the brain. Those that are overused usually become fatigued, and so their sensitivities will be reduced. On the other hand, those that are underused will become rested, and their sensitivities will increase. Thus, fatigue and recovery from fatigue constitute an important short-term means of moderating the sensitivities of the different nervous system circuits, helping to keep them operating in a range of sensitivity that allows effective function.

Long-term Changes in Synaptic Sensitivity Caused by Automatic Downgrading or Upgrading of Synaptic Receptors. Recently it has been learned that the long-term sensitivities of synapses

Figure 46-18. Successive flexor reflexes illustrating fatigue of conduction through the reflex pathway.

can be changed tremendously by downgrading the number of receptor proteins at the synaptic sites when there is overactivity and upgrading the receptors when there is underactivity. The mechanism for this is believed to be the following: Receptor proteins are being formed constantly by the endoplasmic reticular–Golgi apparatus system and are constantly being inserted into the synaptic membrane. However, when the synapses are overused and excesses of transmitter substance combine with the receptor proteins, many of these proteins are inactivated permanently and presumably removed from the synaptic membrane. This is especially true when some of the "modulator" transmitter substances are released at the synapses.

It is indeed fortunate that fatigue and downgrading or upgrading of receptors, as well as other control mechanisms of the nervous system, continually adjust the sensitivity in each circuit to almost the exact level required for proper function. Think for a moment how serious it would be if the sensitivities of only a few of these circuits should be abnormally high; one might then expect almost continual muscle cramps, convulsions, psychotic disturbances, hallucinations, tension, or many other nervous disorders. But the automatic controls normally readjust the sensitivities of the circuits back to a controllable range of reactivity any time the circuits begin to be either too active or too depressed.

REFERENCES

An der Heiden, U.: Analysis of Neural Networks. New York, Springer-Verlag, 1980.

Baldissera, F., et al.: Integration in spinal neuronal systems. In Brooks, V. B. (ed.): Handbook of Physiology. Sec. 1, Vol. II. Bethesda, Md., American Physiological Society, 1981, p. 509.

Bjorklund, A., and Stenevi, U.: Intercerebral neural implants: Neuronal replacement and reconstruction of damaged circuitries. Annu. Rev. Neurosci., 7:279, 1984.

Bousfield, D. (ed.): Neurotransmitters in Action. New York, Elsevier Science Publishing Co., 1985.

Connor, J. A.: Neural pacemakers and rhythmicity. Annu. Rev. Physiol., 47:17, 1985.

Cotman, C. W., et al.: Synapse replacement in the nervous system of adult vertebrates. Physiol. Rev., 61:684, 1981.

Cowan, W. M.: The development of the brain. Sci. Am., 241(3):112, 1979.

Dumont, J. P. C., and Robertson R. M.: Neuronal circuits: An evolutionary perspective. Science, 233:849, 1986.

Faber, D. S., and Korn, H.: Electrical field effects: Their relevance in central neural networks. Physiol. Rev., 69:821, 1989.

Gilbert, C. D.: Microcircuitry of the visual cortex. Annu. Rev. Neurosci., 6:217, 1983.

Hemmings, H. C., Jr., et al.: Role of protein phosphorylation in neuronal signal transduction. FASEB J., 3:1583, 1989.

Henneman, E., and Mendell, L. M.: Functional organization of motoneuron pool and its inputs. In Brooks, V. B. (ed.): Handbook of Physiology. Sec. 1, Vol. II. Bethesda, Md., American Physiological Society, 1981, p. 423.

Hopfield, J. J., and Tank, D. W.: Computing with neural circuits: A model. Science, 233:625, 1986.

Kalia, M. P.: Anatomical organization of central respiratory neurons. Annu. Rev. Physiol., 43:105, 1981.

Laduron, P. M.: Presynaptic heteroreceptors in regulation of neuronal transmission. Biochem. Pharmacol., 34:467, 1985.

Llinas, R. R.: The intrinsic electrophysiological properties of mammalian neurons: Insights into central nervous system function. Science, 242:1654, 1988.

Mendell, L. M.: Modifiability of spinal synapses. Physiol. Rev., 64:260, 1984.

Mountcastle, V. B.: Central nervous mechanisms in mechanoreceptive sensibility. In Darian-Smith, I. (ed.): Handbook of Physiology. Sec. 1, Vol. III. Bethesda, Md., American Physiological Society, 1984, p. 789.

Pinsker, H. M., and Willis, W. D., Jr. (eds.): Information Processing in the Nervous System. New York, Raven Press, 1980.

Purves, D., and Lichtman, J. W.: Specific connections between nerve cells. Annu. Rev. Physiol., 45:553, 1983.

Robinson, D. A.: Integrating with neurons. Annu. Rev. Physiol., 12:33, 1989.

Sachs, M. B.: Neural coding of complex sounds: Speech. Annu. Rev. Physiol., 46:261, 1984.

Sejnowski, T. J., et al.: Computational neuroscience. Science, 241:1299, 1988.

Selverston, A. I., and Moulins, M.: Oscillatory neural networks. Annu. Rev. Physiol., 47:29, 1985.

Sherman, S. M., and Spear, P. D.: Organization of visual pathways in normal and visually deprived cats. Physiol. Rev., 62:738, 1982.

Starke, K., et al.: Modulation of neurotransmitter release by presynaptic autoreceptors. Physiol. Rev., 69:864, 1989.

Sterling, P.: Microcircuitry of the cat retina. Annu. Rev. Neurosci., 6:149, 1983.

Su, C.: Purinergic neurotransmission and neuromodulation. Annu. Rev. Pharmacol. Toxicol., 23:397, 1983.

Turek, F. W.: Circadian neural rhythms in mammals. Annu. Rev. Physiol., 47:49, 1985.

Wong, R. K., et al.: Local circuit interactions in synchronization of cortical neurons. J. Exp. Biol., 112:169, 1984.

CHAPTER 47

• *Somatic Sensations: I. General Organization; the Tactile and Position Senses* •

The *somatic senses* are the nervous mechanisms that collect sensory information from the body. These senses are in contradistinction to the *special senses,* which mean specifically vision, hearing, smell, taste, and equilibrium.

CLASSIFICATION OF SOMATIC SENSES

The somatic senses can be classified into three different physiological types: (1) the *mechanoreceptive somatic senses,* which include both *tactile* and *position* sensations that are stimulated by mechanical displacement of some tissue of the body, (2) the *thermoreceptive senses,* which detect heat and cold, and (3) the *pain sense,* which is activated by any factor that damages the tissues. This chapter deals with the mechanoreceptive tactile and position senses, and the following chapter discusses the thermoreceptive and pain senses.

The tactile senses include *touch, pressure, vibration,* and *tickle* senses, and the position senses include *static position* and *rate of movement* senses.

Other Classifications of Somatic Sensations. Somatic sensations are also often grouped together in other classes that are not necessarily mutually exclusive, as follows:

Exteroreceptive sensations are those from the surface of the body. *Proprioceptive sensations* are those having to do with the physical state of the body, including position sensations, tendon and muscle sensations, pressure sensations from the bottom of the feet, and even the sensation of equilibrium, which is generally considered to be a "special" sensation rather than a somatic sensation.

Visceral sensations are those from the viscera of the body; in using this term one usually refers specifically to sensations from the internal organs.

The *deep sensations* are those that come from the deep tissues, such as from fasciae, muscles, bone, and so forth. These include mainly "deep" pressure, pain, and vibration.

• DETECTION AND TRANSMISSION OF TACTILE SENSATIONS

Interrelationship Between the Tactile Sensations of Touch, Pressure, and Vibration. Though touch, pressure, and vibration are frequently classified as separate sensations, they are all detected by the same types of receptors. There are only three differences among them: (1) touch sensation generally results from stimulation of tactile receptors in the skin or in tissues immediately beneath the skin; (2) pressure sensation generally results from deformation of deeper tissues; (3) vibration sensation results from rapidly repetitive sensory signals, but some of the same types of receptors as those for touch and pressure are utilized — specifically the very rapidly adapting types of receptors.

The Tactile Receptors. At least six entirely different types of tactile receptors are known, but many more similar to these also exist. Some of these receptors were illustrated in Figure 46–1 of Chapter 46, and their special characteristics are the following:

First, some *free nerve endings,* which are found everywhere in the skin and in many other tissues, can detect touch and pressure. For instance, even light contact with the cornea of the eye, which contains no other type of nerve ending besides free nerve endings, can nevertheless elicit touch and pressure sensations.

Second, a touch receptor of special sensitivity is *Meissner's corpuscle,* an elongated encapsulated nerve ending that excites a large (type $A\beta$) myelinated sensory nerve fiber. Inside the capsulation are many whorls of terminal nerve filaments. These receptors are present in the nonhairy parts of the skin (called *glabrous skin*) and are particularly abundant in the fingertips, lips, and other areas of the skin where one's ability to discern spatial characteristics of touch sen-

507

sations is highly developed. Meissner's corpuscles adapt in a fraction of a second after they are stimulated, which means that they are particularly sensitive to movement of very light objects over the surface of the skin and also to low frequency vibration.

Third, the fingertips and other areas that contain large numbers of Meissner's corpuscles also contain large numbers of *expanded tip tactile receptors,* one type of which is *Merkel's discs,* illustrated in Figure 47–1. The hairy parts of the skin also contain moderate numbers of expanded tip receptors, even though they have almost no Meissner's corpuscles. These receptors differ from Meissner's corpuscles in that they transmit an initially strong but partially adapting signal and then a continuing weaker signal that adapts only slowly. Therefore, they are responsible for giving steady state signals that allow one to determine continuous touch of objects against the skin. Merkel's discs are often grouped together in a single receptor organ called the *Iggo dome receptor,* which projects upward against the underside of the epithelium of the skin, as also illustrated in Figure 47–1. This causes the epithelium at this point to protrude outward, thus creating a dome and constituting an extremely sensitive receptor. Also note that the entire group of Merkel's discs is innervated by a single large type of myelinated nerve fiber (type Aβ). These receptors, along with the Meissner's corpuscles discussed above, play extremely important roles in localizing touch sensations to the specific surface areas of the body and also in determining the texture of what is felt.

Fourth, slight movement of any hair on the body stimulates the nerve fiber entwining its base. Thus, each hair and its basal nerve fiber, called the *hair end-organ,* is also a touch receptor. This receptor adapts readily and, therefore, like Meissner's corpuscles, detects mainly movement of objects on the surface of the body or initial contact with the body.

Fifth, located in the deeper layers of the skin and also in deeper tissues are many *Ruffini's end-organs,* which are multibranched, encapsulated endings, as illustrated in the previous chapter. These endings adapt very little and, therefore, are important for signaling continuous states of deformation of the skin and deeper tissues, such as heavy and continuous touch signals and pressure signals. They are also found in joint capsules and help signal the degree of joint rotation.

Sixth, *pacinian corpuscles,* which were discussed in detail in Chapter 46, lie both immediately beneath the skin and also deep in the fascial tissues of the body. These are stimulated only by very rapid movement of the tissues because they adapt in a few hundredths of a second. Therefore, they are particularly important for detecting tissue vibration or other extremely rapid changes in the mechanical state of the tissues.

Transmission of Tactile Sensations in Peripheral Nerve Fibers. Almost all the specialized sensory receptors, such as Meissner's corpuscles, Iggo dome receptors, hair receptors, pacinian corpuscles, and Ruffini's endings, transmit their signals in type Aβ nerve fibers that have transmission velocities of 30 to 70 m/sec. On the other hand, free nerve ending tactile receptors transmit signals mainly via the small type Aδ myelinated fibers that conduct at velocities of 5 to 30 m/sec. Some tactile free nerve endings transmit via type C unmyelinated fibers at velocities from a fraction of a meter up to 2 m/sec; these send signals into the spinal cord and lower brain stem, probably subserving mainly the sensation of tickle. Thus, the more critical types of sensory signals—those that help to determine precise localization on the skin, minute gradations of intensity, or rapid changes in sensory signal intensity—are all transmitted in more rapidly conducting types of sensory nerve fibers. On the other hand, the cruder types of signals, such as crude pressure, poorly localized touch, and especially tickle, are transmitted via much slower nerve fibers

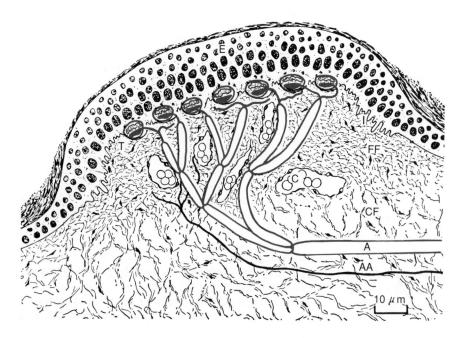

Figure 47–1. The Iggo dome receptor. Note the multiple numbers of Merkel's discs innervated by a single large myelinated fiber and abutting tightly the undersurface of the epithelium. (From Iggo and Muir: *J. Physiol., 200:763, 1969.*)

that require much less space in the nerve bundle than the faster fibers.

DETECTION OF VIBRATION

All the different tactile receptors are involved in detection of vibration, though different receptors detect different frequencies of vibration. Pacinian corpuscles can signal vibrations from 30 to 800 cycles per second, because they respond extremely rapidly to minute and rapid deformations of the tissues, and they also transmit their signals over type Aβ nerve fibers, which can transmit more than 1000 impulses per second.

Low frequency vibrations up to 80 cycles per second, on the other hand, stimulate other tactile receptors — especially Meissner's corpuscles, which are less rapidly adapting than pacinian corpuscles.

TICKLING AND ITCH

Recent neurophysiological studies have demonstrated the existence of very sensitive, rapidly adapting, mechanoreceptive free nerve endings that elicit only the tickle and itch sensation. Furthermore, these endings are found almost exclusively in the superficial layers of the skin, which is also the only tissue from which the tickle and itch sensation usually can be elicited. This sensation is transmitted by very small type C, unmyelinated fibers similar to those that transmit the aching, slow type of pain.

The purpose of the itch sensation is presumably to call attention to mild surface stimuli such as a flea crawling on the skin or a fly about to bite, and the elicited signals then excite the scratch reflex or other maneuvers that rid the host of the irritant.

Itch can be relieved by the process of scratching if this removes the irritant or if the scratch is strong enough to elicit pain. The pain signals are believed to suppress the itch signals in the cord by the process of lateral inhibition that will be described in the following chapter.

● THE TWO SENSORY PATHWAYS FOR TRANSMISSION OF SOMATIC SIGNALS INTO THE CENTRAL NERVOUS SYSTEM

Almost all sensory information from the somatic segments of the body enters the spinal cord through the dorsal roots of the spinal nerves (with the exception of a few very small fibers of questionable importance that enter the ventral roots). However, from the entry point of the cord and then to the brain the sensory signals are carried through one of two alternate sensory pathways: (1) the *dorsal column – lemniscal system* and (2) the *anterolateral system*. These two systems again come together partially at the level of the thalamus.

The dorsal column – lemniscal system, as its name implies, carries signals mainly in the *dorsal columns* of the cord and then, after crossing to the opposite side in the medulla, upward through the brain stem to the thalamus by way of the *medial lemniscus*. On the other hand, signals of the anterolateral system, after originating in the dorsal horns of the spinal gray matter, cross to the opposite side of the cord and ascend through the anterior and lateral white columns to terminate at all levels of the brain stem and also in the thalamus.

The dorsal column – lemniscal system is composed of large, myelinated nerve fibers that transmit signals to the brain at velocities of 30 to 110 m/sec, whereas the anterolateral system is composed of much smaller myelinated fibers (averaging 4 micrometers in diameter) that transmit signals at velocities ranging from a few meters per second up to 40 m/sec.

Another difference between the two systems is that the dorsal column – lemniscal system has a very high degree of spatial orientation of the nerve fibers with respect to their origin on the surface of the body, whereas the anterolateral system has a much smaller degree of spatial orientation.

These differences immediately characterize the types of sensory information that can be transmitted by the two systems. That is, sensory information that must be transmitted rapidly and with temporal and spatial fidelity is transmitted in the dorsal column – lemniscal system, while that which does not need to be transmitted rapidly nor with great spatial fidelity is transmitted mainly in the anterolateral system. On the other hand, the anterolateral system has a special capability that the dorsal system does not have: the ability to transmit a broad spectrum of sensory modalities — pain, warmth, cold, and crude tactile sensations; the dorsal system is limited to the more discrete types of mechanoreceptive sensations alone.

With this differentiation in mind we can now list the types of sensations transmitted in the two systems:

The Dorsal Column – Lemniscal System

1. Touch sensations requiring a high degree of localization of the stimulus.
2. Touch sensations requiring transmission of fine gradations of intensity.
3. Phasic sensations, such as vibratory sensations.
4. Sensations that signal movement against the skin.
5. Position sensations.
6. Pressure sensations having to do with fine degrees of judgment of pressure intensity.

The Anterolateral System

1. Pain.
2. Thermal sensations, including both warm and cold sensations.
3. Crude touch and pressure sensations capable of only crude localizing ability on the surface of the body.

4. Tickle and itch sensations.
5. Sexual sensations.

● TRANSMISSION IN THE DORSAL COLUMN–LEMNISCAL SYSTEM

ANATOMY OF THE DORSAL COLUMN–LEMNISCAL SYSTEM

On entering the spinal cord from the spinal nerve dorsal roots, the large myelinated fibers from specialized mechanoreceptors pass medially into the lateral margin of the dorsal white columns. However, almost immediately each fiber divides to form a *medial branch* and a *lateral branch*, as illustrated by the medial fiber from the dorsal root in Figure 47–2. The medial branch turns upward in the dorsal column and proceeds by way of the dorsal column pathway to the brain.

The lateral branch enters the dorsal horn of the cord gray matter and then divides many times, synapsing with neurons in almost all parts of the intermediate and anterior portions of the cord gray matter. The neurons that are excited, in turn, serve three functions: (1) A few of them give off second-order fibers that re-enter the dorsal column, to make up about 15% of all the dorsal column fibers; and a few other second-order fibers enter the posterolateral column, forming the *spinocervical tract* that rejoins the dorsal column system in the neck and low medulla. (2) Many of the neurons elicit local spinal cord reflexes, which will be discussed in Chapter 54. (3) Others give rise to the spinocerebellar tracts, which we will discuss in Chapter 56 in relation to the function of the cerebellum.

The Dorsal Column–Medial Lemniscal Pathway. Note in Figure 47–3 that the nerve fibers entering the dorsal columns pass uninterrupted up to the medulla, where they synapse in the *dorsal column nuclei* (the *cuneate* and *gracile nuclei*). From here, *second-order neurons* decussate immediately to the opposite side and then continue upward to the thalamus through bilateral pathways called the *medial lemnisci*. In this pathway through the brain stem, the medial lemniscus is joined by additional fibers from the *main sensory nucleus of the trigeminal nerve* and from the *upper portion of its descending nuclei;* these fibers subserve

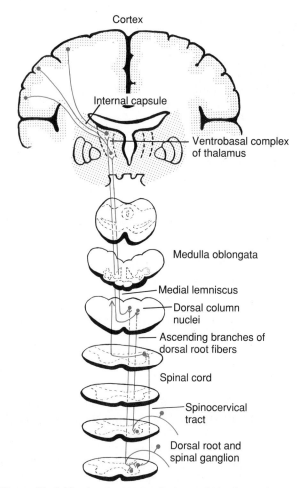

Figure 47–3. The dorsal column and spinocervical pathways for transmitting critical types of tactile signals. (Modified from Ranson and Clark: Anatomy of the Nervous System. Philadelphia, W. B. Saunders Company, 1959.)

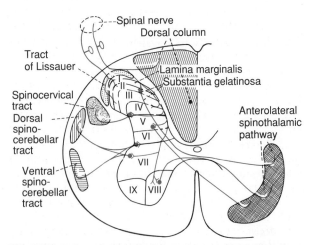

Figure 47–2. Cross-section of the spinal cord, showing the anatomical laminae I through IX of the cord gray matter and the ascending sensory tracts in the white columns of the spinal cord.

the same sensory functions for the head that the dorsal column fibers subserve for the body.

In the thalamus, the medial lemniscal fibers from the dorsal columns terminate in the *ventral posterolateral nucleus,* whereas those from the trigeminal nuclei terminate in the *ventral posteromedial nucleus*. These two nuclei, along with the posterior thalamic nuclei, where some fibers from the anterolateral system terminate, are together called the *ventrobasal complex*. From the ventrobasal complex, *third-order nerve fibers* project, as shown in Figure 47–4, mainly to the *postcentral gyrus* of the *cerebral cortex,* which is called *somatic sensory area I*. In addition, fewer fibers project to the lowermost lateral portion of each parietal lobe, an area called *somatic sensory area II*.

Spatial Orientation of the Nerve Fibers in the Dorsal Column–Lemniscal System

One of the distinguishing features of the dorsal column–lemniscal system is a distinct spatial orientation of nerve fibers from the individual parts of the body that is maintained throughout. For instance, in the dorsal columns, the fibers from the lower parts of the body lie toward the center, while those that enter

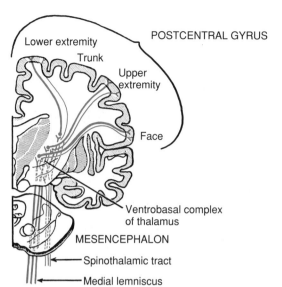

POSTCENTRAL GYRUS

Lower extremity
Trunk
Upper extremity
Face
Ventrobasal complex of thalamus
MESENCEPHALON
— Spinothalamic tract
— Medial lemniscus

Figure 47–4. Projection of the dorsal column–lemniscal system from the thalamus to the somatic sensory cortex. (Modified from Brodal: Neurological Anatomy in Relation to Clinical Medicine. New York, Oxford University Press, 1969.)

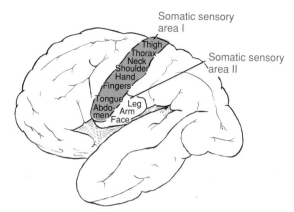

Somatic sensory area I
Somatic sensory area II
Thigh
Thorax
Neck
Shoulder
Hand
Fingers
Tongue
Abdomen
Leg
Arm
Face

Figure 47–6. The two somatic sensory cortical areas, somatic sensory areas I and II.

the spinal cord at progressively higher segmental levels form successive layers laterally.

In the thalamus, the distinct spatial orientation is still maintained, with the tail end of the body represented by the most lateral portions of the ventrobasal complex and the head and face represented in the medial component of the complex. However, because of the crossing of the medial lemnisci in the medulla, the left side of the body is represented in the right side of the thalamus, and the right side of the body is represented in the left side of the thalamus.

THE SOMATIC SENSORY CORTEX

Before discussing the role of the cerebral cortex in somatic sensation, we need to give an orientation to the various areas of the cortex. Figure 47–5 is a map of the human cerebral cortex, showing that it is divided

Central fissure
3 2
4
7a
8
5
7b
9
6
1
46
40
39
10
44
19
45
43 41 42
47
22
11
38
37 18
21
Lateral fissure
20
17

Figure 47–5. Structurally distinct areas, called "Brodmann areas," of the human cerebral cortex. (From Everett: Functional Neuroanatomy. 5th ed. Philadelphia, Lea & Febiger, 1965. Modified from Brodmann.)

into about 50 distinct areas called *Brodmann areas* based on histological structural differences. The map itself is important because it has come to be used by virtually all neurophysiologists and neurologists in referring to the different functional areas of the human cortex.

Note in the figure the large *central fissure* (also called "central sulcus") that extends horizontally across the brain. In general, sensory signals from all modalities of sensation terminate in the cerebral cortex posterior to the central fissure. Most importantly, the *somatic sensory cortex* lies immediately behind the central fissure, located mainly in Brodmann areas 1, 2, 3, 5, 7, and 40. By and large, these constitute the *parietal lobe* of the cortex. In addition, visual signals terminate in the occipital lobe; and auditory signals, in the temporal lobe.

The portion of the cortex anterior to the central fissure is devoted to motor control of the body and to some aspects of analytical thought.

Two distinct and separate areas are known to receive direct afferent nerve fibers from the somesthetic relay nuclei in the ventrobasal complex of the thalamus; these, called *somatic sensory area I* (S-I area) and *somatic sensory area II* (S-II area), are illustrated in Figure 47–6. However, somatic sensory area I is so much more important to the sensory functions of the body than is somatic sensory area II that in popular usage the term "somatic sensory cortex" most often is used to mean area I.

Projection of the Body in Somatic Sensory Area I. Somatic sensory area I lies in the postcentral gyrus of the human cerebral cortex (in Brodmann areas 3, 1, and 2). A distinct spatial orientation exists in this area for reception of nerve signals from the different areas of the body. Figure 47–7 illustrates a cross-section through the brain at the level of the postcentral gyrus, showing the representations of the different parts of the body in separate regions of somatic sensory area I. Note, however, that each side of the cortex receives sensory information exclusively from the opposite side of the body (with the exception of a small amount of sensory information from the same side of the face).

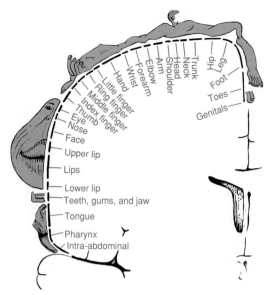

Figure 47–7. Representation of the different areas of the body in the somatic sensory area I of the cortex. (From Penfield and Rasmussen: Cerebral Cortex of Man: A Clinical Study of Localization of Function. New York, Macmillan Company, 1968.)

Figure 47–8. Structure of the cerebral cortex, illustrating *I*, molecular layer; *II*, external granular layer; *III*, layer of pyramidal cells; *IV*, internal granular layer; *V*, large pyramidal cell layer; and *VI*, layer of fusiform or polymorphic cells. (From Ranson and Clark [after Brodmann]: Anatomy of the Nervous System. Philadelphia, W. B. Saunders Company, 1959.)

Some areas of the body are represented by large areas in the somatic cortex—the lips the greatest of all, followed by the face and thumb—whereas the entire trunk and lower part of the body are represented by relatively small areas. The sizes of these areas are directly proportional to the number of specialized sensory receptors in each respective peripheral area of the body. For instance, a great number of specialized nerve endings are found in the lips and thumb, whereas only a few are present in the skin of the trunk.

Note also that the head is represented in the most lateral portion of somatic sensory area I, whereas the lower part of the body is represented medially.

Somatic Sensory Area II. The second cortical area to which thalamic somatic fibers project, somatic sensory area II, is much smaller and lies posterior and inferior to the lateral end of somatic sensory area I, as shown in Figure 47–6. The degree of localization of the different parts of the body is very poor in this area, compared with somatic sensory area I. The face is represented anteriorly, the arms centrally, and the legs posteriorly.

So little is known about the function of somatic sensory area II that it cannot be discussed intelligently. It is known that signals enter this area from both sides of the body, from somatic sensory area I, and also from other sensory areas of the brain, such as visual and auditory signals. Also, in lower animals, ablation of this area makes it difficult for the animal to learn to discriminate different shapes of objects.

The Layers of the Somatic Sensory Cortex and Their Function

The cerebral cortex contains *six* separate layers of neurons, beginning with layer I next to the surface and extending progressively deeper to layer VI, as illustrated in Figure 47–8. As would be expected, the neurons in each layer perform functions different from those in other layers. Some of these functions are as follows:

1. The incoming sensory signal excites mainly neuronal layer IV first; then the signal spreads toward the surface of the cortex and also toward the deeper layers.

2. Layers I and II receive a diffuse, nonspecific input from lower brain centers that can facilitate a whole region of the cortex at once; this system will be described in Chapter 57. This input perhaps controls the overall level of excitability of the region stimulated.

3. The neurons in layers II and III send axons to other closely related portions of the cerebral cortex.

4. The neurons in layers V and VI send axons to more distant parts of the nervous system. Those in layer V are generally larger and project to more distant areas. For instance, many of these pass all the way into the brain stem and spinal cord to provide control signals to these areas. From layer VI, especially large numbers of axons extend to the thalamus, providing feedback signals from the cerebral cortex to the thalamus.

Representation of the Different Sensory Modalities in the Somatic Sensory Cortex—The Vertical Columns of Neurons

Functionally, the neurons of the somatic sensory cortex are arranged in vertical columns extending all the way through the six layers of the cortex, each column having a diameter of 0.3 to 0.5 millimeter and containing perhaps 10,000 neuronal cell bodies. Each of these columns serves a single specific sensory modality, some columns responding to stretch receptors around joints, some to stimulation of tactile hairs, others to discrete localized pressure points on the skin, and so forth. Furthermore, the columns for the different modalities are interspersed among each other. At layer IV, where the signals first enter the cord, the columns of neurons function almost entirely separately from each other. However, at other levels of the columns interactions occur that allow beginning analysis of the meanings of the sensory signals.

In the most anterior portion of the postcentral gyrus, located deep in the central fissure, in Brodmann area 3a, a disproportionately large share of the vertical columns respond to muscle, tendon, or joint stretch receptors. Many of the signals from these in turn spread directly to the motor cortex located immediately anterior to the central fissure and help control muscle function. As one proceeds more posteriorly in somatic sensory cortex I, more and more of the vertical columns respond to the slowly adapting cutaneous receptors, and then still farther posteriorly greater numbers of the columns are sensitive to deep pressure.

In the most posterior portion of somatic sensory area I, about 6 per cent of the vertical columns respond only when a stimulus moves across the skin in a particular direction. Thus, this is a still higher order of interpretation of sensory signals; and the process becomes even more complex still farther posteriorly in the parietal cortex, which is called the *somatic association area,* as we discuss subsequently.

Functions of Somatic Sensory Area I

The functional capabilities of different areas of the somatic sensory cortex have been determined by selective excision of the different portions. Widespread excision of somatic sensory area I causes loss of the following types of sensory judgment:

1. The person is unable to localize discretely the different sensations in the different parts of the body. However, he or she can localize these sensations very crudely, such as to a particular hand, which indicates that the thalamus or parts of the cerebral cortex not normally considered to be concerned with somatic sensations can perform some degree of localization.

2. He is unable to judge critical degrees of pressure against his body.

3. He is unable to judge exactly the weights of objects.

4. He is unable to judge shapes or forms of objects. This is called *astereognosis.*

5. He is unable to judge texture of materials, for this type of judgment depends on highly critical sensations caused by movement of the skin over the surface to be judged.

Note in the list that nothing has been said about loss of pain and temperature sense. However, in the absence of somatic sensory area I, the appreciation of these sensory modalities may be altered either in quality or in intensity. But more important, the pain and temperature sensations that do occur are poorly localized, indicating that both pain and temperature localization probably depend mainly upon simultaneous stimulation of tactile stimuli that uses the topographical map of the body in somatic sensory area I to localize the source.

SOMATIC ASSOCIATION AREAS

Brodmann areas 5 and 7 of the cerebral cortex, which are located in the parietal cortex behind somatic sensory area I and above somatic sensory area II, play important roles in deciphering the sensory information that enters the somatic sensory areas. Therefore, these areas are called the *somatic association areas.*

Electrical stimulation in the somatic association area can occasionally cause a person to experience a complex somatic sensation, sometimes even the "feeling" of an object such as a knife or a ball. Therefore, it seems clear that the somatic association area combines information from multiple points in the somatic sensory area to decipher its meaning. This also fits with the anatomical arrangement of the neuronal tracts that enter the somatic association area, for it receives signals from (1) somatic sensory area I, (2) the ventrobasal nuclei of the thalamus, (3) other areas of the thalamus, (4) the visual cortex, and (5) the auditory cortex.

Effect of Removing the Somatic Association Area — Amorphosynthesis. When the somatic association area is removed, the person loses the ability to recognize complex objects and complex forms by the process of feeling them. In addition, he or she loses most of the sense of form of his or her own body. Especially interesting, the person is mainly oblivious to the opposite side of the body — that is, forgets that it is there. Therefore, he also often forgets to use the other side for motor functions as well. Likewise, when feeling objects, the person will tend to feel only one side of the object and to forget that the other side even exists. This complex sensory deficit is called *amorphosynthesis.*

OVERALL CHARACTERISTICS OF SIGNAL TRANSMISSION AND ANALYSIS IN THE DORSAL COLUMN–LEMNISCAL SYSTEM

Basic Neuronal Circuit and Discharge Cortical "Field" in the Dorsal Column–Lemniscal System. The lower part of Figure 47–9 illustrates the basic organization of the neuronal circuit of the dorsal column pathway, showing that at each synaptic stage divergence occurs. However, the upper part of the figure shows that a single receptor stimulus on the skin does not cause all the cortical neurons with which that

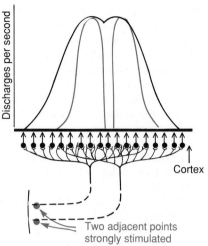

Figure 47–10. Transmission of signals to the cortex from two adjacent pinpoint stimuli. The solid black curve represents the pattern of cortical stimulation without "surround" inhibition, and the two colored curves represent the pattern with "surround" inhibition.

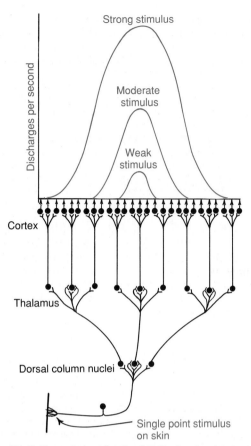

Figure 47–9. Transmission of a pinpoint stimulus signal to the cortex.

receptor connects to discharge at the same rate. Instead, the cortical neurons that discharge to the greatest extent are those in a central part of the cortical "field" for each respective receptor. Thus, a weak stimulus causes only the centralmost neurons to fire. A stronger stimulus causes still more neurons to fire, but those in the center still discharge at a considerably more rapid rate than do those farther away from the center.

Two-Point Discrimination. A method frequently used to test tactile capabilities is to determine a person's so-called "two-point discriminatory ability." In this test, two needles are pressed lightly against the skin, and the subject determines whether two points of stimulus are felt or one point. On the tips of the fingers a person can distinguish two separate points even when the needles are as close together as 1 to 2 millimeters. However, on the person's back, the needles must usually be as far apart as 30 to 70 millimeters before two separate points can be detected. The reason for this difference is the different numbers of specialized tactile receptors in the two areas.

Figure 47–10 illustrates the mechanism by which the dorsal column pathway, and all other sensory pathways as well, transmit two-point discriminatory information. This figure shows two adjacent points on the skin that are strongly stimulated, and it also shows the area of the somatic sensory cortex (greatly enlarged) that is excited by signals from the two stim-

ulated points. The solid black curve shows the spatial pattern of cortical excitation when both skin points are stimulated simultaneously. Note that the resultant zone of excitation has two separate peaks. It is these two peaks, separated by a valley, that allow the sensory cortex to detect the presence of two stimulatory points, rather than a single point. However, the capability of the sensorium to distinguish between two points of stimulation is strongly influenced by another mechanism, *lateral inhibition,* as explained in the following section.

Effect of Lateral Inhibition in Increasing the Degree of Contrast in the Perceived Spatial Pattern. It was pointed out in Chapter 46 that virtually every sensory pathway, when excited, gives rise simultaneously to lateral inhibitory signals; these spread to the sides of the excitatory signal and inhibit adjacent neurons. For instance, consider an excited neuron in a dorsal column nucleus. Aside from the central excitatory signal, short collateral fibers transmit inhibitory signals to the surrounding neurons. Some of these pass through an additional interneuron that secretes an inhibitory transmitter, and others pass directly to presynaptic terminals on the adjacent neurons and inhibit them by the mechanism of presynaptic inhibition.

The importance of *lateral inhibition* is that it blocks lateral spread of the excitatory signals and therefore increases the degree of contrast in the sensory pattern perceived in the cerebral cortex.

In the case of the dorsal column system, lateral inhibitory signals occur at each synaptic level, for instance in the dorsal column nuclei, in the ventrobasal nuclei of the thalamus, and in the cortex itself. At each of these levels, the lateral inhibition helps to block lateral spread of the excitatory signal. As a result, the peaks of excitation stand out, and much of the surrounding diffuse stimulation is blocked. This effect is illustrated by the two colored curves in Figure

47–10, showing complete separation of the peaks when the intensity of the lateral inhibition, also called *surround inhibition,* is very great. Obviously, this mechanism accentuates the contrast between the areas of peak stimulation and the surrounding areas, thus greatly increasing the contrast or sharpness of the perceived spatial pattern.

Transmission of Rapidly Changing and Repetitive Sensations. The dorsal column system is of particular value for apprising the sensorium of rapidly changing peripheral conditions. This system can "follow" changing stimuli up to at least 400 cycles per second and can "detect" changes as high as 700 cycles.

Vibratory Sensation. Vibratory signals are rapidly repetitive and can be detected as vibration up to 700 cycles per second. The higher frequency vibratory signals originate from the pacinian corpuscles, but lower frequency (below about 100 per second) can originate from Meissner's corpuscles as well. These signals are transmitted only in the dorsal column pathway. For this reason, application of vibration with a tuning fork to different peripheral parts of the body is an important tool used by the neurologist for testing the functional integrity of the dorsal columns.

PSYCHIC INTERPRETATION OF SENSORY STIMULUS INTENSITY

The ultimate goal of most sensory stimulation is to apprise the psyche of the state of the body and its surroundings. Therefore, it is important that we discuss briefly some of the principles related to the transmission of sensory stimulus intensity to the higher levels of the nervous system.

The first question that comes to mind is: How is it possible for the sensory system to transmit sensory experiences of tremendously varying intensities? For instance, the auditory system can detect the weakest possible whisper but can also discern the meanings of an explosive sound only a few feet away, even though the sound intensities of these two experiences can vary more than 10 billion times; the eyes can see visual images with light intensities that vary as much as a half million times; or the skin can detect pressure differences of 10,000 to 100,000 times.

As a partial explanation of these effects, Figure 46–4 in the previous chapter showed the relationship of the receptor potential produced by the pacinian corpuscle to the intensity of the sensory stimulus. At low stimulus strength, very slight changes in strength increase the potential markedly, whereas at high levels of stimulus strength further increases in receptor potential are very slight. Thus, the pacinian corpuscle is capable of accurately measuring extremely minute changes in stimulus strength at low intensity levels, but at high intensity levels the change in stimulus strength must be much greater to cause the same amount of change in receptor potential.

The transduction mechanism for detecting sound by the cochlea of the ear illustrates still another method for separating gradations of stimulus intensity. When sound causes vibration at a specific point on the basilar membrane, weak vibration stimulates only those hair cells at the point of maximum vibration. But as the vibration intensity increases, not only do these hair cells become more intensely stimulated, but still many more hair cells in each direction farther away from the maximum vibratory point also become stimulated. Thus, signals transmitted over progressively increasing numbers of nerve fibers is another mechanism by which stimulus strength is transmitted into the central nervous system. This mechanism, plus the direct effect of stimulus strength on impulse rate in each nerve fiber, as well as several other mechanisms, makes it possible for most sensory systems to operate reasonably faithfully at stimulus intensity levels changing more than hundreds of thousands to billions of times.

Importance of the Tremendous Intensity Range of Sensory Reception. Were it not for the tremendous intensity range of sensory reception that we can experience, the various sensory systems would more often than not operate in the wrong range. This is illustrated by the attempts of most persons to adjust the light exposure on a camera without using a light meter. Left to intuitive judgment of light intensity, a person almost always overexposes the film on very bright days and greatly underexposes the film at twilight. Yet that person's eyes are capable of discriminating with great detail the visual objects in both very bright sunlight and at twilight; the camera cannot do this because of the narrow critical range of light intensity required for proper exposure of film.

JUDGMENT OF STIMULUS INTENSITY

Physiopsychologists have evolved numerous methods for testing one's judgment of sensory stimulus intensity, but only rarely do the results from the different methods agree with each other. Yet the basic principle of decreasing intensity discrimination as the sensory intensity increases is applicable to virtually all sensory modalities. Two formulations of this principle are widely discussed in the physiopsychology field of sensory interpretation: the *Weber-Fechner principle* and the *power principle*.

The Weber-Fechner Principle—Detection of "Ratio" of Stimulus Strength. In the mid-1800s, Weber first and Fechner later proposed the principle that *gradations of stimulus strength are discriminated approximately in proportion to the logarithm of stimulus strength.* That is, a typical test of this principle might show that a person can barely detect a 1 gram increase in weight when holding 30 grams or a 10 gram increase when holding 300 grams. Thus, the *ratio* of the change in stimulus strength required for detection of a change remains essentially constant, about 1 to 30, which is what the logarithmic principle means. To express this mathematically,

$$\text{Interpreted signal strength} = \log (\text{Stimulus}) + \text{Constant}$$

More recently it has become evident that the Weber-Fechner principle is quantitatively accurate only for the higher intensities of visual, auditory, and cutaneous sensory experience and applies only poorly to most other types of sensory experience.

Yet the Weber-Fechner principle is still a good one to remember because it emphasizes that the greater the background sensory intensity, the greater also must be the additional change in stimulus strength in order for the psyche to detect the change.

The Power Law. Another attempt by physiopsychologists to find a good mathematical relationship is the following formula, known as the power law:

$$\text{Interpreted signal strength} = K \cdot (\text{Stimulus} - k)^y$$

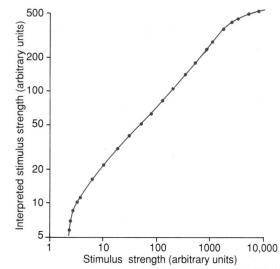

Figure 47–11. Graphical demonstration of the "power law" relationship between actual stimulus strength and strength that the psyche interprets it to be. Note that the power law does not hold at either very weak or very strong stimulus strengths.

In this formula the exponent y and the constants K and k are different for each type of sensation.

When this power law relationship is plotted on a graph using double logarithmic coordinates, as illustrated in Figure 47–11, a linear relationship can be attained between interpreted stimulus strength and actual stimulus strength over a large range for almost any type of sensory perception. However, as illustrated in the figure, even this power law relationship fails to hold satisfactorily at both very low and very high stimulus strengths.

THE POSITION SENSES

The *position senses* are frequently also called *proprioceptive senses*. They can be divided into two subtypes: (1) *static position sense,* which means conscious orientation of the different parts of the body with respect to each other, and (2) *rate of movement sense,* also called *kinesthesia* or *dynamic proprioception.*

The Position Sensory Receptors. Knowledge of position, both static and dynamic, depends upon knowing the degrees of angulation of all joints in all planes and their rates of change. Therefore, multiple different types of receptors help to determine joint angulation and are used together for position sense. Furthermore, both skin tactile receptors and deep receptors near the joints are also used. In the case of the fingers, where skin receptors are in great abundance, as much as half of position recognition is probably detected through the skin receptors. On the other hand, for most of the larger joints of the body, deep receptors are more important.

For determining joint angulation in midranges of motion, the most important receptors are believed to be the *muscle spindles.* These are also exceedingly important in helping to control muscle movement, as we see in Chapter 54. When the angle of a joint is changing, some muscles are being stretched while others are loosened, and the stretch information from the spindles is passed into the computational system of the spinal cord and higher regions of the dorsal column system for deciphering the complex interrelations of joint angulations.

At the extremes of joint angulation, the stretch of the ligaments and deep tissues around the joints is an additional important factor in determining position. Some types of endings used for this are the pacinian corpuscles, Ruffini's endings, and receptors similar to the Golgi tendon receptors found in muscle tendons.

The pacinian corpuscles and muscle spindles are especially adapted for detecting rapid rates of change. Therefore, it is likely that these are the receptors most responsible for detecting rate of movement.

Processing of Position Sense Information in the Dorsal Column–Lemniscal Pathway. Despite the usual faithfulness of transmission of signals from the periphery to the sensory cortex in the dorsal column–lemniscal system, there nevertheless seems to be some processing of position sense signals before they reach the cerebral cortex. For instance individual joint receptors are stimulated maximally at specific degrees of rotation of the joint, with the intensity of stimulation decreasing on either side of the maximal point for each receptor. However, the static position signal for joint rotation is quite different at the level of the thalamus, as can be seen by referring to Figure 47–12. This figure shows that the thalamic neurons that respond to joint rotation are of two types: (1) those maximally stimulated when the joint is at full rotation and (2) those maximally stimulated when the joint is at minimal rotation. In each case, as the degree of rotation changes, the rate of stimulation of the neuron either decreases or increases, depending on the direction in which the joint is being rotated. Furthermore, the intensity of neuronal excitation changes over angles of 40 to 60 degrees of angulation, in contrast to 20 to 30 degrees for the individual joint receptors. Thus, the signals from the individual joint receptors have been integrated in the space domain by the time they reach the thalamic neurons, illustrating some degree of processing of the signals either in the cord or in the thalamus.

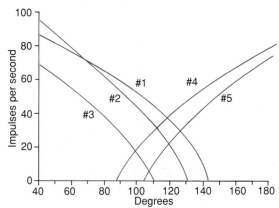

Figure 47–12. Typical responses of five different neurons in the knee joint receptor field of the thalamic ventrobasal complex when the knee joint is moved through its range of motion. (The curves were constructed from data in Mountcastle et al.: *J. Neurophysiol., 26*:807, 1963.)

● TRANSMISSION IN THE ANTEROLATERAL SYSTEM

The anterolateral system, in contrast to the dorsal column system, transmits sensory signals that do not require highly discrete localization of the signal source and also that do not require discrimination of fine gradations of intensity. These include pain, heat, cold, crude tactile, tickle and itch, and sexual sensations. In the following chapter pain and temperature sensations are discussed; the present chapter is still concerned principally with transmission of the tactile sensations, but now with the less acute types.

ANATOMY OF THE ANTEROLATERAL PATHWAY

The anterolateral fibers originate mainly in laminae I, IV, V, and VI (see Fig. 47–2) in the dorsal horns where many of the dorsal root sensory nerve fibers terminate after entering the cord. Then, as illustrated in Figure 47–13, the fibers cross in the anterior commissure of the cord to the opposite anterior and lateral white columns, where they turn upward toward the brain. These fibers ascend rather diffusely throughout the anterolateral columns. However, anatomical studies suggest a partial differentiation of this pathway

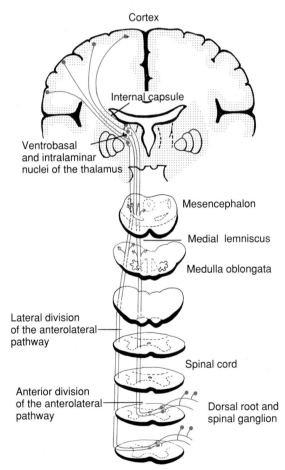

Figure 47–13. The anterior and lateral divisions of the anterolateral pathway.

into an anterior division, called the *anterior spinothalamic tract,* and a lateral division, called the *lateral spinothalamic tract.* Also encompassed in the anterolateral pathway are a *spinoreticular pathway* (to the reticular substance of the brain stem) and a *spinotectal tract* (to the tectum of the mesencephalon). However, it has been difficult to make these differentiations using electrical recording techniques.

The upper terminus of the anterolateral pathway is mainly twofold: (1) throughout the *reticular nuclei of the brain stem* and (2) in two different nuclear complexes of the thalamus, the *ventrobasal complex* and the *intralaminar nuclei.* In general, the tactile signals are transmitted mainly into the ventrobasal complex, terminating in the same *ventral posterior lateral* and *medial* nuclei as the dorsal column system, and this is probably also true for the temperature signals. From here, the tactile signals are transmitted to the somatosensory cortex along with the signals from the dorsal columns. On the other hand, only part of the pain signals project to this complex. Instead, most of these enter the reticular nuclei of the brain stem and via relay from the brain stem to the intralaminar nuclei of the thalamus, as is discussed in greater detail in the following chapter.

Characteristics of Transmission in the Anterolateral Pathway. In general, the same principles apply to transmission in the anterolateral pathway as in the dorsal column–lemniscal system except for the following differences: (1) the velocities of transmission are only one-third to one-half those in the dorsal column–lemniscal system, ranging between 8 and 40 m/sec; (2) the degree of spatial localization of signals is poor, especially in the pain pathways; (3) the gradations of intensities are also far less accurate, most of the sensations being recognized in 10 to 20 gradations of strength, rather than as many as 100 gradations for the dorsal column system; and (4) the ability to transmit rapidly repetitive signals is poor.

Thus, it is evident that the anterolateral system is a cruder type of transmission system than the dorsal column–lemniscal system. Even so, certain modalities of sensation are transmitted only in this system and not at all in the dorsal column–lemniscal system. These are pain, thermal, tickle and itch, and sexual sensations in addition to crude touch and pressure.

● SOME SPECIAL ASPECTS OF SOMATIC SENSORY FUNCTION

Function of the Thalamus in Somatic Sensation

When the somatosensory cortex of a human being is destroyed, that person loses most critical tactile sensibilities, but a slight degree of crude tactile sensibility does return. Therefore, it must be assumed that the thalamus (as well as other lower centers) has a slight ability to discriminate tactile sensation even though the thalamus normally functions mainly to relay this type of information to the cortex.

On the other hand, loss of the somatosensory cortex has little effect on one's perception of pain sensation and only a moderate effect on the perception of temperature. Therefore, there is much reason to believe that the brain stem, the thalamus, and other associated basal regions of the brain play perhaps the dominant role in discrimination of these sensibilities. It is interesting that these sensibilities appeared very early in the phylogenetic development of animalhood, whereas the critical tactile sensibilities were a late development.

Cortical Control of Sensory Sensitivity — "Corticofugal" Signals

In addition to somatic sensory signals transmitted from the periphery to the brain, "corticofugal" signals are transmitted in the backward direction from the cerebral cortex to the lower sensory relay stations of the thalamus, medulla, and spinal cord; these control the sensitivity of the sensory input. The corticofugal signals are inhibitory, so that when the input intensity becomes too great, the corticofugal signals automatically decrease the transmission in the relay nuclei. Obviously, this does two things: First, it decreases lateral spread of the sensory signals into adjacent neurons and therefore increases the contrast of the signal pattern. Second, it keeps the sensory system operating in a range of sensitivity that is not so low that the signals are ineffectual nor so high that the system is swamped beyond its capacity to differentiate sensory patterns.

This principle of corticofugal sensory control is utilized by all of the different sensory systems, not only the somatic system, as we shall see in subsequent chapters.

Segmental Fields of Sensation — The Dermatomes

Each spinal nerve innervates a "segmental field" of the skin called a *dermatome*. The different dermatomes are illustrated in Figure 47–14. However, these are shown as if there were distinct borders between the adjacent dermatomes, which is far from true, because much overlap exists from segment to segment.

The figure shows that the anal region of the body lies in the dermatome of the most distal cord segment. In the embryo, this is the tail region and is the most distal portion of the body. The legs develop from the lumbar and upper sacral segments, rather than from the distal sacral segments, which is evident from the dermatomal map. Obviously, one can use a dermatomal map such as that illustrated in Figure 47–14 to determine the level in the spinal cord at which various cord injuries may have occurred when the peripheral sensations are disturbed.

Figure 47–14. The dermatomes. (Modified from Grinker and Sahs: Neurology. Springfield, Ill., Charles C Thomas, 1966.)

REFERENCES

Akil, H., and Lewis, J. W. (eds.): Neurotransmitters and Pain Control. New York, S. Karger Publishers, Inc., 1987.

Akil, H., et al.: Endogenous opioids: Etiology and function. Annu. Rev. Neurosci., 7:223, 1984.

American Physiological Society: Sensory Processes. Washington, D.C., American Physiological Society, 1984.

Amit, Z., and Galina Z. H.: Stress-induced analgesia: Adaptive pain suppression. Physiol. Rev., 68:1091, 1988.

Basbaum, A. I., and Fields, H. L.: Endogenous pain control systems: Brainstem spinal pathways and endorphin circuitry. Annu. Rev. Neurosci., 7:309, 1984.

Berger, P. A., et al.: Behavioral pharmacology of the endorphins. Annu. Rev. Med., 33:397, 1982.

Besson, J. M., and Chaouch, A.: Peripheral and spinal mechanisms of nociception. Physiol. Rev., 67:67, 1987.

Bond, M. R.: Pain — Its Nature, Analysis, and Treatment. New York, Churchill Livingstone, 1984.

Darian-Smith, I.: The sense of touch: Performance and peripheral neural processes. In Darian-Smith I. (ed.): Handbook of Physiology. Sec. 1, Vol. III. Bethesda, Md., American Physiological Society, 1984, p. 739.

Darian-Smith, I.: Thermal sensibility. In Darian-Smith, I. (ed.): Handbook of Physiology. Sec. 1, Vol. III. Bethesda, Md., American Physiological Society, 1984, p. 879.

Dubner, R., and Bennett, G. J.: Spinal and trigeminal mechanisms of nociception. Annu. Rev. Neurosci., 6:381, 1983.

Emmers, R.: Somesthetic System of the Rat. New York, Raven Press, 1988.

Fields, H. L. (ed.): Pain: Mechanisms and Management. New York, McGraw-Hill Book Co., 1987.

Foreman, R. D., and Blair, R. W.: Central organization of sympathetic cardiovascular response to pain. Annu. Rev. Physiol., 50:607, 1988.

Friedhoff, A. J., and Miller, J. C.: Clinical implications of receptor sensitivity modification. Annu. Rev. Neurosci., 6:121, 1983.

Gelmers, H. J.: Calcium-channel blockers in the treatment of migraine. Am. J. Cardiol., 55:139B, 1985.

Goldman-Rakic, P. S.: Topography of cognition: Parallel distributed networks in primate association cortex. Annu. Rev. Neurosci., 11:137, 1988.

Goldstein, E. B.: Sensation and Perception. Belmont, Calif. Wadsworth Publishing Co., 1980.

Guyton, A. C., and Reeder, R. C.: Pain and contracture in poliomyelitis. Arch. Neurol. Psychiatr., 63:954, 1950.

Haft, J. I. (ed.): Differential Diagnosis of Chest Pain and Other Cardiac Symptoms. Mt. Kisco, N.Y., Futura Publishing Co., 1983.

Han, J. S., and Terenius, L.: Neurochemical basis of acupuncture analgesia. Annu. Rev. Pharmacol. Toxicol., 22:193, 1982.

Hnik, P., et al. (eds.): Mechanoreceptors. Development, Structure and Function. New York, Plenum Publishing Corp., 1988.

Hochberg, J.: Perception. In Darian-Smith, I. (ed.): Handbook of Physiology. Sec. 1, Vol. III. Bethesda, Md., American Physiological Society, 1984, p. 75.

Hyvarinen, J.: Posterior parietal lobe of the primate brain. Physiol. Rev., 62:1060, 1982.

Iggo, A., et al. (eds.): Nociception and Pain. New York, Cambridge University Press, 1986.

Jung, R.: Sensory research in historical perspective: Some philosophical foundations of perception. In Darian-Smith, I. (ed.): Handbook of Physiology. Sec. 1, Vol. III. Bethesda, Md., American Physiological Society, 1984, p. 1.

Kaas, J. H.: What, if anything, is SI? Organization of first somatosensory area of cortex. Physiol. Rev., 63:206, 1983.

Kaas, J. J., et al.: The reorganization of the somatosensory cortex following peripheral nerve damage in adult and developing mammals. Annu. Rev. Neurosci., 6:325, 1983.

Kruger, L. (ed.): Neural Mechanisms of Pain. New York, Raven Press, 1984.

Lewis, R. V., and Stern, A. S.: Biosynthesis of the enkephalins and enkephalin-containing polypeptides. Annu. Rev. Pharmacol. Toxicol., 23:353, 1983.

Loewenstein, E. R.: Excitation and inactivation in the receptor membrane. Ann. N.Y. Acad. Sci., 94:510, 1961.

Lucente, F. E., and Cooper, B. C.: Management of Facial, Head and Neck Pain. Philadelphia, W. B. Saunders Co., 1989.

Lund, J. S. (ed.): Sensory Processing in the Mammalian Brain. New York, Oxford University Press, 1988.

McCloskey, D. I.: Kinesthetic sensibility. Physiol. Rev., 58:763, 1978.

Mountcastle, V. B.: Central nervous mechanisms in mechanoreceptive sensibility. In Darian-Smith, I. (ed.): Handbook of Physiology. Sec. 1, Vol. III. Bethesda, Md., American Physiological Society, 1984, p. 789.

Nakanishi, S.: Substance P precursor and kininogen: Their structures, gene organizations, and regulation. Physiol. Rev., 67:1117, 1987.

Neff, W. D. (ed.): Contributions to Sensory Physiology. New York, Academic Press, 1982.

Paintal, A. S.: The visceral sensations—some basic mechanisms. Prog. Brain Res., 67:3, 1986.

Paris, P. M., and Stewart, R. D.: Pain Management in Emergency Medicine. East Norwalk, Conn., Appleton & Lange, 1988.

Perl, E. R.: Pain and nociception. In Darian-Smith, I. (ed.): Handbook of Physiology. Sec. 1, Vol. III. Bethesda, Md., American Physiological Society, 1984, p. 915.

Porter, R. (ed.): Studies in Neurophysiology. New York, Cambridge University Press, 1978.

Price, D. D.: Psychological and Neural Mechanisms of Pain. New York, Raven Press, 1988.

Saper, J. P. (ed.): Controversies and Clinical Variants of Migraine. New York, Pergamon Press, 1987.

Scheibel, A. B.: The brain stem reticular core and sensory function: In Darian-Smith, I. (ed.): Handbook of Physiology. Sec. 1, Vol. III. Bethesda, Md., American Physiological Society, 1984, p. 213.

Stebbins, W. C., et al.: Sensory function in animals. In Darian-Smith, I. (ed.): Handbook of Physiology. Sec. 1, Vol. III. Bethesda, Md., American Physiological Society, 1984, p. 123.

Tollison, C. D., et al.: Handbook of Chronic Pain Management. Baltimore, Williams & Wilkins, 1988.

Udin, S. B., and Fawcett, J. W.: Formation of topographic maps. Annu. Rev. Neurosci., 11:289, 1988.

Weiss, T. F.: Relation of receptor potentials of cochlear hair cells to spike discharges of cochlear neurons. Annu. Rev. Physiol., 46:247, 1984.

CHAPTER 48

• *Somatic Sensations: II. Pain, Headache, and Thermal Sensations* •

Many, if not most, ailments of the body cause pain. Furthermore, the ability to diagnose different diseases depends to a great extent on a doctor's knowledge of the different qualities of pain. For these reasons, the present chapter is devoted mainly to pain and to the physiologic basis of some of the associated clinical phenomena.

The Purpose of Pain. Pain is a protective mechanism for the body; it occurs whenever any tissues are being damaged, and it causes the individual to react to remove the pain stimulus. Even such simple activities as sitting for a long time on the ischia can cause tissue destruction because of lack of blood flow to the skin where the skin is compressed by the weight of the body. When the skin becomes painful as a result of the ischemia, the person normally shifts weight unconsciously. But a person who has lost the pain sense, such as after spinal cord injury, fails to feel the pain and therefore fails to shift. This very soon results in ulceration at the areas of pressure.

● THE TWO TYPES OF PAIN AND THEIR QUALITIES — FAST PAIN AND SLOW PAIN

Pain has been classified into two different major types: *fast pain* and *slow pain*. Fast pain occurs within about 0.1 second when a pain stimulus is applied, whereas slow pain begins only after a second or more and then increases slowly over many seconds and sometimes even minutes. During the course of this chapter we shall see that the conduction pathways for these two types of pain are different and that each of them has specific qualities.

Fast pain is also described by many alternate names, such as *sharp pain, pricking pain, acute pain, electric pain,* and others. This type of pain is felt when a needle is stuck into the skin or when the skin is cut with a knife, and this pain is also felt when the skin is subjected to electric shock. Fast, sharp pain is not felt in most of the deeper tissues of the body.

Slow pain also goes by multiple additional names such as *burning pain, aching pain, throbbing pain, nauseous pain,* and *chronic pain.* This type of pain is usually associated with *tissue destruction.* It can become excruciating and can lead to prolonged, unbearable suffering. It can occur both in the skin and in almost any deep tissue or organ.

We will learn later that the fast type of pain is transmitted through type Aδ pain fibers, whereas the slow type of pain results from stimulation of the more primitive type C fibers.

● THE PAIN RECEPTORS AND THEIR STIMULATION

All Pain Receptors Are Free Nerve Endings. The pain receptors in the skin and other tissues are all free nerve endings. They are widespread in the superficial layers of the *skin* and also in certain internal tissues, such as the *periosteum,* the *arterial walls,* the *joint surfaces,* and the *falx* and *tentorium* of the cranial vault. Most of the other deep tissues are not extensively supplied with pain endings but are weakly supplied; nevertheless, any widespread tissue damage can still summate to cause the slow-chronic-aching type of pain in these areas.

Three Different Types of Stimuli Excite Pain Receptors — Mechanical, Thermal, and Chemical. Most pain fibers can be excited by multiple types of stimuli. However, some fibers are more likely to respond to excessive mechanical stretch, others to extremes of heat or cold, and still others to specific chemicals in the tissues. These are classified respectively as *mechanical, thermal,* and *chemical pain receptors.* In general, fast pain is elicited by the mechanical and thermal types of receptors, whereas slow pain can be elicited by all three types.

Some of the chemicals that excite the chemical type of pain receptors include *bradykinin, serotonin, histamine, potassium ions, acids, acetylcholine,* and *proteolytic enzymes.* In addition, *prostaglandins* enhance the sensitivity of pain endings, but do not directly excite them. The chemical substances are especially important in stimulating the slow, suffering type of pain that occurs following tissue injury.

Nonadapting Nature of Pain Receptors. In contrast to most other sensory receptors of the body the pain receptors adapt very little and sometimes not at all. In fact, under some conditions, the excitation of the pain fibers becomes progressively greater as the pain stimulus continues. This increase in sensitivity of the pain receptors is called *hyperalgesia.*

One can readily understand the importance of this failure of pain receptors to adapt, for it allows them to keep the person apprised of a damaging stimulus that causes the pain as long as it persists.

RATE OF TISSUE DAMAGE AS A CAUSE OF PAIN

The average person first begins to perceive pain when the skin is heated above 45°C, as illustrated in Figure 48–1. This is also the temperature at which the tissues begin to be damaged by heat; indeed, the tissues are eventually completely destroyed if the temperature remains above this level indefinitely. Therefore, it is immediately apparent that pain resulting from heat is closely correlated with the ability of heat to damage the tissues.

Furthermore, the intensity of pain has also been closely correlated with the rate of tissue damage from causes other than heat — bacterial infection, tissue ischemia, tissue contusion, and so forth.

Special Importance of Chemical Pain Stimuli During Tissue Damage. Extracts from damaged tissues cause intense pain when injected beneath the normal skin. All the chemicals listed above that excite the chemical pain receptors are found in these extracts. However, the chemical that seems to be most painful of all is *bradykinin.* Therefore, many research workers have suggested that bradykinin might be the single agent most responsible for causing the tissue damage type of pain. Also, the intensity of the pain felt correlates with the local increase in potassium ion concentration as well. And it should be remembered, too, that proteolytic enzymes can directly attack the nerve endings and excite pain by making their membranes more permeable to ions.

Release of the various chemical pain excitants not only stimulates the chemosensitive pain endings but also greatly decreases the threshold for stimulation of the mechanosensitive and thermosensitive pain receptors as well. A widely known example of this is the extreme pain caused by slight mechanical or heat stimuli following tissue damage by sunburn.

Tissue Ischemia as a Cause of Pain. When blood flow to a tissue is blocked, the tissue becomes very painful within a few minutes. And the greater the rate of metabolism of the tissue, the more rapidly the pain appears. For instance, if a blood pressure cuff is placed around the upper arm and inflated until the arterial blood flow ceases, exercise of the forearm muscles can cause severe muscle pain within 15 to 20 seconds. In the absence of muscle exercise, the pain will not appear for three to four minutes.

One of the suggested causes of pain in ischemia is accumulation of large amounts of lactic acid in the tissues, formed as a consequence of the anaerobic metabolism (metabolism without oxygen) that occurs during ischemia. However, it is also possible that other chemical agents, such as bradykinin, proteolytic enzymes, and so forth, are formed in the tissues because of cell damage and that these, rather than lactic acid, stimulate the pain nerve endings.

Muscle Spasm as a Cause of Pain. Muscle spasm is also a very common cause of pain, and it is the basis of many clinical pain syndromes. This pain probably results partially from the direct effect of muscle spasm in stimulating mechanosensitive pain receptors. However, it possibly results also from the indirect effect of muscle spasm to compress the blood vessels and cause ischemia. Also, the spasm increases the rate of metabolism in the muscle tissue at the same time, thus making the relative ischemia even greater, creating ideal conditions for release of chemical pain-inducing substances.

● THE DUAL TRANSMISSION OF PAIN SIGNALS INTO THE CENTRAL NERVOUS SYSTEM

Even though all pain endings are free nerve endings, these endings utilize two separate pathways for

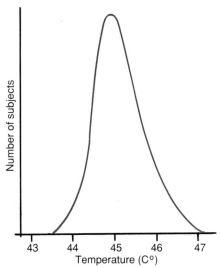

Figure 48–1. Distribution curve obtained from a large number of subjects of the minimal skin temperature that causes pain. (Modified from Hardy: *J. Chronic Dis.,* 4:22, 1956.)

transmitting pain signals into the central nervous system. The two pathways correspond to the two different types of pain, a *fast-sharp pain pathway,* and a *slow-chronic pain pathway.*

The Peripheral Pain Fibers—"Fast" and "Slow" Fibers. The fast-sharp pain signals are transmitted in the peripheral nerves to the spinal cord by small type Aδ fibers at velocities of between 6 and 30 m/sec. On the other hand, the slow-chronic type of pain is transmitted by type C fibers at velocities of between 0.5 and 2 m/sec. When the type Aδ fibers are blocked without blocking the C fibers by moderate compression of the nerve trunk, the fast-sharp pain disappears. On the other hand, when the type C fibers are blocked without blocking the delta fibers by low concentrations of local anesthetic, the slow-chronic-aching type of pain disappears.

Because of this double system of pain innervation, a sudden onset of painful stimulus gives a "double" pain sensation: a fast-sharp pain followed a second or so later by a slow, burning pain. The sharp pain apprises the person very rapidly of a damaging influence and, therefore, plays an important role in making the person react immediately to remove himself or herself from the stimulus. On the other hand, the slow, burning sensation tends to become more and more painful over a period of time. This sensation eventually gives one the intolerable suffering of long-continued pain.

On entering the spinal cord from the dorsal spinal roots, the pain fibers ascend or descend one to three segments in the *tract of Lissauer* that lies immediately posterior to the dorsal horn of the cord gray matter. Then they terminate on neurons in the dorsal horns. However, here again, there are two systems for processing the pain signals on their way to the brain, as illustrated in Figures 48–2 and 48–3. These are as follows:

Dual Pain Pathways in the Cord and Brain Stem—the Neospinothalamic Tract and the Paleospinothalamic Tract

On entering the spinal cord, the pain signals take two different pathways to the brain, through the *neo-*

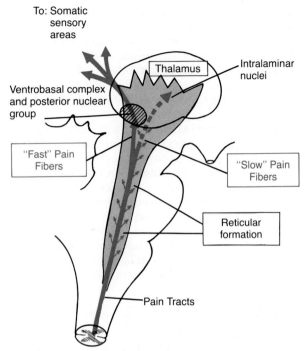

Figure 48 – 3. Transmission of pain signals into the hindbrain, thalamus, and cortex via the fast "pricking pain" pathway and the slow "burning pain" pathway.

spinothalamic tract and through the *paleospinothalamic tract.*

The Neospinothalamic Tract for Fast Pain. The "fast" type Aδ pain fibers transmit mainly mechanical and thermal pain. They terminate mainly in lamina I (lamina marginalis) of the dorsal horns and there excite second-order neurons of the neospinothalamic tract. These give rise to long fibers that cross immediately to the opposite side of the cord through the anterior commissure and then pass upward to the brain in the anterolateral columns.

Termination of the Neospinothalamic Tract in the Brain Stem and Thalamus. A few fibers of the neospinothalamic tract terminate in the reticular areas of the brain stem, but most pass all the way to the thalamus, terminating in the *ventrobasal complex*

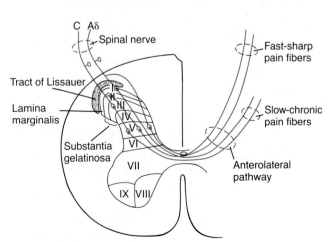

Figure 48 – 2. Transmission of both "acute-sharp" and "slow-chronic" pain signals into and through the spinal cord on the way to the brain stem.

along with the dorsal column–medial lemniscal tract discussed in the previous chapter. A few also terminate in the posterior nuclear group of the thalamus. From these areas the signals are transmitted to other basal areas of the brain and to the somatic sensory cortex.

Capability of the Nervous System to Localize Fast Pain in the Body. The fast-sharp type of pain can be localized much more exactly in the different parts of the body than can slow-chronic pain. However, even fast pain, when only pain receptors are stimulated without simultaneously stimulating tactile receptors, is still quite poorly localized, often only within 10 centimeters or so of the stimulated area. Yet when tactile receptors are also stimulated, the localization can be very exact.

The Paleospinothalamic Pathway for Transmitting Slow-Chronic Pain. The paleospinothalamic pathway is a much older system, and transmits pain mainly carried in the peripheral slow-suffering type C pain fibers, though it does also transmit some signals from type Aδ fibers as well. In this pathway, the peripheral fibers terminate almost entirely in laminas II and III of the dorsal horns, which together are called the *substantia gelatinosa*, as illustrated by the lateralmost dorsal root fiber in Figure 48–2. Most of the signals then pass through one or more additional short fiber neurons within the dorsal horns themselves before entering mainly lamina V, also in the dorsal horn. Here the last neuron in the series gives rise to long axons that mostly join the fibers from the fast pathway, passing through the anterior commissure to the opposite side of the cord, then upward to the brain in the same anterolateral pathway. However, a few of these fibers do not cross but instead pass ipsilaterally to the brain.

Substance P, the Probable Neurotransmitter of the Type C Nerve Endings. Where the type C fibers synapse in the dorsal horns of the spinal cord, they are believed to release substance P as the synaptic transmitter. Substance P is a neuropeptide; and as is true of all neuropeptides, it is slow to build up at the synapse and also slow to be destroyed. Therefore, its concentration at the synapse is believed to increase for at least several seconds, and perhaps much longer, after pain stimulation begins. After the pain is over, the substance P probably persists for many more seconds or perhaps minutes. The importance of this is that it might explain the progressive increase in intensity of slow-chronic pain with time, and it might also explain at least partially the persistence of this type of pain even after the painful stimulus has been removed.

Termination of the Slow-Chronic Pain Signals in the Brain Stem and Thalamus. The slow-chronic pathway terminates very widely in the brain stem, in the large pink-shaded area illustrated in Figure 48–3. Only one-tenth to one-fourth of the fibers pass all the way to the thalamus. Instead, they terminate principally in one of three different areas: (1) the *reticular nuclei* of the medulla, pons, and mesencephalon; (2) the *tectal area* of the mesencephalon deep to the superior and inferior colliculi; and (3) the *periaqueductal gray region* surrounding the aqueduct of Sylvius. These lower regions of the brain appear to be very important in the appreciation of the suffering types of pain, for animals with their brains sectioned above the mesencephalon evince undeniable evidence of suffering when any part of the body is traumatized.

From the reticular area of the brain stem, multiple short-fiber neurons relay the pain signals upward into the intralaminar nuclei of the thalamus and also into certain portions of the hypothalamus and other adjacent regions of the basal brain.

Capability of the Nervous System to Localize Pain Transmitted in the Slow-Chronic Pathway. Localization of pain transmitted in the paleospinothalamic pathway is very poor. In fact, electrophysiological studies suggest that the localization is often only to a major part of the body such as to one limb but not to a detailed point on the limb. This is in keeping with the multisynaptic, diffuse connectivity to the brain. It also explains why patients often have serious difficulty in localizing the source of some chronic types of pain.

Function of the Reticular Formation, Thalamus, and Cerebral Cortex in the Appreciation of Pain. Complete removal of the somatic sensory areas of the cerebral cortex does not destroy one's ability to perceive pain. Therefore, it is likely that pain impulses entering the reticular formation, thalamus, and other lower centers can cause conscious perception of pain. However, this does not mean that the cerebral cortex has nothing to do with normal pain appreciation; indeed, electrical stimulation of the cortical somatic sensory areas causes a person to perceive mild pain in approximately 3 per cent of the different points stimulated. It is believed that the cortex plays an important role in interpreting the quality of pain even though pain perception might be a function of lower centers.

Special Capability of Pain Signals to Arouse the Nervous System. Electrical stimulation in the reticular areas of the brain stem and also in the intralaminar nuclei of the thalamus, the areas where the slow-suffering type of pain terminates, has a strong arousal effect on nervous activity throughout the brain. In fact, these two areas are parts of the brain's principal arousal system, which is discussed in Chapter 59. This explains why a person with severe pain is frequently strongly aroused, and it also explains why it is almost impossible for a person to sleep when he or she is subjected to pain.

Surgical Interruption of Pain Pathways. Often a person has such severe and intractable pain (often resulting from rapidly spreading cancer) that it is necessary to relieve the pain. To do this the pain pathway can be destroyed at any one of several different points. If the pain is in the lower part of the body, a *cordotomy* in the upper thoracic region often relieves the pain for a few weeks to a few months. To do this, the spinal cord on the side opposite the pain is sectioned almost entirely through its anterolateral quadrant, which interrupts the anterolateral sensory pathway.

Unfortunately, though, the cordotomy is not always successful in relieving the pain for two reasons. First, many of the pain fibers from the upper part of the body do not cross to the opposite side of the spinal cord until they have reached the brain, so that the cordotomy does not transect these fibers. Second, pain frequently returns several months later, perhaps caused partly by sensitization of other pain pathways and partly by fibrous tissue from the cordotomy stimulating still remaining fibers. This new pain is often even more objectionable than the original pain.

Another operative procedure to relieve pain is to place lesions in the intralaminar nuclei in the thalamus, which often relieve the suffering type of pain while leaving intact one's appreciation of "acute" pain, an important protective mechanism.

● A PAIN CONTROL ("ANALGESIA") SYSTEM IN THE BRAIN AND SPINAL CORD

The degree to which each person reacts to pain varies tremendously. This results partly from the capability of the brain itself to control the degree of input of pain signals to the nervous system by activation of a pain control system, called an *analgesia system.*

The analgesia system is illustrated in Figure 48–4. It consists of three major components (plus other accessory components): (1) The *periaqueductal gray area* of the mesencephalon and upper pons surrounding the aqueduct of Sylvius. Neurons from this area send their signals to (2) the *raphe magnus nucleus,* a thin midline nucleus located in the lower pons and upper medulla. From here the signals are transmitted down the dorsolateral columns in the spinal cord to (3) a *pain inhibitory complex located in the dorsal horns of the spinal cord.* At this point the analgesia signals can block the pain before it is relayed on to the brain.

Electrical stimulation either in the periaqueductal gray area or in the raphe magnus nucleus can almost completely suppress many very strong pain signals entering by way of the dorsal spinal roots. Also, stimulation of areas at still higher levels of the brain that in turn excite the periaqueductal gray, especially the *periventricular nuclei in the hypothalamus* lying adjacent to the third ventricle and to a lesser extent the *medial forebrain bundle* also in the hypothalamus, can suppress pain, though perhaps not quite so much so.

Several different transmitter substances are involved in the analgesia system; especially involved are *enkephalin* and *serotonin.* Many of the nerve fibers derived from both periventricular nuclei and the periaqueductal gray area secrete enkephalin at their endings. Thus, as shown in Figure 48–4, the endings of many of the fibers in the raphe magnus nucleus release enkephalin. The fibers originating in this nucleus but terminating in the dorsal horns of the spinal cord secrete serotonin at their endings. The serotonin in turn acts on still another set of local cord neurons that are believed to secrete enkephalin. Enkephalin is believed to cause *presynaptic inhibition* of both incoming type C and type Aδ pain fibers where they synapse in the dorsal horns. It probably does this by blocking calcium channels in the membranes of the nerve terminals. Because it is calcium ions that cause release of transmitter at the synapse, such calcium blockage would obviously result in presynaptic inhibition. Furthermore, the blockage appears to last for prolonged periods of time, because after activating the analgesia system, analgesia often lasts for many minutes or even for hours.

Thus, the analgesia system can block pain signals at the initial entry point to the spinal cord. In fact, it can also block many of the local cord reflexes that result from pain signals, especially the withdrawal reflexes that are described in Chapter 54.

It is probable that this analgesia system can also inhibit pain transmission at other points in the pain pathway, especially in the reticular nuclei in the brain stem and in the intralaminar nuclei of the thalamus.

The Brain's Opiate System — The Endorphins and Enkephalins

More than 20 years ago it was discovered that injection of extremely minute quantities of morphine either into the periventricular nucleus around the third ventricle of the diencephalon or into the periaqueduc-

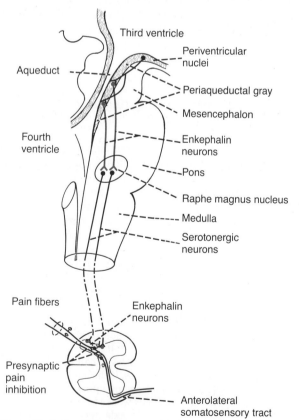

Figure 48–4. The analgesia system of the brain and spinal cord, showing inhibition of incoming pain signals at the cord level.

tal gray area of the brain stem will cause an extreme degree of analgesia. In subsequent studies, it has now been found that morphine acts at still many other points in the analgesia system, including the dorsal horns of the spinal cord. Because most drugs that alter the excitability of neurons do so by acting on synaptic receptors, it was assumed that the "morphine-receptors" of the analgesia system must in fact be receptors for some morphine-like neurotransmitter that is naturally secreted in the brain. Therefore, an extensive search was set into motion for a natural opiate of the brain. About a dozen such opiatelike substances have now been found in different points of the nervous system, but all are breakdown products of three large protein molecules: *proopiomelanocortin, proenkephalin,* and *prodynorphin.* Furthermore, multiple areas of the brain have been shown to have opiate receptors, especially the areas in the analgesia system. Among the more important of the opiate substances are *β-endorphin, met-enkephalin, leu-enkephalin,* and *dynorphin.*

The two enkephalins are found in the portions of the analgesia system described earlier, and *β*-endorphin is present both in the hypothalamus and in the pituitary gland. Dynorphin, though found in only minute quantities in nervous tissue, is important because it is an extremely powerful opiate, having 200 times as much pain-killing effect as morphine when injected directly into the analgesia system.

Thus, although all the fine details of the brain's opiate system are not yet entirely understood, nevertheless activation of the analgesia system either by nervous signals entering the periaqueductal gray area or by morphine-like drugs can totally or almost totally suppress many pain signals entering through the peripheral nerves.

Inhibition of Pain Transmission by Tactile Sensory Signals

Another important landmark in the saga of pain control was the discovery that stimulation of large sensory fibers from the peripheral tactile receptors depresses the transmission of pain signals either from the same area of the body or even from areas sometimes located many segments away. This effect presumably results from a type of local lateral inhibition. It explains why such simple maneuvers as rubbing the skin near painful areas is often very effective in relieving pain. And it probably also explains why liniments are often useful in the relief of pain. This mechanism and simultaneous psychogenic excitation of the central analgesia system are probably also the basis of pain relief by acupuncture.

Treatment of Pain by Electrical Stimulation

Several clinical procedures have been developed recently for suppressing pain by electrical stimulation of large sensory nerve fibers. The stimulating electrodes are placed on selected areas of the skin, or on occasion they have been implanted over the spinal cord to stimulate the dorsal sensory columns.

In a few patients, electrodes have also been placed stereotaxically in the intralaminar nuclei of the thalamus or in the periventricular or periaqueductal area of the diencephalon. The patient can then personally control the degree of stimulation. Dramatic relief has been reported in some instances. Also, the pain relief often lasts for as long as 24 hours after only a few minutes of stimulation.

● REFERRED PAIN

Often a person feels pain in a part of his or her body that is considerably removed from the tissues causing the pain. This pain is called *referred pain.* Usually the pain is initiated in one of the visceral organs and referred to an area on the body surface. Also, pain may be referred to another deep area of the body not exactly coincident with the location of the viscus producing the pain. A knowledge of these different types of referred pain is extremely important in clinical diagnosis because many visceral ailments cause no other signs except referred pain.

Mechanism of Referred Pain. Figure 48–5 illustrates the most likely mechanism by which most pain is referred. In the figure, branches of visceral pain fibers are shown to synapse in the spinal cord with some of the same second-order neurons that receive pain fibers from the skin. When the visceral pain fibers are stimulated, pain signals from the viscera are then conducted through at least some of the same neurons that conduct pain signals from the skin, and the person has the feeling that the sensations actually originate in the skin itself.

● VISCERAL PAIN

In clinical diagnosis, pain from the different viscera of the abdomen and chest is one of the few criteria that can be used for diagnosing visceral inflammation, disease, and other ailments. In general, the viscera have sensory receptors for no other modalities of sensation besides pain. Also, visceral pain differs from surface pain in several important aspects.

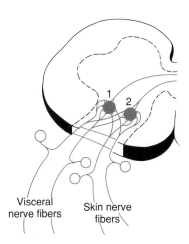

Figure 48–5. Mechanism of referred pain and referred hyperalgesia.

One of the most important differences between surface pain and visceral pain is that highly localized types of damage to the viscera rarely cause severe pain. For instance, a surgeon can cut the gut entirely in two in a patient who is awake without causing significant pain. On the other hand, any stimulus that causes *diffuse stimulation of pain nerve endings* throughout a viscus causes pain that can be extremely severe. For instance, ischemia caused by occluding the blood supply to a large area of gut stimulates many diffuse pain fibers at the same time and can result in extreme pain.

CAUSES OF TRUE VISCERAL PAIN

Any stimulus that excites pain nerve endings in diffuse areas of the viscera causes visceral pain. Such stimuli include ischemia of visceral tissue, chemical damage to the surfaces of the viscera, spasm of the smooth muscle in a hollow viscus, distension of a hollow viscus, or stretching of the ligaments.

Essentially all the true visceral pain originating in the thoracic and abdominal cavities is transmitted through sensory nerve fibers that run in the sympathetic nerves. These fibers are small type C fibers and, therefore, can transmit only the chronic-aching-suffering type of pain.

Ischemia. Ischemia causes visceral pain in exactly the same way that it does in other tissues, presumably because of the formation of acidic metabolic end products or tissue-degenerative products, such as bradykinin, proteolytic enzymes, or others that stimulate the pain nerve endings.

Chemical Stimuli. On occasion, damaging substances leak from the gastrointestinal tract into the peritoneal cavity. For instance, proteolytic acidic gastric juice often leaks through a ruptured gastric or duodenal ulcer. This juice causes widespread digestion of the visceral peritoneum, thus stimulating extremely broad areas of pain fibers. The pain is usually extremely severe.

Spasm of a Hollow Viscus. Spasm of the gut, the gallbladder, a bile duct, the ureter, or any other hollow viscus can cause pain possibly by mechanical stimulation of the pain endings. Or its cause might be diminished blood flow to the muscle combined with increased metabolic need of the muscle for nutrients. Thus, *relative* ischemia could develop, which causes severe pain.

Often, pain from a spastic viscus occurs in the form of *cramps,* the pain increasing to a high degree of severity and then subsiding, this process continuing rhythmically once every few minutes. The rhythmic cycles result from rhythmic contraction of smooth muscle. For instance, each time a peristaltic wave travels along an overly excitable spastic gut, a cramp occurs. The cramping type of pain frequently occurs in gastroenteritis, constipation, menstruation, parturition, gallbladder disease, or ureteral obstruction.

Overdistension of a Hollow Viscus. Extreme overfilling of a hollow viscus also results in pain, presumably because of overstretch of the tissues themselves. However, overdistension can also collapse the blood vessels that encircle the viscus, or that pass into its wall, thus perhaps promoting ischemic pain.

Insensitive Viscera

A few visceral areas are almost entirely insensitive to pain of any type. These include the parenchyma of the liver and the alveoli of the lungs. Yet the liver *capsule* is extremely sensitive to both direct trauma and stretch, and the *bile* ducts are also sensitive to pain. In the lungs, even though the alveoli are insensitive, the *bronchi* and the *parietal pleura* are both very sensitive to pain.

PARIETAL PAIN CAUSED BY VISCERAL DAMAGE

In addition to true visceral pain, pain sensations are also transmitted from the viscera through nonvisceral nerve fibers that innervate the parietal peritoneum, pleura, or pericardium.

When a disease affects a viscus, it often spreads to the parietal wall of the visceral cavity. This wall, like the skin, is supplied with extensive innervation from the spinal nerves, not from the sympathetic nerves. Therefore, pain from the parietal wall overlying the viscus is frequently very sharp. To emphasize the difference between this pain and true visceral pain: a knife incision through the *parietal* peritoneum is very painful, even though a similar cut through the visceral peritoneum or through a gut is not very painful if at all.

LOCALIZATION OF VISCERAL PAIN—THE "VISCERAL" AND THE "PARIETAL" TRANSMISSION PATHWAYS

Pain from the different viscera is frequently difficult to localize for a number of reasons. First, the brain does not know from firsthand experience that the different organs exist, and, therefore, any pain that originates internally can be localized only generally. Second, sensations from the abdomen and thorax are transmitted through two separate pathways to the central nervous system—the *true visceral pathway* and the *parietal pathway*. The true visceral pain is transmitted via sensory fibers of the autonomic nervous system (both sympathetic and parasympathetic), and the sensations are *referred* to surface areas of the body often far from the painful organ. On the other hand, parietal sensations are conducted *directly* into the local spinal nerves from the parietal peritoneum, pleura, or pericardium, and the sensations are usually *localized directly over the painful area.*

Localization of Referred Pain Transmitted in the Visceral Pathways. When visceral pain is referred to the surface of the body, the person generally localizes it in the dermatomal segment from which the visceral organ originated in the embryo. For instance the heart originated in the neck and upper thorax, so that the heart's visceral pain fibers pass into the cord between segments C-3 and T-5. Therefore, as illustrated in Figure 48–6, pain from the heart is referred to the side of the neck, over the shoulder, over the pectoral muscles, down the arm, and into the substernal area of the chest. Most frequently, the pain is on the left side, rather than on the right—because the left side of the heart is much more frequently involved in coronary disease than the right.

The stomach originated approximately from the seventh to the ninth thoracic segments of the embryo. Therefore, stomach pain is referred to the anterior epigastrium above the umbilicus, which is the surface area of the body subserved by the seventh through ninth thoracic segments. And Figure 48–6 shows several other surface areas to which visceral pain is referred from other organs, representing in general the areas in the embryo from which the respective organ originated.

The Parietal Pathway for Transmission of Abdominal and Thoracic Pain. Pain from the viscera is fre-

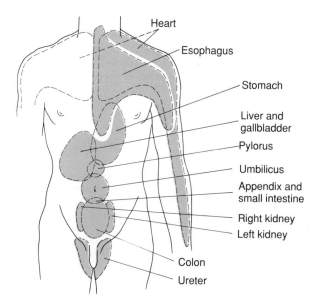

Figure 48-6. Surface areas of referred pain from different visceral organs.

quently localized to two surface areas of the body at the same time because of the dual transmission of pain through the referred visceral pathway and the direct parietal pathway. Thus, Figure 48-7 illustrates dual transmission from an inflamed appendix. Impulses pass from the appendix through the sympathetic visceral pain fibers into the sympathetic chain and then into the spinal cord at approximately T-10 or T-11; this pain is referred to an area around the umbilicus and is of the aching, cramping type. On the other hand, pain impulses also often originate in the parietal peritoneum where the inflamed appendix touches or is adherent to the abdominal wall. These cause pain of the sharp type directly over the irritated peritoneum in the right lower quadrant of the abdomen.

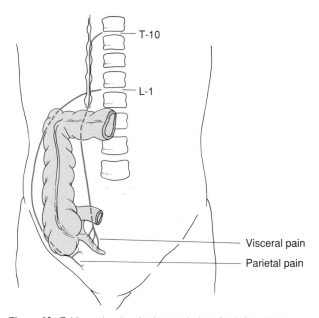

Figure 48-7. Visceral and parietal transmission of pain from the appendix.

● SOME CLINICAL ABNORMALITIES OF PAIN AND OTHER SOMATIC SENSATIONS

HYPERALGESIA

A pain pathway sometimes becomes excessively excitable; this gives rise to *hyperalgesia,* which means hypersensitivity to pain. The basic causes of hyperalgesia are (1) excessive sensitivity of the pain receptors themselves, which is called *primary hyperalgesia,* or facilitation of sensory transmission, which is called *secondary hyperalgesia.*

An example of primary hyperalgesia is the extreme sensitivity of sunburned skin, which is believed to result from sensitization of the pain endings by local tissue products of the burn—perhaps histamine, perhaps prostaglandins, perhaps others. Secondary hyperalgesia frequently results from lesions in the spinal cord or in the thalamus. Several of these are discussed in subsequent sections.

THE THALAMIC SYNDROME

Occasionally the posterolateral branch of the posterior cerebral artery, a small artery supplying the posteroventral portion of the thalamus, becomes blocked by thrombosis, so that the nuclei of this area of the thalamus degenerate, while the medial and anterior nuclei of the thalamus remain intact. The patient suffers a series of abnormalities, as follows: First, loss of almost all sensations from the opposite side of the body occurs because of destruction of the relay nuclei. Second, ataxia (inability to control movements precisely) may be evident because of loss of position and kinesthetic signals normally relayed through the thalamus to the cortex. Third, after a few weeks to a few months some sensory perception in the opposite side of the body returns, but strong stimuli are usually necessary to elicit this. When the sensations do occur, they are poorly—if at all—localized, almost always very painful, sometimes lancinating, regardless of the type of stimulus applied to the body. Fourth, the person is likely to perceive many affective sensations of extreme unpleasantness or, rarely, extreme pleasantness; the unpleasant ones are often associated with emotional tirades.

The medial nuclei of the thalamus are not destroyed by thrombosis of the artery. Therefore, it is believed that these nuclei become facilitated and give rise to the enhanced sensitivity to pain transmitted through the reticular system as well as to the affective perceptions.

HERPES ZOSTER (SHINGLES)

Occasionally a herpes virus infects a dorsal root ganglion. This causes severe pain in the dermatomal segment normally subserved by the ganglion, thus eliciting a segmental type of pain that circles halfway around the body. The disease is called *herpes zoster,* or "shingles" because of the eruption described below.

The cause of the pain is presumably excitation of the neuronal cells of the dorsal root ganglion by the virus infection. Aside from causing pain, the virus is also carried by neuronal cytoplasmic flow outward through the peripheral axons to their cutaneous terminals. Here the virus causes a rash that vesiculates within a few days and within another few days crusts over, all of this occurring within the dermatomal area served by the infected dorsal root.

TIC DOULOUREUX

Lancinating pains occur in some persons over one side of the face in part of the sensory distribution area of the fifth or ninth nerves; this phenomenon is called *tic douloureux* (or *trigeminal neuralgia* or *glossopharyngeal neuralgia*). The pains feel like sudden electric shocks, and they may appear for only a few seconds at a time or they may be almost continuous. Often, they are set off by exceedingly sensitive "trigger areas" on the surface of the face, in the mouth or in the throat—almost always by a mechanoreceptive stimulus instead of a pain stimulus. For instance, when the patient swallows a bolus of food, as the food touches a tonsil it might set off a severe lancinating pain in the mandibular portion of the fifth nerve.

The pain of tic douloureux can usually be blocked by cutting the peripheral nerve from the hypersensitive area. The sensory portion of the fifth nerve is often sectioned immediately inside the cranium, where the motor and sensory roots of the fifth nerve can be separated so that the motor portions, which are needed for many of the jaw movements, are spared while the sensory elements are destroyed. Obviously, this operation leaves the side of the face anesthetic, which in itself may be annoying. Furthermore, it is sometimes unsuccessful, indicating that the lesion that causes the pain is in the sensory nucleus in the brain stem and not in the peripheral nerves.

THE BROWN-SÉQUARD SYNDROME

Obviously, if the spinal cord is transected entirely, all sensations and motor functions distal to the segments of transection are blocked, but if only one half of the spinal cord is transected on a single side, the so-called Brown-Séquard syndrome occurs. The following effects of such a transection occur, and these can be predicted from a knowledge of the cord fiber tracts illustrated in Figure 48–8: All motor functions are blocked on the side of the transection in all segments below the level of the transection. Yet only some of the modalities of sensation are lost on the transected side, and others are lost on the opposite side. The sensations of pain, heat, and cold are lost on the opposite side of the body in all dermatomes two to six segments below the level of the transection. The sensations that are transmitted only in the dorsal and dorsolateral columns—kinesthetic and position sensations, vibration sensation, discrete localization, and two-point discrimination—are lost entirely on the side of the transection in all dermatomes below the level of the transection. Touch is impaired on the side of the transection because the principal pathway for transmission of light touch, the dorsal columns, is transected. Yet "crude touch," which is poorly localized, still persists because of transmission in the opposite ventral spinothalamic tract.

● HEADACHE

Headaches are actually referred pain to the surface of the head from the deep structures. Many headaches result from pain stimuli arising inside the cranium, but others result from pain arising outside the cranium, such as from the nasal sinuses.

HEADACHE OF INTRACRANIAL ORIGIN

Pain-Sensitive Areas in the Cranial Vault. The brain itself is almost totally insensitive to pain. Even cutting or electrically stimulating the somatic sensory areas of the cortex only occasionally causes pain; instead, it causes pins-and-needles types of paresthesias on the area of the body represented by the portion of the sensory cortex stimulated. Therefore, it is likely that much or most of the pain of headache is not caused by damage within the brain itself.

On the other hand, *tugging on the venous sinuses, damaging the tentorium,* or *stretching the dura at the base of the brain* can all cause intense pain that is recognized as headache. Also, almost any type of traumatizing, crushing, or stretching stimulus to the *blood vessels of the dura* can cause headache. A very sensitive structure is the middle meningeal artery, and neurosurgeons are careful to anesthetize this artery specifically when performing brain operations under local anesthesia.

Areas of the Head to Which Intracranial Headache Is Referred. Stimulation of pain receptors in the intracranial vault above the tentorium, including the upper surface of the tentorium itself, initiates impulses in the fifth nerve and, therefore, causes referred headache to the front half of the head in the area supplied by the fifth cranial nerve, as illustrated in Figure 48–9.

On the other hand, pain impulses from beneath the tentorium enter the central nervous system mainly through the

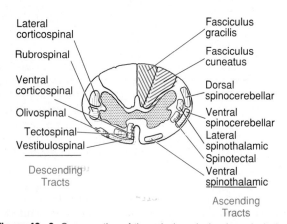

Lateral corticospinal
Rubrospinal
Ventral corticospinal
Olivospinal
Tectospinal
Vestibulospinal
Descending Tracts

Fasciculus gracilis
Fasciculus cuneatus
Dorsal spinocerebellar
Ventral spinocerebellar
Lateral spinothalamic
Spinotectal
Ventral spinothalamic
Ascending Tracts

Figure 48–8. Cross-section of the spinal cord, showing principal ascending tracts on the right and principal descending tracts on the left.

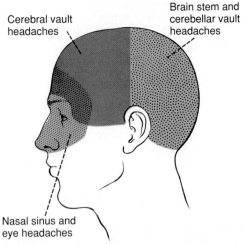

Cerebral vault headaches
Brain stem and cerebellar vault headaches
Nasal sinus and eye headaches

Figure 48–9. Areas of headache resulting from different causes.

second cervical nerve, which also supplies the scalp behind the ear. Therefore, subtentorial pain stimuli cause "occipital headache" referred to the posterior part of the head as shown in Figure 48–9.

Types of Intracranial Headache. *Headache of Meningitis.* One of the most severe headaches of all is that resulting from meningitis, which causes inflammation of all of the meninges, including the sensitive areas of the dura and the sensitive areas around the venous sinuses. Such intense damage as this can cause extreme headache pain referred over the entire head.

Headache Caused by Low Cerebrospinal Fluid Pressure. Removing as little as 20 milliliters of fluid from the spinal canal, particularly if the person remains in the upright position, often causes intense intracranial headache. Removing this quantity of fluid removes the flotation for the brain that is normally provided by the cerebrospinal fluid. Therefore, the weight of the brain stretches and otherwise distorts the various dural surfaces and thereby elicits the pain that causes the headache.

Migraine Headache. Migraine headache is a special type of headache that is thought to result from abnormal vascular phenomena, though the exact mechanism is unknown.

Migraine headaches often begin with various prodromal sensations, such as nausea, loss of vision in part of the field of vision, visual aura, or other types of sensory hallucinations. Ordinarily, the prodromal symptoms begin half an hour to an hour prior to the beginning of the headache itself. Therefore, any theory that explains migraine headache must also explain these prodromal symptoms.

One of the *theories* of the cause of migraine headaches is that prolonged emotion or tension causes reflex vasospasm of some of the arteries of the head, including arteries that supply the brain itself. The vasospasm theoretically produces ischemia of portions of the brain, and this is responsible for the prodromal symptoms. Then, as a result of the intense ischemia, something happens to the vascular wall to allow it to become flaccid and incapable of maintaining vascular tone for 24 to 48 hours. The blood pressure in the vessels causes them to dilate and pulsate intensely, and it is postulated that the excessive stretching of the walls of the arteries—including some extracranial arteries as well, such as the temporal artery—causes the actual pain of migraine headaches. However, it is possible that diffuse aftereffects of ischemia in the brain itself are at least partially if not mainly responsible for this type of headache.

Alcoholic Headache. As many people have experienced, a headache usually follows an alcoholic binge. It is most likely that alcohol, because it is toxic to tissues, directly irritates the meninges and causes intracranial pain.

Headache Caused by Constipation. Constipation causes headache in many persons. Because it has been shown that constipation headache can occur in persons whose spinal cords have been cut, we know that this headache is not caused by nervous impulses from the colon. Therefore, it possibly results from absorbed toxic products or from changes in the circulatory system resulting from loss of fluid into the gut.

EXTRACRANIAL TYPES OF HEADACHE

Headache Resulting from Muscular Spasm. Emotional tension often causes many of the muscles of the head, including especially those muscles attached to the scalp and the neck muscles attached to the occiput, to become spastic, and it is postulated that this is one of the common causes of headache. The pain of the spastic head muscles supposedly is referred to the overlying areas of the head and gives one the same type of headache as do intracranial lesions.

Headache Caused by Irritation of the Nasal and Accessory Nasal Structures. The mucous membranes of the nose and also of all the nasal sinuses are sensitive to pain, but not intensely so. Nevertheless, infection or other irritative processes in widespread areas of the nasal structures usually cause headache that is referred behind the eyes or, in the case of frontal sinus infection, to the frontal surfaces of the forehead and scalp, as illustrated in Figure 48–9. Also, pain from the lower sinuses—such as the maxillary sinuses—can be felt in the face.

Headache Caused by Eye Disorders. Difficulty in focusing one's eyes clearly may cause excessive contraction of the ciliary muscles in an attempt to gain clear vision. Even though these muscles are extremely small, tonic contraction of them can be the cause of retro-orbital headache. Also, excessive attempts to focus the eyes can result in reflex spasm in various facial and extraocular muscles, which is also a possible cause of headache.

A second type of headache originating in the eyes occurs when the eyes are exposed to excessive irradiation by light rays, especially ultraviolet light. Watching the sun or the arc of an arc-welder for even a few seconds may result in headache that lasts from 24 to 48 hours. The headache sometimes results from "actinic" irritation of the conjunctivae, and the pain is referred to the surface of the head or retro-orbitally. However, focusing intense light from an arc or the sun on the retina can actually burn the retina, and this could result in headache.

● THERMAL SENSATIONS

THERMAL RECEPTORS AND THEIR EXCITATION

The human being can perceive different gradations of cold and heat, progressing from *freezing cold* to *cold* to *cool* to *indifferent* to *warm* to *hot* to *burning hot*.

Thermal gradations are discriminated by at least three different types of sensory receptors: the cold receptors, the warmth receptors, and pain receptors. The pain receptors are stimulated only by extreme degrees of heat or cold and therefore are responsible, along with the cold and warmth receptors, for "freezing cold" and "burning hot" sensations.

The cold and warmth receptors are located immediately under the skin at discrete but separated points, each having a stimulatory diameter of about 1 millimeter. In most areas of the body there are three to ten times as many cold receptors as warmth receptors, and the number in different areas of the body varies from as great as 15 to 25 cold points per square centimeter in the lips, to 3 to 5 cold points per square centimeter in the finger, to less than 1 cold point per square centimeter in some broad surface areas of the trunk. There are correspondingly fewer numbers of warmth points.

Although it is quite certain, on the basis of psychological tests, that there are distinctive warmth nerve endings, these have not yet been identified histologically. They are presumed to be free nerve endings because warmth signals are transmitted mainly over type C nerve fibers at transmission velocities of only 0.4 to 2 m/sec.

On the other hand, a definitive cold receptor has been identified. It is a special, small, type Aδ myelinated nerve ending that branches a number of times, the tips of which protrude into the bottom surfaces of basal epidermal cells. Signals are transmitted from these receptors via delta nerve fibers at velocities of up to about 20 m/sec. However, some cold sensations are also transmitted in type C nerve fibers, which suggests that some free nerve endings also might function as cold receptors.

Stimulation of Thermal Receptors—Sensations of Cold, Cool, Indifferent, Warm, and Hot. Figure 48–10 illustrates the effects of different temperatures on the responses of four different nerve fibers: (1) a pain fiber stimulated by cold, (2) a cold fiber, (3) a warmth fiber, and (4) a pain fiber stimulated by heat. Note especially that these fibers respond differently at different levels of temperature. For instance, in the *very* cold region only the pain fibers are stimulated (if the skin becomes even colder so that it nearly freezes or actually does freeze, even these fibers cannot be stimulated). As the temperature rises to 10° to 15° C, pain impulses cease, but the cold receptors begin to be stimulated. Then, above about 30° C the warmth receptors become stimulated, and the cold receptors fade out at about 43° C. Finally, at around 45° C, pain fibers begin to be stimulated by heat.

One can understand from Figure 48–10, therefore, that a person determines the different gradations of thermal sensations by the relative degrees of stimulation of the different types of endings. One can understand also from this figure why extreme degrees of cold or heat can both be painful and why both these sensations, when intense enough, may give almost exactly the same quality of sensation—that is, freezing cold and burning hot sensations feel almost alike; they are both very painful.

Stimulatory Effects of Rising and Falling Temperature—Adaption of Thermal Recep-

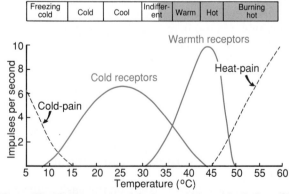

Figure 48–10. Frequencies of discharge of (1) a cold-pain fiber, (2) a cold fiber, (3) a warmth fiber, and (4) a heat-pain fiber. (The responses of these fibers are drawn from original data collected in separate experiments by Zotterman, Hensel, and Kenshalo.)

tors. When a cold receptor is suddenly subjected to an abrupt fall in temperature, it becomes strongly stimulated at first, but this stimulation fades rapidly during the first few seconds and progressively more slowly during the next half hour or more. In other words, the receptor "adapts" to a very great extent but does not appear to adapt 100 per cent.

Thus, it is evident that the thermal senses respond markedly to *changes in temperature* in addition to being able to respond to steady states of temperature. This means, therefore, that when the temperature of the skin is actively falling, a person feels much colder than when the temperature remains at the same level. Conversely, if the temperature is actively rising the person feels much warmer than he or she would at the same temperature if it were constant.

The response to changes in temperature explains the extreme degree of heat that one feels on first entering a tub of hot water and the extreme degree of cold felt on going from a heated room to the out-of-doors on a cold day.

Mechanism of Stimulation of the Thermal Receptors

It is believed that the cold and warmth receptors are stimulated by changes in their metabolic rates, these changes resulting from the fact that temperature alters the rates of intracellular chemical reactions more than twofold for each 10° C change. In other words, thermal detection probably results not from direct physical effects of heat or cold on the nerve endings, but instead from chemical stimulation of the endings as modified by the temperature.

Spatial Summation of Thermal Sensations. Because the number of cold or warmth endings in any one surface area of the body is very slight, it is difficult to judge gradations of temperature when small areas are stimulated. However, when a large area of the body is stimulated all at once, the thermal signals from the entire area summate. For instance, rapid changes in temperature of as little as 0.01° C can be detected if this change affects the entire surface of the body simultaneously. On the other hand, temperature changes 100 times this great might not be detected when the skin surface affected is only a square centimeter or so in size.

TRANSMISSION OF THERMAL SIGNALS IN THE NERVOUS SYSTEM

In general, thermal signals are transmitted in almost parallel, but not the same, pathways as pain signals. On entering the spinal cord, the signals travel for a few segments upward or downward in the *tract of Lissauer* and then terminate mainly in laminae I, II, and III of the dorsal horns—the same as for pain. After a small amount of processing by one or more cord neurons, the signals enter long, ascending thermal fibers that cross to the opposite anterolateral sensory tract and terminate in (1) the reticular areas of the brain stem and (2) the ventrobasal complex of the thalamus. A few thermal signals are also relayed to the somatic sensory cortex from the ventrobasal complex. Occasionally, a neuron in somatic sensory area I has been found by microelectrode studies to be directly responsive to either cold or warm stimuli in specific areas of the skin. Furthermore, it is known that removal of the postcentral gyrus in the human being reduces the ability to distinguish gradations of temperature.

REFERENCES

See references for Chapter 47.

UNIT X

THE NERVOUS SYSTEM: B. THE SPECIAL SENSES

CHAPTER 49

• *The Eye: I. Optics of Vision* •

• PHYSICAL PRINCIPLES OF OPTICS

Before it is possible to understand the optical system of the eye, the student must be thoroughly familiar with the basic physical principles of optics, including the physics of refraction, a knowledge of focusing, depth of focus, and so forth. Therefore, a brief review of these physical principles is first presented, and then the optics of the eye is discussed.

REFRACTION OF LIGHT

The Refractive Index of a Transparent Substance. Light rays travel through air at a velocity of approximately 300,000 km/sec but much slower through transparent solids and liquids. The refractive index of a transparent substance is the *ratio* of the velocity of light in air to the velocity in the substance. Obviously, the refractive index of air itself is 1.00.

If light travels through a particular type of glass at a velocity of 200,000 km/sec, the refractive index of this glass is 300,000 divided by 200,000, or 1.50.

Refraction of Light Rays at an Interface Between Two Media with Different Refractive Indices. When light waves traveling forward in a beam, as shown in the upper part of Figure 49–1, strike an interface that is perpendicular to the beam, the waves enter the second refractive medium without deviating from their course. The only effect that occurs is decreased velocity of transmission and shorter wavelength. On the other hand, as illustrated in the lower part of the figure, if the light waves strike an angulated interface, the light waves bend if the refractive indices of the two media are different from each other. In this particular figure the light waves are leaving air, which has a refractive index of 1.00, and are entering a block of glass having a refractive index of 1.50. When the beam first strikes the angulated interface, the lower edge of the beam enters the glass ahead of the upper edge. The wave front in the upper portion of the beam continues to travel at a velocity of 300,000 km/sec, whereas that which has entered the glass travels at a velocity of 200,000 km/sec. This causes the upper portion of the wave front to move ahead of the lower portion so that the wave front is no longer vertical but is angulated to the right. Because *the direction in which light travels is always perpendicular to the plane of the wave front,* the direction of travel of the light beam now bends downward.

The bending of light rays at an angulated interface is known as *refraction.* Note particularly that the degree of refraction increases as a function of (1) the ratio of the two refractive indices of the two transparent media and (2) the degree of angulation between the interface and the entering wave front.

APPLICATION OF REFRACTIVE PRINCIPLES TO LENSES

The Convex Lens—Focusing of Light Rays. Figure 49–2 shows parallel light rays entering a convex lens. The light rays passing through the center of the lens strike the lens exactly perpendicular to the lens surface and therefore pass through the lens without being refracted at all. Toward either edge of the lens, however, the light rays strike a progressively more angulated interface. Therefore, the outer rays bend more and more toward the center. Half the bending occurs when the rays enter the lens and half as they exit from the opposite side. (At this time the student should pause and analyze why the rays still bend toward the center upon leaving the lens.)

Finally, if the lens is ground with exactly the proper curvature, parallel light rays passing through each part of the lens will be bent exactly enough so that all the rays will pass through a single point, which is called the *focal point.*

The Concave Lens. Figure 49–3 shows the effect of a concave lens on parallel light rays. The rays that enter the very center of the lens strike an interface that is absolutely perpendicular to the beam and, therefore, do not refract at all. The rays at the edge of the lens enter the lens ahead of the rays toward the center. This is opposite to the effect in the convex lens, and it causes the peripheral light rays to *diverge* away from the light rays that pass through the center of the lens.

Thus, the concave lens *diverges* light rays, whereas the convex lens *converges* light rays.

Cylindrical Lenses—Comparison With Spherical Lenses. Figure 49–4 illustrates both a convex *spherical* lens and a convex *cylindrical* lens. Note that the cylindrical lens bends light rays from the two sides of the lens but not from either the top or the bottom. Therefore, parallel light rays are bent to a *focal line.* On the other hand, light rays that pass through the spherical lens are refracted at all edges of the lens toward the central ray, and all the rays come to a *focal point.*

The cylindrical lens is well illustrated by a test tube full of water. If the test tube is placed in a beam of sunlight and a piece of paper is brought progressively closer to the opposite

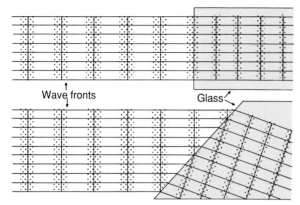

Figure 49 – 1. Wave fronts entering *(top)* a glass surface perpendicular to the light rays and *(bottom)* a glass surface angulated to the light rays. This figure illustrates that the distance between waves after they enter the glass is shortened to approximately two thirds that in air. It also illustrates that light rays striking an angulated glass surface are bent.

side of the tube, a certain distance will be found at which the light rays come to a *focal line*. On the other hand, the spherical lens is illustrated by an ordinary magnifying glass. If such a lens is placed in a beam of sunlight and a piece of paper is brought progressively closer to the lens, the light rays will impinge on a common focal point at an appropriate distance.

Concave cylindrical lenses *diverge* light rays in only one plane in the same manner that *convex* cylindrical lenses *converge* light rays in one plane.

Combination of Two Cylindrical Lenses to Equal a Spherical Lens. Figure 49–5 shows two convex cylindrical lenses at right angles to each other. The vertical cylindrical lens causes convergence of the light rays that pass through the two sides of the lens, and the horizontal lens converges the top and bottom rays. Thus, all the light rays come to a single-point focus. In other words, *two cylindrical lenses crossed at right angles to each other perform the same function as one spherical lens of the same refractive power.*

FOCAL LENGTH OF A LENS

The distance beyond a convex lens at which *parallel* rays converge to a common focal point is called the *focal length* of the lens. The diagram at the top of Figure 49–6 illustrates this focusing of parallel light rays.

In the middle diagram, the light rays that enter the convex lens are not parallel but are diverging because the origin of the light is a point source not far away from the lens itself. Because these rays are diverging outward from the point source, it can be seen from the diagram that they do not come to a point focus at the same distance away from the lens as do parallel rays. In other words, when rays of light

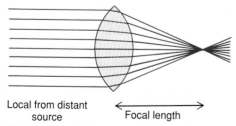

Figure 49 – 2. Bending of light rays at each surface of a convex spherical lens, showing that parallel light rays are focused to a point focus.

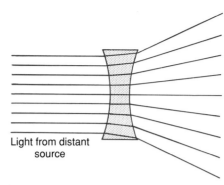

Figure 49 – 3. Bending of light rays at each surface of a concave spherical lens, illustrating that parallel light rays are diverged by a concave lens.

that are already diverging enter a convex lens, the distance of focus on the other side of the lens is farther from the lens than is the case when the entering rays are parallel to each other.

In the lower diagram of Figure 49–6 are shown light rays that are diverging toward a convex lens with far greater curvature than that of the upper two lenses of the figure. In this diagram the distance from the lens at which the light rays come to a focus is exactly the same as that from the lens in the first diagram, in which the lens was less convex but the rays entering it were parallel. This illustrates that both parallel rays and diverging rays can be focused at the same distance behind a lens provided the lens changes its convexity.

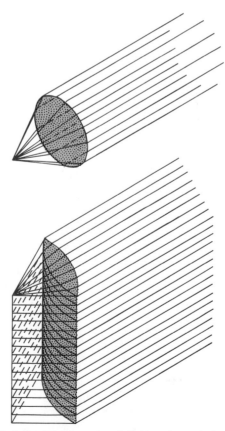

Figure 49 – 4. *Top:* Point focus of parallel light rays by a spherical convex lens. *Bottom:* Line focus of parallel light rays by a cylindrical convex lens.

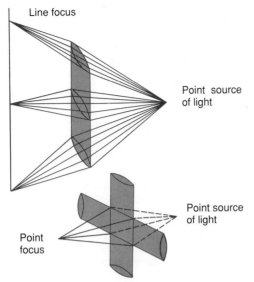

Line focus

Point source
of light

Point source
of light

Point
focus

Figure 49–5. Two cylindrical convex lenses at right angles to each other, illustrating that one lens converges light rays in one plane and the other lens converges light rays in the plane at right angles. The two lenses combined give the same point focus as that obtained with a spherical convex lens.

The relationship of focal length of the lens, distance of the point source of light, and distance of focus is expressed by the following formula:

$$\frac{1}{f} = \frac{1}{a} + \frac{1}{b}$$

in which *f* is the focal length of the lens, *a* the distance of the point source of light from the lens, and *b* the distance of focus from the lens.

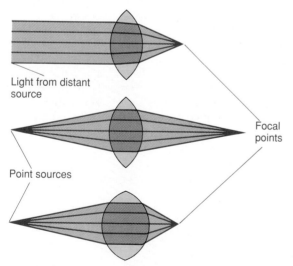

Light from distant
source

Point sources

Focal
points

Figure 49–6. The upper two lenses of this figure have the same strength, but the light rays entering the top lens are parallel, whereas those entering the second lens are diverging; the effect of parallel versus diverging rays on the focal distance is illustrated. The bottom lens has far more refractive power than either of the other two lenses, illustrating that the stronger the lens the nearer to the lens is the point focus.

FORMATION OF AN IMAGE BY A CONVEX LENS

The upper drawing of Figure 49–7 illustrates a convex lens with two point sources of light to the left. Because light rays pass through the center of a convex lens without being refracted in either direction, the light rays from each point source of light are shown to come to a point focus on the opposite side of the lens *directly in line with the point source and the center of the lens.*

Any object in front of the lens is in reality a mosaic of point sources of light. Some of these points are very bright, some are very weak, and they vary in color. And each point source of light on the object comes to a separate point focus on the opposite side of the lens in line with the lens center. Furthermore, all the focal points behind the lens will fall in a common plane a certain distance behind the lens. If a white piece of paper is placed at this distance, one can see an image of the object, as illustrated in the lower portion of Figure 49–7. However, this image is upside down with respect to the original object, and the two lateral sides of the image are reversed as well. This is the method by which the lens of a camera focuses images on the camera film.

MEASUREMENT OF THE REFRACTIVE POWER OF A LENS—THE DIOPTER

The more a lens bends light rays, the greater is its "refractive power." This refractive power is measured in terms of *diopters.* The refractive power of a convex lens is equal to 1 meter divided by its focal length. Thus, a spherical lens that converges parallel light rays to a focal point 1 meter beyond the lens has a refractive power of +1 diopter, as illustrated in Figure 49–8. If the lens is capable of bending parallel light rays twice as much as a lens with a power of +1 diopter, it is said to have a strength of +2 diopters, and the light rays come to a focal point 0.5 meter beyond the lens. A lens capable of converging parallel light rays to a focal point only 10 cm (0.10 meter) beyond the lens has a refractive power of +10 diopters.

The refractive power of concave lenses cannot be stated in terms of the focal distance beyond the lens because the light rays diverge, rather than focusing to a point. However, if a concave lens diverges light rays the same amount that a 1 diopter convex lens converges them, the concave lens is said to have a dioptric strength of −1. Likewise, if the concave lens diverges the light rays as much as a +10 diopter lens converges them, it is said to have a strength of −10 diopters.

Note particularly that concave lenses "neutralize" the refractive power of convex lenses. Thus, placing a 1 diopter concave lens immediately in front of a 1 diopter convex lens results in a lens system with zero refractive power.

The strengths of cylindrical lenses are computed in the same manner as the strengths of spherical lenses. If a cylindrical lens focuses parallel light rays to a line focus 1 meter beyond the lens, it has a strength of +1 diopter. On the other hand, if a cylindrical lens of a concave type *diverges* light rays as much as a +1 diopter cylindrical lens *converges* them, it has a strength of −1 diopter. However, the *axis* of the cylindrical lens must also be stated in addition to its strength.

● THE OPTICS OF THE EYE

THE EYE AS A CAMERA

The eye, illustrated in Figure 49–9, is optically equivalent to the usual photographic camera, for it

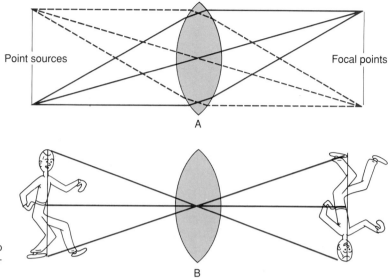

Figure 49–7. *A,* Two point sources of light focused at two separate points on the opposite side of the lens. *B,* Formation of an image by a convex spherical lens.

has a lens system, a variable aperture system (the pupil), and a retina that corresponds to the film. The lens system of the eye is composed of four refractive interfaces: (1) the interface between air and the anterior surface of the cornea, (2) the interface between the posterior surface of the cornea and the aqueous humor, (3) the interface between the aqueous humor and the anterior surface of the crystalline lens of the eye, and (4) the interface between the posterior surface of the lens and the vitreous humor. The refractive index of air is 1; the cornea, 1.38; the aqueous humor, 1.33; the crystalline lens (on the average), 1.40; and the vitreous humor, 1.34.

The Reduced Eye. If all the refractive surfaces of the eye are algebraically added together and then considered to be one single lens, the optics of the normal eye may be simplified and represented schematically as a "reduced eye." This is useful in simple calculations. In the reduced eye, a single refractive surface is considered to exist with its central point 17 mm in front of the retina and to have a total refractive power of approximately 59 diopters when the lens is accommodated for distant vision.

Most of the refractive power of the eye is provided not by the crystalline lens but instead by the anterior surface of the cornea. The principal reason for this is that the refractive index of the cornea is markedly different from that of air.

On the other hand, the total refractive power of the crystalline lens of the eye, as it normally lies in the eye surrounded by fluid on each side, is only 20 diopters, about one-third the total refractive power of the eye's lens system. If this lens were removed from the eye and then surrounded by air, its refractive power would be about six times as great. The reason for this difference is that the fluids surrounding the lens have refractive indices not greatly different from the refractive index of the lens itself, and the smallness of the differences greatly decreases the amount of light refraction at the lens interfaces. But the importance of the crystalline lens is that its curvature can be increased markedly to provide "accommodation," which will be discussed later in the chapter.

Formation of an Image on the Retina. In exactly the same manner that a glass lens can focus an image on a sheet of paper, the lens system of the eye can focus an image on the retina. The image is inverted and reversed with respect to the object. However, the mind perceives objects in the upright position despite the upside-down orientation on the retina because the brain is trained to consider an inverted image as the normal.

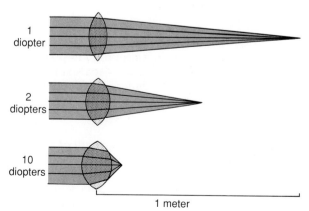

Figure 49–8. Effect of lens strength on the focal distance.

Figure 49–9. The eye as a camera. The numbers are the refractive indices.

THE MECHANISM OF ACCOMMODATION

The refractive power of the crystalline lens of the eye can be voluntarily increased from 20 diopters to approximately 34 diopters in young children; this is a total "accommodation" of 14 diopters. To do this, the shape of the lens is changed from that of a moderately convex lens to that of a very convex lens. The mechanism of this is the following:

In the young person, the lens is composed of a strong elastic capsule filled with viscous, proteinaceous, but transparent fibers. When the lens is in a relaxed state, with no tension on its capsule, it assumes an almost spherical shape, owing entirely to the elasticity of the lens capsule. However, as illustrated in Figure 49–10, approximately 70 ligaments (called *zonules*) attach radially around the lens, pulling the lens edges toward the anterior edges of the retina. These ligaments are constantly tensed by the elastic pull of their attachments to the ciliary body at the anterior border of the choroid. The tension on the ligaments causes the lens to remain relatively flat under normal resting conditions of the eye. At the insertions of the ligaments in the ciliary body is the *ciliary muscle,* which has two sets of smooth muscle fibers, the *meridional fibers* and the *circular fibers.* The meridional fibers extend to the corneoscleral junction. When these muscle fibers contract, the peripheral insertions of the lens ligaments are pulled forward, thereby releasing a certain amount of tension on the lens. The circular fibers are arranged circularly all the way around the eye so that when they contract, a sphincter-like action occurs, decreasing the diameter of the circle of ligament attachments and also allowing the ligaments to pull less on the lens capsule.

Thus, contraction of both sets of smooth muscle fibers in the ciliary muscle relaxes the ligaments to the lens capsule, and the lens assumes a more spherical shape, like that of a balloon, because of the natural elasticity of its capsule. Therefore, when the ciliary muscle is completely relaxed, the dioptric strength of the lens is as weak as it can become. On the other hand, when the ciliary muscle contracts as strongly as possible, the dioptric strength of the lens becomes maximal.

Autonomic Control of Accommodation. The ciliary muscle is controlled almost entirely by the parasympathetic nervous system. Stimulation of the parasympathetic nerves contracts the ciliary muscle, which relaxes the lens ligaments and increases the refractive power. With an increased refractive power, the eye is capable of focusing on objects nearer at hand than when the eye has less refractive power. Consequently, as a distant object moves toward the eye, the number of parasympathetic impulses impinging on the ciliary muscle must be progressively increased for the eye to keep the object constantly in focus. (Sympathetic stimulation has a weak effect in relaxing the ciliary muscle, but this plays almost no role in the normal accommodation mechanism, the neurology of which will be discussed in Chapter 51.)

Presbyopia. As a person grows older, the lens grows larger and thicker, and it also becomes less elastic, partly because of progressive denaturation of the lens proteins. Therefore, the ability of the lens to change shape progressively decreases, and the power of accommodation decreases from approximately 14 diopters in the young child to less than 2 diopters at the age of 45 to 50 and to about zero at age 70. Thereafter, the lens is then almost totally nonaccommodating, a condition known as "presbyopia."

Once a person has reached the state of presbyopia, each eye remains focused permanently at an almost constant distance; this distance depends on the physical characteristics of each individual's eyes. Obviously, the eyes can no longer accommodate for both near and far vision. Therefore, to see clearly both in the distance and nearby, an older person must wear bifocal glasses with the upper segment normally focused for far-seeing and the lower segment focused for near-seeing.

THE PUPILLARY APERTURE

A major function of the iris is to increase the amount of light that enters the eye during darkness and to decrease the light in bright light. The reflexes for controlling this mechanism will be considered in the discussion of the neurology of the eye in Chapter 51. The amount of light that enters the eye through the pupil is proportional to the *area* of the pupil or to the *square of the diameter* of the pupil. The pupil of the human eye can become as small as approximately 1.5 mm and as large as 8 mm in diameter. Therefore, the quantity of light entering the eye may vary approximately 30 times as a result of changes in pupillary aperture.

Depth of Focus of the Lens System of the Eye. Figure 49–11 illustrates two separate eyes that are exactly alike except for the diameters of the pupillary apertures. In the upper eye the pupillary aperture is small, and in the lower eye the aperture is large. In

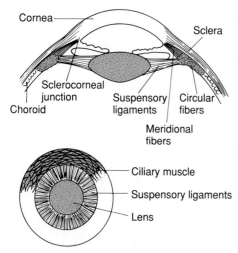

Figure 49–10. Mechanism of accommodation (focusing).

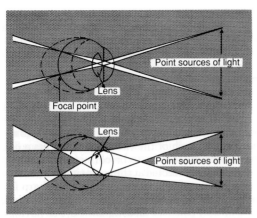

Figure 49–11. Effect of small and large pupillary apertures on the depth of focus.

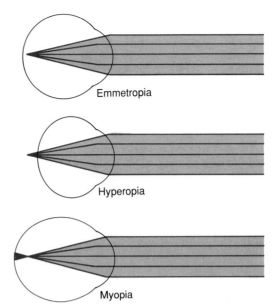

Figure 49–12. Parallel light rays focus on the retina in emmetropia, behind the retina in hyperopia, and in front of the retina in myopia.

front of each of these two eyes are two small point sources of light, and light from each passes through the pupillary aperture and focuses on the retina. Consequently, in both eyes the retina sees two spots of light in perfect focus. It is evident from the diagrams, however, that if the retina is moved forward or backward to an out-of-focus position, the size of each spot will not change much in the upper eye, but in the lower eye the size of each spot will increase greatly, becoming a "blur circle." In other words, the upper lens system has far greater *depth of focus* than the bottom lens system. When a lens system has great depth of focus, the retina can be considerably displaced from the focal plane and still the image remains in sharp focus; whereas when a lens system has a shallow depth of focus, moving the retina only slightly away from the focal plane causes extreme blurring.

The greatest possible depth of focus occurs when the pupil is extremely small. The reason for this is that with a very small aperture all light rays must pass through the center of the lens, and the centralmost rays are always in focus, as was explained earlier.

ERRORS OF REFRACTION

Emmetropia. As shown in Figure 49–12, the eye is considered to be normal, or "emmetropic," if parallel light rays from distant objects are in sharp focus on the retina *when the ciliary muscle is completely relaxed.* This means that the emmetropic eye can see all distant objects clearly, with its ciliary muscle relaxed, but to focus objects at close range it must contract its ciliary muscle and thereby provide various degrees of accommodation.

Hyperopia. Hyperopia, which is also known as "far-sightedness," is usually due either to an eyeball that is too short or occasionally to a lens system that is too weak when the ciliary muscle is relaxed. In this condition, as seen in the middle panel of Figure 49–12, parallel light rays are not bent sufficiently by the lens system to come to a focus by the time they reach the retina. To overcome this abnormality, the ciliary muscle may contract to increase the strength of the lens. Therefore, the far-sighted person is capable, by using his mechanism of accommodation, of focusing distant objects on the retina. If he has used only a small amount of

strength in his ciliary muscle to accommodate for the distant objects, then he still has much accommodative power left, and objects closer and closer to the eye can also be focused sharply until the ciliary muscle has contracted to its limit.

In old age, when the lens becomes presbyopic, the far-sighted person often is not able to accommodate his or her lens sufficiently to focus even distant objects, much less to focus near objects.

Myopia. In myopia, or "near-sightedness," when the ciliary muscle is completely relaxed the light rays coming from distant objects are focused in front of the retina, as shown in the lower panel of Figure 49–12. This is usually due to too long an eyeball but it can occasionally result from too much refractive power of the lens system of the eye.

No mechanism exists by which the eye can decrease the strength of its lens to less than that which exists when the ciliary muscle is completely relaxed. Therefore, the myopic person has no mechanism by which he can ever focus distant objects sharply on his retina. However, as an object comes nearer to his eye, it finally comes near enough that its image will focus. Then, when the object comes still closer to the eye, the person can use his mechanism of accommodation to keep the image focused clearly. Therefore, a myopic person has a definite limiting "far point" for clear vision.

Correction of Myopia and Hyperopia by Use of Lenses. It will be recalled that light rays passing through a concave lens diverge. Therefore, if the refractive surfaces of the eye have too much refractive power, as in *myopia,* some of this excessive refractive power can be neutralized by placing in front of the eye a concave spherical lens, which will diverge rays.

On the other hand, in a person who has *hyperopia*—that is, someone who has too weak a lens system—the abnormal vision can be corrected by adding refractive power with a convex lens in front of the eye. These corrections are illustrated in Figure 49–13.

One usually determines the strength of the concave or convex lens needed for clear vision by "trial and error"—that is, by trying first a strong lens and then a stronger or

DETERMINATION OF DISTANCE OF AN OBJECT FROM THE EYE—DEPTH PERCEPTION

The visual apparatus normally perceives distance by three major means. This phenomenon is known as *depth perception*. These means are (1) the size of the image of known objects on the retina, (2) the phenomenon of moving parallax, and (3) the phenomenon of stereopsis.

Determination of Distance by Sizes of Retinal Images of Known Objects. If one knows that a man whom he is viewing is 6 feet tall, he can determine how far away the man is simply by the size of the man's image on his retina. He does not consciously think about the size, but his brain has learned to calculate automatically from image sizes the distances of objects when the dimensions are known.

Determination of Distance by Moving Parallax. Another important means by which the eyes determine distance is that of moving parallax. If a person looks off into the distance with his eyes completely still, he perceives no moving parallax, but when he moves his head to one side or the other, the images of objects close to him move rapidly across his retinas while the images of distant objects remain rather stationary. For instance, if he moves his head 1 inch and an object is only 1 inch in front of his eye, the image moves almost all the way across his retinas, whereas the image of an object 200 feet away from his eyes does not move perceptibly. Thus, by this mechanism of moving parallax, one can tell the *relative distances* of different objects even though only one eye is used.

Determination of Distance by Stereopsis—Binocular Vision. Another method by which one perceives parallax is that of binocular vision. Because one eye is a little more than 2 inches to one side of the other eye, the images on the two retinas are different one from the other—that is, an object that is 1 inch in front of the bridge of the nose forms an image on the temporal portion of the retina of each eye, whereas a small object 20 feet in front of the nose has its image at closely corresponding points in the middle of each retina. This type of parallax is illustrated in Figure 49–17, which shows the images of a black spot and a square actually reversed on the two retinas because they are at different distances in front of the eyes. This gives a type of parallax that is present all the time when both eyes are being used. It is almost entirely this binocular parallax (or stereopsis) that gives a person with two eyes far greater ability to judge relative distances *when objects are nearby* than a person who has only one eye. However, stereopsis is virtually useless for depth perception at distances beyond 200 feet.

● OPTICAL INSTRUMENTS

THE OPHTHALMOSCOPE

The ophthalmoscope is an instrument through which an observer can look into another person's eye and see the retina with clarity. Though the ophthalmoscope appears to be a relatively complicated instrument, its principles are simple. The basic components are illustrated in Figure 49–18 and may be explained as follows.

If a bright spot of light is on the retina of an *emmetropic eye,* light rays from this spot diverge toward the lens system of the eye, and, after passing through the lens system, they are parallel with each other because the retina is located exactly one focal length distance behind the lens. Then, when these parallel rays pass into an emmetropic eye of another person, they focus back again to a point focus on the retina of the second person because his retina is also one focal length distance behind the lens. Therefore, any spot of light on the retina of the observed eye comes to a focal spot on the retina of the observing eye. Likewise, when the bright spot of light is moved to different points on the observed retina, the focal spot on the retina of the observer also moves an equal amount. Thus, if the retina of one person is made to emit light, the image of his retina will be focused on the retina of the observer provided the two eyes are simply looking into each other. These principles, of course, apply only to completely emmetropic eyes.

To make an ophthalmoscope, one need only devise a means for illuminating the retina to be examined. Then, the reflected light from that retina can be seen by the observer simply by putting the two eyes close to each other. To illuminate the retina of the observed eye, an angulated mirror or a segment of a prism is placed in front of the observed eye in such a manner, as illustrated in Figure 49–18, that light from a bulb is reflected into the observed eye. Thus, the retina is illuminated through the pupil, and the observer sees into the subject's pupil by looking over the edge of the

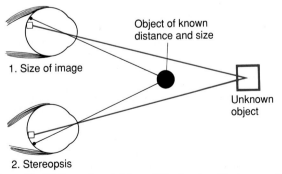

Figure 49–17. Perception of distance (1) by the size of the image on the retina and (2) as a result of stereopsis.

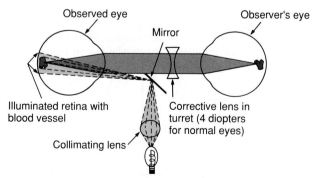

Figure 49–18. The optical system of the ophthalmoscope.

mirror or prism, or *through* an appropriately designed prism so that the light will not have to enter the pupil at an angle.

It was noted above that these principles apply only to persons with completely emmetropic eyes. If the refractive power of either eye is abnormal, it is necessary to correct this refractive power in order for the observer to see a sharp image of the observed retina. Therefore, the usual ophthalmoscope has a series of lenses mounted on a turret so that the turret can be rotated from one lens to another, and the correction for abnormal refractive power of either or both eyes can be made by selecting a lens of appropriate strength. In normal young adults, when the two eyes come close together, a natural accommodative reflex occurs that causes an approximate +2 diopter increase in the strength of the lens of each eye. To correct for this, it is necessary that the lens turret be rotated to an approximately −4 diopter correction.

● THE FLUID SYSTEM OF THE EYE— THE INTRAOCULAR FLUID

The eye is filled with *intraocular fluid,* which maintains sufficient pressure in the eyeball to keep it distended. Figure 49–19 illustrates that this fluid can be divided into two portions, the *aqueous humor,* which lies in front and to the sides of the lens, and the fluid of the *vitreous humor,* which lies between the lens and the retina. The aqueous humor is a freely flowing fluid, whereas the vitreous humor, sometimes called the *vitreous body,* is a gelatinous mass held together by a fine fibrillar network composed primarily of large proteoglycan molecules. Substances can *diffuse* slowly in the vitreous humor, but there is little *flow* of fluid.

Aqueous humor is continually being formed and reabsorbed. The balance between formation and reabsorption of aqueous humor regulates the total volume and pressure of the intraocular fluid.

FORMATION OF AQUEOUS HUMOR BY THE CILIARY BODY

Aqueous humor is formed in the eye *at an average rate of 2 to 3 microliters each minute.* Essentially all of this is se-

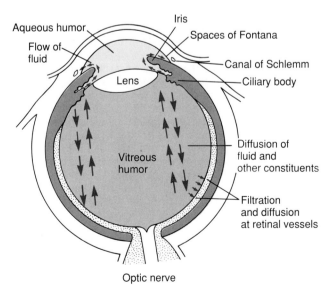

Figure 49–19. Formation and flow of fluid in the eye.

Figure 49–20. Anatomy of the ciliary processes.

creted by the *ciliary processes,* which are linear folds projecting from the *ciliary body* into the space behind the iris where the lens ligaments also attach to the eyeball. A cross-section of these ciliary processes is illustrated in Figure 49–20, and their relationship to the fluid chambers of the eye can be seen in Figure 49–19. Because of their folded architecture, the total surface area of the ciliary processes is approximately 6 square centimeters in each eye — a large area, considering the small size of the ciliary body. The surfaces of these processes are covered by epithelial cells, and immediately beneath these is a highly vascular area.

Aqueous humor is formed almost entirely as an active secretion of the epithelium lining the ciliary processes. Secretion begins with active transport of sodium ions into the spaces between the epithelial cells. The sodium ions in turn pull chloride and bicarbonate ions along with them to maintain electrical neutrality. Then all these ions together cause osmosis of water from the sublying tissue into the same epithelial intercellular spaces, and the resulting solution washes from the spaces onto the surfaces of the ciliary processes. In addition, several nutrients are transported across the epithelium by active transport or facilitated diffusion; these include amino acids, ascorbic acid, and probably also glucose.

OUTFLOW OF AQUEOUS HUMOR FROM THE EYE

After aqueous humor is formed by the ciliary processes, it flows, as shown in Figure 49–19, *between the ligaments of the lens,* then *through the pupil,* and finally *into the anterior chamber of the eye.* Here, the fluid flows into the *angle between the cornea and the iris* and thence through a meshwork of *trabeculae,* finally entering the *canal of Schlemm,* which empties into extraocular veins. Figure 49–21 illustrates the anatomical structures at the iridocorneal angle, showing that the spaces between the trabeculae extend all the way from the anterior chamber to the canal of Schlemm. The canal of Schlemm in turn is a thin-walled vein that extends circumferentially all the way around the eye. Its endothelial membrane is so porous that even large protein molecules, as well as small particulate matter up to the size of red blood cells, can pass from the anterior chamber into the canal of Schlemm. Even though the canal of Schlemm is actually a venous blood vessel, so much aqueous humor nor-

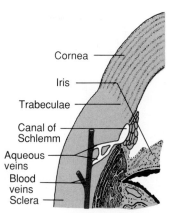

Figure 49–21. Anatomy of the iridocorneal angle, showing the system for outflow of aqueous humor into the conjunctival veins.

mally flows into it that it is filled only with aqueous humor rather than with blood. Also, the small veins that lead from the canal of Schlemm to the larger veins of the eye usually contain only aqueous humor, and these are called *aqueous veins.*

INTRAOCULAR PRESSURE

The average normal intraocular pressure is approximately 15 mm Hg, with a range of from 12 to 20.

Tonometry. Because it is impractical to pass a needle into a patient's eye for measurement of intraocular pressure, this pressure is measured clinically by means of a tonometer, the principle of which is illustrated in Figure 49–22. The cornea of the eye is anesthetized with a local anesthetic, and the footplate of the tonometer is placed on the cornea. A small force is then applied to a central plunger, causing the part of the cornea beneath the plunger to be displaced inward. The amount of displacement is recorded on the scale of the tonometer, and this in turn is calibrated in terms of intraocular pressure.

Regulation of Intraocular Pressure. Intraocular pressure remains very constant in the normal eye, normally within about ±2 mm Hg. The level of this pressure is determined mainly by the resistance to outflow of aqueous humor from the anterior chamber into the canal of Schlemm. This outflow resistance results from a meshwork of trabeculae through which the fluid must percolate on its way from the

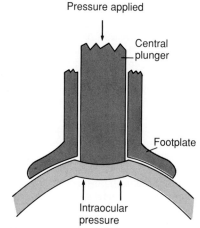

Figure 49–22. Principles of the tonometer.

lateral angles of the anterior chamber to the wall of the canal of Schlemm. These trabeculae have minute openings of only 2 to 3 micrometers. The rate of fluid flow into the canal increases markedly as the pressure rises. At approximately 15 mm Hg in the normal eye the amount of fluid leaving the eye by way of the canal of Schlemm averages 2.5 μl/min and exactly equals the inflow of fluid from the ciliary body. Therefore, normally the pressure remains at approximately this level of 15 mm Hg.

Cleansing of the Trabecular Spaces and of the Intraocular Fluid. When large amounts of debris occur in the aqueous humor, as occurs following hemorrhage into the eye or during intraocular infection, the debris is likely to accumulate in the trabecular spaces leading to the canal of Schlemm, therefore preventing adequate reabsorption of fluid from the anterior chamber and sometimes causing glaucoma, as explained subsequently. However, on the surfaces of the trabecular plates are large numbers of phagocytic cells. Also, immediately outside the canal of Schlemm is a layer of interstitial gel containing large numbers of reticuloendothelial cells that have an extremely high capacity for both engulfing debris and degrading it into small molecular substances that can then be absorbed. Thus, this phagocytic system keeps the trabecular spaces cleaned.

In addition, the surface of the iris and other surfaces of the eye behind the iris are covered with an epithelium that is capable of phagocytizing proteins and small particles from the aqueous humor, thereby helping maintain a perfectly clear fluid.

Glaucoma. Glaucoma is one of the most common causes of blindness. It is a disease of the eye in which the intraocular pressure becomes pathologically high, sometimes rising to as high as 60 to 70 mm Hg. Pressures rising above as little as 20 to 30 mm Hg can cause loss of vision when maintained for long periods of time. And the extremely high pressures can cause blindness within days or even hours. As the pressure rises, the axons of the optic nerve are compressed where they leave the eyeball at the optic disk. This compression is believed to block axonal flow of cytoplasm from the neuronal cell bodies in the retina to the peripheral optic nerve fibers entering the brain. The result is lack of appropriate nutrition, which eventually causes death of the involved neurons. It is possible that compression of the retinal artery, which also enters the eyeball at the optic disk, adds as well to the neuronal damage by reducing nutrition to the retina.

In most cases of glaucoma the abnormally high pressure results from increased resistance to fluid outflow through the trabecular spaces into the canal of Schlemm at the iridocorneal junction. For instance, in acute eye inflammation, white blood cells and tissue debris can block these spaces and cause acute increase in intraocular pressure. In chronic conditions, especially in older age, fibrous occlusion of the trabecular spaces appears to be the likely culprit.

Glaucoma can sometimes be treated by placing drops in the eye containing a drug that diffuses into the eyeball and causes reduced secretion or increased absorption of aqueous humor. However, when drug therapy fails, operative techniques to open the spaces of the trabeculae or to make channels directly between the fluid space of the eyeball and the subconjunctival space outside the eyeball can often effectively reduce the pressure.

REFERENCES

Apple, D. J., et al.: Intraocular Lenses: Evolution, Design, Complications and Pathology. Baltimore, Williams & Wilkins, 1988.

Bentley, P. J.: The crystalline lens of the eye: An optical microcosm. News in Physiol. Sci., 1:195, 1986.

Bill, A.: Blood circulation and fluid dynamics in the eye. Physiol. Rev., 55:383, 1975.

Bill, A.: Circulation in the eye. In Renkin, E. M., and Michel, C. C. (eds.): Handbook of Physiology. Sec. 2, Vol. IV. Bethesda, Md., American Physiological Society, 1984, p. 1001.

Caldwell, D. R.: Cataracts. New York, Raven Press, 1988.

Cavanagh, H. D.: The Cornea: Transactions of the World Congress on the Cornea III. New York, Raven Press, 1988.

Collins, R., and Van der Werff, T. J.: Mathematical Models of the Dynamics of the Human Eye. New York, Springer-Verlag, 1980.

Duncan, G., and Jacob, T. J.: Calcium and the physiology of cataract. Ciba Found. Symp., 106:132, 1984.

Elliot, R. H.: A Treatise on Glaucoma. Huntington, N.Y., R. E. Krieger, 1979.

Fischbarg, J., and Lim, J. J.: Fluid and electrolyte transports across corneal endothelium. Curr. Top. Eye Res., 4:201, 1984.

Guyton, D. L.: Sights and Sounds in Ophthalmology: Ocular Motility and Binocular Vision. St. Louis, C. V. Mosby Co., 1989.

Jaffe, N. S.: Cataract Surgery and Its Complications. St. Louis, C. V. Mosby Co., 1983.

Kavner, R. S., and Dusky, L.: Total Vision. New York, A & W Publishers, 1980.

Koretz, J. F., and Handelman, G. H.: How the human eye focuses. Sci. Am., July, 1988, p. 92.

Kuszak, J. R., et al.: Sutures of the crystalline lens: A review. Scan. Electron Miscrosc. 3:1369, 1984.

Lee, J. R.: Contact Lens Handbook. Philadelphia, W. B. Saunders Co., 1986.

Leydhecker, W., and Krieglstein, G. K. (eds.): Recent Advances in Glaucoma. New York, Springer-Verlag, 1979.

Michaels, D. D.: Basic Refraction Techniques. New York, Raven Press, 1988.

Moses, R. A.: Adler's Physiology of the Eye; Clinical Application. 7th Ed. St. Louis, C. V. Mosby, 1981.

Piatigorsky, J.: Lens crystallins and their genes: diversity and tissue-specific expression. FASEB J., 3:1933, 1989.

Ritch, R., et al. (eds.): The Glaucomas. St. Louis, C. V. Mosby Co., 1989.

Roth, H. W., and Roth-Wittig, M.: Contact Lenses. Hagerstown, MD, Harper & Row, 1980.

Safir, A. (ed.): Refraction and Clinical Optics. Hagerstown, MD, Harper & Row, 1980.

Shields, M. B.: Textbook of Glaucoma, 2nd Ed. Baltimore, Williams & Wilkins, 1986.

Stenson, S. M.: Contact Lenses: Guide to Selection, Fitting, and Management of Complications. East Norwalk, Conn., Appleton & Lange, 1987.

Toates, F. M.: Accommodation function of the human eye. Physiol. Rev., 52:828, 1972.

Whitnall, S. E. The Anatomy of the Human Orbit and Accessory Organs of Vision. Huntington, N. Y., R. E. Kreiger Publishing Co., 1979.

Wiederholt M.: Ion transport by the cornea. News in Physiol. Sci., 3:97, 1988.

Yellott, J. I., Jr., et al.: The beginnings of visual perception: The retinal image and its initial encoding. In Darian-Smith, I. (ed.): Handbook of Physiology. Sec. 1, Vol. III. Bethesda Md., American Physiological Society, 1984, p. 257.

CHAPTER 50

• *The Eye: II. Receptor and Neural Function of the Retina* •

The retina is the light-sensitive portion of the eye, containing the cones, which are responsible for color vision, and the rods, which are mainly responsible for vision in the dark. When the rods and cones are excited, signals are transmitted through successive neurons in the retina itself and finally into the optic nerve fibers and cerebral cortex. The purpose of the present chapter is to explain specifically the mechanisms by which the rods and cones detect both white and colored light and then convert the visual image into nerve impulses.

● ANATOMY AND FUNCTION OF THE STRUCTURAL ELEMENTS OF THE RETINA

The Layers of the Retina. Figure 50–1 shows the functional components of the retina arranged in layers from the outside to the inside as follows: (1) pigment layer, (2) layer of rods and cones projecting into the pigment, (3) outer lining membrane, (4) outer nuclear layer containing the cell bodies of the rods and cones, (5) outer plexiform layer, (6) inner nuclear layer, (7) inner plexiform layer, (8) ganglionic layer, (9) layer of optic nerve fibers, and (10) inner limiting membrane.

After light passes through the lens system of the eye and then through the vitreous humor, it enters the retina from the inside (see Figure 50–1); that is, it passes through the ganglion cells, the plexiform layers, the nuclear layer, and the limiting membranes before it finally reaches the layer of rods and cones located all the way on the outer side of the retina. This distance is a thickness of several hundred micrometers; visual acuity is obviously decreased by this passage through such nonhomogeneous tissue. However, in the central region of the retina, as will be discussed below, the initial layers are pulled aside for prevention of this loss of acuity.

The Foveal Region of the Retina and Its Importance in Acute Vision. A minute area in the center of the retina, illustrated in Figure 50–2, called the *macula* and occupying a total area of less than 1 square millimeter, is especially capable of acute and detailed vision. The central portion of the macula, only 0.4 millimeter in diameter, is called the *fovea;* this area is composed entirely of cones, and the cones have a special structure that aids their detection of detail in the visual image, especially a long slender body, in contradiction to much larger cones located further peripherally in the retina. Also, in this region the blood vessels, the ganglion cells, the inner nuclear layer of cells, and the plexiform layers are all displaced to one side rather than resting directly on top of the cones. This allows light to pass unimpeded to the cones.

The Rods and Cones. Figure 50–3 is a diagrammatic representation of a photoreceptor (either a rod or a cone), though the cones are distinguished by having a conical upper end (the outer segment) as shown in Figure 50–4. In general, the rods are narrower and longer than the cones, but this is not always the case. In the peripheral portions of the retina the rods are 2 to 5 micrometers in diameter, whereas the cones are 5 to 8 micrometers in diameter; in the central part of the retina, in the fovea, the cones have a diameter of only 1.5 micrometers.

To the right in Figure 50–3 are labeled the four major functional segments of either a rod or a cone: (1) the *outer segment,* (2) the *inner segment,* (3) the *nucleus,* and (4) the *synaptic body.* In the outer segment the light-sensitive photochemical is found. In the case of the rods, this is *rhodopsin,* and in the cones it is one of several "color" photochemicals that function almost exactly the same as rhodopsin except for differences in spectral sensitivity.

Note in both Figures 50–3 and 50–4 the large numbers of discs in both the rods and the cones. In the cones, each of the discs is actually an infolded shelf of cell membrane; in the rods this is also true near the base of the rod. However, toward the tip of the rod the discs separate from the membrane and are flat sacs lying totally inside the cell. There are as many as 1000 discs in each rod or cone.

Both rhodopsin and the color photochemicals are conjugated proteins. These are incorporated into the membranes of the discs in the form of transmembrane proteins. The concentrations of these photosensitive pigments in the disc are so great that they constitute approximately 40 per cent of the entire mass of the outer segment.

The inner segment contains the usual cytoplasm of the cell with the usual cytoplasmic organelles. Particularly important are the mitochondria; we see later that the mitochondria in this segment play an important role in providing the energy for function of the photoreceptors.

The synaptic body is the portion of the rod and cone that connects with the subsequent neuronal cells, the horizontal and bipolar cells, that represent the next stages in the vision chain.

OUTSIDE

- - - Pigmented layer

Layer of rods and cones
- Outer limiting membrane

Outer nuclear layer

Outer plexiform layer

Horizontal cell

Inner nuclear layer

Fiber of Müller

Amacrine cell

Inner plexiform layer

Ganglion cells

Ganglionic layer

Stratum opticum

Inner limiting membrane

DIRECTION
OF LIGHT

Figure 50-1. Plan of the retinal neurons. (Modified from Polyak: The Retina. ©1941 by The University of Chicago. All rights reserved.)

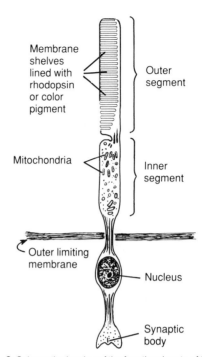

Membrane
shelves
lined with
rhodopsin
or color
pigment

Outer
segment

Mitochondria

Inner
segment

Outer limiting
membrane

Nucleus

Synaptic
body

Figure 50-3. Schematic drawing of the functional parts of the rods and cones.

The Pigment Layer of the Retina. The black pigment *melanin* in the pigment layer prevents light reflection throughout the globe of the eyeball; this is extremely important for clear vision. This pigment performs the same function in the eye as the black coloring inside the bellows of a camera. Without it, light rays would be reflected in all directions within the eyeball and would cause diffuse lighting of the retina rather than the contrast between dark and light spots required for formation of precise images.

The importance of melanin in the pigment layer and choroid is well illustrated by its absence in *albinos,* persons hereditarily lacking in melanin pigment in all parts of their bodies. When an albino enters a bright area, light that im-

pinges on the retina is reflected in all directions by the white unpigmented surfaces so that a single discrete spot of light that would normally excite only a few rods or cones is reflected everywhere and excites many of the receptors. Therefore, the visual acuity of albinos, even with the best of optical correction, is rarely better than 20/100 to 20/200.

The pigment layer also stores large quantities of *vitamin A.* This vitamin A is exchanged back and forth through the membranes of the outer segments of the rods and cones, which themselves are embedded in the pigment layers. We shall see later that vitamin A is an important precursor of the photosensitive pigments and that this interchange of vitamin A is very important for adjustment of the light sensitivity of the receptors.

Figure 50-2. Photomicrograph of the macula and of the fovea in its center. Note that the inner layers of the retina are pulled to the side to decrease the interference with light transmission. (From Fawcett: Bloom and Fawcett: A Textbook of Histology. 11th ed. Philadelphia, W. B. Saunders Company, 1986; courtesy of H. Mizoguchi.)

Figure 50–4. Membranous structures of the outer segments of a rod (left) and a cone (right). (Courtesy of Dr. Richard Young.)

The Blood Supply of the Retina—The Arterial System and the Choroid. The nutrient blood supply for the inner layers of the retina is derived from the central retinal artery, which enters the eyeball along with the optic nerve and then divides to supply the entire inner retinal surface. Thus, to a great extent, the retina has its own blood supply independent of the other structures of the eye.

However, the outer surface of the retina is adherent to the *choroid,* which is a highly vascular tissue between the retina and the sclera. The outer layers of the retina, including the outer segments of the rods and cones, depend mainly on diffusion from the choroid vessels for their nutrition, especially for their oxygen.

Retinal Detachment. The neural retina occasionally detaches from the pigment epithelium. In some instances the cause of such detachment is injury to the eyeball that allows fluid or blood to collect between the retina and the pigment epithelium, but often it is also caused by contracture of fine collagenous fibrils in the vitreous humor, which pull the retina unevenly toward the interior of the globe.

Fortunately, partly because of diffusion across the detachment gap and partly because of the independent blood supply to the retina through the retinal artery, the detached retina can resist degeneration for days and can become functional once again if surgically replaced in its normal relationship with the pigment epithelium. But, if not replaced soon, the retina finally is destroyed and is then unable to function even after surgical repair.

● PHOTOCHEMISTRY OF VISION

Both the rods and cones contain chemicals that decompose on exposure to light and, in the process, excite the nerve fibers leading from the eye. The chemical in the *rods* is called *rhodopsin,* and the light-sensitive chemicals in the *cones* have compositions only slightly different from that of rhodopsin.

In the present section we discuss principally the photochemistry of rhodopsin, but we can apply almost exactly the same principles to the photochemistry of the cones.

THE RHODOPSIN-RETINAL VISUAL CYCLE, AND EXCITATION OF THE RODS

Rhodopsin and Its Decomposition by Light Energy. The outer segment of the rod that projects into the pigment layer of the retina has a concentration of about 40 per cent of the light-sensitive pigment called *rhodopsin,* or *visual purple.* This substance is a combination of the protein *scotopsin* and the carotenoid pigment *retinal* (also called "retinene"). Furthermore, the retinal is a particular type called 11-*cis* retinal. This *cis* form of the retinal is important because only this form can bind with scotopsin to synthesize rhodopsin.

When light energy is absorbed by rhodopsin, the rhodopsin begins within trillionths of a second to decompose, as shown at the top of Figure 50–5. The cause of this is photoactivation of electrons in the retinal portion of the rhodopsin, which leads to an instantaneous change (on the order of trillionths of a second) of the *cis* form of retinal into an all-*trans* form, which still has the same chemical structure as the *cis* form but has a different physical structure—a straight molecule rather than a curved molecule. Be-

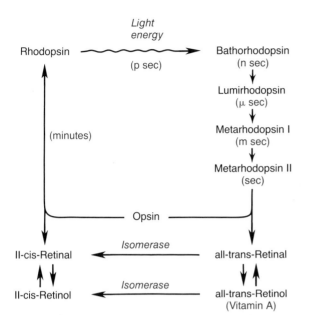

Figure 50–5. The rhodopsin-retinal visual cycle, showing decomposition of rhodopsin during exposure to light and subsequent slow reformation of rhodopsin by the chemical processes of the rod.

cause the three-dimensional orientation of the reactive sites of the all-*trans* retinal no longer fits with that of the reactive sites on the protein scotopsin, it begins to pull away from the scotopsin. The immediate product is *bathorhodopsin,* which is a partially split combination of the all-*trans* retinal and scotopsin. Bathorhodopsin itself is an extremely unstable compound and decays in nanoseconds to *lumirhodopsin.* This then decays in microseconds to *metarhodopsin I,* then in about a millisecond to *metarhodopsin II,* and, finally, much more slowly (in seconds) into the completely split products: *scotopsin* and *all-trans retinal.* It is the metarhodopsin II, also called *activated rhodopsin,* that excites electrical changes in the rods that then transmit the visual image into the central nervous system, as we discuss later.

Reformation of Rhodopsin. The first stage in reformation of rhodopsin, as shown in Figure 50–5, is to reconvert the all-*trans* retinal into 11-*cis* retinal. In the dark this process is catalyzed by the enzyme *retinal isomerase.* Once the 11-*cis* retinal is formed, it automatically recombines with the scotopsin to reform rhodopsin, which then remains stable until its decomposition is again triggered by absorption of light energy.

The Role of Vitamin A in the Formation of Rhodopsin. Note in Figure 50–5 that there is a second chemical route by which all-*trans* retinal can be converted into 11-*cis* retinal. This is by conversion of the all-*trans* retinal first into *all-trans retinol,* which is one form of vitamin A. Then, the all-*trans* retinol is converted into 11-*cis* retinol under the influence of the enzyme isomerase. And, finally, the 11-*cis* retinol is converted into 11-*cis* retinal.

Vitamin A is present both in the cytoplasm of the rods and in the pigment layer of the retina as well. Therefore, vitamin A is normally always available to form new retinal when needed. On the other hand, when there is excess retinal in the retina, the excess is converted back into vitamin A, thus reducing the amount of light-sensitive pigment in the retina. We shall see later that this interconversion between retinal and vitamin A is especially important in long-term adaptation of the retina to different light intensities.

Night Blindness. Night blindness occurs in severe vitamin A deficiency. The simple reason for this is that not enough vitamin A is then available to form adequate quantities of retinal. Therefore, the amounts of rhodopsin that can be formed in the rods, as well as the amounts of color-photosensitive chemicals in the cones, are all depressed. This condition is called night blindness because the amount of light available at night is then too little to permit adequate vision, though in daylight the cones especially can still be excited despite their reduction in photochemical substances.

For night blindness to occur, a person usually must remain on a vitamin A–deficient diet for months, because large quantities of vitamin A are normally stored in the liver and can be made available to the eyes. However, once night blindness does develop, it can sometimes be completely cured in less than an hour by intravenous injection of vitamin A.

Excitation of the Rod When Rhodopsin Is Activated

The Rod Receptor Potential Is Hyperpolarizing, Not Depolarizing. The rod receptor potential is different from the receptor potentials in almost all other sensory receptors. That is, excitation of the rod causes *increased negativity* of the membrane potential, which is a state of *hyperpolarization,* rather than decreased negativity, which is the process of "depolarization" that is characteristic of almost all other sensory receptors.

But how does the activation of rhodopsin cause hyperpolarization? The answer to this is that *when rhodopsin decomposes, it decreases the membrane conductance for sodium ions in the outer segment of the rod.* And this causes hyperpolarization of the entire rod membrane in the following way:

Figure 50–6 illustrates movement of sodium ions in a complete electrical circuit through the inner and outer segments of the rod. The inner segment continually pumps sodium from inside the rod to the outside, thereby creating a negative potential on the inside of the entire cell. However, the membrane of the outer segment, in the *dark* state, is very leaky to sodium. Therefore, sodium continually leaks back to the inside of the rod and thereby neutralizes much of the negativity on the inside of the entire cell. Thus, under normal conditions, when the rod is not excited there is a reduced amount of electronegativity inside the membrane of the rod, normally about −40 millivolts.

When the rhodopsin in the outer segment of the rod is exposed to light and begins to decompose, this *decreases* the outer segment conductance of sodium to the interior of the rod even though sodium ions continue to be pumped out of the inner segment. Thus,

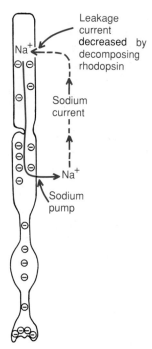

Figure 50–6. Theoretical basis for the generation of a hyperpolarization receptor potential caused by rhodopsin decomposition.

more sodium ions now leave the rod than leak back in. Because these are positive ions, their loss from inside the rod creates increased negativity inside the membrane; and the greater the amount of light energy striking the rod, the greater the electronegativity — that is, the greater the degree of *hyperpolarization.* At maximum light intensity, the membrane potential approaches -70 to -80 millivolts, which is near the equilibrium potential for potassium ions across the membrane.

Duration of the Receptor Potential and Logarithmic Relationship of the Receptor Potential to Light Intensity. When a sudden pulse of light strikes the retina, the transient hyperpolarization that occurs in rods — that is, the receptor potential that occurs — reaches a peak in about 0.3 second and lasts for more than a second. In cones these changes occur four times as fast. Therefore, a visual image impinged on the retina for only a millionth of a second nevertheless can cause the sensation of seeing the image sometimes for longer than a second.

Another characteristic of the receptor potential is that it is approximately proportional to the logarithm of the light intensity. This is exceedingly important, because it allows the eye to discriminate light intensities through a range many thousand times as great as would be possible otherwise.

Mechanism by Which Rhodopsin Decomposition Decreases Membrane Sodium Conductance — The Excitation "Cascade." Under optimal conditions, a single photon of light, the smallest possible quantal unit of light energy, can cause a measurable receptor potential in a rod of about 1 millivolt. Only thirty photons of light will cause half saturation

of the rod. How can such a small amount of light cause such great excitation? The answer is that the photoreceptors have an extremely sensitive chemical cascade that amplifies the stimulatory effects about a millionfold, as follows:

1. The *photon activates an electron* in the *11*-cis *retinal* portion of the rhodopsin; this leads to the formation of *metarhodopsin II,* which is the active form of rhodopsin, as already discussed and illustrated in Figure 50–5.

2. The *activated rhodopsin* functions as an enzyme to activate many molecules of *transducin,* a protein present in an inactive form in the membranes of the discs and cell membrane of the rod.

3. The *activated transducin* in turn activates many more molecules of *phosphodiesterase.*

4. *Activated phosphodiesterase* is another enzyme; it immediately hydrolyzes many, many molecules of *cyclic* guanosine monophosphate (cGMP), thus destroying it. Before being destroyed, the cGMP had been bound with the sodium channel protein in a way to "splint" it in the open state, allowing continued rapid influx of sodium ions during dark conditions. But in light, when phosphodiesterase hydrolyzes the cGMP, this removes the splinting and causes the sodium channels to close. Several hundred channels close for each originally activated molecule of rhodopsin. Because the sodium flux through each of these channels is extremely rapid, flow of more than a million sodium ions is blocked by the channel closure before the channel opens again. This diminishment of sodium ion flow is what excites the rod, as already discussed.

5. Within a small fraction of a second, another enzyme, *rhodopsin kinase,* that is always present in the rod, inactivates the activated rhodopsin, and the entire cascade reverses back to the normal state with open sodium channels.

Thus, the rods have invented an important chemical cascade that amplifies the effect of a single photon of light to cause movement of millions of sodium ions. This explains the extreme sensitivity of the rods under dark conditions.

The cones are about 300 times less sensitive than the rods, but even this allows color vision in any light greater than very dim twilight.

PHOTOCHEMISTRY OF COLOR VISION BY THE CONES

It was pointed out at the outset of this discussion that the photochemicals in the cones have almost exactly the same chemical composition as that of rhodopsin in the rods. The only difference is that the protein portions, the opsins, called *photopsins* in the cones, are different from the scotopsin of the rods. The retinal portion is exactly the same in the cones as in the rods. The color-sensitive pigments of the cones, therefore, are combinations of retinal and photopsins.

In the discussion of color vision later in the chapter, it will become evident that three different types of

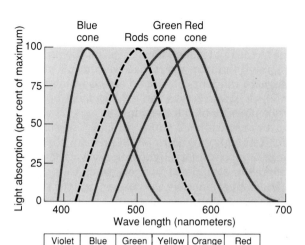

Figure 50–7. Light absorption by the respective pigments of the three color-receptive cones of the human retina. (Drawn from curves recorded by Marks, Dobelle, and MacNichol, Jr.: *Science, 143*:1181, 1964, and by Brown and Wald: *Science, 144*:45, 1964. Copyright 1964 by the American Association for the Advancement of Science.)

photochemicals are present in different cones, thus making these cones selectively sensitive to the different colors of blue, green, and red. These photochemicals are called, respectively, *blue-sensitive pigment, green-sensitive pigment,* and *red-sensitive pigment.* The absorption characteristics of the pigments in the three types of cones show peak absorbancies at light wavelengths, respectively, of 445, 535, and 570 nanometers. These are also the wavelengths for peak light sensitivity for each type of cone, which begins to explain how the retina differentiates the colors. The approximate absorption curves for these three pigments are shown in Figure 50–7. Also shown is the absorption curve for the rhodopsin of the rods, having a peak at 505 nanometers.

AUTOMATIC REGULATION OF RETINAL SENSITIVITY—DARK AND LIGHT ADAPTATION

Relationship of Sensitivity to Pigment Concentration. The sensitivity of rods is approximately proportional to the antilogarithm of the rhodopsin concentration, and it is assumed that this relationship also holds true in the cones. Therefore, the sensitivity of the rods and cones can be altered up or down tremendously by only slight changes in concentrations of the photosensitive chemicals.

Light and Dark Adaptation. If a person has been in bright light for a long time, large proportions of the photochemicals in both the rods and cones have been reduced to retinal and opsins. Furthermore, much of the retinal of both the rods and cones has also been converted into vitamin A. Because of these two effects, the concentrations of the photosensitive chemicals are considerably reduced, and the sensitivity of the eye to light is even more reduced. This is called *light adaptation.*

On the other hand, if the person remains in darkness for a long time, the retinal and opsins in the rods and cones are converted back into the light-sensitive pigments. Furthermore, vitamin A is reconverted back into retinal to give still additional light-sensitive pigments, the final limit being determined by the amount of opsins in the rods and cones. This is called *dark adaptation.*

Figure 50–8 illustrates the course of dark adaptation when a person is exposed to total darkness after having been exposed to bright light for several hours. Note that sensitivity of the retina is very low on first entering the darkness, but within 1 minute the sensitivity has increased tenfold—that is, the retina can respond to light of one-tenth the previously required intensity. At the end of 20 minutes the sensitivity has increased about 6000-fold, and at the end of 40 minutes it has increased about 25,000-fold.

The resulting curve of Figure 50–8 is called the *dark adaptation curve.* Note, however, the inflection in the curve. The early portion of the curve is caused by adaptation of the cones, for all of the chemical events of vision occur about 4 times as rapidly in cones as in rods. On the other hand, the cones do not achieve anywhere near the same degree of sensitivity as the rods. Therefore, despite rapid adaptation by the cones, they cease adapting after only a few minutes, while the slowly adapting rods continue to adapt for many minutes and even hours, their sensitivity increasing tremendously. In addition, a large share of the greater sensitivity of the rods is also caused by convergence of as many as 100 or more rods onto a single ganglion cell in the retina; these rods summate to increase their sensitivity, as will be discussed later in the chapter.

Other Mechanisms of Light and Dark Adaptation. In addition to adaptation caused by changes in concentrations of rhodopsin or color photochemicals, the eye has two other mechanisms for light and dark adaptation. The first of these is a *change in pupillary size,* which was discussed in the previous chapter. This can cause adaptation of approxi-

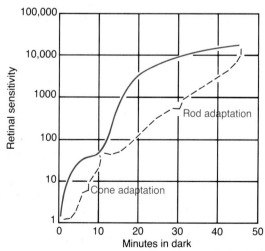

Figure 50–8. Dark adaptation, illustrating the relationship of cone adaptation to rod adaptation.

mately 30-fold because of changes in the amount of light allowed through the pupillary opening.

The other mechanism is *neural adaptation,* involving the neurons in the successive stages of the visual chain in the retina itself. That is, when the light intensity first increases, the intensities of the signals transmitted by the bipolar cells, the horizontal cells, the amacrine cells, and the ganglion cells are all very intense. However, the intensities of most of these signals decrease rapidly. Although the degree of this adaptation is only a fewfold rather than the many thousand-fold that occurs during adaptation of the photochemical system, this neural adaptation occurs in a fraction of a second, in contrast to the many minutes required for full adaptation by the photochemicals.

Value of Light and Dark Adaptation in Vision. Between the limits of maximal dark adaptation and maximal light adaptation, the eye can change its sensitivity to light by as much as 500,000 to 1,000,000 times, the sensitivity automatically adjusting to changes in illumination.

Since the registration of images by the retina requires detection of both dark and light spots in the image, it is essential that the sensitivity of the retina always be adjusted so that the receptors respond to the lighter areas but not to the darker areas. An example of maladjustment of the retina occurs when a person leaves a movie theater and enters the bright sunlight, for even the dark spots in the images then seem exceedingly bright, and, as a consequence, the entire visual image is bleached, having little contrast between its different parts. Obviously, this is poor vision, and it remains poor until the retina has adapted sufficiently for the darker areas of the image no longer to stimulate the receptors excessively.

Conversely, when a person enters darkness, the sensitivity of the retina is usually so slight that even the light spots in the image cannot excite the retina. After dark adaptation, however, the light spots begin to register. As an example of the extremes of light and dark adaptation, the intensity of sunlight is approximately 10 billion times that of starlight; yet the eye can function both in bright sunlight and in starlight.

● COLOR VISION

From the preceding sections, we know that different cones are sensitive to different colors of light. The present section is a discussion of the mechanisms by which the retina detects the different gradations of color in the visual spectrum.

THE TRICOLOR MECHANISM OF COLOR DETECTION

All the theories of color vision are based on the well-known observation that the human eye can detect almost all gradations of colors when red, green, and blue monochromatic lights are appropriately mixed in different combinations.

Spectral Sensitivities of the Three Types of Cones. On the basis of color vision tests, the spectral

sensitivities of the three different types of cones in human beings have been proved to be essentially the same as the light absorption curves for the three types of pigment found in the respective cones. These were illustrated in Figure 50-7 and are also shown in Figure 50-9. These curves can explain most but not all the phenomena of color vision.

Interpretation of Color in the Nervous System. Referring to Figure 50-9, one can see that an orange monochromatic light with a wavelength of 580 nanometers stimulates the red cones to a stimulus value of approximately 99 (99 per cent of the peak stimulation at optimum wavelength), whereas it stimulates the green cones to a stimulus value of approximately 42 and the blue cones not at all. Thus, the ratios of stimulation of the three different types of cones in this instance are 99:42:0. The nervous system interprets this set of ratios as the sensation of orange. On the other hand, a monochromatic blue light with a wavelength of 450 nanometers stimulates the red cones to a stimulus value of 0, the green cones to a value of 0, and the blue cones to a value of 97. This set of ratios—0:0:97—is interpreted by the nervous system as blue. Likewise, ratios of 83:83:0 are interpreted as yellow, and 31:67:36 as green.

Perception of White Light. Approximately equal stimulation of all the red, green, and blue cones gives one the sensation of seeing white. Yet there is no wavelength of light corresponding to white; instead, white is a combination of all the wavelengths of the spectrum. Furthermore, the sensation of white can be achieved by stimulating the retina with a proper combination of only three chosen colors that stimulate the respective types of cones approximately equally.

Failure of Changes in the Color of an Illuminating Light to Alter the Perceived Colors of a Visual Scene —The Phenomenon of Color Constancy. When Edwin Land was developing the Polaroid color camera, he noted that changing the color of a light illuminating a scene altered the hue of a color picture taken by the camera but did not significantly alter the hue of the scene as observed by

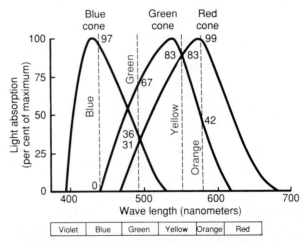

Figure 50-9. Demonstration of the degree of stimulation of the different color-sensitive cones by monochromatic lights of four separate colors: blue, green, yellow, and orange.

the human eye under the same changed light conditions. This phenomenon is called *color constancy;* to this day it has not been completely explained. What is believed to occur is the following:

First, the brain computes from all the colors in the scene the overall hue of the entire vision. This computation is helped when some areas in the picture are known by the person to be white. Using this information of overall hue, the brain then adjusts mathematically for the changed color of the illuminating light, though the exact neural mechanism for doing this has not been explained. Spotted throughout the primary visual cortex of the brain are irregular peg-shaped blocks of cells called by the simple name "blobs" that demonstrate color constancy when the illuminating light changes its wavelength. Therefore, it is believed that somewhere in the neighborhood of these blobs is located the computational machinery that allows for this phenomenon of color constancy.

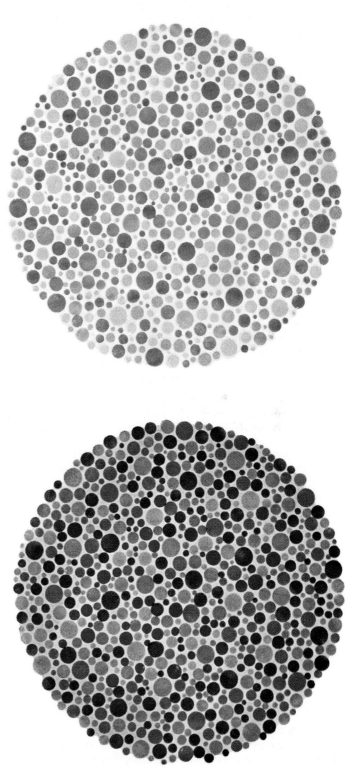

Figure 50–10. Two Ishihara charts. *Upper:* In this chart, the normal person reads "74," whereas the red-green color blind person reads "21." *Lower:* In this chart, the red-blind person (protanope) reads "2," whereas the green-blind person (deuteranope) reads "4." The normal person reads "42." (From Ishihara: Tests for Colour-Blindness. Tokyo, Kanehara and Co.)

Obviously, color constancy is of survival value to the foraging animal when he must distinguish nutritious food from poisonous plants both in bright sunlight and in the rose-colored dawn of day.

COLOR BLINDNESS

Red-Green Color Blindness. When a single group of color receptive cones is missing from the eye, the person is unable to distinguish some colors from others. For instance, one can see in Figure 50–9 that green, yellow, orange, and red colors, which are the colors between the wavelengths of 525 and 675 nanometers, are normally distinguished one from the other entirely by the red and the green cones. If either of these two cones is missing, one no longer can use this mechanism for distinguishing these four colors; the person is especially unable to distinguish red from green and therefore is said to have *red-green color blindness.*

The person with loss of red cones is called a *protanope;* his overall visual spectrum is noticeably shortened at the long wavelength end because of lack of the red cones. The color blind person who lacks green cones is called a *deuteranope;* this person has a perfectly normal visual spectral width because the absent green cones operate in the middle of the spectrum.

Red-green color blindness is a genetic disease in males that is transmitted through the female. That is, genes in the female X chromosome code for the respective cones. Yet color blindness almost never occurs in the female because at least one of her two X chromosomes will almost always have normal genes for all the cones. But the male has only one X chromosome, so that a missing gene will lead to color blindness in him.

Since the X chromosome in the male is always inherited from the mother, never from the father, color blindness is passed from mother to son, and the mother is said to be a *color blindness* carrier; this is the case for about 8 per cent of all women.

Blue Weakness. Only rarely are blue cones missing, though sometimes they are underrepresented, which is a state also genetically inherited, giving rise to the phenomenon called blue weakness.

Color Test Charts. A rapid method for determining color blindness is based on the use of spot-charts such as those illustrated in Figure 50–10. These charts are arranged with a confusion of spots of several different colors. In the top chart, the normal person reads "74," while the red-green color blind person reads "21." In the bottom chart, the normal person reads "42," while the red blind "protanope" reads "2," and the green blind "deuteranope" reads "4."

If one will study these charts while at the same time observing the spectral sensitivity curves of the different cones in Figure 50–9, it can be readily understood how excessive emphasis can be placed on spots of certain colors by color blind persons in comparison with normal persons.

● THE NEURAL FUNCTION OF THE RETINA

THE NEURAL CIRCUITRY OF THE RETINA

The first figure of this chapter, Figure 50–1, illustrated the tremendous complexity of neural organization in the retina. To simplify this, Figure 50–11

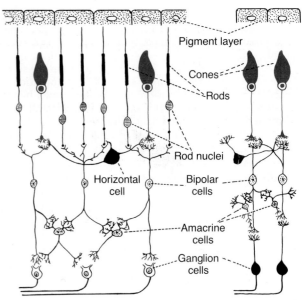

Figure 50–11. Neural organization of the retina: peripheral area to the left, foveal area to the right.

presents the basic essentials of the retina's neural connections. The different neuronal cell types are:

1. The photoreceptors themselves: the *rods* and *cones.*

2. The *horizontal cells,* which transmit signals horizontally in the outer plexiform layer from the rods and cones to the bipolar cell dendrites.

3. The *bipolar cells,* which transmit signals from the rods, cones, and horizontal cells to the inner plexiform layer, where they synapse with either amacrine cells or ganglion cells.

4. The *amacrine cells,* which transmit signals in two directions, either directly from bipolar cells to ganglion cells or horizontally within the inner plexiform layer between the axons of the bipolar cells, the dendrites of the ganglion cells, and/or other amacrine cells.

5. The *ganglion cells,* which transmit output signals from the retina through the optic nerve into the brain.

Still a sixth type of neuronal cell in the retina is the *interplexiform* cell. This cell transmits signals in the retrograde direction from the inner plexiform layer to the outer plexiform layer. These signals are all inhibitory and are believed to control the lateral spread of visual signals by the horizontal cells in the outer plexiform layer. Their role possibly is to control the degree of contrast in the visual image.

The Direct Visual Pathways From the Receptors to the Ganglion Cells. As is true of many of our sensory systems, the retina has both a very old type of vision based on rod vision and a new type of vision based on cone vision. The neurons and nerve fibers that conduct the visual signals for cone vision are considerably larger than those for rod vision, and the signals are conducted to the brain two to five times as rapidly. Also, the circuitries for the two systems are slightly different as follows:

To the far right of Figure 50–11 is illustrated the visual pathway from the foveal portion of the retina, representing the new, fast system. This shows three neurons in the direct pathway: (1) cones, (2) bipolar cells, and (3) ganglion cells. In addition, horizontal cells transmit inhibitory signals laterally in the outer plexiform layer, and amacrine cells transmit signals laterally in the inner plexiform layer.

To the left in Figure 50–11 is illustrated the neural connections for the peripheral retina where both rods and cones are present. Three bipolar cells are shown; the middle of these connects only to rods, representing the old visual system. In this case, the output from the bipolar cell passes only to amacrine cells, and these in turn relay the signals to the ganglion cells. Thus, for pure rod vision there are four neurons in the direct visual pathway: (1) rods, (2) bipolar cells, (3) amacrine cells, and (4) ganglion cells. Also, both horizontal and amacrine cells provide lateral connectivity.

The other two bipolar cells illustrated in the peripheral retinal circuitry of Figure 50–11 connect with both rods and cones; the outputs of these bipolar cells pass both directly to ganglion cells and also by way of amacrine cells.

Neurotransmitters Released by Retinal Neurons. The neurotransmitters employed for synaptic transmission in the retina still have not all been delineated clearly. However, it is believed that both the rods and the cones release *glutamate,* an excitatory transmitter, at their synapses with the bipolar and horizontal cells. And histological and pharmacological studies have shown there to be many different types of amacrine cells secreting at least five different types of transmitter substances: *gamma-aminobutyric acid (GABA), glycine, dopamine, acetylcholine,* and *indolamine,* all of which normally function as inhibitory transmitters. The transmitters of the bipolar, horizontal, and interplexiform cells are still unknown.

Transmission of Most Signals Occurs in the Retina by Electrotonic Conduction, not by Action Potentials. The only retinal neurons that always transmit visual signals by means of action potentials are the ganglion cells; and these send their signals all the way to the brain. Occasionally, though, action potentials have also been recorded in amacrine cells, though the importance of these action potentials is questionable. Otherwise, all the retinal neurons conduct their visual signals by *electrotonic conduction,* which can be explained as follows:

Electrotonic conduction means direct flow of electrical current, not action potentials, in the neuronal cytoplasm from the point of excitation all the way to the output synapses. In fact, even in the rods and cones, conduction from their outer segments where the visual signals are generated to the synaptic bodies is by electrotonic conduction. That is, when hyperpolarization occurs in response to light in the outer segment, approximately the same degree of hyperpolarization is conducted by direct electrical current flow to

the synaptic body, and no action potential at all occurs. Then, when the transmitter from a rod or cone stimulates a bipolar cell or horizontal cell, once again the signal is transmitted from the input to the output by direct electrical current flow, not by action potentials. Electrotonic conduction is also the means of signal transmission in most, if not all, of the different types of amacrine cells as well.

The importance of electrotonic conduction is that it allows *graded conduction* of signal strength. Thus, for the rods and cones, the hyperpolarizing output signal is directly related to the intensity of illumination; the signal is not all-or-none, as would be the case for action potential conduction.

Lateral Inhibition to Enhance Visual Contrast—Function of the Horizontal Cells

The horizontal cells, illustrated in Figure 50–11, connect laterally between the synaptic bodies of the rods and cones and also with the dendrites of the bipolar cells. The outputs of the horizontal cells are always inhibitory. Therefore, this lateral connection provides the same phenomenon of lateral inhibition that is important in all other sensory systems, that is, allowing faithful transmission of visual patterns into the central nervous system. This phenomenon is illustrated in Figure 50–12, which shows a very minute spot of light focused on the retina. The visual pathway from the centralmost area where the light strikes is excited, whereas the area to the side, called the "surround," is inhibited. In other words, instead of the excitatory signal spreading widely in the retina because of the spreading dendritic and axonal trees in the plexiform layers, transmission through the horizontal cells puts a stop to this by providing lateral inhibition in the surrounding area. This is an essential mechanism allowing high visual accuracy in transmitting contrast borders in the visual image. It is

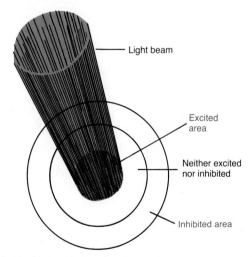

Figure 50–12. Excitation and inhibition of a retinal area caused by a small beam of light.

probable that some of the amacrine cells provide additional lateral inhibition and further enhancement of visual contrast in the inner plexiform layer of the retina as well.

Excitation of Some Bipolar Cells and Inhibition of Others — The Depolarizing and Hyperpolarizing Bipolar Cells

Two different types of bipolar cells provide opposing excitatory and inhibitory signals in the visual pathway, the *depolarizing bipolar cell* and the *hyperpolarizing bipolar cell*. That is, some bipolar cells depolarize when the rods and cones are excited, and others hyperpolarize.

There are two possible explanations for this difference in response of the two different types of bipolar cells. One explanation is that the two bipolar cells are of entirely different types, one responding to the glutamate neurotransmitter released by the rods and cones by depolarizing and the other responding by hyperpolarizing. The other possibility is that one of the bipolar cells receives direct excitation from the rods and cones, whereas the other receives its signal indirectly through a horizontal cell. Because the horizontal cell is an inhibitory cell, this would reverse the polarity of the electrical response.

Regardless of the mechanism for the two different types of bipolar responses, the importance of this phenomenon is that it allows half of the bipolar cells to transmit positive signals and the other half to transmit negative signals. We shall see later that both positive and negative signals are used in transmitting visual information to the brain.

Another importance of this reciprocal relationship between depolarizing and hyperpolarizing bipolar cells is that it provides a second mechanism for lateral inhibition in addition to the horizontal cell mechanism. Since depolarizing and hyperpolarizing bipolar cells lie immediately against each other, this gives an extremely acute mechanism for separating contrast borders in the visual image even when the border lies exactly between two adjacent photoreceptors.

The Amacrine Cells and Their Functions

About 30 different types of amacrine cells have been identified by morphological or histochemical means. The functions of a half dozen different types of amacrine cells have been characterized, and all of these are different from each other. It is probable that other amacrine cells have many additional functions yet to be determined.

One type of amacrine cell is part of the direct pathway for rod vision — that is, from rod to bipolar cells to amacrine cells to ganglion cells.

Another type of amacrine cell responds very strongly at the onset of a visual signal, but the response dies out rapidly. Other amacrine cells respond very strongly at the offset of visual signals, but again the response dies quickly. Finally, still other amacrine cells respond both when a light is turned on or off, signalling simply a change in illumination irrespective of direction.

Still another type of amacrine cell responds to movement of a spot across the retina in a specific direction; therefore, these amacrine cells are said to be *directional sensitive*.

In a sense, then, amacrine cells are types of interneurons that help in the beginning analysis of visual signals before they ever leave the retina.

THE GANGLION CELLS

Connectivity of the Ganglion Cells With Cones in the Fovea and With Rods and Cones in the Peripheral Retina. Each retina contains about 100,000,000 rods and 3,000,000 cones; yet the number of ganglion cells is only about 1,600,000. Thus, an average of 60 rods and 2 cones converge on each optic nerve fiber.

However, major differences exist between the peripheral retina and the central retina. As one approaches the fovea, fewer rods and cones converge on each optic fiber, and the rods and cones both become slenderer. These two effects progressively increase the acuity of vision toward the central retina. And in the very center, in the *fovea* itself, there are only slender cones, about 35,000 of them, and no rods at all. Also, the number of optic nerve fibers leading from this part of the retina is almost equal to the number of cones, as illustrated to the right in Figure 50–11. This mainly explains the high degree of visual acuity in the central retina in comparison with much poorer acuity peripherally.

Another difference between the peripheral and central portions of the retina is a much greater sensitivity of the peripheral retina to weak light. This results partly from the fact that rods are about 300 times more sensitive to light than are cones, but it is further magnified by the fact that as many as 200 rods converge on the same optic nerve fiber in the more peripheral portions of the retina, so that the signals from the rods summate to give even more intense stimulation of the peripheral ganglion cells.

Three Different Types of Retinal Ganglion Cells and Their Respective Fields

There are three distinct groups of ganglion cells designated as W, X, and Y cells. Each of these serves a different function:

Transmission of Rod Vision by the W Cells. The W cells, constituting about 40 per cent of all the ganglion cells, are small, having a diameter less than 10 micrometers and transmitting signals in their optic nerve fibers at the slow velocity of only 8 m/sec. These ganglion cells receive most of their excitation from rods, transmitted by way of small bipolar cells and amacrine cells. They have very broad fields in the retina because their dendrites spread widely in the inner plexiform layer, receiving signals from broad areas.

On the basis of histology as well as physiological experiments, it appears that the W cells are especially sensitive for detecting directional movement anywhere in the field of vision, and they probably also are important for much of our rod vision under dark conditions.

Transmission of the Visual Image and Color by the X Cells. The most numerous of the ganglion cells are the X cells, representing 55 per cent of the total. They are of medium diameter, between 10 and 15 micrometers, and transmit signals in their optic nerve fibers at about 14 m/sec.

The X cells have very small fields because their dendrites do not spread widely in the retina. Because of this, the signals represent rather discrete retinal locations. Therefore, it is through the X cells that the visual image itself is mainly transmitted. Also, because every X cell receives input from at least one cone, X cell transmission is probably responsible for all color vision as well.

Function of the Y Cells to Transmit Instantaneous Changes in the Visual Image. The Y cells are the largest of all, up to 35 micrometers in diameter, and they transmit their signals to the brain faster than 50 m/sec. However, they are also the fewest of all the ganglion cells, representing only 5 per cent of the total. Yet they have very broad dendritic fields, so that signals are picked up by these cells from widespread retinal areas.

The Y ganglion cells respond like many of the amacrine cells to rapid changes in the visual image, either rapid movement or rapid change in light intensity, sending bursts of signals for only a fraction of a second before the signal dies out. Therefore, these ganglion cells undoubtedly apprise the central nervous system almost instantaneously when an abnormal visual event occurs anywhere in the visual field, but without specifying with great accuracy the location of the event other than to give appropriate clues for moving the eyes toward the exciting vision.

EXCITATION OF THE GANGLION CELLS

Spontaneous, Continuous Action Potentials in the Ganglion Cells. It is from the ganglion cells that the long fibers of the optic nerve lead into the brain. Because of the distance involved, the electrotonic method of conduction is no longer appropriate; and, true enough, ganglion cells transmit their signals by means of action potentials instead. Furthermore, even when unstimulated they still transmit continuous impulses at rates varying between 5 and 40 per second, with the larger nerve fibers, in general, firing more rapidly. The visual signals, in turn, are superimposed onto this background ganglion cell firing.

Transmission of Changes in Light Intensity —The On-Off Response. Many ganglion cells are especially excited by *changes* in light intensity. This is illustrated by the records of nerve impulses in Figure

50–13, showing in the upper panel strong excitation for a fraction of a second when a light was first turned on; then in another fraction of a second the level of excitation diminished. The lower tracing is from a ganglion cell located in the dark area lateral to the spot of light; this cell was markedly inhibited when the light was turned on because of lateral inhibition. Then, when the light was turned off, exactly the opposite effects occurred. Thus, these records are called "on-off" and "off-on" responses. The opposite directions of these responses to light are caused, respectively, by the depolarizing and hyperpolarizing bipolar cells, and the transient nature of the responses was probably generated by the amacrine cells, many of which also have similar transient responses themselves.

This capability of the eyes to detect change in light intensity is equally developed in the peripheral retina as in the central retina. For instance, a minute gnat flying across the peripheral field of vision is instantaneously detected. On the other hand, the same gnat sitting quietly remains entirely below the threshold of visual detection.

Transmission of Signals Depicting Contrasts in the Visual Scene—The Role of Lateral Inhibition

Most of the ganglion cells do not respond to the actual level of illumination of the scene; instead they respond mainly to contrast borders in the scene. Since it seems that this is the major means by which the form of the scene is transmitted to the brain, let us explain how this process occurs.

When flat light is applied to the entire retina—that is, when all the photoreceptors are stimulated equally by the incident light—the contrast type of ganglion cell is neither stimulated nor inhibited. The reason for this is that the signals transmitted *directly* from the photoreceptors through the depolarizing bipolar cells are excitatory, whereas the signals transmitted *laterally* through the horizontal cells and hyperpolarizing bipolar cells are inhibitory. Thus, the direct excitatory signal through one pathway is likely to be completely neutralized by the inhibitory signals through

Figure 50 – 13. Responses of ganglion cells to light in (1) an area excited by a spot of light and (2) an area immediately adjacent to the excited spot; the ganglion cells in this area are inhibited by the mechanism of lateral inhibition. (Modified from Granit: Receptors and Sensory Perception: A Discussion of Aims, Means, and Results of Electrophysiological Research Into the Process of Reception. New Haven, Conn., Yale University Press, 1955.)

the lateral pathways. One circuit for this is illustrated in Figure 50–14, which shows three photoreceptors; the central one of these receptors excites a depolarizing bipolar cell. However, the two receptors on either side are connected to the same bipolar cell through inhibitory horizontal cells that neutralize the direct excitatory signal if these receptors are also stimulated by light.

Now, let us examine what happens when a contrast border occurs in the visual scene. Referring again to Figure 50–14, let us assume that the central photoreceptor is stimulated by a bright spot of light while one of the two lateral receptors is in the dark. The bright spot of light will excite the direct pathway through the bipolar cell. Then, in addition, the fact that one of the lateral photoreceptors is in the dark causes one of the horizontal cells to be inhibited. In turn, this cell loses its inhibitory effect on the bipolar cell, and this allows still more excitation of the bipolar cell. Thus, when light is everywhere, the excitatory and inhibitory signals to the bipolar cells mainly neutralize each other, but where contrasts occur the signals through the direct and lateral pathways actually accentuate each other.

Thus, the mechanism of lateral inhibition functions in the eye in the same way that it functions in most other sensory systems as well—that is, to provide contrast detection and enhancement.

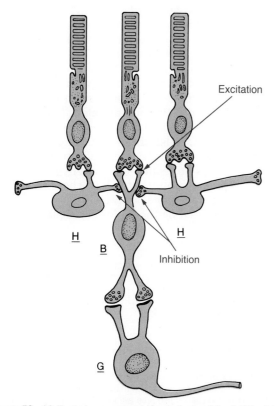

Figure 50–14. Typical arrangement of rods, horizontal cells (H), a bipolar cell (B), and a ganglion cell (G) in the retina, showing excitation at the synapses between the rods and the horizontal cells but inhibition between the horizontal cells and the bipolar cells.

Transmission of Color Signals by the Ganglion Cells

A single ganglion cell may be stimulated by a number of cones or by only a very few. When all three types of cones — the red, blue, and green types — stimulate the same ganglion cell, the signal transmitted through the ganglion cell is the same for any color of the spectrum. Therefore, this signal plays no role in the detection of the different colors. Instead, it is a "white" signal.

On the other hand, some of the ganglion cells are excited by only one color type of cone but inhibited by a second type. For instance, this frequently occurs for the red and green cones, red causing excitation and green causing inhibition—or vice versa, with green causing excitation and red, inhibition. The same type of reciprocal effect also occurs between blue cones on the one hand and a combination of red and green cones on the other hand, giving a reciprocal excitation inhibition relationship between the blue and yellow colors.

The mechanism of this opposing effect of colors is the following: One color-type cone excites the ganglion cell by the direct excitatory route through a depolarizing bipolar cell, while the other color type inhibits the ganglion cell by the indirect inhibitory route through a horizontal cell or a hyperpolarizing bipolar cell.

The importance of these color-contrast mechanisms is that they represent a mechanism by which the retina itself begins to differentiate colors. Thus each color-contrast type of ganglion cell is excited by one color but inhibited by the "opponent color." Therefore, the process of color analysis begins in the retina and is not entirely a function of the brain.

REFERENCES

Allansmith, M. R.: The Eye and Immunology. St. Louis, C. V. Mosby Co., 1983.
Benson, W. E., et al.: Diabetes and Its Ocular Complications. Philadelphia, W. B. Saunders Co., 1988.
Cunha-Vaz, J. G. (ed.): The Blood-Retinal Barriers. New York, Plenum Press, 1980.
Dacey, D. M.: Dopamine-accumulating retinal neurons revealed by in vitro fluorescence display a unique morphology. Science, 240:1196, 1988.
Daw, N. W., et al.: The function of synaptic transmitters in the retina. Annu. Rev. Neurosci., 12:205, 1989.
DeValois, R. L., and Jacobs, G. H.: Neural mechanisms of color vision. In Darian-Smith, I. (ed.): Handbook of Physiology. Sec. 1, Vol. III. Bethesda, Md., American Physiological Society, 1984, p. 525.
Dowling, J. E., and Dubin, M. W.: The vertebrate retina. In Darian-Smith, I. (ed.): Handbook of Physiology. Sec. 1, Vol. III. Bethesda, Md., American Physiological Society, 1984, p. 317.
Fine, B. S., and Yanoff, M.: Ocular Histology: A Text and Atlas. Hagerstown, Md., Harper & Row, 1979.
Finlay, B. L., and Sengelaub, D. R. (eds.): Development of the Vertebrate Retina. New York, Plenum Publishing Corp., 1989.
Gurney, A. M., and Lester, H. A.: Light-flash physiology with synthetic photosensitive compounds. Physiol. Rev., 67:583, 1987.
Hillman, P., et al.: Transduction in invertebrate photoreceptors: Role of pigment bistability. Physiol. Rev., 63:668, 1983.
Huismans, H., et al.: The Photographed Fundus. Baltimore, Williams & Wilkins, 1988.
Hurley, J. G.: Molecular properties of the cGMP cascade of vertebrate photoreceptors. Annu. Rev. Physiol., 49:793, 1987.
Kaneko, A.: Physiology of the retina. Annu. Rev. Neurosci., 2:169, 1979.
Kanski, J. J. (ed.): BIMR Ophthalmology. Vol. 1. Disorders of the Vitreous, Retina, and Choroid. Woburn, Mass., Butterworths, 1983.

Land, E. H.: The retinex theory of color vision. Sci. Am., 237(6):108, 1977.

Liebman, P. A., et al.: The molecular mechanism of visual excitation and its relation to the structure and composition of the rod outer segment. Annu. Rev. Physiol., 49:765, 1987.

MacNichol, E. F., Jr.: Three-pigment color vision. Sci. Am., 211:48, 1964.

Marks, W. B., et al.: Visual pigments of single primate cones. Science, 143:1181, 1964.

Michaelson, I. C.: Textbook of the Fundus of the Eye. New York, Churchill Livingstone, 1980.

Ming, A. L. S., and Constable, I. J.: Colour Atlas of Ophthalmology. Boston, Houghton Mifflin, 1979.

Montgomery, G.: Seeing With the Brain. Discover, December 1988, p. 52.

Newsome, D. A.: Retinal Dystrophies and Degenerations. New York, Raven Press, 1988.

Owen, W. G.: Ionic conductances in rod photoreceptors. Annu. Rev. Physiol., 49:743, 1987.

Pugh, E. N., Jr.: The nature and identity of the internal excitational transmitter of vertebrate phototransduction. Annu. Rev. Physiol., 49:715, 1987.

Rushton, W. A. H.: Visual pigments and color blindness. Sci. Am., 232(3):64, 1975.

Ryan, S. J., et al. (eds.): Retina. St. Louis, C. V. Mosby Co., 1989.

Saibil, H. R.: From photon to receptor potential: The biochemistry of vision. News in Physiol. Sci., 1:122, 1986.

Schepens, C. L.: Retinal Detachment and Allied Diseases. Philadelphia, W. B. Saunders Co., 1983.

Sherman, S. M., and Spear, P. D.: Organization of visual pathways in normal and visually deprived cats. Physiol. Rev., 62:738, 1982.

Spaeth, G. L. (ed.): Ophthalmic Surgery: Principles and Practice, 2nd Ed. Philadelphia, W. B. Saunders Co., 1989.

Stillman, A. J.: Current concepts in photoreceptor physiology. Physiologist, 28:122, 1985.

Wolf, G.: Multiple functions of vitamin A. Physiol. Rev., 64:738, 1982.

Yannuzzi, L. A., et al. (eds.): The Macula: A Comprehensive Text and Atlas. Baltimore, Williams & Wilkins, 1978.

CHAPTER 51

• The Eye: III. Central Neurophysiology of Vision •

THE VISUAL PATHWAYS

Figure 51–1 illustrates the principal visual pathways from the two retinas to the *visual cortex.* After nerve impulses leave the retinas they pass backward through the *optic nerves.* At the *optic chiasm* all the fibers from the nasal halves of the retinas cross to the opposite side, where they join the fibers from the opposite temporal retinas to form the *optic tracts.* The fibers of each optic tract synapse in the *dorsal lateral geniculate nucleus,* and from here the *geniculocalcarine fibers* pass by way of the *optic radiation,* or *geniculocalcarine tract,* to the *primary visual cortex* in the calcarine area of the occipital lobe.

In addition, visual fibers also pass to older precortical areas of the brain: (1) from the optic tracts to the *suprachiasmatic nucleus of the hypothalamus,* presumably for controlling circadian rhythms; (2) into the *pretectal nuclei,* for eliciting some reflex movements of the eyes focused on objects of importance and also for activating the pupillary light reflex; (3) into the *superior colliculus,* for control of rapid directional movements of the two eyes; and (4) into the *ventral lateral geniculate nucleus* of the thalamus and thence into surrounding basal regions of the brain, presumably to help control some of the body's behavioral functions.

Thus, the visual pathways can be divided roughly into an *old system* to the midbrain and base of the forebrain and a *new system* for direct transmission into the visual cortex. The new system is responsible in man for the perception of virtually all aspects of visual form, colors, and other conscious vision. On the other hand, in many lower animals, even visual form is detected by the older system, using the superior colliculus in the same manner that the visual cortex is used in mammals.

Function of the Dorsal Lateral Geniculate Nucleus

The optic nerve fibers of the new visual system all terminate in the *dorsal lateral geniculate nucleus,* located at the dorsal end of the thalamus and frequently also called the *lateral geniculate body.* The dorsal lateral geniculate nucleus serves two principal functions: First, it serves as a relay station to relay visual information from the optic tract to the *visual cortex* by way of the *geniculocalcarine tract.* This relay function is very accurate, so much so that there is exact point-to-point transmission with a high degree of spatial fidelity all the way from the retina to the visual cortex.

It will be recalled that half of the fibers in each optic tract after passing the optic chiasm are derived from one eye and half from the other eye, representing corresponding points on the two retinas. However, the signals from the two eyes are kept apart in the dorsal lateral geniculate nucleus. This nucleus is composed of six nuclear layers. Layers II, III, and V (from ventral to dorsal) receive signals from the temporal portion of the ipsilateral retina, whereas layers I, IV, and VI receive signals from the nasal retina of the opposite eye. The respective retinal areas of the two eyes connect with neurons that are approximately superimposed over each other in the paired layers, and similar parallel transmission is preserved all the way back to the visual cortex.

The second major function of the dorsal lateral geniculate nucleus is to "gate" the transmission of signals to the visual cortex, that is, to control how much of the signal is allowed to pass to the cortex. The nucleus receives gating control signals from two major sources, (1) *corticofugal fibers* returning in a backward direction from the primary visual cortex to the lateral geniculate nucleus and (2) the reticular areas of the mesencephalon. Both of these are inhibitory and, when stimulated, can literally turn off transmission through selected portions of the dorsal lateral geniculate nucleus. Therefore, it is assumed that both of these gating circuits help to control the visual information that is allowed to pass.

Finally, the dorsal lateral geniculate nucleus is divided in another way: (1) Layers I and II are called *magnocellular layers* because they contain very large neurons. These receive their usual input almost entirely from the large type Y retinal ganglion cells. This

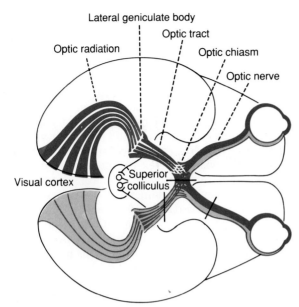

Figure 51–1. The principal visual pathways from the eyes to the visual cortex. (Modified from Polyak: The Retina. ©1941 by The University of Chicago. All rights reserved.)

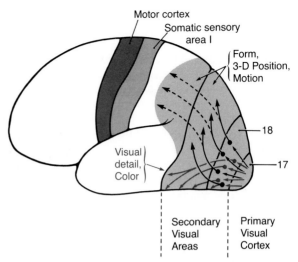

Figure 51–2. Transmission of visual signals from the primary visual cortex into secondary visual areas. Note that the signals representing form, third dimensional position, and motion are transmitted mainly superiorly into the superior portions of the occipital lobe and the posterior parietal lobe. By contrast, the signals for visual detail and color are transmitted mainly into the anteroventral portion of the occipital lobe and ventral portion of the posterior temporal lobe.

magnocellular system provides a very rapidly conducting pathway to the visual cortex. On the other hand, it is "color blind," transmitting only black and white information. Also, its point-to-point transmission is poor, for there are not so many Y ganglion cells, and their dendrites spread widely in the retina. (2) Layers III through VI are called *parvocellular layers* because they contain large numbers of small- to medium-sized neurons. These receive their input almost entirely from the type X retinal ganglion cells that transmit color and also convey accurate point-to-point spatial information but at only a moderate velocity of conduction, rather than high velocity.

● ORGANIZATION AND FUNCTION OF THE VISUAL CORTEX

Figures 51–2 and 51–3 show that the *visual cortex* is located primarily in the occipital lobes. Like the cortical representations of the other sensory systems, the visual cortex is divided into a *primary visual cortex* and *secondary visual areas.*

The Primary Visual Cortex. The primary visual cortex (Figure 51–3) lies in the *calcarine fissure area* and extending to the *occipital pole* on the medial aspect of each occipital cortex. This area is the terminus of the most direct visual signals from the eyes. Signals from the macular area of the retina terminate near the occipital pole, while signals from the more peripheral retina terminate in concentric circles anterior to the pole and along the calcarine fissure. The upper portion of the retina is represented superiorly and the lower portion inferiorly. Note in the figure the especially large area that represents the macula. It is

to this region that the fovea transmits its signals. The fovea is responsible for the highest degree of visual acuity. Based on retinal area, the fovea has several hundred times as much representation in the primary visual cortex as do the peripheral portions of the retina.

The primary visual cortex is coextensive with *Brodmann cortical area 17* (see the diagram of the Brodmann areas in Figure 47–5 in Chapter 47). It is frequently also called *visual area I* or simply V-1. Still another name for the primary visual cortex is the *striate cortex* because this area has a grossly striated appearance.

The Secondary Visual Areas. The secondary visual areas, also called *visual association areas,* lie anterior, superior, and inferior to the primary visual cortex. Secondary signals are transmitted to these areas for further analysis of visual meanings. For instance,

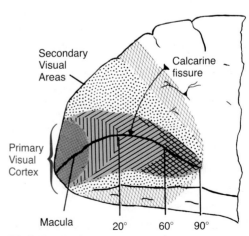

Figure 51–3. The visual cortex.

on all sides of the primary visual cortex is *Brodmann area 18* (Figure 51–2), which is the association area where virtually all signals from the primary visual cortex pass next. Therefore, Brodmann area 18 is called *visual area II* or simply V-2. The other more distant secondary visual areas have specific designations V-3, V-4, and so forth. The importance of all these areas is that various aspects of the visual image are progressively dissected and analyzed in separate ones of them.

The Layered Structure of the Primary Visual Cortex

Like almost all other portions of the cerebral cortex, the primary visual cortex has six distinct layers, as illustrated in Figure 51–4. As is true for the other sensory systems, the geniculocalcarine fibers terminate mainly in layer 4. But this layer, too, is organized in subdivisions. The rapidly conducted signals from the Y retinal ganglion cells terminate in layer 4cα and from here are relayed vertically both outward toward the cortical surface and inward toward deeper levels.

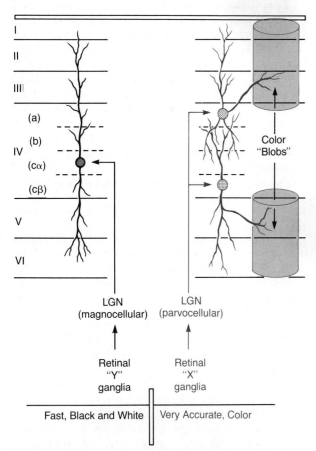

Figure 51–4. The six layers of the primary visual cortex. The connections to the left transmit very rapidly changing black and white visual signals. The pathways to the right transmit signals depicting very accurate detail and also color. Note especially small areas of the visual cortex called "color blobs" that are necessary for detection of color.

The visual signals from the medium-sized optic nerve fibers, derived from the X ganglion cells in the retina, also terminate in layer 4 but at points different from the Y signals, in layers 4a and 4cβ, the shallowest and deepest portions of layer 4. From here, these signals again are transmitted vertically both toward the surface of the cortex and to deeper layers. It is these X ganglion pathways that transmit the very accurate point-to-point type of vision and also color vision.

The Vertical Neuronal Columns in the Visual Cortex. The visual cortex is organized structurally into several million vertical columns of neuronal cells, each column having a diameter of 30 to 50 micrometers. This same vertical columnar organization is found throughout the cerebral cortex. Each column represents a functional unit. One can calculate from rough data that the number of neurons in each of the visual vertical columns is around 1000.

After the optic signals terminate in layer 4, they are further processed as they spread both outward and inward along each vertical column unit. This processing is believed to decipher separate bits of visual information at successive stations along the pathway. The signals that pass outward to layers 1, 2, and 3 eventually transmit higher orders of signals for short distances laterally in the cortex. On the other hand, the signals that pass inward to layers 5 and 6 excite neurons that transmit signals much greater distances.

The "Color Blobs" in the Visual Cortex. Interspersed among the primary visual columns are special column-like areas called *color blobs*. These receive lateral signals from the adjacent visual columns and respond specifically to color signals. Therefore, it is presumed that these blobs are the primary areas for deciphering color. Also, in certain secondary visual areas additional color blobs are found, which presumably perform still higher levels of color deciphering.

Interaction of Visual Signals From the Separate Eyes. Recall that the visual signals from the two separate eyes are relayed through separate neuronal layers in the lateral geniculate nucleus. And these signals still remain separated from each other when they arrive in layer 4 of the primary visual cortex. In fact, layer 4 is interlaced with horizontal zebra-like stripes of neuronal columns, each stripe about 0.5 mm wide; the signals from one eye enter the columns of every other stripe, alternating with the signals from the other eye. However, as the signals spread vertically into the more superficial or deeper layers of the cortex, this separation is lost because of lateral spread of the visual signals. In the meantime the cortex deciphers whether the respective areas of the two visual images are "in register" with each other, that is, whether corresponding points on the two retinas fit with each other. In turn, this deciphered information is used to control the movements of the eyes so that they will fuse with each other (brought into "register"). The information also allows a person to distinguish distances of objects by the mechanism of stereopsis.

Two Major Pathways for Analysis of Visual Information — The Fast "Position" and "Motion" Pathway; The Accurate Color Pathway

Figure 51–2 shows that after leaving the primary visual cortex, the visual information is analyzed in two major pathways in the secondary visual areas.

1. Analysis of Three-Dimensional Position, Gross Form, and Motion of Objects. One of the analytical pathways, illustrated in Figure 51–2 by the broad black arrows, analyzes the third dimensional positions of visual objects in the coordinates of space around the body. From this information, this pathway also analyzes the overall form of the visual scene as well as motion in the scene. In other words, this pathway tells "where" every object is at each instant and whether it is moving. After leaving the primary visual cortex (Brodmann area 17), the signals of this pathway next synapse in visual area 2 (Brodmann area 18), then flow generally into the posterior midtemporal area, and thence upward into the broad occipitoparietal cortex. At the anterior border of this last area, the signals overlap with signals from the posterior somatic association areas that analyze form and three-dimensional aspects of somatic sensory signals. The signals transmitted in this position-form-motion pathway are mainly from the large Y optic nerve fibers of the retinal Y ganglion cells, transmitting rapid signals but only black and white signals.

2. Analysis of Visual Detail and Color. The red arrows in Figure 51–2, passing from the primary visual cortex (Brodmann area 17) into visual area 2 (Brodmann area 18) and thence into the inferior ventral and medial regions of the occipital and temporal cortex, illustrate the principal pathway for analysis of visual detail. Also, separate portions of this pathway specifically dissect out color as well. Therefore, this pathway is concerned with such visual feats as recognizing letters, reading, determining the texture of surfaces, determining detailed colors of objects, and deciphering from all this information "what" the object is and its meaning.

● NEURONAL PATTERNS OF STIMULATION DURING ANALYSIS OF THE VISUAL IMAGE

Analysis of Contrasts in the Visual Image. If a person looks at a blank wall, only a few neurons in the primary visual cortex will be stimulated whether the illumination of the wall is bright or weak. Therefore, the question must be asked, What does the visual cortex detect? To answer this, let us now place on the wall a large solid cross as illustrated to the left in Figure 51–5. To the right is illustrated the spatial pattern of the greater majority of the excited neurons in the visual cortex. *Note that the areas of maximum excitation occur along the sharp borders of the visual*

Retinal image   Cortical stimulation

Figure 51–5. Pattern of excitation occurring in the visual cortex in response to a retinal image of a dark cross.

pattern. Thus, the visual signal in the primary visual cortex is concerned mainly with the *contrasts* in the visual scene, rather than with the flat areas. We saw in the previous chapter that this is true of most of the retinal ganglion cells as well, because equally stimulated adjacent retinal receptors mutually inhibit each other. But at any border in the visual scene where there is a change from dark to light or light to dark, mutual inhibition does not occur, and the intensity of stimulation is proportional to the *gradient of contrast* —that is, the greater the sharpness of contrast and the greater the intensity difference between the light and dark areas, the greater the degree of stimulation.

Detection of Orientation of Lines and Borders — The "Simple" Cells. Not only does the visual cortex detect the existence of lines and borders in the different areas of the retinal image, but it also detects the orientation of each line or border — that is, whether it is vertical or horizontal or lies at some degree of inclination. This is believed to result from linear organizations of mutually inhibiting cells that excite second order neurons when mutual inhibition falls all along a line of cells, that is, where there is a contrast edge. Thus, for each such orientation of a line, a specific neuronal cell is stimulated. And a line oriented in a different direction excites a different cell. These neuronal cells are called *simple cells.* They are found mainly in layer 4 of the primary visual cortex.

Detection of Line Orientation When the Line Is Displaced Laterally or Vertically in the Visual Field — "Complex" Cells. As the signal progresses farther away from layer 4, some neurons now respond to lines still oriented in the same direction but not position-specific. That is, the line can be displaced moderate distances laterally or vertically in any direction in the field, and still the neuron will be stimulated if the line has the same direction. These cells are called *complex cells.*

Detection of Lines of Specific Lengths, Angles, or Other Shapes. Many neurons in the outer layers of the primary visual columns, as well as neurons in some secondary visual areas, are stimulated only by lines or borders of specific lengths, or by specific angulated shapes, or by images having other characteristics. Thus, these neurons detect still higher orders of information from the visual scene; therefore, they are called *hypercomplex cells.*

Thus, as one goes farther into the analytical pathway of the visual cortex, progressively more characteristics of each area of the visual scene are deciphered.

Detection of Color

Color is detected in much the same way that lines are detected: by means of color contrast. The contrasts are between cones that lie immediately adjacent to each other or cones that lie far apart. For instance, a red area is often contrasted against a green area, or a blue area against a red, or a green area against a yellow. All of these colors can also be contrasted against a white area within the visual scene. In fact, it is this contrasting against white that is believed to be mainly responsible for the phenomenon called color constancy that was discussed in the previous chapter; that is, when the color of the illuminating light changes, the color of the "white" changes with the light, and appropriate computation in the brain allows red to be interpreted as red even though the illuminating light has actually changed the color spectrum entering the eyes.

The mechanism of color contrast analysis depends on the fact that contrasting colors, called opponent colors, mutually excite certain of the neuronal cells. It is presumed that the initial details of color contrast are detected by simple cells, whereas more complex contrasts are detected by complex and hypercomplex cells.

Serial Analysis of the Visual Image Versus Parallel Analysis

From the foregoing discussion, it should by now be clear that the visual image is deciphered and analyzed by both serial and parallel pathways. The sequence from simple to complex to hypercomplex cells is serial analysis, with more and more details being deciphered. The transmission of different types of visual information into different brain locations represents parallel processing. It is the combination of both types of these analyses that gives one full interpretation of a visual scene. However, the highest levels of analysis are still mainly beyond present physiological understanding.

Effect of Removing the Primary Visual Cortex

Removal of the primary visual cortex in the human being causes loss of conscious vision. However, psychological studies demonstrate that such persons can still react subconsciously to changes in light intensity, to movement, and even to some gross patterns of vision. These reactions include turning the eyes, turning the head, avoidance, and so on. This vision is believed to be subserved by neuronal pathways that pass from the optic tracts mainly into the superior colliculi and other portions of the older visual system.

● THE FIELDS OF VISION; PERIMETRY

The *field of vision* is the area seen by an eye at a given instant. The area seen to the nasal side is called the *nasal field of vision,* and the area seen to the lateral side is called the *temporal field of vision.*

To diagnose blindness in specific portions of the retinas, one charts the field of vision for each eye by a process known as *perimetry.* This is done by having the subject look with one eye toward a central spot directly in front of the eye. Then a small dot of light or a small object is moved back and forth in all areas of the field of vision, and the person indicates when the spot of light or object can be seen and when it cannot. Thus, the field of vision is plotted as illustrated in Figure 51–6.

In all perimetry charts, a *blind spot* caused by lack of rods and cones in the retina over the *optic disc* is found approximately 15 degrees lateral to the central point of vision, as illustrated in the figure.

Abnormalities in the Fields of Vision. Occasionally blind spots are found in portions of the field of vision other than the optic disc area. Such blind spots are called *scotomata;* they frequently result from allergic reactions in the retina or from toxic conditions, such as lead poisoning or excessive use of tobacco.

Still another condition that can be diagnosed by perimetry is *retinitis pigmentosa.* In this disease, portions of the retina degenerate and excessive melanin pigment deposits in the degenerated areas. Retinitis pigmentosa generally causes blindness in the peripheral field of vision first and then gradually encroaches on the central areas.

Effect of Lesions in the Optic Pathway on the Fields of Vision. Destruction of an entire *optic nerve* obviously causes blindness of the respective eye. Destruction of the *optic chiasm,* as shown by the longitudinal line across the chiasm in Figure 51–1, prevents the passage of impulses from the nasal halves of the two retinae to the opposite optic tracts. Therefore, the nasal halves are both blinded, which means that the person is blind in both temporal fields of vision *because the image of the field of vision is inverted on the retina;* this condition is called *bitemporal hemianopsia.*

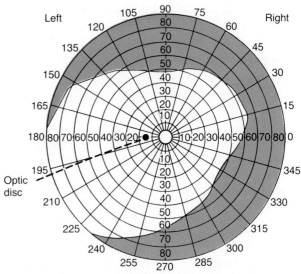

Figure 51–6. A perimetry chart, showing the field of vision for the left eye.

Such lesions frequently result from tumors of the adenohypophysis pressing upward on the optic chiasm.

Interruption of an *optic tract,* which is shown by another line in Figure 51–1, denervates the corresponding half of each retina on the same side as the lesion, and, as a result, neither eye can see objects to the opposite side. This condition is known as *homonymous hemianopsia.* Destruction of the *optic radiation* or the *visual cortex* of one side also causes homonymous hemianopsia. A common condition that destroys the visual cortex is thrombosis of the posterior cerebral artery, which infarcts the occipital cortex except for part of the foveal area, thus often sparing central vision.

One can differentiate a lesion in the optic tract from a lesion in the geniculocalcarine tract or visual cortex by determining whether impulses can still be transmitted into the pretectal nuclei to initiate a pupillary light reflex.

● EYE MOVEMENTS AND THEIR CONTROL

To make use of the abilities of the eye, almost equally as important as the system for interpretation of the visual signals from the eyes is the cerebral control system for directing the eyes toward the object to be viewed.

Muscular Control of Eye Movements. The eye movements are controlled by three separate pairs of muscles, shown in Figure 51–7: (1) the *medial* and *lateral recti,* (2) the *superior* and *inferior recti,* and (3) the *superior* and *inferior obliques.* The medial and lateral recti contract reciprocally mainly to move the eyes from side to side. The superior and inferior recti contract reciprocally to move the eyes mainly upward or downward. And the oblique muscles func-

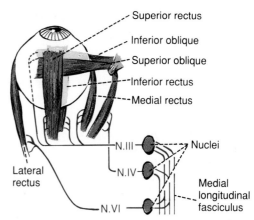

Figure 51–7. The extraocular muscles of the eye and their innervation.

tion mainly to rotate the eyeballs to keep the visual fields in the upright position.

Neural Pathways for Control of Eye Movements. Figure 51–7 also illustrates the nuclei of the third, fourth, and sixth cranial nerves and their innervation of the ocular muscles. Shown, too, are the interconnections among these three nuclei through the *medial longitudinal fasciculus.* Either by way of this fasciculus or by way of other closely associated pathways, each of the three sets of muscles to each eye is *reciprocally* innervated so that one muscle of the pair relaxes while the other contracts.

Figure 51–8 illustrates cortical control of the oculomotor apparatus, showing spread of signals from the occipital visual areas through occipitotectal and occipitocollicular tracts into the pretectal and superior colliculus areas of the

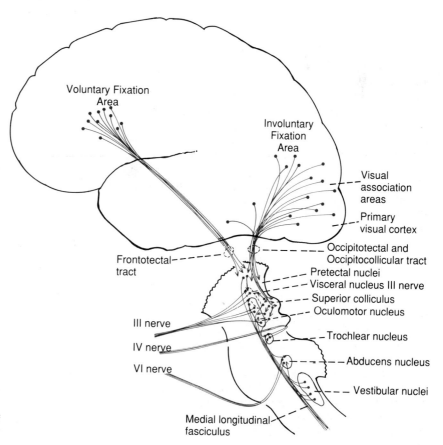

Figure 51–8. Neural pathways for control of conjugate movement of the eyes.

brain stem. In addition, a frontotectal tract passes from the frontal cortex into the pretectal area. From both the pretectal and the superior colliculus areas, the oculomotor control signals then pass to the nuclei of the oculomotor nerves. Strong signals are also transmitted into the oculomotor system from the vestibular nuclei by way of the medial longitudinal fasciculus.

FIXATION MOVEMENTS OF THE EYES

Perhaps the most important movements of the eyes are those that cause the eyes to "fix" on a discrete portion of the field of vision.

Fixation movements are controlled by two different neuronal mechanisms. The first of these allows the person to move his eyes voluntarily to find the object upon which he wishes to fix his vision; this is called the *voluntary fixation mechanism*. The second is an involuntary mechanism that holds the eyes firmly on the object once it has been found; this is called the *involuntary fixation mechanism*.

The voluntary fixation movements are controlled by a small cortical field located bilaterally in the premotor cortical regions of the frontal lobes, as illustrated in Figure 51–8. Bilateral dysfunction or destruction of these areas makes it difficult or almost impossible for the person to "unlock" the eyes from one point of fixation and then move them to another point. It is usually necessary for the person to blink the eyes or put a hand over the eyes for a short time, which then allows the eyes to be moved.

On the other hand, the fixation mechanism that causes the eyes to "lock" on the object of attention once it is found is controlled by *secondary visual areas of the occipital cortex* — mainly Brodmann area 19 located anterior to visual areas V-1 and V-2 (Brodmann areas 17 and 18). When this area is destroyed bilaterally, an animal has difficulty keeping its eyes directed toward a given fixation point or becomes completely unable to do so.

To summarize, the posterior eye fields automatically "lock" the eyes on a given spot of the visual field and thereby prevent movement of the image across the retina. To unlock this visual fixation, voluntary impulses must be transmitted from the "voluntary" eye fields located in the frontal areas.

Mechanism of the Involuntary Locking Fixation — Role of the Superior Colliculi. The involuntary locking type of fixation discussed in the previous section results from a negative feedback mechanism that prevents the object of attention from leaving the foveal portion of the retina. The eyes even normally have three types of continuous but almost imperceptible movements: (1) a *continuous tremor* at a rate of 30 to 80 cycles per second caused by successive contractions of the motor units in the ocular muscles, (2) a *slow drift* of the eyeballs in one direction or another, and (3) sudden *flicking movements* that are controlled by the involuntary fixation mechanism. When a spot of light has become fixed on the foveal region of the retina, the tremorous movements cause the spot to move back and forth at a rapid rate across the cones, and the drifting movements cause it to drift slowly across the cones. However, each time the spot of light drifts as far as the edge of the fovea, a sudden reflex reaction occurs, producing a flicking movement that moves the spot away from this edge back toward the center. Thus, an automatic response moves the image back toward the central portion of the fovea. These drifting and flicking motions are illustrated in Figure 51–9, which shows by the dashed lines the slow drifting across the

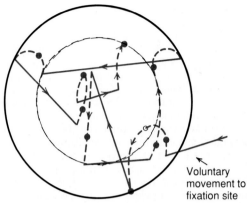

Figure 51–9. Movements of a spot of light on the fovea, showing sudden "flicking" movements to move the spot back toward the center of the fovea whenever it drifts to the foveal edge. (The dashed lines represent slow drifting movements, and the solid lines represent sudden flicking movements.) (Modified from Whitteridge: Handbook of Physiology. Vol. 2, Sec. 1. Baltimore, Williams & Wilkins, 1960.)

retina and by the solid lines the flicks that keep the image from leaving the foveal region.

This involuntary fixation capability is mostly lost when the superior colliculi are destroyed. After the signals for fixation originate in the visual fixation areas of the occipital cortex, they pass to the superior colliculi, probably from there to reticular areas around the oculomotor nuclei, and thence into the motor nuclei themselves.

Saccadic Movement of the Eyes — a Mechanism of Successive Fixation Points. When the visual scene is moving continually before the eyes, such as when a person is riding in a car or turning around, the eyes fix on one highlight after another in the visual field, jumping from one to the next at a rate of two to three jumps per second. The jumps are called *saccades,* and the movements are called *opticokinetic movements.* The saccades occur so rapidly that not more than 10 per cent of the total time is spent in moving the eyes, 90 per cent of the time being allocated to the fixation sites. Also, the brain suppresses the visual image during the saccades so that one is completely unconscious of the movements from point to point.

Saccadic Movements During Reading. During the process of reading, a person usually makes several saccadic movements of the eyes for each line. In this case the visual scene is not moving past the eyes, but the eyes are trained to scan across the visual scene to extract the important information. Similar saccades occur when a person observes a painting, except that the saccades occur in one direction after another from one highlight of the painting to another, then another, and so forth.

Fixation on Moving Objects — "Pursuit Movements." The eyes can also remain fixed on a moving object, which is called *pursuit movement.* A highly developed cortical mechanism automatically detects the course of movement of an object and then gradually develops a similar course of movement of the eyes. For instance, if an object is moving up and down in a wavelike form at a rate of several times per second, the eyes at first may be completely unable to fixate on it. However, after a second or so the eyes begin to jump coarsely in approximately the same pattern of movement as that of the object. Then after a few more seconds, the eyes develop progressively smoother and smoother movements and finally follow the course of movement almost exactly. This represents a high degree of automatic, subconscious computational ability by the cerebral cortex.

The Superior Colliculi Are Mainly Responsible for Turning the Eyes and Head Toward a Visual Disturbance

Even after the visual cortex has been destroyed, a sudden visual disturbance in a lateral area of the visual field will cause immediate turning of the eyes in that direction. This will not occur if the superior colliculi have also been destroyed. To support this function, the various points of the retina are represented topologically in the superior colliculi in the same way as in the primary visual cortex, though with less accuracy. Even so, the principal direction of a flash of light in a peripheral retinal field is mapped by the colliculi, and secondary signals are then transmitted to the oculomotor nuclei to turn the eyes.

The optic nerve fibers from the eyes to the colliculi that are responsible for these rapid turning movements are branches from the rapidly conducting Y fibers, with one branch going to the visual cortex and the other going to the superior colliculi. (The superior colliculi and other regions of the brain stem are also strongly supplied with visual signals transmitted in type W optic nerve fibers. These represent the older visual pathway, but their function is unclear.)

In addition to causing the eyes to turn toward the visual disturbance, signals are also relayed from the superior colliculi through the *medial longitudinal fasciculus* to other levels of the brain stem to cause turning of the whole head and even of the whole body toward the direction of the disturbance. Also, other types of disturbances besides visual, such as strong sounds or even stroking of the side of the body, will cause similar turning of the eyes, head, and body, but only if the superior colliculi are intact. Therefore, the superior colliculi play a global role in orienting the eyes, the head, and the body with respect to external disturbances whether visual, auditory, or somatic.

FUSION OF THE VISUAL IMAGES FROM THE TWO EYES

To make the visual perceptions more meaningful, the visual images in the two eyes normally *fuse* with each other on "corresponding points" of the two retinas.

The visual cortex plays a very important role in fusion. It was pointed out earlier in the chapter that corresponding points of the two retinas transmit visual signals to different neuronal layers of the lateral geniculate body, and these signals in turn are relayed to parallel stripes of neurons in the visual cortex. Interactions occur between the stripes of cortical neurons; these cause *interference patterns of excitation* in some of the local neuronal cells when the two visual images are not precisely "in register"—that is, not precisely fused. This excitation presumably provides the signal that is transmitted to the oculomotor apparatus to cause convergence or divergence or rotation of the eyes so that fusion can be reestablished. Once the corresponding points of the retinas are precisely in register with each other, the excitation of the specific cells in the visual cortex is greatly diminished or disappears.

The Neural Mechanism of Stereopsis for Judging Distances of Visual Objects

In Chapter 49 it was pointed out that because the two eyes are more than 2 inches apart the images on the two retinas are not exactly the same. That is, the right eye sees a little more of the right-hand side of the object and the left eye a little more of the left-hand side, and the closer the object the greater the disparity. Therefore, even when the two eyes are fused with each other, it is still impossible for all corresponding points in the two visual images to be absolutely in register at the same time. Furthermore, the nearer the object is to the eyes, the less the degree of register. This degree of non-register provides the mechanism for *stereopsis*, a very important mechanism for judging distances of visual objects up to distances of about 100 meters.

The neuronal cellular mechanism for stereopsis is based on the fact that some of the fiber pathways from the retinas to the visual cortex stray 1 to 2 degrees on either side of the central pathway. Therefore, some optic pathways from the two eyes will be exactly in register for objects 2 meters away; and still another set of pathways will be in register for objects 75 meters away. Thus, the distance is determined by which set of pathways interact with each other. This phenomenon is called *depth perception*, which is another name for stereopsis.

Strabismus

Strabismus, which is also called *squint* or *cross-eyedness*, means lack of fusion of the eyes in one or more of the coordinates described above. Three basic types of strabismus are illustrated in Figure 51–10: *horizontal strabismus, vertical strabismus,* and *torsional strabismus.* However, combinations of two or even of all three of the different types of strabismus often occur.

Strabismus is often caused by an abnormal "set" of the fusion mechanism of the visual system. That is, in the early efforts of the child to fixate the two eyes on the same object, one of the eyes fixates satisfactorily while the other fails to fixate, or they both fixate satisfactorily but never simultaneously. Soon, the patterns of conjugate movements of the eyes become abnormally "set" so that the eyes never fuse.

Frequently, some abnormality of the eyes contributes to the failure of the two eyes to fixate on the same point. For instance, in hyperopic infants, intense impulses must be transmitted to the ciliary muscles to focus the eyes, and some of these impulses overflow into the oculomotor nuclei to cause simultaneous convergence of the eyes, as will be discussed later. As a result, the child's fusion mechanism becomes "set" for continual inward deviation of the eyes.

Suppression of the Visual Image From a Repressed Eye. In a few patients with strabismus the eyes alternate in fixing on the object of attention. However, in other patients, one eye alone is used all the time while the other eye becomes repressed and is never used for vision. The vision in the repressed eye develops only slightly, usually remaining 20/400 or less. If the dominant eye then becomes blinded, vision in the repressed eye can develop only to a slight extent in the adult but far more in young children. This illustrates that visual acuity is highly dependent on proper development of the central synaptic connections from the

Horizontal strabismus   Torsional strabismus   Vertical strabismus

Figure 51–10. The three basic types of strabismus.

eyes. In fact, even the numbers of neuronal connections diminish in the cortical stripes that receive signals from the repressed eye.

● AUTONOMIC CONTROL OF ACCOMMODATION AND PUPILLARY APERTURE

The Autonomic Nerves to the Eyes. The eye is innervated by both parasympathetic and sympathetic fibers, as illustrated in Figure 51–11. The parasympathetic preganglionic fibers arise in the *Edinger-Westphal nucleus* (the visceral nucleus of the third nerve) and then pass in the *third nerve* to the *ciliary ganglion,* which lies immediately behind the eye. Here the preganglionic fibers synapse with postganglionic parasympathetic neurons that, in turn, send fibers through the *ciliary nerves* into the eyeball. These nerves excite the ciliary muscle and the sphincter of the iris.

The sympathetic innervation of the eye originates in the *intermediolateral horn cells* of the first thoracic segment of the spinal cord. From here, sympathetic fibers enter the sympathetic chain and pass upward to the *superior cervical ganglion,* where they synapse with postganglionic neurons. Fibers from these spread along the carotid artery and successively smaller arteries until they reach the eye. There the sympathetic fibers innervate the radial fibers of the iris as well as several extraocular structures around the eye, which are discussed shortly in relation to Horner's syndrome. Also, they supply very weak innervation to the ciliary muscle.

CONTROL OF ACCOMMODATION (FOCUSING THE EYES)

The accommodation mechanism—that is, the mechanism that focuses the lens system of the eye—is essential

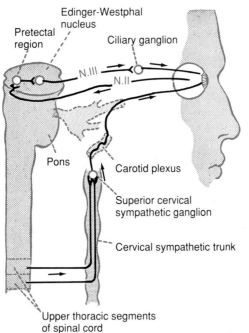

Figure 51–11. Autonomic innervation of the eye, showing also the reflex arc of the light reflex. (Modified from Ranson and Clark: Anatomy of the Nervous System. Philadelphia, W. B. Saunders Company, 1959.)

for a high degree of visual acuity. Accommodation results from contraction or relaxation of the ciliary muscle, contraction causing increased strength of the lens system, as explained in Chapter 49, and relaxation causing decreased strength. The question that must be answered now is, How does a person adjust accommodation to keep the eyes in focus all the time?

Accommodation of the lens is regulated by a negative feedback mechanism that automatically adjusts the focal power of the lens for the highest degree of visual acuity. When the eyes have been fixed on some far object and then suddenly fix on a near object, the lens accommodates for maximum acuity of vision usually within less than 1 second. Though the precise control mechanism that causes this rapid and accurate focusing of the eye is still unclear, some of the known features are the following:

First, when the eyes suddenly change the distance of the fixation point, the lens always changes its strength in the proper direction to achieve a new state of focus. In other words, the lens does not make a mistake and change its strength in the wrong direction in an attempt to find the focus.

Second, different types of clues that can help the lens change its strength in the proper direction include the following: (1) *Chromatic aberration* appears to be important. That is, the red light rays focus slightly posteriorly to the blue light rays. The eyes appear to be able to detect which of these two types of rays is in better focus, and this clue relays information to the accommodation mechanism whether to make the lens stronger or weaker. (2) When the eyes fixate on a near object they also converge toward each other. The neural mechanisms for *convergence cause a simultaneous signal to strengthen the lens of the eye.* (3) *Because the fovea lies in a hollowed-out depression that is deeper than the remainder of the retina, the clarity of focus in the depth of the fovea versus the clarity of focus on the edges will be different.* It has been suggested that this also gives clues as to which way the strength of the lens needs to be changed. (4) It has been found that *the degree of accommodation of the lens oscillates slightly* all of the time, at a frequency up to two times per second. It has been suggested that the visual image becomes clearer when the oscillation of the lens strength is changing in the appropriate direction and poorer when the lens strength is changing in the wrong direction. This could give a rapid cue as to which way the strength of the lens needs to change to provide appropriate focus.

It is presumed that the cortical areas that control accommodation closely parallel those that control fixation movements of the eyes, with final integration of the visual signals in Brodmann areas 18 and 19 and transmission of motor signals to the ciliary muscle through the pretectal area and Edinger-Westphal nucleus.

CONTROL OF THE PUPILLARY DIAMETER

Stimulation of the parasympathetic nerves excites the pupillary sphincter muscle, thereby decreasing the pupillary aperture; this is called *miosis.* On the other hand, stimulation of the sympathetic nerves excites the radial fibers of the iris and causes pupillary dilatation, which is called *mydriasis.*

The Pupillary Light Reflex. When light is shone into the eyes, the pupils constrict, a reaction called the *pupillary light reflex.* The neuronal pathway for this reflex is illustrated in Figure 51–11. When light impinges on the retina, the resulting impulses pass through the optic nerves and optic tracts to the pretectal nuclei. From here, impulses pass to the *Edinger-Westphal nucleus* and finally back through the *parasympathetic nerves* to constrict the sphincter of the iris. In darkness, the reflex becomes inhibited, which results in dilatation of the pupil.

The function of the light reflex is to help the eye adapt extremely rapidly to changing light conditions, as explained in the previous chapter. The limits of pupillary diameter are about 1.5 mm on the small side and 8 mm on the large side. Therefore, the range of light adaptation that can be effected by the pupillary reflex is about 30 to 1.

Pupillary Reflexes in Central Nervous System Disease. Certain central nervous system diseases block the transmission of visual signals from the retinas to the Edinger-Westphal nucleus. Such blocks frequently occur as a result of *central nervous system syphilis, alcoholism, encephalitis,* and so forth. The block usually occurs in the pretectal region of the brain stem, though it can also result from destruction of the small afferent fibers in the optic nerves.

The final nerve fibers in the pathway through the pretectal area to the Edinger-Westphal nucleus are of the inhibitory type. Therefore, when their inhibitory effect is lost, the nucleus becomes chronically active, causing the pupils thereafter to remain partially constricted in addition to their failure to respond to light.

Yet the pupils can still constrict some more if the Edinger-Westphal nucleus is stimulated through some other pathway. For instance, when the eyes fixate on a near object, the signals that cause accommodation of the lens and also those that cause convergence of the two eyes cause a mild degree of pupillary constriction at the same time. This is called the *accommodation reflex.* Such a pupil that fails to respond to light but does respond to accommodation and also is very small (an *Argyll Robertson pupil*) is an important diagnostic sign of central nervous system disease — very often syphilis.

Horner's Syndrome. The sympathetic nerves to the eye are occasionally interrupted, and this interruption frequently occurs in the cervical sympathetic chain. This results in *Horner's syndrome,* which consists of the following effects: First, because of interruption of fibers to the pupillary dilator muscle, the pupil remains persistently constricted to a smaller diameter than that of the pupil of the opposite eye. Second, the superior eyelid droops because this eyelid is normally maintained in an open position during the waking hours partly by contraction of a smooth muscle embedded in the lid and innervated by the sympathetics. Therefore, destruction of the sympathetics makes it impossible to open the superior eyelid nearly as widely as normally. Third, the blood vessels on the corresponding side of the face and head become persistently dilated. And fourth, sweating cannot occur on the side of the face and head affected by Horner's syndrome.

REFERENCES

Andersen, R. A.: Visual and eye movement functions of the posterior parietal cortex. Annu. Rev. Neurosci., 12:377, 1989.

Anderson, D. R.: Testing the Field of Vision. St. Louis, C. V. Mosby Co., 1983.

Bahill, A. T., and Hamm, T. M.: Using open-loop experiments to study physiological systems, with examples from the human eye-movement systems. News Physiol. Sci., 4:104, 1989.

Bishop, P. O.: Processing of visual information within the retinostriate system. In Darian-Smith, I. (ed.): Handbook of Physiology. Sec. 1, Vol. III. Bethesda, Md., American Physiological Society, 1984, p. 341.

Blasdel, G. G.: Visualization of neuronal activity in monkey striate cortex. Annu. Rev. Physiol., 51:561, 1989.

Buttner, E. J. (ed.): Neuroanatomy of the Oculomotor System. New York, Elsevier Science Publishing Co., 1984.

DeValois, R. L., and DeValois, K. K.: Spatial Vision. New York, Oxford University Press, 1988.

DeValois, R. L., and Jacobs, G. H.: Neural mechanisms of color vision. In Darian-Smith, I. (ed.): Handbook of Physiology. Sec. 1, Vol. III. Bethesda, Md., American Physiological Society, 1984, p. 525.

Eckmiller, R.: Neural control of pursuit eye movements. Physiol. Rev., 67:797, 1987.

Fregnac, Y., and Imbert, M.: Development of neuronal selectivity in primary visual cortex of cat. Physiol. Rev., 64:325, 1984.

Gilbert, C. C.: Microcircuitry of the visual cortex. Annu. Rev. Neurosci., 6:217, 1983.

Hubel, D. H., and Wiesel, T. N.: Brain mechanisms of vision. Sci. Am., 241(3):150, 1979.

Hubel, D. H., and Wiesel, T. N.: Cortical and callosal connections concerned with vertical meridian of visual fields in the cat. J. Neurophysiol., 30:1561, 1967.

Hubel, D. H., and Wiesel, T. N.: Receptive fields of cells in striate cortex of very young, visually inexperienced kittens. J. Neurophysiol., 26:994, 1963.

Jones, G. M.: The remarkable vestibuloocular reflex. News Physiol. Sci., 2:85, 1987.

Lennerstrand, G., et al. (eds.): Strabismus and Amblyopia. New York, Plenum Publishing Corp., 1988.

Livingstone, M., and Hubel, D.: Segregation of form, color, movement, and depth: Anatomy, physiology, and perception. Science, 240:740, 1988.

Lund, J. S.: Anatomical organization of Macaque monkey striate visual cortex. Annu. Rev. Neurosci., 11:253, 1988.

Mitchell, D. E., and Timney, B.: Postnatal development of function in the mammalian visual system. In Darian-Smith, I. (ed.): Handbook of Physiology. Sec. 1, Vol. III. Bethesda, Md., American Physiological Society, 1984, p. 507.

Moses, R. A.: Adler's Physiology of the Eye: Clinical Application, 7th Ed. St. Louis, C. V. Mosby Co., 1981.

Peters, A., and Jones, E. G. (eds.): Visual Cortex, New York, Plenum Publishing Corp., 1985.

Poggio, G. F., and Poggio, T.: The analysis of stereopsis. Annu. Rev. Neurosci., 7:379, 1984.

Reinecke, R. D., and Parks, M. M.: Strabismus, 3rd Ed. East Norwalk, Conn., Appleton & Lange, 1987.

Robinson, D. A.: Control of eye movements. In Brooks, V. B. (ed.): Handbook of Physiology. Sec. 1, Vol. II. Bethesda, Md., American Physiological Society, 1981, p. 1275.

Schiller, P. H.: The superior colliculus and visual function. In Darian-Smith, I. (ed.): Handbook of Physiology. Sec. 1, Vol. III. Bethesda, Md., American Physiological Society, 1984, p. 457.

Schor, C. M. (ed.): Vergence Eye Movements: Basic and Clinical Aspects. Woburn, Mass., Butterworth, 1982.

Sherman, S. M., and Spear, P. D.: Organization of visual pathways in normal and visually deprived cats. Physiol. Rev., 62:738, 1982.

Simpson, J. I.: The accessory optic system. Annu. Rev. Neurosci., 7:13, 1984.

Song, P.-S.: Protozoan and related photoreceptors: Molecular aspects. Annu. Rev. Biophys. Bioeng., 12:35, 1983.

Sparks, D. L.: Translation of sensory signals into commands for control of saccadic eye movements: Role of primate superior colliculus. Physiol. Rev., 66:118, 1986.

Sterling, P.: Microcircuitry of the cat retina. Annu. Rev. Neurosci., 6:149, 1983.

Wolfe, J. M. (ed.): The Mind's Eye. New York, W. H. Freeman and Company, 1986.

Woolsey, C. N. (ed.): Cortical Sensory Organization. Multiple Visual Areas. Clifton, N.J., Humana Press, 1981.

Wurtz, R. H., and Albano, J. E.: Visual-motor function of the primate superior colliculus. Annu. Rev. Neurosci., 3:189, 1980.

CHAPTER 52

• *The Sense of*
Hearing •

The purpose of this chapter is to describe and explain the mechanism by which the ear receives sound waves, discriminates their frequencies, and finally transmits auditory information into the central nervous system where its meaning is deciphered.

● THE TYMPANIC MEMBRANE AND THE OSSICULAR SYSTEM

CONDUCTION OF SOUND FROM THE TYMPANIC MEMBRANE TO THE COCHLEA

Figure 52–1 illustrates the *tympanic membrane* (commonly called the *eardrum)* and the *ossicular system,* which conducts sound through the middle ear. The tympanic membrane is cone-shaped, with its concavity facing downward and outward toward the auditory canal. Attached to the very center of the tympanic membrane is the *handle* of the *malleus.* At its other end the malleus is tightly bound to the *incus* by ligaments so that whenever the malleus moves the incus moves with it. The opposite end of the incus in turn articulates with the stem of the *stapes,* and the *faceplate* of the stapes lies against the membranous labyrinth in the opening of the oval window where sound waves are conducted into the inner ear, the *cochlea.*

The ossicles of the middle ear are suspended by ligaments in such a way that the combined malleus and incus act as a single lever having its fulcrum approximately at the border of the tympanic membrane. The large *head* of the malleus, which is on the opposite side of the fulcrum from the handle, almost exactly balances the other end of the lever.

The articulation of the incus with the stapes causes the stapes to push forward on the cochlear fluid every time the handle of the malleus moves inward and to pull backward on the fluid every time the malleus moves outward, which promotes inward and outward motion of the faceplate at the oval window. ⌐

The handle of the malleus is constantly pulled inward by the *tensor tympani muscle,* which keeps the tympanic membrane tensed. This allows sound vibra-

tions on *any* portion of the tympanic membrane to be transmitted to the malleus, which would not be true if the membrane were lax.

Impedance Matching by the Ossicular System. The amplitude of movement of the stapes faceplate with each sound vibration is only three-fourths as much as the amplitude of the handle of the malleus. Therefore, the ossicular lever system does not amplify the movement distance of the stapes, as is commonly believed. Instead, the system actually reduces the amplitude but increases the *force* of movement about 1.3 times. However, the surface area of the tympanic membrane is approximately 55 square millimeters, whereas the surface area of the stapes averages 3.2 square millimeters. This 17-fold difference times the 1.3-fold ratio of the lever system allows energy of a sound wave impinging on the tympanic membrane to be applied to the small faceplate of the stapes, causing approximately 22 times as much *pressure* on the fluid of the cochlea as is exerted by the sound wave against the tympanic membrane. Because fluid has far greater inertia than air, it is easily understood that increased amounts of pressure are needed to cause vibration in the fluid. Therefore, the tympanic membrane and ossicular system provide *impedance matching* between the sound waves in air and the sound vibrations in the fluid of the cochlea. Indeed, the impedance matching is about 50 to 75 per cent of perfect for sound frequencies between 300 and 3000 cycles per second, which allows utilization of most of the energy in the incoming sound waves.

In the absence of the ossicular system and tympanum, sound waves can travel directly through the air of the middle ear and can enter the cochlea at the oval window. However, the sensitivity for hearing is then 15 to 20 decibels less than for ossicular transmission — equivalent to a decrease from a medium voice to a barely perceptible voice level.

Attenuation of Sound by Contraction of the Stapedius and Tensor Tympani Muscles. When loud sounds are transmitted through the ossicular system into the central nervous system, a reflex occurs after a latent period of 40 to 80 milliseconds to cause contraction of the *stapedius* and *tensor tympani muscles.* The tensor tympani muscle pulls the handle of the malleus inward while the stapedius muscle pulls

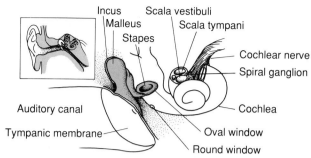

Figure 52–1. The tympanic membrane, the ossicular system of the middle ear, and the inner ear.

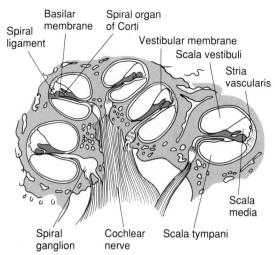

Figure 52–2. The cochlea. (From Goss, C. M. [ed.]: Gray's Anatomy of the Human Body. Philadelphia, Lea & Febiger.)

the stapes outward. These two forces oppose each other and thereby cause the entire ossicular system to develop a high degree of rigidity, thus greatly reducing the ossicular conduction of low frequency sound, mainly frequencies below 1000 cycles per second.

This *attenuation reflex* can reduce the intensity of sound transmission by as much as 30 to 40 decibels, which is about the same difference as that between a loud voice and the sound of a whisper. The function of this mechanism is probably twofold:

1. To *protect* the cochlea from damaging vibrations caused by excessively loud sound.

2. To *mask* low frequency sounds in loud environments. This usually removes a major share of the background noise and allows a person to concentrate on sounds above 1000 cycles per second, where most of the pertinent information in voice communication is transmitted.

Another function of the tensor tympani and stapedius muscles is to decrease a person's hearing sensitivity to his or her own speech. This effect is activated by collateral signals transmitted to these muscles at the same time that the brain activates the voice mechanism.

TRANSMISSION OF SOUND THROUGH BONE

Because the inner ear, the *cochlea,* is embedded in a bony cavity in the temporal bone called the bony labyrinth, vibrations of the entire skull can cause fluid vibrations in the cochlea itself. Therefore, under appropriate conditions, a tuning fork or an electronic vibrator placed on any bony protuberance of the skull, but especially on the mastoid process, causes the person to hear the sound. Unfortunately, the energy available even in very loud sound in the air is not sufficient to cause hearing through the bone except when a special electromechanical sound-transmitting device is applied directly to the bone.

● THE COCHLEA

FUNCTIONAL ANATOMY OF THE COCHLEA

The cochlea is a system of coiled tubes, shown in Figure 52–1 and in cross section in Figures 52–2 and

52–3. It consists of three different tubes coiled side by side: the *scala vestibuli,* the *scala media,* and the *scala tympani.* The scala vestibuli and scala media are separated from each other by *Reissner's membrane* (also called the *vestibular membrane*), shown in Figure 52–3; and the scala tympani and scala media are separated from each other by the *basilar membrane.* On the surface of the basilar membrane lies a structure, the *organ of Corti,* which contains a series of electromechanically sensitive cells, the *hair cells.* These are the receptive end-organs that generate nerve impulses in response to sound vibrations.

Figure 52–4 diagrams the functional parts of the uncoiled cochlea for conduction of sound vibrations. First, note that Reissner's membrane is missing from this figure. This membrane is so thin and so easily moved that it does not obstruct the passage of sound vibrations from the scala vestibuli into the scala media at all. Therefore, so far as the conduction of sound is concerned, the scala vestibuli and scala media are considered to be a single chamber. The importance of Reissner's membrane is to maintain a special fluid in the scala media that is required for normal function of the sound-receptive hair cells, as discussed later in the chapter.

Sound vibrations enter the scala vestibuli from the faceplate of the stapes at the oval window. The faceplate covers this window and is connected with the window's edges by a relatively loose annular ligament so that it can move inward and outward with the sound vibrations. Inward movement causes the fluid to move into the scala vestibuli and scala media, and outward movement causes the fluid to move backward.

The Basilar Membrane and Resonance in the Cochlea. The basilar membrane is a fibrous membrane that separates the scala media and the scala tympani. It contains 20,000 to 30,000 *basilar fibers* that project from the bony center of the cochlea, the *modiolus,* toward the outer wall. These fibers are stiff, elastic, reedlike structures that are fixed at their basal ends in the central bony structure of the cochlea (the

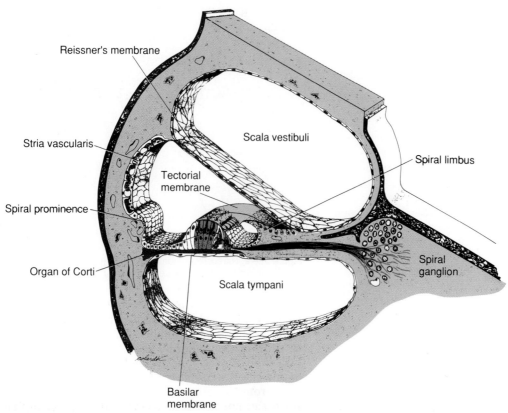

Reissner's membrane

Stria vascularis

Tectorial membrane

Spiral prominence

Organ of Corti

Scala vestibuli

Spiral limbus

Spiral ganglion

Scala tympani

Basilar membrane

Figure 52–3. A section through one of the turns of the cochlea. (Drawn by Sylvia Colard Keene. From Fawcett: A Textbook of Histology, 11th Ed. Philadelphia, W. B. Saunders Company, 1986.)

modiolus) but not fixed at their distal ends, except that the distal ends are embedded in the loose basilar membrane. Because the fibers are stiff and also free at one end, they can vibrate like reeds of a harmonica.

The lengths of the basilar fibers increase progressively as one goes from the base of the cochlea to its apex, from a length of approximately 0.04 millimeters near the oval and round windows to 0.5 millimeters at the tip of the cochlea, a 12-fold increase in length.

The diameters of the fibers, on the other hand, decrease from the base to the helicotrema, so that their overall stiffness decreases more than 100-fold. As a result, the stiff, short fibers near the oval window of the cochlea will vibrate at a high frequency, whereas the long, limber fibers near the tip of the cochlea will vibrate at a low frequency.

Oval window

Stapes

Scala vestibuli and scala media

Round window

Scala tympani

Basilar membrane

Helicotrema

Figure 52–4. Movement of fluid in the cochlea following forward thrust of the stapes.

Thus, high frequency resonance of the basilar membrane occurs near the base, where the sound waves enter the cochlea through the oval window; and low frequency resonance occurs near the apex mainly because of difference in stiffness of the fibers but also because of increased "loading" of the basilar membrane with extra amounts of fluid that must vibrate with the membrane at the apex.

TRANSMISSION OF SOUND WAVES IN THE COCHLEA—THE "TRAVELING WAVE"

If the foot of the stapes moves inward instantaneously, the round window must also bulge outward instantaneously because the cochlea is bounded on all sides by bony walls. Therefore, the initial effect is to cause the basilar membrane at the very base of the cochlea to bulge in the direction of the round window. However, the elastic tension that is built up in the basilar fibers as they bend toward the round window initiates a wave that "travels" along the basilar membrane toward the helicotrema, as illustrated in Figure 52–5. Figure 52–5A shows movement of a high frequency wave down the basilar membrane; Figure 52–5B, a medium frequency wave; and Figure 52–5C, a very low frequency wave. Movement of the wave along the basilar membrane is comparable to the movement of a pressure wave along the arterial walls, which was

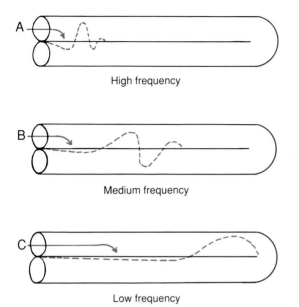

Figure 52–5. "Traveling waves" along the basilar membrane for high, medium, and low frequency sounds.

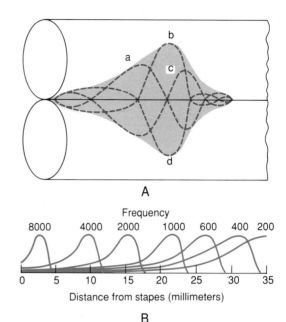

Figure 52–6. *A,* Amplitude pattern of vibration of the basilar membrane for a medium frequency sound. *B,* Amplitude patterns for sounds of all frequencies between 200 and 8000 per second, showing the points of maximum amplitude (the resonance points) on the basilar membrane for the different frequencies.

discussed in Chapter 15, or it is also comparable to the wave that travels along the surface of a pond.

Pattern of Vibration of the Basilar Membrane for Different Sound Frequencies. Note in Figure 52–5 the different patterns of transmission for sound waves of different frequencies. Each wave is relatively weak at the outset but becomes strong when it reaches that portion of the basilar membrane that has a natural resonant frequency equal to the respective sound frequency. At this point the basilar membrane can vibrate back and forth with such great ease that the energy in the wave is completely dissipated. Consequently, the wave dies out at this point and fails to travel the remaining distance along the basilar membrane. Thus, a high frequency sound wave travels only a short distance along the basilar membrane before it reaches its resonant point and dies out; a medium frequency sound wave travels about halfway and then dies out; and finally, a very low frequency sound wave travels the entire distance along the membrane.

Another feature of the traveling wave is that it travels fast along the initial portion of the basilar membrane but progressively more slowly as it goes farther and farther into the cochlea. The cause of this is the high coefficient of elasticity of the basilar fibers near the stapes and a progressively decreasing coefficient farther along the membrane. This rapid initial transmission of the wave allows the high frequency sounds to travel far enough into the cochlea to spread out and separate from each other on the basilar membrane. Without this, all the high frequency waves would be bunched together within the first millimeter or so of the basilar membrane, and their frequencies could not be discriminated one from the other.

Amplitude Pattern of Vibration of the Basilar Membrane. The dashed curves of Figure 52–6A

show the position of a sound wave on the basilar membrane when the stapes (*a*) is all the way inward, (*b*) has moved back to the neutral point, (*c*) is all the way outward, and (*d*) has moved back again to the neutral point but is moving inward. The shaded area around these different waves shows the extent of vibration of the basilar membrane during a complete vibratory cycle. This is the *amplitude pattern of vibration* of the basilar membrane for this particular sound frequency.

Figure 52–6B shows the amplitude patterns of vibration for different frequencies, showing that the maximum amplitude for 8000 cycles occurs near the base of the cochlea, whereas that for frequencies less than 200 cycles per second is all the way at the tip of the basilar membrane near the helicotrema where the scala vestibuli opens into the scala tympani.

The principal method by which sound frequencies, especially those above 200 cycles per second, are discriminated from one another is based on the "place" of maximum stimulation of the nerve fibers from the organ of Corti lying on the basilar membrane, as is explained in the following section.

FUNCTION OF THE ORGAN OF CORTI

The organ of Corti, illustrated in Figures 52–2, 52–3, and 52–7, is the receptor organ that generates nerve impulses in response to vibration of the basilar membrane. Note that the organ of Corti lies on the surface of the basilar fibers and basilar membrane. The actual sensory receptors in the organ of Corti are two types of *hair cells,* a single row of *internal hair cells,* numbering about 3500 and measuring about 12

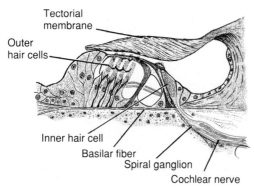

Figure 52–7. The organ of Corti, showing especially the hair cells and the tectorial membrane against the projecting hairs.

micrometers in diameter, and three to four rows of *external hair cells,* numbering about 15,000 and having diameters of only about 8 micrometers. The bases and sides of the hair cells synapse with a network of cochlear nerve endings. These lead to the *spiral ganglion of Corti,* which lies in the modiolus (the center) of the cochlea. The spiral ganglion in turn sends axons into the *cochlear nerve* and thence into the central nervous system at the level of the upper medulla. The relationship of the organ of Corti to the spiral ganglion and to the cochlear nerve is illustrated in Figure 52–2.

Excitation of the Hair Cells. Note in Figure 52–7 that minute hairs, or *stereocilia,* project upward from the hair cells and either touch or are embedded in the surface gel coating of the *tectorial membrane,* which lies above the stereocilia in the scala media. These hair cells are similar to the hair cells found in the macula and cristae ampullaris of the vestibular apparatus, which are discussed in Chapter 55. Bending of the hairs in one direction depolarizes the hair cells, and bending them in the opposite direction hyperpolarizes them. This in turn excites the nerve fibers synapsing with their bases.

Figure 52–8 illustrates the mechanism by which vibration of the basilar membrane excites the hair endings. The upper ends of the hair cells are fixed tightly in a rigid structure composed of a flat plate, called the *reticular lamina,* supported by triangular *rods of Corti,* which in turn are attached tightly to the

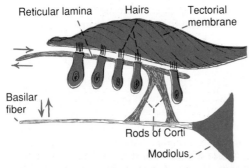

Figure 52–8. Stimulation of the hair cells by the to-and-fro movement of the hairs in the tectorial membrane.

basilar fibers. Therefore, the basilar fiber, the rods of Corti, and the reticular lamina all move as a rigid unit.

Upward movement of the basilar fiber rocks the reticular lamina upward and *inward.* Then, when the basilar membrane moves downward, the reticular lamina rocks downward and *outward.* The inward and outward motion causes the hairs to shear back and forth against the tectorial membrane; or, in the case of the internal hair cells, the hairs of which do not necessarily touch the tectorial membrane, fluid rushes back and forth over the hairs and bends them. Thus, the hair cells are excited whenever the basilar membrane vibrates.

Hair Cell Receptor Potentials and Excitation of Auditory Nerve Fibers. The stereocilia are stiff structures because each of these has a rigid internal structural protein framework as explained for all cilia in Chapter 2. Each hair cell has about 100 stereocilia on its apical border. These become progressively longer on the side away from the modiolus, and the tops of the shorter stereocilia are each attached by a thin filament to the side of its adjacent longer stereocilium. Therefore, whenever the cilia are bent in the direction of the longer ones, the tips of the smaller stereocilia are tugged outward from the surface of the hair cell. This causes a mechanical transduction that opens as many as 200 to 300 cation conducting channels, allowing rapid movement of positively charged potassium ions into the tips of the stereocilia, which in turn causes depolarization of the entire hair cell membrane.

Thus, when the basilar fibers bend toward the scala vestibuli, the hair cells depolarize, and in the opposite direction they hyperpolarize, thus generating an alternating hair cell receptor potential. This in turn stimulates the cochlear nerve endings that synapse with the bases of the hair cells. It is believed that a rapidly acting neurotransmitter is released by the hair cells at these synapses during depolarization. It is possible that the transmitter substance is glutamate, but this is not certain.

The Endocochlear Potential. To explain even more fully the electrical potentials generated by the hair cells, we need to explain another electrical phenomenon called the endocochlear potential: The scala media is filled with a fluid called *endolymph,* in contradistinction to the *perilymph* present in the scala vestibuli and scala tympani. The scala vestibuli and scala tympani communicate directly with the subarachnoid space around the brain, so that the perilymph is almost identical with cerebrospinal fluid. On the other hand, the endolymph that fills the scala media is an entirely different fluid secreted by the *stria vascularis,* a highly vascular area on the outer wall of the scala media. Endolymph contains a very high concentration of potassium and a very low concentration of sodium, which is exactly opposite to the perilymph.

An electrical potential of approximately +80 millivolts exists all the time between the endolymph and the perilymph, with positivity inside the scala media and negativity outside. This is called the *endocochlear potential,* and it is believed to be generated by continual transport of positive potassium ions from the perilymph into the scala media by the stria vascularis.

The importance of the endocochlear potential is that the tops of the hair cells project through the reticular lamina and are bathed by the endolymph of the scala media, whereas perilymph bathes the lower bodies of the hair cells. Furthermore, the hair cells have a negative intracellular potential of −60 millivolts with respect to the perilymph, but −140 millivolts with respect to the endolymph at their upper surfaces, where the hairs project into the endolymph. It is believed that this high electrical potential at the tips of the stereocilia greatly sensitizes the cell, thereby increasing its ability to respond to the slightest sound.

DETERMINATION OF SOUND FREQUENCY — THE "PLACE" PRINCIPLE

From earlier discussions in this chapter it is already apparent that low frequency sounds cause maximal activation of the basilar membrane near the apex of the cochlea, sounds of high frequency activate the basilar membrane near the base of the cochlea, and intermediate frequencies activate the membrane at intermediate distances between these two extremes. Furthermore, there is spatial organization of the nerve fibers in the cochlear pathway all the way from the cochlea to the cerebral cortex. And recording of signals from the auditory tracts in the brain stem and from the auditory receptive fields in the cerebral cortex shows that specific neurons are activated by specific sound frequencies. Therefore, the major method used by the nervous system to detect different frequencies is to determine the position along the basilar membrane that is most stimulated. This is called the *place principle* for determination of frequency (or of sound "pitch").

Yet, referring again to Figure 52–6, one can see that the distal end of the basilar membrane at the helicotrema is stimulated by all sound frequencies below 200 cycles per second. Therefore, it has been difficult to understand from the place principle how one can differentiate between very low sound frequencies, from 200 down to 20. It is postulated that these low frequencies are discriminated mainly by the so-called *volley* or *frequency principle*. That is, low frequency sounds, from 20 up to 2000 to 4000 cycles per second, can cause volleys of impulses at the same low frequencies to be transmitted by the cochlear nerve into the cochlear nuclei. It is believed that the cochlear nuclei then distinguish the different frequencies. In fact, destruction of the entire apical half of the cochlea, which destroys the basilar membrane where all the lower frequency sounds are normally detected, still does not completely eliminate the discrimination of low frequency sounds.

DETERMINATION OF LOUDNESS

Loudness is determined by the auditory system in at least three different ways: First, as the sound becomes louder, the amplitude of vibration of the basilar membrane and hair cells also increases, so that the hair cells excite the nerve endings at more rapid rates.

Second, as the amplitude of vibration increases, it causes more and more of the hair cells on the fringes of the resonating portion of the basilar membrane to become stimulated, thus causing *spatial summation* of impulses — that is, transmission through many nerve fibers, rather than through a few. Third, certain hair cells do not become stimulated until the vibration of the basilar membrane reaches a relatively high intensity, and it is believed that stimulation of these cells in some way apprises the nervous system that the sound is then very loud.

Detection of Changes in Loudness—The Power Law. It was pointed out in Chapter 46 that a person interprets changes in intensity of sensory stimuli approximately in proportion to a power function of the actual intensity. In the case of sound, the interpreted sensation changes approximately in proportion to the cube root of the actual sound intensity. To express this another way, the ear can discriminate differences in sound intensity from the softest whisper to the loudest possible noise, representing an *approximate 1 trillion times* increase in sound energy or 1 million times increase in amplitude of movement of the basilar membrane. Yet the ear interprets this much difference in sound level as approximately a 10,000-fold change. Thus, the scale of intensity is greatly "compressed" by the sound perception mechanisms of the auditory system. This obviously allows a person to interpret differences in sound intensities over an extremely wide range, a far broader range than would be possible were it not for compression of the scale.

The Decibel Unit. Because of the extreme changes in sound intensities that the ear can detect and discriminate, sound intensities are usually expressed in terms of the logarithm of their actual intensities. A 10-fold increase in sound energy (or a $\sqrt{10}$-fold increase in sound pressure, because energy is proportional to the square of pressure) is called 1 *bel,* and 0.1 bel is called 1 *decibel.* One decibel represents an actual increase in sound energy of 1.26 times.

Another reason for using the decibel system in expressing changes in loudness is that, in the usual sound intensity range for communication, the ears can barely distinguish approximately a 1 decibel *change* in sound intensity.

Threshold for Hearing Sound at Different Frequencies. Figure 52–9 shows the pressure thresholds at which sounds of different frequencies can barely be heard by the ear. This figure illustrates that a 3000 cycle per second sound can be heard even when its intensity is as low as 70 decibels below 1 dyne/cm² sound pressure level, which is one ten-millionth microwatt per square centimeter. On the other hand, a 100 cycle per second sound can be detected only if its intensity is 10,000 times as great as this.

Frequency Range of Hearing. The frequencies of sound that a young person can hear, before aging has occurred in the ears, is generally stated to be between 20 and 20,000 cycles per second. However, referring again to Figure 52–9, we see that the sound range depends to a great extent on intensity. If the intensity is 60 decibels below the 1 dyne/cm² sound pressure level, the sound range is 500 to 5000 cycles per second, and only with intense sounds can the

Figure 52–9. Relationship of the threshold of hearing and the threshold of somesthetic perception to the sound energy level at each sound frequency. (Modified from Stevens and Davis: Hearing. New York, John Wiley & Sons.)

complete range of 20 to 20,000 cycles be achieved. In old age, the frequency range falls to 50 to 8000 cycles per second or less, as discussed later in the chapter.

● CENTRAL AUDITORY MECHANISMS

THE AUDITORY PATHWAY

Figure 52–10 illustrates the major auditory pathways. It shows that nerve fibers from the *spiral ganglion of Corti* enter the *dorsal* and *ventral cochlear nuclei* located in the upper part of the medulla. At this point, all the fibers synapse, and second-order neurons pass mainly to the opposite side of the brain stem

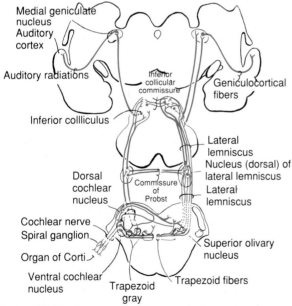

Figure 52–10. The auditory pathway. (Modified from Crosby, E., Humphrey, T., and Lauer, E.: Correlative Anatomy of the Nervous System. New York, Macmillan Publishing Co., 1962. Copyright 1962 by Macmillan Publishing Co. Reprinted with permission.)

through the *trapezoid body* to the *superior olivary nucleus.* However, some second-order fibers also pass ipsilaterally to the superior olivary nucleus on the same side. From the superior olivary nucleus the auditory pathway then passes upward through the *lateral lemniscus;* and some, but not all, of the fibers terminate in the *nucleus of the lateral lemniscus.* Many bypass this nucleus and pass on to the inferior colliculus, where either all or almost all of them terminate. From here, the pathway passes to the *medial geniculate nucleus,* where all the fibers again synapse. And, finally, the auditory pathway proceeds by way of the *auditory radiation* to the *auditory cortex,* located mainly in the superior gyrus of the temporal lobe.

Several points of importance in relation to the auditory pathway should be noted. First, signals from both ears are transmitted through the pathways of both sides of the brain with only slight preponderance of transmission in the contralateral pathway. In at least three different places in the brain stem crossing-over occurs between the two pathways: (1) in the trapezoid body, (2) in the commissure of Probst between the two nuclei of the lateral lemnisci, and (3) in the commissure connecting the two inferior colliculi.

Second, many collateral fibers from the auditory tracts pass directly into the *reticular activating system of the brain stem.* This system projects diffusely upward into the cerebral cortex and downward into the spinal cord and activates the entire nervous system in response to a loud sound. Other collaterals go to the *vermis of the cerebellum,* which is also activated instantaneously in the event of a sudden noise.

Third, a high degree of spatial orientation is maintained in the fiber tracts from the cochlea all the way to the cortex. In fact, there are *three* different spatial representations of sound frequencies in the cochlear nuclei, *two* representations in the inferior colliculi, *one precise* representation for discrete sound frequencies in the auditory cortex, and *at least five other less precise* representations in the auditory cortex and auditory association areas.

Firing Rates at Different Levels of the Auditory Pathway. Single nerve fibers entering the cochlear nuclei from the auditory nerve can fire at rates up to at least 1000 per second, the rate being determined mainly by the loudness of the sound. At sound frequencies up to 2000 to 4000 cycles per second, the auditory nerve impulses are often synchronized with the sound waves, but they do not necessarily occur with every wave.

In the auditory tracts of the brain stem, the firing is usually no longer synchronized with the sound frequency except at sound frequencies below 200 cycles per second. And above the level of the inferior colliculi, even this synchronization is mainly lost. These findings demonstrate that the sound signals are not transmitted unchanged directly from the ear to the higher levels of the brain; instead, information from the sound signals begins to be dissected from the impulse traffic at levels as low as the cochlear nuclei. We will have more to say about this later, especially in relation to perception of direction from which sound comes.

Another significant feature of the auditory pathways is that low rates of impulse firing continue even in the absence of sound all the way from the cochlear nerve fibers to the auditory cortex. When the basilar membrane moves toward the scala vestibuli, the impulse traffic increases; when the basilar membrane moves toward the scala tympani, the impulse traffic decreases. Thus, the presence of this background signal allows information to be transmitted from the basilar membrane when the membrane moves in either direction: positive information in one direction and negative information in the opposite direction. Were it not for the background signal, only the positive half of the information could be transmitted. This type of so-called "carrier wave" method for transmitting information is utilized in many parts of the brain, as discussed in several of the succeeding chapters.

FUNCTION OF THE CEREBRAL CORTEX IN HEARING

The projection areas of the auditory pathway to the cerebral cortex are illustrated in Figure 52–11, which shows that the auditory cortex lies principally on the *supratemporal plane of the superior temporal gyrus* but also extends over the *lateral border of the temporal lobe,* over much of the *insular cortex,* and even into the most lateral portion of the *parietal operculum.*

Two separate areas are shown in Figure 52–11: the *primary auditory cortex* and the *auditory association cortex* (also called the *secondary auditory cortex*). The primary auditory cortex is directly excited by projections from the medial geniculate body, whereas the auditory association areas are excited secondarily by impulses from the primary auditory cortex and by projections from thalamic association areas adjacent to the medial geniculate body.

Sound Frequency Perception in the Primary Auditory Cortex. At least six different *tonotopic maps* have been found in the primary auditory cortex and auditory association areas. In each of these maps,

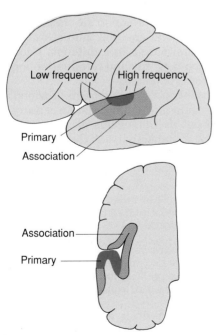

Figure 52–11. The auditory cortex.

high frequency sounds excite neurons at one end of the map, whereas low frequency sounds excite the neurons at the opposite end. In most, the low frequency sounds are located anteriorly, as shown in Figure 52–11; and the high frequency sounds, posteriorly. However, this is not true for all the maps. The question that one must ask is why does the auditory cortex have so many different tonotopic maps? The answer is presumably that each of the separate areas dissects out some specific feature of the sounds. For instance, one of the large maps in the primary auditory cortex almost certainly discriminates the sound frequencies themselves and gives the person the psychic sensation of sound pitches. Another one of the maps probably is used to detect the direction from which the sound comes.

The frequency range to which each individual neuron in the auditory cortex responds is much narrower than that in the cochlear and brain stem relay nuclei. Referring back to Figure 52–6B, we note that the basilar membrane near the base of the cochlea is stimulated by all frequency sounds, and in the cochlear nuclei this same breadth of sound representation is found. Yet by the time the excitation has reached the cerebral cortex, most sound-responsive neurons respond only to a narrow range of frequencies, rather than a broad range. Therefore, somewhere along the pathway, processing mechanisms "sharpen" the frequency response. It is believed that this sharpening effect is caused mainly by the phenomenon of lateral inhibition, which was discussed in Chapter 46 in relation to mechanisms for transmitting information in nerves. That is, stimulation of the cochlea at one frequency causes inhibition of signals caused by sound frequencies on either side of the stimulated frequency, this resulting from collateral fibers angling off the primary signal pathway and exerting inhibitory influences on adjacent pathways. The same effect has also been demonstrated to be important in sharpening patterns of somesthetic images, visual images, and other types of sensations.

A large share of the neurons in the auditory cortex, especially in the auditory association cortex, do not respond to specific sound frequencies in the ear. It is believed that these neurons "associate" different sound frequencies with each other or associate sound information with information from other sensory areas of the cortex. Indeed, the parietal portion of the auditory association cortex partly overlaps somatic sensory area II, which could provide easy opportunity for association of auditory information with somatic sensory information.

Discrimination of Sound "Patterns" by the Auditory Cortex. Complete bilateral removal of the auditory cortex does not prevent a cat or monkey from detecting sounds or reacting in a crude manner to the sounds. However, it does greatly reduce or sometimes even abolish its ability to discriminate different sound pitches and especially *patterns of sound.* For instance, an animal that has been trained to recognize a combination or sequence of tones, one following the other in a particular pattern, loses this ability when the auditory cortex is destroyed; and, furthermore, it cannot relearn this type of response. Therefore, the auditory cortex is important in the discrimination of *tonal* and *sequential sound patterns.*

Total destruction of both primary auditory cortices in the human being is said to reduce greatly one's sensitivity for hearing, which is quite different from the effect in lower animals. However, this information is not clear. On the other hand, destruction of the primary auditory cortex on only one side in the human being has little effect on hearing because of the many crossover connections from side to side in the neural pathway. Yet this does affect one's ability to localize the source of sound, because comparative signals in both cortices are required for this localization function.

Lesions in the human being affecting the auditory association areas but not affecting the primary auditory cortex do not decrease the person's ability to hear and differentiate sound tones and to interpret at least simple patterns of sound. However, he or she will often be unable to interpret the *meaning* of the sound heard. For instance, lesions in the posterior portion of the superior temporal gyrus, which is called Wernicke's area and is also part of the auditory association cortex, often make it impossible for the person to interpret the meanings of words even though he hears them perfectly well and can even repeat them. These functions of the auditory association areas and their relationship to the overall intellectual functions of the brain are discussed in detail in Chapter 57.

DISCRIMINATION OF THE DIRECTION FROM WHICH SOUND EMANATES

A person determines the direction from which sound emanates by two principal mechanisms: (1) by the time lag between the entry of sound into one ear and into the opposite ear and (2) by the difference between the intensities of the sounds in the two ears. The first mechanism functions best at frequencies below 3000 cycles per second, and the intensity mechanism operates best at higher frequencies because the head acts as a sound barrier at these frequencies. The time lag mechanism discriminates direction much more exactly than the intensity mechanism, for the time lag mechanism does not depend on extraneous factors but only on an exact interval of time between two acoustical signals. If a person is looking straight toward the sound, the sound reaches both ears at exactly the same instant, while, if the right ear is closer to the sound than the left ear, the sound signals from the right ear enter the brain ahead of those from the left ear.

Neural Mechanisms for Detecting Sound Direction. Destruction of the auditory cortex on both sides of the brain, in either human beings or lower mammals, causes loss of almost all ability to detect the direction from which sound comes. Yet the mechanism for this detection process begins in the superior

olivary nuclei, even though it requires the neural pathways all the way from these nuclei to the cortex for interpretation of the signals. The mechanism is believed to be the following:

First, the superior olivary nucleus is divided into two sections, (1) the *medial superior olivary nucleus* and (2) the *lateral superior olivary nucleus.* The lateral nucleus is concerned with detecting the direction from which the sound is coming by the *difference in intensities of the sound* reaching the two ears, presumably by simply comparing the two intensities and sending an appropriate signal to the auditory cortex to estimate the direction.

The *medial superior olivary nucleus,* on the other hand, has a very specific mechanism for *detecting the time-lag between acoustic signals entering the two ears.* This nucleus contains large numbers of neurons that have two major dendrites, one projecting to the right and the other to the left. The acoustical signal from the right ear impinges on the right dendrite, and the signal from the left ear impinges on the left dendrite. The intensity of excitation of each of these neurons is highly sensitive to a specific time-lag between the two acoustical signals from the two ears. That is, the neurons near one border of the nucleus respond maximally to a short time-lag; whereas those near the opposite border respond to a very long time-lag; and those between, to intermediate time-lags. Thus, a spatial pattern of the neuronal stimulation develops in the medial superior olivary nucleus, with sound from directly in front of the head stimulating one set of olivary neurons maximally and sounds from different side angles stimulating other sets of neurons on opposite sides of the straight front neurons. This spatial orientation of signals is then transmitted all the way to the auditory cortex, where sound direction is determined by the locus in the cortex that is stimulated maximally. It is believed that the signals for determining sound direction are transmitted through a different pathway and that this pathway terminates in the cerebral cortex in a different locus from the transmission pathway and termination locus for the tonal patterns of sound.

This mechanism for detection of sound direction indicates again how information in sensory signals is dissected out as the signals pass through different levels of neuronal activity. In this case, the "quality" of sound direction is separated from the "quality" of sound tones at the level of the superior olivary nuclei.

CENTRIFUGAL SIGNALS FROM THE CENTRAL NERVOUS SYSTEM TO LOWER AUDITORY CENTERS

Retrograde pathways have been demonstrated at each level of the nervous system from the auditory cortex to the cochlea. The final pathway is mainly from the superior olivary nucleus to the hair cells themselves in the organ of Corti.

These retrograde fibers are inhibitory. Indeed, direct stimulation of discrete points in the olivary nucleus has been shown to inhibit specific areas of the organ of Corti, reducing their sound sensitivities as much as 15 to 20 decibels. One can readily understand how this could allow a person to direct attention to sounds of particular qualities while rejecting sounds of other qualities. This is readily demonstrated when one listens to a single instrument in a symphony orchestra.

● HEARING ABNORMALITIES

TYPES OF DEAFNESS

Deafness is usually divided into two types: first, that caused by impairment of the cochlea or auditory nerve, which is usually classed as "nerve deafness," and, second, that caused by impairment of the mechanisms for transmitting sound into the cochlea, which is usually called "conduction deafness." Obviously, if either the cochlea or the auditory nerve is completely destroyed, the person is permanently deaf. However, if the cochlea and nerve are still intact but the ossicular system has been destroyed or ankylosed ("frozen" in place by fibrosis or calcification), sound waves can still be conducted into the cochlea by means of bone conduction.

The Audiometer. To determine the nature of hearing disabilities, the audiometer is used. Simply an earphone connected to an electronic oscillator capable of emitting pure tones ranging from low frequencies to high frequencies, the instrument is calibrated so that the zero intensity level of sound at each frequency is the loudness that can barely be heard by the normal person, based on previous studies of normal persons. However, a calibrated volume control can increase or decrease the loudness of each tone above or below the zero level. If the loudness of a tone must be increased to 30 decibels above normal before it can be heard, the person is said to have a *hearing loss* of 30 decibels for that particular tone.

In performing a hearing test using an audiometer, one tests approximately 8 to 10 frequencies covering the auditory spectrum, and the hearing loss is determined for each of these frequencies. Then the so-called "audiogram" is plotted as shown in Figures 52–12 and 52–13, depicting the hearing loss for each of the frequencies in the auditory spectrum.

The audiometer, in addition to being equipped with an earphone for testing air conduction by the ear, is also equipped with an electronic vibrator for testing bone conduction from the mastoid process into the cochlea.

Figure 52–12. Audiogram of the old-age type of nerve deafness.

Figure 52–13. Audiogram of deafness resulting from middle ear sclerosis.

The Audiogram in Nerve Deafness. In nerve deafness—this term including damage to the cochlea, to the auditory nerve, or to the central nervous system circuits from the ear—the person has decreased or total loss of ability to hear sound as tested by both air conduction and bone conduction. An audiogram depicting partial nerve deafness is illustrated in Figure 52–12. In this figure the deafness is mainly for high frequency sound. Such deafness could be caused by damage to the base of the cochlea. This type of deafness occurs to some extent in almost all older persons.

Other patterns of nerve deafness frequently occur as follows: (1) deafness for low frequency sounds caused by excessive and prolonged exposure to very loud sounds (the rock band or the jet airplane engine) because low frequency sounds are usually louder and more damaging to the organ of Corti, and (2) deafness for all frequencies caused by drug sensitivity of the organ of Corti, especially sensitivity to some antibiotics such as streptomycin, kanamycin, and chloramphenicol.

The Audiogram in Conduction Deafness. A frequent type of deafness is that caused by fibrosis of the middle ear following repeated infection in the middle ear or fibrosis occurring in the hereditary disease called *otosclerosis.* In this instance the sound waves cannot be transmitted easily through the ossicles from the tympanic membrane to the oval window. Figure 52–13 illustrates an audiogram from a person with "middle ear deafness" of this type. In this case the bone conduction is essentially normal, but air conduction is greatly depressed at all frequencies, more so at the low frequencies. In this type of deafness, the faceplate of the stapes frequently becomes "ankylosed" by bony overgrowth to the edges of the oval window. In this case, the person becomes totally deaf for air conduction but can be made to hear again almost normally by removing the stapes and replacing it with a minute Teflon or metal prosthesis that transmits the sound from the incus to the oval window.

REFERENCES

Aitkin, L. M., et al.: Central neural mechanisms of hearing. In Darian-Smith, I. (ed.): Handbook of Physiology. Sec. 1, Vol. III. Bethesda, Md., American Physiological Society, 1984, p. 675.

Altschuler, R. A., et al.: Neurobiology of Hearing: The Cochlea. New York, Raven Press, 1986.

Ballenger, J. J. (ed.): Diseases of the Nose, Throat, Ear, Head, and Neck. Philadelphia, Lea & Febiger, 1985.

Becker, W.: Atlas of Ear, Nose and Throat Diseases, Including Bronchoesophagology. Philadelphia, W. B. Saunders Co., 1984.

Borg, E., and Counter, S. A.: The middle-ear muscles. Sci. Am., August, 1989, p. 74.

Brugge, J. F., and Geisler, C. D.: Auditory mechanisms of the lower brainstem. Annu. Rev. Neurosci., 1:363, 1978.

Dallos, P.: Peripheral mechanisms of hearing. In Darian-Smith, I. (ed.): Handbook of Physiology. Sec. 1, Vol. III. Bethesda, Md., American Physiological Society, 1984, p. 595.

Fujimura, O.: Vocal Physiology: Voice Production, Mechanisms and Functions. New York, Raven Press, 1988.

Glasscock, M., III: Shambaugh's Surgery of the Ear, 4th Ed. Philadelphia, W. B. Saunders Co., 1989.

Green, D. M., and Wier, C. C.: Auditory perception. In Darian-Smith, I. (ed.): Handbook of Physiology. Sec. 1, Vol. III. Bethesda, Md., American Physiological Society, 1984, p. 557.

Guth, P. S., and Melamed, B.: Neurotransmission in the auditory system: A primer for pharmacologists. Annu. Rev. Pharmacol. Toxicol., 22:383, 1982.

Hawke, M., et al.: Diseases of the Ear: Clinical and Pathologic Aspects. Philadelphia, Lea & Febiger, 1987.

Hudspeth, A. J.: Mechanoelectrical transduction by hair cells in the acousticolateralis sensory system. Annu. Rev. Neurosci., 6:187, 1983.

Hudspeth, A. J.: The cellular basis of hearing: The biophysics of hair cells. Science, 230:745, 1985.

Imig, T. J., and Morel, A.: Organization of the thalamocortical auditory system in the cat. Annu. Rev. Neurosci., 6:95, 1983.

Kay, R. H.: Hearing of modulation in sounds. Physiol. Rev., 62:187, 1983.

Kiang, N. Y. S.: Peripheral neural processing of auditory information. In Darian-Smith, I. (ed.): Handbook of Physiology. Sec. 1, Vol. III. Bethesda, Md., American Physiological Society, 1984, p. 639.

Lee, K. J.: Textbook of Otolaryngology and Head and Neck Surgery. New York, Elsevier Science Publishing Co., 1989.

Lucente, F. E., and Sobol, S. M.: Essentials of Otolaryngology, 2nd Ed. New York, Raven Press, 1988.

Masterton, R. B., and Imig, T. J.: Neural mechanisms of sound localization. Annu. Rev. Physiol., 46:275, 1984.

Patuzzi, R., and Robertson, D.: Tuning in the mammalian cochlea. Physiol. Rev., 68:1009, 1988.

Rhode, W. S.: Cochlear mechanisms. Annu. Rev. Physiol., 46:231, 1984.

Rubel, E. W.: Ontogeny of auditory system function. Annu. Rev. Physiol., 46:213, 1984.

Sachs, M. B.: Neural coding of complex sounds: Speech. Annu. Rev. Physiol., 46:261, 1984.

Sataloff, J., et al.: Hearing Loss, 2nd Ed. Philadelphia, J. B. Lippincott Co., 1980.

Schneiderman, C. R.: Basic Anatomy and Physiology in Speech and Hearing. San Diego, College-Hill Press, 1984.

Simmons, J. A., and Kick, S. A.: Physiological mechanisms for spatial filtering and image enhancement in the sonar of bats. Annu. Rev. Physiol., 46:599, 1984.

Singh, R. P.: Anatomy of Hearing and Speech. New York, Oxford University Press, 1980.

Sterkers, O., et al.: How are inner ear fluids formed? News Physiol. Sci., 2:176, 1987.

Stevens, S. S.: Hearing. Its Psychology and Physiology. New York, Acoustical Society of America, 1983.

Syka, J., and Masterton, R. B. (eds.): Auditory Pathway. Structure and Function. New York, Plenum Publishing Corp., 1988.

Weiss, T. F.: Relation of receptor potentials of cochlear hair cells to spike discharges of cochlear neurons. Annu. Rev. Physiol., 46:247, 1984.

Wever, E. G., and Lawrence, M.: Physiological Acoustics. Princeton, Princeton University Press, 1954.

Woolsey, D. N. (ed.): Cortical Sensory Organization. Multiple Auditory Areas. Clifton, N. J., Humana Press, 1982.

CHAPTER 53

• *The Chemical Senses—Taste and Smell* •

The senses of taste and smell allow us to separate undesirable or even lethal foods from those that are nutritious. And the sense of smell allows animals to recognize the proximity of other animals, or even individuals among animals. Finally, both senses are strongly tied to primitive emotional and behavioral functions of our nervous systems.

● THE SENSE OF TASTE

Taste is mainly a function of the *taste buds* in the mouth, but it is common experience that one's sense of smell also contributes strongly to taste perception. In addition, the texture of food, as detected by tactual senses of the mouth, and the presence of such substances in the food as pepper, which stimulate pain endings, greatly condition the taste experience. The importance of taste lies in the fact that it allows a person to select food in accord with desires and perhaps also in accord with the needs of the tissues for specific nutritive substances.

THE PRIMARY SENSATIONS OF TASTE

The identities of the specific chemicals that excite different taste receptors are still very incomplete. Even so, psychophysiological and neurophysiological studies have identified at least 13 possible or probable chemical receptors in the taste cells as follows: 2 sodium receptors, 2 potassium receptors, 1 chloride receptor, 1 adenosine receptor, 1 inosine receptor, 2 sweet receptors, 2 bitter receptors, 1 glutamate receptor, and 1 hydrogen ion receptor.

For practical analysis of taste, however, the above receptor capabilities have been collected into 4 general categories called the *primary sensations of taste.* These are *sour, salty, sweet,* and *bitter.*

We know, of course, that a person can perceive literally hundreds of different tastes. These are all supposed to be combinations of the elementary sensations in the same manner that all the colors that we can see are combinations of the three primary colors as described in Chapter 50.

The Sour Taste. The sour taste is caused by acids, and the intensity of the taste sensation is approximately proportional to the logarithm of the *hydrogen ion concentration.* That is, the more acidic the acid, the stronger becomes the sensation.

The Salty Taste. The salty taste is elicited by ionized salts. The quality of the taste varies somewhat from one salt to another because the salts also elicit other taste sensations besides saltiness. The cations of the salts are mainly responsible for the salty taste, but the anions also contribute to a lesser extent.

The Sweet Taste. The sweet taste is not caused by any single class of chemicals. A list of some of the types of chemicals that cause this taste includes sugars, glycols, alcohols, aldehydes, ketones, amides, esters, amino acids, sulfonic acids, halogenated acids, and inorganic salts of lead and beryllium. Note specifically that most of the substances that cause a sweet taste are organic chemicals. It is especially interesting that very slight changes in the chemical structure, such as addition of a simple radical, can often change the substance from sweet to bitter.

The Bitter Taste. The bitter taste, like the sweet taste, is not caused by any single type of chemical agent; but, here again, the substances that give the bitter taste are almost entirely organic substances. Two particular classes of substances are especially likely to cause bitter taste sensations: (1) long chain organic substances containing nitrogen and (2) alkaloids. The alkaloids include many of the drugs used in medicines such as quinine, caffeine, strychnine, and nicotine.

Some substances that at first taste sweet have a bitter aftertaste. This is true of saccharin, which makes this substance objectionable to some people.

The bitter taste, when it occurs in high intensity, usually causes the person or animal to reject the food. This is undoubtedly an important purposive function of the bitter taste sensation, because many of the deadly toxins found in poisonous plants are alkaloids, which all cause intensely bitter taste.

Threshold for Taste

The threshold for stimulation of the sour taste by hydrochloric acid averages 0.0009 N; for stimulation of the salty taste by sodium chloride, 0.01 M; for the sweet taste by sucrose, 0.01 M; and for the bitter taste by quinine, 0.000008 M. Note especially how much more sensitive is the bitter taste sense than all the others, which would be expected, because this sensation provides an important protective function.

Table 53-1 gives the relative taste indices (the reciprocals of the taste thresholds) of different substances. In this table, the intensities of the four different primary sensations of taste are referred, respectively, to the intensities of taste of hydrochloric acid, quinine, sucrose, and sodium chloride, each of which is considered to have a taste index of 1.

Taste Blindness. Many persons are taste blind for certain substances, especially for different types of thiourea compounds. A substance used frequently by psychologists for demonstrating taste blindness is *phenylthiocarbamide,* for which approximately 15 to 30 per cent of all people exhibit taste blindness, the exact percentage depending on the method of testing and the concentration of the substance.

THE TASTE BUD AND ITS FUNCTION

Figure 53-1 illustrates a taste bud, which has a diameter of about 1/30 millimeter and a length of about 1/16 millimeter. The taste bud is composed of about 40 modified epithelial cells, some of which are supporting cells called *sustentacular cells* and others are *taste cells.* The taste cells are continually being replaced by mitotic division from the surrounding epithelial cells so that some are young cells and others are mature cells that lie toward the center of the bud and soon break up and dissolve. The life span of each taste cell is about ten days in lower mammals but is unknown for the human being.

The outer tips of the taste cells are arranged around a minute *taste pore,* shown in Figure 53-1. From the tip of each cell, several *microvilli,* or *taste hairs,* protrude outward into the taste pore to approach the cavity of the mouth. These microvilli provide the receptor surface for taste.

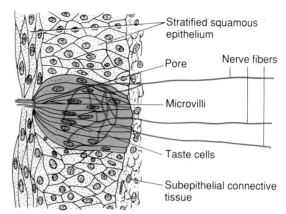

Figure 53-1. The taste bud.

Interwoven among the taste cells is a branching terminal network of several *taste nerve fibers* that are stimulated by the taste receptor cells. Some of these fibers invaginate into folds of the taste cell membranes. Many vesicles form beneath the membrane near the fibers, suggesting that these might secrete a neurotransmitter to excite the nerve fibers in response to taste stimulation.

Location of the Taste Buds. The taste buds are found on three different types of papillae of the tongue, as follows: (1) A large number of taste buds are on the walls of the troughs that surround the circumvallate papillae, which form a V line toward the posterior of the tongue. (2) Moderate numbers of taste buds are on the fungiform papillae over the flat anterior surface of the tongue. (3) Moderate numbers are on the foliate papillae located in the folds along the lateral surfaces of the tongue. Additional taste buds are located on the palate and a few on the tonsillar pillars, the epiglottis, and even the proximal esophagus. Adults have approximately 10,000 taste buds, and children a few more. Beyond the age of 45 years many taste buds rapidly degenerate, causing the taste sensation to become progressively less critical.

Especially important in relation to taste is the tendency for taste buds subserving particular primary sensations of taste to be located in special areas. The sweet and salty tastes are located *principally* on the tip of the tongue, the sour taste on the two lateral sides of the tongue, and the bitter taste on the posterior tongue and soft palate.

Table 53-1 RELATIVE TASTE INDICES OF DIFFERENT SUBSTANCES

Sour Substances	Index	Bitter Substances	Index	Sweet Substances	Index	Salty Substances	Index
Hydrochloric acid	1	Quinine	1	Sucrose	1	NaCl	1
Formic acid	1.1	Brucine	11	1-propoxy-2-amino-		NaF	2
Chloracetic acid	0.9	Strychnine	3.1	4-nitrobenzene	5000	$CaCl_2$	1
Acetyllactic acid	0.85	Nicotine	1.3	Saccharin	675	NaBr	0.4
Lactic acid	0.85	Phenylthiourea	0.9	Chloroform	40	NaI	0.35
Tartaric acid	0.7	Caffeine	0.4	Fructose	1.7	LiCl	0.4
Malic acid	0.6	Veratrine	0.2	Alanine	1.3	NH_4Cl	2.5
Potassium H tartrate	0.58	Pilocarpine	0.16	Glucose	0.8	KCl	0.6
Acetic acid	0.55	Atropine	0.13	Maltose	0.45		
Citric acid	0.46	Cocaine	0.02	Galactose	0.32		
Carbonic acid	0.06	Morphine	0.02	Lactose	0.3		

(From Derma: *Proc. Oklahoma Acad. Sci.,* 27:9, 1947; and Pfaffman: Handbook of Physiology. Sec. I, Vol. I. Baltimore, Williams & Wilkins, 1959, p. 507.)

Specificity of Taste Buds for the Primary Taste Stimuli. Microelectrode studies from single taste buds while they are stimulated successively by the four different primary taste stimuli have shown that most of them can be excited by two, three, or even four of the primary taste stimuli as well as by a few other taste stimuli that do not fit into the "primary" categories. Usually, though, one or two of the taste categories will predominate.

Mechanism of Stimulation of Taste Buds. *The Receptor Potential.* The membrane of the taste cell, like that of other sensory receptor cells, is negatively charged on the inside with respect to the outside. Application of a taste substance to the taste hairs causes partial loss of this negative potential — that is, the taste cell is *depolarized.* The decrease in potential, within a wide range, is approximately proportional to the logarithm of concentration of the stimulating substance. This change in potential in the taste cell is the *receptor potential* for taste.

The mechanism by which the stimulating substance reacts with the taste villi to initiate the receptor potential is believed to be by binding of the taste chemicals to protein receptor molecules that protrude through the villus membrane. This in turn opens ion channels, which allow sodium ions to enter and depolarize the cell. Then the taste chemical is gradually washed away from the taste hair by the saliva, which removes the stimulus. Supposedly, the types of receptors in each taste hair determine the types of taste that will elicit responses.

Generation of Nerve Impulses by the Taste Bud. On first application of the taste stimulus, the rate of discharge of the nerve fibers rises to a peak in a small fraction of a second, but then it adapts within the next 2 seconds back to a lower steady level. Thus, a strong immediate signal is transmitted by the taste nerve, and a weaker continuous signal is transmitted as long as the taste bud is exposed to the taste stimulus.

TRANSMISSION OF TASTE SIGNALS INTO THE CENTRAL NERVOUS SYSTEM

Figure 53–2 illustrates the neuronal pathways for transmission of taste signals from the tongue and pharyngeal region into the central nervous system. Taste impulses from the anterior two thirds of the tongue pass first into the *fifth nerve* and then through the *chorda tympani* into the *facial nerve,* thence into the *tractus solitarius* in the brain stem. Taste sensations from the circumvallate papillae on the back of the tongue and from other posterior regions of the mouth are transmitted through the *glossopharyngeal nerve* also into the *tractus solitarius* but at a slightly lower level. Finally, a few taste signals are transmitted into the *tractus solitarius* from the base of the tongue and other parts of the pharyngeal region by way of the *vagus nerve.*

All taste fibers synapse in the *nuclei of the tractus solitarius* and send second-order neurons to a small

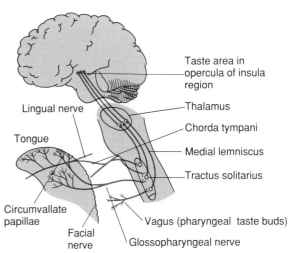

Figure 53–2. Transmission of taste impulses into the central nervous system.

area of the *ventral posterior medial nucleus of the thalamus* located slightly medial to the thalamic terminations of the facial regions of the dorsal column-medial lemniscal system. From the thalamus, third-order neurons are transmitted to the *lower tip of the postcentral gyrus in the parietal cortex,* where it curls deep into the sylvian fissure and also into the adjacent *operculorinsular area,* also in the sylvian fissure. This lies slightly lateral, ventral, and rostral to the tongue area of somatic area I.

From this description of the taste pathways, it immediately becomes evident that they parallel closely the somatic pathways from the tongue.

Taste Reflexes Integrated in the Brain Stem. From the tractus solitarius a large number of impulses are transmitted within the brain stem itself directly into the *superior* and *inferior salivatory nuclei,* and these in turn transmit impulses to the submandibular, sublingual, and parotid glands to help control the secretion of saliva during the ingestion of food.

Adaptation of Taste. Everyone is familiar with the fact that taste sensations adapt rapidly, often with almost complete adaptation within a minute or so of continuous stimulation. Yet, from electrophysiological studies of taste nerve fibers, it is clear that the taste buds themselves adapt enough to account for no more than about one-half of this. Therefore, the extreme degree of adaptation that occurs in the sensation of taste almost certainly occurs in the central nervous system itself, though the mechanism and site of this are not known. At any rate, it is a mechanism different from that of most other sensory systems, which adapt mainly at the receptors.

TASTE PREFERENCE AND CONTROL OF THE DIET

Taste preferences mean simply that an animal will choose certain types of food in preference to others, and it automatically uses this to help control the type of diet it eats. Furthermore, its taste preferences often change in accord with the needs of the body for certain specific substances. The following experiments illustrate this ability of animals to choose food in accord with the needs of its body: First, adrenalectomized animals automatically select drinking water with a high concentration of sodium chloride in preference to pure water, and this in many instances is sufficient to supply the needs of the body and prevent death as a result of salt depletion. Second, an animal given injections of excessive amounts of insulin develops a depleted blood sugar, and it automatically chooses the sweetest food from among many samples. Third, parathyroidectomized animals automatically choose drinking water with a high concentration of calcium chloride.

These same phenomena are also observed in many instances of everyday life. For instance, the salt licks of the desert region are known to attract animals from far and wide, and even the human being rejects any food that has an unpleasant affective sensation, which certainly in many instances protects our bodies from undesirable substances.

The phenomenon of taste preference almost certainly results from some mechanism located in the central nervous system and not from a mechanism in the taste receptors themselves, although it is true that the receptors often do become sensitized to the needed nutrient. An important reason for believing taste preference to be mainly a central phenomenon is that previous experience with unpleasant or pleasant tastes plays a major role in determining one's different taste preferences. For instance, if a person becomes sick soon after eating a particular type of food, the person generally develops a negative taste preference, or *taste aversion,* for that particular food thereafter; the same effect can be demonstrated in animals.

● THE SENSE OF SMELL

Smell is the least understood of our senses. This results partly from the fact that the sense of smell is a subjective phenomenon that cannot be studied with ease in lower animals. Still another complicating problem is that the sense of smell is almost rudimentary in the human being in comparison with that of some lower animals.

THE OLFACTORY MEMBRANE

The olfactory membrane lies in the superior part of each nostril, as illustrated in Figure 53–3. Medially it folds downward over the surface of the septum, and laterally it folds over the superior turbinate and even over a small portion of the upper surface of the middle turbinate. In each nostril the olfactory membrane has a surface area of approximately 2.4 square centimeters.

The Olfactory Cells. The receptor cells for the smell sensation are the *olfactory cells,* which are actually bipolar nerve cells derived originally from the central nervous system itself. There are about 100 million of these cells in the olfactory epithelium interspersed among *sustentacular cells,* as shown in Figure 53–4. The mucosal end of the olfactory cell forms a knob from which 6 to 12 *olfactory hairs,* or *cilia,* 0.3 micrometer in diameter and up to 200 micrometers in length, project into the mucus that coats the inner

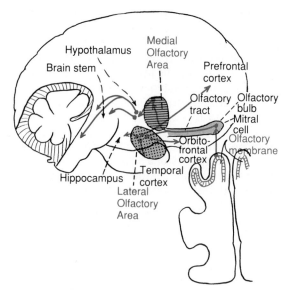

Figure 53–3. Neural connections of the olfactory system.

surface of the nasal cavity. These projecting olfactory cilia form a dense mat in the mucus, and it is these cilia that react to odors in the air and then stimulate the olfactory cells, as discussed later. Spaced among the olfactory cells in the olfactory membrane are many small *glands of Bowman* that secrete mucus onto the surface of the olfactory membrane.

STIMULATION OF THE OLFACTORY CELLS

Mechanism of Excitation of the Olfactory Cells. The portion of the olfactory cells that responds to the olfactory chemical stimuli is the *cilia.* The membranes of the cilia contain large numbers of protein molecules that protrude all the way through the membrane and that can bind with different odorant substances. These proteins are called *odorant binding proteins.* It is presumed that this binding is the necessary stimulus for exciting the olfactory cells.

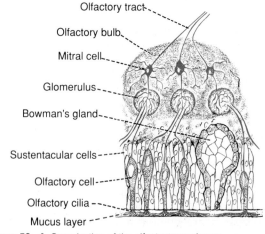

Figure 53–4. Organization of the olfactory membrane.

Two different theories have been proposed for the mechanism of excitation: The simplest theory suggests that the molecules of the odorant binding proteins themselves open up to become ion channels when the odorant binds, allowing mainly large numbers of positively charged sodium ions to flow to the interior of the olfactory cell and depolarize it. The second theory proposes that binding of the odorant causes the odorant binding protein to become an activated adenylate cyclase at its end that protrudes to the interior of the cell. The cyclase in turn catalyzes the formation of cyclic adenosine monophosphate (cAMP), and the cAMP acts on many other membrane proteins to open ion channels through them. This second mechanism would provide an extremely sensitive receptor because of a cascade effect that would occur and allow even the most minute of stimulation to cause a reaction.

Regardless of the basic chemical mechanism by which the olfactory cells are stimulated, several physical factors also affect the degree of stimulation. First, only volatile substances that can be sniffed into the nostrils can be smelled. Second, the stimulating substance must be at least slightly water soluble, so that it can pass through the mucus to reach the olfactory cells. And, third, it must be at least slightly lipid soluble, presumably because the lipid constituents of the cell membrane repel odorants from the membrane receptor proteins.

Membrane Potentials and Action Potentials in Olfactory Cells. The membrane potential of unstimulated olfactory cells, as measured by microelectrodes, averages about -55 millivolts. At this potential, most of the cells generate continuous action potentials at a rate varying from once every 20 seconds up to two to three per second.

Most odorants cause depolarization of the olfactory membrane, decreasing the negative potential in the olfactory cell from -55 down to as low as -30 millivolts or even less. Along with this, the number of action potentials increases to about 20 per second, which is a very high rate for the very minute, fraction-of-a-micrometer olfactory nerve fibers.

A few odorants hyperpolarize the olfactory cell membrane, thus decreasing, instead of increasing, the nerve firing rate.

Over a wide range, the rate of olfactory nerve impulses is approximately proportional to the logarithm of the stimulus strength, which illustrates that the olfactory receptors tend to obey principles of transduction similar to those of other sensory receptors.

Adaptation. The olfactory receptors adapt approximately 50 per cent in the first second or so after stimulation. Thereafter, they adapt very little and very slowly. Yet we all know from our own experience that smell sensations adapt almost to extinction within a minute or so after one enters a strongly odorous atmosphere. Because this psychological adaptation is far greater than the degree of adaptation of the receptors themselves, it is almost certain that most of the adaptation occurs in the central nervous system, which seems also to be true for the adaptation of taste sensations. A postulated neuronal mechanism for this adaptation is the following: Large numbers of centrifugal nerve fibers pass from the olfactory regions of the brain backward along the olfactory tract and terminate on special inhibitory cells in the olfactory bulb, the *granule cells*. It is postulated that after the onset of an olfactory stimulus the central nervous system gradually develops a strong feedback inhibition to suppress relaying of the smell signals through the olfactory bulb.

Search for the Primary Sensations of Smell

Most physiologists are convinced that the many smell sensations are subserved by a few rather discrete primary sensations, in the same way that vision and taste are subserved by a select few sensations. But, thus far, only minor success has been achieved in classifying the primary sensations of smell. Yet, on the basis of psychological tests and action potential studies from various points in the olfactory nerve pathways, it has been postulated that about seven different primary classes of olfactory stimulants preferentially excite separate olfactory cells. These classes of olfactory stimulants are characterized as follows:

1. Camphoraceous.
2. Musky.
3. Floral.
4. Pepperminty.
5. Ethereal.
6. Pungent.
7. Putrid.

However, it is unlikely that this list actually represents the true primary sensations of smell, even though it does illustrate the results of one of the many attempts to classify them. Indeed, several clues in recent years have indicated that there may be as many as *50* or more primary sensations of smell—a marked contrast to only *three* primary sensations of color detected by the eyes and only a few primary sensations of taste detected by the tongue. For instance, persons have been found who have *odor blindness* for single substances; and such discrete odor blindness has been identified for more than 50 different substances. Because it is presumed that odor blindness for each substance represents a lack of the appropriate receptor protein in olfactory cells for that substance, it is possible that the sense of smell might be subserved by 50 or more primary smell sensations.

Affective Nature of Smell. Smell, equally as much as taste, has the affective qualities of either pleasantness or unpleasantness. Because of this, smell is as important as, if not more important than, taste in the selection of food. Indeed, a person who has previously eaten food that has disagreed with him is often nauseated by even the smell of that same type of food on a second occasion. Other types of odors that have proved to be unpleasant in the past may also provoke a disagreeable feeling; on the other hand, perfume of the right quality can wreak havoc with masculine emotions. In addition, in some lower animals odors are the primary excitant of sexual drive.

Threshold for Smell. One of the principal characteristics of smell is the minute quantity of the stimulating agent in the air often required to effect a smell sensation. For instance, the substance *methyl mercaptan* can be smelled when only one 25 billionth of a milligram is present in each milliliter of air. Because of this low threshold, this substance is mixed with natural gas to give the gas an odor that can be detected when it leaks from a gas pipe.

Measurement of Smell Threshold. One of the problems in studying smell has been difficulty in obtaining accurate measurements of the threshold stimulus required to induce smell. The simplest technique is simply to allow a person to sniff different substances in the usual manner of smelling. Indeed, some investigators feel that this is as satisfactory as almost any other procedure. However, to eliminate variations from person to person, more objective methods have been developed: One of these has been to place a box containing the volatilized agent over the subject's head. Appropriate precautions are taken to exclude odors from the person's own body. The person is allowed to breathe naturally, but the volatilized agent is distributed evenly in the air that is breathed.

Gradations of Smell Intensities. Though the threshold concentrations of substances that evoke smell are extremely slight, concentrations only 10 to 50 times above the threshold values for certain substances sometimes evoke maximum intensity of smell. This is in contrast to most other sensory systems of the body, in which the ranges of detection are tremendous — for instance, 500,000 to 1 in the case of the eyes and 1 trillion to 1 in the case of the ears. This perhaps can be explained by the fact that smell is concerned more with detecting the presence or absence of odors than with quantitative detection of their intensities.

TRANSMISSION OF SMELL SIGNALS INTO THE CENTRAL NERVOUS SYSTEM

The olfactory portions of the brain are among its oldest structures, and much of the remainder of the brain developed around these olfactory beginnings. In fact, part of the brain that originally subserved olfaction later evolved into the basal brain structures that in the human being control emotions and other aspects of behavior; this is the system we call the *limbic system,* discussed in Chapter 58.

Transmission of Olfactory Signals Into the Olfactory Bulb. The olfactory bulb, which is also called cranial nerve I, is illustrated in Figure 53–3. Although it looks like a nerve, in reality it is an anterior outgrowth of brain tissue from the base of the brain having a bulbous enlargement, the *olfactory bulb,* at its end that lies over the *cribriform plate* separating the brain cavity from the upper reaches of the nasal cavity. The cribriform plate has multiple small perforations through which an equal number of small nerves enter the olfactory bulb from the olfactory membrane. Figure 53–4 illustrates the close relationship between the *olfactory cells* in the olfactory membrane and the olfactory bulb, showing very short axons terminating in multiple globular structures of the olfactory bulb called *glomeruli.* Each bulb has several thousand such glomeruli, each of which is the

terminus for about 25,000 axons from olfactory cells. Each glomerulus also is the terminus for dendrites from about 25 large *mitral cells* and about 60 smaller *tufted cells* the cell bodies of which lie also in the olfactory bulb superior to the glomeruli. These cells in turn send axons through the olfactory tract into the central nervous system.

Recent research work suggests that different glomeruli respond to different odors. Therefore, it is possible that the specific glomeruli that are stimulated are the real clue to the analysis of different odor signals transmitted into the central nervous system.

The Very Old, the Old, and the Newer Olfactory Pathways Into the Central Nervous System

The olfactory tract enters the brain at the junction between the mesencephalon and cerebrum; there the tract divides into two pathways, one passing medially into the *medial olfactory area* and the other laterally into the *lateral olfactory area.* The medial olfactory area represents a very old olfactory system, while the lateral olfactory area is the input to both a less old olfactory system and a newer system.

The Very Old Olfactory System — The Medial Olfactory Area. The medial olfactory area consists of a group of nuclei located in the midbasal portions of the brain anterior and superior to the hypothalamus. Most conspicuous are the *septal nuclei,* which are midline nuclei that feed into the hypothalamus and other portions of the brain's limbic system, the system that is concerned with basic behavior as described in Chapter 58.

The importance of this medial olfactory area is best understood by considering what happens in animals when the lateral olfactory areas on both sides of the brain are removed and only the medial system remains. The answer is that this hardly affects the more primitive responses to olfaction such as licking the lips, salivation, and other feeding responses caused by the smell of food, or such as primitive emotional drives associated with smell. On the other hand, removal of the lateral areas does abolish the more complicated olfactory conditioned reflexes.

The Old Olfactory System — The Lateral Olfactory Area. The lateral olfactory area is composed mainly of the *prepyriform* and *pyriform cortex* plus the *cortical portion of the amygdaloid nuclei.* From these areas, signal pathways pass into almost all portions of the limbic system, especially into the hippocampus, which is most important in the learning process — in this case presumably learning of likes and dislikes of certain foods depending upon experiences with the foods. For instance, it is this lateral olfactory area and its many connections with the limbic behavioral system that cause a person to develop absolute aversion to foods that have previously caused nausea and vomiting.

An important feature of the lateral olfactory area is that many signal pathways from this area feed directly into an older type of cerebral cortex called the paleocortex in the anteromedial portion of the temporal lobe. This is the only area of the entire cerebral cortex where sensory signals pass directly to the cortex without passing through the thalamus.

The Newer Pathway. Still a newer olfactory pathway has now been found that does indeed pass through the thalamus, passing to the dorsomedial thalamic nucleus and thence to the lateroposterior quadrant of the orbitofrontal cortex. Based on studies in monkeys, this newer system probably helps especially in the conscious analysis of odor.

Thus, there appears to be a very old olfactory system that subserves the basic olfactory reflexes, an old system that provides automatic but learned control of food intake and aversion to toxic and unhealthy foods, and finally a newer system that is comparable to most of the other cortical sensory systems and is used for conscious perception of olfaction.

Centrifugal Control of Activity in the Olfactory Bulb by the Central Nervous System. Many nerve fibers originating in the olfactory portions of the brain pass in the backward direction in the olfactory tract to the olfactory bulb, that is, "centrifugally" from the brain to the periphery. These terminate on a very large number of small *granule cells* located in the center of the bulb. These in turn send short, inhibitory *dendrites* to the mitral and tufted cells. It is believed that this inhibitory feedback to the olfactory bulb might be a means of helping to sharpen one's specific capability of distinguishing one odor from another.

Electrical Activity in the Olfactory Nerves and Tracts. Electrophysiological studies show that the mitral and tufted cells are continually active, the same as is true for the olfactory receptors as discussed earlier. Superimposed on this background are increases or decreases in impulse traffic caused by different odors. Thus, the olfactory stimuli *modulate* the frequency of impulses in the olfactory system and in this way transmit the olfactory information.

REFERENCES

Alberts, J. R.: Producing and interpreting experimental olfactory deficits. Physiol. Behav., 12:657, 1974.

Chanel, J.: The olfactory system as a molecular descriptor. News Physiol. Sci., 2:203, 1987.

Dastoli, F. R.: Taste receptor proteins. Life Sci., 14:1417, 1974.

Denton, D. A.: Salt appetite. In Code, C. F., and Heidel, W. (eds.): Handbook of Physiology. Sec. 6, Vol. 1. Baltimore, Md., Williams & Wilkins, 1967, p. 433.

Douek, E.: The Sense of Smell and Its Abnormalities. New York, Churchill Livingstone, 1974.

Getchell, T. V.: Functional properties of vertebrate olfactory receptor neurons. Physiol. Rev., 66:772, 1986.

Kashara, Y. (ed.): Proceedings of the Seventeenth Japanese Symposium on Taste and Smell. Arlington, VA, IRL Press, 1984.

Lat, J.: Self-selection of dietary components. In Code, C. F., and Heidel, W. (eds.): Handbook of Physiology. Sec. 6, Vol. 1, Baltimore, Md., Williams and Wilkins, 1967, p. 367.

Margolis, F. L., and Getchell, T. V. (eds.): Molecular Neurobiology of the Olfactory System. New York, Plenum Publishing Corp., 1988.

McBurney, D. H.: Taste and olfaction: Sensory discrimination. In Darian-Smith, I. (ed.): Handbook of Physiology. Sec. 1, Vol. III. Bethesda, Md., American Physiological Society, 1984, p. 1067.

Monmaney, T.: Are we led by the nose? Discover, September, 1987, p. 48.

Moulton, D. G., and Beidler, L. M.: Structure and function in the peripheral olfactory system. Physiol. Rev., 47:1, 1967.

Norgren, R.: Central neural mechanisms of taste. In Darian-Smith, I. (ed.): Handbook of Physiology. Sec. 1, Vol. III. Bethesda, Md., American Physiological Society, 1984, p. 1087.

Oakley, B., and Benjamin, R. M.: Neural mechanisms of taste. Physiol. Rev., 46:173, 1966.

Roper, S. D.: The cell biology of vertebrate taste receptors. Annu. Rev. Neurosci., 12:329, 1989.

Schiffman, S. S.: Taste transduction and modulation. News Physiol. Sci., 3:109, 1988.

Shepherd, G. M.: The olfactory bulb: A simple system in the mammalian brain. In Brookhart, J. M., and Mountcastle, V. B. (eds.): Handbook of Physiology. Sec. 1, Vol. I. Baltimore, Md., Williams & Wilkins, 1977, p. 945.

Takagi, S. F.: The olfactory nervous system of the Old World monkey. Jpn. J. Physiol., 34:51, 1984.

Zotterman, Y.: Olfaction and Taste. New York, Macmillan Co., 1963.

UNIT XI

THE NERVOUS SYSTEM: C. MOTOR AND INTEGRATIVE NEUROPHYSIOLOGY

CHAPTER 54

• *Motor Functions of the Spinal Cord; the Cord Reflexes* •

In the discussion of the nervous system thus far, we have considered principally the input of sensory information. In the following chapters we discuss the origin and output of motor signals, the signals that cause muscle contraction, secretory function, and other motor effects throughout the body.

Sensory information is integrated at all levels of the nervous system and causes appropriate motor responses, beginning in the spinal cord with relatively simple reflexes, extending into the brain stem with still more complicated responses, and finally extending to the cerebrum, where the most complicated responses are controlled.

In the present chapter we discuss the control of muscle function by the spinal cord. The spinal cord is not merely a conduit for sensory signals to the brain or for motor signals from the brain back to the periphery. In fact, without the special neuronal circuits of the cord, even the most fundamental motor control systems in the brain cannot cause any purposeful muscle movement. To give an example, there is no neuronal circuit anywhere in the brain that causes the specific to-and-fro movement of the legs that is required in walking. Instead, the circuits for these movements are in the cord, and the brain simply sends *command* signals to set into motion the walking process. Thus, under appropriate conditions, a cat or a dog with its cord transected in the neck can be made to walk.

Yet let us not belittle the role of the brain as well, for the brain gives the sequential directions to the cord activities, to promote turning movements when they are required, to lean the body forward during acceleration, to change the movements from walking to jumping as needed, and to monitor continuously and control equilibrium. All of this is done through "command" signals from above. But it also requires the many neuronal circuits of the spinal cord that are themselves the objects of the commands. These circuits in turn provide all but a small fraction of the direct control of the muscles.

Experimental Preparations for Studying Cord Reflexes—The Spinal Animal and the Decerebrate Animal. Two different types of experimental preparations

have been especially useful in studying spinal cord function: (1) the *spinal animal,* in which the spinal cord is transected, frequently in the neck so that most of the cord still remains functional; and (2) the *decerebrate animal,* in which the brain stem is transected in the lower part of the mesencephalon.

Immediately after preparing a *spinal animal,* most spinal cord function is severely depressed below the level of the transection. However, after a few hours in lower animals, and after a few days to weeks in monkeys, most of the intrinsic spinal cord functions return nearly to normal and provide a suitable experimental preparation for study.

In the *decerebrate animal* the brain stem is transected at the lower mesencephalic level, which blocks the normal inhibitory signals from the higher control centers of the brain to the pontile reticular and vestibular nuclei. This causes these nuclei to become tonically active, transmitting facilitatory signals to most of the spinal cord motor control circuits. The result is that these become easy to activate by even the slightest sensory input signals to the cord. Using this preparation, one can study very easily the intrinsic motor functions of the cord itself.

ORGANIZATION OF THE SPINAL CORD FOR MOTOR FUNCTIONS

The cord gray matter is the integrative area for the cord reflexes and other motor functions. Figure 54–1 shows the typical organization of the cord gray matter in a single cord segment. Sensory signals enter the cord almost entirely through the sensory (posterior) roots. After entering the cord, every sensory signal travels to two separate destinations. First, one branch of the sensory nerve terminates in the gray matter of the cord and elicits local segmental reflexes and other effects. Second, another branch transmits signals to higher levels of the nervous system—to higher levels in the cord itself, to the brain stem, or even to the cerebral cortex, as described in earlier chapters.

Each segment of the spinal cord has several million neurons in its gray matter. Aside from the sensory relay neurons discussed in Chapters 47 and 48, the remainder of these neurons are of two separate types, the *anterior motor neurons* and the *interneurons.*

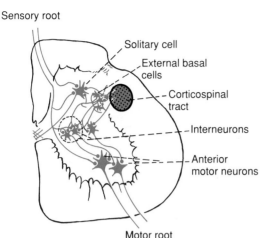

Sensory root

Solitary cell

External basal cells

Corticospinal tract

Interneurons

Anterior motor neurons

Motor root

Figure 54–1. Connections of the sensory fibers and corticospinal fibers with the interneurons and anterior motor neurons of the spinal cord.

The Anterior Motor Neurons. Located in each segment of the anterior horns of the cord gray matter are several thousand neurons that are 50 to 100 per cent larger than most of the others and called *anterior motor neurons*. These give rise to the nerve fibers that leave the cord via the anterior roots and innervate the skeletal muscle fibers. The neurons are of two types, the *alpha motor neurons* and the *gamma motor neurons*.

The Alpha Motor Neurons. The alpha motor neurons give rise to large type A alpha (Aα) nerve fibers ranging from 9 to 20 micrometers in diameter that innervate the large skeletal muscle fibers. Stimulation of a single nerve fiber excites from as few as three to as many as several hundred skeletal muscle fibers, which are collectively called the *motor unit.* Transmission of nerve impulses into skeletal muscles and their stimulation of the muscles were discussed in Chapters 6 and 7.

The Gamma Motor Neurons. In addition to the alpha motor neurons that excite contraction of the skeletal muscle fibers, about one half as many much smaller gamma motor neurons are located along with the alpha motor neurons in the anterior horns. These transmit impulses through type A gamma (Aγ) fibers, averaging 5 micrometers in diameter, to very small, special skeletal muscle fibers called *intrafusal fibers.* These are part of the *muscle spindle,* which is discussed later in the chapter.

The Interneurons. Interneurons are present in all areas of the cord gray matter — in the dorsal horns, in the anterior horns, and in the intermediate areas between these two. These cells are numerous — approximately 30 times as numerous as the anterior motor neurons. They are small and highly excitable, often exhibiting spontaneous activity and capable of firing as rapidly as 1500 times per second. They have many interconnections one with the other, and many of them directly innervate the anterior motor neurons as illustrated in Figure 54–1. The interconnections among the interneurons and anterior motor neurons are responsible for many of the integrative functions of the spinal cord that are discussed in the remainder of this chapter.

Essentially all the different types of neuronal circuits described in Chapter 46 are found in the interneuron pool of cells of the spinal cord, including the *diverging, converging,* and *repetitive-discharge* circuits. In this chapter we see many applications of these different circuits to the performance of specific reflex acts by the spinal cord.

Only a few incoming sensory signals from the spinal nerves or signals from the brain terminate directly on the anterior motor neurons. Instead, most of them are transmitted first through interneurons, where they are appropriately processed. Thus, in Figure 54–1, it is shown that the corticospinal tract terminates almost entirely on interneurons, and it is only after the signals from this tract have been integrated in the interneuron pool with signals from other spinal tracts or from the spinal nerves that they finally impinge on the anterior motor neurons to control muscular function.

The Renshaw Cell Inhibitory System. Located also in the ventral horns of the spinal cord in close association with the motor neurons are a large number of small interneurons called *Renshaw cells.* Almost immediately after the axon leaves the body of the anterior motor neuron, collateral branches from the axon pass to the adjacent Renshaw cells. These in turn are inhibitory cells that transmit inhibitory signals to the nearby motor neurons. Thus, stimulation of each motor neuron tends to inhibit the surrounding motor neurons, an effect called *recurrent inhibition.* This effect is probably important for the following major reason:

It shows that the motor system utilizes the principle of lateral inhibition to focus, or sharpen, its signals in the same way that the sensory system utilizes this principle — that is, to allow unabated transmission of the primary signal while suppressing the tendency for signals to spread to adjacent neurons.

Multisegmental Connections in the Spinal Cord — The Propriospinal Fibers

More than half of all the nerve fibers ascending and descending in the spinal cord are *propriospinal fibers*. These are fibers that run from one segment of the cord to another. In addition, the sensory fibers as they enter the cord branch both up and down the spinal cord, some of the branches transmitting signals only a segment or two in each direction whereas others transmit signals many segments. These ascending and descending fibers of the cord provide pathways for the multisegmental reflexes that are described later in this chapter, including reflexes that coordinate simultaneous movements in the forelimbs and hindlimbs.

● THE MUSCLE RECEPTORS —MUSCLE SPINDLES AND GOLGI TENDON ORGANS— AND THEIR ROLES IN MUSCLE CONTROL

Proper control of muscle function requires not only excitation of the muscle by the anterior motor neu-

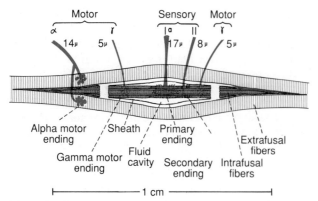

Figure 54–2. The muscle spindle, showing its relationship to the large extrafusal skeletal muscle fibers. Note also both the motor and the sensory innervation of the muscle spindle.

rons but also continuous feedback of information from each muscle to the nervous system, giving the status of the muscle at each instant. That is, what is the length of the muscle, what is its instantaneous tension, and how rapidly is its length or tension changing? To provide this information, the muscles and their tendons are supplied abundantly with two special types of sensory receptors: (1) *muscle spindles,* which are distributed throughout the belly of the muscle and send information to the nervous system about either the muscle length or rate of change of its length, and (2) *Golgi tendon organs,* which are located in the muscle tendons and transmit information about tension or the rate of change of tension.

The signals from these two receptors are either entirely or almost entirely for the purpose of muscle control itself, because they operate almost entirely at a subconscious level. Even so, they transmit tremendous amounts of information into the spinal cord, the cerebellum, and even the cerebral cortex, helping each of these portions of the nervous system in its function for controlling muscle contraction.

RECEPTOR FUNCTION OF THE MUSCLE SPINDLE

Structure and Innervation of the Muscle Spindle. The physiologic organization of the muscle spindle is illustrated in Figure 54–2. Each spindle is built around 3 to 12 small *intrafusal muscle fibers* that are pointed at their ends and are attached to the glycocalyx of the surrounding *extrafusal* skeletal muscle fibers. Each intrafusal fiber is a very small skeletal

muscle fiber. However, the central region of each of these fibers—that is, the area midway between its two ends—has either no or few actin and myosin filaments. Therefore, this central portion does not contract when the ends do. Instead, it functions as a sensory receptor, as we describe later. The end portions are excited by the small *gamma motor nerve fibers* originating from the gamma motor neurons that were described earlier. These fibers are frequently also called *gamma efferent fibers,* in contradistinction to the *alpha efferent fibers* that innervate the extrafusal skeletal muscle.

Excitation of the Spindle Receptors

The receptor portion of the muscle spindle is its central portion, where the intrafusal muscle fibers have no contractile elements. As illustrated in Figure 54–2, and also in more detail in Figure 54–3, sensory fibers originate in this area, and these are stimulated by stretching of this midportion of the spindle. One can readily see that the muscle spindle receptor can be excited in two different ways:

1. Lengthening the whole muscle will obviously stretch the midportion of the spindle and therefore excite the receptor.

2. Even if the length of the entire muscle does not change, contraction of the end portions of the intrafusal fibers will also stretch the midportions of the spindle fibers and therefore excite the receptor.

Two types of sensory endings are found in the receptor area of the muscle spindle. These are:

The Primary Ending. In the very center of the receptor area a large sensory fiber encircles the central portion of each intrafusal fiber, forming the so-called *primary ending* or *annulospiral ending.* This nerve fiber is a type Ia fiber averaging 17 micrometers in diameter, and it transmits sensory signals to the spinal cord at a velocity of 70 to 120 m/sec, as rapidly as any type of sensory nerve fiber in the entire body.

The Secondary Ending. Usually one but sometimes two smaller sensory nerve fibers, type II fibers with an average diameter of 8 micrometers, innervate the receptor region on one side of the primary ending, as illustrated in Figures 54–2 and 54–3. This sensory ending is called the *secondary ending,* or sometimes the *flower spray ending,* because in some preparations it looks like a flower spray even though it mainly encircles the intrafusal fibers in the same way that the type Ia fiber does.

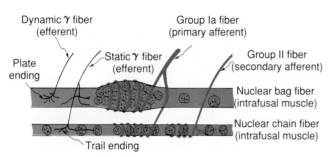

Figure 54–3. Details of nerve connections to the nuclear bag and nuclear chain muscle spindle fibers. (Modified from Stein: *Physiol. Rev.,* 54:225, 1974, and Boyd: *Philios. Trans. R. Soc. Lond. [Biol Sci.],* 245:81, 1962.)

Division of the Intrafusal Fibers Into Nuclear Bag and Nuclear Chain Fibers — Dynamic and Static Responses of the Muscle Spindle. There are also two different types of intrafusal fibers: (1) *nuclear bag fibers* (one to three in each spindle), in which a large number of nuclei are congregated into an expanded bag in the central portion of the receptor area as shown by the top fiber in Figure 54–3, and (2) *nuclear chain fibers* (three to nine), which are about half as large in diameter and half as long as the nuclear bag fibers and have nuclei aligned in a chain throughout the receptor area, as illustrated by the bottom fiber in the figure. The primary ending innervates both the nuclear bag intrafusal fibers *and* the nuclear chain fibers. On the other hand, the secondary ending usually innervates only the nuclear chain fibers. These relationships are all illustrated in Figure 54–3.

Response of Both the Primary and Secondary Endings to the Length of the Receptor — The "Static" Response. When the receptor portion of the muscle spindle is stretched slowly, the number of impulses transmitted from both the primary and the secondary endings increases almost directly in proportion to the degree of stretching, and the endings continue to transmit these impulses for many minutes. This effect is called the *static response* of the spindle receptor, meaning simply that both the primary and the secondary endings continue to transmit their signals for as long as the receptor itself remains stretched. Because only the *nuclear chain* type of intrafusal fiber is innervated by both the primary and the secondary endings, it is believed that these nuclear chain fibers are responsible for the static response.

Response of the Primary Ending (but Not the Secondary Ending) to the Rate of Change of Receptor Length — The "Dynamic" Response. When the length of the spindle receptor increases suddenly, the primary ending (but not the secondary ending) is stimulated especially powerfully, much more powerfully than the stimulus caused by the static response. This excess stimulus of the primary ending is called the *dynamic response,* which means that the primary ending responds extremely actively to a rapid *rate of change* in length. When the length of a spindle receptor increases only a fraction of a micrometer, if this increase occurs in a fraction of a second, the primary receptor transmits tremendous numbers of excess impulses into the Ia fiber, but only *while the length is actually increasing.* As soon as the length has stopped increasing, the rate of impulse discharge returns back to the level of the much smaller static response that is still present in the signal.

Conversely, when the spindle receptor shortens, this change momentarily decreases the impulse output from the primary ending; then, as soon as the receptor area has reached its new shortened length, the impulses reappear in the Ia fiber within a fraction of a second. Thus, the primary ending sends extremely strong signals to the central nervous system to apprise it of any change in length of the spindle receptor area.

Because only the primary endings transmit the dynamic response and the nuclear bag intrafusal fibers have only primary endings, it is assumed that the nuclear bag fibers are responsible for the powerful dynamic response.

Control of the Static and Dynamic Responses by the Gamma Motor Nerves. The gamma motor nerves to the muscle spindle can be divided into two different types: gamma-dynamic (gamma-d) and gamma-static (gamma-s). The first of these excites mainly the nuclear bag intrafusal fibers and the second mainly the nuclear chain intrafusal fibers. When the gamma-d fibers excite the nuclear bag fibers, the dynamic response of the muscle spindle becomes tremendously enhanced, whereas the static response is hardly affected. On the other hand, stimulation of the gamma-s fibers, which excite the nuclear chain fibers, enhances the static response while having little influence on the dynamic response. We shall see in subsequent paragraphs that these two different types of responses of the muscle spindle are exceedingly important in different types of muscle control.

Continuous Discharge of the Muscle Spindles Under Normal Conditions. Normally, particularly when there is a slight amount of gamma motor excitation, the muscle spindles emit sensory nerve impulses continuously. Stretching the muscle spindles increases the rate of firing, while shortening the spindle decreases this rate of firing. Thus, the spindles can send either *positive signals* to the spinal cord — that is, increased numbers of impulses to indicate increased stretch of a muscle, or they can send *negative signals* — decreased numbers of impulses below the normal level — to indicate that the muscle is actually being unstretched.

THE MUSCLE STRETCH REFLEX (ALSO CALLED MYOTATIC REFLEX)

The simplest manifestation of muscle spindle function is the *muscle stretch reflex* — that is, whenever a muscle is stretched, excitation of the spindles causes reflex contraction of the large skeletal muscle fibers that lie around the spindles.

Neuronal Circuitry of the Stretch Reflex. Figure 54–4 illustrates the basic circuit of the muscle spindle stretch reflex, showing a type Ia nerve fiber originating in a muscle spindle and entering the dorsal root of the spinal cord. Then, in contrast to most other nerve fibers entering the cord, one branch of it passes directly to the anterior horn of the cord gray matter and synapses directly with anterior motor neurons that send nerve fibers back to the same muscle from whence the muscle spindle fiber originated. Thus, this is a *monosynaptic pathway* that allows a reflex signal to return with the shortest possible delay back to the same muscle after excitation of the spindle.

Some of the type II fibers from the secondary spin-

Proprioceptor nerve

Motor nerve

Muscle spindle

Stretch reflex

Figure 54–4. Neuronal circuit of the stretch reflex.

dle endings also terminate monosynaptically with the anterior motor neurons. However, most of the type II fibers (as well as many collaterals from the Ia fibers from the primary endings) terminate on multiple interneurons in the cord gray matter, and these in turn transmit more delayed signals to the anterior motor neurons and also serve other functions.

The Dynamic Stretch Reflex Versus the Static Stretch Reflex. The stretch reflex can be divided into two separate components called respectively the dynamic stretch reflex and the static stretch reflex. The *dynamic stretch reflex* is elicited by the potent dynamic signal transmitted from the primary endings of the muscle spindles. That is, when a muscle is suddenly stretched, a strong signal is transmitted to the spinal cord, and this causes an instantaneous, very strong reflex contraction of the same muscle from which the signal originated. Thus, the reflex functions to oppose sudden changes in the length of the muscle, because the muscle contraction opposes the stretch.

The dynamic stretch reflex is over within a fraction of a second after the muscle has been stretched to its new length, but then a weaker *static stretch reflex* continues for a prolonged period of time thereafter. This reflex is elicited by the continuous static receptor signals transmitted by both the primary and secondary endings. The importance of the static stretch reflex is that it continues to cause muscle contraction as long as the muscle is maintained at excessive length. The muscle contraction in turn opposes the force that is causing the excess length.

The Negative Stretch Reflex. When a muscle is suddenly shortened, exactly opposite effects occur because of negative signals from the spindles. If the muscle is already taut, any sudden release of the load on the muscle that allows it to shorten will elicit both dynamic and static reflex *muscle inhibition* rather than reflex excitation. Thus, *this negative stretch reflex* opposes the shortening of the muscle in the same way that the positive stretch reflex opposes lengthening of the muscle. Therefore, one can begin to see that the stretch reflex tends to maintain the status quo for the length of a muscle.

The Damping Function of the Dynamic and Static Stretch Reflexes

An especially important function of the stretch reflex is its ability to prevent some types of oscillation and jerkiness of the body movements. This is a *damping*, or smoothing, function. An example is the following:

Use of the Damping Mechanism in Smoothing Muscle Contraction. Occasionally, signals from other parts of the nervous system are transmitted to a muscle in a very unsmooth form, increasing in intensity for a few milliseconds, then decreasing in intensity, then changing to another intensity level, and so forth. When the muscle spindle apparatus is not functioning satisfactorily, the muscle contraction is very jerky during the course of such a signal. This effect is illustrated in Figure 54–5, which shows an experiment in which a sensory nerve signal entering one side of the cord is transmitted to a motor nerve on the other side of the cord to excite a muscle. In curve A the muscle spindle reflex of the excited muscle is intact. Note that the contraction is relatively smooth even though the sensory nerve is excited at a very slow frequency of 8 per second. Curve B, on the other hand, is the same experiment in an animal whose muscle spindle sensory nerves from the muscle had been sectioned 3 months earlier. Note the very unsmooth muscle contraction. Thus, curve A illustrates very graphically the ability of the damping mechanism of the muscle spindle to smooth muscle contractions even though the input signals to the muscle motor system may themselves be very jerky. This effect can also be called a *signal averaging* function of the muscle spindle reflex.

ROLE OF THE MUSCLE SPINDLE IN VOLUNTARY MOTOR ACTIVITY

To emphasize the importance of the gamma efferent system, one needs to recognize that 31 per cent of

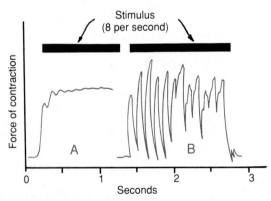

Figure 54–5. Muscle contraction caused by a spinal cord signal under two different conditions: *A*, in a normal muscle, and *B*, in a muscle whose muscle spindles had been denervated by section of the posterior roots of the cord 82 days previously. Note the smoothing effect of the muscle spindle reflex in *A*. (Modified from Creed, R. S., et al.: Reflex Activity of the Spinal Cord. New York, Oxford University Press, 1932.)

all the motor nerve fibers to the muscle are gamma efferent fibers rather than large, type A alpha motor fibers. Whenever signals are transmitted from the motor cortex or from any other area of the brain to the alpha motor neurons, almost always, the gamma motor neurons are stimulated simultaneously, an effect called *coactivation* of the alpha and gamma motor neurons. This causes both the extrafusal and the intrafusal muscle fibers to contract at the same time.

The purpose of contracting the muscle spindle fibers at the same time that the large skeletal muscle fibers contract is probably twofold: First, it keeps the length of the receptor portion of the muscle spindle from changing and therefore keeps the muscle spindle from opposing the muscle contraction. Second, it maintains proper damping function of the muscle spindle regardless of change in muscle length. For instance, if the muscle spindle should not contract and relax along with the large muscle fibers, the receptor portion of the spindle would sometimes be flail and at other times be overstretched, in neither instance operating under optimal conditions for spindle function.

Possible "Servo-Assist" Function of the Muscle Spindle Reflex

It is possible that the muscle spindle reflex also functions as a "servo-assist" mechanism during muscle contraction. But, first, let us explain what is meant by "servo-assist mechanism."

When both the alpha and gamma motor neurons are stimulated simultaneously, if the intra- and extrafusal fibers contract equal amounts, the degree of stimulation of the muscle spindles will not change at all—neither increase nor decrease. However, in case the extrafusal muscle fibers should contract less than the intrafusal fibers (as might occur when the muscle is contracting against a great load), this mismatch would stretch the receptor portions of the spindles and, therefore, elicit a stretch reflex that would provide extra excitation of the extrafusal fibers. This is exactly the same mechanism employed by power steering in an automobile. That is, if the front wheels are resistant to following the movement of the steering wheel, a servo-assist device becomes activated that applies additional force to turn the wheels.

A servo-assist function of the muscle spindle reflex could have several important advantages, as follows:

1. It could allow the brain to cause a muscle contraction against a load without the brain having to expend much extra nervous energy—instead, the spindle reflex would provide most of the nervous energy.

2. It could make the muscle contract almost the desired length even when the load is increased or decreased between successive contractions. In other words, it would make the length of contraction less load-sensitive.

3. It could compensate for fatigue or other abnormalities of the muscle itself, because any failure of the muscle to provide the proper contraction would elicit an additional muscle spindle reflex stimulus to make the contraction occur.

But, unfortunately, we still do not know how important this possible function of the muscle spindle reflex actually is.

Brain Areas for Control of the Gamma Efferent System

The gamma efferent system is excited by the same signals that excite the alpha motor neurons and also by signals from the *bulboreticular facilitatory* region of the brain stem, and secondarily by impulses transmitted into the bulboreticular area from (a) the *cerebellum,* (b) the *basal ganglia,* and even (c) the *cerebral cortex.* Unfortunately, though, little is known about the precise mechanisms of control of the gamma efferent system. However, since the bulboreticular facilitatory area is particularly concerned with antigravity contractions, and also because the antigravity muscles have an especially high density of muscle spindles, emphasis is given to the possible or probable importance of the gamma efferent mechanism in controlling muscle contraction for positioning the different parts of the body and for damping the movements of the different parts.

CLINICAL APPLICATIONS OF THE STRETCH REFLEX

The stretch reflex is tested by the clinician almost every time he performs a physical examination. His purpose is to determine how much background excitation, or "tone," the brain is sending to the spinal cord. This reflex is elicited as follows:

The Knee Jerk and Other Muscle Jerks. Clinically, a method used to determine the sensitivity of the stretch reflexes is to elicit the knee jerk and other muscle jerks. The knee jerk can be elicited by simply striking the patellar tendon with a reflex hammer; this stretches the quadriceps muscle and initiates a *dynamic stretch reflex* that in turn causes the lower leg to jerk forward. The upper part of Figure 54–6 illustrates a myogram from the quadriceps muscle recorded during a knee jerk.

Similar reflexes can be obtained from almost any muscle of the body either by striking the tendon of the muscle or by striking the belly of the muscle itself. In other words, sudden stretch of muscle spindles is all that is required to elicit a stretch reflex.

The muscle jerks are used by neurologists to assess the degree of facilitation of spinal cord centers. When large numbers of facilitatory impulses are being transmitted from the upper regions of the central nervous system into the cord, the muscle jerks are greatly exacerbated. On the other

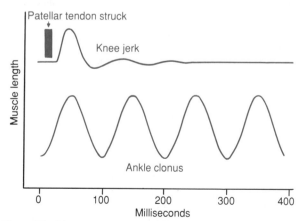

Figure 54–6. Myograms recorded from the quadriceps muscle during elicitation of the knee jerk and from the gastrocnemius muscle during ankle clonus.

hand, if the facilitatory impulses are depressed or abrogated, the muscle jerks are considerably weakened or completely absent. These reflexes are used most frequently in determining the presence or absence of muscle spasticity following lesions in the motor areas of the brain or muscle spasticity in diseases that excite the bulboreticular facilitatory area of the brain stem. Ordinarily, large lesions in the contralateral motor areas of the cerebral cortex, especially those caused by strokes or brain tumors, cause greatly exacerbated muscle jerks.

Clonus. Under appropriate conditions, the muscle jerks can oscillate, a phenomenon called *clonus* (see lower myogram, Fig. 54–6). Oscillation can be explained particularly well in relation to ankle clonus, as follows:

If a man standing on his tiptoes suddenly drops his body downward to stretch one of the gastrocnemius muscles, impulses are transmitted from the muscle spindles into the spinal cord. These reflexly excite the stretched muscle, which lifts the body back up again. After a fraction of a second, the reflex contraction of the muscle dies out and the body falls again, thus stretching the spindles a second time. Again a dynamic stretch reflex lifts the body, but this too dies out after a fraction of a second, and the body falls once more to elicit still a new cycle. In this way, the stretch reflex of the gastrocnemius muscle continues to oscillate, often for long periods of time; this is clonus.

Clonus ordinarily occurs only if the stretch reflex is highly sensitized by facilitatory impulses from the brain. For instance, in the decerebrate animal, in which the stretch reflexes are highly facilitated, clonus develops readily. Therefore, to determine the degree of facilitation of the spinal cord, neurologists test patients for clonus by suddenly stretching a muscle and keeping a steady stretching force applied to the muscle. If clonus occurs, the degree of facilitation is certain to be very high.

THE GOLGI TENDON REFLEX

The Golgi Tendon Organ and Its Excitation. The Golgi tendon organ, illustrated in Figure 54–7, is an encapsulated sensory receptor through which a small bundle of muscle tendon fibers pass immediately beyond their point of fusion with the muscle fibers. An average of 10 to 15 muscle fibers is usually connected in series with each Golgi tendon organ, and the organ is stimulated by the tension produced by this small bundle of muscle fibers. Thus, the major difference between the function of the Golgi tendon organ and the muscle spindle is that the spindle detects muscle length and changes in muscle length, while the tendon organ detects muscle *tension*.

The tendon organ, like the primary receptor of the

muscle spindle, has both a *dynamic response* and a *static response,* responding very intensely when the muscle tension suddenly increases (the dynamic response) but within a small fraction of a second settling down to a lower level of steady-state firing that is almost directly proportional to the muscle tension (the static response). Thus, the Golgi tendon organs provide the nervous system with instantaneous information on the degree of tension in each small segment of each muscle.

Transmission of Impulses from the Tendon Organ Into the Central Nervous System. Signals from the tendon organ are transmitted through large, rapidly conducting type Ib nerve fibers, averaging 16 micrometers in diameter, only slightly smaller than those from the primary ending of the muscle spindle. These fibers, like those from the primary endings, transmit signals both into local areas of the cord and through long fiber pathways such as the spinocerebellar tracts into the cerebellum and through still other tracts to the cerebral cortex. The local cord signal excites a single *inhibitory* interneuron that in turn inhibits the anterior motor neuron. This local circuit directly inhibits the individual muscle without affecting adjacent muscles. The signals to the brain are discussed in Chapter 56.

Inhibitory Nature of the Tendon Reflex and Its Importance

When the Golgi tendon organs of a muscle are stimulated by increased muscle tension, signals are transmitted into the spinal cord to cause reflex effects in the respective muscle. However, this reflex is entirely inhibitory, the exact opposite of the muscle spindle reflex. Thus, this reflex provides a negative feedback mechanism that prevents the development of too much tension on the muscle.

When tension on the muscle and, therefore, on the tendon becomes extreme, the inhibitory effect from the tendon organ can be so great that it leads to a sudden reaction in the spinal cord and instantaneous relaxation of the entire muscle. This effect is called the *lengthening reaction;* it is possibly or even probably a protective mechanism to prevent tearing of the muscle or avulsion of the tendon from its attachments to the bone. We know, for instance, that direct electrical stimulation of muscles in the laboratory, which cannot be opposed by this negative reflex, can frequently cause such destructive effects.

Possible Role of the Tendon Reflex to Equalize Contractile Force Among the Muscle Fibers. Another likely function of the Golgi tendon reflex is to equalize the contractile forces of the separate muscle fibers. That is, those fibers that exert excess tension become inhibited by the reflex while those that exert too little tension become more excited because of absence of reflex inhibition. Obviously this would spread the muscle load over all the fibers and especially would prevent local muscle damage where small numbers of fibers would be overloaded.

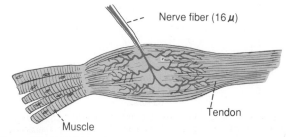

Figure 54–7. Golgi tendon organ.

Nerve fiber (16 μ)

Tendon

Muscle

FUNCTION OF THE MUSCLE SPINDLES AND GOLGI TENDON ORGANS IN CONJUNCTION WITH MOTOR CONTROL FROM HIGHER LEVELS OF THE BRAIN

Although we have emphasized the function of the muscle spindles and Golgi tendon organs in spinal cord control of motor function, these two sensory organs also apprise the higher motor control centers of instantaneous changes taking place in the muscles. For instance, the dorsal spinocerebellar tracts carry instantaneous information from both the muscle spindles and the Golgi tendon organs directly to the cerebellum at conduction velocities approaching 120 m/sec. Additional pathways transmit similar information into the reticular regions of the brain stem and also all the way to the motor areas of the cerebral cortex. We shall learn in the following two chapters that information from these receptors is crucial for feedback control of motor signals originating in all of these areas.

● THE FLEXOR REFLEX (THE WITHDRAWAL REFLEXES)

In the spinal or decerebrate animal, almost any type of cutaneous sensory stimulus on a limb is likely to cause the flexor muscles of the limb to contract, thereby withdrawing the limb from the stimulus. This is called the *flexor reflex.*

In its classic form the flexor reflex is elicited most powerfully by stimulation of pain endings, such as by a pinprick, heat, or some other painful stimulus, for which reason it is also frequently called a *nociceptive reflex,* or simply *pain reflex.* However, stimulation of the touch receptors can also elicit a weaker and less prolonged flexor reflex.

If some part of the body besides one of the limbs is painfully stimulated, this part, in a similar manner, will be *withdrawn from the stimulus,* but the reflex may not be confined entirely to flexor muscles even though it is basically the same type of reflex. Therefore, the many patterns of reflexes of this type in the different areas of the body are called the *withdrawal reflexes.*

Neuronal Mechanism of the Flexor Reflex. The left-hand portion of Figure 54–8 illustrates the neuronal pathways for the flexor reflex. In this instance, a painful stimulus is applied to the hand; as a result, the flexor muscles of the upper arm become reflexly excited, thus withdrawing the hand from the painful stimulus.

The pathways for eliciting the flexor reflex do not pass directly to the anterior motor neurons but, instead, pass first into the interneuron pool of neurons and then to the motor neurons. The shortest possible circuit is a three- or four-neuron arc; however, most of the signals of the reflex traverse many more neurons than this and involve the following basic types of circuits: (1) diverging circuits to spread the reflex to the

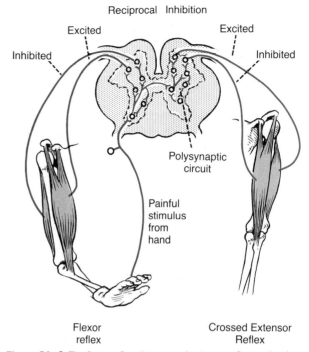

Figure 54–8. The flexor reflex, the crossed extensor reflex, and reciprocal inhibition.

necessary muscles for withdrawal, (2) circuits to inhibit the antagonist muscles, called *reciprocal inhibition circuits,* and (3) circuits to cause a prolonged repetitive afterdischarge even after the stimulus is over.

Figure 54–9 illustrates a typical myogram from a flexor muscle during a flexor reflex. Within a few milliseconds after a pain nerve begins to be stimulated, the flexor response appears. Then, in the next few seconds the reflex begins to *fatigue,* which is characteristic of essentially all of the more complex integrative reflexes of the spinal cord. Then, soon after the stimulus is over, the contraction of the muscle begins to return toward the base line, but, because of *afterdischarge,* will not return all the way for many milliseconds. The duration of the afterdischarge depends on the intensity of the sensory stimulus that had elicited the reflex; a weak tactile stimulus causes almost no afterdischarge in contrast to an afterdischarge lasting for a second or more following a very strong pain stimulus.

The afterdischarge that occurs in the flexor reflex

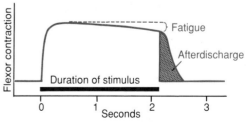

Figure 54–9. Myogram of the flexor reflex, showing rapid onset of the reflex, an interval of fatigue, and finally afterdischarge after the stimulus is over.

almost certainly results from both types of repetitive-discharge circuits that were discussed in Chapter 46. Electrophysiological studies indicate that the immediate afterdischarge, lasting for about 6 to 8 milliseconds, results from repetitive firing of the excited interneurons themselves. However, the prolonged afterdischarge that occurs following strong pain stimuli almost certainly results from recurrent pathways that excite reverberating interneuron circuits, these transmitting impulses to the anterior motor neurons sometimes for several seconds after the incoming sensory signal is completely over.

Thus, the flexor reflex is appropriately organized to withdraw a pained or otherwise irritated part of the body away from the stimulus. Furthermore, because of the afterdischarge, the reflex can still hold the irritated part away from the stimulus for as long as 1 to 3 seconds after the irritation is over. During this time, other reflexes and actions of the central nervous system can move the entire body away from the painful stimulus.

The Pattern of Withdrawal. The pattern of withdrawal that results when the flexor reflex (or the many other types of withdrawal reflexes) is elicited depends on the sensory nerve that is stimulated. Thus, a painful stimulus on the inside of the arm not only elicits a flexor reflex in the arm but also contracts the abductor muscles to pull the arm outward. In other words, the integrative centers of the cord cause those muscles to contract that can most effectively remove the pained part of the body from the object that causes pain. This same principle, which is called the principle of "local sign," applies to any part of the body but especially to the limbs, because they have highly developed flexor reflexes.

● THE CROSSED EXTENSOR REFLEX

Approximately 0.2 to 0.5 second after a stimulus elicits a flexor reflex in one limb, the opposite limb begins to extend. This is called the *crossed extensor reflex.* Extension of the opposite limb obviously can push the entire body away from the object causing the painful stimulus.

Neuronal Mechanism of the Cross Extensor Reflex. The right-hand portion of Figure 54–8 illustrates the neuronal circuit responsible for the crossed extensor reflex, showing that signals from the sensory nerves cross to the opposite side of the cord to cause reactions exactly opposite those that cause the flexor reflex. Because the crossed extensor reflex usually does not begin until 200 to 500 milliseconds following the initial pain stimulus, it is certain that many interneurons are involved in the circuit between the incoming sensory neuron and the motor neurons of the opposite side of the cord responsible for the crossed extension. Furthermore, after the painful stimulus is removed, the crossed extensor reflex continues for an even longer period of afterdischarge than that for the

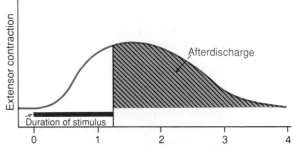

Figure 54–10. Myogram of a crossed extensor reflex, showing slow onset but prolonged afterdischarge.

flexor reflex. Therefore, again, it is presumed that this prolonged afterdischarge results from reverberatory circuits among the internuncial cells.

Figure 54–10 illustrates a typical myogram recorded from a muscle involved in a crossed extensor reflex. This shows the relatively long latency before the reflex begins and also the long afterdischarge following the end of the stimulus. The prolonged afterdischarge obviously would be of benefit in holding the body away from a painful object until other nervous reactions could cause the body to move away.

● RECIPROCAL INHIBITION AND RECIPROCAL INNERVATION

In the foregoing paragraphs we have pointed out several times that excitation of one group of muscles is usually associated with inhibition of another group. For instance, when a stretch reflex excites one muscle, it simultaneously inhibits the antagonist muscles. This is the phenomenon of *reciprocal inhibition,* and the neuronal circuit that causes this reciprocal relationship is called *reciprocal innervation.* Likewise, reciprocal relationships exist between the two sides of the cord, as exemplified by the flexor and extensor reflexes described above.

Figure 54–11 illustrates a typical example of reciprocal inhibition. In this instance, a moderate but prolonged flexor reflex is elicited from one limb of the body; and while this reflex is still being elicited, a still stronger flexor reflex is elicited in the opposite limb. This reflex then sends reciprocal inhibitory signals to

Figure 54–11. Myogram of a flexor reflex, illustrating reciprocal inhibition caused by a stronger flexor reflex in the opposite limb.

the first limb and depresses its degree of flexion. Finally, removal of the stronger reflex allows the original reflex to reassume its previous intensity.

● THE REFLEXES OF POSTURE AND LOCOMOTION

THE POSTURAL AND LOCOMOTIVE REFLEXES OF THE CORD

The Positive Supportive Reaction. Pressure on the footpad of a decerebrate animal causes the limb to extend against the pressure being applied to the foot. Indeed, this reflex is so strong that an animal whose spinal cord has been transected for several months—that is, after the reflexes have become exacerbated—can often be placed on its feet, and the reflex will stiffen the limbs sufficiently to support the weight of the body—the animal will stand in a rigid position. This reflex is called the *positive supportive reaction.*

The positive supportive reaction involves a complex circuit in the interneurons similar to those responsible for the flexor and the cross-extensor reflexes. The locus of the pressure on the pad of the foot determines the position to which the limb will extend; pressure on one side causes extension in that direction, an effect called the *magnet reaction.* This obviously helps keep an animal from falling to that side.

The Cord "Righting" Reflexes. When a spinal cat or even a well-recovered young spinal dog is laid on its side, it will make incoordinate movements that indicate that it is trying to raise itself to the standing position. This is called a *cord righting reflex.* Such a reflex illustrates that relatively complicated reflexes associated with posture are integrated in the spinal cord. Indeed, a puppy with a well-healed transected thoracic cord between the level for the forelimbs and the hindlimbs can completely right itself from the lying position and can even walk on its hindlimbs. And, in the case of the opossum with a similar transection of the thoracic cord, the walking movements of the hindlimbs are hardly different from those in the normal opossum—except that the hindlimb movements are not synchronized with those of the forelimbs as is normally the case.

Stepping and Walking Movements

Rhythmic Stepping Movements of a Single Limb. Rhythmic stepping movements are frequently observed in the limbs of spinal animals. Indeed, even when the lumbar portion of the spinal cord is separated from the remainder of the cord and a longitudinal section is made down the center of the cord to block neuronal connections between the two limbs, each hindlimb can still perform individual stepping functions. Forward flexion of the limb is followed a second or so later by backward extension. Then flexion occurs again, and the cycle is repeated over and over.

This oscillation back and forth between the flexor and extensor muscles can occur even after the sensory nerves have been cut, and it seems to result mainly from mutually reciprocal inhibition circuits that oscillate between the agonist and antagonist muscles within the matrix of the cord itself. That is, the forward flexion of the limb is coincident with reciprocal inhibition of the cord center controlling the extensor muscles. Then, as the flexion begins to die out, *rebound* excitation of the extensors causes the leg to move downward and backward, with simultaneous reciprocal inhibition of the flexor muscles. And the oscillating cycle continues again and again.

The sensory signals from the footpads and from the position sensors around the joints play a strong role in controlling foot pressure and rate of stepping when the foot is allowed to walk along a surface. In fact, the cord mechanism for control of stepping can be still more complex. For instance, if during the forward thrust of the foot the top of the foot encounters an obstruction, the forward thrust will stop temporarily, the foot will be lifted higher, and then the foot will proceed forward to be placed over the obstruction. Thus, the cord is an intelligent walking controller.

Reciprocal Stepping of Opposite Limbs. If the lumbar spinal cord is not split down its center as noted above, every time stepping occurs in the forward direction in one limb, the opposite limb ordinarily steps backward. This effect results from reciprocal innervation between the two limbs.

Diagonal Stepping of All Four Limbs—The "Mark Time" Reflex. If a well-healed spinal animal, with the spinal transection above the forelimb area of the cord, is held up from the floor and its legs are allowed to fall downward as illustrated in Figure 54–12, the stretch on the limbs occasionally elicits stepping reflexes that involve all four limbs. In general, stepping occurs diagonally between the forelimbs and hindlimbs. This diagonal response is another manifestation of reciprocal innervation, this time occurring the entire distance up and down the

Figure 54–12. Diagonal stepping movements exhibited by a spinal animal.

cord between the forelimbs and hindlimbs. Such a walking pattern is called a *mark time reflex.*

The Galloping Reflex. Another type of reflex that occasionally develops in the spinal animal is the galloping reflex, in which both forelimbs move backward in unison while both hindlimbs move forward. This often occurs when stretch or pressure stimuli are applied almost exactly equally to opposite limbs at the same time, whereas unequal stimulation of one side versus the other elicits the diagonal walking reflex. This is in keeping with the normal patterns of walking and of galloping, for, in walking, only one limb at a time is stimulated, and this would predispose to continued walking. Conversely, when the animal strikes the ground during galloping, the limbs on both sides are stimulated approximately equally; this obviously would predispose to further galloping and, therefore, would continue this pattern of motion in contradistinction to the walking pattern.

● THE SCRATCH REFLEX

An especially important cord reflex in some animals is the scratch reflex, which is initiated by the *itch and tickle sensation.* It actually involves two different functions: (1) a *position sense* that allows the paw to find the exact point of irritation on the surface of the body, and (2) a *to-and-fro scratching movement.*

Obviously, the *to-and-fro movement,* like the stepping movements of locomotion, involves reciprocal innervation circuits that cause oscillation, which can still function even when all the sensory roots from the oscillating limb are sectioned, as is true for basic walking movements.

The *position sense* of the scratch reflex is a very highly developed function, for even though a flea might be crawling as far forward as the shoulder of a spinal animal, the hind paw can often find its position even though 19 different muscles in the limb must be contracted simultaneously in a precise pattern to bring the paw to the position of the crawling flea. To make the reflex even more complicated, when the flea crosses the midline, the first paw stops scratching, and the opposite paw begins the to-and-fro motion and eventually finds the flea.

● THE SPINAL CORD REFLEXES THAT CAUSE MUSCLE SPASM

In human beings, local muscle spasm is often observed. The mechanism of this has not been elucidated to complete satisfaction even in experimental animals, but it is known that pain stimuli can cause reflex spasm of local muscles, which presumably is the cause of much if not most of the muscle spasm observed in localized regions of the human body.

Muscle Spasm Resulting from a Broken Bone. One type of clinically important spasm occurs in muscles surrounding a broken bone. This seems to result from the pain impulses initiated from the broken edges of the bone, which cause the muscles surrounding the area to contract powerfully and tonically. Relief of the pain by injection of a local anesthetic relieves the spasm; a general anesthetic also relieves the spasm. One of these procedures is often necessary

before the spasm can be overcome sufficiently for the two ends of the bone to be set back into appropriate positions.

Abdominal Muscle Spasm in Peritonitis. Another type of local spasm caused by cord reflexes is the abdominal spasm resulting from irritation of the parietal peritoneum by peritonitis. Here, again, relief of the pain caused by the peritonitis allows the spastic muscle to relax. Almost the same type of spasm often occurs during surgical operations; pain impulses from the parietal peritoneum cause the abdominal muscles to contract extensively and sometimes actually to extrude the intestines through the surgical wound. For this reason deep surgical anesthesia is usually required for intra-abdominal operations.

Muscle Cramps. Still another type of local spasm is the typical muscle cramp. Electromyographic studies indicate that the cause of at least some muscle cramps is the following:

Any local irritating factor or metabolic abnormality of a muscle — such as severe cold, lack of blood flow to the muscle, or overexercise of the muscle — can elicit pain or other types of sensory impulses that are transmitted from the muscle to the spinal cord, thus causing reflex muscle contraction. The contraction in turn stimulates the same sensory receptors still more, which causes the spinal cord to increase the intensity of contraction still further. Thus, a positive feedback develops so that a small amount of initial irritation causes more and more contraction until a full-blown muscle cramp ensues. Reciprocal inhibition of the muscle can sometimes relieve the cramp. That is, if a person purposefully contracts the muscle on the side of the joint opposite the cramped muscle while at the same time using the other hand or foot to prevent movement of the joint, the reciprocal inhibition that occurs in the cramped muscle can at times relieve the cramp.

● THE AUTONOMIC REFLEXES IN THE SPINAL CORD

Many different types of segmental autonomic reflexes occur in the spinal cord, most of which are discussed in other chapters. Briefly, these include (1) changes in vascular tone, resulting from local skin heat and cold (Chapter 73); (2) sweating, which results from localized heat on the surface of the body (Chapter 73); (3) intestinointestinal reflexes that control some motor functions of the gut (Chapter 62); (4) peritoneointestinal reflexes that inhibit gastric motility in response to peritoneal irritation (Chapter 66); and (5) evacuation reflexes for emptying the bladder (Chapter 31) and the colon (Chapter 63). In addition, all the segmental reflexes can at times be elicited simultaneously in the form of the so-called mass reflex as follows:

The Mass Reflex. In a spinal animal or human being, the spinal cord sometimes suddenly becomes excessively active, causing massive discharge of large portions of the cord. The usual stimulus that causes this is a strong nociceptive stimulus to the skin or excessive filling of a viscus, such as overdistention of the bladder or of the gut. Regardless of the type of stimulus, the resulting reflex, called the *mass reflex,* involves large portions or even all of the cord, and its pattern of reaction is the same. The effects are (1) a major portion of the body goes into strong flexor spasm, (2) the colon and bladder are likely to evacuate, (3) the arterial pressure often rises to maximal values — sometimes to a mean pressure well over 200 mm Hg, and (4) large areas of the body break out into profuse sweating. The mass reflex might be likened to epileptic seizures that involve the central nervous system in which large portions of the brain become massively activated.

The precise neuronal mechanism of the mass reflex is unknown. However, because it lasts for minutes, it presumably results from activation of great masses of reverberating circuits that excite large areas of the cord at once.

● SPINAL CORD TRANSECTION AND SPINAL SHOCK

When the spinal cord is suddenly transected, essentially all cord functions, including the cord reflexes, immediately become depressed to the point of total silence, a reaction called *spinal shock*. The reason for this is that normal activity of the cord neurons depends to a great extent on continual tonic discharges of nerve fibers entering the cord from higher centers, particularly discharges transmitted through the reticulospinal tracts, vestibulospinal tracts, and corticospinal tracts.

After a few hours to a few weeks, the spinal neurons gradually regain their excitability. This seems to be a natural characteristic of neurons everywhere in the nervous system —that is, after they lose their source of facilitatory impulses, they increase their own natural degree of excitability to make up for the loss. But there is also some possibility of sprouting of multiple new nerve endings in the cord, which could also increase excitability. In most nonprimates, the excitability of the cord centers returns essentially to normal within a few hours to a day or so, but in human beings the return is often delayed for several weeks and occasionally is never complete; or, on the other hand, recovery is sometimes excessive, with resultant hyperexcitability of some or all cord functions.

Some of the spinal functions specifically affected during or following spinal shock are the following: (1) The arterial blood pressure falls immediately—sometimes to as low as 40 mm Hg—thus illustrating that sympathetic activity becomes blocked almost to extinction. However, the pressure ordinarily returns to normal within a few days even in the human being. (2) All skeletal muscle reflexes integrated in the spinal cord are completely blocked during the initial stages of shock. In lower animals, a few hours to a few days are required for these reflexes to return to normal, and in human beings 2 weeks to several months are usually required. Sometimes, both in animals and people, some reflexes eventually become hyperexcitable, particularly if a few facilitatory pathways remain intact between the brain and the cord while the remainder of the spinal cord is transected. The first reflexes to return are the stretch reflexes, followed in order by the progressively more complex reflexes: the flexor reflexes, the postural antigravity reflexes, and remnants of stepping reflexes. (3) The sacral reflexes for control of bladder and colon evacuation are completely suppressed in human beings for the first few weeks following cord transection, but they eventually return. These effects are discussed in Chapters 31 and 66.

REFERENCES

Adams, R. D., and Victor, M. (eds.): Principles of Neurology, 4th Ed. New York, McGraw-Hill Book Co., 1989.
Austin, G.: The Spinal Cord. New York, Igaku Shoin Medical Publishers, 1981.
Baldissera, F., et al.: Integration in spinal neuronal systems. In Brooks, V. B. (ed.): Handbook of Physiology. Sec. 1, Vol. II. Bethesda, Md., American Physiological Society, 1981, p. 509.
Brooks, V. B.: The Neural Basis of Motor Control. New York, Oxford University Press, 1986.
Burke, R. E.: Motor units: Anatomy, physiology, and functional organization. In Brooks, V. B. (ed.): Handbook of Physiology. Sec. 1, Vol. II. Bethesda, Md., American Physiological Society, 1981, p. 345.
Creed, R. S., et al.: Reflex Activity of the Spinal Cord. New York, Oxford University Press, 1932.
Emonet-Denand, F., et al.: How muscle spindles signal changes of muscle length. News Physiol. Sci., 3:105, 1988.
Evarts, E. V., et al. (eds.): Motor System in Neurobiology. New York, Elsevier Science Publishing Co., 1986.
Hammond, D. L.: New insights regarding organization of spinal cord pain pathways. News Physiol. Sci., 4:98, 1989.
Hasan, A., and Stuart, D. G.: Animal solutions to problems of movement control: The role of proprioceptors. Annu. Rev. Neurosci., 11:199, 1988.
Henneman, E., and Mendell, L. M.: Functional organization of motoneuron pool and its inputs. In Brooks, V. B. (ed.): Handbook of Physiology. Sec. 1, Vol. II. Bethesda, Md., American Physiological Society, 1981, p. 423.
Hnik, P., et al. (eds.): Mechanoreceptors. Development, Structure, and Function. New York, Plenum Publishing Corp., 1988.
Houk, J. C., and Rymer, W. Z.: Neural control of muscle length and tension. In Brooks, V. B. (ed.): Handbook of Physiology. Sec. 1, Vol. II. Bethesda, Md., American Physiological Society, 1981, p. 257.
Houk, J. C.: Control strategies in physiological systems. FASEB J., 2:97, 1988.
Hunt, C. C., and Perl, E. R.: Spinal reflex mechanisms concerned with skeletal muscle. Physiol. Rev., 40:538, 1960.
Illis, L. S.: Spinal Cord Dysfunction. New York, Oxford University Press, 1988.
Janig, W., and McLachlan, E. M.: Organization of lumbar spinal outflow to distal colon and pelvic organs. Physiol. Rev., 67:1332, 1987.
Kao, C. C. (ed.): Spinal Cord Reconstruction. New York, Raven Press, 1983.
Le Douarin, N. M., et al.: From the neural crest to the ganglia of the peripheral nervous system. Annu. Rev. Physiol., 43:653, 1981.
Matthews, P. B. C.: Muscle spindles: Their messages and their fusimotor supply. In Brooks, V. B. (ed.): Handbook of Physiology. Sec. 1, Vol. II. Bethesda, Md., American Physiological Society, 1981, p. 189.
Mendell, L. M.: Modifiability of spinal synapses. Physiol. Rev., 64:260, 1984.
Rack, P. M. H.: Limitations of somatosensory feedback in control of posture and movement. In Brooks, V. B. (ed.): Handbook of Physiology. Sec. 1, Vol. II. Bethesda, Md., American Physiological Society, 1981, p. 229.
Redman, S. J.: Monosynaptic transmission in the spinal cord. News Physiol. Sci., 1:171, 1986.
Rowell, L. B.: Reflex control of regional circulation in humans. J. Auton. Nerv. Syst., 11:101, 1984.
Sachs, F.: Mechanical transduction: Unification? News Physiol. Sci., 1:98, 1986.
Sherrington, C. S.: The Integrative Action of the Nervous System. New Haven, Conn., Yale University Press, 1911.
Stein, D. G., and Sabel, B. A. (eds.): Pharmacological Approaches to the Treatment of Brain and Spinal Cord Injury. New York, Plenum Publishing Corp., 1988.
Stein, R. B.: Peripheral control of movement. Physiol. Rev., 54:215, 1974.
Stein, R. B., and Lee, R. G.: Tremor and clonus. In Brooks, V. B. (ed.): Handbook of Physiology. Sec. 1, Vol. II. Bethesda, Md., American Physiological Society, 1981, p. 325.
Wiesendanger, M., and Miles, T. S.: Ascending pathway of low-threshold muscle afferents to the cerebral cortex and its possible role in motor control. Physiol. Rev., 62:1234, 1982.
Youmans, J. R. (ed.): Neurological Surgery. Philadelphia, W. B. Saunders Co., 1989.

CHAPTER 55

• *Cortical and Brain Stem Control of Motor Function* •

In this chapter we will discuss the control of the body's movements by the cerebral cortex and brain stem. These two neural areas, along with the basal ganglia and cerebellum, which are discussed in the following chapter, control the very complex movements that the human being and other higher animals have developed for their special purposes.

Virtually all "voluntary" movements involve conscious activity in the cerebral cortex. Yet this does not mean that each contraction of each muscle is willed by the cortex itself. Instead, most control used by the cortex involves the patterns of function in lower brain areas — in the cord, in the brain stem, in the basal ganglia, in the cerebellum — and these lower centers in turn send most of the specific activating signals to the muscles. However, for a few types of movements, the cortex does have almost a direct pathway to the anterior motor neurons of the cord, bypassing other motor centers on the way, especially for control of the very fine dexterous movements of our fingers and hands. It will be the goal of this and the following chapter to explain the interplay among the different motor areas of the brain and spinal cord that provides this overall synthesis of motor function.

● THE MOTOR CORTEX AND THE CORTICOSPINAL TRACT

Figure 55–1 illustrates the functional areas of the cerebral cortex. Anterior to the central sulcus, occupying approximately the posterior third of the frontal lobes, is the *motor cortex.* Posterior to the central sulcus is the *somatic sensory cortex,* an area discussed in detail in earlier chapters that feeds many signals into the motor cortex for control of motor activities.

The motor cortex itself is further divided into three separate subareas, each of which has its own topographical representation of all the muscle groups of the body: (1) the *primary motor cortex,* (2) the *premotor area,* and (3) the *supplemental motor area.*

The Primary Motor Cortex

The primary motor cortex, shown in Figure 55–1, lies in the first convolution of the frontal lobes anterior to the central sulcus. It begins laterally in the sylvian fissure and spreads superiorly to the uppermost portion of the brain, then dips over into the longitudinal fissure. This area is the same as area 4 in Brodmann's classification of the brain cortical areas shown in Figure 47–5 of Chapter 47.

Figure 55–1 lists the topographical representations of the different muscle areas of the body in the primary motor cortex, beginning with the face and mouth region, near the sylvian fissure; the arm and hand area, in the midportions of the primary motor cortex; the trunk, near the apex of the brain; and the leg and foot areas, in that part of the primary motor cortex that dips into the longitudinal fissure. This topographical organization is illustrated even more graphically in Figure 55–2, which shows the degrees of representation of the different muscle areas as mapped by Penfield and Rasmussen. This mapping was done by electrically stimulating the different areas of the motor cortex in human beings who were undergoing neurosurgical operations. Note that more than half of the entire primary motor cortex is concerned with controlling the hands and the muscles of speech. Point stimulations in these motor areas will cause contraction of a single muscle or even a portion of a single muscle. But in those areas with a lesser degree of representation, such as the trunk area, electrical stimulation usually contracts a group of muscles instead.

The Premotor Area

The premotor area, also shown in Figure 55–1, lies immediately anterior to the primary motor cortex, projecting 1 to 3 centimeters anteriorly and extending inferiorly into the sylvian fissure and superiorly into the longitudinal fissure, where it abuts the supplemental motor area. Note that the topographical organization of the premotor cortex is roughly the same as that of the primary motor cortex, with the face area

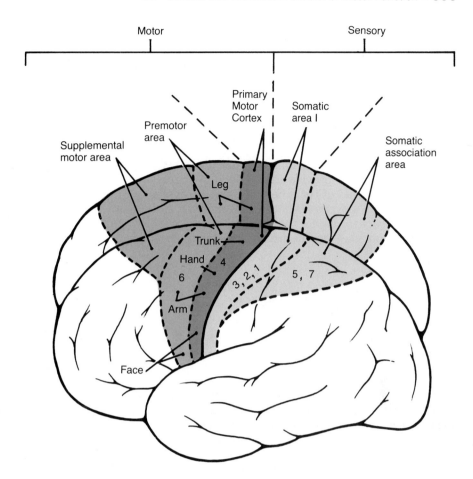

Figure 55–1. The motor and somatic sensory functional areas of the cerebral cortex.

located most laterally and then in the upward direction the arm, trunk, and leg areas. The premotor area is frequently called simply motor area 6 because it occupies a large share of area 6 in the Brodmann classification of brain topology.

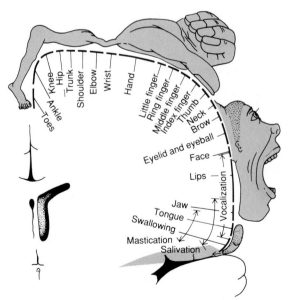

Figure 55–2. Degree of representation of the different muscles of the body in the motor cortex. (From Penfield and Rasmussen: The Cerebral Cortex of Man: A Clinical Study of Localization of Function. New York, Macmillan Co., 1968.)

Most nerve signals generated in the premotor area cause patterns of movement involving groups of muscles that perform specific tasks. For instance, the task may be to position the shoulders and arms so that the hands become properly oriented to perform specific tasks. To achieve these results, the premotor area sends its signals either directly into the primary motor cortex to excite multiple groups of muscles or, more likely, by way of the basal ganglia and then back through the thalamus to the primary motor cortex. Thus, the premotor cortex, the basal ganglia, the thalamus, and the primary motor cortex constitute a complex overall system for control of many of the body's more complex patterns of coordinated muscle activity.

The Supplemental Motor Area

The supplemental motor area has still another topographical organization for control of motor function. It lies immediately superior and anterior to the premotor area, lying mainly in the longitudinal fissure but extending a centimeter or so over the edge onto the superiormost portion of the exposed cortex. Note in Figure 55–1 that the leg area lies most posteriorly and the face most anteriorly.

Considerably stronger electrical stimuli are required in the supplemental motor area to cause muscle contraction than in the other motor areas. However, when contractions are elicited, they are often bilateral rather than unilateral. And stimulation frequently leads to movements such as unilateral grasping of a hand or at other times bilateral grasping of both hands simultaneously; these movements are perhaps rudiments of the hand functions required for climbing. Also, there may be rotation of the hands, movement of the eyes, vocalization, or yawning. But other than these subtle clues to the function of the supplemental motor area, very little else is known. In general, this area probably functions in concert with the premotor area to provide attitudinal movements, fixation movements of the different segments of the body, positional movements of the head and eyes, and so forth, as background for the finer motor control of the hands and feet by the premotor and primary motor cortex.

SOME SPECIALIZED AREAS OF MOTOR CONTROL FOUND IN THE HUMAN MOTOR CORTEX

Neurosurgeons have found a few highly specialized motor regions of the human cerebral cortex, located mainly in the premotor areas as illustrated in Figure 55–3, that control very specific motor functions. These have been localized either by electrical stimulation or by noting the loss of motor function when destructive lesions have occurred in specific cortical areas. Some of the more important of these are the following:

Broca's Area and Speech. Figure 55–3 illustrates a premotor area lying immediately anterior to the primary motor cortex and immediately above the sylvian fissure labeled "word formation." This region is called *Broca's area*. Damage to it does not prevent a person from vocalizing, but it does make it impossible for the person to speak whole words other than simple utterances such as "no" or "yes." A closely associated cortical area also causes appropriate respiratory function so that respiratory activation of vocal cords can occur simultaneously with the movements of the mouth and tongue during speech. Thus, the premotor activities that are related to Broca's area are highly complex.

The "Voluntary" Eye Movement Field. Immediately above Broca's area is a locus for controlling eye movements. Damage to this area prevents a person from voluntarily moving the eyes toward different objects. Instead, the eyes tend to lock on specific objects, an effect controlled by signals from the occipital cortex, as explained in Chapter 51. This frontal area also controls eyelid movements such as blinking.

Head Rotation Area. Still slightly higher in the motor association area, electrical stimulation will elicit head rotation. This area is closely associated with the eye movement field and is presumably related to directing the head toward different objects.

Area for Hand Skills. In the premotor area immediately anterior to the primary motor cortex for the hands and fingers is a region neurosurgeons have called an area for hand skills. That is, when tumors or other lesions cause destruction in this area, the hand movements become incoordinate and nonpurposeful, a condition called *motor apraxia.*

TRANSMISSION OF SIGNALS FROM THE MOTOR CORTEX TO THE MUSCLES

Motor signals are transmitted directly from the cortex to the spinal cord through the *corticospinal tract* and indirectly through multiple accessory pathways that involve the *basal ganglia,* the *cerebellum,* and various *nuclei of the brain stem.* In general, the direct pathways are concerned more with discrete and detailed movements, especially of the distal segments of the limbs, particularly the hands and the fingers.

The Corticospinal Tract (Pyramidal Tract)

The most important output pathway from the motor cortex is the *corticospinal tract,* also called the *pyramidal tract,* which is illustrated in Figure 55–4.

The corticospinal tract originates about 30 per cent from the primary motor cortex, 30 per cent from the premotor and supplementary motor areas, and 40 per cent from the somatic sensory areas posterior to the central sulcus. After leaving the cortex it passes through the posterior limb of the internal capsule (between the caudate nucleus and the putamen of the basal ganglia) and then downward through the brain stem, forming the *pyramids of the medulla.* By far the majority of the pyramidal fibers then cross to the opposite side and descend in the *lateral corticospinal tracts* of the cord, finally terminating principally on the interneurons in the intermediate regions of the cord gray matter but also a few on sensory relay neurons in the dorsal horn, and some directly on the anterior motor neurons.

A few of the fibers do not cross to the opposite side in the medulla but pass ipsilaterally down the cord in the *ventral corticospinal tracts,* but many of these fibers also cross to the

Figure 55–3. Representation of the different muscles of the body in the motor cortex and location of other cortical areas responsible for certain types of motor movements.

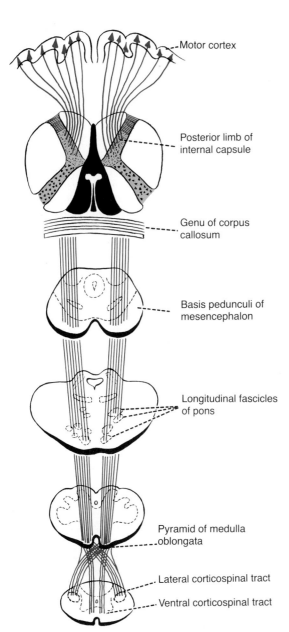

Motor cortex

*Posterior limb of
internal capsule*

*Genu of corpus
callosum*

*Basis pedunculi of
mesencephalon*

*Longitudinal fascicles
of pons*

*Pyramid of medulla
oblongata*

Lateral corticospinal tract

Ventral corticospinal tract

Figure 55-4. The pyramidal tract. (Modified from Ranson and Clark: *Anatomy of the Nervous System.* Philadelphia, W. B. Saunders Co., 1959.)

opposite side of the cord either in the neck or in the upper thoracic region. These fibers are perhaps concerned with control by the supplementary motor area of bilateral postural movements.

The most impressive fibers in the pyramidal tract are a population of large myelinated fibers with mean diameter of 16 micrometers. These originate from the *giant pyramidal cells,* also called *Betz cells,* that are found only in the primary motor cortex. These cells are about 60 micrometers in diameter, and their fibers transmit nerve impulses to the spinal cord at a velocity of about 70 m/sec, the most rapid rate of transmission of any signals from the brain to the cord. There are approximately 34,000 of these large Betz cell fibers in each corticospinal tract. However, the total number of fibers in each corticospinal tract is more than 1,000,000; so these large fibers represent only 3 per cent of all of them. The other 97 per cent are mainly fibers smaller than 4 micrometers in diameter.

Other Fiber Pathways From the Motor Cortex

The motor cortex gives rise to very large numbers of fibers from the cortex or collaterals from the pyramidal tract that go to deeper regions of the cerebrum and also into the brain stem, including the following:

1. The axons from the giant Betz cells send short collaterals back to the cortex itself. It is believed that these collaterals mainly inhibit adjacent regions of the cortex when the Betz cells discharge, thereby "sharpening" the boundaries of the excitatory signal.

2. A large body of fibers passes into the *caudate nucleus* and *putamen.* From here additional pathways extend through several neurons into the brain stem, as discussed in the following chapter.

3. A moderate number of fibers pass to the *red nuclei.* From these, additional fibers pass down the cord through the *rubrospinal tract.*

4. A moderate number of fibers deviate into the *reticular substance* and *vestibular nuclei* of the brain stem; from here signals go to the cord via *reticulospinal* and *vestibulospinal tracts,* and others go to the cerebellum via *reticulocerebellar* and *vestibulocerebellar tracts.*

5. A tremendous number of fibers synapse in the pontile nuclei, which give rise to the *pontocerebellar fibers,* carrying signals into the cerebellar hemispheres.

6. Collaterals also terminate in the *inferior olivary nuclei,* and from here secondary *olivocerebellar fibers* transmit signals to many areas of the cerebellum.

Thus, the basal ganglia, the brain stem, and the cerebellum all receive strong signals from the corticospinal system every time a signal is transmitted down the spinal cord to cause a motor activity.

INCOMING FIBER PATHWAYS TO THE MOTOR CORTEX

The functions of the motor cortex are controlled mainly by the somatic sensory system but also to a lesser extent by the other sensory systems such as hearing and vision. Once the sensory information is derived from these sources, the motor cortex operates in association with the basal ganglia and cerebellum to process the information and to determine the appropriate course of motor action. The more important incoming fiber pathways to the motor cortex are as follows:

1. Subcortical fibers from adjacent regions of the cortex, especially from the somatic sensory areas of the parietal cortex and from the frontal areas as well as subcortical fibers from the visual and auditory cortices.

2. Subcortical fibers that pass through the corpus callosum from the opposite cerebral hemisphere. These fibers connect corresponding areas of the motor cortices in the two sides of the brain.

3. Somatic sensory fibers derived directly from the ventrobasal complex of the thalamus. These transmit mainly cutaneous tactile signals and joint and muscle signals.

4. Tracts from the ventrolateral and ventroanterior nuclei of the thalamus, which in turn receive tracts from the cerebellum and the basal ganglia. These tracts provide signals that are necessary for coordination between the functions of the motor cortex, the basal ganglia, and the cerebellum.

5. Fibers from the intralaminar nuclei of the thalamus. These fibers probably control the general level of excitability of the motor cortex in the same manner that they also control the general level of excitability of most other regions of the cerebral cortex.

THE RED NUCLEUS SERVES AS AN ALTERNATE PATHWAY FOR TRANSMITTING CORTICAL SIGNALS TO THE SPINAL CORD

The *red nucleus,* located in the mesencephalon, functions in close association with the corticospinal tract. As illustrated in Figure 55–5, it receives a large number of direct fibers from the primary motor cortex through the *corticorubral tract* as well as branching fibers from the corticospinal tract as it passes through the mesencephalon. These fibers synapse in the lower portion of the red nucleus, the *magnocellular portion,* that contains large neurons similar to the Betz cells in the motor cortex. These large neurons give rise to the *rubrospinal tract,* which crosses to the opposite side in the lower brain stem and follows a course parallel to the corticospinal tract into the lateral columns of the spinal cord. This tract partially overlaps the corticospinal tract but on the average lies slightly anterior to it. The rubrospinal fibers terminate mainly on the interneurons of the intermediate areas of the cord gray matter along with the corticospinal fibers, but a few of the rubrospinal fibers also terminate on the anterior motor neurons, along with some corticospinal fibers.

The red nucleus also has close connections with the cerebellum, similar to the connections between the motor cortex and the cerebellum.

Function of the Corticorubrospinal System. The magnocellular portion of the red nucleus has a somatographic representation of all the muscles of the body, as is true of the motor cortex. Therefore, stimulation of a single point in this portion of the red nucleus will cause contraction of either a single muscle or small group of muscles. However, the fineness of representation of the different muscles is far less developed than in the motor cortex. This is especially true in human beings who have a relatively small red nucleus.

The corticorubrospinal pathway serves as an accessory route for the transmission of relatively discrete signals from the motor cortex to the spinal cord. When the corticospinal fibers are destroyed without destroying this other pathway, discrete movements can still occur, except that the movements of the fingers and hands are considerably impaired. Wrist movements are still well developed, which is not true when the corticorubrospinal pathway is also blocked. Therefore, the pathway through the red nucleus to the spinal cord is associated far more with the corticospinal system than with the other major brain stem motor pathway, the vestibuloreticulospinal system that controls mainly the axial and girdle muscles of the body, as we shall discuss later in the chapter. Furthermore, the rubrospinal tract lies in the lateral columns of the spinal cord, along with the corticospinal tract, and terminates more on the interneurons and motor neurons that control the distal muscles of the limbs. Therefore, the corticospinal and rubrospinal tracts together are frequently called the *lateral motor system of the cord,* in contradistinction to the vestibuloreticulospinal system that lies mainly medially in the cord and is called the *medial motor system of the cord.*

THE EXTRAPYRAMIDAL SYSTEM

The term *"extrapyramidal motor system"* is widely used in clinical circles to denote all those portions of the brain and brain stem that contribute to motor control that are not part of the direct corticospinal-pyramidal system. This includes pathways through the basal ganglia, the reticular formation of the brain stem, the vestibular nuclei, and often the red nuclei as well. However, this is such an all-inclusive and diverse group of motor control areas that it is difficult to ascribe specific neurophysiological functions to the extrapyramidal system as a whole. For this reason, the term "extrapyramidal" is beginning to have less usage clinically as well as physiologically.

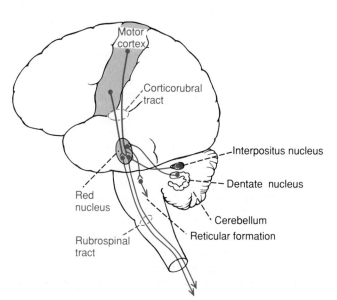

Figure 55–5. The corticorubrospinal pathway for motor control, showing also the relationship of this pathway to the cerebellum.

EXCITATION OF THE SPINAL CORD BY THE PRIMARY MOTOR CORTEX AND THE RED NUCLEUS

Vertical Columnar Arrangement of the Neurons in the Motor Cortex. In Chapters 47 and 51 it is pointed out that the cells in the somatic sensory cortex and visual cortex — and in all other parts of the brain as well — are organized in vertical columns of cells. In a like manner, the cells of the motor cortex are also organized in vertical columns a fraction of a millimeter in diameter and having thousands of neurons in each column.

Each column of cells functions as a unit, stimulating either a single muscle or a group of synergistic muscles. Also, each column is arranged in six distinct layers of cells, like the arrangement throughout almost all the cerebral cortex. The pyramidal cells that give rise to the corticospinal fibers all lie in the fifth layer of cells from the cortical surface, whereas the input signals to the column of cells all enter layers 2 through 4. The sixth layer gives rise mainly to fibers that communicate with other regions of the cerebral cortex itself.

Function of Each Column of Neurons. The neurons of each column operate as an integrative processing system, utilizing information from multiple input sources to determine the output response from the column. In addition, each column can function as an amplifying system to stimulate large numbers of pyramidal fibers to the same muscle or to synergistic muscles simultaneously. This is important because stimulation of a single pyramidal cell can rarely excite a muscle. Instead, as many as 50 to 100 pyramidal cells usually need to be excited simultaneously or in rapid succession to achieve muscle contraction.

Dynamic and Static Signals Transmitted by the Pyramidal Neurons. If a strong signal is sent at first to a muscle to cause initial rapid contraction, then a much weaker signal can maintain the contraction for long periods thereafter. This is the manner in which excitation for causing muscle contractions is usually provided. To do this, each column of cells excites two separate populations of pyramidal cell neurons, one called *dynamic neurons* and the other *static neurons.* The dynamic neurons are excessively excited for a short period of time at the beginning of the contraction, causing the initial *development of force.* Then the static neurons fire at a much slower rate, but they continue at this slow rate indefinitely to *maintain the force* of contraction as long as the contraction is required.

The neurons of the red nucleus have similar dynamic and static characteristics, except that more dynamic neurons are in the red nucleus and more static neurons in the primary motor cortex. This perhaps relates to the fact that the red nucleus is closely allied with the cerebellum, and we shall learn later in this chapter that the cerebellum also plays an important role in the rapid initiation of muscle contraction.

Somatic Sensory Feedback to the Motor Cortex

When nerve signals from the motor cortex cause a muscle to contract, somatic sensory signals return from the activated region of the body to the neurons in the motor cortex that are causing the action. Most of these somatic sensory signals arise in the muscle spindles or in the tactile receptors of the skin overlying the muscles. In general, the somatic signals cause a positive feedback enhancement of the muscle contraction in the following ways: In the case of the muscle spindles, if the fusimotor muscle fibers in the spindles contract more than the large skeletal muscle itself does, then the spindles become excited, and the signals from these spindles stimulate the pyramidal cells in the motor cortex, which further excites the muscle, helping its contraction catch up with the contraction of the spindles. In the case of the tactile receptors, if the muscle contraction causes compression of the skin against an object, such as compression of the fingers around an object that is being grasped, the signals from these receptors cause further excitement of the muscles and therefore increase the muscle contraction — such as increasing the tightness of the grasp of the hand.

Stimulation of the Spinal Motor Neurons

Figure 55–6 shows a segment of the spinal cord, illustrating multiple motor tracts entering the cord from the brain and also showing a representative anterior motor neuron. The corticospinal tract and the rubrospinal tract lie in the dorsal portions of the lateral columns. At most levels of the cord, their fibers terminate mainly on interneurons in the intermediate area of the cord gray matter. However, in the cervical enlargement of the cord where the hands and fingers are represented, moderate numbers of both cortico-

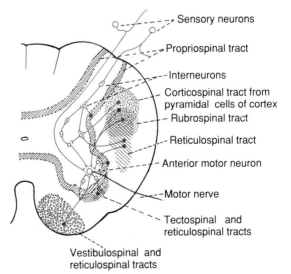

Figure 55–6. Convergence of all the different motor pathways on the anterior motor neurons.

Sensory neurons
Propriospinal tract
Interneurons
Corticospinal tract from pyramidal cells of cortex
Rubrospinal tract
Reticulospinal tract
Anterior motor neuron
Motor nerve
Tectospinal and reticulospinal tracts
Vestibulospinal and reticulospinal tracts

spinal and rubrospinal fibers terminate directly on the anterior motor neurons, thus allowing a direct route from the brain for activating muscle contraction. This is in keeping with the fact that the primary motor cortex has an extremely high degree of representation for fine control of hand, finger, and thumb actions.

Patterns of Movement Elicited by Spinal Cord Centers. From the previous chapter, recall that the spinal cord can provide specific reflex patterns of movement in response to sensory nerve stimulation. Many of these patterns are also important when the anterior motor neurons are excited by signals from the brain. For instance, the stretch reflex is functional at all times, helping to damp the motor movements initiated from the brain and probably providing at least part of the motive power required to cause the muscle contractions, employing the servo-assist mechanism that was described in Chapter 54.

Also, when a brain signal excites an agonist muscle, it is not necessary to transmit an inverse signal to the antagonist at the same time; this transmission will be achieved by the reciprocal innervation circuit that is always present in the cord for coordinating the functions of antagonistic pairs of muscles.

Finally, parts of the other reflex mechanisms, such as withdrawal, stepping and walking, scratching, postural mechanisms, and so forth, can be activated by "command" signals from the brain. Thus, very simple signals from the brain can initiate many of our normal motor activities, particularly for such functions as walking and the attainment of different postural attitudes of the body.

Effect of Lesions in the Motor Cortex or Corticospinal Pathway — The "Stroke"

The motor cortex or corticospinal pathway is frequently damaged, especially by the common abnormality called a "stroke." This is caused either by a ruptured blood vessel that allows hemorrhage into the brain or by thrombosis of one of the major arteries supplying the brain, in either case causing loss of blood supply to the cortex, or very frequently to the corticospinal tract, where it passes through the internal capsule between the caudate nucleus and the putamen. Also, experiments have been performed in animals to remove selectively different parts of the motor cortex.

Removal of the Primary Motor Cortex (the Area Pyramidalis). Removal of a portion of the primary motor cortex — the area that contains the giant Betz pyramidal cells — in a monkey causes varying degrees of paralysis of the represented muscles. If the sublying caudate nucleus and the adjacent premotor area are not damaged, gross postural and limb "fixation" movements can still be performed, but the animal *loses voluntary control of discrete movements of the distal segments of the limbs — especially of the hands and fingers.* This does not mean that the muscles themselves cannot contract, but that the animal's ability to control the fine movements is gone.

From these results one can conclude that the area pyramidalis is essential for voluntary initiation of finely controlled movements, especially of the hands and fingers.

Muscle Spasticity Caused by Lesions That Damage Large Areas Adjacent to the Motor Cortex. Ablation of the primary motor cortex alone causes *hypotonia*, not spasticity, because the primary motor cortex normally exerts a continual tonic stimulatory effect on the motor neurons of the spinal cord; when this is removed, hypotonia results.

On the other hand, most lesions of the motor cortex, especially those caused by a stroke, involve not only the primary motor cortex but also adjacent cortical areas and deeper structures of the cerebrum as well, especially the basal ganglia. In these instances, muscle spasm almost invariably occurs in the afflicted muscle areas on the opposite side of the body (because all the motor pathways cross to the opposite side). Obviously, this spasm is not caused by loss of either the primary motor cortex or blockage of the corticospinal fibers to the cord.

Instead, it is believed to result mainly from damage to accessory pathways from the cortex that normally inhibit the vestibular and reticular brain stem nuclei. When these nuclei lose this inhibition (that is, when they are said to be "disinhibited"), they become spontaneously active and cause excessive spastic tone in the involved areas of the body, as we discuss more fully later in the chapter. This is the spasticity that normally accompanies a "stroke" in the human being.

The Babinski Sign Used As a Clinical Tool in Testing Corticospinal Integrity. Destruction of the foot region of the area pyramidalis or transection of the foot portion of the corticospinal tract causes a peculiar response of the foot called the *Babinski sign.* This response is demonstrated when a firm tactile stimulus is applied to the lateral sole of the foot: The great toe extends upward, and the other toes fan outward. This is in contradistinction to the normal effect in which all the toes bend downward. The Babinski sign does not occur when damage occurs in the noncorticospinal portions of the motor control system without involving the corticospinal tract. Therefore, the sign is used clinically to detect damage specifically in the corticospinal portion of the motor control system.

The cause of the Babinski sign is believed to be the following: The corticospinal tract is a major controller of muscle activity for performance of voluntary, purposeful activity. On the other hand, the noncorticospinal pathways constitute a much older motor control system and are concerned to a great extent with protection of the body from damage. Therefore, when only the noncorticospinal system is functional, stimuli to the bottom of the feet cause a typical withdrawal protective type of reflex, expressed by the upturned great toe and fanning of the other toes. But when the corticospinal system is also fully functional, it suppresses the protective reflex and instead excites a higher order of motor function, including the normal effect of causing downward bending of the toes and foot in response to sensory stimuli from the bottom of the feet, a response that helps us to walk.

● ROLE OF THE BRAIN STEM IN CONTROLLING MOTOR FUNCTION

The brain stem consists of the *medulla, pons,* and *mesencephalon.* In one sense it is an extension of the spinal cord upward into the cranial cavity, because it contains motor and sensory nuclei that perform motor and sensory functions for the face and head regions in the same way that the anterior and posterior gray horns of the spinal cord perform these same functions from the neck down. But in another sense, it is its own master, because it provides many special control functions, such as the following:

1. Control of respiration.
2. Control of the cardiovascular system.
3. Control of gastrointestinal function.
4. Control of many stereotyped movements of the body.
5. Control of equilibrium.
6. Control of eye movement.

Finally, the brain stem serves also as an instrument of higher neural centers that transmit many "command" signals into the brain stem to initiate or modify the brain stem's specific control functions.

In the following sections of this chapter we discuss the role of the brain stem in controlling whole body movement and equilibrium. Especially important for this purpose are the brain stem's *reticular nuclei* and *vestibular nuclei* plus the *vestibular apparatus* that sends most of the equilibrium control signals to the vestibular nuclei and to a lesser extent to the reticular nuclei as well.

SUPPORT OF THE BODY AGAINST GRAVITY—ROLES OF THE RETICULAR AND VESTIBULAR NUCLEI

Excitatory-Inhibitory Antagonism Between Pontine and Medullary Reticular Nuclei

Figure 55–7 illustrates the locations of the reticular and vestibular nuclei. The reticular nuclei are divided

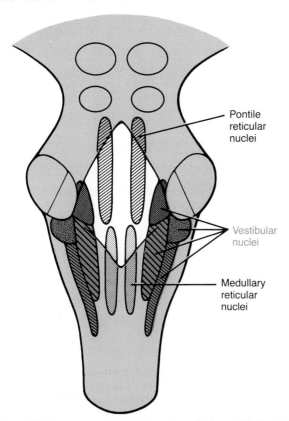

Figure 55–7. Locations of the reticular and vestibular nuclei in the brain stem.

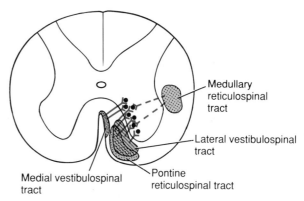

Figure 55–8. The vestibulospinal and reticulospinal tracts descending in the spinal cord to excite (solid lines) or inhibit (dashed lines) the anterior motor neurons that control the body's axial musculature.

into two major groups, (1) the *pontine reticular nuclei,* located mainly in the pons but extending into the mesencephalon as well, lying more laterally in the brain stem, and (2) the *medullary reticular nuclei,* which extend the entire extent of the medulla, lying ventrally and medially near the midline. These two sets of nuclei function mainly antagonistically to each other, the pontine exciting the antigravity muscles and the medullary inhibiting them. The pontine reticular nuclei transmit excitatory signals downward into the cord through the *pontine* (or *medial*) *reticulospinal tract,* illustrated in Figure 55–8. The fibers of this pathway terminate on the medial anterior motor neurons that excite the muscles that support the body against gravity, that is, the muscles of the spinal column and the extensor muscles of the limbs.

The pontine reticular nuclei have a high degree of natural excitability. In addition, they receive excitatory signals from local circuits within the brain stem and especially strong excitatory signals from the vestibular nuclei and also the deep nuclei of the cerebellum. Therefore, when the pontine reticular excitatory system is unopposed by the medullary reticular system, it causes powerful excitation of the antigravity muscles throughout the body, so much so that animals can then stand up against gravity without any signals from the higher levels of the brain.

The Medullary Reticular System. The medullary nuclei, on the other hand, transmit inhibitory signals to the same antigravity anterior motor neurons by way of a different tract, the *medullary* (or *lateral*) *reticulospinal tract,* also illustrated in Figure 55–8. The medullary reticular nuclei receive strong input collaterals from (1) the corticospinal tract, (2) the rubrospinal tract, and (3) other motor pathways. These normally activate the medullary reticular inhibitory system to counterbalance the excitatory signals from the pontine reticular system. Yet other signals from the cerebral, red nucleus, and cerebellar pathways "disinhibit" the medullary system when the brain wishes for excitation by the pontine system to cause standing. Or at other times, excitation of the medullary reticular system can inhibit the antigravity muscles in certain portions of the body to allow those portions to perform other motor activities, which would be impossible if the antigravity muscles opposed the necessary movements.

Therefore, the excitatory and inhibitory reticular nuclei constitute a controllable system that is manipulated by motor signals from the cortex and elsewhere to provide the necessary muscle contractions for standing against gravity and yet to inhibit appropriate groups of muscles as needed so that other functions can be performed as required.

Role of the Vestibular Nuclei in Exciting the Antigravity Muscles

The vestibular nuclei, illustrated in Figure 55–7, function also in association with the pontine reticular nuclei to excite the antigravity muscles. The *lateral vestibular nuclei* (indicated by the heavy dots in the figure), especially, transmit strong excitatory signals by way of both the *lateral* and *medial vestibulospinal tracts* in the anterior column of the spinal cord, as illustrated in Figure 55–8. In fact, without the support of the vestibular nuclei, the pontine reticular system loses much of its force. The specific role of the vestibular nuclei, however, is to control selectively the excitatory signals to the different antigravity muscles to maintain equilibrium in response to signals from the vestibular apparatus. We shall discuss this more fully later in the chapter.

The Decerebrate Animal Develops Spastic Rigidity

When the brain stem is sectioned between the pons and the mesencephalon, leaving both the pontine and medullary reticular systems as well as the vestibular system intact, the animal develops a condition called *decerebrate rigidity*. This rigidity does not occur in all muscles of the body, but instead occurs in the antigravity muscles — the muscles of the neck and the trunk and the extensors of the legs.

The cause of decerebrate *rigidity* is blockage of the normally strong excitatory input to the medullary reticular nuclei from the cerebral cortex, red nuclei, and basal ganglia. As a result, the medullary vestibular inhibitor system becomes nonfunctional because of loss of its usual excitatory drive, thus allowing full overactivity of the pontine excitatory system.

A specific characteristic of decerebrate rigidity is that the antigravity muscles exhibit the phenomenon called *spasticity* as well as rigidity. This means that any attempt to change the position of a limb or other part of the body, especially attempts to stretch the muscles suddenly, is resisted by very powerful stretch reflexes described in the previous chapter. This occurs because the pontine and vestibular antigravity signals to the cord selectively excite the gamma motor neurons in the spinal cord much more than they excite the alpha motor neurons. This tightens the intrafused muscle fibers of the muscle spindles, which in turn strongly sensitizes the stretch reflex feedback loop.

We see later that other types of rigidity occur in other neuromotor diseases, especially in lesions of the basal ganglia. In many of these the rigidity involves all muscles equally without the excessive spastic stretch reflex component.

● VESTIBULAR SENSATIONS AND THE MAINTENANCE OF EQUILIBRIUM

THE VESTIBULAR APPARATUS

The vestibular apparatus is the organ that detects sensations of equilibrium. It is composed of a system of bony tubes and chambers in the petrous portion of the temporal bone called the *bony labyrinth* and within this a system of membranous tubes and chambers called the *membranous labyrinth,* which is the functional part of the apparatus. The top of Figure 55–9 illustrates the membranous labyrinth; it is composed mainly of the *cochlea,* three *semicircular ducts,* and two large chambers known as the *utricle* and the *saccule.* The cochlea is the major sensory area for hearing that was discussed in Chapter 52 and has nothing to do with equilibrium. However, the *utricle,* the *semicircular ducts,* and the *saccule* are all integral parts of the equilibrium mechanism.

Membranous Labyrinth

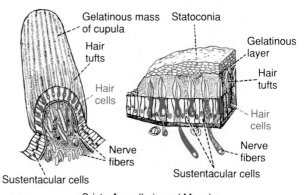

Crista Ampullaris and Macula

Figure 55–9. The membranous labyrinth and organization of the crista ampullaris and the macula. (Modified from Goss: Gray's Anatomy of the Human Body. Philadelphia, Lea & Febiger; modified from Kolmer by Buchanan: Functional Neuroanatomy. Philadelphia, Lea & Febiger.)

The Maculae—The Sensory Organs of the Utricle and the Saccule for Detecting the Orientation of the Head With Respect to Gravity. Located on the inside surface of each utricle and saccule is a small sensory area slightly over two mm in diameter called a *macula*. The macula of the utricle lies mainly in the horizontal plane on the inferior surface of the utricle and plays an important role in determining the normal orientation of the head with respect to the direction of gravitational or acceleratory forces when a person is upright. On the other hand, the macula of the saccule is located mainly in a vertical plane and therefore is important in equilibrium when the person is lying down.

Each macula is covered by a gelatinous layer in which many small calcium carbonate crystals called *statoconia* (or *otoliths*) are imbedded. Also, in the macula are thousands of *hair cells,* one of which is illustrated in Figure 55–10; these project *cilia* up into the gelatinous layer. The bases and sides of the hair cells synapse with sensory endings of the vestibular nerve.

Directional Sensitivity of the Hair Cells—The Kinocilium. Each hair cell has an average of 50 to 70 small cilia called *stereocilia,* plus one very large cilium, the *kinocilium,* as illustrated in the figure. The kinocilium is located always to one side, and the stereocilia become progressively shorter toward the other side of the cell. Very minute filamentous attachments, almost invisible even to the electron microscope, connect the tip of each stereocilium to the next longer stereocilium and finally to the kinocilium. Because of these attachments, when the brush pile of stereocilia and kinocilium is bent in the direction of the kinocilium, the filamentous attachments tug one after the other on the stereocilia, pulling them away from the cell body. This opens several hundred channels in each cilium membrane for conducting positive sodium ions, and large quantities of these positive ions pour into the cell from the surrounding fluids, causing *depolarization.* Conversely, bending the pile of cilia in the opposite direction (away from the kinocilium) reduces the tension on the attachments, and this closes the ion channels, thus causing *hyperpolarization.*

Under normal resting conditions, the nerve fibers leading from the hair cells transmit continuous nerve impulses at rates about 100 per second. When the cilia are bent toward the kinocilium, the impulse traffic can increase to several hundred per second; conversely, bending the cilia in the opposite direction decreases the impulse traffic, often turning it off completely. Therefore, as the orientation of the head in space changes and the weight of the statoconia (whose specific gravity is about three times that of the surrounding tissues) bends the cilia, appropriate signals are transmitted to the brain to control equilibrium.

In each macula the different hair cells are oriented in different directions so that some of them are stimulated when the head bends forward, some when it bends backward, others when it bends to one side, and so forth. Therefore, a different pattern of excitation

Figure 55–10. A hair cell of the membranous labyrinth of the equilibrium apparatus.

occurs in the nerve fibers from the macula for each position of the head; it is this "pattern" that apprises the brain of the head's orientation.

The Semicircular Ducts. The three semicircular ducts in each vestibular apparatus, known respectively as the *anterior, posterior,* and *horizontal semicircular ducts,* are arranged at right angles to each other so that they represent all three planes in space. When the head is bent forward approximately 30 degrees, the horizontal semicircular ducts are located approximately horizontal with respect to the surface of the earth. The anterior ducts are then located in vertical planes that project *forward and 45 degrees outward,* and the posterior ducts are also in vertical planes but project *backward and 45 degrees outward.* Thus, the anterior duct on each side of the head is in a plane parallel to that of the posterior duct on the opposite side of the head, whereas the horizontal ducts on the two sides are located in approximately the same plane.

Each semicircular duct has an enlargement at one of its ends called the *ampulla,* and the ducts are filled with a viscous fluid called *endolymph.* Flow of this fluid from one of the ducts into the ampulla excites the sensory organ of the ampulla in the following manner: Figure 55–11 illustrates in each ampulla a small crest called a *crista ampullaris.* On top of this crista is a gelatinous mass, the *cupula.* When the head begins to rotate in any direction, the inertia of the fluid in one or more of the semicircular ducts will cause the fluid to remain stationary while the semicircular duct rotates with the head. This causes fluid flow from the duct into the ampulla, bending the cupula to one side as illustrated by the position of the shaded cupula in Figure 55–11. Rotation of the head in the opposite direction causes the cupula to bend to the opposite side.

Into the cupula are projected hundreds of cilia from hair cells located along the ampullary crest. The *kinocilium* of each of these hair cells is always directed toward the same side of the cupula as the others, and bending the cupula in that direction causes depolarization of the hair cells, whereas bending it in the opposite direction will hyperpolarize the cells. From the hair cells, appropriate signals are sent by way of the *vestibular nerve* to apprise the central nervous system of changes in the rate and direction of rotation of the head in the three different planes of space.

FUNCTION OF THE UTRICLE AND SACCULE IN THE MAINTENANCE OF STATIC EQUILIBRIUM

It is especially important that the different hair cells are oriented in all different directions in the

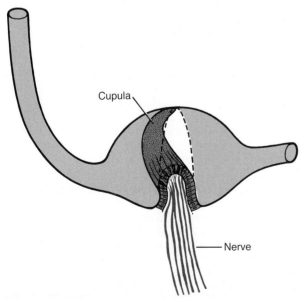

Figure 55–11. Movement of the cupula and its embedded hairs at the onset of rotation.

maculae of the utricles and saccules so that at different positions of the head, different hair cells become stimulated. The "patterns" of stimulation of the different hair cells apprise the nervous system of the position of the head with respect to the pull of gravity. In turn, the vestibular, cerebellar, and reticular motor systems excite by reflex the appropriate muscles to maintain proper equilibrium.

The maculae, especially those in the utricles, function extremely effectively for maintaining equilibrium when the head is in the near-vertical position. Indeed, a person can determine as little as a half-degree of malequilibrium when the head leans from the precise upright position. On the other hand, as the head is leaned farther and farther from the upright, the determination of head orientation by the vestibular sense becomes poorer and poorer. Obviously, extreme sensitivity in the upright position is of major importance for maintenance of precise vertical static equilibrium, which is the most essential function of the vestibular apparatus.

Detection of Linear Acceleration by the Maculae. When the body is suddenly thrust forward—that is, when the body accelerates—the statoconia, which have greater inertia than the surrounding fluids, fall backward on the hair cell cilia, and information of malequilibrium is sent into the nervous centers, causing the individual to feel as though he were falling backward. This automatically causes him to lean his body forward until the anterior shift of the statoconia caused by leaning exactly equals the tendency for the statoconia to fall backward. At this point, the nervous system detects a state of proper equilibrium and therefore leans the body no farther forward. Thus, the maculae operate to maintain equilibrium during linear acceleration in exactly the same manner as they operate in static equilibrium.

The maculae *do not* operate for the detection of linear *velocity.* When runners first begin to run, they must lean far forward to keep from falling over backward because of *acceleration,* but once they have achieved running speed, they would not have to lean forward at all if they were running in a vacuum. When running in air they lean forward to maintain equilibrium only because of the air resistance against their bodies; in this instance it is not the maculae that make them lean but the pressure of the air acting on pressure end-organs in the skin, which initiate the appropriate equilibrium adjustments to prevent falling.

DETECTION OF HEAD ROTATION BY THE SEMICIRCULAR DUCTS

When the head suddenly *begins* to rotate in any direction (this is called angular acceleration), the endolymph in the semicircular ducts, because of its inertia, tends to remain stationary while the semicircular ducts themselves turn. This causes relative fluid flow in the ducts in the direction opposite to the rotation of the head.

Figure 55–12 illustrates a typical discharge signal

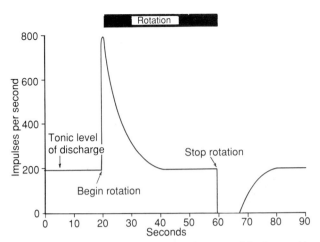

Figure 55–12. Response of a hair cell when a semicircular canal is stimulated first by rotation and then by stopping rotation.

from a single hair cell in the crista ampullaris when an animal is rotated for 40 seconds, showing that (1) even when the cupula is in its resting position the hair cell emits a tonic discharge of approximately 100 impulses per second; (2) when the animal is rotated, the hairs bend to one side and the rate of discharge increases greatly; and (3) with continued rotation, the excess discharge of the hair cell gradually subsides back to the resting level in about 20 seconds.

The reason for this adaptation of the receptor is that within a second or more of rotation, back-pressure from the bent cupula causes the endolymph to rotate as rapidly as the semicircular canal itself; then in an additional 15 to 20 seconds the cupula slowly returns to its resting position in the middle of the ampulla because of its own elastic recoil.

When the rotation suddenly stops, exactly the opposite effects take place: the endolymph continues to rotate while the semicircular duct stops. This time the cupula is bent in the opposite direction, causing the hair cell to stop discharging entirely. After another few seconds, the endolymph stops moving, and the cupula returns gradually to its resting position in about 20 seconds, thus allowing the discharge of the hair cell to return to its normal tonic level as shown to the right in Figure 55–12.

Thus, the semicircular duct transmits a signal of one polarity when the head *begins* to rotate and of opposite polarity when it *stops* rotating. Furthermore, at least some hair cells will always respond to rotation in any plane — horizontal, sagittal, or coronal — for fluid movement always occurs in at least one semicircular duct.

Rate of Angular Acceleration Required to Stimulate the Semicircular Ducts. The angular acceleration required to stimulate the semicircular ducts in the human being averages about 1 degree per second per second. In other words, when one begins to rotate, the velocity of rotation must be as much as 1 degree per second by the end of the first second, 2 degrees per second by the end of the second second, 3 degrees per second by the end of the third second,

and so forth, in order for the person barely to detect that the rate of rotation is increasing.

"Predictive" Function of the Semicircular Ducts in the Maintenance of Equilibrium. Because the semicircular ducts do not detect that the body is off balance in the forward direction, in the side direction, or in the backward direction, one might at first ask: What is the function of the semicircular ducts in the maintenance of equilibrium? All they detect is that the person's head is beginning to rotate or stopping rotation in one direction or another. Therefore, the function of the semicircular ducts is not likely to be to maintain static equilibrium or to maintain equilibrium during linear acceleration or when the person is exposed to steady centrifugal forces. Yet loss of function of the semicircular ducts causes a person to have very poor equilibrium when attempting to perform *rapid* and *intricate* body movements.

We can explain the function of the semicircular ducts best by the following illustration. If a person is running forward rapidly, and then suddenly begins to turn to one side, he falls off balance a fraction of a second later unless appropriate corrections are made *ahead of time.* But, unfortunately, the macula of the utricle cannot detect that he is off balance until *after* this has occurred. On the other hand, the semicircular ducts will have already detected that the person is turning, and this information can easily apprise the central nervous system of the fact that the person *will* fall off balance within the next fraction of a second or so unless some correction is made. In other words, the semicircular duct mechanism *predicts ahead of time* that malequilibrium is going to occur even before it does occur and thereby causes the equilibrium centers to make appropriate preventive adjustments. In this way, the person need not fall off balance before he begins to correct the situation.

Removal of the flocculonodular lobes of the cerebellum prevents normal function of the semicircular ducts but has less effect on the function of the macular receptors. It is especially interesting in this connection that the cerebellum serves as a "predictive" organ for most of the other rapid movements of the body as well as those having to do with equilibrium. These other functions of the cerebellum are discussed in the following chapter.

VESTIBULAR POSTURAL REFLEXES

Sudden changes in the orientation of an animal in space elicit reflexes that help to maintain equilibrium and posture. For instance, if an animal is suddenly pushed to the right, even before it can fall more than a few degrees its right legs extend instantaneously. In other words, this mechanism *anticipates* that the animal will be off balance in a few seconds and makes appropriate adjustments to prevent this.

Another type of vestibular postural reflex occurs when the animal suddenly falls forward. When this occurs, the forepaws extend forward, the extensor muscles tighten, and the muscles in the back of the neck stiffen to prevent the animal's head from striking the ground. This reflex is probably also of importance in locomotion, for, in the case of the galloping horse, the downward thrust of the head can automatically provide reflex thrust of the forelimbs to move the animal forward for the next gallop.

VESTIBULAR MECHANISM FOR STABILIZING THE EYES

When a person changes his direction of movement rapidly, or even leans his head sideways, forward, or backward, it would be impossible for him to maintain a stable image on the retinae of his eyes unless he had some automatic control mechanism to stabilize the direction of gaze of the eyes. In addition, the eyes would be of little use in detecting an image unless they remained "fixed" on each object long enough to gain a clear image. Fortunately, each time the head is suddenly rotated, signals from the semicircular ducts cause the eyes to rotate in a direction equal and opposite to the rotation of the head. This results from reflexes transmitted from the canals through the *vestibular nuclei* and the *medial longitudinal fasciculus* to the *ocular nuclei* that were described in Chapter 51.

OTHER FACTORS CONCERNED WITH EQUILIBRIUM

The Neck Proprioceptors. The vestibular apparatus detects the orientation and movements *only of the head.* Therefore, it is essential that the nervous centers also receive appropriate information depicting the orientation of the head with respect to the body. This information is transmitted from the proprioceptors of the neck and body directly into the vestibular and reticular nuclei of the brain stem and also indirectly by way of the cerebellum.

By far the most important proprioceptive information needed for the maintenance of equilibrium is that derived from the *joint receptors of the neck.* When the head is leaned in one direction by bending the neck, impulses from the neck proprioceptors keep the vestibular apparatus from giving the person a sense of malequilibrium. They do this by transmitting signals that exactly oppose the signals transmitted from the vestibular apparatuses. However, *when the entire body* leans in one direction, the impulses from the vestibular apparatuses *are not opposed* by the neck proprioceptors; therefore, the person in this instance does perceive a change in equilibrium status of the entire body.

The Neck Reflexes. In an animal *whose vestibular apparatuses have been destroyed,* bending the neck causes immediate muscular reflexes called *neck reflexes* occurring especially in the forelimbs. For instance, bending the head forward causes both forelimbs to relax. However, when the vestibular apparatuses are intact, this effect does *not* occur because the vestibular reflexes function in a manner exactly opposite that of the neck reflexes. Because the equilibrium of the entire body and not of the head alone must be maintained, it is easy to understand that the vestibular and neck reflexes must function oppositely.

Proprioceptive and Exteroceptive Information from Other Parts of the Body. Proprioceptive information from other parts of the body besides the neck is also important in the maintenance of equilibrium. For instance, pressure sensations from the footpads can tell one (1) whether weight is distributed equally between the two feet and (2) whether weight is more forward or backward on the feet.

An instance in which exteroceptive information is necessary for maintenance of equilibrium occurs when a person is running. The air pressure against the front of the body signals that a force is opposing the body in a direction different from that caused by gravitational pull; as a result, the person leans forward to oppose this.

Importance of Visual Information in the Maintenance of Equilibrium. After complete destruction of the vestibular apparatus, and even after loss of most proprioceptive information from the body, a person can still use the visual mechanisms effectively for maintaining equilibrium. Even slight linear or rotational movement of the body instantaneously shifts the visual images on the retina, and this information is relayed to the equilibrium centers. Many persons with complete destruction of the vestibular apparatus have almost normal equilibrium as long as their eyes are open and as long as they perform all motions slowly. But when one is moving rapidly or when the eyes are closed, equilibrium is immediately lost.

Neuronal Connections of the Vestibular Apparatus with the Central Nervous System

Figure 55–13 illustrates the central connections of the vestibular nerve. Most of the vestibular nerve fibers end in the vestibular nuclei, which are located approximately at the junction of the medulla and the pons, but some fibers pass without synapsing to the brain stem reticular nuclei and the fastigial nuclei, uvula, and flocculonodular lobes of the cerebellum. The fibers that end in the vestibular nuclei synapse with second-order neurons that also send fibers into these areas of the cerebellum as well as to the cortex of other portions of the cerebellum, into the vestibulospinal tracts, into the medial longitudinal fasciculus, and to other areas of the brain stem, particularly the reticular nuclei.

The primary pathway for the reflexes of equilibrium begins in the vestibular nerves and passes next to both the vestibular nuclei and the cerebellum. Then, along with two-way traffic of impulses between these two, signals are also sent into the reticular nuclei of the brain stem as well as down the spinal cord via the vestibulospinal and reticulospinal tracts. In turn, the signals to the cord control the interplay between facilitation and inhibition of the antigravity muscles, thus automatically controlling equilibrium.

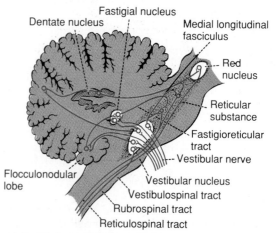

Figure 55–13. Connections of vestibular nerves in the central nervous system.

The *flocculonodular lobes* of the cerebellum seem to be especially concerned with equilibrium functions of the semicircular ducts, because destruction of these lobes gives almost exactly the same clinical symptoms as destruction of the semicircular ducts themselves. That is, severe injury to either of these structures causes loss of equilibrium during *rapid changes in direction of motion* but does not seriously disturb equilibrium under static conditions, as discussed earlier. It is also believed that the *uvula* of the cerebellum plays a similar important role in static equilibrium.

Signals transmitted upward in the brain stem from both the vestibular nuclei and the cerebellum via the *medial longitudinal fasciculus* cause corrective movements of the eyes every time the head rotates, so that the eyes can remain fixed on a specific visual object. Signals also pass upward (either through this same tract or through reticular tracts) to the cerebral cortex, probably terminating in a primary cortical center for equilibrium located in the parietal lobe deep in the sylvian fissure, on the opposite side of the fissure from the auditory area of the superior temporal gyrus. These signals apprise the psyche of the equilibrium status of the body.

The vestibular nuclei on either side of the brain stem are divided into four separate subdivisions: (1 and 2) the *superior* and *medical vestibular nuclei,* which receive signals mainly from the semicircular ducts and in turn send large numbers of nerve signals into the *medial longitudinal fasciculus* to cause corrective movements of the eyes as well as signals through the *medial vestibulospinal tract* to cause appropriate movements of the neck and head; (3) the *lateral vestibular nucleus,* which receives its innervation primarily from the utricle and saccule and in turn transmits outflow signals to the spinal cord through the *lateral vestibulospinal tract* to control body movement; (4) the *inferior vestibular nucleus,* which receives signals from both the semicircular ducts and the utricle and in turn sends signals into both the cerebellum and the reticular formation of the brain stem.

● FUNCTIONS OF SPECIFIC BRAIN STEM NUCLEI IN CONTROLLING SUBCONSCIOUS, STEREOTYPED MOVEMENTS

Rarely, a child called an *anencephalic monster* is born without brain structures above the mesencephalic region, and some of these children have been kept alive for many months. They are able to perform essentially all of the functions of feeding, such as suckling, extrusion of unpleasant food from the mouth, and moving the hands to the mouth to suck the fingers. In addition, they can yawn and stretch. They can cry and follow objects with movements of the eyes and head. Also, placing pressure on the upper anterior parts of their legs will cause them to pull to the sitting position.

Therefore, it is obvious that many of the stereotyped motor functions of the human being are integrated in the brain stem. Unfortunately, the loci of most of the different motor control systems have not been found except for the following:

Stereotyped Body Movements. Most movements of the trunk and head can be classified into several simple movements, such as forward flexion, extension, rotation, and turning movements of the entire body. These types of movements are controlled by special nuclei located mainly in the mesencephalic and lower diencephalic region. For instance, *rotational movements* of the head and eyes are controlled by the *interstitial nucleus.* This nucleus lies in the

mesencephalon in close approximation to the *medial longitudinal fasciculus,* through which it transmits a major portion of its control impulses. The *raising movements* of the head and body are controlled by the *prestitial nucleus,* which is located approximately at the juncture of the diencephalon and mesencephalon. On the other hand, the *flexing movements* of the head and body are controlled by the *nucleus precommissuralis* located at the level of the posterior commissure. Finally, the *turning movements* of the entire body, which are much more complicated, involve both the pontile and mesencephalic reticular nuclei.

REFERENCES

Adams, R. D., and Victor, M. (eds.): Principles of Neurology, 4th Ed. New York, McGraw Hill Book Co., 1989.

Asanuma, H.: The pyramidal tract. In Brooks, V. B. (ed.): Handbook of Physiology. Sec. 1, Vol. II. Bethesda, Md., American Physiological Society, 1981, p. 703.

Bahill, A. T., and Hamm, T. M.: Using open-loop experiments to study physiological systems, with examples from the human eye-movement systems. News Physiol. Sci., 4:104, 1989.

Brooks, V. B.: The Neural Basis of Motor Control. New York, Oxford University Press, 1986.

Carpenter, M. B.: Anatomy of the corpus striatum and brain stem integrating systems. In Brooks, V. B. (ed.): Handbook of Physiology. Sec. 1, Vol. II. Bethesda, Md., American Physiological Society, 1981, p. 947.

Dampney, R. A., et al.: Identification of cardiovascular cell groups in the brain stem. Clin. Exp. Hypertens., 6:205, 1984.

Desmedt, J. E. (ed.): Cerebral Motor Control in Man: Long Loop Mechanisms. New York, S. Karger, 1978.

Dublin, W. B.: Fundamentals of Vestibular Pathology. St. Louis, Warren H. Green, Inc., 1985.

Dutia, M. B.: Mechanisms of head stabilization. News Physiol. Sci., 4:101, 1989.

Elder, H. Y., and Trueman, E. R. (eds.): Aspects of Animal Movement. New York, Cambridge University Press, 1980.

Evarts, E. V., et al. (eds.): Motor System in Neurobiology. New York, Elsevier Science Publishing Co., 1986.

Evarts, E. V.: Role of motor cortex in voluntary movements in primates. In Brooks, V. B. (ed.): Handbook of Physiology. Sec. 1, Vol. II. Bethesda, Md., American Physiological Society, 1981, p. 1083.

Fernstrom, J. D.: Role of precursor availability on control of monoamine biosynthesis in brain. Physiol. Rev., 63:484, 1983.

Fournier, E., and Pierrot-Deseilligny, E.: Changes in transmission in some reflex pathways during movement in humans. News Physiol. Sci., 4:29, 1989.

Goldberg, J. M., and Fernandez, C.: The vestibular system. In Darian-Smith, I. (ed.): Handbook of Physiology, Sec. 1, Vol. III. Bethesda, Md., American Physiological Society, 1984, p. 977.

Grillner, S.: Control of locomotion in bipeds, tetrapods, and fish. In Brooks, V. B. (ed.): Handbook of Physiology. Sec. 1, Vol. II. Bethesda, Md., American Physiological Society, 1981, p. 1179.

Grillner, S.: Locomotion in vertebrates: Central mechanisms and reflex interaction. Physiol. Rev., 55:247, 1975.

Hasan, Z., and Stuart, D. G.: Animal solutions to problems of movement control: The role of proprioceptors. Annu. Rev. Neurosci., 11:199, 1988.

Hobson, J. A., and Brazier, M. A. B. (eds.): The Reticular Formation Revisited: Specifying Function for a Nonspecific System, New York, Raven Press, 1980.

Keele, S. W.: Behavioral analysis of movement. In Brooks, V. B. (ed.): Handbook of Physiology. Sec. 1, Vol. II. Bethesda, Md., American Physiological Society, 1981, p. 1391.

Kuypers, H. B. J. M.: Anatomy of the descending pathways. In Brooks, V. B. (ed.): Handbook of Physiology. Sec. 1, Vol. II. Bethesda, Md., American Physiological Society, 1981, p. 597.

Llinas, R.: Electrophysiology of the cerebellar networks. In Brooks, V. B. (ed.): Handbook of Physiology. Sec. 1, Vol. II. Bethesda, Md., American Physiological Society, 1981, p. 831.

Luschei, E. S., and Goldberg, L. J.: Neural mechanisms of mandibular control: Mastication and voluntary biting. In Brooks, V. B. (ed.): Handbook of Physiology. Sec. 1, Vol. II. Bethesda, Md., American Physiological Society, 1981, p. 1237.

McCloskey, D. I.: Corollary discharges: Motor commands and perception. In Brooks, V. B. (ed.): Handbook of Physiology. Sec. 1, Vol. II. Bethesda, Md., American Physiological Society, 1981, p. 1415.

Oosterveld, W. J. (ed.): Audio-Vestibular System and Facial Nerve. New York, S. Karger, 1977.

Pearson, K.: The control of walking. Sci. Am., 235(6):72, 1976.

Penfield, W., and Rasmussen, T.: The Cerebral Cortex of Man. New York, Macmillan Co., 1950.

Peterson, B. W., and Richmond, F. J. (eds.): Control of Head Movement. New York, Oxford University Press, 1988.

Peterson, B. W.: Reticulospinal projections of spinal motor nuclei. Annu. Rev. Physiol., 41:127, 1979.

Porter, R.: Influences of movement detectors on pyramidal tract neurons in primates. Annu. Rev. Physiol., 38:121, 1976.

Porter, R.: Internal organization of the motor cortex for input-output arrangements. In Brooks, V. B. (ed.): Handbook of Physiology. Sec. 1, Vol. II. Bethesda, Md., American Physiological Society, 1981, p. 1063.

Poulton, E. C.: Human manual control. In Brooks, V. B. (ed.): Handbook of Physiology. Sec. 1, Vol. II. Bethesda, Md., American Physiological Society, 1981, p. 1337.

Precht, W.: Vestibular mechanisms. Annu. Rev. Neurosci., 2:265, 1979.

Rack, P. M. H.: Limitations of somatosensory feedback in control of posture and movement. In Brooks, V. B. (ed.): Handbook of Physiology. Sec. 1, Vol. II. Bethesda, Md., American Physiological Society, 1981, p. 229.

Scheibel, A. B.: The brain stem reticular core and sensory function. In Darian-Smith, I. (ed.): Handbook of Physiology. Sec. 1, Vol. III. Bethesda, Md., American Physiological Society, 1984, p. 213.

Sherrington, C. S.: Decerebrate rigidity and reflex coordination of movements. J. Physiol. (Lond.), 22:319, 1898.

Shik, M. L., and Orlovsky, B. N.: Neurophysiology of locomotor automatism. Physiol. Rev., 56:465, 1976.

Silverman, A. J.: Magnocellular neurosecretory system. Annu. Rev. Neurosci., 6:357, 1983.

Stein, P. S. B.: Motor systems with reference to the control of locomotion. Annu. Rev. Neurosci., 1:61, 1978.

Valentinuzzi, M.: The Organs of Equilibrium and Orientation as a Control System. New York, Harwood Academic Publishers, 1980.

Wiesendanger, M., and Miles, T. S.: Ascending pathway of low-threshold muscle afferents to the cerebral cortex and its possible role in motor control. Physiol. Rev., 62:1234, 1982.

Wilson, V. J., and Peterson, B. W.: Peripheral and central substrates of vestibulospinal reflexes. Physiol. Rev., 58:80, 1978.

CHAPTER 56

• *The Cerebellum, the Basal Ganglia, and Overall Motor Control* •

Aside from the cerebral cortical areas for control of muscle activity, two other brain structures are also essential for normal motor function. These are the *cerebellum* and the *basal ganglia*. Yet neither of these two can initiate muscle function by themselves. Instead, *they always function in association with other systems of motor control.*

Basically, the cerebellum plays major roles in the timing of motor activities and in rapid progression from one movement to the next; it also helps to control the instantaneous interplay between agonist and antagonist muscle groups.

The basal ganglia, on the other hand, help to control complex patterns of muscle movement, controlling the relative intensities of movements, directions of movement, and sequencing of multiple successive and parallel movements for achieving specific motor goals.

It is the purpose of this chapter to explain the basic mechanisms of function of the cerebellum and basal ganglia and also to discuss what we know about the overall brain mechanisms for achieving the intricate coordination of total motor activity.

● THE CEREBELLUM AND ITS MOTOR FUNCTIONS

The cerebellum has long been called a *silent area* of the brain, principally because electrical excitation of this structure does not cause any sensation and rarely any motor movement. However, as we shall see, removal of the cerebellum does cause movement to become highly abnormal. The cerebellum is especially vital to the control of rapid muscular activities such as running, typing, playing the piano, and even talking. Loss of this area of the brain can cause almost total incoordination of these activities even though its loss causes paralysis of no muscles.

But how is it that the cerebellum can be so important when it has no direct ability to cause muscle contraction? The answer to this is that it both helps *sequence the motor activities* and also *monitors and makes corrective adjustments in the motor activities elicited by other parts of the brain.* It receives continuously updated information on the desired program of muscle contractions from the motor control areas of the other parts of the brain; and it receives continuous sensory information from the peripheral parts of the body to determine the sequential changes in the status of each part of the body — its position, its rate of movement, forces acting on it, and so forth. The cerebellum *compares* the actual movements as depicted by the peripheral sensory feedback information with the movements intended by the motor system. If the two do not compare favorably, then appropriate corrective signals are transmitted instantaneously back into the motor system to increase or decrease the levels of activation of the specific muscles.

In addition, the cerebellum aids the cerebral cortex in planning the next sequential movement a fraction of a second in advance while the present movement is still being executed, thus helping one to progress smoothly from one movement to the next. Also, it learns by its mistakes — that is, if a movement does not occur exactly as intended, the cerebellar circuit learns to make a stronger or weaker movement the next time. To do this, long-lasting changes occur in the excitability of the appropriate cerebellar neurons, thus bringing the subsequent contractions into better correspondence with the intended movements.

THE ANATOMICAL FUNCTIONAL AREAS OF THE CEREBELLUM

Anatomically, the cerebellum is divided into three separate lobes by two deep fissures, as shown in Figures 56–1 and 56–2: (1) the *anterior lobe,* (2) the *posterior lobe,* and (3) the *flocculonodular lobe.* The flocculonodular lobe is the oldest of all portions of the cerebellum; it developed along with (and functions with) the vestibular system in controlling equilibrium, as discussed in the previous chapter.

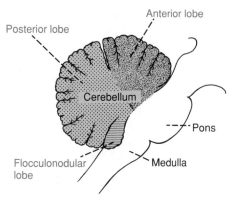

Figure 56–1. The anatomical lobes of the cerebellum as seen from the lateral side.

The Longitudinal Functional Divisions of the Anterior and Posterior Lobes. From a functional point of view, the anterior and posterior lobes are organized not by lobes, but instead along the longitudinal axis, as illustrated in Figure 56–2, which shows the human cerebellum after the lower end of the posterior cerebellum has been rolled downward from its normally hidden position. Note down the center of the cerebellum a narrow band separated from the remainder of the cerebellum by shallow grooves. This is called the *vermis.* In this area most cerebellar control functions for the muscle movements of the axial body, the neck, and the shoulders and hips are located.

To each side of the vermis is a large, laterally protruding *cerebellar hemisphere,* and each of these hemispheres is divided into an *intermediate zone* and *a lateral zone.*

The intermediate zone of the hemisphere is concerned with the control of muscular contractions in the distal portions of the upper and lower limbs, especially of the hands and fingers and feet and toes.

The lateral zone of the hemisphere operates at a much more remote level, for this area joins in the overall planning of sequential motor movements. Without this lateral zone, most discrete motor activities of the body lose their appropriate timing and therefore become incoordinate, as we discuss more fully later.

Topographical Representation of the Body in the Vermis and Intermediate Zones. In the same manner that the sensory cortex, the motor cortex, the basal ganglia, the red nuclei, and the reticular formation all have topographical representations of the different parts of the body, so also is this true for the vermis and intermediate zones of the cerebellum. Figure 56–3 illustrates two separate such representations. Note that the axial portions of the body lie in the vermal part of the cerebellum, whereas the limbs and facial regions lie in the intermediate zones. These topographical representations receive afferent nerve signals from all the respective parts of the body as well as from the corresponding topographical areas of the motor cortex and brain stem motor areas. In turn, they send motor signals into the same respective topographical areas of the motor cortex, the red nucleus, and the reticular formation.

However, note that the large lateral portions of the cerebellar hemispheres *do not* have topographical representations of the body. These areas of the cerebellum connect mainly with corresponding association areas of the cerebral cortex, especially the premotor area of the frontal cortex and the somatic sensory and sensory association areas of the parietal cortex. Presumably this connectivity with the association areas allows the lateral portions of the cerebellar hemispheres to play important roles in planning and coordinating the sequential muscular activities.

The Input Pathways to the Cerebellum

Afferent Pathways From the Brain. The basic input pathways to the cerebellum are illustrated in Figure 56–4. An extensive and important afferent pathway is the *corticopontocerebellar pathway,* which originates mainly in the *motor* and *premotor cortices* but to a lesser extent in the sensory cortex as well and then passes by way of the *pontile nuclei* and *pontocerebellar tracts* to the contralateral hemisphere of the cerebellum.

In addition, important afferent tracts originate in the brain stem; they include (1) an extensive *olivocerebellar* tract, which passes from the *inferior olive* to all parts of the cerebellum and is excited by fibers from the *motor cortex,* the *basal ganglia,* widespread areas of the *reticular formation,* and the *spinal cord;* (2) *vestibulocerebellar fibers,* some of which originate in the vestibular apparatus itself and others from the vestibular nuclei and most of which termi-

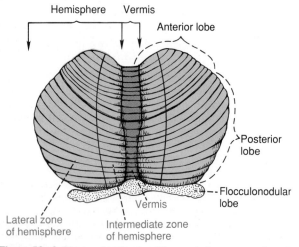

Figure 56–2. The functional parts of the cerebellum as seen from the posteroinferior view, with the inferiormost portion of the cerebellum rolled outward to flatten the surface.

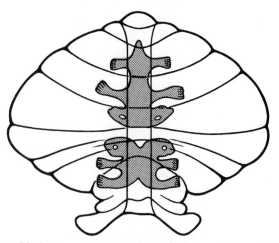

Figure 56–3. The somatic sensory projection areas in the cerebellar cortex.

Figure 56-4. The principal afferent tracts to the cerebellum.

nate in the *flocculonodular lobe* and *fastigial nucleus* of the cerebellum; and (3) *reticulocerebellar fibers,* which originate in different portions of the reticular formation and terminate mainly in the midline cerebellar areas (the vermis).

Afferent Pathways From the Periphery. The cerebellum also receives important sensory signals directly from the peripheral parts of the body through four separate tracts, two of which are located dorsally in the cord and two ventrally. The two most important of these tracts are illustrated in Figure 56-5: the *dorsal spinocerebellar tract* and the *ventral spinocerebellar tract* (plus similar tracts from the neck and facial regions). The dorsal tracts enter the cerebellum through the inferior cerebellar peduncle and terminate in the vermis and intermediate zones of the cerebellum on the same side as their origin. The two ventral tracts enter the same areas of the cerebellum through the superior cerebellar peduncle, but they terminate in both sides of the cerebellum.

The signals transmitted in the dorsal spinocerebellar

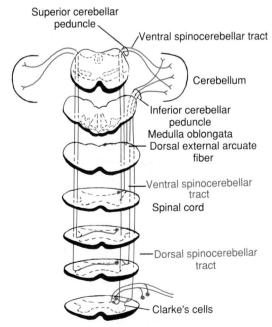

Figure 56-5. The spinocerebellar tracts.

tracts come mainly from the muscle spindles and to a lesser extent from other somatic receptors throughout the body, such as from the Golgi tendon organs, the large tactile receptors of the skin, and the joint receptors. All these signals apprise the cerebellum of the momentary status of muscle contraction, degree of tension on the muscle tendons, positions and rates of movement of the parts of the body, and forces acting on the surfaces of the body.

On the other hand, the ventral spinocerebellar tracts receive less information from the peripheral receptors. Instead, they are excited mainly by the motor signals arriving in the anterior horns of the spinal cord from the brain through the corticospinal and rubrospinal tracts as well as from the internal motor pattern generators in the cord itself. Thus, this ventral fiber pathway tells the cerebellum what motor signals have arrived at the anterior horns; this feedback is called the *efference copy* of the anterior horn motor drive.

The spinocerebellar pathways can transmit impulses at velocities of up to 120 m/sec, which is the most rapid conduction of any pathway in the entire central nervous system. This extremely rapid conduction is important for the instantaneous apprisal of the cerebellum of the changes that take place in peripheral motor actions.

In addition to the signals in the spinocerebellar tracts, other signals are transmitted through the dorsal columns to the dorsal column nuclei of the medulla and then relayed from there to the cerebellum. Likewise, signals are transmitted through the *spinoreticular pathway* to the reticular formation of the brain stem and through the *spino-olivary pathway* to the inferior olivary nucleus and then relayed from both these areas to the cerebellum. Thus, the cerebellum continually collects information about all parts of the body even though it is operating at a subconscious level.

Output Signals From the Cerebellum

The Deep Cerebellar Nuclei and the Efferent Pathways. Located deep in the cerebellar mass are three *deep cerebellar nuclei*—the *dentate, interpositus,* and *fastigial.* The *vestibular nuclei* in the medulla also function in some respects as if they were deep cerebellar nuclei because of their direct connections with the cortex of the flocculonodular lobe. All the deep cerebellar nuclei receive signals from two different sources: (1) the cerebellar cortex and (2) the sensory afferent tracts to the cerebellum. Each time an input signal arrives in the cerebellum, it divides and goes in two directions: (1) directly to one of the deep nuclei and (2) to a corresponding area of the cerebellar cortex overlying the deep nucleus. Then, a short time later, the cerebellar cortex relays its output signals back to the same deep nucleus. Thus, all the input signals that enter the cerebellum eventually end in the deep nuclei, from which output signals are then distributed to other parts of the brain.

Three major efferent pathways lead out of the cerebellum, as illustrated in Figure 56–6:

1. A pathway that originates in the *midline structures of the cerebellum* (the *vermis*) and then passes through the *fastigial nuclei* into the *medullary* and *pontile regions of the brain stem.* This circuit functions in close association with the equilibrium apparatus to help control equilibrium and also, in association with the reticular formation of the brain stem, to help control the postural attitudes of the body. It was discussed in detail in the previous chapter in relation to equilibrium.

2. A pathway that originates in the *intermediate zone of the cerebellar hemisphere*, then passes (a) through the *nucleus interpositus* to the *ventrolateral and ventroanterior nuclei of the thalamus,* and thence to the *cerebral cortex,* (b) to several *midline structures* of the *thalamus* and thence to the *basal ganglia,* and (c) to the *red nucleus* and *reticular formation* of the upper portion of the brain stem. This circuit is believed to coordinate mainly the reciprocal contractions of agonist and antagonist muscles in the peripheral portions of the limbs — especially in the hands, fingers, and thumbs.

3. A pathway that begins in the *cortex of the lateral zone of the cerebellar hemisphere,* then passes to the *dentate nucleus,* next to the *ventrolateral and ventroanterior nuclei of the thalamus,* and finally to the *cerebral cortex.* This pathway plays an important role in helping coordinate sequential motor activities initiated by the cerebral cortex.

THE NEURONAL CIRCUIT OF THE CEREBELLUM

The human cerebellar cortex is actually a large folded sheet, approximately 17 cm wide by 120 cm long, with the folds lying crosswise, as illustrated in Figures 56–2 and 56–3. Each fold is called a *folium.* And lying deep in the folded mass of cortex are the deep nuclei.

The Functional Unit of the Cerebellar Cortex — The Purkinje Cell and the Deep Nuclear Cell.
The cerebellum has approximately 30 million nearly identical functional units, one of which is illustrated to the left in Figure 56–7. This functional unit centers on a single very large *Purkinje cell,* 30 million of which are in the cerebellar cortex.

To the right in Figure 56–7, the three major layers of the cerebellar cortex are illustrated, the *molecular layer,* the *Purkinje cell layer,* and the *granular cell layer.* And far beneath these cortical layers, in the center of the cerebellar mass, are the deep nuclei.

The Neuronal Circuit of the Functional Unit.
As illustrated in the left half of Figure 56–7, the output from the functional unit is from a deep nuclear cell. However, this cell is continually under the influence of both excitatory and inhibitory influences. The excitatory influences arise from direct connections with the afferent fibers that enter the cerebellum from the brain or the periphery. The inhibitory influences arise entirely from the Purkinje cell in the cortex of the cerebellum.

The afferent inputs to the cerebellum are mainly of two types, one called the *climbing fiber type* and the other called the *mossy fiber type.*

The climbing fibers *all originate from the inferior olivary complex of the medulla.* There is one climbing fiber for about 10 Purkinje cells. After sending branches to several deep nuclear cells, the climbing fiber projects all the way to the molecular layer of the cerebellar cortex, where it makes about 300 synapses with the soma and dendrites of each Purkinje cell. This climbing fiber is distinguished by the fact that a single impulse in it will always cause a single, very prolonged (up to 1 second), and peculiar oscillatory type of action potential in each Purkinje cell with which it connects. This action potential is called the *complex spike.*

The mossy fibers are all the other fibers that enter the cerebellum from multiple sources: the higher brain, the brain stem, and the spinal cord. These

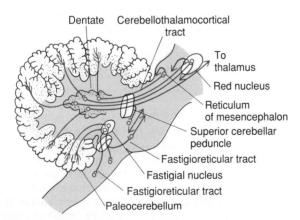

Figure 56–6. Principal efferent tracts from the cerebellum.

Dentate
Cerebellothalamocortical tract
To thalamus
Red nucleus
Reticulum of mesencephalon
Superior cerebellar peduncle
Fastigioreticular tract
Fastigial nucleus
Fastigioreticular tract
Paleocerebellum

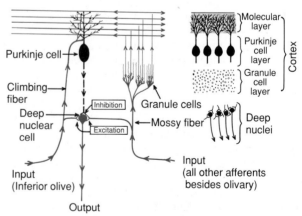

Figure 56–7. The left side of this figure shows the basic neuronal circuit of the cerebellum, with excitatory neurons shown in red. To the right is illustrated the physical relationship of the deep cerebellar nuclei to the cerebellar cortex with its three layers.

Purkinje cell
Climbing fiber
Deep nuclear cell
Input (Inferior olive)
Output
Inhibition
Excitation
Granule cells
Mossy fiber
Input (all other afferents besides olivary)
Molecular layer
Purkinje cell layer
Granule cell layer
Cortex
Deep nuclei

fibers also send collaterals to excite deep nuclear cells. Then they proceed to the granular layer of the cortex, where they synapse with hundreds of *granule cells.* In turn, the granule cells send very small axons, less than 1 micrometer in diameter, up to the outer surface of the cerebellar cortex to enter the molecular layer. Here the axons divide into two branches that extend 1 to 2 millimeters in each direction parallel to the folia. There are literally billions of these *parallel nerve fibers,* for there are some 500 to 1000 granule cells for every Purkinje cell. It is into this molecular layer that the dendrites of the Purkinje cells project, and 80,000 to 200,000 of these parallel fibers synapse with each Purkinje cell. As these fibers pass along their 1 to 2 millimeter course, each of them contacts about 250 to 500 Purkinje cells.

Yet the mossy fiber input to the Purkinje cell is quite different from the climbing fiber input because their synaptic connections are very weak, so that large numbers of mossy fibers must be stimulated simultaneously to alter the activation of the Purkinje cell. Furthermore, this activation usually takes the form of facilitation or excitation that causes repetitive Purkinje cell firing of short-duration action potentials called *simple spikes,* rather than the prolonged complex action potential occurring in response to the climbing fiber input.

Continual Firing of the Cerebellum Purkinje Cells and Deep Nuclear Cells Under Normal Resting Conditions. One of the characteristics of both the Purkinje cells and the deep nuclear cells is that normally they fire continually, the Purkinje cell fires at about 50 to 100 action potentials per second and the deep nuclear cells at still much higher rates. Therefore, the output activity of both these cells can be modulated either upward or downward. For instance, a decrease in the firing rate of the deep nuclear cells below the normal level would actually provide an *inhibitory output signal* to the motor system. On the other hand, any factor that should increase the firing rate above normal would provide an *excitatory output signal.* In this way, the cerebellum can provide either excitation or inhibition as the need arises.

Balance Between Excitation and Inhibition of the Deep Cerebellar Nuclei. Referring again to the circuit of Figure 56–7, *one should note that direct stimulation of the deep nuclear cells by both the climbing and the mossy fibers excites them.* By contrast, *the signals arriving from the Purkinje cells inhibit them.* Normally, the balance between these two effects is slightly in favor of excitation, so that the output from the deep nuclear cell remains relatively constant at a moderate level of continuous stimulation. On the other hand, in the execution of rapid motor movements, the *timing* of the two effects on the deep nuclei is such that the excitation appears before the inhibition. Then a few milliseconds later inhibition occurs. In this way, there is first a very rapid excitatory signal fed back into the motor pathway to modify the motor movement, but this is followed within a few milliseconds by an inhibitory signal. This inhibitory signal resembles a "delay-line" negative feedback signal of the type that is very effective in providing *damping.* That is, when the motor system is excited, a negative feedback signal presumably occurs after a short delay to stop the muscle movement from overshooting its mark, which is the usual cause of oscillation.

Other Inhibitory Cells in the Cerebellar Cortex. In addition to the granule cells and Purkinje cells, three other types of neurons are also located in the cerebellar cortex: *basket cells, stellate cells,* and *Golgi cells.* All of these are inhibitory cells with very short axons. Both the basket cells and the stellate cells are located in the molecular layer of the cortex, lying among and stimulated by the parallel fibers. These cells in turn send their axons at right angles across the parallel fibers and cause *lateral inhibition* of the adjacent Purkinje cells, thus sharpening the signal in the same manner that lateral inhibition sharpens the contrast of signals in many other areas of the nervous system. The Golgi cells, on the other hand, lie beneath the parallel fibers, though their dendrites are also stimulated by the parallel fibers. Their axons then feed back to inhibit the granule cells. The function of this feedback is to limit the duration of the signal transmitted into the cerebellar cortex from the granule cells. That is, within a short fraction of a second after the granule cells are stimulated, their initial burst of excitation is reduced back to a lower level of excitation that is sustained only as long as the input signal lasts.

The Turn-On/Turn-Off and Turn-Off/Turn-On Output Signals From the Cerebellum

The typical function of the cerebellum is to help provide rapid turn-on signals for agonist muscles and simultaneous reciprocal turn-off signals for the antagonist muscles at the onset of a movement. Then, at the termination of the movement, the cerebellum is mainly responsible for timing and executing the turn-off signals to the agonists and turn-on signals to the antagonists. Although the exact means by which the cerebellum achieves these turn-on and turn-off signals are not fully known, one can speculate from the basic cerebellar circuit of Figure 56–7 how this might work as follows:

First, let us suppose that the turn-on/turn-off pattern of agonist/antagonist contraction at the onset of movement begins with signals from the cerebral cortex that pass directly to the agonist muscle to begin the initial contraction. At the same time, parallel signals are also sent by way of the pontile mossy fibers into the cerebellum. One branch of each mossy fiber goes directly to deep nuclear cells in the dentate or other deep nucleus; this instantly sends an excitatory signal back into the corticospinal motor system, either by way of the return signals through the thalamus to the cortex or by way of neuronal circuitry in the brain stem, to support the muscle contraction signal that had already been begun by the cerebral cortex. As a consequence, the turn-on signal, after a few milliseconds, becomes even more powerful than it was at the start because it is now the sum of both the cortical and cerebellar signals. This is the normal effect when the cerebellum is intact, but in the absence of the cerebellum the secondary extra supportive signal is missing. Obviously, this cerebellar support makes the turn-on muscle contraction much stronger than otherwise it would be.

Now, what causes the turn-off signal for the agonist muscles at the termination of the movement? Remember that all mossy fibers have a second branch that transmits signals by way of the granule cells to the cerebellar cortex and eventually to the Purkinje cells, and the Purkinje cells in turn *inhibit* the deep nuclear cells. This pathway passes through some of the smallest nerve fibers known in the entire nervous system, the parallel fibers of the cerebellar cortical molecular layer that have diameters of only a fraction of a millimeter. Also, the signals from these fibers are weak, so that they require a finite period of time to build up enough excitation in the dendrites of the Purkinje cell to excite it. But once the Purkinje cell is excited, it sends *inhibitory* signals to the same deep nuclear cells that had originally turned on the movement. Therefore, theoretically, this could turn off the cerebellar excitation of the agonist muscles.

Thus, one can see how this circuit could cause a rapid turn-on of agonist contraction at the beginning of a movement and yet cause also a precisely timed turn-off of the same agonist contraction after a given period of time.

Now let us speculate on a circuit for the antagonist muscles. Most importantly, remember that throughout the spinal cord, there are reciprocal agonist/antagonist circuits for virtually every movement that the cord can initiate. Therefore, these circuits are probably the major basis for the antagonist turn-off at the onset of movement and turn-on at its termination, always mirroring whatever occurs in the agonist muscles. But we must remember, too, that the cerebellum contains several other types of inhibitory cells besides the Purkinje cells. The functions of some of these are still to be determined; these, too, could play roles in the initial inhibition of the antagonist muscles and then subsequent excitation.

Obviously, these theoretical mechanisms are still mainly speculation. They are presented here only to illustrate possible ways by which the cerebellum could indeed cause reciprocal turn-on and turn-off signals in the agonist and antagonist muscles, and with controlled timing as well.

The Purkinje Cells Can "Learn" to Correct Motor Errors — The Role of the Climbing Fibers

The degree to which the cerebellum supports the onset and offset of muscle contractions, as well as the timing of the contractions, can be learned by the cerebellum itself. Typically, when a person first performs a new motor act, the degree of motor enhancement provided by the cerebellum to the onset agonist contraction, the degree of onset inhibition of the antagonist, the timing of the offset, the extent of inhibition of the agonist at the offset, and the extent of contraction of the antagonist at the offset, all are almost always incorrect for precise performance of the movement. But after the act has been performed many times, these individual events become progressively more precise in performing the movement exactly as desired, sometimes requiring only a few movements before the desired result is achieved but at other times requiring hundreds of movements.

Yet how do these adjustments come about? The exact answer is not known, although it is known that sensitivity levels of cerebellar circuits themselves progressively adapt during the training process. For instance, the sensitivity of the Purkinje cells to respond to the parallel fibers from the granule cells becomes altered. Furthermore, research studies suggest that this sensitivity change is brought about by signals from the climbing fibers entering the cerebellum from the inferior olivary complex. These signals adjust the long-term sensitivity of the Purkinje cells to stimulation by the parallel fibers.

Under resting conditions, the climbing fibers fire about once per second. But each time they do fire, they cause extreme depolarization of the entire dendritic tree of the Purkinje cell, lasting for up to a second. During this time, the Purkinje cell fires with one initial very strong output spike followed by a series of oscillatory waves in the membrane potential. When a person performs a new movement for the first time and the achieved movement does not match the intended movement, the firing by the climbing fibers changes markedly, either greatly increased or decreased as needed, up to a maximum of about 4 per second or all the way down to zero. These changes in stimulatory rate are believed to alter the long-term sensitivity of the Purkinje cells to the subsequent signals from the mossy fiber circuit. That is, the greater or lesser the climbing fiber input, the greater becomes the accumulative change in long-term sensitivity to the mossy fiber input. Over a period of time, this change in sensitivity, along with other possible "learning" functions of the cerebellum, is believed to make the timing and other aspects of cerebellar control of movements approach perfection. When this has been achieved, the climbing fibers no longer send their "error" signals to the cerebellum to cause further change.

Finally, we need to answer how the climbing fibers themselves know to alter their own rate of firing when a performed movement is imperfect. What is known about this is that the inferior olivary complex receives full information from the corticospinal tracts as well as from the motor centers of the brain stem detailing the *intent* of each motor movement; and it also receives full information from the sensory nerve endings in the muscles and surrounding tissues detailing the movement that actually occurs. Therefore, it is presumed that the inferior olivary complex then functions as a *comparator* to test how well the actual performance matches the intended performance. If there is a match, no change in firing of the climbing fibers occurs. But if there is a mismatch, then the climbing fibers are stimulated or inhibited as needed in proportion to the degree of mismatch, thus leading to progressive changes in Purkinje cell sensitivity until no further mismatch occurs — or so the theory goes.

OVERALL FUNCTION OF THE CEREBELLUM IN CONTROLLING MOVEMENTS

It is already clear that the cerebellum functions in motor control only in association with motor activities initiated elsewhere in the nervous system. These activities may originate in the spinal cord, in the brain stem reticular nuclei, or in the cerebral cortex. We discuss first the operation of the cerebellum in association with the spinal cord and brain stem for control of postural movements and equilibrium and then discuss its function in association with the motor cortex for control of voluntary movements.

FUNCTION OF THE CEREBELLUM WITH THE SPINAL CORD AND BRAIN STEM TO CONTROL POSTURAL AND EQUILIBRIUM MOVEMENTS

The cerebellum originated phylogenetically at about the same time that the vestibular apparatus developed. Furthermore, as discussed in the previous chapter, loss of the flocculonodular lobes and portions of the vermis of the cerebellum causes extreme disturbance of equilibrium.

Yet we still must ask the question, what role does the cerebellum play in equilibrium that cannot be provided by the other neuronal machinery of the brain stem? A clue is the fact that in persons with cerebellar dysfunction equilibrium is far more disturbed during performance of rapid motions than during stasis — especially so when the movements involve changes in direction that stimulate the semicircular ducts. This suggests that the cerebellum is especially important in controlling the balance between agonist and antagonist muscle contractions during *rapid changes* in body positions as dictated by the vestibular apparatus.

One of the major problems in controlling balance is the time required to transmit position signals and velocity of movement signals from the different parts of the body to the brain. Even when the most rapidly conducting sensory pathways, up to 120 m/sec, are used, as by the spinocerebellar system, the delay for transmission from the feet to the brain is still 15 to 20 milliseconds. The feet of a person running rapidly can move as much as 10 inches during this time. Therefore, it is never possible for the return signals from the peripheral parts of the body to reach the brain at the same time that the movements actually occur. How, then, is it possible for the brain to know when to stop a movement in order to perform the next sequential act, especially when the movements are performed very rapidly? The answer is that the signals from the periphery tell the brain not only positions of the different parts of the body but also how rapidly and in what

directions they are moving. It is the function of the cerebellum then to *calculate* from these rates and directions where the different parts of the body will be during the next few milliseconds. The results of these calculations are the key to the brain's progression to the next sequential movement.

Thus, during the control of equilibrium, it is presumed that the information from the vestibular apparatus is used in a typical feedback control circuit to provide almost instantaneous correction of postural motor signals as necessary for maintaining equilibrium even during extremely rapid motion, including rapidly changing directions of motion. The feedback signals from the peripheral areas of the body help in this process. Their help is mediated mainly through the *cerebellar vermis* that functions in association with the axial and girdle muscles of the body; it is the role of the cerebellum to compute actual positions of the respective parts of the body at any given time, despite the long delay time from the periphery to the cerebellum.

FUNCTION OF THE CEREBELLUM IN VOLUNTARY MUSCLE CONTROL

In addition to the feedback circuitry between the body periphery and the cerebellum, an almost entirely independent feedback circuitry exists between the motor cortex of the cerebrum and the cerebellum. This circuitry affects only slightly if at all the control of equilibrium and other postural movements of the axial and girdle muscles of the body. Instead, it serves two other principal functions: (1) It helps the cerebral cortex to coordinate patterns of movement involving mostly the distal parts of the limbs — especially the hands, fingers, and feet. The part of the cerebellum involved in this function is mainly the *intermediate zone of the cerebellar cortex and its associated nucleus interpositus.* (2) It helps the cerebral cortex to plan the timing and sequencing of the next successive movement that will be performed after the present movement is completed. The part of the cerebellum involved in this is the large *lateral zone of the cerebellar hemisphere,* along with its associated *dentate nucleus.* Let us discuss each of these two functions separately.

Cerebellar Feedback Control of Distal Limb Movements by Way of the Intermediate Cerebellar Cortex and the Nucleus Interpositus

As illustrated in Figure 56–8, the intermediate zone of each cerebellar hemisphere receives two types of information when a movement is performed: (1) direct information from the motor cortex and red nucleus, telling the cerebellum the sequential *intended plan of movement* for the next few fractions of a second, and (2) feedback information from the peripheral parts of the body, especially from the distal parts of the limbs, telling the cerebellum what *actual movements* result. After the intermediate zone of the cere-

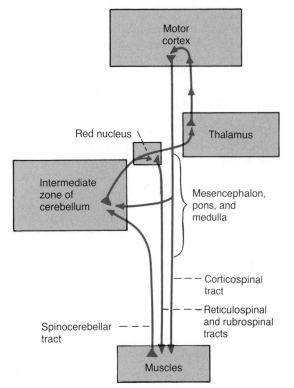

Figure 56–8. Cerebral and cerebellar control of voluntary movements, involving especially the intermediate zone of the cerebellar cortex and its associated nucleus interpositus.

bellum has compared the intended movements with the actual movement, the nucleus interpositus sends *corrective* output signals (a) back to the *motor cortex* through relay nuclei in the *thalamus* and (b) to the *magnocellular portion* (the lower portion) *of the red nucleus,* which gives rise to the *rubrospinal tract.* The rubrospinal tract, in turn, joins the corticospinal tract in innervating the lateralmost motor neurons in the anterior horns of the spinal cord gray matter, the neurons that control the distal parts of the limbs, particularly the hands and fingers.

This part of the cerebellar motor control system provides smooth, coordinate movements of the agonist and antagonist muscles of the distal limbs for the performance of acute purposeful patterned movements. The cerebellum seems to compare the "intentions" of the higher levels of the motor control system, as transmitted to the intermediate cerebellar zone through the corticopontocerebellar tract, with the "performance" by the respective parts of the body as transmitted back to the cerebellum from the periphery. In fact, the ventral spinocerebellar tract even transmits back to the cerebellum an "efference" copy of the actual motor control signals that reach the anterior motor neurons, and this, too, is integrated with the signals arriving from the muscle spindles and other proprioceptor sensory organs. We learned earlier that similar comparator signals also go to the inferior olivary complex; if the signals do not compare favorably, the olivary-Purkinje cell system, along with possible other cerebellar learning mechanisms,

will eventually correct the motions until they perform the desired function.

Once the cerebellum has learned its role in each pattern of movement, it provides rapid turn-on of agonist muscle activity at the onset of each movement while inhibiting the antagonist muscles. Then it continues agonist contraction until near the end of the movement, when the cerebellar circuit again plays the major role in rapid turn-off of the agonist muscles and turn-on of the antagonist muscles. The point at which the reversal of excitation between agonist and antagonist muscles occurs depends on (1) the rate of movement and (2) the previously learned knowledge of the inertia of the system. The faster the movement and the greater the inertia, the earlier the reversal point must occur in the course of movement to stop the movement at the proper point.

Function of the Cerebellum to Prevent Overshoot of Movements and to "Damp" Movements. Almost all movements of the body are "pendular." For instance, when an arm is moved, momentum develops, and the momentum must be overcome before the movement can be stopped. Because of the momentum, all pendular movements have a tendency to *overshoot.* If overshooting does occur in a person whose cerebellum has been destroyed, the conscious centers of the cerebrum eventually recognize this and initiate a movement in the opposite direction to bring the arm to its intended position. But again the arm, by virtue of its momentum, overshoots, and appropriate corrective signals must again be instituted. Thus, the arm oscillates back and forth past its intended point for several cycles before it finally fixes on its mark. This effect is called an *action tremor,* or *intention tremor.*

However, if the cerebellum is intact, appropriate learned, subconscious signals stop the movement precisely at the intended point, thereby preventing the overshoot and also the tremor. This is the basic characteristic of a damping system. All control systems regulating pendular elements that have inertia must have damping circuits built into the mechanisms. In the motor control system of our central nervous system, the cerebellum provides most of this damping function.

Cerebellar Control of Ballistic Movements. Many rapid movements of the body, such as the movements of the fingers in typing, occur so rapidly that it is not possible to receive feedback information either from the periphery to the cerebellum or from the cerebellum back to the motor cortex before the movements are over. These movements are called *ballistic movements,* meaning that the entire movement is preplanned and is set into motion to go a specific distance and then to stop. Another important example is the saccadic movements of the eyes, in which the eyes jump from one position to the next when reading or when looking at successive points along a road as a person is moving in a car.

Much can be understood about the function of the cerebellum by studying the changes that occur in the ballistic movements when the cerebellum is removed.

Three major changes occur: (1) the movements are slow to develop and do not have the extra onset surge that the cerebellum usually gives to an agonist movement, (2) the force development is weak, and (3) the movements are slow to turn off, usually allowing the movement to go well beyond the intended mark. Therefore, in the absence of the cerebellar circuit, the motor cortex has to think extra hard to turn ballistic movements on and again has to think hard and take extra time to turn the movement off. Thus, the automatism of ballistic movements is lost.

If one will consider once again the circuitry of the cerebellum as described earlier in the chapter, one will see that it is beautifully organized to perform this biphasic, first excitatory and then delayed inhibitory, function that is required for ballistic movements. One will also see that the time delay circuits of the cerebellar cortex are fundamental to this particular ability of the cerebellum.

Function of the Large Lateral Zone of the Cerebellar Hemisphere — The "Sequencing" and "Timing" Functions

In human beings, the lateral zones of the two cerebellar hemispheres have become very highly developed and greatly enlarged, along with the human ability to perform intricate sequential patterns of movement, especially with the hands and fingers, and along with the ability to speak. Yet, strangely enough, these large lateral portions of the cerebellar hemispheres have no direct input of information from the peripheral parts of the body. Also, almost all the communication between these lateral cerebellar areas and the cortex is not with the primary motor cortex itself but instead with the premotor area and primary and association somatic sensory areas. Even so, destruction of the lateral portions of the cerebellar hemispheres along with their deep nuclei, the dentate nuclei, can lead to extreme incoordination of the purposeful movements of the hands, fingers, feet, and speech apparatus. This has been difficult to understand because of lack of direct communication between this part of the cerebellum and the primary motor cortex. However, recent experimental studies suggest that these portions of the cerebellum are concerned with two other important aspects of motor control: (1) the planning of sequential movements, and (2) the "timing" of the sequential movements.

The Planning of Sequential Movements. The planning of sequential movements seems to be related to the fact that the lateral hemispheres communicate with the premotor and sensory portions of the cerebral cortex and that there is also two-way communication between these same areas and corresponding areas of the basal ganglia. It seems that the "plan" of the sequential movements is transmitted from the sensory and premotor areas of the cortex to the lateral zones of the cerebellar hemispheres, and two-way traffic between the cerebellum and the cortex is necessary to provide appropriate transition from one movement to the next. An exceedingly interesting observa-

tion that supports this view is that many of the neurons in the dentate nuclei display the activity pattern of the movement that is yet to follow at the same time that the present movement is occurring. Thus, the lateral hemispheres appear to be involved not with what is happening at a given moment, but instead with *what will be happening during the next sequential movement.*

To summarize, one of the most important features of normal motor function is one's ability to progress smoothly from one movement to the next in orderly succession. In the absence of the cerebellar hemispheres, this capability is seriously disturbed, especially for rapid movements.

The Timing Function. Another important function of the lateral cerebellar hemispheres is to provide appropriate timing for each movement. In the absence of these lateral areas, one loses the subconscious ability to predict ahead of time how far the different parts of the body will move in a given time. And without this timing capability, the person becomes unable to determine when the next movement should begin. As a result, the succeeding movement may begin too early or, more likely, too late. Therefore, cerebellar lesions cause complex movements, such as those required for writing, running, or even talking, to become totally incoordinate, lacking completely in the ability to progress in an orderly sequence from one movement to the next. Such cerebellar lesions are said to cause *failure of smooth progression of movements.*

Extramotor Predictive Functions of the Cerebellum. The cerebellum also plays a role in predicting events other than movements of the body. For instance, the rates of progression of both auditory and visual phenomena can be predicted, and both of these require cerebellar participation. As an example, a person can predict from the changing visual scene how rapidly he or she is approaching an object. A striking experiment that demonstrates the importance of the cerebellum in this ability is the removal of the "head" portion of the cerebellum in monkeys. Such a monkey occasionally charges the wall of a corridor and literally bashes its brains out because it is unable to predict when it will reach the wall.

Unfortunately, we are only now beginning to learn about these extramotor predictive functions of the cerebellum. It is quite possible that the cerebellum provides a "time base," perhaps utilizing time-delay circuits, against which signals from other parts of the central nervous system can be compared. It is often stated that the cerebellum is especially important in interpreting *spatiotemporal relationships* in sensory information.

CLINICAL ABNORMALITIES OF THE CEREBELLUM

An important feature of clinical cerebellar abnormalities is that destruction of small portions of the cerebellar *cortex* rarely causes detectable abnormalities in motor function. In fact, several months after as much as half the cerebellar cortex has been removed, if the deep cerebellar nuclei are not removed along with the cortex, the motor functions of an animal appear to be almost entirely normal as long as the animal performs all movements slowly. Thus, the remaining portions of the motor control system are capable of compensating tremendously for loss of parts of the cerebellum.

Therefore, to cause serious and continuing dysfunction of the cerebellum, the cerebellar lesion must usually involve one or more of the deep cerebellar nuclei — the *dentate, interpositus,* and *fastigial nuclei* — as well as the cerebellar cortex.

Dysmetria and Ataxia. Two of the most important symptoms of cerebellar disease are dysmetria and ataxia. It was pointed out earlier that in the absence of the cerebellum the subconscious motor control system cannot predict ahead of time how far movements will go. Therefore, the movements ordinarily overshoot their intended mark, and then the conscious portion of the brain overcompensates in the opposite direction for the succeeding movements. This effect is called *dysmetria,* and it results in incoordinate movements that are called *ataxia.*

Dysmetria and ataxia can also result from lesions in the spinocerebellar tracts, for the feedback information from the moving parts of the body is essential for accurate control of the movements.

Past Pointing. Past pointing means that in the absence of the cerebellum a person ordinarily moves the hand or some other moving part of the body considerably beyond the point of intention. This probably results from the fact that normally the cerebellum provides most of the motor signal that turns off a movement after it has begun; and if the cerebellum is not available to do this, the movement ordinarily goes beyond the intended point. Therefore, past pointing is actually a manifestation of dysmetria.

Failure of Progression. *Dysdiadochokinesia.* When the motor control system fails to predict ahead of time where the different parts of the body will be at a given time, it temporarily "loses" the parts during rapid motor movements. As a result, the succeeding movement may begin much too early or much too late, so that no orderly "progression of movement" can occur. One can demonstrate this readily by having a patient with cerebellar damage turn one hand upward and downward at a rapid rate. The patient rapidly "loses" all perception of the instantaneous position of the hand during any portion of the movement. As a result, a series of jumbled movements occurs instead of the normal coordinate upward and downward motions. This is called *dysdiadochokinesia.*

Dysarthria. Another instance in which failure of progression occurs is in talking, for the formation of words depends on rapid and orderly succession of individual muscular movements in the larynx, mouth, and respiratory system. Lack of coordination between these and inability to predict either the intensity of the sound or the duration of each successive sound cause jumbled vocalization, with some syllables loud, some weak, some held long, some held for short intervals, and resultant speech that is almost completely unintelligible. This is called *dysarthria.*

Intention Tremor. When a person who has lost the cerebellum performs a voluntary act, the movements tend to oscillate, especially when they approach the intended mark, first overshooting the mark and then vibrating back and forth several times before settling on the mark. This reaction is called an *intention tremor* or an *action tremor,* and it results from cerebellar overshooting and failure of the cerebellar system to damp the motor movements.

Cerebellar Nystagmus. Cerebellar nystagmus is a tremor of the eyeballs that occurs usually when one attempts to fixate the eyes on a scene to one side of the head. This off-center type of fixation results in rapid, tremulous movements of the eyes rather than a steady fixation, and it is another manifestation of the failure of damping by the cerebellum. It occurs especially when the flocculonodular lobes are damaged; in this instance it is associated with loss of equilibrium, presumably because of dysfunction of the pathways through the cerebellum from the semicircular ducts.

Rebound. If a person with cerebellar disease is asked to pull upward strongly with his or her arm while the physician holds it back at first and then lets go, the arm will fly back until it strikes the face instead of being automatically stopped. This is called *rebound,* and it results from *loss of the cerebellar component of the stretch reflex.* That is, the normal cerebellum ordinarily instantaneously adds a large amount of additional feedback support to the spinal cord stretch reflex mechanism whenever a portion of the body begins to move unexpectedly in an unwilled direction. Without the cerebellum, strong activation of the muscles fails to occur, thus allowing overmovement of the limb in the unwanted direction.

Hypotonia. Loss of the deep cerebellar nuclei, particularly the dentate and interpositus, causes decreased tone of the peripheral musculature on the side of the lesion, though after several months the cerebral motor cortex usually compensates for this by an increase in its intrinsic activity. The hypotonia results from loss of cerebellar facilitation of the motor cortex and brain stem motor nuclei by the tonic discharge of the deep cerebellar nuclei.

● THE BASAL GANGLIA — THEIR MOTOR FUNCTIONS

The basal ganglia, like the cerebellum, are another accessory motor system that functions not by itself but always in close association with the cerebral cortex and corticospinal system. In fact, the basal ganglia receive almost all their input signals from the cortex itself and in turn return almost all of their output signals back to the cortex.

Figure 56–9 illustrates the anatomical relationships of the basal ganglia to the other structures of the brain. Note that they are located mainly lateral to the thalamus, occupying a large portion of the deeper regions of both cerebral hemispheres. Note also that almost all of the motor and sensory nerve fibers connecting the cerebral cortex and spinal cord pass between the two major masses of the basal ganglia, the *caudate nucleus* and the *putamen.* This mass of nerve fibers is called the *internal capsule* of the brain. It is important to our present discussion because of the intimate association between the basal ganglia and the corticospinal system for motor control.

The Neuronal Circuitry of the Basal Ganglia. The anatomical connections between the basal ganglia and the other elements of motor control are very complex, as illustrated in Figure 56–10. To the left is shown the motor cortex, the thalamus, the corticospinal pathways, and associated brain stem and cerebellar circuitry. To the right is the major circuitry of the basal ganglia system, showing the tremendous number of interconnections among the basal ganglia

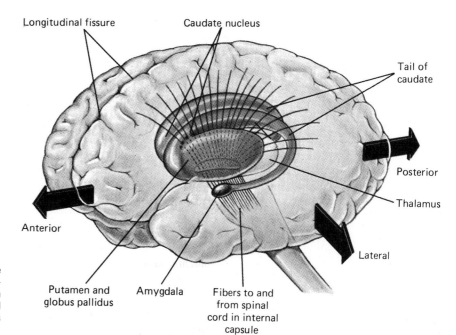

Figure 56–9. Anatomical relationships of the basal ganglia to the cerebral cortex and thalamus, shown in three-dimensional view. (From Guyton: Basic Neuroscience: Anatomy and Physiology. Philadelphia, W. B. Saunders Company, 1987.)

themselves—plus extensive input and output pathways between the motor regions of the cerebral cortex and the basal ganglia.

Anatomists consider the motor portions of the basal ganglia to be the *caudate nucleus,* the *putamen,*

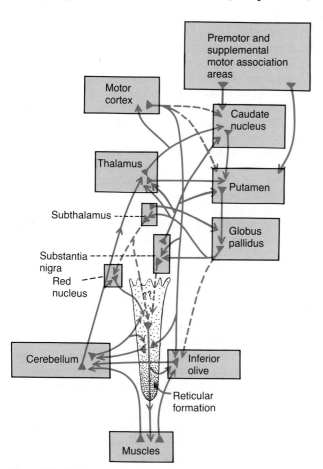

Figure 56–10. Relation of the basal ganglial circuitry to the corticospinal-cerebellar system for movement control.

and the *globus pallidus.* But, physiologically, two other structures that are not normally classified as basal ganglia are intimately involved as well, the *subthalamus* and *substantia nigra,* which are located inferior and posterior to the thalamus in the lower diencephalon and upper mesencephalon. Several specific re-entrant circuits interconnect the subthalamus and substantia nigra with all three of the basal ganglia. Also, both the subthalamus and substantia nigra feed back to the thalamus and thence to the cortical areas for motor control.

From Figure 56–10, it is already clear that the circuitry of the basal ganglial system is intensely complex. However, we try in the next few sections to dissect out the major pathways of action and attempt to describe their functional attributes. We concentrate especially on two major circuits called, respectively, the *putamen circuit* and the *caudate circuit.*

Function of the Basal Ganglia in Executing Patterns of Motor Activity— The Putamen Circuit

One of the principal roles of the basal ganglia in motor control is to function in association with the corticospinal system to control complex patterns of motor activity. An example is the writing of letters of the alphabet. When there is serious damage to the basal ganglia, the cortical system of motor control can no longer provide these patterns. Instead, one's writing becomes crude, as if one were learning for the first time how to write.

Other patterns requiring the basal ganglia are cutting paper with scissors, hammering nails, shooting basketballs through a hoop, passing a football, throwing a baseball, the movements of shoveling dirt, some aspects of vocalization, and virtually any other of our skilled movements.

The Neural Circuit Through the Putamen for Executing Patterns of Movements. Figure 56–11 illustrates the principal pathways through the basal ganglia for executing learned patterns of movement. These begin mainly in the premotor and supplemental motor areas of the motor cortex and also the primary somatic sensory area of the sensory cortex. Next they pass, as shown in the bright red in the figure, to the putamen (mainly bypassing the caudate nucleus), then to the internal portion of the globus pallidus, next to the ventroanterior and ventrolateral nuclei of the thalamus, and then finally return to the primary motor cortex and portions of the premotor and supplemental areas closely associated with the primary motor cortex. Thus, this putamen circuit has its inputs mainly from those parts of the brain adjacent to the primary motor cortex, but not much from the primary motor cortex itself. Then its outputs do go mainly back to the *primary* motor cortex.

Functioning in close association with this primary putamen circuit are three ancillary circuits: (1) from the putamen to the external globus pallidus, to the subthalamus, to the relay nuclei of the thalamus, and back to the motor cortex; (2) from the putamen to the internal globus pallidus, to the substantia nigra, to the relay nuclei of the thalamus, and also returning to the motor cortex; and (3) a local feedback circuit from the external globus pallidus to the subthalamus and returning again to the external globus pallidus.

Athetosis, Hemiballismus, and Chorea. How does the above putamen circuit function in the execution of patterns of movement? The answer is only poorly known. However, when any portion of the circuit is damaged or blocked, certain patterns of movement become severely abnormal. For instance, lesions in the *globus pallidus* frequently lead to spontaneous *writhing movements* of a hand, an arm, the neck, or the face, movements called *athetosis*.

A lesion in the *subthalamus* often leads to sudden *flailing movements* of an entire limb, a condition called *hemiballismus*.

Multiple small lesions in the *putamen* lead to *flicking movements* in the hands, face, and other parts of the body, which is called *chorea*.

And lesions of the *substantia nigra* lead to the common and extremely severe disease of rigidity and tremors known as *Parkinson's disease* that we shall discuss in more detail later.

Role of the Basal Ganglia for Cognitive Control of Sequences of Motor Patterns — The Caudate Circuit

The term cognition means the thinking processes of the brain, utilizing both the sensory input to the brain as well as information already stored in memory. Obviously, most of our motor actions occur as a consequence of thoughts generated in the mind, a process called *cognitive control of motor activity*. The caudate nucleus plays a major role in this cognitive control of motor activity.

The neural connections between the corticospinal motor control system and the caudate nucleus, illustrated in Figure 56–12, are somewhat different from those of the putamen circuit. Part of the reason for this is that the caudate nucleus extends into all lobes of the cerebrum, beginning anteriorly in the frontal lobes, then passing posteriorly through the parietal and occipital lobes, and finally curving forward again like a letter "C" into the temporal lobes. Furthermore, the caudate nucleus receives large amounts of its input from the *association areas* of the cerebral cortex, the areas that integrate the different types of sensory and motor information into usable thought patterns.

After the signals pass from the cerebral cortex to the caudate nucleus, they are transmitted next to the internal globus pallidus, then to the relay nuclei of the ventroanterior and ventrolateral thalamus, and finally back to the prefrontal, premotor, and supplemental motor areas of the cerebral cortex, but with almost none of the returning signals passing directly to the primary motor cortex. Instead, the returning signals go to those accessory motor regions that are concerned with patterns of movement instead of individual muscle movements.

A good example of this would be for a person to see a lion approach and then respond instantaneously and automatically by (1) turning away from the lion, (2) beginning to run, and (3) even attempting to climb a tree. Without the cognitive functions, the person might not have the instinctive knowledge, without thinking for too long a time, to respond quickly and appropriately. Thus, cognitive control of motor activity determines which patterns of movement will be used together and in what sequence to achieve a complex goal.

Figure 56–11. The *putamen circuit* through the basal ganglia for subconscious execution of learned patterns of movement.

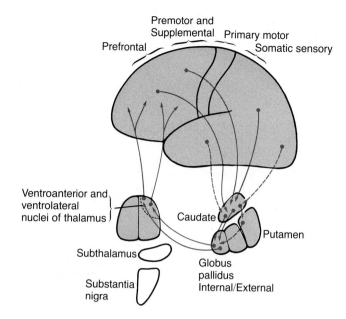

The Caudate Circuit

Figure 56–12. The *caudate circuit* through the basal ganglia for cognitive planning of the combinations of sequential and parallel motor patterns to achieve specific conscious goals.

Function of the Basal Ganglia to Change the Timing and to Scale the Intensity of Movements

Two important capabilities of the brain in controlling movement are (1) to determine how rapidly it is to be performed and (2) to control how large the movement will be. For instance, one may write the letter

Figure 56–13. A typical drawing made by a person who has severe damage in his or her left parietal cortex, where the spatial coordinates of the right side of the body and right field of vision are calculated.

"a" slowly or rapidly. Also, he may write a small "a" or a very large letter "a" on a chalk board. Regardless of his choices, the proportional characteristics of the letter will remain the same. This is also true even though the person might use the fingers for writing the letter in one instance or the whole arm at another time.

In the absence of the basal ganglia, these timing and scaling functions are very poor, in fact almost nonexistent. Of course, here again, the basal ganglia do not function alone; they function in close association with the cerebral cortex as well. One especially important cortical area is the posterior parietal cortex, which is the locus of the spatial coordinates for all parts of the body as well as for the relationship of the body and its parts to all surroundings. Figure 56–13 illustrates the way in which a person lacking a left posterior parietal cortex might draw the face of another human being, providing proper proportions for the right side of the face, but almost ignoring the left side (which is in his right field of vision). Also, such a person will try always to avoid using his right arm, right hand, or other portions of his right body for the performance of tasks, almost not knowing that these parts of his body even exist.

Because it is the caudate circuit of the basal ganglial system that functions mainly with the association areas of the cortex, such as the posterior parietal cortex, presumably the timing and scaling of movements are functions of this caudate cognitive motor control circuit.

FUNCTIONS OF SPECIFIC NEUROTRANSMITTERS IN THE BASAL GANGLIAL SYSTEM

Figure 56–14 illustrates the interplay of some specific neurotransmitters that are known to function

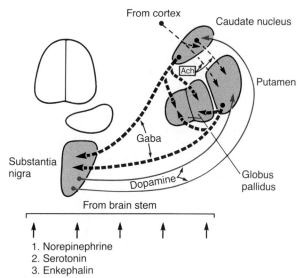

Figure 56 – 14. Neuronal pathways that secrete different types of neurotransmitter substances in the basal ganglia.

within the basal ganglia, showing (1) a *dopamine* pathway from the substantia nigra to the caudate nucleus and putamen; (2) a *gamma-aminobutyric acid (GABA)* pathway from the caudate nucleus and putamen to the globus pallidus and substantia nigra; (3) *acetylcholine* pathways from the cortex to the caudate nucleus and putamen; and (4) multiple general pathways from the brain stem that secrete *norepinephrine, serotonin, enkephalin,* and several other neurotransmitters in the basal ganglia as well as in other parts of the cerebrum. We will have more to say about some of these hormonal systems in the following sections when we discuss diseases of the basal ganglia as well as in subsequent chapters when we discuss behavior, sleep, wakefulness, and functions of the autonomic nervous system.

For the present, it should be remembered that the neurotransmitter GABA always functions as an inhibitory agent. Therefore, the GABA neurons in the feedback loops from the cortex through the basal ganglia and then back to the cortex make virtually all these loops *negative feedback loops,* rather than positive feedback loops, thus lending stability to the motor control systems. Dopamine also functions as an inhibitory neurotransmitter in most parts of the brain, so that it too may function as a stabilizer. Acetylcholine, on the other hand, usually functions as an excitatory transmitter and therefore probably provides many of the positive features of motor action.

CLINICAL SYNDROMES RESULTING FROM DAMAGE TO THE BASAL GANGLIA

Aside from athetosis and hemiballismus, which have already been mentioned in relation to lesions in the globus pallidus and the subthalamus, two other major diseases result from damage in the basal ganglia. These are Parkinson's disease and Huntington's chorea.

Parkinson's Disease

Parkinson's disease, also known as *paralysis agitans,* results from *widespread destruction of that portion of the substantia nigra, the pars compacta, that sends dopamine-secreting nerve fibers to the caudate nucleus and putamen.* The disease is characterized by (1) *rigidity* of much if not most of the musculature of the body, (2) *involuntary tremor* of the involved areas even when the person is resting and always at a fixed rate of 3 to 6 cycles per second, and (3) a serious inability to initiate movement called *akinesia.*

The causes of these abnormal motor effects are almost entirely unknown. However, if the dopamine secreted in the caudate nucleus and putamen functions as an inhibitory transmitter, then destruction of the substantia nigra theoretically would allow these structures to become overly active and possibly cause continuous output of excitatory signals to the corticospinal motor control system. These signals could certainly overly excite many or all muscles of the body, thus leading to *rigidity.* And some of the feedback circuits might easily *oscillate* because of high feedback gains after loss of their inhibition, leading to the *tremor* of Parkinson's disease. This tremor is quite different from that of cerebellar disease, for it occurs during all waking hours and is therefore called an *involuntary tremor,* in contradistinction to cerebellar tremor, which occurs only when the person performs intentionally initiated movements and therefore is called *intention tremor.*

The *akinesia* that occurs in Parkinson's disease is often much more distressing to the patient than are the symptoms of muscle rigidity and tremor, for to perform even the simplest movement in severe Parkinsonism the person must exert the highest degree of concentration. The mental effort, even mental anguish, that is necessary to make the movement "go" is often at the limit of the patient's will power. Then, when the movement does occur, it is stiff and staccato in character instead of occurring smoothly. Unfortunately, the cause of this akinesia is still entirely speculative. It is presumed that loss of dopamine secretion in the caudate nucleus and putamen might lead to loss of balance between the excitatory and inhibitory systems. Since *patterns of movement* require sequential changes between excitation and inhibition, any effect that would lock basal ganglia activity always in one direction would obviously prevent the initiation of and progression through sequential patterns, which is exactly what happens in akinesia.

Treatment With L-Dopa. Administration of the drug L-dopa to patients with Parkinson's disease ameliorates many of the symptoms, especially the rigidity and akinesia, in most patients. The reason for this is believed to be that L-dopa is converted in the brain into dopamine, and the dopamine then restores the normal balance between inhibition and excitation in the caudate nucleus and putamen. Unfortunately, administration of dopamine itself does not have the same effect because dopamine has a chemical structure that will not allow it to pass through the blood-brain barrier, even though the slightly different structure of L-dopa does allow it to pass.

Coagulation of the Ventrolateral and Ventroanterior Nuclei of the Thalamus for Treatment of Parkinson's Disease. Many researchers have also treated Parkinson's disease, with varying degrees of success, by surgically destroying portions of the basal ganglia, the thalamus, or even the motor cortex to block basal ganglial feedback to the cortex. The most widely employed of these has been destruction of the *ventrolateral* and *ventroanterior nuclei of the thalamus,* usually by electrocoagulation. Almost all of the feedback pathways from the basal ganglia to the cerebral cortex pass through these nuclei. It is presumed that blockage of these feedbacks prevents function of the neuronal loops that cause the tremor and some other symptoms of Parkinson's disease.

Huntington's Chorea

Huntington's chorea is a hereditary disorder that usually begins to cause symptoms in the third or fourth decade of life. It is characterized at first by flicking movements at individual joints and then progressive severe distortional movements of the entire body. In addition, severe dementia also develops along with the motor dysfunctions.

The abnormal movements of Huntington's chorea are *believed to be caused by loss of most of the cell bodies of the GABA-secreting neurons in the caudate nucleus and putamen.* The axon terminals of these neurons normally cause inhibition in the globus pallidus and substantia nigra. This loss of inhibition is believed to allow spontaneous outbursts of globus pallidus and substantia nigra activity that cause the distortional movements.

The dementia in Huntington's chorea probably does not result from the loss of GABA neurons but instead from loss of many acetylcholine-secreting neurons at the same time. This loss occurs not only in the basal ganglia but in much of the cerebral cortex as well, which could easily block much of the thinking process.

● INTEGRATION OF ALL PARTS OF THE TOTAL MOTOR CONTROL SYSTEM

Finally, we need to summarize as best we can what is known about overall control of movement. To do this, let us first give a synopsis of the different levels of control:

The Spinal Level

Programmed in the spinal cord are local patterns of movement for all muscle areas of the body — for instance, programmed withdrawal reflexes that pull any part of the body away from a source of pain. And the cord is the locus even of complex patterns of rhythmical motions such as to-and-fro movement of the limbs for walking, plus reciprocal activity of opposite sides of the body, or hind limbs versus forelimbs.

All these programs of the cord can be commanded into action by the higher levels of motor control, or they can be inhibited while the higher levels take over control.

The Hindbrain Level

The hindbrain provides two major functions for general motor control of the body: (1) maintenance of axial tone of the body for the purpose of standing and (2) continuous modification of the different directions of this tone in response to continuous information from the vestibular apparatuses for the purpose of maintaining equilibrium.

The Corticospinal Level

The corticospinal system transmits most of the motor signals from the motor cortex to the spinal cord. It functions partly by issuing commands to set into motion the various cord patterns of motor control. It can also change the intensity of the different patterns or modify their timing or other characteristics. When needed, the corticospinal system can bypass the cord patterns by issuing inhibitory commands and replacing them with higher level patterns from the brain stem or from the cerebral cortex. Usually the cortical patterns are more complex; also, they can be learned by practice, while the cord patterns are mainly set by heredity and are said to be "hard wired."

The Associated Function of the Cerebellum. The cerebellum functions with all levels of muscle control. It functions with the spinal cord especially to enhance the stretch reflex, so that when a contracting muscle meets an unexpectedly heavy load, a long stretch reflex arc through the cerebellum and back again to the cord strongly facilitates the load-resisting effect of the basic stretch reflex.

At the brain stem level, the cerebellum functions to make the postural movements of the body, especially the rapid movements required by the equilibrium system, smooth and continuous and without abnormal oscillations.

At the cerebral cortex level, the cerebellum functions to provide many accessory motor commands, especially to provide extra motor force to turn on muscle contraction very rapidly and forcefully at the start of movements. And near the end of each movement, the cerebellum turns on antagonist muscles at exactly the right time and with proper force to stop the movement at the intended point. Furthermore, there is good physiological evidence that all aspects of this turn-on/turn-off patterning by the cerebellum can be learned with experience.

In addition, the cerebellum functions with the cerebral cortex at still another level of motor planning: it helps to program in advance the muscle contractions that are required for smooth progression from the present movement in one direction to the next movement in another direction. The neural circuit for this passes from the cerebral cortex to the large lateral hemispheres of the cerebellum and then back to the cortex.

It should be noted especially that the cerebellum functions mainly with very rapid movements. Without the cerebellum, slow and calculated movements can still occur, but it is difficult for the corticospinal system to achieve well-controlled rapid intended movements to a particular goal, or especially to progress smoothly from one movement to the next.

The Associated Functions of the Basal Ganglia. The basal ganglia are essential to motor control in ways entirely different from those of the cerebellum. Their two most important functions are (1) to help the cortex execute subconscious but *learned* patterns of movement and (2) to help plan multiple parallel and sequential patterns of movement that the mind must put together to accomplish a purposeful task.

The types of motor patterns that require the basal ganglia include those for writing all the different letters of the alphabet, for throwing a ball, for typing, or so forth. Also, the basal ganglia are required to modify these patterns for slow execution, rapid execution, to write small or write very large — thus controlling both timing and dimensions of the patterns.

At still a higher level of control is another cerebral cortex-basal ganglia circuit, beginning in the thinking processes of the brain and providing the overall sequence of action for responding to each new situation — such as planning one's immediate response to an assailant who hits the person in the face or one's sequential response to an unexpectedly fond embrace.

An important part of all these basal ganglial planning processes is not only the motor cortex and basal ganglia but also the somatic sensory cortex of the parietal lobe, especially the posterior portion where the instantaneous spatial coordinates of all parts of one's body are continuously calculated, and even the spatial coordinates of the relationships of the body parts to the physical surroundings. If one of the two parietal cortices is severely damaged, then the person simply ignores the opposite side of his or her body and even ignores objects on the opposite side; then the movements are planned around use of only the consciously recognized side of the body.

WHAT DRIVES US TO ACTION?

Finally, what is it that arouses us from inactivity and sets into play our trains of movement. Fortunately, we are beginning to learn about the motivational systems of the brain. Basically, the brain has an older core centered beneath, anterior, and lateral to the thalamus — including the hypothalamus, the amygdala, the hippocampus, the septal region anterior to the hypothalamus and thalamus, and even older regions of the thalamus and cerebral cortex themselves — all of which function together to motivate most of the motor and other functional activities of the brain. These areas are collectively called the *limbic system* of the brain. We discuss this system in detail in Chapter 58.

REFERENCES

Atkeson, C. G.: Learning arm kinematics and dynamics. Annu. Rev. Neurosci., 12:157, 1989.

Baldessarini, R. J., and Tarsey, D.: Dopamine and the pathophysiology of dyskinesias induced by antipsychotic drugs. Annu. Rev. Pharmacol. Toxicol., 20:533, 1980.

Bloedel, J. R., and Courville, J.: Cerebellar afferent systems. In Brooks, V. B. (ed.): Handbook of Physiology. Sec. 1, Vol. II. Bethesda, Md., American Physiological Society, 1981, p. 735.

Brooks, V. B.: The Neural Basis of Motor Control. New York, Oxford University Press, 1986.

Brooks, V. B., and Thach, W. T.: Cerebellar control of posture and movement. In Handbook of Physiology. Sec. 1, Vol. II. Bethesda, Md., American Physiological Society, 1981, p. 877.

Carpenter, M. B.: Anatomy of the corpus striatum and brain stem integrating system. In Handbook of Physiology. Sec. 1, Vol. II. Bethesda, Md., American Physiological Society, 1981, p. 947.

Collier, T. J., and Sladek, J. R., Jr.: Neural transplantation in animal models of neurodegenerative disease. News Physiol. Sci., 3:204, 1988.

Courville, J., et al. (eds.): The Inferior Olivary Nucleus. New York, Raven Press, 1980.

DeLong, M., and Georgopoulos, A. P.: Motor functions of the basal ganglia. In Handbook of Physiology. Sec. 1, Vol. II. Bethesda, Md., American Physiological Society, 1981, p. 1017.

Di Chiara, G. (ed.): GABA and the Basal Ganglia. New York, Raven Press, 1981.

Duvoisin, R. C.: Parkinson's Disease. New York, Raven Press, 1978.

Eckmiller, R.: Neural control of pursuit eye movements. Physiol. Rev., 67:797, 1987.

Evarts, E. V.: Role of motor cortex involuntary movements in primates. In Handbook of Physiology. Sec. 1, Vol. II. Bethesda, Md., American Physiological Society, 1981, p. 1083.

Evarts, E. V., et al. (eds.): Motor System in Neurobiology. New York, Elsevier Science Publishing Co., 1986.

Fernstrom, J. D.: Role of precursor availability on control of monoamine biosynthesis in the brain. Physiol. Rev., 63:484, 1983.

Fuster, J. M.: Prefrontal cortex in motor control. In Handbook of Physiology. Sec. 1, Vol. II. Bethesda, Md., American Physiological Society, 1981, p. 1149.

Georgopoulos A. P.: Neural integration of movement: role of motor cortex in reaching. FASEB J., 1:2849, 1988.

Glickstein M., and Yeo, C. (eds.): Cerebellum and Neuronal Plasticity. New York, Plenum Publishing Corp., 1987.

Goldstein, M., et al. (eds.): Central D_1 Dopamine Receptors. New York, Plenum Publishing Corp., 1988.

Grillner, S.: Control of locomotion in bipeds, tetrapods, and fish. In Handbook of Physiology. Sec. 1, Vol. II. Bethesda, Md., American Physiological Society, 1981, p. 1179.

Ito, M.: The Cerebellum and Neural Control. New York, Raven Press, 1984.

Ito, M.: Where are neurophysiologists going? News Physiol. Sci., 1:30, 1986.

Jones, E. G., and Peters, A. (eds.): Sensory-Motor Areas and Aspects of Cortical Connectivity. New York, Plenum Publishing Corp., 1986.

Keele, S. W.: Behavioral analysis of movement. In Handbook of Physiology. Sec. 1, Vol. II. Bethesda, Md., American Physiological Society, 1981, p. 1391.

Kitai, S. T.: Electrophysiology of the corpus striatum and brain stem integrating systems. In Handbook of Physiology. Sec. 1, Vol. II. Bethesda, Md., American Physiological Society, 1981, p. 997.

Kuypers, H. G. J. M.: Anatomy of the descending pathways. In Handbook of Physiology. Sec. 1, Vol. II. Bethesda, Md., American Physiological Society, 1981, p. 597.

Lewin, R.: Brain grafts benefit Parkinson's patients. Science, 236:149, 1987.

Llinas, R.: Eighteenth Bowditch lecture. Motor aspects of cerebellar control. Physiologist, 17:19, 1974.

Llinas, R.: Electrophysiology of the cerebellar networks. In Handbook of Physiology. Sec. 1, Vol. II. Bethesda, Md., American Physiological Society, 1981, p. 831.

McCloskey, D. I., et al.: Sensing position and movements of the fingers. News Physiol. Sci., 2:226, 1987.

Olsen, R. W.: Drug interactions at the GABA receptor-ionophore complex. Annu. Rev. Pharmacol. Toxicol., 22:245, 1982.

Palacios, J. M.: Neurotransmitters, their receptors and the degenerative diseases of the aging brain. Triangle, 25:85, 1986.

Palay, S. L., and Chan-Palay, V.: The Cerebellum — New Vistas. New York, Springer-Verlag, 1982.

Penney, J. B., Jr., and Young, A. B.: Speculations on the functional anatomy of basal ganglia disorders. Annu. Rev. Neurosci., 6:73, 1983.

Peterson, B. W., and Richmond, F. J. (eds.): Control of Head Movement. New York, Oxford University Press, 1988.

Porter, R.: Internal organization of the motor cortex for input-output arrangements. In Handbook of Physiology. Sec. 1, Vol. II. Bethesda, Md., American Physiological Society, 1981, p. 1063.

Poulton, E. C.: Human manual control. In Handbook of Physiology. Sec. 1, Vol. II. Bethesda, American Physiological Society, 1981, p. 1337.

Riklan, M.: L-Dopa and Parkinsonism. Springfield, Ill., Charles C Thomas, 1973.

Robinson, D. A.: The windfalls of technology in the oculomotor system. Inv. Ophthal. Vis. Sci., 28:1912, 1987.

Sandler, M., et al. (eds.): Neurotransmitter Interactions in the Basal Ganglia. New York, Raven Press, 1987.

Scheibel, A. B.: The brain stem reticular core and sensory function. In Handbook of Physiology. Sec. 1, Vol. II. Bethesda, Md., American Physiological Society, 1981, p. 213.

Shik, M. L., and Orlovsky, G. N.: Neurophysiology of locomotor automatism. Physiol. Rev., 56:465, 1976.

Stein, R. B., and Lee, R. G.: Tremor and clonus. In Handbook of Physiology. Sec. 1, Vol. II. Bethesda, Md., American Physiological Society, 1981, p. 325.

Wiesendanger, M.: Organization of secondary motor areas of cerebral cortex. In Handbook of Physiology. Sec. 1, Vol. II. Bethesda, Md., American Physiological Society, 1981, p. 1121.

Wiesendanger, M., and Miles, T. S.: Ascending pathway of low-threshold muscle afferents to the cerebral cortex and its possible role in motor control. Physiol. Rev., 62:1234, 1982.

CHAPTER 57

• The Cerebral Cortex; Intellectual Functions of the Brain and Learning and Memory •

It is ironic that of all the parts of the brain, we know least about the mechanisms of the cerebral cortex, even though it is by far the largest portion of the nervous system. Yet we do know the effects of destruction or of specific stimulation of various portions of the cortex. In the early part of the present chapter the facts known about cortical functions are discussed; then some basic theories of the neuronal mechanisms involved in thought processes, memory, analysis of sensory information, and so forth, are presented briefly.

PHYSIOLOGIC ANATOMY OF THE CEREBRAL CORTEX

The functional part of the cerebral cortex is composed mainly of a thin layer of neurons 2 to 5 millimeters in thickness, covering the surface of all the convolutions of the cerebrum and having a total area of about 0.25 square meter. The total cerebral cortex probably contains 100 billion or more neurons.

Figure 57–1 illustrates the typical structure of the cerebral cortex, showing successive layers of different types of cells. Most of the cells are of three types: *granular* (also called *stellate*), *fusiform,* and *pyramidal,* the latter named for their characteristic pyramidal shape. The *granule cells,* in general, have short axons and therefore function mainly as intracortical interneurons. Some are excitatory, probably releasing the excitatory neurotransmitter *glutamate;* others are inhibitory and release the inhibitory neurotransmitter *gamma-aminobutyric acid (GABA).* The sensory areas of the cortex, as well as the association areas between sensory and motor, have large concentrations of these granule cells, suggesting a high degree of intracortical processing of the incoming sensory signals in the sensory areas and of the cognitive analytical signals in the association areas.

The *pyramidal* and *fusiform cells,* on the other hand, give rise to almost all of the output fibers from the cortex. The pyramidal cells are the larger of the two and are more nu-

merous than the fusiform cells. They are the source of the long, large nerve fibers that go all the way to the spinal cord. They also give rise to most of the large subcortical association fiber bundles that pass from one major part of the brain to the other.

To the right in Figure 57–1 is illustrated the typical organization of nerve fibers within the different layers of the cortex. Note particularly the large number of *horizontal fibers* extending between adjacent areas of the cortex, but note also the *vertical fibers* that extend to and from the cortex to lower areas of the brain and to the spinal cord or to distant regions of the cerebral cortex through the long association bundles.

The functions of the specific layers of the cerebral cortex have been discussed briefly in Chapters 47 and 51. By way of review, let us recall that most incoming specific sensory signals terminate in cortical layer IV. Most of the output signals leave the cortex from neurons located in layers V and VI, the very large fibers to the brain stem and cord arise generally in layer V, and the tremendous numbers of fibers to the thalamus arise in layer VI. Layers I, II, and III perform most of the intracortical association functions, with especially large numbers of neurons in layers II and III making short horizontal connections with adjacent cortical areas.

Anatomical and Functional Relationships of the Cerebral Cortex to the Thalamus and Other Lower Centers. All areas of the cerebral cortex have extensive to-and-fro efferent and afferent connections with the deeper structures of the brain. It is especially important to emphasize the relationship between the cerebral cortex and the thalamus. When the thalamus is damaged along with the cortex, the loss of cerebral function is far greater than when the cortex alone is damaged, for thalamic excitation of the cortex is necessary for almost all cortical activity.

Figure 57–2 shows the areas of the cerebral cortex connected with specific parts of the thalamus. These connections act in *two* directions, both from the thalamus to the cortex and then from the cortex back to

Figure 57–1. Structure of the cerebral cortex, illustrating *I,* molecular layer; *II,* external granular layer; *III,* layer of pyramidal cells: *IV,* internal granular layer: *V,* large pyramidal cell layer; and *VI,* layer of fusiform or polymorphic cells. (From Ranson and Clark (after Brodmann): Anatomy of the Nervous System. Philadelphia, W. B. Saunders Company, 1959.)

essentially the same area of the thalamus. Furthermore, when the thalamic connections are cut, the functions of the corresponding cortical area become entirely abrogated. Therefore, the cortex operates in close association with the thalamus and can almost be considered both anatomically and functionally to be a unit with the thalamus; for this reason the thalamus and the cortex together are sometimes called the *thalamocortical system.* Also, all pathways from the sensory organs to the cortex pass through the thalamus, with the single exception of most sensory pathways of the olfactory tract.

● FUNCTIONS OF SPECIFIC CORTICAL AREAS

Studies in human beings by neurosurgeons, neurologists, and neuropathologists have shown that different cortical areas have their own separate functions. Figure 57–3 is a map of some of these functions as determined by Penfield and Rasmussen from electrical stimulation of the cortex in awake patients or during neurological examination of patients after portions of the cortex had been removed. The electrically stimulated patients either told the surgeons their thoughts evoked by the stimulation or at times experienced a movement or a spontaneously emitted sound or even a word or some other evidence of the stimulation. In the patients in whom portions of the cortex had been removed, the subsequent neurological examinations demonstrated different deficits of brain function.

Information of the type illustrated in Figure 57–3 from many different sources gives a more general map, as illustrated in Figure 57–4. This figure shows the major primary and secondary motor areas of the cortex, as well as the major primary and secondary sensory areas for somatic sensation, vision, and hearing, all of which have been discussed in previous chapters. The primary areas have direct connections with specific muscles or specific sensory receptors, for causing discrete muscle movements or experiencing a sensation—visual, auditory, or somatic—from a minute receptor area. The secondary areas, on the other hand, make sense out of the functions of the primary areas. For instance, the supplemental and premotor areas function along with the primary motor cortex and basal ganglia to provide highly specific patterns of motor activity. On the sensory side, the secondary sensory areas, which are located within a few centimeters of the primary areas, begin to make sense out of the specific sensory signals, such as interpreting the shape or texture of an object in one's hand;

Figure 57–2. Areas of the cerebral cortex that connect with specific portions of the thalamus. (Modified from Elliott: Textbook of the Nervous System. Philadelphia, J. B. Lippincott Company.)

Figure 57–3. Functional areas of the human cerebral cortex as determined by electrical stimulation of the cortex during neurosurgical operations and by neurological examinations of patients with destroyed cortical regions. (From Penfield and Rasmussen: The Cerebral Cortex of Man: A Clinical Study of Localization of Function. New York, Macmillan Company, 1968.)

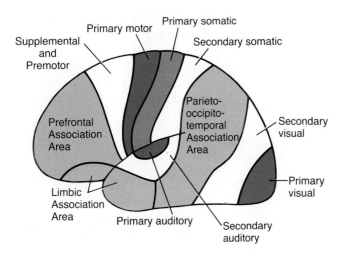

Figure 57–4. Locations of the major association areas of the cerebral cortex, shown in relation to the primary and secondary motor and sensory areas.

the color, the light intensity, the directions of lines and angles, and other aspects of vision; and the combination of tones, sequence of tones, and beginning interpretation of the meanings of auditory signals.

THE ASSOCIATION AREAS

Figure 57–4 also shows several large areas of the cerebral cortex that do not fit into the rigid categories of primary or secondary motor and sensory areas. These are called *association areas* because they receive and analyze signals from multiple regions of the cortex and even subcortical structures. Yet even the association areas have their own specializations, as we shall see. The three most important association areas are (1) the *parieto-occipitotemporal association area,* (2) the *prefrontal association area,* and (3) the *limbic association area.* The functions of these are the following:

The Parieto-occipitotemporal Association Area. This association area lies in the large cortical space between the somatic sensory cortex anteriorly, the visual cortex posteriorly, and the auditory cortex laterally. As would be expected, it provides a high level of interpretive meaning for the signals from all the surrounding sensory areas. However, even the parieto-occipitotemporal association area has its own functional subareas, which are illustrated in Figure 57–5:

1. An area beginning in the *posterior parietal cortex and extending into the superior occipital cortex provides continuous analysis of the spatial coordinates of all parts of the body as well as of the surroundings of the body.* This area receives visual information from the posterior occipital cortex and simultaneous somatic information from the anterior parietal cortex; from this it computes the coordinates. But why does a person need to know these spatial coordinates? The an-

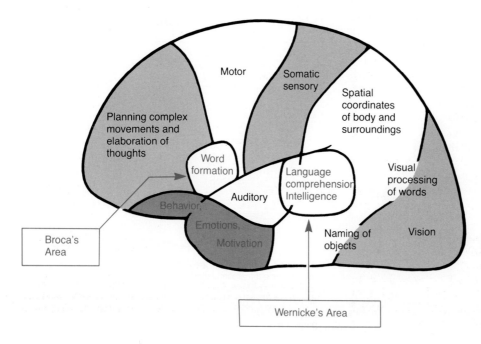

Figure 57–5. Map of specific functional areas in the cerebral cortex, showing especially Wernicke's and Broca's areas for language comprehension and speech production, which in 95 per cent of all persons are located in the left hemisphere.

swer is that to control the body movements, the brain must know at all times where each part of his body is located and also its relation to the surroundings. He also needs this information to analyze incoming somatic sensory signals. In fact, as was illustrated in Figure 56–13 of Chapter 56, a person missing this area of the brain actually loses recognition of the fact that he or she has an opposite side of the body or surroundings and, as a consequence, will fail to consider the existence of the opposite side either for sensation or for planning voluntary movements.

2. The major area for language comprehension, called *Wernicke's area,* lies behind *the primary auditory cortex in the posterior part of the superior temporal lobe.* We discuss this area much more fully later; it is the most important region of the entire brain for higher intellectual functions because almost all intellectual functions are language-based.

3. *Posterior to the language comprehension area, lying mainly in the angular gyrus region of the occipital lobe, is a secondary visual processing area that feeds the visual signals of words read from a page into Wernicke's area, the language comprehension area.* This angular gyrus area is needed to make meaning out of the visually perceived words. In its absence, a person can still have excellent language comprehension through hearing but not through reading.

4. *In the most lateral portions of both the anterior occipital lobe and posterior temporal lobe is an area for naming objects.* The names presumably originate mainly through auditory input, whereas the nature of the objects originates mainly through visual input. In turn, the names are essential for language comprehension and intelligence, functions performed in Wernicke's area, located immediately superior to the "names" region.

The Prefrontal Association Area. In the previous chapter we learned that the prefrontal association area functions in close association with the motor cortex to plan complex patterns and sequences of motor movements. To aid in this function, it receives very strong input through a massive subcortical bundle of fibers connecting the parieto-occipitotemporal association area with the prefrontal association area. Through this bundle the prefrontal cortex receives much preanalyzed sensory information, especially information on the spatial coordinates of the body that is absolutely necessary in the planning of effective movements. Much of the output from the prefrontal area into the motor control system passes through the caudate portion of the basal ganglia-thalamic feedback circuit for motor planning, which provides many of the sequential and parallel components of the movement complex.

The prefrontal association area is also essential to carrying out prolonged thought processes in the mind. This presumably results from some of the same capabilities of the prefrontal cortex that allow it to plan motor activities. That is, it seems to be capable of combining nonmotor information from widespread areas of the brain and therefore to achieve nonmotor

types of thinking as well as motor types. In fact, the prefrontal association area is frequently described simply as important for the *elaboration of thoughts.*

A special region in the frontal cortex, called *Broca's area, provides the neural circuitry for word formation.* This area, illustrated in Figure 57–5, is located partly in the posterior lateral prefrontal cortex and partly in the premotor area. It is here that the plans and motor patterns for the expression of individual words or even short phrases are initiated and executed. This area also works in close association with Wernicke's language comprehension center in the temporal association cortex, as we discuss more fully later in the chapter.

The Limbic Association Area. Figure 57–4 illustrates still another association area called the *limbic area.* This is found in the anterior pole of the temporal lobe, in the ventral portions of the frontal lobes, and in the cingulate gyri on the midsurfaces of the cerebral hemispheres. This region is concerned primarily with *behavior, emotions,* and *motivation,* as illustrated in Figure 57–5. We will learn in the following chapter that the limbic cortex is part of a much more extensive system, the *limbic system,* that includes a complex set of neuronal structures in the midbasal regions of the brain. It is this limbic system that provides most of the drives for setting the other areas of the brain into action and even provides the motivational drive for the process of learning itself.

An Area for Recognition of Faces

An interesting type of brain abnormality called *prosophenosia* is the inability to recognize faces. This occurs in persons who have extensive damage on the medial undersides of both occipital lobes and along the medioventral surfaces of the temporal lobes, as illustrated in Figure 57–6. Loss of these face recogni-

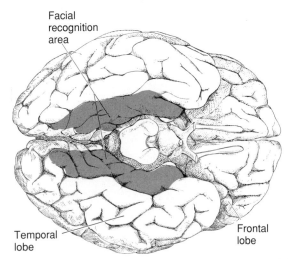

Facial recognition area

Temporal lobe

Frontal lobe

Figure 57–6. Facial recognition areas located on the underside of the brain in the medial occipital and temporal lobes. (From Geschwind: *Sci. Am.,* 241:180, 1979. © 1979 by Scientific American, Inc. All rights reserved.)

tion areas, strangely enough, results in very little other abnormality of brain function.

One wonders why so much of the cerebral cortex should be reserved for the simple task of face recognition. However, when it is remembered that most of our daily tasks involve associations with other people, one can see the importance of this intellectual function.

The occipital portion of this area is contiguous with the visual cortex, and the temporal portion is closely associated with the limbic system that has to do with emotions, brain activation, and control of one's behavioral response to the environment, as we see later in the following chapter.

INTERPRETATIVE FUNCTION OF THE POSTERIOR SUPERIOR TEMPORAL LOBE—WERNICKE'S AREA (A GENERAL INTERPRETATIVE AREA)

The somatic, visual, and auditory secondary and association areas, which can actually be called sensory interpretative areas, all meet one another in the posterior part of the superior temporal lobe, as illustrated in Figure 57–7, where the temporal, parietal, and occipital lobes all come together. This area of confluence of the different sensory interpretative areas is especially highly developed in the *dominant* side of the brain—the *left side* in almost all right-handed persons—and it plays the greatest single role of any part of the cerebral cortex in the higher levels of brain function that we call *intelligence*. Therefore, this region has frequently been called by different names suggestive of the area having almost global importance: the *general interpretative area*, the *gnostic area*, the *knowing area*, the *tertiary association area*, and so forth. However, it is best known as *Wernicke's area* in honor of the neurologist who first described its special significance in intellectual processes.

Following severe damage in Wernicke's area, a person might hear perfectly well and even recognize different words but still be unable to arrange these words into a coherent thought. Likewise, the person may be able to read words from the printed page but be unable to recognize the thought that is conveyed.

Electrical stimulation in Wernicke's area of the conscious patient occasionally causes a highly complex thought. This is particularly true when the stimulatory electrode is passed deep enough into the brain to approach the corresponding connecting areas of the thalamus. The types of thoughts that might be experienced include complicated visual scenes that one might remember from childhood, auditory hallucinations such as a specific musical piece, or even a discourse by a specific person. For this reason it is believed that activation of Wernicke's area can call forth complicated memory patterns involving more than one sensory modality even though many of the memory patterns may be stored elsewhere. This belief is in accord with the importance of Wernicke's area in interpretation of the complicated meanings of different sensory experiences.

The Angular Gyrus—Interpretation of Visual Information. The angular gyrus is the most inferior portion of the posterior parietal lobe, lying immediately behind Wernicke's area and fusing posteriorly into the visual areas of the occipital lobe as well. If this region is destroyed while Wernicke's area in the temporal lobe is still intact, the person can still interpret auditory experiences as usual, but the stream of visual experiences passing into Wernicke's area from the visual cortex is mainly blocked. Therefore, the person may be able to see words and even know they are words but, nevertheless, not be able to interpret their meanings. This is the condition called *dyslexia*, or *word blindness*.

Let us again emphasize the global importance of Wernicke's area for most intellectual functions of the brain. Loss of this area in an adult usually leads thereafter to a lifetime of almost demented existence.

The Concept of the Dominant Hemisphere

The general interpretative functions of Wernicke's area and of the angular gyrus, and also the functions of the speech and motor control areas, are usually much more highly developed in one cerebral hemisphere than in the other. Therefore, this hemisphere is called the *dominant hemisphere*. In about 95 per cent of all persons the left hemisphere is the dominant one. Even at birth, the area of the cortex that will eventually become Wernicke's area is as much as 50 per cent larger in the left hemisphere than in the right in more than one half of newborn babies. Therefore, it is easy to understand why the left side of the brain might become dominant over the right side. However, if for some reason this left side area is damaged or removed in early childhood, the opposite side of the brain can develop full dominant characteristics.

A theory that can explain the capability of one hemisphere to dominate the other hemisphere is the following:

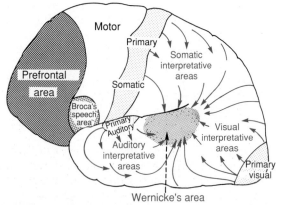

Figure 57–7. Organization of the somatic, auditory, and visual association areas into a general mechanism for interpretation of sensory experience. All these feed also into *Wernicke's area*, located in the posterosuperior portion of the temporal lobe. Note also the prefrontal area and Broca's speech area.

The attention of the "mind" seems to be directed to one portion of the brain at a time. Presumably, because its size is usually larger at birth, the left temporal lobe normally begins to be used to a greater extent than the right, and, thenceforth, because of the tendency to direct one's attention to the better developed region, the rate of learning in the cerebral hemisphere that gains the first start increases rapidly while that in the opposite side remains slight. Therefore, in the normal human being, one side becomes dominant over the other.

In about 95 per cent of all persons the left temporal lobe and angular gyrus become dominant, and in the remaining 5 per cent either both sides develop simultaneously to have dual dominance, or, more rarely, the right side alone becomes highly developed.

Usually associated with the dominant temporal lobe and angular gyrus is dominance of certain portions of the somatic sensory cortex and motor cortex for control of voluntary motor functions. For instance, as is discussed later in the chapter, the prefrontal and premotor speech area (Broca's area), located far laterally in the intermediate frontal lobe, is almost always dominant also on the left side of the brain. This speech area causes the formation of words by exciting simultaneously the laryngeal muscles, the respiratory muscles, and the muscles of the mouth.

Also, the motor areas for controlling the hands are dominant on the left side of the brain in about nine of ten persons, thus causing "right-handedness" in most people.

Although the interpretative areas of the temporal lobe and angular gyrus, as well as many of the motor areas, are highly developed in only a single hemisphere, they are capable of receiving sensory information from both hemispheres and are also capable of controlling motor activities in both hemispheres, utilizing mainly fiber pathways in the *corpus callosum* for communication between the two hemispheres. This unitary, cross-feeding organization prevents interference between the functions of the two sides of the brain; such interference, obviously, could create havoc with both thoughts and motor responses.

Role of Language in the Function of Wernicke's Area and in Intellectual Functions

A major share of our sensory experience is converted into its language equivalent before being stored in the memory areas of the brain and before being processed for other intellectual purposes. For instance, when we read a book, we do not store the visual images of the printed words but, instead, store the words themselves in language form. Also, the information conveyed by the words is usually converted to language form before its meaning is discerned.

The sensory area of the dominant hemisphere for interpretation of language is Wernicke's area, and this is very closely associated with both the primary hearing area and the secondary auditory areas of the temporal lobe. This very close relationship probably results from the fact that the first introduction to language is by way of hearing. Later in life, when visual perception of language through the medium of reading develops, the visual information is then presumably channeled into the already developed language regions of the dominant temporal lobe.

FUNCTIONS OF THE PARIETO-OCCIPITOTEMPORAL CORTEX IN THE NONDOMINANT HEMISPHERE

When Wernicke's area in the dominant hemisphere is destroyed, the person normally loses almost all intellectual functions associated with language or verbal symbolism, such as ability to read, ability to perform mathematical operations, and even the ability to think through logical problems. However, many other types of interpretative capabilities, some of which utilize the temporal lobe and angular gyrus regions of the opposite hemisphere, are retained. Psychological studies in patients with damage to their nondominant hemispheres have suggested that this hemisphere may be especially important for understanding and interpreting music, nonverbal visual experiences (especially visual patterns), spatial relationships between the person and the surroundings, the significance of "body language" and intonations of persons' voices, and probably also many somatic experiences related to use of the limbs and hands.

Thus, even though we speak of the "dominant" hemisphere, this dominance is primarily for language- or verbal symbolism-related intellectual functions; the opposite hemisphere is actually dominant for some other types of intelligence.

THE HIGHER INTELLECTUAL FUNCTIONS OF THE PREFRONTAL ASSOCIATION AREA

For years it has been taught that the prefrontal cortex is the locus of the higher intellect in the human being, principally because the main difference between the brain of monkeys and of human beings is the great prominence of the human prefrontal areas. Yet efforts to show that the prefrontal cortex is more important in higher intellectual functions than other portions of the brain have not been entirely successful. Indeed, destruction of the language comprehension area in the posterior superior temporal lobe (Wernicke's area) and the angular gyrus region in the dominant hemisphere causes infinitely more harm to the intellect than does destruction of the prefrontal area. The prefrontal areas do, however, have less definable but nevertheless very important intellectual functions of their own. These can be explained best by describing what happens to patients in whom the prefrontal lobes have become nonfunctional as follows:

Several decades ago, before the advent of modern drugs for treating psychiatric conditions, it was found that some patients could receive significant relief from severe psychotic depression by severing of the neuronal connections between the prefrontal areas of the brain and the remainder of the brain, that is, by a procedure called *prefrontal lobotomy*. This was done by inserting a blunt, thin-bladed knife through small openings in the lateral frontal skull on both sides and slicing the brain from top to bottom. Subsequent studies in these patients showed the following mental changes:

1. The patients lost their ability to solve complex problems.

2. They became unable to string together sequential tasks to reach specific goals, and in general lost all ambition.

3. They became unable to learn to do several parallel tasks at the same time.

4. Their level of aggressiveness was decreased, sometimes markedly.

5. Their social responses were often inappropriate for the occasion, including loss of morals and little embarrassment in relation to sex and excretion.

6. The patients could still talk and comprehend language, but they were unable to carry through any long trains of thought, and their moods changed rapidly from sweetness to rage to exhilaration to madness.

7. The patients could also still perform most of the usual patterns of motor function that they had performed throughout life, but often without purpose.

From this information, let us try to piece together a coherent understanding of the function of the prefrontal association areas.

Decreased Aggressiveness and Inappropriate Social Responses. These two characteristics probably result from loss of the ventral parts of the frontal lobes on the underside of the brain. As explained earlier and illustrated in Figure 57–4, this area is considered to be part of the limbic association cortex, rather than the prefrontal association cortex. This limbic area helps to control behavior.

Inability to Progress Toward Goals or to Carry Through With Sequential Thoughts. We learned earlier in the chapter that the prefrontal association areas appear to have the capability of calling forth information from widespread areas of the brain and then using it in deeper thought patterns for attaining goals. If these goals include motor action, so be it. If they do not, then the thought processes attain intellectual analytical goals. Although persons without prefrontal cortices can still think, they show little concerted thinking in logical sequence for longer than a few seconds or a few minutes at most. One of the results is that persons without prefrontal cortices are *easily distracted from the central theme of the thought,* whereas persons with functioning prefrontal cortices can drive themselves to completion of their thought goals irrespective of distractions.

Elaboration of Thought, Prognostication, and Performance of Higher Intellectual Functions by the Prefrontal Areas. Another function that has been ascribed to the prefrontal areas by psychologists and neurologists is *elaboration of thought.* This means simply an increase in depth and abstractness of the different thoughts. Psychological tests have shown that prefrontal lobectomized lower animals presented with successive bits of sensory information fail to keep track of these bits even in temporary memory — probably because they are distracted

so easily that they cannot hold thoughts long enough for storage to take place. This ability of the prefrontal areas to keep track of many bits of information simultaneously, and then to cause recall of this information bit by bit as it is needed for subsequent thoughts, could well explain the many functions of the brain that we associate with higher intelligence, such as the abilities to (1) prognosticate, (2) plan for the future, (3) delay action in response to incoming sensory signals so that the sensory information can be weighed until the best course of response is decided, (4) consider the consequences of motor actions even before these are performed, (5) solve complicated mathematical, legal, or philosophical problems, (6) correlate all avenues of information in diagnosing rare diseases, and (7) control one's activities in accord with moral laws.

● FUNCTION OF THE BRAIN IN COMMUNICATION

One of the most important differences between the human being and lower animals is the facility with which human beings can communicate with one another. Furthermore, because neurological tests can easily assess the ability of a person to communicate with others, we know more about the sensory and motor systems related to communication than about any other segment of cortical function. Therefore, we will review, with the help of the anatomical maps of neural pathways in Figure 57–8, the function of the cortex in communication, and from this one can see immediately how the principles of sensory analysis and motor control apply to this art.

There are two aspects to communication: first, the *sensory aspect,* involving the ears and eyes, and, second, the *motor aspect,* involving vocalization and its control.

Sensory Aspects of Communication

We noted earlier in the chapter that destruction of portions of the *auditory* and *visual association areas* of the cortex can result in inability to understand the spoken word or the written word. These effects are called, respectively, *auditory receptive aphasia* and *visual receptive aphasia* or, more commonly, *word deafness* and *word blindness* (also called *dyslexia*).

Wernicke's Aphasia and Global Aphasia. Some persons are perfectly capable of understanding either the spoken word or the written word but are *unable to interpret the thought* that is expressed. This results most frequently when *Wernicke's area* in the *posterior portion of the dominant hemisphere superior temporal gyrus* is damaged or destroyed. Therefore, this type of aphasia is generally called *Wernicke's aphasia.*

When the lesion in Wernicke's area is widespread and extends (1) backward into the angular gyrus region, (2) inferiorly into the lower areas of the temporal lobe, and (3) superiorly into the superior border of the sylvian fissure, the person is likely to be almost totally demented and therefore is said to have *global aphasia.*

Motor Aspects of Communication

The process of speech involves two principal stages of mentation: (1) formation in the mind of thoughts to be ex-

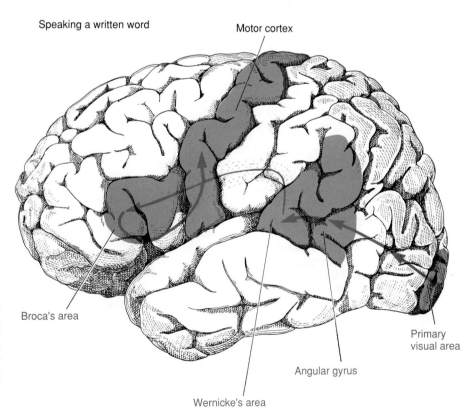

Figure 57–8. Brain pathways for *(top)* perception of the heard word and then speaking the same word and *(bottom)* perception of the written word and then speaking the same word. (From Geschwind: *Sci. Am.,* 241:180, 1979. © 1979 by Scientific American, Inc. All rights reserved.)

pressed and choice of words to be used, then (2) motor control of vocalization and the actual act of vocalization itself. The formation of thoughts and even most choices of words are the function of the sensory areas of the brain. Again, it is Wernicke's area in the posterior part of the superior tem-poral gyrus that is most important for this ability. Therefore, persons with either Wernicke's aphasia or global aphasia are unable to formulate the thoughts that are to be communicated. Or, if the lesion is less severe, the person may be able to formulate thoughts but yet be unable to put together the appropriate sequence of words to express the thought. Often, the person is very fluent in words but the words are jumbled.

Motor Aphasia. Often a person is perfectly capable of deciding what he wishes to say, and he is capable of vocalizing, but he simply cannot make his vocal system emit words instead of noises. This effect, called *motor aphasia*, results from damage to *Broca's speech area*, which lies in the *prefrontal* and *premotor* facial region of the cortex—about 95 per cent of the time in the left hemisphere, as illustrated in Figures 57–5 and 57–8. Therefore, we assume that the *skilled motor patterns* for control of the larynx, lips, mouth, respiratory system, and other accessory muscles of articulation are all initiated from this area.

Articulation. Finally, we have the act of articulation itself, which means the muscular movements of the mouth, tongue, larynx, and so forth, that are responsible for the actual emission of sound. The *facial and laryngeal regions of the motor cortex* activate these muscles, and the *cerebellum, basal ganglia,* and *sensory cortex* all help control the muscle contractions by feedback mechanisms described in Chapters 55 and 56. Destruction of these regions can cause either total or partial inability to speak distinctly.

Summary

Figure 57–8 illustrates two principal pathways for communication. The upper half of the figure shows the pathway involved in hearing and speaking. This sequence is the following: (1) reception in the primary auditory area of the sound signals that encode the words; (2) interpretation of the words in Wernicke's area; (3) determination, also in Wernicke's area, of the thoughts and the words to be spoken; (4) transmission of signals from Wernicke's area to Broca's area via the *arcuate fasciculus;* (5) activation of the skilled motor programs in Broca's area for control of word formation; and (6) transmission of appropriate signals into the motor cortex to control the speech muscles.

The lower figure illustrates the comparable steps in reading and then speaking in response. The initial receptive area for the words is in the primary visual area rather than in the primary auditory area. Then the information passes through early stages of interpretation in the *angular gyrus region* and finally reaches its full level of recognition in Wernicke's area. From here, the sequence is the same as for speaking in response to the spoken word.

● FUNCTION OF THE CORPUS CALLOSUM AND ANTERIOR COMMISSURE TO TRANSFER THOUGHTS, MEMORIES, TRAINING, AND OTHER INFORMATION TO THE OPPOSITE HEMISPHERE

Fibers in the *corpus callosum* connect most of the respective cortical areas of the two hemispheres with each other except for the anterior portions of the temporal lobes; these temporal areas, including especially the *amygdala*, are interconnected by fibers that pass through the *anterior commissure*. Because of the tremendous number of fibers in the corpus callosum, it was assumed from the beginning that this massive structure must have some important function to correlate activities of the two cerebral hemispheres. However, after cutting the corpus callosum in experimental animals, it was difficult to discern deficits in brain function. Therefore, for a long time the function of the corpus callosum was a mystery.

Yet properly designed psychological experiments have now demonstrated the extremely important functions of the corpus callosum and anterior commissure. These can be explained best by recounting one of the experiments. A monkey is first prepared by cutting the corpus callosum and splitting the optic chiasm longitudinally so that signals from each eye can go only to the cerebral hemisphere on the side of the eye. Then the monkey is taught to recognize different types of objects with its right eye while its left eye is covered. Next, the right eye is covered and the monkey is tested to determine whether or not its left eye can recognize the same object. The answer to this is that the left eye *cannot* recognize the object. Yet, on repeating the same experiment in another monkey with the optic chiasm split but the corpus callosum intact, it is found invariably that recognition in one hemisphere of the brain creates recognition also in the opposite hemisphere.

Thus, one of the functions of the corpus callosum and the anterior commissure is to make information stored in the cortex of one hemisphere available to cortical areas of the opposite hemisphere. Three important examples of such cooperation between the two hemispheres are the following:

1. Cutting of the corpus callosum blocks transfer of information from Wernicke's area of the dominant hemisphere to the motor cortex on the opposite side of the brain. Therefore, the intellectual functions of the brain, located primarily in the dominant hemisphere, lose their control over the right motor cortex and therefore also of the voluntary motor functions of the left hand and arm even though the usual subconscious movements of the left hand and arm are completely normal.

2. Cutting of the corpus callosum prevents transfer of somatic and visual information from the right hemisphere into Wernicke's area of the dominant hemisphere. Therefore, somatic and visual information from the left side of the body frequently fails to reach this general interpretative area of the brain and therefore cannot be used for decision making.

3. Finally, persons whose corpus callosum is completely sectioned are found to have two entirely separate conscious portions of the brain. For example, in a recently studied teen-age boy with a sectioned corpus callosum, only the left half of his brain could understand the spoken word, because it was the dominant hemisphere. On the other hand, the right side of the brain could understand the written word and could elicit a motor response to it without the left side of the brain ever knowing why the response was performed. Yet the effect was quite different when an emotional response was evoked in the right side of the brain: in this case a subconscious emotional response occurred in the left side of the brain as well. This undoubtedly occurred because the areas of the two sides of the brain for emotions, the anterior temporal cortices and adjacent areas, were still communicating with each other through the anterior commissure that was not sectioned. For instance, when the command "kiss" was written for the right half of his brain to see, the boy immediately and with full emotion, said "No way!" This response obviously required function of Wernicke's area and the motor areas for speech in the left hemispheres. But, when questioned why he said this, the boy could not explain. Thus, the two halves of the brain have independent capabilities for consciousness, memory storage, communication, and control of motor activities. The corpus callosum is required for the two sides to operate cooperatively, and the anterior commissure plays an important additional role in unifying the emotional responses of the two sides of the brain.

THOUGHTS, CONSCIOUSNESS, AND MEMORY

Our most difficult problem in discussing consciousness, thoughts, memory, and learning is that we do not know the neural mechanism of a thought. We know that destruction of large portions of the cerebral cortex does not prevent a person from having thoughts, but it usually does reduce the *degree* of awareness of the surroundings.

Each thought almost certainly involves simultaneous signals in many portions of the cerebral cortex, thalamus, limbic system, and reticular formation of the brain stem. Some crude thoughts probably depend almost entirely on lower centers; the thought of pain is probably a good example, for electrical stimulation of the human cortex rarely elicits anything more than the mildest degree of pain, whereas stimulation of certain areas of the hypothalamus and mesencephalon often causes excruciating pain. On the other hand, a type of thought pattern that requires mainly the cerebral cortex is that of vision, because loss of the visual cortex causes complete inability to perceive visual form or color.

Therefore, we might formulate a definition of a thought in terms of neural activity as follows: A thought results from the "pattern" of stimulation of many different parts of the nervous system at the same time and in definite sequence, probably involving most importantly the cerebral cortex, the thalamus, the limbic system, and the upper reticular formation of the brain stem. This is called the *holistic theory* of thoughts. The stimulated areas of the limbic system, thalamus, and reticular formation are believed to determine the general nature of the thought, giving it such qualities as pleasure, displeasure, pain, comfort, crude modalities of sensation, localization to gross areas of the body, and other general characteristics. On the other hand, the stimulated areas of the cerebral cortex determine the discrete characteristics of the thought such as specific localization of sensations on the body and of objects in the fields of vision, discrete patterns of sensation such as the rectangular pattern of a concrete block wall or the texture of a rug, and other individual characteristics that enter into the overall awareness of a particular instant.

Consciousness can perhaps be described as our continuing stream of awareness of either our surroundings or our sequential thoughts.

MEMORY—ROLES OF SYNAPTIC FACILITATION AND SYNAPTIC INHIBITION

Physiologically, memories are caused by changes in the capability of synaptic transmission from one neuron to the next as a result of previous neural activity. These changes in turn cause new pathways to develop for transmission of signals through the neural circuits of the brain. The new pathways are called *memory traces*. They are important because, once established, they can be activated by the thinking mind to reproduce the memories.

Experiments in lower animals have demonstrated that memory traces can occur at all levels of the nervous system. Even spinal cord reflexes can change at least slightly in response to repetitive cord activation, which is part of the memory process. Also, even some long-term memories result from changed synaptic conduction in the lower brain centers. To give an example, the blinking reflex is a learned function involving neuronal circuits in the cerebellum.

Yet there is much reason to believe that most of the memory that we associate with intellectual processes is based on memory traces mainly in the cerebral cortex.

Positive and Negative Memory—"Sensitization" or "Habituation" of Synaptic Transmission. Although we often think of memories as being positive recollections of previous thoughts or experiences, probably the greater share of our memories is negative memories, not positive. That is, our brain is inundated with sensory information from all of our senses. If our minds attempted to remember all of this information, the memory capacity of the brain would be exceeded within minutes. Fortunately, though, the brain has the peculiar capacity to learn to ignore information that is of no consequence. This results from *inhibition* of the synaptic pathways for this type of information, and the resulting effect is called *habituation*. This is, in a sense, a type of negative memory.

On the other hand, for those types of incoming information that cause important consequences, such as pain or pleasure, the brain also has the automatic capability of enhancing and storing the memory traces. Obviously, this is positive memory. It results from *facilitation* of the synaptic pathways, and the process is called *memory sensitization*. We learn later that special areas in the basal limbic regions of the brain determine whether information is important or unimportant and make the subconscious decision whether to store the thought as an enhanced memory trace or to suppress it.

Classification of Memories. We all know that some memories last only a few seconds, and others hours, days, months, or years. For the purpose of discussing these, let us use a common classification of memories that divides memories into (1) *immediate memory*, which includes memories that last for seconds or at most minutes unless they are converted into short-term memories; (2) *short-term memories*, which last for days to weeks but eventually are lost; and (3) *long-term memory*, which, once stored, can be recalled up to years or even a lifetime later.

IMMEDIATE MEMORY

Immediate memory is typified by one's memory of up to seven to ten telephone numbers at a time (or other discrete facts) for a few seconds to a few minutes at a time, but lasting only so long as the person continues to think about the numbers or facts.

Many physiologists have suggested that immediate memory is caused by continual neural activity resulting from nerve signals that travel around and around in a temporary memory trace through a *circuit of reverberating neurons*. Unfortunately, it has not yet been possible to prove this theory.

Another possible explanation of immediate memory is *presynaptic facilitation or inhibition*. This occurs at synapses that lie on presynaptic terminals, not on the subsequent neuron. The neurotransmitters secreted at such terminals frequently cause prolonged facilitation or inhibition (depending upon the type of transmitter secreted) for as long as seconds or even several minutes. Obviously, circuits of this type could lead to immediate memory.

A final possibility for explaining immediate memory is synaptic potentiation, which can enhance synaptic conduction. It can result from the accumulation of large amounts of calcium ions in the presynaptic terminals. That is, when a train of impulses passes through a presynaptic terminal, the amount of calcium ions increases with each impulse. When the amount of calcium ions becomes greater than the mitochondria and endoplasmic reticulum can absorb, the excess calcium then causes prolonged release of transmitter substance at the synapse. Thus, this, too, could be a mechanism for immediate memory.

SHORT-TERM MEMORY

Now we come to short-term memories that may last for many minutes or even weeks. Yet these will eventually be lost unless the memory traces become more permanent, then are classified as long-term memories. Recent experiments in primitive animals have demonstrated that memories of this type can result from temporary chemical or physical changes, or both, in either the presynaptic terminals or in the postsynaptic membrane, changes that can persist for up to several weeks. These mechanisms are so important that they deserve special description.

Memory Based on Chemical and Physical Changes in the Presynaptic Terminal or Postsynaptic Neuronal Membrane

Figure 57–9 illustrates a mechanism of memory studied especially by Kandel and his colleagues that can cause memories lasting for up to 3 weeks in the large snail *Aplysia*. In this figure there are two separate presynaptic terminals. One terminal is from a primary input sensory neuron and terminates on the surface of the neuron that is to be stimulated; this is called the *sensory terminal*. The other terminal lies on the surface of the sensory terminal and is called the *facilitator terminal*. When the sensory terminal is stimulated repeatedly but without stimulating the facilitator terminal, signal transmission at first is very

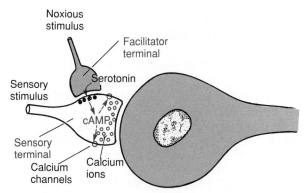

Figure 57–9. A memory system that has been discovered in the snail *Aplysia*.

great, but this becomes less and less intense with repeated stimulation until transmission almost ceases. This phenomenon is called *habituation*. It is a type of memory that causes the neuronal circuit to lose its response to repeated events that are insignificant.

On the other hand, if a noxious stimulus excites the facilitator terminal at the same time that the sensory terminal is stimulated, then, instead of the transmitted signal becoming progressively weaker, the ease of transmission becomes much stronger and will remain strong for hours, days, or, with more intense training, up to about 3 weeks even without further stimulation of the facilitator terminal. Thus, the noxious stimulus causes the memory pathway to become facilitated for days or weeks thereafter. It is especially interesting that once habituation has occurred, the pathway can be converted to a facilitated pathway with only a few noxious stimuli.

At the molecular level, the habituation effect in the sensory terminal results from progressive closure of calcium channels of the terminal membrane, although the cause of this is not fully known. Nevertheless, much smaller than normal amounts of calcium can then diffuse into this terminal when action potentials occur, and much less transmitter is therefore released because calcium entry is the stimulus for transmitter release (as was discussed in Chapter 45).

In the case of facilitation, the molecular mechanism is believed to be the following:

1. Stimulation of the facilitator neuron at the same time that the sensory neuron is stimulated causes serotonin release at the facilitator synapse on the sensory presynaptic terminal.

2. The serotonin acts on *serotonin receptors* in the sensory terminal membrane, and these activate the enzyme *adenylate cyclase* inside the membrane. This causes the formation of *cyclic adenosine monophosphate (cAMP)* inside the sensory presynaptic terminal.

3. The cAMP activates a *protein kinase* that causes phosphorylation of a protein that is part of the potassium channels in the sensory terminal membrane. This blocks these channels for potassium conductance. This blockage of the potassium channels can last for minutes up to several weeks.

4. Lack of potassium conductance causes a greatly prolonged action potential in the presynaptic terminal because the flow of potassium ions out of the terminal is necessary for recovery from the action potential.

5. The prolonged action potential causes prolonged activation of the calcium pores, allowing tremendous quantities of calcium ions to enter the sensory terminal. These calcium ions then cause greatly increased transmitter release, thereby greatly facilitating synaptic transmission.

Thus, in a very indirect way the associative effect of stimulating the facilitator neuron at the same time that the sensory neuron is stimulated causes a prolonged change in the sensory terminal that produces the memory trace.

In addition, recent studies by Byrne and his colleagues, also in the snail *Aplysia,* have suggested still another mechanism of cellular memory; their studies have shown that stimuli from two separate sources acting on a single neuron can, under appropriate conditions, cause long-term changes in the membrane properties of the entire postsynaptic neuron. Thus, this is another possible mechanism of short-term memory.

LONG-TERM MEMORY

There is no real demarcation between the more prolonged types of short-term memory and long-term memory. The distinction is one of degree. However, long-term memory is generally believed to result from actual *structural changes* at the synapses that enhance or suppress signal conduction. Again, let us recall experiments in primitive animals (where the nervous systems are much easier to study) that have aided immensely in understanding possible mechanisms of long-term memory.

Structural and Other Physical Changes in Synapses During the Development of Long-Term Memory

If the reader will refer again to Figure 7–2 in Chapter 7, he or she will see that the vesicles in a presynaptic terminal release their transmitter substance into the synaptic cleft through a special release site. When extra amounts of calcium enter the terminal, those vesicles near the release site attach to receptors at the site; then it is this attachment that causes vesicular exocytosis of the transmitter substance into the synaptic cleft.

Electron-microscopic pictures in invertebrate animals have now demonstrated that the total area of this vesicular release site increases in the presynaptic terminal during the development of long-term memory traces. Conversely, during long periods of synaptic inactivity, the release site diminishes and may actually disappear. Furthermore, the growth of the site depends upon activation of specific genetic control

mechanisms for synthesizing proteins that are required for assembling the release structures.

An intriguing characteristic of this mechanism for learning is that increased areas of vesicular release sites can be seen experimentally within hours after initiating the training sessions. Thus, it is entirely possible that much of what we now consider to be the longer types of short-term memory are actually early stages of this purely anatomically based long-term memory.

Therefore, at least in these primitive animals, at last we are beginning to understand a physical, structural basis for the development of long-term memory.

Other Possible Physical, Anatomical Mechanisms for Long-Term Memory

Aside from increasing the physical capability for release of neurotransmitter from the presynaptic terminals, memory development is also associated with an increase in the number of transmitter vesicles in the presynaptic terminals. And in some instances, the number of terminals themselves increases. Both of these effects could account for enhanced signal transmission. In fact, as a child grows and learns, the number of synapses in the brain increases greatly. Conversely, if a newborn animal is prevented from seeing with one of its eyes, those stripes of the visual cortex that are normally connected to the blinded eye fail to develop the same profusion of synapses as occurs for the other eye. Furthermore, taking the cover off the blinded eye reveals that the eye has not learned to see and has even lost much of its future capacity for learning to see.

Finally, in addition to changed synaptic conduction as a basis for learning, there is also the possibility of changing numbers of neurons in the used circuits. However, in this case, the process seems to be one of negative selection, for the brain has its greatest peak number of neurons at the time of birth or shortly after birth. Then, during the subsequent period of most rapid learning, those neurons that are excited appear to flourish; whereas those that remain unexcited may disappear altogether.

Therefore, it is likely that the human brain utilizes several different methods for enhancing or suppressing neuronal transmission when establishing memories.

CONSOLIDATION OF MEMORY

For an immediate memory to be converted into either a more prolonged short-term memory or into a long-term memory that can be recalled weeks or years later, it must become "consolidated." That is, the memory must in some way initiate the chemical, physical, and anatomical changes in the synapses that are responsible for the long-term type of memory. This process requires 5 to 10 minutes for minimal consolidation and an hour or more for maximal consolidation. For instance, if a strong sensory impression is made on the brain but is then followed within a minute or so by an electrically induced brain convulsion, the sensory experience will not be remembered at all. Likewise, brain concussion, sudden application of deep general anesthesia, or any other effect that temporarily blocks the dynamic function of the brain can prevent consolidation.

However, if the strong electrical shock is delayed for more than 5 to 10 minutes, at least part of the memory trace will have become established. If the shock is delayed for an hour, the memory will have become even much more fully consolidated.

The process of consolidation and the time required for consolidation can probably be explained by the phenomenon of *rehearsal* of the immediate memory as follows:

Role of Rehearsal in Transference of Immediate Memory into Longer Memory. Psychological studies have shown that rehearsal of the same information again and again accelerates and potentiates the degree of transfer of immediate memory into longer-term memory, and therefore also accelerates and potentiates the process of consolidation. The brain has a natural tendency to rehearse newfound information, and especially to rehearse newfound information that catches the mind's attention. Therefore, over a period of time the important features of sensory experiences become progressively more and more fixed in the secondary memory stores. This explains why a person can remember small amounts of information studied in depth far better than large amounts of information studied only superficially. It also explains why a person who is wide awake will consolidate memories far better than a person who is in a state of mental fatigue.

Codifying of Memories During the Process of Consolidation. One of the most important features of the process of consolidation is that memories placed permanently into the longer-term memory storehouse are codified into different classes of information. During this process similar information is recalled from the memory storage bins and is used to help process the new information. The new and old are compared for similarities and for differences, and part of the storage process is to store the information about these similarities and differences, rather than simply to store the information unprocessed. Thus, during the process of consolidation, the new memories are not stored randomly in the brain, but instead are stored in direct association with other memories of the same type. This is obviously necessary if one is to be able to "search" the memory store at a later date to find the required information.

Role of Specific Parts of the Brain in the Memory Process

Role of the Hippocampus for Storage of Memories — Anterograde Amnesia Following Hippocampal Lesions. The hippocampus is the most medial portion of the temporal lobe cortex where it folds underneath the brain and then upward into the lower surface of the lateral ventricle. The two hippocampi have been removed for the treatment of epilepsy in a number of patients. This procedure does not seriously affect the person's memory for information stored in the brain prior to removal of the hippocampi. However, after removal, these persons have very little capacity for storing *verbal and symbolic types* of memories in long-term memory, or even in short-term memory lasting longer than a few minutes. Therefore, these persons are unable to establish new long-term memories of those types of information that are the basis of intelligence. This is called *anterograde amnesia.*

But why is the hippocampus so important in helping the brain to store new memories? The probable answer is that the hippocampus is one of the important output pathways from the "reward" and "punishment" areas of the limbic system. These areas are found in many of the basal regions of the brain, and they feed into the hippocampus. Those sensory stimuli, or even thoughts, that cause pain or aversion excite the *punishment centers,* whereas those stimuli that cause pleasure, happiness, or a sense of reward excite the *reward centers.* All of these together provide the background mood and motivations of the person. Among these motivations is the drive in the brain to remember those experiences and thoughts that are either pleasant or unpleasant. The hippocampus especially and to a lesser degree the dorsal medial nuclei of the thalamus, another limbic structure, have proved especially important in making a decision about which of our thoughts are important enough on a basis of reward or punishment to be worthy of memory.

Lesions in other parts of the temporal lobes besides the hippocampi, especially of the amygdalas, are also frequently associated with reduced ability to store new memories. This probably results from two factors: (1) the association of the other parts of the temporal lobes with the hippocampi and therefore failure of the usual consolidation process for memories, and (2) the fact that Wernicke's area, which is the major locus of intellectual operations of the brain, is located in the temporal lobe. The reason that lesions affecting Wernicke's area might diminish memory storage is probably that consolidation of memories requires analysis of the memory so that it can be stored in association with other memories of like kind.

Retrograde Amnesia. *Retrograde amnesia* means inability to recall memories from the past — that is, from the long-term memory storage bins — even though the memories are known to be still there. When retrograde amnesia occurs, the degree of amnesia for recent events is likely to be much greater than for events of the distant past. The reason for this difference is probably that the distant memories have been rehearsed so many times that the memory traces are deeply engrained so that elements of these memories are stored in widespread areas of the brain.

In some persons who have hippocampal lesions,

some degree of retrograde amnesia occurs along with anterograde amnesia just discussed, which suggests that these two types of amnesia are at least partially related and that hippocampal lesions can cause both. However, it has also been claimed that damage in some thalamic areas can lead specifically to retrograde amnesia without causing significant anterograde amnesia. A possible explanation of this is that the thalamus might play a role in helping the person "search" the memory storehouses and thus "read out" the memories. That is, the memory process not only requires the storing of memories but also the ability to search and find the memory at a later date. The possible function of the thalamus in this process is discussed in the following chapter.

Lack of Importance of the Hippocampi in Reflexive Learning. It should be noted, however, that persons with either temporal lobe or hippocampal lesions usually do not have difficulty in learning physical skills that do not involve verbalization or symbolic types of intelligence. For instance, these persons can still learn hand and physical skills such as those required in many types of sports. This type of learning is called *reflexive learning;* it depends on physically repeating the required tasks over and over again, rather than on symbolical rehearsing in the mind.

REFERENCES

Avoli, M., et al. (eds.): Neurotransmitters and Cortical Function. New York, Plenum Publishing Corp., 1988.

Baddeley, A.: Working Memory. New York, Oxford University Press, 1987.

Bear, M. F., et al.: A physiological basis for a theory of synapse modification. Science, 237:42, 1987.

Benson, D. F.: Aphasia, Alexia, and Agraphia. New York, Churchill Livingstone, 1979.

Brown, T. H., et al.: Long-term synaptic potentiation. Science, 242:724, 1988.

Byrne, J. H.: Can learning and memory be understood? News Physiol. Sci., 1:182, 1986.

Byrne, J. H.: Cellular analysis of associative learning. Physiol. Rev., 67:329, 1987.

Cotman, C. W., et al.: The role of the NMDA receptor in central nervous system plasticity and pathology. J. NIH Res., 1:65, 1989.

Damasio, A. R., and Geschwind, N.: The neural basis of language. Annu. Rev. Neurosci., 7:127, 1984.

De Valois, R. L., and De Valois, K. K.: Spatial Vision. New York, Oxford University Press, 1988.

De Wied, D.: Neuroendocrine aspects of learning and memory processes. News Physiol. Sci., 4:32, 1989.

DeFelipe, J., and Jones, E. G. (eds.): Cajal on the Cerebral Cortex. New York, Oxford University Press, 1988.

Geschwind, N.: Specializations of the human brain. Sci. Am., 241(3):180, 1979.

Gold, P. E.: Sweet memories. Am. Sci., 75:151, 1987.

Goldman-Rakic, P. S.: Topography of cognition: Parallel distributed networks in primate association cortex. Annu. Rev. Neurosci., 11:137, 1988.

Gregory, R. L. (ed.): The Oxford Companion to the Mind. Oxford University Press, 1987.

Heppenheimer, T. A.: Nerves of silicon. Discover, February 1988, p. 70.

Hixon, T. J., et al. (eds.): Introduction to Communicative Disorders. Englewood Cliffs, N. J., Prentice-Hall, 1980.

Hubel, D. H.: The brain. Sci. Am., 241(3):44, 1979.

Hundert, E. M.: Philosophy, Psychiatry and Neuroscience—Three Approaches to the Mind. New York, Oxford University Press, 1989.

Hyvarinen, J.: Posterior parietal lobe of the primate brain. Physiol. Rev., 62:1060, 1982.

Ito, M.: Long-term depression. Annu. Rev. Neurosci., 12:85, 1989.

John, E. R., et al.: Double-labeled metabolic maps of memory. Science, 233:1167, 1986.

Kandel, E. R.: A Cell-Biological Approach to Learning. Bethesda, Md., Society for Neuroscience, 1977, p. 1137.

Kandel, E. R.: Neuronal plasticity and the modification of behavior. In Brookhart, J. M., and Mountcastle, V. B. (eds.): Handbook of Physiology. Sec. 1, Vol. I. Baltimore, Williams & Wilkins, 1977, p. 1137.

Kolata, G.: Associations or rules in learning language? Science, 237:133, 1987.

Kosslyn, S. M.: Aspects of a cognitive neuroscience of mental imagery. Science, 240:1621, 1988.

Laduron, P. M.: Presynaptic heteroreceptors in regulation of neuronal transmission. Biochem. Pharmacol., 34:467, 1985.

Lieke, E. E., et al.: Optical imaging of cortical activity. Annu. Rev. Physiol., 51:543, 1989.

McCloskey, D. I.: Corollary discharges: Motor commands and perception. In Brooks, V. B. (ed.): Handbook of Physiology. Sec. 1, Bol. II. Bethesda, Md., American Physiological Society, 1981, p. 1415.

McNeil, M. R. (ed.): The Dysarthrias. Physiology, Acoustics, and Perception Management. San Diego, College-Hill Press, 1984.

Mitzdorf, U.: Current source-density method and application in cat cerebral cortex: Investigation of evoked potentials and EEG phenomena. Physiol. Rev., 65:37, 1985.

Peters, A., and Jones, E. G. (eds.): Development and Maturation of Cerebral Cortex. New York, Plenum Publishing Corp., 1988.

Plum, F. (ed.): Language, Communication, and the Brain. New York, Raven Press, 1988.

Quinn, W. G., and Greenspan, R. J.: Learning and courtship in Drosophila: Two stories with mutants. Annu. Rev. Neurosci., 7:67, 1984.

Rakic, P.: Specification of cerebral cortical areas. Science, 241:170, 1988.

Rosenbek, J. C. (ed.): Apraxia of Speech. Physiology, Acoustics, Linguistics, Management. San Diego, College-Hill Press, 1984.

Squire, L. R.: Mechanisms of memory. Science, 232:1612, 1986.

Squire, L. R.: Memory and Brain. New York, Oxford University Press, 1987.

Sutcliffe, J. G.: mRNA in the mammalian central nervous system. Annu. Rev. Neurosci., 11:157, 1988.

Thompson, R. F., et al.: Cellular processes of learning and memory in the mammalian CNS. Annu. Rev. Neurosci., 6:447, 1983.

Trevarthen, C.: Hemispheric specialization. In Darian-Smith, I. (ed.): Handbook of Physiology. Sec. 1, Vol. III. Bethesda, Md., American Physiological Society, 1984, p. 1129.

Truman, J. W.: Cell death in invertebrate nervous systems. Annu. Rev. Neurosci., 7:171, 1984.

Waldrop, M. M.: Soar: A unified theory of cognition? Science, 241:296, 1988.

Walters, E. T., and Byrne, J. H.: Associative conditioning of single sensory neurons suggest a cellular mechanism for learning. Science, 219:405, 1983.

Weiskrantz, L. (ed.): Thought without Language. New York, Oxford University Press, 1988.

Wong, R. K., et al.: Local circuit interactions in synchronization of cortical neurones. J. Exp. Biol., 112:169, 1984.

Woody, C. D., et al. (eds.): Cellular Mechanisms of Conditioning and Behavioral Plasticity. New York, Plenum Publishing Corp., 1988.

Zucker, R. S.: Short-term synaptic plasticity. Annu. Rev. Neurosci., 12:13, 1989.

CHAPTER 58

• *Behavioral and Motivational Mechanisms of the Brain — The Limbic System and the Hypothalamus* •

The control of behavior is a function of the entire nervous system. Even the discrete cord reflexes are an element of behavior, and the wakefulness and sleep cycle discussed in the following chapter is certainly one of the most important of our behavioral patterns. However, in this chapter we deal first with those mechanisms that control the levels of activity in the different parts of the brain. Then we discuss the bases of motivational drives, especially the motivational control of the learning process and the feelings of pleasure or punishment. These functions of the nervous system are performed mainly by the basal regions of the brain, which together are loosely called the *limbic*, meaning "border," *system.*

● THE ACTIVATING-DRIVING SYSTEMS OF THE BRAIN

In the absence of continuous transmission of nerve signals from the brain stem into the cerebrum, the brain becomes useless. In fact, severe compression of the brain stem at the juncture between the mesencephalon and cerebrum, often resulting from a pineal tumor, usually causes the person to go into unremitting coma lasting for the remainder of the person's life.

Nerve signals in the brain stem activate the cerebral part of the brain in two different ways: (1) by directly stimulating the background level of activity in wide areas of the brain and (2) by activating neurohormonal systems that release specific facilitory or inhibitory hormonal substances into selected areas of the brain. These two activating systems always function together and cannot be distinguished entirely from each other; nevertheless, let us discuss each as a separate entity.

CONTROL OF CEREBRAL ACTIVITY BY CONTINUOUS EXCITATORY SIGNALS FROM THE BRAIN STEM

The Reticular Excitatory Area of the Brain Stem

Figure 58–1 illustrates a general system for controlling the level of activity of the brain. The central driving component of this system is an excitatory area called the *bulboreticular facilitory area.* This lies in the reticular substance of the middle and lateral pons and mesencephalon. Actually, we have already discussed this area in Chapter 55, for it is the same brain stem reticular area that transmits facilitory signals downward to the spinal cord to maintain tone in the antigravity muscles and also to control the level of activity of the spinal cord reflexes. In addition to these downward signals, this area sends a profusion of signals in the upward direction as well. Most of these synapse in the thalamus and are distributed from there to all regions of the cerebral cortex, although others go to most of the other subcortical structures besides the thalamus as well.

The signals passing through the thalamus are of two types. One type is rapidly transmitted action potentials that excite the cerebrum for only a few milliseconds. These originate from very large neuronal cell bodies that lie throughout the reticular area. Their nerve endings release the neurotransmitter substance *acetylcholine,* which serves as the excitatory agent, lasting for only a few milliseconds before it is destroyed.

The second type of excitatory signal originates from large numbers of very small neurons spread throughout the reticular excitatory area. Again, most of these pass to the thalamus, but this time through

Figure 58 – 1. The excitatory-activating system of the brain. Also shown is an inhibitory area in the medulla that can inhibit or depress the activating system.

small, very slowly conducting fibers, and synapsing mainly in the intralaminar nuclei of the thalamus and reticular nuclei over the surface of the thalamus. From here, additional very small fibers are distributed everywhere in the cerebral cortex. The excitatory effect caused by this system of fibers can build up progressively for many seconds to a minute or more, which suggests that its signals are especially important for controlling the longer-term background excitability level of the brain.

Excitation of the Brain Stem Excitatory Area by Peripheral Sensory Signals. The level of activity of the brain stem excitatory area, and therefore the level of activity of the entire brain, is determined to a great extent by the sensory signals that enter the excitatory area from the periphery. Pain signals, in particular, increase the activity in this area and therefore strongly excite the brain to attention.

The importance of sensory signals in activating the excitatory area is illustrated by the effect of cutting the brain stem above the point where the bilateral fifth nerves enter the pons. These are the highest nerves that transmit significant numbers of somatosensory signals into the brain. When all of these signals are gone, the level of activity in the excitatory area diminishes abruptly, and the brain proceeds instantly to greatly reduced activity, actually approaching a permanent state of coma. Yet when the brain stem is transected below the fifth nerves, which leaves much input of the sensory signals from the facial and oral regions, the coma is averted.

Increased Activity of the Brain Stem Excitatory Area Caused by Feedback Signals From the Cerebrum. Not only do excitatory signals pass to the cerebrum from the bulboreticular excitatory area of the brain stem, but signals in turn return from the cerebrum back to the bulbar regions. Therefore, any time the cerebral cortex becomes activated by either thinking or motor processes, reverse signals are sent back to the brain stem excitatory areas; this obviously helps to maintain the level of excitation of the cerebral cortex or even to enhance it. Thus, this is a general mechanism of *positive feedback* that allows any beginning activity in the cerebrum to support still more activity, thus leading to an awake mind.

The Thalamus Is a Distribution Center That Controls Activity in Specific Regions of the Cortex. It was pointed out in the previous chapter, and illustrated in Figure 57 – 2, that almost every area of the cerebral cortex connects with its own highly specific area in the thalamus. Therefore, electrical stimulation of a specific point in the thalamus will activate a specific small region of the cortex. Furthermore, signals regularly reverberate back and forth between the thalamus and the cerebral cortex, the thalamus exciting the cortex and the cortex then re-exciting the thalamus by way of return fibers. It has been suggested that the part of the thinking process that helps us to establish long-term memories might result from just such back-and-forth reverberation of signals.

However, can the thalamus also function internally within the brain to call forth specific memories or to activate specific thought processes? The answer to this is not known. Yet, certainly, the thalamus does have the appropriate neuronal circuitry to do this.

An Inhibitory Reticular Area Located in the Lower Brain Stem

Figure 58 – 1 illustrates still another area that is important in controlling brain activity. This is the *reticular inhibitory area* located medially and ventrally in the medulla. In Chapter 55 we saw that this area can inhibit the reticular facilitory area of the upper brain stem and thereby reduce the tonic nerve signals transmitted through the spinal cord to the antigravity muscles. Likewise, this same inhibitory area, when excited, will decrease activity in the superior portions of the brain as well. One of the mechanisms that it uses is to excite serotonergic neurons; these, in turn, secrete the inhibitory neurohormone serotonin at crucial points in the brain; we discuss this in more detail later.

NEUROHORMONAL CONTROL OF BRAIN ACTIVITY

Aside from direct control of brain activity by specific transmission of nerve signals from the lower brain areas to the cortical regions of the brain, still another method is also used to control brain activity.

This is the release of excitatory or inhibitory neurotransmitter hormonal agents into the substance of the brain. These neurohormones often persist for minutes or even hours and thereby provide long periods of control rather than instantaneous activation or inhibition.

Figure 58–2 illustrates three neurohormonal systems that have been mapped in detail in the rat brain, a *norepinephrine system,* a *dopamine system,* and a *serotonin system.* Usually norepinephrine functions as an excitatory hormone, while serotonin is usually inhibitory, and dopamine is excitatory in some areas but inhibitory in others. Therefore, as would be expected, these three different systems have different effects on the levels of excitability in different parts of the brain. The norepinephrine system spreads to virtually every area of the brain, whereas the serotonin and dopamine systems are directed to much more specific brain regions, the dopamine system mainly into the basal ganglial regions and the serotonin system more into the midline structures.

Neurohormonal Systems in the Human Brain. Figure 58–3 illustrates the brain stem areas in the human brain for activating four different neurohormonal systems, the same three discussed above for the rat and one other, the *acetylcholine system.* Some of the specific functions of these are as follows:

1. *The locus ceruleus and the norepinephrine system.* The *locus ceruleus* is a small area located bilaterally and posteriorly at the juncture between the pons and the mesencephalon. Nerve fibers from this area spread throughout the brain, the same as illustrated for the rat in the top frame of Figure 58–2, and they secrete *norepinephrine.* The norepinephrine excites the brain to generalized increased activity. However, it has inhibitory effects in a few areas because of inhibitory receptors at certain neuronal synapses. In the following chapter we see that this system probably plays a very important role in causing a dreaming type of sleep called REM sleep.

2. *The substantia nigra and the dopamine system.* The *substantia nigra* is discussed in Chapter 56 in relation to the basal ganglia. It lies anteriorly in the superior mesencephalon, and its neurons send nerve endings mainly to the caudate nucleus and putamen, where they secrete *dopamine.* Other neurons located in adjacent regions also secrete dopamine, but these send their endings into the ventral areas of the cerebrum, especially to the hypothalamus and the limbic system. The dopamine is believed to act as an inhibitory transmitter in the basal ganglia, but in some of the other areas of the brain it is possibly excitatory. Also, remember from Chapter 56 that destruction of the dopaminergic neurons in the substantia nigra is the basic cause of Parkinson's disease.

3. *The raphe nuclei and the serotonin system.* In the

Figure 58–2. Three neurohormonal systems that have been mapped in the rat brain: a *norepinephrine system,* a *dopamine system,* and a *serotonin system.* (Adapted from Kelly, J. P., after Cooper, Bloom, and Roth, in Kandel and Schwartz: Principles of Neural Science, 2nd Ed. New York, Elsevier, 1985.)

Figure 58–3. Multiple centers in the brain stem, the neurons of which secrete different transmitter substances. These neurons send control signals upward into the diencephalon and cerebrum and downward into the spinal cord.

midline of the lower pons and medulla are several very thin nuclei called the *raphe nuclei.* Many of the neurons in these nuclei secrete *serotonin.* They send many fibers into the diencephalon and fewer fibers to the cerebral cortex; still many others descend to the spinal cord. The cord fibers have the ability to suppress pain, which is discussed in Chapter 48. The serotonin released in the diencephalon and cerebrum almost certainly plays an essential inhibitory role to help cause normal sleep, as we discuss in the following chapter.

4. *The gigantocellular neurons of the reticular excitatory area and the acetylcholine system.* Earlier, we discussed the gigantocellular neurons (the *giant cells*) in the reticular excitatory area of the pons and mesencephalon. The fibers from these large cells divide immediately into two branches, one passing upward to the higher levels of the brain and the other passing downward through the reticulospinal tracts into the spinal cord. The neurohormone secreted at their terminals is *acetylcholine.* In most places, the acetylcholine functions as an excitatory neurotransmitter at specific synapses.

Still other acetylcholine-secreting neurons are present in some regions of the diencephalon; some psychiatric disorders of the brain have been found to be associated with decreased function or even destruction of some of these neurons.

Other Neurotransmitters and Neurohormonal Substances Secreted in the Brain. Without describing their function, the following is a list of still other neurohormonal substances that, among others, function either at synapses or by release into the fluids of the brain: enkephalins, gamma-aminobutyric acid (GABA), glutamate, vasopressin, adrenocorticotropic hormone, epinephrine, endorphins, angiotensin II, neurotensin.

Thus, there are multiple neurohormonal systems in the brain, the activation of each of which plays its own role in controlling a different quality of brain function.

● THE LIMBIC SYSTEM

The word "limbic" means "border." Originally, the term "limbic" was used to describe the border structures around the basal regions of the cerebrum, but as we have learned more about the functions of the limbic system, the term *"limbic system"* has been expanded to mean the entire neuronal circuitry that controls emotional behavior and motivational drives.

A major part of the limbic system is the *hypothalamus,* with its related structures. In addition to their roles in behavioral control, these areas also control many internal conditions of the body as well, such as body temperature, osmolality of the body fluids, the drive to eat and drink and control body weight, and so forth. These internal functions are collectively called *vegetative functions* of the brain, and their control is obviously closely related to behavior.

● FUNCTIONAL ANATOMY OF THE LIMBIC SYSTEM; ITS RELATION ALSO TO THE HYPOTHALAMUS

Figure 58–4 illustrates the anatomical structures of the limbic system, showing these to be an interconnected complex of basal brain elements. Located in the midst of all these is the *hypothalamus,* which is considered by some anatomists to be a structure separate from the remainder of the limbic system but which, from a physiological point of view, is one of the central elements of the system. Figure 58–5 illustrates schematically this key position of the hypothalamus in the limbic system and shows that surrounding it are the other subcortical structures of the limbic system, including the *septum,* the *paraolfactory area,* the *epithalamus,* the *anterior nucleus of the thalamus, portions of the basal ganglia,* the *hippocampus,* and the *amygdala.*

Surrounding the subcortical limbic areas is the *limbic cortex,* composed of a ring of cerebral cortex (1) beginning in the *orbitofrontal area* on the ventral surface of the frontal lobes, (2) extending upward in the *subcallosal gyrus* beneath the anterior limb of the corpus callosum, (3) over the top of the corpus callosum onto the medial aspect of the cerebral hemisphere in the *cingulate gyrus,* and finally (4) passing behind the corpus callosum and downward onto the ventromedial surface of the temporal lobe to the *parahippocampal gyrus* and *uncus.* Thus, on the medial and ventral surfaces of each cerebral hemisphere is a ring mostly of *paleocortex* that surrounds a group of deep structures intimately associated with overall behavior and with emotions. In turn, this ring of limbic cortex functions as a two-way communication and association linkage between the *neocortex* and the lower limbic structures.

It is also important to recognize that many of the behavioral functions elicited from the hypothalamus and other limbic structures are mediated through the reticular nuclei and their associated nuclei in the brain stem. It is pointed out in Chapter 55 and earlier in this chapter that stimulation of the excitatory portion of the reticular formation can cause high degrees of somatic and cerebral cortical excitability; in Chapter 60 we see that most of the hypothalamic signals for control of the autonomic nervous system also are transmitted through nuclei located in the brain stem.

A very important route of communication between the limbic system and the brain stem is the *medial forebrain bundle* that extends from the septal and orbitofrontal cortical regions downward through the middle of the hypothalamus to the brain stem reticular formation. This bundle carries fibers in both directions, forming a trunk line communication system. A second route of communication is through short pathways among the reticular formation of the brain stem, the thalamus, the hypothalamus, and most of the other contiguous areas of the basal brain.

● THE HYPOTHALAMUS, A MAJOR OUTPUT PATHWAY OF THE LIMBIC SYSTEM

The hypothalamus has communicating pathways with all levels of the limbic system. In turn, it and its closely allied structures send output signals in three directions: (1) downward to the brain stem, mainly into the reticular areas of the mesencephalon, pons, and medulla; (2) upward toward many higher areas of

Figure 58–4. Anatomy of the limbic system illustrated by the shaded areas of the figure. (From Warwick and Williams: Gray's Anatomy. 35th Br. Ed. London, Longman Group, Ltd., 1973.)

the diencephalon and cerebrum, especially to the anterior thalamus and the limbic cortex; and (3) into the infundibulum to control most of the secretory functions of both the posterior and anterior pituitary glands.

Thus, the hypothalamus, which represents less than 1 per cent of the brain mass, nevertheless is one of the most important of the motor output pathways of the limbic system. It controls most of the vegetative and endocrine functions of the body as well as many aspects of emotional behavior. Let us discuss first the vegetative and endocrine control functions and then return to the behavioral functions of the hypothalamus to see how all these operate together.

VEGETATIVE AND ENDOCRINE CONTROL FUNCTIONS OF THE HYPOTHALAMUS

The different hypothalamic mechanisms for controlling the vegetative and endocrine functions of the body are discussed in many different chapters throughout this text. For instance, the role of the hypothalamus in arterial pressure regulation is discussed in Chapter 18, thirst and water conservation in Chapter 29, temperature regulation in Chapter 73, and endocrine control in Chapter 75. However, to illustrate the organization of the hypothalamus as a functional unit, let us summarize the more important of its vegetative and endocrine functions here as well.

Figures 58–6 and 58–7 show enlarged coronal and sagittal views of the hypothalamus, which represents only a small area in Figure 58–4. Please take a few minutes to study these diagrams, especially to read in Figure 58–6 the multiple activities that are excited or inhibited when respective hypothalamic nuclei are stimulated. In addition to those centers that are illustrated in Figure 58–6, a large *lateral hypothalamic* area overlies the illustrated areas on each side of the

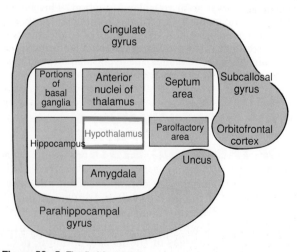

Figure 58–5. The limbic system.

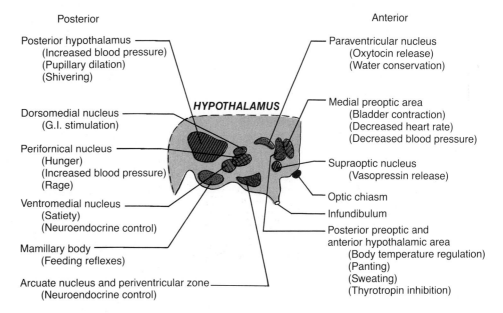

Figure 58-6. Control centers of the hypothalamus.

Posterior

Posterior hypothalamus
(Increased blood pressure)
(Pupillary dilation)
(Shivering)

HYPOTHALAMUS

Dorsomedial nucleus
(G.I. stimulation)

Perifornical nucleus
(Hunger)
(Increased blood pressure)
(Rage)

Ventromedial nucleus
(Satiety)
(Neuroendocrine control)

Mamillary body
(Feeding reflexes)

Arcuate nucleus and periventricular zone
(Neuroendocrine control)

Anterior

Paraventricular nucleus
(Oxytocin release)
(Water conservation)

Medial preoptic area
(Bladder contraction)
(Decreased heart rate)
(Decreased blood pressure)

Supraoptic nucleus
(Vasopressin release)

Optic chiasm

Infundibulum

Posterior preoptic and
anterior hypothalamic area
(Body temperature regulation)
(Panting)
(Sweating)
(Thyrotropin inhibition)

Lateral hypothalamic area (not shown)
(Thirst and hunger)

hypothalamus. The lateral areas are especially important in controlling thirst, hunger, and many of the emotional drives.

A word of caution must be issued for studying these diagrams, however, for the areas that cause specific activities are not nearly so accurately localized as suggested in the figure. Also, it is not known whether the effects noted in the figure result from stimulation of specific control nuclei or whether they result merely from activation of fiber tracts leading from control nuclei located elsewhere. With this caution in mind, we can give the following general description of the vegetative and control functions of the hypothalamus.

Cardiovascular Regulation. Stimulation of different areas throughout the hypothalamus can cause every known type of neurogenic effect on the cardiovascular system, including increased arterial pressure, decreased arterial pressure, increased heart rate, and decreased heart rate. In general, stimulation in the *posterior* and *lateral hypothalamus* increases the arterial pressure and heart rate, whereas stimulation in the *preoptic area* often has opposite effects, causing a decrease in both heart rate and arterial pressure. These effects are transmitted mainly through the cardiovascular control centers in the reticular regions of the medulla and pons.

Regulation of Body Temperature. The anterior portion of the hypothalamus, especially the *preoptic area,* is concerned with regulation of body temperature. An increase in the temperature of the blood flowing through this area increases the activity of temperature-sensitive neurons, whereas a decrease in temperature decreases their activity. In turn, these neurons control the mechanisms for increasing or decreasing body temperature, as discussed in Chapter 73.

Regulation of Body Water. The hypothalamus regulates body water in two separate ways: (1) by creating the sensation of thirst, which makes an animal drink water, and (2) by controlling the excretion of water into the urine. An area called the *thirst center* is located in the lateral hypothalamus. When the electrolytes inside the neurons either of this center or allied areas of the hypothalamus become too concentrated, the animal develops an intense desire to drink water; it will search out the nearest source of water and drink enough to return the electrolyte concentration of the thirst center neurons to normal.

Control of renal excretion of water is vested mainly in the *supraoptic* nucleus. When the body fluids become too concentrated, the neurons of this area become stimulated. The nerve fibers from these neurons project downward through the infundibulum into the posterior pituitary gland, where they secrete a hormone called *antidiuretic hormone* (also called *vasopressin*). This hormone is then absorbed into the blood and acts on the collecting ducts of the kidneys to cause massive reabsorption of water, thereby decreasing the loss of water into the urine. These functions are presented in Chapter 28.

Regulation of Uterine Contractility and of Milk Ejection by the Breasts. Stimulation of the *paraventricular nucleus* causes its neuronal cells to secrete the hormone

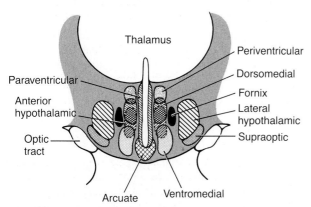

Figure 58-7. Coronal view of the hypothalamus, showing the mediolateral positions of the respective hypothalamic nuclei.

Thalamus

Periventricular

Dorsomedial

Fornix

Lateral
hypothalamic

Supraoptic

Paraventricular

Anterior
hypothalamic

Optic
tract

Arcuate   Ventromedial

oxytocin. This in turn causes increased contractility of the uterus and also contraction of the myoepithelial cells that surround the alveoli of the breasts, which then causes the alveoli to empty the milk through the nipples. At the end of pregnancy, especially large quantities of oxytocin are secreted, and this secretion helps to promote labor contractions that expel the baby. Also, when a baby suckles the mother's breast, a reflex signal from the nipple to the hypothalamus causes oxytocin release, and the oxytocin then performs the necessary function of expelling the milk through the nipples so that the baby can nourish itself. These functions are discussed in Chapter 82.

Gastrointestinal and Feeding Regulation. Stimulation of several areas of the hypothalamus causes an animal to experience extreme hunger, a voracious appetite, and an intense desire to search for food. The area most associated with hunger is the *lateral hypothalamic area.* On the other hand, damage to this causes the animal to lose desire for food, sometimes causing lethal starvation.

A center that opposes the desire for food, called the *satiety center,* is located in the *ventromedial nucleus.* When this center is stimulated, an animal that is eating food suddenly stops eating and shows complete indifference to food. On the other hand, if this area is destroyed bilaterally, the animal cannot be satiated; instead, its hypothalamic hunger centers become overactive, so that it has a voracious appetite, resulting in tremendous obesity.

Another area closely associated with the hypothalamus that enters into the overall control of gastrointestinal activity is the *mamillary bodies;* these control the patterns of many feeding reflexes, such as licking the lips and swallowing.

Hypothalamic Control of the Anterior Pituitary Gland

Stimulation of certain areas of the hypothalamus also causes the *anterior* pituitary gland to secrete its hormones. This subject is discussed in detail in Chapter 74 in relation to the neural control of the endocrine glands. Briefly, the basic mechanisms are the following:

The anterior pituitary gland receives its blood supply mainly from venous blood that flows into the anterior pituitary sinuses after having passed first through the lower part of the hypothalamus. As the blood courses through the hypothalamus before reaching the anterior pituitary, *releasing* and *inhibitory hormones* are secreted into the blood by various hypothalamic nuclei. These hormones are then transported in the blood to the anterior pituitary, where they act on the glandular cells to control the release of the anterior pituitary hormones.

The cell bodies of the neurons that secrete these releasing hormones and inhibitory hormones are located mainly in the medial basal nuclei of the hypothalamus, especially in the *paraventricular zone,* the *arcuate nucleus,* and part of the *ventromedial nucleus.* However, the axons from these nuclei then project to the *median eminence,* which is an enlarged area of the infundibulum where it arises from the inferior border of the hypothalamus. It is here that the nerve terminals actually secrete their releasing and inhibitory hormones. These hormones are then absorbed into the blood capillaries in the median eminence and carried in the venous blood down along the infundibulum to the anterior pituitary gland.

Summary. A number of areas of the hypothalamus control specific vegetative functions. However, these areas are still poorly delimited, so much so that the above specification of different areas for different hypothalamic functions is still tentative.

BEHAVIORAL FUNCTIONS OF THE HYPOTHALAMUS AND ASSOCIATED LIMBIC STRUCTURES

Aside from the vegetative and endocrine functions of the hypothalamus, stimulation of or lesions in the hypothalamus often have profound effects on the emotional behavior of animals or human beings.

In animals, some of the behavioral effects of stimulation are the following:

1. Stimulation in the *lateral hypothalamus* not only causes thirst and eating as discussed above but also increases the general level of activity of the animal, sometimes leading to overt rage and fighting, as is discussed subsequently.

2. Stimulation in the *ventromedial nucleus* and surrounding areas mainly causes effects opposite to those caused by lateral hypothalamic stimulation — that is, a sense of *satiety, decreased eating,* and *tranquility.*

3. Stimulation of a *thin zone of the periventricular nucleus,* located immediately adjacent to the third ventricle (or also stimulation of the central gray area of the mesencephalon that is continuous with this portion of the hypothalamus) usually leads to *fear* and *punishment reactions.*

4. *Sexual drive* can be stimulated from several areas of the hypothalamus, especially the most anterior and most posterior portions of the hypothalamus.

Lesions in the hypothalamus, in general, cause the opposite effects. For instance:

1. Bilateral lesions in the lateral hypothalamus will decrease drinking and eating almost to zero, often leading to lethal starvation. These lesions cause extreme *passivity* of the animal as well, with loss of most of its overt drives.

2. Bilateral lesions of the ventromedial areas of the hypothalamus cause effects that are mainly opposite to those caused by lesions of the lateral hypothalamus: excessive drinking and eating as well as hyperactivity and often continuous savagery along with frequent bouts of extreme rage on the slightest provocation.

Stimulation or lesions in other regions of the limbic system, especially the amygdala, the septal area, and areas in the mesencephalon, often cause effects similar to those elicited from the hypothalamus. We discuss some of these in more detail later.

THE REWARD AND PUNISHMENT FUNCTION OF THE LIMBIC SYSTEM

From the preceding discussion, it is already clear that several limbic structures, including the hypothalamus, are particularly concerned with the *affective* nature of sensory sensations — that is, whether the sensations are *pleasant* or *unpleasant.* These affective qualities are also called *reward* or *punishment,* or *satisfaction* or *aversion.* Electrical stimulation of certain regions pleases or satisfies the animal, whereas electrical stimulation of other regions causes terror, pain, fear, defense, escape reactions, and all the other elements of punishment. Obviously, these two oppositely responding systems greatly affect the behavior of the animal.

Reward Centers

Figure 58–8 illustrates a technique that has been used for localizing specific reward and punishment areas of the brain. In this figure a lever is placed at the side of the cage and is arranged so that depressing the lever makes electrical contact with a stimulator. Electrodes are placed successively at different areas in the brain so that the animal can stimulate the area by pressing the lever. If stimulating the particular area gives the animal a sense of reward, then it will press the lever again and again, sometimes as much as thousands of times per hour. Furthermore, when offered the choice of eating some delectable food as opposed to the opportunity to stimulate the reward center, it often chooses the electrical stimulation.

By using this procedure, the major reward centers have been found to be located *along the course of the medial forebrain bundle,* especially in the *lateral* and *ventromedial nucleus of the hypothalamus.* It is strange that the lateral nucleus should be included among the reward areas — indeed, they are one of the most potent of all — because even stronger stimuli in this area can cause rage. But this is true in many areas, with weaker stimuli giving a sense of reward and stronger ones a sense of punishment.

Less potent reward centers, which are perhaps secondary to the major ones in the hypothalamus, are found in the septum, the amygdala, certain areas of

Figure 58–8. Technique for localizing reward and punishment centers in the brain of a monkey.

the thalamus and basal ganglia, and finally extending downward into the basal tegmentum of the mesencephalon.

Punishment Centers

The apparatus illustrated in Figure 58–8 can also be connected so that pressing the lever turns off, rather than turning on, an electrical stimulus. In this case, the animal will not turn the stimulus off when the electrode is in one of the reward areas; but when it is in certain other areas, it immediately learns to turn it off. Stimulation in these areas causes the animal to show all the signs of displeasure, fear, terror, and punishment. Furthermore, prolonged stimulation for 24 hours or more can cause the animal to become severely sick and actually lead to death.

By means of this technique, the most potent areas for punishment and escape tendencies have been found in the *central gray area surrounding the aqueduct of Sylvius in the mesencephalon* and extending upward into the *periventricular zones of the hypothalamus and thalamus.* Also, less potent punishment areas are found in the *amygdala* and the *hippocampus.*

It is particularly interesting that stimulation in the punishment centers can frequently inhibit the reward and pleasure centers completely, illustrating that punishment and fear can take precedence over pleasure and reward.

Importance of Reward and Punishment in Behavior

Almost everything that we do is related in some way to reward and punishment. If we are doing something that is rewarding, we continue to do it; if it is punishing, we cease to do it. Therefore, the reward and punishment centers undoubtedly constitute one of the most important of all the controllers of our bodily activities, our drives, our aversions, our motivations.

Effect of Tranquilizers on the Reward and Punishment Centers. Administration of a tranquilizer, such as chlorpromazine, inhibits both the reward and punishment centers, thereby greatly decreasing the affective reactivity of the animal. Therefore, it is presumed that tranquilizers function in psychotic states by suppressing many of the important behavioral areas of the hypothalamus and its associated regions of the brain, a subject that we discuss more fully later.

Importance of Reward and Punishment in Learning and Memory — Habituation or Reinforcement

Animal experiments have shown that a sensory experience causing neither reward nor punishment is remembered hardly at all. Electrical recordings show that new and novel sensory stimuli always excite the cerebral cortex. But repetition of the stimulus over and over leads to almost complete extinction of the cortical response if the sensory experience does not elicit either a sense of reward or punishment. Thus, the animal becomes *habituated* to the sensory stimulus and thereafter ignores it.

However, if the stimulus causes either reward or punishment rather than indifference, the cortical response becomes progressively more and more intense with repeated stimulation instead of fading away, and the response is said to be *reinforced*. Thus, an animal builds up strong memory traces for sensations that are either rewarding or punishing but, on the other hand, develops complete habituation to indifferent sensory stimuli. Therefore, it is evident that the reward and punishment centers of the limbic system have much to do with selecting the information that we learn.

RAGE

An emotional pattern that involves the hypothalamus and many other limbic structures, and has also been well characterized, is the *rage pattern*. This can be described as follows:

Strong stimulation of the punishment centers of the brain, especially in the *periventricular zone of the hypothalamus* or in the *lateral hypothalamus,* causes the animal to (1) develop a defense posture, (2) extend its claws, (3) lift its tail, (4) hiss, (5) spit, (6) growl, and (7) develop piloerection, wide-open eyes, and dilated pupils. Furthermore, even the slightest provocation causes an immediate savage attack. This is approximately the behavior that one would expect from an animal being severely punished, and it is a pattern of behavior that is called *rage*.

Stimulation of the more rostral areas of the punishment areas — in the midline preoptic areas — causes mainly fear and anxiety, associated with a tendency for the animal to run away.

In the normal animal the rage phenomenon is held in check mainly by counterbalancing activity of the ventromedial nucleus of the hypothalamus. In addition, the hippocampus, the amygdala, and the anterior portions of the limbic cortex, especially the limbic cortex of the anterior cingulate gyrus and the subcallosal gyrus, help suppress the rage phenomenon. Conversely, if these portions of the limbic system are damaged or destroyed, the animal (also the human being) becomes far more susceptible to bouts of rage.

Placidity and Tameness. Exactly the opposite emotional behavior patterns occur when the reward centers are stimulated: placidity and tameness.

● SPECIFIC FUNCTIONS OF OTHER PARTS OF THE LIMBIC SYSTEM

FUNCTIONS OF THE AMYGDALA

The amygdala is a complex of nuclei located immediately beneath the cortex of the medial anterior pole of each temporal lobe. It has abundant bidirectional connections with the hypothalamus.

In lower animals, the amygdala is concerned to a great extent with association of olfactory stimuli with stimuli from other parts of the brain. Indeed, it is pointed out in Chapter 53 that one of the major divisions of the olfactory tract leads directly to a portion of the amygdala called the *corticomedial nuclei* that lies immediately beneath the cortex in the pyriform area of the temporal lobe. However, in the human being, another portion of the amygdala, the *basolateral nuclei,* has become much more highly developed than this olfactory portion and plays exceedingly important roles in many behavioral activities not generally associated with olfactory stimuli.

The amygdala receives neuronal signals from all portions of the limbic cortex as well as from the neocortex of the temporal, parietal, and occipital lobes, especially from the auditory and visual association areas. Because of these multiple connections, the amygdala has been called the "window" through which the limbic system sees the place of the person in the world. In turn, the amygdala transmits signals (1) back into these same cortical areas, (2) into the hippocampus, (3) into the septum, (4) into the thalamus, and (5) especially into the hypothalamus.

Effects of Stimulating the Amygdala. In general, stimulation in the amygdala can cause almost all the same effects as those elicited by stimulation of the hypothalamus, plus still other effects. The effects that are mediated through the hypothalamus include (1) increases or decreases in arterial pressure, (2) increases or decreases in heart rate, (3) increases or decreases in gastrointestinal motility and secretion, (4) defecation and micturition, (5) pupillary dilatation or, rarely, constriction, (6) piloerection, (7) secretion of the various anterior pituitary hormones, especially the gonadotropins and adrenocorticotropic hormone.

Aside from these effects mediated through the hypothalamus, amygdala stimulation can also cause different types of involuntary movement. These include (1) tonic movements, such as raising the head or bending the body, (2) circling movements, (3) occasionally clonic, rhythmic movements, and (4) different types of movements associated with olfaction and eating, such as licking, chewing, and swallowing.

In addition, stimulation of certain amygdaloid nuclei can, rarely, cause a pattern of rage, escape, punishment, and fear similar to the rage pattern elicited from the hypothalamus as described earlier. And stimulation of other nuclei can give reactions of reward and pleasure.

Finally, excitation of still other portions of the amygdala can cause sexual activities that include erection, copulatory movements, ejaculation, ovulation, uterine activity, and premature labor.

Effects of Bilateral Ablation of the Amygdala — The Klüver-Bucy Syndrome. When the anterior portions of both temporal lobes are destroyed in a monkey, this not only removes the temporal cortex but also the amygdalas that lie deep in these parts of the temporal lobes. This causes a combination of changes in behavior called the Klüver-Bucy syndrome, which includes (1) excessive tendency to examine objects orally, (2) loss of fear, (3) decreased aggressiveness, (4) tameness, (5) changes in dietary habits, even to the extent that a herbivorous animal frequently becomes carnivorous, (6) sometimes psychic blindness, and (7) often excessive sex drive. The characteristic picture is of an animal that is not afraid of anything, has extreme curiosity about everything, forgets very rapidly, has a tendency to place everything in its mouth and sometimes even tries to eat solid objects, and, finally, often has a sex drive so strong that it attempts to copulate with immature animals, animals of the wrong sex, or animals of a different species.

Though similar lesions in human beings are rare, afflicted persons respond in a manner not too different from that of the monkey.

Overall Function of the Amygdala. The amygdala seems to be a behavioral awareness area that operates at a semiconscious level. It also seems to project into the limbic system one's present status in relation both to surroundings and thoughts. On the basis of this information, the amygdala is believed to help pattern the person's behavioral response so that it is appropriate for each occasion.

FUNCTIONS OF THE HIPPOCAMPUS

The hippocampus is the elongated, medial portion of the temporal cortex that folds upward and inward to form the ventral surface of the inferior horn of the lateral ventricle. One end of the hippocampus abuts the amygdaloid nuclei, and it also fuses along one of its borders with the parahippocampal gyrus, which is the cortex of the ventromedial surface of the temporal lobe.

The hippocampus has numerous but mainly indirect connections with many portions of the cerebral cortex as well as with the basic structures of the limbic system—the amygdala, the hypothalamus, the septum, and the mamillary bodies. Almost any type of sensory experience causes activation of at least some part of the hippocampus, and the hippocampus in turn distributes many outgoing signals to the anterior thalamus, the hypothalamus, and other parts of the limbic system, especially through the *fornix,* its major output pathway. Thus, the hippocampus, like the amygdala, is an additional channel through which incoming sensory signals can lead to appropriate behavioral reactions, but perhaps for different purposes, as can be seen later.

As in other limbic structures, stimulation of different areas in the hippocampus can cause almost any one of different behavioral patterns, such as rage, passivity, excess sex drive, and so forth.

Another feature of the hippocampus is that very weak electrical stimuli can cause local epileptic seizures that persist for many seconds after the stimulation is over, suggesting that the hippocampus can perhaps give off prolonged output signals even under normal functioning conditions. During hippocampal seizures, the person experiences various psychomotor effects, including olfactory, visual, auditory, tactile, and other types of hallucinations that cannot be suppressed even though the person has not lost consciousness and knows these hallucinations to be unreal. Probably one of the reasons for this hyperexcitability of the hippocampus is that it is composed of a different type of cortex from that elsewhere in the cerebrum, having only three nerve cell layers instead of the six layers found elsewhere.

Role of the Hippocampus in Learning

Effect of Bilateral Removal of the Hippocampi—Inability to Learn. The hippocampi have been surgically removed bilaterally in a few human beings for the treatment of epilepsy. These persons can recall most previously learned memories satisfactorily. However, they can learn essentially no new information that is based on verbal symbolism. In fact, they cannot even learn the names of persons with whom they come in contact every day. Yet they can remember for a moment or so what transpires during the course of their activities. Thus, they are capable of the type of very short-term memory called "immediate memory" even though their ability to establish secondary memories lasting longer than a few minutes is either completely or almost completely abolished, which is the phenomenon called *anterograde amnesia* discussed in the previous chapter.

Destruction of the hippocampi also causes some deficit in previously learned memories (retrograde amnesia), a little more so for memories up to the past year or so than for memories of the distant past.

Theoretical Function of the Hippocampus in Learning. The hippocampus originated as part of the olfactory cortex. In the very lowest animals it plays essential roles in determining whether the animal will eat a particular food, whether the smell of a particular object suggests danger, and whether the odor is sexually inviting and in making other decisions that are of life-or-death importance. Thus, very early in the development of the brain, the hippocampus presumably became a critical decision-making neuronal mechanism, determining the importance and type of importance of the incoming sensory signals. Once this critical decision-making capability had been established, presumably the remainder of the brain began to call on it for the same decision-making. If the hippocampus says that a neuronal signal is important, it is likely to be committed to memory.

Earlier in this chapter (and also in the previous chapter), it was pointed out that reward and punishment play a major role in determining the importance of information and especially whether or not the information will be stored in memory. A person rapidly becomes habituated to indifferent stimuli but learns assiduously any sensory experience that causes either pleasure or punishment. Yet, what is the mechanism by which this occurs? It has been suggested that the hippocampus provides the drive that causes translation of immediate memory into secondary memory—that is, it transmits some type of signal or signals that seem to make the mind rehearse over and over the new information until permanent storage takes place.

Whatever the mechanism, without the hippocampi *consolidation* of long-term memories of verbal or symbolic type does not take place.

FUNCTION OF THE LIMBIC CORTEX

Probably the most poorly understood portion of the entire limbic system is the ring of cerebral cortex called the *limbic cortex* that surrounds the subcortical limbic structures. This cortex functions as a transitional zone through which signals are transmitted from the remainder of the cortex into the limbic system. Therefore, it is presumed that the limbic cortex functions as a cerebral *association area for control of behavior.*

Stimulation of the different regions of the limbic cortex has failed to give any real idea of their functions. However, as is true of so many other portions of the limbic system, essentially all the behavioral patterns that have already been described can also be elicited by stimulation in different portions of the limbic cortex. Likewise, ablation of a few limbic cortical areas can cause persistent changes in an animal's behavior, as follows:

Ablation of the Temporal Cortex. When the anterior temporal cortex is ablated bilaterally, the amygdala is almost invariably damaged as well. This was discussed earlier, and it was pointed out that the Klüver-Bucy syndrome occurs. The animal especially develops consummatory behavior, investigates any and all objects, has intense sex drives toward inappropriate animals or even inanimate objects, and loses all fear—thus develops tameness as well.

Ablation of the Posterior Orbital Frontal Cortex. Bilateral removal of the posterior portion of the orbital frontal cortex often causes an animal to develop insomnia and an intense degree of motor restlessness, becoming unable to sit still but moving about continually.

Ablation of the Anterior Cingulate Gyri and Subcallosal Gyri. The anterior cingulate gyri and the subcallosal gyri are the portions of the limbic cortex that communicate between the prefrontal cerebral cortex and the subcortical limbic structures. Destruction of these gyri bilaterally releases the rage centers of the septum and hypothalamus from any prefrontal inhibitory influence. Therefore, the animal can become vicious and much more subject to fits of rage than normally.

Summary. Until further information is available, it is perhaps best to state that the cortical regions of the limbic system occupy intermediate associative positions between the functions of the remainder of the cerebral cortex and the functions of the subcortical limbic structures for control of behavioral patterns. Thus, in the anterior temporal cortex one especially finds gustatory and olfactory associations. In the parahippocampal gyri there is a tendency for complex auditory associations and also complex thought associations derived from Wernicke's area of the posterior temporal lobe. In the middle and posterior cingulate cortex, there is reason to believe that sensorimotor associations occur.

REFERENCES

Aoki, C., and Siekevitz, P.: Plasticity in brain development. Sci. Am., December, 1988, p. 56.

Ashton, H.: Brain Systems, Disorders and Psychotropic Drugs. New York, Oxford University Press, 1987.

Avoli, M., et al. (eds.): Neurotransmitters and Cortical Function. New York, Plenum Publishing Corp., 1988.

Barnes, D. M.: The biological tangle of drug addiction. Science, 241:415, 1988.

Barnes, D. M.: Neural models yield data on learning. Science, 236:1628, 1987.

Berger, P. A., et al.: Behavioral pharmacology of the endorphins. Annu. Rev. Med., 33:397, 1982.

Borbely, A. A., and Tobler, I.: Endogenous sleep-promoting substances and sleep regulation. Physiol. Rev., 69:605, 1989.

Burchfield, S. R. (ed.): Stress, Physiological and Psychological Interactions. Washington, D.C., Hemisphere Publishing Corp., 1985.

Buzsaki, G.: Feed-forward inhibition in the hippocampal formation. Prog. Neurobiol., 22:131, 1984.

Byrne, J. H.: Cellular analysis of associative learning. Physiol. Rev., 67:329, 1987.

Chiba, A., et al.: Synaptic rearrangement during postembryonic development in the cricket. Science, 240:901, 1988.

Chrousos, G. P., et al. (eds.): Mechanisms of Physical and Emotional Stress. New York, Plenum Publishing Corp., 1988.

Clynes, M., and Panksepp, J. (eds.): Emotions and Psychopathology. New York, Plenum Publishing Corp., 1988.

Cohen, D. H., and Randall, D. C.: Classical conditioning of cardiovascular responses. Annu. Rev. Physiol., 46:187, 1984.

De Wied, D., and Jolles, J.: Neuropeptides derived from pro-opiocortin: Behavioral, physiological, and neurochemical effects. Physiol. Rev., 62:976, 1982.

Doane, B. K., and Livingston, K. E. (eds.): The Limbic System. New York, Raven Press, 1986.

Doris, P. A.: Vasopressin and central integrative processes. Neuroendocrinology, 38:75, 1984.

Engel, B. T., and Schneiderman, N.: Operant conditioning and the modulation of cardiovascular function. Annu. Rev. Physiol., 46:199, 1984.

Engel, J., et al. (eds.): Brain Reward Systems and Abuse. New York, Raven Press, 1987.

Fink, M.: Convulsive and drug therapies of depression. Annu. Rev. Med., 32:405, 1981.

Foote, S. L., et al.: Nucleus locus ceruleus: New evidence of anatomical and physiological specificity. Physiol. Rev., 63:844, 1983.

Ganong, W. F.: The brain renin-angiotensin system. Annu. Rev. Physiol., 46:17, 1984.

Givens, J. R.: The Hypothalamus in Health and Disease. Chicago, Year Book Medical Publishers, 1984.

Goldstein, G. (ed.): Neuropsychology Review. New York, Plenum Publishing Corp., 1989.

Iversen, L. L.: Nonopioid neuropeptides in mammalian CNS. Annu. Rev. Pharmacol. Toxicol., 23:1, 1983.

Jones, E. G., and Peters, A. (eds.): Further Aspects of Cortical Function, Including Hippocampus. New York, Plenum Publishing Corp., 1987.

Kandel, E. R.: Molecular Neurobiology in Neurology and Psychiatry. New York, Raven Press, 1987.

Klimov, P. K.: Behavior of the organs of the digestive system. Neurosci. Behav. Physiol., 14:333, 1984.

Malick, J. B. (ed.): Anxiolytics, Neurochemical, Behavioral, and Clinical Perspectives. New York, Raven Press, 1983.

Mariani, J., and Delhaye-Bouchaud, N.: Elimination of functional synapses during development of the nervous system. News Physiol. Sci., 2:93, 1987.

Miller, J. A.: En route to thought: Recognition and recall. Sci. News, 128:373, 1985.

Mishkin, M., and Appenzeller, T.: The Anatomy of Memory. Sci. Am., Special Report, 1987, p. 2.

Nerozzi, D., et al. (eds.): Hypothalamic Dysfunction in Neuropsychiatric Disorders. New York, Raven Press, 1987.

Rescorla, R. A.: Behavioral studies of Pavlovian conditioning. Annu. Rev. Neurosci., 11:329, 1988.

Russell, R. W.: Cholinergic system in behavior: The search for mechanisms of action. Annu. Rev. Pharmacol. Toxicol., 22:435, 1982.

Schatzberg, A. F., and Nemeroff, C. B. (eds.): The Hypothalamic-Pituitary-Adrenal Axis. New York, Raven Press, 1988.

Shepherd, G. M.: Neurobiology, 2nd Ed. New York, Oxford University Press, 1987.

Siddle, D.: Orienting and Habituation. Perspectives in Human Research. New York, John Wiley & Sons, 1982.

Skodol, A. E., and Spitzer, R. L.: The development of reliable diagnostic criteria in psychiatry. Annu. Rev. Med., 33:317, 1982.

Smith, O. A., and DeVito, J. L.: Central neural integration for the control of autonomic responses associated with emotion. Annu. Rev. Neurosci., 7:43, 1984.

Stephenson, R. B.: Modification of reflex regulation of blood pressure by behavior. Annu. Rev. Physiol., 46:133, 1984.

Steriade, M., and Llinas, R. R.: The functional states of the thalamus and the associated neuronal interplay. Physiol. Rev., 68:649, 1988.

Swanson, L. W., and Sawchenko, P. E.: Hypothalamic integration: Organization of the paraventricular and supraoptic nuclei. Annu. Rev. Neurosci., 6:269, 1983.

Tucek, S.: Regulation of acetylcholine synthesis in the brain. J. Neurochem., 44:11, 1985.

Tyrer, P. (ed.): Psychopharmacology of Anxiety. New York, Oxford University Press, 1989.

Usdin, E.: Stress. The Role of Catecholamines and Other Neurotransmitters. New York, Gordon Press Publishers, 1984.

Verrier, R. L., and Lown, B.: Behavioral stress and cardiac arrhythmias. Annu. Rev. Physiol., 46:155, 1984.

Wise, S. P., and Desimone, R.: Behavioral neurophysiology: Insights into seeing and grasping. Science, 242:736, 1988.

Wolman, B. B.: Psychosomatic Disorders. New York, Plenum Publishing Corp., 1988.

Woody, C. D., et al. (eds.): Cellular Mechanisms of Conditioning and Behavioral Plasticity. New York, Plenum Publishing Corp., 1988.

CHAPTER 59

• States of Brain Activity — Sleep; Brain Waves; Epilepsy; Psychoses •

All of us are aware of the many different states of brain activity, including sleep, wakefulness, extreme excitement, and even different levels of mood such as exhilaration, depression, and fear. All these states result from different activating or inhibiting forces generated usually within the brain itself. In the last chapter, we began a partial discussion of this subject when we described different systems that are capable of activating either large or isolated portions of the brain. In this chapter, we present brief surveys of what is known about other states of brain activity, beginning with sleep.

• SLEEP

Sleep is defined as unconsciousness from which the person can be aroused by sensory or other stimuli. It is to be distinguished from *coma,* which is unconsciousness from which the person cannot be aroused. However, there are multiple stages of sleep, from very light sleep to very deep sleep, and most sleep researchers even divide sleep into two different types of sleep that have different qualities, as follows:

Two Different Types of Sleep—(1) Slow Wave Sleep and (2) REM Sleep. During each night a person goes through stages of two different types of sleep that alternate with each other. These are called (1) *slow wave sleep,* because in this type of sleep the brain waves are very slow, as we discuss later; and (2) *REM sleep,* which stands for *rapid eye movement* sleep, because in this type of sleep the eyes undergo rapid movements despite the fact that the person is still asleep.

Most sleep during each night is of the slow wave variety; this is the deep, restful type of sleep that the person experiences during the first hour of sleep after having been kept awake for many hours. Episodes of REM sleep occur periodically during sleep and occupy about 25 per cent of the sleep time of the young adult; they normally recur about every 90 minutes. This type of sleep is not so restful, and it is usually associated with dreaming, as we discuss later.

SLOW WAVE SLEEP

Most of us can understand the characteristics of deep slow wave sleep by remembering the last time that we were kept awake for more than 24 hours and then remembering the deep sleep that occurred during the first hour after going to sleep. This sleep is exceedingly restful and is associated with a decrease in both peripheral vascular tone and many other vegetative functions of the body as well. In addition, there is a 10 to 30 per cent decrease in blood pressure, respiratory rate, and basal metabolic rate.

Though slow wave sleep is frequently called "dreamless sleep," dreams do occur often during slow wave sleep, and nightmares even occur during this type of sleep. However, the difference between the dreams occurring in slow wave sleep and those in REM sleep is that those of REM sleep are remembered, whereas those of slow wave sleep usually are not. That is, during slow wave sleep the process of consolidation of the dreams in memory does not occur.

REM SLEEP (PARADOXICAL SLEEP, DESYNCHRONIZED SLEEP)

In a normal night of sleep, bouts of REM sleep lasting 5 to 30 minutes usually appear on the average every 90 minutes, the first such period occurring 80 to 100 minutes after the person falls asleep. When the person is extremely sleepy, the duration of each bout of REM sleep is very short, and it may even be absent.

On the other hand, as the person becomes more rested through the night, the duration of the REM bouts greatly increases.

There are several very important characteristics of REM sleep:

1. It is usually associated with active dreaming.

2. The person is even more difficult to arouse by sensory stimuli than during deep slow wave sleep, and yet persons usually awaken in the morning during an episode of REM sleep, not from slow wave sleep.

3. The muscle tone throughout the body is exceedingly depressed, indicating strong inhibition of the spinal projections from the excitatory areas of the brain stem.

4. The heart rate and respiration usually become irregular, which is characteristic of the dream state.

5. Despite the extreme inhibition of the peripheral muscles, a few irregular muscle movements occur. These include, in particular, rapid movements of the eyes; this is the origin of the acronym REM, for "rapid eye movements."

6. The brain is highly active in REM sleep, and the overall brain metabolism may be increased as much as 20 per cent. Also, the electroencephalogram shows a pattern of brain waves similar to those that occur during wakefulness. Therefore, this type of sleep is also frequently called *paradoxical sleep* because it is a paradox that a person can still be asleep despite marked activity in the brain.

In summary, REM sleep is a type of sleep in which the brain is quite active. However, the brain activity is not channeled in the proper direction for persons to be fully aware of their surroundings and therefore to be awake.

BASIC THEORIES OF SLEEP

The Active Theory of Sleep. An earlier theory of sleep was that the excitatory areas of the upper brain stem, which was called the *reticular activating system,* and other parts of the brain simply fatigued over the period of a waking day and therefore became inactive as a result. This was called the *passive theory of sleep.* However, an important experiment changed this view to the current belief that *sleep is probably caused by an active inhibitory process.* This was the experiment in which it was discovered that transecting the brain stem in the midpontile region leads to a brain that never goes to sleep. In other words, there seems to be some center or centers located below the midpontile level of the brain stem that actively cause sleep by inhibiting other parts of the brain. This is called the *active theory* of sleep.

Neuronal Centers, Neurohumoral Substances, and Mechanisms That Can Cause Sleep — A Possible Specific Role for Serotonin

Stimulation of several specific areas of the brain can produce sleep with characteristics very near those of natural sleep. Some of these are the following:

1. The most conspicuous stimulation area for causing almost natural sleep is the raphe nuclei in the lower half of the pons and in the medulla. These are a thin sheet of nuclei located in the midline. Nerve fibers from these nuclei spread widely in the reticular formation and also upward into the thalamus, neocortex, hypothalamus, and most areas of the limbic system. In addition, they extend downward into the spinal cord, terminating in the posterior horns where they can inhibit incoming pain signals, as was discussed in Chapter 48. It is also known that many of the endings of fibers from these raphe neurons secrete *serotonin.* Also, when a drug that blocks the formation of serotonin is administered to an animal, the animal often cannot sleep for the next several days. Therefore, it is assumed that serotonin is the major transmitter substance associated with production of sleep.

2. Stimulation of some areas in the *nucleus of the tractus solitarius,* which is the sensory region of the medulla and pons for the visceral sensory signals entering the brain via the vagus and glossopharyngeal nerves, will also promote sleep. However, this will not occur if the raphe nuclei have been destroyed. Therefore, these regions probably act by exciting the raphe nuclei and the serotonin system.

3. Stimulation of several regions in the diencephalon can also help promote sleep, including (a) the rostral part of the hypothalamus, mainly in the suprachiasmal area, and (b) an occasional area in the diffuse nuclei of the thalamus.

Effect of Lesions in the Sleep-Promoting Centers. Discrete lesions in the raphe nuclei lead to a high state of wakefulness. This is also true of bilateral lesions in the mediorostral suprachiasmal portion of the anterior hypothalamus. In both instances, the excitatory reticular nuclei of the mesencephalon and upper pons seem to become released from inhibition. Indeed, the lesions of the anterior hypothalamus can sometimes cause such intense wakefulness that the animal actually dies of exhaustion.

Other Possible Transmitter Substances Related to Sleep. Experiments have shown that the cerebrospinal fluid and also the blood and urine of animals that have been kept awake for several days contain a substance or substances that will cause sleep when injected into the ventricular system of an animal. One of these substances has been identified as *muramyl peptide,* a low molecular weight substance that accumulates in the cerebrospinal fluid and in the urine in animals kept awake for several days. When only micrograms of this sleep-producing substance are injected into the third ventricle, almost natural sleep occurs within a few minutes, and the animal may then stay asleep for several hours. Another substance that has similar effects in causing sleep is a nonapeptide isolated from the blood of sleeping animals. And still a third and different sleep factor has been isolated from the neuronal tissues of the brain stem of animals kept awake for days. Therefore, it is possible that prolonged wakefulness causes progressive accumulation of a sleep factor in the brain stem or in the cerebrospinal fluid that leads to sleep.

Possible Causes of REM Sleep

Why slow wave sleep is broken periodically by REM sleep is not understood. However, a lesion in the *locus ceruleus* on each side of the brain stem can reduce REM sleep, and if the lesion includes other contiguous areas of the brain stem, REM sleep can be prevented altogether. Therefore, it has been postulated that when stimulated, the norepinephrine-secreting nerve fibers that originate in the locus ceruleus can activate many portions of the brain. This theoretically causes the excess activity that occurs in certain regions of the brain in REM sleep, but the signals are not channeled appropriately in the brain to cause normal conscious awareness that is characteristic of wakefulness.

The Cycle Between Sleep and Wakefulness

The preceding discussions have merely identified neuronal areas, transmitters, and mechanisms that are related to sleep. However, they have not explained the cyclic, reciprocal operation of the sleep-wakefulness cycle. There is, as yet, no explanation. Therefore, we can let our imaginations run wild and suggest the following possible mechanism for causing the rhythmicity of the sleep-wakefulness cycle:

When the sleep centers are not activated, the release from inhibition of the mesencephalic and upper pontile reticular nuclei allows this region to become spontaneously active. This in turn excites both the cerebral cortex and the peripheral nervous system, both of which then send numerous positive feedback signals back to the same reticular nuclei to activate them still further. Thus, once wakefulness begins, it has a natural tendency to sustain itself because of all this positive feedback activity.

However, after the brain remains activated for many hours, even the neurons within the activating system presumably will become fatigued to some extent, and other factors presumably activate the sleep centers. Consequently, the positive feedback cycle between the mesencephalic reticular nuclei and the cortex, and also that between these and the periphery, will fade, and the inhibitory effects of the sleep centers (as well as inhibition by possible sleep-producing chemical transmitter substances) will take over, leading to rapid transition from the wakefulness state to the sleep state.

Then, one could postulate that during sleep the excitatory neurons of the reticular activating system gradually become more and more excitable because of the prolonged rest, while the inhibitory neurons of the sleep centers become less excitable because of their overactivity, thus leading to a new cycle of wakefulness.

This theory can explain the rapid transitions from sleep to wakefulness and from wakefulness to sleep. It can also explain arousal, the insomnia that occurs when a person's mind becomes preoccupied with a thought, the wakefulness that is produced by bodily activity, and many other conditions that affect the person's state of sleep or wakefulness.

PHYSIOLOGICAL EFFECTS OF SLEEP

Sleep causes two major types of physiological effects: first, effects on the nervous system itself and, second, effects on other structures of the body. The first of these seems to be by far the more important, for any person who has a transected spinal cord in the neck shows no harmful effects in the body beneath the level of transection that can be attributed to a sleep and wakefulness cycle; that is, lack of this sleep-wakefulness cycle in the nervous system at any point below the brain causes neither harm to the bodily organs nor even any deranged function. On the other hand, lack of sleep certainly does affect the functions of the central nervous system.

Prolonged wakefulness is often associated with progressive malfunction of the mind and sometimes even causes abnormal behavioral activities of the nervous system. We are all familiar with the increased sluggishness of thought that occurs toward the end of a prolonged wakeful period, but in addition, a person can become irritable or even psychotic following forced wakefulness for prolonged periods of time. Therefore, we can assume that sleep in some way not presently understood restores both normal levels of activity and normal "balance" among the different parts of the central nervous system. This might be likened to the "rezeroing" of electronic analog computers after prolonged use, for all computers of this type gradually lose their "baseline" of operation; it is reasonable to assume that the same effect occurs in the central nervous system, because overuse of some brain areas during wakefulness could easily throw these out of balance with the remainder of the nervous system. Therefore, in the absence of any definitely demonstrated functional value of sleep, we might postulate that the principal value of sleep is to restore the natural balance among the neuronal centers.

Even though, as we pointed out earlier, wakefulness and sleep have not been shown to be harmful to the somatic functions of the body, the cycle of enhanced and depressed nervous excitability that follows the cycle of wakefulness and sleep does have moderate physiological effects on the peripheral body. For instance, there is enhanced sympathetic activity during wakefulness and also enhanced numbers of impulses to the skeletal musculature to increase muscle tone. Conversely, during sleep, sympathetic activity decreases while parasympathetic activity increases. Therefore, arterial blood pressure falls, pulse rate decreases, skin vessels dilate, activity of the gastrointestinal tract sometimes increases, muscles fall into a mainly relaxed state, and the overall basal metabolic rate of the body falls by 10 to 30 per cent.

● BRAIN WAVES

Electrical recordings from the surface of the brain or from the outer surface of the head demonstrate continuous electrical activity in the brain. Both the intensity and patterns of this electrical activity are determined to a great extent by the overall level of excitation of the brain resulting from *sleep, wakefulness,* and brain diseases such as *epilepsy* and even some *psychoses.* The undulations in the recorded electrical potentials, shown in Figure 59–1, are called *brain waves,* and the entire record is called an *electroencephalogram* (EEG).

The intensities of the brain waves on the surface of the scalp range from 0 to 200 microvolts, and their frequencies range from once every few seconds to 50 or more per second. The character of the waves is highly dependent on the degree of activity of the cerebral cortex, and the waves change markedly between the states of wakefulness and sleep and coma.

Much of the time, the brain waves are irregular, and no general pattern can be discerned in the EEG. However, at other times, distinct patterns do appear. Some of these are characteristic of specific abnormalities of the brain, such as epilepsy, which is discussed later. Others occur even in normal persons and can be classified as *alpha, beta, theta,* and *delta waves,* which are all illustrated in Figure 59–1.

Alpha waves are rhythmic waves occurring at a frequency of between 8 and 13 per second and are found in the EEGs of almost all normal adult persons when they are awake in a quiet, resting state of cerebration. These waves occur most intensely in the occipital region but can also be recorded from the parietal and frontal regions of the scalp. Their voltage usually is about 50 microvolts. During deep sleep the alpha waves disappear entirely; and when the awake person's attention is directed to some specific type of mental activity, the alpha waves are replaced by asynchronous, higher-frequency but lower-voltage beta waves. Figure 59–2 illustrates the effect on the alpha waves of simply opening the eyes in bright light and then closing the eyes again. Note that the visual sensations cause immediate cessation of the alpha waves and that these are replaced by low-voltage, asynchronous beta waves.

Beta waves occur at frequencies of more than 14 cycles per second and as high as 25 and rarely 50 cycles per second. These are most frequently recorded from the parietal and frontal regions of the scalp during activation of the central nervous system or during tension.

Theta waves have frequencies of between 4 and 7 cycles per second. These occur mainly in the parietal and temporal regions in children, but they also occur during emotional stress in some adults, particularly during disappointment

Figure 59–2. Replacement of the alpha rhythm by an asynchronous discharge on opening the eyes.

and frustration. Theta waves also occur in many brain disorders.

Delta waves include all the waves of the EEG below 3.5 cycles per second and sometimes as low as 1 cycle every 2 to 3 seconds. These occur in very deep sleep, in infancy, and in serious organic brain disease. They also occur in the cortex of animals that have had subcortical transections separating the cerebral cortex from the thalamus. Therefore, delta waves can occur strictly in the cortex independent of activities in lower regions of the brain.

ORIGIN IN THE BRAIN OF THE BRAIN WAVES

The discharge of a single neuron or single nerve fiber in the brain can never be recorded from the surface of the head. Instead, many thousands or even millions of neurons or fibers *must fire synchronously;* only then will the potentials from the individual neurons or fibers summate enough to be recorded all the way through the skull. Thus, the intensity of the brain waves from the scalp is determined mainly by the number of neurons and fibers that fire in synchrony with each other, not by the total level of electrical activity in the brain, for even very strong *nonsynchronous* nerve signals will actually nullify each other and cause only weak waves. This was illustrated in Figure 59–2, which showed, when the eyes were closed, synchronous discharge of many neurons in the cerebral cortex at a frequency of about 12 per second, which gave the *alpha waves.* Then, when the eyes were opened, the activity of the brain increased greatly, but the synchronization of the signals decreased to so little that the brain waves decreased to weak waves of generally higher but irregular frequency, called *beta waves.*

The brain wave electrical potentials are generated mainly from cortical layers I and II, especially from the large matt of dendrites that extend into these surface areas from neuronal cells located deeper in the cortex. The potentials generated in the tissue fluids surrounding these dendrites can be either positive or negative, for the following reasons: When the neuronal cell bodies in the deeper layers discharge, negative charges leak out of the cell bodies and cause negativity in the deeper cortical fluids; at the same time this loss of negative charges leaves the interior of the neuronal cell membranes positive. This positivity is conducted electrotonically upward through the dendrites to the surface layers of the brain and then transmitted by a capacitative effect across the dendrite membranes to the fluids surrounding these dendrites. Therefore, stimulation of neurons deep in the cerebral cortex usually causes initial positivity on the surface of the brain. On the other hand, other cortical synapses lie not on the deep cell bodies but instead on the surface dendrites themselves, especially in cortical layers II and III. When these synapses are excited, local depolarization occurs in the dendrites themselves, allowing negative charges to leak outward; then the electrical waves recorded from the surface of the scalp are negative. This difference between positivity and negativity is important because it sometimes allows one to distinguish the depth in the cortex of the neuronal discharges that produce specific types of waves.

Figure 59–1. Different types of normal electroencephalographic waves.

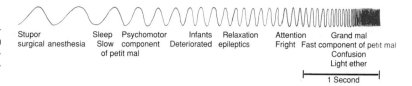

Figure 59-3. Effect of varying degrees of cerebral activity on the basic rhythm of the electroencephalogram (EEG). (From Gibbs and Gibbs: Atlas of Electroencephalography, 2nd. Ed. Vol. I. Reading, Mass., Addison-Wesley, 1974. Reprinted by permission.)

Stupor
surgical anesthesia

Sleep
Slow component
of petit mal

Psychomotor
component

Infants
Deteriorated

Relaxation
epileptics

Attention
Fright

Grand mal
Fast component of petit mal
Confusion
Light ether

1 Second

Origin of Alpha Waves. Alpha waves will *not* occur in the cortex without connections with the thalamus. Also, stimulation in the nonspecific thalamic nuclei often sets up waves in the generalized thalamocortical system at a frequency of between 8 and 13 per second, the natural frequency of the alpha waves. Therefore, it is likely that the alpha waves result from spontaneous activity in the nonspecific thalamocortical system, which causes both the periodicity of the alpha waves and the synchronous activation of literally millions of cortical neurons during each wave.

Origin of Delta Waves. Transection of the fiber tracts from the thalamus to the cortex, which blocks the thalamic activation of the cortex and eliminates the alpha waves, nevertheless causes delta waves in the cortex. This indicates that some synchronizing mechanism can occur in the cortical neurons themselves—entirely independently of lower structures in the brain—to cause the delta waves.

Delta waves also occur in very deep "slow wave" sleep; and this suggests that the cortex then might be released from the activating influences of the lower centers.

EFFECT OF VARYING DEGREES OF CEREBRAL ACTIVITY ON THE BASIC FREQUENCY OF THE ELECTROENCEPHALOGRAM

There is a general relationship between the degree of cerebral activity and the average frequency of the electroencephalographic rhythm, the average frequency increasing progressively with higher degrees of activity. This is illustrated in Figure 59-3, which shows the existence of delta waves in stupor, surgical anesthesia, and sleep; theta waves in psychomotor states and in infants; alpha waves during relaxed states; and beta waves during periods of intense mental activity. *However, during periods of mental activity the waves usually become asynchronous rather than synchronous, so that the voltage falls considerably, despite increased cortical activity,* as illustrated in Figure 59-2.

ELECTROENCEPHALOGRAPHIC CHANGES IN THE DIFFERENT STAGES OF WAKEFULNESS AND SLEEP

Figure 59-4 illustrates the electroencephalogram from a typical person in different stages of wakefulness and sleep. Alert wakefulness is characterized by high-frequency *beta waves,* whereas quiet wakefulness is usually associated with *alpha waves,* as illustrated by the first two electroencephalograms of the figure.

Slow wave sleep is generally divided into four stages. In the first stage, a stage of very light sleep, the voltage of the electroencephalographic waves becomes very low; but this is broken by *"sleep spindles,"* that is, short spindle-shaped bursts of alpha waves that occur periodically. In stages 2, 3, and 4 of slow wave sleep the frequency of the electroencephalogram becomes progressively slower until it reaches a frequency of only 2 to 3 waves per second; these are typical *delta waves.*

Finally, the bottom record in Figure 59-4 illustrates the electroencephalogram during REM sleep. It is often difficult to tell a difference between this brain wave pattern and that of an alert awake person. The voltage of these waves is considerably lower than the voltage in deep stage 4 slow wave sleep, and the waves are themselves irregular high-frequency beta waves, which is normally suggestive of excess but desynchronized nervous activity as found in the awake state. Therefore, REM sleep is frequently called *desynchronized sleep* because there is a lack of synchrony in the firing of the neurons despite very significant brain activity.

● EPILEPSY

Epilepsy is characterized by uncontrolled excessive activity of either a part or all of the central nervous system. A person who is predisposed to epilepsy has attacks when the basal level of excitability of the nervous system (or of the part that is susceptible to the epileptic state) rises above a certain critical threshold. But as long as the degree of excitability is held below this threshold, no attack occurs.

Basically, epilepsy can be classified into three major types: *grand mal epilepsy, petit mal epilepsy,* and *focal epilepsy.*

GRAND MAL EPILEPSY

Grand mal epilepsy is characterized by extreme neuronal discharges in all areas of the brain—in the cortex, in the deeper parts of the cerebrum, and even in the brain stem and thalamus. Also, discharges into the spinal cord cause

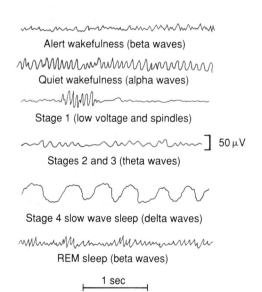

Alert wakefulness (beta waves)

Quiet wakefulness (alpha waves)

Stage 1 (low voltage and spindles)

Stages 2 and 3 (theta waves)] 50 μV

Stage 4 slow wave sleep (delta waves)

REM sleep (beta waves)

1 sec

Figure 59-4. Progressive change in the characteristics of the brain waves during different stages of wakefulness and sleep.

generalized *tonic convulsions* of the entire body, followed toward the end of the attack by alternating tonic and then spasmotic muscular contractions called *tonic-clonic convulsions*. Often the person bites or "swallows" the tongue and usually has difficulty in breathing, sometimes to the extent of developing cyanosis. Also, signals to the viscera frequently cause urination and defecation.

The grand mal seizure lasts from a few seconds to as long as three to four minutes and is characterized by postseizure depression of the entire nervous system; the person remains in stupor for one to many minutes after the attack is over and then often remains severely fatigued or even asleep for many hours thereafter.

The top recording of Figure 59–5 illustrates a typical electroencephalogram from almost any region of the cortex during the tonic phase of a grand mal attack. This illustrates that high-voltage, synchronous discharges occur over the entire cortex. Furthermore, the same type of discharge occurs on both sides of the brain at the same time, illustrating that the abnormal neuronal circuitry responsible for the attack strongly involves the basal regions of the brain that drive the cortex.

In experimental animals or even in human beings, grand mal attacks can be initiated by administering neuronal stimulants, such as the drug Metrazol, or they can be caused by insulin hypoglycemia or by the passage of alternating electrical current directly through the brain. Electrical recordings from the thalamus and also from the reticular formation of the brain stem during the grand mal attack show typical high-voltage activity in both of these areas similar to that recorded from the cerebral cortex.

Presumably, therefore, a grand mal attack is caused by abnormal activation in the lower portions of the brain activating system itself.

What Initiates a Grand Mal Attack? Most persons who have grand mal attacks have a hereditary predisposition to epilepsy, a predisposition that occurs in about 1 of every 50 to 100 people. In such persons, some of the factors that can increase the excitability of the abnormal "epileptogenic" circuitry enough to precipitate attacks are (1) strong emotional stimuli, (2) alkalosis caused by overbreathing, (3) drugs, (4) fever, and (5) loud noises or flashing lights. Also, even in persons not genetically predisposed, traumatic lesions in almost any part of the brain can cause excess excitability of local brain areas, as we discuss shortly; and these, too, can transmit signals into the basal activating systems of the brain to elicit grand mal seizures.

Figure 59–5. Electroencephalograms in different types of epilepsy.

What Stops the Grand Mal Attack? The cause of the extreme neuronal overactivity during a grand mal attack is presumed to be massive activation of many reverberating pathways throughout the brain. Presumably, also, the major factor, or at least one of the major factors, that stops the attack after a few minutes is the phenomenon of neuronal *fatigue*. However, a second factor is probably *active inhibition* by inhibitory neurons that have also been activated by the attack. The stupor and total body fatigue that occur after a grand mal seizure is over are believed to result from the intense fatigue of the neuronal synapses following their intensive activity during the grand mal attack.

PETIT MAL EPILEPSY

Petit mal epilepsy is closely allied to grand mal epilepsy in that it too almost certainly involves the basic brain activating system. It is usually characterized by 3 to 30 seconds of unconsciousness during which the person has several twitchlike contractions of the muscles, usually in the head region — especially blinking of the eyes; this is followed by return of consciousness and resumption of previous activities. The patient may have one such attack in many months or in rare instances may have a rapid series of attacks, one following the other. However, the usual course is for the petit mal attacks to appear in late childhood and then to disappear entirely by the age of 30. On occasion, a petit mal epileptic attack will initiate a grand mal attack.

The brain wave pattern in petit mal epilepsy is illustrated by the middle record of Figure 59–5, which is typified by a *spike and dome pattern*. The spike portion of this recording is almost identical to the spikes that occur in grand mal epilepsy, but the dome portion is distinctly different. The spike and dome can be recorded over most or all of the cerebral cortex, illustrating that the seizure involves the entire activating system of the brain.

FOCAL EPILEPSY

Focal epilepsy can involve almost any part of the brain, either localized regions of the cerebral cortex or deeper structures of both the cerebrum and brain stem. And almost always, focal epilepsy results from some localized organic lesion or functional abnormality, such as a scar that pulls on the neuronal tissue, a tumor that compresses an area of the brain, a destroyed area of brain tissue, or congenitally deranged local circuitry. Lesions such as these can promote extremely rapid discharges in the local neurons; and when the discharge rate rises above approximately 1000 per second, synchronous waves begin to spread over the adjacent cortical regions. These waves presumably result from *localized reverberating circuits* that gradually recruit adjacent areas of the cortex into the discharge zone. The process spreads to adjacent areas at a rate as slow as a few millimeters a minute to as fast as several centimeters per second. When such a wave of excitation spreads over the motor cortex, it causes a progressive "march" of muscular contractions throughout the opposite side of the body, beginning most characteristically in the mouth region and marching progressively downward to the legs, but at other times marching in the opposite direction. This is called *jacksonian epilepsy*.

A focal epileptic attack may remain confined to a single area of the brain, but in many instances the strong signals from the convulsing cortex or other part of the brain excite the mesencephalic portion of the brain activating system so greatly that a grand mal epileptic attack ensues as well.

Another type of focal epilepsy is the so-called *psychomotor seizure,* which may cause (1) a short period of amnesia; (2) an attack of abnormal rage; (3) sudden anxiety, discomfort, or fear; (4) a moment of incoherent speech or mumbling of some trite phrase; or (5) a motor act to attack someone, to rub the face with the hand, and so forth. Sometimes the person cannot remember his activities during the attack, but at other times he will have been conscious of everything that he had been doing but unable to control it. Attacks of this type characteristically involve part of the limbic portion of the brain, such as the hippocampus, the amygdala, the septum, and the temporal cortex.

The lower tracing of Figure 59–5 illustrates a typical electroencephalogram during a psychomotor attack, showing a low-frequency rectangular wave with a frequency of between 2 and 4 per second and with superimposed 14-per-second waves.

The electroencephalogram can often be used to localize abnormal spiking waves originating in areas of organic brain disease that might predispose to focal epileptic attacks. Once such a focal point is found, surgical excision of the focus frequently prevents future attacks.

● PSYCHOTIC BEHAVIOR AND DEMENTIA — ROLES OF SPECIFIC NEUROTRANSMITTER SYSTEMS

Clinical studies of patients with different psychoses and also some types of dementia have suggested that many if not most of these conditions result from diminished function of classes of neurons that secrete specific neurotransmitters. Use of appropriate drugs to counteract the loss of the respective transmitters has been quite successful in treating some patients.

In Chapter 56 we discussed the cause of Parkinson's disease, the loss of the neurons in the substantia nigra whose axons secrete dopamine in the caudate nucleus and putamen. The loss of acetylcholine-secreting neurons in the basal ganglia is associated with the abnormal motor patterns of Huntington's chorea, as well as with the dementia that develops later in the same patients. In the present section, we extend this concept to other abnormalities and other classes of neurons that lead to additional types of psychotic behavior or dementia.

Depression and Manic-Depressive Psychoses — Decreased Activity of the Norepinephrine and Serotonin Neurotransmitter Systems

In the past few years much evidence has accumulated suggesting that the *mental depression psychosis,* which afflicts about 8 million people in the United States at any one time, might be caused by diminished formation of either norepinephrine or serotonin or both. These patients experience symptoms of grief, unhappiness, despair, and misery. In addition, they lose their appetite and sex drive and also have severe insomnia. And associated with all these is a state of psychomotor agitation despite the depression.

In the previous chapter it was pointed out that large numbers of *norepinephrine-secreting neurons* are located in the brain stem, especially in the *locus ceruleus,* and that these send fibers upward to most parts of the limbic system, the thalamus, and the cerebral cortex. Also, many *serotonin-producing neurons* are located in the *midline raphe nuclei* of

the lower pons and medulla and also project fibers to many areas of the limbic system and to some other areas of the brain as well.

A principal reason for believing that depression is caused by diminished activity of the norepinephrine and serotonin systems is that drugs that block the secretion of norepinephrine and serotonin, such as the drug reserpine, frequently cause depression. Conversely, about 70 per cent of depressive patients can be treated very effectively with one of two types of drugs that increase especially the excitatory effects of norepinephrine at the nerve endings, and perhaps of serotonin as well: (1) *monoamine oxidase inhibitors,* which block destruction of norepinephrine and serotonin once they are formed; and (2) *tricyclic antidepressants,* which block reuptake of norepinephrine and serotonin by the nerve endings, so that these transmitters remain active for longer periods of time after secretion.

Mental depression can also be treated effectively by electroconvulsive therapy — commonly called "shock therapy." In this therapy an electric shock is used to cause a generalized seizure similar to that of an epileptic attack. This has also been shown to enhance norepinephrine transmission efficiency.

Some patients with mental depression alternate between depression and mania, which is called *manic-depressive psychosis,* and a few persons exhibit only mania without the depressive episodes. Drugs that diminish the formation or action of norepinephrine and serotonin, such as lithium compounds, can be effective in treating the manic condition.

Therefore, it is presumed that the norepinephrine system especially and perhaps the serotonin system as well normally function to provide motor drive to the limbic system to increase a person's sense of well-being, to create happiness, contentment, good appetite, appropriate sex drive, and psychomotor balance, although too much of a good thing can cause mania. In support of this concept is the fact that the pleasure and reward centers of the hypothalamus and surrounding areas receive large numbers of nerve endings from the norepinephrine system.

Schizophrenia — Exaggerated Function of Part of the Dopamine System

Schizophrenia comes in many different varieties. One of the most common is the person who hears voices and has delusions of grandeur, or intense fear, or other types of feelings that are unreal. Schizophrenics are often highly paranoid, with a sense of persecution from outside sources; they may develop incoherent speech, dissociation of ideas, and abnormal sequences of thought; and they are often withdrawn, sometimes with abnormal posture and even rigidity.

There is reason to believe that schizophrenia results from excessive functioning of a group of neurons that secrete dopamine. These neurons are located in the ventral tegmentum of the mesencephalon, medial and superior to the substantia nigra. They give rise to the so-called *mesolimbic dopaminergic system,* which projects nerve fibers mainly into the medial and anterior portions of the limbic system, especially into the nucleus accumbens, the amygdala, the anterior caudate nucleus, and the anterior cingulate gyrus of the cortex, all of which are powerful behavioral control centers.

Some of the reasons for believing the mesolimbic dopaminergic system to be related to schizophrenia are the following: When Parkinson's disease patients are treated with

L-dopa, which releases dopamine in the brain, the parkinsonian patient sometimes develops schizophrenic-like symptoms, indicating that excess dopaminergic activity can cause dissociation of a person's drives and thought patterns. However, an even more compelling reason for believing that schizophrenia might be caused by excess production of dopamine is that those drugs that are effective in treating schizophrenia, such as chlorpromazine, haloperidol, and thiothixene, all decrease the secretion of dopamine by the dopaminergic nerve endings or decrease the effect of dopamine on the subsequent neurons.

Almost certainly there are other factors in schizophrenia besides excess secretion of dopamine; nevertheless, the symptoms of schizophrenia are similar to the behavioral effects of excessive dopamine.

Alzheimer's Disease — Loss of Acetylcholine-Secreting Neurons

Alzheimer's disease is defined as premature aging of the brain, usually beginning in midadult life and progressing rapidly to extreme loss of mental powers as usually seen in very, very old age. These patients usually require continuous care within a few years after the disease begins.

A consistent finding in Alzheimer's disease is about 75 per cent loss of the neurons in the *nucleus basalis of Meynert,* located beneath the globus pallidus in the substantia innominata. The neurons of this nucleus send *acetylcholine-secreting* fibers to a large share of the neocortex. It is presumed that the acetylcholine in some way activates the neuronal mechanisms for storing and recalling memories. The nucleus basalis in turn receives input signals from multiple portions of the limbic system, which provides the motivational drive for the memory process that was discussed in the previous chapter.

Other neurotransmitters that have also been found deficient in Alzheimer's disease are *somatostatin* and *substance P.* Therefore, the basic cause of the disease might be more global, rather than simply a loss of one specific set of acetylcholine-secreting neurons.

REFERENCES

Ashton, H.: Brain Systems, Disorders and Psychotropic Drugs. New York, Oxford University Press, 1987.

Barnes, D. M.: Biological issues in schizophrenia. Science, 235:430, 1987.

Barnes, D. M.: Debate about epilepsy: What initiates seizures? Science, 234:938, 1986.

Bindman, L.: The Neurophysiology of the Cerebral Cortex. Austin, University of Texas Press, 1981.

Chrousos, G. P., et al. (eds.): Mechanisms of Physical and Emotional Stress. New York, Plenum Publishing Corp., 1988.

Conn, D. K., et al. (eds.): Psychiatric Consequences of Brain Disease in the Elderly. New York, Plenum Publishing Corp., 1989.

Cooper, A. J. L., and Plum, F.: Biochemistry and physiology of brain ammonia. Physiol. Rev., 67:440, 1987.

Dichter, M. A. (ed.): Mechanisms of Epileptogenesis. New York, Plenum Publishing Corp., 1988.

Dichter, M. A., and Ayala, G. F.: Cellular mechanisms of epilepsy: A status report. Science, 237:157, 1987.

DiDonato, S., et al. (eds.): Molecular Genetics of Neurological and Neuromuscular Disease. New York, Raven Press, 1988.

Georgotas, A., and Cancro, R. (eds.): Depression and Mania. New York, Elsevier Science Publishing Co., 1988.

Glaser, G. H., et al. (eds.): Antiepileptic Drugs. New York, Raven Press, 1980.

Hansen, A. J.: Effect of anoxia on ion distribution in the brain. Physiol. Rev., 65:101, 1985.

Hobson, J. A., and Brazier, M. A. B. (eds.): The Reticular Formation Revisited: Specifying Function for a Nonspecific System. New York, Raven Press, 1980.

Hyvarinen, J.: Posterior parietal lobe of the primate brain. Physiol. Rev., 62:1060, 1982.

Jacobs, B. L.: How hallucinogenic drugs work. Am. Sci., 75:386, 1987.

Jones, E. G.: Organization of the thalamocortical complex and its relation to sensory processes. In Darian-Smith, I. (ed.): Handbook of Physiology, Sec. 1, Vol. III. Bethesda, Md., American Physiological Society, 1984, p. 149.

Kaplan, H. I., and Sadock, B. J.: Synopsis of Psychiatry: Behavioral Sciences Clinical Psychiatry, 5th Ed. Baltimore, Williams & Wilkins, 1988.

Klee, M. (ed.): Physiology and Pharmacology of Epileptogenic Phenomena. New York, Raven Press, 1982.

Kryger, M. H., et al.: Principles and Practice of Sleep Medicine. Philadelphia, W. B. Saunders Co., 1986.

Livingston, R. B.: Sensory Processing, Perception, and Behavior. New York, Raven Press, 1978.

Mann, J. J. (ed.): The Phenomenology of Depressive Illness. New York, Plenum Publishing Corp., 1988.

McKinney, W. T.: Models of Mental Disorders. New York, Plenum Publishing Corp., 1988.

Meijer, J. H., and Rietveld, W. J.: Neurophysiology of the suprachiasmatic circadian pacemaker in rodents. Physiol. Rev., 69:671, 1989.

Mesulam, M. M.: The cholinergic connection in Alzheimer's disease. News Physiol. Sci., 1:107, 1986.

Mitzdorf, U.: Current source-density method and application in cat cerebral cortex: Investigation of evoked potentials and EEG phenomena. Physiol. Rev., 65:37, 1985.

Nappi, G., et al. (eds.): Neurodegenerative Disorders. New York, Raven Press, 1988.

Nerozzi, D., et al. (eds.): Hypothalamic Dysfunction in Neuropsychiatric Disorders. New York, Raven Press, 1987.

Newmark, M. E., and Penry, J. K.: Genetics of Epilepsy: A Review. New York, Churchill Livingstone, 1980.

Plum, F., and Posner, J. B.: The Diagnosis of Stupor and Coma, 3rd Ed. Philadelphia, F. A. Davis Co., 1980.

Pollack, M. H., et al.: Propranolol and depression revisited: Three cases and a review. J. Nerv. Ment. Dis., 173:118, 1985.

Stern, R. M., et al.: Psychophysiological Recording. New York, Oxford University Press, 1980.

Strong, R., et al. (eds.): Central Nervous System Disorders of Aging. New York, Raven Press, 1988.

Terry, R. D. (ed.): Aging and the Brain. New York, Raven Press, 1988.

Tucek, S.: Regulation of acetylcholine synthesis in the brain. J. Neurochem., 44:11, 1985.

Wauquier, A., et al.: Slow Wave Sleep: Physiological, Pathophysiological, and Functional Aspects. New York, Raven Press, 1989.

Wolman, B. B.: Psychosomatic Disorders. New York, Plenum Publishing Corp., 1988.

CHAPTER 60

• *The Autonomic Nervous System; The Adrenal Medulla* •

The portion of the nervous system that controls the visceral functions of the body is called the *autonomic nervous system.* This system helps control arterial pressure, gastrointestinal motility and secretion, urinary bladder emptying, sweating, body temperature, and many other activities, some of which are controlled almost entirely and some only partially by the autonomic nervous system.

One of the most striking characteristics of the autonomic nervous system is the rapidity and intensity with which it can change visceral functions. For instance, within 3 to 5 seconds it can increase the heart rate to two times normal, and the arterial pressure can be doubled within as little as 10 to 15 seconds; or, at the other extreme, the arterial pressure can be decreased low enough within 4 to 5 seconds to cause fainting. Sweating can begin within seconds, and the bladder may empty involuntarily, also within seconds. It is these extremely rapid changes that are measured by the lie detector polygraph, reflecting the innermost feelings of a person.

● GENERAL ORGANIZATION OF THE AUTONOMIC NERVOUS SYSTEM

The autonomic nervous system is activated mainly by centers located in the *spinal cord, brain stem,* and *hypothalamus.* Also, portions of the cerebral cortex, especially of the limbic cortex, can transmit impulses to the lower centers and in this way influence autonomic control. Often the autonomic nervous system also operates by means of *visceral reflexes.* That is, sensory signals entering the autonomic ganglia, cord, brain stem, or hypothalamus can elicit appropriate reflex responses back to the visceral organs to control their activities.

The efferent autonomic signals are transmitted to the body through two major subdivisions called the *sympathetic nervous system* and the *parasympathetic nervous system,* the characteristics and functions of which follow.

PHYSIOLOGIC ANATOMY OF THE SYMPATHETIC NERVOUS SYSTEM

Figure 60-1 illustrates the general organization of the sympathetic nervous system, showing one of the two *paravertebral sympathetic chains of ganglia* that lie to the two sides of the spinal column, two *prevertebral ganglia* (the *celiac* and *hypogastric),* and nerves extending from the ganglia to the different internal organs. The sympathetic nerves originate in the spinal cord between the segments T-1 and L-2 and pass from here first into the sympathetic chain and thence to the tissues and organs that are stimulated by the sympathetic nerves.

Preganglionic and Postganglionic Sympathetic Neurons

The sympathetic nerves are different from skeletal motor nerves in the following way: Each sympathetic pathway from the cord to the stimulated tissue is composed of two neurons, a *preganglionic neuron* and a *postganglionic neuron,* in contrast to only a single neuron in the skeletal motor pathway. The cell body of each preganglionic neuron lies in the *intermediolateral horn* of the spinal cord; and its fiber passes, as illustrated in Figure 60-2, through an *anterior root* of the cord into the corresponding *spinal nerve.*

Immediately after the spinal nerve leaves the spinal column, the preganglionic sympathetic fibers leave the nerve and pass through the *white ramus* into one of the *ganglia* of the *sympathetic chain.* Then the course of the fibers can be one of the following three: (1) It can synapse with postgan-

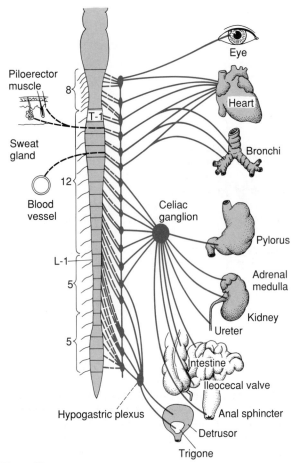

Figure 60–1. The sympathetic nervous system. Dashed lines represent postganglionic fibers in the gray rami leading into the spinal nerves for distribution to blood vessels, sweat glands, and piloerector muscles.

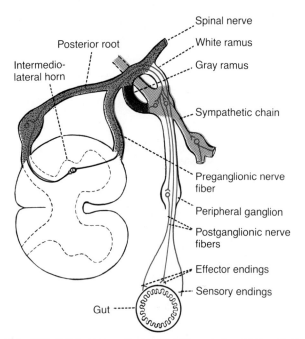

Figure 60–2. Nerve connections between the spinal cord, sympathetic chain, spinal nerves, and peripheral sympathetic nerves.

glionic neurons in the ganglion that it enters. (2) It can pass upward or downward in the chain and synapse in one of the other ganglia of the chain. Or (3) it can pass for variable distances through the chain and then through one of the *sympathetic nerves* radiating outward from the chain, finally terminating in one of the *prevertebral ganglia*.

The postganglionic neuron then originates either in one of the sympathetic chain ganglia or in one of the prevertebral ganglia. From either of these two sources, the postganglionic fibers travel to their destinations in the various organs.

Sympathetic Nerve Fibers in the Skeletal Nerves. Some of the postganglionic fibers pass back from the sympathetic chain into the spinal nerves through *gray rami* at all levels of the cord, illustrated by the dashed fiber in Figure 60–2. These pathways are made up of type C fibers that extend to all parts of the body in the skeletal nerves. They control the blood vessels, sweat glands, and piloerector muscles of the hairs. Approximately 8 per cent of the fibers in the average skeletal nerve are sympathetic fibers, a fact that indicates their importance.

Segmental Distribution of the Sympathetic Nerves. The sympathetic pathways originating in the different segments of the spinal cord are not necessarily distributed to the same part of the body as the spinal nerve fibers from the same segments. Instead, the *sympathetic fibers from cord segment T-1 generally pass up the sympathetic chain to the head; from T-2 into the neck; from T-3, T-4, T-5, and T-6 into the thorax; from T-7, T-8, T-9, T-10, and T-11 into the abdomen; and from T-12, L-1, and L-2 into the legs.* This distribution is only approximate and overlaps greatly.

The distribution of sympathetic nerves to each organ is determined partly by the position in the embryo where the organ originates. For instance, the heart receives many sympathetic nerve fibers from the neck portion of the sympathetic chain because the heart originates in the neck of the embryo. Likewise, the abdominal organs receive their sympathetic innervation from the lower thoracic segments because most of the primitive gut originates in this area.

Special Nature of the Sympathetic Nerve Endings in the Adrenal Medullae. Preganglionic sympathetic nerve fibers pass, without synapsing, all the way from the intermediolateral horn cells of the spinal cord, through the sympathetic chains, through the splanchnic nerves, and finally into the adrenal medullae. There they end directly on modified neuronal cells that secrete *epinephrine* and *norepinephrine* into the blood stream. These secretory cells embryologically are derived from nervous tissue and are analogous to postganglionic neurons; indeed, they even have rudimentary nerve fibers, and it is these fibers that secrete the hormones.

PHYSIOLOGIC ANATOMY OF THE PARASYMPATHETIC NERVOUS SYSTEM

The parasympathetic nervous system is illustrated in Figure 60–3, showing that parasympathetic fibers leave the central nervous system through cranial nerves III, VII, IX, and X; the second and third sacral spinal nerves; and occasionally the first and fourth sacral nerves. About 75 per cent of all parasympathetic nerve fibers are in the vagus nerves, passing to the entire thoracic and abdominal regions of the body. Therefore, a physiologist speaking of the parasympathetic nervous system often thinks mainly of the two vagus nerves. The vagus nerves supply parasympathetic nerves to

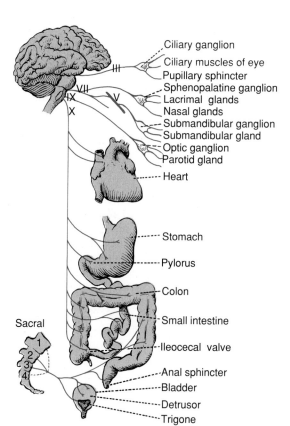

Figure 60-3. The parasympathetic nervous system.

the heart, the lungs, the esophagus, the stomach, the entire small intestine, the proximal half of the colon, the liver, the gallbladder, the pancreas, and the upper portions of the ureters.

Parasympathetic fibers in the *third nerve* flow to the pupillary sphincters and ciliary muscles of the eye. Fibers from the *seventh nerve* pass to the lacrimal, nasal, and submandibular glands, and fibers from the *ninth nerve* pass to the parotid gland.

The sacral parasympathetic fibers congregate in the form of the *nervi erigentes,* also called the *pelvic nerves,* which leave the sacral plexus on each side of the cord and distribute their peripheral fibers to the descending colon, rectum, bladder, and lower portions of the ureters. Also, this sacral group of parasympathetics supplies fibers to the external genitalia to cause sexual stimulation.

Preganglionic and Postganglionic Parasympathetic Neurons. The parasympathetic system, like the sympathetic, has both preganglionic and postganglionic neurons. However, except in the case of a few cranial parasympathetic nerves, the *preganglionic fibers* pass uninterrupted all the way to the organ that is to be controlled. Then, in the wall of the organ are located the *postganglionic neurons.* The preganglionic fibers synapse with these, and short postganglionic fibers, 1 millimeter to several centimeters in length, leave the neurons to spread through the substance of the organ. This location of the parasympathetic postganglionic neurons in the visceral organ itself is quite different from the arrangement of the sympathetic ganglia, for the cell bodies of the sympathetic postganglionic neurons are almost always located in ganglia of the sympathetic chain or in various other discrete ganglia in the abdomen, rather than in the excited organ itself.

● BASIC CHARACTERISTICS OF SYMPATHETIC AND PARASYMPATHETIC FUNCTION

CHOLINERGIC AND ADRENERGIC FIBERS — SECRETION OF ACETYLCHOLINE OR NOREPINEPHRINE

The sympathetic and parasympathetic nerve fibers all secrete one of the two synaptic transmitter substances, *acetylcholine* or *norepinephrine.* Those that secrete acetylcholine are said to be *cholinergic.* Those that secrete norepinephrine are said to be *adrenergic,* a term derived from *adrenalin,* which is the British name for epinephrine.

All *preganglionic neurons* are *cholinergic* in both the sympathetic and parasympathetic nervous systems. Therefore, acetylcholine or acetylcholine-like substances, when applied to the ganglia, will excite both sympathetic and parasympathetic postganglionic neurons.

The *postganglionic neurons of the parasympathetic system are* also *all cholinergic.*

On the other hand, *most of the postganglionic sympathetic neurons are adrenergic,* though this is not entirely true, because the postganglionic sympathetic nerve fibers to the sweat glands, to the piloerector muscles, and to a few blood vessels are cholinergic.

Thus, the terminal nerve endings of the parasympathetic system *all* secrete *acetylcholine,* and *most* of the sympathetic nerve endings secrete *norepinephrine.* These hormones, in turn, act on the different organs to cause the respective parasympathetic and sympathetic effects. Therefore, they are often called, respectively, *parasympathetic* and *sympathetic transmitters.*

These are the molecular structures of acetylcholine and norepinephrine:

$$CH_3-\underset{\underset{O}{\|}}{C}-O-CH_2-CH_2-\overset{+}{\underset{\underset{CH_3}{|}}{N}}\overset{CH_3}{\diagdown}CH_3$$

Acetylcholine

$$HO-\underset{HO}{\bigcirc}-\underset{\underset{OH}{|}}{CH}-CH_2-NH_2$$

Norepinephrine

Mechanisms of Transmitter Secretion and Removal at the Postganglionic Endings

Secretion of Acetylcholine and Norepinephrine by Postganglionic Nerve Endings. Some of the postganglionic autonomic nerve endings, especially those of the parasympathetic nerves, are similar to but much smaller in size than those of the skeletal neuromuscular junction. However, most of the sympathetic nerve fibers merely touch the effector cells of the organs that they innervate as they pass by; and in some instances they terminate in connective tissue located adjacent to the cells that are to be stimulated. Where these filaments pass over or near the effector cells, they usually have bulbous enlargements called *varicosities;* it is in these varicosities that the transmitter vesicles of acetylcholine or norepinephrine are found. Also in the varicosities are large numbers of mitochondria to supply the adenosine triphosphate (ATP) required to energize acetylcholine or norepinephrine synthesis.

When an action potential spreads over the terminal fibers, the depolarization process increases the permeability of the fiber membrane to calcium ions, allowing these to diffuse into the nerve terminals. There they interact with the vesicles that are adjacent to the membrane, causing them to fuse with the membrane and to empty their contents to the exterior. Thus, the transmitter substance is secreted.

Synthesis of Acetylcholine, Its Destruction After Secretion, and Duration of Action. Acetylcholine is synthesized in the terminal endings of cholinergic nerve fibers. Most of this synthesis occurs in the axoplasm outside the vesicles, and then the acetylcholine is transported to the interior of the vesicles, where it is stored in a highly concentrated form until it is released. The basic chemical reaction of this synthesis is the following:

$$\text{Acetyl-CoA} + \text{Choline} \xrightarrow{\text{choline acetyl-transferase}} \text{Acetylcholine}$$

Once the acetylcholine has been secreted by the cholinergic nerve ending, it persists in the tissue for a few seconds; then most of it is split into an acetate ion and choline by the enzyme *acetylcholinesterase* bound with collagen and glycosaminoglycans in the local connective tissue. Thus, this is the same mechanism of acetylcholine destruction that occurs at the neuromuscular junctions of skeletal nerve fibers. The choline that is formed is in turn transported back into the terminal nerve ending, where it is used again for synthesis of new acetylcholine.

Synthesis of Norepinephrine, Its Removal, and Duration of Action. Synthesis of norepinephrine begins in the axoplasm of the terminal nerve endings of adrenergic nerve fibers but is completed inside the vesicles. The basic steps are the following:

1. $\text{Tyrosine} \xrightarrow{\text{hydroxylation}} \text{DOPA}$

2. $\text{DOPA} \xrightarrow{\text{decarboxylation}} \text{Dopamine}$

3. Transport of dopamine into the vesicles

4. $\text{Dopamine} \xrightarrow{\text{hydroxylation}} \text{Norepinephrine}$

In the adrenal medulla this reaction goes still one step further to transform about 80 per cent of the norepinephrine into epinephrine, as follows:

5. $\text{Norepinephrine} \xrightarrow{\text{methylation}} \text{Epinephrine}$

After secretion of norepinephrine by the terminal nerve endings, it is removed from the secretory site in three different ways: (1) reuptake into the adrenergic nerve endings themselves by an active transport process — accounting for removal of 50 to 80 per cent of the secreted norepinephrine; (2) diffusion away from the nerve endings into the surrounding body fluids and thence into the blood — accounting for removal of most of the remainder of the norepinephrine; and (3) destruction by enzymes to a slight extent (one of these enzymes is *monoamine oxidase,* which is found in the nerve endings themselves, and another is *catechol-O-methyl transferase,* which is present diffusely in all tissues).

Ordinarily, the norepinephrine secreted directly into a tissue remains active for only a few seconds, illustrating that its reuptake and diffusion away from the tissue are rapid. However, the norepinephrine and epinephrine secreted into the blood by the adrenal medullae remain active until they diffuse into some tissue where they are destroyed by catechol-O-methyl transferase; this occurs mainly in the liver. Therefore, when secreted into the blood, both norepinephrine and epinephrine remain very active for 10 to 30 seconds; additional decreasing activity follows for 1 to several minutes.

RECEPTORS OF THE EFFECTOR ORGANS

Before the acetylcholine, norepinephrine, or epinephrine transmitter secreted at the autonomic nerve endings can stimulate the effector organ, it must first bind with highly specific *receptors* of the effector cells. The receptor is usually on the outside of the cell membrane, bound as a prosthetic group to a protein molecule that penetrates all the way through the cell membrane. When the transmitter binds with the receptor, this generally causes a conformational change in the structure of the protein molecule. In turn, the altered protein molecule excites or inhibits the cell, most often by (1) causing a change in the cell membrane permeability to one or more ions or (2) activating or inactivating an enzyme attached to the other end of the receptor protein where it protrudes into the interior of the cell.

Excitation or Inhibition of the Effector Cell by Changing Its Membrane Permeability. Because the receptor protein is an integral part of the cell membrane, a conformational change in the structures of many of these proteins opens or closes *ion channels,* thus altering the permeability of the cell membrane to various ions. For instance, sodium and/or calcium ion channels frequently become opened and allow rapid influx of the respective ions into the cell, usually depolarizing the cell membrane and exciting the cell. At other times, potassium channels are opened, allowing potassium ions to diffuse out of the cell, and this usually inhibits it. Also, in some cells, the ions will cause an internal cell action, such as the direct effect of calcium ions in promoting smooth muscle contraction.

Receptor Action by Altering Intracellular Enzymes. Another way the receptor functions is in activating or inactivating an enzyme (or other intracellular chemical) inside the cell. The enzyme usually is attached to the receptor protein where it protrudes into the interior of the cell. For instance, binding of epinephrine with its receptor on the outside of many cells increases the activity of the enzyme *adenylcyclase* on the inside of the cell, and this then causes the formation of *cyclic adenosine monophosphate (cAMP)*. The cAMP in turn can initiate any one of many different intracellular actions, the exact effect depending on the chemical machinery of the effector cell.

Therefore, it is easy to understand how an autonomic transmitter substance can cause inhibition in some organs or excitation in others. This is usually determined by the nature of the receptor protein in the cell membrane and the effect of receptor binding on its conformational state. In each organ, the resulting effects are likely to be entirely different from those in other organs.

The Acetylcholine Receptors — Muscarinic and Nicotinic Receptors

Acetylcholine activates two different types of receptors. These are called *muscarinic* and *nicotinic* receptors. The reason for these names is that muscarine, a poison from toadstools, activates only the muscarinic receptors but will not activate the nicotinic receptors, whereas nicotine will activate only nicotinic receptors; acetylcholine activates both of them.

The muscarinic receptors are found in all effector cells stimulated by the postganglionic neurons of the parasympathetic nervous system, as well as those stimulated by the postganglionic cholinergic neurons of the sympathetic system.

The nicotinic receptors are found in the synapses between the pre- and postganglionic neurons of both the sympathetic and parasympathetic systems and also in the membranes of skeletal muscle fibers at the neuromuscular junction (discussed in Chapter 7).

An understanding of the two different types of receptors is especially important because specific drugs are frequently used in the practice of medicine to stimulate or to block one or the other of the two types of receptors.

The Adrenergic Receptors — Alpha and Beta Receptors

Research experiments using different drugs that mimic the action of norepinephrine on sympathetic effector organs (called *sympathomimetic drugs*) have shown that there are two major types of adrenergic receptors, *alpha receptors* and *beta receptors*. (The beta receptors in turn are divided into *beta_1* and *beta_2* receptors because certain drugs affect some beta receptors but not all. Also, there is a less distinct division of alpha receptors into alpha_1 and alpha_2 receptors.)

Norepinephrine and epinephrine, both of which are secreted by the adrenal medulla, have somewhat different effects in exciting the alpha and beta receptors. Norepinephrine excites mainly alpha receptors but excites the beta receptors to a slight extent as well. On the other hand, epinephrine excites both types of receptors approximately equally. Therefore, the relative effects of norepinephrine and epinephrine on different effector organs is determined by the types of receptors in the organs. Obviously, if they are all beta receptors, epinephrine will be the more effective excitant.

Table 60-1 gives the distribution of alpha and beta receptors in some of the organs and systems controlled by the sympathetics. Note that certain alpha functions are excitatory while others are inhibitory. Likewise, certain beta functions are excitatory and others are inhibitory. Therefore, alpha and beta receptors are not necessarily associated with excitation or inhibition but simply with the affinity of the hormone for the receptors in the given effector organ.

A synthetic hormone chemically similar to epinephrine and norepinephrine, *isopropyl norepinephrine,* has an extremely strong action on beta receptors but essentially no action on alpha receptors.

EXCITATORY AND INHIBITORY ACTIONS OF SYMPATHETIC AND PARASYMPATHETIC STIMULATION

Table 60-2 lists the effects on different visceral functions of the body caused by stimulating the parasympathetic and sympathetic nerves. From this table it can be seen again that *sympathetic stimulation causes excitatory effects in some organs but inhibitory effects in others. Likewise, parasympathetic stimulation causes excitation in some but inhibition in others.*

Table 60-1 ADRENERGIC RECEPTORS AND FUNCTION

Alpha Receptor	Beta Receptor
Vasoconstriction	Vasodilatation (β_2)
Iris dilatation	Cardioacceleration (β_1)
Intestinal relaxation	Increased myocardial strength (β_1)
Intestinal sphincter contraction	Intestinal relaxation (β_2)
Pilomotor contraction	Uterus relaxation (β_2)
Bladder sphincter contraction	Bronchodilatation (β_2)
	Calorigenesis (β_2)
	Glycogenolysis (β_2)
	Lipolysis (β_1)
	Bladder wall relaxation (β_2)

Table 60–2 AUTONOMIC EFFECTS ON VARIOUS ORGANS OF THE BODY

Organ	Effect of Sympathetic Stimulation	Effect of Parasympathetic Stimulation
Eye		
Pupil	Dilated	Constricted
Ciliary muscle	Slight relaxation (far vision)	Constricted (near vision)
Glands	Vasoconstriction and slight secretion	Stimulation of copious secretion (containing many enzymes for enzyme-secreting glands)
Nasal		
Lacrimal		
Parotid		
Submandibular		
Gastric		
Pancreatic		
Sweat glands	Copious sweating (cholinergic)	Sweating on palms of hands
Apocrine glands	Thick, odoriferous secretion	None
Heart		
Muscle	Increased rate Increased force of contraction	Slowed rate Decreased force of contraction (especially of atria)
Coronaries	Dilated (β_2); constricted (α)	Dilated
Lungs		
Bronchi	Dilated	Constricted
Blood vessels	Mildly constricted	? Dilated
Gut		
Lumen	Decreased peristalsis and tone	Increased peristalsis and tone
Sphincter	Increased tone (most times)	Relaxed (most times)
Liver	Glucose released	Slight glycogen synthesis
Gallbladder and bile ducts	Relaxed	Contracted
Kidney	Decreased output and renin secretion	None
Bladder		
Detrusor	Relaxed (slight)	Contracted
Trigone	Contracted	Relaxed
Penis	Ejaculation	Erection
Systemic arterioles		
Abdominal viscera	Constricted	None
Muscle	Constricted (adrenergic α) Dilated (adrenergic β_2) Dilated (cholinergic)	None
Skin	Constricted	None
Blood		
Coagulation	Increased	None
Glucose	Increased	None
Lipids	Increased	None
Basal metabolism	Increased up to 100%	None
Adrenal medullary secretion	Increased	None
Mental activity	Increased	None
Piloerector muscles	Contracted	None
Skeletal muscle	Increased glycogenolysis Increased strength	None
Fat cells	Lipolysis	None

Also, when sympathetic stimulation excites a particular organ, parasympathetic stimulation sometimes inhibits it, illustrating that the two systems occasionally act reciprocally to each other. However, most organs are dominantly controlled by one or the other of the two systems.

There is no generalization one can use to explain whether sympathetic or parasympathetic stimulation will cause excitation or inhibition of a particular organ. Therefore, to understand sympathetic and parasympathetic function, one must learn the functions of these two nervous systems as listed in Table 60–2. Some of these functions need to be clarified in still greater detail as follows:

EFFECTS OF SYMPATHETIC AND PARASYMPATHETIC STIMULATION ON SPECIFIC ORGANS

The Eye. Two functions of the eye are controlled by the autonomic nervous system. These are the pupillary opening and the focus of the lens. Sympathetic stimulation contracts the meridional *fibers of the iris* that dilate the pupil, whereas parasympathetic stimulation contracts the *circular*

muscle of the iris to constrict the pupil. The parasympathetics that control the pupil are reflexly stimulated when excess light enters the eyes, which is explained in Chapter 51; this reflex reduces the pupillary opening and decreases the amount of light that strikes the retina. On the other hand, the sympathetics become stimulated during periods of excitement and, therefore, increase the pupillary opening at these times.

Focusing of the lens is controlled almost entirely by the parasympathetic nervous system. The lens is normally held in a flattened state by intrinsic elastic tension of its radial ligaments. Parasympathetic excitation contracts the *ciliary muscle,* which releases this tension and allows the lens to become more convex. This causes the eye to focus on objects near at hand. The focusing mechanism is discussed in Chapters 49 and 51 in relation to the function of the eyes.

The Glands of the Body. The *nasal, lacrimal, salivary,* and many *gastrointestinal glands* are all strongly stimulated by the parasympathetic nervous system, usually resulting in copious quantities of secretion. The glands of the alimentary tract most strongly stimulated by the parasympathetics are those of the upper tract, especially those of the mouth and stomach. The glands of the small and large intestines are controlled principally by local factors in the intestinal tract itself and not by the autonomic nerves.

Sympathetic stimulation has a slight direct effect on glandular cells in causing formation of a concentrated secretion. However, it also causes vasoconstriction of the blood vessels supplying the glands and in this way often reduces their rates of secretion.

The *sweat glands* secrete large quantities of sweat when the sympathetic nerves are stimulated, but no effect is caused by stimulating the parasympathetic nerves. However, the sympathetic fibers to most sweat glands are *cholinergic* (except for a few adrenergic fibers to palms of the hand and the soles of the feet), in contrast to almost all other sympathetic fibers, which are adrenergic. Furthermore, the sweat glands are stimulated primarily by centers in the hypothalamus that are usually considered to be parasympathetic centers. Therefore, sweating could be called a parasympathetic function.

The *apocrine glands* in the axillae secrete a thick, odoriferous secretion as a result of sympathetic stimulation, but they do not react to parasympathetic stimulation. Instead, the apocrine glands, despite their close embryological relationship to sweat glands, are controlled by adrenergic fibers rather than by cholinergic fibers and are controlled by the sympathetic centers of the central nervous system rather than by the parasympathetic centers.

The Gastrointestinal System. The gastrointestinal system has its own intrinsic set of nerves known as the *intramural plexus.* However, both parasympathetic and sympathetic stimulation can affect gastrointestinal activity. Parasympathetic stimulation, in general, increases the overall degree of activity of the gastrointestinal tract by promoting peristalsis and relaxing the sphincters, thus allowing rapid propulsion of contents along the tract. This propulsive effect is associated with simultaneous increases in rates of secretion by many of the gastrointestinal glands, which was described earlier.

Normal function of the gastrointestinal tract is not very dependent on sympathetic stimulation. However, strong sympathetic stimulation inhibits peristalsis and increases the tone of the sphincters. The net result is greatly slowed propulsion of food through the tract and sometimes decreased secretion as well.

The Heart. In general, sympathetic stimulation increases the overall activity of the heart. This is accomplished by increasing both the rate and force of heart contraction. Parasympathetic stimulation causes mainly the opposite effects. To express these effects in another way, sympathetic stimulation increases the effectiveness of the heart as a pump, whereas parasympathetic stimulation decreases its pumping capability.

Systemic Blood Vessels. Most systemic blood vessels, especially those of the abdominal viscera and the skin of the limbs, are constricted by sympathetic stimulation. Parasympathetic stimulation generally has almost no effects on blood vessels but does dilate vessels in certain restricted areas such as in the blush area of the face. Under some conditions, the beta function of the sympathetics causes vascular dilatation, especially when drugs have paralyzed the sympathetic alpha vasoconstrictor effects, which are usually by far dominant over the beta effects.

Effect of Sympathetic and Parasympathetic Stimulation on Arterial Pressure. The arterial pressure is determined by two factors, the propulsion of blood by the heart and the resistance to flow of this blood through the blood vessels. Sympathetic stimulation increases both propulsion by the heart and resistance to flow, which usually causes the pressure to increase greatly.

On the other hand, parasympathetic stimulation decreases the pumping by the heart but has virtually no effect on total peripheral resistance. The usual effect is a slight fall in pressure. Yet very strong vagal parasympathetic stimulation can occasionally stop the heart entirely and cause loss of all arterial pressure.

Effects of Sympathetic and Parasympathetic Stimulation on Other Functions of the Body. Because of the great importance of the sympathetic and parasympathetic control systems, these are discussed many times in this text in relation to a myriad of body functions that are not considered in detail here. In general, most of the entodermal structures, such as the ducts of the liver, the gallbladder, the ureter, the bladder, and the bronchi, are inhibited by sympathetic stimulation but excited by parasympathetic stimulation. Also, sympathetic stimulation has metabolic effects, causing release of glucose from the liver, increase in blood glucose concentration, increase in glycogenolysis in both liver and muscle, increase in muscle strength, increase in basal metabolic rate, and increase in mental activity. Finally, the sympathetics and parasympathetics are involved in the execution of the male and female sexual acts, as are explained in Chapters 80 and 81.

FUNCTION OF THE ADRENAL MEDULLAE

Stimulation of the sympathetic nerves to the adrenal medullae causes large quantities of epinephrine and norepinephrine to be released into the circulating blood, and these two hormones in turn are carried in the blood to all tissues of the body. On the average, approximately 80 per cent of the secretion is epinephrine and 20 per cent is norepinephrine, though the relative proportions can change considerably under different physiological conditions.

The circulating epinephrine and norepinephrine have almost the same effects on the different organs as those caused by direct sympathetic stimulation, except that *the effects last 5 to 10 times as long* because these hormones are removed from the blood slowly.

The circulating norepinephrine causes constriction

of essentially all the blood vessels of the body; it causes increased activity of the heart, inhibition of the gastrointestinal tract, dilation of the pupils of the eyes, and so forth.

Epinephrine causes almost the same effects as those caused by norepinephrine, but the effects differ in the following respects: First, epinephrine, because of its greater effect in stimulating the beta receptors, has a greater effect on cardiac stimulation than norepinephrine. Second, epinephrine causes only weak constriction of the blood vessels of the muscles, in comparison with the much stronger constriction caused by norepinephrine. Because the muscle vessels represent a major segment of the vessels of the body, this difference is of special importance because norepinephrine greatly increases the total peripheral resistance and thereby greatly elevates arterial pressure, whereas epinephrine raises the arterial pressure to a lesser extent but increases the cardiac output considerably more because of its excitatory effect on the heart.

A third difference between the actions of epinephrine and norepinephrine relates to their effects on tissue metabolism. Epinephrine has up to five to ten times as great a metabolic effect as norepinephrine. Indeed, the epinephrine secreted by the adrenal medullae can increase the metabolic rate of the body often to as much as 100 per cent above normal, in this way increasing the activity and excitability of the whole body. It also increases the rate of other metabolic activities, such as glycogenolysis in the liver and muscle and glucose release into the blood.

In summary, stimulation of the adrenal medullae causes the release of hormones that have almost the same effects throughout the body as direct sympathetic stimulation, except that the effects are greatly prolonged, up to a minute or two after the stimulation is over. The only significant differences are caused by the beta effects of the epinephrine in the secretion, which increase the rate of metabolism and cardiac output to a greater extent than is caused by direct sympathetic stimulation.

Value of the Adrenal Medullae to the Function of the Sympathetic Nervous System. Epinephrine and norepinephrine are almost always released by the adrenal medullae at the same time that the different organs are stimulated directly by generalized sympathetic activation. Therefore, the organs are actually stimulated in two different ways simultaneously, directly by the sympathetic nerves and indirectly by the medullary hormones. The two means of stimulation support each other, and either can usually substitute for the other. For instance, destruction of the direct sympathetic pathways to the organs does not abrogate excitation of the organs, because norepinephrine and epinephrine are still released into the circulating blood and indirectly cause stimulation. Likewise, total loss of the two adrenal medullae usually has little effect on the operation of the sympathetic nervous system because the direct pathways can still perform almost all the necessary duties. Thus, the dual mechanism of sympathetic stimula-

tion provides a safety factor, one mechanism substituting for the other when it is missing.

Another important value of the adrenal medullae is the capability of epinephrine and norepinephrine to stimulate structures of the body that are not innervated by direct sympathetic fibers. For instance, the metabolic rate of every cell of the body is increased by these hormones, especially by epinephrine, even though only a small proportion of all the cells in the body are innervated directly by sympathetic fibers.

RELATIONSHIP OF STIMULUS RATE TO DEGREE OF SYMPATHETIC AND PARASYMPATHETIC EFFECT

A special difference between the autonomic nervous system and the skeletal nervous system is that only a very low frequency of stimulation is required for full activation of autonomic effectors. In general, only one nerve impulse every second or so suffices to maintain normal sympathetic or parasympathetic effect, and full activation occurs when the nerve fibers discharge 10 to 20 times per second. This compares will full activation in the skeletal nervous system at 50 to 500 or more impulses per second.

SYMPATHETIC AND PARASYMPATHETIC "TONE"

The sympathetic and parasympathetic systems are continually active, and the basal rates of activity are known, respectively, as *sympathetic tone* or *parasympathetic tone*.

The value of tone is that *it allows a single nervous system to increase or to decrease the activity of a stimulated organ.* For instance, sympathetic tone normally keeps almost all of the systemic arterioles constricted to approximately half their maximum diameter. By increasing the degree of sympathetic stimulation, these vessels can be constricted even more; but, on the other hand, by inhibiting the normal tone, they can be dilated. If it were not for the continual sympathetic tone, the sympathetic system could cause only vasoconstriction, never vasodilatation.

Another interesting example of tone is that of the parasympathetics in the gastrointestinal tract. Surgical removal of the parasympathetic supply to most of the gut by cutting the vagus nerves can cause serious and prolonged gastric and intestinal "atony" with resulting blockage of gastrointestinal propulsion and consequent serious constipation, thus illustrating that parasympathetic tone to the gut is normally very strong. This tone can be decreased by the brain, thereby inhibiting gastrointestinal motility, or it can be increased, thereby promoting increased gastrointestinal activity.

Tone Caused by Basal Secretion of Epinephrine and Norepinephrine by the Adrenal Medullae. The normal resting rate of secretion by the adrenal medullae is about 0.2 μg/kg/min of epinephrine and about 0.05 μg/kg/min of norepinephrine. These quantities are considerable — indeed, enough to maintain the blood pressure almost up to the normal value even if all direct sympathetic pathways to the cardiovascular system are removed. Therefore, it is obvious that much of the overall tone of the sympathetic nervous system results from basal secretion of epinephrine and norepinephrine in addition to the tone resulting from direct sympathetic stimulation.

Effect of Loss of Sympathetic or Parasympathetic Tone Following Denervation. Immediately after a sympathetic or parasympathetic nerve is cut, the innervated organ loses its sympathetic or parasympathetic tone. In the case of the blood vessels, for instance, cutting the sympathetic nerves results immediately in almost maximal vasodilatation. However, over minutes, hours, days, or weeks, *intrinsic tone* in the smooth muscle of the vessels increases, usually eventually restoring almost normal vasoconstriction.

Essentially the same events occur in most effector organs whenever sympathetic or parasympathetic tone is lost. That is, intrinsic compensation soon develops to return the function of the organ almost to its normal basal level. However, in the parasympathetic system, the compensation sometimes requires many months. For instance, loss of parasympathetic tone to the heart increases the heart rate to 160 beats per minute in a dog, and this will still be partially elevated 6 months later.

DENERVATION SUPERSENSITIVITY OF SYMPATHETIC AND PARASYMPATHETIC ORGANS FOLLOWING DENERVATION

During the first week or so after a sympathetic or parasympathetic nerve is destroyed, the innervated organ becomes more and more sensitive to injected norepinephrine or acetylcholine, respectively. This effect is illustrated in Figure 60–4, showing the blood flow in the forearm before removal of the sympathetics to be about 200 ml/min; a test dose of norepinephrine causes only a slight depression in flow. Then the stellate ganglion is removed, and normal sympathetic tone is lost. At first, the blood flow rises markedly because of the lost vascular tone, but over a period of days to weeks the blood flow returns almost to normal because of progressive increase in intrinsic tone of the vascular musculature itself, thus compensating for the loss of sympathetic tone. Another test dose of norepinephrine is then administered; and the blood flow decreases much more

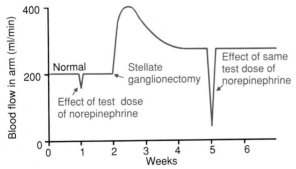

Figure 60–4. Effect of sympathectomy on blood flow in the arm and the effect of a test dose of norepinephrine before and after sympathectomy, showing sensitization of the vasculature to norepinephrine.

than before, illustrating that the blood vessels become about two to four times as responsive to norepinephrine as previously. This phenomenon is called *denervation supersensitivity*. It occurs in both sympathetic and parasympathetic organs and to a far greater extent in some organs than in others, often increasing the response more than tenfold.

Mechanism of Denervation Supersensitivity. The cause of denervation supersensitivity is only partially known. Part of the answer is that the number of receptors in the postsynaptic membranes of the effector cells increases —sometimes manyfold—when norepinephrine or acetylcholine is no longer released at the synapses, a process called "up-regulation" of the receptors. Therefore, when these hormones are injected temporarily into the circulating blood, the effector reaction is vastly enhanced.

● THE AUTONOMIC REFLEXES

Many of the visceral functions of the body are regulated by *autonomic reflexes*. Throughout this text the functions of these reflexes are discussed in relation to individual organ systems; but to illustrate their importance, a few are presented here briefly.

Cardiovascular Autonomic Reflexes. Several reflexes in the cardiovascular system help to control especially the arterial blood pressure and the heart rate. One of these is the *baroreceptor reflex*, which is described in Chapter 18 along with other cardiovascular reflexes. Briefly, stretch receptors called *baroreceptors* are located in the walls of the major arteries, including the carotid arteries and the aorta. When these become stretched by high pressure, signals are transmitted to the brain stem, where they inhibit the sympathetic impulses to the heart and blood vessels, which allows the arterial pressure to fall back toward normal.

Gastrointestinal Autonomic Reflexes. The uppermost part of the gastrointestinal tract and also the rectum are controlled principally by autonomic reflexes. For instance, the smell of appetizing food or the presence of food in the mouth initiates signals from the nose and mouth to the vagal, glossopharyngeal, and salivary nuclei of the brain stem. These in turn transmit signals through the parasympathetic nerves to the secretory glands of the mouth and stomach, causing secretion of digestive juices even before food enters the mouth. And when fecal matter fills the rectum at the other end of the alimentary canal, sensory impulses initiated by stretching the rectum are sent to the sacral portion of the spinal cord, and a reflex signal is retransmitted through the parasympathetics to the distal parts of the colon; these result in strong peristaltic contractions that empty the bowel.

Other Autonomic Reflexes. Emptying of the bladder is controlled in the same way as emptying the rectum; stretching of the bladder sends impulses to the sacral cord, and this in turn causes contraction of the bladder as well as relaxation of the urinary sphincters, thereby promoting micturition.

Also important are the sexual reflexes, which are initiated both by psychic stimuli from the brain and stimuli from the sexual organs. Impulses from these sources converge on the sacral cord and, in the male, result, first, in erection, mainly a parasympathetic function, and then in ejaculation, a sympathetic function.

Other autonomic reflexes include reflex contributions to the regulation of pancreatic secretion, gallbladder emptying, kidney excretion of urine, sweating, blood glucose concentration, and many other visceral functions, all of which are discussed in detail at other points in this text.

● STIMULATION OF DISCRETE ORGANS IN SOME INSTANCES AND MASS STIMULATION IN OTHER INSTANCES BY THE SYMPATHETIC AND PARASYMPATHETIC SYSTEMS

The Sympathetic System. In many instances, the sympathetic nervous system discharges almost as a complete unit, a phenomenon called *mass discharge.* This frequently occurs when the hypothalamus is activated by fright or fear or severe pain. The result is a widespread reaction throughout the body called the *alarm* or *stress response,* which we shall discuss shortly.

However, at other times sympathetic activation occurs in isolated portions of the system. The most important of these are the following: (1) In the process of heat regulation, the sympathetics control sweating and blood flow in the skin without affecting other organs innervated by the sympathetics. (2) During muscular activity in some animals, specific cholinergic vasodilator fibers of the skeletal muscles are stimulated independently, apart from the remainder of the sympathetic system. (3) Many "local reflexes" involving sensory afferent fibers that travel centrally in the sympathetic nerves to the sympathetic ganglia and spinal cord cause highly localized reflex responses. For instance, heating a local skin area causes local vasodilatation and enhanced local sweating, whereas cooling causes the opposite effects. (4) Many of the sympathetic reflexes that control gastrointestinal functions are very discrete, operating sometimes by way of nerve pathways that do not even enter the spinal cord, merely passing from the gut to the sympathetic ganglia, mainly the prevertebral ganglia, and then back to the gut through the sympathetic nerves to control motor or secretory activity.

The Parasympathetic System. In contrast to the sympathetic system, control functions of the parasympathetic system are much more likely to be highly specific. For instance, parasympathetic cardiovascular reflexes usually act only on the heart to increase or decrease its rate of beating. Likewise, other parasympathetic reflexes cause secretion mainly in the mouth glands, whereas in other instances secretion mainly in the stomach glands. Finally, the rectal emptying reflex does not affect other parts of the bowel to a major extent.

Yet there is often association between closely allied parasympathetic functions. For instance, though salivary secretion can occur independently of gastric secretion, these two often also occur together, and pancreatic secretion frequently occurs at the same time. Also, the rectal emptying reflex often initiates a blad-der emptying reflex, resulting in simultaneous emptying of both the bladder and rectum. Conversely, the bladder emptying reflex can help initiate rectal emptying.

"ALARM" OR "STRESS" RESPONSE OF THE SYMPATHETIC NERVOUS SYSTEM

When large portions of the sympathetic nervous system discharge at the same time—that is, a *mass discharge*—this increases in many different ways the ability of the body to perform vigorous muscle activity. Let us quickly summarize these ways:

1. Increased arterial pressure.
2. Increased blood flow to active muscles concurrent with decreased blood flow to organs such as the gastrointestinal tract and the kidneys that are not needed for rapid motor activity.
3. Increased rates of cellular metabolism throughout the body.
4. Increased blood glucose concentration.
5. Increased glycolysis in the liver and in muscle.
6. Increased muscle strength.
7. Increased mental activity.
8. Increased rate of blood coagulation.

The sum of these effects permits the person to perform far more strenuous physical activity than would otherwise be possible. Because it is mental or physical *stress* that usually excites the sympathetic system, it is frequently said that the purpose of the sympathetic system is to provide extra activation of the body in states of stress: this is often called the sympathetic *stress response.*

The sympathetic system is especially strongly activated in many emotional states. For instance, in the state of *rage,* which is elicited mainly by stimulating the hypothalamus, signals are transmitted downward through the reticular formation and spinal cord to cause massive sympathetic discharge, and all of the sympathetic events listed above ensue immediately. This is called the sympathetic *alarm reaction.* It is also frequently called the *fight or flight reaction* because an animal in this state decides almost instantly whether to stand and fight or to run. In either event, the sympathetic alarm reaction makes the animal's subsequent activities vigorous.

MEDULLARY, PONTINE, AND MESENCEPHALIC CONTROL OF THE AUTONOMIC NERVOUS SYSTEM

Many areas in the reticular substance of the medulla, pons, and mesencephalon, as well as many special nuclei (Fig. 60–5), control different autonomic functions such as arterial pressure, heart rate, glandular secretion in the upper part of the gastrointestinal tract, gastrointestinal peristalsis, the degree of contraction of the urinary bladder, and many others. The control of each of these is discussed at appropriate points in this text. Suffice it to point out here that the most important factors controlled in the lower

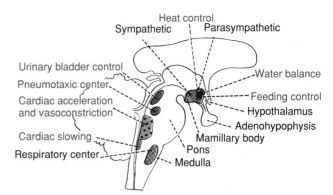

Figure 60–5. Autonomic control areas of the brain stem and hypothalamus.

brain stem are arterial pressure, heart rate, and respiration. Indeed, transection of the brain stem at the midpontine level allows normal basal control of arterial pressure to continue as before but prevents its modulation by higher nervous centers, particularly the hypothalamus. On the other hand, transection immediately below the medulla causes the arterial pressure to fall to about one-half normal for several hours or several days after the transection.

Closely associated with the cardiovascular regulatory centers in the medulla are the medullary and pontine centers for regulation of respiration, which are discussed in Chapter 41. Although this is not considered to be an autonomic function, it is one of the *involuntary* functions of the body.

Control of Lower Brain Stem Autonomic Centers by Higher Areas. Signals from the hypothalamus and even from the cerebrum can affect the activities of almost all the lower brain stem autonomic control centers. For instance, stimulation in appropriate areas of the hypothalamus can activate the medullary cardiovascular control centers strongly enough to increase the arterial pressure to more than double normal. Likewise, other hypothalamic centers can control body temperature, increase or decrease salivation and gastrointestinal activity, or cause bladder emptying. To some extent, therefore, the autonomic centers in the lower brain stem act as relay stations for control activities initiated at higher levels of the brain.

In the previous two chapters it is pointed out that many of our behavioral responses are mediated through the hypothalamus, the reticular areas of the brain stem, and the autonomic nervous system. Indeed, the higher areas of the brain can alter the function of the whole autonomic nervous system or of portions of it strongly enough to cause severe autonomic-induced disease, such as peptic ulcer, constipation, heart palpitation, and even heart attacks.

● PHARMACOLOGY OF THE AUTONOMIC NERVOUS SYSTEM

DRUGS THAT ACT ON ADRENERGIC EFFECTOR ORGANS—THE SYMPATHOMIMETIC DRUGS

From the foregoing discussion, it is obvious that intravenous injection of norepinephrine causes essentially the same effects throughout the body as sympathetic stimulation. Therefore, norepinephrine is called a *sympathomimetic, or adrenergic, drug. Epinephrine* and *methoxamine* are also sympathomimetic drugs, and there are many others. These differ from each other in the degree to which they stimulate different sympathetic effector organs and in their durations of action. Norepinephrine and epinephrine have actions as short as 1 to 2 minutes, whereas the actions of most other commonly used sympathomimetic drugs last 30 minutes to 2 hours.

Important drugs that stimulate specific adrenergic receptors but not the others are *phenylephrine*—alpha receptors; *isoproterenol*—beta receptors; and *albuterol*—only beta$_2$ receptors.

Drugs That Cause Release of Norepinephrine From Nerve Endings. Certain drugs have an indirect sympathomimetic action rather than directly exciting adrenergic effector organs. These drugs include *ephedrine, tyramine,* and *amphetamine.* Their effect is to cause release of norepinephrine from its storage vesicles in the sympathetic nerve endings. The released norepinephrine in turn causes the sympathetic effects.

Drugs That Block Adrenergic Activity. Adrenergic activity can be blocked at several different points in the stimulatory process as follows:

1. The synthesis and storage of norepinephrine in the sympathetic nerve endings can be prevented. The best known drug that causes this effect is *reserpine.*

2. Release of norepinephrine from the sympathetic endings can be blocked. This is caused by *guanethidine.*

3. The *alpha* receptors can be blocked. Two drugs that cause this effect are *phenoxybenzamine* and *phentolamine.*

4. The beta receptors can be blocked. A drug that blocks all beta receptors is *propranolol.* One that blocks only beta$_1$ receptors is *metoprolol.*

5. Sympathetic activity can be blocked by drugs that block transmission of nerve impulses through the autonomic ganglia. These are discussed in a later section, but the most important drug for blockade of both sympathetic and parasympathetic transmission through the ganglia is *hexamethonium.*

DRUGS THAT ACT ON CHOLINERGIC EFFECTOR ORGANS

Parasympathomimetic Drugs (Muscarinic Drugs). Acetylcholine injected intravenously usually does not cause exactly the same effects throughout the body as parasympathetic stimulation because the acetylcholine is destroyed by cholinesterase in the blood and body fluids before it can reach all the effector organs. Yet a number of other drugs that are not so rapidly destroyed can produce typical parasympathetic effects, and these are called *parasympathomimetic drugs.*

Two commonly used parasympathomimetic drugs are *pilocarpine* and *methacholine.* These act directly on the muscarinic type of cholinergic receptors.

Parasympathomimetic drugs act on the effector organs of cholinergic *sympathetic* fibers also. For instance, these drugs cause profuse sweating. Also, they cause vascular dilatation in some organs, this effect occurring even in vessels not innervated by cholinergic fibers.

Drugs That Have a Parasympathetic Potentiating Effect — Anticholinesterase Drugs. Some drugs do not have a direct effect on parasympathetic effector organs but do potentiate the effects of the naturally secreted acetylcholine at the parasympathetic endings. These are the same drugs as those discussed in Chapter 7 that potentiate the effect of acetylcholine at the neuromuscular junction — that is, *neostigmine, pyridostigmine,* and *ambenonium.* These inhibit acetylcholinesterase, thus preventing rapid destruction of the acetylcholine liberated by the parasympathetic nerve endings. As a consequence, the quantity of acetylcholine acting on the effector organs progressively increases with successive stimuli, and the degree of action also increases.

Drugs That Block Cholinergic Activity at Effector Organs — Antimuscarinic Drugs. *Atropine* and similar drugs, such as *homatropine* and *scopolamine,* block the action of acetylcholine on the muscarinic type of cholinergic effector organs. However, these drugs do not affect the nicotinic action of acetylcholine on the postganglionic neurons or on skeletal muscle.

DRUGS THAT STIMULATE OR BLOCK SYMPATHETIC AND PARASYMPATHETIC POSTGANGLIONIC NEURONS

Drugs That Stimulate Autonomic Ganglia. The preganglionic neurons of both the parasympathetic and sympathetic nervous systems secrete acetylcholine at their endings, and this acetylcholine in turn stimulates the postganglionic neurons. Furthermore, injected acetylcholine can also stimulate the postganglionic neurons of both systems, thereby causing at the same time both sympathetic and parasympathetic effects throughout the body. *Nicotine* is a drug that can also stimulate postganglionic neurons in the same manner as acetylcholine because the membranes of these neurons all contain the *nicotinic type of acetylcholine receptor.* Therefore, drugs that cause autonomic effects by stimulating the postganglionic neurons are frequently called *nicotinic drugs.* Some drugs, such as *acetylcholine* itself and *methacholine,* have both nicotinic and muscarinic actions, whereas pilocarpine has only muscarinic actions.

Nicotine excites both the sympathetic and parasympathetic postganglionic neurons at the same time, resulting in strong sympathetic vasoconstriction in the abdominal organs and limbs, but at the same time resulting in parasympathetic effects, such as increased gastrointestinal activity and, sometimes, slowing of the heart.

Ganglionic Blocking Drugs. Many important drugs block impulse transmission from the preganglionic neurons to the postganglionic neurons, including *tetraethyl ammonium ion, hexamethonium ion,* and *pentolinium.* These inhibit impulse transmission in both the sympathetic and parasympathetic systems simultaneously. They are often used for blocking sympathetic activity but rarely for blocking parasympathetic activity, because the sympathetic blockade usually far overshadows the effects of parasympathetic blockade. The ganglionic blocking drugs can especially reduce the arterial pressure in patients with hypertension, but these drugs are not very useful for this purpose because their effects are difficult to control.

REFERENCES

Abboud, F. M. (ed.): Disturbances in Neurogenic Control of the Circulation. Baltimore, Williams & Wilkins, 1981.

Bannister, Sir R. (ed.): Autonomic Failure. New York, Oxford University Press, 1988.

Buckley, J. P., et al. (eds.): Brain Peptides and Catecholamines in Cardiovascular Regulation. New York, Raven Press, 1987

Burattini, R., and Borgdorff, P.: Closed-loop baroreflex control of total peripheral resistance in the cat: Identification of gains by aid of a model. Cardiovasc. Res., 18:715, 1984.

Burchfield, S. R. (ed.): Stress. Physiological and Psychological Interactions. Washington, D.C., Hemisphere Publishing Corp., 1985.

Christensen, N. J., and Galbo, H.: Sympathetic nervous activity during exercise. Annu. Rev. Physiol., 45:139, 1983.

Cotman, C. W., et al. (eds.): The Neuro-Immune-Endocrine Connection. New York, Raven Press, 1987.

Davies, A. O., and Lefkowitz, R. J.: Regulation of B-adrenergic receptors by steroid hormones. Annu. Rev. Physiol., 46:119, 1984.

Donald, D. E., and Shepherd, J. T.: Autonomic regulation of the peripheral circulation. Annu. Rev. Physiol., 42:429, 1980.

Francis, G. S., and Cohn, J. N.: Catecholamines in Cardiovascular Disease. Current Concepts, February, 1988.

Gillis, C. N., and Pitt, B. R.: The fate of circulating amines within the pulmonary circulation. Annu. Rev. Physiol., 44:269, 1982.

Givens, J. R.: The Hypothalamus in Health and Disease. Chicago, Year Book Medical Publishers, 1984.

Goldstein, D. S., and Eisenhofer, G.: Plasma catechols — What do they mean? News Physiol. Sci., 3:138, 1988.

Guyton, A. C., and Gillespie, W. M., Jr.: Constant infusion of epinephrine: Rate of epinephrine secretion and destruction in the body. Am. J. Physiol., 164:319, 1951.

Guyton, A. C., and Reeder, R. C.: Quantitative studies on the autonomic actions of curare. J. Pharmacol. Exp. Ther., 98:188, 1950.

Hirst, G. D. S., and Edwards, F. R.: Sympathetic neuroeffector transmission in arteries and arterioles. Physiol. Rev., 69:546, 1989.

Hoffman, B. B., and Lefkowitz, R. J.: Radioligand binding studies of adrenergic receptors: New insights into molecular and physiological regulation. Annu. Rev. Pharmacol. Toxicol., 20:581, 1980.

Janig, W.: Pre- and postganglionic vasoconstrictor neurons: Differentiation, types, and discharge properties. Annu. Rev. Physiol., 50:525, 1988.

Kobilka, B. K., et al.: Chimeric α_2, β_2 adrenergic receptors: Delineation of domains involved in effector coupling and ligand binding specificity. Science, 240:1310, 1988.

Kreulen, D. L.: Integration in autonomic ganglia. Physiologist, 27:49, 1984.

Levitzki, A.: β-adrenergic receptors and their mode of coupling to adenylate cyclase. Physiol. Rev., 66:819, 1986.

Livett, B. G.: Adrenal medullary chromaffin cells in vitro. Physiol. Rev., 64:1103, 1984.

Ludbrook, J., and Evans, R.: Posthemorrhagic syncope. News Physiol. Sci., 4:120, 1989.

Lundberg, J. M., et al.: Neuropeptide Y: Sympathetic cotransmitter and modulator? News Physiol. Sci., 4:13, 1989.

Robinson, R.: Tumours That Secrete Catecholamines: A Study of Their Natural History and Their Diagnosis. New York, John Wiley & Sons, 1980.

Rowell, L. B.: Reflex control of regional circulations in humans. J. Auton. Nerv. Syst., 11:101, 1984.

Simon, P. (ed.): Neurotransmitters. New York, Pergamon Press, 1979.

Stiles, G. L., et al.: β-Adrenergic receptors: Biochemical mechanisms of physiological regulation. Physiol. Rev., 64:661, 1984.

Stjarne, L.: New paradigm: Sympathetic neurotransmission by lateral interaction between secretory units? News Physiol. Sci., 1:103, 1986.

Tauc, L.: Nonvesicular release of neurotransmitter. Physiol. Rev., 62:857, 1982.

Torretti, J.: Sympathetic control of renin release. Annu. Rev. Pharmacol. Toxicol., 22:167, 1982.

Ungar, A., and Phillips, J. H.: Regulation of the adrenal medulla. Physiol. Rev., 63:787, 1983.

Usdin, E.: Stress. The Role of Catecholamines and Other Neurotransmitters. New York, Gordon Press Publishers, 1984.

Vanhoutte, P. M.: Vasodilatation: Vascular Smooth Muscle, Peptides, Autonomic Nerves, and Endothelium. New York, Raven Press, 1988.

Von Euler, U. S.: Noradrenaline. Springfield, Ill., Charles C Thomas, 1956.

Youmans, J. R. (ed.): Neurological Surgery. Philadelphia, W. B. Saunders Co., 1989.

CHAPTER 61

• Cerebral Blood Flow, the Cerebrospinal Fluid, and Brain Metabolism •

Thus far, we have discussed the function of the brain as if it were independent of its blood flow, its metabolism, and its fluids. However, this is far from the truth, for abnormalities of any of these can profoundly affect brain function. For instance, total cessation of blood flow to the brain causes unconsciousness within 5 to 10 seconds. This is true because lack of oxygen delivery to the brain cells shuts down most of their metabolism. Also, on a longer time scale, abnormalities of the cerebrospinal fluid, either in its composition or in its fluid pressure, can have equally severe effects on brain function.

● CEREBRAL BLOOD FLOW

NORMAL RATE OF CEREBRAL BLOOD FLOW

The normal blood flow through the brain tissue of the adult averages 50 to 55 milliliters per 100 grams of brain per minute. For the entire brain, this is approximately 750 ml/min, or 15 per cent of the total resting cardiac output.

REGULATION OF CEREBRAL BLOOD FLOW

Metabolic Control of Flow

As in most other vascular areas of the body, cerebral blood flow is highly related to the metabolism of the cerebral tissue. At least three different metabolic factors have potent effects in controlling cerebral blood flow. These are carbon dioxide concentration, hydrogen ion concentration, and oxygen concentration. An *increase* in either the carbon dioxide or the hydrogen ion concentration increases cerebral blood flow, whereas a *decrease* in oxygen concentration increases the flow.

Regulation of Cerebral Blood Flow in Response to Excess Carbon Dioxide or Hydrogen Ion Concentration. An increase in carbon dioxide concentration in the arterial blood perfusing the brain greatly increases cerebral blood flow. This is illustrated in Figure 61–1, which shows that a 70 per cent increase in arterial Pco_2 approximately doubles the blood flow.

Carbon dioxide is believed to increase cerebral blood flow almost entirely by combining first with water in the body fluids to form carbonic acid, with subsequent dissociation to form hydrogen ions. The hydrogen ions then cause vasodilatation of the cerebral vessels—the dilatation being almost directly proportional to the increase in hydrogen ion concentration.

Any other substance that increases the acidity of the brain tissue, and therefore also increases the hydrogen ion concentration, increases blood flow as well. Such substances include lactic acid, pyruvic acid, and any other acidic material formed during the course of metabolism.

Importance of the Carbon Dioxide and Hydrogen Control of Cerebral Blood Flow. Increased hydrogen ion concentration greatly depresses neuronal activity. Therefore, it is fortunate that an increase in hydrogen ion concentration causes an increase in blood flow, which in turn carries both carbon dioxide and other acidic substances away from the brain tissues. Loss of the carbon dioxide removes carbonic acid from the tissues; and this, along with the removal of other acids, reduces the hydrogen ion concentration back toward normal. Thus, this mechanism helps maintain a constant hydrogen ion concentration in the cerebral fluids and thereby helps to maintain the normal level of neuronal activity.

Oxygen Deficiency as a Regulator of Cerebral Blood Flow. Except during periods of intense brain activity, the utilization of oxygen by the brain tissue remains within very narrow limits—within a few percentage points of 3.5 milliliters of oxygen per 100 grams of brain tissue per minute. If the blood flow to the brain ever becomes insufficient and cannot supply this needed amount of oxygen, the oxygen deficiency mechanism for causing vasodilatation, discussed in Chapter 17, which functions in essentially all tissues of the body, immediately causes vasodilatation, returning the blood flow and transport of oxygen to the cerebral tissues to near normal. Thus, this local blood flow regulatory mechanism is much the same in the brain as in the coronary and skeletal muscle circulation and in many other circulatory areas of the body.

Experiments have shown that a decrease in cerebral *tissue* Po_2 below approximately 30 mm Hg (normal value is 35 to 40 mm Hg) will begin to increase cerebral blood flow. This is very fortuitous because brain function becomes deranged at not much lower values of Po_2, especially so at Po_2 levels below 20 mm Hg. Even coma can result at these low levels. Thus, the oxygen mechanism for local regulation of cerebral blood flow is also a very important protective re-

679

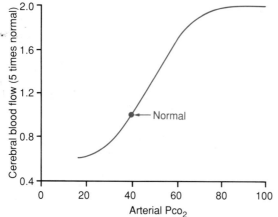

Figure 61–1. Relationship between arterial PCO₂ and cerebral blood flow.

Figure 61–2. Increase in blood flow to the occipital regions of the brain when a light is flashed in the eyes of an animal.

sponse against diminished cerebral neuronal activity and, therefore, against derangement of mental capability.

Measurement of Cerebral Blood Flow and Effect of Cerebral Activity on the Flow. A method has recently been developed to record blood flow in as many as two hundred fifty-six isolated segments of the human cerebral cortex simultaneously. A radioactive substance, usually radioactive xenon, is injected into the carotid artery; then the radioactivity of each segment of the cortex is recorded as the radioactive substance passes through the brain tissue. For this to be done, 256 small radioactive scintillation detectors are focused on the same number of separate parts of the cortex; the rapidity of decay of the radioactivity after it once peaks in each tissue segment is a direct measure of the rate of blood flow through the segment.

With this technique, it has become clear that the blood flow in each individual segment of the brain changes within seconds in response to changes in local neuronal activity. For instance, simply making a fist of the hand causes an immediate increase in blood flow in the motor cortex of the opposite side of the brain. Or, reading a book increases the blood flow in multiple areas of the brain, especially in the occipital cortex and in the language areas of the temporal cortex. This measuring procedure can also be used for localizing the origin of epileptic attacks, for the blood flow increases acutely and markedly at the focal point of the attack at its onset.

Illustrating the effect of local neuronal activity on cerebral blood flow, Figure 61–2 shows an increase in occipital blood flow recorded in a cat when intense light was shone into its eyes for a period of 0.5 minute.

Autoregulation of Cerebral Blood Flow When the Arterial Pressure Is Changed. Cerebral blood flow is autoregulated extremely well between the pressure limits of 60 and 140 mm Hg. That is, the arterial pressure can be decreased acutely to as low as 60 mm Hg or increased to as high as 140 mm Hg without a significant change in cerebral blood flow. In persons who have hypertension, this autoregulatory range shifts to even higher pressure levels, up to as high as 180 to 200 mm Hg. This effect is illustrated in Figure 61–3, which shows cerebral blood flows measured both in normal human beings and in hypertensive patients. Note the extreme constancy of cerebral blood flow between the limits of 60 and 180 mm Hg mean arterial pressure. On the other hand, if the arterial pressure does fall below 60 mm Hg, cerebral blood flow then becomes severely compro-

mised, and if the pressure rises above the upper limit of autoregulation, the blood flow rises rapidly and can cause severe overstretching of the cerebral blood vessels, sometimes resulting in serious brain edema.

Role of the Sympathetic Nervous System in Regulating Cerebral Blood Flow

The cerebral circulatory system has a strong sympathetic innervation that passes upward from the superior cervical sympathetic ganglia along with the cerebral arteries. This innervation supplies both the large superficial arteries and the small arteries that penetrate into the substance of the brain. However, neither transection of these sympathetic nerves nor mild to moderate stimulation of them normally causes significant change in the cerebral blood flow. Therefore, it has long been stated that the sympathetic nerves play essentially no role in regulating cerebral blood flow.

However, recent experiments have shown that cerebral sympathetic stimulation can, under some conditions, become activated strongly enough to constrict the cerebral arteries markedly. The reason that this usually does not occur is that the local blood flow autoregulatory mechanism is so powerful that it normally compensates almost entirely for the effects of the sympathetic stimulation. Yet in those conditions in which the autoregulatory mechanism fails to compensate enough, sympathetic control of cerebral blood

Figure 61–3. Relationship of mean arterial pressure to cerebral blood flow in normotensive, hypotensive, and hypertensive persons. (Modified from Lassen: *Physiol. Rev.,* 39:183, 1959.)

flow becomes quite important. For instance, when the arterial pressure rises to a very high level during strenuous exercise and during other states of excessive circulatory activity, the sympathetic nervous system constricts the large and intermediate-sized arteries and prevents the very high pressures from ever reaching the smaller blood vessels. Experiments have shown that this is important in preventing the occurrence of a vascular hemorrhage into the brain — that is, for preventing the occurrence of cerebral stroke.

Also, sympathetic reflexes are believed to cause vasospasm in the intermediate and large arteries in some instances of brain damage, such as after a cerebral stroke has occurred or in patients with a subdural hematoma or brain tumor.

THE CEREBRAL MICROCIRCULATION

As in almost all other tissues of the body, the density of the blood capillaries in the brain is greatest where the metabolic needs are greatest. The overall metabolic rate of the brain gray matter, where the neuronal cell bodies lie, is about four times as great as that of white matter; correspondingly, the number of capillaries and rate of blood flow are also about four times as great in the gray matter.

Another important structural characteristic of the brain capillaries is that they are much less "leaky" than the capillaries in almost any other tissue of the body. Most importantly, the capillaries are supported on all sides by "glial feet," which are small projections from the surrounding glia that abut against all surfaces of the capillaries and provide physical support to prevent overstretching of the capillaries in case of high pressure. In addition, the walls of the small arterioles leading to the brain capillaries become greatly thickened in persons who develop high blood pressure, and these arterioles remain significantly constricted all of the time to prevent transmission of the high pressure to the capillaries. We shall see later in the chapter that whenever these systems for protecting against transudation of fluid into the brain break down, serious brain edema ensues, which can lead rapidly to coma and death.

● THE CEREBROSPINAL FLUID SYSTEM

The entire cavity enclosing the brain and spinal cord has a volume of approximately 1600 milliliters, and about 150 milliliters of this volume is occupied by cerebrospinal fluid. This fluid, as shown in Figure 61–4, is found in the *ventricles of the brain,* in the *cisterns around the brain,* and in the *subarachnoid space around both the brain and the spinal cord.* All these chambers are connected with each other, and the pressure of the fluid is regulated at a constant level.

CUSHIONING FUNCTION OF THE CEREBROSPINAL FLUID

A major function of the cerebrospinal fluid is to cushion the brain within its solid vault. Fortunately, the brain and the cerebrospinal fluid have approximately the same specific gravity (only about 4 per cent different), so that the brain simply floats in the fluid. Therefore, a blow to the head moves the entire brain simultaneously, causing no one portion of the brain to be momentarily contorted by the blow.

Contrecoup. When a blow to the head is extremely severe, it usually does not damage the brain on the side of the

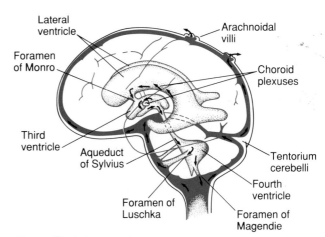

Figure 61–4. Pathway of cerebrospinal fluid flow from the choroid plexuses in the lateral ventricles to the arachnoidal villi protruding into the dural sinuses.

head where the blow is struck, but, instead, on the opposite side. This phenomenon is known as "contrecoup," and the reason for this effect is the following: When the blow is struck, the fluid on the struck side is so incompressible that, as the skull moves, the fluid pushes the brain at the same time. However, on the opposite side, the sudden movement of the skull causes it to pull away from the brain momentarily because of the brain's inertia, creating for a split second a vacuum space in the cranial vault at this point. Then, when the skull is no longer being accelerated by the blow, the vacuum suddenly collapses and the brain strikes the inner surface of the skull. Because of this effect, the damage to the brain of a boxer usually does not occur in the frontal regions but, instead, in the occipital regions.

FORMATION, FLOW, AND ABSORPTION OF CEREBROSPINAL FLUID

Cerebrospinal fluid is formed at a rate of approximately 500 milliliters each day, which is about three times as much as the total volume of fluid in the entire cerebrospinal fluid system. Probably two thirds or more of this fluid originates as a secretion from the choroid plexuses in the four ventricles, mainly in the two lateral ventricles. Additional amounts of fluid are secreted by all the ependymal surfaces of the ventricles and the arachnoidal membranes, and a small amount comes from the brain itself through the perivascular spaces that surround the blood vessels entering the brain.

The arrows in Figure 61–4 show the main channel of fluid flow from the choroid plexuses and then through the cerebrospinal fluid system. The fluid secreted in the lateral ventricles and the third ventricle passes along the *aqueduct of Sylvius* into the fourth ventricle, where a small amount of additional fluid is added. It then passes out of the fourth ventricle through three small openings, two lateral *foramina of Luschka* and a midline *foramen of Magendie,* entering the *cisterna magna,* a large fluid space that lies behind the medulla and beneath the cerebellum. The cisterna magna is continuous with the *subarachnoid space* that surrounds the entire brain and spinal cord. Almost all of the cerebrospinal fluid then flows upward through this space toward the cerebrum. From the cerebral subarachnoid spaces, the fluid flows into multiple *arachnoidal villi* that project into the large sagittal venous sinus and other venous sinuses. Finally, the fluid empties into the venous blood through the surfaces of these villi.

Secretion by the Choroid Plexus. The choroid plexus, which is illustrated in Figure 61–5, is a cauliflower-like growth of blood vessels covered by a thin layer of epithelial cells. This plexus projects into (1) the temporal horn of each lateral ventricle, (2) the posterior portion of the third ventricle, and (3) the roof of the fourth ventricle.

The secretion of fluid by the choroid plexus depends mainly on active transport of sodium ions through the epithelial cells that line the outsides of the plexus. The sodium ions in turn pull along large amounts of chloride ions as well because the positive charge of the sodium ion attracts the chloride ion's negative charge. The two of these together increase the quantity of osmotically active substances in the cerebrospinal fluid, which then causes almost immediate osmosis of water through the membrane, thus providing the fluid of the secretion. Less important transport processes move small amounts of glucose into the cerebrospinal fluid and both potassium and bicarbonate ions out of the cerebrospinal fluid into the capillaries. Therefore, the resulting characteristics of the cerebrospinal fluid become approximately the following: osmotic pressure, approximately equal to that of plasma; sodium ion concentration, also approximately equal to that of plasma; chloride, about 15 per cent greater than in plasma; potassium, approximately 40 per cent less; and glucose, about 30 per cent less.

Absorption of Cerebrospinal Fluid Through the Arachnoidal Villi. The *arachnoidal villi* are microscopic fingerlike projections of the arachnoidal membrane through the walls of the venous sinuses. Large conglomerates of these villi are usually found together and form macroscopic structures called *arachnoidal granulations* that can be seen protruding into the sinuses. The endothelial cells covering the villi have been shown by electron microscopy to have large vesicular holes directly through the bodies of the cells. It has been proposed that these are large enough to allow relatively free flow of cerebrospinal fluid, protein molecules, and even particles as large as red blood cells into the venous blood.

The Perivascular Spaces and Cerebrospinal Fluid. The blood vessels entering the substance of the brain pass first along the surface of the brain and then penetrate inward, carrying a layer of *pia mater,* the membrane that covers the brain, with them, as shown in Figure 61–6. The pia is only loosely adherent to the vessels, so that a space, the *perivascular space,* exists between it and each vessel. Perivascular spaces follow both the arteries and the veins into the brain as far as the arterioles and venules but not to the capillaries.

The Lymphatic Function of the Perivascular Spaces. As is true elsewhere in the body, a small amount of protein leaks out of the parenchymal capillaries into the interstitial spaces of the brain; and because no true lymphatics are present in brain tissue, this protein leaves the tissue mainly through the perivascular spaces but partly also by direct diffusion through the pia mater into the subarachnoid spaces. On reaching the subarachnoid spaces, the protein flows along with the cerebrospinal fluid to be absorbed through the *arachnoidal villi* into the cerebral veins. Therefore, the perivascular spaces, in effect, are a modified lymphatic system for the brain.

In addition to transporting fluid and proteins, the perivascular spaces also transport extraneous particulate matter from the brain into the subarachnoid space. For instance, whenever infection occurs in the brain, dead white blood cells are carried away through the perivascular spaces.

CEREBROSPINAL FLUID PRESSURE

The normal pressure in the cerebrospinal fluid system when one is lying in a horizontal position averages 130 millimeters of water (10 mm Hg), though this may be as low as 70 millimeters of water or as high as 180 millimeters of water even in the normal person. These values are considerably more positive than the −3 to −5 mm Hg pressure in the interstitial spaces of the subcutaneous tissue.

Regulation of Cerebrospinal Fluid Pressure by the Arachnoidal Villi. The cerebrospinal fluid pressure is regulated almost entirely by absorption of the fluid through the arachnoidal villi. The reason for this is that the rate of cerebrospinal fluid formation is very constant, so that this is rarely a factor in pressure control. On the other hand, the

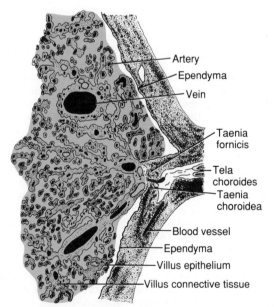

Figure 61–5. The choroid plexus. (Modified from Clara: Das Nervensystem des Menschen. Barth.)

Artery
Ependyma
Vein
Taenia fornicis
Tela choroides
Taenia choroidea
Blood vessel
Ependyma
Villus epithelium
Villus connective tissue

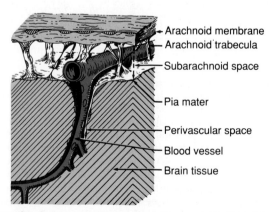

Figure 61–6. Drainage of the perivascular spaces into the subarachnoid space. (From Ranson and Clark: Anatomy of the Nervous System. Philadelphia, W. B. Saunders Company, 1959.)

Arachnoid membrane
Arachnoid trabecula
Subarachnoid space
Pia mater
Perivascular space
Blood vessel
Brain tissue

villi function like "valves" that allow the fluid and its contents to flow readily into the venous blood of the sinuses while not allowing the blood to flow backward in the opposite direction. Normally, this valve action of the villi allows cerebrospinal fluid to begin to flow into the blood when its pressure is about 1.5 mm Hg greater than the pressure of the blood in the sinuses. Then, as the cerebrospinal fluid pressure rises still higher, the valves open very widely so that, under normal conditions, the pressure almost never rises more than a few millimeters of mercury more than the pressure in the venous sinuses.

On the other hand, in disease states the villi sometimes become blocked by large particulate matter, by fibrosis, or even by excesses of plasma protein molecules that have leaked into the cerebrospinal fluid in brain diseases. Such blockage can cause very high cerebrospinal fluid pressure, as we discuss later.

Cerebrospinal Fluid Pressure in Pathological Conditions of the Brain. Often a large *brain tumor* elevates the cerebrospinal fluid pressure by decreasing the rate of absorption of fluid. For instance, if the tumor is above the tentorium and becomes so large that it compresses the brain downward, the upward flow of fluid through the subarachnoid space around the brain stem where it passes through the tentorial opening may become blocked and the absorption of fluid by the cerebral arachnoidal villi greatly curtailed. As a result, the cerebrospinal fluid pressure can rise to as high as 500 millimeters of water (37 mm Hg) or more.

The pressure also rises considerably when *hemorrhage* or *infection* occurs in the cranial vault. In both of these conditions, large numbers of cells suddenly appear in the cerebrospinal fluid, and these can cause serious blockage of the small channels for absorption through the arachnoidal villi. This sometimes elevates the cerebrospinal fluid pressure to as high as 400 to 600 millimeters of water (about four times normal).

Occasionally babies are born with high cerebrospinal fluid pressure. This is usually caused by abnormally high resistance to fluid reabsorption through the arachnoidal villi, resulting either from too few arachnoidal villi or villi with abnormal absorptive properties. This is discussed later in connection with *hydrocephalus.*

Measurement of Cerebrospinal Fluid Pressure. The usual procedure for measuring cerebrospinal fluid pressure is the following: First, the subject lies exactly horizontally on his side so that the spinal fluid pressure is equal to the pressure in the cranial vault. A spinal needle is then inserted into the lumbar spinal canal below the lower end of the cord and is connected to a glass tube. The spinal fluid is allowed to rise in the tube as high as it will. If it rises to a level 136 millimeters above the level of the needle, the pressure is said to be 136 millimeters of water pressure or, dividing this by 13.6, which is the specific gravity of mercury, about 10 mm Hg pressure.

Effect of High Cerebrospinal Fluid Pressure on the Optic Disc—Papilledema. Anatomically, the dura of the brain extends as a sheath around the optic nerve and then connects with the sclera of the eye. When the pressure rises in the cerebrospinal fluid system, it also rises in the optic nerve sheath. The retinal artery and vein pierce this sheath a few millimeters behind the eye and then pass with the optic nerve into the eye itself. The high pressure in the optic sheath pushes fluid along the optic nerve fibers to the interior of the eyeball. Also, the pressure in the sheath impedes the flow of blood in the retinal vein, thereby also increasing the retinal capillary pressure throughout the eye, which results in additional retinal edema. The tissues of the optic disc are much more distensible than those of the remainder of the retina, so that the disc becomes far more edematous than the remainder of the retina and swells into the cavity of the eye. The swelling of the disc, which can be observed with an ophthalmoscope, is called *papilledema,* and neurologists can estimate the cerebrospinal fluid pressure level by assessing the extent to which the optic disc protrudes into the eyeball.

OBSTRUCTION TO THE FLOW OF CEREBROSPINAL FLUID

Hydrocephalus. "Hydrocephalus" means excess water in the cranial vault. This condition is frequently divided into *communicating hydrocephalus* and *noncommunicating hydrocephalus.* In communicating hydrocephalus fluid flows readily from the ventricular system into the subarachnoid space; whereas in noncommunicating hydrocephalus fluid flow out of one or more of the ventricles is blocked.

Usually the *noncommunicating* type of hydrocephalus is caused by a block in the aqueduct of Sylvius, resulting from *atresia* (closure) before birth in many babies or from a brain tumor at any age. As fluid is formed by the choroid plexuses in the two lateral and the third ventricles, the volumes of these three ventricles increase greatly. This flattens the brain into a thin shell against the skull. In newborn babies the increased pressure also causes the whole head to swell because the skull bones still have not fused.

The *communicating* type of hydrocephalus is usually caused by blockage of fluid flow in the subarachnoid space around the basal regions of the brain or blockage of the arachnoidal villi themselves. Fluid therefore collects both inside the ventricles and on the outside of the brain, causing the head to swell tremendously if it occurs in infants when the skull is still pliable and can be stretched, and usually damaging the brain severely at any age.

The most effective therapy for hydrocephalus is surgical institution of a silicone rubber tube shunt all the way from one of the ventricles to the peritoneal cavity, where the fluid can then be absorbed through the peritoneum.

THE BLOOD-CEREBROSPINAL FLUID AND BLOOD-BRAIN BARRIERS

It has already been pointed out that the constituents of the cerebrospinal fluid are not exactly the same as those of the extracellular fluid elsewhere in the body. Furthermore, many large molecular substances hardly pass at all from the blood into the cerebrospinal fluid or into the interstitial fluids of the brain even though these same substances pass readily into the usual interstitial fluids of the body. Therefore, it is said that barriers, called the *blood-cerebrospinal fluid barrier* and the *blood-brain barrier,* exist between the blood and the cerebrospinal fluid and brain fluid, respectively. These barriers exist both in the choroid plexus and at the tissue capillary membranes in essentially all areas of the brain parenchyma *except in some areas of the hypothalamus,* the *pineal gland,* and the *area postrema,* where substances diffuse with ease into the tissue spaces. This ease of diffusion is very important because these areas of the brain have sensory organs that respond to different changes in the body fluids, such as changes in osmolality, glucose concentration, and so forth; these responses provide the signals for feedback regulation of each of the factors.

In general, the blood-cerebrospinal fluid and blood-brain barriers are highly permeable to water, carbon dioxide, oxygen, and most lipid-soluble substances such as alcohol and most anesthetics; slightly permeable to the electrolytes, such as sodium, chloride, and potassium; and almost totally impermeable to plasma proteins and many large organic molecules. Therefore, the blood-cerebrospinal fluid and blood-brain barriers often make it impossible to achieve effective concentrations of either protein antibodies or some non–lipid-soluble drugs in the cerebrospinal fluid or parenchyma of the brain.

The cause of the low permeability of the blood-cerebrospinal fluid and blood-brain barriers is the manner in which the endothelial cells of the capillaries are joined to each other. They are joined by so-called *tight junctions.* That is, the membranes of the adjacent endothelial cells are almost fused with each other, rather than having slit-pores between them, as is the case in most other capillaries of the body.

Diffusion Between the Cerebrospinal Fluid and the Brain Interstitial Fluid. The surfaces of the ventricles are lined with a thin cuboidal epithelium called the *ependyma* and the cerebrospinal fluid on the outer surfaces of the brain is separated from the brain tissue by a thin membrane called the *pia mater.* Both the ependyma and the pia mater are extremely permeable so that almost all substances that enter the cerebrospinal fluid can also diffuse readily into the surface areas of the brain interstitial fluid. Or, likewise, substances in the interstitial fluid can diffuse in the other direction as well. Therefore, some drugs that have no effect at all on the brain when introduced into the blood stream nevertheless can have important effects on the brain when injected into the cerebrospinal fluid.

BRAIN EDEMA

One of the most serious complications of abnormal cerebral hemodynamics and fluid dynamics is the development of brain edema. Because the brain is encased in a solid vault, the accumulation of edema fluid compresses the blood vessels, with eventual depression of blood flow and destruction of brain tissue.

The usual cause of brain edema is either greatly increased capillary pressure or damage to the capillary wall. One cause of excessively high capillary pressure is a sudden increase in the cerebral blood pressure to levels too high for the autoregulatory mechanism to cope with. However, the most common cause is brain concussion, in which the brain tissues and capillaries are traumatized and capillary fluid leaks into the traumatized tissues. Once brain edema begins, it often initiates two vicious circles because of the following positive feedback: (1) edema compresses the vasculature. This in turn decreases the blood flow and causes brain ischemia. The ischemia causes arteriolar dilatation with increased capillary pressure. The increased capillary pressure then causes more edema fluid, so that the edema becomes progressively worse. (2) The decreased blood flow decreases oxygen delivery. This increases the permeability of the capillaries, allowing more fluid leakage. It also turns off the sodium pumps of the tissue cells, thus allowing these to swell.

Once these two vicious circles have begun, heroic measures must be used to prevent total destruction of the brain. One such measure is to infuse intravenously a concentrated osmotic substance, such as a very concentrated mannitol solution. This pulls fluid by osmosis from the brain tissue and breaks up the vicious circle. Another procedure is to remove fluid quickly from the lateral ventricles of the brain via ventricular puncture, thereby relieving the intracerebral pressure.

● BRAIN METABOLISM

Like other tissues, the brain requires oxygen and solid nutrients to supply its metabolic needs. However, there are special peculiarities of brain metabolism that need to be mentioned.

Total Metabolic Rate and Metabolic Rate of Neurons. Under resting conditions, the metabolism of the brain accounts for about 15 per cent of the total metabolism in the body, even though the mass of the brain is only 2 per cent of the total body mass. Therefore, under resting conditions brain metabolism is about 7½ times the average metabolism in the rest of the body.

Most of this excess metabolism of the brain occurs in the neurons, not in the glial supportive tissues. The major need for metabolism in the neurons is to pump ions through their membranes, mainly to transport sodium and calcium ions to the outside of the neuronal membrane and potassium and chloride ions to the interior. Each time a neuron conducts an action potential, these ions move through the membranes, increasing the need for membrane transport to restore the proper ionic concentrations. Therefore, during excessive brain activity, neuronal metabolism can increase several times.

Special Requirement of the Brain for Oxygen — Lack of Significant Anaerobic Metabolism. Most tissues of the body can go without oxygen for several minutes and some as long as a half hour. During this time the tissue cells obtain their energy through the processes of anaerobic metabolism, which means the release of energy by partial breakdown of glucose and glycogen but without combining with oxygen. This delivers energy only at the expense of consuming tremendous amounts of glucose and glycogen. However, it does keep the tissues functioning.

Unfortunately, the brain is not capable of much anaerobic metabolism. One of the reasons for this is the very high metabolic rate of the neurons, so that far more energy is required by each brain cell than is required in most tissues. An additional reason is that the amount of glycogen stored in the neurons is very slight. The stores of oxygen in brain tissues are also slight. Therefore, neuronal activity depends on second-by-second delivery of oxygen from the blood.

Putting these various factors together, one can understand why a sudden cessation of blood flow to the brain or a sudden lack of oxygen in the blood can cause unconsciousness within 5 to 10 seconds.

Under Normal Conditions, Most Brain Energy Is Supplied by Glucose. Under normal conditions, almost all of the energy used by the brain cells is supplied by glucose derived from the blood. As is true for oxygen, most of this is derived minute by minute and second by second from the capillary blood, with a total of only about a 2 minute supply of glucose normally stored as glycogen in the neurons at any given time.

A special feature of glucose delivery to the neurons is that its transport into the neurons through the cell membrane is not dependent on insulin, as is true for almost all other tissue cells. Therefore, even in patients who have serious diabetes with essentially zero secretion of insulin, glucose still diffuses readily into the neurons — which is most fortunate in preventing loss of mental function in diabetic patients. However, when a diabetic patient is overtreated with insulin, the blood glucose concentration can sometimes fall extremely low, because the excess insulin causes almost all of the glucose in the blood to be transported rapidly into the insulin-sensitive nonneural cells throughout the body. When this happens, not enough glucose is left in the blood to supply the neurons, and mental function does then become seriously deranged, leading sometimes to coma but more often to mental imbalances and psychotic disturbances.

REFERENCES

Angerson, W. J., et al. (eds.): Blood Flow in the Brain. New York, Oxford University Press, 1989.

Bevan, J. A., et al.: Sympathetic control of cerebral arteries: specialization in receptor type, reserve, affinity, and distribution. FASEB J., 1:193, 1987.

Cserr, H. F.: Physiology of the choroid plexus. Physiol. Rev., 51:273, 1971.

Daveson, H.: The Physiology of the Cerebrospinal Fluid. Boston, Little, Brown, 1967.

Fenstermacher, J. D., and Rapoport, S. I.: Blood-brain barrier. In Renkin, E. M., and Michel, C. C. (eds.): Handbook of Physiology. Sec. 2, Vol. IV. Bethesda, Md., American Physiological Society, 1984, p. 969.

Finger, S., et al. (eds.): Brain Injury and Recovery. New York, Plenum Publishing Corp., 1988.

Guyton, A. C., et al.: Circulatory Physiology. II. Dynamics and Control of the Body Fluids. Philadelphia, W. B. Saunders Co., 1975.

Hibbard, L. S., et al.: Three-dimensional representation and analysis of brain energy metabolism. Science, 236:1641, 1987.

Hochwald, G. M.: Animal models of hydrocephalus: Recent developments. Proc. Soc. Exp. Biol. Med., 178:1, 1985.

Kazemi, H., and Johnson, D. C.: Regulation of cerebrospinal fluid acid-base balance. Physiol. Rev., 66:953, 1986.

Levin, H. S., and Eisenberg, H. S. (eds.): Mild Head Injury. New York, Oxford University Press, 1989.

Mayhan, W. G., et al.: Cerebral microcirculation. News Physiol. Sci., 3:164, 1988.

McLaurin, R. L., et al. (eds.): Pediatric Neurosurgery, 2nd Ed. Philadelphia, W. B. Saunders Co., 1989.

Neuwelt, E. A. (ed.): Implications of the Blood-Brain Barrier and Its Manipulation. New York, Plenum Publishing Corp., 1989.

Oldendorf, W. H.: Blood-brain barrier permeability to drugs. Annu. Rev. Pharmacol., 14:239, 1974.

Rescigno, A., and Boicelli, A.: Cerebral Blood Flow. New York, Plenum Publishing Corp., 1988.

Saunders, N. R., and Milgard, K.: Development of the blood-brain barrier. J. Dev. Physiol., 6:45, 1984.

Shulman, K. (ed.): Intracranial Pressure IV. New York, Springer-Verlag, 1980.

Siesjo, B. K.: Cerebral circulation and metabolism. J. Neurosurg., 60:883, 1984.

Somjen, G. (ed.): Mechanisms of Cerebral Hypoxia and Stroke. New York, Plenum Publishing Corp., 1988.

Wood, J. H. (ed.): Cerebral Blood Flow: Physiologic and Clinical Aspects. New York, McGraw-Hill Book Co., 1987.

GASTROINTESTINAL PHYSIOLOGY

CHAPTER 62

•General Principles of Gastrointestinal Function — Motility, Nervous Control, and Blood Circulation •

The alimentary tract provides the body with a continual supply of water, electrolytes, and nutrients. To achieve this requires (1) movement of food through the alimentary tract; (2) secretion of digestive juices and digestion of the food; (3) absorption of the digestive products, water, and the various electrolytes; (4) circulation of blood through the gastrointestinal organs to carry away the absorbed substances; and (5) control of all of these functions by the nervous and hormonal systems.

Figure 62-1 illustrates the entire alimentary tract. Each part is adapted to its specific functions, some to simple passage of food, such as the esophagus; others to storage of food, such as the stomach; and others to digestion and absorption, such as the small intestine. In the present chapter we discuss the basic principles of the function of the entire alimentary tract; and in the following chapters, the specific functions of the different segments of the tract.

● GENERAL PRINCIPLES OF GASTROINTESTINAL MOTILITY

CHARACTERISTICS OF THE GASTROINTESTINAL WALL

Figure 62-2 illustrates a typical section of the intestinal wall, showing the following layers from the outer surface inward: (1) the *serosa*, (2) a *longitudinal muscle layer*, (3) a *circular muscle layer*, (4) the *submucosa*, and (5) the *mucosa*. In addition, a sparse layer of smooth muscle fibers, the *muscularis mucosae*, lies in the deeper layers of the mucosa. The motor func-

tions of the gut are performed by the different layers of smooth muscle.

The general characteristics of smooth muscle and its function are discussed in Chapter 8. However, the specific characteristics of smooth muscle in the gut are the following:

The Gastrointestinal Smooth Muscle — Its Function as a Syncytium. The individual smooth muscle fibers in the gastrointestinal tract are between 200 and 500 micrometers in length and 2 and 10 micrometers in diameter, and they are arranged in bundles of as many as 1000 parallel fibers. In the longitudinal muscle layer, these bundles extend longitudinally down the intestinal tract; in the circular muscle layer they extend around the gut. Within each bundle the muscle fibers are electrically connected with each other through large numbers of *gap junctions* that allow low resistance movement of ions from one cell to the next. Therefore, electrical signals can travel readily from one fiber to the next.

Each bundle of smooth muscle fibers is separated from the next by loose connective tissue, but the bundles fuse with each other at many points so that in reality each muscle layer represents a branching latticework of smooth muscle bundles. Therefore, each muscle layer functions as a *syncytium* — that is, when an action potential is elicited anywhere within the muscle mass, it generally travels in all directions in the muscle. However, the distance that it travels depends upon the excitability of the muscle; sometimes it stops after only a few millimeters and at other times travels many centimeters or even the entire length of the intestinal tract. Also, a few connections exist between the longitudinal and circular muscle layers so that excitation of one of these layers usually excites the other as well.

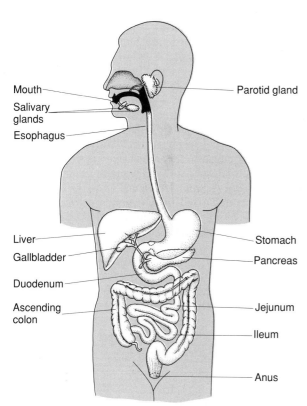

Figure 62-1. The alimentary tract.

Electrical Activity of Gastrointestinal Smooth Muscle

The smooth muscle of the gastrointestinal tract undergoes almost continual but slow electrical activity. This activity tends to have two basic types of electrical waves: (1) *slow waves* and (2) *spikes,* both of which are illustrated in Figure 62-3. In addition, the voltage of the resting membrane potential of the gastrointestinal smooth muscle can change to different levels without the generation of waves, and this too

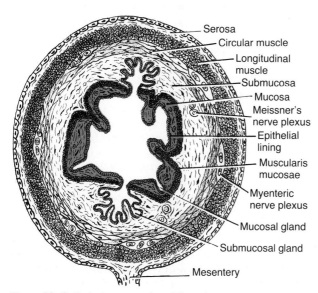

Figure 62-2. Typical cross-section of the gut.

Figure 62-3. Membrane potentials in intestinal smooth muscle. Note the slow waves, the spike potentials, total depolarization, and hyperpolarization, all of which occur under different physiological conditions of the intestine.

can have important effects in controlling motor activity of the gastrointestinal tract.

The Slow Waves. Most gastrointestinal contractions occur rhythmically, and this rhythm is determined mainly by the frequency of the so-called "slow waves" in the smooth muscle membrane potential. These waves, illustrated in Figure 62-3, are not action potentials. Instead, they are slow, undulating changes in the resting membrane potential. Their intensity usually varies between 5 and 15 millivolts, and their frequency ranges in different parts of the human gastrointestinal tract between 3 and 12 per minute: about 3 in the body of the stomach, as much as 12 in the duodenum, and changing to about 8 or 9 in the terminal ileum. Therefore, the rhythm of contraction of the body of the stomach usually is about 3 per minute, of the duodenum about 12 per minute, and the ileum 8 to 9 per minute.

The cause of the slow waves is not known. However, as discussed in Chapter 8, it is believed that they might be caused by a slow undulation of the pumping activity of the sodium-potassium pump.

The slow waves themselves usually *do not directly cause* muscle contraction. However, they do control the appearance of intermittent spike potentials, and the spike potentials in turn actually cause the muscle contraction.

The Spike Potentials. The spike potentials are true action potentials. They occur automatically when the resting membrane potential of the gastrointestinal smooth muscle becomes more positive than about −40 millivolts (the normal resting membrane potential is between −50 and −60 millivolts). Thus, note in Figure 62-3 that each time the peaks of the slow waves rise temporarily above the −40 millivolt level—that is, become less negative than −40 millivolts—spike potentials appear on these peaks. And the higher the slow wave potential rises above this level, the greater the frequency of the spike potentials, usually ranging between 1 and 10 spikes per second.

The spike potentials last 10 to 40 times as long in the gastrointestinal muscle as the action potentials in large nerve fibers, each lasting as long as 10 to 20 milliseconds. Another very important difference between the action potentials of the gastrointestinal smooth muscle and those of nerve fibers is the manner in which they are generated. In nerve fibers, the action potentials are caused almost entirely by rapid entry of sodium ions through sodium channels to the interior of the fibers. In the gastrointestinal smooth muscle, the channels responsible for the action potentials are quite different; they allow especially large numbers of calcium ions to enter along with smaller numbers of sodium ions and therefore are called *calcium-sodium channels.* These channels are much slower to open and also to close than are the rapid sodium channels, which accounts for the longer duration of the action potentials. Also, the movement of large amounts of calcium ions to the interior of the muscle fiber during the action potential plays a special role in causing the intestinal smooth muscle to contract, as we discuss shortly.

Changes in the Voltage of the Resting Membrane Potential. In addition to the slow waves and spike potentials, the voltage level of the resting membrane potential can also change. Under normal conditions the resting membrane potential averages about −56 millivolts, but multiple factors can change this level. When the potential becomes more positive, which is called *depolarization* of the membrane, the muscle fiber becomes more excitable. When the potential becomes more negative, which is called *hyperpolarization,* the fiber becomes less excitable.

Factors that depolarize the membrane — that is, make it more excitable — are (1) stretching of the muscle, (2) stimulation by acetylcholine, (3) stimulation by the parasympathetic nerves that secrete acetylcholine at their endings, and (4) stimulation by several specific gastrointestinal hormones.

Important factors that make the membrane potential more negative — that is, hyperpolarize the membrane and make the muscle fiber less excitable — are (1) the effect of norepinephrine or epinephrine on the muscle membrane and (2) stimulation of the sympathetic nerves that secrete norepinephrine at their endings.

Calcium Ions and Muscle Contraction. Muscle contraction occurs in response to the entry of calcium into the muscle fiber. As was explained in Chapter 8, it is believed that in smooth muscle the calcium ions, acting through a calmodulin control mechanism, activate the myosin filaments in the fiber, causing attractive forces between these and the actin filaments and thereby causing the muscle to contract.

The slow waves do not cause calcium ions to enter the smooth muscle fiber but instead only sodium ions. Therefore, the slow waves by themselves usually cause no contraction or very little. Instead, it is during the spike potentials that are generated at the peaks of the slow waves that large quantities of calcium ions do enter the fibers and cause most contraction.

Tonic Contraction of Gastrointestinal Smooth Muscle. The smooth muscle of the gastrointestinal tract usually exhibits *tonic contraction* as well as rhythmical contractions. Tonic contraction is continuous, not associated with the basic electrical rhythm of the slow waves, but instead often lasting several minutes or even several hours. The tonic contraction occasionally increases or decreases in intensity but nevertheless continues. Tonic contraction is sometimes caused by repetitive series of spike potentials — the greater the frequency, the greater the degree of contraction. However, at other times tonic contraction is caused by hormones or other factors that bring about continuous depolarization of the smooth muscle membrane without causing action potentials. And still a third cause of tonic contraction is continuous entry of calcium into the interior of the cell, brought about in ways not at all associated with changes in the membrane potential. Unfortunately, the details of these mechanisms are still very unclear.

● NEURAL CONTROL OF GASTROINTESTINAL FUNCTION

The gastrointestinal tract has a nervous system all its own called the *enteric nervous system.* It lies entirely in the wall of the gut, beginning in the esophagus and extending all the way to the anus. The number of neurons in this enteric system is about 100,000,000, almost exactly equal to the number in the entire spinal cord; this illustrates the importance of the enteric system for controlling gastrointestinal function. It especially controls gastrointestinal movements and secretion.

The enteric system is composed mainly of two plexuses, as illustrated in Figure 62–2: an outer plexus lying between the longitudinal and circular muscular layers, called the *myenteric plexus* or *Auerbach's plexus;* and (2) an inner plexus, called the *submucosal plexus* or *Meissner's plexus,* that lies in the submucosa. The nervous connections within and between these two plexuses are illustrated in Figure 62–4. The myenteric plexus controls mainly the gastrointestinal movements, and the submucosal plexus controls mainly gastrointestinal secretion and local blood flow.

Note in Figure 62–4 the sympathetic and parasympathetic fibers that connect with both the myenteric and submucosal plexuses. Although the enteric nervous system can function on its own, independently of these extrinsic nerves, stimulation of the parasympathetic and sympathetic systems can further activate or inhibit gastrointestinal functions, as we discuss later.

Also shown in Figure 62–4 are sensory nerve endings that originate in the gastrointestinal epithelium or gut wall and then send afferent fibers to both plexuses of the enteric system and also to the prevertebral ganglia of the sympathetic nervous system, some

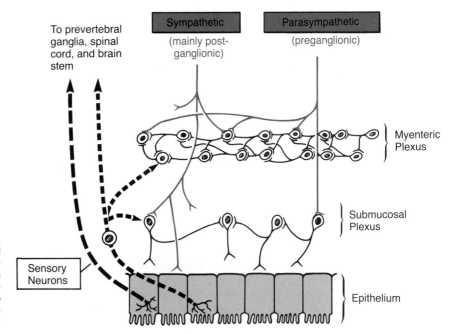

Figure 62–4. Neural control of the gut wall, showing (1) the myenteric and submucosal plexuses; (2) extrinsic control of these plexuses by the sympathetic and parasympathetic nervous systems; and (3) sensory fibers passing from the luminal epithelium and gut wall to the enteric plexuses and from there to the prevertebral ganglia, spinal cord, and brain stem.

traveling in the sympathetic nerves to the spinal cord and others traveling in the vagi all the way to the brain stem. These sensory nerves elicit local reflexes within the gut itself and also reflexes that are relayed back to the gut from either the prevertebral ganglia or the central nervous system.

Differences Between the Characteristics of the Myenteric and Submucosal Plexuses

The myenteric plexus consists mostly of linear chains of many interconnecting neurons that extend the entire length of the gastrointestinal tract. One of these linear chains is illustrated in Figure 62–4; similar chains are located parallel to each other a few millimeters apart all the way around the gut wall. The separate chains also have lateral fiber connections from one to the other as well as connections with the deeper lying submucosal plexus.

Because the myenteric plexus is a linear plexus extending all the way down the intestinal wall, and because it lies between the longitudinal and circular masses of intestinal smooth muscle, it is concerned mainly with controlling motor activity along the length of the gut. When stimulated, its principal effects are (1) increased tonic contraction, or "tone," of the gut wall, (2) increased intensity of the rhythmical contractions, (3) slightly increased rate of the rhythm of contraction, and (4) increased velocity of conduction of excitatory waves along the gut wall, causing more rapid movement of the peristaltic waves.

However, the myenteric plexus must not be considered entirely excitatory, because some of the neurons instead are inhibitory; their fiber endings secrete an inhibitory transmitter, possibly vasoactive intestinal polypeptide (VIP) or some other peptide. The result-

ing inhibitory signals are especially useful for inhibiting some of the contracted intestinal sphincter muscles that impede movement of food between successive segments of the gastrointestinal tract, such as the *pyloric sphincter*, which controls the emptying of the stomach, and the *sphincter of the ileocecal valve*, which controls the emptying of the small intestine into the cecum.

The submucosal plexus, in contrast to the myenteric plexus, is mainly concerned with controlling function within the inner wall of each minute segment of the intestine. For instance, many sensory signals originate from the gastrointestinal epithelium and are then integrated in the submucosal plexus to help control local intestinal secretion, local absorption, and local contraction of the submucosal muscle that causes various degrees of infolding of the stomach mucosa.

Types of Neurotransmitters Secreted by the Enteric Neurons

In an attempt to understand better the multiple functions of the enteric nervous system, research workers the world over have recently identified a dozen or more different neurotransmitter substances that are released by the nerve endings of different types of enteric neurons. Two of these with which we are already familiar are (1) *acetylcholine* and (2) *norepinephrine*. Others are (3) *adenosine triphosphate,* (4) *serotonin,* (5) *dopamine,* (6) *cholecystokinin,* (7) *substance P,* (8) *vasoactive intestinal polypeptide,* (9) *somatostatin,* (10) *leu-enkephalin,* (11) *met-enkephalin,* and (12) *bombesin.* The specific functions of most of these are not yet known well enough to justify extensive discussion, other than to point out the following: Acetylcholine most often excites gastrointestinal activity. Norepinephrine, on the other hand, almost always inhibits gastrointestinal activity. This is also true of epinephrine which reaches the gastrointestinal tract by way of the blood after it is secreted by the adrenal medullae into the circulation. The other transmitter substances listed above are a mixture of excitatory and inhibitory agents, but their importance and even their functions are still mainly to be determined.

Autonomic Control of the Gastrointestinal Tract

Parasympathetic Innervation. The parasympathetic supply to the gut is divided into *cranial* and *sacral divisions,* which are discussed in Chapter 60. Except for a few parasympathetic fibers to the mouth and pharyngeal regions of the alimentary tract, the cranial parasympathetics are transmitted almost entirely in the *vagus nerves.* These fibers provide extensive innervation to the esophagus, stomach, pancreas, and first half of the large intestine (but rather little innervation to the small intestine). The sacral parasympathetics originate in the second, third, and fourth sacral segments of the spinal cord and pass through the *pelvic nerves* to the distal half of the large intestine. The sigmoidal, rectal, and anal regions of the large intestine are considerably better supplied with parasympathetic fibers than are the other portions. These fibers function especially in the defecation reflexes, which are discussed in the following chapter.

The *postganglionic neurons* of the parasympathetic system are located in the myenteric and submucosal plexuses, and stimulation of the parasympathetic nerves causes a general increase in activity of the entire enteric nervous system. This in turn enhances the activity of most gastrointestinal functions, but not all, for some of the enteric neurons are inhibitory and therefore inhibit certain of the functions.

Sympathetic Innervation. The sympathetic fibers to the gastrointestinal tract originate in the spinal cord between the segments T-5 and L-2. The preganglionic fibers, after leaving the cord, enter the sympathetic chains and pass through the chains to outlying ganglia, such as the *celiac ganglion* and various *mesenteric ganglia.* Here, most of the *postganglionic neuron bodies* are located, and postganglionic fibers spread from them along with the blood vessels to all parts of the gut, terminating principally on neurons in the enteric nervous system. The sympathetics innervate essentially all portions of the gastrointestinal tract, rather than being more extensively supplied to the most orad and most analward portions, as is true of the parasympathetics. The sympathetic nerve endings secrete *norepinephrine.*

In general, stimulation of the sympathetic nervous system inhibits activity in the gastrointestinal tract, causing effects essentially opposite to those of the parasympathetic system. It exerts its effects in two different ways: (1) to a slight extent by direct effect of the norepinephrine on the smooth muscle to inhibit this (except the muscularis mucosa, which it excites), and (2) to a major extent by an inhibitory effect of the norepinephrine on the neurons of the enteric nervous system. Thus, strong stimulation of the sympathetic system can totally block movement of food through the gastrointestinal tract.

Afferent Nerve Fibers From the Gut

Many afferent nerve fibers arise in the gut. Some of these have their cell bodies in the enteric nervous system itself. These nerves can be stimulated by (1) irritation of the gut mucosa, (2) excessive distention of the gut, or (3) the presence of specific chemical substances in the gut. Signals transmitted through these fibers can cause excitation or, under some conditions, inhibition of intestinal movements or intestinal secretion.

In addition to the afferent fibers that terminate in the enteric nervous system, still two other types of afferent fibers are associated with this system. One of these has its cell bodies in the enteric nervous system but sends its axons through the autonomic nerves to terminate in the *prevertebral sympathetic ganglia,* that is, in the *celiac, mesenteric,* and *hypogastric ganglia.* The other type of afferent fibers has its cell bodies in the dorsal root ganglia of the spinal cord or in the cranial nerve ganglia; these fibers transmit their signals directly into the spinal cord or brain stem, traveling in the same nerve trunks along with the sympathetic or parasympathetic nerve fibers. For example, 80 per cent of the nerve fibers in the vagus nerves are afferent rather than efferent. These fibers transmit afferent signals into the medulla, which in turn initiates many vagal reflex signals returning to the gastrointestinal tract to control many of its functions.

The Gastrointestinal Reflexes

The anatomical arrangement of the enteric nervous system and its connections with the sympathetic and parasympathetic systems supports three different types of gastrointestinal reflexes that are essential to gastrointestinal control. These are the following:

1. *Reflexes that occur entirely within the enteric nervous system.* These include reflexes that control gastrointestinal secretion, peristalsis, mixing contractions, local inhibitory effects, and so forth.

2. *Reflexes from the gut to the prevertebral sympathetic ganglia and then back to the gastrointestinal tract.* These reflexes transmit signals for long distances in the gastrointestinal tract, such as signals from the stomach to cause evacuation of the colon (the *gastrocolic reflex*), signals from the colon and small intestine to inhibit stomach motility and stomach secretion (the *enterogastric reflexes*), and reflexes from the colon to inhibit emptying of ileal contents into the colon (the *colonoileal reflex*).

3. Reflexes from the gut to the spinal cord or brain stem and then back to the gastrointestinal tract. These include especially (a) reflexes from the stomach and duodenum to the brain stem and back to the stomach to control gastric motor and secretory activity; (b) pain reflexes that cause general inhibition of the entire gastrointestinal tract; and (c) defecation reflexes that travel to the spinal cord and back again to produce the powerful colonic, rectal, and abdominal contractions required for defecation (the defecation reflexes).

HORMONAL CONTROL OF GASTROINTESTINAL MOTILITY

In Chapter 64 we discuss the extreme importance of several hormones for controlling gastrointestinal secretion. Most of these same hormones also affect the motility of some parts of the gastrointestinal tract. Though the motility effects are less important than the secretory effects of the hormones, some of the more important of them are the following:

Cholecystokinin, which is secreted mainly by the mucosa of the jejunum in response to the presence of fatty substances in the intestinal contents, has a very potent effect in increasing contractility of the gallbladder, thus expelling bile into the small intestine, where the bile then plays important roles in emulsifying fatty substances, allowing them to be digested and absorbed. Cholecystokinin also inhibits stomach motility moderately. Therefore, at the same time that this hormone causes emptying of the gallbladder, it also slows the emptying of food from the stomach to give adequate time for digestion of the fats in the upper intestinal tract.

Secretin, which is secreted by the mucosa of the duodenum in response to acidic gastric juice emptied from the stomach through the pylorus, has a mild inhibitory effect on the motility of most of the gastrointestinal tract.

Gastric inhibitory peptide, which is secreted by the mucosa of the upper small intestine, mainly in response to fat but to a lesser extent in response to carbohydrate, has a mild effect in decreasing motor activity of the stomach and therefore slowing the emptying of gastric contents into the duodenum when the upper small intestine is already oversupplied with food products.

● FUNCTIONAL TYPES OF MOVEMENTS IN THE GASTROINTESTINAL TRACT

Two basic types of movements occur in the gastrointestinal tract: (1) *propulsive movements,* which cause food to move forward along the tract at an appropriate rate for digestion and absorption, and (2) *mixing movements,* which keep the intestinal contents thoroughly mixed at all times.

The Propulsive Movements — Peristalsis

The basic propulsive movement of the gastrointestinal tract is *peristalsis,* which is illustrated in Figure 62–5. A contractile ring appears around the gut and then moves forward; this is analogous to putting one's fingers around a thin distended tube, then constricting the fingers and sliding them forward along the tube. Obviously, any material in front of the contractile ring is moved forward.

Peristalsis is an inherent property of many syncytial smooth muscle tubes; stimulation at any point can cause a contractile ring to appear and then spread along the tube. Thus, peristalsis occurs in the gastrointestinal tract, the bile ducts, other glandular ducts throughout the body, the ureters, and many other smooth muscle tubes of the body.

The usual stimulus for peristalsis is *distention.* That is, if a large amount of food collects at any point in the gut, the stretching of the gut wall stimulates the gut 2 to 3 centimeters above this point, and a contractile ring appears that initiates a peristaltic movement. Other stimuli that can initiate peristalsis include irritation of the epithelium lining the gut and extrinsic nervous signals, particularly parasympathetic, that excite the gut.

Function of the Myenteric Plexus in Peristalsis. Peristalsis occurs only weakly, if at all, in any portion of the gastrointestinal tract that has congenital absence of the myenteric plexus. Also, it is greatly depressed or completely blocked in the entire gut when the person is treated with atropine to paralyze the cholinergic nerve endings of the myenteric plexus. Therefore, *effectual* peristalsis requires an active myenteric plexus.

Directional Movement of Peristaltic Waves Toward the Anus. Peristalsis, theoretically, can occur in either direction from a stimulated point, but it normally dies out rapidly in the orad direction while continuing for a considerable distance analward. The exact cause of this directional transmission of peristalsis has never been ascertained, though it probably results mainly from the fact that the myenteric plexus itself is "polarized" in the anal direction, which can be explained as follows:

The Peristaltic Reflex and the "Law of the Gut." When a segment of the intestinal tract is ex-

Figure 62–5. Peristalsis.

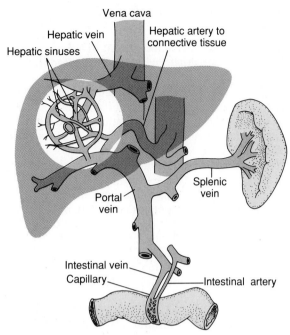

Figure 62–6. The splanchnic circulation.

cited by distention and thereby initiates peristalsis, the contractile ring for causing the peristalsis normally begins slightly on the orad side of the distended segment; then it moves toward the distended segment, thus pushing the intestinal contents in the anal direction. At the same time, the gut sometimes relaxes several centimeters downstream toward the anus, which is called "receptive relaxation," thus allowing the food to be propelled more easily analward than in the orad direction.

This complex pattern does not occur in the absence of the myenteric plexus. Therefore, the complex is frequently called the *myenteric reflex,* or it is also called simply the *peristaltic reflex.* And the peristaltic reflex plus the analward direction of movement of the peristalsis is called the "law of the gut."

The Mixing Movements

The mixing movements are quite different in different parts of the alimentary tract. In some areas, the peristaltic contractions themselves cause most of the mixing. This is especially true when forward progression of the intestinal contents is blocked by a

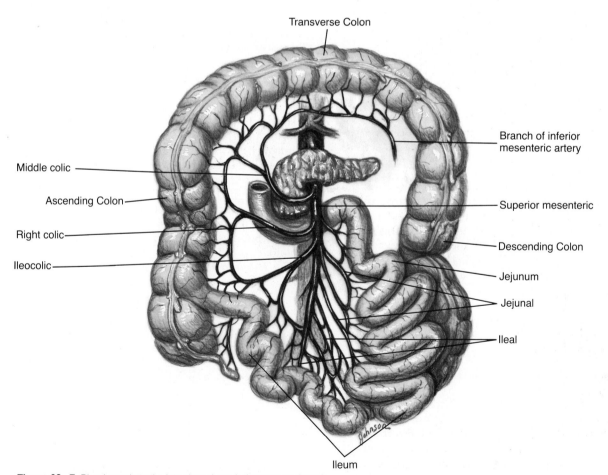

Figure 62–7. Blood supply to the intestines through the mesenteric web.

sphincter, so that a peristaltic wave can then only churn the intestinal contents, rather than propelling them forward. At other times, *local constrictive contractions* occur every few centimeters in the gut wall. These constrictions usually last only a few seconds; then new constrictions occur at other points in the gut, thus "chopping" the contents first here and then there. These peristaltic and constrictive movements are modified in different parts of the gastrointestinal tract for proper propulsion and mixing, as are discussed for each portion of the tract in the following chapter.

● GASTROINTESTINAL BLOOD FLOW

The blood vessels of the gastrointestinal system are part of a more extensive system called the *splanchnic circulation,* illustrated in Figure 62–6. It includes the blood flow through the alimentary tract itself plus the blood flow through the spleen, the pancreas, and the liver. The design of this system is such that all of the blood that courses through the gut, spleen, and pancreas then flows immediately into the liver by way of the portal vein. In the liver, the blood passes through millions of fine liver sinusoids and finally leaves the liver by way of the hepatic veins that empty into the vena cava of the general circulation. This secondary flow of blood through the liver allows the *reticuloendothelial cells* lining the liver sinusoids to remove bacteria and other particulate matter that might enter the blood from the gastrointestinal tract, thus preventing direct access of potentially harmful agents into the remainder of the body.

Many of the nutrients absorbed from the gut are also transported in the portal venous blood to the same liver sinusoids. Here, both the reticuloendothelial cells and the principal parenchymal cells of the liver, the *hepatic cells,* absorb from the blood and store temporarily from one half to three quarters of all the absorbed nutrients. In addition, much intermediary processing of these nutrients occurs in the liver as well. We discuss these nutritional functions of the liver in Chapters 67 through 71.

Anatomy of the Gastrointestinal Blood Supply

Figure 62–7 illustrates the general plan of the blood supply to the gut, showing the superior mesenteric and inferior mesenteric arteries supplying the walls of the small and large intestines by way of an arching arterial system. Not shown in the figure is the celiac artery that provides a similar blood supply to the stomach.

On entering the wall of the gut the arteries branch and send smaller arteries circling in both directions around the gut, with the tips of these meeting on the side of the gut wall opposite to the mesenteric attachment. From the circling arteries, still much smaller

arteries penetrate into the intestinal wall and spread (1) along the muscle bundles, (2) into the villi, and (3) into submucosal vessels beneath the epithelium to serve the secretory and absorptive functions of the gut.

Figure 62–8 illustrates the special organization of the blood flow through an intestinal villus, showing a small arteriole and venule that interconnect with a system of multiple looping capillaries. The walls of the arterioles are highly muscular, and they are also highly active in controlling villus blood flow.

Effect of Gut Activity and Metabolic Factors on Gastrointestinal Blood Flow

Under normal conditions the blood flow in each area of the gastrointestinal tract, as well as in each layer of the gut wall, is directly related to the level of local activity. For instance, during active absorption, blood flow in the villi and adjacent regions of the submucosa is greatly increased. Likewise, blood flow in the muscle layers of the intestinal wall increases with increased motor activity in the gut. For instance, after a meal the motor activity, secretory activity, and ab-

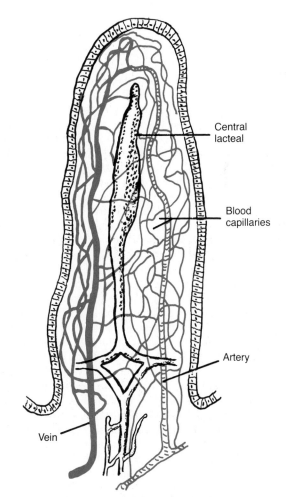

Figure 62–8. Microvasculature of the villus, showing a countercurrent arrangement of blood flow in the arterioles and venules.

sorptive activity all increase, and likewise the blood flow increases as much as 100 to 150 per cent, usually lasting for 3 to 6 hours.

Possible Causes of the Increased Blood Flow During Activity. Although the precise cause or causes of the increased blood flow during increased gastrointestinal activity are still unclear, some facts are known.

First, several different vasodilator substances are released from the mucosa of the intestinal tract during the digestive process. Most of these are peptide hormones, including *cholecystokinin, vasoactive intestinal peptide, gastrin,* and *secretin.* These same hormones are also important in controlling certain specific motor and secretory activities of the gut, as we see in the next two chapters.

Second, some of the gastrointestinal glands also release into the gut wall two kinins, *kallidin* and *bradykinin,* at the same time that they secrete their secretions into the lumen. These kinins are powerful vasodilators that are believed by some researchers to cause much of the increased mucosal vasodilatation that occurs along with secretion.

Third, *decreased oxygen concentration* in the gut wall can increase intestinal blood flow as much as 50 per cent; therefore, the increased metabolic rate during gut activity probably lowers the oxygen concentration enough to cause much of the vasodilatation. Maybe it is the lack of enough oxygen to supply the metabolism of the smooth muscle in the vessel walls that is the cause of this vasodilatation. A lack of oxygen can also lead to the release of *adenosine,* by as much as fourfold, which is a well-known vasodilator that could be responsible for much of the increased flow.

Thus, the answer to the increased blood flow during increased gastrointestinal activity is not clear. It is probably a combination of all or many of the above factors plus still others yet undiscovered.

A Countercurrent Blood Flow Mechanism in the Villi. Note in Figure 62–8 that the arterial flow into the villus and the venous flow out of the villus are in directions opposite to one another, and the vessels lie in close apposition to one another. Because of this vascular arrangement, much of the blood oxygen diffuses out of the arterioles directly into the adjacent venules without ever being carried in the blood to the tips of the villi. As much as 80 per cent of the oxygen may take this short-circuit route and therefore not be available for the local metabolic functions of the villi. The reader will recognize that this type of countercurrent mechanism in the villi is analogous to the countercurrent mechanism in the vasa recta of the kidney medulla, discussed in detail in Chapter 28.

Under normal conditions, this shunting of oxygen from the arterioles to the venules is not harmful to the villi, but in disease conditions in which blood flow to the gut becomes greatly curtailed, such as in circulatory shock, the oxygen deficit in the tips of the villi can become so great that the villus tip or even the whole villus suffers ischemic death and can actually disintegrate. Therefore, in many gastrointestinal diseases the villi become seriously blunted, leading to greatly diminished gastrointestinal absorptive capacity.

Nervous Control of Gastrointestinal Blood Flow

Stimulation of the parasympathetic nerves to the *stomach* and *lower colon* increases local blood flow at the same time that it increases glandular secretion. However, this increased flow probably results secondarily from the increased glandular activity and not as a direct effect of nervous stimulation.

Sympathetic stimulation, in contrast, has a direct effect on essentially all of the gastrointestinal tract by causing intense vasoconstriction of the arterioles with greatly decreased blood flow. However, after a few minutes of this vasoconstriction, the flow returns almost to normal via a mechanism called "autoregulatory escape." That is, the local metabolic vasodilator mechanisms that are elicited by ischemia become prepotent over the sympathetic vasoconstriction and, therefore, redilate the arterioles, thus causing return of necessary nutrient blood flow to the gastrointestinal glands and muscle.

Importance of Nervous Control When Other Parts of the Body Need Extra Blood Flow. A major value of sympathetic vasoconstriction in the gut is that it allows the shutting off of gastrointestinal and other splanchnic blood flow for short periods of time during heavy exercise when increased flow is needed by the skeletal muscle and heart. Also, in circulatory shock when all the vital tissues are in danger of cellular death for lack of blood flow — especially the brain and the heart — sympathetic stimulation can almost completely block splanchnic blood flow for as long as an hour.

Sympathetic stimulation also causes especially strong vasoconstriction of the intestinal and mesenteric *veins.* Furthermore, this vasoconstriction does not "escape." Instead, it decreases the volume of these veins and thereby displaces large amounts of blood into other parts of the circulation. In hemorrhagic shock or other states of low blood volume, this mechanism can provide probably 200 to 300 milliliters of extra blood to sustain the general circulation.

REFERENCES

Bofak-Gyovai, L. A., and Manzione, J. V.: Oral Medicine: Patient Evaluation and Management. Baltimore, Williams & Wilkins, 1980.

Bongiovanni, G. L. (ed.): Essentials of Clinical Gastroenterology, 2nd Ed. New York, McGraw-Hill Book Co., 1988.

Bulbring, E. (ed.): Smooth Muscle. An Assessment of Current Knowledge. Austin, University of Texas Press, 1981.

Chou, C. C.: Relationship between intestinal blood flow and motility. Annu. Rev. Physiol., 44:29, 1982.

Costa, M., et al.: Chemical coding of enteric neurons. Prog. Brain Res., 68:217, 1986.

Costa, M., et al.: Histochemistry of the enteric nervous system. In: Johnson, L. R. (ed.) Physiology of the Gastrointestinal Tract, 2nd Ed. New York, Raven Press, 1987.

Donald, D. E.: Splanchnic circulation. In Shephard, J. T., and Abboud, F. M. (eds.): Handbook of Physiology. Sec. 2, Vol. III. Bethesda, Md., American Physiological Society, 1983, p. 219.

Figlewicz, D. P., et al.: Gastroenteropancreatic peptides and the central nervous system. Annu. Rev. Physiol., 49:383, 1987.

Forte, J. G.: Gastrointestinal physiology Annu. Rev. Physiol., 57:173, 1985.

Forte, J. G.: Gastrointestinal physiology. Annu. Rev. Physiol., 48:73, 1986.

Gabella, G.: Structural apparatus for force transmission in smooth muscle. Physiol. Rev., 64:455, 1984.

Gonella, J., et al.: Extrinsic nervous control of motility of small and large intestines and related sphincters. Physiol. Rev. 67:902, 1987.

Grossman, M. I.: Neural and hormonal regulation of gastrointestinal function: An overview. Annu. Rev. Physiol., 41:27, 1979.

Hunt, J. N.: Mechanisms and disorders of gastric emptying. Annu. Rev. Med., 34:219, 1983.

Janig, W., and McLachlan E. M.: Organization of lumbar spinal outflow to distal colon and pelvic organs. Physiol. Rev., 67:1332, 1987.

Jodal, M. and Lundgren, O.: Countercurrent mechanisms in the mammalian gastrointestinal tract. Gastroenterology, 91:225, 1986.

Johnson, L. R., et al.: Physiology of the Gastrointestinal Tract, 2nd Ed. New York, Raven Press, 1987.

Kirsner, J. B., and Shorter, R. G. (eds.): Diseases of the Colon, Rectum, and Anal Canal. Baltimore, Williams & Wilkins, 1988.

Klimov, P. K.: Behavior of the organs of the digestive system. Neurosci. Behav. Physiol., 14:333, 1984.

Levitt, M. D., and Bond, J. H.: Flatulence. Annu. Rev. Med., 31:127, 1980.

Loewenstein, W. R.: Junctional intercellular communication: The cell-to-cell membrane channel. Physiol. Rev., 61:829, 1981.

Luschei, E. S., and Goldberg, L. J.: Neural mechanism of mandibular control: Mastication and voluntary biting. In Brooks, V. B. (ed.): Handbook of Physiology. Sec. 1, Vol. II. Bethesda, Md., American Physiological Society, 1981, p. 1237.

Miller, A. J. Deglutition. Physiol. Rev., 62:129, 1982.

Sernka, T. J., and Jacobson, E. D.: Gastrointestinal Physiology; The Essentials. Baltimore, Williams & Wilkins, 1979.

Shaffer, E., and Thomson, A. B. R. (eds.): Modern Concepts in Gastroenterology. New York, Plenum Publishing Corp., 1989.

Sjovall, H.: Sympathetic control of jejunal fluid and electrolyte transport. Acta Physiol. Scand., Suppl. 535:1, 1984.

Sternini, C.: Structural and chemical organization of the myenteric plexus. Annu. Rev. Physiol., 50:81, 1988.

Szurszewski, J. H.: Physiology of mammalian prevertebral ganglia. Annu. Rev. Physiol., 43:53, 1981.

Weems, W. A.: The intestine as a fluid propelling system. Annu. Rev. Physiol., 43:9, 1981.

Weisbrodt, N. W.: Patterns of intestinal motility. Annu. Rev. Physiol., 43:21, 1981.

Williamson, J. R., and Monck, J. R.: Hormone effects on cellular Ca^{2+} fluxes. Annu. Rev. Physiol., 51:107, 1989.

Wood, J. D.: Intrinsic neural control of intestinal motility. Annu. Rev. Physiol., 43:33, 1981.

CHAPTER 63

• *Transport and Mixing of Food in the Alimentary Tract* •

For food to be processed optimally in the alimentary tract, the time that it remains in each part of the tract is critical. Also, appropriate mixing must be provided. Yet because the requirements for mixing and propulsion are quite different at each stage of processing, multiple automatic nervous and hormonal feedback mechanisms control each aspect of these so that they will both occur optimally, not too rapidly, not too slowly.

The purpose of the present chapter is to discuss these movements, and especially to discuss the automatic mechanisms that control them.

● INGESTION OF FOOD

The amount of food that a person ingests is determined principally by the intrinsic desire for food called *hunger.* The type of food that a person preferentially seeks is determined by *appetite.* These mechanisms in themselves are extremely important automatic regulatory systems for maintaining an adequate nutritional supply for the body, and they are discussed in Chapter 71 in relation to nutrition of the body. The present discussion is confined to the mechanical aspects of food ingestion, especially *mastication* and *swallowing.*

MASTICATION (CHEWING)

The teeth are admirably designed for chewing, the anterior teeth (incisors) providing a strong cutting action and the posterior teeth (molars) a grinding action. All the jaw muscles working together can close the teeth with a force as great as 55 pounds on the incisors and 200 pounds on the molars.

Most of the muscles of chewing are innervated by the motor branch of the fifth cranial nerve, and the chewing process is controlled by nuclei in the brain stem. Stimulation of the reticular formation near the brain stem centers for taste can cause continual rhythmic chewing movements. Also, stimulation of areas in the hypothalamus, amygdala, and even in the

cerebral cortex near the sensory areas for taste and smell can cause chewing.

Much of the chewing process is caused by the *chewing reflex,* which may be explained as follows: The presence of a bolus of food in the mouth causes reflex inhibition of the muscles of mastication, which allows the lower jaw to drop. The drop in turn initiates a stretch reflex of the jaw muscles that leads to *rebound* contraction. This automatically raises the jaw to cause closure of the teeth, but it also compresses the bolus again against the linings of the mouth, which inhibits the jaw muscles once again, allowing the jaw to drop and rebound another time; and this is repeated again and again.

Chewing of the food is important for digestion of all foods, but it is especially important for most fruits and raw vegetables, because these have undigestible cellulose membranes around their nutrient portions which must be broken before the food can be utilized. Chewing aids in the digestion of food for the following simple reason: Because the *digestive enzymes act only on the surfaces of food particles,* the rate of digestion is highly dependent on the total surface area exposed to the intestinal secretions. Also, grinding the food to a very fine particulate consistency prevents excoriation of the gastrointestinal tract and increases the ease with which food is emptied from the stomach into the small intestine and thence into all succeeding segments of the gut.

SWALLOWING (DEGLUTITION)

Swallowing is a complicated mechanism, principally because the pharynx most of the time subserves several other functions besides swallowing and is converted for only a few seconds at a time into a tract for propulsion of food. It is especially important that respiration not be compromised because of swallowing.

In general, swallowing can be divided into (1) the *voluntary stage,* which initiates the swallowing process, (2) the *pharyngeal stage,* which is involuntary and constitutes the passage of food through the pharynx into the esophagus, and (3) the *esophageal stage,* another involuntary phase that promotes passage of food from the pharynx to the stomach.

Voluntary Stage of Swallowing. When the food is ready for swallowing, it is "voluntarily" squeezed or rolled posteriorly into the pharynx by pressure of the tongue upward and backward against the palate, as shown in Figure 63–1. From here on, the process of swallowing becomes entirely, or almost entirely, automatic and ordinarily cannot be stopped.

Pharyngeal Stage of Swallowing. As the bolus of food enters the pharynx, it stimulates *swallowing receptor areas* all around the opening of the pharynx, especially on the tonsillar pillars, and impulses from these pass to the brain stem to initiate a series of automatic pharyngeal muscular contractions as follows:

1. The soft palate is pulled upward to close the posterior nares, in this way preventing reflux of food into the nasal cavities.

2. The palatopharyngeal folds on either side of the pharynx are pulled medially to approximate each other. In this way these folds form a sagittal slit through which the food must pass into the posterior pharynx. This slit performs a selective action, allowing food that has been masticated properly to pass with ease while impeding the passage of large objects. Because this stage of swallowing lasts less than 1 second, any large object is usually impeded too much to pass through the pharynx into the esophagus.

3. The vocal cords of the larynx are strongly approximated, and the larynx is pulled upward and anteriorly by the neck muscles. This action, combined with the presence of ligaments that prevent upward movement of the epiglottis, causes the epiglottis to swing backward over the opening of the larynx. Both effects prevent passage of food into the trachea. Especially important is the approximation of the vocal

cords, but the epiglottis helps to prevent food from ever getting as far as the vocal cords. Destruction of the vocal cords or of the muscles that approximate them can cause strangulation. On the other hand, removal of the epiglottis usually does not cause serious debility in swallowing.

4. The upward movement of the larynx also enlarges the opening of the esophagus. At the same time, the upper 3 to 4 centimeters of the esophageal muscular wall, an area called the *upper esophageal sphincter* or the *pharyngoesophageal sphincter,* relaxes, thus allowing food to move easily and freely from the posterior pharynx into the upper esophagus. This sphincter, between swallows, remains strongly contracted, thereby preventing air from going into the esophagus during respiration. The upward movement of the larynx also lifts the glottis out of the main stream of food flow so that the food usually passes on either side of the epiglottis rather than over its surface; this adds still another protection against entry of food into the trachea.

5. At the same time that the larynx is raised and the pharyngoesophageal sphincter is relaxed, the entire muscular wall of the pharynx contracts, beginning in the superior part of the pharynx and spreading downward as a rapid peristaltic wave over the middle and inferior pharyngeal muscles and thence into the esophagus, which propels the food into the esophagus.

To summarize the mechanics of the pharyngeal stage of swallowing: the trachea is closed, the esophagus is opened, and a fast peristaltic wave originating in the pharynx forces the bolus of food into the upper esophagus, the entire process occurring in 1 to 2 seconds.

Nervous Control of the Pharyngeal Stage of Swallowing. The most sensitive tactile areas of the pharynx for initiation of the pharyngeal stage of swallowing lie in a ring around the pharyngeal opening, with greatest sensitivity in the tonsillar pillars. Impulses are transmitted from these areas through the sensory portions of the trigeminal and glossopharyngeal nerves into a region of the medulla oblongata closely associated with the *tractus solitarius,* which receives essentially all sensory impulses from the mouth.

The successive stages of the swallowing process are then automatically controlled in orderly sequence by neuronal areas distributed throughout the reticular substance of the medulla and lower portion of the pons. The sequence of the swallowing reflex is the same from one swallow to the next, and the timing of the entire cycle also remains constant from one swallow to the next. The areas in the medulla and lower pons that control swallowing are collectively called the *deglutition* or *swallowing center.*

The motor impulses from the swallowing center to the pharynx and upper esophagus that cause swallowing are transmitted by the fifth, ninth, tenth, and twelfth cranial nerves and even a few of the superior cervical nerves.

In summary, the pharyngeal stage of swallowing is

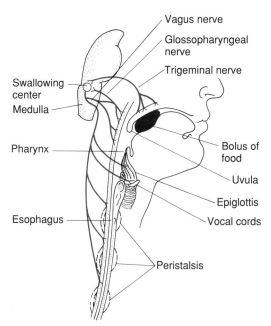

Figure 63–1. The swallowing mechanism.

Vagus nerve
Glossopharyngeal nerve
Trigeminal nerve
Swallowing center
Medulla
Pharynx
Bolus of food
Uvula
Epiglottis
Vocal cords
Esophagus
Peristalsis

principally a reflex act. It is almost never initiated by direct stimuli to the swallowing center from higher regions of the central nervous system. Instead, it is almost always initiated by voluntary movement of food into the back of the mouth, which, in turn, elicits the swallowing reflex.

Effect of the Pharyngeal Stage of Swallowing on Respiration. The entire pharyngeal stage of swallowing occurs in less than 1 to 2 seconds, thereby interrupting respiration for only a fraction of a usual respiratory cycle. The swallowing center specifically inhibits the respiratory center of the medulla during this time, halting respiration at any point in its cycle to allow swallowing to proceed. Yet even while a person is talking, swallowing interrupts respiration for such a short time that it is hardly noticeable.

Esophageal Stage of Swallowing

The esophagus functions primarily to conduct food from the pharynx to the stomach, and its movements are organized specifically for this function.

Normally the esophagus exhibits two types of peristaltic movements—*primary peristalsis* and *secondary peristalsis.* Primary peristalsis is simply a continuation of the peristaltic wave that begins in the pharynx and spreads into the esophagus during the pharyngeal stage of swallowing. This wave passes all the way from the pharynx to the stomach in approximately 8 to 10 seconds. However, food swallowed by a person who is in the upright position is usually transmitted to the lower end of the esophagus even more rapidly than the peristaltic wave itself, in about 5 to 8 seconds, because of the additional effect of gravity pulling the food downward. If the primary peristaltic wave fails to move all the food that has entered the esophagus into the stomach, *secondary peristaltic waves* result from distention of the esophagus by the retained food, and they continue until all the food has emptied into the stomach. These secondary waves are initiated partly by intrinsic neural circuits in the esophageal enteric nervous system and partly by reflexes that are transmitted through *vagal afferent fibers* from the esophagus to the medulla and then back again to the esophagus through *vagal efferent fibers.*

The musculature of the pharynx and the upper quarter of the esophagus is *striated muscle.* Therefore, the peristaltic waves in these regions are controlled only by skeletal nerve impulses in the glossopharyngeal and vagus nerves. In the lower two thirds of the esophagus, the musculature is smooth, but this portion of the esophagus is also strongly controlled by the vagus nerves acting through their connections with the enteric nervous system. However, when the vagus nerves to the esophagus are sectioned, the myenteric nerve plexus of the esophagus becomes excitable enough after several days to cause strong secondary peristaltic waves even without support from the vagal reflexes. Therefore, following paralysis of the swallowing reflex, food forced in some other way into the upper esophagus and then pulled by gravity to the lower esophagus still passes readily into the stomach.

Receptive Relaxation of the Stomach. As the esophageal peristaltic wave passes toward the stomach, a wave of relaxation, transmitted through myenteric inhibitory neurons, precedes the constriction. Furthermore, the entire stomach and, to a lesser extent, even the duodenum become relaxed as this wave reaches the lower end of the esophagus and thus are prepared ahead of time to receive the food propelled down the esophagus during the swallowing act.

Function of the Lower Esophageal Sphincter (Gastroesophageal Sphincter)

At the lower end of the esophagus, extending from about 2 to 5 cm above its juncture with the stomach, the esophageal circular muscle is slightly thickened and functions as a *lower esophageal sphincter* or *gastroesophageal sphincter.* Anatomically this sphincter is no different from the remainder of the esophagus. However, physiologically, it normally remains tonically constricted, in contrast to the mid and upper portions of the esophagus, which normally remain completely relaxed. Yet when a peristaltic swallowing wave passes down the esophagus, "receptive relaxation" relaxes the lower esophageal sphincter ahead of the peristaltic wave and allows easy propulsion of the swallowed food into the stomach. Rarely, the sphincter does not relax satisfactorily, resulting in a condition called *achalasia,* which is discussed in Chapter 66.

The stomach contents are highly acidic and contain many proteolytic enzymes. The esophageal mucosa, except in the lower eighth of the esophagus, is not capable of resisting for long the digestive action of gastric secretions. Fortunately, the tonic constriction of the lower esophageal sphincter helps to prevent significant reflux of stomach contents into the esophagus except under abnormal conditions.

Additional Prevention of Reflux by Valvelike Closure of the Distal End of the Esophagus. Another factor that prevents reflux is a valvelike mechanism of that short portion of the esophagus that lies immediately beneath the diaphragm before reaching the stomach. Increased intra-abdominal pressure caves the esophagus inward at this point at the same time that this pressure also increases the intragastric pressure. Thus, this valvelike closure of the lower esophagus prevents the high abdominal pressure from forcing stomach contents into the esophagus. Otherwise, every time we walked, coughed, or breathed hard, we might expel acid into the esophagus.

● MOTOR FUNCTIONS OF THE STOMACH

The motor functions of the stomach are threefold: (1) storage of large quantities of food until it can be accommodated in the duodenum, (2) mixing of this food with gastric secretions until it forms a semifluid mixture called *chyme,* and (3) slow emptying of the food from the stomach into the small intestine at a rate suitable for proper digestion and absorption by the small intestine.

Figure 63–2 illustrates the basic anatomy of the stomach. Physiologically, the stomach can be divided into two major parts: (1) the *corpus,* or *body,* and (2) the *antrum.* In addition, the *fundus,* located at the upper end of the body of the stomach, is considered by some anatomists to be a separate entity from the body, but physiologically the fundus functions mainly as part of the body.

STORAGE FUNCTION OF THE STOMACH

As food enters the stomach, it forms concentric circles in the body and fundus of the stomach, the newest food lying closest to the esophageal opening and the oldest food lying nearest the wall of the stomach. Normally, when food enters the stomach, a vagal reflex greatly reduces the tone in the muscular wall of the body of the stomach so that the wall can bulge progressively outward, accommodating greater and greater quantities of food up to a limit of about 1.5 liters. The pressure in the stomach remains low until this limit is approached.

MIXING AND PROPULSION OF FOOD IN THE STOMACH—THE BASIC ELECTRICAL RHYTHM OF THE STOMACH

The digestive juices of the stomach are secreted by the *gastric glands,* which cover almost the entire wall of the body of the stomach except along a strip on the lesser curvature of the stomach. These secretions come immediately into contact with that portion of the stored food lying against the mucosal surface of the stomach; when the stomach is filled, weak peristaltic *constrictor waves,* also called *mixing waves,* move toward the antrum along the stomach wall approximately once every 20 seconds. These waves are

initiated by the *basic electrical rhythm* (called *BER*) consisting of electrical "slow waves" that occur spontaneously in the stomach wall. In most parts of the gastrointestinal tract, these waves are not strong enough to cause contractions unless they first elicit action potentials, but in the stomach their positive peaks do rise above the threshold for excitation even without action potentials.

As the slow waves move down the stomach, they not only cause the secretions to mix with the outer portions of the stored food but also provide weak propulsion to move the food toward the antrum. When the stomach is full, these mixing contractions usually begin near the midpoint of the stomach; but, as the stomach empties, the contractions become stronger and also originate farther back up the stomach wall, thus propelling the last vestiges of stored food into the stomach antrum. Then when completely empty, the stomach becomes mainly quiescent until new food enters.

As the constrictor waves progress from the body of the stomach into the antrum, they become more intense, some becoming extremely intense and providing powerful *peristaltic constrictor rings* that force the antral contents under high pressure toward the pylorus. These constrictor rings also play an exceedingly important role in mixing the stomach contents in the following way: Each time a peristaltic wave passes over the antrum toward the pylorus, it digs deeply into the contents of the antrum. Yet the opening of the pylorus is small enough that only a few milliliters of antral contents are expelled into the duodenum with each peristaltic wave. Also, as each peristaltic wave approaches the pylorus, the pyloric muscle itself contracts, which further impedes emptying through the pylorus. Therefore, most of the antral contents are squirted backward through the peristaltic ring toward the body of the stomach. Thus, the moving peristaltic constrictive ring, combined with this squirting action, called "retropulsion," is an exceedingly important mixing mechanism of the stomach.

Chyme. After the food has become mixed with the stomach secretions, the resulting mixture that passes on down the gut is called *chyme.* The degree of fluidity of chyme depends on the relative amounts of food and stomach secretions and on the degree of digestion that has occurred. The appearance of chyme is that of a murky, milky semifluid or paste.

Hunger Contractions. Besides the peristaltic contractions that occur when food is present in the stomach, another type of intense contractions, called *hunger contractions,* often occurs when the stomach has been empty for a long time. These are rhythmic peristaltic contractions in the *body* of the stomach. However, when they become extremely strong, they often fuse together to cause a continuing tetanic contraction lasting for as long as 2 to 3 minutes.

Hunger contractions are usually most intense in young, healthy persons with high degrees of gastrointestinal tonus; and they are also greatly increased by a low level of blood sugar.

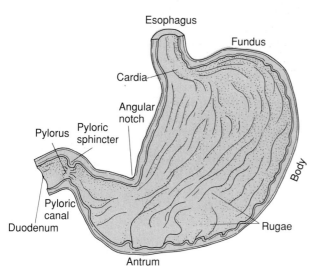

Figure 63–2. Physiologic anatomy of the stomach.

When hunger contractions occur in the stomach, the person sometimes experiences a sensation of pain in the pit of the stomach, called *hunger pangs.* Hunger pangs usually do not begin until 12 to 24 hours after the last ingestion of food; in starvation they reach their greatest intensity in 3 to 4 days and then gradually weaken in succeeding days.

Hunger contractions are often associated with a feeling of hunger and therefore are perhaps an important means by which the alimentary tract intensifies the animal drive to acquire food when a person is in a state of incipient starvation.

EMPTYING OF THE STOMACH

Mainly, stomach emptying is promoted by the intensity of the peristaltic contractions of the stomach antrum. At the same time emptying is opposed by varying degrees of resistance to the passage of chyme at the pylorus.

Intense Antral Peristaltic Contractions During Stomach Emptying — The Pyloric Pump. Most of the time the antral peristaltic contractions are weak and function mainly to cause mixing of the food and gastric secretions. However, about 20 per cent of the time while food is in the stomach, the antral contractions become very intense, beginning at the incisura angularis of the stomach and then spreading through the antrum no longer as weak mixing contractions but instead as strong peristaltic, ringlike constrictions. As the stomach becomes progressively more and more empty, these constrictions begin farther and farther up the body of the stomach, gradually pinching off the lowermost portions of the stored food and adding this food to the chyme in the antrum. These intense peristaltic contractions often create as much as 50 to 70 centimeters of water pressure, which is about six times as powerful as the usual mixing type of peristaltic waves. Thus, the intensity of this antral peristalsis is the principal factor that determines the rate of stomach emptying.

When pyloric tone is normal, each strong antral peristaltic wave forces several milliliters of chyme into the duodenum. Thus, the peristaltic waves provide a pumping action that is frequently called the "pyloric pump."

Role of the Pylorus in Controlling Stomach Emptying. The distal opening of the stomach is the *pylorus.* Here the thickness of the circular muscle becomes 50 to 100 per cent greater than in the earlier portions of the stomach antrum, and it also remains slightly tonically contracted almost all the time. Therefore, the pyloric circular muscle is frequently called the *pyloric sphincter.*

Despite the tonic contraction of the pyloric sphincter, the pylorus usually remains slightly open, large enough for water and other fluids normally to empty from the stomach with ease. On the other hand, the constriction usually prevents passage of most larger food particles until they have become mixed in the chyme to an almost fluid consistency.

However, the degree of constriction of the pylorus can be increased or decreased under the influence of nervous and humoral signals from both the stomach and the duodenum, as is discussed shortly. Thus, the pylorus as well as the stomach antrum also enters into the control of stomach emptying.

REGULATION OF STOMACH EMPTYING

The rate at which the stomach empties is regulated by signals both from the stomach and from the duodenum. The stomach signals are mainly twofold: (1) nervous signals caused by distention of the stomach by food and (2) the hormone *gastrin* released from the antral mucosa in response to the presence of certain types of food in the stomach. Both these signals mainly increase pyloric pumping force and at the same time slightly inhibit the pylorus, thus promoting stomach emptying.

On the other hand, signals from the duodenum depress the pyloric pump and usually increase pyloric tone at the same time. In general, when an excess volume of chyme or excessive amounts of certain types of chyme enter the duodenum, strong *negative* feedback signals, both *nervous* and *hormonal,* depress the pyloric pump and enhance pyloric sphincter tone. Obviously, these feedback signals allow chyme to enter the duodenum only as rapidly as it can be processed by the small intestine.

Gastric Factors That Promote Emptying

Effect of Gastric Food Volume on Rate of Emptying. It is very easy to see how increased food volume in the stomach could promote increased emptying from the stomach. However, this increased emptying does not occur for the reasons that one would expect. It is not increased pressure in the stomach that causes the increased emptying, because in the usual normal range of volume the increase in volume does not increase the pressure significantly. On the other hand, stretching of the stomach wall does elicit vagal and local myenteric reflexes in the wall that greatly excite the activity of the pyloric pump and at the same time slightly inhibit the pylorus.

Effect of the Hormone Gastrin on Stomach Emptying. In the following chapter we shall see that stretch, as well as the presence of certain types of foods in the stomach — particularly meat — elicits release of a hormone called *gastrin* from the antral mucosa, and this has potent effects on causing secretion of highly acidic gastric juice by the stomach fundic glands. Gastrin also has moderate stimulatory effects on motor functions of the stomach as well. Most important, it enhances the activity of the pyloric pump. Thus, it, too, has an important influence in promoting stomach emptying.

Duodenal Factors That Inhibit Emptying

The Inhibitory Effect of Enterogastric Nervous Reflexes From the Duodenum. When food enters the duodenum, multiple nervous reflexes are initiated from the duodenum wall that pass back to the stomach and slow or even stop stomach emptying if the volume of chyme in the duodenum has become too much. These reflexes are mediated by three different routes: (1) directly from the duodenum to the stomach through the enteric nervous system in the gut wall, (2) through extrinsic nerves that go to the prevertebral sympathetic ganglia and then back through inhibitory sympathetic nerve fibers to the stomach, and (3) through the vagus nerves all the way to the brain stem, where they inhibit the normal excitatory signals transmitted to the stomach through the vagi. All these parallel reflexes have two effects on stomach emptying: first, they strongly inhibit the antral propulsive contractions; and second, they increase slightly to moderately the tone of the pyloric sphincter.

The types of factors that are continually monitored in the duodenum and that can excite the enterogastric reflexes include

1. The degree of distention of the duodenum.
2. The presence of any degree of irritation of the duodenal mucosa.
3. The degree of acidity of the duodenal chyme.
4. The degree of osmolality of the chyme.
5. The presence of certain breakdown products in the chyme, especially breakdown products of proteins and perhaps to a lesser extent of fats.

The enterogastric reflexes are especially sensitive to the presence of irritants and acids in the duodenal chyme. For instance, whenever the pH of the chyme in the duodenum falls below approximately 3.5 to 4, the reflexes frequently block entirely further release of acidic stomach contents into the duodenum until the duodenal chyme can be neutralized by pancreatic and other secretions.

Breakdown products of protein digestion will also elicit these reflexes; by slowing the rate of stomach emptying, sufficient time is ensured for adequate protein digestion in the upper portion of the small intestine.

Finally, either hypo- or hypertonic fluids (especially hypertonic) will elicit the reflexes. These then prevent too rapid flow of nonisotonic fluids into the small intestine, thereby preventing rapid changes in electrolyte balance of the body fluids during absorption of the intestinal contents.

Hormonal Feedback From the Duodenum in Inhibiting Gastric Emptying — Role of Fats. Not only do nervous reflexes from the duodenum to the stomach inhibit stomach emptying, but hormones released from the upper intestine do so as well. The stimulus for producing the hormones is mainly fats entering the duodenum, though other types of foods can increase the hormones to a lesser degree.

On entering the duodenum, the fats extract several different hormones from the duodenal and jejunal epithelium, either by binding with "receptors" in the epithelial cells or in some other way. In turn, the hormones are carried by way of the blood to the stomach, where they (1) inhibit the activity of the pyloric pump and at the same time (2) increase slightly the strength

of contraction of the pyloric sphincter. These effects are important because fats are much slower to be digested than are most other foods.

Unfortunately, what the precise hormones are that cause the hormonal feedback inhibition of the stomach is not fully clear. However, the most potent appears to be *cholecystokinin (CCK)*, which is released from the mucosa of the jejunum in response to fatty substances in the chyme. This hormone acts as a competitive inhibitor to block the increased stomach motility caused by gastrin. Another is the hormone *secretin*, which is released mainly from the duodenal mucosa in response to gastric acid released from the stomach through the pylorus. This hormone has a general but only weak effect of decreasing gastrointestinal motility. Finally, a hormone called *gastric inhibitory peptide (GIP)*, which is released from the upper small intestine in response mainly to fat in the chyme but to carbohydrates as well, is known also to inhibit gastric motility under some conditions. (However, its effect at physiological concentrations is probably mainly to stimulate the secretion of insulin by the pancreas.) All these hormones will be discussed at greater length elsewhere in this text, especially in the following chapter, in relation to the control of gallbladder emptying and pancreatic secretion.

In summary, several different hormones are known that could serve as hormonal mechanisms for inhibiting gastric emptying when excess quantities of chyme, especially acidic or fatty chyme, enter the duodenum from the stomach. CCK is probably the most important.

Summary of the Control of Stomach Emptying

Emptying of the stomach is controlled to a moderate degree by stomach factors, such as the degree of filling in the stomach and the excitatory effect of gastrin on antral peristalsis. However, probably the more important control of stomach emptying resides in feedback signals from the duodenum, including both the enterogastric feedback reflexes and hormonal feedback. These two feedback inhibitory mechanisms work together to slow the rate of emptying when (1) too much chyme is already in the small intestine or (2) the chyme is excessively acid, contains too much unprocessed protein or fat, is hypotonic or hypertonic, or is irritating. In this way the rate of stomach emptying is limited to that amount of chyme that the small intestine can process.

● MOVEMENTS OF THE SMALL INTESTINE

The movements of the small intestine, as elsewhere in the gastrointestinal tract, can be divided into *mixing contractions* and *propulsive contractions*. However, to a great extent this separation is artificial because essentially all movements of the small intestine cause at least some degree of both mixing and propulsion. Yet the usual classification of these processes is the following:

MIXING CONTRACTIONS (SEGMENTATION CONTRACTIONS)

When a portion of the small intestine becomes distended with chyme, the stretch of the intestinal wall elicits localized concentric contractions spaced at intervals along the intestine. The longitudinal length of each one of the contractions is only about 1 cm, so that each set of contractions causes "segmentation" of the small intestine, as illustrated in Figure 63–3, dividing the intestine into spaced segments that have the appearance of a chain of sausages. As one set of segmentation contractions relaxes, a new set begins, but the contractions this time occur at new points between the previous contractions. These segmentation contractions "chop" the chyme as often as 8 to 12 times a minute, in this way promoting progressive mixing of the solid food particles with the secretions of the small intestine.

The maximum frequency of the segmentation contractions in the small intestine is determined by the frequency of the *slow waves* in the intestinal wall, which is the basic electrical rhythm (BER) as explained earlier. Since this frequency is about 12 per minute in the duodenum and proximal jejunum, the maximum frequency of the segmentation contractions in these areas is also about 12 per minute. However, in the terminal ileum, the maximum frequency is usually 8 to 9 contractions per minute.

The segmentation contractions become exceedingly weak when the excitatory activity of the enteric nervous system is blocked by atropine. Therefore, even though it is the slow waves in the smooth muscle itself that control the segmentation contractions, these contractions are not really effective without background excitation by the enteric nervous system, especially by the myenteric plexis.

PROPULSIVE MOVEMENTS

Peristalsis in the Small Intestine. Chyme is propelled through the small intestine by *peristaltic*

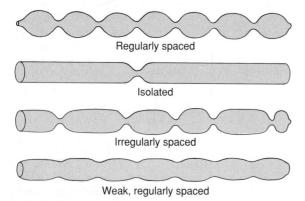

Figure 63–3. Segmentation movements of the small intestine.

waves. These can occur in any part of the small intestine, and they move analward at a velocity of 0.5 to 2 cm/sec, much faster in the proximal intestine and much slower in the terminal intestine. However, they are normally very weak and usually die out after traveling only 3 to 5 centimeters, very rarely farther than 10 centimeters, so that movement of the chyme is also very poor, so poor in fact that the *net* movement of the chyme along the small intestine averages only 1 cm/min. This means that normally 3 to 5 hours are required for passage of chyme from the pylorus to the ileocecal valve.

Control of Peristalsis by Nervous and Hormonal Signals. Peristaltic activity of the small intestine is greatly increased after a meal. This is caused partly by the beginning entry of chyme into the duodenum but also by a so-called *gastroenteric reflex* that is initiated by distention of the stomach and conducted principally through the myenteric plexus from the stomach down along the wall of the small intestine.

In addition to the nervous signals that affect small intestinal peristalsis, several hormonal factors also affect peristalsis. These include *gastrin, cholecystokinin, insulin,* and *serotonin,* all of which enhance intestinal motility and are secreted during various phases of food processing. On the other hand, *secretin* and *glucagon* inhibit small intestinal motility. Unfortunately, the quantitative importance of each of these hormonal factors for controlling motility is still questionable.

The function of the peristaltic waves in the small intestine is not only to cause progression of the chyme toward the ileocecal valve but also to spread out the chyme along the intestinal mucosa. As the chyme enters the intestine from the stomach and causes initial distention of the proximal intestine, the elicited peristaltic waves begin immediately to spread the chyme along the intestine, and this process intensifies as additional chyme enters the duodenum. On reaching the ileocecal valve, the chyme is sometimes blocked for several hours until the person eats another meal, when a new *gastroenteric* (also called *gastroileal*) reflex intensifies the peristalsis in the ileum and forces the remaining chyme through the ileocecal valve into the cecum.

The Propulsive Effect of the Segmentation Movements. The segmentation movements, though they last for only a few seconds, also travel in the anal direction and help propel the food down the intestine. Therefore, the difference between the segmentation and the peristaltic movements is not as great as might be implied by their separation into these two classifications.

The Peristaltic Rush. Though peristalsis in the small intestine is normally very weak, intense irritation of the intestinal mucosa, as occurs in some severe cases of infectious diarrhea, can cause both very powerful and rapid peristalsis called the *peristaltic rush.* This is initiated partly by extrinsic nervous reflexes to the brain stem and back again to the gut and partly by direct enhancement of the myenteric plexus reflexes. The powerful peristaltic contractions then travel long distances in the small intestine within minutes, sweeping the contents of the intestine into the colon and thereby relieving the small intestine of either irritative chyme or excessive distention.

Peristalsis in the Fasting Human Being— The "Migrating Motor Complex"

After a person or an animal eats a meal, the nature of the gastrointestinal motor functions is determined mainly by the stimulating effects of the food in the gastrointestinal tract itself. However, late at night, many hours after a meal, or at other times when the human being is fasting, a different pattern of activity recurs approximately every 1.5 to 2 hours in the stomach and small intestine, called the *migrating motor complex*. This migrating complex causes moderately active peristaltic waves to sweep slowly downward along the stomach and small intestine, sweeping the excess digestive secretions into the colon and therefore preventing their accumulation in the upper gastrointestinal tract.

The migrating complex begins in the body of the stomach and spreads all the way through the ileum. At any one time only about 40 centimeters of the intestinal tract is actively engaged in the peristaltic waves, and this lasts for only 6 to 10 minutes; but this 40 centimeter area moves slowly along the intestinal tract at a velocity of 6 to 12 cm/min. Then approximately at the time that one migrating complex reaches the end of the ileum, a new one begins in the stomach.

Movements Caused by the Muscularis Mucosae and Muscle Fibers of the Villi

The muscularis mucosae can cause short or long folds to appear in the intestinal mucosa and also cause the folds to move to progressively newer areas of mucosa. Also, individual fibers from this muscle extend into the intestinal villi and cause them to contract intermittently. The mucosal folds increase the surface area exposed to the chyme, thereby increasing the rate of absorption. The contractions of the villi—shortening, elongating, and shortening again —"milk" the villi so that lymph flows freely from the central lacteals into the lymphatic system. Both of these types of contraction also agitate the fluids surrounding the villi, so that progressively new areas of fluid become exposed to absorption.

These mucosal and villous contractions are initiated by local nervous reflexes in the submucosal plexus that occur in response to chyme in the small intestine.

FUNCTION OF THE ILEOCECAL VALVE

A principal function of the ileocecal valve is to prevent backflow of fecal contents from the colon into the small intestine. As illustrated in Figure 63–4, the lips of the ileocecal valve protrude into the lumen of the cecum and therefore are forcefully closed when any excess pressure builds up in the cecum and tries to push the cecal contents backwards against the lips. Usually the valve can resist reverse pressure of as much as 50 to 60 centimeters of water.

In addition, the wall of the ileum for several centi-

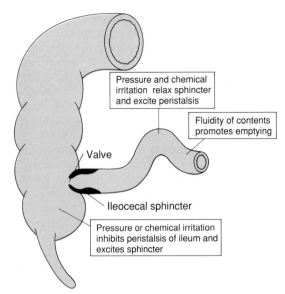

Figure 63–4. Emptying at the ileocecal valve.

meters immediately preceding the ileocecal valve has a thickened muscular coat called the *ileocecal sphincter*. This normally remains mildly constricted and slows the emptying of ileal contents into the cecum except immediately after a meal, when a gastroileal reflex (described earlier) intensifies the peristalsis in the ileum. Also, the hormone *gastrin*, which is liberated from the stomach mucosa in response to food in the stomach, increases ileal contractions and relaxes the ileocecal sphincter.

The resistance to emptying at the ileocecal valve prolongs the stay of chyme in the ileum and, therefore, facilitates absorption. Only about 1500 milliliters of chyme empty into the cecum each day.

Feedback Control of the Ileocecal Sphincter. The degree of contraction of the ileocecal sphincter, as well as the intensity of peristalsis in the terminal ileum, is also controlled strongly by reflexes from the cecum. Whenever the cecum is distended, the contraction of the ileocecal sphincter is intensified while ileal peristalsis is inhibited, which greatly delays emptying of additional chyme from the ileum. Also, any irritant in the cecum delays emptying. For instance, when a person has an inflamed appendix, the irritation of this vestigial remnant of the cecum can cause such intense spasm of the ileocecal sphincter and paralysis of the ileum that this completely blocks emptying of the ileum. These reflexes from the cecum to the ileocecal sphincter and ileum are mediated both by way of the myenteric plexus in the gut wall itself and also through extrinsic nerves, especially reflexes by way of the prevertebral sympathetic ganglia.

● MOVEMENTS OF THE COLON

The principal functions of the colon are (1) absorption of water and electrolytes from the chyme and (2) storage of fecal matter until it can be expelled. The proximal half of the colon, illustrated in Figure 63–5, is concerned principally with absorption, and the distal half with storage. Because intense movements are not required for these functions, the movements of the colon are normally sluggish. Yet in a sluggish manner, the movements still have characteristics similar to those of the small intestine and can be divided once again into mixing movements and propulsive movements.

Mixing Movements—Haustrations. In the same manner that segmentation movements occur in the small intestine, large circular constrictions also occur in the large intestine. At each of these constriction points, about 2.5 centimeters of the circular muscle contracts, sometimes constricting the lumen of the colon to almost complete occlusion. At the same time, the longitudinal muscle of the colon, which is aggregated into three longitudinal strips called the *teneae coli*, contract. These combined contractions of the circular and longitudinal smooth muscle cause the unstimulated portion of the large intestine to bulge outward into baglike sacs called *haustrations*. The haustral contractions, once initiated, usually reach peak intensity in about 30 seconds and then disappear during the next 60 seconds. They also at times move slowly anally during their period of contraction, especially in the cecum and ascending colon, and therefore provide a minor amount of forward propulsion of the colonic contents. After another few minutes, new haustral contractions occur in other areas nearby. Therefore, the fecal material in the large intestine is slowly *dug into and rolled over* in much the same manner that one spades the earth. In this way, all the fecal material is gradually exposed to the surface of the large intestine, and fluid and dissolved substances are progressively absorbed until only 80 to 200 milliliters of the 1500 milliliter daily load of chyme is lost in the feces.

Propulsive Movements—"Mass Movements." Peristaltic waves of the type seen in the small intestine only rarely occur in most parts of the colon. Instead, most propulsion occurs by (1) the slow analward movement of the *haustral contractions* just discussed and (2) *mass movements*.

Much of the propulsion in the cecum and ascending colon results from the slow but persistent haustral contractions, requiring as many as 8 to 15 hours to move the chyme only from the ileocecal valve through the transverse colon, while the chyme itself becomes fecal in quality and also becomes a semisolid slush instead of a semifluid.

From the beginning of the transverse colon to the sigmoid, mass movements mainly take over the propulsive role. These movements usually occur only a few times each day, most abundantly for about 15 minutes during the first hour after eating breakfast.

A mass movement is a modified type of peristalsis characterized by the following sequence of events: First, a constrictive ring occurs at a distended or irritated point in the colon, usually in the transverse colon, and then rapidly thereafter the 20 or more centimeters of colon *distal* to the constriction lose their haustrations and instead contract as a unit, forcing the fecal material in this segment *en masse* down the colon. The contraction develops progressively more force for about 30 seconds, and relaxation then occurs during the next 2 to 3 minutes before another mass movement occurs, this time perhaps farther along the colon. But the whole series of mass movements will usually persist for only 10 minutes to half an hour, and they will then return perhaps a half day or even a day later. When they have forced a mass of feces into the rectum, the desire for defecation is felt.

Initiation of Mass Movements by the Gastrocolic and Duodenocolic Reflexes. The appearance of mass movements after meals is facilitated by *gastrocolic* and *duodenocolic reflexes*. These reflexes result from distention of the stomach and duodenum. They can take place only weakly when the extrinsic nerves are removed; therefore, it is probable that only weak reflexes are transmitted through the myenteric plexus, whereas reflexes conducted through the extrinsic nerves of the autonomic nervous system probably provide most of the intensity of the gastrocolic and duodenocolic reflexes.

Irritation in the colon can also initiate intense mass movements. For instance, a person who has an ulcerated condition of the colon *(ulcerative colitis)* frequently has mass movements that persist almost all the time.

Mass movements can also be initiated by intense stimulation of the parasympathetic nervous system or simply by overdistention of a segment of the colon.

DEFECATION

Most of the time, the rectum is empty of feces. This results partly from the fact that a weak functional sphincter exists approximately 20 centimeters from the anus at the juncture between the sigmoid and the rectum. There is also a sharp angulation here that contributes additional resistance to filling of the rectum. However, when a mass movement forces feces

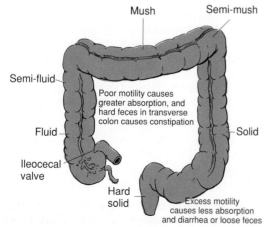

Mush
Semi-mush
Semi-fluid
Fluid
Ileocecal valve
Hard solid
Solid

Poor motility causes greater absorption, and hard feces in transverse colon causes constipation

Excess motility causes less absorption and diarrhea or loose feces

Figure 63–5. Absorptive and storage functions of the large intestine.

into the rectum, the desire for defecation is normally initiated, including reflex contraction of the rectum and relaxation of the anal sphincters.

Continual dribble of fecal matter through the anus is prevented by tonic constriction of (1) the *internal anal sphincter,* a thickening of the intestinal circular smooth muscle that lies immediately inside the anus, and (2) the *external anal sphincter,* composed of striated voluntary muscle that both surrounds the internal sphincter and also extends distal to it; the external sphincter is controlled by nerve fibers in the pudendal nerve, which is part of the somatic nervous system and therefore is under *voluntary, conscious control.*

The Defecation Reflexes. Ordinarily, defecation is initiated by *defecation reflexes.* One of these reflexes is an *intrinsic reflex* mediated by the local enteric nervous system. This can be described as follows: When the feces enter the rectum, distention of the rectal wall initiates afferent signals that spread through the *myenteric plexus* to initiate peristaltic waves in the descending colon, sigmoid, and rectum, forcing feces toward the anus. As the peristaltic wave approaches the anus, the internal anal sphincter is relaxed by inhibitory signals from the myenteric plexus; and if the external anal sphincter is voluntarily relaxed at the same time, defecation will occur.

However, the intrinsic defecation reflex itself is weak; and to be effective in causing defecation it usually must be fortified by another type of defecation reflex, a *parasympathetic defecation reflex* that involves the sacral segments of the spinal cord, as illustrated in Figure 63–6. When the nerve endings in the rectum are stimulated, signals are transmitted into the spinal cord and thence, reflexly, back to the descending colon, sigmoid, rectum, and anus by way of parasympathetic nerve fibers in the *pelvic nerves.* These parasympathetic signals greatly intensify the peristaltic waves as well as relaxing the internal anal sphincter and thus convert the intrinsic defecation

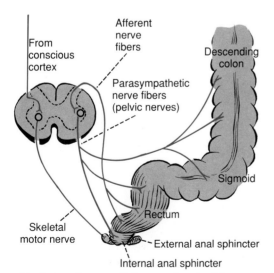

Figure 63–6. The afferent and efferent pathways of the parasympathetic mechanism for enhancing the defecation reflex.

reflex from an ineffectual weak movement into a powerful process of defecation that is sometimes effective in emptying the large bowel in one movement all the way from the splenic flexure of the colon to the anus. Also, the afferent signals entering the spinal cord initiate other effects, such as taking a deep breath, closure of the glottis, contraction of the abdominal wall muscles to force the fecal contents of the colon downward, and at the same time cause the pelvic floor to extend downward and pull outward on the anal ring to evaginate the feces.

However, despite the defecation reflexes, other effects are also necessary before actual defecation occurs. In the toilet-trained human being, relaxation of the internal sphincter and forward movement of feces toward the anus normally initiate an instantaneous contraction of the external sphincter, which still temporarily prevents defecation. Except in babies and mentally inept persons, the conscious mind then takes over voluntary control of the external sphincter and either relaxes it to allow defecation to occur or further contracts it if the moment is not socially acceptable for defecation. If the external sphincter is kept contracted, the defecation reflexes die out after a few minutes, and they remain quiescent for several hours or until additional amounts of feces enter the rectum.

When it becomes convenient for the person to defecate, the defecation reflexes can sometimes be excited by taking a deep breath to move the diaphragm downward and then contracting the abdominal muscles to increase the pressure in the abdomen, thus forcing fecal contents into the rectum to elicit new reflexes. Unfortunately, reflexes initiated in this way are almost never as effective as those that arise naturally, for which reason people who too often inhibit their natural reflexes are likely to become severely constipated.

In the newborn baby and in some persons with transected spinal cords, the defecation reflexes cause automatic emptying of the lower bowel without the normal control exercised through contraction of the external anal sphincter.

● OTHER AUTONOMIC REFLEXES AFFECTING BOWEL ACTIVITY

Aside from the duodenocolic, gastrocolic, gastroileal, enterogastric, and defecation reflexes that have been discussed in this chapter, several other important nervous reflexes can affect the overall degree of bowel activity. These are the peritoneointestinal reflex, renointestinal reflex, vesicointestinal reflex, and somatointestinal reflex. All of these reflexes are initiated by sensory signals that pass to the prevertebral sympathetic ganglia or to the spinal cord and then are transmitted through the sympathetic nervous system back to the gut. And they all *inhibit* gastrointestinal activity.

The *peritoneointestinal reflex* results from irritation of the peritoneum; it strongly inhibits the excitatory enteric nerves and thereby causes intestinal paralysis. The *renoin-*

testinal and *vesicointestinal reflexes* inhibit intestinal activity as a result of kidney or bladder irritation. Finally, the *somatointestinal reflex* causes intestinal inhibition when the skin over the abdomen is irritatingly stimulated.

REFERENCES

Berk, J. E., et al.: Bockus Gastroenterology, 4th Ed. Philadelphia, W. B. Saunders Co., 1985.

Bofak-Gyovai, L. A., and Manzione, J. V.: Oral Medicine; Patient Evaluation and Management. Baltimore, Williams & Wilkins, 1980.

Chou, C. C.: Relationship between intestinal blood flow and motility. Annu. Rev. Physiol., 44:29, 1982.

Cohen, S., et al.: Gastrointestinal motility. In Crane, R. K. (ed.): International Review of Physiology: Gastrointestinal Physiology III. Vol. 19. Baltimore, University Park Press, 1979, p. 107.

Davenport, H. W.: A Digest of Digestion, 2nd Ed. Chicago, Year Book Medical Publishers, 1978.

Dickray, G.: Comparative biochemistry and physiology of gut hormones. Annu. Rev. Physiol., 41:83, 1979.

Duthie, H. S. (ed.): Gastrointestinal Motility in Health and Disease. Baltimore, University Press, 1978.

Grossman, M. I.: Neural and hormonal regulation of gastrointestinal function: An overview. Annu. Rev. Physiol., 41:27, 1979.

Hunt, J. N.: Mechanisms and disorders of gastric emptying. Annu. Rev. Med., 34:219, 1983.

Hurwitz, A. L., et al.: Disorders of Esophageal Motility. Philadelphia, W. B. Saunders Co., 1979.

Jenkins, G. N.: The Physiology and Biochemistry of the Mouth. Philadelphia, J. B. Lippincott Co., 1978.

Johnson, L. R., et al.: Physiology of the Gastrointestinal Tract, 2nd Ed. New York, Raven Press, 1987.

Klimov, P. K.: Behavior of the organs of the digestive system. Neurosci. Behav. Physiol., 14:333, 1984.

Levitt, M. D., and Bond, J. H.: Flatulence. Annu. Rev. Med., 31:127, 1980.

Luschei, E. S., and Goldberg, L. J.: Neural mechanisms of mandibular control: Mastication and voluntary biting. In Brooks, V. B. (ed.): Handbook of Physiology. Sec. 1, Vol. II. Bethesda, Md., American Physiological Society, 1981, p. 1237.

Magee, D. F.: Interdigestive activity in the gastrointestinal tract. News in Physiol. Sci., 2:101, 1987.

Mei, N.: Intestinal chemosensitivity. Physiol. Rev., 65:211, 1985.

Miller, A. J.: Deglutition. Physiol. Rev., 62:129, 1982.

Murphy, R. A.: Muscle cells of hollow organs. News Physiol. Sci., 3:124, 1988.

Phillips, S. F., and Devroede, G. J.: Functions of the large intestine. In Crane, R. K. (ed.): International Review of Physiology: Gastrointestinal Physiology III. Vol. 19. Baltimore, University Park Press, 1979, p. 263.

Rehfeld, J. F.: Gastrointestinal hormones, In Crane, R. K. (ed.): International Review of Physiology: Gastrointestinal Physiology III. Vol. 19., Baltimore, University Park Press, 1979, p. 291.

Sant'Ambrogio, G., and Mathew, O. P.: Control of upper airway muscles. News Physiol. Sci., 3:167, 1988.

Sernka, T. J., and Jacobson, E. D.: Gastrointestinal Physiology: The Essentials. Baltimore, Williams & Wilkins, 1979.

Sternini, C.: Structural and chemical organization of the myenteric plexus. Annu. Rev. Physiol., 50:81, 1988.

Thompson, J. C., et al. (eds.): Gastrointestinal Endocrinology. New York, McGraw-Hill Book Co., 1987.

Weems, W. A.: The intestine as a fluid propelling system. Annu. Rev. Physiol., 43:9, 1981.

Weisbrodt, N. W.: Patterns of intestinal motility. Annu. Rev. Physiol., 43:21, 1981.

CHAPTER 64

• *Secretory Functions of the Alimentary Tract* •

Throughout the gastrointestinal tract secretory glands subserve two primary functions: First, digestive enzymes are secreted in most areas from the mouth to the distal end of the ileum. Second, mucous glands, from the mouth to the anus, provide mucus for lubrication and protection of all parts of the alimentary tract.

Most digestive secretions are formed only in response to the presence of food in the alimentary tract, and the quantity secreted in each segment of the tract is almost exactly the amount needed for proper digestion. Furthermore, in some portions of the gastrointestinal tract even the types of enzymes and other constituents of the secretions are varied in accordance with the types of food present. The purpose of the present chapter, therefore, is to describe the different alimentary secretions, their functions, and regulation of their production.

● GENERAL PRINCIPLES OF ALIMENTARY TRACT SECRETION

ANATOMICAL TYPES OF GLANDS

Several types of glands provide the different types of secretions in the alimentary tract. First, on the surface of the epithelium in most parts of the gastrointestinal tract are literally billions of *single cell mucous glands* called simply *mucous cells* or sometimes *goblet cells*. These function mainly by themselves in response to local stimulation of the epithelium, and they simply extrude their mucus directly onto the epithelial surface to act as a lubricant.

Second, many surface areas of the gastrointestinal tract are lined by pits that represent invaginations of the epithelium into the submucosa. In the small intestine these pits, called *crypts of Lieberkühn,* are deep and contain specialized secretory cells. One of these is illustrated in Figure 64–14.

Third, in the stomach and upper duodenum are found large numbers of deep *tubular glands.* A typical tubular gland is illustrated in Figure 64–4, which shows an acid- and pepsinogen-secreting gland of the stomach.

Fourth, also associated with the alimentary tract are several complex glands—the *salivary glands,* the *pancreas,* and the *liver*—which provide secretions for digestion or emulsification of food. The liver has a highly specialized structure that will be discussed in Chapter 70. The salivary glands and the pancreas are compound acinous glands of the type illustrated in Figure 64–2. These glands lie completely outside the walls of the alimentary tract and, in this, differ from all other alimentary glands. These glands contain millions of *acini* lined with secreting glandular cells; these feed into a system of ducts that finally empty into the alimentary tract itself.

BASIC MECHANISMS OF STIMULATION OF THE ALIMENTARY TRACT GLANDS

Effect of Local Contact and Enteric Nervous Stimuli. The mechanical presence of food in a particular segment of the gastrointestinal tract usually causes the glands of that region, and often of adjacent regions, to secrete moderate to large quantities of digestive juices. Part of this local effect results from direct contact stimulation of the surface glandular cells by contact with the food. In addition, epithelial stimulation also activates the *enteric nervous system* of the gut wall. The types of stimuli that do this are (1) tactile stimulation, (2) chemical irritation, and (3) distention of the gut wall. The resulting nervous reflexes stimulate both the mucous cells on the epithelial surface and the deeper glands in the mucosa to increase their secretion.

Autonomic Stimulation of Secretion. *Parasympathetic Stimulation.* Stimulation of the parasympathetic nerves to the alimentary tract almost invariably increases the rates of glandular secretion. This is especially true of the glands in the upper portion of the tract, innervated by the vagus and other cranial parasympathetic nerves—including the salivary glands, esophageal glands, gastric glands, pancreas, and Brunner's glands in the duodenum—and also of the glands of the distal portion of the large intestine, innervated by the pelvic parasympathetic nerves. Secretion in the remainder of the small intestine and in the first two thirds of the large intestine occurs mainly in response to segmental neural and hormonal stimuli.

Sympathetic Stimulation. Stimulation of sympathetic nerves in some parts of the gastrointestinal tract causes a slight to moderate increase in secretion by the respective glands. On the other hand, sympathetic stimulation also results in constriction of the blood vessels supplying the glands. Therefore, sympathetic stimulation can have a dual effect: First, sympathetic stimulation alone usually slightly increases secretion. But, second, if parasympathetic or hormonal stimulation is causing copious secretion by the glands, superimposed sympathetic stimulation usually reduces the secretion, sometimes very significantly, mainly because of reduced blood supply.

Regulation of Glandular Secretion by Hormones. In the stomach and intestine several different *gastrointestinal hormones* help regulate the volume and character of the secretions. These hormones are liberated from the gastrointestinal mucosa in response to the presence of foods in the lumen of the gut. They then are absorbed into the blood and are carried to the glands, where they stimulate secretion. This type of stimulation is particularly valuable in increasing the output of gastric juice and pancreatic juice when food enters the stomach or duodenum. Also, hormonal stimulation of the gallbladder wall causes it to empty its stored bile into the duodenum. Other hormones that are still of doubtful value have been postulated to stimulate secretion by the glands of the small intestine.

Chemically, the gastrointestinal hormones are polypeptides or polypeptide derivatives.

BASIC MECHANISM OF SECRETION BY GLANDULAR CELLS

Secretion of Organic Substances. Although all of the basic mechanisms by which glandular cells form different secretions and then extrude these to the exterior are not known, experimental evidence points to the following basic principles of secretion by glandular cells, as illustrated in Figure 64–1.

1. The nutrient material needed for formation of the secretion must diffuse or be actively transported from the capillary into the base of the glandular cell.

2. Many *mitochondria* located inside the cell near its base utilize oxidative energy for formation of adenosine triphosphate (ATP).

3. Energy from the ATP, along with appropriate substrates, is then used for synthesis of the organic substances; this synthesis occurs almost entirely in the *endoplasmic reticulum* and *Golgi complex*. The *ribosomes* adherent to this reticulum are specifically responsible for formation of the proteins that are to be secreted.

4. The secretory materials are transported through the tubules of the endoplasmic reticulum, passing in about 20 minutes all the way to the vesicles of the Golgi complex, which lies near the secretory ends of the cells.

5. In the Golgi complex the materials are modified, added to, concentrated, and discharged into the cyto-

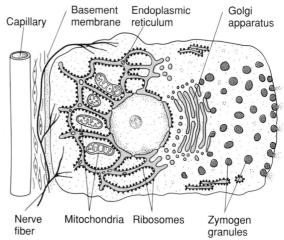

Figure 64–1. Typical function of a glandular cell in formation and secretion of enzymes or other secretory substances.

plasm in the form of *secretory vesicles,* which are stored in the apical ends of the secretory cells.

6. These vesicles remain stored until nervous or hormonal control signals cause them to extrude their contents through the cell's surface. This probably occurs in the following way: The control signal first increases the cell membrane permeability to calcium, and calcium enters the cell. The calcium in turn causes many of the vesicles to fuse with the cell membrane and then to break open on their outer surfaces, thus emptying their contents to the exterior; this process is called *exocytosis.*

Water and Electrolyte Secretion. A second necessity for glandular secretion is sufficient water and electrolytes to be secreted along with the organic substances. The following is a postulated method by which nervous stimulation causes water and salts to pass through the glandular cells in great profusion, which washes the organic substances through the secretory border of the cells at the same time:

1. Nerve stimulation has a specific effect on the *basal* portion of the cell membrane, causing active transport of chloride ions to the interior.

2. The resulting increase in electronegativity inside the cell then causes positive ions also to move to the interior of the cell.

3. The excess of both of these ions inside the cell creates an osmotic force that pulls water to the interior, thereby increasing the hydrostatic pressure inside the cell and causing the cell itself to swell.

4. The pressure in the cell then results in minute ruptures of the secretory border of the cell and causes flushing of water, electrolytes, and organic materials out of the secretory end of the glandular cell and into the lumen of the gland.

In support of this theory have been the following findings: First, the nerve endings on glandular cells are principally on the bases of the cells. Second, microelectrode studies show that the normal electrical potential across the membrane at the base of the cell is between 30 and 40 millivolts, with negativity on the interior and positivity on the exterior. Parasympathetic stimulation increases this polarization voltage to values some 10 and 20 millivolts more negative than normal. This increase in polarization occurs a second or more after the nerve signal has arrived, indicating that it is caused by movement of negative ions (presumably chloride ions) through the membrane to the interior of the cell.

Although this mechanism for secretion is still partly theoretical, it does explain how it would be possible for nerve impulses to regulate secretion. Obviously, hormonal effects on the cell membrane could cause similar results.

LUBRICATING AND PROTECTIVE PROPERTIES OF MUCUS AND ITS IMPORTANCE IN THE GASTROINTESTINAL TRACT

Mucus is a thick secretion composed mainly of water, electrolytes, and a mixture of several glycoproteins, which themselves are composed of large polysaccharides bound with much smaller quantities of protein. Mucus is slightly different in different parts of the gastrointestinal tract, but everywhere it has several important characteristics that make it both an excellent lubricant and a protectant for the wall of the gut. *First,* mucus has adherent qualities that make it adhere tightly to the food or other particles and also to spread as a thin film over the surfaces. *Second,* it has sufficient *body* that it coats the wall of the gut and prevents actual contact of the food particles with the mucosa. *Third,* mucus has a low resistance for slippage so that the particles can slide along the epithelium with great ease. *Fourth,* mucus causes fecal particles to adhere to each other to form the fecal masses that are expelled during a bowel movement. *Fifth,* mucus is strongly resistant to digestion by the gastrointestinal enzymes. And, *sixth,* the glycoproteins of mucus have amphoteric properties, which means that they are capable of buffering small amounts of either acids or alkalies; also, mucus often contains moderate quantities of bicarbonate ions, which specifically neutralize acids.

In summary, mucus has the ability to allow easy slippage of food along the gastrointestinal tract and also to prevent excoriative or chemical damage to the epithelium. A person becomes acutely aware of the lubricating qualities of mucus when the salivary glands fail to secrete saliva, for under these circumstances it is extremely difficult to swallow solid food even when it is taken with large amounts of water.

● SECRETION OF SALIVA

The Salivary Glands; Characteristics of Saliva. The principal glands of salivation are the *parotid, submandibular,* and *sublingual* glands; in addition, there are many small *buccal* glands. The daily secretion of saliva normally ranges between about 800 and 1500 milliliters, as shown in Table 64–1.

Saliva contains two major types of protein secretion: (1) a *serous secretion* containing *ptyalin* (an α-amylase), which is an enzyme for digesting starches, and (2) *mucous secretion* containing *mucin* for lubricating purposes. The parotid glands secrete entirely the serous type, and the submandibular and sublin-

Table 64–1 DAILY SECRETION OF INTESTINAL JUICES

	Daily Volume (ml)	pH
Saliva	1000	6.0–7.0
Gastric secretion	1500	1.0–3.5
Pancreatic secretion	1000	8.0–8.3
Bile	1000	7.8
Small intestinal secretion	1800	7.5–8.0
Brunner's gland secretion	200	8.0–8.9
Large intestinal secretion	200	7.5–8.0
Total	6700	

gual glands secrete both the serous type and mucus. The buccal glands secrete only mucus. Saliva has a pH between 6.0 and 7.4, a favorable range for the digestive action of ptyalin.

Secretion of Ions in the Saliva. Saliva contains especially large quantities of potassium and bicarbonate ions. On the other hand, the concentrations of both sodium and chloride ions are considerably less in saliva than in plasma. One can understand these special concentrations of ions in the saliva from the following description of the mechanism for secretion of saliva.

Figure 64–2 illustrates secretion by the submaxillary gland, a typical compound gland containing both *acini* and *salivary ducts.* Salivary secretion is a two-stage operation: the first stage involves the acini; and the second, the salivary ducts. The acini secrete a *primary secretion* that contains ptyalin and/or mucin in a solution of ions in concentrations not greatly different from those of typical extracellular fluid. However, as the primary secretion flows through the ducts, two major active transport processes take place that markedly modify the ionic composition of the saliva.

First, *sodium ions* are actively reabsorbed from all of the salivary ducts; and *potassium ions* are actively secreted, but at a slower rate, in exchange for the sodium. Therefore, the sodium concentration of the

Primary secretion
1. Ptyalin
2. Mucus
3. Extracellular fluid

Na$^+$ active absorption
Cl$^-$ passive absorption
K$^+$ active secretion
HCO$_3^-$ secretion

Saliva

Figure 64–2. Formation and secretion of saliva by a salivary gland.

saliva becomes greatly reduced, whereas the potassium ion concentration becomes increased. The great excess of sodium reabsorption over potassium secretion creates negativity of about −70 millivolts in the salivary ducts, and this causes chloride ions to be reabsorbed passively; therefore, the chloride ion concentration falls to a very low level along with the decrease in sodium ion concentration.

Second, *bicarbonate ions* are secreted by the ductal epithelium into the lumen of the duct. This is at least partly caused by exchange of bicarbonate for chloride ions, but it may also result partly from an active secretory process.

The net result of these active transport processes is that *under resting conditions,* the concentrations of sodium and chloride ions in the saliva are only about 15 mEq/liter each, approximately one-seventh to one-tenth their concentrations in plasma. On the other hand, the concentration of potassium ions is about 30 mEq/liter, seven times as great as its concentration in plasma; and the concentration of bicarbonate ions is 50 to 70 mEq/liter, about two to three times that of plasma.

During maximal salivation, the salivary ionic concentrations change considerably because the rate of formation of primary secretion by the acini can increase as much as 20-fold. As a result, this secretion then flows through the ducts so rapidly that the ductal reconditioning of the secretion is considerably reduced. Therefore, when copious quantities of saliva are secreted, the sodium chloride concentration rises to about one-half to two-thirds that of plasma, whereas the potassium concentration falls to only four times that of plasma.

In the presence of excess aldosterone secretion, the sodium and chloride reabsorption and the potassium secretion become greatly increased, so that the sodium chloride concentration in the saliva is sometimes reduced almost to zero while the potassium concentration increases still more.

Because of the high potassium ion concentration of saliva, in any abnormal state in which the saliva is lost to the exterior of the body for long periods of time, a person can develop serious depletion of potassium ions in the body, leading eventually to serious hypokalemia and paralysis.

Function of Saliva for Oral Hygiene. Under basal conditions, about 0.5 ml/min of saliva, almost entirely of the mucous type, is secreted all the time except during sleep, when the secretion becomes very little. This secretion plays an exceedingly important role in maintaining healthy oral tissues. The mouth is loaded with pathogenic bacteria that can easily destroy tissues and can also cause dental caries. However, saliva helps prevent the deteriorative processes in several ways: *First,* the flow of saliva itself helps wash away the pathogenic bacteria, as well as the food particles that provide their metabolic support. *Second,* the saliva also contains several factors that actually destroy bacteria. One of these is *thiocyanate ions* and another is several *proteolytic enzymes*—most importantly, *lysozyme*—that (1) attack the bacteria; (2) aid the thiocyanate ions in entering the bacteria, where they in turn become bactericidal; and (3) digest food particles, thus helping further to remove the

bacterial metabolic support. And, *third,* saliva often contains significant amounts of protein antibodies that can destroy oral bacteria, including those that cause dental caries.

Therefore, in the absence of salivation, the oral tissues become ulcerated and otherwise infected, and caries of the teeth becomes rampant.

Nervous Regulation of Salivary Secretion. Figure 64–3 illustrates the parasympathetic nervous pathways for regulation of salivation, showing that the salivary glands are controlled mainly by *parasympathetic nervous signals* from the *salivatory nuclei.* The salivatory nuclei are located approximately at the juncture of the medulla and pons and are excited by both taste and tactile stimuli from the tongue and other areas of the mouth. Many taste stimuli, especially the sour taste, elicit copious secretion of saliva —often as much as 5 to 8 ml/min, or 8 to 20 times the basal rate of secretion. Also, certain tactile stimuli, such as the presence of smooth objects in the mouth (a pebble, for instance), cause marked salivation; whereas rough objects cause less salivation and occasionally even inhibit salivation.

Salivation can also be stimulated or inhibited by impulses arriving in the salivatory nuclei from higher centers of the central nervous system. For instance, when a person smells or eats favorite foods, salivation is greater than when disliked food is smelled or eaten. The *appetite area* of the brain, which partially regulates these effects, is located in close proximity to the parasympathetic centers of the anterior hypothalamus, and it functions to a great extent in response to signals from the taste and smell areas of the cerebral cortex or amygdala.

Salivation also occurs in response to reflexes originating in the stomach and upper intestines—particularly when very irritating foods are swallowed or when a person is nauseated because of some gastrointestinal abnormality. The swallowed saliva presum-

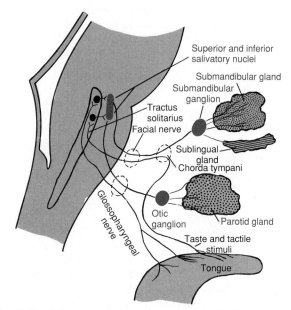

Figure 64–3. Parasympathetic nervous regulation of salivary secretion.

ably helps to remove the irritating factor in the gastrointestinal tract by diluting or neutralizing the irritating substances.

Sympathetic stimulation can also increase salivation a moderate amount, but much less so than does parasympathetic stimulation. The sympathetic nerves originate from the superior cervical ganglia and then travel along the blood vessels to the salivary glands.

A secondary factor that also affects secretion is the *blood supply to the glands,* for secretion always requires adequate nutrition. The process of salivation indirectly dilates the blood vessels, thus providing increased nutrition as it is needed. Part of the vasodilator effect is caused by *kallikrein* secreted by the activated salivary cells, which in turn is converted in the blood into *bradykinin,* a strong vasodilator.

Figure 64–4. An oxyntic gland from the body or fundus of the stomach.

● ESOPHAGEAL SECRETION

The esophageal secretions are entirely mucoid in character and principally provide lubrication for swallowing. The main body of the esophagus is lined with many *simple mucous glands;* but at the gastric end, and to a lesser extent in the initial portion of the esophagus, there are many *compound mucous glands.* The mucus secreted by the compound glands in the upper esophagus prevents mucosal excoriation by the newly entering food, whereas the compound glands near the esophagogastric junction protect the esophageal wall from digestion by gastric juices that flow back into the lower esophagus. Despite this protection, a peptic ulcer at times may occur at the gastric end of the esophagus.

● GASTRIC SECRETION

CHARACTERISTICS OF THE GASTRIC SECRETIONS

In addition to mucus-secreting cells that line the entire surface of the stomach, the stomach mucosa has two important types of tubular glands: the *oxyntic* (or *gastric) glands* and the *pyloric glands.* The oxyntic (acid-forming) glands secrete *hydrochloric acid, pepsinogen, intrinsic factor,* and *mucus,* and the pyloric glands secrete mainly *mucus* for protection of the pyloric mucosa but also some *pepsinogen* and, very important, the hormone *gastrin.* The oxyntic glands are located on the inside surfaces of the body and fundus of the stomach, comprising the proximal 80 per cent of the stomach; and the pyloric glands are located in the antral portion of the stomach.

The Secretions From the Oxyntic Glands

A typical oxyntic gland is shown in Figure 64–4. It is composed of three different types of cells: the *mucous neck cells,* which secrete mainly mucus but also some pepsinogen; the *peptic* (or *chief) cells,* which secrete large quantities of pepsinogen; and the *parietal* (or *oxyntic) cells,* which secrete hydrochloric acid and intrinsic factor. A postulated mechanism for se-

cretion of mucus and pepsinogen by the mucous neck cells and the peptic cells was given earlier in the chapter and was depicted in Figure 64–1. However, secretion of hydrochloric acid by the parietal cells involves special mechanisms as follows:

Basic Mechanism of Hydrochloric Acid Secretion. When stimulated, the parietal cells secrete an acid solution containing about 160 millimoles of hydrochloric acid per liter, which is almost exactly isotonic with the body fluids. The pH of this acid is approximately 0.8, illustrating its extreme acidity. At this pH the hydrogen ion concentration is about 3 million times that of the arterial blood. And to concentrate the hydrogen ions this tremendous amount requires over 1500 calories of energy per liter of gastric juice, as discussed in Chapter 4 in relation to membrane transport mechanisms.

Figure 64–5 illustrates the basic structure of a parietal cell, showing that it contains several large *intracellular canaliculi* that empty directly into the lumen of the oxyntic gland. The hydrochloric acid is formed

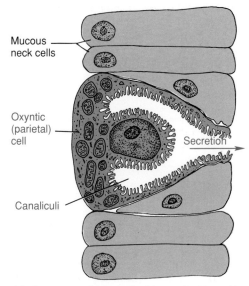

Figure 64–5. Anatomy of the canaliculi in a parietal (oxyntic) cell.

at the villuslike membranes of these canaliculi and then conducted to the exterior.

Different suggestions for the precise mechanism of hydrochloric acid formation have been offered. One of these is illustrated in Figure 64–6 and consists of the following steps:

1. Chloride ion is actively transported from the cytoplasm of the parietal cell into the lumen of the canaliculus, and sodium ions are actively transported out of the lumen. These two effects together create a negative potential of −40 to −70 millivolts in the canaliculus, which in turn causes passive diffusion of positively charged potassium ions and a small number of sodium ions from the cell cytoplasm also into the canaliculus. Thus, in effect, potassium chloride and much smaller amounts of sodium chloride enter the canaliculus.

2. Water is dissociated into hydrogen ions and hydroxyl ions in the cell cytoplasm. The hydrogen ions are then actively secreted into the canaliculus in exchange for potassium ions, this active exchange process being catalyzed by H^+,K^+-ATPase. In addition, the sodium ions are actively reabsorbed by a separate sodium pump. Thus, most of the potassium and sodium ions that had diffused into the canaliculus are reabsorbed, and hydrogen ions take their place, giving a very strong solution of hydrochloric acid.

3. Water passes through the cell and into the canaliculus by osmosis. Thus the final secretion entering the canaliculus contains hydrochloric acid in a concentration of 155 mEq/liter, potassium chloride in a concentration of 15 mEq/liter, and a very small amount of sodium chloride.

4. Finally, carbon dioxide, either formed during metabolism in the cell or entering the cell from the blood, combines under the influence of *carbonic anhydrase* with the hydroxyl ions (formed in step 1 when water was dissociated) to form bicarbonate ions. This then diffuses out of the cell into the extracellular fluid in exchange for chloride ions that enter the cell and later will be secreted into the canaliculus. The importance of carbon dioxide in the chemical reactions for formation of hydrochloric acid is illustrated by the fact that carbonic anhydrase inhibition by the drug *acetazolamide* diminishes the formation of hydrochloric acid.

Secretion and Activation of Pepsinogen. Several different types of pepsinogen are secreted by the peptic and mucous cells of the gastric glands. Even so, all the pepsinogens perform essentially the same functions. When the pepsinogens are first secreted, they have no digestive activity. However, as soon as they come in contact with hydrochloric acid, and especially when they come in contact with previously formed pepsin plus the hydrochloric acid, they are immediately activated to form active *pepsin*. In this process, the pepsinogen molecule, having a molecular weight of about 42,500, is split to form the pepsin molecule, having a molecular weight of about 35,000.

Pepsin is an active proteolytic enzyme in a highly acid medium (optimum pH 1.8 to 3.5), but above a pH of about 5 it has little proteolytic activity and even becomes completely inactivated in a short time. Therefore, hydrochloric acid is as necessary as pepsin for protein digestion in the stomach; this is discussed in the following chapter.

Secretion of Other Enzymes. Small quantities of other enzymes are also secreted in the stomach juices, including *gastric lipase, gastric amylase,* and a *gelatinase.* Gastric lipase is of little quantitative importance and is actually a *tributyrase,* for its principal activity is on tributyrin, which is butterfat; it has almost no lipolytic activity on the other fats. Gastric amylase plays a very minor role in digestion of starches, and the gelatinase helps to liquefy some of the proteoglycans in meats.

Secretion of Intrinsic Factor. The substance *intrinsic factor,* essential for absorption of vitamin B_{12} in the ileum, is secreted by the *parietal cells* along with the secretion of hydrochloric acid. Therefore, when the acid-producing cells of the stomach are destroyed, which frequently occurs in chronic gastritis, the person not only develops *achlorhydria* but also develops *pernicious anemia* because of failure of maturation of the red blood cells in the absence of vitamin B_{12} stimulation of the bone marrow. This was discussed in detail in Chapter 32.

Figure 64–6. Postulated mechanism for the secretion of hydrochloric acid.

Secretion of Mucus in the Stomach. The pyloric glands are structurally similar to the oxyntic glands but contain few peptic cells and almost no parietal cells. Instead, they contain mostly mucous cells that are identical with the mucous neck cells of the gastric glands. These cells secrete a small amount of pepsinogen, as discussed earlier, and an especially large amount of thin mucus that helps to protect the stomach wall from digestion by the gastric enzymes. The pyloric glands also secrete the hormone *gastrin,* which plays a key role in controlling gastric secretion, as we discuss shortly.

In addition, the entire surface of the stomach mucosa between glands has a continuous layer of mucous cells that secrete large quantities of a far more *viscid and alkaline mucus* that is mainly insoluble and coats the mucosa with a gel layer of mucus often more than a millimeter thick, thus providing a major shell of protection for the stomach wall as well as contributing to lubrication of food transport. Even the slightest contact with food or especially any irritation of the mucosa directly stimulates the mucous cells to secrete copious quantities of this thick, viscid mucus.

REGULATION OF GASTRIC SECRETION BY NERVOUS AND HORMONAL MECHANISMS

Basic Factors That Stimulate Gastric Secretion: Acetylcholine, Gastrin, and Histamine

The basic neurotransmitters or hormones that directly stimulate secretion by the gastric glands are *acetylcholine, gastrin,* and *histamine.* All of these function by binding first with receptors on the secretory cells. Then the receptors activate the secretory processes. Acetylcholine excites secretion by all of the secretory cell types in the gastric glands, including secretion of *pepsinogen* by the *peptic cells, hydrochloric acid* by the *parietal cells, mucus* by the *mucus cells,* and *gastrin* by the *gastrin cells.* On the other hand, both *gastrin* and *histamine* stimulate very strongly the *secretion of acid* by the parietal cells but have much less effect in stimulating the other cells.

A few other substances also stimulate the gastric secretory cells, such as circulating amino acids, caffeine, and possibly alcohol. However, the stimulatory effects of these are slight in comparison with acetylcholine, gastrin, and histamine.

Stimulation of Acid Secretion

Nervous Stimulation. About half of the nerve signals to the stomach that cause gastric secretion originate in the *dorsal motor nuclei of the vagi* and pass via the *vagus nerves* first to the *enteric nervous system* of the stomach wall and then to the gastric glands. The other half of the secretory signals are generated by local reflexes that occur entirely within the enteric nervous system itself. All the secretory nerves release acetylcholine as the neurotransmitter at their endings on the glandular cells, with one exception: for those signals that go to the gastrin-secreting cells in the pyloric glands, an intermediate neuron serves as the final path and secretes *gastrin-releasing peptide (GRP)* as the neurotransmitter.

Nerve stimulation of gastric secretion can be initiated by signals that originate either in the brain, especially in the limbic system, or in the stomach itself. And the stomach-initiated signals can activate two different types of reflexes: (1) *long reflexes* that are transmitted from the stomach mucosa all the way to the brain stem and then back to the stomach through the vagus nerves and (2) *short reflexes* that originate locally and are transmitted entirely through the local enteric nervous system.

The types of stimuli that can initiate the reflexes are (1) distention of the stomach, (2) tactile stimuli on the surface of the stomach mucosa, and (3) chemical stimuli, including especially *amino acids* and *peptides* derived from protein foods or *acid* that has already been secreted by the gastric glands.

Stimulation of Acid Secretion by Gastrin. Both the nerve signals from the vagus nerves and those from the local enteric reflexes, aside from causing direct stimulation of glandular secretion of stomach juices, also cause the mucosa in the stomach antrum to secrete the hormone *gastrin.* This hormone is secreted by the *gastrin cells,* also called *G cells,* in the pyloric glands. Gastrin is a large peptide secreted in two forms, a large form called G-34, containing 34 amino acids, and a smaller form, G-17, containing 17 amino acids. Although both of these are important, the smaller is more abundant.

Gastrin is absorbed into the blood and carried to the *oxyntic glands* in the body of the stomach; there it stimulates the *parietal cells* very strongly and perhaps the peptic cells as well, but to a lesser extent. Thus, the really important effect is to increase the rate of hydrochloric acid secretion, often as much as eight times. In turn, the hydrochloric acid excites still additional enteric reflex activity that not only further increases hydrochloric acid secretion but also stimulates secondarily the secretion of enzymes by the peptic cells to increase as much as two to four times.

Role of Histamine in Controlling Gastric Secretion. *Histamine,* an amino acid derivative, also stimulates acid secretion by the *parietal cells.* A small amount of histamine is formed continually in the gastric mucosa, either in response to acid in the stomach or for other reasons. This amount, acting by itself, causes very little acid secretion. However, whenever acetylcholine or gastrin stimulates the parietal cells at the same time, then even the small normal amounts of histamine greatly enhance acid secretion. We know this to be true because when the action of histamine is blocked by an appropriate antihistaminic drug such as cimetidine, neither acetylcholine nor gastrin can then cause significant amounts of acid secretion. Thus, histamine is a necessary *cofactor* for exciting significant acid secretion.

However, it should be noted that the histamine receptors on the parietal cells are of the H_2 type, not of the H_1 type. Therefore, only antihistaminic drugs that block the action of histamine H_2 receptors are effective in blocking stomach acid secretion. The first important drug of this type was *cimetidine,* but others are now becoming available.

Multiplicative Effect of Acetylcholine, Gastrin, and Histamine in Stimulating Acid Secretion. Because no one of the primary stimulators of the acid-secreting parietal cells — acetylcholine, gastrin, or histamine — is effective in causing secretion of more than slight amounts of acid when functioning alone, it has been postulated that all three of the receptors to these separate transmitter-hormonal substances must be activated simultaneously to give a truly effective stimulus for gastric acid secretion. The histamine seems to be always present under normal conditions in small amounts. Then when the vagi are stimulated, acetylcholine is released at the parasympathetic nerve endings, and the GRP neurons stimulated by the vagi cause simultaneous release of gastrin by the gastrin cells. Therefore, all three stimuli become available, and copious amounts of acid are secreted. When food in the stomach elicits enteric reflexes, this, too, causes both gastrin and acetylcholine secretion, once again promoting the flow of tremendous quantities of acid.

Regulation of Pepsinogen Secretion

Regulation of *pepsinogen* secretion is much less complex than that of acid secretion; it occurs in response to two types of signals: (1) stimulation of the *peptic cells* by *acetylcholine* released from the *vagus nerves* or other enteric nerves and (2) stimulation of peptic secretion in response to acid in the stomach. The acid probably does not stimulate the peptic cells directly, but instead elicits additional enteric reflexes, thus further supporting the original nervous signals to the peptic cells. It is possible, too, that the gastrin released during acid secretion has an additional slight direct effect in stimulating the peptic cells, although this is not clear. Nevertheless, the rate of secretion of *pepsinogen,* the precursor of pepsin that causes protein digestion, is strongly influenced by the amount of acid in the stomach. In persons who have lost the ability to secrete large amounts of acid, the secretion of pepsinogen is very little, even though the peptic cells may still be normal.

Feedback Inhibition of Gastric Secretion by Excess Acid. When the acidity of the gastric juices increases to a pH of 3.0, the gastrin mechanism for stimulating gastric secretion becomes blocked. This effect probably results from two different factors. First, greatly enhanced acidity depresses or blocks the secretion of gastrin itself by the G cells. Second, the acid seems to cause an inhibitory nervous reflex that inhibits gastric secretion.

Obviously, this feedback inhibition of the gastric glands plays an important role in protecting the stomach against excessive acidity, which would promote peptic ulceration. In addition to this protective effect, the feedback mechanism is also important in maintaining optimal pH for function of the peptic enzymes in the digestive process, which is a pH of about 3.0.

The Three Phases of Gastric Secretion

Gastric secretion is said to occur in three separate phases (as illustrated in Fig. 64–7): a *cephalic phase,* a *gastric phase,* and an *intestinal phase.* However, as will be apparent in the following discussion, these three phases in reality fuse together.

The Cephalic Phase. The cephalic phase of gastric secretion occurs even before food enters the stomach or while it is being eaten. It results from the sight, smell, thought, or taste of food; and the greater the appetite, the more intense is the stimulation. Neurogenic signals causing the cephalic phase of secretion can originate in the cerebral cortex or in the appetite centers of the amygdala or hypothalamus.

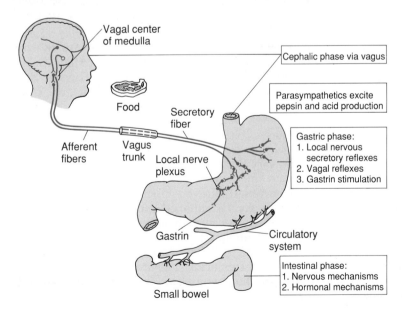

Figure 64–7. The phases of gastric secretion and their regulation.

They are transmitted through the dorsal motor nuclei of the vagi to the stomach. This phase of secretion normally accounts for less than one fifth of the gastric secretion associated with eating a meal.

The Gastric Phase. Once the food enters the stomach, it excites the long vagovagal reflexes, the local enteric reflexes, and the gastrin mechanism, which in turn cause secretion of gastric juice that continues throughout the several hours that the food remains in the stomach.

The gastric phase of secretion accounts for at least two thirds of the total gastric secretion associated with eating a meal and, therefore, accounts for most of the total daily gastric secretion of about 1500 milliliters.

The Intestinal Phase. The presence of food in the upper portion of the small intestine, particularly in the duodenum, can cause the stomach to secrete small amounts of gastric juice, probably partly because of the small amounts of gastrin that are also released by the duodenal mucosa in response to distention or chemical stimuli of the same type as those that stimulate the stomach gastrin mechanism. In addition, amino acids absorbed into the blood, as well as several other hormones or reflexes, play minor roles in causing secretion of gastric juice.

However, several intestinal factors can also inhibit gastric secretion, and these often completely override the excitatory factors.

Inhibition of Gastric Secretion by Intestinal Factors

Although chyme stimulates gastric secretion during the intestinal phase of secretion, it paradoxically often inhibits secretion during the gastric phase. This inhibition results from at least two different influences:

1. The presence of food in the small intestine initiates an *enterogastric reflex,* transmitted through the enteric nervous system as well as through the extrinsic sympathetic and vagus nerves, that inhibits stomach secretion. This reflex can be initiated by distention of the small bowel, the presence of acid in the upper intestine, the presence of protein breakdown products, or irritation of the mucosa. This is part of the complex mechanism discussed in the preceding chapter for slowing down stomach emptying when the intestines are already filled.

2. The presence of acid, fat, protein breakdown products, hyper- or hypo-osmotic fluids, or any irritating factor in the upper small intestine causes the release of several intestinal hormones. Two of these are *secretin* and *cholecystokinin.* Both of these are especially important for control of pancreatic secretion, and cholecystokinin is also important in promoting emptying of the gallbladder. However, in addition to having these effects, they both oppose stomach secretion. Also, three other hormones, *gastric inhibitory peptide, vasoactive intestinal polypeptide,* and *somatostatin* all have slight to moderate effects in inhibiting gastric secretion.

The functional purpose of the inhibition of gastric secretion by intestinal factors is probably to slow the release of chyme from the stomach when the small intestine is already filled. In fact, the enterogastric reflex and also these inhibitory hormones usually reduce stomach motility at the same time that they reduce gastric secretion, as was discussed in the previous chapter.

Secretion During the Interdigestive Period. The stomach secretes only a few milliliters of gastric juice per hour during the "interdigestive period," when little or no digestion is occurring anywhere in the gut. Furthermore, the secretion that does occur is almost entirely of the so-called nonoxyntic type, meaning that it is composed mainly of mucus containing very little pepsin and almost no acid. In fact, on occasion, the fluid is actually slightly alkaline, containing moderate quantities of sodium bicarbonate. However, strong emotional stimuli frequently increase the interdigestive secretion to 50 milliliters or more of highly peptic and highly acidic gastric juice per hour, in very much the same manner that the cephalic phase of gastric secretion excites secretion at the onset of a meal. This increase of secretion during the presence of emotional stimuli is believed to be one of the factors in the development of peptic ulcers, as is discussed in Chapter 66.

CHEMICAL COMPOSITION OF GASTRIN AND OTHER GASTROINTESTINAL HORMONES

Figure 64–8 illustrates the amino acid compositions of *gastrin-17* and also of *cholecystokinin* and *secretin,* which will be discussed later in the chapter. Note that all are polypeptides and that the terminal five amino acids in the gastrin and cholecystokinin molecular chains are exactly the same. The activity of gastrin resides in the terminal four amino acids and in the terminal eight amino acids for cholecystokinin; all of the amino acids in the secretin molecule are essential. A synthetic gastrin, composed of the terminal four amino acids of natural gastrins plus the amino acid alanine, has all the same physiological properties as the natural gastrins. This synthetic product is called *pentagastrin.*

Gastrin:
Glu- Gly- Pro- Trp- Leu- Glu- Glu- Glu- Glu- Glu- Ala- Tyr- Gly- Trp- Met- Asp- Phe- NH_2
                                        |
                                   HSO_3

Cholecystokinin:
Lys- (Ala, Gly, Pro, Ser)- Arg- Val- (Ile, Met, Ser)- Lys- Asn- (Asn, Gln, His, Leu_2,
               Pro, Ser_2)- Arg- Ile- (Asp, Ser)- Arg- Asp- Tyr- Met- Gly- Trp- Met- Asp- Phe- NH_2
                                                             |
                                                             HSO_3

Secretin:
His- Ser- Asp- Gly- Thr- Phe- Thr- Ser- Glu- Leu- Ser- Arg- Leu- Arg- Asp- Ser-
                               Ala- Arg- Leu- Gln- Arg- Leu- Leu- Gln- Gly- Leu- Val- NH_2

Figure 64–8. Amino acid composition of gastrin-17, cholecystokinin, and secretin.

● PANCREATIC SECRETION

The pancreas, which lies parallel to and beneath the stomach, is a large compound gland with an internal structure similar to that of the salivary glands. In addition to secreting insulin by the islets of Langerhans in the pancreas, digestive enzymes, also, are secreted by the pancreatic acini, and large volumes of sodium bicarbonate solution are secreted by both the small ductules and the larger ducts leading from the acini. The combined product then flows through a long pancreatic duct that usually joins the hepatic duct immediately before it empties into the duodenum through the sphincter of Oddi. Pancreatic juice is secreted most abundantly in response to the presence of chyme in the upper portions of the small intestine, and the characteristics of the pancreatic juice are determined to some extent by the types of food in the chyme.

Characteristics of Pancreatic Juice

Pancreatic juice contains enzymes for digesting all three major types of food: proteins, carbohydrates, and fats. It also contains large quantities of bicarbonate ions, which play an important role in neutralizing the acid chyme emptied by the stomach into the duodenum.

The more important of the proteolytic enzymes are *trypsin, chymotrypsin,* and *carboxypolypeptidase.* Less important are several *elastases* and *nucleases.* By far the most abundant of these is trypsin. The trypsin and chymotrypsin split whole and partially digested proteins into peptides of various sizes, but these do not cause the release of individual amino acids. On the other hand, carboxypolypeptidase splits the peptides into the individual amino acids, thus completing the digestion of much of the proteins all the way to the amino acid state.

The digestive enzyme for carbohydrates is *pancreatic amylase,* which hydrolyzes starches, glycogen, and most other carbohydrates (except cellulose) to form disaccharides and a few trisaccharides.

The main enzymes for fat digestion are *pancreatic lipase,* which is capable of hydrolyzing neutral fat into fatty acids and monoglycerides, *cholesterol esterase,* which causes hydrolysis of cholesterol esters, and *phospholipase,* which splits fatty acids from phospholipids.

When synthesized in the pancreatic cells, the proteolytic enzymes are in the inactive forms *trypsinogen, chymotrypsinogen,* and *procarboxypolypeptidase,* which are all enzymatically inactive. These become activated only after they are secreted into the intestinal tract. Trypsinogen is activated by an enzyme called *enterokinase,* which is secreted by the intestinal mucosa when chyme comes in contact with the mucosa. Also, trypsinogen can be autocatalytically activated by trypsin that has already been formed. Chymotrypsinogen is activated by trypsin to form chymotrypsin, and procarboxypolypeptidase is activated in a similar manner.

Secretion of Trypsin Inhibitor. It is important that the proteolytic enzymes of the pancreatic juice not become activated until they have been secreted into the intestine, for the trypsin and other enzymes would digest the pancreas itself. Fortunately, the same cells that secrete the proteolytic enzymes into the acini of the pancreas secrete simultaneously another substance called *trypsin inhibitor.* This substance is stored in the cytoplasm of the glandular cells surrounding the enzyme granules, and it prevents activation of trypsin both inside the secretory cells and in the acini and ducts of the pancreas. Because it is trypsin that activates the other pancreatic proteolytic enzymes, trypsin inhibitor also prevents the subsequent activation of all of these as well.

However, when the pancreas becomes severely damaged or when a duct becomes blocked, large quantities of pancreatic secretion become pooled in the damaged areas of the pancreas. Under these conditions, the effect of trypsin inhibitor is sometimes overwhelmed, in which case the pancreatic secretions rapidly become activated and literally digest the entire pancreas within a few hours, giving rise to the condition called *acute pancreatitis.* This often is lethal because of accompanying shock; and even if not lethal, it leads to a lifetime of pancreatic insufficiency.

Secretion of Bicarbonate Ions

Although the enzymes of the pancreatic juice are secreted entirely by the acini of the pancreatic glands, the other two important components of pancreatic juice, bicarbonate ions and water, are secreted in large amounts mainly by the epithelial cells of the ductules and ducts leading from the acini. We shall see later that the stimulatory mechanisms for (1) enzyme production and (2) production of bicarbonate ions and water are also quite different. When the pancreas is stimulated to secrete copious quantities of pancreatic juice, the bicarbonate ion concentration can rise to as high as 145 mEq/liter, a value approximately five times that of bicarbonate ions in the plasma. Obviously, this provides a large quantity of alkali in the pancreatic juice that serves to neutralize acid emptied into the duodenum from the stomach.

The basic steps in the cellular mechanism for secreting sodium bicarbonate solution into the pancreatic ductules and ducts are illustrated in Figure 64–9. They are the following:

1. Carbon dioxide diffuses to the interior of the cell from the blood and combines with water under the influence of carbonic anhydrase to form carbonic acid. The carbonic acid in turn dissociates into bicarbonate ions and hydrogen ions. Then the bicarbonate ions are actively transported through the *luminal* border of the cell into the lumen of the duct.

2. The hydrogen ions formed by dissociation of carbonic acid inside the cell are exchanged for sodium ions through the *blood* border of the cell, also by an

Figure 64–9. Secretion of iso-osmotic sodium bicarbonate solution by the pancreatic ductules and ducts.

active transport process. The sodium ions in turn either diffuse through or are actively transported through the *luminal* border into the pancreatic duct to provide electrical neutrality for the bicarbonate ions.

3. The movement of sodium and bicarbonate ions from the blood to the lumen creates an osmotic gradient that causes osmosis of water also into the pancreatic duct, thus forming the bicarbonate solution.

REGULATION OF PANCREATIC SECRETION

Basic Stimuli of Pancreatic Secretion

Four basic stimuli are important in causing pancreatic secretion:

1. *Acetylcholine,* which is released from the parasympathetic vagus nerve endings as well as from other cholinergic nerves in the enteric nervous system.

2. *Gastrin,* which is liberated in copious quantities during the gastric phase of stomach secretion.

3. *Cholecystokinin (CCK),* which is secreted by the duodenal and upper jejunal mucosa when food enters the small intestine.

4. *Secretin,* which is secreted by the same duodenal and jejunal mucosa when highly acid food enters the small intestine.

The first three of the above stimuli, acetylcholine, gastrin, and cholecystokinin, all stimulate the acinar cells of the pancreas much more than the ductal cells. Therefore, they cause the production of large quantities of digestive enzymes but relatively small quantities of fluid to go with the enzymes. Without the fluid, most of the enzymes remain temporarily stored in the acini and ducts until more fluid secretion comes along to wash them into the duodenum.

Secretin, in contrast to the other three basic stimuli, mainly stimulates the secretion of large quantities of sodium bicarbonate solution by the ductal epithelium, but almost no stimulation of enzyme secretion.

Multiplicative Effects of the Different Stimuli. When all the different stimuli of pancreatic secretion occur at once, the secretion is far greater than the sum of the secretions caused by each one separately. Therefore, the various stimuli are said to "multiply," or "potentiate," each other. Thus, pancreatic secretion normally results from the combined effects of multiple basic stimuli, not one alone.

The Phases of Pancreatic Secretion

Pancreatic secretion occurs in three different phases, the same as for gastric secretion: The *cephalic* phase, the *gastric* phase, and the *intestinal phase.* Their characteristics are as follows:

Cephalic and Gastric Phases. During the cephalic phase of pancreatic secretion, the same nervous signals that cause secretion in the stomach also cause acetylcholine release by the vagal nerve endings in the pancreas. This causes moderate amounts of enzymes to be secreted into the pancreatic acini. However, little secretion flows out the pancreatic ducts into the intestine because only small amounts of water and electrolytes are secreted along with the enzymes.

During the gastric phase, the nervous stimulation of enzyme secretion continues; in addition, the large quantities of gastrin formed in the stomach stimulate still more enzyme secretion. Still only small amounts reach the duodenal lumen because of continued lack of significant quantities of fluid secretion.

Intestinal Phase. After chyme enters the small intestine, pancreatic secretion becomes copious, mainly in response to the hormone *secretin.* In addition, *cholecystokinin* causes still much more increase in the secretion of enzymes.

Stimulation of Secretion of Copious Quantities of Bicarbonate by Secretin — Neutralization of the Acidic Chyme. Secretin is a polypeptide containing 27 amino acids (molecular weight about 3400) that is present in so-called *S cells* in the mucosa of the upper small intestine (duodenum and upper jejunum) in an inactive form, *prosecretin.* When acid chyme enters the intestine, it causes the release and activation of secretin, which is subsequently absorbed into the blood. The one truly potent constituent of chyme that causes secretin release is hydrochloric acid.

Secretin causes the pancreas to secrete large quantities of fluid containing a high concentration of bicarbonate ion (up to 145 mEq/liter) but a low concentration of chloride ion. However, this fluid contains very little enzymes when the pancreas is stimulated only by secretin because secretin does not stimulate the acinar cells.

The secretin mechanism is especially important for two reasons: *First,* secretin begins to be released from the mucosa of the small intestine when the pH of the duodenal contents falls below 4.5, and its release increases very greatly as the pH falls to 3.0 and as more and more acid reaches deeper into the duodenum and jejunum. This immediately causes large quantities of

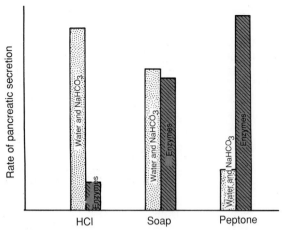

Figure 64–10. Sodium bicarbonate and enzyme secretion by the pancreas, caused respectively by the presence of acid, fat (soap), or peptone solutions in the duodenum.

pancreatic juice containing abundant amounts of sodium bicarbonate to be secreted, which results in the following reaction in the duodenum:

$$HCl + NaHCO_3 \rightarrow NaCl + H_2CO_3$$

The carbonic acid immediately dissociates into carbon dioxide and water, and the carbon dioxide is absorbed into the blood and expired through the lungs, thus leaving a neutral solution of sodium chloride in the duodenum. In this way, the acid contents emptied into the duodenum from the stomach become neutralized, and the peptic activity of the gastric juices is immediately blocked. Because the mucosa of the small intestine cannot withstand the digestive action of acid gastric juice, this is a highly important and even essential protective mechanism against the development of duodenal ulcers, discussed in further detail in Chapter 66.

Second, bicarbonate secretion by the pancreas provides an appropriate pH for action of the pancreatic enzymes. All of these function optimally in a slightly alkaline or neutral medium. The pH of the sodium bicarbonate secretion averages 8.0.

Cholecystokinin—Control of Enzyme Secretion by the Pancreas. The presence of food in the upper small intestine also causes a second hormone, cholecystokinin, a polypeptide containing 33 amino acids, to be released from a still different group of cells, called I cells, in the mucosa of the duodenum and upper jejunum. This results especially from the presence of *proteoses* and *peptones,* which are products of partial protein digestion, and of *long chain fatty acids;* however, acids will also cause its release in smaller quantities. Cholecystokinin, like secretin, passes by way of the blood to the pancreas but, instead of causing sodium bicarbonate secretion, causes secretion of large quantities of digestive enzymes by the acinar cells, an effect similar to the effects of vagal stimulation and gastrin but even more pronounced.

The differences between the stimulatory effects of secretin and cholecystokinin are shown in Figure 64–10, which illustrates (1) intense sodium bicarbonate secretion in response to acid in the duodenum, (2) a dual effect in response to soap (a fat), and (3) intense enzyme secretion in response to peptones.

Figure 64–11 summarizes the more important factors in the regulation of pancreatic secretion. The total amount secreted each day is about 1000 milliliters.

● SECRETION OF BILE AND FUNCTIONS OF THE BILIARY TREE

One of the many functions of the liver is to secrete bile, normally between 600 and 1200 ml/day. Bile serves two important functions: *First,* it plays a very important role in fat digestion and absorption, not because of any enzymes in the bile that cause fat digestion, but instead because the bile acids contained in the bile do two things: (1) they help to emulsify the large fat particles of the food into many minute particles that can be attacked by the lipases secreted in pancreatic juice, and (2) they aid in the transport and absorption of the digested fat end products to and through the intestinal mucosal membrane. *Second,* bile serves as a means for excretion of several important waste products from the blood. These include especially *bilirubin,* an end product of hemoglobin destruction, and excesses of *cholesterol* synthesized by the liver cells.

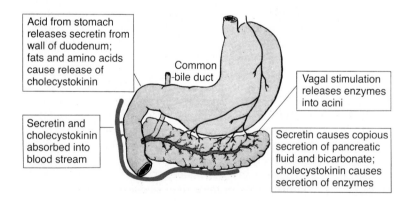

Acid from stomach releases secretin from wall of duodenum; fats and amino acids cause release of cholecystokinin

Common bile duct

Vagal stimulation releases enzymes into acini

Secretin and cholecystokinin absorbed into blood stream

Secretin causes copious secretion of pancreatic fluid and bicarbonate; cholecystokinin causes secretion of enzymes

Figure 64–11. Regulation of pancreatic secretion.

Physiologic Anatomy of Biliary Secretion

Bile is secreted in two stages by the liver: (1) Initially bile is secreted by the liver hepatocytes; this initial secretion contains large amounts of bile acids, cholesterol, and other organic constituents, secreted into the minute *bile canaliculi* that lie between the hepatic cells in the hepatic plates. (2) Next, the bile flows peripherally toward the interlobular septa, where the canaliculi empty into *terminal bile ducts,* then into progressively larger ducts, finally reaching the *hepatic duct* and *common bile duct,* from which the bile either empties directly into the duodenum or is diverted through the *cystic duct* into the gallbladder, illustrated in Figure 64–12. In its course through these bile ducts, a secondary secretion is added to the initial bile. This additional secretion is a watery solution of sodium and bicarbonate ions, and it sometimes increases the total quantity of bile by as much as an additional 100 per cent. The secondary secretion is stimulated by secretin, thus causing increased quantities of bicarbonate ions that supplement the pancreatic secretions in neutralizing acid from the stomach.

Storage of Bile in the Gallbladder. The bile secreted continually by the liver cells is normally stored in the gallbladder until needed in the duodenum. The maximum volume of the gallbladder is only 20 to 60 milliliters. Nevertheless, as much as 12 hours' bile secretion (usually about 450 milliliters) can be stored in the gallbladder because water, sodium, chloride, and most other small electrolytes are continually absorbed by the gallbladder mucosa, concentrating

the other bile constituents, including the bile salts, cholesterol, lecithin, and bilirubin. Most of this absorption is caused by active transport of sodium through the gallbladder epithelium. Bile is normally concentrated about 5-fold, but it can be concentrated up to a maximum of 12- to 20-fold.

Composition of Bile. Table 64–2 gives the composition of bile when it is first secreted by the liver and then after it has been concentrated in the gallbladder. This table shows that by far the most abundant substance secreted in the bile is the *bile salts,* accounting for about half of the total solutes of bile, but also secreted or excreted in large concentrations are *bilirubin, cholesterol, lecithin,* and the usual *electrolytes* of plasma. In the concentrating process in the gallbladder, water and large portions of the electrolytes (ex-

Table 64–2 COMPOSITION OF BILE

	Liver Bile		Gallbladder Bile	
Water	97.5	gm/dl	92	gm/dl
Bile salts	1.1	gm/dl	6	gm/dl
Bilirubin	0.04	gm/dl	0.3	gm/dl
Cholesterol	0.1	gm/dl	0.3 to 0.9	gm/dl
Fatty acids	0.12	gm/dl	0.3 to 1.2	gm/dl
Lecithin	0.04	gm/dl	0.3	gm/dl
Na^+	145	mEq/liter	130	mEq/liter
K^+	5	mEq/liter	12	mEq/liter
Ca^+	5	mEq/liter	23	mEq/liter
Cl^-	100	mEq/liter	25	mEq/liter
HCO_3^-	28	mEq/liter	10	mEq/liter

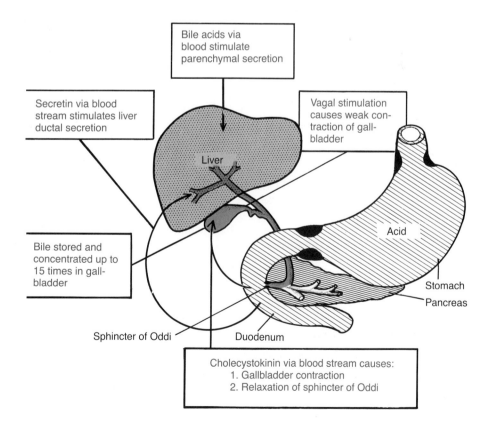

Figure 64–12. Liver secretion and gallbladder emptying.

cept calcium ions) are reabsorbed by the gallbladder mucosa; but essentially all the other constituents, including especially the bile salts and the lipid substances cholesterol and lecithin, are not reabsorbed and therefore become highly concentrated in the gallbladder bile.

Emptying of the Gallbladder — Role of Cholecystokinin. When food begins to be digested in the upper gastrointestinal tract, the gallbladder also begins to empty, especially as fatty foods enter the duodenum. The basic cause of the emptying is rhythmic contractions of the wall of the gallbladder, but effective emptying also requires simultaneous relaxation of the *sphincter of Oddi* that guards the exit of the common bile duct into the duodenum.

By far the most potent stimulus for causing the gallbladder contractions is the hormone *cholecystokinin.* This is the same cholecystokinin that causes increased secretion of enzymes by the acinar cells of the pancreas. And the stimulus for its release into the blood from the duodenal mucosa is mainly fatty foods entering the duodenum.

However, in addition to cholecystokinin, the gallbladder is also stimulated less strongly by cholinergic nerve fibers from both the vagi and the enteric nervous system. These are the same nerves that promote motility and secretion in other parts of the upper gastrointestinal tract.

Yet even with relatively strong contractions of the gallbladder, emptying can still be difficult because the sphincter of Oddi normally remains tonically contracted. Therefore, before emptying will occur, it, too, must be relaxed. At least three different factors help in this: First, cholecystokinin, instead of stimulating the sphincter of Oddi, has a weak relaxing effect; but this effect is usually not sufficient by itself to allow significant emptying. Second, the rhythmic contractions of the gallbladder transmit peristaltic waves down the common bile duct to the sphincter of Oddi, causing a leading wave of relaxation that partially inhibits the sphincter in advance of the peristaltic wave. But this, too, is usually not enough to allow large amounts of emptying. Third, when intestinal peristaltic waves travel over the wall of the duodenum itself, the relaxation phase of each of these waves strongly relaxes the sphincter of Oddi along with the relaxation of the muscle of the gut wall. This seems to be by far the most potent of all of the relaxant effects on the sphincter of Oddi. As a result, bile usually enters the duodenum in the form of squirts that are synchronized with the duodenal peristaltic contractions.

In summary, the gallbladder empties its store of concentrated bile into the duodenum mainly in response to the cholecystokinin stimulus. When fat is not in the meal, the gallbladder empties poorly; but when adequate quantities of fat are present, the gallbladder empties completely in about 1 hour.

Figure 64 – 12 summarizes the secretion of bile, its storage in the gallbladder, and its release from the bladder to the gut.

THE BILE SALTS AND THEIR FUNCTION

The liver cells form about 0.5 gram of *bile salts* daily. The precursor of the bile salts is *cholesterol,* which is either supplied in the diet or synthesized in the liver cells during the course of fat metabolism and then converted to *cholic acid* or *chenodeoxycholic acid* in about equal quantities. These acids then combine principally with glycine and to a lesser extent with taurine to form *glyco-* and *tauro-conjugated bile acids.* The salts of these acids are secreted in the bile.

The bile salts have two important actions in the intestinal tract. First, they have a detergent action on the fat particles in the food, which decreases the surface tension of the particles and allows the agitation in the intestinal tract to break the fat globules into minute sizes. This is called the *emulsifying* or *detergent function* of bile salts. Second, and even more important than the emulsifying function, bile salts help in the absorption of fatty acids, monoglycerides, cholesterol, and other lipids from the intestinal tract. They do this by forming minute complexes with these lipids; the complexes are called *micelles,* and they are highly soluble because of the electrical charges of the bile salts. The lipids are "ferried" in this form to the mucosa, where they are then absorbed; this mechanism is described in detail in the following chapter. Without the presence of bile salts in the intestinal tract, up to 40 per cent of the lipids are lost into the stools, and the person often develops a metabolic deficit due to this nutrient loss.

Enterohepatic Circulation of Bile Salts. Approximately 94 per cent of the bile salts are reabsorbed by an *active transport* process through the intestinal mucosa in the distal ileum. They enter the portal blood and pass to the liver. On reaching the liver these salts are absorbed almost totally on the first passage through the venous sinusoids into the hepatic cells and then resecreted into the bile. In this way about 94 per cent of all the bile salts are recirculated into the bile, so that on the average these salts make the entire circuit some 18 times before being carried out in the feces. The small quantities of bile salts lost into the feces are replaced by new amounts formed continually by the liver cells. This recirculation of the bile salts is called the *enterohepatic circulation.*

The quantity of bile secreted by the liver each day is highly dependent on the availability of bile salts — the greater the quantity of bile salts in the enterohepatic circulation (usually a total of about 2.5 grams), the greater the rate of bile secretion. Indeed, ingestion of an excess of bile salts can increase bile secretion by several hundred milliliters per day.

If a bile fistula empties the bile salts to the exterior for several days to several weeks so that they cannot be reabsorbed from the ileum, the liver increases its production of bile salts as much as tenfold, which increases the rate of bile secretion approximately back to normal. This demonstrates that the daily rate of bile salt secretion is actively controlled by the availability (or lack of availability) of bile salts in the enterohepatic circulation.

Role of Secretin in Controlling Bile Secretion. In addition to the strong stimulating effect of bile acids on bile secretion, the hormone *secretin* also increases bile secretion, sometimes more than doubling the secretion rate for several hours after a meal. However, this increase in secretion represents mainly secretion of a bicarbonate-rich watery solution by the epithelial cells of the bile ductules and ducts and not increased secretion of bile acids by the liver parenchymal cells. The bicarbonate in turn passes into the small intestine and joins the bicarbonate from the pancreas in neutralizing the acid from the stomach. Thus, the secretin feedback mechanism for neutralizing duodenal acid operates not only through its effects on pancreatic secretion but also through its effect on the secretion by the liver ductules and ducts.

SECRETION OF CHOLESTEROL; GALLSTONE FORMATION

Bile salts are formed in the hepatic cells from cholesterol, and in the process of secreting the bile salts about one-tenth as much cholesterol is also secreted into the bile. This amounts to a total of 1 to 2 grams a day. No specific function is known for the cholesterol in the bile, and it is presumed that it is simply a byproduct of bile salt formation and secretion.

Cholesterol is almost insoluble in pure water, but the bile salts and lecithin in bile combine physically with the cholesterol to form ultramicroscopic *micelles* that are soluble, as is explained in more detail in the following chapter. When the bile becomes concentrated in the gallbladder, the bile salts and lecithin become concentrated along with the cholesterol, which keeps the cholesterol in solution.

Nevertheless, under abnormal conditions the cholesterol may precipitate, resulting in the formation of *cholesterol gallstones,* as shown in Figure 64–13. The different conditions that can cause cholesterol precipitation are (1) too much absorption of water from the bile, (2) too much absorption of bile salts and lecithin from the bile, (3) too much secretion of cholesterol in the bile, and (4) inflammation of the epithelium of the gallbladder. The latter two of these require special explanation.

The amount of cholesterol in the bile is determined partly by the quantity of fat that the person eats, for the hepatic cells synthesize cholesterol as one of the products of fat metabolism in the body. For this reason, persons on a high fat diet over a period of many years are prone to the development of gallstones.

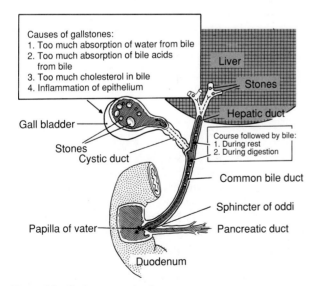

Figure 64–13. Formation of gallstones.

Inflammation of the gallbladder epithelium often results from low grade chronic infection; this changes the absorptive characteristics of the gallbladder mucosa, sometimes allowing excessive absorption of water, bile salts, or other substances that are necessary to keep the cholesterol in solution. As a result, cholesterol begins to precipitate, usually forming many small crystals of cholesterol on the surface of the inflamed mucosa or on small precipitated particles of bilirubin. These, in turn, act as nidi for further precipitation of cholesterol, and the crystals grow larger. Occasionally tremendous numbers of sandlike stones develop, but much more frequently these coalesce to form a few large gallstones, or even a single stone that fills the entire gallbladder. Also, calcium ions, which are usually concentrated fivefold or more in the gallbladder, often precipitate in the gallstones, making the stones x-ray–opaque so that they can be seen in x-rays of the abdomen.

Medical Therapy for Dissolving Gallstones. Simple cholesterol gallstones in many patients can be dissolved over a period of 1 to 2 years by feeding the patient 1 to 1.5 grams of *chenodeoxycholic acid* daily. This is one of the naturally secreted bile acids, and its exogenous administration adds greatly to the enterohepatic pool of bile acids. This causes dissolution and reabsorption of the gallstones in the following ways: (1) The increased quantity of bile acids increases the volume of bile formed and therefore decreases the concentration of cholesterol in the bile. (2) The increase of bile acids in the bile makes the cholesterol that is present more soluble. (3) The exogenous administration of bile acids decreases the formation of bile acids by the liver, which at the same time reduces the secretion of cholesterol.

● SECRETIONS OF THE SMALL INTESTINE

Secretion of Mucus by Brunner's Glands

An extensive array of compound mucous glands, called *Brunner's glands,* is located in the first few centimeters of the duodenum, mainly between the pylorus and the papilla of Vater, where the pancreatic juices and bile empty into the duodenum. These glands secrete mucus in response to (1) tactile stimuli or irritating stimuli of the overlying mucosa; (2) vagal stimulation, which causes secretion concurrently with increase in stomach secretion; and (3) gastrointestinal hormones, especially secretin. The function of the mucus secreted by Brunner's glands is to protect the duodenal wall from digestion by the gastric juice, and their rapid and intense response to irritating stimuli is especially geared to this purpose.

Brunner's glands are inhibited by sympathetic stimulation; therefore, such stimulation is likely to leave the duodenal bulb unprotected and is perhaps one of the factors that cause this area of the gastrointestinal tract to be the site of peptic ulcers in about 50 per cent of all cases.

SECRETION OF THE INTESTINAL DIGESTIVE JUICES — THE CRYPTS OF LIEBERKÜHN

Located on the entire surface of the small intestine are small pits called *crypts of Lieberkühn,* one of which is illustrated in Figure 64–14. The intestinal secretions are formed by the epithelial cells in these crypts at a rate of about 1800 ml/day. The secretions are almost pure extracellular fluid and have a slightly alkaline pH in the range of 7.5 to 8.0. They are rapidly reabsorbed by the villi. This circulation of fluid from the crypts to the villi obviously supplies a watery vehicle for absorption of substances from the chyme as it comes in contact with the villi, which is one of the primary functions of the small intestine.

Mechanism of Secretion of the Watery Fluid. The exact mechanism causing the marked secretion of watery fluid by the crypts of Lieberkühn is not known. However, it is believed to involve at least two active secretory processes: (1) active secretion of chloride ions into the crypts and (2) active secretion of bicarbonate ions. The secretion of these ions, especially the chloride ions, causes electrical drag of sodium ions through the membrane as well. Finally, all these ions together cause osmotic movement of water.

Effect of Cholera Toxin on Intestinal Secretion. Cholera toxin causes an extreme rate of fluid secretion into the crypts of Lieberkühn, especially in the jejunal region of the small intestine. During the first day of a cholera attack, as much as 5 to 10 liters of diarrhea fluid can be lost from the bowels, often leading to circulatory shock caused by dehydration and culminating in death within a few hours. Fortunately, though, most cases of cholera can be treated successfully simply by administering large amounts of saline and glucose solutions, either by mouth or intravenously.

The cholera toxin seems to have a specific effect in increasing active transport of chloride ions into the crypts of Lieberkühn; this in turn leads to the massive loss of fluid into the intestinal tract. Several other bacterial diseases of the intestines can also cause increased chloride transport similar to the effect of cholera, though usually they are less severe; these include infections caused by the colon bacillus, the dysentery bacillus, and others.

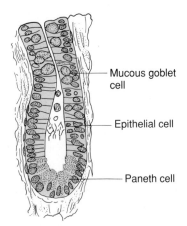

Figure 64–14. A crypt of Lieberkühn, found in all parts of the small intestine between the villi, which secretes almost pure extracellular fluid.

Mucous goblet cell

Epithelial cell

Paneth cell

Enzymes in the Small Intestinal Secretion. When secretions of the small intestine are collected without cellular debris, they have almost no enzymes. However, the epithelial cells of the mucosa, especially those covering the villi, do contain digestive enzymes that digest some food substances *while* they are being absorbed through the epithelium. These enzymes are the following: (1) several different *peptidases* for splitting small peptides into amino acids, (2) four enzymes for splitting disaccharides into monosaccharides—*sucrase, maltase, isomaltase,* and *lactase,* and (3) small amounts of *intestinal lipase* for splitting neutral fats into glycerol and fatty acids. Most, if not all, of these enzymes are mainly in the brush border of the epithelial cells. They are believed to catalyze hydrolysis of the foods on the outside surfaces of the microvilli prior to absorption of the end products.

The epithelial cells deep in the crypts of Lieberkühn continually undergo mitosis, and the new cells gradually migrate along the basement membrane upward out of the crypts toward the tips of the villi, where they are finally shed into the intestinal secretions. The life cycle of an intestinal epithelial cell is approximately 5 days. This rapid growth of new cells allows rapid repair of excoriations that occur in the mucosa.

REGULATION OF SMALL INTESTINAL SECRETION

Local Stimuli. By far the most important means for regulating small intestinal secretion are various local nervous reflexes, especially reflexes initiated by tactile or irritative stimuli. Therefore, for the most part, secretion in the small intestine occurs simply in response to the presence of chyme in the intestine—the greater the amount of chyme, the greater the secretion.

Hormonal Regulation. Some of the same hormones that promote secretion elsewhere in the gastrointestinal tract also increase small intestinal secretion, especially secretin and cholecystokinin. Also, some experiments suggest that other hormonal substances extracted from the small intestinal mucosa by the chyme might help to control secretion. However, in general, the local enteric reflex mechanisms probably play the dominant role.

● SECRETIONS OF THE LARGE INTESTINE

Mucus Secretion. The mucosa of the large intestine, like that of the small intestine, is lined with crypts of Lieberkühn; but in this mucosa, unlike that of the small intestine, there are no villi. Also, the epithelial cells contain almost no enzymes. Instead, they are lined almost entirely by mucous cells that secrete only mucus. Additionally, on the surface epithelium of the large intestine are large numbers of mucous cells dispersed among the other epithelial cells.

Therefore, the great preponderance of secretion in the large intestine is mucus. This mucus contains large amounts of bicarbonate ions caused by active transport of these ions by the epithelial cells. The rate of secretion of mucus is regulated principally by direct, tactile stimulation of the mucous cells on the surface of the mucosa and by local nervous reflexes to the mucous cells in the crypts of Lieberkühn. However, stimulation of the pelvic nerves, which carry the parasympathetic innervation to the distal one half to two thirds of the large intestine, also causes marked increase in the secretion of mucus. This occurs along with an increase in motility, which was discussed in the preceding chapter. Therefore, during extreme parasympathetic stimulation, often caused by emotional disturbances, so much mucus may be secreted into the large intestine that the person has a bowel movement of ropy mucus as often as every 30 minutes; the mucus contains little or no fecal material.

Mucus in the large intestine obviously protects the wall against excoriation; but, in addition, it provides the adherent medium for holding fecal matter together. Furthermore, it protects the intestinal wall from the great amount of bacterial activity that takes place inside the feces; and it, plus the alkalinity of the secretion (pH of 8.0 caused by large amounts of sodium bicarbonate), also provides a barrier to keep acids formed deep in the feces from attacking the intestinal wall.

Secretion of Water and Electrolytes in Response to Irritation. Whenever a segment of the large intestine becomes intensely irritated, as occurs when bacterial infection becomes rampant during *enteritis,* the mucosa then secretes large quantities of water and electrolytes in addition to the normal viscid solution of alkaline mucus. This acts to dilute the irritating factors and to cause rapid movement of the feces toward the anus. The usual result is *diarrhea,* with loss of large quantities of water and electrolytes, which promotes earlier recovery from the disease than would otherwise occur.

REFERENCES

Ackerman, S. H.: The developmental physiology of the stomach: Possible contributions to the regulatory disturbances of duodenal ulcer disease. Proc. Soc. Exp. Biol. Med., 175:119, 1984.

Berglindh, T.: The mammalian gastric parietal cell in vitro. Annu. Rev. Physiol., 46:377, 1984.

Burnham, D. B., and Williams, J. A.: Stimulus-secretion coupling in pancreatic acinar cells. J. Pediatr. Gastroenterol. Nutr., 3 (Suppl. 1):S1, 1984.

Cheli, R., et al.: Gastric Protection. New York, Raven Press, 1988.

Cheli, R.: Gastric Secretion: A Physiological and Pharmacological Approach. New York, Raven Press, 1986.

Cooke, H. J.: Role of the "little brain" in the gut in water and electrolyte homeostasis. FASEB J., 3:127, 1989.

Daugherty, D., and Yamada, T.: Posttranslational processing of gastrin. Physiol. Rev., 69:482, 1989.

DeLisle, R. C., and Williams, J. A.: Regulation of membrane fusion in secretory exocytosis. Annu. Rev. Physiol., 48:225, 1986.

Flemstrom, G., and Garner A.: Some characteristics of duodenal epithelium, Ciba Found. Symp., 109:94, 1984.

Forte, J. G., et al.: Mechanisms of gastric H^+ and Cl^- transport. Annu. Rev. Physiol., 42:111, 1980.

Fushiki, T., and Iwai, K.: Two hypotheses on the feedback regulation of pancreatic enzyme secretion. FASEB J., 3:121, 1989.

Go, V. L. W., et al.: The Exocrine Pancreas: Biology, Pathobiology, and Diseases. New York, Raven Press, 1986.

Guth, P. H.: Stomach blood flow and acid secretion. Annu. Rev. Physiol., 44:3, 1982.

Henning, S. J.: Ontogeny of enzymes in the small intestine. Annu. Rev. Physiol., 47:231, 1985.

Hersey, S. J., et al.: Cellular control of pepsinogen secretion, Annu. Rev. Physiol., 46:393, 1984.

Hopfer, U., and Liedtke, C. M.: Proton and bicarbonate transport mechanisms in the intestine. Annu. Rev. Physiol., 49:51, 1987.

Jerzy Glass, G. B.: Gastrointestinal Hormones. New York, Raven Press, 1980.

Johnson, L. R., et al.: Physiology of the Gastrointestinal Tract, 2nd Ed. New York, Raven Press, 1987.

Karlstrom, L.: Mechanisms in bile salt-induced secretion in the small intestine. Acta Physiol. Scand., 126 (Suppl. 549):1, 1986.

Kuijpers, G. A. J., et al.: Role of proton and bicarbonate transport in pancreatic cell function. Annu. Rev. Physiol., 49:87, 1987.

Lauger, P.: Dynamics of ion transport systems in membranes. Physiol. Rev., 67:1296, 1987.

Lipkin, M.: Growth and development of the stomach. Annu. Rev. Physiol., 47:175, 1985.

Lundgren, O.: Microcirculation of the gastrointestinal tract and pancreas. In Renkin, E. M., and Michel, C. C. (eds.): Handbook of Physiology. Sec. 2, Vol. IV. Bethesda, Md., American Physiological Society, 1984, p. 799.

Machen, T. E., and Paradiso, A. M.: Regulation of intracellular pH in the stomach. Annu. Rev. Physiol., 49:19, 1987.

Malinowska, D. H., and Sachs, G.: Cellular mechanisms of acid secretion. Clin. Gastroenterol., 13:309, 1984.

Muallem, S.: Calcium transport pathways of pancreatic acinar cells. Annu. Rev. Physiol., 51:83, 1989.

Petersen, O. H., and Findlay, I.: Electrophysiology of the pancreas. Physiol. Rev., 67:1054, 1987.

Petersen, O. H., and Gallacher, D. V.: Electrophysiology of pancreatic and salivary acinar cells. Annu. Rev. Physiol., 50:65, 1988.

Putney, J. W., Jr.: Identification of cellular activation mechanisms associated with salivary secretion. Annu. Rev. Physiol., 48:75, 1986.

Rehfeld, J. F., and Amdrup, E. (eds.): Gastrins and the Vagus. New York, Academic Press, 1979.

Rehfeld, J. F.: Gastrointestinal hormones. In Crane, R. K. (ed.): International Review of Physiology: Gastrointestinal Physiology III. Vol. 19. Baltimore, University Park Press, 1979, p. 291.

Reuss, L., and Stoddard, J. S.: Role of H^+ and HCO_3^- in salt transport in gallbladder epithelium. Annu. Rev. Physiol., 49:35, 1987.

Sachs, G., et al.: Gastric secretion. Int. Rev. Physiol., 12:127, 1977.

Salen, G., and Shefer, S.: Bile acid synthesis. Annu. Rev. Physiol., 45:679, 1983

Satir, B.: The final steps in secretion. Sci. Am. 233(4):28, 1975.

Schulz, I., and Stolze, H. H.: The exocrine pancreas: The role of secretagogues, cyclic nucleotides and calcium in enzyme secretion. Annu. Rev. Physiol., 42:127, 1980.

Shepherd, A. P.: Local control of intestinal oxygenation and blood flow. Annu. Rev. Physiol., 44:13, 1982.

Strange, R. C.: Hepatic bile flow. Physiol. Rev., 64:1055, 1984.

Streebny, L. M.: The Salivary System. Boca Raton, CRC Press Inc., 1987.

Szursezewski, J. H.: Physiology of mammalian prevertebral ganglia. Annu. Rev. Physiol., 43:53, 1981.

Tache, Y.: CNS peptides and regulation of gastric acid secretion. Annu. Rev. Physiol., 50:19, 1988.

Thompson, J. C., et al.: Gastrointestinal Endocrinology. New York, McGraw-Hill Book Co., 1987.

Walsh, J. H.: Peptides as regulators of gastric acid secretion. Annu. Rev. Physiol., 50:41, 1988.

Williams, J. A.: Regulatory mechanisms in pancreas and salivary acini. Annu. Rev. Physiol., 46:361, 1984.

Willingham, M. C., and Pastan, I.: Endocytosis and exocytosis: Current concepts of vesicle traffic in animal cells. Int. Rev. Cytol., 92:51, 1984.

Young, J. A.: Salivary secretion of inorganic electrolytes. In Crane, R. K. (ed.): International Review of Physiology: Gastrointestinal Physiology III. Vol. 19. Baltimore, University Park Press, 1979, p. 1.

CHAPTER 65

• *Digestion and Absorption in the Gastrointestinal Tract* •

The foods on which the body lives, with the exception of small quantities of substances such as vitamins and minerals, can be classified as carbohydrates, fats, and proteins. However, these generally cannot be absorbed in their natural forms through the gastrointestinal mucosa and, for this reason, are useless as nutrients without the preliminary process of digestion. Therefore, the present chapter discusses, first, the processes by which carbohydrates, fats, and proteins are digested into small enough compounds for absorption and, second, the mechanisms by which the digestive end products, as well as water, electrolytes, and other substances, are absorbed.

● DIGESTION OF THE VARIOUS FOODS

Hydrolysis as the Basic Process of Digestion. Almost all the carbohydrates of the diet are large *polysaccharides* or *disaccharides,* which are combinations of *monosaccharides* bound to each other by the process of *condensation.* This means that a hydrogen ion has been removed from one of the monosaccharides, while a hydroxyl ion has been removed from the next one. The two monosaccharides then are combined with each other at these sites of removal, and the hydrogen and hydroxyl ions combine to form water. When the carbohydrates are digested back into monosaccharides, specific enzymes return the hydrogen and hydroxyl ions to the polysaccharides and thereby separate the monosaccharides from each other. This process, called *hydrolysis,* is the following:

$$R'' - R' + H_2O \xrightarrow[enzyme]{digestive} R''OH + R'H$$

Almost the entire fat portion of the diet consists of triglycerides (neutral fats), which are combinations of three *fatty acid* molecules condensed with a single *glycerol* molecule. In the process of condensation, three molecules of water had been removed. Digestion of the triglycerides consists of the reverse process, the fat-digesting enzymes returning molecules of water to the triglyceride molecule and thereby splitting the fatty acid molecules away from the glycerol. Here again, the digestive process is one of *hydrolysis.*

Finally, proteins are formed from *amino acids* that are bound together by *peptide linkages.* In this linkage a hydroxyl ion is removed from one amino acid, while a hydrogen ion is removed from the succeeding one; thus, the amino acids in the protein chain are bound together by a process of condensation, and digestion occurs by the reverse effect of hydrolysis, the proteolytic enzymes returning the water to the protein molecules to split them into their constituent amino acids.

Therefore, the chemistry of digestion is really simple, for in the case of all three major types of food, the same basic process of *hydrolysis* is involved. The only difference lies in the enzymes required to promote the reactions for each type of food.

All the digestive enzymes are proteins. Their secretion by the different gastrointestinal glands was discussed in the preceding chapter.

DIGESTION OF CARBOHYDRATES

The Carbohydrate Foods of the Diet. Only three major sources of carbohydrates exist in the normal human diet. These are sucrose, which is the disaccharide known popularly as cane sugar; lactose, which is a disaccharide in milk; and starches, which are large polysaccharides present in almost all nonanimal foods and particularly in the grains. Other carbohydrates ingested to a slight extent are amylose, glycogen, alcohol, lactic acid, pyruvic acid, pectins, dextrins, and minor quantities of other carbohydrate derivatives in meats. The diet also contains a large amount of cellulose, which is a carbohydrate. However, no enzymes capable of hydrolyzing cellulose are secreted by the human digestive tract. Consequently, cellulose cannot be considered to be a food for the human being.

Digestion of Carbohydrates in the Mouth and Stomach. When food is chewed, it is mixed with the saliva, which contains the enzyme *ptyalin* (an α-amylase) secreted mainly by the parotid glands. This enzyme hydrolyzes starch into the disaccharide *maltose* and other small polymers of glucose containing three to nine glucose molecules (such as *maltotriose* and α *limit dextrins* that are the branch points of the starch molecule), as shown in Figure 65–1. But the food remains in the mouth only a short time, and probably not more than 3 to 5 per cent of all the starches that are eaten will have become hydrolyzed by the time the food is swallowed. However, digestion continues in the body and fundus of the stomach for as long as an hour until the food is mixed with the stomach secretions. Then the activity of the salivary amylase is blocked by the acid of the gastric secretions, for it is essentially nonactive as an enzyme once the pH of the medium falls below approximately 4.0. Nevertheless, on the average, before the food becomes completely mixed with the gastric secretions, as much as 30 to 40 per cent of the starches will have been hydrolyzed mainly to maltose.

Digestion of Carbohydrates in the Small Intestine.
Digestion by Pancreatic Amylase. Pancreatic secretion, like saliva, contains a large quantity of α-amylase that is almost identical in its function with the α-amylase of saliva but is several times as powerful. Therefore, within 15 to 30 minutes after the chyme empties from the stomach into the duodenum and mixes with pancreatic juice, virtually all of the starches are digested. In general, the starches are almost totally converted into *maltose* and *other very small glucose polymers* before they have passed beyond the duodenum or upper jejunum.

Hydrolysis of Disaccharides and Small Glucose Polymers Into Monosaccharides by the Intestinal Epithelial Enzymes. The epithelial cells lining the small intestine contain the four enzymes *lactase, sucrase, maltase,* and α-*dextrinase,* which are capable of splitting the disaccharides lactose, sucrose, and maltose, as well as the other small glucose polymers, into their constituent monosaccharides. These enzymes are located *in the membranes of the microvilli brush border* of the cells lining the lumen of the intestine, and the disaccharides are digested as they come in contact with these membranes. Lactose splits into a molecule of *galactose* and a molecule of *glucose*. Sucrose splits into a molecule of *fructose* and a molecule of *glucose*. Maltose and the other small glucose polymers all split into *molecules of glucose*. Thus, the final products of carbohydrate digestion are all monosaccharides, and these are absorbed immediately into the portal blood.

In the ordinary diet, which contains far more starches than either sucrose or lactose, glucose represents more than 80 per cent of the final products of carbohydrate digestion, and galactose and fructose each rarely represent more than 10 per cent of the products of carbohydrate digestion.

The major steps in carbohydrate digestion are summarized in Figure 65-1.

DIGESTION OF PROTEINS

The Proteins of the Diet. The dietary proteins are formed of long chains of amino acids bound together by *peptide linkages*. A typical linkage is the following:

$$NH_2 \quad\quad H$$
$$R—CH—C +OH + H—N—CH—COOH →$$
$$\quad\quad\; \| \quad\quad\quad\quad\quad\quad\quad\quad |$$
$$\quad\quad\; O \quad\quad\quad\quad\quad\quad\quad\quad R$$

$$NH_2 \quad\quad H$$
$$R—CH—C—N—CH—COOH + H_2O$$
$$\quad\quad\; \| \quad\quad\quad |$$
$$\quad\quad\; O \quad\quad\quad R$$

The characteristics of each type of protein are determined by the types of amino acids in the protein molecule and by the arrangement of these amino acids. The physical and chemical characteristics of the different proteins are discussed in Chapter 69.

Digestion of Proteins in the Stomach. *Pepsin,* the important peptic enzyme of the stomach, is most active at a pH of 2 to 3 and is completely inactive at a pH above approximately 5. Consequently, for this enzyme to cause any digestive action on protein, the stomach juices must be acidic. It will be recalled from Chapter 64 that the gastric glands secrete a large quantity of hydrochloric acid. This hydrochloric acid is secreted by the parietal cells at a pH of about 0.8, but, by the time it is mixed with the stomach contents and with the secretions from the nonparietal glandular cells of the stomach, the pH ranges around 2 to 3, a highly favorable range of acidity for pepsin activity.

One of the important features of pepsin digestion is its ability to digest collagen, an albuminoid that is affected little by other digestive enzymes. Collagen is a major constituent of the intercellular connective tissue of meats, and for the digestive enzymes of the digestive tract to penetrate meats and digest the cellular proteins it is first necessary that the collagen fibers be digested. Consequently, in persons lacking peptic activity in the stomach, the ingested meats are less well penetrated by the digestive enzymes and, therefore, may be poorly digested.

As illustrated in Figure 65-2, pepsin only begins the process of protein digestion, usually providing only 10 to 20 per cent of the total protein digestion. This splitting of proteins is a process of hydrolysis occurring at the peptide linkages between the amino acids.

Digestion of Proteins by Pancreatic Secretions.
Most protein digestion occurs principally in the upper small intestine, in the duodenum and jejunum, under the influence of the proteolytic enzymes of the pancreatic secretion. When the proteins leave the stomach, they ordinarily are mainly in the form of proteoses, peptones, and large polypeptides. Immediately upon entering the small intestine, the partial breakdown products are attacked by the major

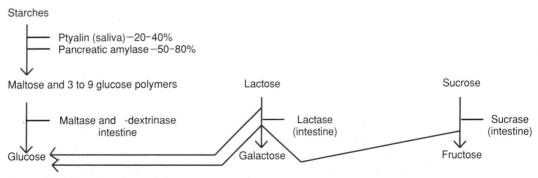

Figure 65-1. Digestion of carbohydrates.

Figure 65–2. Digestion of proteins.

proteolytic pancreatic enzymes *trypsin, chymotrypsin,* and *carboxypolypeptidase,* as illustrated in Figure 65–2. Both trypsin and chymotrypsin can split protein molecules into small polypeptides; carboxypolypeptidase then cleaves individual amino acids from the carboxyl ends of the polypeptides. However, only a small percentage of the proteins are digested all the way to their constituent amino acids by the pancreatic juices. Instead, most remain as dipeptides, tripeptides, and some even larger.

Digestion of Peptides by Peptidases in the Epithelial Cells Lining the Small Intestinal Villi. The last digestion of the proteins in the intestinal lumen is achieved by the epithelial cells that line the villi of the small intestine, mainly in the duodenum and jejunum. These epithelial cells have a *brush border* that consists literally of hundreds of microvilli projecting from the surface of each cell. In the cell membrane surrounding each of these microvilli are multiple *peptidases* that protrude through the membranes to the exterior, where they come in contact with the intestinal fluids. Two types of peptidase enzymes are especially important, *aminopolypeptidase* and several *dipeptidases.* These succeed in splitting the remaining larger polypeptides into tripeptides and dipeptides and a few all the way to amino acids. Both the amino acids and the dipeptides and tripeptides are easily transported through the microvillar membrane to the interior of the epithelial cell.

Finally, inside the cytosol of the epithelial cell are multiple other peptidases that are specific for the remaining types of linkages between the amino acids. Within minutes virtually all of the last dipeptides and tripeptides are digested to the final stage of single amino acids; these then pass on through the opposite side of the epithelial cell membrane into the blood. More than 99 per cent of the final protein digestive products that are absorbed are individual amino acids, with only rare absorption of peptides, and very, very rare absorption of whole protein molecules. Unfortunately, even these very few molecules of protein can sometimes cause serious allergic or immunologic disturbances, as discussed in Chapter 34.

DIGESTION OF FATS

The Fats of the Diet. By far the most abundant fats of the diet are the neutral fats, also known as *triglycerides,* each molecule of which is composed of a glycerol nucleus and three fatty acids, as illustrated in Figure 65–3. Neutral fat is a major constituent in food of animal origin and much less so of plant origin.

In the usual diet are also small quantities of phospholipids, cholesterol, and cholesterol esters. The phospholipids and cholesterol esters contain fatty acid, and, therefore, can be considered to be fats themselves. Cholesterol, on the other hand, is a sterol compound containing no fatty acid, but it does exhibit some of the physical and chemical characteristics of fats; it is derived from fats, and it is metabolized similarly to fats. Therefore, cholesterol is considered from a dietary point of view to be a fat.

Digestion of Fats in the Intestine. A small amount of short chain triglycerides of butterfat origin is digested in the stomach by gastric lipase *(tributyrase).* However, the amount of digestion is so slight that it is unimportant. Instead, essentially all fat digestion occurs in the small intestine as follows:

Emulsification of Fat by Bile Acids and Lecithin. The first step in fat digestion is to break the fat globules into small sizes so that the water-soluble digestive enzymes can act on the globule surfaces. This process is called emulsification of the fat, and it is achieved under the influence of *bile,* the secretion of the liver that does not contain any digestive enzymes. However, bile does contain a large quantity of *bile salts* and the phospholipid lecithin mainly in the form of ionized sodium salts, both of which are extremely important for the emulsification of fat. The polar parts of the bile salt and lecithin molecules are highly soluble in water, whereas most of the remaining portions of their molecules are highly soluble in fat. Therefore, the fat-soluble portions dissolve in the surface layer of the fat globule but

$$CH_3-(CH_2)_{16}-\overset{\overset{\displaystyle O}{\|}}{C}-O-CH_2$$

$$CH_3-(CH_2)_{16}-\overset{\overset{\displaystyle O}{\|}}{C}-O-CH + 2H_2O \xrightarrow{\text{lipase}}$$

$$CH_3-(CH_2)_{16}-\overset{\overset{\displaystyle O}{\|}}{C}-O-CH_2$$

(Tristearin)

$$CH_3-(CH_2)_{16}-\overset{\overset{\displaystyle O}{\|}}{C}-O-\overset{\overset{\displaystyle HO-CH_2}{|}}{\underset{\underset{\displaystyle HO-CH_2}{|}}{CH}} + 2CH_3-(CH_2)_{16}-\overset{\overset{\displaystyle O}{\|}}{C}-OH$$

(2-Monoglyceride)   (Stearic acid)

Figure 65–3. Hydrolysis of neutral fat catalyzed by lipase.

with the polar portions projecting outward and soluble in the surrounding fluids; this effect greatly decreases the interfacial tension of the fat.

When the interfacial tension of a globule of nonmiscible fluid is low, this nonmiscible fluid, on agitation, can be broken up into many minute particles far more easily than it can when the interfacial tension is great. Consequently, a major function of the bile salts and lecithin in the bile is to make the fat globules readily fragmentable by agitation in the small bowel. This action is the same as that of many detergents that are used widely in most household cleansers for removing grease.

Each time the diameters of the fat globules are decreased by a factor of 2 as a result of agitation in the small intestine, the total surface area of the fat increases two times. In other words, the total surface area of the fat particles in the intestinal contents is inversely proportional to the diameters of the particles. Since the average size of the emulsified fat particles in the intestine is less than 1 micrometer, this represents an increase of as much as 1000-fold in the total surface area of the fats caused by the emulsification process.

The lipases are water-soluble compounds and can attack the fat globules only on their surfaces. Consequently, it can be readily understood how important this detergent function of bile salts is for the digestion of fats.

Digestion of Triglycerides by Pancreatic Lipase. By far the most important enzyme for the digestion of triglycerides is *pancreatic lipase* in the pancreatic juice. This is present in enormous quantities in pancreatic juice, enough to digest all triglycerides that it can reach within a few minutes. In addition, the epithelial cells of the small intestine contain a minute quantity of lipase known as *enteric lipase*, but this is usually unimportant.

End Products of Fat Digestion. Most of the triglycerides of the diet are split into *free fatty acids* and *2-monoglycerides,* as illustrated in Figure 65-4. However, small portions remain in the diglyceride state.

Role of Bile Salts in Accelerating Fat Digestion — Formation of Micelles. The hydrolysis of triglycerides is a highly reversible process; therefore, accumulation of monoglycerides and free fatty acids in the vicinity of digesting fats very quickly blocks further digestion. Fortunately, the bile salts play an important role in removing the monoglycerides and free fatty acids from the vicinity of the digesting fat globules almost as rapidly as these end products of digestion are formed. This occurs in the following way:

Bile salts have the propensity to form *micelles,* which are small spherical globules about 3 to 4 nanometers in diameter composed of 20 to 40 molecules of bile salt. These develop because each bile salt molecule is composed of a sterol nucleus, most of which is highly fat-soluble, and a polar group that is highly water-soluble. The sterol nuclei of the 20 to 40 bile salt molecules of the micelle aggregate together to form a small fat globule in the middle of the micelle. This aggregation causes the polar groups to project outward to cover the surface of the micelle. Because these polar groups are negatively charged, they allow the entire micelle globule to become dissolved in the water of the digestive fluids and to remain in stable solution despite the very large size of the micelle.

During triglyceride digestion, as rapidly as the monoglycerides and free fatty acids are formed they become dissolved in the central fatty portion of the micelles, which immediately reduces the concentrations of these end products of digestion in the vicinity of the digesting fat globules. Consequently, the digestive process can proceed unabated.

The bile salt micelles also act as a transport medium to carry the monoglycerides and the free fatty acids, both of which would otherwise be relatively insoluble, to the brush borders of the epithelial cells. There the monoglycerides and free fatty acids are absorbed, as is discussed later. On delivery of these substances to the brush border, the bile salts are again released back into the chyme to be used again and again for this "ferrying" process.

Digestion of Cholesterol Esters and Phospholipids. Most of the cholesterol in the diet is in the form of cholesterol esters, which are combinations of free cholesterol and one molecule of fatty acid. Phospholipids also contain fatty acid chains within their molecules. Both the cholesterol esters and the phospholipids are hydrolyzed by lipases in the pancreatic secretion that free the fatty acids — the enzyme *cholesterol ester hydrolase* to hydrolyze the cholesterol ester and *phospholipase A_2* to hydrolyze the phospholipid.

The bile salt micelles play the same role in "ferrying" free cholesterol and also the remaining portions of the digested phospholipid molecules as they play in "ferrying" monoglycerides and free fatty acids. Indeed, this role of the micelles is absolutely essential to the absorption of cholesterol because essentially no cholesterol can be absorbed without the function of the micelles. On the other hand, as much as 60 per cent of the triglycerides can be digested and absorbed even in the absence of the bile salt micelles.

● BASIC PRINCIPLES OF GASTROINTESTINAL ABSORPTION

ANATOMICAL BASIS OF ABSORPTION

The total quantity of fluid that must be absorbed each day is equal to the ingested fluid (about 1.5 liters) plus that secreted in the various gastrointestinal secretions (about 7 liters). This comes to a total of approximately 8 to 9 liters. All but about 1.5 liters of this is absorbed in the small intestine, leaving only this 1.5 liters to pass through the ileocecal valve into the colon each day.

The stomach is a poor absorptive area of the gastrointestinal tract because it lacks the typical villus-type of absorptive membrane and also because the junctions between the epithelial cells are tight junctions. Only a few highly lipid-soluble substances, such as alcohol and some drugs like aspirin, can be absorbed in small quantities.

The Absorptive Surface of the Intestinal Mucosa — The Villi. Figure 65-5 illustrates the absorptive surface of the intestinal mucosa, showing many folds called *valvulae conniventes* (or *folds of Kerckring*), which increase the surface area of the absorptive mucosa about threefold. These folds extend circularly most of the way around the intestine and are especially well developed in the duodenum and

Fat $\xrightarrow{\text{(Bile + Agitation)}}$ Emulsified fat

Emulsified fat $\xrightarrow{\text{Pancreatic lipase}}$ Fatty acids and 2-monoglycerides

Figure 65-4. Digestion of fats.

Figure 65–7. Brush border of the gastrointestinal epithelial cell, showing, also, pinocytic vesicles, mitochondria, and endoplasmic reticulum lying immediately beneath the brush border. (Courtesy of Dr. William Lockwood.)

Figure 65–5. A longitudinal section of the small intestine, showing the valvulae conniventes covered by villi.

jejunum, where they often protrude as much as 8 millimeters into the lumen.

Located over the entire surface of the small intestine, from approximately the point at which the common bile duct empties into the duodenum down to the ileocecal valve, are literally millions of small *villi,* which project about 1 millimeter from the surface of the mucosa, as shown on the surfaces of the valvulae conniventes in Figure 65–5 and in more detail in Figure 65–6. These villi lie so close to each other in the upper small intestine that they actually touch in most areas, but their distribution is less profuse in the distal small intestine. The presence of villi on the mucosal surface enhances the absorptive area another tenfold.

Finally, each intestinal epithelial cell is characterized by a brush border, consisting of about 600 *microvilli* 1 micrometer in length and 0.1 micrometer in diameter protruding into the intestinal chyme; these are illustrated in the electron micrograph in Figure 65–7. This increases the surface area exposed to the intestinal materials another 20-fold. Thus, the combination of the folds of Kerckring, the villi, and the microvilli increases the absorptive area of the mucosa about 600-fold, making a tremendous total area of about 250 square meters for the entire small intestine —about the surface area of a tennis court.

Figure 65–6A illustrates the general organization of the villus, emphasizing especially the advantageous arrangement of the vascular system for absorption of fluid and dissolved material into the portal blood, and the arrangement of the *central lacteal* for absorption into the lymph. Figure 65–6B shows the cross-section of the villus, and Figure 65–7 shows many small *pinocytic* vesicles, which are pinched-off portions of infolded epithelium surrounding extracellular materials that have been entrapped inside the cells. Small amounts of substances are absorbed by this physical process of *pinocytosis,* though, as noted later in the

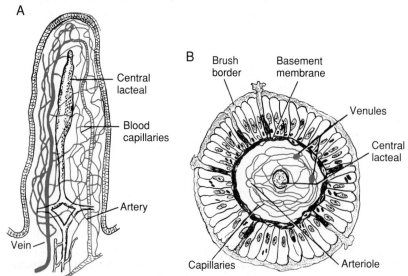

Figure 65–6. Functional organization of the villus. *A,* Longitudinal section. *B,* Cross-section showing the epithelial cells and basement membrane.

chapter, most absorption occurs by means of single molecular transfer. Also, extending linearly into each microvillus of the brush border are multiple actin filaments that are believed to contract intermittently and cause continual movement of the microvilli, keeping them constantly exposed to new quantities of intestinal fluid.

BASIC MECHANISMS OF ABSORPTION

Absorption through the gastrointestinal mucosa occurs by *active transport* and by *diffusion*, as is also true for other membranes. The physical principles of these processes were explained in Chapter 4.

Briefly, active transport imparts energy to the substance as it is being transported for the purpose of concentrating it on the other side of the membrane or for moving it against an electrical potential. On the other hand, the term "diffusion" means simply transport of substances through the membrane as a result of molecular movement *along*, rather than against, an electrochemical gradient.

● ABSORPTION IN THE SMALL INTESTINE

Normally, absorption from the small intestine each day consists of several hundred grams of carbohydrates, 100 or more grams of fat, 50 to 100 grams of amino acids, 50 to 100 grams of ions, and 7 to 8 liters of water. However, the absorptive *capacity* of the normal small intestine is far greater than this: as much as several kilograms of carbohydrates per day, 500 to 1000 grams of fat per day, 500 to 700 grams of proteins per day, and 20 or more liters of water per day. In addition, the *large* intestine can absorb still more water and ions, though almost no nutrients.

ABSORPTION OF WATER

Isosmotic Absorption. Water is transported through the intestinal membrane entirely by the process of *diffusion*. Furthermore, this diffusion obeys the usual laws of osmosis. Therefore, when the chyme is dilute, water is absorbed through the intestinal mucosa into the blood of the villi by osmosis.

On the other hand, water can also be transported in the opposite direction from the plasma into the chyme. This occurs especially when hyperosmotic solutions are discharged from the stomach into the duodenum. Usually within minutes, sufficient water is transferred by osmosis to make the chyme isosmotic with the plasma.

As dissolved substances are absorbed from the lumen of the gut into the blood, the absorption tends to decrease the osmotic pressure of the chyme; but water diffuses so readily through the intestinal membrane (because of large 0.7 to 1.5 nanometer intercellular pores through the so-called "tight junctions" be-

tween the epithelial cells) that it almost instantaneously "follows" the absorbed substances into the blood. Therefore, as ions and nutrients are absorbed, so also is an isosmotic equivalent of water absorbed.

ABSORPTION OF IONS

Active Transport of Sodium. Twenty to 30 grams of sodium are secreted into the intestinal secretions each day. In addition, the normal person eats 5 to 8 grams of sodium each day. Combining these two, the small intestine must absorb 25 to 35 grams of sodium each day, which is equal to about one seventh of all the sodium present in the body. One can well understand that whenever the intestinal secretions are lost to the exterior, as in extreme diarrhea, the sodium reserves of the body can be depleted to a lethal level within hours. Normally, though, less than 0.5 per cent of the intestinal sodium is lost in the feces each day. The sodium plays an important role in the absorption of sugars and amino acids, as we shall see in subsequent discussions.

The basic mechanism of sodium absorption from the intestine is illustrated in Figure 65-8. The principles of this mechanism, which were discussed in Chapter 4, are also essentially the same as those for absorption of sodium from the gallbladder and from the renal tubules, as discussed in Chapter 27. The motive power for the sodium absorption is provided by active transport of sodium from inside the epithelial cells through the basal and side walls of these cells into the intercellular spaces. This is illustrated by the heavy black arrows in Figure 65-8. This active transport obeys the usual laws of active transport: it requires energy, and it is catalyzed by appropriate ATPase carrier enzymes in the cell membrane (see Chapter 4). Part of the sodium is absorbed along with chloride ions that are passively "dragged" along by the positive electrical charges of the sodium ion. However, other sodium ions are absorbed while either potassium or hydrogen ions are transported in the opposite direction in exchange for the sodium ions.

The active transport of sodium through the baso-

Figure 65-8. Absorption of sodium through the intestinal epithelium. Note also the osmotic absorption of water—that is, the water "follows" the sodium through the epithelial membrane.

lateral membranes of the cell reduces its concentration inside the cell to a low value (about 50 mEq/liter), as also illustrated in Figure 65–8. Since the sodium concentration in the chyme is normally about 142 mEq/liter (that is, approximately equal to that in the plasma), sodium moves down a steep electrochemical gradient from the chyme through the brush border of the epithelial cell into the epithelial cell cytoplasm. This replaces the sodium that is actively transported out of the epithelial cells into the intercellular spaces.

The next step in the transport process is osmosis of water into the intercellular spaces. This is caused by the osmotic gradient created by the elevated concentration of ions in the intercellular space. Most of this osmosis occurs through the "tight junctions" between the apical borders of the epithelial cells, as discussed earlier; but a smaller proportion occurs through the cells themselves. The osmotic movement of water creates a flow of fluid into the intercellular space, then through the basement membrane of the epithelium, and finally into the circulating blood of the villi.

Effect of Aldosterone in Enhancing Greatly Sodium Absorption. When a person becomes dehydrated, very large amounts of aldosterone almost always are secreted by the adrenal glands. Within 1 to 3 hours the excess aldosterone greatly enhances all the enzyme and transport mechanisms for all aspects of sodium absorption by the intestinal epithelial cells. The increased sodium absorption then causes secondary increases also in absorption of chloride ions, water, and some other substances. This effect of aldosterone is especially important in the colon because it allows virtually no loss of sodium chloride in the feces and very little water loss. This absorptive mechanism in the intestinal tract is the same as that activated by aldosterone in the renal tubules, which also serves to conserve salt and water in the body when a person becomes dehydrated.

Absorption of Chloride Ions in the Duodenum and Jejunum. In the upper part of the small intestine chloride absorption is very rapid and mainly by passive diffusion. The absorption of sodium ions through the epithelium creates slight electronegativity in the chyme and electropositivity on the basal side of the epithelial cells. Then chloride ions move along this electrical gradient to "follow" the sodium ions.

Absorption of Bicarbonate Ions in the Duodenum and Jejunum. Often, large quantities of bicarbonate ions must be reabsorbed from the upper small intestine because of the large amounts of bicarbonate ions in both the pancreatic secretion and bile. However, the bicarbonate ion is absorbed in an indirect way as follows: When sodium ions are absorbed, moderate amounts of hydrogen ions are secreted into the lumen of the gut in exchange for some of the sodium, as explained earlier. These hydrogen ions in turn combine with the bicarbonate ions to form carbonic acid (H_2CO_3), which then dissociates to form H_2O and CO_2. The water remains as part of the chyme in the intestines, but the carbon dioxide is readily absorbed into the blood and subsequently expired through the lungs. Thus, this is a so-called "active" absorption of bicarbonate ions. It is the same mechanism that occurs in the tubules of the kidneys.

Secretion of Bicarbonate Ions in the Ileum and Large Intestine — Simultaneous Absorption of Chloride Ions

The epithelial cells on the surfaces of the villi in the ileum as well as on all surfaces of the large intestine have a special capability of secreting bicarbonate ions in exchange for absorption of chloride ions. This is important because it provides alkaline bicarbonate ions that are used to neutralize acid products formed by bacteria, especially so in the large intestine. The entire mechanism of this exchange is not clear, but basically it depends on an exchange protein in the luminal membrane of the epithelial cell that forcibly exchanges bicarbonate ions formed inside the cell for chloride ions in the intestinal lumen. Then the excess chloride in the cell is transported by facilitated diffusion through the basolateral membrane of the epithelial cell, thus completing the chloride absorption.

Extreme Secretion of Chloride Ions, Sodium Ions, and Water From the Crypts of Lieberkühn in Cholera and Some Other Types of Diarrhea. Deep in the crypts of Lieberkühn are immature epithelial cells that continually divide to form new epithelial cells that then spread outward over the luminal surfaces of the intestines. These new cells, while still in the crypts, have properties different from those of the mature cells already on the outer luminal surfaces. Even normally, they secrete small quantities of sodium chloride and water into the intestinal lumen, but this secretion is immediately reabsorbed by the older epithelial cells, thus providing a watery solution for absorbing intestinal digestates. However, the toxins of cholera and some other types of diarrheal bacteria can stimulate the crypt secretion so greatly that this secretion totally inundates the reabsorption, thus often causing loss of as much as 5 to 10 liters of water and salt as diarrhea each day. Within 1 to 5 days many severely affected patients die of this loss of fluid alone.

This extreme secretion is initiated by attachment of the cholera toxin to chloride channel proteins in the apical membranes of the crypt cells. This allows chloride ions to flow rapidly from inside the cell into the crypts; in turn, this is believed to activate a sodium pump that pumps sodium ions into the crypts to go along with the chloride ions. Finally, all this extra sodium chloride causes extreme osmosis of water into the crypts as well, thus providing the very rapid flow of fluid along with the salt. Initially, all of this excess fluid washes away the bacteria and is of value in combating the disease, but too much of a good thing can be lethal because of the serious dehydration of the body that may ensue.

In most instances, the life of the cholera victim can be saved by simple administration of tremendous amounts of sodium chloride solution to make up for the loss.

Absorption of Other Ions. Calcium ions are actively absorbed, especially from the duodenum, and calcium ion absorption is exactly controlled in relation to the need of the body for calcium. One important factor controlling calcium absorption is parathyroid hormone secreted by the parathyroid glands, and another is vitamin D. The parathyroid hormone activates vitamin D in the kidneys, and the activated vitamin D in turn greatly enhances calcium absorption. These effects are discussed in Chapter 79.

Iron ions are also actively absorbed from the small intestine. The principles of iron absorption and the regulation of its absorption in proportion to the body's need for iron were discussed in Chapter 32.

Potassium, magnesium, phosphate, and probably still other ions can also be actively absorbed through the mucosa. In general, the monovalent ions are absorbed with ease and in great quantities. On the other hand, bivalent ions are normally absorbed in only small amounts; for instance, the maximum absorption of calcium ions is only $\frac{1}{50}$ as great as the normal absorption of sodium ions. Fortunately, only small quantities of the bivalent ions are normally needed by the body.

ABSORPTION OF NUTRIENTS

Absorption of Carbohydrates

Essentially all of the carbohydrates are absorbed in the form of monosaccharides, only a small fraction of a per cent being absorbed as disaccharides, and almost none as larger carbohydrate compounds. Furthermore, little carbohydrate absorption results from simple diffusion, for the pores of the mucosa through which diffusion occurs are essentially impermeable to water-soluble solutes with molecular weights greater than 100.

The transport of most monosaccharides through the intestinal membrane can occur against large concentration gradients and therefore requires an active source of energy. Also, the transport is selective: certain monosaccharides are transported, whereas others are not. The order of preference for the transport of different monosaccharides and their relative rates of transport in comparison with that of glucose are as follows:

Galactose	1.1
Glucose	1.0
Fructose	0.4
Mannose	0.2
Xylose	0.15
Arabinose	0.1

There is a maximum rate of transport for each type of monosaccharide. The most rapidly transported monosaccharide is galactose, with glucose running a close second. Fructose, which is also one of the three important monosaccharides for nutrition, is absorbed less than half as rapidly as either galactose or glucose; also, its mechanism of absorption is different, as is explained below.

Mechanism of Glucose and Galactose Absorption — The Role of Sodium. Glucose and galactose transport either ceases or is greatly reduced wherever active sodium transport is blocked. The rea-

son is that the energy required for transport of these two monosaccharides is actually provided secondarily by the sodium transport system in the following way: A carrier protein for transport of glucose (which is the carrier for galactose as well) is present in the brush border of the epithelial cell. However, this carrier will not transport the glucose molecule in the absence of sodium transport. Therefore, it is believed that the carrier protein has receptor sites for both a glucose molecule and a sodium ion, and that it will not transport either of these to the interior of the epithelial cell until both receptor sites are simultaneously filled. The energy to cause movement of both the sodium ion and the glucose molecule from the exterior of the membrane to the interior is derived from the difference in sodium concentration between the outside and inside; the sodium and glucose are coupled in such a way that they must move together. That is, as sodium diffuses to the inside of the cell it "drags" the glucose along with it, thus providing the energy for transport of the glucose.

For obvious reasons, this explanation is called the *sodium co-transport theory* for glucose transport; it is also called *secondary active transport* of glucose. This sodium co-transport of glucose obviously moves the glucose only to the interior of the cell. However, this increases the intracellular glucose concentration to a higher than normal level, and the glucose then diffuses, by facilitated diffusion, through the basolateral membrane of the epithelial cell into the extracellular fluid.

Subsequently, we will see that sodium transport is also required for transport of many, if not all, amino acids, utilizing a similar co-transport mechanism for amino acid transport.

Absorption of Fructose. Transport of fructose is slightly different from that of most other monosaccharides. It is not blocked by some of the same metabolic poisons — specifically, phlorhizin — and it does not require metabolic energy for transport, even though it does require a specific carrier. Therefore, it is transported by *facilitated diffusion* rather than active transport. Also, it is partly converted into glucose inside the epithelial cell before entering the portal blood, the fructose first becoming phosphorylated, then converted to glucose, and finally released from the epithelial cell into the blood.

Absorption of Proteins

As explained earlier in the chapter, most proteins are absorbed through the luminal membranes of the epithelial cells in the form of dipeptides, tripeptides, and a few free amino acids. The energy for most of this transport is supplied by a sodium co-transport mechanism in exactly the same way that sodium co-transport of glucose occurs. That is, most peptide or amino acid molecules bind with a specific transport protein that also requires sodium binding before transport can occur. The sodium ion then moves down its electrochemical gradient to the interior of the cell and pulls the amino acid or peptide along with it. Therefore, this is called *co-transport* or *secondary active transport of the amino acids or peptides.*

However, a few amino acids do not require this sodium co-transport mechanism, but instead are transported by special membrane transport proteins in the same way that fructose is transported, probably by a process of facilitated diffusion.

At least five different types of amino acid and peptide transport proteins have been characterized in the luminal membrane of intestinal epithelial cells. This multiplicity of transport proteins is required because of the diverse binding properties of the different amino acids and peptides.

Absorption of Fats

Earlier in this chapter it was pointed out that as fats are digested to form monoglycerides and free fatty acids, both of these digestive end products become dissolved in the central lipid portion of the bile acid micelles. Because of the molecular dimensions of these micelles, only 3 nanometers in diameter, and also because of their highly charged exterior, they are soluble in the chyme. In this form the monoglycerides and the fatty acids are carried to the surfaces of the microvilli in the brush border, even penetrating into the recesses among the moving, agitating microvilli. Here, both the monoglycerides and the fatty acids diffuse first into the local fluids and then immediately through the epithelial membrane, because they are equally as soluble in this membrane as in the micelles. This leaves the bile acid micelles still in the chyme. The micelles then diffuse back through the chyme and absorb still more monoglycerides and fatty acids and similarly carry these also to the epithelial cells. Thus, the bile acids perform a "ferrying" function, which is highly important for fat absorption. In the presence of an abundance of bile acids, approximately 97 per cent of the fat is absorbed; in the absence of bile acids, only 50 to 60 per cent is normally absorbed.

The mechanism for absorption of the monoglycerides and fatty acids through the brush border is based entirely on the fact that both these substances are highly lipid-soluble. Therefore, they become dissolved in the membrane and simply diffuse to the interior of the cell.

The undigested triglycerides and the diglycerides are both also highly soluble in the lipid membrane of the epithelial cell. However, only small quantities of these are normally absorbed because the bile acid micelles will not dissolve either triglycerides or diglycerides and therefore will not ferry them to the epithelial membrane.

After entering the epithelial cell, the fatty acids and monoglycerides are taken up by the smooth endoplasmic reticulum, and here they are mainly recombined to form new triglycerides. However, a few of the monoglycerides are further digested into glycerol and fatty acids by an epithelial cell lipase. Then, these free fatty acids, too, are reconstituted by the smooth endoplasmic reticulum into triglycerides, using for this purpose new glycerol that is synthesized de novo from alpha-glycerophosphate, this synthesis requiring both energy from ATP and a complex of enzymes to catalyze the reactions.

Formation of Chylomicrons. Once formed, the reconstituted triglycerides aggregate within the endoplasmic reticulum into globules along with absorbed cholesterol, absorbed phospholipids, and small amounts of newly synthesized cholesterol and phospholipids. The phospholipids arrange themselves in these globules with the fatty portion of the phospholipid toward the center and the polar portions located on the surface. This provides an electrically charged surface that makes these globules miscible with the fluids of the cell. In addition, small amounts of *β-lipoprotein,* also synthesized by the endoplasmic reticulum, coat part of the surface of each globule. In this form the globule diffuses to the side of the epithelial cell and is excreted by the process of cellular *exocytosis* into the space between the cells; from there it passes into the lymph in the central lacteal of the villus. These globules are then called *chylomicrons.*

The *β*-lipoprotein is essential for cellular exocytosis of the chylomicrons to occur, because this protein provides a means for attaching the fatty globule to the cell membrane before it is extruded. In persons who have a genetic inability to form this *β*-lipoprotein, the epithelial cells become engorged with fatty products that cannot proceed the rest of the way to be absorbed.

Transport of the Chylomicrons in the Lymph. From the sides of the epithelial cells the chylomicrons wend their way into the central lacteals of the villi and from here are propelled, along with the lymph, by the lymphatic pump upward through the thoracic duct to be emptied into the great veins of the neck. Between 80 and 90 per cent of all fat absorbed from the gut is absorbed in this manner and is transported to the blood by way of the thoracic lymph in the form of chylomicrons.

Direct Absorption of Fatty Acids into the Portal Blood. Small quantities of short chain fatty acids, such as those from butterfat, are absorbed directly into the portal blood rather than being converted into triglycerides and absorbed into the lymphatics. The cause of this difference between short and long chain fatty acid absorption is that the shorter chain fatty acids are more water-soluble and mostly are not reconverted into triglycerides by the endoplasmic reticulum. This allows direct diffusion of these fatty acids from the epithelial cells into the capillary blood of the villi.

● ABSORPTION IN THE LARGE INTESTINE: FORMATION OF THE FECES

Approximately 1500 milliliters of chyme pass through the ileocecal valve into the large intestine each day. Most of the water and electrolytes in this are absorbed in the colon, usually leaving less than 100 milliliters of fluid to be excreted in the feces. Also, essentially all the ions are also absorbed, leaving only 1 to 5 milliequivalents each of sodium and chloride ions to be lost in the feces.

Most of the absorption in the large intestine occurs in the proximal half of the colon, giving this portion the name *absorbing colon,* whereas the distal colon functions principally for storage and is therefore called the *storage colon.*

Absorption and Secretion of Electrolytes and Water. The mucosa of the large intestine, like that of the small intestine, has a high capability for active absorption of sodium, and the electrical potential created by the absorption of the sodium causes chloride absorption as well. The "tight junctions" between the epithelial cells of the large intestinal epithelium are much tighter than those of the small intestine. This prevents significant amounts of back-diffusion of ions through these junctions, thus allowing the large intestinal mucosa to absorb sodium ions far more completely — that is, against a much higher concentration gradient — than can occur in the small intestine. This is especially true when larger quantities of aldosterone are available, for aldosterone greatly enhances the sodium transport capability.

In addition, as in the distal portion of the small intestine, the mucosa of the large intestine secretes *bicarbonate ions* while it simultaneously absorbs an equal number of chloride ions in an exchange transport process that has already been described. The bicarbonate helps neutralize the acidic end products of bacterial action in the colon.

The absorption of sodium and chloride ions creates an osmotic gradient across the large intestinal mucosa, which in turn causes absorption of water.

Maximum Absorption Capacity of the Large Intestine. The large intestine can absorb a maximum of about 5 to 7 liters of fluid and electrolytes each day. When the total quantity entering the large intestine through the ileocecal valve or by way of large intestinal secretion exceeds this amount, the excess appears in the feces as diarrhea. As noted earlier in the chapter, the toxins from cholera or certain other bacterial infections often cause the crypts of Lieberkühn in the terminal ileum and in the large intestine to secrete as much as 12 or more liters of fluid each day, leading to severe and sometimes even lethal diarrhea.

Bacterial Action in the Colon. Numerous bacteria, especially colon bacilli, are present even normally in the absorbing colon. These are capable of digesting small amounts of cellulose, in this way providing a few calories of nutrition to the body each day. In herbivorous animals this source of energy is very significant, though it is of negligible importance in the human being. Other substances formed as a result of bacterial activity are vitamin K, vitamin B_{12}, thiamin, riboflavin, and various gases that contribute to *flatus* in the colon — especially carbon dioxide, hydrogen gas, and methane. Vitamin K is especially important, for the amount of this vitamin in the ingested foods is normally insufficient to maintain adequate blood coagulation.

Composition of the Feces. The feces normally are about three-fourths water and one-fourth solid matter that itself is composed of about 30 per cent dead bacteria, 10 to 20 per cent fat, 10 to 20 per cent inorganic matter, 2 to 3 per cent protein, and 30 per cent undigested roughage of the food and dried constituents of digestive juices, such as bile pigment and sloughed epithelial cells. The large amount of fat derives mainly from fat formed by bacteria and fat in the sloughed epithelial cells.

The brown color of feces is caused by *stercobilin* and *urobilin,* which are derivatives of bilirubin. The odor is caused principally by the products of bacterial action; these vary from one person to another, depending on each person's colonic bacterial flora and on the type of food eaten. The actual odoriferous products include *indole, skatole, mercaptans,* and *hydrogen sulfide.*

REFERENCES

Anderson , P. O.: Vascular control in the colon and rectum. Scand. J. Gastroenterol., 93:65, 1984.

Baker, P. F., and Knight, D. E.: Chemiosmotic hypotheses of exocytosis: A critique. Biosci. Rep., 4:285, 1984.

Bickel, H. (ed.): Digestion and Absorption of Nutrients. Ft. Lee, N.J., J. K. Burgess, 1983.

Buddington, R. K., and Diamond, J. M.: Ontogenetic development of intestinal nutrient transporters. Annu. Rev. Physiol., 51:601, 1989.

Case, R. M.: The role of Ca^{2+} stores in secretion. Cell-Calcium, 5:89, 1984.

Christensen, H. N.: The regulation of amino acid and sugar absorption by diet. Nutr. Rev., 42:237, 1984.

Cummings, J. H.: Colonic absorption: The importance of short chain fatty acids in man. Scand. J. Gastroenterol. (Suppl), 93:89, 1984.

Donowitz, M., et al.: Cytosol free Ca^{++} in the regulation of active intestinal Na and Cl transport. KROC Found. Ser., 17:171, 1984.

Donowitz, M., and Welsh, M. J.: Ca^{2+} and cyclic AMP in regulation of intestinal Na, K, and Cl transport. Annu. Rev. Physiol., 48:135, 1986.

Ferraris, R. P., and Diamond, J. M.: Specific regulation of intestinal nutrient transporters by their dietary substrates. Annu. Rev. Physiol., 51:125, 1989.

Flemstrom, G., and Garner, A.: Some characteristics of duodenal epithelium. Ciba Found. Symp., 109:94, 1984.

Gardner, M. L.: Intestinal assimilation of intact peptides and proteins from the diet — a neglected field? Biol. Rev., 59:289, 1984.

Hopfer, U., and Liedtke, C. M.: Proton and bicarbonate transport mechanisms in the intestine. Annu. Rev. Physiol., 49:51, 1987.

Horl, W. H., and Heidland, A. (eds.): Proteases. New York, Plenum Publishing Corp., 1988.

Johnson, L. R.: Regulation of gastrointestinal mucosal growth. Physiol. Rev., 68:456, 1988.

Johnson, L., et al.: Physiology of the Gastrointestinal Tract, 2nd Ed. New York, Raven Press, 1987.

Kenny, A. J., and Maroux, S.: Topology of microvillar membrane hydrolases of kidney and intestine. Physiol. Rev., 62:91, 1982.

Liedtke, C. M.: Regulation of chloride transport in epithelia. Annu. Rev. Physiol., 51:143, 1989.

Mailman, D.: Relationships between intestinal absorption and hemodynamics. Annu. Rev. Physiol., 44:43, 1982.

Norum, K. R., et al.: Transport of cholesterol. Physiol. Rev., 63:1343, 1983.

Ockner, R. K., and Isselbacher, K. J.: Recent concepts of intestinal fat absorption. Rev. Physiol. Biochem. Pharmacol., 71:107, 1984.

Petersen, O. H.: Potassium channels and fluid secretion. News Physiol. Sci., 1:92, 1986.

Schultz, S. G.: A cellular model for active sodium absorption by mammalian colon. Annu. Rev. Physiol., 46:435, 1984.

Schultz, S. G., and Hudson, R. L.: How do sodium-absorbing cells do their job and survive? News in Physiol. Sci., 1:185, 1986.

Setchell, K. D. R., et al. (eds.): The Bile Acids. New York, Plenum Publishing Corp., 1988.

Sjovall, H., et al.: Sympathetic control of intestinal fluid and electrolyte transport. News Physiol. Sci., 2:214, 1987.

Smith, M. W.: Expression of digestive and absorptive function in differentiating enterocytes. Annu. Rev. Physiol., 47:247, 1985.

Smith, P. L., and McCabe, R. D.: Mechanisms and regulation of transcellular potassium transport by the colon. Am. J. Physiol., 247:G445, 1984.

Stevens, B. R., et al.: Intestinal transport of amino acids and sugars: Advances using membrane vesicles. Annu. Rev. Physiol., 46:417, 1984.

Thompson, J. C., et al. (eds.): Gastrointestinal Endocrinology. New York, McGraw-Hill Book Co., 1987.

Ugolev, A. M.: Membrane transport and hydrolytic enzymes under physiological vs. acute experimental conditions. News Physiol. Sci., 2:186, 1987.

Watson, D. W., and Sodeman, W. A., Jr.: The small intestine. In Sodeman, W. A., Jr., and Sodeman, T. M. (eds.): Pathologic Physiology: Mechanisms of Disease, 6th Ed. Philadelphia, W. B. Saunders Co., 1979, p. 824.

Wheeler, T. J., and Hinkle, P. C.: The glucose transporter of mammalian cells. Annu. Rev. Physiol., 47:503, 1985.

CHAPTER 66

• *Physiology of Gastrointestinal Disorders* •

The logical therapy of most gastrointestinal disorders depends on a basic knowledge of gastrointestinal physiology. The purpose of this chapter, therefore, is to discuss a few representative types of malfunction that have special physiological bases or consequences.

● DISORDERS OF SWALLOWING AND OF THE ESOPHAGUS

Paralysis of the Swallowing Mechanism. Damage to the 5th, 9th, or 10th nerve can cause paralysis of significant portions of the swallowing mechanism. Also, a few diseases, such as poliomyelitis or encephalitis, can prevent normal swallowing by damaging the swallowing center in the brain stem. Finally, malfunction of the swallowing muscles, as occurs in *muscle dystrophy* or in failure of neuromuscular transmission in *myasthenia gravis* or *botulism,* can also prevent normal swallowing.

When the swallowing mechanism is partially or totally paralyzed, the abnormalities that can occur include (1) complete abrogation of the swallowing act so that swallowing cannot occur at all, (2) failure of the glottis to close so that food passes into the lungs instead of the esophagus, or (3) failure of the soft palate and uvula to close the posterior nares so that food refluxes into the nose during swallowing.

One of the most serious instances of paralysis of the swallowing mechanism occurs when patients are under deep anesthesia. Often they vomit large quantities of materials from the stomach into the pharynx; then, instead of swallowing the materials again, they simply suck them into the trachea because the anesthetic has blocked the reflex mechanism of swallowing. As a result, such patients occasionally choke to death on their own vomitus.

Achalasia and Megaesophagus. Achalasia is a condition in which the lower esophageal sphincter fails to relax during the swallowing mechanism. As a result, food transmission from the esophagus into the stomach is impeded or prevented. Pathological studies have suggested the physiological basis of this condition to be either pathology of or nonfunction of the myenteric plexus in the lower two thirds of the esophagus. The musculature of the lower esophagus instead remains incoordinately contracted, and the myenteric plexus has lost the ability to transmit a signal to cause "receptive relaxation" of the gastroesophageal sphincter as food approaches this area during the swallowing process.

When achalasia becomes severe, the esophagus may not empty the swallowed food into the stomach for many hours, instead of within a few seconds, which is the normal time. Over months and years, the esophagus becomes tremendously enlarged until it often can hold as much as one liter of food, which becomes putridly infected during the long periods of esophageal stasis. The infection may also cause ulceration of the esophageal mucosa, sometimes leading to severe substernal pain or even rupture and death. Fortunately, considerable benefit can be achieved by stretching the lower end of the esophagus by means of a balloon inflated on the end of a swallowed esophageal tube. Antispasmotic drugs (drugs that relax smooth muscle), such as hydralazine and nifedipine, can also be helpful.

● DISORDERS OF THE STOMACH

Gastritis. Gastritis means inflammation of the gastric mucosa. Mild to moderate chronic gastritis is exceedingly common in the population as a whole, especially in the later years of adult life.

The inflammation of gastritis can be only superficial and therefore not very harmful, or it can penetrate deeply into the gastric mucosa and in many long-standing cases cause almost complete atrophy of the gastric mucosa. In a few cases, gastritis can be very acute and severe, with ulcerative excoriation of the stomach mucosa by the stomach's own peptic secretions.

The cause of gastritis in most instances is not known. In the past this was ascribed mainly to irritant foods, but the present clinical belief is that almost no food can be as irritant to the gastric mucosa as the normal acid-pepsin gastric juices themselves. Therefore, patients with gastritis are usually told to eat almost any food that will not cause nausea or burning epigastric pain. Yet a few substances that are often ingested can be very damaging to the protective gastric mucosal barrier — that is, to the mucous glands and the tight epithelial junctions between the gastric lining cells — often leading to severe acute or chronic gastritis. The two most common of these substances are *alcohol* and *aspirin.*

The Gastric Barrier and Its Penetration in Gastritis. Absorption from the stomach is normally very slight. This low level of absorption is mainly caused by two specific features of the gastric mucosa: (1) it is lined with highly resistant mucous cells that secrete a very viscid and adherent mucus, and (2) it has very tight junctions between the adjacent epithelial cells. These two together, plus other impediments to gastric absorption, are called the "gastric barrier." Normally this barrier is so resistant to diffusion that even the highly concentrated hydrogen ions of the gastric juice, averaging about 100,000 times the concentration of the hydrogen ions in the plasma, barely diffuse through the epithelial membrane. However, in gastritis, this barrier becomes inflamed, and its permeability is greatly increased. The hydrogen ions do then diffuse into the stomach epithelium, creating additional havoc and leading to a vicious circle of progressive stomach mucosal damage and atrophy. It also makes the mucosa susceptible to peptic digestion, thus frequently resulting in gastric ulcer.

Gastric Atrophy. In many persons who have chronic gastritis, the mucosa gradually becomes atrophic until little or no gastric gland activity remains. It is also believed that some persons develop autoimmunity against the gastric mucosa, this also leading eventually to gastric atrophy. Loss of the stomach secretions in gastric atrophy leads to achlorhydria and, occasionally, to *pernicious anemia.*

Achlorhydria (and Hypochlorhydria). Achlorhydria means simply that the stomach fails to secrete hydrochloric acid, and it is diagnosed when the pH of the gastric secretions fails to decrease below 6.5 following maximal stimulation. Hypochlorhydria means diminished acid secretion. Usually, when acid is not secreted; pepsin also is not secreted; even when it is, the lack of acid prevents it from functioning, because pepsin requires an acid medium for activity.

Although achlorhydria is associated with depressed or even no digestive capability by the stomach, the overall digestion of food in the entire gastrointestinal tract is still almost normal. This is because trypsin and other enzymes secreted by the pancreas are capable of digesting virtually all of the protein in the diet.

Pernicious Anemia in Gastric Atrophy. Pernicious anemia is a common accompaniment of achlorhydria and gastric atrophy. The normal gastric secretions contain a glycoprotein called *intrinsic factor,* which is secreted by the same parietal cells that secrete the hydrochloric acid; intrinsic factor must be present for adequate absorption of vitamin B_{12} from the ileum. The intrinsic factor combines with vitamin B_{12} in the stomach and thereafter protects it from being destroyed as it passes through the gastrointestinal tract. Then, when the intrinsic factor–vitamin B_{12} complex reaches the ileum, the intrinsic factor binds with receptors on the ileal epithelial surface. This in turn makes it possible for the vitamin B_{12} to be absorbed. In the absence of intrinsic factor, an adequate amount of vitamin B_{12} is not made available from the foods. As a result, maturation failure occurs in the bone marrow, resulting in pernicious anemia. This subject was discussed in more detail in Chapter 32.

Pernicious anemia also occurs frequently when most of the stomach has been removed for treatment of either stomach ulcer or gastric cancer, or when the ileum, where vitamin B_{12} is almost entirely absorbed, is removed.

PEPTIC ULCER

A peptic ulcer is an excoriated area of the mucosa caused by the digestive action of gastric juice. Figure 66–1 illustrates the points in the gastrointestinal tract at which peptic ulcers frequently occur, showing that by far the most frequent site is in the first few centimeters of the duodenum. In addition, peptic ulcers frequently occur along the lesser

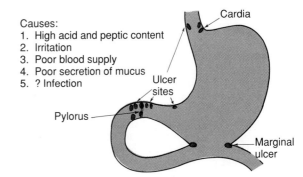

Causes:
1. High acid and peptic content
2. Irritation
3. Poor blood supply
4. Poor secretion of mucus
5. ? Infection

Figure 66–1. Peptic ulcer.

curvature of the antral end of the stomach or, more rarely, in the lower end of the esophagus where stomach juices frequently reflux. A peptic ulcer called a *marginal ulcer* also frequently occurs wherever a surgical opening, such as a gastrojejunostomy, is made between the stomach and some portion of the small intestine.

Basic Cause of Peptic Ulceration. The usual cause of peptic ulceration is an *imbalance* between the rate of secretion of gastric juice and the degree of protection afforded by the gastroduodenal mucosal barrier as well as the neutralization of the gastric acid by duodenal juices. It will be recalled that all areas normally exposed to gastric juice are well supplied with mucous glands, beginning with the compound mucous glands of the lower esophagus, then including the mucous cell coating of the stomach mucosa, the mucous neck cells of the gastric glands, the deep pyloric glands that secrete mainly mucus, and, finally, the glands of Brunner of the upper duodenum, which secrete a highly alkaline mucus.

In addition to the mucus protection of the mucosa, the duodenum is also protected by the alkalinity of the small intestinal secretions. Especially important is pancreatic secretion, which contains large quantities of sodium bicarbonate that neutralize the hydrochloric acid of the gastric juice, thus inactivating the pepsin to prevent digestion of the mucosa. Two additional mechanisms ensure that this neutralization of gastric juices is complete:

1. When excess acid enters the duodenum, it reflexly inhibits gastric secretion and peristalsis in the stomach, both nervously and hormonally, thereby decreasing the rate of gastric emptying.

2. The presence of acid in the small intestine liberates secretin from the intestinal mucosa, which then passes by way of the blood to the pancreas to promote rapid secretion of pancreatic juice containing a high concentration of sodium bicarbonate, thus making more sodium bicarbonate available for neutralization of the acid.

Therefore, a peptic ulcer can be caused in either of two ways: (1) *excess secretion of acid and pepsin* by the gastric mucosa or (2) *diminished ability of the gastroduodenal mucosal barrier to protect against the digestive properties of the acid-pepsin complex.*

Experimental Peptic Ulcer

Experimental peptic ulcers have been created in dogs and other animals in the following ways: (a) *Feeding ground glass* to an animal causes excoriation of the pyloric wall and allows the peptic juices to begin digesting the deeper layers of the mucosa. (b) *Transplantation of the pancreatic duct to the ileum* removes the normal neutralizing effect of pancreatic secretion in the duodenum and, therefore, allows the gastric juice to attack the mucosa of the upper duodenum. (c) *Repeated injection of histamine* causes excessive secretion of gastric juice. (d) *Continual infusion of hydrochloric acid* through a tube into the stomach causes direct damage to the stomach mucosa and also prevents full neutralization of the gastric juices by the pancreatic and other secretions of the small intestine. Therefore, peptic activity is enhanced, and ulcers appear in either the stomach or duodenum. (e) A portion of the *stomach is anastomosed directly to the small intestine* so that gastric juice can pass directly and rapidly into the small intestine. The mucosa of the small intestine, except at the uppermost part of the duodenum, is not sufficiently resistant to gastric juice to prevent peptic digestion. (f) *Obstruction of the blood flow* or even reduction of the blood flow to an area of the stomach or upper duodenum will cause ulcers to develop because the local epithelium degenerates or cannot produce appropriate protective secretions.

In summary, any factor that (1) *increases the rate of production of gastric juice* or (2) *blocks the normal protective mechanisms* against this juice can produce peptic ulcers. The same general principles apply to the development of peptic ulcers in the human being.

Causes of Peptic Ulcer in the Human Being

About 50 per cent of the patients with peptic ulcer *of the duodenum* secrete up to two times as much gastric acid following injection of a test dose of pentagastrin as do normal persons. Therefore, it is believed that a large percentage of *duodenal* ulcers are caused by excessive secretion of acid and pepsin by the gastric glands. In the remainder of the patients, it is presumed that one of four other abnormalities might be the cause: (a) possible secretion of an abnormal mucus that has less than normal protective value, (b) diminished secretion of mucus, (c) failure of the normal duodenal-gastric feedback mechanisms to limit the rate of gastric emptying into the duodenum, or (d) failure of the secretin-pancreatic and secretin–bile duct feedback mechanisms to cause the secretion of enough alkaline pancreatic juice and bile to neutralize the gastric juice as it enters the duodenum.

The development of duodenal peptic ulcers is strongly hereditary. For instance, the offspring of persons who secrete excessive amounts of gastric acid, like their parents, tend to secrete excessive amounts of acid.

In addition to hereditary factors, psychogenic factors that cause stress anxiety can also play a role in peptic ulceration, though this probably occurs mainly in persons who already have a hereditary basis for developing ulcers. The number of persons with peptic ulcer increased greatly during the air raids of London in World War II. Therefore, it is believed that many instances of duodenal peptic ulcer in the human being are precipitated by excessive stimulation of the dorsal motor nucleus of the vagus by signals originating elsewhere in the brain, with this in turn greatly increasing the stomach secretions.

Paradoxically, *gastric* ulcers, in contradistinction to duodenal ulcers, usually occur in patients who have normal or even low secretion of hydrochloric acid. However, these patients frequently have an associated, often atrophic, gastritis, indicating that ulceration in the stomach most often results from reduced resistance of the stomach mucosa to digestion rather than from excess secretion of gastric juice. Stomach ulceration occurs especially frequently in patients who have ingested substances such as aspirin or alcohol that reduce the mucosal resistance.

Physiology of Treatment

The usual medical treatment for peptic ulcer is a combination of (1) reduction of stressful situations that might lead to excessive acid secretion, (2) administration of antacid drugs to neutralize much of the acid in the stomach secretions, (3) administration of the type of antihistamic drug that blocks the effect of histamine on H_2-receptors and thereby reduces gastric acid secretion by 70 to 80 per cent, such drugs as *cimetidine, ranitidine,* and *famotidine,* (4) interdiction of smoking because statistical studies have shown that smokers are several times as prone to have peptic ulcers as are nonsmokers, and (5) removal of such ulcer-causing factors as alcohol, aspirin, or other substances that might break down the gastroduodenal mucosal barrier.

Surgical treatment of peptic ulceration is usually by one or both of two procedures: (1) *vagotomy* or (2) *removal of a portion of the stomach.* Fortunately, however, since the advent of the drug cimetidine, the more drastic surgical procedures are employed far less frequently than in the past.

Vagotomy means section of the vagus nerves to the stomach; this temporarily blocks almost all secretion of acid and pepsin by the stomach and often cures the ulcer or ulcers within a week after the operation is performed. Unfortunately, though, some of the basal stomach secretion returns after a few months, and in many patients the ulcer also returns. Also, *gastric atony* usually follows section of both vagus trunks; this can be very distressing, since stomach motility is often reduced so much that gastric emptying becomes minimal, leading to partial or sometimes almost total pyloric obstruction. Several modified procedures prevent this, such as performing along with the vagotomy a *pyloroplasty,* which is an operation to increase the size of the opening of the pylorus from the stomach into the duodenum or removing the stomach antrum and pylorus and then connecting the body of the stomach directly to the proximal end of the duodenum.

● DISORDERS OF THE SMALL INTESTINE

Abnormal Digestion of Food in the Small Intestine — Pancreatic Failure

A common cause of abnormal digestion is failure of the pancreas to secrete its juice into the small intestine. Lack of pancreatic secretion frequently occurs (1) in *pancreatitis,* which is discussed later, (2) when the *pancreatic duct is blocked* by a gallstone at the papilla of Vater, or (3) after the *head of the pancreas has been removed* because of malignancy. Loss of pancreatic juice means loss of trypsin, chymotrypsin, carboxypolypeptidase, pancreatic amylase, pancreatic lipase, and still a few other digestive enzymes. Without these enzymes, most of the fat entering the small intestine may go unabsorbed and as much as one third to one half of the proteins and carbohydrates as well. As a result, large portions of the ingested food are not utilized for nutrition; and copious, fatty feces are excreted.

Pancreatitis. Pancreatitis means inflammation of the pancreas, and this can occur in the form of either *acute pancreatitis* or *chronic pancreatitis.* The commonest cause of pancreatitis is alcohol and the second most common cause is blockage of the papilla of Vater by a gallstone; the two together account for over 90 per cent of all cases. When a gallstone blocks the papilla of Vater, this blocks the main secretory duct from the pancreas as well as the common bile duct. The pancreatic enzymes are then dammed up in the ducts and acini of the pancreas. Eventually, so much trypsinogen accumulates that it *overcomes the trypsin inhibitor* in the secretions, and a small quantity of trypsinogen becomes activated to form trypsin. Once this happens, the trypsin activates still more trypsinogen as well as chymotrypsinogen and carboxypolypeptidase, resulting in a vicious circle until most of the proteolytic enzymes in the pancreatic ducts and acini become activated. These rapidly digest large portions of the pancreas itself, sometimes completely and permanently destroying the ability of the pancreas to secrete digestive enzymes.

Malabsorption by the Small Intestinal Mucosa — Sprue

Occasionally, nutrients are not adequately absorbed from the small intestine even though the food is well digested. Several different diseases can cause decreased absorption by the mucosa; these are often classified together under the general heading of *"sprue."* Obviously, also, malabsorption can occur when large portions of the small intestine have been removed.

Nontropical Sprue. One type of sprue, called variously by the names *idiopathic sprue, celiac disease* (in children), or *gluten enteropathy*, results from the toxic effects of *gluten* present in certain types of grains, especially wheat and rye. Only some persons are susceptible to this effect, but in these persons gluten has a direct destructive effect on the intestinal epithelial cells, or perhaps as the result of an immunological or allergic reaction. In milder forms of the disease, the microvilli of the absorbing epithelial cells on the villi are destroyed, thus decreasing the absorptive surface area as much as twofold. In the more severe forms, the villi themselves become blunted or disappear altogether, thus still further reducing the absorptive area of the gut. Removal of wheat and rye flour from the diet frequently results in a miraculous cure within weeks, especially in children with this disease.

Tropical Sprue. A different type of sprue called *tropical sprue* frequently occurs in the tropics and can often be treated with antibacterial agents. Even though no bacterium has been implicated as the cause, it is believed that this variety of sprue is usually caused by inflammation of the intestinal mucosa resulting from yet unidentified infectious agents.

Malabsorption in Sprue. In the early stages of sprue, the absorption of fats is more impaired than the absorption of other digestive products. The fat appearing in the stools is almost entirely in the form of salts of fatty acids rather than undigested neutral fat, illustrating that the problem is one of absorption and not of digestion. In this stage of sprue, the condition is frequently called *idiopathic steatorrhea*, which means simply excess fats in the stools as a result of unknown causes.

In more severe cases of sprue the absorption of proteins, carbohydrates, calcium, vitamin K, folic acid, and vitamin B_{12}, as well as many other important substances, becomes greatly impaired. As a result, the person suffers (1) severe nutritional deficiency, often developing severe wasting of the body, (2) osteomalacia (demineralization of the bones because of calcium lack), (3) inadequate blood coagulation due to lack of vitamin K, and (4) macrocytic anemia of the pernicious anemia type, owing to diminished vitamin B_{12} and folic acid absorption.

● DISORDERS OF THE LARGE INTESTINE

CONSTIPATION

Constipation means slow movement of feces through the large intestine; it is often associated with large quantities of dry, hard feces in the descending colon which accumulate because of the long time available for absorption of fluid.

A frequent cause of constipation is irregular bowel habits that have developed through a lifetime of inhibition of the normal defecation reflexes. Newborn children are rarely constipated, but part of their training in the early years of life requires that they learn to control defecation, and this control is effected by inhibiting the natural defecation reflexes. Clinical experience shows that if one fails to allow defecation to occur when the defecation reflexes are excited or if one overuses laxatives to take the place of natural bowel function, the reflexes themselves become progressively less strong over a period of time, and the colon becomes *atonic*. For this reason, if a person establishes regular bowel habits early in life, usually defecating in the morning after breakfast when the gastrocolic and duodenocolic reflexes cause mass movements in the large intestine, the development of constipation in later life can generally be prevented.

Constipation can also result from spasm of a small segment of the sigmoid colon. It should be recalled that motility, even normally, is weak in the large intestine, so that even a slight degree of spasm is often capable of causing serious constipation. After the constipation has continued for several days and excessive feces have accumulated above the spastic sigmoid colon, excessive colonic secretions can lead to a day or so of diarrhea. Following this, the cycle begins again, with repeated bouts of alternating constipation and diarrhea.

Megacolon

Occasionally, constipation develops that is so severe that bowel movements occur only once every week or so. Obviously, this allows tremendous quantities of fecal matter to accumulate in the colon, causing the colon sometimes to distend to a diameter as great as 3 to 4 inches. The condition is called megacolon, or *Hirschsprung's disease.*

The most frequent cause of megacolon is lack of or deficiency of ganglion cells in the myenteric plexus in a segment of the sigmoid colon. As a consequence, neither defecation reflexes nor peristaltic motility can occur through this area of the large intestine. The sigmoid itself becomes small and almost spastic while feces accumulate proximal to this area, causing megacolon. Thus, the diseased portion of the large intestine appears normal by x-ray, whereas the originally normal portion of the large intestine appears greatly enlarged.

DIARRHEA

Diarrhea results from rapid movement of fecal matter through the large intestine. Several causes of diarrhea with important physiological overtones are the following:

Enteritis. Enteritis means infection caused either by a virus or by bacteria in the intestinal tract. In the usual infectious diarrhea, the infection is most extensive in the large intestine and the distal end of the ileum. Everywhere that the infection is present, the mucosa becomes extensively irritated, and its rate of secretion becomes greatly enhanced. In addition, the motility of the intestinal wall usually increases manyfold. As a result, large quantities of fluid are made available for washing the infectious agent toward the anus, and at the same time strong propulsive movements propel this fluid forward. Obviously, this is an important mechanism for ridding the intestinal tract of the debilitating infection.

Of special interest is the diarrhea caused by *cholera* (and sometimes by other bacteria such as pathogenic colon bacilli). As explained in the previous chapter, cholera toxin directly stimulates excessive secretion of electrolytes and fluid from the crypts of Lieberkühn in the distal ileum and colon. The amount can be as great as 12 liters per day, and the colon can usually reabsorb a maximum of 6 liters per day. Therefore, the loss of fluid and electrolytes can be so debilitating within a day or so that death ensues. Therefore, the most important physiological basis of therapy is simply to replace the fluid and electrolytes as rapidly as they are lost, mainly by giving the patient saline and glucose solutions. With proper therapy of this type, along with the use of antibiotics, almost no cholera patients die, but without therapy 50 per cent or more do.

Psychogenic Diarrhea. Everyone is familiar with the diarrhea that accompanies periods of nervous tension, such as during examination time or when a soldier is about to go into battle. This type of diarrhea, called psychogenic or emotional diarrhea, is caused by excessive stimulation of the parasympathetic nervous system, which greatly excites both motility and secretion of mucus in the distal colon. These two effects added together can cause marked diarrhea.

Ulcerative Colitis. Ulcerative colitis is a disease in which extensive areas of the walls of the large intestine become inflamed and ulcerated. The motility of the ulcerated colon is often so great that mass movements occur most of the time, rather than the usual 10 to 20 minutes per day. Also, the colon's secretions are greatly enhanced. As a result, the patient has repeated diarrheal bowel movements.

The cause of ulcerative colitis is unknown. Some clinicians believe that it results from an allergic or immune destructive effect. But whatever the cause, there is a strong hereditary tendency for susceptibility to ulcerative colitis. Once the condition has progressed very far, the ulcers are perpetuated by superimposed bacterial infection and rarely will heal until an ileostomy is performed to allow the intestinal contents to drain to the exterior rather than to flow through the colon. Even then the ulcers sometimes fail to heal, and the only solution is removal of the colon itself.

PARALYSIS OF DEFECATION IN SPINAL CORD INJURIES

From Chapter 63 it will be recalled that defecation is normally initiated by the movement of feces into the rectum, which causes a cord-mediated defecation reflex passing from the rectum to the spinal cord and then back to the descending colon, sigmoid, rectum, and anus. This reflex adds greatly to the activity of the intrinsic defecation reflex mediated through the myenteric plexus of the sigmoid and colon wall itself. Frequently, this *cord defecation reflex* is blocked or altered by spinal cord injuries. For instance, destruction of the *conus medullaris* of the spinal cord destroys

the sacral centers in which the cord reflex is integrated and therefore almost completely paralyzes defecation. In such instances defecation requires extensive supportive measures, such as cathartics and large enemas.

But, more frequently, the spinal cord is injured somewhere between the conus medullaris and the brain, in which case the voluntary portion of the defecation act is blocked while the basic cord reflex for defecation is still intact. Nevertheless, loss of the voluntary aid to defecation—that is, loss of the increased abdominal pressure and the relaxation of the voluntary anal sphincter—often makes defecation a difficult process in the person with this type of cord injury. Yet, because the cord defecation reflex can still occur, a small enema to excite the action of this reflex, usually given in the morning shortly after a meal, can often cause adequate defecation. In this way persons with spinal cord injuries can usually control their bowel movements each day.

● GENERAL DISORDERS OF THE GASTROINTESTINAL TRACT

VOMITING

Vomiting is the means by which the gastrointestinal tract rids itself of its contents when almost any part of the upper gastrointestinal tract becomes excessively irritated, overdistended, or even overexcitable. Excessive distention or irritation of the duodenum provides the strongest stimulus for vomiting. Impulses are transmitted, as illustrated in Figure 66–2, by both vagal and sympathetic afferents to the bilateral *vomiting center* of the medulla, which lies near the tractus solitarius at approximately the level of the dorsal motor nucleus of the vagus. Appropriate motor reactions are then instituted to cause the vomiting act. The motor impulses that cause the actual vomiting are transmitted from the vomiting center through the 5th, 7th, 9th, 10th, and 12th cranial nerves to the upper gastrointestinal tract and through the spinal nerves to the diaphragm and abdominal muscles.

Antiperistalsis, the Prelude to Vomiting. In the early stages of excessive gastrointestinal irritation or overdistention, *antiperistalsis* begins to occur, often many minutes before vomiting appears. The antiperistalsis may begin as far down in the intestinal tract as the ileum, and the antiperistaltic wave travels backward up the intestine at a rate of 2 to 3 cm/sec; this process can actually push the intestinal contents all the way back to the duodenum and stomach within 3 to 5 minutes. Then, as these upper portions of the gastrointestinal tract, especially the duodenum, become overly distended, this distention becomes the exciting factor that initiates the actual vomiting act. During the vomiting, strong intrinsic contractions occur in both the duodenum and the stomach, along with beginning relaxation of the lower esophageal sphincter, thus allowing the vomitus to begin moving into the esophagus. From here a specific vomiting act involving the abdominal muscles expels the vomitus to the exterior, as explained in the following paragraph.

The Vomiting Act. Once the vomiting center has been sufficiently stimulated and the vomiting act instituted, the first effects are (1) a deep breath, (2) raising of the hyoid bone and the larynx to pull the upper esophageal sphincter open, (3) closing of the glottis, and (4) lifting of the soft palate to close the posterior nares. Next comes a strong downward contraction of the diaphragm along with simul-

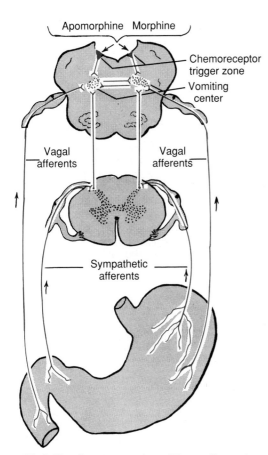

Figure 66-2. The afferent connections of the vomiting center.

taneous contraction of all the abdominal wall muscles. This obviously squeezes the stomach between the two sets of muscles, building the intragastric pressure to a high level. Finally, the lower esophageal sphincter relaxes completely, allowing expulsion of the gastric contents upward through the esophagus.

Thus, the vomiting act results from a squeezing action of the muscles of the abdomen associated with opening of the esophageal sphincters so that the gastric contents can be expelled.

The Chemoreceptor Trigger Zone of the Medulla for Initiation of Vomiting by Drugs or by Motion Sickness. Aside from the vomiting initiated by irritative stimuli in the gastrointestinal tract itself, vomiting can also be caused by nervous signals arising in areas of the brain outside the vomiting center. This is particularly true of a small area located bilaterally on the floor of the fourth ventricle in or above the area postrema and called the *chemoreceptor trigger zone.* Electrical stimulation of this area will initiate vomiting; more importantly, administration of certain drugs, including apomorphine, morphine, and some of the digitalis derivatives, can directly stimulate the chemoreceptor trigger zone and initiate vomiting. Destruction of this area blocks this type of vomiting but does not block vomiting resulting from irritative stimuli in the gastrointestinal tract itself.

Also, it is well known that rapidly changing directions of motion of the body cause certain people to vomit. The mechanism for this is the following: the motion stimulates the receptors of the labyrinth, and impulses are transmitted mainly by way of the vestibular nuclei into the cerebellum,

then to the chemoreceptor trigger zone, and finally to the vomiting center to cause vomiting.

Cerebral Excitation of Vomiting. Various psychic stimuli, including disquieting scenes, noisome odors, and other similar psychological factors, can also cause vomiting. Stimulation of certain areas of the hypothalamus also causes vomiting. The precise neuronal connections for these effects are not known, although it is probable that the impulses pass directly to the vomiting center and do not involve the chemoreceptor trigger zone.

NAUSEA

Everyone has experienced the sensation of nausea and knows that it is often a prodrome of vomiting. Nausea is the conscious recognition of subconscious excitation in an area of the medulla closely associated with or part of the vomiting center, and it can be caused by irritative impulses coming from the gastrointestinal tract, impulses originating in the lower brain associated with motion sickness, or impulses from the cerebral cortex to initiate vomiting. However, vomiting occasionally occurs without the prodromal sensation of nausea, indicating that only certain portions of the vomiting centers are associated with the sensation of nausea.

GASTROINTESTINAL OBSTRUCTION

The gastrointestinal tract can become obstructed at almost any point along its course, as illustrated in Figure 66-3; some common causes of obstruction are *cancer, fibrotic constriction resulting from ulceration or from peritoneal adhesions, spasm of a segment of the gut,* or *paralysis of a segment of the gut.*

The abnormal consequences of obstruction depend on the point in the gastrointestinal tract that becomes obstructed. If the obstruction occurs at the pylorus, which results often from fibrotic constriction following peptic ulceration, persistent vomiting of stomach contents occurs. This obviously depresses bodily nutrition, and it also causes excessive loss of hydrogen ions from the body and can result in various degrees of alkalosis.

If the obstruction is beyond the stomach, antiperistaltic reflux from the small intestine causes the intestinal juices to flow backward into the stomach, and these are vomited along with the stomach secretions. In this instance the person loses large amounts of water and electrolytes, so that he

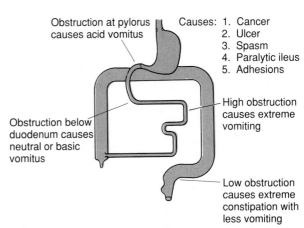

Causes: 1. Cancer
2. Ulcer
3. Spasm
4. Paralytic ileus
5. Adhesions

Obstruction at pylorus causes acid vomitus

High obstruction causes extreme vomiting

Obstruction below duodenum causes neutral or basic vomitus

Low obstruction causes extreme constipation with less vomiting

Figure 66-3. Obstruction in different parts of the gastrointestinal tract.

becomes severely dehydrated; but the loss of acids and bases may be approximately equal, so that little change in acid-base balance occurs. If the obstruction is near the lower end of the small intestine, then it is actually possible to vomit more basic than acidic substances; in this case acidosis may result. In addition, the vomitus, after a few days of obstruction, becomes fecal in character.

Also important in obstruction of the small intestine is marked distention of the intestine proximal to the obstructed point. Large quantities of fluid and electrolytes continue to be secreted into the lumen of the small intestine; and even large amounts of proteins are lost from the blood stream, partly into the intestinal lumen and partly into the gut wall, which becomes edematous as a result of the excessive distention. The plasma volume diminishes because of the protein loss, and severe circulatory shock often ensues. One might immediately ask: Why does the small intestine not reabsorb these fluids and electrolytes? The answer is that distention of the gut usually stimulates the secretory activity of the gut but does not correspondingly increase the rate of absorption. Normally this would act to flush the chyme farther down the small intestine and therefore relieve the distention. But if obstruction is present, this normal mechanism cannot occur, and instead a vicious circle of more and more distention occurs.

If the obstruction is near the distal end of the large intestine, feces can accumulate in the colon for several weeks. The patient develops an intense feeling of constipation, but in the first stages of the obstruction vomiting is not severe. After the large intestine has become completely filled and it finally becomes impossible for additional chyme to move from the small intestine into the large intestine, vomiting does then begin severely. Prolonged obstruction of the large intestine will finally cause rupture of the intestine itself or dehydration and circulatory shock resulting from the severe vomiting.

GASES IN THE GASTROINTESTINAL TRACT AND FLATUS

Gases can enter the gastrointestinal tract from three different sources: (1) swallowed air, (2) gases formed as a result of bacterial action, and (3) gases that diffuse from the blood into the gastrointestinal tract.

Most gases in the stomach are nitrogen and oxygen derived from swallowed air, and in the normal person most of these are expelled by belching.

Only small amounts of gas are usually present in the small intestine, and much of these is air that passes from the stomach into the intestinal tract. In addition, considerable carbon dioxide also often occurs because the reaction between acidic gastric juice and bicarbonate in pancreatic juice sometimes is too rapid for the liberated CO_2 to be absorbed.

In the large intestine, the greater proportion of the gases is derived from bacterial action, including especially *carbon dioxide, methane,* and *hydrogen.* These occur along with varying amounts of oxygen and nitrogen from swallowed air. When the methane and hydrogen become suitably mixed with oxygen from swallowed air, an actual explosive mixture is occasionally formed; and use of the electric cautery during sigmoidoscopy has on rare occasions caused colon explosions.

Certain foods are known to cause greater expulsion of flatus through the anus than others—beans, cabbage, onion, cauliflower, corn, and certain irritant foods such as vinegar. Some of these foods serve as a suitable medium for gas-forming bacteria, especially because of unabsorbed fermentable types of carbohydrates (for instance, beans contain an undigestible sugar that passes into the colon and becomes a superior food for the colonic bacteria), but in other instances excess gas results from irritation of the large intestine, which promotes rapid peristaltic expulsion of the gases before they can be absorbed.

The amount of gases entering or forming in the large intestine each day averages 7 to 10 liters, whereas the average amount expelled through the anus is usually only about 0.6 liter. The remainder is absorbed through the intestinal mucosa.

REFERENCES

Ackerman, S. H.: The developmental physiology of the stomach: Possible contributions to the regulatory disturbances of duodenal ulcer disease. Proc. Soc. Exp. Biol. Med., 175:119, 1984.

Awouters, F., et al.: Pharmacology of antidiarrheal drugs. Annu. Rev. Pharmacol. Toxicol., 23:279, 1983.

Baillie, J., and Soltis, R. D.: Systemic complications of inflammatory bowel disease. Geriatrics, 40:53, 59, 1985.

Berglinch, T.: The mammalian gastric parietal cell in vitro. Annu. Rev. Physiol., 46:377, 1984.

Berk, J. E., et al.: Bockus Gastroenterology, 4th Ed. Philadelphia, W. B. Saunders Co., 1985.

Bongiovanni, G. L. (ed.): Essentials of Clinical Gastroenterology, 2nd Ed. New York, McGraw-Hill Book Co., 1988.

Castell, D. O.: Calcium-channel blocking agents for gastrointestinal disorders. Am J. Cardiol., 55:210B, 1985.

Giannella, R. A.: Pathogenesis of acute bacterial diarrheal disorders. Annu. Rev. Med., 32:341, 1981.

Gitnick, G. (ed.): Handbook of Gastrointestinal Emergencies, 2nd Ed. New York Elsevier Science Publishing Co., 1988.

Guth, P. H.: Pathogenesis of gastric mucosal injury. Annu. Rev. Med., 33:183, 1982.

Hayworth, M. F., and Jones, A. L.: Immunology of the Gastrointestinal Tract and Liver. New York, Raven Press, 1988.

Johnson, L. R., et al.: Physiology of the Gastrointestinal Tract, 2nd Ed. New York, Raven Press, 1987.

Kestenbaum, D., and Behar, J.: Pathogenesis, diagnosis, and management of reflux esophagitis. Annu. Rev. Med., 32:443, 1981.

Kirsner, J. B., and Shorter, R. G. (eds.): Diseases of the Colon, Rectum, and Anal Canal. Baltimore, Williams & Wilkins, 1988.

Kirsner, J. B., and Shorter, R. G. (eds.): Inflammatory Bowel Disease, 2nd Ed. Philadelphia, Lea & Febiger, 1980.

McCarthy, D. M.: Zollinger-Ellison syndrome. Annu. Rev. Med., 33:197, 1982.

Misiewicz, J. J., et al.: Atlas of Clinical Gastroenterology. Philadelphia, Lea & Febiger, 1987.

Nelson, R. L., and Nyhus, L. M.: Surgery of the Small Intestine. East Norwalk, CT, Appleton & Lange, 1987.

Rodolfo, C., et al. (eds.): Gastric Protection. New York, Raven Press, 1988.

Shaffer, E., and Thomson, A. B. R.: Modern Concepts in Gastroenterology. New York, Plenum Publishing Corp., 1989.

Shearman, D. J. (ed.): Diseases of the Gastrointestinal Tract and Liver, New York, Churchill Livingstone, 1982.

Silva, J., Jr., and Fekety, R.: Clostridia and antimicrobial enterocolitis. Annu. Rev. Med., 32:327, 1981.

Snape, W. J., Jr. (ed.): Pathogenesis of Functional Bowel Disease. New York, Plenum Publishing Corp., 1989.

Strickland, R. G., and Jewell, D. P.: Immunoregulatory mechanisms in non-specific inflammatory bowel disease. Annu. Rev. Med., 34:195, 1983.

UNIT XIII

METABOLISM AND TEMPERATURE REGULATION

CHAPTER 67

• Metabolism of Carbohydrates and Formation of Adenosine Triphosphate •

The next few chapters deal with metabolism in the body, which means the chemical processes that make it possible for the cells to continue living. It is not the purpose of this textbook, however, to present the chemical details of all the various cellular reactions, for this lies in the discipline of biochemistry. Instead, these chapters are devoted to (1) a review of the principal chemical processes of the cell, and (2) an analysis of their physiological implications, especially in relation to the manner in which they fit into the overall concept of homeostasis.

RELEASE OF ENERGY FROM FOODS AND THE CONCEPT OF "FREE ENERGY"

A great proportion of the chemical reactions in the cells is concerned with making the energy in foods available to the various physiological systems of the cell. For instance, energy is required for muscular activity, secretion by the glands, maintenance of membrane potentials by the nerve and muscle fibers, synthesis of substances in the cells, absorption of foods from the gastrointestinal tract, and many other functions.

Coupled Reactions. All the energy foods—carbohydrates, fats, and proteins—can be oxidized in the cells, and in this process large amounts of energy are released. These same foods can also be burned with pure oxygen outside the body in an actual fire, again releasing large amounts of energy. However, this time the energy is released suddenly, all in the form of heat. The energy needed by the physiological processes of the cells is not heat, but energy to cause mechanical movement in the case of muscle function, to concentrate solutes in the case of glandular secretion, and to effect other functions. To provide this energy, the chemical reactions must be "coupled" with the systems responsible for these physiological functions. This coupling is accomplished by special cellular enzyme and energy transfer systems, some of which will be explained in this and subsequent chapters.

"Free Energy." The amount of energy liberated by complete oxidation of a food is called *the free energy of oxidation of the food,* and this is generally represented by the symbol ΔG. Free energy is usually expressed in terms of

calories per mole of substance. For instance, the amount of free energy liberated by oxidation of 1 mole of glucose (180 grams of glucose) is 686,000 calories.

ROLE OF ADENOSINE TRIPHOSPHATE IN METABOLISM

Adenosine triphosphate (ATP) is a labile chemical compound that is present in all cells and has the chemical structure shown in Figure 67–1.

From this formula it can be seen that ATP is a combination of adenine, ribose, and three phosphate radicals. The last two phosphate radicals are connected with the remainder of the molecule by so-called *high-energy bonds,* which are indicated by the symbol $\sim$. The amount of free energy in each of these high-energy bonds per mole of ATP is approximately 7300 calories under standard conditions and about 12,000 calories under the conditions of temperature and concentrations of the reactants in the body. Therefore, removal of each phosphate radical liberates 12,000 calories of energy. After loss of one phosphate radical from ATP, the compound becomes *adenosine diphosphate* (ADP), and after loss of the second phosphate radical the compound becomes *adenosine monophosphate* (AMP). The interconversions between ATP, ADP, and AMP are the following:

$$\text{ATP} \underset{+\,12,000\text{ cal}}{\overset{-\,12,000\text{ cal}}{\rightleftarrows}} \left\{ \begin{matrix} \text{ADP} \\ + \\ \text{PO}_3 \end{matrix} \right\} \underset{+\,12,000\text{ cal}}{\overset{-\,12,000\text{ cal}}{\rightleftarrows}} \left\{ \begin{matrix} \text{AMP} \\ + \\ 2\text{PO}_3 \end{matrix} \right\}$$

ATP is present everywhere in the cytoplasm and nucleoplasm of all cells, and essentially all of the physiological mechanisms that require energy for operation obtain this directly from the ATP (or some other similar high-energy compound—guanosine triphosphate [GTP], for example). In turn, the food in the cells is gradually oxidized, and the released energy is used to re-form the ATP, thus always maintaining a supply of this substance; all these energy transfers take place by means of coupled reactions.

In summary, ATP is an intermediary compound that has

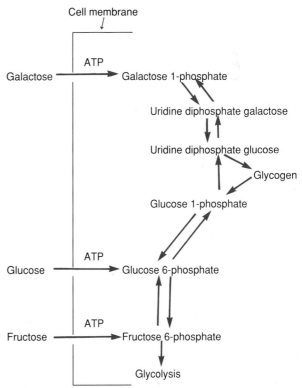

Figure 67-1. Chemical structure of adenosine triphosphate (ATP).

the peculiar ability to enter into many coupled reactions — reactions with the food to extract energy and reactions in many physiological mechanisms to provide energy for their operation. For this reason, ATP has been called the energy *currency* of the body that can be gained and spent again and again.

The principal purpose of the present chapter is to explain how the energy from carbohydrates can be used to form ATP in the cells. Normally, 90 per cent or more of all the carbohydrates utilized by the body are used for this purpose.

● CENTRAL ROLE OF GLUCOSE IN CARBOHYDRATE METABOLISM

From Chapter 65 it will be recalled that the final products of carbohydrate digestion in the alimentary tract are almost entirely glucose, fructose, and galactose — with glucose representing, on the average, about 80 per cent of these. After absorption from the intestinal tract, much of the fructose and almost all of the galactose is also rapidly converted into glucose. Therefore, very little fructose and galactose is present in the circulating blood. Glucose thus becomes the final common pathway for transport of almost all carbohydrates to the tissue cells.

In liver cells, appropriate enzymes are available to promote interconversions among the monosaccharides, as shown in Figure 67-2. Furthermore, the dynamics of the reactions are such that when the liver releases the monosaccharides back into the blood, the final product is almost entirely glucose. The reason for this is that the liver cells contain large amounts of *glucose phosphatase*. Therefore, glucose 6-phosphate can be degraded back to glucose and phosphate, and the glucose can be transported back into the blood.

Therefore, once again it should be emphasized that usually more than 95 per cent of all the monosaccharides that circulate in the blood are the final conversion product, glucose.

● TRANSPORT OF GLUCOSE THROUGH THE CELL MEMBRANE

Before glucose can be used by the cells, it must be transported through the cell membrane into the cellular cyto-

plasm. However, glucose cannot diffuse through the pores of the cell membrane because the maximum molecular weight of particles that can do this is about 100, whereas glucose has a molecular weight of 180. Yet glucose does pass to the interior of the cells with a reasonable degree of freedom by the mechanism of *facilitated diffusion*. The principles of this type of transport were discussed in Chapter 4. Basically, they are the following: In the lipid matrix of the cell membrane are large numbers of protein *carrier* molecules that can bind with glucose. In this bound form the glucose can be transported by the carrier from one side of the membrane to the other side and then released. Therefore, if the concentration of glucose is greater on one side of the membrane than on the other side, more glucose will be

Figure 67-2. Interconversions of the three major monosaccharides — glucose, fructose, and galactose — in liver cells.

transported from the high concentration area than in the opposite direction.

It should be noted that the transport of glucose through the membranes of most tissue cells is quite different from that which occurs through the gastrointestinal membrane or through the epithelium of the renal tubules. In both these cases, the glucose is transported by the mechanism of *active sodium-glucose cotransport,* in which active transport of sodium provides energy for absorbing glucose *against a concentration difference.* This sodium cotransport mechanism functions only in certain special epithelial cells that are specifically adapted for active absorption of glucose. At all other cell membranes, glucose is transported only from higher concentration toward lower concentration by the process of facilitated diffusion, made possible by the special binding properties of the membrane glucose carrier protein.

Effect of Insulin to Increase Facilitated Diffusion of Glucose

The rate of glucose transport, and also transport of some other monosaccharides, is greatly increased by insulin. When large amounts of insulin are secreted by the pancreas, the rate of glucose transport into most cells increases to 10 or more times the rate of transport when no insulin at all is secreted. Conversely, the amounts of glucose that can diffuse to the insides of most cells of the body in the absence of insulin, with the exceptions of the liver and brain cells, are far too little to supply anywhere near the amount of glucose normally required for energy metabolism. Therefore, in effect, the rate of carbohydrate utilization by most cells is controlled by the rate of insulin secretion from the pancreas. The functions of insulin and its control of carbohydrate metabolism will be discussed in detail in Chapter 78.

PHOSPHORYLATION OF GLUCOSE

Immediately upon entry into the cells, glucose combines with a phosphate radical in accordance with the following reaction:

$$\text{Glucose} \xrightarrow[\text{+ ATP}]{\text{glucokinase or hexokinase}} \text{Glucose 6-phosphate}$$

This phosphorylation is promoted by the enzyme *glucokinase* in the liver or hexokinase in most other cells.

The phosphorylation of glucose is almost completely irreversible except in the liver cells, the renal tubular epithelium, and the intestinal epithelial cells, in which *glucose phosphatase* is also available for reversing the reaction. Therefore, in most tissues of the body phosphorylation serves to *capture* the glucose in the cell—once *in* the cell the glucose will not diffuse back out except from those special cells that have the phosphatase.

● STORAGE OF GLYCOGEN IN LIVER AND MUSCLE

After absorption into the cells, glucose can be used immediately for release of energy to the cells or it can be stored in the form of *glycogen,* which is a large polymer of glucose.

All cells of the body are capable of storing at least some glycogen, but certain cells can store large amounts, especially the liver cells, which can store up to 5 to 8 per cent of

their weight as glycogen, and muscle cells, which can store up to 1 to 3 per cent glycogen. The glycogen molecules can be polymerized to almost any molecular weight, the average molecular weight being 5 million or greater; most of the glycogen precipitates in the form of solid granules. This conversion of the monosaccharides into a high molecular weight, precipitated compound makes it possible to store large quantities of carbohydrates without significantly altering the osmotic pressure of the intracellular fluids. Obviously, high concentrations of low molecular weight, soluble monosaccharides would play havoc with the osmotic relationships between intracellular and extracellular fluids.

GLYCOGENESIS

Glycogenesis is the process of glycogen formation, the chemical reactions for which are illustrated in Figure 67–3. From this figure it can be seen that *glucose 6-phosphate* first becomes *glucose 1-phosphate;* then this is converted to *uridine diphosphate glucose,* which is then converted into glycogen. Several specific enzymes are required to cause these conversions, and any monosaccharide that can be converted into glucose obviously can enter into the reactions. Certain smaller compounds, including *lactic acid, glycerol, pyruvic acid,* and *some deaminated amino acids,* can also be converted into glucose or closely allied compounds and thence into glycogen.

REMOVAL OF STORED GLYCOGEN— GLYCOGENOLYSIS

Glycogenolysis means the breakdown of glycogen to re-form glucose in the cells. The glucose can then be used to provide energy. Glycogenolysis does not occur by reversal of the same chemical reactions that serve to form glycogen; instead, each succeeding glucose molecule on each branch of the glycogen polymer is split away by a process of *phosphorylation,* catalyzed by the enzyme *phosphorylase* (several other enzymes split the glycogen molecule at the branching points).

Under resting conditions, the phosphorylase is in an inactive form, so that glycogen can be stored but not reconverted into glucose. When it is necessary to re-form glucose

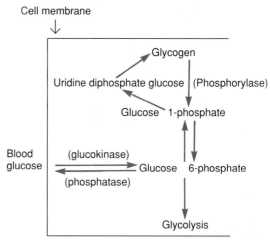

Figure 67–3. The chemical reactions of glycogenesis and glycogenolysis, showing also the interconversions between blood glucose and liver glycogen. (The phosphatase required for release of glucose from the cell is present in liver cells but not in most other cells.)

from glycogen, therefore, the phosphorylase must first be activated. This can be accomplished in several ways, including the following two.

Activation of Phosphorylase by Epinephrine and Glucagon. Two hormones, *epinephrine* and *glucagon,* can specifically activate phosphorylase and thereby cause rapid glycogenolysis. The initial effect of each of these hormones is to cause the formation of *cyclic adenosine monophosphate (cAMP)* in the cells. This substance then initiates a cascade of chemical reactions that activates the phosphorylase. This is discussed in more detail in Chapter 78.

Epinephrine is released by the adrenal medullae when the sympathetic nervous system is stimulated. Therefore, one of the functions of the sympathetic nervous system is to increase the availability of glucose for rapid energy metabolism. This function of epinephrine occurs markedly both in liver cells and in muscle, thereby contributing, along with other effects of sympathetic stimulation, to preparing the body for action, as discussed in Chapter 60.

Glucagon is a hormone secreted by the *alpha cells* of the pancreas when the blood glucose concentration falls too low. It stimulates the formation of cAMP mainly in the liver cells. Therefore, its effect is primarily to convert liver glycogen into glucose and to release this into the blood, thereby elevating blood glucose concentration. The function of glucagon in blood glucose regulation is discussed in Chapter 78.

● RELEASE OF ENERGY FROM THE GLUCOSE MOLECULE BY THE GLYCOLYTIC PATHWAY

Because complete oxidation of 1 gram-molecule of glucose releases 686,000 calories of energy, and only 12,000 calories of energy are required to form 1 gram-molecule of adenosine triphosphate (ATP), it would be extremely wasteful of energy if glucose should be decomposed all the way into water and carbon dioxide at once while forming only a single ATP molecule. Fortunately, all cells contain an extensive series of different protein enzymes that cause the glucose molecule to split a little at a time in many suc-

cessive steps, with its energy released in small packets to form one molecule of ATP at a time, forming a total of 38 moles of ATP for each mole of glucose utilized by the cells.

The purpose of the present section is to describe the basic principles by which the glucose molecule is progressively dissected and its energy released to form ATP.

GLYCOLYSIS AND THE FORMATION OF PYRUVIC ACID

By far the most important means by which energy is released from the glucose molecule is by the process of *glycolysis* and then *oxidation of the end products of glycolysis.* Glycolysis means splitting of the glucose molecule to form *two molecules of pyruvic acid.* This occurs by 10 successive steps of chemical reactions, as illustrated in Figure 67–4. Each step is catalyzed by at least one specific protein enzyme. Note that glucose is first converted into fructose 1,6-phosphate and then split into two three-carbon atom molecules, each of which is then converted through five successive steps into pyruvic acid.

Formation of Adenosine ATP During Glycolysis. Despite the many chemical reactions in the glycolytic series, only a small portion of the free energy in the glucose molecule is released. However, between the 1,3-diphosphoglyceric acid and the 3-phosphoglyceric acid stages, and again between the phosphopyruvic acid and the pyruvic acid stages, the packets of energy released are greater than 12,000 calories per mole, the amount required to form ATP, and the reactions are coupled in such a way that ATP is formed. Thus, a total of 4 moles of ATP are formed for each mole of fructose 1,6-phosphate that is split into pyruvic acid.

Yet 2 moles of ATP had been required to phosphorylate the original glucose to form fructose 1,6-phosphate before glycolysis could begin. Therefore, the net gain in ATP molecules by the entire glycolytic process is only 2 moles for each mole of glucose utilized. This amounts to 24,000 calories of energy stored in the form of ATP, but during glycolysis a total of 56,000 calories of energy is lost from the original glucose, giving an overall *efficiency* for ATP formation of 43

Figure 67–4. The sequence of chemical reactions responsible for glycolysis.

Net reaction:

Glucose + 2ADP + 2PO$_4$ $^{---}$ ⟶ 2 Pyruvic acid + 2ATP + 4H per cent. The remaining 57 per cent of the energy is lost in the form of heat.

CONVERSION OF PYRUVIC ACID TO ACETYLCOENZYME A

The next stage in the degradation of glucose is (1) facilitated transport of the two derivative pyruvic acid molecules into the matrix of the mitochondria, and then (2) conversion of these into two molecules of *acetylcoenzyme A* (acetyl-CoA) in accordance with the following reaction:

$$2\ CH_3-\overset{\overset{O}{\|}}{C}-COOH\ +\ 2\ CoA-SH \rightarrow$$
$$\text{(Pyruvic acid)}\qquad\text{(Coenzyme A)}$$

$$2\ CH_3-\overset{\overset{O}{\|}}{C}-S-CoA\ +\ 2CO_2\ +\ 4H$$
$$\text{(Acetyl-CoA)}$$

From this reaction it can be seen that two carbon dioxide molecules and four hydrogen atoms are released, while the remainders of the two pyruvic acid molecules combine with coenzyme A, a derivative of the vitamin pantothenic acid, to form two molecules of acetyl-CoA. In this conversion, no ATP is formed, but up to six molecules of ATP are formed when the four hydrogen atoms are later oxidized, as is discussed in a later section.

THE CITRIC ACID CYCLE

The next stage in the degradation of the glucose molecule is called the *citric acid cycle* (also called the *tricarboxylic acid cycle* or *Krebs cycle*). This is a sequence of chemical reactions in which the acetyl portion of acetyl-CoA is degraded to carbon dioxide and hydrogen atoms. These reactions all occur in the matrix of the mitochondrion. The released hydrogen atoms are subsequently oxidized, as is discussed later, releasing tremendous amounts of energy to form ATP.

Figure 67–5 shows the different stages of the chemical reactions in the citric acid cycle. The substances to the left are added during the chemical reactions, and the products of the chemical reactions are shown to the right. Note at the top of the column that the cycle begins with *oxaloacetic acid,* and then at the bottom of the chain of reactions *oxaloacetic acid* is formed once again. Thus, the cycle can continue over and over.

In the initial stage of the citric acid cycle, *acetyl-CoA* combines with *oxaloacetic acid* to form *citric acid.* The coenzyme A portion of the acetyl-CoA is released and can be used again and again for the formation of still more quantities of acetyl-CoA from pyruvic acid. The acetyl portion, however, becomes an integral part of the citric acid molecule. During the successive stages of the citric acid cycle, several molecules of water are added, and *carbon dioxide* and *hydrogen atoms* are released at various stages in the cycle, as shown on the right in the figure.

The net results of the entire citric acid cycle are given in the legend of Figure 67–5, illustrating that for each molecule of glucose originally metabolized, two acetyl-CoA molecules enter into the citric acid cycle along with six molecules of water. These then are degraded into four carbon dioxide molecules, 16 hydrogen atoms, and two molecules of coenzyme A.

Figure 67–5. The chemical reactions of the citric acid cycle, showing the release of carbon dioxide and a number of hydrogen atoms during the cycle.

Net reaction per molecule of glucose:

$$2\ \text{Acetyl-CoA} + 6H_2O + 2ADP \longrightarrow 4CO_2 + 16H + 2CoA + 2ATP$$

Formation of ATP in the Citric Acid Cycle. Not a great amount of energy is released during the citric acid cycle itself; in only one of the chemical reactions — during the change from α-ketoglutaric acid to succinic acid — is a molecule of ATP formed. Thus, for each molecule of glucose metabolized, two acetyl-CoA molecules pass through the citric acid cycle, each forming a molecule of ATP; or a total of two molecules of ATP is formed.

Function of the Dehydrogenases and Nicotinamide Adenine Dinucleotide (NAD⁺) in Causing Release of Hydrogen Atoms. As noted at several points in this discussion, hydrogen atoms are released during different chemical reactions — 4 hydrogen atoms during glycolysis, 4 during the formation of acetyl-CoA from pyruvic acid, and 16 in the citric acid cycle; this makes a total of 24 hydrogen atoms. However, the hydrogen atoms are not simply turned loose in the intracellular fluid. Instead, they are released in packets of two, and in each instance the release is catalyzed by a specific protein enzyme called a *dehydrogenase.* Twenty of the 24 hydrogen atoms immediately combine with nicotinamide adenine dinucleotide (NAD⁺), a derivative of the vitamin niacin, in accordance with the following reaction:

$$NADH + H^+ + Substrate$$

This reaction will not occur without the initial intermediation of the specific dehydrogenase nor without the availability of NAD⁺ to act as a hydrogen carrier. Both the free hydrogen ion and the hydrogen bound with NAD⁺ subsequently enter into the oxidative chemical reactions that form tremendous quantities of ATP, as will be discussed later.

The remaining four hydrogen atoms released during the breakdown of glucose — the four released during the citric acid cycle between the succinic and fumaric acid stages — combine with a specific dehydrogenase but are not then subsequently released to NAD⁺. Instead, they pass directly from the dehydrogenase into the oxidative process.

Function of Decarboxylases in Causing Release of Carbon Dioxide. Referring again to the chemical reactions of the citric acid cycle and also to those for formation of acetyl-CoA from pyruvic acid, we find that there are three stages in which carbon dioxide is released. To cause the carbon dioxide release, other specific protein enzymes, called *decarboxylases,* split the carbon dioxide away from the substrate. The carbon dioxide in turn becomes dissolved in the fluids of the cells and is transported to the lungs, where it is expired from the body (see Chapter 40).

FORMATION OF ATP BY OXIDATIVE PHOSPHORYLATION

Despite all the complexities of (1) glycolysis, (2) the citric acid cycle, (3) dehydrogenation, and (4) decarboxylation, pitifully small amounts of ATP are formed during all these processes — only two ATP molecules in the glycolysis scheme and another two in the citric acid cycle. Instead, 90 per cent of the total ATP formed by glucose metabolism is

formed during subsequent oxidation of the hydrogen atoms that were released during these earlier stages of glucose degradation. Indeed, the principal function of all these earlier stages is to make the hydrogen of the glucose molecule available in forms that can be utilized for oxidation.

Oxidation of hydrogen is accomplished by a series of enzymatically catalyzed reactions *in the mitochondria* that (1) split each hydrogen atom into a hydrogen ion and an electron, and (2) use the electrons eventually to change the dissolved oxygen of the fluids into hydroxyl ions. Then the hydrogen and hydroxyl ions combine with each other to form water. During the sequence of oxidative reactions, tremendous quantities of energy are released to form ATP. Formation of ATP in this manner is called *oxidative phosphorylation.* This occurs entirely in the mitochondria by a highly specialized process called the chemiosmotic mechanism.

The Chemiosmotic Mechanism of the Mitochondria to Form ATP

Ionization of Hydrogen, the Electron Transport Chain, and Formation of Water. The first step in oxidative phosphorylation in the mitochondria is to ionize the hydrogen atoms that have been removed from the food substrates. As described earlier, these hydrogen atoms are removed in pairs: one immediately becomes a hydrogen ion, H⁺; and the other combines with NAD⁺ to form NADH. The upper portion of Figure 67–6 shows in color the subsequent fate of the NADH and H⁺. The initial effect is to release the other hydrogen atom from the NADH to form another hydrogen ion, H⁺; this process also reconstitutes NAD⁺ that will be reused again and again.

The electrons that are removed from the hydrogen atoms to cause the ionization immediately enter an *electron transport chain* that is an integral part of the inner membrane (the shelf membrane) of the mitochondrion. This transport chain consists of a series of electron acceptors that can be reversibly reduced or oxidized by accepting or giving up electrons. The important members of this electron transport chain include *flavoprotein, several iron sulfide proteins, ubiquinone,* and *cytochromes B, C_1, C, A, and A_3.* Each electron is shuttled from one of these acceptors to the next until it finally reaches cytochrome A_3, which is called *cytochrome oxidase* because it is capable, by giving up two electrons, of reducing elemental oxygen, which then combines with hydrogen ions to form water.

Thus, Figure 67–6 illustrates transport of electrons through the electron chain and then ultimate use of these by cytochrome oxidase to cause the formation of water molecules. During the transport of these electrons through the electron transport chain, energy is released that is later used to cause synthesis of ATP.

Pumping of Hydrogen Ions Into the Outer Chamber of the Mitochondrion, Caused by the Electron Transport Chain. As the electrons pass through the electron transport chain, large amounts of energy are released. This energy is used to pump hydrogen ions from the inner matrix of the mitochondrion into the outer chamber between the inner and outer mitochondrial membranes. This creates a high concentration of hydrogen ions in this chamber, and it also creates a strong negative electrical potential in the inner matrix.

Formation of ATP. The next step in oxidative phosphorylation is to convert ADP into ATP. This occurs in conjunction with a large protein molecule that protrudes all

Figure 67–6. The chemiosmotic mechanism of oxidative phosphorylation for forming great quantities of ATP.

the way through the inner mitochondrial membrane and projects with a knoblike head into the inner matrix. This molecule is an ATPase, the physical nature of which is illustrated in Figure 67–6. It is called *ATP synthetase*. The high concentration of the positively charged hydrogen ions in the outer chamber and the large electrical potential difference across the inner membrane cause the hydrogen ions to flow into the mitochondrial matrix *through the substance of the ATPase molecule*. In doing so, energy derived from this hydrogen ion flow is utilized by the ATPase to convert ADP into ATP by combining ADP with inorganic phosphate (Pi), thus adding an additional high-energy phosphate bond to the molecule.

The final step in the process is transfer of the ATP from the inside of the mitochondrion back to the cytoplasm. This occurs by facilitated diffusion outward through the inner membrane and then by simple diffusion through the very permeable outer mitochondrial membrane. In turn, ADP is continually transferred in the other direction and in like manner back into the mitochondrial matrix for continual conversion into ATP.

For each two electrons that pass through the entire electron transport chain (representing the ionization of two hydrogen atoms), up to three ATP molecules are synthesized.

SUMMARY OF ATP FORMATION DURING THE BREAKDOWN OF GLUCOSE

We can now determine the total number of ATP molecules that, under optimal conditions, can be formed by the energy from one molecule of glucose.

1. During glycolysis, four molecules of ATP are formed, whereas two are expended to cause the initial phosphorylation of glucose to start the process going. This gives a net gain of *two molecules of ATP*.

2. During each revolution of the citric acid cycle, one molecule of ATP is formed. However, because each glucose molecule splits into two pyruvic acid molecules, there are two revolutions of the cycle for each molecule of glucose metabolized, giving a net production of *two more molecules of ATP*.

3. During the entire schema of glucose breakdown, a total of 24 hydrogen atoms is released during glycolysis and during the citric acid cycle. Twenty of these atoms are oxidized in conjunction with the chemiosmotic mechanism shown in Figure 67–6, with the release of up to three ATP molecules per two atoms of hydrogen metabolized. This gives an additional *30 ATP molecules*.

4. The remaining four hydrogen atoms are released by their dehydrogenase into the chemiosmotic oxidative schema in the mitochondrion beyond the first stage of Figure 67–6, so that for these four hydrogen atoms only two ATP molecules are usually released for each two hydrogen atoms oxidized, giving a total of *four more ATP molecules*.

Now, adding all the ATP molecules formed, we find a maximum of *38 ATP molecules* formed for each molecule of glucose degraded to carbon dioxide and water. Thus, 456,000 calories of energy can be stored in the form of ATP, whereas 686,000 calories are released during the complete oxidation of each gram-molecule of glucose. This represents an overall maximum *efficiency* of energy transfer of 66 per cent. The remaining 34 per cent of the energy becomes heat and therefore cannot be used by the cells to perform specific functions.

EFFECT OF ATP AND ADP CONCENTRATIONS IN THE CELL TO CONTROL THE RATE OF GLYCOLYSIS

Continual release of energy from glucose when the energy is not needed by the cells would be an extremely wasteful process. Fortunately, glycolysis and the subsequent oxidation of hydrogen atoms are continually controlled in accordance with the needs of the cells for ATP. This control is accomplished by multiple feedback control mechanisms within the chemical schemata. However, among the more important of these are the effects of both ADP and ATP in controlling the rates of chemical reactions in the energy metabolism sequence.

An important way in which ATP helps control energy metabolism is its effect in causing an *allosteric inhibition of the enzyme phosphofructokinase*. Because this enzyme promotes the formation of fructose 1,6-phosphate, one of the initial steps in the glycolytic series of reactions, the net effect of excess ATP, therefore, is to stop glycolysis, which in turn stops most carbohydrate metabolism. Conversely, ADP (and AMP as well) causes the opposite allosteric change in this enzyme, greatly increasing its activity. Therefore, whenever ATP is used by the tissues for energy, this reduces the ATP inhibition of the enzyme but at the same time increases its activity as a result of the ADP that is formed, and the glycolytic process is set in motion. Soon the total cellular store of ATP is replenished.

Another control linkage is the *citrate ion* formed in the citric acid cycle. An excess of this ion also *strongly inhibits phosphofructokinase*, thus preventing the glycolytic process from getting ahead of the ability of the citric acid cycle to utilize the pyruvic acid formed during glycolysis.

A third way by which the ATP-ADP-AMP system controls carbohydrate metabolism, as well as controlling energy release from fats and proteins, is the following: Referring back to the various chemical reactions for energy release, we see that at many different stages ADP is converted into ATP. If ADP is not available, the reactions at each of these stages simply cannot occur, and thus the degradation of the glucose molecule, fat molecule, or protein molecule is stopped. Then, when ATP is used to energize the different physiological functions in the cell, the resulting ADP and AMP turn on the energy processes again, and the ADP and AMP are almost instantly returned to the ATP state. In this way, essentially a full store of ATP is automatically maintained all the time, except during extreme cellular activity as might occur in very heavy exercise.

ANAEROBIC RELEASE OF ENERGY— "ANAEROBIC GLYCOLYSIS"

Occasionally, oxygen becomes either unavailable or insufficient, so that oxidative phosphorylation cannot take place. Yet even under these conditions, a small amount of energy can still be released to the cells by the glycolysis stage of carbohydrate degradation, for the chemical reactions in the glycolytic breakdown of glucose to pyruvic acid do not require oxygen. Unfortunately, this process is extremely wasteful of glucose because only 24,000 calories of energy are used to form ATP for each molecule of glucose utilized, which represents only a little over 3 per cent of the total energy in the glucose molecule. Nevertheless, this release of glycolytic energy to the cells, which is called *anaerobic energy,* can be a life-saving measure for up to a few minutes' time when oxygen becomes unavailable.

Formation of Lactic Acid During Anaerobic Glycolysis

Allows the Release of Extra Anaerobic Energy. The *law of mass action* states that as the end products of a chemical reaction build up in a reacting medium, the rate of the reaction approaches zero. The two end products of the glycolytic reactions (see Fig. 67–4) are (1) pyruvic acid and (2) hydrogen atoms combined with NAD^+ to form NADH and H^+. The buildup of either or both of these would stop the glycolytic process and prevent further formation of ATP. Fortunately, when their quantities begin to be excessive, these two end products react with each other to form lactic acid in accordance with the following equation:

$$
\underset{\text{(Pyruvic acid)}}{CH_3-\overset{\overset{\textstyle O}{\|}}{C}-COOH} + NADH + H^+ \xrightleftharpoons{\overset{\textstyle lactic}{dehydrogenase}}
$$

$$
\underset{\text{(Lactic acid)}}{CH_3-\overset{\overset{\textstyle OH}{|}}{\underset{\underset{\textstyle H}{|}}{C}}-COOH} + NAD^+
$$

Thus, under anaerobic conditions, by far the major portion of the pyruvic acid is converted into lactic acid, which diffuses readily out of the cells into the extracellular fluids and even into the intracellular fluids of other less active cells. Therefore, lactic acid represents a type of "sinkhole" into which the glycolytic end products can disappear, thus allowing glycolysis to proceed far longer than would otherwise be possible. Indeed, glycolysis could proceed for only a few seconds without this conversion. Instead, it can proceed for several minutes, supplying the body with considerable

quantities of ATP even in the absence of respiratory oxygen.

Reconversion of Lactic Acid When Oxygen Becomes Available Again. When a person begins to breathe oxygen again after a period of anaerobic metabolism, the lactic acid is rapidly reconverted to pyruvic acid and NADH plus H^+. Large portions of these are immediately oxidized to form large quantities of ATP. This excess ATP then causes as much as three fourths of the remaining excess pyruvic acid to be converted back into glucose.

Thus, the great amount of lactic acid that forms during anaerobic glycolysis does not become lost from the body, for when oxygen is again available, the lactic acid either can be reconverted to glucose or can be used directly for energy. By far the greater portion of this reconversion occurs in the liver, but a small amount can also occur in other tissues.

Use of Lactic Acid by the Heart for Energy. Heart muscle is especially capable of converting lactic acid to pyruvic acid and then utilizing it for energy. This occurs to a great extent in very heavy exercise, during which large amounts of lactic acid are released into the blood from the skeletal muscles and then are consumed as an extra energy source by the heart.

● RELEASE OF ENERGY FROM GLUCOSE BY THE PENTOSE PHOSPHATE PATHWAY

Though, in muscle, essentially all the carbohydrates utilized for energy are first degraded to pyruvic acid by glycolysis and then oxidized, this glycolytic schema is not the only means by which glucose can be degraded and then used to provide energy. A second important schema for breakdown and oxidation of glucose is called the *pentose phosphate pathway* (or *phosphogluconate pathway*), which is responsible for as much as 30 per cent of the glucose breakdown *in the liver and even more than this in fat cells.* It is especially important because it can provide energy independently of all the enzymes of the citric acid cycle and therefore is an alternate pathway for energy metabolism when some enzymatic abnormalities occur in cells. Also, it has a special capacity for providing energy to multiple cellular synthetic processes, as we shall see.

Release of Carbon Dioxide and Hydrogen by Means of the Pentose Phosphate Pathway. Figure 67–7 illustrates most of the basic chemical reactions in the pentose phosphate pathway. It shows that glucose, after several stages of conversion, releases one molecule of carbon dioxide and four atoms of hydrogen, with resultant formation of a five-carbon sugar, D-ribulose. This substance in turn can change progressively into several other five-, four-, seven-, and three-carbon sugars. Finally, various combinations of these sugars can resynthesize glucose. However, *only five molecules of glucose are resynthesized for every six molecules of glucose that initially enter into the reactions.* That is, the pentose phosphate pathway is a cyclic process in which one molecule of glucose is metabolized for each revolution of the cycle. Thus, by revolution of the cycle again and again, all the glucose can eventually be converted into carbon dioxide and hydrogen, and the hydrogen in turn can enter the oxidative phosphorylation pathway to form ATP, or more often it is used for synthesis of fat or other substances as follows:

Use of Hydrogen to Synthesize Fat; The Function of Nicotinamide Adenine Dinucleotide Phosphate ($NADP^+$). The hydrogen released during the pentose phos-

Glucose   6-phosphate

6-Phosphoglucono-δ-lactone

6-Phosphogluconic acid

3-Keto-6-phosphogluconic acid

D-Ribulose 5-phosphate

{ D-Xylulose 5-phosphate
  +
  D-Ribose 5-phosphate }

{ D-Sedoheptulose 7-phosphate
  +
  D-Glyceraldehyde 3-phosphate }

{ Fructose 6-phosphate
  +
  Erythrose 4-phosphate }

Net reaction:

$$Glucose + 12NADP^+ + 6H_2O \rightarrow 6CO_2 + 12H + 12NADPH$$

Figure 67–7. The pentose phosphate pathway for glucose metabolism.

phate cycle does not combine with NAD^+ as in the glycolytic pathway, but combines with $NADP^+$, which is almost identical to NAD^+ except for an extra phosphate radical. Yet this difference is extremely significant because only hydrogen bound with $NADP^+$ in the form of NADPH can be used for synthesis of fats from carbohydrates, which is discussed in the following chapter, as well as synthesis of some other substances. When glycolysis becomes slowed because of cellular inactivity, the pentose phosphate pathway still remains operative (mainly in the liver) to break down any excess glucose that continues to be transported into the cells, and NADPH becomes abundant to help convert acetyl radicals into long fatty acid chains. This is another way in which the stored energy in the glucose molecule is used besides the formation of ATP.

GLUCOSE CONVERSION TO GLYCOGEN OR FAT

When glucose is not immediately required for energy, the extra glucose that continually enters the cells either is stored as glycogen or is converted into fat. Glucose is preferentially stored as glycogen until the cells have stored as much glycogen as they can—an amount sufficient to supply the energy needs of the body for only 12 to 24 hours. When the cells (primarily liver and muscle cells) approach saturation with glycogen, the additional glucose is then converted into fat in the liver and in fat cells and then is stored in the fat cells. Other steps in the chemistry of this conversion are discussed in the following chapter.

● FORMATION OF CARBOHYDRATES FROM PROTEINS AND FATS— "GLUCONEOGENESIS"

When the body's stores of carbohydrates decrease below normal, moderate quantities of glucose can be formed from *amino acids* and from the *glycerol* portion of fat. This pro-

cess is called *gluconeogenesis*. Approximately 60 per cent of the amino acids in the body proteins can be converted easily into carbohydrates; the remaining 40 per cent have chemical configurations that make this difficult. Each amino acid is converted into glucose by a slightly different chemical process. For instance, alanine can be converted directly into pyruvic acid simply by deamination; the pyruvic acid then is converted into glucose or stored glycogen. Several of the more complicated amino acids can be converted into different sugars containing three-, four-, five-, or seven-carbon atoms; these can then enter the phosphogluconate pathway and eventually form glucose. Thus, by means of deamination plus several simple interconversions, many of the amino acids become glucose. Similar interconversions can change glycerol into carbohydrates.

Regulation of Gluconeogenesis. Diminished carbohydrates in the cells and decreased blood sugar are the basic stimuli that set off an increase in the rate of gluconeogenesis. The diminished carbohydrates can directly cause reversal of many of the glycolytic and phosphogluconate reactions, thus allowing conversion of deaminated amino acids and glycerol into carbohydrates. However, in addition, cortisol, one of the hormones, is especially important in this regulation, as follows:

Effect of Corticotropin and Glucocorticoids on Gluconeogenesis. When normal quantities of carbohydrates are not available to the cells, the adenohypophysis, for reasons not completely understood at present, begins to secrete increased quantities of *corticotropin*, which stimulate the adrenal cortex to produce large quantities of *glucocorticoid hormones,* especially *cortisol*. In turn, cortisol mobilizes proteins from essentially all cells of the body, making these available in the form of amino acids in the body fluids. A high proportion of these immediately become deaminated in the liver and provide ideal substrates for conversion into glucose. Thus, one of the most important means by which gluconeogenesis is promoted is through the release of glucocorticoids from the adrenal cortex.

● BLOOD GLUCOSE

The normal blood glucose concentration in a person who has not eaten a meal within the past 3 to 4 hours is approximately 90 mg/dl. After a meal containing large amounts of carbohydrates, this rarely rises above 140 mg/dl unless the person has diabetes mellitus, which will be discussed in Chapter 78.

The regulation of blood glucose concentration is intimately related to the hormones insulin and glucagon; this subject will be discussed in detail in Chapter 78 in relation to the functions of these hormones.

REFERENCES

Brazy, P. C., and Mandel, L. J.: Does availability of inorganic phosphate regulate cellular oxidative metabolism? News Physiol. Sci., 1:100, 1986.

Butler, T. M., and Davies, R. E.: High-energy phosphates in smooth muscle. In Bohr, D. F., et al. (eds.): Handbook of Physiology. Sec. 2, Vol. II. Baltimore, Williams & Wilkins, 1980, p. 237.

Cornish-Bowden, A.: Fundamentals of Enzyme Kinetics. Boston, Butterworths, 1979.

Dickerson, R. E.: Cytochrome C and the evolution of energy metabolism. Sci. Am., 242(3):136, 1980.

Felig, P.: Disorders of carbohydrate metabolism. In Bondy, P. K., and Rosenberg, L. E. (eds.): Metabolic Control and Disease, 8th Ed. Philadelphia, W. B. Saunders Co., 1980, p. 276.

Foster, R. L.: The Nature of Enzymology. New York, John Wiley & Sons, 1979.

Friedmann, H. C. (ed.): Enzymes. Stroudsburg, Pa., Dowden, Hutchinson & Ross, 1980.

Frohman, L. A.: CNS peptides and glucoregulation. Annu. Rev. Physiol., 45:95, 1983.

Golinick, P. D.: Metabolism of substrates: Energy substrate metabolism during exercise and as modified by training. Fed. Proc., 44:353, 1985.

Hems, D. A., and Whitton, P. D.: Control of hepatic glycogenolysis. Physiol. Rev., 60:1, 1980.

Hetenyi, G., Jr., et al.: Turnover and precursor-product relationships of non-lipid metabolites. Physiol. Rev., 63:606, 1983.

Homsher, E., and Kean, C. J.: Skeletal muscle energetics and metabolism. Annu. Rev. Physiol., 40:93, 1978.

Howell, S. L., and Tyhurst, M.: Insulin secretion: The effector system. Experimentia, 40:1098, 1984.

Jacquez, J. A.: Red blood cell as glucose carrier: Significance for placental and cerebral glucose transfer. Am. J. Physiol., 246:R289, 1984.

Jequier, E., and Flatt, J.-P.: Recent advances in human energetics. News Physiol. Sci., 1:112, 1986.

Kraus-Friedmann, N.: Hormonal regulation of hepatic gluconeogenesis. Physiol. Rev., 64:170, 1984.

Krebs, H. A.: The tricarboxylic acid cycle. Harvey Lectures, 44:165, 1948–1949.

Lehninger, A. L.: Principles of Biochemistry. New York, Worth Publishers, Inc., 1982.

Lemasters, J. J., et al. (eds.): Integration of Mitochondrial Function. New York, Plenum Publishing Corp., 1988.

Lund-Andersen, H.: Transport of glucose from blood to brain. Physiol. Rev., 549:305, 1979.

McGilvery, R. W.: Biochemistry: A Functional Approach. Philadelphia, W. B. Saunders Co., 1983.

Oomura, Y., and Yoshimatsu, H.: Neural network of glucose monitoring system. J. Auton. Nerv. Syst., 10:359, 1984.

Sairam, M. R.: Role of carbohydrates in glycoprotein hormone signal transduction. FASEB J., 3:1915, 1989.

Senior, A. E.: ATP synthesis by oxidative phosphorylation. Physiol. Rev., 68:177, 1988.

Storlien, L. H.: The role of the ventromedial hypothalamic area in periprandial glucoregulation. Life-Sci., 36:505, 1985.

Stryer, L.: Biochemistry. New York, W. H. Freeman Co., 1988.

Wang, J. H.: Coupling of proton flux to the hydrolysis and synthesis of ATP. Annu. Rev. Biophys. Bioeng., 12:21, 1983.

CHAPTER 68

• *Lipid Metabolism* •

A number of different chemical compounds in the food and in the body are classified as *lipids*. These include (1) *neutral fat,* known also as *triglycerides,* (2) the *phospholipids,* (3) *cholesterol,* and (4) a few others of less importance. Chemically, the basic lipid moiety of the triglycerides and the phospholipids is *fatty acids,* which are simply long chain hydrocarbon organic acids. A typical fatty acid, palmitic acid, is the following:

$$CH_3(CH_2)_{14}COOH$$

Although cholesterol does not contain fatty acid, its sterol nucleus is synthesized from degradation products of fatty acid molecules, thus giving it many of the physical and chemical properties of other lipid substances.

The triglycerides are used in the body mainly to provide energy for the different metabolic processes; this function they share almost equally with the carbohydrates. However, some lipids, especially cholesterol, the phospholipids, and derivatives of these, are used throughout the body to perform other cellular functions.

Basic Chemical Structure of Triglycerides (Neutral Fat). Because most of this chapter deals with utilization of triglycerides for energy, the following basic structure of the triglyceride molecule must be understood:

$$CH_3-(CH_2)_{16}-COO-CH_2$$
$$CH_3-(CH_2)_{16}-COO-CH$$
$$CH_3-(CH_2)_{16}-COO-CH_2$$

Tristearin

Note that three long chain fatty acid molecules are bound with one molecule of glycerol. In the human body, the three fatty acids most commonly present in the triglycerides are (1) *stearic acid* (illustrated above), which has an 18-carbon chain and is fully saturated with hydrogen atoms; (2) *oleic acid,* which also has an 18-carbon chain but has one double bond in the middle of the chain; and (3) *palmitic acid,* which has 16 carbon atoms and is fully saturated.

● TRANSPORT OF LIPIDS IN THE BODY FLUIDS

Transport of Triglycerides and Other Lipids by the Lymph From the Gastrointestinal Tract — the Chylomicrons

It will be recalled from Chapter 65 that almost all the fats of the diet are absorbed from the intestines into the lymph.

During digestion, most triglycerides are split into monoglycerides and fatty acids. Then, while passing through the intestinal epithelial cells, these are resynthesized into new molecules of triglycerides that aggregate and enter the lymph as minute, dispersed droplets called *chylomicrons,* having sizes between 0.08 and 0.5 micrometer. A small amount of the protein apoprotein B adsorbs to the outer surfaces of the chylomicrons; this increases their suspension stability in the fluid of the lymph and prevents their adherence to the lymphatic vessel walls.

Most of the cholesterol and phospholipids absorbed from the gastrointestinal tract, as well as small amounts of phospholipids that are continually synthesized by the intestinal mucosa, also enter the chylomicrons. Thus, chylomicrons are composed principally of triglycerides, but they also contain approximately 9 per cent phospholipids, 3 per cent cholesterol, and 1 per cent apoprotein B as well. The chylomicrons are then transported up the thoracic duct and emptied into the venous blood at the juncture of the jugular and subclavian veins.

Removal of the Chylomicrons From the Blood

An hour or so after a meal that contains large quantities of fat, the chylomicron concentration in the plasma may rise to as high as 1 to 2 per cent; and because of the large sizes of the chylomicrons, the plasma appears turbid and sometimes yellow. However, the chylomicrons have a half-life of less than an hour, so that the plasma becomes clear again within a few hours. The fat of the chylomicrons is removed mainly in the following way:

Hydrolysis of the Chylomicron Triglycerides by Lipoprotein Lipase; Fat Storage in the Fat and Liver Cells. Most of the chylomicrons are removed from the circulating blood as they pass through the capillaries of adipose tissue and the liver. Both adipose tissue and the liver contain large quantities of the enzyme called *lipoprotein lipase.* This enzyme is active in the capillary endothelium, where it hydrolyzes the triglycerides of chylomicrons that stick to the endothelial wall, releasing fatty acids and glycerol. The fatty acids, being highly miscible with the membranes of the cells, immediately diffuse into the fat cells of adipose tissue and the liver cells. Once within these cells, the fatty acids are resynthesized into triglycerides, new glycerol being supplied by the metabolic processes of the cells, as is discussed later in the chapter. The lipase also causes hydrolysis of phospholipids, this, too, releasing fatty acids to be stored in the cells in the same way. Thus, most of the mass of the chylomicrons is removed from the circulating blood; then the remnants are taken up mainly by the liver.

TRANSPORT OF FATTY ACIDS IN THE BLOOD IN COMBINATION WITH ALBUMIN —"FREE FATTY ACID"

When the fat that has been stored in the adipose tissue is to be used elsewhere in the body, usually to provide energy, it must first be transported to the other tissues. It is transported mainly in the form of *free fatty acid.* This is achieved by hydrolysis of the triglycerides back into fatty acids and glycerol. At least two classes of stimuli play important roles in promoting this hydrolysis. First, when the availability of glucose to the fat cell falls too low, one of its breakdown products, α-glycerophosphate, also becomes too low. This substance is required to form the glycerol portion of newly synthesized triglycerides, and in its absence the equilibrium shifts in favor of hydrolysis. Second, a *hormone-sensitive cellular lipase* can be activated by many different hormones, and this promotes rapid hydrolysis of the triglyceride. This also is discussed later in the chapter.

On leaving the fat cells, the fatty acids ionize strongly in the plasma and immediately combine with albumin molecules of the plasma proteins. The fatty acid bound in this manner is called *free fatty acid* or *nonesterified fatty acid* (or simply *FFA* or *NEFA*) to distinguish it from other fatty acids in the plasma that exist in the form of esters of glycerol, cholesterol, or other substances.

The concentration of free fatty acid in the plasma under resting conditions is about 15 mg/dl, which is a total of only 0.45 gram of fatty acids in the entire circulatory system. Yet, strangely enough, even this small amount accounts for almost all the transport of fatty acids from one part of the body to another for the following reasons:

(1) Despite the minute amount of free fatty acid in the blood, its rate of "turnover" is extremely rapid; *half the plasma fatty acid is replaced by new fatty acid every 2 to 3 minutes.* One can calculate that at this rate almost all the normal energy requirements of the body can be provided by oxidation of the transported free fatty acid without using any carbohydrates or proteins for energy. (2) All conditions that increase the rate of utilization of fat for cellular energy also increase the free fatty acid concentration in the blood; this concentration sometimes increases as much as five- to eightfold. This increase occurs especially in starvation and in diabetes, in both of which conditions the person derives little or no energy from carbohydrates.

Under normal conditions, about 3 molecules of fatty acid combine with each molecule of albumin, but as many as 30 fatty acid molecules can combine with a single albumin molecule when the need for fatty acid transport is extreme. This shows how variable the rate of lipid transport can be with different physiological needs.

THE LIPOPROTEINS—THEIR SPECIAL FUNCTION IN TRANSPORTING CHOLESTEROL AND PHOSPHOLIPIDS

In the postabsorptive state — that is, after all of the chylomicrons have been removed from the blood — over 95 per cent of all the lipids in the plasma are in the form of *lipoproteins,* which are small particles much smaller than chylomicrons but qualitatively similar in composition, containing *triglycerides, cholesterol, phospholipids,* and *protein.* The protein averages about one fourth to one third of the total constituents, and the remainder is lipids. The total concentration of lipoproteins in the plasma averages about 700 mg/dl, and this can be broken down into the following average concentrations of the individual constituents:

	mg/dl of plasma
Cholesterol	180
Phopholipids	160
Triglycerides	160
Protein	200

Types of Lipoproteins. Aside from the chylomicrons, which are themselves very large lipoproteins, there are four other major classes of lipoproteins classified by their densities as measured in the ultracentrifuge: (1) *very low density lipoproteins,* which contain high concentrations of triglycerides and moderate concentrations of both cholesterol and phospholipids; (2) *intermediate density lipoproteins,* which are very low density lipoproteins from which a large share of the triglycerides has been removed so that the concentrations of cholesterol and phospholipids are increased; (3) *low density lipoproteins,* which are intermediate density lipoproteins from which almost all of the triglycerides have been removed, leaving an especially high concentration of cholesterol and moderately high concentration of phospholipids; and (4) *high density lipoproteins,* which contain a very high concentration of protein, about 50 per cent, but smaller concentrations of cholesterol and phospholipids.

Formation and Function of the Lipoproteins. Almost all of the lipoproteins are formed in the liver, which is where most of the plasma cholesterol, phospholipids, and triglycerides (except those absorbed from the intestines in the chylomicrons) are synthesized. However, small quantities of high density lipoproteins are also synthesized in the intestinal epithelium during the absorption of fatty acids from the intestines.

The primary function of the lipoproteins is to transport their lipid components in the blood. The very low density lipoproteins transport triglycerides synthesized in the liver mainly to the adipose tissue, whereas the other lipoproteins are especially important in the different stages of phospholipid and cholesterol transport from the liver to the peripheral tissues or from the periphery back to the liver. Later in the chapter, we will discuss in detail special problems of cholesterol transport in relation to the disease *atherosclerosis.*

● THE FAT DEPOSITS

ADIPOSE TISSUE

Large quantities of fat are stored in two major tissues of the body, the adipose tissue and the liver. The adipose tissue is usually called the *fat deposits,* or simply the *fat depots.*

The major function of adipose tissue is storage of triglycerides until these are needed to provide energy elsewhere in the body. However, a subsidiary function is to provide heat insulation for the body, as will be discussed in Chapter 73.

The Fat Cells. The fat cells of adipose tissue are modified fibroblasts that are capable of storing almost pure triglycerides in quantities equal to 80 to 95 per cent of their volume. The triglycerides are generally in a liquid form, and when the tissues of the skin are exposed to prolonged cold, the fatty acid chains of the triglycerides, over a period of weeks, become either shorter or more unsaturated to decrease their melting point, thereby always allowing the fat to remain in a liquid state. This is particularly important because only liquid fat can be hydrolyzed and transported from the cells.

Fat cells can synthesize very small amounts of fatty acids and triglycerides from carbohydrates, this function supplementing the synthesis of fat in the liver, as discussed later in the chapter.

Exchange of Fat Between the Adipose Tissue and the Blood — Tissue Lipases. As discussed above, large quantities of lipases are present in adipose tissue. Some of these enzymes catalyze the deposition of triglycerides from the chylomicrons and very low density lipoproteins. Others, when activated by hormones, cause splitting of the triglycerides of the fat cells to release free fatty acids. Because of rapid exchanges of the fatty acids, the triglycerides in the fat cells are renewed approximately once every 2 to 3 weeks, which means that the fat stored in the tissues today is not the same fat that was stored last month, thus emphasizing the dynamic state of the storage fat.

THE LIVER LIPIDS

The principal functions of the liver in lipid metabolism are (1) to degrade fatty acids into small compounds that can be used for energy, (2) to synthesize triglycerides, mainly from carbohydrates but to a lesser extent from proteins as well, and (3) to synthesize other lipids from fatty acids, especially cholesterol and phospholipids.

Large quantities of triglycerides appear in the liver (a) during starvation, (b) in diabetes mellitus, or (c) in any other condition in which fat is being utilized rapidly for energy. In these conditions, large quantities of triglycerides are mobilized from the adipose tissue, transported as free fatty acids in the blood, and then redeposited as triglycerides in the liver, where the initial stages of much of the fat degradation begin. Thus, under normal physiological conditions the total amount of triglycerides in the liver is controlled to a great extent by the overall rate at which lipids are being utilized for energy.

The liver cells, in addition to containing triglycerides, contain large quantities of phospholipids and cholesterol, which are continually synthesized by the liver. Also, the liver cells are much more capable than other tissues of desaturating fatty acids so that the liver triglycerides normally are much more unsaturated than the triglycerides of the adipose tissue. This capability of the liver to desaturate fatty acids is functionally important to all the tissues of the body, because many of the structural members of all cells contain reasonable quantities of unsaturated fats, and their principal source is the liver. This desaturation is accomplished by a dehydrogenase in the liver cells.

● USE OF TRIGLYCERIDES FOR ENERGY, AND FORMATION OF ADENOSINE TRIPHOSPHATE (ATP)

Approximately 40 to 45 per cent of the calories in the normal American diet are derived from fats, which is about equal to the calories derived from carbohydrates. Therefore, the use of fats by the body for energy is as important as the use of carbohydrates. In addition, much of the carbohydrates ingested with each meal is converted into triglycerides, then stored, and later utilized as triglycerides for energy.

Hydrolysis of the Triglycerides. The first stage in the utilization of triglycerides for energy is hydrolysis of these compounds into fatty acids and glycerol and subsequent transport of both these products to the active tissues where they will be oxidized to give energy. Almost all cells, with some degree of exception for brain tissue, can use fatty acids almost interchangeably with glucose for energy.

The glycerol, on entering the active tissue, is immediately changed by intracellular enzymes into *glycerol 3-phosphate*, which enters the glycolytic pathway for glucose breakdown and in this way is used for energy. However, before the fatty acids can be used for energy, they must be processed further in the following way:

Entry of Fatty Acids Into the Mitochondria. The degradation and oxidation of fatty acids occur only in the mitochondria. Therefore, the first step for utilization of the fatty acids is their transport into the mitochondria. This is a carrier-mediated process that employs *carnitine* as the carrier substance. Once inside the mitochondria, the fatty acid splits away from the carnitine and is then degraded and oxidized.

Degradation of Fatty Acid to Acetylcoenzyme A by Beta Oxidation. The fatty acid molecule is degraded in the mitochondria by progressive release of two-carbon segments in the form of acetylcoenzyme A (acetyl-CoA). This process, which is illustrated in Figure 68–1, is called the *beta oxidation* process for degradation of fatty acids. Let us summarize the results of the equations in this figure:

1. The beta carbon of the fatty acid molecule is oxidized by adding a water molecule, then removing two hydrogen ions, leaving the oxygen from the water bound with the beta carbon.

2. The molecule then splits between the alpha and beta carbons, releasing one molecule of acetyl-CoA. Thus, the fatty acid molecule becomes two carbon atoms shorter.

3. Then the process repeats itself again and again until

(1) $RCH_2CH_2CH_2COOH + CoA + ATP \xrightleftharpoons{\text{Thiokinase}} RCH_2CH_2CH_2COCoA + AMP + \text{Pyrophosphate}$
 (Fatty acid) (Fatty acyl CoA)

(2) $RCH_2CH_2CH_2COCoA + FAD \xrightarrow{\text{Acyl dehydrogenase}} RCH_2CH=CHCOCoA + FADH_2$
 (Fatty acyl CoA)

(3) $RCH_2CH=CHCOCoA + H_2O \xrightleftharpoons{\text{Enoyl hydrase}} RCH_2CHOHCH_2COCoA$

(4) $RCH_2CHOHCH_2COCoA + NAD^+ \xrightleftharpoons[\text{Dehydrogenase}]{\beta\text{-Hydroxyacyl}} RCH_2COCH_2COCoA + NADH + H^+$

(5) $RCH_2COCH_2COCoA + CoA \xrightarrow{\text{Thiolase}} RCH_2COCoA + CH_3COCoA$
 (Fatty acyl CoA)(Acetyl CoA)

Figure 68–1. Beta oxidation of fatty acids to yield acetylcoenzyme A.

the entire fatty acid chain has been split into multiple acetyl-CoA molecules.

4. For each acetyl-CoA molecule split from the fatty acid, a total of four hydrogen atoms is released. These are oxidized in the mitochondria to form large amounts of ATP, as we shall discuss subsequently.

Oxidation of Acetyl-CoA. The acetyl-CoA molecules formed by beta oxidation of fatty acids enter into the *citric acid cycle* as explained in the preceding chapter, combining first with oxaloacetic acid to form citric acid, which then is degraded into carbon dioxide and hydrogen atoms. The hydrogen is subsequently oxidized by the *chemiosmotic oxidative system of the mitochondria,* which is explained in the preceding chapter. The net reaction for each molecule of acetyl-CoA is the following:

$$CH_3COCo\text{-}A + Oxaloacetic\ acid + 3H_2O + ADP$$

$$\xrightarrow{\textit{Citric acid cycle}}$$

$$2CO_2 + 8H + HCo\text{-}A + ATP + Oxaloacetic\ acid$$

Thus, after the initial degradation of fatty acids to acetyl-CoA, their final breakdown is precisely the same as that of the acetyl-CoA formed from pyruvic acid during the metabolism of glucose. Then the hydrogen is oxidized by the same *chemiosmotic oxidative system of the mitochondria* that is employed in carbohydrate oxidation, liberating tremendous amounts of ATP.

ATP Formed by Oxidation of Fatty Acid. In Figure 68–1 note also that four hydrogen atoms are released each time a molecule of acetyl-CoA is formed from the fatty acid chain. Therefore, for every stearic acid molecule that is split, a total of 32 hydrogen atoms is removed. In addition, for each acetyl-CoA degraded by the citric acid cycle, 8 hydrogen atoms are removed, making an additional 72 hydrogens for each molecule of stearic acid metabolized. This, added to the above 32 hydrogen atoms, makes a total of 104 hydrogen atoms. Of this group 34 are removed from the degrading fatty acid by flavoproteins and 70 are removed by NAD^+ as NADH and H^+. These two groups of hydrogen atoms are oxidized in the mitochondria, as discussed in the preceding chapter, but they enter the oxidative system at different points, so that up to 1 molecule of ATP is synthesized for each of the 34 flavoprotein hydrogens and up to 1.5 molecule of ATP is synthesized for each of the 70 NADH and H^+ hydrogens. This makes 34 plus 105, or a total of 139 molecules of ATP formed by the oxidation of hydrogen derived from each molecule of stearic acid. And another nine molecules of ATP are formed in the citric acid cycle, one for each of the nine acetyl-CoA molecules metabolized. Thus, a total of 148 molecules of ATP is formed during the complete oxidation of one molecule of stearic acid. However, two high energy bonds are consumed in the initial combination of coenzyme A with the fatty acid molecule, making a net gain of 146 molecules of ATP.

FORMATION OF ACETOACETIC ACID IN THE LIVER AND ITS TRANSPORT IN THE BLOOD

A large share of the initial degradation of fatty acids occurs in the liver, especially when excessive amounts of lipids are being mobilized for energy. However, the liver uses only a small proportion of the fatty acids for its own intrinsic metabolic processes. Instead, when the fatty acid

chains have been split into acetyl-CoA, two molecules of acetyl-CoA condense to form one molecule of acetoacetic acid, which is then transported in the blood to the other cells throughout the body, where it is used for energy. The chemical processes are the following:

$$2CH_3COCo\text{-}A + H_2O \underset{\text{other cells}}{\overset{\text{liver cells}}{\rightleftharpoons}}$$
$$\text{\scriptsize Acetyl-CoA}$$

$$CH_3COCH_2COOH + 2HCo\text{-}A$$
$$\text{\scriptsize Acetoacetic acid}$$

A large part of the acetoacetic acid is also converted into *β-hydroxybutyric acid* and minute quantities into *acetone* in accord with the following reactions:

The acetoacetic acid and β-hydroxybutyric acid diffuse freely through the liver cell membranes and are transported by the blood to the peripheral tissues. Here they again diffuse into the cells, where reverse reactions occur and acetyl-CoA molecules are formed. These in turn enter the citric acid cycle and are oxidized for energy, as explained.

Normally, the acetoacetic acid and β-hydroxybutyric acid that enter the blood are transported so rapidly to the tissues that their combined concentration in the plasma rarely rises above 3 mg/dl. Yet despite the small quantities in the blood, large amounts are actually transported; this is analogous to the high rate of free fatty acid transport. The rapid transport of both these substances depends on their high solubility in the membranes of the target cells, which allows rapid diffusion through the cell membranes.

Ketosis and Its Occurrence in Starvation, Diabetes, and Other Diseases

The concentrations of acetoacetic acid, β-hydroxybutyric acid, and acetone occasionally rise very high in the blood and interstitial fluids; this condition is called *ketosis* because acetoacetic acid is a keto acid, and the three compounds are called *ketone bodies.* Ketosis occurs especially in starvation, in diabetes mellitus, and sometimes even when a person's diet is composed almost entirely of fat. In all these states, essentially no carbohydrates are metabolized—in starvation and following a high fat diet because carbohydrates are not available and in diabetes because insulin is not available to cause glucose transport into the cells.

When carbohydrates are not utilized for energy, almost all the energy of the body must come from metabolism of fats. We see later in the chapter that lack of availability of carbohydrates automatically increases the rate of removal of fatty acids from adipose tissues; in addition, several hormonal factors—such as increased secretion of corticotropin by the adenohypophysis, increased secretion of glucocorticoids by the adrenal cortex, and decreased secretion of insulin by the pancreas—further enhance the removal of fatty acids from the fat tissues. As a result, tremendous quantities of fatty acids become available to the liver for degradation.

The ketone bodies in turn pour out of the liver to be carried to the cells. Yet the cells are limited in the amount of ketone bodies that can be oxidized for several reasons, the most important of which is the following: One of the products of carbohydrate metabolism is the *oxaloacetate* that is required to bind with acetyl-CoA before it can be processed in the citric acid cycle. Therefore, deficiency of oxaloacetate limits the entry of acetyl-CoA into the citric acid cycles, and because of the simultaneous outpouring of tremendous quantities of acetoacetic acid and the other ketone bodies from the liver, the blood concentration of acetoacetic acid and β-hydroxybutyric acid sometimes rises to as high as 20 to 30 times normal, thus leading to extreme acidosis, as explained in Chapter 30.

The acetone that is formed during ketosis is a volatile substance some of which is blown off in small quantities in the expired air of the lungs, often giving the breath an acetone smell. This smell is frequently used as a diagnostic criterion of ketosis.

Adaptation to a High Fat Diet. Upon changing slowly from a carbohydrate diet to an almost completely fat diet, a person's body adapts to the utilization of far more acetoacetic acid than usual, and, in this instance, ketosis does not occur. For instance, the Eskimos, who sometimes live almost entirely on a fat diet, do not develop ketosis. Undoubtedly several factors enhance the rate of acetoacetic acid metabolism by the cells. Even the brain cells, after a few weeks, can then derive as much as 50 to 75 per cent of their energy from fats.

SYNTHESIS OF TRIGLYCERIDES FROM CARBOHYDRATES

Whenever a greater quantity of carbohydrates enters the body than can be used immediately for energy or stored in the form of glycogen, the excess is rapidly converted into triglycerides and stored in this form in the adipose tissue. In human beings, most triglyceride synthesis occurs in the liver, but small quantities are also synthesized in the adipose tissue. The triglycerides formed in the liver are transported at least partly by the very low density lipoproteins to the adipose tissue.

Conversion of Acetyl-CoA Into Fatty Acids. The first step in the synthesis of triglycerides is conversion of carbohydrates into acetyl-CoA. It will be recalled from the preceding chapter that this occurs during the normal degradation of glucose by the glycolytic system. It will also be

remembered from earlier in this chapter that fatty acids are actually large polymers of acetic acid. Therefore, it is easy to understand how acetyl-CoA could be coverted into fatty acids.

However, synthesis of the fatty acids from acetyl-CoA is not achieved by simply reversing the oxidative degradation that was described above. Instead, this occurs by the two-step process shown in Figure 68–2, using *malonyl CoA* and NADPH as the principal intermediates in the polymerization process.

Combination of Fatty Acids With α-Glycerophosphate to Form Triglycerides. Once the synthesized fatty acid chains have grown to contain 14 to 18 carbon atoms, they then become bound to glycerol to form triglycerides. The enzymes that cause this conversion are highly specific for fatty acids with chain lengths of 14 carbon atoms or greater, a factor that actually controls the physical quality of the triglycerides stored in the body.

As illustrated in Figure 68–3, the glycerol portion of the triglyceride is furnished by α-glycerophosphate, which is a product derived from the glycolytic scheme of glucose degradation. The mechanism of this was discussed in Chapter 67.

Efficiency of Carbohydrate Conversion Into Fat. During triglyceride synthesis, only about 15 per cent of the original energy in the glucose is lost in the form of heat, while the remaining 85 per cent is transferred to the stored triglycerides.

Importance of Fat Synthesis and Storage. Fat synthesis from carbohydrates is especially important for two reasons: (1) The ability of the different cells of the body to store carbohydrates in the form of glycogen is generally slight; a maximum of only a few hundred grams of glycogen can be stored in the liver, the skeletal muscles, and all other tissues of the body put together. In contrast, many kilograms of fat can be stored. Therefore, fat synthesis provides a means by which the energy of excess ingested carbohydrates (and proteins, too) can be stored for later use. Indeed, the average person has almost 150 times as much energy stored in the form of fat as stored in the form of carbohydrate. (2) Each gram of fat contains 2.25 times as many calories of energy as each gram of glycogen. Therefore, for a given weight gain, a person can store far more energy in the form of fat than in the form of carbohydrate, which is exceedingly important when an animal must be highly motile to survive.

Failure to Synthesize Fats From Carbohydrates in the Absence of Insulin. When no insulin is available, as occurs in very serious diabetes mellitus, fats are poorly, if at all, synthesized, for the following reasons: First, when insulin is not available, glucose does not enter the fat and liver

Step 1:

$$CH_3COCoA + CO_2 + ATP$$

$\Updownarrow$ (Acetyl CoA carboxylase)

$$\begin{array}{l} COOH \\ || \\ CH_2 \\ | \\ O=C-CoA \end{array} + ADP + PO_4^{---}$$

Malonyl CoA

Step 2:

1 Acetyl CoA + 8 Malonyl CoA + 16 NADPH + 16H$^+$ $\longrightarrow$

1 Stearic acid + 8CO$_2$ + 9CoA + 16NADP$^+$ + 7H$_2$O

Figure 68–2. Synthesis of fatty acids.

Figure 68–3. An overall schema for synthesis of triglycerides from glucose.

cells satisfactorily, so that little of the acetyl-CoA and NADPH needed for fat synthesis can be derived from glucose. Second, lack of glucose in the fat cells greatly reduces the availability of α-glycerophosphate, which also makes it difficult for the tissues to form triglycerides.

SYNTHESIS OF TRIGLYCERIDES FROM PROTEINS

Many amino acids can be converted into acetyl-CoA, as is discussed in the following chapter. Obviously, this can be synthesized into triglycerides. Therefore, when people have more proteins in their diets than their tissues can use as proteins, a large share of the excess is stored as fat.

● REGULATION OF ENERGY RELEASE FROM TRIGLYCERIDES

Fat-Sparing Effect of Carbohydrates; Control of Fat Synthesis From Carbohydrates. When excess quantities of carbohydrates are available in the body, carbohydrates are utilized preferentially over triglycerides for energy. There are several reasons for this "fat-sparing" effect of carbohydrate. One of the most important is the following: The fats in adipose tissue cells are present in two different forms, the stored triglycerides and small quantities of free fatty acids. These are in constant equilibrium with each other. When excess quantities of α-glycerophosphate are present, the equilibrium between free fatty acids and triglycerides shifts toward the stored triglycerides; as a result, only minute quantities of fatty acids are then available to be utilized for energy. Since α-glycerophosphate is an important product of glucose metabolism, the availability of large amounts of glucose automatically inhibits the use of fatty acids for energy.

Second, when carbohydrates are available in excess, fatty acids are synthesized more rapidly than they are degraded. This effect is caused partially by the large quantities of acetyl-CoA formed from the carbohydrates and by the low concentration of free fatty acids in the adipose tissue, thus creating conditions appropriate for conversion of acetyl-CoA into fatty acids. However, an even more important effect that promotes conversion of carbohydrates to fats is the following: The first step, and the rate-limiting step, in the synthesis of fatty acids is carboxylation of acetyl-CoA to form malonyl-CoA, as was discussed previously. The rate of this reaction is controlled primarily by the enzyme *acetyl-CoA carboxylase,* the activity of which is accelerated in the presence of the intermediates of the citric acid cycle. When excess carbohydrates are being utilized, these intermediates increase, thus automatically causing increased synthesis of fatty acids. Thus, an excess of carbohydrates in the diet not only acts as a fat-sparer but also increases the fat in the fat stores.

Acceleration of Fat Utilization for Energy in the Absence of Carbohydrates. All the fat-sparing effects of carbohydrates are lost and are actually reversed when carbohydrates are not available or are available in short supply. Therefore, the equilibrium now shifts in the opposite direction, and fat is mobilized from the adipose cells and used for energy in place of the absent carbohydrates.

But also important are several hormonal changes that take place to promote rapid fatty acid mobilization from adipose tissue. Among the most important of these is a marked decrease in pancreatic secretion of insulin caused by the absence of carbohydrates. This not only reduces the rate of glucose utilization by the tissues but also decreases fat synthesis, which further shifts the equilibrium in favor of fat utilization in place of carbohydrates.

Hormonal Regulation of Fat Utilization. At least seven of the hormones secreted by the endocrine glands have significant or even very large effects on fat utilization. In addition to the important effect of *insulin lack* discussed earlier, some other important hormonal effects on fat metabolism are listed here.

Probably the most dramatic increase that occurs in fat utilization is that observed during heavy exercise. This results almost entirely from release of *epinephrine* and *norepinephrine* by the adrenal medullae during exercise, as a result of sympathetic stimulation. These two hormones directly activate *hormone-sensitive triglyceride lipase* that is present in abundance in the fat cells, and this causes very rapid breakdown of triglycerides and mobilization of fatty acids. Sometimes the free fatty acid concentration in the blood rises as much as eightfold, and the use of these fatty acids by the muscles for energy is correspondingly increased. Other types of stress that activate the sympathetic nervous system can also increase fatty acid mobilization and utilization in a similar manner.

Stress also causes large quantities of *corticotropin* to be released by the anterior pituitary gland, and this, in turn, causes the adrenal cortex to secrete excessive quantities of *glucocorticoids*. Both the corticotropin and glucocorticoids activate either the same hormone-sensitive triglyceride lipase as that activated by epinephrine and norepinephrine or a similar lipase. Therefore, this is still another mechanism for increasing the release of fatty acids from fat tissue. When corticotropin and glucocorticoids are secreted in excessive amounts for long periods of time, as occurs in the endocrine disease called Cushing's disease, fats are frequently mobilized to such a great extent that ketosis results. Corticotropin and glucocorticoids are then said to have a *ketogenic effect.*

Growth hormone has an effect similar to but less than that of corticotropin and glucocorticoids in activating the hormone-sensitive lipase. Therefore, growth hormone can also have a mild ketogenic effect.

Finally, *thyroid hormone* causes rapid mobilization of fat, which is believed to result indirectly from an increased overall rate of energy metabolism in all cells of the body under the influence of this hormone. The resulting reduction in acetyl-CoA and other intermediates of both fat and carbohydrate metabolism in the cells would then be a stimulus to cause fat mobilization.

The effects of the different hormones on metabolism are discussed further in the chapters dealing with each of them.

OBESITY

Obesity means deposition of excess fat in the body. This subject is discussed in detail in relation to dietary balances in Chapter 71, but briefly it is caused by ingestion of greater amounts of food than can be utilized by the body for energy. The excess food, whether fats, carbohydrates, or proteins, is then stored as fat in the adipose tissue to be used later for energy.

Strains of rats have been found in which *hereditary obesity* occurs. In at least one of these, the obesity is caused by ineffective mobilization of fat from the adipose tissue by tissue lipase while synthesis and storage of fat continue normally. Obviously, such a one-way process causes progressive enhancement of the fat stores, resulting in severe obesity.

● PHOSPHOLIPIDS AND CHOLESTEROL

PHOSPHOLIPIDS

The three major types of body phospholipids are *lecithins, cephalins,* and *sphingomyelins,* typical examples of which are shown in Figure 68–4.

Phospholipids always contain one or more fatty acid molecules and one phosphoric acid radical, and they usually contain a nitrogenous base. Although the chemical structures of phospholipids are somewhat variant, their physical properties are similar, for they are all lipid soluble, are transported together in lipoproteins, and are utilized in similar ways throughout the body for various structural purposes such as for use in cell membranes and intracellular membranes.

Formation of Phospholipids. Phospholipids are formed in essentially all cells of the body, though certain cells have a special ability to form them. Probably 90 per cent are formed in the liver cells; reasonable quantities are also formed by the intestinal epithelial cells during lipid absorption from the gut.

The rate of phospholipid formation is governed to some extent by the usual factors that control the rate of fat metabolism, for when triglycerides are deposited in the liver, the rate of phospholipid formation increases. Also, certain specific chemical substances are needed for formation of some phospholipids. For instance, *choline,* either in the diet or synthesized in the body, is needed for the formation of lecithin because choline is the nitrogenous base of the lecithin molecule. Also, *inositol* is needed for the formation of some cephalins.

Specific Use of Phospholipids. Several isolated functions of the phospholipids are the following: (1) Phospholipids are an important constituent of lipoproteins and are essential for the formation and function of some of these; in

Figure 68–4. Typical phospholipids.

their absence serious abnormalities of cholesterol and other lipid transport can occur. (2) Thromboplastin, which is needed to initiate the clotting process, is composed mainly of one of the cephalins. (3) Large quantities of sphingomyelin are present in the nervous system; this substance acts as an insulator in the myelin sheath around nerve fibers. (4) Phospholipids are donors of phosphate radicals when these are needed for different chemical reactions in the tissues. (5) Perhaps the most important of all the functions of phospholipids is participation in the formation of structural elements — mainly membranes — within the cells throughout the body, as is discussed later in connection with a similar function for cholesterol.

CHOLESTEROL

Cholesterol, the formula of which is illustrated in Figure 68–5, is present in the diet of all persons, and it can be absorbed slowly from the gastrointestinal tract into the in-

Figure 68-5. Cholesterol.

testinal lymph. It is highly fat soluble but only slightly soluble in water, and it is capable of forming esters with fatty acids. Indeed, approximately 70 per cent of the cholesterol in the lipoproteins of the plasma is in the form of cholesterol esters.

Formation of Cholesterol. Besides the cholesterol absorbed each day from the gastrointestinal tract, which is called *exogenous cholesterol,* an even greater quantity is also formed in the cells of the body, called *endogenous cholesterol.* Essentially all of the endogenous cholesterol that circulates in the lipoproteins of the plasma is formed by the liver, but all of the other cells of the body form at least some cholesterol, which is consistent with the fact that many of the membranous structures of all cells are partially composed of this substance.

As illustrated by the formula of cholesterol, its basic structure is a sterol nucleus. This is synthesized entirely from multiple molecules of acetyl-CoA. In turn, the sterol nucleus can be modified by means of various side chains to form (a) cholesterol; (b) cholic acid, which is the basis of the bile acids formed in the liver; and (c) many important steroid hormones secreted by the adrenal cortex, the ovaries, and the testes (these are discussed in later chapters).

Factors That Affect Plasma Cholesterol Concentration—Feedback Control of Body Cholesterol. Among the important factors that affect plasma cholesterol concentration are the following:

1. An increase in the *amount of cholesterol ingested each day* increases the plasma concentration slightly. However, when cholesterol is ingested, the rising concentration of cholesterol inhibits the most essential enzyme for endogenous synthesis of cholesterol, 3-hydroxy-3-methylglutaryl CoA reductase, thus providing an intrinsic feedback control system to regulate plasma cholesterol concentration. As a result, plasma cholesterol concentration *usually* is not changed upward or downward more than ±15 per cent by altering the amount of cholesterol in the diet, though the response of individuals differs markedly.

2. A *highly saturated* fat diet increases blood cholesterol concentration as much as 15 to 25 per cent. This results from increased fat deposition in the liver, which then provides increased quantities of acetyl-CoA in the liver cells for production of cholesterol. Therefore, to decrease the blood cholesterol concentration, it is usually even more important to maintain a diet low in saturated fat than to maintain a diet low in cholesterol concentration.

3. Ingestion of fat containing highly *unsaturated fatty acids* usually depresses the blood cholesterol concentration a slight to moderate amount. Unfortunately, the mechanism of this effect is unknown despite the fact that this observation is the basis of much present-day dietary strategy.

4. *Lack of insulin* or *thyroid hormone* increases the blood cholesterol concentration, whereas excess thyroid hormone decreases the concentration. These effects are probably caused mainly by changes in the activities of specific enzymes responsible for the metabolism of lipid substances.

Specific Uses of Cholesterol. By far the most abundant use of cholesterol in the body is to form cholic acid in the liver. As much as 80 per cent of the cholesterol is converted into cholic acid. As explained in Chapter 70, this is conjugated with other substances to form bile salts, which promote digestion and absorption of fats, as has already been discussed in connection with fat digestion.

A small quantity of cholesterol is used by (a) the adrenal glands to form *adrenocortical hormones,* (b) the ovaries to form *progesterone* and *estrogen,* and (c) the testes to form *testosterone.* However, these glands can also synthesize their own sterols and then form their hormones from these, as is discussed in the chapters on endocrinology later in the text.

A large amount of cholesterol is precipitated in the corneum of the skin. This, along with other lipids, makes the skin highly resistant to the absorption of water-soluble substances and also to the action of many chemical agents, for cholesterol and the other lipids are highly inert to acids and different solvents that might otherwise easily penetrate the body. Also, these lipid substances help to prevent water evaporation from the skin; without this protection the amount of evaporation (as occurs in burn patients who have lost their skin) can be as much as 5 to 10 liters per day instead of the usual 300 to 400 milliliters.

CELLULAR STRUCTURAL FUNCTIONS OF PHOSPHOLIPIDS AND CHOLESTEROL— ESPECIALLY FOR MEMBRANES

The above-listed uses of phospholipids and cholesterol are of only minor importance in comparison with their importance for general structural purposes throughout the cells of the body, mainly for formation of membranes.

In Chapter 2 it was pointed out that large quantities of phospholipids and cholesterol are present in both the cell membrane and the membranes of the internal organelles of all cells. It is also known that the *ratio* of membrane cholesterol to phospholipids is especially important in determining the fluidity of the cell membranes.

For membranes to be formed, substances that are not soluble in water must be available. In general, the only substances in the body that are not soluble in water (besides the inorganic substances of bone) are the lipids and some proteins. Thus, the physical integrity of cells throughout the body is based mainly on phospholipids, cholesterol, and certain insoluble proteins. The polar changes on the phospholipids in particular also reduce the interfacial tension between the membranes and the surrounding fluids.

Another fact that indicates the importance of phospholipids and cholesterol for the formation of structural elements of the cells is the slow turnover rates of these substances in most nonhepatic tissues—turnover rates measured in months or years. For instance, their function in the cells of the brain is mainly related to their indestructible physical properties.

● ATHEROSCLEROSIS

Atherosclerosis is a disease of the arteries in which fatty lesions called *atheromatous plaques* develop in the inside arterial wall. These plaques begin by deposition of minute crystals of cholesterol in the intima and sublying smooth muscle. With time, the crystals grow larger and coalesce to form very large matlike crystals. In addition, the surrounding fibrous and smooth muscle tissues proliferate to form additional layers for still further growth of larger and larger plaques. The plaque plus the cellular proliferation can become so large that the inside surface of the vessel bulges deeply into the lumen and greatly reduces blood flow, sometimes even to complete vessel occlusion. Even without occlusion, the fibroblasts of the plaque eventually deposit such extensive amounts of dense connective tissue that *sclerosis* (fibrosis) becomes so great that the arteries become stiff and unyielding. Still later, calcium salts often precipitate with the cholesterol and other lipids of the plaques, leading to bony-hard calcifications that make the arteries at times completely rigid tubes. Both of these latter stages of the disease are called "hardening of the arteries."

Obviously, arteriosclerotic arteries lose most of their distensibility, and because of the degenerative areas in their walls they are easily ruptured. Also, where the plaques protrude into the flowing blood, the roughness of their surfaces causes blood clots to develop, with resultant thrombus or embolus formation (see Chapter 36), thus blocking all blood flow in the artery suddenly. Almost half of all persons in the United States and Europe die of arteriosclerosis. Approximately two thirds of these deaths are caused by thrombosis of one or more coronary arteries. The remaining third are caused by thrombosis or hemorrhage of vessels in other organs of the body—especially the brain, causing strokes, but also in the kidneys, liver, gastrointestinal tract, limbs, and so forth.

Basic Causes of Atherosclerosis—The Role of Cholesterol and Lipoproteins

The factor most important in causing atherosclerosis is a high blood plasma concentration of cholesterol in the form of low density lipoproteins. As explained earlier in this chapter, the plasma concentration of these high cholesterol low density lipoproteins is directly increased by increased saturated fat in the daily diet. To a lesser extent, it is also increased by increased cholesterol in the diet. Therefore, both or either of these dietary factors can contribute to the development of atherosclerosis. An interesting example of this occurs in rabbits that normally have very low plasma cholesterol concentrations because of their vegetarian diet. Simply feeding these animals large quantities of cholesterol as part of their daily nutrition will lead to serious atherosclerotic plaques all through their arterial systems.

The Lipoprotein System Controls Cholesterol Deposition in All the Body's Tissues. Three of the lipoproteins are especially concerned with controlling cholesterol transport to the deposition sites in the tissue cells. These are the *very low density lipoproteins,* the *intermediate density lipoproteins,* and the *low density lipoproteins,* all of which work together in a common system as follows: Of these three lipoproteins, only one is initially formed in the liver, the *very low density lipoproteins.* These contain large quantities of liver-formed triglycerides in addition to less cholesterol and phospholipids. However, as these lipoproteins circulate in the blood, the *lipoprotein lipase* in the walls of the tissue capillaries (especially in the adipose tissue) hydrolyzes a large share of the triglycerides into glycerol and fatty acids, releasing these to be stored in the fat tissue as new triglycerides or to be used for energy as explained earlier.

After loss of much of the triglycerides, the density of what had been the very low density lipoproteins becomes slightly greater, and they are then called *intermediate density lipoproteins.* At this stage, many of these intermediate density lipoproteins are attracted back to the liver cells because of receptors on the liver cell membranes for a surface protein on the lipoprotein, called *apoprotein B.* Normally, the liver removes about half of these intermediate density lipoproteins. Those that remain in the blood lose almost all of their remaining triglycerides by further hydrolysis in the capillaries under the influence of the capillary lipoprotein lipase. As a result, the density of the lipoproteins becomes still greater, and their cholesterol and phosphospholipid reach their greatest concentrations. The lipoproteins are then called *low density lipoproteins.* The structure of the low density lipoproteins is illustrated in Figure 68–6. The center of this lipoprotein is composed almost entirely of *fat-soluble esterified cholesterol.* Most of the surface, on the other hand, is composed of *phospholipids* and *nonesterified cholesterol.* Both of these surface substances have electrically charged prosthetic groups that protrude outward and provide a negative electrical charge over the surface of the lipoprotein; it is this negative charge that allows the lipoprotein to remain soluble in the plasma. However, at one pole of the low density lipoprotein is a single large molecule of apoprotein B (molecular weight 400,000), a protein that provides a recognition site for the low density lipoprotein receptors on the cell membranes of virtually all cells of the body. Attachment of this protein to the receptors causes the entire lipoprotein to be transported to the inside of the cell by the process of pinocytosis; then it is digested internally to release the lipoprotein constituents. In this way, the low density lipoproteins deliver cholesterol and phospholipids to virtually all cells of the body to be used for cellular structural purposes.

Control of Intracellular Cholesterol Concentration. When the concentration of cholesterol inside a cell becomes too great, decreased production of that cell's low density lipoprotein receptors occurs; this now reduces the absorption of additional low density lipoproteins. In this way, each cell controls its own internal concentration of cholesterol.

Control of Cholesterol Synthesis by the Liver. The liver cells, like all other cells of the body, participate in the attraction and ingestion of the low density lipoproteins. In addition, the liver cells also ingest about half of the intermediate density lipoproteins as already discussed. This return of both of these types of lipoproteins to the liver cells maintains a high concentration of cholesterol in the liver cells; the greater the return, the greater the concentration. In turn, this excess cholesterol inhibits the liver enzyme

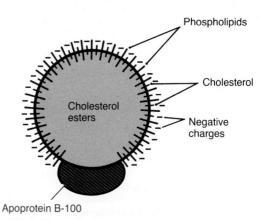

Figure 68–6. Structure and composition of a low density lipoprotein.

system for formation of new cholesterol. Therefore, when the tissue cells throughout the body do not utilize cholesterol, the remainder returns to the liver, and further synthesis of cholesterol is inhibited.

Familial Hypercholesterolemia. *Familial hypercholesterolemia* is a hereditary disease in which the person inherits defective genes for the formation of low density lipoprotein receptors on the membrane surfaces of the body's cells. In the absence of these receptors, the liver cannot reabsorb either the intermediate density lipoproteins or the low density lipoproteins. Without this normal redelivery of cholesterol to the liver cells, the cholesterol machinery of the liver cells goes on a rampage of producing new cholesterol, no longer responsive to the feedback inhibition of too much plasma cholesterol. As a result, the number of very low density lipoproteins released into the plasma increases immensely, and the patient with full-blown familial hypocholesterolemia will have a blood cholesterol concentration of 600 to 1000 mg/dl, a level 4 to 6 times normal. Almost all of these persons die before the age of 20, some as early as the age of 4 to 6, because of myocardial infarction or other sequelae of atherosclerotic blockage of blood vessels throughout the body.

Role of High Density Lipoproteins in Helping to Prevent Atherosclerosis. Much less is known about the function of high density lipoproteins than about that of low density lipoproteins. However, these are formed mainly in the liver but to a lesser extent also in the intestinal epithelium during the absorption of fat from the gut. An especially large portion of these lipoproteins is composed of two types of apoproteins, apoprotein A-I and apoprotein A-II, which lie on the outer surface of the lipoprotein. These apoproteins have affinities for entirely different receptors on the tissue cells than those for the apoprotein B of the low density lipoproteins, and the high density lipoproteins also have entirely different functions. It is believed that the high density lipoproteins can actually absorb cholesterol crystals that are beginning to deposit in the arterial walls and can transfer the cholesterol to intermediate density and low density lipoproteins to be carried back to the liver. Whether this mechanism is true or not, nevertheless, the high density lipoproteins do help to protect against the development of atherosclerosis. Consequently, when a person has a high *ratio* of high density to low density lipoproteins, the likelihood of developing atherosclerosis is considerably reduced.

Other Factors That Lead to Atherosclerosis

In some persons with perfectly normal levels of cholesterol and lipoproteins atherosclerosis will still develop. In most instances, the reasons are not known. Yet special factors that do predispose to atherosclerosis are (1) diabetes mellitus, (2) hypothyroidism, and (3) smoking. In addition, the human male in his early and midadult years of life is several times more likely to have atherosclerosis than the female, suggesting that the male sex hormones might be atherogenic or, conversely, that the female sex hormones might be protective. Some of these factors cause atherosclerosis by increasing the concentration of low density lipoproteins in the plasma. Others lead to atherosclerosis by causing local changes in the vascular tissues that in turn predispose to cholesterol deposition.

Prevention of Atherosclerosis

By far the most important preventive measure against the development of atherosclerosis is to eat a low fat diet containing mainly unsaturated fat with low cholesterol content. However, two types of drug therapy have also proved in recent years to be valuable:

First, most of the cholesterol formed in the liver is converted into bile acids and secreted in this form into the duodenum; then over 90 per cent of these same bile acids are reabsorbed in the terminal ileum and are used over and over again in the bile. Therefore, any agent that will combine with the bile acids in the gastrointestinal tract and prevent their return to the circulation can decrease the total bile acid pool in the circulating blood. This in turn causes far more of the liver cholesterol to be converted into new bile acids. Thus, eating *oat bran,* which binds bile acids and is a constituent of many breakfast cereals, will increase the proportion of the liver cholesterol that forms new bile acids, rather than forming new low density lipoproteins.

Second, a drug *mevinolin,* inhibits the rate-limiting enzyme for cholesterol synthesis in the liver, *3-hydroxy-3-methylglutaryl CoA reductase.* Therefore, it greatly reduces the rate of cholesterol synthesis, sometimes reducing the person's low density lipoprotein cholesterol level as much as 25 to 45 per cent.

In general, one can state that for each 1 mg/dl decrease in low density lipoprotein cholesterol in the plasma there is approximately a 2 per cent decrease in mortality from atherosclerotic heart disease. Therefore, preventive measures will probably prove to be very valuable in decreasing the incidence of heart attacks.

REFERENCES

Bierman, E. L.: Hyperlipoproteinemia. Current Concepts. Upjohn Company, May 1984.
Birdi, K. S.: Lipid and Biopolymer Monolayers at Liquid Interfaces. New York, Plenum Publishing Corp., 1989.
Bisgaier, C., and Glickman, R. M.: Intestinal synthesis, secretion, and transport of lipoproteins. Annu. Rev. Physiol., 45:625, 1983.
Bremer, J.: Carnitine—metabolism and functions. Physiol Rev., 63:1420, 1983.
Breslow, J. L.: Apolipoprotein genetic variation and human disease. Physiol. Rev., 68:85, 1988.
Brown, M. S., and Goldstein, J. L.: A receptor-mediated pathway for cholesterol homeostasis. Science, 232:34, 1986.
Brownell, K. D.: The psychology and physiology of obesity: Implications for screening and treatment. J. Am. Diet. Assoc., 84:406, 1984.
Bukowiecki, L. J.: Mechanisms of stimulus-calorigenesis coupling in brown adipose tissue. Can. J. Biochem. Cell. Biol., 62:623, 1984.
Campbell, J. H., and Campbell, G. R.: Potential role of heparinase in atherosclerosis. News Physiol. Sci., 4:9, 1989.
Carey, M. C., et al.: Lipid digestion and absorption. Annu. Rev. Physiol., 45:651, 1983.
Catapano, A. L., et al.: High-Density Lipoproteins: Physiopathological Aspects and Clinical Significance. New York, Raven Press, 1987.
Chan, L.: Hormonal control of apolipoprotein synthesis. Annu. Rev. Physiol., 45:615, 1983.
Dickerson, R. M.: Gastric exclusion surgery in the management of morbid obesity. Annu. Rev. Med., 33:263, 1982.
Forte, T. M.: Primary hepatocytes in monolayer culture: A model for studies on lipoprotein metabolism. Annu. Rev. Physiol., 46:403, 1984.
Gaber, B. P., and Schnur, J. M.: Biotechnological Applications of Lipid Microstructures. New York, Plenum Publishing Corp., 1989.
Glatz, J. F. C., et al.: Fatty acid-binding proteins and their physiological significance. News Physiol. Sci., 3:41, 1988.
Golinick, P. D.: Metabolism of substrates: Energy substrate metabolism during exercise and as modified by training. Fed. Proc., 44:353, 1985.
Grundy, S. M.: Absorption and metabolism of dietary cholesterol. Annu. Rev. Nutr., 3:71, 1983.
Guthrie, H. A.: Introductory Nutrition, 7th Ed. St. Louis, C. V. Mosby Co., 1988.
Havel, R. J., and Kane, J. P.: Therapy of hyperlipidemic states. Annu. Rev. Med., 33:417, 1982.
Havel, R. J., et al.: Lipoproteins and lipid transport. In Bondy, P. K., and Rosenberg, L. E. (eds.): Metabolic Control and Disease. 8th Ed. Philadelphia, W. B. Saunders Company, 1980, p. 393.
Havel, R. J.: Functional activities of hepatic lipoprotein receptors. Annu. Rev. Physiol., 48:119, 1986.

Hidalgo, C. (ed.): Physical Properties of Biological Membranes and Their Functional Implications. New York, Plenum Publishing Corp., 1988.

Hietanen, E.: Regulation of Serum Lipids by Physical Exercise. Boca Raton, FL, CRC Press, 1982.

Hilderson, H. J. (ed.): Fluorescence Studies on Biological Membranes. New York, Plenum Publishing Corp., 1988.

Holtzman, E.: Lysosomes. New York, Plenum Publishing Corp., 1989.

Horrocks, L. A. (ed.): Phospholipids in the Nervous System. Metabolism. New York, Raven Press, 1982.

Kane, J. P.: Apolipoprotein B: Structural and metabolic heterogeneity. Annu. Rev. Physiol., 45:637, 1983.

Levy, R. I., et al.: Lipoproteins and Atherosclerosis. New York, Raven Press, 1988.

Mahley, R. W.: Apolipoprotein E: cholesterol transport protein with expanding role in cell biology. Science, 240:611, 1988.

Miller, G. J.: High density lipoproteins and atherosclerosis. Annu. Rev. Med., 31:97, 1980.

Paige, D. M.: Clinical Nutrition. St. Louis, C. V. Mosby Co., 1988.

Parmley, W. W., et al.: Modification of experimental atherosclerosis by calcium-channel blockers. Am. J. Cardiol., 55:165B, 1985.

Peng, S. K., and Taylor, C. B.: Cholesterol autoxidation, health and arteriosclerosis. A review on situations in developed countries. World Rev. Nutr. Diet., 44:117, 1984.

Redgrave, T. G.: A new approach to the physiology of lipid transport. News Physiol. Sci., 3:10, 1988.

Robinson, A. M., and Williamson, D. H.: Physiological roles of ketone bodies as substrates and signals in mammalian tissues. Physiol. Rev., 60:143, 1980.

Rosell, S.: Microcirculation and transport in adipose tissue. In Renkin, E. M., and Michel, C. C. (eds.): Handbook of Physiology. Sec. 2, Vol. IV. Bethesda, Md. American Physiological Society, 1984, p. 949.

Rosenberg, R. D.: Role of heparin and heparin-like molecules in thrombosis and atherosclerosis. Fed. Proc., 44:404, 1985.

Royce, S. M., et al.: The influence of dietary isomeric and saturated fatty acids on atherosclerosis and eicosanoid synthesis in swine. Am. J. Clin. Nutr., 39:215, 1984.

Samuel, P., et al.: The role of diet in the etiology and treatment of atherosclerosis. Annu. Rev. Med., 34:179, 1983.

Stallones, R. A.: Ischemic heart disease and lipids in blood and diet. Annu. Rev. Nutr., 3:155, 1983.

Stokes, J. I., and Mancini, M.: Hypercholesterolemia: Clinical and Therapeutic Implications. New York, Raven Press, 1988.

Stryer, L.: Biochemistry. New York, W. H. Freeman Co., 1988.

Waite, M.: The Phospholipases. New York, Plenum Publishing Corp., 1987.

White, F. N., and Somero, G.: Acid-base regulation and phospholipid adaptations to temperature: Time courses and physiological significance of modifying the milieu for protein function. Physiol. Rev., 62:40, 1982.

Wight, T. N.: Proteoglycans in pathological conditions: Atherosclerosis. Fed. Proc., 44:381, 1985.

Williams, S. R.: Nutrition and Diet Therapy, 6th Ed. St. Louis, C. V. Mosby Co., 1989.

Wong, P. Y.-K., and Serhan, C. N. (eds.): Lipoxins. New York, Plenum Publishing Corp., 1988.

CHAPTER 69

• *Protein Metabolism* •

About three quarters of the body solids are proteins. These include *structural proteins, enzymes, nucleoproteins, proteins that transport oxygen, proteins of the muscle that cause contraction,* and many other types that perform specific functions both intracellularly and extracellularly throughout the body.

The basic chemical properties of proteins that explain their diverse functions are so extensive that they constitute a major portion of the entire discipline of biochemistry. For this reason, the present discussion is confined to a few specific aspects of protein metabolism that are important as background for other discussions in this text.

● BASIC PROPERTIES

THE AMINO ACIDS

The principal constituents of proteins are amino acids, 20 of which are present in the body proteins in significant quantities. Figure 69-1 illustrates the chemical formulas of these 20 amino acids, showing that they all have two features in common: Each amino acid has an acidic group (—COOH) and a nitrogen radical attached to the molecule near the acidic radical, usually represented by the amino group (—NH$_2$).

Peptide Linkages and Peptide Chains. In proteins, the amino acids are aggregated into long chains by means of *peptide linkages,* one of which is illustrated by the following reaction:

$$NH_2 \qquad\qquad H_{\backslash}NH$$
$$|\qquad\qquad\qquad |$$
$$R - CH - CO\backslash OH / + R' - CH - COOH \rightarrow$$
$$NH_2$$
$$|$$
$$R - CH - CO$$
$$|$$
$$NH + H_2O$$
$$|$$
$$R' - CH - COOH$$

Note in this reaction that the nitrogen of the amino radical of one amino acid bonds with the carbon of the carboxyl radical of the other amino acid. A hydrogen atom is released from the amino radical, and a hydroxyl radical is released from the carboxyl radical; these two combine to form a molecule of water. After the peptide linkage has been formed, an amino radical and a carboxyl radical are still in the new molecule, both of which are capable of combining with additional amino acids to form a *peptide chain.* Some complicated protein molecules have many thousand amino acids combined together principally by peptide linkages, and even the smallest protein usually has more than 20 amino acids combined together by peptide linkages. The average is about 400 amino acids.

Other Linkages in Protein Molecules. Some protein molecules are composed of several peptide chains rather than a single chain, and these in turn are bound with each other by other linkages, often by *hydrogen bonding* between the CO and NH radicals of the peptides as follows:

$$\begin{array}{ccc}
 & C{=}O\text{-----------}H{-}N & \\
R{-}HC & & CH{-}R' \\
 & N{-}H\text{-----------}O{=}C &
\end{array}$$

Also, many peptide chains are coiled or folded, and the successive coils or folds are held in a tight spiral or in other shapes by similar hydrogen bonding and other forces. In addition to hydrogen bonds, separate peptide chains can be held together by hydrophobic bonds, electrostatic forces, and sulfhydryl, phenolic, and salt linkages, as well as by others.

● TRANSPORT AND STORAGE OF AMINO ACIDS

THE BLOOD AMINO ACIDS

The normal concentration of amino acids in the blood is between 35 and 65 mg/dl. This is an average of about 2 mg/dl for each of the 20 amino acids, though some are present in far greater concentrations than others. Because the amino acids are relatively strong acids, they exist in the blood principally in the ionized state and account for 2 to 3 milliequivalents of the negative ions in the blood. The precise distribution of the different amino acids in the blood depends to some extent on the types of proteins ingested, but the concentrations of at least some individual amino acids are regulated by selective synthesis in the different cells.

Fate of Amino Acids Absorbed From the Gastrointestinal Tract. It will be recalled from Chapter 65 that the products of protein digestion and absorption in the gastrointestinal tract are almost entirely amino acids and that only rare polypeptide or whole protein molecules are absorbed into the blood. Immediately after a meal, the amino acid concentration in the blood rises, but the rise is usually

Figure 69–1. The amino acids, showing the 10 essential amino acids, which cannot be synthesized at all or in sufficient quantity in the body.

only a few milligrams per deciliter for two reasons: First, protein digestion and absorption are usually extended over 2 to 3 hours, which allows only small quantities of amino acids to be absorbed at a time. Second, after entering the blood, the excess amino acids are absorbed within 5 to 10 minutes by cells throughout the entire body, especially by the liver. Therefore, almost never do large concentrations of amino acids accumulate in the blood. Nevertheless, the turnover rate of the amino acids is so rapid that many grams of proteins can be carried from one part of the body to another in the form of amino acids each hour.

Active Transport of Amino Acids Into the Cells. The molecules of essentially all of the amino acids are much too large to diffuse through the pores of the cell membranes. Therefore, significant quantities of amino acids can be transported through the membrane only by facilitated or active transport utilizing carrier mechanisms. The nature of some of the carrier mechanisms is still poorly understood, but some are discussed in Chapter 4.

Renal Threshold for Amino Acids. One of the special functions of carrier transport of amino acids is preventing loss of these in the urine. All the different amino acids can be *actively transported* through the proximal tubular epithelium, which removes them from the glomerular filtrate and returns them to the blood. However, as is true of other active transport mechanisms in the renal tubules, there is an upper limit to the rate at which each type of amino acid can be transported. For this reason, when a particular type of amino acid rises to too high a concentration in the plasma and glomerular filtrate, the excess above that which can be actively reabsorbed is lost into the urine.

In Chapter 31 it was pointed out that appropriate carrier systems for active reabsorption of certain amino acids are often deficient or lacking in the renal tubular epithelium. Under these conditions the plasma threshold at which each affected amino acid begins to be lost in the urine is greatly reduced.

STORAGE OF AMINO ACIDS AS PROTEINS IN THE CELLS

Almost immediately after entry into the cells, amino acids are combined by peptide linkages, under the direction of the messenger RNA and ribosomal system, to form cellular proteins. Therefore, the concentrations of free amino acids inside the cells usually remain low. Thus, storage of large quantities of amino acids as such probably does not occur in the cells; instead, they are mainly stored in the form of actual proteins. Yet many intracellular proteins can be rapidly decomposed again into amino acids under the influence of intracellular lysosomal digestive enzymes, and these amino acids in turn can be transported back out of the cell into the blood. However, the proteins in the chromosomes of the nucleus and the structural proteins such as collagen and muscle contractile proteins do not participate significantly in this reversible storage of amino acids.

Some tissues of the body participate in the storage of amino acids to a greater extent than others. For instance, the liver, which is a large organ and also has special systems for processing amino acids, can store large quantities of proteins; this is also true of the kidney and the intestinal mucosa.

Release of Amino Acids From the Cells and Regulation of Plasma Amino Acid Concentration. Whenever the plasma amino acid concentrations fall below their normal levels, amino acids are transported out of the cells to replenish the supply in the plasma. In this way plasma concentration of each type of amino acid is maintained at a reasonably constant value. Later it will be noted that the various hormones secreted by the endocrine glands are able to alter the balance between tissue proteins and circulating amino acids. Growth hormone and insulin increase the formation of tissue proteins, and the adrenocortical glucocorticoid hormones increase the concentration of circulating amino acids.

Reversible Equilibrium Between the Proteins of Different Parts of the Body. Because cellular proteins in the liver (and to a much lesser extent in other tissues) can be synthesized rapidly from plasma amino acids and many of these in turn can be degraded and returned to the plasma almost equally as rapidly, there is constant equilibrium between the plasma amino acids and the labile proteins in the cells of the body. Therefore, it follows that there is also equilibrium between the proteins from one type of cell and another. For instance, if any particular tissue requires proteins, it can synthesize new proteins from the amino acids of the blood; in turn, these are replenished by degradation of proteins from other cells of the body, especially from the liver cells. These effects are particularly noticeable in relation to protein synthesis in cancer cells. Cancer cells are prolific users of amino acids; therefore, the proteins of the other cells become markedly depleted.

Upper Limit to the Storage of Proteins. Each particular type of cell has an upper limit to the amount of proteins that it can store. After all the cells have reached their limits, the excess amino acids in the circulation are then degraded into other products and used for energy, as is discussed subsequently, or they are converted to fat or glycogen and stored in these forms.

● METABOLIC ROLES OF THE PLASMA PROTEINS

The three major types of protein present in the plasma are *albumin, globulin,* and *fibrinogen.* The principal function of albumin is to provide *colloid osmotic pressure* in the plasma, which in turn prevents plasma loss from the capillaries, as discussed in Chapter 16. The globulins perform a number of enzymatic functions in the plasma; but, equally important, they are principally responsible for both the natural and acquired immunity that a person has against invading organisms, which was discussed in Chapter 34. The fibrinogen polymerizes into long fibrin threads during blood coagulation, thereby forming blood clots that help to repair leaks in the circulatory system, which was discussed in Chapter 36.

Formation of the Plasma Proteins. Essentially all the albumin and fibrinogen of the plasma proteins, as well as 50 to 80 per cent of the globulins, are formed in the liver. The remainder of the globulins are formed in the lymphoid tissues. These are mainly the gamma globulins that constitute the antibodies.

The rate of plasma protein formation by the liver can be extremely high, as much as 30 grams per day. Certain disease conditions often cause rapid loss of plasma proteins; severe burns that denude large surface areas of the skin cause loss of many liters of plasma through the denuded areas each day. The rapid production of plasma proteins by the liver is obviously valuable in preventing death in such states. Furthermore, occasionally, a person with severe renal disease loses as much as 20 grams of plasma protein in the urine each day for months, and these are continually replaced.

Use of Plasma Proteins by the Tissues. When the tissues become depleted of proteins, the plasma proteins can act as a source for rapid replacement of the tissue proteins. Indeed, whole plasma proteins can be imbibed *in toto* by the tissue macrophages by the process of pinocytosis; then, once in the cells, these are split into amino acids that are transported back into the blood and utilized throughout the body to build cellular proteins. In this way, the plasma proteins function as a labile protein storage medium and represent a rapidly available source of amino acids whenever a particular tissue requires these.

Reversible Equilibrium Between the Plasma Proteins and the Tissue Proteins. There is a constant state of equilibrium, as illustrated in Figure 69–2, between the plasma proteins, the amino acids of the blood, and the tissue proteins. It has been estimated from radioactive tracer studies that about 400 grams of body protein are synthesized and degraded each day as part of the continual state of flux of amino acids. This illustrates once again the general principle of reversible exchange of amino acids among the different proteins of the body. Even during starvation or during severe debilitating diseases, the ratio of total tissue proteins to total plasma proteins in the body remains relatively constant at about 33 to 1.

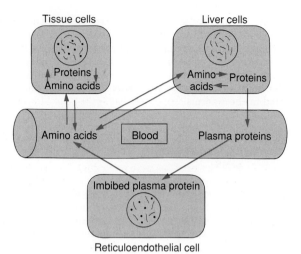

Figure 69–2. Reversible equilibrium among the tissue proteins, plasma proteins, and plasma amino acids.

Because of this reversible equilibrium between plasma proteins and the other proteins of the body, one of the most effective of all therapies for severe acute protein deficiency is intravenous administration of plasma protein. Within a few days or sometimes within hours, the amino acids of the administered protein become distributed throughout the cells of the body to form proteins where they are needed.

ESSENTIAL AND NONESSENTIAL AMINO ACIDS

Ten of the amino acids normally present in animal proteins can be synthesized in the cells, whereas the other ten either cannot be synthesized at all or are synthesized in quantities too small to supply the body's needs. This second group of amino acids is called *essential amino acids.* Use of the word "essential" does not mean that the other ten "nonessential" amino acids are not equally as essential for formation of the proteins, but only that these others are *not essential in the diet* because they can be synthesized in the body.

Synthesis of the nonessential amino acids depends mainly on the formation first of appropriate α-keto acids, which are the precursors of the respective amino acids. For instance, *pyruvic acid,* which is formed in large quantities during the glycolytic breakdown of glucose, is the keto acid precursor of the amino acid *alanine.* Then, by the process of *transamination,* an amino radical is transferred to the α-keto acid while the keto oxygen is transferred to the donor of the amino radical. This reaction is illustrated in Figure 69–3. Note in this reaction that the amino radical is transferred to the pyruvic acid from another chemical that is closely allied to the amino acids, *glutamine.* Glutamine is present in the tissues in large quantities, and one of its principal functions is to serve as an amino radical store-

house. In addition, amino radicals can be transferred from *asparagine, glutamic acid,* and *aspartic acid.*

Transamination is promoted by several different enzymes among which are the *aminotransferases,* which are derivatives of pyridoxine, one of the B vitamins (B_6). Without this vitamin, the nonessential amino acids are synthesized only poorly, and, therefore, protein formation cannot proceed normally.

USE OF PROTEINS FOR ENERGY

Once the cells are filled to their limits for protein storage, any additional amino acids in the body fluids are degraded and used for energy or stored as fat. This degradation occurs almost entirely in the liver, and it begins with the process known as *deamination.* To initiate this process, the excess amino acids in the cells, especially in the liver, induce the production of large quantities of *aminotransferases,* the enzymes responsible for initiating most deamination.

Deamination. Deamination means removal of the amino groups from the amino acids. This occurs mainly by transamination, which means transfer of the amino group to some acceptor substance, which is the reverse of the transamination explained above in relation to the synthesis of amino acids.

The greatest amount of deamination occurs by the following transamination schema:

α–Ketoglutaric acid + Amino acid

Glutamic acid + α–Keto acid

+ NAD^+ + H_2O

$NADH$ + H^+ + NH_3

Note from this schema that the amino group from the amino acid is transferred to α-ketoglutaric acid, which then becomes glutamic acid. The glutamic acid can then transfer the amino group to still other substances or can release it in the form of ammonia. In the process of losing the amino group, the glutamic acid once again becomes α-ketoglutaric acid, so that the cycle can be repeated again and again.

Urea Formation by the Liver. The ammonia released during deamination is removed from the blood almost entirely by conversion into urea; two molecules of ammonia and one molecule of carbon dioxide combine in accordance with the following net reaction:

$$2NH_3 + CO_2 \rightarrow H_2N-\underset{\underset{O}{\|}}{C}-NH_2 + H_2O$$

Essentially all urea formed in the human body is synthe-

NH₂—C—CH₂—CH₂—CH—COOH   CH₃—C—COOH
‖ | ‖
O   NH₂   O
(Glutamine)   (Pyruvic acid)

→ Transaminase →

NH₂—C—CH₂—CH₂—C—COOH   CH₃—CH—COOH
‖   ‖ |
O   O   NH
(α—Ketoglutamic acid)   (Alanine)

Figure 69–3. Synthesis of alanine from pyruvic acid by transamination.

sized in the liver. In the absence of the liver or in serious liver disease, ammonia accumulates in the blood. This in turn is extremely toxic, especially to the brain, often leading to a state called *hepatic coma.*

The stages in the formation of urea are essentially the following:

After its formation, the urea diffuses from the liver cells into the body fluids and is excreted by the kidneys.

Oxidation of Deaminated Amino Acids. Once amino acids have been deaminated, the resulting keto acid products can in most instances be oxidized to release energy for metabolic purposes. This usually involves two successive processes: (1) the keto acid is changed into an appropriate chemical substance that can enter the citric acid cycle, and (2) this substance is then degraded by this cycle and used for energy in the same manner that acetyl coenzyme A (acetyl-CoA) derived from carbohydrate and lipid metabolism is used.

In general, the amount of adenosine triphosphate formed for each gram of protein that is oxidized is slightly less than that formed for each gram of glucose oxidized.

Gluconeogenesis and Ketogenesis. Certain deaminated amino acids are similar to the substrates normally used by the cells, mainly the liver cells, to synthesize glucose or fatty acids. For instance, deaminated alanine is pyruvic acid. Obviously, this can be converted into glucose or glycogen. Or it can be converted into acetyl-CoA, which can then be polymerized into fatty acids. Also, two molecules of acetyl-CoA can condense to form acetoacetic acid, which is one of the ketone bodies, as explained in the preceding chapter.

The conversion of amino acids into glucose or glycogen is called *gluconeogenesis,* and the conversion of amino acids into keto acids or fatty acids is called *ketogenesis.* Eighteen of 20 of the deaminated amino acids have chemical structures that allow them to be converted into glucose, and 19 can be converted into fatty acids.

OBLIGATORY DEGRADATION OF PROTEINS

When a person eats no proteins, a certain proportion of that person's own body proteins continues to be degraded into amino acids, then deaminated and oxidized. This involves 20 to 30 grams of protein each day, which is called the *obligatory loss* of proteins. Therefore, to prevent a net loss of protein from the body, one must ingest a minimum of 20 to 30 grams of protein each day, and to be on the safe side a minimum of 60 to 75 grams is usually recommended.

The ratios of the different amino acids in the dietary protein must be about the same as the ratios in the body tissues if the entire protein is to be fully usable. If one particular type of essential amino acid is low in concentration, the others become unusable as well because cells form either whole proteins or none at all, as was explained in Chapter 3 in relation to protein synthesis. The unusable amino acids are then deaminated and oxidized. A protein that has a ratio of amino acids different from that of the average body protein is called a *partial protein* or *incomplete protein,* and such a protein is obviously less valuable for nutrition than is the *complete protein.*

Effect of Starvation on Protein Degradation. Except for the excess protein in the diet or the 20 to 30 grams of obligatory protein degradation each day, the body uses almost entirely carbohydrates or fats for energy as long as these are available. However, after several weeks of starvation, when the quantity of stored fats begins to run out, the amino acids of the blood begin to be rapidly deaminated and oxidized for energy. From this point on, the proteins of the tissues degrade rapidly — as much as 125 grams daily — and the cellular functions deteriorate precipitously.

Because carbohydrate and fat utilization for energy occurs in preference to protein utilization, carbohydrates and fats are called *protein sparers.*

● HORMONAL REGULATION OF PROTEIN METABOLISM

Growth Hormone. Growth hormone increases the rate of synthesis of cellular proteins, causing the tissue proteins to increase. The precise mechanism by which growth hormone increases the rate of protein synthesis is not known, but it is believed mainly to enhance the transport of amino acids through the cell membranes and/or to accelerate the DNA and RNA transcription and translation processes for protein synthesis. Part of the action might also result from the effect of growth hormone on fat metabolism, for this hormone causes increased rate of fat liberation from the fat depots, making this available for energy. This in turn reduces the rate of oxidation of amino acids and consequently makes increased quantities of amino acids available to the tissues to be synthesized into proteins.

Insulin. Lack of insulin reduces protein synthesis almost to zero. The mechanism by which this hormone affects protein metabolism is also unknown, but insulin does accelerate amino acid transport into the cells, which could be the stimulus to protein synthesis. Also, insulin increases the availability of glucose to the cells so that the use of amino acids for energy becomes correspondingly reduced.

Glucocorticoids. The glucocorticoids secreted by the adrenal cortex *decrease* the quantity of protein in most tissues while increasing the amino acid concentration in the plasma. However, contrary to elsewhere in the body, these hormones *increase* both the liver proteins and the plasma proteins. It is believed that the glucocorticoids act by increasing the rate of breakdown of extrahepatic proteins, thereby making increased quantities of amino acids available in the body fluids. This in turn supposedly allows the liver to synthesize increased quantities of hepatic cellular proteins and plasma proteins.

The effects of glucocorticoids on protein metabolism are especially important in promoting ketogenesis and gluconeogenesis; in the absence of these hormones insufficient quantities of amino acids are usually available in the plasma to allow either of these processes to occur significantly.

Testosterone. Testosterone, the male sex hormone, causes increased deposition of protein in the tissues throughout the body, including especially an increase in the contractile proteins of the muscles. The mechanism by which this effect comes about is unknown, but it is definitely different from the effect of growth hormone in the following way: Growth hormone causes tissues to continue growing almost indefinitely, whereas testosterone causes the muscles and other protein tissues to enlarge only for several months. Beyond that time, despite continued administration of testosterone, further protein deposition ceases.

Estrogen, the principal female sex hormone, also causes slight deposition of protein, but its effect is relatively insignificant in comparison with that of testosterone.

Thyroxine. Thyroxine increases the rate of metabolism of all cells, and, as a result, indirectly affects protein metabolism. If insufficient carbohydrates and fats are available for energy, thyroxine causes rapid degradation of proteins and uses these for energy. On the other hand, if adequate quantities of carbohydrates and fats are available and excesses of amino acids are also available in the extracellular fluid, thyroxine can actually increase the rate of protein synthesis.

Conversely, in growing animals deficiency of thyroxine causes growth to be greatly inhibited because of lack of protein synthesis. In essence, it is believed that thyroxine has little specific direct effect on protein metabolism but does have an important general effect in increasing the rates of both normal anabolic and normal catabolic protein reactions.

REFERENCES

Benevenga, N. J., and Steele, R. D.: Adverse effects of excessive consumption of amino acids. Annu. Rev. Nutr., 4:157, 1984.

Borum, P. R.: Carnitine. Annu. Rev. Nutr., 3:233, 1983.

Christensen, H. N.: Interorgan amino acid nutrition. Physiol. Rev., 62:1193, 1982.

Coleman, J. E.: Metabolic interrelationships between carbohydrates, lipids and proteins. In Bondy, P. K., and Rosenberg, L. E. (eds.): Metabolic Control and Disease, 8th Ed. Philadelphia, W. B. Saunders Co., 1980, p. 161.

DeMartino, G. N., and Croall, D. E.: Calcium-dependent proteases: A prevalent proteolytic system of uncertain function. News Physiol. Sci., 2:82, 1987.

Dohm, G. L., et al.: Protein metabolism during endurance exercise. Fed Proc., 44:348, 1985.

Fasman, G. D. (ed.): Prediction of Protein Structure and the Principles of Protein Conformation. New York, Plenum Publishing Corp., 1989.

Friedmann, H. C. (ed.): Enzymes. Stroudsburg, Pa., Dowden, Hutchinson & Ross, 1980.

Galjaard, H.: Genetic Metabolic Diseases: Early Diagnosis and Prenatal Analysis. New York, Elsevier/North-Holland, 1980.

Gevers, W.: Protein metabolism of the heart. J. Mol. Cell Cardiol., 16:3, 1984.

Guthrie, H. A.: Introductory Nutrition, 7th Ed. St. Louis, C. V. Mosby Co., 1988.

Harper, A. E., et al.: Branched-chain amino acid metabolism. Annu. Rev. Nutr., 4:409, 1984.

Kinney, J. M., and Elwyn, D. H.: Protein metabolism and injury. Annu. Rev. Nutr., 3:433, 1983.

Lehminger, A. L.: Principles of Biochemistry. New York, Worth Publishers, Inc., 1982.

Oriordan, J. L.: Recent Advances in Endocrinology and Metabolism. New York, Churchill Livingstone, 1982.

Rosenberg, L. E., and Scriver, C. R.: Disorders of amino acid metabolism. In Bondy, P. K., and Rosenberg, L. E. (eds.): Metabolic Control and Disease, 8th Ed. Philadelphia, W. B. Saunders Co., 1980, p. 583.

Rothschild, M. A.: Albumin synthesis. In Javitt, N. B. (ed.): International Review of Physiology: Liver and Biliary Tract Physiology I. Vol. 21. Baltimore, University Park Press, 1980, p. 249.

Rothschild, M. A., et al.: Effect of albumin concentration on albumin synthesis in the perfused liver. Am. J. Physiol., 216:1127, 1969.

Seegmiller, J. E.: Diseases of purine and pyrimidine metabolism. In Bondy, P. K., and Rosenberg, L. E. (eds.): Metabolic Control and Disease, 8th Ed. Philadelphia, W. B. Saunders Co., 1980, p. 777.

Stryer, L.: Biochemistry. New York, W. H. Freeman Co., 1988.

Sugden, P. H.: The effects of hormonal factors on cardiac protein turnover. Adv. Myocardial., 5:105, 1985.

Visek, W. J.: An update of concepts of essential amino acids. Annu. Rev. Neurosci., 5:137, 1982.

Waterlow, J. C. (ed.): Nitrogen Metabolism in Man. New York, Elsevier Science Publishing Co., 1982.

Weinsier, R. L., et al.: Handbook of Clinical Nutrition: Clinician's Manual for the Diagnosis and Management of Nutritional Problems. St. Louis, C. V. Mosby Co., 1988.

Williams, S. R.: Basic Nutrition and Diet Therapy, 8th Ed. St. Louis, C. V. Mosby Co., 1989.

Zak, R., et al.: Assessment of protein turnover by use of radioisotopic tracers. Physiol. Rev., 59:407, 1979.

Zamyatnin, A. A.: Amino acid, peptide, and protein volume in solution. Annu. Rev. Biophys. Bioeng., 13:145, 1984.

CHAPTER 70

• *The Liver as an Organ* •

In many previous chapters we have encountered the liver, and each time it is performing a different function. Yet we often forget that the liver is a discrete organ and that many of its various functions interrelate to each other. This becomes particularly evident in clinical abnormalities of the liver because many of its functions are disturbed simultaneously in different combinations, depending upon the nature of the disorder. Therefore, the purpose of the present chapter is to summarize the different functions of the liver and to show how the liver operates as an individual organ.

The basic functions of the liver can be divided into (1) its vascular functions for storage and filtration of blood, (2) its metabolic functions concerned with the majority of the metabolic systems of the body, and (3) its secretory and excretory functions that are responsible for forming the bile that flows through the bile ducts into the gastrointestinal tract.

PHYSIOLOGIC ANATOMY OF THE LIVER

The basic functional unit of the liver is the *liver lobule*, which is a cylindrical structure several millimeters in length and 0.8 to 2 millimeters in diameter. The human liver contains 50,000 to 100,000 individual lobules.

The liver lobule is constructed around a *central vein* that empties into the hepatic veins and thence into the vena cava. The lobule itself is composed principally of many *hepatic cellular plates* (two of which are illustrated in Figure 70–1) that radiate centrifugally from the central vein like spokes in a wheel. Each hepatic plate is one to two cells thick, and between the adjacent cells lie small *bile canaliculi* that empty into *bile ducts* in the fibrous septa separating the adjacent liver lobules.

Also in the septa are small *portal venules* that receive their blood mainly from the venous outflow of the gastrointestinal tract by way of the portal veins. From these venules blood flows into flat, branching *hepatic sinusoids* that lie between the hepatic plates, and thence into the central vein. Thus, the hepatic cells are exposed continuously to portal venous blood.

In addition to the portal venules, *hepatic arterioles* are also present in the interlobular septa. These arterioles supply arterial blood to the septal tissues, and many of the small arterioles also empty directly into the hepatic sinusoids, most frequently emptying into these about one third of the distance away from the interlobular septa, as shown in Figure 70–1.

The venous sinusoids are lined by two types of cells: (1) typical *endothelial cells* and (2) large *Kupffer cells*, which are tissue macrophages (also called reticuloendothelial cells) capable of phagocytizing bacteria and other foreign matter in the blood. The endothelial lining of the venous sinusoids has extremely large pores, some of which are almost 1 micrometer in diameter. Beneath this lining, lying between the endothelial cells and the hepatic cells, is a very narrow tissue space called the *space of Disse*. The millions of spaces of Disse in turn connect with lymphatic vessels in the interlobular septa. Therefore, excess fluid in these spaces is removed through the lymphatics. Because of the large pores in the endothelium, substances in the plasma move freely into the space of Disse. Even large portions of the plasma proteins diffuse freely into this space.

● FUNCTION OF THE HEPATIC VASCULAR SYSTEM

The function of the hepatic vascular system was discussed in Chapter 15 in connection with the portal veins. Briefly, this can be summarized as follows:

Blood Flow Through the Liver

About 1100 milliliters of blood flows from the portal vein into the liver sinusoids each minute, and approximately an additional 350 milliliters flows into the sinusoids from the hepatic artery, the total averaging about 1450 ml/min. This amounts to about 29 per cent of the resting cardiac output.

Measurement of Liver Blood Flow. Total hepatic blood flow per minute can be measured by a modified Fick procedure in which the dye *indocyanine green,* a dye that is removed from the blood only by the liver, is infused continuously into the circulatory system until its concentration in the arterial blood remains constant. When this is achieved, one knows the rate of excretion of the dye because it is then exactly equal to the rate of infusion. Next, the concentration of the dye is measured in the arterial blood and also in venous blood collected from the hepatic vein by means of a catheter. From these measurements, one can calculate the arteriovenous difference (A-V difference) of the dye. The liver blood flow is then calculated by the usual Fick formula, as follows:

$$\text{Hepatic blood flow (ml/min)} = \frac{\text{Rate of dye excretion (mg/min)}}{\text{A-V difference in dye (mg/ml)}}$$

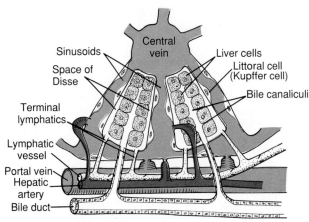

Figure 70-1. Basic structure of a liver lobule, showing the hepatic cellular plates, the blood vessels, the bile-collecting system, and the lymph flow system composed of the spaces of Disse and the interlobular lymphatics. (Reprinted from Guyton, Taylor, and Granger [as modified from Elias]: Circulatory Physiology Two: Dynamics and Control of the Body Fluids. Philadelphia, W. B. Saunders Company, 1975.)

Pressures and Resistance in the Hepatic Vessels

The pressure in the portal vein leading into the liver averages about 9 mm Hg, and the pressure in the hepatic vein leading from the liver into the vena cava averages almost exactly 0 mm Hg. This small pressure difference shows that the resistance to blood flow through the liver sinusoids is normally low, especially when one considers that about 1.45 liter of blood flows by this route each minute.

Blockage of the Portal System—"Cirrhosis" of the Liver. Frequently, extreme amounts of fibrous tissue develop within the liver structure, destroying many of the parenchymal cells and eventually contracting around the blood vessels, thereby greatly impeding the flow of portal blood through the liver. This disease process is known as *cirrhosis of the liver*. It results most frequently from alcoholism, but it can also follow ingestion of poisons such as carbon tetrachloride, virus diseases such as infectious hepatitis, or infectious processes in the bile ducts.

The portal system is also occasionally blocked by a large clot developing in the portal vein or in its major branches. When the portal system is suddenly blocked, the return of blood from the intestines and spleen to the systemic circulation is tremendously impeded, the capillary pressure rising as much as 15 to 20 mm Hg; and the patient often dies within a few hours because of excessive loss of fluid from the capillaries into the lumens and walls of the intestines.

Reservoir Function of the Liver

Because the liver is an expandable organ, large quantities of blood can be stored in its blood vessels. Its normal blood volume, including both that in the hepatic veins and hepatic sinuses, is 450 milliliters, or almost 10 per cent of the total blood volume. However, when high pressure in the right atrium causes back pressure in the liver, the liver expands, and as much as 0.5 to 1 liter of extra blood occasionally is thereby stored in the hepatic veins and sinuses. This occurs especially in cardiac failure with peripheral congestion, which was discussed in Chapter 22.

Thus, in effect, the liver is a large, expandable, venous organ capable of acting as a valuable blood reservoir in times of excess blood volume and capable of supplying extra blood in times of diminished blood volume.

Lymph Flow From the Liver

Because the pores in the hepatic sinusoids are very, very permeable and allow ready passage of both fluid and proteins into the spaces of Disse, the lymph draining from the liver usually has a protein concentration of about 6 gm/dl, which is only slightly less than the protein concentration of plasma. Also, the extreme permeability of the liver sinusoids allows large quantities of lymph to form. Therefore, about one half of all the lymph formed in the body under resting conditions arises in the liver.

Effects of High Hepatic Vascular Pressures in Causing Fluid Transudation From the Liver and Portal Capillaries—Ascites. When the hepatic venous pressure where the liver veins empty into the vena cava rises only 3 to 7 mm Hg above normal, excessive amounts of fluid begin to transude into the lymph and also to leak through the outer surface of the liver capsule directly into the abdominal cavity. This fluid is almost pure plasma, containing 80 to 90 per cent as much protein as normal plasma. At still higher vena caval pressures, at 10 to 15 mm Hg, liver lymph flow increases to as much as 20 times normal, and the "sweating" from the surface of the liver can be so great that it causes large amounts of free fluid in the abdominal cavity, which is called *ascites.*

Blockage of portal flow through the liver also causes very high capillary pressures in the entire gastrointestinal tract, resulting in edema of the gut wall and transudation of fluid through the serosa of the gut into the abdominal cavity. This, too, can cause ascites but is less likely to do so than is sweating from the liver surface, because collateral vascular channels develop rapidly from the portal veins to the systemic veins, decreasing the intestinal capillary pressure back to a safe value.

The Hepatic Macrophage System—The Blood-Cleansing Function of the Liver

Blood flowing through the intestinal capillaries picks up many bacteria from the intestines. Indeed, a sample of blood from the portal system almost always grows colon bacilli when cultured, whereas growth of colon bacilli from blood in the systemic circulation is extremely rare. Special high-speed motion pictures of the action of Kupffer cells, the large phagocytic macrophages that line the hepatic sinuses, have demonstrated that these cells can cleanse blood extremely efficiently as it passes through the sinuses; when a bacterium comes into momentary contact with a Kupffer cell, in less than 0.01 second the bacterium passes inward through the wall of the Kupffer cell to become permanently lodged therein until it is digested. Probably not over 1 per cent of the bacteria entering the portal blood from the intestines succeeds in passing through the liver into the systemic circulation.

● METABOLIC FUNCTIONS OF THE LIVER

The liver cells all together are a large chemically reactant pool having a very high rate of metabolism, sharing substrates and energy from one metabolic system to another, processing and synthesizing multiple substances that are transported to other areas of the body, and performing a myriad of other metabolic functions. For all of these reasons, a major share of the entire discipline of biochemistry is devoted to the metabolic reactions in the liver. But, here, let us summarize those metabolic functions that are especially important in understanding the integrated physiology of the body.

CARBOHYDRATE METABOLISM

In carbohydrate metabolism the liver performs the following specific functions: (1) storage of glycogen, (2) conversion of galactose and fructose to glucose, (3) gluconeogenesis, and (4) formation of many important chemical compounds from the intermediate products of carbohydrate metabolism.

The liver is especially important for maintaining a normal blood glucose concentration. For instance, storage of glycogen allows the liver to remove excess glucose from the blood, store it, and then return it to the blood when the blood glucose concentration begins to fall too low. This is called the *glucose buffer function* of the liver. As an example, immediately after a meal containing large amounts of carbohydrates, the blood glucose concentration rises about three times as much in a person with a nonfunctional liver as in a person with a normal liver.

Gluconeogenesis in the liver is also concerned with maintaining a normal blood glucose concentration, for gluconeogenesis occurs to a significant extent only when the glucose concentration begins to fall below normal. In such a case, large amounts of amino acids are converted into glucose, thereby contributing still another means for maintaining a relatively normal blood glucose concentration.

FAT METABOLISM

Although some fat metabolism can take place in almost all cells of the body, certain aspects of fat metabolism occur mainly in the liver. Specific functions of the liver in fat metabolism are (1) a very high rate of oxidation of fatty acids to supply energy for other bodily functions, (2) formation of most of the lipoproteins, (3) synthesis of large quantities of cholesterol and phospholipids, and (4) conversion of large quantities of carbohydrates and proteins to fat.

To derive energy from neutral fats, the fat is first split into glycerol and fatty acids; then the fatty acids are split by *beta oxidation* into two-carbon acetyl radicals that then form *acetylcoenzyme A* (acetyl-CoA). This in turn can enter the citric acid cycle and be oxidized to liberate tremendous amounts of energy. Beta oxidation can take place in all cells of the body, but it occurs especially rapidly in the hepatic cells. Yet the liver itself cannot utilize all the acetyl-CoA that is formed; instead, this is converted by condensation of two molecules of acetyl-CoA into *acetoacetic acid,* which is a highly soluble acid that passes from the liver cells into the extracellular fluids and then is transported throughout the body to be absorbed by the other tissues. These tissues in turn reconvert the acetoacetic acid into acetyl-CoA and then oxidize it in the usual manner. In these ways the liver is responsible for a major part of the metabolism of fats.

About 80 per cent of the cholesterol synthesized in the liver is converted into bile salts, which in turn are secreted into the bile; the remainder is transported in the lipoproteins, which are carried by the blood to the tissue cells everywhere in the body. Phospholipids are likewise synthesized in the liver and transported principally in the lipoproteins. Both the cholesterol and the phospholipids are used by the cells to form membranes, intracellular structures, and multiple derived chemical substances that are important to cellular function.

Most of the fat synthesis in the body from carbohydrates and proteins also occurs in the liver. After fat is synthesized in the liver it is transported in the lipoproteins to the adipose tissue to be stored.

PROTEIN METABOLISM

Even though a large proportion of the processes for carbohydrate and fat metabolism occurs in the liver, the body could probably dispense with many of these functions of the liver and still survive. On the other hand, the body cannot dispense with the services of the liver in protein metabolism for more than a few days without death ensuing. The most important functions of the liver in protein metabolism are (1) deamination of amino acids, (2) formation of urea for removal of ammonia from the body fluids, (3) formation of plasma proteins, and (4) interconversions among the different amino acids and other compounds important to the metabolic processes of the body.

Deamination of the amino acids is required before these can be used for energy or before they can be converted into carbohydrates or fats. A small amount of deamination can occur in the other tissues of the body, especially in the kidney, but the percentage of deamination occurring extrahepatically is so small that it is almost completely unimportant.

Formation of urea by the liver removes ammonia from the body fluids. Large amounts of ammonia are formed by the deamination process, and still additional amounts are continually formed in the gut by bacteria and are then absorbed into the blood. Therefore, in the absence of this function of the liver to form urea, the plasma ammonia concentration rises rapidly and results in *hepatic coma* and death. Indeed, even greatly decreased blood flow through the liver — as occurs occasionally when a shunt develops between the portal vein and the vena cava — can also cause excessive ammonia in the blood, an exceedingly toxic condition.

Essentially all the plasma proteins, with the exception of part of the gamma globulins, are formed by the hepatic cells. This accounts for about 90 per cent of all of the plasma proteins. The remaining gamma globulins are the antibodies formed mainly by plasma cells in the lymph tissue of the body. The liver can form plasma proteins at a maximum rate of 15 to 50 grams a day. Therefore, after loss of as much as half the plasma proteins from the body, these can be replenished in a week or two. It is particularly interesting that plasma protein depletion causes rapid mitosis of the hepatic cells and actual growth of the liver to a larger size; these effects are coupled with rapid output of plasma proteins until the plasma concentration returns to normal.

Among the most important functions of the liver is its ability to synthesize certain amino acids and also to synthesize other important chemical compounds from amino acids. For instance, the so-called nonessential amino acids can all be synthesized in the liver. To do this, a keto acid having the same chemical composition (except at the keto oxygen) as that of the amino acid to be formed is first synthesized. Then an amino radical is transferred through several stages of *transamination* from an available amino acid to the keto acid to take the place of the keto oxygen.

MISCELLANEOUS METABOLIC FUNCTIONS OF THE LIVER

Storage of Vitamins. The liver has a particular propensity for storing vitamins and has long been known as an excellent source of certain vitamins in treating patients. The single vitamin stored in greatest quantity in the liver is vitamin A, but large quantities of vitamin D and vitamin B_{12}

Fevery, J., and Heirwegh, K. P. M.: Bilirubin metabolism. In Javitt, N. B. (ed.): International Review of Physiology: Liver and Biliary Tract Physiology I. Vol. 21. Baltimore, University Park Press, 1980, p. 171.

Gitnick, G. (ed.): Modern Concepts of Acute and Chronic Hepatitis. New York, Plenum Publishing Corp., 1989.

Gitnick, G., et al. (eds.): Principles and Practice of Gastroenterology and Hepatology. New York, Elsevier Science Publishing Co., 1988.

Goresky, C. A., and Groom, A. C.: Microcirculatory events in the liver and the spleen. In Renkin, E. M., and Michel, C. C. (eds.): Handbook of Physiology. Sec. 2, Vol. IV. Bethesda, Md., American Physiological Society, 1984, p. 689.

Gustafsson, J. A.: Sex steroid induced changes in hepatic enzymes. Annu. Rev. Physiol., 45:51, 1983.

Havel, R. J.: Functional activities of hepatic lipoprotein receptors. Annu. Rev. Physiol., 48:119, 1986.

Henderson, J. R., and Daniel, P. M.: Capillary beds and portal circulations. In Renkin, E. M., and Michel, C. C. (eds.): Handbook of Physiology. Sec. 2, Vol. IV. Bethesda, Md., American Physiological Society, 1984. p. 1035.

Heyworth, M. F., and Jones, A. L.: Immunology of the Gastrointestinal Tract and Liver. New York, Raven Press, 1988.

Huebers, H. A., and Finch, C. A.: The physiology of transferrin and transferrin receptors. Physiol. Rev., 67:520, 1987.

Johnson, L. R., et al.: Physiology of the Gastrointestinal Tract, 2nd Ed. New York, Raven Press, 1987.

Jungermann, K., and Katz, N.: Functional specialization of different hepatocyte populations. Physiol. Rev., 69:708, 1989.

Kraus-Friedmann, N.: Hormonal regulation of hepatic gluconeogenesis. Physiol. Rev., 64:170, 1984.

LaRusso, N. F.: Proteins in bile: How they get there and what they do. Am. J. Physiol., 247:G199, 1984.

Rappaport, A. M.: Hepatic blood flow: Morphologic aspects and physiologic regulation. In Javitt, N. B. (ed.): International Review of Physiology: Liver and Biliary Tract Physiology I. Vol. 21. Baltimore, University Park Press, 1980, p. 1.

Reichen, J., and Paumgartner, G.: Excretory function of the liver. In Javitt, N. B. (ed.): International Review of Physiology: Liver and Biliary Tract Physiology I. Vol. 21. Baltimore, University Park Press, 1980, p. 103.

Richardson, P. D. I., and Withrington, P. G.: Physiological regulation of the hepatic circulation. Annu. Rev. Physiol., 44:57, 1982.

Salen, G., and Shefer, S.: Bile acid synthesis. Annu. Rev. Physiol., 45:679, 1983.

Schachter D.: Fluidity and function of hepatocyte plasma membranes. Hepatology, 4:140, 1984.

Scharschmidt, B. F., and Van Dyke, R. W.: Proton transport by hepatocyte organelles and isolated membrane vesicles. Annu. Rev. Physiol., 49:69, 1987.

Seeff, L. B., and Lewis, J. H. (eds.): Current Perspectives in Hepatology. New York, Plenum Publishing Corp., 1989.

Setchell, K. D. R., et al. (eds.): The Bile Acids. New York, Plenum Publishing Corp., 1988.

Shearman, D. J. (ed.): Diseases of the Gastrointestinal Tract and Liver. New York, Churchill Livingstone, 1982.

Soloway, R. D. (ed.): Chronic Active Liver Disease. New York, Churchill Livingstone, 1983.

Strange, R. C.: Hepatic bile flow. Physiol. Rev., 64:1055, 1984.

Stryer, L.: Biochemistry. New York, W. H. Freeman Co., 1988.

CHAPTER 71

• *Dietary Balances; Regulation of Feeding; Obesity and Starvation; Vitamins and Minerals* •

The intake of food must always be sufficient to supply the metabolic needs of the body and yet not so much as to cause obesity. Also, since different foods contain different proportions of proteins, carbohydrates, fats, minerals, and vitamins, appropriate balance must be maintained among these so that all segments of the body's metabolic systems can be supplied with the requisite materials. This chapter, therefore, discusses the problems of balance among the types of food and also the mechanisms by which the intake of food is regulated in accordance with the metabolic needs of the body.

● DIETARY BALANCES

ENERGY AVAILABLE IN FOODS

The energy liberated from each gram of carbohydrate as it is oxidized to carbon dioxide and water is 4.1 Calories, and that liberated from fat is 9.3 Calories. The energy liberated from metabolism of the average protein of the diet as each gram is oxidized to carbon dioxide, water, and urea is 4.35 Calories. Also, these different substances vary in the average percentages that are absorbed from the gastrointestinal tract: approximately 98 per cent of the carbohydrate, 95 per cent of the fat, and 92 per cent of the protein. Therefore, in round figures the average *physiologically available energy* in each gram of the three different foodstuffs in the diet is:

	Calories
Carbohydrates	4.0
Fat	9.0
Protein	4.0

Average Americans receive approximately 15 per cent of their energy from protein, about 40 per cent from fat, and 45 per cent from carbohydrates. In most other parts of the world the quantity of energy derived from carbohydrates far exceeds that derived from both proteins and fats. Indeed, in Mongolia the energy received from fats and proteins combined is said to be no greater than 15 to 20 per cent.

Table 71–1 gives the compositions of selected foods, illustrating especially the high proportions of fats and proteins in meat products and the high proportions of carbohydrates in most vegetable and grain products. Fat is deceptive in the diet, for it usually exists as 100 per cent fat, whereas both proteins and carbohydrates are mixed in watery media so that each of these normally represents less than 25 per cent of the weight. Therefore, the fat of one pat of butter mixed with an entire helping of potato sometimes contains as much energy as the potato itself.

Daily Requirement for Protein. Twenty to 30 grams of the body proteins are degraded and used for producing other body chemicals daily. Therefore, all cells must continue to form new proteins to take the place of those that are being destroyed, and a supply of protein is needed in the diet for this purpose. An average person can maintain normal stores of protein provided the *daily intake is above 30 to 55 grams.*

However, as was discussed in Chapter 69, some proteins have inadequate quantities of certain essential amino acids and therefore cannot be used to form the body proteins. Such proteins are called *partial proteins,* and when they are present in large quantities in the diet, the daily requirement of protein is much greater than normally. In general, proteins derived from animal foodstuffs are more nearly complete than are proteins derived from vegetable and grain sources. A special example of dietary deficiency resulting from partial proteins occurs in the diet of many African natives who subsist primarily on corn meal. The protein of corn has almost no tryptophan; therefore, for practical purposes, the entire diet of these natives is almost completely protein deficient. As a result, these natives, especially the children, develop the protein deficiency syndrome called *kwashiorkor,* which consists of failure to grow, lethargy, depressed mentality, and hypoprotein edema.

Table 71–1 PROTEIN, FAT, AND CARBOHYDRATE
CONTENT OF DIFFERENT FOODS

Food	% Protein	% Fat	% Carbohydrate	Fuel Value per 100 Grams, Calories
Apples	0.3	0.4	14.9	64
Asparagus	2.2	0.2	3.9	26
Bacon, fat	6.2	76.0	0.7	712
broiled	25.0	55.0	1.0	599
Beef, medium	17.5	22.0	1.0	268
Beets, fresh	1.6	0.1	9.6	46
Bread, white	9.0	3.6	49.8	268
Butter	0.6	81.0	0.4	733
Cabbage	1.4	0.2	5.3	29
Carrots	1.2	0.3	9.3	45
Cashew nuts	19.6	47.2	26.4	609
Cheese, Cheddar, American	23.9	32.3	1.7	393
Chicken, total edible	21.6	2.7	1.0	111
Chocolate	(5.5)	52.9	(18.0)	570
Corn (maize), entire	10.0	4.3	73.4	372
Haddock	17.2	0.3	0.5	72
Lamb, leg, intermediate	18.0	17.5	1.0	230
Milk, fresh whole	3.5	3.9	4.9	69
Molasses, medium	0.0	0.0	(60.0)	240
Oatmeal, dry, uncooked	14.2	7.4	68.2	396
Oranges	0.9	0.2	11.2	50
Peanuts	26.9	44.2	23.6	600
Peas, fresh	6.7	0.4	17.7	101
Pork, ham, medium	15.2	31.0	1.0	340
Potatoes	2.0	0.1	19.1	85
Spinach	2.3	0.3	3.2	25
Strawberries	0.8	0.6	8.1	41
Tomatoes	1.0	0.3	4.0	23
Tuna, canned	24.2	10.8	0.5	194
Walnuts, English	15.0	64.4	15.6	702

Carbohydrates and Fats as "Protein Sparers." It was also noted in Chapters 67 and 68 that when the diet contains an abundance of carbohydrates and fats, almost all the body's energy is derived from these two substances and very little is derived from proteins. Therefore, both carbohydrates and fats are said to be *protein sparers.* On the other hand, in starvation, after the carbohydrates and fats have been depleted, the body's protein stores are then consumed rapidly for energy, sometimes at rates approaching several hundred grams per day rather than at the normal daily rate of 30 to 55 grams.

Clinical and Experimental Methods for Determining Metabolic Utilization of Proteins, Carbohydrates, and Fats

In animal experiments, and also in human beings who have metabolic problems, it is often important to determine the relative rates of utilization of proteins, carbohydrates, and fats. Two principal methods are used for this purpose:

Determination of the Rate of Protein Metabolism in the Body. The average protein of the diet contains approximately 16 per cent nitrogen. During metabolism of the protein about 90 per cent of this nitrogen is excreted in the urine in the form of urea, uric acid, creatinine, and other less important nitrogen products. The remaining 10 per cent is excreted in the feces. Therefore, it is a simple matter to estimate quite accurately the rate of protein breakdown in the body by simply measuring the amount of nitrogen in the urine, then adding another 11 per cent for the nitrogen excreted in the feces, and finally multiplying by 6.25 to determine the total amount of protein metabolism in grams per day. Thus, excretion of 8 grams of nitrogen in the urine each day means that there has been about 55 grams of protein breakdown.

If the daily intake of protein is less than the daily breakdown of protein, the person is said to have a *negative nitrogen balance,* which means that his or her body stores of protein are decreasing daily.

Relative Utilization of Fat and Carbohydrates— The "Respiratory Quotient." When carbohydrates are metabolized with oxygen, for each molecule of oxygen consumed exactly 1 carbon dioxide molecule is formed. This ratio of carbon dioxide output to oxygen usage is called the *respiratory quotient,* and therefore the respiratory quotient for carbohydrates is 1.0.

On the other hand, when fat is oxidized in the body's cells, on average only 70 carbon dioxide molecules are formed for each 100 molecules of oxygen consumed. Therefore, the respiratory quotient for the metabolism of fat averages 0.70. In a similar manner, when proteins are oxidized by the cells, the average respiratory quotient is 0.80. The reason for these lower respiratory quotients for fat and protein than for carbohydrates is that a large share of the oxygen metabolized with these foods is required to combine with the excess hydrogen atoms present in their molecules so that less carbon dioxide is formed.

Now let us see how one can utilize this respiratory quotient to determine the relative utilization of different foods by the body. First, it will be recalled from Chapter 39 that the output of carbon dioxide by the lungs divided by the uptake of oxygen during the same period of time is called the *respiratory exchange ratio.* Over a period of an hour or more, the respiratory exchange ratio exactly equals the average respiratory quotient of the metabolic reactions throughout the body. And it is obvious that if a person has a respiratory quotient of 1.0, he will probably be metabolizing carbohydrates, because the respiratory quotients for both fat and protein metabolism are considerably less than 1.0. Likewise, when the respiratory quotient is approximately 0.70, the body is metabolizing fat almost entirely, to the exclusion of carbohydrates and proteins. And, finally, if we ignore the normally small amount of protein metabolism, respiratory quotients between 0.70 and 1.0 describe the approximate ratios of carbohydrate to fat metabolism. To be more exact, though, one can first determine the protein utilization by measuring nitrogen excretion and then, using appropriate mathematical formulation, calculate almost exactly the utilization of the three different foodstuffs.

Some of the important findings from studies of respiratory quotient are the following:

1. Immediately after a meal almost all the food that is metabolized is carbohydrates, so that the respiratory quotient at that time approaches 1.0.

2. Approximately 8 to 10 hours after a meal, the body has already used up most of its readily available carbohydrate, and the respiratory quotient approaches that for fat metabolism, approximately 0.70.

3. In diabetes mellitus, very little carbohydrate can be utilized by the body's cells under any conditions because insulin is required for this. Therefore, when diabetes is very severe, the respiratory quotient remains most of the time very near to that for fat metabolism, 0.70.

● REGULATION OF FOOD INTAKE

Hunger. The term "hunger" means a craving for food, and it is associated with a number of objective sensations. For instance, in a person who has not had food for many hours, the stomach undergoes intense rhythmic contractions called *hunger contractions.* These cause a tight or gnawing feeling in the pit of the stomach and sometimes actually cause pain called *hunger pangs.* However, even after the stomach is completely removed, the psychic sensations of hunger still occur, and craving for food still makes the person search for an adequate food supply.

Appetite. The term "appetite" is often used in the same sense as hunger except that it usually implies desire for specific types of food instead of food in general. Therefore, appetite helps a person choose the quality of food to eat.

Satiety. Satiety is the opposite of hunger. It means a feeling of fulfillment in the quest for food. Satiety usually results from a filling meal, particularly when the person's nutritional storage depots, the adipose tissue and glycogen stores, are already filled.

NEURAL CENTERS FOR REGULATION OF FOOD INTAKE

Hunger and Satiety Centers. Stimulation of the *lateral hypothalamus* causes an animal to eat voraciously, which is called *hyperphagia.* On the other hand, stimulation of the *ventromedial nuclei of the hypothalamus* causes complete satiety; and even in the presence of highly appetizing food, the animal will still refuse to eat, which is *aphagia.* Conversely, destructive lesions of the two areas cause results exactly opposite to those caused by stimulation. That is, ventromedial lesions cause voracious and continued eating until the animal becomes extremely obese, sometimes as large as four times normal size. And lesions of the lateral nuclei on the two sides of the hypothalamus cause complete lack of desire for food and progressive inanition of the animal. Therefore, we can label the lateral nuclei of the hypothalamus as a *hunger center* or *feeding center,* and we can label the ventromedial nuclei of the hypothalamus as a *satiety center.*

The feeding center operates by directly exciting the emotional drive to search for food (while also stimulating other emotional drives as well — see Chapters 58 and 60). On the other hand, it is believed that the satiety center operates primarily by inhibiting the feeding center.

Other Neural Centers That Enter Into Feeding. If the brain is sectioned below the hypothalamus but above the mesencephalon, the animal can still perform the basic mechanical features of the feeding process. It can salivate, lick its lips, chew food, and swallow. Therefore, *the actual mechanics of feeding are controlled by centers in the brain stem.* The function of the hypothalamus in feeding, then, is to control the quantity of food intake and to excite the lower centers to activity.

Higher centers than the hypothalamus also play important roles in the control of feeding, particularly in the control of appetite. These centers include especially the *amygdala* and the *prefrontal cortex,* which are closely coupled with the hypothalamus. It will be recalled from the discussion of the sense of smell that portions of the amygdala are a major part of the olfactory nervous system. Destructive lesions in the amygdala have demonstrated that some of its areas greatly increase feeding, whereas others inhibit feeding. In addition, stimulation of some areas of the amygdala elicits the mechanical act of feeding. However, the most important effect of destruction of the amygdala on both sides of the brain is a "psychic blindness" in the choice of foods. In other words, the animal (and presumably the human being as well) loses or at least partially loses the mechanism of appetite control of the type and quality of food that it eats.

FACTORS THAT REGULATE FOOD INTAKE

We can divide the regulation of food into (1) *nutritional regulation* (or *long-term regulation*), which is concerned primarily with long-term maintenance of normal quantities of nutrient stores in the body, and (2) *alimentary regulation* (or *short-term regulation*), which is concerned primarily with preventing overeating at the time of each meal.

Nutritional Regulation (Long-Term Regulation)

An animal that has been starved for a long time and is then presented with unlimited food eats a far greater quantity than does an animal that has been on a regular diet. Conversely, an animal that has been force-fed for several weeks eats little when allowed to eat according to its own desires. Thus, the feeding control mechanism of the body is geared to the nutritional status of the body. Some of the nutritional factors that control the degree of activity of the feeding center are the following:

Effect of Blood Concentrations of Glucose, Amino Acids, and Lipids on Hunger and Feeding — The Glucostatic, Aminostatic, and Lipostatic Theories. It has long been known that a decrease in blood glucose concentration causes hunger, which has led to the so-called *glucostatic theory of hunger and feeding regulation.* Similar studies have more recently demonstrated the same effect for blood amino acid concentration and blood concentration of breakdown products of lipids such as the keto acids and some fatty acids, leading to the *aminostatic* and *lipostatic* theories of regulation. That is, when the availability of any of the three major types of food decreases, the animal automatically increases its feeding, which eventually returns the blood metabolite concentrations back toward normal.

Neurophysiological studies of function in the hypothalamus have also substantiated the glucostatic, aminostatic, and lipostatic theories by the following observations: (1) an increase in blood glucose level *increases* the rate of firing of *glucoreceptor neurons* in the *satiety center in the ventromedial nucleus of the hypothalamus.* (2) The same increase in blood glucose level simultaneously *decreases* the firing of neurons called *glucosensitive neurons* in the *hunger center of the lateral hypothalamus.* In addition, some amino acids and lipid substances also affect the rates of firing of these same neurons.

Still other neurons, found in the *dorsomedial nuclei of the hypothalamus,* respond to the rate of utilization of all the foodstuffs that provide energy for the cells. This has led to a more global theory of hunger and feeding regulation based on *power generation* within these cells.

Interrelationship Between Body Temperature and Food Intake. When an animal is exposed to cold, it tends to overeat; when exposed to heat, it tends to undereat. This is caused by interaction within the hypothalamus between the temperature-regulating system (see Chapter 73) and the food intake–regulating system. It is important because increased food intake in the cold animal (1) increases its metabolic rate and (2) provides increased fat for insulation, both of which tend to correct the cold state.

Summary of Long-Term Regulation. Even though our information on the different feedback factors in long-term feeding regulation is imprecise, we can make the following general statement: When the nutrient stores of the body fall below normal, the feeding center of the hypothalamus becomes highly active, and the person exhibits increased hunger; on the other hand, when the nutrient stores are abundant, the person loses the hunger and develops a state of satiety.

Alimentary Regulation of Feeding (Short-Term Regulation)

When a person is driven by hunger to eat, what turns off the eating when he or she has eaten enough? It is not the nutritional feedback mechanisms that we have discussed above, because all of them take an hour to several hours before enough quantities of the nutritional factors are absorbed into the blood to cause the necessary inhibition of eating. Yet it is very important that the person not overeat and even that he eat an amount of food that approximates his nutritional needs. The following are several different types of signals that are important for this purpose:

Gastrointestinal Filling. When the gastrointestinal tract becomes distended, especially the stomach and the duodenum, inhibitory signals are transmitted mainly by way of the vagi to suppress the feeding center, thereby reducing the desire for food.

Humoral and Hormonal Factors That Suppress Feeding — Cholecystokinin, Glucagon, and Insulin. The gastrointestinal hormone *cholecystokinin,* released mainly in response to fat entering the duodenum, has a strong direct effect on the feeding center to reduce further eating. In addition, for reasons that are not entirely understood, the presence of food in the stomach and duodenum causes the pancreas to secrete significant quantities of both *glucagon* and *insulin,* both of which also suppress the hypothalamic feeding center.

Metering of Food by Oral Receptors. When a person with an esophageal fistula is fed large quantities of food, even though this food is immediately lost again to the exterior, the degree of hunger is decreased after a reasonable quantity of food has passed through the mouth. This effect occurs despite the fact that the gastrointestinal tract does not become the least bit filled. Therefore, it is postulated that various "oral factors" relating to feeding, such as chewing, salivation, swallowing, and tasting, "meter" the food as it passes through the mouth, and after a certain amount has passed, the hypothalamic feeding center becomes inhibited. However, the inhibition caused by this metering mechanism is considerably less intense and less lasting, usually lasting only 20 to 40 minutes, than is the inhibition caused by gastrointestinal filling.

Importance of Having Both Long- and Short-Term Regulatory Systems for Feeding

The long-term regulatory system for feeding, which includes all the metabolite feedback mechanisms, obviously helps to maintain constant stores of nutrients in the tissues, preventing these from becoming too low or becoming too high. The short-term regulatory stimuli, on the other hand, serve two other purposes. First, they make the person or animal eat smaller quantities at a time, thus allowing food to pass through the gastrointestinal tract at a steadier pace so that its digestive and absorptive mechanisms can work at more optimal rates rather than becoming excessively overburdened only when the animal needs food. Second, they prevent the person or animal from eating amounts of food at each meal that would be too much for the metabolic storage systems once the food has all been absorbed.

● OBESITY

Energy Input Versus Energy Output. When greater quantities of energy (in the form of food) enter the body than are expended, the body weight increases. Therefore, obesity is obviously caused by excess energy input over energy output. For each 9.3 Calories of excess energy entering the body, 1 gram of fat is stored.

Excess energy input occurs *only during the developing phase of obesity.* Once a person has become obese, all that is required to remain obese is that the energy input equal the energy output. For the person to reduce in weight, the input must be *less* than the output. Indeed, studies of obese persons have shown that the intake of food of most of them in the static stage of obesity (after the obesity has already been attained) is approximately the same as that for normal persons.

Effect of Muscular Activity on Energy Output. About one third of the energy used each day by the normal person goes into muscular activity, and in the laborer as much as two thirds or occasionally three fourths is used in this way. Because muscular activity is by far the most important means by which energy is expended in the body, it is frequently said that obesity in the otherwise normal person results from *too high a ratio of food intake to daily exercise.*

Abnormal Feeding Regulation as a Pathological Cause of Obesity

We have already emphasized that the rate of feeding is normally regulated in proportion to the nutrient stores in the body. When these stores begin to approach an optimal level in a normal person, feeding is automatically reduced to prevent overstorage. However, in many obese persons this is not true, for feeding does not slacken until body weight is far above normal. Therefore, in effect, obesity is often caused by an abnormality of the feeding regulatory mechanism. This can result from either psychogenic factors that affect the regulation or actual abnormalities of the hypothalamus itself.

Psychogenic Obesity. Studies of obese patients show that a large proportion of obesity results from psychogenic factors. Perhaps the most common psychogenic factor contributing to obesity is the prevalent idea that healthy eating habits require three meals a day and that each meal must be filling. Many young children are forced into this habit by overly solicitous parents, and the children continue to practice it throughout life. In addition, persons are known often to gain large amounts of weight during or after stressful situations, such as the death of a parent, a severe illness, or even mental depression. It seems that eating is often a means of release from tension.

Hypothalamic Abnormalities as a Cause of Obesity. In the preceding discussion of feeding regulation, it was pointed out that lesions in the ventromedial nuclei of the hypothalamus cause an animal to eat excessively and become obese. It has also been discovered that such lesions cause excess insulin production, which in turn increases fat deposition. In addition, many persons with hypophysial tumors that encroach on the hypothalamus develop progressive obesity, illustrating that obesity in the human being, too, can definitely result from damage to the hypothalamus.

Yet in the normal obese person hypothalamic damage is almost never found. Nevertheless, it is possible that the functional organization of the feeding center is different in the obese person from that of the nonobese person. For instance, a normally obese person who has reduced to normal weight by strict dietary measures usually develops intense hunger that is demonstrably far greater than that of the normal person. This indicates that the "set-point" of the obese person's feeding center is at a much higher level of nutrient storage than that of the normal person.

Genetic Factors in Obesity. Obesity definitely runs in families. Furthermore, identical twins usually maintain weight levels within 2 pounds of each other throughout life if they live under similar conditions, or within 5 pounds of each other if their conditions of life differ markedly. This might result partly from eating habits engendered during childhood, but it is generally believed that this close similarity between twins is genetically controlled.

The genes can direct the degree of feeding in several different ways, including (1) a genetic abnormality of the feeding center to set the level of nutrient storage high or low and (2) abnormal hereditary psychic factors that either whet the appetite or cause the person to eat as a "release" mechanism.

Genetic abnormalities in the *chemistry of fat storage* are also known to cause obesity in certain strains of rats and mice. In one strain of rats, fat is easily stored in the adipose tissue, but the quantity of hormone-sensitive lipase in the adipose tissue is greatly reduced, so that little of the fat can be removed. This obviously results in a one-way path, the fat continually being deposited but never released. In a strain of obese mice that has been studied, there is an excess of fatty acid synthetase, which causes excess synthesis of fatty acids. Therefore, similar genetic mechanisms are possible causes of obesity in human beings.

Childhood Overnutrition as a Possible Cause of Obesity. The rate of formation of new fat cells is especially rapid in the first few years of life, and the greater the rate of fat storage, the greater also becomes the number of fat cells. In obese children the number of fat cells is often as much as three times that in normal children. However, after adolescence, the number of fat cells remains almost identically the same throughout the remainder of life. Therefore, it has been suggested that overfeeding children, especially in infancy and to a lesser extent during the older years of childhood, can lead to a lifetime of obesity. The person who has excess fat cells is thought to have a higher setting of fat storage by the hypothalamic feedback autoregulatory mechanism for control of adipose tissues.

In less obese persons, especially those who become obese in middle or old age, most of the obesity results from hypertrophy of already existing fat cells. This type of obesity is far more susceptible to treatment than is the lifelong type.

TREATMENT OF OBESITY

Treatment of obesity depends simply on decreasing energy input below energy expenditure. In other words, this means partial starvation. For this purpose, most diets are designed to contain large quantities of "bulk," which, in general, are made up of nonnutritive cellulose substances. This bulk distends the stomach and thereby partially appeases the hunger. In most lower animals such a procedure simply makes the animal increase its food intake still further, but human beings can often fool themselves because their food intake is sometimes controlled as much by habit as by hunger. As pointed out later in connection with starvation, it is important to prevent vitamin deficiencies during the dieting period.

Various *drugs for decreasing the degree of hunger* have been used in the treatment of obesity. The most important of these is *amphetamine* (or amphetamine derivatives), which directly inhibits the feeding center in the lateral hypothalamus. However, there is danger in using this drug because it simultaneously overexcites the central nervous system, making the person nervous and elevating the blood pressure. Also, a person soon adapts to the drug so that weight reduction is usually no greater than 5 to 10 per cent.

Finally, the more exercise one takes, the greater is the daily energy expenditure and the more rapidly the obesity disappears. Therefore, forced exercise is often an essential part of the treatment for obesity.

● INANITION

Inanition is the exact opposite of obesity. In addition to inanition caused by inadequate availability of food, both psychogenic and hypothalamic abnormalities can on occasion cause greatly decreased feeding. One such condition, *anorexia nervosa,* is an abnormal psychic state in which a person loses all desire for food and even becomes nauseated by food; as a result, severe inanition occurs. Also, destructive lesions of the hypothalamus, often caused by vascular thrombosis, frequently cause a condition called *cachexia;* the term simply means severe inanition.

● STARVATION

Depletion of Food Stores in the Body Tissues During Starvation. Even though the tissues preferentially use carbohydrate for energy over both fat and protein, the quantity of carbohydrate stores of the body is only a few hundred grams (mainly glycogen in the liver and muscles), and it can supply the energy required for body function for perhaps half a day. Therefore, except for the first few hours of starvation, the major effects are progressive depletion of tissue fat and protein. Because fat is the prime source of energy, its rate of depletion continues unabated, as illustrated in Figure 71–1, until most of the fat stores in the body are gone.

Protein undergoes three different phases of depletion: rapid depletion at first, then greatly slowed depletion, and finally, rapid depletion again shortly before death. The initial rapid depletion is caused by use of easily mobilizable protein for direct metabolism or for conversion to glucose and then metabolism of the glucose mainly by the brain. However, after the readily mobilizable protein stores have been depleted during the early phase of starvation, the remaining protein is not so easily removed. At this time, the rate of gluconeogenesis decreases to one-third to one-fifth its previous rate, and the rate of depletion of protein becomes greatly decreased. The lessened availability of glucose then initiates a series of events that often leads to *ketosis,* which was discussed in Chapter 68. Fortunately, the ketone bodies, like glucose, can cross the blood-brain barrier and can be utilized by the brain cells for energy. Therefore, approximately two thirds of the brain's energy now is derived from these ketone bodies, principally from beta-hydroxybutyrate. This sequence of events thus leads to at least partial preservation of the protein stores of the body.

However, there finally comes a time when the fat stores also are almost totally depleted, and the only remaining source of energy is proteins. At that time, the protein stores once again enter a stage of rapid depletion. Since the proteins are essential for maintenance of cellular function, death ordinarily ensues when the proteins of the body have been depleted to approximately one-half their normal level.

Vitamin Deficiencies in Starvation. The stores of some of the vitamins, especially the water-soluble vitamins —the vitamin B group and vitamin C—do not last long during starvation. Consequently, after a week or more of starvation mild vitamin deficiencies usually begin to appear, and after several weeks severe vitamin deficiencies can occur. Obviously, these can add to the debility that leads to death.

● VITAMINS

Daily Requirements of Vitamins. A vitamin is an organic compound needed in small quantities for operation of normal bodily metabolism and that cannot be manufactured in the cells of the body. When lacking in the diet, vitamins can cause specific metabolic deficits.

Table 71–2 lists the amounts of important vitamins required daily by the average adult man. These requirements vary considerably, depending on such factors as body size, rate of growth, amount of exercise, pregnancy, and so forth.

Storage of Vitamins in the Body. Vitamins are stored to a slight extent in all cells. However, some vitamins are stored to a major extent in the liver. For instance, the quantity of vitamin A stored in the liver may be sufficient to maintain a person without any intake of vitamin A for up to 10 months, and ordinarily the quantity of vitamin D stored in the liver is sufficient to maintain a person for 2 to 4 months without any additional intake of vitamin D.

The storage of most water-soluble vitamins is relatively slight. This applies especially to the vitamin B compounds, for when a person's diet is deficient in vitamin B compounds, clinical symptoms of the deficiency can sometimes be recognized within a few days (except for vitamin B_{12}, which can last in the liver for a year or longer). Absence of vitamin C, another water-soluble vitamin, can cause symptoms within a few weeks and can cause death from scurvy in 20 to 30 weeks.

VITAMIN A

Vitamin A occurs in animal tissues as *retinol,* the formula of which is illustrated. This vitamin does not occur in foods

Figure 71–1. Effect of starvation on the food stores of the body.

Table 71–2 REQUIRED DAILY AMOUNTS OF THE VITAMINS

A	5000 IU
Thiamine	1.5 mg
Riboflavin	1.8 mg
Niacin	20 mg
Ascorbic acid	45 mg
D	400 IU
E	15 IU
K	None
Folic acid	0.4 mg
B_{12}	3 μg
Pyridoxine	2 mg
Pantothenic acid	Unknown

of vegetable origin, but *provitamins* for the formation of vitamin A do occur in abundance in many vegetable foods. These are the yellow and red *carotenoid pigments,* which, because they have chemical structures similar to that of vitamin A, can be changed into vitamin A in the liver.

CH₃ structure

Vitamin A₁

The basic function of vitamin A in the metabolism of the body is not known except in relation to its use in the formation of the retinal pigments of the eye, which is discussed in Chapter 50. Nevertheless, some of the effects of vitamin A lack have been well documented. In addition to the need for vitamin A to form the visual pigments and therefore to prevent night blindness, it is also necessary for normal growth of most cells of the body and especially for normal growth and proliferation of the different types of epithelial cells. When vitamin A is lacking, the epithelial structures of the body tend to become stratified and keratinized. Vitamin A deficiency manifests itself by (1) scaliness of the skin and sometimes acne; (2) failure of growth of young animals, including cessation of skeletal growth; (3) failure of reproduction, associated especially with atrophy of the germinal epithelium of the testes and sometimes with interruption of the female sexual cycle; and (4) keratinization of the cornea with resultant corneal opacity and blindness.

Also, the damaged epithelial structures often become infected, for example, in the eyes, the kidneys, or the respiratory passages. Therefore, vitamin A has been called an "anti-infection" vitamin.

THIAMINE (VITAMIN B₁)

Thiamine operates in the metabolic systems of the body principally as *thiamine pyrophosphate;* this compound functions as a *cocarboxylase,* operating mainly in conjunction with a protein decarboxylase for decarboxylation of pyruvic acid and other α-keto acids, as discussed in Chapter 67.

Thiamine chloride

Thiamine deficiency causes decreased utilization of pyruvic acid and some amino acids by the tissues but increased utilization of fats. Thus, thiamine is specifically needed for final metabolism of carbohydrates and many amino acids. The decreased utilization of these nutrients is the responsible factor for many debilities associated with thiamine deficiency.

Thiamine Deficiency and the Nervous System. The central nervous system normally depends almost entirely on the metabolism of carbohydrates for its energy. In thia-

mine deficiency the utilization of glucose by nervous tissue may be decreased as much as 50 to 60 per cent. The neuronal cells of the central nervous system frequently show chromatolysis and swelling during thiamine deficiency, changes that are characteristic of neuronal cells with poor nutrition. Obviously, such changes as these can disrupt communication in many different portions of the central nervous system.

Also, thiamine deficiency can cause *degeneration of myelin sheaths* of nerve fibers both in the peripheral nerves and in the central nervous system. The lesions in the peripheral nerves frequently cause these nerves to become extremely irritable, resulting in "polyneuritis" characterized by pain radiating along the course of one or more peripheral nerves. Also, fiber tracts in the cord can degenerate to such an extent that *paralysis* occasionally results; and, even in the absence of paralysis, the muscles atrophy, with resultant severe weakness.

Thiamine Deficiency and the Cardiovascular System. Thiamine deficiency also weakens the heart muscle, so that a person with severe thiamine deficiency eventually develops *cardiac failure.* Furthermore, the return of blood to the heart may be increased to as much as two times normal. This occurs because thiamine deficiency causes *peripheral vasodilation* throughout the circulatory system, possibly as a result of decreased release of metabolic energy in the tissues and this leading to local vascular dilation. Therefore, the cardiac effects of thiamine deficiency are due partly to excessive load of blood to the heart and partly to primary weakness of the cardiac muscle. *Peripheral edema* and *ascites* also occur to a major extent in some persons with thiamine deficiency because of the cardiac failure.

Thiamine Deficiency and the Gastrointestinal Tract. Among the gastrointestinal symptoms of thiamine deficiency are indigestion, severe constipation, anorexia, gastric atony, and hypochlorhydria. All these effects presumably result from failure of the smooth muscle and glands of the gastrointestinal tract to derive sufficient energy from carbohydrate metabolism.

The overall picture of thiamine deficiency, including polyneuritis, cardiovascular symptoms, and gastrointestinal disorders, is frequently referred to as "beriberi" — especially when the cardiovascular symptoms predominate.

NIACIN

Niacin, also called *nicotinic acid,* functions in the body as coenzymes in the forms of nicotinamide adenine dinucleotide (NAD) and nicotinamide adenine dinucleotide phosphate (NADP). These coenzymes are hydrogen acceptors; they combine with hydrogen atoms as they are removed from food substrates by many different types of dehydrogenases. Typical operation of both of them is presented in Chapter 67. When a deficiency of niacin exists, the normal rate of dehydrogenation cannot be maintained; therefore, oxidative delivery of energy from the foodstuffs to the functioning elements of all cells cannot occur at normal rates.

—COOH

Niacin

In the early stages of niacin deficiency simple physiological changes such as muscular weakness and poor glandular secretion may occur, but in severe niacin deficiency actual death of tissues ensues. Pathological lesions appear in many parts of the central nervous system, and permanent dementia or any of many different types of psychoses may result. Also, the skin develops a cracked, pigmented scaliness in areas that are exposed to mechanical irritation or sun irradiation; thus, it seems as though the skin were unable to repair the different types of irritative damage.

Niacin deficiency causes intense irritation and inflammation of the mucous membranes of the mouth and other portions of the gastrointestinal tract, thus instituting many digestive abnormalities, leading in severe cases to widespread gastrointestinal hemorrhage. It is possible that this results from generalized depression of metabolism in the gastrointestinal epithelium and failure of appropriate epithelial repair.

The clinical entity called *pellagra* and the canine disease called black tongue are caused mainly by niacin deficiency. Pellagra is greatly exacerbated in persons on a corn diet (such as many of the natives of Africa) because corn is very deficient in the amino acid tryptophan, which can be converted in limited quantities to niacin in the body.

RIBOFLAVIN (VITAMIN B₂)

Riboflavin normally combines in the tissues with phosphoric acid to form two coenzymes, *flavin mononucleotide (FMN)*, and *flavin adenine dinucleotide (FAD)*. These in turn operate as hydrogen carriers in the important oxidative systems of the mitochondria. Usually, NAD, operating in association with specific dehydrogenases, accepts hydrogen removed from various food substrates and then passes the hydrogen to FMN or FAD; finally, the hydrogen is released as an ion into the mitochondrial matrix to become oxidized by oxygen, the system for which is described in Chapter 67.

Riboflavin

Deficiency of riboflavin in lower animals causes severe *dermatitis, vomiting, diarrhea, muscular spasticity* that finally becomes muscular weakness, and then *death* preceded by coma and decline in body temperature. Thus, severe riboflavin deficiency can cause many of the same effects as a lack of niacin in the diet; presumably the debilities that result in each instance are due to generally depressed oxidative processes within the cells.

In the human being riboflavin deficiency has never been known to be severe enough to cause the marked debilities noted in animal experiments, but mild riboflavin deficiency is probably common. Such deficiency causes digestive disturbances, burning sensations of the skin and eyes, cracking at the corners of the mouth, headaches, mental depression, forgetfulness, and so on.

Though the manifestations of riboflavin deficiency are usually relatively mild, this deficiency frequently occurs in association with deficiency of thiamine or niacin. There-fore, many deficiency syndromes, including *pellagra, beriberi, sprue,* and *kwashiorkor,* are probably due to a combined deficiency of a number of the vitamins, as well as to other aspects of malnutrition.

VITAMIN B₁₂

Several different *cobalamin* compounds which possess the common prosthetic group illustrated below exhibit so-called "vitamin B₁₂" activity.

Note that this prosthetic group contains cobalt, which has coordination bonds similar to those of iron in the hemoglobin molecule. It is likely that the cobalt atom functions in much the same way that the iron atom functions to combine reversibly with other substances.

Vitamin B₁₂ performs several metabolic functions, acting as a hydrogen acceptor coenzyme. Its most important function is to act as a coenzyme for reducing ribonucleotides to deoxyribonucleotides, a step that is necessary in the replication of genes. This could explain the two major functions of vitamin B₁₂: (1) promotion of growth and (2) promotion of red blood cell formation and maturation. This latter function was described in detail in Chapter 32 in relation to pernicious anemia, a type of anemia caused by failure of red blood cell maturation when vitamin B₁₂ is deficient.

A special effect of vitamin B₁₂ deficiency is often demyelination of the large nerve fibers of the spinal cord, especially of the posterior columns and occasionally of the lateral columns. As a result, persons with pernicious anemia frequently have much loss of peripheral sensation and, in severe cases, even become paralyzed.

The usual cause of vitamin B₁₂ deficiency is not lack of this vitamin in the food but instead deficiency of formation of *intrinsic factor,* which is normally secreted by the parietal cells of the gastric glands and is essential for absorption of vitamin B₁₂ by the ileal mucosa. This was discussed in Chapter 32 and in Chapter 66.

FOLIC ACID (PTEROYLGLUTAMIC ACID)

Several different pteroylglutamic acids, one of which is illustrated below, exhibit the "folic acid effect." Folic acid functions as a carrier of hydroxymethyl and formyl groups. Perhaps its most important use in the body is in the synthesis of purines and thymine, which are required for formation of deoxyribonucleic acid. Therefore, folic acid, like vitamin B₁₂, is also required for replication of the cellular genes. This perhaps explains one of the most important functions of folic acid—that is, to promote growth. Indeed, when it is absent from the diet, an animal will grow very little.

Folic acid is an even more potent growth promoter than vitamin B₁₂, and, like vitamin B₁₂, is also important for the maturation of red blood cells, as discussed in Chapter 32. However, vitamin B₁₂ and folic acid each perform specific and different functions in promoting growth and maturation of red blood cells. One of the very significant effects of folic acid deficiency is development of macrocytic anemia almost identical to that which occurs in pernicious anemia. This often can be treated effectively with folic acid alone.

Folic acid (pteroylglutamic acid)

PYRIDOXINE (VITAMIN B₆)

Pyridoxine exists in the form of *pyridoxal phosphate* in the cells and functions as a coenzyme for many different chemical reactions relating to amino acid and protein metabolism. Its most important role is that of coenzyme in the transamination process for the synthesis of amino acids. As a result, pyridoxine plays many key roles in metabolism — especially in protein metabolism. Also, it is believed to act in the transport of some amino acids across cell membranes.

Pyridoxine

Dietary lack of pyridoxine in lower animals can cause dermatitis, decreased rate of growth, development of fatty liver, anemia, and evidence of mental deterioration. Rarely, in children, pyridoxine deficiency has been known to cause convulsions, dermatitis, and gastrointestinal disturbances such as nausea and vomiting.

PANTOTHENIC ACID

Pantothenic acid mainly is incorporated in the body into *coenzyme A* (CoA), which has many metabolic roles in the cells. Two of these discussed at length in Chapters 67 and 68 are (1) conversion of decarboxylated pyruvic acid into acetyl-CoA prior to its entry into the citric acid cycle, and (2) degradation of fatty acid molecules into multiple molecules of acetyl-CoA. Thus, lack of pantothenic acid can lead to depressed metabolism of both carbohydrates and fats.

Pantothenic acid

Deficiency of pantothenic acid in lower animals can cause retarded growth, failure of reproduction, graying of the hair, dermatitis, fatty liver, and hemorrhagic adrenal cortical necrosis. In the human being no definite deficiency syndrome has been proved, presumably because of the wide occurrence of this vitamin in almost all foods and because small amounts of the vitamin can probably be synthesized in the body. Nevertheless, this does not mean that pantothenic acid is not of value in the metabolic systems of the body; indeed, it is perhaps as necessary as any other vitamin.

ASCORBIC ACID (VITAMIN C)

Ascorbic acid is essential for activating the enzyme *prolyl hydroxylase,* which promotes the hydroxylation step in the formation of hydroxyproline, an integral constituent of collagen. Without ascorbic acid the collagen that is formed is defective and weak. Therefore, this vitamin is essential for growth of subcutaneous tissue, cartilage, bone, and teeth.

Ascorbic acid
(vitamin C)

Deficiency of ascorbic acid for 20 to 30 weeks, which occurred frequently during long sailing voyages in the past, causes *scurvy,* some effects of which are the following:

One of the most important effects of scurvy is *failure of wounds to heal.* This is caused by failure of the cells to deposit collagen fibrils and intercellular cement substances. As a result, healing of a wound may require several months instead of the several days ordinarily necessary.

Lack of ascorbic acid causes *cessation of bone growth.* The cells of the growing epiphyses continue to proliferate, but no new collagen is laid down between the cells, and the bones fracture easily at the point of growth because of failure to ossify. Also, when an already ossified bone fractures in a person with ascorbic acid deficiency, the osteoblasts cannot form new bone matrix. Consequently, the fractured bone does not heal.

The *blood vessel walls become extremely fragile* in scurvy, because of failure of the endothelial cells to be cemented together properly and failure to form the collagen fibrils normally present in vessel walls. The capillaries especially are likely to rupture, and as a result many small petechial hemorrhages occur throughout the body. The hemorrhages beneath the skin cause purpuric blotches, sometimes over the entire body. To test for ascorbic acid deficiency, one can produce such petechial hemorrhages by inflating a blood pressure cuff over the upper arm; this occludes the venous return of blood, the capillary pressure rises, and red blotches occur in the skin of the forearm if there is a sufficiently severe ascorbic acid deficiency.

In extreme scurvy the muscle cells sometimes fragment; lesions of the gums with loosening of the teeth occur; infections of the mouth develop; vomiting of blood, bloody stools, and cerebral hemorrhage can all occur; and, finally, high fever often develops before death.

VITAMIN D

Vitamin D increases calcium absorption from the gastrointestinal tract and also helps control calcium deposition in the bone. The mechanism by which vitamin D increases calcium absorption is to promote active transport of calcium through the epithelium of the ileum. It especially increases the formation of a calcium-binding protein in the intestinal epithelial cells that aids in calcium absorption. The specific functions of vitamin D in relation to overall body calcium metabolism and to bone formation are presented in Chapter 79.

$$CH_3-CH-(CH_2)_3-CH \overset{\displaystyle CH_3}{\underset{\displaystyle CH_3}{<}}$$

Vitamin D₃ (cholecalciferol)

VITAMIN E

Several related compounds, one of which is illustrated below, exhibit so-called "vitamin E activity." Only rare instances of vitamin E deficiency have occurred in human beings. In lower animals, lack of vitamin E can cause degeneration of the germinal epithelium in the testis and therefore can cause male sterility. Lack of vitamin E can also cause resorption of a fetus after conception in the female. Because of these effects of vitamin E deficiency, vitamin E is sometimes called the "antisterility vitamin."

Vitamin E (alpha-tocopherol)

Also, as is true of almost all the vitamins, deficiency of vitamin E prevents normal growth, and it sometimes causes degeneration of the renal tubular cells and also muscle cells.

Vitamin E is believed to function mainly in relation to unsaturated fatty acids, providing a protective role to prevent oxidation of the unsaturated fats. In the absence of vitamin E, the quantity of unsaturated fats in the cells becomes diminished, causing abnormal structure and function of such cellular organelles as the mitochondria, the lysosomes, and even the cell membrane.

VITAMIN K

Vitamin K is necessary for the formation by the liver of prothrombin, Factor VII (proconvertin), Factor IX, and Factor X, all of which are important in blood coagulation. Therefore, when vitamin K deficiency occurs, blood clotting

Table 71–3 CONTENT IN GRAMS OF A 70 KILOGRAM MAN

Water	41,400
Fat	12,600
Protein	12,600
Carbohydrate	300
Na	63
K	150
Ca	1,160
Mg	21
Cl	85
P	670
S	112
Fe	3
I	0.014

is retarded. The function of this vitamin and its relationships with some of the anticoagulants, such as dicumarol, have been presented in greater detail in Chapter 35.

Several different compounds, both natural and synthetic, exhibit vitamin K activity. The chemical formula for one of the natural vitamin K compounds is illustrated below. Because vitamin K is synthesized by bacteria in the colon, a dietary source for this vitamin is not usually necessary; but when the bacteria of the colon are destroyed by administration of large quantities of antibiotic drugs, vitamin K deficiency occurs rapidly because of the paucity of this compound in the normal diet.

● MINERAL METABOLISM

The functions of many of the minerals, such as sodium, potassium, chloride, and so forth, have been presented at appropriate points in the text. Therefore, only specific functions of minerals not covered elsewhere are mentioned here.

The body content of the most important minerals is listed in Table 71–3, and the daily requirements of these are given in Table 71–4.

Magnesium. Magnesium is approximately one-sixth as plentiful in cells as potassium. Magnesium is especially required as a catalyst for many intracellular enzymatic reactions, particularly those relating to carbohydrate metabolism.

The extracellular magnesium concentration is slight, only 1.8 to 2.5 mEq/liter. Increased extracellular concentration of magnesium depresses activity in the nervous system and also depresses skeletal muscle contraction. This latter effect can be blocked by administration of calcium. Low magnesium concentration causes increased irritability of the nervous system, peripheral vasodilatation, and cardiac arrhythmias — especially so after acute myocardial infarction.

Table 71–4 REQUIRED DAILY AMOUNTS OF MINERALS

Na	3.0 gm
K	1.0 gm
Cl	3.5 gm
Ca	1.2 gm
P	1.2 gm
Fe	18.0 mg
I	150.0 μg
Mg	0.4 gm
Co	Unknown
Cu	Unknown
Mn	Unknown
Zn	15 mg

Vitamin K$_1$ (2-methyl-3-phytyl-1,4-naphthoquinone)

Calcium. Calcium is present in the body mainly in the form of calcium phosphate in the bone. This subject is discussed in detail in Chapter 79, as is also the calcium content of the extracellular fluid.

Excess quantities of calcium ions in the extracellular fluids can cause the heart to stop in systole and can act as a mental depressant. At the other extreme, low levels of calcium can cause spontaneous discharge of nerve fibers, resulting in tetany. This, too, is discussed in Chapter 79.

Phosphorus. Phosphate is the major anion of intracellular fluids. Phosphates have the ability to combine reversibly with many coenzyme systems and also with multiple other compounds that are necessary for operation of the metabolic processes. Many important reactions of phosphates have been catalogued at other points in this text, especially in relation to the functions of ATP, ADP, phosphocreatine, and so forth. Also, bone contains a tremendous amount of calcium phosphate, which is discussed in Chapter 79.

Iron. The function of iron in the body, especially in relation to the formation of hemoglobin, was discussed in Chapter 32. Two thirds of the iron in the body is in the form of hemoglobin, though smaller quantities are present in other forms, especially in the liver and in the bone marrow. Electron carriers containing iron (especially the cytochromes) are present in the mitochondria of all cells of the body and are essential for most of the oxidation that occurs in the cells. Therefore, iron is absolutely essential both for transport of oxygen to the tissues and for operation of oxidative systems within the tissue cells, without which life would cease within a few seconds.

Important Trace Elements in the Body. A few elements are present in the body in such small quantities that they are called *trace elements.* Usually, the amounts of these in the foods are also minute. Yet without any one of them a specific deficiency syndrome is likely to develop. Three of the most important of these are:

Iodine. The best known of the trace elements is iodine. This element is discussed in Chapter 76 in connection with thyroid hormone; as illustrated in Table 71–3, the entire body contains an average of only 14 milligrams. Iodine is essential for the formation of *thyroxine* and *triiodothyronine,* the two thyroid hormones that are essential for maintenance of normal metabolic rates in all of the cells.

Zinc. Zinc is an integral part of many enzymes, one of the most important of which is *carbonic anhydrase,* present in especially high concentration in the red blood cells. This enzyme is responsible for rapid combination of carbon dioxide with water in the red blood cells of the peripheral capillary blood and for rapid release of carbon dioxide from the pulmonary capillary blood into the alveoli. Carbonic anhydrase is also present to a major extent in the gastrointestinal mucosa, in the tubules of the kidney, and in the epithelial cells of many glands of the body. Consequently, zinc in small quantities is essential for the performance of many reactions relating to carbon dioxide metabolism.

Zinc is also a component of *lactic dehydrogenase* and, therefore, is important for the interconversions between pyruvic acid and lactic acid. Finally, zinc is a component part of some *peptidases* and therefore is important for digestion of proteins in the gastrointestinal tract.

Fluorine. Fluorine does not seem to be a necessary element for metabolism, but the presence of a small quantity of fluorine in the body during the period of life when the teeth are being formed subsequently protects against carious teeth. Fluorine does not make the teeth themselves stronger but, instead, has a poorly understood effect in suppressing the cariogenic process. It has been suggested that fluorine is deposited in the hydroxyapatite crystals of the tooth enamel and combines with and therefore blocks the functions of various trace metals that are necessary for activation of the bacterial enzymes. Therefore, the enzymes remain inactive and cause no caries.

Excessive intake of fluorine causes *fluorosis,* which is manifest in its mild state by mottled teeth and in a more severe state by enlarged bones. It has been postulated that in this condition fluorine combines with trace metals in some of the metabolic enzymes, including the phosphatases, so that various metabolic systems become partially inactivated. According to this theory, the mottled teeth and enlarged bones are due to abnormal enzyme systems in the odontoblasts and osteoblasts. Even though mottled teeth are highly resistant to the development of caries, the structural strength of these teeth is considerably lessened by the mottling process.

REFERENCES

Alderton, P.: The Vitamin, Mineral Connection. Louisville, Booksworld, 1985.

Altura, B. M., and Altura, B. T.: Magnesium, electrolyte transport and coronary vascular tone. Drugs, 28 (Suppl. 1):120, 1984.

Anderson, K. E., et al.: Nutrient regulation of chemical metabolism in humans. Fed. Proc., 44:130, 1985.

Atkins, F. M., and Metcalfe, D. D.: The diagnosis and treatment of food allergy. Annu. Rev. Nutr., 4:233, 1984.

Bendich, A., and Olson, J. A.: Biological actions of carotenoids. FASEB J., 3:1927, 1989.

Berger, H.: Vitamins and Minerals in Pregnancy and Lactation. New York, Raven Press, 1988.

Brownell, K. D.: The psychology and physiology of obesity: Implications for screening and treatment. J. Am. Diet. Assoc., 84:406, 1984.

Bukowiecki, L. J.: Mechanisms of stimulus-calorigenesis coupling in brown adipose tissue. Can. J. Biochem. Cell. Biol., 62:623, 1984.

Burk, R. F.: Biological activity of selenium. Annu. Rev. Nutr., 3:53, 1983.

Christensen, H. N.: The regulation of amino acid and sugar absorption by diet. Nutr. Rev., 42:237, 1984.

Combs, G. F., Jr., and Combs, S. B.: The nutritional biochemistry of selenium. Annu. Rev. Nutr., 4:257, 1984.

Cousins, R. J.: Absorption, transport, and hepatic metabolism of copper and zinc with special reference to metallothionein and ceruloplasmin. Physiol. Rev., 65:238, 1985.

Cunningham, J. J.: Introduction to Nutritional Physiology. Philadelphia, G. F. Stickley, 1983.

De Weck, A. L.: Pathophysiologic mechanisms of allergic and pseudoallergic reactions to foods, food additives and drugs. Ann. Allergy, 53:583, 1984.

DiSilvestro, R. A., and Cousins, R. J.: Physiological ligands for copper and zinc. Annu. Rev. Nutr., 3:261, 1983.

Flint, D. J., et al.: Can obesity be controlled? News Physiol. Sci., 2:1, 1987.

Forbes, R. M., and Erdman, J. W., Jr.: Bioavailability of trace mineral elements. Annu. Rev. Nutr., 3:213, 1983.

Gershwin, M. E.: Nutrition and Immunity. New York, Academic Press, 1985.

Guengerich, F. P.: Effects of nutritive factors on metabolic processes involving bioactivation and detoxication of chemicals. Annu. Rev. Nutr., 4:207, 1984.

Guthrie, H. A.: Introductory Nutrition, 7th Ed. St. Louis, C. V. Mosby Co., 1988.

Hanson, L. A.: Biology of Human Milk. New York, Raven Press, 1988.

Henderson, L. M.: Niacin. Annu. Rev. Nutr., 3:289, 1983.

Howard, L., and Michalek, A. V.: Home parenteral nutrition (HPN). Annu. Rev. Nutr., 4:69, 1984.

Katch, F. I., and McArdle, W. D.: Nutrition, Weight Control, and Exercise. Philadelphia, Lea & Febiger, 1988.

Kinney, J. M., et al.: Nutrition and Metabolism in Patient Care. Philadelphia, W. B. Saunders Co., 1988.

Luigi, B., et al.: Nutrition in Gastrointestinal Disease. New York, Raven Press, 1987.

MacLean: Pediatric Nutrition in Clinical Practice. Menlo Park, Addison-Wesley Publishing Co., 1984.

Magnen, J. L.: Body energy balance and food intake: A neuroendocrine regulatory mechanism. Physiol. Rev., 63:314, 1983.

Maho, Y. L., et al: Starvation as a treatment for obesity: The need to conserve body protein. News Physiol. Sci., 3:21, 1988.

Marshall, C. W.: Vitamins and Minerals: Help or Harm? Philadelphia, G. F. Stickley, 1983.

Morr, M. L.: Introductory Foods. New York, Macmillan Publishing Co., 1985.

Muno, H. N., and Danford, D. E.: Nutrition, Aging, and the Elderly. New York, Plenum Publishing Corp., 1989.

Nicolaidis, S.: What determines food intake? The ischymetric theory. News Physiol. Sci., 2:104, 1987.

Nielsen, F. H.: Ultratrace elements in nutrition. Annu. Rev. Nutr., 4:21, 1984.

Olson, R. E.: The function and metabolism of vitamin K. Annu. Rev. Nutr., 4:281, 1984.

Oomura, Y., and Yoshimatsu, H.: Neural network of glucose monitoring system. J. Auton. Nerv. Syst., 10:359, 1984.

Oomura, Y.: Regulation of feeding by neural responses to endogenous factors. News Physiol. Sci., 2:199, 1987.

Paige, D. M.: Clinical Nutrition. St. Louis, C. V. Mosby Co., 1988.

Peng, S. K., and Taylor, C. B.: Cholesterol autoxidation, health and arteriosclerosis. A review of situations in developed countries. World Rev. Nutr. Diet., 44:117, 1984.

Prohaska J. R.: Functions of trace elements in brain metabolism. Physiol Rev., 67:858, 1987.

Shils, M. E., and Young, V. R.: Modern Nutrition in Health and Disease, 7th Ed. Philadelphia, Lea & Febiger, 1988.

Storlien, L. H.: The role of the ventromedial hypothalamic area in periprandial glucoregulation. Life. Sci., 360:505, 1985.

Stryer, L.: Biochemistry. New York, W. H. Freeman Co., 1988.

Sullivan, A. C., and Gruen, R. K.: Mechanisms of appetite modulation by drugs. Fed. Proc., 44:139, 1985.

Swan, P. B.: Food consumption by individuals in the United States: Two major surveys. Annu. Rev. Nutr., 3:413, 1983.

Tielsch, J. M., and Sommer, A.: The epidemiology of vitamin A deficiency and xerophthalmia. Annu. Rev. Nutr., 4:183, 1984.

Tresguerres J. A. F., et al.: Dietary essential polyunsaturated fatty acids and blood pressure. News Physiol. Sci., 4:64, 1989.

Weinsier, R. L., et al.: Handbook of Clinical Nutrition: Clinician's Manual for the Diagnosis and Management of Nutritional Problems. St. Louis, C. V. Mosby Co., 1988.

Weir, D. G., et al.: Folate metabolism, the enterohepatic circulation and alcohol. Biochem. Pharmacol., 34:1, 1985.

Williams, S. R.: Nutrition and Diet Therapy, 6th Ed. St. Louis, C. V. Mosby Co., 1989.

Wolf, G.: Multiple functions of vitamin A. Physiol. Rev., 64:873, 1984.

CHAPTER 72

• *Energetics and Metabolic Rate* •

● IMPORTANCE OF ADENOSINE TRIPHOSPHATE IN METABOLISM

In the last few chapters it has been pointed out that carbohydrates, fats, and proteins can all be used by cells to synthesize large quantities of adenosine triphosphate (ATP), and that in turn the ATP can be used as an energy source for many other cellular functions. For these reasons, ATP has been called an energy "currency" that can be created and expended. Indeed, the cells can transfer energy from the different foodstuffs to most functional systems of the cells only through this medium of ATP(or the very similar nucleotide guanosine triphosphate [GTP]). Many of the attributes of ATP were presented in Chapter 2, but others require discussion at this point.

An attribute of ATP that makes it highly valuable as a means of energy currency is the large quantity of free energy (about 7300 calories per mole under standard conditions but as much as 12,000 calories under physiological conditions) vested in each of its two high-energy phosphate bonds. The amount of energy in each bond, when liberated by decomposition of ATP, is enough to cause almost any step of any chemical reaction in the body to take place if appropriate transfer of the energy is achieved. Some chemical reactions that require ATP energy use only a few hundred of the available 12,000 calories, and the remainder of this energy is then lost in the form of heat. Yet even this inefficiency in the utilization of energy is better than not being able to energize the necessary chemical reactions at all.

Use of ATP for Synthesis of Important Cellular Components. Among the most important of the intracellular processes that require ATP is the formation of peptide linkages between amino acids during the synthesis of proteins. The different peptide linkages, depending on which types of amino acids are linked together, require from 500 to 5000 calories of energy per mole. It will be recalled from the discussion of protein synthesis in Chapter 3 that four separate high-energy phosphate bonds are expended during the cascade of reactions required for forming each peptide linkage. This provides a total of 48,000 calories of energy, which is far more than the 500 to 5000 calories eventually stored in each of the different peptide linkages.

Also, it will be recalled from the preceding chapters that ATP is utilized in the synthesis of glucose from lactic acid and in the synthesis of fatty acids from acetyl coenzyme A (acetyl-CoA). In addition, ATP is utilized for synthesis of cholesterol, phospholipids, the hormones, and almost all other substances of the body. Even the urea excreted by the kidneys requires ATP to cause its formation from ammonia.

One might wonder at the advisability of expending energy to form urea, which then is simply thrown away from the body. However, remembering the extreme toxicity of ammonia in the body fluids, one can see the value of this reaction, which keeps the ammonia concentration of the body fluids always at a low level.

Use of ATP for Muscular Contraction. Muscular contraction will not occur without energy from ATP. Myosin, one of the important contractile proteins of the muscle fiber, acts as an enzyme to cause breakdown of ATP into adenosine diphosphate (ADP), thus causing the release of energy. Only a small amount of ATP is normally degraded in muscles when muscular contraction is not occurring, but this rate of ATP usage can rise to at least 150 times the resting level during short bursts of maximal contraction. The postulated mechanism by which ATP is utilized to cause muscle contraction was discussed in Chapter 6.

Use of ATP for Active Transport Across Membranes. In Chapters 4, 27, and 65, active transport of electrolytes and various nutrients across cell membranes and from the renal tubules and gastrointestinal tract was discussed. In each instance, it was noted that active transport of most electrolytes and other substances such as glucose, amino acids, and acetoacetate can occur against an electrochemical gradient, even though the natural diffusion of the substances would be in the opposite direction. Obviously, to oppose the electrochemical gradient requires energy, as was discussed in Chapter 4. This energy is provided by ATP.

Energy for Glandular Secretion. The same principles apply to glandular secretion as to the absorption of substances against concentration gradients, for energy is required to concentrate substances as they are secreted by the glandular cell. In addition, energy is also required to synthesize the organic compounds to be secreted.

Energy for Nerve Conduction. The energy utilized during propagation of a nerve impulse is derived from the potential energy stored in the form of concentration differences of ions across the membranes. That is, a high concentration of potassium inside the fiber and a low concentration outside the fiber constitute a type of energy storage. Likewise, a high concentration of sodium on the outside of the membrane and a low concentration on the inside represent another store of energy. The energy needed to pass each action potential along the fiber membrane is derived from this energy storage, with small amounts of potassium transferring out of the cell and sodium into the cell during each of the action potentials. However, active transport systems, energized by ATP, then retransport the ions back through the membrane to their former positions.

PHOSPHOCREATINE AS A STORAGE DEPOT FOR ENERGY AND FOR BUFFERING THE CONCENTRATION OF ATP

Despite the paramount importance of ATP as a coupling agent for energy transfer, this substance is not the most abundant store of high-energy phosphate bonds in the cells. On the contrary, *phosphocreatine,* which also contains high-energy phosphate bonds, is three to eight times as abundant. Also, the high-energy bond of phosphocreatine contains about 8500 calories per mole under standard conditions, and as much as 13,000 calories per mole under conditions in the body (37°C and low concentrations of the reactants). This is somewhat greater than the 12,000 calories per mole in each of the two high-energy phosphate bonds of ATP. The formula for creatine phosphate is the following:

$$HOOC-CH_2-\underset{\underset{H}{|}}{N}-\underset{\overset{\|}{NH}}{C}-\underset{\overset{|}{H}}{N}\sim\underset{\underset{H}{\overset{|}{O}}}{\overset{\overset{\|}{O}}{P}}-OH$$

Phosphocreatine cannot act in the same manner as ATP as a coupling agent for transfer of energy between the foods and the functional cellular systems. But it can transfer energy interchangeably with ATP. When extra amounts of ATP are available in the cell, much of its energy is utilized to synthesize phosphocreatine, thus building up this storehouse of energy. Then when the ATP begins to be used up, the energy in the phosphocreatine is transferred rapidly back to ATP and from this to the functional systems of the cells. This reversible interrelationship between ATP and creatine phosphate is illustrated by the following equation:

$$Phosphocreatine \ + \ ADP$$
$$\Updownarrow$$
$$ATP \ + \ Creatine$$

Note particularly that the higher energy level of the high-energy phosphate bond in phosphocreatine — 1000 to 1500 calories per mole greater than that in ATP — causes the reaction between phosphocreatine and ADP to proceed very rapidly toward the formation of new ATP every time even the slightest amount of ATP releases energy and forms ADP. Therefore, the slightest utilization of ATP by the cells calls forth the energy from the phosphocreatine to synthesize new ATP. This effect keeps the concentration of ATP at an almost constant high level as long as any phosphocreatine remains. For this reason we can call the ATP-phosphocreatine system an ATP "buffer" system. One can readily understand the importance of keeping the concentration of ATP very nearly constant, because the rates of almost all the reactions in the body depend on this constancy.

ANAEROBIC VERSUS AEROBIC ENERGY

Anaerobic energy means energy that can be derived from foods without the simultaneous utilization of oxygen; *aerobic energy* means energy that can be derived from foods only by oxidative metabolism. In the discussions in Chapters 67 through 69, it was noted that carbohydrates, fats, and proteins can all be oxidized to cause synthesis of ATP. However, carbohydrates are the only significant foods that can be utilized to provide energy without utilization of oxygen; this energy release occurs during glycolytic breakdown of glucose or glycogen to pyruvic acid. For each mole of glucose that is split into pyruvic acid, 2 moles of ATP are formed. However, when glycogen is split to pyruvic acid, each mole of glucose in the glycogen gives rise to 3 moles of ATP. The reason for this difference is that free glucose entering the cell must be phosphorylated by 1 mole of ATP before it can begin to be split, whereas this is not true of glucose derived from glycogen because it comes from the glycogen already in the phosphorylated state without additional expenditure of ATP. Thus, the best source of energy under anaerobic conditions is the stored glycogen of the cells.

Anaerobic Energy During Hypoxia. One of the prime examples of anaerobic energy utilization occurs in acute hypoxia. When a person stops breathing, there is already a small amount of oxygen stored in the lungs and an additional amount stored in the hemoglobin of the blood. However, these are sufficient to keep the metabolic processes functioning only for about 2 minutes. Continued life beyond this time requires an additional source of energy. This can be derived for another minute or so from glycolysis, the glycogen of the cells splitting into pyruvic acid and the pyruvic acid in turn becoming lactic acid, which diffuses out of the cells as described in Chapter 67.

Anaerobic Energy Usage During Strenuous Bursts of Activity. It is common knowledge that muscles can perform extreme feats of strength for a few seconds but are much less capable during prolonged activity. The energy used during strenuous activity is derived from (1) ATP already present in the muscle cells, (2) phosphocreatine in the cells, (3) anaerobic energy released by glycolytic breakdown of glycogen to lactic acid, and (4) oxidative energy released continuously by oxidative processes in the cells. The speed of the oxidative processes cannot approach that which is required to supply all the energy demands during strenuous bursts of activity. Therefore, the other three sources of energy are called upon to their maximum extent.

The maximum amount of ATP in muscle is only about 5 millimoles per liter of intracellular fluid, and this amount can maintain maximum muscle contraction for no more than a second or so. The amount of phosphocreatine in the cells is three to eight times this amount, but even by utilization of all the phosphocreatine, the amount of time that maximum contraction can be maintained is still only a few more seconds. Release of energy by glycolysis can occur much more rapidly than can oxidative release of energy. Consequently, most of the extra energy required during strenuous activity that lasts for more than 5 to 10 seconds but less than 1 to 2 minutes is derived from anaerobic glycolysis. As a result, the glycogen content of muscles during strenuous bouts of exercise becomes reduced, whereas the lactic acid concentration of the blood rises. Then, immediately after the exercise is over, oxidative metabolism is used to reconvert about four fifths of the lactic acid into glucose; the remainder becomes pyruvic acid and is degraded and oxidized in the citric acid cycle. The reconversion to glucose occurs principally in the liver cells, and the glucose is then transported in the blood back to the muscles, where it is stored once more in the form of glycogen.

Oxygen Debt. After a period of strenuous exercise, a person continues to breathe hard and to consume excessive amounts of oxygen for at least a few minutes and sometimes for up to as long as an hour thereafter. This excess oxygen is used (1) to reconvert the lactic acid that has accumulated during exercise back into glucose, (2) to reconvert the decomposed ATP and creatine phosphate to their original states, (3) to re-establish normal concentrations of oxygen bound with hemoglobin and myoglobin, and (4) to raise the concentration of oxygen in the lungs up to its normal level. This excess consumption of oxygen after the exercise is over is called the *oxygen debt*.

The principle of oxygen debt will be discussed further in Chapter 84 in relation to sports physiology; the ability of a person to build up an oxygen debt is especially important in many types of athletics.

SUMMARY OF ENERGY UTILIZATION BY THE CELLS

With the background of the past few chapters and of the preceding discussion, we can now synthesize a composite picture of overall energy utilization by the cells as illustrated in Figure 72-1. This figure shows the anaerobic utilization of glycogen and glucose to form ATP and also the aerobic utilization of compounds derived from carbohydrates, fats, proteins, and other substances for the formation of still additional ATP. In turn, ATP is in reversible equilibrium with phosphocreatine in the cells; and because larger quantities of phosphocreatine are present in the cell than ATP, much of the stored energy of the cell is in this energy storehouse.

Energy from ATP can be utilized by the different functioning systems of the cells to provide for synthesis and growth, muscular contraction, glandular secretion, nerve impulse conduction, active absorption, and other cellular activities. If greater amounts of energy are called forth for cellular activities than can be provided by oxidative metabolism, the phosphocreatine storehouse is first utilized, and then anaerobic breakdown of glycogen follows rapidly. Thus, oxidative metabolism cannot deliver energy to the cells nearly so rapidly as can the anaerobic processes; but in contrast, it is quantitatively almost inexhaustible.

● CONTROL OF ENERGY RELEASE IN THE CELL

Rate Control of Enzyme-Catalyzed Reactions. Before it is possible to discuss the control of energy release in the cell, it is necessary to consider the basic principles of *rate control* of enzymatically catalyzed chemical reactions, which are the types of reactions that occur almost universally throughout the body.

The mechanism by which an enzyme catalyzes a chemical reaction is for the enzyme first to combine loosely with one of the substrates of the reaction. This alters the bonding forces on the substrate sufficiently that it can then react with other substances. Therefore, the rate of the overall chemical reaction is determined by both the concentration of the enzyme and the concentration of the substrate that binds with the enzyme. The basic equation expressing this concept is as follows:

$$\text{Rate of reaction} = \frac{K_1 \cdot \text{Enzyme} \cdot \text{Substrate}}{K_2 + \text{Substrate}}$$

This is called the *Michaelis-Menten equation*. Figure 72-2 illustrates the application of this equation.

Role of Enzyme Concentration in the Regulation of Metabolic Reactions. Figure 72-2 shows that *when the substrate is present in reasonably high concentration*, as shown in the right half of the figure, the rate of a chemical reaction is then determined almost entirely by the concentration of the enzyme. Thus, as the enzyme concentration increases from an arbitrary value of 1 up to 2, 4, or 8, the rate of the reaction increases proportionately, as illustrated by the rising levels of the curves. As an example, when large quantities of glucose enter the renal tubules in diabetes mellitus, the rate of reabsorption of the glucose is determined almost entirely by the concentration of the glucose transport enzymes in the proximal tubular cells because the substrate is then present in excess.

Role of Substrate Concentration in Regulation of Metabolic Reactions. Note also in Figure 72-2 that when the substrate concentration becomes low enough that only a

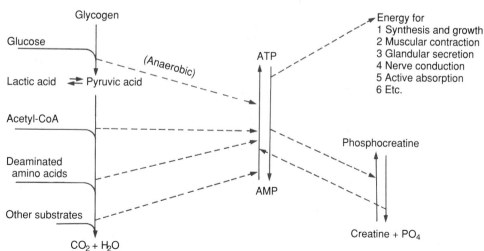

Figure 72-1. Overall schema of energy transfer from foods to the adenylic acid system and then to the functional elements of the cells. (Modified from Soskin and Levine: Carbohydrate Metabolism. Chicago, University of Chicago Press. © 1946, 1952 by The University of Chicago. All rights reserved.)

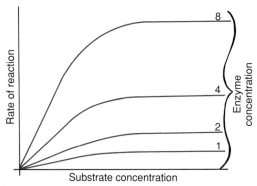

Figure 72–2. Effect of substrate and enzyme concentrations on the rate of enzyme-catalyzed reaction.

small portion of the enzyme is required in the reaction, the rate of the reaction then becomes directly proportional to the substrate concentration as well as enzyme concentration. This is the effect seen in the absorption of substances from the intestinal tract and renal tubules when their concentrations have become very low.

Rate Limitation in a Series of Reactions. Almost all chemical reactions of the body occur in series, the product of one reaction acting as a substrate for the next reaction, and so on. It is immediately obvious also that the overall rate of a complex series of chemical reactions is determined mainly by the rate of reaction of the slowest step in the series. This is called the *rate-limiting step* in the entire series.

Adenosine Diphosphate (ADP) Concentration as a Rate-Controlling Factor in Energy Release. Under *resting* conditions, the concentration of ADP in the cells is extremely slight, so that the chemical reactions that depend on ADP as one of the substrates likewise are very slow. These include all the oxidative metabolic pathways as well as essentially all other pathways for release of energy in the body. Thus, *ADP is a major rate-limiting factor for almost all energy metabolism of the body.*

When the cells become active, regardless of the type of activity, ATP is converted into ADP, increasing the concentration of ADP in direct proportion to the degree of activity of the cell. This automatically increases the rates of reactions of the ADP rate-limiting steps in the metabolic release of energy. Thus, by this simple process, the amount of energy released in the cell is controlled by the degree of activity of the cell. In the absence of cellular activity, the release of energy stops because all of the ADP soon becomes ATP.

● THE METABOLIC RATE

The *metabolism* of the body means simply all the chemical reactions in all the cells of the body, and the *metabolic rate* is normally expressed in terms of the rate of heat liberation during the chemical reactions.

Heat as the End Product of Almost All of the Energy Released in the Body. In discussing many of the metabolic reactions of the preceding chapters, we have noted that not all the energy in the foods is transferred to ATP; instead, a large portion of this energy becomes heat. On the average, 35 per cent of the energy in the foods becomes heat during ATP formation. Then still more energy becomes heat as it is transferred from ATP to the functional

systems of the cells, so that even under the best of conditions not more than 27 per cent of all the energy from the food is finally utilized by the functional systems.

Even though 27 per cent of the energy should reach the functional systems of the cells, almost all of this also eventually becomes heat for the following reasons: We might first consider the synthesis of protein and other growing elements of the body. When proteins are synthesized, large portions of ATP are used to form the peptide linkages, and this stores energy in these linkages. But we also noted in our discussions of proteins in Chapter 69 that there is continuous turnover of proteins, some being degraded while others are being formed. When the proteins are degraded, the energy stored in the peptide linkages is released in the form of heat into the body.

Now let us consider the energy used for muscle activity. Much of this energy simply overcomes the viscosity of the muscles themselves or of the tissues so that the limbs can move. The viscous movement in turn causes friction within the tissues, which generates heat.

We might also consider the energy expended by the heart in pumping blood. The blood distends the arterial system, the distention in itself representing a reservoir of potential energy. However, as the blood flows through the peripheral vessels, the friction of the different layers of blood flowing over each other and the friction of the blood against the walls of the vessels turn all of this energy into heat.

Therefore, essentially all of the energy expended by the body is eventually converted into heat. The only significant exception to this occurs when the muscles are used to perform some form of work outside the body. For instance, when the muscles elevate an object to a height or carry the person's body up steps, a type of potential energy is thus created by raising a mass against gravity. But when external expenditure of energy is not taking place, it is safe to consider that all the energy released by the metabolic processes eventually becomes body heat.

The Calorie. To discuss the metabolic rate and related subjects intelligently, it is necessary to use some unit for expressing the quantity of energy released from the different foods or expended by the different functional processes of the body. Most often, the *Calorie* is the unit used for this purpose. It will be recalled that 1 *calorie,* spelled with a small "c" and often called a *gram calorie,* is the quantity of heat required to raise the temperature of 1 gram of water 1°C. The calorie is much too small a unit for ease of expression in speaking of energy in the body. Consequently the large Calorie, sometimes spelled with a capital C and often called a *kilocalorie,* which is equivalent to 1000 calories, is the unit ordinarily used in discussing energy metabolism.

MEASUREMENT OF THE METABOLIC RATE

Direct Calorimetry. Because a person ordinarily is not performing any external work, the metabolic rate can be determined by simply measuring the total quantity of heat liberated from the body in a given time. This method is called *direct calorimetry.*

In determining the metabolic rate by direct calorimetry, one measures the quantity of heat liberated from the body in a large, specially constructed *calorimeter.* The subject is placed in an air chamber that is so well insulated that no heat can leak through the walls of the chamber. Heat formed by the subject's body warms the air of the chamber. However, the air temperature within the chamber is maintained at a constant level by forcing the air through pipes in a cool water bath. The rate of heat gain by the water bath, which can be measured with an accurate thermometer, is equal to the rate at which heat is liberated by the subject's body.

Obviously, direct calorimetry is physically difficult to perform and therefore is used only for research purposes.

Indirect Calorimetry—The "Energy Equivalent" of Oxygen. Because more than 95 per cent of the energy expended in the body is derived from reactions of oxygen with the different foods, the metabolic rate can also be calculated with a high degree of accuracy from the rate of oxygen utilization. When 1 liter of oxygen is metabolized with glucose, 5.01 Calories of energy are released; when metabolized with starches, 5.06 Calories are released; with fat, 4.70 calories; and with protein, 4.60 Calories.

From these figures, it is striking how nearly equivalent are the quantities of energy liberated per liter of oxygen regardless of the type of food that is being metabolized. For the average diet, the *quantity of energy liberated per liter of oxygen utilized in the body averages approximately 4.825 Calories.* This is called the *energy equivalent* of oxygen; using this energy equivalent, one can calculate with a high degree of precision the rate of heat liberation in the body from the quantity of oxygen utilized in a given period of time.

If a person should metabolize only carbohydrates during the period of the metabolic rate determination, the calculated quantity of energy liberated, based on the value for the average energy equivalent of oxygen (4.825 Calories per liter), would be approximately 4 per cent too little. On the other hand, if the person were obtaining most energy from fat, the calculated value would be approximately 4 per cent too great.

The Metabolator. Figure 72–3 illustrates the metabolator usually used for indirect calorimetry. This apparatus contains a floating drum, containing an oxygen chamber connected to a mouthpiece through two rubber tubes. A valve in one of these rubber tubes allows air to pass from the oxygen chamber into the mouth, while air passing from the mouth back to the chamber is directed by means of another valve through the second tube. Before the expired air from the mouth enters the upper portion of the oxygen chamber, it flows through a lower chamber containing pellets of soda lime, which combine chemically with the carbon dioxide in the expired air. Therefore, as oxygen is used by the person's body and the carbon dioxide is absorbed by the soda lime, the floating oxygen chamber, which is precisely balanced by a weight, gradually sinks in the water, owing to the oxygen loss. This chamber is coupled to a pen that records on a moving drum the rate at which the chamber sinks in the water and thereby records the rate at which the body utilizes oxygen.

FACTORS THAT AFFECT THE METABOLIC RATE

Factors that increase the chemical activity in the cells also increase the metabolic rate. Some of these are the following:

Exercise. The factor that causes by far the most dramatic effect on metabolic rate is strenuous exercise. Short bursts of maximal muscle contraction in any single muscle can liberate as much as a hundred times its normal resting amount of heat for a few seconds at a time. In considering the entire body, however, maximal muscle exercise can increase the overall heat production of the body for a few seconds to about 50 times normal or sustained for several minutes to about 20 times normal in the well-trained athlete, which is an increase in metabolic rate to 2000 per cent of normal.

Energy Requirements for Daily Activities. When an average man of 70 kilograms lies in bed all day, he utilizes approximately 1650 Calories of energy. The process of eating and digesting food increases the amount of energy utilized each day by an additional 200 or more Calories, so that the same man lying in bed and also eating a reasonable diet requires a dietary intake of approximately 1850 Calories per day. If he sits in a chair all day, his total energy requirement reaches 2000 to 2250 Calories. Therefore, in round figures, it can be assumed that the daily energy requirements simply for existing (that is, performing essential functions only) is about 2000 Calories.

Effects of Different Types of Work on Daily Energy Requirements. Table 72–1 illustrates the rates of energy utilization while one performs different types of activities. Note that walking up stairs requires approximately 17 times as much energy as lying in bed asleep. In general, over a 24-hour period a laborer can achieve a maximum rate of energy utilization as great as 6000 to 7000 Calories—in other words, as much as 3.5 times the basal rate of metabolism.

Specific Dynamic Action of Protein. After a meal is ingested, the metabolic rate increases. This is believed to

Figure 72–3. The metabolator.

Table 72-1 ENERGY EXPENDITURE PER HOUR DURING DIFFERENT TYPES OF ACTIVITY FOR A 70 KILOGRAM MAN

Form of Activity	Calories per Hour
Sleeping	65
Awake lying still	77
Sitting at rest	100
Standing relaxed	105
Dressing and undressing	118
Tailoring	135
Typewriting rapidly	140
"Light" exercise	170
Walking slowly (2.6 miles per hour)	200
Carpentry, metal working, industrial painting	240
"Active" exercise	290
"Severe" exercise	450
Sawing wood	480
Swimming	500
Running (5.3 miles per hour)	570
"Very severe" exercise	600
Walking very fast (5.3 miles per hour)	650
Walking up stairs	1100

Extracted from data compiled by Professor M. S. Rose.

result to a very slight extent from the different chemical reactions associated with digestion, absorption, and storage of food in the body. However, it mainly results from a stimulatory effect on the cellular chemical processes by certain of the amino acids derived from the proteins of the ingested food.

After a meal containing a large quantity of carbohydrates or fats, the metabolic rate usually increases only about 4 per cent. However, after a meal containing large quantities of protein, the metabolic rate usually begins rising within 1 hour, reaches a maximum about 30 per cent above normal, and lasts for as long as 3 to 12 hours. This effect of protein on the metabolic rate is called the *specific dynamic action* of protein.

Age. The metabolic rate of a young child in relation to its size is almost two times that of an old person. This is illus-

trated in Figure 72-4, which shows the declining metabolic rates of both males and females from birth until very old age. The high metabolic rate of young children results from high rates of cellular reactions, including partly the rapid synthesis of cellular materials and growth of the body, which require moderate quantities of energy.

Thyroid Hormone. When the thyroid gland secretes maximal quantities of thyroxine, the metabolic rate sometimes rises to as high as 50 to 100 per cent above normal. On the other hand, total loss of thyroid secretion decreases the metabolic rate to as low as 40 to 60 per cent of normal. These effects can readily be explained by the basic function of thyroxine of increasing the rates of activity of almost all the chemical reactions in all cells of the body. This relationship between thyroxine and metabolic rate is discussed in much greater detail in Chapter 76 in connection with thyroid function.

Sympathetic Stimulation—"Nonshivering Thermogenesis." Stimulating the sympathetic nervous system, with liberation of norepinephrine and epinephrine, increases the metabolic rate of many tissues of the body. These hormones have a direct effect on muscle and liver cells in causing glycogenolysis; and this, along with other intracellular effects, increases cellular activity. However, even more important is the effect of sympathetic stimulation on a certain type of fat tissue called *brown fat* in causing marked liberation of heat. This type of fat contains large numbers of mitochondria and many small globules of fat instead of one large fat globule. In these cells, the process of oxidative phosphorylation in the mitochondria is mainly "uncoupled." That is, when the cells are stimulated by the sympathetic nerves, their mitochondria produce a large amount of heat but almost no ATP, so that almost all the released oxidative energy immediately becomes heat. The newborn child has a considerable number of such fat cells, and maximal sympathetic stimulation can increase the child's metabolism more than 100 per cent. This is called *nonshivering thermogenesis*. The magnitude of this effect in the adult human being is in question—probably less than 10 to 15 per cent—though this might increase following cold adaptation.

Male Sex Hormone. The male sex hormone can in-

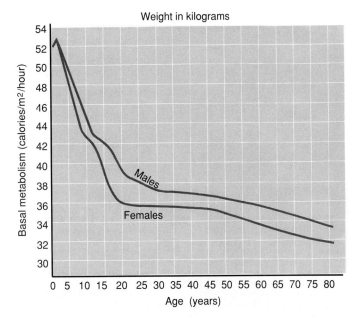

Figure 72-4. Normal basal metabolic rates at different ages for each sex.

crease the basal metabolic rate about 10 to 15 per cent, and the female sex hormone perhaps a small amount but usually not enough to be of significance. The difference in metabolic rates of males and females is illustrated in Figure 72-4.

Growth Hormone. Growth hormone can increase the basal metabolic rate as much as 15 to 20 per cent as a result of direct stimulation of cellular metabolism.

Fever. Fever, regardless of its cause, increases the metabolic rate. This is because all chemical reactions, either in the body or in the test tube, increase their rates of reaction an average of about 120 per cent for every 10°C rise in temperature. However, the body's temperature control system diminishes this effect somewhat, as we discuss in the following chapter.

Climate. Studies of metabolic rates of persons living in the different geographic zones have shown metabolic rates as much as 10 to 20 per cent lower in tropical regions than in arctic regions. This difference is caused to some extent by adaptation of the thyroid gland, with increased secretion in cold climates and decreased secretion in hot climates. Indeed, far more persons develop hyperthyroidism in cold regions of the earth than in tropical regions.

Sleep. The metabolic rate falls 10 to 15 per cent below normal during sleep. This fall is presumably due to two principal factors: (1) decreased tone of the skeletal musculature during sleep and (2) decreased activity of the sympathetic nervous system.

Malnutrition. Prolonged malnutrition can decrease the metabolic rate as much as 20 to 30 per cent; this decrease is presumably caused by paucity of the necessary food substances in the cells.

In the final stages of many disease conditions, the inanition that accompanies the disease causes a marked premortem decrease in metabolic rate, even to the extent that the body temperature may fall a number of degrees shortly before death.

THE BASAL METABOLIC RATE

The Basal Metabolic Rate as a Method for Comparing Metabolic Rates Between Individuals. It has been extremely important to establish a procedure that will measure the inherent activity of the tissues independently of exercise and other extraneous factors that would make it impossible to compare one person's metabolic rate with that of another person. To do this, the metabolic rate is usually measured under so-called *basal conditions,* and the metabolic rate then measured is called the *basal metabolic rate* (BMR).

Basal Conditions. The basal metabolic rate means the rate of energy utilization in the body during absolute rest but while the person is awake. The following basal conditions are necessary for measuring the basal metabolic rate:

1. The person must not have eaten any food for at least 12 hours.

2. The basal metabolic rate is determined after a night of restful sleep.

3. No strenuous exercise is performed during the preceding hour or more.

4. All psychic and physical factors that cause excitement must be eliminated.

5. The temperature of the air must be comfortable and be somewhere between the limits of 68° and 80°F.

Usual Technique for Determining the Basal Metabolic Rate. The usual method for determining basal meta-

bolic rate is to measure the rate of oxygen utilization using a metabolator of the type illustrated in Figure 72-3. Then the basal metabolic rate is calculated in terms of *Calories per hour.* This normally averages about 60 Calories per hour for the average young man and about 53 Calories per hour in young women.

Expressing the Basal Metabolic Rate in Terms of Surface Area. Obviously, if one subject is much larger than another, the total amount of energy utilized by the two subjects will be considerably different simply because of differences in body size. Experimentally, among normal persons, the average basal metabolic rate varies approximately *in proportion to the body surface area,* and the surface area can be determined from height and weight tables.

Therefore, to compare basal metabolic rates between persons, these are expressed as *Calories per hour per square meter.* The normal values for both males and females at different ages are shown in Figure 72-4.

Expressing the Basal Metabolic Rate in Percentage Above or Below Normal. Finally, because of the changing basal metabolic rates at different ages, it has become the practice to compare the actual rate with the normal rate shown in Figure 72-4. Then, the rate is expressed as a percentage above or below normal. Thus, the basal metabolic rate, usually called simply the BMR, is expressed as *plus 25* when it is 25 per cent too high, or *minus 15* when 15 per cent too low.

Constancy of the Metabolic Rate in the Same Person. Basal metabolic rates have been measured in many subjects at repeated intervals for as long as 20 or more years. As long as a subject remains healthy, almost invariably the BMR, when expressed as a percentage of normal, does not vary more than 5 to 10 per cent except for the age-related changes.

Constancy of the Basal Metabolic Rate From Person to Person. When the basal metabolic rate is measured in a wide variety of different persons and comparisons are made within single age, weight, and sex groups, 85 per cent of normal persons have been found to have BMRs within 10 per cent of the mean. Thus, it is obvious that measurements of metabolic rates performed under basal conditions offer a reasonable means for comparing the rate of metabolism from one person to another.

REFERENCES

Alexander, G.: Cold thermogenesis. In Robertshaw, D. (ed.): International Review of Physiology: Environmental Physiology III. Vol. 20. Baltimore, University Park Press, 1979, p. 43.

Baldwin, R. L., and Bywater, A. C.: Nutritional energetics of animals. Annu. Rev. Nutr., 4:101, 1984.

Becker, D. J.: The endocrine responses to protein-calorie malnutrition. Annu. Rev. Nutr., 3:187, 1983.

Bray, G. A.: Regulation of energy balance. Physiologist, 28:186, 1985.

Bukowiecki, L. J.: Mechanisms of stimulus-calorigenesis coupling in brown adipose tissue. Can. J. Biochem. Cell. Biol., 62:623, 1984.

Calder, W. A. III: Scaling energetics of homeothermic vertebrates: An operational allometry. Annu. Rev. Physiol., 49:107, 1987.

Durnin, J. V.: Energy balance in childhood and adolescence. Proc. Nutr. Soc., 43:271, 1984.

Golinick, P. D.: Metabolism of substrates: Energy substrate metabolism during exercise and as modified by training. Fed. Proc., 44:353, 1985.

Guyton, A. C., and Farish, C. A.: A rapidly responding continuous oxygen consumption recorder. J. Appl. Physiol., 14:143, 1959.

Homsher, E., and Kean, C. J.: Skeletal muscle energetics and metabolism. Annu. Rev. Physiol., 40:93, 1978.

Jequier, E., and Flatt, J. -P.: Recent advances in human energetics. News Physiol. Sci., 1:112, 1986.

Kim, C. H., et al. (eds.): Advances in Membrane Biochemistry and Bioenergetics. New York, Plenum Publishing Corp., 1989.

Magnen, J. L.: Body energy balance and food intake: A neuroendocrine regulatory mechanism. Physiol. Rev., 63:314, 1983.

Nicholls, D. G., and Locke, R. M.: Thermogenic mechanisms in brown fat. Physiol. Rev., 64:1, 1984.

Oppenheimer, J. H.: Thyroid hormone action at the nuclear level. Ann. Intern. Med., 102:374, 1985.

Plowman, P. N.: Endocrinology and Metabolic Diseases. New York, Elsevier Science Publishing Co., 1987.

Storlein, L. H.: The role of the ventromedial hypothalamic area in periprandial glucoregulation. Life Sci., 36:505, 1985.

Stryer, L.: Biochemistry. New York, W. H. Freeman Co., 1988.

Van der Laarse, W. J., and Woledge, R. C.: Energetics at the single cell level. News Physiol. Sci., 4:91, 1989.

Wyndham, C. H., and Loots, H.: Responses to cold during a year in Antarctica. J. Appl. Physiol., 27:696, 1969.

CHAPTER 73

• Body Temperature, Temperature Regulation, and Fever •

NORMAL BODY TEMPERATURES

Core Temperature and Skin Temperature. The temperature of the deep tissues of the body — the "core" — remains almost exactly constant, within ±1°F, day in and day out except when a person develops a febrile illness. Indeed, the nude person can be exposed to temperatures as low as 55°F or as high as 140°F in *dry* air and still maintain an almost constant internal body temperature. Obviously, the mechanisms for control of body temperature represent a beautifully designed control system. The purpose of this chapter is to discuss this system as it operates in health and in disease.

The *skin temperature,* in contrast to the *core temperature,* rises and falls with the temperature of the surroundings. This is the temperature that is important when we refer to the ability of the skin to lose heat to the surroundings.

The Normal Core Temperature. No single temperature level can be considered to be normal, for measurements on many normal persons have shown a *range* of normal temperatures, as illustrated in Figure 73–1, from less than 97°F (36°C) to over 99.5°F (37.5°C). When measured by rectum, the values are approximately 1°F greater than the oral temperatures. The average normal temperature is generally considered to be between 98.0°F and 98.6°F (36.7°C and 37°C) when measured orally and approximately 1°F or 0.6°C higher when measured rectally.

The body temperature varies somewhat with exercise and with extremes of temperature of the surroundings, because the temperature regulatory mechanisms are not 100 per cent perfect. When excessive heat is produced in the body by strenuous exercise, the rectal temperature can rise to as high as 101° to 104°F. On the other hand, when the body is exposed to cold, the rectal temperature can often fall to values below 97°F.

• BODY TEMPERATURE IS CONTROLLED BY BALANCING HEAT PRODUCTION AGAINST HEAT LOSS

When the rate of heat production in the body is greater than the rate at which heat is being lost, heat obviously builds up in the body and the body temperature rises. Conversely, when heat loss is greater, both body heat and body temperature decrease. Therefore, most of the remainder of this chapter will be concerned with this balance between heat production and heat loss and the mechanisms by which the body controls each of these.

HEAT PRODUCTION

Heat production is one of the principal byproducts of metabolism. In the previous chapter, which summarized body energetics, we discussed the different factors that determine the rate of heat production, called the *metabolic rate of the body.* The most important of these are listed again here: (1) basal rate of metabolism of all the cells of the body; (2) extra rate of metabolism caused by muscle activity, including muscle contractions caused by shivering; (3) extra metabolism caused by the effect of thyroxin on the cells; (4) extra metabolism caused by the effect of epinephrine, norepinephrine, and sympathetic stimulation on the cells; and (5) extra metabolism caused by increased temperature of the body cells themselves.

HEAT LOSS

Most of the heat produced in the body is generated in the deep organs, especially in the liver, the brain, the heart, and the skeletal muscles — especially so

Figure 73–1. Estimated range of body temperature in normal persons. (From DuBois: Fever. Springfield, Ill., Charles C Thomas.)

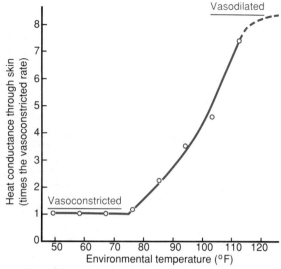

Figure 73–3. Effect of changes in the environmental temperature on the heat conductance from the body core to the skin surface. (Modified from Benzinger: Heat and Temperature: Fundamentals of Medical Physiology. New York, Dowden, Hutchinson and Ross, 1980.)

during exercise. Then this heat is transferred from the deeper organs and tissues to the skin, where it is lost to the air and other surroundings. Therefore, the rate at which heat is lost is determined almost entirely by two factors: (1) how rapidly heat can be conducted from the core to the skin, and (2) how rapidly heat can then be transferred from the skin to the surroundings. Let us begin by discussing the insulator system that insulates the core from the skin surface.

The Insulator System of the Body

The skin, the subcutaneous tissues, and the fat of the subcutaneous tissues are a heat insulator for the body. The fat is especially important because it conducts heat only *one-third* as readily as other tissues. When no blood is flowing from the heated internal organs to the skin, the insulating properties of the normal male body are approximately equal to three-quarters the insulating properties of a usual suit of clothes. In women this insulation is still better.

The insulation beneath the skin is an effective means of maintaining normal internal core temperature, even though it allows the temperature of the skin to approach the temperature of the surroundings.

"Radiator" System of the Body: Flow of Blood to the Skin From the Body Core

Blood vessels penetrate the fatty subcutaneous insulator tissues and are distributed profusely immediately beneath the skin. Especially important is a continuous venous plexus that is supplied by inflow of blood from the skin capillaries, illustrated in Figure 73–2. In the most exposed areas of the body — the hands, feet, and ears — blood is also supplied to the plexus directly from the small arteries through highly muscular venous *arteriovenous anastomoses*. The rate of blood flow into the venous plexus can vary tremendously — from barely above zero to as great as 30 per cent of the total cardiac output. A high rate of blood flow causes heat to be conducted from the core of the body to the skin with great efficiency, whereas reduction in the rate of blood flow decreases the efficiency of heat conduction from the core. Figure 73–3 shows quantitatively the effect of skin blood flow on

Figure 73–2. The skin circulation.

conductance of heat from the body core to the skin surface, illustrating an approximate eightfold increase in heat conductance between the fully vasoconstricted state and the fully vasodilated state.

Obviously, therefore, the skin is an effective "radiator" system, and the flow of blood to the skin is a most effective mechanism of heat transfer from the body core to the skin.

Control of Heat Conduction to the Skin by the Sympathetic Nervous System. Heat conduction to the skin by the blood is controlled by the degree of vasoconstriction of the arterioles and arteriovenous anastomoses that supply blood to the venous plexus of the skin. This vasoconstriction in turn is controlled almost entirely by the sympathetic nervous system in response to changes in the body core temperature and changes in the environmental temperature. This will be discussed later in the chapter in connection with the control of body temperature by the hypothalamus.

The Basic Physics of How Heat Is Lost from the Skin Surface

The various methods by which heat is lost from the skin to the surroundings are pictured in Figure 73–4. These include *radiation, conduction,* and *evaporation* and may be explained as follows:

Radiation. As illustrated in Figure 73–4, a nude person in a room at normal room temperature loses about 60 per cent of the total heat loss by radiation.

Loss of heat by radiation means loss in the form of infrared heat rays, a type of electromagnetic wave. Most infrared heat rays radiating from the body have wavelengths of 5 to 20 micrometers, 10 to 30 times the wavelengths of light rays. All objects that are not at absolute zero temperature radiate such rays. Therefore, the human body radiates heat rays in all directions. However, heat rays are also being radiated from the walls and other objects toward the body. If the temperature of the body is greater than the temperature of the surroundings, a greater quantity of heat is radiated from the body than is radiated to the body.

Conduction. As illustrated in Figure 73–4, only minute quantities of heat are normally lost from the body by direct conduction from the surface of the body *to other objects,* such as a chair or a bed. On the other hand, loss of heat by *conduction to air* does represent a sizable proportion of the body's heat loss even under

normal conditions. It will be recalled that heat is actually the kinetic energy of molecular motion, and the molecules of the skin are continually undergoing vibratory motion. Much of the energy of this motion can be transferred to the air if the air is colder than the skin, thus increasing the velocity of motion of the air molecules. However, once the temperature of the air immediately adjacent to the skin equals the temperature of the skin, no further loss of heat from the body to the air can occur. Therefore, conduction of heat from the body to the air is self-limited *unless the heated air moves away from the skin,* so that new, unheated air is continually brought in contact with the skin, a phenomenon called air convection.

Convection. The removal of heat from the body by convection air currents is commonly called heat loss by convection. Actually, the heat must first be *conducted* to the air and then carried away by the convection currents.

A small amount of convection almost always occurs around the body because of the tendency for the air adjacent to the skin to rise as it becomes heated. Therefore, a nude person seated in a comfortable room without gross air movement still loses about 12 per cent of his or her heat by conduction to the air and then by air convection away from the body.

Cooling Effect of Wind. When the body is exposed to wind, the layer of air immediately adjacent to the skin is replaced by new air much more rapidly than normally, and heat loss by convection increases accordingly. The cooling effect of wind at low velocities is approximately proportional to the *square root of the wind velocity.* For instance, a wind of 4 miles per hour is about two times as effective for cooling as a wind of 1 mile per hour.

Conduction and Convection of Heat From a Person Exposed to Water. Water has a specific heat several thousand times as great as that of air, so that each unit portion of water adjacent to the skin can absorb far greater quantities of heat than can air. Also, the conductivity of water for heat is marked in comparison with that of air. Consequently, it is impossible for the body to heat a thin layer of water next to the body to form an "insulator zone" as occurs in air. Therefore, the rate of heat loss to water at moderate temperatures is many times as great as the rate of heat loss to air of the same temperature. However, when both the water and air are extremely cold, the rate of heat loss to air becomes almost as great as to water, for both water and air are then capable of carrying away essentially all the heat that can diffuse through the subcutaneous insulation of the skin.

Evaporation. When water evaporates from the body surface, 0.58 Calorie (kilocalorie) of heat is lost for each gram of water that evaporates. Even when a person is not sweating, water still evaporates *insensibly* from the skin and lungs at a rate of about 600 milliliters per day. This causes continual heat loss at a rate of 12 to 16 Calories per hour. Unfortunately, this insensible evaporation through the skin and lungs cannot be controlled for purposes of temperature regulation because it results from continual diffusion of water molecules through the skin and respiratory surfaces regardless of body temperature. However, loss of heat by evaporation of sweat can be controlled by

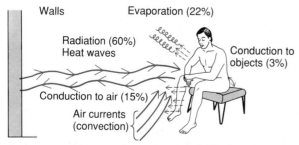

Walls   Evaporation (22%)

Radiation (60%)
Heat waves   Conduction to
                                    objects (3%)

Conduction to air (15%)

Air currents
(convection)

Figure 73–4. Mechanisms of heat loss from the body.

regulating the rate of sweating, which is discussed later in the chapter.

Evaporation as a Necessary Refrigeration Mechanism at Very High Air Temperatures. As long as skin temperature is greater than the temperature of the surroundings, heat can be lost by radiation and conduction; but when the temperature of the surroundings is greater than that of the skin, instead of losing heat, the body gains heat by radiation and conduction. Under these conditions, *the only means by which the body can rid itself of heat is evaporation.* Therefore, any factor that prevents adequate evaporation when the surrounding temperatures are higher than skin temperature will cause the body temperature to rise. This occurs occasionally in human beings who are born with congenital absence of sweat glands. These persons can stand cold temperatures as well as can normal persons, but they are likely to die of heat stroke in tropical zones, for without the evaporative refrigeration system, they cannot prevent a rise in body temperature when the air temperature is above that of the body.

Effect of Clothing on Heat Loss. *Effect on Conductive Heat Loss.* Clothing entraps air next to the skin and in the weave of the cloth, thereby increasing the thickness of the so-called "private zone" of air adjacent to the skin and decreasing the flow of convection air currents. Consequently, the rate of heat loss from the body by conduction and convection is greatly depressed. A usual suit of clothes decreases the rate of heat loss to about half that from a nude body, whereas arctic-type clothing can decrease this heat loss to as little as one sixth.

About half the heat transmitted from the skin to the clothing is radiated to the clothing instead of being conducted across the small intervening space. Therefore, coating the inside of clothing with a thin layer of gold, which reflects radiant heat back to the body, makes the insulating properties of clothing far more effective than otherwise. Using this technique, clothing for use in the arctic can be decreased in weight by about half.

The effectiveness of clothing in maintaining body temperature is almost completely lost when it becomes wet, because the high conductivity of water increases the rate of heat transmission as much as 20-fold or more. One of the most important factors for protecting the body against cold in arctic regions is extreme caution against allowing the clothing to become wet. Indeed, one must be careful not to overheat oneself even temporarily, for sweating in one's clothes makes them much less effective thereafter as an insulator.

Sweating and Its Regulation by the Autonomic Nervous System

Stimulation of the anterior hypothalamus–preoptic area either electrically or by excess heat causes sweating. The impulses from this area that cause sweating are transmitted in the autonomic pathways to the cord and thence through the sympathetic outflow to the skin everywhere in the body.

It should be recalled from the discussion of the autonomic nervous system in Chapter 60 that the sweat glands are innervated by sympathetic *cholinergic*

nerve fibers. However, these glands can also be stimulated by epinephrine or norepinephrine circulating in the blood, even though the glands themselves, in most parts of the body, do not have adrenergic innervation. This is important during exercise, when these hormones are secreted by the adrenal medullae and the body needs to lose the extra heat produced by the active muscles.

Mechanism of Sweat Secretion. Figure 73–5 illustrates a sweat gland, showing it to be a tubular structure consisting of two parts: (1) a deep subdermal *coiled portion* that secretes the sweat, and (2) a *duct portion* passing outward through the dermis and epidermis of the skin. As is true of so many other glands, the secretory portion of the sweat gland secretes a fluid called the *primary secretion* or *precursor secretion;* then the concentrations of the constituents in the fluid are modified as the fluid flows through the duct.

The precursor secretion is an active secretory product of the epithelial cells lining the coiled portion of the sweat gland. Cholinergic sympathetic nerve fibers ending on or near the glandular cells elicit the secretion.

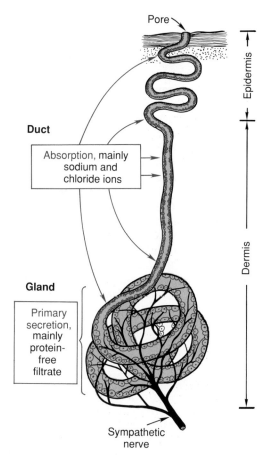

Figure 73–5. A sweat gland innervated by an acetylcholine secreting sympathetic nerve. A *primary secretion* is formed by the glandular portion, but much if not most of the electrolytes are reabsorbed in the duct, leaving a dilute, watery secretion.

The composition of the precursor secretion is very similar to that of plasma except that it does not contain the plasma proteins. The concentration of sodium is about 142 mEq/liter and chloride about 104 mEq/liter, with much smaller concentrations of the other solutes of plasma. However, as this precursor solution flows through the duct portion of the gland, it is modified by reabsorption of most of the sodium and chloride ions. The degree of this reabsorption depends upon the rate of sweating as follows:

When the sweat glands are stimulated only slightly, the precursor fluid passes through the duct very slowly. In this instance, essentially all the sodium and chloride ions are reabsorbed, and the concentration of each of these falls to as low as 5 mEq/liter. This reduces the osmotic pressure of the fluid to such a low level that most of the water is also reabsorbed, which concentrates most of the other constituents. Therefore, at low rates of sweating, such constituents as urea, lactic acid, and potassium ions are usually very concentrated.

On the other hand, when the sweat glands are strongly stimulated by the sympathetic nervous system, very large amounts of precursor secretion are formed, and the duct may now reabsorb only slightly more than one half of the sodium chloride; the concentrations of the sodium and chloride ions then usually rise (in the *unacclimatized* person) to maximum levels of about 60 mEq/liter, slightly less than half the concentrations in plasma. Furthermore, the water reabsorption is greatly reduced because the sweat flows on through the glandular tubules too rapidly. Therefore, the other dissolved constituents of the sweat besides sodium chloride are only moderately increased in concentration: urea is then about two times that in the plasma; lactic acid, about four times; and potassium, about 1.2 times.

Note especially the large loss of sodium chloride in the sweat when the person is unacclimatized. However, this is quite different once the person has become acclimatized to heat as follows:

Acclimatization of the Sweating Mechanism. Though a normal, unacclimatized person can rarely produce more than about 700 ml of sweat per hour, when exposed to hot weather for 1 to 6 weeks the person sweats progressively more profusely, often increasing the maximum sweat production to as much as 2 liters/hr. Evaporation of this much sweat can remove heat from the body at a rate *more than ten times* the normal basal rate of heat production. This increased effectiveness of the sweating mechanism is caused by a direct increase in the sweating capability of the sweat glands themselves.

Also associated with the acclimatization is decreased concentration of sodium chloride in the sweat, which allows progressively better conservation of salt. Most of this effect is caused by *increased secretion of aldosterone,* which, in turn, results from a slight decrease in the level of sodium chloride in the body. An unacclimatized person who sweats profusely often loses as much as 15 to 30 grams of salt each day for the first few days. But after 4 to 6 weeks of acclimatization, the loss will usually be as little as 3 to 5 grams a day.

Loss of Heat by Panting

Many lower animals have very little ability to lose heat from the surfaces of their bodies for two reasons: first, the surfaces are usually covered with fur; and second, the skin of most lower animals is not supplied with sweat glands, which prevents much evaporative loss of heat from the skin. Therefore, a substitute mechanism, the *panting* mechanism, is used by many lower animals as a means of dissipating heat.

The phenomenon of panting is "turned on" by the thermoregulator centers of the brain that we discuss later. When the blood becomes overheated, the hypothalamus initiates neurogenic signals to decrease the body temperature. One of these signals initiates panting. The actual panting process is then controlled by a *panting center* that is closely related to the pneumotaxic respiratory center in the pons.

When an animal pants, it breathes in and out rapidly, so that large quantities of new air from the exterior come in contact with the upper portions of the respiratory passages; this cools the blood in the mucosa as a result of water evaporation from the mucosal surfaces, especially evaporation of saliva from the tongue. Yet panting does not increase the alveolar ventilation more than is required for proper control of the blood gases, because each breath is extremely shallow; therefore, most of the air that enters the alveoli is dead space air.

● REGULATION OF BODY TEMPERATURE—ROLE OF THE HYPOTHALAMUS

Figure 73–6 illustrates approximately what happens to the temperature of the nude body after a few hours' exposure to *dry* air ranging from 30° to 170°F.

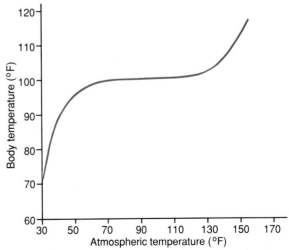

Figure 73–6. Effect of high and low atmospheric temperature for several hours' duration on the internal body temperature, showing that the internal body temperature remains stable despite wide changes in atmospheric temperature.

Obviously, the precise dimensions of this curve depend on the movement of the air, the amount of moisture in the air, and even the nature of the surroundings. However, in general, between approximately 60° and 130°F in dry air the nude body is capable of maintaining a normal body core temperature somewhere between 97° and 100°F.

The temperature of the body is regulated almost entirely by nervous feedback mechanisms, and almost all of these operate through *temperature-regulating centers* located in the *hypothalamus*. However, for these feedback mechanisms to operate, there must also exist temperature detectors, discussed next, to determine when the body temperature becomes either too hot or too cold.

Thermostatic Detection of Temperature in the Hypothalamus — Role of the Anterior Hypothalamus – Preoptic Area

In recent years, experiments have been performed in which minute areas in the brain have been either heated or cooled by use of a so-called *thermode*. This small, needle-like device is heated by electrical means or by passing hot water through it, or it is cooled by cold water. The principal area in the brain in which heat from a thermode affects body temperature control consists of the preoptic and anterior hypothalamic nuclei of the hypothalamus.

Using the thermode, the anterior hypothalamic – preoptic area has been found to contain large numbers of heat-sensitive neurons and about a third as many cold-sensitive neurons that seem to function as temperature sensors for controlling body temperature. The heat-sensitive neurons increase their firing rate as the temperature rises, twofold to tenfold with an increase in body temperature of 10°C. The cold-sensitive neurons, by contrast, increase their firing rate when the body temperature falls.

When the preoptic area is heated, the skin immediately breaks out into a profuse sweat while at the same time the skin blood vessels over the entire body become greatly vasodilated. Thus, this is an immediate reaction to cause the body to lose heat, thereby helping to return the body temperature toward the normal level. In addition, excess body heat production is inhibited. Therefore, it is clear that the preoptic area of the hypothalamus has the capability of serving as a thermostatic body temperature control center.

Detection of Temperature by Receptors in the Skin and Deep Body Tissues

Though the signals generated by the temperature receptors of the hypothalamus are extremely powerful in controlling body temperature, receptors in other parts of the body also play important roles in temperature regulation. This is especially true of temperature receptors in the skin and in a few specific deep tissues of the body.

It will be recalled from the discussion of sensory receptors in Chapter 48 that the skin is endowed with both *cold* and *warmth* receptors. However, there are far more cold receptors than warmth receptors; in fact, ten times as many in many parts of the skin. Therefore, peripheral detection of temperature mainly concerns detecting cool and cold instead of warm temperatures.

When the skin is chilled over the entire body, immediate reflex effects are invoked to increase the temperature of the body in several ways: (1) by providing a strong stimulus to cause shivering, with resultant increase in the rate of body heat production, (2) by inhibiting the process of sweating if this should be occurring, and (3) by promoting skin vasoconstriction to diminish the transfer of body heat to the skin.

Deep body temperature receptors are also found in certain parts of the body, mainly in the *spinal cord,* in the *abdominal viscera,* and in or around the *great veins*. However, these deep receptors function differently from the skin receptors, for they are exposed to the body core temperature rather than the body surface temperature. Yet, like the skin temperature receptors, they mainly detect cold rather than warmth. It is probable that both the skin and the deep body receptors are concerned with preventing hypothermia — that is, preventing low body temperatures.

Role of the Posterior Hypothalamus in Integrating Peripheral and Central Temperature Signals

Even though a large share of the signals for temperature detection arise in peripheral receptors, these signals help control body temperature mainly through the hypothalamus. However, the area of the hypothalamus that they stimulate is not the anterior hypothalamus – preoptic area but instead an area located bilaterally in the posterior hypothalamus approximately at the level of the mammary bodies. The thermostatic signals from the anterior hypothalamus – preoptic area are also transmitted into this posterior hypothalamus area. Here the signals from the preoptic area and the signals from the body periphery are combined to provide mainly the heat-producing and heat-conserving reactions of the body.

NEURONAL EFFECTOR MECHANISMS THAT DECREASE OR INCREASE BODY TEMPERATURE

When the hypothalamic temperature centers detect that the body temperature is either too hot or too cold, they institute appropriate temperature-decreasing or temperature-increasing procedures. The student is familiar with most of these from personal experience, but special features are the following:

Temperature-Decreasing Mechanisms When the Body Is Too Hot

The temperature control system employs three important mechanisms to reduce body heat when the body temperature becomes too great:

1. *Vasodilatation.* In almost all areas of the body the skin blood vessels become intensely dilated. This is caused by *inhibition of the sympathetic centers in the posterior hypothalamus that cause vasoconstriction.* Full vasodilatation can increase the rate of heat transfer to the skin as much as eightfold.

2. *Sweating.* The effect of increased temperature on causing sweating is illustrated by the solid curve in Figure 73–7, which shows a sharp increase in the rate of evaporative heat loss resulting from sweating when the body core temperature rises above the critical temperature level of 37°C (98.6° F). An additional 1°C increase in body temperature causes enough sweating to remove ten times the basal rate of body heat production.

3. *Decrease in heat production.* Those mechanisms that cause excess heat production, such as shivering and chemical thermogenesis, are strongly inhibited.

Temperature-Increasing Mechanisms When the Body Is Too Cold

When the body is too cold, the temperature control system institutes exactly opposite procedures. These are:

1. *Skin vasoconstriction throughout the body.* This is caused by stimulation of the posterior hypothalamic sympathetic centers.

2. *Piloerection.* Piloerection means hairs "standing on end." Sympathetic stimulation causes the arrector pili muscles attached to the hair follicles to contract, which brings the hairs to an upright stance. This is not important in the human being, but in lower animals upright projection of the hairs allows them to entrap a thick layer of "insulator air" next to the skin so that the transfer of heat to the surroundings is greatly depressed.

3. *Increase in heat production.* Heat production by the metabolic systems is increased by promoting (a) shivering, (b) sympathetic excitation of heat production, and (c) thyroxine secretion. These require additional explanation, as follows:

Hypothalamic Stimulation of Shivering. Located in the dorsomedial portion of the posterior hypothalamus near the wall of the third ventricle is an area called the *primary motor center for shivering.* This area is normally inhibited by signals from the heat center in the anterior hypothalamic–preoptic area but is excited by cold signals from the skin and spinal cord. Therefore, as illustrated by the dashed curve in Figure 73–7, this center becomes activated when the body temperature falls even a fraction of a degree below a critical temperature level. It then transmits signals that cause shivering through bilateral tracts down the brain stem, into the lateral columns of the spinal cord, and, finally, to the anterior motor neurons. These signals are nonrhythmic and do not cause the actual muscle shaking. Instead, they increase the tone of the skeletal muscles throughout the body. When the tone rises above a certain critical level, shivering begins. This probably results from feedback oscillation of the muscle spindle stretch reflex mechanism, which was discussed in Chapter 54. During maximum shivering, body heat production can rise to as high as four to five times normal.

Sympathetic "Chemical" Excitation of Heat Production. It was pointed out in Chapter 72 that either sympathetic stimulation or circulating norepinephrine and epinephrine in the blood can cause an immediate increase in the rate of cellular metabolism; this effect is called *chemical thermogenesis,* and it results at least partially from the ability of norepinephrine and epinephrine to *uncouple* oxidative phosphorylation, which means that excess foodstuffs are oxidized and thereby release energy in the form of heat but do not cause adenosine triphosphate to be formed.

The degree of chemical thermogenesis that occurs in an animal is almost directly proportional to the amount of *brown* fat that exists in the animal's tissues. This is a type of fat that contains large numbers of special mitochondria where the uncoupled oxidation occurs, as described in Chapter 72; and these cells are supplied by a strong sympathetic innervation.

The process of acclimatization greatly affects the intensity of chemical thermogenesis; some animals that have been exposed for several weeks to a very cold environment exhibit as much as a 100 to 500 per cent increase in heat production when acutely exposed to cold, in contrast to the unacclimatized animal, which responds with an increase of perhaps one-third as much.

In adult human beings, who have almost no brown fat, it is rare that chemical thermogenesis increases

Figure 73–7. Effect of hypothalamic temperature on (1) evaporative heat loss from the body and (2) heat production caused primarily by muscular activity and shivering. This figure demonstrates the extremely critical temperature level at which increased heat loss begins and increased heat production stops. (Drawn from data in Benzinger, Kitzinger, and Pratt, in Hardy [ed.]: Temperature, Part 3, p. 637. Reinhold Publishing Corp.)

the rate of heat production more than 10 to 15 per cent. However, in infants, who *do* have a small amount of brown fat in the interscapular space, chemical thermogenesis can increase the rate of heat production as much as 100 per cent, which is probably a very important factor in maintaining normal body temperature in the neonate.

Increased Thyroxine Output as a Cause of Increased Heat Production. Cooling the anterior hypothalamic–preoptic area of the hypothalamus also increases the production of the neurosecretory hormone *thyrotropin-releasing hormone* by the hypothalamus. This hormone is carried by way of the hypothalamic portal veins to the anterior pituitary gland, where it stimulates the secretion of *thyroid-stimulating hormone*. Thyroid-stimulating hormone, in turn, stimulates increased output of *thyroxine* by the thyroid gland, as is explained in Chapter 76. The increased thyroxine increases the rate of cellular metabolism throughout the body, which is yet another mechanism of *chemical thermogenesis*. However, this increase in metabolism does not occur immediately but requires several weeks for the thyroid gland to hypertrophy before it reaches its new level of thyroxine secretion.

Exposure of animals to extreme cold for several weeks can cause their thyroid glands to increase in size as much as 20 to 40 per cent. However, human beings rarely allow themselves to be exposed to the same degree of cold as that to which animals have been subjected. Therefore, we still do not know, quantitatively, how important the thyroid method of adaptation to cold is in the human being. Yet isolated measurements have shown that military personnel residing for several months in the Arctic develop increased metabolic rates; Eskimos also have abnormally high basal metabolic rates. Also, the continuous stimulatory effect of cold on the thyroid gland can probably explain the much higher incidence of toxic thyroid goiters in persons living in colder climates than in those living in warmer climates.

THE CONCEPT OF A "SET-POINT" FOR TEMPERATURE CONTROL

In the example of Figure 73–7, it is clear that at a very critical body core temperature, at a level of almost exactly 37.1°C, drastic changes occur in the rates of both heat loss and heat production. At temperatures above this level, the rate of heat loss is greater than that of heat production, so that the body temperature falls and reapproaches the 37.1°C level. At temperatures below this level, the rate of heat production is greater than heat loss, so that now the body temperature rises and again approaches the 37.1°C level. Therefore, this critical temperature level is called the "set-point" of the temperature control mechanism. That is, all the temperature control mechanisms continually attempt to bring the body temperature back to this set-point level.

The Feedback Gain for Body Temperature Control. Let us recall for a moment the discussion of feedback gain of control systems presented in Chapter 1. Feedback gain is a measure of the effectiveness of a control system. In the case of body temperature control, it is important for the internal core temperature to change as little as possible despite marked changes in the environmental temperature, and the gain of the temperature control system is approximately equal to the ratio of the change in environmental temperature to the change in body temperature that this causes. Experiments have shown that the body temperature of man changes about 1°C for each 25° to 30°C change in environmental temperature. Therefore, the feedback gain of the total mechanism for control of body temperature averages about 27, which is an extremely high gain for a biological control system (the baroreceptor arterial pressure control system, for instance, has a gain of less than 2).

Effect of Skin and Deep Body Temperatures in Altering the Temperature Level of the Set-Point

The critical temperature set-point in the hypothalamus above which sweating begins and below which shivering begins is determined mainly by the degree of activity of the heat temperature receptors in the anterior hypothalamic–preoptic area of the hypothalamus. However, we have already pointed out that temperature signals from the peripheral areas of the body, especially from the skin and certain deep body tissues (the spinal cord and the abdominal viscera), also contribute very significantly to body temperature regulation. But how do they contribute? The answer is that they alter the set-point of the hypothalamic temperature control center. This effect is illustrated in Figures 73–8 and 73–9.

Figure 73–8 illustrates the effect of different skin temperatures on the set-point, showing that it increases as the skin temperature decreases. Thus, in the person represented in this figure, the hypothalamic set-point increased from 36.7°C when the skin temperature was higher than 33°C to a set-point of 37.4°C when the skin temperature had fallen to 29°C. Therefore, when the skin temperature was very high, sweating began at a much lower hypothalamic temperature than when the skin temperature was low. One can readily understand the logic of a system such as this, for it is important that sweating be inhibited when the skin temperature is low; otherwise, the combined effect of a low skin temperature and sweating at the same time could cause far too much loss of body heat.

A similar but opposite effect occurs in shivering, as illustrated in Figure 73–9. That is, when the skin becomes cold, it drives the hypothalamic centers to the shivering threshold even when the hypothalamic temperature itself is still quite hot. Here again, one can well understand the logic of the control system, because a cold skin temperature will soon lead to

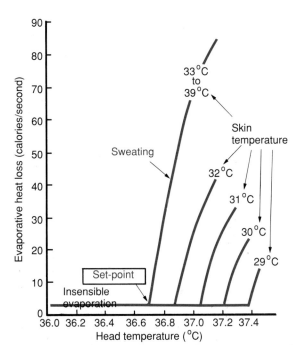

Figure 73 – 8. Effect of changes in the internal head temperature on the rate of evaporative heat loss from the body. Note also that the skin temperature determines the set-point level at which sweating begins. (Courtesy of Dr. T. H. Benzinger.)

deeply depressed body temperature unless heat production is increased. Thus, this effect of a cold skin temperature on increasing heat production actually "anticipates" a possible fall in internal body temperature and prevents its occurrence.

Figure 73 – 9. Effect of changes in the internal head temperature on the rate of heat production by the body. Note also that the skin temperature determines the set-point level at which shivering begins. (Courtesy of Dr. T. H. Benzinger.)

BEHAVIORAL CONTROL OF BODY TEMPERATURE

Aside from the subconscious mechanisms for body temperature control, the body has still another temperature-controlling mechanism that is even more potent. This is *behavioral control of temperature,* which can be explained as follows: Whenever the internal body temperature becomes too high, signals from the brain temperature controlling areas give the person a psychic sensation of being overheated. Conversely, whenever the body becomes too cold, signals from the skin and probably also from the deep body receptors elicit the feeling of cold discomfort. Therefore, the person makes appropriate environmental adjustments to re-establish comfort. This is a much more powerful system of body temperature control than most physiologists have recognized in the past. Indeed, for people, this is the only really effective mechanism for body heat control in severely cold environs.

LOCAL SKIN TEMPERATURE REFLEXES

When a person places a foot under a hot lamp and leaves it there for a short time, *local vasodilatation* and mild *local sweating* occur. Conversely, placing the foot in cold water causes local vasoconstriction and local cessation of sweating. These reactions are caused by local effects of temperature directly on the blood vessels and sweat glands and also by local cord reflexes conducted from the skin receptors to the spinal cord and back to the same skin area. However, the *intensity* of these local effects is controlled by the central brain temperature controller, so that the overall effect is approximately proportional to the hypothalamic heat control signal *times* the local signal. Such reflexes can help prevent excessive heat exchange from locally cooled or heated portions of the body.

Regulation of Internal Body Temperature After Cutting the Spinal Cord. After cutting the spinal cord in the neck above the sympathetic outflow from the cord, regulation of body temperature becomes extremely poor, for the hypothalamus can then no longer control either skin blood flow or the degree of sweating anywhere in the body. This is true even though the local temperature reflexes originating in the skin, spinal cord, and intra-abdominal receptors still exist. Unfortunately, these reflexes are not powerful. In persons with this condition, body temperature must be regulated principally by the patients' psychic response to cold and hot sensations in the head region — that is, by behavioral control of clothing and environment.

● ABNORMALITIES OF BODY TEMPERATURE REGULATION

FEVER

Fever, which means a body temperature above the usual range of normal, may be caused by abnormalities in the brain itself or by toxic substances that affect the temperature-regulating centers. Some causes of fever are presented in Figure 73 – 10. These include bacterial diseases, brain tumors, and environmental conditions that may terminate in heat stroke.

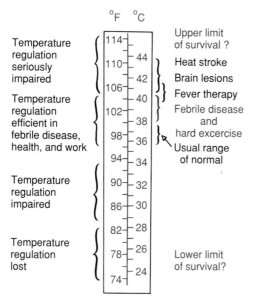

Figure 73–10. Body temperatures under different conditions. (From DuBois: Fever. Springfield, Ill., Charles C Thomas.)

Resetting the Hypothalamic Temperature Regulating Center in Febrile Diseases — Effect of Pyrogens

Many proteins, breakdown products of proteins, and certain other substances, especially lipopolysaccharide toxins secreted by bacteria, can cause the set-point of the hypothalamic thermostat to rise. Substances that cause this effect are called *pyrogens*. It is pyrogens secreted by toxic bacteria or pyrogens released from degenerating tissues of the body that cause fever during disease conditions. When the set-point of the hypothalamic temperature regulating center becomes increased to a higher level than normal, all the mechanisms for raising the body temperature are brought into play, including heat conservation and increased heat production. Within a few hours after the set-point has been increased to a higher level, the body temperature also approaches this level.

Mechanism of Action of Pyrogens in Causing Fever — Role of Interleukin-1. Experiments in animals have shown that some pyrogens, when injected into the hypothalamus, can act directly on the hypothalamic temperature regulating center to increase its set-point, though still other pyrogens function indirectly and also may require several hours of latency before causing their effects. This is true of many of the bacterial pyrogens, especially the *endotoxins* from gram-negative bacteria, as follows:

When bacteria or breakdown products of bacteria are present in the tissues or in the blood, these are *phagocytized by the blood leukocytes, the tissue macrophages,* and the *large granular killer lymphocytes.* All of these cells in turn digest the bacterial products and then release into the body fluids the *substance interleukin-1,* which is also called *leukocyte pyrogen* or *endogenous pyrogen.* The interleukin-1, on reaching the hypothalamus, immediately produces fever, increasing the body temperature in as little as 8 to 10 minutes. As little as one ten millionth of a gram of endotoxin lipopolysaccharide acting in this manner in concert with the blood leukocytes, tissue macrophages, and killer lymphocytes can cause fever. The amount of interleukin-1 that is formed in response to the lipopolysaccharide to cause the fever is only a few nanograms.

Several recent experiments have suggested that interleukin-1 causes fever by first inducing the formation of one of the prostaglandins or a similar substance and this in turn acting in the hypothalamus to elicit the fever reaction. When prostaglandin formation is blocked by drugs, the fever is either completely abrogated or at least reduced. In fact, this may be the explanation for the manner in which aspirin reduces the degree of fever, because aspirin impedes the formation of prostaglandins from arachidonic acid. It also would explain why aspirin does not lower the body temperature in a normal person, because a normal person does not have any interleukin-1. Drugs such as aspirin that reduce the level of fever are called *antipyretics.*

Fever Caused by Brain Lesions. When a brain surgeon operates in the region of the hypothalamus, severe fever almost always occurs; rarely, however, the opposite effect occurs, thus illustrating both the potency of the hypothalamic mechanisms for body temperature control and also the ease with which abnormalities of the hypothalamus can alter the set-point of temperature control. Another condition that frequently causes prolonged high temperature is compression of the hypothalamus by brain tumors.

Characteristics of Febrile Conditions

Chills. When the set-point of the hypothalamic temperature control center is suddenly changed from the normal level to higher than normal as a result of tissue destruction, pyrogenic substances, or dehydration, the body temperature usually takes several hours to reach the new temperature set-point. Figure 73–11 illustrates this, showing the effect of suddenly increasing the set-point to a level of

Figure 73–11. Effects of changing the set-point of the hypothalamic temperature controller.

103°F. Because the blood temperature is less than the set-point of the hypothalamic temperature controller, the usual responses that cause elevation of body temperature occur. During this period the person experiences chills and feels extremely cold, even though his or her body temperature may already be above normal. Also, the skin is cold because of vasoconstriction, and the person shivers. Chills continue until the body temperature reaches the hypothalamic setting of 103°F. Then the person no longer experiences chills but instead feels neither cold nor hot. As long as the factor that is causing the hypothalamic temperature controller to be set at this high value continues, the body temperature is regulated more or less in the normal manner but at the high temperature set-point level.

The Crisis, or "Flush." If the factor that is causing the high temperature is suddenly removed, the set-point of the hypothalamic temperature controller is suddenly reduced to a lower value—perhaps even back to the normal level, as illustrated in Figure 73–11. In this instance, the body temperature is still 103°F, but the hypothalamus is attempting to regulate the temperature to 98.6°F. This situation is analogous to excessive heating of the anterior hypothalamic–preoptic area, which causes intense sweating and sudden development of a hot skin because of vasodilatation everywhere. This sudden change of events in a febrile disease is known as the "crisis" or, more appropriately, the "flush." In the days before the advent of antibiotics, the crisis was always awaited, for once this occurred, the doctor knew immediately that the patient's temperature would soon be falling.

Heat Stroke

The limits of extreme heat that one can stand depend almost entirely on whether the heat is dry or wet. If the air is completely dry and sufficient convection air currents are flowing to promote rapid evaporation from the body, a person can withstand several hours of air temperature at 130°F. On the other hand, if the air is 100 per cent humidified or if the body is in water, the body temperature begins to rise whenever the environmental temperature rises above approximately 94°F. If the person is performing very heavy work, this critical temperature level may be as low as 85° to 90°F.

Unfortunately, there is a limit to the rate at which the body can lose heat even with maximal sweating. Furthermore, when the hypothalamus becomes excessively heated, its heat-regulating ability becomes greatly depressed and sweating diminishes. As a result, a high body temperature tends to perpetuate itself unless measures are taken specifically to decrease body heat.

When the body temperature rises beyond a critical temperature, into the range of 106° to 108°F, the person is likely to develop *heat stroke*. The symptoms include dizziness, abdominal distress, sometimes delirium, and eventually loss of consciousness if the body temperature is not soon decreased. Some of these symptoms are exacerbated by a mild degree of *circulatory shock* brought on by excessive loss of fluid and electrolytes in the sweat. However, the hyperpyrexia itself is also exceedingly damaging to the body tissues, especially to the brain, and therefore is responsible for many of the effects. In fact, even a few minutes of very high body temperature can sometimes be fatal. For this reason, many authorities recommend immediate treatment of heat stroke by placing the person in an ice water bath. However, because this often induces uncontrollable shivering with considerable increase in rate of heat production, others have suggested that sponge or spray cooling of the skin is likely to be more effective for rapidly decreasing the body core temperature.

Harmful Effects of the High Temperature. The pathological findings in a person who dies of hyperpyrexia are local hemorrhages and parenchymatous degeneration of cells throughout the entire body, but especially in the brain. Unfortunately, once neuronal cells are destroyed, they can never be replaced. Damage to the liver, kidneys, and other body organs can often be great enough that failure of one or more of these eventually causes death, sometimes not until several days after the heat stroke.

Acclimatization to Heat. It is often extremely important to acclimatize persons to extreme heat; some examples are acclimatization of soldiers for tropical duty and acclimatization of miners for work in the 2-mile-deep gold mines of South Africa, where the temperature approaches body temperature and the humidity approaches 100 per cent. Exposure of a person to heat for several hours each day while working a reasonably heavy work load will develop increased tolerance to hot and humid conditions in 1 to 3 weeks. Among the most important physiological changes that occur during this acclimatization process are an increase in the maximum rate of sweating, sometimes to more than twofold, an increase in the plasma volume, and diminished loss of salt in the sweat and urine; the latter two of these effects result from increased secretion of aldosterone.

EXPOSURE OF THE BODY TO EXTREME COLD

Unless treated immediately, a person exposed to ice water for approximately 20 to 30 minutes ordinarily dies because of heart standstill or heart fibrillation. By that time, the internal body temperature will have fallen to about 77°F. Yet if warmed rapidly by application of external heat, the person's life can often be saved.

Loss of Temperature Regulation at Low Temperatures. As noted in Figure 73–10, once the body temperature has fallen below about 85°F, the ability of the hypothalamus to regulate temperature is completely lost; it is greatly impaired even when the body temperature falls below approximately 94°F. Part of the reason for this loss of temperature regulation is that the rate of chemical heat production in each cell is depressed almost twofold for each 10°F decrease in body temperature. Also, sleepiness and even coma are likely to develop, which depress the activity of the central nervous system heat control mechanisms and also prevent shivering.

Frostbite. When the body is exposed to extremely low temperatures, surface areas can actually freeze; the freezing is called *frostbite*. This occurs especially in the lobes of the ears and in the digits of the hands and feet. If the freeze has been sufficient to cause extensive formation of ice crystals in the cells, permanent damage may result, such as permanent circulatory impairment as well as local tissue damage. Often gangrene follows thawing, and the frostbitten areas are lost.

Cold-Induced Vasodilation as a Protection Against Frostbite. When the temperature of tissues falls almost to freezing, the smooth muscle in the vascular wall becomes paralyzed because of the cold itself, and sudden vasodilation occurs, often manifested by a flush of the skin. This mechanism, fortunately, helps prevent frostbite by delivering warm blood to the skin. Unfortunately, it is a far less well-developed mechanism in humans than in most lower animals that live in the cold all the time.

Artificial Hypothermia. It is very easy to decrease the temperature of a person by first administering a strong sedative to depress the reactivity of the hypothalamic temperature controller and then cooling the person with ice, with cooling blankets, or otherwise until the temperature falls. The temperature can then be maintained below 90°F for several days to a week or more by continual sprinkling of cool water or alcohol on the body. Such artificial cooling is often used during heart surgery so that the heart can be stopped artificially for many minutes at a time. Cooling to this extent does not cause severe physiological results. It does slow the heart and greatly depresses body metabolism.

REFERENCES

Benzinger, T. H.: Heat regulation: Homeostasis of central temperature in man. Physiol. Rev., 49:671, 1969.

Boulant, J. A., and Dean, J. B.: Temperature receptors in the central nervous system. Annu. Rev. Physiol., 48:639, 1986.

Boulant, J. A., and Scott, I. M.: Effects of leukocytic pyrogen on hypothalamic neurons in tissue slices. Environment, Drugs and Thermoregulation, 5th Int. Symp. Pharmacol. Thermoregulation, Saint-Paul-de-Vence, 1982, p. 125.

Brengelmann, G. L.: Circulatory adjustments to exercise and heat stress. Annu. Rev. Physiol., 45:191, 1983.

Bukowiecki, L. J.: Mechanisms of stimulus-calorigenesis coupling in brown adipose tissue. Can. J. Biochem. Cell. Biol., 62:623, 1984.

Calder, W. A., III: Scaling of physiological processes in homeothermic animals. Annu. Rev. Physiol., 43:301, 1981.

Cena, K., and Clark, J. A.: Transfer of heat through animal coats and clothing. In Robertshaw, D. (ed.): International Review of Physiology: Environmental Physiology III. Vol. 20. Baltimore, University Park Press, 1979, p. 1.

Crawshaw, L. I.: Temperature regulation in vertebrates. Annu. Rev. Physiol., 42:473, 1980.

Felig, P., et al. (eds.): Endocrinology and Metabolism, 2nd Ed. New York, McGraw-Hill Book Co., 1987.

Galanter, E.: Detection and discrimination and environmental change. In Darian-Smith, I. (ed.): Handbook of Physiology, Sec. 1, Vol. III. Bethesda, Md., American Physiological Society, 1984, p. 103.

Gordon, C. J., and Heath, J. E.: Integration and central processing in temperature regulation. Annu. Rev. Physiol., 48:595, 1986.

Hales, J. E. (ed.): Thermal Physiology. New York, Raven Press, 1984.

Hardy, J. D.: Physiology of temperature regulation. Physiol. Rev., 41:521, 1961.

Harrison, M. H.: Effects of thermal stress and exercise on blood volume in humans. Physiol. Rev., 65:149, 1985.

Hellon, R.: Thermoreceptors. In Shepherd, J. T., and Abboud, F. M. (eds.): Handbook of Physiology. Sec. 1, Vol. III. Bethesda, Md., American Physiological Society, 1983, p. 659.

Hensel, H.: Neural processes in thermoregulation. Physiol. Rev., 53:948, 1973.

Hensel, H.: Thermoreceptors. Annu. Rev. Physiol., 36:233, 1974.

Hong, S. K., et al.: Humans can acclimatize to cold: A lesson from Korean women divers. News Physiol. Sci., 2:79, 1987.

Kelso, S. R., et al.: Thermosensitive single-unit activity of in vitro hypothalamic slices. Am. J. Physiol., 242:R77, 1982.

Kluger, M. J.: Fever: A hot topic. News Physiol. Sci., 1:25, 1986.

Kluger, M. J.: Temperature regulation, fever, and disease. In Robertshaw, D. (ed.): International Review of Physiology: Environmental Physiology III. Vol. 20. Baltimore, University Park Press, 1979, p. 209.

Lipton, J. M., and Clark, W. G.: Neurotransmitters in temperature control. Annu. Rev. Physiol., 48:613, 1986.

Mitchell, D., and Laburn, H. P.: Pathophysiology of temperature regulation. Physiologist, 28:507, 1985.

Myers, R. D.: Neurochemistry of thermoregulation. Physiologist, 27:41, 1984.

Nicholls, D. G., and Locke, R. M.: Thermogenic mechanisms in brown fat. Physiol. Rev., 64:1, 1984.

Prosser, C. L., and Nelson, D. O.: The role of nervous systems in temperature adaptation of poikilotherms. Annu. Rev. Physiol., 43:281, 1981.

Quinton, P. M.: Sweating and its disorders. Annu. Rev. Med., 34:453, 1983.

Robertshaw, D.: Role of the adrenal medulla in thermoregulation. Int. Rev. Physiol., 15:189, 1977.

Rowell, L. B.: Cardiovascular adjustments to thermal stress. In Shepherd, J. T., and Abboud, F. M. (eds.): Handbook of Physiology. Sec. 2, Vol. III. Bethesda, American Physiological Society, 1983, p. 967.

Schmidt-Nielsen, K.: Animal Physiology: Adaptation and Environment. London, Cambridge University Press, 1975.

Scott, I. M., and Boulant, J. A.: Dopamine effects on thermosensitive neurons in hypothalamic tissue slices. Brain Res., 306:157, 1984.

Simon, E., et al.: Central and peripheral thermal control of effectors in homeothermic temperature regulation. Physiol. Rev., 66:235, 1986.

Spray, D. C.: Cutaneous temperature receptors. Annu. Rev. Physiol., 48:625, 1986.

Underwood, L. S., and Tieszen, L. L. (eds.): Comparative Mechanism of Cold Adaptation. New York, Academic Press, 1979.

Wang, L. C. H., and Hudson, J. W. (eds.): Strategies in Cold: Natural Torpidity and Thermogenesis. New York, Academic Press, 1978.

Wolkomir, R.: Chilling out for science. Discover, February, 1988, p. 44.

Wyndham, C. H.: The physiology of exercise under heat stress. Annu. Rev. Physiol., 35:193, 1973.

UNIT XIV

ENDOCRINOLOGY AND REPRODUCTION

tex (*cortisol* and *aldosterone*), (b) the ovaries (*estrogen* and *progesterone*), (c) the testes (*testosterone*), and (d) the placenta (*estrogen* and *progesterone*).

2. *Derivatives of the amino acid tyrosine:* Two groups of hormones are derivatives of the amino acid tyrosine. The two metabolic thyroid hormones, *thyroxine* and *triiodothyronine,* are iodinated forms of tyrosine derivatives. And the two principal hormones of the adrenal medullae, *epinephrine* and *norepinephrine,* are both catecholamines, also derived from tyrosine.

3. *Proteins or peptides:* All the remaining important endocrine hormones are either proteins, peptides, or immediate derivatives of these. The anterior pituitary hormones are either proteins or large polypeptides; the posterior pituitary hormones, antidiuretic hormone and oxytocin, are peptides, each containing only eight amino acids. And insulin, glucagon, and parathormone are all large polypeptides.

STORAGE AND SECRETION OF HORMONES

We shall see in the following chapters that there is no single way in which all of the endocrine glands store and secrete their hormones. However, several general patterns occur for most of the hormones. For instance, all of the protein hormones are formed by the granular endoplasmic reticulum of the glandular cells in the same manner that other secretory proteins are formed as described in Chapter 3. However, the initial protein formed by the endoplasmic reticulum almost never is the final hormone itself. Instead, it is larger than the active hormone and is called a *preprohormone.* Then this large protein is further cleaved, usually while still in the endoplasmic reticulum, to form a smaller protein called the *prohormone.* This in turn is transported in the ER transport vesicles to the Golgi apparatus, where still another section of the protein is cleaved; in this way the final active protein hormone is formed. The Golgi apparatus usually also compacts the hormone molecules into small membrane-encapsulated vesicles called *secretory vesicles* or *secretory granules.* These then remain stored in the cytoplasmic compartment of the endocrine cell until a specific signal, such as a nerve signal, another hormonal signal, or a local chemical or physical signal, comes along to cause secretion.

The two groups of hormones derived from tyrosine, the thyroid and adrenal medullary hormones, are both formed by the actions of enzymes in the cytoplasmic compartments of the glandular cells. In the case of the adrenal medullary hormones norepinephrine and epinephrine, these are then absorbed into preformed vesicles and stored until they are to be secreted. The thyroid metabolic hormones, thyroxine and triiodothyronine, on the other hand, are formed as component parts of a large protein molecule called thyroglobulin, and this is then stored in large follicles within the thyroid gland. When the thyroid hormones are to be secreted, specific enzyme systems within the thyroid glandular cells cleave the thyroglobulin molecule, thereby allowing the thyroid hormones to be released into the blood.

For the steroid hormones secreted by the adrenal cortex, the ovaries, or the testes, the amounts stored in the glandular cells are usually quite small, but large amounts of precursor molecules, especially cholesterol and various intermediates between cholesterol and the final hormones, are present in the cells. Upon appropriate stimulation, enzymes in these cells can within minutes cause the chemical conversions to the final hormones, followed almost immediately by secretion.

Onset of Hormone Secretion Following a Stimulus, and Durations of Action of Different Hormones

Some hormones are secreted within seconds after the gland is stimulated, and they may develop full action within another few seconds to minutes; the actions of other hormones may require months for full effect. To give examples, the two adrenal medullary hormones, epinephrine and norepinephrine, begin to be secreted in response to sympathetic nerve stimuli within the first second of stimulation and usually reach maximum concentrations within a minute after the onset of stimulation. However, these hormones are also destroyed rapidly by local tissue enzymes, or they are absorbed into cells. Therefore, their duration of action is usually no more than 1 to 3 minutes at most after the stimulation is over.

At the other extreme, the thyroid hormones are stored in the form of thyroglobulin in the thyroid follicles, sometimes for several months, before finally being secreted. Then, after secretion, several hours to several days are required before even initial activity begins; but their effect on enhancement of tissue metabolism can last as long as 6 weeks.

Thus, each of the different hormones has its own characteristic onset and duration of action—each tailored to perform its specific control function.

Concentrations of Hormones in the Circulating Blood and Hormonal Secretion Rates. The quantitative amounts of hormones that are required to control most metabolic and endocrine functions are incredibly small. Their concentrations in the blood range from as little as 1 picogram (which is one millionth of a millionth of a gram) in each milliliter of blood up to at most a few micrograms (one millionth of a gram) per milliliter of blood. Similarly, the rates of secretion of the various hormones are also extremely small, usually measured in micrograms or milligrams per day. We shall see later in this chapter that highly specialized mechanisms are available in the target tissues that allow even these minute quantities of hormones to exert powerful control over the physiologic systems.

Control of Hormone Secretion Rate— The Role of Negative Feedback

Without exception, the rate of secretion of every hormone that has ever been studied is itself controlled very exactly by some internal control system. In most instances, this control is exerted through a negative feedback mechanism as follows:

1. The endocrine gland has a natural tendency to *oversecrete* its hormone.

2. Because of this tendency, the hormone exerts more and more of its control effect on the target organ.

3. The target organ in turn performs its function.

4. But when too much function occurs, usually some factor about the function then *feeds back* to the endocrine gland and causes a *negative* effect on the gland to decrease its secretory rate. Thus, the function of the hormone is monitored, and this information in turn provides negative feedback control of the secretory rate by the gland.

Careful consideration of the above negative feedback mechanism will show that the important factor to be controlled usually is not the secretory rate of the hormone but instead the degree of activity of the target organ. Therefore, only when the target organ activity rises to an appropriate level will the feedback to the gland become powerful enough to slow further secretion of the hormone. If the target organ responds poorly to the hormone, the endocrine gland will almost always secrete still more and more of its hormone until the target organ eventually reaches the appropriate level of activity, but at the expense of excessive secretion of the controlling hormone.

HORMONE RECEPTORS AND THEIR ACTIVATION

The endocrine hormones almost never act directly on the intracellular machinery to control the different cellular chemical reactions; instead, they almost invariably first combine with *hormone receptors* on the surfaces of the cells or inside the cells. The combination of hormone and receptor then usually initiates a *cascade of reactions* in the cell, with each stage of reaction in the cascade becoming much more powerfully activated than the previous stage, so that owing to a very small initiating hormonal stimulus a very large final effect occurs.

Either all or almost all hormonal receptors are very large proteins, and each cell usually has some 2000 to 100,000 receptors.

Also, each receptor is usually highly specific for a single hormone; this determines the type of hormone that will act on a particular tissue. Obviously, the target tissues that are affected by a hormone are those that contain its specific receptors.

The locations of the receptors for the different types of hormones are generally the following:

1. *In or on the surface of the cell membrane.* The membrane receptors are specific mostly to the protein, peptide, and catecholamine (epinephrine and norepinephrine) hormones.

2. *In the cell cytoplasm.* The receptors for the different steroid hormones are found almost entirely in the cytoplasm.

3. *In the cell nucleus.* The receptors for the meta-bolic thyroid hormones (thyroxine and triiodothyronine) are found in the nucleus, believed to be located in direct association with one or more of the chromosomes.

Regulation of the Number of Receptors. The number of receptors in a target cell usually does not remain constant from day to day, or even from minute to minute, for the receptor proteins themselves are often destroyed and new ones are manufactured by the protein-manufacturing mechanism of the cell. For instance, binding of a hormone with its target cell receptors often, if not usually, causes the number of receptors to decrease, either because of inactivation of some of the receptor molecules or because of decreased production of the molecules. In either event, this is called *down-regulation* of the receptors. Obviously, this decreases the responsiveness of the target tissue to the hormone as the number of receptors decreases.

In a few instances, hormones cause *up-regulation* of receptors; that is, the stimulating hormone induces the formation of more receptor molecules than normally by the protein-manufacturing machinery of the target cell. In this instance, the target tissue becomes progressively more sensitive to the stimulating effects of the hormone.

● MECHANISMS OF HORMONAL ACTION

The Hormonal Receptors Play a Major Role in Hormonal Action

Almost without exception, a hormone affects its target tissues by first activating target receptors in the tissue cells. This alters the function of the receptor itself, and this receptor is then the direct cause of the hormonal effects. To explain this, let us give a few examples:

1. Change in Membrane Permeability. Virtually all of the neurotransmitter substances, which are themselves local hormones, combine with receptors in the postsynaptic membrane. Almost always this causes a conformational change in the protein structure of the receptor, usually opening or closing a channel for one or more ions. Some receptors provide open (or closed) channels for sodium ions, others for potassium ions, others for calcium ions, and so forth. Then, it is the altered movement of these ions through the channels that causes the subsequent effects on the postsynaptic cells.

A few of the general hormones have similar effects in opening or closing membrane ion channels. This is most notably true for many of the actions of the adrenal medullary secretions, norepinephrine and epinephrine. For instance, norepinephrine and epinephrine have especially powerful effects in changing the membrane potentials of smooth muscle cells, causing excitation when the receptor is an excitatory type or inhibition when it is an inhibitory type.

2. Activation of an Intracellular Enzyme When a Hormone Combines With a Membrane Receptor. Another common effect of membrane receptor binding is activation (or occasionally inactivation) of an enzyme immediately inside the cell membrane. A good example of this is the effect of insulin. Insulin binds with the portion of its membrane receptor that protrudes to the exterior of the cell. This produces a structural change in the receptor molecule itself, causing the portion of the molecule that protrudes to the inside to become an activated kinase. This kinase then promotes phosphorylation of many different substances inside the cell. Most of the actions of insulin on the cell then result secondarily from these phosphorylation processes.

A second example, one widely used in hormonal control of cell function, is for the hormone to bind with a special transmembrane receptor that then becomes the activated enzyme *adenyl cyclase* at its end that protrudes to the interior of the cell. This cyclase in turn causes the formation of the substance *cyclic adenosine monophosphate (cAMP)*. And cAMP has a multitude of effects inside the cell to control cell activity, as is discussed in greater detail later. The cAMP is called a *second messenger*, for it is not the hormone itself that directly institutes the intracellular changes; instead, it is the cAMP that serves as a "second messenger" to cause these effects.

In a few instances, *cyclic guanosine monophosphate (cGMP)*, which is only slightly different from cAMP, serves in a similar manner as a "second messenger."

3. Activation of Genes by Binding With Intracellular Receptors. Several hormones, especially the steroid hormones and thyroid hormones, bind with protein receptors inside the cell, not in the cell membrane. For both the steroid and thyroid hormones, the activated hormone–receptor complex then binds with or activates specific points on the DNA strands of the cell nucleus, which initiates transcription of specific genes to form messenger RNA. Therefore, many minutes, hours, or even days after the hormone has entered the cell, newly formed proteins appear in the cell and become the controllers of new or increased cellular functions.

Other hormones enhance the translation of messenger RNA in the cytoplasm. This is believed to be especially true of growth hormone and perhaps also of insulin.

SECOND MESSENGER MECHANISMS FOR MEDIATING INTRACELLULAR HORMONAL FUNCTIONS

We noted previously that one of the principal means by which hormones exert intracellular actions is causing the "second messenger" cAMP to be formed inside the cell membrane. Then the cAMP in turn causes all or most of the intracellular effects of the hormone. Thus, the only direct effect that the hormone has on the cell is to activate a single type of membrane receptor. The second messenger does the rest.

Cyclic AMP is not the only second messenger system utilized by the different hormones. Two other especially important ones are (a) calcium ions and associated *calmodulin,* and (b) products of membrane phospholipid breakdown.

Intracellular Mechanism of the cAMP "Second Messenger" System

The cAMP mechanism has been shown to be the way in which all the following hormones (and many more) often stimulate their target tissues:

1. Adrenocorticotropin.
2. Thyroid-stimulating hormone.
3. Luteinizing hormone.
4. Follicle-stimulating hormone.
5. Vasopressin.
6. Parathyroid hormone.
7. Glucagon.
8. Catecholamines.
9. Secretin.
10. The hypothalamic releasing hormones.

Figure 74–2 illustrates the function of the cyclic AMP mechanism in more detail. The stimulating hormone first binds with a specific "receptor" for that hormone on the membrane surface of the target cell. The specificity of the receptor determines which hormone will affect the target cell. After binding with the membrane receptor, the portion of the receptor that protrudes to the interior of the cell membrane is activated to become the protein enzyme *adenyl cyclase.* This enzyme, in turn, causes immediate *conversion of much of the cytoplasmic ATP into cAMP,* which is the compound *cyclic 3',5'-adenosine monophosphate.*

Once cAMP is formed inside the cell, it activates still other enzymes. In fact, it usually activates a cas-

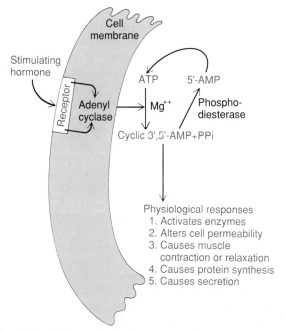

Figure 74–2. The cyclic 3′,5′-adenosine monophosphate (cAMP) mechanism by which many hormones exert their control of cell function.

cade of enzymes. That is, a first enzyme is activated and then activates another enzyme, which activates still a third, and so forth. The importance of this mechanism is that only a few molecules of activated adenyl cyclase in the cell membrane can cause many more molecules of the next enzyme to be activated, which can cause still many times that many molecules of the third enzyme to be activated, and so forth. In this way, even the slightest amount of hormone acting on the cell surface can initiate a very powerful cascading activating force for the entire cell.

The specific action that occurs in response to CAMP in each type of target cell depends upon the nature of the intracellular machinery, some cells having one set of enzymes and other cells having other enzymes. Therefore, different functions are elicited in different target cells—such functions as initiating synthesis of specific intracellular chemicals, causing muscle contraction or relaxation, initiating secretion by the cells, altering the cell permeability, and many other effects.

Thus, a thyroid cell stimulated by cAMP forms the metabolic hormones thyroxine and triiodothyronine, whereas the same cAMP in an adrenocortical cell causes secretion of the adrenocortical steroid hormones. On the other hand, cAMP affects epithelial cells of the renal tubules by increasing their permeability to water.

Role of Calcium Ions and "Calmodulin" as a Second Messenger System

Another second messenger system operates in response to the entry of calcium ions into cells. The calcium entry may be initiated by changes in the membrane electrical potential that open membrane calcium channels or by hormones interacting with membrane receptors that similarly open calcium channels.

On entering the cell, the calcium ions bind with a protein called *calmodulin.* This protein has four separate calcium sites; when as many as three or four of these sites have bound with calcium, a conformational change occurs that activates the calmodulin, causing multiple effects inside the cell in the same way that cAMP functions. For instance, it activates a set of enzymes different from those activated by cAMP, thus causing an additional set of intracellular metabolic reactions. One of the specific functions of calmodulin is to activate myosin kinase that then acts directly on the myosin of smooth muscle to cause smooth muscle contraction.

The normal calcium ion concentration in most cells of the body is about 10^{-7} to 10^{-8} mol/liter, which is not enough to activate the calmodulin system. But when the calcium ion concentration rises to as high as 10^{-6} to 10^{-5} mol/liter, enough binding then occurs to cause all the intracellular actions of calmodulin. This is almost exactly the same amount of calcium ion change that is required in skeletal muscle to activate troponin C, which in turn causes skeletal muscle contraction, as explained in Chapter 7. It is interesting

that troponin C is very similar to calmodulin both in function and in protein structure.

Membrane Phospholipid Breakdown Products as "Second Messengers"

Some hormones activate transmembrane receptors that then become activated *phospholipase C.* This enzyme causes some of the phospholipids in the cell membrane itself to split into smaller substances that have widespread "second messenger" intracellular effects. The types of hormones that cause this effect are mainly local hormones, most notably hormonal factors released by tissue immune and allergic reactions as explained in Chapter 34.

The most important membrane phospholipid broken down in this way is *phosphatidyl inositol bisphosphate.* And the most important products that serve as second messengers are *inositol trisphosphate* and *diacylglycerol.* The inositol trisphosphate especially mobilizes calcium ions from both the mitochondria and the endoplasmic reticulum, and the calcium ions then promote all their own second messenger effects such as smooth muscle contraction, changes in secretion by secreting cells, changes in ciliary action, and so forth.

The diacylglycerol, the other lipid second messenger, activates the enzyme *protein kinase C.* This activation is further enhanced by the increased calcium ions that have been released in response to inositol trisphosphate. In turn, the activated protein kinase C plays an especially important role in promoting cell division and cell proliferation.

In addition, the lipid portion of diacylglycerol is *arachidonic acid,* which is the precursor for the *prostaglandins* and other local hormones that cause a wide variety of local effects in tissues throughout the body.

Action of Steroid Hormones on the Genes to Cause Protein Synthesis

A second major means by which hormones act — specifically the steroid hormones secreted by the adrenal cortex, the ovaries, and the testes — is to cause synthesis of proteins in the target cells; these proteins then function as enzymes, transport proteins, or structural proteins that in turn provide other functions of the cells.

The sequence of events in steroid function is the following:

1. The steroid hormone enters the cytoplasm of the cell, where it binds with a specific *receptor protein.*

2. The combined receptor protein/hormone then diffuses into or is transported into the nucleus.

3. The combination now binds at specific points to the DNA strands in the chromosomes, which activates the transcription process of specific genes to form messenger RNA.

4. The messenger RNA diffuses into the cytoplasm, where it promotes the translation process at the ribosomes to form new proteins.

To give an example, *aldosterone,* one of the hormones secreted by the adrenal cortex, enters the cytoplasm of renal tubular cells, which contain its specific receptor protein. Therefore, in these cells the above sequence of events ensues. After about 45 minutes, proteins begin to appear in the renal tubular cells that promote sodium reabsorption from the tubules and potassium secretion into the tubules. Thus, there is a characteristic delay in the beginning action of the steroid hormone of at least 45 minutes and up to several hours or even days for full action, which is in marked contrast to the almost instantaneous action of some of the peptide and amino acid–derived hormones, such as vasopressin and norepinephrine.

Action of the Thyroid Hormones in the Cell Nucleus

The thyroid hormones *thyroxine* and *triiodothyronine* cause increased transcription by certain genes in the nucleus. To accomplish this, these hormones first bind directly with receptor proteins in the nucleus itself; these receptors are probably protein molecules located within the chromosomal complex, and they probably control the function of the genetic promotors or operators as explained in Chapter 3.

Two important features of thyroid hormone function in the nucleus are the following:

1. They activate the genetic mechanisms for formation of many different types of intracellular proteins—probably a hundred or more. Many of these are enzymes that promote enhanced intracellular metabolic activity in virtually all cells of the body.

2. Once bound to the intranuclear receptors, the thyroid hormones can continue to express their control functions for days or even weeks.

● MEASUREMENT OF HORMONE CONCENTRATIONS IN THE BLOOD

Most hormones are present in the blood in extremely minute quantities, some in concentrations as low as one millionth of a milligram (1 picogram) per milliliter. Therefore, except in a few instances, it has been almost impossible to measure these concentrations by the usual chemical means. Fortunately, though, an extremely sensitive method was developed about 25 years ago that revolutionized the measurement of hormones, their precursors, and their metabolic end products. This is the method of *radioimmunoassay*.

Radioimmunoassay

The principle of radioimmunoassay is the following:

First, an antibody is made in large quantities in some lower animal that is highly specific for the hormone to be measured.

Second, a small quantity of this antibody is mixed with (1) a quantity of fluid from the animal containing the hormone to be measured and (2) an appropriate amount of purified standard hormone that has been tagged with a radioactive isotope. However, one specific condition must be met: there must be too little antibody to bind completely

both the radioactively tagged hormone and the hormone in the fluid to be assayed. Therefore, the natural hormone in the assay fluid and the radioactive standard hormone *compete for the binding sites* on the antibody. In the process of competing, the quantity of each of the two hormones, the natural and the radioactive, that binds is proportional to its concentration.

Third, after binding has reached equilibrium, the antibody-hormone complex is separated from the remainder of the solution, and the quantity of radioactive hormone bound with antibody is measured by radioactive counting techniques. If a *large amount of radioactive hormone* has bound with the antibody, then it is clear that there was only a *small amount of natural hormone* to compete with the radioactive hormone, and therefore the concentration of the natural hormone in the assayed fluid was small. Conversely, if only a small amount of radioactive hormone has bound, it is clear that there was a very large amount of natural hormone to compete for the binding sites.

Fourth, to make the assay highly quantitative, the radioimmunoassay procedure is performed also for "standard" solutions of untagged hormone at several different concentration levels. Then, a "standard curve" is plotted as illustrated in Figure 74–3. By comparing the radioactive counts recorded from the original assay procedures with the standard curve, one can determine within an error of ± 10 to 15 per cent the concentration of the hormone in the assayed fluid. As little as one trillionth of a gram of hormone can often be assayed in this way.

Other Competitive Binding Assay Procedures. Several other competitive binding techniques for assay of minute quantities of hormones have also been employed. One of these is to use in place of the antibody a specific carrier globulin of the plasma that is a natural binding agent for some specific hormone. For instance, the plasma-binding globulin for the adrenocortical hormone cortisol is highly specific. Therefore, this globulin is substituted for the antibody in the assay process for cortisol; the assay is then carried out in exactly the same way as the radioimmunoassay procedure.

Figure 74–3. A "standard curve" for radioimmunoassay of aldosterone. (Courtesy of Dr. Manis Smith.)

Measurement of the Metabolic Clearance Rate of Hormones

Two different factors can increase or decrease the concentration of a hormone in the blood. One of these, obviously, is the rate of secretion of the hormone. The second is the rate of removal of the hormone from the blood, which is called the *metabolic clearance rate*. This is expressed in terms of the number of milliliters of plasma cleared of the hormone per minute. To calculate this clearance rate, one makes the following two measurements:

1. The rate of disappearance of the hormone from the plasma per minute.

2. The concentration of the hormone in each milliliter of plasma.

Then, the metabolic clearance rate is calculated by the following formula:

$$\text{Metabolic clearance rate} = \frac{\left\{ \begin{array}{c} \text{Rate of disappearance of} \\ \text{hormone from the plasma} \end{array} \right\}}{\left\{ \begin{array}{c} \text{Concentration of hormone in} \\ \text{each milliliter of plasma} \end{array} \right\}}$$

The usual procedure for making this measurement is the following: A purified solution of the hormone to be measured is tagged with a radioactive substance. Then the radioactive hormone is infused at a constant rate into the blood stream until the radioactive concentration in the plasma becomes steady. At this time, the rate of disappearance of the radioactive hormone from the plasma obviously equals the rate at which it is infused, which gives one the rate of disappearance. At the same time, the plasma concentration of the radioactive hormone is measured with the use of a standard radioactive counting procedure. Then, with the formula just cited, the metabolic clearance rate is calculated.

Hormones are "cleared" from the plasma in several ways, including (1) metabolic destruction by the tissues, (2) binding with the tissues, (3) excretion by the liver into the bile, and (4) excretion by the kidneys into the urine. For certain hormones, a decreased metabolic clearance rate frequently causes an excessively high concentration of the hormone in the circulating body fluids. For instance, this occurs for several of the steroid hormones when the liver is diseased, because these hormones are mainly conjugated in the liver and then "cleared" into the bile.

Measurement of the Rate of Hormone Secretion

A simple method for estimating the rate of hormone secretion is, first, to measure the concentration of the natural hormone in the plasma by means of a radioimmunoassay procedure. Next, the metabolic clearance rate is measured. Then, by multiplying the concentration of the natural hormone times the metabolic clearance rate, one derives a value that is equal to the steady-state rate of hormone production.

However, hormone production often increases and decreases very rapidly. In such cases, one can measure the changing rates of secretion only by collecting samples of both the arterial blood entering the gland and the venous blood leaving the gland, and simultaneously measuring the rate of blood flow through the gland. Then, by multiplying the rate of blood flow times the venous increase in hormone concentration over that in the arterial blood, one can derive the instantaneous secretion rate. With methods such as this, it has been possible to show that many endocrine glands can be stimulated to extremely high secretory rates within minutes, and many glands also secrete hormones very intermittently, rather than steadily. This is especially true for the pituitary hormones as well as for cortisol from the adrenal gland.

REFERENCES

Almers, W., and Stirling, C.: Distribution of transport proteins over animal cell membranes. J. Membr. Biol., 77:169, 1984.

Barnes, D. M.: How cells respond to signals. Science, 234:286, 1986.

Blalock, J. E.: A molecular basis for bidirectional communication between the immune and neuroendocrine systems. Physiol. Rev., 69:1, 1989.

Cheung, W. Y.: Calmodulin: Its potential role in cell proliferation and heavy metal toxicity. Fed. Proc., 43:2995, 1984.

Conn, P. M. (ed.): Neuroendocrine Peptide Methodology. San Diego, Calif., Academic Press, 1988.

Daly, J. W.: Forskolin, adenylate cyclase, and cell physiology: an overview. Adv. Cyclic Nucleotide Res., 17:81, 1984.

De Crombrugghe, B., et al.: Cyclic AMP receptor protein: Role in transcription activation. Science, 224:831, 1984.

DeGroot, L. J., et al.: Endocrinology, 2nd Ed. Philadelphia, W. B. Saunders Co., 1989.

Douglass, J., et al.: Polyprotein gene expression: Generation of diversity of neuroendocrine peptides. Annu. Rev. Biochem., 53:665, 1984.

Ebels, I., and Balemans, M. G. M.: Physiological aspects of pineal functions in mammals. Physiol. Rev., 66:581, 1986.

Evans, R. M.: The steroid and thyroid hormone receptor superfamily. Science, 240:889, 1988.

Ganong, W. F., and Martini, L.: Frontiers in Neuroendocrinology. New York, Raven Press, 1986.

Glick, B.: Interrelation of the avian immune and neuroendocrine systems. J. Exp. Zool., 232:671, 1984.

Godfraind, T.: Calcium modulators: Tools for cell biology. Adv. Cyclic Nucleotide Res., 17:551, 1984.

Goldbeter, A.: Periodic signaling as an optimal mode of intercellular communication. News Physiol. Sci. 3:103, 1988.

Goodman, H. M.: Basic Medical Endocrinology. New York, Raven Press, 1988.

Greengard, P., and Alan R. G.: Advances in Second Messenger and Phosphoprotein Research. New York, Raven Press, 1988.

Gustafsson, J. A., et al.: Central control of prolactin and estrogen receptors in rat liver—Expression of a novel endocrine system, the hypothalamo-pituitary-liver axis. Annu. Rev. Pharmacol. Toxicol., 23:259, 1983.

Hall, P. F.: The role of the cytoskeleton in hormone action. Can. J. Biochem. Cell. Biol., 62:653, 1984.

Howlett, T. A., and Rees, L. H.: Endogenous opioid peptides and hypothalamo-pituitary function. Annu. Rev. Physiol., 48:527, 1986.

Hughes, J. P., and Friesen, H. G.: The nature and regulation of the receptors for pituitary growth hormone. Annu. Rev. Physiol., 47:469, 1985.

Isaksson, O. G. P., et al.: Mode of action of pituitary growth hormone on target cells. Annu. Rev. Physiol., 47:483, 1985.

Karsch, F. J.: Central actions of ovarian steroids in the feedback regulation of pulsatile secretion in leuteinizing hormone. Annu. Rev. Physiol., 49:365, 1987.

Martini, L., and Ganong, W. F., (eds.): Frontiers in Neuroendocrinology. New York, Raven Press, 1988.

Mazzaferri, E. L., (ed.): Textbook of Endocrinology. New York, Elsevier Science Publishing Co., 1986.

McCloskey, M., and Poo, M. M.: Protein diffusion in cell membranes: Some biological implications. Int. Rev. Cytol., 87:19, 1984.

Means, A. R., et al.: Physiological implications of the presence, distribution, and regulation of calmodulin in eukaryotic cells. Physiol. Rev., 62:1, 1982.

Muller, E. E.: Neural control of somatotropic function. Physiol. Rev., 67:962, 1987.

Nishizuka, Y.: Studies and perspectives of protein kinase C. Science, 233:305, 1986.

Norman, A. W., and Litwak, G.: Hormones. San Diego, Calif., Academic Press, 1987.

Olson, T. S., and Lane, M. D.: A common mechanism for postranslational activation of plasma membrane receptors. FASEB J., 3:1618, 1989.

Ozawa, S., and Sand, O.: Electrophysiology of excitable endocrine cells. Physiol. Rev., 66:887, 1986.

Pardridge, W. M.: Neuropeptides and the blood-brain barrier. Annu. Rev. Physiol., 45:73, 1983.

Patel, Y. C., and Srikant, C. B.: Somatostatin mediation of adenohypophysial secretion. Annu. Rev. Physiol., 48:551, 1986.

Poste, G.: New insights into receptor regulation. J. Appl. Physiol., 57:1297, 1984.

Prentki, M., and Matschinsky, F. M.: Ca²⁺-cAMP, and phospholipid-derived messengers in coupling mechanisms of insulin secretion. Physiol. Rev., 67:1185, 1987.

Rasmussen, H., and Barrett, P. Q.: Calcium messenger system: An integrated view. Physiol. Rev., 64:938, 1984.

Rodbard, D. (ed.): Computers in Endocrinology. New York, Raven Press, 1984.

Rousseau, G. G.: Structure and regulation of the glucocorticoid hormone receptor. Mol. Cell. Endocrinol., 38:1, 1984.

Shenoklkar, S.: Protein phosphorylation: hormones, drugs, and bioregulation. FASEB J., 2:2753, 1988.

Sherman, M. R., and Stevens, J.: Structure of mammalian steroid receptors: Evolving concepts and methodological developments. Annu. Rev. Physiol., 46:83, 1984.

Simonsen, M., et al.: Compound receptors in the cell membrane: Ruminations from the borderland of immunology and physiology. Prog. Allergy, 36:151, 1985.

Sowers, J. R., and Felicetta, J. V.: Endocrinology of Aging. New York, Raven Press, 1988.

Stiles, G. L., et al.: β-Adrenergic receptors: Biochemical mechanisms of physiological regulation. Physiol. Rev., 64:661, 1984.

Weiss, E. R., et al.: Receptor activation of G proteins. FASEB J., 2:2841, 1988.

Yalow, R. W.: Radioimmunoassay. Annu. Rev. Biophys. Bioeng., 9:327, 1980.

CHAPTER 75

• *The Pituitary Hormones and Their Control by the Hypothalamus* •

● THE PITUITARY GLAND AND ITS RELATION TO THE HYPOTHALAMUS

The *pituitary gland* (Fig. 75–1), also called the *hypophysis,* is a small gland—about 1 centimeter in diameter and 0.5 to 1 gram in weight—that lies in the *sella turcica* at the base of the brain and is connected with the hypothalamus by the *pituitary* (or *hypophysial*) stalk. Physiologically, the pituitary gland is divisible into two distinct portions: the *anterior pituitary,* also known as the *adenohypophysis,* and the *posterior pituitary,* also known as the *neurohypophysis.* Between these is a small, relatively avascular zone called the *pars intermedia,* which is almost absent in the human being while much larger and much more functional in some lower animals.

Embryologically, the two portions of the pituitary originate from different sources, the anterior pituitary from *Rathke's pouch,* which is an embryonic invagination of the pharyngeal epithelium, and the posterior pituitary from an outgrowth of the hypothalamus. The origin of the anterior pituitary from the pharyngeal epithelium explains the epithelioid nature of its cells, while the origin of the posterior pituitary from neural tissue explains the presence of large numbers of glial-type cells in this gland.

Six very important hormones plus several less important ones are secreted by the *anterior* pituitary, and two important hormones are secreted by the *posterior* pituitary. The hormones of the anterior pituitary play major roles in the control of metabolic functions throughout the body, as shown in Figure 75–2. (1) *Growth hormone* promotes growth of the animal by affecting protein formation, cell multiplication, and cell differentiation. (2) *Adrenocorticotropin (corticotropin)* controls the secretion of some of the adrenocortical hormones, which in turn affect the metabolism of glucose, proteins, and fats. (3) *Thyroid-stimulating hormone (thyrotropin)* controls the rate of secretion of thyroxine by the thyroid gland,

and thyroxine in turn controls the rates of most chemical reactions of the entire body. (4) *Prolactin* promotes the mammary gland development and milk production. And two separate gonadotropic hormones, (5) *follicle-stimulating hormone* and (6) *luteinizing hormone,* control growth of the gonads as well as their reproductive activities.

The two hormones secreted by the posterior pituitary play other roles. (1) *Antidiuretic hormone* (also called *vasopressin)* controls the rate of water excretion into the urine and in this way helps control the concentration of water in the body fluids. (2) *Oxytocin* (a) helps deliver milk from the glands of the breast to the nipples during suckling and (b) possibly helps in the delivery of the baby at the end of gestation.

CELL TYPES IN THE ANTERIOR PITUITARY GLAND

The anterior pituitary gland contains many different types of secretory cells, as illustrated in Figure 75–3. Usually, there is one cell type for each major hormone formed in this gland. With special stains attached to high affinity antibodies that bind with the distinctive hormones, at least five cell types can be differentiated one from another, as follows:

1. *Somatotropes*—human growth hormone (hGH).
2. *Corticotropes*—adrenocorticotropin (ACTH).
3. *Thyrotropes*—thyroid-stimulating hormone (TSH).
4. *Gonadotropes*—gonadotropic hormones, which include both luteinizing hormone (LH) and follicle-stimulating hormone (FSH).
5. *Lactotropes*—prolactin (PRL).

About 30 to 40 per cent of the anterior pituitary cells are somatotropes that secrete growth hormone, and about 20 per cent are corticotropes that secrete ACTH. The other cell types each number only 3 to 5 per cent of the total; nevertheless, they secrete the extremely powerful hormones for controlling thyroid function, sexual functions, and milk secretion by the breasts.

Somatotropes stain very strongly with acid dyes and therefore are called *acidophils.* Thus, pituitary tumors that secrete large quantities of human growth hormone are called *acidophilic tumors.*

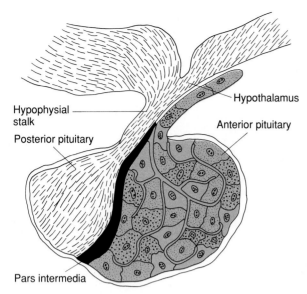

Figure 75-1. The pituitary gland.

The cell bodies of the cells that secrete the posterior pituitary hormones are not located in the posterior pituitary gland itself but instead are large neurons located in the *supraoptic* and *paraventricular nuclei* of the hypothalamus; the hormones are then transported to the posterior pituitary glands in the axoplasm of the neurons' nerve fibers passing from the hypothalamus to the posterior pituitary gland. This will be discussed more fully later in the chapter.

● CONTROL OF PITUITARY SECRETION BY THE HYPOTHALAMUS

Almost all secretion by the pituitary is controlled by either hormonal or nervous signals from the hypothalamus. Indeed, when the pituitary gland is re-

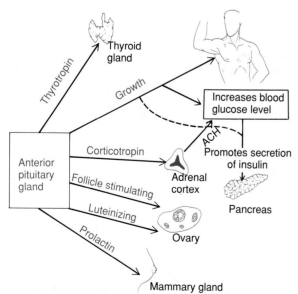

Figure 75-2. Metabolic functions of the anterior pituitary hormones.

moved from its normal position beneath the hypothalamus and transplanted to some other part of the body, its rates of secretion of the different hormones (except for prolactin) fall to low levels — in the case of some of the hormones, to zero.

Secretion from the posterior pituitary is controlled by nerve signals originating in the hypothalamus and terminating in the posterior pituitary. In contrast, secretion by the anterior pituitary is controlled by hormones called *hypothalamic releasing* and *inhibitory hormones* (or *factors*) secreted within the hypothalamus itself and then conducted to the anterior pituitary through minute blood vessels called *hypothalamic-hypophysial portal vessels*. In the anterior pituitary these releasing and inhibitory hormones act on the glandular cells to control their secretion. This system of control will be discussed in detail later in the chapter.

The hypothalamus in turn receives signals from almost all possible sources in the nervous system. Thus, when a person is exposed to pain, a portion of the pain signal is transmitted into the hypothalamus. Likewise, when a person experiences some powerful depressing or exciting thought, a portion of the signal is transmitted into the hypothalamus. Olfactory stimuli denoting pleasant or unpleasant smells transmit strong signal components directly and through the amygdaloid nuclei into the hypothalamus. *Even the concentrations of nutrients, electrolytes, water, and various hormones* in the blood excite or inhibit various portions of the hypothalamus. Thus, the hypothalamus is a collecting center for information concerned with the internal well-being of the body, and in turn much of this information is used to control secretions of the many globally important pituitary hormones.

THE HYPOTHALAMIC-HYPOPHYSIAL PORTAL SYSTEM

The anterior pituitary is a highly vascular gland with extensive capillary sinuses among the glandular cells. Almost all the blood that enters these sinuses passes first through another capillary bed in the lower tip of the hypothalamus and then through small *hypothalamic-hypophysial portal vessels* into the anterior pituitary sinuses. Thus, Figure 75-4 illustrates a small artery supplying the lowermost portion of the hypothalamus called the *median eminence* that connects inferiorly with the pituitary stalk. Small blood vessels penetrate into the substance of the median eminence and then return to its surface, coalescing to form the hypothalamic-hypophysial portal vessels. These in turn pass downward along the pituitary stalk to supply blood to the anterior pituitary sinuses.

Secretion of Hypothalamic Releasing and Inhibitory Hormones Into the Median Eminence. Special neurons in the hypothalamus synthesize and secrete the *hypothalamic releasing* and *inhibitory hormones* (or *releasing* and *inhibitory factors*) that control the secretion of the anterior pituitary hormones.

Sinusoid   Alpha (*a*) cell   Epsilon (ε) acidophil cell

Gamma (γ) cell

Delta (δ) basophil cell

Beta (β) cell

Figure 75–3. Cellular structure of the anterior pituitary gland. (From Guyton: Physiology of the Human Body. 6th Ed. Philadelphia, Saunders College Publishing, 1984.)

These neurons originate in various parts of the hypothalamus and send their nerve fibers into the median eminence and the tuber cinereum, an extension of hypothalamic tissue that extends into the pituitary stalk. The endings of these fibers are different from most endings in the central nervous system in that their function is not to transmit signals from one neuron to another but merely to secrete the hypothalamic releasing and inhibitory hormones (factors) into the tissue fluids. These hormones are immediately absorbed into the hypothalamic-hypophysial portal system and carried directly to the sinuses of the anterior pituitary gland.

(To avoid confusion, the student needs to know the difference between a "factor" and a "hormone." A substance that has the actions of a hormone but that has not been purified and identified as a distinct chemical compound is called a *factor*. Once it has been so identified, it is thereafter known as a *hormone* instead of simply a factor.)

Function of the Releasing and Inhibitory Hormones. The function of the releasing and inhibitory hormones is to control the secretion of the anterior pituitary hormones. For most of the anterior pituitary hormones it is the releasing hormones that are important; but, for prolactin, an inhibitory hormone probably exerts most control. The hypothalamic releasing and inhibitory hormones (or factors) that are of major importance are the following:

1. *Thyroid-stimulating hormone releasing hormone* (TRH), which causes release of thyroid-stimulating hormone.

2. *Corticotropin-releasing hormone* (CRH), which causes release of adrenocorticotropin.

3. *Growth hormone releasing hormone* (GHRH), which causes release of growth hormone, and *growth hormone inhibitory hormone* (GHIH), which is the same as the hormone *somatostatin* and which inhibits the release of growth hormone.

4. *Gonadotropin-releasing hormone* (GnRH), which causes release of the two gonadotropic hormones, luteinizing hormone and follicle-stimulating hormone.

5. *Prolactin inhibitory factor* (PIF), which causes inhibition of prolactin secretion.

In addition to these more important hypothalamic hormones, still another probably excites the secretion of prolactin, and several possibly inhibit some of the other anterior pituitary hormones. Each of the more important hypothalamic hormones will be discussed in detail at the time that the specific hormonal system

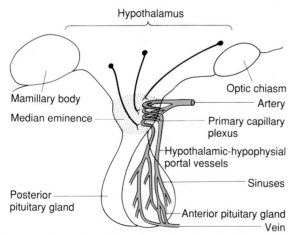

Hypothalamus

Optic chiasm

Mamillary body
Artery

Median eminence
Primary capillary plexus

Hypothalamic-hypophysial portal vessels

Sinuses

Posterior pituitary gland

Anterior pituitary gland
Vein

Figure 75–4. The hypothalamic-hypophysial portal system.

controlled by them is presented in this and subsequent chapters.

Specific Areas in the Hypothalamus That Control Secretion of Specific Hypothalamic Releasing and Inhibitory Factors. All or most of the hypothalamic hormones are secreted at nerve endings in the median eminence before being transported to the anterior pituitary gland. Electrical stimulation of this region excites these nerve endings and therefore causes release of essentially all the hypothalamic hormones. However, the neuronal cell bodies that give rise to these median eminence nerve endings are located in other discrete areas of the hypothalamus or in closely related areas of the basal brain. Unfortunately, the specific loci of the neuronal cell bodies that form the different hypothalamic releasing or inhibitory hormones are so incompletely known that it would be misleading to attempt a delineation here.

● PHYSIOLOGICAL FUNCTIONS OF GROWTH HORMONE

All the major anterior pituitary hormones besides growth hormone exert their principal effects by stimulating target glands — the thyroid gland, the adrenal cortex, the ovaries, the testicles, and the mammary glands. The functions of each of these pituitary hormones are so intimately concerned with the functions of the respective target glands that, except for growth hormone, their functions and the function of their corresponding hypothalamic releasing and inhibitory hormones will be discussed in subsequent chapters along with the target glands. Growth hormone, in contrast to other hormones, does not function through a target gland but instead exerts its effects on all or almost all tissues of the body.

EFFECT ON GROWTH OF GROWTH HORMONE

Growth hormone (GH), also called *somatotropic hormone* (SH) or *somatotropin*, is a small protein molecule containing 191 amino acids in a single chain and having a molecular weight of 22,005. It causes growth of almost all tissues of the body that are capable of growing. It promotes increased sizes of the cells, increased mitosis with development of increased numbers of cells, and specific differentiation of certain types of cells such as bone growth cells and early muscle cells.

Figure 75–5 illustrates weight charts of two growing rats, one of which received daily injections of growth hormone, compared with a littermate that did not receive growth hormone. This figure shows marked exacerbation of growth by growth hormone — both in the early days of life and even after the two rats had reached adulthood. In the early stages of development, all organs of the treated rat increased proportionately in size, but after adulthood was

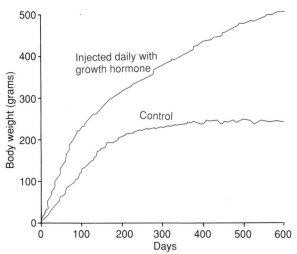

Figure 75–5. Comparison of weight gain of a rat injected daily with growth hormone with that of a normal rat.

reached, most of the bones ceased lengthening while the soft tissues continued to grow. This results from the fact that once the epiphyses of the long bones have united with the shafts, further growth of bone length cannot occur even though most other tissues of the body can continue to grow throughout life.

METABOLIC EFFECTS OF GROWTH HORMONE

Aside from its general effect in causing growth, growth hormone has many specific metabolic effects as well, including especially (1) increased rate of protein synthesis in all cells of the body; (2) increased mobilization of fatty acids from adipose tissue and increased use of the fatty acids for energy; and (3) decreased rate of glucose utilization throughout the body.

Thus, in effect, growth hormone enhances the body protein, uses up the fat stores, and conserves carbohydrate.

Role of Growth Hormone in Promoting Protein Deposition

Although the most important cause of the increased protein deposition caused by growth hormone is not known, a series of different effects are known, all of which can lead to enhanced protein:

1. Enhancement of Amino Acid Transport Through the Cell Membranes. Growth hormone directly enhances transport of at least some and perhaps most amino acids through the cell membranes to the interior of the cells. This increases the concentrations of the amino acids in the cells and is presumed to be at least partly responsible for the increased protein synthesis. This control of amino acid transport is similar to the effect of insulin in controlling glucose transport through the membrane, as discussed in Chapters 67 and 78.

2. Enhancement of RNA Translation to Promote Protein Synthesis by the Ribosomes. Even when the amino acids are not increased in the cells, growth hormone still excites increased ribonucleic acid (RNA) translation, causing protein to be synthesized in increased amounts by the ribosomes in the cytoplasm.

3. Increased Nuclear Transcription of DNA to Form RNA. Over more prolonged periods of time (24 to 48 hours), growth hormone also stimulates the transcription of deoxyribonucleic acid (DNA) in the nucleus, causing formation of increased quantities of RNA. This in turn promotes more protein synthesis and also promotes growth if sufficient energy, amino acids, vitamins, and other necessities for growth are available. In the long run, this perhaps is the most important of all the functions of growth hormone.

4. Decreased Catabolism of Protein and Amino Acids. In addition to the increase in protein synthesis, there is a decrease in the breakdown of cell protein. A probable reason for this is that growth hormone also mobilizes large quantities of free fatty acids from the adipose tissue, and these in turn are used to supply most of the energy for the body cells, thus acting as a potent "protein sparer."

Summary. Growth hormone enhances almost all facets of amino acid uptake and protein synthesis by cells, while at the same time reducing the breakdown of proteins.

Effect of Growth Hormone in Enhancing Fat Utilization for Energy

Growth hormone has a specific effect in causing release of fatty acids from adipose tissue and, therefore, increasing the concentration of fatty acids in the body fluids. In addition, in the tissues throughout the body it enhances the conversion of fatty acids to acetylcoenzyme A (acetyl-CoA) with subsequent utilization of this for energy. Therefore, under the influence of growth hormone, fat is utilized for energy in preference to both carbohydrates and proteins.

Some research workers have considered this fat mobilization effect of growth hormone to be one of its most important functions and have also considered the protein-sparing effect to be a major factor that promotes protein deposition and growth. However, growth hormone mobilization of fat requires hours to occur, whereas enhancement of cellular protein synthesis can begin in minutes under the influence of growth hormone.

Ketogenic Effect of Growth Hormone. Under the influence of excessive amounts of growth hormone, fat mobilization from adipose tissue can sometimes become so great that large quantities of acetoacetic acid are formed by the liver and released into the body fluids, thus causing *ketosis*. This excessive mobilization of fat from the adipose tissue also frequently causes a fatty liver.

Effect of Growth Hormone on Carbohydrate Metabolism

Growth hormone has four major effects on cellular metabolism of glucose: (1) decreased use of glucose for

energy, (2) enhancement of glycogen deposition in the cells, (3) diminished uptake of glucose by the cells, and (4) increased insulin secretion and decreased sensitivity to insulin.

1. Decreased Use of Glucose for Energy. Unfortunately, we do not know the precise mechanism by which growth hormone decreases glucose utilization by the cells. However, the decrease probably results partially from the increased mobilization and utilization of fatty acids for energy caused by growth hormone. That is, the fatty acids form large quantities of acetyl-CoA that in turn initiate feedback effects to block the glycolytic breakdown of glucose and glycogen.

2. Enhancement of Glycogen Deposition in the Cells. Because glucose and glycogen cannot easily be utilized for energy, the glucose that does enter the cells is rapidly polymerized into glycogen and deposited. Therefore, the cells rapidly become saturated with glycogen and can store no more.

3. Diminished Uptake of Glucose by the Cells and Increased Blood Glucose Concentration — "Pituitary Diabetes." When growth hormone is first administered to an animal, the cellular uptake of glucose is enhanced, and the blood glucose concentration falls slightly. However, this effect lasts for only 30 minutes to an hour or so and is then followed by exactly the opposite effect, namely, decreased transport of glucose into the cell. This perhaps results from the fact that the cells have already taken up an excess of glucose that they are having difficulty using. Without normal cellular uptake and use, the blood concentration of glucose increases, sometimes to as high as 50 to 100 per cent above normal, and the condition is called *pituitary diabetes*. When this diabetes is treated with insulin, it is "insulin insensitive," requiring excessive amounts of insulin for therapy.

4. Increased Insulin Secretion — Diabetogenic Effect of Growth Hormone. The increase in blood glucose concentration caused by growth hormone stimulates the beta cells of the islets of Langerhans to secrete extra insulin. In addition, the growth hormone also has a direct stimulatory effect on the beta cells as well. The combination of these two effects sometimes so greatly overstimulates insulin secretion by the beta cells that they literally "burn out." When this occurs diabetes mellitus develops, a disease that will be discussed in detail in Chapter 78. Therefore, growth hormone is said to have a *diabetogenic effect*.

Necessity of Insulin and Carbohydrate for the Growth-Promoting Action of Growth Hormone. Growth hormone fails to cause growth in an animal lacking a pancreas, and it also fails to cause growth if carbohydrates are excluded from the diet. This shows that adequate insulin activity as well as adequate availability of carbohydrates is necessary for growth hormone to be effective. Part of this requirement for carbohydrates and insulin is to provide the energy needed for the metabolism of growth. But there seem to be other effects as well. Especially important is the specific effect of insulin in enhancing amino acid transport into cells in the same way that it enhances glucose transport.

STIMULATION OF CARTILAGE AND BONE GROWTH

Although growth hormone stimulates increased deposition of proteins and increased growth in almost all tissues of the body, its most obvious effect is to increase growth of the skeletal frame. This results from multiple effects of growth hormone on bone, including (1) increased deposition of protein by the chondrocytic and osteogenic cells that cause bone growth, (2) increased rate of reproduction of these cells as well, and (3) the specific effect of converting chondrocytes into osteogenic cells, thus causing specific deposition of new bone.

There are two principal mechanisms of bone growth: In one of these, the long bones grow in length at the epiphyseal cartilages, where the epiphyses at the ends of the bone are separated from the shaft. This growth causes deposition of new cartilage, followed by conversion of this into new bone, thus elongating the shaft and pushing the epiphyses farther and farther apart. However, at the same time, the epiphyseal cartilage itself is progressively used up so that by late adolescence no additional epiphyseal cartilage remains to provide for further growth. At this time bony fusion occurs between the shaft and each end epiphysis so that no further lengthening of the long bone can occur thereafter. Growth hormone stimulates all of these processes of epiphyseal cartilage growth and growth of the long bones. However, once the epiphyses have united with the shafts, growth hormone has no further ability to lengthen the bones.

For the second mechanism of bone growth, *osteoblasts* in the bone periosteum and in some bone cavities deposit new bone on the surfaces of older bone. Simultaneously, *osteoclasts* in the bone (to be discussed in detail in Chapter 79) remove old bone. When the rate of deposition is greater than that of resorption, the thickness of the bone increases. Growth hormone strongly stimulates the osteoblasts. Therefore, the bones can continue to enlarge throughout life under the influence of growth hormone; this is especially true for the membranous bones. For instance, the jaw bones can be stimulated to grow even after adolescence, causing forward protrusion of the chin and lower teeth. Likewise, the bones of the skull grow in thickness and also give rise to bony protrusions over the eyes.

GROWTH HORMONE EXERTS MUCH OF ITS EFFECTS THROUGH INTERMEDIATE SUBSTANCES CALLED SOMATOMEDINS

When growth hormone is supplied directly to cartilage chondrocytes cultured outside the body, proliferation or enlargement of the chondrocytes usually fails to occur. Yet growth hormone injected into the intact animal does cause such proliferation and growth of the same cells.

In brief, it has been found that growth hormone causes the liver (and to much less extent other tissues) to form several small proteins called *somatomedins* that in turn have the very potent effect of increasing all aspects of bone growth. At least four different somatomedins have been isolated, but by far the most important of these is *somatomedin-C,* the molecular weight of which is approximately 7500. The concentration of *somatomedin-C* in the plasma normally follows closely the rate of secretion of growth hormone. The pygmies of Africa have a congenital inability to synthesize significant amounts of somatomedin-C. Therefore, even though their plasma concentration of growth hormone is either normal or high, there remain diminished amounts of somatomedin-C in the plasma, thus apparently accounting for the small stature of these people. Some other dwarfs (the Levi-Lorain dwarf) also have this problem.

Therefore, it has been postulated that most if not nearly all of the growth effects of growth hormone result from somatomedin-C and other somatomedins, rather than from direct effects of growth hormone on the bones and other peripheral tissues. Even so, recent experiments have demonstrated that injection of growth hormone directly into the epiphyseal cartilages of bones will cause specific growth of these cartilage areas alone, and the amount of growth hormone required for this is minute. Therefore, some aspects of the somatomedin hypothesis are still questionable. One possibility is that growth hormone can also cause formation of enough somatomedin-C in the local tissue to cause the local growth. However, it is also possible that growth hormone itself is directly responsible for increased growth in some tissues and that the somatomedin mechanism is an alternative means of increasing growth but not always a necessary one.

(Somatomedin-C also has insulin-like effects, promoting glucose transport through membranes. Therefore, it has also been called insulin-like growth factor, or IGF-G).

The Short Duration of Action of Growth Hormone but Prolonged Action of Somatomedin-C. Growth hormone is only weakly attached to the plasma proteins in the blood. Therefore, it is released from the blood into the tissues rapidly, having a half-time in the blood of less than 20 minutes. By contrast, somatomedin-C attaches very strongly to a carrier protein in the blood that, like somatomedin-C, is itself produced in response to growth hormone. As a result, the somatomedin-C is released only very slowly to the tissues with a half-time of about 20 hours. This obviously greatly prolongs the growth-promoting effects of the bursts of growth hormone secretion illustrated in Figure 75–6.

REGULATION OF GROWTH HORMONE SECRETION

For many years it was believed that growth hormone was secreted primarily during the period of

Figure 75-6. Typical variations in growth hormone secretion throughout the day, illustrating the especially powerful effect on secretion caused by strenuous exercise and illustrating also the high rate of growth hormone secretion that occurs during the first few hours of deep sleep.

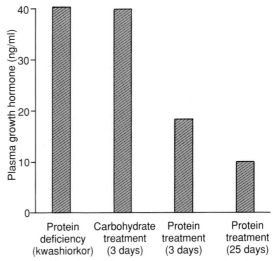

Figure 75-7. Effect of extreme protein deficiency on the concentration of growth hormone in the plasma in the disease kwashiorkor. The figure also shows the failure of carbohydrate treatment but the effectiveness of protein treatment in lowering growth hormone concentration. (Drawn from data in Pimstone: *Am. J. Clin. Nutr.*, 21:482, 1968.)

growth but then disappeared from the blood at adolescence. However, this has proved not to be true, because after adolescence, secretion decreases only slowly with aging, finally falling to about 25 per cent of the adolescent level in very old age.

The rate of growth hormone secretion increases and decreases within minutes, as illustrated in Figure 75-6, sometimes for reasons that are not at all understood, but at other times definitely in relation to the person's state of nutrition or stress, such as during (1) *starvation,* (2) *hypoglycemia* or *low concentration of fatty acids in the blood,* (3) *exercise,* (4) *excitement,* and (5) *trauma.* And it characteristically increases during the first 2 hours of *deep sleep,* as also is shown in Figure 75-6.

The normal concentration of growth hormone in the plasma of an adult is between 1.5 and 3 ng/ml and in a child or adolescent about 6 ng/ml. However, these values often increase to as high as 50 ng/ml after depletion of the body stores of proteins or carbohydrates. Under acute conditions, hypoglycemia is a far more potent stimulator of growth hormone secretion than is an acute decrease in protein intake.

On the other hand, in chronic conditions, growth hormone secretion seems to correlate more with the degree of cellular protein depletion than with the degree of glucose insufficiency. For instance, the extremely high levels of growth hormone that occur during starvation are very closely related to the amount of protein depletion. Figure 75-7 illustrates this relationship: the first column shows growth hormone levels in children with extreme protein deficiency during the malnutrition disease called *kwashiorkor*; the second column shows the levels in the same children after 3 days of treatment with more than adequate quantities of carbohydrates in their diets, illustrating that the carbohydrates did not lower the plasma growth hormone concentration; the third and fourth columns show the levels after treatment with protein supplement in their diet for 3 and 25 days, with concomitant decrease in the hormone. These results demonstrate that under very severe conditions of protein malnutrition, adequate calories alone are not sufficient to correct the excess production of growth hormone. Instead, the protein defi-

ciency must also be corrected before the growth hormone concentration will return to normal.

Role of the Hypothalamus, Growth Hormone Releasing Hormone, and Somatostatin in the Control of Growth Hormone Secretion

From the preceding description of the many different factors that can affect growth hormone secretion, one can readily understand the perplexity that has faced physiologists in their attempt to unravel the mysteries of the regulation of growth hormone secretion. However, it is known that growth hormone secretion is controlled almost entirely in response to two factors secreted in the hypothalamus and then transported to the anterior pituitary gland through the hypothalamic-hypophysial portal vessels. These are *growth hormone releasing hormone* (GHRH) and *growth hormone inhibitory hormone* (GHIH), also called *somatostatin.*

The hypothalamic nucleus that causes secretion of growth hormone releasing hormone is the ventromedial nucleus, the same area of the hypothalamus that is known to be sensitive to hypoglycemia and to cause hunger in hypoglycemic states. The secretion of somatostatin is controlled by other nearby areas of the hypothalamus. Therefore, it is reasonable to believe that some of the same signals that modify a person's behavioral feeding instincts also alter the rate of growth hormone secretion. In a similar manner, hypothalamic signals depicting emotions, stress, and trauma can all affect hypothalamic control of growth hormone secretion. In fact, definitive experiments have shown that catecholamines, dopamine, and serotonin, each of which is released by a different neuronal system in the hypothalamus, will increase the rate of growth hormone secretion.

Most of the control of growth hormone secretion is probably mediated through growth hormone releasing hormone rather than through the inhibitory hormone somatostatin. However, it is important to note that somatostatin is also secreted by the delta cells of the islets of Langerhans in the pancreas, and that it can inhibit the secretion of insulin and glucagon by the beta and alpha cells in the islets of Langerhans in the same way that it inhibits the secretion of growth hormone by the anterior pituitary gland. For this reason, somatostatin might have a widespread role in modulating the functions of multiple hormonal systems.

When growth hormone is administered to an animal over a period of hours, the rate of endogenous growth hormone secretion becomes decreased. This illustrates that growth hormone secretion, as is true of that of essentially all other hormones, is subject to typical negative feedback control. However, the nature of this feedback mechanism and whether it is mediated through inhibition of growth hormone releasing hormone or enhancement of somatostatin are still uncertain.

In summary, our present knowledge of the regulation of growth hormone secretion is not sufficient to describe a composite picture. Yet, because of its undoubted long-term effect on promoting protein synthesis and growth of tissues, one would like to propose that the major long-term controller of growth hormone secretion is the state of nutrition of the tissues themselves, especially their level of protein nutrition. That is, nutritional deficiency or excess tissue need for cellular proteins — for instance, after a severe bout of exercise when the muscles' nutritional status has been taxed — would in some way increase the rate of growth hormone secretion. The growth hormone in turn would promote synthesis of new proteins while at the same time conserving the proteins already present in the cells.

ABNORMALITIES OF GROWTH HORMONE SECRETION

Panhypopituitarism. This term means decreased secretion of all the anterior pituitary hormones. The decrease in secretion may be congenital (present from birth), or it may occur suddenly or slowly at any time during the life of the individual.

Dwarfism. Most instances of dwarfism result from generalized deficiency of anterior pituitary secretion (panhypopituitarism) during childhood. In general, the features of the body develop in appropriate proportion to each other, but the rate of development is greatly decreased. A child who has reached the age of 10 years may have the bodily development of a child of 4 to 5 years, and the same person on reaching the age of 20 years may have the bodily development of a child of 7 to 10 years.

The panhypopituitary dwarf does not pass through puberty and never secretes a sufficient quantity of gonadotropic hormones to develop adult sexual functions. In one third of the dwarfs, however, the deficiency is of growth hormone alone; these individuals do mature sexually and occasionally do reproduce.

In one type of dwarfism (the African pygmy and the Levi-Lorain dwarf), the rate of growth hormone secretion is actually normal or high, but there is a hereditary inability to form somatomedin-C in response to the growth hormone.

Treatment With Human Growth Hormone. The growth hormones of different species of animals are sufficiently different from each other that they will function to cause growth only in the one animal species or at most very closely allied species. For this reason, growth hormone prepared from lower animals (except to some extent from primates) is not effective in human beings. Therefore, the growth hormone of the human being is called *human growth hormone* (hGH) to distinguish it from the others.

Unfortunately, in the past it has been very difficult to obtain sufficient quantities of human growth hormone to treat patients with growth hormone deficiency except on an experimental basis because it has had to be prepared from human pituitary glands. However, human growth hormone can now be synthesized by *Escherichia coli* bacteria as a result of successful application of recombinant DNA technology. Therefore, this hormone is now beginning to become available in sufficient quantities for treatment purposes. Dwarfs who have pure growth hormone deficiency can be completely cured. Also, human growth hormone might prove to be beneficial in other metabolic disorders, because of its widespread metabolic functions.

Panhypopituitarism in the Adult. Panhypopituitarism occurring in adulthood frequently results from one of three common abnormalities: Two tumorous conditions, craniopharyngiomas and chromophobe tumors, may compress the pituitary gland until the functioning anterior pituitary cells are totally or almost totally destroyed. The third cause is thrombosis of the pituitary blood vessels. This occurs occasionally when a mother develops circulatory shock following the birth of a baby.

The effects of adult panhypopituitarism, in general, are (1) hypothyroidism, (2) depressed production of glucocorticoids by the adrenal glands, and (3) suppressed secretion of the gonadotropic hormones to the point at which sexual functions are lost. Thus, the picture is that of a lethargic person (from lack of thyroid hormones) who is gaining weight because of lack of fat mobilization by growth, adrenocorticotropic, adrenocortical, and thyroid hormones and who has lost all sexual functions. Except for the abnormal sexual functions, the patient can usually be treated satisfactorily by administration of adrenocortical and thyroid hormones.

Gigantism. Occasionally, the acidophilic, growth hormone–producing cells of the anterior pituitary become excessively active, and sometimes even acidophilic tumors occur in the gland. As a result, large quantities of growth hormone are produced. All body tissues grow rapidly, including the bones, and if the condition occurs before adolescence, that is, before the epiphyses of the long bones have not become fused with the shafts, height increases so that the person becomes a giant 8 to 9 feet tall.

The giant ordinarily has *hyperglycemia,* and the beta cells of the islets of Langerhans in the pancreas are prone to degenerate, partially because they become overactive owing to the hyperglycemia and partially because of a direct overstimulating effect of growth hormone on the islet cells. Consequently, in about 10 per cent of giants finally *full-blown diabetes mellitus* develops.

In most giants, unfortunately, panhypopituitarism eventually develops if they remain untreated, because the gigantism is usually caused by a tumor of the pituitary gland that grows until the gland itself is destroyed. This general deficiency of pituitary hormones usually causes death in early adulthood. However, once gigantism is diagnosed, further development can often be blocked by microsurgical removal of the tumor from the pituitary gland or, if not in this way, by irradiation of the gland.

Acromegaly. If an acidophilic tumor occurs after adolescence — that is, after the epiphyses of the long bones have fused with the shafts — the person cannot grow taller; but the soft tissues can continue to grow, and the bones can grow in thickness. This condition, which is illustrated in Figure 75-8, is known as *acromegaly.* Enlargement is especially marked in the bones of the hands and feet and in the *membranous bones,* including the cranium, the nose, the bosses on the forehead, the supraorbital ridges, the lower jawbone, and portions of the vertebrae, for their growth does not cease at adolescence. Consequently, the jaw protrudes forward, sometimes as much as a half inch, the forehead slants forward because of excess development of the supraorbital ridges, the nose increases to as much as twice normal size, the foot requires a size 14 or larger shoe, and the fingers become extremely thickened so that the hand develops a size almost twice normal. In addition to these effects, changes in the vertebrae ordinarily cause a hunched back, which is known clinically as *kyphosis.* Finally, many soft tissue organs, such as the tongue, liver, and especially the kidneys, become greatly enlarged.

Possible Role of Decreased Growth Hormone Secretion in Causing Aging

In persons who have lost their ability to secrete growth hormone, the aging process accelerates. For instance, a person at age 50 who has been without growth hormone for many years will probably have the appearance of a person aged 65. The aging seems to result mainly from decreased protein deposition in most tissues of the body and in its place increased deposition of fat. The physical and physio-

logical effects are increased wrinkling of the skin, diminished rates of function of some of the organs, and diminished muscular mass and muscle strength.

As one ages, the average plasma concentration of growth hormone changes approximately as follows:

	ng/ml
5 to 20 years	6
20 to 40 years	3
40 to 70 years	1.6

Thus, it is highly possible that part of the aging effects in normal older life result from diminished growth hormone secretion.

● THE POSTERIOR PITUITARY GLAND AND ITS RELATION TO THE HYPOTHALAMUS

The *posterior pituitary gland,* also called the *neurohypophysis,* is composed mainly of glial-like cells called *pituicytes.* However, the pituicytes do not secrete hormones; they act simply as a supporting structure for large numbers of *terminal nerve fibers* and *terminal nerve endings* from nerve tracts that originate in the *supraoptic* and *paraventricular nuclei* of the hypothalamus, as shown in Figure 75-9. These tracts pass to the neurohypophysis through the *pituitary stalk* (hypophysial stalk). The nerve endings are bulbous knobs containing many secretory granules that lie on the surfaces of capillaries onto which they secrete the two posterior pituitary hormones: (1) *anti-*

Figure 75-8. An acromegalic patient.

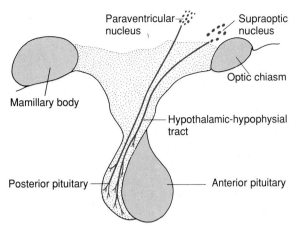

Figure 75–9. Hypothalamic control of the posterior pituitary.

diuretic hormone (ADH), also called *vasopressin;* and (2) *oxytocin.*

If the pituitary stalk is cut above the pituitary gland but the entire hypothalamus is left intact, the posterior pituitary hormones continue, after a transient decrease for a few days, to be secreted almost normally; but they are then secreted by the cut ends of the fibers within the hypothalamus and not by the nerve endings in the posterior pituitary. The reason for this is that the hormones are initially synthesized in the cell bodies of the supraoptic and paraventricular nuclei and are then transported in combination with "carrier" proteins called *neurophysins* down to the nerve endings in the posterior pituitary gland, requiring several days to reach the gland.

ADH is formed primarily in the supraoptic nuclei, whereas oxytocin is formed primarily in the paraventricular nuclei. However, each of these two nuclei can synthesize approximately one-sixth as much of the second hormone as of its primary hormone.

When nerve impulses are transmitted downward along the fibers from the supraoptic or paraventricular nuclei, the hormone is immediately released from the secretory granules in the nerve endings by the usual secretory mechanism of *exocytosis* and is absorbed into adjacent capillaries. Both the neurophysin and the hormone are secreted together; but because they are only loosely bound to each other, the hormone separates almost immediately. The neurophysin has no known function after leaving the nerve terminals.

CHEMICAL NATURE OF ANTIDIURETIC HORMONE (VASOPRESSIN) AND OXYTOCIN

Both oxytocin and ADH (vasopressin) are polypeptides containing nine amino acids. The amino acid sequences of these are the following:

Vasopressin:
    Cys-Tyr-Phe-Gln-Asn-Cys-Pro-Arg-GlyNH$_2$
Oxytocin:
    Cys-Tyr-Ile-Gln-Asn-Cys-Pro-Leu-GlyNH$_2$

Note that these two hormones are almost identical except that in vasopressin phenylalanine and arginine replace isoleucine and leucine of the oxytocin molecule. The similarity of the molecules explains their partial functional similarities.

PHYSIOLOGICAL FUNCTIONS OF ANTIDIURETIC HORMONE (VASOPRESSIN)

Extremely minute quantities of antidiuretic hormone (ADH) — as small as 2 nanograms — when injected into a person can cause antidiuresis, that is, decreased excretion of water by the kidneys. This antidiuretic effect was discussed in detail in Chapter 28. Briefly, in the absence of ADH, the collecting tubules and ducts are almost totally impermeable to water, which prevents significant reabsorption of water and therefore allows extreme loss of water into the urine. On the other hand, in the presence of ADH the permeability of the collecting ducts and tubules to water increases greatly and allows most of the water to be reabsorbed as the tubular fluid passes through these ducts, thereby conserving water in the body.

The precise mechanism by which ADH acts on the ducts to increase their permeability is only partially known. The hormone causes special structural changes in the apical membranes of the tubular epithelial cells. This provides temporarily many new pores that allow free diffusion of water between the tubular and peritubular fluids. Water is then absorbed from the collecting tubules and ducts by osmosis, as explained in relation to the concentrating mechanism of the kidney in Chapter 28.

Regulation of ADH Production

Osmotic Regulation. When a concentrated electrolyte solution is injected into the artery supplying the hypothalamus, the ADH neurons in the supraoptic and paraventricular nuclei immediately transmit impulses into the posterior pituitary to release large quantities of ADH into the circulating blood. Conversely, injection of a dilute solution into this artery causes complete cessation of the impulses and essentially complete cessation of ADH secretion. The ADH that is already in the tissues is destroyed at a rate of approximately one half every 15 to 20 minutes. Thus, the concentration of ADH in the body fluids can change from small amounts to large amounts, or vice versa, in only a few minutes.

The precise way that the osmotic concentration of the extracellular fluids controls ADH secretion is still not clear. Yet somewhere in or near the hypothalamus are modified neuron receptors called *osmoreceptors.* When the extracellular fluid becomes too concentrated, fluid is pulled by osmosis out of the osmoreceptor cell, decreasing its size and initiating appropriate signals in the hypothalamus to cause additional ADH secretion. Conversely, when the extracellular fluid becomes too dilute, water moves by osmosis in the opposite direction into the cell, and this decreases the signal for ADH secretion. Although some research workers place these osmoreceptors in the hypothalamus itself, others believe that they are located in the *organum vasculosum,* a highly vascular structure located in the anteroventral wall of the third ventricle. Regardless of the mechanism, concentrated body fluids do stimulate the supraoptic nuclei, whereas dilute body fluids inhibit them. Therefore, a feedback control system is available to control the total osmotic pressure of the body fluids, operating as follows:

When the body fluids become highly concentrated, the supraoptic nuclei become excited, impulses are transmitted to the posterior pituitary, and ADH is secreted. This passes by way of the blood to the kidneys, where it increases the permeability of the collecting ducts to water. As a result, most of the water is then reabsorbed from the tubular fluid, while electrolytes continue to be lost into the urine. This dilutes the extracellular fluid, returning it to a reasonably normal osmotic composition.

Further details on the function of antidiuretic hormone to control renal function and body fluid osmolality are presented in Chapter 28.

Vasoconstrictor and Pressor Effects of ADH, and Increased ADH Secretion Caused by Low Blood Volume

Aside from the effect of minute concentrations of ADH in causing increased water conservation by the kidneys, higher concentrations of ADH have the very potent effect of constricting the arterioles everywhere in the body and therefore of increasing the arterial pressure. For this reason, ADH also has another name, *vasopressin.*

One of the stimuli for causing very intense ADH (vasopressin) secretion is decreased blood volume. This occurs especially strongly when the blood volume decreases 15 to 20 per cent, with the secretory rate then rising sometimes to 20 to 50 times normal. The cause of this is the following:

The atria, especially the right atrium, have stretch receptors that are excited by overfilling. When excited, they send signals to the brain to inhibit ADH secretion. Conversely, when unexcited as a result of underfilling, they send opposite signals to increase ADH secretion greatly. Furthermore, aside from the atrial stretch receptors, decreased stretch of the baroreceptors of the carotid, aortic, and pulmonary regions also participates in increasing ADH secretion.

For further details about this blood volume-pressure feedback mechanism, please refer again to Chapter 28, in which the ADH-vasopressin mechanisms are discussed more fully.

OXYTOCIC HORMONE

Effect on the Uterus and on Birth. An oxytocic substance is one that causes contraction of the pregnant uterus. The hormone *oxytocin,* in accordance with its name, powerfully stimulates the pregnant uterus, especially so toward the end of gestation. Therefore, many obstetricians believe that this hormone is at least partially responsible for effecting birth. This is supported by the following facts: (1) In a hypophysectomized animal, the duration of labor is prolonged, thus indicating a possible effect of oxytocin during delivery. (2) The amount of oxytocin in the plasma increases during labor, especially during the last stage. (3) Stimulation of the cervix in a pregnant animal elicits nervous signals that pass to the hypothalamus and cause increased secretion of oxytocin. These effects and this possible mechanism for aiding the birth of the baby will be discussed in Chapter 82.

Effect of Oxytocin on Milk Ejection. Oxytocin plays an especially important role in the process of lactation, a role that is far more certain than its possible role in delivery of the baby. In lactation, it causes milk to be expressed from the alveoli into the ducts so that the baby can obtain it by suckling. This mechanism works as follows: The suckling stimuli on the nipple of the breast cause signals to be transmitted through the sensory nerves to the brain. The signals finally reach the oxytocin neurons in the paraventricular and supraoptic nuclei in the hypothalamus, to cause release of oxytocin. The oxytocin then is carried by the blood to the breasts, where it causes contraction of *myoepithelial cells* that lie outside of and form a latticework surrounding the alveoli of the mammary glands. In less than a minute after the beginning of suckling, milk begins to flow. Therefore, this mechanism is called *milk letdown* or *milk ejection.* This process is discussed further in Chapter 82 in relation to lactation.

REFERENCES

Almers, W., and Stirling, C.: Distribution of transport proteins over animal cell membranes. J. Membr. Biol., 77:169, 1984.

Arad, Z., and Skadhauge, E.: Plasma hormones (arginine vasotocin, prolactin, aldosterone, and corticosterone) in relation to hydratoin state, NaCl intake, and egg laying in fowls. J. Exp. Zool., 232:707, 1984.

Bercu, B. B. (ed.): Basic and Clinical Aspects of Growth Hormone. New York, Plenum Publishing Corp., 1988.

Brownstein, M. J.: Biosynthesis of vasopressin and oxytocin. Annu. Rev. Physiol., 45:129, 1983.

Campion, D. R., et al. (eds.): Animal Growth Regulation. New York, Plenum Publishing Corp., 1989.

Chrousos, G. P., et al.: NIH Conference: Clinical applications of corticotropin-releasing factor. Ann. Intern. Med., 102:344, 1985.

Collu, R., et al.: Pediatric Endocrinology, 2nd Ed. New York, Raven Press, 1989.

Conn, P. M., et al.: Gonadotropin-releasing hormone: Molecular and cell biology, physiology, and clinical applications. Fed. Proc., 43:2351, 1984.

DeGroot, L. J. (ed.): Endocrinology, 2nd Ed. Philadelphia, W. B. Saunders Co., 1989.

Doris, P. A.: Vasopressin and central integrative processes. Neuroendocrinology, 38:75, 1984.

Dubocovish, M. L.: Pharmacology and function of melatonin receptors. FASEB J., 2:2765, 1988.

Farner, D. S.: Annual Rhythms. Annu. Rev. Physiol., 47:65, 1985.

Felig, P., et al. (eds.): Endocrinology and Metabolism, 2nd Ed. New York, McGraw-Hill Book Co., 1987.

Fiddes, J. C., and Talmadge, K.: Structure, expression, and evolution of the genes for the human glycoprotein hormones. Recent Prog. Horm. Res., 40:43, 1984.

Gann, D. S., et al.: Neural interaction in control of adrenocorticotropin. Fed. Proc., 44:161, 1985.

Gash, D. M., and Boer, G. J. (eds.): Vasopressin. New York, Plenum Publishing Corp., 1987.

Goodman, H. M.: Basic Medical Endocrinology. New York, Raven Press, 1988.

Handler, J. S., and Orloff, J.: Antidiuretic hormone. Annu. Rev. Physiol., 43:611, 1981.

Homo-Delarche, F.: Glucocorticoid receptors and steroid sensitivity in normal and neoplastic human lymphoid tissues: A review. Cancer Res., 44:431, 1984.

Hseuh, A. J. W., and Jones, P. B. C.: Gonadotropin releasing hormone: Extrapituitary actions and paracrine control mechanisms. Annu. Rev. Physiol., 45:83, 1983.

Hughes, J. P., and Friesen, H. G.: The nature and regulation of the receptors for pituitary growth hormone. Annu. Rev. Physiol., 47:469, 1985.

Isaksson, O. G. P., et al.: Mode of action of pituitary growth hormone on target cells. Annu. Rev. Physiol., 47:483, 1985.

Kannan, C. R.: The Pituitary Gland. New York, Plenum Publishing Corp., 1987.

Keller-Wood, M. E., and Dallman, M. F.: Corticosteroid inhibition of ACTH secretion. Endocr. Rev., 5:1, 1984.

Kolata, G.: New growth industry in human growth hormone? Science, 234:22, 1986.

Kudlow, J. E., et al. (eds.): Biology of Growth Factors. New York, Plenum Publishing Corp., 1988.

Lewis, U. J.: Variants of growth hormone and prolactin and their posttranslational modifications. Annu. Rev. Physiol., 46:33, 1984.

Menninger, R. P.: Current concepts of volume receptor regulation of vasopressin release. Fed. Proc., 4:55, 1985.

Muller, E. E.: Neural control of somatotropic function. Physiol. Rev., 67:962, 1987.

Nishimoto, I., and Kojima I.: Calcium signalling system triggered by insulin-like growth factor II. News Physiol. Sci., 4:94, 1989.

Oriordan, J. L.: Recent Advances in Endocrinology and Metabolism. New York, Churchill Livingstone, 1982.

Poste, G.: New insights into receptor regulation. J. Appl. Physiol., 57:1297, 1984.

Robbins, R. J., and Melmed, S. (eds.): Acromegaly. New York, Plenum Publishing Corp., 1987.

Sawyer, W. H., and Manning, M.: The use of antagonists of vasopressin in studies of its physiological functions. Fed. Proc., 44:78, 1985.

Silva, N. L., and Boulant, J. A.: Effects of osmotic pressure, glucose, and temperature on neurons in preoptic tissue slices. Am. J. Physiol. 247:R335, 1984.

Sklar, A. H., and Schrier, R. W.: Central nervous system mediators of vasopressin release. Physiol. Rev., 63:1243, 1983.

Swanson, L. W., and Sawchenko, P. E.: Hypothalamic integration: Organization of the paraventricular and supraoptic nuclei. Ann. Rev. Neurosci., 6:269, 1983.

Wass, J. A. H., and Besser, G. M.: The medical management of hormone secreting tumors of the pituitary. Annu. Rev. Med., 34:283, 1983.

CHAPTER 76

• *The Thyroid Metabolic Hormones* •

The thyroid gland, which is located immediately below the larynx on either side of and anterior to the trachea, secretes two significant hormones, *thyroxine* and *triiodothyronine*, commonly called T_4 and T_3, that have the profound effect of increasing the metabolic rate of the body. It also secretes *calcitonin*, an important hormone for calcium metabolism that is considered in detail in Chapter 79. Complete lack of thyroid secretion usually causes the basal metabolic rate to fall about 40 per cent below normal, and extreme excesses of thyroid secretion can cause the basal metabolic rate to rise as high as 60 to 100 per cent above normal. Thyroid secretion is controlled primarily by thyroid-stimulating hormone (TSH) secreted by the anterior pituitary gland.

The purpose of this chapter is to discuss the formation and secretion of the thyroid hormones, their functions in the metabolic scheme of the body, and regulation of their secretion.

• FORMATION AND SECRETION OF THE THYROID HORMONES

About 90 per cent of the hormone secreted by the thyroid gland is *thyroxine*, and 10 per cent is *triiodothyronine*. However, most of the thyroxine is eventually converted to triiodothyronine in the tissues, so that both are important functionally. The functions of these two hormones are qualitatively the same, but they differ in rapidity and intensity of action. Triiodothyronine is about four times as potent as thyroxine, but it is present in the blood in much smaller quantities and persists for a much shorter time than does thyroxine.

Physiologic Anatomy of the Thyroid Gland. The thyroid gland is composed, as shown in Figure 76–1, of large numbers of closed *follicles* (150 to 300 micrometers in diameter) filled with a secretory substance called *colloid* and lined with *cuboidal epithelioid cells* that secrete into the interior of the follicles. The major constituent of colloid is the large glycoprotein *thyroglobulin*, which contains the thyroid

hormones within its molecule. Once the secretion has entered the follicles, it must be absorbed back through the follicular epithelium into the blood before it can function in the body. The thyroid gland has a blood flow about five times the weight of the gland each minute, which is a blood supply as rich as that of any other area of the body, with the possible exception of the adrenal cortex.

IODINE REQUIREMENTS FOR FORMATION OF THYROXINE

To form normal quantities of thyroxine, approximately 50 milligrams of ingested iodine in the form of iodides are required *each year*, or approximately *1 mg/week*. To prevent iodine deficiency, common table salt is iodized with one part sodium iodide to every 100,000 parts sodium chloride.

Fate of Ingested Iodides. Iodides ingested orally are absorbed from the gastrointestinal tract into the blood in approximately the same manner as chlorides. However, most of them are rapidly excreted by the kidneys, but only after about one fifth are selectively removed from the circulating blood by the cells of the thyroid gland and used for synthesis of the thyroid hormones.

THE IODIDE PUMP (IODIDE TRAPPING)

The first stage in the formation of thyroid hormones, shown in Figure 76–2, is transport of iodides from the extracellular fluid into the thyroid glandular cells and follicles. The basal membrane of the thyroid cell has the specific ability to pump the iodide actively to the interior of the cell. This is called *iodide trapping*. In a normal gland, the iodide pump concentrates the iodide to about 30 times its concentration in the blood. However, when the thyroid gland becomes maximally active, the concentration ratio can rise to as high as 250 times.

THYROGLOBULIN AND CHEMISTRY OF THYROXINE AND TRIIODOTHYRONINE FORMATION

Formation and Secretion of Thyroglobulin by the Thyroid Cells. The thyroid cells are typical pro-

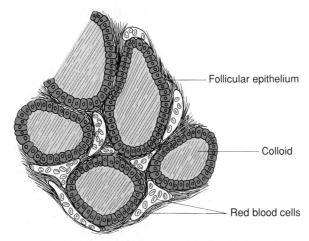

Figure 76-1. Microscopic appearance of the thyroid gland, showing the secretion of thyroglobulin into the follicles.

tein-secreting glandular cells, as illustrated in Figure 76-2. The endoplasmic reticulum and Golgi apparatus synthesize and secrete into the follicles a large glycoprotein molecule called *thyroglobulin,* with a molecular weight of 670,000.

Each molecule of thyroglobulin contains 140 tyrosine amino acids, and these are the major substrates that combine with iodine to form the thyroid hormones. These hormones form *within* the thyroglobulin molecule. That is, the thyroxine and triiodothyronine hormones formed from the tyrosine amino acids remain a part of the thyroglobulin molecule during synthesis of the thyroid hormones and even afterward as stored hormones in the follicular colloid.

In addition to secreting the thyroglobulin, the glandular cells also process the iodine and provide the enzymes and other substances necessary for thyroid hormone synthesis.

Oxidation of the Iodide Ion. The first essential step in the formation of the thyroid hormones is con-

version of the iodide ions to an *oxidized form of iodine,* either nascent iodine (I^0) or I_3^-, that is then capable of combining directly with the amino acid tyrosine.

This oxidation of iodine is promoted by the enzyme *peroxidase* and its accompanying *hydrogen peroxide,* which provide a potent system capable of oxidizing iodides. The peroxidase is located either in the apical membrane of the cell or attached to it, thus providing the oxidized iodine at exactly the point in the cell where the thyroglobulin molecule issues forth from the Golgi apparatus and thence through the membrane into the stored colloid. When the peroxidase system is blocked, or when it is hereditarily absent from the cells, the rate of formation of thyroid hormones falls to zero.

Iodination of Tyrosine and Formation of the Thyroid Hormones—"Organification" of Thyroglobulin. The binding of iodine with the thyroglobulin molecule is called *organification* of the thyroglobulin. Oxidized iodine even in the molecular form will bind directly but slowly with the amino acid tyrosine, but in the thyroid cells the oxidized iodine is associated with an *iodinase* enzyme that causes the process to occur within seconds or minutes. Therefore, almost as rapidly as the thyroglobulin molecule is released from the Golgi apparatus, or as it is secreted through the apical cell membrane into the follicle, iodine binds with about one sixth of the tyrosine amino acids within the thyroglobulin molecule.

Figure 76-3 illustrates the successive stages of iodination of tyrosine and the final formation of the two important thyroid hormones, thyroxine and triiodothyronine. Tyrosine is first iodized to *monoiodotyrosine* and then to *diiodotyrosine.* Then during the next few minutes, hours, and even days, more and more of the diiodotyrosine residues become *coupled* with each other (though the mechanism of the coupling is not understood). The product of the coupling reaction is the molecule of *thyroxine* that also remains part of the thyroglobulin molecule. Or one molecule of monoiodotyrosine couples with one molecule of diiodotyrosine to form *triiodothyronine.*

Storage of Thyroglobulin. After synthesis of the thyroid hormones has run its course, each thyroglobu-

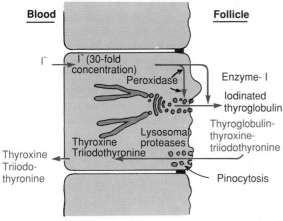

Figure 76-2. Thyroid cellular mechanisms for iodine transport, thyroxine (and triiodothyronine) formation, and thyroxine (and triiodothyronine) release into the blood.

Figure 76-3. Chemistry of thyroxine and triiodothyronine formation.

lin molecule contains 1 to 3 thyroxine molecules and an average of 1 triiodothyronine molecule for every 10 molecules of thyroxine. In this form the thyroid hormones are stored in the follicles in an amount sufficient to supply the body with its normal requirements of thyroid hormones for 2 to 3 months. Therefore, even when synthesis of thyroid hormone ceases entirely, the effects of deficiency will not be observed for several months.

RELEASE OF THYROXINE AND TRIIODOTHYRONINE FROM THE THYROID GLAND

Thyroglobulin itself is not released into the circulating blood in measurable amounts; instead, the thyroxine and triiodothyronine are first cleaved from the thyroglobulin molecule, and then these free hormones are released. This process occurs as follows: The apical surface of the thyroid cells sends out pseudopod extensions that close around small portions of the colloid to form and ingest *pinocytic vesicles.* Then *lysosomes* immediately fuse with these vesicles to form digestive vesicles containing the digestive enzymes from the lysosomes mixed with the colloid. The *proteinases* among these enzymes digest the thyroglobulin molecules and release the thyroxine and triiodothyronine, which then diffuse through the base of the thyroid cell into the surrounding capillaries. Thus, the thyroid hormones are released into the blood.

About three quarters of the iodinated tyrosine in the thyroglobulin never becomes thyroid hormones but instead remains monoiodotyrosine or diiodotyrosine. During the digestion of the thyroglobulin molecule to cause release of thyroxine and triiodothyronine, these iodinated tyrosines also are freed from the thyroid cells. However, they are not secreted into the blood. Instead, their iodine is cleaved from them by a *deiodinase enzyme* that makes most of this iodine available again for recycling within the gland for forming additional thyroid hormones. In the congenital absence of this deiodinase enzyme, persons frequently become iodine-deficient because of failure of this recycling process.

Daily Rate of Secretion of Thyroxine and Triiodothyronine. Over 90 per cent of the thyroid hormone released from the thyroid gland is normally thyroxine, and slightly less than 10 per cent is triiodothyronine. However, during the ensuing few days, as these hormones come in contact with their target tissues, most of the thyroxine is slowly deiodinated to form additional triiodothyronine. Therefore, the hormone finally delivered to and used by the tissues is mainly triiodothyronine, a total of about 35 micrograms of triiodothyronine per day. (Another 35 micrograms of so-called reverse triiodothyronine is formed each day by removal of one of the iodines of thyroxine near its carboxyl end. However, reverse triiodothyronine is almost totally inactive, and it is eventually destroyed.)

TRANSPORT OF THYROXINE AND TRIIODOTHYRONINE TO THE TISSUES

Binding of Thyroxine and Triiodothyronine with Plasma Proteins. On entering the blood, all but minute portions of the thyroxine and triiodothyronine combine immediately with several of the plasma proteins. They combine approximately as follows: 80 per cent with *thyroxine-binding globulin;* 10 to 15 per cent with *thyroxine-binding prealbumin;* and the remainder with *albumin.* The quantity of thyroxine-binding globulin in the blood is only 1 to 1.5 milligrams per deciliter of plasma, but its affinity for the thyroid hormones is so great that it still binds most of the hormones. Its binding affinity (and that of the other plasma proteins) is over 6 times as great for thyroxine as for triiodothyronine. This difference, and the fact that the concentration of thyroxine in the plasma is considerably greater than that of triiodothyronine, causes the total amount of protein-bound thyroxine to be about 60 times as great as the protein-bound triiodothyronine.

Slow Release of Thyroxine and Triiodothyronine to the Tissue Cells. Because of the very high affinity of the plasma-binding proteins for the thyroid hormones, these substances—in particular, thyroxine—are released to the tissue cells only very slowly. Half of the thyroxine in the blood is released to the tissue cells approximately every 6 days, whereas half of the triiodothyronine—because of its lower affinity—is released to the cells in approximately one day.

On entering the cells, both of these hormones again bind with intracellular proteins, the thyroxine once again binding more strongly than the triiodothyronine. Therefore, they are again stored, but this time in the functional cells themselves; and they are used slowly over a period of days or weeks.

Latency and Duration of Action of the Thyroid Hormones. After injection of a large quantity of thyroxine into a human being, essentially no effect on the metabolic rate can be discerned for 2 to 3 days, thereby illustrating that there is a *long latent* period before thyroxine activity begins. Once activity does begin, it increases progressively and reaches a maximum in 10 to 12 days, as shown in Figure 76–4. Thereafter, it decreases with a half-life of about 15 days. Some of the activity still persists as long as 6 weeks to 2 months later.

The actions of triiodothyronine occur about four times as rapidly as those of thyroxine, with the latent period as short as 6 to 12 hours and maximum cellular activity occurring within 2 to 3 days.

Most of the latency and prolonged period of action of these hormones is probably caused by their binding with proteins both in the plasma and in the tissue cells, followed by their slow release. However, we shall see in subsequent discussions that part of the latent period also results from the manner in which these hormones perform their functions in the cells themselves.

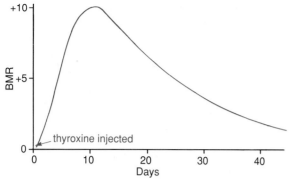

Figure 76 – 4. Approximate prolonged effect on the basal metabolic rate caused by administering a single large dose of thyroxine.

● FUNCTIONS OF THE THYROID HORMONES IN THE TISSUES

EFFECT OF THYROID HORMONES IN INCREASING THE TRANSCRIPTION OF LARGE NUMBERS OF GENES

The general effect of thyroid hormone is to cause wholesale nuclear transcription of large numbers of genes. Therefore, in virtually all cells of the body great numbers of protein enzymes, structural proteins, transport proteins, and other substances increase. The net result of all this is a generalized increase in functional activity throughout the body.

Conversion of Thyroxine (T_4) Into Triiodothyronine (T_3), and Activation of Nuclear Receptors. Before acting on the genes to increase genetic transcription, almost all of the thyroxine is deiodinated by one iodide ion, thus forming triiodothyronine. This in turn has a very high binding affinity to the cellular thyroid hormone receptors. Consequently, about 90 per cent of the thyroid hormone molecules that bind with the receptors is triiodothyronine, and only 10 per cent, thyroxine.

The thyroid hormone receptors are either attached to the DNA genetic strands or in close proximity to them. Upon binding with thyroid hormone, they become activated and initiate the transcription process. Then large numbers of different types of messenger RNA are formed, followed within another few minutes and hours by RNA translation on the cytoplasmic ribosomes to form hundreds of new types of proteins. However, not all proteins are increased by similar percentages — some only slightly and others at least as much as six-fold. It is believed that most, if not all, of the actions of thyroid hormone result from the enzymatic and other functions of these new proteins.

Important Types of Increased Cellular Metabolic Activity

The thyroid hormones increase the metabolic activities of all or almost all of the tissues of the body. The basal metabolic rate can increase to as much as 60 to 100 per cent above normal when large quantities of the hormones are secreted. The rate of utilization of foods for energy is greatly accelerated. Although the rate of protein synthesis is increased, at the same time the rate of protein catabolism is also increased. The growth rate of young persons is greatly accelerated. The mental processes are excited, and the activity of most of the endocrine glands is increased.

Effect of Thyroid Hormones on Mitochondria. When thyroxine or triiodothyronine is given to an animal, the mitochondria in most cells of the body increase in size and also in number. Furthermore, the total membrane surface area of the mitochondria increases almost directly in proportion to the increased metabolic rate of the whole animal. Therefore, it seems almost to be an obvious deduction that one of the principal functions of thyroxine might be simply to increase the number and activity of mitochondria, and these in turn increase the rate of formation of adenosine triphosphate (ATP) to energize cellular function. Unfortunately, though, the increase in the number and activity of mitochondria could be the *result* of increased activity of the cells as well as the cause of the increase.

When *extremely* high concentrations of thyroid hormone are administered, the mitochondria swell inordinately, and there is then uncoupling of the oxidative phosphorylation process with production of large amounts of heat but little ATP. However, under natural conditions, it is questionable whether the concentration of thyroid hormones ever becomes high enough to cause this effect even in persons who have thyrotoxicosis.

Effect of Thyroid Hormone in Increasing Active Transport of Ions Through Cell Membranes. One of the enzymes that becomes increased in response to thyroid hormone is *Na,K-ATPase*. This in turn increases the rate of transport of both sodium and potassium through the cell membranes of some tissues. Because this process utilizes energy and also increases the amount of heat produced in the body, it has also been suggested that this might be one of the mechanisms by which thyroid hormone increases the body's metabolic rate. In fact, thyroid hormone also causes the cell membranes of most cells to become leaky to sodium ions, therefore further activating the sodium pump and further increasing heat production.

Effect of Thyroid Hormone on Growth

Thyroid hormone has both general and specific effects on growth. For instance, it has long been known that thyroid hormone is essential for the metamorphic change of the tadpole into the frog. In the human being, the effect of thyroid hormone on growth is manifest mainly in growing children. In those who are hypothyroid, the rate of growth is greatly retarded. In those who are hyperthyroid, excessive skeletal growth often occurs, causing the child to become considerably taller at an earlier age. However, the bones also mature more rapidly, and the epiphyses close at an early age so that the duration of growth, and the eventual height of the adult, may actually be shortened.

An important effect of thyroid hormone is to promote growth and development of the brain during fetal life and for the first few years of postnatal life. If the fetus does not secrete sufficient quantities of thyroid hormone, growth and maturation of the brain both before birth and afterward are greatly retarded. Without specific thyroid therapy within days or weeks after birth, the child without a thyroid gland will remain mentally deficient throughout life. This is discussed more fully later in the chapter.

EFFECTS OF THYROID HORMONE ON SPECIFIC BODILY MECHANISMS

Effect on Carbohydrate Metabolism. Thyroid hormone stimulates almost all aspects of carbohydrate metabolism, including rapid uptake of glucose by the cells, enhanced glycolysis, enhanced gluconeogenesis, increased rate of absorption from the gastrointestinal tract, and even increased insulin secretion with its resultant secondary effects on carbohydrate metabolism. All these effects probably result from the overall increase in enzymes caused by thyroid hormone.

Effect on Fat Metabolism. Essentially all aspects of fat metabolism are also enhanced under the influence of thyroid hormone. However, because fats are the major source of long-term energy supplies, the fat stores of the body are depleted to a greater extent than are most of the other tissue elements. In particular, lipids are mobilized from the fat tissue, which increases the free fatty acid concentration in the plasma; and thyroid hormone also greatly accelerates the oxidation of free fatty acids by the cells.

Effect on Plasma and Liver Fats. Increased thyroid hormone *decreases* the quantity of cholesterol, phospholipids, and triglycerides in the plasma, even though it *increases* the free fatty acids. On the other hand, decreased thyroid secretion greatly increases the concentrations of cholesterol, phospholipids, and triglycerides and almost always causes excessive deposition of fat in the liver. The large increase in circulating plasma cholesterol in prolonged hypothyroidism is often associated with severe arteriosclerosis, discussed in Chapter 68.

One of the mechanisms by which thyroid hormone decreases the plasma cholesterol concentration is to increase significantly the rate of cholesterol secretion in the bile and consequent loss in the feces. A possible mechanism for the increased cholesterol secretion is that thyroid hormone induces increased numbers of low density lipoprotein receptors on the liver cells, leading to rapid removal of low density lipoproteins from the plasma and subsequent secretion of the lipoprotein cholesterol by the liver cells.

Effect on Vitamin Metabolism. Because thyroid hormone increases the quantities of many of the different enzymes and because vitamins are essential parts of some of the enzymes or coenzymes, thyroid hormone causes increased need for vitamins. Therefore, a relative vitamin deficiency can occur when excess thyroid hormone is secreted, unless at the same time increased quantities of vitamins are available.

Effect on Basal Metabolic Rate. Because thyroid hormone increases metabolism in almost all cells of the body,

excessive quantities of the hormone can occasionally increase the basal metabolic rate to as much as 60 to 100 per cent above normal. On the other hand, when no thyroid hormone is produced, the basal metabolic rate falls almost to half normal; that is, the basal metabolic rate becomes −30 to −50, as discussed in Chapter 72. Figure 76−5 shows the approximate relationship between the daily supply of thyroid hormones and the basal metabolic rate. Extreme amounts of the hormones are required to cause very high basal metabolic rates.

Effect on Body Weight. Greatly increased thyroid hormone production almost always decreases the body weight, and greatly decreased production almost always increases the body weight; but these effects do not always occur, because thyroid hormone increases the appetite, and this may overbalance the change in the metabolic rate.

Effect on the Cardiovascular System. *Blood Flow and Cardiac Output.* Increased metabolism in the tissues causes more rapid utilization of oxygen than normal and causes greater than normal quantities of metabolic end products to be released from the tissues. These effects cause vasodilatation in most of the body tissues, thus increasing blood flow. The rate of blood flow in the skin especially increases because of the increased need for heat elimination.

As a consequence of the increased blood flow, the cardiac output also increases, sometimes rising to 60 per cent or more above normal when excessive thyroid hormone is present, and falling to only 50 per cent of normal in severe hypothyroidism.

Heart Rate. The heart rate increases considerably more under the influence of thyroid hormone than would be expected from the increase in cardiac output. Therefore, thyroid hormone probably has a direct effect on the excitability of the heart, which in turn increases the heart rate. This effect is of particular importance because the heart rate is one of the sensitive physical signs that the clinician uses in determining whether a patient has excessive or diminished thyroid hormone production.

Strength of Heartbeat. The increased enzymatic activity caused by increased thyroid hormone production ap-

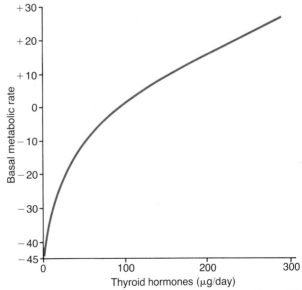

Figure 76−5. Approximate relationship of thyroid hormone (T$_4$ and T$_3$) daily rate of secretion to the basal metabolic rate.

parently increases the strength of the heart when only a slight excess of thyroid hormone is secreted. This is analogous to the increase in strength of heart beat that occurs in mild fevers and during exercise. However, when thyroid hormone is increased markedly, the heart muscle strength becomes depressed because of excessive protein catabolism. Indeed, some severely thyrotoxic patients die of cardiac decompensation secondary to myocardial failure and increased cardiac load imposed by the increased output.

Blood Volume. Thyroid hormone causes the blood volume to increase slightly. The effect probably results at least partly from vasodilatation, which allows increased quantities of blood to collect in the circulatory system.

Arterial Pressure. The *mean* arterial pressure usually is unchanged. However, because of the increased stroke volume with each heartbeat and increased blood flow through the tissues between heartbeats, the pulse pressure is increased, with the systolic pressure elevated in hyperthyroidism 10 to 15 mm Hg, and the diastolic pressure correspondingly reduced.

Effect on Respiration. The increased rate of metabolism increases the utilization of oxygen and the formation of carbon dioxide; these effects activate all the mechanisms that increase the rate and depth of respiration.

Effect on the Gastrointestinal Tract. In addition to increased appetite and food intake, which has been discussed, thyroid hormone increases both the rate of secretion of the digestive juices and the motility of the gastrointestinal tract. Often, diarrhea results. Lack of thyroid hormone causes constipation.

Effect on the Central Nervous System. In general, thyroid hormone increases the rapidity of cerebration but also often dissociates this; on the other hand, lack of thyroid hormone decreases this function. The hyperthyroid individual is likely to have extreme nervousness and many psychoneurotic tendencies, such as anxiety complexes, extreme worry, or paranoia.

Effect on the Function of the Muscles. Slight increase in thyroid hormone usually makes the muscles react with vigor, but when the quantity of hormone becomes excessive, the muscles become weakened because of excess protein catabolism. On the other hand, lack of thyroid hormone causes the muscles to become extremely sluggish, and they relax slowly after a contraction.

Muscle Tremor. One of the most characteristic signs of hyperthyroidism is a fine muscle tremor. This is not the coarse tremor that occurs in Parkinson's disease or in shivering, for it occurs at the rapid frequency of 10 to 15 times per second. The tremor can be observed easily by placing a sheet of paper on the extended fingers and noting the degree of vibration of the paper. This tremor is believed to be caused by increased reactivity of the neuronal synapses in the areas of the cord that control muscle tone. The tremor is an important means for assessing the degree of thyroid hormone effect on the central nervous system.

Effect on Sleep. Because of the exhausting effect of thyroid hormone on the musculature and on the central nervous system, the hyperthyroid subject often has a feeling of constant tiredness; but because of the excitable effects of thyroid hormone on the synapses, it is difficult to sleep. On the other hand, extreme somnolence is characteristic of hypothyroidism.

Effect on Other Endocrine Glands. Increased thyroid hormone increases the rates of secretion of most other endocrine glands, but it also increases the need of the tissues for the hormones. For instance, increased thyroxine secretion increases the rate of glucose metabolism everywhere in the body and therefore causes a corresponding need for increased insulin secretion by the pancreas. Also, thyroid hormone increases many metabolic activities related to bone formation and, as a consequence, increases the need for parathyroid hormone. Finally, thyroid hormone increases the rate at which adrenal glucocorticoids are inactivated by the liver. This leads to feedback increase in adrenocorticotropic hormone production by the anterior pituitary and, therefore, increased rate of glucocorticoid secretion by the adrenal glands also.

Effect of Thyroid Hormone on Sexual Function. For normal sexual function, thyroid secretion needs to be approximately normal. In men, lack of thyroid hormone is likely to cause complete loss of libido; on the other hand, great excesses of the hormone frequently cause impotence. In women, lack of thyroid hormone often causes *menorrhagia* and *polymenorrhea,* that is, respectively, excessive and frequent menstrual bleeding. Yet, strangely enough, in other women thyroid lack may cause irregular periods and, occasionally, even total *amenorrhea.* A hypothyroid woman, like a man, is also likely to have greatly decreased libido. To make the picture still more confusing, in the hyperthyroid woman, *oligomenorrhea,* which means greatly reduced bleeding, is common, and occasionally amenorrhea results.

The action of thyroid hormone on the gonads cannot be pinpointed to a specific function but probably results from a combination of direct metabolic effects on the gonads and of excitatory and inhibitory effects operating through the anterior pituitary hormones that control the sexual functions.

● REGULATION OF THYROID HORMONE SECRETION

To maintain normal levels of metabolic activity in the body, precisely the right amount of thyroid hormone must be secreted at all times; and to provide this, specific feedback mechanisms operate through the hypothalamus and anterior pituitary gland to control the rate of thyroid secretion. These mechanisms can be explained as follows:

Effects of Thyroid-Stimulating Hormone on Thyroid Secretion. Thyroid-stimulating hormone (TSH), also known as *thyrotropin,* is an anterior pituitary hormone, a glycoprotein with a molecular weight of about 28,000, discussed in Chapter 74; it increases the secretion of thyroxine and triiodothyronine by the thyroid gland. Its specific effects on the thyroid gland are

1. Increased proteolysis of the thyroglobulin that has already been stored in the follicles, with the resultant release of the thyroid hormones into the circulating blood and diminishment of the follicular substance itself.

2. Increased activity of the iodide pump, which increases the rate of "iodide trapping" in the glandular cells, sometimes increasing the ratio of intracellular to extracellular iodide concentration to as much as eight times normal.

3. Increased iodination of tyrosine and increased coupling to form the thyroid hormones.

4. Increased size and increased secretory activity of the thyroid cells.

5. Increased number of thyroid cells, plus a change from cuboidal to columnar cells and much infolding of the thyroid epithelium into the follicles.

In summary, TSH *increases all the known activities of the thyroid glandular cells.*

The most important early effect following administration of TSH is proteolysis of the thyroglobulin, which causes release of thyroxine and triiodothyronine into the blood within 30 minutes. The other effects require hours or even days and weeks to develop fully.

Role of Cyclic Adenosine Monophosphate in the Stimulatory Effect of TSH. In the past it was difficult to explain the many and varied effects of thyroid-stimulating hormone on the thyroid cell. However, it is now clear that at least most of these result from activation of the "second messenger" *cyclic adenosine monophosphate (cAMP)* system of the cell. The first event in this activation is binding of the thyroid-stimulating hormone with specific TSH receptors on the basal membrane surfaces of the cell. This then activates *adenylcyclase* in the membrane, which increases the formation of cAMP in the cell. Finally, the cAMP acts as a *second messenger* to activate protein kinase, which causes multiple phosphorylations throughout the cell. The result is both an immediate increase in secretion of thyroid hormones and prolonged growth of the thyroid glandular tissue itself. This method for control of thyroid cell activity is similar to the function of cAMP in many other target tissues of the body.

Hypothalamic Regulation of TSH Secretion by the Anterior Pituitary — Thyrotropin-Releasing Hormone

Anterior pituitary secretion is controlled by a hypothalamic hormone, *thyrotropin-releasing hormone* (TRH), which is secreted by nerve endings in the median eminence of the hypothalamus and then transported from there to the anterior pituitary in the hypothalamic-hypophysial portal blood, as explained in Chapter 74. The precise nuclei of the hypothalamus that are responsible for causing TRH secretion in the median eminence are not known. However, injection of radioactive antibodies that attach specifically to TRH have shown this hormone to be present in many different hypothalamic loci, including the (1) *dorsomedial nucleus*, (2) *suprachiasmatic nucleus*, (3) *ventromedial nucleus*, (4) *anterior hypothalamus*, (5) *preoptic area*, and (6) *paraventricular nucleus*.

TRH has been obtained in pure form. It is a very simple substance, a tripeptide amide — *pyroglutamyl-histidyl-proline-amide*. TRH directly affects the anterior pituitary gland cells to increase their output of TSH. When the portal system from the hypothalamus to the anterior pituitary gland is completely blocked, the rate of secretion of TSH by the anterior pituitary is greatly decreased but not reduced to zero.

The molecular mechanism by which TRH causes the TSH-secreting cells of the anterior pituitary to produce TSH is first to bind with TRH receptors in the pituitary cell membrane. This in turn *activates the phospholipid second messenger* system to produce large amounts of phospholipid C, followed by many other second messenger products, including calcium ions and diacyl glycerol, which eventually leads to TSH release.

Effects of Cold and Other Neurogenic Stimuli on TRH and TSH Secretion. One of the best-known stimuli for increasing the rate of TRH secretion by the hypothalamus, and therefore TSH secretion by the anterior pituitary, is exposure of an animal to cold. This effect almost certainly results from excitation of the anterior hypothalamus and the preoptic area, which are the hypothalamic center for body temperature control. Exposure of rats for several weeks to very severe cold increases the output of thyroid hormones sometimes more than 100 per cent and can increase the basal metabolic rate as much as 50 per cent. Indeed, people moving to arctic regions have been known to develop basal metabolic rates 15 to 20 per cent above normal; however, the behavioral propensity of human beings to protect themselves from cold usually prevents such effects.

Various emotional reactions can also affect the output of TRH and TSH and can, therefore, indirectly affect the secretion of thyroid hormones. On the other hand, excitement and anxiety — conditions that greatly stimulate the sympathetic nervous system — cause acute decrease in secretion of TSH, perhaps because these states increase the metabolic rate and the body heat and exert an inverse effect on the heat control center.

Neither these emotional effects nor the effect of cold is observed after the hypophysial stalk has been cut, illustrating that both these effects are mediated by way of the hypothalamus.

Feedback Effect of Thyroid Hormone in Decreasing Anterior Pituitary Secretion of TSH — Feedback Regulation of Thyroid Secretion

Increased thyroid hormone in the body fluids decreases the secretion of TSH by the anterior pituitary. When the rate of thyroid hormone secretion rises to about 1.75 times normal, the rate of TSH secretion falls essentially to zero. Almost all of this feedback depressant effect occurs even when the anterior pituitary has been completely separated from the hypothalamus. Therefore, as illustrated in Figure 76–6, it is probable that increased thyroid hormone inhibits anterior pituitary secretion of TSH mainly by a direct effect on the anterior pituitary itself, though perhaps secondarily by much weaker effects acting through the hypothalamus.

One mechanism that has been suggested for the

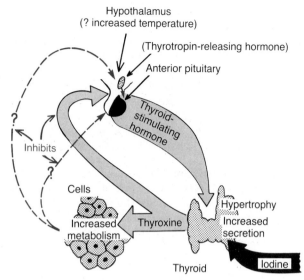

Figure 76-6. Regulation of thyroid secretion.

feedback effect on the anterior pituitary gland is that thyroid hormone reduces the number of TRH receptors on the cells that secrete TSH. Therefore, the stimulating effect on these cells by TRH from the hypothalamus is greatly reduced.

Regardless of the mechanism of the feedback, its effect is to maintain an almost constant concentration of free thyroid hormones in the circulating body fluids.

If there is a feedback effect through the hypothalamus in addition to the direct feedback to the pituitary gland itself, this probably operates very slowly and might be caused at least partly by changes in body heat and its effect on the body temperature-controlling centers of the hypothalamus, which are known to have a significant effect in controlling the thyroid hormone system.

ANTITHYROID SUBSTANCES

Drugs that suppress thyroid secretion are called antithyroid substances. The three best known of these are thiocyanate, propylthiouracil, and high concentrations of inorganic iodides. The mechanism by which each of these blocks thyroid secretion is different from the others, and they can be explained as follows:

Decreased Iodide Trapping Caused by Thiocyanate Ions. The same active pump that pumps iodide ions into the thyroid cells can also pump thiocyanate ions, perchlorate ions, and nitrate ions. Therefore, administration of thiocyanate (or one of the other ions as well) in high enough concentration can cause competitive inhibition of iodide transport into the cell—that is, inhibition of the iodide-trapping mechanism.

The decreased availability of iodide in the glandular cells does not stop the formation of thyroglobulin; it merely prevents the thyroglobulin that is formed from becoming iodinated and therefore from forming the thyroid hormones. This deficiency of the hormones in turn leads to increased secretion of TSH by the anterior pituitary gland, which causes overgrowth of the thyroid gland even though the

gland still does not form adequate quantities of thyroid hormones. Therefore, use of thiocyanates and some other ions to block thyroid secretion, or the presence of these ions in certain foods, can lead to the development of an enlarged thyroid gland, which is called a *goiter*.

Depression of Thyroid Hormone Formation by Propylthiouracil. Propylthiouracil (and other similar compounds such as methimazole and carbimazole) prevents formation of thyroid hormone from iodides and tyrosine. The mechanism of this is partly to block the peroxidase enzyme that is required for iodination of tyrosine but partly also to block the coupling of two iodinated tyrosines to form thyroxine or triiodothyronine.

Propylthiouracil, like thiocyanate, does not prevent formation of thyroglobulin, but the absence of thyroxine and triiodothyronine in the thyroglobulin can lead to tremendous feedback enhancement of TSH secretion by the anterior pituitary gland, thus promoting growth of the glandular tissue and forming a goiter.

Decrease in Thyroid Activity Caused by Iodides. When iodides are present in the blood *in high concentration* (100 times the normal plasma level), most activities of the thyroid gland are decreased, but often they remain decreased only for a few weeks. The effect is to reduce the rate of iodide trapping, and the rate of iodination of tyrosine to form thyroid hormones is also decreased. Even more important, the normal endocytosis of colloid from the follicles by the thyroid glandular cells is paralyzed by the high iodide concentrations. Because this is the first step in release of the thyroid hormones from the storage colloid, there is almost immediate shutdown of thyroid hormone secretion into the blood.

Because iodides in high concentrations decrease all phases of thyroid activity, they also decrease the size of the thyroid gland and especially decrease its blood supply, in contradistinction to the opposite effects caused by most of the other antithyroid agents. For this reason, iodides are frequently administered to patients for 2 or 3 weeks prior to surgical removal of the thyroid gland to decrease the necessary amount of surgery.

● DISEASES OF THE THYROID

HYPERTHYROIDISM

Most effects of hyperthyroidism are obvious from the preceding discussion of the various physiological effects of thyroid hormone. However, some specific effects should be mentioned in connection especially with the development, diagnosis, and treatment of hyperthyroidism.

Causes of Hyperthyroidism (Toxic Goiter, Thyrotoxicosis, Graves' Disease). In most patients with hyperthyroidism the entire thyroid gland is increased to two to three times normal size, with tremendous hyperplasia and folding of the follicular cell lining into the follicles, so that the number of cells is increased several more times than the size of the gland is increased. Also, each cell increases its rate of secretion severalfold; radioactive iodine uptake studies indicate that some of these hyperplastic glands secrete thyroid hormone at rates as great as 5 to 15 times normal.

The changes in the thyroid gland are similar to those caused by excessive TSH. However, radioimmunoassay studies have shown the plasma TSH concentrations to be less than normal rather than enhanced in almost all patients, and often to be essentially zero. On the other hand, other substances that have actions similar to that of TSH are found in the blood of almost all these patients. These substances are immunoglobulin antibodies that bind with the same membrane receptors that bind TSH. They induce continual activation of the cAMP system of the cells, with resultant development of hyperthyroidism. These antibodies are called simply *thyroid-stimulating antibodies* and are designated "TSAb." They have a prolonged stimulating effect on the thyroid gland, lasting as long as 12 hours in contrast to a little over 1 hour for TSH. The high level of thyroid hormone secretion caused by TSAb in turn suppresses anterior pituitary formation of TSH.

The antibodies that cause hyperthyroidism almost certainly develop as the result of autoimmunity that has developed against thyroid tissue. Presumably, at some time in the history of the person an excess of thyroid cell antigens has been released from the thyroid cells, and this has resulted in the formation of antibodies against the thyroid gland itself.

Thyroid Adenoma. Aside from the aforementioned variety of hyperthyroidism, the condition rarely also results from a localized adenoma (a tumor) that develops in the thyroid tissue and secretes large quantities of thyroid hormone. This is different from the more usual type of hyperthyroidism in that it usually is not associated with evidence of autoimmune disease. An interesting effect of the adenoma is that as long as it continues to secrete large quantities of thyroid hormone, function in the remainder of the thyroid gland is almost totally inhibited because the thyroid hormone from the adenoma depresses the production of TSH by the pituitary gland.

Symptoms of Hyperthyroidism

The symptoms of hyperthyroidism are obvious from the preceding discussion of the physiology of the thyroid hormones — intolerance to heat, increased sweating, mild to extreme weight loss (sometimes as much as 100 pounds), varying degrees of diarrhea, muscular weakness, nervousness or other psychic disorders, extreme fatigue but inability to sleep, and tremor of the hands.

Exophthalmos. Most, but not all, persons with hyperthyroidism develop some degree of protrusion of the eyeballs, as illustrated in Figure 76-7. This condition is called *exophthalmos.* A major degree of exophthalmos occurs in about one third of hyperthyroid patients, and the condition

Figure 76-7. Patient with exophthalmic hyperthyroidism. Note protrusion of the eyes and retraction of the superior eyelids. The basal metabolic rate was +40. (Courtesy of Dr. Leonard Posey.)

sometimes becomes severe enough that the eyeball protrusion stretches the optic nerve enough to damage vision. Much more often, the eyes are damaged because the eyelids do not close completely when the person blinks or is asleep. As a result, the epithelial surfaces of the eyes become dry and irritated and often infected, resulting in ulceration of the cornea.

The cause of the protruding eyes is edematous swelling of the retro-orbital tissues and degenerative changes in the extraocular muscles. The factor or factors that initiate these changes are still in serious dispute. In most patients, antibodies can be found in the blood that react with the eye muscles. Furthermore, the concentration of these antibodies is highest in patients who have very high concentrations of thyroid-stimulating antibodies. Therefore, there is much reason to believe that exophthalmos, like hyperthyroidism itself, is an autoimmune process. Usually, the exophthalmos is greatly ameliorated with treatment of the hyperthyroidism.

Diagnostic Tests for Hyperthyroidism. For the usual case of hyperthyroidism, the most accurate diagnostic test is direct measurement of the concentration of "free" thyroxine (and sometimes triiodothyronine) in the plasma, using appropriate radioimmunoassay procedures.

Other tests that are sometimes used are as follows:

1. The basal metabolic rate is usually increased to +30 or +60 in severe hyperthyroidism.

2. The concentration of TSH in the plasma is measured by radioimmunoassay. In the usual type of thyrotoxicosis, the anterior pituitary secretion of TSH is so completely suppressed by the large amounts of circulating thyroxine and triiodothyronine that there will be almost no plasma TSH.

3. The concentration of TSAb is measured by radioimmunoassay. This is normally high in the usual type of thyrotoxicosis but low in thyroid adenoma.

Physiology of Treatment in Hyperthyroidism. The most direct treatment for hyperthyroidism is surgical removal of most of the thyroid gland. In general, it is desirable to prepare the patient for surgical removal of the gland prior to the operation. This is done by administering propylthiouracil, usually for several weeks, until the basal metabolic rate of the patient has returned to normal. Then administration of high concentrations of iodides for 1 to 2 weeks immediately prior to operation causes the gland itself to recede in size and its blood supply to diminish. By using these preoperative procedures, the operative mortality is less than 1 in 1000 in the better hospitals, whereas prior to development of these procedures the operative mortality was as great as 1 in 25.

Treatment of the Hyperplastic Thyroid Gland with Radioactive Iodine. As much as 80 to 90 per cent of an injected dose of iodide is absorbed by the hyperplastic, toxic thyroid gland within a day after injection. If this injected iodine is radioactive, it can destroy internally the secretory cells of the thyroid gland. Usually 5 millicuries of radioactive iodine is given to the patient, whose condition is reassessed several weeks later. If the patient is still hyperthyroid, additional doses are repeated until normal thyroid status is reached.

HYPOTHYROIDISM

The effects of hypothyroidism in general are opposite to those of hyperthyroidism; but here again, a few physiological mechanisms peculiar to hypothyroidism alone are involved.

Hypothyroidism, like hyperthyroidism, probably also results in most instances from autoimmunity against the thyroid gland, but immunity that destroys the gland, rather than stimulating it. The thyroid glands of most of these patients first have "thyroiditis," which means thyroid inflammation. This causes progressive deterioration and finally fibrosis of the gland, with resultant diminished or absent secretion of thyroid hormone. However, several other types of hypothyroidism also occur, often associated with development of enlarged thyroid glands, called thyroid goiter.

Endemic Colloid Goiter. The term goiter means a greatly enlarged thyroid gland. As pointed out in the discussion of iodine metabolism, about 50 milligrams of iodine is necessary each year for the formation of adequate quantities of thyroid hormone. In certain areas of the world, notably in the Swiss Alps, in the Andes, and in the Great Lakes region of the United States, insufficient iodine is present in the soil for the foodstuffs to contain even this minute quantity of iodine. Therefore, in days prior to iodized table salt, in many persons living in these areas extremely large thyroid glands developed, called *endemic goiters*.

The mechanism for development of the large endemic goiters is the following: Lack of iodine prevents production of both thyroxine and triiodothyronine but does not stop the formation of thyroglobulin. As a result, no hormone is available to inhibit production of TSH by the anterior pituitary; this allows the pituitary to secrete excessively large quantities of TSH. The TSH then causes the thyroid cells to secrete tremendous amounts of thyroglobulin (colloid) into the follicles, and the gland grows larger and larger. But unfortunately, because of lack of iodine, still thyroxine and triiodothyronine production does not occur within the thyroglobulin and therefore does not cause the normal suppression of TSH production by the anterior pituitary. The follicles become tremendous in size, and the thyroid gland may increase to as large as 300 to 500 grams or more.

Idiopathic Nontoxic Colloid Goiter. Enlarged thyroid glands similar to those of endemic colloid goiter frequently also occur in persons who do not have iodine deficiency. These goitrous glands may secrete normal quantities of thyroid hormones; but more frequently the secretion of hormone is depressed, as in endemic colloid goiter.

The exact cause of the enlarged thyroid gland in patients with idiopathic colloid goiter is not known, but most of these patients show signs of mild thyroiditis; therefore, it has been suggested that the thyroiditis causes slight hypothyroidism, which then leads to increased TSH secretion and progressive growth of the noninflamed portions of the gland. This could explain why these glands usually are very nodular, with some portions of the gland growing while other portions are being destroyed by thyroiditis.

In some persons with colloid goiter, the thyroid gland has an abnormality of the enzyme system required for the formation of the thyroid hormones. Among the abnormalities often encountered are as follows:

1. Deficient iodide-trapping mechanism, in which iodine is not pumped adequately into the thyroid cells.

2. Deficient peroxidase system, in which the iodides are not oxidized to the iodine state.

3. Deficient coupling of iodinated tyrosines in the thyroglobulin molecule, so that the final thyroid hormones cannot be formed.

4. Deficiency of the deiodinase enzyme; this prevents recovery of iodine from the iodinated tyrosines that are not coupled to form the thyroid hormones (this is about two thirds of the iodine), thus leading to iodine deficiency.

Finally, some foods contain *goitrogenic substances* that have a propylthiouracil-type of antithyroid activity, thus also leading to TSH-stimulated enlargement of the thyroid gland. Such goitrogenic substances are found especially in some varieties of turnips and cabbages.

Characteristics of Hypothyroidism. Whether hypothyroidism is due to thyroiditis, endemic colloid goiter, idiopathic colloid goiter, destruction of the thyroid gland by irradiation, or surgical removal of the thyroid gland, the physiological effects are the same. These include fatigue and extreme somnolence with sleeping up to 14 to 16 hours a day, extreme muscular sluggishness, slowed heart rate, decreased cardiac output, decreased blood volume, sometimes increased weight, constipation, mental sluggishness, failure of many trophic functions in the body evidenced by depressed growth of hair and scaliness of the skin, development of a froglike husky voice, and, in severe cases, development of an edematous appearance throughout the body called myxedema.

Myxedema. *Myxedema* develops in the patient with almost total lack of thyroid function. Figure 76–8 shows such a patient, illustrating bagginess under the eyes and swelling of the face. In this condition, for reasons not yet explained, greatly increased quantities of proteins mixed with hyaluronic acid and chondroitin sulfate form excessive quantities of tissue gel in the interstitial spaces, and this causes the total quantity of interstitial fluid also to increase. Because of the gel nature of the excess fluid, it is relatively immobile, and the edema is nonpitting in type.

Arteriosclerosis in Hypothyroidism. As pointed out earlier, lack of thyroid hormone increases the quantity of blood cholesterol because of diminished liver excretion of cholesterol in the bile. The increase in blood cholesterol is usually associated with increased atherosclerosis and arteriosclerosis. Therefore, in many hypothyroid patients, particularly those with myxedema, arteriosclerosis develops, which results in peripheral vascular disease, deafness, and

Figure 76–8. Patient with myxedema. (Courtesy of Dr. Herbert Langford.)

often extreme coronary sclerosis with consequent early death.

Diagnostic Tests in Hypothyroidism. The tests already described for diagnosis of hyperthyroidism give the opposite results in hypothyroidism. The free thyroxine in the blood is low. The basal metabolic rate in myxedema ranges between −30 and −50. And the secretion of TSH by the anterior pituitary when a test dose of TRH is administered is usually greatly increased (except in those rare instances of hypothyroidism caused by depressed response of the pituitary gland to the TRH).

Treatment of Hypothyroidism. Figure 76–4 shows the effect of thyroxine on the basal metabolic rate, illustrating that the hormone normally has a duration of action of more than 1 month. Consequently, it is easy to maintain a steady level of thyroid hormone activity in the body by daily oral ingestion of a tablet or so containing thyroxine. Furthermore, proper treatment of the hypothyroid patient results in such complete normality that formerly myxedematous patients properly treated have lived into their 90s after treatment for over 50 years.

Cretinism

Cretinism is the condition caused by extreme hypothyroidism during fetal life, infancy, and childhood. It is characterized especially by failure of growth and mental retardation. Cretinism results from congenital lack of a thyroid gland *(congenital cretinism)*, from failure of the thyroid gland to produce thyroid hormone because of a genetic defect of the gland, or from iodine lack in the diet *(endemic cretinism)*. The severity of endemic cretinism varies greatly, depending on the amount of iodine in the diet, and whole populaces of an endemic area have been known to have cretinoid tendencies.

A newborn baby without a thyroid gland may have absolutely normal appearance and function because he had been supplied with some (but usually not enough) thyroid hormone by the mother while *in utero,* but a few weeks after birth its movements become sluggish, and both its physical and mental growth are greatly retarded. Treatment of the cretin at any time usually causes normal return of physical growth, but, unless the cretin is treated within a few weeks after birth, its mental growth will be permanently retarded. This results from retardation of the growth, branching, and myelination of the neuronal cells of the central nervous system at this very critical time in the normal development of the mental powers.

Skeletal growth in the cretin is characteristically more inhibited than is soft tissue growth. As a result of this disproportionate rate of growth, the soft tissues are likely to enlarge excessively, giving the cretin the appearance of an obese and stocky, short child. Occasionally the tongue becomes so large in relation to the skeletal growth that it obstructs swallowing and breathing, inducing a characteristic guttural breathing that sometimes chokes the baby.

REFERENCES

Bayliss, R. I.: Thyroid Disease. New York, Oxford University Press, 1982.
Burrow, G. N., et al.: Thyroid Function and Disease. Philadelphia, W. B. Saunders Co., 1989.
Chopra, I. J., and Solomon, D. H.: Pathogenesis of hyperthyroidism. Annu. Rev. Med., 34:267, 1983.
Delange, F., et al. (eds.): Research in Congenital Hypothyroidism. New York, Plenum Publishing Corp., 1989.
DeLong, G. R., et al. (eds.): Iodine and the Brain. New York, Plenum Publishing Corp., 1989.
Dussault, J. H., and Ruel, J.: Thyroid hormones and brain development. Annu. Rev. Physiol., 49:321, 1987.
Felig, P., et al. (eds.): Endocrinology and Metabolism, 2nd Ed. New York, McGraw-Hill Book Co., 1987.
Gershengorn, M. C.: Mechanism of thyrotropin releasing hormone stimulation of pituitary hormone secretion. Annu. Rev. Physiol., 48:515, 1986.
Green, W. L. (ed.): The Thyroid. New York, Elsevier Science Publishing Co., 1987.
Ho, K. L.: Basophilic degeneration of skeletal muscle in hypothyroid myopathy. Histochemical and ultrastructural studies. Arch. Pathol. Lab. Med., 108:239, 1984.
King, D. B., and May, J. D.: Thyroidal influence on body growth. J. Exp. Zool., 232:453, 1984.
Kourides, I. A., et al.: The regulation and organization of thyroid stimulating hormone genes. Recent Prog. Horm. Res., 40:79, 1984.
Kreiger, D. T. (ed.): Current Therapy in Endocrinology, 1983–1984. St. Louis, C. V. Mosby, 1983.
Lenzen, S., and Bailey, C. J.: Thyroid hormones, gonadal and adrenocortical steroids and the function of the islets of Langerhans. Endocrinol. Rev., 5:411, 1984.
LiVolsi, V. A.: Pathology of the Thyroid. Philadelphia, W. B. Saunders Co., 1989.
Matovinovic, J.: Endemic goiter and cretinism at the dawn of the third millennium. Annu. Rev. Nutr., 3:341, 1983.
McClung, M. R., and Greer, M. A.: Treatment of hyperthyroidism. Annu. Rev. Med., 31:385, 1980.
Mederiros-Neto, G. A., and Gaitan, E. (eds.): Frontiers in Thyroidology. New York, Plenum Publishing Corp., 1987.
Oppenheimer, J. H.: Thyroid hormone action at the nuclear level. Ann. Intern. Med., 102:374, 1985.
Pinchera, A., et al. (eds.): Thyroid Autoimmunity. New York, Plenum Publishing Corp., 1987.
Robbins, J., et al.: The thyroid and iodine metabolism. In Bondy, P. K., and Rosenberg, L. E. (eds.): Metabolic Control and Disease, 8th Ed. Philadelphia, W. B. Saunders Co., 1980, p. 1325.
Samuels, H. H., et al.: Regulation of gene expression by thyroid hormone. Annu. Rev. Physiol., 51:623, 1989.
Silva, J. E.: Brown adipose tissue: An extrathyroidal source of triiodothyronine. News Physiol. Sci., 1:119, 1986.
Stanbury, J. B. (ed.): Endemic Goiter and Endemic Cretinism. New York, John Wiley & Sons, 1980.
Wilber, J. F.: Human pituitary thyrotropin. In DeGroot, L. J., et al. (eds.): Endocrinology. Vol. 1. New York, Grune & Stratton, 1979, p. 141.

CHAPTER 77

• *The Adrenocortical Hormones* •

The *adrenal glands,* each of which weighs about 4 grams, lie at the superior poles of the two kidneys. And, as illustrated in Figure 77–1, each gland is composed of two distinct parts, the *adrenal medulla* and the *adrenal cortex.* The adrenal medulla, the central 20 per cent of the gland, is functionally related to the sympathetic nervous system; it secretes the hormones *epinephrine* and *norepinephrine* in response to sympathetic stimulation. In turn, these hormones cause almost the same effects as direct stimulation of the sympathetic nerves in all parts of the body. These hormones and their effects are discussed in detail in Chapter 60 in relation to the sympathetic nervous system.

The adrenal cortex secretes an entirely different group of hormones, called *corticosteroids.* These hormones are all synthesized from the steroid cholesterol, and they all have similar chemical formulas. However, very slight differences in their molecular structures give them several very different but very important functions.

The Corticosteroids — Mineralocorticoids, Glucocorticoids, and Androgens. Two major types of adrenocortical hormones, the *mineralocorticoids* and the *glucocorticoids,* are secreted by the adrenal cortex. In addition to these, small amounts of sex hormones are secreted, especially *androgenic hormones,* which exhibit approximately the same effects in the body as the male sex hormone testosterone. These are normally of only slight importance, though in certain abnormalities of the adrenal cortices extreme quantities can be secreted (which is discussed later in the chapter) and can then result in masculinizing effects.

The *mineralocorticoids* have gained this name because they especially affect the electrolytes of the extracellular fluids — sodium and potassium, in particular. The *glucocorticoids* have gained their name because they exhibit an important effect in increasing blood glucose concentration. However, the glucocorticoids have additional effects on both protein and fat metabolism that probably are as important to body function as their effects on carbohydrate metabolism, if not more important.

Over 30 different steroids have been isolated from

the adrenal cortex, but only two of these are of exceptional importance to the normal endocrine function of the human body — *aldosterone,* which is the principal mineralocorticoid, and *cortisol,* which is the principal glucocorticoid.

● CHEMISTRY OF ADRENOCORTICAL SECRETION

The Layers of the Adrenal Cortex, and Hormone Formation. Figure 77–1 shows that the adrenal cortex is composed of three relatively distinct layers. *Aldosterone* is secreted by the *zona glomerulosa,* the outermost and very thin layer of cells on the surface. *Cortisol* and several other glucocorticoids are secreted by both the *zona fasciculata,* the middle layer, and the *zona reticularis,* the deep layer, with probably more secretion of these hormones by the zona fasciculata than by the reticularis. The *adrenal androgens* are also secreted by both these layers.

Those conditions that increase the output of aldosterone cause hypertrophy of the zona glomerulosa while not affecting the other two zones. On the other hand, those factors that cause increased secretion of cortisol and adrenal androgens cause hypertrophy of the zona fasciculata and zona reticularis while having very little or no effect on the zona glomerulosa; this is especially true of stimulation of the gland by adrenocorticotropic hormone (ACTH), as is discussed later in this chapter.

Chemistry of the Adrenocortical Hormones. All the adrenocortical hormones are steroid compounds. They are formed mainly from cholesterol absorbed directly from the circulating blood by endocytosis through the cell membrane. This membrane has specific receptors for the low density lipoproteins that contain very high concentrations of cholesterol, and attachment of these lipoproteins to the membrane promotes the endocytotic process. Small amounts of cholesterol are also synthesized within the cortical cells from acetyl coenzyme A; this, too, can be used for forming the adrenocortical hormones.

Figure 77–2 gives the principal steps in the formation of the three important steroid products of the adrenal cortex: aldosterone, cortisol, and the androgens. Essentially all these steps occur in two of the organelles of the cell, the *mitochondria* and the *endoplasmic reticulum,* some steps occurring in one of these organelles and some in the other. Each step is catalyzed by a specific enzyme system. A change in even a single enzyme in the schema can cause

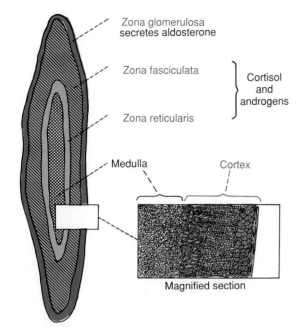

Figure 77–1. Secretion of adrenocortical hormones by the different zones of the adrenal cortex.

vastly different types of hormones to be formed, such as especially large quantities of masculinizing or, very rarely, feminizing sex hormones or other steroid compounds not normally present in the blood but that have either mineralocorticoid or glucocorticoid actions or a combination of both.

Figure 77–3 illustrates the chemical formulas of aldosterone and cortisol. It is the oxygen atom bound at the number 18 carbon of the cholesterol nucleus that is most important in providing the mineralocorticoid activity of aldosterone. The glucocorticoid activity of cortisol is provided principally by the presence of the keto-oxygen on carbon number 3 and the hydroxylation of carbon numbers 11 and 21.

In addition to aldosterone and cortisol, which are, respectively, the principal mineralocorticoid and glucocorticoid hormones, still other steroids having one or both of these activities are secreted in small amounts even normally by the adrenal cortex. And several additional potent steroid hormones not normally formed in the adrenal glands have been synthesized and are used in various forms of therapy. The more important of the adrenocortical hormones are the following:

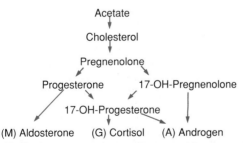

Figure 77–2. The major steps in the synthesis of the three principal adrenal steroids. The physiologic characteristics are designated (M) mineralocorticoid effect, (G) glucocorticoid effect, and (A) androgenic effect.

Mineralocorticoids:
  Aldosterone (very potent, accounts for nearly 90 per cent of mineralocorticoid activity)
  Desoxycorticosterone (one-fifteenth as potent as aldosterone, very small quantities secreted)
  Corticosterone (slight activity)
  9α-Fluorocortisol (synthetic, slightly more potent than aldosterone)
  Cortisol (very slight activity but large quantity secreted)
  Cortisone (synthetic, slight activity)

Glucocorticoids:
  Cortisol (very potent, accounts for about 95 per cent of all glucocorticoid activity)
  Corticosterone (about 4 per cent of total glucocorticoid activity, but much less potent than cortisol)
  Cortisone (synthetic, almost as potent as cortisol)
  Prednisone (synthetic, four times as potent as cortisol)
  Methylprednisone (synthetic, five times as potent as cortisol)
  Dexamethasone (synthetic, 30 times as potent as cortisol)

It is clear from this list that some of these hormones have both glucocorticoid and mineralocorticoid activities. It is especially significant that cortisol has a small amount of mineralocorticoid activity because some syndromes of excess cortisol secretion can cause significant mineralocorticoid effects, along with its much more potent glucocorticoid effects.

The very intense glucocorticoid activity of dexamethasone, which has almost zero mineralocorticoid activity, makes this an especially important drug for stimulating specific glucocorticoid activity.

Transport and Fate of the Adrenal Hormones. Cortisol combines with a globulin called *cortisol-binding globulin* or *transcortin* and, to a lesser extent, with albumin—about 94 per cent is normally transported in the bound form and about 6 per cent free. On the other hand, aldosterone combines only loosely with the plasma proteins so that about 50 per cent is in the free form. In both the combined and free forms the hormones are transported throughout the extracellular fluid compartment. In general, the hormones become fixed in the target tissues within an hour or two for cortisol and within about 30 minutes for aldosterone.

The adrenal steroids are degraded mainly in the liver and conjugated especially to form glucuronides and, to a lesser extent, sulfates. About 25 per cent of these are excreted in the bile and then in the feces and the remaining 75 per cent in the urine. The conjugated forms of these hormones are inactive.

The normal concentration of aldosterone in blood is about 6 nanograms (6 billionths of a gram) per deciliter, and the secretory rate is 150 to 250 μg/day.

The concentration of cortisol in the blood averages 12 μg/dl, and the secretory rate averages 15 to 20 mg/day.

● FUNCTIONS OF THE MINERALOCORTICOIDS— ALDOSTERONE

Total loss of adrenocortical secretion usually causes death within 3 days to 2 weeks unless the person receives extensive salt therapy or mineralocorti-

Figure 77-3. The two important corticosteroids.

coid therapy. Without mineralocorticoids, the potassium ion concentration of the extracellular fluid rises markedly, the sodium and chloride concentrations decrease, and the total extracellular fluid volume and blood volume also become greatly reduced. The person soon develops diminished cardiac output, which proceeds to a shocklike state followed by death. This entire sequence can be prevented by the administration of aldosterone or some other mineralocorticoid. Therefore, the mineralocorticoids are said to be the acute "life-saving" portion of the adrenocortical hormones; the glucocorticoids are equally necessary, allowing the person to resist the destructive effects of life's intermittent physical and mental "stresses," as is discussed later in the chapter.

Aldosterone exerts nearly 90 per cent of the mineralocorticoid activity of the adrenocortical secretion, but cortisol, the major glucocorticoid secreted by the adrenal cortex, also provides a significant amount of mineralocorticoid activity — its mineralocorticoid activity is only 1/400 that of aldosterone, but about 80 times as much is secreted! Other adrenal steroids secreted in small amounts that have mineralocorticoid effects are *corticosterone*, which also exerts glucocorticoid effects, and *deoxycorticosterone*, which has almost the same effects as aldosterone but with a potency 1/30 that of aldosterone.

RENAL AND CIRCULATORY EFFECTS OF ALDOSTERONE

By far the most important function of aldosterone is to promote transport of sodium and potassium through some portions of the renal tubular walls and, to a lesser extent, transport of hydrogen ions. The mechanisms of these effects are discussed in detail in Chapters 27 through 30. However, let us summarize briefly the renal and body fluid effects of aldosterone.

Effect on Tubular Reabsorption of Sodium and Tubular Secretion of Potassium. It will be recalled from Chapter 27 that aldosterone causes increased exchange transport of sodium and potassium — that is, absorption of sodium and simultaneous excretion of potassium by the tubular epithelial cells — especially in the collecting tubule but also to a lesser extent in the distal tubule and collecting duct. Therefore, aldosterone causes sodium to be conserved

in the extracellular fluid while more potassium is excreted into the urine.

A high concentration of aldosterone in the plasma can decrease the sodium loss into the urine to as little as a few milligrams a day. At the same time, potassium loss into the urine increases manyfold. Conversely, total lack of aldosterone secretion can cause loss of as much as 20 grams of sodium in the urine a day, an amount equal to one fifth of all the sodium in the body. At the same time, potassium is conserved tenaciously in the extracellular fluid.

Therefore, the net effect of excess aldosterone in the plasma is to increase the total quantity of sodium in the extracellular fluid while decreasing the potassium.

Effect on Extracellular Fluid Volume and Arterial Pressure. Even though aldosterone has a potent effect in decreasing the rate of sodium ion excretion by the kidneys, the concentration of sodium in the extracellular fluid rises very little. The reason for this is that when sodium is reabsorbed by the tubules, there is simultaneous osmotic absorption of almost equivalent amounts of water. Therefore, the extracellular fluid volume increases almost exactly as much as the retained sodium but without much change in sodium concentration.

A persistent increase in extracellular fluid volume eventually leads to an increase in arterial pressure, as explained in Chapter 19. The increase in arterial pressure leads then to greatly increased kidney excretion of both water and salt, which is the phenomenon called pressure diuresis. Thus, in a roundabout manner, after the extracellular fluid volume has increased to about 10 to 15 per cent above normal in response to great excesses of aldosterone, the arterial pressure also will have increased some 15 to 25 mm Hg, and this returns the renal output of water and salt to normal despite the excess aldosterone. This secondary increase in water and salt excretion by the kidneys as a result of pressure diuresis is called *aldosterone escape* because the net gain of salt and water by the body thereafter is zero.

Conversely, when aldosterone secretion becomes zero, very large amounts of salt are lost in the urine, not only diminishing the amount of sodium chloride in the extracellular fluid but also decreasing the extracellular fluid volume. The result is severe extracellular fluid dehydration and low blood volume, leading to *circulatory shock.* Without immediate therapy, this usually causes death within a few days after the adrenal glands suddenly stop secreting aldosterone.

Hypokalemia and Muscle Weakness; Hyperkalemia and Cardiac Toxicity. The excessive loss of potassium ions from the extracellular fluid into the urine under the influence of aldosterone causes a serious decrease in the plasma potassium concentration, often from the normal value of 4.5 mEq/liter to as low as 1 to 2 mEq/liter. This condition is called *hypokalemia.* When the potassium ion concentration falls below approximately one-half normal, severe muscle weakness often develops. This is caused by alteration of the electrical properties of the nerve and muscle fiber membranes (see Chapter 5), which prevents transmission of action potentials.

On the other hand, when aldosterone is deficient, the extracellular fluid potassium ion concentration can rise far above normal. When it rises to 60 to 100 per cent above normal, serious cardiac toxicity, including weakness of heart contraction and development of arrhythmia, becomes evident; a still higher concentration of potassium leads inevitably to cardiac death.

Effect of Aldosterone on Increasing Tubular Hydrogen Ion Secretion, With Resultant Mild Alkalosis. Though aldosterone mainly causes potassium to be secreted into the tubules in exchange for sodium reabsorption, to a much smaller extent it also causes tubular secretion of hydrogen ions in exchange for sodium. The obvious effect of this is to decrease the hydrogen ion concentration in the extracellular fluid. However, this effect is not a strong one, usually causing only a mild degree of alkalosis.

EFFECTS OF ALDOSTERONE ON SWEAT GLANDS, SALIVARY GLANDS, AND INTESTINAL ABSORPTION

Aldosterone has almost the same effects on sweat glands and salivary glands that it has on the renal tubules. Both these glands form a primary secretion that contains large quantities of sodium chloride, but much of the sodium chloride on passing through the excretory ducts is reabsorbed, whereas potassium and bicarbonate ions are secreted. Aldosterone greatly increases the reabsorption of sodium chloride and the secretion of potassium by the ducts. The effect on the sweat glands is important to conserve body salt in hot environments, and the effect on the salivary glands is necessary to conserve salt when excessive quantities of saliva are lost.

Aldosterone also greatly enhances sodium absorption by the intestines, especially in the colon, which obviously prevents loss of sodium in the stools. On the other hand, in the absence of aldosterone, sodium absorption can be very poor, leading to failure to absorb anions and water as well. The unabsorbed sodium chloride and water then lead to diarrhea, with further loss of salt from the body.

CELLULAR MECHANISM OF ALDOSTERONE ACTION

Although for many years we have known the overall effects of mineralocorticoids on the body, the basic action of aldosterone on the tubular cells to increase transport of sodium is still only partly understood. The sequence of events that leads to increased sodium reabsorption seems to be the following:

First, because of its lipid solubility in the cellular membranes aldosterone diffuses readily to the interior of the tubular epithelial cells.

Second, in the cytoplasm of the tubular cells aldosterone combines with a highly specific cytoplasmic *receptor protein,* a protein that has a stereomolecular configuration that will allow only aldosterone or extremely similar compounds to combine.

Third, the aldosterone-receptor complex diffuses into the nucleus, where it may undergo further alterations; then it induces specific portions of the DNA to form a type or types of messenger RNA related to the process of sodium and potassium transport.

Fourth, the messenger RNA diffuses back into the cytoplasm, where, operating in conjunction with the ribosomes, it causes protein formation. The proteins formed are a mixture of (1) one or more enzymes, (2) receptor proteins, and (3) membrane transport proteins that collectively are required for sodium, potassium, and hydrogen transport through the cell membrane. One of the enzymes especially increased is *sodium-potassium adenosine triphosphatase,* which serves as the principal part of the pump for sodium and potassium exchange at the *basolateral membranes* of the renal tubular cells. Another protein, perhaps equally important, is a channel protein inserted into the *luminal membrane* of the same tubular cells that allows rapid diffusion of sodium ions from the tubular lumen into the cell; then the sodium is pumped the rest of the way by the sodium-potassium pump located in the basolateral membranes of the cell.

Thus, aldosterone does not have an immediate effect on sodium transport but must await the sequence of events that leads to the formation of the specific intracellular substance or substances required for sodium transport. Approximately 30 minutes is required before new RNA appears in the cells, and approximately 45 minutes is required before the rate of sodium transport begins to increase; the effect reaches maximum only after several hours.

REGULATION OF ALDOSTERONE SECRETION

The regulation of aldosterone secretion is so deeply intertwined with the regulation of extracellular fluid electrolyte concentrations, extracellular fluid volume, blood volume, arterial pressure, and many special aspects of renal function that it is not possible to discuss the regulation of aldosterone secretion independently of all these other factors. This subject is presented in great detail in Chapters 28 and 29, to which the reader is referred. However, it is important to list here as well the most important points of aldosterone secretion control.

Let us note first that aldosterone is secreted by the *zona glomerulosa,* a very thin zone of cells located on the surface of the adrenal cortex immediately beneath the capsule. These cells function almost entirely independently of the deeper cells in the zona reticularis and zona fasciculata, which secrete cortisol and the androgens. Furthermore, the regulation of aldosterone secretion is almost entirely independent of the regulation of these other hormones.

Four different factors are presently known to play essential roles in the regulation of aldosterone. In the probable order of their importance these are:

1. Potassium ion concentration of the extracellular fluid.
2. Renin-angiotensin system.
3. Sodium ion concentration in the extracellular fluid.
4. Adrenocorticotropic hormone (ACTH).

Of the above factors, *potassium ion concentration* and the *renin-angiotensin system* are by far the most potent in regulating aldosterone secretion. A very low percentage change in potassium concentration can cause a several-fold change in aldosterone secretion. Likewise, activation of the renin-angiotensin system, usually in response to diminished blood flow to the kidney, can also cause a several-fold change in aldosterone secretion. In turn, the aldosterone acts on the kidneys to help correct the abnormal potassium ion concentration in the extracellular fluid or to correct abnormal circulatory function, thus returning the renin-angiotensin system also back toward its normal level of activity. These feedback control mechanisms are essential for maintaining life, and the reader is referred again to Chapters 28 and 29 for a full understanding of their functions.

Figure 77–4 illustrates the effect on plasma aldosterone concentration of continuous infusion of angiotensin II at two different infusion rates for 2 weeks. Note especially the very large acute effect, which allows the renin-angiotensin-aldosterone control system to correct abnormalities of extracellular fluid electrolytes as rapidly as possible.

Figure 77–4. Effects on plasma aldosterone concentration caused by continuous infusion of angiotensin II at two different infusion rates for a period of two weeks. Note the very marked acute effect but the much weaker chronic effect. (Drawn from data in Cowley and McCaa, *Circ. Res.*, 39:788, 1976.)

By contrast, the effects of sodium ion concentration and of ACTH in controlling aldosterone secretion are usually minor. Nevertheless, a 10 to 20 per cent decrease in extracellular fluid sodium ion concentration, which occurs on rare occasions, can perhaps double aldosterone secretion. In the case of ACTH, if there is even a small amount of ACTH, it is usually enough to permit the adrenal glands to secrete whatever amount of aldosterone is required; but total absence of ACTH can significantly reduce aldosterone secretion.

● FUNCTIONS OF THE GLUCOCORTICOIDS

Even though mineralocorticoids can save the life of an acutely adrenalectomized animal, the animal still is far from normal. Instead, its metabolic systems for utilization of proteins, carbohydrates, and fats are considerably deranged. Furthermore, the animal cannot resist different types of physical or even mental stress, and minor illnesses such as respiratory tract infections can lead to death. Therefore, the glucocorticoids have functions just as important to the long-continued life of the animal as those of the mineralocorticoids. These are explained in the following sections.

At least 95 per cent of the glucocorticoid activity of the adrenocortical secretions results from the secretion of *cortisol,* known also as *hydrocortisone.* In addition to this, a small but significant amount of glucocorticoid activity is provided by *corticosterone.*

EFFECTS OF CORTISOL ON CARBOHYDRATE METABOLISM

Stimulation of Gluconeogenesis. By far the best-known metabolic effect of cortisol and other glucocorticoids on metabolism is their ability to stimulate gluconeogenesis (formation of carbohydrate from proteins and some other substances) by the liver, often increasing the rate of gluconeogenesis as much as six- to tenfold. This results mainly from two different effects of cortisol:

First, all the enzymes required to convert amino acids into glucose are increased in the liver cells. This results from the effect of the glucocorticoids to activate DNA transcription in the liver cell nuclei, in the same way that aldosterone functions in the renal tubular cells, with formation of messenger RNAs that in turn lead to the array of enzymes required for gluconeogenesis.

Second, cortisol causes mobilization of amino acids from the extrahepatic tissues, mainly from muscle. As a result, more amino acids become available in the plasma to enter into the gluconeogenesis process of the liver and thereby to promote the formation of glucose.

One of the effects of increased gluconeogenesis is a marked increase in glycogen storage in the liver cells.

Decreased Glucose Utilization by the Cells. Cortisol also causes a moderate decrease in the rate of glucose utilization by the cells everywhere in the body. Though the cause of this decrease is unknown, most physiologists believe that somewhere between the point of entry of glucose into the cells and its final degradation cortisol directly delays the rate of glucose utilization. A suggested mechanism is based on the observation that glucocorticoids depress the oxidation of nicotinamide-adenine dinucleotide (NADH). Because NADH must be oxidized to allow rapid glycolysis, this effect could account for the diminished utilization of glucose by the cells.

Elevated Blood Glucose Concentration and Adrenal Diabetes. Both the increased rate of gluconeogenesis and the moderate reduction in the rate of glucose utilization by the cells cause the blood glucose concentration to rise. The increase in concentration is occasionally great enough—50 per cent or more above normal—that the condition is called *adrenal diabetes,* and it has many similarities to pituitary diabetes, discussed in Chapter 75. Administration of insulin lowers the blood glucose concentration only a moderate amount in adrenal diabetes, not nearly so much as it does in pancreatic diabetes. On the other hand, insulin causes greater decrease in blood glucose concentration in adrenal diabetes than in pituitary diabetes. Therefore, *pituitary diabetes is said to be weakly insulin sensitive, adrenal diabetes moderately insulin sensitive, and pancreatic diabetes strongly insulin sensitive.*

EFFECTS OF CORTISOL ON PROTEIN METABOLISM

Reduction in Cellular Protein. One of the principal effects of cortisol on the metabolic systems of the body is reduction of the protein stores in essentially all body cells except those of the liver. This is caused by both decreased protein synthesis and increased catabolism of protein already in the cells. Both these effects may possibly result from decreased amino acid transport into extrahepatic tissues, as is discussed later; but this probably is not the only cause, because cortisol also depresses the formation of RNA in many extrahepatic tissues, especially muscle and lymphoid tissue.

In the presence of great excesses of cortisol, the muscles can become so weak that the person cannot rise from the squatting position. And the immunity functions of the lymphoid tissue can be decreased to a small fraction of normal.

Increased Liver Protein and Plasma Proteins Caused by Cortisol. Coincidentally with the reduced proteins elsewhere in the body, the liver proteins become enhanced. Furthermore, the plasma proteins (which are produced by the liver and then released into the blood) are also increased. Therefore, these are exceptions to the protein depletion that occurs elsewhere in the body. It is believed that this difference results from a possible effect of cortisol in enhancing amino acid transport into liver cells (but not into most other cells) and of enhancement of the liver enzymes required for protein synthesis.

Increased Blood Amino Acids, Diminished Transport of Amino Acids Into Extrahepatic Cells, and Enhanced Transport Into Hepatic Cells. Recent studies in isolated tissue have demonstrated that cortisol depresses amino acid transport into muscle cells and perhaps into other extrahepatic cells.

Obviously, the decreased transport of amino acids into extrahepatic cells decreases their intracellular amino acid concentrations and as a consequence decreases the synthesis of protein. Yet catabolism of proteins in the cells continues to release amino acids from the already existing proteins, and these diffuse out of the cells to increase the plasma amino acid concentration. Therefore, *cortisol mobilizes amino acids from the nonhepatic tissues* and in doing so diminishes the tissue stores of protein.

The increased plasma concentration of amino acids, plus the fact that cortisol enhances transport of amino acids into the hepatic cells, could also account for enhanced utilization of amino acids by the liver to cause such effects as (1) increased rate of deamination of amino acids by the liver, (2) increased protein synthesis in the liver, (3) increased formation of plasma proteins by the liver, and (4) increased conversion of amino acids to glucose—that is, enhanced gluconeogenesis.

Thus, it is possible that many of the effects of cortisol on the metabolic systems of the body result mainly from this ability of cortisol to mobilize amino acids from the peripheral tissues while at the same time increasing the liver enzymes required for the hepatic effects by causing the formation of appropriate messenger RNAs.

EFFECTS OF CORTISOL ON FAT METABOLISM

Mobilization of Fatty Acids. In much the same manner that cortisol promotes amino acid mobilization from muscle, it also promotes mobilization of fatty acids from adipose tissue. This increases the concentration of free fatty acids in the plasma, which also increases their utilization for energy. Cortisol moderately enhances the oxidation of fatty acids in the cells as well; this effect perhaps results secondarily from the reduced availability of glycolytic products for metabolism.

The mechanism by which cortisol promotes fatty acid mobilization is not yet understood. However, part of the effect probably results from diminished transport of glucose into the fat cells. It will be remembered that α-glycerophosphate, which is derived from glucose, is required for both deposition of and maintenance of triglycerides in these cells, and in its absence the fat cells begin to release fatty acids.

The increased mobilization of fats by cortisol, combined with increased oxidation of fatty acids in the cells, helps shift the metabolic systems of the cells in times of starvation or other stresses from utilization of glucose for energy to utilization of fatty acids instead. This cortisol mechanism, however, requires several hours to become fully developed—not nearly so rapid or nearly so powerful an effect as a similar shift elicited by a decrease in insulin. Nevertheless, it is probably an important factor for long-term conservation of body glucose and glycogen.

Ketogenic Effects of Cortisol. It is frequently said that cortisol has a *ketogenic effect* because ketosis usually will not develop unless cortisol is available to cause fat mobilization. However, this ketogenic effect occurs only under certain conditions, such as when insulin also is deficient.

Obesity Caused by Cortisol. Despite the fact that cortisol can cause a moderate degree of fatty acid mobilization from adipose tissue, persons with excess cortisol secretion frequently develop a peculiar type of obesity, with excess deposition of fat in the chest and head regions of the body, giving a buffalo-like torso and a rounded face, a "moon face." Experiments have suggested that this obesity results from excess stimulation of food intake so that fat is generated in some tissues of the body at a rate even more rapidly than it is mobilized and oxidized.

OTHER EFFECTS OF CORTISOL

Function of Cortisol in Different Types of Stress

It is amazing that almost any type of stress, whether physical or neurogenic, will cause an immediate and marked increase in ACTH secretion by the anterior pituitary gland, followed within minutes by greatly increased adrenocortical secretion of cortisol. This is illustrated dramatically by the experiment shown in Figure 77–5, in which corticosteroid formation and secretion increased sixfold in a rat within 4 to 20 minutes after fracture of two leg bones.

Some of the different types of stress that increase cortisol release are the following:

1. Trauma of almost any type.
2. Infection.
3. Intense heat or cold.
4. Injection of norepinephrine and other sympathomimetic drugs.
5. Surgical operations.
6. Injection of necrotizing substances beneath the skin.
7. Restraining an animal so that it cannot move.
8. Almost any debilitating disease.

Thus, a wide variety of nonspecific stimuli can cause marked increase in the rate of cortisol secretion by the adrenal cortex.

Yet even though we know that cortisol secretion often increases greatly in stressful situations, we still are not sure why this is of significant benefit to the animal. One guess, which is probably as good as any other, is that the glucocorticoids cause rapid mobilization of amino acids and fats from their cellular

Figure 77–5. Rapid reaction of the adrenal cortex of a rat to stress caused by fracture of the tibia and fibula. (Courtesy of Drs. Guillemin, Dear, and Lipscomb.)

stores, making these available both for energy and for synthesis of other compounds, including glucose, needed by the different tissues of the body. Indeed, it has been shown in a few instances that damaged tissues that are momentarily severely depleted of proteins can utilize the newly available amino acids to form new proteins that are essential to the lives of the cells. Also, the amino acids are perhaps used to synthesize such essential intracellular substances as purines, pyrimidines, and creatine phosphate, which are necessary for maintenance of cellular life and reproduction of new cells.

But all this is mainly supposition. It is supported only by the fact that cortisol usually does not mobilize the basic functional proteins of the cells until almost all other proteins have been released. This preferential effect of cortisol in mobilizing labile proteins could make amino acids available to needy cells to synthesize substances essential to life.

Anti-inflammatory Effects of Cortisol

When tissues are damaged by trauma, by infection with bacteria, or in almost any other way, they almost always become inflamed. In some conditions the inflammation is more damaging than the trauma or disease itself. Administration of large amounts of cortisol can usually block this inflammation or even reverse many of its effects once it has begun. Before attempting to explain the way in which cortisol functions to block inflammation, let us first review the basic steps in the inflammation process, discussed in more detail in Chapter 33.

Basically, there are five main stages of inflammation: (1) release from the damaged tissue cells of chemical substances that activate the inflammation process—chemicals such as histamine, bradykinin, proteolytic enzymes, and so forth; (2) an increase in blood flow in the inflamed area caused by some of the released products from the tissues, a process called *erythema;* (3) leakage of large quantities of almost pure plasma out of the capillaries into the damaged areas, followed by clotting of the tissue fluid, thus causing a *nonpitting type of edema;* (4) infiltration of the area by leukocytes; and, finally, (5) tissue healing, which is often accomplished at least partially by ingrowth of fibrous tissue.

When very large amounts of cortisol are secreted or injected into a person, the cortisol has two basic *anti-inflammatory effects:* (1) it blocks the early stages of the inflammation process; and (2) if inflammation has already begun, it causes rapid resolution of the inflammation and increased rapidity of healing. These effects are explained further as follows:

Prevention of the Development of Inflammation — Lysosome Stabilization and Other Effects. Cortisol has the following effects in preventing inflammation:

1. One of the most important anti-inflammatory effects of cortisol is its ability to cause *stabilization of the lysosomal membranes.* That is, cortisol makes it much more difficult than is normal for the membranes of the intracellular lysosomes to rupture. Therefore, most of the proteolytic enzymes that are usually released by damaged cells to cause inflammation, which are mainly stored in the lysosomes, are released in greatly decreased quantity.

2. Cortisol decreases the permeability of the capillaries, probably as a secondary effect of the reduced release of proteolytic enzymes. This prevents loss of plasma into the tissues.

3. Cortisol decreases both migration of white blood cells into the inflamed area and phagocytosis of the damaged cells. These effects probably result from the fact that cortisol diminishes release of lysosomal enzymes and other substances from the damaged cells because of its ability to stabilize both the cell and lysosome membranes.

4. Cortisol suppresses the immune system, causing lymphocyte reproduction to decrease markedly. The T lymphocytes are especially suppressed. In turn, reduced amounts of T cells and antibodies in the inflamed area lessen the tissue reactions that otherwise would promote further the inflammation process.

5. Cortisol lowers fever, mainly because it reduces the release of interleukin-1 from the white blood cells, which is one of the principal excitants to the hypothalamic temperature control system. The decreased temperature in turn reduces the degree of vasodilatation.

Thus, cortisol has an almost systemic effect in reducing all aspects of the inflammatory process. How much of this results from the simple effect of cortisol in stabilizing lysosomal and cell membranes versus other possible effects of cortisol is still unknown.

Effect of Cortisol in Causing Resolution of Inflammation. Even after inflammation has become well established, administration of cortisol can often reduce inflammation within hours to a few days. The immediate effect is to block most of the factors that are promoting the inflammation. Then the rate of healing is also enhanced. This probably results from those same, mainly undefined factors that allow the body to resist many other types of physical stress when large quantities of cortisol are secreted: perhaps this results from the mobilization of amino acids and use of these to repair the damaged tissues; perhaps it results from increased amounts of glucose and fatty acids available for cellular energy; or perhaps it depends on some effect of cortisol for inactivating or removing inflammatory products.

Regardless of the precise mechanisms by which the anti-inflammatory effect occurs, this effect of cortisol can play a major role in combating certain types of diseases, such as rheumatoid arthritis, rheumatic fever, and acute glomerulonephritis. All these are characterized by severe local inflammation, and the harmful effects on the body are caused mainly by the inflammation itself and not by other aspects of the disease. When cortisol or other glucocorticoids are administered to patients with these diseases, almost invariably the inflammation subsides within 24 to 48 hours. And even though the cortisol does not correct the basic disease condition, merely preventing the damaging effects of the inflammatory response, this alone can be a lifesaving measure.

Effect on Allergy

Cortisol blocks the inflammatory response to allergic reactions in exactly the same way that it blocks other types of inflammatory responses. The basic allergic reaction between antigen and antibody is not affected by cortisol, and even some of the secondary effects of the allergic reaction still occur. However, because the inflammatory response is responsible for many of the serious and sometimes lethal effects of allergic reactions, administration of cortisol, followed by its effect in reducing inflammation and the release of inflammatory products, can be lifesaving. For instance, cortisol effectively prevents shock or death in anaphylaxis, which otherwise kills many persons, as explained in Chapter 34.

Effect on Blood Cells and on Immunity

Cortisol decreases the number of eosinophils and lymphocytes in the blood; this effect begins within a few minutes after the injection of cortisol and becomes marked within a few hours. Indeed, a finding of lymphocytopenia or eosinopenia is an important diagnostic criterion for overproduction of cortisol by the adrenal gland.

Likewise, administration of large doses of cortisol causes significant atrophy of all the lymphoid tissue throughout the body, which in turn decreases the output of both T cells and antibodies from the lymphoid tissue. As a result, the level of immunity for almost all foreign invaders of the body is decreased. This can occasionally lead to fulminating infection and death from diseases that otherwise would not be lethal, such as fulminating tuberculosis in a person whose disease had previously been arrested. On the other hand, this ability of cortisol and other glucocorticoids to suppress immunity makes these among the most useful of all drugs to prevent immunologic rejection of transplanted hearts, kidneys, and other tissues.

Cortisol increases the production of red blood cells, the cause of which is unknown. When excess cortisol is secreted by the adrenal glands, polycythemia often results, and, conversely, when the adrenal glands secrete no cortisol, anemia often results.

REGULATION OF CORTISOL SECRETION—ADRENOCORTICOTROPIC HORMONE

Control of Cortisol Secretion by ACTH. Unlike aldosterone secretion by the zona glomerulosa, which is controlled mainly by potassium and angiotensin acting directly on the adrenocortical cells, *almost no stimuli have direct effects* on the adrenal cells to control cortisol secretion. Instead, *secretion of cortisol is controlled almost entirely by adrenocorticotropic hormone (ACTH) secreted by the anterior pituitary gland.* This hormone, also called *corticotropin* or *adrenocorticotropin,* also enhances the production of adrenal androgens. As pointed out earlier in the chapter, small amounts of ACTH are also required for aldosterone secretion, providing a permissive role that allows the other, more important factors to exert their more powerful controls.

Chemistry of ACTH. ACTH has been isolated in pure form from the anterior pituitary. It is a large polypeptide having a chain length of 39 amino acids. A smaller polypeptide, a digested product of ACTH having a chain length of 24 amino acids, also has all the effects of the total molecule.

Control of ACTH Secretion by the Hypothalamus—Corticotropin-Releasing Factor. In the same way that other pituitary hormones are controlled by releasing hormones or factors from the hypothalamus, so also does an important releasing factor control ACTH secretion. This is called *corticotropin-releasing factor* (CRF). It is secreted into the primary capillary plexus of the hypophysial portal system in the median eminence of the hypothalamus and then carried to the anterior pituitary gland, where it induces ACTH secretion. CRF is a peptide composed of 41 amino acids. The cell bodies of the neurons that secrete CRF are located mainly in the paraventricular nucleus of the hypothalamus. This nucleus in turn receives many nervous connections from the limbic system and lower brain stem.

The anterior pituitary gland can secrete only minute quantities of ACTH in the absence of CRF. Instead, most conditions that cause high ACTH secretory rates initiate this secretion by signals that begin in the basal regions of the brain, including the hypothalamus, and are then transmitted by CRF to the anterior pituitary gland.

Mechanism by Which ACTH Activates Adrenocortical Cells to Produce Steroids—Function of Cyclic

Adenosine Monophosphate (cAMP). The principal effect of ACTH on the adrenocortical cells is to activate *adenyl cyclase* in the cell membrane. This then induces the formation of *cAMP* in the cell cytoplasm, reaching its maximum effect in about 3 minutes. The cAMP in turn activates the intracellular enzymes that cause formation of the adrenocortical hormones. Thus, this is another example of cAMP as a *second messenger* hormone.

The most important of all of the ACTH-stimulated steps for controlling adrenocortical secretion is activation of the enzyme *desmolase,* which *causes initial conversion of cholesterol to pregnenolone.* This initial conversion is the "rate-limiting" step for all the adrenocortical hormones, which explains why ACTH normally is necessary for any adrenocortical hormones to be formed. Long-term stimulation of the adrenal cortex by ACTH not only increases secretory activity but also causes hypertrophy and proliferation of the adrenocortical cells, especially in the zona fasciculata and reticularis, where cortisol and the androgens are secreted.

Effect of Physiological Stress on ACTH Secretion

It was pointed out earlier in the chapter that almost any type of physical or even mental stress can lead within minutes to greatly enhanced secretion of ACTH and consequently of cortisol as well, often increasing cortisol secretion as much as 20-fold. This effect was illustrated forcefully by the curves in Figure 77–5, which shows a great increase in plasma corticosterone concentration in a rat within minutes after the tibia and fibula had been broken (corticosterone is the principal glucocorticoid secreted by the rat adrenal instead of cortisol). It is believed that pain stimuli caused by the stress are first transmitted upward through the brain stem to the perifornical area of the hypothalamus and from here into the paraventricular nucleus of the hypothalamus and eventually to the median eminence, as shown in Figure 77–6, where CRF is secreted into the hypophysial portal system.

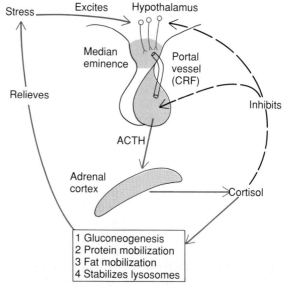

Figure 77–6. Mechanism for regulation of glucocorticoid secretion.

Within minutes the entire control sequence leads to large quantities of the glucocorticoids in the blood.

Mental stress can also cause equally as rapid increase in ACTH secretion. This is believed to result from increased activity in the limbic system, especially in the region of the amygdala and hippocampus, both of these then transmitting signals to the posterior medial hypothalamus.

Inhibitory Effect of Cortisol on the Hypothalamus and on the Anterior Pituitary to Cause Decreased ACTH Secretion. Cortisol has direct negative feedback effects on (1) the hypothalamus to decrease the formation of CRF and (2) the anterior pituitary gland to decrease the formation of ACTH. These feedbacks help regulate the plasma concentration of cortisol. That is, whenever the concentration becomes too great, the feedbacks automatically reduce the ACTH toward a normal control level.

Summary of the Control System

Figure 77–6 illustrates the overall system for control of cortisol secretion. The central key to this control is the excitation of the hypothalamus by different types of stress. These activate the entire system to cause rapid release of cortisol, and the cortisol in turn initiates a series of metabolic effects directed toward relieving the damaging nature of the stressful state. In addition, there is also direct feedback of the cortisol to the hypothalamus and anterior pituitary gland to stabilize the concentration of cortisol in the plasma at times when the body is not experiencing stress. However, the stress stimuli are the prepotent ones; they can always break through this direct inhibitory feedback of cortisol, causing periodic exacerbations of cortisol secretion at multiple times during the day (Fig. 77–7).

Circadian Rhythm of Glucocorticoid Secretion. The secretory rates of CRH, ACTH, and cortisol are all high in the early morning but low in the late evening, as illustrated in Figure 77–7; the plasma cortisol level ranges between a high of about 20 μg/dl an hour before arising in the morning and a low of about 5 μg/dl around midnight. This effect results from a 24-hour cyclic alteration in the signals from the hypothalamus that cause cortisol secretion. When a person changes daily sleeping habits, the cycle changes correspondingly. One of the reasons the cycle is so important is that measurements of blood cortisol levels are meaningful only when expressed in terms of the time in the cycle at which the measurements are made.

Secretion of Melanocyte-Stimulating Hormone, Lipotropin, and Endorphin in Association With ACTH

When ACTH is secreted by the anterior pituitary gland, several other hormones that have similar chemical structures are secreted simultaneously. The reason for this is that the RNA molecule that causes ACTH formation initially causes the formation of a considerably larger protein molecule, a preprohormone, that contains ACTH as only one of its subunits. This same preprohormone also contains *melanocyte-stimulating hormone (MSH)*, *beta-lipotropin*, and *beta-endorphin*. Under normal conditions, none of these is known to be secreted in enough quantity to have a significant effect on the body, but this may not be true, especially for MSH, when the rate of secretion of ACTH is very high, as occurs in some types of Addison's disease, which is discussed later.

MSH causes the *melanocytes,* which are located in abundance between the dermis and the epidermis of the skin, to form the black pigment *melanin* and to disperse this in the cells of the epidermis. Injection of melanocyte-stimulating hormone into a person over a period of 8 to 10 days can cause intense darkening of the skin. However, the effect is much greater in persons who have genetically dark skins than in light-skinned persons.

In some lower animals, an intermediate "lobe" of the pituitary gland, called the *pars intermedia,* is highly developed, lying between the anterior and the posterior pituitary lobes. This lobe secretes an especially large amount of melanocyte-stimulating hormone. Furthermore, this secretion is independently controlled by the hypothalamus in response to the amount of light to which the animal is exposed or in response to other environmental factors. For instance, some arctic animals develop darkened fur in the summer and yet have entirely white fur in the winter.

ACTH, because of its similarity to MSH, has about one-thirtieth as much melanocyte-stimulating effect as MSH. Furthermore, because the quantities of pure MSH secreted in the human being are extremely small, whereas those of ACTH are large, it is likely that ACTH is considerably more important than MSH in determining the amount of melanin in the skin.

● THE ADRENAL ANDROGENS

Several moderately active male sex hormones called *adrenal androgens* (the most important of which is *dehydroepiandrosterone*) are continually secreted by the adrenal cortex, especially so during fetal life, as discussed more fully in Chapter 83. Also, progesterone and estrogens, which are female sex hormones, are secreted in very minute quantities.

Normally, in the human being, the adrenal androgens have only weak effects. However, it is possible that part of the early development of the male sex organs results from childhood secretion of adrenal androgens. The adrenal an-

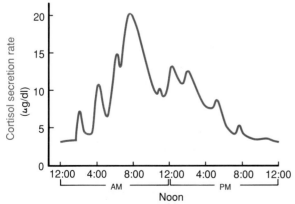

Figure 77–7. A typical pattern of cortisol secretion during the 24 hour day. Note the oscillations in secretion as well as a daily secretory surge an hour or so after awaking in the morning.

drogens also exert mild effects in the female, not only before puberty but also throughout life. Much of the growth of the pubic and axillary hair in the female results from the action of these hormones. Some of the adrenal androgens are converted to testosterone, the major male sex hormone, in the extra-adrenal tissues, which probably accounts for much of their androgenic activity. The physiological effects of androgens are discussed in Chapter 80 in relation to male sexual function.

● ABNORMALITIES OF ADRENOCORTICAL SECRETION

HYPOADRENALISM — ADDISON'S DISEASE

Addison's disease results from failure of the adrenal cortices to produce adrenocortical hormones, and this in turn is most frequently caused by *primary atrophy* of the adrenal cortices, resulting in about 80 per cent of the cases from autoimmunity against the cortices, but also frequently caused by tuberculous destruction of the adrenal glands or invasion of the adrenal cortices by cancer. Basically, the disturbances in Addison's disease are as follows:

Mineralocorticoid Deficiency. Lack of aldosterone secretion greatly decreases sodium reabsorption and consequently allows sodium ions, chloride ions, and water to be lost into urine in great profusion. The net result is a greatly decreased extracellular fluid volume. Furthermore, hyperkalemia and mild acidosis develop because of failure of potassium and hydrogen ions to be secreted in exchange for sodium reabsorption.

As the extracellular fluid becomes depleted, the plasma volume falls, the red blood cell concentration rises markedly, the cardiac output decreases, and the patient dies in shock, death usually occurring in the untreated patient 4 days to 2 weeks after complete cessation of mineralocorticoid secretion.

Glucocorticoid Deficiency. Loss of cortisol secretion makes it impossible for the person with Addison's disease to maintain normal blood glucose concentration between meals because he cannot synthesize significant quantities of glucose by gluconeogenesis. Furthermore, lack of cortisol reduces the mobilization of both proteins and fats from the tissues, thereby depressing many other metabolic functions of the body. This sluggishness of energy mobilization when cortisol is not available is one of the major detrimental effects of glucocorticoid lack. However, even when excess quantities of glucose and other nutrients are available, the person's muscles are still weak, indicating that glucocorticoids are needed to maintain other metabolic functions of the tissues in addition to energy metabolism.

Lack of adequate glucocorticoid secretion also makes the person with Addison's disease highly susceptible to the deteriorating effects of different types of stress, and even a mild respiratory infection can sometimes cause death.

Melanin Pigmentation. Another characteristic of most persons with Addison's disease is melanin pigmentation of the mucous membranes and skin. This melanin is not always deposited evenly but occasionally in blotches and especially in the thin skin areas, such as the mucous membranes of the lips and the thin skin of the nipples.

The cause of the melanin deposition is believed to be the following: When cortisol secretion is depressed, the normal negative feedback to the hypothalamus and anterior pituitary is also depressed, therefore allowing tremendous rates of ACTH secretion as well as simultaneous secretion of increased amounts of MSH. Probably the tremendous amounts of ACTH cause most of the pigmenting effect because these can stimulate formation of melanin by the melanocytes in the same way that MSH does. Even though MSH has 30 times as much melanocyte-stimulating effect as ACTH, the amounts secreted by the human being are extremely small.

Treatment of Persons With Addison's Disease. The untreated person with total adrenal destruction dies within a few days to a few weeks because of consuming weakness and eventual circulatory shock. Yet such a person can usually live for years if small quantities of mineralocorticoids and glucocorticoids are administered daily.

The Addisonian Crisis. As noted earlier in the chapter, great quantities of glucocorticoids are occasionally secreted in response to different types of physical or mental stress. In the person with Addison's disease, the output of glucocorticoids does not increase during stress. Yet whenever different types of trauma, disease, or other stresses such as surgical operations supervene, a person is likely to have an acute need for excessive amounts of glucocorticoids and must be given as much as 10 or more times the normal quantities of glucocorticoids in order to prevent death.

This critical need for extra glucocorticoids and the associated severe debility in times of stress is called an *addisonian crisis.*

HYPERADRENALISM — CUSHING'S SYNDROME

Hypersecretion of cortisol by the adrenal cortex causes a complex of hormonal effects called *Cushing's disease,* and this results from either a cortisol-secreting tumor of one adrenal cortex or general hyperplasia of both adrenal cortices. The hyperplasia in turn is usually caused by increased secretion of ACTH by the anterior pituitary or by "ectopic secretion" of ACTH by a tumor elsewhere in the body, such as an abdominal carcinoma. Most abnormalities of Cushing's syndrome are ascribable to abnormal amounts of cortisol, but secretion of androgens is also of significance.

A special characteristic of Cushing's disease is mobilization of fat from the lower part of the body, with concomitant extra deposition of fat in the thoracic and upper abdominal regions, giving rise to a so-called "buffalo" torso. The excess secretion of steroids also leads to an edematous appearance of the face, and the androgenic potency of some of the hormones sometimes causes acne and hirsutism (excess growth of facial hair). The total appearance of the face is frequently described as a "moon face," as illustrated to the left in Figure 77–8, which shows a patient with Cushing's syndrome prior to treatment. About 80 per cent of the patients have hypertension, presumably because of the slight mineralocorticoid effects of cortisol.

Effects on Carbohydrate and Protein Metabolism. The abundance of cortisol secreted in Cushing's syndrome can cause increased blood glucose concentration, sometimes to values as high as 200 mg/dl after meals, more than two times normal. This results mainly from enhanced gluconeogenesis. If this "adrenal diabetes" lasts for many months, the beta cells in the islets of Langerhans in the pancreas occasionally "burn out" because the high blood glucose greatly overstimulates them to secrete insulin. The destruction of these cells then causes frank pancreatic diabetes mellitus, which is permanent for the remainder of life.

The effects of glucocorticoids on protein catabolism are

Figure 77–8. A person with Cushing's disease before subtotal adrenalectomy (left) and after subtotal adrenalectomy (right). (Courtesy of Dr. Leonard Posey.)

often profound in Cushing's syndrome, causing greatly decreased tissue proteins almost everywhere in the body with the exceptions of the liver and the plasma proteins. The loss of protein from the muscles in particular causes severe weakness. The loss of protein synthesis in the lymphoid tissues leads to a suppressed immune system, so that many of these patients die of infections. Even the collagen fibers in the subcutaneous tissue are diminished so that the subcutaneous tissues tear easily, resulting in development of large *purplish striae* where the subcutaneous tissues have torn apart. In addition, lack of protein deposition in the bones often causes very severe *osteoporosis* with consequent weakness of the bones.

Treatment of Cushing's Disease. Treatment in Cushing's disease consists of removing an adrenal tumor if this is the cause, or of decreasing the secretion of ACTH, if this is possible. Hypertrophied pituitary glands or even small tumors in the pituitary that oversecrete ACTH can be surgically or microsurgically removed or destroyed by radiation. If ACTH secretion cannot easily be decreased, the only satisfactory treatment is usually bilateral partial (or even total) adrenalectomy, followed by administration of adrenal steroids to make up for any insufficiency that develops.

PRIMARY ALDOSTERONISM

Occasionally a small tumor of the zona glomerulosa cells occurs and secretes large amounts of aldosterone. In a few instances, hyperplastic adrenal cortices secrete aldosterone rather than cortisol. The effects of the excess aldosterone are those discussed in detail earlier in the chapter. The most important effects are hypokalemia, slight increase in extracellular fluid volume and blood volume, very slight increase in plasma sodium concentration (usually not over a 2 to 3 per cent increase), and almost always hypertension. Especially interesting in primary aldosteronism are occasional periods of muscular paralysis caused by the hypokalemia. The paralysis is caused by a depressant effect of hypokale-

mia on action potential transmission by the nerve fibers, as explained in Chapter 5.

One of the diagnostic criteria of primary aldosteronism is a decreased plasma renin concentration. This results from feedback suppression of renin secretion caused by the excess aldosterone or by the excess extracellular fluid volume and arterial pressure resulting from the aldosteronism.

Treatment of primary aldosteronism is usually surgical removal of the tumor or of most of the adrenal tissue when hyperplasia is the cause. In about 30 per cent of the treated patients, the arterial pressure fails to return to normal even though plasma aldosterone concentrations do become normal; this presumably results from some long-term damage to the kidneys.

ADRENOGENITAL SYNDROME

An occasional adrenocortical tumor secretes excessive quantities of androgens that cause intense masculinizing effects throughout the body. If this occurs in a female, she develops virile characteristics, including growth of a beard, a much deeper voice, occasionally baldness if she also has the genetic inheritance for baldness, masculine distribution of hair on the body and on the pubis, growth of the clitoris to resemble a penis, and deposition of proteins in the skin and especially in the muscles to give typical masculine characteristics.

In the prepubertal male a virilizing adrenal tumor causes the same characteristics as in the female, plus rapid development of the male sexual organs and creation of male sexual desires. Typical development of the male sexual organs in a 4 year old boy with the adrenogenital syndrome is shown in Figure 77–9.

In the adult male, the virilizing characteristics of the adrenogenital syndrome are usually completely obscured by the normal virilizing characteristics of the testosterone secreted by the testes. Therefore, it is often difficult to make a diagnosis of adrenogenital syndrome in the adult male.

Figure 77-9. Adrenogenital syndrome in a 4 year old boy. (Courtesy of Dr. Leonard Posey.)

In the adrenogenital syndrome, the excretion of 17-keto-steroids (which are derived from androgens) in the urine may be as much as 10 to 15 times normal.

REFERENCES

Burnstein, K. L., and Cidlowski, J. A.: Regulation of gene expression by glucocorticoids. Annu. Rev. Physiol., 51:683, 1989.

Chrousos, G. P., et al.: NIH Conference: Clinical applications of corticotropin-releasing factor. Ann. Intern. Med., 102:344, 1985.

D'Agata, R., and Chrousos, G. P.: Recent Advances in Adrenal Regulation and Function. New York, Raven Press, 1987.

DeGroot, L. J. (ed.): Endocrinology, 2nd Ed. Philadelphia, W. B. Saunders Co., 1989.

Felig, P., et al. (eds.): Endocrinology and Metabolism, 2nd Ed. New York, McGraw-Hill Book Co., 1987.

Funder, J. W., and Sheppard, K.: Adrenocortical steroids and the brain. Annu. Rev. Physiol., 49:397, 1987.

Funder, J. W.: Adrenal steroids: New answers, new questions. Science, 237:236, 1987.

Gann, D. S., et al.: Neural interaction in control of adrenocorticotropin. Fed. Proc., 44:161, 1985.

Hall, J. E., et al.: Control of arterial pressure and renal function during glucocorticoid excess in dogs. Hypertension, 2:139, 1980.

Jones, M. T., and Gilham, B.: Factors involved in the regulation of adreno-corticotropic hormone/β-lipotropic hormone. Physiol. Rev., 68:743, 1988.

Kannan, C. R.: The Adrenal Gland. New York, Plenum Publishing Corp., 1988.

Keller-Wood, M. E., and Dallman, M. F.: Corticosteroid inhibition of ACTH secretion. Endocrinol. Rev., 5:1, 1984.

Kraus-Friedmann, N.: Hormonal regulation of hepatic gluconeogenesis. Physiol. Rev., 64:170, 1984.

Lan, N. C., et al.: Mechanisms of glucocorticoid hormone action. J. Steroid Biochem., 20:77, 1984.

McDougall, J. G.: The physiology of aldosterone secretion. News Physiol. Sci., 2:126, 1987.

McEwen, B. S., et al.: Adrenal steroid receptors and actions in the nervous system. Physiol. Rev., 66:1121, 1986.

Meyer, J. S.: Biochemical effects of corticosteroids on neural tissues. Physiol. Rev., 65:946, 1985.

Miller, M.: Assessment of hormonal disorders of water metabolism. Clin. Lab. Med., 4:729, 1984.

Morel, F., and Doucet, A.: Hormonal control of kidney functions at the cell level. Physiol. Rev., 66:377, 1986.

Moudgil, V. K. (ed.): Steroid Receptors in Health and Disease. New York, Plenum Publishing Corp., 1988.

Mulrow, P. J. (ed.): The Adrenal Gland. New York, Elsevier Science Publishing Co., 1986.

Ozawa, S., and Sand, O.: Electrophysiology of excitable endocrine cells. Physiol. Rev., 66:887, 1986.

Parker, L.: Adrenal Androgens in Clinical Medicine. San Diego, Calif., Academic Press, 1988.

Quinn, S. J., and Williams, G. H.: Regulation of aldosterone secretion. Annu. Rev. Physiol., 50:409, 1988.

Rivier, C. L., and Plotsky, P. M.: Mediation by corticotropin releasing factor (CRF) of adenohypophysial hormone secretion. Annu. Rev. Physiol., 48:475, 1986.

Rousseau, G. G.: Control of gene expression by glucocorticoid hormones. Biochem. J., 224:1, 1984.

Rousseau, G. G.: Structure and regulation of the glucocorticoid hormone receptor. Mol. Cell. Endocrinol., 38:1, 1984.

Schatzberg, A. F., and Nemeroff, C. B. (eds.): The Hypothalamic-Pituitary-Adrenal Axis. New York, Raven Press, 1988.

Schmidt, T. J., and Litwack, G.: Activation of the glucocorticoid-receptor complex. Physiol. Rev., 62:1131, 1982.

Schneider, E. G., et al.: Effect of osmolality on aldosterone secretion. Endocrinology 116:1621, 1985.

Schneider, E. G., et al.: Effect of sodium concentration on aldosterone secretion by isolated perfused canine adrenal glands. Endocrinology, 115:2195, 1984.

Seldin, D. W., and Giebisch, G.: The Regulation of Potassium Balance. New York, Raven Press, 1989.

Seldin, D. W., and Giebisch, G.: The Regulation of Sodium and Chloride Balance. New York, Raven Press, 1989.

Sherman, M. R., and Stevens, J.: Structure of mammalian steroid receptors: Evolving concepts and methodological developments. Annu. Rev. Physiol., 46:83, 1984.

Smith, P. L., and McCabe, R. D.: Mechanism and regulation of transcellular potassium transport by the colon. Am. J. Physiol., 247:G445, 1984.

Young, D. B., and Guyton, A. C.: Steady state aldosterone dose response relationships. Circ. Res., 40(2):138, 1977.

CHAPTER 78

• *Insulin, Glucagon, and Diabetes Mellitus* •

The pancreas, in addition to its digestive functions, secretes two important hormones, *insulin* and *glucagon.* The purpose of this chapter is to discuss the functions of these hormones in regulating glucose, lipid, and protein metabolism, as well as to discuss briefly the two diseases—*diabetes mellitus* and *hyperinsulinism*—caused, respectively, by hyposecretion of insulin and excess secretion of insulin.

Physiologic Anatomy of the Pancreas. The pancreas is composed of two major types of tissues, as shown in Figure 78–1: (1) the *acini,* which secrete digestive juices into the duodenum; and (2) the *islets of Langerhans,* which do not have any means for emptying their secretions externally but instead secrete insulin and glucagon directly into the blood. The digestive secretions of the pancreas are discussed in Chapter 64.

The pancreas of the human being has 1 to 2 million islets of Langerhans, each only about 0.3 millimeters in diameter and organized around small capillaries into which its cells secrete their hormones. The islets contain three major types of cells, *alpha, beta,* and *delta* cells, which are distinguished from one another by their morphologic and staining characteristics. The beta cells, constituting about 60 per cent of all the cells, lie mainly in the middle of each islet and secrete *insulin.* The alpha cells, about 25 per cent of the total, secrete *glucagon.* And the delta cells, about 10 per cent of the total, secrete *somatostatin.* In addition, at least one other type of cell, the PP cell, is present in small numbers in the islets and secretes a hormone of uncertain function called *pancreatic polypeptide.*

The close interrelationships among these different cell types in the islets of Langerhans allow direct control of secretion of some of the hormones by the other hormones. For instance, insulin inhibits glucagon secretion, and somatostatin inhibits the secretion of both insulin and glucagon.

● INSULIN AND ITS METABOLIC EFFECTS

Insulin was first isolated from the pancreas in 1922 by Banting and Best, and almost overnight the outlook for the severely diabetic patient changed from one of rapid decline and death to that of a nearly normal person.

Historically, insulin has been associated with "blood sugar," and, true enough, insulin has profound effects on carbohydrate metabolism. Yet, in addition, abnormalities of fat metabolism, causing such conditions as acidosis and arteriosclerosis, are the usual causes of death of a diabetic patient. Also, in patients with prolonged diabetes, the inability to synthesize proteins leads to wasting of the tissues as well as many cellular functional disorders. Therefore, it is clear that insulin affects fat and protein metabolism almost as much as it does carbohydrate metabolism.

Insulin Is a Hormone Associated with Energy Abundance

As we discuss insulin in the next few pages, it will become apparent that insulin secretion is associated with energy abundance. That is, when there is great abundance of energy-giving foods in the diet, insulin is secreted in great quantity. This is especially true for excess carbohydrates, less for excess proteins, and slightly even for fats.

In turn, the insulin plays an important role in storing the excess energy substances. In the case of excess carbohydrates, it causes these to be stored as glycogen mainly in the liver and muscles. It causes fat storage in the adipose tissue. Also, all the excess carbohydrates that cannot be stored as glycogen are converted under the stimulus of insulin into fats and also stored in the adipose tissue. In the case of proteins, insulin has a direct effect in promoting amino acid uptake by cells and conversion of these into protein. In addition, it inhibits the breakdown of the proteins that are already in the cells.

Chemistry of Insulin

Insulin is a small protein, and human insulin has a molecular weight of 5808. It is composed of two amino acid chains, illustrated in Figure 78–2, connected to each other by disulfide linkages. When the two amino acid chains are split apart, the functional activity of the insulin molecule is lost.

855

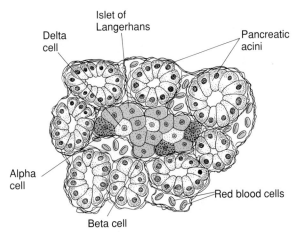

Figure 78–1. Physiologic anatomy of the pancreas.

Insulin is synthesized in the beta cells by the usual cell machinery for protein synthesis, as explained in Chapter 3, beginning with translation of the insulin RNA by ribosomes attached to the endoplasmic reticulum to form an *insulin preprohormone.* This initial preprohormone has a molecular weight of about 11,500, but it is then cleaved in the endoplasmic reticulum to form a *proinsulin* with a molecular weight of about 9000; most of this is further cleaved in the Golgi apparatus to form insulin before being packaged in the secretory granules. However, about one sixth of the final secreted product is still in the form of proinsulin. Unfortunately, the proinsulin has no insulin activity.

When insulin is secreted into the blood, it circulates almost entirely in an unbound form; it has a plasma half-life averaging only about 6 minutes, so that it is mainly cleared from the circulation within 10 to 15 minutes. Except for that portion of the insulin that combines with receptors in the target cells, the remainder is mainly degraded by the enzyme insulinase in the liver and to a lesser extent in the kidneys. This rapid removal from the plasma is important because at times it is equally as important to turn off rapidly as to turn on the control functions of insulin.

Activation of Target Cell Receptors by Insulin, and the Resulting Cellular Effects

To initiate its effects on target cells, insulin first binds with and activates a membrane receptor protein having a molecular weight of about 300,000. It is the activated receptor, not the insulin, that causes the subsequent effects.

The insulin receptor is a combination of four separate subunits held together by disulfide linkages, *two alpha subunits* that lie entirely outside the cell membrane and *two beta subunits* that penetrate through the membrane with an end of each of these protruding into the cell cytoplasm. The insulin binds with the alpha subunits on the outside of the cell; but because of the linkages with the beta subunits, the portions of the beta subunits protruding into the cell become autophosphorylated. This makes them become an activated enzyme, a local *protein kinase,* which in turn causes phosphorylation of multiple other cytosol enzymes. The net effect is to activate some of these enzymes while inactivating others. Thus, in this roundabout way, insulin directs the intracellular metabolic machinery to produce the desired effects. Unfortunately, from this point on, the molecular mechanisms are almost entirely unknown.

Nevertheless, the end effects of insulin stimulation are clear. These are basically the following:

1. Within seconds after insulin binds with its membrane receptors, the membranes of muscle cells, adipose cells, and many other types of cells in the body — constituting about 80 per cent of all cells — become highly permeable to glucose. This allows rapid entry of glucose into the cells. Inside the cell, the glucose is immediately phosphorylated and becomes a substrate for all the usual carbohydrate metabolic functions. The increased glucose transport is believed to result from opening of gates in a glucose transport protein, a membrane protein with a molecular weight of about 55,000.

2. In addition to increased membrane permeability for glucose, the cell membrane also becomes more permeable for many of the amino acids, potassium ions, magnesium ions, and phosphate ions.

3. Slower effects occur during the next 10 to 15 minutes to change the activity levels of still many more intracellular metabolic enzymes. These effects result mainly from the changed states of phosphorylation of the enzymes.

4. Still much slower effects continue to occur for hours and even for several days. These result from changed rates of translation of messenger RNAs at the ribosomes to form new proteins and still slower effects from changed rates of transcription of DNA in the cell nucleus. In this way, insulin remolds much of the cellular enzymatic machinery to achieve its metabolic goals.

Gly•Ileu•Val•Glu•Glu•Cy•Cy•Thr•Ser•Ileu•Cy•Ser•Leu•Tyr•Glu•Leu•Glu•Asp•Tyr•Cy•Asp

Phe•Val•Asp•Glu•His•Leu•Cy•Gly•Ser•His•Leu•Val•Glu•Ala•Leu•Tyr•Leu•Val•Cy•Gly•Glu•Arg•Gly•Phe•Phe•Tyr•Thr•Pro•Lys•Thr

Figure 78–2. The human insulin molecule.

EFFECT OF INSULIN ON CARBOHYDRATE METABOLISM

Immediately after a high carbohydrate meal, the glucose that is absorbed into the blood causes rapid secretion of insulin, which is discussed in detail later in the chapter. The insulin in turn causes rapid uptake, storage, and use of glucose by almost all tissues of the body, but especially by the muscles, adipose tissue, and liver.

Effect of Insulin in Promoting Glucose Metabolism in Muscle

During much of the day, muscle tissue depends not on glucose for its energy but instead on fatty acids. The principal reason for this is that the normal *resting muscle* membrane is only slightly permeable to glucose except when the muscle fiber is stimulated by insulin; and between meals, the amount of insulin that is secreted is too small to promote significant amounts of glucose entry into the muscle cells.

However, under two conditions the muscles do utilize large amounts of glucose. One of these is during periods of moderate to heavy exercise. This usage of glucose does not require large amounts of insulin because exercising muscle fibers, for reasons not understood, become highly permeable to glucose even in the absence of insulin because of the contraction process itself.

The second condition for muscle usage of large amounts of glucose is during the few hours after a meal. At this time the blood glucose concentration is high; also, the pancreas is secreting large quantities of insulin, and the extra insulin causes rapid transport of glucose into the muscle cells. This causes the muscle cell during this period of time to utilize carbohydrates, and to do so preferentially over fatty acids because the flow of fatty acids from the adipose tissue is strongly inhibited by insulin, as we discuss later.

Storage of Glycogen in Muscle. If the muscles are not exercising during the period following a meal and yet glucose is transported into the muscle cells in abundance, then most of the glucose is stored in the form of muscle glycogen instead of being used for energy, up to a limit of about 2 per cent concentration. The glycogen can later be used for energy by the muscle. It is especially useful for short periods of extreme energy use by the muscles and even to provide spurts of anaerobic energy for a few minutes at a time by glycolytic breakdown of the glycogen to lactic acid, which can occur even in the absence of oxygen.

Insulin Facilitation of Glucose Transport Through the Muscle Cell Membrane. Insulin has a direct effect on the muscle cell membrane to *facilitate glucose transport*. This is illustrated by the experimental results depicted in Figure 78–3. The lower curve labeled "control" shows the concentration of free glucose measured inside the cell, illustrating that the glucose concentration remained almost exactly zero despite increased extracellular glucose concen-

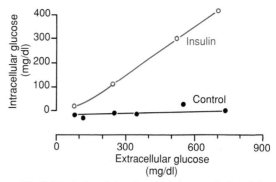

Figure 78–3. Effect of insulin in enhancing the concentration of glucose inside muscle cells. Note that in the absence of insulin (control) the intracellular glucose concentration remained near zero despite very high extracellular glucose concentrations of glucose. (From Park, Morgan, Kaji, and Smith, *in* Eisenstein [ed.]: The Biochemical Aspects of Hormone Action. Boston, Little, Brown & Co.)

tration up to as high as 750 mg/dl. In contrast, the curve labeled "insulin" illustrates that the intracellular glucose concentration rose to as high as 400 milligrams when insulin was added. Thus, it is clear that insulin can increase the rate of transport of glucose into the resting muscle cell by at least 10- to 20-fold.

Effect of Insulin on Promoting Liver Uptake, Storage, and Use of Glucose

One of the most important of all the effects of insulin is to cause most of the glucose absorbed after a meal to be stored almost immediately in the liver in the form of glycogen. Then, between meals, when food is not available and the blood glucose concentration begins to fall, the liver glycogen is split back into glucose, which is released back into the blood to keep the blood glucose concentration from falling too low.

The mechanism by which insulin causes glucose uptake and storage in the liver includes several almost simultaneous steps:

1. Insulin *inhibits liver phosphorylase*, the enzyme that causes liver glycogen to split into glucose. This obviously prevents breakdown of the glycogen that is already in the liver cells.

2. Insulin causes *enhanced uptake of glucose* from the blood by the liver cells. It does this by *increasing the activity of the enzyme glucokinase*, which is the enzyme that causes the initial phosphorylation of glucose after it diffuses into the liver cells. Once phosphorylated, the glucose is *temporarily* trapped inside the liver cells, because phosphorylated glucose cannot diffuse back through the cell membrane.

3. Insulin also increases the activities of the enzymes that promote glycogen synthesis, including *phosphofructokinase*, which causes the second stage in the phosphorylation of the glucose molecule and *glycogen synthetase*, which is responsible for polymerization of the monosaccharide units to form the glycogen molecules.

The net effect of all these actions is to increase the amount of glycogen in the liver. The glycogen can increase to a total of about 5 to 6 per cent of the liver mass, which is equivalent to almost 100 grams of stored glycogen.

Release of Glucose From the Liver Between Meals. After the meal is over and the blood glucose level begins to fall to a low level, several events now transpire that cause the liver to release glucose back into the circulating blood:

1. The decreasing blood glucose causes the pancreas to decrease its insulin secretion.

2. The lack of insulin then reverses all the effects listed above for glycogen storage, essentially stopping further synthesis of glycogen in the liver and preventing further uptake of glucose by the liver from the blood.

3. The lack of insulin (along with increase of glucagon, which will be discussed later) activates the enzyme *phosphorylase*, which causes the splitting of glycogen into *glucose phosphate.*

4. The enzyme *glucose phosphatase,* which has been inhibited by insulin, now becomes activated by the insulin lack and causes the phosphate radical to split away from the glucose; and this allows the free glucose to diffuse back into the blood.

Thus, the liver removes glucose from the blood when it is present in excess after a meal and returns it to the blood when it is needed between meals. Ordinarily, about 60 per cent of the glucose in the meal is stored in this way in the liver and then returned later.

Other Effects of Insulin on Carbohydrate Metabolism in the Liver. When the quantity of glucose entering the liver cells is more than can be stored as glycogen, *insulin promotes the conversion of all of this excess glucose into fatty acids.* These fatty acids are subsequently packaged as triglycerides in very low density lipoproteins and transported to the adipose tissue and deposited as fat.

Insulin also *inhibits gluconeogenesis.* It does this mainly by decreasing the quantities and activities of the liver enzymes required for gluconeogenesis. However, part of the effect is caused by an action of insulin that decreases the release of amino acids from muscle and other extrahepatic tissues and in turn the availability of the necessary precursors required for gluconeogenesis. This will be discussed further in relation to the effect of insulin on protein metabolism.

Lack of Effect of Insulin on Glucose Uptake and Usage by the Brain

The brain is quite different from most other tissues of the body in that insulin has either little or no effect on uptake or use of glucose. Instead, *the brain cells are permeable to glucose without the intermediation of insulin.*

The brain cells are also quite different from most other cells of the body in that they normally use only glucose for energy. Therefore, it is essential that the blood glucose level be maintained always above a critical level, which is one of the important functions of the blood glucose control system. When the blood glu-

cose does fall too low, into the range of 20 to 50 mg/dl, symptoms of *hypoglycemic shock* develop, characterized by progressive nervous irritability that leads to fainting, convulsions, and even coma.

Effect of Insulin on Carbohydrate Metabolism in Other Cells

Insulin increases glucose transport into and glucose usage by most other cells of the body (with the exception of the liver and brain cells, as noted) in the same way that it affects glucose transport through the muscle cell membrane. The transport of glucose into adipose cells is essential for providing the glycerol portion of the fat molecule for deposition of fat in these cells.

EFFECT OF INSULIN ON FAT METABOLISM

Though not quite as visible as the acute effects of insulin on carbohydrate metabolism, insulin also affects fat metabolism in ways that, in the long run, are equally as important. Especially dramatic is the long-term effect of insulin lack in causing extreme atherosclerosis, often leading to heart attacks, cerebral strokes, and other vascular accidents. But, first, let us discuss the acute effects of insulin on fat metabolism.

Effect of Insulin Excess on Fat Synthesis and Storage

Insulin has several different effects that lead to fat storage in adipose tissue. Insulin increases the utilization of glucose by most of the body's tissues, which automatically decreases the utilization of fat, thus functioning as a "fat sparer." However, insulin also promotes fatty acid synthesis. Almost all of this synthesis occurs in the liver cells, and the fatty acids are then transported in the lipoproteins to the adipose cells to be stored. However, a minute part of the synthesis occurs in the fat cells themselves. The different factors that lead to increased fatty acid synthesis in the liver include the following:

1. Insulin increases the transport of glucose into the liver cells. After the liver glycogen concentration reaches 5 to 6 per cent, this in itself inhibits further glycogen synthesis. Then all the additional glucose entering the liver cells becomes available to form fat. The glucose is first split to pyruvate in the glycolytic pathway, and the pyruvate subsequently is converted to acetyl coenzyme A (acetyl-CoA), the substrate from which fatty acids are synthesized.

2. An excess of *citrate* and *isocitrate ions* is formed by the citric acid cycle when excess amounts of glucose are being used for energy. These ions then have a direct effect in activating *acetyl-CoA carboxylase,* the enzyme required to carboxylate acetyl-CoA to form malonyl-CoA, the first stage of fatty acid synthesis.

3. Most of the fatty acids are then synthesized within the liver itself and used to form triglycerides,

the usual form of storage fat. These are released from the liver cells to the blood in the lipoproteins. Insulin activates *lipoprotein lipase* in the capillary walls of the adipose tissue, which splits the triglycerides again into fatty acids, a requirement for them to be absorbed into the adipose cells, where they are again converted to triglycerides and stored.

Storage of Fat in the Adipose Cells. Insulin also has two other essential effects that are required for fat storage in adipose cells:

1. Insulin *inhibits the action of hormone-sensitive lipase.* This is the enzyme that causes hydrolysis of the triglycerides already stored in the fat cells. Therefore, the release of fatty acids into the circulating blood is inhibited.

2. Insulin *promotes glucose transport through the cell membrane into the fat cells* in exactly the same way that it promotes glucose transport into muscle cells. Some of this glucose is then utilized to synthesize small amounts of fatty acids; but, more important, it also forms large quantities of the substance *α-glycero-phosphate.* This substance supplies the *glycerol* that combines with fatty acids to form the triglycerides that are the storage form of fat in adipose cells. Therefore, when insulin is not available, even storage of the large amounts of fatty acids transported from the liver in the lipoproteins is almost totally blocked.

Increased Metabolic Use of Fat Caused by Insulin Lack

All aspects of fat breakdown and use for providing energy are greatly *enhanced in the absence of insulin.* This occurs even normally between meals when secretion of insulin is minimal, but it becomes extreme in diabetes mellitus when secretion of insulin is almost zero. The resulting effects are as follows:

Lipolysis of Storage Fat and Release of Free Fatty Acids During Insulin Lack. In the absence of insulin, all the effects of insulin noted above causing storage of fat are reversed. The most important effect is that the enzyme *hormone-sensitive lipase* in the fat cells becomes strongly activated. This causes hydrolysis of the stored triglycerides, releasing large quantities of fatty acids and glycerol into the circulating blood. Consequently, the plasma concentration of free fatty acids begins to rise within minutes. This free fatty acid then becomes the main energy substrate used by essentially all tissues of the body besides the brain. Figure 78–4 illustrates the effect of insulin lack on the plasma concentration of free fatty acids, glucose, and acetoacetic acid. Note that immediately after removal of the pancreas the free fatty acid concentration in the plasma begins to rise, rising considerably more rapidly even than the concentration of glucose.

Effect of Insulin Lack on Plasma Cholesterol and Phospholipid Concentrations. The excess of fatty acids in the plasma also promotes liver conversion of some of the fatty acids into phospholipids and cholesterol, two of the major products of fat metabo-

Figure 78–4. Effect of removing the pancreas on the concentrations of blood glucose, plasma free fatty acids, and acetoacetic acid.

lism. These two substances, along with excess triglycerides formed at the same time in the liver, are then discharged into the blood in the lipoproteins. Occasionally, the plasma lipoproteins increase as much as threefold in the absence of insulin, giving a total concentration of plasma lipids of several per cent rather than the normal 0.6 per cent. This high lipid concentration—especially the high concentration of cholesterol—leads to rapid development of atherosclerosis in persons with serious diabetes.

Ketogenic and Acidotic Effects of Insulin Lack. Insulin lack also causes excessive amounts of *acetoacetic acid* to be formed in the liver cells. This results from the following effect: In the absence of insulin but in the presence of excess fatty acids in the liver cells, the carnitine transport mechanism for transporting fatty acids into the mitochondria becomes increasingly activated. In the mitochondria, beta oxidation of the fatty acids then proceeds exceedingly rapidly, releasing extreme amounts of acetyl-CoA. A large part of this excess acetyl-CoA is then condensed to form acetoacetic acid, which, in turn, is released into the circulating blood. Most of this passes to the peripheral cells, where it is again converted into acetyl-CoA and used for energy in the usual manner.

However, the absence of insulin also depresses the utilization of acetoacetic acid in the peripheral tissues. At the same time, so much acetoacetic acid is released from the liver that it cannot all be metabolized by the tissues. Therefore, as illustrated in Figure 78–4, its concentration rises during the days following cessation of insulin secretion, sometimes reaching concentrations as high as 10 mEq/liter or more. As explained in Chapter 68, some of the acetoacetic acid is also converted into β-hydroxybutyric acid and *acetone.* These two substances, along with the acetoacetic acid, are called *ketone bodies,* and their presence in large quantities in the body fluids is called *ketosis.* We see later that in severe diabetes the acetoacetic acid and the β-hydroxybutyric acid can cause severe *acidosis* and *coma,* which often leads to death.

EFFECT OF INSULIN ON PROTEIN METABOLISM AND GROWTH

Effect of Insulin on Protein Synthesis and Storage. During the few hours following a meal when excess quantities of nutrients are available in the circulating blood, not only carbohydrates and fats but proteins as well are stored in the tissues; insulin is required for this to occur. The manner in which insulin causes protein storage is not as well understood as the mechanisms for both glucose and fat storage. Some of the facts follow:

1. Insulin causes active transport of many of the amino acids into the cells. Among the amino acids most strongly transported are *valine, leucine, isoleucine, tyrosine,* and *phenylalanine.* Thus, insulin shares with growth hormone the capability of increasing the uptake of amino acids into cells. However, the amino acids affected are not necessarily the same ones.

2. Insulin has a direct effect on the ribosomes in *increasing the translation of messenger RNA,* thus forming new proteins. In some unexplained way, insulin "turns on" the ribosomal machinery. In the absence of insulin the ribosomes simply stop working, almost as if insulin operates an "on-off" mechanism.

3. Over a longer period of time insulin also *increases the rate of transcription of selected DNA genetic sequences* in the cell nuclei, thus forming increased quantities of RNA and still more protein synthesis—especially promoting a vast array of enzymes for storage of carbohydrates, fats, and proteins.

4. Insulin also *inhibits the catabolism of proteins,* thus decreasing the rate of amino acid release from the cells, especially from the muscle cells. Presumably this results from some ability of the insulin to diminish the normal degradation of proteins by the cellular lysosomes.

5. In the liver, insulin *depresses the rate of gluconeogenesis.* It does this by decreasing the activity of the enzymes that promote gluconeogenesis. Because the substrates most used for synthesis of glucose by the process of gluconeogenesis are the plasma amino acids, this suppression of gluconeogenesis conserves the amino acids in the protein stores of the body.

In summary, insulin promotes protein formation and also prevents the degradation of proteins.

Protein Depletion and Increased Plasma Amino Acids Caused by Insulin Lack. Virtually all protein storage comes to a complete halt when insulin is not available. The catabolism of proteins increases, protein synthesis stops, and large quantities of amino acids are dumped into the plasma. The plasma amino acid concentration rises considerably, and most of the excess amino acids are either used directly for energy or as substrates for gluconeogenesis. This degradation of the amino acids also leads to enhanced urea excretion in the urine. The resulting protein wasting is one of the most serious of all of the effects of severe diabetes mellitus. It can lead to extreme weakness as well as to many deranged functions of the organs.

Effect of Insulin on Growth—Its Synergistic Effect With Growth Hormone. Because insulin is required for the synthesis of proteins, it is equally as essential for growth of an animal as is growth hormone. This is illustrated in Figure 78–5, which shows that a depancreatized, hypophysectomized rat without therapy hardly grows at all. Furthermore, administration of either growth hormone or insulin one at a time causes almost no growth. Yet a combination of both of these hormones does cause dramatic growth. Thus it appears that the two hormones function synergistically to promote growth, each performing a specific function that is separate from that of the other. Perhaps part of this necessity for both hormones results from the fact that each promotes cellular uptake of a different selection of amino acids, all of which are required if growth is to be achieved.

CONTROL OF INSULIN SECRETION

Formerly, it was believed that insulin secretion is controlled almost entirely by the blood glucose concentration. However, as more has been learned about the metabolic functions of insulin for protein and fat metabolism, it has been learned that blood amino acids and other factors also play important roles in controlling insulin secretion.

Stimulation of Insulin Secretion by Blood Glucose. At the normal fasting level of blood glucose of 80 to 90 mg/dl, the rate of insulin secretion is minimal—on the order of 25 ng/min/kg of body weight. If the blood glucose concentration is suddenly increased to a level two to three times normal and is kept at this high level thereafter, insulin secretion increases markedly in two stages, as shown by the changes in plasma insulin concentration seen in Figure 78–6:

1. Plasma insulin concentration increases almost tenfold within 3 to 5 minutes after acute elevation of the blood glucose; this results from immediate dumping of preformed insulin from the beta cells of the islets of Langerhans. However, the initial high rate of

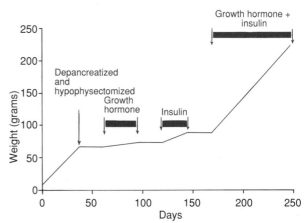

Figure 78–5. Effect of growth hormone, insulin, and growth hormone plus insulin on growth in a depancreatized and hypophysectomized rat.

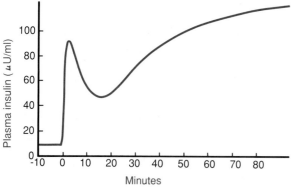

Figure 78–6. Increase in plasma insulin concentration following a sudden increase in blood glucose to two to three times the normal range. Note an initial rapid surge in insulin concentration and then a delayed but higher and continuing increase in concentration beginning 15 to 20 minutes later.

secretion is not maintained; instead, the insulin concentration decreases about halfway back toward normal in another 5 to 10 minutes.

2. After about 15 minutes, insulin secretion rises a second time and reaches a new plateau in 2 to 3 hours, this time usually at a rate of secretion even greater than that in the initial phase. This secretion results both from additional release of preformed insulin and from activation of the enzyme system that synthesizes and releases new insulin from the cells.

Feedback Relationship Between Blood Glucose Concentration and Insulin Secretion Rate. As the concentration of blood glucose rises above 100 mg/dl of blood, the rate of insulin secretion rises rapidly, reaching a peak some 10 to 25 times the basal level at blood glucose concentrations between 400 and 600 mg/dl, as illustrated in Figure 78–7. Thus, the increase in insulin secretion under a glucose stimulus is dramatic both in its rapidity and in the tremendous level of secretion achieved. Furthermore, the turn-off of insulin secretion is almost equally as rapid, occurring within minutes after reduction in blood glucose concentration back to the fasting level.

This response of insulin secretion to an elevated blood glucose concentration provides an extremely important feedback mechanism for regulating blood glucose concentration. That is, any rise in blood glucose increases insulin secretion, and the insulin in turn causes transport of glucose into the liver, muscle, and other cells, thereby reducing the blood glucose concentration back toward the normal value.

OTHER FACTORS THAT STIMULATE INSULIN SECRETION

Amino Acids. In addition to excess glucose stimulating insulin secretion, many of the amino acids have a similar effect. The most potent of these are *arginine* and *lysine*. However, this effect differs from glucose stimulation of insulin secretion in the following way: Amino acids administered in the absence of a rise in blood glucose cause only a small increase in insulin secretion. However, when administered at the same time that the blood glucose concentration is elevated, the glucose-induced secretion of insulin may be as much as doubled in the presence of the excess amino acids. Thus, the amino acids very strongly potentiate the glucose stimulus for insulin secretion.

The stimulation of insulin secretion by amino acids seems to be a purposeful response because the insulin in turn promotes transport of the amino acids into the tissue cells and also promotes intracellular formation of protein. That is, the insulin is important for proper utilization of excess amino acids in the same way that it is important for the utilization of carbohydrates.

Gastrointestinal Hormones. A mixture of several important gastrointestinal hormones — *gastrin, secretin, cholecystokinin,* and *gastric inhibitory peptide* (which seems to be the most potent of them all) — will cause a moderate increase in insulin secretion. These hormones are released in the gastrointestinal tract after a person eats a meal. They then cause an "anticipatory" increase in blood insulin in preparation for the glucose and amino acids to be absorbed from the meal. These gastrointestinal hormones can almost double the rate of insulin secretion as the blood glucose level rises.

Other Hormones and the Autonomic Nervous System. Other hormones that either directly increase insulin secretion or potentiate the glucose stimulus for insulin secretion include *glucagon, growth hormone, cortisol,* and to a lesser extent *progesterone* and *estrogen.* The importance of the stimulatory effects of these hormones is that prolonged secretion of any one of them in large quantities can occasionally lead to exhaustion of the beta cells of the islets of Langerhans and thereby cause diabetes mellitus. Indeed, diabetes often occurs in persons maintained on high pharmacological doses of some of these hormones. Diabetes is particularly common in giants or acromegalic persons with growth hormone – secreting tumors or in persons whose adrenal glands or adrenal gland tumors secrete excess glucocorticoids.

Under some conditions, stimulation of either the parasympathetic or the sympathetic nerves to the pancreas can also increase insulin secretion. However, it is doubtful that either of these effects is of physiological significance for regulating insulin secretion.

ROLE OF INSULIN (AND OTHER HORMONES) IN "SWITCHING" BETWEEN CARBOHYDRATE AND LIPID METABOLISM

From the preceding discussions it should already be clear that insulin promotes the utilization of carbohy-

Figure 78–7. Approximate increase in insulin secretion at different plasma glucose levels.

drates for energy, whereas it depresses the utilization of fats. Conversely, lack of insulin causes fat utilization mainly to the exclusion of glucose utilization, except by brain tissue. Furthermore, the signal that controls this switching mechanism is principally the blood glucose concentration. When the glucose concentration is low, insulin secretion is suppressed, and fat is utilized almost exclusively for energy everywhere except in the brain. When the glucose concentration is high, insulin secretion is stimulated, and carbohydrate is utilized instead of fat, and the excess blood glucose is stored in the form of liver glycogen or muscle glycogen. Therefore, one of the most important functional roles of insulin in the body is to control which of these two foods from moment to moment will be utilized by the cells for energy.

At least four other known hormones also play important roles in this switching mechanism: *growth hormone* from the anterior pituitary gland, *cortisol* from the adrenal cortex, *epinephrine* from the adrenal medulla, and *glucagon* from the alpha cells of the islets of Langerhans in the pancreas. Glucagon is discussed in the following section of this chapter. Both growth hormone and cortisol are secreted in response to hypoglycemia, and both inhibit cellular utilization of glucose while promoting fat utilization. However, the effects of both of these hormones develop very slowly, usually requiring many hours to obtain maximal levels.

Epinephrine is especially important in increasing plasma glucose concentration during periods of stress when the sympathetic nervous system is excited. However, epinephrine acts differently from the other hormones in that it increases the plasma fatty acid concentration at the same time. The reasons for these effects are as follows: (1) epinephrine has the very potent effect of causing glycogenolysis in the liver, thus releasing large quantities of glucose into the blood; and (2) it also has a direct lipolytic effect on the adipose cells because it activates adipose tissue hormone-sensitive lipase, thus greatly enhancing the blood concentration of fatty acids as well. Quantitatively, the enhancement of fatty acids is far greater than the enhancement of blood glucose. Therefore, epinephrine especially enhances the utilization of fat in such stressful states as exercise, circulatory shock, anxiety, and so forth.

● GLUCAGON AND ITS FUNCTIONS

Glucagon, a hormone secreted by the *alpha cells* of the islets of Langerhans when the blood glucose concentration falls, has multiple functions that are diametrically opposed to those of insulin. Most important of these is to increase the blood glucose concentration, an effect which is exactly the opposite of that of insulin.

Like insulin, glucagon is a large polypeptide. It has a molecular weight of 3485 and is composed of a chain of 29 amino acids. On injection of purified glucagon into an animal, a profound *hyper*glycemic effect occurs. Only 1 microgram per kilogram of glucagon can elevate the blood glucose concentration approximately 20 mg/dl of blood in about 20 minutes. For this reason, glucagon is also called the *hyperglycemic hormone*.

EFFECTS ON GLUCOSE METABOLISM

The two major effects of glucagon on glucose metabolism are (1) breakdown of liver glycogen (*glycogenolysis*) and (2) increased *gluconeogenesis* in the liver. Both of these effects greatly enhance the availability of glucose to the other organs of the body.

Glycogenolysis and Increased Blood Glucose Concentration Caused by Glucagon. The most dramatic effect of glucagon is its ability to cause glycogenolysis in the liver, which in turn increases the blood glucose concentration within minutes.

It does this by the following complex cascade of events:

1. Glucagon activates *adenyl cyclase* in the hepatic cell membrane,
2. Which causes the formation of *cyclic AMP*,
3. Which activates *protein kinase regulator protein*,
4. Which activates *protein kinase*,
5. Which activates *phosphorylase b kinase*,
6. Which converts *phosphorylase b* into *phosphorylase a*,
7. Which promotes the degradation of glycogen into glucose-1-phosphate,
8. Which then is dephosphorylated; and the glucose is released from the liver cells.

This sequence of events is exceedingly important for several reasons. First, it is one of the most thoroughly studied of all the *second messenger* functions of cyclic adenosine monophosphate. Second, it illustrates a cascading system in which *each succeeding product is produced in greater quantity than the preceding product*. Therefore, it represents a potent *amplifying* mechanism. This explains how only a few micrograms of glucagon can cause the blood glucose level to double or rise even more within a few minutes.

Infusion of glucagon for about 4 hours can cause such intensive liver glycogenolysis that all the liver stores of glycogen become totally depleted.

Gluconeogenesis Caused by Glucagon. Even after all the glycogen in the liver has been exhausted under the influence of glucagon, continued infusion of this hormone still causes continued hyperglycemia. This results from the effect of glucagon of increasing the rate of gluconeogenesis in the liver cells. This is achieved by activating multiple enzymes that are required for gluconeogenesis, especially activation of the enzyme system for converting pyruvate to phosphoenolpyruvate, a rate-limiting step in gluconeogenesis. Also, glucagon increases the extraction of amino acids from the blood by the liver cells, thus making a greater quantity of these available to be converted into glucose.

Other Effects of Glucagon

Most other effects of glucagon occur only when its concentration rises well above that normally in the blood. Perhaps the most important effect is that glucagon *activates adipose cell lipase,* making increased quantities of fatty acids available to the energy systems of the body. It also inhibits the storage of triglycerides in the liver, which prevents the liver from removing fatty acids from the blood; this also helps make additional amounts of fatty acids available for the other tissues of the body.

Glucagon in very large concentrations also (1) enhances the strength of the heart, (2) enhances bile secretion, and (3) inhibits gastric acid secretion. However, all these effects are probably unimportant in the normal function of the body.

REGULATION OF GLUCAGON SECRETION

Inhibitory Effect of Blood Glucose. The blood glucose concentration is by far the most potent factor controlling glucagon secretion. Note specifically, however, that *the effect of blood glucose concentration on glucagon secretion is in exactly the opposite direction from the effect of glucose on insulin secretion.* This is illustrated in Figure 78–8, showing that a *decrease* in the blood glucose concentration from its normal fasting level of about 90 mg/dl of blood down to hypoglycemic levels can increase the plasma concentration of glucagon severalfold. On the other hand, increasing the blood glucose to hyperglycemic levels decreases plasma glucagon. Thus, in hypoglycemia, glucagon is secreted in large amounts; it then greatly increases the output of glucose from the liver and thereby serves the very important function of correcting the hypoglycemia.

Excitatory Effect of Amino Acids. High concentrations of amino acids, as occur in the blood after a protein meal (especially the amino acids *alanine* and *arginine*), stimulate the secretion of glucagon. This is the same effect that amino acids have in stimulating insulin secretion. Thus, in this instance the glucagon and insulin responses are not opposites.

The importance of amino acid stimulation of glucagon secretion is that the glucagon then promotes

rapid conversion of the amino acids to glucose, thus making even more glucose available to the tissues.

Excitatory Effect of Exercise. In exhaustive exercise the blood concentration of glucagon often increases as much as four- to fivefold. However, what causes this is not understood, because the blood glucose concentration does not necessarily fall. Nevertheless, a beneficial effect of the glucagon is that it prevents a decrease in blood glucose. One of the factors that might increase glucagon secretion in exercise is increased circulating amino acids. However, other factors, such as nervous stimulation of the islets of Langerhans, could also play a role.

● SOMATOSTATIN—ITS EFFECT ON INHIBITION OF GLUCAGON AND INSULIN SECRETION

The *delta cells* of the islets of Langerhans secrete the hormone *somatostatin,* which is a polypeptide containing only 14 amino acids that has an extremely short half-life in the circulating blood of only three minutes. Almost all factors related to the ingestion of food will stimulate somatostatin secretion. These include (1) increased blood glucose, (2) increased amino acids, (3) increased fatty acids, and (4) increased concentrations of several of the gastrointestinal hormones released from the upper gastrointestinal tract in response to food intake.

In turn, somatostatin has multiple inhibitory effects as follows:

1. Somatostatin acts locally within the islets of Langerhans themselves to depress the secretion of both insulin and glucagon.

2. Somatostatin decreases the motility of the stomach, the duodenum, and the gallbladder.

3. Somatostatin decreases both secretion and absorption in the gastrointestinal tract.

Putting all this information together, it has been suggested that the principal role of somatostatin is to extend the period of time over which the food nutrients are assimilated into the blood. At the same time, the effect of somatostatin to depress insulin and glucagon secretion decreases the utilization of the absorbed nutrients by the tissues, thus preventing rapid exhaustion of the food and therefore making it available over a longer period of time.

It should also be recalled that somatostatin is the same chemical substance as *growth hormone inhibitory hormone,* which is secreted in the hypothalamus and suppresses anterior pituitary gland growth hormone secretion.

● SUMMARY OF BLOOD GLUCOSE REGULATION

In the normal person the blood glucose concentration is very narrowly controlled, usually in a range between 80 and 90 mg/dl of blood in the fasting person each morning before breakfast. This concentration increases to 120 to 140 mg/dl during the first hour or so following a meal, but the feedback systems for control of blood glucose return the glucose concentration very rapidly back to the control level, usually within 2 hours after the last absorption of carbohydrates. Con-

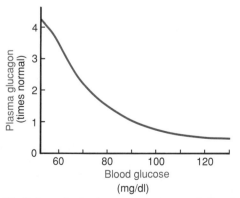

Figure 78–8. Approximate plasma glucagon concentration at different blood glucose levels.

versely, in starvation the gluconeogenesis function of the liver provides the glucose that is required to maintain the fasting blood glucose level.

The mechanisms for achieving this high degree of control have been presented in this chapter. Let us summarize these briefly:

1. The liver functions as a very important *blood glucose buffer system.* That is, when the blood glucose rises to a very high concentration following a meal and the rate of insulin secretion also increases, as much as two thirds of the glucose absorbed from the gut is almost immediately stored in the liver in the form of glycogen. Then, during the succeeding hours, when both the blood glucose concentration and the rate of insulin secretion fall, the liver releases the glucose back into the blood. In this way, the liver decreases the fluctuations in blood glucose concentration to about one-third what they otherwise would be. In fact, in patients with severe liver disease, it becomes almost impossible to maintain a narrow range of blood glucose concentration.

2. It is very clear that both insulin and glucagon function as important feedback control systems for maintaining a normal blood glucose concentration. When the concentration rises to a level too high, insulin is secreted; the insulin in turn causes the blood glucose concentration to decrease toward normal. Conversely, a decrease in blood glucose stimulates glucagon secretion; the glucagon then functions in the opposite direction to increase the glucose up toward normal. Under most normal conditions, the insulin feedback mechanism is much more important than the glucagon mechanism, but in instances of diminished glucose intake or excessive utilization of glucose during exercise and other stressful situations, the glucagon mechanism then also becomes very valuable.

3. Also, in hypoglycemia, a direct effect of low blood glucose on the hypothalamus stimulates the sympathetic nervous system. In turn, the epinephrine secreted by the adrenal glands causes still further release of glucose from the liver. This, too, helps protect against severe hypoglycemia.

4. And, finally, over a period of hours and days, both growth hormone and cortisol are secreted in response to prolonged hypoglycemia, and they both decrease the rate of glucose utilization by most cells of the body. This, too, helps return the blood glucose concentration toward normal.

Importance of Blood Glucose Regulation. One might ask the question: Why is it so important to maintain a constant blood glucose concentration, particularly since most tissues can shift to utilization of fats and proteins for energy in the absence of glucose? The answer is that glucose is the only nutrient that normally can be utilized by the *brain, retina,* and *germinal epithelium of the gonads* in sufficient quantities to supply them with their required energy. Therefore, it is important to maintain the blood glucose concentration at a sufficiently high level to provide this necessary nutrition.

Most of the glucose formed by gluconeogenesis during the interdigestive period is used for metabolism in the brain. Indeed, it is important that the pancreas not secrete any insulin during this time, for otherwise the scant supplies of glucose that are available would all go into the muscles and other peripheral tissues, leaving the brain without a nutritive source.

On the other hand, it is also important that the blood glucose concentration not rise too high for three reasons: First, glucose exerts a large amount of osmotic pressure in the extracellular fluid; and, if the glucose concentration rises to excessive values, this can cause considerable cellular dehydration. Second, an excessively high level of blood glucose concentration causes loss of glucose in the urine. Third, this causes osmotic diuresis by the kidneys, which can deplete the body of its fluids and electrolytes.

● DIABETES MELLITUS

In most instances, diabetes mellitus results from diminished secretion of insulin by the beta cells of the islets of Langerhans. Heredity usually plays a major role in determining in whom diabetes will develop and in whom it will not. Sometimes it does this by increasing the susceptibility of the beta cells to viruses, or by favoring the development of autoimmune antibodies against the beta cells, thus leading to their destruction. In other instances there appears to be a simple hereditary tendency for beta cell degeneration.

Obesity also plays a role in the development of clinical diabetes. One reason is that in obesity the beta cells of the islets of Langerhans become less responsive to stimulation by increased blood glucose; therefore, the blood insulin levels do not increase when needed. Another reason is that obesity decreases the number of insulin receptors in the insulin target cells throughout the body, thus making the amount of insulin that is available even less effective in promoting its usual metabolic effects.

PATHOLOGICAL PHYSIOLOGY OF DIABETES MELLITUS

Most of the pathologic features of diabetes mellitus can be attributed to one of the following three major effects of the lack of insulin: (1) decreased utilization of glucose by the body cells, with a resultant increase in blood glucose concentration to as high as 300 to 1200 mg/dl; (2) markedly increased mobilization of fats from the fat storage areas, causing abnormal fat metabolism as well as deposition of cholesterol in arterial walls, causing atherosclerosis; and (3) depletion of protein in the tissues of the body.

In addition, some special pathophysiological problems occur in diabetes mellitus that are not so readily apparent.

Loss of Glucose in the Urine in Diabetes. Whenever the quantity of glucose entering the kidney tubules in the glomerular filtrate rises above a critical level, a significant proportion of the excess glucose then cannot be reabsorbed and instead spills into the urine. This normally occurs when the blood glucose concentration rises over 180 mg/dl, a level that is called the blood "threshold" for the appearance of glucose in the urine. When the blood glucose level rises to 300 to 500 mg/dl—common values in persons with severe untreated diabetes—100 or more grams of glucose can be lost into the urine each day.

Dehydrating Effect of Elevated Blood Glucose Levels in Diabetes. Blood glucose levels as high as 1200 mg/dl, 12 times normal, can occur under certain conditions in extreme untreated diabetes, and levels of 300 to 400 mg/dl are common. Yet the only significant effect of the elevated glucose is dehydration of the tissue cells, for glucose does not diffuse easily through the pores of the cell membrane, and the increased osmotic pressure in the extracellular fluids causes osmotic transfer of water out of the cells.

In addition to the direct cellular dehydrating effect of excessive glucose, the loss of glucose in the urine causes *osmotic diuresis* because the osmotic effect of glucose in the tubules greatly decreases tubular reabsorption of fluid. The overall effect is dehydration of the extracellular fluid, which also causes compensatory dehydration of the intracellular fluid for reasons discussed in Chapter 26. Thus, one of the important features of diabetes is a tendency for both extracellular and intracellular dehydration to develop, and these can contribute to the development of *circulatory shock.*

Acidosis and Coma in Diabetes. The shift from carbohydrate to fat metabolism in diabetes has already been discussed. When the body depends almost entirely on fat for energy, the level of the keto acids acetoacetic acid and β-hydroxybutyric acid in the body fluids may rise from 1 mEq/liter to as high as 10 mEq/liter. All of this extra acid, obviously, is likely to result in acidosis.

A second effect, which is usually even more important in causing acidosis than is the direct increase in keto acids, is a decrease in sodium concentration caused by the following: Keto acids have a low threshold for excretion by the kidneys; therefore, when the keto acid concentration rises in diabetes, as much as 100 to 200 grams of keto acids can be excreted in the urine each day. Because these are strong acids, having a pK averaging 4.0 or less, very little of them can be excreted in the acidic form but instead is excreted combined with sodium derived from the extracellular fluid. As a result, the sodium concentration in the extracellular fluid usually decreases, and part of the sodium is replaced by increased quantities of hydrogen ions, thus adding greatly to the acidosis.

Obviously, all the usual reactions that occur in metabolic acidosis take place in diabetic acidosis. These include *rapid and deep breathing* called "Kussmaul respiration," which causes excessive expiration of carbon dioxide, and *marked decrease in bicarbonate content of the extracellular fluids.* Although these extreme effects occur only in the most severe instances of uncontrolled diabetes, they can lead to *acidotic coma* and death within hours when the pH of the blood falls below about 7.0. The overall changes in the electrolytes of the blood as a result of severe diabetic acidosis are illustrated in Figure 78–9.

Relationship of Other Diabetic Symptoms to the Pathological Physiology of Lack of Insulin. *Polyuria* (excessive elimination of urine), *polydipsia* (excessive drinking of water), *polyphagia* (excessive eating), *loss of weight,* and *asthenia* (lack of energy) are the earliest symptoms of diabetes. As explained, the polyuria is due to the osmotic diuretic effect of glucose in the kidney tubules. In turn, the polydipsia is due to dehydration resulting from polyuria. The failure of glucose (and protein) utilization by the body causes loss of weight and a tendency toward polyphagia. The asthenia apparently also is caused mainly by loss of body protein.

PHYSIOLOGY OF DIAGNOSIS

The usual methods for diagnosing diabetes are based on various chemical tests of the urine and the blood.

Urinary Sugar. Simple office tests or more complicated quantitative laboratory tests may be used for determining

Figure 78–9. Changes in blood constituents in diabetic coma, showing normal values (light bars) and diabetic values (dark bars).

the quantity of glucose lost in the urine. In general, the normal person loses undetectable amounts of glucose, whereas the diabetic loses glucose in small to large amounts, in proportion to the severity of disease and the intake of carbohydrates.

The Fasting Blood Glucose Level. The fasting blood sugar level in the early morning is normally 80 to 90 mg/dl, and 110 mg/dl is generally considered to be the upper limit of normal. A fasting blood sugar level above this value often indicates diabetes mellitus or, less commonly, either pituitary diabetes or adrenal diabetes.

The Glucose Tolerance Test. As illustrated by the bottom curve in Figure 78–10, called a "glucose tolerance curve," when a normal, fasting person ingests 1 gram of glucose per kilogram of body weight, the blood glucose level rises from approximately 90 mg/dl to 120 to 140 mg/dl and falls back to below normal within about 2 hours.

In the diabetic person the fasting blood glucose concentration is almost always above 100 mg/dl and often above 140 mg/dl. Also, the glucose tolerance test is almost always abnormal. On ingestion of glucose, these persons exhibit a much greater than normal rise in blood glucose level, as illustrated by the upper curve, and the glucose level falls back to the control value only after 4 to 6 hours; furthermore, it fails to fall below the control level. This slow fall of

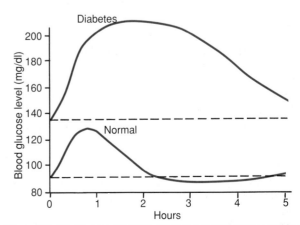

Figure 78–10. Glucose tolerance curve in a normal person and in a diabetic person.

the curve and its failure to fall below the control level illustrates that the normal increase in insulin secretion following glucose ingestion does not occur in the diabetic person, and a diagnosis of diabetes mellitus can usually be definitely established on the basis of such a curve.

Acetone Breath. As pointed out in Chapter 68, small quantities of acetoacetic acid, which increase greatly in severe diabetes, can be converted to acetone, which is volatile and is vaporized into the expired air. Consequently, one frequently can make a diagnosis of diabetes mellitus simply by smelling acetone on the breath of a patient. Also, keto acids can be detected by chemical means in the urine, and their quantitation aids in determining the severity of the diabetes.

TREATMENT OF DIABETES

The theory of treatment of diabetes mellitus is to administer enough insulin so that the patient will have as nearly normal carbohydrate, fat, and protein metabolism as possible. Insulin is available in several different forms. "Regular" insulin has a duration of action lasting from 3 to 8 hours, whereas other forms of insulin (precipitated with zinc or with various protein derivatives) are absorbed slowly from the injectate site and therefore have effects that last as long as 10 to 48 hours. Ordinarily, the severely diabetic patient is given a single dose of one of the longer-acting insulins each day to increase overall carbohydrate metabolism throughout the day. Then additional quantities of regular insulin are given at those times of the day when the blood glucose level tends to rise too high, such as at meal times. Thus, each patient is established on an individualized pattern of treatment.

Relationship of Treatment to Arteriosclerosis. Diabetic patients, mainly because of their high levels of circulating cholesterol and other lipids, develop atherosclerosis, arteriosclerosis, severe coronary heart disease, and multiple microcirculatory lesions far more easily than do normal persons. Indeed, those who have relatively poorly controlled diabetes throughout childhood are likely to die of heart disease in their twenties.

In the early days of treating diabetes it was the tendency to reduce severely the carbohydrates in the diet so that the insulin requirements would be minimized. This procedure kept the blood sugar level down to normal values and prevented loss of glucose in the urine, but it did not prevent many of the abnormalities of fat metabolism. Consequently, there is a tendency at present to allow the patient an almost normal carbohydrate diet and to give simultaneously large enough quantities of insulin to metabolize the carbohydrates. This depresses the rate of fat metabolism and also depresses the high level of blood cholesterol.

Because the complications of diabetes — such as atherosclerosis, greatly increased susceptibility to infection, diabetic retinopathy, cataracts, hypertension, and chronic renal disease — are more closely associated with the level of the blood lipids than with the level of blood glucose, it is the object of some clinics treating diabetes to administer sufficient glucose and insulin so that the quantity of blood lipids becomes normal.

● HYPERINSULINISM

Though much rarer than diabetes, increased insulin production, known as *hyperinsulinism*, does occasionally occur. This usually results from an adenoma of an islet of Langerhans. About 10 to 15 per cent of these adenomas are malignant, and occasionally metastases from the islets of Langerhans spread throughout the body, causing tremendous production of insulin by both the primary and the metastatic cancers. Indeed, in order to prevent hypoglycemia, in some of these patients more than 1000 grams of glucose have had to be administered every 24 hours.

Diagnosis of hyperinsulinism is made with assurance by measuring very high levels of plasma insulin by means of radioimmunoassay — especially when the insulin remains high constantly throughout the day without rising significantly with increased carbohydrate intake.

Insulin Shock and Hypoglycemia. As already emphasized, the central nervous system normally derives essentially all its energy from glucose metabolism, and insulin is not necessary for this use of glucose. However, if insulin causes the level of blood glucose to fall to low values, the metabolism of the central nervous system becomes depressed. Consequently, in patients with hyperinsulinism, or in diabetic patients who administer too much insulin to themselves, the syndrome called *insulin shock* may occur as follows:

As the blood sugar level falls into the range of 50 to 70 mg/dl, the central nervous system usually becomes quite excitable, for this degree of hypoglycemia facilitates neuronal activity. Sometimes various forms of hallucinations result, but more often the patient simply experiences extreme nervousness, trembles all over, and breaks out in a sweat. As the blood glucose level falls to 20 to 50 mg/dl, clonic convulsions and loss of consciousness are likely to occur. As the glucose level falls still lower, the convulsions cease, and only a state of coma remains. Indeed, at times it is difficult to distinguish between diabetic coma as a result of insulin lack and coma due to hypoglycemia caused by excess insulin. However, the acetone breath and the rapid, deep breathing of diabetic coma are not present in hypoglycemic coma.

Obviously, proper treatment for a patient who has hypoglycemic shock or coma is immediate intravenous administration of large quantities of glucose. This usually brings the patient out of shock within a minute or more. Also, administration of glucagon (or, less effectively, epinephrine) can cause glycogenolysis in the liver and thereby increase the blood glucose level extremely rapidly.

If treatment is not effected immediately, permanent damage to the neuronal cells of the central nervous system often occurs; this happens especially in prolonged hyperinsulinism due to pancreatic tumors.

REFERENCES

Baskin, D. G., et al.: Insulin in the brain. Annu. Rev. Physiol., 49:335, 1987.
Bennett, P. H.: The diagnosis of diabetes: New international classification and diagnostic criteria. Annu. Rev. Med., 34:295, 1983.
Bergeron, J. J. M., et al.: Uptake of insulin and other ligands into receptor-rich endocytic components of target cells: The endosomal apparatus. Annu. Rev. Physiol., 47:383, 1985.
Bliss, M.: Who discovered insulin? News Physiol. Sci., 1:31, 1986.
Bonner-Weir, S., and Weir, G. C.: The islets of Langerhans and diabetes mellitus. Current Concepts. Upjohn, November 1986.
Camerini-Davalos, R. A., and Cole, H. S.: Prediabetes. New York, Plenum Publishing Corp., 1988.
Cheng, K., and Larner, J.: Intracellular mediators of insulin action. Annu. Rev. Physiol., 47:405, 1985.
Cudworth, A. G., et al.: Etiology of Type I diabetes mellitus: Heterogeneity and immunological events leading to clinical onset. Annu. Rev. Med., 34:13, 1983.
Czech, M. P.: New perspectives on the mechanism of insulin action. Recent Prog. Horm. Res., 40:347, 1984.
Czech, M. P.: The nature and regulation of the insulin receptor: Structure and function. Annu. Rev. Physiol., 47:357, 1985.

DeGroot, L. J. (ed.): Endocrinology, 2nd Ed. Philadelphia, W. B. Saunders Co., 1989.

Fain, J. N.: Insulin secretion and action. Metabolism, 33:672, 1984.

Fein, F. S., and Sonnenblick, E. H.: Diabetic cardiomyopathy. Prog. Cardiovasc. Dis., 27:255, 1985.

Felig, P., et al. (eds.): Endocrinology and Metabolism, 2nd Ed. New York, McGraw-Hill Book Co., 1987.

Flier, J. S.: Insulin receptors and insulin resistance. Annu. Rev. Med., 34:145, 1983.

Gammeltoft, S.: Insulin receptors: Binding kinetics and structure-function relationship of insulin. Physiol. Rev., 64:1321, 1984.

Goren, H. J., et al.: Insulin Action and Diabetes. New York, Raven Press, 1988.

Guyton, J. R., et al.: A model of glucose-insulin homeostasis in man that incorporates the heterogeneous fast pool theory of pancreatic insulin release. Diabetes, 27:1027, 1978.

Hedeskov, C. J.: Mechanism of glucose-induced insulin secretion. Physiol. Rev., 60:442, 1980.

Hellerstrom, C.: The life story of the pancreatic β cell. Diabetologia, 26:393, 1984.

Howell, S. L., and Tyhurst, M.: Insulin secretion: The effector system. Experientia, 40:1098, 1984.

Jackson, R. L., and Guthrie, R. A.: The Physiological Management of Diabetes in Children. New York, Elsevier Science Publishing Co., 1986.

Jacobs, S., and Cuatrecasas, P.: Insulin receptors. Annu. Rev. Pharmacol. Toxicol., 23:461, 1983.

Krall, L. P., and Beaser, R.: Joslin diabetes manual, 12th Ed. Philadelphia, Lea & Febiger, 1989.

Kraus-Friedmann, N.: Hormonal regulation of hepatic gluconeogenesis. Physiol. Rev., 64:170, 1984.

Lenzen, S., and Bailey, C. J.: Thyroid hormones, gonadal and adrenocortical steroids and the function of the islets of Langerhans. Endocr. Rev., 5:411, 1984.

Malaisse, W. J., et al.: Coupling factors in nutrient-induced insulin release. Experientia, 40:1035, 1984.

Meisler, M. H., and Howard, G.: Effects of insulin on gene transcription. Annu. Rev. Physiol., 51:701, 1989.

Nishimoto, I., Kojima, I.: Calcium signalling system triggered by insulin-like growth factor II. News Physiol. Sci., 4:94, 1989.

Nujima, A., and Mei, N.: Glucose sensors in viscera and control of blood glucose level. News Physiol. Sci., 2:164, 1987.

Park, C. R., et al.: The action of insulin on the transport of glucose through the cell membrane. Am. J. Med., 26:674, 1959.

Petersen, O. H., and Findlay, I.: Electrophysiology of the pancreas. Physiol. Rev., 67:1054, 1987.

Pipeleers, D.: Islet cell interactions with pancreatic β-cells. Experientia, 40:1114, 1984.

Post, R. L., et al.: Regulation of glucose uptake in muscle III. The interaction of membrane transport and phosphorylation in the control of glucose uptake. J. Biol. Chem., 236:269, 1961.

Prentki, M., and Matschinsky, F. M.: Ca^{2+}-cAMP, and phospholipid-derived messengers in coupling mechanisms of insulin secretion. Physiol. Rev., 67:1185, 1987.

Reaven, G. M., and Reaven, E. P.: Aging, glucose intolerance, and non-insulin-dependent diabetes mellitus. Current Concepts. Upjohn, June 1986.

Reichlin, S. (ed.):Somatostatin. New York, Plenum Publishing Corp., 1987.

Rosen, O. M.: After insulin binds. Science, 237:1452, 1987.

Sonne, O.: Receptor-mediated endocytosis and degradation of insulin. Physiol. Rev., 68:1129, 1988.

Standaert, M. L., and Pollet, R. J.: Insulin-glycerolipid mediators and gene expression. FASEB J., 2:2453, 1988.

Ward, W. K., et al.: Pathophysiology of insulin secretion in noninsulin-dependent diabetes mellitus. Diabetes Care, 7:491, 1984.

Wollheim, C. B., and Sharp, G. W. G.: Regulation of insulin release by calcium. Physiol. Rev., 61:914, 1981.

CHAPTER 79

• *Parathyroid Hormone, Calcitonin, Calcium and Phosphate Metabolism, Vitamin D, Bone, and Teeth* •

The physiology of calcium and phosphate metabolism, the function of vitamin D, and the formation of bone and teeth are all tied together in a common system along with the two regulatory hormones *parathyroid hormone* and *calcitonin*. Therefore, they are discussed together in this chapter.

● CALCIUM AND PHOSPHATE IN THE EXTRACELLULAR FLUID AND PLASMA— FUNCTION OF VITAMIN D

ABSORPTION AND EXCRETION OF CALCIUM AND PHOSPHATE

Intestinal Absorption of Calcium and Phosphate. Calcium is poorly absorbed from the intestinal tract because of the relative insolubility of many of its compounds and also because bivalent cations are poorly absorbed anyway. On the other hand, phosphate is easily absorbed most of the time except when excess calcium is in the diet; the calcium tends to form almost insoluble calcium phosphate compounds that fail to be absorbed but instead pass on through the bowels to be excreted in the feces.

Excretion of Calcium in Feces and Urine; Net Rate of Absorption. About nine tenths of the daily intake of calcium is excreted in the feces, and the remaining tenth in the urine. The approximate daily turnover rates for calcium in the adult follow:

	mg
Intake	1000
Intestinal absorption	350
Secretion in gastrointestinal juices	250
Net absorption over secretion	100
Loss in the feces	900
Excretion in the urine	100

Excretion of calcium in the urine conforms to much the same principles as sodium excretion. About two thirds of the calcium in the glomerular filtrate is reabsorbed in the proximal tubules. Then, in the ascending limbs of the loops of Henle and in the distal tubules and collecting ducts, reabsorption of the remaining calcium is very selective, depending upon the calcium ion concentration in the blood. When low, this reabsorption is very great, so that almost no calcium is lost in the urine. On the other hand, even a minute increase in calcium ion concentration above normal increases calcium excretion markedly. We shall see later in the chapter that the most important factor controlling this reabsorption of calcium in the distal portions of the nephron, and therefore controlling the rate of calcium excretion, is parathyroid hormone.

Intestinal and Urinary Excretion of Phosphate. Except for the portion of phosphate that is excreted in the feces in combination with calcium, almost all of the dietary phosphate is absorbed into the blood from the gut and later excreted in the urine.

Phosphate is a *threshold substance;* that is, when its concentration in the plasma is below the critical value of approximately 1 mmol/liter, no phosphate at all is lost into the urine; but above this critical concentration, the rate of phosphate loss is directly proportional to the additional increase. Thus, the kidney regulates the phosphate concentration in the extracellular fluid by altering the rate of phosphate excretion in accordance with the plasma phosphate concentration.

However, as discussed later in the chapter, phosphate excretion by the kidneys is greatly increased by parathyroid hormone, thereby playing an important role in the control of plasma phosphate concentration.

VITAMIN D AND ITS ROLE IN CALCIUM AND PHOSPHATE ABSORPTION

Vitamin D has a potent effect on increasing calcium absorption from the intestinal tract; it also has important effects on both bone deposition and bone reabsorption, as will be discussed later in the chapter. However, vitamin D itself is not the active substance that actually causes these effects. Instead, the vitamin D must first be converted through a succession of reactions in the liver and the kidney to the final active product, *1,25-dihydroxycholecalciferol*, also called 1,25-(OH)$_2$-D$_3$. Figure 79–1 illustrates the succession of steps that leads to the formation of this substance from vitamin D. Let us discuss these steps.

The Vitamin D Compounds. Several different compounds derived from sterols belong to the vitamin D family, and all of these perform more or less the same functions. The most important of these, called vitamin D$_3$, is *cholecalciferol*. Most of this substance is formed in the skin as a result of irradiation of *7-dehydrocholesterol* by ultraviolet rays from the sun. Consequently, appropriate exposure to the sun prevents vitamin D deficiency. The vitamin D compounds in the food are identical to cholecalciferol except for the substitution of one or more atoms that do not affect their function.

Conversion of Cholecalciferol to 25-Hydroxycholecalciferol in the Liver and Its Feedback Control. The first step in the activation of cholecalciferol is to convert it to 25-hydroxycholecalciferol; this occurs in the liver. The process, however, is itself a limited one because the 25-hydroxycholecalciferol

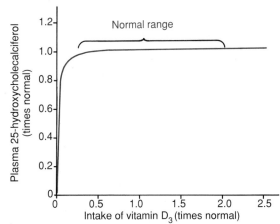

Figure 79–2. Effect of increasing vitamin D$_3$ intake on the plasma concentration of 25-hydroxycholecalciferol. This figure shows that tremendous changes in vitamin D intake have little effect on the final quantity of activated vitamin D that is formed.

has a feedback inhibitory effect on the conversion reactions. This feedback effect is extremely important for two reasons:

First, the feedback mechanism regulates very precisely the concentration of 25-hydroxycholecalciferol in the plasma, an effect that is illustrated in Figure 79–2. Note that the intake of vitamin D$_3$ can change many times, and yet the concentration of 25-hydroxycholecalciferol still remains within a few per cent of its normal mean value. Obviously, this high degree of feedback control prevents excessive action of vitamin D when it is present in too great a quantity.

Second, this controlled conversion of vitamin D$_3$ to 25-hydroxycholecalciferol conserves the vitamin D for future use, because once it is converted, it persists in the body for only a few weeks thereafter, whereas in the vitamin D form it can be stored in the liver for many months.

Formation of 1,25-Dihydroxycholecalciferol in the Kidneys and Its Control by Parathyroid Hormone. Figure 79–1 also illustrates the conversion in the proximal tubules of the kidneys of 25-hydroxycholecalciferol to *1,25-dihydroxycholecalciferol*. This latter substance is by far the most active form of vitamin D, for the previous products in the scheme of Figure 79–1 have less than 1/1000 as much vitamin D effect. Therefore, in the absence of the kidneys vitamin D loses almost all of its effectiveness.

Note also in Figure 79–1 that the conversion of 25-hydroxycholecalciferol to 1,25-dihydroxycholecalciferol requires parathyroid hormone. In the absence of this hormone, either none or almost none of the 1,25-dihydroxycholecalciferol is formed. Therefore, parathyroid hormone exerts a potent effect in determining the functional effects of vitamin D in the body, specifically its effects on calcium absorption in the intestines and its effects on bone.

Effect of Calcium Ion Concentration on Control of the Formation of 1,25-Dihydroxycholecalciferol. Figure 79–3 illustrates that the plasma

Figure 79–1. Activation of vitamin D$_3$ to form *1,25-dihydroxycholecalciferol*; and the role of vitamin D in controlling the plasma calcium concentration.

Figure 79–3. Effect of plasma calcium concentration on the plasma concentration of 1,25-dihydroxycholecalciferol. This figure shows that a very slight decrease in calcium concentration below normal causes marked formation of activated vitamin D, which in turn leads to greatly increased absorption of calcium from the intestine.

concentration of 1,25-dihydroxycholecalciferol is inversely affected by the concentration of calcium in the plasma. There are two reasons for this. First, the calcium ion itself has a slight effect in prevention of the conversion of 25-hydroxycholecalciferol to 1,25-dihydroxycholecalciferol. Second, and even more important, as we shall see later in the chapter, the rate of secretion of parathyroid hormone is greatly suppressed when the plasma calcium ion concentration rises above 10 mg/dl. Parathyroid hormone promotes the conversion of 25-hydroxycholecalciferol to 1,25-dihydroxycholecalciferol in the kidney; however, when parathyroid hormone is suppressed, the 25-hydroxycholecalciferol is converted to a slightly different compound — 24,25-dihydroxycholecalciferol — that has very little vitamin D effect.

Therefore, when the plasma calcium concentration is already too high, the formation of 1,25-dihydroxycholecalciferol is greatly depressed. Lack of this in turn decreases the absorption of calcium from the intestines, from the bones, and from the renal tubules, thus causing the calcium ion concentration to fall back toward its normal level.

"Hormonal" Effect of 1,25-Dihydroxycholecalciferol on the Intestinal Epithelium in Promoting Calcium Absorption. 1,25-Dihydroxycholecalciferol itself functions as a "hormone" to promote intestinal absorption of calcium. It does this principally by increasing over a period of about 2 days formation of a *calcium-binding protein* in the intestinal epithelial cells. This protein functions in the brush border of these cells to transport calcium into the cell cytoplasm, and the calcium then moves through the basolateral membrane of the cell by facilitated diffusion. The rate of calcium absorption seems to be directly proportional to the quantity of this calcium-binding protein. Furthermore, this protein remains in the cells for several weeks after the 1,25-di-

hydroxycholecalciferol has been removed from the body, thus causing a prolonged effect on calcium absorption.

Other effects of this "hormone," 1,25-dihydroxycholecalciferol, that might play a role in promoting calcium absorption are (1) the formation of a calcium-stimulated ATPase in the brush border of the epithelial cells and (2) the formation of an alkaline phosphatase in the epithelial cells. Unfortunately, the precise details of all of these effects are still unknown.

Effect of Vitamin D on Phosphate Absorption. Much less is known about the effect of vitamin D on phosphate absorption than on calcium absorption. Also, this is much less important because phosphate is usually absorbed relatively easily anyway. However, phosphate flux through the gastrointestinal epithelium is enhanced by vitamin D. It is believed that this results from a direct effect of 1,25-dihydroxycholecalciferol, but it is possible that it results secondarily from this hormone's action on calcium absorption, the calcium in turn acting as a transport mediator for the phosphate.

THE CALCIUM IN THE PLASMA AND INTERSTITIAL FLUID

The concentration of calcium in the plasma averages approximately 9.4 mg/dl, normally varying between 9.0 and 10.0 mg/dl. This is equivalent to approximately 2.4 mmol/liter. It is apparent that the calcium level in the plasma is regulated within very narrow limits — and mainly by parathyroid hormone, as discussed later in the chapter.

The calcium in the plasma is present in three different forms, as shown in Figure 79–4. (1) Approximately 40 per cent (1.0 mmol/liter) of the calcium is combined with the plasma proteins and in this form is

Figure 79–4. Distribution of ionic calcium (Ca^{++}), diffusible but un-ionized calcium (Ca X), and nondiffusible calcium proteinate (Ca Prot) in blood plasma.

nondiffusible through the capillary membrane. (2) Approximately 10 per cent of the calcium (0.2 mmol/liter) is diffusible through the capillary membrane but is combined with other substances of the plasma and interstitial fluids (citrate and phosphate, for instance) in such a manner that it is not ionized. (3) The remaining 50 per cent of the calcium in the plasma is both diffusible through the capillary membrane and ionized. Thus, the plasma and interstitial fluids have a normal *calcium ion concentration of approximately 1.2 mmol/liter, or 2.4 m Eq/liter,* a level only one-half the total plasma calcium concentration. This ionic calcium is important for most functions of calcium in the body, including the effect of calcium on the heart, on the nervous system, and on bone formation.

THE INORGANIC PHOSPHATE IN THE EXTRACELLULAR FLUIDS

Inorganic phosphate in the plasma is mainly in two forms: HPO_4^{--} and $H_2PO_4^{-}$. The concentration of HPO_4^{--} is approximately 1.05 mmol/liter and the concentration of $H_2PO_4^{-}$ approximately 0.26 mmol/liter. When the total quantity of phosphate in the extracellular fluid rises, so does the quantity of each of these two types of phosphate ions. Furthermore, when the pH of the extracellular fluid becomes more acid, there is a relative increase in $H_2PO_4^{-}$ and decrease in the HPO_4^{--}, whereas the opposite occurs when the extracellular fluid becomes alkaline. These relationships were presented in the discussion of acid-base balance in Chapter 30.

Because it is difficult to determine chemically the exact quantities of HPO_4^{--} and $H_2PO_4^{-}$ in the blood, ordinarily the total quantity of phosphate is expressed in terms of milligrams of *phosphorus* per deciliter of blood. The average total quantity of inorganic phosphorus represented by both phosphate ions is about 4 mg/dl, varying between normal limits of 3 to 4 mg/dl in adults and 4 to 5 mg/dl in children.

EFFECTS OF ALTERED CALCIUM AND PHOSPHATE CONCENTRATIONS IN THE BODY FLUIDS

Changing the level of phosphate in the extracellular fluid from far below normal to as high as three to four times normal does not cause significant immediate effects on the body.

On the other hand, elevation or depletion of calcium ion in the extracellular fluid causes extreme immediate effects. Both chronic hypocalcemia and hypophosphatemia greatly decrease bone mineralization, as explained later in the chapter.

Tetany Resulting From Hypocalcemia. When the extracellular fluid concentration of calcium ions falls below normal, the nervous system becomes progressively more and more excitable because this causes increased neuronal membrane permeability to

sodium ions, allowing easy initiation of action potentials. At plasma calcium ion concentrations about 50 per cent below normal, the peripheral nerve fibers especially become so excitable that they begin to discharge spontaneously, initiating trains of nerve impulses that pass to the peripheral skeletal muscles and elicit tetanic muscle contraction. Consequently, hypocalcemia causes tetany. But it also occasionally causes convulsions because of its action of increasing excitability in the brain as well.

Figure 79–5 illustrates tetany in the hand, which usually occurs before tetany develops in most other parts of the body. This is called "carpopedal spasm."

Tetany ordinarily occurs when the blood concentration of calcium falls from its normal level of 9.4 mg/dl to approximately 6 mg/dl, which is only 35 per cent below the normal calcium concentration, and it is usually lethal at about 4 mg/dl.

In experimental animals, in which the level of calcium can be reduced beyond the normal lethal stage, extreme hypocalcemia can cause still other effects that rarely are evident in patients; these effects are marked dilatation of the heart, changes in cellular enzyme activities, increased cell membrane permeability in other cells in addition to nerve cells, and impaired blood clotting.

Hypercalcemia. When the level of calcium in the body fluids rises above normal, the nervous system is depressed, and reflex activities of the central nervous system become sluggish. Also, increased calcium ion concentration decreases the QT interval of the heart, and it causes constipation and lack of appetite, probably because of depressed contractility of the muscle walls of the gastrointestinal tract.

The depressive effects of increased calcium level begin to appear when the blood level of calcium rises above approximately 12 mg/dl, and they can become marked as the calcium level rises above 15 mg/dl. When the level of calcium rises above approximately 17 mg/dl in the body fluids, calcium phosphate is likely to precipitate throughout the body; this condition is discussed shortly in connection with parathyroid poisoning.

Figure 79–5. Hypocalcemic tetany in the hand, called "carpopedal spasm." (Courtesy of Dr. Herbert Langford.)

● BONE AND ITS RELATIONSHIPS TO EXTRACELLULAR CALCIUM AND PHOSPHATES

Bone is composed of a tough *organic matrix* that is greatly strengthened by deposits of *calcium salts.* Average *compact bone* contains by weight approximately 30 per cent matrix and 70 per cent salts. However, *newly formed bone* may have a considerably higher percentage of matrix in relation to salts.

The Organic Matrix of Bone. The organic matrix of bone is 90 to 95 per cent *collagen fibers,* and the remainder is a homogeneous medium called *ground substance.* The collagen fibers extend primarily along the lines of tensional force. These fibers give bone its powerful tensile strength.

The ground substance is composed of extracellular fluid plus *proteoglycans,* especially *chondroitin sulfate* and *hyaluronic acid.* The precise function of these is not known, though perhaps they help to control the deposition of calcium salts.

The Bone Salts. The crystalline salts deposited in the organic matrix of bone are composed principally of *calcium* and *phosphate,* and the formula for the major crystalline salt, known as *hydroxyapatite,* is the following:

$$Ca_{10}(PO_4)_6(OH)_2$$

Each crystal—about 400 angstroms long, 10 to 30 angstroms thick, and 100 angstroms wide—is shaped like a long, flat plate. The relative ratio of calcium to phosphorus can vary markedly under different nutritional conditions, the Ca/P ratio on a weight basis varying between 1.3 and 2.0.

Magnesium, sodium, potassium, and *carbonate* ions are also present among the bone salts, though x-ray diffraction studies fail to show definite crystals formed by these. Therefore, they are believed to be conjugated to the hydroxyapatite crystals rather than organized into distinct crystals of their own. This ability of many different types of ions to conjugate to bone crystals extends to many ions normally foreign to bone, such as *strontium, uranium, plutonium, the other transuranic elements, lead, gold, other heavy metals,* and *at least 9 of 14 of the major radioactive products released by explosion of the hydrogen bomb.* Deposition of radioactive substances in the bone can cause prolonged irradiation of the bone tissues, and, if a sufficient amount is deposited, an osteogenic sarcoma (bone cancer) almost invariably eventually develops.

Tensile and Compressional Strength of Bone. Each collagen fiber of *compact* bone is composed of repeating periodic segments every 640 angstroms along its length; hydroxyapatite crystals lie adjacent to each segment of the fiber, bound tightly to it. This intimate bonding prevents "shear" in the bone; that is, it prevents the crystals and collagen fibers from slipping out of place, which is essential in providing strength to the bone. In addition, the segments of adjacent collagen fibers overlap each other, also causing hydroxyapatite crystals to be overlapped like bricks keyed to each other in a brick wall.

The collagen fibers of bone, like those of tendons, have great tensile strength; whereas the calcium salts, which are similar in physical properties to marble, have great compressional strength. These combined properties, plus the degree of bondage between the collagen fibers and the crystals, provide a bony structure that has both extreme tensile and compressional strength. Thus, bones are constructed in exactly the same way that reinforced concrete is constructed. The steel of reinforced concrete provides the tensile strength, while the cement, sand, and rock provide the compressional strength. Indeed, the compressional strength of bone is greater than that of even the best reinforced concrete, and the tensile strength approaches that of reinforced concrete.

PRECIPITATION AND ABSORPTION OF CALCIUM AND PHOSPHATE IN BONE— EQUILIBRIUM WITH THE EXTRACELLULAR FLUIDS

Supersaturated State of Calcium and Phosphate Ions in the Extracellular Fluids With Respect to Hydroxyapatite. The concentrations of calcium and phosphate ions in extracellular fluid are considerably greater than those required to cause precipitation of hydroxyapatite. However, inhibitors are present in most tissues of the body, as well as in plasma, to prevent such precipitation; one such inhibitor is pyrophosphate. Therefore, hydroxyapatite crystals fail to precipitate in normal tissues except in bone despite the state of supersaturation of the ions.

Mechanism of Bone Calcification. The initial stage in bone production is the secretion of *collagen molecules* (called collagen monomers) and *ground substance* (mainly proteoglycans) by *osteoblasts.* The collagen monomers polymerize rapidly to form collagen fibers; and the resultant tissue becomes *osteoid,* a cartilage-like material but differing from cartilage in that calcium salts precipitate in it. As the osteoid is formed, some of the osteoblasts become entrapped in the osteoid and then are called *osteocytes.*

Within a few days after the osteoid is formed, calcium salts begin to precipitate on the surfaces of the collagen fibers. The precipitates first appear at intervals along each collagen fiber, forming minute nidi that rapidly multiply and grow over a period of days and weeks into the finished product, *hydroxyapatite crystals.*

The initial calcium salts to be deposited are not hydroxyapatite crystals but, instead, amorphous compounds (noncrystalline), a probable mixture of such salts as $CaHPO_4 \cdot 2H_2O$, $Ca_3(PO_4)_2 \cdot 3H_2O$, and others. Then by a process of substitution and addition of atoms, or reabsorption and reprecipitation, these salts are converted into the hydroxyapatite crystals over a period of weeks or months. Yet as much as 20 to 30 per cent may remain permanently in the amorphous form. This is important, because these salts can be absorbed rapidly when there is need for extra calcium in the extracellular fluid.

It is still not known what causes calcium salts to be deposited in osteoid. One theory holds that at the time of formation the collagen fibers are specially constituted in advance for causing precipitation of calcium salts. The osteoblasts supposedly also secrete a substance into the osteoid to neutralize an inhibitor (perhaps pyrophosphate) that normally prevents hydroxyapatite crystallization. Once the pyrophosphate has been neutralized, then the natural affinity of the collagen fibers for calcium salts supposedly causes the precipitation. In support of this theory is the fact that properly prepared collagen fibers from other tissues of the body besides bone will also cause precipitation of hydroxyapatite crystals from plasma.

Precipitation of Calcium in Nonosseous Tissues Under Abnormal Conditions. Although calcium salts almost never precipitate in normal tissues besides bone, under abnormal conditions they do precipitate. For instance, they precipitate in arterial walls in the condition called arteriosclerosis and cause the arteries to become bonelike tubes. Likewise, calcium salts frequently deposit in degenerating tissues or in old blood clots. Presumably, in these instances, the inhibitor factors that normally prevent deposition of calcium salts disappear from the tissues, thereby allowing precipitation.

EXCHANGEABLE CALCIUM

If soluble calcium salts are injected intravenously, the calcium ion concentration can be made to increase immediately to very high levels. However, within a half hour to an hour or more, the calcium ion concentration returns to normal. Likewise, if large quantities of calcium ions are removed from the circulating body fluids, the calcium ion concentration again returns to normal within a half hour to an hour or so. These effects result partly from the fact that the body contains a type of *exchangeable* calcium that is always in equilibrium with the calcium ions in the extracellular fluids. A small portion of this exchangeable calcium is that calcium found in all tissue cells, especially in highly permeable types of cells such as those of the liver and the gastrointestinal tract. However, most of the exchangeable calcium, as shown by studies using radioactively tagged calcium, is in the bone, and it normally amounts to about 0.4 to 1.0 per cent of the total bone calcium. Most of this calcium is probably deposited in the bones in the form of readily mobilizable salts such as $CaHPO_4$ and the other amorphous salts.

The importance of exchangeable calcium to the body is that it provides a rapid buffering mechanism to keep the calcium ion concentration in the extracellular fluids from rising to excessive levels or falling to very low levels under transient conditions of excess or hypoavailability of calcium.

DEPOSITION AND ABSORPTION OF BONE —REMODELING OF BONE

Deposition of Bone by the Osteoblasts. Bone is continually being deposited by *osteoblasts,* and it is continually being absorbed where *osteoclasts* are active. Osteoblasts are found on the outer surfaces of the bones and in the bone cavities. A small amount of osteoblastic activity occurs continually in all living bones (on about 4 per cent of all surfaces at any given time, as shown in Figure 79–6) so that at least some new bone is being formed constantly.

Absorption of Bone — Function of the Osteoclasts. Bone is also being continually absorbed in the presence of osteoclasts, which are large phagocytic, multinucleated cells (as many as 50 nuclei), derivatives of monocytes or monocyte-like cells formed in the bone marrow. The osteoclasts are normally active on less than 1 per cent of the bone surfaces. Later in the chapter we see that parathyroid hormone controls the bone absorptive activity of osteoclasts.

Histologically, bone absorption occurs immediately adjacent to the osteoclasts, as illustrated in Figure 79–6. The mechanism of this absorption is believed to be the following: The osteoclasts send out villus-like projections toward the bone, forming a so-called ruffled border adjacent to the bone. The villi secrete two types of substances: (1) proteolytic enzymes, released from the lysosomes of the osteoclasts; and (2) several acids, including citric acid and lactic acid, released mainly from the mitochondria. The enzymes presumably digest or dissolve the organic matrix of the bone, and the acids cause solution of the bone salts.

Equilibrium Between Bone Deposition and Absorption. Normally, except in growing bones, the rates of bone deposition and absorption are equal to each other, so that the total mass of bone remains constant. Usually, osteoclasts exist in small but concentrated masses; and once a mass of osteoclasts begins to develop, it usually eats away at

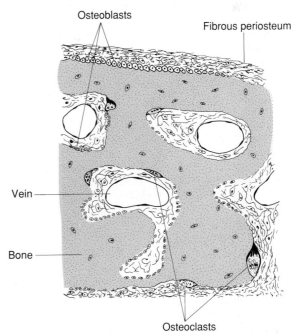

Figure 79–6. Osteoblastic and osteoclastic activity in the same bone.

the bone for about 3 weeks, eating out a tunnel that may be as large as 1 millimeter in diameter and several millimeters in length. At the end of this time the osteoclasts disappear, and the tunnel is invaded by osteoblasts instead; then new bone begins to develop. Bone deposition then continues for several months, the new bone being laid down in successive layers of concentric circles (*lamellae*) on the inner surfaces of the cavity until the tunnel is filled. Deposition of new bone ceases when the bone begins to encroach on the blood vessels supplying the area. The canal through which these vessels run, called the *haversian canal,* therefore, is all that remains of the original cavity. Each new area of bone deposited in this way is called an *osteon,* as shown in Figure 79–7.

Value of Continual Remodeling of Bone. The continual deposition and absorption of bone has a number of physiologically important functions. First, bone ordinarily adjusts its strength in proportion to the degree of bone stress. Consequently, bones thicken when subjected to heavy loads. Second, even the shape of the bone can be rearranged for proper support of mechanical forces by deposition and absorption of bone in accordance with stress patterns. Third, since old bone becomes relatively weak and brittle, new organic matrix is needed as the old organic matrix degenerates. In this manner the normal toughness of bone is maintained. Indeed, the bones of children, in whom the rates of deposition and absorption are rapid, show little brittleness in comparison with the bones of old age, at which time the rates of deposition and absorption are slow.

Control of the Rate of Bone Deposition by Bone "Stress." Bone is deposited in proportion to the compressional load that the bone must carry. For instance, the bones of athletes become considerably heavier than those of nonathletes. Also, if a person has one leg in a cast but continues to walk on the opposite leg, the bone of the leg in the cast becomes thin and as much as 30 per cent decalcified within a few weeks, while the opposite bone remains thick and normally calcified. Therefore, continual physical stress stimulates calcification and osteoblastic deposition of bone.

Bone stress also determines the shape of bones under certain circumstances. For instance, if a long bone of the leg breaks in its center and then heals at an angle, the compression stress on the inside of the angle causes increased deposition of bone, and increased absorption occurs on the outer side of the angle where the bone is not compressed. After many years of increased deposition on the inner side of the angulated bone and absorption on the outer side, the bone can become almost straight, especially in children because of the rapid remodeling of bone at younger ages.

The deposition of bone at points of compressional stress has been suggested to be caused by a *piezoelectric* effect, as follows: Compression of bone causes a negative potential in the compressed areas and a positive potential elsewhere in the bone. It has been shown that minute quantities of current flowing in bone cause osteoblastic activity at the negative end of the current flow, which could explain the increased bone deposition at compression sites. On the other hand, usual osteoclastic activity could account for reabsorption of bone elsewhere.

Repair of a Fracture. Fracture of a bone in some way maximally activates all the periosteal and intraosseous osteoblasts involved in the break. Also, immense numbers of new osteoblasts are formed almost immediately from so-called *osteoprogenitor cells,* which are bone stem cells. Therefore, within a short time a large bulge of osteoblastic tissue and new organic bone matrix, followed shortly by the deposition of calcium salts, develops between the two broken ends of the bone. This is called a *callus.*

Many bone surgeons utilize the phenomenon of bone stress to accelerate the rate of fracture healing. This is done by use of special mechanical fixation apparatuses for holding the ends of the broken bone together so that the patient can use the bone immediately. This obviously causes stress on the opposed ends of the broken bones, which accelerates osteoblastic activity at the break and often shortens convalescence.

Blood Alkaline Phosphatase as an Indication of the Rate of Bone Deposition. The osteoblasts secrete large quantities of alkaline phosphatase when they are actively depositing bone matrix. This phosphatase is believed either to increase the local concentration of inorganic phosphate or to activate the collagen fibers in such a way that they cause the deposition of calcium salts. Because some alkaline phosphatase diffuses into the blood, the blood level of alkaline phosphatase is usually a good indicator of the rate of bone formation.

● PARATHYROID HORMONE

For many years it has been known that increased activity of the parathyroid gland causes rapid absorption of calcium salts from the bones, with resultant *hypercalcemia* in the extracellular fluid; conversely, hypofunction of the parathyroid glands causes *hypocalcemia,* often with resultant tetany. Also, parathyroid hormone is important in phosphate metabolism as well as in calcium metabolism.

Physiologic Anatomy of the Parathyroid Glands. Normally there are four parathyroid glands in the human being; these are located immediately behind the thyroid gland—one behind each of the upper and each of the lower poles of the thyroid. Each parathyroid gland is approximately 6 millimeters

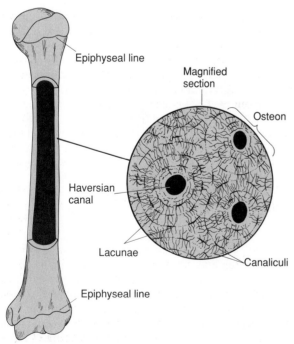

Figure 79–7. The structure of bone.

Epiphyseal line

Magnified section

Osteon

Haversian canal

Lacunae

Canaliculi

Epiphyseal line

long, 3 millimeters wide, and 2 millimeters thick and has a macroscopic appearance of dark brown fat; therefore, the parathyroid glands are difficult to locate during thyroid operations. For this reason, before the importance of these glands was generally recognized, total or subtotal thyroidectomy frequently resulted in total removal of the parathyroid glands as well.

Removal of half the parathyroid glands usually causes little physiological abnormality. However, removal of three of four normal glands usually causes transient hypoparathyroidism. But even a small quantity of remaining parathyroid tissue is usually capable of hypertrophying satisfactorily to perform the function of all of the glands.

The parathyroid gland of the adult human being, illustrated in Figure 79–8, contains mainly *chief cells* and *oxyphil cells,* but oxyphil cells are absent in many animals and in young human beings. The chief cells secrete most of the parathyroid hormone. The function of the oxyphil cells is not certain; they are believed to be modified or depleted chief cells that no longer secrete hormone.

Chemistry of Parathyroid Hormone. Parathyroid hormone has been isolated in a pure form. It is first synthesized on the ribosomes in the form of a preprohormone, a polypeptide chain of 110 amino acids. This is cleaved first to a prohormone and finally to the hormone itself by the endoplasmic reticulum and Golgi apparatus, and then packaged in secretory granules in the cytoplasm of the cells. The final hormone has a molecular weight of approximately 9500 and is composed of 84 amino acids. Smaller compounds with as few as 34 amino acids adjacent to the N terminus of the molecule have also been isolated from the parathyroid glands that exhibit full parathyroid hormone activity.

EFFECT OF PARATHYROID HORMONE ON CALCIUM AND PHOSPHATE CONCENTRATIONS IN THE EXTRACELLULAR FLUID

Figure 79–9 illustrates the effect on the blood calcium and phosphate concentrations caused by suddenly beginning to infuse parathyroid hormone into

Figure 79–9. Approximate changes in calcium and phosphate concentrations during the first 5 hours of parathyroid hormone infusion at a moderate rate.

an animal and continuing this for a number of hours. Note that at the onset of infusion the calcium ion concentration begins to rise and reaches a plateau level in about 4 hours. On the other hand, the phosphate concentration falls more rapidly and also reaches a depressed level within an hour or two. The rise in calcium concentration is caused principally by two effects: (1) an effect of parathyroid hormone in causing calcium and phosphate absorption from the bone and (2) a rapid effect of parathyroid hormone in decreasing the excretion of calcium by the kidneys. The decline in phosphate concentration, on the other hand, is caused by a very strong effect of parathyroid hormone on the kidney in causing excessive renal phosphate excretion, an effect that is usually great enough to override increased phosphate absorption from the bone.

Calcium and Phosphate Absorption From the Bone Caused by Parathyroid Hormone

Parathyroid hormone seems to have two separate effects on bone in causing absorption of calcium and phosphate. One is a very rapid phase that takes place in minutes and probably results from activation of the already existing bone cells to promote calcium and phosphate absorption. The second phase is a much slower one, requiring several days or even weeks to become fully developed; and it results from proliferation of the osteoclasts, followed by greatly increased osteoclastic reabsorption of the bone itself, not merely absorption of the calcium phosphate salts from the bone.

The Rapid Phase of Calcium and Phosphate Absorption—Osteolysis. When large quantities of parathyroid hormone are injected, the calcium ion concentration in the blood begins to rise within minutes, long before any new bone cells can be developed. Histological and physiological studies have shown that the parathyroid hormone causes removal of bone salts from the bone matrix in the vicinity of the osteocytes lying within the bone itself and also in the vicinity of the osteoblasts along the bone surface. Yet, strangely enough, one does not usually think of either osteoblasts or osteocytes functioning to cause bone

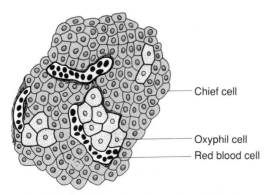

Chief cell

Oxyphil cell
Red blood cell

Figure 79–8. Histological structure of a parathyroid gland.

salt absorption, because both these types of cells are osteoblastic in nature and are normally associated with bone deposition and its calcification. However, recent studies have shown that the osteoblasts and osteocytes form a system of interconnected cells that spreads over all the bone surfaces except the small surface areas that are adjacent to the osteoclasts. Also, long filmy processes extend from osteocyte to osteocyte throughout the bone structure, and these processes also connect with the surface osteocytes and osteoblasts. This extensive system is called the *osteocytic membrane system,* and it is believed to provide a membrane that separates the bone itself from the extracellular fluid. Between the osteocytic membrane and the bone is a small amount of fluid called simply *bone fluid.* Indirect experiments indicate that the osteocytic membrane pumps calcium ions from the bone fluid into the extracellular fluid, creating a calcium ion concentration in the bone fluid only one-third that in the extracellular fluid. When the osteocytic pump becomes excessively activated, the bone fluid calcium concentration falls even lower, and calcium phosphate salts are then absorbed from the bone. This effect is called *osteolysis,* and it occurs without absorption of the bone matrix. When the pump is inactivated, the bone fluid calcium concentration rises to a higher level, and calcium phosphate salts are then redeposited in the matrix.

But where does parathyroid hormone fit into this picture? First, the cell membranes of both the osteoblasts and the osteocytes have receptor proteins for binding parathyroid hormone. It seems that parathyroid hormone can activate the calcium pump strongly, thereby causing rapid removal of calcium phosphate salts from the amorphous bone crystals that lie near the cells. The parathyroid hormone is believed to stimulate this pump by increasing the calcium permeability of the bone fluid side of the osteocytic membrane, thus allowing calcium ions to diffuse into the membrane cells from the bone fluid. Then the calcium pump on the other side of the cell membrane transfers the calcium ions the rest of the way into the extracellular fluid.

The Slow Phase of Bone Absorption and Calcium Phosphate Release — Activation of the Osteoclasts. A much better known effect of parathyroid hormone, and one for which the evidence is also much clearer, is its activation of the osteoclasts. Yet, strangely enough, the osteoclasts do not themselves have membrane receptor proteins for parathyroid hormone. Instead, it is believed that the activated osteoblasts and osteocytes send a secondary but unknown "signal" to the osteoclasts, causing them in turn to set about their usual task of gobbling up the bone.

Activation of the osteoclastic system occurs in two stages: (1) immediate activation of the osteoclasts that are already formed and (2) formation of new osteoclasts. Usually several days of excess parathyroid hormone cause the osteoclastic system to become well developed, but it can continue to grow for literally months under the influence of very strong parathyroid hormone stimulation.

After a few weeks to a few months, the osteoclastic resorption of bone leads to weakened bones and secondary further stimulation of the osteoblasts. Therefore, the late effect is actually to enhance both osteoblastic and osteoclastic activity. Still, even in the late stages, there is more bone absorption than bone deposition.

Bone contains such great amounts of calcium in comparison with the total amount in all the extracellular fluids (about 1000 times as much) that even when parathyroid hormone causes a great rise in calcium concentration in the fluids, it is impossible to discern any immediate effect on the bones. Yet prolonged administration or secretion of parathyroid hormone — over a period of many months or even years — finally results in evident absorption in all the bones with development of large cavities filled with very large, multinucleated osteoclasts.

Effect of Parathyroid Hormone on Phosphate and Calcium Excretion by the Kidneys

Administration of parathyroid hormone causes immediate and rapid loss of phosphate in the urine, due to diminished proximal tubular reabsorption of phosphate ions.

Parathyroid hormone also increases tubular reabsorption of calcium at the same time that it diminishes phosphate reabsorption. Moreover, it also increases the rate of reabsorption of magnesium ions and hydrogen ions while it decreases the reabsorption of sodium, potassium, and amino acid ions in much the same way that it affects phosphate. However, the increased calcium absorption occurs in the *ascending limbs of the loops of Henle, distal tubules,* and *collecting ducts* instead of the proximal tubules.

Were it not for the effect of parathyroid hormone on the kidneys to increase calcium reabsorption, continual loss of calcium into the urine would eventually deplete the bones of this mineral.

Effect of Parathyroid Hormone on Intestinal Absorption of Calcium and Phosphate

At this point we should be reminded again that parathyroid hormone greatly enhances both calcium and phosphate absorption from the intestines by increasing the formation in the kidneys of 1,25-dihydroxycholecalciferol from vitamin D, as was discussed earlier in the chapter.

Effect of Vitamin D on Bone and Its Relation to Parathyroid Hormone Activity

Vitamin D plays important roles in both bone absorption and bone deposition. Administration of extreme quantities of vitamin D causes absorption of bone in much the same way that administration of parathyroid hormone does. Also, in the absence of vitamin D, the effect of parathyroid hormone in causing bone absorption is greatly reduced or even prevented. The mechanism of this action of vitamin D is not known, but it is believed to result from the effect of 1,25-dihydroxycholecalciferol (the principal active product of vitamin D) in increasing calcium transport through cellular membranes.

Vitamin D in smaller quantities promotes bone calcification. Obviously, one of the ways in which it does this is to increase calcium and phosphate absorption from the intestines. However, even in the absence of such increase, it still enhances the mineralization of bone. Here again, the mechanism of the effect is unknown, but it probably also results from the ability of 1,25-dihydroxycholecalciferol to cause transport of calcium ions through cell membranes—but perhaps in the opposite direction through the osteoblastic or osteocytic cell membranes.

Role of Cyclic Adenosine Monophosphate as a Mediator of Parathyroid Stimulation. A large share of the effect of parathyroid hormone on its target organs is almost certainly mediated by the cyclic adenosine monophosphate (cAMP) *second messenger* mechanism. Within a few minutes after parathyroid hormone administration, the concentration of cAMP increases in the osteoclasts and other target cells. This cAMP, in turn, is probably responsible for such functions as osteoclastic secretion of enzymes and acids to cause bone reabsorption, formation of 1,25-dihydroxycholecalciferol in the kidneys, and so forth. However, there are probably other direct effects of parathyroid hormone that function independently of the second messenger mechanism.

CONTROL OF PARATHYROID SECRETION BY THE CALCIUM ION CONCENTRATION

Even the slightest decrease in calcium ion concentration in the extracellular fluid causes the parathyroid glands to increase their rate of secretion within minutes; and if the decreased calcium concentration persists, the glands will hypertrophy, sometimes fivefold or more. For instance, the parathyroid glands become greatly enlarged in *rickets,* in which the level of calcium is usually depressed only a small amount; also, they become greatly enlarged in *pregnancy,* even though the decrease in calcium ion concentration in the mother's extracellular fluid is hardly measurable; and they are greatly enlarged during *lactation* because calcium is used for milk formation.

On the other hand, any condition that increases the calcium ion concentration causes decreased activity and reduced size of the parathyroid glands. Such conditions include (1) excess quantities of calcium in the diet; (2) increased vitamin D in the diet; and (3) bone absorption caused by factors other than parathyroid hormone (for example, bone absorption caused by disuse of the bones).

Figure 79–10. Approximate effect of plasma calcium concentration on the plasma concentrations of parathyroid hormone and calcitonin. Note especially that long-term, chronic changes in calcium concentration can cause as much as a 100 per cent change in parathyroid hormone concentration for only a few per cent change in calcium concentration.

Figure 79–10 illustrates quantitatively the approximate relationship between plasma calcium concentration and plasma parathyroid hormone concentration. The solid red curve shows the acute relationship when the calcium concentration is changed over a period of a few hours. This shows that a decrease in calcium concentration from the normal value of 9.4 to 8.4 mg/dl doubles or triples the plasma parathyroid hormone. On the other hand, the approximate chronic relationship that one finds when the calcium ion concentration changes over a period of many weeks, thus allowing time for the glands to hypertrophy, is illustrated by the dashed red line; this illustrates that as little as a decrease of a fraction of a milligram per deciliter in plasma calcium concentration can double parathyroid hormone secretion. Obviously, this is the basis of the body's extremely potent feedback system for control of plasma calcium ion concentration.

● CALCITONIN

About 30 years ago, a new hormone that has *weak* effects on blood calcium opposite to those of parathyroid hormone was discovered. This hormone, named *calcitonin* because it reduces the blood calcium ion concentration, is secreted in the human being not by the parathyroid glands but instead by the thyroid gland. In fish, amphibians, reptiles, and birds calcitonin is secreted by the *ultimobranchial glands;* it plays an especially important role in helping control the blood calcium ion concentration when these animals change their habitat from fresh water to sea water, where there are great excesses of calcium. Furthermore, its concentration in these ultimobranchial glands is extremely great. In the human being, ultimobranchial glands do not exist as such but have become incorporated into the thyroid gland. The so-called *parafollicular cells*, or C cells, in the interstitial tissue between the follicles of the human thyroid gland are remnants of the ultimobranchial glands of lower animals, constituting only 0.1 per cent of the thyroid gland; and it is these cells that secrete the calcitonin.

Calcitonin is a large polypeptide with a molecular weight of approximately 3400 and a chain of 32 amino acids.

Effect of Calcitonin in Decreasing Plasma Calcium Concentration. In some young animals, but very little in the human being, calcitonin decreases blood calcium ion concentration very rapidly, beginning within minutes after injection of the calcitonin. Thus, the effect of calcitonin on blood calcium ion concentration is exactly opposite that of parathyroid hormone, and it occurs several times as rapidly.

Calcitonin reduces plasma calcium concentration in at least two separate ways:

1. The immediate effect is to decrease the absorptive activities of the osteoclasts and probably also the osteolytic effect of the osteocytic membrane throughout the bone, thus shifting the balance in favor of deposition of calcium in the rapidly exchangeable pool of bone calcium salts. This effect is especially significant in young animals because of the rapid interchange of absorbed and deposited calcium.

2. The second and more prolonged effect of calcitonin is to decrease the formation of new osteoclasts. Also, since osteoclastic resorption of bone leads secondarily to osteoblastic activity, the depressed numbers of osteoclasts are followed by depressed numbers of osteoblasts as well. Therefore, over a long period of time the net result is simply greatly reduced osteoclastic and osteoblastic activity; consequently, there is not a significant prolonged effect on plasma calcium ion concentration. That is, the effect on plasma calcium is mainly a transient one, lasting for a few hours to a few days at most.

Calcitonin also has minor effects on calcium handling in the kidney tubules and intestinal tract. Again, the effects are opposite those of parathyroid hormone, but quantitatively they appear to be of such little import that they are rarely considered.

Importance of the Calcitonin Effect on Plasma Calcium Concentration. Calcitonin has only a very weak effect on plasma calcium concentration in the adult human being. The reason for this is twofold. First, any initial reduction of the calcium ion concentration caused by calcitonin leads within hours to a powerful stimulation of parathyroid hormone secretion, which almost completely overrides the calcitonin effect. Second, in the adult, the daily rates of absorption and deposition of calcium are very small, and even after the rate of absorption is slowed and deposition increased by calcitonin, this still has only a small effect on plasma calcium ion concentration. On the other hand, the effect in children is much more marked because bone remodeling occurs rapidly in children, with absorption and deposition of calcium as great as 5 or more grams per day — equal to five to ten times the total calcium in all the extracellular fluid. Also, in certain bone diseases such as Paget's disease in which osteoclastic activity is greatly accelerated, calcitonin then has the potent effect of reducing the calcium absorption.

Effect of Plasma Calcium Concentration on the Secretion of Calcitonin

In young animals, but much less so in older animals and in human beings, an increase in plasma calcium concentration of about 10 per cent causes an immediate three- to sixfold increase in the rate of secretion of calcitonin, which is illustrated by the dot-dash line of Figure 79–10. This provides a second hormonal feedback mechanism for controlling the plasma calcium ion concentration, but one that works in a way exactly opposite that of the parathyroid hormone system.

However, there are two major differences between the calcitonin and the parathyroid feedback systems. First, the calcitonin mechanism operates more rapidly, reaching peak activity in less than an hour, in contrast to the 3 to 4 hours required for peak activity to be attained following the onset of parathyroid secretion.

The second difference is that the calcitonin mechanism acts mainly as a short-term regulator of calcium ion concentration because it is very rapidly overridden by the much more powerful parathyroid control mechanism. Therefore, over a prolonged period of time it is almost entirely the parathyroid system that sets the long-term level of calcium ions in the extracellular fluid. Yet for short periods, such as for an hour or so after a high calcium meal, calcitonin does seem to play a significant role in some young animals in decreasing the rise in the calcium ion concentration that otherwise would occur.

When the thyroid gland is removed and calcitonin is no longer secreted, the long-term blood calcium ion concentration is not measurably altered — which again demonstrates the overriding effect of the parathyroid hormonal system of control.

● OVERALL CONTROL OF CALCIUM ION CONCENTRATION

At times the amount of calcium absorbed into or lost from the body fluids is as much as 0.3 gram in an hour. For instance, in cases of diarrhea, several grams of calcium can be secreted in the intestinal juices, passed into the intestinal tract, and lost into the feces each day. Conversely, after ingestion of large quantities of calcium, particularly when there is also an excess of vitamin D activity, a person may absorb as much as 0.3 gram in an hour. This figure compares with a *total quantity of calcium in all the extracellular fluid of about 1 gram.* The addition or subtraction of 0.3 gram to or from such a small amount of calcium in the extracellular fluid would obviously cause serious hyper- or hypocalcemia. However, there is a first line of defense to prevent this from occurring even before the parathyroid and calcitonin hormone feedback systems have a chance to act.

Buffer Function of the Exchangeable Calcium in Bones. The exchangeable calcium salts in the bones, which were discussed earlier in this chapter, are amorphous calcium phosphate compounds, probably mainly $CaHPO_4$ or some similar compound loosely bound in the bone and in reversible equilibrium with the calcium and phosphate ions in the extracellular fluid. The quantity of these salts that is available for exchange is about 0.5 to 1 per cent of the total calcium salts of the bone, a total of 5 to 10 grams of calcium. Because of the ease of deposition of these exchangeable salts and their ease of resolubility, an increase in the concentrations of extracellular fluid calcium and phosphate ions above normal causes immediate deposition of exchangeable salt. Conversely, a decrease in these concentrations causes immediate absorption of exchangeable salt. This reaction is so rapid that a single passage through a bone of blood containing a high concentration of calcium will remove almost all of the excess calcium. This rapid effect results from the fact that the amorphous bone crystals are extremely small, and their total surface area exposed to the fluids of the bone is perhaps an acre or more. Also, about 5 per cent of all the blood flows through the bones each minute — that is, about 1 per cent of all the extracellular fluid each minute. Therefore, about half of any excess calcium that appears in the extracellular fluid is removed by this buffer function of the bones in approximately 70 minutes.

In addition to the buffer function of the bones, the mitochondria of many of the tissues of the body, especially of the liver and intestine, also contain a reasonable amount of exchangeable calcium (a total of about 10 grams in the whole body) that provides an additional buffer system for helping to maintain constancy of the extracellular fluid calcium ion concentration.

Hormonal Control of Calcium Ion Concentration, the Second Line of Defense. At the same time that the exchangeable calcium mechanism in the bones is "buffering" the calcium in the extracellular fluid, both the parathyroid and calcitonin hormonal systems are also beginning to act. Within 3 to 5 minutes after an acute increase in the calcium ion concentration, the rate of parathyroid hormone secretion decreases. As already explained, this sets into play multiple mechanisms for reducing the calcium ion concentration back toward normal. However, this is a slow process. On the other hand, at the same time that parathyroid hormone decreases, calcitonin increases. In young animals, and possibly in young children as well (but probably very little in adults), the calcitonin causes rapid deposition of calcium in the bones, and

perhaps also in some cells of other tissues. Therefore, in very young animals, excess calcitonin can cause a high calcium ion concentration to return to normal perhaps considerably more rapidly than can be achieved by the exchangeable calcium-buffering mechanism alone.

In prolonged calcium excess or prolonged calcium deficiency, only the parathyroid hormonal mechanism seems to be really important in maintaining a normal plasma calcium ion concentration. When a person has a continuing deficiency of calcium in the diet, parathyroid hormone often can stimulate enough calcium absorption from the bones to maintain a normal plasma calcium ion concentration for a year or more; but eventually even the bones will run out of calcium. Thus, in effect, the bones are a large buffer-reservoir of calcium that can be manipulated by parathyroid hormone, but the ultimate very long-term control of calcium ion concentration resides in the roles of parathyroid hormone and vitamin D in controlling calcium absorption from the gut and calcium excretion in the urine

● PHYSIOLOGY OF PARATHYROID AND BONE DISEASES

HYPOPARATHYROIDISM

When the parathyroid glands do not secrete sufficient parathyroid hormone, the osteocytic reabsorption of exchangeable calcium decreases, and the osteoclasts also become almost totally inactive. As a result, calcium reabsorption from the bones is so depressed that the level of calcium in the body fluids decreases. Yet, because calcium and phosphates are not being absorbed from the bone, the bone usually remains strong.

When the parathyroid glands are suddenly removed, the calcium level in the blood falls from the normal of 9.4 to 6 to 7 mg/dl within 2 to 3 days, and the blood phosphate concentration may double. When this low calcium level is reached, the usual signs of tetany develop. Among the muscles of the body especially sensitive to tetanic spasm are the laryngeal muscles. Spasm of these obstructs respiration, which is the usual cause of death in tetany unless appropriate treatment is applied.

Treatment of Hypoparathyroidism. *Parathyroid Hormone (Parathormone).* Parathyroid hormone is occasionally used for treating hypoparathyroidism. However, because of the expense of this hormone, because its effect lasts a few hours at most, and because the tendency of the body to develop immune bodies against it makes it progressively less and less effective, treatment of hypoparathyroidism with parathyroid hormone is rare in present-day therapy.

Vitamin D and Calcium Therapy. In most patients, administration of extremely large quantities of vitamin D, to as high as 100,000 units per day, along with intake of 1 to 2 grams of calcium, will suffice to keep the calcium ion concentration in a normal range. At times it might be necessary to administer 1,25-dihydroxycholecalciferol instead of the nonactivated form of vitamin D because of its much more potent and much more rapid action. However, this can also cause unwanted effects, because it is sometimes difficult to prevent overactivity by this activated form of vitamin D.

HYPERPARATHYROIDISM

The cause of hyperparathyroidism ordinarily is a tumor of one of the parathyroid glands; such tumors occur much more frequently in women than in men or children, presumably because pregnancy and lactation stimulate the parathyroid glands and therefore predispose to the development of such a tumor.

Hyperparathyroidism causes extreme osteoclastic activity in the bones. This elevates the calcium ion concentration in the extracellular fluid while usually (but not always) depressing the concentration of phosphate ions because of increased renal excretion of phosphate.

Bone Disease in Hyperparathyroidism. Although in mild hyperparathyroidism new bone can be deposited rapidly enough to compensate for the increased osteoclastic reabsorption of bone, in severe hyperparathyroidism the osteoclastic absorption soon far outstrips osteoblastic deposition, and the bone may be eaten away almost entirely. Indeed, the reason a hyperparathyroid person comes to the doctor is often a broken bone. Radiographs of the bone show extensive decalcification and occasionally large punched-out cystic areas of the bone that are filled with osteoclasts in the form of so-called giant cell "tumors." Obviously, multiple fractures of the weakened bones can result from only slight trauma, especially where cysts develop. The cystic bone disease of hyperparathyroidism is called *osteitis fibrosa cystica.*

Osteoblastic activity in the bones also increases greatly in a vain attempt to form enough new bone to make up for the old bone absorbed by the osteoclastic activity. As explained earlier in the chapter, when the osteoblasts become active, they secrete large quantities of alkaline phosphatase. Therefore, one of the important diagnostic findings in hyperparathyroidism is a very high level of plasma alkaline phosphatase.

Effects of Hypercalcemia in Hyperparathyroidism. Hyperparathyroidism can at times cause the plasma calcium level to rise to as high as 12 to 15 mg/dl and rarely even higher. The effects of such elevated calcium levels, as detailed earlier in the chapter, are depression of the central and peripheral nervous systems, muscular weakness, constipation, abdominal pain, peptic ulcer, lack of appetite, and depressed relaxation of the heart during diastole.

Parathyroid Poisoning and Metastatic Calcification. When, on rare occasions, extreme quantities of parathyroid hormones are secreted, the level of calcium in the body fluids rises rapidly to very high values. Even the extracellular fluid phosphate concentration also often rises markedly instead of falling as is usually the case, probably because the kidneys cannot excrete rapidly enough all of the phosphate being absorbed from the bone. Therefore, the calcium and phosphate in the body fluids become greatly supersaturated, so that calcium phosphate ($CaHPO_4$) crystals begin to deposit in the alveoli of the lungs, in the tubules of the kidneys, in the thyroid gland, in the acid-producing area of the stomach mucosa, and in the walls of the arteries throughout the body. This extensive *metastatic* deposition of calcium phosphate can develop within a few days.

Ordinarily, the level of calcium in the blood must rise above 17 mg/dl before there is danger of parathyroid poisoning, but once such elevation develops along with concurrent elevation of phosphate, death can occur in only a few days.

Formation of Kidney Stones in Hyperparathyroidism. Most patients with mild hyperparathyroidism show few signs of bone disease and few general abnormalities as a result of elevated calcium. Yet they do have an extreme tendency to form kidney stones. The reason is that all the excess calcium and phosphate absorbed from the intestines or mobilized from the bones in hyperparathyroidism is excreted by the kidneys, causing a proportionate increase in the concentrations of these substances in the urine. As a result, crystals of calcium phosphate tend to precipitate in the kidney, forming calcium phosphate stones. Also, calcium oxalate stones develop because even normal levels of oxalate will cause calcium precipitation at high calcium levels. Because the solubility of most renal stones is slight in alkaline media, the tendency for formation of renal calculi is considerably greater in alkaline urine than in acid urine. For this reason, acidotic diets and acidic drugs are frequently used for treating renal calculi.

RICKETS

Rickets occurs mainly in children as a result of calcium or phosphate deficiency in the extracellular fluid. Yet, ordinarily rickets is due to lack of vitamin D, rather than a dietary lack of calcium or phosphate. If the child is properly exposed to sunlight, the 7-dehydrocholesterol in the skin becomes activated by the ultraviolet rays and forms vitamin D_3, which prevents rickets by promoting calcium and phosphate absorption from the intestines, as discussed earlier in the chapter.

Children who remain indoors through the winter in general do not receive adequate quantities of vitamin D without some supplementary therapy in the diet. Rickets tends to occur especially in the spring months because vitamin D formed during the preceding summer is stored in the liver and is still available for use during the early winter months. Also, calcium and phosphate absorption from the bones can prevent clinical signs of rickets for the first few months of vitamin D deficiency.

Plasma Concentrations of Calcium and Phosphate in Rickets. Ordinarily, the plasma calcium concentration in rickets is only slightly depressed, but the level of phosphate is greatly depressed. This is because the parathyroid glands prevent the calcium level from falling by promoting bone absorption every time the calcium level begins to fall. On the other hand, there is no good regulatory system for preventing a falling level of phosphate, and the increased parathyroid activity actually increases the excretion of phosphates in the urine.

Effect of Rickets on the Bones. During prolonged rickets, the marked compensatory increase in parathyroid hormone secretion causes extreme osteoclastic absorption of the bone; this in turn causes the bone to become progressively weaker and imposes marked physical stress on the bone, resulting in rapid osteoblastic activity. The osteoblasts lay down large quantities of osteoid, which does not become calcified because of insufficient calcium and phosphate ions. Consequently, the newly formed, uncalcified, and very weak osteoid gradually takes the place of the older bone that is being reabsorbed.

Tetany in Rickets. In the early stages of rickets, tetany almost never occurs because the parathyroid glands continually stimulate osteoclastic absorption of bone and therefore maintain an almost normal level of calcium in the extracellular fluid. However, when the bones finally become exhausted of calcium, the level of calcium may fall rapidly. As the blood level of calcium falls below 7 mg/dl the usual signs of tetany develop, and the child may die of tetanic respiratory spasm unless intravenous calcium is administered, which relieves the tetany immediately.

Treatment. The treatment of rickets, obviously, depends on supplying adequate calcium and phosphate in the diet and, equally important, on administering large amounts of vitamin D. If vitamin D is not administered, little calcium and phosphate are absorbed from the gut.

Osteomalacia

Osteomalacia is rickets in adults and is frequently called "adult rickets."

Normal adults rarely have a serious *dietary* deficiency of vitamin D or calcium because large quantities of calcium are not needed for bone growth as in children. However, a serious deficiency of both vitamin D and calcium occasionally occurs as a result of steatorrhea (failure to absorb fat), for vitamin D is fat-soluble, and calcium tends to form insoluble soaps with fat; consequently, in steatorrhea both vitamin D and calcium tend to pass into the feces. Under these conditions an adult occasionally has such poor calcium and phosphate absorption that adult rickets can occur, though this almost never proceeds to the stage of tetany — but very often is a cause of severe bone disability.

Osteomalacia and Rickets Caused by Renal Disease. "Renal rickets" is a type of osteomalacia resulting from prolonged kidney damage. The cause of this condition is mainly failure of the damaged kidneys to form 1,25-dihydroxycholecalciferol, the active form of vitamin D. In patients whose kidneys have been completely removed or destroyed and who are being treated by hemodialysis, the problem of renal rickets is often a very severe one.

Another type of renal disease that leads to rickets and osteomalacia is *congenital hypophosphatemia* resulting from congenitally reduced reabsorption of phosphates by the renal tubules. This type of rickets must be treated with phosphate compounds instead of calcium and vitamin D, and it is called *vitamin D – resistant* rickets.

OSTEOPOROSIS

Osteoporosis is the most common of all bone diseases in adults, especially in old age. It is a different disease from osteomalacia and rickets, for it results from diminished organic bone matrix rather than abnormal bone calcification. Usually, in osteoporosis the osteoblastic activity in the bone is less than normal, and consequently the rate of bone osteoid deposition is depressed. But occasionally, as in hyperparathyroidism, the cause of the diminished bone is excess osteoclastic activity.

The many common causes of osteoporosis are (1) lack of physical stress on the bones because of inactivity; (2) malnutrition to the extent that sufficient protein matrix cannot be formed; (3) lack of vitamin C, which is necessary for the secretion of intercellular substances by all cells, including formation of osteoid by the osteoblasts; (4) postmenopausal lack of estrogen secretion, for estrogens have an osteoblast-stimulating activity; (5) old age, in which growth hormone and other growth factors diminish greatly, plus the fact that many of the protein anabolic functions are poor anyway, so that bone matrix cannot be deposited satisfactorily; and (6) Cushing's disease, because massive quantities of glucocorticoids cause decreased deposition of protein throughout the body, cause increased catabolism of protein, and also have the specific effect of depressing osteoblastic activity. Thus, many different diseases or deficiencies of protein metabolism can cause osteoporosis.

● PHYSIOLOGY OF THE TEETH

The teeth cut, grind, and mix the food eaten. To perform these functions the jaws have powerful muscles capable of providing an occlusive force between the front teeth of as much as 50 to 100 pounds and as much as 150 to 200 pounds for the jaw teeth. Also, the upper and lower teeth are provided with projections and facets that interdigitate so that each set of teeth fits with the other. This fitting is called *occlusion*, and it allows even small particles of food to be caught and ground between the tooth surfaces.

FUNCTION OF THE DIFFERENT PARTS OF THE TEETH

Figure 79–11 illustrates a sagittal section of a tooth, showing its major functional parts: the *enamel, dentine, cementum,* and *pulp.* The tooth can also be divided into the *crown,* which is the portion that protrudes out from the gum into the mouth, and the *root,* which is the portion that protrudes into the bony socket of the jaw. The collar between the crown and the root where the tooth is surrounded by the gum is called the *neck.*

Dentine. The main body of the tooth is composed of dentine, which has a strong, bony structure. Dentine is made up principally of hydroxyapatite crystals similar to those in the bone, but much more dense. These are embedded in a strong meshwork of collagen fibers. In other words, the principal constituents of dentine are very much the same as those of bone. The major difference is its histological organization, for dentine does not contain any osteoblasts, osteocytes, osteoclasts, or spaces for blood vessels or nerves. Instead, it is deposited and nourished by a layer of cells called *odontoblasts,* which line its inner surface along the wall of the pulp cavity.

The calcium salts in dentine make it extremely resistant to compressional forces, while the collagen fibers make it

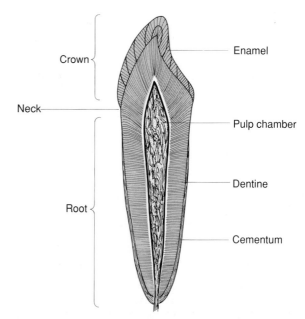

Figure 79–11. Functional parts of a tooth.

tough and resistant to tensional forces that might result when the teeth are struck by solid objects.

Enamel. The outer surface of the tooth is covered by a layer of enamel that is formed prior to eruption of the tooth by special epithelial cells called *ameloblasts.* Once the tooth has erupted, no more enamel is formed. Enamel is composed of very large and very dense crystals of hydroxyapatite with adsorbed carbonate, magnesium, sodium, potassium, and other ions embedded in a fine meshwork of strong and almost completely insoluble protein fibers that are similar in physical characteristics (but not chemically identical) to the keratin of hair. The crystalline structure of the salts makes the enamel extremely hard, much harder than the dentine. Also, the special protein fiber meshwork, though constituting only about 1 per cent of the enamel mass, nevertheless makes enamel very resistant to acids, enzymes, and other corrosive agents because this protein is one of the most insoluble and resistant proteins known.

Cementum. Cementum is a bony substance secreted by cells of the *periodontal membrane,* which lines the tooth socket. Many collagen fibers pass directly from the bone of the jaw, through the periodontal membrane, and then into the cementum. These collagen fibers and the cementum hold the tooth in place. When the teeth are exposed to excessive strain, the layer of cementum becomes thicker and stronger. Also, it increases in thickness and strength with age, causing the teeth to become progressively more firmly seated in the jaws as one reaches adulthood and older.

Pulp. The pulp cavity of each tooth is filled with *pulp,* which in turn is composed of connective tissue with an abundant supply of nerve fibers, blood vessels, and lymphatics. The cells lining the surface of the pulp cavity are the odontoblasts, which, during the formative years of the tooth, lay down the dentine but at the same time encroach more and more on the pulp cavity, making it smaller. In later life the dentine stops growing and the pulp cavity remains essentially constant in size. However, the odontoblasts are still viable and send projections into small *dentinal tubules* that penetrate all the way through the dentine; these are of importance for providing nutrition and for exchange of calcium, phosphate, and other minerals.

DENTITION

Each human being and most other mammals develop two sets of teeth during a lifetime. The first teeth are called the *deciduous teeth,* or *milk teeth,* and they number 20 in the human being. These erupt between the seventh month and second year of life, and they last until the sixth to the thirteenth year. After each deciduous tooth is lost, a permanent tooth replaces it, and an additional 8 to 12 molars appear posteriorly in the jaws, making the total number of permanent teeth 28 to 32, depending on whether the four *wisdom teeth* finally appear, which does not occur in everyone.

Formation of the Teeth. Figure 79–12 illustrates the formation and eruption of teeth. Figure 79–12A shows invagination of the oral epithelium into the *dental lamina;* this is followed by the development of a tooth-producing organ. The epithelial cells above form ameloblasts, which form the enamel on the outside of the tooth. The epithelial cells below invaginate upward into the middle of the tooth to form the pulp cavity and also the odontoblasts that secrete dentine. Thus, enamel is formed on the outside of the tooth, and dentine on the inside, giving rise to an early tooth as illustrated in Figure 79–12B.

Eruption of Teeth. During early childhood, the teeth begin to protrude upward from the jaw bone through the oral epithelium into the mouth. The cause of "eruption" is unknown, though several theories have been offered in an attempt to explain this phenomenon. The most likely theory is that growth of the tooth root as well as of the bone underneath the tooth progressively shoves the tooth forward.

Development of the Permanent Teeth. During embryonic life, a tooth-forming organ also develops in the dental lamina for each permanent tooth that will be needed after the deciduous teeth are gone. These tooth-producing organs slowly form the permanent teeth throughout the first 6 to 20 years of life. When each permanent tooth becomes fully formed, it, like the deciduous tooth, pushes upward through the bone of the jaw. In so doing it erodes the root of the deciduous tooth and eventually causes it to loosen and fall out. Soon thereafter, the permanent tooth erupts to take the place of the original one.

Metabolic Factors in Development of the Teeth. The rate of development and the speed of eruption of teeth can be accelerated by both thyroid and growth hormones. Also, the deposition of salts in the early forming teeth is affected considerably by various factors of metabolism, such as the availability of calcium and phosphate in the diet, the amount of vitamin D present, and the rate of parathyroid hormone secretion. When all these factors are normal, the dentine and enamel will be correspondingly healthy; but

Enamel organ of milk tooth

Primordium of enamel organ of permanent tooth

Oral epithelium

Enamel
Dentine

Mesenchymal primordium of pulp

Alveolar bone

A B C

Figure 79–12. *A,* Primordial tooth organ. *B,* The developing tooth. *C,* The erupting tooth.

when they are deficient, the calcification of the teeth also may be defective, so that the teeth will be abnormal throughout life.

MINERAL EXCHANGE IN TEETH

The salts of teeth, like those of bone, are composed basically of hydroxyapatite with adsorbed carbonates and various cations bound together in a hard crystalline substance. Also, new salts are constantly being deposited while old salts are being reabsorbed from the teeth, as also occurs in bone. However, experiments indicate that deposition and reabsorption occur mainly in the dentine and cementum, and very little occurs in the enamel. Most of that which does occur in the enamel occurs by exchange of minerals with the saliva instead of with the fluids of the pulp cavity.

The rate of absorption and deposition of minerals in the cementum is approximately equal to that in the surrounding bone of the jaw, while the rate of deposition and absorption of minerals in the dentine is only one-third that of bone. The cementum has characteristics almost identical to those of usual bone, including the presence of osteoblasts and osteoclasts; whereas dentine does not have these characteristics, as explained above. This difference undoubtedly explains the different rates of mineral exchange.

The mechanism by which minerals are deposited and reabsorbed from the dentine is not clear. It is probable that the thin, tubular processes of the odontoblasts that protrude into the tubules of the dentine are capable of absorbing salts and then of providing new salts to take the place of the old.

In summary, rapid mineral exchange occurs in the dentine and cementum of teeth, although the mechanism of this exchange in dentine is unclear. On the other hand, enamel exhibits extremely slow mineral exchange, so that it maintains most of its original mineral complement throughout life.

DENTAL ABNORMALITIES

The two most common dental abnormalities are *caries* and *malocclusion*. Caries means erosions of the teeth, whereas malocclusion means failure of the projections of the upper and lower teeth to interdigitate properly.

Caries and the Role of Fluorine. It is generally agreed by all research investigators of dental caries that caries results from the action of bacteria on the teeth, the most common of which is *Streptococcus mutans*. The first event in the development of caries is the deposit of *plaque*, a film of precipitated products of saliva and food, on the teeth. Large numbers of bacteria inhabit this plaque and are readily available to cause caries. However, these bacteria depend to a great extent on carbohydrates for their food. When carbohydrates are available, their metabolic systems are strongly activated and they multiply. In addition, they form acids, particularly lactic acid, and proteolytic enzymes. The acids are the major culprit in the causation of caries, because the calcium salts of teeth are slowly dissolved in a highly acid medium. And once the salts have become absorbed, the remaining organic matrix is rapidly digested by the proteolytic enzymes.

Enamel is far more resistant to demineralization by acids than is dentine, primarily because the crystals of enamel are very dense and also are about 200 times as large in volume as the dentine crystals. Therefore, the enamel of the tooth is the primary barrier to the development of caries. Once the carious process has penetrated through the enamel to the dentine, it then proceeds many times as rapidly because of the high degree of solubility of the dentine salts.

Because of the dependence of the caries bacteria on carbohydrates, it has frequently been taught that eating a diet high in carbohydrate content will lead to excessive development of caries. However, it is not the quantity of carbohydrate ingested but instead the frequency with which it is eaten that is important. If eaten in many small parcels throughout the day, such as in the form of candy, the bacteria are supplied with their preferential metabolic substrate for many hours of the day, and the development of caries is extreme. If eaten in large amounts only at mealtimes, the extensiveness of the caries is greatly reduced.

Some teeth are more resistant to caries than others. Studies show that teeth formed in children who drink water containing small amounts of fluorine develop enamel that is more resistant to caries than the enamel in children who drink water not containing fluorine. Fluorine does not make the enamel harder than usual, but instead it displaces hydroxyl ions in the hydroxyapatite crystals, which, in turn, makes the enamel several times less soluble. It is also believed that the fluorine is toxic to the bacteria as well. Finally, when small pits do develop in the enamel, fluorine is believed to promote deposition of calcium phosphate to "heal" the enamel surface. Regardless of the precise means by which fluorine protects the teeth, it is known that small amounts of fluorine deposited in enamel make teeth about three times as resistant to caries as are teeth without fluorine.

Malocclusion. Malocclusion is usually caused by a hereditary abnormality that causes the teeth of one jaw to grow to abnormal positions. In malocclusion, the teeth cannot perform their normal grinding or cutting action adequately. Occasionally malocclusion also results in abnormal displacement of the lower jaw in relation to the upper jaw, causing such undesirable effects as pain in the mandibular joint or deterioration of the teeth.

The orthodontist can usually correct malocclusion by applying prolonged gentle pressure against the teeth with appropriate braces. The gentle pressure causes absorption of alveolar jaw bone on the compressed side of the tooth and deposition of new bone on the tensional side of the tooth. In this way the tooth gradually moves to a new position as directed by the applied pressure.

REFERENCES

Albertson, D. A., et al.: Subtotal parathyroidectomy versus total parathyroidectomy with autotransplantation for secondary hyperparathyroidism. Am. Surg., 51:16, 1985.

Albright, J. A., and Brand, R. A.: The Scientific Basis of Orthopaedics, 2nd Ed. East Norwalk, Conn., Appleton & Lange, 1987.

Allen, D. L., et al.: Perodontics for the Dental Hygienist, 4th Ed. Philadelphia, Lea & Febiger, 1983.

Avioli, L. V.: Calcium and osteoporosis. Annu. Rev. Nutr., 4:471, 1984.

Avioli, L., and Krane, S. M.: Metabolic Bone Disease, 2nd Ed. Philadelphia, W. B. Saunders Co., 1989.

Barnes, D. M.: Close encounters with an osteoclast. Science, 236:914, 1987.

Barnes, D. M.: New tumor factor may disrupt calcium levels. Science, 237:363, 1987.

Berridge, M. J., and Galione, A.: Cytosolic calcium oscillators. FASEB J., 2:3074, 1988.

Berridge, M. J.: Calcium: A universal second messenger. Triangle, 24:79, 1985.

Canalis, E.: Bone-related growth factors. Triangle, 27:11, 1988.

Carney, S. L., and Muir, H.: The structure and function of cartilage proteoglycans. Physiol. Rev., 68:858, 1988.

Chambers, T. J.: The effect of calcitonin on the osteoclast. Triangle, 27:53, 1988.

Copp, D. H.: Calcitonin: Comparative endocrinology. In DeGroot, L. J., et al. (eds.): Endocrinology, Vol. 2. New York, Grune & Stratton, 1979, p. 637.

Dee, R., et al. (eds.): Principles of Orthopaedic Practice. New York, McGraw-Hill Book Co., 1989.

DeLuca, H. F.: The vitamin D story: a collaborative effort of basic science and clinical medicine. FASEB J., 2:224, 1988.

Einhorn, T. A.: Biomechanical properties of bone. Triangle, 27:27, 1988.

Felig, P., et al. (eds.): Endocrinology and Metabolism, 2nd Ed. New York, McGraw-Hill Book Co., 1987.

Fiskum, G. (ed.): Cell Calcium Metabolism. New York, Plenum Publishing Corp., 1989.

Gallacher, D. V.: Control of calcium influx in cells without action potentials. News Physiol. Sci., 3:244, 1988.

Garel, J. -M.: Hormonal control of calcium metabolism during the reproductive cycle in mammals. Physiol. Rev., 67:1, 1987.

Grossman, L. I., et al: Endodontic Practice, 11th Ed. Philadelphia, Lea & Febiger, 1988.

Habener, J. F., et al.: Parathyroid hormone: Biochemical aspects of biosynthesis, secretion, action, and metabolism. Physiol. Rev., 64:985, 1984.

Hauschka, P. V., et al.: Osteocalcin and matrix Gla protein: Vitamin K-dependent proteins in bone. Physiol. Rev., 69:990, 1989.

Heaney, R. P., and Barger-Lux, M. J.: Calcium, bone metabolism, and structural failure. Triangle, 24:91, 1985.

Henry, H. L., and Norman, A. W.: Vitamin D: Metabolism and biological actions. Annu. Rev. Nutr., 4:493, 1984.

Herzig, J. W.: Calcium ions and the electromechanical coupling of the heart. Triangle, 24:115, 1985.

Inesi, G.: Mechanism of calcium transport. Annu. Rev. Physiol., 47:573, 1985.

Kelly, P. J.: Pathways of transport in bone. In Shepherd, J. T., and Abboud, F. M. (eds.): Handbook of Physiology. Sec. 2, Vol. III. Bethesda, Md., American Physiological Society, 1983, p. 371.

Kumar, R.: Metabolism of 1,25-dihydroxyvitamin D_3. Physiol. Rev., 64:478, 1984.

Martin, R. B., and Burr, D. B.: Structure, Function, and Adaptation of Compact Bone. New York, Raven Press, 1989.

Minghetti, P. P., and Norman, A. W.: 1.25$(OH)_2$-vitamin D_3 receptors: gene regulation and genetic circuitry. FASEB J., 2:3043, 1988.

Newman, H. N.: Dental Plaque: The Ecology of the Flora on Human Teeth. Springfield, Ill., Charles C Thomas, 1980.

Nijweide, P. J., et al.: Cells of bone: Proliferation, differentiation, and hormonal regulation. Physiol. Rev., 66:855, 1986.

Norman, A. W., et al. (eds.): Calcium-Binding Proteins in Health and Disease. San Diego, Academic Press, 1987.

Ostrowski, K., and Dziedzic-Goclawska, A.: Uses of radiation-induced paramagnetic centers in bone. News Physiol. Sci., 4:112, 1989.

Pitkin, R. M.: Calcium metabolism in pregnancy and the perinatal period: A review. Am. J. Obstet. Gynecol., 151:99, 1985.

Poole-Wilson, P. A., et al.: Calcium out of control. J. Mol. Cell. Cardiol., 16:175, 1984.

Provenza, D. V.: Fundamentals of Oral Histology and Embryology, 2nd Ed. Philadelphia, Lea & Febiger, 1988.

Raisz, L. G.: Bone metabolism and its hormonal regulation: An update. Triangle, 27:5, 1988.

Rao, G. S.: Dietary intake and bioavailability of fluoride. Annu. Rev. Nutr., 4:115, 1984.

Richmond, V. L.: Thirty years of fluoridation: A review. Am. J. Clin. Nutr., 41:129, 1985.

Riggs, B. L., and Melton, L. J., III: Osteoporosis: Etiology, Diagnosis, and Management. New York, Raven Press, 1988.

Ryan, K. J.: Postmenopausal estrogen use. Annu. Rev. Med., 33:171, 1982.

Stevenson, J. C.: Pathophysiology of osteoporosis. Triangle, 27:47, 1988.

Talmage, R. V., and Cooper, C. W.: Physiology and mode of action of calcitonin. In DeGroot, L. J., et al. (eds.): Endocrinology. Vol. 2. New York, Grune & Stratton, 1979, p. 647.

Tam, C. S., et al. (eds.): Metabolic Bone Disease: Cellular and Tissue Mechanisms. Boca Raton, CRC Press, Inc., 1988.

Thaller, S. R., and Montgomery, W. W. (eds.): Guide to Dental Problems for Physicians and Surgeons. Baltimore, Williams & Wilkins, 1988.

Uhthoff, H. K., and Wiley, J. J.: Behavior of the Growth Plate. New York, Raven Press, 1988.

Wasserman, R. H., and Fullmer, C. S.: Calcium transport proteins, calcium absorption, and vitamin D. Annu. Rev. Physiol., 45:375, 1983.

Wozney, J. M., et al.: Novel regulators of bone formation: molecular clones and activities. Science, 242:1528, 1988.

CHAPTER 80

• Reproductive and Hormonal Functions of the Male •

The reproductive functions of the male can be divided into three major subdivisions: first, spermatogenesis, which means simply the formation of sperm; second, performance of the male sexual act; and third, regulation of male reproductive functions by the various hormones. Associated with these reproductive functions are the effects of the male sex hormones on the accessory sexual organs, on cellular metabolism, on growth, and on other functions of the body.

Physiologic Anatomy of the Male Sexual Organs. Figure 80–1A illustrates the various portions of the male reproductive system, and Figure 80–1B gives a more detailed structure of the testis and epididymis. The testis is composed of up to 900 coiled *seminiferous tubules,* each averaging more than half a meter long, in which the sperm are formed. The sperm then empty into the *epididymis,* another coiled tube approximately 6 meters long. The epididymis leads into the *vas deferens,* which enlarges into the *ampulla of the vas deferens* immediately before the vas enters the body of the *prostate gland.* A *seminal vesicle,* one located on each side of the prostate, empties into the prostatic end of the ampulla, and the contents from both the ampulla and the seminal vesicle pass into an *ejaculatory duct* leading through the body of the prostate gland to empty into the *internal urethra. Prostatic ducts* in turn empty from the prostate gland into the ejaculatory duct. Finally, the *urethra* is the last connecting link from the testis to the exterior. The urethra is supplied with mucus derived from a large number of minute *urethral glands* located along its entire extent and even more so from bilateral *bulbourethral glands* (Cowper's glands) located near the origin of the urethra.

● SPERMATOGENESIS

Spermatogenesis occurs in all the seminiferous tubules during active sexual life, beginning at an average age of 13 years as the result of stimulation by anterior pituitary gonadotropic hormones and continuing throughout the remainder of life.

THE STEPS OF SPERMATOGENESIS

The seminiferous tubules, a cross-section of one of which is illustrated in Figure 80–2A, contain large numbers of germinal epithelial cells called *spermatogonia,* located in two to three layers along the outer border of the tubular epithelium. These continually proliferate to replenish themselves, and a portion of them differentiate through definite stages of development to form sperm, as shown in Figure 80–2B.

In the first stage of spermatogenesis, primordial spermatogonia located immediately adjacent to the basement membrane of the germinal epithelium, called *type A spermatogonia,* divide four times to form 16 slightly more differentiated cells, the *type B spermatogonia.*

At this stage, the spermatogonia migrate centrally toward and among the *Sertoli cells.* These cells are very large, with overflowing cytoplasmic envelopes that extend from the spermatogonial cell layers all the way to the central lumen of the tubule. The membranes of the Sertoli cells are tightly adherent to each other at their bases, forming a barrier that prevents penetration from the capillaries surrounding the tubules of large protein molecules such as immunoglobulins that might interfere with further development of the spermatogonia into spermatozoa. Yet the spermatogonia do penetrate this barrier and become totally enveloped within the enfolding cytoplasmic processes of the Sertoli cells. This close relationship with the Sertoli cells continues throughout the entire remainder of spermatozoon development.

Meiosis. For a period averaging 24 days, each spermatogonium that crosses the barrier into the Sertoli cell layer becomes progressively modified and enlarged to form a very large *primary spermatocyte.* At the end of the 24 days, each of these then divides to form two *secondary spermatocytes.* However, this division is not a normal division. Instead it is called the *first meiotic division.* In the initial stage of this division, all of the DNA in the 46 chromosomes is replicated. In this process, each of the 46 chromosomes becomes two *chromatids* that remain bound together at the centromeres, the two chromatids having duplicate genes of that chromosome. It is at this time that the primary spermatocyte divides into two secondary spermatocytes, each pair of chromosomes separating so that 23 chromosomes, each containing two chromatids, go to one of the secondary spermatocytes while the other 23 chromosomes go to the other secondary spermatocyte. Within 2 to 3 days, a *second meiotic division* occurs, now forming four *spermatids* for each original primary spermatocyte. This division also is not a normal division. Instead, the two chromatids in each of the 23 chromosomes splits apart at the centromeres, forming

885

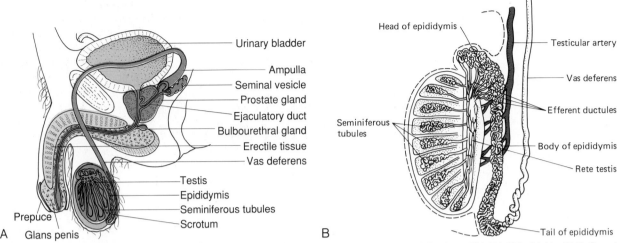

Figure 80-1. *A*, The male reproductive system. (Modified from Bloom and Fawcett: A Textbook of Histology. 10th Ed. Philadelphia, W. B. Saunders Company, 1975.) *B*, Internal structure of the testicle and relationship of the testicle to the epididymis. (From Guyton: Anatomy and Physiology. Philadelphia, Saunders College Publishing, 1985.)

two sets of 23 chromosomes. During the division one set of 23 chromosomes passes into one daughter spermatid and the other set into the second daughter spermatid.

The importance of these two meiotic divisions is that each spermatid that is finally formed carries only 23 chromosomes, having only half the genes of the original sperma-

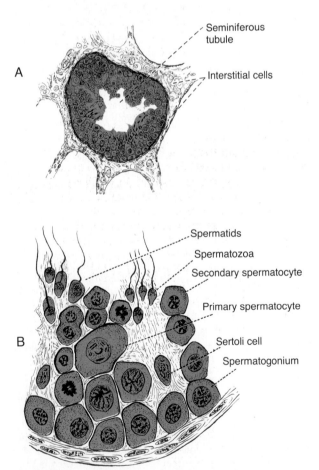

Figure 80-2. Cross-section of a seminiferous tubule, and the stages in the development of sperm from spermatogonia.

togonium. Therefore, the eventual spermatozoon that fertilizes the female ovum will provide half of the genetic material to the fertilized ovum, and the ovum will provide the other half.

During the following few weeks, each spermatid is nursed and physically reshaped by its enveloping Sertoli cell, changing it slowly into a spermatozoon (a sperm) by (1) losing some of its cytoplasm, (2) reorganizing the chromatin material of its nucleus to form a compact head, and (3) collecting the remaining cytoplasm and cell membranes at one end of the cell to form a tail.

All of the stages of the final conversion of the spermatocytes into sperm occur with the spermatocytes and spermatids actually embedded in the Sertoli cells. The Sertoli cells nurture and control the spermatogenesis process. The entire period of spermatogenesis from germinal cell to sperm takes about 74 days.

The Sex Chromosomes. In each spermatogonium one of the 23 pairs of chromosomes carries the genetic information that determines the sex of the eventual offspring. This pair is composed of one X chromosome, which is called the *female chromosome,* and one Y chromosome, the *male chromosome.* During meiotic division the sex-determining chromosomes divide among the secondary spermatocytes so that half the sperm become *male sperm* containing the Y chromosome and the other half *female sperm* containing the X chromosome. The sex of the offspring is determined by which of these two types of sperm fertilizes the ovum. This is discussed further in Chapter 82.

Formation of Sperm. When the spermatids are first formed, they still have the usual characteristics of epithelioid cells, but soon each spermatid begins to elongate into a spermatozoon, illustrated in Figure 80-3, composed of a *head* and a *tail.* The head is composed of the condensed nucleus of the cell with only a very thin cytoplasmic and cell membrane layer around its surface. However, on the outside of the anterior two thirds of the head is a thick cap called the acrosome that has been formed mainly from the Golgi apparatus. This contains a number of enzymes similar to those found in lysosomes of the typical cell, including *hyaluronidase,* which can digest proteoglycan filaments of tissues, and powerful *proteolytic enzymes,* which can digest proteins. These enzymes play important roles in allowing the sperm to fertilize the ovum.

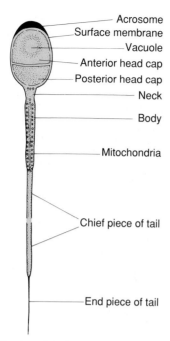

Acrosome
Surface membrane
Vacuole
Anterior head cap
Posterior head cap
Neck
Body
Mitochondria
Chief piece of tail
End piece of tail

Figure 80-3. Structure of the human spermatozoon.

The tail of the sperm, called the *flagellum*, has three major components: (1) a central skeleton constructed of 11 microtubules, collectively called the *axoneme* — the structure of this is similar to that of cilia described in Chapter 2; (2) a thin cell membrane covering the axoneme; and (3) a collection of mitochondria surrounding the axoneme in the proximal portion of the tail (called the *body of the tail*).

To-and-fro movement of the tail (flagellar movement) provides motility for the sperm. This movement results from a rhythmic longitudinal sliding motion between the anterior and posterior tubules that make up the axoneme. The energy for this process is supplied in the form of adenosine triphosphate synthesized by the mitochondria in the body of the tail. Normal sperm move in a straight line at a velocity of 1 to 4 mm/min. This allows them to move through the female genital tract in quest of the ovum.

Hormonal Factors That Stimulate Spermatogenesis

We shall speak more about the role of hormones in reproduction later, but at this point we need to note that several hormones play essential roles in spermatogenesis. Some of these are as follows:

1. *Testosterone*, secreted by the Leydig cells located in the interstitium of the testis, is essential for growth and division of the germinal cells in forming sperm.

2. *Luteinizing hormone (LH)*, secreted by the anterior pituitary gland, stimulates the Leydig cells to secrete testosterone.

3. *Follicle-stimulating hormone (FSH)*, also secreted by the anterior pituitary gland, stimulates the Sertoli cells; without this stimulation, the conversion of the spermatids to sperm (the process of spermiogenesis) will not occur.

4. *Estrogens*, formed from testosterone by the Ser-

toli cells when they are stimulated by FSH, are probably also essential for spermiogenesis. The Sertoli cells also secrete an *androgen-binding protein* that binds both testosterone and estrogens and carries these into the fluid in the seminiferous tubular lumen, making both these hormones available to the maturing sperm.

5. *Growth hormone* (as well as most of the other hormones) is necessary for controlling background metabolic functions of the testes. Growth hormone specifically promotes early division of the spermatogonia themselves; in its absence, as in pituitary dwarfs, spermatogenesis is severely deficient or entirely absent.

Maturation of Sperm in the Epididymis

Following formation in the seminiferous tubules, the sperm require several days to pass through the 6 meter long *epididymis*. Sperm removed from the seminiferous tubules and from the early portions of the epididymis are completely nonmotile, and they cannot fertilize an ovum. However, after the sperm have been in the epididymis for some 18 to 24 hours, they develop the capability of motility, even though several inhibitory proteins in the epididymal fluid still prevent actual motility until after ejaculation. The sperm also become capable of fertilizing the ovum if placed in the uterus, a process called *maturation*. The Sertoli cells and the epithelium of the epididymis secrete a special nutrient fluid containing hormones (both testosterone and estrogens), enzymes, and special nutrients that may be important or even essential for sperm maturation.

Storage of Sperm. The two testes of the young adult form about 120 million sperm each day. A small quantity of these can be stored in the epididymis, but most of them are stored in the vas deferens and the ampulla of the vas deferens. They can remain stored, maintaining their fertility, in the genital ducts for at least a month. During all this time they are kept in a deeply suppressed inactive state because of multiple inhibitory substances in the secretions of the ducts. On the other hand, with excessive sexual activity storage may be no longer than a few days at most.

Physiology of the Mature Sperm. The usual motile and fertile sperm are capable of flagellated movement through the fluid media at a rate of approximately 1 to 4 mm/min. Furthermore, *normal* sperm tend to travel in a straight line, rather than with a circuitous movement. The activity of sperm is greatly enhanced in neutral and slightly alkaline media as exists in the ejaculated semen, but it is greatly depressed in mildly acid media, and strong acid media can cause rapid death of sperm. The activity of sperm increases markedly with increasing temperature, but so does the rate of metabolism, causing the life of the sperm to be considerably shortened. Though sperm can live for many weeks in the genital ducts of the testes, the life of sperm in the female genital tract is only 1 to 2 days.

FUNCTION OF THE SEMINAL VESICLES

From early anatomical studies of the seminal vesicles it was erroneously believed that sperm were stored in these, whence came the name "seminal vesicles." However, these structures are only secretory glands, not sperm storage areas.

Each seminal vesicle is a tortuous, loculated tube lined with a secretory epithelium that secretes a mucoid material containing an abundance of *fructose* and other nutrient substances, as well as large quantities of *prostaglandins* and *fibrinogen*. During the process of emission each seminal vesicle empties its contents into the ejaculatory duct shortly after the vas deferens empties the sperm. This adds greatly to the bulk of the ejaculated semen, and the fructose and other substances in the seminal fluid are of considerable nutrient value for the ejaculated sperm until one of them fertilizes the ovum. The prostaglandins are believed to aid fertilization in two ways: (1) by reacting with the cervical mucus to make it more receptive to sperm and (2) possibly causing reverse peristaltic contractions in the uterus and fallopian tubes to move the sperm toward the ovaries (a few sperm reach the upper end of the fallopian tubes within 5 minutes).

FUNCTION OF THE PROSTATE GLAND

The prostate gland secretes a thin, milky, alkaline fluid containing citrate ion, calcium, acid phosphate, a clotting enzyme, and a profibrinolysin. During emission, the capsule of the prostate gland contracts simultaneously with the contractions of the vas deferens so that the thin, milky fluid of the prostate gland adds further to the bulk of the semen. The alkaline characteristic of the prostatic fluid may be quite important for successful fertilization of the ovum, because the fluid of the vas deferens is relatively acidic owing to the presence of metabolic end products of the sperm and, consequently, helps to inhibit sperm fertility. Also, the vaginal secretions of the female are acidic (pH of 3.5 to 4.0). Sperm do not become optimally motile until the pH of the surrounding fluids rises to approximately 6 to 6.5. Consequently, it is probable that prostatic fluid neutralizes the acidity of these other fluids after ejaculation and greatly enhances the motility and fertility of the sperm.

SEMEN

Semen, which is ejaculated during the male sexual act, is composed of the fluid and sperm from the vas deferens (about 10 per cent of the total), fluid from the seminal vesicles (about 60 per cent), fluid from the prostate gland (about 30 per cent), and small amounts from the mucous glands, especially the bulbourethral glands. Thus, the bulk of the semen is seminal vesicle fluid, which is the last to be ejaculated and serves to wash the sperm out of the ejaculatory duct and urethra. The average pH of the combined semen is approximately 7.5, the alkaline prostatic fluid having neutralized the mild acidity of the other portions of the semen. The prostatic fluid gives the semen a milky appearance, while fluid from the seminal vesicles and from the mucous glands gives the semen a mucoid consistency. Also, the clotting enzyme of the prostatic fluid causes the fibrinogen of the seminal vesicle fluid to form a weak coagulum that holds the semen in the deeper regions of the vagina where the uterine cervix lies. The coagulum then dissolves during the next 15 to 30 minutes because of lysis by fibrinolysin formed from the prostatic profibrinolysin. In the early minutes after ejaculation, the sperm remain relatively immobile, possibly because of the viscosity of the coagulum. However, as the coagulum dissolves, the sperm simultaneously become highly motile.

Although sperm can live for many weeks in the male genital ducts, once they are ejaculated in the semen their maximal life span is only 24 to 48 hours at body temperature. At lowered temperatures, however, semen may be stored for several weeks; and when frozen at temperatures below $-100°C$, sperm have been preserved for years.

Capacitation of the Spermatozoa — Making It Possible for Them to Penetrate the Ovum

Although the spermatozoa are said to be "mature" when they leave the epididymis, nevertheless, their activity is still held in check by multiple inhibitory factors secreted by the genital duct epithelia. Therefore, when they are first expelled in the semen, they are still unable to perform their duties in fertilizing the ovum. However, on coming in contact with the fluids of the female genital tract, multiple changes occur that activate the sperm for the final processes of fertilization. These collective changes are called *capacitation of the spermatozoa*. This normally requires from 1 to 10 hours. Some changes that are known to occur are the following:

1. The uterine and fallopian tube fluids wash away the various inhibitory factors that had suppressed sperm activity in the male genital ducts.

2. While the spermatozoa had remained in the fluid of the male genital ducts, they were continually exposed to many floating vesicles from the seminiferous tubules containing large amounts of cholesterol. This cholesterol was continually donated to the cellular membrane covering the sperm acrosome, toughening this membrane and preventing release of its enzymes. After ejaculation, the sperm that are deposited in the vagina swim away from the cholesterol vesicles upward into the uterine fluid, and they gradually lose much of their excess cholesterol through the next few hours. In so doing the membrane at the head of the sperm becomes much weaker.

3. The membrane of the sperm head also becomes much more permeable to calcium ions, so that calcium now enters the sperm in abundance and changes the activity of the flagellum, giving it a powerful whiplash motion in contrast to its previously weak undulating

motion. In addition, the calcium ions probably also cause changes in the intracellular membrane covering the leading edge of the acrosome, making it possible for the acrosome to release its enzymes very rapidly and easily as the sperm penetrates the granulosa cell mass surrounding the ovum and even more so as it attempts to penetrate the zona pellucida of the ovum itself.

Thus, multiple changes occur during capacitation. Without these, the sperm cannot make its way to the interior of the ovum to cause fertilization.

The Acrosome Enzymes, the "Acrosome Reaction," and Penetration of the Ovum

Stored in the acrosomes of the sperm are large quantities of *hyaluronidase* and *proteolytic enzymes.* Hyaluronidase depolymerizes the hyaluronic acid polymers in the intercellular cement that holds the granulosa cells together. The proteolytic enzymes digest proteins in the structural elements of the tissues.

When the ovum is expelled from the ovarian follicle into the abdominal cavity and fallopian tube, it carries with it multiple layers of granulosal cells. Before a sperm can fertilize the ovum, it must first pass through the granulosa cell layer, and then it must penetrate through the thick covering of the ovum itself, the *zona pellucida.* In the sperm capacitation process, the anterior membrane of the acrosome fuses with the cell membrane at the head of the sperm. This allows beginning release of small amounts of the acrosome enzymes. It is believed that the hyaluronidase among these enzymes is especially important in opening pathways between the granulosa cells so that the sperm can reach the ovum.

On reaching the zona pellucida of the ovum, the anterior membrane of the sperm binds specifically with a receptor protein in the zona pellucida. Then, rapidly, the entire anterior membrane of the acrosome dissolves, and all the acrosomal enzymes are immediately released. Within minutes, these open a penetrating pathway for passage of the sperm head through the zona pellucida. The head at first enters the *perivitelline space* lying beneath the zona pellucida but outside the membrane of the underlying oocyte. Within 30 minutes, the membranes of the sperm head and of the oocyte fuse, and the sperm genetic material enters the oocyte to cause fertilization; then the embryo begins to develop, as discussed in Chapter 82.

Why Does Only One Sperm Enter the Oocyte? With as many sperm as there are, why does only one enter the oocyte? The reason is not entirely known, but some of the facts are the following:

First, only a few sperm ever get as far as the zona pellucida, so that it might be as long as 10, 20, or even 30 minutes before a second sperm arrives.

Second, within a few minutes after the first sperm penetrates the zona pellucida, calcium ions diffuse through the oocyte membrane and cause multiple cortical granules to be released by exocytosis from the oocyte into the perivitelline space. These granules contain substances that permeate all portions of the zona pellucida and prevent binding of additional sperm and even cause the sperm that have already bound to fall off.

And, third, changes in the oocyte membrane after its fusion with the sperm are believed to cause electrical depolarization; this, too, may play a role in fending off subsequent sperm.

At any rate, almost never does more than one sperm enter the oocyte during the process of fertilization.

ABNORMAL SPERMATOGENESIS AND MALE FERTILITY

The seminiferous tubular epithelium can be destroyed by a number of different diseases. For instance, bilateral orchitis resulting from mumps causes sterility in a large percentage of afflicted males. Also, many male infants are born with degenerate tubular epithelium as a result of strictures in the genital ducts or as a result of genetic abnormalities. Finally, another cause of sterility, usually temporary, is excessive temperature of the testes, as follows:

Effect of Temperature on Spermatogenesis. Increasing the temperature of the testes can prevent spermatogenesis by causing degeneration of most cells of the seminiferous tubules besides the spermatogonia.

It has often been stated that the reason the testicles are located in the dangling scrotum is to maintain the temperature of these glands below the internal temperature of the body, although usually only about 2°C below the internal temperature. On cold days scrotal reflexes cause the musculature of the scrotum to contract, pulling the testicles close to the body. Thus the scrotum theoretically acts as a cooling mechanism for the testicles, without which spermatogenesis is said to be deficient during hot weather.

Cryptorchidism. Cryptorchidism means failure of a testis to descend from the abdomen into the scrotum. During the development of the male fetus, the testes are derived from the genital ridges in the abdomen. However, during the late stages of gestation, the testes descend through the inguinal canals into the scrotum. Occasionally this descent does not occur at all or occurs incompletely, so that one or both testes remain in the abdomen, in the inguinal canal, or elsewhere along the route of descent.

A testicle that remains throughout life in the abdominal cavity is incapable of forming sperm. The tubular epithelium is completely degenerate, leaving only the interstitial structures of the testis. It has been claimed that even the few degrees' higher temperature in the abdomen than in the scrotum is sufficient to cause degeneration of the tubular epithelium and consequently to cause sterility, although this is questionable. Nevertheless, for this reason operations to relocate the cryptorchid testes from the abdominal cavity into the scrotum prior to the beginning of adult sexual life are frequently performed on boys who have undescended testes.

Testosterone secretion by the fetal testes themselves is the normal stimulus that causes the testes to descend into the scrotum from the abdomen. Therefore, many, if not most, instances of cryptorchidism are caused by abnormally formed testes that are unable to secrete enough testosterone. Obviously, the surgical operation for cryptorchidism in these patients is unlikely to be successful.

Effect of Sperm Count on Fertility. The usual quantity of semen ejaculated at each coitus averages approxi-

Figure 80–4. Abnormal infertile sperm, compared with a normal sperm on the right.

mately 3.5 milliliters, and in each milliliter of semen is an average of approximately 120 million sperm, though even in "normal" persons this can vary from 35 to 200 million. This means an average total of 400 million sperm is usually present in each ejaculate. When the number of sperm in each milliliter falls below approximately 20 million, the person is likely to be infertile. Thus, even though only a single sperm is necessary to fertilize the ovum, for reasons not yet completely understood, the ejaculate usually must contain a tremendous number of sperm for one to fertilize the ovum.

Effect of Sperm Morphology and Motility on Fertility. Occasionally a man has a completely normal number of sperm but is still infertile. When this occurs, as many as half the sperm are sometimes found to be abnormal physically, having two heads, abnormally shaped heads, or abnormal tails, as illustrated in Figure 80–4; at other times, the sperm appear to be structurally normal but for reasons not understood are either entirely nonmotile or relatively nonmotile. Whenever the majority of the sperm are morphologically abnormal or are found to be nonmotile, the person is likely to be infertile even though the remainder of the sperm appear to be normal.

● THE MALE SEXUAL ACT

NEURONAL STIMULUS FOR PERFORMANCE OF THE MALE SEXUAL ACT

The most important source of impulses for initiating the male sexual act is the glans penis, for the glans contains an especially sensitive sensory end-organ system that transmits into the central nervous system that special modality of sensation called *sexual sensation.* The massaging action of intercourse on the glans stimulates the sensory end-organs, and the sexual sensations in turn pass through the pudendal nerve, thence through the sacral plexus into the sacral portion of the spinal cord, and finally up the cord to undefined areas of the cerebrum. Impulses may also enter the spinal cord from areas adjacent to the penis to aid in stimulating the sexual act. For instance, stimulation of the anal epithelium, the scrotum, and perineal structures in general can all send signals into the cord that add to the sexual sensation. Sexual sensations can even originate in internal structures, such as irritated areas of the urethra, the bladder, the prostate, the seminal vesicles, the testes, and the vas deferens. Indeed, one of the causes of "sexual drive" is filling of the sexual organs with secretions. Infection and inflammation of these sexual organs sometimes cause almost continual sexual desire, and "aphrodisiac" drugs, such as cantharides, increase the sexual desire by irritating the bladder and urethral mucosa.

The Psychic Element of Male Sexual Stimulation. Appropriate psychic stimuli can greatly enhance the ability of a person to perform the sexual act. Simply thinking sexual thoughts or even dreaming that the act of intercourse is being performed can cause the male sexual act to occur and to culminate in ejaculation. Indeed, *nocturnal emissions* during dreams occur in many males during some stages of sexual life, especially during the teens.

Integration of the Male Sexual Act in the Spinal Cord. Though psychic factors usually play an important part in the male sexual act and can actually initiate or inhibit it, the cerebrum is probably not absolutely necessary for its performance, because appropriate genital stimulation can cause ejaculation in some animals and occasionally in a human being after their spinal cords have been cut above the lumbar region. Therefore, the male sexual act results from inherent reflex mechanisms integrated in the sacral and lumbar spinal cord, and these mechanisms can be initiated by either psychic stimulation or actual sexual stimulation but most likely a combination of both.

STAGES OF THE MALE SEXUAL ACT

Erection; Role of the Parasympathetic Nerves. Erection is the first effect of male sexual stimulation, and the degree of erection is proportional to the degree of stimulation, whether psychic or physical.

Erection is caused by parasympathetic impulses that pass from the sacral portion of the spinal cord through the *nervi erigentes* to the penis. These parasympathetic impulses dilate the arteries of the penis, thus allowing arterial blood to build up under high pressure in the *erectile tissue* of the *corpus cavernosum* and *corpus spongiosum* in the shaft of the penis, illustrated in Figure 80–5. This erectile tissue is nothing more than large, cavernous, venous sinusoids, which are normally relatively empty but which become dilated tremendously when arterial blood flows rapidly into them under pressure, since the venous outflow is partially occluded. Also, the erectile bodies, especially the two corpora cavernosa, are surrounded by strong fibrous coats; therefore, high pressure within the sinusoids causes ballooning of the erectile tissue to such an extent that the penis becomes hard and elongated.

Lubrication, a Parasympathetic Function. During sexual stimulation, the parasympathetic impulses, in addition to promoting erection, cause the urethral glands and the bulbourethral glands to se-

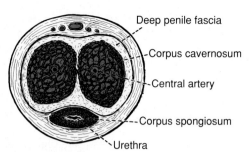

Figure 80-5. Erectile tissue of the penis.

crete mucus. This mucus flows through the urethra during intercourse to aid in the lubrication of coitus. However, most of the lubrication of coitus is provided by the female sexual organs rather than by the male. Without satisfactory lubrication, the male sexual act is rarely successful because unlubricated intercourse causes grating, painful sensations that inhibit rather than excite sexual sensations.

Emission and Ejaculation; Function of the Sympathetic Nerves. Emission and ejaculation are the culmination of the male sexual act. When the sexual stimulus becomes extremely intense, the reflex centers of the spinal cord begin to emit *sympathetic impulses* that leave the cord at L-1 and L-2 and pass to the genital organs through the hypogastric and pelvic plexuses to initiate emission, the forerunner of ejaculation.

Emission begins with contraction of the vas deferens and the ampulla to cause expulsion of sperm into the internal urethra. Then, contractions of the muscular coat of the prostate gland followed finally by contraction of the seminal vesicles expel prostatic fluid and seminal fluid, forcing the sperm forward. All of these fluids mix in the internal urethra with the mucus already secreted by the bulbourethral glands to form the semen. The process to this point is *emission.*

The filling of the internal urethra then elicits sensory signals that are transmitted through the pudendal nerves to the sacral regions of the cord, giving the feeling of sudden fullness in the internal genital organs. Also, these sensory signals further excite the rhythmic contraction of the internal genital organs and also cause contraction of the ischiocavernosus and bulbocavernosus muscles that compress the bases of the penile erectile tissue. These effects together cause rhythmic, wavelike increases in pressure in the genital ducts and urethra, which "ejaculate" the semen from the urethra to the exterior. The process is called *ejaculation.* At the same time, rhythmic contractions of the pelvic muscles and even of some of the muscles of the body trunk cause thrusting movements of the pelvis and penis, which also help propel the semen into the deepest recesses of the vagina and perhaps even slightly into the cervix of the uterus.

This entire period of emission and ejaculation is called the *male orgasm.* At its termination, the male sexual excitement disappears almost entirely within 1 to 2 minutes, and erection ceases.

● TESTOSTERONE AND OTHER MALE SEX HORMONES

SECRETION, METABOLISM, AND CHEMISTRY OF THE MALE SEX HORMONE

Secretion of Testosterone by the Interstitial Cells of Leydig in the Testes. The testes secrete several male sex hormones, which are collectively called *androgens,* including *testosterone, dihydrotestosterone,* and *androstenedione.* However, testosterone is so much more abundant than the others that one can consider it to be the significant testicular hormone, although as we shall see, much, if not most, of the testosterone is converted into the more active hormone dihydrotesterone in the target tissues.

Testosterone is formed by the *interstitial cells of Leydig,* which lie in the interstices between the seminiferous tubules and constitute about 20 per cent of the mass of the adult testes, as illustrated in Figure 80-6. Leydig cells are almost nonexistent in the testes during childhood, when the testes secrete almost no testosterone, but they *are* numerous in the newborn male infant and also in the adult male any time after puberty; at both these times the testes secrete large quantities of testosterone. Furthermore, when tumors develop from the interstitial cells of Leydig, great quantities of testosterone are secreted. Finally, when the germinal epithelium of the testes is destroyed by x-ray treatment or by excessive heat, the Leydig cells, which are less easily destroyed, continue to produce testosterone.

Secretion of Androgens Elsewhere in the Body. The term "androgen" means any steroid hormone that has masculinizing effects, including, of course, testosterone itself; but it also includes male sex hormones produced elsewhere in the body besides the testes. For instance, the adrenal glands secrete at least five different androgens, though

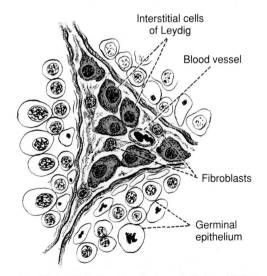

Figure 80-6. Interstitial cells of Leydig, the cells that secrete testosterone, located in the interstices between the seminiferous tubules.

the total masculinizing activity of all these is normally so slight that they do not cause significant masculine characteristics even in women, except for causing growth of pubic and axillary hair. But when an adrenal tumor of the adrenal androgen–producing cells occurs, the quantity of androgenic hormones may then become great enough to cause all the usual male secondary sexual characteristics. These effects are described in connection with the adrenogenital syndrome in Chapter 77.

Rarely, embryonic rest cells in the ovary can develop into a tumor that produces excessive quantities of androgens in women; one such tumor is the *arrhenoblastoma.* The normal ovary also produces minute quantities of androgens, but these are not significant.

Chemistry of the Androgens. All androgens are steroid compounds, as illustrated by the formulas in Figure 80–7 for *testosterone* and *dihydrotestosterone.* Both in the testes and in the adrenals, the androgens can be synthesized either from cholesterol or directly from acetyl coenzyme A.

Metabolism of Testosterone. After secretion by the testes, most of the testosterone becomes loosely bound with plasma albumin or more tightly with a beta globulin called *gonadal steroid-binding globulin* and circulates in the blood for about 30 minutes to an hour or so. By that time it either becomes fixed to the tissues or degraded into inactive products that are subsequently excreted.

Much of the testosterone that becomes fixed to the tissues is converted within the cells to *dihydrotestosterone,* and lesser quantities into 5-α-androstanediol, especially in certain target organs such as the prostate gland in the adult and in the external genitalia of the fetal male. Some actions of testosterone are dependent on this conversion, whereas other actions are not. The intracellular functions are discussed later in the chapter.

Degradation and Excretion of Testosterone. The testosterone that does not become fixed to the tissues is rapidly converted, mainly by the liver, into *androsterone* and *dehydroepiandrosterone* and simultaneously conjugated either as glucuronides or sulfates (glucuronides, particularly). These are excreted either into the gut in the bile or into the urine.

Production of Estrogen in the Male. In addition to testosterone, small amounts of estrogens are formed in the male (about one-fifth the amount in the nonpregnant female), and a reasonable quantity of these can be recovered from a man's urine.

The exact source of the estrogens in the male is also still doubtful, but the following are known: (1) The concentration of estrogens in the fluid of the seminiferous tubules is quite high and probably plays an important role in spermiogenesis. This estrogen is believed to be formed by the Sertoli cells by converting testosterone mainly to estradiol. (2) Estrogens are formed from testosterone and androstanediol in other tissues of the body, especially the liver, probably accounting for as much as 80 per cent of the total male estrogen production.

FUNCTIONS OF TESTOSTERONE

In general, testosterone is responsible for the distinguishing characteristics of the masculine body. Even during fetal life, the testes are stimulated by chorionic gonadotropin from the placenta to produce moderate quantities of testosterone throughout the entire period of fetal development and for up to 10 or more weeks after birth; thereafter, essentially no testosterone is produced during childhood until approximately the age of 10 to 13 years. Then testosterone production increases rapidly under the stimulus of anterior pituitary gonadotropic hormones at the onset of puberty and lasts throughout most of the remainder of life, as illustrated in Figure 80–8, dwindling rapidly beyond the age of 50 to become 20 to 50 per cent of the peak value by the age of 80.

Functions of Testosterone During Fetal Development

Testosterone begins to be elaborated by the male fetal testes at about the seventh week of embryonic life. Indeed, one of the major functional differences between the female and the male sex chromosome is that the male chromosome causes the newly developing genital ridge to secrete testosterone, whereas the female chromosome causes this ridge to secrete estrogens. Injection of large quantities of male sex hormone into pregnant animals causes development of male sexual organs even though the fetus is female. Also, removal of the testes in a male fetus causes development of female sexual organs. Therefore, testosterone secreted first by the genital ridges and later by

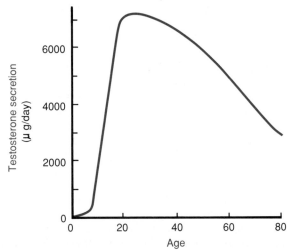

Figure 80–8. Approximate rates of testosterone secretion at different ages.

Figure 80–7. Testosterone and dihydrotestosterone.

Testosterone   Dihydrotestosterone

the fetal testes is responsible for the development of the male body characteristics, including the formation of a penis and a scrotum rather than the formation of a clitoris and a vagina. Also, it causes formation of the prostate gland, the seminal vesicles, and the male genital ducts, while at the same time suppressing the formation of female genital organs.

Effect on the Descent of the Testes. The testes usually descend into the scrotum during the last 2 months of gestation, when the testes are secreting reasonable quantities of testosterone. If a male child is born with undescended but otherwise normal testes, administration of testosterone can still cause the testes to descend in the usual manner if the inguinal canals are large enough to allow the testes to pass. Administration of gonadotropic hormones, which stimulate the Leydig cells of the testes to produce testosterone, can also cause the testes to descend. Thus, the stimulus for descent of the testes is testosterone, indicating again that testosterone is an important hormone for male sexual development during fetal life.

Effect of Testosterone on Development of Adult Primary and Secondary Sexual Characteristics

After puberty, the reinitiation of testosterone secretion causes the penis, the scrotum, and the testes all to enlarge about eightfold before the age of 20 years. In addition, testosterone causes the "secondary sexual characteristics" of the male to develop at the same time, beginning at puberty and ending at maturity. These secondary sexual characteristics, in addition to the sexual organs themselves, distinguish the male from the female as follows:

Effect on the Distribution of Body Hair. Testosterone causes growth of hair (1) over the pubis, (2) upward along the linea alba sometimes to the umbilicus and above, (3) on the face, (4) usually on the chest, and (5) less often on other regions of the body, such as the back. It also causes the hair on most other portions of the body to become more prolific.

Baldness. Testosterone decreases the growth of hair on the top of the head; a man who does not have functional testes does not become bald. However, many virile men never become bald, for baldness is a result of two factors: first, a *genetic background* for the development of baldness and, second, superimposed on this genetic background, *large quantities of androgenic hormones.* A woman who has the appropriate genetic background and who develops a long-sustained androgenic tumor becomes bald in the same manner as does a man.

Effect on the Voice. Testosterone secreted by the testes or injected into the body causes hypertrophy of the laryngeal mucosa and enlargement of the larynx. The effects cause at first a relatively discordant, "cracking" voice, but this gradually changes into the typical adult masculine bass voice.

Effect on the Skin, and Development of Acne. Testosterone increases the thickness of the skin over the entire body and increases the ruggedness of the subcutaneous tissues.

Testosterone increases the rate of secretion by some or perhaps all the sebaceous glands. Especially important is the excessive secretion by the sebaceous glands of the face, for oversecretion of these can result in *acne.* Therefore, acne is one of the most common features of adolescence when the male body is first becoming introduced to increased testosterone. After several years of testosterone secretion, the skin adapts itself to the testosterone in some way that allows it to overcome the acne.

Effect on Protein Formation and Muscular Development. One of the most important male characteristics is the development of increasing musculature following puberty, averaging about a 50 per cent increase in muscle mass over that in the female. This is associated with increased protein in other parts of the body as well. Many of the changes in the skin are also due to deposition of proteins in the skin, and the changes in the voice probably even result from this protein anabolic function of testosterone.

Because of the very great effect that testosterone has on the body musculature, it (or more usually a synthetic androgen instead) is widely used by athletes to improve their muscular performance. This practice is to be severely deprecated because of prolonged harmful effects of excess testosterone, as we discuss in Chapter 84 in relation to sports physiology. Testosterone is also used in old age as a "youth hormone" to improve muscle strength and vigor.

Effect on Bone Growth and Calcium Retention. Following puberty or following prolonged injection of testosterone, the bones grow considerably in thickness and also deposit considerable additional calcium salts. Thus, testosterone increases the total quantity of bone matrix, and it also causes calcium retention. The increase in bone matrix is believed to result from the general protein anabolic function of testosterone, and the deposition of calcium salts to result secondarily to the increased bone matrix.

Finally, testosterone has a specific effect on the pelvis to (1) narrow the pelvic outlet, (2) lengthen it, (3) cause a funnel-like shape instead of the broad ovoid shape of the female pelvis, and (4) greatly increase the strength of the entire pelvis for load-bearing. In the absence of testosterone, the male pelvis develops into one very similar to that of the female.

Because of the ability of testosterone to increase the size and strength of bones, testosterone is often also used in old age to treat osteoporosis.

When great quantities of testosterone (or any other androgen) are secreted in the still-growing child, the rate of bone growth increases markedly, causing a spurt in total body height as well. However, the testosterone also causes the epiphyses of the long bones to unite with the shafts of the bones at an early age in life. Therefore, despite the rapidity of growth, this early uniting of the epiphyses prevents the person from growing as tall as he would have grown had testosterone not been secreted at all. Even in normal men the final adult height is slightly less than that which would have been attained had the person been castrated prior to puberty.

Effect on Basal Metabolism. Injection of large quantities of testosterone can increase the basal metabolic rate by as much as 15 per cent, and it is believed that even the usual quantity of testosterone secreted by the testes during adolescence and early adult life increases the rate of metabolism some 5 to 10 per cent above the value that it would be were the testes not active. This increased rate of metabolism is possibly an indirect result of the effect of testosterone on protein anabolism, the increased quantity of proteins—the enzymes especially—increasing the activities of all cells.

Effect on the Red Blood Cells. When normal quantities of testosterone are injected into a castrated adult, the number of red blood cells per cubic millimeter of blood increases 15 to 20 per cent. Also, the average man has about 700,000 more red blood cells per cubic millimeter than the average woman. However, this difference may be due partly to the increased metabolic rate following testosterone administration rather than to a direct effect of testosterone on red blood cell production.

Effect on Electrolyte and Water Balance. As pointed out in Chapter 77, many different steroid hormones can increase the reabsorption of sodium in the distal tubules of the kidneys. Testosterone has such an effect but only to a minor degree in comparison with the adrenal mineralocorticoids. Nevertheless, after puberty the blood and extracellular fluid volumes of the male in relation to his weight increase to some extent.

BASIC INTRACELLULAR MECHANISM OF ACTION OF TESTOSTERONE

Probably all or almost all the effects just listed result from increased rate of protein formation in the target cells. This has been studied extensively in the prostate gland, one of the organs that is most affected by testosterone. In this gland, testosterone enters the cells within a few minutes after secretion, is there converted to *dihydrotestosterone,* and binds with a cytoplasmic "receptor protein." This combination then migrates to the nucleus where it binds with a nuclear protein and induces the DNA-RNA transcription process. Within 30 minutes RNA polymerase has become activated and the concentration of RNA begins to increase in the cells; this is followed by progressive increase in cellular protein. After several days the quantity of DNA in the gland has also increased, and there has been a simultaneous increase in the number of prostatic cells.

Therefore, it is assumed that testosterone greatly stimulates production of proteins in general, though increasing more specifically those proteins in "target" organs or tissues responsible for the development of secondary sexual characteristics.

Some important target tissues do not have appropriate enzymes in their cells to convert testosterone into dihydrotestosterone. In these tissues, testosterone functions directly, though usually with only about half the potency, to induce the formation of cellular proteins. For instance, this direct effect of testosterone is essential in the male fetus for the development of the epididymis, the vas deferens, and the seminal vesicles. The direct effect is probably also responsible for much of the effects of testosterone on the musculature, the skeleton, and the larynx and also to promote spermatogenesis.

CONTROL OF MALE SEXUAL FUNCTIONS BY HORMONES FROM THE HYPOTHALAMUS AND ANTERIOR PITUITARY GLAND

A major share of the control of sexual functions in both the male and female begins with the secretion of *gonadotropin-releasing hormone* (GnRH) by the hypothalamus. This hormone in turn stimulates the anterior pituitary gland to secrete two other hormones called gonadotropic hormones: (1) *luteinizing hormone (LH)* and (2) *follicle-stimulating hormone (FSH).* In turn, luteinizing hormone is the primary stimulus for the secretion of testosterone by the testes, and FSH helps to stimulate spermatogenesis.

Gondotropin-Releasing Hormone and Its Effect in Increasing the Secretion of LH and FSH

GnRH is a 10 amino acid peptide secreted by neurons whose cell bodies are located mainly in the arcuate nuclei of the hypothalamus. The endings of these neurons terminate mainly in the median eminence of the hypothalamus, where they release GnRH into the hypothalamic-hypophysial portal vascular system. Then the GnRH is transported to the anterior pituitary gland in the portal blood and stimulates the release of the two gonadotropins, LH and FSH.

GnRH is secreted intermittently a few minutes at a time once every 1 to 3 hours. The intensity of this hormone's stimulus is determined in two ways: (1) by the frequency of these cycles of secretion and (2) by the quantity of GnRH released with each cycle. The secretion of LH by the anterior pituitary gland is also cyclic, with LH following fairly faithfully the pulsatile release of GnRH. On the other hand, FSH secretion increases and decreases only slightly with the fluctuating GnRH; instead, it changes more slowly over a period of many hours in response to longer term changes in GnRH. Because of the much closer relationship between GnRH secretion and LH hormone secretion, GnRH in the past has also been known widely as *LH-releasing hormone (LHRH).*

The Gonadotropic Hormones: LH and FSH

Both of the gonadotropic hormones, LH and FSH, are secreted by the same cells, called *gonadotropes,* in the anterior pituitary gland. In the total absence of GnRH from the hypothalamus, the gonadotropes secrete almost no LH or FSH.

LH and FSH are *glycoproteins;* however, the quantity of carbohydrate bound with the protein in the molecules varies considerably under different conditions, which may change the potencies of activity.

Both LH and FSH exert their effects on their target tissues in the testes by *activating the cyclic adenosine monophosphate (cAMP) second messenger system,* which in turn activates specific enzyme systems in the respective target cells.

Testosterone — Regulation of Its Production by LH. *Testosterone* is secreted by the *interstitial cells of Leydig* in the testes, but only when they are stimulated by LH from the pituitary gland. Furthermore, the quantity of testosterone secreted increases approximately in direct proportion to the amount of LH available.

Mature Leydig cells are not normally found in the child's testes, except for a few weeks after birth, until after the age of approximately 10 years. However, either injection of purified LH into a child at any age or secretion of LH at puberty causes cells that look like fibroblasts in the interstitial areas of the testes to evolve into interstitial cells of Leydig.

Reciprocal Inhibition of Anterior Pituitary Secretion of LH and FSH by Testosterone — Negative Feedback Control of Testosterone Secretion. The testosterone secreted by the testes in response to LH has the reciprocal effect of turning off anterior pituitary secretion of LH. It does this in two ways:

1. By far the greater part of the inhibition results from the direct effect of testosterone on the hypothalamus of decreasing the secretion of GnRH. This in turn causes a corresponding decrease in the secretion of both LH and FSH by the anterior pituitary, and the decrease in LH decreases the secretion of testosterone by the testes. Thus, whenever the secretion of testosterone becomes too great, this automatic negative feedback effect, operating through the hypothalamus and anterior pituitary gland, reduces the testosterone secretion back toward its normal operating level. Conversely, too little testosterone allows the hypothalamus to secrete large amounts of GnRH, with a corresponding increase in anterior pituitary LH and FSH secretion and an increase in testicular testosterone secretion.

2. Testosterone probably also has a weak negative feedback effect, acting directly on the anterior pituitary gland to diminish LH secretion. Consequently, a small degree of regulation of testosterone secretion is believed to occur in this way as well.

Regulation of Spermatogenesis by FSH and Testosterone

FSH binds with specific FSH receptors attached to the Sertoli cells in the seminiferous tubules. This causes these cells to grow and secrete various sper-

matogenic substances. Simultaneously, testosterone diffusing into the tubules from the Leydig cells in the interstitial spaces also has a strong tropic effect on spermatogenesis. To initiate spermatogenesis, both FSH and testosterone are necessary, although once the initial stimulation has occurred, testosterone alone can maintain spermatogenesis for a long time thereafter.

Negative Feedback Control of Seminiferous Tubule Activity — Role of Inhibin. When the seminiferous tubules fail to produce sperm, the secretion of FSH by the anterior pituitary gland increases markedly. Conversely, when spermatogenesis proceeds too rapidly, the secretion of FSH diminishes. The cause of this negative feedback effect on the anterior pituitary is believed to be secretion by the Sertoli cells of still another hormone called *inhibin*. This hormone has a strong direct effect on the anterior pituitary gland in inhibiting the secretion of FSH; it possibly also has a slight effect on the hypothalamus as well in inhibiting secretion of GnRH.

Inhibin is a glycoprotein, like both LH and FSH, having a molecular weight between 10,000 and 30,000. It has been isolated from cultured Sertoli cells. Its potent inhibitory feedback effect on the anterior pituitary gland is believed to provide an important negative feedback mechanism for control of spermatogenesis, operating simultaneously with and in parallel to the negative feedback mechanism for control of testosterone secretion.

Psychic Factors That Affect Gonadotropin Secretion and Sexual Activity

Many psychic factors, feeding especially from the limbic system of the brain into the hypothalamus, can affect the rate of secretion of GnRH by the hypothalamus and therefore also affect most other aspects of sexual and reproductive functions in both the male and the female. For instance, transporting a prize bull in a rough truck is said to inhibit fertility — and the human male is hardly different.

Human Chorionic Gonadotropin and Its Effect on the Fetal Testes

During pregnancy, still another hormone, *human chorionic gonadotropin* (hCG), is secreted by the placenta. This hormone has almost exactly the same effects on the sexual organs as LH, capable of stimulating the secretion of testosterone by the Leydig cells in the testes and therefore causing all the male sexual effects caused by testosterone.

During pregnancy, if the fetus is a male, chorionic gonadotropin from the placenta causes the testes to secrete testosterone. This testosterone is critical for promoting formation of the male sexual organs, as was pointed out earlier. We discuss hCG and its functions during pregnancy in greater detail in Chapter 82.

Puberty and Regulation of Its Onset

Initiation of the onset of puberty has long been a mystery. In the earliest history of humanity, the belief was simply that the testicles "ripened" at this time. With the discovery of the gonadotropins, ripening of the anterior pituitary gland was then considered responsible. But, now it is known from experiments in which both testicular and pituitary tissues have been transplanted from infant animals into adult animals that both the testes and the anterior pituitary of the infant are capable of performing adult functions if appropriately stimulated. Therefore, it is now certain that *during childhood the hypothalamus simply does not secrete significant amounts of GnRH.* One of the reasons for this is that the slightest secretion of sex steroid hormones exerts a very strong inhibitory effect on hypothalamic secretion of GnRH.

The Male Adult Sexual Life and the Male Climacteric. Following puberty, gonadotropic hormones are produced by the male pituitary gland for the remainder of life, and at least some spermatogenesis usually continues until death. Most men, however, begin to exhibit slowly decreasing sexual functions in their late 40s or 50s, and one study showed that the average age for terminating intersexual relations was 68, though the variation was great. This decline in sexual function is related to decrease in testosterone secretion, as depicted in Figure 80–8. The decrease in male sexual function is called the *male climacteric*. Occasionally, the male climacteric is associated with symptoms of hot flashes, suffocation, and psychic disorders similar to the menopausal symptoms of the female. These symptoms can be abrogated by administration of testosterone, synthetic androgens, or even estrogens, which are used for treatment of menopausal symptoms in the female.

● ABNORMALITIES OF MALE SEXUAL FUNCTION

THE PROSTATE GLAND AND ITS ABNORMALITIES

The prostate gland remains relatively small throughout childhood and begins to grow at puberty under the stimulus of testosterone. This gland reaches an almost stationary size by the age of about 20 years and remains at this size up to the age of approximately 50 years. At that time in some men it begins to involute, along with the decreased production of testosterone by the testes. However, a benign prostatic fibroadenoma frequently develops in the prostate in older men and causes urinary obstruction. This hypertrophy is not caused by testosterone.

Cancer of the prostate gland is a common cause of death, resulting in approximately 2 to 3 per cent of all male deaths.

Once cancer of the prostate gland does occur, the cancerous cells are usually stimulated to more rapid growth by testosterone and are inhibited by removal of the testes, so that testosterone cannot be formed. Also, prostatic cancer usually can be inhibited by administration of estrogens. Even some patients who have prostatic cancer that has already metastasized to almost all of the bones of the body can be successfully treated for a few months to years by removal of the testes, by estrogen therapy, or by both; after this therapy the metastases degenerate and the bones heal. This treatment does not completely stop the cancer but does slow it and greatly diminishes the severe bone pain.

HYPOGONADISM IN THE MALE

When the testes are nonfunctional during fetal life or when there is genetic absence of androgen receptors in the target cells, none of the male sexual organs develop. Instead, normal female organs are formed. The reason for this is that the basic genetic characteristic of the fetus, whether male or female, is to form female sexual organs if there are no sex hormones. But in the presence of testosterone, formation of the female sexual organs is suppressed, and instead male organs are induced.

When a boy loses his testes prior to puberty, a state of *eunuchism* ensues in which he continues to have infantile sexual characteristics throughout life. The height of the adult eunuch is slightly greater than that of the normal man, though the bones are quite thin, the muscles are considerably weaker than those of the normal man, and, obviously, the sexual organs and secondary sexual characteristics remain those of a child rather than those of an adult. The voice is childlike, there is no loss of hair on the head, and the normal masculine hair distribution on the face and elsewhere does not occur.

When a man is castrated after puberty, some male secondary sexual characteristics revert to those of a child, and others remain of masculine character. The sexual organs regress slightly in size but not to a childlike state, and the voice regresses from the bass quality only slightly. On the other hand, there is loss of masculine hair production, loss of the thick masculine bones, and loss of the musculature of the virile male.

In the castrated adult male, sexual desires are decreased but not totally lost, provided sexual activities have been practiced previously. Erection can still occur as before though with less ease; but it is rare that ejaculation can take place, primarily because the semen-forming organs become degenerate, and there has been a loss of the testosterone-driven psychic desire.

Some instances of hypogonadism are caused by the genetic inability of the hypothalamus to secrete normal amounts of GnRH. Often this is associated with a simultaneous abnormality of the feeding center of the hypothalamus, causing the person to greatly overeat. Consequently, severe obesity occurs along with eunuchism. A patient with this condition is illustrated in Figure 80–9; the condition is called *adiposogenital syndrome, Fröhlich's syndrome,* or *hypothalamic eunuchism.*

TESTICULAR TUMORS AND HYPERGONADISM IN THE MALE

Interstitial Leydig cell tumors develop rarely in the testes; but when they do develop, they sometimes produce as much as 100 times the normal quantities of testosterone. When such tumors develop in young children, they cause rapid growth of the musculature and bones but also early uniting of the epiphyses, so that the eventual adult height actually is less than that which would have been achieved otherwise. Obviously, such interstitial cell tumors cause excessive development of the sexual organs, of all muscles, and of all of the other secondary sexual characteristics. In the adult male, small interstitial cell tumors are difficult to diagnose because masculine features are already present.

Much more common than the interstitial Leydig cell

Figure 80–9. Adiposogenital syndrome in an adolescent male. Note the obesity and the childlike sexual organs. (Courtesy of Dr. Leonard Posey.)

tumors are tumors of the germinal epithelium. Because germinal cells are capable of differentiating into almost any type of cell, many of these tumors contain multiple tissues, such as placental tissue, hair, teeth, bone, skin, and so forth, all found together in the same tumorous mass called a *teratoma*. Often these tumors secrete no hormones, but if a significant quantity of placental tissue develops in the tumor, it may secrete large quantities of chorionic gonadotropin with functions very similar to those of LH. Also, estrogenic hormones are frequently secreted by these tumors and cause the condition called *gynecomastia*, overgrowth of the breasts.

● THE PINEAL GLAND—ITS FUNCTION IN CONTROLLING SEASONAL FERTILITY

For as long as the pineal gland has been known to exist, to it has been ascribed a myriad of different functions, including being the seat of the soul. It is known from comparative anatomy that the pineal gland is a vestigial remnant of what was a third eye high in the back of the head in lower animals. Many physiologists have been content with the idea that this gland is also a nonfunctional remnant, but others have claimed for many years that it plays important roles in the control of sexual activities and reproduction, functions that still others said were nothing more than the fanciful imaginings of physiologists preoccupied with sexual delusions.

But now, after years of dispute, it looks as though the sex advocates might have won, and that the pineal gland does indeed play an important regulatory role in sexual and reproductive function. For, in lower animals that bear their

young at certain seasons of the year and in which the pineal gland has been removed or the nervous circuits to the pineal gland have been sectioned, the normal annual periods of seasonal fertility are lost. To these animals such seasonal fertility is very important because it allows birth of the offspring in the spring and summer months, when survival is most likely. Unfortunately, the mechanism of this effect is still not entirely clear, but it seems to be the following:

First, the pineal gland is controlled by the amount of light seen by the eyes each day. For instance, in the hamster, a number of hours greater than 13 of *darkness* each day activates the pineal gland, while less than that amount of darkness fails to activate it, with a very critical balance between activation and nonactivation. The nervous pathway involved is this: passage of light signals from the eyes to the suprachiasmal nucleus of the hypothalamus and thence to the pineal gland activates pineal secretion.

Second, the pineal gland secretes *melatonin* and several other similar substances. Either melatonin or one of the other substances then passes either by way of the blood or through the fluid of the third ventricle to the anterior pituitary gland to control gonadotropic hormone secretion.

In the presence of pineal gland secretion, the gonadotropic hormone secretion is suppressed in at least some species of animals, and the gonads become inhibited and even involuted. This is what presumably occurs during the early winter months when there is increasing darkness. But after about 4 months of dysfunction, the gonadotropic hormone secretion breaks through the inhibitory effect of the pineal gland, and the gonads become functional once more, ready for a full springtime of activity.

But does the pineal gland have a similar function for control of reproduction in man? The answer to this is still far from known. However, tumors often occur in the region of the pineal gland. Some of these are pineal tumors that secrete excessive quantities of pineal hormones, whereas others are tumors of surrounding tissue and press on the pineal gland to destroy it. Both types of tumors are often associated with serious hypo- or hypergonadal function. So perhaps the pineal gland does play at least some role in controlling sexual drive and reproduction in man.

REFERENCES

Berger, F. G., and Watson, G: Androgen-regulated gene expression. Annu. Rev. Physiol., 51:51, 1989.
Beyer, C., and Feder, H. H.: Sex steroids and afferent input: Their roles in brain sexual differentiation. Annu. Rev. Physiol., 49:349, 1987.
Brooks, D. E.: Metabolic activity in the epididymis and its regulation by androgens. Physiol. Rev., 61:515, 1981.
Byskov, A. G.: Differentiation of mammalian embryonic gonad. Physiol. Rev., 66:71, 1986.
Conn, P. M., et al.: Gonadotropin-releasing hormone: Molecular and cell biology, physiology, and clinical applications. Fed. Proc., 43:2351, 1984.
Conn, P. M., et al.: Mechanism of action of gonadotropin releasing hormone. Annu. Rev. Physiol., 48:495, 1986.
Cooke, B. A., and Sharpe, R. M.: The Molecular and Cellular Endocrinology of the Testis. New York, Raven Press, 1988.
Darszon, A., et al.: Ionic channels in sea urchin sperm physiology. News Physiol. Sci., 3:181, 1988.
DeFelice, L. J., and Kell, M. J.: Sperm-activated currents in ascidian oocytes. Develop. Biol., 119:123, 1987.
DeGroot, L. J., et al. (eds.): Endocrinology, 2nd Ed. Philadelphia, W. B. Saunders Co., 1989.
Diczfalusy, E., and Bygdeman, M.: Fertility Regulation Today and Tomorrow. New York, Raven Press, 1987.
Donahoe, P. K., et al.: Mechanism of action of mullerian inhibiting substance. Annu. Rev. Physiol., 46:53, 1984.
Habenicht, U. F.: Hormonal Regulation of Testicular Descent. New York, Springer-Verlag, 1983.
Hinton, B. T., and Turner, T. T.: Is the epididymis a kidney analogue? News Physiol. Sci., 3:28, 1988.

Figure 81–1. The female reproductive organs.

Figure 81–3. Approximate plasma concentrations of the gonadotropins and ovarian hormones during the normal female sexual cycle.

rates of secretion of the female hormones and corresponding changes in the ovaries and sexual organs as well. This rhythmic pattern is called the *female sexual cycle* (or less accurately, the *menstrual cycle*). The duration of the cycle averages 28 days. It may be as short as 20 days or as long as 45 days even in completely normal women, though abnormal cycle length is frequently associated with decreased fertility.

The two significant results of the female sexual cycle are: First, only a *single* mature ovum is normally released from the ovaries each month, so that only a single fetus can begin to grow at a time. Second, the uterine endometrium is prepared for implantation of the fertilized ovum at the required time of the month.

THE GONADOTROPIC HORMONES AND THEIR EFFECTS ON THE OVARIES

The ovarian changes during the sexual cycle depend completely on the gonadotropic hormones, *FSH* and *LH*, secreted by the anterior pituitary gland. Ovaries that are not stimulated by these hormones remain completely inactive, which is essentially the case throughout childhood, when almost no gonadotropic hormones are secreted. However, at the age of 9 to 10 years, the pituitary begins to secrete progressively more FSH and LH, which culminates in the initiation of monthly sexual cycles between the ages

of 11 and 16 years. This period of change is called *puberty*, and the first menstrual cycle is called *menarche.*

(The ovaries also function during fetal life because of stimulation by another gonadotropic hormone, *chorionic gonadotropin*, secreted by the placenta as we discuss in Chapter 83. But within a few weeks after birth this stimulus is lost, and the ovaries become almost totally dormant until the prepubertal period.)

Both FSH and LH are small glycoproteins having molecular weights of about 30,000. The only significant effects of FSH and LH are on the testes in the male, as explained in the previous chapter, and on the ovaries in the female.

During each month of the female sexual cycle, there is a cyclic increase and decrease of both FSH and LH, as illustrated in Figure 81–3. These cyclic variations in turn cause the cyclic ovarian changes, which are explained in the following sections.

Both FSH and LH stimulate their ovarian target cells by combining with highly specific FSH and LH receptors in the cell membranes. The activated receptors in turn increase both the rates of secretion of these cells and growth and proliferation of the cells. Almost all of these stimulatory effects result from *activation of the cyclic adenosine monophosphate*

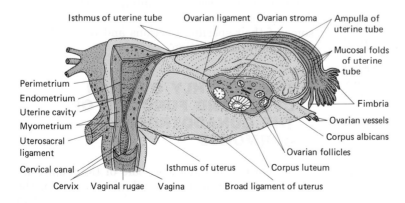

Figure 81–2. The internal structures of the uterus, ovary, and a uterine tube. (From Guyton: Physiology of the Human Body. 6th ed. Philadelphia, Saunders College Publishing, 1984.)

(cAMP) second messenger system in the cell cytoplasm, which in turn causes the formation of protein kinase, then multiple phosphorylations of key enzymes that excite many intracellular functions, as explained in Chapter 74. However, some effects of the hormones cannot be attributed entirely to the cAMP system.

OVARIAN FOLLICLE GROWTH—THE "FOLLICULAR" PHASE OF THE OVARIAN CYCLE

Figure 81–4 depicts the various stages of follicular growth in the ovaries. At birth in the female child, each of the ova is surrounded by a single layer of granulosa cells, and the ovum, with its granulosa cell sheath, is called a *primordial follicle,* as illustrated in the figure. Throughout childhood the granulosa cells are believed to provide nourishment for the ovum and also to secrete an *oocyte maturation-inhibiting factor* that keeps the ovum in its primordial state, suspended during this entire time in the prophase stage of meiotic division. Then, after puberty, when FSH and LH from the anterior pituitary gland begin to be secreted in large quantity, the entire ovaries, together with some of the follicles within them, begin to grow.

The first stage of follicular growth is moderate enlargement of the ovum itself, which increases in diameter two- to threefold. Then follows growth of additional layers of granulosa cells, and the follicle becomes known as a *primary follicle.* At least some of the development to this stage can occur even in the absence of FSH and LH, but development beyond this point is not possible without these two hormones.

Development of Antral and Vesicular Follicles. During the first few days after the beginning of menstruation, the concentrations of FSH and LH increase slightly to moderately, with the increase in FSH preceding that of LH by a few days. These hormones, especially the FSH, cause accelerated growth of 6 to 12 primary follicles each month. The initial effect is rapid proliferation of the granulosa cells, giving rise to many more layers of granulosa cells. In addition, many spindle cells derived from the ovarian interstitium also collect in several layers outside the granulosa cells, giving rise to a second class of cells called the *theca.* This is divided into two sublayers: in the *theca interna,* the cells take on epithelioid characteristics similar to those of the granulosa cells and develop an ability to secrete steroid hormones, similar to the ability of the granulosa cells to secrete slightly different hormones. The outer sublayer, the *theca externa,* is a highly vascular connective tissue capsule. This becomes the capsule of the developing follicle.

After the early proliferative phase of growth, lasting for a few days, the mass of granulosa cells secretes a *follicular fluid* that contains a high concentration of estrogen, one of the important female sex hormones that is discussed later. The accumulation of this fluid causes an *antrum* to appear within the mass of granulosa cells, as illustrated in Figure 81–4. Once the atrum is formed, the granulosa and theca cells proliferate even more rapidly, the rate of secretion accelerates, and each of the growing follicles becomes an *antral follicle.*

The early growth of the primary follicle up to the antral stage is stimulated mainly by FSH alone. Then greatly accelerated growth occurs in the antral follicles, leading to much larger follicles called *vesicular follicles.* This accelerated growth is caused by the following: (1) Estrogen is secreted into the follicle and causes the granulosa cells to form increasing numbers of FSH receptors; this causes a positive feedback effect because it makes the granulosa cells more sensitive than ever to the FSH from the anterior pituitary gland. (2) The pituitary FSH and the estrogens combine to promote LH receptors as well on the original granulosa cells, thus allowing LH stimulation of these cells in addition to the FSH stimulation and creating a very rapid increase in follicular secretion. (3) The increasing estrogens from the follicle plus the increasing LH from the anterior pituitary gland act together to cause proliferation of the follicular thecal cells and increase their secretion as well. Therefore, once the antral follicles begin to grow, their further growth occurs very rapidly.

As the follicle enlarges, the ovum itself remains imbedded in a mass of granulosa cells located at one pole of the follicle, as illustrated in Figure 81–4. The ovum, together with its surrounding granulosa cells, is called the *cumulus oophorus.*

Full Maturation of Only One Follicle, Atresia of the Remainder. After a week or more of growth —but before ovulation occurs—one of the follicles begins to outgrow all the others; the remainder begin to involute (a process called *atresia*), and these follicles are said to become *atretic.* The cause of the atresia is unknown, but it has been postulated to be the fol-

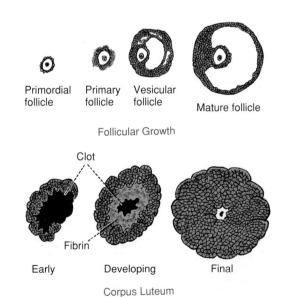

Follicular Growth

Primordial follicle   Primary follicle   Vesicular follicle   Mature follicle

Clot

Fibrin

Early   Developing   Final

Corpus Luteum

Figure 81–4. Stages of follicular growth in the ovary, showing also formation of the corpus luteum. (Modified from Arey: Developmental Anatomy. 7th ed. Philadelphia, W. B. Saunders Company, 1974.)

lowing: The one follicle that becomes more highly developed than the others also secretes more estrogen. Furthermore, this estrogen causes a *positive* feedback effect in that single local follicle, because the FSH (1) enhances both granulosa cell and thecal cell proliferation, which leads to further estrogen production and still a new cycle of cellular proliferation — thus the positive feedback; and (2) the combination of FSH and estrogens also causes increasing numbers of both FSH and LH receptors on the granulosa cells and to a lesser extent on the thecal cells, thus promoting another positive feedback cycle. These effects together cause an explosive increase in the rate of secretion of fluid and hormones into this one rapidly developing follicle. At the same time, the large amounts of estrogen from this follicle act on the hypothalamus to depress further enhancement of secretion of FSH and LH by the anterior pituitary gland, believed in this way to block further growth of the less well developed follicles that have not yet initiated their own intrinsic positive feedback stimulation. Therefore, the largest follicle continues to grow because of its intrinsic positive feedback effects while all the other follicles stop growing and, indeed, involute.

This process of atresia obviously is important in that it allows only one of the follicles to grow large enough to ovulate. The single follicle reaches a size of 1 to 1.5 centimeters at the time of ovulation and is called the *mature follicle*.

Ovulation

Ovulation in a woman who has a normal 28 day female sexual cycle occurs 14 days after the onset of menstruation.

Shortly before ovulation, the protruding outer wall of the follicle swells rapidly, and a small area in the center of the capsule, called the *stigma*, protrudes like a nipple. In another half hour or so, fluid begins to ooze from the follicle through the stigma. About 2 minutes later, as the follicle becomes smaller because of loss of fluid, the stigma ruptures widely, and a more viscous fluid that has occupied that central portion of the follicle is evaginated outward into the abdomen. This viscous fluid carries with it the ovum surrounded by several thousand small granulosa cells called the *corona radiata*.

Need for LH in Ovulation — Ovulatory Surge of LH. LH is necessary for final follicular growth and ovulation. Without this hormone, even though large quantities of FSH are available, the follicle will not progress to the stage of ovulation.

Approximately 2 days before ovulation, for reasons that are not completely known at present but that are discussed in more detail later in the chapter, the rate of secretion of LH by the anterior pituitary gland increases markedly, rising six- to tenfold and peaking about 16 hours before ovulation. FSH also increases about two- to threefold at the same time, and the two hormones act synergistically to cause rapid swelling of the follicle during the several days before ovulation. The LH also has the specific effect on the granulosa and theca cells of converting them more to progesterone-secreting cells and less estrogen secretion. Therefore, the rate of secretion of estrogen begins to fall approximately 1 day prior to ovulation, while small amounts of progesterone begin to be secreted.

It is in this environment of (1) very rapid growth of the follicle, (2) diminishing estrogen secretion after a prolonged phase of excessive estrogen secretion, and (3) beginning secretion of progesterone that ovulation occurs. Without the initial preovulatory surge of LH, ovulation will not take place.

Initiation of Ovulation. Figure 81–5 gives a schema for the initiation of ovulation. It shows especially the role of the large quantity of LH secreted by the anterior pituitary gland. The LH in turn causes rapid secretion of follicular steroid hormones containing a small amount of progesterone for the first time. Within a few hours two events occur, both of which are necessary for ovulation: (1) The theca externa (the capsule of the follicle) begins to release proteolytic enzymes from lysosomes that cause dissolution of the capsular wall and consequent weakening of the wall, resulting in further swelling of the entire follicle and degeneration at the stigma. (2) Simultaneously, there is rapid growth of new blood vessels into the follicle wall, and at the same time prostaglandins (local hormones that cause vasodilation) are secreted in the follicular tissues. These two effects in turn cause plasma transudation into the follicle, which also contributes to follicle swelling. Finally, the combined follicle swelling and simultaneous degeneration of the stigma cause follicle rupture with discharge of the ovum.

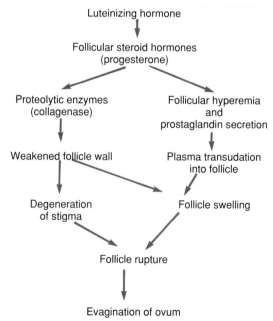

Figure 81–5. Postulated mechanism of ovulation. (Based primarily on the research studies of H. Lipner.)

THE CORPUS LUTEUM—THE "LUTEAL" PHASE OF THE OVARIAN CYCLE

During the first few hours after expulsion of the ovum from the follicle, the remaining granulosa cells change rapidly into *lutein cells.* These enlarge to a diameter two or more times that of the granulosa cells, and they become filled with lipid inclusions that give them a yellowish appearance. This process is called *luteinization,* and the total mass of cells together is called the *corpus luteum,* which is illustrated in the bottom of Figure 81–4. The lutein cells develop a very extensive smooth endoplasmic reticulum that forms large amounts of the female sex hormones progesterone and estrogen, but more so of progesterone. A well-developed vascular supply also grows into the corpus luteum.

Incorporated in the corpus luteum are both granulosa cells and theca cells. The theca cells form mainly the androgens *androstenedione* and *testosterone* rather than female sex hormones. However, most of these are then converted by the granulosa cells into the female hormones.

In the normal female, the corpus luteum grows to approximately 1.5 centimeters in diameter, reaching this stage of development approximately 7 or 8 days after ovulation. Then it begins to involute and eventually loses its secretory function, as well as its lipid characteristics, approximately 12 days after ovulation, becoming the so-called *corpus albicans;* during the ensuing few weeks this is replaced by connective tissue.

Luteinizing Function of LH. The change of granulosa cells into lutein cells is mainly dependent on the LH secreted by the anterior pituitary gland. In fact, this function gave LH its name "luteinizing." However, luteinization of the granulosa cells also depends upon extrusion of the ovum from the follicle. A yet undiscovered local hormone in the follicular fluid, called *luteinization-inhibiting factor,* seems to hold the luteinization process mainly in check until after ovulation. Also for this reason, a corpus luteum only rarely develops in a follicle that does not ovulate.

Secretion by the Corpus Luteum: Function of LH. The corpus luteum is a highly secretory organ, secreting large amounts of both *progesterone* and *estrogen.* Once LH (mainly that which has been secreted during the ovulatory surge) has acted on the granulosa cells to cause luteinization, the newly formed lutein cells seem to be programmed to go through a preordained sequence of (1) proliferation, (2) enlargement, and (3) secretion, then to be followed by (4) degeneration. Even in the absence of further secretion of LH by the anterior pituitary gland, this process will still occur, but it will last only 4 to 8 days. On the other hand, in the presence of LH the degree of growth of the corpus luteum is enhanced, its secretion is greater, and its life is extended, usually to about 12 days. We shall see in the discussion of pregnancy in the next chapter that another hormone that has almost exactly the same properties as LH, *chorionic gonadotropin,* which is secreted by the placenta, can also act on the corpus luteum to prolong its life—usually maintaining it for at least the first 2 to 4 months of pregnancy.

Involution of the Corpus Luteum and Onset of the Next Ovarian Cycle. The estrogen, in particular, and progesterone, to a lesser extent, secreted by the corpus luteum during the luteal phase of the ovarian cycle have the strong feedback effect on the anterior pituitary gland of decreasing the secretion of both FSH and LH. In addition, the luteal cells also secrete small amounts of the hormone *inhibin,* the same as the inhibin secreted by the Sertoli cells of the male testes. This hormone, too, inhibits secretion by the anterior pituitary gland, especially FSH secretion. As a result, both FSH and LH in the blood fall to low concentrations, and loss of these causes the corpus luteum to degenerate completely, a process called *involution* of the corpus luteum. This occurs at the end of almost exactly 12 days of corpus luteum life, which is approximately on the 26th day of the normal female sexual cycle, 2 days before menstruation begins. Now, the lack of secretion of the estrogen, progesterone, and inhibin by the corpus luteum removes the feedback inhibition of the anterior pituitary gland, allowing it again to begin secreting increasing amounts of FSH and a few days later increased quantities of LH as well. The FSH and LH initiate growth of new follicles to begin a new ovarian cycle. At the same time, the paucity of secretion of progesterone and estrogen leads to menstruation by the uterus, as is explained later.

SUMMARY

Approximately each 28 days, gonadotropic hormones from the anterior pituitary gland cause new follicles to begin to grow in the ovaries. One of the follicles finally becomes completely "mature" and ovulates at the 14th day of the cycle. During growth of the follicles, mainly estrogen is secreted.

Following ovulation, the secretory cells of the follicle develop into a corpus luteum that secretes large quantities of the female hormones progesterone and estrogen. After another 2 weeks the corpus luteum degenerates, whereupon the ovarian hormones estrogen and progesterone decrease greatly and menstruation begins. A new ovarian cycle then follows.

● FUNCTIONS OF THE OVARIAN HORMONES— ESTRADIOL AND PROGESTERONE

The two types of ovarian sex hormones are the *estrogens* and the *progestins.* By far the most important of the estrogens is the hormone *estradiol,* and by far the most important progestin is *progesterone.* The estrogens mainly promote proliferation and growth of specific cells in the body and are responsible for development of most secondary sexual characteristics of the female. On the other hand, the progestins are concerned almost entirely with final preparation of the uterus for pregnancy and the breasts for lactation.

CHEMISTRY OF THE SEX HORMONES

The Estrogens. In the normal nonpregnant female, estrogens are secreted in major quantities only by the ovaries, though minute amounts are also secreted by the adrenal cortices. In pregnancy, tremendous quantities are also secreted by the placenta, as we shall discuss in the following chapter.

Only three estrogens are present in significant quantities in the plasma of the human female: *β-estradiol, estrone,* and *estriol,* the formulas for which are illustrated in Figure 81–6. The principal estrogen secreted by the ovaries is *β*-estradiol. Small amounts of estrone are also secreted, but most of this is formed in the peripheral tissues from androgens secreted by the adrenal cortices and by the ovarian thecal cells. Estriol is a very weak estrogenic oxidative product derived from both estradiol and estrone, the conversion occurring mainly in the liver.

The estrogenic potency of *β*-estradiol is 12 times that of estrone and 80 times that of estriol. Considering these relative potencies, one can see that the total estrogenic effect of *β*-estradiol is usually many times that of the other two together. For this reason *β*-estradiol is considered to be the major estrogen, though the estrogenic effects of estrone are far from negligible.

The Progestins. By far the most important of the progestins is progesterone. However, small amounts of another progestin, 17-α-hydroxyprogesterone, also are secreted along with progesterone and have essentially the same effects. Yet for practical purposes it is usually proper to consider progesterone to be the single important progestin.

In the normal nonpregnant female, progesterone is secreted in significant amounts only during the latter half of each ovarian cycle, when it is secreted by the corpus luteum. Only minute amounts of progesterone appear in the plasma during the first half of the ovarian cycle, secreted approximately equally by the ovaries and the adrenal cortices. Yet, as we shall see in the following chapter, very large amounts of progesterone are also secreted by the placenta during pregnancy, especially after the fourth month of gestation.

Synthesis of the Estrogens and Progestins. Note from the chemical formulas of the estrogens and progesterone in Figure 81–6 that all these are steroids. They are synthesized in the ovaries mainly from cholesterol derived from the blood, but to a slight extent also from acetyl coenzyme A, multiple molecules of which can combine to form the appropriate steroid nucleus. During synthesis, progesterone and the male sex hormone testosterone are mainly synthesized first; then, during the follicular phase of the ovarian cycle, before these can leave the ovaries, almost all the testosterone and much of the progesterone are con-

Figure 81–6. Chemical formulas of the principal female hormones.

verted into estrogens by the granulosa cells. During the luteal phase of the cycle, far too much progesterone is formed for all of it to be converted, which accounts for the large progesterone secretion at this time. Even so, about one-fifteenth as much testosterone is secreted into the plasma of the female by the ovaries as is secreted into the plasma of the male by the testes.

Transport of Estrogens and Progesterone in the Blood. Estrogens and progesterone are both transported in the blood bound mainly with plasma albumin and specific estrogen- and progesterone-binding globulins. However, the binding between these hormones and the plasma proteins is loose enough that they are rapidly released to the tissues over a period of 30 minutes or so.

Fate of the Estrogens; Function of the Liver in Estrogen Degradation. The liver conjugates the estrogens to form glucuronides and sulfates, and about one fifth of these conjugated products are excreted in the bile while most of the remainder are excreted in the urine. Also, the liver converts the potent estrogens estradiol and estrone into the almost totally impotent estrogen estriol. Therefore, diminished liver function actually *increases* the activity of estrogens in the body, sometimes causing *hyperestrinism*.

Fate of Progesterone. Within a few minutes after secretion, almost all the progesterone is degraded to other steroids that have no progesteronic effect. Here, as is also true with the estrogens, the liver is especially important for this metabolic degradation.

The major end product of progesterone degradation is *pregnanediol.* Approximately 10 per cent of the original progesterone is excreted in the urine in this form. Therefore, one can estimate the rate of progesterone formation in the body from the rate of this excretion.

FUNCTIONS OF THE ESTROGENS — THEIR EFFECTS ON THE PRIMARY AND SECONDARY FEMALE SEX CHARACTERISTICS

The principal function of the estrogens is to cause cellular proliferation and growth of the tissues of the sex organs and of other tissues related to reproduction.

Effect on the Uterus and External Sex Organs. During childhood, estrogens are secreted only in minute quantities, but at puberty the quantity of estrogens secreted under the influence of the pituitary gonadotropic hormones increases some 20-fold or more. At this time the female sex organs change from those of a child to those of an adult. The ovaries, fallopian tubes, uterus, and vagina all increase several times in size. Also, the external genitalia enlarge, with deposition of fat in the mons pubis and labia majora and with enlargement of the labia minora.

In addition, estrogens change the vaginal epithelium from a cuboidal into a stratified type, which is considerably more resistant to trauma and infection than is the prepubertal epithelium. Infections in children, such as gonorrheal vaginitis, can actually be cured by administration of estrogens simply because of the resulting increased resistance of the vaginal epithelium.

During the few years following puberty, the size of the uterus increases two- to threefold. More important, however, than increases in size are the changes that take place in the endometrium under the influence of estrogens, for estrogens cause marked proliferation of the endometrial stroma and greatly increased development of the endometrial glands that will later be used to aid in nutrition of the implanting ovum. These effects are discussed later in the chapter in connection with the endometrial cycle.

Effect on the Fallopian Tubes. The estrogens have an effect on the mucosal lining of the fallopian tubes similar to that on the uterine endometrium: They cause the glandular tissues to proliferate; and, especially important, they cause the number of ciliated epithelial cells that line the fallopian tubes to increase. Also, the activity of the cilia is considerably enhanced, these always beating toward the uterus. This helps propel the fertilized ovum toward the uterus.

Effect on the Breasts. The primordial breasts of both female and male are exactly alike, and under the influence of appropriate hormones, the masculine breast, at least during the first two decades of life, can develop sufficiently to produce milk in the same manner as the female breast.

Estrogens cause (1) development of the stromal tissues of the breasts, (2) growth of an extensive ductile system, and (3) deposition of fat in the breasts. The lobules and alveoli of the breast develop to a slight extent, but it is progesterone and prolactin that cause the determinative growth and function of these structures. In summary, the estrogens initiate growth of the breasts and the breasts' milk-producing apparatus, and they are also responsible for the characteristic external appearance of the mature female breast, but they do not complete the job of converting the breasts into milk-producing organs.

Effect on the Skeleton. Estrogens cause increased osteoblastic activity. Therefore, at puberty, when the female enters her reproductive years, her growth rate becomes rapid for several years. However, estrogens have another potent effect on skeletal growth — that is, they cause early uniting of the epiphyses with the shafts of the long bones. This effect is much stronger in the female than is a similar effect of testosterone in the male. As a result, growth of the female usually ceases several years earlier than growth of the male. The female eunuch who is completely devoid of estrogen production usually grows several inches taller than the normal mature female because her epiphyses do not unite early.

Osteoporosis of the Bones Caused by Estrogen Deficiency in Old Age. After the menopause, almost no estrogens are secreted by the ovaries. This estrogen deficiency leads to (1) diminished osteoblastic activity in the bones, (2) decreased bone matrix, and (3) decreased deposition of bone calcium and phosphate. In some women, this effect is extremely severe, and the resulting condition is *osteoporosis,* described in Chapter 79. Because this can greatly weaken the bones and lead to bone fracture, especially fracture of the vertebrae, a large share of postmenopausal women are treated continually with substitutive estrogens.

Effect on Protein Deposition. Estrogens cause a slight increase in total body protein, which is evidenced by a slight positive nitrogen balance when estrogens are administered. This probably results from the growth-promoting effect of estrogen on the sexual organs, the bones, and a few other tissues of the body. The enhanced protein deposition caused by testosterone is much more general and many times as powerful as that caused by estrogens.

Effect on Metabolism and Fat Deposition. Estrogens increase the metabolic rate slightly but only about a third as much as the male sex hormone testosterone. They also cause deposition of increased quantities of fat in the subcutaneous tissues. As a result, the overall specific gravity of the female body, as judged by flotation in water, is considerably less than that of the male body, which contains more protein and less fat. In addition to deposition of fat in the breasts and subcutaneous tissues, estrogens cause the deposition of fat in the buttocks and thighs that is characteristic of the feminine figure.

Effect on Hair Distribution. Estrogens do not greatly affect hair distribution. However, hair develops in the pubic region and in the axillae after puberty. Androgens formed by the adrenal glands are mainly responsible for this.

Effect on the Skin. Estrogens cause the skin to develop a texture that is soft and usually smooth but nevertheless thicker than that of the child or the female castrate. Also, estrogens cause the skin to become more vascular than normal; this effect is often associated with increased warmth of the skin, and it often results in greater bleeding of cut surfaces than is observed in men.

The adrenal androgens, which are secreted in increased quantities after puberty, cause increased secretion by the axillary sweat glands and also often cause acne.

Effect on Electrolyte Balance. The chemical similarity of estrogenic hormones to adrenocortical hormones has been pointed out; and estrogens, like aldosterone and some other adrenocortical hormones, cause sodium and water retention by the kidney tubules. However, this effect of estrogens is slight and rarely of significance except in pregnancy, as discussed in the following chapter.

Intracellular Functions of Estrogens. Thus far we have discussed the gross effects of estrogens on the body. The cellular mechanism behind these effects is: Estrogens circulate in the blood for only a few minutes before they are delivered to the target cells. On entry into these cells, they combine within 10 to 15 seconds with "receptor" protein molecules in the cytoplasm and then, in combination with this protein, interact with specific portions of the chromosomal DNA. This immediately initiates the process of transcription; therefore, RNA begins to be produced within a few minutes. In addition, over many hours new DNA may also be produced, resulting eventually in division of

the cell. The RNA diffuses to the cytoplasm, where it causes greatly increased protein formation and subsequently altered cellular function.

One of the principal differences between the protein anabolic effect of the estrogens and that of testosterone is that estrogen causes its effect almost exclusively in a few specific target organs, such as the uterus, the breasts, the skeleton, and certain fatty areas of the body, whereas testosterone has a more generalized effect throughout the body.

FUNCTIONS OF PROGESTERONE

Effect on the Uterus. By far the most important function of progesterone is *to promote secretory changes in the uterine endometrium* during the latter half of the female sexual cycle, thus preparing the uterus for implantation of the fertilized ovum. This function is discussed later in connection with the endometrial cycle of the uterus.

In addition to this effect on the endometrium, progesterone decreases the frequency of uterine contractions, thereby helping to prevent expulsion of the implanted ovum, an effect discussed in the following chapter.

Effect on the Fallopian Tubes. Progesterone also promotes secretory changes in the mucosal lining of the fallopian tubes. These secretions are necessary for nutrition of the fertilized, dividing ovum as it traverses the fallopian tube prior to implantation.

Effect on the Breasts. Progesterone promotes development of the lobules and alveoli of the breasts, causing the alveolar cells to proliferate, to enlarge, and to become secretory in nature. However, progesterone does not cause the alveoli actually to secrete milk, for, as discussed in the following chapter, milk is secreted only after the prepared breast is further stimulated by prolactin from the anterior pituitary.

Progesterone also causes the breasts to swell. Part of this swelling is due to the secretory development in the lobules and alveoli, but part also seems to result from increased fluid in the subcutaneous tissue itself.

Effect on Electrolyte Balance. Progesterone in very large quantity, like estrogens, testosterone, and adrenocortical hormones, can enhance sodium, chloride, and water reabsorption from the distal tubules of the kidney. Yet, strangely enough, progesterone more often causes increased sodium and water excretion. The cause of this is competition between progesterone and aldosterone, which probably occurs as follows: These two substances combine with the same receptor proteins in the epithelial cells of the tubules. When progesterone combines with these, aldosterone cannot combine. Yet progesterone exerts many hundred times less sodium transport effect than does aldosterone. Therefore, despite the fact that under appropriate conditions progesterone can weakly promote sodium and water retention by the renal tubules, it blocks the far more potent effect of aldosterone, thus usually resulting in net loss of sodium and water from the body.

THE ENDOMETRIAL CYCLE AND MENSTRUATION

Associated with the cyclic production of estrogens and progesterone by the ovaries is an endometrial cycle operating through the following stages: first, proliferation of the uterine endometrium; second, secretory changes in the endometrium; and third, desquamation of the endometrium, which is known as *menstruation.* The various phases of the endometrial cycle are illustrated in Figure 81–7.

Proliferative Phase (Estrogen Phase) of the Endometrial Cycle. At the beginning of each monthly cycle, most of the endometrium is desquamated by the process of menstruation. After menstruation, only a thin layer of endometrial stroma remains at the base of the original endometrium, and the only epithelial cells that are left are those located in the remaining deep portions of the glands and crypts of the endometrium. *Under the influence of estrogens,* secreted in increasing quantities by the ovary during the first part of the ovarian cycle, the stromal cells and the epithelial cells proliferate rapidly. The endometrial surface is re-epithelialized within 4 to 7 days after the beginning of menstruation. For the first 2 weeks of the cycle — that is, until ovulation — the endometrium increases greatly in thickness, owing to increasing numbers of stromal cells and to progressive growth of the endometrial glands and blood vessels into the endometrium. At the time of ovulation the endometrium is approximately 3 to 4 millimeters thick. At this time, the endometrial glands, especially those of the cervical region, secrete a thin, stringy mucus. The mucous strings align themselves along the length of the cervical canal, forming channels that help guide the sperm in the proper direction into the uterus.

Secretory Phase (Progestational Phase) of the Endometrial Cycle. During the latter half of the monthly cycle, progesterone and estrogen are secreted in large quantity by the corpus luteum. The estrogens cause slight additional cellular proliferation in the endometrium during this phase of the endometrial cycle, whereas progesterone causes marked swelling and secretory development of the endometrium. The glands increase in tortuosity; an excess of secretory substances accumulates in the glandular ep-

ithelial cells. Also, the cytoplasm of the stromal cells increases; lipid and glycogen deposits increase greatly in the stromal cells; and the blood supply to the endometrium further increases in proportion to the developing secretory activity, the blood vessels becoming highly tortuous. At the end of the secretory phase the endometrium has a thickness of 5 to 6 millimeters.

The whole purpose of all these endometrial changes is to produce a highly secretory endometrium containing large amounts of stored nutrients that can provide appropriate conditions for implantation of a fertilized ovum during the latter half of the monthly cycle. From the time a fertilized ovum enters the uterine cavity from the fallopian tube until the time the ovum implants, the uterine secretions, called "uterine milk," provide nutrition for the early dividing ovum. Then, once the ovum implants in the endometrium, the trophoblastic cells on the surface of the blastocyst begin to digest the endometrium and to absorb the endometrial stored substances, thus making still far greater quantities of nutrients available to the early embryo.

Menstruation. Approximately 2 days before the end of the monthly cycle, the ovarian hormones, estrogens and progesterone, decrease sharply to low levels of secretion, as illustrated in Figure 81–3, and menstruation follows.

Menstruation is caused by this sudden reduction of the estrogens and progesterone, especially the progesterone, at the end of the monthly ovarian cycle. The first effect is decreased stimulation of the endometrial cells by these two hormones, followed rapidly by involution of the endometrium itself to about 65 per cent of its previous thickness. During the 24 hours preceding the onset of menstruation, the tortuous blood vessels leading to the mucosal layers of the endometrium become vasospastic, presumably because of some effect of the involution, such as release of a vasoconstrictor material — possibly one of the vasoconstrictor types of prostaglandins that are present in abundance at that time. The vasospasm and loss of hormonal stimulation cause beginning necrosis in the endometrium, especially of blood vessels in the stratum vasculare. As a result, blood seeps into the vascular layer of the endometrium, and the hemorrhagic areas grow over a period of 24 to 36 hours. Gradually, the necrotic outer layers of the endometrium separate from the uterus at the site of the hemorrhages, until, approximately 48 hours following the onset of menstruation, all the superficial layers of the endometrium have desquamated. The mass of desquamated tissue and blood in the uterine cavity, possibly plus contractile effects of the prostaglandins, initiate uterine contractions that expel the uterine contents.

During normal menstruation, approximately 40 milliliters of blood and an additional 35 milliliters of serous fluid are lost. This menstrual fluid is normally nonclotting, because a *fibrinolysin* is released along with the necrotic endometrial material. However, if excessive bleeding occurs from the uterine surface, the quantity of fibrinolysin may not be sufficient to

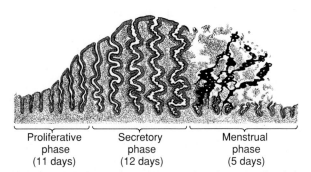

| Proliferative phase (11 days) | Secretory phase (12 days) | Menstrual phase (5 days) |

Figure 81–7. Phases of endometrial growth and menstruation during each monthly female sexual cycle.

prevent clotting. The presence of clots during menstruation ordinarily is clinical evidence of uterine pathology.

Within 4 to 7 days after menstruation starts, the loss of blood ceases, for by this time the endometrium has become completely re-epithelialized.

Leukorrhea During Menstruation. During menstruation tremendous numbers of leukocytes are released along with the necrotic material and blood. It is probable that some substance liberated by the endometrial necrosis causes this outflow of leukocytes. As a result of these many leukocytes and possible other factors, the uterus is highly resistant to infection during menstruation even though the endometrial surfaces are denuded. Obviously, this is of extreme protective value.

● REGULATION OF THE FEMALE MONTHLY RHYTHM —INTERPLAY BETWEEN THE OVARIAN AND HYPOTHALAMIC-PITUITARY HORMONES

Now that we have presented the major cyclic changes that occur during the female sexual cycle, we can attempt to explain the basic rhythmic mechanism that causes these cyclic variations.

Function of the Hypothalamus in the Regulation of Gonadotropin Secretion — Gonadotropin-Releasing Hormone

As pointed out in Chapter 74, secretion of most of the anterior pituitary hormones is controlled by releasing hormones formed in the hypothalamus and then transported to the anterior pituitary gland by way of the hypothalamic-hypophysial portal system. In the case of the gonadotropins, at least one releasing hormone, *gonadotropin-releasing hormone (GnRH)*, also called *luteinizing hormone–releasing hormone (LHRH)*, is important. This has been purified and has been found to be a decapeptide having the following formula:

$$\text{Glu-His-Trp-Ser-Tyr-Gly-Leu-Arg-}$$
$$\text{Pro-Gly-NH}_2$$

Intermittent, Pulsatile Secretion of GnRH by the Hypothalamus — and Pulsatile Release of LH From the Anterior Pituitary. Recent experiments have demonstrated that the hypothalamus does not secrete GnRA continuously, but instead secretes it in pulses lasting several minutes that occur every 1 to 3 hours. Furthermore, when GnRH is infused continuously so that it is available all of the time rather than in these pulses, its effects in causing release of LH and FSH by the anterior pituitary gland are completely lost. Therefore, for reasons unknown, the pulsatile nature of GnRH release is absolutely essential to its function.

The pulsatile release of GnRH also causes pulsatile output of LH. This is illustrated in Figure 81–8, which shows in the upper tracing pulsatile release of LH in a monkey. The lower tracing illustrates the number of neuronal action potentials recorded from the mediobasal hypothalamus, the locus of the neurons that secrete GnRH. Note that the hypothalamic pulse occurs a few minutes in advance of the increased LH secretion by the anterior pituitary gland.

To a slight extent, the secretion of FSH is also modulated upward and downward by the hypothalamic pulses of GnRH, but there is a more important prolonged effect on FSH secretion that persists for many hours rather than changing greatly from one pulse to the next.

Hypothalamic Centers for Release of GnRH. The neuronal activity that causes pulsatile release of GnRH occurs primarily in the mediobasal hypothala-

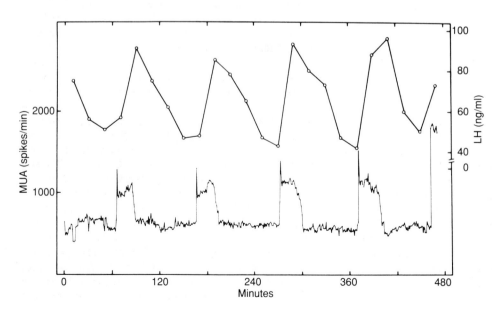

Figure 81–8. *Lower curve:* Changes in multiunit electrical activity (MUA) recorded from mediobasal hypothalamus. *Upper curve:* Luteinizing hormone (LH) pulses in peripheral circulation in a pentobarbital-anesthetized ovariectomized rhesus monkey. (From Wilson, Kesner, Kaufman, Uemura, Akema, and Knobil: *Neuroendocrinology* 39:256, 1984.)

mus, especially in the arcuate nuclei of this area. Therefore, it is believed that these arcuate nuclei control most female sexual activity, though other neurons located in the preoptic area of the anterior hypothalamus also secrete GnRH in moderate amounts, the function of which is unclear. Multiple neuronal centers in the brain's limbic system transmit signals into the arcuate nuclei to modify both the intensity of GnRH release and the frequency of the pulses, thus providing a possible explanation of why psychic factors very often modify female sexual function.

Negative Feedback Effect of Estrogen, Progesterone, and Inhibin on Secretion of Follicle-Stimulating Hormone and Luteinizing Hormone

Estrogen in small amounts and progesterone in large amounts inhibit the production of FSH and LH. These feedback effects seem to operate mainly directly on the anterior pituitary gland but to a lesser extent on the hypothalamus to decrease secretion of GnRH, especially by altering the frequency of the GnRH pulses.

In addition to the feedback effects of estrogen and progesterone, still another hormone also seems to be involved. This is *inhibin,* which is secreted along with the steroid sex hormones by the corpus luteum in the same way that the Sertoli cells secrete this same hormone in the male testes. This inhibin has the same effect in the female as in the male of inhibiting the secretion of FSH by the anterior pituitary gland and LH to a lesser extent as well. Therefore, it is believed that inhibin might be especially important in causing the decrease in secretion of FSH and LH toward the end of the female sexual month.

Positive Feedback Effect of Estrogen Before Ovulation — The Preovulatory Luteinizing Hormone Surge

For reasons not completely understood, the anterior pituitary gland secretes greatly increased amounts of LH for a period of 1 to 2 days beginning 24 to 48 hours before ovulation. This effect is illustrated in Figure 81–3, and the figure shows a much smaller preovulatory surge of FSH as well.

Experiments have shown that infusion of estrogen into a female above a critical rate for a period of 2 to 3 days during the first half of the ovarian cycle will cause rapidly accelerating growth of the follicles and also rapidly accelerating secretion of ovarian estrogens. During this period the secretion of both FSH and LH by the anterior pituitary gland is at first suppressed slightly. Then abruptly the secretion of LH increases six- to eightfold, and the secretion of FSH increases about twofold. The cause of this abrupt increase in secretion of the gonadotropins is not known. However, several possible causes are as follows: (1) It has been suggested that estrogen at this point in the cycle has a peculiar *positive feedback effect* to stimulate pituitary secretion of the gonadotropins; this is in sharp contrast to its normal negative feedback effect that occurs during the remainder of the female monthly cycle. (2) The granulosa cells of the follicles begin to secrete small but increasing quantities of progesterone a day or so prior to the preovulatory LH surge, and it has been suggested that this might be the factor that stimulates the excess LH secretion.

Without this normal preovulatory surge of LH, ovulation will not occur.

FEEDBACK OSCILLATION OF THE HYPOTHALAMIC-PITUITARY-OVARIAN SYSTEM

Now, after discussing much of the known information about the interrelationships of the different components of the female hormonal system, we can digress from the area of proven fact into the realm of speculation and attempt to explain the feedback oscillation that controls the rhythm of the female sexual cycle. It seems to operate in approximately the following sequence of three successive events:

1. The Postovulatory Secretion of the Ovarian Hormones and Depression of Gonadotropins. The easiest part of the cycle to explain is the events that occur during the postovulatory phase — between ovulation and the beginning of menstruation. During this time the corpus luteum secretes large quantities of both progesterone and estrogen and probably the hormone inhibin as well. All these hormones together have a combined negative feedback effect on the anterior pituitary gland and the hypothalamus to cause suppression of both FSH and LH, decreasing these to their lowest levels at about three to four days before the onset of menstruation. These effects are illustrated in Figure 81–3.

2. The Follicular Growth Phase. Two to three days before menstruation the corpus luteum involutes, and the secretion of estrogen, progesterone, and inhibin decreases to a low ebb. This releases the hypothalamus and anterior pituitary from the feedback effect of these hormones; and a day or so later, at about the time that menstruation begins, FSH increases as much as twofold; then several days after menstruation begins, LH secretion increases as much as twofold as well. These hormones initiate new follicular growth and progressive increase in the secretion of estrogen, reaching a peak estrogen secretion at about 12.5 to 13 days after the onset of menstruation. During the first 11 to 12 days of this follicular growth the rates of secretion of the gonadotropins FSH and LH decrease very slightly because of the negative feedback effect mainly of estrogen on the anterior pituitary gland. Then comes a sudden increase in secretion of both of these hormones, leading to the preovulatory surge of LH, followed by ovulation.

3. Preovulatory Surge of LH and FSH; Ovulation. At approximately 11.5 to 12 days after the onset of menstruation, the decline in secretion of FSH and LH comes to an abrupt halt. It is believed that the high level of estrogens at this time (or the beginning secretion of progesterone by the follicles) causes a positive feedback effect principally on the anterior pituitary, as explained earlier, which leads to a terrific surge of secretion of LH and to a lesser extent of FSH. Whatever the cause of this preovulatory LH and FSH surge, the LH leads to both ovulation and subsequent secretion by the corpus luteum. Thus, the hormonal system begins a new round of the cycle until the next ovulation.

Anovulatory Cycles — The Sexual Cycles at Puberty

If the preovulatory surge of LH is not of sufficient magnitude, ovulation will not occur, and the cycle is then said to be "anovulatory." Most of the cyclic variations of the sexual cycle continue, but they are altered in the following ways: First, lack of ovulation causes failure of development of the corpus luteum and consequently there is almost no secretion of progesterone during the latter portion of the cycle. Second, the cycle is shortened by several days, but the rhythm continues. Therefore, it is likely that progesterone is not required for maintenance of the cycle itself though it can alter its rhythm.

Anovulatory cycles are usual during the first few cycles during puberty and for several years prior to menopause, presumably because the LH surge is not potent enough at these times to cause ovulation.

PUBERTY AND MENARCHE

Puberty means the onset of adult sexual life, and menarche means the onset of menstruation. The period of puberty is caused by a gradual increase in gonadotropic hormone secretion by the pituitary, beginning approximately in the eighth year of life, as illustrated in Figure 81–9, and usually culminating in the onset of menstruation between the ages of 11 and 16 years (average, 13 years).

Figure 81–9. Total rates of secretion of gonadotropic hormones throughout the sexual lives of females and males, showing an especially abrupt increase in gonadotropic hormones at the menopause in the female.

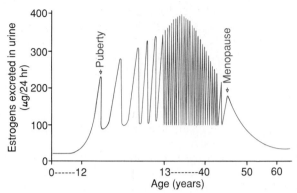

Figure 81–10. Estrogen secretion throughout sexual life.

In the female, as in the male, the infantile pituitary gland and ovaries are capable of full function if appropriately stimulated. However, as is also true in the male and for reasons not yet understood, the hypothalamus does not secrete significant quantities of GnRH during childhood. Experiments have shown that the hypothalamus itself is perfectly capable of secreting this hormone, but there is lack of the appropriate signal from some other brain area to cause the secretion. Therefore, it is now believed that the onset of puberty is initiated by some maturation process occurring elsewhere in the brain, perhaps somewhere in the limbic system.

Figure 81–10 illustrates (1) the increasing levels of estrogen secretion at puberty, (2) the cyclic variation during the monthly sexual cycles, (3) the further increase in estrogen secretion during the first few years of reproductive life, (4) then progressive decrease in estrogen secretion toward the end of reproductive life, and (5) finally almost no estrogen secretion beyond the menopause.

THE MENOPAUSE

At the age of 40 to 50 years the sexual cycles usually become irregular, and ovulation fails to occur during many of the cycles. After a few months to a few years, the cycles cease altogether, as illustrated in Figure 81–10. This period during which the cycles cease and the female sex hormones diminish rapidly to almost none at all is called the *menopause*.

The cause of the menopause is "burning out" of the ovaries. Throughout a woman's reproductive life about 400 of the primordial follicles grow into vesicular follicles and ovulate, while literally hundreds of thousands of the ova degenerate. At the age of about 45 years only a few primordial follicles still remain to be stimulated by FSH and LH, and the production of estrogens by the ovary decreases as the number of primordial follicles approaches zero (also illustrated in Figure 81–10). When estrogen production falls below a critical value, the estrogens can no longer inhibit the production of FSH and LH; nor can they cause an ovulatory surge of LH and FSH to cause ovulatory cycles. Instead, as illustrated in Figure

81-9, FSH and LH (mainly FSH) are produced thereafter in large and continuous quantities. Estrogens are produced in subcritical quantities for a short time after the menopause; but over a few years, as the final remaining primordial follicles become atretic, the production of estrogens by the ovaries falls almost to zero.

At the time of the menopause a woman must readjust her life from one that has been physiologically stimulated by estrogen and progesterone production to one devoid of these hormones. The loss of the estrogens often causes marked physiological changes in the function of the body, including (1) "hot flushes" characterized by extreme flushing of the skin, (2) psychic sensations of dyspnea, (3) irritability, (4) fatigue, (5) anxiety, and (6) occasionally various psychotic states. These symptoms are of sufficient magnitude in approximately 15 per cent of women to warrant treatment. If counseling fails, daily administration of an estrogen in small quantities will reverse the symptoms, and, by gradually decreasing the dose, the postmenopausal woman is likely to avoid severe symptoms; unfortunately, such treatment can prolong the symptoms.

● ABNORMALITIES OF SECRETION BY THE OVARIES

Hypogonadism. Less than normal secretion by the ovaries can result from poorly formed ovaries, lack of ovaries, or genetically abnormal ovaries that secrete the wrong hormones because of missing enzymes in the secretory cells. When ovaries are absent from birth or when they become nonfunctional before puberty, *female eunuchism* occurs. In this condition the usual secondary sexual characteristics do not appear, and the sexual organs remain infantile. Especially characteristic of this condition is prolonged growth of the long bones because the epiphyses do not unite with the shafts of these bones at as early an age as in the normal adolescent woman. Consequently, the female eunuch is essentially as tall as, or perhaps even slightly taller than, her male counterpart of similar genetic background.

When the ovaries of a fully developed woman are removed, the sexual organs regress to some extent so that the uterus becomes almost infantile in size, the vagina becomes smaller, and the vaginal epithelium becomes thin and easily damaged. The breasts atrophy and become pendulous, and the pubic hair becomes thinner. These same changes occur in women after menopause.

Irregularity of Menses and Amenorrhea Due to Hypogonadism. As pointed out in the preceding discussion of the menopause, the quantity of estrogens produced by the ovaries must rise above a critical value if they are to be able to cause rhythmical sexual cycles. Consequently, in hypogonadism or when the gonads are secreting small quantities of estrogens as a result of other factors, such as *hypothyroidism,* the ovarian cycle likely will not occur normally. Instead, several months may elapse between menstrual periods, or menstruation may cease altogether (amenorrhea). Prolonged ovarian cycles are frequently associated with failure of ovulation, presumably due to insufficient secretion of LH, which is necessary for ovulation.

Hypersecretion by the Ovaries. Extreme hypersecre-

tion of ovarian hormones by the ovaries is a rare clinical entity, for excessive secretion of estrogens automatically decreases the production of gonadotropins by the pituitary, and this in turn limits the production of the ovarian hormones. Consequently, hypersecretion of feminizing hormones is recognized clinically only when a feminizing tumor develops.

A rare granulosa cell tumor can develop in an ovary, occurring more often after menopause than before. These tumors secrete large quantities of estrogens, which exert the usual estrogenic effects, including hypertrophy of the uterine endometrium and irregular bleeding from this endometrium. In fact, bleeding is often the first and only indication that such a tumor exists.

● THE FEMALE SEXUAL ACT

Stimulation of the Female Sexual Act. As is true in the male sexual act, successful performance of the female sexual act depends on both psychic stimulation and local sexual stimulation.

As is also true in the male, the thinking of erotic thoughts can lead to female sexual desire, and this aids greatly in the performance of the female sexual act. Such desire is probably based as much on one's background training as on physiological drive, though sexual desire does increase in proportion to the level of secretion of the sex hormones. Desire also changes during the sexual month, reaching a peak near the time of ovulation, probably because of the high levels of estrogen secretion during the preovulatory period.

Local sexual stimulation in women occurs in more or less the same manner as in men, for massage, irritation, or other types of stimulation of the vulva, vagina, and other perineal regions, and even of the urinary tract, create sexual sensations. The glans of the *clitoris* is especially sensitive for initiating sexual sensations. As in the male, the sexual sensory signals are mediated to the sacral segments of the spinal cord through the pudendal nerve and sacral plexus. Once these signals have entered the spinal cord, they are transmitted thence to the cerebrum. Also, local reflexes integrated in the sacral and lumbar spinal cord are at least partly responsible for female sexual reactions.

Female Erection and Lubrication. Located around the introitus and extending into the clitoris is erectile tissue almost identical with the erectile tissue of the penis. This erectile tissue, like that of the penis, is controlled by the parasympathetic nerves that pass through the nervi erigentes from the sacral plexus to the external genitalia. In the early phases of sexual stimulation, parasympathetic signals dilate the arteries of the erectile tissues, and this allows rapid accumulation of blood in the erectile tissue so that the introitus tightens around the penis; this aids the male greatly in his attainment of sufficient sexual stimulation for ejaculation to occur.

Parasympathetic signals also pass to the bilateral Bartholin's glands located beneath the labia minora to cause secretion of mucus immediately inside the introitus. This mucus is responsible for much of the lubrication during sexual intercourse, though much is also provided by mucus secreted by the vaginal epithelium as well and a small amount from the male urethral glands. The lubrication in turn is necessary for establishing during intercourse a satisfactory massaging sensation rather than an irritative sensation, which may be provoked by a dry vagina. A massaging sensation constitutes the optimal type of sensation for evoking the appropriate reflexes that culminate in both the male and female climaxes.

The Female Orgasm. When local sexual stimulation reaches maximum intensity, and especially when the local sensations are supported by appropriate psychic conditioning signals from the cerebrum, reflexes are initiated that cause the female orgasm, also called the *female climax.* The female orgasm is analogous to emission and ejaculation in the male, and it perhaps helps promote fertilization of the ovum. Indeed, the human female is known to be somewhat more fertile when inseminated by normal sexual intercourse rather than by artificial methods, thus indicating an important function of the female orgasm. Possible reasons for this are as follows:

First, during the orgasm the perineal muscles of the female contract rhythmically, which results from spinal cord reflexes similar to those that cause ejaculation in the male. It is possible that these same reflexes increase uterine and fallopian tube motility during the orgasm, thus helping transport the sperm toward the ovum, but information on this subject is scanty. Also, the orgasm seems to cause dilation of the cervical canal for up to half an hour, thus allowing easy transport of the sperm.

Second, in many lower animals, copulation causes the posterior pituitary gland to secrete oxytocin; this effect is probably mediated through the amygdaloid nuclei and then through the hypothalamus to the pituitary. The oxytocin in turn causes increased rhythmical contractions of the uterus, which have been postulated to cause rapid transport of the sperm. Sperm have been shown to traverse the entire length of the fallopian tube in the cow in approximately 5 minutes, a rate at least 10 times as fast as that which the swimming motions of the sperm themselves could possibly achieve. Whether or not this occurs in the human female is unknown.

In addition to the possible effects of the orgasm on fertilization, the intense sexual sensations that develop during the orgasm also pass to the cerebrum and cause intense muscle tension throughout the body. But after culmination of the sexual act, this gives way during the succeeding minutes to a sense of satisfaction characterized by relaxed peacefulness, an effect called *resolution.*

● FEMALE FERTILITY

The Fertile Period of Each Sexual Cycle. The ovum remains viable and capable of being fertilized after it is expelled from the ovary probably no longer than 24 hours. Therefore, sperm must be available soon after ovulation if fertilization is to take place. On the other hand, a few sperm can remain fertile in the female reproductive tract for up to 72 hours, though most of them for not more than 24 hours. Therefore, for fertilization to take place, intercourse usually must occur some time between 1 day before ovulation up to 1 day after ovulation. Thus, the period of female fertility during each sexual cycle is short.

The Rhythm Method of Contraception. One of the often practiced methods of contraception is to avoid intercourse near the time of ovulation. The difficulty with this method of contraception is the impossibility of predicting the exact time of ovulation. Yet the interval from ovulation until the next succeeding onset of menstruation is almost always between 13 and 15 days. Therefore, if the menstrual cycle is regular, with a periodicity of 28 days, ovulation usually occurs within 1 day of the 14th day of the cycle. If, on the other hand, the periodicity of the cycle is 40 days, ovulation usually occurs within 1 day of the 26th day of the cycle. Finally, if the periodicity of the cycle is 21 days, ovulation usually occurs within 1 day of the 7th day of the cycle. Therefore, it is usually stated that avoidance of intercourse for 4 days prior to the calculated day of ovulation and 3 days afterward prevents conception. But such a method of contraception can be used only when the periodicity of the menstrual cycle is regular.

Hormonal Suppression of Fertility — "The Pill." It has long been known that administration of either estrogen or progesterone, if given in appropriate quantity during the first half of the monthly female cycle, can inhibit ovulation. The reason for this is that appropriate administration of either of these can prevent the preovulatory surge of LH secretion by the pituitary gland, which, it will be recalled, is essential in causing ovulation.

The reason the administration of estrogen or progesterone prevents the preovulatory surge of LH secretion is not fully understood. However, experimental work has suggested that immediately before the surge occurs there is probably a sudden depression of estrogen secretion by the ovarian follicles, and that this might be the necessary signal for causing the subsequent positive feedback effect that leads to the surge. Obviously, administration of the sex hormones could prevent the initial hormonal depression that might be the initiating signal for ovulation.

The problem in devising methods for hormonal suppression of ovulation has been in developing appropriate combinations of estrogens and progestins that will suppress ovulation but that will not cause unwanted effects of these two hormones. For instance, too much of either of the hormones can cause abnormal menstrual bleeding patterns. However, use of certain synthetic progestins in place of progesterone, especially the 19-norsteroids, along with small amounts of estrogens will usually prevent ovulation and yet also allow almost a normal pattern of menstruation. Therefore, almost all "pills" used for control of fertility consist of some combination of synthetic estrogens and synthetic progestins. The main reason for using synthetic estrogens and synthetic progestins is that the *natural* hormones are almost entirely destroyed by the liver within a short time after they are absorbed from the gastrointestinal tract into the portal circulation. However, many of the *synthetic* hormones can resist this destructive propensity of the liver, thus allowing oral administration.

Two of the most commonly used synthetic estrogens are *ethynyl estradiol* and *mestranol.* Among the most commonly used progestins are *norethindrone, norethynodrel, ethyno-*

diol, and *norgestrel.* The medication is usually begun in the early stages of the monthly cycle and continued beyond the time that ovulation normally would have occurred. Then the medication is stopped, allowing menstruation to occur and a new cycle to begin.

Abnormal Conditions Causing Female Sterility. Approximately 1 of every 6 to 8 marriages is infertile; in about 60 per cent of these, the infertility is due to female sterility.

Occasionally, no abnormality whatever can be discovered in the female genital organs, in which case it must be assumed that the infertility is due either to abnormal physiological function of the genital system or to abnormal genetic development of the ova themselves.

However, probably by far the most common cause of female sterility is failure to ovulate. This can result from hyposecretion of gonadotropic hormones, in which case the intensity of the hormonal stimuli simply is not sufficient to cause ovulation; or it can result from abnormal ovaries that will not allow ovulation. For instance, thick capsules occasionally exist on the outside of the ovaries that make ovulation difficult.

Because of the high incidence of anovulation in sterile women, special methods are often utilized to determine whether or not ovulation occurs. These are all based on the effects of progesterone on the body, for the normal increase in progesterone secretion usually does not occur during the latter half of anovulatory cycles. In the absence of progesteronic effects, the cycle can be assumed to be anovulatory. One of these tests is simply to analyze the urine for a surge in pregnanediol, the end product of progesterone metabolism, during the latter half of the sexual cycle, the lack of which indicates failure of ovulation. However, another common test is for the woman to chart her body temperature throughout the cycle. Secretion of progesterone during the latter half of the cycle raises the body temperature about 0.5°F, the temperature rise coming abruptly at the time of ovulation. Such a temperature chart, showing the point of ovulation, is illustrated in Figure 81–11.

Lack of ovulation caused by hyposecretion of the pituitary gonadotropic hormones can sometimes be treated by appropriately timed administration of *human chorionic gonadotropin,* a hormone that is discussed in the following chapter and that is extracted from the human placenta. This hormone, although secreted by the placenta, has almost exactly the same effects as LH and, therefore, is a powerful stimulator of ovulation. However, excess use of this hormone can cause ovulation from many follicles simultaneously; and this results in multiple births, an effect that has caused as many as seven babies to be born to mothers treated for infertility with this hormone.

One of the most common causes of female sterility is *endometriosis,* a very common condition in which endometrial tissue almost identical to that of the uterine endometrium grows and even menstruates in the pelvic cavity surrounding the uterus, fallopian tubes, and ovaries. Endometriosis causes fibrosis throughout the pelvis; and this fibrosis sometimes so enshrouds the ovaries that an ovum cannot be released into the abdominal cavity. Often, also, endometriosis occludes the fallopian tubes, either at the fimbriated ends or elsewhere along their extent. Another common cause of female infertility is salpingitis, that is, inflammation of the fallopian tubes; this causes fibrosis in the tubes, thereby occluding them. In past years, such inflammation was extremely common as a result of gonococcal infection, but with modern therapy this is becoming a less prevalent cause of female infertility.

Finally, still another cause of infertility is secretion of abnormal mucus by the uterine cervix. Ordinarily, at the time of ovulation, the hormonal environment of estrogen causes secretion of a thin mucus with special characteristics that will allow rapid mobility of sperm into the uterus and will actually guide the sperm up along mucous "threads." Abnormalities of the cervix itself, such as low-grade infection or inflammation, or abnormal hormonal stimulation of the cervix can lead to a viscous mucous plug that will prevent fertilization.

REFERENCES

Beato, M. (ed.): Steroid-Induced Uterine Proteins. New York, Elsevier/North-Holland, 1980.

Beyer, C., and Feder, H. H.: Sex steroids and afferent input: Their roles in brain sexual differentiation. Annu. Rev. Physiol., 49:349, 1987.

Chambon, P., et al.: Promoter elements of genes coding for proteins and modulation of transcription by estrogens and progesterone. Recent Prog. Horm. Res., 40:1, 1984.

Conn, P. M., et al: Gonadotropin-releasing hormone: Molecular and cell biology, physiology, and clinical applications. Fed. Proc., 43:2351, 1984.

Coutts, S. R. (ed.): Functional Morphology of the Human Ovary. Baltimore, University Park Press, 1981.

DeGroot, L. J., et al. (eds.): Endocrinology, 2nd Ed. Philadelphia, W. B. Saunders Co., 1989.

DeJong, F. H.: Inhibin. Physiol. Rev., 68:555, 1988.

Diczfausy, E., and Bygdeman, M.: Fertility Regulation Today and Tomorrow. New York, Raven Press, 1987.

Dufau, M. L.: Endocrine regulation and communicating functions of the Leydig cell. Annu. Rev. Physiol., 50:483, 1988.

Fishman, J.: Aromatic hydroxylation of estrogens. Annu. Rev. Physiol., 45:61, 1983.

Freeman, M. E.: The nonpregnant uterus as an endocrine organ. News Physiol. Sci., 3:31, 1988.

Garel, J.-M.: Hormonal control of calcium metabolism during the reproductive cycle in mammals. Physiol. Rev., 67:1, 1987.

Gruhn, J. G., and Kazer, R. R.: Hormonal Regulation of the Menstrual Cycle. New York, Plenum Publishing Corp., 1989.

Hammond, C. B., and Maxson, W. S.: Physiology of the menopause. Current Concepts, Upjohn, February, 1983.

Jones, H. W., III, et al. (eds.): Novak's Textbook of Gynecology, 11th Ed. Baltimore, Williams & Wilkins, 1988.

Karsch, F. J.: Central actions of ovarian steroids in the feedback regulation of pulsatile secretion of luteinizing hormone. Annu. Rev. Physiol., 49:365, 1987.

Keyes, P. L., and Wiltbank, M. C.: Endocrine regulation of the corpus luteum. Annu. Rev. Physiol., 50:465, 1988.

Knobil, E., et al. (eds.): The Physiology of Reproduction. New York, Raven Press, 1988.

Knobil, E.: A hypothalamic pulse generator governs mammalian reproduction. News Physiol. Sci., 2:42, 1987.

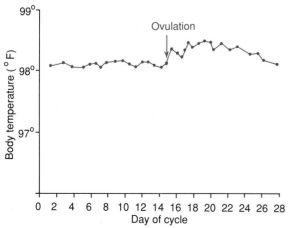

Figure 81–11. Elevation in body temperature shortly after ovulation.

Leung, P. C. K., and Armstrong, D. T.: Interactions of steroids and gonadotropins in the control of steroidogenesis on the ovarian follicle. Annu. Rev. Physiol., 42:71, 1980.

Leung, P. C. K., et al. (eds.): Endocrinology and Physiology of Reproduction. New York, Plenum Publishing Corp., 1987.

Mahesh, V. B., et al. (eds.): Regulation of Ovarian and Testicular Function. New York, Plenum Publishing Corp., 1987.

Mahesh, V. B.: The dynamic interaction between steroids and gonadotropins in the mammalian ovulatory cycle. Neurosci. Biobehav. Rev., 9(2):245, 1985.

McCann, S. M.: Physiology and pharmacology of LHRH and somatostatin. Annu. Rev. Pharmacol. Toxicol., 22:491, 1982.

McEwen, B. S., and Parsons, B.: Gonadal steroid action on the brain: Neurochemistry and neuropharmacology. Annu. Rev. Pharmacol. Toxicol., 22:555, 1982.

Meschia, G.: Circulation to female reproductive organs. In Shepherd, J. T., and Abboud, F. M. (eds.): Handbook of Physiology. Sec. 2, Vol. III. Bethesda, Md., American Physiological Society, 1983, p. 241.

Millar, R. P., and King, J. A.: Evolution of gonadotropin-releasing hormone: Multiple usage of a peptide. News Physiol. Sci., 3:49, 1988.

Mishell, D. R., Jr. (ed.): Long-Acting Steroid Contraception. New York, Raven Press, 1983.

Molinatti, G. M. (ed.): Androgenization in Women. Pathophysiology and Clinical Aspects. New York, Raven Press, 1983.

Newton, M., and Newton, E. R.: Complications of Gynecologic and Obstetric Management. Philadelphia, W. B. Saunders Co., 1988.

Ochiai, K. (ed.): Endocrine Correlates of Reproduction. New York, Springer-Verlag, 1984.

Pohl, C. R., and Knobil, E.: The role of the central nervous system in the control of ovarian function in higher primates. Annu. Rev. Physiol., 44:583, 1982.

Reiter, E. O., and Grumbach, M. M.: Neuroendocrine control mechanisms and the onset of puberty. Annu. Rev. Physiol., 44:595, 1982.

Richards, J. S., and Hedin, L.: Molecular aspects of hormone action in ovarian folicular development, ovulation, and luteinization. Annu. Rev. Physiol., 50:441, 1988.

Richards, J. S.: Maturation of ovarian follicles: Actions and interactions of pituitary and ovarian hormones on follicular cell differentiation. Physiol. Rev., 60:51, 1980.

Rories, C., and Spelsberg, T. C.: Ovarian steroid action on gene expression: Mechanisms and Models. Annu. Rev. Physiol., 51:653, 1989.

Schwartz, N. B. (ed.): Dynamics of Ovarian Function. New York, Raven Press, 1981.

Serra, G. B. (ed.): The Ovary. New York, Raven Press, 1983.

Soules, M. F.: Problems in Reproductive Endocrinology and Infertility. New York, Elsevier Science Publishing Co., 1989.

Stouffer, R. L. (ed.): The Primate Ovary. New York, Plenum Publishing Corp., 1987.

Wassarman, P. M.: Eggs, sperm, and sugar: A recipe for fertilization. News Physiol. Sci., 3:120, 1988.

Weiss, G.: Relaxin. Annu. Rev. Physiol., 46:43, 1984.

Wynn, R. M., and Jollie, W. (eds.): The Biology of the Uterus, 2nd Ed. New York, Plenum Publishing Corp., 1989.

Yen, S. S. C., and Jaffe, R.: Reproductive Endocrinology: Physiology, Pathophysiology and Clinical Management, 2nd Ed. Philadelphia, W. B. Saunders Co., 1986.

CHAPTER 82

• *Pregnancy and Lactation* •

In the preceding two chapters the sexual functions of the male and female are described to the point of fertilization of the ovum. If the ovum becomes fertilized, a completely new sequence of events called *gestation,* or *pregnancy,* takes place; and the fertilized ovum eventually develops into a full-term fetus. The purpose of the present chapter is to discuss the early stages of ovum development after fertilization and then to discuss the physiology of pregnancy. In the following chapter some special problems of fetal and early childhood physiology are discussed.

MATURATION OF THE OVUM

Shortly before the *primary oocyte* (the original stage of the ovum in the ovary) is released from the follicle, its nucleus divides by meiosis and a so-called *first polar body* is expelled from the nucleus of the oocyte, and this then becomes the *secondary oocyte.* In this process, each of the 23 pairs of chromosomes loses one of the partners that becomes incorporated in the polar body that is expelled, so that 23 *unpaired* chromosomes remain in the secondary oocyte. A few hours after the sperm enters the oocyte, the nucleus divides again; and a *second polar body* is expelled, thus forming the *mature ovum,* which still contains 23 unpaired chromosomes.

One of the 23 chromosomes is always a female chromosome, called an *X chromosome.* When this combines with a sperm that carries an X chromosome, giving an XX combination, a female child is formed, as explained in Chapter 80. And when the X chromosome of the ovum is paired with a sperm that carries a Y chromosome, giving an XY combination, a male child is formed.

TRANSPORT, FERTILIZATION, AND IMPLANTATION OF THE DEVELOPING OVUM

Entry of the Ovum Into the Fallopian Tube. When ovulation occurs, the ovum, along with a hundred or more attached granulosa cells, the *corona radiata,* is expelled directly into the peritoneal cavity and must then enter one of the fallopian tubes. The fimbriated ends of each fallopian tube fall naturally around the ovaries; the inner surfaces of the fimbriated tentacles are lined with ciliated epithelium, and the *cilia,* increased in number and activated by estrogen from the ovaries, continually beat toward the *abdominal ostium* of the fallopian tube. One can actually see a slow fluid current flowing toward the ostium. By this means the ovum enters one or the other fallopian tube.

It would seem likely that many ova might fail to enter the fallopian tubes. However, on the basis of conception studies, it is probable that as many as 98 per cent succeed in this task. Indeed, cases are on record in which women with one ovary removed and the opposite fallopian tube removed have had as many as four children with relative ease of conception, thus illustrating that ova can even enter the opposite fallopian tube.

Fertilization of the Ovum. After ejaculation, a few sperm are transported through the uterus to the *ampullae* in the ovarian ends of the fallopian tubes within 5 to 10 minutes, probably aided by contractions of the uterus and fallopian tubes stimulated by prostaglandins in the seminal fluid and oxytocin released from the posterior pituitary gland during the female orgasm. However, of the half billion sperm deposited in the vagina, only 1000 to 3000 succeed in actually reaching the ampulla.

Fertilization of the ovum normally takes place soon after the ovum enters the ampulla. However, before a sperm can enter the ovum, it must first penetrate the multiple layers of granulosa cells attached to the outside of the ovum, called the *corona radiata,* and must bind to and penetrate the *zona pellucida* surrounding the ovum itself. The mechanisms employed by the sperm for these purposes are presented in Chapter 80.

Once a sperm has entered the ovum, its head swells rapidly to form a *male pronucleus,* which is illustrated in Figure 82–1. Later, after the nucleus of the ovum has discharged the second polar body, the 23 unpaired chromosomes of the male pronucleus and the 23 unpaired chromosomes of the *female pronucleus* align themselves to reform a complete complement of 46 chromosomes (23 pairs) in the fertilized ovum.

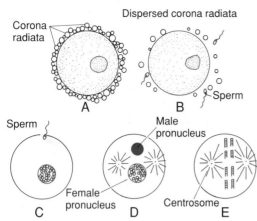

Figure 82–1. Fertilization of the ovum, showing *A,* the mature ovum surrounded by the corona radiata, *B,* dispersal of the corona radiata, *C,* entry of the sperm, *D,* formation of the male and female pronuclei, and *E,* reorganization of a full complement of chromosomes and beginning division of the ovum. (Modified from Arey: Developmental Anatomy. 7th ed. Philadelphia, W. B. Saunders Company, 1974.)

Figure 82–2. Implantation of the early human embryo, showing trophoblastic digestion and invasion of the endometrium. (Courtesy of Dr. Arthur Hertig.)

Transport in the Fallopian Tubes

After fertilization has occurred, an additional 3 to 4 days is normally required for transport of the ovum through the fallopian tube into the cavity of the uterus. This transport is effected mainly by a feeble fluid current in the tube resulting from epithelial secretion plus action of the ciliated epithelium that lines the tube, the cilia always beating toward the uterus. It is possible also that weak contractions of the fallopian tube aid in the passage of the ovum.

The fallopian tubes are lined with a rugged, cryptoid surface that actually impedes the passage of the ovum despite the fluid current. Also, the *isthmus* of the fallopian tube (the last 2 centimeters before the uterus is entered) remains spastically contracted for the first 3 days following ovulation. After this time, the rapidly increasing progesterone first promotes increasing progesterone receptors on the smooth muscle cells and then activates these, exerting a relaxing effect that allows entry of the ovum into the uterus.

The delayed transport of the ovum through the fallopian tube allows several stages of division to occur before the dividing ovum, now called a *blastocyst* with about 100 cells, enters the uterus. During this time, large quantities of secretions are formed by secretory cells that alternate with ciliated cells lining the fallopian tube. These secretions are for nutrition of the blastocyst.

Implantation of the Blastocyst in the Uterus

After reaching the uterus, the developing blastocyst usually remains in the uterine cavity an additional 2 to 4 days before it implants in the endometrium; thus, implantation ordinarily occurs on about the seventh day after ovulation. Before implantation, the blastocyst obtains its nutrition from the endometrial secretions called "uterine milk."

Implantation results from the action of *trophoblast cells* that develop over the surface of the blastocyst. These cells secrete proteolytic enzymes that digest and liquefy the cells of the endometrium. The fluid and nutrients thus released are then actively transported by the same trophoblast cells into the blastocyst, adding still further sustenance for growth. Figure 82–2 shows an early implanted human blastocyst, with a very small embryo illustrated in color.

Once implantation has taken place, the trophoblast and other sublying cells proliferate rapidly, forming the placenta and the various membranes of pregnancy.

● EARLY INTRAUTERINE NUTRITION OF THE EMBRYO

In the previous chapter it was pointed out that the progesterone secreted during the latter half of each sexual cycle has a special effect on the endometrium to convert the endometrial stromal cells into large swollen cells that contain extra quantities of glycogen, proteins, lipids, and even some necessary minerals for development of the conceptus. Then when the conceptus implants in the endometrium, the continued secretion of progesterone causes the endometrial cells to swell still more and to store even more nutrients. These cells are now called *decidual cells,* and the total mass of cells is called the *decidua.*

As the trophoblast cells invade the decidua, digesting and imbibing it, the stored nutrients in the decidua are used by the embryo for appropriate growth and development. During the first week after implantation, this is the only means by which the embryo can obtain nutrients, and the embryo continues to obtain a large measure of its total nutrition in this way for at least 8 weeks, though the placenta also begins to provide slight amounts of nutrition after approximately the 16th day beyond fertilization (a little over a week

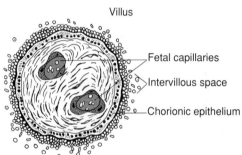

Figure 82–4. *Above,* Organization of the mature placenta. *Below,* Relationship of the fetal blood in the villus capillaries to the mother's blood in the intervillous spaces. (Modified from Gray and Goss: Anatomy of the Human Body. Philadelphia, Lea & Febiger; and from Arey: Developmental Anatomy. 7th ed. Philadelphia, W. B. Saunders Company, 1974.)

Figure 82–3. Nutrition of the fetus, illustrating that most of the early nutrition is due to trophoblastic digestion and absorption of nutrients from the endometrial decidua and that essentially all the later nutrition results from diffusion through the placental membrane.

after implantation). Figure 82–3 depicts this trophoblastic period of nutrition, which gradually gives way to placental nutrition.

● FUNCTION OF THE PLACENTA

DEVELOPMENTAL AND PHYSIOLOGIC ANATOMY OF THE PLACENTA

While the trophoblastic cords from the blastocyst are attaching to the uterus, blood capillaries grow into the cords from the vascular system of the embryo; and, by the 16th day of fertilization, blood begins to flow. Simultaneously, *blood sinuses* supplied with blood from the mother develop around the trophoblastic cords. The trophoblast cells send out more and more projections, which become the *placental villi* into which fetal capillaries grow. Thus, the villi, carrying fetal blood, are surrounded by sinuses containing maternal blood.

The final structure of the placenta is illustrated in Figure 82–4. Note that the fetus' blood flows through two *umbilical arteries,* then to the capillaries of the villi, and thence back through the *umbilical vein* into the fetus. At the same time, the mother's blood flows from the *uterine arteries* into large *maternal sinuses* surrounding the villi and then back into the *uterine veins* of the mother.

The lower part of Figure 82–4 illustrates the relationship between the fetal blood of the villus and the blood of the mother in the fully developed placenta. The capillaries of the villus are lined with an extremely thin endothelium and are surrounded by a layer of *mesenchymal tissue* that is covered on the outside of the villus by a layer of *syncytial trophoblast* cells. During the first 16 weeks of pregnancy, still an additional layer of cells is present immediately beneath the syncytial trophoblast layer. This layer is composed of distinct cuboidal cells called *cytotrophoblast cells,* or *Langerhans' cells.*

The total surface area of all the villi of the mature placenta is only a few square meters — many times less than the area of the pulmonary membrane. Also, remember that even at full maturity the placental membrane is still several cell layers thick, and the minimum distance between the maternal blood and the fetal blood is 3.5 micrometers, al-

most ten times the distance across the alveolar membranes of the lung. Nevertheless, many nutrients and other substances pass through the placental membrane by diffusion in very much the same manner that diffusion occurs through the alveolar membranes of the lungs and the capillary membranes elsewhere in the body.

Placental Permeability and Membrane Diffusion Conductance

The major function of the placenta is to provide for *diffusion of foodstuffs* from the mother's blood into the fetus's blood and also *diffusion of excretory products* from the fetus back into the mother. Therefore, it is important to know the permeability and total membrane diffusion conductance of the placental membrane. Permeability is expressed as the net quantity of a substance that will diffuse through a unit surface area of placental membrane for a given concentration difference, and *total* membrane conductance is the permeability times the total surface area.

In the early months of pregnancy, the placental membrane is still thick because it is not fully developed. Therefore, its permeability is low. Furthermore, the surface area is also very slight because the placenta simply has not grown significantly. Therefore, the total diffusion conductance is miniscule at first. On the other hand, in later pregnancy, the permeability increases because of thinning of the membrane diffusion layers; and the surface area becomes enormous because of growth, thus giving the tremendous increase in placental conductance illustrated in Figure 82–3.

Rarely, "breaks" occur in the placental membrane, which allow fetal blood cells to pass into the mother or, more rarely, the mother's cells to pass into the fetus. Indeed, there are instances in which the fetus bleeds severely into the mother's circulation because of a ruptured placental membrane.

Diffusion of Oxygen Through the Placental Membrane. Almost exactly the same principles are applicable for the diffusion of oxygen through the placental membrane as through the pulmonary membrane; these principles are discussed in Chapter 39. The dissolved oxygen in the blood of the large maternal sinuses simply passes into the fetal blood because of an oxygen pressure gradient from the mother's blood to the fetus' blood. The mean Po_2 in the mother's blood in the maternal sinuses is approximately 50 mm Hg toward the end of pregnancy, and the mean Po_2 in the fetal blood after it becomes "oxygenated" is about 30 mm Hg. Therefore, the mean pressure gradient for diffusion of oxygen through the placental membrane is about 20 mm Hg.

One might wonder how it is possible for a fetus to obtain sufficient oxygen when the fetal blood leaving the placenta has a Po_2 of only 30 mm Hg. However, there are three different reasons why even this low Po_2 is capable of allowing the fetal blood to transport almost as much oxygen to the fetal tissues as is transported by the mother's blood to her tissues:

First, the hemoglobin of the fetus is mainly *fetal hemoglobin,* a type of hemoglobin synthesized in the fetus prior to birth. Figure 82–5 illustrates the comparative oxygen dissociation curves of maternal hemoglobin and fetal hemoglobin, showing that the curve for fetal hemoglobin is shifted to the left of that for maternal hemoglobin. This means that at a given Po_2, the fetal hemoglobin can carry as much as 20 to 30 per cent more oxygen than can maternal hemoglobin.

Second, the *hemoglobin concentration of the fetus is about 50 per cent greater than that of the mother;* therefore, this is an even more important factor in enhancing the amount of oxygen transported to the fetal tissues.

Third, the *Bohr effect,* which is explained in relation to the exchange of carbon dioxide and oxygen in the lung in Chapter 40, provides another factor that enhances the transport of oxygen by the fetal blood. That is, hemoglobin can carry more oxygen at a low Pco_2 than it can at a high Pco_2. The fetal blood entering the placenta carries large amounts of carbon dioxide, but much of this carbon dioxide diffuses from the fetal blood into the maternal blood. Loss of the carbon dioxide makes the fetal blood more alkaline, whereas the increased carbon dioxide in the maternal blood makes this more acidic. These changes cause the combining capacity of fetal blood for oxygen to become increased and that of the maternal blood to become decreased. This forces more oxygen from the maternal blood while enhancing the oxygen in the fetal blood. Thus, the Bohr shift operates in one direction in the maternal blood and in the other in the fetal blood, these two effects making the Bohr shift twice as important here as it is for oxygen exchange in the lungs; therefore, it is actually called the *double Bohr effect.*

By these three means, the fetus is capable of receiving more than adequate oxygen through the placenta despite the fact that the fetal blood leaving the placenta has a Po_2 of only 30 mm Hg.

The total *diffusing capacity* of the entire placenta for oxygen at term is about 1.2 milliliters of oxygen per minute per millimeter of oxygen difference across the membrane. This compares favorably with that of the lungs of the newborn baby.

Diffusion of Carbon Dioxide Through the Placental Membrane. Carbon dioxide is continually formed in the tissues of the fetus in the same way that it is formed in maternal tissues. And the only means for excreting the carbon dioxide is through the placenta into the mother's blood. The Pco_2 of the fetal blood is 2 to 3 mm Hg higher than that of the maternal blood. This small pressure gradient for carbon dioxide across the membrane is sufficient to allow adequate diffusion of carbon dioxide because the extreme solubility of carbon dioxide in the placental membrane allows carbon dioxide to diffuse about 20 times as rapidly as oxygen.

The Pco_2 of the blood in the maternal sinuses is usually less than the normal value of 40 mm Hg found in the arterial blood of the nonpregnant woman because the estrogens and progesterone of pregnancy overdrive the mother's breathing, thus causing her to blow off excess CO_2 from her lungs. This obviously helps keep the Pco_2 in the fetus' blood at a low level as well, usually very near to the normal value of 40 mm Hg.

Figure 82–5. Oxygen-hemoglobin dissociation curves for maternal *(M)* and fetal *(F)* bloods, illustrating the ability of the fetal blood to carry a much greater quantity of oxygen than can maternal blood for a given blood Po_2. (From Metcalfe, Moli, and Bartels: *Fed. Proc., 23:*775, 1964.)

Diffusion of Foodstuffs Through the Placental Membrane. Other metabolic substrates needed by the fetus diffuse into the fetal blood in the same manner as oxygen does. For instance, in the late stages of pregnancy the fetus often utilizes more glucose than does the entire body of the mother. To provide this much glucose, the trophoblast cells lining the placental villi provide for *facilitated diffusion* of glucose through the placental membrane. That is, the glucose is transported by carrier molecules in the trophoblast cell membrane. Even so, the glucose level in the fetal blood is still 20 to 30 per cent lower than that in the maternal blood.

Because of the high solubility of fatty acids in cell membranes, these also diffuse from the maternal blood into the fetal blood but more slowly than glucose, so that glucose is preferentially used by the fetus for nutrition. Also, such substances as ketone bodies and potassium, sodium, and chloride ions diffuse from the maternal blood into the fetal blood.

Active Absorption by the Placental Membrane. As pointed out previously, early nutrition of the embryo depends on phagocytosis of fallopian tube and uterine secretions and even on phagocytosis of the endometrial decidua. The trophoblast cells that line the outer surface of the villi can also *actively absorb* certain nutrients from the maternal blood in the placenta, at least during the first half of pregnancy and perhaps even throughout the entire pregnancy. For instance, the measured *amino acid* content of fetal blood is greater than that of maternal blood, and *calcium* and *inorganic phosphate* occur in greater concentration in fetal blood than in maternal blood, whereas *ascorbic acid* is as much as three times as concentrated in the fetal blood. These findings indicate that the placental membrane has the ability to absorb actively at least small amounts of certain substances even during the latter part of pregnancy.

Excretion Through the Placental Membrane. In the same manner that carbon dioxide diffuses from the fetal blood into the maternal blood, other excretory products formed in the fetus diffuse in the opposite direction into the maternal blood and then are excreted along with the excretory products of the mother. These include especially the *nonprotein nitrogens* such as *urea, uric acid,* and *creatinine.* The level of urea in the fetal blood is only slightly greater than that in maternal blood because urea diffuses through the placental membrane with considerable ease. On the other hand, creatinine, which does not diffuse as easily, has a concentration gradient of a considerably higher percentage. Therefore, excretion from the fetus occurs mainly, if not entirely, as a result of diffusion gradients across the placental membrane—that is, because of higher concentrations of the excretory products in the fetal blood than in the maternal blood.

● HORMONAL FACTORS IN PREGNANCY

In pregnancy, the placenta forms especially large quantities of *human chorionic gonadotropin, estrogens, progesterone, and human chorionic somatomammotropin,* the first three of which, and perhaps the fourth as well, are essential to a normal pregnancy.

HUMAN CHORIONIC GONADOTROPIN AND ITS EFFECT IN CAUSING PERSISTENCE OF THE CORPUS LUTEUM AND IN PREVENTING MENSTRUATION

Menstruation normally occurs approximately 14 days after ovulation, at which time most of the endometrium of the uterus sloughs away from the uterine wall and is expelled to the exterior. If this should happen after an ovum has implanted, the pregnancy would terminate. However, this is prevented by the secretion of human chorionic gonadotropin in the following manner:

Coincidentally with the development of the trophoblast cells from an early fertilized ovum, the hormone *human chorionic gonadotropin* is secreted by the syncytial trophoblast cells into the fluids of the mother, as illustrated in Figure 82-6. The secretion of this hormone can first be measured in the blood 8 to 9 days after ovulation, shortly after the blastocyst implants in the endometrium. Then the rate of secretion rises rapidly to reach a maximum approximately 10 to 12 weeks after ovulation and decreases to a much lower value by 16 to 20 weeks after ovulation.

Function of Human Chorionic Gonadotropin. Human chorionic gonadotropin is a glycoprotein having a molecular weight of about 39,000 and very much the same molecular structure and function as luteinizing hormone secreted by the pituitary. By far its most important function is to prevent the normal involution of the corpus luteum at the end of the female sexual cycle. Instead, it causes the corpus luteum to secrete even larger quantities of its usual sex hormones, progesterone and estrogens. These sex hormones cause the endometrium to continue growing and to store large amounts of nutrients rather than pass them in the menstruum. As a result, the *decidua-like cells* that develop in the endometrium during the normal female sexual cycle become actual, greatly swollen, and nutritious *decidual cells* soon after the blastocyst implants.

Under the influence of human chorionic gonadotropin, the corpus luteum grows to about two times its

Figure 82-6. Rates of secretion of estrogens, progesterone, and chorionic gonadotropin at different stages of pregnancy.

initial size by a month or so after pregnancy begins, and its continued secretion of estrogens and progesterone maintains the decidual nature of the uterine endometrium, which is necessary for the early development of the fetus. If the corpus luteum is removed before approximately the 7th week of pregnancy, spontaneous abortion almost always occurs, sometimes even up to the 12th week. After that time the placenta itself secretes sufficient quantities of progesterone and estrogens to maintain pregnancy for the remainder of the gestation period. The corpus luteum involutes slowly after the 13th to 17th week of gestation.

Effect of Human Chorionic Gonadotropin on the Fetal Testes. Human chorionic gonadotropin also exerts an *interstitial cell–stimulating effect* on the testes, thus resulting in the production of testosterone in male fetuses. This small secretion of testosterone during gestation is the factor that causes the fetus to grow male sex organs. Near the end of pregnancy, the testosterone secreted by the fetal testes also causes the testicles to descend into the scrotum.

SECRETION OF ESTROGENS BY THE PLACENTA

The placenta, like the corpus luteum, secretes both estrogens and progesterone. Both histochemical and physiological studies indicate that these two hormones, like most other placental hormones, are also secreted by the *syncytial trophoblast cells.*

Figure 82–6 shows that the daily production of placental estrogens increases, in terms of estrogenic activity, to about 30 times normal toward the end of pregnancy. However, the secretion of estrogens by the placenta is quite different from the secretion by the ovaries in several ways, as follows: First a large share of the secreted estrogens is estriol, which, as will be recalled from Chapter 81, is formed in only small amounts in the nongravid female. Yet because of the very low estrogenic potency of estriol, this accounts for little of the total estrogenic activity. Second, the estrogens secreted by the placenta are not synthesized *de novo* from basic substrates in the placenta. Instead, the androgenic steroid compounds *dehydroepiandrosterone* and *16-hydroxydehydroepiandrosterone,* which are formed in the mother's adrenal glands and also in the adrenal glands of the fetus, are transported by the blood to the placenta and converted into estradiol, estrone, and estriol. (The cortices of the fetal adrenal glands are extremely large, composed almost entirely of a so-called *fetal zone,* the primary function of which seems to be to secrete the dehydroepiandrosterone.)

Function of Estrogen in Pregnancy. In the discussions of estrogens in the preceding chapter it was pointed out that these hormones exert mainly a proliferative function on most reproductive and associated organs of the mother. During pregnancy, the extreme quantities of estrogens cause (1) enlargement of the uterus, (2) enlargement of the breasts and growth of the breast ductal structure, and (3) enlargement of the female external genitalia.

The estrogens also relax the various pelvic ligaments so that the sacroiliac joints become relatively limber and the symphysis pubis becomes elastic. These changes obviously make for easier passage of the fetus through the birth canal.

There is much reason to believe that estrogens also affect some general aspects of the development of the fetus during pregnancy, for example, by affecting the rate of cell reproduction in the early embryo.

SECRETION OF PROGESTERONE BY THE PLACENTA

Progesterone is also a hormone essential for pregnancy, in fact equally as important as estrogen. In addition to being secreted in moderate quantities by the corpus luteum at the beginning of pregnancy, it is secreted in tremendous quantities by the placenta, averaging about one-quarter gram per day toward the end of pregnancy. Indeed, the rate of progesterone secretion increases by as much as tenfold during the course of pregnancy, as illustrated in Figure 82–6.

The special effects of progesterone that are essential for normal progression of pregnancy and even for its continuance are the following:

1. As pointed out earlier, progesterone causes decidual cells to develop in the uterine endometrium, and these then play an important role in the nutrition of the early embryo.

2. Progesterone has a special effect on decreasing the contractility of the gravid uterus, thus preventing uterine contractions from causing spontaneous abortion.

3. Progesterone also contributes to the development of the conceptus even prior to implantation, for it specifically increases the secretions of the fallopian tubes and uterus to provide appropriate nutritive matter for the developing *morula* and *blastocyst.* There is reason to believe, too, that progesterone even affects cell cleavage in the early developing embryo.

4. The progesterone secreted during pregnancy also helps the estrogen prepare the breasts for lactation, which is discussed later in the chapter.

HUMAN CHORIONIC SOMATOMAMMOTROPIN

More recently, a new placental hormone called *human chorionic somatomammotropin* was discovered. This is a protein, having a molecular weight of about 38,000, that begins to be secreted about the fifth week of pregnancy and increases progressively throughout the remainder of pregnancy in direct proportion to the weight of the placenta. Although the functions of chorionic somatomammotropin are still uncertain, this hormone is secreted in quantities several times as great as that of all other pregnancy hormones combined. It has several possible important effects:

First, when administered to several different types of lower animals, human chorionic somatomammotropin causes at least partial development of the breasts and in some instances causes lactation. Because this was the first function of the hormone discovered, it was at first named *placental lactogen* and was believed to have functions similar to those of prolactin. However, attempts to promote lactation in the human being with its use have not been successful.

Second, this hormone has weak actions similar to those of growth hormone, causing deposition of protein tissues in the same way as growth hormone does. It also has a chemical structure similar to that of growth hormone, but 100 times as much human chorionic somatomammotropin as growth hormone is required to promote growth.

Third, human chorionic somatomammotropin causes decreased insulin sensitivity and decreased utilization of glucose in the mother, thereby making larger quantities of glucose available to the fetus. Since glucose is the major substrate used by the fetus to energize its growth, the importance of this hormonal effect is obvious. Furthermore, the hormone promotes release of free fatty acids from the fat stores of the mother, thus providing this alternative source of energy for her metabolism.

Therefore, it is beginning to appear that human chorionic somatomammotropin is a general metabolic hormone that has specific nutritional implications for both the mother and the fetus.

OTHER HORMONAL FACTORS IN PREGNANCY

Almost all of the nonsexual endocrine glands of the mother react markedly to pregnancy. This results mainly from the increased metabolic load on the mother but also to some extent from effects of placental hormones on the pituitary and other glands. In addition, some of the most notable effects are the following:

Pituitary Secretion. The anterior pituitary gland enlarges at least 50 per cent during pregnancy and increases its production of *corticotropin, thyrotropin,* and *prolactin.* On the other hand, follicle-stimulating hormone and luteinizing hormone are greatly suppressed as a result of the inhibitory effects of estrogens and progesterone from the placenta.

Corticosteroid Secretion. The rate of adrenocortical secretion of the *glucocorticoids* is moderately increased throughout pregnancy. It is possible that the glucocorticoids help mobilize amino acids from the mother's tissues so that these can be used for synthesis of tissues in the fetus.

Pregnant women also usually have about a twofold increase in secretion of *aldosterone*, reaching the peak at the end of gestation. This, along with the actions of the estrogens, causes a tendency for even the normal pregnant woman to reabsorb excess sodium from the renal tubules and therefore to retain fluid, occasionally leading to hypertension.

Secretions by the Thyroid Gland. The thyroid gland ordinarily enlarges up to 50 per cent during pregnancy and increases its production of thyroxine a corresponding amount. The increased thyroxine production is caused at least partly by a thyrotropic effect of *human chorionic go-*

nadotropin and also by small quantities of a specific thyroid-stimulating hormone, *human chorionic thyrotropin,* secreted by the placenta.

Secretion by the Parathyroid Glands. The parathyroid glands also usually enlarge during pregnancy; this is especially true if the mother is on a calcium-deficient diet. Enlargement of these glands causes calcium absorption from the mother's bones, thereby maintaining normal calcium ion concentration in the mother's extracellular fluids as the fetus removes calcium for ossifying its own bones. This secretion of parathyroid hormone is even more intensified during lactation following the birth of the baby, because the baby then requires many times more calcium than the fetus.

Secretion of "Relaxin" by the Ovaries and Placenta. An additional substance besides the estrogens and progesterone, a hormone called relaxin, is secreted by the corpus luteum of the ovary and also by the placenta. Its secretion by the corpus luteum is increased by human chorionic gonadotropin at the same time that the corpus luteum also secretes large quantities of estrogen and progesterone.

Relaxin is a polypeptide having a molecular weight of about 9000. This hormone, when injected, causes relaxation of the ligaments of the symphysis pubis in the estrous rat and guinea pig. However, this effect is very poor in the pregnant woman. Instead, this role is probably played mainly by the estrogens, which also cause relaxation of the pelvic ligaments.

It has also been claimed that relaxin softens the cervix of the pregnant woman at the time of delivery.

● RESPONSE OF THE MOTHER'S BODY TO PREGNANCY

Most apparent among the many reactions of the mother to the fetus and to the excessive hormones of pregnancy is the increased size of the various sexual organs. For instance, the uterus increases from about 50 grams to about 1100 grams, and the breasts approximately double in size. At the same time the vagina enlarges, and the introitus opens more widely. Also, the various hormones can cause marked changes in the appearance of the woman, sometimes resulting in the development of edema, acne, and masculine or acromegalic features.

Weight Gain in the Pregnant Woman

The average weight gain during pregnancy is about 24 pounds, most of this gain occurring during the last two trimesters. Of this, approximately 7 pounds is fetus, and 4 pounds is amniotic fluid, placenta, and fetal membranes. The uterus increases approximately 2 pounds, and the breasts another 2 pounds, still leaving an average increase in weight of the woman's body of 9 pounds. About 6 pounds of this is extra fluid in the blood and extracellular fluid, and the remaining 3 pounds in general is fat accumulation. The extra fluid is excreted in the urine during the first few days after birth, that is, after loss of the fluid-retaining hormones of the placenta.

Often during pregnancy a woman has a greatly increased desire for food, partly as a result of fetal removal of food substrates from the mother's blood and partly because of hormonal factors. Without appropriate prenatal care the mother's weight gain can be as great as 75 pounds.

Metabolism During Pregnancy

As a consequence of the increased secretion of many different hormones during pregnancy, including thyroxine, adrenocortical hormones, and the sex hormones, the basal metabolic rate of the pregnant woman increases about 15 per cent during the latter half of the pregnancy. As a result, she frequently has sensations of becoming overheated. Also, owing to the extra load that she is carrying, greater amounts of energy than normal must be expended for muscular activity.

Nutrition During Pregnancy. By far the greatest growth of the fetus occurs during the last trimester of pregnancy; its weight almost doubles during the last two months of pregnancy. Ordinarily, the mother does not absorb sufficient protein, calcium, phosphates, and iron from the gastrointestinal tract during the last months of pregnancy to supply the fetus. However, from the beginning of pregnancy her body has been storing these substances, some in the placenta but most in the normal storage depots of the mother.

If appropriate nutritional elements are not present in the pregnant woman's diet, a number of maternal deficiencies can occur. Deficiencies often occur in calcium, phosphates, iron, and the vitamins. For example, approximately 375 milligrams of iron is needed by the fetus to form its blood, and an additional 600 milligrams is needed by the mother to form her own extra blood. The normal store of nonhemoglobin iron in the mother at the outset of pregnancy is often only 100 or so milligrams and almost never over 700 milligrams. Therefore, without sufficient iron in her food, a pregnant woman usually develops anemia. Also, it is especially important that she receive vitamin D, for although the total quantity of calcium utilized by the fetus is small, calcium even normally is poorly absorbed by the gastrointestinal tract. Finally, shortly before birth of the baby, vitamin K is often added to the mother's diet so that the baby will have sufficient prothrombin to prevent hemorrhage, particularly brain hemorrhage, caused by the birth process.

Changes in the Maternal Circulatory System During Pregnancy

Blood Flow Through the Placenta and Cardiac Output During Pregnancy. About 625 milliliters of blood flows through the maternal circulation of the placenta each minute during the latter phases of pregnancy. This, plus the general increase in metabolism, increases the mother's cardiac output to 30 to 40 per cent above normal by the 27th week of pregnancy, but then, for reasons unexplained, the cardiac output falls to only a little above normal during the last 8 weeks of pregnancy, despite the high uterine blood flow.

Blood Volume During Pregnancy. The maternal blood volume shortly before term is approximately 30 per cent above normal. This increase occurs mainly during the latter half of pregnancy, as illustrated by the curve of Figure 82–7. The cause of the increased volume is mainly hormonal, for both aldosterone and estrogens, which are greatly increased in pregnancy, cause increased fluid retention by the kidneys. Also, the bone marrow becomes increasingly active and produces some extra red blood cells to go with the excess fluid volume. Therefore, at the time of birth of the baby, the mother has approximately 1 to 2 liters of extra blood in her circulatory system. Only about one fourth of this amount is normally lost during delivery of the

Figure 82–7. Effect of pregnancy on blood volume.

baby, thereby allowing a considerable safety factor for the mother.

Respiration During Pregnancy

Because of the increased basal metabolic rate of the pregnant woman and also because of her increase in size, the total amount of oxygen utilized by the mother shortly before birth of the baby is approximately 20 per cent above normal, and a commensurate amount of carbon dioxide is formed. These effects cause the minute ventilation to increase. It is also believed that the high levels of progesterone during pregnancy increase the minute ventilation still more, because progesterone increases the sensitivity of the respiratory center to carbon dioxide. The net result is an increase in minute ventilation of approximately 50 per cent and a decrease in arterial Pco_2 to several millimeters of mercury below that of the normal woman. Simultaneously, the growing uterus presses upward against the abdominal contents, and these in turn press upward against the diaphragm, so that the total excursion of the diaphragm is decreased. Consequently, the respiratory rate is increased to maintain adequate ventilation.

Function of the Maternal Urinary System During Pregnancy

The rate of urine formation by the pregnant woman is usually slightly increased because of increased fluid intake and also increased load of excretory products. But, in addition, several special alterations of urinary function occur:

First, reabsorption of sodium, chloride, and water by the renal tubules is increased as much as 50 per cent as a consequence of increased production of steroid hormones by the placenta and adrenal cortex.

Second, the glomerular filtration rate also increases as much as 50 per cent during pregnancy, which tends to increase the rate of water and electrolyte loss in the urine. This factor normally almost balances the first so that the mother ordinarily accumulates only about 6 pounds of extra water and salt except when she develops *preeclampsia;* this condition is discussed later in the chapter.

The Amniotic Fluid and Its Formation

Normally, the volume of amniotic fluid is between 500 milliliters and 1 liter, but it can be only a few milliliters or as much as several liters. Studies with isotopes on the rate of formation of amniotic fluid show that on the average the water in amniotic fluid is completely replaced once every 3 hours, and the electrolytes sodium and potassium are replaced once every 15 hours. A portion of the fluid is derived from renal excretion by the fetus. Likewise, a certain amount of absorption occurs by way of the gastrointestinal tract and lungs of the fetus. However, even after the death of a fetus, the rate of turnover of the amniotic fluid is still one-half as great as it is when the fetus is normal, which indicates that much of the fluid is formed and absorbed directly through the amniotic membranes. The total volume of amniotic fluid is probably regulated mainly by the amniotic membranes, but the volume also increases when the fetal excretion of urine increases, and decreases when there is no urinary output.

PREECLAMPSIA AND ECLAMPSIA

Approximately 4 per cent of all pregnant women experience a rapid rise in arterial blood pressure associated with loss of large amounts of protein in the urine during the latter few months of pregnancy. This condition is called *preeclampsia* or *toxemia of pregnancy*. It is often also characterized by salt and water retention by the kidneys, weight gain, and development of edema. In addition, arterial spasm occurs in many parts of the body, most significantly in the kidneys, brain, and liver. Both the renal blood flow and the glomerular filtration rate are decreased, which is exactly opposite to the changes that occur in the normal pregnant woman. The renal effects are caused by thickened glomerular tufts that contain a protein deposit in the basement membranes.

Various attempts have been made to prove that preeclampsia is caused by excessive secretion of placental or adrenal hormones, but proof of a hormonal basis is still lacking. Indeed, another plausible theory is that preeclampsia results from some type of autoimmunity or allergy resulting from the presence of the fetus. In support of this, the acute symptoms disappear within a few days after birth of the baby.

The severity of preeclampsia symptoms is closely associated with the increase in arterial pressure. In fact, an increasing pressure seems to set off a vicious circle that intensifies the arterial spasm and other pathological effects of preeclampsia.

Eclampsia is an extreme degree of the same effects as those observed in preeclampsia, characterized by extreme vascular spasm throughout the body, clonic convulsions sometimes followed by coma, greatly decreased kidney output, malfunction of the liver, often extreme hypertension, and a generalized toxic condition of the body. Usually, it occurs shortly before parturition. Without treatment, a very high percentage of eclamptic patients die. However, with optimal and immediate use of rapidly acting vasodilating drugs to reduce the arterial pressure to normal, followed by immediate termination of pregnancy—by cesarean section if necessary—the mortality has been reduced to 1 per cent or less.

● PARTURITION

INCREASED UTERINE EXCITABILITY NEAR TERM

Parturition means simply the process by which the baby is born. Toward the end of pregnancy the uterus becomes progressively more excitable until finally it begins strong rhythmic contractions with such force that the baby is expelled. The exact cause of the increased activity of the uterus is not known, but at least two major categories of effects lead up to the culminating contractions responsible for parturition: first, progressive hormonal changes that cause increased excitability of the uterine musculature; and, second, progressive mechanical changes.

Hormonal Factors That Cause Increased Uterine Contractility. *Ratio of Estrogens to Progesterone.* Progesterone inhibits uterine contractility during pregnancy, thereby helping to prevent expulsion of the fetus. On the other hand, estrogens have a definite tendency to increase the degree of uterine contractility, at least partly because estrogens increase the number of gap junctions between the adjacent uterine smooth muscle cells. Both progesterone and estrogen are secreted in progressively greater quantities throughout most of pregnancy, but from the seventh month onward estrogen secretion continues to increase while progesterone secretion remains constant or perhaps even decreases slightly. Therefore, it has been postulated that the *estrogen-to-progesterone ratio* increases sufficiently toward the end of pregnancy to be at least partly responsible for the increased contractility of the uterus.

Effect of Oxytocin on the Uterus. Oxytocin is a hormone secreted by the neurohypophysis that specifically causes uterine contraction (see Chapter 75). There are four reasons for believing that oxytocin might be important in increasing the contractility of the uterus near term: (1) The uterus increases its responsiveness to a given dose of oxytocin during the latter few months of pregnancy. (2) The rate of oxytocin secretion by the neurohypophysis is considerably increased at the time of labor. (3) Though hypophysectomized animals and human beings can still deliver their young at term, labor is prolonged. (4) Experiments in animals indicate that irritation or stretching of the uterine cervix, as occurs during labor, can cause a neurogenic reflex that causes the posterior pituitary gland to increase its secretion of oxytocin.

Effect of Fetal Hormones on the Uterus. The fetus' pituitary gland also secretes increasing quantities of oxytocin that could possibly play a role in exciting the uterus, and its adrenal gland secretes large quantities of cortisol that are also a possible uterine stimulant. In addition, the fetal membranes release prostaglandins in high concentration at the time of labor. These, too, can increase the intensity of the uterine contractions.

Mechanical Factors That Increase the Contractility of the Uterus. *Stretch of the Uterine Musculature.* Simply stretching smooth muscle organs usually increases their contractility. Furthermore, intermittent stretch, as occurs repetitively in the uterus because of movements of the fetus, can also elicit smooth muscle contraction.

Note especially that twins are born on the average *19 days* earlier than a single child, which emphasizes the probable importance of mechanical stretch in eliciting uterine contractions.

Stretch or Irritation of the Cervix. There is much reason to believe that stretching or irritation of the uterine cervix is particularly important in eliciting uterine contractions. For instance, the obstetrician frequently induces labor by rupturing the membranes so that the head of the baby stretches the cervix more forcefully than usual or irritates it in other ways.

The mechanism by which cervical irritation excites the body of the uterus is not known. It has been suggested that stretching or irritation of nerves in the cervix initiates reflexes to the body of the uterus, but the effect could also result simply from myogenic transmission of signals from the cervix to the body of the uterus.

ONSET OF LABOR—A POSITIVE FEEDBACK THEORY FOR ITS INITIATION

During most of the months of pregnancy the uterus undergoes periodic episodes of weak and slow rhythmic contractions called *Braxton-Hicks contractions.* These become progressively stronger toward the end of pregnancy; then they change rather suddenly, within hours, to become exceptionally strong contractions that start stretching the cervix and later force the baby through the birth canal, thereby causing parturition. This process is called *labor,* and the strong contractions that result in final parturition are called *labor contractions.*

Yet, strangely enough, we do not know what suddenly changes the slow and weak rhythmicity of the uterus into the strong labor contractions. However, on the basis of experience during the past few years with other types of physiological control systems, a theory has been proposed for explaining the onset of labor based on positive feedback. This theory suggests that stretching of the cervix by the fetus' head finally becomes great enough to elicit strong reflex increase in contractility of the uterine body. This pushes the baby forward, which stretches the cervix still more and initiates still more positive feedback to the uterine body. Thus, the process continues again and again until the baby is expelled. This theory is illustrated in Figure 82–8, and the data supporting it are the following:

First, labor contractions obey all the principles of positive feedback. That is, once the strength of uterine contraction becomes greater than a critical value, each contraction leads to subsequent contractions that become stronger and stronger until maximum effect is achieved. Referring to the discussion in Chapter 1 of positive feedback in control systems, one will see that this is the precise nature of all positive feedback mechanisms when the feedback gain becomes greater than a critical value.

Second, two known types of positive feedback increase uterine contractions during labor: (1) Stretching of the cervix causes the entire body of the uterus to contract, and this contraction stretches the cervix still more because of the downward thrust of the baby's head. (2) Cervical stretching also causes the pituitary gland to secrete oxytocin, which is still another means for increasing uterine contractility.

1. Baby's head stretches cervix...
2. Cervical stretch excites fundic contraction
3. Fundic contraction pushes baby down and stretches cervix some more...
4. Cycle repeats over and over again...

Figure 82–8. Theory for the onset of intensely strong contractions during labor.

To summarize the theory, we can assume that multiple factors increase the contractility of the uterus toward the end of pregnancy. Eventually, a uterine contraction becomes strong enough to irritate the uterus and increase its contractility still more because of positive feedback, resulting in a second contraction stronger than the first, a third stronger than the second, and so forth. Once these contractions become strong enough to cause this type of feedback, with each succeeding contraction greater than the one preceding, the process proceeds to completion—all simply *because positive feedback initiates a vicious circle when the gain of the feedback is greater than a critical level.*

One might immediately ask about the many instances of false labor in which the contractions become stronger and stronger and then fade away. Remember that for a vicious circle to continue, *each* new cycle of the positive feedback must be stronger than the previous one. If, at any time after labor starts, some contractions fail to re-excite the uterus sufficiently, the positive feedback could go into a retrograde decline, and the labor contractions would fade away.

ABDOMINAL MUSCLE CONTRACTIONS DURING LABOR

Once the uterine contractions become very strong during labor, pain signals then originate both from the uterus itself and from the birth canal. These signals, in addition to causing suffering, also elicit neurogenic reflexes from the spinal cord to the abdominal muscles, causing intense contractions of these muscles. The abdominal contractions in turn add greatly to the force that causes expulsion of the baby.

MECHANICS OF PARTURITION

The uterine contractions during labor begin at the top of the uterine fundus and spread downward over the body of the uterus. Also, the intensity of contraction is great in the top and body of the uterus but weak in the lower segment of the uterus adjacent to the cervix. Therefore, each uterine contraction tends to force the baby downward toward the cervix.

In the early part of labor, the contractions might occur only once every 30 minutes. As labor progresses, the contractions finally appear as often as once every 1 to 3 minutes, and the intensity of contraction increases greatly, with only a short period of relaxation between contractions.

The combined contractions of the uterine and abdominal musculature during delivery of the baby cause downward force on the fetus of approximately 25 pounds during each strong contraction.

It is fortunate that the contractions of labor occur intermittently, because strong contractions impede or sometimes even stop blood flow through the placenta and would cause death of the fetus were the contractions continuous. Indeed, in clinical use of various uterine stimulants, such as oxytocin, overuse of the drugs can cause uterine spasm rather than rhythmic contractions and can lead to death of the fetus.

In 19 of 20 births the head is the first part of the baby to be expelled, and in most of the remaining instances the buttocks are presented first. The head acts as a wedge to open the structures of the birth canal as the fetus is forced downward.

The first major obstruction to expulsion of the fetus is the uterine cervix. Toward the end of pregnancy the cervix becomes soft, which allows it to stretch when labor contractions begin in the uterus. The so-called *first stage of labor* is the period of progressive cervical dilatation, lasting until the opening is as large as the head of the fetus. This stage usually lasts 8 to 24 hours in the first pregnancy but often only a few minutes after many pregnancies.

Once the cervix has dilated fully, the fetus' head moves rapidly into the birth canal, and, with additional force from above, continues to wedge its way through the canal until delivery is effected. This is called the *second stage of labor,* and it may last from as little as a minute after many pregnancies up to half an hour or more in the first pregnancy.

SEPARATION AND DELIVERY OF THE PLACENTA

During the succeeding 10 to 45 minutes after birth of the baby, the uterus contracts to a very small size, which causes a *shearing* effect between the walls of the uterus and the placenta, thus separating the placenta from its implantation site. Obviously, separation of the placenta opens the placental sinuses and causes bleeding. However, the amount of bleeding is limited to an average of 350 milliliters by the following mechanism: The smooth muscle fibers of the uterine musculature are arranged in figures of 8 around the blood vessels as they pass through the uterine wall. Therefore, contraction of the uterus following delivery of the baby constricts the vessels that had previously supplied blood to the placenta. In addition, it is believed that vasoconstrictor prostaglandins formed at the placental separation site cause additional blood vessel spasm.

LABOR PAINS

With each uterine contraction the mother experiences considerable pain. The pain in early labor is probably caused mainly by hypoxia of the uterine muscle resulting from compression of the blood vessels to the uterus. This pain is not felt when the *hypogastric nerves,* which carry the visceral sensory fibers leading from the uterus, have been sectioned. However, during the second stage of labor, when the fetus is being expelled through the birth canal, much more severe pain is caused by cervical stretching, perineal stretching, and stretching or tearing of structures in the vaginal canal itself. This pain is conducted by somatic nerves instead of by the visceral sensory hypogastric nerves.

INVOLUTION OF THE UTERUS

During the first 4 to 5 weeks following parturition, the uterus involutes. Its weight becomes less than one-half its immediate postpartum weight within a week; and in 4 weeks, if the mother lactates, the uterus may be as small as it had been prior to pregnancy. This effect of lactation results from the suppression of gonadotropin and ovarian hormone secretion during the first few months of lactation, as is discussed later. During early involution of the uterus the placental site on the endometrial surface autolyzes, causing a vaginal discharge known as "lochia," which is first bloody and then serous in nature, continuing in all for approximately a week and a half. After this time, the endometrial surface will have become re-epithelialized and ready for normal, nongravid sex life again.

● LACTATION

DEVELOPMENT OF THE BREASTS

The breasts, illustrated in Figure 82-9, begin to develop at puberty; this development is stimulated by the estrogens of the monthly sexual cycles; these stimulate growth of the breast's *mammary gland* plus deposition of fat to give mass to the breasts. In addition, still far greater growth occurs during pregnancy, and the glandular tissue only then becomes completely developed for actual production of milk.

Growth of the Ductal System — Role of the Estrogens. All through pregnancy, the tremendous quantities of estrogens secreted by the placenta cause the ductal system of the breasts to grow and to branch. Simultaneously, the stroma of the breasts also increases in quantity, and large quantities of fat are laid down in the stroma.

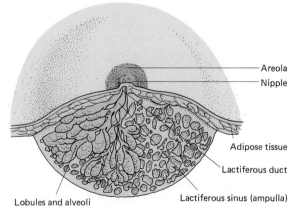

Figure 82-9. The breast and its mammary gland. (From Guyton: Physiology of the Human Body. 6th ed. Philadelphia, Saunders College Publishing, 1984.)

However, also important in growth of the ductal system are at least four other hormones: *growth hormone, prolactin,* the *adrenal glucocorticoids,* and *insulin.* Each of these is known to play at least some role in protein metabolism, which presumably explains their function in the development of the breasts.

Development of the Lobule-Alveolar System — Role of Progesterone. Final development of the breasts into milk-secreting organs also requires progesterone. Once the ductal system has developed, progesterone, acting synergistically especially with estrogen but also with all the other hormones just mentioned, causes growth of the lobules, budding of alveoli, and development of secretory characteristics in the cells of the alveoli. These changes are analogous to the secretory effects of progesterone on the endometrium of the uterus during the latter half of the female menstrual cycle.

INITIATION OF LACTATION — FUNCTION OF PROLACTIN

Though estrogen and progesterone are essential for the physical development of the breasts during pregnancy, a specific effect of both these hormones is to inhibit *the actual secretion of milk.* On the other hand, the hormone *prolactin* has exactly the opposite effect, promotion of milk secretion. This hormone is secreted by the mother's pituitary gland, and its concentration in her blood rises steadily from the fifth week of pregnancy until birth of the baby, at which time it has risen to about 10 times the normal nonpregnant level. This very high level of prolactin at the end of pregnancy is illustrated in Figure 82–10. In addition, the placenta secretes large quantities of *human chorionic somatomammotropin,* which also has mild lactogenic properties, thus supporting the prolactin from the mother's pituitary. Even so, because of the suppressive effects of estrogen and progesterone, never more than a few milliliters of fluid are secreted each day until after the baby is born. The fluid that is secreted the last few days or weeks before parturition is called *colostrum;* it contains essentially the same concentrations of proteins and lactose as milk but almost no fat, and its maximum rate of production is about 1/100 the subsequent rate of milk production.

Immediately after the baby is born, the sudden loss of both estrogen and progesterone secretion by the placenta now allows the lactogenic effect of the prolactin from the mother's pituitary gland to assume its natural milk-promoting role, and over the next 1 to 7 days the breasts begin to secrete copious quantities of milk instead of colostrum. This secretion of milk requires an adequate background secretion of most of the mother's other hormones as well, but most important of all are *growth hormone, cortisol,* and *parathyroid hormone.* These hormones are necessary to provide the amino acids, fatty acids, glucose, and calcium that are required for milk formation.

Following birth of the baby, the *basal level* of prolactin secretion returns during the next few weeks to the nonpregnant level, as shown in Figure 82–10. However, each time the mother nurses her baby, nervous signals from the nipples to the hypothalamus cause a 10- to 20-fold surge in prolactin secretion lasting about 1 hour, which is also shown in the figure. This prolactin in turn acts on the breasts to keep the mammary glands functional for the subsequent nursing periods. If this prolactin surge is absent, if it is blocked as a result of hypothalamic or pituitary damage, or if nursing does not continue, the breasts lose their ability to produce milk within a week or so. However, milk production can continue for several years if the child continues to suckle, though the rate of milk formation normally decreases considerably within 7 to 9 months.

Hypothalamic Control of Prolactin Secretion. The hypothalamus plays an essential role in controlling prolactin secretion, as it does also for the control of secretion of almost all the other anterior pituitary hormones. However, this control is different in one aspect: the hypothalamus mainly *stimulates* the production of all the other hormones, but it mainly *inhibits* prolactin production. Consequently, damage to the hypothalamus or blockage of the hypothalamic-hypophysial portal system increases prolactin secretion while it depresses secretion of the other anterior pituitary hormones. Yet under special conditions, such as when the baby suckles, it seems that a different type of signal from the hypothalamus can then increase the secretion of prolactin.

Therefore, it is believed that two different factors formed in the hypothalamus are transported to the anterior pituitary through the hypothalamic-hypophysial portal system to control prolactin release by the anterior pituitary gland. These are called *prolactin inhibitory hormone* (PIH), which is the dominant hormone under most normal conditions, and *prolactin-releasing factor* (PRF), which can intermittently increase prolactin secretion. PIH is almost certainly the catecholamine *dopamine,* which is known to be secreted in the hypothalamus and which can decrease prolactin secretion as much as tenfold.

Figure 82–10. Changes in rates of secretion of estrogens, progesterone, and prolactin for 8 weeks prior to parturition and for 36 weeks thereafter. Note especially the decrease of prolactin secretion back to basal levels within a few weeks, but also the intermittent periods of marked prolactin secretion (for about 1 hour at a time) during and after periods of nursing.

Suppression of the Female Ovarian Cycles During Nursing.

In about half of nursing mothers, the ovarian cycle and ovulation do not resume until a few weeks after cessation of nursing the baby. The cause of this seems to be that the same nervous signals from the breasts to the hypothalamus that cause prolactin secretion during suckling simultaneously inhibit secretion of gonadotropin-releasing hormone by the hypothalamus, which in turn suppresses the formation of the pituitary gonadotropic hormones, luteinizing hormone and follicle-stimulating hormone. Yet after several months of lactation, in about half of mothers the pituitary begins again to secrete sufficient gonadotropic hormones to reinstate the monthly sexual cycle.

THE EJECTION OR "LET DOWN" PROCESS IN MILK SECRETION— FUNCTIONS OF OXYTOCIN

Milk is secreted continuously into the alveoli of the breasts, but milk does not flow easily from the alveoli into the ductal system and therefore does not continually leak from the breast nipples. Instead, the milk must be ejected, or "let down," from the alveoli to the ducts before the baby can obtain it. This process is caused by a combined neurogenic and hormonal reflex involving the posterior pituitary hormone *oxytocin.*

When the baby first suckles, it receives virtually no milk. Instead, sensory impulses are transmitted through somatic nerves from the nipples to the spinal cord and then to the hypothalamus, there causing *oxytocin* secretion at the same time that they cause prolactin secretion. The oxytocin is then carried in the blood to the breasts, where it causes the *myoepithelial cells* that surround the outer walls of the alveoli to contract, thereby expressing the milk from the alveoli into the ducts at a pressure of plus 10 to 20 mm Hg, and then the baby's suckling does become effective in removing the milk. Thus, within 30 seconds to a minute after a baby begins to suckle, milk begins to flow. This process is called milk ejection or milk let-down.

Suckling on one breast causes milk flow not only in that breast but also in the opposite breast. It is especially interesting that fondling of the baby by the mother or hearing the baby crying also often gives enough of a signal to her hypothalamus to cause milk ejection.

Inhibition of Milk Ejection. A particular problem in nursing the baby comes from the fact that many psychogenic factors or generalized sympathetic stimulation throughout the body can inhibit oxytocin secretion and consequently depress milk ejection. For this reason, many mothers must have an undisturbed puerperium if they are to be successful in nursing their babies.

MILK COMPOSITION AND THE METABOLIC DRAIN ON THE MOTHER CAUSED BY LACTATION

Table 82–1 lists the contents of human milk and cow's milk. The concentration of lactose in human

Table 82–1 PERCENTAGE COMPOSITION OF MILK

	Human Milk	Cow's Milk
Water	88.5	87
Fat	3.3	3.5
Lactose	6.8	4.8
Casein	0.9	2.7
Lactalbumin and other protein	0.4	0.7
Ash	0.2	0.7

milk is approximately 50 per cent greater than in cow's milk, but on the other hand the concentration of protein in cow's milk is ordinarily two or more times as great as in human milk. Finally, the ash, which contains the calcium and other minerals, is only one-third as much in human milk as in cow's milk.

At the height of lactation 1.5 liters of milk may be formed each day. With this degree of lactation great quantities of metabolic substrates are drained from the mother. For instance, approximately 50 grams of fat then enter the milk each day; and approximately 100 grams of lactose, which must be derived from glucose, are lost from the mother each day. Also, 2 to 3 grams of calcium phosphate may be lost each day; and, unless the mother is drinking large quantities of milk and has an adequate intake of vitamin D, the output of calcium and phosphate by the lactating mammae will be much greater than the intake of these substances. To supply the needed calcium and phosphate, the parathyroid glands enlarge greatly, and the bones become progressively decalcified. The problem of decalcification is usually not very great during pregnancy, but it can be a distinct problem during lactation.

REFERENCES

Aynsley-Green, A.: Metabolic and endocrine interrelations in the human fetus and neonate. Am. J. Clin. Nutr., 41:399, 1985.

Beckman, D. A., and Brent, R. L.: Mechanisms of teratogenesis. Annu. Rev. Pharmacol. Toxicol., 24:483, 1984.

Ben-Jonathan, N., et al.: Suckling-induced rise in prolactin: Mediation by prolactin-releasing factor from posterior pituitary. News Physiol. Sci., 3:172, 1988.

Conn, P. M., et al.: Gonadotropin-releasing hormone: Molecular and cell biology, physiology, and clinical applications. Fed. Proc., 43:2351, 1984.

Cummings, A. M., and Yochim, J. M.: Differentiation of the uterus in preparation for gestation: A model for the action of progesterone. J. Theor. Biol., 106:353, 1984.

DeGroot, L. J., et al. (eds.): Endocrinology, 2nd Ed. Philadelphia, W. B. Saunders Co., 1989.

Gould, M., and Stephano, J. L.: Electrical responses of eggs to acrosomal protein similar to those induced by sperm. Science, 235:1634, 1987.

Hankins, G. D., et al.: Myocardial infarction during pregnancy: A review. Obstet. Gynecol., 65:139, 1985.

Hanson, L. A.: Biology of Human Milk. New York, Raven Press, 1988.

Jaffe, L. A., and Cross, N. L.: Electrical regulation of sperm-egg fusion. Annu. Rev. Physiol., 48:191, 1986.

Jones, H. W. Jr., et al. (eds.): In Vitro Fertilization. Baltimore, Williams & Wilkins, 1986.

Kaufmann, P., and Miller, R. K., (eds.): Placental Vascularization and Blood Flow. New York, Plenum Publishing Corp., 1988.

Keyes, P. L., and Wiltbank, M. C.: Endocrine regulation of the corpus luteum. Annu. Rev. Physiol., 50:465, 1988.

Knobil, E., et al. (eds.): The Physiology of Reproduction. New York, Raven Press, 1988.

Leong, D. A., et al.: Neuroendocrine control of prolactin secretion. Annu. Rev. Physiol., 45:109, 1983.

Leung, P. C. K., et al. (eds.): Endocrinology and Physiology of Reproduction. New York, Plenum Publishing Corp., 1987.

Li, C. H. (ed.): The Chemistry of Prolactin. New York, Academic Press, 1980.

Lindheimer, M. D., and Katz, A. K.: Pathophysiology of preeclampsia. Annu. Rev. Med., 32:273, 1981.

Lotgering, F. K., et al.: Maternal and fetal responses to exercise during pregnancy. Physiol. Rev., 65:1, 1985.

Munro, H. N., et al.: The placenta in nutrition. Annu. Rev. Nutr., 3:97, 1983.

Neville, M. C., and Daniel, C. W. (eds.): The Mammary Gland. New York, Plenum Publishing Corp., 1987.

Neville, M. C.: Mammary cultures on floating gels: A model system for mammary function. News Physiol. Sci., 2:107, 1987.

Ochiai, K. (ed.): Endocrine Correlates of Reproduction. New York, Springer-Verlag, 1984.

Olsson, K.: Pregnancy—A challenge to water balance. News Physiol. Sci., 1:131, 1986.

Peaker, M., and Wilde, C. J.: Milk secretion: Autocrine control. News Physiol. Sci., 2:124, 1987.

Pitkin, R. M.: Calcium metabolism in pregnancy and the perinatal period: A review. Am J. Obstet. Gynecol., 151:99, 1985.

Sampson, D. A., and Jansen, G. R.: Protein and energy nutrition during lactation. Annu. Rev. Nutr., 4:43, 1984.

Simpson, E. R., and MacDonald, P. C.: Endocrine physiology of the placenta. Annu. Rev. Physiol., 43:163, 1981.

Stern, L., et al. Physiologic Foundations of Perinatal Care. New York, Elsevier Science Publishing Co., 1989.

Thorburn, G. D., and Challis, J. R. G.: Endocrine control of parturition. Physiol. Rev., 59:863, 1979.

Veeck, L. L.: Atlas of the Human Oocyte and Early Conceptus. Baltimore, Williams & Wilkins, 1986.

Vonderhaar, B. K., and Ziska, S. E.: Hormonal regulation of milk protein gene expression. Annu. Rev. Physiol., 51:641, 1989.

Wassarman, P. M.: Fertilization in mammals. Sci. Am., December 1988, p. 78.

Wassarman, P. M.: The biology and chemistry of fertilization. Science, 235:553, 1987.

Wintour, E. M.: Amniotic fluid—Our first environment. News Physiol. Sci., 1:95, 1986.

Wolf, D. P., et al. (eds.): In Vitro Fertilization and Embryo Transfer. New York, Plenum Publishing Corp., 1988.

Yen, S. S. C., and Jaffe, R.: Reproductive Endocrinology: Physiology, Pathophysiology and Clinical Management, 2nd Ed. Philadelphia, W. B. Saunders Co., 1986.

CHAPTER 83

• *Fetal and Neonatal Physiology* •

A complete discussion of fetal development, functioning of the child immediately after birth, and growth and development through the early years of life lies within the province of formal courses in obstetrics and pediatrics. However, many aspects of these are strictly physiological problems, some of which relate to the physiological principles that we have discussed for the adult and some of which are peculiar to the infant itself. The present chapter discusses some of the most important of these special problems.

● GROWTH AND FUNCTIONAL DEVELOPMENT OF THE FETUS

Early development of the placenta and of the fetal membranes occurs far more rapidly than development of the fetus itself. During the first 2 to 3 weeks the fetus remains almost microscopic in size, but thereafter, as illustrated in Figure 83–1, the dimensions of the fetus increase almost in proportion to age. At 12 weeks the length of the fetus is approximately 10 centimeters; at 20 weeks, 25 centimeters; and at term (40 weeks), 53 centimeters (about 21 inches). Because the weight of the fetus is proportional to the cube of the length, the weight increases approximately in proportion to the cube of the age of the fetus. Note in Figure 83–1 that the weight remains miniscule during the first 12 weeks and reaches 1 pound only at 23 weeks (5½ months) of gestation. Then, during the last trimester of pregnancy, the fetus gains tremendously, so that 2 months prior to birth the weight averages 3 pounds, 1 month prior to birth, 4.5 pounds, and at birth, 7 pounds — the birth weight varying from as low as 4.5 pounds to as high as 11 pounds in completely normal infants with completely normal gestational periods.

DEVELOPMENT OF THE ORGAN SYSTEMS

Within 1 month after fertilization of the ovum, all of the different organs of the fetus have already been "blocked out"; and during the next 2 to 3 months, most of the details of the different organs are established. Beyond the fourth month, the organs of the fetus are grossly the same as those of the newborn child. However, cellular development in each organ is usually far from complete at this time and requires the full remaining 5 months of pregnancy for complete development. Even at birth certain structures, particularly the nervous system, the kidneys, and the liver, still lack full development, which will be discussed in more detail later in the chapter.

The Circulatory System

The human heart begins beating during the fourth week following fertilization, contracting at the rate of about 65 beats per minute. This increases steadily as the fetus grows and reaches a rate of approximately 140 beats per minute immediately before birth.

Formation of Blood Cells. Nucleated red blood cells begin to be formed in the yolk sac and mesothelial layers of the placenta at about the third week of fetal development. This is followed a week later (at 4 to 5 weeks) by the formation of nonnucleated red blood cells by the fetal mesenchyme and by the endothelium of the fetal blood vessels. Then at approximately 6 weeks, the liver begins to form blood cells, and in the third month the spleen and other lymphoid tissues of the body also begin forming blood cells. Finally, from approximately the third month on, the bone marrow also forms red and white blood cells. During the midportion of fetal life, the extramarrow areas are the major sources of the fetus' blood cells; but during the latter 3 months of fetal life, the bone marrow gradually takes over while the other blood cell–forming areas lose completely their ability to form blood cells.

The Respiratory System

Obviously, respiration cannot occur during fetal life. However, respiratory movements do take place beginning at the end of the first trimester of pregnancy. Tactile stimuli or fetal asphyxia especially cause respiratory movements.

However, during the latter 3 to 4 months of pregnancy, the respiratory movements of the fetus are mainly inhibited, for reasons unknown. This could possibly result from (1) special chemical conditions in the body fluids of the fetus, (2) the presence of fluid in the fetal lungs, or (3) other possible unexplored stimuli.

The inhibition of respiration during the latter months of fetal life prevents filling of the lungs with debris from the meconium excreted by the gastrointestinal tract into the amniotic fluid. Also, fluid is secreted into the lungs by the alveolar epithelium up until the moment of birth, thus keeping the pulmonary spaces filled with this clean secretion.

The Nervous System

Most of the skin reflexes of the fetus are present by the third to fourth month of pregnancy. However, most higher functions of the central nervous system that involve the

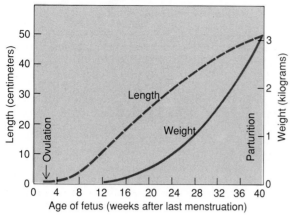

Figure 83–1. Growth of the fetus.

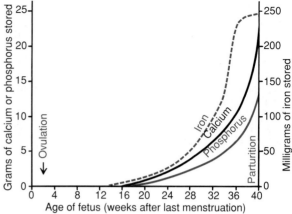

Figure 83–2. Calcium, phosphorus, and iron storage in the fetus at different stages of gestation.

cerebral cortex are still undeveloped even at birth. Indeed, myelinization of some major tracts of the central nervous system becomes complete only after approximately a year of postnatal life.

The Gastrointestinal Tract

By midpregnancy the fetus ingests and absorbs large quantities of amniotic fluid; and during the last 2 to 3 months, gastrointestinal function approaches that of the normal newborn infant. Small quantities of *meconium* are continually formed in the gastrointestinal tract and excreted from the bowels into the amniotic fluid. Meconium is composed partly of unabsorbed residue of amniotic fluid and partly of excretory products from the gastrointestinal mucosa and glands.

The Kidneys

The fetal kidneys are capable of excreting urine during at least the latter half of pregnancy, and urination occurs normally *in utero.* However, the renal control systems for regulation of extracellular fluid electrolyte balances and especially acid-base balance are almost nonexistent until after midfetal life and do not reach full development until a few months after birth.

Fetal Metabolism

The fetus utilizes mainly glucose for energy, and it has a high rate of storage of fat and protein, much if not most of the fat being synthesized from glucose, rather than being absorbed from the mother's blood. Aside from these generalities, there are some special problems of fetal metabolism in relation to calcium, phosphate, iron, and some vitamins.

Metabolism of Calcium and Phosphate. Figure 83–2 illustrates the rates of calcium and phosphate accumulation in the fetus, showing that approximately 22.5 grams of calcium and 13.5 grams of phosphorus are accumulated in the average fetus during gestation. About half of this accumulates during the last 4 weeks of gestation, which is also coincident with the period of rapid ossification of the fetal bones as well as with the period of rapid weight gain of the fetus.

During the earlier part of fetal life, the bones are relatively unossified and have mainly a cartilaginous matrix.

Indeed, x-ray pictures ordinarily will not show any ossification until approximately the fourth month of pregnancy.

Note especially that the total amounts of calcium and phosphate needed by the fetus during gestation represent only about 1/50 the quantities of these substances in the mother's bones. Therefore, this is a minimal drain from the mother. However, a much greater drain occurs after birth during lactation.

Accumulation of Iron. Figure 83–2 also shows that iron accumulates in the fetus somewhat more rapidly than calcium and phosphates. Most of the iron is in the form of hemoglobin, which begins to be formed as early as the third week after fertilization of the ovum.

Small amounts of iron are concentrated in the progestational endometrium even prior to implantation of the ovum; this iron is ingested into the embryo by the trophoblastic cells for early formation of the red blood cells.

Approximately one-third of the iron in a fully developed fetus is normally stored in the liver. This iron can then be used for several months after birth by the newborn infant for formation of additional hemoglobin.

Utilization and Storage of Vitamins. The fetus needs vitamins equally as much as the adult and in some instances to a far greater extent. In general, the vitamins function the same in the fetus as in the adult as discussed in Chapter 71. Special functions of several vitamins should be mentioned, however.

The B vitamins, especially vitamin B_{12} and folic acid, are necessary for formation of red blood cells and also for overall growth of the fetus.

Vitamin C is necessary for appropriate formation of intercellular substances, especially the bone matrix and fibers of connective tissue.

Vitamin D probably is needed for normal bone growth in the fetus; but even more importantly, the mother needs it for adequate absorption of calcium from her gastrointestinal tract. If the mother has plenty of this vitamin in her body fluids, large quantities of the vitamin will be stored by the fetal liver to be used by the newborn child for several months after birth.

Vitamin E, although its precise functions are not clear, is necessary for normal development of the early embryo. In its absence in experimental animals, spontaneous abortion usually occurs at an early age.

Vitamin K is used by the fetal liver for formation of Factor VII, prothrombin, and several other blood coagulation factors. When vitamin K is insufficient in the mother, Factor VII and prothrombin become deficient in the child as well as in the mother. Since most vitamin K absorbed into the blood is vitamin K formed by bacterial action in the colon, the newborn child has no adequate source of vitamin K for the first week or so of life until normal colonic bacterial flora become established. Therefore, prenatal storage of at least small amounts of vitamin K in the fetal liver is helpful in preventing hemorrhage—particularly hemorrhage in the brain when the head is traumatized by squeezing through the birth canal.

● ADJUSTMENTS OF THE INFANT TO EXTRAUTERINE LIFE

ONSET OF BREATHING

The most obvious effect of birth on the baby is loss of the placental connection with the mother and therefore loss of this means of metabolic support. By far the most important immediate adjustment required of the infant is to begin breathing.

Cause of Breathing at Birth. After completely normal delivery from a mother who has not been depressed by anesthetics, the child ordinarily begins to breathe immediately and has a completely normal respiratory rhythm within seconds after birth. The promptness with which the fetus begins to breathe indicates that breathing is initiated by sudden exposure to the exterior world, probably resulting from a slightly asphyxiated state incident to the birth process but also from sensory impulses originating in the suddenly cooled skin. In an infant who does not breathe immediately, the body becomes progressively more hypoxic and hypercapnic, which provides additional stimulus to the respiratory center and usually causes breathing within an additional few seconds to few minutes after birth.

Delayed and Abnormal Breathing at Birth— Danger of Hypoxia. If the mother has been depressed by a general anesthetic during delivery, which at least partially anesthetizes the child as well, respiration is likely to be delayed for several minutes, thus illustrating the importance of using as little obstetrical anesthesia as feasible. Also, many infants who have had head trauma during delivery or who undergo very prolonged delivery are slow to breathe or sometimes will not breathe at all. This can result from two possible effects: First, in a few infants, intracranial hemorrhage or brain contusion causes a concussion syndrome with a greatly depressed respiratory center. Second, and probably much more important, prolonged fetal hypoxia during delivery also causes serious depression of the respiratory center. Hypoxia frequently occurs during delivery because of (1) compression of the umbilical cord; (2) premature separation of the placenta; (3) excessive contraction of the uterus, which cuts off the blood flow to the placenta; or (4) excessive anesthesia of the mother, which depresses the oxygenation even of her blood.

Degree of Hypoxia That an Infant Can Tolerate. In the adult, failure to breathe for only 4 minutes often causes death, but a newborn infant often survives as long as 10 to 15 minutes of failure to breathe after birth. Unfortunately, though, permanent and very evident brain impairment often ensues if breathing is delayed more than 8 to 10 minutes. Indeed, actual lesions develop mainly in the thalamus, the inferior colliculi, and in other brain stem areas, thus affecting many of the motor functions of the body.

Expansion of the Lungs at Birth

At birth, the walls of the alveoli are kept collapsed by the surface tension of the viscid fluid that fills them. More than 25 mm Hg of negative pressure is required to oppose the effects of this surface tension when opening the alveoli for the first time. But once the alveoli are open, further respiration can be effected with relatively weak respiratory movements. Fortunately, the first inspirations of the newborn infant are extremely powerful, usually capable of creating as much as 60 mm Hg negative pressure in the intrapleural space.

Figure 83–3 illustrates the tremendous forces required to open the lungs at the onset of breathing. To the left is shown the pressure-volume curve (compliance curve) for the first breath after birth. Observe, first, the lowermost curve beginning at the zero pressure point and moving to the right. The curve shows that the volume of air in the lungs remains almost exactly zero until the negative pressure has reached −40 centimeters water (−30 mm Hg). Then, as the negative pressure increases to −60 centimeters of water, about 40 milliliters of air enters the lungs. To deflate the lungs, considerable positive pressure, about +40 centimeters of water, is required because of the viscous resistance offered by the fluid in the bronchioles.

Note that the second breath is much easier. However, breathing does not become completely normal until about 40 minutes after birth, as shown by the third compliance curve, the shape of which compares favorably with that for the normal adult, as shown in Chapter 38.

Respiratory Distress Syndrome Caused by Lack of Surfactant Secretion. A small number of infants, especially premature infants and infants born of diabetic mothers, develop severe respiratory distress a few hours to several days after birth and frequently succumb within the next day or so. The alveoli of these infants at death contain large quantities of proteinaceous fluid, almost as if pure plasma had leaked out of the capillaries into the alveoli. The fluid also contains desquamated alveolar epithelial cells. This condition is also called *hyaline membrane disease* because microscopic slides of the lung show this alveolar material to look like a hyaline membrane.

One of the most characteristic findings in respiratory distress syndrome is failure to secrete adequate quantities of *surfactant,* a substance normally secreted into the alveoli that decreases the surface tension of the alveolar fluid, therefore allowing the alveoli to open easily. The surfactant-secreting cells (the type II alveolar epithelial cells) do not begin to secrete surfactant until the last 1 to 3 months of gestation. Therefore, many premature babies and some full-term babies are born without the capability of secreting surfactant, which therefore causes both a collapse tendency of the alveoli and development of pulmonary edema. The

Figure 83–3. Pressure-volume curves of the lungs (compliance curves) of a newborn baby immediately after birth, showing *(a)* the extreme forces required for breathing during the first two breaths of life and *(b)* development of a nearly normal compliance within 40 minutes after birth. (From Smith: The first breath. *Sci. Amer.,* 209:32, 1963, © 1963 by Scientific American, Inc. All rights reserved.)

role of surfactant in preventing these effects is discussed in Chapter 37.

CIRCULATORY READJUSTMENTS AT BIRTH

Equally as essential as the onset of breathing at birth are the immediate circulatory adjustments that allow adequate blood flow through the lungs. Also, circulatory adjustments during the first few hours of life shunt more and more blood through the liver as well. To describe these readjustments, we must first consider briefly the anatomical structure of the fetal circulation.

Specific Anatomical Structure of the Fetal Circulation. Because the lungs are mainly nonfunctional during fetal life and because the liver is only partially functional, it is not necessary for the fetal heart to pump much blood through either the lungs or the liver. On the other hand, the fetal heart must pump large quantities of blood through the placenta. Therefore, special anatomical arrangements cause the fetal circulatory system to operate considerably differently from that of the adult. First, as illustrated in Figure 83–4, blood returning from the placenta through the umbilical vein passes through the *ductus venosus,* mainly bypassing the liver. Then most of the blood entering the right atrium from the inferior vena cava is directed in a straight pathway across the posterior aspect of the right atrium and thence through the *foramen ovale* directly into the left atrium. Thus, this well-oxygenated blood from the placenta enters mainly the left side of the heart, rather than the right side, and is pumped by the left ventricle mainly into the vessels of the head and forelimbs.

The blood entering the right atrium from the superior vena cava is directed downward through the tricuspid valve into the right ventricle. This blood is mainly deoxygenated blood from the head region of the fetus, and it is pumped by the right ventricle into the pulmonary artery, then mainly through the *ductus arteriosus* into the descending aorta and through the two umbilical arteries into the placenta, where the deoxygenated blood becomes oxygenated.

Figure 83–5 gives the relative percentages of the total blood pumped by the heart that passes through the different vascular circuits of the fetus. This figure shows that 55 per cent of all the blood goes through the placenta, leaving only 45 per cent to pass through all the tissues of the fetus. Furthermore, during fetal life, only 12 per cent of the blood flows through the lungs; immediately after birth virtually all of the blood flows through the lungs, a manyfold increase at birth.

Changes in the Fetal Circulation at Birth. The basic changes in the fetal circulation at birth are discussed in Chapter 23 in relation to congenital anomalies of the ductus arteriosus and foramen ovale that persist throughout life. Briefly, these changes are the following:

Primary Changes in Pulmonary and Systemic Vascular Resistance at Birth. The primary changes in the circulation at birth are, first, loss of the tremendous blood flow through the placenta, which *approximately doubles the systemic vascular resistance at birth.* This obviously *increases the aortic pressure* as well as the pressures in the left ventricle and left atrium.

Second, the *pulmonary vascular resistance greatly decreases* as a result of expansion of the lungs. In the unexpanded fetal lungs, the blood vessels are compressed because of the small volume of the lungs. Immediately upon expansion these vessels are no longer compressed, and the resistance to blood flow decreases severalfold. Also, in fetal life the hypoxia of the lungs causes considerable tonic vasoconstriction of the lung blood vessels, but vasodilation takes place when aeration of the lungs eliminates the hypoxia. All these changes together reduce the resistance to blood flow through the lungs as much as fivefold, which obviously *reduces the pulmonary arterial pressure,* the right ventricular pressure, and the right atrial pressure.

Closure of the Foramen Ovale. The *low right atrial pressure* and the *high left atrial pressure* that occur secondarily to the changes in pulmonary and systemic resistance

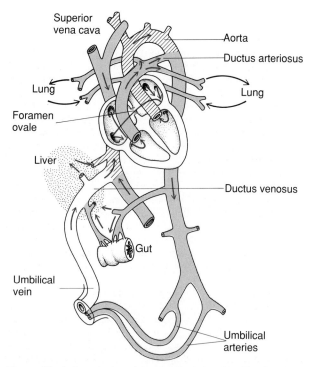

Figure 83–4. Organization of the fetal circulation. (Modified from Arey: Developmental Anatomy. 7th ed. Philadelphia, W. B. Saunders Company, 1974.)

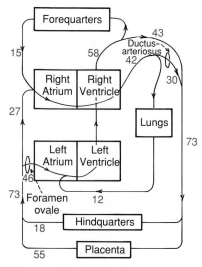

Figure 83–5. Diagram of the fetal circulatory system, showing relative distribution of blood flow to the different vascular areas. The numerals represent per cent of the total cardiac output flowing through the particular area.

at birth cause blood now to attempt to flow backward through the foramen ovale from the left atrium into the right atrium, rather than in the other direction, as occurred during fetal life. Consequently, the small valve that lies over the foramen ovale on the left side of the atrial septum closes over this opening, thereby preventing further flow through the foramen ovale. In two thirds of all persons the valve becomes adherent over the foramen ovale within a few months to a few years and forms a permanent closure. But even if permanent closure does not occur, the left atrial pressure throughout life remains 2 to 4 mm Hg greater than the right atrial pressure, and the back pressure keeps the valve closed.

Closure of the Ductus Arteriosus. The ductus arteriosus also closes, but for different reasons. First, the increased systemic resistance *elevates the aortic pressure* while the decreased pulmonary resistance *reduces the pulmonary arterial pressure*. As a consequence, within a few hours after birth, blood begins to flow backward from the aorta into the pulmonary artery through the ductus arteriosus, rather than in the other direction as in fetal life. However, after only a few hours the muscular wall of the ductus arteriosus constricts markedly, and within 1 to 8 days the constriction is sufficient to stop all blood flow. This is called *functional closure* of the ductus arteriosus. Then, during the next 1 to 4 months the ductus arteriosus ordinarily becomes anatomically *occluded* by growth of fibrous tissue into its lumen.

The cause of ductus closure relates to the increased oxygenation of the blood flowing through the ductus. In fetal life the P_{O_2} of the ductus blood is only 15 to 20 mm Hg, but it increases to about 100 mm Hg within a few hours after birth. Furthermore, many experiments have shown that the degree of contraction of the smooth muscle in the ductus wall is highly related to the availability of oxygen.

In one of several thousand infants the ductus fails to close, resulting in a *patent ductus arteriosus,* the consequences of which are discussed in Chapter 23.

Closure of the Ductus Venosus. In fetal life, the portal blood from the fetus' abdomen joins the blood from the umbilical vein; and these together pass through the *ductus venosus* directly into the vena cava, thus bypassing the liver. Immediately after birth, blood flow through the umbilical vein ceases; but most of the portal blood still flows through the ductus venosus, with only a small amount passing through the channels of the liver. However, within 1 to 3 hours the muscular wall of the ductus venosus contracts strongly and closes this avenue of flow. As a consequence, the portal venous pressure rises from near 0 mm Hg to 6 to 10 mm Hg, which is enough to force blood flow through the liver sinuses. Although the ductus venosus almost never fails to close, unfortunately we know almost nothing about what causes the closure.

NUTRITION OF THE NEWBORN INFANT

The fetus obtains almost all its energy from glucose obtained from the mother's blood. Immediately after birth, the amount of glucose stored in the infant's body in the form of glycogen is sufficient to supply the infant's needs for only a few hours, and unfortunately the liver of the newborn infant is still far from functionally adequate at birth, which prevents significant gluconeogenesis. Therefore, the infant's blood glucose concentration frequently falls the first day to as low as 30 to 40 milligrams per deciliter of plasma, less than half the normal value. Fortunately, appropriate mechanisms are available for the infant to utilize its stored fats and proteins for metabolism until mother's milk can be provided 2 to 3 days later.

Special problems are also frequently associated with getting an adequate fluid supply to the newborn infant because the infant's rate of body fluid turnover averages seven times that of an adult, and the mother's milk supply requires several days to develop. Ordinarily, the infant's weight decreases 5 to 10 per cent and sometimes as much as 20 per cent within the first 2 to 3 days of life. Most of this weight loss is loss of fluid rather than of body solids.

● SPECIAL FUNCTIONAL PROBLEMS IN THE NEONATAL INFANT

The most important characteristic of the newborn infant is instability of the various hormonal and neurogenic control systems. This results partly from the immature development of the different organs of the body and partly from the fact that the control systems simply have not become adjusted to the completely new way of life.

The Respiratory System

The normal rate of respiration in the newborn is about 40 breaths per minute, and tidal air with each breath averages 16 milliliters. This gives a total minute respiratory volume of 640 ml/min, which is about two times as great in relation to the body weight as that of an adult. *The functional residual capacity of the infant is only half that of an adult in relation to body weight.* This allows cyclic changes in blood gas concentration when the respiration becomes slowed because it is the residual air in the lungs that smooths out the blood gas variations in the adult.

The Circulation

Blood Volume. The blood volume of a newborn infant immediately after birth averages about 300 milliliters; but if the infant is left attached to the placenta for a few minutes after birth or if the umbilical cord is stripped to force blood out of its vessels into the baby, an additional 75 milliliters of blood enters the infant, to make a total of 375 milliliters. Then, during the ensuing few hours, fluid is lost into the tissue spaces from this blood, which increases the hematocrit but returns the blood volume once again to the normal value of about 300 milliliters. Some pediatricians feel that this extra blood volume in some instances causes mild pulmonary edema with some degree of respiratory distress.

Cardiac Output. The cardiac output of the newborn infant averages 550 ml/min, which, like respiration and body metabolism, is about two times as much in relation to body weight as in the adult. Occasionally a child is born with an especially low cardiac output caused by hemorrhage of much of its blood volume through the placental membrane into the mother's body prior to birth.

Arterial Pressure. The arterial pressure during the first day after birth averages about 70/50; this increases slowly during the next several months to approximately 90/60. Then there is a much slower rise during the subsequent years until the adult pressure of 115/70 is attained at adolescence.

Blood Characteristics. The red blood cell count in the newborn infant averages about 4 million per microliter. If blood is stripped from the cord into the infant, the red blood cell count rises an additional 0.5 to 0.75 million during the first few hours of life, giving a red blood cell count of about 4.75 million per microliter, as illustrated in Figure 83-6. Subsequent to this, however, few new red blood cells are formed in the infant during the first few weeks of life, presumably because the hypoxic stimulus of fetal life is no longer present to stimulate red cell production. Thus, as shown in Figure 83-6, the average red blood cell count falls to 3.25 million per microliter by about 8 to 10 weeks of age. From that time on, increasing activity by the fetus provides the appropriate stimulus for returning the red blood cell count to normal within another 2 to 3 months.

Immediately after birth, the white blood cell count of the infant is about 45,000 per microliter, which is about five times as great as that of the normal adult.

Neonatal Jaundice and Erythroblastosis Fetalis. Bilirubin formed in the fetus can cross the placenta and be excreted through the liver of the mother; but immediately after birth the only means for ridding the infant of bilirubin is through the infant's own liver, which for the first week or so of life still functions poorly and is incapable of conjugating significant quantities of bilirubin with glucuronic acid for excretion into the bile. Consequently, the plasma bilirubin concentration rises from a normal value of less than 1 mg/dl to an average of 5 mg/dl during the first 3 days of life, then gradually falls back to normal as the liver becomes functional. This condition, called *physiologic hyperbilirubinemia*, is illustrated in Figure 83-6, and it is associated with a mild *jaundice* (yellowness) of the infant's skin and especially of the sclerae of its eyes.

However, by far the most important cause of serious neonatal jaundice is *erythroblastosis fetalis,* which is discussed in detail in Chapter 32 in relation to Rh factor incompatibility between the infant and mother. Briefly, the erythroblastic baby usually inherits Rh positive red cells from the father while the mother is Rh negative. The mother then becomes immunized against the fetus' blood cells, and her antibodies in turn destroy the fetal red cells, releasing extreme quantities of bilirubin into the plasma. Before the advent of modern obstetrics, this condition occurred either mildly or seriously in 1 of every 50 to 100 newborn infants.

Fluid Balance, Acid-Base Balance, and Renal Function

The rate of fluid intake and fluid excretion in the infant is seven times as great in relation to weight as in the adult,

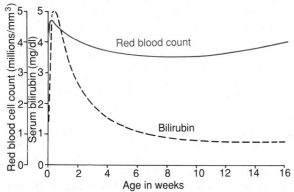

Figure 83-6. Changes in the red blood cell count and in the serum bilirubin concentration during the first 16 weeks of life, showing "physiologic anemia" at 6 to 12 weeks of life and "physiologic hyperbilirubinemia" during the first 2 weeks of life.

which means that even a slight alteration of fluid balance can cause rapidly developing abnormalities. Second, the rate of metabolism in the infant is two times as great in relation to body mass as in the adult, which means that two times as much acid is normally formed, which leads to a tendency toward acidosis in the infant. Third, functional development of the kidneys is not complete until the end of approximately the first month of life. For instance, the kidneys of the newborn can concentrate urine to only 1.5 times the osmolality of the plasma instead of the normal three to four times in the adult.

Therefore, considering the immaturity of the kidney, together with the marked fluid turnover in the infant and rapid formation of acid, one can readily understand that among the most important problems of infancy are acidosis, dehydration, and some instances of overhydration.

Liver Function

During the first few days of life, liver function may be quite deficient, as evidenced by the following effects:

1. The liver of the newborn conjugates bilirubin with glucuronic acid poorly and therefore excretes bilirubin only slightly during the first few days of life.

2. The liver of the newborn is deficient in forming plasma proteins, so that the plasma protein concentration falls to 15 to 20 per cent less than that for older children. Occasionally, the protein concentration falls so low that the infant actually develops hypoproteinemic edema.

3. The gluconeogenesis function of the liver is particularly deficient. As a result, the blood glucose level of the unfed newborn infant falls to about 30 to 40 mg/dl, and the infant must depend on its stored fats for energy until feeding can occur.

4. The liver of the newborn usually also forms too little of the factors needed for normal blood coagulation.

Digestion, Absorption, Metabolism of Energy Foods, and Nutrition

In general, the ability of the newborn infant to digest, absorb, and metabolize foods is not different from that of the older child, with the following three exceptions:

First, secretion of pancreatic amylase in the newborn infant is deficient, so that the infant utilizes starches less adequately than do older children.

Second, absorption of fats from the gastrointestinal tract is somewhat less than that in the older child. Consequently, milk with a high fat content, such as cow's milk, is frequently inadequately utilized.

Third, because the liver functions imperfectly during at least the first week of life, the glucose concentration in the blood is unstable and also low.

The neonate is especially capable of synthesizing and storing proteins. Indeed, with a completely adequate diet, as much as 90 per cent of the ingested amino acids is utilized for formation of body proteins. This is a much higher percentage than in adults.

Metabolic Rate and Body Temperature. The normal metabolic rate of the neonate in relation to body weight is about two times that of the adult, which accounts also for the two times as great cardiac output and two times as great minute respiratory volume in the infant.

However, because the body surface area is very large in relation to body mass, heat is readily lost from the body. As a result, the body temperature of the newborn infant, particularly of premature infants, falls easily. Figure 83-7

Figure 83-7. Fall in body temperature of the infant immediately after birth and instability of body temperature during the first few days of life.

shows that the body temperature of even the normal infant often falls several degrees during the first few hours after birth but returns to normal in 7 to 10 hours. Still, the body temperature regulatory mechanisms remain poor during the early days of life, allowing marked deviations in temperature, which are also illustrated in Figure 83-7.

Nutritional Needs During the Early Weeks of Life. At birth, a newborn infant is usually in complete nutritional balance, provided the mother has had an adequate diet. Furthermore, function of the gastrointestinal system is usually more than adequate to digest and assimilate all the nutritional needs of the infant if these are provided in the diet. However, three specific problems do occur in the early nutrition of the infant:

Need for Calcium and Vitamin D. The newborn infant is in a stage of rapid ossification of its bones at birth, so that a ready supply of calcium throughout infancy is needed. This is ordinarily supplied adequately by the usual diet of milk. Yet absorption of calcium by the gastrointestinal tract is poor in the absence of vitamin D. Therefore, the vitamin D-deficient infant can develop severe rickets in only a few weeks. This is particularly true in premature babies because their gastrointestinal tracts absorb calcium even less effectively than those of normal infants.

Necessity for Iron in the Diet. If the mother has had adequate amounts of iron in her diet, the liver of the infant usually has stored enough iron to keep forming blood cells for 4 to 6 months after birth. But if the mother has had insufficient iron in her diet, severe anemia is likely to occur in the infant after about 3 months of life. To prevent this possibility, early feeding of egg yolk, which contains reasonably large quantities of iron, or administration of iron in some other form is desirable by the second or third month of life.

Vitamin C Deficiency in Infants. Ascorbic acid (vitamin C) is not stored in significant quantities in the fetal tissues; yet it is required for proper formation of cartilage, bone, and other intercellular structures of the infant. Furthermore, milk has poor supplies of ascorbic acid, especially cow's milk, which has only one-fourth as much as mother's milk. For this reason, orange juice or other sources of ascorbic acid are usually prescribed by the third week of life.

Immunity

Fortunately, the neonate inherits much immunity from the mother because many antibodies diffuse from the mother's blood through the placenta into the fetus. However, the neonate does not form antibodies of its own to a significant extent. By the end of the first month, the baby's gamma globulins, which contain the antibodies, have decreased to less than one half the original level, with corresponding decrease in immunity. Thereafter, the baby's own immunity system begins to form antibodies, and the gamma globulin concentration returns essentially to normal by the age of 12 to 20 months.

Despite the decrease in gamma globulins soon after birth, the antibodies inherited from the mother still protect the infant for about 6 months against most major childhood infectious diseases, including diphtheria, measles, and polio. Therefore, immunization against these diseases before 6 months is usually unnecessary. On the other hand, the inherited antibodies against whooping cough are normally insufficient to protect the neonate; therefore, for full safety the infant requires immunization against this disease within the first month or so of life.

Allergy. Fortunately, the newborn infant is rarely subject to allergy. Several months later, however, when the neonate's own antibodies first begin to form, extreme allergic states can develop, often resulting in serious eczema, gastrointestinal abnormalities, or even anaphylaxis. As the child grows older and still higher degrees of immunity develop, these allergic manifestations usually disappear. This relationship of immunity to allergy is discussed in Chapter 34.

Endocrine Problems

Ordinarily the endocrine system of the infant is highly developed at birth, and the infant rarely exhibits any immediate endocrine abnormalities. However, there are special instances in which the endocrinology of infancy is important.

1. If a pregnant mother bearing a female child is treated with an androgenic hormone or if an androgenic tumor develops during pregnancy, the child will be born with a high degree of masculinization of her sexual organs, thus resulting in a type of *hermaphroditism.*

2. The sex hormones secreted by the placenta and by the mother's glands during pregnancy occasionally cause the neonate's breasts to form milk during the first few days of life. Sometimes the breasts then become inflamed, or infectious mastitis develops.

3. An infant born of a diabetic mother will have considerable hypertrophy and hyperfunction of the islets of Langerhans. As a consequence, the infant's blood glucose concentration may fall to lower than 20 mg/dl shortly after birth. Fortunately, in the newborn infant, unlike the adult, insulin shock or coma from this low level of blood glucose concentration only rarely develops.

Because of metabolic deficits in the diabetic mother, the fetus is often stunted in growth, and growth and tissue maturation of the newborn infant are often impaired. Also, there is a high rate of intrauterine mortality, and among those fetuses that do come to term, there is still a high mortality rate. Two thirds of the infants who die succumb to the respiratory distress syndrome, described earlier in the chapter.

4. Occasionally, a child is born with hypofunctional adrenal cortices, often resulting from *agenesis* of the glands or *exhaustion atrophy,* which can occur when the adrenal glands have been overstimulated.

5. If a pregnant woman has hyperthyroidism or is treated with excess thyroid hormone, the infant is likely to be born with a temporarily hyposecreting thyroid gland. On the other hand, if prior to pregnancy a woman had had her thyroid gland removed, her pituitary may secrete great

quantities of thyrotropin during gestation, and the child might be born with temporary hyperthyroidism.

6. In a fetus lacking thyroid hormone secretion, the bones grow poorly, and there is also mental retardation. This causes the condition called *cretin dwarfism*, discussed in Chapter 76.

● SPECIAL PROBLEMS OF PREMATURITY

All the problems just noted for neonatal life are especially exacerbated in prematurity. These can be categorized under the following two headings: (1) immaturity of certain organ systems and (2) instability of the different homeostatic control systems. Because of these effects, a premature baby rarely lives if it is born more than 2.5 to 3 months before term.

IMMATURE DEVELOPMENT OF THE PREMATURE INFANT

Almost all the organ systems of the body are immature in the premature infant, but some require particular attention if the life of the premature baby is to be saved.

Respiration. The respiratory system is especially likely to be underdeveloped in the premature infant. The vital capacity and the functional residual capacity of the lungs are especially small in relation to the size of the infant. Also, surfactant secretion is depressed or absent. As a consequence, the respiratory distress syndrome is a common cause of death. Also, the low functional residual capacity in the premature infant is often associated with periodic breathing of the Cheyne-Stokes type.

Gastrointestinal Function. Another major problem of the premature infant is to ingest and absorb adequate food. If the infant is more than 2 months premature, the digestive and absorptive systems are almost always inadequate. The absorption of fat is so poor that the premature infant must have a low fat diet. Furthermore, the premature infant has unusual difficulty in absorbing calcium and therefore can develop severe rickets before one recognizes the difficulty. For this reason, special attention must be paid to adequate calcium and vitamin D intake.

Function of Other Organs. Immaturity of other organ systems that frequently causes serious difficulties in the premature infant includes: (1) immaturity of the liver, which results in poor intermediary metabolism and often also a bleeding tendency as a result of poor formation of coagulation factors; (2) immaturity of the kidneys, which are particularly deficient in their ability to rid the body of acids, thereby predisposing to acidosis as well as to serious fluid balance abnormalities; (3) immaturity of the blood-forming mechanism of the bone marrow, which allows rapid development of anemia; and (4) depressed formation of gamma globulin by the lymphoid system, which is often associated with serious infection.

INSTABILITY OF THE CONTROL SYSTEMS IN THE PREMATURE INFANT

Immaturity of the different organ systems in the premature infant creates a high degree of instability in the homeostatic systems of the body. For instance, the acid-base balance can vary tremendously, particularly when the food intake varies from time to time. Likewise, the blood protein concentration is usually somewhat low because of immature liver development, often leading to *hypoproteinemic edema*. And inability of the infant to regulate its calcium ion concentration frequently brings on hypocalcemic tetany. Also, the blood glucose concentration can vary between the extremely wide limits of 20 to over 100 mg/dl, depending principally on the regularity of feeding. It is no wonder, then, with these extreme variations in the internal environment of the premature infant, that mortality is high.

Instability of Body Temperature. One of the particular problems of the premature infant is inability to maintain normal body temperature. Its temperature tends to approach that of its surroundings. At normal room temperature the infant's temperature may stabilize in the low 90s or even in the 80s. Statistical studies show that a body temperature maintained below 96°F (35.5°C) is associated with a particularly high incidence of death, which explains the common use of the incubator in the treatment of prematurity.

DANGER OF BLINDNESS CAUSED BY OXYGEN THERAPY IN THE PREMATURE INFANT

Because the premature infant frequently develops respiratory distress, oxygen therapy has often been used in treating prematurity. However, it has been discovered that use of high oxygen concentrations in treating premature infants, especially in early prematurity, causes ingrowth of a complex of blood vessels into the vitreous humor of the eyes *when the infant is later withdrawn from the oxygen*. This is followed still later by fibrosis. This condition, known as *retrolental fibroplasia*, causes permanent blindness. For this reason, it is particularly important to avoid treatment of premature infants with high concentrations of respiratory oxygen. Physiological studies indicate that the premature infant is usually safe with up to 40 per cent oxygen, but some child physiologists believe that complete safety can be achieved only at normal oxygen concentration.

● GROWTH AND DEVELOPMENT OF THE CHILD

The major physiological problems of the child beyond the neonatal period are related to special metabolic needs for growth, which have been fully covered in the sections on metabolism and endocrinology.

Figure 83–8 illustrates the changes in heights of boys and girls from the time of birth until the age of 20 years. Note especially that these parallel each other almost exactly until the end of the first decade of life. Between the ages of 11 and 13 years the female estrogens cause rapid growth but early uniting of the epiphyses at about the 14th to 16th year of life, so that growth in height ceases. This contrasts with the effect of testosterone in the male, which causes growth at a slightly later age—mainly between ages 13 and 17 years. The male, however, undergoes much more prolonged growth so that his final height is considerably greater than that of the female.

BEHAVIORAL GROWTH

Behavioral growth is principally a problem of maturity of the nervous system. Here, it is extremely difficult to dissoci-

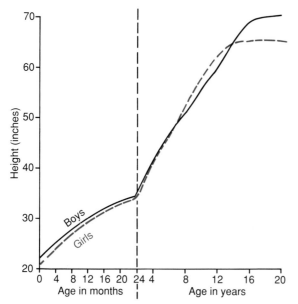

Figure 83–8. Height of boys and girls from infancy to 20 years of age.

ate maturity of the anatomical structures of the nervous system from maturity caused by training. Anatomical studies show that certain major tracts in the central nervous system are not completely myelinated until the end of the first year of life. For this reason it is frequently stated that the nervous system is not fully functional at birth. The brain cortex and its associated mechanisms such as vision seem to require several months after birth for significant functional development.

At birth, the brain size is only 26 per cent of the adult size and 55 per cent at 1 year but reaches almost adult proportions by the end of the second year. This is also associated with closure of the fontanels and sutures of the skull, which

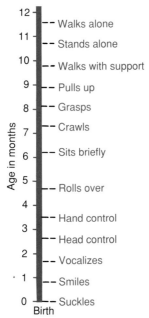

Figure 83–9. Behavioral development of the infant during the first year of life.

allows only 20 per cent additional growth of the brain beyond the first two years of life.

Figure 83–9 illustrates a normal progress chart for the infant during the first year of life. Comparison of this with the baby's actual development is used for clinical assessment of mental and behavioral growth.

REFERENCES

Aynsley-Green, A.: Metabolic and endocrine interrelations in the human fetus and neonate. Am. J. Clin. Nutr., 41:399, 1985.

Battaglia, F. C., and Meschia, G.: Principal substrates of fetal metabolism. Physiol. Rev., 58:499, 1978.

Bissonnette, J. M., et al.: Regulation of cerebral blood flow in the fetus. J. Dev. Physiol., 6:275, 1984.

Buckley, N. M., et al.: Circulatory effects of splanchnic nerve stimulation in developing swine. Am. J. Physiol., 248:H69, 1985.

Byskov, A. G.: Differentiation of mammalian embryonic gonad. Physiol. Rev., 66:71, 1986.

Caveney, S.: The role of gap junctions in development. Annu. Rev. Physiol., 47:319, 1985.

Cooke, J.: The early embryo and the formation of body pattern. Am. Sci., 76:35, 1988.

Dallman, P. R.: Anemia of prematurity. Annu. Rev. Med., 32:143, 1981.

Durnin, J. V. Energy balance in childhood and adolescence. Proc. Nutr. Soc., 43:271, 1984.

Fanaroff, A. A.: Behrman's Neonatal-Perinatal Medicine. Diseases of the Fetus and Infant. St. Louis, C. V. Mosby Co., 1983.

Fenichel, G. M.: Neonatal Neurology. New York, Churchill Livingstone, 1980.

Gaull, G. E., et al.: Significance of growth modulators in human milk. Pediatrics, 75:142, 1984.

Gold, J. J., and Josimovich, J. B. (eds.): Gynecologic Endocrinology, 4th Ed. New York, Plenum Publishing Corp., 1987.

Gomella, T. L., and Cunningham, M. D.: Neonatology. East Norwalk, Conn., Appleton & Lange, 1988.

Gootman, N., and Gootman, P. M. (eds.): Neural regulation of cardiovascular function in the perinatal period. In Perinatal Cardiovascular Function. New York, Marcel Dekker, 1983, p. 265.

Gootman, P. M., et al.: Characteristics of spontaneous efferent splanchnic discharge in neonatal swine. In Catecholamines: Basic and Peripheral Mechanisms. New York, Alan R. Liss, 1984, p. 369.

Gootman, P. M., et al.: Localization of vasoactive sites in the thalamus of newborn swine. London, Academic Press, 1985, p. 599.

Gootman, P. M., et al.: Postnatal maturation of the respiratory rhythm generator. In The Physiological Development of the Fetus and Newborn. London, Academic Press, 1985, p. 223.

Haller, J. O., and Schneider, M.: Pediatric Ultrasound. Chicago, Year Book Medical Publishers, 1980.

Harding, R.: Function of the larynx in the fetus and newborn. Annu. Rev. Physiol., 46:645, 1984.

Heymann, M. A., et al.: Factors affecting changes in the neonatal systemic circulation. Annu. Rev. Physiol., 43:371, 1981.

Jones, C. T., and Rolph, T. P.: Metabolism during fetal life: A functional assessment of metabolic development. Physiol. Rev., 65:357, 1985.

Josso, N., and Picard, J.-Y.: Anti-müllerian hormone. Physiol. Rev., 66:1038, 1986.

King, D. B., and May, J. D.: Thyroidal influence on body growth. J Exp. Zool., 232:453, 1984.

Knobil, E., et al. (eds.): The Physiology of Reproduction. New York, Raven Press, 1988.

Lauer, R. M., and Shekelle, R. B. (eds.): Childhood Prevention of Atherosclerosis and Hypertension. New York, Raven Press, 1980.

Lotgering, F. K., et al.: Maternal and fetal responses to exercise during pregnancy. Physiol Rev., 65:1, 1985.

Lumbers, E. R.: Renal function during intrauterine life. News Physiol. Sci., 2:220, 1987.

Mortola, J. P.: Dynamics of breathing in newborn mammals. Physiol. Rev., 67:187, 1987.

Mott, J. C. and Walker, D. W.: Neural and endocrine regulation of circulation in the fetus and newborn. In Shepherd J. T., and Abboud, F. M. (eds.): Handbook of Physiology. Sec. 2, Vol. III. Bethesda, Md., American Physiological Society, 1983, p. 837.

Naeye, R. L.: Diseases of the Fetus and Newborn: Clinical Correlations and Medical-Legal Considerations. St. Louis, C. V. Mosby Co., 1989.

Nathan, D. G., and Oski, F. A. (eds.): Hematology of Infancy and Childhood, 2nd Ed. Philadelphia, W. B. Saunders Co., 1980.

Nissley, S. P., and Rechler, M. M.: Somatomedin/insulin-like growth factor tissue receptors. Clin. Endocrinol. Metab., 13:43, 1984.

Olley, P. M., and Coceani, F.: Prostaglandins and the ductus arteriosus. Annu. Rev. Med., 32:375, 1981.

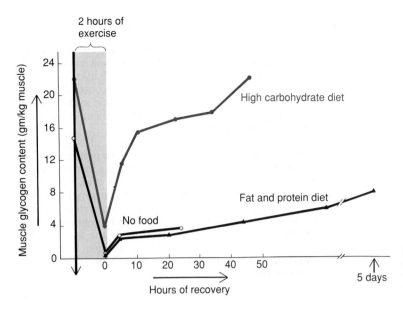

Figure 84–3. Effect of diet on the rate of muscle glycogen replenishment following prolonged exercise. (Reprinted from Fox: Sports Physiology. Philadelphia, Saunders College Publishing, 1979.)

fatty acids and *acetoacetic acid* (see Chapter 68) and also use to a much less extent proteins in the form of *amino acids*. In fact, even under the best conditions, in those endurance athletic events that last longer than 4 to 5 hours, the glycogen stores of the muscle become depleted and are then of little further use for energizing muscle contraction. Instead, the muscle now depends upon energy from other sources, mainly from fats.

Figure 84–4 illustrates the approximate relative usage of carbohydrates and fat for energy during prolonged exhaustive exercise under three different dietary conditions: high carbohydrate diet, mixed diet, and high fat diet. Note that most of the energy is derived from carbohydrate during the first few seconds or minutes of the exercise, but at the time of exhaustion, as much as 60 to 85 per cent of the energy is being derived from fats, rather than carbohydrates.

Not all the energy from carbohydrates comes from the stored muscle glycogen. In many persons almost as much glycogen is stored in the liver as in the muscles; and this can be released into the blood in the form of glucose, then taken up by the muscles as an energy source. In addition, glucose solutions given to an athlete to drink during the course of an athletic event (in optimal concentrations of 2 to 2.5 per cent) can provide as much as 30 to 40 per cent of the energy required during prolonged events such as marathon races.

In essence, then, if muscle glycogen and blood glucose are available, these are the energy nutrients of choice for intense muscle activity. Even so, for a real endurance event one can expect fat to supply more than 50 per cent of the required energy after about the first 3 to 4 hours.

EFFECT OF ATHLETIC TRAINING ON MUSCLES AND MUSCLE PERFORMANCE

Importance of Maximum Resistance Training. One of the cardinal principles of muscle development during athletic training is the following: Muscles that function under no load, even if they are exercised for hours upon end, increase little in strength. At the other extreme, muscles that contract at or near their maximal force of contraction will develop strength very rapidly even if the contractions are performed only a few times each day. Utilizing this principle, experiments on muscle building have shown that 6 maximal or nearly maximal muscle contractions performed in three separate sets 3 days a week give approximately optimal increase in muscle strength and without producing chronic muscle fatigue. The upper curve in Figure 84–5 illustrates the approximate percentage increase in strength that can be achieved in the previously untrained person by this optimal resistive training program, showing that the muscle strength increases about 30 per cent during the first 6 to 8 weeks but almost plateaus after that time. Along with this increase in strength is an approximately equal percentage increase in muscle mass, which is called *muscle hypertrophy*.

Muscle Hypertrophy. The basic size of a person's muscles is determined mainly by heredity plus the level of testosterone secretion, which, in men, causes considerably larger muscles than in women. However, with training, the muscles can become hypertrophied perhaps an additional 30 to 60 per cent. Most of this hypertrophy results from increased diameter of the muscle fibers; but this is not entirely true, because a very few greatly enlarged muscle fibers

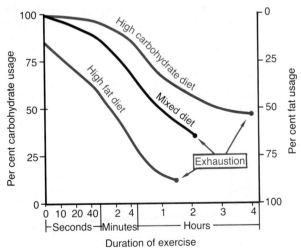

Figure 84–4. Effect of duration of exercise as well as type of diet on relative percentages of carbohydrate or fat used for energy by muscles. (Based partly on data in Fox: Sports Physiology. Philadelphia, Saunders College Publishing, 1979.)

Figure 84–5. Approximate effect of optimal resistive exercise training on increase in muscle strength over a training period of 10 weeks.

are believed to split down the middle along their entire length to form entirely new fibers, thus increasing the numbers of fibers slightly.

The changes that occur inside the hypertrophied muscle fibers themselves include (1) increased numbers of myofibrils, proportionate to the degree of hypertrophy; (2) up to 120 per cent increase in mitochondrial enzymes; (3) as much as 60 to 80 per cent increase in the components of the phosphagen metabolic system, including both ATP and phosphocreatine; (4) as much as 50 per cent increase in stored glycogen; and (5) as much as 75 to 100 per cent increase in stored triglyceride (fat). Because of all these changes, the capabilities of both the anaerobic and the aerobic metabolic systems are increased, increasing especially the maximum oxidation rate and efficiency of the oxidative metabolic system as much as 45 per cent.

Fast Twitch and Slow Twitch Muscle Fibers

In the human being, all muscles have varying percentages of *fast twitch* and *slow twitch muscle fibers.* For instance, the gastrocnemius muscle has a higher preponderance of fast twitch fibers, which gives it the capability of very forceful and rapid contraction of the type used in jumping. On the other hand, the soleus muscle has a higher preponderance of slow twitch muscle fibers and therefore is used to a greater extent for prolonged lower leg muscle activity.

The basic differences between the fast twitch and the slow twitch fibers are the following:

1. Fast twitch fibers are about two times as large in diameter.

2. The enzymes that promote rapid release of energy from the phosphagen and glycogen–lactic acid energy systems are two to three times as active in fast twitch fibers as in slow twitch fibers, thus making the maximal power that can be achieved by fast twitch fibers as great as two times that of slow twitch fibers.

3. Slow twitch fibers are mainly organized for endurance, especially for generation of aerobic energy. They have far more mitochondria than the fast twitch fibers. In addition, they contain considerably more myoglobin, a hemoglobin-like protein that combines with oxygen within the muscle fiber; and even more important, myoglobin increases the rate of diffusion of oxygen throughout the fiber by shuttling oxygen from one molecule of myoglobin to the next. In addition, the enzymes of the aerobic metabolic system are con-

siderably more active in slow twitch fibers than in fast twitch fibers.

4. The number of capillaries per mass of fibers is greater in the vicinity of slow twitch fibers than in the vicinity of fast twitch fibers.

In summary, fast twitch fibers can deliver extreme amounts of power for a few seconds to a minute or so. On the other hand, slow twitch fibers provide endurance, delivering prolonged strength of contraction over many minutes to hours.

Hereditary Differences Among Athletes for Fast Twitch Versus Slow Twitch Muscle Fibers. Some persons have considerably more fast twitch than slow twitch fibers, and others have more slow twitch fibers; this obviously could determine to some extent the athletic capabilities of different individuals. Unfortunately, athletic training has not been shown to change the relative proportions of fast twitch and slow twitch fibers however much an athlete might wish to develop one type of athletic prowess over another. Instead, this is determined almost entirely by genetic inheritance, and this in turn helps determine which area of athletics is most suited to each person: some people are born to be marathoners; others are born to be sprinters and jumpers. For example, the following are recorded percentages of fast twitch versus slow twitch fiber in the quadriceps muscles of different types of athletes:

	Fast Twitch	*Slow Twitch*
Marathoners	18	82
Swimmers	26	74
Average man	55	45
Weight lifters	55	45
Sprinters	63	37
Jumpers	63	37

● RESPIRATION IN EXERCISE

Although one's respiratory ability is of relatively little concern in the performance of sprint types of athletics, it is critical for maximal performance in endurance athletics.

Oxygen Consumption and Pulmonary Ventilation in Exercise. Normal oxygen consumption for a young adult man at rest is about 250 ml/min. However, under maximal conditions this can be increased to approximately the following average levels:

	ml/min
Untrained average male	3600
Athletically trained average male	4000
Male marathon runners	5100

Figure 84–6 illustrates the relationship between oxygen consumption at different degrees of exercise and *total pulmonary ventilation*. It is clear from this figure, as would be expected, that there is a linear relationship. In round numbers, both oxygen consumption and total pulmonary ventilation increase about 20-fold between the resting state and maximum intensity of exercise.

The Limits of Pulmonary Ventilation. How severely do we stress our respiratory systems during exercise? This can be answered by the following comparison for the normal man:

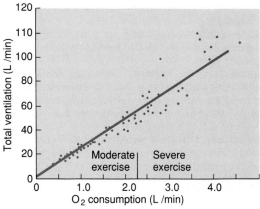

Figure 84–6. Effect of exercise on oxygen consumption and ventilatory rate. (From Gray: Pulmonary Ventilation and Its Physiological Regulation. Springfield, Ill., Charles C Thomas.)

	Liters/min
Pulmonary ventilation at maximal exercise	100 to 110
Maximal breathing capacity	150 to 170

Thus, the maximal breathing capacity is about 50 per cent greater than the actual pulmonary ventilation during maximal exercise. This obviously provides an element of safety for athletes, giving them extra ventilation that can be called on in such conditions as (1) exercise at high altitudes, (2) exercise under very hot conditions, and (3) abnormalities in the respiratory system.

The important point is that the respiratory system is not normally the most limiting factor in the delivery of oxygen to the muscles during maximal muscle aerobic metabolism. We shall see shortly that the ability of the heart to pump blood to the muscles is usually a greater limiting factor.

Effect of Training on $\dot{V}O_2$ Max. The abbreviation for the rate of oxygen usage under maximal aerobic metabolism is $\dot{V}O_2$ Max. Figure 84–7 illustrates the progressive effect of athletic training on $\dot{V}O_2$ Max recorded in a group of subjects beginning at the level of no training and then pursuing the training program for 7 to 13 weeks. In this study, it is surprising that the $\dot{V}O_2$ Max increased only about 10 per cent. Furthermore, the frequency of training, whether two times

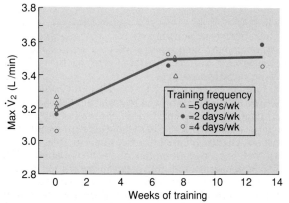

Figure 84–7. Increase in $\dot{V}O_2$ Max over a period of 7 to 13 weeks of athletic training. (Reprinted from Fox: Sports Physiology. Philadelphia, Saunders College Publishing, 1979.)

or five times per week, had little effect on the increase in $\dot{V}O_2$ Max. Yet, as was pointed out earlier, the $\dot{V}O_2$ Max of marathoners is about 45 per cent greater than that of the untrained person. Part of this greater $\dot{V}O_2$ Max of the marathoner is genetically determined; that is, those persons who have greater chest sizes in relation to body size and stronger respiratory muscles select themselves to become marathoners. However, it is also very likely that the very prolonged training of the marathoner does increase the $\dot{V}O_2$ Max by values considerably greater than the 10 per cent that has been recorded in short-term experiments such as that in Figure 84–7.

The O_2 Diffusing Capacity of Athletes. The O_2 *diffusing capacity* is a measure of the rate at which oxygen can diffuse from the alveoli into the blood. This is expressed in terms of *milliliters of oxygen that will diffuse for each millimeter of mercury difference between alveolar partial pressure of oxygen and pulmonary blood oxygen pressure.* That is, if the partial pressure of oxygen in the alveoli is 91 mm Hg while the pressure in the blood is 90 mm Hg, the amount of oxygen that diffuses through the respiratory membrane each minute is the diffusing capacity. The following are measured values for different diffusing capacities:

	ml/min
Nonathlete at rest	23
Nonathlete during maximum exercise	48
Speed skaters during maximum exercise	64
Swimmers during maximum exercise	71
Oarsmen during maximum exercise	80

The most startling fact about these results is the almost threefold increase in diffusing capacity between the resting state and the state of maximum exercise. This results mainly from the fact that blood flow through many of the pulmonary capillaries is very sluggish or even dormant in the resting state, whereas in exercise increased blood flow through the lungs causes all the pulmonary capillaries to be perfused at their maximum level, thus providing far greater surface area through which oxygen can diffuse into the pulmonary capillary blood.

It is also clear from these values that those athletes who require greater amounts of oxygen per minute have higher diffusing capacities. Is this because persons with naturally greater diffusing capacities choose these types of sports, or is it because something about the training procedures increases the diffusing capacity? The answer is not known; but one must believe that training does play some role, particularly endurance training.

The Blood Gases During Exercise. Because of the great usage of oxygen by the muscles in exercise, one would expect the oxygen pressure of the arterial blood to decrease markedly during strenuous athletics and the carbon dioxide pressure of the venous blood to increase far above normal. However, this normally is not the case. Both of these values remain nearly normal, illustrating the extreme ability of the respiratory system to provide very adequate aeration of the blood even in heavy exercise, which illustrates another very important point. The blood gases do not have to become abnormal for respiration to be stimulated in exercise. Instead, respiration is stimulated mainly by neurogenic mechanisms in exercise, as discussed in Chapter 41. Part of this stimulation results from direct stimulation of the respiratory center by the same nervous signals that are transmitted from the brain to the muscles to cause the exercise. Part is believed to result from sensory signals transmitted into the respiratory center from the contracting muscles and moving joints. All this nervous stimulation of respiration is normally sufficient to provide almost exactly the proper increase in pulmonary ventilation to keep the blood respiratory gases — the oxygen and the carbon dioxide — almost normal.

Effect of Smoking on Pulmonary Ventilation in Exercise. It is widely stated that smoking can decrease an athlete's "wind." This is true for many reasons. First, one effect of nicotine is constriction of the terminal bronchioles of the lungs, which increases the resistance of air flow into and out of the lungs. Second, the irritating effects of smoke cause increased fluid secretion in the bronchial tree, as well as some swelling of the epithelial linings. Third, nicotine paralyzes the cilia on the surfaces of the respiratory epithelial cells that normally beat continuously to remove excess fluids and foreign particles from the respiratory tract. As a result, much debris accumulates in the respiratory passageways and adds further to the difficulty of breathing. Putting all these factors together, even the light smoker will feel respiratory strain during maximal exercise, and the level of performance obviously may be reduced.

Much more severe are the effects of chronic smoking. There are very few chronic smokers in whom some degree of emphysema does not develop. In this disease, the following occur: (1) chronic bronchitis, (2) obstruction of many of the terminal bronchioles, and (3) destruction of many alveolar walls. In severe emphysema, as much as four fifths of the respiratory membrane can be destroyed; then even the slightest exercise can cause respiratory distress. In fact, many such patients cannot even perform the athletic feat of walking across the floor of a single room without gasping for breath. Such is the indictment of smoking.

● THE CARDIOVASCULAR SYSTEM IN EXERCISE

Muscle Blood Flow. The final common denominator of cardiovascular function in exercise is to deliver oxygen and other nutrients to the muscles. For this purpose, the muscle blood flow increases drastically during exercise. Figure 84–8 illustrates a recording of muscle blood flow in the calf of a person for a period of 6 minutes during strong intermittent contraction. Note not only the great increase in flow —

Figure 84–8. Effects of muscle exercise on blood flow in the calf of a leg during strong rhythmic contraction. The blood flow was much less during contraction than between contractions. (From Barcroft and Dornhorst: *J. Physiol.*, 109:402, 1949.)

about 13-fold — but also that the flow decreased during each muscle contraction. Two points can be made from this study: (1) The actual contractile process itself temporarily decreases muscle blood flow because the contracting muscle compresses the intramuscular blood vessels; therefore, strong tonic contractions can cause rapid muscle fatigue because of lack of delivery of enough oxygen and nutrients during the continuous contraction. (2) The blood flow to muscles during exercise can increase markedly. The following comparison illustrates the maximum increase in blood flow that can occur in the well-trained athlete.

	ml/100 gm Muscle/min
Resting blood flow	3.6
Blood flow during maximal exercise	90

Thus, muscle blood flow can increase a maximum of about 25-fold during the most strenuous exercise. About half this increase in flow results from intramuscular vasodilation caused by the direct effects of increased muscle metabolism, as was explained in Chapter 21. The other half results from multiple factors, the most important of which is probably the moderate increase in arterial blood pressure that occurs in exercise, usually about a 30 per cent increase. The increase in pressure not only forces more blood through the blood vessels but also stretches the walls of the arterioles and further reduces the vascular resistance. Therefore, a 30 per cent increase in blood pressure can often more than double the blood flow; this is added to the great increase in flow already caused by the metabolic vasodilation.

Work Output, Oxygen Consumption, and Cardiac Output During Exercise. Figure 84–9 illustrates the interrelationships among work output, oxygen consumption, and cardiac output during exercise. It is not surprising that all these are directly related to each other, as shown by the linear functions, because the muscle work output increases oxygen consumption, and oxygen consumption in turn dilates the muscle blood vessels, thus increasing venous return and cardiac output. Typical cardiac outputs at several levels of exercise are the following:

	Liters/min
Average young man at rest	5.5
Maximum output during exercise in young untrained man	23
Maximum output during exercise in male marathoner	30

Thus, the normal untrained person can increase cardiac output a little over fourfold, and the well-trained athlete can increase output about sixfold. Individual marathoners have been clocked at cardiac outputs as great as 35 to 40 liters/min.

Effect of Training on Heart Hypertrophy and on Cardiac Output. From the foregoing data, it is clear that marathoners can achieve maximum cardiac outputs about 40 per cent greater than that achieved by the untrained person. This results mainly from the fact that the heart chambers of marathoners enlarge about 40 per cent; along with enlargement of the chambers, the heart mass increases 40 per cent or more as well. Therefore, not only do the skeletal muscles hypertrophy during athletic training but the heart does also. However, heart enlargement and increased pumping capacity occur only in the endurance types, not in the sprint types, of athletic training.

Figure 84–9. Relationship between cardiac output and work output (solid line) and between oxygen consumption and work output (dashed line) during different levels of exercise. (From Guyton, Jones, and Coleman: Circulatory Physiology: Cardiac Output and Its Regulation. Philadelphia, W. B. Saunders Company, 1973.)

Figure 84–10. Approximate stroke volume output and heart rate at different levels of cardiac output in a marathon athlete.

Even though the heart of the marathoner is considerably larger than that of the normal person, resting cardiac output is almost exactly the same as that in the normal person. However, this normal cardiac output is achieved by a large stroke volume at a reduced heart rate. Table 84–2 compares stroke volume and heart rate in the untrained person and the marathoner.

Thus, the heart-pumping effectiveness of each heart beat is 40 to 50 per cent greater in the highly trained athlete than in the untrained person, but there is a corresponding decrease in heart rate at rest.

Role of Stroke Volume and Heart Rate in Increasing the Cardiac Output. Figure 84–10 illustrates the approximate changes in stroke volume and heart rate as the cardiac output increases from its resting level of about 5.5 liters/min to 30 liters/min in the marathon runner. The *stroke volume* increases from 105 to 162 milliliters, an increase of about 50 per cent; whereas the heart rate increases from 50 to 185 beats per minute, an increase of 270 per cent. Therefore, the heart rate increase accounts by far for a greater proportion of the increase in cardiac output than does the increase in stroke volume during strenuous exercise. The stroke volume normally reaches its maximum by the time the cardiac output has increased only halfway to its maximum. Any further increase in cardiac output must occur by increasing the heart rate.

Relationship of Cardiovascular Performance to $\dot{V}o_2$ Max. During maximal exercise, both the heart rate and the stroke volume are increased to about 95 per cent of their maximal levels. Since the cardiac output is equal to stroke volume *times* heart rate, one finds that the cardiac output is about 90 per cent of the maximum that the person can achieve. This is in contrast to about 65 per cent of maximum for pulmonary ventilation. Therefore, one can readily see that the cardiovascular system is normally much more limiting on $\dot{V}o_2$ Max than is the respiratory system. For this reason, it is frequently stated that the performance that can be achieved by the marathoner mainly depends on his or her heart, for this is the most limiting link in the delivery of adequate oxygen to the exercising muscles. Therefore, the 40 per cent advantage in maximum cardiac output that the marathoner has over the average untrained male is probably the single most important physiological benefit of the marathoner's training program.

Effect of Heart Disease and Old Age on Athletic Performance. Because of the critical limitation that the cardiovascular system places on maximal performance in endurance athletics, one can readily understand that any type of heart disease that reduces the maximum cardiac output will cause an almost corresponding decrease in achievable muscle power. Therefore, a person with congestive heart failure frequently has difficulty achieving even the muscle power required to climb out of bed, much less to walk across the floor.

The maximum cardiac output of older persons also decreases considerably—there is as much as a 50 per cent decrease between the teens and the age of eighty years. Again, one finds that the maximum achievable muscle power is greatly reduced.

● BODY HEAT IN EXERCISE

Almost all the energy released by the internal metabolism of nutrients is eventually converted into body heat. This applies even to the energy that causes muscle contraction, for the following reasons: First, the maximum efficiency for conversion of nutrient energy into muscle work, even under the best of conditions, is only 20 to 25 per cent; the remainder of the nutrient energy is converted into heat during the course of the intracellular chemical reactions. Second, almost all the energy that does go into creating muscle work still becomes body heat because all but a small portion of this energy is used for (1) overcoming viscous resistance to the movement of the muscles and joints, (2) overcoming the friction of the blood flowing through the blood vessels, and

Table 84–2 COMPARISON OF CARDIAC OUTPUT BETWEEN MARATHONER AND NONATHLETE

	Stroke Volume (ml)	Heart Rate (beats/min)
Resting		
Nonathlete	75	75
Marathoner	105	50
Maximum		
Nonathlete	110	195
Marathoner	162	185

(3) other similar effects—all of which convert the muscle contractile energy into heat.

Now, recognizing that the oxygen consumption by the body can increase as much as 20-fold in the well-trained athlete and that the amount of heat liberated in the body is directly proportional to the oxygen consumption (as discussed in Chapter 72), one quickly realizes that tremendous amounts of heat are injected into the internal body tissues during endurance athletic events.

Next, with a vast rate of heat flow into the body on a very hot and humid day, so that the sweating mechanism cannot eliminate the heat, an intolerable and even lethal condition called *heat stroke* can easily develop in the athlete.

Heat Stroke. During endurance athletics, even under normal environmental conditions the body temperature often rises from its normal level of 98.6°F to 102°F to 103°F (37°C to 40°C). With very hot and humid conditions or excess clothing, the body temperature can easily rise to as high as 106°F to 108°F (41°C to 42°C). At this level the elevated temperature itself becomes destructive to tissue cells, especially destructive to brain cells. When this happens, multiple symptoms begin to appear, including extreme weakness, exhaustion, headache, dizziness, nausea, profuse sweating, confusion, staggering gait, collapse, and unconsciousness.

This whole complex is called heat stroke, and failure to treat it immediately can lead to death. In fact, even though the person has stopped exercising, the temperature does not easily decrease by itself. One of the reasons for this is that at these high temperatures the temperature-regulating mechanism itself often fails (see Chapter 73). A second reason is that the high temperature approximately doubles the rates of all intracellular chemical reactions, thus liberating still far more heat.

The treatment of heat stroke is to reduce the body temperature as rapidly as possible. The most practical way to do this is to remove all clothing, maintain a spray of water on all surfaces of the body or continually sponge the body, and blow air over the body with a strong fan. Experiments have shown that this treatment can reduce the temperature either as rapidly or almost as rapidly as any other procedure, though some physicians prefer total immersion of the body in ice water containing a mush of crushed ice if available.

● BODY FLUIDS AND SALT IN EXERCISE

As much as a 5 to 10 pound weight loss has been recorded in athletes in a period of 1 hour during endurance athletic events under hot and humid conditions. Essentially all of this weight loss results from loss of sweat. Loss of enough sweat to decrease body weight only 3 per cent can significantly diminish a person's performance, and a 5 to 10 per cent rapid decrease in weight can often be very serious, leading to muscle cramps, nausea, and other effects. Therefore, it is essential to replace fluid as it is lost.

Replacement of Salt and Potassium. Sweat contains a large amount of salt, for which reason it has long been stated that all athletes should take salt (sodium chloride) tablets when performing exercise on hot and humid days. Unfortunately, overuse of salt tablets has often done more harm than good. Furthermore, if an athlete becomes acclimatized to the heat by progressive increase in athletic exposure over a period of 1 to 2 weeks rather than performing maximal athletic feats on the first day, the sweat glands also become acclimatized, so that the amount of salt lost in the sweat is only a small fraction of that prior to acclimatiza-tion. This sweat gland acclimatization results mainly from increased aldosterone secretion by the adrenal cortex. The aldosterone in turn has a direct effect on the sweat glands, increasing the reabsorption of sodium chloride from the sweat before it issues forth from the sweat gland tubules onto the surface of the skin. Once the athlete is acclimatized, only rarely do salt supplements need to be considered during athletic events.

On the other hand, recent experience by military units suddenly exposed to heavy exercise in the desert has demonstrated still another electrolyte problem—the problem of potassium loss. Potassium loss results partly from the increased secretion of aldosterone during heat acclimatization, which increases the loss of potassium in the urine as well as in the sweat. As a consequence of these new findings, some of the newer supplemental fluids for athletics are beginning to contain properly proportioned amounts of potassium and sodium, usually in the form of fruit juices.

● DRUGS AND ATHLETES

Without belaboring this issue, let us list some of the effects of drugs in athletics:

First, *caffeine* can increase athletic performance. In one experiment on a marathon runner, his running time for the marathon was reduced by 7 per cent by judicious use of caffeine in amounts similar to those found in one to three cups of coffee.

Second, use of *male sex hormones (androgens)* to increase muscle strength undoubtedly can increase athletic performance under some conditions, especially in women and in men who are poorly endowed with normal testosterone secretion. Unfortunately, some of the synthetic testosterone analog preparations can cause liver damage and even liver cancer. In men, any type of male sex hormone preparation also leads to decreased testicular function, including both decreased formation of sperm and decreased secretion of the person's own natural testosterone, with residual effects lasting at least for many months and perhaps indefinitely. In a woman, even more dire effects can occur because she is not normally adapted to the male sex hormone—hair on the face, a bass voice, ruddy skin, cessation of menses.

Other drugs, such as *amphetamines* and *cocaine,* have been reputed to increase one's athletic performance. However, it is equally true that overuse of these drugs can lead to deterioration of performance. Furthermore, experiments have failed to prove the value of these drugs except as a psychic stimulant. Some athletes have been known to die during athletic events because of interaction between such drugs and the norepinephrine and epinephrine released by the sympathetic nervous system during exercise. One of the causes of death under these conditions is overexcitability of the heart, leading to ventricular fibrillation, which is lethal within seconds.

REFERENCES

Armstrong, R. B.: Mechanism of exercise-induced delayed onset muscular soreness: A brief review. Med. Sci. Sports Exerc., 16:529, 1984.

Barrow, H. M., and Brown, J. D.: Man and Movement: Principles of Physical Education, 4th Ed. Philadelphia, Lea & Febiger, 1988.

Blomqvist, C. G., and Saltin, B.: Cardiovascular adaptations to physical training. Annu. Rev. Physiol., 45:169, 1983.

Brengelmann, G. L.: Circulatory adjustments to exercise and heat stress. Annu. Rev. Physiol., 45:191, 1983.

Bye, P. T. P., et al.: Respiratory factors limiting exercise. Annu. Rev. Physiol., 45:439, 1983.

Daniels, L., and Worthingham, C.: Muscle Testing: Techniques of Manual Examination, 5th Ed. Philadelphia, W. B. Saunders Co., 1986.

Dehn, M. M., et al.: Clinical exercise performance. Clin. Sports Med., 3:319, 1984.

Dohm, G. L., et al.: Protein metabolism during endurance exercise. Fed. Proc., 44:348, 1985.

Fisher, A. G., and Jensen, C. R.: Scientific Basis of Athletic Conditioning, 3rd Ed. Philadelphia, Lea & Febiger, 1989.

Franklin, B. A., et al. (eds.): Exercise in Modern Medicine. Baltimore, Williams & Wilkins, 1988.

Friedman, M. J., and Ferkel, R. D. (eds.): Prosthetic Ligament Reconstruction of the Knee. Philadelphia, W. B. Saunders Co., 1988.

Goldspink, G.: Malleability of the motor system: A comparative approach. J. Exp. Biol., 115:375, 1985.

Golinick, P. D.: Metabolism of substrates: Energy substrate metabolism during exercise and as modified by training. Fed. Proc., 44:353, 1985.

Gowitzke, B. A., and Milner, M.: Scientific Bases of Human Movement, 3rd Ed. Baltimore, Williams & Wilkins, 1988.

Guyton, A. C., et al.: Circulatory Physiology: Cardiac Output and Its Regulation. 2nd Ed. Philadelphia, W. B. Saunders Company, 1973.

Hammond, H. K., and Froelicher, V. F.: Normal and abnormal heart rate responses to exercise. Prog. Cardiovasc. Dis., 27:271, 1985.

Harrison, M. H.: Effects of thermal stress and exercise on blood volume in humans. Physiol. Rev., 65:149, 1985.

Harrison, M. H.: Effects on thermal stress and exercise on blood volume in humans. Physiol. Rev., 65:149, 1985.

Hockey, R.: Stress and Fatigue in Human Performance. New York, John Wiley & Sons, 1983.

Jones, N. L.: Clinical Exercise Testing, 3rd Ed. Philadelphia, W. B. Saunders Co., 1988.

Killian, K. J., and Campbell, E. J. M.: Dyspnea and exercise. Annu. Rev. Physiol., 45:465, 1983.

Lamb, D. R.: Physiology of Exercises. Responses and Adaptations. New York, Macmillan Publishing Co., 1984.

Lind, A. R.: Cardiovascular adjustments to isometric contractions: Static effort. In Shepherd, J. T., and Abboud, F. M. (eds.): Handbook of Physiology. Sec. 2, Vol. III. Bethesda, Md., American Physiological Society, 1983, p. 947.

Lotgering, F. K., et al.: Maternal and fetal responses to exercise during pregnancy. Physiol. Rev., 65:1, 1985.

McArdle, W. D., et al.: Exercise Physiology: Energy, Nutrition, and Human Performance, 2nd Ed. Philadelphia, Lea & Febiger, 1986.

Mitchell, J. H., et al.: The exercise pressor reflex: Its cardiovascular effects, afferent mechanisms, and control pathways. Annu. Rev. Physiol., 45:229, 1983.

Nadel, E. R.: Physiological adaptations to aerobic training. Am. Sci., 73:334, 1985.

Nordin, M., and Frankel, V. H. (eds.): Basic Biomechanics of the Musculoskeletal System, 2nd Ed. Philadelphia, Lea & Febiger, 1989.

Partridge, L. D., and Benton, L. A.: Muscle, the motor. In Brooks, V. B. (ed.): Handbook of Physiology. Sec. 1, Vol. II. Bethesda, Md., American Physiological Society, 1981, p. 43.

Pearson, M. L. (ed.): Muscle Development. Molecular and Cellular Control. Cold Spring Harbor, N. Y., Cold Spring Harbor Institute, 1982.

Rasch, P. J. (ed.): Kinesiology and Applied Anatomy, 7th Ed. Philadelphia, Lea & Febiger, 1989.

Rash, J.: Neuromuscular Atlas. New York, Praeger Publishers, 1984.

Rigotti, N. A., et al.: Exercise and coronary heart disease. Annu. Rev. Med. 34:391, 1983.

Shepherd, J. T.: Circulation to skeletal muscle. In Shepherd, J. T., and Abboud, F. M. (eds.): Handbook of Physiology, Sec. 2, Vol. III. Bethesda, Md., American Physiological Society, 1983, p. 319.

Smith, N. J., and Stanitski, C. L.: Sports Medicine: A Practical Guide. Philadelphia, W. B. Saunders Co., 1987.

Stone, H. L.: Control of the coronary circulation during exercise. Annu. Rev. Physiol., 45:213, 1983.

Strauss, R. H. (ed.): Sports Medicine and Physiology. Philadelphia, W. B. Saunders Company, 1979.

Strauss, R. H.: Drugs and Performance in Sports. Philadelphia, W. B. Saunders Co., 1987.

Stray-Gundersen, J.: Unethical alterations of oxygen-carrying capacity in endurance athletes. News Physiol. Sci., 3:241, 1988.

Taylor, C. R.: Carrying loads: The cost of generating muscular force. News Physiol. Sci., 1:153, 1986.

Thomas, J. A. (ed.): Drugs, Athletes, and Physical Performance. New York, Plenum Publishing Corp., 1988.

Wagner, P. D.: The lungs during exercise. News Physiol. Sci., 2:6, 1987.

Winter, D. A.: Biomechanics of Human Movement. New York, John Wiley & Sons, 1979.

Index

Note: Page numbers in *italics* refer to illustrations; page numbers followed by t refer to tables.

Diffusion *(Continued)*
concentration gradient in, 42–44, *42*
definition of, 38–39
distance of, vs. oxygen metabolic use, 439
electrical potential in, 44, *44*
facilitated, 42–43, *42–43*
definition of, 39, *39*
in fructose absorption, 733
in glucose transport, 745–746
in kidney tubules, 299
in placenta, 919
in artificial kidney, 347–348, *348*
in capillaries, 3–4, *3*, 172, 172t
in gastrointestinal absorption, 731–734, *731*
in interstitial gel, 173–174
in kidney tubules, 300–301
kinetics of, 38–39, 45–46
membrane permeability and, 43
membrane potentials from, 51–53, *52–53*
net (osmosis), 40t, 44–46, *44–45*
of lipid-insoluble substances, 40, 40t, 172, 172t
of lipid-soluble substances, 39–40, 40t, 172
of neurotransmitters from synaptic cleft, 486
of water, 40, 40t, 44–46, *44–45*
of water-soluble substances, 172, 172t
osmosis and, 40t, 44–46, *44–45*
pressure difference in, 44, *44*
rate of, maximum (V_{max}), 42–43, *42*
net, factors affecting, 43–44, *44*
vs. concentration difference, 173
vs. lipid solubility, 40, 40t
simple, definition of, 39, *39*
through lipid bilayer, 39–40, 40t
through protein channels, 40–42, *41–42*
vs. facilitated diffusion, 42–43, *42*
through capillary membrane, 172–173, *172*, 172t, 179
through capillary wall, 171
through cell membrane, 11–12, *11*
through cellular sheet, 49, *49*
through lymph capillary membrane, 181
through placental membrane, 917–919, *918*
through skin, water loss in, 274, 275t
vs. active transport, 38, *39*
vs. filtration, 179
with active transport, 49, *49*
Diffusion coefficient, 43
of gases, 424, 429
Diffusion potential, 51–52, *52*, 54–55, *54*
Digestion (at cellular level), 17–18, *17*
by lysosomes, 13
of phagocytized particles, 367–368
Digestion (gastrointestinal). See also *Gastrointestinal secretion.*
chewing role in, 698
disorders of, 738–739
gallbladder emptying in, *721*, 722
hydrolysis in, 726
in neonate, 934
of carbohydrates, 718, 726–727, *727*
of fats, 703, 718, 726, 728–729, *728–729*
of proteins, 703, 718, 726–728, *728*
Digitalis, in cardiac failure, 248–249, 252–253, *252*
toxic effects of, 136–137, *136*, 143
Dihydrotestosterone, 891–892, *892*, 894
1,25-Dihydroxycholecalciferol, 869–870, *869–870*, 877, 879–880

24,25-Dihydroxycholecalciferol, 870
Diiodotyrosine, 832–833, *832*
Diisopropyl fluorophosphate, 83
Diluting segment, of distal tubule, 302
Dilution method, in body fluid compartment measurement, 276, *276*
in cardiac output measurement, 232, *232*
in functional residual capacity measurement, 408–409
Diopter, of lens, 536, *537*
Dipalmitoyl lecithin, in alveolar surfactant, 405
Dipeptidase, in protein digestion, 728
2,3-Diphosphoglycerate, oxygen-hemoglobin dissociation curve and, 438, *438*
1,3-Diphosphoglyceric acid, in glycolysis, 747, *747*
Dipole layer, in cell membrane, 53
Disaccharides, digestion of, 726–727, *727*
Disc(s), in cones, 546, *547–548*
in rods, 546, *547–548*
intercalcated, in cardiac muscle, 98, *99*, 114
Discharge zone, of nerve fiber, 502, *502*
Disse, space of, 771, *772*
Disseminated intravascular coagulation, 269, 398
Dissociation, in buffer systems, 331–332
of carbonic acid, 331–332, 336–337, *336440–441, 440*
Dissociation curves. See also *Hemoglobin-oxygen dissociation curve.*
carbon dioxide, 441–442, *441–442*
Distensibility, vascular, 159–160, *160*
Distention, gastrointestinal, peristalsis and, 693–694
Distractability, prefrontal association cortex and, 640
Diuresis, osmotic, in diabetes mellitus, 865
pressure, 205–207, *206–207*. See also *Renal-body fluid system.*
aldosterone and, 844
arterial pressure effects on, *293*, 295
in blood volume control, 320, *321*
water, 314, *314*
Diuretics, action of, 350–351
in cardiac failure, 250
in hypertension, 218, 250
loop, 351
metabolic alkalosis from, 341
osmotic, 351
Divergence, of light rays, 534, *535*
of nerve signals, 502, *502*
Diver's paralysis, 473–474, *473*
Diving, carbon dioxide toxicity in, 473
compressed air sickness after, 473–474, *473*
decompression after, 473–474, *473*
in submarines, 475
nitrogen narcosis in, 471
oxygen pressure control in, hemoglobin buffer system in, 437
oxygen toxicity in, 471–473, *472*
pressure vs. depth and, 471, *472*
"saturation" (helium-oxygen mixtures), 474
Scuba, 474–475, *475*
sea level volume of dissolved nitrogen, vs. depth, 473
volume of gas vs. depth and, 471
Dysbarism, 473–474, *473*
DNA. See *Deoxyribonucleic acid.*
DNA ligase, 32
DNA polymerase, 32
Dominant hemisphere, of cerebral cortex, 638–639

Donnan effect, on colloid osmotic pressure, 177, *177*
L-Dopa, in Parkinson's disease, 630
Dopamine, action of, 485
in basal ganglia, 630, *630*
in brain activity control, 650, *650*
in enteric nervous system, 691
in norepinephrine synthesis, 670
in Parkinson's disease, 630
in prolactin release, 926
in retina, 555
in schizophrenia, 665–666
secretion of, 485
Doppler flowmeter, in blood flow measurement, 153–154, *154*
Dorsal column-lemniscal system, in signal transmission, 509
anatomy of, 510–511, *510–511*
in two-point discrimination, 514, *514*
lateral inhibition in, 514–515, *514*
neuronal circuits of, 513–514, *514*
Dorsal respiratory group of neurons, 444–445, *445*
Dorsomedial nucleus, of hypothalamus, *653*
Double Bohr effect, in placental diffusion, 918
Down-regulation, of hormone receptors, 813
dP/dt, in cardiac contractility assessment, 108–109, *109*
DPG (2,3-diphosphoglycerate), oxygen-hemoglobin dissociation curve and, 438, *438*
Dreams, 659–660
Drinking, in diabetes, 865
thirst relief from, 315
threshold for, 315–316
urinary output after, 314, *314*
Drugs. See also specific drug.
in athletics, 949
metabolic alkalosis from, 341
metabolism of, in liver, 774
neuromuscular transmission and, 83
synaptic transmission effects of, 493
Ductus arteriosus, 932–933, *932*
closure of, 933
patent, 259–260, *259*
phonocardiogram in, *256*
pulse in, 161, *161*
Ductus venosus, 932–933, *932*
Duodenocolic reflex, 706
Duodenum, bicarbonate absorption in, 732
chloride absorption in, 732
distention of, in feeding regulation, 780
peptic ulcer of, 737–738, *737*
stomach emptying and, 702–703
Dural sinus, negative pressure in, 166
Dwarfism, 824, 826
Dye dilution method, in blood flow measurement, 771
in body fluid compartment measurement, 276, *276*
in cardiac output measurement, 232, *232*
Dynamic proprioception, 516
Dynein, in ciliary movement, 23
Dynorphin, 525
Dysarthria, in cerebellar dysfunction, 626
Dysbarism, 473–474, *473*
Dysdiadochokinesia, 626
Dyslexia, 638, 640
Dysmetria, in cerebellar dysfunction, 626
Dyspnea, 460–461
in asthma, 459
in cardiac reserve depletion, 251
in hypercapnia, 460
J receptors role in, 452